Mycotoxins in Plants and Plant Products

Martin Weidenbörner

Mycotoxins in Plants and Plant Products

Cocoa, Coffee, Fruits and Fruit Products, Medicinal Plants, Nuts, Spices, Wine

Springer

Martin Weidenbörner
Bonn, Germany

ISBN 978-3-319-92849-4 ISBN 978-3-319-92850-0 (eBook)
https://doi.org/10.1007/978-3-319-92850-0

Library of Congress Control Number: 2018947379

Printed on acid-free paper

This Springer imprint is published by the registered company Springer Nature Switzerland AG
The registered company address is: Gewerbestrasse 11, 6330 Cham, Switzerland

Hem Chandra Jha

Preface

Mycotoxins in Plants and Plant Products—Cocoa, Coffee, Fruits and Fruit Products, Medicinal Plants, Nuts, Spices, Wine (Volume II) deals with the mycotoxin contamination of all the remaining plants and plant products, which have not been covered in the first volume *Mycotoxins in Plants and Plant Products—Cereals and Cereal Products.* Although not being mentioned in the title of the book, mycotoxin contamination of, e.g., beans, beverages, chocolate, figs, jams, oils, pastes, peanut butter, sauces, seeds, vine fruit, and vinegar is also considered. Extraordinary and exotic items, usually not associated with mycotoxins like halva, honey, ice cream, kubeba, kuru-kuru, margarine, marzipan, muhallebi, pekmez, shiitake, tahini, and tea, form also an essential part of the book.

The foodstuffs are listed in alphabetical order. Terms within brackets are followed by terms separated by a comma. The single items are ordered and grouped together normally according to their botanical origin. However, because of the easier use of the book, some items have been arranged due to the current language. All types of, e.g., flours, fruits, juices, meals, nuts, oils, pastes, products, spices, and wines are put together to get a quicker overview about mycotoxins involved in the corresponding foodstuffs. This was done as far as possible. Data of more than 280 new publications are added to Volume II. In the end, data of more than 900 articles have been considered in the book. Each item contains the relevant information as far as possible. The present book will be followed by an additional volume, dealing with the mycotoxin contamination of animal products.

Since mankind exist, the need of food is indispensable. But food supplies can be attacked, including by microorganisms. If conditions are suitable, microorganisms can grow and subsequently spoil the food. Thereby, filamentous fungi are of special interest. Once they grow up, they do not even spoil the food but can also contaminate it with mycotoxins. The presence of these invisible mycotoxins in food is undesirable and can produce serious problems. At high levels mycotoxins cause even death shortly after exposure. At lower levels, they can cause disorders in various organs and/or impair immunity. In the end, mycotoxins are a cost factor in human health.

In contrast to bacterial contamination of food, the enhanced contamination by fungi becomes more visual. Each consumer knows about the appearance of molded food, and the awareness of the public concerning the health hazard of foods is great. Furthermore, consumers demand for high-quality food. As a prerequisite for producing such food, the whole food industry must know about fungi and their corresponding mycotoxins. In this case the second volume *Mycotoxins in Plants and Plant Products—Cocoa, Coffee, Fruits and Fruit Products, Medicinal Plants, Nuts, Spices, Wine* gives very good information about the main mycotoxins occurring in this kind of food.

Mycotoxin contamination of food occurs as a result of crop invasion by field fungi, e.g., *Fusarium* spp. and *Alternaria* spp. Drought of plants, close planting, competition from weeds, reduced fertilization, and other factors are causing stress to plants which enhance growth of these fungi. Furthermore, the growth of storage fungi (*Aspergillus* spp. and *Penicillium* spp.) in improperly stored crops or after processing plants into food products causes spoilage and possible mycotoxin contamination. Specialists of these genera can also grow when food is preserved, e.g., in a refrigerator.

In the worst case, these fungi produce secondary metabolites, the so-called mycotoxins. They can be very harmful to humans and animals. In 1960, turkey X disease was caused by a toxin (aflatoxin) produced by molds belonging to the *Aspergillus flavus* group which caused the deaths of 100,000 turkeys in the UK. Hence, the study of mycotoxins began. Approximately 400 known mycotoxins are known today. They have various negative effects on one or several organs in humans or animals. It is beyond the scope of this book listing all mycotoxins and microbial metabolites contaminating plants and plant products in extra entries. Briefly this was done in Chapter 2: Further Mycotoxins and Microbial Metabolites. This chapter comprises the remaining mycotoxins and microbial metabolites which are not enumerated under their items. Usually, substances which belong to the masked but not real mycotoxins such as deoxynivalenol-3-glucoside, zearalenone-4-glucoside, and zearalenone-4-sulfate or do not count as main mycotoxins, e.g., alternariol-3-sulfate, fumonisin B_4, and tentoxin, have been listed in Chapter 2: Further Mycotoxins and Microbial Metabolites.

This book provides a basic overview of the most well-known mycotoxins in plants and plant products except cereals and cereal products. Their degree of contamination and the range of concentration of mycotoxins, as well as the mean of the positive samples (as far as possible), are shown in each case. These basic details are intended for readers interested in this topic.

For detailed information, the more interested reader is referred to the points "Co-contamination" and "Further contamination." Co-contaminations of a sample with two or more mycotoxins are listed as described in the corresponding article. If "For detailed information see the article" occurs behind "Co-contamination," the corresponding publication dealt with this co-contamination mostly from a general point of view. All entries contain the point "Further contamination," which gives information about additional mycotoxin contamination of food of plant origin except cereals and cereal products. Besides, if there is written "see also ..." in the end of "Further contamination," a look into Chapter 2: Further Mycotoxins and Microbial Metabolites, e.g., Dagwa, aflatoxin B_1, and literature[659], is appropriate. Here, the remaining mycotoxins/microbial metabolites of a plant/plant product are listed. Finally, the book bears plenty of information if totally utilized.

The book also shows whether a food of plant origin except cereals or cereal products is predisposed to mycotoxin contamination, documented by the number of different mycotoxins and the number of citations for the corresponding food item. This was done as far as possible for each case of mycotoxin contamination. Documenting this, more than 900 publications have been used which are all available at German libraries. The plants and their products are listed in alphabetical order. As far as possible, food and other items have been considered if food of plant origin has been used in their preparation. All items dealing with food of plant origin appear either in the present Volume II or in *Mycotoxins in Plants and Plant Products—Cereals and Cereal Products* (Volume I). Items having an animal product background will occur in the third volume dealing with animal products.

The main part of the book represents Chapter 1: Mycotoxin Contamination of Plants and Plant Products (Except Cereals and Cereal Products). It contains the mycotoxin contamination of single plants as well as plant products in detail. This is followed by Chapter 2: Further Mycotoxins and Microbial Metabolites, Chapter 3: Mycotoxins and Their Plant and Plant Product Spectrum (Except Cereals and Cereal Products), Chapter 4: Plants and Plant Products and Their Mycotoxins (Except Cereals and Cereal Products), and Chapter 5: Mycotoxin Contamination in Conventional and Organic Farming. This chapter lists articles with their index numbers which compare the mycotoxin contamination in conventional and organic production. Finally, the numerical and alphabetical bibliography follows.

The chief publications considered are in English language. Articles in other languages have been chosen only if comprehensive summaries or tables, listing detailed results, were given in English. Articles that unequivocally describe the mycotoxin contamination of plants and plant products (best known already apparent in the title) have been selected. There are exceptions, however. Also, the most cited publications that satisfied the aforementioned requirements have been considered for the book. Determination methods of mycotoxins are quite beyond the scope of this book (see the title of the book). The focus lies on natural mycotoxin contamination of plants and plant products without cereals and cereal products. Publications containing data about mycotoxins due to artificial contamination (infected by direct fungal inoculation) have not been included in this volume. Results of field trials have been taken into account only to a minor extent. The treated publications are mainly dealing with natural mycotoxin contamination of marketed plants and plant products. *Mycotoxins in Plants and Plant Products—Cocoa, Coffee, Fruits and Fruit Products, Medicinal Plants, Nuts, Spices, Wine* will contribute to more information and transparency about plants and plant products consumed by humans in relation to mycotoxin contamination. It is a suitable reference text for mycotoxin contamination in this kind of foodstuff.

This fundamental book is especially written for scientists and researchers who are interested in the mycotoxin contamination of food of plant origin other than cereals and cereal products for human consumption and will be especially suitable for those working in food microbiology, food technology, and the food industry (e.g., food producers, supervisors of food, food traders), as well as ministries, offices, and departments of farming and environmental regulation at national and international levels, bureaus, associations, agricultural bodies, mycologists, mycotoxicologists, toxicologists, biologists, chemists, biochemists, plant pathologists, supervisors in food quality control, lawyers and experts in food law, students of the respective fields, and other interested groups.

Bonn, Germany Martin Weidenbörner

How to Handle the Book

The different mycotoxins in plants and plant products (except cereals and cereal products) are classified by the fungal genera which produce them. However, also other fungal genera may be able to produce a special mycotoxin, but only main producers or relevant species interesting for this book are chosen. For a quick overview, each mycotoxin can furthermore be looked up for its presence in plants and plant products (except cereals and cereal products) at the end of the book (Chapter 3: Mycotoxins and Their Plant and Plant Product Spectrum, Tables 1–6) and vice versa (Chapter 4: Plants and Plant Products and Their Mycotoxins).

In Chapter 1: "Mycotoxin Contamination of Plants and Plant Products—Cocoa, Coffee, Fruits and Fruit Products, Medicinal Plants, Nuts, Spices, Wine (Volume II)," mycotoxins are arranged in the following way:

Alternaria Toxins
Altenuene (ALT), alternariol (AOH), alternariol methyl ether (AME), altertoxin I (ATX-I), tenuazonic acid (TA)

Aspergillus Toxins
Aflatoxin/s: Aflatoxin B_1 (AFB_1), aflatoxin B_2 (AFB_2), aflatoxin G_1 (AFG_1), aflatoxin G_2 (AFG_2), aflatoxin M_1 (AFM_1), aflatoxin M_2 (AFM_2), aflatoxin/s (AF/AFS)
Other ***Aspergillus*** Toxins: Gliotoxin (GT), sterigmatocystin (STG)

Aspergillus and ***Penicillium*** Toxins
3-Nitropropionic acid (3-NPA), citreoviridin (CTV), citrinin (CIT), cyclopiazonic acid (CPA), kojic acid (KA), ochratoxin A (OTA), ochratoxin B (OTB), penicillic acid (PA), viomellein (VIOM)

Fusarium Toxins
Depsipeptides: Beauvericin (BEA), enniatin A (ENA), enniatin A_1 (ENA_1), enniatin B (ENB), enniatin B_1 (ENB_1), enniatin B_2 (ENB_2), enniatin B_3 (ENB_3), enniatin B_4 (ENB_4), enniatin/s (EN/ENS)
Fumonisin/s: Fumonisin B_1 (FB_1), fumonisin B_2 (FB_2), fumonisin B_3 (FB_3), fumonisin/s (FB/FBS)
Type A Trichothecenes: Diacetoxyscirpenol (DAS), HT-2 toxin (HT-2), monoacetoxyscirpenol (MAS), neosolaniol (NEO), scirpentriol (STO), T-2 toxin (T-2), T-2 tetraol (T-2TET), T-2 triol (T-2TRI)
Type B Trichothecenes: Deoxynivalenol (DON), 3-acetyldeoxynivalenol (3-AcDON), 15-acetyldeoxynivalenol (15-AcDON) fusarenon X (FUS-X), nivalenol (NIV)
Other ***Fusarium*** Toxins: Fusaproliferin (FP), fusaric acid (FA), moniliformin (MON), zearalenone (ZEA)

Penicillium Toxins
Mycophenolic acid (MA), roquefortine (ROQ), roquefortine-C (ROQ-C), rubratoxin (RBT)

Each declaration of the mycotoxin contamination of plants and plant products comprises seven main pieces of information:

incidence:	3/7 = 3 positives for aflatoxin contamination in relation to 7 investigated
concentration range:	minimum and maximum residue values or only the maximum residue value
Ø concentration:	Ø mycotoxin contamination for the positive samples (as far as possible)
sample origin:	location, where the sample/s come/s from (as far as possible)
sample year:	in which month/year the sample was collected/investigated (as far as possible)
country:	origin of publication
literature number:	Japan[732]

Usually, the highest mycotoxin value or the lowest and the highest value of mycotoxin contamination is given. In general, mean values of the mycotoxin-positive samples are stated. The presented concentrations (measuring units) occur always in µg/kg or µg/l. If a variant of a trial is not listed, "no contamination" was determined in the corresponding article. In some cases, a variant may be stated although mycotoxin concentration has not been detected. In this case "no contamination" is written.

Each entry is marked with an index number, which is located behind the name of the involved country/countries where the publication has been carried out. This index number stands for the article where the presented results can be checked. It occurs again in the numerical bibliography. The index number refers to the title of the corresponding article. In the alphabetical bibliography, the literature is additionally arranged according to the author´s first name of a publication. Behind each alphabetical literature, there is a number in brackets giving information about the position of the literature in the numerical bibliography, e.g., (620) in the alphabetical bibliography denotes the 620th position in the numerical bibliography.

For easier finding of further mycotoxin-contaminated plants or plant products in an entry, the reader has to keep in mind the following scheme:

"Further contamination": **Yams, FB_1, literature**[774]

- title of item: **Yams**
- mycotoxin: **FB_1**
- index number: **literature**[774]

With these facts, further contaminated/co-contaminated plants or plant products presented in the original publication can be found in this book. Moreover, all mycotoxins, contaminating the same plant or plant product (listed in "Further contamination"), are quoted in this book. Additionally, the corresponding data in "Further contamination" are marked with an index number. This number can be related to the article, which is presented in the numerical bibliography.

Acknowledgments

My family, especially my wife, is thanked for the patience, valuable advice, and support given to me in writing this book. I also wish to thank Mrs. Beate Frohn and Mrs. Ursula Kleinheyer-Thomas as well as the whole staff from the Bran Library for Medicine, Science and Agricultural Science, Bonn University, for very good support in searching and finding the corresponding publications. I also appreciate the support of my daughter Anna as well as Dr. Ursula Monnerjahn-Karbach for the valuable advice.

Contents

Abbreviations

15-AcDON	15-Acetyldeoxynivalenol
3-AcDON	3-Acetyldeoxynivalenol
3-NPA	3-Nitropropionic acid
AC	Agroclavine
AFB_1	Aflatoxin B_1
AFB_2	Aflatoxin B_2
AFG_1	Aflatoxin G_1
AFG_2	Aflatoxin G_2
AFL-I	Aflatoxicol I
AF/S	Aflatoxin/s
ALT	Altenuene
AME	Alternariol methyl ether
AOH	Alternariol
ATX-I	Altertoxin I
AVE-Y	Avenacein Y
BEA	Beauvericin
BEN	Balkan endemic nephropathy
BGY fluorescence	Bright greenish-yellow fluorescence
CAOBISCO	Association of the Chocolate, Biscuit and Confectionery Industries of the EU
CEN	Comité Européen de Normalisation
CHAE	Chaetoglobosin A
CIT	Citrinin
CL-ELISA	Chemiluminescent enzyme-linked immunosorbent assay
COL-ELISA	Colorimetric enzyme-linked immunosorbent assay
CNV	Chanoclavine
CPA	Cyclopiazonic acid
DAS	Diacetoxyscirpenol
DOCa	Denominación de Origen Calificada
DON	Deoxynivalenol
DPDON	Deepoxydeoxynivalenol
EN	Endemic nephropathy
ENA	Enniatin A
ENA_1	Enniatin A_1
ENB	Enniatin B
ENB_1	Enniatin B_1
ENB_2	Enniatin B_2
ENB_3	Enniatin B_3
EN/S	Enniatin/s
EMOD	Emodin
ESM	Electronic sorting machine
EU	European Union
EVENA	Estación de Viticulture y Enología de Navarra
FA	Fusaric acid
FB_1	Fumonisin B_1

Iso-FB_1	Iso-fumonisin B_1
FB_2	Fumonisin B_2
Iso-$FB_{2,3}$	Iso-fumonisin $B_{2,3}$
FB_3	Fumonisin B_3
3-epi-FB_3	3-Epi-fumonisin B_3
FB_4	Fumonisin B_4
3-epi-FB_4	3-Epi-fumonisin B_4
FB_5	Fumonisin B_5
Iso-FB_5	Iso-fumonisin B_5
FB/S	Fumonisin/s
FDA	US Food and Drug Administration
FP	Fusaproliferin
FUS-X	Fusarenon-X
GLI	Gliotoxin
GRIS	Griseofulvin
HACCP	Hazard analysis and critical control points
HCC	Human hepatocellular carcinoma
HPLC	High-performance liquid chromatography
HT-2	HT-2 toxin
KA	Kojic acid
KSU	Kahramanmaras Sutcu Iman University
LOD	Limit of detection
LOQ	Limit of quantification
MA	Mycophenolic acid
MAS	Monoacetoxyscirpenol
MeOTA	Methylochratoxin A
MeOTB	Methylochratoxin B
MON	Moniliformin
NEO	Neosolaniol
NIV	Nivalenol
NSW	New South Wales
NWFP	North-West Frontier Province
OTA	Ochratoxin A
OTB	Ochratoxin B
PA	Penicillic acid
PAT	Patulin
PEA	Penitrem A
pr	Present
QbA	Quality wine produced in specified regions
RBT	Rubratoxin
ROQ-C	Roquefortine C
ROR	Roridin A
sa	Sample/s
SCIRP	Scirpentriol
STG	Sterigmatocystin
STO	Scirpentriol
SUL	Sulochrin
T-2	T-2 toxin
T-2TET	T-2 tetraol
T-2TRI	T-2 triol
TA	Tenuazonic acid
TCM	Traditional Chinese medicine
TEN	Tentoxin
TLC	Thin-layer chromatography
TRFIA	Time-resolved fluorescence immunoassay

VER	Verruculogen
VIOM	Viomellein
WHO	World Health Organization
ZEA	Zearalenone
ZEAA	Zearalanone
ZEAOL	Zearalenol
α-ZEL	α-Zearalenol
ß-ZEL	ß-Zearalenol

Mycotoxin Contamination of Plants and Plant Products (Except Cereals and Cereal Products)

Alcoholic drink see Drinks (alcoholic drink)

Almond paste see Marzipan (almond paste)

Almonds see Nuts (almond)

Ammi see Spices (ammi)

Amontillado see Wine (white)

Anise see Spices (anise)

Aperitif see Wine, miscellaneous

Apple see Fruits (apple)

Apple cider may contain the following mycotoxins:

Aspergillus and ***Penicillium*** Toxins

Patulin

incidence: 1/2, conc.: <LOQ, sample origin: local supermarkets or organic food shops, Portugal, sample year: unknown, country: Portugal[27]

- Co-contamination: not reported
- Further contamination of plants and plant products other than cereals and their products reported in the present book/article[27]: Baby food, Fruits (apple), Fruits (quince), Jam (quince jam), Juices (apple juice), PAT, literature[27]

incidence: 2/8*, conc. range: 5–10 µg/l, ∅ conc.: 7.5 µg/l, sample origin: retail outlets in Cape Town (city), South Africa, sample year: 1996–1998, country: South Africa[37], *apple cider, alcoholic

- Co-contamination: not reported
- Further contamination of plants and plant products other than cereals and their products reported in the present book/article[37]: Juices, Juices (apple juice), Products (apple product), PAT, literature[37]

incidence: 2/5, conc. range: tr (1sa.), LOQ-2.8 µg/l (1 sa.), sample origin: supermarkets and handicraft producers, Belgium, sample year: 2001, country: Belgium[38]

incidence: 1/2, conc.: 6.1 µg/l, sample origin: imports from France, Germany, and Switzerland purchased in supermarkets, Belgium, sample year: 2001, country: Belgium[38]

- Co-contamination: not reported
- Further contamination of plants and plant products other than cereals and their products reported in the present book/article[38]: Juices (apple juice), PAT, literature[38]

incidence: 5/5*, conc. range: 244–3,993 µg/l, ∅ conc.: 1,902 µg/l, sample origin: roadside stands or directly from producers, middle and north Georgia (state), USA, sample year: unknown, country: USA[39], *apple cider, pasteurized

- Co-contamination: not reported
- Further contamination of plants and plant products other than cereals and their products reported in the present book/article[39]: not reported

incidence: 9*/100, conc. range: ≤45,000 µg/l, sample origin: cider mills, USA, sample year: 1971, country: USA[40], *5 organic apple ciders contained ≤45,000 µg/l and 4 conventional apple ciders contained ≤25,000 µg/l

- Co-contamination: not reported
- Further contamination of plants and plant products other than cereals and their products reported in the present book/article[40]: not reported

incidence: 55/324*, conc. range: 4.8–329.8 µg/l, sample origin: cider mills and retail establishments, Michigan (state), USA, sample year: fall seasons in 2002–2004, country: USA[41], *food safety control measures: none

incidence: 27/95*, conc. range: 4.6–467.4 µg/l, sample origin: cider mills and retail establishments, Michigan (state), USA, sample year: fall seasons in 2002–2004, country: USA[41], *food safety control measures: thermal pasteurization

incidence: 10/74*, conc. range: 5.5–59.1 µg/l, sample origin: cider mills and retail establishments, Michigan (state), USA, sample year: fall seasons in 2002–2004, country: USA[41], *food safety control measures: UV light radiation

- Co-contamination: not reported
- Further contamination of plants and plant products other than cereals and their products reported in the present book/article[41]: Juices (apple juice), PAT, literature[41]

For detailed information see the article.

incidence: 2/2, conc. range: 12.3–48.3 µg/l, ∅ conc.: 30.3 µg/l, sample origin: Czech retail market, Czech Republic, sample year: unknown, country: Czech Republic[911]

- Co-contamination: not reported
- Further contamination of plants and plant products other than cereals and their products reported in the present book/article[911]: Baby food, Fruit salad, Fruits (apple), Fruits (pear), Juices (apple juice), Juices (pear juice), Pulp (apple pulp), PAT, literature[911]

Apple concentrate see Juices (apple concentrate)

Apple flavor products see Products (apple flavor product)

Apple juice see Juices (apple juice)

Apple juice concentrate see Juices (apple concentrate)

Apple leather may contain the following mycotoxins:

Aspergillus and ***Penicillium*** Toxins

Patulin

incidence: 35/35, conc. range: <10–2,559 µg/kg, ∅ conc.: 620 µg/kg, sample origin: retail markets, Iran, sample year: 2010/2011, country: Iran[795]

- Co-contamination: not reported
- Further contamination of plants and plant products other than cereals and their products reported in the present book/article[795]: not reported

M. Weidenbörner, *Mycotoxins in Plants and Plant Products*, https://doi.org/10.1007/978-3-319-92850-0_1

Apple marmalade see **Marmalade (apple marmalade)**

Apple products see **Products (apple product)**

Apple puree see **Puree**

Apple rings may contain the following mycotoxins:

Aspergillus and ***Penicillium*** Toxins

Patulin
incidence: 2/8*, conc. range: ~10 μg/kg, sample origin: industrial sa., South Africa, sample year: unknown, country: South Africa[51], *apple rings, dried
- Co-contamination: not reported
- Further contamination of plants and plant products other than cereals and their products reported in the present book/article[51]: not reported

Apricot kernels see **Kernels (apricot)**

Apricots see **Fruits, dried**

Apulian Denominazione di origine controllata (DOC) wines see **Wine (red)**

Areca nuts see **Betel nuts** and **Medicinal plants**

Aromatic herbs see **Herbs** and **Medicinal plants**

Asparagus see **Vegetables (asparagus)**

Asure may contain the following mycotoxins:

Aspergillus Toxins

Aflatoxin/s
incidence: 2/4, conc. range: ≤1.773 μg/kg*, sample origin: groceries and supermarkets, Turkey, sample year: 2003, country: Turkey[119], *AFS total
- Co-contamination: not reported
- Further contamination of plants and plant products other than cereals and their products reported in the present book/article[119]: Beverages, Cocoa, Cream, Ice cream, Jello, Keskul, Muhallebi, Pudding, Salep, Sauces (chocolate sauce), Soups, AF/S, literature[119]

Asure is a Turkish dessert. Ingredients were sugar, potato starch, dehydrated beans, chickpeas, dried grapes, and currants.

"Auslese" see **Wine**

Baby food may contain the following mycotoxins:

Aspergillus Toxins

Aflatoxin B_1
incidence: 1/3* **, conc.: 0.003 μg/kg, sample origin: organic product retail outlets in Lisbon (capital), Portugal, sample year: May/June 2007, country: Portugal/Netherlands[88], *organic, **sa. consisted of soy proteins or milk and vegetable oils
- Co-contamination: 1 sa. co-contaminated with AFB_1 and OTA
- Further contamination of plants and plant products other than cereals and their products reported in the present book/article[88]: Baby food, AFM_1 and OTA, literature[88]; Biscuits, OTA, literature[88]; Flour, AFB_1 and OTA, literature[88]

incidence: 0/4* **, conc.: no contamination, sample origin: supermarkets in Lisbon (capital), Portugal, sample year: May/June 2007, country: Portugal/Netherlands[88], *conventional, **sa. consisted of milk and one or more ingredient(s) (vegetable or vegetal fish oils in each sa.)
- Co-contamination: not reported
- Further contamination of plants and plant products other than cereals and their products reported in the present book/article[88]: Baby food, AFM_1 and OTA, literature[88]; Biscuits, OTA, literature[88]; Flour, AFB_1 and OTA, literature[88]

For detailed information see the article.

incidence: 2/7*, conc. range: 181.6–4,806 μg/l, Ø conc.: 2,494 μg/l, sample origin: locally-made and from markets in Ibadan (city), Nigeria, sample year: unknown, country: Nigeria[105], *weaning food
- Co-contamination: 1 sa. co-contaminated with AFB_1 and AFB_2; 1 sa. contaminated solely with AFB_1
- Further contamination of plants and plant products other than cereals and their products reported in the present book/article[105]: Baby food, AFB_2, AFG_1, AFG_2, and AFM_1, literature[105]

For detailed information see the article (authors emphasized that there was an extremely high AFS-conc. in the pos. sa.).

Aflatoxin B_2
incidence: 4/7*, conc. range: 103–8,290 μg/l, Ø conc.: 2,311 μg/l, sample origin: locally made and from markets in Ibadan (city), Nigeria, sample year: unknown, country: Nigeria[105], *weaning food
- Co-contamination: 1 sa. co-contaminated with AFB_2, AFG_1, and AFM_1; 2 sa. co-contaminated with AFB_2, AFG_2, and AFM_1; 1 sa. co-contaminated with AFB_1 and AFB_2
- Further contamination of plants and plant products other than cereals and their products reported in the present book/article[105]: Baby food, AFB_1, AFG_1, AFG_2, and AFM_1, literature[105]

For detailed information see the article (authors emphasized that there was an extremely high AFS-conc. in the pos. sa.).

Aflatoxin G_1
incidence: 1/7*, conc.: 144 μg/l, sample origin: locally made and from markets in Ibadan (city), Nigeria, sample year: unknown, country: Nigeria[105], *weaning food
- Co-contamination: 1 sa. co-contaminated with AFB_2, AFG_1, and AFM_1
- Further contamination of plants and plant products other than cereals and their products reported in the present book/article[105]: Baby food, AFB_1, AFB_2, AFG_2, and AFM_1, literature[105]

For detailed information see the article (authors emphasized that there was an extremely high AFS-conc. in the pos. sa.).

Aflatoxin G_2
incidence: 2/7*, conc. range: 1,169–1,699 μg/l, Ø conc.: 1,434 μg/l, sample origin: locally made and from markets in Ibadan (city), Nigeria, sample year: unknown, country: Nigeria[105], *weaning food
- Co-contamination: 2 sa. co-contaminated with AFB_2, AFG_2, and AFM_1
- Further contamination of plants and plant products other than cereals and their products reported in the present book/article[105]: Baby food, AFB_1, AFB_2, AFG_1, and AFM_1, literature[105]

For detailed information see the article (authors emphasized that there was an extremely high AFS-conc. in the pos. sa.).

Aflatoxin M_1

incidence: 2/3* **, conc. range: 0.007–0.011 µg/kg, ∅ conc.: 0.009 µg/kg, sample origin: organic product retail outlets in Lisbon (capital), Portugal, sample year: May/June 2007, country: Portugal/Netherlands[88], *organic, **sa. consisted of soy proteins or milk and vegetable oils

- Co-contamination: 1 sa. co-contaminated with AFM_1 and OTA; 1 sa. contaminated solely with AFM_1
- Further contamination of plants and plant products other than cereals and their products reported in the present book/article[88]: Baby food, Flour, AFB_1 and OTA, literature[88]; Biscuits, OTA, literature[88]

incidence: 4/4* **, conc. range: 0.005–0.041 µg/kg, ∅ conc.: 0.017 µg/kg, sample origin: supermarkets in Lisbon (capital), Portugal, sample year: May/June 2007, country: Portugal/Netherlands[88], *conventional, **sa. consisted of milk and one or more ingredient(s) (vegetable or vegetal fish oils in each sa.)

- Co-contamination: not reported
- Further contamination of plants and plant products other than cereals and their products reported in the present book/article[88]: Baby food, Flour, AFB_1 and OTA, literature[88]; Biscuits, OTA, literature[88]

For detailed information see the article.

incidence: 1/6*, conc.: 0.016 µg/kg or µg/l, sample origin: supermarkets and drug stores in Corum (city), Turkey, sample year: January/February 2011, country: Turkey[89], *infant formula (for infants aged up to 4–6 months)

- Co-contamination: not reported
- Further contamination of plants and plant products other than cereals and their products reported in the present book/article[89]: Baby food, OTA, literature[89]

incidence: 2/36*, conc. range: 0.016–0.020 µg/kg or µg/l, ∅ conc.: 0.018 µg/kg or µg/l, sample origin: supermarkets and drug stores in Corum (city), Turkey, sample year: January/February 2011, country: Turkey[89], *follow-on formula (for infants aged from 6 months)

- Co-contamination: not reported
- Further contamination of plants and plant products other than cereals and their products reported in the present book/article[89]: Baby food, OTA, literature[89]

incidence: 2/20*, conc. range: 0.017–0.022 µg/kg or µg/l, ∅ conc.: 0.020 µg/kg or µg/l, sample origin: supermarkets and drug stores in Corum (city), Turkey, sample year: January/February 2011, country: Turkey[89], *toddler formula (for young children aged 12–36 months)

- Co-contamination: 1 sa. co-contaminated with AFM_1 and OTA; 1 sa. contaminated solely with AFM_1
- Further contamination of plants and plant products other than cereals and their products reported in the present book/article[89]: Baby food, OTA, literature[89]

incidence: 3/7*, conc. range: 4.6–530 µg/l, ∅ conc.: 221 µg/l, sample origin: locally made and from markets in Ibadan (city), Nigeria, sample year: unknown, country: Nigeria[105], *weaning food

- Co-contamination: 1 sa. co-contaminated with AFB_2, AFG_1, and AFM_1; 2 sa. co-contaminated with AFB_2, AFG_2, and AFM_1
- Further contamination of plants and plant products other than cereals and their products reported in the present book/article[105]: Baby food, AFB_1, AFB_2, AFG_1, and AFG_2, literature[105]

For detailed information see the article (authors emphasized that there was an extremely high AFS-conc. in the pos. sa.).

Aflatoxin/s

incidence: 1/17*, conc.: 19.0 µg/kg**, sample origin: retailers, factories, and stores, Libya, sample year: unknown, country: UK/Libya[663], *cereal with nuts, **AFB_1, AFB_2, AFG_1, and AFG_2

incidence: 1/24*, conc.: 70.1 µg/kg**, sample origin: retailers, factories, and stores, Libya, sample year: unknown, country: UK/Libya[663], *cereal with milk, vegetable, and fruits, **AFB_1, AFB_2, AFG_1, and AFG_2

- Co-contamination: not reported
- Further contamination of plants and plant products other than cereals and their products reported in the present book/article[663]: not reported

For detailed information see the article.

Aspergillus and ***Penicillium*** Toxins

Ochratoxin A

incidence: 2/3* **, conc. range: 0.011–0.136 µg/kg, ∅ conc.: 0.074 µg/kg, sample origin: organic product retail outlets in Lisbon (capital), Portugal, sample year: May/June 2007, country: Portugal/Netherlands[88], *organic, **sa. consisted of soy proteins or milk and vegetable oils

- Co-contamination: 1 sa. co-contaminated with AFM_1 and OTA; 1 sa. contaminated solely with OTA
- Further contamination of plants and plant products other than cereals and their products reported in the present book/article[88]: Baby food, AFB_1 and AFM_1, literature[88]; Biscuits, OTA, literature[88]; Flour, AFB_1 and OTA, literature[88]

incidence: 1/4* **, conc.: 0.135 µg/kg, sample origin: supermarkets in Lisbon (capital), Portugal, sample year: May/June 2007, country: Portugal/Netherlands[88], *conventional, **sa. consisted of milk and one or more ingredient(s) (vegetable or vegetal fish oils in each sa.)

- Co-contamination: not reported
- Further contamination of plants and plant products other than cereals and their products reported in the present book/article[88]: Baby food, AFB_1 and AFM_1, literature[88]; Biscuits, OTA, literature[88]; Flour, AFB_1 and OTA, literature[88]

For detailed information see the article.

incidence: 0/6*, conc.: no contamination, sample origin: supermarkets and drug stores in Corum (city), Turkey, sample year: January/February 2011, country: Turkey[89], *infant formula (for infants aged up to 4–6 months)

- Co-contamination: not reported
- Further contamination of plants and plant products other than cereals and their products reported in the present book/article[89]: Baby food, AFM_1, literature[89]

incidence: 2/36*, conc. range: 0.017–0.029 µg/kg or µg/l, ∅ conc.: 0.023 µg/kg or µg/l, sample origin: supermarkets and drug stores in Corum (city), Turkey, sample year: January/February 2011, country: Turkey[89], *follow-on formula (for infants aged from 6 months)

- Co-contamination: not reported
- Further contamination of plants and plant products other than cereals and their products reported in the present book/article[89]: Baby food, AFM_1, literature[89]

incidence: 10/20*, conc. range: 0.024–0.184 µg/kg or µg/l, ∅ conc.: 0.119 µg/kg or µg/l, sample origin: supermarkets and drug stores in Corum (city), Turkey, sample year: January/February 2011, country: Turkey[89], *toddler formula (for young children aged 12–36 months)

- Co-contamination: 1 sa. co-contaminated with AFM_1 and OTA; 9 sa. contaminated solely with OTA
- Further contamination of plants and plant products other than cereals and their products reported in the present book/article[89]: Baby food, AFM_1, literature[89]

incidence: 1/1*, conc.: 0.01 µg/kg, sample origin: supermarkets and fast-food chains in Quebec (city), Canada, sample year: September/October 2008, country: Canada[90], *formulas, soy-based

incidence: 1/1*, conc.: 0.02 µg/kg, sample origin: supermarkets and fast-food chains in Calgary (city), Canada, sample year: September/October 2009, country: Canada[90], *formulas, soy-based

- Co-contamination: not reported
- Further contamination of plants and plant products other than cereals and their products reported in the present book/article[90]: Beans, Beverages, Butter (peanut butter), Candies, Chocolate, Coffee, Extracts, Herbs, Ice cream, Nuts, Pies, Pizza, Products, Pudding, Raisins, Seeds, Wine, OTA, literature[90]

incidence: 63/97, conc. range: LOD/LOQ-0.9 µg/kg (60 sa.), ≤2.13 µg/kg (3 sa.), sample origin: Germany, sample year: 1995–1998, country: EU[698]

incidence: 6/6, conc. range: <5 µg/kg (6 sa.), sample origin: Ireland, sample year: 1999, country: EU[698]

- Co-contamination: not reported
- Further contamination of plants and plant products other than cereals and their products reported in the present book/article[698]: Beans (soybean), Chocolate, Cocoa, Coffee, Coffee beans, Currants, Fruits, dried, Herbs, Jam, Juices, Juices (grape juice), Legumes, Nuts, Oil seeds, Oils (olive oil), Peas, Peas (chickpea), Products (plant product), Raisins, Spices, Spices (cayenne pepper, chilli), Spices (nutmeg), Spices (paprika), Spices (pepper), Sultanas, Tea, Vinegar, Wine, Wine (red), Wine (rosé), Wine (white), Wine, miscellaneous, OTA, literature[698]

Patulin

incidence: 1*/2, conc.: 9.1 µg/kg**, sample origin: local supermarkets or organic food shops in Porto (city), Portugal, sample year: unknown, country: Portugal[27], *75 and 100%**apple-based (100%) baby food

- Co-contamination: not reported
- Further contamination of plants and plant products other than cereals and their products reported in the present book/article[27]: Apple cider, Fruits (apple), Fruits (quince), Jam (quince jam), Juices (apple juice), PAT, literature[27]

incidence: 3/12*, conc. range: 6.25–10 µg/kg (3 sa., maximum: 9.2 µg/kg), ∅ conc.: 7.5 µg/kg, sample origin: hypermarkets and supermarkets in 12 main cities, Catalonia (autonomous community), Spain, sample year: June/July 2008, country: Spain[42], *apple juice

incidence: 15/36*, conc. range: 6.25–10 µg/kg (15 sa., maximum: 9.6 µg/kg), ∅ conc.: 7.4 µg/kg, sample origin: hypermarkets and supermarkets in 12 main cities, Catalonia (autonomous community), Spain, sample year: June/July 2008, country: Spain[42], *apple compote

incidence: 24/76*, conc. range: 6.25–10 µg/kg (24 sa., maximum: 8.6 µg/kg), ∅ conc.: 6.9 µg/kg, sample origin: hypermarkets and supermarkets in 12 main cities, Catalonia (autonomous community), Spain, sample year: June/July 2008, country: Spain[42], *multifruit compote

- Co-contamination: not reported
- Further contamination of plants and plant products other than cereals and their products reported in the present book/article[42]: Juices (apple juice), Products (apple products), PAT, literature[42]

incidence: 1/11*, conc.: 10 µg/kg**, sample origin: unknown, sample year: November 1993, country: UK[107], *baby apple and orange juices/drinks, **apple and blackcurrant drink

- Co-contamination: not reported
- Further contamination of plants and plant products other than cereals and their products reported in the present book/article[107]: not reported

incidence: 3/3*, conc. range: 4–6.9 µg/kg, ∅ conc.: 5.10 µg/kg, sample origin: Greek market, Greece, sample year: October 2004–June 2006, country: Greece[108], *baby fruit juice containing apple pulp

- Co-contamination: not reported
- Further contamination of plants and plant products other than cereals and their products reported in the present book/article[108]: Drinks (apple drink), Juices, Juices (apple concentrate), Juices (apple juice), Juices (apricot concentrate), Juices (orange juice), Juices (pineapple concentrate), Juices (rosehip juice), PAT, literature[108]

incidence: 78/120* **, conc. range: 3–6 µg/kg (37 sa.), 6–8 µg/kg (15 sa.), 8–9 µg/kg (26 sa.), ∅ conc.: 6.28 µg/kg, sample origin: Italian supermarkets, Italy, sample year: 2008/2009, country: Italy[109], *homogenized fruit-based (60 apple tastes and 60 mixed tastes sa.) baby food products, **80 conventional and 40 organic sa.

- Co-contamination: not reported
- Further contamination of plants and plant products other than cereals and their products reported in the present book/article[109]: not reported

incidence: 15/30*, conc. range: ≤67.3 µg/kg, sample origin: supermarkets and stores in Changchun (city), China, sample year: 2009, country: China[890], *baby food made from apple products

- Co-contamination: not reported
- Further contamination of plants and plant products other than cereals and their products reported in the present book/article[890]: Juices, Juices (apple concentrate), Juices (apple juice), PAT, literature[890]

incidence: 7/25, conc. range: ≤165 µg/l sample origin: major supermarkets and stores, Tunisia, sample year: 2011?, country: Tunisia[895], *baby food made from apple products

- Co-contamination: not reported
- Further contamination of plants and plant products other than cereals and their products reported in the present book/article[895]: Juices, Juices (apple juice), PAT, literature[895]

incidence: 2/10*, conc. range: 2.1–5.0 µg/kg, ∅ conc.: 3.6 µg/kg, sample origin: Czech retail market, Czech Republic, sample year: unknown, country: Czech Republic[911], *apple-based baby food

- Co-contamination: not reported
- Further contamination of plants and plant products other than cereals and their products reported in the present book/article[911]: Apple cider, Fruit salad, Fruits (apple), Fruits (pear), Juices (apple juice), Juices (pear juice), Pulp (apple pulp), PAT, literature[911]

see also **Biscuits**, **Flour** and **Butter (peanut butter)**

Barbera see Wine

Bars may contain the following mycotoxins:

Aspergillus **Toxins**

Aflatoxin B$_1$

incidence: 4/41*, conc. range: 0.1–4.8 µg/kg, ∅ conc.: 1.4 µg/kg, sample origin: available in New Zealand, sample year: 2008–2010, country: New Zealand[277], *snack bars

- Co-contamination: not reported
- Further contamination of plants and plant products other than cereals and their products reported in the present book/article[277]: Bars, AF/S, DON, NIV, and OTA, literature[277]; Butter (peanut butter), Confectionery, Food, Fruits, dried, Nuts, Nuts (peanut), Nuts (pistachio nut), Sauces (peanut sauce), AFB$_1$ and AF/S, literature[277]; Coffee, Dates, Spices (nutmeg), Spices (pepper), Vine fruits, dried, OTA, literature[277]; Figs, Spices (cayenne pepper, chilli), Spices (curry), Spices (ginger), Spices (paprika), AFB$_1$, AF/S, and OTA, literature[277]; Nuts (Brazil nut, paranut), AFB$_1$, literature[277]

Aflatoxin/s

incidence: 4/41*, conc. range: 0.2–7.7 µg/kg**, ∅ conc.: 2.1 µg/kg**, sample origin: available in New Zealand, sample year: 2008–2010, country: New Zealand[277], *snack bars, **AFS total

- Co-contamination: not reported
- Further contamination of plants and plant products other than cereals and their products reported in the present book/article[277]: Bars, AFB$_1$, DON, NIV, and OTA, literature[277]; Butter (peanut butter), Confectionery, Food, Fruits, dried, Nuts, Nuts (peanut), Nuts (pistachio nut), Sauces (peanut sauce), AFB$_1$ and AF/S, literature[277]; Coffee, Dates, Spices (nutmeg), Spices (pepper), Vine fruits, dried, OTA, literature[277]; Figs, Spices (cayenne pepper, chilli), Spices (curry), Spices (ginger), Spices (paprika), AFB$_1$, AF/S, and OTA, literature[277]; Nuts (Brazil nut, paranut), AFB$_1$, literature[277]

Aspergillus **and** ***Penicillium*** **Toxins**

Ochratoxin A

incidence: 40/47*, conc. range: ≤3.6 µg/kg, sample origin: regions around Detmold, Giessen, Hamburg, Jena, Karlsruhe, Kiel, Kulmbach, Trier (cities), Germany, sample year: unknown, country: Germany[262], *nut bars

- Co-contamination: not reported
- Further contamination of plants and plant products other than cereals and their products reported in the present book/article[262]: Chocolate, Drinks (chocolate drink), Figs, Fruits, dried, Nut crème, Nuts, Nuts (hazelnut), Nuts (peanut), Powder (cocoa powder), Pudding, Raisins, Seeds (linseed), Seeds (sesame), Seeds (sunflower), OTA, literature[262]

incidence: 2/8*, conc. range: 0.21–0.30 µg/kg, ∅ conc.: 0.255 µg/kg, sample origin: available in New Zealand, sample year: 2011, country: New Zealand[277], *snack bars

- Co-contamination: not reported
- Further contamination of plants and plant products other than cereals and their products reported in the present book/article[277]: Bars, AFB$_1$, AF/S, DON, and NIV, literature[277]; Butter (peanut butter), Confectionery, Food, Fruits, dried, Nuts, Nuts (peanut), Nuts (pistachio nut), Sauces (peanut sauce), AFB$_1$ and AF/S, literature[277]; Coffee, Dates, Spices (nutmeg), Spices (pepper), Vine fruits, dried, OTA, literature[277]; Figs, Spices (cayenne pepper, chilli), Spices (curry), Spices (ginger), Spices (paprika), AFB$_1$, AF/S, and OTA, literature[277]; Nuts (Brazil nut, paranut), AFB$_1$, literature[277]

Fusarium **Toxins**

Type B Trichothecenes

Deoxynivalenol

incidence: 1/8*, conc.: 2.5 µg/kg, sample origin: available in New Zealand, sample year: 2009, country: New Zealand[277], *snack bars

- Co-contamination: not reported
- Further contamination of plants and plant products other than cereals and their products reported in the present book/article[277]: Bars, AFB$_1$, AF/S, NIV, and OTA, literature[277]; Butter (peanut butter), Confectionery, Food, Fruits, dried, Nuts, Nuts (peanut), Nuts (pistachio nut), Sauces (peanut sauce), AFB$_1$ and AF/S, literature[277]; Coffee, Dates, Spices (nutmeg), Spices (pepper), Vine fruits, dried, OTA, literature[277]; Figs, Spices (cayenne pepper, chilli), Spices (curry), Spices (ginger), Spices (paprika), AFB$_1$, AF/S, and OTA, literature[277]; Nuts (Brazil nut, paranut), AFB$_1$, literature[277]

Nivalenol

incidence: 1/8*, conc.: 2.5 µg/kg, sample origin: available in New Zealand, sample year: 2009, country: New Zealand[277], *snack bars

- Co-contamination: not reported
- Further contamination of plants and plant products other than cereals and their products reported in the present book/article[277]: Bars, AFB$_1$, AF/S, DON, and OTA, literature[277]; Butter (peanut butter), Confectionery, Food, Fruits, dried, Nuts, Nuts (peanut), Nuts (pistachio nut), Sauces (peanut sauce), AFB$_1$ and AF/S, literature[277]; Coffee, Dates, Spices (nutmeg), Spices (pepper), Vine fruits, dried, OTA, literature[277]; Figs, Spices (cayenne pepper, chilli), Spices (curry), Spices (ginger), Spices (paprika), AFB$_1$, AF/S, and OTA, literature[277]; Nuts (Brazil nut, paranut), AFB$_1$, literature[277]

Bean curd see Curd (bean curd)

Beans may contain the following mycotoxins:

Aspergillus **Toxins**

Aflatoxin B$_1$

incidence: 4?/381*, conc. range: 1.4–254 µg/kg, sample origin: unknown, sample year: unknown, country: Japan[97], *beans for bean jam

- Co-contamination: not reported
- Further contamination of plants and plant products other than cereals and their products reported in the present book/article[97]: Beans, AFB$_2$, literature[97]; Jam (bean jam), Spices, AFB$_1$ and AFB$_2$, literature[97]; Butter (peanut butter), Oils (peanut oil), Spices (pepper), AFB$_1$, AFB$_2$, and AFG$_1$, literature[97]; Spices (nutmeg), AFB$_1$, AFB$_2$, AFG$_1$, and AFG$_2$, literature[97]

incidence: 4?/293*, conc. range: 4.7–52.3 µg/kg, ∅ conc.: 17.9 µg/kg, sample origin: imports from Myanmar, commercially available market basket sa. of foods and foodstuffs in Tokyo (capital), Japan, sample year: 1986–1990, country: Japan[99], *beans for bean jam

- Co-contamination: not reported
- Further contamination of plants and plant products other than cereals and their products reported in the present book/

article[99]: Beans, AFB_2, literature[99]; Seeds (sesame), Sugar, AFB_1 and AFB_2, literature[99]; Food, Jam (bean jam), Paste (peanut paste), AFB_1, literature[99]; Nuts (Brazil nut, paranut), Nuts (peanut), Nuts (pistachio nut), Spices (nutmeg), AFB_1, AFB_2, AFG_1, and AFG_2, literature[99]; Spices, Spices (cayenne pepper, chilli), Spices (pepper), AFB_1, AFB_2, and AFG_1, literature[99]; Spices (paprika), AFB_1 and AFG_1, literature[99]

incidence: 2/145, conc. range: 1–10 µg/kg, sample origin: unknown, sample year: 1990–1996, country: Cuba[126]
- Co-contamination: not reported
- Further contamination of plants and plant products other than cereals and their products reported in the present book/ article[126]: Beans (soybean), Cocoa beans, Nuts (peanut), Peas, AFB_1, literature[126]

incidence: 1/231*, conc.: 1.4 µg/kg, sample origin: commercial sources in Tokyo (capital), Japan, sample year: November 1977–June 1982, country: Japan[129], *small red beans as raw material for bean jam
- Co-contamination: 1 sa. co-contaminated with AFB_1 and AFB_2
- Further contamination of plants and plant products other than cereals and their products reported in the present book/ article[129]: Beans, AFB_2, literature[129]; Beans (butter beans), Beans (kidney bean), Beans (saltani saltapya bean), AFB_1 and AFB_2, literature[129]; Nuts (pistachio nut), AFB_1, AFB_2, and AFM_1, literature[129]; see also Nuts (pistachio nut), chapter: Further Mycotoxins and Microbial Metabolites

incidence: 5/30*, conc. range: 0.21–0.29 µg/kg, Ø conc.: 0.24 µg/kg, sample origin: Mashhad (city), Khorasan (historical region), Iran, sample year: unknown, country: Iran[131]
- Co-contamination: not reported
- Further contamination of plants and plant products other than cereals and their products reported in the present book/ article[131]: Beans, OTA, literature[131]

Aflatoxin B_2

incidence: 4?/381*, conc. range: 1.2–8.5 µg/kg, sample origin: unknown, sample year: unknown, country: Japan[97], *beans for bean jam
- Co-contamination: not reported
- Further contamination of plants and plant products other than cereals and their products reported in the present book/article[97]: Beans, AFB_1, literature[97]; Jam (bean jam), Spices (AFB_1 and AFB_2), literature[97]; Butter (peanut butter), Oils (peanut oil), Spices (pepper), AFB_1, AFB_2, and AFG_1, literature[97]; Spices (nutmeg), AFB_1, AFB_2, AFG_1, and AFG_2, literature[97]

incidence: 4?/293*, conc. range: 0.4–1.5 µg/kg, Ø conc.: 0.7 µg/kg, sample origin: imports from Myanmar, commercially available market basket sa. of foods and foodstuffs in Tokyo (capital), Japan, sample year: 1986–1990, country: Japan[99], *beans for bean jam
- Co-contamination: not reported
- Further contamination of plants and plant products other than cereals and their products reported in the present book/ article[99]: Beans, AFB_1, literature[99]; Seeds (sesame), Sugar, AFB_1 and AFB_2, literature[99]; Food, Jam (bean jam), Paste (peanut paste), AFB_1, literature[99]; Nuts (Brazil nut, paranut), Nuts (peanut), Nuts (pistachio nut), Spices (nutmeg), AFB_1, AFB_2, AFG_1, and AFG_2, literature[99]; Spices, Spices (cayenne pepper, chilli), Spices (pepper), AFB_1, AFB_2, and AFG_1, literature[99]; Spices (paprika), AFB_1 and AFG_1, literature[99]

incidence: 1/231*, conc.: 0.4 µg/kg, sample origin: commercial sources in Tokyo (capital), Japan, sample year: November 1977–June 1982, country: Japan[129], *small red beans as raw material for bean jam
- Co-contamination: 1 sa. co-contaminated with AFB_1 and AFB_2
- Further contamination of plants and plant products other than cereals and their products reported in the present book/ article[129]: Beans, AFB_1, literature[129]; Beans (butter beans), Beans (kidney bean), Beans (saltani saltapya bean), AFB_1 and AFB_2, literature[129]; Nuts (pistachio nut), AFB_1, AFB_2, and AFM_1, literature[129]; see also Nuts (pistachio nut), chapter: Further Mycotoxins and Microbial Metabolites

Aflatoxin G_1

incidence: 2/20*, conc. range: 20**–50*** µg/kg, Ø conc.: 35 µg/kg, sample origin: markets around Bangkok (capital) area, Thailand, sample year: July 1983–April 1985, country: Thailand[133], *different kinds of beans, **lantao bean, ***fava bean
- Co-contamination: not reported
- Further contamination of plants and plant products other than cereals and their products reported in the present book/ article[133]: Beans (soybean), Curd (bean curd), Nuts (peanut), Sauces (soy sauce), AFB_1 and AFG_1, literature[133]

Aflatoxin/s

incidence: 233/416*, Ø conc.: 19.1 µg/kg**, sample origin: greater Manila (capital) area, Philippines, sample year: unknown, country: Philippines[1], *bean and bean products, **AFS of pos sa.?
- Co-contamination: not reported
- Further contamination of plants and plant products other than cereals and their products reported in the present book/ article[1]: Butter (peanut butter), Cassava, Cocoa, Coffee, Drinks (alcoholic drink), Nuts (cashew nut), Nuts (peanut), Products (coconut product), Rootcrops, Spices (cayenne pepper, chilli), Sweet potatoes, Yams, AF/S, literature[1]

incidence: 2/29, conc. range: >30–≤86 µg/kg*, sample origin: markets in or near Manila (capital), Philippines, sample year: 1967–1969, country: USA[134], *AF
- Co-contamination: not reported
- Further contamination of plants and plant products other than cereals and their products reported in the present book/article[134]: Butter (peanut butter), Nuts (peanuts), Products (coconut product), Products (peanut product), Tubers, AF/S, literature[134]

incidence: 10/160, Ø conc.: 34.4 µg/kg*, sample origin: Highveld, Middleveld, Lowveld, and Lubombo Region, Swaziland, sample year: July 1982–June 1983, country: South Africa/France/Kenya/ Netherlands[135], *AFS total
- Co-contamination: not reported
- Further contamination of plants and plant products other than cereals and their products reported in the present book/ article[135]: Beans (cowpea), Beans (soybean), Nuts (peanut), AF/S, literature[135]

incidence: 46/64, conc. range: 1–100 µg/kg* (30 sa.), 100–1,000 µg/kg* (11 sa.), >1,000 µg/kg* (5 sa.), sample origin: natives homes and markets, Uganda, sample year: September 1966–June 1967, country: Uganda/USA[137], *AFB_1, AFB_2, AFG_1, and AFG_2
- Co-contamination: 15 sa. contained AFB_1, 42 sa. contained AFB_2, 11 sa. contained AFG_1, 1 sa. contained AFG_2
- Further contamination of plants and plant products other than cereals and their products reported in the present book/ article[137]: Cassava, Nuts (peanut), Peas, AF/S, literature[137]

incidence: 10/322*, conc. range: ≤1,620 µg/kg**, ∅ conc.: 213 µg/kg**, sample origin: market retailers, street vendors, wholesalers, and distributors, Thailand, sample year: September 1967–July 1969, country: Thailand[139], **Phaseolus aureus* Roxb., **AFB_1, AFB_2, AFG_1, and AFG_2

- Co-contamination: not reported
- Further contamination of plants and plant products other than cereals and their products reported in the present book/article[139]: Beans (mung bean), Food, Nuts (peanut), Seeds (sesame), Spices (cayenne pepper, chilli), Spices (garlic/onion), Starch (cassava starch), Vegetables, AF/S, literature[139]

For detailed information see the article.

incidence: 4/79, conc. range: >6.8–10 µg/kg* (2 sa.), –15 µg/kg* (2 sa.), sample origin: different Ecuadorian regions, Ecuador, sample year: April 1992–October 1994, country: Switzerland[140], *AFB_1, AFB_2, AFG_1, and AFG_2

- Co-contamination: For detailed information see the article.
- Further contamination of plants and plant products other than cereals and their products reported in the present book/article[140]: Nuts (peanut), AF/S, literature[140]

incidence: 38/51*, conc. range: 0.025–0.42 µg/kg**, sample origin: different households from employees of the University of São Paulo at Pirassununga (municipality), Brazil, sample year: June 2011–March 2012, country: Brazil[752], **in natura*, **AFB_1, AFB_2, AFG_1, and AFG_2

- Co-contamination: not reported
- Further contamination of plants and plant products other than cereals and their products reported in the present book/article[752]: not reported

incidence: 0/3*, conc.: no contamination, sample origin: purchases and donations in Bamenda (city), Cameroon, sample year: December 2004–April 2005, country: South Africa/Bulgaria[804], **Phaseolus vulgaris* L.

incidence: 0/2*, conc.: no contamination, sample origin: purchases and donations in Douala (city), Cameroon, sample year: December 2004–April 2005, country: South Africa/Bulgaria[804], **Phaseolus vulgaris* L.

incidence: 5/10*, conc. range: 0.2–6.2 µg/kg**, ∅ conc.: 2.4 µg/kg**, sample origin: purchases and donations in Kumbo (city), Cameroon, sample year: December 2004–April 2005, country: South Africa/Bulgaria[804], **Phaseolus vulgaris* L., **AFB_1, AFB_2, AFG_1, and AFG_2

incidence: 0/0, conc.: no sa. analyzed, sample origin: purchases and donations in Yaounde (capital), Cameroon, sample year: 2005, country: South Africa/Bulgaria[804]

- Co-contamination: For detailed information see the article.
- Further contamination of plants and plant products other than cereals and their products reported in the present book/article[804]: Beans, DON, FB_1, and ZEA, literature[804]; Beans (soybean), AF/S, DON, and FB_1, literature[804]; Food, AF/S, DON, and ZEA, literature[804]; Nuts (peanut), AF/S, DON, FB_1, OTA, and ZEA, literature[804]

Aspergillus and ***Penicillium*** Toxins

Ochratoxin A

incidence: 2/2*, conc. range: 0.01–0.53 µg/kg, sample origin: supermarkets and fast-food chains in Quebec (city), Canada, sample year: September/October 2008, country: Canada[90], *beans, baked and canned as well as beans, string

incidence: 0/2*, conc.: no contamination, sample origin: supermarkets and fast-food chains in Calgary (city), Canada, sample year: September/October 2009, country: Canada[90], *beans, baked and canned as well as beans, string

- Co-contamination: not reported
- Further contamination of plants and plant products other than cereals and their products reported in the present book/article[90]: Baby food, Beverages, Butter (peanut butter), Candies, Chocolate, Coffee, Extracts, Herbs, Ice cream, Nuts, Pies, Pizza, Products, Pudding, Raisins, Seeds, Wine, OTA, literature[90]

incidence: 5/30*, conc. range: 0.23–0.39 µg/kg, ∅ conc.: 0.29 µg/kg, sample origin: Mashhad (city), Khorasan (historical region), Iran, sample year: unknown, country: Iran[131]

- Co-contamination: not reported
- Further contamination of plants and plant products other than cereals and their products reported in the present book/article[131]: Beans, AFB_1, literature[131]

incidence: 2/28, conc. range: 0.7–19.3 µg/kg, ∅ conc.: 10 µg/kg, sample origin: individual households in BEN control area, Bulgaria, sample year: harvest 1981–February/March 1982, country: Bulgaria/France[141]

incidence: 4/24, conc. range: 4.8–33.9 µg/kg, sample origin: individual households in BEN area, Bulgaria, sample year: harvest 1981–February/March 1982, country: Bulgaria/France[141]

- Co-contamination: not reported
- Further contamination of plants and plant products other than cereals and their products reported in the present book/article[141]: not reported

incidence: 4/42, conc. range: 12.0–22.4 µg/kg, ∅ conc.: 14.7 µg/kg, sample origin: markets or farmers, Egypt, sample year: unknown, country: Egypt[142]

- Co-contamination: not reported
- Further contamination of plants and plant products other than cereals and their products reported in the present book/article[142]: Coffee beans, OTA, literature[142]

incidence: 1*/50, conc.: 0.3 µg/kg, sample origin: retail outlets across UK, sample year: August 1996, country: UK[143], *beans, baked (cut and normal price)

- Co-contamination: not reported
- Further contamination of plants and plant products other than cereals and their products reported in the present book/article[143]: Legumes, Powder (cocoa powder), Wine (red), OTA, literature[143]

incidence: 17/45*, conc. range: 0.25–0.92 µg/kg, ∅ conc.: 0.41 µg/kg, sample origin: farmers in 13 counties, Croatia, sample year: fall 2001, country: Croatia[145], **Phaseolus vulgaris* L.

- Co-contamination: not reported
- Further contamination of plants and plant products other than cereals and their products reported in the present book/article[145]: not reported

incidence: 1/12, conc. range: ≤107 µg/kg, sample origin: 12 regions, Czech Republic, sample year: 2011–2013, country: Czech Republic[811]

- Co-contamination: not reported
- Further contamination of plants and plant products other than cereals and their products reported in the present book/article[811]: Chocolate, Coffee, Confectionery, Figs, Herbs and

Spices, Juices, Licorice, Powder (cocoa powder), Products (ginger product), Raisins, Spices (cayenne pepper, chilli), Spices (caraway), Spices (coriander), Spices (ginger), Spices (mustard), Spices (nutmeg), Spices (paprika), Spices (pepper), Tea, Wine (red), Wine (white), OTA, literature[811]

incidence: 1/3, conc.: 3.7 µg/kg, sample origin: local markets, China, sample year: unknown, country: China[853]
- Co-contamination: not reported
- Further contamination of plants and plant products other than cereals and their products reported in the present book/article[853]: Oils (maize oil), AFB_1, AFB_2, and ZEA, literature[853]; Oils (peanut oil), AFB_1 and AFB_2, literature[853]

Penicillic Acid
incidence: 5/20*, conc. range: 11–179 µg/kg, sample origin: North Central States, USA, sample year: spring 1973, country: USA[102], *beans, dried
- Co-contamination: not reported
- Further contamination of plants and plant products other than cereals and their products reported in the present book/article[102]: not reported

Fusarium Toxins

Fumonisin/s

Fumonisin B_1
incidence: 1/3*, conc.: 746 µg/kg, sample origin: purchases and donations in Bamenda (city), Cameroon, sample year: December 2004–April 2005, country: South Africa/Bulgaria[804], **Phaseolus vulgaris* L.

incidence: 0/2*, conc.: no contamination, sample origin: purchases and donations in Douala (city), Cameroon, sample year: December 2004–April 2005, country: South Africa/Bulgaria[804], **Phaseolus vulgaris* L.

incidence: 2/10*, conc. range: 28–1,350.8 µg/kg, ∅ conc.: 690 µg/kg, sample origin: purchases and donations in Kumbo (city), Cameroon, sample year: December 2004–April 2005, country: South Africa/Bulgaria[804], **Phaseolus vulgaris* L.

incidence: 0/0, conc.: no sa. analyzed, sample origin: purchases and donations in Yaounde (capital), Cameroon, sample year: 2005, country: South Africa/Bulgaria[804]
- Co-contamination: For detailed information see the article.
- Further contamination of plants and plant products other than cereals and their products reported in the present book/article[804]: Beans, AF/S, DON, and ZEA, literature[804]; Beans (soybean), AF/S, DON, and FB_1, literature[804]; Food, AF/S, DON, and ZEA, literature[804]; Nuts (peanut), AF/S, DON, FB_1, OTA, and ZEA, literature[804]

Type A Trichothecenes

T-2 Toxin
incidence: 2/5*, conc. range: 1,500–1,900 µg/kg, ∅ conc.: 1,700 µg/kg, sample origin: market in Istanbul (capital), Turkey, sample year: unknown, country: Turkey[150], *beans, dried
- Co-contamination: not reported
- Further contamination of plants and plant products other than cereals and their products reported in the present book/article[150]: Lentils, Peas (chickpea), T-2, literature[150]

Type B Trichothecenes

Deoxynivalenol
incidence: 5/22*, conc. range: 2–160 µg/kg, ∅ conc.: 42.8 µg/kg, sample origin: individual household storage in EN region, Croatia, sample year: unknown, country: Croatia/Japan[149], *ncac
- Co-contamination: 1 sa. co-contaminated with DON and NIV; 1 sa. co-contaminated with DON and ZEA; 3 sa. contaminated solely with DON
- Further contamination of plants and plant products other than cereals and their products reported in the present book/article[149]: Beans, NIV and ZEA, literature[149]

incidence: 0/3*, conc.: no contamination, sample origin: purchases and donations in Bamenda (city), Cameroon, sample year: December 2004–April 2005, country: South Africa/Bulgaria[804], **Phaseolus vulgaris* L.

incidence: 1/2*, conc.: 24 µg/kg, sample origin: purchases and donations in Douala (city), Cameroon, sample year: December 2004–April 2005, country: South Africa/Bulgaria[804], **Phaseolus vulgaris* L.

incidence: 6/10*, conc. range: 13–35 µg/kg, ∅ conc.: 26 µg/kg, sample origin: purchases and donations in Kumbo (city), Cameroon, sample year: December 2004–April 2005, country: South Africa/Bulgaria[804], **Phaseolus vulgaris* L.

incidence: 0/0, conc.: no sa. analyzed, sample origin: purchases and donations in Yaounde (capital), Cameroon, sample year: 2005, country: South Africa/Bulgaria[804]
- Co-contamination: For detailed information see the article.
- Further contamination of plants and plant products other than cereals and their products reported in the present book/article[804]: Beans, AF/S, FB_1, and ZEA, literature[804]; Beans (soybean), AF/S, DON, and FB_1, literature[804]; Food, AF/S, DON, and ZEA, literature[804]; Nuts (peanut), AF/S, DON, FB_1, OTA, and ZEA, literature[804]

Nivalenol
incidence: 1/22*, conc.: 24 µg/kg, sample origin: individual household storage in EN region, Croatia, sample year: unknown, country: Croatia/Japan[149], *ncac
- Co-contamination: 1 sa. co-contaminated with DON and NIV
- Further contamination of plants and plant products other than cereals and their products reported in the present book/article[149]: Beans, DON and ZEA, literature[149]

Other ***Fusarium*** Toxins

Zearalenone
incidence: 1/22*, conc.: 3 µg/kg, sample origin: individual household storage in EN region, Croatia, sample year: unknown, country: Croatia/Japan[149], *ncac
- Co-contamination: 1 sa. co-contaminated with DON and ZEA
- Further contamination of plants and plant products other than cereals and their products reported in the present book/article[149]: Beans, DON and NIV, literature[149]

incidence: 1/5, conc.: 7 µg/kg, sample origin: food stores, Germany, sample year: 2000/2001, country: Germany[151]
- Co-contamination: not reported
- Further contamination of plants and plant products other than cereals and their products reported in the present

book/article[151]: Nuts (hazelnut), Seeds (sunflower), HT-2, T-2, and ZEA, literature[151]; Products (potato product), DAS, HT-2, MAS, SCIRP, and ZEA, literature[151]; Seeds (pumpkin), HT-2 and ZEA, literature[151]; Spices (cayenne pepper, chilli), ZEA, literature[151]

incidence: 1/3*, conc.: 57 µg/kg, sample origin: purchases and donations in Bamenda (city), Cameroon, sample year: December 2004–April 2005, country: South Africa/Bulgaria[804], **Phaseolus vulgaris* L.

incidence: 0/2*, conc.: no contamination, sample origin: purchases and donations in Douala (city), Cameroon, sample year: December 2004–April 2005, country: South Africa/Bulgaria[804], **Phaseolus vulgaris* L.

incidence: 4/10*, conc. range: 27–81 µg/kg, ∅ conc.: 39 µg/kg, sample origin: purchases and donations in Kumbo (city), Cameroon, sample year: December 2004–April 2005, country: South Africa/Bulgaria[804], **Phaseolus vulgaris* L.

incidence: 0/0, conc.: no sa. analyzed, sample origin: purchases and donations in Yaounde (capital), Cameroon, sample year: 2005, country: South Africa/Bulgaria[804]

- Co-contamination: For detailed information see the article.
- Further contamination of plants and plant products other than cereals and their products reported in the present book/article[804]: Beans, AF/S, DON, and FB_1, literature[804]; Beans (soybean), AF/S, DON, and FB_1, literature[804]; Food, AF/S, DON, and ZEA, literature[804]; Nuts (peanut), AF/S, DON, FB_1, OTA, and ZEA, literature[804]

Beans (black bean) may contain the following mycotoxins:

Aspergillus Toxins

Aflatoxin/s

incidence: 4/4*, conc. range: 6–20 µg/kg** (4 sa., maximum: 20 µg/kg**), sample origin: silos, homes, and small shops in different villages on the southern coast, Guatemala, sample year: 1976, country: Guatemala[136], *sa. stored for 6 months during dry season, **AFB_1, AFB_2, and AFG_1

incidence: 2/5*, conc. range: 6–20 µg/kg** (1 sa.), 30 µg/kg** (1 sa.), sample origin: silos, homes, and small shops in different villages on the southern coast, Guatemala, sample year: 1976, country: Guatemala[136], *sa. collected within 20 days after harvest (June) middle of the rainy season, **AFB_1, AFB_2, AFG_1, and AFG_2

- Co-contamination: not reported
- Further contamination of plants and plant products other than cereals and their products reported in the present book/article[136]: Coffee, Seeds (sesame), AFB_1, literature[136]

incidence: 1/17*, conc.: 22 µg/kg**, sample origin: storage facilities, silos sacks, small shops, and fields in zones with different climatic conditions, Guatemala, sample year: 1975/1976–?, country: Guatemala[138], *sa. collected at the end of rainy season, **AFB_1, AFB_2, AFG_1, and AFG_2

- Co-contamination: not reported
- Further contamination of plants and plant products other than cereals and their products reported in the present book/article[138]: Coffee beans, Nuts (peanuts), AF/S, literature[138]

Fusarium Toxins

Type A Trichothecenes

HT-2 Toxin

incidence: 2/15, conc. range: 21.1–22.3 µg/kg, ∅ conc.: 21.7 µg/kg, sample origin: Jinju, Masan, and Sancheon/Samcheon-po (cities) in Gosung and Ham-an (counties), Gyeongnam (province), South Korea, sample year: unknown country: South Korea/Tanzania[762]

- Co-contamination: not reported
- Further contamination of plants and plant products other than cereals and their products reported in the present book/article[762]: Beans (black beans), T-2, literature[762]

T-2 Toxin

incidence: 4/15, conc. range: 10–100 µg/kg (3 sa.), 199.3 µg/kg (1 sa.), sample origin: Jinju, Masan, and Sancheon/Samcheon-po (cities) in Gosung and Ham-an (counties), Gyeongnam (province), South Korea, sample year: unknown country: South Korea/Tanzania[762]

- Co-contamination: not reported
- Further contamination of plants and plant products other than cereals and their products reported in the present book/article[762]: Beans (black beans), HT-2, literature[762]

Beans (broad bean) may contain the following mycotoxins:

Aspergillus Toxins

Aflatoxin B_1

incidence: 1/30* **, conc.: 125.0 µg/kg, sample origin: different storage facilities, Egypt, sample year: November 1978–April 1979, country: Egypt[4], *faba beans, **sa. moldy with mechanical or insect damage

- Co-contamination: 1 sa. co-contaminated with AFB_1 and AFB_2
- Further contamination of plants and plant products other than cereals and their products reported in the present book/article[4]: Beans (broad bean), AFB_2, literature[4]; Lentils, Nuts (almond), AFB_1, literature[4]; Nuts (peanut), AFB_1, AFB_2, AFG_1, and AFG_2, literature[4]

incidence: 1/6*, conc.: 1.7 µg/kg**, sample origin: Alexandria, Luxor (cities), and Cairo (capital), Egypt, sample year: unknown, country: Egypt/USA[127], *3 sa. freshly produced and 3 sa. stored for 12 months, **stored sa.

- Co-contamination: 1 sa. co-contaminated with AFB_1 and AFB_2
- Further contamination of plants and plant products other than cereals and their products reported in the present book/article[127]: Beans (broad bean), AFB_2, literature[127]; Lentils, AFB_1, literature[127]; Nuts (peanut), AFB_1 and AFB_2, literature[127]; Spices (fenugreek), AFB_1, AFB_2, and AFG_1, literature[127]

Aflatoxin B_2

incidence: 1/30*, conc.: 12.5 µg/kg**, sample origin: different storage facilities, Egypt, sample year: November 1978–April 1979, country: Egypt[4], *faba beans, **sa. moldy with mechanical or insect damage

- Co-contamination: 1 sa. co-contaminated with AFB_1 and AFB_2
- Further contamination of plants and plant products other than cereals and their products reported in the present book/article[4]: Beans (broad bean), Lentils, Nuts (almond), AFB_1, literature[4]; Nuts (peanut), AFB_1, AFB_2, AFG_1, and AFG_2, literature[4]

incidence: 1/6*, conc.: 1.5 μg/kg**, sample origin: Alexandria, Luxor (cities), and Cairo (capital), Egypt, sample year: unknown, country: Egypt/USA[127], *3 sa. freshly produced and 3 sa. stored for 12 months, **stored sa.

- Co-contamination: 1 sa. co-contaminated with AFB_1 and AFB_2
- Further contamination of plants and plant products other than cereals and their products reported in the present book/article[127]: Beans (broad bean), Lentils, AFB_1, literature[127]; Nuts (peanut), AFB_1 and AFB_2, literature[127]; Spices (fenugreek), AFB_1, AFB_2, and AFG_1, literature[127]

Beans (brown bean) may contain the following mycotoxins:

Aspergillus and *Penicillium* Toxins

Ochratoxin A

incidence: 2/20, conc. range: 0.1–5 μg/kg, sample origin: ethnic stores, health-food stores, and local stores as well as supermarkets in the four largest cities, Sweden, sample year: autumn 1996–summer 1998, country: Sweden[13]

- Co-contamination: not reported
- Further contamination of plants and plant products other than cereals and their products reported in the present book/article[13]: Juices (apple juice), PAT, literature[13]; Nuts (almond), AFB_1, literature[13]; Nuts (Brazil nut, paranut), Nuts (pistachio nut), Oil seeds, Products (peanut product), AFB_1 and AF/S, literature[13]; Peas, Peas (chickpea), OTA, literature[13]

incidence: 4/31, conc. range: 10–442 μg/kg, Ø conc.: 124 μg/kg, sample origin: retail outlets, Sweden, sample year: 1976, country: Sweden[144]

incidence: 2/22, conc. range: 139–277 μg/kg, Ø conc.: 208 μg/kg, sample origin: retail outlets, Sweden, sample year: 1977, country: Sweden[144]

- Co-contamination: not reported
- Further contamination of plants and plant products other than cereals and their products reported in the present book/article[144]: Peas, OTA, PA, and ZEA, literature[144]

Beans (butter bean) may contain the following mycotoxins:

Aspergillus Toxins

Aflatoxin B_1

incidence: 5/99*, conc. range: 1.5–12.0 μg/kg, Ø conc.: 4.5 μg/kg, sample origin: commercial sources in Tokyo (capital), Japan, sample year: November 1977–June 1982, country: Japan[129], *raw material for bean jam

- Co-contamination: 5 sa. co-contaminated with AFB_1 and AFB_2
- Further contamination of plants and plant products other than cereals and their products reported in the present book/article[129]: Beans, Beans (butter beans), AFB_2, literature[129]; Beans (kidney bean), Beans (saltani saltapya bean), AFB_1 and AFB_2, literature[129]; Nuts (pistachio nut), AFB_1, AFB_2, and AFM_1, literature[129]; see also Nuts (pistachio nut), chapter: Further Mycotoxins and Microbial Metabolites

Aflatoxin B_2

incidence: 5/99*, conc. range: 0.7–2.2 μg/kg, Ø conc.: 1.24 μg/kg, sample origin: commercial sources in Tokyo (capital), Japan, sample year: November 1977–June 1982, country: Japan[129], *raw material for bean jam

- Co-contamination: 5 sa. co-contaminated with AFB_1 and AFB_2
- Further contamination of plants and plant products other than cereals and their products reported in the present book/article[129]: Beans, Beans (butter beans), AFB_1, literature[129]; Beans (kidney bean), Beans (saltani saltapya bean), AFB_1 and AFB_2, literature[129]; Nuts (pistachio nut), AFB_1, AFB_2, and AFM_1, literature[129]; see also Nuts (pistachio nut), chapter: Further Mycotoxins and Microbial Metabolites

Beans (cowpea) may contain the following mycotoxins:

Aspergillus Toxins

Aflatoxin B_1

incidence: 5/16, conc. range: 26–68 μg/kg, sample origin: villages of Khagaria and Saharsa (districts) of North Bihar (state), India, sample year: August–October 1987, country: India[128]

- Co-contamination: not reported
- Further contamination of plants and plant products other than cereals and their products reported in the present book/article[128]: Beans (mung bean), Field mustard, Lentils, Seeds (linseed), Peas (chickpea), AFB_1, literature[128]

incidence: 6/92*, conc. range: 0.40–8.04 μg/kg, Ø conc.: 3.58 μg/kg, sample origin: farmer stocks in Forest Mosaic Savannah, Southern Guinea Savannah, Northern Guinea Savannah, and Sudan Savannah, Benin, sample year: July 2005–September 2006, country: Benin/Ghana/Denmark[324], *collected after 3 months of storage

- Co-contamination: not reported
- Further contamination of plants and plant products other than cereals and their products reported in the present book/article[324]: Beans (cowpea), FB_1, literature[324]

For detailed information see the article.

Aflatoxin/s

incidence: 7/101, Ø conc.: 14.1 μg/kg*, sample origin: Highveld, Middleveld, Lowveld, and Lumbobo Region, Swaziland, sample year: July 1982–June 1983, country: South Africa/France/Kenya/Netherlands[135], *AFS total

- Co-contamination: not reported
- Further contamination of plants and plant products other than cereals and their products reported in the present book/article[135]: Beans, Beans (soybean), Nuts (peanut), AF/S, literature[135]

incidence: 4/20*, conc. range: 45–112 μg/kg**, Ø conc.: 75.8 μg/kg**, sample origin: International Institute of Tropical Agriculture, Ibadan (city), Nigeria and Egyptian Agricultural Organization, Egypt, sample year: 1987/1988, country: Egypt[325], *ncac, **AFB_1, AFB_2, AFG_1, and AFG_2

- Co-contamination: not reported
- Further contamination of plants and plant products other than cereals and their products reported in the present book/article[325]: not reported

incidence: 2/10, conc. range: 1.8–2.6 μg/kg*, Ø conc.: 2.2 μg/kg*, sample origin: retail shops and local markets, Pakistan, sample year: January–November 2009, country: Pakistan[782], *AFB_1, AFB_2, AFG_1, and AFG_2

- Co-contamination: not reported
- Further contamination of plants and plant products other than cereals and their products reported in the present book/article[782]: Beans (kidney bean), Beans (soybean), Peas, Peas (chickpea), AF/S, literature[782]

Fusarium Toxins

Fumonisin/s

Fumonisin B_1
incidence: 3/92*, conc. range: 10–68 µg/kg, Ø conc.: 30 µg/kg, sample origin: farmer stocks in Forest Mosaic Savannah, Southern Guinea Savannah, Northern Guinea Savannah, and Sudan Savannah, Benin, sample year: July 2005–September 2006, country: Benin/Ghana/Denmark[324], *collected at harvest
- Co-contamination: not reported
- Further contamination of plants and plant products other than cereals and their products reported in the present book/article[324]: Beans (cowpea), AFB_1, literature[324]

For detailed information see the article.

incidence: 4/4*, conc. range: 120–610 µg/kg, Ø conc.: 255 µg/kg, sample origin: Agricultural Research Council—Grain Crops Institute in Potchefstroom (city), South Africa, sample year: unknown, country: South Africa[326], *ncac
- Co-contamination: not reported
- Further contamination of plants and plant products other than cereals and their products reported in the present book/article[326]: not reported

Beans (green bean) may contain the following mycotoxins:

Aspergillus and ***Penicillium*** Toxins

Ochratoxin A
incidence: 3/11, conc. range: 0.14–0.76 µg/kg, Ø conc.: 0.45 µg/kg, sample origin: local supermarkets and small retail shops across Japan, sample year: summer 2004–winter 2005, country: Japan[146]
- Co-contamination: not reported
- Further contamination of plants and plant products other than cereals and their products reported in the present book/article[146]: Butter (peanut butter), AFB_1, AFB_2, AFG_1, and AFG_2, literature[146]; Chocolate, AFB_1 and OTA, literature[146]; Coffee, Raisins, Wine, OTA, literature[146]

Beans (kidney bean) may contain the following mycotoxins:

Aspergillus Toxins

Aflatoxin B_1
incidence: 2/2*, conc. range: 6.3–26.9 µg/kg, Ø conc.: 16.6 µg/kg, sample origin: commercial sources in Tokyo (capital), Japan, sample year: November 1977–June 1982, country: Japan[129], *raw material for bean jam (red beans)
- Co-contamination: 2 sa. co-contaminated with AFB_1 and AFB_2
- Further contamination of plants and plant products other than cereals and their products reported in the present book/article[129]: Beans, Beans (butter beans), Beans (saltani saltapya bean), AFB_1 and AFB_2, literature[129]; Beans (kidney bean), AFB_2, literature[129]; Nuts (pistachio nut), AFB_1, AFB_2, and AFM_1, literature[129]; see also Nuts (pistachio nut), chapter: Further Mycotoxins and Microbial Metabolites

incidence: 1/3*, conc.: 52 µg/kg, sample origin: retail stores including large supermarkets and small neighborhood grocery shops in Campinas (municipality), Brazil, sample year: May 1985–June 1986, country: Brazil[130], *Mulatinho beans, dried
- Co-contamination: 1 sa. co-contaminated with AFB_1, AFG_1, and AFG_2
- Further contamination of plants and plant products other than cereals and their products reported in the present book/article[130]: Beans (kidney bean), AFG_1 and AFG_2, literature[130]; Beans (pinto bean), AFB_1, AFG_1, AFG_2, and OTA, literature[130]; Flour (cassava flour), OTA, literature[130]

incidence: 6/20*, conc. range: 3.6–21.2 µg/kg, Ø conc.: 9.93 µg/kg, sample origin: small markets in several urban Shanghai (city) areas, China, sample year: April 1988–April 1989, country: China/USA[132], *red beans, dried
- Co-contamination: not reported
- Further contamination of plants and plant products other than cereals and their products reported in the present book/article[132]: Butter (peanut butter), Curd (bean curd), Nuts (peanut), Oils (rapeseed oil), Oils (soybean oil), Paste (bean paste), Sauces (soy sauce), Tofu, Wine, miscellaneous, AFB_1, literature[132]

Aflatoxin B_2
incidence: 2/2*, conc. range: 3.5–6.9 µg/kg, Ø conc.: 5.2 µg/kg, sample origin: commercial sources in Tokyo (capital), Japan, sample year: November 1977–June 1982, country: Japan[129], *raw material for bean jam (red beans)
- Co-contamination: 2 sa. co-contaminated with AFB_1 and AFB_2
- Further contamination of plants and plant products other than cereals and their products reported in the present book/article[129]: Beans, Beans (butter beans), Beans (saltani saltapya bean), AFB_1 and AFB_2, literature[129]; Beans (kidney bean), AFB_1, literature[129]; Nuts (pistachio nut), AFB_1, AFB_2, and AFM_1, literature[129]; see also Nuts (pistachio nut), chapter: Further Mycotoxins and Microbial Metabolites

Aflatoxin G_1
incidence: 1/3*, conc.: 31 µg/kg, sample origin: retail stores including large supermarkets and small neighborhood grocery shops in Campinas (municipality), Brazil, sample year: May 1985-June 1986, country: Brazil[130], *Mulatinho beans, dried
- Co-contamination: 1 sa. co-contaminated with AFB_1, AFG_1, and AFG_2
- Further contamination of plants and plant products other than cereals and their products reported in the present book/article[130]: Beans (kidney bean), AFB_1 and AFG_2, literature[130]; Beans (pinto bean), AFB_1, AFG_1, AFG_2, and OTA, literature[130]; Flour (cassava flour), OTA, literature[130]

Aflatoxin G_2
incidence: 1/3*, conc.: 8 µg/kg, sample origin: retail stores including large supermarkets and small neighborhood grocery shops in Campinas (municipality), Brazil, sample year: May 1985–June 1986, country: Brazil[130], *Mulatinho beans, dried
- Co-contamination: 1 sa. co-contaminated with AFB_1, AFG_1, and AFG_2
- Further contamination of plants and plant products other than cereals and their products reported in the present book/article[130]: Beans (kidney bean), AFB_1 and AFG_1, literature[130]; Beans (pinto bean), AFB_1, AFG_1, AFG_2, and OTA, literature[130]; Flour (cassava flour), OTA, literature[130]

Aflatoxin/s
incidence: 2/10*, conc. range: 2.6–7.4 µg/kg**, Ø conc.: 5.0 µg/kg**, sample origin: retail shops and local markets, Pakistan, sample

year: January–November 2009, country: Pakistan[782], *red kidney beans, **AFB_1, AFB_2, AFG_1, and AFG_2

- Co-contamination: not reported
- Further contamination of plants and plant products other than cereals and their products reported in the present book/article[782]: Beans (cowpea), Beans (soybean), Peas, Peas (chickpea), AF/S, literature[782]

***Aspergillus* and *Penicillium* Toxins**

Ochratoxin A

incidence: 1/1*, conc.: 4.64 µg/kg**, sample origin: stores, Taiwan, sample year: unknown, country: Taiwan/Russia[147], *red kidney bean, **CL-ELISA

- Co-contamination: not reported
- Further contamination of plants and plant products other than cereals and their products reported in the present book/article[147]: Berries (grape), Coffee, Raisins, OTA, literature[147]

Beans (locust bean) may contain the following mycotoxins:

***Aspergillus* and *Penicillium* Toxins**

3-Nitropropionic Acid

incidence: 1/9, conc.: 5,500.5 µg/kg, sample origin: markets in Lagos, Ogun, and Oyo (states), Southwest Nigeria, sample year: March 2016, country: Nigeria/South Africa/Austria/Ireland[361]

- Co-contamination: not reported
- Further contamination of plants and plant products other than cereals and their products reported in the present book/article[361]: Beans (locust bean), BEA and CIT, literature[361]; Iru, CIT, OTA, and STG, literature[361]; Ogiri, 3-NPA, AFB_1, BEA, CIT, and STG, literature[361]; Seeds (melon), 3-NPA, AFB_1, AFB_2, BEA, and STG, literature[361]

Citrinin

incidence: 2/9, conc. range: 0.9–6.6 µg/kg, ∅ conc.: 3.7 µg/kg, sample origin: markets in Lagos, Ogun, and Oyo (states), Southwest Nigeria, sample year: March 2016, country: Nigeria/South Africa/Austria/Ireland[361]

- Co-contamination: not reported
- Further contamination of plants and plant products other than cereals and their products reported in the present book/article[361]: Beans (locust bean), 3-NPA and BEA, literature[361]; Iru, CIT, OTA, and STG, literature[361]; Ogiri, 3-NPA, AFB_1, BEA, CIT, and STG, literature[361]; Seeds (melon), 3-NPA, AFB_1, AFB_2, BEA, and STG, literature[361]

***Fusarium* Toxins**

Depsipeptides

Beauvericin

incidence: 6/9, conc. range: 0.12–29.3 µg/kg, ∅ conc.: 6 µg/kg, sample origin: markets in Lagos, Ogun, and Oyo (states), Southwest Nigeria, sample year: March 2016, country: Nigeria/South Africa/Austria/Ireland[361]

- Co-contamination: not reported
- Further contamination of plants and plant products other than cereals and their products reported in the present book/article[361]: Beans (locust bean), 3-NPA and CIT, literature[361]; Iru, CIT, OTA, and STG, literature[361]; Ogiri, 3-NPA, AFB_1, BEA, CIT, and STG, literature[361]; Seeds (melon), 3-NPA, AFB_1, AFB_2, BEA, and STG, literature[361]

Beans (mung bean) may contain the following mycotoxins:

***Aspergillus* Toxins**

Aflatoxin B1

incidence: 15/25, conc. range: tr-2,800 µg/kg, sample origin: storage centers in Bhagalpur (city) and adjoining areas, India, sample year: unknown, country: India[125]

- Co-contamination: 1 sa. co-contaminated with AFB_1 and AFB_2; 14 sa. contaminated solely with AFB_1
- Further contamination of plants and plant products other than cereals and their products reported in the present book/article[125]: Beans (mung bean), AFB_2, literature[125]

incidence: 14/27, conc. range: 21–100 µg/kg (4 sa.), 101–500 µg/kg (7 sa.), 501–1,000 µg/kg (1 sa.), >1,000 µg/kg (2 sa., maximum: 1,040 µg/kg), sample origin: villages of Khagaria and Saharsa (districts) of North Bihar (state), India, sample year: August–October 1987, country: India[128]

- Co-contamination: not reported
- Further contamination of plants and plant products other than cereals and their products reported in the present book/article[128]: Beans (cowpea), Field mustard, Lentils, Seeds (linseed), Peas (chickpea), AFB_1, literature[128]

incidence: 2/8, conc. range: 1.6–1.7 µg/kg, ∅ conc.: 1.65 µg/kg, sample origin: cultivated in Indonesia, Myanmar, and Taiwan purchased in Taiwanese stores, Taiwan, sample year: unknown, country: Taiwan/Russia[893]

- Co-contamination: not reported
- Further contamination of plants and plant products other than cereals and their products reported in the present book/article[893]: not reported

Aflatoxin B2

incidence: 1/25, conc.: pr, sample origin: storage centers in Bhagalpur (city) and adjoining areas, India, sample year: unknown, country: India[125]

- Co-contamination: 1 sa. co-contaminated with AFB_1 and AFB_2
- Further contamination of plants and plant products other than cereals and their products reported in the present book/article[125]: Beans (mung bean), AFB_1, literature[125]

Aflatoxin/s

incidence: 7/140, conc. range: ≤112 µg/kg*, ∅ conc.: 16 µg/kg*, sample origin: market retailers, street vendors, wholesalers, and distributors, Hong Kong (British colony) and Thailand, sample year: September 1967–July 1969, country: Thailand[139], *AFB_1, AFB_2, AFG_1, and AFG_2

- Co-contamination: not reported
- Further contamination of plants and plant products other than cereals and their products reported in the present book/article[139]: Beans, Food, Nuts (peanut), Seeds (sesame), Spices (cayenne pepper, chilli), Spices (garlic/onion), Starch (cassava starch), Vegetables, AF/S, literature[139]

For detailed information see the article.

***Fusarium* Toxins**

Type A Trichothecenes

HT-2 Toxin

incidence: 28/40*, conc. range: ≤45.7 µg/kg, ∅ conc.: 9.9 µg/kg, sample origin: local supermarkets and small retail shops, Japan, sample year: summer 2010–winter 2012, country: Japan[892], *azuki beans

- Co-contamination: not reported
- Further contamination of plants and plant products other than cereals and their products reported in the present book/article[892]: Beans (mung bean), DON, T-2, and ZEA, literature[892]; Beans (soybean), HT-2 and T-2, literature[892]; Seeds (sesame), T-2 and ZEA, literature[892]

T-2 Toxin

incidence: 28/40*, conc. range: ≤48.4 µg/kg, ∅ conc.: 11.1 µg/kg, sample origin: local supermarkets and small retail shops, Japan, sample year: summer 2010–winter 2012, country: Japan[892], *azuki beans

- Co-contamination: not reported
- Further contamination of plants and plant products other than cereals and their products reported in the present book/article[892]: Beans (mung bean), DON, HT-2, and ZEA, literature[892]; Beans (soybean), HT-2 and T-2, literature[892]; Seeds (sesame), T-2 and ZEA, literature[892]

Type B Trichothecenes

Deoxynivalenol

incidence: 15/40*, conc. range: ≤38.2 µg/kg, ∅ conc.: 12.2 µg/kg, sample origin: local supermarkets and small retail shops, Japan, sample year: summer 2010–winter 2012, country: Japan[892], *azuki beans

- Co-contamination: not reported
- Further contamination of plants and plant products other than cereals and their products reported in the present book/article[892]: Beans (mung bean), HT-2, T-2, and ZEA, literature[892]; Beans (soybean), HT-2 and T-2, literature[892]; Seeds (sesame), T-2 and ZEA, literature[892]

Other *Fusarium* Toxins

Zearalenone

incidence: 29/40*, conc. range: ≤125 µg/kg, ∅ conc.: 44.6 µg/kg, sample origin: local supermarkets and small retail shops, Japan, sample year: summer 2010–winter 2012, country: Japan[892], *azuki beans

- Co-contamination: not reported
- Further contamination of plants and plant products other than cereals and their products reported in the present book/article[892]: Beans (mung bean), DON, HT-2, and T-2, literature[892]; Beans (soybean), HT-2 and T-2, literature[892]; Seeds (sesame), T-2 and ZEA, literature[892]

Beans (pinto bean) may contain the following mycotoxins:

Aspergillus Toxins

Aflatoxin B_1

incidence: 1/13*, conc.: 39 µg/kg**, sample origin: retail stores including large supermarkets and small neighborhood grocery shops in Campinas (municipality), Brazil, sample year: May 1985–June 1986, country: Brazil[130], *Carioquinha** and *Rosinha beans, dried

- Co-contamination: 1 sa. co-contaminated with AFB_1, AFG_1, and AFG_2; no further information available
- Further contamination of plants and plant products other than cereals and their products reported in the present book/article[130]: Beans (kidney bean), AFB_1, AFG_1, and AFG_2, literature[130]; Beans (pinto bean), AFG_1, AFG_2, and OTA, literature[130]; Flour (cassava flour), OTA, literature[130]

Aflatoxin G_1

incidence: 1/13*, conc.: 21 µg/kg**, sample origin: retail stores including large supermarkets and small neighborhood grocery shops in Campinas (municipality), Brazil, sample year: May 1985–June 1986, country: Brazil[130], *Carioquinha** and *Rosinha beans, dried

- Co-contamination: 1 sa. co-contaminated with AFB_1, AFG_1, and AFG_2; no further information available
- Further contamination of plants and plant products other than cereals and their products reported in the present book/article[130]: Beans (pinto bean), AFB_1, AFG_2, and OTA, literature[130]; Beans (kidney bean), AFB_1, AFG_1, and AFG_2, literature[130]; Flour (cassava flour), OTA, literature[130]

Aflatoxin G_2

incidence: 1/13*, conc.: 4 µg/kg**, sample origin: retail stores including large supermarkets and small neighborhood grocery shops in Campinas (municipality), Brazil, sample year: May 1985–June 1986, country: Brazil[130], *Carioquinha** and *Rosinha beans, dried

- Co-contamination: 1 sa. co contaminated with AFB_1, AFG_1, and AFG_2; no further information available
- Further contamination of plants and plant products other than cereals and their products reported in the present book/article[130]: Beans (kidney bean), AFB_1, AFG_1, and AFG_2, literature[130]; Beans (pinto bean), AFB_1, AFG_1, and OTA, literature[130]; Flour (cassava flour), OTA, literature[130]

Aspergillus and *Penicillium* Toxins

Ochratoxin A

incidence: 2/13*, conc. range: 94**–160*** µg/kg, ∅ conc.: 127 µg/kg, sample origin: retail stores including large supermarkets and small neighborhood grocery shops in Campinas (municipality), Brazil, sample year: May 1985–June 1986, country: Brazil[130], *Carioquinha** and *Rosinha*** beans, dried

- Co-contamination: not reported
- Further contamination of plants and plant products other than cereals and their products reported in the present book/article[130]: Beans (kidney bean), Beans (pinto bean), AFB_1, AFG_1, and AFG_2, literature[130]; Flour (cassava flour), OTA, literature[130]

Beans (saltani saltapya bean) may contain the following mycotoxins:

Aspergillus Toxins

Aflatoxin B_1

incidence: 3/37*, conc. range: 1.3–11.0 µg/kg, ∅ conc.: 4.56 µg/kg, sample origin: commercial sources in Tokyo (capital), Japan, sample year: November 1977–June 1982, country: Japan[129], *raw material for bean jam

- Co-contamination: 3 sa. co-contaminated with AFB_1 and AFB_2
- Further contamination of plants and plant products other than cereals and their products reported in the present book/article[129]: Beans, Beans (butter beans), Beans (kidney bean), AFB_1 and AFB_2, literature[129]; Beans (saltani saltapya bean), AFB_2, literature[129]; Nuts (pistachio nut), AFB_1, AFB_2, and AFM_1, literature[129]; see also Nuts (pistachio nut), chapter: Further Mycotoxins and Microbial Metabolites

Aflatoxin B_2

incidence: 3/37*, conc. range: 0.4–3.0 µg/kg, ∅ conc.: 2.0 µg/kg, sample origin: commercial sources in Tokyo (capital), Japan, sample year: November 1977–June 1982, country: Japan[129], *raw material for bean jam

- Co-contamination: 3 sa. co-contaminated with AFB_1 and AFB_2
- Further contamination of plants and plant products other than cereals and their products reported in the present book/article[129]: Beans, Beans (butter beans), Beans (kidney bean), AFB_1 and AFB_2, literature[129]; Beans (saltani saltapya bean), AFB_1, literature[129]; Nuts (pistachio nut), AFB_1, AFB_2, and AFM_1, literature[129]; see also Nuts (pistachio nut), chapter: Further Mycotoxins and Microbial Metabolites

Beans (soybean) may contain the following mycotoxins:

Alternaria **Toxins**

Alternariol

incidence: 23/50, conc. range: 25–211 µg/kg, ∅ conc.: 71.3 µg/kg, sample origin: local grain storage company, Argentina, sample year: 2006–2008, country: Argentina[631]

- Co-contamination: 15 sa. co-contaminated with AME and AOH; 8 sa. contaminated solely with AOH
- Further contamination of plants and plant products other than cereals and their products reported in the present book/article[631]: Beans (soybean), AME, literature[631]

Alternariol Methyl Ether

incidence: 22/50, conc. range: 62–1,153 µg/kg, ∅ conc.: 334 µg/kg, sample origin: local grain storage company, Argentina, sample year: 2006–2008, country: Argentina[631]

- Co-contamination: 15 sa. co-contaminated with AME and AOH; 7 sa. contaminated solely with AME
- Further contamination of plants and plant products other than cereals and their products reported in the present book/article[631]: Beans (soybean), AOH, literature[631]

Aspergillus **Toxins**

Aflatoxin B_1

incidence: 12/260, conc. range: 1–10 µg/kg (1 sa.), >50–100 µg/kg (11 sa.), sample origin: unknown, sample year: 1990–1996, country: Cuba[126]

- Co-contamination: not reported
- Further contamination of plants and plant products other than cereals and their products reported in the present book/article[126]: Beans, Cocoa beans, Nuts (peanut), Peas, AFB_1, literature[126]

incidence: 3/10*, conc. range: 10–20 µg/kg, ∅ conc.: 13.3 µg/kg, sample origin: markets around Bangkok (capital) area, Thailand, sample year: July 1983–April 1985, country: Thailand[133], *fermented soybean (tao-chiew)

- Co-contamination: 2 sa. co-contaminated with AFB_1 and AFG_1; 1 sa. contaminated solely with AFB_1
- Further contamination of plants and plant products other than cereals and their products reported in the present book/article[133]: Beans, Beans (soybean), AFG_1, literature[133]; Curd (bean curd), Nuts (peanut), Sauces (soy sauce), AFB_1 and AFG_1, literature[133]

incidence: 1/1*, conc.: 97 µg/kg, sample origin: unknown, sample year: unknown, country: UK[384], *ncac

- Co-contamination: 1 sa. co-contaminated with AFB_1, AFB_2, AFG_1, and AFG_2
- Further contamination of plants and plant products other than cereals and their products reported in the present book/article[384]: Beans (soybean), AFB_2, AFG_1, and AFG_2, literature[384]; Nuts (peanut), AFB_1, AFB_2, AFG_1, and AFG_2, literature[384]

incidence: 1/1, conc.: 4.8 µg/kg, sample origin: Nepal, sample year: unknown, country: Japan/Nepal[403]

- Co-contamination: 1 sa. co-contaminated with AFB_1 and AFG_1 as well as other aflatoxins
- Further contamination of plants and plant products other than cereals and their products reported in the present book/article[403]: Beans (soybean), AFG_1, literature[403]; Nuts (peanut), AFB_1 and AFG_1, literature[403]

incidence: 4/40, conc. range: ≤24 µg/kg, ∅ conc.: 13.00 µg/kg, sample origin: several villages and markets in Kichha (town) as well as Haldwani, Rudrapur, and Bareilly (cities) in Nainital (district), Uttar Pradesh (state), India, sample year: unknown, country: India[632]

- Co-contamination: not reported
- Further contamination of plants and plant products other than cereals and their products reported in the present book/article[632]: not reported

incidence: 1/4, conc.: 6.1 µg/kg, sample origin: supermarkets and wholesale markets of agro-products in Yangtze Delta region, China, sample year: April 2010, country: China[748]

- Co-contamination: 1 sa. co-contaminated with AFB_1, AF/S, and ZEA
- Further contamination of plants and plant products other than cereals and their products reported in the present book/article[748]: Beans (soybean), AF/S and ZEA; Oils, AFB_1, AF/S, and ZEA, literature[748]

Aflatoxin B_2

incidence: 1/1*, conc.: 20 µg/kg, sample origin: unknown, sample year: unknown, country: UK[384], *ncac

- Co-contamination: 1 sa. co-contaminated with AFB_1, AFB_2, AFG_1, and AFG_2
- Further contamination of plants and plant products other than cereals and their products reported in the present book/article[384]: Beans (soybean), AFB_1, AFG_1, and AFG_2, literature[384]; Nuts (peanut), AFB_1, AFB_2, AFG_1, and AFG_2, literature[384]

Aflatoxin G_1

incidence: 3/10*, conc. range: 20–100 µg/kg, ∅ conc.: 53.3 µg/kg, sample origin: markets around Bangkok (capital) area, Thailand, sample year: July 1983–April 1985, country: Thailand[133], *fermented soybean (tao-chiew)

- Co-contamination: 2 sa. co-contaminated with AFB_1 and AFG_1; 1 sa. contaminated solely with AFG_1
- Further contamination of plants and plant products other than cereals and their products reported in the present book/article[133]: Beans, AFG_1, literature[133]; Beans (soybean), AFB_1, literature[133]; Curd (bean curd), Nuts (peanut), Sauces (soy sauce), AFB_1 and AFG_1, literature[133]

incidence: 1/1*, conc.: 11 µg/kg, sample origin: unknown, sample year: unknown, country: UK[384], *ncac

- Co-contamination: 1 sa. co-contaminated with AFB_1, AFB_2, AFG_1, and AFG_2
- Further contamination of plants and plant products other than cereals and their products reported in the present book/

article[384]: Beans (soybean), AFB_1, AFB_2, and AFG_2, literature[384]; Nuts (peanut), AFB_1, AFB_2, AFG_1, and AFG_2, literature[384]

incidence: 1/1, conc.: 5.0 µg/kg, sample origin: Nepal, sample year: unknown, country: Japan/Nepal[403]
- Co-contamination: 1 sa. co-contaminated with AFB_1 and AFG_1 as well as other aflatoxins
- Further contamination of plants and plant products other than cereals and their products reported in the present book/article[403]: Beans (soybean), AFB_1, literature[403]; Nuts (peanut), AFB_1 and AFG_1, literature[403]

Aflatoxin G_2
incidence: 1/1*, conc.: 3.5 µg/kg, sample origin: unknown, sample year: unknown, country: UK[384], *ncac
- Co-contamination: 1 sa. co-contaminated with AFB_1, AFB_2, AFG_1, and AFG_2
- Further contamination of plants and plant products other than cereals and their products reported in the present book/article[384]: Beans (soybean), AFB_1, AFB_2, and AFG_1, literature[384]; Nuts (peanut), AFB_1, AFB_2, AFG_1, and AFG_2, literature[384]

Aflatoxin/s
incidence: 11/133, ∅ conc.: 22.8 µg/kg*, sample origin: Highveld, Middleveld, Lowveld, and Lumbobo Region, Swaziland, sample year: July 1982–June 1983, country: South Africa/France/Kenya/Netherlands[135], *AFS total
- Co-contamination: not reported
- Further contamination of plants and plant products other than cereals and their products reported in the present book/article[135]: Beans, Beans (cowpea), Nuts (peanut), AF/S, literature[135]

incidence: 35/100, conc. range: 5–35 µg/kg*, sample origin: localities in Egyptian governorates, Egypt, sample year: unknown, country: Egypt[714], *5 sa. contained only AFB_1 and AFB_2 (5–13 µg/kg), while the remaining 30 sa. contained AFB_1, AFB_2, AFG_1, and AFB_2 (9–35 µg/kg)
- Co-contamination: not reported
- Further contamination of plants and plant products other than cereals and their products reported in the present book/article[714]: not reported

incidence: 1/4, conc.: 7.6 µg/kg, sample origin: supermarkets and wholesale markets of agro-products in Yangtze Delta region, China, sample year: April 2010, country: China[748]
- Co-contamination: 1 sa. co-contaminated with AFB_1, AF/S, and ZEA
- Further contamination of plants and plant products other than cereals and their products reported in the present book/article[748]: Beans (soybean), AFB_1 and ZEA; Oils, AFB_1, AF/S, and ZEA, literature[748]

incidence: 3/20, conc. range: 3.2–10.5 µg/kg*, ∅ conc.: 6.4 µg/kg*, sample origin: retail shops and local markets, Pakistan, sample year: January–November 2009, country: Pakistan[782], *AFB_1, AFB_2, AFG_1, and AFG_2
- Co-contamination: not reported
- Further contamination of plants and plant products other than cereals and their products reported in the present book/article[782]: Beans (cowpea), Beans (kidney bean), Peas, Peas (chickpea), AF/S, literature[782]

incidence: 1/1, conc.: 3.9 µg/kg*, sample origin: purchases and donations in Bamenda (city), Cameroon, sample year: December 2004–April 2005, country: South Africa/Bulgaria[804], *AFB_1, AFB_2, AFG_1, and AFG_2

incidence: 0/2, conc.: no contamination, sample origin: purchases and donations in Douala (city), Cameroon, sample year: December 2004–April 2005, country: South Africa/Bulgaria[804]

incidence: 1/2, conc.: 0.2 µg/kg*, sample origin: purchases and donations in Kumbo (city), Cameroon, sample year: December 2004–April 2005, country: South Africa/Bulgaria[804], *AFB_1, AFB_2, AFG_1, and AFG_2

incidence: 0/0, conc.: no sa. analyzed, sample origin: purchases and donations in Yaounde (capital), Cameroon, sample year: 2005, country: South Africa/Bulgaria[804]
- Co-contamination: For detailed information see the article.
- Further contamination of plants and plant products other than cereals and their products reported in the present book/article[804]: Beans, AF/S, DON, FB_1, and ZEA, literature[804]; Beans (soybean), DON and FB_1, literature[804]; Food, AF/S, DON and ZEA, literature[804]; Nuts (peanut), AF/S, DON, FB_1, OTA, and ZEA, literature[804]

Aspergillus and ***Penicillium*** Toxins

Ochratoxin A
incidence: 26/31, conc. range: ≤0.100 µg/kg, sample origin: retail stores in Detmold, Jena, Karlsruhe, Kiel, Kulmbach, and Trier (cities), as well as Berlin (capital), Germany, sample year: 1996–1998, country: Germany[597]
- Co-contamination: not reported
- Further contamination of plants and plant products other than cereals and their products reported in the present book/article[597]: Legumes, OTA, literature[597]

incidence: 5/31, conc. range: LOD/LOQ-0.09 µg/kg (5 sa.), sample origin: Germany, sample year: 1995–1998, country: EU[698]
- Co-contamination: not reported
- Further contamination of plants and plant products other than cereals and their products reported in the present book/article[698]: Baby food, Chocolate, Cocoa, Coffee, Coffee beans, Currants, Fruits, dried, Herbs, Jam, Juices, Juices (grape juice), Legumes, Nuts, Oil seeds, Oils (olive oil), Peas, Peas (chickpea), Products (plant product), Raisins, Spices, Spices (cayenne pepper, chilli), Spices (nutmeg), Spices (paprika), Spices (pepper), Sultanas, Tea, Vinegar, Wine, Wine (red), Wine (rosé), Wine (white), Wine, miscellaneous, OTA, literature[698]

Fusarium Toxins

Fumonisin/s

Fumonisin B_1
incidence: 13/82, conc. range: ≤8.0 µg/kg, ∅ conc.: 4.5 µg/kg, sample origin: local supermarkets and small retail shops in seven major cities, Japan, sample year: summer 2004–winter 2007, country: Japan[85]
- Co-contamination: not reported
- Further contamination of plants and plant products other than cereals and their products reported in the present book/article[85]: Beans (soybean), FB_2, literature[85]; Beverages, Chocolate, Cocoa, Coffee, Coffee beans, Raisins, Spices (coriander), Wine, OTA, literature[85]; Vegetables (asparagus), FB_1 and FB_2, literature[85]

incidence: 4/39, conc. range: 138–1,495 µg/kg, ∅ conc.: 543 µg/kg, sample origin: shops and food stores in Rio Grande do Sul and Santa Catarina (states), southern Brazil, sample year: March–October 2013, country: Brazil[392]

- Co-contamination: 2 sa. co-contaminated with FB_1 and FB_2; 2 sa. contaminated solely with FB_1
- Further contamination of plants and plant products other than cereals and their products reported in the present book/article[392]: Beans (soybean), FB_2, literature[392]

incidence: 1/1*, conc.: 8,700 µg/kg, sample origin: agricultural cooperatives and factories, Spain, sample year: 1994–1996, country: Spain[635], *ncac

- Co-contamination: not reported
- Further contamination of plants and plant products other than cereals and their products reported in the present book/article[635]: not reported

incidence: 0/1, conc.: no contamination, sample origin: purchases and donations in Bamenda (city), Cameroon, sample year: December 2004–April 2005, country: South Africa/Bulgaria[804]

incidence: 1/2, conc.: 25 µg/kg, sample origin: purchases and donations in Douala (city), Cameroon, sample year: December 2004–April 2005, country: South Africa/Bulgaria[804]

incidence: 1/2, conc.: 365 µg/kg, sample origin: purchases and donations in Kumbo (city), Cameroon, sample year: December 2004–April 2005, country: South Africa/Bulgaria[804]

incidence: 0/0, conc.: no sa. analyzed, sample origin: purchases and donations in Yaounde (capital), Cameroon, sample year: 2005, country: South Africa/Bulgaria[804]

- Co-contamination: For detailed information see the article.
- Further contamination of plants and plant products other than cereals and their products reported in the present book/article[804]: Beans, AF/S, DON, FB_1, and ZEA, literature[804]; Beans (soybean), AF/S and DON, literature[804]; Food, AF/S, DON and ZEA, literature[804]; Nuts (peanut), AF/S, DON, FB_1, OTA, and ZEA, literature[804]

Fumonisin B_2

incidence: 3/82, conc. range: ≤4.8 µg/kg, ∅ conc.: 4.3 µg/kg, sample origin: local supermarkets and small retail shops in seven major cities, Japan, sample year: summer 2004–winter 2007, country: Japan[85]

- Co-contamination: not reported
- Further contamination of plants and plant products other than cereals and their products reported in the present book/article[85]: Beans (soybean), FB_1, literature[85]; Beverages, Chocolate, Cocoa, Coffee, Coffee beans, Raisins, Spices (coriander), Wine, OTA, literature[85]; Vegetables (asparagus), FB_1 and FB_2, literature[85]

incidence: 2/39, conc. range: 178–552 µg/kg, ∅ conc.: 365 µg/kg, sample origin: shops and food stores in Rio Grande do Sul and Santa Catarina (states), southern Brazil, sample year: March–October 2013, country: Brazil[392]

- Co-contamination: 2 sa. co-contaminated with FB_1 and FB_2
- Further contamination of plants and plant products other than cereals and their products reported in the present book/article[392]: Beans (soybean), FB_1, literature[392]

Type A Trichothecenes

HT-2 Toxin

incidence: 2/44*, conc. range: 99–189 µg/kg, ∅ conc.: 144 µg/kg, sample origin: Cordoba (province), Argentina, sample year: 2008/2009, country: Argentina/Italy[636], *ncac

- Co-contamination: 2 sa. co-contaminated with HT-2 and T-2
- Further contamination of plants and plant products other than cereals and their products reported in the present book/article[636]: Beans (soybean), T-2, literature[636]

incidence: 1/36, conc.: 3.1 µg/kg, sample origin: local supermarkets and small retail shops, Japan, sample year: summer 2010–winter 2012, country: Japan[892]

- Co-contamination: not reported
- Further contamination of plants and plant products other than cereals and their products reported in the present book/article[892]: Beans (mung bean), DON, HT-2, T-2, and ZEA, literature[892]; Beans (soybean), T-2, literature[892]; Seeds (sesame), T-2 and ZEA, literature[892]

T-2 Toxin

incidence: 2/44*, conc. range: 110–184 µg/kg, ∅ conc.: 147 µg/kg, sample origin: Cordoba (province), Argentina, sample year: 2008/2009, country: Argentina/Italy[636], *ncac

- Co-contamination: 2 sa. co-contaminated with HT-2 and T-2
- Further contamination of plants and plant products other than cereals and their products reported in the present book/article[636]: Beans (soybean), HT-2, literature[636]

incidence: 1/36, conc.: 1.4 µg/kg, sample origin: local supermarkets and small retail shops, Japan, sample year: summer 2010–winter 2012, country: Japan[892]

- Co-contamination: not reported
- Further contamination of plants and plant products other than cereals and their products reported in the present book/article[892]: Beans (mung bean), DON, HT-2, T-2, and ZEA, literature[892]; Beans (soybean), HT-2, literature[892]; Seeds (sesame), T-2 and ZEA, literature[892]

Type B Trichothecenes

Deoxynivalenol

incidence: 1/3*, conc.: 36 µg/kg, sample origin: imported and purchased in supermarkets in Port Moresby (city), Papua New Guinea, sample year: 1991, country: Papua New Guinea/Japan[633], *ncac

- Co-contamination: 1 sa. co-contaminated with DON and NIV
- Further contamination of plants and plant products other than cereals and their products reported in the present book/article[633]: Beans (soybean), NIV, literature[633]

incidence: 1/13* **, conc.: 110 µg/kg, sample origin: different locations, Serbia, sample year: 2004, country: Serbia[634], *ncac, **soybean and soybean meal

incidence: 1/11* **, conc.: 100 µg/kg, sample origin: different locations, Serbia, sample year: 2005, country: Serbia[634], *ncac, **soybean and soybean meal

- Co-contamination: not reported
- Further contamination of plants and plant products other than cereals and their products reported in the present book/article[633]: Seeds (sunflower), DON, literature[634]

incidence: 9/53*, conc. range: 100–500 µg/kg (5 sa.), 500–1,000 µg/kg (4 sa.), sample origin: homegrown as well as imported and purchased in retail outlets and from Portuguese cereal traders, Portugal, sample year: unknown, country: Portugal[787], *for food and feed

- Co-contamination: For detailed information see the article.
- Further contamination of plants and plant products other than cereals and their products reported in the present book/article[787]: Beans (soybean), ZEA, literature[787]; Seeds (sunflower), DON and ZEA, literature[787]

incidence: 0/1, conc.: no contamination, sample origin: purchases and donations in Bamenda (city), Cameroon, sample year: December 2004–April 2005, country: South Africa/Bulgaria[804]

incidence: 0/2, conc.: no contamination, sample origin: purchases and donations in Douala (city), Cameroon, sample year: December 2004–April 2005, country: South Africa/Bulgaria[804]

incidence: 2/2, conc. range: 13–207 µg/kg, ∅ conc.: 110 µg/kg, sample origin: purchases and donations in Kumbo (city), Cameroon, sample year: December 2004–April 2005, country: South Africa/Bulgaria[804]

incidence: 0/0, conc.: no sa. analyzed, sample origin: purchases and donations in Yaounde (capital), Cameroon, sample year: 2005, country: South Africa/Bulgaria[804]

- Co-contamination: For detailed information see the article.
- Further contamination of plants and plant products other than cereals and their products reported in the present book/article[804]: Beans, AF/S, DON, FB_1, and ZEA, literature[804]; Beans (soybean), AF/S and FB_1, literature[804]; Food, AF/S, DON and ZEA, literature[804]; Nuts (peanut), AF/S, DON, FB_1, OTA, and ZEA, literature[804]

Nivalenol

incidence: 1/3*, conc.: 50 µg/kg, sample origin: imported and purchased in supermarkets in Port Moresby (city), Papua New Guinea, sample year: 1991, country: Papua New Guinea/Japan[633], *ncac

- Co-contamination: 1 sa. co-contaminated with DON and NIV
- Further contamination of plants and plant products other than cereals and their products reported in the present book/article[633]: Beans (soybean), DON, literature[633]

incidence: 3/9, conc. range: 8.7–12.0 µg/kg, ∅ conc.: 10.6 µg/kg, sample origin: imports to Japan, sample year: 1983, country: Japan[640]

- Co-contamination: not reported
- Further contamination of plants and plant products other than cereals and their products reported in the present book/article[640]: not reported

Other *Fusarium* Toxins

Zearalenone

incidence: 2/17* **, conc. range: >200 µg/kg**, sample origin: imported and domestic sa. from Uruguay, sample year: 1993–1995, country: Uruguay[269], *ncac, **soybeans and by-products

- Co-contamination: not reported
- Further contamination of plants and plant products other than cereals and their products reported in the present book/article[269]: Cocoa beans, AF/S, literature[269]; Fruits, dried, Oil seeds, Vegetables, dried, AF/S and ZEA, literature[269]

incidence: 7/75, conc. range: 1.01–15.26 µg/kg, ∅ conc.: 7.70 µg/kg, sample origin: eight provinces in Korea, sample year: March–September 1988, country: Korea/USA[637]

- Co-contamination: not reported
- Further contamination of plants and plant products other than cereals and their products reported in the present book/article[637]: not reported

incidence: 2/4, conc. range: 11–130 µg/kg, ∅ conc.: 71 µg/kg, sample origin: supermarkets and wholesale markets of agro-products in Yangtze Delta region, China, sample year: April 2010, country: China[748]

- Co-contamination: 1 sa. co-contaminated with AFB_1, AF/S, and ZEA; 1 sa. contaminated solely with ZEA
- Further contamination of plants and plant products other than cereals and their products reported in the present book/article[748]: Beans (soybean), AFB_1 and AF/S; Oils, AFB_1, AF/S, and ZEA, literature[748]

incidence: 21/53*, conc. range: 5–100 µg/kg (21 sa.), sample origin: homegrown as well as imported and purchased in retail outlets and from Portuguese cereal traders, Portugal, sample year: unknown, country: Portugal[787], *for food and feed

- Co-contamination: For detailed information see the article.
- Further contamination of plants and plant products other than cereals and their products reported in the present book/article[787]: Beans (soybean), DON, literature[787]; Seeds (sunflower), DON and ZEA, literature[787]

see also **Flour (soybean flour)**

Bee pollen may contain the following mycotoxins:

Aspergillus Toxins

Aflatoxin/s

incidence: 5?/5*, conc. range: 15.80–16.20 µg/kg**, ∅ conc.: 16.00 µg/kg**, sample origin: beekeepers, west Slovakia, sample year: spring 2009, country: Slovakia[153], *poppy bee pollen, frozen, **AFS total

- Co-contamination: 5 sa. co-contaminated with AFS, DON, OTA, T-2, and ZEA
- Further contamination of plants and plant products other than cereals and their products reported in the present book/article[153]: Bee pollen, DON, FB/S, OTA, T-2, and ZEA, literature[153]

incidence: 5?/5*, conc. range: 3.50–3.70 µg/kg**, ∅ conc.: 3.58 µg/kg**, sample origin: beekeepers, west Slovakia, sample year: spring 2009, country: Slovakia[153], *rape bee pollen, frozen, **AFS total

- Co-contamination: 5 sa. co-contaminated with AFS, DON, OTA, T-2, and ZEA
- Further contamination of plants and plant products other than cereals and their products reported in the present book/article[153]: Bee pollen, DON, FB/S, OTA, T-2, and ZEA, literature[153]

incidence: 5?/5*, conc. range: 1.20–1.60 µg/kg**, ∅ conc.: 1.40 µg/kg**, sample origin: beekeepers, west Slovakia, sample year: spring 2009, country: Slovakia[153], *sunflower bee pollen, frozen, **AFS total

- Co-contamination: 5 sa. co-contaminated with AFS, FBS, DON, OTA, T-2, and ZEA
- Further contamination of plants and plant products other than cereals and their products reported in the present book/article[153]: Bee pollen, DON, FB/S, OTA, T-2, and ZEA, literature[153]

Dried and UV-radiated bee pollen was also investigated. For detailed information see the article.

Aspergillus **and** ***Penicillium*** **Toxins**

Ochratoxin A

incidence: 5?/5*, conc. range: 6.12–6.41 µg/kg, ∅ conc.: 6.21 µg/kg, sample origin: beekeepers, west Slovakia, sample year: spring 2009, country: Slovakia[153], *poppy bee pollen, frozen

- Co-contamination: 5 sa. co-contaminated with AFS, DON, OTA, T-2, and ZEA
- Further contamination of plants and plant products other than cereals and their products reported in the present book/article[153]: Bee pollen, AF/S, DON, FB/S, T-2, and ZEA, literature[153]

incidence: 5?/5*, conc. range: 9.70–9.87 µg/kg, ∅ conc.: 9.81 µg/kg, sample origin: beekeepers, west Slovakia, sample year: spring 2009, country: Slovakia[153], *rape bee pollen, frozen

- Co-contamination: 5 sa. co-contaminated with AFS, DON, OTA, T-2, and ZEA
- Further contamination of plants and plant products other than cereals and their products reported in the present book/article[153]: Bee pollen, AF/S, DON, FB/S, T-2, and ZEA, literature[153]

incidence: 5?/5*, conc. range: 6.89–6.93 µg/kg, ∅ conc.: 6.91 µg/kg, sample origin: beekeepers, west Slovakia, sample year: spring 2009, country: Slovakia[153], *sunflower bee pollen, frozen

- Co-contamination: 5 sa. co-contaminated with AFS, FBS, DON, OTA, T-2, and ZEA
- Further contamination of plants and plant products other than cereals and their products reported in the present book/article[153]: Bee pollen, AF/S, DON, FB/S, T-2, and ZEA, literature[153]

Dried and UV-radiated bee pollen was also investigated. For detailed information see the article.

Fusarium **Toxins**

Fumonisin/s

incidence: 0/5*, conc.: no contamination, sample origin: beekeepers, west Slovakia, sample year: spring 2009, country: Slovakia[153], *poppy bee pollen, frozen

- Co-contamination: not reported
- Further contamination of plants and plant products other than cereals and their products reported in the present book/article[153]: Bee pollen, AF/S, DON, OTA, T-2, and ZEA, literature[153]

incidence: 0/5*, conc.: no contamination, sample origin: beekeepers, west Slovakia, sample year: spring 2009, country: Slovakia[153], *rape bee pollen, frozen

- Co-contamination: not reported
- Further contamination of plants and plant products other than cereals and their products reported in the present book/article[153]: Bee pollen, AF/S, DON, OTA, T-2, and ZEA, literature[153]

incidence: 5?/5*, conc. range: 9.40–9.70 µg/kg**, ∅ conc.: 9.60 µg/kg**, sample origin: beekeepers, west Slovakia, sample year: spring 2009, country: Slovakia[153], *sunflower bee pollen, frozen, **FBS total

- Co-contamination: 5 sa. co-contaminated with AFS, DON, OTA, T-2, and ZEA
- Further contamination of plants and plant products other than cereals and their products reported in the present book/article[153]: Bee pollen, AF/S, DON, OTA, T-2, and ZEA, literature[153]

Dried and UV-radiated bee pollen was also investigated. For detailed information see the article.

Type A Trichothecenes

Neosolaniol

incidence: 2/15, conc. range: 22–30 µg/kg, ∅ conc.: 26 µg/kg, sample origin: supermarkets, Valencia (city) metropolitan area, Spain, sample year: unknown, country: Spain[154]

- Co-contamination: 2 sa. co-contaminated with NEO and NIV
- Further contamination of plants and plant products other than cereals and their products reported in the present book/article[154]: Bee pollen, NIV, literature[154]

T-2 Toxin

incidence: 5?/5*, conc. range: 299.00–299.60 µg/kg, ∅ conc.: 299.38 µg/kg, sample origin: beekeepers, west Slovakia, sample year: spring 2009, country: Slovakia[153], *poppy bee pollen, frozen

- Co-contamination: 5 sa. co-contaminated with AFS, DON, OTA, T-2, and ZEA
- Further contamination of plants and plant products other than cereals and their products reported in the present book/article[153]: Bee pollen, AF/S, DON, FB/S, OTA, and ZEA, literature[153]

incidence: 5?/5*, conc. range: 265.20–265.70 µg/kg, ∅ conc.: 265.40 µg/kg, sample origin: beekeepers, west Slovakia, sample year: spring 2009, country: Slovakia[153], *rape bee pollen, frozen

- Co-contamination: 5 sa. co-contaminated with AFS, DON, OTA, T-2, and ZEA
- Further contamination of plants and plant products other than cereals and their products reported in the present book/article[153]: Bee pollen, AF/S, DON, FB/S, OTA, and ZEA, literature[153]

incidence: 5?/5*, conc. range: 364.60–364.90 µg/kg, ∅ conc.: 364.72 µg/kg, sample origin: beekeepers, west Slovakia, sample year: spring 2009, country: Slovakia[153], *sunflower bee pollen, frozen

- Co-contamination: 5 sa. co-contaminated with AFS, DON, FBS, OTA, T-2, and ZEA
- Further contamination of plants and plant products other than cereals and their products reported in the present book/article[153]: Bee pollen, AF/S, DON, FB/S, OTA, and ZEA, literature[153]

Dried and UV-radiated bee pollen was also investigated. For detailed information see the article.

Type B Trichothecenes

Deoxynivalenol

incidence: 5?/5*, conc. range: 183.10–183.70 µg/kg, ∅ conc.: 183.38 µg/kg, sample origin: beekeepers, west Slovakia, sample year: spring 2009, country: Slovakia[153], *poppy bee pollen, frozen

- Co-contamination: 5 sa. co-contaminated with AFS, DON, OTA, T-2, and ZEA
- Further contamination of plants and plant products other than cereals and their products reported in the present book/article[153]: Bee pollen, AF/S, FB/S, OTA, T-2, and ZEA, literature[153]

incidence: 5?/5*, conc. range: 215.00–215.70 µg/kg, ∅ conc.: 215.32 µg/kg, sample origin: beekeepers, west Slovakia, sample year: spring 2009, country: Slovakia[153], *rape bee pollen, frozen

- Co-contamination: 5 sa. co-contaminated with AFS, DON, OTA, T-2, and ZEA
- Further contamination of plants and plant products other than cereals and their products reported in the present book/article[153]: Bee pollen, AF/S, FB/S, OTA, T-2, and ZEA, literature[153]

incidence: 5?/5*, conc. range: 203.10–203.50 µg/kg, ∅ conc.: 203.24 µg/kg, sample origin: beekeepers, west Slovakia, sample year: spring 2009, country: Slovakia[153], *sunflower bee pollen, frozen

- Co-contamination: 5 sa. co-contaminated with AFS, DON, OTA, T-2, and ZEA
- Further contamination of plants and plant products other than cereals and their products reported in the present book/article[153]: Bee pollen, AF/S, FB/S, OTA, T-2, and ZEA, literature[153]

Dried and UV-radiated bee pollen was also investigated. For detailed information see the article.

Nivalenol

incidence: 2/15, conc. range: ~1 µg/kg, sample origin: supermarkets, Valencia (city) metropolitan area, Spain, sample year: unknown, country: Spain[154]

- Co-contamination: 2 sa. co-contaminated with NEO and NIV
- Further contamination of plants and plant products other than cereals and their products reported in the present book/article[154]: Bee pollen, NEO, literature[154]

Other *Fusarium* Toxins

Zearalenone

incidence: 5?/5*, conc. range: 361.30–361.90 µg/kg, ∅ conc.: 361.58 µg/kg, sample origin: beekeepers, west Slovakia, sample year: spring 2009, country: Slovakia[153], *poppy bee pollen, frozen

- Co-contamination: 5 sa. co-contaminated with AFS, DON, OTA, T-2, and ZEA
- Further contamination of plants and plant products other than cereals and their products reported in the present book/article[153]: Bee pollen, AF/S, DON, FB/S, OTA, and T-2, literature[153]

incidence: 5?/5*, conc. range: 181.40–181.60 µg/kg, ∅ conc.: 181.50 µg/kg, sample origin: beekeepers, west Slovakia, sample year: spring 2009, country: Slovakia[153], *rape bee pollen, frozen

- Co-contamination: 5 sa. co-contaminated with AFS, DON, OTA, T-2, and ZEA
- Further contamination of plants and plant products other than cereals and their products reported in the present book/article[153]: Bee pollen, AF/S, DON, FB/S, OTA, and T-2, literature[153]

incidence: 5?/5*, conc. range: 147.10–147.40 µg/kg, ∅ conc.: 147.26 µg/kg, sample origin: beekeepers, west Slovakia, sample year: spring 2009, country: Slovakia[153], *sunflower bee pollen, frozen

- Co-contamination: 5 sa. co-contaminated with AFS, DON, FBS, OTA, T-2, and ZEA
- Further contamination of plants and plant products other than cereals and their products reported in the present book/article[153]: Bee pollen, AF/S, DON, FB/S, OTA, and T-2, literature[153]

Dried and UV-radiated bee pollen was also investigated. For detailed information see the article.

"Beerenauslese" see Wine and Wine (white)

Berries (blueberry) may contain the following mycotoxins:

Aspergillus and *Penicillium* Toxins

Patulin

incidence: 1/12, conc.: 21 µg/kg (pulp), sample origin: Uppsala (city) and Stockholm (capital) area, Sweden, sample year: summer 1976, country: Sweden[155]

- Co-contamination: not reported
- Further contamination of plants and plant products other than cereals and their products reported in the present book/article[155]: Berries (lingonberry), Fruits (peach), Fruits (plum), PAT, literature[155]

incidence: 3/16*, conc. range: ≤190 µg/kg, sample origin: Malmö (city), Sweden, sample year: 1975, country: Sweden[156], *blueberries, frozen

- Co-contamination: not reported
- Further contamination of plants and plant products other than cereals and their products reported in the present book/article[156]: not reported

Berries (grape) may contain the following mycotoxins:

Aspergillus and *Penicillium* Toxins

Ochratoxin A

incidence: 1/1*, conc.: 15.6 µg/kg**, sample origin: stores, Taiwan, sample year: unknown, country: Taiwan/Russia[147], *black grape, **COL-ELISA

- Co-contamination: not reported
- Further contamination of plants and plant products other than cereals and their products reported in the present book/article[147]: Beans (kidney beans), Coffee, Raisins, OTA, literature[147]

incidence: 0/12*, conc.: no contamination, sample origin: vineyards in Apulia (region) area, southern Italy, sample year: harvest season 2001, country: Italy[157], *Apulian wine grapes rotten (intact berries)

incidence: 2/12*, conc. range: 0.06–0.08 µg/kg, ∅ conc.: 0.07 µg/kg, sample origin: vineyards in Apulia (region) area, southern Italy, sample year: harvest season 2001, country: Italy[157], *Apulian wine grapes rotten by *Aspergillus* spp.

incidence: 12/12*, conc. range: 0.06–681 µg/kg, ∅ conc.: 108.46 µg/kg, sample origin: vineyards in Apulia (region) area, southern Italy, sample year: harvest season 2001, country: Italy[157], *Apulian wine grapes rotten by *Aspergillus* spp. and damaged by larvae of *Lobesia botrana*

incidence: 3/5*, conc. range: 0.02–0.03 µg/kg, ∅ conc.: 0.02 µg/kg, sample origin: vineyards in Apulia (region) area, southern Italy, sample year: harvest season 2002, country: Italy[157], *Apulian wine grapes (intact berries)

incidence: 5/5* **, conc. range: 0.03–0.36 µg/kg, ∅ conc.: 0.19 µg/kg, sample origin: vineyards in Apulia (region) area, southern Italy, sample year: harvest season 2002, country: Italy[157], *Apulian wine grapes rotten by *Aspergillus* spp.

incidence: 5/5* **, conc. range: 0.09–9.59 µg/kg, ∅ conc.: 5.22 µg/kg, sample origin: vineyards in Apulia (region) area, southern Italy, sample year: harvest season 2002, country: Italy[157], *Apulian wine grapes rotten by *Aspergillus* spp. and damaged by larvae of *Lobesia botrana*

- Co-contamination: not reported
- Further contamination of plants and plant products other than cereals and their products reported in the present book/article[157]: not reported

incidence: 21/23*, conc. range: 0.10–0.50 µg/l (2 sa.), 0.50–1.00 µg/l (6 sa.), 1.00–2.00 µg/l (3 sa.), >2.00–15.62 µg/l (10 sa.), sample origin: unknown, sample year: unknown, country: Spain[158], *grapes dried by sun

incidence: 1/3*, conc.: 1.74 µg/l, sample origin: unknown, sample year: unknown, country: Spain[158], *grapes dried in a warm chamber

incidence: 4/8*, conc. range: 0.50–1.00 µg/l (1 sa.), 1.00–2.00 µg/l (1 sa.), >2.00–4.79 µg/l (2 sa.), sample origin: unknown, sample year: unknown, country: Spain[158], *grapes dried in a fresh chamber

- Co-contamination: not reported
- Further contamination of plants and plant products other than cereals and their products reported in the present book/article[158]: Must, Wine, miscellaneous, OTA, literature[158]

incidence: 15/50*, conc. range: 0.50–2.69 µg/kg, ∅ conc.: 0.962 µg/kg, sample origin: producers and markets across Greece, sample year: August–October 2005, country: Greece[159], *grapes of different varieties for winemaking and/or table grapes

- Co-contamination: not reported
- Further contamination of plants and plant products other than cereals and their products reported in the present book/article[159]: not reported

For detailed information see the article.

incidence: 2/10, conc. range: 0.23–0.39 µg/kg, ∅ conc.: 0.31 µg/kg, sample origin: supermarkets and small shops in the northeast of China, sample year: January–June 2010, country: China[302]

- Co-contamination: not reported
- Further contamination of plants and plant products other than cereals and their products reported in the present book/article[302]: Coffee, Wine (red), OTA, literature[302]

incidence: 3/4*, conc. range: 0.0095–0.1156 µg/l, ∅ conc.: 0.0731 µg/l, sample origin: Dão (region) and Madeira (island) Portugal, sample year: September/October 2003, country: Portugal[841], *grapes for wine production

- Co-contamination: not reported
- Further contamination of plants and plant products other than cereals and their products reported in the present book/article[841]: not reported

incidence: 11/63, conc. range: 0.003–0.020 µg/l, sample origin: northern Italy, sample year: 2003–2007, country: Italy[914]

incidence: 1/30, conc. range: 0.003–0.020 µg/l, sample origin: central Italy, sample year: 2003–2007, country: Italy[914]

incidence: 50/111, conc. range: 0.003–0.020 µg/l (23 sa.), 0.02–0.20 µg/l (16 sa.), 0.20–2.00 µg/l (6 sa.), >2.00 µg/l (5 sa.), sample origin: southern Italy, sample year: 2003–2007, country: Italy[914]

- Co-contamination: not reported
- Further contamination of plants and plant products other than cereals and their products reported in the present book/article[914]: not reported

Berries (lingonberry) may contain the following mycotoxins:

Aspergillus and ***Penicillium*** Toxins

Patulin

incidence: 1/2, conc.: 265 µg/kg (pulp), sample origin: Uppsala (city) and Stockholm (capital) area, Sweden, sample year: summer 1976, country: Sweden[155]

- Co-contamination: not reported
- Further contamination of plants and plant products other than cereals and their products reported in the present book/article[155]: Berries (blueberry), Fruits (peach), Fruits (plum), PAT, literature[155]

Betel nuts may contain the following mycotoxins:

Aspergillus Toxins

Aflatoxin B_1

incidence: 10/10*, conc. range: 2.1–10.2 µg/kg, sample origin: imported from Asian countries, produced from different commercial outlets in greater Durban (city) metropolitan area, South Africa, sample year: unknown, country: South Africa[56], *betel nuts, raw and sliced

incidence: 2/20*, conc. range: ≤0.1 µg/kg, sample origin: imported from Asian countries, produced from different commercial outlets in greater Durban (city) metropolitan area, South Africa, sample year: unknown, country: South Africa[56], *betel nuts, boiled

- Co-contamination: not reported
- Further contamination of plants and plant products other than cereals and their products reported in the present book/article[56]: Betel nuts, AFB_2, AFG_1, and AFG_2, literature[56]

incidence: 12/32, conc. range: 18–208 µg/kg, ∅ conc.: 94.2 µg/kg, sample origin: markets of some important north Indian cities, India, sample year: unknown, country: India[57]

- Co-contamination: 3 sa. co-contaminated with AFB_1, AFB_2, AFG_1, and AFG_2; 3 sa. co-contaminated with AFB_1, AFB_2, and AFG_1; 3 sa. co-contaminated with AFB_1 and AFG_1; 3 sa. contaminated solely with AFB_1
- Further contamination of plants and plant products other than cereals and their products reported in the present book/article[57]: Betel nuts, AFB_2, AFG_1, and AFG_2, literature[57]

incidence: 21/21, conc. range: 10–20 µg/kg (2 sa.), ≤262.0 µg/kg (19 sa.), ∅ conc.: 92.5 µg/kg, sample origin: imported from India collected at the port of Karachi (city), Pakistan, sample year: 2010/2011, country: Pakistan[681]

incidence: 41/51, conc. range: 2–10 µg/kg (19 sa.), 10–20 µg/kg (19 sa.), ≤39.2 µg/kg (3 sa.), sample origin: imported from Indonesia collected at the port of Karachi (city), Pakistan, sample year: 2010/2011, country: Pakistan[681]

incidence: 25/34, conc. range: 2–10 µg/kg (3 sa.), 10–20 µg/kg (9 sa.), ≤103.4 µg/kg (13 sa.), sample origin: imported from Sri Lanka collected at the port of Karachi (city), Pakistan, sample year: 2010/2011, country: Pakistan[681]

incidence: 52/172, conc. range: 2–10 µg/kg (50 sa.), ≤77.0 µg/kg (2 sa.), sample origin: imported from Indonesia collected at the

port of Karachi (city), Pakistan, sample year: 2010/2011, country: Pakistan[681]
- Co-contamination: not reported
- Further contamination of plants and plant products other than cereals and their products reported in the present book/article[681]: not reported

Aflatoxin B_2
incidence: 10/10*, conc. range: 0.2–1.3 µg/kg, sample origin: imported from Asian countries, produced from different commercial outlets in greater Durban (city) metropolitan area, South Africa, sample year: unknown, country: South Africa[56], *betel nuts, raw and sliced

incidence: 0/20*, conc.: no contamination, sample origin: imported from Asian countries, produced from different commercial outlets in greater Durban (city) metropolitan area, South Africa, sample year: unknown, country: South Africa[56], *betel nuts, boiled
- Co-contamination: not reported
- Further contamination of plants and plant products other than cereals and their products reported in the present book/article[56]: Betel nuts, AFB_1, AFG_1, and AFG_2, literature[56]

incidence: 6/32, conc. range: 10.0–52.0 µg/kg, ∅ conc.: 24.3 µg/kg, sample origin: markets of some important north Indian cities, India, sample year: unknown, country: India[57]
- Co-contamination: 3 sa. co-contaminated with AFB_1, AFB_2, AFG_1, and AFG_2; 3 sa. co-contaminated with AFB_1, AFB_2, and AFG_1
- Further contamination of plants and plant products other than cereals and their products reported in the present book/article[57]: Betel nuts, AFB_1, AFG_1, and AFG_2, literature[57]

Aflatoxin G_1
incidence: 10/10*, conc. range: 0.2–16.6 µg/kg, sample origin: imported from Asian countries, produced from different commercial outlets in greater Durban (city) metropolitan area, South Africa, sample year: unknown, country: South Africa[56], *betel nuts, raw and sliced

incidence: 0/20*, conc.: no contamination, sample origin: imported from Asian countries, produced from different commercial outlets in greater Durban (city) metropolitan area, South Africa, sample year: unknown, country: South Africa[56], *betel nuts, boiled
- Co-contamination: not reported
- Further contamination of plants and plant products other than cereals and their products reported in the present book/article[56]: Betel nuts, AFB_1, AFB_2, and AFG_2, literature[56]

incidence: 9/32, conc. range: tr-81.0 µg/kg, sample origin: markets of some important north Indian cities, India, sample year: unknown, country: India[57]
- Co-contamination: 3 sa. co-contaminated with AFB_1, AFB_2, AFG_1, and AFG_2; 3 sa. co-contaminated with AFB_1, AFB_2, and AFG_1; 3 sa. co-contaminated with AFB_1 and AFG_1
- Further contamination of plants and plant products other than cereals and their products reported in the present book/article[57]: Betel nuts, AFB_1, AFB_2, and AFG_2, literature[57]

Aflatoxin G_2
incidence: 10/10*, conc. range: 0.1–1.8 µg/kg, sample origin: imported from Asian countries, produced from different commercial outlets in greater Durban (city) metropolitan area, South Africa, sample year: unknown, country: South Africa[56], *betel nuts, raw and sliced

incidence: 0/20*, conc.: no contamination, sample origin: imported from Asian countries, produced from different commercial outlets in greater Durban (city) metropolitan area, South Africa, sample year: unknown, country: South Africa[56], *betel nuts, boiled
- Co-contamination: not reported
- Further contamination of plants and plant products other than cereals and their products reported in the present book/article[56]: Betel nut, AFB_1, AFB_2, and AFG_1, literature[56]

incidence: 3/32, conc. range: tr-<20.0 µg/kg, sample origin: markets of some important north Indian cities, India, sample year: unknown, country: India[57]
- Co-contamination: 3 sa. co-contaminated with AFB_1, AFB_2, AFG_1, and AFG_2
- Further contamination of plants and plant products other than cereals and their products reported in the present book/article[57]: Betel nut, AFB_1, AFB_2, and AFG_1, literature[57]

Aflatoxin/s
incidence: 20/80, conc. range: <30 µg/kg*, sample origin: 16 districts of eastern Nepal, sample year: 1995–2003, country: Nepal[58], *AF
- Co-contamination: not reported
- Further contamination of plants and plant products other than cereals and their products reported in the present book/article[58]: Butter (peanut butter), Nuts (peanut), AFB_1, literature[58]

Beverages may contain the following mycotoxins:

Alternaria Toxins

Alternariol
incidence: 1/1*, conc.: 0.27 µg/kg, sample origin: commercially available, sample year: unknown, country: Germany[160], *fruit punch
- Co-contamination: 1 sa. co-contaminated with AME and AOH
- Further contamination of plants and plant products other than cereals and their products reported in the present book/article[160]: Juices (apple juice), Juices (grape juice), Juices (orange juice), Juices (tomato juice), Juices (vegetable juice), Wine (red), Wine (white), Wine, miscellaneous, AME and AOH, literature[160]; Beverages, AME, literature[160]

Alternariol Methyl Ether
incidence: 1/1*, conc.: 0.04 µg/kg, sample origin: commercially available, sample year: unknown, country: Germany[160], *fruit punch
- Co-contamination: 1 sa. co-contaminated with AME and AOH
- Further contamination of plants and plant products other than cereals and their products reported in the present book/article[160]: Juices (apple juice), Juices (grape juice), Juices (orange juice), Juices (tomato juice), Juices (vegetable juice), Wine (red), Wine (white), Wine, miscellaneous, AME and AOH, literature[160]; Beverages, AOH, literature[160]

Aspergillus Toxins

Aflatoxin B_1
incidence: 9/17*, conc. range: 10–50 µg/kg, ∅ conc.: 23.3 µg/kg, sample origin: brewers in Dar es Salaam (city), Tanzania, sample year: 1988, country: Finland/Tanzania[161], *brewed or distilled beverages included 15 locally made beers produced from grain, fruits, or sugarcane and honey
- Co-contamination: not reported
- Further contamination of plants and plant products other than cereals and their products reported in the present book/article[161]: not reported

For detailed information see the article.

incidence: 3/25*, conc. range: 1.2–3.1 µg/l, ∅ conc.: 2.4 µg/l, sample origin: supermarkets, street vendors, juice bars, and ice cream parlors, Valencia (city), Spain, sample year: unknown, country: Spain[162], *tigernut beverage

- Co-contamination: 3 sa. co-contaminated with AFB_1, AFB_2, and AFG_2
- Further contamination of plants and plant products other than cereals and their products reported in the present book/article[162]: Beverages, AFB_2 and AFG_2, literature[162]; Nuts (tigernut), AFB_1, AFB_2, AFG_1, and AFG_2, literature[162]

incidence: 12/190*, conc. range: 0.8–1.7 µg/l, sample origin: supermarkets, street vendors, juice bars, and ice cream parlors, Valencia (city), Spain, sample year: March 2009–March 2010, country: Spain[163], *included sterilized, concentrated, pasteurized, and fresh serving horchata (tigernut beverage)

- Co-contamination: For detailed information see the article.
- Further contamination of plants and plant products other than cereals and their products reported in the present book/article[163]: Beverages, AFB_2, AFG_2, and OTA, literature[163]; Nuts (tigernut), AFB_1, AFB_2, AFG_2, and OTA, literature[163]

For detailed information see the article.

incidence: 5/5*, conc. range: 20–30 µg/l, sample origin: shops and markets, Assiut (governorate), Egypt, sample year: unknown, country: Egypt[443], *apple beverage

- Co-contamination: 5 sa. co-contaminated with AFB_1 and AFG_1
- Further contamination of plants and plant products other than cereals and their products reported in the present book/article[443]: Beverages, AFG_1, literature[443]; Juices (guava juice), AFB_1, literature[443]

incidence: 1/3*, conc.: <LOQ, sample origin: local supermarkets, Spain, sample year: unknown, country: Spain[793], *soy beverage

- Co-contamination: 1 sa. co-contaminated with AFB_1, AFG_1, AFG_2, OTA, and ZEA
- Further contamination of plants and plant products other than cereals and their products reported in the present book/article[793]: Beverages, AFG_1, AFG_2, OTA, and ZEA, literature[793]

Aflatoxin B_2

incidence: 3/25*, conc. range: 1.3–2.2 µg/l, ∅ conc.: 1.8 µg/l, sample origin: supermarkets, street vendors, juice bars, and ice cream parlors, Valencia (city), Spain, sample year: unknown, country: Spain[162], *tigernut beverage

- Co-contamination: 3 sa. co-contaminated with AFB_1, AFB_2, and AFG_2
- Further contamination of plants and plant products other than cereals and their products reported in the present book/article[162]: Beverages, AFB_1 and AFG_2, literature[162]; Nuts (tigernut), AFB_1, AFB_2, AFG_1, and AFG_2, literature[162]

incidence: 2/190*, conc. range: 1.1–1.6 µg/l, ∅ conc.: 1.35 µg/l, sample origin: supermarkets, street vendors, juice bars, and ice cream parlors, Valencia (city), Spain, sample year: March 2009–March 2010, country: Spain[163], *included sterilized, concentrated, pasteurized, and fresh serving horchata (tigernut beverage)

- Co-contamination: For detailed information see the article.
- Further contamination of plants and plant products other than cereals and their products reported in the present book/article[163]: Beverages, AFB_1, AFG_2, and OTA, literature[163]; Nuts (tigernut), AFB_1, AFB_2, AFG_2, and OTA, literature[163]

For detailed information see the article.

Aflatoxin G_1

incidence: 5/5*, conc. range: 20–30 µg/l, sample origin: shops and markets, Assiut (governorate), Egypt, sample year: unknown, country: Egypt[443], *apple beverage

- Co-contamination: 5 sa. co-contaminated with AFB_1 and AFG_1
- Further contamination of plants and plant products other than cereals and their products reported in the present book/article[443]: Beverages, AFB_1, literature[443]; Juices (guava juice), AFB_1, literature[443]

incidence: 2/3*, conc.: <LOQ, sample origin: local supermarkets, Spain, sample year: unknown, country: Spain[793], *soy beverage

- Co-contamination: 1 sa. co-contaminated with AFB_1, AFG_1, AFG_2, OTA, and ZEA; 1 sa. co-contaminated with AFG_1, AFG_2, and OTA
- Further contamination of plants and plant products other than cereals and their products reported in the present book/article[793]: Beverages, AFB_1, AFG_2, OTA, and ZEA, literature[793]

Aflatoxin G_2

incidence: 3/25*, conc. range: 1.3–1.4 µg/l, ∅ conc.: 1.36 µg/l, sample origin: supermarkets, street vendors, juice bars, and ice cream parlors, Valencia (city), Spain, sample year: unknown, country: Spain[162], *tigernut beverage

- Co-contamination: 3 sa. co-contaminated with AFB_1, AFB_2, and AFG_2
- Further contamination of plants and plant products other than cereals and their products reported in the present book/article[162]: Beverages, AFB_1 and AFB_2, literature[162]; Nuts (tigernut), AFB_1, AFB_2, AFG_1, and AFG_2, literature[162]

incidence: 1?/190*, conc. range: 1.2–2.3 µg/l, sample origin: supermarkets, street vendors, juice bars, and ice cream parlors, Valencia (city), Spain, sample year: March 2009–March 2010, country: Spain[163], *included sterilized, concentrated, pasteurized, and fresh serving horchata (tigernut beverage)

- Co-contamination: For detailed information see the article.
- Further contamination of plants and plant products other than cereals and their products reported in the present book/article[163]: Beverages, AFB_1, AFB_2, and OTA, literature[163]; Nuts (tigernut), AFB_1, AFB_2, AFG_2, and OTA, literature[163]

For detailed information see the article.

incidence: 2/3*, conc.: <LOQ, sample origin: local supermarkets, Spain, sample year: unknown, country: Spain[793], *soy beverage

- Co-contamination: 1 sa. co-contaminated with AFB_1, AFG_1, AFG_2, OTA, and ZEA; 1 sa. co-contaminated with AFG_1, AFG_2, and OTA
- Further contamination of plants and plant products other than cereals and their products reported in the present book/article[793]: Beverages, AFB_1, AFG_1, OTA, and ZEA, literature[793]

Aflatoxin/s

incidence: 4/4*, conc. range: 0.772–5.937 µg/kg**, ∅ conc.: 2.572 µg/kg**, sample origin: groceries and supermarkets, Turkey, sample year: 2003, country: Turkey[119], *instant fruit flavored drink, **AFS total

- Co-contamination: not reported
- Further contamination of plants and plant products other than cereals and their products reported in the present book/article[119]: Asure, Cocoa, Cream, Ice cream, Jello, Keskul, Muhallebi, Pudding, Salep, Sauces (chocolate sauce), Soups, AF/S, literature[119]

Instant fruit flavored drink ingredients contained sugar, acidity regulator, fruit flavor, thickener agent, sweetener, and coloring agents.

***Aspergillus* and *Penicillium* Toxins**

Ochratoxin A

incidence: 11/41*, conc. range: ≤0.039 µg/kg, Ø conc.: 0.026 µg/kg, sample origin: local supermarkets and small retail shops in seven major cities, Japan, sample year: 2006–winter 2007, country: Japan[85], *coffee beverage

- Co-contamination: not reported
- Further contamination of plants and plant products other than cereals and their products reported in the present book/article[85]: Beans (soybean), Vegetables (asparagus), FB_1 and FB_2, literature[85]; Chocolate, Cocoa, Coffee, Coffee beans, Raisins, Spices (coriander), Wine, OTA, literature[85]

incidence: 1/1*, conc.: 0.05 µg/l, sample origin: supermarkets and fast-food chains in Quebec (city), Canada, sample year: September/October 2008, country: Canada[90], *soy beverage, fortified

incidence: 1/1*, conc.: 0.02 µg/l, sample origin: supermarkets and fast-food chains in Calgary (city), Canada, sample year: September/October 2009, country: Canada[90], *soy beverage, fortified

- Co-contamination: not reported
- Further contamination of plants and plant products other than cereals and their products reported in the present book/article[90]: Baby food, Beans, Butter (peanut butter), Candies, Chocolate, Coffee, Extracts, Herbs, Ice cream, Nuts, Pies, Pizza, Products, Pudding, Raisins, Seeds, Wine, OTA, literature[90]

incidence: 2/190*, conc. range: 2–2.4 µg/l, Ø conc.: 2.2 µg/l, sample origin: supermarkets, street vendors, juice bars, and ice cream parlors, Valencia (city), Spain, sample year: March 2009–March 2010, country: Spain[163], *included sterilized, concentrated, pasteurized, and fresh serving horchata (tigernut beverage)

- Co-contamination: For detailed information see the article.
- Further contamination of plants and plant products other than cereals and their products reported in the present book/article[163]: Beverages, AFB_1, AFB_2, and AFG_2, literature[163]; Nuts (tigernut), AFB_1, AFB_2, AFG_2, and OTA, literature[163]

For detailed information see the article.

incidence: 2/3*, conc.: <LOQ–0.1 µg/kg, sample origin: local supermarkets, Spain, sample year: unknown, country: Spain[793], *soy beverage

- Co-contamination: 1 sa. co-contaminated with AFB_1, AFG_1, AFG_2, OTA, and ZEA; 1 sa. co-contaminated with AFG_1, AFG_2, and OTA
- Further contamination of plants and plant products other than cereals and their products reported in the present book/article[793]: Beverages, AFB_1, AFG_1, AFG_2, and ZEA, literature[793]

Patulin

incidence: 29/66*, conc. range: 2.5–27 µg/l, sample origin: partly Swedish and partly imported, Sweden, sample year: unknown, country: Sweden[451], *apple beverages

- Co-contamination: not reported
- Further contamination of plants and plant products other than cereals and their products reported in the present book/article[451]: not reported

***Fusarium* Toxins**

Zearalenone

incidence: 1/3*, conc.: <LOQ, sample origin: local supermarkets, Spain, sample year: unknown, country: Spain[793], *soy beverage

- Co-contamination: 1 sa. co-contaminated with AFB_1, AFG_1, AFG_2, OTA, and ZEA
- Further contamination of plants and plant products other than cereals and their products reported in the present book/article[793]: Beverages, AFB_1, AFG_1, AFG_2, and OTA, literature[793]

see also **Juices (apple juice)**

Biscuits may contain the following mycotoxins:

***Aspergillus* Toxins**

Aflatoxin B_1

incidence: 3/25*, conc. range: 0.91–1.06 µg/kg, Ø conc.: 0.99 µg/kg, sample origin: local markets and supermarkets in Novi Sad (city), Serbia, sample year: February–June 2014, country: Serbia[903], *biscuits with fig filling

incidence: 2/14*, conc. range: 0.95–1.92 µg/kg, Ø conc.: 1.48 µg/kg, sample origin: local markets and supermarkets in Novi Sad (city), Serbia, sample year: February–June 2014, country: Serbia[903], *biscuits with cherry filling

- Co-contamination: not reported
- Further contamination of plants and plant products other than cereals and their products reported in the present book/article[903]: Biscuits, OTA, T-2, and ZEA, literature[903]; Jam (fig jam), AFB_1, AFB_2, OTA, and T-2, literature[903]

***Aspergillus* and *Penicillium* Toxins**

Ochratoxin A

incidence: 1/1* **, conc.: 0.052 µg/kg, sample origin: organic product retail outlets in Lisbon (capital), Portugal, sample year: May/June 2007, country: Portugal/Netherlands[88], *organic baby food, **biscuits consisted of wheat 60%, grape juice 28%, sunflower oil 2%, and ginger 1%

- Co-contamination: not reported
- Further contamination of plants and plant products other than cereals and their products reported in the present book/article[88]: Baby food, AFB_1, AFM_1, and OTA, literature[88]; Flour, AFB_1 and OTA, literature[88]

For detailed information see the article.

incidence: 14/25*, conc. range: 1.58–11.6 µg/kg, Ø conc.: 5.68 µg/kg, sample origin: local markets and supermarkets in Novi Sad (city), Serbia, sample year: February–June 2014, country: Serbia[903], *biscuits with fig filling

incidence: 3/14*, conc. range: 1.75–3.75 µg/kg, Ø conc.: 2.50 µg/kg, sample origin: local markets and supermarkets in Novi Sad (city), Serbia, sample year: February–June 2014, country: Serbia[903], *biscuits with cherry filling

- Co-contamination: not reported
- Further contamination of plants and plant products other than cereals and their products reported in the present book/article[903]: Biscuits, AFB_1, T-2, and ZEA, literature[903]; Jam (fig jam), AFB_1, AFB_2, OTA, and T-2, literature[903]

Fusarium Toxins

Type A Trichothecenes

T-2 Toxin
incidence: 11/25*, conc. range: 2.75–26.7 μg/kg, ∅ conc.: 8.72 μg/kg, sample origin: local markets and supermarkets in Novi Sad (city), Serbia, sample year: February–June 2014, country: Serbia[903], *biscuits with fig filling

incidence: 3/14*, conc. range: 2.90–11.8 μg/kg, ∅ conc.: 5.94 μg/kg, sample origin: local markets and supermarkets in Novi Sad (city), Serbia, sample year: February–June 2014, country: Serbia[903], *biscuits with cherry filling

- Co-contamination: not reported
- Further contamination of plants and plant products other than cereals and their products reported in the present book/article[903]: Biscuits, AFB_1, OTA, and ZEA, literature[903]; Jam (fig jam), AFB_1, AFB_2, OTA, and T-2, literature[903]

Other ***Fusarium*** Toxins

Zearalenone
incidence: 14/25*, conc. range: 1.11–4.15 μg/kg, ∅ conc.: 2.28 μg/kg, sample origin: local markets and supermarkets in Novi Sad (city), Serbia, sample year: February–June 2014, country: Serbia[903], *biscuits with fig filling

incidence: 6/14*, conc. range: 1.05–8.86 μg/kg, ∅ conc.: 2.78 μg/kg, sample origin: local markets and supermarkets in Novi Sad (city), Serbia, sample year: February–June 2014, country: Serbia[903], *biscuits with cherry filling

- Co-contamination: not reported
- Further contamination of plants and plant products other than cereals and their products reported in the present book/article[903]: Biscuits, AFB_1, OTA, and T-2, literature[903]; Jam (fig jam), AFB_1, AFB_2, OTA, and T-2, literature[903]

Biryani masala see **Spices (biryani masala)**

Black gram see **Products (legume product)**

Black radish see **Vegetables (black radish)**

Black tea see **Tea**

Bondakadledkai may contain the following mycotoxins:

Aspergillus Toxins

Aflatoxin B_1
incidence: 22/54, conc. range: 3–1,500 μg/kg, sample origin: Mysore (city) market, India, sample year: October 1991–January 1992, country: India[199]

- Co-contamination: 1 sa. co-contaminated with AFB_1, AFB_2, AFG_1, and AFG_2; no further information available
- Further contamination of plants and plant products other than cereals and their products reported in the present book/article[199]: Bondakadledkai, AFB_2, literature[199]; Congressbele, Toffees, AFB_1 and AFB_2, literature[199]

Aflatoxin B_2
incidence: 14/54, conc. range: 3–370 μg/kg, sample origin: Mysore (city) market, India, sample year: October 1991–January 1992, country: India[199]

- Co-contamination: 1 sa. co-contaminated with AFB_1, AFB_2, AFG_1, and AFG_2; no further information available
- Further contamination of plants and plant products other than cereals and their products reported in the present book/article[199]: Bondakadledkai, AFB_1, literature[199]; Congressbele, Toffees, AFB_1 and AFB_2, literature[199]

Bondakaledkai is a groundnut-based (whole seeds are used), spiced snack product.

Brazil nuts see **Nuts (Brazil nut, paranut)**

Breadfruit may contain the following mycotoxins:

Aspergillus Toxins

Aflatoxin B_1
incidence: 9/9*, conc. range: 40.06–48.59 μg/kg, ∅ conc.: 45.37 μg/kg, sample origin: markets in the Niger Delta region, Nigeria, sample year: November 2004–March 2005, country: Nigeria[210]

- Co-contamination: not reported
- Further contamination of plants and plant products other than cereals and their products reported in the present book/article[210]: Nuts (peanut), AFB_1, literature[210]

Breakfast cereals may contain the following mycotoxins:

Aspergillus Toxins

Aflatoxin B_1
incidence: 2/4* **, conc. range: LOD-2 μg/kg (1 sa.), >2 μg/kg (1 sa.), sample origin: supermarkets in Athens (capital), Greece, sample year: 2006/2007, country: Greece[91], *breakfast cereals+dried fruits, **conventional

incidence: 0/0* **, no sa. investigated, supermarkets in Athens (capital), Greece, sample year: 2006/2007, country: Greece[91], *breakfast cereals+dried fruits, **organic

incidence: 1/1* **, conc.: >2 μg/kg (1 sa.), sample origin: supermarkets in Athens (capital), Greece, sample year: 2006/2007, country: Greece[91], *breakfast cereals+dried fruits+cinnamon, **conventional

incidence: 0/0* **, no sa. investigated, supermarkets in Athens (capital), Greece, sample year: 2006/2007, country: Greece[91], *breakfast cereals+dried fruits+cinnamon, **organic

incidence: 1/3* **, conc.: LOD-2 μg/kg (1 sa.), sample origin: supermarkets in Athens (capital), Greece, sample year: 2006/2007, country: Greece[91], *breakfast cereals, multicereal+nuts, **conventional

incidence: 1/1* **, conc.: 0.59 μg/kg, sample origin: supermarkets in Athens (capital), Greece, sample year: 2006/2007, country: Greece[91], *breakfast cereals, multicereal+nuts, **organic

incidence: 2/6* **, conc. range: LOD-2 μg/kg (2 sa., maximum: 1.90 μg/kg), sample origin: supermarkets in Athens (capital), Greece, sample year: 2006/2007, country: Greece[91], *breakfast cereals+dried fruits+nuts, **conventional

incidence: 0/0* **, no sa. investigated, supermarkets in Athens (capital), Greece, sample year: 2006/2007, country: Greece[91], *breakfast cereals+dried fruits+nuts, **organic

incidence: 0/0* **, no sa. investigated, supermarkets in Athens (capital), Greece, sample year: 2006/2007, country: Greece[91], *breakfast cereals, (wheat+maize-based+cocoa), **conventional incidence: 0/1* **, conc.: no contamination, sample origin: supermarkets in Athens (capital), Greece, sample year: 2006/2007, country: Greece[91], *breakfast cereals, (wheat+maize-based+cocoa), **organic

incidence: 0/0* **, no sa. investigated, supermarkets in Athens (capital), Greece, sample year: 2006/2007, country: Greece[91], *breakfast cereals, multicereal formula+cocoa (maize+barley-based), **conventional

incidence: 1/1* **, conc.: 0.88 µg/kg, sample origin: supermarkets in Athens (capital), Greece, sample year: 2006/2007, country: Greece[91], *breakfast cereals, multicereal formula+cocoa (maize+barley-based), **organic

- Co-contamination: For detailed information see the article.
- Further contamination of plants and plant products other than cereals and their products reported in the present book/article[91]: Breakfast cereals, OTA, literature[91]

incidence: 18/26*, conc. range: ≤0.130 µg/kg, sample origin: supermarkets in Lisbon (capital) region, Portugal, sample year: unknown, country: Portugal/Brazil[537], *breakfast cereals in part with chocolate

- Co-contamination: For detailed information see the article.
- Further contamination of plants and plant products other than cereals and their products reported in the present book/article[537]: Breakfast cereals, AFB_2, AFG_1, AFM_1, DON, FB_1, FB_2, NIV, OTA, and ZEA, literature[537]

Aflatoxin B_2

incidence: 7/26*, conc. range: ≤0.011 µg/kg, sample origin: supermarkets in Lisbon (capital) region, Portugal, sample year: unknown, country: Portugal/Brazil[537], *breakfast cereals in part with chocolate

- Co-contamination: For detailed information see the article.
- Further contamination of plants and plant products other than cereals and their products reported in the present book/article[537]: Breakfast cereals, AFB_1, AFG_1, AFM_1, DON, FB_1, FB_2, NIV, OTA, and ZEA, literature[537]

Aflatoxin G_1

incidence: 1/26*, conc.: 0.014 µg/kg, sample origin: supermarkets in Lisbon (capital) region, Portugal, sample year: unknown, country: Portugal/Brazil[537], *breakfast cereals in part with chocolate

- Co-contamination: For detailed information see the article.
- Further contamination of plants and plant products other than cereals and their products reported in the present book/article[537]: Breakfast cereals, AFB_1, AFB_2, AFM_1, DON, FB_1, FB_2, NIV, OTA, and ZEA, literature[537]

Aflatoxin M_1

incidence: 3/26*, conc. range: ≤0.024 µg/kg, sample origin: supermarkets in Lisbon (capital) region, Portugal, sample year: unknown, country: Portugal/Brazil[537], *breakfast cereals in part with chocolate

- Co-contamination: For detailed information see the article.
- Further contamination of plants and plant products other than cereals and their products reported in the present book/article[537]: Breakfast cereals, AFB_1, AFB_2, AFG_1, DON, FB_1, FB_2, NIV, OTA, and ZEA, literature[537]

Aspergillus and ***Penicillium*** Toxins

Citrinin

incidence: 8/45*, conc. range: <LOQ-42 µg/kg, sample origin: retail outlets, France, sample year: unknown, country: France[94], *breakfast cereals in part with fruits, raisins, and/or cocoa

- Co-contamination: 6 sa. co-contaminated with CIT, FB_1, and OTA; 2 sa. co-contaminated with CIT and OTA
- Further contamination of plants and plant products other than cereals and their products reported in the present book/article[94]: Breakfast cereals, FB_1 and OTA, literature[94]

For detailed information see the article.

Ochratoxin A

incidence: 1/5* **, conc.: >2 µg/kg, sample origin: supermarkets in Athens (capital), Greece, sample year: 2006/2007, country: Greece[91], *breakfast cereals, wheat-based+chocolate, **conventional

incidence: 0/0* **, no sa. investigated, supermarkets in Athens (capital), Greece, sample year: 2006/2007, country: Greece[91], *breakfast cereals, wheat-based+chocolate, **organic

incidence: 2/4* **, conc. range: LOD-3 µg/kg (2 sa., maximum: 0.87 µg/kg), sample origin: supermarkets in Athens (capital), Greece, sample year: 2006/2007, country: Greece[91], *breakfast cereals+dried fruits, **conventional

incidence: 0/0* **, no sa. investigated, supermarkets in Athens (capital), Greece, sample year: 2006/2007, country: Greece[91], *breakfast cereals+dried fruits, **organic

incidence: 1/3* **, conc.: LOD-3 µg/kg, sample origin: supermarkets in Athens (capital), Greece, sample year: 2006/2007, country: Greece[91], *breakfast cereals, multicereal+nuts, **conventional

incidence: 1/1* **, conc.: 0.10 µg/kg, sample origin: supermarkets in Athens (capital), Greece, sample year: 2006/2007, country: Greece[91], *breakfast cereals, multicereal+nuts, **organic

incidence: 5/6* **, conc. range: LOD-3 µg/kg (5 sa.), sample origin: supermarkets in Athens (capital), Greece, sample year: 2006/2007, country: Greece[91], *breakfast cereals+dried fruits+nuts, **conventional

incidence: 0/0* **, no sa. investigated, supermarkets in Athens (capital), Greece, sample year: 2006/2007, country: Greece[91], *breakfast cereals+dried fruits+nuts, **organic

incidence: 0/0* **, no sa. investigated, supermarkets in Athens (capital), Greece, sample year: 2006/2007, country: Greece[91], *breakfast cereals, multicereal formula+cocoa (maize+barley-based), **conventional

incidence: 0/1* **, conc.: no contamination, sample origin: supermarkets in Athens (capital), Greece, sample year: 2006/2007, country: Greece[91], *breakfast cereals, multicereal formula+cocoa (maize+barley-based), **organic

- Co-contamination: For detailed information see the article.
- Further contamination of plants and plant products other than cereals and their products reported in the present book/article[91]: Breakfast cereals, AFB_1, literature[91]

incidence: 31/45*, conc. range: <LOQ-8.8 µg/kg, sample origin: retail outlets, France, sample year: unknown, country: France[94], *breakfast cereals in part with fruits, raisins, and/or cocoa

- Co-contamination: 13 sa. co-contaminated with FB_1 and OTA; 6 sa. co-contaminated with CIT, FB_1, and OTA; 2 sa. co-contaminated with CIT and OTA; 10 sa. contaminated solely with OTA
- Further contamination of plants and plant products other than cereals and their products reported in the present book/article[94]: Breakfast cereals, CIT and FB_1, literature[94]

For detailed information see the article.

incidence: 1/2*, conc.: 127.5 µg/kg, sample origin: supermarkets and pharmacies in Rabat (capital), Morocco, sample year: unknown, country: Morocco/Spain/Italy[95], *breakfast cereals with fruit rings
- Co-contamination: not reported
- Further contamination of plants and plant products other than cereals and their products reported in the present book/article[95]: not reported

incidence: 18/26*, conc. range: ≤0.100 µg/kg, sample origin: supermarkets in Lisbon (capital) region, Portugal, sample year: unknown, country: Portugal/Brazil[537], *breakfast cereals in part with chocolate
- Co-contamination: For detailed information see the article.
- Further contamination of plants and plant products other than cereals and their products reported in the present book/article[537]: Breakfast cereals, AFB_1, AFB_2, AFG_1, AFM_1, DON, FB_1, FB_2, NIV, and ZEA, literature[537]

incidence: 1/2*, conc.: 0.53 µg/kg, sample origin: supermarkets in Novi Sad (city), Serbia, sample year: 2012, country: Serbia[900], *multigrain + cinnamon

incidence: 2/5*, conc. range: 0.08–0.09 µg/kg, Ø conc.: 0.085 µg/kg, sample origin: supermarkets in Novi Sad (city), Serbia, sample year: 2012, country: Serbia[900], *multigrain + chocolate

incidence: 0/3*, conc.: no contamination, sample origin: supermarkets in Novi Sad (city), Serbia, sample year: 2012, country: Serbia[900], *multigrain + nuts

incidence: 3/14*, conc. range: 0.08–11.81 µg/kg, Ø conc.: 4.01 µg/kg, sample origin: supermarkets in Novi Sad (city), Serbia, sample year: 2012, country: Serbia[900], *multigrain + dried fruits

incidence: 4/13*, conc. range: 0.11–8.14 µg/kg, Ø conc.: 4.04 µg/kg, sample origin: supermarkets in Novi Sad (city), Serbia, sample year: 2012, country: Serbia[900], *multigrain + nuts + dried fruit

incidence: 0/1*, conc.: no contamination, sample origin: supermarkets in Novi Sad (city), Serbia, sample year: 2015, country: Serbia[900], *multigrain + cinnamon

incidence: 0/1*, conc.: no contamination, sample origin: supermarkets in Novi Sad (city), Serbia, sample year: 2015, country: Serbia[900], *multigrain + chocolate

incidence: 1/2*, conc.: 0.14 µg/kg, sample origin: supermarkets in Novi Sad (city), Serbia, sample year: 2015, country: Serbia[900], *multigrain + nuts

incidence: 3/21*, conc. range: 0.09–0.20 µg/kg, Ø conc.: 0.14 µg/kg, sample origin: supermarkets in Novi Sad (city), Serbia, sample year: 2015, country: Serbia[900], *multigrain + dried fruits

incidence: 2/10*, conc. range: 0.38–2.33 µg/kg, Ø conc.: 1.36 µg/kg, sample origin: supermarkets in Novi Sad (city), Serbia, sample year: 2015, country: Serbia[900], *multigrain + nuts + dried fruit
- Co-contamination: For detailed information see the article.
- Further contamination of plants and plant products other than cereals and their products reported in the present book/article[900]: not reported

Fusarium Toxins

Fumonisin/s

Fumonisin B_1

incidence: 30/32*, conc. range: <LOQ-1,113 µg/kg, sample origin: retail outlets, France, sample year: unknown, country: France[94], *breakfast cereals in part with fruits, raisins, and/or cocoa
- Co-contamination: 13 sa. co-contaminated with FB_1 and OTA; 6 sa. co-contaminated with CIT, FB_1, and OTA; 11 sa. contaminated solely with FB_1
- Further contamination of plants and plant products other than cereals and their products reported in the present book/article[94]: Breakfast cereals, CIT and OTA, literature[94]

For detailed information see the article.

incidence: 15/26*, conc. range: ≤67.0 µg/kg, sample origin: supermarkets in Lisbon (capital) region, Portugal, sample year: unknown, country: Portugal/Brazil[537], *breakfast cereals in part with chocolate
- Co-contamination: For detailed information see the article.
- Further contamination of plants and plant products other than cereals and their products reported in the present book/article[537]: Breakfast cereals, AFB_1, AFB_2, AFG_1, AFM_1, DON, FB_2, NIV, OTA, and ZEA, literature[537]

incidence: 1/2*, conc.: 24.8 µg/kg, sample origin: supermarkets and pharmacies in Rabat-Salé (area), Morocco, sample year: unknown, country: Morocco/Spain[656], *breakfast cereals with fruit rings
- Co-contamination: 1 sa. co-contaminated with FB_1, FB_2, and FB_3
- Further contamination of plants and plant products other than cereals and their products reported in the present book/article[656]: Breakfast cereals, FB_2 and FB_3, literature[656]

Fumonisin B_2

incidence: 10/26*, conc. range: ≤14.0 µg/kg, sample origin: supermarkets in Lisbon (capital) region, Portugal, sample year: unknown, country: Portugal/Brazil[537], *breakfast cereals in part with chocolate
- Co-contamination: For detailed information see the article.
- Further contamination of plants and plant products other than cereals and their products reported in the present book/article[537]: Breakfast cereals, AFB_1, AFB_2, AFG_1, AFM_1, DON, FB_1, NIV, OTA, and ZEA, literature[537]

incidence: 1/2*, conc.: 20.4 µg/kg, sample origin: supermarkets and pharmacies in Rabat-Salé (area), Morocco, sample year: unknown, country: Morocco/Spain[656], *breakfast cereals with fruit rings
- Co-contamination: 1 sa. co-contaminated with FB_1, FB_2, and FB_3
- Further contamination of plants and plant products other than cereals and their products reported in the present book/article[656]: Breakfast cereals, FB_1 and FB_3, literature[656]

Fumonisin B_3

incidence: 1/2*, conc.: 2.7 µg/kg, sample origin: supermarkets and pharmacies in Rabat-Salé (area), Morocco, sample year: unknown, country: Morocco/Spain[656], *breakfast cereals with fruit rings
- Co-contamination: 1 sa. co-contaminated with FB_1, FB_2, and FB_3
- Further contamination of plants and plant products other than cereals and their products reported in the present book/article[656]: Breakfast cereals, FB_1 and FB_2, literature[656]

Type B Trichothecenes

Deoxynivalenol

incidence: 16/26*, conc. range: ≤207.8 μg/kg, sample origin: supermarkets in Lisbon (capital) region, Portugal, sample year: unknown, country: Portugal/Brazil[537], *breakfast cereals in part with chocolate

- Co-contamination: For detailed information see the article.
- Further contamination of plants and plant products other than cereals and their products reported in the present book/article[537]: Breakfast cereals, AFB_1, AFB_2, AFG_1, AFM_1, FB_1, FB_2, NIV, OTA, and ZEA, literature[537]

Nivalenol

incidence: 1/26*, conc.: 27.1 μg/kg, sample origin: supermarkets in Lisbon (capital) region, Portugal, sample year: unknown, country: Portugal/Brazil[537], *breakfast cereals in part with chocolate

- Co-contamination: For detailed information see the article.
- Further contamination of plants and plant products other than cereals and their products reported in the present book/article[537]: Breakfast cereals, AFB_1, AFB_2, AFG_1, AFM_1, DON, FB_1, FB_2, OTA, and ZEA, literature[537]

Other ***Fusarium*** Toxins

Zearalenone

incidence: 19/26*, conc. range: ≤5.6 μg/kg, sample origin: supermarkets in Lisbon (capital) region, Portugal, sample year: unknown, country: Portugal/Brazil[537], *breakfast cereals in part with chocolate

- Co-contamination: For detailed information see the article.
- Further contamination of plants and plant products other than cereals and their products reported in the present book/article[537]: Breakfast cereals, AFB_1, AFB_2, AFG_1, AFM_1, DON, FB_1, FB_2, NIV, and OTA, literature[537]

Bsissa may contain the following mycotoxins:

Aspergillus and ***Penicillium*** Toxins

Ochratoxin A

incidence: 11/11*, conc. range: 0.4–12,770 μg/kg, sample origin: Tunisia, sample year: unknown, country: Tunisia/France[345], *bsissa sa. composed of wheat, barley, chickpea, and a kind of lentil

- Co-contamination: not reported
- Further contamination of plants and plant products other than cereals and their products reported in the present book/article[345]: Olives, Vegetables, dried, OTA, literature[345]

Bukolo see Sauces (flour sauce)

Bush mango seeds see Seeds (bush mango)

Butter (cocoa butter) may contain the following mycotoxins:

Aspergillus Toxins

Aflatoxin B_1

incidence: 7/25, conc. range: ≤0.38 μg/kg, sample origin: different stages of manufacture at processing plants, Brazil, sample year: unknown, country: Brazil[211]

- Co-contamination: not reported
- Further contamination of plants and plant products other than cereals and their products reported in the present book/article[211]: Butter (cocoa butter), AFB_2, literature[211]; Powder (cocoa powder), AFB_1, AFB_2, and AFG_1, literature[211]; Products (cocoa product), AFB_1, AFB_2, AFG_1, and AFG_2, literature[211]

Aflatoxin B_2

incidence: 1/25, conc.: 0.04 μg/kg, sample origin: different stages of manufacture at processing plants, Brazil, sample year: unknown, country: Brazil[211]

- Co-contamination: not reported
- Further contamination of plants and plant products other than cereals and their products reported in the present book/article[211]: Butter (cocoa butter), AFB_1, literature[211]; Powder (cocoa powder), AFB_1, AFB_2, and AFG_1, literature[211]; Products (cocoa product), AFB_1, AFB_2, AFG_1, and AFG_2, literature[211]

Aspergillus and ***Penicillium*** Toxins

Ochratoxin A

incidence: 20/25, conc. range: ≤0.06 μg/kg, sample origin: processing plants in the region of Ilhéus (city) in Bahia (state) and São Paulo (city) in São Paulo (state), Brazil, sample year: unknown, country: Brazil[212]

- Co-contamination: not reported
- Further contamination of plants and plant products other than cereals and their products reported in the present book/article[212]: Powder (cocoa powder), Products (cocoa product), OTA, literature[212]

incidence: 1?/5, conc.: 0.08 μg/kg, sample origin: Canada, sample year: unknown, country: Canada[864]

- Co-contamination: For detailed information see the article.
- Further contamination of plants and plant products other than cereals and their products reported in the present book/article[864]: Chocolate, Cocoa beans, Products (cocoa product), AFB_1, AFB_2, AFG_1, AFG_2, and OTA, literature[864]; Powder (cocoa powder), AFB_1, AFB_2, AFG_1, and OTA, literature[864]

incidence: 24/36, conc. range: ≤1.88 μg/kg, sample origin: Kumba? (city), Cameroon, sample year: 2005–2007, country: Cameroon/France[907]

- Co-contamination: not reported
- Further contamination of plants and plant products other than cereals and their products reported in the present book/article[907]: Chocolate, Powder (cocoa powder), OTA, literature[907]

Butter (nut butter) may contain the following mycotoxins:

Aspergillus Toxins

Aflatoxin B_1

incidence: 2/6*, conc. range: 0.2 μg/kg (1 sa.), 2.5 μg/kg** (1 sa.), ∅ conc.: 1.35 μg/kg, sample origin: retail outlets, UK, sample year: September 2001, country: UK[203], *included almond, cashew, hazelnut, and three nut butter**

- Co-contamination: 1 sa. co-contaminated with AFB_1, AFB_2, AFG_1, AFG_2, and OTA; 1 sa. contaminated solely with AFB_1
- Further contamination of plants and plant products other than cereals and their products reported in the present book/article[203]: Butter (nut butter), AFB_1, AFB_2, AFG_2, and OTA, literature[203]; Butter (peanut butter), AFB_1, AFB_2, AFG_1, AFG_2, and

OTA, literature[203]; Dates, AFB_1 and OTA, literature[203]; Figs, AFB_1, AFB_2, AFG_2, and OTA, literature[203]; Fruits, dried, Nuts (pistachio nut), AFB_1, AFB_2, and OTA, literature[203]; Nuts (Brazil nut, paranut), AFB_1, AFB_2, and AFG_1, literature[203]; Nuts (peanut), AFB_1 and AFB_2, literature[203]

incidence: 1/1*, conc.: 2.9 μg/kg, sample origin: retail outlets, UK, sample year: September 2001, country: UK[203], *crunchy three nut butter

- Co-contamination: 1 sa. co-contaminated with AFB_1, AFB_2, AFG_1, and AFG_2
- Further contamination of plants and plant products other than cereals and their products reported in the present book/article[203]: Butter (nut butter), AFB_2, AFG_1, AFG_2, and OTA, literature[203]; Butter (peanut butter), AFB_1, AFB_2, AFG_1, AFG_2, and OTA, literature[203]; Dates, AFB_1 and OTA, literature[203]; Figs, AFB_1, AFB_2, AFG_2, and OTA, literature[203]; Fruits, dried, Nuts (pistachio nut), AFB_1, AFB_2, and OTA, literature[203]; Nuts (Brazil nut, paranut), AFB_1, AFB_2, and AFG_1, literature[203]; Nuts (peanut), AFB_1 and AFB_2, literature[203]

Aflatoxin B_2

incidence: 1/6*, conc.: 0.6 μg/kg, sample origin: retail outlets, UK, sample year: September 2001, country: UK[203], *included almond, cashew, hazelnut, and three nut butter

- Co-contamination: 1 sa. co-contaminated with AFB_1, AFB_2, AFG_1, AFG_2, and OTA
- Further contamination of plants and plant products other than cereals and their products reported in the present book/article[203]: Butter (nut butter), AFB_1, AFB_2, AFG_2, and OTA, literature[203]; Butter (peanut butter), AFB_1, AFB_2, AFG_1, AFG_2, and OTA, literature[203]; Dates, AFB_1 and OTA, literature[203]; Figs, AFB_1, AFB_2, AFG_2, and OTA, literature[203]; Fruits, dried, Nuts (pistachio nut), AFB_1, AFB_2, and OTA, literature[203]; Nuts (Brazil nut, paranut), AFB_1, AFB_2, and AFG_1, literature[203]; Nuts (peanut), AFB_1 and AFB_2, literature[203]

incidence: 1/1*, conc.: 0.6 μg/kg, sample origin: retail outlets, UK, sample year: September 2001, country: UK[203], *crunchy three nut butter

- Co-contamination: 1 sa. co-contaminated with AFB_1, AFB_2, AFG_1, and AFG_2
- Further contamination of plants and plant products other than cereals and their products reported in the present book/article[203]: Butter (nut butter), AFB_2, AFG_1, AFG_2, and OTA, literature[203]; Butter (peanut butter), AFB_1, AFB_2, AFG_1, AFG_2, and OTA, literature[203]; Dates, AFB_1 and OTA, literature[203]; Figs, AFB_1, AFB_2, AFG_2, and OTA, literature[203]; Fruits, dried, Nuts (pistachio nut), AFB_1, AFB_2, and OTA, literature[203]; Nuts (Brazil nut, paranut), AFB_1, AFB_2, and AFG_1, literature[203]; Nuts (peanut), AFB_1 and AFB_2, literature[203]

Aflatoxin G_1

incidence: 1/6*, conc.: 0.8 μg/kg, sample origin: retail outlets, UK, sample year: September 2001, country: UK[203], *included almond, cashew, hazelnut, and three nut butter

- Co-contamination: 1 sa. co-contaminated with AFB_1, AFB_2, AFG_1, AFG_2, and OTA
- Further contamination of plants and plant products other than cereals and their products reported in the present book/article[203]: Butter (nut butter), AFB_1, AFB_2, AFG_2, and OTA, literature[203]; Butter (peanut butter), AFB_1, AFB_2, AFG_1, AFG_2, and OTA, literature[203]; Dates, AFB_1 and OTA, literature[203]; Figs, AFB_1, AFB_2, AFG_2, and OTA, literature[203]; Fruits, dried, Nuts (pistachio nut), AFB_1, AFB_2, and OTA, literature[203]; Nuts (Brazil nut, paranut), AFB_1, AFB_2, and AFG_1, literature[203]; Nuts (peanut), AFB_1 and AFB_2, literature[203]

incidence: 1/1*, conc.: 1 μg/kg, sample origin: retail outlets, UK, sample year: September 2001, country: UK[203], *crunchy 3 nut butter

- Co-contamination: 1 sa. co-contaminated with AFB_1, AFB_2, AFG_1, and AFG_2
- Further contamination of plants and plant products other than cereals and their products reported in the present book/article[203]: Butter (nut butter), AFB_2, AFG_1, AFG_2, and OTA, literature[203]; Butter (peanut butter), AFB_1, AFB_2, AFG_1, AFG_2, and OTA, literature[203]; Dates, AFB_1 and OTA, literature[203]; Figs, AFB_1, AFB_2, AFG_2, and OTA, literature[203]; Fruits, dried, Nuts (pistachio nut), AFB_1, AFB_2, and OTA, literature[203]; Nuts (Brazil nut, paranut), AFB_1, AFB_2, and AFG_1, literature[203]; Nuts (peanut), AFB_1 and AFB_2, literature[203]

Aflatoxin G_2

incidence: 1/6*, conc.: 0.3 μg/kg, sample origin: retail outlets, UK, sample year: September 2001, country: UK[203], *included almond, cashew, hazelnut, and 3 nut butter

- Co-contamination: 1 sa. co-contaminated with AFB_1, AFB_2, AFG_1, AFG_2, and OTA
- Further contamination of plants and plant products other than cereals and their products reported in the present book/article[203]: Butter (nut butter), AFB_1, AFB_2, AFG_2, and OTA, literature[203]; Butter (peanut butter), AFB_1, AFB_2, AFG_1, AFG_2, and OTA, literature[203]; Dates, AFB_1 and OTA, literature[203]; Figs, AFB_1, AFB_2, AFG_2, and OTA, literature[203]; Fruits, dried, Nuts (pistachio nut), AFB_1, AFB_2, and OTA, literature[203]; Nuts (Brazil nut, paranut), AFB_1, AFB_2, and AFG_1, literature[203]; Nuts (peanut), AFB_1 and AFB_2, literature[203]

incidence: 1/1*, conc.: 0.3 μg/kg, sample origin: retail outlets, UK, sample year: September 2001, country: UK[203], *crunchy 3 nut butter

- Co-contamination: 1 sa. co-contaminated with AFB_1, AFB_2, AFG_1, and AFG_2
- Further contamination of plants and plant products other than cereals and their products reported in the present book/article[203]: Butter (nut butter), AFB_2, AFG_1, AFG_2, and OTA, literature[203]; Butter (peanut butter), AFB_1, AFB_2, AFG_1, AFG_2, and OTA, literature[203]; Dates, AFB_1 and OTA, literature[203]; Figs, AFB_1, AFB_2, AFG_2, and OTA, literature[203]; Fruits, dried, Nuts (pistachio nut), AFB_1, AFB_2, and OTA, literature[203]; Nuts (Brazil nut, paranut), AFB_1, AFB_2, and AFG_1, literature[203]; Nuts (peanut), AFB_1 and AFB_2, literature[203]

Aspergillus and *Penicillium* Toxins

Ochratoxin A

incidence: 2/6*, conc. range: 0.7–2.0 μg/kg, ∅ conc.: 1.35 μg/kg, sample origin: retail outlets, UK, sample year: September 2001, country: UK[203], *included almond, cashew, hazelnut, and 3 nut butter

- Co-contamination: 1 sa. co-contaminated with AFB_1, AFB_2, AFG_1, AFG_2, and OTA; 1 sa. contaminated solely with OTA
- Further contamination of plants and plant products other than cereals and their products reported in the present book/article[203]: Butter (nut butter), AFB_1, AFB_2, AFG_2, and OTA,

literature[203]; Butter (peanut butter), AFB_1, AFB_2, AFG_1, AFG_2, and OTA, literature[203]; Dates, AFB_1 and OTA, literature[203]; Figs, AFB_1, AFB_2, AFG_2, and OTA, literature[203]; Fruits, dried, Nuts (pistachio nut), AFB_1, AFB_2, and OTA, literature[203]; Nuts (Brazil nut, paranut), AFB_1, AFB_2, and AFG_1, literature[203]; Nuts (peanut), AFB_1 and AFB_2, literature[203]

incidence: 0/1*, conc.: no contamination, sample origin: retail outlets, UK, sample year: September 2001, country: UK[203], *crunchy 3 nut butter

- Co-contamination: For detailed information see the article.
- Further contamination of plants and plant products other than cereals and their products reported in the present book/article[203]: Butter (nut butter), AFB_1, AFB_2, AFG_1, and AFG_2, literature[203]; Butter (peanut butter), AFB_1, AFB_2, AFG_1, AFG_2, and OTA, literature[203]; Dates, AFB_1 and OTA, literature[203]; Figs, AFB_1, AFB_2, AFG_2, and OTA, literature[203]; Fruits, dried, Nuts (pistachio nut), AFB_1, AFB_2, and OTA, literature[203]; Nuts (Brazil nut, paranut), AFB_1, AFB_2, and AFG_1, literature[203]; Nuts (peanut), AFB_1 and AFB_2, literature[203]

Butter (peanut butter) may contain the following mycotoxins:

Aspergillus Toxins

Aflatoxin B_1

incidence: 7/16* **, conc. range: 2–5 µg/kg (6 sa.), 7 µg/kg (1 sa.), sample origin: retail outlets in 20 different regions, England, sample year: January 1982, country: UK[3], *regular food, **peanut butter, smooth

incidence: 5/16* **, conc. range: 2–5 µg/kg (3 sa.), 6–10 µg/kg (1 sa.), 12 µg/kg (1 sa.), sample origin: retail outlets in 20 different regions, England, sample year: January 1982, country: UK[3], *regular food, **peanut butter, crunchy

incidence: 4/11* **, conc. range: 6–10 µg/kg (2 sa.), 31–100 µg/kg (2 sa., maximum: 49 µg/kg), sample origin: retail outlets in 20 different regions, England, sample year: January 1982, country: UK[3], *health food, **peanut butter, smooth

incidence: 10/14* **, conc. range: 2–5 µg/kg (2 sa.), 6–10 µg/kg (1 sa.), 11–30 µg/kg (1 sa.), 31–100 µg/kg (1 sa.), >100 µg/kg (5 sa., maximum: 318 µg/kg), sample origin: retail outlets in 20 different regions, England, sample year: January 1982, country: UK[3], *health food, **peanut butter, crunchy

incidence: 5/6* **, conc. range: 11–30 µg/kg (3 sa.), 31–100 µg/kg (2 sa., maximum: 76 µg/kg), sample origin: retail outlets in 20 different regions, sample year: June 1983, country: UK[3], *health food, **peanut butter, smooth

incidence: 7/9* **, conc. range: 2–5 µg/kg (5 sa.), 6–10 µg/kg (1 sa.), 58 µg/kg (1 sa.), sample origin: retail outlets in 20 different regions, England, sample year: June 1983, country: UK[3], *health food, **peanut butter, crunchy

incidence: 1/4* **, conc.: 13 µg/kg, sample origin: retail outlets in 20 different regions, England, sample year: December 1984, country: UK[3], *health food, **peanut butter, smooth

incidence: 7/15* **, conc. range: 6–10 µg/kg (1 sa.), 11–30 µg/kg (3 sa.), 31–100 µg/kg (3 sa., maximum: 73 µg/kg), sample origin: retail outlets in 20 different regions, England, sample year: December 1984, country: UK[3], *health food, **peanut butter, crunchy

- Co-contamination: not reported
- Further contamination of plants and plant products other than cereals and their products reported in the present book/article[3]: Butter (peanut butter), AF/S, literature[3]; Nuts (almond), Nuts (Brazil nut, paranut), Nuts (hazelnut), Nuts (peanut), AFB_1 and AF/S, literature[3]

incidence: 8/8, ∅ conc.: 1.3 µg/kg, sample origin: shops, supermarkets, and wholesalers, France, sample year: 1989/1990, country: France[5]

- Co-contamination: not reported
- Further contamination of plants and plant products other than cereals and their products reported in the present book/article[5]: Spices (nutmeg), AFB_1, literature[5]

incidence: 22?/26, conc. range: 0.09–2.65 µg/kg, sample origin: unknown, sample year: unknown, country: Korea[9]

- Co-contamination: not reported
- Further contamination of plants and plant products other than cereals and their products reported in the present book/article[9]: Butter (peanut butter), AFB_2, AFG_1, and AFG_2, literature[9]; Nuts (almond), Nuts (peanut), AFB_1 and AFB_2, literature[9]; Products (nut product), Tea, AFB_1, literature[9]

incidence: 5/5, conc. range: 1.30–6.44 µg/kg, ∅ conc.: 3.60 µg/kg, sample origin: grocery markets in Busan, Daegu, Daejeon, Gangneung, and Gwangju (cities) as well as Seoul (capital), Korea, sample year: May 2004–June 2005, country: Korea[10]

- Co-contamination: not reported
- Further contamination of plants and plant products other than cereals and their products reported in the present book/article[10]: Nuts, Nuts (almonds), Nuts (peanuts), Nuts (pistachio nuts), Paste (soybean paste), Sauces (soy sauce), Spices (cayenne pepper, chilli), Spices (curry), Spices (pepper), AFB_1, literature[10]

incidence: 2/2, conc. range: 3.5–5.2 µg/kg, ∅ conc.: 4.4 µg/kg, sample origin: local marketing institutions, GDR, sample year: unknown, country: GDR[29]

- Co-contamination: 2 sa. co-contaminated with AFB_1, AFB_2, AFG_1, and AFG_2
- Further contamination of plants and plant products other than cereals and their products reported in the present book/article[29]: Butter (peanut butter), AFB_2, AFG_1, and AFG_2, literature[29]; Coffee, OTA, literature[29]; Fruits, AFB_1, AFG_1, and AFG_2, literature[29]; Fruits (apple), Fruits (peach), Juices (apple juice), PAT, literature[29]; Nuts (peanut), Products (peanut product), Seeds (sunflower), AFB_1, AFB_2, AFG_1, and AFG_2, literature[29]

incidence: 20*/101**, conc. range: 64–1,736 µg/kg, sample origin: 16 districts of eastern Nepal, sample year: 1995–2003, country: Nepal[58], *>30 µg/kg, **peanut butter/vegetable oil

- Co-contamination: not reported
- Further contamination of plants and plant products other than cereals and their products reported in the present book/article[58]: Betel nuts, AF/S, literature[58]; Nuts (peanut), AFB_1, literature[58]

incidence: 3?/6, conc. range: 0.6–2.4 µg/kg, sample origin: unknown, sample year: unknown, country: Japan[97]

- Co-contamination: not reported
- Further contamination of plants and plant products other than cereals and their products reported in the present book/article[97]: Beans, Jam (bean jam), Spices, AFB_1 and AFB_2, literature[97];

Butter (peanut butter), AFB_2 and AFG_1, literature[97]; Oils (peanut oil), Spices (pepper), AFB_1, AFB_2, and AFG_1, literature[97]; Spices (nutmeg), AFB_1, AFB_2, AFG_1, and AFG_2, literature[97]

incidence: 7/8, conc. range: 2.0–4.0 μg/kg (1 sa.), >4.0–≤249.0 μg/kg (6 sa.), sample origin: supermarkets and/or traditional markets in Yogyakarta (city), Indonesia, sample year: 1998/1999, country: Indonesia/Austria/Canada[106]

- Co-contamination: not reported
- Further contamination of plants and plant products other than cereals and their products reported in the present book/article[106]: Cake (peanut), Food, Nuts (peanut), Products, Sauces (peanut-chilli sauce), AFB_1, literature[106]

incidence: 1/13, conc.: 20 μg/kg, sample origin: main kitchen store, California (state), USA, sample year: August 1985, country: USA[113]

- Co-contamination: not reported
- Further contamination of plants and plant products other than cereals and their products reported in the present book/article[113]: not reported

incidence: 40/62, conc. range: 0.5–54.6 μg/kg, Ø conc.: 7.10 μg/kg, sample origin: small markets in several urban Shanghai (city) areas, China, sample year: April 1988–April 1989, country: China/USA[132]

- Co-contamination: not reported
- Further contamination of plants and plant products other than cereals and their products reported in the present book/article[132]: Beans (kidney bean), Curd (bean curd), Nuts (peanut), Oils (rapeseed oil), Oils (soybean oil), Paste (bean paste), Sauces (soy sauce), Tofu, Wine, miscellaneous, AFB_1, literature[132]

incidence: 10*/21, conc. range: 0.17–2.59 μg/kg, Ø conc.: 1.07 μg/kg, sample origin: local supermarkets and small retail shops across Japan, sample year: summer 2004–winter 2005, country: Japan[146], *domestic and imported products

- Co-contamination: 4 sa. co-contaminated with AFB_1, AFB_2, AFG_1, and AFG_2; 3 sa. co-contaminated with AFB_1 and AFB_2; 3 sa. contaminated solely with AFB_1
- Further contamination of plants and plant products other than cereals and their products reported in the present book/article[146]: Beans (green bean), Coffee, Raisins, Wine, OTA, literature[146]; Butter (peanut butter), AFB_2, AFG_1, and AFG_2, literature[146]; Chocolate, AFB_1 and OTA, literature[146]

incidence: 3/4, conc. range: 0.6–1.4 μg/kg, Ø conc.: 1.0 μg/kg, sample origin: retail stores in Osaka (prefecture), Japan, sample year: 1988–1992, country: Japan[173]

- Co-contamination: 2 sa. co-contaminated with AFB_1, AFB_2, and AFG_1; 1 sa. co-contaminated with AFB_1 and AFB_2
- Further contamination of plants and plant products other than cereals and their products reported in the present book/article[173]: Butter (peanut butter), AFB_2 and AFG_1, literature[173]; Nuts (peanut), Spices (cayenne pepper, chilli), Spices (pepper), AFB_1, literature[173]; Spices (nutmeg), AFB_1, AFB_2, and AFG_1, literature[173]

incidence: 2/2, conc. range: 0.76–5.90 μg/kg, Ø conc.: 3.33 μg/kg, sample origin: local stores in Munich (city), Germany, sample year: unknown, country: Germany[181]

- Co-contamination: 1 sa. co-contaminated with AFB_1, AFB_2, and AFG_1; 1 sa. co-contaminated with AFB_1 and AFB_2
- Further contamination of plants and plant products other than cereals and their products reported in the present book/article[181]: Butter (peanut butter), AFB_2, and AFG_1, literature[181]; Nuts (hazelnut), Spices (paprika), AFB_1 and AFG_1, literature[181]; Nuts (pistachio nut), Spices (pepper), AFB_1, literature[181]; Spices (nutmeg), AFB_1, AFB_2, and AFG_1, literature[181]

incidence: 21?/74, conc. range: 1.2–73 μg/kg, Ø conc.: 40.6 μg/kg, sample origin: sites of primary storage or processing (domestic products) and sites of importation, Cyprus, sample year: 1992–1996, country: Cyprus[200]

- Co-contamination: not reported
- Further contamination of plants and plant products other than cereals and their products reported in the present book/article[200]: Butter (peanut butter), AFB_2, AFG_1, and AFG_2, literature[200]; Figs, Nuts (peanut), AFB_1, AFB_2, AFG_1, and AFG_2, literature[200]; Nuts (Brazil nut, paranut), Nuts (walnut), AFB_1, AFB_2, and AFG_1, literature[200]; Nuts (chestnut), Products (sesame product), Spices, AFB_1, literature[200]; Nuts (pistachio nut), AFB_1 and AFB_2, literature[200]; Seeds (sesame), AFB_1, AFG_1, and AFG_2, literature[200]

incidence: 19/29, conc. range: LOQ-2 μg/kg (13 sa.), >2–5 μg/kg (5 sa.), 7.5 μg/kg (1 sa.), Ø conc.: 1.82 μg/kg, sample origin: retail outlets, UK, sample year: September 2001, country: UK[203]

- Co-contamination: 2 sa. co-contaminated with AFB_1, AFB_2, AFG_1, AFG_2, and OTA; 3 sa. co-contaminated with AFB_1, AFB_2, AFG_1, and AFG_2; 3 sa. co-contaminated with AFB_1, AFB_2, and AFG_1; 2 sa. co-contaminated with AFB_1 and AFB_2; 4 sa. co-contaminated with AFB_1 and AFG_1; 1 sa. co-contaminated with AFB_1 and OTA; 4 sa. contaminated solely with AFB_1
- Further contamination of plants and plant products other than cereals and their products reported in the present book/article[203]: Butter (nut butter), AFB_1, AFB_2, AFG_1, AFG_2, and OTA, literature[203]; Butter (peanut butter), AFB_2, AFG_1, AFG_2, and OTA, literature[203]; Dates, AFB_1 and OTA, literature[203]; Figs, AFB_1, AFB_2, AFG_2, and OTA, literature[203]; Fruits, dried, Nuts (pistachio nut), AFB_1, AFB_2, and OTA, literature[203]; Nuts (Brazil nut, paranut), AFB_1, AFB_2, and AFG_1, literature[203]; Nuts (peanut), AFB_1 and AFB_2, literature[203]

incidence: 16/27, conc. range: LOQ-2 μg/kg (14 sa.), 2.5 μg/kg (1 sa.), 2.9 μg/kg (1 sa.), Ø conc.: 1.14 μg/kg, sample origin: retail outlets, UK, sample year: September 2001, country: UK[203]

- Co-contamination: 2 sa. co-contaminated with AFB_1, AFB_2, AFG_1, and AFG_2; 1 sa. co-contaminated with AFB_1, AFB_2, and AFG_1; 2 sa. co-contaminated with AFB_1, AFB_2, AFG_1, and OTA; 1 sa. co-contaminated with AFB_1 and AFB_2; 3 sa. co-contaminated with AFB_1 and AFG_1; 7 sa. contaminated solely with AFB_1
- Further contamination of plants and plant products other than cereals and their products reported in the present book/article[203]: Butter (nut butter), AFB_1, AFB_2, AFG_1, AFG_2, and OTA, literature[203]; Butter (peanut butter), AFB_2, AFG_1, AFG_2, and OTA, literature[203]; Dates, AFB_1 and OTA, literature[203]; Figs, AFB_1, AFB_2, AFG_2, and OTA, literature[203]; Fruits, dried, Nuts (pistachio nut), AFB_1, AFB_2, and OTA, literature[203]; Nuts (Brazil nut, paranut), AFB_1, AFB_2, and AFG_1, literature[203]; Nuts (peanut), AFB_1 and AFB_2, literature[203]

incidence: 6/50* **, conc. range: 2.5–5 μg/kg (2 sa.), 11 to ≤28.6 μg/kg (4 sa.), sample origin: retail health-food outlets in 25 different geographical locations, UK, sample year: 1986, country: UK[213], *health food, **peanut butter, smooth

incidence: 9/79* **, conc. range: 2.5–5 μg/kg (2 sa.), 5.1–10 μg/kg (3 sa.), 11–30 μg/kg (3 sa.), 41.0 μg/kg (1 sa.), sample origin: retail health-food outlets in 25 different geographical locations, UK, sample year: 1986, country: UK[213], *health food, **peanut butter, crunchy

- Co-contamination: not reported
- Further contamination of plants and plant products other than cereals and their products reported in the present book/article[213]: Butter (peanut butter), AF/S, literature[213]

incidence: 14/30, conc. range: 5–117 µg/kg, ∅ conc.: 25 µg/kg, sample origin: markets in 16 areas (cities and counties), Taiwan, sample year: 1977/1978, country: Taiwan[214]

- Co-contamination: not reported
- Further contamination of plants and plant products other than cereals and their products reported in the present book/article[214]: Butter (peanut butter), AFB_2, AFG_1, and AFG_2, literature[214]; Cake (peanut), Nuts (peanut), Oils (peanut oil), AFB_1, AFB_2, AFG_1, and AFG_2, literature[214]

incidence: 10/63, conc. range: 2–20 µg/kg, ∅ conc.: 7 µg/kg, sample origin: supermarkets in East Lansing (city), USA, sample year: unknown, country: USA[215]

- Co-contamination: not reported
- Further contamination of plants and plant products other than cereals and their products reported in the present book/article[215]: not reported

incidence: 9/16, conc. range: 20–730 µg/kg, ∅ conc.: 218 µg/kg, sample origin: USA and Taiwan, sample year: unknown, country: Taiwan[216]

- Co-contamination: not reported
- Further contamination of plants and plant products other than cereals and their products reported in the present book/article[216]: Nuts (peanut), Oils (peanut oil), AFB_1, literature[216]

incidence: 2/2, conc. range: 1.8–5.5 µg/kg, ∅ conc.: 3.7 µg/kg, sample origin: unknown, sample year: unknown, country: Denmark[217]

- Co-contamination: 2 sa. co-contaminated with AFB_1, AFB_2, AFG_1, and AFG_2
- Further contamination of plants and plant products other than cereals and their products reported in the present book/article[217]: Butter (peanut butter), AFB_2, AFG_1, and AFG_2, literature[217]; Figs, Spices (cayenne pepper, chilli), AFB_1, AFB_2, AFG_1, and AFG_2, literature[217]; Meal (pistachio meal), AFB_1 and AFB_2, literature[217]

incidence: 15?/21, conc. range: 3.2–16.0 µg/kg, sample origin: storage depots and retail outlets, Botswana, sample year: January 1996–December 1997, country: Botswana[218]

- Co-contamination: not reported
- Further contamination of plants and plant products other than cereals and their products reported in the present book/article[218]: Butter (peanut butter), AFB_2, AFG_1, and AFG_2, literature[218]; Nuts (peanut), AFB_1, AFB_2, AFG_1, and AFG_2, literature[218]

incidence: 14/14*, conc. range: 29–128 µg/kg, ∅ conc.: 63.9 µg/kg, sample origin: street sellers, Khartoum (capital, state) area, Sudan, sample year: 2005/2006, country: Sudan[219], *traditionally processed peanut butter (homemade)

incidence: 10/10*, conc. range: 21–131 µg/kg, ∅ conc.: 54.5 µg/kg, sample origin: street sellers, Khartoum (capital, state) north area, Sudan, sample year: 2005/2006, country: Sudan[219], *traditionally processed peanut butter (homemade)

incidence: 12/12*, conc. range: 17–170 µg/kg, ∅ conc.: 101 µg/kg, sample origin: street sellers, Omdurman (city) area, Sudan, sample year: 2005/2006, country: Sudan[219], *traditionally processed peanut butter (homemade)

incidence: 10/14, conc. range: 1–57 µg/kg, ∅ conc.: 14.5 µg/kg, sample origin: retail stores, Khartoum (state), Sudan, sample year: 2005/2006, country: Sudan[219]

- Co-contamination: not reported
- Further contamination of plants and plant products other than cereals and their products reported in the present book/article[219]: not reported

incidence: 2/2, conc. range: 5.82–6.44 µg/kg, ∅ conc.: 6.13 µg/kg, sample origin: supermarkets, local stores, and markets in Busan, Daejeon, Gwangju (cities), and Seoul (capital), Korea, sample year: June/July 2004, country: Korea[220]

- Co-contamination: not reported
- Further contamination of plants and plant products other than cereals and their products reported in the present book/article[220]: Butter (peanut butter), AFB_2, literature[220]; Nuts, AFB_1 and AFB_2, literature[220]; Nuts (peanut), AFB_1, AFB_2, AFG_1, and AFG_2, literature[220]; Nuts (pistachio nut), AFB_1, literature[220]

incidence: 19/20, conc. range: 2.06–63.72 µg/kg, ∅ conc.: 16.6 µg/kg, sample origin: markets in Ankara (capital), Turkey, sample year: unknown, country: Turkey[221]

- Co-contamination: 19 sa. co-contaminated with AFB_1, AFB_2, and AFG_1
- Further contamination of plants and plant products other than cereals and their products reported in the present book/article[221]: Butter (peanut butter), AFB_2 and AFG_1, literature[221]; Seeds (sesame), AFG_1, literature[221]

incidence: 5/5, conc. range: 0.62–14.3 µg/kg, ∅ conc.: 4.6 µg/kg, sample origin: unknown, sample year: unknown, country: USA[222]

- Co-contamination: 4 sa. co-contaminated with AFB_1, AFB_2, AFG_1, and AFG_2; 1 sa. co-contaminated with AFB_1 and AFB_2
- Further contamination of plants and plant products other than cereals and their products reported in the present book/article[222]: Butter (peanut butter), AFB_2, AFG_1, and AFG_2, literature[222]

incidence: 1/1, conc.: 26.3 µg/kg, sample origin: market in Khartoum (capital), Sudan, sample year: February 1981–1984?, country: UK/Sudan[223]

- Co-contamination: 1 sa. co-contaminated with AFB_1, AFB_2, and AFG_1
- Further contamination of plants and plant products other than cereals and their products reported in the present book/article[223]: Butter (peanut butter), AFB_2 and AFG_1, literature[223]; Nuts (peanut), AFB_1, AFB_2, AFG_1, and AFG_2, literature[223]; Okra, AFG_2, literature[223]; Peas (chickpea), AFB_1, literature[223]

incidence: 41/50*, conc. range: 0.39–68.51 µg/kg, sample origin: retail markets in Changchun, Chengdu, Shanghai, Shijiazhuang, Zhengzhou (cities), and Beijing (capital), China, sample year: 2007, country: China[224], *manufactured domestically

- Co-contamination: 41 sa. co-contaminated with AFB_1 and AFB_2; no further information available
- Further contamination of plants and plant products other than cereals and their products reported in the present book/article[224]: Butter (peanut butter), AFB_2, AFG_1, and AFG_2, literature[224]; Paste (sesame paste), AFB_1, AFB_2, AFG_1, and AFG_2, literature[224]

incidence: 7?/12, conc. range: 13.3–56.6 µg/kg, sample origin: supermarkets and village stores in five districts of Penang (state) including Penang Island and Wellesley (province), Malaysia, sample year: unknown, country: Malaysia[225]
- Co-contamination: not reported
- Further contamination of plants and plant products other than cereals and their products reported in the present book/article[225]: Butter (peanut butter), AFB_2, literature[225]; Cake (peanut), Nuts (peanut), AFB_1, AFB_2, and AFG_2, literature[225]; Confectionery, Nuts (walnut), Products (nut product), AFB_1 and AFB_2, literature[225]

incidence: 4/10, conc. range: 5–61 µg/kg, sample origin: traditional markets and supermarkets, Indonesia, sample year: 2001/2002, country: Austria/Indonesia/Thailand[226]
- Co-contamination: not reported
- Further contamination of plants and plant products other than cereals and their products reported in the present book/article[226]: Butter (peanut butter), AF/S, literature[226]; Cake (peanut), Nuts (peanut), Sauces (peanut-chilli sauce), Snacks, AFB_1 and AF/S, literature[226]

incidence: 31?/33, conc. range: ≤51.4 µg/kg, sample origin: local stores in Hangzhou, Ningbo, and Quzhou (cities), Zhejiang (province), China, sample year: unknown, country: China[227]
- Co-contamination: not reported
- Further contamination of plants and plant products other than cereals and their products reported in the present book/article[227]: Butter (peanut butter), AFB_2, AFG_1, AFG_2, AFM_1, and AFM_2, literature[227]; Nuts (peanut), AFB_1, AFB_2, AFG_1, AFG_2, AFM_1, and AFM_2, literature[227]

incidence: 31/75, conc. range: 0.1–2.1 µg/kg, ∅ conc.: 0.5 µg/kg, sample origin: available in New Zealand, sample year: 2010, country: New Zealand[277]
- Co-contamination: not reported
- Further contamination of plants and plant products other than cereals and their products reported in the present book/article[277]: Bars, AFB_1, AF/S, DON, NIV, and OTA, literature[277]; Confectionery, Food, Fruits, dried, Nuts, Nuts (peanut), Nuts (pistachio nut), Sauces (peanut sauce), AFB_1 and AF/S, literature[277]; Butter (peanut butter), AF/S, literature[277]; Coffee, Dates, Spices (nutmeg), Spices (pepper), Vine fruits, dried, OTA, literature[277]; Figs, Spices (cayenne pepper, chilli), Spices (curry), Spices (ginger), Spices (paprika), AFB_1, AF/S, and OTA, literature[277]; Nuts (Brazil nut, paranut), AFB_1, literature[277]

incidence: 28/43, conc. range: 73.9–534 µg/kg, ∅ conc.: 223 µg/kg, sample origin: households, small factories, supermarkets, and grocery shops in Khartoum State, Sudan, sample year: January 2010, country: Sudan[718]
- Co-contamination: not reported
- Further contamination of plants and plant products other than cereals and their products reported in the present book/article[718]: Butter (peanut butter), AFB_2, AFG_1, and AFG_2, literature[718]

incidence: 14/14*, conc. range: 13.2–40.6 µg/kg, sample origin: market in Lilongwe (city), Malawi, sample year: December 2012, country: Malawi[790], *peanut butter, local
- Co-contamination: 14 sa. co-contaminated with AFB_1, AFB_2, AFG_1, and AFG_2
- Further contamination of plants and plant products other than cereals and their products reported in the present book/article[790]: Butter (peanut butter), AFB_2, AFG_1, and AFG_2, literature[790]; Food, Nuts (peanut), AFB_1, AFB_2, AFG_1, and AFG_2, literature[790]

incidence: 8/11*, conc. range: 0.5–1.4 µg/kg, sample origin: market in Lilongwe (city), Malawi, sample year: December 2012, country: Malawi[790], *peanut butter, imported
- Co-contamination: not reported
- Further contamination of plants and plant products other than cereals and their products reported in the present book/article[790]: Butter (peanut butter), AFB_2, AFG_1, and AFG_2, literature[790]; Food, Nuts (peanut), AFB_1, AFB_2, AFG_1, and AFG_2, literature[790]

incidence: 10/11, conc. range: 6.1–191 µg/kg, ∅ conc.: 55.7 µg/kg, sample origin: retail shops and vendors at informal markets in Bulawayo metropolitan area, Zimbabwe, sample year: October/November 2011, country: Botswana/South Africa[797]
- Co-contamination: 1 sa. co-contaminated with AFB_1, AFB_2, AFG_1, and AFG_2; 3 sa. co-contaminated with AFB_1, AFB_2, and AFG_1; 1 sa. co-contaminated with AFB_1 and AFG_1; 5 sa. contaminated solely with AFB_1
- Further contamination of plants and plant products other than cereals and their products reported in the present book/article[797]: Butter (peanut butter), AFB_2, AFG_1, and AFG_2, literature[797]; Nuts (peanut), AFB_1 and AF/S, literature[797]

Aflatoxin B_2

incidence: 22?/26, conc. range: 0.04–0.65 µg/kg, sample origin: unknown, sample year: unknown, country: Korea[9]
- Co-contamination: not reported
- Further contamination of plants and plant products other than cereals and their products reported in the present book/article[9]: Butter (peanut butter), AFB_1, AFG_1, and AFG_2, literature[9]; Nuts (almond), Nuts (peanut), AFB_1 and AFB_2, literature[9]; Products (nut product), Tea, AFB_1, literature[9]

incidence: 2/2, conc. range: 0.5–0.6 µg/kg, ∅ conc.: 0.55 µg/kg, sample origin: local marketing institutions, GDR, sample year: unknown, country: GDR[29]
- Co-contamination: 2 sa. co-contaminated with AFB_1, AFB_2, AFG_1, and AFG_2
- Further contamination of plants and plant products other than cereals and their products reported in the present book/article[29]: Butter (peanut butter), AFB_1, AFG_1, and AFG_2, literature[29]; Coffee, OTA, literature[29]; Fruits, AFB_1, AFG_1, and AFG_2, literature[29]; Fruits (apple), Fruits (peach), Juices (apple juice), PAT, literature[29]; Nuts (peanut), Products (peanut product), Seeds (sunflower), AFB_1, AFB_2, AFG_1, and AFG_2, literature[29]

incidence: 1/6, conc.: 0. 4 µg/kg, sample origin: unknown, sample year: unknown, country: Japan[97]
- Co-contamination: not reported
- Further contamination of plants and plant products other than cereals and their products reported in the present book/article[97]: Beans, Jam (bean jam), Spices, AFB_1 and AFB_2, literature[97]; Butter (peanut butter), AFB_1 and AFG_1, literature[97]; Oils (peanut oil), Spices (pepper), AFB_1, AFB_2, and AFG_1, literature[97]; Spices (nutmeg), AFB_1, AFB_2, AFG_1, and AFG_2, literature[97]

incidence: 7*/21, conc. range: 0.16–0.52 µg/kg, ∅ conc.: 0.27 µg/kg, sample origin: local supermarkets and small retail shops across Japan, sample year: summer 2004–winter 2005, country: Japan[146], *domestic and imported products

- Co-contamination: 4 sa. co-contaminated with AFB_1, AFB_2, AFG_1, and AFG_2; 3 sa. co-contaminated with AFB_1 and AFB_2
- Further contamination of plants and plant products other than cereals and their products reported in the present book/article[146]: Beans (green bean), Coffee, Raisins, Wine, OTA, literature[146]; Butter (peanut butter), AFB_1, AFG_1, and AFG_2, literature[146]; Chocolate, AFB_1 and OTA, literature[146]

incidence: 3/4, conc. range: 0.1–0.3 µg/kg, ∅ conc.: 0.2 µg/kg, sample origin: retail stores in Osaka (prefecture), Japan, sample year: 1988–1992, country: Japan[173]

- Co-contamination: 2 sa. co-contaminated with AFB_1, AFB_2, and AFG_1; 1 sa. co-contaminated with AFB_1 and AFB_2
- Further contamination of plants and plant products other than cereals and their products reported in the present book/article[173]: Butter (peanut butter), AFB_1 and AFG_1, literature[173]; Nuts (peanut), Spices (cayenne pepper, chilli), Spices (pepper), AFB_1, literature[173]; Spices (nutmeg), AFB_1, AFB_2, and AFG_1, literature[173]

incidence: 2/2, conc. range: 0.12–1.38 µg/kg, ∅ conc.: 0.75 µg/kg, sample origin: local stores in Munich (city), Germany, sample year: unknown, country: Germany[181]

- Co-contamination: 1 sa. co-contaminated with AFB_1, AFB_2, and AFG_1; 1 sa. co-contaminated with AFB_1 and AFB_2
- Further contamination of plants and plant products other than cereals and their products reported in the present book/article[181]: Butter (peanut butter), Nuts (hazelnut), Spices (paprika), AFB_1 and AFG_1, literature[181]; Nuts (pistachio nut), Spices (pepper), AFB_1, literature[181]; Spices (nutmeg), AFB_1, AFB_2, and AFG_1, literature[181]

incidence: 21?/74, conc. range: 0.3–9 µg/kg, ∅ conc.: 6.4 µg/kg, sample origin: sites of primary storage or processing (domestic products) and sites of importation, Cyprus, sample year: 1992–1996, country: Cyprus[200]

- Co-contamination: not reported
- Further contamination of plants and plant products other than cereals and their products reported in the present book/article[200]: Butter (peanut butter), AFB_1, AFG_1, and AFG_2, literature[200]; Figs, Nuts (peanut), AFB_1, AFB_2, AFG_1, and AFG_2, literature[200]; Nuts (Brazil nut, paranut), Nuts (walnut), AFB_1, AFB_2, and AFG_1, literature[200]; Nuts (chestnut), Products (sesame product), Spices, AFB_1, literature[200]; Nuts (pistachio nut), AFB_1 and AFB_2, literature[200]; Seeds (sesame), AFB_1, AFG_1, and AFG_2, literature[200]

incidence: 10/29, conc. range: LOQ-2 µg/kg (10 sa., maximum: 1.8 µg/kg), ∅ conc.: 0.72 µg/kg, sample origin: retail outlets, UK, sample year: September 2001, country: UK[203]

- Co-contamination: 2 sa. co-contaminated with AFB_1, AFB_2, AFG_1, AFG_2, and OTA; 3 sa. co-contaminated with AFB_1, AFB_2, AFG_1, and AFG_2; 3 sa. co-contaminated with AFB_1, AFB_2, and AFG_1; 2 sa. co-contaminated with AFB_1 and AFB_2
- Further contamination of plants and plant products other than cereals and their products reported in the present book/article[203]: Butter (nut butter), AFB_1, AFB_2, AFG_1, AFG_2, and OTA, literature[203]; Butter (peanut butter), AFB_2, AFG_1, AFG_2, and OTA, literature[203]; Dates, AFB_1 and OTA, literature[203]; Figs, AFB_1, AFB_2, AFG_2, and OTA, literature[203]; Fruits, dried, Nuts (pistachio nut), AFB_1, AFB_2, and OTA, literature[203]; Nuts (Brazil nut, paranut), AFB_1, AFB_2, and AFG_1, literature[203]; Nuts (peanut), AFB_1 and AFB_2, literature[203]

incidence: 6/27, conc. range: LOQ-2 µg/kg (6 sa., maximum: 0.6 µg/kg), ∅ conc.: 0.3 µg/kg, sample origin: retail outlets, UK, sample year: September 2001, country: UK[203]

- Co-contamination: 2 sa. co-contaminated with AFB_1, AFB_2, AFG_1, and AFG_2; 1 sa. co-contaminated with AFB_1, AFB_2, and AFG_1; 2 sa. co-contaminated with AFB_1, AFB_2, AFG_1, and OTA; 1 sa. co-contaminated with AFB_1 and AFB_2
- Further contamination of plants and plant products other than cereals and their products reported in the present book/article[203]: Butter (nut butter), AFB_1, AFB_2, AFG_1, AFG_2, and OTA, literature[203]; Butter (peanut butter), AFB_2, AFG_1, AFG_2, and OTA, literature[203]; Dates, AFB_1 and OTA, literature[203]; Figs, AFB_1, AFB_2, AFG_2, and OTA, literature[203]; Fruits, dried, Nuts (pistachio nut), AFB_1, AFB_2, and OTA, literature[203]; Nuts (Brazil nut, paranut), AFB_1, AFB_2, and AFG_1, literature[203]; Nuts (peanut), AFB_1 and AFB_2, literature[203]

incidence: 5/30, conc. range: 5–50 µg/kg, ∅ conc.: 18 µg/kg, sample origin: markets in 16 areas (cities and counties), Taiwan, sample year: 1977/1978, country: Taiwan[214]

- Co-contamination: not reported
- Further contamination of plants and plant products other than cereals and their products reported in the present book/article[214]: Butter (peanut butter), AFB_1, AFG_1, and AFG_2, literature[214]; Cake (peanut), Nuts (peanut), Oils (peanut oil), AFB_1, AFB_2, AFG_1, and AFG_2, literature[214]

incidence: 2/2, conc. range: 0.4–1.0 µg/kg, ∅ conc.: 0.7 µg/kg, sample origin: unknown, sample year: unknown, country: Denmark[217]

- Co-contamination: 2 sa. co-contaminated with AFB_1, AFB_2, AFG_1, and AFG_2
- Further contamination of plants and plant products other than cereals and their products reported in the present book/article[217]: Butter (peanut butter), AFB_1, AFG_1, and AFG_2, literature[217]; Figs, Spices (cayenne pepper, chilli), AFB_1, AFB_2, AFG_1, and AFG_2, literature[217]; Meal (pistachio meal), AFB_1 and AFB_2, literature[217]

incidence: 15?/21, conc. range: 1.6–20.0 µg/kg, sample origin: storage depots and retail outlets, Botswana, sample year: January 1996–December 1997, country: Botswana[218]

- Co-contamination: not reported
- Further contamination of plants and plant products other than cereals and their products reported in the present book/article[218]: Butter (peanut butter), AFB_1, AFG_1, and AFG_2, literature[218]; Nuts (peanut), AFB_1, AFB_2, AFG_1, and AFG_2, literature[218]

incidence: 1/2, conc.: 1.15 µg/kg, sample origin: supermarkets, local stores, and markets in Busan, Daejeon, Gwangju (cities), and Seoul (capital), Korea, sample year: June/July 2004, country: Korea[220]

- Co-contamination: not reported
- Further contamination of plants and plant products other than cereals and their products reported in the present book/article[220]: Butter (peanut butter), Nuts (pistachio nut), AFB_1, literature[220]; Nuts, AFB_1 and AFB_2, literature[220]; Nuts (peanut), AFB_1, AFB_2, AFG_1, and AFG_2, literature[220]

incidence: 20/20, conc. range: 0.14–4.68 µg/kg, ∅ conc.: 1.23 µg/kg, sample origin: markets in Ankara (capital), Turkey, sample year: unknown, country: Turkey[221]

- Co-contamination: 19 sa. co-contaminated with AFB_1, AFB_2, and AFG_1; 1 sa. co-contaminated with AFB_2 and AFG_1
- Further contamination of plants and plant products other than cereals and their products reported in the present book/article[221]: Butter (peanut butter), AFB_1 and AFG_1, literature[221]; Seeds (sesame), AFG_1, literature[221]

incidence: 5/5, conc. range: 0.14–3.3 µg/kg, ∅ conc.: 1.0 µg/kg, sample origin: unknown, sample year: unknown, country: USA[222]

- Co-contamination: 4 sa. co-contaminated with AFB_1, AFB_2, AFG_1, and AFG_2; 1 sa. co-contaminated with AFB_1 and AFB_2
- Further contamination of plants and plant products other than cereals and their products reported in the present book/article[222]: Butter (peanut butter), AFB_1, AFG_1, and AFG_2, literature[222]

incidence: 1/1, conc.: 9.72 µg/kg, sample origin: market in Khartoum (capital), Sudan, sample year: February 1981–1984?, country: UK/Sudan[223]

- Co-contamination: 1 sa. co-contaminated with AFB_1, AFB_2, and AFG_1
- Further contamination of plants and plant products other than cereals and their products reported in the present book/article[223]: Butter (peanut butter), AFB_1 and AFG_1, literature[223]; Nuts (peanut), AFB_1, AFB_2, AFG_1, and AFG_2, literature[223]; Okra, AFG_2, literature[223]; Peas (chickpea), AFB_1, literature[223]

incidence: 41/50*, conc. range: ≤5.52 µg/kg, sample origin: retail markets in Changchun, Chengdu, Shanghai, Shijiazhuang, Zhengzhou (cities), and Beijing (capital), China, sample year: 2007, country: China[224], *manufactured domestically

- Co-contamination: 41 sa. co-contaminated with AFB_1 and AFB_2; no further information available
- Further contamination of plants and plant products other than cereals and their products reported in the present book/article[224]: Butter (peanut butter), AFB_1, AFG_1, and AFG_2, literature[224]; Paste (sesame paste), AFB_1, AFB_2, AFG_1, and AFG_2, literature[224]

incidence: 7?/12, conc. range: 3.31–10.8 µg/kg, sample origin: supermarkets and village stores in five districts of Penang (state) including Penang Island and Wellesley (province), Malaysia, sample year: unknown, country: Malaysia[225]

- Co-contamination: not reported
- Further contamination of plants and plant products other than cereals and their products reported in the present book/article[225]: Butter (peanut butter), AFB_1, literature[225]; Cake (peanut), Nuts (peanut), AFB_1, AFB_2, and AFG_2, literature[225]; Confectionery, Nuts (walnut), Products (nut product), AFB_1 and AFB_2, literature[225]

incidence: 31?/33, conc. range: ≤14.0 µg/kg, sample origin: local stores in Hangzhou, Ningbo, and Quzhou (cities), Zhejiang (province), China, sample year: unknown, country: China[227]

- Co-contamination: not reported
- Further contamination of plants and plant products other than cereals and their products reported in the present book/article[227]: Butter (peanut butter), AFB_1, AFG_1, AFG_2, AFM_1, and AFM_2, literature[227]; Nuts (peanut), AFB_1, AFB_2, AFG_1, AFG_2, AFM_1, and AFM_2, literature[227]

incidence: 42/43, conc. range: 0.18–23.9 µg/kg, ∅ conc.: 3.20 µg/kg, sample origin: households, small factories, supermarkets, and grocery shops in Kharthoum State, Sudan, sample year: January 2010, country: Sudan[718]

- Co-contamination: not reported
- Further contamination of plants and plant products other than cereals and their products reported in the present book/article[718]: Butter (peanut butter), AFB_1, AFG_1, and AFG_2, literature[718]

incidence: 14/14*, conc. range: 1.7–7.2 µg/kg, sample origin: market in Lilongwe (city), Malawi, sample year: December 2012, country: Malawi[790], *peanut butter, local

- Co-contamination: 14 sa. co-contaminated with AFB_1, AFB_2, AFG_1, and AFG_2
- Further contamination of plants and plant products other than cereals and their products reported in the present book/article[790]: Butter (peanut butter), AFB_1, AFG_1, and AFG_2, literature[790]; Food, Nuts (peanut), AFB_1, AFB_2, AFG_1, and AFG_2, literature[790]

incidence: 8/11*, conc. range: 0.2–0.4 µg/kg, sample origin: market in Lilongwe (city), Malawi, sample year: December 2012, country: Malawi[790], *peanut butter, imported

- Co-contamination: not reported
- Further contamination of plants and plant products other than cereals and their products reported in the present book/article[790]: Butter (peanut butter), AFB_1, AFG_1, and AFG_2, literature[790]; Food, Nuts (peanut), AFB_1, AFB_2, AFG_1, and AFG_2, literature[790]

incidence: 4/11, conc. range: 6.2–25.7 µg/kg, ∅ conc.: 15.8 µg/kg, sample origin: retail shops and vendors at informal markets in Bulawayo metropolitan area, Zimbabwe, sample year: October/November 2011, country: Botswana/South Africa[797]

- Co-contamination: 1 sa. co-contaminated with AFB_1, AFB_2, AFG_1, and AFG_2; 3 sa. co-contaminated with AFB_1, AFB_2, and AFG_1
- Further contamination of plants and plant products other than cereals and their products reported in the present book/article[797]: Butter (peanut butter), AFB_1, AFG_1, and AFG_2, literature[797]; Nuts (peanut), AFB_1 and AF/S, literature[797]

Aflatoxin G_1

incidence: 22?/26, conc. range: 0.13–2.21 µg/kg, sample origin: unknown, sample year: unknown, country: Korea[9]

- Co-contamination: not reported
- Further contamination of plants and plant products other than cereals and their products reported in the present book/article[9]: Butter (peanut butter), AFB_1, AFB_2, and AFG_2, literature[9]; Nuts (almond), Nuts (peanut), AFB_1 and AFB_2, literature[9]; Products (nut product), Tea, AFB_1, literature[9]

incidence: 2/2, conc. range: 3.5–5.2 µg/kg, ∅ conc.: 4.4 µg/kg, sample origin: local marketing institutions, GDR, sample year: unknown, country: GDR[29]

- Co-contamination: 2 sa. co-contaminated with AFB_1, AFB_2, AFG_1, and AFG_2
- Further contamination of plants and plant products other than cereals and their products reported in the present book/article[29]: Butter (peanut butter), AFB_1, AFB_2, and AFG_2, literature[29]; Coffee, OTA, literature[29]; Fruits, AFB_1, AFG_1, and AFG_2, literature[29]; Fruits (apple), Fruits (peach), Juices (apple juice), PAT, literature[29]; Nuts (peanut), Products (peanut product), Seeds (sunflower), AFB_1, AFB_2, AFG_1, and AFG_2, literature[29]

incidence: 3?/6, conc. range: 0.1–0.4 µg/kg, sample origin: unknown, sample year: unknown, country: Japan[97]

- Co-contamination: not reported
- Further contamination of plants and plant products other than cereals and their products reported in the present book/article[97]: Beans, Butter (peanut butter), Jam (bean jam), Spices, AFB_1 and AFB_2, literature[97]; Oils (peanut oil), Spices (pepper), AFB_1, AFB_2, and AFG_1, literature[97]; Spices (nutmeg), AFB_1, AFB_2, AFG_1, and AFG_2, literature[97]

incidence: 4*/21, conc. range: 0.17–0.81 µg/kg, ∅ conc.: 0.4 µg/kg, sample origin: local supermarkets and small retail shops across Japan, sample year: summer 2004–winter 2005, country: Japan[146], *domestic and imported products

- Co-contamination: 4 sa. co-contaminated with AFB_1, AFB_2, AFG_1, and AFG_2
- Further contamination of plants and plant products other than cereals and their products reported in the present book/article[146]: Beans (green bean), Coffee, Raisins, Wine, OTA, literature[146]; Butter (peanut butter), AFB_1, AFB_2, and AFG_2, literature[146]; Chocolate, AFB_1 and OTA, literature[146]

incidence: 2/4, conc. range: 0.3 µg/kg, ∅ conc.: 0.3 µg/kg, sample origin: retail stores in Osaka (prefecture), Japan, sample year: 1988–1992, country: Japan[173]

- Co-contamination: 2 sa. co-contaminated with AFB_1, AFB_2, and AFG_1
- Further contamination of plants and plant products other than cereals and their products reported in the present book/article[173]: Butter (peanut butter), AFB_1 and AFB_2, literature[173]; Nuts (peanut), Spices (cayenne pepper, chilli), Spices (pepper), AFB_1, literature[173]; Spices (nutmeg), AFB_1, AFB_2, and AFG_1, literature[173]

incidence: 1/2, conc.: 0.52 µg/kg, sample origin: local stores in Munich (city), Germany, sample year: unknown, country: Germany[181]

- Co-contamination: 1 sa. co-contaminated with AFB_1, AFB_2, and AFG_1
- Further contamination of plants and plant products other than cereals and their products reported in the present book/article[181]: Butter (peanut butter), AFB_1 and AFB_2, literature[181]; Nuts (hazelnut), Spices (paprika), AFB_1 and AFG_1, literature[181]; Nuts (pistachio nut), Spices (pepper), AFB_1, literature[181]; Spices (nutmeg), AFB_1, AFB_2, and AFG_1, literature[181]

incidence: 21?/74, conc. range: <0.4–0.9 µg/kg, ∅ conc.: 0.6 µg/kg, sample origin: sites of primary storage or processing (domestic products) and sites of importation, Cyprus, sample year: 1992–1996, country: Cyprus[200]

- Co-contamination: not reported
- Further contamination of plants and plant products other than cereals and their products reported in the present book/article[200]: Butter (peanut butter), AFB_1, AFB_2, and AFG_2, literature[200]; Figs, Nuts (peanut), AFB_1, AFB_2, AFG_1, and AFG_2, literature[200]; Nuts (Brazil nut, paranut), Nuts (walnut), AFB_1, AFB_2, and AFG_1, literature[200]; Nuts (chestnut), Products (sesame product), Spices, AFB_1, literature[200]; Nuts (pistachio nut), AFB_1 and AFB_2, literature[200]; Seeds (sesame), AFB_1, AFG_1, and AFG_2, literature[200]

incidence: 13/29, conc. range: LOQ-2 µg/kg (13 sa., maximum: 1.4 µg/kg), ∅ conc.: 0.79 µg/kg, sample origin: retail outlets, UK, sample year: September 2001, country: UK[203]

- Co-contamination: 2 sa. co-contaminated with AFB_1, AFB_2, AFG_1, AFG_2, and OTA; 3 sa. co-contaminated with AFB_1, AFB_2, AFG_1, and AFG_2; 3 sa. co-contaminated with AFB_1, AFB_2, and AFG_1; 4 sa. co-contaminated with AFB_1 and AFG_1; 1 sa. contaminated solely with AFG_1
- Further contamination of plants and plant products other than cereals and their products reported in the present book/article[203]: Butter (nut butter), AFB_1, AFB_2, AFG_1, AFG_2, and OTA, literature[203]; Butter (peanut butter), AFB_2, AFG_1, AFG_2, and OTA, literature[203]; Dates, AFB_1 and OTA, literature[203]; Figs, AFB_1, AFB_2, AFG_2, and OTA, literature[203]; Fruits, dried, Nuts (pistachio nut), AFB_1, AFB_2, and OTA, literature[203]; Nuts (Brazil nut, paranut), AFB_1, AFB_2, and AFG_1, literature[203]; Nuts (peanut), AFB_1 and AFB_2, literature[203]

incidence: 8/27, conc. range: LOQ-2 µg/kg (8 sa., maximum: 1 µg/kg), ∅ conc.: 0.44 µg/kg, sample origin: retail outlets, UK, sample year: September 2001, country: UK[203]

- Co-contamination: 2 sa. co-contaminated with AFB_1, AFB_2, AFG_1, and AFG_2; 1 sa. co-contaminated with AFB_1, AFB_2, and AFG_1; 2 sa. co-contaminated with AFB_1, AFB_2, AFG_1, and OTA; 3 sa. co-contaminated with AFB_1 and AFG_1
- Further contamination of plants and plant products other than cereals and their products reported in the present book/article[203]: Butter (nut butter), AFB_1, AFB_2, AFG_1, AFG_2, and OTA, literature[203]; Butter (peanut butter), AFB_1, AFB_2, AFG_2, and OTA, literature[203]; Dates, AFB_1 and OTA, literature[203]; Figs, AFB_1, AFB_2, AFG_2, and OTA, literature[203]; Fruits, dried, Nuts (pistachio nut), AFB_1, AFB_2, and OTA, literature[203]; Nuts (Brazil nut, paranut), AFB_1, AFB_2, and AFG_1, literature[203]; Nuts (peanut), AFB_1 and AFB_2, literature[203]

incidence: 5/30, conc. range: 6–135 µg/kg, ∅ conc.: 54 µg/kg, sample origin: markets in 16 areas (cities and counties), Taiwan, sample year: 1977/1978, country: Taiwan[214]

- Co-contamination: not reported
- Further contamination of plants and plant products other than cereals and their products reported in the present book/article[214]: Butter (peanut butter), AFB_1, AFB_2, and AFG_2, literature[214]; Cake (peanut), Nuts (peanut), Oils (peanut oil), AFB_1, AFB_2, AFG_1, and AFG_2, literature[214]

incidence: 2/2, conc. range: 0.6–1.2 µg/kg, ∅ conc.: 0.9 µg/kg, sample origin: unknown, sample year: unknown, country: Denmark[217]

- Co-contamination: 2 sa. co-contaminated with AFB_1, AFB_2, AFG_1, and AFG_2
- Further contamination of plants and plant products other than cereals and their products reported in the present book/article[217]: Butter (peanut butter), AFB_1, AFB_2, and AFG_2, literature[217]; Figs, Spices (cayenne pepper, chilli), AFB_1, AFB_2, AFG_1, and AFG_2, literature[217]; Meal (pistachio meal), AFB_1 and AFB_2, literature[217]

incidence: 15?/21, conc. range: 3.2–20.0 µg/kg, sample origin: storage depots and retail outlets, Botswana, sample year: January 1996–December 1997, country: Botswana[218]

- Co-contamination: not reported
- Further contamination of plants and plant products other than cereals and their products reported in the present book/article[218]: Butter (peanut butter), AFB_1, AFB_2, and AFG_2, literature[218]; Nuts (peanut), AFB_1, AFB_2, AFG_1, and AFG_2, literature[218]

incidence: 20/20, conc. range: 1.66–32.78 µg/kg, ∅ conc.: 10.6 µg/kg, sample origin: markets in Ankara (capital), Turkey, sample year: unknown, country: Turkey[221]

- Co-contamination: 19 sa. co-contaminated with AFB_1, AFB_2, and AFG_1; 1 sa. co-contaminated with AFB_2 and AFG_1
- Further contamination of plants and plant products other than cereals and their products reported in the present book/article[221]: Butter (peanut butter), AFB_1 and AFB_2, literature[221]; Seeds (sesame), AFG_1, literature[221]

incidence: 4/5, conc. range: 0.11–2.01 µg/kg, ∅ conc.: 0.7 µg/kg, sample origin: unknown, sample year: unknown, country: USA[222]

- Co-contamination: 4 sa. co-contaminated with AFB_1, AFB_2, AFG_1, and AFG_2
- Further contamination of plants and plant products other than cereals and their products reported in the present book/article[222]: Butter (peanut butter), AFB_1, AFB_2, and AFG_2, literature[222]

incidence: 1/1, conc.: 84.5 µg/kg, sample origin: market in Khartoum (capital), Sudan, sample year: February 1981–1984?, country: UK/Sudan[223]

- Co-contamination: 1 sa. co-contaminated with AFB_1, AFB_2, and AFG_1
- Further contamination of plants and plant products other than cereals and their products reported in the present book/article[223]: Butter (peanut butter), AFB_1 and AFB_2, literature[223]; Nuts (peanut), AFB_1, AFB_2, AFG_1, and AFG_2, literature[223]; Okra, AFG_2, literature[223]; Peas (chickpea), AFB_1, literature[223]

incidence: 40/50*, conc. range: ≤21.22 µg/kg, sample origin: retail markets in Changchun, Chengdu, Shanghai, Shijiazhuang, Zhengzhou (cities), and Beijing (capital), China, sample year: 2007, country: China[224], *manufactured domestically

- Co-contamination: not reported
- Further contamination of plants and plant products other than cereals and their products reported in the present book/article[224]: Butter (peanut butter), AFB_1, AFB_2, and AFG_2, literature[224]; Paste (sesame paste), AFB_1, AFB_2, AFG_1, and AFG_2, literature[224]

incidence: 31?/33, conc. range: ≤20.8 µg/kg, sample origin: local stores in Hangzhou, Ningbo, and Quzhou (cities), Zhejiang (province), China, sample year: unknown, country: China[227]

- Co-contamination: not reported
- Further contamination of plants and plant products other than cereals and their products reported in the present book/article[227]: Butter (peanut butter), AFB_1, AFB_2, AFG_2, AFM_1, and AFM_2, literature[227]; Nuts (peanut), AFB_1, AFB_2, AFG_1, AFG_2, AFM_1, and AFM_2, literature[227]

incidence: 43/43, conc. range: 26.5–401 µg/kg, ∅ conc.: 137 µg/kg, sample origin: households, small factories, supermarkets, and grocery shops in Khartoum State, Sudan, sample year: January 2010, country: Sudan[718]

- Co-contamination: not reported
- Further contamination of plants and plant products other than cereals and their products reported in the present book/article[718]: Butter (peanut butter), AFB_1, AFB_2, and AFG_2, literature[718]

incidence: 14/14*, conc. range: 14.8–65.0 µg/kg, sample origin: market in Lilongwe (city), Malawi, sample year: December 2012, country: Malawi[790], *peanut butter, local

- Co-contamination: 14 sa. co-contaminated with AFB_1, AFB_2, AFG_1, and AFG_2
- Further contamination of plants and plant products other than cereals and their products reported in the present book/article[790]: Butter (peanut butter), AFB_1, AFB_2, and AFG_2, literature[790]; Food, Nuts (peanut), AFB_1, AFB_2, AFG_1, and AFG_2, literature[790]

incidence: 7/11*, conc. range: 0.5–1.8 µg/kg, sample origin: market in Lilongwe (city), Malawi, sample year: December 2012, country: Malawi[790], *peanut butter, imported

- Co-contamination: not reported
- Further contamination of plants and plant products other than cereals and their products reported in the present book/article[790]: Butter (peanut butter), AFB_1, AFB_2, and AFG_2, literature[790]; Food, Nuts (peanut), AFB_1, AFB_2, AFG_1, and AFG_2, literature[790]

incidence: 5/11, conc. range: 9.3–47.1 µg/kg, ∅ conc.: 25.5 µg/kg, sample origin: retail shops and vendors at informal markets in Bulawayo metropolitan area, Zimbabwe, sample year: October/November 2011, country: Botswana/South Africa[797]

- Co-contamination: 1 sa. co-contaminated with AFB_1, AFB_2, AFG_1, and AFG_2; 3 sa. co-contaminated with AFB_1, AFB_2, and AFG_1; 1 sa. co-contaminated with AFB_1 and AFG_1
- Further contamination of plants and plant products other than cereals and their products reported in the present book/article[797]: Butter (peanut butter), AFB_1, AFB_2, and AFG_2, literature[797]; Nuts (peanut), AFB_1 and AF/S, literature[797]

Aflatoxin G_2

incidence: 2/26, conc. range: 0.21–0.30 µg/kg, ∅ conc.: 0.255 µg/kg, sample origin: unknown, sample year: unknown, country: Korea[9]

- Co-contamination: not reported
- Further contamination of plants and plant products other than cereals and their products reported in the present book/article[9]: Butter (peanut butter), AFB_1, AFB_2, and AFG_1, literature[9]; Nuts (almond), Nuts (peanut), AFB_1 and AFB_2, literature[9]; Products (nut product), Tea, AFB_1, literature[9]

incidence: 2/2, conc. range: 1.3–1.7 µg/kg, ∅ conc.: 1.5 µg/kg, sample origin: local marketing institutions, GDR, sample year: unknown, country: GDR[29]

- Co-contamination: 2 sa. co-contaminated with AFB_1, AFB_2, AFG_1, and AFG_2
- Further contamination of plants and plant products other than cereals and their products reported in the present book/article[29]: Butter (peanut butter), AFB_1, AFB_2, and AFG_1, literature[29]; Coffee, OTA, literature[29]; Fruits, AFB_1, AFG_1, and AFG_2, literature[29]; Fruits (apple), Fruits (peach), Juices (apple juice), PAT, literature[29]; Nuts (peanut), Products (peanut product), Seeds (sunflower), AFB_1, AFB_2, AFG_1, and AFG_2, literature[29]

incidence: 4*/21, conc. range: 0.12–0.46 µg/kg, ∅ conc.: 0.21 µg/kg, sample origin: local supermarkets and small retail shops across Japan, sample year: summer 2004–winter 2005, country: Japan[146], *domestic and imported products

- Co-contamination: 4 sa. co-contaminated with AFB_1, AFB_2, AFG_1, and AFG_2
- Further contamination of plants and plant products other than cereals and their products reported in the present book/article[146]: Beans (green bean), Coffee, Raisins, Wine, OTA, literature[146]; Butter (peanut butter), AFB_1, AFB_2, and AFG_1, literature[146]; Chocolate, AFB_1 and OTA, literature[146]

incidence: 1/74, conc.: 0.3 µg/kg, sample origin: sites of primary storage or processing (domestic products) and sites of importation, Cyprus, sample year: 1992–1996, country: Cyprus[200]
- Co-contamination: not reported
- Further contamination of plants and plant products other than cereals and their products reported in the present book/article[200]: Butter (peanut butter), AFB_1, AFB_2, and AFG_1, literature[200]; Figs, Nuts (peanut), AFB_1, AFB_2, AFG_1, and AFG_2, literature[200]; Nuts (Brazil nut, paranut), Nuts (walnut), AFB_1, AFB_2, and AFG_1, literature[200]; Nuts (chestnut), Products (sesame product), Spices, AFB_1, literature[200]; Nuts (pistachio nut), AFB_1 and AFB_2, literature[200]; Seeds (sesame), AFB_1, AFG_1, and AFG_2, literature[200]

incidence: 5/29, conc. range: LOQ-2 µg/kg (5 sa., maximum: 0.5 µg/kg), ∅ conc.: 0.38 µg/kg, sample origin: retail outlets, UK, sample year: September 2001, country: UK[203]
- Co-contamination: 2 sa. co-contaminated with AFB_1, AFB_2, AFG_1, AFG_2, and OTA; 3 sa. co-contaminated with AFB_1, AFB_2, AFG_1, and AFG_2
- Further contamination of plants and plant products other than cereals and their products reported in the present book/article[203]: Butter (nut butter), AFB_1, AFB_2, AFG_1, AFG_2, and OTA, literature[203]; Butter (peanut butter), AFB_1, AFB_2, AFG_1, and OTA, literature[203]; Dates, AFB_1 and OTA, literature[203]; Figs, AFB_1, AFB_2, AFG_2, and OTA, literature[203]; Fruits, dried, Nuts (pistachio nut), AFB_1, AFB_2, and OTA, literature[203]; Nuts (Brazil nut, paranut), AFB_1, AFB_2, and AFG_1, literature[203]; Nuts (peanut), AFB_1 and AFB_2, literature[203]

incidence: 2/27, conc. range: LOQ-2 µg/kg (2 sa., maximum: 0.3 µg/kg), ∅ conc.: 0.25 µg/kg, sample origin: retail outlets, UK, sample year: September 2001, country: UK[203]
- Co-contamination: 2 sa. co-contaminated with AFB_1, AFB_2, AFG_1, and AFG_2
- Further contamination of plants and plant products other than cereals and their products reported in the present book/article[203]: Butter (nut butter), AFB_1, AFB_2, AFG_1, AFG_2, and OTA, literature[203]; Butter (peanut butter), AFB_1, AFG_1, and OTA, lit erature[203]; Dates, AFB_1 and OTA, literature[203]; Figs, AFB_1, AFB_2, AFG_2, and OTA, literature[203]; Fruits, dried, Nuts (pistachio nut), AFB_1, AFB_2, and OTA, literature[203]; Nuts (Brazil nut, paranut), AFB_1, AFB_2, and AFG_1, literature[203]; Nuts (peanut), AFB_1 and AFB_2, literature[203]

incidence: 3/30, conc. range: 33–124 µg/kg, ∅ conc.: 68 µg/kg, sample origin: markets in 16 areas (cities and counties), Taiwan, sample year: 1977/1978, country: Taiwan[214]
- Co-contamination: not reported
- Further contamination of plants and plant products other than cereals and their products reported in the present book/article[214]: Butter (peanut butter), AFB_1, AFB_2, and AFG_1, literature[214]; Cake (peanut), Nuts (peanut), Oils (peanut oil), AFB_1, AFB_2, AFG_1, and AFG_2, literature[214]

incidence: 2/2, conc. range: 0.2 µg/kg, ∅ conc.: 0.2 µg/kg, sample origin: unknown, sample year: unknown, country: Denmark[217]
- Co-contamination: 2 sa. co-contaminated with AFB_1, AFB_2, AFG_1, and AFG_2
- Further contamination of plants and plant products other than cereals and their products reported in the present book/article[217]: Butter (peanut butter), AFB_1, AFB_2, and AFG_1, literature[217]; Figs, Spices (cayenne pepper, chilli), AFB_1, AFB_2, AFG_1, and AFG_2, literature[217]; Meal (pistachio meal), AFB_1 and AFB_2, literature[217]

incidence: 15?/21, conc. range: 1.6–20.0 µg/kg, sample origin: storage depots and retail outlets, Botswana, sample year: January 1996–December 1997, country: Botswana[218]
- Co-contamination: not reported
- Further contamination of plants and plant products other than cereals and their products reported in the present book/article[218]: Butter (peanut butter), AFB_1, AFB_2, and AFG_1, literature[218]; Nuts (peanut), AFB_1, AFB_2, AFG_1, and AFG_2, literature[218]

incidence: 4/5, conc. range: 0.11–0.44 µg/kg, ∅ conc.: 0.2 µg/kg, sample origin: unknown, sample year: unknown, country: USA[222]
- Co-contamination: 4 sa. co-contaminated with AFB_1, AFB_2, AFG_1, and AFG_2
- Further contamination of plants and plant products other than cereals and their products reported in the present book/article[222]: Butter (peanut butter), AFB_1, AFB_2, and AFG_1, literature[222]

incidence: 37/50*, conc. range: ≤6.36 µg/kg, sample origin: retail markets in Changchun, Chengdu, Shanghai, Shijiazhuang, Zhengzhou (cities), and Beijing (capital), China, sample year: 2007, country: China[224], *manufactured domestically
- Co-contamination: not reported
- Further contamination of plants and plant products other than cereals and their products reported in the present book/article[224]: Butter (peanut butter), AFB_1, AFB_2, and AFG_1, literature[224]; Paste (sesame paste), AFB_1, AFB_2, AFG_1, and AFG_2, literature[224]

incidence: 31?/33, conc. range: ≤4.5 µg/kg, sample origin: local stores in Hangzhou, Ningbo, and Quzhou (cities), Zhejiang (province), China, sample year: unknown, country: China[227]
- Co-contamination: not reported
- Further contamination of plants and plant products other than cereals and their products reported in the present book/article[227]: Butter (peanut butter), AFB_1, AFB_2, AFG_1, AFM_1, and AFM_2, literature[227]; Nuts (peanut), AFB_1, AFB_2, AFG_1, AFG_2, AFM_1, and AFM_2, literature[227]

incidence: 4/43, conc. range: 8.60–30.1 µg/kg, ∅ conc.: 18.5 µg/kg, sample origin: households, small factories, supermarkets, and grocery shops in Khartoum State, Sudan, sample year: January 2010, country: Sudan[718]
- Co-contamination: not reported
- Further contamination of plants and plant products other than cereals and their products reported in the present book/article[718]: Butter (peanut butter), AFB_1, AFB_2, and AFG_1, literature[718]

incidence: 14/14*, conc. range: 1.7–6.4 µg/kg, sample origin: market in Lilongwe (city), Malawi, sample year: December 2012, country: Malawi[790], *peanut butter, local
- Co-contamination: 14 sa. co-contaminated with AFB_1, AFB_2, AFG_1, and AFG_2
- Further contamination of plants and plant products other than cereals and their products reported in the present book/article[790]: Butter (peanut butter), AFB_1, AFB_2, and AFG_1, literature[790]; Food, Nuts (peanut), AFB_1, AFB_2, AFG_1, and AFG_2, literature[790]

incidence: 6/11*, conc. range: 0.2–0.7 µg/kg, sample origin: market in Lilongwe (city), Malawi, sample year: December 2012, country: Malawi[790], *peanut butter, imported

- Co-contamination: not reported
- Further contamination of plants and plant products other than cereals and their products reported in the present book/article[790]: Butter (peanut butter), AFB_1, AFB_2, and AFG_1, literature[790]; Food, Nuts (peanut), AFB_1, AFB_2, AFG_1, and AFG_2, literature[790]

incidence: 1/11, conc.: 8.8 µg/kg, sample origin: retail shops and vendors at informal markets in Bulawayo metropolitan area, Zimbabwe, sample year: October/November 2011, country: Botswana/South Africa[797]

- Co-contamination: 1 sa. co-contaminated with AFB_1, AFB_2, AFG_1, and AFG_2
- Further contamination of plants and plant products other than cereals and their products reported in the present book/article[797]: Butter (peanut butter), AFB_1, AFB_2, and AFG_1, literature[797]; Nuts (peanut), AFB_1 and AF/S, literature[797]

Aflatoxin M_1

incidence: 31?/33, conc. range: ≤4.2 µg/kg, sample origin: local stores in Hangzhou, Ningbo, and Quzhou (cities), Zhejiang (province), China, sample year: unknown, country: China[227]

- Co-contamination: not reported
- Further contamination of plants and plant products other than cereals and their products reported in the present book/article[227]: Butter (peanut butter), AFB_1, AFB_2, AFG_1, AFG_2, and AFM_2, literature[227]; Nuts (peanut), AFB_1, AFB_2, AFG_1, AFG_2, AFM_1, and AFM_2, literature[227]

Aflatoxin M_2

incidence: 31?/33, conc. range: ≤1.8 µg/kg, sample origin: local stores in Hangzhou, Ningbo, and Quzhou (cities), Zhejiang (province), China, sample year: unknown, country: China[227]

- Co-contamination: not reported
- Further contamination of plants and plant products other than cereals and their products reported in the present book/article[227]: Butter (peanut butter), AFB_1, AFB_2, AFG_1, AFG_2, and AFM_1, literature[227]; Nuts (peanut), AFB_1, AFB_2, AFG_1, AFG_2, AFM_1, and AFM_2, literature[227]

Aflatoxin/S

incidence: 478/483, Ø conc.: 143.6 µg/kg*, sample origin: greater Manila (capital) area, Philippines, sample year: unknown, country: Philippines[1], *AF of pos sa.?

- Co-contamination: not reported
- Further contamination of plants and plant products other than cereals and their products reported in the present book/article[1]: Beans, Cassava, Cocoa, Coffee, Drinks (alcoholic drink), Nuts (cashew nut), Nuts (peanut), Products (coconut product), Rootcrops, Spices (cayenne pepper, chilli), Sweet potatoes, Yams, AF/S, literature[1]

incidence: 7/16* **, conc. range: 2–5 µg/kg*** (6 sa.), 8 µg/kg*** (1 sa.), sample origin: retail outlets in 20 different regions, England, sample year: January 1982, country: UK[3], *regular food, **peanut butter, smooth, ***AFB_1, AFB_2, AFG_1, and AFG_2

incidence: 6/16* **, conc. range: 2–5 µg/kg*** (4 sa.), 6–10 µg/kg*** (1 sa.), 14 µg/kg*** (1 sa.), sample origin: retail outlets in 20 different regions, England, sample year: January 1982, country: UK[3], *regular food, **peanut butter, crunchy, ***AFB_1, AFB_2, AFG_1, and AFG_2

incidence: 6/11* **, conc. range: 2–5 µg/kg*** (1 sa.), 6–10 µg/kg*** (2 sa.), 11–30 µg/kg*** (1 sa.), 31–100 µg/kg*** (2 sa., maximum: 85 µg/kg***), sample origin: retail outlets in 20 different regions, England, sample year: January 1982, country: UK[3], *health food, **peanut butter, smooth, ***AFB_1, AFB_2, AFG_1, and AFG_2

incidence: 10/14* **, conc. range: 2–5 µg/kg*** (1 sa.), 6–10 µg/kg*** (2 sa.), 11–30 µg/kg*** (1 sa.), 31–100 µg/kg*** (1 sa.), >100 µg/kg*** (5 sa., maximum: 345 µg/kg***), sample origin: retail outlets in 20 different regions, England, sample year: January 1982, country: UK[3], *health food, **peanut butter, crunchy, ***AFB_1, AFB_2, AFG_1, and AFG_2

incidence: 6/6* **, conc. range: 6–10 µg/kg*** (1 sa.), 11–30 µg/kg*** (1 sa.), 31–100 µg/kg*** (3 sa.), 175 µg/kg*** (1 sa.), sample origin: retail outlets in 20 different regions, sample year: June 1983, country: UK[3], *health food, **peanut butter, smooth, ***AFB_1, AFB_2, AFG_1, and AFG_2

incidence: 7/9* **, conc. range: 2–5 µg/kg*** (3 sa.), 6–10 µg/kg*** (2 sa.), 11–30 µg/kg*** (1 sa.), 211 µg/kg*** (1 sa.), sample origin: retail outlets in 20 different regions, England, sample year: June 1983, country: UK[3], *health food, **peanut butter, crunchy, ***AFB_1, AFB_2, AFG_1, and AFG_2

incidence: 1/4* **, conc.: 27 µg/kg***, sample origin: retail outlets in 20 different regions, England, sample year: December 1984, country: UK[3], *health food, **peanut butter, smooth, ***AFB_1, AFB_2, AFG_1, and AFG_2

incidence: 7/15* **, conc. range: 6–10 µg/kg*** (1 sa.), 11–30 µg/kg*** (1 sa.), 31–100 µg/kg*** (2 sa.), >100 µg/kg*** (3 sa., maximum: 147 µg/kg***), sample origin: retail outlets in 20 different regions, England, sample year: December 1984, country: UK[3], *health food, **peanut butter, crunchy, ***AFB_1, AFB_2, AFG_1, and AFG_2

- Co-contamination: not reported
- Further contamination of plants and plant products other than cereals and their products reported in the present book/article[3]: Butter (peanut butter), AFB_1, literature[3]; Nuts (almond), Nuts (Brazil nut, paranut), Nuts (hazelnut), Nuts (peanut), AFB_1 and AF/S, literature[3]

incidence: 98/2,092, conc. range: 5.0–9.9 µg/kg* (47 sa.), 10.0–14.9 µg/kg* (36 sa.), 15.0–19.9 µg/kg* (12 sa.), >25 µg/kg* (3 sa.), sample origin: brokers, distributors, growers, processors, and retail trade, Canada, sample year: 1970–1975, country: Canada[15], *AFB_1, AFB_2, AFG_1, and AFG_2

- Co-contamination: not reported
- Further contamination of plants and plant products other than cereals and their products reported in the present book/article[15]: Cocoa beans, Nuts (almond), Nuts (cashew nut), Nuts (peanut), Nuts (pecan nut), Nuts (pistachio nut), Nuts (walnut), Seeds (sunflower), AF/S, literature[15]

incidence: 17/104, conc. range: ≤20 µg/kg* (13 sa.), >20 µg/kg* (4 sa., maximum: 27 µg/kg*), sample origin: FDA districts, USA, sample year: 1986, country: USA[19], *AFB_1, AFB_2, AFG_1, and AFG_2

incidence: 1/3, conc.: 43 µg/kg*, sample origin: imported to USA, sample year: 1986, country: USA[19], *AFB_1, AFB_2, AFG_1, and AFG_2

- Co-contamination: not reported
- Further contamination of plants and plant products other than cereals and their products reported in the present book/article[19]: Candies (peanut candy), Candies (pistachio candy), Nuts, Nuts (almond), Nuts (Brazil nut, paranut), Nuts (peanut), Nuts (pecan nut), Nuts (pistachio nut), Nuts (walnut), Oils (peanut oil), Paste (peanut paste), Paste (pipian paste), Products (peanut product), Seeds (melon), Spices (cayenne pepper, chilli), Spices (ginger), Spices (nutmeg), AF/S, literature[19]

incidence: 29/29, conc. range: >30–8,600* µg/kg**, sample origin: markets in or near Manila (capital), Philippines, sample year: 1967/1968, country: USA[134], *baby food, **AF

- Co-contamination: not reported
- Further contamination of plants and plant products other than cereals and their products reported in the present book/ article[134]: Beans, Nuts (peanut), Products (coconut product), Products (peanut product), Tubers, AF/S, literature[134]

incidence: 0/3, conc.: no contamination, sample origin: imported from USA purchased on markets in or near Manila (capital), Philippines, sample year: 1967/1968, country: USA[134]

incidence: 13/42, conc.: 1.0–3.9 µg/kg* (8 sa.), 4.0–10 µg/kg* (4 sa.), 21 µg/kg* (1 sa.), sample origin: retail sa. from Norwich area and health food stores, market stalls, and supermarkets across the UK, sample year: unknown, country: UK[206], *AFB_1, AFB_2, AFG_1, and AFG_2

- Co-contamination: not reported
- Further contamination of plants and plant products other than cereals and their products reported in the present book/ article[206]: Figs, Nuts (Brazil nut, paranut), Nuts (peanut), Nuts (pistachio nut), Paste (fig paste), AF/S, literature[206]

incidence: 6/50* **, conc. range: 2.5–5 µg/kg*** (2 sa.), 11–30 µg/kg*** (3 sa.), 40.1 µg/kg*** (1 sa.), sample origin: retail health-food outlets in 25 different geographical locations, UK, sample year: 1986, country: UK[213], *health food, **peanut butter, smooth, *** AFB_1, AFB_2, AFG_1, and AFG_2

incidence: 12/79* **, conc. range: 2.5–5 µg/kg*** (4 sa.), 5.1–10 µg/kg*** (4 sa.), 11–30 µg/kg*** (2 sa.), 31 to ≤53.3 µg/kg*** (2 sa.), sample origin: retail health-food outlets in 25 different geographical locations, UK, sample year: 1986, country: UK[213], *health food, **peanut butter, crunchy, *** AFB_1, AFB_2, AFG_1, and AFG_2

- Co-contamination: not reported
- Further contamination of plants and plant products other than cereals and their products reported in the present book/ article[213]: Butter (peanut butter), AFB_1, literature[213]

incidence: 4/10, conc. range: 7–228 µg/kg*, sample origin: traditional markets and supermarkets, Indonesia, sample year: 2001/2002, country: Austria/Indonesia/Thailand[226], *AFB_1, AFB_2, AFG_1, and AFG_2

- Co-contamination: not reported
- Further contamination of plants and plant products other than cereals and their products reported in the present book/ article[226]: Butter (peanut butter), AFB_1, literature[226]; Cake (peanut), Nuts (peanut), Sauces (peanut-chilli sauce), Snacks, AFB_1 and AF/S, literature[226]

incidence: 10/18, conc. range: 1 to 1.0 to <5.0 µg/kg* (1 sa.), 5.0 to <10.0 µg/kg* (3 sa.), 10.0 to <50.0 µg/kg* (4 sa.), 50.0 to <100.0 µg/kg* (1 sa.), 775.0 µg/kg* (1 sa.), sample origin: commercially available, sample year: unknown, country: UK[228], *AFB_1, AFB_2, AFG_1, and AFG_2

- Co-contamination: not reported
- Further contamination of plants and plant products other than cereals and their products reported in the present book/ article[228]: not reported

incidence: 7/7, conc. range: 0.32–13.26 µg/kg*, sample origin: sales outlets, Qatar, sample year: October 2002, country: Qatar[229], *AFB_1, AFB_2, AFG_1, and AFG_2

- Co-contamination: not reported
- Further contamination of plants and plant products other than cereals and their products reported in the present book/ article[229]: Coffee, Raisins, OTA, literature[229]; Figs, Nuts (peanut), Nuts (pistachio nut), AF/S, literature[229]; Spices, Spices (cayenne pepper, chilli), AF/S and OTA, literature[229]

incidence: 256/400, conc. range: 32–54 µg/kg*, Ø conc.: 43.5 µg/kg*, sample origin: local market and processing centers, central region of Sudan, sample year: unknown, country: Sudan/Saudi Arabia[230], *AFB_1, AFB_2, AFG_1, and AFG_2

- Co-contamination: not reported
- Further contamination of plants and plant products other than cereals and their products reported in the present book/ article[230]: Nuts (peanut), AF/S, literature[230]

incidence: 4/4, conc. range: 6.50–14.90 µg/kg*, Ø conc.: 10.7 µg/kg*, sample origin: unknown, sample year: unknown, country: USA[231], *AFB_1, AFB_2, AFG_1, and AFG_2

- Co-contamination: not reported
- Further contamination of plants and plant products other than cereals and their products reported in the present book/ article[231]: Paste (peanut paste), AF/S, literature[231]

incidence: 27/95, conc. range: <20 µg/kg*, Ø conc.: 3.5 µg/kg*, sample origin: market basket sa. commercially available in Virginia (state), USA, sample year: 1982, country: USA[232], *AFB_1, AFB_2, AFG_1, and AFG_2

incidence: 273/355, conc. range: <20 µg/kg* (242 sa.), ≥20 µg/kg* (31 sa.), Ø conc.: 7.7 µg/kg*, sample origin: market basket sa. commercially available in Virginia (state), USA, sample year: 1983, country: USA[232], *AFB_1, AFB_2, AFG_1, and AFG_2

incidence: 196/258, conc. range: <20 µg/kg* (189 sa.), ≥20 µg/kg* (7 sa.), Ø conc.: 6.6 µg/kg*, sample origin: market basket sa. commercially available in Virginia (state), USA, sample year: 1984, country: USA[232], *AFB_1, AFB_2, AFG_1, and AFG_2

incidence: 184/235, conc. range: <20 µg/kg* (179 sa.), ≥20 µg/kg* (5 sa.), Ø conc.: 5.7 µg/kg*, sample origin: market basket sa. commercially available in Virginia (state), USA, sample year: 1985, country: USA[232]

incidence: 435/465, conc. range: <20 µg/kg* (375 sa.), ≥20 µg/kg* (60 sa., maximum: 76.6 µg/kg*), Ø conc.: 9.6 µg/kg*, sample origin: market basket sa. commercially available in Virginia (state), USA, sample year: 1986, country: USA[232], *AFB_1, AFB_2, AFG_1, and AFG_2

incidence: 404/449, conc. range: <20 µg/kg* (273 sa.), 20–50 µg/kg* (24 sa.), 50–100 µg/kg* (70 sa.), >100 µg/kg* (37 sa., maximum: 215 µg/kg*), Ø conc.: 33.0 µg/kg*, sample origin: market basket sa. commercially available in Virginia (state), USA, sample year: 1987, country: USA[232], *AFB_1, AFB_2, AFG_1, and AFG_2

incidence: 187/322, conc. range: <20 µg/kg* (175 sa.), ≥20 µg/kg* (12 sa.), Ø conc.: 8.9 µg/kg*, sample origin: market basket sa. commercially available in Virginia (state), USA, sample year: 1988, country: USA[232], *AFB_1, AFB_2, AFG_1, and AFG_2

incidence: 144/331, conc. range: <20 µg/kg*, Ø conc.: 2.5 µg/kg*, sample origin: market basket sa. commercially available in Virginia (state), USA, sample year: 1989, country: USA[232], *AFB_1, AFB_2, AFG_1, and AFG_2

- Co-contamination: not reported
- Further contamination of plants and plant products other than cereals and their products reported in the present book/ article[232]: not reported

incidence: 16/35, conc. range: 0.1–5 µg/kg* (5 sa.), 6–10 µg/kg* (6 sa.), 11–20 µg/kg* (3 sa.), 21–35 µg/kg* (2 sa.), sample origin:

commercial and imported food from retail outlets and points of entries, Malaysia, sample year: 1995–1999, country: Malaysia[233], *AF not specified

- Co-contamination: not reported
- Further contamination of plants and plant products other than cereals and their products reported in the present book/article[233]: Candies (peanut candy), Nuts, Nuts (peanut), Products (peanut product), Spices (cayenne pepper, chilli), AF/S, literature[233]

incidence: 31/75, conc. range: 0.1–3.4 µg/kg*, ∅ conc.: 0.7 µg/kg*, sample origin: available in New Zealand, sample year: 2010, country: New Zealand[277], *AFS total

- Co-contamination: not reported
- Further contamination of plants and plant products other than cereals and their products reported in the present book/article[277]: Bars, AFB_1, AF/S, DON, NIV, and OTA, literature[277]; Confectionery, Food, Fruits, dried, Nuts, Nuts (peanut), Nuts (pistachio nut), Sauces (peanut sauce), AFB_1 and AF/S, literature[277]; Butter (peanut butter), AFB_1, literature[277]; Coffee, Dates, Spices (nutmeg), Spices (pepper), Vine fruits, dried, OTA, literature[277]; Figs, Spices (cayenne pepper, chilli), Spices (curry), Spices (ginger), Spices (paprika), AFB_1, AF/S, and OTA, literature[277]; Nuts (Brazil nut, paranut), AFB_1, literature[277]

incidence: 10/11, conc. range: LOD-20 µg/kg* (1 sa.), 20–100 µg/kg* (3 sa.), 100–1,000 µg/kg* (4 sa.), ≤2,720 µg/kg* (2 sa.), sample origin: open-air markets in Port-au-Prince (capital), Haiti, sample year: July 2012, country: USA[376], *AFB_1, AFB_2, AFG_1, and AFG_2

incidence: 21/21, conc. range: 20–100 µg/kg* (2 sa.), 100–1,000 µg/kg* (17 sa.), ≤1,850 µg/kg* (2 sa.), sample origin: open-air markets in Cap Haitien, Haiti, sample year: July 2013, country: USA[376], *AFB_1, AFB_2, AFG_1, and AFG_2

- Co-contamination: not reported
- Further contamination of plants and plant products other than cereals and their products reported in the present book/article[376]: Nuts (peanut), AF/S, literature[376]

incidence: 75/142, conc. range: 0.2–32.5 µg/kg*, sample origin: traditional markets, supermarkets, grocery stores, wholesale stores, and convenience stores in 25 counties, Taiwan, sample year: 1997–2011 (with exceptions), country: Taiwan[540], *AFB_1, AFB_2, AFG_1, and AFG_2

- Co-contamination: For detailed information see the article.
- Further contamination of plants and plant products other than cereals and their products reported in the present book/article[540]: Candies (peanut candy), Flour (peanut flour), Nuts (peanut), AF/S, literature[540]

For detailed information see the article.

incidence: 16/32, conc. range: ≤32.2 µg/kg*, sample origin: markets, superstores, malls, and venders of major districts of Punjab (province), Pakistan, sample year: unknown, country: Pakistan[573], *AFB_1, AFB_2, AFG_1, and AFG_2

- Co-contamination: not reported
- Further contamination of plants and plant products other than cereals and their products reported in the present book/article[573]: Cookies, Nuts (peanut), Products (peanut product), AF/S, literature[573]

incidence: 5/160, conc. range: 2–20 µg/kg*, sample origin: imported to New Zealand, sample year: 1997–1999, 2001–2004, country: New Zealand[700], *AFB_1, AFB_2, AFG_1, and AFG_2

- Co-contamination: not reported
- Further contamination of plants and plant products other than cereals and their products reported in the present book/article[700]: Nuts (peanut), Nuts (pistachio nut), AF/S, literature[700]

incidence: 31/75, conc. range: ≤3.4 µg/kg*, sample origin: New Zealand, sample year: 2010, country: New Zealand[701], *AFB_1, AFB_2, AFG_1, and AFG_2

- Co-contamination: not reported
- Further contamination of plants and plant products other than cereals and their products reported in the present book/article[701]: Nuts (Brazil nut, paranut), AFB_1, literature[701]; Nuts (peanut), Nuts (pistachio nut), Products (nut product), Products (peanut product), AF/S, literature[701]

incidence: 14/14*, conc. range: 34.2–115.6 µg/kg**, sample origin: supermarkets in Lilongwe (capital), Malawi, sample year: December 2012, country: Belgium/Malawi[722], *manufactured in Malawi, **AFB_1, AFB_2, AFG_1, and AFG_2

incidence: 8/11*, conc. range: ≤4.3 µg/kg**, sample origin: supermarkets in Lilongwe (capital), Malawi, sample year: December 2012, country: Belgium/Malawi[722], *imported from South Africa, **AFB_1, AFB_2, AFG_1, and AFG_2

- Co-contamination: not reported
- Further contamination of plants and plant products other than cereals and their products reported in the present book/article[722]: Nuts (peanut), AF/S, literature[722]

Aspergillus and ***Penicillium*** Toxins

Ochratoxin A

incidence: 1/1, conc.: 0.04 µg/kg, sample origin: supermarkets and fast-food chains in Quebec (city), Canada, sample year: September/October 2008, country: Canada[90]

incidence: 0/1, conc.: no contamination, sample origin: supermarkets and fast-food chains in Calgary (city), Canada, sample year: September/October 2009, country: Canada[90]

- Co-contamination: not reported
- Further contamination of plants and plant products other than cereals and their products reported in the present book/article[90]: Baby food, Beans, Beverages, Candies, Chocolate, Coffee, Extracts, Herbs, Ice cream, Nuts, Pies, Pizza, Products, Pudding, Raisins, Seeds, Wine, OTA, literature[90]

incidence: 3/29, conc. range: 0.4–5.4 µg/kg, ∅ conc.: 2.6 µg/kg, sample origin: retail outlets, UK, sample year: September 2001, country: UK[203]

- Co-contamination: 2 sa. co-contaminated with AFB_1, AFB_2, AFG_1, AFG_2, and OTA; 1 sa. co-contaminated with AFB_1 and OTA
- Further contamination of plants and plant products other than cereals and their products reported in the present book/article[203]: Butter (nut butter), AFB_1, AFB_2, AFG_1, AFG_2, and OTA, literature[203]; Butter (peanut butter), AFB_1, AFB_2, AFG_1, and AFG_2, literature[203]; Dates, AFB_1 and OTA, literature[203]; Figs, AFB_1, AFB_2, AFG_2, and OTA, literature[203]; Fruits, dried, Nuts (pistachio nut), AFB_1, AFB_2, and OTA, literature[203]; Nuts (Brazil nut, paranut), AFB_1, AFB_2, and AFG_1, literature[203]; Nuts (peanut), AFB_1 and AFB_2, literature[203]

incidence: 2/27, conc. range: LOQ-2 µg/kg (2 sa., maximum: 0.4 µg/kg), ∅ conc.: 0.35 µg/kg, sample origin: retail outlets, UK, sample year: September 2001, country: UK[203]

- Co-contamination: 2 sa. co-contaminated with AFB_1, AFB_2, AFG_1, and OTA

- Further contamination of plants and plant products other than cereals and their products reported in the present book/article[203]: Butter (nut butter), AFB_1, AFB_2, AFG_1, AFG_2, and OTA, literature[203]; Butter (peanut butter), AFB_1, AFB_2, AFG_1, and AFG_2, literature[203]; Dates, AFB_1 and OTA, literature[203]; Figs, AFB_1, AFB_2, AFG_2, and OTA, literature[203]; Fruits, dried, Nuts (pistachio nut), AFB_1, AFB_2, and OTA, literature[203]; Nuts (Brazil nut, paranut), AFB_1, AFB_2, and AFG_1, literature[203]; Nuts (peanut), AFB_1 and AFB_2, literature[203]

see also **Food**

Butter (sesame butter) may contain the following mycotoxins:

Aspergillus Toxins

Aflatoxin/s

incidence: 1/14*, conc.: >176 µg/kg**, sample origin: retail markets, Turkey, sample year: unknown, country: Turkey[234], *tahini, **AFB_1, AFB_2, AFG_1, and AFG_2

- Co-contamination: not reported
- Further contamination of plants and plant products other than cereals and their products reported in the present book/article[234]: not reported

Cabernet Sauvignon see **Wine (red)**

Cake (date) may contain the following mycotoxins:

Aspergillus Toxins

Aflatoxin/s

incidence: 3/19, conc. range: ≤8.15 µg/kg*, sample origin: markets and local shops in big cities of Punjab (province), Pakistan, sample year: November 2011–March 2012, country: Pakistan/Malaysia[574], *AFB_1, AFB_2, AFG_1, and AFG_2

- Co-contamination: not reported
- Further contamination of plants and plant products other than cereals and their products reported in the present book/article[574]: Cookies, Dates, Halva, AF/S, literature[574]

Cake (peanut) may contain the following mycotoxins:

Aspergillus Toxins

Aflatoxin B_1

incidence: 4/6*, conc. range: 2.0 to ≤3.0 µg/kg, sample origin: supermarkets and/or traditional markets in Yogyakarta (city), Indonesia, sample year: 1998/1999, country: Indonesia/Austria/Canada[106], *"ampyang" (peanut-brown sugar cake)

- Co-contamination: not reported
- Further contamination of plants and plant products other than cereals and their products reported in the present book/article[106]: Butter (peanut butter), Food, Nuts (peanut), Products, Sauces (peanut-chilli sauce), AFB_1, literature[106]

incidence: 8/33, conc. range: 6–52 µg/kg, ∅ conc.: 22 µg/kg, sample origin: markets in 16 areas (cities and counties), Taiwan, sample year: 1977/1978, country: Taiwan[214]

- Co-contamination: not reported
- Further contamination of plants and plant products other than cereals and their products reported in the present book/article[214]: Butter (peanut butter), Nuts (peanut), Oils (peanut oil), AFB_1, AFB_2, AFG_1, and AFG_2, literature[214]; Cake (peanut), AFB_2, AFG_1, and AFG_2, literature[214]

incidence: 2/3*, conc. range: 47.2–48.9 µg/kg, ∅ conc.: 48.1 µg/kg, sample origin: supermarkets and village stores in five districts of Penang (state) including Penang Island and Wellesley (province), Malaysia, sample year: unknown, country: Malaysia[225], *gung tang

- Co-contamination: 1 sa. co-contaminated with AFB_1, AFB_2, and AFG_2; 1 sa. co-contaminated with AFB_1 and AFB_2
- Further contamination of plants and plant products other than cereals and their products reported in the present book/article[225]: Butter (peanut butter), Confectionery, Nuts (walnut), Products (nut product), AFB_1 and AFB_2, literature[225]; Cake (peanut), AFB_2 and AFG_2, literature[225]; Nuts (peanut), AFB_1, AFB_2, and AFG_2, literature[225]

incidence: 3/10, conc. range: 4–117 µg/kg, sample origin: traditional markets and supermarkets, Indonesia, sample year: 2001/2002, country: Austria/Indonesia/Thailand[226]

- Co-contamination: not reported
- Further contamination of plants and plant products other than cereals and their products reported in the present book/article[226]: Butter (peanut butter), Nuts (peanut), Sauces (peanut-chilli sauce), Snacks, AFB_1 and AF/S, literature[226]; Cake (peanut), AF/S, literature[226]

incidence: 15/15*, conc. range: <LOQ-183 µg/kg, sample origin: different agroecological zones, Benin, sample year: unknown, country: Belgium[235], *"kulikuli"

- Co-contamination: 6 sa. co-contaminated with AFB_1, AFB_2, AFG_1, AFG_2, and OTA; 3 sa. co-contaminated with AFB_1, AFB_2, AFG_1, and AFG_2; 1 sa. co-contaminated with AFB_1, AFB_2, AFG_2, and OTA; 2 sa. co-contaminated with AFB_1, AFG_1, AFG_2, and OTA; 3 sa. co-contaminated with AFB_1, AFG_1, and AFG_2
- Further contamination of plants and plant products other than cereals and their products reported in the present book/article[235]: Cake (peanut), AFB_2, AFG_1, AFG_2, and OTA, literature[235]; Flour (cassava flour), AFB_1, AFB_2, DAS, FB_1, and ZEA, literature[235]

incidence: 7/10*, conc. range: <4 µg/kg (2 sa.), 4–20 µg/kg (2 sa.), ≤39.1 µg/kg (3 sa.), sample origin: Babile (town), eastern parts of Ethiopia, sample year: 2014/2015, country: Ethiopia/USA[389], *"Halawa" (local cake)

incidence: 8/10*, conc. range: <4 µg/kg (2 sa.), 4–20 µg/kg (1 sa.), 20–100 µg/kg (3 sa.), ≤158.1 µg/kg (2 sa.), sample origin: Dire Dawa (city), eastern parts of Ethiopia, sample year: 2014/2015, country: Ethiopia/USA[389], *"Halawa" (local cake)

incidence: 10/10*, conc. range: 2.8 µg/kg (1 sa.), ≤9.4 µg/kg (9 sa.), sample origin: Jijiga (city), eastern parts of Ethiopia, sample year: 2014/2015, country: Ethiopia/USA[389], *"Halawa" (local cake)

- Co-contamination: not reported
- Further contamination of plants and plant products other than cereals and their products reported in the present book/article[389]: Cake (peanut), AFB_2, literature[389]; Nuts (peanut), AFB_1, AFB_2, AFG_1, and AFG_2, literature[389]

incidence: 29/29, conc. range: 12.6–2,824 µg/kg, ∅ conc.: 561 µg/kg, sample origin: markets in five states, Nigeria, sample year: unknown, country: Nigeria/Austria[738]

- Co-contamination: For detailed information see the article.
- Further contamination of plants and plant products other than cereals and their products reported in the present book/article[738]: Cake (peanut), 3-NPA, AFB_2, AFG_1, AFG_2, AFM_1, BEA, KA, OTA, and STG, literature[738]; see also Cake (peanut), chapter: Further Mycotoxins and Microbial Metabolites

incidence: 5/15, conc. range: 120–340 µg/kg, ∅ conc.: 219 µg/kg, sample origin: Cameroon, sample year: December 2010, country: Belgium/Russia[805]

- Co-contamination: 5 sa. co-contaminated with AFB_1 and OTA
- Further contamination of plants and plant products other than cereals and their products reported in the present book/article[805]: Cake (peanut), OTA, literature[805]; Flour (cassava flour), AFB_1, FB_1, and OTA, literature[805]

Aflatoxin B_2

incidence: 9/33, conc. range: 8–66 µg/kg, ∅ conc.: 25 µg/kg, sample origin: markets in 16 areas (cities and counties), Taiwan, sample year: 1977/1978, country: Taiwan[214]

- Co-contamination: not reported
- Further contamination of plants and plant products other than cereals and their products reported in the present book/article[214]: Butter (peanut butter), Nuts (peanut), Oils (peanut oil), AFB_1, AFB_2, AFG_1, and AFG_2, literature[214]; Cake (peanut), AFB_1, AFG_1, and AFG_2, literature[214]

incidence: 2/3*, conc. range: 13.0–22.3 µg/kg, ∅ conc.: 17.7 µg/kg, sample origin: supermarkets and village stores in five districts of Penang (state) including Penang Island and Wellesley (province), Malaysia, sample year: unknown, country: Malaysia[225], *gung tang

- Co-contamination: 1 sa. co-contaminated with AFB_1, AFB_2, and AFG_2; 1 sa. co-contaminated with AFB_1 and AFB_2
- Further contamination of plants and plant products other than cereals and their products reported in the present book/article[225]: Butter (peanut butter), Confectionery, Nuts (walnut), Products (nut product), AFB_1 and AFB_2, literature[225]; Cake (peanut), AFB_1 and AFG_2, literature[225]; Nuts (peanut), AFB_1, AFB_2, and AFG_2, literature[225]

incidence: 10/15*, conc. range: <LOQ-31 µg/kg, sample origin: different agroecological zones, Benin, sample year: unknown, country: Belgium[235], *"kulikuli"

- Co-contamination: 6 sa. co-contaminated with AFB_1, AFB_2, AFG_1, AFG_2, and OTA; 3 sa. co-contaminated with AFB_1, AFB_2, AFG_1, and AFG_2; 1 sa. co-contaminated with AFB_1, AFB_2, AFG_2, and OTA
- Further contamination of plants and plant products other than cereals and their products reported in the present book/article[235]: Cake (peanut), AFB_1, AFG_1, AFG_2, and OTA, literature[235]; Flour (cassava flour), AFB_1, AFB_2, DAS, FB_1, and ZEA, literature[235]

incidence: 7/10*, conc. range: <4 µg/kg (6 sa.), ≤4.6 µg/kg (1 sa.), sample origin: Babile (town), eastern parts of Ethiopia, sample year: 2014/2015, country: Ethiopia/USA[389], *"Halawa" (local cake)

incidence: 8/10*, conc. range: <4 µg/kg (6 sa.), ≤15.3 µg/kg (2 sa.), sample origin: Dire Dawa (city), eastern parts of Ethiopia, sample year: 2014/2015, country: Ethiopia/USA[389], *"Halawa" (local cake)

incidence: 10/10*, conc. range: 0.2–0.8 µg/kg (10 sa.), sample origin: Jijiga (city), eastern parts of Ethiopia, sample year: 2014/2015, country: Ethiopia/USA[389], *"Halawa" (local cake)

- Co-contamination: not reported
- Further contamination of plants and plant products other than cereals and their products reported in the present book/article[389]: Cake (peanut), AFB_1, literature[389]; Nuts (peanut), AFB_1, AFB_2, AFG_1, and AFG_2, literature[389]

incidence: 26/29, conc. range: 5–533 µg/kg, ∅ conc.: 117 µg/kg, sample origin: markets in five states, Nigeria, sample year: unknown, country: Nigeria/Austria[738]

- Co-contamination: For detailed information see the article.
- Further contamination of plants and plant products other than cereals and their products reported in the present book/article[738]: Cake (peanut), 3-NPA, AFB_1, AFG_1, AFG_2, AFM_1, BEA, KA, OTA, and STG, literature[738]; see also Cake (peanut), chapter: Further Mycotoxins and Microbial Metabolites

Aflatoxin G_1

incidence: 7/33, conc. range: 8–120 µg/kg, ∅ conc.: 54 µg/kg, sample origin: markets in 16 areas (cities and counties), Taiwan, sample year: 1977/1978, country: Taiwan[214]

- Co-contamination: not reported
- Further contamination of plants and plant products other than cereals and their products reported in the present book/article[214]: Butter (peanut butter), Nuts (peanut), Oils (peanut oil), AFB_1, AFB_2, AFG_1, and AFG_2, literature[214]; Cake (peanut), AFB_1, AFB_2, and AFG_2, literature[214]

incidence: 14/15*, conc. range: <LOQ-79 µg/kg, sample origin: different agroecological zones, Benin, sample year: unknown, country: Belgium[235], *"kulikuli"

- Co-contamination: 6 sa. co-contaminated with AFB_1, AFB_2, AFG_1, AFG_2, and OTA; 3 sa. co-contaminated with AFB_1, AFB_2, AFG_1, and AFG_2; 2 sa. co-contaminated with AFB_1, AFG_1, AFG_2, and OTA; 3 sa. co-contaminated with AFB_1, AFG_1, and AFG_2
- Further contamination of plants and plant products other than cereals and their products reported in the present book/article[235]: Cake (peanut), AFB_1, AFB_2, AFG_2, and OTA, literature[235]; Flour (cassava flour), AFB_1, AFB_2, DAS, FB_1, and ZEA, literature[235]

incidence: 24/29, conc. range: 5.8–477 µg/kg, ∅ conc.: 75 µg/kg, sample origin: markets in five states, Nigeria, sample year: unknown, country: Nigeria/Austria[738]

- Co-contamination: For detailed information see the article.
- Further contamination of plants and plant products other than cereals and their products reported in the present book/article[738]: Cake (peanut), 3-NPA, AFB_1, AFB_2, AFG_2, AFM_1, BEA, KA, OTA, and STG, literature[738]; see also Cake (peanut), chapter: Further Mycotoxins and Microbial Metabolites

Aflatoxin G_2

incidence: 7/33, conc. range: 18–84 µg/kg, ∅ conc.: 40 µg/kg, sample origin: markets in 16 areas (cities and counties), Taiwan, sample year: 1977/1978, country: Taiwan[214]

- Co-contamination: not reported
- Further contamination of plants and plant products other than cereals and their products reported in the present book/article[214]: Butter (peanut butter), Nuts (peanut), Oils (peanut oil), AFB_1, AFB_2, AFG_1, and AFG_2, literature[214]; Cake (peanut), AFB_1, AFB_2, and AFG_1, literature[214]

incidence: 1/3*, conc.: 14.5 µg/kg, sample origin: supermarkets and village stores in five districts of Penang (state) including Penang Island and Wellesley (province), Malaysia, sample year: unknown, country: Malaysia[225], *gung tang

- Co-contamination: 1 sa. co-contaminated with AFB_1, AFB_2, and AFG_2
- Further contamination of plants and plant products other than cereals and their products reported in the present book/

article[225]: Butter (peanut butter), Cake (peanut), Confectionery, Nuts (walnut), Products (nut product), AFB_1 and AFB_2, literature[225]; Nuts (peanut), AFB_1, AFB_2, and AFG_2, literature[225]

incidence: 15/15*, conc. range: 6–96 µg/kg, ∅ conc.: 43 µg/kg, sample origin: different agroecological zones, Benin, sample year: unknown, country: Belgium[235], *"kulikuli"

- Co-contamination: 6 sa. co-contaminated with AFB_1, AFB_2, AFG_1, AFG_2, and OTA; 3 sa. co-contaminated with AFB_1, AFB_2, AFG_1, and AFG_2; 1 sa. co-contaminated with AFB_1, AFB_2, AFG_2, and OTA; 2 sa. co-contaminated with AFB_1, AFG_1, AFG_2, and OTA; 3 sa. co-contaminated with AFB_1, AFG_1, and AFG_2
- Further contamination of plants and plant products other than cereals and their products reported in the present book/article[235]: Cake (peanut), AFB_1, AFB_2, AFG_1, and OTA, literature[235]; Flour (cassava flour), AFB_1, AFB_2, DAS, FB_1, and ZEA, literature[235]

incidence: 11/29, conc. range: 4.3–113 µg/kg, ∅ conc.: 23 µg/kg, sample origin: markets in five states, Nigeria, sample year: unknown, country: Nigeria/Austria[738]

- Co-contamination: For detailed information see the article.
- Further contamination of plants and plant products other than cereals and their products reported in the present book/article[738]: Cake (peanut), 3-NPA, AFB_1, AFB_2, AFG_1, AFM_1, BEA, KA, OTA, and STG, literature[738]; see also Cake (peanut), chapter: Further Mycotoxins and Microbial Metabolites

Aflatoxin M_1

incidence: 19/29, conc. range: 4.4–126 µg/kg, ∅ conc.: 37 µg/kg, sample origin: markets in five states, Nigeria, sample year: unknown, country: Nigeria/Austria[738]

- Co-contamination: For detailed information see the article.
- Further contamination of plants and plant products other than cereals and their products reported in the present book/article[738]: Cake (peanut), 3-NPA, AFB_1, AFB_2, AFG_1, AFG_2, BEA, KA, OTA, and STG, literature[738]; see also Cake (peanut), chapter: Further Mycotoxins and Microbial Metabolites

Aflatoxin/s

incidence: 3/10, conc. range: 5–302 µg/kg, sample origin: traditional markets and supermarkets, Indonesia, sample year: 2001/2002, country: Austria/Indonesia/Thailand[226]

- Co-contamination: not reported
- Further contamination of plants and plant products other than cereals and their products reported in the present book/article[226]: Butter (peanut butter), Nuts (peanut), Sauces (peanut-chilli sauce), Snacks, AFB_1 and AF/S, literature[226]; Cake (peanut), AFB_1, literature[226]

Other ***Aspergillus*** Toxins

Sterigmatocystin

incidence: 18/29, conc. range: 2.2–22 µg/kg, ∅ conc.: 8 µg/kg, sample origin: markets in five states, Nigeria, sample year: unknown, country: Nigeria/Austria[738]

- Co-contamination: For detailed information see the article.
- Further contamination of plants and plant products other than cereals and their products reported in the present book/article[738]: Cake (peanut), 3-NPA, AFB_1, AFB_2, AFG_1, AFG_2, AFM_1, BEA, KA, and OTA, literature[738]; see also Cake (peanut), chapter: Further Mycotoxins and Microbial Metabolites

Aspergillus and ***Penicillium*** Toxins

3-Nitropropionic Acid

incidence: 26/29, conc. range: 24.5–471 µg/kg, ∅ conc.: 181 µg/kg, sample origin: markets in five states, Nigeria, sample year: unknown, country: Nigeria/Austria[738]

- Co-contamination: For detailed information see the article.
- Further contamination of plants and plant products other than cereals and their products reported in the present book/article[738]: Cake (peanut), AFB_1, AFB_2, AFG_1, AFG_2, AFM_1, BEA, KA, OTA, and STG, literature[738]; see also Cake (peanut), chapter: Further Mycotoxins and Microbial Metabolites

Kojic Acid

incidence: 28/29, conc. range: 86–7,478 µg/kg, ∅ conc.: 1,356 µg/kg, sample origin: markets in five states, Nigeria, sample year: unknown, country: Nigeria/Austria[738]

- Co-contamination: For detailed information see the article.
- Further contamination of plants and plant products other than cereals and their products reported in the present book/article[738]: Cake (peanut), 3-NPA, AFB_1, AFB_2, AFG_1, AFG_2, AFM_1, BEA, OTA, and STG, literature[738]; see also Cake (peanut), chapter: Further Mycotoxins and Microbial Metabolites

Ochratoxin A

incidence: 9/15*, conc. range: <LOQ-2 µg/kg, sample origin: different agroecological zones, Benin, sample year: unknown, country: Belgium[235], *"kulikuli"

- Co-contamination: 6 sa. co-contaminated with AFB_1, AFB_2, AFG_1, AFG_2, and OTA; 1 sa. co-contaminated with AFB_1, AFB_2, AFG_2, and OTA; 2 sa. co-contaminated with AFB_1, AFG_1, AFG_2, and OTA
- Further contamination of plants and plant products other than cereals and their products reported in the present book/article[235]: Cake (peanut), AFB_1, AFB_2, AFG_1, and AFG_2, literature[235]; Flour (cassava flour), AFB_1, AFB_2, DAS, FB_1, and ZEA, literature[235]

incidence: 4/29, conc. range: 5–13 µg/kg, ∅ conc.: 9 µg/kg, sample origin: markets in five states, Nigeria, sample year: unknown, country: Nigeria/Austria[738]

- Co-contamination: For detailed information see the article.
- Further contamination of plants and plant products other than cereals and their products reported in the present book/article[738]: Cake (peanut), 3-NPA, AFB_1, AFB_2, AFG_1, AFG_2, AFM_1, BEA, KA, and STG, literature[738]; see also Cake (peanut), chapter: Further Mycotoxins and Microbial Metabolites

incidence: 5/15, conc. range: 2–15 µg/kg, ∅ conc.: 5.66 µg/kg, sample origin: Cameroon, sample year: December 2010, country: Belgium/Russia[805]

- Co-contamination: 5 sa. co-contaminated with AFB_1 and OTA
- Further contamination of plants and plant products other than cereals and their products reported in the present book/article[805]: Cake (peanut), AFB_1, literature[805]; Flour (cassava flour), AFB_1, FB_1, and OTA, literature[805]

Fusarium Toxins

Depsipeptides

Beauvericin

incidence: 29/29, conc. range: 0.05–3.36 µg/kg, ∅ conc.: 0.92 µg/kg, sample origin: markets in five states, Nigeria, sample year: unknown, country: Nigeria/Austria[738]

- Co-contamination: For detailed information see the article.

- Further contamination of plants and plant products other than cereals and their products reported in the present book/article[738]: Cake (peanut), 3-NPA, AFB_1, AFB_2, AFG_1, AFG_2, AFM_1, KA, OTA, and STG, literature[738]; see also Cake (peanut), chapter: Further Mycotoxins and Microbial Metabolites

Candies may contain the following mycotoxins:

Aspergillus and *Penicillium* Toxins

Ochratoxin A

incidence: 1/1, conc.: 0.01 µg/kg, sample origin: supermarkets and fast-food chains in Quebec (city), Canada, sample year: September/October 2008, country: Canada[90]

incidence: 0/1, conc.: no contamination, sample origin: supermarkets and fast-food chains in Calgary (city), Canada, sample year: September/October 2009, country: Canada[90]

- Co-contamination: not reported
- Further contamination of plants and plant products other than cereals and their products reported in the present book/article[90]: Baby food, Beans, Beverages, Butter (peanut butter), Chocolate, Coffee, Extracts, Herbs, Ice cream, Nuts, Pies, Pizza, Products, Pudding, Raisins, Seeds, Wine, OTA, literature[90]

Candies (coconut candy) may contain the following mycotoxins:

Aspergillus Toxins

Aflatoxin B_1

incidence: 4/6, conc. range: 20 µg/kg, sample origin: local markets, oil mills, and houses, Andhra Pradesh (state), India, sample year: 1982, country: India[236]

- Co-contamination: not reported
- Further contamination of plants and plant products other than cereals and their products reported in the present book/article[236]: Candies (coconut candy), AFB_2, AFG_1, and AFG_2, literature[236]; Copra, Meal (copra meal), AFB_1, AFB_2, AFG_1, and AFG_2, literature[236]; Oils (coconut oil), AFB_1, AFB_2, and AFG_1, literature[236]

Aflatoxin B_2

incidence: 4/6, conc. range: 20 µg/kg, sample origin: local markets, oil mills, and houses, Andhra Pradesh (state), India, sample year: 1982, country: India[236]

- Co-contamination: not reported
- Further contamination of plants and plant products other than cereals and their products reported in the present book/article[236]: Candies (coconut candy), AFB_1, AFG_1, and AFG_2, literature[236]; Copra, Meal (copra meal), AFB_1, AFB_2, AFG_1, and AFG_2, literature[236]; Oils (coconut oil), AFB_1, AFB_2, and AFG_1, literature[236]

Aflatoxin G_1

incidence: 4/6, conc. range: 40–100 µg/kg, sample origin: local markets, oil mills, and houses, Andhra Pradesh (state), India, sample year: 1982, country: India[236]

- Co-contamination: not reported
- Further contamination of plants and plant products other than cereals and their products reported in the present book/article[236]: Candies (coconut candy), AFB_1, AFB_2, and AFG_2, literature[236]; Copra, Meal (copra meal), AFB_1, AFB_2, AFG_1, and AFG_2, literature[236]; Oils (coconut oil), AFB_1, AFB_2, and AFG_1, literature[236]

Aflatoxin G_2

incidence: 4/6, conc. range: 40–100 µg/kg, sample origin: local markets, oil mills, and houses, Andhra Pradesh (state), India, sample year: 1982, country: India[236]

- Co-contamination: not reported
- Further contamination of plants and plant products other than cereals and their products reported in the present book/article[236]: Candies (coconut candy), Oils (coconut oil), AFB_1, AFB_2, and AFG_1, literature[236]; Copra, Meal (copra meal), AFB_1, AFB_2, AFG_1, and AFG_2, literature[236]

Candies (peanut candy) may contain the following mycotoxins:

Aspergillus Toxins

Aflatoxin B_1

incidence: 39/48*, conc. range: 0.50–43.29 µg/kg, ∅ conc.: 4.33 µg/kg, sample origin: Araras, Leme, Pirassununga, and Porto Ferreira (cities), northeast of São Paulo (state), Brazil, sample year: June 2006–May 2007, country: Brazil[237], *Paçoca (candy, ground)

incidence: 21/48*, conc. range: 0.52–32.85 µg/kg, ∅ conc.: 1.78 µg/kg, sample origin: Araras, Leme, Pirassununga, and Porto Ferreira (cities), northeast of São Paulo (state), Brazil, sample year: June 2006–May 2007, country: Brazil[237], *dragee, salty

incidence: 17/48*, conc. range: 0.59–13.98 µg/kg, ∅ conc.: 1.47 µg/kg, sample origin: Araras, Leme, Pirassununga, and Porto Ferreira (cities), northeast of São Paulo (state), Brazil, sample year: June 2006–May 2007, country: Brazil[237], *dragee, sweet

- Co-contamination: not reported
- Further contamination of plants and plant products other than cereals and their products reported in the present book/article[237]: Candies (peanut candy), AFB_2, AFG_1, and AFG_2, literature[237]; Nuts (peanut), AFB_1, AFB_2, AFG_1, and AFG_2, literature[237]

incidence: 12/27, conc. range: LOD-20 µg/kg (10 sa.), >20 µg/kg (2 sa., maximum: 69.6 µg/kg), sample origin: retail outlets in Paraná (state), Brazil, sample year: April 2006–May 2007, country: Brazil[784]

- Co-contamination: For detailed information see the article.
- Further contamination of plants and plant products other than cereals and their products reported in the present book/article[784]: Candies (peanut candy), AFB_2, AFG_1, and AFG_2, literature[784]; Nuts (peanut), AFB_1, AFB_2, AFG_1, and AFG_2, literature[784]

Aflatoxin B_2

incidence: 31/48*, conc. range: 0.52–11.72 µg/kg, ∅ conc.: 1.30 µg/kg, sample origin: Araras, Leme, Pirassununga, and Porto Ferreira (cities), northeast of São Paulo (state), Brazil, sample year: June 2006–May 2007, country: Brazil[237], *Paçoca (candy, ground)

incidence: 11/48*, conc. range: <LOQ-2.55 µg/kg, ∅ conc.: <LOQ, sample origin: Araras, Leme, Pirassununga, and Porto Ferreira (cities), northeast of São Paulo (state), Brazil, sample year: June 2006–May 2007, country: Brazil[237], *dragee, salty

incidence: 13/48*, conc. range: <LOQ-3.40 µg/kg, ∅ conc.: <LOQ, sample origin: Araras, Leme, Pirassununga, and Porto Ferreira (cities), northeast of São Paulo (state), Brazil, sample year: June 2006–May 2007, country: Brazil[237], *dragee, sweet

- Co-contamination: not reported
- Further contamination of plants and plant products other than cereals and their products reported in the present book/article[237]: Candies (peanut candy), AFB_1, AFG_1, and AFG_2, literature[237]; Nuts (peanut), AFB_1, AFB_2, AFG_1, and AFG_2, literature[237]

incidence: 11/27, conc. range: LOD-20 µg/kg (11 sa., maximum: 10.6 µg/kg), sample origin: retail outlets in Paraná (state), Brazil, sample year: April 2006–May 2007, country: Brazil[784]

- Co-contamination: For detailed information see the article.
- Further contamination of plants and plant products other than cereals and their products reported in the present book/article[784]: Candies (peanut candy), AFB_1, AFG_1, and AFG_2, literature[784]; Nuts (peanut), AFB_1, AFB_2, AFG_1, and AFG_2, literature[784]

Aflatoxin G_1

incidence: 35/48*, conc. range: 0.52–38.92 µg/kg, Ø conc.: 2.63 µg/kg, sample origin: Araras, Leme, Pirassununga, and Porto Ferreira (cities), northeast of São Paulo (state), Brazil, sample year: June 2006–May 2007, country: Brazil[237], *Paçoca (candy, ground)

incidence: 19/48*, conc. range: 0.54–3.65 µg/kg, Ø conc.: 0.83 µg/kg, sample origin: Araras, Leme, Pirassununga, and Porto Ferreira (cities), northeast of São Paulo (state), Brazil, sample year: June 2006–May 2007, country: Brazil[237], *dragee, salty

incidence: 15/48*, conc. range: 0.50–2.29 µg/kg, Ø conc.: 0.90 µg/kg, sample origin: Araras, Leme, Pirassununga, and Porto Ferreira (cities), northeast of São Paulo (state), Brazil, sample year: June 2006–May 2007, country: Brazil[237], *dragee, sweet

- Co-contamination: not reported
- Further contamination of plants and plant products other than cereals and their products reported in the present book/article[237]: Candies (peanut candy), AFB_1, AFB_2, and AFG_2, literature[237]; Nuts (peanut), AFB_1, AFB_2, AFG_1, and AFG_2, literature[237]

incidence: 7/27, conc. range: LOD-20 µg/kg (7 sa., maximum: 2.7 µg/kg), sample origin: retail outlets in Paraná (state), Brazil, sample year: April 2006–May 2007, country: Brazil[784]

- Co-contamination: For detailed information see the article.
- Further contamination of plants and plant products other than cereals and their products reported in the present book/article[784]: Candies (peanut candy), AFB_1, AFB_2, and AFG_2, literature[784]; Nuts (peanut), AFB_1, AFB_2, AFG_1, and AFG_2, literature[784]

Aflatoxin G_2

incidence: 30/48*, conc. range: 0.57–9.85 µg/kg, Ø conc.: 0.70 µg/kg, sample origin: Araras, Leme, Pirassununga, and Porto Ferreira (cities), northeast of São Paulo (state), Brazil, sample year: June 2006–May 2007, country: Brazil[237], *Paçoca (candy, ground)

incidence: 17/48*, conc. range: <LOQ-2.45 µg/kg, Ø conc.: <LOQ, sample origin: Araras, Leme, Pirassununga, and Porto Ferreira (cities), northeast of São Paulo (state), Brazil, sample year: June 2006–May 2007, country: Brazil[237], *dragee, salty

incidence: 11/48*, conc. range: 0.50–7.19 µg/kg, Ø conc.: 0.73 µg/kg, sample origin: Araras, Leme, Pirassununga, and Porto Ferreira (cities), northeast of São Paulo (state), Brazil, sample year: June 2006–May 2007, country: Brazil[237], *dragee, sweet

- Co-contamination: not reported
- Further contamination of plants and plant products other than cereals and their products reported in the present book/article[237]: Candies (peanut candy), AFB_1, AFB_2, and AFG_1, literature[237]; Nuts (peanut), AFB_1, AFB_2, AFG_1, and AFG_2, literature[237]

incidence: 3/27, conc. range: LOD-20 µg/kg (3 sa., maximum: 1.8 µg/kg), sample origin: retail outlets in Paraná (state), Brazil, sample year: April 2006–May 2007, country: Brazil[784]

- Co-contamination: For detailed information see the article.
- Further contamination of plants and plant products other than cereals and their products reported in the present book/article[784]: Candies (peanut candy), AFB_1, AFB_2, and AFG_1, literature[784]; Nuts (peanut), AFB_1, AFB_2, AFG_1, and AFG_2, literature[784]

Aflatoxin/s

incidence: 10/18, conc. range: ≤20 µg/kg*, sample origin: imported to the USA, sample year: 1986, country: USA[19], *AFB_1, AFB_2, AFG_1, and AFG_2

- Co-contamination: not reported
- Further contamination of plants and plant products other than cereals and their products reported in the present book/article[19]: Butter (peanut butter), Candies (pistachio candy), Nuts, Nuts (almond), Nuts (peanut), Nuts (pecan nut), Nuts (pistachio nut), Nuts (walnut), Oils (peanut oil), Paste (peanut paste), Paste (pipian paste), Products (peanut product), Seeds (melon), Spices (cayenne pepper, chilli), Spices (ginger), Spices (nutmeg), AF/S, literature[19]

incidence: 44/63*, conc. range: 1–5 µg/kg** (2 sa.), 6–10 µg/kg** (14 sa.), 11–20 µg/kg** (6 sa.), 21–35 µg/kg** (8 sa.), 36–40 µg/kg** (3 sa.), 41–60 µg/kg** (4 sa.), 61–80 µg/kg** (2 sa.), 81–150 µg/kg** (2 sa.), >150 µg/kg** (3 sa.), sample origin: commercial and imported food from retail outlets and points of entries, Malaysia, sample year: 1995–1999, country: Malaysia[233], *kacang tumbuk (finely ground peanut and sugar), **AF not specified

- Co-contamination: not reported
- Further contamination of plants and plant products other than cereals and their products reported in the present book/article[233]: Butter (peanut butter), Nuts, Nuts (peanut), Products (peanut products), Spices (cayenne pepper, chilli), AF/S, literature[233]

incidence: 338/1031, conc. range: 0.2–513.4 µg/kg*, sample origin: traditional markets, supermarkets, grocery stores, wholesale stores, and convenience stores in 25 counties, Taiwan, sample year: 1997–2011 (with exceptions), country: Taiwan[540], *AFB_1, AFB_2, AFG_1, and AFG_2

- Co-contamination: For detailed information see the article.
- Further contamination of plants and plant products other than cereals and their products reported in the present book/article[540]: Butter (peanut butter), Flour (peanut flour), Nuts (peanut), AF/S, literature[540]

For detailed information see the article.

Candies (pistachio candy) may contain the following mycotoxins:

Aspergillus Toxins

Aflatoxin/s

incidence: 1/1, conc.: 78 µg/kg*, sample origin: imported to the USA, sample year: 1986, country: USA[19], *AFB_1, AFB_2, AFG_1, and AFG_2

- Co-contamination: not reported
- Further contamination of plants and plant products other than cereals and their products reported in the present book/article[19]: Butter (peanut butter), Candies (peanut candy), Nuts, Nuts (almond), Nuts (Brazil nut, paranut), Nuts (peanut), Nuts (pecan nut), Nuts (pistachio nut), Nuts (walnut), Oils (peanut oil), Paste (peanut paste), Paste (pipian paste), Products (peanut product), Seeds (melon), Spices (cayenne pepper, chilli), Spices (ginger), Spices (nutmeg), AF/S, literature[19]

Cashew nuts see Nuts (cashew nut)

Cassava may contain the following mycotoxins:

Alternaria Toxins

Alternariol

incidence: 114/405*, conc. range: ≤1,051 µg/kg, sample origin: households and markets, Tanzania, sample year: June–September 2012, country: Austria/Tanzania/Rwanda[823], *cassava, processed and stored

incidence: 10/222*, conc. range: ≤76.8 µg/kg, sample origin: households and markets, Rwanda, sample year: July 2013–April 2014, country: Austria/Tanzania/Rwanda[823], *cassava, processed and stored

- Co-contamination: not reported
- Further contamination of plants and plant products other than cereals and their products reported in the present book/article[823]: Cassava, 3-NPA, AFB_1, AFB_2, AFG_1, AFG_2, AFM_1, AME, ATX-I, BEA, CIT, CPA, CTV, DAS, DON, ENA, ENA_1, ENB, ENB_1, ENB_2, ENB_3, FA, FB_1, FB_2, FB_3, FP, HT-2, KA, MA, MAS, MON, NEO, NIV, OTA, OTB, PA, ROQ-C, STG, T-2, T-2TET, TA, and ZEA, literature[823]; see also Cassava, chapter: Further Mycotoxins and Microbial Metabolites

Alternariol Methyl Ether

incidence: 202/405*, conc. range: ≤1,430 µg/kg, sample origin: households and markets, Tanzania, sample year: June–September 2012, country: Austria/Tanzania/Rwanda[823], *cassava, processed and stored

incidence: 57/222*, conc. range: ≤51.8 µg/kg, sample origin: households and markets, Rwanda, sample year: July 2013–April 2014, country: Austria/Tanzania/Rwanda[823], *cassava, processed and stored

- Co-contamination: not reported
- Further contamination of plants and plant products other than cereals and their products reported in the present book/article[823]: Cassava, 3-NPA, AFB_1, AFB_2, AFG_1, AFG_2, AFM_1, AOH, ATX-I, BEA, CIT, CPA, CTV, DAS, DON, ENA, ENA_1, ENB, ENB_1, ENB_2, ENB_3, FA, FB_1, FB_2, FB_3, FP, HT-2, KA, MA, MAS, MON, NEO, NIV, OTA, OTB, PA, ROQ-C, STG, TA, T-2, T-2TET, and ZEA, literature[823]; see also Cassava, chapter: Further Mycotoxins and Microbial Metabolites

Altertoxin I

incidence: 17/405*, conc. range: ≤25.8 µg/kg, sample origin: households and markets, Tanzania, sample year: June–September 2012, country: Austria/Tanzania/Rwanda[823], *cassava, processed and stored

incidence: 1/222*, conc.: 0.4 µg/kg, sample origin: households and markets, Rwanda, sample year: July 2013–April 2014, country: Austria/Tanzania/Rwanda[823], *cassava, processed and stored

- Co-contamination: not reported
- Further contamination of plants and plant products other than cereals and their products reported in the present book/article[823]: Cassava, 3-NPA, AFB_1, AFB_2, AFG_1, AFG_2, AFM_1, AME, AOH, BEA, CIT, CPA, CTV, DAS, DON, ENA, ENA_1, ENB, ENB_1, ENB_2, ENB_3, FA, FB_1, FB_2, FB_3, FP, HT-2, KA, MA, MAS, MON, NEO, NIV, OTA, OTB, PA, ROQ-C, STG, T-2, T-2TET, TA, and ZEA, literature[823]; see also Cassava, chapter: Further Mycotoxins and Microbial Metabolites

Tenuazonic Acid

incidence: 3/27*, conc. range: 20–340 µg/kg, ∅ conc.: 130 µg/kg, sample origin: households and markets, Ghana, sample year: unknown, country: UK/Ghana[238], *kokonte (cassava, dried)

- Co-contamination: 2 sa. co-contaminated with STG and TA; 1 sa. contaminated solely with TA
- Further contamination of plants and plant products other than cereals and their products reported in the present book/article[238]: Cassava, CPA, PA, PAT, and STG, literature[238]

incidence: 158/405*, conc. range: ≤2,642 µg/kg, sample origin: households and markets, Tanzania, sample year: June–September 2012, country: Austria/Tanzania/Rwanda[823], *cassava, processed and stored

incidence: 26/222*, conc. range: ≤562.1 µg/kg, sample origin: households and markets, Rwanda, sample year: July 2013–April 2014, country: Austria/Tanzania/Rwanda[823], *cassava, processed and stored

- Co-contamination: not reported
- Further contamination of plants and plant products other than cereals and their products reported in the present book/article[823]: Cassava, 3-NPA, AFB_1, AFB_2, AFG_1, AFG_2, AFM_1, AME, AOH, ATX-I, BEA, CIT, CPA, CTV, DAS, DON, ENA, ENA_1, ENB, ENB_1, ENB_2, ENB_3, FA, FB_1, FB_2, FB_3, FP, HT-2, KA, MA, MAS, MON, NEO, NIV, OTA, OTB, PA, ROQ-C, STG, T-2, T-2TET, and ZEA, literature[823]; see also Cassava, chapter: Further Mycotoxins and Microbial Metabolites

Aspergillus Toxins

Aflatoxin B_1

incidence: 43/405*, conc. range: ≤76.6 µg/kg, sample origin: households and markets, Tanzania, sample year: June–September 2012, country: Austria/Tanzania/Rwanda[823], *cassava, processed and stored

incidence: 0/222*, conc.: no contamination, sample origin: households and markets, Rwanda, sample year: July 2013–April 2014, country: Austria/Tanzania/Rwanda[823], *cassava, processed and stored

- Co-contamination: not reported
- Further contamination of plants and plant products other than cereals and their products reported in the present book/article[823]: Cassava, 3-NPA, AFB_2, AFG_1, AFG_2, AFM_1, AME, AOH, ATX-I, BEA, CIT, CPA, CTV, DAS, DON, ENA, ENA_1, ENB, ENB_1, ENB_2, ENB_3, FA, FB_1, FB_2, FB_3, FP, HT-2, KA, MA, MAS, MON, NEO, NIV, OTA, OTB, PA, ROQ-C, STG, T-2, T-2TET, TA, and ZEA, literature[823]; see also Cassava, chapter: Further Mycotoxins and Microbial Metabolites

Aflatoxin B_2

incidence: 11/405*, conc. range: ≤2.2 µg/kg, sample origin: households and markets, Tanzania, sample year: June–September 2012, country: Austria/Tanzania/Rwanda[823], *cassava, processed and stored

incidence: 0/222*, conc.: no contamination, sample origin: households and markets, Rwanda, sample year: July 2013–April 2014, country: Austria/Tanzania/Rwanda[823], *cassava, processed and stored

- Co-contamination: not reported
- Further contamination of plants and plant products other than cereals and their products reported in the present book/article[823]: Cassava, 3-NPA, AFB_1, AFG_1, AFG_2, AFM_1, AME, AOH,

ATX-I, BEA, CIT, CPA, CTV, DAS, DON, ENA, ENA_1, ENB, ENB_1, ENB_2, ENB_3, FA, FB_1, FB_2, FB_3, FP, HT-2, KA, MA, MAS, MON, NEO, NIV, OTA, OTB, PA, ROQ-C, STG, T-2, T-2TET, TA, and ZEA, literature[823]; see also Cassava, chapter: Further Mycotoxins and Microbial Metabolites

Aflatoxin G_1

incidence: 39/405*, conc. range: ≤60.7 µg/kg, sample origin: households and markets, Tanzania, sample year: June–September 2012, country: Austria/Tanzania/Rwanda[823], *cassava, processed and stored

incidence: 2/222*, conc. range: ≤1.4 µg/kg, sample origin: households and markets, Rwanda, sample year: July 2013–April 2014, country: Austria/Tanzania/Rwanda[823], *cassava, processed and stored

- Co-contamination: not reported
- Further contamination of plants and plant products other than cereals and their products reported in the present book/article[823]: Cassava, 3-NPA, AFB_1, AFB_2, AFG_2, AFM_1, AME, AOH, ATX-I, BEA, CIT, CPA, CTV, DAS, DON, ENA, ENA_1, ENB, ENB_1, ENB_2, ENB_3, FA, FB_1, FB_2, FB_3, FP, HT-2, KA, MA, MAS, MON, NEO, NIV, OTA, OTB, PA, ROQ-C, STG, T-2, T-2TET, TA, and ZEA, literature[823]; see also Cassava, chapter: Further Mycotoxins and Microbial Metabolites

Aflatoxin G_2

incidence: 1/405*, conc.: 4.3 µg/kg, sample origin: households and markets, Tanzania, sample year: June–September 2012, country: Austria/Tanzania/Rwanda[823], *cassava, processed and stored

incidence: 0/222*, conc.: no contamination, sample origin: households and markets, Rwanda, sample year: July 2013–April 2014, country: Austria/Tanzania/Rwanda[823], *cassava, processed and stored

- Co-contamination: not reported
- Further contamination of plants and plant products other than cereals and their products reported in the present book/article[823]: Cassava, 3-NPA, AFB_1, AFB_2, AFG_1, AFM_1, AME, AOH, ATX-I, BEA, CIT, CPA, CTV, DAS, DON, ENA, ENA_1, ENB, ENB_1, ENB_2, ENB_3, FA, FB_1, FB_2, FB_3, FP, HT-2, KA, MA, MAS, MON, NEO, NIV, OTA, OTB, PA, ROQ-C, STG, T-2, T-2TET, TA, and ZEA, literature[823]; see also Cassava, chapter: Further Mycotoxins and Microbial Metabolites

Aflatoxin M_1

incidence: 2/405*, conc. range: ≤0.3 µg/kg, sample origin: households and markets, Tanzania, sample year: June–September 2012, country: Austria/Tanzania/Rwanda[823], *cassava, processed and stored

incidence: 0/222*, conc.: no contamination, sample origin: households and markets, Rwanda, sample year: July 2013–April 2014, country: Austria/Tanzania/Rwanda[823], *cassava, processed and stored

- Co-contamination: not reported
- Further contamination of plants and plant products other than cereals and their products reported in the present book/article[823]: Cassava, 3-NPA, AFB_1, AFB_2, AFG_1, AFG_2, AME, AOH, ATX-I, BEA, CIT, CPA, CTV, DAS, DON, ENA, ENA_1, ENB, ENB_1, ENB_2, ENB_3, FA, FB_1, FB_2, FB_3, FP, HT-2, KA, MA, MAS, MON, NEO, NIV, OTA, OTB, PA, ROQ-C, STG, T-2, T-2TET, TA, and ZEA, literature[823]; see also Cassava, chapter: Further Mycotoxins and Microbial Metabolites

Aflatoxin/s

incidence: 142/142*, Ø conc.: 467.5 µg/kg**, sample origin: greater Manila (capital) area, Philippines, sample year: unknown, country: Philippines[1], *cassava and cassava products, **AFS of pos sa.?

- Co-contamination: not reported
- Further contamination of plants and plant products other than cereals and their products reported in the present book/article[1]: Beans, Butter (peanut butter), Cocoa, Coffee, Drinks (alcoholic drink), Nuts (cashew nut), Nuts (peanut), Products (coconut product), Rootcrops, Spices (cayenne pepper, chilli), Sweet potatoes, Yams, AF/S, literature[1]

incidence: 4/34, conc. range: 100–1,000 µg/kg* (2 sa.), >1,000 µg/kg* (2 sa.), sample origin: natives homes and markets, Uganda, sample year: September 1966–June 1967, country: Uganda/USA[137], *AFB_1, AFB_2, AFG_1, and AFG_2

- Co-contamination: 2 sa. contained AFB_1 and AFG_1, 4 sa. contained AFB_2 (2 sa. co-contaminated with AFB_1 and AFB_2, 2 sa. contaminated solely with AFB_2); no further information available
- Further contamination of plants and plant products other than cereals and their products reported in the present book/article[137]: Beans, Nuts (peanut), Peas, AF/S, literature[137]

incidence: 1/1, conc.: 1,700 µg/kg*, sample origin: Africa, sample year: unknown, country: USA/Thailand[240], *AF

- Co-contamination: not reported
- Further contamination of plants and plant products other than cereals and their products reported in the present book/article[240]: not reported

Other *Aspergillus* Toxins

Sterigmatocystin

incidence: 10/27*, conc. range: 130–1,670 µg/kg, Ø conc.: 545 µg/kg, sample origin: households and markets, Ghana, sample year: unknown, country: UK/Ghana[238], *kokonte (cassava, dried)

- Co-contamination: 2 sa. co-contaminated with TA and STG; 1 sa. co-contaminated with CPA, PA, and STG; 1 sa. co-contaminated with CPA and STG; 1 sa. co-contaminated with PA, PAT, and STG; 1 sa. co-contaminated with PA and STG; 4 sa. contaminated solely with STG
- Further contamination of plants and plant products other than cereals and their products reported in the present book/article[238]: Cassava, CPA, PA, PAT, and TA, literature[238]

incidence: 90/405*, conc. range: ≤144 µg/kg, sample origin: households and markets, Tanzania, sample year: June–September 2012, country: Austria/Tanzania/Rwanda[823], *cassava, processed and stored

incidence: 28/222*, conc. range: ≤16.2 µg/kg, sample origin: households and markets, Rwanda, sample year: July 2013–April 2014, country: Austria/Tanzania/Rwanda[823], *cassava, processed and stored

- Co-contamination: not reported
- Further contamination of plants and plant products other than cereals and their products reported in the present book/article[823]: Cassava, 3-NPA, AFB_1, AFB_2, AFG_1, AFG_2, AFM_1, AME, AOH, ATX-I, BEA, CIT, CPA, CTV, DAS, DON, ENA, ENA_1, ENB, ENB_1, ENB_2, ENB_3, FA, FB_1, FB_2, FB_3, FP, HT-2, KA, MA, MAS, MON, NEO, NIV, OTA, OTB, PA, ROQ-C, T-2, T-2TET, TA, and ZEA, literature[823]; see also Cassava, chapter: Further Mycotoxins and Microbial Metabolites

Aspergillus and ***Penicillium*** Toxins

3-Nitropropionic Acid

incidence: 260/405*, conc. range: ≤5,159 μg/kg, sample origin: households and markets, Tanzania, sample year: June–September 2012, country: Austria/Tanzania/Rwanda[823], *cassava, processed and stored

incidence: 66/222*, conc. range: ≤2,598 μg/kg, sample origin: households and markets, Rwanda, sample year: July 2013–April 2014, country: Austria/Tanzania/Rwanda[823], *cassava, processed and stored

- Co-contamination: not reported
- Further contamination of plants and plant products other than cereals and their products reported in the present book/article[823]: Cassava, AFB_1, AFB_2, AFG_1, AFG_2, AFM_1, AME, AOH, ATX-I, BEA, CIT, CPA, CTV, DAS, DON, ENA, ENA_1, ENB, ENB_1, ENB_2, ENB_3, FA, FB_1, FB_2, FB_3, FP, HT-2, KA, MA, MAS, MON, NEO, NIV, OTA, OTB, PA, ROQ-C, STG, TA, T-2, T-2TET, and ZEA, literature[823]; see also Cassava, chapter: Further Mycotoxins and Microbial Metabolites

Citreoviridin

incidence: 1/405*, conc.: 18.6 μg/kg, sample origin: households and markets, Tanzania, sample year: June–September 2012, country: Austria/Tanzania/Rwanda[823], *cassava, processed and stored

incidence: 0/222*, conc.: no contamination, sample origin: households and markets, Rwanda, sample year: July 2013–April 2014, country: Austria/Tanzania/Rwanda[823], *cassava, processed and stored

- Co-contamination: not reported
- Further contamination of plants and plant products other than cereals and their products reported in the present book/article[823]: Cassava, 3-NPA, AFB_1, AFB_2, AFG_1, AFG_2, AFM_1, AME, AOH, ATX-I, BEA, CIT, CPA, DAS, DON, ENA, ENA_1, ENB, ENB_1, ENB_2, ENB_3, FA, FB_1, FB_2, FB_3, FP, HT-2, KA, MA, MAS, MON, NEO, NIV, OTA, OTB, PA, ROQ-C, STG, T-2, T-2TET, TA, and ZEA, literature[823]; see also Cassava, chapter: Further Mycotoxins and Microbial Metabolites

Citrinin

incidence: 9/405*, conc. range: ≤462 μg/kg, sample origin: households and markets, Tanzania, sample year: June–September 2012, country: Austria/Tanzania/Rwanda[823], *cassava, processed and stored

incidence: 0/222*, conc.: no contamination, sample origin: households and markets, Rwanda, sample year: July 2013–April 2014, country: Austria/Tanzania/Rwanda[823], *cassava, processed and stored

- Co-contamination: not reported
- Further contamination of plants and plant products other than cereals and their products reported in the present book/article[823]: Cassava, 3-NPA, AFB_1, AFB_2, AFG_1, AFG_2, AFM_1, AME, AOH, ATX-I, BEA, CPA, CTV, DAS, DON, ENA, ENA_1, ENB, ENB_1, ENB_2, ENB_3, FA, FB_1, FB_2, FB_3, FP, HT-2, KA, MA, MAS, MON, NEO, NIV, OTA, OTB, PA, ROQ-C, STG, T-2, T-2TET, TA, and ZEA, literature[823]; see also Cassava, chapter: Further Mycotoxins and Microbial Metabolites

Cyclopiazonic Acid

incidence: 4/27*, conc. range: 80–720 μg/kg, ∅ conc.: 430 μg/kg, sample origin: households and markets, Ghana, sample year: unknown, country: UK/Ghana[238], *kokonte (cassava, dried)

- Co-contamination: 1 sa. co-contaminated with CPA, PA, and PAT; 1 sa. co-contaminated with CPA, PAT, and STG; 1 sa. co-contaminated with CPA and STG; 1 sa. contaminated solely with CPA
- Further contamination of plants and plant products other than cereals and their products reported in the present book/article[238]: Cassava, PA, PAT, STG, and TA, literature[238]

incidence: 23/405*, conc. range: ≤560 μg/kg, sample origin: households and markets, Tanzania, sample year: June–September 2012, country: Austria/Tanzania/Rwanda[823], *cassava, processed and stored

incidence: 1/222*, conc.: 28.0 μg/kg, sample origin: households and markets, Rwanda, sample year: July 2013–April 2014, country: Austria/Tanzania/Rwanda[823], *cassava, processed and stored

- Co-contamination: not reported
- Further contamination of plants and plant products other than cereals and their products reported in the present book/article[823]: Cassava, 3-NPA, AFB_1, AFB_2, AFG_1, AFG_2, AFM_1, AME, AOH, ATX-I, BEA, CIT, CTV, DAS, DON, ENA, ENA_1, ENB, ENB_1, ENB_2, ENB_3, FA, FB_1, FB_2, FB_3, FP, HT-2, KA, MA, MAS, MON, NEO, NIV, OTA, OTB, PA, ROQ-C, STG, T-2, T-2TET, TA, and ZEA, literature[823]; see also Cassava, chapter: Further Mycotoxins and Microbial Metabolites

Kojic Acid

incidence: 345/405*, conc. range: ≤650,379 μg/kg, sample origin: households and markets, Tanzania, sample year: June–September 2012, country: Austria/Tanzania/Rwanda[823], *cassava, processed and stored

incidence: 110/222*, conc. range: ≤93,654 μg/kg, sample origin: households and markets, Rwanda, sample year: July 2013–April 2014, country: Austria/Tanzania/Rwanda[823], *cassava, processed and stored

- Co-contamination: not reported
- Further contamination of plants and plant products other than cereals and their products reported in the present book/article[823]: Cassava, 3-NPA, AFB_1, AFB_2, AFG_1, AFG_2, AFM_1, AME, AOH, ATX-I, BEA, CIT, CPA, CTV, DAS, DON, ENA, ENA_1, ENB, ENB_1, ENB_2, ENB_3, FA, FB_1, FB_2, FB_3, FP, HT-2, MA, MAS, MON, NEO, NIV, OTA, OTB, PA, ROQ-C, STG, T-2, T-2TET, TA, and ZEA, literature[823]; see also Cassava, chapter: Further Mycotoxins and Microbial Metabolites

Ochratoxin A

incidence: 21/405*, conc. range: ≤271 μg/kg, sample origin: households and markets, Tanzania, sample year: June–September 2012, country: Austria/Tanzania/Rwanda[823], *cassava, processed and stored

incidence: 2/222*, conc. range: ≤24.7 μg/kg, sample origin: households and markets, Rwanda, sample year: July 2013–April 2014, country: Austria/Tanzania/Rwanda[823], *cassava, processed and stored

- Co-contamination: not reported
- Further contamination of plants and plant products other than cereals and their products reported in the present book/article[823]: Cassava, 3-NPA, AFB_1, AFB_2, AFG_1, AFG_2, AFM_1, AME, AOH, ATX-I, BEA, CIT, CPA, CTV, DAS, DON, ENA, ENA_1, ENB, ENB_1, ENB_2, ENB_3, FA, FB_1, FB_2, FB_3, FP, HT-2, KA, MA, MAS, MON, NEO, NIV, OTB, PA, ROQ-C, STG, T-2, T-2TET, TA, and ZEA, literature[823]; see also Cassava, chapter: Further Mycotoxins and Microbial Metabolites

Ochratoxin B

incidence: 2/405*, conc. range: ≤61.2 µg/kg, sample origin: households and markets, Tanzania, sample year: June–September 2012, country: Austria/Tanzania/Rwanda[823], *cassava, processed and stored

incidence: 1/222*, conc.: 23.5 µg/kg, sample origin: households and markets, Rwanda, sample year: July 2013–April 2014, country: Austria/Tanzania/Rwanda[823], *cassava, processed and stored

- Co-contamination: not reported
- Further contamination of plants and plant products other than cereals and their products reported in the present book/article[823]: Cassava, 3-NPA, AFB_1, AFB_2, AFG_1, AFG_2, AFM_1, AME, AOH, ATX-I, BEA, CIT, CPA, CTV, DAS, DON, ENA, ENA_1, ENB, ENB_1, ENB_2, ENB_3, FA, FB_1, FB_2, FB_3, FP, HT-2, KA, MA, MAS, MON, NEO, NIV, OTA, PA, ROQ-C, STG, T-2, T-2TET, TA, and ZEA, literature[823]; see also Cassava, chapter: Further Mycotoxins and Microbial Metabolites

Patulin

incidence: 4/27*, conc. range: 550–850 µg/kg, Ø conc.: 683 µg/kg, sample origin: households and markets, Ghana, sample year: unknown, country: UK/Ghana[238], *kokonte (cassava, dried)

- Co-contamination: 1 sa. co-contaminated with CPA, PA, and PAT; 1 sa. co-contaminated with CPA, PAT, and STG; 1 sa. co-contaminated with PA, PAT, and STG; 1 sa. co-contaminated with PA and PAT
- Further contamination of plants and plant products other than cereals and their products reported in the present book/article[238]: Cassava, CPA, PA, STG, and TA, literature[238]

Penicillic Acid

incidence: 5/27*, conc. range: 110–850 µg/kg, Ø conc.: 302 µg/kg, sample origin: households and markets, Ghana, sample year: unknown, country: UK/Ghana[238], *kokonte (cassava, dried)

- Co-contamination: 1 sa. co-contaminated with CPA, PA, and PAT; 1 sa. co-contaminated with PA and STG; 1 sa. co-contaminated with PA, PAT, and STG; 1 sa. co-contaminated with PA and PAT; 1 sa. contaminated solely with PA
- Further contamination of plants and plant products other than cereals and their products reported in the present book/article[238]: Cassava, CPA, PAT, STG, and TA, literature[238]

incidence: 3/405*, conc. range: ≤17.5 µg/kg, sample origin: households and markets, Tanzania, sample year: June–September 2012, country: Austria/Tanzania/Rwanda[823], *cassava, processed and stored

incidence: 3/222*, conc. range: ≤40.3 µg/kg, sample origin: households and markets, Rwanda, sample year: July 2013–April 2014, country: Austria/Tanzania/Rwanda[823], *cassava, processed and stored

- Co-contamination: not reported
- Further contamination of plants and plant products other than cereals and their products reported in the present book/article[823]: Cassava, 3-NPA, AFB_1, AFB_2, AFG_1, AFG_2, AFM_1, AME, AOH, ATX-I, BEA, CIT, CPA, CTV, DAS, DON, ENA, ENA_1, ENB, ENB_1, ENB_2, ENB_3, FA, FB_1, FB_2, FB_3, FP, HT-2, KA, MA, MAS, MON, NEO, NIV, OTA, OTB, ROQ-C, STG, T-2, T-2TET, TA, and ZEA, literature[823]; see also Cassava, chapter: Further Mycotoxins and Microbial Metabolites

Fusarium Toxins

Depsipeptides

Beauvericin

incidence: 329/405*, conc. range: ≤527 µg/kg, sample origin: households and markets, Tanzania, sample year: June–September 2012, country: Austria/Tanzania/Rwanda[823], *cassava, processed and stored

incidence: 199/222*, conc. range: ≤280 µg/kg, sample origin: households and markets, Rwanda, sample year: July 2013–April 2014, country: Austria/Tanzania/Rwanda[823], *cassava, processed and stored

- Co-contamination: not reported
- Further contamination of plants and plant products other than cereals and their products reported in the present book/article[823]: Cassava, 3-NPA, AFB_1, AFB_2, AFG_1, AFG_2, AFM_1, AME, AOH, ATX-I, CIT, CPA, CTV, DAS, DON, ENA, ENA_1, ENB, ENB_1, ENB_2, ENB_3, FA, FB_1, FB_2, FB_3, FP, HT-2, KA, MA, MAS, MON, NEO, NIV, OTA, OTB, PA, ROQ-C, STG, T-2, T-2TET, TA, and ZEA, literature[823]; see also Cassava, chapter: Further Mycotoxins and Microbial Metabolites

Enniatin A

incidence: 10/405*, conc. range: ≤17.9 µg/kg, sample origin: households and markets, Tanzania, sample year: June–September 2012, country: Austria/Tanzania/Rwanda[823], *cassava, processed and stored

incidence: 22/222*, conc. range: ≤46.6 µg/kg, sample origin: households and markets, Rwanda, sample year: July 2013–April 2014, country: Austria/Tanzania/Rwanda[823], *cassava, processed and stored

- Co-contamination: not reported
- Further contamination of plants and plant products other than cereals and their products reported in the present book/article[823]: Cassava, 3-NPA, AFB_1, AFB_2, AFG_1, AFG_2, AFM_1, AME, AOH, ATX-I, BEA, CIT, CPA, CTV, DAS, DON, ENA_1, ENB, ENB_1, ENB_2, ENB_3, FA, FB_1, FB_2, FB_3, FP, HT-2, KA, MA, MAS, MON, NEO, NIV, OTA, OTB, PA, ROQ-C, STG, T-2, T-2TET, TA, and ZEA, literature[823]; see also Cassava, chapter: Further Mycotoxins and Microbial Metabolites

Enniatin A_1

incidence: 20/405*, conc. range: ≤125 µg/kg, sample origin: households and markets, Tanzania, sample year: June–September 2012, country: Austria/Tanzania/Rwanda[823], *cassava, processed and stored

incidence: 42/222*, conc. range: ≤165.2 µg/kg, sample origin: households and markets, Rwanda, sample year: July 2013–April 2014, country: Austria/Tanzania/Rwanda[823], *cassava, processed and stored

- Co-contamination: not reported
- Further contamination of plants and plant products other than cereals and their products reported in the present book/article[823]: Cassava, 3-NPA, AFB_1, AFB_2, AFG_1, AFG_2, AFM_1, AME, AOH, ATX-I, BEA, CIT, CPA, CTV, DAS, DON, ENA, ENB, ENB_1, ENB_2, ENB_3, FA, FB_1, FB_2, FB_3, FP, HT-2, KA, MA, MAS, MON, NEO, NIV, OTA, OTB, PA, ROQ-C, STG, T-2, T-2TET, TA, and ZEA, literature[823]; see also Cassava, chapter: Further Mycotoxins and Microbial Metabolites

Enniatin B

incidence: 126/405*, conc. range: ≤29.5 µg/kg, sample origin: households and markets, Tanzania, sample year: June–September 2012, country: Austria/Tanzania/Rwanda[823], *cassava, processed and stored

incidence: 68/222*, conc. range: ≤50.4 µg/kg, sample origin: households and markets, Rwanda, sample year: July 2013–April 2014, country: Austria/Tanzania/Rwanda[823], *cassava, processed and stored

- Co-contamination: not reported
- Further contamination of plants and plant products other than cereals and their products reported in the present book/ article[823]: Cassava, 3-NPA, AFB_1, AFB_2, AFG_1, AFG_2, AFM_1, AME, AOH, ATX-I, BEA, CIT, CPA, CTV, DAS, DON, ENA, ENA_1, ENB_1, ENB_2, ENB_3, FA, FB_1, FB_2, FB_3, FP, HT-2, KA, MA, MAS, MON, NEO, NIV, OTA, OTB, PA, ROQ-C, STG, T-2, T-2TET, TA, and ZEA, literature[823]; see also Cassava, chapter: Further Mycotoxins and Microbial Metabolites

Enniatin B_1

incidence: 103/405*, conc. range: ≤146.1 µg/kg, sample origin: households and markets, Tanzania, sample year: June–September 2012, country: Austria/Tanzania/Rwanda[823], *cassava, processed and stored

incidence: 37/222*, conc. range: ≤163 µg/kg, sample origin: households and markets, Rwanda, sample year: July 2013–April 2014, country: Austria/Tanzania/Rwanda[823], *cassava, processed and stored

- Co-contamination: not reported
- Further contamination of plants and plant products other than cereals and their products reported in the present book/ article[823]: Cassava, 3-NPA, AFB_1, AFB_2, AFG_1, AFG_2, AFM_1, AME, AOH, ATX-I, BEA, CIT, CPA, CTV, DAS, DON, ENA, ENA_1, ENB, ENB_2, ENB_3, FA, FB_1, FB_2, FB_3, FP, HT-2, KA, MA, MAS, MON, NEO, NIV, OTA, OTB, PA, ROQ-C, STG, T-2, T-2TET, TA, and ZEA, literature[823]; see also Cassava, chapter: Further Mycotoxins and Microbial Metabolites

Enniatin B_2

incidence: 3/405*, conc. range: ≤0.1 µg/kg, sample origin: households and markets, Tanzania, sample year: June–September 2012, country: Austria/Tanzania/Rwanda[823], *cassava, processed and stored

incidence: 0/222*, conc.: no contamination, sample origin: households and markets, Rwanda, sample year: July 2013–April 2014, country: Austria/Tanzania/Rwanda[823], *cassava, processed and stored

- Co-contamination: not reported
- Further contamination of plants and plant products other than cereals and their products reported in the present book/ article[823]: Cassava, 3-NPA, AFB_1, AFB_2, AFG_1, AFG_2, AFM_1, AME, AOH, ATX-I, BEA, CIT, CPA, CTV, DAS, DON, ENA, ENA_1, ENB, ENB_1, ENB_3, FA, FB_1, FB_2, FB_3, FP, HT-2, KA, MA, MAS, MON, NEO, NIV, OTA, OTB, PA, ROQ-C, STG, T-2, T-2TET, TA, and ZEA, literature[823]; see also Cassava, chapter: Further Mycotoxins and Microbial Metabolites

Enniatin B_3

incidence: 55/405*, conc. range: ≤309.5 µg/kg, sample origin: households and markets, Tanzania, sample year: June–September 2012, country: Austria/Tanzania/Rwanda[823], *cassava, processed and stored

incidence: 0/222*, conc.: no contamination, sample origin: households and markets, Rwanda, sample year: July 2013–April 2014, country: Austria/Tanzania/Rwanda[823], *cassava, processed and stored

- Co-contamination: not reported
- Further contamination of plants and plant products other than cereals and their products reported in the present book/ article[823]: Cassava, 3-NPA, AFB_1, AFB_2, AFG_1, AFG_2, AFM_1, AME, AOH, ATX-I, BEA, CIT, CPA, CTV, DAS, DON, ENA, ENA_1, ENB, ENB_1, ENB_2, FA, FB_1, FB_2, FB_3, FP, HT-2, KA, MA, MAS, MON, NEO, NIV, OTA, OTB, PA, ROQ-C, STG, T-2, T-2TET, TA, and ZEA, literature[823]; see also Cassava, chapter: Further Mycotoxins and Microbial Metabolites

Fumonisin/s

Fumonisin B_1

incidence: 3/59*, conc. range: 4.8–41.4 µg/kg, ∅ conc.: 21.8 µg/kg, sample origin: five Colombian Amazon communities, Colombia, sample year: May–October 2013, country: Colombia/Austria[709], *cassava root, processed

- Co-contamination: not reported
- Further contamination of plants and plant products other than cereals and their products reported in the present book/ article[709]: Cassava, FB_2, HT-2, MON, T-2, and ZEA, literature[709]

incidence: 27/405*, conc. range: ≤115 µg/kg, sample origin: households and markets, Tanzania, sample year: June–September 2012, country: Austria/Tanzania/Rwanda[823], *cassava, processed and stored

incidence: 0/222*, conc.: no contamination, sample origin: households and markets, Rwanda, sample year: July 2013–April 2014, country: Austria/Tanzania/Rwanda[823], *cassava, processed and stored

- Co-contamination: not reported
- Further contamination of plants and plant products other than cereals and their products reported in the present book/ article[823]: Cassava, 3-NPA, AFB_1, AFB_2, AFG_1, AFG_2, AFM_1, AME, AOH, ATX-I, BEA, CIT, CPA, CTV, DAS, DON, ENA, ENA_1, ENB, ENB_1, ENB_2, ENB_3, FA, FB_2, FB_3, FP, HT-2, KA, MA, MAS, MON, NEO, NIV, OTA, OTB, PA, ROQ-C, STG, T-2, T-2TET, TA, and ZEA, literature[823]; see also Cassava, chapter: Further Mycotoxins and Microbial Metabolites

Fumonisin B_2

incidence: 2/59*, conc. range: 3.9–7.5 µg/kg, ∅ conc.: 7.5 µg/kg, sample origin: five Colombian Amazon communities, Colombia, sample year: May–October 2013, country: Colombia/Austria[709], *processed cassava root

- Co-contamination: not reported
- Further contamination of plants and plant products other than cereals and their products reported in the present book/ article[709]: Cassava, FB_1, HT-2, MON, T-2, and ZEA, literature[709]

incidence: 17/405*, conc. range: ≤160 µg/kg, sample origin: households and markets, Tanzania, sample year: June–September 2012, country: Austria/Tanzania/Rwanda[823], *cassava, processed and stored

incidence: 0/222*, conc.: no contamination, sample origin: households and markets, Rwanda, sample year: July 2013–April

2014, country: Austria/Tanzania/Rwanda[823], *cassava, processed and stored

- Co-contamination: not reported
- Further contamination of plants and plant products other than cereals and their products reported in the present book/article[823]: Cassava, 3-NPA, AFB_1, AFB_2, AFG_1, AFG_2, AFM_1, AME, AOH, ATX-I, BEA, CIT, CPA, CTV, DAS, DON, ENA, ENA_1, ENB, ENB_1, ENB_2, ENB_3, FA, FB_1, FB_3, FP, HT-2, KA, MA, MAS, MON, NEO, NIV, OTA, OTB, PA, ROQ-C, STG, T-2, T-2TET, TA, and ZEA, literature[823]; see also Cassava, chapter: Further Mycotoxins and Microbial Metabolites

Fumonisin B_3

incidence: 9/405*, conc. range: ≤17.7 µg/kg, sample origin: households and markets, Tanzania, sample year: June–September 2012, country: Austria/Tanzania/Rwanda[823], *cassava, processed and stored

incidence: 0/222*, conc.: no contamination, sample origin: households and markets, Rwanda, sample year: July 2013–April 2014, country: Austria/Tanzania/Rwanda[823], *cassava, processed and stored

- Co-contamination: not reported
- Further contamination of plants and plant products other than cereals and their products reported in the present book/article[823]: Cassava, 3-NPA, AFB_1, AFB_2, AFG_1, AFG_2, AFM_1, AME, AOH, ATX-I, BEA, CIT, CPA, CTV, DAS, DON, ENA, ENA_1, ENB, ENB_1, ENB_2, ENB_3, FA, FB_1, FB_2, FP, HT-2, KA, MA, MAS, MON, NEO, NIV, OTA, OTB, PA, ROQ-C, STG, T-2, T-2TET, TA, and ZEA, literature[823]; see also Cassava, chapter: Further Mycotoxins and Microbial Metabolites

Type A Trichothecenes

Diacetoxyscirpenol

incidence: 82/405*, conc. range: ≤80.7 µg/kg, sample origin: households and markets, Tanzania, sample year: June–September 2012, country: Austria/Tanzania/Rwanda[823], *cassava, processed and stored

incidence: 35/222*, conc. range: ≤32.9 µg/kg, sample origin: households and markets, Rwanda, sample year: July 2013–April 2014, country: Austria/Tanzania/Rwanda[823], *cassava, processed and stored

- Co-contamination: not reported
- Further contamination of plants and plant products other than cereals and their products reported in the present book/article[823]: Cassava, 3-NPA, AFB_1, AFB_2, AFG_1, AFG_2, AFM_1, AME, AOH, ATX-I, BEA, CIT, CPA, CTV, DON, ENA, ENA_1, ENB, ENB_1, ENB_2, ENB_3, FA, FB_1, FB_2, FB_3, FP, HT-2, KA, MA, MAS, MON, NEO, NIV, OTA, OTB, PA, ROQ-C, STG, T-2, T-2TET, TA, and ZEA, literature[823]; see also Cassava, chapter: Further Mycotoxins and Microbial Metabolites

HT-2 Toxin

incidence: 6/59*, conc. range: 37.5–251.4 µg/kg, ∅ conc.: 121.1 µg/kg, sample origin: five Colombian Amazon communities, Colombia, sample year: May–October 2013, country: Colombia/Austria[709], *cassava root, processed

- Co-contamination: not reported
- Further contamination of plants and plant products other than cereals and their products reported in the present book/article[709]: Cassava, FB_1, FB_2, MON, T-2, and ZEA, literature[709]

incidence: 3/405*, conc. range: ≤26.2 µg/kg, sample origin: households and markets, Tanzania, sample year: June–September 2012, country: Austria/Tanzania/Rwanda[823], *cassava, processed and stored

incidence: 0/222*, conc.: no contamination, sample origin: households and markets, Rwanda, sample year: July 2013–April 2014, country: Austria/Tanzania/Rwanda[823], *cassava, processed and stored

- Co-contamination: not reported
- Further contamination of plants and plant products other than cereals and their products reported in the present book/article[823]: Cassava, 3-NPA, AFB_1, AFB_2, AFG_1, AFG_2, AFM_1, AME, AOH, ATX-I, BEA, CIT, CPA, CTV, DAS, DON, ENA, ENA_1, ENB, ENB_1, ENB_2, ENB_3, FA, FB_1, FB_2, FB_3, FP, KA, MA, MAS, MON, NEO, NIV, OTA, OTB, PA, ROQ-C, STG, T-2, T-2TET, TA, and ZEA, literature[823]; see also Cassava, chapter: Further Mycotoxins and Microbial Metabolites

Monoacetoxyscirpenol

incidence: 59/405*, conc. range: ≤93.2 µg/kg, sample origin: households and markets, Tanzania, sample year: June–September 2012, country: Austria/Tanzania/Rwanda[823], *cassava, processed and stored

incidence: 27/222*, conc. range: ≤48.9 µg/kg, sample origin: households and markets, Rwanda, sample year: July 2013–April 2014, country: Austria/Tanzania/Rwanda[823], *cassava, processed and stored

- Co-contamination: not reported
- Further contamination of plants and plant products other than cereals and their products reported in the present book/article[823]: Cassava, 3-NPA, AFB_1, AFB_2, AFG_1, AFG_2, AFM_1, AME, AOH, ATX-I, BEA, CIT, CPA, CTV, DAS, DON, ENA, ENA_1, ENB, ENB_1, ENB_2, ENB_3, FA, FB_1, FB_2, FB_3, FP, HT-2, KA, MA, MON, NEO, NIV, OTA, OTB, PA, ROQ-C, STG, T-2, T-2TET, TA, and ZEA, literature[823]; see also Cassava, chapter: Further Mycotoxins and Microbial Metabolites

Neosolaniol

incidence: 51/405*, conc. range: ≤20.3 µg/kg, sample origin: households and markets, Tanzania, sample year: June–September 2012, country: Austria/Tanzania/Rwanda[823], *cassava, processed and stored

incidence: 12/222*, conc. range: ≤53.5 µg/kg, sample origin: households and markets, Rwanda, sample year: July 2013–April 2014, country: Austria/Tanzania/Rwanda[823], *cassava, processed and stored

- Co-contamination: not reported
- Further contamination of plants and plant products other than cereals and their products reported in the present book/article[823]: Cassava, 3-NPA, AFB_1, AFB_2, AFG_1, AFG_2, AFM_1, AME, AOH, ATX-I, BEA, CIT, CPA, CTV, DAS, DON, ENA, ENA_1, ENB, ENB_1, ENB_2, ENB_3, FA, FB_1, FB_2, FB_3, FP, HT-2, KA, MA, MAS, MON, NIV, OTA, OTB, PA, ROQ-C, STG, T-2, T-2TET, TA, and ZEA, literature[823]; see also Cassava, chapter: Further Mycotoxins and Microbial Metabolites

T-2 Toxin

incidence: 3/59*, conc. range: 3.9–4.9 µg/kg, ∅ conc.: 4.4 µg/kg, sample origin: five Colombian Amazon communities, Colombia, sample year: May–October 2013, country: Colombia/Austria[709], *processed cassava root

- Co-contamination: not reported
- Further contamination of plants and plant products other than cereals and their products reported in the present book/article[709]: Cassava, FB_1, FB_2, HT-2, MON, and ZEA, literature[709]

incidence: 14/405*, conc. range: ≤147 µg/kg, sample origin: households and markets, Tanzania, sample year: June–September 2012, country: Austria/Tanzania/Rwanda[823], *cassava, processed and stored

incidence: 0/222*, conc.: no contamination, sample origin: households and markets, Rwanda, sample year: July 2013–April 2014, country: Austria/Tanzania/Rwanda[823], *cassava, processed and stored

- Co-contamination: not reported
- Further contamination of plants and plant products other than cereals and their products reported in the present book/article[823]: Cassava, 3-NPA, AFB_1, AFB_2, AFG_1, AFG_2, AFM_1, AME, AOH, ATX-I, BEA, CIT, CPA, CTV, DAS, DON, ENA, ENA_1, ENB, ENB_1, ENB_2, ENB_3, FA, FB_1, FB_2, FB_3, FP, HT-2, KA, MA, MAS, MON, NEO, NIV, OTA, OTB, PA, ROQ-C, STG, T-2TET, TA, and ZEA, literature[823]; see also Cassava, chapter: Further Mycotoxins and Microbial Metabolites

T-2 Tetraol

incidence: 7/405*, conc. range: ≤433 µg/kg, sample origin: households and markets, Tanzania, sample year: June–September 2012, country: Austria/Tanzania/Rwanda[823], *cassava, processed and stored

incidence: 2/222*, conc. range: ≤138 µg/kg, sample origin: households and markets, Rwanda, sample year: July 2013–April 2014, country: Austria/Tanzania/Rwanda[823], *cassava, processed and stored

- Co-contamination: not reported
- Further contamination of plants and plant products other than cereals and their products reported in the present book/article[823]: Cassava, 3-NPA, AFB_1, AFB_2, AFG_1, AFG_2, AFM_1, AME, AOH, ATX-I, BEA, CIT, CPA, CTV, DAS, DON, ENA, ENA_1, ENB, ENB_1, ENB_2, ENB_3, FA, FB_1, FB_2, FB_3, FP, HT-2, KA, MA, MAS, MON, NEO, NIV, OTA, OTB, PA, ROQ-C, STG, T-2, TA, and ZEA, literature[823]; see also Cassava, chapter: Further Mycotoxins and Microbial Metabolites

Type B Trichothecenes

Deoxynivalenol

incidence: 22/405*, conc. range: ≤157 µg/kg, sample origin: households and markets, Tanzania, sample year: June–September 2012, country: Austria/Tanzania/Rwanda[823], *cassava, processed and stored

incidence: 12/222*, conc. range: ≤27.4 µg/kg, sample origin: households and markets, Rwanda, sample year: July 2013–April 2014, country: Austria/Tanzania/Rwanda[823], *cassava, processed and stored

- Co-contamination: not reported
- Further contamination of plants and plant products other than cereals and their products reported in the present book/article[823]: Cassava, 3-NPA, AFB_1, AFB_2, AFG_1, AFG_2, AFM_1, AME, AOH, ATX-I, BEA, CIT, CPA, CTV, DAS, ENA, ENA_1, ENB, ENB_1, ENB_2, ENB_3, FA, FB_1, FB_2, FB_3, FP, HT-2, KA, MA, MAS, MON, NEO, NIV, OTA, OTB, PA, ROQ-C, STG, T-2, T-2TET, TA, and ZEA, literature[823]; see also Cassava, chapter: Further Mycotoxins and Microbial Metabolites

Nivalenol

incidence: 74/405*, conc. range: ≤376 µg/kg, sample origin: households and markets, Tanzania, sample year: June–September 2012, country: Austria/Tanzania/Rwanda[823], *cassava, processed and stored

incidence: 33/222*, conc. range: ≤44.1 µg/kg, sample origin: households and markets, Rwanda, sample year: July 2013–April 2014, country: Austria/Tanzania/Rwanda[823], *cassava, processed and stored

- Co-contamination: not reported
- Further contamination of plants and plant products other than cereals and their products reported in the present book/article[823]: Cassava, 3-NPA, AFB_1, AFB_2, AFG_1, AFG_2, AFM_1, AME, AOH, ATX-I, BEA, CIT, CPA, CTV, DAS, DON, ENA, ENA_1, ENB, ENB_1, ENB_2, ENB_3, FA, FB_1, FB_2, FB_3, FP, HT-2, KA, MA, MAS, MON, NEO, OTA, OTB, PA, ROQ-C, STG, T-2, T-2TET, TA, and ZEA, literature[823]; see also Cassava, chapter: Further Mycotoxins and Microbial Metabolites

Other *Fusarium* Toxins

Fusaric Acid

incidence: 88/405*, conc. range: ≤6,207 µg/kg, sample origin: households and markets, Tanzania, sample year: June–September 2012, country: Austria/Tanzania/Rwanda[823], *cassava, processed and stored

incidence: 58/222*, conc. range: ≤3,102 µg/kg, sample origin: households and markets, Rwanda, sample year: July 2013–April 2014, country: Austria/Tanzania/Rwanda[823], *cassava, processed and stored

- Co-contamination: not reported
- Further contamination of plants and plant products other than cereals and their products reported in the present book/article[823]: Cassava, 3-NPA, AFB_1, AFB_2, AFG_1, AFG_2, AFM_1, AME, AOH, ATX-I, BEA, CIT, CPA, CTV, DAS, DON, ENA, ENA_1, ENB, ENB_1, ENB_2, ENB_3, FB_1, FB_2, FB_3, FP, HT-2, KA, MA, MAS, MON, NEO, NIV, OTA, OTB, PA, ROQ-C, STG, T-2, T-2TET, TA, and ZEA, literature[823]; see also Cassava, chapter: Further Mycotoxins and Microbial Metabolites

Fusaproliferin

incidence: 37/405*, conc. range: ≤479 µg/kg, sample origin: households and markets, Tanzania, sample year: June–September 2012, country: Austria/Tanzania/Rwanda[823], *cassava, processed and stored

incidence: 1/222*, conc.: 197 µg/kg, sample origin: households and markets, Rwanda, sample year: July 2013–April 2014, country: Austria/Tanzania/Rwanda[823], *cassava, processed and stored

- Co-contamination: not reported
- Further contamination of plants and plant products other than cereals and their products reported in the present book/article[823]: Cassava, 3-NPA, AFB_1, AFB_2, AFG_1, AFG_2, AFM_1, AME, AOH, ATX-I, BEA, CIT, CPA, CTV, DAS, DON, ENA, ENA_1, ENB, ENB_1, ENB_2, ENB_3, FA, FB_1, FB_2, FB_3, HT-2, KA, MA, MAS, MON, NEO, NIV, OTA, OTB, PA, ROQ-C, STG, T-2, T-2TET, TA, and ZEA, literature[823]; see also Cassava, chapter: Further Mycotoxins and Microbial Metabolites

Moniliformin

incidence: 1/59*, conc.: 313.9 μg/kg, sample origin: five Colombian Amazon communities, Colombia, sample year: May–October 2013, country: Colombia/Austria[709], *processed cassava root

- Co-contamination: not reported
- Further contamination of plants and plant products other than cereals and their products reported in the present book/article[709]: Cassava, FB_1, FB_2, HT-2, T-2, and ZEA, literature[709]

incidence: 131/405*, conc. range: ≤73.2 μg/kg, sample origin: households and markets, Tanzania, sample year: June–September 2012, country: Austria/Tanzania/Rwanda[823], *cassava, processed and stored

incidence: 61/222*, conc. range: ≤1,426 μg/kg, sample origin: households and markets, Rwanda, sample year: July 2013–April 2014, country: Austria/Tanzania/Rwanda[823], *cassava, processed and stored

- Co-contamination: not reported
- Further contamination of plants and plant products other than cereals and their products reported in the present book/article[823]: Cassava, 3-NPA, AFB_1, AFB_2, AFG_1, AFG_2, AFM_1, AME, AOH, ATX-I, BEA, CIT, CPA, CTV, DAS, DON, ENA, ENA_1, ENB, ENB_1, ENB_2, ENB_3, FA, FB_1, FB_2, FB_3, FP, HT-2, KA, MA, MAS, NEO, NIV, OTA, OTB, PA, ROQ-C, STG, T-2, T-2TET, TA, and ZEA, literature[823]; see also Cassava, chapter: Further Mycotoxins and Microbial Metabolites

Zearalenone

incidence: 3/59*, conc. range: 0.9–3.2 μg/kg, ∅ conc.: 1.7 μg/kg, sample origin: five Colombian Amazon communities, Colombia, sample year: May–October 2013, country: Colombia/Austria[709], *processed cassava root

- Co-contamination: not reported
- Further contamination of plants and plant products other than cereals and their products reported in the present book/article[709]: Cassava, FB_1, FB_2, HT-2, MON, and T-2, literature[709]

incidence: 154/405*, conc. range: ≤8,493 μg/kg, sample origin: households and markets, Tanzania, sample year: June–September 2012, country: Austria/Tanzania/Rwanda[823], *cassava, processed and stored

incidence: 104/222*, conc. range: ≤2,826 μg/kg, sample origin: households and markets, Rwanda, sample year: July 2013–April 2014, country: Austria/Tanzania/Rwanda[823], *cassava, processed and stored

- Co-contamination: not reported
- Further contamination of plants and plant products other than cereals and their products reported in the present book/article[823]: Cassava, 3-NPA, AFB_1, AFB_2, AFG_1, AFG_2, AFM_1, AME, AOH, ATX-I, BEA, CIT, CPA, CTV, DAS, DON, ENA, ENA_1, ENB, ENB_1, ENB_2, ENB_3, FA, FB_1, FB_2, FB_3, FP, HT-2, KA, MA, MAS, MON, NEO, NIV, OTA, OTB, PA, ROQ-C, STG, T-2, T-2TET, and TA, literature[823]; see also Cassava, chapter: Further Mycotoxins and Microbial Metabolites

Penicillium Toxins

Mycophenolic Acid

incidence: 11/405*, conc. range: ≤147 μg/kg, sample origin: households and markets, Tanzania, sample year: June–September 2012, country: Austria/Tanzania/Rwanda[823], *cassava, processed and stored

incidence: 50/222*, conc. range: ≤2,279.7 μg/kg, sample origin: households and markets, Rwanda, sample year: July 2013–April 2014, country: Austria/Tanzania/Rwanda[823], *cassava, processed and stored

- Co-contamination: not reported
- Further contamination of plants and plant products other than cereals and their products reported in the present book/article[823]: Cassava, 3-NPA, AFB_1, AFB_2, AFG_1, AFG_2, AFM_1, AME, AOH, ATX-I, BEA, CIT, CPA, CTV, DAS, DON, ENA, ENA_1, ENB, ENB_1, ENB_2, ENB_3, FA, FB_1, FB_2, FB_3, FP, HT-2, KA, MAS, MON, NEO, NIV, OTA, OTB, PA, ROQ-C, STG, T-2, T-2TET, TA, and ZEA, literature[823]; see also Cassava, chapter: Further Mycotoxins and Microbial Metabolites

Roquefortine C

incidence: 6/405*, conc. range: ≤138.2 μg/kg, sample origin: households and markets, Tanzania, sample year: June–September 2012, country: Austria/Tanzania/Rwanda[823], *cassava, processed and stored

incidence: 59/222*, conc. range: ≤1,455 μg/kg, sample origin: households and markets, Rwanda, sample year: July 2013–April 2014, country: Austria/Tanzania/Rwanda[823], *cassava, processed and stored

- Co-contamination: not reported
- Further contamination of plants and plant products other than cereals and their products reported in the present book/article[823]: Cassava, 3-NPA, AFB_1, AFB_2, AFG_1, AFG_2, AFM_1, AME, AOH, ATX-I, BEA, CIT, CPA, CTV, DAS, DON, ENA, ENA_1, ENB, ENB_1, ENB_2, ENB_3, FA, FB_1, FB_2, FB_3, FP, HT-2, KA, MA, MAS, MON, NEO, NIV, OTA, OTB, PA, STG, T-2, T-2TET, TA, and ZEA, literature[823]; see also Cassava, chapter: Further Mycotoxins and Microbial Metabolites

Cassava flour see Flour (cassava flour)

Cassia see Spices (cassia)

Cassis may contain the following mycotoxins:

Aspergillus and ***Penicillium*** Toxins

Ochratoxin A

incidence: 1/1*, conc.: 0.38 μg/kg, sample origin: unknown, sample year: unknown, country: Japan[112], *cassis, dry

- Co-contamination: not reported
- Further contamination of plants and plant products other than cereals and their products reported in the present book/article[112]: Cocoa, Coffee, Food, Vinegar, OTA and OTB, literature[112]; Juices, Raisins, Wine, OTA, literature[112]; Powder (cocoa powder), AFB_1, AFB_2, AFG_1, OTA, and OTB, literature[112]

Cayenne pepper see Spices (cayenne pepper, chilli)

Chardonnay see Wine (white)

Chenin Blanc see Wine (white)

Chestnuts see Nuts (chestnut)

Chickpeas see Peas (chickpea)

Chilli see Spices (cayenne pepper, chilli)

Chilli pickles see Pickles

Chilli sauce see Sauces (chilli sauce)

Chinese rice wine see Wine, miscellaneous

Chips (cassava chips) may contain the following mycotoxins:

Aspergillus **Toxins**

Aflatoxin B_1

incidence: 6/6, conc. range: 0.4–4.38 µg/kg, sample origin: markets in Brazzaville (capital), Congo, sample year: unknown, country: Nigeria/Benin/Tanzania[256]

incidence: 7?/7*, conc. range: ≤1.58 µg/kg, sample origin: Mtwara, Mkongi, Ugunja, and Zanzibar Island (regions), Tanzania, sample year: unknown, country: Nigeria/Benin/Tanzania[256], *fresh sa.

incidence: 13/13*, conc. range: 0.12–2.08 µg/kg, sample origin: Mtwara, Mkongi, Ugunja, and Zanzibar Island (regions), Tanzania, sample year: unknown, country: Nigeria/Benin/Tanzania[256], *stored sa.

incidence: 5/5*, conc. range: 0.9–1.44 µg/kg, sample origin: Mtwara, Mkongi, Ugunja, and Zanzibar Island (regions), Tanzania, sample year: unknown, country: Nigeria/Benin/Tanzania[256], *smoked sa.

incidence: 6/6*, conc. range: 0.86–6.98 µg/kg, sample origin: Mtwara, Mkongi, Ugunja, and Zanzibar Island (regions), Tanzania, sample year: unknown, country: Nigeria/Benin/Tanzania[256], *fresh sa.

incidence: 24/24*, conc. range: 0.1–17.11 µg/kg, sample origin: Mtwara, Mkongi, Ugunja, and Zanzibar Island (regions), Tanzania, sample year: unknown, country: Nigeria/Benin/Tanzania[256], *stored sa. for 3–7 days

incidence: 9/9*, conc. range: 2.3–33.8 µg/kg, sample origin: Mtwara, Mkongi, Ugunja, and Zanzibar Island (regions), Tanzania, sample year: unknown, country: Nigeria/Benin/Tanzania[256], *stored sa. for 1–4 months

- Co-contamination: not reported
- Further contamination of plants and plant products other than cereals and their products reported in the present book/article[256]: Flour (cassava flour), AFB_1, literature[256]

Aflatoxin/s

incidence: 11/36*, conc. range: 5.8–11.4 µg/kg, ∅ conc.: 8.0 µg/kg, sample origin: Yaoundé (capital), Cameroon, sample year: June–August 1999, country: Nigeria/Cameroon/Denmark/Ethiopia[719], *cassava chips (pellets), stored

incidence: 7/36*, conc. range: 5.2–14.5 µg/kg, ∅ conc.: 8.3 µg/kg, sample origin: Ebolowa (city), Cameroon, sample year: June–August 1999, country: Nigeria/Cameroon/Denmark/Ethiopia[719], *cassava chips (balls), stored

- Co-contamination: not reported
- Further contamination of plants and plant products other than cereals and their products reported in the present book/article[719]: not reported

For detailed information see the article.

incidence: 18/60, conc. range: ≤4.5 µg/kg, sample origin: Kumi (district), Uganda, sample year: unknown, country: Uganda[756]

- Co-contamination: not reported
- Further contamination of plants and plant products other than cereals and their products reported in the present book/article[756]: not reported

For detailed information see the article.

Chips (yam chips) may contain the following mycotoxins:

Aspergillus **Toxins**

Aflatoxin B_1

incidence: 10/44, conc. range: 10.3–65.7 µg/kg, ∅ conc.: 28.1 µg/kg, sample origin: markets in Ogun (state) in the rain forest belt of Nigeria, sample year: unknown, country: Nigeria[257]

incidence: 7/32, conc. range: 8.5–80.4 µg/kg, ∅ conc.: 25.6 µg/kg, sample origin: markets in Oyo (state) in the rain forest belt of Nigeria, sample year: unknown, country: Nigeria[257]

- Co-contamination: not reported
- Further contamination of plants and plant products other than cereals and their products reported in the present book/article[257]: not reported

incidence: 52/96*, conc. range: 4–186 µg/kg, ∅ conc.: 23 µg/kg, sample origin: markets in the rainforest belt, Nigeria, sample year: June–September 2001, country: Nigeria[258], *yam chips, dried

- Co-contamination: not reported
- Further contamination of plants and plant products other than cereals and their products reported in the present book/article[258]: Chips (yam chips), AFB_2, AFG_1, and AFG_2, literature[258]

incidence: 10/10*, conc. range: 11–27 µg/kg, ∅ conc.: 19 µg/kg, sample origin: Bodija (market) in Ibadan (city), Oyo State, Nigeria, sample year: unknown, country: Nigeria[660], *yam chips, dried

incidence: 9/10*, conc. range: 14–21 µg/kg, sample origin: Ojaoba (market) in Ibadan (city), Oyo State, Nigeria, sample year: unknown, country: Nigeria[660], *yam chips, dried

incidence: 10/10*, conc. range: 169–216 µg/kg, ∅ conc.: 190 µg/kg, sample origin: Oje (market) in Ibadan (city), Oyo State, Nigeria, sample year: unknown, country: Nigeria[660], *yam chips, dried

incidence: 10/10*, conc. range: 141–196 µg/kg, ∅ conc.: 180 µg/kg, sample origin: Omiadio (market) in Ibadan (city), Oyo State, Nigeria, sample year: unknown, country: Nigeria[660], *yam chips, dried

- Co-contamination: Further aflatoxins are found. For detailed information see the article.
- Further contamination of plants and plant products other than cereals and their products reported in the present book/article[660]: see co-contamination

Aflatoxin B_2

incidence: 31/96*, conc. range: 2–55 µg/kg, ∅ conc.: 11 µg/kg, sample origin: markets in the rainforest belt, Nigeria, sample year: June–September 2001, country: Nigeria[258], *yam chips, dried

- Co-contamination: not reported
- Further contamination of plants and plant products other than cereals and their products reported in the present book/article[258]: Chips (yam chips), AFB_1, AFG_1, and AFG_2, literature[258]

Aflatoxin G_1

incidence: 5/96*, conc. range: 4–18 µg/kg, ∅ conc.: 6 µg/kg, sample origin: markets in the rainforest belt, Nigeria, sample year: June–September 2001, country: Nigeria[258], *yam chips, dried

- Co-contamination: not reported
- Further contamination of plants and plant products other than cereals and their products reported in the present book/article[258]: Chips (yam chips), AFB_1, AFB_2, and AFG_2, literature[258]

Aflatoxin G_2

incidence: 2/96*, conc. range: 5–11 µg/kg, Ø conc.: 8 µg/kg, sample origin: markets in the rainforest belt, Nigeria, sample year: June–September 2001, country: Nigeria[258], *yam chips, dried

- Co-contamination: not reported
- Further contamination of plants and plant products other than cereals and their products reported in the present book/article[258]: Chips (yam chips), AFB_1, AFB_2, and AFG_1, literature[258]

Aflatoxin/s

incidence: 86/107*, conc. range: 4–15 µg/kg* (61 sa.), >15 µg/kg* (25 sa., maximum: 200 µg/kg*), sample origin: central and southern part of Benin, sample year: July/August 2000 and May/June 2001, country: France/Benin[259], *AFB_1, AFB_2, AFG_1, and AFG_2

- Co-contamination: not reported
- Further contamination of plants and plant products other than cereals and their products reported in the present book/article[259]: not reported

Chocolate may contain the following mycotoxins:

***Aspergillus* Toxins**

Aflatoxin B_1

incidence: 22*/42**, conc. range: 0.11–0.60 µg/kg, Ø conc.: 0.18 µg/kg, sample origin: local supermarkets and small retail shops across Japan, sample year: summer 2004–winter 2005, country: Japan[146], *domestic products, **bitter chocolate

- Co-contamination: not reported
- Further contamination of plants and plant products other than cereals and their products reported in the present book/article[146]: Beans (green bean), Chocolate, Raisins, Wine, OTA, literature[146]; Butter (peanut butter), AFB_1, AFB_2, AFG_1, and AFG_2, literature[146]

incidence: 24/25*, conc. range: ≤0.96 µg/kg, sample origin: supermarkets in Campinas and São Paulo (municipalities), Brazil, sample year: unknown, country: Brazil[260], *powdered chocolate

- Co-contamination: 24 sa. co-contaminated with AFB_1 and OTA (only co-contamination with OTA stated); no further information available
- Further contamination of plants and plant products other than cereals and their products reported in the present book/article[260]: Chocolate, AFB_2, AFG_1, AFG_2, and OTA, literature[260]

incidence: 25/25*, conc. range: 0.10–0.87 µg/kg, Ø conc.: 0.33 µg/kg, sample origin: supermarkets in Campinas and São Paulo (municipalities), Brazil, sample year: unknown, country: Brazil[260], *bitter chocolate

- Co-contamination: 25 sa. co-contaminated with AFB_1 and OTA (only co-contamination with OTA stated); no further information available
- Further contamination of plants and plant products other than cereals and their products reported in the present book/article[260]: Chocolate, AFB_2, AFG_1, AFG_2, and OTA, literature[260]

incidence: 25/25*, conc. range: 0.04–0.91 µg/kg, Ø conc.: 0.43 µg/kg, sample origin: supermarkets in Campinas and São Paulo (municipalities), Brazil, sample year: unknown, country: Brazil[260], *dark chocolate

- Co-contamination: 25 sa. co-contaminated with AFB_1 and OTA (only co-contamination with OTA stated); no further information available
- Further contamination of plants and plant products other than cereals and their products reported in the present book/article[260]: Chocolate, AFB_2, AFG_1, and OTA, literature[260]

incidence: 18/25*, conc. range: ≤0.27 µg/kg, sample origin: supermarkets in Campinas and São Paulo (municipalities), Brazil, sample year: unknown, country: Brazil[260], *milk chocolate

- Co-contamination: 18 sa. co-contaminated with AFB_1 and OTA (only co-contamination with OTA stated); no further information available
- Further contamination of plants and plant products other than cereals and their products reported in the present book/article[260]: Chocolate, AFB_2, AFG_1, and OTA, literature[260]

incidence: 5/25*, conc. range: ≤0.10 µg/kg, sample origin: supermarkets in Campinas and São Paulo (municipalities), Brazil, sample year: unknown, country: Brazil[260], *white chocolate

- Co-contamination: 5 sa. co-contaminated with AFB_1 and OTA
- Further contamination of plants and plant products other than cereals and their products reported in the present book/article[260]: Chocolate, AFB_2, AFG_1, AFG_2, and OTA, literature[260]

incidence: 7*/9**, conc. range: ≤0.53 µg/kg, sample origin: Canada, sample year: unknown, country: Canada[864], *n>LOQ, **baking chocolate

incidence: 16*/20**, conc. range: ≤0.63 µg/kg, sample origin: Canada, sample year: unknown, country: Canada[864], *n>LOQ, **dark chocolate

incidence: 7*/10**, conc. range: ≤0.18 µg/kg, sample origin: Canada, sample year: unknown, country: Canada[864], *n>LOQ, **milk chocolate

- Co-contamination: For detailed information see the article.
- Further contamination of plants and plant products other than cereals and their products reported in the present book/article[864]: Butter (cocoa butter), OTA, literature[864]; Chocolate, AFB_2, AFG_1, AFG_2, and OTA, literature[864]; Cocoa beans, Products (cocoa product), AFB_1, AFB_2, AFG_1, AFG_2, and OTA, literature[864]; Powder (cocoa powder), AFB_1, AFB_2, AFG_1, and OTA, literature[864]

Aflatoxin B_2

incidence: 20/25*, conc. range: ≤0.60 µg/kg, sample origin: supermarkets in Campinas and São Paulo (municipalities), Brazil, sample year: unknown, country: Brazil[260], *powdered chocolate

- Co-contamination: 20 sa. co-contaminated with AFB_2 and OTA (only co-contamination with OTA stated); no further information available
- Further contamination of plants and plant products other than cereals and their products reported in the present book/article[260]: Chocolate, AFB_1, AFG_1, AFG_2, and OTA, literature[260]

incidence: 19/25*, conc. range: ≤0.15 µg/kg, sample origin: supermarkets in Campinas and São Paulo (municipalities), Brazil, sample year: unknown, country: Brazil[260], *bitter chocolate

- Co-contamination: 19 sa. co-contaminated with AFB_1, AFB_2, and OTA (only co-contamination with OTA stated); no further information available

- Further contamination of plants and plant products other than cereals and their products reported in the present book/article[260]: Chocolate, AFB_1, AFG_1, AFG_2, and OTA, literature[260]

incidence: 22/25*, conc. range: ≤0.11 μg/kg, sample origin: supermarkets in Campinas and São Paulo (municipalities), Brazil, sample year: unknown, country: Brazil[260], *dark chocolate

- Co-contamination: 22 sa. co-contaminated with AFB_1, AFB_2, and OTA (only co-contamination with OTA stated); no further information available
- Further contamination of plants and plant products other than cereals and their products reported in the present book/article[260]: Chocolate, AFB_1, AFG_1, and OTA, literature[260]

incidence: 15/25*, conc. range: ≤0.07 μg/kg, sample origin: supermarkets in Campinas and São Paulo (municipalities), Brazil, sample year: unknown, country: Brazil[260], *milk chocolate

- Co-contamination: 15 sa. co-contaminated with AFB_2 and OTA (only co-contamination with OTA stated); no further information available
- Further contamination of plants and plant products other than cereals and their products reported in the present book/article[260]: Chocolate, AFB_1, AFG_1, and OTA, literature[260]

incidence: 0/25*, conc.: no contamination, sample origin: supermarkets in Campinas and São Paulo (municipalities), Brazil, sample year: unknown, country: Brazil[260], *white chocolate

- Co-contamination: not reported
- Further contamination of plants and plant products other than cereals and their products reported in the present book/article[260]: Chocolate, AFB_1, AFG_1, AFG_2, and OTA, literature[260]

incidence: 7*/9**, conc. range: ≤0.1 μg/kg, sample origin: Canada, sample year: unknown, country: Canada[864], *n>LOQ, **baking chocolate

incidence: 13*/20**, conc. range: ≤0.11 μg/kg, sample origin: Canada, sample year: unknown, country: Canada[864], *n>LOQ, **dark chocolate

incidence: 7*/10**, conc. range: ≤0.05 μg/kg, sample origin: Canada, sample year: unknown, country: Canada[864], *n>LOQ, **milk chocolate

- Co-contamination: For detailed information see the article.
- Further contamination of plants and plant products other than cereals and their products reported in the present book/article[864]: Butter (cocoa butter), OTA, literature[864]; Chocolate, AFB_1, AFG_1, AFG_2, and OTA, literature[864]; Cocoa beans, Products (cocoa product), AFB_1, AFB_2, AFG_1, AFG_2, and OTA, literature[864]; Powder (cocoa powder), AFB_1, AFB_2, AFG_1, and OTA, literature[864]

Aflatoxin G_1

incidence: 11/25*, conc. range: ≤0.48 μg/kg, sample origin: supermarkets in Campinas and São Paulo (municipalities), Brazil, sample year: unknown, country: Brazil[260], *powdered chocolate

- Co-contamination: 11 sa. co-contaminated with AFG_1 and OTA (only co-contamination with OTA stated); no further information available
- Further contamination of plants and plant products other than cereals and their products reported in the present book/article[260]: Chocolate, AFB_1, AFB_2, AFG_2, and OTA, literature[260]

incidence: 24/25*, conc. range: ≤0.63 μg/kg, sample origin: supermarkets in Campinas and São Paulo (municipalities), Brazil, sample year: unknown, country: Brazil[260], *bitter chocolate

- Co-contamination: 24 sa. co-contaminated with AFB_1, AFG_1, and OTA (only co-contamination with OTA stated); no further information available
- Further contamination of plants and plant products other than cereals and their products reported in the present book/article[260]: Chocolate, AFB_1, AFB_2, AFG_2, and OTA, literature[260]

incidence: 2/25*, conc. range: ≤0.06 μg/kg, sample origin: supermarkets in Campinas and São Paulo (municipalities), Brazil, sample year: unknown, country: Brazil[260], *dark chocolate

- Co-contamination: 2 sa. co-contaminated with AFB_1, AFG_1, and OTA (only co-contamination with OTA stated); no further information available
- Further contamination of plants and plant products other than cereals and their products reported in the present book/article[260]: Chocolate, AFB_1, AFB_2, and OTA, literature[260]

incidence: 5/25*, conc. range: ≤0.07 μg/kg, sample origin: supermarkets in Campinas and São Paulo (municipalities), Brazil, sample year: unknown, country: Brazil[260], *milk chocolate

- Co-contamination: 5 sa. co-contaminated with AFG_1 and OTA (only co-contamination with OTA stated); no further information available
- Further contamination of plants and plant products other than cereals and their products reported in the present book/article[260]: Chocolate, AFB_1, AFB_2, and OTA, literature[260]

incidence: 0/25*, conc.: no contamination, sample origin: supermarkets in Campinas and São Paulo (municipalities), Brazil, sample year: unknown, country: Brazil[260], *white chocolate

- Co-contamination: not reported
- Further contamination of plants and plant products other than cereals and their products reported in the present book/article[260]: Chocolate, AFB_1, AFB_2, AFG_2, and OTA, literature[260]

incidence: 1?*/9**, conc.: 0.05 μg/kg, sample origin: Canada, sample year: unknown, country: Canada[864], *n>LOQ, **baking chocolate

incidence: 1?*/20**, conc.: 0.09 μg/kg, sample origin: Canada, sample year: unknown, country: Canada[864], *n>LOQ, **dark chocolate

incidence: 1*/10**, conc.: 0.28 μg/kg, sample origin: Canada, sample year: unknown, country: Canada[864], *n>LOQ, **milk chocolate

- Co-contamination: For detailed information see the article.
- Further contamination of plants and plant products other than cereals and their products reported in the present book/article[864]: Butter (cocoa butter), OTA, literature[864]; Chocolate, AFB_1, AFB_2, AFG_2, and OTA, literature[864]; Cocoa beans, Products (cocoa product), AFB_1, AFB_2, AFG_1, AFG_2, and OTA, literature[864]; Powder (cocoa powder), AFB_1, AFB_2, AFG_1, and OTA, literature[864]

Aflatoxin G_2

incidence: 3/25*, conc. range: ≤0.10 μg/kg, sample origin: supermarkets in Campinas and São Paulo (municipalities), Brazil, sample year: unknown, country: Brazil[260], *powdered chocolate

- Co-contamination: 3 sa. co-contaminated with AFG_2 and OTA (only co-contamination with OTA stated); no further information available
- Further contamination of plants and plant products other than cereals and their products reported in the present book/article[260]: Chocolate, AFB_1, AFB_2, AFG_1, and OTA, literature[260]

incidence: 5/25*, conc. range: ≤0.03 µg/kg, sample origin: supermarkets in Campinas and São Paulo (municipalities), Brazil, sample year: unknown, country: Brazil[260], *bitter chocolate

- Co-contamination: 5 sa. co-contaminated with AFB_1, AFG_2, and OTA (only co-contamination with OTA stated); no further information available
- Further contamination of plants and plant products other than cereals and their products reported in the present book/ article[260]: Chocolate, AFB_1, AFB_2, AFG_1, and OTA, literature[260]

incidence: 0/25*, conc.: no contamination, sample origin: supermarkets in Campinas and São Paulo (municipalities), Brazil, sample year: unknown, country: Brazil[260], *dark chocolate

- Co-contamination: not reported
- Further contamination of plants and plant products other than cereals and their products reported in the present book/ article[260]: Chocolate, AFB_1, AFB_2, AFG_1, and OTA, literature[260]

incidence: 0/25*, conc.: no contamination, sample origin: supermarkets in Campinas and São Paulo (municipalities), Brazil, sample year: unknown, country: Brazil[260], *milk chocolate

- Co-contamination: not reported
- Further contamination of plants and plant products other than cereals and their products reported in the present book/ article[260]: Chocolate, AFB_1, AFB_2, AFG_1, and OTA, literature[260]

incidence: 0/25*, conc.: no contamination, sample origin: supermarkets in Campinas and São Paulo (municipalities), Brazil, sample year: unknown, country: Brazil[260], *white chocolate

- Co-contamination: not reported
- Further contamination of plants and plant products other than cereals and their products reported in the present book/ article[260]: Chocolate, AFB_1, AFB_2, AFG_1, and OTA, literature[260]

incidence: 0*/9**, conc.: no contamination, sample origin: Canada, sample year: unknown, country: Canada[864], *n>LOQ, **baking chocolate

incidence: 1*/20**, conc.: 0.08 µg/kg, sample origin: Canada, sample year: unknown, country: Canada[864], *n>LOQ, **dark chocolate

incidence: 1?*/10**, conc.: 0.04 µg/kg, sample origin: Canada, sample year: unknown, country: Canada[864], *n>LOQ, **milk chocolate

- Co-contamination: For detailed information see the article.
- Further contamination of plants and plant products other than cereals and their products reported in the present book/ article[864]: Butter (cocoa butter), OTA, literature[864]; Chocolate, AFB_1, AFB_2, AFG_1, and OTA, literature[864]; Cocoa beans, Products (cocoa product), AFB_1, AFB_2, AFG_1, AFG_2, and OTA, literature[864]; Powder (cocoa powder), AFB_1, AFB_2, AFG_1, and OTA, literature[864]

Aspergillus and *Penicillium* Toxins

Ochratoxin A

incidence: 84/115, conc. range: ≤1.75 µg/kg, ∅ conc.: 0.31 µg/kg, sample origin: local supermarkets and small retail shops in seven major cities, Japan, sample year: 2005–winter 2007, country: Japan[85]

- Co-contamination: not reported
- Further contamination of plants and plant products other than cereals and their products reported in the present book/ article[85]: Beans (soybean), Vegetables (asparagus), FB_1 and FB_2, literature[85]; Beverages, Cocoa, Coffee, Coffee beans, Raisins, Spices (coriander), Wine, OTA, literature[85]

incidence: 1/1*, conc.: 0.30 µg/kg, sample origin: supermarkets and fast-food chains in Quebec (city), Canada, sample year: September/October 2008, country: Canada[90], *chocolate bars

incidence: 1/1*, conc.: 0.13 µg/kg, sample origin: supermarkets and fast-food chains in Calgary (city), Canada, sample year: September/October 2009, country: Canada[90], *chocolate bars

incidence: 1/1*, conc.: 0.02 µg/l, sample origin: supermarkets and fast-food chains in Quebec (city), Canada, sample year: September/ October 2008, country: Canada[90], *milk, chocolate, 1%

incidence: 1/1*, conc.: 0.01 µg/l, sample origin: supermarkets and fast-food chains in Calgary (city), Canada, sample year: September/October 2009, country: Canada[90], *milk, chocolate, 1%

- Co-contamination: not reported
- Further contamination of plants and plant products other than cereals and their products reported in the present book/ article[90]: Baby food, Beans, Beverages, Butter (peanut butter), Candies, Coffee, Extracts, Herbs, Ice cream, Nuts, Pies, Pizza, Products, Pudding, Raisins, Seeds, Wine, OTA, literature[90]

incidence: 27*/41**, conc. range: 0.10–0.94 µg/kg, ∅ conc.: 0.35 µg/kg, sample origin: local supermarkets and small retail shops across Japan, sample year: summer 2004–winter 2005, country: Japan[146], *domestic products, **bitter chocolate

- Co-contamination: not reported
- Further contamination of plants and plant products other than cereals and their products reported in the present book/ article[146]: Beans (green bean), Coffee, Raisins, Wine, OTA, literature[146]; Butter (peanut butter), AFB_1, AFB_2, AFG_1, and AFG_2, literature[146]; Chocolate, AFB_1, literature[146]

incidence: 25/25*, conc. range: 0.03–0.92 µg/kg, ∅ conc.: 0.39 µg/kg, sample origin: supermarkets in Campinas and São Paulo (municipalities), Brazil, sample year: unknown, country: Brazil[260], *powdered chocolate

- Co-contamination: 24 sa. co-contaminated with AFB_1 and OTA; 20 sa. co-contaminated with AFB_2 and OTA; 11 sa. co-contaminated with AFG_1 and OTA; 3 sa. co-contaminated with AFG_2 and OTA (only co-contamination with OTA stated); no further information available
- Further contamination of plants and plant products other than cereals and their products reported in the present book/ article[260]: Chocolate, AFB_1, AFB_2, AFG_1, and AFG_2, literature[260]

incidence: 25/25*, conc. range: 0.06–0.60 µg/kg, ∅ conc.: 0.31 µg/kg, sample origin: supermarkets in Campinas and São Paulo (municipalities), Brazil, sample year: unknown, country: Brazil[260], *bitter chocolate

- Co-contamination: 25 sa. co-contaminated with AFB_1 and OTA; 19 sa. co-contaminated with AFB_2 and OTA; 24 sa. co-contaminated with AFG_1 and OTA; 5 sa. co-contaminated with AFG_2 and OTA (only co-contamination with OTA stated); no further information available
- Further contamination of plants and plant products other than cereals and their products reported in the present book/ article[260]: Chocolate, AFB_1, AFB_2, AFG_1, and AFG_2, literature[260]

incidence: 25/25*, conc. range: 0.09–0.87 µg/kg, ∅ conc.: 0.34 µg/kg, sample origin: supermarkets in Campinas and São Paulo (municipalities), Brazil, sample year: unknown, country: Brazil[260], *dark chocolate

- Co-contamination: 25 sa. co-contaminated with AFB_1 and OTA; 22 sa. co-contaminated with AFB_2 and OTA; 2 sa. co-contaminated with AFG_1 and OTA (only co-contamination with OTA stated); no further information available
- Further contamination of plants and plant products other than cereals and their products reported in the present book/article[260]: Chocolate, AFB_1, AFB_2, and AFG_1, literature[260]

incidence: 25/25*, conc. range: 0.08–0.45 µg/kg, ∅ conc.: 0.15 µg/kg, sample origin: supermarkets in Campinas and São Paulo (municipalities), Brazil, sample year: unknown, country: Brazil[260], *milk chocolate

- Co-contamination: 18 sa. co-contaminated with AFB_1 and OTA; 15 sa. co-contaminated with AFB_2 and OTA; 5 sa. co-contaminated with AFG_1 and OTA (only co-contamination with OTA stated); no further information available
- Further contamination of plants and plant products other than cereals and their products reported in the present book/article[260]: Chocolate, AFB_1, AFB_2, and AFG_1, literature[260]

incidence: 23/25*, conc. range: ≤0.05 µg/kg, sample origin: supermarkets in Campinas and São Paulo (municipalities), Brazil, sample year: unknown, country: Brazil[260], *white chocolate

- Co-contamination: 5 sa. co-contaminated with AFB_1 and OTA (only co-contamination with OTA stated); no further information available
- Further contamination of plants and plant products other than cereals and their products reported in the present book/article[260]: Chocolate, AFB_1, AFB_2, AFG_1, and AFG_2, literature[260]

incidence: 8/11*, conc. range: 0.1–1.59 µg/kg, ∅ conc.: 0.63 µg/kg, sample origin: Cameroon, Guinea, Ivory Coast, and Nigeria, sample year: unknown, country: Spain[261], *chocolate and chocolate cream

- Co-contamination: not reported
- Further contamination of plants and plant products other than cereals and their products reported in the present book/article[261]: Cocoa beans, Powder (cocoa powder), Products (cocoa product), OTA, literature[261]

incidence: 36/39*, conc. range: ≤0.410 µg/kg, sample origin: regions around Detmold, Giessen, Hamburg, Jena, Karlsruhe, Kiel, Kulmbach, Trier (cities), Germany, sample year: unknown, country: Germany[262], *whole milk chocolate

incidence: 78/78*, conc. range: 0.020–0.660 µg/kg, sample origin: regions around Detmold, Giessen, Hamburg, Jena, Karlsruhe, Kiel, Kulmbach, Trier (cities), Germany, sample year: unknown, country: Germany[262], *half bitter/bitter chocolate

incidence: 31/35*, conc. range: ≤0.160 µg/kg, sample origin: regions around Detmold, Giessen, Hamburg, Jena, Karlsruhe, Kiel, Kulmbach, Trier (cities), Germany, sample year: unknown, country: Germany[262], *chocolate with nuts

incidence: 55/58*, conc. range: ≤0.340 µg/kg, sample origin: regions around Detmold, Giessen, Hamburg, Jena, Karlsruhe, Kiel, Kulmbach, Trier (cities), Germany, sample year: unknown, country: Germany[262], *chocolate, filled

- Co-contamination: not reported
- Further contamination of plants and plant products other than cereals and their products reported in the present book/article[262]: Bars, Drinks (chocolate drink), Figs, Fruits, dried, Nut crème, Nuts, Nuts (hazelnut), Nuts (peanut), Powder (cocoa powder), Pudding, Raisins, Seeds (linseed), Seeds (sesame), Seeds (sunflower), OTA, literature[262]

incidence: 30/40, conc. range: 0.1 µg/kg (12 sa.), 0.2–4.0 µg/kg (18 sa., maximum: 0.6 µg/kg), sample origin: 13 regions in UK, sample year: May 1998, country: UK[263]

- Co-contamination: not reported
- Further contamination of plants and plant products other than cereals and their products reported in the present book/article[263]: Currants, Figs, Juices (grape juice), Powder (cocoa powder), Raisins, Sultanas, Wine (red), OTA, literature[263]

incidence: 86/87, conc. range: 0.025–4.289* µg/kg, sample origin: different countries, sample year: 2001–2003, country: Spain[264], *chocolate containing raisins

incidence: 169/169*, conc. range: 0.012–0.693 µg/kg, ∅ conc.: 0.139 µg/kg, sample origin: different countries, sample year: 2001–2003, country: Spain[264], *milk chocolate

incidence: 14/14*, conc. range: 0.016–0.190 µg/kg, ∅ conc.: 0.050 µg/kg, sample origin: different countries, sample year: 2001–2003, country: Spain[264] *white chocolate

- Co-contamination: not reported
- Further contamination of plants and plant products other than cereals and their products reported in the present book/article[264]: Powder (cocoa powder), OTA, literature[264]

incidence: 7/7*, conc. range: 0.12–1.4 µg/kg, ∅ conc.: 0.63 µg/kg, sample origin: sa. available in Canada, sample year: unknown, country: Canada[265] *baking chocolate

incidence: 14/14*, conc. range: 0.17–0.88 µg/kg, ∅ conc.: 0.38 µg/kg, sample origin: sa. available in Canada, sample year: unknown, country: Canada[265], *dark chocolate

incidence: 5/7*, conc. range: ≤0.19 µg/kg, sample origin: sa. available in Canada, sample year: unknown, country: Canada[265], *milk chocolate

- Co-contamination: not reported
- Further contamination of plants and plant products other than cereals and their products reported in the present book/article[265]: Powder (cocoa powder), OTA, literature[265]

incidence: 92/120*, conc. range: ≤0.74 µg/kg, sample origin: discount stores in south, center, and north Italy, sample year: unknown, country: Italy[266], *dark chocolate bars

incidence: 21/78*, conc. range: ≤0.26 µg/kg, sample origin: discount stores in south, center, and north Italy, sample year: unknown, country: Italy[266], *milk chocolate bars

incidence: 21/47*, conc. range: ≤0.42 µg/kg, sample origin: discount stores in south, center, and north Italy, sample year: unknown, country: Italy[266], *chocolate candies

incidence: 5/15*, conc. range: ≤0.50 µg/kg, sample origin: discount stores in south, center, and north Italy, sample year: unknown, country: Italy[266], *easter eggs

- Co-contamination: not reported
- Further contamination of plants and plant products other than cereals and their products reported in the present book/article[266]: Powder (cocoa powder), OTA, literature[266]

incidence: 18/18*, conc. range: ≤1.66 µg/kg, ∅ conc.: 0.377 µg/kg, sample origin: supermarkets and fresh provision shops, Hong Kong, China, sample year: August/September 2005, country: China[697], *chocolate and its products

- Co-contamination: not reported
- Further contamination of plants and plant products other than cereals and their products reported in the present book/

article[697]: Coffee, Fruits, dried, Juices, Legumes, Spices, OTA, literature[697]

incidence: 297/352*, conc. range: LOD/LOQ-0.9 μg/kg (295 sa.), ≤3.60 μg/kg (2 sa.), sample origin: Germany, sample year: 1995–1998, country: EU[698], *chocolate and sweets

- Co-contamination: not reported
- Further contamination of plants and plant products other than cereals and their products reported in the present book/article[698]: Baby food, Beans (soybean), Cocoa, Coffee, Coffee beans, Currants, Fruits, dried, Herbs, Jam, Juices, Juices (grape juice), Legumes, Nuts, Oil seeds, Oils (olive oil), Peas, Peas (chickpea), Products (plant product), Raisins, Spices, Spices (cayenne pepper, chilli), Spices (nutmeg), Spices (paprika), Spices (pepper), Sultanas, Tea, Vinegar, Wine, Wine (red), Wine (rosé), Wine (white), Wine, miscellaneous, OTA, literature[698]

incidence: 5/12*, conc. range: ≤1.01 μg/kg, sample origin: 12 regions, Czech Republic, sample year: 2011–2013, country: Czech Republic[811], *bitter chocolate

incidence: 2/12*, conc. range: ≤0.33 μg/kg, sample origin: 12 regions, Czech Republic, sample year: 2011–2013, country: Czech Republic[811], *filled chocolate

incidence: 4/12*, conc. range: ≤0.40 μg/kg, sample origin: 12 regions, Czech Republic, sample year: 2011–2013, country: Czech Republic[811], *milk chocolate

- Co-contamination: not reported
- Further contamination of plants and plant products other than cereals and their products reported in the present book/article[811]: Beans, Coffee, Confectionery, Figs, Herbs and Spices, Juices, Licorice, Powder (cocoa powder), Products (ginger product), Raisins, Spices (cayenne pepper, chilli), Spices (caraway), Spices (coriander), Spices (ginger), Spices (mustard), Spices (nutmeg), Spices (paprika), Spices (pepper), Tea, Wine (red), Wine (white), OTA, literature[811]

incidence: 9*/9**, conc. range: ≤0.91 μg/kg, sample origin: Canada, sample year: unknown, country: Canada[864], *n>LOQ, **baking chocolate

incidence: 20*/20**, conc. range: ≤0.65 μg/kg, sample origin: Canada, sample year: unknown, country: Canada[864], *n>LOQ, **dark chocolate

incidence: 7*/10**, conc. range: ≤0.33 μg/kg, sample origin: Canada, sample year: unknown, country: Canada[864], *n>LOQ, **milk chocolate

- Co-contamination: For detailed information see the article.
- Further contamination of plants and plant products other than cereals and their products reported in the present book/article[864]: Butter (cocoa butter), OTA, literature[864]; Chocolate, AFB_1, AFB_2, AFG_1, and AFG_2, literature[864]; Cocoa beans, Products (cocoa product), AFB_1, AFB_2, AFG_1, AFG_2, and OTA, literature[864]; Powder (cocoa powder), AFB_1, AFB_2, AFG_1, and OTA, literature[864]

incidence: 24/36, conc. range: ≤16.15 μg/kg, sample origin: Kumba? (city), Cameroon, sample year: 2005–2007, country: Cameroon/France[907]

- Co-contamination: not reported
- Further contamination of plants and plant products other than cereals and their products reported in the present book/article[907]: Butter (cocoa butter), Powder (cocoa powder), OTA, literature[907]

Chocolate bars see Chocolate

Chocolate sauce see Sauces (chocolate sauce)

Cigarettes see Tobacco

Cinnamon see Spices (cinnamon)

Cocoa may contain the following mycotoxins:

Aspergillus Toxins

Aflatoxin B_1

incidence: 1/40*, conc.: 5 μg/kg, sample origin: stores in Oslo (capital) area, Norway, sample year: autumn 1973, country: Norway[201], *cocoa and cocoa products

- Co-contamination: 1 sa. co-contaminated with AFB_1, AFB_2, AFG_1, and AFG_2
- Further contamination of plants and plant products other than cereals and their products reported in the present book/article[201]: Cocoa, AFB_2, AFG_1, and AFG_2, literature[201]; Nuts (Brazil nut, paranut), Nuts (cashew nut), Nuts (peanut), Nuts (walnut), AFB_1, AFB_2, AFG_1, and AFG_2, literature[201]; Nuts, AFB_2, AFG_1, and AFG_2, literature[201]

Aflatoxin B_2

incidence: 1/40*, conc.: tr, sample origin: stores in Oslo (capital) area, Norway, sample year: autumn 1973, country: Norway[201], *cocoa and cocoa products

- Co-contamination: 1 sa. co-contaminated with AFB_1, AFB_2, AFG_1, and AFG_2
- Further contamination of plants and plant products other than cereals and their products reported in the present book/article[201]: Cocoa, AFB_1, AFG_1, and AFG_2, literature[201]; Nuts (Brazil nut, paranut), Nuts (cashew nut), Nuts (peanut), Nuts (walnut), AFB_1, AFB_2, AFG_1, and AFG_2, literature[201]; Nuts, AFB_2, AFG_1, and AFG_2, literature[201]

Aflatoxin G_1

incidence: 1/40*, conc.: 4 μg/kg, sample origin: stores in Oslo (capital) area, Norway, sample year: autumn 1973, country: Norway[201], *cocoa and cocoa products

- Co-contamination: 1 sa. co-contaminated with AFB_1, AFB_2, AFG_1, and AFG_2
- Further contamination of plants and plant products other than cereals and their products reported in the present book/article[201]: Cocoa, AFB_1, AFB_2, and AFG_2, literature[201]; Nuts (Brazil nut, paranut), Nuts (cashew nut), Nuts (peanut), Nuts (walnut), AFB_1, AFB_2, AFG_1, and AFG_2, literature[201]; Nuts, AFB_2, AFG_1, and AFG_2, literature[201]

Aflatoxin G_2

incidence: 1/40, conc.: tr, sample origin: stores in Oslo (capital) area, Norway, sample year: autumn 1973, country: Norway[201], *cocoa and cocoa products

- Co-contamination: 1 sa. co-contaminated with AFB_1, AFB_2, AFG_1, and AFG_2
- Further contamination of plants and plant products other than cereals and their products reported in the present book/article[201]: Cocoa, AFB_1, AFB_2, and AFG_1, literature[201]; Nuts (Brazil nut, paranut), Nuts (cashew nut), Nuts (peanut), Nuts (walnut), AFB_1, AFB_2, AFG_1, and AFG_2, literature[201]; Nuts, AFB_2, AFG_1, and AFG_2, literature[201]

Aflatoxin/s

incidence: 19/27*, Ø conc.: 50.6 µg/kg**, sample origin: greater Manila (capital) area, Philippines, sample year: unknown, country: Philippines[1], *cocoa and products, **AF of pos sa.?

- Co-contamination: not reported
- Further contamination of plants and plant products other than cereals and their products reported in the present book/article[1]: Beans, Butter (peanut butter), Cassava, Coffee, Drinks (alcoholic drink), Nuts (cashew nut), Nuts (peanut), Products (coconut product), Rootcrops, Spices (cayenne pepper, chilli), Sweet potatoes, Yams, AF/S, literature[1]

incidence: 6/6, conc. range: 2.090–25.753 µg/kg*, Ø conc.: 7.483 µg/kg*, sample origin: groceries and supermarkets, Turkey, sample year: 2003, country: Turkey[119], *AFS total

- Co-contamination: not reported
- Further contamination of plants and plant products other than cereals and their products reported in the present book/article[119]: Asure, Beverages, Cream, Ice cream, Jello, Keskul, Muhallebi, Pudding, Salep, Sauces (chocolate sauce), Soups, AF/S, literature[119]

incidence: 2/20, conc. range: 12.6–21.7 µg/kg*, Ø conc.: 17.6 µg/kg*, sample origin: markets from Aswan, Assiut, El-Minya, Qena, and Sohag, (governorates), Egypt, sample year: unknown, country: Egypt[196], *AFB_1 and AFB_2

- Co-contamination: not reported
- Further contamination of plants and plant products other than cereals and their products reported in the present book/article[196]: Tea, AF/S, literature[196]

incidence: 49/130* conc. range: ≤17.795 µg/kg**, sample origin: southern region of Bahia (state), Brazil, sample year: unknown, country: Brazil[615], *cocoa clones, **AFB_1, AFB_2, AFG_1, and AFG_2

- Co-contamination: For detailed information see the article.
- Further contamination of plants and plant products other than cereals and their products reported in the present book/article[615]: Cocoa, OTA, literature[615]

For detailed information see the article.

incidence: 2/2, conc. range: 0.2–0.7 µg/kg, Ø conc.: 0.45 µg/kg, sample origin: unknown, sample year: unknown, country: Germany[673]

- Co-contamination: 1 sa. co-contaminated with AFS and OTA; 1 sa. contaminated solely with AFS
- Further contamination of plants and plant products other than cereals and their products reported in the present book/article[673]: Cocoa, OTA, literature[673]; Raisins, Sultanas, CIT and OTA, literature[673]

Aspergillus and ***Penicillium*** Toxins

Ochratoxin A

incidence: 37/38, conc. range: ≤3.45 µg/kg, Ø conc.: 0.89 µg/kg, sample origin: local supermarkets and small retail shops in seven major cities, Japan, sample year: 2006–winter 2007, country: Japan[85]

- Co-contamination: not reported
- Further contamination of plants and plant products other than cereals and their products reported in the present book/article[85]: Beans (soybean), Vegetables (asparagus), FB_1 and FB_2, literature[85]; Beverages, Chocolate, Coffee, Coffee beans, Raisins, Spices (coriander), Wine, OTA, literature[85]

incidence: 5?/8, conc. range: 0.11–0.35 µg/kg, sample origin: unknown, sample year: unknown, country: Japan[112]

- Co-contamination: For detailed information see the article.
- Further contamination of plants and plant products other than cereals and their products reported in the present book/article[112]: Cassis, Juices, Raisins, Wine, OTA, literature[112]; Cocoa, OTB, literature[112]; Coffee, Food, Vinegar, OTA and OTB, literature[112]; Powder (cocoa powder), AFB_1, AFB_2, AFG_1, OTA, and OTB, literature[112]

incidence: 23/130*, conc. range: ≤274.900 µg/kg, sample origin: southern region of Bahia (state), Brazil, sample year: unknown, country: Brazil[615], *cocoa clones

- Co-contamination: For detailed information see the article.
- Further contamination of plants and plant products other than cereals and their products reported in the present book/article[615]: Cocoa, AF/S, literature[615]; literature[615]

For detailed information see the article.

incidence: 1/2, conc.: 1.4 µg/kg, sample origin: unknown, sample year: unknown, country: Germany[673]

- Co-contamination: 1 sa. co-contaminated with AFS and OTA
- Further contamination of plants and plant products other than cereals and their products reported in the present book/article[673]: Cocoa, AF/S, literature[673]; Raisins, Sultanas, CIT and OTA, literature[673]

incidence: 91/96, conc. range: LOD/LOQ-0.9 µg/kg (88 sa.), ≤1.80 µg/kg (3 sa.), sample origin: Germany, sample year: 1995–1998, country: EU[698]

- Co-contamination: not reported
- Further contamination of plants and plant products other than cereals and their products reported in the present book/article[698]: Baby food, Beans (soybean), Chocolate, Coffee, Coffee beans, Currants, Fruits, dried, Herbs, Jam, Juices, Juices (grape juice), Legumes, Nuts, Oil seeds, Oils (olive oil), Peas, Peas (chickpea), Products (plant product), Raisins, Spices, Spices (cayenne pepper, chilli), Spices (nutmeg), Spices (paprika), Spices (pepper), Sultanas, Tea, Vinegar, Wine, Wine (red), Wine (rosé), Wine (white), Wine, miscellaneous, OTA, literature[698]

Ochratoxin B

incidence: 1/8, conc.: 0.10 µg/kg, sample origin: unknown, sample year: unknown, country: Japan[112]

- Co-contamination: For detailed information see the article.
- Further contamination of plants and plant products other than cereals and their products reported in the present book/article[112]: Cassis, Cocoa, Juices, Raisins, Wine, OTA, literature[112]; Coffee, Food, Vinegar, OTA and OTB, literature[112]; Powder (cocoa powder), AFB_1, AFB_2, AFG_1, OTA, and OTB, literature[112]

see also **Coffee**

Cocoa beans may contain the following mycotoxins:

Aspergillus Toxins

Aflatoxin B_1

incidence: 14/200, conc. range: 1–10 µg/kg (14 sa.), sample origin: unknown, sample year: 1990–1996, country: Cuba[126]

- Co-contamination: not reported
- Further contamination of plants and plant products other than cereals and their products reported in the present book/

article[126]: Beans, Beans (soybean), Nuts (peanut), Peas, AFB_1, literature[126]

incidence: 1/1, conc.: 0.6 μg/kg, sample origin: imported sa. collected by the Mycotoxin Research Association, Japan, sample year: 1985, country: Japan[171]

- Co-contamination: not reported
- Further contamination of plants and plant products other than cereals and their products reported in the present book/article[171]: Nuts (pistachio nut), Spices (cayenne pepper, chilli), Spices (pepper), AFB_1, literature[171]

incidence: 4/151*, conc. range: ≤2.6 μg/kg, sample origin: imported to USSR, sample year: 1985–1987, country: USSR[267], *cocoa beans and products

- Co-contamination: not reported
- Further contamination of plants and plant products other than cereals and their products reported in the present book/article[267]: Coffee beans, Legumes, Nuts, Nuts (peanut), AFB_1, literature[267]

incidence: 0/25*, conc.: no contamination, sample origin: farms in Bahia (state), Brazil, sample year: unknown, country: Brazil[268], *before fermentation

incidence: 2/51*, conc. range: ≤0.1 μg/kg, sample origin: farms in Bahia (state), Brazil, sample year: unknown, country: Brazil[268], *during fermentation

incidence: 11/85*, conc. range: ≤6.66 μg/kg, sample origin: farms in Bahia (state), Brazil, sample year: unknown, country: Brazil[268], *during sun drying

incidence: 3/65*, conc. range: ≤0.14 μg/kg, sample origin: farms in Bahia (state), Brazil, sample year: unknown, country: Brazil[268], *during storage

- Co-contamination: not reported
- Further contamination of plants and plant products other than cereals and their products reported in the present book/article[268]: Cocoa beans, AFB_2 and AFG_1, literature[268]

incidence: 14*/15**, conc. range: ≤2.60 μg/kg, sample origin: Canada, sample year: unknown, country: Canada[864], *n>LOQ, **cocoa, natural

- Co-contamination: For detailed information see the article.
- Further contamination of plants and plant products other than cereals and their products reported in the present book/article[864]: Butter (cocoa butter), OTA, literature[864]; Chocolate, Products (cocoa product), AFB_1, AFB_2, AFG_1, AFG_2, and OTA, literature[864]; Cocoa beans, AFB_2, AFG_1, AFG_2, and OTA, literature[864]; Powder (cocoa powder), AFB_1, AFB_2, AFG_1, and OTA, literature[864]

Aflatoxin B_2

incidence: 0/25*, conc.: no contamination, sample origin: farms in Bahia (state), Brazil, sample year: unknown, country: Brazil[268], *before fermentation

incidence: 1/51*, conc.: 0.04 μg/kg, sample origin: farms in Bahia (state), Brazil, sample year: unknown, country: Brazil[268], *during fermentation

incidence: 4/85*, conc. range: ≤0.37 μg/kg, sample origin: farms in Bahia (state), Brazil, sample year: unknown, country: Brazil[268], *during sun drying

incidence: 0/65*, conc.: no contamination, sample origin: farms in Bahia (state), Brazil, sample year: unknown, country: Brazil[268], *during storage

- Co-contamination: not reported
- Further contamination of plants and plant products other than cereals and their products reported in the present book/article[268]: Cocoa beans, AFB_1 and AFG_1, literature[268]

incidence: 13*/15**, conc. range: ≤0.44 μg/kg, sample origin: Canada, sample year: unknown, country: Canada[864], *n>LOQ, **cocoa, natural

- Co-contamination: For detailed information see the article.
- Further contamination of plants and plant products other than cereals and their products reported in the present book/article[864]: Butter (cocoa butter), OTA, literature[864]; Chocolate, Products (cocoa product), AFB_1, AFB_2, AFG_1, AFG_2, and OTA, literature[864]; Cocoa beans, AFB_1, AFG_1, AFG_2, and OTA, literature[864]; Powder (cocoa powder), AFB_1, AFB_2, AFG_1, and OTA, literature[864]

Aflatoxin G_1

incidence: 0/25*, conc.: no contamination, sample origin: farms in Bahia (state), Brazil, sample year: unknown, country: Brazil[268], *before fermentation

incidence: 1/51*, conc.: 0.06 μg/kg, sample origin: farms in Bahia (state), Brazil, sample year: unknown, country: Brazil[268], *during fermentation

incidence: 0/85*, conc.: no contamination, sample origin: farms in Bahia (state), Brazil, sample year: unknown, country: Brazil[268], *during sun drying

incidence: 0/65*, conc.: no contamination, sample origin: farms in Bahia (state), Brazil, sample year: unknown, country: Brazil[268], *during storage

- Co-contamination: not reported
- Further contamination of plants and plant products other than cereals and their products reported in the present book/article[268]: Cocoa beans, AFB_1 and AFB_2, literature[268]

incidence: 7*/15**, conc. range: ≤0.43 μg/kg, sample origin: Canada, sample year: unknown, country: Canada[864], *n>LOQ, **cocoa, natural

- Co-contamination: For detailed information see the article.
- Further contamination of plants and plant products other than cereals and their products reported in the present book/article[864]: Butter (cocoa butter), OTA, literature[864]; Chocolate, Products (cocoa product), AFB_1, AFB_2, AFG_1, AFG_2, and OTA, literature[864]; Cocoa beans, AFB_1, AFB_2, AFG_2, and OTA, literature[864]; Powder (cocoa powder), AFB_1, AFB_2, AFG_1, and OTA, literature[864]

Aflatoxin G_2

incidence: 2*/15**, conc. range: ≤0.07 μg/kg, sample origin: Canada, sample year: unknown, country: Canada[864], *n>LOQ, **cocoa, natural

- Co-contamination: For detailed information see the article.
- Further contamination of plants and plant products other than cereals and their products reported in the present book/article[864]: Butter (cocoa butter), OTA, literature[864]; Chocolate, Products (cocoa product), AFB_1, AFB_2, AFG_1, AFG_2, and OTA, literature[864]; Cocoa beans, Powder (cocoa powder), AFB_1, AFB_2, AFG_1, and OTA, literature[864]

Aflatoxin/s

incidence: 2/47, conc. range: 5.0–9.9 μg/kg* (2 sa.), sample origin: brokers, distributors, growers, processors, and retail trade, Canada, sample year: 1970–1975, country: Canada[15], *AFB_1, AFB_2, AFG_1, and AFG_2

- Co-contamination: not reported
- Further contamination of plants and plant products other than cereals and their products reported in the present book/article[15]: Butter (peanut butter), Nuts (almond), Nuts (cashew nut), Nuts (peanut), Nuts (pecan nut), Nuts (pistachio nut), Nuts (walnut), Seeds (sunflower), AF/S, literature[15]

incidence: 3/91*, conc. range: 2–20 µg/kg** (1 sa.), >20 µg/kg** (2 sa.), sample origin: imported and domestic sa. from Uruguay, sample year: 1993–1995, country: Uruguay[269], *cocoa beans and by-products, **AFB_1, AFB_2, AFG_1, and AFG_2

- Co-contamination: not reported
- Further contamination of plants and plant products other than cereals and their products reported in the present book/article[269]: Beans (soybean), ZEA, literature[269]; Fruits, dried, Oil seeds, Vegetables, dried, AF/S and ZEA, literature[269]

Aspergillus and ***Penicillium*** Toxins

Ochratoxin A

incidence: 16/21, conc. range: 0.1–3.5 µg/kg, ∅ conc.: 0.45 µg/kg, sample origin: Cameroon, Guinea, Ivory Coast, Nigeria, and Senegal, sample year: unknown, country: Spain[261]

- Co-contamination: not reported
- Further contamination of plants and plant products other than cereals and their products reported in the present book/article[261]: Chocolate, Powder (cocoa powder), Products (cocoa product), OTA, literature[261]

incidence: 24/33, conc. range: 0.041–14.8 µg/kg, ∅ conc.: 1.85 µg/kg, sample origin: Ivory Coast, sa. imported from a national import factory, Spain, sample year: 2003/2004, country: Spain[270]

incidence: 3/7, conc. range: 0.055–3.88 µg/kg, ∅ conc.: 1.55 µg/kg, sample origin: Cameroon, sa. imported from a national import factory, Spain, sample year: 2003/2004, country: Spain[270]

incidence: 2/6, conc. range: 0.107–0.416 µg/kg, ∅ conc.: 0.26 µg/kg, sample origin: Equatorial Guinea, sa. imported from a national import factory, Spain, sample year: 2003/2004, country: Spain[270]

- Co-contamination: not reported
- Further contamination of plants and plant products other than cereals and their products reported in the present book/article[270]: not reported

incidence: 0/25*, conc.: no contamination, sample origin: farms in Bahia (state), Brazil, sample year: 2006–2008, country: Brazil/Australia[272], *before fermentation

incidence: 14/51*, conc. range: ≤1.7 µg/kg**, sample origin: farms in Bahia (state), Brazil, sample year: 2006–2008, country: Brazil/Australia[272], *sa. remained for fermentation in wooden boxes (1–6 days), **only 3 sa. above 0.10 µg/kg

incidence: 41/8*, conc. range: ≤5.54 µg/kg**, sample origin: farms in Bahia (state), Brazil, sample year: 2006–2008, country: Brazil/Australia[272], *sa. remained on wooden floor platforms with movable roofs for sun drying (1–12 days), **30 sa. lower than 0.10 µg/kg

incidence: 33/65*, conc. range: ≤4.64 µg/kg**, sample origin: farms in Bahia (state), Brazil, sample year: 2006–2008, country: Brazil/Australia[272], *dried cocoa bean sa. in storage, **only 1 sa. with the high value of 4.64 µg/kg

- Co-contamination: not reported
- Further contamination of plants and plant products other than cereals and their products reported in the present book/article[272]: not reported

incidence: 5/9*, conc. range: >LOD<LOQ (5 sa.), sample origin: south of Bahia (state), Brazil, sample year: unknown, country: Brazil/France[705], *conventional production, early crop season

incidence: 5/9*, conc. range: >LOD<LOQ (3 sa.), 0.2–2.0 µg/kg (2 sa.), sample origin: south of Bahia (state), Brazil, sample year: unknown, country: Brazil/France[705], *conventional production, harvest season

incidence: 7/9*, conc. range: >LOD<LOQ (3 sa.), 0.2–2.0 µg/kg (4 sa.), sample origin: south of Bahia (state), Brazil, sample year: unknown, country: Brazil/France[705], *organic production, early crop season

incidence: 7/9*, conc. range: >LOD<LOQ (5 sa.), 0.2–2.0 µg/kg (2 sa.), sample origin: south of Bahia (state), Brazil, sample year: unknown, country: Brazil/France[705], *organic production, harvest season

incidence: 9/9*, conc. range: 0.2–2.0 µg/kg (5 sa.), >2.0 µg/kg (4 sa.), sample origin: south of Bahia (state), Brazil, sample year: unknown, country: Brazil/France[705], *fine and flavor production, early crop season

incidence: 5/9*, conc. range: >LOD<LOQ (2 sa.), 0.2–2.0 µg/kg (3 sa.), sample origin: south of Bahia (state), Brazil, sample year: unknown, country: Brazil/France[705], *fine and flavor production, harvest season

- Co-contamination: not reported
- Further contamination of plants and plant products other than cereals and their products reported in the present book/article[705]: not reported

incidence: 15*/15**, conc. range: ≤4.72 µg/kg, sample origin: Canada, sample year: unknown, country: Canada[864], *n>LOQ, **cocoa, natural

- Co-contamination: For detailed information see the article.
- Further contamination of plants and plant products other than cereals and their products reported in the present book/article[864]: Butter (cocoa butter), OTA, literature[864]; Chocolate, Products (cocoa product), AFB_1, AFB_2, AFG_1, AFG_2, and OTA, literature[864]; Cocoa beans, AFB_1, AFB_2, AFG_1, and AFG_2, literature[864]; Powder (cocoa powder), AFB_1, AFB_2, AFG_1, and OTA, literature[864]

Cocoa hazelnut creme see **Nut creme**

Cocoa products see **Products (cocoa product)**

Coconut oil see **Oils (coconut oil)**

Coconut products see **Products (coconut product)**

Coconuts may contain the following mycotoxins:

Aspergillus Toxins

Aflatoxin B_1

incidence: 5/9, conc. range: 28–260 µg/kg, sample origin: major local retail outlets at Nainital (district), India, sample year: 1983–1985, country: India[7]

- Co-contamination: Often more than one mycotoxin was present in a sa.
- Further contamination of plants and plant products other than cereals and their products reported in the present book/article[7]: Coconuts, AFB_2, AFG_1, AFG_2, and CIT, literature[7];

Dates, AFB_1, AFB_2, AFG_1, and OTA, literature[7]; Grogannuts, AFB_1, AFB_2, AFG_1, AFG_2, and CIT, literature[7]; Nuts (almond), AFB_1, AFB_2, AFG_1, RBT, STG, and ZEA, literature[7]; Nuts (cashew nut), AFB_1, AFB_2, AFG_1, AFG_2, CIT, and OTA, literature[7]; Nuts (peanut), Raisins, AFB_1, AFB_2, AFG_1, AFG_2, and OTA, literature[7]; Nuts (pistachio nut), AFB_1, AFB_2, AFG_1, AFG_2, and STG, literature[7]; Nuts (walnut), AFB_1, AFB_2, AFG_1, AFG_2, and RBT, literature[7]

incidence: 5/25, conc. range: 15–25 µg/kg, sample origin: markets of different sanitation levels in the Upper Egyptian Governorates, Egypt, sample year: March 1992, country: Egypt[275]

- Co-contamination: not reported
- Further contamination of plants and plant products other than cereals and their products reported in the present book/article[275]: Coconuts, OTA, literature[275]

incidence: 1/1, conc.: 0.65 µg/kg, sample origin: local supermarket in Okayama (prefecture), Japan, sample year: unknown, country: Japan[339]

- Co-contamination: 1 sa. co-contaminated with AFB_1 and AFG_2
- Further contamination of plants and plant products other than cereals and their products reported in the present book/article[339]: Coconuts, Raisins, AFG_2, literature[339]; Figs, AFB_2 and AFG_2, literature[339]; Fruits, dried, Seeds (sesame), AFB_1 and AFG_2, literature[339]; Flour (soybean flour), AFB_1, literature[339]; Nuts, Spices, AFB_1, AFB_2, AFG_1, and AFG_2, literature[339]

Aflatoxin B_2

incidence: 5/9, conc. range: 15–109 µg/kg, sample origin: major local retail outlets at Nainital (district), India, sample year: 1983–1985, country: India[7]

- Co-contamination: Often more than one mycotoxin was present in a sa.
- Further contamination of plants and plant products other than cereals and their products reported in the present book/article[7]: Coconuts, AFB_1, AFG_1, AFG_2, and CIT, literature[7]; Dates, AFB_1, AFB_2, AFG_1, and OTA, literature[7]; Grogannuts, AFB_1, AFB_2, AFG_1, AFG_2, and CIT, literature[7]; Nuts (almond), AFB_1, AFB_2, AFG_1, RBT, STG, and ZEA, literature[7]; Nuts (cashew nut), AFB_1, AFB_2, AFG_1, AFG_2, CIT, and OTA, literature[7]; Nuts (peanut), Raisins, AFB_1, AFB_2, AFG_1, AFG_2, and OTA, literature[7]; Nuts (pistachio nut), AFB_1, AFB_2, AFG_1, AFG_2, and STG, literature[7]; Nuts (walnut), AFB_1, AFB_2, AFG_1, AFG_2, and RBT, literature[7]

Aflatoxin G_1

incidence: 5/9, conc. range: 12–75 µg/kg, sample origin: major local retail outlets at Nainital (district), India, sample year: 1983–1985, country: India[7]

- Co-contamination: Often more than one mycotoxin was present in a sa.
- Further contamination of plants and plant products other than cereals and their products reported in the present book/article[7]: Coconuts, AFB_1, AFB_2, AFG_2, and CIT, literature[7]; Dates, AFB_1, AFB_2, AFG_1, and OTA, literature[7]; Grogannuts, AFB_1, AFB_2, AFG_1, AFG_2, and CIT, literature[7]; Nuts (almond), AFB_1, AFB_2, AFG_1, RBT, STG, and ZEA, literature[7]; Nuts (cashew nut), AFB_1, AFB_2, AFG_1, AFG_2, CIT, and OTA, literature[7]; Nuts (peanut), Raisins, AFB_1, AFB_2, AFG_1, AFG_2, and OTA, literature[7]; Nuts (pistachio nut), AFB_1, AFB_2, AFG_1, AFG_2, and STG, literature[7]; Nuts (walnut), AFB_1, AFB_2, AFG_1, AFG_2, and RBT, literature[7]

Aflatoxin G_2

incidence: 2/9, conc. range: 20–68 µg/kg, Ø conc.: 44 µg/kg, sample origin: major local retail outlets at Nainital (district), India, sample year: 1983–1985, country: India[7]

- Co-contamination: Often more than one mycotoxin was present in a sa.
- Further contamination of plants and plant products other than cereals and their products reported in the present book/article[7]: Coconuts, AFB_1, AFB_2, AFG_1, and CIT, literature[7]; Dates, AFB_1, AFB_2, AFG_1, and OTA, literature[7]; Grogannuts, AFB_1, AFB_2, AFG_1, AFG_2, and CIT, literature[7]; Nuts (almond), AFB_1, AFB_2, AFG_1, RBT, STG, and ZEA, literature[7]; Nuts (cashew nut), AFB_1, AFB_2, AFG_1, AFG_2, CIT, and OTA, literature[7]; Nuts (peanut), Raisins, AFB_1, AFB_2, AFG_1, AFG_2, and OTA, literature[7]; Nuts (pistachio nut), AFB_1, AFB_2, AFG_1, AFG_2, and STG, literature[7]; Nuts (walnut), AFB_1, AFB_2, AFG_1, AFG_2, and RBT, literature[7]

incidence: 1/1, conc.: 1.06 µg/kg, sample origin: local supermarket in Okayama (prefecture), Japan, sample year: unknown, country: Japan[339]

- Co-contamination: 1 sa. co-contaminated with AFB_1 and AFG_2
- Further contamination of plants and plant products other than cereals and their products reported in the present book/article[339]: Coconuts, Flour (soybean flour), AFB_1, literature[339]; Figs, AFB_2 and AFG_2, literature[339]; Fruits, dried, Seeds (sesame), AFB_1 and AFG_2, literature[339]; Nuts, Spices, AFB_1, AFB_2, AFG_1, and AFG_2, literature[339] Raisins, AFG_2, literature[339]

Aspergillus and ***Penicillium*** Toxins

Citrinin

incidence: 4/9, conc. range: 5–60 µg/kg, sample origin: major local retail outlets at Nainital (district), India, sample year: 1983–1985, country: India[7]

- Co-contamination: Often more than one mycotoxin was present in a sa.
- Further contamination of plants and plant products other than cereals and their products reported in the present book/article[7]: Coconuts, AFB_1, AFB_2, AFG_1, and AFG_2, literature[7]; Dates, AFB_1, AFB_2, AFG_1, and OTA, literature[7]; Grogannuts, AFB_1, AFB_2, AFG_1, AFG_2, and CIT, literature[7]; Nuts (almond), AFB_1, AFB_2, AFG_1, RBT, STG, and ZEA, literature[7]; Nuts (cashew nut), AFB_1, AFB_2, AFG_1, AFG_2, CIT, and OTA, literature[7]; Nuts (peanut), Raisins, AFB_1, AFB_2, AFG_1, AFG_2, and OTA, literature[7]; Nuts (pistachio nut), AFB_1, AFB_2, AFG_1, AFG_2, and STG, literature[7]; Nuts (walnut), AFB_1, AFB_2, AFG_1, AFG_2, and RBT, literature[7]

Ochratoxin A

incidence: 3/25, conc. range: 50–205 µg/kg, sample origin: markets of different sanitation levels in the Upper Egyptian Governorates, Egypt, sample year: March 1992, country: Egypt[275]

- Co-contamination: not reported
- Further contamination of plants and plant products other than cereals and their products reported in the present book/article[275]: Coconuts, AFB_1, literature[275]

Coffee may contain the following mycotoxins:

Aspergillus Toxins

Aflatoxin B_1

incidence: 2/4*, conc. range: 21–100 µg/kg (2 sa., maximum: 37 µg/kg), sample origin: silos, homes, and small shops in different

villages on the southern coast, Guatemala, sample year: 1976, country: Guatemala[136], *sa. stored for 6 months during dry season
- Co-contamination: not reported
- Further contamination of plants and plant products other than cereals and their products reported in the present book/article[136]: Beans (black bean), AF/S, literature[136]; Seeds (sesame), AFB_1, literature[136]

incidence: 9/30*, conc. range: 0.36–3.01 µg/kg, ∅ conc.: 1.4 µg/kg, sample origin: local markets in Giza (city) and Cairo (capital), Egypt, sample year: unknown, country: Egypt[279], * coffee (beans), ground and roasted
- Co-contamination: 2 sa. co-contaminated with AFB_1, AFB_2, AFG_1, and AFG_2; 1 sa. co-contaminated with AFB_1, AFB_2, and AFG_1; 1 sa. co-contaminated with AFB_1, AFB_2, and AFG_2; 3 sa. co-contaminated with AFB_1 and AFB_2; 1 sa. co-contaminated with AFB_1 and AFG_1; 1 sa. co-contaminated with AFB_1 and AFG_2
- Further contamination of plants and plant products other than cereals and their products reported in the present book/article[279]: Coffee, AFB_2, AFG_1, and AFG_2, literature[279]; Coffee beans, AFB_1, AFB_2, AFG_1, and AFG_2, literature[279]

incidence: 37/169*, conc. range: 0.75–1.14 µg/kg, ∅ conc.: 0.91 µg/kg, sample origin: supermarkets in Valencia (city), Spain, sample year: unknown, country: Spain[731], *coffee, brewed
- Co-contamination: not reported
- Further contamination of plants and plant products other than cereals and their products reported in the present book/article[731]: Coffee, AFB_2, AFG_1, AFG_2, BEA, DAS, DON, 3-AcDON, 15-AcDON, ENA, ENA_1, ENB, ENB_1, FB_1, FB_2, HT-2, NEO, NIV, OTA, STG, and T-2, literature[731]

For detailed information see the article.

Aflatoxin B_2

incidence: 7/30*, conc. range: 0.52–1.52 µg/kg, ∅ conc.: 0.8 µg/kg, sample origin: local markets in Giza (city) and Cairo (capital), Egypt, sample year: unknown, country: Egypt[279], *coffee (beans), ground and roasted
- Co-contamination: 2 sa. co-contaminated with AFB_1, AFB_2, AFG_1, and AFG_2; 1 sa. co-contaminated with AFB_1, AFB_2, and AFG_1; 1 sa. co-contaminated with AFB_1, AFB_2, and AFG_2; 3 sa. co-contaminated with AFB_1 and AFB_2
- Further contamination of plants and plant products other than cereals and their products reported in the present book/article[279]: Coffee, AFB_1, AFG_1, and AFG_2, literature[279]; Coffee beans, AFB_1, AFB_2, AFG_1, and AFG_2, literature[279]

incidence: 17/169*, conc. range: 0.51–1.57 µg/kg, ∅ conc.: 0.69 µg/kg, sample origin: supermarkets in Valencia (city), Spain, sample year: unknown, country: Spain[731], *coffee, brewed
- Co-contamination: not reported
- Further contamination of plants and plant products other than cereals and their products reported in the present book/article[731]: Coffee, AFB_1, AFG_1, AFG_2, BEA, DAS, DON, 3-AcDON, 15-AcDON, ENA, ENA_1, ENB, ENB_1, FB_1, FB_2, HT-2, NEO, NIV, OTA, STG, and T-2, literature[731]

For detailed information see the article.

Aflatoxin G_1

incidence: 5/30*, conc. range: 0.80–2.91 µg/kg, ∅ conc.: 1.53 µg/kg, sample origin: local markets in Giza (city) and Cairo (capital), Egypt, sample year: unknown, country: Egypt[279], *coffee (beans), ground and roasted
- Co-contamination: 2 sa. co-contaminated with AFB_1, AFB_2, AFG_1, and AFG_2; 1 sa. co-contaminated with AFB_1, AFB_2, and AFG_1; 1 sa. co-contaminated with AFB_1 and AFG_1; 1 sa. co-contaminated with AFG_1 and AFG_2
- Further contamination of plants and plant products other than cereals and their products reported in the present book/article[279]: Coffee, AFB_1, AFB_2, and AFG_2, literature[279]; Coffee beans, AFB_1, AFB_2, AFG_1, and AFG_2, literature[279]

incidence: 25/169*, conc. range: 2.54–50.80 µg/kg, ∅ conc.: 3.05 µg/kg?, sample origin: supermarkets in Valencia (city), Spain, sample year: unknown, country: Spain[731], *coffee, brewed
- Co-contamination: not reported
- Further contamination of plants and plant products other than cereals and their products reported in the present book/article[731]: Coffee, AFB_1, AFB_2, AFG_2, BEA, DAS, DON, 3-AcDON, 15-AcDON, ENA, ENA_1, ENB, ENB_1, FB_1, FB_2, HT-2, NEO, NIV, OTA, STG, and T-2, literature[731]

For detailed information see the article.

Aflatoxin G_2

incidence: 7/30*, conc. range: 0.13–2.68 µg/kg, ∅ conc.: 1.2 µg/kg, sample origin: local markets in Giza (city) and Cairo (capital), Egypt, sample year: unknown, country: Egypt[279], *coffee (beans), ground and roasted
- Co-contamination: 2 sa. co-contaminated with AFB_1, AFB_2, AFG_1, and AFG_2; 1 sa. co-contaminated with AFB_1, AFB_2, and AFG_2; 1 sa. co-contaminated with AFB_1 and AFG_2; 1 sa. co-contaminated with AFG_1 and AFG_2; 1 sa. contaminated solely with AFG_2
- Further contamination of plants and plant products other than cereals and their products reported in the present book/article[279]: Coffee, AFB_1, AFB_2, and AFG_1, literature[279]; Coffee beans, AFB_1, AFB_2, AFG_1, and AFG_2, literature[279]

incidence: 42/169*, conc. range: 0.99–25.87 µg/kg, ∅ conc.: 4.03 µg/kg, sample origin: supermarkets in Valencia (city), Spain, sample year: unknown, country: Spain[731], *coffee, brewed
- Co-contamination: not reported
- Further contamination of plants and plant products other than cereals and their products reported in the present book/article[731]: Coffee, AFB_1, AFB_2, AFG_1, BEA, DAS, DON, 3-AcDON, 15-AcDON, ENA, ENA_1, ENB, ENB_1, FB_1, FB_2, HT-2, NEO, NIV, OTA, STG, and T-2, literature[731]

For detailed information see the article.

Aflatoxin/s

incidence: 9/17*, ∅ conc.: 11.2 µg/kg**, sample origin: greater Manila (capital) area, Philippines, sample year: unknown, country: Philippines[1], *instant coffee, **AF of pos sa.?
- Co-contamination: not reported
- Further contamination of plants and plant products other than cereals and their products reported in the present book/article[1]: Beans, Butter (peanut butter), Cassava, Cocoa, Drinks (alcoholic drink), Nuts (cashew nut), Nuts (peanut), Products (coconut product), Rootcrops, Spices (cayenne pepper, chilli), Sweet potatoes, Yams, AF/S, literature[1]

Other *Aspergillus* Toxins

Sterigmatocystin

incidence: 27/169*, conc. range: 7.65–63.19 µg/kg, ∅ conc.: 22.63 µg/kg, sample origin: supermarkets in Valencia (city), Spain, sample year: unknown, country: Spain[731], *coffee, brewed

- Co-contamination: not reported
- Further contamination of plants and plant products other than cereals and their products reported in the present book/article[731]: Coffee, AFB_1, AFB_2, AFG_1, AFG_2, BEA, DAS, DON, 3-AcDON, 15-AcDON, ENA, ENA_1, ENB, ENB_1, FB_1, FB_2, HT-2, NEO, NIV, OTA, and T-2, literature[731]

For detailed information see the article.

Aspergillus and ***Penicillium*** Toxins

Ochratoxin A

incidence: 2/4*, conc. range: 10–90 µg/kg, ∅ conc.: 50 µg/kg, sample origin: local marketing institutions, GDR, sample year: unknown, country: GDR[29], *moldy

- Co-contamination: not reported
- Further contamination of plants and plant products other than cereals and their products reported in the present book/article[29]: Butter (peanut butter), Nuts (peanut), Products (peanut product), Seeds (sunflower), AFB_1, AFB_2, AFG_1, and AFG_2, literature[29]; Fruits, AFB_1, AFG_1, and AFG_2, literature[29]; Fruits (apple), Fruits (peach), Juices (apple juice), PAT, literature[29]

incidence: 18/49*, conc. range: ≤2.75 µg/kg, ∅ conc.: 0.55 µg/kg, sample origin: local supermarkets and small retail shops in seven major cities, Japan, sample year: summer 2004–winter 2007, country: Japan[85], *coffee beans, roasted

incidence: 63/66*, conc. range: ≤4.23 µg/kg, ∅ conc.: 0.72 µg/kg, sample origin: local supermarkets and small retail shops in seven major cities, Japan, sample year: 2005–winter 2007, country: Japan[85], *instant coffee

- Co-contamination: not reported
- Further contamination of plants and plant products other than cereals and their products reported in the present book/article[85]: Beans (soybean), Vegetables (asparagus), FB_1 and FB_2, literature[85]; Beverages, Chocolate, Cocoa, Coffee beans, Raisins, Spices (coriander), Wine, OTA, literature[85]

incidence: 1/1, conc.: 0.02 µg/l, sample origin: supermarkets and fast-food chains in Quebec (city), Canada, sample year: September/October 2008, country: Canada[90]

incidence: 1/1, conc.: 0.01 µg/l, sample origin: supermarkets and fast-food chains in Calgary (city), Canada, sample year: September/October 2009, country: Canada[90]

- Co-contamination: not reported
- Further contamination of plants and plant products other than cereals and their products reported in the present book/article[90]: Baby food, Beans, Beverages, Butter (peanut butter), Candies, Chocolate, Extracts, Herbs, Ice cream, Nuts, Pies, Pizza, Products, Pudding, Raisins, Seeds, Wine, OTA, literature[90]

incidence: 2/10*, conc. range: 0.51–0.81 µg/kg, ∅ conc.: 0.66 µg/kg, sample origin: unknown, sample year: unknown, country: Japan[112], *coffee, raw

incidence: 3/23*, conc. range: 0.25–0.43 µg/kg, sample origin: unknown, sample year: unknown, country: Japan[112], *coffee, roasted

incidence: 5/7*, conc. range: 0.16–1.1 µg/kg, sample origin: unknown, sample year: unknown, country: Japan[112], *instant coffee

- Co-contamination: not reported
- Further contamination of plants and plant products other than cereals and their products reported in the present book/article[112]: Cassis, Juices, Raisins, Wine, OTA, literature[112]; Cocoa, Food, Vinegar, OTA and OTB, literature[112]; Coffee, OTB, literature[112]; Powder (cocoa powder), AFB_1, AFB_2, AFG_1, OTA, and OTB, literature[112]

incidence: 12/12*, conc. range: 0.03–0.6 µg/kg, ∅ conc.: 0.27 µg/kg, sample origin: Warsaw (capital) market or food producers, Poland, sample year: 2004, country: Poland[116], *coffee, roasted

incidence: 6/6*, conc. range: 0.02–11.90 µg/kg, ∅ conc.: 3.11 µg/kg, sample origin: Warsaw (capital) market or from food producers, Poland, sample year: 2005, country: Poland[116], *coffee and cocoa

- Co-contamination: not reported
- Further contamination of plants and plant products other than cereals and their products reported in the present book/article[116]: Food, AFB_1 and AFG_1, literature[116]; Nuts, Spices, AFB_1, AFB_2, AFG_1, and AFG_2, literature[116]; Raisins, OTA, literature[116]

incidence: 3/9*, conc. range: 0.11–0.33 µg/kg, ∅ conc.: 0.22 µg/kg, sample origin: local supermarkets and small retail shops across Japan, sample year: summer 2004–winter 2005, country: Japan[146], *coffee, roasted

- Co-contamination: not reported
- Further contamination of plants and plant products other than cereals and their products reported in the present book/article[146]: Beans (green bean), Raisins, Wine, OTA, literature[146]; Butter (peanut butter), AFB_1, AFB_2, AFG_1, and AFG_2, literature[146]; Chocolate, AFB_1 and OTA, literature[146]

incidence: 2/2, conc. range: 15.2*–53.6** µg/kg***, sample origin: stores, Taiwan, sample year: unknown, country: Taiwan/Russia[147], *instant and **roasted coffee, ***COL-ELISA

incidence: 2/3, conc. range: 0.96*–3.36** µg/kg***, sample origin: stores, Taiwan, sample year: unknown, country: Taiwan/Russia[147], *unroasted and **instant coffee, ***CL-ELISA

- Co-contamination: not reported
- Further contamination of plants and plant products other than cereals and their products reported in the present book/article[147]: Beans (kidney beans), Berries (grape), Raisins, OTA, literature[147]

incidence: 64/80*, conc. range: LOD/LOQ-4.9 µg/kg (61 sa.), 5.5–9.9 µg/kg (3 sa., maximum: 8.0 µg/kg), sample origin: UK, sample year: 1994, country: EU[194], *coffee, soluble

incidence: 17/20*, conc. range: LOD/LOQ-4.9 µg/kg (maximum: 2.1 µg/kg), sample origin: UK, sample year: 1994, country: EU[194], *coffee, roasted and grounded

- Co-contamination: not reported
- Further contamination of plants and plant products other than cereals and their products reported in the present book/article[194]: Coffee beans, Spices (cayenne pepper, chilli), Spices (curry), Spices (majoran), Spices (pepper), OTA, literature[194]

incidence: 52/113*, conc. range: ≤6.32 µg/kg, ∅ conc.: 1.15 µg/kg, sample origin: retail shops, Germany, sample year: unknown, country: Germany[197], *coffee, roasted and ground

incidence: 32/67*, conc. range: ≤3.34 µg/kg, ∅ conc.: 1.01 µg/kg, sample origin: retail shops, Germany, sample year: unknown, country: Germany[197], *coffee, roasted, ground and decaffeinated

incidence: 21/60*, conc. range: ≤4.75 µg/kg, ∅ conc.: 1.0 µg/kg, sample origin: retail shops, Germany, sample year: unknown, country: Germany[197], *coffee, mild, roasted, and ground

incidence: 46/52*, conc. range: ≤9.47 µg/kg, ∅ conc.: 2.05 µg/kg, sample origin: retail shops, Germany, sample year: unknown, country: Germany[197], *instant coffee incidence: 19/32*, conc. range: ≤1.8 µg/kg, ∅ conc.: 0.89 µg/kg, sample origin: retail shops, Germany, sample year: unknown, country: Germany[197], *instant coffee, decaffeinated

incidence: 5/33*, conc. range: ≤0.96 µg/kg, ∅ conc.: 0.65 µg/kg, sample origin: retail shops, Germany, sample year: unknown, country: Germany[197], *malt coffee

- Co-contamination: not reported
- Further contamination of plants and plant products other than cereals and their products reported in the present book/article[197]: Tea, OTA, literature[197]

incidence: 1/3*, conc.: 0.20 µg/kg, sample origin: sales outles, Qatar, sample year: October 2002, country: Qatar[229], *instant coffee

- Co-contamination: not reported
- Further contamination of plants and plant products other than cereals and their products reported in the present book/article[229]: Butter (peanut butter), Figs, Nuts (peanut), Nuts (pistachio nut), AF/S, literature[229]; Raisins, OTA, literature[229]; Spices, Spices (cayenne pepper, chilli), AF/S and OTA, literature[229]

incidence: 5/8*, conc. range: 0.28–1.48 µg/kg, sample origin: available in New Zealand, sample year: 2007, country: New Zealand[277], *coffee, various

incidence: 1/8*, conc.: 0.09 µg/kg, sample origin: available in New Zealand, sample year: 2011, country: New Zealand[277], *coffee, brewed

- Co-contamination: not reported
- Further contamination of plants and plant products other than cereals and their products reported in the present book/article[277]: Bars, AFB_1, AF/S, DON, NIV, and OTA, literature[277]; Butter (peanut butter), Confectionery, Food, Fruits, dried, Nuts, Nuts (peanut), Nuts (pistachio nut), Sauces (peanut sauce), AFB_1 and AF/S, literature[277]; Dates, Spices (nutmeg), Spices (pepper), Vine fruits, dried, OTA, literature[277]; Figs, Spices (cayenne pepper, chilli), Spices (curry), Spices (ginger), Spices (paprika), AFB_1, AF/S and OTA, literature[277]; Nuts (Brazil nut, paranut), AFB_1, literature[277]

incidence: 16/16*, conc. range: 0.5–5.1 µg/kg, ∅ conc.: 2.2 µg/kg, sample origin: food retail stores in Campinas (city), Brazil, sample year: unknown, country: Brazil[281], *instant coffee

incidence: 23/34* **, conc. range: 0.3–6.5 µg/kg, ∅ conc.: 1.4 µg/kg, sample origin: food retail stores in Campinas (city), Brazil, sample year: unknown, country: Brazil[281], *included 2 sa. decaffeinated coffee (not contaminated), **coffee, roasted and ground

- Co-contamination: not reported
- Further contamination of plants and plant products other than cereals and their products reported in the present book/article[281]: not reported

incidence: 2/2*, conc. range: 3.8–23.0 µg/kg, ∅ conc.: 13.4 µg/kg, sample origin: Ivory Coast and Uganda, sample year: unknown, country: Italy[282], *coffee, roasted

- Co-contamination: not reported
- Further contamination of plants and plant products other than cereals and their products reported in the present book/article[282]: Coffee beans, OTA, literature[282]

incidence: 22/38*, conc. range: 0.17–0.91 µg/kg, ∅ conc.: 0.5 µg/kg, sample origin: food stores in Debrecen (city) and Budapest (capital), Hungary, sample year: 2001, country: Hungary[283], *coffee, roasted

incidence: 11/12*, conc. range: 0.34–1.3 µg/kg, ∅ conc.: 0.72 µg/kg, sample origin: food stores in Debrecen (city) and Budapest (capital), Hungary, sample year: 2001, country: Hungary[283], *real coffee blended with "Ersatz coffee" = roasted barley (substitute)

- Co-contamination: not reported
- Further contamination of plants and plant products other than cereals and their products reported in the present book/article[283]: not reported

incidence: 5/16*, conc. range: 0.1–0.3 µg/kg (2 sa.), 0.3–0.5 µg/kg (3 sa.), sample origin: supermarkets in eight counties, Taiwan, sample year: spring and fall 2003, country: Taiwan[284], *instant coffee

incidence: 8/19*, conc. range: 0.1–0.3 µg/kg (5 sa.), 0.3–0.5 µg/kg (3 sa.), sample origin: supermarkets in eight counties, Taiwan, sample year: spring and fall 2003, country: Taiwan[284], *coffee, ready-to-drink

- Co-contamination: not reported
- Further contamination of plants and plant products other than cereals and their products reported in the present book/article[284]: Wine (red), OTA, literature[284]

incidence: 11/11*, conc. range: LOD-0.99 µg/kg (10 sa.), 3.2 µg/kg (1 sa.), sample origin: retail shops, Denmark, sample year: 1995, country: Denmark[285], *coffee, roasted

- Co-contamination: not reported
- Further contamination of plants and plant products other than cereals and their products reported in the present book/article[285]: not reported

incidence: 38/59*, conc. range: <0.1–2.3 µg/kg, ∅ conc.: 0.63 µg/kg, sample origin: retail location across Canada, sample year: unknown, country: Canada[287], *coffee, non-decaffeinated, roasted and ground

incidence: 4/12*, conc. range: <0.1–1.3 µg/kg, ∅ conc.: 0.55 µg/kg, sample year: unknown, country: Canada[287], *coffee, decaffeinated, roasted and ground

incidence: 15/21*, conc. range: <0.1–3.1 µg/kg, ∅ conc.: 1.11 µg/kg, sample year: unknown, country: Canada[287], *instant coffee, non-decaffeinated

incidence: 5/9*, conc. range: <0.1–2.5 µg/kg, ∅ conc.: 0.9 µg/kg, sample year: unknown, country: Canada[287], *instant coffee, decaffeinated

- Co-contamination: not reported
- Further contamination of plants and plant products other than cereals and their products reported in the present book/article[287]: not reported

incidence: 191/419*, conc. range: 0.21–12.1 µg/kg, sample origin: retail stores and coffee roasters, Germany, sample year: 1995–1999, country: Germany[288], *coffee, roasted

incidence: 26/71* **, conc. range: 0.15–2.7 µg/kg, sample origin: retail stores and coffee roasters, Germany sample year: 1995–1999, country: Germany[288], *coffee, decaffeinated and roasted, as well as low-acid coffee, decaffeinated and roasted

incidence: 29/41*, conc. range: 0.28–4.8 µg/kg, sample origin: retail stores and coffee roasters, Germany sample year: 1995–1999, country: Germany[288], *instant coffee

- Co-contamination: not reported
- Further contamination of plants and plant products other than cereals and their products reported in the present book/article[288]: Coffee beans, OTA, literature[288]

incidence: 11/12*, conc. range: 0.2–8.0 µg/kg, sample origin: retail outlets across the UK, sample year: 1995, country: UK[289], *coffee powder, decaffeinated and non-decaffeinated

incidence: 31/36*, conc. range: 0.1–4.9 µg/kg, sample origin: retail outlets across the UK, sample year: 1995, country: UK[289], *coffee granules, decaffeinated and non-decaffeinated

incidence: 22/32* **, conc. range: 0.2–3.0 µg/kg, sample origin: retail outlets across the UK, sample year: 1995, country: UK[289], *freeze-dried coffee, decaffeinated and non-decaffeinated

incidence: 17/20*, conc. range: 0.2–2.1 µg/kg, sample origin: retail outlets across the UK, sample year: 1995, country: UK[289], *coffee, roasted and ground, non-decaffeinated

- Co-contamination: not reported
- Further contamination of plants and plant products other than cereals and their products reported in the present book/ article[289]: not reported

incidence: 75/101*, conc. range: 0.2–6.5 µg/kg, sample origin: different countries, sample year: unknown, country: Switzerland[290], *coffee, pure soluble

incidence: 15/15*, conc. range: 1.2–15.9 µg/kg, ∅ conc.: 5.9 µg/kg, sample origin: Czech Republic, Hungary, and Russia, sample year: unknown, country: Switzerland[290], *coffee, adulterated soluble

- Co-contamination: not reported
- Further contamination of plants and plant products other than cereals and their products reported in the present book/ article[290]: not reported

incidence: 334/633*, conc. range: ≤27.2 µg/kg, sample origin: retail markets in different European countries, sample year: 1995, country: Netherlands/Germany/Switzerland/Italy[291], *partly roasted, ground, decaffeinated, instant, or mixed coffee sa.

- Co-contamination: not reported
- Further contamination of plants and plant products other than cereals and their products reported in the present book/ article[291]: not reported

For detailed information see the article.

incidence: 9/13*, conc. range: 0.1–1.2 µg/kg (9 sa.), ∅ conc.: 0.41 µg/kg, sample origin: coffee importer (imported from South America), sample year: unknown, country: USA[292], *coffee, roasted

- Co-contamination: not reported
- Further contamination of plants and plant products other than cereals and their products reported in the present book/ article[292]: Coffee beans, OTA, literature[292]

incidence: 5/68, conc. range: 3.2–17.0 µg/kg*, ∅ conc.: 7.6 µg/kg*, sample origin: imports from different countries purchased on a market in Nagoya (city), Japan, sample year: January/February 1987, country: Japan[293], *contaminated sa. from Indonesia and Yemen

- Co-contamination: not reported
- Further contamination of plants and plant products other than cereals and their products reported in the present book/ article[293]: not reported

incidence: 81/82*, conc. range: 0.16–0.50 µg/kg (13 sa.), 0.51–1.00 µg/kg (20 sa.), 1.01–1.50 µg/kg (22 sa.), 1.51–2.00 µg/kg (18 sa.), 2.01–2.50 µg/kg (3 sa.), 2.51–5.00 µg/kg (3 sa.), 6.29 µg/kg (1 sa.), sample origin: supermarkets in São Paulo (city), Brazil, sample year: August–December 2004, country: Brazil[294], *instant coffee

- Co-contamination: not reported
- Further contamination of plants and plant products other than cereals and their products reported in the present book/ article[294]: not reported

incidence: 11/24*, conc. range: 0.3–9.4 µg/kg, ∅ conc.: 1.75 µg/kg, sample origin: commercially available, sample year: unknown, country: France[295], *coffee, roasted and ground

- Co-contamination: not reported
- Further contamination of plants and plant products other than cereals and their products reported in the present book/ article[295]: not reported

incidence: 28/28*, conc. range: 0.22–5.64 µg/kg, ∅ conc.: 0.88 µg/kg, sample origin: coffee roasters, Spain, sample year: 1997, country: Spain[296], *coffee, roasted

incidence: 9/9*, conc. range: 0.19–1.08 µg/kg, ∅ conc.: 0.50 µg/kg, sample origin: coffee roasters, Spain, sample year: 1997, country: Spain[296], *coffee, soluble

- Co-contamination: not reported
- Further contamination of plants and plant products other than cereals and their products reported in the present book/ article[296]: not reported

incidence: 43/45*, conc. range: 1.30–5.24 µg/kg, ∅ conc.: 2.07 µg/kg, sample origin: hypermarkets and supermarkets in 12 cities of Catalonia (autonomous community), Spain, sample year: November 2008, country: Spain[297], *coffee, ground (different brands)

incidence: 35/72*, conc. range: 1.21–4.21 µg/kg, ∅ conc.: 2.17 µg/kg, sample origin: hypermarkets and supermarkets in 12 cities of Catalonia (autonomous community), Spain, sample year: November 2008, country: Spain[297], *coffee, ground

- Co-contamination: not reported
- Further contamination of plants and plant products other than cereals and their products reported in the present book/ article[297]: not reported

For detailed information see the article.

incidence: 23/25*, conc. range: 0.1–1.0 µg/kg (19 sa.), 1.1–2.0 µg/kg (3 sa.), 4.1–5.0 µg/kg (1 sa.), sample origin: M/s. Fresh & Honest Café Limited, Chennai (city) and M/s. CCL Products (India) Limited, Hyderabad (city), India, sample year: unknown, country: India[298], *special coffee (monsooned)

incidence: 4/4*, conc. range: 0.1–1.0 µg/kg (4 sa.), sample origin: M/s. Fresh & Honest Café Limited, Chennai (city) and M/s. CCL Products (India) Limited, Hyderabad (city), India, sample year: unknown, country: India[298], *special coffee (organic)

incidence: 0/2*, conc.: no contamination, sample origin: M/s. Fresh & Honest Café Limited, Chennai (city) and M/s. CCL Products (India) Limited, Hyderabad (city), India, sample year: unknown, country: India[298], *coffee, roasted and ground

incidence: 13/17*, conc. range: 0.1–1.0 µg/kg (10 sa.), 1.1–2.0 µg/kg (1 sa.), 2.1–3.0 µg/kg (2 sa.), sample origin: M/s. Fresh & Honest Café Limited, Chennai (city) and M/s. CCL Products (India) Limited, Hyderabad (city), India, sample year: unknown, country: India[298], *coffee, soluble

- Co-contamination: not reported
- Further contamination of plants and plant products other than cereals and their products reported in the present book/ article[298]: Coffee beans, OTA, literature[298]

incidence: 29/30*, conc. range: ≤15.08 μg/kg**, sample origin: supermarkets, France, sample year: unknown, country: France[299], *coffee, ground, **evaluated with CEN 15141-1 Entwilse
- Co-contamination: not reported
- Further contamination of plants and plant products other than cereals and their products reported in the present book/article[299]: not reported

incidence: 9/36*, conc. range: <1.0 μg/kg (2 sa.), 1.0–2.0 μg/kg (2 sa.), >2.0 μg/kg (5 sa., maximum: 12.4 μg/kg), sample origin: commercial outlets in Karnataka (state) and Hyderabad (city) regions, India, sample year: unknown, country: India[301], *roast coffee (roast, ground, and instant coffee)
- Co-contamination: not reported
- Further contamination of plants and plant products other than cereals and their products reported in the present book/article[301]: Coffee beans, OTA, literature[301]

incidence: 1/10*, conc.: 0.5 μg/kg, sample origin: supermarkets and small shops in the northeast of China, sample year: January–June 2010, country: China[302], *coffee, soluble
- Co-contamination: not reported
- Further contamination of plants and plant products other than cereals and their products reported in the present book/article[302]: Berries (grape), Wine (red), OTA, literature[302]

incidence: 1/1*, conc.: 8.6 μg/kg, sample origin: Zaire, sample year: unknown, country: Italy[311], *Robusta (green coffee beans)

incidence: 1/1*, conc.: tr, sample origin: Zaire, sample year: unknown, country: Italy[311], *Robusta, simulated light roasted (same sa. than above)

incidence: 1/1*, conc.: 0.2 μg/kg, sample origin: Zaire, sample year: unknown, country: Italy[311], *Robusta, simulated medium roasted (same sa. than above)

incidence: 1/1*, conc.: 0.2 μg/kg, sample origin: Zaire, sample year: unknown, country: Italy[311], *Robusta, simulated dark roasted (same sa. than above)

incidence: 1/1*, conc.: 4.0 μg/kg, sample origin: Brazil, sample year: unknown, country: Italy[311], *Arabica conillon (green coffee beans)

incidence: 1/1*, conc.: 0.2 μg/kg, sample origin: Brazil, sample year: unknown, country: Italy[311], *Arabica conillon, simulated light roasted (same sa. than above)

incidence: 1/1*, conc.: 0.3 μg/kg, sample origin: Brazil, sample year: unknown, country: Italy[311], *Arabica conillon, simulated medium roasted (same sa. than above)

incidence: 1/1*, conc.: 0.3 μg/kg, sample origin: Brazil, sample year: unknown, country: Italy[311], *Arabica conillon, simulated dark roasted (same sa. than above)
- Co-contamination: not reported
- Further contamination of plants and plant products other than cereals and their products reported in the present book/article[311]: Coffee beans, OTA, literature[311]

incidence: 26/32*, conc. range: 0.24–4.27 μg/kg, ∅ conc.: 0.90 μg/kg, sample origin: conventional supermarkets and stores in Ho Chi Minh City and Can Tho City, Vietnam, sample year: 2011/2012, country: Japan[653], *coffee, roasted (Robusta)
- Co-contamination: 11 sa. co-contaminated with OTA and OTB; 15 sa. contaminated solely with OTA
- Further contamination of plants and plant products other than cereals and their products reported in the present book/article[653]: Coffee beans, OTB, literature[653]

incidence: 4/30*, conc. range: 0.40–1.00 μg/kg, ∅ conc.: 0.66 μg/kg, sample origin: Chiang Mai City, Thailand, sample year: 2012/2013, country: Japan[653], *coffee, roasted (Arabica)
- Co-contamination: not reported
- Further contamination of plants and plant products other than cereals and their products reported in the present book/article[653]: Coffee beans, OTB, literature[653]

incidence: 10/38*, conc. range: 1.28–6.20 μg/kg, ∅ conc.: 2.90 μg/kg, sample origin: Chiang Mai City and Bangkok (capital), Thailand, sample year: 2012/2013, country: Japan[653], *instant coffee (mainly Robusta)
- Co-contamination: not reported
- Further contamination of plants and plant products other than cereals and their products reported in the present book/article[653]: Coffee, OTB, literature[653]

incidence: 3/44*, conc. range: <1–4.61 μg/kg, sample origin: supermarkets in Bangkok (capital), Thailand, sample year: February/March and November/December 2005, country: Thailand/Japan[667], *imported and local roasted coffee (Arabica-Robusta mix)
- Co-contamination: not reported
- Further contamination of plants and plant products other than cereals and their products reported in the present book/article[667]: not reported

For detailed information see the article.

incidence: 10/10*, conc. range: <0.15–11.8 μg/kg, sample origin: north of Portugal, sample year: 2012, country: Portugal[691], *plain soluble coffee

incidence: 13/13*, conc. range: <0.15–6.5 μg/kg, sample origin: north of Portugal, sample year: 2012, country: Portugal[691], *substitutes with coffee

incidence: 12/17*, conc. range: <0.15–1.3 μg/kg, sample origin: north of Portugal, sample year: 2012, country: Portugal[691], *substitutes without coffee
- Co-contamination: not reported
- Further contamination of plants and plant products other than cereals and their products reported in the present book/article[691]: not reported

For detailed information see the article.

incidence: 6/24*, conc. range: ≤0.97 μg/kg, sample origin: supermarkets and fresh provision shops, Hong Kong, China, sample year: August/September 2005, country: China[697], coffee and tea
- Co-contamination: not reported
- Further contamination of plants and plant products other than cereals and their products reported in the present book/article[697]: Chocolate, Fruits, dried, Juices, Legumes, Spices, OTA, literature[697]

incidence: 5/15*, conc. range: LOD/LOQ-0.9 μg/kg (5 sa.), sample origin: Finland, sample year: 1997, country: EU[698], *coffee, roasted

incidence: 9/10*, conc. range: LOD/LOQ-0.9 μg/kg (4 sa.), 1.0–2.9 μg/kg (4 sa.), 3.0 μg/kg (1 sa.), sample origin: Finland, sample year: 1998, country: EU[698], *coffee, roasted incidence: 3/11*, conc. range: LOD/LOQ-≤0.7 µg/kg (3 sa.), sample origin: Finland, sample year: 1999, country: EU[698], *coffee, roasted

incidence: 15/25*, conc. range: LOD/LOQ-0.9 µg/kg (5 sa.), 1.0–2.9 µg/kg (12 sa.), 3.0–4.9 µg/kg (2 sa.), 5.0 µg/kg (1 sa.), sample origin: Finland, sample year: 1997, country: EU[698], *instant coffee

incidence: 6/7*, conc. range: LOD/LOQ-0.9 µg/kg (3 sa.), 1.0–2.9 µg/kg (1 sa.), 3.0–4.9 µg/kg (1 sa.), 6.0 µg/kg (1 sa.), sample origin: Finland, sample year: 1998, country: EU[698], *instant coffee

incidence: 6/8*, conc. range: LOD/LOQ-0.9 µg/kg (1 sa.), 1.0–2.9 µg/kg (4 sa.), 7.0 µg/kg (1 sa.), sample origin: Finland, sample year: 1999, country: EU[698], *instant coffee

incidence: 3/34*, conc. range: LOD/LOQ-0.9 µg/kg (1 sa.), ≤1.6 µg/kg (2 sa.), sample origin: France, sample year: 1998/1999, country: EU[698], *coffee, roasted

incidence: 3/13*, conc. range: 3.0–4.9 µg/kg (2 sa.), 6.4 µg/kg (1 sa.), sample origin: France, sample year: 1998/1999, country: EU[698], *instant coffee

incidence: 76/183*, conc. range: LOD/LOQ-0.9 µg/kg (51 sa.), 1.0–2.9 µg/kg (20 sa.), 3.0–4.9 µg/kg (3 sa.), ≤6.32 µg/kg (2 sa.), sample origin: Germany, sample year: 1995–1998, country: EU[698], *coffee, roasted

incidence: 46/55*, conc. range: LOD/LOQ-0.9 µg/kg (16 sa.), 1.0–2.9 µg/kg (19 sa.), 3.0–4.9 µg/kg (8 sa.), ≤9.47 µg/kg (3 sa.), sample origin: Germany, sample year: 1995–1998, country: EU[698], *instant coffee

incidence: 45/94*, conc. range: LOD/LOQ-0.9 µg/kg (23 sa.), 1.0–2.9 µg/kg (21 sa.), 3.34 µg/kg (1 sa.), sample origin: Germany, sample year: 1995–1998, country: EU[698], *coffee, decaffeinated

incidence: 5/33*, conc. range: LOD/LOQ-≤0.96 µg/kg (5 sa.), sample origin: Germany, sample year: 1995–1998, country: EU[698], *malt coffee

incidence: 0/5*, conc.: no contamination, sample origin: Greece, sample year: 1998, country: EU[698], *coffee, roasted

incidence: 10/11*, conc. range: 1.0–2.9 µg/kg (7 sa.), 3.0–4.9 µg/kg (1 sa.), ≤7.2 µg/kg (2 sa.), sample origin: Greece, sample year: 1999, country: EU[698], *coffee, roasted

incidence: 3/3*, conc. range: ≤2.6 µg/kg (3 sa.), sample origin: Greece, sample year: 1999, country: EU[698], *instant coffee

incidence: 0/2*, conc.: no contamination, sample origin: Greece, sample year: 1999, country: EU[698], *coffee, decaffeinated

incidence: 6/6*, conc. range: <20 µg/kg (6 sa.), sample origin: Ireland, sample year: 1998, country: EU[698], *coffee, roasted

incidence: 15/15*, conc. range: LOD/LOQ-0.9 µg/kg (13 sa.), 1.0–2.9 µg/kg (1 sa.), 10 µg/kg (1 sa.), sample origin: Ireland, sample year: 1999, country: EU[698], *coffee, roasted

incidence: 46/198*, conc. range: LOD/LOQ-0.9 µg/kg (18 sa.), 1.0–2.9 µg/kg (19 sa.), 3.0–4.9 µg/kg (8 sa.), 11.5 µg/kg (1 sa.), sample origin: Italy, sample year: 1999, country: EU[698], *coffee, roasted

incidence: 3/13*, conc. range: LOD/LOQ-0.9 µg/kg (2 sa.), 1.6 µg/kg (1 sa.), sample origin: Italy, sample year: 1999, country: EU[698], *instant coffee

incidence: 22?/50*, conc. range: LOD/LOQ-0.9 µg/kg (29 sa.), 1.0–2.9 µg/kg (2 sa.), 4.1 µg/kg (1 sa.), sample origin: Norway, sample year: 1997, country: EU[698], *coffee, roasted

incidence: 5/8*, conc. range: ≤2.20 µg/kg (5 sa.), sample origin: Portugal, sample year: 1998, country: EU[698], *coffee, roasted

incidence: 1/25*, conc.: 2.70 µg/kg (1 sa.), sample origin: Portugal, sample year: 1999, country: EU[698], *coffee, roasted

incidence: 29/29*, conc. range: LOD/LOQ-0.9 µg/kg (15 sa.), 1.0–2.9 µg/kg (13 sa.), 5.64 µg/kg (1 sa.), sample origin: Spain, sample year: 1997, country: EU[698], *coffee, roasted

incidence: 9/9*, conc. range: LOD/LOQ-0.9 µg/kg (8 sa.), 1.08 µg/kg (1 sa.), sample origin: Spain, sample year: 1997, country: EU[698], *instant coffee

incidence: 42/77*, conc. range: LOD/LOQ-0.9 µg/kg (3 sa.), 1.0–2.9 µg/kg (36 sa.), ≤3.90 µg/kg (3 sa.), sample origin: Netherlands, sample year: 1995, country: EU[698], *coffee, roasted

incidence: 15/67*, conc. range: LOD/LOQ-0.9 µg/kg (12 sa.), 1.0–2.9 µg/kg (2 sa.), 5.0 µg/kg (1 sa.), sample origin: Netherlands, sample year: 1997–1999, country: EU[698], *coffee, roasted

incidence: 7/14*, conc. range: LOD/LOQ-0.9 µg/kg (6 sa.), 3.30 µg/kg (1 sa.), sample origin: Netherlands, sample year: 1998/1999, country: EU[698], *coffee, roasted

incidence: 5/13*, conc. range: 1.0–2.9 µg/kg (4 sa.), 3.60 µg/kg (1 sa.), sample origin: Netherlands, sample year: 1995, country: EU[698], *instant coffee

incidence: 1/2*, conc.: 4.50 µg/kg (1 sa.), sample origin: Netherlands, sample year: 1998/1999, country: EU[698], *instant coffee

incidence: 4/10*, conc. range: LOD/LOQ-0.9 µg/kg (2 sa.), ≤2.30 µg/kg (2 sa.), sample origin: Netherlands, sample year: 1995, country: EU[698], *coffee, roasted and decaffeinated

incidence: 1/5*, conc.: 0.20 µg/kg (1 sa.), sample origin: Netherlands, sample year: 1998/1999, country: EU[698], *coffee, roasted and decaffeinated

- Co-contamination: not reported
- Further contamination of plants and plant products other than cereals and their products reported in the present book/article[698]: Baby food, Beans (soybean), Chocolate, Cocoa, Coffee beans, Currants, Fruits, dried, Herbs, Jam, Juices, Juices (grape juice), Legumes, Nuts, Oil seeds, Oils (olive oil), Peas, Peas (chickpea), Products (plant product), Raisins, Spices, Spices (cayenne pepper, chilli), Spices (nutmeg), Spices (paprika), Spices (pepper), Sultanas, Tea, Vinegar, Wine, Wine (red), Wine (rosé), Wine (white), Wine, miscellaneous, OTA, literature[698]

incidence: 1/8*, conc.: 0.09 µg/kg, sample origin: New Zealand, sample year: 2011, country: New Zealand[703], *coffee, brewed

- Co-contamination: not reported
- Further contamination of plants and plant products other than cereals and their products reported in the present book/article[703]: not reported

incidence: 25/25, conc. range: 0.587–4.5 µg/kg, sample origin: five governates, Kuwait, sample year: unknown, country: Kuwait[704]

- Co-contamination: not reported
- Further contamination of plants and plant products other than cereals and their products reported in the present book/article[704]: Nuts, Spices, AF/S, literature[704]

incidence: 24/24*, conc. range: 0.30–0.84 µg/kg, ∅ conc.: 0.47 µg/kg, sample origin: specialized stores, markets, and supermarkets in Biobío, Metropolitana, and Valparaíso (regions), Chile, sample year: unknown, country: Chile[730], *coffee, roasted incidence: 39/39*, conc. range: 0.28–5.58 μg/kg, ∅ conc.: 1.80 μg/kg, sample origin: specialized stores, markets, and supermarkets in Biobío, Metropolitana, and Valparaíso (regions), Chile, sample year: unknown, country: Chile[730], *instant coffee

- Co-contamination: not reported
- Further contamination of plants and plant products other than cereals and their products reported in the present book/article[730]: not reported

incidence: 61/169*, conc. range: 0.67–4.70 μg/kg, ∅ conc.: 1.87 μg/kg, sample origin: supermarkets in Valencia (city), Spain, sample year: unknown, country: Spain[731], *coffee, brewed

- Co-contamination: not reported
- Further contamination of plants and plant products other than cereals and their products reported in the present book/article[731]: Coffee, AFB_1, AFB_2, AFG_1, AFG_2, BEA, DAS, DON, 3-AcDON, 15-AcDON, ENA, ENA_1, ENB, ENB_1, FB_1, FB_2, HT-2, NEO, NIV, STG, and T-2, literature[731]

For detailed information see the article.

incidence: 26/57*, conc. range: ≤21 μg/kg, ∅ conc.: 2.3 μg/kg, sample origin: purchased from retail markets in Denmark, sample year: unknown, country: Denmark/Brazil[803], *coffee, roasted

incidence: 14/25*, conc. range: ≤8.3 μg/kg, ∅ conc.: 4.5 μg/kg, sample origin: purchased from retail markets in Kenya, Tanzania, and Uganda, sample year: unknown, country: Denmark/Brazil[803], *instant coffee

- Co-contamination: not reported
- Further contamination of plants and plant products other than cereals and their products reported in the present book/article[803]: Coffee beans, OTA and FB_2, literature[803]; see also Coffee beans, chapter: Further Mycotoxins and Microbial Metabolites

incidence: 7/12*, conc. range: ≤1.04 μg/kg, sample origin: 12 regions, Czech Republic, sample year: 2011–2013, country: Czech Republic[811], *coffee, ground and roasted

incidence: 11/12*, conc. range: ≤4.91 μg/kg, sample origin: 12 regions, Czech Republic, sample year: 2011–2013, country: Czech Republic[811], *instant coffee

- Co-contamination: not reported
- Further contamination of plants and plant products other than cereals and their products reported in the present book/article[811]: Beans, Chocolate, Confectionery, Figs, Herbs and Spices, Juices, Licorice, Powder (cocoa powder), Products (ginger product), Raisins, Spices (caraway), Spices (cayenne pepper, chilli), Spices (coriander), Spices (ginger), Spices (mustard), Spices (nutmeg), Spices (paprika), Spices (pepper), Tea, Wine (red), Wine (white), OTA, literature[811]

incidence: 8/30*, conc. range: LOD-2.0 μg/kg (7 sa.), 2.0–5.0 μg/kg (1 sa.), ∅ conc.: 1.03 μg/kg, sample origin: marketed in Italy, sample year: unknown, country: Italy[820], *coffee, roasted

- Co-contamination: not reported
- Further contamination of plants and plant products other than cereals and their products reported in the present book/article[820]: Spices (cayenne pepper, chilli), Wine (red), OTA, literature[820]

incidence: 13/24*, conc. range: 0.11–5.78 μg/kg, sample origin: markets in Lujan and Merlo (cities), Buenos Aires (province), Argentina, sample year: unknown, country: Argentina[866], *coffee, ground and roasted

incidence: 17/22*, conc. range: 0.22–13.66 μg/kg, sample origin: markets in Lujan and Merlo (cities), Buenos Aires (province), Argentina, sample year: unknown, country: Argentina[866], *coffee, soluble

- Co-contamination: not reported
- Further contamination of plants and plant products other than cereals and their products reported in the present book/article[866]: Coffee beans, OTA, literature[866]

incidence: 10/10*, conc. range: <0.2–2.04 μg/kg, sample origin: supermarkets and retail stores, Italy, sample year: 2011, country: Italy[869], *instant coffee, decaffeinated and granulated

incidence: 34/34*, conc. range: <0.2–6.40 μg/kg, sample origin: supermarkets and retail stores, Italy, sample year: 2011, country: Italy[869], *instant coffee, granulated or powdered

incidence: 4/4*, conc. range: 0.33–0.52 μg/kg, ∅ conc.: 0.4 μg/kg, sample origin: supermarkets and retail stores, Italy, sample year: 2011, country: Italy[869], *coffee products (coffee:barley = ca. 30:70) powdered

incidence: 2/2*, conc. range: 0.51 μg/kg, ∅ conc.: 0.51 μg/kg, sample origin: supermarkets and retail stores, Italy, sample year: 2011, country: Italy[869], *coffee products (coffee:chicory = ca. 30:70) granulated or powdered

- Co-contamination: not reported
- Further contamination of plants and plant products other than cereals and their products reported in the present book/article[869]: not reported

Ochratoxin B

incidence: 1/10*, conc.: 0.10 μg/kg, sample origin: unknown, sample year: unknown, country: Japan[112], *coffee, raw

incidence: 0/23*, conc.: no contamination, sample origin: unknown, sample year: unknown, country: Japan[112], *coffee, roasted

incidence: 0/7*, conc.: no contamination, sample origin: unknown, sample year: unknown, country: Japan[112], *instant coffee

- Co-contamination: not reported
- Further contamination of plants and plant products other than cereals and their products reported in the present book/article[112]: Cassis, Coffee, Juices, Raisins, Wine, OTA, literature[112]; Cocoa, Food, Vinegar, OTA and OTB, literature[112]; Powder (cocoa powder), AFB_1, AFB_2, AFG_1, OTA, and OTB, literature[112]

incidence: 11/32*, conc. range: 0.21–0.64 μg/kg, ∅ conc.: 0.40 μg/kg, sample origin: conventional supermarkets and stores in Ho Chi Minh City and Can Tho City, Vietnam, sample year: 2011/2012, country: Japan[653], *coffee, roasted (Robusta)

- Co-contamination: 11 sa. co-contaminated with OTA and OTB
- Further contamination of plants and plant products other than cereals and their products reported in the present book/article[653]: Coffee, OTA, literature[653]

incidence: 1/30*, conc.: 0.56 μg/kg, sample origin: Chiang Mai City, Thailand, sample year: 2012/2013, country: Japan[653], *coffee, roasted (Arabica)

- Co-contamination: not reported
- Further contamination of plants and plant products other than cereals and their products reported in the present book/article[653]: Coffee, OTA, literature[653]

incidence: 0/38*, conc.: no contamination, sample origin: Chiang Mai City and Bangkok (capital), Thailand, sample year: 2012/2013, country: Japan[653], *instant coffee (mainly Robusta)

- Co-contamination: not reported
- Further contamination of plants and plant products other than cereals and their products reported in the present book/article[653]: Coffee, OTA, literature[653]

Fusarium Toxins

Depsipeptides

Beauvericin

incidence: 9/169*, conc. range: 0.52–3.07 μg/kg, Ø conc.: 0.97 μg/kg, sample origin: supermarkets in Valencia (city), Spain, sample year: unknown, country: Spain[731], *coffee, brewed

- Co-contamination: not reported
- Further contamination of plants and plant products other than cereals and their products reported in the present book/article[731]: Coffee, AFB_1, AFB_2, AFG_1, AFG_2, DAS, DON, 3-AcDON, 15-AcDON, ENA, ENA_1, ENB, ENB_1, FB_1, FB_2, HT-2, NEO, NIV, OTA, STG, and T-2, literature[731]

For detailed information see the article.

Enniatin A

incidence: 100/169*, conc. range: 5.88–2,875.57 μg/kg, Ø conc.: 282.89 μg/kg, sample origin: supermarkets in Valencia (city), Spain, sample year: unknown, country: Spain[731], *coffee, brewed

- Co-contamination: not reported
- Further contamination of plants and plant products other than cereals and their products reported in the present book/article[731]: Coffee, AFB_1, AFB_2, AFG_1, AFG_2, BEA, DAS, DON, 3-AcDON, 15-AcDON, ENA_1, ENB, ENB_1, FB_1, FB_2, HT-2, NEO, NIV, OTA, STG, and T-2, literature[731]

For detailed information see the article.

Enniatin A_1

incidence: 112/169*, conc. range: 2.22–12,039.12 μg/kg, Ø conc.: 246.69 μg/kg, sample origin: supermarkets in Valencia (city), Spain, sample year: unknown, country: Spain[731], *coffee, brewed

- Co-contamination: not reported
- Further contamination of plants and plant products other than cereals and their products reported in the present book/article[731]: Coffee, AFB_1, AFB_2, AFG_1, AFG_2, BEA, DAS, DON, 3-AcDON, 15-AcDON, ENA, ENB, ENB_1, FB_1, FB_2, HT-2, NEO, NIV, OTA, STG, and T-2, literature[731]

For detailed information see the article.

Enniatin B

incidence: 157/169*, conc. range: 4.68–4,147.96 μg/kg, Ø conc.: 176.26 μg/kg, sample origin: supermarkets in Valencia (city), Spain, sample year: unknown, country: Spain[731], *coffee, brewed

- Co-contamination: not reported
- Further contamination of plants and plant products other than cereals and their products reported in the present book/article[731]: Coffee, AFB_1, AFB_2, AFG_1, AFG_2, BEA, DAS, DON, 3-AcDON, 15-AcDON, ENA, ENA_1, ENB_1, FB_1, FB_2, HT-2, NEO, NIV, OTA, STG, and T-2, literature[731]

For detailed information see the article.

Enniatin B_1

incidence: 127/169*, conc. range: 5.16–3,932.47 μg/kg, Ø conc.: 138.57 μg/kg, sample origin: supermarkets in Valencia (city), Spain, sample year: unknown, country: Spain[731], *coffee, brewed

- Co-contamination: not reported
- Further contamination of plants and plant products other than cereals and their products reported in the present book/article[731]: Coffee, AFB_1, AFB_2, AFG_1, AFG_2, BEA, DAS, DON, 3-AcDON, 15-AcDON, ENA, ENA_1, ENB, FB_1, FB_2, HT-2, NEO, NIV, OTA, STG, and T-2, literature[731]

For detailed information see the article.

Fumonisin/s

Fumonisin B_1

incidence: 7/169*, conc. range: 15.52–18.25 μg/kg, Ø conc.: 15.55 μg/kg?, sample origin: supermarkets in Valencia (city), Spain, sample year: unknown, country: Spain[731], *coffee, brewed

- Co-contamination: not reported
- Further contamination of plants and plant products other than cereals and their products reported in the present book/article[731]: Coffee, AFB_1, AFB_2, AFG_1, AFG_2, BEA, DAS, DON, 3-AcDON, 15-AcDON, ENA, ENA_1, ENB, ENB_1, FB_2, HT-2, NEO, NIV, OTA, STG, and T-2, literature[731]

For detailed information see the article.

Fumonisin B_2

incidence: 7/169*, conc. range: 20.02–23.99 μg/kg, Ø conc.: 22.46 μg/kg, sample origin: supermarkets in Valencia (city), Spain, sample year: unknown, country: Spain[731], *coffee, brewed

- Co-contamination: not reported
- Further contamination of plants and plant products other than cereals and their products reported in the present book/article[731]: Coffee, AFB_1, AFB_2, AFG_1, AFG_2, BEA, DAS, DON, 3-AcDON, 15-AcDON, ENA, ENA_1, ENB, ENB_1, FB_2, HT-2, NEO, NIV, OTA, STG, and T-2, literature[731]

For detailed information see the article.

Type A Trichothecenes

Diacetoxyscirpenol

incidence: 12/169*, conc. range: 195.91–402.23 μg/kg, Ø conc.: 196.20 μg/kg?, sample origin: supermarkets in Valencia (city), Spain, sample year: unknown, country: Spain[731], *coffee, brewed

- Co-contamination: not reported
- Further contamination of plants and plant products other than cereals and their products reported in the present book/article[731]: Coffee, AFB_1, AFB_2, AFG_1, AFG_2, BEA, DON, 3-AcDON, 15-AcDON, ENA, ENA_1, ENB, ENB_1, FB_1, FB_2, HT-2, NEO, NIV, OTA, STG, and T-2, literature[731]

For detailed information see the article.

HT-2 Toxin

incidence: 15/169*, conc. range: 8.41–37.01 μg/kg, Ø conc.: 12.20 μg/kg, sample origin: supermarkets in Valencia (city), Spain, sample year: unknown, country: Spain[731], *coffee, brewed

- Co-contamination: not reported
- Further contamination of plants and plant products other than cereals and their products reported in the present book/article[731]: Coffee, AFB_1, AFB_2, AFG_1, AFG_2, BEA, DAS, DON,

3-AcDON, 15-AcDON, ENA, ENA_1, ENB, ENB_1, FB_1, FB_2, NEO, NIV, OTA, STG, and T-2, literature[731]

For detailed information see the article.

Neosolaniol

incidence: 22/169*, conc. range: 29.06–313.84 µg/kg, ∅ conc.: 86.73 µg/kg, sample origin: supermarkets in Valencia (city), Spain, sample year: unknown, country: Spain[731], *coffee, brewed

- Co-contamination: not reported
- Further contamination of plants and plant products other than cereals and their products reported in the present book/article[731]: Coffee, AFB_1, AFB_2, AFG_1, AFG_2, BEA, DAS, DON, 3-AcDON, 15-AcDON, ENA, ENA_1, ENB, ENB_1, FB_1, FB_2, HT-2, NIV, OTA, STG, and T-2, literature[731]

For detailed information see the article.

T-2 Toxin

incidence: 49/169*, conc. range: 0.74–21.17 µg/kg, ∅ conc.: 2.37 µg/kg, sample origin: supermarkets in Valencia (city), Spain, sample year: unknown, country: Spain[731], *coffee, brewed

- Co-contamination: not reported
- Further contamination of plants and plant products other than cereals and their products reported in the present book/article[731]: Coffee, AFB_1, AFB_2, AFG_1, AFG_2, BEA, DAS, DON, 3-AcDON, 15-AcDON, ENA, ENA_1, ENB, ENB_1, FB_1, FB_2, HT-2, NEO, NIV, OTA, and STG, literature[731]

For detailed information see the article.

Type B Trichothecenes

Deoxynivalenol

incidence: 73/169*, conc. range: 16.58–954.91 µg/kg, ∅ conc.: 102.22 µg/kg, sample origin: supermarkets in Valencia (city), Spain, sample year: unknown, country: Spain[731], *coffee, brewed

- Co-contamination: not reported
- Further contamination of plants and plant products other than cereals and their products reported in the present book/article[731]: Coffee, AFB_1, AFB_2, AFG_1, AFG_2, BEA, DAS, 3-AcDON, 15-AcDON, ENA, ENA_1, ENB, ENB_1, FB_1, FB_2, HT-2, NEO, NIV, OTA, STG, and T-2, literature[731]

For detailed information see the article.

3-Acetyldeoxynivalenol

incidence: 34/169*, conc. range: 6.31–24.57 µg/kg, ∅ conc.: 10.45 µg/kg, sample origin: supermarkets in Valencia (city), Spain, sample year: unknown, country: Spain[731], *coffee, brewed

- Co-contamination: not reported
- Further contamination of plants and plant products other than cereals and their products reported in the present book/article[731]: Coffee, AFB_1, AFB_2, AFG_1, AFG_2, BEA, DAS, DON, 15-AcDON, ENA, ENA_1, ENB, ENB_1, FB_1, FB_2, HT-2, NEO, NIV, OTA, STG, and T-2, literature[731]

For detailed information see the article.

15-Acetyldeoxynivalenol

incidence: 69/169*, conc. range: 61.04–884.25 µg/kg, ∅ conc.: 94.86 µg/kg, sample origin: supermarkets in Valencia (city), Spain, sample year: unknown, country: Spain[731], *coffee, brewed

- Co-contamination: not reported
- Further contamination of plants and plant products other than cereals and their products reported in the present book/article[731]: Coffee, AFB_1, AFB_2, AFG_1, AFG_2, BEA, DAS, DON, 3-AcDON, ENA, ENA_1, ENB, ENB_1, FB_1, FB_2, HT-2, NEO, NIV, OTA, STG, and T-2, literature[731]

For detailed information see the article.

Nivalenol

incidence: 73/169*, conc. range: 0.12–37.83 µg/kg, ∅ conc.: 1.68 µg/kg, sample origin: supermarkets in Valencia (city), Spain, sample year: unknown, country: Spain[731], *coffee, brewed

- Co-contamination: not reported
- Further contamination of plants and plant products other than cereals and their products reported in the present book/article[731]: Coffee, AFB_1, AFB_2, AFG_1, AFG_2, BEA, DAS, DON, 3-AcDON, 15-AcDON, ENA, ENA_1, ENB, ENB_1, FB_1, FB_2, HT-2, NEO, OTA, STG, and T-2, literature[731]

For detailed information see the article.

see also **Coffee beans**

Coffee beans may contain the following mycotoxins:

Aspergillus Toxins

Aflatoxin B_1

incidence: 6/57*, conc. range: ≤3.5 µg/kg, sample origin: imported to USSR, sample year: 1985–1987, country: USSR[267], *green coffee

- Co-contamination: not reported
- Further contamination of plants and plant products other than cereals and their products reported in the present book/article[267]: Cocoa beans, Legumes, Nuts, Nuts (peanut), AFB_1, literature[267]

incidence: 13/30*, conc. range: 0.57–6.00 µg/kg; ∅ conc.: 2.75 µg/kg, sample origin: local markets in Giza (city) and Cairo (capital), Egypt, sample year: unknown, country: Egypt[279], *green coffee beans

- Co-contamination: 2 sa. co-contaminated with AFB_1, AFB_2, AFG_1, and AFG_2; 6 sa. co-contaminated with AFB_1, AFB_2, and AFG_1; 1 sa. co-contaminated with AFB_1, AFB_2, and AFG_2; 2 sa. co-contaminated with AFB_1 and AFB_2; 1 sa. co-contaminated with AFB_1 and AFG_1; 1 sa. contaminated solely with AFB_1
- Further contamination of plants and plant products other than cereals and their products reported in the present book/article[279]: Coffee, AFB_1, AFB_2, AFG_1, and AFG_2, literature[279]; Coffee beans, AFB_2, AFG_1, and AFG_2, literature[279]

incidence: 5/10*, conc. range: 0.0024–0.0202 µg/kg, ∅ conc.: 0.0076 µg/kg, sample origin: imported from Yemen collected in Nagoya (city), Japan, sample year: 1988–1993, country: Japan[303], *commercial green coffee beans (Arabica)

- Co-contamination: 3 sa. co-contaminated with AFB_1 and OTA; 2 sa. contaminated solely with AFB_1
- Further contamination of plants and plant products other than cereals and their products reported in the present book/article[303]: Coffee beans, OTA, literature[303]

incidence: 3/9*, conc. range: 0.002–0.004 µg/kg, ∅ conc.: 0.0027 µg/kg, sample origin: imported from Tanzania collected in Nagoya (city), Japan, sample year: 1988–1993, country: Japan[303], *commercial green coffee beans (Arabica)

- Co-contamination: 2 sa. co-contaminated with AFB_1 and OTA; 1 sa. contaminated solely with AFB_1
- Further contamination of plants and plant products other than cereals and their products reported in the present book/article[303]: Coffee beans, OTA, literature[303]

incidence: 7/9*, conc. range: 0.0025–0.0329 µg/kg, ∅ conc.: 0.0010 µg/kg, sample origin: imported from Indonesia collected in Nagoya (city), Japan, sample year: 1988–1993, country: Japan[303], *commercial green coffee beans (Robusta)

- Co-contamination: 2 sa. co-contaminated with AFB_1 and OTA; 5 sa. contaminated solely with AFB_1
- Further contamination of plants and plant products other than cereals and their products reported in the present book/article[303]: Coffee beans, OTA, literature[303]

Aflatoxin B_2

incidence: 11/30*, conc. range: 0.13–1.56 µg/kg, ∅ conc.: 0.72 µg/kg, sample origin: local markets in Giza (city) and Cairo (capital), Egypt, sample year: unknown, country: Egypt[279], *green coffee beans

- Co-contamination: 2 sa. co-contaminated with AFB_1, AFB_2, AFG_1, and AFG_2; 6 sa. co-contaminated with AFB_1, AFB_2, and AFG_1; 1 sa. co-contaminated with AFB_1, AFB_2, and AFG_2; 2 sa. co-contaminated with AFB_1 and AFB_2
- Further contamination of plants and plant products other than cereals and their products reported in the present book/article[279]: Coffee, AFB_1, AFB_2, AFG_1, and AFG_2, literature[279]; Coffee beans, AFB_1, AFG_1, and AFG_2, literature[279]

Aflatoxin G_1

incidence: 9/30*, conc. range: 0.19–2.72 µg/kg, ∅ conc.: 1.23 µg/kg, sample origin: local markets in Giza (city) and Cairo (capital), Egypt, sample year: unknown, country: Egypt[279], *green coffee beans

- Co-contamination: 2 sa. co-contaminated with AFB_1, AFB_2, AFG_1, and AFG_2; 6 sa. co-contaminated with AFB_1, AFB_2, and AFG_1; 1 sa. co-contaminated with AFB_1 and AFG_1
- Further contamination of plants and plant products other than cereals and their products reported in the present book/article[279]: Coffee, AFB_1, AFB_2, AFG_1, and AFG_2, literature[279]; Coffee beans, AFB_1, AFB_2, and AFG_2, literature[279]

Aflatoxin G_2

incidence: 3/30*, conc. range: 0.13–0.39 µg/kg, ∅ conc.: 0.26 µg/kg, sample origin: local markets in Giza (city) and Cairo (capital), Egypt, sample year: unknown, country: Egypt[279], *green coffee beans

- Co-contamination: 2 sa. co-contaminated with AFB_1, AFB_2, AFG_1, and AFG_2; 1 sa. co-contaminated with AFB_1, AFB_2, and AFG_2
- Further contamination of plants and plant products other than cereals and their products reported in the present book/article[279]: Coffee, AFB_1, AFB_2, AFG_1, and AFG_2, literature[279]; Coffee beans, AFB_1, AFB_2, and AFG_1, literature[279]

Aflatoxin/s

incidence: 1/11* **, conc.: 8 µg/kg***, sample origin: storage facilities, silos sacks, small shops, and fields in zones with different climatic conditions, Guatemala, sample year: 1975/1976–?, country: Guatemala[138], *coffee beans, unroasted, **sa. collected at the end of rainy season, ***AFB_1, AFB_2, and AFG_1

- Co-contamination: not reported
- Further contamination of plants and plant products other than cereals and their products reported in the present book/article[138]: Beans (black beans), Nuts (peanut), AF/S, literature[138]

incidence: 2/201*, conc. range: 3–12 µg/kg**, ∅ conc.: 7.5 µg/kg**, sample origin: FDA in New Orleans (city), USA, sample year: unknown, country: USA[860], *green coffee, **AF

- Co-contamination: not reported
- Further contamination of plants and plant products other than cereals and their products reported in the present book/article[860]: Coffee beans, OTA, literature[860]

Other ***Aspergillus*** Toxins

Sterigmatocystin

incidende: 1/2* **, conc.: 1,143 µg/kg, sample origin: unknown, sample year: unknown, country: South Africa[280], *coffee beans, **sa. condemned as unfit for human consumption

- Co-contamination: not reported
- Further contamination of plants and plant products other than cereals and their products reported in the present book/article[280]: not reported

incidence: 1*/502**, conc.: 12,000 µg/kg*, sample origin: sa. from South America, Africa, and Asia collected at the port of Trieste (city), Italy, sample year: unknown, country: Italy[305], *not marketable, **green coffee

- Co-contamination: not reported
- Further contamination of plants and plant products other than cereals and their products reported in the present book/article[305]: not reported

Aspergillus and ***Penicillium*** Toxins

Ochratoxin A

incidence: 5/21*, conc. range: ≤0.76 µg/kg, ∅ conc.: 0.40 µg/kg, sample origin: local supermarkets and small retail shops in seven major cities, Japan, sample year: summer 2004–winter 2007, country: Japan[85], *green coffee beans

- Co-contamination: not reported
- Further contamination of plants and plant products other than cereals and their products reported in the present book/article[85]: Beans (soybean), Vegetables (asparagus), FB_1 and FB_2, literature[85]; Beverages, Chocolate, Cocoa, Coffee, Raisins, Spices (coriander), Wine, OTA, literature[85]

incidence: 3/20*, conc. range: 13.7–124 µg/kg, ∅ conc.: 17.4 µg/kg, sample origin: markets or farmers, Egypt, sample year: unknown, country: Egypt[142]

- Co-contamination: not reported
- Further contamination of plants and plant products other than cereals and their products reported in the present book/article[142]: Beans, OTA, literature[142]

incidence: 1/1*, conc.: 8 µg/kg, sample origin: Austria, sample year: 1994, country: EU[194], *raw green coffee

incidence: 2/2*, conc. range: LOD/LOQ-4.9 µg/kg (2 sa., maximum: 2.2 µg/kg), sample origin: Ireland, sample year: 1995, country: EU[194], *raw green coffee

incidence: 2/14*, conc. range: 5.0–9.9 µg/kg (2 sa., maximum: 7 µg/kg), sample origin: Netherlands, sample year: 1995, country: EU[194], *green coffee beans

- Co-contamination: not reported
- Further contamination of plants and plant products other than cereals and their products reported in the present book/article[194]: Coffee, Spices (cayenne pepper, chilli), Spices (curry), Spices (majoran), Spices (pepper), OTA, literature[194]

incidence: 9/40*, conc. range: 0.5–23.0 µg/kg, ∅ conc.: 4.3 µg/kg, sample origin: different coffee-producing countries, sample year: unknown, country: Italy[282], *green coffee beans

- Co-contamination: not reported
- Further contamination of plants and plant products other than cereals and their products reported in the present book/article[282]: Coffee, OTA, literature[282]

incidence: 22/82*, conc. range: 0.23–24.5 μg/kg, sample origin: retail stores and coffee roasters, Germany sample year: 1995–1999, country: Germany[288], *green coffee

- Co-contamination: not reported
- Further contamination of plants and plant products other than cereals and their products reported in the present book/article[288]: Coffee, OTA, literature[288]

incidence: 9/19*, conc. range: 0.1–4.6 μg/kg (9 sa.), ∅ conc.: 1.41 μg/kg, sample origin: coffee importer (imported from South America), sample year: unknown, country: USA[292], *green coffee

- Co-contamination: not reported
- Further contamination of plants and plant products other than cereals and their products reported in the present book/article[292]: Coffee, OTA, literature[292]

incidence: 36/45*, conc. range: 0.1–1.0 μg/kg (34 sa.), 1.1–2.0 μg/kg (1 sa.), 2.1–3.0 μg/kg (1 sa.), sample origin: coffee curer/trader and local auction platforms, India, sample year: unknown, country: India[298], *regular green coffee: Arabica parchment

incidence: 17/18*, conc. range: 0.1–1.0 μg/kg (11 sa.), 1.1–2.0 μg/kg (1 sa.), 2.1–3.0 μg/kg (3 sa.), 4.1–5.0 μg/kg (1 sa.), 9.8 μg/kg (1 sa.), sample origin: coffee curer/trader and local auction platforms, India, sample year: unknown, country: India[298], *regular green coffee: Arabica cherry

incidence: 22/30*, conc. range: 0.1–1.0 μg/kg (20 sa.), 1.1–2.0 μg/kg (2 sa.), sample origin: coffee curer/trader and local auction platforms, India, sample year: unknown, country: India[298], *regular green coffee: Robusta parchment

incidence: 7/7*, conc. range: 0.1–1.0 μg/kg (2 sa.), 1.1–2.0 μg/kg (1 sa.), 2.1–3.0 μg/kg (1 sa.), 3.1–4.0 μg/kg (1 sa.), 5.1–15 μg/kg (2 sa., maximum: 11.7 μg/kg), sample origin: coffee curer/trader and local auction platforms, India, sample year: unknown, country: India[298], *regular green coffee: Robusta cherry

- Co-contamination: not reported
- Further contamination of plants and plant products other than cereals and their products reported in the present book/article[298]: Coffee, OTA, literature[298]

incidence: 14/20*, conc. range: <1.0 μg/kg (6 sa.), 1.0–5.0 μg/kg (5 sa.), >10 μg/kg (3 sa.), sample origin: different coffee growing regions in Karnataka (state), India, sample year: unknown, country: India[301], *Robusta cherry

incidence: 8/21*, conc. range: <1.0 μg/kg (6 sa.), 1.0–5.0 μg/kg (2 sa.), sample origin: different coffee growing regions in Karnataka (state), India, sample year: unknown, country: India[301], *Robusta parchment

incidence: 5/7*, conc. range: <1.0 μg/kg (1 sa.), 1.0–5.0 μg/kg (3 sa.), >10 μg/kg (1 sa.), sample origin: different coffee growing regions in Karnataka (state), India, sample year: unknown, country: India[301], *Arabica cherry

incidence: 35/74*, conc. range: <1.0 μg/kg (21 sa.), 1.0–5.0 μg/kg (11 sa.), 5–10 μg/kg (1 sa.), >10 μg/kg (2 sa.), sample origin: different coffee growing regions in Karnataka (state), India, sample year: unknown, country: India[301], *Arabica parchment

incidence: 16/16*, conc. range: <1.0 μg/kg (7 sa.), 1.0–5.0 μg/kg (8 sa.), >10 μg/kg (1 sa.), sample origin: different coffee growing regions in Karnataka (state), India, sample year: unknown, country: India[301], *cherry with hull

incidence: 17/20*, conc. range: <1.0 μg/kg (6 sa.), 1.0–5.0 μg/kg (8 sa.), 5–10 μg/kg (1 sa.), >10 μg/kg (2 sa.), sample origin: different coffee growing regions in Karnataka (state), India, sample year: unknown, country: India[301], *monsooned coffee

incidence: 6/7*, conc. range: 0.1–0.8 μg/kg, sample origin: different coffee growing regions in Karnataka (state), India, sample year: unknown, country: India[301], *low grade/defective coffee beans (plantation blacks)

incidence: 11/13*, conc. range: 0.2–19.0 μg/kg, sample origin: different coffee growing regions in Karnataka (state), India, sample year: unknown, country: India[301], *low grade/defective coffee beans (rejected beans)

incidence: 13/14*, conc. range: 0.4–103.0 μg/kg, sample origin: different coffee growing regions in Karnataka (state), India, sample year: unknown, country: India[301], *low grade/defective coffee beans (gleanings = windfall cherries)

- Co-contamination: not reported
- Further contamination of plants and plant products other than cereals and their products reported in the present book/article[301]: Coffee, OTA, literature[301]

incidence: 7/10*, conc. range: 0.7–17.4 μg/kg, ∅ conc.: 4.0 μg/kg, sample origin: imported from Yemen collected in Nagoya (city), Japan, sample year: 1988–1993, country: Japan[303], *commercial green coffee beans (Arabica)

- Co-contamination: 3 sa. co-contaminated with AFB_1 and OTA; 4 sa. contaminated solely with OTA
- Further contamination of plants and plant products other than cereals and their products reported in the present book/article[303]: Coffee beans, AFB_1, literature[303]

incidence: 5/9*, conc. range: 0.1–7.2 μg/kg, ∅ conc.: 1.58 μg/kg, sample origin: imported from Tanzania collected in Nagoya (city), Japan, sample year: 1988–1993, country: Japan[303], *commercial green coffee beans (Arabica)

- Co-contamination: 2 sa. co-contaminated with AFB_1 and OTA; 3 sa. contaminated solely with AFB_1
- Further contamination of plants and plant products other than cereals and their products reported in the present book/article[303]: Coffee beans, AFB_1, literature[303]

incidence: 2/9*, conc. range: 0.2–1.0 μg/kg, ∅ conc.: 0.6 μg/kg, sample origin: imported from Indonesia collected in Nagoya (city), Japan, sample year: 1988–1993, country: Japan[303], *commercial green coffee beans (Robusta)

- Co-contamination: 2 sa. co-contaminated with AFB_1 and OTA
- Further contamination of plants and plant products other than cereals and their products reported in the present book/article[303]: Coffee beans, AFB_1, literature[303]

incidence: 29/30*, conc. range: 0.7–72.0 μg/kg, ∅ conc.: 8.44 μg/kg, sample origin: different countries and unknown, sample year: unknown, country: Germany[304], *green coffee

- Co-contamination: not reported
- Further contamination of plants and plant products other than cereals and their products reported in the present book/article[304]: not reported

incidence: 27/132*, conc. range: 0.7–47.8 μg/kg, Ø conc.: 7.1 μg/kg, sample origin: sa. from five coffee-producing states collected at sales points in Londrina and Santos (cities), Brazil, sample year: unknown, country: Brazil[306], *green coffee

- Co-contamination: not reported
- Further contamination of plants and plant products other than cereals and their products reported in the present book/article[306]: not reported

incidence: 13/25*, conc. range: 1.2–56.0 μg/kg, Ø conc.: 10.3 μg/kg, sample origin: different countries provided by various coffee companies, sample year: unknown, country: Switzerland[307], *green beans

- Co-contamination: not reported
- Further contamination of plants and plant products other than cereals and their products reported in the present book/article[307]: not reported

incidence: 2/13*, conc. range: 58–168 μg/kg, Ø conc.: 113 μg/kg, sample origin: different countries, sample year: unknown, country: Belgium[308], *green coffee beans (Arabica)

incidence: 3/8*, conc. range: 4–27 μg/kg, Ø conc.: 15.7 μg/kg, sample origin: different countries, sample year: unknown, country: Belgium[308], *green coffee beans (Robusta)

- Co-contamination: not reported
- Further contamination of plants and plant products other than cereals and their products reported in the present book/article[308]: not reported

incidence: 41/41*, conc. range: <5 μg/kg (22 sa.), 5–10 μg/kg (10 sa.), 10–20 μg/kg (6 sa.), 31.5 μg/kg (3 sa.), sample origin: different countries collected from various factories, sample year: unknown, country: Spain[309], *green coffee beans (Arabica)

incidence: 16/16*, conc. range: <5 μg/kg (8 sa.), 5–10 μg/kg (7 sa.), 23.3 μg/kg (1 sa.), sample origin: different countries, sample year: unknown, country: Spain[309], *green coffee beans (Robusta)

- Co-contamination: not reported
- Further contamination of plants and plant products other than cereals and their products reported in the present book/article[309]: not reported

incidence: 19/267*, conc. range: tr-360 μg/kg, sample origin: different countries, sample year: unknown, country: USA[310], *hand-cleaned green coffee beans

incidence: 3/68*, conc. range: tr-80** μg/kg, sample origin: different countries, sample year: unknown, country: USA[310], *commercial green coffee beans, **a 2nd subsample from the batch which contained tr-80 μg OTA/kg did not show any OTA contamination anymore

- Co-contamination: not reported
- Further contamination of plants and plant products other than cereals and their products reported in the present book/article[310]: not reported

incidence: 13/22*, conc. range: tr 5.5 μg/kg, sample origin: Brazil, Colombia, Costa Rica, Kenya, and Mexico, sample year: unknown, country: Italy[311], *green coffee beans (Arabica)

incidence: 6/7*, conc. range: tr-15.0 μg/kg, sample origin: Cameroon, Ivory Coast, and Zaire, sample year: unknown, country: Italy[311], *green coffee beans (Robusta)

- Co-contamination: not reported
- Further contamination of plants and plant products other than cereals and their products reported in the present book/article[311]: Coffee, OTA, literature[311]

incidence: 76/84*, conc. range: 0.5–48 μg/kg, Ø conc.: 5.2 μg/kg, sample origin: Africa, sample year: unknown, country: Italy[312], *green coffee beans

incidence: 19/60*, conc. range: 0.1–7.7 μg/kg, Ø conc.: 1.3 μg/kg, sample origin: America, sample year: unknown, country: Italy[312], *green coffee beans

incidence: 11/18*, conc. range: 0.2–4.9 μg/kg, Ø conc.: 1.6 μg/kg, sample origin: Asia, sample year: unknown, country: Italy[312], *green coffee beans

- Co-contamination: not reported
- Further contamination of plants and plant products other than cereals and their products reported in the present book/article[312]: not reported

incidence: 31/153*, conc. range: 0.26–1.0 μg/kg (19 sa.), 1.0–5.0 μg/kg (8 sa.), 5.0–10.0 μg/kg (4 sa., maximum 9.0 μg/kg), sample origin: different countries, sample year: unknown, country: UK[313], *green coffee beans (Arabica)

incidence: 55/75*, conc. range: 0.26–1.0 μg/kg (34 sa.), 1.0–5.0 μg/kg (15 sa.), 5.0–10.0 μg/kg (3 sa.), >10.0 μg/kg (3 sa., maximum 27.3 μg/kg), sample origin: different countries, sample year: unknown, country: UK[313], *green coffee beans (Robusta)

incidence: 24/63*, conc. range: 0.26–1.0 μg/kg (14 sa.), 1.0–5.0 μg/kg (7 sa.), 5.0–10.0 μg/kg (3 sa.), sample origin: different countries, sample year: unknown, country: UK[313], *green coffee beans

- Co-contamination: not reported
- Further contamination of plants and plant products other than cereals and their products reported in the present book/article[313]: not reported

incidence: 7/11*, conc. range: 2.19–6.57 μg/kg, Ø conc.: 4.09 μg/kg, sample origin: supermarkets and small shops, Jordan, sample year: September 2008–March 2009, country: Jordan[315], *green coffee beans

- Co-contamination: not reported
- Further contamination of plants and plant products other than cereals and their products reported in the present book/article[315]: Nuts, Seeds (sunflower), OTA, literature[315]

incidence: 75/75*, conc. range: tr (6 sa.), <1 μg/kg (47 sa.), 1–9.9 μg/kg (22 sa., maximum: 7.48 μg/kg), sample origin: Kenya, sample year: unknown, country: France/Kenya/Italy[316], *sound beans not roasted

incidence: 62/62*, conc. range: tr (8 sa), <1 μg/kg (19 sa.), 1–9.9 μg/kg (21 sa.), 10–99.9 μg/kg (11 sa.), ≥100 μg/kg (3 sa., maximum: 152.6 μg/kg), sample origin: Kenya, sample year: unknown, country: France/Kenya/ Italy[316], *diseased beans not roasted

incidence: 32/32*, conc. range: tr (1 sa.), <1 μg/kg (11 sa.), 1–9.9 μg/kg (16 sa.), 10–99.9 μg/kg (2 sa.), ≥100 μg/kg (2 sa., maximum: 130 μg/kg), sample origin: Kenya, sample year: unknown, country: France/Kenya/ Italy[316], *black beans not roasted

incidence: 16/16*, conc. range: tr (3 sa.), <1 μg/kg (4 sa.), 1–9.9 μg/kg (4 sa.), 10–99.9 μg/kg (4 sa.), 499.3 μg/kg (1 sa.), sample origin: Kenya, sample year: unknown, country: France/Kenya/ Italy[316], *insect damaged beans not roasted incidence: 16?/16*, conc. range: tr (2 sa.), <1 µg/kg (2 sa.), 1–9.9 µg/kg (4 sa.), 10–99.9 µg/kg (3 sa.), ≥100 µg/kg (4 sa., maximum: 727 µg/kg), sample origin: Kenya, sample year: unknown, country: France/Kenya/ Italy[316], *foxy beans not roasted

incidence: 10/10*, conc. range: <1 µg/kg (1 sa.), 1–9.9 µg/kg (8 sa.), 32.1 µg/kg (1 sa.), sample origin: Kenya, sample year: unknown, country: France/Kenya/ Italy[316], *stinkers not roasted

incidence: 14/14*, conc. range: tr (1 sa.), <1 µg/kg (10 sa.), 1–9.9 µg/kg (2 sa.), 29.6 µg/kg (1 sa.), sample origin: Kenya, sample year: unknown, country: France/Kenya/ Italy[316], *hulled ears not roasted

incidence: 4/4*, conc. range: 1–9.9 µg/kg (3 sa.), 2.5 µg/kg? (1 sa.), sample origin: Kenya, sample year: unknown, country: France/ Kenya/ Italy[316], *coffee, unroasted

- Co-contamination: not reported
- Further contamination of plants and plant products other than cereals and their products reported in the present book/ article[316]: not reported

incidence: 20/60*, conc. range: 0.2–1.0 µg/kg (10 sa.), 1.1–5.0 µg/kg (6 sa.), 5.3–7.3 µg/kg (4 sa.), ∅ conc.: 2.38 µg/kg, sample origin: Brazil, sample year: 2000, country: Portugal/USA[317], *green coffee beans (Arabica)

- Co-contamination: not reported
- Further contamination of plants and plant products other than cereals and their products reported in the present book/ article[317]: not reported

incidence: 22/54*, conc. range: 0.3–160** µg/kg, sample origin: Bom Jardim area in the highlands of Rio de Janeiro (state), Brazil, sample year: May–November 1999, country: Brazil[318], *green coffee (Arabica), **varrição (swept from the ground) coffee

- Co-contamination: not reported
- Further contamination of plants and plant products other than cereals and their products reported in the present book/ article[318]: not reported

incidence: 10/21*, conc. range: 0.2–1.4 µg/kg, ∅ conc.: 0.816 µg/kg, sample origin: Chikmagalur, Coorg, and Hassan (districts) in Karnataka (state) as well as Dindigul and Salem (districts) in Tamil Nadu (state), India, sample year: 2006/2007, country: India[319], *Arabica parchment (green coffee beans, destined for export)

incidence: 15/18*, conc. range: 0.5–4.3 µg/kg, ∅ conc.: 1.97 µg/kg, sample origin: Chikmagalur, Coorg, and Hassan (districts) in Karnataka (state) as well as Dindigul and Salem (districts) in Tamil Nadu (state), India, sample year: 2006/2007, country: India[319], *Arabica cherry (green coffee beans, destined for export)

incidence: 9/14*, conc. range: 0.5–1.4 µg/kg, ∅ conc.: 0.91 µg/kg, sample origin: Chikmagalur, Coorg, and Hassan (districts) in Karnataka (state) as well as Wayand (district) in Kerala (state), India, sample year: 2006/2007, country: India[319], *Robusta parchment (green coffee beans, destined for export)

incidence: 25/27*, conc. range: 0.5–13.5 µg/kg, ∅ conc.: 3.3 µg/kg, sample origin: Chikmagalur, Coorg, and Hassan (districts) in Karnataka (state) as well as Wayand and Palakkad (districts) in Kerala (state), sample year: 2006/2007, country: India[319], *Robusta cherry (green coffee beans, destined for export)

- Co-contamination: not reported
- Further contamination of plants and plant products other than cereals and their products reported in the present book/ article[319]: not reported

For detailed information see the article.

incidence: 1/9*, conc.: 23.70 µg/kg**, sample origin: imported from Brazil, Costa Rica, Ghana, India, Indonesia, and Uganda to Italy, sample year: January 2008–December 2009, country: Italy[320], *green coffee, **contaminated sa. from Costa Rica

- Co-contamination: not reported
- Further contamination of plants and plant products other than cereals and their products reported in the present book/ article[320]: Kernels (apricot), AFB_1, AFB_2, and AFG_1, literature[320]; Nuts (hazelnut), Paste (hazelnut paste), AFB_1, AFB_2, AFG_1, and AFG_2, literature[320]; Nuts (peanut), Nuts (pistachio nut), AFB_1 and AFB_2, literature[320]; Vine fruits, dried, OTA, literature[320]

incidence: 1/13*, conc.: 6.3 µg/kg, sample origin: Buon Ma Thuot at Dac Lac plateau in the central highlands, Vietnam, sample year: October 2010, country: Japan[657], *green coffee beans (Robusta)

- Co-contamination: 1 sa. co-contaminated with FB_2 and OTA
- Further contamination of plants and plant products other than cereals and their products reported in the present book/ article[657]: Coffee beans, FB_2, literature[657]

incidence: 1/3*, conc.: 2.3 µg/kg, sample origin: Buon Ma Thuot at Dac Lac plateau in the central highlands, Vietnam, sample year: October 2010, country: Japan[657], *green coffee beans (Arabica)

- Co-contamination: not reported
- Further contamination of plants and plant products other than cereals and their products reported in the present book/ article[657]: Coffee beans, FB_2, literature[657]

incidence: 21/99*, conc. range: LOD/LOQ-0.9 µg/kg (4 sa.), 1.0–2.9 µg/kg (10 sa.), 5.0–9.9 µg/kg (5 sa.), ≤23.0 µg/kg (2 sa.), sample origin: Finland, sample year: 1995, country: EU[698], *green coffee

incidence: 82/202*, conc. range: LOD/LOQ-0.9 µg/kg (8 sa.), 1.0–2.9 µg/kg (36 sa.), 3.0–4.9 µg/kg (14 sa.), 5.0–9.9 µg/kg (18 sa.), ≤19.0 µg/kg (6 sa.), sample origin: Finland, sample year: 1996, country: EU[698], *green coffee

incidence: 66/148*, conc. range: LOD/LOQ-0.9 µg/kg (28 sa.), 1.0–2.9 µg/kg (22 sa.), 3.0–4.9 µg/kg (5 sa.), 5.0–9.9 µg/kg (6 sa.), 10.0–24.9 µg/kg (2 sa.), ≤80.0 µg/kg (3 sa.), sample origin: Finland, sample year: 1997, country: EU[698], *green coffee

incidence: 64/119*, conc. range: LOD/LOQ-0.9 µg/kg (29 sa.), 1.0–2.9 µg/kg (16 sa.), 3.0–4.9 µg/kg (12 sa.), 5.0–9.9 µg/kg (6 sa.), 24 µg/kg (1 sa.), sample origin: Finland, sample year: 1998, country: EU[698], *green coffee

incidence: 27/97*, conc. range: LOD/LOQ-0.9 µg/kg (13 sa.), 1.0–2.9 µg/kg (10 sa.), 3.0–4.9 µg/kg (2 sa.), ≤7.0 µg/kg (2 sa.), sample origin: Finland, sample year: 1999, country: EU[698], *green coffee

incidence: 20/42*, conc. range: LOD/LOQ-0.9 µg/kg (6 sa.), 1.0–2.9 µg/kg (2 sa.), 3.0–4.9 µg/kg (5 sa.), 5.0–9.9 µg/kg (1 sa.), 10.0–24.9 µg/kg (1 sa.), ≤65.5 µg/kg (5 sa.), sample origin: France, sample year: 1998/1999, country: EU[698], *green coffee

incidence: 19/21*, conc. range: 1.0–2.9 µg/kg (4 sa.), 5.0–9.9 µg/kg (4 sa.), 10.0–24.9 µg/kg (5 sa.), ≤200.9 µg/kg (6 sa.), sample origin: Greece, sample year: 1997, country: EU[698], *green coffee

incidence: 5/6*, conc. range: 1.0–2.9 µg/kg (2 sa.), 3.0–4.9 µg/kg (1 sa.), ≤13.3 µg/kg (2 sa.), sample origin: Greece, sample year: 1998, country: EU[698], *green coffee incidence: 3/17*, conc. range: 3.0–4.9 µg/kg (2 sa.), 7.4 µg/kg (1 sa.), sample origin: Greece, sample year: 1999, country: EU[698], *green coffee

incidence: 132/564*, conc. range: LOD/LOQ-0.9 µg/kg (37 sa.), 1.0–2.9 µg/kg (31 sa.), 3.0–4.9 µg/kg (32 sa.), 5.0–9.9 µg/kg (21 sa.), 10.0–24.9 µg/kg (7 sa.), ≤42 µg/kg (4 sa.), sample origin: Italy, sample year: 1999, country: EU[698], *green coffee

incidence: 81/98*, conc. range: LOD/LOQ-0.9 µg/kg (53 sa.), 1.0–2.9 µg/kg (21 sa.), 3.0–4.9 µg/kg (4 sa.), ≤7.9 µg/kg (3 sa.), sample origin: Norway, sample year: 1999, country: EU[698], *green coffee

- Co-contamination: not reported
- Further contamination of plants and plant products other than cereals and their products reported in the present book/article[698]: Baby food, Beans (soybean), Chocolate, Cocoa, Coffee, Currants, Fruits, dried, Herbs, Jam, Juices, Juices (grape juice), Legumes, Nuts, Oil seeds, Oils (olive oil), Peas, Peas (chickpea), Products (plant product), Raisins, Spices, Spices (cayenne pepper, chilli), Spices (nutmeg), Spices (paprika), Spices (pepper), Sultanas, Tea, Vinegar, Wine, Wine (red), Wine (rosé), Wine (white), Wine, miscellaneous, OTA, literature[698]

incidence: 7/18*, conc. range: ≤2.8 µg/kg, ∅ conc.: 1.7 µg/kg, sample origin: purchased from retail markets in Kenya, Tanzania, and Uganda, sample year: unknown, country: Denmark/Brazil[803], *green coffee

- Co-contamination: not reported
- Further contamination of plants and plant products other than cereals and their products reported in the present book/article[803]: Coffee, OTA, literature[803]; Coffee beans, FB_2, literature[803]; see also Coffee beans, chapter: Further Mycotoxins and Microbial Metabolites

incidence: ?/6*, conc. range: ≤0.4 µg/kg, sample origin: farms of Alta Paulista, Sorocabana, and Alta Mogiana (regions) in São Paulo (state), as well as Cerrado Mineiro in Minas Gerais (state), Brazil, sample year: 1999/2000, country: Brazil/Australia[856], **Coffea arabica*, cherries from tree

incidence: 0/16*, conc.: no contamination, sample origin: farms of Alta Paulista, Sorocabana, and Alta Mogiana (regions) in São Paulo (state), as well as Cerrado Mineiro in Minas Gerais (state), Brazil, sample year: 1999/2000, country: Brazil/Australia[856], * *Coffea arabica*, overripe from tree

incidence: 2/25*, conc. range: 6.0–37 µg/kg, ∅ conc.: 21.5 µg/kg, sample origin: farms of Alta Paulista, Sorocabana, and Alta Mogiana (regions) in São Paulo (state), as well as Cerrado Mineiro in Minas Gerais (state), Brazil, sample year: 1999/2000, country: Brazil/Australia[856], * *Coffea arabica*, overripe from soil

incidence: 3/40*, conc. range: 5.0–48 µg/kg, ∅ conc.: 24 µg/kg, sample origin: farms of Alta Paulista, Sorocabana, and Alta Mogiana (regions) in São Paulo (state), as well as Cerrado Mineiro in Minas Gerais (state), Brazil, sample year: 1999/2000, country: Brazil/Australia[856], * *Coffea arabica*, drying yard

incidence: 5/48*, conc. range: 9.6–109 µg/kg, ∅ conc.: 31.3 µg/kg, sample origin: farms of Alta Paulista, Sorocabana, and Alta Mogiana (regions) in São Paulo (state), as well as Cerrado Mineiro in Minas Gerais (state), Brazil, sample year: 1999/2000, country: Brazil/Australia[856], * *Coffea arabica*, storage

- Co-contamination: not reported
- Further contamination of plants and plant products other than cereals and their products reported in the present book/article[856]: not reported

For detailed information see the article.

incidence: 2/201*, conc. range: 24–96 µg/kg, ∅ conc.: 60 µg/kg, sample origin: FDA in New Orleans (city), USA, sample year: unknown, country: USA[860], *green coffee

- Co-contamination: not reported
- Further contamination of plants and plant products other than cereals and their products reported in the present book/article[860]: Coffee beans, AF/S, literature[860]

incidence: 5/5*, conc. range: 0.20–20.30 µg/kg, ∅ conc.: 5.77 µg/kg, sample origin: manufacture company, Argentina, sample year: unknown, country: Argentina[866], *green coffee

- Co-contamination: not reported
- Further contamination of plants and plant products other than cereals and their products reported in the present book/article[866]: Coffee, OTA, literature[866]

see also **Coffee**

Fusarium Toxins

Fumonisin/s

Fumonisin B_2

incidence: 1/13*, conc.: 19 µg/kg, sample origin: Buon Ma Thuot at Dac Lac plateau in the central highlands, Vietnam, sample year: October 2010, country: Japan[657], *green coffee beans (Robusta)

- Co-contamination: 1 sa. co-contaminated with FB_2 and OTA
- Further contamination of plants and plant products other than cereals and their products reported in the present book/article[657]: Coffee beans, OTA, literature[657]

incidence: 1/3*, conc.: 55 µg/kg, sample origin: Buon Ma Thuot at Dac Lac plateau in the central highlands, Vietnam, sample year: October 2010, country: Japan[657], *green coffee beans (Arabica)

- Co-contamination: not reported
- Further contamination of plants and plant products other than cereals and their products reported in the present book/article[657]: Coffee beans, OTA, literature[657]

incidence: 8/18*, conc. range: ≤134 µg/kg, ∅ conc.: 25 µg/kg, sample origin: purchased from retail markets in Kenya, Tanzania, and Uganda, sample year: unknown, country: Denmark/Brazil[803], *green coffee

- Co-contamination: not reported
- Further contamination of plants and plant products other than cereals and their products reported in the present book/article[803]: Coffee, Coffee beans, OTA, literature[803]; see also Coffee beans, chapter: Further Mycotoxins and Microbial Metabolites

Coffee beverage see **Beverages**

Compote see **Products (fruit product)**

Confectionery may contain the following mycotoxins:

Aspergillus Toxins

Aflatoxin B_1

incidence: 2/16, conc. range: 12.6–13.8 µg/kg, ∅ conc.: 13.2 µg/kg, sample origin: supermarkets and village stores in five districts of

Penang (state) including Penang Island and Wellesley (province), Malaysia, sample year: unknown, country: Malaysia[225]

- Co-contamination: 2 sa. co-contaminated with AFB_1, AFB_2, and AFG_2
- Further contamination of plants and plant products other than cereals and their products reported in the present book/article[225]: Butter (peanut butter), Nuts (walnut), Products (nut product), AFB_1 and AFB_2, literature[225]; Cake (peanut), Nuts (peanut), AFB_1, AFB_2, and AFG_2, literature[225]; Confectionery, AFB_2, literature[225]

incidence: 3/17*, conc. range: 0.1–2.1 μg/kg, ∅ conc.: 1.0 μg/kg, sample origin: available in New Zealand, sample year: 2010, country: New Zealand[277], *peanut confectionery

- Co-contamination: not reported
- Further contamination of plants and plant products other than cereals and their products reported in the present book/article[277]: Bars, AFB_1, AF/S, DON, NIV, and OTA, literature[277]; Butter (peanut butter), Food, Fruits, dried, Nuts, Nuts (peanut), Nuts (pistachio nut), Sauces (peanut sauce), AFB_1 and AF/S, literature[277]; Confectionery, AF/S, literature[277]; Coffee, Dates, Spices (nutmeg), Spices (pepper), Vine fruits, dried, OTA, literature[277]; Figs, Spices (cayenne pepper, chilli), Spices (curry), Spices (ginger), Spices (paprika), AFB_1, AF/S, and OTA, literature[277]; Nuts (Brazil nut, paranut), AFB_1, literature[277]

Aflatoxin B_2

incidence: 2/16, conc. range: 4.44–7.62 μg/kg, ∅ conc.: 6.03 μg/kg, sample origin: supermarkets and village stores in five districts of Penang (state) including Penang Island and Wellesley (province), Malaysia, sample year: unknown, country: Malaysia[225]

- Co-contamination: 2 sa. co-contaminated with AFB_1, AFB_2, and AFG_2
- Further contamination of plants and plant products other than cereals and their products reported in the present book/article[225]: Butter (peanut butter), Nuts (walnut), Products (nut product), AFB_1 and AFB_2, literature[225]; Cake (peanut), Nuts (peanut), AFB_1, AFB_2, and AFG_2, literature[225]; Confectionery, AFB_1, literature[225]

Aflatoxin/s

incidence: 3/17*, conc. range: 0.5–2.3 μg/kg**, ∅ conc.: 1.1 μg/kg**, sample origin: available in New Zealand, sample year: 2010, country: New Zealand[277], *peanut confectionery, **AFS total

- Co-contamination: not reported
- Further contamination of plants and plant products other than cereals and their products reported in the present book/article[277]: Bars, AFB_1, AF/S, DON, NIV, and OTA, literature[277]; Butter (peanut butter), Food, Fruits, dried, Nuts, Nuts (peanut), Nuts (pistachio nut), Sauces (peanut sauce), AFB_1 and AF/S, literature[277]; Coffee, Dates, Spices (nutmeg), Spices (pepper), Vine fruits, dried, OTA, literature[277]; Confectionery, AFB_1, literature[277]; Figs, Spices (cayenne pepper, chilli), Spices (curry), Spices (ginger), Spices (paprika), AFB_1, AF/S, and OTA, literature[277]; Nuts (Brazil nut, paranut), AFB_1, literature[277]

Aspergillus and *Penicillium* Toxins

Ochratoxin A

incidence: 6/12*, conc. range: ≤1.16 μg/kg, sample origin: 12 regions, Czech Republic, sample year: 2011–2013, country: Czech Republic[811], *chocolate coated confectionery

incidence: 10/12*, conc. range: ≤1.78 μg/kg, sample origin: 12 regions, Czech Republic, sample year: 2011–2013, country: Czech Republic[811], *confectionery, sweet

- Co-contamination: not reported
- Further contamination of plants and plant products other than cereals and their products reported in the present book/article[811]: Beans, Chocolate, Coffee, Figs, Herbs and Spices, Juices, Licorice, Powder (cocoa powder), Products (ginger product), Raisins, Spices (caraway), Spices (cayenne pepper, chilli), Spices (coriander), Spices (ginger), Spices (mustard), Spices (nutmeg), Spices (paprika), Spices (pepper), Tea, Wine (red), Wine (white), OTA, literature[811]

Congressbele may contain the following mycotoxins:

Aspergillus Toxins

Aflatoxin B_1

incidence: 9/41, conc. range: 6–1,100 μg/kg, sample origin: Mysore (city) market, India, sample year: October 1991–January 1992, country: India[199]

- Co-contamination: not reported
- Further contamination of plants and plant products other than cereals and their products reported in the present book/article[199]: Bondakadledkai, Toffees, AFB_1 and AFB_2, literature[199]; Congressbele, AFB_2, literature[199]

Aflatoxin B_2

incidence: 5/41, conc. range: 4–700 μg/kg, sample origin: Mysore (city) market, India, sample year: October 1991–January 1992, country: India[199]

- Co-contamination: not reported
- Further contamination of plants and plant products other than cereals and their products reported in the present book/article[199]: Bondakadledkai, Toffees, AFB_1 and AFB_2, literature[199]; Congressbele, AFB_1, literature[199]

Congressbele is a groundnut-based (only the cotyledons are used), spiced snack product.

Cookies may contain the following mycotoxins:

Aspergillus Toxins

Aflatoxin/s

incidence: 10/24*, conc. range: ≤31.8 μg/kg**, sample origin: markets, superstores, malls, and venders of major districts of Punjab (province), Pakistan, sample year: unknown, country: Pakistan[573], *peanut cookies, **AFB_1, AFB_2, AFG_1, and AFG_2

- Co-contamination: not reported
- Further contamination of plants and plant products other than cereals and their products reported in the present book/article[573]: Butter (peanut butter), Nuts (peanut), Products (peanut product), AF/S, literature[573]

incidence: 6/22*, conc. range: ≤9.85 μg/kg**, sample origin: markets and local shops in big cities of Punjab (province), Pakistan, sample year: November 2011–March 2012, country: Pakistan/Malaysia[574], *date cookies, **AFB_1, AFB_2, AFG_1, and AFG_2

- Co-contamination: not reported
- Further contamination of plants and plant products other than cereals and their products reported in the present book/article[574]: Cake (date), Dates, Halva, AF/S, literature[574]

Copra may contain the following mycotoxins:

Aspergillus Toxins

Aflatoxin B_1

incidence: 40/62*, conc. range: 10–4,000 µg/kg, sample origin: local markets, oil mills, and houses, Andhra Pradesh (state), India, sample year: 1982, country: India[236], *copra, dry

- Co-contamination: not reported
- Further contamination of plants and plant products other than cereals and their products reported in the present book/article[236]: Candies (coconut candy), Meal (copra meal), AFB_1, AFB_2, AFG_1, and AFG_2, literature[236]; Copra, AFB_2, AFG_1, and AFG_2, literature[236]; Oils (coconut oil), AFB_1, AFB_2, and AFG_1, literature[236]

Aflatoxin B_2

incidence: 20/62*, conc. range: 10–1,000 µg/kg, sample origin: local markets, oil mills, and houses, Andhra Pradesh (state), India, sample year: 1982, country: India[236], *copra, dry

- Co-contamination: not reported
- Further contamination of plants and plant products other than cereals and their products reported in the present book/article[236]: Candies (coconut candy), Meal (copra meal), AFB_1, AFB_2, AFG_1, and AFG_2, literature[236]; Copra, AFB_1, AFG_1, and AFG_2, literature[236]; Oils (coconut oil), AFB_1, AFB_2, and AFG_1, literature[236]

Aflatoxin G_1

incidence: 20/62*, conc. range: 20–416 µg/kg, sample origin: local markets, oil mills, and houses, Andhra Pradesh (state), India, sample year: 1982, country: India[236], *copra, dry

- Co-contamination: not reported
- Further contamination of plants and plant products other than cereals and their products reported in the present book/article[236]: Candies (coconut candy), Meal (copra meal), AFB_1, AFB_2, AFG_1, and AFG_2, literature[236]; Copra, AFB_1, AFB_2, and AFG_2, literature[236]; Oils (coconut oil), AFB_1, AFB_2, and AFG_1, literature[236]

Aflatoxin G_2

incidence: 16/62*, conc. range: 20–400 µg/kg, sample origin: local markets, oil mills, and houses, Andhra Pradesh (state), India, sample year: 1982, country: India[236], *copra, dry

- Co-contamination: not reported
- Further contamination of plants and plant products other than cereals and their products reported in the present book/article[236]: Candies (coconut candy), Meal (copra meal), AFB_1, AFB_2, AFG_1, and AFG_2, literature[236]; Copra, Oils (coconut oil), AFB_1, AFB_2, and AFG_1, literature[236]

Aflatoxin/s

incidence: 2/2, conc. range: 4 µg/kg*, Ø conc.: 4 µg/kg*, sample origin: copra mills in part from outlying islands, Tonga, sample year: unknown, country: Fiji/Zambia[321], *AF

- Co-contamination: not reported
- Further contamination of plants and plant products other than cereals and their products reported in the present book/article[321]: Meal (copra meal), Nuts (peanut), AF/S, literature[321]

Coriander see Spices (coriander)

Cortese see Wine

Costers del Segre see Wine (red) and Wine (white)

Cowpea see Beans (cowpea)

Cream may contain the following mycotoxins:

Aspergillus Toxins

Aflatoxin/s

incidence: 5/5*, conc. range: 0.065–4.905 µg/kg**, Ø conc.: 1.550 µg/kg**, sample origin: groceries and supermarkets, Turkey, sample year: 2003, country: Turkey[119], *cream chantilly, **AFS total

- Co-contamination: not reported
- Further contamination of plants and plant products other than cereals and their products reported in the present book/article[119]: Asure, Beverages, Cocoa, Ice cream, Jello, Keskul, Muhallebi, Pudding, Salep, Sauces (chocolate sauce), Soups, AF/S, literature[119]

Cream chantilly is whipped cream. Ingredients were sugar, milk powder, modified starch, emulgator, thickener agent, and vanilla flavor.

Crianza see Wine (red)

Cumin see Spices (cumin)

Curd (bean curd) may contain the following mycotoxins:

Aspergillus Toxins

Aflatoxin B_1

incidence: 15/45*, conc. range: 0.3–4.3 µg/kg, Ø conc.: 1.10 µg/kg, sample origin: small markets in several urban Shanghai (city) areas, China, sample year: April 1988–April 1989, country: China/USA[132], *fermented bean curd

- Co-contamination: not reported
- Further contamination of plants and plant products other than cereals and their products reported in the present book/article[132]: Beans (kidney bean), Butter (peanut butter), Nuts (peanut), Oils (rapeseed oil), Oils (soybean oil), Paste (bean paste), Sauces (soy sauce), Tofu, Wine, miscellaneous, AFB_1, literature[132]

incidence: 1/10*, conc.: 210 µg/kg, sample origin: markets around Bangkok (capital) area, Thailand, sample year: July 1983–April 1985, country: Thailand[133], *salted bean curd (tao-hoo-yee)

- Co-contamination: 1 sa. co-contaminated with AFB_1 and AFG_1
- Further contamination of plants and plant products other than cereals and their products reported in the present book/article[133]: Beans, Curd (bean curd), AFG_1, literature[133]; Beans (soybean), Nuts (peanut), Sauces (soy sauce), AFB_1 and AFG_1, literature[133]

Aflatoxin G_1

incidence: 2/10*, conc. range: 110–270 µg/kg, Ø conc.: 190 µg/kg, sample origin: markets around Bangkok (capital) area, Thailand, sample year: July 1983–April 1985, country: Thailand[133], *salted bean curd (tao-hoo-yee)

- Co-contamination: 1 sa. co-contaminated with AFB_1 and AFG_1; 1 sa. contaminated solely with AFG_1
- Further contamination of plants and plant products other than cereals and their products reported in the present book/article[133]: Beans, AFG_1, literature[133]; Beans (soybean), Nuts (peanut), Sauces (soy sauce), AFB_1 and AFG_1, literature[133]; Curd (bean curd), AFB_1, literature[133]

Currants may contain the following mycotoxins:

Aspergillus Toxins

Aflatoxin B_1

incidence: 2/18, conc. range: 0.20–0.63 μg/kg, Ø conc.: 0.41 μg/kg, sample origin: supermarkets, shops, and market stalls in Mashhad (city), Khorasan (province), Iran, sample year: 2009–2011, country: Iran[726]
- Co-contamination: not reported
- Further contamination of plants and plant products other than cereals and their products reported in the present book/article[726]: Currants, AFB_2, AFG_1, and OTA, literature[726]; Raisins, AFB_1 and OTA literature[726]

incidence: 3/8, conc. range: 0.05–0.1 μg/kg, Ø conc.: 0.07 μg/kg, sample origin: Corinth (city), Peloponnes (peninsula), Greece, sample year: 2012, country: Greece[767]
- Co-contamination: 3 sa. co-contaminated with AFB_1 and OTA
- Further contamination of plants and plant products other than cereals and their products reported in the present book/article[767]: Currants, OTA, literature[767]; Raisins, Sultanas, AFB_1 and OTA, literature[767]

Aflatoxin B_2

incidence: 1/18, conc.: 0.33 μg/kg, sample origin: supermarkets, shops, and market stalls in Mashhad (city), Khorasan (province), Iran, sample year: 2009–2011, country: Iran[726]
- Co-contamination: not reported
- Further contamination of plants and plant products other than cereals and their products reported in the present book/article[726]: Currants, AFB_1, AFG_1, and OTA, literature[726]; Raisins, AFB_1 and OTA, literature[726]

Aflatoxin G_1

incidence: 1/18, conc.: 0.49 μg/kg, sample origin: supermarkets, shops, and market stalls in Mashhad (city), Khorasan (province), Iran, sample year: 2009–2011, country: Iran[726]
- Co-contamination: not reported
- Further contamination of plants and plant products other than cereals and their products reported in the present book/article[726]: Currants, AFB_1, AFB_2, and OTA, literature[726]; Raisins, AFB_1 and OTA, literature[726]

Aspergillus and ***Penicillium*** Toxins

Ochratoxin A

incidence: 96/100, conc. range: 0.2–4.0 μg/kg (62 sa.), 4.1–10 μg/kg (22 sa.), 10.1–20 μg/kg (7 sa.), 20.1–30 μg/kg (4 sa.), 40.8 μg/kg (1 sa.), sample origin: 13 regions in the UK, sample year: October/November 1998, country: UK[263]
- Co-contamination: not reported
- Further contamination of plants and plant products other than cereals and their products reported in the present book/article[263]: Chocolate, Figs, Juices (grape juice), Powder (cocoa powder), Raisins, Sultanas, Wine (red), OTA, literature[263]

incidence: 9/12, conc. range: ≤12.4 μg/kg, sample origin: Peloponissos (peninsula), Greece, sample year: 1998, country: Greece[327]

incidence: 24/32, conc. range: ≤13.8 μg/kg, sample origin: Peloponissos (peninsula), Greece, sample year: 1999, country: Greece[327]

incidence: 10/10, conc. range: 0.6–4.9 μg/kg, Ø conc.: 2.3 μg/kg, sample origin: Peloponissos (peninsula), Greece, sample year: 2000, country: Greece[327]
- Co-contamination: not reported
- Further contamination of plants and plant products other than cereals and their products reported in the present book/article[327]: Sultanas, Wine (red), Wine (rosé), Wine (white), Wine, miscellaneous, OTA, literature[327]

incidence: 13/20, conc. range: 2.9–74.7 μg/kg, Ø conc.: 12.2 μg/kg, sample origin: imported from Greece purchased in shops in Norwich (city) and supermarkets, small shops, etc. across the UK, sample year: unknown, country: UK[328]
- Co-contamination: not reported
- Further contamination of plants and plant products other than cereals and their products reported in the present book/article[328]: Raisins, Sultanas, OTA, literature[328]

incidence: 2/2, conc. range: 0.8–4.85 μg/kg, Ø conc.: 2.81 μg/kg, sample origin: South Africa and unknown collected in retail locations across Canada, sample year: 1998/1999, country: Canada[329]
- Co-contamination: not reported
- Further contamination of plants and plant products other than cereals and their products reported in the present book/article[329]: Raisins, Sultanas, OTA, literature[329]

incidence: 2/22, conc. range: 0.77–2.59 μg/kg, Ø conc.: 2.07 μg/kg, sample origin: different supermarkets and traditional bazaars in Istanbul (capital), Turkey, sample year: 2008/2009, country: Turkey[665]
- Co-contamination: not reported
- Further contamination of plants and plant products other than cereals and their products reported in the present book/article[665]: Juices (grape juice), Pekmez, Sultanas, OTA, literature[665]

incidence: 9/20, conc. range: 1.0–2.9 μg/kg (7 sa.), 3.0–4.9 μg/kg (1 sa.), 8.4 μg/kg (1 sa.), sample origin: Greece, sample year: 1997, country: EU[698]

incidence: 5/15, conc. range: 1.0–2.9 μg/kg (4 sa.), 3.5 μg/kg (1 sa.), sample origin: Greece, sample year: 1998, country: EU[698]

incidence: 9/12, conc. range: LOD/LOQ-0.9 μg/kg (1 sa.), 1.0–2.9 μg/kg (4 sa.), 3.0–4.9 μg/kg (2 sa.), 5.0–9.9 μg/kg (1 sa.), 12.38 μg/kg (1 sa.), sample origin: Greece, sample year: 1999, country: EU[698]
- Co-contamination: not reported
- Further contamination of plants and plant products other than cereals and their products reported in the present book/article[698]: Baby food, Beans (soybean), Chocolate, Cocoa, Coffee, Coffee beans, Fruits, dried, Herbs, Jam, Juices, Juices (grape juice), Legumes, Nuts, Oil seeds, Oils (olive oil), Peas, Peas (chickpea), Products (plant product), Raisins, Spices, Spices (cayenne pepper, chilli), Spices (nutmeg), Spices

(paprika), Spices (pepper), Sultanas, Tea, Vinegar, Wine, Wine (red), Wine (rosé), Wine (white), Wine, miscellaneous, OTA, literature[698]

incidence: 1/18, conc.: 2.99 µg/kg, sample origin: supermarkets, shops, and market stalls in Mashhad (city), Khorasan (province), Iran, sample year: 2009–2011, country: Iran[726]

- Co-contamination: not reported
- Further contamination of plants and plant products other than cereals and their products reported in the present book/article[726]: Currants, AFB_1, AFB_2, and AFG_1, literature[726]; Raisins, AFB_1 and OTA, literature[726]

incidence: 23/40, conc. range: 0.16–5 µg/kg (20 sa.), 5–10 µg/kg (3 sa., maximum: 7.10 µg/kg), sample origin: dried grape factories in Hamadan (province), Iran, sample year: October 2012–March 2013, country: Iran[742]

- Co-contamination: not reported
- Further contamination of plants and plant products other than cereals and their products reported in the present book/article[742]: Raisins, Sultanas, OTA, literature[742]

incidence: 2/2, conc. range: 10–100 µg/kg, sample origin: Crete (island, administrative unit), Greece, sample year: 2012, country: Greece[767]

incidence: 9/9, conc. range: <0.8–10 µg/kg (1 sa.), 10–100 µg/kg (4 sa.), >100 µg/kg (4 sa., maximum: 138.3 µg/kg), sample origin: Aigio and Corinth (town and city), Peloponnes (peninsula), Greece, sample year: 2012, country: Greece[767]

incidence: 1/1, conc.: 10–100 µg/kg, sample origin: American sa. imported to Greece, sample year: 2012, country: Greece[767]

- Co-contamination: 3 sa. co-contaminated with AFB_1 and OTA; no further information available
- Further contamination of plants and plant products other than cereals and their products reported in the present book/article[767]: Currants, AFB_1, literature[767]; Raisins, Sultanas, AFB_1 and OTA, literature[767]

incidence: 14/24, conc. range: ≤8.22 µg/kg, sample origin: retail markets, China, sample year: 2012, country: China[897]

- Co-contamination: not reported
- Further contamination of plants and plant products other than cereals and their products reported in the present book/article[897]: Sultanas, OTA, literature[897]

For detailed information see the article.
see also **Fruits, dried, Raisins,** and **Sultanas**

Curry see Spices (curry)

Dagwa may contain the following mycotoxins:

Alternaria Toxins

Alternariol Methyl Ether

incidence: 5/8, conc. range: 0.4–6 µg/kg, ∅ conc.: 3.9 µg/kg, sample origin: Bamenda (region, western highlands), Cameroon, sample year: July–September 2011, country: Cameroon/Austria/South Africa[659]

- Co-contamination: For detailed information see the article.
- Further contamination of plants and plant products other than cereals and their products reported in the present book/article[659]: Dagwa, AFB_1, AFB_2, AFG_1, AFG_2, BEA, CIT, DON, ENA, ENA_1, ENB, ENB_1, FA, FB_1, FB_2, FB_3, KA, MA, NIV, OTA, OTB, STG, VIOM, and ZEA, literature[659]; Flour (soybean flour), AFB_1, AFG_1, BEA, DON, ENB, ENB_1, FA, FB_1, FB_2, FB_3, FUS-X, KA, NIV, and ZEA, literature[659]; Kuru-kuru, AFB_1, AFB_2, AFG_1, AFG_2, BEA, ENA, ENA_1, ENB, ENB_1, FA, FB_1, FB_3, KA, MA, OTA, OTB, ROQ-C, STG, VIOM, and ZEA, literature[659]; Paste (peanut paste), AFB_1, AFB_2, AFG_1, AFG_2, AME, BEA, ENA, ENA_1, ENB, ENB_1, FA, FB_1, FB_2, FB_3, KA, MA, OTA, OTB, STG, VIOM, and ZEA, literature[659]; Soups, AFB_1, AFB_2, AFG_1, AFG_2, AME, AOH, BEA, DON, ENA, ENA_1, ENB, ENB_1, FB_1, FB_2, FB_3, KA, OTA, OTB, and STG, literature[659]; see also Dagwa, Flour (soybean flour), Kuru-kuru, Paste (peanut paste), and Soups, chapter: Further Mycotoxins and Microbial Metabolites

Aspergillus Toxins

Aflatoxin B_1

incidence: 8/8, conc. range: 2–224 µg/kg, ∅ conc.: 45 µg/kg, sample origin: Bamenda (region, western highlands), Cameroon, sample year: July–September 2011, country: Cameroon/Austria/South Africa[659]

- Co-contamination: For detailed information see the article.
- Further contamination of plants and plant products other than cereals and their products reported in the present book/article[659]: Dagwa, AFB_2, AFG_1, AFG_2, AME, BEA, CIT, DON, ENA, ENA_1, ENB, ENB_1, FA, FB_1, FB_2, FB_3, KA, MA, NIV, OTA, OTB, STG, VIOM, and ZEA, literature[659]; Flour (soybean flour), AFB_1, AFG_1, BEA, DON, ENB, ENB_1, FA, FB_1, FB_2, FB_3, FUS-X, KA, NIV, and ZEA, literature[659]; Kuru-kuru, AFB_1, AFB_2, AFG_1, AFG_2, BEA, ENA, ENA_1, ENB, ENB_1, FA, FB_1, FB_3, KA, MA, OTA, OTB, ROQ-C, STG, VIOM, and ZEA, literature[659]; Paste (peanut paste), AFB_1, AFB_2, AFG_1, AFG_2, AME, BEA, ENA, ENA_1, ENB, ENB_1, FA, FB_1, FB_2, FB_3, KA, MA, OTA, OTB, STG, VIOM, and ZEA, literature[659]; Soups, AFB_1, AFB_2, AFG_1, AFG_2, AME, AOH, BEA, DON, ENA, ENA_1, ENB, ENB_1, FB_1, FB_2, FB_3, KA, OTA, OTB, and STG, literature[659]; see also Dagwa, Flour (soybean flour), Kuru-kuru, Paste (peanut paste), and Soups, chapter: Further Mycotoxins and Microbial Metabolites

Aflatoxin B_2

incidence: 6/8, conc. range: <LOQ-14 µg/kg, ∅ conc.: 6 µg/kg, sample origin: Bamenda (region, western highlands), Cameroon, sample year: July–September 2011, country: Cameroon/Austria/South Africa[659]

- Co-contamination: For detailed information see the article.
- Further contamination of plants and plant products other than cereals and their products reported in the present book/article[659]: Dagwa, AFB_1, AFG_1, AFG_2, AME, BEA, CIT, DON, ENA, ENA_1, ENB, ENB_1, FA, FB_1, FB_2, FB_3, KA, MA, NIV, OTA, OTB, STG, VIOM, and ZEA, literature[659]; Flour (soybean flour), AFB_1, AFG_1, BEA, DON, ENB, ENB_1, FA, FB_1, FB_2, FB_3, FUS-X, KA, NIV, and ZEA, literature[659]; Kuru-kuru, AFB_1, AFB_2, AFG_1, AFG_2, BEA, ENA, ENA_1, ENB, ENB_1, FA, FB_1, FB_3, KA, MA, OTA, OTB, ROQ-C, STG, VIOM, and ZEA, literature[659]; Paste (peanut paste), AFB_1, AFB_2, AFG_1, AFG_2, AME, BEA, ENA, ENA_1, ENB, ENB_1, FA, FB_1, FB_2, FB_3, KA, MA, OTA, OTB, STG, VIOM, and ZEA, literature[659]; Soups, AFB_1, AFB_2, AFG_1, AFG_2, AME, AOH, BEA, DON, ENA, ENA_1, ENB, ENB_1, FB_1, FB_2, FB_3, KA, OTA, OTB, and STG, literature[659]; see also Dagwa, Flour (soybean flour), Kuru-kuru, Paste (peanut paste), and Soups, chapter: Further Mycotoxins and Microbial Metabolites

Aflatoxin G_1

incidence: 8/8, conc. range: 2–149 µg/kg, Ø conc.: 24 µg/kg, sample origin: Bamenda (region, western highlands), Cameroon, sample year: July–September 2011, country: Cameroon/Austria/South Africa[659]

- Co-contamination: For detailed information see the article.
- Further contamination of plants and plant products other than cereals and their products reported in the present book/article[659]: Dagwa, AFB_1, AFB_2, AFG_2, AME, BEA, CIT, DON, ENA, ENA_1, ENB, ENB_1, FA, FB_1, FB_2, FB_3, KA, MA, NIV, OTA, OTB, STG, VIOM, and ZEA, literature[659]; Flour (soybean flour), AFB_1, AFG_1, BEA, DON, ENB, ENB_1, FA, FB_1, FB_2, FB_3, FUS-X, KA, NIV, and ZEA, literature[659]; Kuru-kuru, AFB_1, AFB_2, AFG_1, AFG_2, BEA, ENA, ENA_1, ENB, ENB_1, FA, FB_1, FB_3, KA, MA, OTA, OTB, ROQ-C, STG, VIOM, and ZEA, literature[659]; Paste (peanut paste), AFB_1, AFB_2, AFG_1, AFG_2, AME, BEA, ENA, ENA_1, ENB, ENB_1, FA, FB_1, FB_2, FB_3, KA, MA, OTA, OTB, STG, VIOM, and ZEA, literature[659]; Soups, AFB_1, AFB_2, AFG_1, AFG_2, AME, AOH, BEA, DON, ENA, ENA_1, ENB, ENB_1, FB_1, FB_2, FB_3, KA, OTA, OTB, and STG, literature[659]; see also Dagwa, Flour (soybean flour), Kuru-kuru, Paste (peanut paste), and Soups, chapter: Further Mycotoxins and Microbial Metabolites

Aflatoxin G_2

incidence: 5/8, conc. range: <LOQ-12 µg/kg, Ø conc.: 7 µg/kg, sample origin: Bamenda (region, western highlands), Cameroon, sample year: July–September 2011, country: Cameroon/Austria/South Africa[659]

- Co-contamination: For detailed information see the article.
- Further contamination of plants and plant products other than cereals and their products reported in the present book/article[659]: Dagwa, AFB_1, AFB_2, AFG_1, AME, BEA, CIT, DON, ENA, ENA_1, ENB, ENB_1, FA, FB_1, FB_2, FB_3, KA, MA, NIV, OTA, OTB, STG, VIOM, and ZEA, literature[659]; Flour (soybean flour), AFB_1, AFG_1, BEA, DON, ENB, ENB_1, FA, FB_1, FB_2, FB_3, FUS-X, KA, NIV, and ZEA, literature[659]; Kuru-kuru, AFB_1, AFB_2, AFG_1, AFG_2, BEA, ENA, ENA_1, ENB, ENB_1, FA, FB_1, FB_3, KA, MA, OTA, OTB, ROQ-C, STG, VIOM, and ZEA, literature[659]; Paste (peanut paste), AFB_1, AFB_2, AFG_1, AFG_2, AME, BEA, ENA, ENA_1, ENB, ENB_1, FA, FB_1, FB_2, FB_3, KA, MA, OTA, OTB, STG, VIOM, and ZEA, literature[659]; Soups, AFB_1, AFB_2, AFG_1, AFG_2, AME, AOH, BEA, DON, ENA, ENA_1, ENB, ENB_1, FB_1, FB_2, FB_3, KA, OTA, OTB, and STG, literature[659]; see also Dagwa, Flour (soybean flour), Kuru-kuru, Paste (peanut paste), and Soups, chapter: Further Mycotoxins and Microbial Metabolites

Other *Aspergillus* Toxins

Sterigmatocystin

incidence: 6/8, conc. range: <LOQ-3 µg/kg, Ø conc.: 1.4 µg/kg, sample origin: Bamenda (region, western highlands), Cameroon, sample year: July–September 2011, country: Cameroon/Austria/South Africa[659]

- Co-contamination: For detailed information see the article.
- Further contamination of plants and plant products other than cereals and their products reported in the present book/article[659]: Dagwa, AFB_1, AFB_2, AFG_1, AFG_2, AME, BEA, CIT, DON, ENA, ENA_1, ENB, ENB_1, FA, FB_1, FB_2, FB_3, KA, MA, NIV, OTA, OTB, VIOM, and ZEA, literature[659]; Flour (soybean flour), AFB_1, AFG_1, BEA, DON, ENB, ENB_1, FA, FB_1, FB_2, FB_3, FUS-X, KA, NIV, and ZEA, literature[659]; Kuru-kuru, AFB_1, AFB_2, AFG_1, AFG_2, BEA, ENA, ENA_1, ENB, ENB_1, FA, FB_1, FB_3, KA, MA, OTA, OTB, ROQ-C, STG, VIOM, and ZEA, literature[659]; Paste (peanut paste), AFB_1, AFB_2, AFG_1, AFG_2, AME, BEA, ENA, ENA_1, ENB, ENB_1, FA, FB_1, FB_2, FB_3, KA, MA, OTA, OTB, STG, VIOM, and ZEA, literature[659]; Soups, AFB_1, AFB_2, AFG_1, AFG_2, AME, AOH, BEA, DON, ENA, ENA_1, ENB, ENB_1, FB_1, FB_2, FB_3, KA, OTA, OTB, and STG, literature[659]; see also Dagwa, Flour (soybean flour), Kuru-kuru, Paste (peanut paste), and Soups, chapter: Further Mycotoxins and Microbial Metabolites

Aspergillus and *Penicillium* Toxins

Citrinin

incidence: 2/8, conc. range: <LOQ-119 µg/kg, Ø conc.: 119 µg/kg, sample origin: Bamenda (region, western highlands), Cameroon, sample year: July–September 2011, country: Cameroon/Austria/South Africa[659]

- Co-contamination: For detailed information see the article.
- Further contamination of plants and plant products other than cereals and their products reported in the present book/article[659]: Dagwa, AFB_1, AFB_2, AFG_1, AFG_2, AME, BEA, DON, ENA, ENA_1, ENB, ENB_1, FA, FB_1, FB_2, FB_3, KA, MA, NIV, OTA, OTB, STG, VIOM, and ZEA, literature[659]; Flour (soybean flour), AFB_1, AFG_1, BEA, DON, ENB, ENB_1, FA, FB_1, FB_2, FB_3, FUS-X, KA, NIV, and ZEA, literature[659]; Kuru-kuru, AFB_1, AFB_2, AFG_1, AFG_2, BEA, ENA, ENA_1, ENB, ENB_1, FA, FB_1, FB_3, KA, MA, OTA, OTB, ROQ-C, STG, VIOM, and ZEA, literature[659]; Paste (peanut paste), AFB_1, AFB_2, AFG_1, AFG_2, AME, BEA, ENA, ENA_1, ENB, ENB_1, FA, FB_1, FB_2, FB_3, KA, MA, OTA, OTB, STG, VIOM, and ZEA, literature[659]; Soups, AFB_1, AFB_2, AFG_1, AFG_2, AME, AOH, BEA, DON, ENA, ENA_1, ENB, ENB_1, FB_1, FB_2, FB_3, KA, OTA, OTB, and STG, literature[659]; see also Dagwa, Flour (soybean flour), Kuru-kuru, Paste (peanut paste), and Soups, chapter: Further Mycotoxins and Microbial Metabolites

Kojic Acid

incidence: 8/8, conc. range: 421–3,549 µg/kg, Ø conc.: 960 µg/kg, sample origin: Bamenda (region, western highlands), Cameroon, sample year: July–September 2011, country: Cameroon/Austria/South Africa[659]

- Co-contamination: For detailed information see the article.
- Further contamination of plants and plant products other than cereals and their products reported in the present book/article[659]: Dagwa, AFB_1, AFB_2, AFG_1, AFG_2, AME, BEA, CIT, DON, ENA, ENA_1, ENB, ENB_1, FA, FB_1, FB_2, FB_3, MA, NIV, OTA, OTB, STG, VIOM, and ZEA, literature[659]; Flour (soybean flour), AFB_1, AFG_1, BEA, DON, ENB, ENB_1, FA, FB_1, FB_2, FB_3, FUS-X, KA, NIV, and ZEA, literature[659]; Kuru-kuru, AFB_1, AFB_2, AFG_1, AFG_2, BEA, ENA, ENA_1, ENB, ENB_1, FA, FB_1, FB_3, KA, MA, OTA, OTB, ROQ-C, STG, VIOM, and ZEA, literature[659]; Paste (peanut paste), AFB_1, AFB_2, AFG_1, AFG_2, AME, BEA, ENA, ENA_1, ENB, ENB_1, FA, FB_1, FB_2, FB_3, KA, MA, OTA, OTB, STG, VIOM, and ZEA, literature[659]; Soups, AFB_1, AFB_2, AFG_1, AFG_2, AME, AOH, BEA, DON, ENA, ENA_1, ENB, ENB_1, FB_1, FB_2, FB_3, KA, OTA, OTB, and STG, literature[659]; see also Dagwa, Flour (soybean flour), Kuru-kuru, Paste (peanut paste), and Soups, chapter: Further Mycotoxins and Microbial Metabolites

Ochratoxin A

incidence: 8/8, conc. range: <LOQ-6 µg/kg, Ø conc.: 3 µg/kg, sample origin: Bamenda (region, western highlands), Cameroon, sample year: July–September 2011, country: Cameroon/Austria/South Africa[659]

- Co-contamination: For detailed information see the article.
- Further contamination of plants and plant products other than cereals and their products reported in the present book/article[659]: Dagwa, AFB_1, AFB_2, AFG_1, AFG_2, AME, BEA, CIT, DON, ENA, ENA_1, ENB, ENB_1, FA, FB_1, FB_2, FB_3, KA, MA, NIV, OTB, STG, VIOM, and ZEA, literature[659]; Flour (soybean flour), AFB_1, AFG_1, BEA, DON, ENB, ENB_1, FA, FB_1, FB_2, FB_3, FUS-X, KA, NIV, and ZEA, literature[659]; Kuru-kuru, AFB_1, AFB_2, AFG_1, AFG_2, BEA, ENA, ENA_1, ENB, ENB_1, FA, FB_1, FB_3, KA, MA, OTA, OTB, ROQ-C, STG, VIOM, and ZEA, literature[659]; Paste (peanut paste), AFB_1, AFB_2, AFG_1, AFG_2, AME, BEA, ENA, ENA_1, ENB, ENB_1, FA, FB_1, FB_2, FB_3, KA, MA, OTA, OTB, STG, VIOM, and ZEA, literature[659]; Soups, AFB_1, AFB_2, AFG_1, AFG_2, AME, AOH, BEA, DON, ENA, ENA_1, ENB, ENB_1, FB_1, FB_2, FB_3, KA, OTA, OTB, and STG, literature[659]; see also Dagwa, Flour (soybean flour), Kuru-kuru, Paste (peanut paste), and Soups, chapter: Further Mycotoxins and Microbial Metabolites

Ochratoxin B

incidence: 6/8, conc. range: <LOQ-1 µg/kg, Ø conc.: 1 µg/kg, sample origin: Bamenda (region, western highlands), Cameroon, sample year: July–September 2011, country: Cameroon/Austria/South Africa[659]

- Co-contamination: For detailed information see the article.
- Further contamination of plants and plant products other than cereals and their products reported in the present book/article[659]: Dagwa, AFB_1, AFB_2, AFG_1, AFG_2, AME, BEA, CIT, DON, ENA, ENA_1, ENB, ENB_1, FA, FB_1, FB_2, FB_3, KA, MA, NIV, OTA, STG, VIOM, and ZEA, literature[659]; Flour (soybean flour), AFB_1, AFG_1, BEA, DON, ENB, ENB_1, FA, FB_1, FB_2, FB_3, FUS-X, KA, NIV, and ZEA, literature[659]; Kuru-kuru, AFB_1, AFB_2, AFG_1, AFG_2, BEA, ENA, ENA_1, ENB, ENB_1, FA, FB_1, FB_3, KA, MA, OTA, OTB, ROQ-C, STG, VIOM, and ZEA, literature[659]; Paste (peanut paste), AFB_1, AFB_2, AFG_1, AFG_2, AME, BEA, ENA, ENA_1, ENB, ENB_1, FA, FB_1, FB_2, FB_3, KA, MA, OTA, OTB, STG, VIOM, and ZEA, literature[659]; Soups, AFB_1, AFB_2, AFG_1, AFG_2, AME, AOH, BEA, DON, ENA, ENA_1, ENB, ENB_1, FB_1, FB_2, FB_3, KA, OTA, OTB, and STG, literature[659]; see also Dagwa, Flour (soybean flour), Kuru-kuru, Paste (peanut paste), and Soups, chapter: Further Mycotoxins and Microbial Metabolites

Viomellein

incidence: 7/8, conc. range: 20–100 µg/kg, Ø conc.: 56 µg/kg, sample origin: Bamenda (region, western highlands), Cameroon, sample year: July–September 2011, country: Cameroon/Austria/South Africa[659]

- Co-contamination: For detailed information see the article.
- Further contamination of plants and plant products other than cereals and their products reported in the present book/article[659]: Dagwa, AFB_1, AFB_2, AFG_1, AFG_2, AME, BEA, CIT, DON, ENA, ENA_1, ENB, ENB_1, FA, FB_1, FB_2, FB_3, KA, MA, NIV, OTA, OTB, STG, and ZEA, literature[659]; Flour (soybean flour), AFB_1, AFG_1, BEA, DON, ENB, ENB_1, FA, FB_1, FB_2, FB_3, FUS-X, KA, NIV, and ZEA, literature[659]; Kuru-kuru, AFB_1, AFB_2, AFG_1, AFG_2, BEA, ENA, ENA_1, ENB, ENB_1, FA, FB_1, FB_3, KA, MA, OTA, OTB, ROQ-C, STG, VIOM, and ZEA, literature[659]; Paste (peanut paste), AFB_1, AFB_2, AFG_1, AFG_2, AME, BEA, ENA, ENA_1, ENB, ENB_1, FA, FB_1, FB_2, FB_3, KA, MA, OTA, OTB, STG, VIOM, and ZEA, literature[659]; Soups, AFB_1, AFB_2, AFG_1, AFG_2, AME, AOH, BEA, DON, ENA, ENA_1, ENB, ENB_1, FB_1, FB_2, FB_3, KA, OTA, OTB, and STG, literature[659]; see also Dagwa, Flour (soybean flour), Kuru-kuru, Paste (peanut paste), and Soups, chapter: Further Mycotoxins and Microbial Metabolites

Fusarium Toxins

Depsipeptides

Beauvericin

incidence: 8/8, conc. range: 3.4–31 µg/kg, Ø conc.: 15 µg/kg, sample origin: Bamenda (region, western highlands), Cameroon, sample year: July–September 2011, country: Cameroon/Austria/South Africa[659]

- Co-contamination: For detailed information see the article.
- Further contamination of plants and plant products other than cereals and their products reported in the present book/article[659]: Dagwa, AFB_1, AFB_2, AFG_1, AFG_2, AME, CIT, DON, ENA, ENA_1, ENB, ENB_1, FA, FB_1, FB_2, FB_3, KA, MA, NIV, OTA, OTB, STG, VIOM, and ZEA, literature[659]; Flour (soybean flour), AFB_1, AFG_1, BEA, DON, ENB, ENB_1, FA, FB_1, FB_2, FB_3, FUS-X, KA, NIV, and ZEA, literature[659]; Kuru-kuru, AFB_1, AFB_2, AFG_1, AFG_2, BEA, ENA, ENA_1, ENB, ENB_1, FA, FB_1, FB_3, KA, MA, OTA, OTB, ROQ-C, STG, VIOM, and ZEA, literature[659]; Paste (peanut paste), AFB_1, AFB_2, AFG_1, AFG_2, AME, BEA, ENA, ENA_1, ENB, ENB_1, FA, FB_1, FB_2, FB_3, KA, MA, OTA, OTB, STG, VIOM, and ZEA, literature[659]; Soups, AFB_1, AFB_2, AFG_1, AFG_2, AME, AOH, BEA, DON, ENA, ENA_1, ENB, ENB_1, FB_1, FB_2, FB_3, KA, OTA, OTB, and STG, literature[659]; see also Dagwa, Flour (soybean flour), Kuru-kuru, Paste (peanut paste), and Soups, chapter: Further Mycotoxins and Microbial Metabolites

Enniatin A

incidence: 5/8, conc. range: <LOQ-0.2 µg/kg, Ø conc.: 0.1 µg/kg, sample origin: Bamenda (region, western highlands), Cameroon, sample year: July–September 2011, country: Cameroon/Austria/South Africa[659]

- Co-contamination: For detailed information see the article.
- Further contamination of plants and plant products other than cereals and their products reported in the present book/article[659]: Dagwa, AFB_1, AFB_2, AFG_1, AFG_2, AME, BEA, CIT, DON, ENA_1, ENB, ENB_1, FA, FB_1, FB_2, FB_3, KA, MA, NIV, OTA, OTB, STG, VIOM, and ZEA, literature[659]; Flour (soybean flour), AFB_1, AFG_1, BEA, DON, ENB, ENB_1, FA, FB_1, FB_2, FB_3, FUS-X, KA, NIV, and ZEA, literature[659]; Kuru-kuru, AFB_1, AFB_2, AFG_1, AFG_2, BEA, ENA, ENA_1, ENB, ENB_1, FA, FB_1, FB_3, KA, MA, OTA, OTB, ROQ-C, STG, VIOM, and ZEA, literature[659]; Paste (peanut paste), AFB_1, AFB_2, AFG_1, AFG_2, AME, BEA, ENA, ENA_1, ENB, ENB_1, FA, FB_1, FB_2, FB_3, KA, MA, OTA, OTB, STG, VIOM, and ZEA, literature[659]; Soups, AFB_1, AFB_2, AFG_1, AFG_2, AME, AOH, BEA, DON, ENA, ENA_1, ENB, ENB_1, FB_1, FB_2, FB_3, KA, OTA, OTB, and STG, literature[659]; see also Dagwa, Flour (soybean flour), Kuru-kuru, Paste (peanut paste), and Soups, chapter: Further Mycotoxins and Microbial Metabolites

Enniatin A_1

incidence: 3/8, conc. range: <LOQ-0.04 µg/kg, Ø conc.: 0.04 µg/kg, sample origin: Bamenda (region, western highlands), Cameroon, sample year: July–September 2011, country: Cameroon/Austria/South Africa[659]

- Co-contamination: For detailed information see the article.
- Further contamination of plants and plant products other than cereals and their products reported in the present book/

article[659]: Dagwa, AFB_1, AFB_2, AFG_1, AFG_2, AME, BEA, CIT, DON, ENA, ENB, ENB_1, FA, FB_1, FB_2, FB_3, KA, MA, NIV, OTA, OTB, STG, VIOM, and ZEA, literature[659]; Flour (soybean flour), AFB_1, AFG_1, BEA, DON, ENB, ENB_1, FA, FB_1, FB_2, FB_3, FUS-X, KA, NIV, and ZEA, literature[659]; Kuru-kuru, AFB_1, AFB_2, AFG_1, AFG_2, BEA, ENA, ENA_1, ENB, ENB_1, FA, FB_1, FB_3, KA, MA, OTA, OTB, ROQ-C, STG, VIOM, and ZEA, literature[659]; Paste (peanut paste), AFB_1, AFB_2, AFG_1, AFG_2, AME, BEA, ENA, ENA_1, ENB, ENB_1, FA, FB_1, FB_2, FB_3, KA, MA, OTA, OTB, STG, VIOM, and ZEA, literature[659]; Soups, AFB_1, AFB_2, AFG_1, AFG_2, AME, AOH, BEA, DON, ENA, ENA_1, ENB, ENB_1, FB_1, FB_2, FB_3, KA, OTA, OTB, and STG, literature[659]; see also Dagwa, Flour (soybean flour), Kuru-kuru, Paste (peanut paste), and Soups, chapter: Further Mycotoxins and Microbial Metabolites

Enniatin B

incidence: 7/8, conc. range: <LOQ-0.1 µg/kg, ∅ conc.: 0.02 µg/kg, sample origin: Bamenda (region, western highlands), Cameroon, sample year: July–September 2011, country: Cameroon/Austria/South Africa[659]

- Co-contamination: For detailed information see the article.
- Further contamination of plants and plant products other than cereals and their products reported in the present book/article[659]: Dagwa, AFB_1, AFB_2, AFG_1, AFG_2, AME, BEA, CIT, DON, ENA, ENA_1, ENB_1, FA, FB_1, FB_2, FB_3, KA, MA, NIV, OTA, OTB, STG, VIOM, and ZEA, literature[659]; Flour (soybean flour), AFB_1, AFG_1, BEA, DON, ENB, ENB_1, FA, FB_1, FB_2, FB_3, FUS-X, KA, NIV, and ZEA, literature[659]; Kuru-kuru, AFB_1, AFB_2, AFG_1, AFG_2, BEA, ENA, ENA_1, ENB, ENB_1, FA, FB_1, FB_3, KA, MA, OTA, OTB, ROQ-C, STG, VIOM, and ZEA, literature[659]; Paste (peanut paste), AFB_1, AFB_2, AFG_1, AFG_2, AME, BEA, ENA, ENA_1, ENB, ENB_1, FA, FB_1, FB_2, FB_3, KA, MA, OTA, OTB, STG, VIOM, and ZEA, literature[659]; Soups, AFB_1, AFB_2, AFG_1, AFG_2, AME, AOH, BEA, DON, ENA, ENA_1, ENB, ENB_1, FB_1, FB_2, FB_3, KA, OTA, OTB, and STG, literature[659]; see also Dagwa, Flour (soybean flour), Kuru-kuru, Paste (peanut paste), and Soups, chapter: Further Mycotoxins and Microbial Metabolites

Enniatin B_1

incidence: 8/8, conc. range: 0.01–0.8 µg/kg, ∅ conc.: 0. 2 µg/kg, sample origin: Bamenda (region, western highlands), Cameroon, sample year: July–September 2011, country: Cameroon/Austria/South Africa[659]

- Co-contamination: For detailed information see the article.
- Further contamination of plants and plant products other than cereals and their products reported in the present book/article[659]: Dagwa, AFB_1, AFB_2, AFG_1, AFG_2, AME, BEA, CIT, DON, ENA, ENA_1, ENB, FA, FB_1, FB_2, FB_3, KA, MA, NIV, OTA, OTB, STG, VIOM, and ZEA, literature[659]; Flour (soybean flour), AFB_1, AFG_1, BEA, DON, ENB, ENB_1, FA, FB_1, FB_2, FB_3, FUS-X, KA, NIV, and ZEA, literature[659]; Kuru-kuru, AFB_1, AFB_2, AFG_1, AFG_2, BEA, ENA, ENA_1, ENB, ENB_1, FA, FB_1, FB_3, KA, MA, OTA, OTB, ROQ-C, STG, VIOM, and ZEA, literature[659]; Paste (peanut paste), AFB_1, AFB_2, AFG_1, AFG_2, AME, BEA, ENA, ENA_1, ENB, ENB_1, FA, FB_1, FB_2, FB_3, KA, MA, OTA, OTB, STG, VIOM, and ZEA, literature[659]; Soups, AFB_1, AFB_2, AFG_1, AFG_2, AME, AOH, BEA, DON, ENA, ENA_1, ENB, ENB_1, FB_1, FB_2, FB_3, KA, OTA, OTB, and STG, literature[659]; see also Dagwa, Flour (soybean flour), Kuru-kuru, Paste (peanut paste), and Soups, chapter: Further Mycotoxins and Microbial Metabolites

Fumonisin/s

Fumonisin B_1

incidence: 8/8, conc. range: 47–132 µg/kg, ∅ conc.: 90 µg/kg, sample origin: Bamenda (region, western highlands), Cameroon, sample year: July–September 2011, country: Cameroon/Austria/South Africa[659]

- Co-contamination: For detailed information see the article.
- Further contamination of plants and plant products other than cereals and their products reported in the present book/article[659]: Dagwa, AFB_1, AFB_2, AFG_1, AFG_2, AME, BEA, CIT, DON, ENA, ENA_1, ENB, ENB_1, FA, FB_2, FB_3, KA, MA, NIV, OTA, OTB, STG, VIOM, and ZEA, literature[659]; Flour (soybean flour), AFB_1, AFG_1, BEA, DON, ENB, ENB_1, FA, FB_1, FB_2, FB_3, FUS-X, KA, NIV, and ZEA, literature[659]; Kuru-kuru, AFB_1, AFB_2, AFG_1, AFG_2, BEA, ENA, ENA_1, ENB, ENB_1, FA, FB_1, FB_3, KA, MA, OTA, OTB, ROQ-C, STG, VIOM, and ZEA, literature[659]; Paste (peanut paste), AFB_1, AFB_2, AFG_1, AFG_2, AME, BEA, ENA, ENA_1, ENB, ENB_1, FA, FB_1, FB_2, FB_3, KA, MA, OTA, OTB, STG, VIOM, and ZEA, literature[659]; Soups, AFB_1, AFB_2, AFG_1, AFG_2, AME, AOH, BEA, DON, ENA, ENA_1, ENB, ENB_1, FB_1, FB_2, FB_3, KA, OTA, OTB, and STG, literature[659]; see also Dagwa, Flour (soybean flour), Kuru-kuru, Paste (peanut paste), and Soups, chapter: Further Mycotoxins and Microbial Metabolites

Fumonisin B_2

incidence: 8/8, conc. range: 15–37 µg/kg, ∅ conc.: 23 µg/kg, sample origin: Bamenda (region, western highlands), Cameroon, sample year: July–September 2011, country: Cameroon/Austria/South Africa[659]

- Co-contamination: For detailed information see the article.
- Further contamination of plants and plant products other than cereals and their products reported in the present book/article[659]: Dagwa, AFB_1, AFB_2, AFG_1, AFG_2, AME, BEA, CIT, DON, ENA, ENA_1, ENB, ENB_1, FA, FB_1, FB_3, KA, MA, NIV, OTA, OTB, STG, VIOM, and ZEA, literature[659]; Flour (soybean flour), AFB_1, AFG_1, BEA, DON, ENB, ENB_1, FA, FB_1, FB_2, FB_3, FUS-X, KA, NIV, and ZEA, literature[659]; Kuru-kuru, AFB_1, AFB_2, AFG_1, AFG_2, BEA, ENA, ENA_1, ENB, ENB_1, FA, FB_1, FB_3, KA, MA, OTA, OTB, ROQ-C, STG, VIOM, and ZEA, literature[659]; Paste (peanut paste), AFB_1, AFB_2, AFG_1, AFG_2, AME, BEA, ENA, ENA_1, ENB, ENB_1, FA, FB_1, FB_2, FB_3, KA, MA, OTA, OTB, STG, VIOM, and ZEA, literature[659]; Soups, AFB_1, AFB_2, AFG_1, AFG_2, AME, AOH, BEA, DON, ENA, ENA_1, ENB, ENB_1, FB_1, FB_2, FB_3, KA, OTA, OTB, and STG, literature[659]; see also Dagwa, Flour (soybean flour), Kuru-kuru, Paste (peanut paste), and Soups, chapter: Further Mycotoxins and Microbial Metabolites

Fumonisin B_3

incidence: 7/8, conc. range: <LOQ-11 µg/kg, ∅ conc.: 7 µg/kg, sample origin: Bamenda (region, western highlands), Cameroon, sample year: July–September 2011, country: Cameroon/Austria/South Africa[659]

- Co-contamination: For detailed information see the article.
- Further contamination of plants and plant products other than cereals and their products reported in the present book/article[659]: Dagwa, AFB_1, AFB_2, AFG_1, AFG_2, AME, BEA, CIT, DON, ENA, ENA_1, ENB, ENB_1, FA, FB_1, FB_2, KA, MA, NIV, OTA, OTB, STG, VIOM, and ZEA, literature[659]; Flour (soybean flour), AFB_1, AFG_1, BEA, DON, ENB, ENB_1, FA, FB_1, FB_2, FB_3, FUS-X, KA, NIV, and ZEA, literature[659]; Kuru-kuru, AFB_1, AFB_2, AFG_1,

AFG_2, BEA, ENA, ENA_1, ENB, ENB_1, FA, FB_1, FB_3, KA, MA, OTA, OTB, ROQ-C, STG, VIOM, and ZEA, literature[659]; Paste (peanut paste), AFB_1, AFB_2, AFG_1, AFG_2, AME, BEA, ENA, ENA_1, ENB, ENB_1, FA, FB_1, FB_2, FB_3, KA, MA, OTA, OTB, STG, VIOM, and ZEA, literature[659]; Soups, AFB_1, AFB_2, AFG_1, AFG_2, AME, AOH, BEA, DON, ENA, ENA_1, ENB, ENB_1, FB_1, FB_2, FB_3, KA, OTA, OTB, and STG, literature[659]; see also Dagwa, Flour (soybean flour), Kuru-kuru, Paste (peanut paste), and Soups, chapter: Further Mycotoxins and Microbial Metabolites

Type B Trichothecenes

Deoxynivalenol

incidence: 8/8, conc. range: 12–116 µg/kg, Ø conc.: 52 µg/kg, sample origin: Bamenda (region, western highlands), Cameroon, sample year: July–September 2011, country: Cameroon/Austria/South Africa[659]

- Co-contamination: For detailed information see the article.
- Further contamination of plants and plant products other than cereals and their products reported in the present book/article[659]: Dagwa, AFB_1, AFB_2, AFG_1, AFG_2, AME, BEA, CIT, ENA, ENA_1, ENB, ENB_1, FA, FB_1, FB_2, FB_3, KA, MA, NIV, OTA, OTB, STG, VIOM, and ZEA, literature[659]; Flour (soybean flour), AFB_1, AFG_1, BEA, DON, ENB, ENB_1, FA, FB_1, FB_2, FB_3, FUS-X, KA, NIV, and ZEA, literature[659]; Kuru-kuru, AFB_1, AFB_2, AFG_1, AFG_2, BEA, ENA, ENA_1, ENB, ENB_1, FA, FB_1, FB_3, KA, MA, OTA, OTB, ROQ-C, STG, VIOM, and ZEA, literature[659]; Paste (peanut paste), AFB_1, AFB_2, AFG_1, AFG_2, AME, BEA, ENA, ENA_1, ENB, ENB_1, FA, FB_1, FB_2, FB_3, KA, MA, OTA, OTB, STG, VIOM, and ZEA, literature[659]; Soups, AFB_1, AFB_2, AFG_1, AFG_2, AME, AOH, BEA, DON, ENA, ENA_1, ENB, ENB_1, FB_1, FB_2, FB_3, KA, OTA, OTB, and STG, literature[659]; see also Dagwa, Flour (soybean flour), Kuru-kuru, Paste (peanut paste), and Soups, chapter: Further Mycotoxins and Microbial Metabolites

Nivalenol

incidence: 7/8, conc. range: 51–155 µg/kg, Ø conc.: 91 µg/kg, sample origin: Bamenda (region, western highlands), Cameroon, sample year: July–September 2011, country: Cameroon/Austria/South Africa[659]

- Co-contamination: For detailed information see the article.
- Further contamination of plants and plant products other than cereals and their products reported in the present book/article[659]: Dagwa, AFB_1, AFB_2, AFG_1, AFG_2, AME, BEA, CIT, DON, ENA, ENA_1, ENB, ENB_1, FA, FB_1, FB_2, FB_3, KA, MA, OTA, OTB, STG, VIOM, and ZEA, literature[659]; Flour (soybean flour), AFB_1, AFG_1, BEA, DON, ENB, ENB_1, FA, FB_1, FB_2, FB_3, FUS-X, KA, NIV, and ZEA, literature[659]; Kuru-kuru, AFB_1, AFB_2, AFG_1, AFG_2, BEA, ENA, ENA_1, ENB, ENB_1, FA, FB_1, FB_3, KA, MA, OTA, OTB, ROQ-C, STG, VIOM, and ZEA, literature[659]; Paste (peanut paste), AFB_1, AFB_2, AFG_1, AFG_2, AME, BEA, ENA, ENA_1, ENB, ENB_1, FA, FB_1, FB_2, FB_3, KA, MA, OTA, OTB, STG, VIOM, and ZEA, literature[659]; Soups, AFB_1, AFB_2, AFG_1, AFG_2, AME, AOH, BEA, DON, ENA, ENA_1, ENB, ENB_1, FB_1, FB_2, FB_3, KA, OTA, OTB, and STG, literature[659]; see also Dagwa, Flour (soybean flour), Kuru-kuru, Paste (peanut paste), and Soups, chapter: Further Mycotoxins and Microbial Metabolites

Other *Fusarium* Toxins

Fusaric Acid

incidence: 8/8, conc. range: 3.8–53.4 µg/kg, Ø conc.: 30.4 µg/kg, sample origin: Bamenda (region, western highlands), Cameroon, sample year: July–September 2011, country: Cameroon/Austria/South Africa[659]

- Co-contamination: For detailed information see the article.
- Further contamination of plants and plant products other than cereals and their products reported in the present book/article[659]: Dagwa, AFB_1, AFB_2, AFG_1, AFG_2, AME, BEA, CIT, DON, ENA, ENA_1, ENB, ENB_1, FB_1, FB_2, FB_3, KA, MA, NIV, OTA, OTB, STG, VIOM, and ZEA, literature[659]; Flour (soybean flour), AFB_1, AFG_1, BEA, DON, ENB, ENB_1, FA, FB_1, FB_2, FB_3, FUS-X, KA, NIV, and ZEA, literature[659]; Kuru-kuru, AFB_1, AFB_2, AFG_1, AFG_2, BEA, ENA, ENA_1, ENB, ENB_1, FA, FB_1, FB_3, KA, MA, OTA, OTB, ROQ-C, STG, VIOM, and ZEA, literature[659]; Paste (peanut paste), AFB_1, AFB_2, AFG_1, AFG_2, AME, BEA, ENA, ENA_1, ENB, ENB_1, FA, FB_1, FB_2, FB_3, KA, MA, OTA, OTB, STG, VIOM, and ZEA, literature[659]; Soups, AFB_1, AFB_2, AFG_1, AFG_2, AME, AOH, BEA, DON, ENA, ENA_1, ENB, ENB_1, FB_1, FB_2, FB_3, KA, OTA, OTB, and STG, literature[659]; see also Dagwa, Flour (soybean flour), Kuru-kuru, Paste (peanut paste), and Soups, chapter: Further Mycotoxins and Microbial Metabolites

Zearalenone

incidence: 8/8, conc. range: 6–57 µg/kg, Ø conc.: 32 µg/kg, sample origin: Bamenda (region, western highlands), Cameroon, sample year: July–September 2011, country: Cameroon/Austria/South Africa[659]

- Co-contamination: For detailed information see the article.
- Further contamination of plants and plant products other than cereals and their products reported in the present book/article[659]: Dagwa, AFB_1, AFB_2, AFG_1, AFG_2, AME, BEA, CIT, DON, ENA, ENA_1, ENB, ENB_1, FA, FB_1, FB_2, FB_3, KA, MA, NIV, OTA, OTB, STG, and VIOM, literature[659]; Flour (soybean flour), AFB_1, AFG_1, BEA, DON, ENB, ENB_1, FA, FB_1, FB_2, FB_3, FUS-X, KA, NIV, and ZEA, literature[659]; Kuru-kuru, AFB_1, AFB_2, AFG_1, AFG_2, BEA, ENA, ENA_1, ENB, ENB_1, FA, FB_1, FB_3, KA, MA, OTA, OTB, ROQ-C, STG, VIOM, and ZEA, literature[659]; Paste (peanut paste), AFB_1, AFB_2, AFG_1, AFG_2, AME, BEA, ENA, ENA_1, ENB, ENB_1, FA, FB_1, FB_2, FB_3, KA, MA, OTA, OTB, STG, VIOM, and ZEA, literature[659]; Soups, AFB_1, AFB_2, AFG_1, AFG_2, AME, AOH, BEA, DON, ENA, ENA_1, ENB, ENB_1, FB_1, FB_2, FB_3, KA, OTA, OTB, and STG, literature[659]; see also Dagwa, Flour (soybean flour), Kuru-kuru, Paste (peanut paste), and Soups, chapter: Further Mycotoxins and Microbial Metabolites

Penicillium Toxins

Mycophenolic Acid

incidence: 5/8, conc. range: 0.6–5 µg/kg, Ø conc.: 1.9 µg/kg, sample origin: Bamenda (region, western highlands), Cameroon, sample year: July–September 2011, country: Cameroon/Austria/South Africa[659]

- Co-contamination: For detailed information see the article.
- Further contamination of plants and plant products other than cereals and their products reported in the present book/article[659]: Dagwa, AFB_1, AFB_2, AFG_1, AFG_2, AME, BEA, CIT, DON, ENA, ENA_1, ENB, ENB_1, FA, FB_1, FB_2, FB_3, KA, NIV, OTA, OTB, STG, VIOM, and ZEA, literature[659]; Flour (soybean flour), AFB_1, AFG_1, BEA, DON, ENB, ENB_1, FA, FB_1, FB_2, FB_3, FUS-X, KA, NIV, and ZEA, literature[659]; Kuru-kuru, AFB_1, AFB_2, AFG_1, AFG_2, BEA, ENA, ENA_1, ENB, ENB_1, FA, FB_1, FB_3, KA, MA, OTA, OTB, ROQ-C, STG, VIOM, and ZEA, literature[659]; Paste (peanut paste), AFB_1, AFB_2, AFG_1, AFG_2, AME, BEA, ENA, ENA_1, ENB, ENB_1, FA, FB_1, FB_2, FB_3, KA, MA, OTA, OTB, STG, VIOM, and ZEA, literature[659]; Soups, AFB_1, AFB_2, AFG_1, AFG_2, AME, AOH, BEA, DON, ENA, ENA_1, ENB, ENB_1, FB_1, FB_2, FB_3, KA, OTA,

OTB, and STG, literature[659]; see also Dagwa, Flour (soybean flour), Kuru-kuru, Paste (peanut paste), and Soups, chapter: Further Mycotoxins and Microbial Metabolites

Dagwa is a solid, dried, fried paste containing milled peanuts and maize.

Dates may contain the following mycotoxins:

Alternaria Toxins

Tenuazonic Acid

incidence: 18/53*, conc. range: 9.6–4,411.4 µg/kg, ∅ conc.: 873.2 µg/kg, sample origin: principally Xinjiang (province), purchased in supermarkets and local markets in Beijing (capital), China, sample year: 2016/2017, country: China[815], *dates, dried

- Co-contamination: For detailed information see the article.
- Further contamination of plants and plant products other than cereals and their products reported in the present book/article[815]: Dates, MA and PA, literature[815]; Fruits, dried, AME, AOH, MA, PAT, and TA, literature[815]; Raisins, AME, AOH, MA, OTA, OTB, and TA, literature[815]; see also Dates and, Fruits, dried, chapter: Further Mycotoxins and Microbial Metabolites

Aspergillus Toxins

Aflatoxin B_1

incidence: 1/6*, conc.: 19 µg/kg, sample origin: major local retail outlets at Nainital (district), India, sample year: 1983–1985, country: India[7], *date palm

- Co-contamination: Often more than one mycotoxin was present in a sa.
- Further contamination of plants and plant products other than cereals and their products reported in the present book/article[7]: Coconuts, AFB_1, AFB_2, AFG_1, AFG_2, and CIT, literature[7]; Dates, AFB_2, AFG_1, and OTA, literature[7]; Grogannuts, AFB_1, AFB_2, AFG_1, AFG_2, and CIT, literature[7]; Nuts (almond), AFB_1, AFB_2, AFG_1, RBT, STG, and ZEA, literature[7]; Nuts (cashew nut), AFB_1, AFB_2, AFG_1, AFG_2, CIT, and OTA, literature[7]; Nuts (peanut), Raisins, AFB_1, AFB_2, AFG_1, AFG_2, and OTA, literature[7]; Nuts (pistachio nut), AFB_1, AFB_2, AFG_1, AFG_2, and STG, literature[7]; Nuts (walnut), AFB_1, AFB_2, AFG_1, AFG_2, and RBT, literature[7]

incidence: 1/12, conc.: 0.7 µg/kg, sample origin: retail outlets, UK, sample year: September 2001, country: UK[203]

- Co-contamination: not reported
- Further contamination of plants and plant products other than cereals and their products reported in the present book/article[203]: Butter (nut butter), AFB_1, AFB_2, AFG_1, AFG_2, and OTA, literature[203]; Butter (peanut butter), AFB_1, AFB_2, AFG_1, AFG_2, and OTA, literature[203]; Dates, OTA, literature[203]; Figs, AFB_1, AFB_2, AFG_2, and OTA, literature[203]; Fruits, dried, Nuts (pistachio nut), AFB_1, AFB_2, and OTA, literature[203]; Nuts (Brazil nut, paranut), AFB_1, AFB_2, and AFG_1, literature[203]; Nuts (peanut), AFB_1 and AFB_2, literature[203]

Aflatoxin B_2

incidence: 1/6*, conc.: 15 µg/kg, sample origin: major local retail outlets at Nainital (district), India, sample year: 1983–1985, country: India[7], *date palm

- Co-contamination: Often more than one mycotoxin was present in a sa.
- Further contamination of plants and plant products other than cereals and their products reported in the present book/article[7]: Coconuts, AFB_1, AFB_2, AFG_1, AFG_2, and CIT, literature[7]; Dates, AFB_1, AFG_1, and OTA, literature[7]; Grogannuts, AFB_1, AFB_2, AFG_1, AFG_2, and CIT, literature[7]; Nuts (almond), AFB_1, AFB_2, AFG_1, RBT, STG, and ZEA, literature[7]; Nuts (cashew nut), AFB_1, AFB_2, AFG_1, AFG_2, CIT, and OTA, literature[7]; Nuts (peanut), Raisins, AFB_1, AFB_2, AFG_1, AFG_2, and OTA, literature[7]; Nuts (pistachio nut), AFB_1, AFB_2, AFG_1, AFG_2, and STG, literature[7]; Nuts (walnut), AFB_1, AFB_2, AFG_1, AFG_2, and RBT, literature[7]

incidence: 16/75, conc. range: 1.1–1.3 µg/kg*, ∅ conc.: 1.14 µg/kg*, sample origin: local and from Algeria, Chile, Iran, Israel, and Turkey, purchased from markets in Valencia (city), Spain, and Tunis (capital), Tunisia, sample year: 2012/2013, country: Spain[720], *purchased in Tunisia

- Co-contamination: For detailed information see the article.
- Further contamination of plants and plant products other than cereals and their products reported in the present book/article[720]: Dates, AFG_1, AFG_2, DAS, ENA_1, ENB, ENB_1, and OTA, literature[720]; Figs, AFG_1, BEA, DAS, ENA_1, ENB, ENB_1, HT-2, and OTA, literature[720]; Fruits, dried, AFG_2, DAS, ENA_1, ENB, ENB_1, HT-2, and OTA, literature[720]; Vine fruits, dried, AFB_1, AFG_2, ENA, ENA_1, ENB, ENB_1, HT-2, and OTA, literature[720]

For detailed information see the article.

Aflatoxin G_1

incidence: 1/6*, conc.: 14 µg/kg, sample origin: major local retail outlets at Nainital (district), India, sample year: 1983–1985, country: India[7], *date palm

- Co-contamination: Often more than one mycotoxin was present in a sa.
- Further contamination of plants and plant products other than cereals and their products reported in the present book/article[7]: Coconuts, AFB_1, AFB_2, AFG_1, AFG_2, and CIT, literature[7]; Dates, AFB_1, AFB_2, and OTA, literature[7]; Grogannuts, AFB_1, AFB_2, AFG_1, AFG_2, and CIT, literature[7]; Nuts (almond), AFB_1, AFB_2, AFG_1, RBT, STG, and ZEA, literature[7]; Nuts (cashew nut), AFB_1, AFB_2, AFG_1, AFG_2, CIT, and OTA, literature[7]; Nuts (peanut), Raisins, AFB_1, AFB_2, AFG_1, AFG_2, and OTA, literature[7]; Nuts (pistachio nut), AFB_1, AFB_2, AFG_1, AFG_2, and STG, literature[7]; Nuts (walnut), AFB_1, AFB_2, AFG_1, AFG_2, and RBT, literature[7]

incidence: 8/75, conc. range: <LOQ-1.8 µg/kg*, ∅ conc.: 1.4 µg/kg*, sample origin: local and from Algeria, Chile, Iran, Israel, and Turkey, purchased from markets in Valencia (city), Spain, and Tunis (capital), Tunisia, sample year: 2012/2013, country: Spain[720], *purchased in Tunisia

- Co-contamination: For detailed information see the article.
- Further contamination of plants and plant products other than cereals and their products reported in the present book/article[720]: Dates, AFB_2, AFG_2, DAS, ENA_1, ENB, ENB_1, and OTA, literature[720]; Figs, AFG_1, BEA, DAS, ENA_1, ENB, ENB_1, HT-2, and OTA, literature[720]; Fruits, dried, AFG_2, DAS, ENA_1, ENB, ENB_1, HT-2, and OTA, literature[720]; Vine fruits, dried, AFB_1, AFG_2, ENA, ENA_1, ENB, ENB_1, HT-2, and OTA, literature[720]

For detailed information see the article.

Aflatoxin G_2

incidence: 15/75, conc. range: <LOQ-2.2 µg/kg*, ∅ conc.: 1.7 µg/kg*, sample origin: local and from Algeria, Chile, Iran, Israel, and Turkey, purchased from markets in Valencia (city), Spain, and Tunis (capital),

Tunisia, sample year: 2012/2013, country: Spain[720], *purchased in Tunisia

- Co-contamination: For detailed information see the article.
- Further contamination of plants and plant products other than cereals and their products reported in the present book/article[720]: Dates, AFB_2, AFG_1, DAS, ENA_1, ENB, ENB_1, and OTA, literature[720]; Figs, AFG_1, BEA, DAS, ENA_1, ENB, ENB_1, HT-2, and OTA, literature[720]; Fruits, dried, AFG_2, DAS, ENA_1, ENB, ENB_1, HT-2, and OTA, literature[720]; Vine fruits, dried, AFB_1, AFG_2, ENA, ENA_1, ENB, ENB_1, HT-2, and OTA, literature[720]

For detailed information see the article.

Aflatoxin/s

incidence: 2/20, conc. range: 2.1–2.9 µg/kg*, ∅ conc.: 2.5 µg/kg*, sample origin: retail shops and local markets in NWFP and northern areas, Pakistan, sample year: January–November 2009, country: Pakistan[17], *AFB_1, AFB_2, AFG_1, and AFG_2

- Co-contamination: not reported
- Further contamination of plants and plant products other than cereals and their products reported in the present book/article[17]: Figs, Fruits, dried, Kernels (apricot), Nuts (almond), Nuts (peanut), Nuts (pine nut), Nuts (pistachio nut), Nuts (walnut), Raisins, AF/S, literature[17]

incidence: 38/96*, conc. range: ≤26.60 µg/kg**, sample origin: southern areas of Punjab and Khyber Pakhtunkhwa (provinces), Pakistan, sample year: November 2011–March 2012, country: Pakistan/Malaysia[574], *different varieties, **AFB_1, AFB_2, AFG_1, and AFG_2

- Co-contamination: not reported
- Further contamination of plants and plant products other than cereals and their products reported in the present book/article[574]: Cake (date), Cookies, Halva, AF/S, literature[574]

Aspergillus and *Penicillium* Toxins

Ochratoxin A

incidence: 1/6*, conc.: pr, sample origin: major local retail outlets at Nainital (district), India, sample year: 1983–1985, country: India[7], *date palm

- Co-contamination: Often more than one mycotoxin was present in a sa.
- Further contamination of plants and plant products other than cereals and their products reported in the present book/article[7]: Coconuts, AFB_1, AFB_2, AFG_1, AFG_2, and CIT, literature[7]; Dates, AFB_1, AFB_2, and AFG_1, literature[7]; Grogannuts, AFB_1, AFB_2, AFG_1, AFG_2, and CIT, literature[7]; Nuts (almond), AFB_1, AFB_2, AFG_1, RBT, STG, and ZEA, literature[7]; Nuts (cashew nut), AFB_1, AFB_2, AFG_1, AFG_2, CIT, and OTA, literature[7]; Nuts (peanut), Raisins, AFB_1, AFB_2, AFG_1, AFG_2, and OTA, literature[7]; Nuts (pistachio nut), AFB_1, AFB_2, AFG_1, AFG_2, and STG, literature[7]; Nuts (walnut), AFB_1, AFB_2, AFG_1, AFG_2, and RBT, literature[7]

incidence: 1/12, conc.: 0.9 µg/kg, sample origin: retail outlets, UK, sample year: September 2001, country: UK[203]

- Co-contamination: not reported
- Further contamination of plants and plant products other than cereals and their products reported in the present book/article[203]: Butter (nut butter), AFB_1, AFB_2, AFG_1, AFG_2, and OTA, literature[203]; Butter (peanut butter), AFB_1, AFB_2, AFG_1, AFG_2, and OTA, literature[203]; Dates, AFB_1, literature[203]; Figs, AFB_1, AFB_2, AFG_2, and OTA, literature[203]; Fruits, dried, Nuts (pistachio nut), AFB_1, AFB_2, and OTA, literature[203]; Nuts (Brazil nut, paranut), AFB_1, AFB_2, and AFG_1, literature[203]; Nuts (peanut), AFB_1 and AFB_2, literature[203]

incidence: 1/2, conc.: 1.02 µg/kg, sample origin: available in New Zealand, sample year: 2007, country: New Zealand[277]

- Co-contamination: not reported
- Further contamination of plants and plant products other than cereals and their products reported in the present book/article[277]: Bars, AFB_1, AF/S, DON, NIV, and OTA, literature[277]; Butter (peanut butter), Confectionery, Food, Fruits, dried, Nuts, Nuts (peanut), Nuts (pistachio nut), Sauces (peanut sauce), AFB_1 and AF/S, literature[277]; Coffee, Spices (nutmeg), Spices (pepper), Vine fruits, dried, OTA, literature[277]; Figs, Spices (cayenne pepper, chilli), Spices (curry), Spices (ginger), Spices (paprika), AFB_1, AF/S, and OTA, literature[277]; Nuts (Brazil nut, paranut), AFB_1, literature[277]

incidence: 2/22*, conc. range: 0.1–5.0 µg/kg, sample origin: different markets in Campinas (city) and São Paulo (city/state), Brazil, sample year: 2002/2003, country: Brazil[331], *dates, dried

- Co-contamination: not reported
- Further contamination of plants and plant products other than cereals and their products reported in the present book/article[331]: Figs, Fruits, dried, Sultanas, OTA, literature[331]

incidence: 1/52*, conc.: 0.39 µg/kg, sample origin: supermarkets and stores, USA, sample year: 2012–2014, country: USA[599], *conventional

incidence: 0/43*, conc.: no contamination, sample origin: supermarkets and stores, USA, sample year: 2012–2014, country: USA[599], *organic

- Co-contamination: not reported
- Further contamination of plants and plant products other than cereals and their products reported in the present book/article[599]: Figs, Nuts (pistachio nut), Raisins, OTA, literature[599]

incidence: 18/75, conc. range: 0.57–3.3 µg/kg*, ∅ conc.: 1.3 µg/kg*, sample origin: local and from Algeria, Chile, Iran, Israel, and Turkey, purchased from markets in Valencia (city), Spain, and Tunis (capital), Tunisia, sample year: 2012/2013, country: Spain[720], *purchased in Tunisia

- Co-contamination: For detailed information see the article.
- Further contamination of plants and plant products other than cereals and their products reported in the present book/article[720]: Dates, AFB_2, AFG_1, AFG_2, DAS, ENA_1, ENB, and, ENB_1, literature[720]; Figs, AFG_1, BEA, DAS, ENA_1, ENB, ENB_1, HT-2, and OTA, literature[720]; Fruits, dried, AFG_2, DAS, ENA_1, ENB, ENB_1, HT-2, and OTA, literature[720]; Vine fruits, dried, AFB_1, AFG_2, ENA, ENA_1, ENB, ENB_1, HT-2, and OTA, literature[720]

For detailed information see the article.

Penicillic Acid

incidence: 14/53*, conc. range: 20.1–85.6 µg/kg, ∅ conc.: 39.1 µg/kg, sample origin: principally Xinjiang (province), purchased in supermarkets and local markets in Beijing (capital), China, sample year: 2016/2017, country: China[815], *dates, dried

- Co-contamination: For detailed information see the article.
- Further contamination of plants and plant products other than cereals and their products reported in the present book/article[815]: Dates, MA and TA, literature[815]; Fruits, dried, AME,

AOH, MA, PAT, and TA, literature[815]; Raisins, AME, AOH, MA, OTA, OTB, and TA, literature[815]; see also Dates, and Fruits, dried, chapter: Further Mycotoxins and Microbial Metabolites

Fusarium Toxins

Depsipeptides

Enniatin A_1

incidence: 28/75, conc. range: <LOQ-8.5 µg/kg*, ∅ conc.: 3.8 µg/kg*, sample origin: local and from Algeria, Chile, Iran, Israel, and Turkey, purchased from markets in Valencia (city), Spain, and Tunis (capital), Tunisia, sample year: 2012/2013, country: Spain[720], *purchased in Spain and Tunisia

- Co-contamination: For detailed information see the article.
- Further contamination of plants and plant products other than cereals and their products reported in the present book/article[720]: Dates, AFB_2, AFG_1, AFG_2, DAS, ENB, ENB_1, and OTA, literature[720]; Figs, AFG_1, BEA, DAS, ENA_1, ENB, ENB_1, HT-2, and OTA, literature[720]; Fruits, dried, AFG_2, DAS, ENA_1, ENB, ENB_1, HT-2, and OTA, literature[720]; Vine fruits, dried, AFB_1, AFG_2, ENA, ENA_1, ENB, ENB_1, HT-2, and OTA, literature[720]

For detailed information see the article.

Enniatin B

incidence: 53/75, conc. range: <LOQ-190 µg/kg*, ∅ conc.: 101.2 µg/kg*, sample origin: local and from Algeria, Chile, Iran, Israel, and Turkey, purchased from markets in Valencia (city), Spain, and Tunis (capital), Tunisia, sample year: 2012/2013, country: Spain[720], *purchased in Spain and Tunisia

- Co-contamination: For detailed information see the article.
- Further contamination of plants and plant products other than cereals and their products reported in the present book/article[720]: Dates, AFB_2, AFG_1, AFG_2, DAS, ENA_1, ENB_1, and OTA, literature[720]; Figs, AFG_1, BEA, DAS, ENA_1, ENB, ENB_1, HT-2, and OTA, literature[720]; Fruits, dried, AFG_2, DAS, ENA_1, ENB, ENB_1, HT-2, and OTA, literature[720]; Vine fruits, dried, AFB_1, AFG_2, ENA, ENA_1, ENB, ENB_1, HT-2, and OTA, literature[720]

For detailed information see the article.

Enniatin B_1

incidence: 3/75, conc. range: 10.7–88.8 µg/kg*, ∅ conc.: 39.8 µg/kg*, sample origin: local and from Algeria, Chile, Iran, Israel, and Turkey, purchased from markets in Valencia (city), Spain, and Tunis (capital), Tunisia, sample year: 2012/2013, country: Spain[720], *purchased in Spain and Tunisia

- Co-contamination: For detailed information see the article.
- Further contamination of plants and plant products other than cereals and their products reported in the present book/article[720]: Dates, AFB_2, AFG_1, AFG_2, DAS, ENA_1, ENB, and OTA, literature[720]; Figs, AFG_1, BEA, DAS, ENA_1, ENB, ENB_1, HT-2, and OTA, literature[720]; Fruits, dried, AFG_2, DAS, ENA_1, ENB, ENB_1, HT-2, and OTA, literature[720]; Vine fruits, dried, AFB_1, AFG_2, ENA, ENA_1, ENB, ENB_1, HT-2, and OTA, literature[720]

For detailed information see the article.

Type A Trichothecenes

Diacetoxyscirpenol

incidence: 2/75, conc. range: <LOQ*, sample origin: local and from Algeria, Chile, Iran, Israel, and Turkey, purchased from markets in Valencia (city), Spain, and Tunis (capital), Tunisia, sample year: 2012/2013, country: Spain[720], *purchased in Tunisia

- Co-contamination: For detailed information see the article.
- Further contamination of plants and plant products other than cereals and their products reported in the present book/article[720]: Dates, AFB_2, AFG_1, AFG_2, ENA_1, ENB, ENB_1, and OTA, literature[720]; Figs, AFG_1, BEA, DAS, ENA_1, ENB, ENB_1, HT-2, and OTA, literature[720]; Fruits, dried, AFG_2, DAS, ENA_1, ENB, ENB_1, HT-2, and OTA, literature[720]; Vine fruits, dried, AFB_1, AFG_2, ENA, ENA_1, ENB, ENB_1, HT-2, and OTA, literature[720]

For detailed information see the article.

Penicillium Toxins

Mycophenolic Acid

incidence: 1/53*, conc.: 7.5 µg/kg, sample origin: principally Xinjiang (province), purchased in supermarkets and local markets in Beijing (capital), China, sample year: 2016/2017, country: China[815], *dates, dried

- Co-contamination: For detailed information see the article.
- Further contamination of plants and plant products other than cereals and their products reported in the present book/article[815]: Dates, PA and TA, literature[815]; Fruits, dried, AME, AOH, MA, PAT, and TA, literature[815]; Raisins, AME, AOH, MA, OTA, OTB, and TA, literature[815]; see also Dates and, Fruits, dried, chapter: Further Mycotoxins and Microbial Metabolites

Dessert wine see Wine, miscellaneous

Dietary plants may contain the following mycotoxins:

Fusarium Toxins

Fumonisin/s

Fumonisin B_1

incidence: 4/19*, conc. range: 34–524 µg/kg, ∅ conc.: 183 µg/kg, sample origin: different locations, South Africa, sample year: November 2002, country: South Africa[81], *dietary wild plants

- Co-contamination: not reported
- Further contamination of plants and plant products other than cereals and their products reported in the present book/article[81]: Medicinal plants, FB_1, literature[81]

For detailed information see the article.

Dietary supplements may contain the following mycotoxins:

Alternaria Toxins

Alternariol

incidence: 29/32*, conc. range: 16–4,560 µg/kg, sample origin: Czech and US retail markets, sample year: unknown, country: Czech Republic/USA[286], *dietary supplements (milk thistle-based supplements)

incidence: 4/9*, conc. range: 82–223 µg/kg, sample origin: Czech and US retail markets, sample year: unknown, country: Czech Republic/USA[286], *dietary supplements for menopause effects treatment

incidence: 8/28*, conc. range: 14–1,460 µg/kg, sample origin: Czech and US retail markets, sample year: unknown, country: Czech Republic/USA[286], *dietary supplements for the general health improvement

- Co-contamination: For detailed information see the article.
- Further contamination of plants and plant products other than cereals and their products reported in the present book/ article[286]: Dietary supplements, AME, BEA, DAS, DON, 3-AcDON, ENA, ENA_1, ENB, ENB_1, FB_1, FB_2, FB_3, FUS-X, HT-2, MA, NEO, OTA, PA, PAT, STG, T-2, TA, and ZEA, literature[286]; see also Dietary supplements, chapter: Further Mycotoxins and Microbial Metabolites

For detailed information see the article.

Alternariol Methyl Ether

incidence: 31/32*, conc. range: 6–3,200 µg/kg, sample origin: Czech and US retail markets, sample year: unknown, country: Czech Republic/USA[286], *dietary supplements (milk thistle-based supplements)

incidence: 6/9*, conc. range: 5–216 µg/kg, sample origin: Czech and US retail markets, sample year: unknown, country: Czech Republic/USA[286], *dietary supplements for menopause effects treatment

incidence: 17/28*, conc. range: 5–1,080 µg/kg, sample origin: Czech and US retail markets, sample year: unknown, country: Czech Republic/USA[286], *dietary supplements for the general health improvement

- Co-contamination: For detailed information see the article.
- Further contamination of plants and plant products other than cereals and their products reported in the present book/ article[286]: Dietary supplements, AOH, BEA, DAS, DON, 3-AcDON, ENA, ENA_1, ENB, ENB_1, FB_1, FB_2, FB_3, FUS-X, HT-2, MA, NEO, OTA, PA, PAT, STG, T-2, TA, and ZEA, literature[286]; see also Dietary supplements, chapter: Further Mycotoxins and Microbial Metabolites

For detailed information see the article.

Tenuazonic Acid

incidence: 7/32*, conc. range: 729–2,140 µg/kg, sample origin: Czech and US retail markets, sample year: unknown, country: Czech Republic/USA[286], *dietary supplements (milk thistle-based supplements)

incidence: 1/9*, conc.: 1,210 µg/kg, sample origin: Czech and US retail markets, sample year: unknown, country: Czech Republic/USA[286], *dietary supplements for menopause effects treatment

incidence: 1/28*, conc.: 6,780 µg/kg, sample origin: Czech and US retail markets, sample year: unknown, country: Czech Republic/USA[286], *dietary supplements for the general health improvement

- Co-contamination: For detailed information see the article.
- Further contamination of plants and plant products other than cereals and their products reported in the present book/ article[286]: Dietary supplements, AME, AOH, BEA, DAS, DON, 3-AcDON, ENA, ENA_1, ENB, ENB_1, FB_1, FB_2, FB_3, FUS-X, HT-2, MA, NEO, OTA, PA, PAT, STG, T-2, and ZEA, literature[286]; see also Dietary supplements, chapter: Further Mycotoxins and Microbial Metabolites

For detailed information see the article.

Aspergillus Toxins

Sterigmatocystin

incidence: 2/32*, conc. range: 7–11 µg/kg, Ø conc.: 9 µg/kg, sample origin: Czech and US retail markets, sample year: unknown, country: Czech Republic/USA[286], *dietary supplements (milk thistle-based supplements)

incidence: 1/9*, conc.: 24 µg/kg, sample origin: Czech and US retail markets, sample year: unknown, country: Czech Republic/ USA[286], *dietary supplements for menopause effects treatment

incidence: 6/28*, conc.: 5–42 µg/kg, sample origin: Czech and US retail markets, sample year: unknown, country: Czech Republic/ USA[286], *dietary supplements for the general health improvement

- Co-contamination: For detailed information see the article.
- Further contamination of plants and plant products other than cereals and their products reported in the present book/ article[286]: Dietary supplements, AME, AOH, BEA, DAS, DON, 3-AcDON, ENA, ENA_1, ENB, ENB_1, FB_1, FB_2, FB_3, FUS-X, HT-2, MA, NEO, OTA, PA, PAT, T-2, TA, and ZEA, literature[286]; see also Dietary supplements, chapter: Further Mycotoxins and Microbial Metabolites

For detailed information see the article.

Aspergillus and *Penicillium* Toxins

Ochratoxin A

incidence: 0/32*, conc.: no contamination, sample origin: Czech and US retail markets, sample year: unknown, country: Czech Republic/ USA[286], *dietary supplements (milk thistle-based supplements)

incidence: 0/9*, conc.: no contamination, sample origin: Czech and US retail markets, sample year: unknown, country: Czech Republic/ USA[286], *dietary supplements for menopause effects treatment

incidence: 1/28*, conc.: 956 µg/kg, sample origin: Czech and US retail markets, sample year: unknown, country: Czech Republic/ USA[286], *dietary supplements for the general health improvement

- Co-contamination: For detailed information see the article.
- Further contamination of plants and plant products other than cereals and their products reported in the present book/ article[286]: Dietary supplements, AME, AOH, BEA, DAS, DON, 3-AcDON, ENA, ENA_1, ENB, ENB_1, FB_1, FB_2, FB_3, FUS-X, HT-2, MA, NEO, PA, PAT, STG, T-2, TA, and ZEA, literature[286]; see also Dietary supplements, chapter: Further Mycotoxins and Microbial Metabolites

For detailed information see the article.

incidence: 18/50*, conc. range: 2.7–136.9 µg/kg, Ø conc.: 37.2 µg/kg, sample origin: internet vendors, sample year: unknown, country: USA[865], *dietary supplements containing green coffee bean extracts

- Co-contamination: 7 sa. co-contaminated with OTA and OTB; 5 sa. co-contaminated with OTA, OTB, and MA; 3 sa. co-contaminated with OTA, OTB, MA and FB_1; 1 sa. co-contaminated with OTA, OTB, and FB_1; 1 sa. co-contaminated with OTA and FB_1; 1 sa. contaminated solely with OTA
- Further contamination of plants and plant products other than cereals and their products reported in the present book/ article[865]: Dietary supplements, OTB, FB_1, and MA, literature[865]

Ochratoxin B

incidence: 16/50*, conc. range: 3.5–20.2 µg/kg, Ø conc.: 6.68 µg/kg, sample origin: internet vendors, sample year: unknown, country: USA[865], *dietary supplements containing green coffee bean extracts

- Co-contamination: 7 sa. co-contaminated with OTA and OTB; 5 sa. co-contaminated with OTA, OTB, and MA; 3 sa. co-contaminated with OTA, OTB, MA and FB_1; 1 sa. co-contaminated with OTA, OTB, and FB_1

- Further contamination of plants and plant products other than cereals and their products reported in the present book/ article[865]: Dietary supplements, OTA, FB_1, and MA, literature[865]

Patulin

incidence: 0/32*, conc.: no contamination, sample origin: Czech and US retail markets, sample year: unknown, country: Czech Republic/ USA[286], *dietary supplements (milk thistle-based supplements)

incidence: 0/9*, conc.: no contamination, sample origin: Czech and US retail markets, sample year: unknown, country: Czech Republic/USA[286], *dietary supplements for menopause effects treatment

incidence: 1/28*, conc.: 380 µg/kg, sample origin: Czech and US retail markets, sample year: unknown, country: Czech Republic/ USA[286], *dietary supplements for the general health improvement

- Co-contamination: For detailed information see the article.
- Further contamination of plants and plant products other than cereals and their products reported in the present book/ article[286]: Dietary supplements, AME, AOH, BEA, DAS, DON, 3-AcDON, ENA, ENA_1, ENB, ENB_1, FB_1, FB_2, FB_3, FUS-X, HT-2, MA, NEO, OTA, PA, STG, T-2, TA, and ZEA, literature[286]; see also Dietary supplements, chapter: Further Mycotoxins and Microbial Metabolites

For detailed information see the article.

Penicillic Acid

incidence: 0/32*, conc.: no contamination, sample origin: Czech and US retail markets, sample year: unknown, country: Czech Republic/ USA[286], *dietary supplements (milk thistle-based supplements)

incidence: 1/9*, conc.: 93 µg/kg, sample origin: Czech and US retail markets, sample year: unknown, country: Czech Republic/ USA[286], *dietary supplements for menopause effects treatment

incidence: 0/28*, conc.: no contamination, sample origin: Czech and US retail markets, sample year: unknown, country: Czech Republic/USA[286], *dietary supplements for the general health improvement

- Co-contamination: For detailed information see the article.
- Further contamination of plants and plant products other than cereals and their products reported in the present book/ article[286]: Dietary supplements, AME, AOH, BEA, DAS, DON, 3-AcDON, ENA, ENA_1, ENB, ENB_1, FB_1, FB_2, FB_3, FUS-X, HT-2, MA, NEO, OTA, PAT, STG, T-2, TA, and ZEA, literature[286]; see also Dietary supplements, chapter: Further Mycotoxins and Microbial Metabolites

For detailed information see the article.

Fusarium Toxins

Depsipeptides

Beauvericin

incidence: 29/32*, conc. range: 5–2,730 µg/kg, sample origin: Czech and US retail markets, sample year: unknown, country: Czech Republic/USA[286], *dietary supplements (milk thistle-based supplements)

incidence: 6/9*, conc. range: 8–131 µg/kg, sample origin: Czech and US retail markets, sample year: unknown, country: Czech Republic/USA[286], *dietary supplements for menopause effects treatment

incidence: 9/28*, conc. range: 12–215 µg/kg, sample origin: Czech and US retail markets, sample year: unknown, country: Czech Republic/USA[286], *dietary supplements for the general health improvement

- Co-contamination: For detailed information see the article.
- Further contamination of plants and plant products other than cereals and their products reported in the present book/ article[286]: Dietary supplements, AME, AOH, DAS, DON, 3-AcDON, ENA, ENA_1, ENB, ENB_1, FB_1, FB_2, FB_3, FUS-X, HT-2, MA, NEO, OTA, PA, PAT, STG, T-2, TA, and ZEA, literature[286]; see also Dietary supplements, chapter: Further Mycotoxins and Microbial Metabolites

For detailed information see the article.

Enniatin A

incidence: 27/32*, conc. range: 5–8,340 µg/kg, sample origin: Czech and US retail markets, sample year: unknown, country: Czech Republic/USA[286], *dietary supplements (milk thistle-based supplements)

incidence: 6/9*, conc. range: 5–89 µg/kg, sample origin: Czech and US retail markets, sample year: unknown, country: Czech Republic/USA[286], *dietary supplements for menopause effects treatment

incidence: 5/28*, conc. range: 5–13 µg/kg, sample origin: Czech and US retail markets, sample year: unknown, country: Czech Republic/USA[286], *dietary supplements for the general health improvement

- Co-contamination: For detailed information see the article.
- Further contamination of plants and plant products other than cereals and their products reported in the present book/ article[286]: Dietary supplements, AME, AOH, BEA, DAS, DON, 3-AcDON, ENA_1, ENB, ENB_1, FB_1, FB_2, FB_3, FUS-X, HT-2, MA, NEO, OTA, PA, PAT, STG, T-2, TA, and ZEA, literature[286]; see also Dietary supplements, chapter: Further Mycotoxins and Microbial Metabolites

For detailed information see the article.

Enniatin A_1

incidence: 27/32*, conc. range: 5–2,340 µg/kg, sample origin: Czech and US retail markets, sample year: unknown, country: Czech Republic/USA[286], *dietary supplements (milk thistle-based supplements)

incidence: 6/9*, conc. range: 12–224 µg/kg, sample origin: Czech and US retail markets, sample year: unknown, country: Czech Republic/ USA[286], *dietary supplements for menopause effects treatment

incidence: 3/28*, conc. range: 15–56 µg/kg, sample origin: Czech and US retail markets, sample year: unknown, country: Czech Republic/USA[286], *dietary supplements for the general health improvement

- Co-contamination: For detailed information see the article.
- Further contamination of plants and plant products other than cereals and their products reported in the present book/ article[286]: Dietary supplements, AME, AOH, BEA, DAS, DON, 3-AcDON, ENA, ENB, ENB_1, FB_1, FB_2, FB_3, FUS-X, HT-2, MA, NEO, OTA, PA, PAT, STG, T-2, TA, and ZEA, literature[286]; see also Dietary supplements, chapter: Further Mycotoxins and Microbial Metabolites

For detailed information see the article.

Enniatin B

incidence: 28/32*, conc. range: 16–9,260 μg/kg, sample origin: Czech and US retail markets, sample year: unknown, country: Czech Republic/USA[286], *dietary supplements (milk thistle-based supplements)

incidence: 7/9*, conc. range: 5–1,230 μg/kg, sample origin: Czech and US retail markets, sample year: unknown, country: Czech Republic/USA[286], *dietary supplements for menopause effects treatment

incidence: 7/28*, conc. range: 5–136 μg/kg, sample origin: Czech and US retail markets, sample year: unknown, country: Czech Republic/USA[286], *dietary supplements for the general health improvement

- Co-contamination: For detailed information see the article.
- Further contamination of plants and plant products other than cereals and their products reported in the present book/article[286]: Dietary supplements, AME, AOH, BEA, DAS, DON, 3-AcDON, ENA, ENA_1, ENB_1, FB_1, FB_2, FB_3, FUS-X, HT-2, MA, NEO, OTA, PA, PAT, STG, T-2, TA, and ZEA, literature[286]; see also Dietary supplements, chapter: Further Mycotoxins and Microbial Metabolites

For detailed information see the article.

Enniatin B$_1$

incidence: 29/32*, conc. range: 6–10,900 μg/kg, sample origin: Czech and US retail markets, sample year: unknown, country: Czech Republic/USA[286], *dietary supplements (milk thistle-based supplements)

incidence: 6/9*, conc. range: 28–447 μg/kg, sample origin: Czech and US retail markets, sample year: unknown, country: Czech Republic/USA[286], *dietary supplements for menopause effects treatment

incidence: 4/28*, conc. range: 5–71 μg/kg, sample origin: Czech and US retail markets, sample year: unknown, country: Czech Republic/USA[286], *dietary supplements for the general health improvement

- Co-contamination: For detailed information see the article.
- Further contamination of plants and plant products other than cereals and their products reported in the present book/article[286]: Dietary supplements, AME, AOH, BEA, DAS, DON, 3-AcDON, ENA, ENA_1, ENB, FB_1, FB_2, FB_3, FUS-X, HT-2, MA, NEO, OTA, PA, PAT, STG, T-2, TA, and ZEA, literature[286]; see also Dietary supplements, chapter: Further Mycotoxins and Microbial Metabolites

For detailed information see the article.

Fumonisin/s

Fumonisin B$_1$

incidence: 0/32*, conc.: no contamination, sample origin: Czech and US retail markets, sample year: unknown, country: Czech Republic/USA[286], *dietary supplements (milk thistle-based supplements)

incidence: 0/9*, conc.: no contamination, sample origin: Czech and US retail markets, sample year: unknown, country: Czech Republic/USA[286], *dietary supplements for menopause effects treatment

incidence: 2/28*, conc. range: 110–141 μg/kg, Ø conc.: 126 μg/kg, sample origin: Czech and US retail markets, sample year: unknown, country: Czech Republic/USA[286], *dietary supplements for the general health improvement

- Co-contamination: For detailed information see the article.
- Further contamination of plants and plant products other than cereals and their products reported in the present book/article[286]: Dietary supplements, AME, AOH, BEA, DAS, DON, 3-AcDON, ENA, ENA_1, ENB, ENB_1, FB_2, FB_3, FUS-X, HT-2, MA, NEO, OTA, PA, PAT, STG, T-2, TA, and ZEA, literature[286]; see also Dietary supplements, chapter: Further Mycotoxins and Microbial Metabolites

For detailed information see the article.

incidence: 5/50*, conc. range: 110.0–415.0 μg/kg, Ø conc.: 295 μg/kg, sample origin: internet vendors, sample year: unknown, country: USA[865], *dietary supplements containing green coffee bean extracts

- Co-contamination: 3 sa. co-contaminated with OTA, OTB, MA, and FB_1; 1 sa. co-contaminated with OTA, OTB, and FB_1; 1 sa. co-contaminated with OTA and FB_1
- Further contamination of plants and plant products other than cereals and their products reported in the present book/article[865]: Dietary supplements, OTA, OTB, and MA, literature[865]

Fumonisin B$_2$

incidence: 0/32*, conc.: no contamination, sample origin: Czech and US retail markets, sample year: unknown, country: Czech Republic/USA[286], *dietary supplements (milk thistle-based supplements)

incidence: 1/9*, conc.: 23 μg/kg, sample origin: Czech and US retail markets, sample year: unknown, country: Czech Republic/USA[286], *dietary supplements for menopause effects treatment

incidence: 4/28*, conc. range: 7–191 μg/kg, sample origin: Czech and US retail markets, sample year: unknown, country: Czech Republic/USA[286], *dietary supplements for the general health improvement

- Co-contamination: For detailed information see the article.
- Further contamination of plants and plant products other than cereals and their products reported in the present book/article[286]: Dietary supplements, AME, AOH, BEA, DAS, DON, 3-AcDON, ENA, ENA_1, ENB, ENB_1, FB_1, FB_3, FUS-X, HT-2, MA, NEO, OTA, PA, PAT, STG, T-2, TA, and ZEA, literature[286]; see also Dietary supplements, chapter: Further Mycotoxins and Microbial Metabolites

For detailed information see the article.

Fumonisin B$_3$

incidence: 1/32*, conc.: 13 μg/kg, sample origin: Czech and US retail markets, sample year: unknown, country: Czech Republic/USA[286], *dietary supplements (milk thistle-based supplements)

incidence: 1/9*, conc.: 5 μg/kg, sample origin: Czech and US retail markets, sample year: unknown, country: Czech Republic/USA[286], *dietary supplements for menopause effects treatment

incidence: 0/28*, conc.: no contamination, sample origin: Czech and US retail markets, sample year: unknown, country: Czech Republic/USA[286], *dietary supplements for the general health improvement

- Co-contamination: For detailed information see the article.
- Further contamination of plants and plant products other than cereals and their products reported in the present book/article[286]: Dietary supplements, AME, AOH, BEA, DAS, DON,

3-AcDON, ENA, ENA_1, ENB, ENB_1, FB_1, FB_2, FUS-X, HT-2, MA, NEO, OTA, PA, PAT, STG, T-2, TA, and ZEA, literature[286]; see also Dietary supplements, chapter: Further Mycotoxins and Microbial Metabolites

For detailed information see the article.

Type A Trichothecenes

Diacetoxyscirpenol

incidence: 0/32*, conc.: no contamination, sample origin: Czech and US retail markets, sample year: unknown, country: Czech Republic/USA[286], *dietary supplements (milk thistle-based supplements)

incidence: 0/9*, conc.: no contamination, sample origin: Czech and US retail markets, sample year: unknown, country: Czech Republic/USA[286], *dietary supplements for menopause effects treatment

incidence: 1/28*, conc.: 89 µg/kg, sample origin: Czech and US retail markets, sample year: unknown, country: Czech Republic/USA[286], *dietary supplements for the general health improvement

- Co-contamination: For detailed information see the article.
- Further contamination of plants and plant products other than cereals and their products reported in the present book/article[286]: Dietary supplements, AME, AOH, BEA, DON, 3-AcDON, ENA, ENA_1, ENB, ENB_1, FB_1, FB_2, FB_3, FUS-X, HT-2, MA, NEO, OTA, PA, PAT, STG, T-2, TA, and ZEA, literature[286]; see also Dietary supplements, chapter: Further Mycotoxins and Microbial Metabolites

For detailed information see the article.

HT-2 Toxin

incidence: 21/32*, conc. range: 59–1,530 µg/kg, sample origin: Czech and US retail markets, sample year: unknown, country: Czech Republic/USA[286], *dietary supplements (milk thistle-based supplements)

incidence: 0/9*, conc.: no contamination, sample origin: Czech and US retail markets, sample year: unknown, country: Czech Republic/USA[286], *dietary supplements for menopause effects treatment

incidence: 0/28*, conc.: no contamination, sample origin: Czech and US retail markets, sample year: unknown, country: Czech Republic/USA[286], *dietary supplements for the general health improvement

- Co-contamination: For detailed information see the article.
- Further contamination of plants and plant products other than cereals and their products reported in the present book/article[286]: Dietary supplements, AME, AOH, BEA, DAS, DON, 3-AcDON, ENA, ENA_1, ENB, ENB_1, FB_1, FB_2, FB_3, FUS-X, MA, NEO, OTA, PA, PAT, STG, T-2, TA, and ZEA, literature[286]; see also Dietary supplements, chapter: Further Mycotoxins and Microbial Metabolites

For detailed information see the article.

Neosolaniol

incidence: 8/32*, conc. range: 20–126 µg/kg, sample origin: Czech and US retail markets, sample year: unknown, country: Czech Republic/USA[286], *dietary supplements (milk thistle-based supplements)

incidence: 0/9*, conc.: no contamination, sample origin: Czech and US retail markets, sample year: unknown, country: Czech Republic/USA[286], *dietary supplements for menopause effects treatment

incidence: ?/28*, conc. range: 8–? µg/kg, sample origin: Czech and US retail markets, sample year: unknown, country: Czech Republic/USA[286], *dietary supplements for the general health improvement

- Co-contamination: For detailed information see the article.
- Further contamination of plants and plant products other than cereals and their products reported in the present book/article[286]: Dietary supplements, AME, AOH, BEA, DAS, DON, 3-AcDON, ENA, ENA_1, ENB, ENB_1, FB_1, FB_2, FB_3, FUS-X, HT-2, MA, OTA, PA, PAT, STG, T-2, TA, and ZEA, literature[286]; see also Dietary supplements, chapter: Further Mycotoxins and Microbial Metabolites

For detailed information see the article.

T-2 Toxin

incidence: 25/32*, conc. range: 69–1,870 µg/kg, sample origin: Czech and US retail markets, sample year: unknown, country: Czech Republic/USA[286], *dietary supplements (milk thistle-based supplements)

incidence: 1/9*, conc.: 169 µg/kg, sample origin: Czech and US retail markets, sample year: unknown, country: Czech Republic/USA[286], *dietary supplements for menopause effects treatment

incidence: 0/28*, conc.: no contamination, sample origin: Czech and US retail markets, sample year: unknown, country: Czech Republic/USA[286], *dietary supplements for the general health improvement

- Co-contamination: For detailed information see the article.
- Further contamination of plants and plant products other than cereals and their products reported in the present book/article[286]: Dietary supplements, AME, AOH, BEA, DAS, DON, 3-AcDON, ENA, ENA_1, ENB, ENB_1, FB_1, FB_2, FB_3, FUS-X, HT-2, MA, NEO, OTA, PA, PAT, STG, TA, and ZEA, literature[286]; see also Dietary supplements, chapter: Further Mycotoxins and Microbial Metabolites

For detailed information see the article.

Type B Trichothecenes

Deoxynivalenol

incidence: 16/32*, conc. range: 153–2,890 µg/kg, sample origin: Czech and US retail markets, sample year: unknown, country: Czech Republic/USA[286], *dietary supplements (milk thistle-based supplements)

incidence: 0/9*, conc.: no contamination, sample origin: Czech and US retail markets, sample year: unknown, country: Czech Republic/USA[286], *dietary supplements for menopause effects treatment

incidence: 0/28*, conc.: no contamination, sample origin: Czech and US retail markets, sample year: unknown, country: Czech Republic/USA[286], *dietary supplements for the general health improvement

- Co-contamination: For detailed information see the article.
- Further contamination of plants and plant products other than cereals and their products reported in the present book/

article[286]: Dietary supplements, AME, AOH, BEA, DAS, 3-AcDON, ENA, ENA_1, ENB, ENB_1, FB_1, FB_2, FB_3, FUS-X, HT-2, MA, NEO, OTA, PA, PAT, STG, T-2, TA, and ZEA, literature[286]; see also Dietary supplements, chapter: Further Mycotoxins and Microbial Metabolites

For detailed information see the article.

3-Acetyldeoxynivalenol

incidence: 5/32*, conc. range: 130–265 µg/kg, sample origin: Czech and US retail markets, sample year: unknown, country: Czech Republic/USA[286], *dietary supplements (milk thistle-based supplements)

incidence: 0/9*, conc.: no contamination, sample origin: Czech and US retail markets, sample year: unknown, country: Czech Republic/USA[286], *dietary supplements for menopause effects treatment

incidence: 0/28*, conc.: no contamination, sample origin: Czech and US retail markets, sample year: unknown, country: Czech Republic/USA[286], *dietary supplements for the general health improvement

- Co-contamination: For detailed information see the article.
- Further contamination of plants and plant products other than cereals and their products reported in the present book/article[286]: Dietary supplements, AME, AOH, BEA, DAS, DON, ENA, ENA_1, ENB, ENB_1, FB_1, FB_2, FB_3, FUS-X, HT-2, MA, NEO, OTA, PA, PAT, STG, T-2, TA, and ZEA, literature[286]; see also Dietary supplements, chapter: Further Mycotoxins and Microbial Metabolites

For detailed information see the article.

Fusarenon-X (4-Acetylnivalenol)

incidence: 4/32*, conc. range: 120–193 µg/kg, sample origin: Czech and US retail markets, sample year: unknown, country: Czech Republic/USA[286], *dietary supplements (milk thistle-based supplements)

incidence: 0/9*, conc.: no contamination, sample origin: Czech and US retail markets, sample year: unknown, country: Czech Republic/USA[286], *dietary supplements for menopause effects treatment

incidence: 0/28*, conc.: no contamination, sample origin: Czech and US retail markets, sample year: unknown, country: Czech Republic/USA[286], *dietary supplements for the general health improvement

- Co-contamination: For detailed information see the article.
- Further contamination of plants and plant products other than cereals and their products reported in the present book/article[286]: Dietary supplements, AME, AOH, BEA, DAS, DON, 3-AcDON, ENA, ENA_1, ENB, ENB_1, FB_1, FB_2, FB_3, HT-2, MA, NEO, OTA, PA, PAT, STG, T-2, TA, and ZEA, literature[286]; see also Dietary supplements, chapter: Further Mycotoxins and Microbial Metabolites

For detailed information see the article.

Other ***Fusarium*** Toxins

Zearalenone

incidence: 25/32*, conc. range: 13–751 µg/kg, sample origin: Czech and US retail markets, sample year: unknown, country: Czech Republic/USA[286], *dietary supplements (milk thistle-based supplements)

incidence: 3/9*, conc. range: 27–227 µg/kg, sample origin: Czech and US retail markets, sample year: unknown, country: Czech Republic/USA[286], *dietary supplements for menopause effects treatment

incidence: 8/28*, conc. range: 5–824 µg/kg, sample origin: Czech and US retail markets, sample year: unknown, country: Czech Republic/USA[286], *dietary supplements for the general health improvement

- Co-contamination: For detailed information see the article.
- Further contamination of plants and plant products other than cereals and their products reported in the present book/article[286]: Dietary supplements, AME, AOH, BEA, DAS, DON, 3-AcDON, ENA, ENA_1, ENB, ENB_1, FB_1, FB_2, FB_3, FUS-X, HT-2, MA, NEO, OTA, PA, PAT, STG, T-2, and TA, literature[286]; see also Dietary supplements, chapter: Further Mycotoxins and Microbial Metabolites

For detailed information see the article.

Penicillium Toxins

Mycophenolic Acid

incidence: 3/32*, conc. range: 15–1,710 µg/kg, sample origin: Czech and US retail markets, sample year: unknown, country: Czech Republic/USA[286], *dietary supplements (milk thistle-based supplements)

incidence: 6/9*, conc. range: 76–2,590 µg/kg, sample origin: Czech and US retail markets, sample year: unknown, country: Czech Republic/USA[286], *dietary supplements for menopause effects treatment

incidence: 5/28*, conc. range: 70–3,260 µg/kg, sample origin: Czech and US retail markets, sample year: unknown, country: Czech Republic/USA[286], *dietary supplements for the general health improvement

- Co-contamination: For detailed information see the article.
- Further contamination of plants and plant products other than cereals and their products reported in the present book/article[286]: Dietary supplements, AME, AOH, BEA, DAS, DON, 3-AcDON, ENA, ENA_1, ENB, ENB_1, FB_1, FB_2, FB_3, FUS-X, HT-2, NEO, OTA, PA, PAT, STG, T-2, TA, and ZEA, literature[286]; see also Dietary supplements, chapter: Further Mycotoxins and Microbial Metabolites

For detailed information see the article.

incidence: 8/50*, conc. range: 43.1–395.0 µg/kg, ∅ conc.: 145 µg/kg, sample origin: internet vendors, sample year: unknown, country: USA[865], *dietary supplements containing green coffee bean extracts

- Co-contamination: 5 sa. co-contaminated with OTA, OTB, and MA; 3 sa. co-contaminated with OTA, OTB, MA, and FB_1
- Further contamination of plants and plant products other than cereals and their products reported in the present book/article[865]: Dietary supplements, OTA, OTB, and FB_1, literature[865]

Dinners see Products

Dolcetto see Wine

Dried fruits see Fruits, dried

Dried vegetables see Vegetables, dried

Drinking water may contain the following mycotoxins:

Aspergillus **Toxins**

Aflatoxin B_1

incidence: 1/27*, conc.: 0.00046 µg/l, sample origin: local stores, Portugal, sample year: June 2012–June 2013, country: Portugal[789], *commercially bottled water sa.

- Co-contamination: not reported
- Further contamination of plants and plant products other than cereals and their products reported in the present book/article[789]: Drinking water, AFB_2, AFG_1, and OTA, literature[789]

Aflatoxin B_2

incidence: 11/27*, conc. range: 0.00016–0.00048 µg/l, sample origin: local stores, Portugal, sample year: June 2012–June 2013, country: Portugal[789], *commercially bottled water sa.

- Co-contamination: not reported
- Further contamination of plants and plant products other than cereals and their products reported in the present book/article[789]: Drinking water, AFB_1, AFG_1, and OTA, literature[789]

Aflatoxin G_1

incidence: 1/27*, conc.: 0.00051 µg/l, sample origin: local stores, Portugal, sample year: June 2012–June 2013, country: Portugal[789], *commercially bottled water sa.

- Co-contamination: not reported
- Further contamination of plants and plant products other than cereals and their products reported in the present book/article[789]: Drinking water, AFB_1, AFB_2, and OTA, literature[789]

Aspergillus **and** ***Penicillium*** **Toxins**

Ochratoxin A

incidence: 1/27*, conc.: 0.00024 µg/l, sample origin: local stores, Portugal, sample year: June 2012–June 2013, country: Portugal[789], *commercially bottled water sa.

- Co-contamination: not reported
- Further contamination of plants and plant products other than cereals and their products reported in the present book/article[789]: Drinking water, AFB_1, AFB_2, and AFG_1, literature[789]

Drinks may contain the following mycotoxins:

Aspergillus **and** ***Penicillium*** **Toxins**

Ochratoxin A

incidence: 2/2*, conc. range: 0.10–0.30 µg/kg (dry matter), Ø conc.: 0.2 µg/kg (dry matter), sample origin: unknown, sample year: unknown, country: Switzerland[333], *malt-based breakfast drink

- Co-contamination: not reported
- Further contamination of plants and plant products other than cereals and their products reported in the present book/article[333]: Wine, OTA, literature[333]

Drinks (alcoholic drink) may contain the following mycotoxins:

Aspergillus **Toxins**

Aflatoxin/s

incidence: 26/55, Ø conc.: 1.9 µg/kg*, sample origin: greater Manila (capital) area, Philippines, sample year: unknown, country: Philippines[1], *AF of pos sa.?

- Co-contamination: not reported
- Further contamination of plants and plant products other than cereals and their products reported in the present book/article[1]: Beans, Butter (peanut butter), Cassava, Cocoa, Coffee, Nuts (cashew nut), Nuts (peanut), Products (coconut product), Rootcrops, Spices (cayenne pepper, chilli), Sweet potatoes, Yams, AF/S, literature[1]

Drinks (apple drink) may contain the following mycotoxins:

Aspergillus **and** ***Penicillium*** **Toxins**

Patulin

incidence: 1/1*, conc.: 3.2 µg/kg, sample origin: Greek market, Greece, sample year: October 2004–June 2006, country: Greece[108], *soft drink

- Co-contamination: not reported
- Further contamination of plants and plant products other than cereals and their products reported in the present book/article[108]: Baby food, Juices, Juices (apple concentrate), Juices (apple juice), Juices (apricot concentrate), Juices (orange juice), Juices (pineapple concentrate), Juices (rosehip juice), PAT, literature[108]

Drinks (chocolate drink) may contain the following mycotoxins:

Aspergillus **and** ***Penicillium*** **Toxins**

Ochratoxin A

incidence: 51/56*, conc. range: ≤0.630 µg/kg, sample origin: regions around Detmold, Giessen, Hamburg, Jena, Karlsruhe, Kiel, Kulmbach, and Trier (cities), Germany, sample year: unknown, country: Germany[262], *chocolate drinks

incidence: 34/34*, conc. range: ≤0.050 µg/kg, sample origin: regions around Detmold, Giessen, Hamburg, Jena, Karlsruhe, Kiel, Kulmbach, and Trier (cities), Germany, sample year: unknown, country: Germany[262], *ready-mixed cocoa drinks

- Co-contamination: not reported
- Further contamination of plants and plant products other than cereals and their products reported in the present book/article[262]: Bars, Chocolate, Figs, Fruits, dried, Nut crème, Nuts, Nuts (hazelnut), Nuts (peanut), Powder (cocoa powder), Pudding, Raisins, Seeds (linseed), Seeds (sesame), Seeds (sunflower), OTA, literature[262]

Drinks (coffee drink) may contain the following mycotoxins:

Aspergillus **and** ***Penicillium*** **Toxins**

Ochratoxin A

incidence: 30/30*, conc. range: tr-0.037 µg/kg, sample origin: local supermarkets and small retail outlets, Japan, sample year: 2007, country: Japan[343], *canned and bottled ready-to-drink coffee (each sa. corresponds to a different brand)

- Co-contamination: not reported
- Further contamination of plants and plant products other than cereals and their products reported in the present book/article[343]: not reported

Drinks (tiger nut drink) may contain the following mycotoxins:

Aspergillus **Toxins**

Aflatoxin B$_1$

incidence: 1/22*, conc.: 0.06 μg/l, sample origin: supermarkets in Belgium and Spain, sample year: 2004, country: Belgium/Germany[342], *soft drink ("horchata")

- Co-contamination: not reported
- Further contamination of plants and plant products other than cereals and their products reported in the present book/article[342]: not reported

Eggplant may contain the following mycotoxins:

Aspergillus **Toxins**

Aflatoxin B$_1$

incidence: 32/50*, conc. range: ≤0.71 μg/kg, sample origin: local markets in Izmir (province), Turkey, sample year: February/March 2015, country: Turkey[449], *eggplants, dried

- Co-contamination: not reported
- Further contamination of plants and plant products other than cereals and their products reported in the present book/article[449]: Eggplant, AFB$_2$, AFG$_1$, AFG$_2$, and OTA, literature[449]; Spices (paprika), AFB$_1$, AFB$_2$, AFG$_1$, AFG$_2$, and OTA, literature[449]

Aflatoxin B$_2$

incidence: 50/50*, conc. range: 0.12–0.67 μg/kg, sample origin: local markets in Izmir (province), Turkey, sample year: February/March 2015, country: Turkey[449], *eggplants, dried

- Co-contamination: not reported
- Further contamination of plants and plant products other than cereals and their products reported in the present book/article[449]: Eggplant, AFB$_1$, AFG$_1$, AFG$_2$, and OTA, literature[449]; Spices (paprika), AFB$_1$, AFB$_2$, AFG$_1$, AFG$_2$, and OTA, literature[449]

Aflatoxin G$_1$

incidence: 50/50*, conc. range: 0.32–1.35 μg/kg, sample origin: local markets in Izmir (province), Turkey, sample year: February/March 2015, country: Turkey[449], *eggplants, dried

- Co-contamination: not reported
- Further contamination of plants and plant products other than cereals and their products reported in the present book/article[449]: Eggplant, AFB$_1$, AFB$_2$, AFG$_2$, and OTA, literature[449]; Spices (paprika), AFB$_1$, AFB$_2$, AFG$_1$, AFG$_2$, and OTA, literature[449]

Aflatoxin G$_2$

incidence: 50/50*, conc. range: 0.10–0.23 μg/kg, sample origin: local markets in Izmir (province), Turkey, sample year: February/March 2015, country: Turkey[449], *eggplants, dried

- Co-contamination: not reported
- Further contamination of plants and plant products other than cereals and their products reported in the present book/article[449]: Eggplant, AFB$_1$, AFB$_2$, AFG$_1$, and OTA, literature[449]; Spices (paprika), AFB$_1$, AFB$_2$, AFG$_1$, AFG$_2$, and OTA, literature[449]

Aspergillus **and** ***Penicillium*** **Toxins**

Ochratoxin A

incidence: 50/50*, conc. range: 8.88–21.35 μg/kg, ∅ conc.: 17.67 μg/kg, sample origin: local markets in Izmir (province), Turkey, sample year: February/March 2015, country: Turkey[449], *eggplants, dried

- Co-contamination: not reported
- Further contamination of plants and plant products other than cereals and their products reported in the present book/article[449]: Eggplant, AFB$_1$, AFB$_2$, AFG$_1$, and AFG$_2$, literature[449]; Spices (paprika), AFB$_1$, AFB$_2$, AFG$_1$, AFG$_2$, and OTA, literature[449]

Egusi see **Food**

Egusi meal see **Meal (egusi meal)**

Egusi melon seeds see **Seeds (melon)**

Emu aran may contain the following mycotoxins:

Aspergillus **Toxins**

Aflatoxin/s

incidence: 2/2, conc. range: 83*–86** μg/kg***, sample origin: Ugbowo in Benin City, Nigeria, sample year: unknown, country: Nigeria[346], *stale and **fresh emu aran, ***AFB

- Co-contamination: not reported
- Further contamination of plants and plant products other than cereals and their products reported in the present book/article[346]: Meal (egusi meal), Ogbono, Ogili-ugba, Ogoro, AF/S, literature[346]

Emu aran is a beverage from fermented sap of the Raphia palm.

Entrees see **Products**

Erbaluce see **Wine**

Extracts may contain the following mycotoxins:

Aspergillus **and** ***Penicillium*** **Toxins**

Ochratoxin A

incidence: 0/1*, conc.: no contamination, sample origin: supermarkets and fast-food chains in Quebec (city), Canada, sample year: September/October 2008, country: Canada[90], *vanilla extract

incidence: 1/1*, conc.: 0.02 μg/l, sample origin: supermarkets and fast-food chains in Calgary (city), Canada, sample year: September/October 2009, country: Canada[90], *vanilla extract

- Co-contamination: not reported
- Further contamination of plants and plant products other than cereals and their products reported in the present book/article[90]: Baby food, Beans, Beverages, Butter (peanut butter), Candies, Chocolate, Coffee, Herbs, Ice cream, Nuts, Pies, Pizza, Products, Pudding, Seeds, Raisins, Wine, OTA, literature[90]

Fennel see **Spices (fennel)**

Field mustard may contain the following mycotoxins:

Aspergillus **Toxins**

Aflatoxin B$_1$

incidence: 17/41*, conc. range: <20 μg/kg (4 sa.), 21–100 μg/kg (8 sa.), 101–500 μg/kg (4 sa.), 780 μg/kg (1 sa.), sample origin: villages of Khagaria and Saharsa (districts) of North Bihar (state), India, sample year: August–October 1987, country: India[128], **Brassica campestris* L.

- Co-contamination: not reported
- Further contamination of plants and plant products other than cereals and their products reported in the present book/

article[128]: Beans (cowpea), Beans (mung bean), Lentils, Seeds (linseed), Peas (chickpea), AFB_1, literature[128]

incidence: 26/148*, conc. range: 8–572 µg/kg, sample origin: mustard growing areas (16 districts) of Bihar (state), India, sample year: February/March 1987, country: India[545], **Brassica campestris* L. (yellow-seeded, *Sarson*)

- Co-contamination: 7 sa. co-contaminated with AFB_1 and AFB_2; 19 sa. contaminated solely with AFB_1
- Further contamination of plants and plant products other than cereals and their products reported in the present book/article[545]: Field mustard, AFB_2, literature[545]; Spices (mustard), AFB_1, AFB_2 and AFG_1, literature[545]

Aflatoxin B$_2$

incidence: 7/148*, conc. range: pr, sample origin: mustard growing areas (16 districts) of Bihar (state), India, sample year: February/March 1987, country: India[545], **Brassica campestris* L. (yellow-seeded, *Sarson*)

- Co-contamination: 7 sa. co-contaminated with AFB_1 and AFB_2
- Further contamination of plants and plant products other than cereals and their products reported in the present book/article[545]: Field mustard, AFB_1, literature[545]; Spices (mustard), AFB_1, AFB_2 and AFG_1, literature[545]

Figpie see Figs

Figs may contain the following mycotoxins:

Alternaria Toxins

Alternariol

incidence: 1/14, conc.: 8.7 µg/kg, sample origin: retail stores, Netherlands, sample year: autumn 2014, country: Netherlands/Germany[906]

- Co-contamination: 1 sa. co-contaminated with AOH and TA
- Further contamination of plants and plant products other than cereals and their products reported in the present book/article[906]: Figs, TA, literature[906]; Fruits (tomato), Juices (tomato juice), AOH and TA, literature[906]; Oils (sunflower oil), AME, literature[906]; Paste (tomato paste), Sauces (tomato sauce), Seeds (sunflower), Tomato ketchup, AME, AOH, and TA, literature[906]

Tenuazonic Acid

incidence: 5/5*, conc. range: 25–2,345 µg/kg, ∅ conc.: 1,043 µg/kg, sample origin: Dutch survey program (2013) and retail shops across The Netherlands, sample year: 2013?, country: Netherlands[568], *figs, dried

- Co-contamination: not reported
- Further contamination of plants and plant products other than cereals and their products reported in the present book/article[568]: Fruits (apple), AOH, literature[568]; Olives, Seeds (sunflower), TA, literature[568]; Sauces (tomato sauce), AME, AOH, and TA, literature[568]; Wine (red), AOH and TA, literature[568]; see also Seeds (sunflower), chapter: Further Mycotoxins and Microbial Metabolites

incidence: 14/14, conc. range: 41 (81?)–1,728 µg/kg, sample origin: retail stores, Netherlands, sample year: autumn 2014, country: Netherlands/Germany[906]

- Co-contamination: 1 sa. co-contaminated with AOH and TA; 13 sa. contaminated solely with TA
- Further contamination of plants and plant products other than cereals and their products reported in the present book/article[906]: Figs, AOH, literature[906]; Fruits (tomato), Juices (tomato juice), AOH and TA, literature[906]; Oils (sunflower oil), AME, literature[906]; Paste (tomato paste), Sauces (tomato sauce), Seeds (sunflower), Tomato ketchup, AME, AOH, and TA, literature[906]

Aspergillus Toxins

Aflatoxin B$_1$

incidence: 2/98*, conc. range: ≤1.02 µg/kg, sample origin: different export companies, Turkey, sample year: unknown, country: Turkey[55], *figs, dried

- Co-contamination: 2 sa. co-contaminated with AFB_1 and OTA (all together 7 sa. contaminated with AFS thereof 2 sa. with AFB_1; see also **Figs, AF/S, literature**[55])
- Further contamination of plants and plant products other than cereals and their products reported in the present book/article[55]: Figs, AF/S and OTA, literature[55]; Fruits, dried, Sultanas, OTA, literature[55]

incidence: 4?/24*, conc. range: 1.4–6 µg/kg, ∅ conc.: 3.7 µg/kg, sample origin: sites of primary storage or processing (domestic products) and sites of importation, Cyprus, sample year: 1992–1996, country: Cyprus[200], *figs and figpie

- Co-contamination: not reported
- Further contamination of plants and plant products other than cereals and their products reported in the present book/article[200]: Butter (peanut butter), Nuts (peanut), AFB_1, AFB_2, AFG_1, and AFG_2, literature[200]; Figs, AFB_2, AFG_1, and AFG_2, literature[200]; Nuts (Brazil nut, paranut), Nuts (walnut), AFB_1, AFB_2, and AFG_1, literature[200]; Nuts (chestnut), Products (sesame product), Spices, AFB_1, literature[200]; Nuts (pistachio nut), AFB_1 and AFB_2, literature[200]; Seeds (sesame), AFB_1, AFG_1, and AFG_2, literature[200]

incidence: 2/20, conc. range: 0.4–0.7 µg/kg, ∅ conc.: 0.55 µg/kg, sample origin: retail outlets, UK, sample year: September 2001, country: UK[203]

- Co-contamination: 1 sa. co-contaminated with AFB_1 and AFB_2; 1 sa. co-contaminated with AFB_1 and AFG_2
- Further contamination of plants and plant products other than cereals and their products reported in the present book/article[203]: Butter (nut butter), AFB_1, AFB_2, AFG_1, AFG_2, and OTA, literature[203]; Butter (peanut butter), AFB_1, AFB_2, AFG_1, AFG_2, and OTA, literature[203]; Dates, AFB_1 and OTA, literature[203]; Figs, AFB_2, AFG_2, and OTA, literature[203]; Fruits, dried, Nuts (pistachio nut), AFB_1, AFB_2, and OTA, literature[203]; Nuts (Brazil nut, paranut), AFB_1, AFB_2, and AFG_1, literature[203]; Nuts (peanut), AFB_1 and AFB_2, literature[203]

incidence: 4/4, conc. range: 1.1–4.4 µg/kg, ∅ conc.: 3.5 µg/kg, sample origin: unknown, sample year: unknown, country: Denmark[217]

- Co-contamination: 1 sa. co-contaminated with AFB_1, AFB_2, AFG_1, and AFG_2; 3 sa. co-contaminated with AFB_1, AFB_2, and AFG_1
- Further contamination of plants and plant products other than cereals and their products reported in the present book/

article[217]: Butter (peanut butter), Spices (cayenne pepper, chilli), AFB_1, AFB_2, AFG_1, and AFG_2, literature[217]; Figs, AFB_2, AFG_1, and AFG_2, literature[217]; Meal (pistachio meal), AFB_1 and AFB_2, literature[217]

incidence: 3/10*, conc. range: 0.1–3.2 µg/kg, ∅ conc.: 1.7 µg/kg, sample origin: available in New Zealand, sample year: 2009, country: New Zealand[277], *figs, dried

- Co-contamination: not reported
- Further contamination of plants and plant products other than cereals and their products reported in the present book/article[277]: Bars, AFB_1, AF/S, DON, NIV, and OTA, literature[277]; Butter (peanut butter), Confectionery, Food, Fruits, dried, Nuts, Nuts (peanut), Nuts (pistachio nut), Sauces (peanut sauce), AFB_1 and AF/S, literature[277]; Coffee, Dates, Spices (nutmeg), Spices (pepper), Vine fruits, dried, OTA, literature[277]; Figs, AF/S and OTA, literature[277]; Nuts (Brazil nut, paranut), AFB_1, literature[277]; Spices (cayenne pepper, chilli), Spices (curry), Spices (ginger), Spices (paprika), AFB_1, AF/S, and OTA, literature[277]

incidence: 2/4*, conc. range: 2.5–11.8 µg/kg, ∅ conc.: 7.2 µg/kg, sample origin: retail outlets in Lattakia (city), Syria, sample year: January 1989, country: Italy[347], *figs, dried

- Co-contamination: not reported
- Further contamination of plants and plant products other than cereals and their products reported in the present book/article[347]: Lentils, AFB_1, literature[347]; Nuts (peanut), AFB_1 and AFB_2, literature[347]

incidence: 94/386*, conc. range: 0.2–1 µg/kg (48 sa.), 1–5 µg/kg (37 sa.), 5–10 µg/kg (7 sa.), 10–20 µg/kg (1 sa.), 20–30 µg/kg (1 sa.), sample origin: Turkey, sample year: 1985, country: Switzerland[348], *figs, dried and randomly selected

incidence: 8/16*, conc. range: 0.2–1 µg/kg (3 sa.), 1–5 µg/kg (3 sa.), 5–10 µg/kg (1 sa.), 10–20 µg/kg (1 sa.), sample origin: Turkey, sample year: 1985, country: Switzerland[348], *figs, dried and discolored (dark)

incidence: 52/62*, conc range: 0.2–1 µg/kg (4 sa.), 1–10 µg/kg (14 sa.), 10–100 µg/kg (8 sa.), 100–1,000 µg/kg (14 sa.), 1,000–10,000 µg/kg (11 sa.), >10,000 µg/kg (1 sa.), sample origin: Turkey, sample year: 1985, country: Switzerland[348], *figs, dried and showing BGY fluorescence

- Co-contamination: not reported
- Further contamination of plants and plant products other than cereals and their products reported in the present book/article[348]: Figs, AFG_1, literature[348]

incidence: 9/58*, conc. range: <0.1–1.0 µg/kg (2 sa.), 1.0–5.0 µg/kg (4 sa.), 5.0–10.0 µg/kg (1 sa.), 10.0–15.0 µg/kg (1 sa.), 35.1 µg/kg (1 sa.), sample origin: Turkish fig exporters, Turkey, sample year: 2003, country: Turkey[349], *figs, dried

- Co-contamination: 2 sa. co-contaminated with AFB_1 and OTA; 7 sa. contaminated solely with AFB_1
- Further contamination of plants and plant products other than cereals and their products reported in the present book/article[349]: Figs, OTA, literature[349]

incidence: 17/41*, conc. range: <0.1–1.0 µg/kg (8 sa.), 1.0–5.0 µg/kg (7 sa.), 5.0–10.0 µg/kg (1 sa.), 20.6 µg/kg (1 sa.), sample origin: Turkish fig exporters, Turkey, sample year: 2004, country: Turkey[349], *figs, dried

- Co-contamination: 2 sa. co-contaminated with AFB_1 and OTA; 15 sa. contaminated solely with AFB_1
- Further contamination of plants and plant products other than cereals and their products reported in the present book/article[349]: Figs, OTA, literature[349]

incidence: 0/20*, conc.: no contamination, sample origin: local markets, Turkey, sample year: 2004, country: Turkey[349], *figs, dried

- Co-contamination: not reported
- Further contamination of plants and plant products other than cereals and their products reported in the present book/article[349]: Figs, OTA, literature[349]

incidence: 37/52*, conc. range: 5–76,000 µg/kg, ∅ conc.: 6,341 µg/kg, sample origin: Turkey, sample year: 1991/1992, country: Switzerland[350], *selected figs (dried) with BGY fluorescence

- Co-contamination: 3 sa. co-contaminated with AFB_1, AFG_1, KA, and OTA; 11 sa. co-contaminated with AFB_1, AFG_1, and KA; 4 sa. co-contaminated with AFB_1, KA, and OTA; 19 sa. co-contaminated with AFB_1 and KA
- Further contamination of plants and plant products other than cereals and their products reported in the present book/article[350]: Figs, AFG_1, KA, and OTA, literature[350]

incidence: 9?/24*, conc. range: 3.7–60.0 µg/kg, sample origin: orchards, farmer store houses, and fig processing plants, Turkey, sample year: 1988, country: Turkey[351], *figs from the tree

incidence: 4?/19*, conc. range: 1.8–63.0 µg/kg, sample origin: orchards, farmer store houses, and fig processing plants, Turkey, sample year: 1988, country: Turkey[351], *figs on the ground

incidence: 2?/8*, conc. range: 0.5–3.7 µg/kg, sample origin: orchards, farmer store houses, and fig processing plants, Turkey, sample year: 1988, country: Turkey[351], *figs during drying (sun)

incidence: 1/3*, conc.: 0.5 µg/kg, sample origin: orchards, farmer store houses, and fig processing plants, Turkey, sample year: 1988, country: Turkey[351], *figs in farmer store house

incidence: 7?/12*, conc. range: 0.5–12.5 µg/kg, sample origin: orchards, farmer store houses, and fig processing plants, Turkey, sample year: 1988, country: Turkey[351], *figs before processing

incidence: 4?/32*, conc. range: 0.5–4.2 µg/kg, sample origin: orchards, farmer store houses, and fig processing plants, Turkey, sample year: 1988, country: Turkey[351], *figs during processing

- Co-contamination: not reported
- Further contamination of plants and plant products other than cereals and their products reported in the present book/article[351]: Figs, AFB_2, AFG_1, AFG_2, and OTA, literature[351]; Paste (fig paste), AFB_1, AFB_2, AFG_1, AFG_2, and OTA, literature[351]

incidence: 1/20*, conc.: 0.28 µg/kg, sample origin: retail shops and local markets in Rabat-Salé (region), Morocco, sample year: January–October 2006, country: Spain/Morocco[352], *figs, dried

- Co-contamination: not reported
- Further contamination of plants and plant products other than cereals and their products reported in the present book/article[352]: Figs, AF/S, literature[352]; Raisins, Nuts (peanut), Nuts (pistachio nut), Nuts (walnut), AFB_1 and AF/S, literature[352]

incidence: 11/48*, conc. range: 0.1–696.3 µg/kg, ∅ conc.: 114 µg/kg, sample origin: orchards in the Aegean Region, Turkey, sample year: unknown, country: Turkey[353], *figs (dried) positive for *Aspergillus* section *Flavi* contamination

- Co-contamination: 1 sa. co-contaminated with AFB_1, AFB_2, AFG_1, AFG_2, and CPA; 4 sa. co-contaminated with AFB_1, AFB_2,

AFG_1, and CPA; 4 sa. co-contaminated with AFB_1, AFB_2, and CPA; 2 sa. co-contaminated with AFB_1 and CPA
- Further contamination of plants and plant products other than cereals and their products reported in the present book/article[353]: Figs, AFB_2, AFG_1, AFG_2, and CPA, literature[353]

For detailed information see the article.

incidence: 49/49*, conc. range: 0.7–2,221 µg/kg, ∅ conc.: 285 µg/kg, sample origin: commercial processing facility, Turkey, sample year: 2005, country: Turkey/UK/Netherlands[354], *49 individual figs unsuitable for food
- Co-contamination: 3 sa. co-contaminated with AFB_1, AFB_2, AFG_1, AFG_2, and OTA; 11 sa. co-contaminated with AFB_1, AFB_2, AFG_1, and OTA; 1 sa. co-contaminated with AFB_1, AFG_1, AFG_2, and OTA; 6 sa. co-contaminated with AFB_1, AFB_2, and AFG_1; 7 sa. co-contaminated with AFB_1, AFB_2, and OTA; 4 sa. co-contaminated with AFB_1, AFG_1, and OTA; 3 sa. co-contaminated with AFB_1 and AFB_2; 4 sa. co-contaminated with AFB_1 and AFG_1; 5 sa. co-contaminated with AFB_1 and OTA; 5 sa. contaminated solely with AFB_1
- Further contamination of plants and plant products other than cereals and their products reported in the present book/article[354]: Figs, AFB_2, AFG_1, AFG_2, and OTA, literature[354]

incidence: 104/2,550*, conc. range: <0.1–1.0 µg/kg (76 sa.), 1.0–2.0 µg/kg (10 sa.), 2.0–10.0 µg/kg (13 sa.), 10.0–20.0 µg/kg (3 sa.), >20.0 µg/kg (2 sa., maximum: 32.0 µg/kg), sample origin: Turkish fig exporters, Turkey, sample year: 2003, country: Turkey[355], *figs, dried

incidence: 386/3,096*, conc. range: <0.1–1.0 µg/kg (261 sa.), 1.0–2.0 µg/kg (61 sa.), 2.0–10.0 µg/kg (54 sa.), 10.0–20.0 µg/kg (10 sa.), >20.0 µg/kg (14 sa., maximum: 115.3 µg/kg), sample origin: Turkish fig exporters, Turkey, sample year: 2004, country: Turkey[355], *figs, dried

incidence: 740/3,920*, conc. range: <0.1–1.0 µg/kg (460 sa.), 1.0–2.0 µg/kg (99 sa.), 2.0–10.0 µg/kg (127 sa.), 10.0–20.0 µg/kg (31 sa.), >20.0 µg/kg (23 sa., maximum: 288.2 µg/kg), sample origin: Turkish fig exporters, Turkey, sample year: 2005, country: Turkey[355], *figs, dried

incidence: 132/830*, conc. range: <0.1–1.0 µg/kg (99 sa.), 1.0–2.0 µg/kg (13 sa.), 2.0–10.0 µg/kg (17 sa.), 10.0–20.0 µg/kg (3 sa.), sample origin: Turkish fig exporters, Turkey, sample year: 2006, country: Turkey[355], *figs, dried
- Co-contamination: not reported
- Further contamination of plants and plant products other than cereals and their products reported in the present book/article[355]: Figs, AF/S, literature[355]

incidence: 8/206*, conc. range: 3.6–320.0 µg/kg**, ∅ conc.: 102 µg/kg**, sample origin: from fields during drying, warehouses, and processing units in ten regions in the Aegean area, Turkey, sample year: 1986, country: Turkey[717], *figs (dried), **low grade figs collected from the drying stage
- Co-contamination: 2 sa. co-contaminated with AFB_1, AFB_2, and AFG_1; 2 sa. co-contaminated with AFB_1 and AFB_2; 1 sa. co-contaminated with AFB_1 and AFG_1; 3 sa. contaminated solely with AFB_1
- Further contamination of plants and plant products other than cereals and their products reported in the present book/article[717]: Figs, AFB_2 and AFG_1, literature[717]

Aflatoxin B_2

incidence: 4?/24*, conc. range: 0.9–1.5 µg/kg, ∅ conc.: 1.2 µg/kg, sample origin: sites of primary storage or processing (domestic products) and sites of importation, Cyprus, sample year: 1992–1996, country: Cyprus[200], *figs and figpie
- Co-contamination: not reported
- Further contamination of plants and plant products other than cereals and their products reported in the present book/article[200]: Butter (peanut butter), Nuts (peanut), AFB_1, AFB_2, AFG_1, and AFG_2, literature[200]; Figs, Seeds (sesame), AFB_1, AFG_1, and AFG_2, literature[200]; Nuts (Brazil nut, paranut), Nuts (walnut), AFB_1, AFB_2, and AFG_1, literature[200]; Nuts (chestnut), Products (sesame product), Spices, AFB_1, literature[200]; Nuts (pistachio nut), AFB_1 and AFB_2, literature[200]

incidence: 1/20, conc.: 0.2 µg/kg, sample origin: retail outlets, UK, sample year: September 2001, country: UK[203]
- Co-contamination: 1 sa. co-contaminated with AFB_1 and AFB_2
- Further contamination of plants and plant products other than cereals and their products reported in the present book/article[203]: Butter (nut butter), AFB_1, AFB_2, AFG_1, AFG_2, and OTA, literature[203]; Butter (peanut butter), AFB_1, AFB_2, AFG_1, AFG_2, and OTA, literature[203]; Dates, AFB_1 and OTA, literature[203]; Figs, AFB_1, AFG_2, and OTA, literature[203]; Fruits, dried, Nuts (pistachio nut), AFB_1, AFB_2, and OTA, literature[203]; Nuts (Brazil nut, paranut), AFB_1, AFB_2, and AFG_1, literature[203]; Nuts (peanut), AFB_1 and AFB_2, literature[203]

incidence: 4/4, conc. range: 0.5–1.1 µg/kg, ∅ conc.: 0.73 µg/kg, sample origin: unknown, sample year: unknown, country: Denmark[217]
- Co-contamination: 1 sa. co-contaminated with AFB_1, AFB_2, AFG_1, and AFG_2; 3 sa. co-contaminated with AFB_1, AFB_2, and AFG_1
- Further contamination of plants and plant products other than cereals and their products reported in the present book/article[217]: Butter (peanut butter), Spices (cayenne pepper, chilli), AFB_1, AFB_2, AFG_1, and AFG_2, literature[217]; Figs, AFB_1, AFG_1, and AFG_2, literature[217]; Meal (pistachio meal), AFB_1 and AFB_2, literature[217]

incidence: 1/1, conc.: 0.94 µg/kg, sample origin: local supermarket in Okayama (prefecture), Japan, sample year: unknown, country: Japan[339]
- Co-contamination: not reported
- Further contamination of plants and plant products other than cereals and their products reported in the present book/article[339]: Coconuts, Fruits, dried, Seeds (sesame), AFB_1 and AFG_2, literature[339]; Figs, Flour (soybean flour), Raisins, AFG_2, literature[339]; Nuts, Spices, AFB_1, AFB_2, AFG_1, and AFG_2, literature[339]

incidence: 9?/24*, conc. range: 9.0–37.7 µg/kg, sample origin: orchards, farmer store houses, and fig processing plants, Turkey, sample year: 1988, country: Turkey[351], *figs from the tree

incidence: 0/19*, conc.: no contamination, sample origin: orchards, farmer store houses, and fig processing plants, Turkey, sample year: 1988, country: Turkey[351], *figs on the ground

incidence: 0/8*, conc.: no contamination, sample origin: orchards, farmer store houses, and fig processing plants, Turkey, sample year: 1988, country: Turkey[351], *figs during drying (sun)

incidence: 0/3*, conc.: no contamination, sample origin: orchards, farmer store houses, and fig processing plants, Turkey, sample year: 1988, country: Turkey[351], *figs in farmer store house incidence: 7?/12*, conc. range: 0.5–4.16 μg/kg, sample origin: orchards, farmer store houses, and fig processing plants, Turkey, sample year: 1988, country: Turkey[351], *figs before processing

incidence: 4?/32*, conc. range: 0.5–2.2 μg/kg, sample origin: orchards, farmer store houses, and fig processing plants, Turkey, sample year: 1988, country: Turkey[351], *figs during processing

- Co-contamination: not reported
- Further contamination of plants and plant products other than cereals and their products reported in the present book/article[351]: Figs, AFB_1, AFG_1, AFG_2, and OTA, literature[351]; Paste (fig paste), AFB_1, AFB_2, AFG_1, AFG_2, and OTA, literature[351]

incidence: 9/48*, conc. range: 0.1–16.7 μg/kg, Ø conc.: 2.92 μg/kg, sample origin: orchards in the Aegean Region, Turkey, sample year: unknown, country: Turkey[353], *figs (dried) positive for *Aspergillus* section *Flavi* contamination

- Co-contamination: 1 sa. co-contaminated with AFB_1, AFB_2, AFG_1, AFG_2, and CPA; 4 sa. co-contaminated with AFB_1, AFB_2, AFG_1, and CPA; 4 sa. co-contaminated with AFB_1, AFB_2, and CPA
- Further contamination of plants and plant products other than cereals and their products reported in the present book/article[353]: Figs, AFB_1, AFG_1, AFG_2, and CPA, literature[353]

For detailed information see the article.

incidence: 30/49*, conc. range: 0.1–74.4 μg/kg, Ø conc.: 6.18 μg/kg, sample origin: commercial processing facility, Turkey, sample year: 2005, country: Turkey/UK/Netherlands[354], *49 individual figs unsuitable for food

- Co-contamination: 3 sa. co-contaminated with AFB_1, AFB_2, AFG_1, AFG_2, and OTA; 11 sa. co-contaminated with AFB_1, AFB_2, AFG_1, and OTA; 6 sa. co-contaminated with AFB_1, AFB_2, and AFG_1; 7 sa. co-contaminated with AFB_1, AFB_2, and OTA; 3 sa. co-contaminated with AFB_1 and AFB_2
- Further contamination of plants and plant products other than cereals and their products reported in the present book/article[354]: Figs, AFB_1, AFG_1, AFG_2, and OTA, literature[354]

incidence: 1/206*, conc. range: 23.5–71.8 μg/kg**, Ø conc.: 50.6 μg/kg**, sample origin: from fields during drying, warehouses, and processing units in ten regions in the Aegean area, Turkey, sample year: 1986, country: Turkey[717], *figs, dried, **low-grade figs collected from the drying stage

- Co-contamination: 2 sa. co-contaminated with AFB_1, AFB_2, and AFG_1; 2 sa. co-contaminated with AFB_1 and AFB_2
- Further contamination of plants and plant products other than cereals and their products reported in the present book/article[717]: Figs, AFB_1 and AFG_1, literature[717]

Aflatoxin G_1

incidence: 4?/24*, conc. range: 0.8–2.1 μg/kg, Ø conc.: 1.4 μg/kg, sample origin: sites of primary storage or processing (domestic products) and sites of importation, Cyprus, sample year: 1992 1996, country: Cyprus[200], *figs and figpie

- Co-contamination: not reported
- Further contamination of plants and plant products other than cereals and their products reported in the present book/article[200]: Butter (peanut butter), Nuts (peanut), AFB_1, AFB_2, AFG_1, and AFG_2, literature[200]; Figs, AFB_1, AFB_2, and AFG_2, literature[200]; Nuts (Brazil nut, paranut), Nuts (walnut), AFB_1, AFB_2, and AFG_1, literature[200]; Nuts (chestnut), Products (sesame product), Spices, AFB_1, literature[200]; Nuts (pistachio nut), AFB_1 and AFB_2, literature[200]; Seeds (sesame), AFB_1, AFG_1, and AFG_2, literature[200]

incidence: 4/4, conc. range: 0.9–1.3 μg/kg, Ø conc.: 1.1 μg/kg, sample origin: unknown, sample year: unknown, country: Denmark[217]

- Co-contamination: 1 sa. co-contaminated with AFB_1, AFB_2, AFG_1, and AFG_2; 3 sa. co-contaminated with AFB_1, AFB_2, and AFG_1
- Further contamination of plants and plant products other than cereals and their products reported in the present book/article[217]: Butter (peanut butter), Spices (cayenne pepper, chilli), AFB_1, AFB_2, AFG_1, and AFG_2, literature[217]; Figs, AFB_1, AFB_2, and AFG_2, literature[217]; Meal (pistachio meal), AFB_1 and AFB_2, literature[217]

incidence: 49/386*, conc. range: 0.2–1 μg/kg (28 sa.), 1–5 μg/kg (17 sa.), 5–10 μg/kg (2 sa.), 20–30 μg/kg (2 sa.), sample origin: Turkey, sample year: 1985, country: Switzerland[348], *figs, dried and randomly selected

incidence: 3/16*, conc. range: 0.2–1 μg/kg (1 sa.), 1–5 μg/kg (2 sa.), sample origin: Turkey, sample year: 1985, country: Switzerland[348], *figs, dried and discolored (dark)

incidence: 21/62*, conc range: 0.2–1 μg/kg (2 sa.), 1–10 μg/kg (5 sa.), 10–100 μg/kg (4 sa.), 100–1,000 μg/kg (4 sa.), 1,000–10,000 μg/kg (4 sa.), >10,000 μg/kg (2 sa.), sample origin: Turkey, sample year: 1985, country: Switzerland[348], *figs, dried and showing BGY fluorescence

- Co-contamination: not reported
- Further contamination of plants and plant products other than cereals and their products reported in the present book/article[348]: Figs, AFB_1, literature[348]

incidence: 14/52* conc. range: 24–180,000 μg/kg, Ø conc.: 15,518 μg/kg, sample origin: Turkey, sample year: 1991/1992, country: Switzerland[350], *selected figs (dried) with BGY fluorescence

- Co-contamination: 3 sa. co-contaminated with AFB_1, AFG_1, KA, and OTA; 11 sa. co-contaminated with AFB_1, AFG_1, and KA
- Further contamination of plants and plant products other than cereals and their products reported in the present book/article[350]: Figs, AFB_1, KA, and OTA, literature[350]

incidence: 9?/21*, conc. range: 3.2–37.7 μg/kg, sample origin: orchards, farmer store houses, and fig processing plants, Turkey, sample year: 1988, country: Turkey[351], *figs from the tree

incidence: 4?/19*, conc. range: 15.2–78.3 μg/kg, sample origin: orchards, farmer store houses, and fig processing plants, Turkey, sample year: 1988, country: Turkey[351], *figs on the ground

incidence: 2?/8*, conc. range: 0.5–4.0 μg/kg, sample origin: orchards, farmer store houses, and fig processing plants, Turkey, sample year: 1988, country: Turkey[351], *figs during drying (sun)

incidence: 0/3*, conc.: no contamination, sample origin: orchards, farmer store houses, and fig processing plants, Turkey, sample year: 1988, country: Turkey[351], *figs in farmer store house

incidence: 7?/12*, conc. range: 0.5–16.6 μg/kg, sample origin: orchards, farmer store houses, and fig processing plants, Turkey, sample year: 1988, country: Turkey[351], *figs before processing

incidence: 4?/32*, conc. range: 0.5–4.2 μg/kg, sample origin: orchards, farmer store houses, and fig processing plants, Turkey, sample year: 1988, country: Turkey[351], *figs during processing

- Co-contamination: not reported
- Further contamination of plants and plant products other than cereals and their products reported in the present book/article[351]: Figs, AFB_1, AFB_2, AFG_2, and OTA, literature[351]; Paste (fig paste), AFB_1, AFB_2, AFG_1, AFG_2, and OTA, literature[351]

incidence: 5/48*, conc. range: 0.4–192.8 µg/kg, ∅ conc.: 52 µg/kg, sample origin: orchards in the Aegean Region, Turkey, sample year: unknown, country: Turkey[353], *figs (dried) positive for *Aspergillus* section *Flavi* contamination

- Co-contamination: 1 sa. co-contaminated with AFB_1, AFB_2, AFG_1, AFG_2, and CPA; 4 sa. co-contaminated with AFB_1, AFB_2, AFG_1, and CPA
- Further contamination of plants and plant products other than cereals and their products reported in the present book/article[353]: Figs, AFB_1, AFB_2, AFG_2, and CPA, literature[353]

For detailed information see the article.

incidence: 29/49*, conc. range: 0.2–734.2 µg/kg, ∅ conc.: 54 µg/kg, sample origin: commercial processing facility, Turkey, sample year: 2005, country: Turkey/UK/Netherlands[354], *49 individual figs unsuitable for food

- Co-contamination: 3 sa. co-contaminated with AFB_1, AFB_2, AFG_1, AFG_2, and OTA; 11 sa. co-contaminated with AFB_1, AFB_2, AFG_1, and OTA; 6 sa. co-contaminated with AFB_1, AFB_2, and AFG_1; 1 sa. co-contaminated with AFB_1, AFG_1, AFG_2, and OTA; 4 sa. co-contaminated with AFB_1, AFG_1, and OTA; 4 sa. co-contaminated with AFB_1 and AFG_1
- Further contamination of plants and plant products other than cereals and their products reported in the present book/article[354]: Figs, AFB_1, AFB_2, AFG_2, and OTA, literature[354]

incidence: 3/206*, conc. range: 12.4–97.5 µg/kg**, ∅ conc.: 61.4 µg/kg**, sample origin: from fields during drying, warehouses, and processing units in ten regions in the Aegean area, Turkey, sample year: 1986, country: Turkey[717], *figs, dried, **low grade figs collected from the drying stage

- Co-contamination: 2 sa. co-contaminated with AFB_1, AFB_2, and AFG_1; 1 sa. co-contaminated with AFB_1 and AFG_1
- Further contamination of plants and plant products other than cereals and their products reported in the present book/article[717]: Figs, AFB_1 and AFB_2, literature[717]

incidence: 4/28, conc. range: 3.96–6.38 µg/kg*, ∅ conc.: 5 µg/kg*, sample origin: local and of Algeria, Chile, Iran, Israel, and Turkey purchased from markets in Valencia (city), Spain and Tunis (capital), Tunisia, sample year: 2012/2013, country: Spain[720], *purchased in Tunisia

- Co-contamination: For detailed information see the article.
- Further contamination of plants and plant products other than cereals and their products reported in the present book/article[720]: Dates, AFB_2, AFG_1, AFG_2, DAS, ENA_1, ENB, ENB_1, and OTA, literature[720]; Figs, BEA, DAS, ENA_1, ENB, ENB_1, HT-2, and OTA, literature[720]; Fruits, dried, AFG_2, DAS, ENA_1, ENB, ENB_1, HT-2, and OTA, literature[720]; Vine fruits, dried, AFB_1, AFG_2, ENA, ENA_1, ENB, ENB_1, HT-2, and OTA, literature[720]

For detailed information see the article.

Aflatoxin G_2

incidence: 1/24*, conc.: 4.2 µg/kg, sample origin: sites of primary storage or processing (domestic products) and sites of importation, Cyprus, sample year: 1992–1996, country: Cyprus[200], *figs and figpie

- Co-contamination: not reported
- Further contamination of plants and plant products other than cereals and their products reported in the present book/article[200]: Butter (peanut butter), Nuts (peanut), AFB_1, AFB_2, AFG_1, and AFG_2, literature[200]; Figs, AFB_1, AFB_2, and AFG_1, literature[200]; Nuts (Brazil nut, paranut), Nuts (walnut), AFB_1, AFB_2, and AFG_1, literature[200]; Nuts (chestnut), Products (sesame product), Spices, AFB_1, literature[200]; Nuts (pistachio nut), AFB_1 and AFB_2, literature[200]; Seeds (sesame), AFB_1, AFG_1, and AFG_2, literature[200]

incidence: 4/20, conc. range: LOQ-2 µg/kg (4 sa., maximum: 1.1 µg/kg), ∅ conc.: 0.9 µg/kg, sample origin: retail outlets, UK, sample year: September 2001, country: UK[203]

- Co-contamination: 1 sa. co-contaminated with AFB_1 and AFG_2; 3 sa. contaminated solely with AFG_2
- Further contamination of plants and plant products other than cereals and their products reported in the present book/article[203]: Butter (nut butter), AFB_1, AFB_2, AFG_1, AFG_2, and OTA, literature[203]; Butter (peanut butter), AFB_1, AFB_2, AFG_1, AFG_2, and OTA, literature[203]; Dates, AFB_1 and OTA, literature[203]; Figs, AFB_1, AFB_2, and OTA, literature[203]; Fruits, dried, Nuts (pistachio nut), AFB_1, AFB_2, and OTA, literature[203]; Nuts (Brazil nut, paranut), AFB_1, AFB_2, and AFG_1, literature[203]; Nuts (peanut), AFB_1 and AFB_2, literature[203]

incidence: 1/4, conc.: 0.7 µg/kg, sample origin: unknown, sample year: unknown, country: Denmark[217]

- Co-contamination: 1 sa. co-contaminated with AFB_1, AFB_2, AFG_1, and AFG_2
- Further contamination of plants and plant products other than cereals and their products reported in the present book/article[217]: Butter (peanut butter), Spices (cayenne pepper, chilli), AFB_1, AFB_2, AFG_1, and AFG_2, literature[217]; Figs, AFB_1, AFB_2, and AFG_1, literature[217]; Meal (pistachio meal), AFB_1 and AFB_2, literature[217]

incidence: 1/1, conc.: 7.89 µg/kg, sample origin: local supermarket in Okayama (prefecture), Japan, sample year: unknown, country: Japan[339]

- Co-contamination: not reported
- Further contamination of plants and plant products other than cereals and their products reported in the present book/article[339]: Coconuts, Fruits, dried, Seeds (sesame), AFB_1 and AFG_2, literature[339]; Figs, AFB_2, literature[339]; Flour (soybean flour), AFB_1, literature[339]; Nuts, Spices, AFB_1, AFB_2, AFG_1, and AFG_2, literature[339]; Raisins, AFG_2, literature[339]

incidence: 9?/24*, conc. range: 0.5–15.0 µg/kg, sample origin: orchards, farmer store houses, and fig processing plants, Turkey, sample year: 1988, country: Turkey[351], *figs from the tree

incidence: 4?/19*, conc. range: 0.5–8.6 µg/kg, sample origin: orchards, farmer store houses, and fig processing plants, Turkey, sample year: 1988, country: Turkey[351], *figs on the ground

incidence: 0/8*, conc.: no contamination, sample origin: orchards, farmer store houses, and fig processing plants, Turkey, sample year: 1988, country: Turkey[351], *figs during drying (sun)

incidence: 0/3*, conc.: no contamination, sample origin: orchards, farmer store houses, and fig processing plants, Turkey, sample year: 1988, country: Turkey[351], *figs in farmer store house

incidence: 7?/12*, conc. range: 0.5–12.5 µg/kg, sample origin: orchards, farmer store houses, and fig processing plants, Turkey, sample year: 1988, country: Turkey[351], *figs before processing incidence: 4?/32*, conc. range: 0.5–2.0 μg/kg, sample origin: orchards, farmer store houses, and fig processing plants, Turkey, sample year: 1988, country: Turkey[351], *figs during processing

- Co-contamination: not reported
- Further contamination of plants and plant products other than cereals and their products reported in the present book/article[351]: Figs, AFB_1, AFB_2, AFG_1, and OTA, literature[351]; Paste (fig paste), AFB_1, AFB_2, AFG_1, AFG_2, and OTA, literature[351]

incidence: 1/48*, conc.: 0.3 μg/kg, sample origin: orchards in the Aegean Region, Turkey, sample year: unknown, country: Turkey[353], *figs (dried) positive for *Aspergillus* section *Flavi* contamination

- Co-contamination: 1 sa. co-contaminated with AFB_1, AFB_2, AFG_1, AFG_2, and CPA
- Further contamination of plants and plant products other than cereals and their products reported in the present book/article[353]: Figs, AFB_1, AFB_2, AFG_1, and CPA, literature[353]

For detailed information see the article.

incidence: 4/49*, conc. range: 0.2–22.3 μg/kg, ∅ conc.: 7.07 μg/kg, sample origin: commercial processing facility, Turkey, sample year: 2005, country: Turkey/UK/Netherlands[354], *49 individual figs unsuitable for food

- Co-contamination: 3 sa. co-contaminated with AFB_1, AFB_2, AFG_1, AFG_2, and OTA; 1 sa. co-contaminated with AFB_1, AFG_1, AFG_2, and OTA
- Further contamination of plants and plant products other than cereals and their products reported in the present book/article[354]: Figs, AFB_1, AFB_2, AFG_1, and OTA, literature[354]

Aflatoxin/s

incidence: 5/10*, conc. range: 0.8–12.5 μg/kg**, ∅ conc.: 5.81 μg/kg**, sample origin: retail shops and local markets in NWFP and northern areas, Pakistan, sample year: January–November 2009, country: Pakistan[17], *figs, dried, **AFB_1, AFB_2, AFG_1, and AFG_2

- Co-contamination: not reported
- Further contamination of plants and plant products other than cereals and their products reported in the present book/article[17]: Dates, Fruits, dried, Kernels (apricot), Nuts (almond), Nuts (peanut), Nuts (pine nut), Nuts (pistachio nut), Nuts (walnut), Raisins, AF/S, literature[17]

incidence: 7/98*, conc. range: 0.23–4.28 μg/kg**, sample origin: different export companies, Turkey, sample year: unknown, country: Turkey[55], *figs, dried, **AFB_1, AFB_2, AFG_1, and AFG_2

- Co-contamination: see **Figs, AFB_1, literature**[55] and the article
- Further contamination of plants and plant products other than cereals and their products reported in the present book/article[55]: Figs, AFB_1 and OTA, literature[55]; Fruits, dried, Sultanas, OTA, literature[55]

incidence: 1/49*, conc.: 0.6 μg/kg**, sample origin: supermarkets and hypermarkets in 12 main cities of Catalonia (autonomous community), Spain, sample year: 2008/2009, country: Spain[93], *figs, dried, **AFB_1, AFB_2, AFG_1, and AFG_2

- Co-contamination: not reported
- Further contamination of plants and plant products other than cereals and their products reported in the present book/article[93]: Nuts (peanut), Nuts (pistachio nut), Spices (cayenne pepper, chilli), AF/S, literature[93]

incidence: 16/20*, conc. range: 1.0–3.9 μg/kg** (10 sa.), 10.1–20.0 μg/kg** (3 sa.), 20.1–50.0 μg/kg** (2 sa.), 89 μg/kg** (1 sa.), sample origin: port sa. from Felixstowe (town), Manchester (city), and Kings Lynn (market town) Port Health Authorities, UK, sample year: unknown, country: UK[206], *figs, dried, **AFB_1, AFB_2, AFG_1, and AFG_2

incidence: 1/29, conc.: 1.0–3.9 μg/kg*, sample origin: retail sa. from Norwich area and health food stores, market stalls, and supermarkets across the UK, sample year: unknown, country: UK[206], *AFB_1, AFB_2, AFG_1, and AFG_2

- Co-contamination: not reported
- Further contamination of plants and plant products other than cereals and their products reported in the present book/article[206]: Butter (peanut butter), Nuts (Brazil nut, paranut), Nuts (peanut), Nuts (pistachio nut), Paste (fig paste), AF/S, literature[206]

incidence: 2/6*, conc. range: 0.70–11.8 μg/kg**, ∅ conc.: 6.25 μg/kg**, sample origin: sales outles, Qatar, sample year: October 2002, country: Qatar[229], *figs, dried, **AFB_1, AFB_2, AFG_1, and AFG_2

- Co-contamination: not reported
- Further contamination of plants and plant products other than cereals and their products reported in the present book/article[229]: Butter (peanut butter), Nuts (peanut), Nuts (pistachio nut), AF/S, literature[229]; Coffee, Raisins, OTA, literature[229]; Spices, Spices (cayenne pepper, chilli), AF/S and OTA, literature[229]

incidence: 3/10*, conc. range: 0.1–6.7 μg/kg, ∅ conc.: 4.2 μg/kg, sample origin: available in New Zealand, sample year: 2009, country: New Zealand[277], *figs, dried

- Co-contamination: not reported
- Further contamination of plants and plant products other than cereals and their products reported in the present book/article[277]: Bars, AFB_1, AF/S, DON, NIV, and OTA, literature[277]; Butter (peanut butter), Confectionery, Food, Fruits, dried, Nuts, Nuts (peanut), Nuts (pistachio nut), Sauces (peanut sauce), AFB_1 and AF/S, literature[277]; Coffee, Dates, Spices (nutmeg), Spices (pepper), Vine fruits, dried, OTA, literature[277]; Figs, AFB_1 and OTA, literature[277]; Nuts (Brazil nut, paranut), AFB_1, literature[277]; Spices (cayenne pepper, chilli), Spices (curry), Spices (ginger), Spices (paprika), AFB_1, AF/S, and OTA, literature[277]

incidence: 6/20*, conc. range: ≤32.9 μg/kg**, ∅ conc.: 8.70 μg/kg**, sample origin: retail shops and local markets in Rabat-Salé (region), Morocco, sample year: January–October 2006, country: Spain/Morocco[352], *figs, dried, **AFB_1, AFB_2, AFG_1, and AFG_2

- Co-contamination: not reported
- Further contamination of plants and plant products other than cereals and their products reported in the present book/article[352]: Figs, AFB_1, literature[352]; Nuts (peanut), Nuts (pistachio nut), Nuts (walnut), Raisins, AFB_1 and AF/S, literature[352]

incidence: 462/2,550*, conc. range: <0.1–1.0 μg/kg** (338 sa.), 1.0–2.0 μg/kg** (44 sa.), 2.0–10.0 μg/kg** (51 sa.), 10.0–20.0 μg/kg** (16 sa.), >20.0 μg/kg** (13 sa., maximum: 200 μg/kg**), sample origin: Turkish fig exporters, Turkey, sample year: 2003, country: Turkey[355], *figs, dried, **AFB_1, AFB_2, AFG_1, and AFG_2

incidence: 518/3,096*, conc. range: <0.1–1.0 μg/kg** (340 sa.), 1.0–2.0 μg/kg** (65 sa.), 2.0–10.0 μg/kg** (76 sa.), 10.0–20.0 μg/kg** (18 sa.), >20.0 μg/kg** (19 sa., maximum: 155 μg/kg**), sample origin: Turkish fig exporters, Turkey, sample year: 2004, country: Turkey[355], *figs, dried, **AFB_1, AFB_2, AFG_1, and AFG_2

incidence: 897/3,920*, conc. range: <0.1–1.0 µg/kg** (542 sa.), 1.0–2.0 µg/kg** (118 sa.), 2.0–10.0 µg/kg** (153 sa.), 10.0–20.0 µg/kg** (49 sa.), >20.0 µg/kg** (35 sa., maximum: 316 µg/kg**), sample origin: Turkish fig exporters, Turkey, sample year: 2005, country: Turkey[355], *figs, dried, **AFB_1, AFB_2, AFG_1, and AFG_2

incidence: 166/830*, conc. range: <0.1–1.0 µg/kg** (107 sa.), 1.0–2.0 µg/kg** (33 sa.), 2.0–10.0 µg/kg** (18 sa.), 10.0–20.0 µg/kg** (5 sa.), >20.0 µg/kg** (3 sa., maximum: 33 µg/kg**), sample origin: Turkish fig exporters, Turkey, sample year: 2006, country: Turkey[355], *figs, dried, **AFB_1, AFB_2, AFG_1, and AFG_2

- Co-contamination: not reported
- Further contamination of plants and plant products other than cereals and their products reported in the present book/article[355]: Figs, AFB_1, literature[355]

incidence: 11/19*, conc. range: 0.3–2.0 µg/kg** (10 sa.), 1,500 µg AFB_1/kg (1 sa.), sample origin: imported from Argentina, Chile, Iran, and Turkey purchased from markets in Campinas (city) and São Paulo (city, state), Brazil, sample year: 2002/2003, country: Brazil[356], *figs, dried, **AFB_1 and AFB_2

- Co-contamination: not reported
- Further contamination of plants and plant products other than cereals and their products reported in the present book/article[356]: Sultanas, AF/S, literature[356]

incidence: 24/30, conc. range: 28–241 µg/kg*, ∅ conc.: 104 µg/kg*, sample origin: imported from France, Greece, Turkey, and unknown origin and purchased in a vegetable shop and health food shops, UK, sample year: unknown, country: UK[357], *AFB_1, AFB_2, AFG_1, and AFG_2

- Co-contamination: not reported
- Further contamination of plants and plant products other than cereals and their products reported in the present book/article[357]: not reported

For detailed information see the article.

incidence: 1575/4917* **, conc. range: 0.2–259.46 µg/kg***, sample origin: exporting companies in Aydin (province), Turkey, sample year: September–December 2007, country: Turkey/USA[358], *figs, dried, **destined for export, ***AFB_1, AFB_2, AFG_1, and AFG_2

- Co-contamination: 94 sa. co-contaminated with AFB_1, AFB_2, AFG_1, and AFG_2; 152 sa. co-contaminated with AFB_1, AFB_2, and AFG_1; 6 sa. co-contaminated with AFB_1, AFG_1, and AFG_2; 134 sa. co-contaminated with AFB_1 and AFB_2; 444 sa. co-contaminated with AFB_1 and AFG_1; 700 sa. contaminated solely with AFB_1; 45 sa. contaminated solely with AFG_1
- Further contamination of plants and plant products other than cereals and their products reported in the present book/article[358]: not reported

For detailed information see the article.

incidence: 11/29*, conc. range: 1.0–10.0 µg/kg** (6 sa.), >10.0 µg/kg** (5 sa., maximum: 96 µg/kg**), sample origin: importers and retailers, UK, sample year: November/December 1988, country: UK[359], *Turkish figs, dried, **AFS total

incidence: 27/57*, conc. range: 1.0–10.0 µg/kg** (21 sa.), >10.0 µg/kg** (6 sa., maximum: 34 µg/kg**), sample origin: containers at port of entry, UK, sample year: April–December 1989, country: UK[359], *Turkish figs, dried, **AFS total

incidence: 18/36*, conc. range: 1.0–10.0 µg/kg** (16 sa.), >10.0 µg/kg** (2 sa., maximum: 25 µg/kg**), sample origin: containers at port of entry, UK, sample year: January–July 1990, country: UK[359], *Turkish figs, dried, **AFS total

- Co-contamination: not reported
- Further contamination of plants and plant products other than cereals and their products reported in the present book/article[359]: Paste (fig paste), AF/S, literature[359]

incidence: 313/2643* **, conc. range: 0.2–162.76 µg/kg***, sample origin: exporting companies in Aydin (province), Turkey, sample year: January–August 2007, country: Turkey[360], *figs, dried, **destined for export, *** AFB_1, AFB_2, AFG_1, and AFG_2

- Co-contamination: 14 sa. co-contaminated with AFB_1, AFB_2, AFG_1, and AFG_2; 13 sa. co-contaminated with AFB_1, AFB_2, and AFG_1; 20 sa. co-contaminated with AFB_1 and AFB_2; 85 sa. co-contaminated with AFB_1 and AFG_1; 159 sa. contaminated solely with AFB_1; 22 sa. contaminated solely with AFG_1
- Further contamination of plants and plant products other than cereals and their products reported in the present book/article[360]: Nuts (hazelnut), Nuts (peanut), Nuts (pistachio nut), Spices (paprika), AF/S, literature[360]

incidence: 12/115*, conc. range: 0.1–763.2 µg/kg**, sample origin: orchards in Erbeyli, Germencik, Incirliova, Ortaklar, Selcuk, Soke, and Torbali (districts) in Aegean (region), Turkey, sample year: unknown, country: Turkey[362], *figs, dried, **AFB_1, AFB_2, AFG_1, and AFG_2

- Co-contamination: 5 sa. co-contaminated with AFS and OTA; 4 sa. contaminated with 4 mycotoxins; 6 sa. contaminated with 3 mycotoxins; 54 sa. contaminated with 2 mycotoxins; 38 sa. contaminated with 1 mycotoxin; no further information available
- Further contamination of plants and plant products other than cereals and their products reported in the present book/article[362]: Figs, CPA, FB_1, and OTA, literature[362]

incidence: 3/10*, conc. range: ≤6.7 µg/kg**, sample origin: New Zealand, sample year: 2009, country: New Zealand[702], *figs, dried, **AFB_1, AFB_2, AFG_1, and AFG_2

- Co-contamination: not reported
- Further contamination of plants and plant products other than cereals and their products reported in the present book/article[702]: Figs, Spices (pepper), Vine fruits, dried, OTA, literature[702]; Fruits, dried, AF/S, literature[702]; Spices (cayenne pepper, chilli), Spices (curry), Spices (ginger), Spices (paprika), AF/S and OTA, literature[702]

Aspergillus and ***Penicillium*** Toxins

Cyclopiazonic Acid

incidence: 28/48*, conc. range: 25–187 µg/kg, ∅ conc.: 58.3 µg/kg, sample origin: orchards in the Aegean Region, Turkey, sample year: unknown, country: Turkey[353], *figs (dried) positive for *Aspergillus* section *Flavi* contamination

- Co-contamination: 1 sa. co-contaminated with AFB_1, AFB_2, AFG_1, AFG_2, and CPA; 4 sa. co-contaminated with AFB_1, AFB_2, AFG_1, and CPA; 4 sa. co-contaminated with AFB_1, AFB_2, and CPA; 2 sa. co-contaminated with AFB_1 and CPA; 17 sa. contaminated solely with CPA
- Further contamination of plants and plant products other than cereals and their products reported in the present book/article[353]: Figs, AFB_1, AFB_2, AFG_1, and AFG_2, literature[353]

For detailed information see the article.

incidence: 28/115*, conc. range: 25.0–187.0 µg/kg**, sample origin: orchards in Erbeyli, Germencik, Incirliova, Ortaklar, Selcuk, Soke, and Torbali (districts) in Aegean (region), Turkey, sample year: unknown, country: Turkey[362], *figs, dried, **AFB_1, AFB_2, AFG_1, and AFG_2

- Co-contamination: 17 sa. co-contaminated with CPA and FB_1; 4 sa. contaminated with 4 mycotoxins; 6 sa. contaminated with 3 mycotoxins; 54 sa. contaminated with 2 mycotoxins; 38 sa. contaminated with 1 mycotoxin; no further information available
- Further contamination of plants and plant products other than cereals and their products reported in the present book/article[362]: Figs, AF/S, FB_1, and OTA, literature[362]

Kojic Acid

incidence: 52/52* conc. range: 8,000–6,900,000 µg/kg, ∅ conc.: 1,336,000 µg/kg, sample origin: Turkey, sample year: 1991/1992, country: Switzerland[350], *selected figs (dried) with BGY fluorescence

- Co-contamination: 3 sa. co-contaminated with AFB_1, AFG_1, KA, and OTA; 11 sa. co-contaminated with AFB_1, AFG_1, and KA; 4 sa. co-contaminated with AFB_1, KA, and OTA; 19 sa. co-contaminated with AFB_1 and KA; 5 sa. co-contaminated with KA and OTA; 10 sa. contaminated solely with KA
- Further contamination of plants and plant products other than cereals and their products reported in the present book/article[350]: Figs, AFB_1, AFG_1, and OTA, literature[350]

Ochratoxin A

incidence: 4/4*, conc. range: 60–120 µg/kg, sample origin: shops and markets of different sanitation levels, Assiut Governorate, Egypt, sample year: March 1992, country: Egypt[54], *figs, dried

- Co-contamination: not reported
- Further contamination of plants and plant products other than cereals and their products reported in the present book/article[54]: Fruits, dried, OTA, literature[54]

incidence: 18/98*, conc. range: 0.87–24.37 µg/kg, sample origin: different export companies, Turkey, sample year: unknown, country: Turkey[55], *figs, dried

- Co-contamination: 2 sa. co-contaminated with AFB_1 and OTA
- Further contamination of plants and plant products other than cereals and their products reported in the present book/article[55]: Figs, AFB_1, AF/S, literature[55]; Fruits, dried, Sultanas, OTA, literature[55]

incidence: 4/20, conc. range: 0.4–151 µg/kg, ∅ conc.: 38.9 µg/kg, sample origin: retail outlets, UK, sample year: September 2001, country: UK[203]

- Co-contamination: not reported
- Further contamination of plants and plant products other than cereals and their products reported in the present book/article[203]: Butter (nut butter), AFB_1, AFB_2, AFG_1, AFG_2, and OTA, literature[203]; Butter (peanut butter), AFB_1, AFB_2, AFG_1, AFG_2, and OTA, literature[203]; Dates, AFB_1 and OTA, literature[203]; Figs, AFB_1, AFB_2, and AFG_2, literature[203]; Fruits, dried, Nuts (pistachio nut), AFB_1, AFB_2, and OTA, literature[203]; Nuts (Brazil nut, paranut), AFB_1, AFB_2, and AFG_1, literature[203]; Nuts (peanut), AFB_1 and AFB_2, literature[203]

incidence: 27/34*, conc. range: ≤3.95 µg/kg, sample origin: regions around Detmold, Giessen, Hamburg, Jena, Karlsruhe, Kiel, Kulmbach, Trier (cities), Germany, sample year: unknown, country: Germany[262], *figs, dried

- Co-contamination: not reported
- Further contamination of plants and plant products other than cereals and their products reported in the present book/article[262]: Bars, Chocolate, Drinks (chocolate drink), Fruits, dried, Nut crème, Nuts, Nuts (hazelnut), Nuts (peanut), Powder (cocoa powder), Pudding, Raisins, Seeds (linseed), Seeds (sesame), Seeds (sunflower), OTA, literature[262]

incidence: 2/20, conc. range: 0.2–0.8 µg/kg, ∅ conc.: 0.5 µg/kg, sample origin: 13 regions in the UK, sample year: May 1998, country: UK[263]

- Co-contamination: not reported
- Further contamination of plants and plant products other than cereals and their products reported in the present book/article[263]: Chocolate, Currants, Juices (grape juice), Powder (cocoa powder), Raisins, Sultanas, Wine (red), OTA, literature[263]

incidence: 1/10*, conc.: 73 µg/kg, sample origin: available in New Zealand, sample year: 2008/2009, country: New Zealand[277], *figs, dried

- Co-contamination: not reported
- Further contamination of plants and plant products other than cereals and their products reported in the present book/article[277]: Bars, AFB_1, AF/S, DON, NIV, and OTA, literature[277]; Butter (peanut butter), Confectionery, Figs, Food, Fruits, dried, Nuts, Nuts (peanut), Nuts (pistachio nut), Sauces (peanut sauce), AFB_1 and AF/S, literature[277]; Coffee, Dates, Spices (nutmeg), Spices (pepper), Vine fruits, dried, OTA, literature[277]; Nuts (Brazil nut, paranut), AFB_1, literature[277]; Spices (cayenne pepper, chilli), Spices (curry), Spices (ginger), Spices (paprika), AFB_1, AF/S, and OTA, literature[277]

incidence: 18/19*, conc. range: 0.1–5.0 µg/kg (13 sa.), 5.1–10.0 µg/kg (2 sa.), 10.1–20.0 µg/kg (2 sa.), 20.1–30.0 µg/kg (1 sa.), sample origin: different markets in Campinas (city) and São Paulo (city/state), Brazil, sample year: 2002/2003, country: Brazil[331], *figs, dried

- Co-contamination: not reported
- Further contamination of plants and plant products other than cereals and their products reported in the present book/article[331]: Dates, Fruits, dried, Sultanas, OTA, literature[331]

incidence: 7/58*, conc. range: <0.1–1.0 µg/kg (2 sa.), 1.0–5.0 µg/kg (1 sa.), 5.0–10.0 µg/kg (1 sa.), 10.0 15.0 µg/kg (3 sa., maximum: 13.0 µg/kg), sample origin: Turkish fig exporters, Turkey, sample year: 2003, country: Turkey[349], *figs, dried

- Co-contamination: 2 sa. co-contaminated with AFB_1 and OTA; 5 sa. contaminated solely with OTA
- Further contamination of plants and plant products other than cereals and their products reported in the present book/article[349]: Figs, AFB_1, literature[349]

incidence: 6/41*, conc. range: <0.1–1.0 µg/kg (2 sa.), 1.0–5.0 µg/kg (1 sa.), >15.0 µg/kg (3 sa., maximum: 26.3 µg/kg), sample origin: Turkish fig exporters, Turkey, sample year: 2004, country: Turkey[349], *figs, dried

- Co-contamination: 2 sa. co-contaminated with AFB_1 and OTA; 4 sa. contaminated solely with OTA
- Further contamination of plants and plant products other than cereals and their products reported in the present book/article[349]: Figs, AFB_1, literature[349]

incidence: 1/20*, conc.: 2.0 µg/kg, sample origin: local markets, Turkey, sample year: 2004, country: Turkey[349], *figs, dried

- Co-contamination: not reported
- Further contamination of plants and plant products other than cereals and their products reported in the present book/article[349]: Figs, AFB_1, literature[349]

incidence: 12/52* conc. range: 5–12,300 μg/kg, Ø conc.: 1,680 μg/kg, sample origin: Turkey, sample year: 1991/1992, country: Switzerland[350], *selected figs (dried) with BGY fluorescence

- Co-contamination: 3 sa. co-contaminated with AFB_1, AFG_1, KA, and OTA; 4 sa. co-contaminated with AFB_1, KA, and OTA; 5 sa. co-contaminated with KA and OTA
- Further contamination of plants and plant products other than cereals and their products reported in the present book/article[350]: Figs, AFB_1, AFG_1, and KA, literature[350]

incidence: 0/24*, conc.: no contamination, sample origin: orchards, farmer store houses, and fig processing plants, Turkey, sample year: 1988, country: Turkey[351], *figs from the tree

incidence: 1/19*, conc.: 7.5 μg/kg, sample origin: orchards, farmer store houses, and fig processing plants, Turkey, sample year: 1988, country: Turkey[351], *figs on the ground

incidence: 0/8*, conc.: no contamination, sample origin: orchards, farmer store houses, and fig processing plants, Turkey, sample year: 1988, country: Turkey[351], *figs during drying (sun)

incidence: 0/3*, conc.: no contamination, sample origin: orchards, farmer store houses, and fig processing plants, Turkey, sample year: 1988, country: Turkey[351], *figs in farmer store house

incidence: 1/12*, conc.: 8.3 μg/kg, sample origin: orchards, farmer store houses, and fig processing plants, Turkey, sample year: 1988, country: Turkey[351], *figs before processing

incidence: 0/32*, conc.: no contamination, sample origin: orchards, farmer store houses, and fig processing plants, Turkey, sample year: 1988, country: Turkey[351], *figs during processing

- Co-contamination: not reported
- Further contamination of plants and plant products other than cereals and their products reported in the present book/article[351]: Figs, AFB_1, AFB_2, AFG_1, and AFG_2, literature[351]; Paste (fig paste), AFB_1, AFB_2, AFG_1, AFG_2, and OTA, literature[351]

incidence: 32/50*, conc. range: 0.5–1,710 μg/kg, Ø conc.: 195 μg/kg, sample origin: commercial processing facility, Turkey, sample year: 2005, country: Turkey/UK/Netherlands[354], *50 individual figs unsuitable for food

- Co-contamination: 3 sa. co-contaminated with AFB_1, AFB_2, AFG_1, AFG_2, and OTA; 11 sa. co-contaminated with AFB_1, AFB_2, AFG_1, and OTA; 1 sa. co-contaminated with AFB_1, AFG_1, AFG_2, and OTA; 7 sa. co-contaminated with AFB_1, AFB_2, and OTA; 4 sa. co-contaminated with AFB_1, AFG_1, and OTA; 5 sa. co-contaminated with AFB_1 and OTA; 1 sa. contaminated solely with OTA
- Further contamination of plants and plant products other than cereals and their products reported in the present book/article[354]: Figs, AFB_1, AFB_2, AFG_1, and AFG_2, literature[354]

incidence: 55/115*, conc. range: 0.1–15.3 μg/kg, sample origin: orchards in Erbeyli, Germencik, Incirliova, Ortaklar, Selcuk, Soke, and Torbali (districts) in Aegean (region), Turkey, sample year: unknown, country: Turkey[362], *figs, dried

- Co-contamination: 5 sa. co-contaminated with AFS and OTA; 4 sa. contaminated with 4 mycotoxins; 6 sa. contaminated with 3 mycotoxins; 54 sa. contaminated with 2 mycotoxins; 38 sa. contaminated with 1 mycotoxin; no further information available
- Further contamination of plants and plant products other than cereals and their products reported in the present book/article[362]: Figs, AF/S, CPA, and FB_1, literature[362]

incidence: 1/9, conc.: 160 μg/kg, sample origin: unknown, sample year: unknown, country: Switzerland[363]

- Co-contamination: not reported
- Further contamination of plants and plant products other than cereals and their products reported in the present book/article[363]: not reported

incidence: 13/20*, conc. range: 0.03–1.42 μg/kg, Ø conc.: 0.33 μg/kg, sample origin: retail shops in Salé (city) and Rabat (capital), Morocco, sample year: July/August 2005, country: Morocco/Spain[364], *figs, dried

- Co-contamination: not reported
- Further contamination of plants and plant products other than cereals and their products reported in the present book/article[364]: Nuts (peanut), Nuts (walnut), Raisins, OTA, literature[364]

incidence: 33/71*, conc. range: 0.12–1.00 μg/kg (31 sa.), >8.01 μg/kg (2 sa., maximum: 15.31 μg/kg), sample origin: orchards in Aegean (region), sample year: 2003 harvest, country: Turkey[365], *figs, dried

incidence: 22/44*, conc. range: 0.12–1.00 μg/kg (20 sa.), >8.01 μg/kg (2 sa., maximum: 11.56 μg/kg), sample origin: orchards in Aegean (region), sample year: 2004 harvest, country: Turkey[365], *figs, dried

- Co-contamination: not reported
- Further contamination of plants and plant products other than cereals and their products reported in the present book/article[365]: not reported

incidence: 31/35*, conc. range: ≥0.06–0.49 μg/kg (25 sa.), 0.5–0.99 μg/kg (3 sa.), >9.9 μg/kg (3 sa.**, maximum: 275.7 μg/kg), sample origin: purchased in local supermarkets and retail shops in Spain: Spanish (29 sa.), Turkish (5 sa.), and French products (1 sa.), sample year: unknown, country: Spain[366], *figs, dried, **2 sa. from Spain and 1 sa. from Turkey

- Co-contamination: not reported
- Further contamination of plants and plant products other than cereals and their products reported in the present book/article[366]: not reported

incidence: 1/49*, conc.: 0.51 μg/kg, sample origin: supermarkets and stores, USA, sample year: 2012–2014, country: USA[599], *conventional

incidence: 3/39*, conc. range: 1.46–3.33 μg/kg, Ø conc.: 2.20 μg/kg, sample origin: supermarkets and stores, USA, sample year: 2012–2014, country: USA[599], *organic

- Co-contamination: not reported
- Further contamination of plants and plant products other than cereals and their products reported in the present book/article[599]: Dates, Nuts (pistachio nut), Raisins, OTA, literature[599]

incidence: 1/10*, conc.: 73 μg/kg, sample origin: New Zealand, sample year: 2009, country: New Zealand[702], *figs, dried

- Co-contamination: not reported
- Further contamination of plants and plant products other than cereals and their products reported in the present book/article[702]: Figs, Fruits, dried, AF/S, literature[702]; Spices (cayenne pepper, chilli), Spices (curry), Spices (ginger), Spices (paprika), AF/S and OTA, literature[702]; Spices (pepper), Vine fruits, dried, OTA, literature[702]

incidence: 5/28, conc. range: <LOQ*, sample origin: local and of Algeria, Chile, Iran, Israel, and Turkey purchased from markets in Valencia (city), Spain and Tunis (capital), Tunisia, sample year: 2012/2013, country: Spain[720], *purchased in Tunisia

- Co-contamination: For detailed information see the article.

- Further contamination of plants and plant products other than cereals and their products reported in the present book/article[720]: Dates, AFB_2, AFG_1, AFG_2, DAS, ENA_1, ENB, ENB_1, and OTA, literature[720]; Figs, AFG_1, BEA, DAS, ENA_1, ENB, ENB_1, and HT-2, literature[720]; Fruits, dried, AFG_2, DAS, ENA_1, ENB, ENB_1, HT-2, and OTA, literature[720]; Vine fruits, dried, AFB_1, AFG_2, ENA, ENA_1, ENB, ENB_1, HT-2, and OTA, literature[720]

For detailed information see the article.

incidence: 1/12*, conc.: 0.99 µg/kg, sample origin: 12 regions, Czech Republic, sample year: 2012/2013, country: Czech Republic[811], *figs, dried

- Co-contamination: not reported
- Further contamination of plants and plant products other than cereals and their products reported in the present book/article[811]: Beans, Chocolate, Coffee, Confectionery, Herbs and Spices, Juices, Licorice, Powder (cocoa powder), Products (ginger product), Raisins, Spices (caraway), Spices (cayenne pepper, chilli), Spices (coriander), Spices (ginger), Spices (mustard), Spices (nutmeg), Spices (paprika), Spices (pepper), Tea, Wine (red), Wine (white), OTA, literature[811]

Fusarium Toxins

Depsipeptides

Beauvericin

incidence: 9/28, conc. range: 5.19–3,620 µg/kg*, Ø conc.: 822 µg/kg*, sample origin: local and of Algeria, Chile, Iran, Israel, and Turkey purchased from markets in Valencia (city), Spain, and Tunis (capital), Tunisia, sample year: 2012/2013, country: Spain[720], *purchased in Spain

- Co-contamination: For detailed information see the article.
- Further contamination of plants and plant products other than cereals and their products reported in the present book/article[720]: Dates, AFB_2, AFG_1, AFG_2, DAS, ENA_1, ENB, ENB_1, and OTA, literature[720]; Figs, AFG_1, DAS, ENA_1, ENB, ENB_1, HT-2, and OTA, literature[720]; Fruits, dried, AFG_2, DAS, ENA_1, ENB, ENB_1, HT-2, and OTA, literature[720]; Vine fruits, dried, AFB_1, AFG_2, ENA, ENA_1, ENB, ENB_1, HT-2, and OTA, literature[720]

For detailed information see the article.

Enniatin A_1

incidence: 14/28, conc. range: <LOQ-14.7 µg/kg*, Ø conc.: 4.1 µg/kg*, sample origin: local and of Algeria, Chile, Iran, Israel, and Turkey purchased from markets in Valencia (city), Spain, and Tunis (capital), Tunisia, sample year: 2012/2013, country: Spain[720], *purchased in Spain and Tunisia

- Co-contamination: For detailed information see the article.
- Further contamination of plants and plant products other than cereals and their products reported in the present book/article[720]: Dates, AFB_2, AFG_1, AFG_2, DAS, ENA_1, ENB, ENB_1, and OTA, literature[720]; Figs, AFG_1, BEA, DAS, ENB, ENB_1, HT-2, and OTA, literature[720]; Fruits, dried, AFG_2, DAS, ENA_1, ENB, ENB_1, HT-2, and OTA, literature[720]; Vine fruits, dried, AFB_1, AFG_2, ENA, ENA_1, ENB, ENB_1, HT-2, and OTA, literature[720]

For detailed information see the article.

Enniatin B

incidence: 14/28, conc. range: <LOQ-251 µg/kg*, Ø conc.: 132 µg/kg*, sample origin: local and of Algeria, Chile, Iran, Israel, and Turkey purchased from markets in Valencia (city), Spain, and Tunis (capital), Tunisia, sample year: 2012/2013, country: Spain[720], *purchased in Spain and Tunisia

- Co-contamination: For detailed information see the article.
- Further contamination of plants and plant products other than cereals and their products reported in the present book/article[720]: Dates, AFB_2, AFG_1, AFG_2, DAS, ENA_1, ENB, ENB_1, and OTA, literature[720]; Figs, AFG_1, BEA, DAS, ENA_1, ENB_1, HT-2, and OTA, literature[720]; Fruits, dried, AFG_2, DAS, ENA_1, ENB, ENB_1, HT-2, and OTA, literature[720]; Vine fruits, dried, AFB_1, AFG_2, ENA, ENA_1, ENB, ENB_1, HT-2, and OTA, literature[720]

For detailed information see the article.

Enniatin B_1

incidence: 14/28, conc. range: <LOQ-34.8 µg/kg*, Ø conc.: 10.2 µg/kg*, sample origin: local and of Algeria, Chile, Iran, Israel, and Turkey purchased from markets in Valencia (city), Spain, and Tunis (capital), Tunisia, sample year: 2012/2013, country: Spain[720], *purchased in Spain and Tunisia

- Co-contamination: For detailed information see the article.
- Further contamination of plants and plant products other than cereals and their products reported in the present book/article[720]: Dates, AFB_2, AFG_1, AFG_2, DAS, ENA_1, ENB, ENB_1, and OTA, literature[720]; Figs, AFG_1, BEA, DAS, ENA_1, ENB, HT-2, and OTA, literature[720]; Fruits, dried, AFG_2, DAS, ENA_1, ENB, ENB_1, HT-2, and OTA, literature[720]; Vine fruits, dried, AFB_1, AFG_2, ENA, ENA_1, ENB, ENB_1, HT-2, and OTA, literature[720]

For detailed information see the article.

Fumonisin/s

Fumonisin B_1

incidence: 86/115*, conc. range: 50–3,650 µg/kg, sample origin: orchards in Erbeyli, Germencik, Incirliova, Ortaklar, Selcuk, Soke, and Torbali (districts) in Aegean (region), Turkey, sample year: unknown, country: Turkey[362], *figs, dried

- Co-contamination: 17 sa. co-contaminated with CPA and FB_1; 4 sa. contaminated with 4 mycotoxins; 6 sa. contaminated with 3 mycotoxins; 54 sa. contaminated with 2 mycotoxins; 38 sa. contaminated with 1 mycotoxin; no further information available
- Further contamination of plants and plant products other than cereals and their products reported in the present book/article[362]: Figs, AF/S, CPA, and OTA, literature[362]

incidence: 64/125*, conc. range: LOD ≤60 µg/kg (41 sa.), 70–80 µg/kg (14 sa.), 90–150 µg/kg (8 sa.), 163 µg/kg (1 sa.), sample origin: Aegean region (Small and Big Meander basin), Turkey, sample year: 2004, country: Turkey[367], *figs, dried

incidence: 111/137*, conc. range: LOD ≤60 µg/kg (45 sa.), 70–80 µg/kg (20 sa.), 90–150 µg/kg (32 sa.), 160–200 µg/kg (11 sa.), 210–400 µg/kg (3 sa., maximum: 332 µg/kg), sample origin: Aegean region (Small and Big Meander basin), Turkey, sample year: 2005, country: Turkey[367], *figs, dried

- Co-contamination: not reported
- Further contamination of plants and plant products other than cereals and their products reported in the present book/article[367]: Figs, FB_2, literature[367]

For detailed information see the article.

Fumonisin B_2

incidence: 12/125*, conc. range: LOD ≤60 µg/kg (10 sa.), 90–150 µg/kg (2 sa., maximum: 145 µg/kg), sample origin: Aegean region (Small and Big Meander basin), Turkey, sample year: 2004, country: Turkey[367], *figs, dried

incidence: 44/137*, conc. range: LOD ≤60 µg/kg (38 sa.), 70–80 µg/kg (1 sa.), 90–150 µg/kg (4 sa.), 198 µg/kg (1 sa.), sample origin:

Aegean region (Small and Big Meander basin), Turkey, sample year: 2005, country: Turkey[367], *figs, dried

- Co-contamination: not reported
- Further contamination of plants and plant products other than cereals and their products reported in the present book/article[367]: Figs, FB_1, literature[367]

For detailed information see the article.

Type A Trichothecenes

Diacetoxyscirpenol

incidence: 3/28, conc. range: <LOQ-77.5 µg/kg*, ∅ conc.: 77.5 µg/kg*?, sample origin: local and of Algeria, Chile, Iran, Israel, and Turkey purchased from markets in Valencia (city), Spain, and Tunis (capital), Tunisia, sample year: 2012/2013, country: Spain[720], *purchased in Spain

- Co-contamination: For detailed information see the article.
- Further contamination of plants and plant products other than cereals and their products reported in the present book/article[720]: Dates, AFB_2, AFG_1, AFG_2, DAS, ENA_1, ENB, ENB_1, and OTA, literature[720]; Figs, AFG_1, BEA, ENA_1, ENB, ENB_1, HT-2, and OTA, literature[720]; Fruits, dried, AFG_2, DAS, ENA_1, ENB, ENB_1, HT-2, and OTA, literature[720]; Vine fruits, dried, AFB_1, AFG_2, ENA, ENA_1, ENB, ENB_1, HT-2, and OTA, literature[720]

For detailed information see the article.

HT-2 Toxin

incidence: 2/28, conc. range: 3.22–7.76 µg/kg*, ∅ conc.: 5.5 µg/kg*, sample origin: local and of Algeria, Chile, Iran, Israel, and Turkey purchased from markets in Valencia (city), Spain, and Tunis (capital), Tunisia, sample year: 2012/2013, country: Spain[720], *purchased in Spain

- Co-contamination: For detailed information see the article.
- Further contamination of plants and plant products other than cereals and their products reported in the present book/article[720]: Dates, AFB_2, AFG_1, AFG_2, DAS, ENA_1, ENB, ENB_1, and OTA, literature[720]; Figs, AFG_1, BEA, DAS, ENA_1, ENB, ENB_1, and OTA, literature[720]; Fruits, dried, AFG_2, DAS, ENA_1, ENB, ENB_1, HT-2, and OTA, literature[720]; Vine fruits, dried, AFB_1, AFG_2, ENA, ENA_1, ENB, ENB_1, HT-2, and OTA, literature[720]

For detailed information see the article.

Filberts see Nuts (hazelnut)

Fino see Wine (white)

Flakes (cassava) may contain the following mycotoxins:

Aspergillus **and** ***Penicillium*** **Toxins**

Ochratoxin A

incidence: 18/18*, conc. range: 3.28–22.73 µg/kg, ∅ conc.: 7.63 µg/kg, sample origin: markets in Kaffin Koro, Minna, and Pasiko (towns), Niger State, Nigeria, sample year: 2010, country: Nigeria[785], *garri

- Co-contamination: not reported
- Further contamination of plants and plant products other than cereals and their products reported in the present book/article[785]: Seeds (sesame), OTA, literature[785]

Flour may contain the following mycotoxins:

Aspergillus **Toxins**

Aflatoxin B_1

incidence: 2/3* **, conc. range: 0.002–0.009 µg/kg, ∅ conc.: 0.0055 µg/kg, sample origin: organic product retail outlets in Lisbon (capital), Portugal, sample year: May/June 2007, country: Portugal/Netherlands[88], *organic baby food, **sa. consisted of cereal ingredients and quinoa flour or apple and nuts

- Co-contamination: 1 sa. co-contaminated with AFB_1 and OTA; 1 sa. contaminated solely with AFB_1
- Further contamination of plants and plant products other than cereals and their products reported in the present book/article[88]: Baby food, AFB_1, AFM_1 and OTA, literature[88]; Biscuits, Flour, OTA, literature[88];

incidence: 0/1* **, conc.: no contamination, sample origin: supermarkets in Lisbon (capital), Portugal, sample year: May/June 2007, country: Portugal/Netherlands[88], *conventional baby food, **sa. consisted of cereal ingredients and fruits

- Co-contamination: not reported
- Further contamination of plants and plant products other than cereals and their products reported in the present book/article[88]: Baby food, AFB_1, AFM_1 and OTA, literature[88]; Biscuits, Flour, OTA, literature[88]

For detailed information see the article.

Aspergillus **and** ***Penicillium*** **Toxins**

Ochratoxin A

incidence: 1/3* **, conc.: 0.212 µg/kg, sample origin: organic product retail outlets in Lisbon (capital), Portugal, sample year: May/June 2007, country: Portugal/Netherlands[88], *organic baby food, **sa. consisted of cereal ingredients and quinoa flour or apple and nuts

- Co-contamination: 1 sa. co-contaminated with AFB_1 and OTA
- Further contamination of plants and plant products other than cereals and their products reported in the present book/article[88]: Baby food, AFB_1, AFM_1 and OTA, literature[88]; Biscuits, OTA, literature[88]; Flour, AFB_1, literature[88]

incidence: 0/1* **, conc.: no contamination, sample origin: supermarkets in Lisbon (capital), Portugal, sample year: May/June 2007, country: Portugal/Netherlands[88], *conventional baby food, **sa. consisted of cereal ingredients and fruits

- Co-contamination: not reported
- Further contamination of plants and plant products other than cereals and their products reported in the present book/article[88]: Baby food, AFB_1, AFM_1 and OTA, literature[88]; Biscuits, OTA, literature[88]; Flour, AFB_1, literature[88]

For detailed information see the article.

Flour (cassava flour) may contain the following mycotoxins:

Aspergillus **Toxins**

Aflatoxin B_1

incidence: 3/4, conc. range: <LOQ-9 µg/kg, sample origin: different agroecological zones, Benin, sample year: unknown, country: Belgium[235]

- Co-contamination: 1 sa. co-contaminated with AFB_1, AFB_2, DAS, FB_1, and ZEA; 1 sa. co-contaminated with AFB_1, DAS, FB_1, and ZEA; 1 sa. co-contaminated with AFB_1, FB_1, and ZEA
- Further contamination of plants and plant products other than cereals and their products reported in the present book/article[235]: Cake (peanut), AFB_1, AFB_2, AFG_1, AFG_2, and OTA, literature[235]; Flour (cassava flour), AFB_2, DAS, FB_1, and ZEA, literature[235]

incidence: 3/3, conc. range: 0.32–1.64 µg/kg, sample origin: markets in Brazzaville (capital), Congo, sample year: unknown, country: Nigeria/Benin/Tanzania[256]

- Co-contamination: not reported
- Further contamination of plants and plant products other than cereals and their products reported in the present book/article[256]: Chips (cassava chips), AFB_1, literature[256]

incidence: 5/15, conc. range: 41–160 µg/kg, ∅ conc.: 87 µg/kg, sample origin: Cameroon, sample year: December 2010, country: Belgium/Russia[805]

- Co-contamination: 2 sa. co-contaminated with AFB_1, FB_1, and OTA; 3 sa. co-contaminated with AFB_1 and OTA
- Further contamination of plants and plant products other than cereals and their products reported in the present book/article[805]: Cake (peanut), AFB_1 and OTA, literature[805]; Flour (cassava flour), FB_1 and OTA, literature[805]

Aflatoxin B_2

incidence: 2/4, conc. range: <LOQ-8 µg/kg, sample origin: different agroecological zones, Benin, sample year: unknown, country: Belgium[235]

- Co-contamination: 1 sa. co-contaminated with AFB_1, AFB_2, DAS, FB_1, and ZEA; 1 sa. co-contaminated with AFB_2, DAS, and FB_1
- Further contamination of plants and plant products other than cereals and their products reported in the present book/article[235]: Cake (peanut), AFB_1, AFB_2, AFG_1, AFG_2, and OTA, literature[235]; Flour (cassava flour), AFB_1, DAS, FB_1, and ZEA, literature[235]

Aspergillus and ***Penicillium*** Toxins

Ochratoxin A

incidence: 2/33*, conc. range: 32–65 µg/kg, ∅ conc.: 48.5 µg/kg, sample origin: retail stores including large supermarkets and small neighborhood grocery shops in Campinas (municipality), Brazil, sample year: May 1985–June 1986, country: Brazil[130], *raw cassava flour

- Co-contamination: not reported
- Further contamination of plants and plant products other than cereals and their products reported in the present book/article[130]: Beans (kidney bean), AFB_1, AFG_1, and AFG_2, literature[130]; Beans (pinto bean), AFB_1, AFG_1, AFG_2, and OTA, literature[130]

incidence: 5/15, conc. range: 3–20 µg/kg, ∅ conc.: 10 µg/kg, sample origin: Cameroon, sample year: December 2010, country: Belgium/Russia[805]

- Co-contamination: 2 sa. co-contaminated with AFB_1, FB_1, and OTA; 3 sa. co-contaminated with AFB_1 and OTA
- Further contamination of plants and plant products other than cereals and their products reported in the present book/article[805]: Cake (peanut), Flour (cassava flour), AFB_1 and OTA, literature[805]; Flour (cassava flour), AFB_1 and FB_1, literature[805]

Fusarium Toxins

Fumonisin/s

Fumonisin B_1

incidence: 4/4, conc. range: 4–21 µg/kg, ∅ conc.: 9.5 µg/kg, sample origin: different agroecological zones, Benin, sample year: unknown, country: Belgium[235]

- Co-contamination: 1 sa. co-contaminated with AFB_1, AFB_2, DAS, FB_1, and ZEA; 1 sa. co-contaminated with AFB_1, DAS, FB_1, and ZEA; 1 sa. co-contaminated with AFB_1, FB_1, and ZEA; 1 sa. co-contaminated with AFB_2, DAS, and FB_1
- Further contamination of plants and plant products other than cereals and their products reported in the present book/article[235]: Cake (peanut), AFB_1, AFB_2, AFG_1, AFG_2, and OTA, literature[235]; Flour (cassava flour), AFB_1, AFB_2, DAS, and ZEA, literature[235]

incidence: 2/15, conc. range: LOQ, sample origin: Cameroon, sample year: December 2010, country: Belgium/Russia[805]

- Co-contamination: 2 sa. co contaminated with AFB_1, FB_1, and OTA
- Further contamination of plants and plant products other than cereals and their products reported in the present book/article[805]: Cake (peanut), Flour (cassava flour), AFB_1 and OTA, literature[805]

Type A Trichothecenes

Diacetoxyscirpenol

incidence: 3/4, conc. range: 1–9 µg/kg, ∅ conc.: 5 µg/kg, sample origin: different agroecological zones, Benin, sample year: unknown, country: Belgium[235]

- Co-contamination: 1 sa. co-contaminated with AFB_1, AFB_2, DAS, FB_1, and ZEA; 1 sa. co-contaminated with AFB_1, DAS, FB_1, and ZEA; 1 sa. co-contaminated with AFB_2, DAS, and FB_1
- Further contamination of plants and plant products other than cereals and their products reported in the present book/article[235]: Cake (peanut), AFB_1, AFB_2, AFG_1, AFG_2, and OTA, literature[235]; Flour (cassava flour), AFB_1, AFB_2, FB_1, and ZEA, literature[235]

Type B Trichothecenes

Deoxynivalenol

incidence: 1/1, conc.: 48 µg/kg, sample origin: local supermarkets in Porto (city), Portugal, sample year: unknown, country: Portugal[96]

- Co-contamination: 1 sa. co-contaminated with DON and ZEA
- Further contamination of plants and plant products other than cereals and their products reported in the present book/article[96]: Flour (cassava flour), ZEA, literature[96]

Other ***Fusarium*** Toxins

Zearalenone

incidence: 1/1, conc.: 14 µg/kg, sample origin: local supermarkets in Porto (city), Portugal, sample year: unknown, country: Portugal[96]

- Co-contamination: 1 sa. co-contaminated with DON and ZEA
- Further contamination of plants and plant products other than cereals and their products reported in the present book/article[96]: Flour (cassava flour), DON, literature[96]

incidence: 3/4, conc. range: 6–12 µg/kg, ∅ conc.: 9 µg/kg, sample origin: different agroecological zones, Benin, sample year: unknown, country: Belgium[235]

- Co-contamination: 1 sa. co-contaminated with AFB_1, AFB_2, DAS, FB_1, and ZEA; 1 sa. co-contaminated with AFB_1, DAS, FB_1, and ZEA; 1 sa. co-contaminated with AFB_1, FB_1, and ZEA
- Further contamination of plants and plant products other than cereals and their products reported in the present book/article[235]: Cake (peanut), AFB_1, AFB_2, AFG_1, AFG_2, and OTA, literature[235]; Flour (cassava flour), AFB_1, AFB_2, DAS, and FB_1, literature[235]

Flour (chestnut flour) may contain the following mycotoxins:

Aspergillus Toxins

Aflatoxin B_1

incidence: 23/25, conc. range: ≤58.6 μg/kg, sample origin: retail outlets in Campania, Emilia-Romagna, Piedmont, and Tuscany (regions), northern Italy, sample year: January 2012–June 2013, country: Italy[687]

- Co-contamination: For detailed information see the article.
- Further contamination of plants and plant products other than cereals and their products reported in the present book/article[687]: Flour (chestnut flour), AFB_2, AFG_1, AFG_2, CIT, MA, OTA, and ROQ-C, literature[687]; Nuts (chestnut), AFB_1, AFB_2, AFG_1, CIT, MA, OTA, and ROQ-C, literature[687]

incidence: 23/37, conc. range: ≤67.88 μg/kg, sample origin: retail outlets principally in Campania, Emilia-Romagna, Piedmont, and Tuscany (regions), northern Italy, sample year: January–June 2011, country: Italy[813]

- Co-contamination: For detailed information see the article.
- Further contamination of plants and plant products other than cereals and their products reported in the present book/article[813]: Flour (chestnut flour), AFB_2, AFG_1, AFG_2, and OTA, literature[813]; Nuts (chestnut), AFB_1, AFB_2, AFG_1, AFG_2, and OTA, literature[813]

Aflatoxin B_2

incidence: 16/25, conc. range: ≤3.30 μg/kg, sample origin: retail outlets in Campania, Emilia-Romagna, Piedmont, and Tuscany (regions), northern Italy, sample year: January 2012–June 2013, country: Italy[687]

- Co-contamination: For detailed information see the article.
- Further contamination of plants and plant products other than cereals and their products reported in the present book/article[687]: Flour (chestnut flour), AFB_1, AFG_1, AFG_2, CIT, MA, OTA, and ROQ-C, literature[687]; Nuts (chestnut), AFB_1, AFB_2, AFG_1, CIT, MA, OTA, and ROQ-C, literature[687]

incidence: 11/37, conc. range: ≤3.19 μg/kg, sample origin: retail outlets principally in Campania, Emilia-Romagna, Piedmont, and Tuscany (regions), northern Italy, sample year: January–June 2011, country: Italy[813]

- Co-contamination: For detailed information see the article.
- Further contamination of plants and plant products other than cereals and their products reported in the present book/article[813]: Flour (chestnut flour), AFB_1, AFG_1, AFG_2, and OTA, literature[813]; Nuts (chestnut), AFB_1, AFB_2, AFG_1, AFG_2, and OTA, literature[813]

Aflatoxin G_1

incidence: 18/25, conc. range: ≤73.8 μg/kg, sample origin: retail outlets in Campania, Emilia-Romagna, Piedmont, and Tuscany (regions), northern Italy, sample year: January 2012–June 2013, country: Italy[687]

- Co-contamination: For detailed information see the article.
- Further contamination of plants and plant products other than cereals and their products reported in the present book/article[687]: Flour (chestnut flour), AFB_1, AFB_2, AFG_2, CIT, MA, OTA, and ROQ-C, literature[687]; Nuts (chestnut), AFB_1, AFB_2, AFG_1, CIT, MA, OTA, and ROQ-C, literature[687]

incidence: 15/37, conc. range: ≤116.96 μg/kg, sample origin: retail outlets principally in Campania, Emilia-Romagna, Piedmont, and Tuscany (regions), northern Italy, sample year: January–June 2011, country: Italy[813]

- Co-contamination: For detailed information see the article.
- Further contamination of plants and plant products other than cereals and their products reported in the present book/article[813]: Flour (chestnut flour), AFB_1, AFB_2, AFG_2, and OTA, literature[813]; Nuts (chestnut), AFB_1, AFB_2, AFG_1, AFG_2, and OTA, literature[813]

Aflatoxin G_2

incidence: 8/25, conc. range: ≤2.54 μg/kg, sample origin: retail outlets in Campania, Emilia-Romagna, Piedmont, and Tuscany (regions), northern Italy, sample year: January 2012–June 2013, country: Italy[687]

- Co-contamination: For detailed information see the article.
- Further contamination of plants and plant products other than cereals and their products reported in the present book/article[687]: Flour (chestnut flour), Nuts (chestnut), AFB_1, AFB_2, AFG_1, CIT, MA, OTA, and ROQ-C, literature[687]

incidence: 7/37, conc. range: ≤2.60 μg/kg, sample origin: retail outlets principally in Campania, Emilia-Romagna, Piedmont, and Tuscany (regions), northern Italy, sample year: January–June 2011, country: Italy[813]

- Co-contamination: For detailed information see the article.
- Further contamination of plants and plant products other than cereals and their products reported in the present book/article[813]: Flour (chestnut flour), AFB_1, AFB_2, AFG_1, and OTA, literature[813]; Nuts (chestnut), AFB_1, AFB_2, AFG_1, AFG_2, and OTA, literature[813]

Aspergillus and ***Penicillium*** Toxins

Citrinin

incidence: 22/25, conc. range: ≤3.81 μg/kg, sample origin: retail outlets in Campania, Emilia-Romagna, Piedmont, and Tuscany (regions), northern Italy, sample year: January 2012–June 2013, country: Italy[687]

- Co-contamination: For detailed information see the article.
- Further contamination of plants and plant products other than cereals and their products reported in the present book/article[687]: Flour (chestnut flour), AFB_1, AFB_2, AFG_1, AFG_2, MA, OTA, and ROQ-C, literature[687]; Nuts (chestnut), AFB_1, AFB_2, AFG_1, CIT, MA, OTA, and ROQ-C, literature[687]

Ochratoxin A

incidence: 17/25, conc. range: ≤9.99 μg/kg, sample origin: retail outlets in Campania, Emilia-Romagna, Piedmont, and Tuscany (regions), northern Italy, sample year: January 2012–June 2013, country: Italy[687]

- Co-contamination: For detailed information see the article.
- Further contamination of plants and plant products other than cereals and their products reported in the present book/

article[687]: Flour (chestnut flour), AFB_1, AFB_2, AFG_1, AFG_2, CIT, MA, and ROQ-C, literature[687]; Nuts (chestnut), AFB_1, AFB_2, AFG_1, CIT, MA, OTA, and ROQ-C, literature[687]

incidence: 37/37, conc. range: 1.02–50.37 µg/kg, Ø conc.: 12.38 µg/kg, sample origin: retail outlets principally in Campania, Emilia-Romagna, Piedmont, and Tuscany (regions), northern Italy, sample year: January–June 2011, country: Italy[813]

- Co-contamination: For detailed information see the article.
- Further contamination of plants and plant products other than cereals and their products reported in the present book/article[813]: Flour (chestnut flour), AFB_1, AFB_2, AFG_1, and AFG_2, literature[813]; Nuts (chestnut), AFB_1, AFB_2, AFG_1, AFG_2, and OTA, literature[813]

Penicillium Toxins

Mycophenolic Acid

incidence: 25/25, conc. range: ≤850 µg/kg, Ø conc.: 237 µg/kg, sample origin: retail outlets in Campania, Emilia-Romagna, Piedmont, and Tuscany (regions), northern Italy, sample year: January 2012–June 2013, country: Italy[687]

- Co-contamination: For detailed information see the article.
- Further contamination of plants and plant products other than cereals and their products reported in the present book/article[687]: Flour (chestnut flour), AFB_1, AFB_2, AFG_1, AFG_2, CIT, OTA, and ROQ-C, literature[687]; Nuts (chestnut), AFB_1, AFB_2, AFG_1, CIT, MA, OTA, and ROQ-C, literature[687]

Roquefortine C

incidence: 7/25, conc. range: ≤34.8 µg/kg, sample origin: retail outlets in Campania, Emilia-Romagna, Piedmont, and Tuscany (regions), northern Italy, sample year: January 2012–June 2013, country: Italy[687]

- Co-contamination: For detailed information see the article.
- Further contamination of plants and plant products other than cereals and their products reported in the present book/article[687]: Flour (chestnut flour), AFB_1, AFB_2, AFG_1, AFG_2, CIT, MA, and OTA, literature[687]; Nuts (chestnut), AFB_1, AFB_2, AFG_1, CIT, MA, OTA, and ROQ-C, literature[687]

Flour (gram flour) may contain the following mycotoxins:

Aspergillus Toxins

Aflatoxin B_1

incidence: 36/127, conc. range: <20 µg/kg (4 sa.), 21–80 µg/kg (3 sa.), 81–120 µg/kg (1 sa.), 121–250 µg/kg (18 sa.), 251–500 µg/kg (5 sa.), 501–750 µg/kg (5 sa., maximum: 682 µg/kg), sample origin: markets and milling factories in the north-east zones of Bihar (state), India, sample year: June–November 1987, 1988, and 1989, country: India[100]

- Co-contamination: 9 sa. co-contaminated with AFB_1 and AFB_2; 27 sa. contaminated solely with AFB_1
- Further contamination of plants and plant products other than cereals and their products reported in the present book/article[100]: Flour (gram flour), AFB_2, literature[100]

Aflatoxin B_2

incidence: 9/127, conc. range: 21–80 µg/kg (2 sa.), 81–120 µg/kg (2 sa.), 121–250 µg/kg (5 sa.), sample origin: markets and milling factories in the north-east zones of Bihar (state), India, sample year: June–November 1987, 1988, and 1989, country: India[100]

- Co-contamination: 9 sa. co-contaminated with AFB_1 and AFB_2
- Further contamination of plants and plant products other than cereals and their products reported in the present book/article[100]: Flour (gram flour), AFB_1, literature[100]

Flour (lentil flour) may contain the following mycotoxins:

Aspergillus Toxins

Aflatoxin B_1

incidence: 1/1, conc.: 0.57 µg/kg, sample origin: supermarkets and traditional bazaars in Ankara (capital), Turkey, sample year: 2002/2003, country: Turkey[368]

- Co-contamination: 1 sa. co-contaminated with AFB_1, AFB_2, AFG_1, AFG_2, and OTA
- Further contamination of plants and plant products other than cereals and their products reported in the present book/article[368]: Flour (lentil flour), AFB_2, AFG_1, AFG_2, and OTA, literature[368]; Flour (potato flour), AFB_2, AFG_1, and OTA, literature[368]; Flour (soybean flour), AFB_1, AFB_2, AFG_1, and OTA, literature[368]; Flour (vetch flour), AFG_1, and OTA, literature[368]

Aflatoxin B_2

incidence: 1/1, conc.: 0.05 µg/kg, sample origin: supermarkets and traditional bazaars in Ankara (capital), Turkey, sample year: 2002/2003, country: Turkey[368]

- Co-contamination: 1 sa. co-contaminated with AFB_1, AFB_2, AFG_1, AFG_2, and OTA
- Further contamination of plants and plant products other than cereals and their products reported in the present book/article[368]: Flour (lentil flour), AFB_1, AFG_1, AFG_2, and OTA, literature[368]; Flour (potato flour), AFB_2, AFG_1, and OTA, literature[368]; Flour (soybean flour), AFB_1, AFB_2, AFG_1, and OTA, literature[368]; Flour (vetch flour), AFG_1, and OTA, literature[368]

Aflatoxin G_1

incidence: 1/1, conc.: 0.03 µg/kg, sample origin: supermarkets and traditional bazaars in Ankara (capital), Turkey, sample year: 2002/2003, country: Turkey[368]

- Co-contamination: 1 sa. co-contaminated with AFB_1, AFB_2, AFG_1, AFG_2, and OTA
- Further contamination of plants and plant products other than cereals and their products reported in the present book/article[368]: Flour (lentil flour), AFB_1, AFB_2, AFG_2, and OTA, literature[368]; Flour (potato flour), AFB_2, AFG_1, and OTA, literature[368]; Flour (soybean flour), AFB_1, AFB_2, AFG_1, and OTA, literature[368]; Flour (vetch flour), AFG_1, and OTA, literature[368]

Aflatoxin G_2

incidence: 1/1, conc.: 0.06 µg/kg, sample origin: supermarkets and traditional bazaars in Ankara (capital), Turkey, sample year: 2002/2003, country: Turkey[368]

- Co-contamination: 1 sa. co-contaminated with AFB_1, AFB_2, AFG_1, AFG_2, and OTA
- Further contamination of plants and plant products other than cereals and their products reported in the present book/article[368]: Flour (lentil flour), Flour (soybean flour), AFB_1, AFB_2, AFG_1, and OTA, literature[368]; Flour (potato flour), AFB_2, AFG_1, and OTA, literature[368]; Flour (vetch flour), AFG_1, and OTA, literature[368]

Aspergillus **and** ***Penicillium*** **Toxins**

Ochratoxin A

incidence: 1/1, conc.: 0.83 µg/kg, sample origin: supermarkets and traditional bazaars in Ankara (capital), Turkey, sample year: 2002/2003, country: Turkey[368]

- Co-contamination: 1 sa. co-contaminated with AFB_1, AFB_2, AFG_1, AFG_2, and OTA
- Further contamination of plants and plant products other than cereals and their products reported in the present book/article[368]: Flour (lentil flour), AFB_1, AFB_2, AFG_1, and AFG_2, literature[368]; Flour (potato flour), AFB_2, AFG_1, and OTA, literature[368]; Flour (soybean flour), AFB_1, AFB_2, AFG_1, and OTA, literature[368]; Flour (vetch flour), AFG_1, and OTA, literature[368]

Flour (peanut flour) may contain the following mycotoxins:

Aspergillus **Toxins**

Aflatoxin/s

incidence: 142/322, conc. range: 0.2–53.8 µg/kg*, sample origin: traditional markets, supermarkets, grocery stores, wholesale stores, and convenience stores in 25 counties, Taiwan, sample year: 1998–2011 (with exceptions), country: Taiwan[540], *AFB_1, AFB_2, AFG_1, and AFG_2

- Co-contamination: For detailed information see the article.
- Further contamination of plants and plant products other than cereals and their products reported in the present book/article[540]: Butter (peanut butter), Candies (peanut candy), Nuts (peanut), AF/S, literature[540]

For detailed information see the article.

Flour (potato flour) may contain the following mycotoxins:

Aspergillus **Toxins**

Aflatoxin B_2

incidence: 1/1, conc.: 0.11 µg/kg, sample origin: supermarkets and traditional bazaars in Ankara (capital), Turkey, sample year: 2002/2003, country: Turkey[368]

- Co-contamination: 1 sa. co-contaminated with AFB_2, AFG_1, and OTA
- Further contamination of plants and plant products other than cereals and their products reported in the present book/article[368]: Flour (lentil flour), AFB_1, AFB_2, AFG_1, AFG_2, and OTA, literature[368]; Flour (potato flour), AFG_1, and OTA, literature[368]; Flour (soybean flour), AFB_1, AFB_2, AFG_1, and OTA, literature[368]; Flour (vetch flour), AFG_1, and OTA, literature[368]

Aflatoxin G_1

incidence: 1/1, conc.: 0.03 µg/kg, sample origin: supermarkets and traditional bazaars in Ankara (capital), Turkey, sample year: 2002/2003, country: Turkey[368]

- Co-contamination: 1 sa. co-contaminated with AFB_2, AFG_1, and OTA
- Further contamination of plants and plant products other than cereals and their products reported in the present book/article[368]: Flour (lentil flour), AFB_1, AFB_2, AFG_1, AFG_2, and OTA, literature[368]; Flour (potato flour), AFB_2, and OTA, literature[368]; Flour (soybean flour), AFB_1, AFB_2, AFG_1, and OTA, literature[368]; Flour (vetch flour), AFG_1, and OTA, literature[368]

Aspergillus **and** ***Penicillium*** **Toxins**

Ochratoxin A

incidence: 1/1, conc.: 0.32 µg/kg, sample origin: supermarkets and traditional bazaars in Ankara (capital), Turkey, sample year: 2002/2003, country: Turkey[368]

- Co-contamination: 1 sa. co-contaminated with AFB_2, AFG_1, and OTA
- Further contamination of plants and plant products other than cereals and their products reported in the present book/article[368]: Flour (lentil flour), AFB_1, AFB_2, AFG_1, AFG_2, and OTA, literature[368]; Flour (potato flour), AFB_2 and AFG_1, literature[368]; Flour (soybean flour), AFB_1, AFB_2, AFG_1, and OTA, literature[368]; Flour (vetch flour), AFG_1 and OTA, literature[368]

Flour (soybean flour) may contain the following mycotoxins:

Aspergillus **Toxins**

Aflatoxin B_1

incidence: 1/1, conc.: 2.08 µg/kg, sample origin: local supermarket in Okayama (prefecture), Japan, sample year: unknown, country: Japan[339]

- Co-contamination: not reported
- Further contamination of plants and plant products other than cereals and their products reported in the present book/article[339]: Coconuts, Fruits, dried, Seeds (sesame), AFB_1 and AFG_2, literature[339]; Figs, AFB_2 and AFG_2, literature[339]; Nuts, Spices, AFB_1, AFB_2, AFG_1, and AFG_2, literature[339]; Raisins, AFG_2, literature[339]

incidence: 1/1, conc.: 0.94 µg/kg, sample origin: supermarkets and traditional bazaars in Ankara (capital), Turkey, sample year: 2002/2003, country: Turkey[368]

- Co-contamination: 1 sa. co-contaminated with AFB_1, AFB_2, AFG_1, and OTA
- Further contamination of plants and plant products other than cereals and their products reported in the present book/article[368]: Flour (lentil flour), AFB_1, AFB_2, AFG_1, AFG_2, and OTA, literature[368]; Flour (potato flour), AFB_2, AFG_1, and OTA, literature[368]; Flour (soybean flour), AFB_2, AFG_1, and OTA, literature[368]; Flour (vetch flour), AFG_1 and OTA, literature[368]

incidence: 10/10, conc. range: 1–3 µg/kg, ∅ conc.: 2.6 µg/kg, sample origin: Bamenda (region, western highlands), Cameroon, sample year: July–September 2011, country: Cameroon/Austria/South Africa[659]

- Co-contamination: For detailed information see the article.
- Further contamination of plants and plant products other than cereals and their products reported in the present book/article[659]: Dagwa, AFB_1, AFB_2, AFG_1, AFG_2, AME, BEA, CIT, DON, ENA, ENA_1, ENB, ENB_1, FA, FB_1, FB_2, FB_3, KA, MA, NIV, OTA, OTB, STG, VIOM, and ZEA, literature[659]; Flour (soybean flour), AFG_1, BEA, DON, ENB, ENB_1, FA, FB_1, FB_2, FB_3, FUS-X, KA, NIV, and ZEA, literature[659]; Kuru-kuru, AFB_1, AFB_2, AFG_1, AFG_2, BEA, ENA, ENA_1, ENB, ENB_1, FA, FB_1, FB_3, KA, MA, OTA, OTB, ROQ-C, STG, VIOM, and ZEA, literature[659]; Paste (peanut paste), AFB_1, AFB_2, AFG_1, AFG_2, AME, BEA, ENA, ENA_1, ENB, ENB_1, FA, FB_1, FB_2, FB_3, KA, MA, OTA, OTB, STG, VIOM, and ZEA, literature[659]; Soups, AFB_1, AFB_2, AFG_1, AFG_2, AME, AOH, BEA, DON, ENA, ENA_1, ENB, ENB_1, FB_1, FB_2, FB_3, KA, OTA, OTB, and STG, literature[659]; see also Dagwa, Flour (soybean flour), Kuru-kuru, Paste (peanut paste), Soups, chapter: Further Mycotoxins and Microbial Metabolites

incidence: 1/1, conc.: 0.15 µg/kg, sample origin: Thrace, Northwest Anatolia, Central Anatolia, West Anatolia (regions), Turkey, sample year: 2008/2009, country: Turkey[707]

- Co-contamination: not reported
- Further contamination of plants and plant products other than cereals and their products reported in the present book/article[707]: not reported

Aflatoxin B$_2$

incidence: 1/1, conc.: 0.18 µg/kg, sample origin: supermarkets and traditional bazaars in Ankara (capital), Turkey, sample year: 2002/2003, country: Turkey[368]

- Co-contamination: 1 sa. co-contaminated with AFB$_1$, AFB$_2$, AFG$_1$, and OTA
- Further contamination of plants and plant products other than cereals and their products reported in the present book/article[368]: Flour (lentil flour), AFB$_1$, AFB$_2$, AFG$_1$, AFG$_2$, and OTA, literature[368]; Flour (potato flour), AFB$_2$, AFG$_1$, and OTA, literature[368]; Flour (soybean flour), AFB$_1$, AFG$_1$, and OTA, literature[368]; Flour (vetch flour), AFG$_1$ and OTA, literature[368]

Aflatoxin G$_1$

incidence: 1/1, conc.: 0.03 µg/kg, sample origin: supermarkets and traditional bazaars in Ankara (capital), Turkey, sample year: 2002/2003, country: Turkey[368]

- Co-contamination: 1 sa. co-contaminated with AFB$_1$, AFB$_2$, AFG$_1$, and OTA
- Further contamination of plants and plant products other than cereals and their products reported in the present book/article[368]: Flour (lentil flour), AFB$_1$, AFB$_2$, AFG$_1$, AFG$_2$, and OTA, literature[368]; Flour (potato flour), AFB$_2$, AFG$_1$, and OTA, literature[368]; Flour (soybean flour), AFB$_1$, AFB$_2$, and OTA, literature[368]; Flour (vetch flour), AFG$_1$ and OTA, literature[368]

incidence: 8/10, conc. range: <LOQ-3 µg/kg, Ø conc.: 3 µg/kg, sample origin: Bamenda (region, western highlands), Cameroon, sample year: July–September 2011, country: Cameroon/Austria/South Africa[659]

- Co-contamination: For detailed information see the article.
- Further contamination of plants and plant products other than cereals and their products reported in the present book/article[659]: Dagwa, AFB$_1$, AFB$_2$, AFG$_1$, AFG$_2$, AME, BEA, CIT, DON, ENA, ENA$_1$, ENB, ENB$_1$, FA, FB$_1$, FB$_2$, FB$_3$, KA, MA, NIV, OTA, OTB, STG, VIOM, and ZEA, literature[659]; Flour (soybean flour), AFB$_1$, BEA, DON, ENB, ENB$_1$, FA, FB$_1$, FB$_2$, FB$_3$, FUS-X, KA, NIV, and ZEA, literature[659]; Kuru-kuru, AFB$_1$, AFB$_2$, AFG$_1$, AFG$_2$, BEA, ENA, ENA$_1$, ENB, ENB$_1$, FA, FB$_1$, FB$_3$, KA, MA, OTA, OTB, ROQ-C, STG, VIOM, and ZEA, literature[659]; Paste (peanut paste), AFB$_1$, AFB$_2$, AFG$_1$, AFG$_2$, AME, BEA, ENA, ENA$_1$, ENB, ENB$_1$, FA, FB$_1$, FB$_2$, FB$_3$, KA, MA, OTA, OTB, STG, VIOM, and ZEA, literature[659]; Soups, AFB$_1$, AFB$_2$, AFG$_1$, AFG$_2$, AME, AOH, BEA, DON, ENA, ENA$_1$, ENB, ENB$_1$, FB$_1$, FB$_2$, FB$_3$, KA, OTA, OTB, and STG, literature[659]; see also Dagwa, Flour (soybean flour), Kuru-kuru, Paste (peanut paste), Soups, chapter: Further Mycotoxins and Microbial Metabolites

Aspergillus and *Penicillium* Toxins

Kojic Acid

incidence: 10/10, conc. range: 86–615 µg/kg, Ø conc.: 148 µg/kg, sample origin: Bamenda (region, western highlands), Cameroon, sample year: July–September 2011, country: Cameroon/Austria/South Africa[659]

- Co-contamination: For detailed information see the article.
- Further contamination of plants and plant products other than cereals and their products reported in the present book/article[659]: Dagwa, AFB$_1$, AFB$_2$, AFG$_1$, AFG$_2$, AME, BEA, CIT, DON, ENA, ENA$_1$, ENB, ENB$_1$, FA, FB$_1$, FB$_2$, FB$_3$, KA, MA, NIV, OTA, OTB, STG, VIOM, and ZEA, literature[659]; Flour (soybean flour), AFB$_1$, AFG$_1$, BEA, DON, ENB, ENB$_1$, FA, FB$_1$, FB$_2$, FB$_3$, FUS-X, NIV, and ZEA, literature[659]; Kuru-kuru, AFB$_1$, AFB$_2$, AFG$_1$, AFG$_2$, BEA, ENA, ENA$_1$, ENB, ENB$_1$, FA, FB$_1$, FB$_3$, KA, MA, OTA, OTB, ROQ-C, STG, VIOM, and ZEA, literature[659]; Paste (peanut paste), AFB$_1$, AFB$_2$, AFG$_1$, AFG$_2$, AME, BEA, ENA, ENA$_1$, ENB, ENB$_1$, FA, FB$_1$, FB$_2$, FB$_3$, KA, MA, OTA, OTB, STG, VIOM, and ZEA, literature[659]; Soups, AFB$_1$, AFB$_2$, AFG$_1$, AFG$_2$, AME, AOH, BEA, DON, ENA, ENA$_1$, ENB, ENB$_1$, FB$_1$, FB$_2$, FB$_3$, KA, OTA, OTB, and STG, literature[659]; see also Dagwa, Flour (soybean flour), Kuru-kuru, Paste (peanut paste), Soups, chapter: Further Mycotoxins and Microbial Metabolites

Ochratoxin A

incidence: 1/1, conc.: 0.57 µg/kg, sample origin: supermarkets and traditional bazaars in Ankara (capital), Turkey, sample year: 2002/2003, country: Turkey[368]

- Co-contamination: 1 sa. co-contaminated with AFB$_1$, AFB$_2$, AFG$_1$, and OTA
- Further contamination of plants and plant products other than cereals and their products reported in the present book/article[368]: Flour (lentil flour), AFB$_1$, AFB$_2$, AFG$_1$, AFG$_2$, and OTA, literature[368]; Flour (potato flour), AFB$_2$, AFG$_1$, and OTA, literature[368]; Flour (soybean flour), AFB$_1$, AFB$_2$, and AFG$_1$, literature[368]; Flour (vetch flour), AFG$_1$ and OTA, literature[368]

Fusarium Toxins

Depsipeptides

Beauvericin

incidence: 10/10, conc. range: 12–19 µg/kg, Ø conc.: 14 µg/kg, sample origin: Bamenda (region, western highlands), Cameroon, sample year: July–September 2011, country: Cameroon/Austria/South Africa[659]

- Co-contamination: For detailed information see the article.
- Further contamination of plants and plant products other than cereals and their products reported in the present book/article[659]: Dagwa, AFB$_1$, AFB$_2$, AFG$_1$, AFG$_2$, AME, BEA, CIT, DON, ENA, ENA$_1$, ENB, ENB$_1$, FA, FB$_1$, FB$_2$, FB$_3$, KA, MA, NIV, OTA, OTB, STG, VIOM, and ZEA, literature[659]; Flour (soybean flour), AFB$_1$, AFG$_1$, DON, ENB, ENB$_1$, FA, FB$_1$, FB$_2$, FB$_3$, FUS-X, KA, NIV, and ZEA, literature[659]; Kuru-kuru, AFB$_1$, AFB$_2$, AFG$_1$, AFG$_2$, BEA, ENA, ENA$_1$, ENB, ENB$_1$, FA, FB$_1$, FB$_3$, KA, MA, OTA, OTB, ROQ-C, STG, VIOM, and ZEA, literature[659]; Paste (peanut paste), AFB$_1$, AFB$_2$, AFG$_1$, AFG$_2$, AME, BEA, ENA, ENA$_1$, ENB, ENB$_1$, FA, FB$_1$, FB$_2$, FB$_3$, KA, MA, OTA, OTB, STG, VIOM, and ZEA, literature[659]; Soups, AFB$_1$, AFB$_2$, AFG$_1$, AFG$_2$, AME, AOH, BEA, DON, ENA, ENA$_1$, ENB, ENB$_1$, FB$_1$, FB$_2$, FB$_3$, KA, OTA, OTB, and STG, literature[659]; see also Dagwa, Flour (soybean flour), Kuru-kuru, Paste (peanut paste), Soups, chapter: Further Mycotoxins and Microbial Metabolites

Enniatin B

incidence: 1/10, conc.: <LOQ, sample origin: Bamenda (region, western highlands), Cameroon, sample year: July–September 2011, country: Cameroon/Austria/South Africa[659]

- Co-contamination: For detailed information see the article.
- Further contamination of plants and plant products other than cereals and their products reported in the present book/article[659]: Dagwa, AFB_1, AFB_2, AFG_1, AFG_2, AME, BEA, CIT, DON, ENA, ENA_1, ENB, ENB_1, FA, FB_1, FB_2, FB_3, KA, MA, NIV, OTA, OTB, STG, VIOM, and ZEA, literature[659]; Flour (soybean flour), AFB_1, AFG_1, BEA, DON, ENB_1, FA, FB_1, FB_2, FB_3, FUS-X, KA, NIV, and ZEA, literature[659]; Kuru-kuru, AFB_1, AFB_2, AFG_1, AFG_2, BEA, ENA, ENA_1, ENB, ENB_1, FA, FB_1, FB_3, KA, MA, OTA, OTB, ROQ-C, STG, VIOM, and ZEA, literature[659]; Paste (peanut paste), AFB_1, AFB_2, AFG_1, AFG_2, AME, BEA, ENA, ENA_1, ENB, ENB_1, FA, FB_1, FB_2, FB_3, KA, MA, OTA, OTB, STG, VIOM, and ZEA, literature[659]; Soups, AFB_1, AFB_2, AFG_1, AFG_2, AME, AOH, BEA, DON, ENA, ENA_1, ENB, ENB_1, FB_1, FB_2, FB_3, KA, OTA, OTB, and STG, literature[659]; see also Dagwa, Flour (soybean flour), Kuru-kuru, Paste (peanut paste), Soups, chapter: Further Mycotoxins and Microbial Metabolites

Enniatin B_1

incidence: 5/10, conc. range: 0.01–0.04 µg/kg, ∅ conc.: 0.02 µg/kg, sample origin: Bamenda (region, western highlands), Cameroon, sample year: July–September 2011, country: Cameroon/Austria/South Africa[659]

- Co-contamination: For detailed information see the article.
- Further contamination of plants and plant products other than cereals and their products reported in the present book/article[659]: Dagwa, AFB_1, AFB_2, AFG_1, AFG_2, AME, BEA, CIT, DON, ENA, ENA_1, ENB, ENB_1, FA, FB_1, FB_2, FB_3, KA, MA, NIV, OTA, OTB, STG, VIOM, and ZEA, literature[659]; Flour (soybean flour), AFB_1, AFG_1, BEA, DON, ENB, FA, FB_1, FB_2, FB_3, FUS-X, KA, NIV, and ZEA, literature[659]; Kuru-kuru, AFB_1, AFB_2, AFG_1, AFG_2, BEA, ENA, ENA_1, ENB, ENB_1, FA, FB_1, FB_3, KA, MA, OTA, OTB, ROQ-C, STG, VIOM, and ZEA, literature[659]; Paste (peanut paste), AFB_1, AFB_2, AFG_1, AFG_2, AME, BEA, ENA, ENA_1, ENB, ENB_1, FA, FB_1, FB_2, FB_3, KA, MA, OTA, OTB, STG, VIOM, and ZEA, literature[659]; Soups, AFB_1, AFB_2, AFG_1, AFG_2, AME, AOH, BEA, DON, ENA, ENA_1, ENB, ENB_1, FB_1, FB_2, FB_3, KA, OTA, OTB, and STG, literature[659]; see also Dagwa, Flour (soybean flour), Kuru-kuru, Paste (peanut paste), Soups, chapter: Further Mycotoxins and Microbial Metabolites

Fumonisin/s

Fumonisin B_1

incidence: 10/10, conc. range: 38–69 µg/kg, ∅ conc.: 49 µg/kg, sample origin: Bamenda (region, western highlands), Cameroon, sample year: July–September 2011, country: Cameroon/Austria/South Africa[659]

- Co-contamination: For detailed information see the article.
- Further contamination of plants and plant products other than cereals and their products reported in the present book/article[659]: Dagwa, AFB_1, AFB_2, AFG_1, AFG_2, AME, BEA, CIT, DON, ENA, ENA_1, ENB, ENB_1, FA, FB_1, FB_2, FB_3, KA, MA, NIV, OTA, OTB, STG, VIOM, and ZEA, literature[659]; Flour (soybean flour), AFB_1, AFG_1, BEA, DON, ENB, ENB_1, FA, FB_2, FB_3, FUS-X, KA, NIV, and ZEA, literature[659]; Kuru-kuru, AFB_1, AFB_2, AFG_1, AFG_2, BEA, ENA, ENA_1, ENB, ENB_1, FA, FB_1, FB_3, KA, MA, OTA, OTB, ROQ-C, STG, VIOM, and ZEA, literature[659]; Paste (peanut paste), AFB_1, AFB_2, AFG_1, AFG_2, AME, BEA, ENA, ENA_1, ENB, ENB_1, FA, FB_1, FB_2, FB_3, KA, MA, OTA, OTB, STG, VIOM, and ZEA, literature[659]; Soups, AFB_1, AFB_2, AFG_1, AFG_2, AME, AOH, BEA, DON, ENA, ENA_1, ENB, ENB_1, FB_1, FB_2, FB_3, KA, OTA, OTB, and STG, literature[659]; see also Dagwa, Flour (soybean flour), Kuru-kuru, Paste (peanut paste), Soups, chapter: Further Mycotoxins and Microbial Metabolites

Fumonisin B_2

incidence: 10/10, conc. range: 7–19 µg/kg, ∅ conc.: 11 µg/kg, sample origin: Bamenda (region, western highlands), Cameroon, sample year: July–September 2011, country: Cameroon/Austria/South Africa[659]

- Co-contamination: For detailed information see the article.
- Further contamination of plants and plant products other than cereals and their products reported in the present book/article[659]: Dagwa, AFB_1, AFB_2, AFG_1, AFG_2, AME, BEA, CIT, DON, ENA, ENA_1, ENB, ENB_1, FA, FB_1, FB_2, FB_3, KA, MA, NIV, OTA, OTB, STG, VIOM, and ZEA, literature[659]; Flour (soybean flour), AFB_1, AFG_1, BEA, DON, ENB, ENB_1, FA, FB_1, FB_3, FUS-X, KA, NIV, and ZEA, literature[659]; Kuru-kuru, AFB_1, AFB_2, AFG_1, AFG_2, BEA, ENA, ENA_1, ENB, ENB_1, FA, FB_1, FB_3, KA, MA, OTA, OTB, ROQ-C, STG, VIOM, and ZEA, literature[659]; Paste (peanut paste), AFB_1, AFB_2, AFG_1, AFG_2, AME, BEA, ENA, ENA_1, ENB, ENB_1, FA, FB_1, FB_2, FB_3, KA, MA, OTA, OTB, STG, VIOM, and ZEA, literature[659]; Soups, AFB_1, AFB_2, AFG_1, AFG_2, AME, AOH, BEA, DON, ENA, ENA_1, ENB, ENB_1, FB_1, FB_2, FB_3, KA, OTA, OTB, and STG, literature[659]; see also Dagwa, Flour (soybean flour), Kuru-kuru, Paste (peanut paste), Soups, chapter: Further Mycotoxins and Microbial Metabolites

Fumonisin B_3

incidence: 10/10, conc. range: 2.3–14 µg/kg, ∅ conc.: 7.5 µg/kg, sample origin: Bamenda (region, western highlands), Cameroon, sample year: July–September 2011, country: Cameroon/Austria/South Africa[659]

- Co-contamination: For detailed information see the article.
- Further contamination of plants and plant products other than cereals and their products reported in the present book/article[659]: Dagwa, AFB_1, AFB_2, AFG_1, AFG_2, AME, BEA, CIT, DON, ENA, ENA_1, ENB, ENB_1, FA, FB_1, FB_2, FB_3, KA, MA, NIV, OTA, OTB, STG, VIOM, and ZEA, literature[659]; Flour (soybean flour), AFB_1, AFG_1, BEA, DON, ENB, ENB_1, FA, FB_1, FB_2, FUS-X, KA, NIV, and ZEA, literature[659]; Kuru-kuru, AFB_1, AFB_2, AFG_1, AFG_2, BEA, ENA, ENA_1, ENB, ENB_1, FA, FB_1, FB_3, KA, MA, OTA, OTB, ROQ-C, STG, VIOM, and ZEA, literature[659]; Paste (peanut paste), AFB_1, AFB_2, AFG_1, AFG_2, AME, BEA, ENA, ENA_1, ENB, ENB_1, FA, FB_1, FB_2, FB_3, KA, MA, OTA, OTB, STG, VIOM, and ZEA, literature[659]; Soups, AFB_1, AFB_2, AFG_1, AFG_2, AME, AOH, BEA, DON, ENA, ENA_1, ENB, ENB_1, FB_1, FB_2, FB_3, KA, OTA, OTB, and STG, literature[659]; see also Dagwa, Flour (soybean flour), Kuru-kuru, Paste (peanut paste), Soups, chapter: Further Mycotoxins and Microbial Metabolites

Type B Trichothecenes

Deoxynivalenol

incidence: 10/10, conc. range: 56–75 µg/kg, ∅ conc.: 67 µg/kg, sample origin: Bamenda (region, western highlands), Cameroon, sample year: July–September 2011, country: Cameroon/Austria/South Africa[659]

- Co-contamination: For detailed information see the article.
- Further contamination of plants and plant products other than cereals and their products reported in the present book/

article[659]: Dagwa, AFB_1, AFB_2, AFG_1, AFG_2, AME, BEA, CIT, DON, ENA, ENA_1, ENB, ENB_1, FA, FB_1, FB_2, FB_3, KA, MA, NIV, OTA, OTB, STG, VIOM, and ZEA, literature[659]; Flour (soybean flour), AFB_1, AFG_1, BEA, ENB, ENB_1, FA, FB_1, FB_2, FB_3, FUS-X, KA, NIV, and ZEA, literature[659]; Kuru-kuru, AFB_1, AFB_2, AFG_1, AFG_2, BEA, ENA, ENA_1, ENB, ENB_1, FA, FB_1, FB_3, KA, MA, OTA, OTB, ROQ-C, STG, VIOM, and ZEA, literature[659]; Paste (peanut paste), AFB_1, AFB_2, AFG_1, AFG_2, AME, BEA, ENA, ENA_1, ENB, ENB_1, FA, FB_1, FB_2, FB_3, KA, MA, OTA, OTB, STG, VIOM, and ZEA, literature[659]; Soups, AFB_1, AFB_2, AFG_1, AFG_2, AME, AOH, BEA, DON, ENA, ENA_1, ENB, ENB_1, FB_1, FB_2, FB_3, KA, OTA, OTB, and STG, literature[659]; see also Dagwa, Flour (soybean flour), Kuru-kuru, Paste (peanut paste), Soups, chapter: Further Mycotoxins and Microbial Metabolites

Fusarenon-X (4-acetylnivalenol)

incidence: 10/10, conc. range: 33–42 µg/kg, Ø conc.: 37 µg/kg, sample origin: Bamenda (region, western highlands), Cameroon, sample year: July–September 2011, country: Cameroon/Austria/South Africa[659]

- Co-contamination: For detailed information see the article.
- Further contamination of plants and plant products other than cereals and their products reported in the present book/article[659]: Dagwa, AFB_1, AFB_2, AFG_1, AFG_2, AME, BEA, CIT, DON, ENA, ENA_1, ENB, ENB_1, FA, FB_1, FB_2, FB_3, KA, MA, NIV, OTA, OTB, STG, VIOM, and ZEA, literature[659]; Flour (soybean flour), AFB_1, AFG_1, BEA, DON, ENB, ENB_1, FA, FB_1, FB_2, FB_3, KA, NIV, and ZEA, literature[659]; Kuru-kuru, AFB_1, AFB_2, AFG_1, AFG_2, BEA, ENA, ENA_1, ENB, ENB_1, FA, FB_1, FB_3, KA, MA, OTA, OTB, ROQ-C, STG, VIOM, and ZEA, literature[659]; Paste (peanut paste), AFB_1, AFB_2, AFG_1, AFG_2, AME, BEA, ENA, ENA_1, ENB, ENB_1, FA, FB_1, FB_2, FB_3, KA, MA, OTA, OTB, STG, VIOM, and ZEA, literature[659]; Soups, AFB_1, AFB_2, AFG_1, AFG_2, AME, AOH, BEA, DON, ENA, ENA_1, ENB, ENB_1, FB_1, FB_2, FB_3, KA, OTA, OTB, and STG, literature[659]; see also Dagwa, Flour (soybean flour), Kuru-kuru, Paste (peanut paste), Soups, chapter: Further Mycotoxins and Microbial Metabolites

Nivalenol

incidence: 9/10, conc. range: 0.3–0.5 µg/kg, Ø conc.: 0.4 µg/kg, sample origin: Bamenda (region, western highlands), Cameroon, sample year: July–September 2011, country: Cameroon/Austria/South Africa[659]

- Co-contamination: For detailed information see the article.
- Further contamination of plants and plant products other than cereals and their products reported in the present book/article[659]: Dagwa, AFB_1, AFB_2, AFG_1, AFG_2, AME, BEA, CIT, DON, ENA, ENA_1, ENB, ENB_1, FA, FB_1, FB_2, FB_3, KA, MA, NIV, OTA, OTB, STG, VIOM, and ZEA, literature[659]; Flour (soybean flour), AFB_1, AFG_1, BEA, DON, ENB, ENB_1, FA, FB_1, FB_2, FB_3, FUS-X, KA, and ZEA, literature[659]; Kuru-kuru, AFB_1, AFB_2, AFG_1, AFG_2, BEA, ENA, ENA_1, ENB, ENB_1, FA, FB_1, FB_3, KA, MA, OTA, OTB, ROQ-C, STG, VIOM, and ZEA, literature[659]; Paste (peanut paste), AFB_1, AFB_2, AFG_1, AFG_2, AME, BEA, ENA, ENA_1, ENB, ENB_1, FA, FB_1, FB_2, FB_3, KA, MA, OTA, OTB, STG, VIOM, and ZEA, literature[659]; Soups, AFB_1, AFB_2, AFG_1, AFG_2, AME, AOH, BEA, DON, ENA, ENA_1, ENB, ENB_1, FB_1, FB_2, FB_3, KA, OTA, OTB, and STG, literature[659]; see also Dagwa, Flour (soybean flour), Kuru-kuru, Paste (peanut paste), Soups, chapter: Further Mycotoxins and Microbial Metabolites

Other *Fusarium* Toxins

Fusaric Acid

incidence: 10/10, conc. range: 33–66 µg/kg, Ø conc.: 47 µg/kg, sample origin: Bamenda (region, western highlands), Cameroon, sample year: July–September 2011, country: Cameroon/Austria/South Africa[659]

- Co-contamination: For detailed information see the article.
- Further contamination of plants and plant products other than cereals and their products reported in the present book/article[659]: Dagwa, AFB_1, AFB_2, AFG_1, AFG_2, AME, BEA, CIT, DON, ENA, ENA_1, ENB, ENB_1, FA, FB_1, FB_2, FB_3, KA, MA, NIV, OTA, OTB, STG, VIOM, and ZEA, literature[659]; Flour (soybean flour), AFB_1, AFG_1, BEA, DON, ENB, ENB_1, FB_1, FB_2, FB_3, FUS-X, KA, NIV, and ZEA, literature[659]; Kuru-kuru, AFB_1, AFB_2, AFG_1, AFG_2, BEA, ENA, ENA_1, ENB, ENB_1, FA, FB_1, FB_3, KA, MA, OTA, OTB, ROQ-C, STG, VIOM, and ZEA, literature[659]; Paste (peanut paste), AFB_1, AFB_2, AFG_1, AFG_2, AME, BEA, ENA, ENA_1, ENB, ENB_1, FA, FB_1, FB_2, FB_3, KA, MA, OTA, OTB, STG, VIOM, and ZEA, literature[659]; Soups, AFB_1, AFB_2, AFG_1, AFG_2, AME, AOH, BEA, DON, ENA, ENA_1, ENB, ENB_1, FB_1, FB_2, FB_3, KA, OTA, OTB, and STG, literature[659]; see also Dagwa, Flour (soybean flour), Kuru-kuru, Paste (peanut paste), Soups, chapter: Further Mycotoxins and Microbial Metabolites

Zearalenone

incidence: 10/10, conc. range: 12–18 µg/kg, Ø conc.: 15 µg/kg, sample origin: Bamenda (region, western highlands), Cameroon, sample year: July–September 2011, country: Cameroon/Austria/South Africa[659]

- Co-contamination: For detailed information see the article.
- Further contamination of plants and plant products other than cereals and their products reported in the present book/article[659]: Dagwa, AFB_1, AFB_2, AFG_1, AFG_2, AME, BEA, CIT, DON, ENA, ENA_1, ENB, ENB_1, FA, FB_1, FB_2, FB_3, KA, MA, NIV, OTA, OTB, STG, VIOM, and ZEA, literature[659]; Flour (soybean flour), AFB_1, AFG_1, BEA, DON, ENB, ENB_1, FA, FB_1, FB_2, FB_3, FUS-X, KA, and NIV, literature[659]; Kuru-kuru, AFB_1, AFB_2, AFG_1, AFG_2, BEA, ENA, ENA_1, ENB, ENB_1, FA, FB_1, FB_3, KA, MA, OTA, OTB, ROQ-C, STG, VIOM, and ZEA, literature[659]; Paste (peanut paste), AFB_1, AFB_2, AFG_1, AFG_2, AME, BEA, ENA, ENA_1, ENB, ENB_1, FA, FB_1, FB_2, FB_3, KA, MA, OTA, OTB, STG, VIOM, and ZEA, literature[659]; Soups, AFB_1, AFB_2, AFG_1, AFG_2, AME, AOH, BEA, DON, ENA, ENA_1, ENB, ENB_1, FB_1, FB_2, FB_3, KA, OTA, OTB, and STG, literature[659]; see also Dagwa, Flour (soybean flour), Kuru-kuru, Paste (peanut paste), Soups, chapter: Further Mycotoxins and Microbial Metabolites

see also **Beans (soybean)**

Flour (vetch flour) may contain the following mycotoxins:

Aspergillus Toxins

Aflatoxin G_1

incidence: 1/1*, conc.: 0.03 µg/kg, sample origin: supermarkets and traditional bazaars in Ankara (capital), Turkey, sample year: 2002/2003, country: Turkey[368], **Lathyrus sativus* L. flour

- Co-contamination: 1 sa. co-contaminated with AFG_1 and OTA
- Further contamination of plants and plant products other than cereals and their products reported in the present book/article[368]: Flour (lentil flour), AFB_1, AFB_2, AFG_1, AFG_2, and OTA, literature[368]; Flour (potato flour), AFB_2, AFG_1, and OTA, literature[368]; Flour (soybean flour), AFB_1, AFB_2, AFG_1, and OTA, literature[368]; Flour (vetch flour), OTA, literature[368]

Aspergillus **and** ***Penicillium*** **Toxins**

Ochratoxin A

incidence: 1/1*, conc.: 0.71 µg/kg, sample origin: supermarkets and traditional bazaars in Ankara (capital), Turkey, sample year: 2002/2003, country: Turkey[368], **Lathyrus sativus* L. flour

- Co-contamination: 1 sa. co-contaminated with AFG_1 and OTA
- Further contamination of plants and plant products other than cereals and their products reported in the present book/article[368]: Flour (lentil flour), AFB_1, AFB_2, AFG_1, AFG_2, and OTA, literature[368]; Flour (potato flour), AFB_2, AFG_1, and OTA, literature[368]; Flour (soybean flour), AFB_1, AFB_2, AFG_1, and OTA, literature[368]; Flour (vetch flour), AFG_1, literature[368]

Flour sauce see Sauces (flour sauce)

Fondillón see Wine, miscellaneous

Food may contain the following mycotoxins:

Alternaria **Toxins**

Alternariol

incidence: 0/30*, conc.: no contamination, sample origin: local markets, cereal traders, food/feed processing enterprises, grain mills, and small-scale farmers in villages and cities, Burkina Faso, sample year: April 2010, country: Austria/Nigeria[117], *included dried fruits, grain based processed foods, millet, rice, sesame, sorghum, and wheat

incidence: 2/7*, conc. range: 8.0–28.4 µg/kg, Ø conc.: 18.2 µg/kg, sample origin: local markets, cereal traders, food/feed processing enterprises, grain mills, and small-scale farmers in the local capital and rural villages in Nampula (province), northern Mozambique, sample year: May 2010, country: Austria/Nigeria[117], *included millet, soy, and waste product (intended for disposal)

- Co-contamination: For detailed information see the article.
- Further contamination of plants and plant products other than cereals and their products reported in the present book/article[117]: Food, 3-NPA, AFB_1, AFB_2, AFG_1, AFG_2, AFM_1, AME, ATX-I, BEA, CIT, CPA, DON, ENA, ENA_1, ENB, ENB_1, ENB_2, FB_1, FB_2, FB_3, MON, NIV, OTA, STG, and ZEA, literature[117]; Nuts (peanut), 3-NPA, AFB_1, AFB_2, AFG_1, BEA, CPA, ENB_1, and STG, literature[117]; see also Food, chapter: Further Mycotoxins and Microbial Metabolites

For detailed information see the article.

Alternariol Methyl Ether

incidence: 3/30*, conc. range: 12.4–17.2 µg/kg, sample origin: local markets, cereal traders, food/feed processing enterprises, grain mills, and small-scale farmers in villages and cities, Burkina Faso, sample year: April 2010, country: Austria/Nigeria[117], *included dried fruits, grain based processed foods, millet, rice, sesame, sorghum, and wheat

incidence: 3/7*, conc. range: 9.0–44.5 µg/kg, sample origin: local markets, cereal traders, food/feed processing enterprises, grain mills, and small-scale farmers in the local capital and rural villages in Nampula (province), northern Mozambique, sample year: May 2010, country: Austria/Nigeria[117], *included millet, soy, and waste product (intended for disposal)

- Co-contamination: For detailed information see the article.
- Further contamination of plants and plant products other than cereals and their products reported in the present book/article[117]: Food, 3-NPA, AFB_1, AFB_2, AFG_1, AFG_2, AFM_1, AOH, ATX-I, BEA, CIT, CPA, DON, ENA, ENA_1, ENB, ENB_1, ENB_2, FB_1, FB_2, FB_3, MON, NIV, OTA, STG, and ZEA, literature[117]; Nuts (peanut), 3-NPA, AFB_1, AFB_2, AFG_1, BEA, CPA, ENB_1, and STG, literature[117]; see also Food, chapter: Further Mycotoxins and Microbial Metabolites

For detailed information see the article.

Altertoxin I

incidence: 1/30*, conc.: 3.1 µg/kg, sample origin: local markets, cereal traders, food/feed processing enterprises, grain mills, and small-scale farmers in villages and cities, Burkina Faso, sample year: April 2010, country: Austria/Nigeria[117], *included dried fruits, grain based processed foods, millet, rice, sesame, sorghum, and wheat

incidence: 1/7*, conc.: 10.1 µg/kg, sample origin: local markets, cereal traders, food/feed processing enterprises, grain mills, and small-scale farmers in the local capital and rural villages in Nampula (province), northern Mozambique, sample year: May 2010, country: Austria/Nigeria[117], *included millet, soy, and waste product (intended for disposal)

- Co-contamination: For detailed information see the article.
- Further contamination of plants and plant products other than cereals and their products reported in the present book/article[117]: Food, 3-NPA, AFB_1, AFB_2, AFG_1, AFG_2, AFM_1, AME, AOH, BEA, CIT, CPA, DON, ENA, ENA1, ENB, ENB_1, ENB_2, FB_1, FB_2, FB_3, MON, NIV, OTA, STG, and ZEA, literature[117]; Nuts (peanut), 3-NPA, AFB_1, AFB_2, AFG_1, BEA, CPA, ENB_1, and STG, literature[117]; see also Food, chapter: Further Mycotoxins and Microbial Metabolites

For detailed information see the article.

Aspergillus **Toxins**

Aflatoxin B_1

incidence: 1/48, conc.: 0.4 µg/kg, sample origin: commercially available market basket sa. of foods and foodstuffs in Tokyo (capital), Japan, sample year: 1986–1990, country: Japan[99]

- Co-contamination: not reported
- Further contamination of plants and plant products other than cereals and their products reported in the present book/article[99]: Beans, Seeds (sesame), Sugar, AFB_1 and AFB_2, literature[99]; Jam (bean jam), Paste (peanut paste), AFB_1, literature[99]; Nuts (Brazil nut, paranut), Nuts (peanut), Nuts (pistachio nut), Spices (nutmeg), AFB_1, AFB_2, AFG_1, and AFG_2, literature[99]; Spices, Spices (cayenne pepper, chilli), Spices (pepper), AFB_1, AFB_2, and AFG_1, literature[99]; Spices (paprika), AFB_1 and AFG_1, literature[99]

incidence: 3/6*, conc. range: >4 µg/kg (3 sa., maximum: 7 µg/kg), sample origin: supermarkets and/or traditional markets in Yogyakarta (city), Indonesia, sample year: 1998/1999, country: Indonesia/Austria/Canada[106], *baby food products

- Co-contamination: not reported
- Further contamination of plants and plant products other than cereals and their products reported in the present book/article[106]: Butter (peanut butter), Cake (peanut), Nuts (peanut), Products, Sauces (peanut-chilli sauce), AFB_1, literature[106]

incidence: 175/2,183*, Ø conc.: 38.1 µg/kg, sample origin: Inhambane Province, Mozambique, sample year: 1969–1974, country: South Africa/UK[114], *prepared food sa. (meals)

- Co-contamination: not reported

- Further contamination of plants and plant products other than cereals and their products reported in the present book/article[114]: Food, AFB_2, AFG_1, and AFG_2, literature[114]

incidence: 34/711* **, conc. range: >0.2–2.0 µg/kg (27 sa.), >2.0–10.0 µg/kg (6 sa.), >10.0–20.0 µg/kg (1 sa.), sample origin: different exporting companies, Turkey, sample year: 2007, country: Turkey[115], *commercial Turkish food, **high sugar and nuts

incidence: 19/275* **, conc. range: >0.2–2.0 µg/kg (12 sa.), >2.0–10.0 µg/kg (6 sa.), >10.0–20.0 µg/kg (1 sa.), sample origin: different exporting companies, Turkey, sample year: 2008, country: Turkey[115], *commercial Turkish food, **high sugar and nuts

incidence: 16/43* **, conc. range: >0.2–2.0 µg/kg (12 sa.), >2.0–10.0 µg/kg (2 sa.), >10.0–20.0 µg/kg (2 sa.), sample origin: different exporting companies, Turkey, sample year: 2009, country: Turkey[115], *commercial Turkish food, **high sugar and nuts

incidence: 87?/617* **, conc. range: >0.2–2.0 µg/kg (38 sa.), >2.0–10.0 µg/kg (26 sa.), >10.0–20.0 µg/kg (8 sa.), >20.0–50.0 µg/kg (6 sa.), >50.0 µg/kg (8 sa.), sample origin: different exporting companies, Turkey, sample year: 2007, country: Turkey[115], *commercial Turkish food, **nuts and seeds

incidence: 139/573* **, conc. range: >0.2–2.0 µg/kg (71 sa.), >2.0–10.0 µg/kg (30 sa.), >10.0–20.0 µg/kg (6 sa.), >20.0–50.0 µg/kg (18 sa.), >50.0 µg/kg (14 sa.), sample origin: different exporting companies, Turkey, sample year: 2008, country: Turkey[115], *commercial Turkish food, **nuts and seeds

incidence: 26/122* **, conc. range: >0.2–2.0 µg/kg (11 sa.), >2.0–10.0 µg/kg (6 sa.), >10.0–20.0 µg/kg (1 sa.), >20.0–50.0 µg/kg (5 sa.), >50.0 µg/kg (3 sa.), sample origin: different exporting companies, Turkey, sample year: 2009, country: Turkey[115], *commercial Turkish food, **nuts and seeds

incidence: 1/49* **, conc. range: >0.2–2.0 µg/kg (1 sa.), sample origin: different exporting companies, Turkey, sample year: 2007, country: Turkey[115], *commercial Turkish food, **dried fruits and vegetables

incidence: 1/61* **, conc. range: >0.2–2.0 µg/kg (1 sa.), sample origin: different exporting companies, Turkey, sample year: 2008, country: Turkey[115], *commercial Turkish food, **dried fruits and vegetables

incidence: 0/37* **, conc.: no contamination, sample origin: different exporting companies, Turkey, sample year: 2009, country: Turkey[115], *commercial Turkish food, **dried fruits and vegetables

- Co-contamination: not reported
- Further contamination of plants and plant products other than cereals and their products reported in the present book/article[115]: Products (cocoa products), Spices, AFB_1, literature[115]

incidence: 4/22*, conc. range: 0.02–0.12 µg/kg, sample origin: Warsaw (capital) market or from food producers, Poland, sample year: 2005, country: Poland[116], *vegetables (dried), fruits (dried), and raisins

- Co-contamination: not reported
- Further contamination of plants and plant products other than cereals and their products reported in the present book/article[116]: Coffee, Raisins, OTA, literature[116]; Food, AFG_1, literature[116]; Nuts, Spices, AFB_1, AFB_2, AFG_1, and AFG_2, literature[116]

incidence: 4/30*, conc. range: 3.1–19.1 µg/kg, sample origin: local markets, cereal traders, food/feed processing enterprises, grain mills, and small-scale farmers in villages and cities, Burkina Faso, sample year: April 2010, country: Austria/Nigeria[117], *included dried fruits, grain based processed foods, millet, rice, sesame, sorghum, and wheat

incidence: 3/7*, conc. range: 3.8–427 µg/kg, sample origin: local markets, cereal traders, food/feed processing enterprises, grain mills, and small-scale farmers in the local capital and rural villages in Nampula (province), northern Mozambique, sample year: May 2010, country: Austria/Nigeria[117], *included millet, soy, and waste product (intended for disposal)

- Co-contamination: For detailed information see the article.
- Further contamination of plants and plant products other than cereals and their products reported in the present book/article[117]: Food, 3-NPA, AFB_2, AFG_1, AFG_2, AFM_1, AME, AOH, ATX-I, BEA, CIT, CPA, DON, ENA, ENA_1, ENB, ENB_1, ENB_2, FB_1, FB_2, FB_3, MON, NIV, OTA, STG, and ZEA, literature[117]; Nuts (peanut), 3-NPA, AFB_1, AFB_2, AFG_1, BEA, CPA, ENB_1, and STG, literature[117]; see also Food, chapter: Further Mycotoxins and Microbial Metabolites

For detailed information see the article.

incidence: 3/19*, conc. range: 0.1 µg/kg, Ø conc.: 0.1 µg/kg, sample origin: available in New Zealand, sample year: 2009, country: New Zealand[277], *fruits and nuts, dried

- Co-contamination: not reported
- Further contamination of plants and plant products other than cereals and their products reported in the present book/article[277]: Bars, AFB_1, AF/S, DON, NIV, and OTA, literature[277]; Butter (peanut butter), Confectionery, Fruits, dried, Nuts, Nuts (peanut), Nuts (pistachio nut), Sauces (peanut sauce), AFB_1 and AF/S, literature[277]; Coffee, Dates, Spices (nutmeg), Spices (pepper), Vine fruits, dried, OTA, literature[277]; Figs Spices (cayenne pepper, chilli), Spices (curry), Spices (ginger), Spices (paprika),, AFB_1, AF/S, and OTA, literature[277]; Food, AF/S, literature[277]; Nuts (Brazil nut, paranut), AFB_1, literature[277]

incidence: 14/17*, conc. range: 4–74 µg/kg, Ø conc.: 24 µg/kg, sample origin: El Atarin, Dokki Square, and Giza Market (districts) in Cairo (capital), Egypt, sample year: unknown, country: USA/Egypt[311], *included hommos (garbanzo beans), peanuts in-shell, peanuts off-shell, pumpkin seeds, and watermelon seeds (all mycotoxin-contaminated to a different degree)

- Co-contamination: not reported
- Further contamination of plants and plant products other than cereals and their products reported in the present book/article[344]: Medicinal plants, Spices, Vegetables, dried, AFB_1, literature[344]

incidence: 3/14*, conc. range: 0.15–0.26 µg/kg, Ø conc.: 0.20 µg/kg, sample origin: Chinese markets in Beijing (capital city), Anhui, Hebei, Ningxia, and Sichuan (provinces), China, sample year: unknown, country: China/Italy[768], *functional food

- Co-contamination: 1 sa. co-contaminated with AFB_1 and AFB_2; 2 sa. contaminated solely with AFB_1
- Further contamination of plants and plant products other than cereals and their products reported in the present book/article[768]: Food, AFB_2 and OTA, literature[768]; Spices, AFB_1, literature[768]; Spices (fennel), Spices (star anise), AFG_2, literature[768]

For detailed information see the article.

incidence: 6/6*, conc. range: 0.5–1.2 µg/kg, sample origin: market in Lilongwe (city), Malawi, sample year: December 2012, country: Malawi[790], *peanut-based therapeutic food

- Co-contamination: 6 sa. co-contaminated with AFB_1, AFB_2, AFG_1, and AFG_2
- Further contamination of plants and plant products other than cereals and their products reported in the present book/article[790]: Butter (peanut butter), Nuts (peanut), AFB_1, AFB_2, AFG_1, and AFG_2, literature[790]; Food, AFB_2, AFG_1, and AFG_2, literature[790]

incidence: 15/31*, conc. range: >20–8,000 µg/kg, sample origin: households in Makueni, Kitui, and Thika (districts), Kenya, sample year: May 2004, country: Kenya[808], *included maize flour, maize grain, dry maize cobs, muthokoi, millet sorghum, and beans

- Co-contamination: not reported
- Further contamination of plants and plant products other than cereals and their products reported in the present book/article[808]: not reported

incidence: 2/30*, conc. range: 0.6–3.7 µg/kg, Ø conc.: 2.15 µg/kg, sample origin: different agro-ecological zones in Tamil Nadu (state), India, sample year: 2005–2007, country: India[871], *included hollow corn, groundnut candies, and raw groundnuts

- Co-contamination: not reported
- Further contamination of plants and plant products other than cereals and their products reported in the present book/article[871]: not reported

Aflatoxin B_2

incidence: 61/2,183*, Ø conc.: 7.4 µg/kg, sample origin: Mozambique, sample year: 1969–1974, country: South Africa/UK[114], *prepared food sa. (meals)

- Co-contamination: not reported
- Further contamination of plants and plant products other than cereals and their products reported in the present book/article[114]: Food, AFB_1, AFG_1, and AFG_2, literature[114]

incidence: 0/30*, conc.: no contamination, sample origin: local markets, cereal traders, food/feed processing enterprises, grain mills, and small-scale farmers in villages and cities, Burkina Faso, sample year: April 2010, country: Austria/Nigeria[117], *included dried fruits, grain based processed foods, millet, rice, sesame, sorghum, and wheat

incidence: 1/7*, conc.: 51.3 µg/kg, sample origin: local markets, cereal traders, food/feed processing enterprises, grain mills, and small-scale farmers in the local capital and rural villages in Nampula (province), northern Mozambique, sample year: May 2010, country: Austria/Nigeria[117], *included millet, soy, and waste product (intended for disposal)

- Co-contamination: For detailed information see the article.
- Further contamination of plants and plant products other than cereals and their products reported in the present book/article[117]: Food, 3-NPA, AFB_1, AFG_1, AFG_2, AFM_1, AME, AOH, ATX-I, BEA, CIT, CPA, DON, ENA, ENA_1, ENB, ENB_1, ENB_2, FB_1, FB_2, FB_3, MON, NIV, OTA, STG, and ZEA, literature[117]; Nuts (peanut), 3-NPA, AFB_1, AFB_2, AFG_1, BEA, CPA, ENB_1, and STG, literature[117]; see also Food, chapter: Further Mycotoxins and Microbial Metabolites

For detailed information see the article.

incidence: 1/14*, conc.: <LOQ, sample origin: Chinese markets in Beijing (capital), Anhui, Hebei, Ningxia, and Sichuan (provinces), China, sample year: unknown, country: China/Italy[768], *functional food

- Co-contamination: 1 sa. co-contaminated with AFB_1 and AFB_2
- Further contamination of plants and plant products other than cereals and their products reported in the present book/article[768]: Food, AFB_1 and OTA, literature[768]; Spices, AFB_1, literature[768]; Spices (fennel), Spices (star anise), AFG_2, literature[768]

For detailed information see the article.

incidence: 6/6*, conc. range: 0.1–0.2 µg/kg, sample origin: market in Lilongwe (city), Malawi, sample year: December 2012, country: Malawi[790], *peanut-based therapeutic food

- Co-contamination: 6 sa. co-contaminated with AFB_1, AFB_2, AFG_1, and AFG_2
- Further contamination of plants and plant products other than cereals and their products reported in the present book/article[790]: Butter (peanut butter), Nuts (peanut), AFB_1, AFB_2, AFG_1, and AFG_2, literature[790]; Food, AFB_1, AFG_1, and AFG_2, literature[790]

Aflatoxin G_1

incidence: 18/2,183*, Ø conc.: 13.7 µg/kg, sample origin: Mozambique, sample year: 1969–1974, country: South Africa/UK[114], *prepared food sa. (meals)

- Co-contamination: not reported
- Further contamination of plants and plant products other than cereals and their products reported in the present book/article[114]: Food, AFB_1, AFB_2, and AFG_2, literature[114]

incidence: 2/22*, conc. range: 0.29–0.48 µg/kg, Ø conc.: 0.385 µg/kg, sample origin: Warsaw (capital) market or from food producers, Poland, sample year: 2005, country: Poland[116], *vegetables (dried), fruits (dried), and raisins

- Co-contamination: not reported
- Further contamination of plants and plant products other than cereals and their products reported in the present book/article[116]: Coffee, Raisins, OTA, literature[116]; Food, AFB_1, literature[116]; Nuts, Spices, AFB_1, AFB_2, AFG_1, and AFG_2, literature[116]

incidence: 0/30*, conc.: no contamination, sample origin: local markets, cereal traders, food/feed processing enterprises, grain mills, and small-scale farmers in villages and cities, Burkina Faso, sample year: April 2010, country: Austria/Nigeria[117], *included dried fruits, grain based processed foods, millet, rice, sesame, sorghum, and wheat

incidence: 1/7*, conc.: 382 µg/kg, sample origin: local markets, cereal traders, food/feed processing enterprises, grain mills, and small-scale farmers in the local capital and rural villages in Nampula (province), northern Mozambique, sample year: May 2010, country: Austria/Nigeria[117], *included millet, soy, and waste product (intended for disposal)

- Co-contamination: For detailed information see the article.
- Further contamination of plants and plant products other than cereals and their products reported in the present book/article[117]: Food, 3-NPA, AFB_1, AFB_2, AFG_2, AFM_1, AME, AOH, ATX-I, BEA, CIT, CPA, DON, ENA, ENA_1, ENB, ENB_1, ENB_2, FB_1, FB_2, FB_3, MON, NIV, OTA, STG, and ZEA, literature[117]; Nuts (peanut), 3-NPA, AFB_1, AFB_2, AFG_1, BEA, CPA, ENB_1, and STG, literature[117]; see also Food, chapter: Further Mycotoxins and Microbial Metabolites

For detailed information see the article.

incidence: 6/6*, conc. range: 0.9–1.4 µg/kg, sample origin: market in Lilongwe (city), Malawi, sample year: December 2012, country: Malawi[790], *peanut-based therapeutic food

- Co-contamination: 6 sa. co-contaminated with AFB_1, AFB_2, AFG_1, and AFG_2
- Further contamination of plants and plant products other than cereals and their products reported in the present book/article[790]: Butter (peanut butter), Nuts (peanut), AFB_1, AFB_2, AFG_1, and AFG_2, literature[790]; Food, AFB_1, AFB_2, and AFG_2, literature[790]

Aflatoxin G_2

incidence: 15/2,183*, Ø conc.: 4.4 µg/kg, sample origin: Mozambique, sample year: 1969–1974, country: South Africa/UK[114], *prepared food sa. (meals)

- Co-contamination: not reported
- Further contamination of plants and plant products other than cereals and their products reported in the present book/article[114]: Food, AFB_1, AFB_2, and AFG_1, literature[114]

incidence: 0/30*, conc.: no contamination, sample origin: local markets, cereal traders, food/feed processing enterprises, grain mills, and small-scale farmers in villages and cities, Burkina Faso, sample year: April 2010, country: Austria/Nigeria[117], *included dried fruits, grain based processed foods, millet, rice, sesame, sorghum, and wheat

incidence: 1/7*, conc.: 48.6 µg/kg, sample origin: local markets, cereal traders, food/feed processing enterprises, grain mills, and small-scale farmers in the local capital and rural villages in Nampula (province), northern Mozambique, sample year: May 2010, country: Austria/Nigeria[117], *included millet, soy, and waste product (intended for disposal)

- Co-contamination: For detailed information see the article.
- Further contamination of plants and plant products other than cereals and their products reported in the present book/article[117]: Food, 3-NPA, AFB_1, AFB_2, AFG_1, AFM_1, AME, AOH, ATX-I, BEA, CIT, CPA, DON, ENA, ENA_1, ENB, ENB_1, ENB_2, FB_1, FB_2, FB_3, MON, NIV, OTA, STG, and ZEA, literature[117]; Nuts (peanut), 3-NPA, AFB_1, AFB_2, AFG_1, BEA, CPA, ENB_1, and STG, literature[117]; see also Food, chapter: Further Mycotoxins and Microbial Metabolites

For detailed information see the article.

incidence: 6/6*, conc. range: 0.1–0.2 µg/kg, sample origin: market in Lilongwe (city), Malawi, sample year: December 2012, country: Malawi[790], *peanut-based therapeutic food

- Co contamination: 6 sa. co-contaminated with AFB_1, AFB_2, AFG_1, and AFG_2
- Further contamination of plants and plant products other than cereals and their products reported in the present book/article[790]: Butter (peanut butter), Nuts (peanut), AFB_1, AFB_2, AFG_1, and AFG_2, literature[790]; Food, AFB_1, AFB_2, and AFG_1, literature[790]

Aflatoxin M_1

incidence: 0/30*, conc.: no contamination, sample origin: local markets, cereal traders, food/feed processing enterprises, grain mills, and small-scale farmers in villages and cities, Burkina Faso, sample year: April 2010, country: Austria/Nigeria[117], *included dried fruits, grain based processed foods, millet, rice, sesame, sorghum, and wheat

incidence: 1/7*, conc.: 6.4 µg/kg, sample origin: local markets, cereal traders, food/feed processing enterprises, grain mills, and small-scale farmers in the local capital and rural villages in Nampula (province), northern Mozambique, sample year: May 2010, country: Austria/Nigeria[117], *included millet, soy, and waste product (intended for disposal)

- Co-contamination: For detailed information see the article.
- Further contamination of plants and plant products other than cereals and their products reported in the present book/article[117]: Food, 3-NPA, AFB_1, AFB_2, AFG_1, AFG_2, AME, AOH, ATX-I, BEA, CIT, CPA, DON, ENA, ENA_1, ENB, ENB_1, ENB_2, FB_1, FB_2, FB_3, MON, NIV, OTA, STG, and ZEA, literature[117]; Nuts (peanut), 3-NPA, AFB_1, AFB_2, AFG_1, BEA, CPA, ENB_1, and STG, literature[117]; see also Food, chapter: Further Mycotoxins and Microbial Metabolites

For detailed information see the article.

Aflatoxin/s

incidence: 8?/8*, conc. range: 0.2–1.4 µg/kg**, sample origin: specialist food shops, UK, sample year: unknown, country: UK[16], *food, canned, **AFB_1, AFB_2, AFG_1, and AFG_2

- Co-contamination: not reported
- Further contamination of plants and plant products other than cereals and their products reported in the present book/article[16]: Food, DON, FUS-X, NIV, OTA, and ZEA, literature[16]; Nuts (almond), AF/S, literature[16]; Oils, Spices (fennel), AF/S and ZEA, literature[16]; Oils (sesame oil), OTA, literature[16]; Paste (curry paste), AF/S, FB/S, NIV, OTA, and ZEA, literature[16]; Pickles, AF/S, FB/S, OTA, and ZEA, literature[16]; Sauces (chilli sauce), NIV, OTA, and ZEA, literature[16]; Spices, AF/S, FB/S, OTA, and ZEA, literature[16]; Spices (cayenne pepper, chilli), AF/S, DAS, DON, HT-2, OTA, and ZEA, literature[16]; Spices (coriander), AF/S, DON, OTA, and ZEA, literature[16]; Spices (curry), AF/S, DAS, FB/S, FUS-X, HT-2, NEO, NIV, OTA, T-2, and ZEA, literature[16]; Spices (garlic), DON, FUS-X, and NIV, literature[16]; Spices (ginger), AF/S, DON, NEO, NIV, OTA, and T-2, literature[16]; Spices (tandoori), AF/S, FB/S, NIV, OTA, and T-2, literature[16]

incidence: 8/30, conc. range: 1–5 µg/kg* (3 sa.), 6–15 µg/kg* (2 sa.), 16–25 µg/kg* (1 sa.), 26–50 µg/kg* (2 sa.), sample origin: Keneba (village), Gambia, sample year: October 1988, country: France/England/Germany/USA/Gambia[118], *AF

- Co-contamination: not reported
- Further contamination of plants and plant products other than cereals and their products reported in the present book/article[118]: Nuts (peanut), Sauces (flour sauce), Sauces (leaf-sauce), AF/S, literature[118]

incidence: 22/364*, conc. range: ≤3,904 µg/kg**, Ø conc.: 510 µg/kg**, sample origin: market retailers, street vendors, wholesalers, and distributors, Thailand, sample year: September 1967–July 1969, country: Thailand[139], *prepared food, **AFB_1, AFB_2, AFG_1, and AFG_2

- Co-contamination: not reported
- Further contamination of plants and plant products other than cereals and their products reported in the present book/article[139]: Beans, Beans (mung bean), Nuts (peanut), Seeds (sesame), Spices (cayenne pepper, chilli), Spices (garlic/onion), Starch (cassava starch), Vegetables, AF/S, literature[139]

For detailed information see the article.

incidence: 3/19*, conc. range: 0.1–0.7 µg/kg**, Ø conc.: 0.4 µg/kg**, sample origin: available in New Zealand, sample year: 2009, country: New Zealand[277], *fruits and nuts, dried, **AFS total

- Co-contamination: not reported
- Further contamination of plants and plant products other than cereals and their products reported in the present book/article[277]: Bars, AFB_1, AF/S, DON, NIV, and OTA, literature[277]; Butter (peanut butter), Confectionery, Fruits, dried, Nuts, Nuts (peanut), Nuts (pistachio nut), Sauces (peanut sauce),

AFB_1 and AF/S, literature[277]; Coffee, Dates, Spices (nutmeg), Spices (pepper), Vine fruits, dried, OTA, literature[277]; Figs, Spices (cayenne pepper, chilli), Spices (curry), Spices (ginger), Spices (paprika), AFB_1, AF/S, and OTA, literature[277]; Food, AFB_1, literature[277]; Nuts (Brazil nut, paranut), AFB_1, literature[277]

incidence: 1/3*, conc.: 0.3 μg/kg**, sample origin: purchases and donations in Bamenda (city), Cameroon, sample year: December 2004–April 2005, country: South Africa/Bulgaria[804], *included rice, pumpkin seeds (egusi), fermented cassava flakes (garri), and fermented cassava flour (nkum nkum), **AFB_1, AFB_2, AFG_1, and AFG_2

incidence: 0/2*, conc.: no contamination, sample origin: purchases and donations in Douala (city), Cameroon, sample year: December 2004–April 2005, country: South Africa/Bulgaria[804], *included rice, pumpkin seeds (egusi), fermented cassava flakes (garri), and fermented cassava flour (nkum nkum)

incidence: 0/1*, conc.: no contamination, sample origin: purchases and donations in Kumbo (city), Cameroon, sample year: December 2004–April 2005, country: South Africa/Bulgaria[804], *included rice, pumpkin seeds (egusi), fermented cassava flakes (garri), and fermented cassava flour (nkum nkum)

incidence: 0/0, conc.: no sa. analyzed, sample origin: purchases and donations in Yaounde (capital), Cameroon, sample year: 2005, country: South Africa/Bulgaria[804]

- Co-contamination: For detailed information see the article.
- Further contamination of plants and plant products other than cereals and their products reported in the present book/article[804]: Beans, AF/S, DON, FB_1, and ZEA, literature[804]; Beans (soybean), AF/S, DON, and FB_1, literature[804]; Food, DON and ZEA, literature[804]; Nuts (peanut), AF/S, DON, FB_1, OTA, and ZEA, literature[804]

Other ***Aspergillus*** Toxins

Sterigmatocystin

incidence: 2/30*, conc. range: 4.8–8.6 μg/kg, ∅ conc.: 6.7 μg/kg, sample origin: local markets, cereal traders, food/feed processing enterprises, grain mills, and small-scale farmers in villages and cities, Burkina Faso, sample year: April 2010, country: Austria/Nigeria[117], *included dried fruits, grain based processed foods, millet, rice, sesame, sorghum, and wheat

incidence: 2/7*, conc. range: 3.0–49.2 μg/kg, ∅ conc.: 26.1 μg/kg, sample origin: local markets, cereal traders, food/feed processing enterprises, grain mills, and small-scale farmers in the local capital and rural villages in Nampula (province), northern Mozambique, sample year: May 2010, country: Austria/Nigeria[117], *included millet, soy, and waste product (intended for disposal)

- Co-contamination: For detailed information see the article.
- Further contamination of plants and plant products other than cereals and their products reported in the present book/article[117]: Food, 3-NPA, AFB_1, AFB_2, AFG_1, AFG_2, AFM_1, AME, AOH, ATX-I, BEA, CIT, CPA, DON, ENA, ENA_1, ENB, ENB_1, ENB_2, FB_1, FB_2, FB_3, MON, NIV, OTA, and ZEA, literature[117]; Nuts (peanut), 3-NPA, AFB_1, AFB_2, AFG_1, BEA, CPA, ENB_1, and STG, literature[117]; see also Food, chapter: Further Mycotoxins and Microbial Metabolites

For detailed information see the article.

Aspergillus and ***Penicillium*** Toxins

3-Nitropropionic Acid

incidence: 6/30*, conc. range: 85.6–1,629 μg/kg, sample origin: local markets, cereal traders, food/feed processing enterprises, grain mills, and small-scale farmers in villages and cities, Burkina Faso, sample year: April 2010, country: Austria/Nigeria[117], *included dried fruits, grain based processed foods, millet, rice, sesame, sorghum, and wheat

incidence: 2/7*, conc. range: 95.0–3,228 μg/kg, ∅ conc.: 1,662 μg/kg, sample origin: local markets, cereal traders, food/feed processing enterprises, grain mills, and small-scale farmers in the local capital and rural villages in Nampula (province), northern Mozambique, sample year: May 2010, country: Austria/Nigeria[117], *included millet, soy, and waste product (intended for disposal)

- Co-contamination: For detailed information see the article.
- Further contamination of plants and plant products other than cereals and their products reported in the present book/article[117]: Food, AFB_1, AFB_2, AFG_1, AFG_2, AFM_1, AME, AOH, ATX-I, BEA, CIT, CPA, DON, ENA, ENA_1, ENB, ENB_1, ENB_2, FB_1, FB_2, FB_3, MON, NIV, OTA, STG, and ZEA, literature[117]; Nuts (peanut), 3-NPA, AFB_1, AFB_2, AFG_1, BEA, CPA, ENB_1, and STG, literature[117]; see also Food, chapter: Further Mycotoxins and Microbial Metabolites

For detailed information see the article.

Citrinin

incidence: 0/30*, conc.: no contamination, sample origin: local markets, cereal traders, food/feed processing enterprises, grain mills, and small-scale farmers in villages and cities, Burkina Faso, sample year: April 2010, country: Austria/Nigeria[117], *included dried fruits, grain based processed foods, millet, rice, sesame, sorghum, and wheat

incidence: 1/7*, conc.: 7,061 μg/kg, sample origin: local markets, cereal traders, food/feed processing enterprises, grain mills, and small-scale farmers in the local capital and rural villages in Nampula (province), northern Mozambique, sample year: May 2010, country: Austria/Nigeria[117], *included millet, soy, and waste product (intended for disposal)

- Co-contamination: For detailed information see the article.
- Further contamination of plants and plant products other than cereals and their products reported in the present book/article[117]: Food, 3-NPA, AFB_1, AFB_2, AFG_1, AFG_2, AFM_1, AME, AOH, ATX-I, BEA, CPA, DON, ENA, ENA_1, ENB, ENB_1, ENB_2, FB_1, FB_2, FB_3, MON, NIV, OTA, STG, and ZEA, literature[117]; Nuts (peanut), 3-NPA, AFB_1, AFB_2, AFG_1, BEA, CPA, ENB_1, and STG, literature[117]; see also Food, chapter: Further Mycotoxins and Microbial Metabolites

For detailed information see the article.

incidence: 10/23, conc. range: 22–105 μg/kg*, sample origin: shops and health food stores in Ingolstadt and Munich (cities), Germany, sample year: unknown, country: Germany[124], *eight vegetarian "meat-like" products and two Asian food contaminated

- Co-contamination: not reported
- Further contamination of plants and plant products other than cereals and their products reported in the present book/article[124]: Food colorant, CIT, literature[124]

Cyclopiazonic Acid

incidence: 0/30*, conc.: no contamination, sample origin: local markets, cereal traders, food/feed processing enterprises, grain mills, and small-scale farmers in villages and cities, Burkina Faso, sample year: April 2010, country: Austria/Nigeria[117], *included dried fruits, grain based processed foods, millet, rice, sesame, sorghum, and wheat

incidence: 1/7*, conc.: 789 µg/kg, sample origin: local markets, cereal traders, food/feed processing enterprises, grain mills, and small-scale farmers in the local capital and rural villages in Nampula (province), northern Mozambique, sample year: May 2010, country: Austria/Nigeria[117], *included millet, soy, and waste product (intended for disposal)

- Co-contamination: For detailed information see the article.
- Further contamination of plants and plant products other than cereals and their products reported in the present book/article[117]: Food, 3-NPA, AFB_1, AFB_2, AFG_1, AFG_2, AFM_1, AME, AOH, ATX-I, BEA, CIT, DON, ENA, ENA_1, ENB, ENB_1, ENB_2, FB_1, FB_2, FB_3, MON, NIV, OTA, STG, and ZEA, literature[117]; Nuts (peanut), 3-NPA, AFB_1, AFB_2, AFG_1, BEA, CPA, ENB_1, and STG, literature[117]; see also Food, chapter: Further Mycotoxins and Microbial Metabolites

For detailed information see the article.

Ochratoxin A

incidence: 8?/8*, conc. range: 0.1–0.3 µg/kg, sample origin: specialist food shops, UK, sample year: unknown, country: UK[16], *food, canned

- Co-contamination: not reported
- Further contamination of plants and plant products other than cereals and their products reported in the present book/article[16]: Food, AF/S, DON, FUS-X, NIV, and ZEA, literature[16]; Nuts (almond), AF/S, literature[16]; Oils, Spices (fennel), AF/S and ZEA, literature[16]; Oils (sesame oil), OTA, literature[16]; Pastes (curry paste), AF/S, FB/S, NIV, OTA, and ZEA, literature[16]; Pickles, AF/S, FB/S, OTA, and ZEA, literature[16]; Sauces (chilli sauce), NIV, OTA, and ZEA, literature[16]; Spices, AF/S, FB/S, OTA, and ZEA, literature[16]; Spices (cayenne pepper, chilli), AF/S, DAS, DON, HT-2, OTA, and ZEA, literature[16]; Spices (coriander), AF/S, DON, OTA, and ZEA, literature[16]; Spices (curry), AF/S, DAS, FB/S, FUS-X, HT-2, NEO, NIV, OTA, T-2, and ZEA, literature[16]; Spices (garlic), DON, FUS-X, and NIV, literature[16]; Spices (ginger), AF/S, DON, NEO, NIV, OTA, and T-2, literature[16]; Spices (tandoori), AF/S, FB/S, NIV, OTA, and T-2, literature[16]

incidence: 1/3*, conc.: 4.0 µg/kg, sample origin: unknown, sample year: unknown, country: Japan[112], *grape soaked in syrup, puree, and liquor

- Co-contamination: not reported
- Further contamination of plants and plant products other than cereals and their products reported in the present book/article[112]: Cassis, Juices, Raisins, Wine, OTA, literature[112]; Cocoa, Coffee, Vinegar, OTA and OTB, literature[112]; Food, OTB, literature[112]; Powder (cocoa powder), AFB_1, AFB_2, AFG_1, OTA, and OTB, literature[112]

incidence: 1/30*, conc.: 13.8 µg/kg, sample origin: local markets, cereal traders, food/feed processing enterprises, grain mills, and small-scale farmers in villages and cities, Burkina Faso, sample year: April 2010, country: Austria/Nigeria[117], *included dried fruits, grain based processed foods, millet, rice, sesame, sorghum, and wheat

incidence: 1/7*, conc.: 5.7 µg/kg, sample origin: local markets, cereal traders, food/feed processing enterprises, grain mills, and small-scale farmers in the local capital and rural villages in Nampula (province), northern Mozambique, sample year: May 2010, country: Austria/Nigeria[117], *included millet, soy, and waste product (intended for disposal)

- Co-contamination: For detailed information see the article.
- Further contamination of plants and plant products other than cereals and their products reported in the present book/article[117]: Food, 3-NPA, AFB_1, AFB_2, AFG_1, AFG_2, AFM_1, AME, AOH, ATX-I, BEA, CIT, CPA, DON, ENA, ENA_1, ENB, ENB_1, ENB_2, FB_1, FB_2, FB_3, MON, NIV, STG, and ZEA, literature[117]; Nuts (peanut), 3-NPA, AFB_1, AFB_2, AFG_1, BEA, CPA, ENB_1, and STG, literature[117]; see also Food, chapter: Further Mycotoxins and Microbial Metabolites

For detailed information see the article.

incidence: 23/69*, conc. range: 2–30 µg/kg (23 sa.), sample origin: stores in individual homes, Yugoslavia, sample year: autumn and winter 1972–1978, country: Yugoslavia/Sweden/USA[120], *beans, paprika, etc.

- Co-contamination: not reported
- Further contamination of plants and plant products other than cereals and their products reported in the present book/article[120]: not reported

incidence: 10/20*, conc. range: ≤0.263 µg/kg, sample origin: supermarkets in Lisbon (capital) region, Portugal, sample year: 2015, country: Portugal/France[551], *cereal-based food with fruits (10 sa.) for children

- Co-contamination: 8 sa. co-contaminated with OTA and PAT; 2 sa. contaminated solely with OTA
- Further contamination of plants and plant products other than cereals and their products reported in the present book/article[551]: Food, PAT, literature[551]

incidence: 1/14*, conc.: 5 µg/kg, sample origin: Chinese markets in Beijing (capital), Anhui, Hebei, Ningxia, and Sichuan (provinces), China, sample year: unknown, country: China/Italy[768], *functional food

- Co-contamination: not reported
- Further contamination of plants and plant products other than cereals and their products reported in the present book/article[768]: Food, AFB_1 and AFB_2, literature[768]; Spices, AFB_1, literature[768]; Spices (fennel), Spices (star anise), AFG_2, literature[768]

For detailed information see the article.

Ochratoxin B

incidence: 1/3*, conc.: 1.8 µg/kg, sample origin: unknown, sample year: unknown, country: Japan[112], *grape soaked in syrup, puree, and liquor

- Co-contamination: not reported
- Further contamination of plants and plant products other than cereals and their products reported in the present book/article[112]: Cassis, Juices, Food, Raisins, Wine, OTA, literature[112]; Cocoa, Coffee, Vinegar, OTA and OTB, literature[112]; Powder (cocoa powder), AFB_1, AFB_2, AFG_1, OTA, and OTB, literature[112]

Patulin

incidence: 15/20*, conc. range: ≤4.50 µg/kg, sample origin: supermarkets in Lisbon (capital) region, Portugal, sample year: 2015, country: Portugal/France[551], *cereal-based food with fruits (10 sa.) for children

- Co-contamination: 8 sa. co-contaminated with OTA and PAT; 7 sa. contaminated solely with PAT
- Further contamination of plants and plant products other than cereals and their products reported in the present book/ article[551]: Food, OTA, literature[551]

Fusarium Toxins

Depsipeptides

Beauvericin

incidence: 16/30*, conc. range: 0.1–47.0 µg/kg, sample origin: local markets, cereal traders, food/feed processing enterprises, grain mills, and small-scale farmers in villages and cities, Burkina Faso, sample year: April 2010, country: Austria/Nigeria[117], *included dried fruits, grain based processed foods, millet, rice, sesame, sorghum, and wheat

incidence: 7/7*, conc. range: 3.5–486 µg/kg, sample origin: local markets, cereal traders, food/feed processing enterprises, grain mills, and small-scale farmers in the local capital and rural villages in Nampula (province), northern Mozambique, sample year: May 2010, country: Austria/Nigeria[117], *included millet, soy, and waste product (intended for disposal)

- Co-contamination: For detailed information see the article.
- Further contamination of plants and plant products other than cereals and their products reported in the present book/ article[117]: Food, 3-NPA, AFB_1, AFB_2, AFG_1, AFG_2, AFM_1, AME, AOH, ATX-I, CIT, CPA, DON, ENA, ENA_1, ENB, ENB_1, ENB_2, FB_1, FB_2, FB_3, MON, NIV, OTA, STG, and ZEA, literature[117]; Nuts (peanut), 3-NPA, AFB_1, AFB_2, AFG_1, BEA, CPA, ENB_1, and STG, literature[117]; see also Food, chapter: Further Mycotoxins and Microbial Metabolites

For detailed information see the article.

Enniatin A

incidence: 6/30*, conc. range: 0.3–1.4 µg/kg, sample origin: local markets, cereal traders, food/feed processing enterprises, grain mills, and small-scale farmers in villages and cities, Burkina Faso, sample year: April 2010, country: Austria/Nigeria[117], *included dried fruits, grain based processed foods, millet, rice, sesame, sorghum, and wheat

incidence: 2/7*, conc. range: 0.2–2.0 µg/kg, ∅ conc.: 1.1 µg/kg, sample origin: local markets, cereal traders, food/feed processing enterprises, grain mills, and small-scale farmers in the local capital and rural villages in Nampula (province), northern Mozambique, sample year: May 2010, country: Austria/Nigeria[117], *included millet, soy, and waste product (intended for disposal)

- Co-contamination: For detailed information see the article.
- Further contamination of plants and plant products other than cereals and their products reported in the present book/ article[117]: Food, 3-NPA, AFB_1, AFB_2, AFG_1, AFG_2, AFM_1, AME, AOH, ATX-I, BEA, CIT, CPA, DON, ENA_1, ENB, ENB_1, ENB_2, FB_1, FB_2, FB_3, MON, NIV, OTA, STG, and ZEA, literature[117]; Nuts (peanut), 3-NPA, AFB_1, AFB_2, AFG_1, BEA, CPA, ENB_1, and STG, literature[117]; see also Food, chapter: Further Mycotoxins and Microbial Metabolites

For detailed information see the article.

Enniatin A_1

incidence: 8/30*, conc. range: 0.2–9.1 µg/kg, sample origin: local markets, cereal traders, food/feed processing enterprises, grain mills, and small-scale farmers in villages and cities, Burkina Faso, sample year: April 2010, country: Austria/Nigeria[117], *included dried fruits, grain based processed foods, millet, rice, sesame, sorghum, and wheat

incidence: 2/7*, conc. range: 0.2–4.1 µg/kg, ∅ conc.: 2.15 µg/kg, sample origin: local markets, cereal traders, food/feed processing enterprises, grain mills, and small-scale farmers in the local capital and rural villages in Nampula (province), northern Mozambique, sample year: May 2010, country: Austria/Nigeria[117], *included millet, soy, and waste product (intended for disposal)

- Co-contamination: For detailed information see the article.
- Further contamination of plants and plant products other than cereals and their products reported in the present book/ article[117]: Food, 3-NPA, AFB_1, AFB_2, AFG_1, AFG_2, AFM_1, AME, AOH, ATX-I, BEA, CIT, CPA, DON, ENA, ENB, ENB_1, ENB_2, FB_1, FB_2, FB_3, MON, NIV, OTA, STG, and ZEA, literature[117]; Nuts (peanut), 3-NPA, AFB_1, AFB_2, AFG_1, BEA, CPA, ENB_1, and STG, literature[117]; see also Food, chapter: Further Mycotoxins and Microbial Metabolites

For detailed information see the article.

Enniatin B

incidence: 8/30*, conc. range: 1.2–16.4 µg/kg, sample origin: local markets, cereal traders, food/feed processing enterprises, grain mills, and small-scale farmers in villages and cities, Burkina Faso, sample year: April 2010, country: Austria/Nigeria[117], *included dried fruits, grain based processed foods, millet, rice, sesame, sorghum, and wheat

incidence: 1/7*, conc.: 0.9 µg/kg, sample origin: local markets, cereal traders, food/feed processing enterprises, grain mills, and small-scale farmers in the local capital and rural villages in Nampula (province), northern Mozambique, sample year: May 2010, country: Austria/Nigeria[117], *included millet, soy, and waste product (intended for disposal)

- Co-contamination: For detailed information see the article.
- Further contamination of plants and plant products other than cereals and their products reported in the present book/ article[117]: Food, 3-NPA, AFB_1, AFB_2, AFG_1, AFG_2, AFM_1, AME, AOH, ATX-I, BEA, CIT, CPA, DON, ENA, ENA_1, ENB_1, ENB_2, FB_1, FB_2, FB_3, MON, NIV, OTA, STG, and ZEA, literature[117]; Nuts (peanut), 3-NPA, AFB_1, AFB_2, AFG_1, BEA, CPA, ENB_1, and STG, literature[117]; see also Food, chapter: Further Mycotoxins and Microbial Metabolites

For detailed information see the article.

Enniatin B_1

incidence: 8/30*, conc. range: 0.9–21.4 µg/kg, sample origin: local markets, cereal traders, food/feed processing enterprises, grain mills, and small-scale farmers in villages and cities, Burkina Faso, sample year: April 2010, country: Austria/Nigeria[117], *included dried fruits, grain based processed foods, millet, rice, sesame, sorghum, and wheat incidence: 1/7*, conc.: 4.1 µg/kg, sample origin: local markets, cereal traders, food/feed processing enterprises, grain mills, and small-scale farmers in the local capital and rural villages in Nampula (province), northern Mozambique, sample year: May 2010, country: Austria/Nigeria[117], *included millet, soy, and waste product (intended for disposal)

- Co-contamination: For detailed information see the article.
- Further contamination of plants and plant products other than cereals and their products reported in the present book/article[117]: Food, 3-NPA, AFB_1, AFB_2, AFG_1, AFG_2, AFM_1, AME, AOH, ATX-I, BEA, CIT, CPA, DON, ENA, ENA_1, ENB, ENB_2, FB_1, FB_2, FB_3, MON, NIV, OTA, STG, and ZEA, literature[117]; Nuts (peanut), 3-NPA, AFB_1, AFB_2, AFG_1, BEA, CPA, ENB_1, and STG, literature[117]; see also Food, chapter: Further Mycotoxins and Microbial Metabolites

For detailed information see the article.

Enniatin B_2

incidence: 4/30*, conc. range: 0.2–0.8 µg/kg, sample origin: local markets, cereal traders, food/feed processing enterprises, grain mills, and small-scale farmers in villages and cities, Burkina Faso, sample year: April 2010, country: Austria/Nigeria[117], *included dried fruits, grain based processed foods, millet, rice, sesame, sorghum, and wheat

incidence: 0/7*, conc.: no contamination, sample origin: local markets, cereal traders, food/feed processing enterprises, grain mills, and small-scale farmers in the local capital and rural villages in Nampula (province), northern Mozambique, sample year: May 2010, country: Austria/Nigeria[117], *included millet, soy, and waste product (intended for disposal)

- Co-contamination: For detailed information see the article.
- Further contamination of plants and plant products other than cereals and their products reported in the present book/article[117]: Food, 3-NPA, AFB_1, AFB_2, AFG_1, AFG_2, AFM_1, AME, AOH, ATX-I, BEA, CIT, CPA, DON, ENA, ENA_1, ENB, ENB_1, FB_1, FB_2, FB_3, MON, NIV, OTA, STG, and ZEA, literature[117]; Nuts (peanut), 3-NPA, AFB_1, AFB_2, AFG_1, BEA, CPA, ENB_1, and STG, literature[117]; see also Food, chapter: Further Mycotoxins and Microbial Metabolites

For detailed information see the article.

Fumonisin/s

Fumonisin B_1

incidence: 1/30*, conc.: 73.8 µg/kg, sample origin: local markets, cereal traders, food/feed processing enterprises, grain mills, and small-scale farmers in villages and cities, Burkina Faso, sample year: April 2010, country: Austria/Nigeria[117], *included dried fruits, grain based processed foods, millet, rice, sesame, sorghum, and wheat

incidence: 3/7*, conc. range: 273–45,450 µg/kg, sample origin: local markets, cereal traders, food/feed processing enterprises, grain mills, and small-scale farmers in the local capital and rural villages in Nampula (province), northern Mozambique, sample year: May 2010, country: Austria/Nigeria[117], *included millet, soy, and waste product (intended for disposal)

- Co-contamination: For detailed information see the article.
- Further contamination of plants and plant products other than cereals and their products reported in the present book/article[117]: Food, 3-NPA, AFB_1, AFB_2, AFG_1, AFG_2, AFM_1, AME, AOH, ATX-I, BEA, CIT, CPA, DON, ENA, ENA_1, ENB, ENB_1, ENB_2, FB_2, FB_3, MON, NIV, OTA, STG, and ZEA, literature[117]; Nuts (peanut), 3-NPA, AFB_1, AFB_2, AFG_1, BEA, CPA, ENB_1, and STG, literature[117]; see also Food, chapter: Further Mycotoxins and Microbial Metabolites

For detailed information see the article.

Fumonisin B_2

incidence: 1/30*, conc.: 28.2 µg/kg, sample origin: local markets, cereal traders, food/feed processing enterprises, grain mills, and small-scale farmers in villages and cities, Burkina Faso, sample year: April 2010, country: Austria/Nigeria[117], *included dried fruits, grain based processed foods, millet, rice, sesame, sorghum, and wheat

incidence: 4/7*, conc. range: 11.5–15,254 µg/kg, sample origin: local markets, cereal traders, food/feed processing enterprises, grain mills, and small-scale farmers in the local capital and rural villages in Nampula (province), northern Mozambique, sample year: May 2010, country: Austria/Nigeria[117], *included millet, soy, and waste product (intended for disposal)

- Co-contamination: For detailed information see the article.
- Further contamination of plants and plant products other than cereals and their products reported in the present book/article[117]: Food, 3-NPA, AFB_1, AFB_2, AFG_1, AFG_2, AFM_1, AME, AOH, ATX-I, BEA, CIT, CPA, DON, ENA, ENA_1, ENB, ENB_1, ENB_2, FB_1, FB_3, MON, NIV, OTA, STG, and ZEA, literature[117]; Nuts (peanut), 3-NPA, AFB_1, AFB_2, AFG_1, BEA, CPA, ENB_1, and STG, literature[117]; see also Food, chapter: Further Mycotoxins and Microbial Metabolites

For detailed information see the article.

Fumonisin B_3

incidence: 0/30*, conc.: no contamination, sample origin: local markets, cereal traders, food/feed processing enterprises, grain mills, and small-scale farmers in villages and cities, Burkina Faso, sample year: April 2010, country: Austria/Nigeria[117], *included dried fruits, grain based processed foods, millet, rice, sesame, sorghum, and wheat

incidence: 3/7*, conc. range: 74.8–5,115 µg/kg, sample origin: local markets, cereal traders, food/feed processing enterprises, grain mills, and small-scale farmers in the local capital and rural villages in Nampula (province), northern Mozambique, sample year: May 2010, country: Austria/Nigeria[117], *included millet, soy, and waste product (intended for disposal)

- Co-contamination: For detailed information see the article.
- Further contamination of plants and plant products other than cereals and their products reported in the present book/article[117]: Food, 3-NPA, AFB_1, AFB_2, AFG_1, AFG_2, AFM_1, AME, AOH, ATX-I, BEA, CIT, CPA, DON, ENA, ENA_1, ENB, ENB_1, ENB_2, FB_1, FB_2, MON, NIV, OTA, STG, and ZEA, literature[117]; Nuts (peanut), 3-NPA, AFB_1, AFB_2, AFG_1, BEA, CPA, ENB_1, and STG, literature[117]; see also Food, chapter: Further Mycotoxins and Microbial Metabolites

For detailed information see the article.

Type A Trichothecenes

Diacetoxyscirpenol

incidence: 1/45*, conc.: 21 µg/kg, sample origin: food stores and health food stores, southwestern Germany, sample year: unknown, country: Germany[121], *soy food included whole beans, soy nuts (roasted), flakes and flour (not defatted), flakes (partially defatted and products (crisp)), soy protein (textured), tofu, proteinisolate including infant formulas, and soy sauce

- Co-contamination: not reported
- Further contamination of plants and plant products other than cereals and their products reported in the present book/article[121]: Food, DON, 3-AcDON, 15-AcDON, HT-2, MAS, STO, T-2TET, and ZEA, literature[121]; see also Food, chapter: Further Mycotoxins and Microbial Metabolites

For detailed information see the article.

HT-2 Toxin

incidence: 3/45*, conc. range: 5–11 µg/kg, Ø conc.: 7 µg/kg, sample origin: food stores and health food stores, southwestern Germany, sample year: unknown, country: Germany[121], *soy food included whole beans, soy nuts (roasted), flakes and flour (not defatted), flakes (partially defatted and products (crisp)), soy protein (textured), tofu, protein isolate including infant formulas, and soy sauce

- Co-contamination: 2 sa. co-contaminated with DON, HT-2, MAS, ZEA, and α-ZEL; 1 sa. co-contaminated with DON, HT-2, MAS, STO, ZEA, α-ZEL, and ß-ZEL
- Further contamination of plants and plant products other than cereals and their products reported in the present book/article[121]: Food, DAS, DON, 3-AcDON, 15-AcDON, HT-2, STO, T-2TET, and ZEA, literature[121]; see also Food, chapter: Further Mycotoxins and Microbial Metabolites

For detailed information see the article.

Monoacetoxyscirpenol

incidence: 4/45*, conc. range: 5–34 µg/kg, Ø conc.: 12.8 µg/kg, sample origin: food stores and health food stores, southwestern Germany, sample year: unknown, country: Germany[121], *soy food included whole beans, soy nuts (roasted), flakes and flour (not defatted), flakes (partially defatted and products (crisp)), soy protein (textured), tofu, protein isolate including infant formulas, and soy sauce

- Co-contamination: 1 sa. co-contaminated with DON, MAS, STO, T-2TET, ZEA, and α-ZEL; 1 sa. co-contaminated with DON, HT-2, MAS, STO, ZEA, α-ZEL and ß-ZEL; 1 sa. co-contaminated with DON, HT-2, MAS, ZEA, and α-ZEL; 1 sa. co-contaminated with DON, HT-2, MAS, STO, ZEA, α-ZEL and ß-ZEL
- Further contamination of plants and plant products other than cereals and their products reported in the present book/article[121]: Food, DAS, DON, 3-AcDON, 15-AcDON, HT-2, STO, T-2TET, and ZEA, literature[121]; see also Food, chapter: Further Mycotoxins and Microbial Metabolites

For detailed information see the article.

Scirpentriol

incidence: 2/45*, conc. range: 25–108 µg/kg, Ø conc.: 66.5 µg/kg, sample origin: food stores and health food stores, southwestern Germany, sample year: unknown, country: Germany[121], *soy food included whole beans, soy nuts (roasted), flakes and flour (not defatted), flakes (partially defatted and products (crisp), soy protein (textured), tofu, protein isolate including infant formulas, and soy sauce

- Co-contamination: 1 sa. co-contaminated with DON, MAS, STO, T-2TET, ZEA, and α-ZEL; 1 sa. co-contaminated with DON, HT-2, MAS, STO, ZEA, α-ZEL and ß-ZEL
- Further contamination of plants and plant products other than cereals and their products reported in the present book/article[121]: Food, DAS, DON, 3-AcDON, 15-AcDON, HT-2, MAS, T-2TET, and ZEA, literature[121]; see also Food, chapter: Further Mycotoxins and Microbial Metabolites

For detailed information see the article.

T-2 Tetraol

incidence: 1/45*, conc.: 32 µg/kg, sample origin: food stores and health food stores, southwestern Germany, sample year: unknown, country: Germany[121], *soy food included whole beans, soy nuts (roasted), flakes and flour (not defatted), flakes (partially defatted and products (crisp)), soy protein (textured), tofu, protein isolate including infant formulas, and soy sauce

- Co-contamination: 1 sa. co-contaminated with DON, MAS, STO, T-2TET, ZEA, and α-ZEL
- Further contamination of plants and plant products other than cereals and their products reported in the present book/article[121]: Food, DAS, DON, 3-AcDON, 15-AcDON, HT-2, MAS, STO, and ZEA, literature[121]; see also Food, chapter: Further Mycotoxins and Microbial Metabolites

For detailed information see the article.

Type B Trichothecenes

Deoxynivalenol

incidence: 8?/8*, conc. range: 4–9 µg/kg, sample origin: specialist food shops, UK, sample year: unknown, country: UK[16], *food, canned

- Co-contamination: not reported
- Further contamination of plants and plant products other than cereals and their products reported in the present book/article[16]: Food, AF/S, FUS-X, NIV, OTA, and ZEA, literature[16]; Nuts (almond), AF/S, literature[16]; Oils, Spices (fennel), AF/S and ZEA, literature[16]; Oils (sesame oil), OTA, literature[16]; Paste (curry paste), AF/S, FB/S, NIV, OTA, and ZEA, literature[16]; Pickles, AF/S, FB/S, OTA, and ZEA, literature[16]; Sauces (chilli sauce), NIV, OTA, and ZEA, literature[16]; Spices, AF/S, FB/S, OTA, and ZEA, literature[16]; Spices (cayenne pepper, chilli), AF/S, DAS, DON, HT-2, OTA, and ZEA, literature[16]; Spices (coriander), AF/S, DON, OTA, and ZEA, literature[16]; Spices (curry), AF/S, DAS, FB/S, FUS-X, HT-2, NEO, NIV, OTA, T-2, and ZEA, literature[16]; Spices (garlic), DON, FUS-X, and NIV, literature[16]; Spices (ginger), AF/S, DON, NEO, NIV, OTA, and T-2, literature[16]; Spices (tandoori), AF/S, FB/S, NIV, OTA, and T-2, literature[16]

incidence: 10/30*, conc. range: 22.3–250 µg/kg, sample origin: local markets, cereal traders, food/feed processing enterprises, grain mills, and small-scale farmers in villages and cities, Burkina Faso, sample year: April 2010, country: Austria/Nigeria[117], *included dried fruits, grain based processed foods, millet, rice, sesame, sorghum, and wheat

incidence: 1/7*, conc.: 145 µg/kg, sample origin: local markets, cereal traders, food/feed processing enterprises, grain mills, and small-scale farmers in the local capital and rural villages in Nampula (province), northern Mozambique, sample year: May

2010, country: Austria/Nigeria[117], *included millet, soy, and waste product (intended for disposal)

- Co-contamination: For detailed information see the article.
- Further contamination of plants and plant products other than cereals and their products reported in the present book/article[117]: Food, 3-NPA, AFB_1, AFB_2, AFG_1, AFG_2, AFM_1, AME, AOH, ATX-I, BEA, CIT, CPA, ENA, ENA_1, ENB, ENB_1, ENB_2, FB_1, FB_2, FB_3, MON, NIV, OTA, STG, and ZEA, literature[117]; Nuts (peanut), 3-NPA, AFB_1, AFB_2, AFG_1, BEA, CPA, ENB_1, and STG, literature[117]; see also Food, chapter: Further Mycotoxins and Microbial Metabolites

For detailed information see the article.

incidence: 6/45*, conc. range: 11–260 µg/kg, Ø conc.: 67.2 µg/kg, sample origin: food stores and health food stores, southwestern Germany, sample year: unknown, country: Germany[121], *soy food included whole beans, soy nuts (roasted), flakes and flour (not defatted), flakes (partially defatted and products (crisp)), soy protein (textured), tofu, protein isolate including infant formulas, and soy sauce

- Co-contamination: 2 sa. co-contaminated with DON, HT-2, MAS, ZEA, and α-ZEL; 1 sa. co-contaminated with DON, MAS, STO, T-2TET, ZEA, and α-ZEL; 1 sa. co-contaminated with DON, HT-2, MAS, STO, ZEA, α-ZEL and ß-ZEL; 1 sa. co-contaminated with DON and 3-AcDON; 1 sa. contaminated solely with DON
- Further contamination of plants and plant products other than cereals and their products reported in the present book/article[121]: Food, DAS, 3-AcDON, 15-AcDON, HT-2, MAS, STO, T-2TET, and ZEA, literature[121]; see also Food, chapter: Further Mycotoxins and Microbial Metabolites

For detailed information see the article.

incidence: 2/3*, conc. range: 13–18 µg/kg, Ø conc.: 15 µg/kg, sample origin: purchases and donations in Bamenda (city), Cameroon, sample year: December 2004–April 2005, country: South Africa/Bulgaria[804], *included rice, pumpkin seeds (egusi), fermented cassava flakes (garri), and fermented cassava flour (nkum nkum)

incidence: 0/2*, conc.: no contamination, sample origin: purchases and donations in Douala (city), Cameroon, sample year: December 2004–April 2005, country: South Africa/Bulgaria[804], *included rice, pumpkin seeds (egusi), fermented cassava flakes (garri), and fermented cassava flour (nkum nkum)

incidence: 1/1*, conc.: 35 µg/kg, sample origin: purchases and donations in Kumbo (city), Cameroon, sample year: December 2004–April 2005, country: South Africa/Bulgaria[804], *included rice, pumpkin seeds (egusi), fermented cassava flakes (garri), and fermented cassava flour (nkum nkum)

incidence: 0/0, conc.: no sa. analyzed, sample origin: purchases and donations in Yaounde (capital), Cameroon, sample year: 2005, country: South Africa/Bulgaria[804]

- Co-contamination: For detailed information see the article.
- Further contamination of plants and plant products other than cereals and their products reported in the present book/article[804]: Beans, AF/S, DON, FB_1, and ZEA, literature[804]; Beans (soybean), AF/S, DON, and FB_1, literature[804]; Food, AF/S and ZEA, literature[804]; Nuts (peanut), AF/S, DON, FB_1, OTA, and ZEA, literature[804]

15-Acetyldeoxynivalenol

incidence: 1/45*, conc.: 11 µg/kg, sample origin: food stores and health food stores, southwestern Germany, sample year: unknown, country: Germany[121], *soy food included whole beans, soy nuts (roasted), flakes and flour (not defatted), flakes (partially defatted and products (crisp)), soy protein (textured), tofu, protein isolate including infant formulas, and soy sauce

- Co-contamination: not reported
- Further contamination of plants and plant products other than cereals and their products reported in the present book/article[121]: Food, DAS, DON, 3-AcDON, HT-2, MAS, STO, T-2TET, and ZEA, literature[121]; see also Food, chapter: Further Mycotoxins and Microbial Metabolites

For detailed information see the article.

3-Acetyldeoxynivalenol

incidence: 1/45*, conc.: 14 µg/kg, sample origin: food stores and health food stores, southwestern Germany, sample year: unknown, country: Germany[121], *soy food included whole beans, soy nuts (roasted), flakes and flour (not defatted), flakes partially defatted and products (crisp), soy protein (textured), tofu, protein isolate including infant formulas, and soy sauce

- Co-contamination: 1 sa. co-contaminated with DON and 3-AcDON
- Further contamination of plants and plant products other than cereals and their products reported in the present book/article[121]: Food, DAS, DON, 15-AcDON, HT-2, MAS, STO, T-2TET, and ZEA, literature[121]; see also Food, chapter: Further Mycotoxins and Microbial Metabolites

For detailed information see the article.

Fusarenon-X (4-Acetylnivalenol)

incidence: 1/8*, conc.: 15 µg/kg, sample origin: specialist food shops, UK, sample year: unknown, country: UK[16], *food, canned

- Co-contamination: not reported
- Further contamination of plants and plant products other than cereals and their products reported in the present book/article[16]: Food, AF/S, DON, NIV, OTA, and ZEA, literature[16]; Nuts (almond), AF/S, literature[16]; Oils, Spices (fennel), AF/S and ZEA, literature[16]; Oils (sesame oil), OTA, literature[16]; Paste (curry paste), AF/S, FB/S, NIV, OTA, and ZEA, literature[16]; Pickles, AF/S, FB/S, OTA, and ZEA, literature[16]; Sauces (chilli sauce), NIV, OTA, and ZEA, literature[16]; Spices, AF/S, FB/S, OTA, and ZEA, literature[16]; Spices (cayenne pepper, chilli), AF/S, DAS, DON, HT-2, OTA, and ZEA, literature[16]; Spices (coriander), AF/S, DON, OTA, and ZEA, literature[16]; Spices (curry), AF/S, DAS, FB/S, FUS-X, HT-2, NEO, NIV, OTA, T-2, and ZEA, literature[16]; Spices (garlic), DON, FUS-X, and NIV, literature[16]; Spices (ginger), AF/S, DON, NEO, NIV, OTA, and T-2, literature[16]; Spices (tandoori), AF/S, FB/S, NIV, OTA, and T 2, literature[16]

Nivalenol

incidence: 1/8*, conc.: 18 µg/kg, sample origin: specialist food shops, UK, sample year: unknown, country: UK[16], *food, canned

- Co-contamination: not reported
- Further contamination of plants and plant products other than cereals and their products reported in the present book/article[16]: Food, AF/S, DON, FUS-X, OTA, and ZEA, literature[16];

Nuts (almond), AF/S, literature[16]; Oils, Spices (fennel), AF/S and ZEA, literature[16]; Oils (sesame oil), OTA, literature[16]; Paste (curry paste), AF/S, FB/S, NIV, OTA, and ZEA, literature[16]; Pickles, AF/S, FB/S, OTA, and ZEA, literature[16]; Sauces (chilli sauce), NIV, OTA, and ZEA, literature[16]; Spices, AF/S, FB/S, OTA, and ZEA, literature[16]; Spices (cayenne pepper, chilli), AF/S, DAS, DON, HT-2, OTA, and ZEA, literature[16]; Spices (coriander), AF/S, DON, OTA, and ZEA, literature[16]; Spices (curry), AF/S, DAS, FB/S, FUS-X, HT-2, NEO, NIV, OTA, T-2, and ZEA, literature[16]; Spices (garlic), DON, FUS-X, and NIV, literature[16]; Spices (ginger), AF/S, DON, NEO, NIV, OTA, and T-2, literature[16]; Spices (tandoori), AF/S, FB/S, NIV, OTA, and T-2, literature[16]

incidence: 1/30*, conc.: 40.2 µg/kg, sample origin: local markets, cereal traders, food/feed processing enterprises, grain mills, and small-scale farmers in villages and cities, Burkina Faso, sample year: April 2010, country: Austria/Nigeria[117], *included dried fruits, grain based processed foods, millet, rice, sesame, sorghum, and wheat

incidence: 2/7*, conc. range: 76.8–113 µg/kg, Ø conc.: 94.9 µg/kg, sample origin: local markets, cereal traders, food/feed processing enterprises, grain mills, and small-scale farmers in the local capital and rural villages in Nampula (province), northern Mozambique, sample year: May 2010, country: Austria/Nigeria[117], *included millet, soy, and waste product (intended for disposal)
- Co-contamination: For detailed information see the article.
- Further contamination of plants and plant products other than cereals and their products reported in the present book/article[117]: Food, 3-NPA, AFB_1, AFB_2, AFG_1, AFG_2, AFM_1, AME, AOH, ATX-I, BEA, CIT, CPA, DON, ENA, ENA_1, ENB, ENB_1, ENB_2, FB_1, FB_2, FB_3, MON, OTA, STG, and ZEA, literature[117]; Nuts (peanut), 3-NPA, AFB_1, AFB_2, AFG_1, BEA, CPA, ENB_1, and STG, literature[117]; see also Food, chapter: Further Mycotoxins and Microbial Metabolites

For detailed information see the article.

Other ***Fusarium*** Toxins

Moniliformin

incidence: 5/30*, conc. range: 70.2–320 µg/kg, sample origin: local markets, cereal traders, food/feed processing enterprises, grain mills, and small-scale farmers in villages and cities, Burkina Faso, sample year: April 2010, country: Austria/Nigeria[117], *included dried fruits, grain based processed foods, millet, rice, sesame, sorghum, and wheat

incidence: 4/7*, conc. range: 46.8–1,923 µg/kg, sample origin: local markets, cereal traders, food/feed processing enterprises, grain mills, and small-scale farmers in the local capital and rural villages in Nampula (province), northern Mozambique, sample year: May 2010, country: Austria/Nigeria[117], *included millet, soy, and waste product (intended for disposal)
- Co-contamination: For detailed information see the article.
- Further contamination of plants and plant products other than cereals and their products reported in the present book/article[117]: Food, 3-NPA, AFB_1, AFB_2, AFG_1, AFG_2, AFM_1, AME, AOH, ATX-I, BEA, CIT, CPA, DON, ENA, ENA_1, ENB, ENB_1, ENB_2, FB_1, FB_2, FB_3, MON, NIV, OTA, and STG, literature[117]; Nuts (peanut), 3-NPA, AFB_1, AFB_2, AFG_1, BEA, CPA, ENB_1, and STG, literature[117]; see also Food, chapter: Further Mycotoxins and Microbial Metabolites

For detailed information see the article.

Zearalenone

incidence: 1/8*, conc.: 6.1 µg/kg, sample origin: specialist food shops, UK, sample year: unknown, country: UK[16], *food, canned
- Co-contamination: not reported
- Further contamination of plants and plant products other than cereals and their products reported in the present book/article[16]: Food, AF/S, DON, FUS-X, NIV, and OTA, literature[16]; Nuts (almond), AF/S, literature[16]; Oils, Spices (fennel), AF/S and ZEA, literature[16]; Oils (sesame oil), OTA, literature[16]; Paste (curry paste), AF/S, FB/S, NIV, OTA, and ZEA, literature[16]; Pickles, AF/S, FB/S, OTA, and ZEA, literature[16]; Sauces (chilli sauce), NIV, OTA, and ZEA, literature[16]; Spices, AF/S, FB/S, OTA, and ZEA, literature[16]; Spices (cayenne pepper, chilli), AF/S, DAS, DON, HT-2, OTA, and ZEA, literature[16]; Spices (coriander), AF/S, DON, OTA, and ZEA, literature[16]; Spices (curry), AF/S, DAS, FB/S, FUS-X, HT-2, NEO, NIV, OTA, T-2, and ZEA, literature[16]; Spices (garlic), DON, FUS-X, and NIV, literature[16]; Spices (ginger), AF/S, DON, NEO, NIV, OTA, and T-2, literature[16]; Spices (tandoori), AF/S, FB/S, NIV, OTA, and T-2, literature[16]

incidence: 8/30*, conc. range: 12.3–17.0 µg/kg, sample origin: local markets, cereal traders, food/feed processing enterprises, grain mills, and small-scale farmers in villages and cities, Burkina Faso, sample year: April 2010, country: Austria/Nigeria[117], *included dried fruits, grain based processed foods, millet, rice, sesame, sorghum, and wheat

incidence: 3/7*, conc. range: 78.8–318 µg/kg, sample origin: local markets, cereal traders, food/feed processing enterprises, grain mills, and small-scale farmers in the local capital and rural villages in Nampula (province), northern Mozambique, sample year: May 2010, country: Austria/Nigeria[117], *included millet, soy, and waste product (intended for disposal)
- Co-contamination: For detailed information see the article.
- Further contamination of plants and plant products other than cereals and their products reported in the present book/article[117]: Food, 3-NPA, AFB_1, AFB_2, AFG_1, AFG_2, AFM_1, AME, AOH, ATX-I, BEA, CIT, CPA, DON, ENA, ENA_1, ENB, ENB_1, ENB_2, FB_1, FB_2, FB_3, MON, NIV, OTA, and STG, literature[117]; Nuts (peanut), 3-NPA, AFB_1, AFB_2, AFG_1, BEA, CPA, ENB_1, and STG, literature[117]; see also Food, chapter: Further Mycotoxins and Microbial Metabolites

For detailed information see the article.

incidence: 7/45*, conc. range: 2–214 µg/kg, Ø conc.: 47 µg/kg, sample origin: food stores and health food stores, southwestern Germany, sample year: unknown, country: Germany[121], *soy food included whole beans, soy nuts (roasted), flakes and flour (not defatted), flakes (partially defatted and products (crisp), soy protein (textured), tofu, protein isolate including infant formulas, and soy sauce
- Co-contamination: 2 sa. co-contaminated with DON, HT-2, MAS, ZEA, and α-ZEL; 1 sa. co-contaminated with DON, MAS, STO, T-2TET, ZEA, and α-ZEL; 1 sa. co-contaminated with ZEA, α-ZEL and ß-ZEL; 1 sa. co-contaminated with DON, HT-2, MAS, STO, ZEA, α-ZEL and ß-ZEL; 2 sa. contaminated solely with ZEA
- Further contamination of plants and plant products other than cereals and their products reported in the present book/article[121]: Food, DAS, DON, 3-AcDON, 15-AcDON, HT-2, MAS, STO, and T-2TET, literature[121]; see also Food, chapter: Further Mycotoxins and Microbial Metabolites

For detailed information see the article.

incidence: 12/100*, conc. range: 5.3–295.8 μg/kg, ∅ conc.: 92.5 μg/kg, sample origin: Beijing, Guizhou, Hainan, Hebei, Jiangsu, Jiangxi, and Zhejiang (provinces), China, sample year: unknown, country: China[122], *foods and medicinal plants

- Co-contamination: not reported
- Further contamination of plants and plant products other than cereals and their products reported in the present book/article[122]: not reported

For detailed information see the article.

incidence: 0/3*, conc.: no contamination, sample origin: purchases and donations in Bamenda (city), Cameroon, sample year: December 2004–April 2005, country: South Africa/Bulgaria[804], *included rice, pumpkin seeds (egusi), fermented cassava flakes (garri), and fermented cassava flour (nkum nkum)

incidence: 0/2*, conc.: no contamination, sample origin: purchases and donations in Douala (city), Cameroon, sample year: December 2004–April 2005, country: South Africa/Bulgaria[804], *included rice, pumpkin seeds (egusi), fermented cassava flakes (garri), and fermented cassava flour (nkum nkum)

incidence: 1/1*, conc.: 67 μg/kg, sample origin: purchases and donations in Kumbo (city), Cameroon, sample year: December 2004–April 2005, country: South Africa/Bulgaria[804], *included rice, pumpkin seeds (egusi), fermented cassava flakes (garri), and fermented cassava flour (nkum nkum)

incidence: 0/0, conc.: no sa. analyzed, sample origin: purchases and donations in Yaounde (capital), Cameroon, sample year: 2005, country: South Africa/Bulgaria[804]

- Co-contamination: For detailed information see the article.
- Further contamination of plants and plant products other than cereals and their products reported in the present book/article[804]: Beans, AF/S, DON, FB_1, and ZEA, literature[804]; Beans (soybean), AF/S, DON, and FB_1, literature[804]; Food, AF/S and DON, literature[804]; Nuts (peanut), AF/S, DON, FB_1, OTA, and ZEA, literature[804]

incidence: 8/170*, conc. range: 12.64–26.54 μg/kg, ∅ conc.: 20.0 μg/kg, sample origin: China, sample year: unknown, country: China[876], *functional food included bulbus, flos, folium, fructus, herba, poria, radix, rhizoma, and semen

- Co-contamination: not reported
- Further contamination of plants and plant products other than cereals and their products reported in the present book/article[876]: not reported

For detailed information see the article.

Food colorant may contain the following mycotoxins:

Aspergillus and *Penicillium* Toxins

Citrinin

incidence: 2/2, conc. range: 157–2,800 μg/kg, ∅ conc.: 1,478 μg/kg, sample origin: companies, Germany, sample year: unknown, country: Germany[124]

- Co-contamination: not reported
- Further contamination of plants and plant products other than cereals and their products reported in the present book/article[124]: Food, CIT, literature[124]

Ochratoxin A

incidence: 9/13*, conc. range: <LOQ-32.00 μg/kg, sample origin: internet or retail stores in Bari (province), Italy, sample year: unknown, country: Italy[845], *derived from *Vitis vinifera* (liquid and solid sa.)

- Co-contamination: not reported
- Further contamination of plants and plant products other than cereals and their products reported in the present book/article[845]: Grape pomace and wine lees, Plant food supplements, OTA, literature[845]

For detailed information see the article.

Fritillaria see Medicinal plants

Fruit juices see Juices

Fruit products see Products (fruit product)

Fruit punch see Beverages

Fruit salad may contain the following mycotoxins:

Aspergillus and *Penicillium* Toxins

Patulin

incidence: 1/1*, conc.: 14.0 μg/kg, sample origin: Czech retail market, Czech Republic, sample year: unknown, country: Czech Republic[911], *mixed fruit salad

- Co-contamination: not reported
- Further contamination of plants and plant products other than cereals and their products reported in the present book/article[911]: Apple cider, Baby food, Fruits (apple), Fruits (pear), Juices (apple juice), Juices (pear juice), Pulp (apple pulp), PAT, literature[911]

Fruit sauce see Fruits (apple)

Fruits may contain the following mycotoxins:

Aspergillus Toxins

Aflatoxin B_1

incidence: 1*/32**, conc.: 4,600 μg/kg, sample origin: local marketing institutions, GDR, sample year: unknown, country: GDR[29], *moldy, **included fruits and fruit products

- Co-contamination: not reported
- Further contamination of plants and plant products other than cereals and their products reported in the present book/article[29]: Butter (peanut butter), Nuts (peanut), Products (peanut product), Seeds (sunflower), AFB_1, AFB_2, AFG_1, and AFG_2, literature[29]; Coffee, OTA, literature[29]; Fruits, AFG_1 and AFG_2, literature[29]; Fruits (apple), Fruits (peach), Juices (apple juice), PAT, literature[29]

Aflatoxin G_1

incidence: 1*/32**, conc.: 21.5 μg/kg, sample origin: local marketing institutions, GDR, sample year: unknown, country: GDR[29], *moldy, **included fruits and fruit products

- Co-contamination: not reported

- Further contamination of plants and plant products other than cereals and their products reported in the present book/article[29]: Butter (peanut butter), Nuts (peanut), Products (peanut product), Seeds (sunflower), AFB_1, AFB_2, AFG_1, and AFG_2, literature[29]; Coffee, OTA, literature[29]; Fruits, AFB_1 and AFG_2, literature[29]; Fruits (apple), Fruits (peach), Juices (apple juice), PAT, literature[29]

Aflatoxin G_2

incidence: 1*/32**, conc.: 1,200 µg/kg, sample origin: different district hygiene institutes, GDR, sample year: unknown, country: GDR[29], *moldy, **included fruits and fruit products

- Co-contamination: not reported
- Further contamination of plants and plant products other than cereals and their products reported in the present book/article[29]: Butter (peanut butter), Nuts (peanut), Products (peanut product), Seeds (sunflower), AFB_1, AFB_2, AFG_1, and AFG_2, literature[29]; Coffee, OTA, literature[29]; Fruits, AFB_1 and AFG_1, literature[29]; Fruits (apple), Fruits (peach), Juices (apple juice), PAT, literature[29]

Fruits (apple) may contain the following mycotoxins:

Alternaria Toxins

Altenuene

incidence: 5/8, conc. range: ≤500 µ/kg, sample origin: cold-storage facilities in central Pennsylvania (state), USA, sample year: unknown, country: USA[23]

- Co-contamination: 3 sa. co-contaminated with ALT, AME, AOH, ATX-I, and TA; 2 sa. co-contaminated with ALT, AME, AOH, and TA
- Further contamination of plants and plant products other than cereals and their products reported in the present book/article[23]: Fruits (apple), AME, AOH, ATX-I, and TA, literature[23]; Fruits (tomato), ALT, AME, AOH, and TA, literature[23]

Alternariol

incidence: 7/8, conc. range: ≤58,800 µg/kg, sample origin: cold-storage facilities in central Pennsylvania (state), USA, sample year: unknown, country: USA[23]

- Co-contamination: 3 sa. co-contaminated with ALT, AME, AOH, ATX-I, and TA; 2 sa. co-contaminated with ALT, AME, AOH, and TA; 2 sa. co-contaminated with AME, AOH, ATX-I, and TA
- Further contamination of plants and plant products other than cereals and their products reported in the present book/article[23]: Fruits (apple), ALT, AME, ATX-I, and TA, literature[23]; Fruits (tomato), ALT, AME, AOH, and TA, literature[23]

incidence: 1/22*, conc.: 160 µg/kg, sample origin: unknown, sample year: unknown, country: Germany[24], *moldy

- Co-contamination: not reported
- Further contamination of plants and plant products other than cereals and their products reported in the present book/article[24]: Fruits (apple), AME, literature[24]

incidence: 1/11*, conc.: 29 µg/kg, sample origin: Dutch survey program (2013) and retail shops across The Netherlands, sample year: 2013?, country: Netherlands[568], *apples, fresh

- Co-contamination: not reported
- Further contamination of plants and plant products other than cereals and their products reported in the present book/article[568]: Figs, Olives, Seeds (sunflower), TA, literature[568]; Sauces (tomato sauce), AME, AOH, and TA, literature[568]; Wine (red), AOH and TA, literature[568]; see also Seeds (sunflower), chapter: Further Mycotoxins and Microbial Metabolites

incidence: 1/1*, conc.: 50 µg/kg, sample origin: private households of staff members, Austria, sample year: unknown, country: Austria[850], *moldy

- Co-contamination: 1 sa. co-contaminated with AME and AOH
- Further contamination of plants and plant products other than cereals and their products reported in the present book/article[850]: Fruits (apple), AME, literature[850]; Fruits (lemon), Puree, MA, literature[850]; Jam, AOH and MA, literature[850]; Nuts (chestnut), AME and AOH, literature[850]; Spices (garlic), BEA, FB_1, FB_2, and FB_3, literature[850]; Wine (red), AME, AOH, and MA, literature[850]; see also Fruits (lemon), Jam, Nuts (chestnut), Wine (red), chapter: Further Mycotoxins and Microbial Metabolites

Alternariol Methyl Ether

incidence: 8/8, conc. range: ≤2,300 µg/kg, sample origin: cold-storage facilities in central Pennsylvania (state), USA, sample year: unknown, country: USA[23]

- Co-contamination: 3 sa. co-contaminated with ALT, AME, AOH, ATX-I, and TA; 2 sa. co-contaminated with ALT, AME, AOH, and TA; 2 sa. co-contaminated with AME, AOH, ATX-I, and TA; 1 sa. co-contaminated with AME and TA
- Further contamination of plants and plant products other than cereals and their products reported in the present book/article[23]: Fruits (apple), ALT, AOH, ATX-I, and TA, literature[23]; Fruits (tomato), ALT, AME, AOH, and TA, literature[23]

incidence: 1/22*, conc.: 250 µg/kg, sample origin: unknown, sample year: unknown, country: Germany[24], *moldy

- Co-contamination: not reported
- Further contamination of plants and plant products other than cereals and their products reported in the present book/article[24]: Fruits (apple), AOH, literature[24]

incidence: 1/1*, conc.: 370 µg/kg, sample origin: private households of staff members, Austria, sample year: unknown, country: Austria[850], *moldy

- Co-contamination: 1 sa. co-contaminated with AME and AOH
- Further contamination of plants and plant products other than cereals and their products reported in the present book/article[850]: Fruits (apple), AOH, literature[850]; Fruits (lemon), Puree, MA, literature[850]; Jam, AOH and MA, literature[850]; Nuts (chestnut), AME and AOH, literature[850]; Spices (garlic), BEA, FB_1, FB_2, and FB_3, literature[850]; Wine (red), AME, AOH, and MA, literature[850]; see also Fruits (lemon), Jam, Nuts (chestnut), Wine (red), chapter: Further Mycotoxins and Microbial Metabolites

Altertoxin I

incidence: 5/8, conc. range: pr, sample origin: cold-storage facilities in central Pennsylvania (state), USA, sample year: unknown, country: USA[23]

- Co-contamination: 3 sa. co-contaminated with ALT, AME, AOH, ATX-I, and TA; 2 sa. co-contaminated with AME, AOH, ATX-I, and TA
- Further contamination of plants and plant products other than cereals and their products reported in the present book/article[23]: Fruits (apple), ALT, AME, AOH, and TA, literature[23]; Fruits (tomato), ALT, AME, AOH, and TA, literature[23]

Tenuazonic Acid

incidence: 8/8, conc. range: 100–500 µg/kg, ∅ conc.: 200 µg/kg, sample origin: cold-storage facilities in central Pennsylvania (state), USA, sample year: unknown, country: USA[23]

- Co-contamination: 3 sa. co-contaminated with ALT, AME, AOH, ATX-I, and TA; 2 sa. co-contaminated with ALT, AME, AOH, and TA; 2 sa. co-contaminated with AME, AOH, ATX-I, and TA; 1 sa. co-contaminated with AME and TA
- Further contamination of plants and plant products other than cereals and their products reported in the present book/article[23]: Fruits (apple), ALT, AME, AOH, and ATX-I, literature[23]; Fruits (tomato), ALT, AME, AOH, and TA, literature[23]

Aspergillus Toxins

Aflatoxin B_1

incidence: 1/15*, conc.: 35 µg/kg, sample origin: unknown, sample year: unknown, country: Germany[239], *apples, moldy

- Co-contamination: not reported
- Further contamination of plants and plant products other than cereals and their products reported in the present book/article[239]: Fruits (cherry), Fruits (lemon), Fruits (orange), Fruits (peach), Fruits (tomato), Nuts (almond), Nuts (cashew nut), Nuts (hazelnut), Nuts (walnut), Seeds (sesame), Spices (nutmeg), AFB_1, literature[239]; Nuts (peanut), AFB_1, AFB_2, AFG_1, and AFG_2, literature[239]

Aspergillus and *Penicillium* Toxins

Citrinin

incidence: 83/351*, Ø conc.: 601 µg/kg, sample origin: groceries and supermarkets, Portugal, sample year: unknown, country: Portugal[25], *apples of different varieties with different-sized rotten spots

- Co-contamination: 69 sa. co-contaminated with CIT and PT; 14 sa. contaminated solely with CIT
- Further contamination of plants and plant products other than cereals and their products reported in the present book/article[25]: Fruits (apple), PAT, literature[25]

For detailed information see the article.

Ochratoxin A

incidence: 2/4*, conc. range: 0.23–0.41 µg/kg, Ø conc.: 0.32 µg/kg, sample origin: households and markets, Germany, sample year: unknown, country: Germany[26], *damaged or moldy

- Co-contamination: not reported
- Further contamination of plants and plant products other than cereals and their products reported in the present book/article[26]: Fruits (cherry), Fruits (peach), Fruits (tomato), Potatoes, Strawberries, OTA, literature[26]

Patulin

incidence: 310/351*, Ø conc.: 18,800 µg/kg, sample origin: groceries and supermarkets, Portugal, sample year: unknown, country: Portugal[25], *apples of different varieties with rotten spots

- Co-contamination: 69 sa. co-contaminated with CIT and PAT; 241 sa. contaminated solely with PAT
- Further contamination of plants and plant products other than cereals and their products reported in the present book/article[25]: Fruits (apple), CIT, literature[25]

For detailed information see the article.

incidence: 4/4*, conc. range: <LOQ-1,500.0 µg/kg, sample origin: supermarket, Portugal, sample year: unknown, country: Portugal[27], *sa. with different-sized brown rotten areas

- Co-contamination: not reported
- Further contamination of plants and plant products other than cereals and their products reported in the present book/article[27]: Apple cider, Baby food, Fruits (quince), Jam (quince jam), Juices (apple juice), PAT, literature[27]

incidence: 5/12*, conc. range: 300–42,000 µg/kg, sample origin: local marketing institutions, GDR, sample year: unknown, country: GDR[29], *apples with rotten spots

- Co-contamination: not reported
- Further contamination of plants and plant products other than cereals and their products reported in the present book/article[29]: Butter (peanut butter), Nuts (peanut), Products (peanut product), Seeds (sunflower), AFB_1, AFB_2, AFG_1, and AFG_2, literature[29]; Coffee, OTA, literature[29]; Fruits, AFB_1, AFG_1, and AFG_2, literature[29]; Fruits (peach), Juices (apple juice), PAT, literature[29]

incidence: 21/21*, conc. range: 2–113,343 µg/kg, Ø conc.: 12,464 µg/kg, sample origin: unknown, sample year: unknown, country: Italy[30], *rotten area

incidence: 17/21*, conc. range: 0.05–1,166.00 µg/kg, Ø conc.: 152 µg/kg, sample origin: unknown, sample year: unknown, country: Italy[30], *unaffected area with peel

incidence: 7/21*, conc. range: 0.44–93.00 µg/kg, Ø conc.: 16.3 µg/kg, sample origin: unknown, sample year: unknown, country: Italy[30], *unaffected area without peel

- Co-contamination: not reported
- Further contamination of plants and plant products other than cereals and their products reported in the present book/article[30]: see the article

incidence: 6/6*, conc. range: 150–267 µg/kg, Ø conc.: 220 µg/kg, sample origin: groceries, markets, and supermarkets, Campinas (city), São Paulo (state), Brazil, sample year: 1992/1993 and 1995, country: Brazil[32], *fruits, spoiled

- Co-contamination: not reported
- Further contamination of plants and plant products other than cereals and their products reported in the present book/article[32]: Fruits (peach), Fruits (pear), Juices (apple juice), PAT, literature[32]

incidence: 2/7*, conc. range: 55 µg/kg, sample origin: orchard, Finland, sample year: March–December 1976, country: Finland[33], *moldy

- Co-contamination: not reported
- Further contamination of plants and plant products other than cereals and their products reported in the present book/article[33]: Jam (apple jam), Juices (apple concentrate), Juices (apple juice), Products (apple flavor product), PAT, literature[33]

incidence: 0/6*, conc.: no contamination, sample origin: farms representative of the cultivation form, France, sample year: unknown, country: France[34], *conventional

incidence: 2/6*, conc. range: ≤1,240 µg/kg, sample origin: farms representative of the cultivation form, France, sample year: unknown, country: France[34], *organic

- Co-contamination: not reported
- Further contamination of plants and plant products other than cereals and their products reported in the present book/article[34]: not reported

incidence: 7/13*, conc. range: >LOQ<10 µg/kg (3 sa.), >10<25 µg/kg (2 sa.), >25<50 µg/kg (1 sa.), 81.7 µg/kg (1 sa.), sample origin:

markets, supermarkets, hypermarkets or organic food shops, Italy, sample year: November 2003–February 2004, country: Italy[35], *conventional

incidence: 3/9*, conc. range: >LOQ<10 μg/kg (1 sa.), >50 μg/kg (2 sa., maximum: 50.8 μg/kg), sample origin: markets, supermarkets, hypermarkets or organic food shops, Italy, sample year: November 2003–February 2004, country: Italy[35], *organic

- Co-contamination: not reported
- Further contamination of plants and plant products other than cereals and their products reported in the present book/article[35]: Juices, Juices (apple juice), Juices (pear juice), Puree, Vinegar (apple), PAT, literature[35]

incidence: 16/16*, conc. range: 254.6–653.4 μg/kg, ∅ conc.: 443 μg/kg, sample origin: juice producer industry, Rio Grande do Sul (state), Brazil, sample year: 2006/2007, country: Brazil[909], *many damaged apples

- Co-contamination: not reported
- Further contamination of plants and plant products other than cereals and their products reported in the present book/article[909]: Juices (apple juice), PAT, literature[909]

incidence: 1/3*, conc.: 415.2 μg/kg, sample origin: Czech retail market, Czech Republic, sample year: unknown, country: Czech Republic[911], *apples (fresh fruits)

- Co-contamination: not reported
- Further contamination of plants and plant products other than cereals and their products reported in the present book/article[911]: Apple cider, Baby food, Fruit salad, Fruits (pear), Juices (apple juice), Juices (pear juice), Pulp (apple pulp), PAT, literature[911]

Fusarium Toxins

Depsipeptides

Enniatin/s (A, A_1, B, B_1)

incidence: 13/20*, conc. range: tr-3,890 μg/kg, sample origin: Agricultural Institute of Slovenia, Brdo pri Lukovici (small settlement), Slovenia, sample year: unknown, country: Denmark/Slovenia[36], *apple sa. showing wet apple core rot symptoms from trees

- Co-contamination: For detailed information see the article.
- Further contamination of plants and plant products other than cereals and their products reported in the present book/article[36]: Fruits (apple), MON, literature[36]; see also Fruits (apple), chapter: Further Mycotoxins and Microbial Metabolites

Other ***Fusarium*** Toxins

Moniliformin

incidence: 14/20*, conc. range: tr-2,870 μg/kg, sample origin: Agricultural Institute of Slovenia, Brdo pri Lukovici (small settlement), Slovenia, sample year: unknown, country: Denmark/Slovenia[36], *apple sa. showing wet apple core rot symptoms from trees

- Co-contamination: For detailed information see the article.
- Further contamination of plants and plant products other than cereals and their products reported in the present book/article[36]: Fruits (apple), EN/S, literature[36]; see also Fruits (apple), chapter: Further Mycotoxins and Microbial Metabolites

Fruits (cherry) may contain the following mycotoxins:

Aspergillus Toxins

Aflatoxin B_1

incidence: 1/8*, conc.: 5 μg/kg, sample origin: unknown, sample year: unknown, country: Germany[239], *cherries, moldy

- Co-contamination: not reported
- Further contamination of plants and plant products other than cereals and their products reported in the present book/article[239]: Fruits (apple), Fruits (lemon), Fruits (orange), Fruits (peach), Fruits (tomato), Nuts (almond), Nuts (cashew nut), Nuts (hazelnut), Nuts (walnut), Seeds (sesame), Spices (nutmeg), AFB_1, literature[239]; Nuts (peanut), AFB_1, AFB_2, AFG_1, and AFG_2, literature[239]

Aspergillus and ***Penicillium*** Toxins

Ochratoxin A

incidence: 6/6*, conc. range: tr-2.71 μg/kg, sample origin: households and markets, Germany, sample year: unknown, country: Germany[26], *damaged or moldy

- Co-contamination: not reported
- Further contamination of plants and plant products other than cereals and their products reported in the present book/article[26]: Fruits (apple), Fruits (peach), Fruits (tomato), Potatoes, Strawberries, OTA, literature[26]

Fruits (goji) may contain the following mycotoxins:

Aspergillus and ***Penicillium*** Toxins

Ochratoxin A

incidence: 3/5, conc. range: 1.11–2.08* μg/kg, ∅ conc.: 1.67 μg/kg, sample origin: China, sa. purchased in local shops in Poland, sample year: unknown, country: Poland[664], *sa. from organic farming

- Co-contamination: not reported
- Further contamination of plants and plant products other than cereals and their products reported in the present book/article[664]: Raisins, AFB_1 and OTA, literature[664]

Fruits (hawthorn) may contain the following mycotoxins:

Aspergillus and ***Penicillium*** Toxins

Patulin

incidence: 8/25*, conc. range: 4.56–38.77 μg/kg, ∅ conc.: 16.26 μg/kg, sample origin: local pharmacies in Jinan (city), China, sample year: 2010, country: China[369], *medicinal hawthorn slices including dried, baked, and charred ones of different grades

- Co-contamination: not reported
- Further contamination of plants and plant products other than cereals and their products reported in the present book/article[369]: not reported

Fruits (lemon) may contain the following mycotoxins:

Aspergillus Toxins

Aflatoxin B_1

incidence: 3/18*, conc. range: 20–60 μg/kg, sample origin: unknown, sample year: unknown, country: Germany[239], *lemons, moldy

- Co-contamination: not reported

- Further contamination of plants and plant products other than cereals and their products reported in the present book/article[239]: Fruits (apple), Fruits (cherry), Fruits (orange), Fruits (peach), Fruits (tomato), Nuts (almond), Nuts (cashew nut), Nuts (hazelnut), Nuts (walnut), Seeds (sesame), Spices (nutmeg), AFB_1, literature[239]; Nuts (peanut), AFB_1, AFB_2, AFG_1, and AFG_2, literature[239]

incidence: 19?/40*, ∅ conc.: 195 μg/kg, sample origin: retail shops, Tirupati? (city), India, sample year: unknown, country: India[370], *pickled in salt, stored in polythene bags

- Co-contamination: not reported
- Further contamination of plants and plant products other than cereals and their products reported in the present book/article[370]: Fruits (lemon), AFB_2, AFG_1, and AFG_2, literature[370]; Fruits (mango), AFB_1, AFB_2, AFG_1, and AFG_2, literature[370]

Aflatoxin B_2

incidence: 19?/40*, ∅ conc.: 42 μg/kg, sample origin: retail shops, Tirupati? (city), India, sample year: unknown, country: India[370], *pickled in salt, stored in polythene bags

- Co-contamination: not reported
- Further contamination of plants and plant products other than cereals and their products reported in the present book/article[370]: Fruits (lemon), AFB_1, AFG_1, and AFG_2, literature[370]; Fruits (mango), AFB_1, AFB_2, AFG_1, and AFG_2, literature[370]

Aflatoxin G_1

incidence: 19?/40*, ∅ conc.: 110 μg/kg, sample origin: retail shops, Tirupati? (city), India, sample year: unknown, country: India[370], *pickled in salt, stored in polythene bags

- Co-contamination: not reported
- Further contamination of plants and plant products other than cereals and their products reported in the present book/article[370]: Fruits (lemon), AFB_1, AFB_2, and AFG_2, literature[370]; Fruits (mango), AFB_1, AFB_2, AFG_1, and AFG_2, literature[370]

Aflatoxin G_2

incidence: 19?/40*, ∅ conc.: 25 μg/kg, sample origin: retail shops, Tirupati? (city), India, sample year: unknown, country: India[370], *pickled in salt, stored in polythene bags

- Co-contamination: not reported
- Further contamination of plants and plant products other than cereals and their products reported in the present book/article[370]: Fruits (lemon), AFB_2, AFG_1, and AFG_1, literature[370]; Fruits (mango), AFB_1, AFB_2, AFG_1, and AFG_2, literature[370]

Penicillium Toxins

Mycophenolic Acid

incidence: 1/1*, conc.: 17 μg/kg, sample origin: private households of staff members, Austria, sample year: unknown, country: Austria[850], *moldy

- Co-contamination: 1 sa. co-contaminated with MA and ROR
- Further contamination of plants and plant products other than cereals and their products reported in the present book/article[850]: Fruits (apple), Nuts (chestnut), AME and AOH, literature[850]; Jam, AOH and MA, literature[850]; Puree, MA, literature[850]; Spices (garlic), BEA, FB_1, FB_2, and FB_3, literature[850]; Wine (red), AME, AOH, and MA, literature[850]; see also Fruits (lemon), Jam, Nuts (chestnut), Wine (red), chapter: Further Mycotoxins and Microbial Metabolites

Fruits (mandarin) may contain the following mycotoxins:

Alternaria Toxins

Alternariol

incidence: 2*/3, conc. range: 1,000–5,200 μg/kg, ∅ conc: 3,100 μg/kg, sample origin: several localities in Apulia and Basilicata (regions), Italy, sample year: winter, country: Italy[164], *visibly affected by *Alternaria* rot

- Co-contamination: 2 sa. co-contaminated with AME, AOH, and TA
- Further contamination of plants and plant products other than cereals and their products reported in the present book/article[164]: Fruits (mandarin), AME and TA, literature[164]; Fruits (tomato), Spices (pepper), AME, AOH, and TA, literature[164]; Fruits (melon), AME and TA, literature[164]; Olives, ALT, AME, AOH, and TA, literature[164]; Seeds (sunflower), AME and AOH, literature[164]

Alternariol Methyl Ether

incidence: 2*/3, conc. range: 550–1,400 μg/kg, ∅ conc.: 975 μg/kg, sample origin: several localities in Apulia and Basilicata (regions), Italy, sample year: winter, country: Italy[164], *visibly affected by *Alternaria* rot

- Co-contamination: 2 sa. co-contaminated with AME, AOH, and TA
- Further contamination of plants and plant products other than cereals and their products reported in the present book/article[164]: Fruits (mandarin), AOH and TA, literature[164]; Fruits (tomato), Spices (pepper), AME, AOH, and TA, literature[164]; Fruits (melon), AME and TA, literature[164]; Olives, ALT, AME, AOH, and TA, literature[164]; Seeds (sunflower), AME and AOH, literature[164]

Tenuazonic Acid

incidence: 3*/3, conc. range: 21,000–173,900 μg/kg, ∅ conc.: 94,033 μg/kg, sample origin: several localities in Apulia and Basilicata (regions), Italy, sample year: winter, country: Italy[164], *visibly affected by *Alternaria* rot

- Co-contamination: 2 sa. co-contaminated with AME, AOH, and TA; 1 sa. contaminated solely with TA
- Further contamination of plants and plant products other than cereals and their products reported in the present book/article[164]: Fruits (mandarin), Seeds (sunflower), AME and AOH, literature[164]; Fruits (tomato), Spices (pepper), AME, AOH, and TA, literature[164]; Fruits (melon), AME and TA, literature[164]; Olives, ALT, AME, AOH, and TA, literature[164]

Fruits (mango) may contain the following mycotoxins:

Aspergillus Toxins

Aflatoxin B_1

incidence: 4?/8*, ∅ conc.: 52 μg/kg, sample origin: retail shops, Tirupati? (city), India, sample year: unknown, country: India[370], *pickled in salt, stored in bottles

incidence: 26?/40*, ∅ conc.: 210 μg/kg, sample origin: retail shops, Tirupati? (city), India, sample year: unknown, country: India[370], *pickled in salt, stored in polythene bags

- Co-contamination: not reported
- Further contamination of plants and plant products other than cereals and their products reported in the present book/

article[370]: Fruits (lemon), AFB_1, AFB_2, AFG_1, and AFG_2, literature[370]; Fruits (mango), AFB_2, AFG_1, and AFG_2, literature[370]

Aflatoxin B$_2$

incidence: 4?/8*, Ø conc.: 5 µg/kg, sample origin: retail shops, Tirupati? (city), India, sample year: unknown, country: India[370], *pickled in salt, stored in bottles

incidence: 26?/40*, Ø conc.: 32 µg/kg, sample origin: retail shops, Tirupati? (city), India, sample year: unknown, country: India[370], *pickled in salt, stored in polythene bags

- Co-contamination: not reported
- Further contamination of plants and plant products other than cereals and their products reported in the present book/article[370]: Fruits (lemon), AFB_1, AFB_2, AFG_1, and AFG_2, literature[370]; Fruits (mango), AFB_1, AFG_1, and AFG_2, literature[370]

Aflatoxin G$_1$

incidence: 4?/8*, Ø conc.: 24 µg/kg, sample origin: retail shops, Tirupati? (city), India, sample year: unknown, country: India[370], *pickled in salt, stored in bottles

incidence: 26?/40*, Ø conc.: 184 µg/kg, sample origin: retail shops, Tirupati? (city), India, sample year: unknown, country: India[370], *pickled in salt, stored in polythene bags

- Co-contamination: not reported
- Further contamination of plants and plant products other than cereals and their products reported in the present book/article[370]: Fruits (lemon), AFB_1, AFB_2, AFG_1, and AFG_2, literature[370]; Fruits (mango), AFB_1, AFB_2, and AFG_2, literature[370]

Aflatoxin G$_2$

incidence: 4?/8*, Ø conc.: tr, sample origin: retail shops, Tirupati? (city), India, sample year: unknown, country: India[370], *pickled in salt, stored in bottles

incidence: 26?/40*, Ø conc.: 15 µg/kg, sample origin: retail shops, Tirupati? (city), India, sample year: unknown, country: India[370], *pickled in salt, stored in polythene bags

- Co-contamination: not reported
- Further contamination of plants and plant products other than cereals and their products reported in the present book/article[370]: Fruits (lemon), AFB_1, AFB_2, AFG_1, and AFG_2, literature[370]; Fruits (mango), AFB_1, AFB_2, and AFG_1, literature[370]

Fruits (melon) may contain the following mycotoxins:

Alternaria **Toxins**

Alternariol Methyl Ether

incidence: 1*/1, conc.: 51 µg/kg**, sample origin: several localities in Apulia and Basilicata (regions), Italy, sample year: late summer, country: Italy[164], *visibly affected by *Alternaria* rot, **toxin concentrations are referred to dry weight

- Co-contamination: 1 sa. co-contaminated with AME and TA
- Further contamination of plants and plant products other than cereals and their products reported in the present book/article[164]: Fruits (mandarin), Fruits (tomato), Spices (pepper), AME, AOH, and TA, literature[164]; Fruits (melon), TA, literature[164]; Olives, ALT, AME, AOH, and TA, literature[164]; Seeds (sunflower), AME and AOH, literature[164]

Tenuazonic Acid

incidence: 1*/1, conc.: 80 µg/kg**, sample origin: several localities in Apulia and Basilicata (regions), Italy, sample year: late summer, country: Italy[164], *visibly affected by *Alternaria* rot, **toxin concentrations are referred to dry weight

- Co-contamination: 1 sa. co-contaminated with AME and TA
- Further contamination of plants and plant products other than cereals and their products reported in the present book/article[164]: Fruits (mandarin), Fruits (tomato), Spices (pepper), AME, AOH, and TA, literature[164]; Fruits (melon), AME, literature[164]; Olives, ALT, AME, AOH, and TA, literature[164]; Seeds (sunflower), AME and AOH, literature[164]

Fruits (orange) may contain the following mycotoxins:

Aspergillus **Toxins**

Aflatoxin B$_1$

incidence: 4/14*, conc. range: 5–50 µg/kg, sample origin: unknown, sample year: unknown, country: Germany[239], *oranges, moldy

- Co-contamination: not reported
- Further contamination of plants and plant products other than cereals and their products reported in the present book/article[239]: Fruits (apple), Fruits (cherry), Fruits (lemon), Fruits (peach), Fruits (tomato), Nuts (almond), Nuts (cashew nut), Nuts (hazelnut), Nuts (walnut), Seeds (sesame), Spices (nutmeg), AFB_1, literature[239]; Nuts (peanut), AFB_1, AFB_2, AFG_1, and AFG_2, literature[239]

Fruits (peach) may contain the following mycotoxins:

Aspergillus **Toxins**

Aflatoxin B$_1$

incidence: 9/20*, conc. range: 5–15 µg/kg, sample origin: unknown, sample year: unknown, country: Germany[239], *peaches, moldy

- Co-contamination: not reported
- Further contamination of plants and plant products other than cereals and their products reported in the present book/article[239]: Fruits (apple), Fruits (cherry), Fruits (lemon), Fruits (orange), Fruits (tomato), Nuts (almond), Nuts (cashew nut), Nuts (hazelnut), Nuts (walnut), Seeds (sesame), Spices (nutmeg), AFB_1, literature[239]; Nuts (peanut), AFB_1, AFB_2, AFG_1, and AFG_2, literature[239]

Aspergillus **and** ***Penicillium*** **Toxins**

Ochratoxin A

incidence: 1/9*, conc.: 0.59 µg/kg, sample origin: households and markets, Germany, sample year: unknown, country: Germany[26], *damaged or moldy

- Co-contamination: not reported
- Further contamination of plants and plant products other than cereals and their products reported in the present book/article[26]: Fruits (apple), Fruits (cherry), Fruits (tomato), Potatoes, Strawberries, OTA, literature[26]

Patulin

incidence: 2/4* **, conc. range: 200–400 µg/kg, Ø conc.: 300 µg/kg, sample origin: local marketing institutions, GDR, sample year: unknown, country: GDR[29], *moldy, **peaches, stewed

- Co-contamination: not reported
- Further contamination of plants and plant products other than cereals and their products reported in the present book/article[29]: Butter (peanut butter), Nuts (peanut), Products (peanut product), Seeds (sunflower), AFB_1, AFB_2, AFG_1, and AFG_2,

literature[29]; Fruits, AFB_1, AFG_1, and AFG_2, literature[29]; Fruits (apple), Juices (apple juice), PAT, literature[29]

incidence: 4/4, conc. range: 92–174 µg/kg, Ø conc.: 120 µg/kg, sample origin: groceries, markets, and supermarkets, Campinas (city), São Paulo (state), Brazil, sample year: 1992/1993 and 1995, country: Brazil[32]

- Co-contamination: not reported
- Further contamination of plants and plant products other than cereals and their products reported in the present book/article[32]: Fruits (apple), Fruits (pear), Juices (apple juice), PAT, literature[32]

incidence: 1/8, conc.: 6 µg/kg (pulp), sample origin: Uppsala (city) and Stockholm (capital) area, Sweden, sample year: summer 1976, country: Sweden[155]

- Co-contamination: not reported
- Further contamination of plants and plant products other than cereals and their products reported in the present book/article[155]: Berries (blueberry), Berries (lingonberry), Fruits (plum), PAT, literature[155]

Fruits (pear) may contain the following mycotoxins:

Aspergillus and *Penicillium* Toxins

Patulin

incidence: 4/4, conc. range: 134–245 µg/kg, Ø conc.: 196 µg/kg, sample origin: groceries, markets, and supermarkets, Campinas (city), São Paulo (state), Brazil, sample year: 1992/1993 and 1995, country: Brazil[32]

- Co-contamination: not reported
- Further contamination of plants and plant products other than cereals and their products reported in the present book/article[32]: Fruits (apple), Fruits (peach), Juices (apple juice), PAT, literature[32]

incidence: 1/3*, conc.: 41.6 µg/kg, sample origin: Czech retail market, Czech Republic, sample year: unknown, country: Czech Republic[911], *pears (fresh fruits)

- Co-contamination: not reported
- Further contamination of plants and plant products other than cereals and their products reported in the present book/article[911]: Apple cider, Baby food, Fruit salad, Fruits (apple), Juices (apple juice), Juices (pear juice), Pulp (apple pulp), PAT, literature[911]

see also **Fruits (apple)**

Fruits (plum) may contain the following mycotoxins:

Aspergillus and *Penicillium* Toxins

Patulin

incidence: 1/6, conc.: 4 µg/kg (pulp), sample origin: Uppsala (city) and Stockholm (capital) area, Sweden, sample year: summer 1976, country: Sweden[155]

- Co-contamination: not reported
- Further contamination of plants and plant products other than cereals and their products reported in the present book/article[155]: Berries (blueberry), Berries (lingonberry), Fruits (peach), PAT, literature[155]

Fruits (quince) may contain the following mycotoxins:

Aspergillus Toxins

Aflatoxin B_1

incidence: 25/25*, conc. range: 96–8,164 µg/kg, Ø conc.: 1,531.9 µg/kg, sample origin: markets in Jammu and Kashmir (state), India, sample year: unknown, country: India[371], *dry fruit slices

- Co-contamination: not reported
- Further contamination of plants and plant products other than cereals and their products reported in the present book/article[371]: not reported

Aspergillus and *Penicillium* Toxins

Ochratoxin A

incidence: 8/25*, conc. range: 500–1,630 µg/kg, Ø conc.: 946 µg/kg, sample origin: markets in Jammu and Kashmir (state), India, sample year: unknown, country: India[372], *dry fruit slices

- Co-contamination: not reported
- Further contamination of plants and plant products other than cereals and their products reported in the present book/article[372]: not reported

Patulin

incidence: 4/4*, conc. range: <LOQ-118.3 µg/kg, sample origin: supermarket, Portugal, sample year: unknown, country: Portugal[27], *sa. from 0 to about 75% brown areas

- Co-contamination: not reported
- Further contamination of plants and plant products other than cereals and their products reported in the present book/article[27]: Apple cider, Baby food, Fruits (apple), Jam (quince jam), Juices (apple juice), PAT, literature[27]

Fruits (tangerine) may contain the following mycotoxins:

Alternaria Toxins

Alternariol

incidence: 2/4*, conc. range: 2.5–3.7 µg/kg, Ø conc.: 3.1 µg/kg, sample origin: farm near Aguaí (city) in the center of São Paulo (state) and local market in Recife (city) in Pernambuco (state), Brazil, sample year: April 2004, country: Brazil[652], *healthy flavedo tissue of tangerines

- Co-contamination: 2 sa. co-contaminated with AME and AOH
- Further contamination of plants and plant products other than cereals and their products reported in the present book/article[652]: Fruits (tangerine), AME, literature[652]

incidence: 4/4*, conc. range: 13.1–17.4 µg/kg, Ø conc.: 14.5 µg/kg, sample origin: farm near Aguaí (city) in the center of São Paulo (state), Brazil, sample year: April 2004, country: Brazil[652], *flavedo tissue of tangerines with symptoms of *Alternaria* brown spot

- Co-contamination: 4 sa. co-contaminated with AME and AOH
- Further contamination of plants and plant products other than cereals and their products reported in the present book/article[652]: Fruits (tangerine), AME, literature[652]

Alternariol Methyl Ether

incidence: 2/4*, conc. range: ≤0.9 µg/kg, sample origin: farm near Aguaí (city) in the center of São Paulo (state) and local market in Recife (city) in Pernambuco (state), Brazil, sample year: April 2004, country: Brazil[652], *healthy flavedo tissue of tangerines

- Co-contamination: 2 sa. co-contaminated with AME and AOH
- Further contamination of plants and plant products other than cereals and their products reported in the present book/article[652]: Fruits (tangerine), AOH, literature[652]

incidence: 4/4*, conc. range: 2.5–3.5 µg/kg, ∅ conc.: 2.85 µg/kg, sample origin: farm near Aguaí (city) in the center of São Paulo (state), Brazil, sample year: April 2004, country: Brazil[652], *flavedo tissue of tangerines with symptoms of *Alternaria* brown spot

- Co-contamination: 4 sa. co-contaminated with AME and AOH
- Further contamination of plants and plant products other than cereals and their products reported in the present book/article[652]: Fruits (tangerine), AOH, literature[652]

Fruits (tomato) may contain the following mycotoxins:

Alternaria Toxins

Altenuene

incidence: 4/19, conc. range: ≤1,100 µg/kg, sample origin: commercial outlets in New York metropolitan area, USA, sample year: unknown, country: USA[23]

- Co-contamination: 1 sa. co-contaminated ALT, AME, AOH, and TA; 1 sa. co-contaminated with ALT, AME, and TA; 1 sa. co-contaminated with ALT and AME; 1 sa. co-contaminated with ALT and TA
- Further contamination of plants and plant products other than cereals and their products reported in the present book/article[23]: Fruits (apple), ALT, AME, AOH, ATX-I, and TA, literature[23]; Fruits (tomato), AME, AOH, and TA, literature[23]

Alternariol

incidence: 6/19, conc. range: ≤5,300 µg/kg, sample origin: commercial outlets in New York metropolitan area, USA, sample year: unknown, country: USA[23]

- Co-contamination: 1 sa. co-contaminated ALT, AME, AOH, and TA; 1 sa. co-contaminated with AME and AOH; 2 sa. co-contaminated with AOH and TA; 2 sa. contaminated solely with AOH
- Further contamination of plants and plant products other than cereals and their products reported in the present book/article[23]: Fruits (apple), ALT, AME, AOH, ATX-I, and TA, literature[23]; Fruits (tomato), ALT, AME, and TA, literature[23]

incidence: 0/4*, conc.: no contamination, sample origin: Swiss market, Switzerland, sample year: autumn and winter 2010, country: Switzerland[47], *tomatoes, fresh and whole

incidence: 3/13*, conc. range: 4–7 µg/kg, sample origin: Swiss market, Switzerland, sample year: autumn and winter 2010, country: Switzerland[47], *tomatoes, peeled and minced

- Co-contamination: For detailed information see the article.
- Further contamination of plants and plant products other than cereals and their products reported in the present book/article[47]: Fruits (tomato), AME and TA, literature[47]; Fruits, dried, ALT, AME, AOH, and TA, literature[47]; Puree, Sauces (tomato sauce), Tomato ketchup, AME, AOH, and TA, literature[47]; see also Fruits (tomato), Fruits, dried, Puree, chapter: Further Mycotoxins and Microbial Metabolites

incidence: 1*/2, conc.: 1,274 µg/kg, sample origin: several localities in Apulia and Basilicata (regions), Italy, sample year: late summer, country: Italy[164], *visibly affected by *Alternaria* rot

- Co-contamination: 1 sa. co-contaminated with AME, AOH, and TA
- Further contamination of plants and plant products other than cereals and their products reported in the present book/article[164]: Fruits (mandarin), Spices (pepper), AME, AOH, and TA, literature[164]; Fruits (melon), Fruits (tomato), AME and TA, literature[164]; Olives, ALT, AME, AOH, and TA, literature[164]; Seeds (sunflower), AME and AOH, literature[164]

incidence: 4/20*, conc. range: LOQ, sample origin: local supermarkets in Valencia (city), Spain, sample year: unknown, country: Spain[415], *tomato, fresh

- Co-contamination: 1 sa. co-contaminated AME and AOH; 3 sa. contaminated solely with AOH
- Further contamination of plants and plant products other than cereals and their products reported in the present book/article[415]: Fruits (tomato), AME, literature[415]

incidence: 3/10*, conc. range: ≤15 µg/kg, sample origin: retail stores, Netherlands, sample year: autumn 2014, country: Netherlands/Germany[906], *sieved, chopped, cubes

- Co-contamination: not reported
- Further contamination of plants and plant products other than cereals and their products reported in the present book/article[906]: Figs, Juices (tomato juice), AOH and TA, literature[906]; Fruits (tomato), TA, literature[906]; Oils (sunflower oil), AME, literature[906]; Paste (tomato paste), Sauces (tomato sauce), Seeds (sunflower), Tomato ketchup, AME, AOH, and TA, literature[906]

Alternariol Methyl Ether

incidence: 5/19, conc. range: ≤800 µg/kg, sample origin: commercial outlets in New York metropolitan area, USA, sample year: unknown, country: USA[23]

- Co-contamination: 1 sa. co-contaminated ALT, AME, AOH, and TA; 1 sa. co-contaminated with ALT, AME, and TA; 1 sa. co-contaminated with ALT and AME; 1 sa. co-contaminated with AME and AOH; 1 sa. co-contaminated with AME and TA
- Further contamination of plants and plant products other than cereals and their products reported in the present book/article[23]: Fruits (apple), ALT, AME, AOH, ATX-I, and TA, literature[23]; Fruits (tomato), ALT, AOH, and TA, literature[23]

incidence: 0/4*, conc.: no contamination, sample origin: Swiss market, Switzerland, sample year: autumn and winter 2010, country: Switzerland[47], *tomatoes, fresh and whole

incidence: 1/13*, conc.: 1 µg/kg, sample origin: Swiss market, Switzerland, sample year: autumn and winter 2010, country: Switzerland[47], *tomatoes, peeled and minced

- Co-contamination: For detailed information see the article.
- Further contamination of plants and plant products other than cereals and their products reported in the present book/article[47]: Fruits (tomato), AOH and TA, literature[47]; Fruits, dried, ALT, AME, AOH, and TA, literature[47]; Puree, Sauces (tomato sauce), Tomato ketchup, AME, AOH, and TA, literature[47]; see also Fruits (tomato), Fruits, dried, Puree, chapter: Further Mycotoxins and Microbial Metabolites

incidence: 2*/2, conc. range: 37–268 µg/kg, ∅ conc.: 153 µg/kg, sample origin: several localities in Apulia and Basilicata (regions),

Italy, sample year: late summer, country: Italy[164], *visibly affected by *Alternaria* rot

- Co-contamination: 1 sa. co-contaminated with AME, AOH, and TA; 1 sa. co-contaminated with AME and TA
- Further contamination of plants and plant products other than cereals and their products reported in the present book/ article[164]: Fruits (mandarin), Spices (pepper), AME, AOH, and TA, literature[164]; Fruits (melon), AME and TA, literature[164]; Fruits (tomato), AOH and TA, literature[164]; Olives, ALT, AME, AOH, and TA, literature[164]; Seeds (sunflower), AME and AOH, literature[164]

incidence: 2/20*, conc. range: LOD-LOQ, sample origin: local supermarkets in Valencia (city), Spain, sample year: unknown, country: Spain[415], *tomato, fresh

- Co-contamination: 1 sa. co-contaminated AME and AOH; 1 sa. contaminated solely with AME
- Further contamination of plants and plant products other than cereals and their products reported in the present book/ article[415]: Fruits (tomato), AOH, literature[415]

Tenuazonic Acid

incidence: 11/19, conc. range: ≤139,000 µg/kg, sample origin: commercial outlets in New York metropolitan area, USA, sample year: unknown, country: USA[23]

- Co-contamination: 1 sa. co-contaminated ALT, AME, AOH, and TA; 1 sa. co-contaminated with ALT, AME, and TA; 1 sa. co-contaminated with ALT and TA; 1 sa. co-contaminated with AME and TA; 2 sa. co-contaminated with AOH and TA; 5 sa. contaminated solely with TA
- Further contamination of plants and plant products other than cereals and their products reported in the present book/ article[23]: Fruits (apple), ALT, AME, AOH, ATX-I, and TA, literature[23]; Fruits (tomato), ALT, AME, and AOH, literature[23]

incidence: 1/4*, conc.: 37 µg/kg, sample origin: Swiss market, Switzerland, sample year: autumn and winter 2010, country: Switzerland[47], *tomatoes, fresh and whole

incidence: 13/13*, conc. range: 25–200 µg/kg, Ø conc.: 81 µg/kg, sample origin: Swiss market, Switzerland, sample year: autumn and winter 2010, country: Switzerland[47], *tomatoes, peeled and minced

- Co-contamination: For detailed information see the article.
- Further contamination of plants and plant products other than cereals and their products reported in the present book/ article[47]: Fruits (tomato), AME and AOH, literature[47]; Fruits, dried, ALT, AME, AOH, and TA, literature[47]; Puree, Sauces (tomato sauce), Tomato ketchup, AME, AOH, and TA, literature[47]; see also Fruits (tomato), Fruits, dried, Puree, chapter: Further Mycotoxins and Microbial Metabolites

incidence: 2*/2, conc. range: 24–7,210 µg/kg, Ø conc.: 3,617 µg/kg, sample origin: several localities in Apulia and Basilicata (regions), Italy, sample year: late summer, country: Italy[164], *visibly affected by *Alternaria* rot

- Co-contamination: 1 sa. co-contaminated with AME, AOH, and TA; 1 sa. co-contaminated with AME and TA
- Further contamination of plants and plant products other than cereals and their products reported in the present book/ article[164]: Fruits (mandarin), Spices (pepper), AME, AOH, and TA, literature[164]; Fruits (melon), AME and TA, literature[164]; Fruits (tomato), Seeds (sunflower), AME and AOH, literature[164]; Olives, ALT, AME, AOH, and TA, literature[164]

incidence: 73/146*, conc. range: 400–69,700 µg/kg**, Ø conc.: 4,940 µg/kg**, sample origin: catsup processing plants in Midwestern and East Coast, as well as California (state), USA, sample year: 1983, country: USA[655], *fresh tomatoes used for catsup production, **µg/kg of moldy tissue

- Co-contamination: not reported
- Further contamination of plants and plant products other than cereals and their products reported in the present book/ article[655]: not reported

incidence: 6/10*, conc. range: ≤100 µg/kg, sample origin: retail stores, Netherlands, sample year: autumn 2014, country: Netherlands/Germany[906], *sieved, chopped, cubes

- Co-contamination: not reported
- Further contamination of plants and plant products other than cereals and their products reported in the present book/ article[906]: Figs, Juices (tomato juice), AOH and TA, literature[906]; Fruits (tomato), AOH, literature[906]; Oils (sunflower oil), AME, literature[906]; Paste (tomato paste), Sauces (tomato sauce), Seeds (sunflower), Tomato ketchup, AME, AOH, and TA, literature[906]

Aspergillus Toxins

Aflatoxin B_1

incidence: 1/8*, conc.: 5 µg/kg, sample origin: unknown, sample year: unknown, country: Germany[239], *tomatoes, moldy

- Co-contamination: not reported
- Further contamination of plants and plant products other than cereals and their products reported in the present book/article[239]: Fruits (apple), Fruits (cherry), Fruits (lemon), Fruits (orange), Fruits (peach), Nuts (almond), Nuts (cashew nut), Nuts (hazelnut), Nuts (walnut), Seeds (sesame), Spices (nutmeg), AFB_1, literature[239]; Nuts (peanut), AFB_1, AFB_2, AFG_1, and AFG_2, literature[239]

Aspergillus and *Penicillium* Toxins

Ochratoxin A

incidence: 6/11*, conc. range: tr-1.44 µg/kg, sample origin: households and markets, Germany, sample year: unknown, country: Germany[26], *damaged or moldy

- Co-contamination: not reported
- Further contamination of plants and plant products other than cereals and their products reported in the present book/ article[26]: Fruits (apple), Fruits (cherry), Fruits (peach), Potatoes, Strawberries, OTA, literature[26]

Fruits, dried may contain the following mycotoxins:

Alternaria Toxins

Altenuene

incidence: 2/8*, conc. range: 2 µg/kg, sample origin: Swiss market, Switzerland, sample year: autumn and winter 2010, country: Switzerland[47], *tomatoes, dried

- Co-contamination: For detailed information see the article.
- Further contamination of plants and plant products other than cereals and their products reported in the present book/ article[47]: Fruits (tomato), Fruits, dried, Puree, Sauces (tomato sauce), Tomato ketchup, AME, AOH, and TA, literature[47]; see also Fruits (tomato), Fruits, dried, Puree, chapter: Further Mycotoxins and Microbial Metabolites

Alternariol

incidence: 1/8*, conc.: 4 µg/kg, sample origin: Swiss market, Switzerland, sample year: autumn and winter 2010, country: Switzerland[47], *tomatoes, dried

- Co-contamination: For detailed information see the article.

- Further contamination of plants and plant products other than cereals and their products reported in the present book/ article[47]: Fruits (tomato), Puree, Sauces (tomato sauce), Tomato ketchup, AME, AOH, and TA, literature[47]; Fruits, dried, ALT, AME, and TA, literature[47]; see also Fruits (tomato), Fruits, dried, Puree, chapter: Further Mycotoxins and Microbial Metabolites

incidence: 0/56*, conc.: no contamination, sample origin: principally Xinjiang (province), purchased in supermarkets and local markets in Beijing (capital), China, sample year: 2016/2017, country: China[815], *apricots, dried

incidence: 2/54*, conc. range: 5.9–27.4 µg/kg, ∅ conc.: 16.6 µg/kg, sample origin: principally Ningxia (province), purchased in supermarkets and local markets in Beijing (capital), China, sample year: 2016/2017, country: China[815], *wolfberries, dried

- Co-contamination: For detailed information see the article.
- Further contamination of plants and plant products other than cereals and their products reported in the present book/ article[815]: Dates, MA, PA, and TA, literature[815]; Fruits, dried, AME, MA, PAT, and TA, literature[815]; Raisins, AME, AOH, MA, OTA, OTB, and TA, literature[815]; see also Dates, Fruits, dried, chapter: Further Mycotoxins and Microbial Metabolites

Alternariol Methyl Ether

incidence: 3/8*, conc. range: 2–7 µg/kg, sample origin: Swiss market, Switzerland, sample year: autumn and winter 2010, country: Switzerland[47], *tomatoes, dried

- Co-contamination: For detailed information see the article.
- Further contamination of plants and plant products other than cereals and their products reported in the present book/ article[47]: Fruits (tomato), Puree, Sauces (tomato sauce), Tomato ketchup, AME, AOH, and TA, literature[47]; Fruits, dried, ALT, AOH, and TA, literature[47]; see also Fruits (tomato), Fruits, dried, Puree, chapter: Further Mycotoxins and Microbial Metabolites

incidence: 3/56*, conc. range: 0.5–2.1 µg/kg, ∅ conc.: 1.3 µg/kg, sample origin: principally Xinjiang (province), purchased in supermarkets and local markets in Beijing (capital), China, sample year: 2016/2017, country: China[815], *apricots, dried

incidence: 4/54*, conc. range: 0.2–15.0 µg/kg, ∅ conc.: 3.9 µg/kg, sample origin: principally Ningxia (province), purchased in supermarkets and local markets in Beijing (capital), China, sample year: 2016/2017, country: China[815], *wolfberries, dried

- Co-contamination: For detailed information see the article.
- Further contamination of plants and plant products other than cereals and their products reported in the present book/ article[815]: Dates, MA, PA, and TA, literature[815]; Fruits, dried, AOH, MA, PAT, and TA, literature[815]; Raisins, AME, AOH, MA, OTA, OTB, and TA, literature[815]; see also Dates, Fruits, dried, chapter: Further Mycotoxins and Microbial Metabolites

Tenuazonic Acid

incidence: 7/8*, conc. range: ≤166 µg/kg, sample origin: Swiss market, Switzerland, sample year: autumn and winter 2010, country: Switzerland[47], *tomatoes, dried

- Co-contamination: For detailed information see the article.
- Further contamination of plants and plant products other than cereals and their products reported in the present book/ article[47]: Fruits (tomato), Puree, Sauces (tomato sauce), Tomato ketchup, AME, AOH, and TA, literature[47]; Fruits, dried, ALT, AME, and AOH, literature[47]; see also Fruits (tomato), Fruits, dried, Puree, chapter: Further Mycotoxins and Microbial Metabolites

incidence: 21/56*, conc. range: 10.4–1,231.8 µg/kg, ∅ conc.: 237.1 µg/kg, sample origin: principally Xinjiang (province), purchased in supermarkets and local markets in Beijing (capital), China, sample year: 2016/2017, country: China[815], *apricots, dried

incidence: 35/54*, conc. range: 23.8–5,665.3 µg/kg, ∅ conc.: 574.8 µg/kg, sample origin: principally Ningxia (province), purchased in supermarkets and local markets in Beijing (capital), China, sample year: 2016/2017, country: China[815], *wolfberries, dried

- Co-contamination: For detailed information see the article.
- Further contamination of plants and plant products other than cereals and their products reported in the present book/ article[815]: Dates, MA, PA, and TA, literature[815]; Fruits, dried, AME, AOH, MA, and PAT, literature[815]; Raisins, AME, AOH, MA, OTA, OTB, and TA, literature[815]; see also Dates, Fruits, dried, chapter: Further Mycotoxins and Microbial Metabolites

Aspergillus Toxins

Aflatoxin B_1

incidence: 9/30*, conc. range: 0.21–5.33 µg/kg, ∅ conc.: 0.88 µg/kg, sample origin: Mashhad (city), Khorasan (province), Iran, sample year: unknown, country: Iran[53], *apricots, dried

incidence: 2/15*, conc. range: 0.23–1.17 µg/kg, ∅ conc.: 0.70 µg/kg, sample origin: Mashhad (city), Khorasan (province), Iran, sample year: unknown, country: Iran[53], *prunes, dried

- Co-contamination: not reported
- Further contamination of plants and plant products other than cereals and their products reported in the present book/ article[53]: Fruits, dried, AFB_2, AFG_1, and OTA, literature[53]

incidence: 2/16*, conc. range: 0.4 µg/kg, ∅ conc.: 0.4 µg/kg, sample origin: retail outlets, UK, sample year: September 2001, country: UK[203], *prunes, dried

- Co-contamination: 1 sa. co-contaminated with AFB_1 and AFB_2; 1 sa. co-contaminated with AFB_1 and OTA
- Further contamination of plants and plant products other than cereals and their products reported in the present book/ article[203]: Butter (nut butter), AFB_1, AFB_2, AFG_1, AFG_2, and OTA, literature[203]; Butter (peanut butter), AFB_1, AFB_2, AFG_1, AFG_2, and OTA, literature[203]; Dates, AFB_1 and OTA, literature[203]; Figs, AFB_1, AFB_2, AFG_2, and OTA, literature[203]; Fruits, dried, AFB_2, and OTA, literature[203]; Nuts (Brazil nut, paranut), AFB_1, AFB_2, and AFG_1, literature[203]; Nuts (peanut), AFB_1 and AFB_2, literature[203]; Nuts (pistachio nut), AFB_1, AFB_2, and OTA, literature[203]

incidence: 1/5*, conc.: 0.1 µg/kg, sample origin: available in New Zealand, sample year: 2010, country: New Zealand[277], *prunes, dried

incidence: 1/10*, conc.: 0.2 µg/kg, sample origin: available in New Zealand, sample year: 2010, country: New Zealand[277], *apricots, dried

- Co-contamination: not reported
- Further contamination of plants and plant products other than cereals and their products reported in the present book/article[277]: Bars, AFB_1, AF/S, DON, NIV, and OTA, literature[277]; Butter (peanut butter), Confectionery, Food, Nuts, Nuts (peanut), Nuts (pistachio nut), Sauces (peanut sauce), AFB_1 and AF/S, literature[277]; Coffee, Dates, Spices (nutmeg), Spices (pepper), Vine fruits, dried, OTA, literature[277]; Figs, Spices (cayenne pepper, chilli), Spices

(curry), Spices (ginger), Spices (paprika), AFB_1, AF/S, and OTA, literature[277]; Fruits, dried, AF/S, literature[277]; Nuts (Brazil nut, paranut), AFB_1, literature[277]

incidence: 9/20, conc. range: 0.10–34.5 µg/kg, sample origin: local markets, Tunisia, sample year: 2005/2006, country: Tunisia[338]

- Co-contamination: 8 sa. co-contaminated with AFB_1, AFB_2, and AFG_1; 1 sa. contaminated solely with AFB_1
- Further contamination of plants and plant products other than cereals and their products reported in the present book/article[338]: Fruits, dried, AFB_2 and AFG_1, literature[338]; Spices, AFB_1, AFB_2, AFG_1, and AFG_2, literature[338]

incidence: 3/7*, conc. range: 0.08–0.58 µg/kg, ∅ conc.: 0.33 µg/kg, sample origin: local supermarket in Okayama (prefecture), Japan, sample year: unknown, country: Japan[339], *included apple**, cranberry**, kiwifruit, mango, papaya, pineapple**, and prune (all dried), **contaminated

- Co-contamination: 3 sa. co-contaminated with AFB_1 and AFG_2
- Further contamination of plants and plant products other than cereals and their products reported in the present book/article[339]: Coconuts, Seeds (sesame), AFB_1 and AFG_2, literature[339]; Figs, AFB_2 and AFG_2, literature[339]; Flour (soybean flour), AFB_1, literature[339]; Fruits, dried, Raisins, AFG_2, literature[339]; Nuts, Spices, AFB_1, AFB_2, AFG_1, and AFG_2, literature[339]

For detailed information see the article.

Aflatoxin B_2

incidence: 1/30*, conc.: 0.32 µg/kg, sample origin: Mashhad (city), Khorasan (province), Iran, sample year: unknown, country: Iran[53], * apricots, dried

incidence: 0/15*, conc: no contamination, sample origin: Mashhad (city), Khorasan (province) Iran, sample year: unknown, country: Iran[53], *prunes, dried

- Co-contamination: not reported
- Further contamination of plants and plant products other than cereals and their products reported in the present book/article[53]: Fruits, dried, AFB_1, AFG_1, and OTA, literature[53]

incidence: 1/16*, conc.: 0.2 µg/kg, sample origin: retail outlets, UK, sample year: September 2001, country: UK[203], *prunes, dried

- Co-contamination: 1 sa. co-contaminated with AFB_1 and AFB_2
- Further contamination of plants and plant products other than cereals and their products reported in the present book/article[203]: Butter (nut butter), AFB_1, AFB_2, AFG_1, AFG_2, and OTA, literature[203]; Butter (peanut butter), AFB_1, AFB_2, AFG_1, AFG_2, and OTA, literature[203]; Dates, AFB_1 and OTA, literature[203]; Figs, AFB_1, AFB_2, AFG_2, and OTA, literature[203]; Fruits, dried, AFB_1 and OTA, literature[203]; Nuts (Brazil nut, paranut), AFB_1, AFB_2, and AFG_1, literature[203]; Nuts (peanut), AFB_1 and AFB_2, literature[203]; Nuts (pistachio nut), AFB_1, AFB_2, and OTA, literature[203]

incidence: 8/20, conc. range: ≤6.1 µg/kg, sample origin: local markets, Tunisia, sample year: 2005/2006, country: Tunisia[338]

- Co-contamination: 8 sa. co-contaminated with AFB_1, AFB_2 and AFG_1
- Further contamination of plants and plant products other than cereals and their products reported in the present book/article[338]: Fruits, dried, AFB_1 and AFG_1, literature[338]; Spices, AFB_1, AFB_2, AFG_1, and AFG_2, literature[338]

Aflatoxin G_1

incidence: 1/30*, conc.: 0.20 µg/kg, sample origin: Mashhad (city), Khorasan (province), Iran, sample year: unknown, country: Iran[53], * apricots, dried

incidence: 1/15*, conc.: 0.37 µg/kg, sample origin: Mashhad (city), Khorasan (province), Iran, sample year: unknown, country: Iran[53], *prunes, dried

- Co-contamination: not reported
- Further contamination of plants and plant products other than cereals and their products reported in the present book/article[53]: Fruits, dried, AFB_1, AFB_2, and OTA, literature[53]

incidence: 1/20, conc.: 0.13 µg/kg, sample origin: local markets, Tunisia, sample year: 2005/2006, country: Tunisia[338]

- Co-contamination: 1 sa. co-contaminated with AFB_1, AFB_2 and AFG_1
- Further contamination of plants and plant products other than cereals and their products reported in the present book/article[338]: Fruits, dried, AFB_1 and AFB_2, literature[338]; Spices, AFB_1, AFB_2, AFG_1, and AFG_2, literature[338]

Aflatoxin G_2

incidence: 6/7*, conc. range: 0.08–0.43 µg/kg, ∅ conc.: 0.23 µg/kg, sample origin: local supermarket in Okayama (prefecture), Japan, sample year: unknown, country: Japan[339], *included apple**, cranberry**, kiwifruit**, mango**, papaya**, pineapple**, and prune (all dried), **contaminated

- Co-contamination: 3 sa. co-contaminated with AFB_1 and AFG_2; 3 sa. contaminated solely with AFG_2
- Further contamination of plants and plant products other than cereals and their products reported in the present book/article[339]: Coconuts, Seeds (sesame), AFB_1 and AFG_2, literature[339]; Figs, AFB_2 and AFG_2, literature[339]; Flour (soybean flour), AFB_1, literature[339]; Fruits, dried, AFB_1, literature[339]; Nuts, Spices, AFB_1, AFB_2, AFG_1, and AFG_2, literature[339]; Raisins, AFG_2, literature[339]

For detailed information see the article.

incidence: 3/27*, conc. range: <LOQ µg/kg**, sample origin: local and of Algeria, Chile, Iran, Israel, and Turkey purchased from markets in Valencia (city), Spain, and Tunis (capital), Tunisia, sample year: 2012/2013, country: Spain[720], *apricots, dried, **purchased in Tunisia

incidence: 7/27*, conc. range: 11.1–15.6 µg/kg**, ∅ conc.: 12.6 µg/kg**, sample origin: local and of Algeria, Chile, Iran, Israel, and Turkey purchased from markets in Valencia (city), Spain, and Tunis (capital), Tunisia, sample year: 2012/2013, country: Spain[720], *plums, dried, **purchased in Spain and Tunisia

- Co-contamination: For detailed information see the article.
- Further contamination of plants and plant products other than cereals and their products reported in the present book/article[720]: Dates, AFB_2, AFG_1, AFG_2, DAS, ENA_1, ENB, ENB_1, and OTA, literature[720]; Figs, AFG_1, BEA, DAS, ENA_1, ENB, ENB_1, HT-2, and OTA, literature[720]; Fruits, dried, DAS, ENA_1, ENB, ENB_1, HT-2, and OTA, literature[720]; Vine fruits, dried, AFB_1, AFG_2, ENA, ENA_1, ENB, ENB_1, HT-2, and OTA, literature[720]

For detailed information see the article.

Aflatoxin/s

incidence: 4/20*, conc. range: 1.5–10.8 µg/kg**, ∅ conc.: 4.55 µg/kg**, sample origin: retail shops and local markets in NWFP and northern areas, Pakistan, sample year: January–November 2009, country: Pakistan[17], * apricots, dried, **AFB_1, AFB_2, AFG_1, and AFG_2

incidence: 4/15*, conc. range: 1.0–3.5 µg/kg**, ∅ conc.: 2.22 µg/kg**, sample origin: retail shops and local markets in NWFP and northern areas, Pakistan, sample year: January–November 2009, country: Pakistan[17], *mulberries, dried, **AFB_1, AFB_2, AFG_1, and AFG_2

- Co-contamination: not reported
- Further contamination of plants and plant products other than cereals and their products reported in the present book/article[17]: Dates, Figs, Kernels (apricot), Nuts (almond), Nuts (peanut), Nuts (pine nut), Nuts (pistachio nut), Nuts (walnut), Raisins, AF/S, literature[17]

incidence: 6/157, conc. range: 2–20 µg/kg*, sample origin: imported and domestic sa. from Uruguay, sample year: 1993–1995, country: Uruguay[269], *AFB_1, AFB_2, AFG_1, and AFG_2

- Co-contamination: not reported
- Further contamination of plants and plant products other than cereals and their products reported in the present book/article[269]: Beans (soybean), Fruits, dried, ZEA, literature[269]; Cocoa beans, AF/S, literature[269]; Oil seeds, Vegetables, dried, AF/S and ZEA, literature[269]

incidence: 1/5*, conc.: 0.5 µg/kg**, sample origin: available in New Zealand, sample year: 2010, country: New Zealand[277], *prunes, dried, **AFS total

incidence: 1/10*, conc.: 0.9 µg/kg**, sample origin: available in New Zealand, sample year: 2010, country: New Zealand[277], *apricots, dried, **AFS total

- Co-contamination: not reported
- Further contamination of plants and plant products other than cereals and their products reported in the present book/article[277]: Bars, AFB_1, AF/S, DON, NIV, and OTA, literature[277]; Butter (peanut butter), Confectionery, Food, Nuts, Nuts (peanut), Nuts (pistachio nut), Sauces (peanut sauce), AFB_1 and AF/S, literature[277]; Coffee, Dates, Spices (nutmeg), Spices (pepper), Vine fruits, dried, OTA, literature[277]; Figs, Spices (cayenne pepper, chilli), Spices (curry), Spices (ginger), Spices (paprika), AFB_1, AF/S, and OTA, literature[277]; Fruits, dried, Nuts (Brazil nut, paranut), AFB_1, literature[277]

incidence: 1/10*, conc.: 0.9 µg/kg**, sample origin: New Zealand, sample year: 2010, country: New Zealand[702], *apricots, dried, **AFB_1, AFB_2, AFG_1, and AFG_2

incidence: 1/10*, conc.: 0.5 µg/kg**, sample origin: New Zealand, sample year: 2010, country: New Zealand[702], *prunes, dried, **AFB_1, AFB_2, AFG_1, and AFG_2

- Co-contamination: not reported
- Further contamination of plants and plant products other than cereals and their products reported in the present book/article[702]: Figs, Spices (cayenne pepper, chilli), Spices (curry), Spices (ginger), Spices (paprika), AF/S and OTA, literature[702]; Vine fruits, dried, Spices (pepper), OTA, literature[702]

incidence: 4/16*, conc. range: 0.92–4.53 µg/kg**, sample origin: Tehran (capital), Iran, sample year: unknown, country: Iran[828], *apricots, dried, **total AFS

incidence: 5/16*, conc. range: 0.96–3.28 µg/kg**, sample origin: Tehran (capital), Iran, sample year: unknown, country: Iran[828], *prunes, dried, **total AFS

incidence: 3/16*, conc. range: 1.36–2.51 µg/kg**, sample origin: Tehran (capital), Iran, sample year: unknown, country: Iran[828], *cherries, dried, **total AFS

- Co-contamination: not reported
- Further contamination of plants and plant products other than cereals and their products reported in the present book/article[828]: Raisins, AF/S, literature[828]

For detailed information see the article.

Aspergillus and ***Penicillium*** Toxins

Ochratoxin A

incidence: 1/30*, conc.: 2.83 µg/kg, sample origin: Mashhad (city), Khorasan (province), Iran, sample year: unknown, country: Iran[53], * apricots, dried

incidence: 3/15*, conc. range: 0.22–2.62 µg/kg, ∅ conc.: 1.28 µg/kg, ample origin: Mashhad (city), Khorasan (province), Iran, sample year: unknown, country: Iran[53], *prunes, dried

- Co-contamination: not reported
- Further contamination of plants and plant products other than cereals and their products reported in the present book/article[53]: Fruits, dried, AFB_1, AFB_2, and AFG_1, literature[53]

incidence: 3/3*, conc. range: 50–110 µg/kg, sample origin: shops and markets of different sanitation levels, Assiut Governorate, Egypt, sample year: March 1992, country: Egypt[54], * apricots, dried

incidence: 3/3*, conc. range: 210–280 µg/kg, sample origin: shops and markets of different sanitation levels, Assiut Governorate, Egypt, sample year: March 1992, country: Egypt[54], *plums, dried

- Co-contamination: not reported
- Further contamination of plants and plant products other than cereals and their products reported in the present book/article[54]: Figs, OTA, literature[54]

incidence: 1/20*, conc.: 0.97 µg/kg, sample origin: different export companies, Turkey, sample year: unknown, country: Turkey[55], *apricots, dried

- Co-contamination: not reported
- Further contamination of plants and plant products other than cereals and their products reported in the present book/article[55]: Figs, AFB_1, AF/S, and OTA, literature[55]; Sultanas, OTA, literature[55]

incidence: 1/16*, conc.: 0.6 µg/kg, sample origin: retail outlets, UK, sample year: September 2001, country: UK[203], *prunes, dried

- Co-contamination: 1 sa. co-contaminated with AFB_1 and OTA
- Further contamination of plants and plant products other than cereals and their products reported in the present book/article[203]: Butter (nut butter), AFB_1, AFB_2, AFG_1, AFG_2, and OTA, literature[203]; Butter (peanut butter), AFB_1, AFB_2, AFG_1, AFG_2, and OTA, literature[203]; Dates, AFB_1 and OTA, literature[203]; Figs, AFB_1, AFB_2, AFG_2, and OTA, literature[203]; Fruits, dried, AFB_1 and AFB_2, literature[203]; Nuts (Brazil nut, paranut), AFB_1, AFB_2, and AFG_1, literature[203]; Nuts (peanut), AFB_1 and AFB_2, literature[203]; Nuts (pistachio nut), AFB_1, AFB_2, and OTA, literature[203]

incidence: 26/31*, conc. range: ≤0.070 µg/kg, sample origin: regions around Detmold, Giessen, Hamburg, Jena, Karlsruhe, Kiel, Kulmbach, Trier (cities), Germany, sample year: unknown, country: Germany[262], *prunes, dried incidence: 26/49*, conc. range: ≤0.090 μg/kg, sample origin: regions around Detmold, Giessen, Hamburg, Jena, Karlsruhe, Kiel, Kulmbach, Trier (cities), Germany, sample year: unknown, country: Germany[262], *different kinds of dried fruits

- Co-contamination: not reported
- Further contamination of plants and plant products other than cereals and their products reported in the present book/article[262]: Bars, Chocolate, Drinks (chocolate drink), Figs, Nut crème, Nuts, Nuts (hazelnut), Nuts (peanut), Powder (cocoa powder), Pudding, Raisins, Seeds (linseed), Seeds (sesame), Seeds (sunflower), OTA, literature[262]

incidence: 0/14*, conc.: no contamination, sample origin: different markets in Campinas (city) and São Paulo (city/state), Brazil, sample year: 2002/2003, country: Brazil[331], *apricots, dried

incidence: 1/19*, conc.: 0.1–5.0 μg/kg, sample origin: different markets in Campinas (city) and São Paulo (city/state), Brazil, sample year: 2002/2003, country: Brazil[331], *plums, dried

- Co-contamination: not reported
- Further contamination of plants and plant products other than cereals and their products reported in the present book/article[331]: Dates, Figs, Sultanas, OTA, literature[331]

incidence: 6/20, conc. range: 0.46–29.9 μg/kg, sample origin: local markets and traditional family reserves, Tunisia, sample year: 2005/2006, country: Tunisia[340]

- Co-contamination: not reported
- Further contamination of plants and plant products other than cereals and their products reported in the present book/article[340]: Spices, OTA, literature[340]

incidence: 6/8*, conc. range: 0.02–0.05 μg/kg, ∅ conc.: 0.035 μg/kg, sample origin: popular markets in Rabat (capital), Morocco, sample year: unknown, country: Morocco/France[341], *prunes, dried

- Co-contamination: not reported
- Further contamination of plants and plant products other than cereals and their products reported in the present book/article[341]: Olives, Raisins, OTA, literature[341]

incidence: 8/21, conc. range: ≤2.09 μg/kg, sample origin: supermarkets and fresh provision shops, Hong Kong, China, sample year: August/September 2005, country: China[697]

- Co-contamination: not reported
- Further contamination of plants and plant products other than cereals and their products reported in the present book/article[697]: Chocolate, Coffee, Juices, Legumes, Spices, OTA, literature[697]

incidence: 2/33, conc. range: LOD/LOQ-0.9 μg/kg (1 sa.), 1.6 μg/kg (1 sa.), sample origin: France, sample year: 1998/1999, country: EU[698]

incidence: 75/114, conc. range: LOD/LOQ-0.9 μg/kg (74 sa.), 3.95 μg/kg (1 sa.), sample origin: Germany, sample year: 1995–1998, country: EU[698]

- Co-contamination: not reported
- Further contamination of plants and plant products other than cereals and their products reported in the present book/article[698]: Baby food, Beans (soybean), Chocolate, Cocoa, Coffee, Coffee beans, Currants, Herbs, Jam, Juices, Juices (grape juice), Legumes, Nuts, Oil seeds, Oils (olive oil), Peas, Peas (chickpea), Products (plant product), Raisins, Spices, Spices (cayenne pepper, chilli), Spices (nutmeg), Spices (paprika), Spices (pepper), Sultanas, Tea, Vinegar, Wine, Wine (red), Wine (rosé), Wine (white), Wine, miscellaneous, OTA, literature[698]

incidence: 3/27*, conc. range: 0.25–0.3 μg/kg**, ∅ conc.: 0.28 μg/kg**, sample origin: local and of Algeria, Chile, Iran, Israel, and Turkey purchased from markets in Valencia (city), Spain and Tunis (capital), Tunisia, sample year: 2012/2013, country: Spain[720], *apricots, dried, **purchased in Tunisia

incidence: 7/27*, conc. range: LOQ-0.2 μg/kg**, ∅ conc.: 0.16 μg/kg**, sample origin: local and of Algeria, Chile, Iran, Israel, and Turkey purchased from markets in Valencia (city), Spain, and Tunis (capital), Tunisia, sample year: 2012/2013, country: Spain[720], *plums, dried, **purchased in Spain and Tunisia

- Co-contamination: For detailed information see the article.
- Further contamination of plants and plant products other than cereals and their products reported in the present book/article[720]: Dates, AFB_2, AFG_1, AFG_2, DAS, ENA_1, ENB, ENB_1, and OTA, literature[720]; Figs, AFG_1, BEA, DAS, ENA_1, ENB, ENB_1, HT-2, and OTA, literature[720]; Fruits, dried, AFG_1, DAS, ENA_1, ENB, ENB_1, and HT-2, literature[720]; Vine fruits, dried, AFB_1, AFG_2, ENA, ENA_1, ENB, ENB_1, HT-2, and OTA, literature[720]

For detailed information see the article.

Patulin

incidence: 1/56*, conc.: 30.6 μg/kg, sample origin: principally Xinjiang (province), purchased in supermarkets and local markets in Beijing (capital), China, sample year: 2016/2017, country: China[815], *apricots, dried

incidence: 0/54*, conc.: no contamination, sample origin: principally Ningxia (province), purchased in supermarkets and local markets in Beijing (capital), China, sample year: 2016/2017, country: China[815], *wolfberries, dried

- Co-contamination: For detailed information see the article.
- Further contamination of plants and plant products other than cereals and their products reported in the present book/article[815]: Dates, MA, PA, and TA, literature[815]; Fruits, dried, AME, AOH, MA, and TA, literature[815]; Raisins, AME, AOH, MA, OTA, OTB, and TA, literature[815]; see also Dates, Fruits, dried, chapter: Further Mycotoxins and Microbial Metabolites

Fusarium Toxins

Depsipeptides

Enniatin A_1

incidence: 4/27*, conc. range: 1.6–5.7 μg/kg**, ∅ conc.: 2.9 μg/kg**, sample origin: local and of Algeria, Chile, Iran, Israel, and Turkey purchased from markets in Valencia (city), Spain, and Tunis (capital), Tunisia, sample year: 2012/2013, country: Spain[720], *apricots, dried, **purchased in Spain

incidence: 1/27*, conc.: 0.21 μg/kg**, sample origin: local and of Algeria, Chile, Iran, Israel, and Turkey purchased from markets in Valencia (city), Spain, and Tunis (capital), Tunisia, sample year: 2012/2013, country: Spain[720], *plums, dried, **purchased in Tunisia

- Co-contamination: For detailed information see the article.
- Further contamination of plants and plant products other than cereals and their products reported in the present book/article[720]: Dates, AFB_2, AFG_1, AFG_2, DAS, ENA_1, ENB, ENB_1, and OTA, literature[720]; Figs, AFG_1, BEA, DAS, ENA_1, ENB, ENB_1, HT-2, and OTA, literature[720]; Fruits, dried, AFG_1, DAS, ENB, ENB_1, HT-2, and OTA, literature[720]; Vine fruits, dried, AFB_1, AFG_2, ENA, ENA_1, ENB, ENB_1, HT-2, and OTA, literature[720]

For detailed information see the article.

Enniatin B

incidence: 8/27*, conc. range: <LOQ-795 µg/kg**, ∅ conc.: 133 µg/kg**, sample origin: local and of Algeria, Chile, Iran, Israel, and Turkey purchased from markets in Valencia (city), Spain, and Tunis (capital), Tunisia, sample year: 2012/2013, country: Spain[720], *apricots, dried, **purchased in Spain and Tunisia

incidence: 4/27*, conc. range: 0.3–1.57 µg/kg**, ∅ conc.: 0.66 µg/kg**, sample origin: local and of Algeria, Chile, Iran, Israel, and Turkey purchased from markets in Valencia (city), Spain, and Tunis (capital), Tunisia, sample year: 2012/2013, country: Spain[720], *plums, dried, **purchased in Tunisia

- Co-contamination: For detailed information see the article.
- Further contamination of plants and plant products other than cereals and their products reported in the present book/article[720]: Dates, AFB_2, AFG_1, AFG_2, DAS, ENA_1, ENB, ENB_1, and OTA, literature[720]; Figs, AFG_1, BEA, DAS, ENA_1, ENB, ENB_1, HT-2, and OTA, literature[720]; Fruits, dried, AFG_1, DAS, ENA_1, ENB_1, HT-2, and OTA, literature[720]; Vine fruits, dried, AFB_1, AFG_2, ENA, ENA_1, ENB, ENB_1, HT-2, and OTA, literature[720]

For detailed information see the article.

Enniatin B₁

incidence: 3/27*, conc. range: 19.7–57.1 µg/kg**, ∅ conc.: 34.2 µg/kg**, sample origin: local and of Algeria, Chile, Iran, Israel, and Turkey purchased from markets in Valencia (city), Spain, and Tunis (capital), Tunisia, sample year: 2012/2013, country: Spain[720], *apricots, dried, **purchased in Spain and Tunisia

incidence: 1/27*, conc.: 2.69 µg/kg**, sample origin: local and of Algeria, Chile, Iran, Israel, and Turkey purchased from markets in Valencia (city), Spain, and Tunis (capital), Tunisia, sample year: 2012/2013, country: Spain[720], *plums, dried, **purchased in Tunisia

- Co-contamination: For detailed information see the article.
- Further contamination of plants and plant products other than cereals and their products reported in the present book/article[720]: Dates, AFB_2, AFG_1, AFG_2, DAS, ENA_1, ENB, ENB_1, and OTA, literature[720]; Figs, AFG_1, BEA, DAS, ENA_1, ENB, ENB_1, HT-2, and OTA, literature[720]; Fruits, dried, AFG_1, DAS, ENA_1, ENB, HT-2, and OTA, literature[720]; Vine fruits, dried, AFB_1, AFG_2, ENA, ENA_1, ENB, ENB_1, HT-2, and OTA, literature[720]

For detailed information see the article.

Type A Trichothecenes

Diacetoxyscirpenol

incidence: 0/27*, conc.: no contamination, sample origin: local and of Algeria, Chile, Iran, Israel, and Turkey purchased from markets in Valencia (city), Spain, and Tunis (capital), Tunisia, sample year: 2012/2013, country: Spain[720], *apricots, dried

incidence: 3/27*, conc. range: LOQ-135 µg/kg**, ∅ conc.: 110 µg/kg**, sample origin: local and of Algeria, Chile, Iran, Israel, and Turkey purchased from markets in Valencia (city), Spain, and Tunis (capital), Tunisia, sample year: 2012/2013, country: Spain[720], *plums, dried, **purchased in Spain

- Co-contamination: For detailed information see the article.
- Further contamination of plants and plant products other than cereals and their products reported in the present book/article[720]: Dates, AFB_2, AFG_1, AFG_2, DAS, ENA_1, ENB, ENB_1, and OTA, literature[720]; Figs, AFG_1, BEA, DAS, ENA_1, ENB, ENB_1, HT-2, and OTA, literature[720]; Fruits, dried, AFG_1, ENA_1, ENB, ENB_1, HT-2, and OTA, literature[720]; Vine fruits, dried, AFB_1, AFG_2, ENA, ENA_1, ENB, ENB_1, HT-2, and OTA, literature[720]

For detailed information see the article.

HT-2 Toxin

incidence: 8/27*, conc. range: <LOQ-7.5 µg/kg**, ∅ conc.: 5.4 µg/kg**, sample origin: local and of Algeria, Chile, Iran, Israel, and Turkey purchased from markets in Valencia (city), Spain, and Tunis (capital), Tunisia, sample year: 2012/2013, country: Spain[720], *apricots, dried, **purchased in Spain and Tunisia

incidence: 7/27*, conc. range: LOQ-10 µg/kg**, ∅ conc.: 7.8 µg/kg**, sample origin: local and of Algeria, Chile, Iran, Israel, and Turkey purchased from markets in Valencia (city), Spain, and Tunis (capital), Tunisia, sample year: 2012/2013, country: Spain[720], *plums, dried, **purchased in Spain and Tunisia

- Co-contamination: For detailed information see the article.
- Further contamination of plants and plant products other than cereals and their products reported in the present book/article[720]: Dates, AFB_2, AFG_1, AFG_2, DAS, ENA_1, ENB, ENB_1, and OTA, literature[720]; Figs, AFG_1, BEA, DAS, ENA_1, ENB, ENB_1, HT-2, and OTA, literature[720]; Fruits, dried, AFG_1, DAS, ENA_1, ENB, ENB_1, and OTA, literature[720]; Vine fruits, dried, AFB_1, AFG_2, ENA, ENA_1, ENB, ENB_1, HT-2, and OTA, literature[720]

For detailed information see the article.

Other ***Fusarium*** Toxins

Zearalenone

incidence: 6/154, conc. range: 100–200 µg/kg (3 sa.), >200 µg/kg (3 sa.), sample origin: imported and domestic sa. from Uruguay, sample year: 1993–1995, country: Uruguay[269]

- Co-contamination: not reported
- Further contamination of plants and plant products other than cereals and their products reported in the present book/article[269]: Beans (soybean), ZEA, literature[269]; Cocoa beans, Fruits, dried, AF/S, literature[269]; Oil seeds, Vegetables, dried, AF/S and ZEA, literature[269]

Penicillium Toxins

Mycophenolic Acid

incidence: 15/56*, conc. range: 1.0–119.7 µg/kg**, ∅ conc.: 26.7 µg/kg**, sample origin: principally Xinjiang (province), purchased in supermarkets and local markets in Beijing (capital), China, sample year: 2016/2017, country: China[815], *apricots, dried

incidence: 0/54*, conc.: no contamination, sample origin: principally Ningxia (province), purchased in supermarkets and local markets in Beijing (capital), China, sample year: 2016/2017, country: China[815], *wolfberries, dried

- Co-contamination: For detailed information see the article.
- Further contamination of plants and plant products other than cereals and their products reported in the present book/article[815]: Dates, MA, PA, and TA, literature[815]; Fruits, dried, AME, AOH, PAT, and TA, literature[815]; Raisins, AME, AOH, MA, OTA, OTB, and TA, literature[815]; see also Dates, Fruits, dried, chapter: Further Mycotoxins and Microbial Metabolites

see also **Dates, Figs, Raisins,** and **Sultanas**

Fruits and nuts, dried see **Food**

Fruits and vegetables may contain the following mycotoxins:

Aspergillus Toxins

Aflatoxin B_1
incidence: 3/7, conc. range: 40–140 μg/kg*, Ø conc.: 80 μg/kg, sample origin: market in Bhagalpur? (city), India, sample year: unknown, country: India[374], **Artocarpus lakoocha* Roxb., *Momordica charantia* L., and *Psidium guajava* L.

- Co-contamination: not reported
- Further contamination of plants and plant products other than cereals and their products reported in the present book/ article[374]: not reported

Functional Food see **Food**

Gabi see **Tubers**

Garlic see **Spices (garlic)**

Garlic pickles see **Pickles**

Garnacha see **Must**

Garri see **Flakes (cassava)** and **Food**

"Gewürztraminer" see **Wine (white)**

Ginger see **Spices (ginger)**

Ginkgo may contain the following mycotoxins:

Aspergillus Toxins

Aflatoxin B_1
incidence: 2/9*, conc. range: 5–54 μg/kg, Ø conc.: 29.5 μg/kg, sample origin: USA, pharmacies in Krakow (city), Poland, and local supermarkets in Almeria (city), Spain, sample year: unknown, country: Spain[64], *ginkgo biloba nutraceutical products

- Co-contamination: not reported
- Further contamination of plants and plant products other than cereals and their products reported in the present book/ article[64]: Ginkgo, AFB_2 and T-2, literature[64]

Aflatoxin B_2
incidence: 2/9*, conc. range: 4–300 μg/kg, Ø conc.: 152 μg/kg, sample origin: USA, pharmacies in Krakow (city), Poland, and local supermarkets in Almeria (city), Spain, sample year: unknown, country: Spain[64], *ginkgo biloba nutraceutical products

- Co-contamination: not reported
- Further contamination of plants and plant products other than cereals and their products reported in the present book/ article[64]: Ginkgo, AFB_1 and T-2, literature[64]

Fusarium Toxins

Type A Trichothecenes

T-2 Toxin
incidence: 2/9*, conc. range: 18–20 μg/kg, Ø conc.: 19 μg/kg, sample origin: USA, pharmacies in Krakow (city), Poland, and local supermarkets in Almeria (city), Spain, sample year: unknown, country: Spain[64], *ginkgo biloba nutraceutical products

- Co-contamination: not reported
- Further contamination of plants and plant products other than cereals and their products reported in the present book/ article[64]: Ginkgo, AFB_1 and AFB_2, literature[64]

Ginseng see **Medicinal plant**

Glavinuša see **Must**

Goji fruits see **Fruits (goji)**

Gram see **Peas (chickpea)**

Grape extract may contain the following mycotoxins:

Aspergillus and ***Penicillium*** Toxins

Ochratoxin A
incidence: 1/1, conc.: 1.3 μg/l, sample origin: unknown, sample year: unknown, country: Italy[729]

- Co-contamination: not reported
- Further contamination of plants and plant products other than cereals and their products reported in the present book/ article[729]: not reported

Grape juice see **Juices (grape juice)**

Grape pomace and wine lees may contain the following mycotoxins:

Aspergillus and ***Penicillium*** Toxins

Ochratoxin A
incidence: 8/8*, conc. range: <LOQ-240.93 μg/kg, sample origin: processing industries in Puglia (region), Italy, sample year: unknown, country: Italy[845], *starting material for the production of ethyl alcohol, tartrates, dried grape seeds, and dried grape skins (solid sa.)

- Co-contamination: not reported
- Further contamination of plants and plant products other than cereals and their products reported in the present book/ article[845]: Food colorants, Plant food supplements, OTA, literature[845]

For detailed information see the article.

Grape pulp see **Pulp (grape pulp)**

Grapes see **Berries (grape)**

Greater cardamom see **Spices (cardamom)**

Green gram see **Products (legume product)**

Green tea see **Tea**

Grogannuts may contain the following mycotoxins:

Aspergillus Toxins

Aflatoxin B_1
incidence: 3/9, conc. range: 20–150 μg/kg, sample origin: major local retail outlets at Nainital (district), India, sample year: 1983–1985, country: India[7]

- Co-contamination: Often more than one mycotoxin was present in a sa.

- Further contamination of plants and plant products other than cereals and their products reported in the present book/article[7]: Coconuts, AFB_1, AFB_2, AFG_1, AFG_2, and CIT, literature[7]; Dates, AFB_1, AFB_2, AFG_1, and OTA, literature[7]; Grogannuts, AFB_2, AFG_1, AFG_2, and CIT, literature[7]; Nuts (almond), AFB_1, AFB_2, AFG_1, RBT, STG, and ZEA, literature[7]; Nuts (cashew nut), AFB_1, AFB_2, AFG_1, AFG_2, CIT, and OTA, literature[7]; Nuts (peanut), Raisins, AFB_1, AFB_2, AFG_1, AFG_2, and OTA, literature[7]; Nuts (pistachio nut), AFB_1, AFB_2, AFG_1, AFG_2, and STG, literature[7]; Nuts (walnut), AFB_1, AFB_2, AFG_1, AFG_2, and RBT, literature[7]

Aflatoxin B_2

incidence: 2/9, conc. range: 15–75 µg/kg, Ø conc.: 45 µg/kg, sample origin: major local retail outlets at Nainital (district), India, sample year: 1983–1985, country: India[7]

- Co-contamination: Often more than one mycotoxin was present in a sa.
- Further contamination of plants and plant products other than cereals and their products reported in the present book/article[7]: Coconuts, AFB_1, AFB_2, AFG_1, AFG_2, and CIT, literature[7]; Dates, AFB_1, AFB_2, AFG_1, and OTA, literature[7]; Grogannuts, AFB_1, AFG_1, AFG_2, and CIT, literature[7]; Nuts (almond), AFB_1, AFB_2, AFG_1, RBT, STG, and ZEA, literature[7]; Nuts (cashew nut), AFB_1, AFB_2, AFG_1, AFG_2, CIT, and OTA, literature[7]; Nuts (peanut), Raisins, AFB_1, AFB_2, AFG_1, AFG_2, and OTA, literature[7]; Nuts (pistachio nut), AFB_1, AFB_2, AFG_1, AFG_2, and STG, literature[7]; Nuts (walnut), AFB_1, AFB_2, AFG_1, AFG_2, and RBT, literature[7]

Aflatoxin G_1

incidence: 2/9, conc. range: 16–76 µg/kg, Ø conc.: 46 µg/kg, sample origin: major local retail outlets at Nainital (district), India, sample year: 1983–1985, country: India[7]

- Co-contamination: Often more than one mycotoxin was present in a sa.
- Further contamination of plants and plant products other than cereals and their products reported in the present book/article[7]: Coconuts, AFB_1, AFB_2, AFG_1, AFG_2, and CIT, literature[7]; Dates, AFB_1, AFB_2, AFG_1, and OTA, literature[7]; Grogannuts, AFB_1, AFB_2, AFG_2, and CIT, literature[7]; Nuts (almond), AFB_1, AFB_2, AFG_1, RBT, STG, and ZEA, literature[7]; Nuts (cashew nut), AFB_1, AFB_2, AFG_1, AFG_2, CIT, and OTA, literature[7]; Nuts (peanut), Raisins, AFB_1, AFB_2, AFG_1, AFG_2, and OTA, literature[7]; Nuts (pistachio nut), AFB_1, AFB_2, AFG_1, AFG_2, and STG, literature[7]; Nuts (walnut), AFB_1, AFB_2, AFG_1, AFG_2, and RBT, literature[7]

Aflatoxin G_2

incidence: 1/9, conc.: 10 µg/kg, sample origin: major local retail outlets at Nainital (district), India, sample year: 1983–1985, country: India[7]

- Co-contamination: Often more than one mycotoxin was present in a sa.
- Further contamination of plants and plant products other than cereals and their products reported in the present book/article[7]: Coconuts, AFB_1, AFB_2, AFG_1, AFG_2, and CIT, literature[7]; Dates, AFB_1, AFB_2, AFG_1, and OTA, literature[7]; Grogannuts, AFB_1, AFB_2, AFG_1, and CIT, literature[7]; Nuts (almond), AFB_1, AFB_2, AFG_1, RBT, STG, and ZEA, literature[7]; Nuts (cashew nut), AFB_1, AFB_2, AFG_1, AFG_2, CIT, and OTA, literature[7]; Nuts (peanut), Raisins, AFB_1, AFB_2, AFG_1, AFG_2, and OTA, literature[7]; Nuts (pistachio nut), AFB_1, AFB_2, AFG_1, AFG_2, and STG, literature[7]; Nuts (walnut), AFB_1, AFB_2, AFG_1, AFG_2, and RBT, literature[7]

Aspergillus and ***Penicillium*** Toxins

Citrinin

incidence: 2/9, conc. range: 4–20 µg/kg, Ø conc.: 12 µg/kg, sample origin: major local retail outlets at Nainital (district), India, sample year: 1983–1985, country: India[7]

- Co-contamination: Often more than one mycotoxin was present in a sa.
- Further contamination of plants and plant products other than cereals and their products reported in the present book/article[7]: Coconuts, AFB_1, AFB_2, AFG_1, AFG_2, and CIT, literature[7]; Dates, AFB_1, AFB_2, AFG_1, and OTA, literature[7]; Grogannuts, AFB_1, AFB_2, AFG_1, and AFG_2, literature[7]; Nuts (almond), AFB_1, AFB_2, AFG_1, RBT, STG, and ZEA, literature[7]; Nuts (cashew nut), AFB_1, AFB_2, AFG_1, AFG_2, CIT, and OTA, literature[7]; Nuts (peanut), Raisins, AFB_1, AFB_2, AFG_1, AFG_2, and OTA, literature[7]; Nuts (pistachio nut), AFB_1, AFB_2, AFG_1, AFG_2, and STG, literature[7]; Nuts (walnut), AFB_1, AFB_2, AFG_1, AFG_2, and RBT, literature[7]

Groundnut sauces see **Sauces (peanut sauce)**

Groundnuts see **Nuts (peanut)**

Gums, food-grade may contain the following mycotoxins:

Aspergillus Toxins

Aflatoxin B_1

incidence: 2/20, conc. range: <LOQ*, sample origin: market sa., USA, sample year: unknown, country: USA[896], *Guar gum and Tragacanth gum

- Co-contamination: not reported
- Further contamination of plants and plant products other than cereals and their products reported in the present book/article[896]: Gums, food grade, ZEA, literature[896]

Fusarium Toxins

Zearalenone

incidence: 1/20, conc.: 6 µg/kg*, sample origin: market sa., USA, sample year: unknown, country: USA[896], *xanthan gum

- Co-contamination: not reported
- Further contamination of plants and plant products other than cereals and their products reported in the present book/article[896]: Gums, food grade, AFB_1, literature[896]

Halva may contain the following mycotoxins:

Aspergillus Toxins

Aflatoxin B_1

incidence: 0/34*, conc.: no contamination, sample origin: halva-factories and supermarkets in Adana (city), Turkey, sample year: summer and winter 2004, country: Turkey[437], *halva

incidence: 0/34*, conc.: no contamination, sample origin: halva-factories and supermarkets in Adana (city), Turkey, sample year: summer and winter 2004, country: Turkey[437], *halva with cacao

incidence: 8/34*, conc. range: 1.5–18 µg/kg, Ø conc.: 7.5 µg/kg, sample origin: halva-factories and supermarkets in Adana (city), Turkey, sample year: summer and winter 2004, country: Turkey[437], *halva with pistachio

- Co-contamination: not reported

- Further contamination of plants and plant products other than cereals and their products reported in the present book/article[437]: not reported

incidence: 12/22*, conc. range: ≤5.94 µg/kg, sample origin: retail outlets, cafeterias, and hotels from major districts of Punjab (state), Pakistan, sample year: unknown, country: Pakistan[750], *halva without pistachio and almonds

incidence: 20/34*, conc. range: ≤19.56 µg/kg, sample origin: retail outlets, cafeterias, and hotels from major districts of Punjab (state), Pakistan, sample year: unknown, country: Pakistan[750], *halva with pistachio, almonds, and milk

incidence: 21/39*, conc. range: ≤19.80 µg/kg, sample origin: retail outlets, cafeterias, and hotels from major districts of Punjab (state), Pakistan, sample year: unknown, country: Pakistan[750], *halva puri

- Co-contamination: not reported
- Further contamination of plants and plant products other than cereals and their products reported in the present book/article[750]: Halva, AFM_1, literature[750]; Nuts (almond), Nuts (pistachio nut), Raisins, Spices (cardamom), AFB_1, literature[750]

Aflatoxin M_1

incidence: 0/22*, conc.: no contamination, sample origin: retail outlets, cafeterias, and hotels from major districts of Punjab (state), Pakistan, sample year: unknown, country: Pakistan[750], *halva without pistachio and almonds

incidence: 15/34*, conc. range: ≤0.08 µg/kg, sample origin: retail outlets, cafeterias, and hotels from major districts of Punjab (state), Pakistan, sample year: unknown, country: Pakistan[750], *halva with pistachio, almonds, and milk

- Co-contamination: not reported
- Further contamination of plants and plant products other than cereals and their products reported in the present book/article[750]: Halva, Nuts (almond), Nuts (pistachio nut), Raisins, Spices (cardamom), AFB_1, literature[750]

Aflatoxin/s

incidence: 9/16*, conc. range: ≤16.70 µg/kg**, sample origin: markets and local shops in big cities of Punjab (province), Pakistan, sample year: November 2011–March 2012, country: Pakistan/Malaysia[574], *date halva, **AFB_1, AFB_2, AFG_1, and AFG_2

- Co-contamination: not reported
- Further contamination of plants and plant products other than cereals and their products reported in the present book/article[574]: Cake (date), Cookies, Dates, AF/S, literature[574]

Halva is a traditional Turkish food (mixture of tahin: milled dehulled roasted sesame seeds, sugar, citric or tartaric acid, and *Saponaria officinalis* root extract).

Hazelnuts see Nuts (hazelnut)

Helva see Halva

Herbal drugs see Medicinal plants

Herbal medicines see Medicinal plants

Herbs may contain the following mycotoxins:

Aspergillus and *Penicillium* Toxins

Ochratoxin A

incidence: 1/1*, conc.: 0.08 µg/kg, sample origin: supermarkets and fast-food chains in Quebec (city), Canada, sample year: September/October 2008, country: Canada[90], *herbs and spices

incidence: 1/1*, conc.: 0.07 µg/kg, sample origin: supermarkets and fast-food chains in Calgary (city), Canada, sample year: September/October 2009, country: Canada[90], *herbs and spices

- Co-contamination: not reported
- Further contamination of plants and plant products other than cereals and their products reported in the present book/article[90]: Baby food, Beans, Beverages, Butter (peanut butter), Candies, Chocolate, Coffee, Extracts, Ice cream, Nuts, Pies, Pizza, Products, Pudding, Raisins, Seeds, Wine, OTA, literature[90]

incidence: 2/13, conc. range: LOD/LOQ-≤0.11 µg/kg (2 sa.), sample origin: Germany, sample year: unknown, country: EU[698]

- Co-contamination: not reported
- Further contamination of plants and plant products other than cereals and their products reported in the present book/article[698]: Baby food, Beans (soybean), Chocolate, Cocoa, Coffee, Coffee beans, Currants, Fruits, dried, Jam, Juices, Juices (grape juice), Legumes, Nuts, Oil seeds, Oils (olive oil), Peas, Peas (chickpea), Products (plant product), Raisins, Spices, Spices (cayenne pepper, chilli), Spices (nutmeg), Spices (paprika), Spices (pepper), Sultanas, Tea, Vinegar, Wine, Wine (red), Wine (rosé), Wine (white), Wine, miscellaneous, OTA, literature[698]

Fusarium Toxins

Fumonisin/s

Fumonisin B_1

incidence: 0/2*, conc.: no contamination, sample origin: different markets, China, sample year: 2009/2010, country: China[82], *moldy aromatic herbs

incidence: 1/9*, conc.: 72.2 µg/kg, sample origin: different markets, China, sample year: 2009/2010, country: China[82], *non-moldy aromatic herbs

- Co-contamination: not reported
- Further contamination of plants and plant products other than cereals and their products reported in the present book/article[82]: Herbs, FB_2, literature[82]; Medicinal plants, FB/S, literature[82]

For detailed information see the article.

Fumonisin B_2

incidence: 1/2*, conc.: 139.7 µg/kg, sample origin: different markets, China, sample year: 2009/2010, country: China[82], *moldy aromatic herbs

incidence: 1/9*, conc.: 449.3 µg/kg, sample origin: different markets, China, sample year: 2009/2010, country: China[82], *non-moldy aromatic herbs

- Co-contamination: not reported
- Further contamination of plants and plant products other than cereals and their products reported in the present book/article[82]: Herbs, FB_1, literature[82]; Medicinal plants, FB/S, literature[82]

For detailed information see the article.

see also **Herbs and Spices, Medicinal plants,** and **Spices**

Herbs (camomile) may contain the following mycotoxins:

Aspergillus Toxins

Aflatoxin B_1

incidence: 7/7, conc. range: 2.0–16.6 μg/kg, sample origin: retail sales points in different regions, Iran, sample year: March–May 2015, country: Iran[491]

- Co-contamination: not reported
- Further contamination of plants and plant products other than cereals and their products reported in the present book/article[491]: Herbs (sage), AFB_1, literature[491]; Spices (cayenne pepper, chilli), Spices (curcuma), Spices (ginger), Spices (pepper), Spices (thyme), AFB_1, literature[491]

incidence: 10/10*, conc. range: 3.4–38.9 μg/kg, ∅ conc.: 28.7 μg/kg, sample origin: organic markets and organic shops, Turkey, sample year: May 2010–May 2011, country: Turkey[859], *organic camomile sa.

- Co-contamination: not reported
- Further contamination of plants and plant products other than cereals and their products reported in the present book/article[859]: Herbs (linden), Herbs (rosehip), Herbs (sage), Spices (anise), Spices (basil), Spices (cayenne pepper, chilli), Spices (cinnamon), Spices (coriander), Spices (cumin), Spices (fennel), Spices (ginger), Spices (laurel leaf), Spices (pepper), Spices (peppermint), Spices (poppy seed), Spices (rosemary), Spices (sumac), AFB_1, literature[859]

Herbs (linden) may contain the following mycotoxins:

Aspergillus Toxins

Aflatoxin B_1

incidence: 3/5*, conc. range: 0.05–40.6 μg/kg, ∅ conc.: 14 μg/kg, sample origin: organic markets and organic shops, Turkey, sample year: May 2010–May 2011, country: Turkey[859], *organic linden sa.

- Co-contamination: not reported
- Further contamination of plants and plant products other than cereals and their products reported in the present book/article[859]: Herbs (camomile), Herbs (rosehip), Herbs (sage), Spices (anise), Spices (basil), Spices (cayenne pepper, chilli), Spices (cinnamon), Spices (coriander), Spices (cumin), Spices (fennel), Spices (ginger), Spices (laurel leaf), Spices (pepper), Spices (peppermint), Spices (poppy seed), Spices (rosemary), Spices (sumac), AFB_1, literature[859]

Herbs (rosehip) may contain the following mycotoxins:

Aspergillus Toxins

Aflatoxin B_1

incidence: 4/6*, conc. range: 20.7–52.5 μg/kg, ∅ conc.: 44.5 μg/kg, sample origin: organic markets and organic shops, Turkey, sample year: May 2010–May 2011, country: Turkey[859], *organic rosehip sa.

- Co-contamination: not reported
- Further contamination of plants and plant products other than cereals and their products reported in the present book/article[859]: Herbs (camomile), Herbs (linden), Herbs (sage), Spices (anise), Spices (basil), Spices (cayenne pepper, chilli), Spices (cinnamon), Spices (coriander), Spices (cumin), Spices (fennel), Spices (ginger), Spices (laurel leaf), Spices (pepper), Spices (peppermint), Spices (poppy seed), Spices (rosemary), Spices (sumac), AFB_1, literature[859]

Herbs (sage) may contain the following mycotoxins:

Aspergillus Toxins

Aflatoxin B_1

incidence: 1/2, conc.: 2.5 μg/kg, sample origin: retail sales points in different regions, Iran, sample year: March–May 2015, country: Iran[491]

- Co-contamination: not reported
- Further contamination of plants and plant products other than cereals and their products reported in the present book/article[491]: Herbs (camomile), AFB_1, literature[491]; Spices (cayenne pepper, chilli), Spices (curcuma), Spices (ginger), Spices (pepper), Spices (thyme), AFB_1, literature[491]

incidence: 9/9*, conc. range: 0.2–32.2 μg/kg, ∅ conc.: 8.9 μg/kg, sample origin: organic markets and organic shops, Turkey, sample year: May 2010–May 2011, country: Turkey[859], *organic sage sa.

- Co-contamination: not reported
- Further contamination of plants and plant products other than cereals and their products reported in the present book/article[859]: Herbs (camomile), Herbs (linden), Herbs (rosehip), Spices (anise), Spices (basil), Spices (cayenne pepper, chilli), Spices (cinnamon), Spices (coriander), Spices (cumin), Spices (fennel), Spices (ginger), Spices (laurel leaf), Spices (pepper), Spices (peppermint), Spices (poppy seed), Spices (rosemary), Spices (sumac), AFB_1, literature[859]

Herbs and spices may contain the following mycotoxins:

Aspergillus and ***Penicillium*** Toxins

Ochratoxin A

incidence: 8/12*, conc. range: ≤9.40 μg/kg, sample origin: 12 regions, Czech Republic, sample year: 2011–2013, country: Czech Republic[811], *mixed herbs and spices

incidence: 1/12*, conc.: 0.40 μg/kg, sample origin: 12 regions, Czech Republic, sample year: 2011–2013, country: Czech Republic[811], *mixed herbs and spices

- Co-contamination: not reported
- Further contamination of plants and plant products other than cereals and their products reported in the present book/article[811]: Beans, Chocolate, Coffee, Confectionery, Figs, Juices, Licorice, Powder (cocoa powder), Products (ginger product), Raisins, Spices (caraway), Spices (cayenne pepper, chilli), Spices (coriander), Spices (ginger), Spices (mustard), Spices (nutmeg), Spices (paprika), Spices (pepper), Tea, Wine (red), Wine (white), OTA, literature[811]

Honey may contain the following mycotoxins:

Aspergillus Toxins

Aflatoxin/s

incidence: 21/21*, conc. range: 0.5–22 μg/kg**, ∅ conc.: 12.1 μg/kg**, sample origin: different regions of the West Bank, Palestine, sample year: April–August 2009, country: Palestine[851], *fresh honey, **AFS, total

- Co-contamination: not reported
- Further contamination of plants and plant products other than cereals and their products reported in the present book/article[851]: not reported

Ice cream may contain the following mycotoxins:

Aspergillus **Toxins**

Aflatoxin/s

incidence: 5/6, conc. range: ≤3.677 µg/kg*, sample origin: groceries and supermarkets, Turkey, sample year: 2003, country: Turkey[119], *AFS total

- Co-contamination: not reported
- Further contamination of plants and plant products other than cereals and their products reported in the present book/article[119]: Asure, Beverages, Cocoa, Cream, Jello, Keskul, Muhallebi, Pudding, Salep, Sauces (chocolate sauce), Soups, AF/S, literature[119]

Ice cream ingredients were sugar, glucose, emulgator, milk protein, stabilizer, fruit flavor, maltodextrin, thickener agent, and sweetener.

Aspergillus **and** ***Penicillium*** **Toxins**

Ochratoxin A

incidence: 1/1, conc.: 0.02 µg/l, sample origin: supermarkets and fast-food chains in Quebec (city), Canada, sample year: September/October 2008, country: Canada[90]

incidence: 1/1, conc.: 0.03 µg/l, sample origin: supermarkets and fast-food chains in Calgary (city), Canada, sample year: September/October 2009, country: Canada[90]

- Co-contamination: not reported
- Further contamination of plants and plant products other than cereals and their products reported in the present book/article[90]: Baby food, Beans, Beverages, Butter (peanut butter), Candies, Chocolate, Coffee, Extracts, Herbs, Nuts, Pies, Pizza, Products, Pudding, Raisins, Seeds, Wine, OTA, literature[90]

Infant food may contain the following mycotoxins:

Aspergillus **Toxins**

Aflatoxin B_1

incidence: 30/30*, conc. range: 0.004–0.161 µg/kg, sample origin: retail level across Canada, sample year: 2002–2005, country: Canada[92], *soy-based infant food

- Co-contamination: not reported
- Further contamination of plants and plant products other than cereals and their products reported in the present book/article[92]: Infant food, AFB_2, literature[92]

incidence: 57/91, conc. range: <LOD (15 sa.), ≤3.11* µg/kg (42 sa.), sample origin: households or purchased in pharmacies and specialized organic feeding shops in Pamplona (city), Navarra (chartered community), Spain, sample year: unknown, country: Spain[98], *outlier

- Co-contamination: not reported
- Further contamination of plants and plant products other than cereals and their products reported in the present book/article[98]: Infant food, AFB_2, AFG_1, and AFG_2, literature[98]

For detailed information see the article.

incidence: 19/49*, conc. range: ≤29.0 µg/kg, sample origin: Ouagadougou (capital), Burkina Faso, sample year: June 2013–December 2014, country: Burkina Faso/France[913], *included flours, oilseeds, and various cereals

- Co-contamination: For detailed information see the article.
- Further contamination of plants and plant products other than cereals and their products reported in the present book/article[913]: Infant food, AF/S, FB/S, and OTA, literature[913]

Aflatoxin B_2

incidence: 20/30*, conc. range: 0.002–0.019 µg/kg, sample origin: retail level across Canada, sample year: 2002–2005, country: Canada[92], *soy-based infant food

- Co-contamination: not reported
- Further contamination of plants and plant products other than cereals and their products reported in the present book/article[92]: Infant food, AFB_1, literature[92]

incidence: 36/91, conc. range: ≤0.41* µg/kg (36 sa.), sample origin: households or purchased in pharmacies and specialized organic feeding shops in Pamplona (city), Navarra (chartered community), Spain, sample year: unknown, country: Spain[98], *outlier

- Co-contamination: not reported
- Further contamination of plants and plant products other than cereals and their products reported in the present book/article[98]: Infant food, AFB_1, AFG_1, and AFG_2, literature[98]

For detailed information see the article.

Aflatoxin G_1

incidence: 60/91, conc. range: <LOD (29 sa.), ≤0.42* µg/kg (31 sa.), sample origin: households or purchased in pharmacies and specialized organic feeding shops in Pamplona (city), Navarra (chartered community), Spain, sample year: unknown, country: Spain[98], *outlier

- Co-contamination: not reported
- Further contamination of plants and plant products other than cereals and their products reported in the present book/article[98]: Infant food, AFB_1, AFB_2, and AFG_2, literature[98]

For detailed information see the article.

Aflatoxin G_2

incidence: 39/91, conc. range: <LOD (29 sa.), ≤0.07* µg/kg (10 sa.), sample origin: households or purchased in pharmacies and specialized organic feeding shops in Pamplona (city), Navarra (chartered community), Spain, sample year: unknown, country: Spain[98], *outlier

- Co-contamination: not reported
- Further contamination of plants and plant products other than cereals and their products reported in the present book/article[98]: Infant food, AFB_1, AFB_2, and AFG_1, literature[98]

For detailed information see the article.

Aflatoxin/s

incidence: 17/49*, conc. range: ≤45.6 µg/kg, sample origin: Ouagadougou (capital), Burkina Faso, sample year: June 2013–December 2014, country: Burkina Faso/France[913], *included flours, oilseeds, and various cereals

- Co-contamination: For detailed information see the article.
- Further contamination of plants and plant products other than cereals and their products reported in the present book/article[913]: Infant food, AFB_1, FB/S, and OTA, literature[913]

Aspergillus **and** ***Penicillium*** **Toxins**

Ochratoxin A

incidence: 7/22*, conc. range: ≤0.9 µg/kg, ∅ conc.: 0.47 µg/kg, sample origin: retail outlets across Canada, sample year: 1997–1999, country: Canada[123], *soy-based infant cereal food

- Co-contamination: not reported
- Further contamination of plants and plant products other than cereals and their products reported in the present book/article[123]: Infant food, DON, FB/S, and ZEA, literature[123]

incidence: ?/49*, conc. range: ≤0.2 µg/kg, sample origin: Ouagadougou (capital), Burkina Faso, sample year: June 2013–December 2014, country: Burkina Faso/France[913], *included flours, oilseeds, and various cereals

- Co-contamination: For detailed information see the article.
- Further contamination of plants and plant products other than cereals and their products reported in the present book/article[913]: Infant food, AFB_1, AF/S, and FB/S, literature[913]

Fusarium Toxins

Fumonisin/s

incidence: 18/24*, conc. range: ≤130 µg/kg**, ∅ conc.: 44 µg/kg**, sample origin: retail outlets across Canada, sample year: 1997–1999, country: Canada[123], *soy-based infant cereal food, **FB_1 and FB_2

- Co-contamination: not reported
- Further contamination of plants and plant products other than cereals and their products reported in the present book/article[123]: Infant food, DON, OTA, and ZEA, literature[123]

incidence: ?/49*, conc. range: ≤138.0 µg/kg**, sample origin: Ouagadougou (capital), Burkina Faso, sample year: June 2013–December 2014, country: Burkina Faso/France[913], *included flours, oilseeds, and various cereals, **FB_1 and FB_2

- Co-contamination: For detailed information see the article.
- Further contamination of plants and plant products other than cereals and their products reported in the present book/article[913]: Infant food, AFB_1, AF/S, and OTA, literature[913]

Type B Trichothecenes

Deoxynivalenol

incidence: 8/8*, conc. range: ≤240 µg/kg, ∅ conc.: 116 µg/kg, sample origin: retail outlets across Canada, sample year: 1997–1999, country: Canada[123], *soy-based infant cereal food

- Co-contamination: not reported
- Further contamination of plants and plant products other than cereals and their products reported in the present book/article[123]: Infant food, FBS, OTA, and ZEA, literature[123]

Other ***Fusarium*** Toxins

Zearalenone

incidence: 23/30*, conc. range: ≤35 µg/kg, ∅ conc.: 18.4 µg/kg, sample origin: retail outlets across Canada, sample year: 1997–1999, country: Canada[123], *soy-based infant cereal food

- Co-contamination: not reported
- Further contamination of plants and plant products other than cereals and their products reported in the present book/article[123]: Infant food, DON, FBS, and OTA, literature[123]

see also **Baby food**

Infant formula see **Infant food**

Instant coffee see **Coffee**

Iru may contain the following mycotoxins:

Alternaria Toxins

Alternariol Methyl Ether

incidence: 6/60*, conc. range: 19–77 µg/kg, ∅ conc.: 38 µg/kg, sample origin: fermented food sellers, Southwest Nigeria, sample year: February 2015–July 2016, country: South Africa/Nigeria/Belgium/Kenya[330], *fermented locust bean

- Co-contamination: For detailed information see the article.
- Further contamination of plants and plant products other than cereals and their products reported in the present book/article[330]: Iru, AFB_1, AFG_1, AFG_1, AFG_2, DON, 3-AcDON, FB_1, FB_2, FB_3, FUS-X, HT-2, OTA, ROQ-C, T-2, and ZEA, literature[330]; Ogiri, AFB_1, AFB_2, AFG_2, AME, DAS, DON, HT-2, NEO, OTA, STG, and ZEA, literature[330]; Ugba, AFB_1, AFB_2, AME, DON, ENB, HT-2, OTA, STG, T-2, and ZEA, literature[330]

Aspergillus Toxins

Aflatoxin B_1

incidence: 1/60*, conc.: 6 µg/kg, sample origin: fermented food sellers, Southwest Nigeria, sample year: February 2015–July 2016, country: South Africa/Nigeria/Belgium/Kenya[330], *fermented locust bean

- Co-contamination: For detailed information see the article.
- Further contamination of plants and plant products other than cereals and their products reported in the present book/article[330]: Iru, AFB_2, AFG_1, AFG_2, AME, DON, 3-AcDON, FB_1, FB_2, FB_3, FUS-X, HT-2, OTA, ROQ-C, T-2, and ZEA, literature[330]; Ogiri, AFB_1, AFB_2, AFG_2, AME, DAS, DON, HT-2, NEO, OTA, STG, and ZEA, literature[330]; Ugba, AFB_1, AFB_2, AME, DON, ENB, HT-2, OTA, STG, T-2, and ZEA, literature[330]

Aflatoxin B_2

incidence: 1/60*, conc.: <LOQ, sample origin: fermented food sellers, Southwest Nigeria, sample year: February 2015–July 2016, country: South Africa/Nigeria/Belgium/Kenya[330], *fermented locust bean

- Co-contamination: For detailed information see the article.
- Further contamination of plants and plant products other than cereals and their products reported in the present book/article[330]: Iru, AFB_1, AFG_1, AFG_2, AME, DON, 3-AcDON, FB_1, FB_2, FB_3, FUS-X, HT-2, OTA, ROQ-C, T-2, and ZEA, literature[330]; Ogiri, AFB_1, AFB_2, AFG_2, AME, DAS, DON, HT-2, NEO, OTA, STG, and ZEA, literature[330]; Ugba, AFB_1, AFB_2, AME, DON, ENB, HT-2, OTA, STG, T-2, and ZEA, literature[330]

Aflatoxin G_1

incidence: 2/60*, conc. range: 8 µg/kg, ∅ conc.: 8 µg/kg, sample origin: fermented food sellers, Southwest Nigeria, sample year: February 2015–July 2016, country: South Africa/Nigeria/Belgium/Kenya[330], *fermented locust bean

- Co-contamination: For detailed information see the article.
- Further contamination of plants and plant products other than cereals and their products reported in the present book/article[330]: Iru, AFB_1, AFB_2, AFG_2, AME, DON, 3-AcDON, FB_1, FB_2, FB_3, FUS-X, HT-2, OTA, ROQ-C, T-2, and ZEA, literature[330]; Ogiri, AFB_1, AFB_2, AFG_2, AME, DAS, DON, HT-2, NEO, OTA, STG, and ZEA, literature[330]; Ugba, AFB_1, AFB_2, AME, DON, ENB, HT-2, OTA, STG, T-2, and ZEA, literature[330]

Aflatoxin G_2

incidence: 3/60*, conc. range: 3–6 µg/kg, ∅ conc.: 4 µg/kg, sample origin: fermented food sellers, Southwest Nigeria, sample year: February 2015–July 2016, country: South Africa/Nigeria/Belgium/Kenya[330], *fermented locust bean

- Co-contamination: For detailed information see the article.
- Further contamination of plants and plant products other than cereals and their products reported in the present book/article[330]: Iru, AFB_1, AFG_1, AFG_1, AME, DON, 3-AcDON, FB_1, FB_2, FB_3, FUS-X, HT-2, OTA, ROQ-C, T-2, and ZEA, literature[330]; Ogiri, AFB_1, AFB_2, AFG_2, AME, DAS, DON, HT-2, NEO, OTA,

STG, and ZEA, literature[330]; Ugba, AFB_1, AFB_2, AME, DON, ENB, HT-2, OTA, STG, T-2, and ZEA, literature[330]

Other ***Aspergillus*** Toxins

Sterigmatocystin

incidence: 1/9, conc.: 1.71 µg/kg, sample origin: markets in Lagos, Ogun, and Oyo (states), Southwest Nigeria, sample year: March 2016, country: Nigeria/South Africa/Austria/Ireland[361]

- Co-contamination: not reported
- Further contamination of plants and plant products other than cereals and their products reported in the present book/article[361]: Beans (locust bean), 3-NPA. BEA, and CIT, literature[361]; Iru, CIT and OTA, literature[361]; Ogiri, 3-NPA, AFB_1, BEA, CIT, and STG, literature[361]; Seeds (melon), 3-NPA, AFB_1, AFB_2, BEA, and STG, literature[361]

Aspergillus and ***Penicillium*** Toxins

Citrinin

incidence: 1/9, conc.: 2.23 µg/kg, sample origin: markets in Lagos, Ogun, and Oyo (states), Southwest Nigeria, sample year: March 2016, country: Nigeria/South Africa/Austria/Ireland[361]

- Co-contamination: not reported
- Further contamination of plants and plant products other than cereals and their products reported in the present book/article[361]: Beans (locust bean), 3-NPA. BEA, and CIT, literature[361]; Iru, OTA and STG, literature[361]; Ogiri, 3-NPA, AFB_1, BEA, CIT, and STG, literature[361]; Seeds (melon), 3-NPA, AFB_1, AFB_2, BEA, and STG, literature[361]

Ochratoxin A

incidence: 7/60*, conc. range: ≤21 µg/kg, Ø conc.: 9 µg/kg, sample origin: fermented food sellers, Southwest Nigeria, sample year: February 2015–July 2016, country: South Africa/Nigeria/Belgium/Kenya[330], *fermented locust bean

- Co-contamination: For detailed information see the article.
- Further contamination of plants and plant products other than cereals and their products reported in the present book/article[330]: Iru, AFB_1, AFG_1, AFG_1, AFG_2, AME, DON, 3-AcDON, FB_1, FB_2, FB_3, FUS-X, HT-2, ROQ-C, T-2, and ZEA, literature[330]; Ogiri, AFB_1, AFB_2, AFG_2, AME, DAS, DON, HT-2, NEO, OTA, STG, and ZEA, literature[330]; Ugba, AFB_1, AFB_2, AME, DON, ENB, HT-2, OTA, STG, T-2, and ZEA, literature[330]

incidence: 1/9, conc.: 7.04 µg/kg, sample origin: markets in Lagos, Ogun, and Oyo (states), Southwest Nigeria, sample year: March 2016, country: Nigeria/South Africa/Austria/Ireland[361]

- Co-contamination: not reported
- Further contamination of plants and plant products other than cereals and their products reported in the present book/article[361]: Beans (locust bean), 3-NPA. BEA, and CIT, literature[361]; Iru, CIT and STG, literature[361]; Ogiri, 3-NPA, AFB_1, BEA, CIT, and STG, literature[361]; Seeds (melon), 3-NPA, AFB_1, AFB_2, BEA, and STG, literature[361]

Fusarium Toxins

Fumonisin/s

Fumonisin B_1

incidence: 4/60*, conc. range: 61–167 µg/kg, Ø conc.: 113 µg/kg, sample origin: fermented food sellers, Southwest Nigeria, sample year: February 2015–July 2016, country: South Africa/Nigeria/Belgium/Kenya[330], *fermented locust bean

- Co-contamination: For detailed information see the article.
- Further contamination of plants and plant products other than cereals and their products reported in the present book/article[330]: Iru, AFB_1, AFG_1, AFG_1, AFG_2, AME, DON, 3-AcDON, FB_2, FB_3, FUS-X, HT-2, OTA, ROQ-C, T-2, and ZEA, literature[330]; Ogiri, AFB_1, AFB_2, AFG_2, AME, DAS, DON, HT-2, NEO, OTA, STG, and ZEA, literature[330]; Ugba, AFB_1, AFB_2, AME, DON, ENB, HT-2, OTA, STG, T-2, and ZEA, literature[330]

Fumonisin B_2

incidence: 4/60*, conc. range: 32–42 µg/kg, Ø conc.: 38 µg/kg, sample origin: fermented food sellers, Southwest Nigeria, sample year: February 2015–July 2016, country: South Africa/Nigeria/Belgium/Kenya[330], *fermented locust bean

- Co-contamination: For detailed information see the article.
- Further contamination of plants and plant products other than cereals and their products reported in the present book/article[330]: Iru, AFB_1, AFG_1, AFG_1, AFG_2, AME, DON, 3-AcDON, FB_1, FB_3, FUS-X, HT-2, OTA, ROQ-C, T-2, and ZEA, literature[330]; Ogiri, AFB_1, AFB_2, AFG_2, AME, DAS, DON, HT-2, NEO, OTA, STG, and ZEA, literature[330]; Ugba, AFB_1, AFB_2, AME, DON, ENB, HT-2, OTA, STG, T-2, and ZEA, literature[330]

Fumonisin B_3

incidence: 4/60*, conc. range: 76–89 µg/kg, Ø conc.: 84 µg/kg, sample origin: fermented food sellers, Southwest Nigeria, sample year: February 2015–July 2016, country: South Africa/Nigeria/Belgium/Kenya[330], *fermented locust bean

- Co-contamination: For detailed information see the article.
- Further contamination of plants and plant products other than cereals and their products reported in the present book/article[330]: Iru, AFB_1, AFG_1, AFG_1, AFG_2, AME, DON, 3-AcDON, FB_1, FB_2, FUS-X, HT-2, OTA, ROQ-C, T-2, and ZEA, literature[330]; Ogiri, AFB_1, AFB_2, AFG_2, AME, DAS, DON, HT-2, NEO, OTA, STG, and ZEA, literature[330]; Ugba, AFB_1, AFB_2, AME, DON, ENB, HT-2, OTA, STG, T-2, and ZEA, literature[330]

Type A Trichothecenes

HT-2 Toxin

incidence: 9/60*, conc. range: 17–51 µg/kg, Ø conc.: 33 µg/kg, sample origin: fermented food sellers, Southwest Nigeria, sample year: February 2015–July 2016, country: South Africa/Nigeria/Belgium/Kenya[330], *fermented locust bean

- Co-contamination: For detailed information see the article.
- Further contamination of plants and plant products other than cereals and their products reported in the present book/article[330]: Iru, AFB_1, AFG_1, AFG_1, AFG_2, AME, DON, 3-AcDON, FB_1, FB_2, FB_3, FUS-X, OTA, ROQ-C, T-2, and ZEA, literature[330]; Ogiri, AFB_1, AFB_2, AFG_2, AME, DAS, DON, HT-2, NEO, OTA, STG, and ZEA, literature[330]; Ugba, AFB_1, AFB_2, AME, DON, ENB, HT-2, OTA, STG, T-2, and ZEA, literature[330]

T-2 Toxin

incidence: 7/60*, conc. range: 28–31 µg/kg, Ø conc.: 29 µg/kg, sample origin: fermented food sellers, Southwest Nigeria, sample year: February 2015–July 2016, country: South Africa/Nigeria/Belgium/Kenya[330], *fermented locust bean

- Co-contamination: For detailed information see the article.
- Further contamination of plants and plant products other than cereals and their products reported in the present book/article[330]: Iru, AFB_1, AFG_1, AFG_1, AFG_2, AME, DON, 3-AcDON,

FB_1, FB_2, FB_3, FUS-X, HT-2, OTA, ROQ-C, and ZEA, literature[330]; Ogiri, AFB_1, AFB_2, AFG_2, AME, DAS, DON, HT-2, NEO, OTA, STG, and ZEA, literature[330]; Ugba, AFB_1, AFB_2, AME, DON, ENB, HT-2, OTA, STG, T-2, and ZEA, literature[330]

Type B Trichothecenes

Deoxynivalenol

incidence: 4/60*, conc. range: ≤118 μg/kg, ∅ conc.: 62 μg/kg, sample origin: fermented food sellers, Southwest Nigeria, sample year: February 2015–July 2016, country: South Africa/Nigeria/Belgium/Kenya[330], *fermented locust bean

- Co-contamination: For detailed information see the article.
- Further contamination of plants and plant products other than cereals and their products reported in the present book/article[330]: Iru, AFB_1, AFB_2, AFG_1, AFG_2, AME, 3-AcDON, FB_1, FB_2, FB_3, FUS-X, HT-2, OTA, ROQ-C, T-2, and ZEA, literature[330]; Ogiri, AFB_1, AFB_2, AFG_2, AME, DAS, DON, HT-2, NEO, OTA, STG, and ZEA, literature[330]; Ugba, AFB_1, AFB_2, AME, DON, ENB, HT-2, OTA, STG, T-2, and ZEA, literature[330]

3-Acetyldeoxynivalenol

incidence: 1/60*, conc.: ? μg/kg, sample origin: fermented food sellers, Southwest Nigeria, sample year: February 2015–July 2016, country: South Africa/Nigeria/Belgium/Kenya[330], *fermented locust bean

- Co-contamination: For detailed information see the article.
- Further contamination of plants and plant products other than cereals and their products reported in the present book/article[330]: Iru, AFB_1, AFB_2, AFG_1, AFG_2, AME, DON, FB_1, FB_2, FB_3, FUS-X, HT-2, OTA, ROQ-C, T-2, and ZEA, literature[330]; Ogiri, AFB_1, AFB_2, AFG_2, AME, DAS, DON, HT-2, NEO, OTA, STG, and ZEA, literature[330]; Ugba, AFB_1, AFB_2, AME, DON, ENB, HT-2, OTA, STG, T-2, and ZEA, literature[330]

Fusarenon-X (4-Acetylnivalenol)

incidence: 3/60*, conc. range: 40–76 μg/kg, ∅ conc.: 62 μg/kg, sample origin: fermented food sellers, Southwest Nigeria, sample year: February 2015–July 2016, country: South Africa/Nigeria/Belgium/Kenya[330], *fermented locust bean

- Co-contamination: For detailed information see the article.
- Further contamination of plants and plant products other than cereals and their products reported in the present book/article[330]: Iru, AFB_1, AFB_2, AFG_1, AFG_2, AME, DON, 3-AcDON, FB_1, FB_2, FB_3, HT-2, OTA, ROQ-C, T-2, and ZEA, literature[330]; Ogiri, AFB_1, AFB_2, AFG_2, AME, DAS, DON, HT-2, NEO, OTA, STG, and ZEA, literature[330]; Ugba, AFB_1, AFB_2, AME, DON, ENB, HT-2, OTA, STG, T-2, and ZEA, literature[330]

Other ***Fusarium*** Toxins

Zearalenone

incidence: 5/60*, conc. range: 11–33 μg/kg, ∅ conc.: 18 μg/kg, sample origin: fermented food sellers, Southwest Nigeria, sample year: February 2015–July 2016, country: South Africa/Nigeria/Belgium/Kenya[330], *fermented locust bean

- Co-contamination: For detailed information see the article.
- Further contamination of plants and plant products other than cereals and their products reported in the present book/article[330]: Iru, AFB_1, AFG_1, AFG_1, AFG_2, AME, DON, 3-AcDON, FB_1, FB_2, FB_3, FUS-X, HT-2, OTA, ROQ-C, and T-2, literature[330]; Ogiri, AFB_1, AFB_2, AFG_2, AME, DAS, DON, HT-2, NEO, OTA, STG, and ZEA, literature[330]; Ugba, AFB_1, AFB_2, AME, DON, ENB, HT-2, OTA, STG, T-2, and ZEA, literature[330]

Penicillium Toxins

Roquefortine C

incidence: 2/60*, conc. range: 10–14 μg/kg, ∅ conc.: 12 μg/kg, sample origin: fermented food sellers, Southwest Nigeria, sample year: February 2015–July 2016, country: South Africa/Nigeria/Belgium/Kenya[330], *fermented locust bean

- Co-contamination: For detailed information see the article.
- Further contamination of plants and plant products other than cereals and their products reported in the present book/article[330]: Iru, AFB_1, AFG_1, AFG_1, AFG_2, AME, DON, 3-AcDON, FB_1, FB_2, FB_3, FUS-X, HT-2, OTA, T-2, and ZEA, literature[330]; Ogiri, AFB_1, AFB_2, AFG_2, AME, DAS, DON, HT-2, NEO, OTA, STG, and ZEA, literature[330]; Ugba, AFB_1, AFB_2, AME, DON, ENB, HT-2, OTA, STG, T-2, and ZEA, literature[330]

Irvingia gabonensis see **Seeds (bush mango)**

Isoflavones see **Soy isoflavones**

Jaggery may contain the following mycotoxins:

Aspergillus Toxins

Aflatoxin/s

incidence: 2/25, conc. range: 0.5–1.0 μg/kg*, ∅ conc.: 0.75 μg/kg*, sample origin: local industries at the nearby regions of Mandya and Mysore (districts), Karnataka (state), India, sample year: June/July 2012, country: India[453], *AF

incidence: 2/41, conc. range: 0.5 μg/kg*, sample origin: local vendors at markets in Mysore (district), Karnataka (state), India, sample year: June/July 2012, country: India[453], *AF

incidence: 2/30, conc. range: 0.5–1.0 μg/kg*, sample origin: local vendors at markets in Mandya (district), Karnataka (state), India, sample year: June/July 2012, country: India[453], *AF

- Co-contamination: not reported
- Further contamination of plants and plant products other than cereals and their products reported in the present book/article[453]: Juices (sugarcane juice), AF/S, literature[453]

Jaggery is a sweetener on natural basis.

Jam may contain the following mycotoxins:

Alternaria Toxins

Alternariol

incidence: 1/2*, conc.: 14 μg/kg, sample origin: private households of staff members, Austria, sample year: unknown, country: Austria[850], *moldy

- Co-contamination: 1 sa. co-contaminated with AOH, CHAE, EMOD, and MA
- Further contamination of plants and plant products other than cereals and their products reported in the present book/article[850]: Fruits (apple), Nuts (chestnut), AME and AOH, literature[850]; Fruits (lemon), Jam, Puree, MA, literature[850]; Spices (garlic), BEA, FB_1, FB_2, and FB_3, literature[850]; Wine (red), AME, AOH, and MA, literature[850]; see also Fruits (lemon), Jam, Nuts (chestnut), and Wine (red), chapter: Further Mycotoxins and Microbial Metabolites

Aspergillus **and** ***Penicillium*** **Toxins**

Ochratoxin A

incidence: 7/12*, conc. range: 0.09–14.33 µg/kg, ∅ conc.: 2.51 µg/kg, sample origin: households, Germany, sample year: unknown, country: Germany[152], *different jams, moldy

- Co-contamination: not reported
- Further contamination of plants and plant products other than cereals and their products reported in the present book/article[152]: not reported

incidence: 4/75, conc. range: LOD/LOQ-≤0.27 µg/kg (4 sa.), sample origin: Germany, sample year: 1995–1998, country: EU[698]

- Co-contamination: not reported
- Further contamination of plants and plant products other than cereals and their products reported in the present book/article[698]: Baby food, Beans (soybean), Chocolate, Cocoa, Coffee, Coffee beans, Currants, Fruits, dried, Herbs, Juices, Juices (grape juice), Legumes, Nuts, Oil seeds, Oils (olive oil), Peas, Peas (chickpea), Products (plant product), Raisins, Spices, Spices (cayenne pepper, chilli), Spices (nutmeg), Spices (paprika), Spices (pepper), Sultanas, Tea, Vinegar, Wine, Wine (red), Wine (rosé), Wine (white), Wine, miscellaneous, OTA, literature[698]

Penicillium **Toxins**

Mycophenolic Acid

incidence: 2/2*, conc. range: 34–850 µg/kg, ∅ conc.: 442 µg/kg, sample origin: private households of staff members, Austria, sample year: unknown, country: Austria[850], *moldy

- Co-contamination: 1 sa. co-contaminated with AOH, CHAE, EMOD, and MA
- Further contamination of plants and plant products other than cereals and their products reported in the present book/article[850]: Fruits (apple), Nuts (chestnut), AME and AOH, literature[850]; Fruits (lemon), Puree, MA, literature[850]; Jam, AOH, literature[850]; Spices (garlic), BEA, FB_1, FB_2, and FB_3, literature[850]; Wine (red), AME, AOH, and MA, literature[850]; see also Fruits (lemon), Jam, Nuts (chestnut), Wine (red), chapter: Further Mycotoxins and Microbial Metabolites

Jam (apple jam) may contain the following mycotoxins:

Aspergillus **and** ***Penicillium*** **Toxins**

Patulin

incidence: 1/2*, conc.: 1,390 µg/kg, sample origin: Finnish customs authorities, Finland, sample year: March–December 1976, country: Finland[33], *moldy

- Co-contamination: not reported
- Further contamination of plants and plant products other than cereals and their products reported in the present book/article[33]: Fruits (apple), Juices (apple concentrate), Juices (apple juice), Products (apple flavor product), PAT, literature[33]

incidence: 5/15, conc. range: 4.7–554 µg/l, ∅ conc.: 302 µg/l, sample origin: local supermarkets, Tunisia, sample year: 2013, country: Tunisia[888]

- Co-contamination: not reported
- Further contamination of plants and plant products other than cereals and their products reported in the present book/article[888]: Jam (pear jam), Juices, Juices (apple juice), Juices (pear juice), Products (fruit product), PAT, literature[888]

Jam (bean jam) may contain the following mycotoxins:

Aspergillus **Toxins**

Aflatoxin B_1

incidence: 5/41*, conc. range: 0.1–0.7 µg/kg, sample origin: unknown, sample year: unknown, country: Japan[97], *white bean jam

incidence: 1/24*, conc.: 0.8 µg/kg, sample origin: unknown, sample year: unknown, country: Japan[97], *red bean jam

- Co-contamination: not reported
- Further contamination of plants and plant products other than cereals and their products reported in the present book/article[97]: Beans, Spices, AFB_1 and AFB_2, literature[97]; Butter (peanut butter), Oils (peanut oil), Spices (pepper), AFB_1, AFB_2, and AFG_1, literature[97]; Jam (bean jam), AFB_2, literature[97]; Spices (nutmeg), AFB_1, AFB_2, AFG_1, and AFG_2, literature[97]

incidence: 1/68, conc.: 0.1 µg/kg, sample origin: commercially available market basket sa. of foods and foodstuffs in Tokyo (capital), Japan, sample year: 1986–1990, country: Japan[99]

- Co-contamination: not reported
- Further contamination of plants and plant products other than cereals and their products reported in the present book/article[99]: Beans, Seeds (sesame), Sugar, AFB_1 and AFB_2, literature[99]; Food, Paste (peanut paste), AFB_1, literature[99]; Nuts (Brazil nut, paranut), Nuts (peanut), Nuts (pistachio nut), Spices (nutmeg), AFB_1, AFB_2, AFG_1, and AFG_2, literature[99]; Spices, Spices (cayenne pepper, chilli), Spices (pepper), AFB_1, AFB_2, and AFG_1, literature[99]; Spices (paprika), AFB_1 and AFG_1, literature[99]

Aflatoxin B_2

incidence: 1/24*, conc.: 0.1 µg/kg, sample origin: unknown, sample year: unknown, country: Japan[97], *red bean jam

- Co-contamination: not reported
- Further contamination of plants and plant products other than cereals and their products reported in the present book/article[97]: Beans, Spices, AFB_1 and AFB_2, literature[97]; Butter (peanut butter), Oils (peanut oil), Spices (pepper), AFB_1, AFB_2, and AFG_1, literature[97]; Jam (bean jam), AFB_1, literature[97]; Spices (nutmeg), AFB_1, AFB_2, AFG_1, and AFG_2, literature[97]

Jam (fig jam) may contain the following mycotoxins:

Aspergillus **Toxins**

Aflatoxin B_1

incidence: 3/10, conc. range: 0.93–4.17 µg/kg, ∅ conc.: 2.00 µg/kg, sample origin: local markets and supermarkets in Novi Sad (city), Serbia, sample year: February–June 2014, country: Serbia[903]

- Co-contamination: not reported
- Further contamination of plants and plant products other than cereals and their products reported in the present book/article[903]: Biscuits, AFB_1, OTA, T-2, and ZEA, literature[903]; Jam (fig jam), AFB_2, OTA, and T-2, literature[903]

Aflatoxin B_2

incidence: 1/10, conc.: 1.15 µg/kg, sample origin: local markets and supermarkets in Novi Sad (city), Serbia, sample year: February–June 2014, country: Serbia[903]

- Co-contamination: not reported
- Further contamination of plants and plant products other than cereals and their products reported in the present book/article[903]: Biscuits, AFB_1, OTA, T-2, and ZEA, literature[903]; Jam (fig jam), AFB_1, OTA, and T-2, literature[903]

Aspergillus **and** ***Penicillium*** Toxins

Ochratoxin A

incidence: 3/10, conc. range: 12.0–25.0 µg/kg, ∅ conc.: 17.7 µg/kg, sample origin: local markets and supermarkets in Novi Sad (city), Serbia, sample year: February–June 2014, country: Serbia[903]

- Co-contamination: not reported
- Further contamination of plants and plant products other than cereals and their products reported in the present book/article[903]: Biscuits, AFB_1, OTA, T-2, and ZEA, literature[903]; Jam (fig jam), AFB_1, AFB_2, and T-2, literature[903]

Fusarium Toxins

Type A Trichothecenes

T-2 Toxin

incidence: 2/10, conc. range: 3.82–4.92 µg/kg, ∅ conc.: 4.37 µg/kg, sample origin: local markets and supermarkets in Novi Sad (city), Serbia, sample year: February–June 2014, country: Serbia[903]

- Co-contamination: not reported
- Further contamination of plants and plant products other than cereals and their products reported in the present book/article[903]: Biscuits, AFB_1, OTA, T-2, and ZEA, literature[903]; Jam (fig jam), AFB_1, AFB_2, and OTA, literature[903]

Jam (pear jam) may contain the following mycotoxins:

Aspergillus **and** ***Penicillium*** Toxins

Patulin

incidence: 7/16, conc. range: 17–325 µg/l, ∅ conc.: 123.7 µg/l, sample origin: local supermarkets, Tunisia, sample year: 2013, country: Tunisia[888]

- Co-contamination: not reported
- Further contamination of plants and plant products other than cereals and their products reported in the present book/article[888]: Jam (apple jam), Juices, Juices (apple juice), Juices (pear juice), Products (fruit product), PAT, literature[888]

Jam (quince jam) may contain the following mycotoxins:

Aspergillus **and** ***Penicillium*** Toxins

Patulin

incidence: 10/10*, conc. range: <LOQ-28.7 µg/kg, sample origin: local supermarkets in Porto (city), Portugal, sample year: unknown, country: Portugal[27], *sa. with different portions of fruit and sugar

- Co-contamination: not reported
- Further contamination of plants and plant products other than cereals and their products reported in the present book/article[27]: Apple cider, Baby food, Fruits (apple), Fruits (quince), Juices (apple juice), PAT, literature[27]

Jellies may contain the following mycotoxins:

Aspergillus **and** ***Penicillium*** Toxins

Patulin

incidence: 2/4* conc. range: <LOQ-3.22 µg/kg, sample origin: supermarkets and organic shops, Portugal, sample year: unknown, country: Portugal[905], *tomato jellies

- Co-contamination: not reported
- Further contamination of plants and plant products other than cereals and their products reported in the present book/article[905]: Juices (apple juice), Juices (tomato juice), Puree, PAT, literature[905]

For detailed information see the article.
see also **Fruits (apple)**

Jello may contain the following mycotoxins:

Aspergillus Toxins

Aflatoxin/s

incidence: 2/3, conc. range: ≤1.727 µg/kg*, sample origin: groceries and supermarkets, Turkey, sample year: 2003, country: Turkey[119], *AFS total

- Co-contamination: not reported
- Further contamination of plants and plant products other than cereals and their products reported in the present book/article[119]: Asure, Beverages, Cocoa, Cream, Ice cream, Keskul, Muhallebi, Pudding, Salep, Sauces (chocolate sauce), Soups, AF/S, literature[119]

Jello is a gelatin dessert. Ingredients were sugar, gelatin, starch, glucose, fruit flavor, and vanilla.

Juices may contain the following mycotoxins:

Alternaria Toxins

Altenuene

incidence: 1/23, conc.: >LOD, sample origin: retail stores in the area of Arnsberg and Münster (cities), Germany, sample year: unknown, country: Germany[743]

- Co-contamination: not reported
- Further contamination of plants and plant products other than cereals and their products reported in the present book/article[743]: Juices, Oils (vegetable oil), Products (tomato product), AME, AOH, and TA, literature[743]; Seeds (sunflower), ALT, AME, AOH, ATX-I, and TA, literature[743]; see also Juices, Oils (vegetable oil), and Seeds (sunflower), chapter: Further Mycotoxins and Microbial Metabolites

Alternariol

incidence: 13/23, conc. range: >LOD (6 sa.), 0.65–16 µg/kg (7 sa.), sample origin: retail stores in the area of Arnsberg and Münster (cities), Germany, sample year: unknown, country: Germany[743]

- Co-contamination: not reported
- Further contamination of plants and plant products other than cereals and their products reported in the present book/article[743]: Juices, ALT, AME, and TA, literature[743]; Oils (vegetable oil), Products (tomato product), AME, AOH, and TA, literature[743]; Seeds (sunflower), ALT, AME, AOH, ATX-I, and TA, literature[743]; see also Juices, Oils (vegetable oil), and Seeds (sunflower), chapter: Further Mycotoxins and Microbial Metabolites

Alternariol Methyl Ether

incidence: 10/23, conc. range: >LOD (7 sa.), 0.14–4.9 µg/kg (3 sa.), sample origin: retail stores in the area of Arnsberg and Münster (cities), Germany, sample year: unknown, country: Germany[743]

- Co-contamination: not reported

- Further contamination of plants and plant products other than cereals and their products reported in the present book/article[743]: Juices, ALT, AOH, and TA, literature[743]; Oils (vegetable oil), Products (tomato product), AME, AOH, and TA, literature[743]; Seeds (sunflower), ALT, AME, AOH, ATX-I, and TA, literature[743]; see also Juices, Oils (vegetable oil), and Seeds (sunflower), chapter: Further Mycotoxins and Microbial Metabolites

Tenuazonic Acid

incidence: 7/7*, conc. range: 0.5–10 μg/kg, sample origin: unknown, sample year: unknown, country: Germany[166], *included cranberry, blueberry, fruit, litchi, passion peach, and raspberry juice

- Co-contamination: not reported
- Further contamination of plants and plant products other than cereals and their products reported in the present book/article[166]: Juices (apple juice), Juices (apricot juice), Juices (banana juice), Juices (blackcurrant juice), Juices (cherry juice), Juices (grape juice), Juices (orange juice), Juices (pear juice), Juices (pomegranate juice), Juices (strawberry juice), Spices, Spices (coriander), Spices (cumin), Spices (curcuma), Spices (curry), Spices (ginger), Spices (paprika), Spices (pepper), TA, literature[166]

incidence: 12/23, conc. range: >LOD (1 sa.), 21–250 μg/kg (11 sa.), sample origin: retail stores in the area of Arnsberg and Münster (cities), Germany, sample year: unknown, country: Germany[743]

- Co-contamination: not reported
- Further contamination of plants and plant products other than cereals and their products reported in the present book/article[743]: Juices, ALT, AME, and AOH, literature[743]; Oils (vegetable oil), Products (tomato product), AME, AOH, and TA, literature[743]; Seeds (sunflower), ALT, AME, AOH, ATX-I, and TA, literature[743]; see also Juices, Oils (vegetable oil), Seeds (sunflower), chapter: Further Mycotoxins and Microbial Metabolites

Aspergillus and ***Penicillium*** Toxins

Ochratoxin A

incidence: 1/11, conc.: 0.11 μg/kg, sample origin: unknown, sample year: unknown, country: Japan[112]

- Co-contamination: not reported
- Further contamination of plants and plant products other than cereals and their products reported in the present book/article[112]: Cassis, Raisins, Wine, OTA, literature[112]; Cocoa, Coffee, Food, Vinegar, OTA and OTB, literature[112]; Powder (cocoa powder), AFB_1, AFB_2, AFG_1, OTA, and OTB, literature[112]

incidence: 1/14*, conc.: 1.16 μg/l**, sample origin: supermarkets in Rabat (capital), Morocco, sample year: unknown, country: Morocco/France[493], *included clementine, cocktail, grapefruit**, mango, orange, peach, and pineapple juice

- Co-contamination: not reported
- Further contamination of plants and plant products other than cereals and their products reported in the present book/article[493]: Wine (red), Wine (rosé), Wine (white), OTA, literature[493]

incidence: 2/12, conc. range: ≤0.22 μg/kg, sample origin: supermarkets and fresh provision shops, Hong Kong, China, sample year: August/September 2005, country: China[697]

- Co-contamination: not reported
- Further contamination of plants and plant products other than cereals and their products reported in the present book/article[697]: Chocolate, Coffee, Fruits, dried, Legumes, Spices, OTA, literature[697]

incidence: 2/3*, conc. range: LOD/LOQ-≤0.14 μg/kg (2 sa.), sample origin: Finland, sample year: 1998, country: EU[698], *miscellaneous fruit juices

incidence: 1/19*, conc.: 3.45 μg/kg (1 sa.), sample origin: France, sample year: 1998/1999, country: EU[698], *fruit juices

incidence: 8/162*, conc. range: LOD/LOQ-≤0.18 μg/kg (8 sa.), sample origin: Germany, sample year: 1995–1998, country: EU[698], *miscellaneous fruit juices

incidence: 3/19*, conc. range: LOD/LOQ-≤0.006 μg/kg (3 sa.), sample origin: Norway, sample year: 1999, country: EU[698], *fruit juices

incidence: 4/4*, conc. range: LOD/LOQ-≤0.016 μg/kg (4 sa.), sample origin: Norway, sample year: 1999, country: EU[698], *miscellaneous fruit juices

- Co-contamination: not reported
- Further contamination of plants and plant products other than cereals and their products reported in the present book/article[698]: Baby food, Beans (soybean), Chocolate, Cocoa, Coffee, Coffee beans, Currants, Fruits, dried, Herbs, Jam, Juices (grape juice), Legumes, Nuts, Oil seeds, Oils (olive oil), Peas, Peas (chickpea), Products (plant product), Raisins, Spices, Spices (cayenne pepper, chilli), Spices (nutmeg), Spices (paprika), Spices (pepper), Sultanas, Tea, Vinegar, Wine, Wine (red), Wine (rosé), Wine (white), Wine, miscellaneous, OTA, literature[698]

incidence: 7/12*, conc. range: ≤0.059 μg/kg, sample origin: 12 regions, Czech Republic, sample year: 2011–2013, country: Czech Republic[811], *fruit juices

- Co-contamination: not reported
- Further contamination of plants and plant products other than cereals and their products reported in the present book/article[811]: Beans, Chocolate, Coffee, Confectionery, Figs, Herbs and Spices, Licorice, Powder (cocoa powder), Products (ginger product), Raisins, Spices (caraway), Spices (cayenne pepper, chilli), Spices (coriander), Spices (ginger), Spices (mustard), Spices (nutmeg), Spices (paprika), Spices (pepper), Tea, Wine (red), Wine (white), OTA, literature[811]

Patulin

incidence: 8/45*, conc. range: >LOQ<10 μg/kg (8 sa., maximum: 1.7 μg/kg), sample origin: markets, supermarkets, hypermarkets or organic food shops, Italy, sample year: November 2003–February 2004, country: Italy[35], *conventional

incidence: 6/12*, conc. range: >LOQ<10 μg/kg (5 sa.), 13.5 μg/kg (1 sa.), sample origin: markets, supermarkets, hypermarkets or organic food shops, Italy, sample year: November 2003–February 2004, country: Italy[35], *organic

- Co-contamination: not reported
- Further contamination of plants and plant products other than cereals and their products reported in the present book/article[35]: Fruits (apple), Juices (apple juice), Juices (pear juice), Puree, Vinegar (apple), PAT, literature[35]

incidence: 2/6*, conc. range: 5 µg/l ∅ conc.: 5 µg/l sample origin: retail outlets, Cape Town (city), South Africa, sample year: 1996–1998, country: South Africa[37], *fruit juices, mixed

incidence: 4/7*, conc. range: 5–20 µg/l ∅ conc.: 13.8 µg/l, sample origin: retail outlets, Cape Town (city), South Africa, sample year: 1996–1998, country: South Africa[37], *mixed fruit juices for infants

- Co-contamination: not reported
- Further contamination of plants and plant products other than cereals and their products reported in the present book/article[37]: Apple cider, Juices (apple juice), Products (apple product), PAT, literature[37]

incidence: 12/12*, conc. range: 2.8–11.2 µg/kg, ∅ conc.: 5.60 µg/kg, sample origin: Greek market, Greece, sample year: October 2004–June 2006, country: Greece[108], *mixed fruit juices (apple included)

- Co-contamination: not reported
- Further contamination of plants and plant products other than cereals and their products reported in the present book/article[108]: Baby food, Drinks (apple drink), Juices (apple concentrate), Juices (apple juice), Juices (apricot concentrate), Juices (orange juice), Juices (pineapple concentrate), Juices (rosehip juice), PAT, literature[108]

incidence: 35/98*, conc. range: <10 µg/kg (20 sa.), 10–50 µg/kg (15 sa., maximum: 44.89 µg/kg), sample origin: supermarkets and organic food shops, Italy, sample year: April–November 2005, country: Italy[438], *conventional mixed and apple juices

incidence: 12/37*, conc. range: <10 µg/kg (4 sa.), 10–50 µg/kg (7 sa.), 55.41 µg/kg (1 sa.), sample origin: supermarkets and organic food shops, Italy, sample year: April–November 2005, country: Italy[438], *organic mixed and apple juices

- Co-contamination: not reported
- Further contamination of plants and plant products other than cereals and their products reported in the present book/article[438]: not reported

For detailed information see the article.

incidence: 2/12*, conc. range: 5–6 µg/l, ∅ conc.: 5.5 µg/l, sample origin: domestic sa. from retail outlets, Tunisia, sample year: unknown, country: Tunisia/France[483], *fruit juice, mixed

incidence: 1/8*, conc.: 5 µg/l, sample origin: imported sa. from retail outlets, Tunisia, sample year: unknown, country: Tunisia/France[483], *fruit juice, mixed

- Co-contamination: not reported
- Further contamination of plants and plant products other than cereals and their products reported in the present book/article[483]: Juices (apple concentrate), Juices (apple juice), PAT, literature[483]

For detailed information see the article.

incidence: 32/35*, conc. range: 1–45 µg/l, sample origin: European and Italian sa. purchased in retail outlets, Italy, sample year: 2008, country: Italy[688], *fruit juices, mixed

- Co-contamination: not reported
- Further contamination of plants and plant products other than cereals and their products reported in the present book/article[688]: Juices (apple juice), Juices (pear juice), PAT, literature[688]

For detailed information see the article.

incidence: 2/10*, conc. range: <30 µg/l (1 sa.), 70 µg/l (1 sa.), ∅ conc.: 46 µg/l, sample origin: local retail markets and juice joints in Lucknow (city), India, sample year: unknown, country: India[838], *branded mixed juices and orange as well as grape juices

- Co-contamination: not reported
- Further contamination of plants and plant products other than cereals and their products reported in the present book/article[838]: Juices (apple juice), PAT, literature[838]

incidence: 9/29*, conc. range: <10 µg/kg (3 sa.), >10 µg/kg (6 sa., maximum: 25.4 µg/kg), sample origin: supermarkets and organic food shops, Northwestern Italy, sample year: January–September 2007, country: Italy[847], *juices, mixed

incidence: 35/98*, conc. range: <10 µg/kg (21 sa.), >10 µg/kg (14 sa., maximum: 33.4 µg/kg), sample origin: supermarkets and organic food shops, Northwestern Italy, sample year: January–September 2007, country: Italy[847], *conventional

incidence: 8/27*, conc. range: <10 µg/kg (3 sa.), >10 µg/kg (5 sa., maximum: 23.5 µg/kg), sample origin: supermarkets and organic food shops, Northwestern Italy, sample year: January–September 2007, country: Italy[847], *organic

- Co-contamination: not reported
- Further contamination of plants and plant products other than cereals and their products reported in the present book/article[847]: Juices (apricot juice), Juices (peach juice), Juices (pear juice), PAT, literature[847]

For detailed information see the article.

incidence: 24/30*, conc. range: 4.5–889 µg/l, ∅ conc.: 158.1 µg/l, sample origin: Tunisian fruit juice company, Tunisia, sample year: 2013, country: Tunisia[888], *juice, concentrated

incidence: 17/34*, conc. range: 10–55.7 µg/l, ∅ conc.: 28.5 µg/l, sample origin: local supermarkets, Tunisia, sample year: 2013, country: Tunisia[888], *juice, mixed

- Co-contamination: not reported
- Further contamination of plants and plant products other than cereals and their products reported in the present book/article[888]: Jam (apple jam), Jam (pear jam), Juices (apple juice), Juices (pear juice), Products (fruit product), PAT, literature[888]

incidence: 24/30*, conc. range: ≤91.8 µg/kg, sample origin: supermarkets and stores in Changchun (city), China, sample year: 2009, country: China[890], *juice, mixed

- Co-contamination: not reported
- Further contamination of plants and plant products other than cereals and their products reported in the present book/article[890]: Baby food, Juices (apple concentrate), Juices (apple juice), PAT, literature[890]

incidence: 12/30, conc. range: ≤125 µg/l sample origin: major supermarkets and stores, Tunisia, sample year: 2011?, country: Tunisia[895]

- Co-contamination: not reported
- Further contamination of plants and plant products other than cereals and their products reported in the present book/article[895]: Baby food, Juices (apple juice), PAT, literature[895]

incidence: 14/36*, conc. range: 1–5 µg/kg (13 sa.), 6.3 µg/kg (1 sa.), ∅ conc.: 3.4 µg/kg, sample origin: supermarkets and retail shops in Novi Sad (city), Vojvodina (province), Serbia, sample year: 2013–2015, country: Serbia/Portugal[904], *apple-based multifruit juice for children incidence: 26/66*, conc. range: 1–10 μg/kg (23 sa.), 10–25 μg/kg (2 sa.), 30.2 μg/kg (1 sa.), ∅ conc.: 5.4 μg/kg, sample origin: supermarkets and retail shops in Novi Sad (city), Vojvodina (province), Serbia, sample year: 2013–2015, country: Serbia/Portugal[904], *apple-based multifruit juice (small package juices with straw)

- Co-contamination: not reported
- Further contamination of plants and plant products other than cereals and their products reported in the present book/article[904]: Juices (apple juice), Puree, PAT, literature[904]

For detailed information see the article.

Juices (apple concentrate) may contain the following mycotoxins:

Alternaria Toxins

Alternariol

incidence: 13/18, conc. range: tr (1 sa.), 1.35–3.56 μg/l (12 sa.), sample origin: processing plant in Lleida (city), Spain, sample year: 1993, country: Spain[488]

incidence: 4/14, conc. range: 2.99–5.42 μg/l, sample origin: processing plant in Lleida (city), Spain, sample year: 1994, country: Spain[488]

- Co-contamination: not reported
- Further contamination of plants and plant products other than cereals and their products reported in the present book/article[488]: Juices (apple concentrate), AME, literature[488]

Alternariol Methyl Ether

incidence: 13/18, conc. range: tr, sample origin: processing plant in Lleida (city), Spain, sample year: 1993, country: Spain[488]

incidence: 4/14, conc. range: tr (3 sa.), 1.71 μg/l (1 sa.), sample origin: processing plant in Lleida (city), Spain, sample year: 1994, country: Spain[488]

- Co-contamination: not reported
- Further contamination of plants and plant products other than cereals and their products reported in the present book/article[488]: Juices (apple concentrate), AOH, literature[488]

Aspergillus and ***Penicillium*** Toxins

Patulin

incidence: 8/16*, conc. range: 5–50 μg/l (6 sa.), >50 μg/l (2 sa., maximum: 646 μg/l), sample origin: retail outlets and directly from producers in Sydney (city) and NSW (state), Australia, sample year: August 1989–May 1990, country: Australia[31], *apple juice concentrate

- Co-contamination: not reported
- Further contamination of plants and plant products other than cereals and their products reported in the present book/article[31]: Juices (apple juice), Products (apple product), PAT, literature[31]

incidence: 13/64, conc. range: 50–690 μg/l, sample origin: imported sa. Finnish customs authorities, Finland, sample year: March–December 1976, country: Finland[33]

incidence: 2/7, conc. range: 460–1,450 μg/l, sample origin: Finland, sample year: March–December 1976, country: Finland[33]

- Co-contamination: not reported
- Further contamination of plants and plant products other than cereals and their products reported in the present book/article[33]: Fruits (apple), Jam (apple jam), Juices (apple juice), Products (apple flavor product), PAT, literature[33]

incidence: 23/23, conc. range: 2.6–36.8 μg/kg, ∅ conc.: 11.60 μg/kg, sample origin: imports from Turkey, Greece, sample year: October 2004–June 2006, country: Greece[108]

incidence: 14/14, conc. range: 3.8–16.2 μg/kg, ∅ conc.: 8.50 μg/kg, sample origin: imports from China, Greece, sample year: October 2004–June 2006, country: Greece[108]

- Co-contamination: not reported
- Further contamination of plants and plant products other than cereals and their products reported in the present book/article[108]: Baby food, Drinks (apple drink), Juices, Juices (apple juice), Juices (apricot concentrate), Juices (orange juice), Juices (pineapple concentrate), Juices (rosehip juice), PAT, literature[108]

incidence: 122/234, conc. range: >50–376 μg/l, sample origin: manufacturers, Turkey, sample year: 1996, country: Turkey[446]

incidence: 41/119, conc. range: >50–153 μg/l, sample origin: manufacturers, Turkey, sample year: 1997, country: Turkey[446]

incidence: 5/67, conc. range: >50–103 μg/l, sample origin: manufacturers, Turkey, sample year: 1998, country: Turkey[446]

incidence: 5/62, conc. range: >50–119 μg/l, sample origin: manufacturers, Turkey, sample year: 1999, country: Turkey[446]

- Co-contamination: not reported
- Further contamination of plants and plant products other than cereals and their products reported in the present book/article[446]: not reported

incidence: 5/39*, conc. range: tr, sample origin: domestic sa. from imported apple juice concentrate purchased in local supermarkets in Tohoku (district), Japan, sample year: March–October 2003, country: Japan[471]

- Co-contamination: not reported
- Further contamination of plants and plant products other than cereals and their products reported in the present book/article[471]: Juices (apple juice), PAT, literature[471]

incidence: 18/23, conc. range: >15–50 μg/l (5 sa.), >50 μg/l (13 sa., maximum: 148.8 μg/l), sample origin: retail outlets and producers, Iran, sample year: 2001/2002, country: Iran[476]

- Co-contamination: not reported
- Further contamination of plants and plant products other than cereals and their products reported in the present book/article[476]: Juices (apple juice), PAT, literature[476]

incidence: 1/8, conc.: 5 μg/l, sample origin: domestic sa. from retail outlets, Tunisia, sample year: unknown, country: Tunisia/France[483]

incidence: 0/7, conc.: no contamination, sample origin: imported sa. from retail outlets, Tunisia, sample year: unknown, country: Tunisia/France[483]

- Co-contamination: not reported
- Further contamination of plants and plant products other than cereals and their products reported in the present book/article[483]: Juices, Juices (apple juice), PAT, literature[483]

For detailed information see the article.

incidence: 16/16, conc. range: 56–231 μg/l, ∅ conc.: 107 μg/l, sample origin: juice producer in Rio Grande do Sul (state), Brazil, sample year: unknown, country: Brazil[489]

- Co-contamination: not reported
- Further contamination of plants and plant products other than cereals and their products reported in the present book/article[489]: not reported

incidence: 215/215, conc. range: 7–50 µg/l (117 sa.), >50 µg/l (98 sa., maximum: 376 µg/l), sample origin: companies, Turkey, sample year: September 1996–January 1997, country: Turkey[490]

- Co-contamination: not reported
- Further contamination of plants and plant products other than cereals and their products reported in the present book/article[490]: not reported

incidence: 14/33*, conc. range: LOD-10 µg/kg (1 sa.), 10–50 µg/kg (8 sa.), >50 µg/kg (5 sa., maximum: 74.4 µg/kg), ∅ conc.: 36.8 µg/kg, sample origin: local producers of fruit concentrates in Valencia? (city), Spain, sample year: 2008, country: Spain[786], *non-reconstituted apple concentrate

- Co-contamination: not reported
- Further contamination of plants and plant products other than cereals and their products reported in the present book/article[786]: Juices (apple juice), Juices (peach concentrate), Juices (pear concentrate), PAT, literature[786]

incidence: 5/28, conc. range: 15.1–30 µg/kg (4 sa.), 41.3 µg/kg (1 sa.), ∅ conc.: 24.0 µg/kg, sample origin: retail outlets and producers of Azerbaijan, Chaharmahal va Bakhtiari, Esfahan, Gillan, Khuzestan, and Tehran (provinces), Iran, sample year: April–November 2013, country: Iran[827]

- Co-contamination: not reported
- Further contamination of plants and plant products other than cereals and their products reported in the present book/article[827]: Juices (apple juice), Juices (grape juice), Juices (peach concentrate), Juices (pear juice), Juices (pomegranate juice), PAT, literature[827]

incidence: 8/20, conc. range: ≤94.7 µg/kg, sample origin: supermarkets and stores in Changchun (city), China, sample year: 2009, country: China[890]

- Co-contamination: not reported
- Further contamination of plants and plant products other than cereals and their products reported in the present book/article[890]: Baby food, Juices, Juices (apple juice), PAT, literature[890]

incidence: 549/568, conc. range: ≤29.0 µg/kg, sample origin: major manufacturers of Shaanxi (province), China, sample year: 2006/2007, country: China[910]

incidence: 824/845, conc. range: ≤78.0 µg/kg*, sample origin: major manufacturers of Shaanxi (province), China, sample year: 2007/2008, country: China[910], *only 4 sa. >50 µg/kg

incidence: 413/419, conc. range: ≤22.7 µg/kg, sample origin: major manufacturers of Shaanxi (province), China, sample year: 2008/2009, country: China[910]

incidence: 155/155, conc. range: ≤20.0 µg/kg, sample origin: major manufacturers of Shaanxi (province), China, sample year: 2009/2010, country: China[910]

- Co-contamination: not reported
- Further contamination of plants and plant products other than cereals and their products reported in the present book/article[910]: not reported

see also **Juices (apple juice)**

Juices (apple juice) may contain the following mycotoxins:

Alternaria Toxins

Altenuene

incidence: 1/10, conc.: 45.6 µg/kg, sample origin: Italian supermarkets, Italy, sample year: unknown, country: Italy[822]

- Co-contamination: not reported
- Further contamination of plants and plant products other than cereals and their products reported in the present book/article[822]: Juices (apple juice), TA, literature[822]; Products (tomato product), ALT and AOH, literature[822]; see also Products (tomato product), chapter: Further Mycotoxins and Microbial Metabolites

Alternariol

incidence: 3/4, conc. range: 0.16–0.22 µg/kg, sample origin: commercially available, sample year: unknown, country: Germany[160]

- Co-contamination: not reported
- Further contamination of plants and plant products other than cereals and their products reported in the present book/article[160]: Beverages, Juices (grape juice), Juices (orange juice), Juices (tomato juice), Juices (vegetable juice), Wine (red), Wine (white), Wine, miscellaneous, AME and AOH, literature[160]; Juices (apple juice), AME, literature[160]

incidence: 9/44, conc. range: 1.7–3.5 µg/l, ∅ conc.: 2.4 µg/l, sample origin: retail shops, Hesse (federal state), Germany, sample year: 2009/2010, country: Germany[439]

- Co-contamination: not reported
- Further contamination of plants and plant products other than cereals and their products reported in the present book/article[439]: Juices (tomato juice), Paste (tomato paste), Sauces (apple sauce), Tomato ketchup, Wine (white), AOH, literature[439]

incidence: 11/11, conc. range: 0.04–2.40 µg/l, ∅ conc.: 0.98 µg/l, sample origin: unknown, sample year: unknown, country: Canada[440]

- Co-contamination: 10 sa. co-contaminated with AME and AOH; 1 sa. contaminated solely with AOH
- Further contamination of plants and plant products other than cereals and their products reported in the present book/article[440]: Juices (apple juice), AME, literature[440]; Juices (grape juice), Nectars (cranberry nectar), Nectars (prune nectar), AME and AOH, literature[440]; Juices (raspberry juice), Wine (red), AOH, literature[440]

incidence: 8/15, conc. range: 0.10–7.94 µg/l, ∅ conc.: 2.57 µg/l, sample origin: local market, China, sample year: unknown, country: China[724]

- Co-contamination: 2 sa. co-contaminated with AME, AOH, and TA; 4 sa. co-contaminated with AME and AOH; 2 sa. contaminated solely with AOH
- Further contamination of plants and plant products other than cereals and their products reported in the present book/article[724]: Juices (apple juice), AME and TA, literature[724]; Wine, miscellaneous, AME, AOH, and TA, literature[724]

Alternariol Methyl Ether

incidence: 3/4, conc. range: pr., sample origin: commercially available, sample year: unknown, country: Germany[160]

- Co-contamination: not reported
- Further contamination of plants and plant products other than cereals and their products reported in the present book/article[160]: Beverages, Juices (grape juice), Juices (orange juice), Juices (tomato juice), Juices (vegetable juice), Wine (red), Wine (white), Wine, miscellaneous, AME and AOH, literature[160]; Juices (apple juice), AOH, literature[160]

incidence: 10/11, conc. range: tr-0.43 µg/l, sample origin: unknown, sample year: unknown, country: Canada[440]

- Co-contamination: 10 sa. co-contaminated with AME and AOH
- Further contamination of plants and plant products other than cereals and their products reported in the present book/article[440]: Juices (apple juice), AOH, literature[440]; Juices (grape juice), Nectars (cranberry nectar), Nectars (prune nectar), AME and AOH, literature[440]; Juices (raspberry juice), Wine (red), AOH, literature[440]

incidence: 1/7, conc.: 0.85 µg/l, sample origin: local stores, Spain or Canada, sample year: unknown, country: Spain/Canada[487]

- Co-contamination: not reported
- Further contamination of plants and plant products other than cereals and their products reported in the present book/article[487]: not reported

incidence: 7/15, conc. range: 0.03–0.88 µg/l, ∅ conc.: 0.3 µg/l, sample origin: local market, China, sample year: unknown, country: China[724]

- Co-contamination: 2 sa. co-contaminated with AME, AOH, and TA; 4 sa. co-contaminated with AME and AOH; 1 sa. co-contaminated with AME and TA
- Further contamination of plants and plant products other than cereals and their products reported in the present book/article[724]: Juices (apple juice), AOH and TA, literature[724]; Wine, miscellaneous, AME, AOH, and TA, literature[724]

Tenuazonic Acid

incidence: 5/5, conc. range: 0.6–6 µg/kg, sample origin: unknown, sample year: unknown, country: Germany[166]

- Co-contamination: not reported
- Further contamination of plants and plant products other than cereals and their products reported in the present book/article[166]: Juices, Juices (apricot juice), Juices (banana juice), Juices (blackcurrant juice), Juices (cherry juice), Juices (grape juice), Juices (orange juice), Juices (pear juice), Juices (pomegranate juice), Juices (strawberry juice), Spices, Spices (coriander), Spices (cumin), Spices (curcuma), Spices (curry), Spices (ginger), Spices (paprika), Spices (pepper), TA, literature[166]

incidence: 1/7, conc.: 58 µg/l, sample origin: retail shops in the area of Giessen (town), Germany, sample year: unknown, country: Germany/Indonesia[442]

- Co-contamination: not reported
- Further contamination of plants and plant products other than cereals and their products reported in the present book/article[442]: Juices (tomato juice), Tomato ketchup, TA, literature[442]

incidence: 2/24, conc. range: <LOQ-7.9 µg/kg, sample origin: local supermarkets, Belgium, sample year: February 2013–February 2015, country: Belgium/Austria/Germany[514]

- Co-contamination: For detailed information see the article.
- Further contamination of plants and plant products other than cereals and their products reported in the present book/article[514]: Juices (grape juice), TA, literature[514]; Juices (tomato juice), Sauces (tomato sauce), Tomato concentrate, ALT, AME, AOH, and TA, literature[514]; see also Juices (tomato juice), Sauces (tomato sauce), Tomato concentrate, chapter: Further Mycotoxins and Microbial Metabolites

For detailed information see the article.

incidence: 6/15, conc. range: 1.75–49.61 µg/l, ∅ conc.: 15.1 µg/l, sample origin: local market, China, sample year: unknown, country: China[724]

- Co-contamination: 2 sa. co-contaminated with AME, AOH, and TA; 1 sa. co-contaminated with AME and TA; 3 sa. contaminated solely with TA
- Further contamination of plants and plant products other than cereals and their products reported in the present book/article[724]: Juices (apple juice), AME and AOH, literature[724]; Wine, miscellaneous, AME, AOH, and TA, literature[724]

incidence: 2/10, conc. range: 24.3–45.3 µg/kg, ∅ conc.: 34.8 µg/kg, sample origin: Italian supermarkets, Italy, sample year: unknown, country: Italy[822]

- Co-contamination: not reported
- Further contamination of plants and plant products other than cereals and their products reported in the present book/article[822]: Juices (apple juice), ALT, literature[822]; Products (tomato product), ALT and AOH, literature[822]; see also Products (tomato product), chapter: Further Mycotoxins and Microbial Metabolites

Aspergillus and ***Penicillium*** Toxins

Citrinin

incidence: 1/35, conc.: 0.13 µg/l, sample origin: local retail stores, Germany, sample year: unknown, country: Germany[711]

- Co-contamination: not reported
- Further contamination of plants and plant products other than cereals and their products reported in the present book/article[711]: Juices (blackcurrant juice), Juices (cherry juice), Juices (tomato juice), CIT, literature[711]

Ochratoxin A

incidence: 15/51*, conc. range: ≤200 µg/l, ∅ conc.: 123 µg/l, sample origin: retail outlets in Jeddah (city), Saudi Arabia, sample year: 2008, country: Saudi Arabia[444], *local and imported apple juices (both kinds contaminated)

- Co-contamination: not reported
- Further contamination of plants and plant products other than cereals and their products reported in the present book/article[444]: Juices (apple juice), PAT, literature[444]

Patulin

incidence: 5/39, conc. range: 2–50 µg/kg, sample origin: ethnic stores, health food stores, and local stores as well as supermarkets in the four largest cities, Sweden, sample year: autumn 1996–summer 1998, country: Sweden[13]

- Co-contamination: not reported
- Further contamination of plants and plant products other than cereals and their products reported in the present book/article[13]: Beans (brown bean), Peas, Peas (chickpea), OTA, literature[13]; Nuts (almond), AFB_1, literature[13]; Nuts (Brazil nut,

paranut), Nuts (pistachio nut), Oil seeds, Products (peanut product), AFB_1 and AF/S, literature[13]

incidence: 9/9* **, conc. range: <LOQ-12.6 µg/kg, sample origin: local supermarkets or organic food shops in Porto (city), Portugal, sample year: unknown, country: Portugal[27], *100% apple juice (pasteurized or high pressure processing), **conventional

incidence: 10/14* **, conc. range: <LOQ-9.9 µg/kg, sample origin: local supermarkets or organic food shops in Porto (city), Portugal, sample year: unknown, country: Portugal[27], *different percentage of apple juice (pasteurized), **conventional

incidence: 6/6* **, conc. range: <LOQ-8.9 µg/kg, sample origin: local supermarkets or organic food shops in Porto (city), Portugal, sample year: unknown, country: Portugal[27], *different kinds of apple juice (pasteurized), **organic

- Co-contamination: not reported
- Further contamination of plants and plant products other than cereals and their products reported in the present book/article[27]: Apple cider, Baby food, Fruits (apple), Fruits (quince), Jam (quince jam), PAT, literature[27]

incidence: 7/36*, conc. range: 20–300 µg/kg, sample origin: local marketing institutions, GDR, sample year: unknown, country: GDR[29], *trade apple juice

incidence: 5/10*, conc. range: 60–50,000 µg/kg, sample origin: local marketing institutions, GDR, sample year: unknown, country: GDR[29], *moldy

- Co-contamination: not reported
- Further contamination of plants and plant products other than cereals and their products reported in the present book/article[29]: Butter (peanut butter), Nuts (peanut), Products (peanut product), Seeds (sunflower), AFB_1, AFB_2, AFG_1, and AFG_2, literature[29]; Fruits, AFB_1, AFG_1, and AFG_2, literature[29]; Fruits (apple), Fruits (peach), PAT, literature[29]

incidence: 140/241*, conc. range: 5–50 µg/l (69 sa.), >50 µg/l (71 sa., maximum: 1,113 µg/l), sample origin: retail outlets and directly from producers in Sydney (city) and NSW (state), Australia, sample year: August 1989–May 1990, country: Australia[31], *apple juice and mixed fruit juices

- Co-contamination: not reported
- Further contamination of plants and plant products other than cereals and their products reported in the present book/article[31]: Juices (apple concentrate), Products (apple product), PAT, literature[31]

incidence: 1/30, conc.: 17 µg/l, sample origin: groceries, markets, and supermarkets, Campinas (city), São Paulo (state), Brazil, sample year: 1992/1993 and 1995, country: Brazil[32]

- Co-contamination: not reported
- Further contamination of plants and plant products other than cereals and their products reported in the present book/article[32]: Fruits (apple), Fruits (peach), Fruits (pear), PAT, literature[32]

incidence: 8/20*, conc. range: 30–16,400 µg/l, sample origin: Finnish customs authorities, Finland, sample year: March–December 1976, country: Finland[33], *homemade apple juice

- Co-contamination: not reported
- Further contamination of plants and plant products other than cereals and their products reported in the present book/article[33]: Fruits (apple), Jam (apple jam), Juices (apple concentrate), Products (apple flavor product), PAT, literature[33]

incidence: 16/33*, conc. range: >LOQ<10 µg/kg (14 sa.), >10<25 µg/kg (1 sa.), 53.4 µg/kg (1 sa.), sample origin: markets, supermarkets, hypermarkets or organic food shops, Italy, sample year: November 2003–February 2004, country: Italy[35], *conventional

incidence: 12/24*, conc. range: >LOQ<10 µg/kg (10 sa.), >50 µg/kg (2 sa., maximum: 69.3 µg/kg), sample origin: markets, supermarkets, hypermarkets, or organic food shops, Italy, sample year: November 2003–February 2004, country: Italy[35], *organic

- Co-contamination: not reported
- Further contamination of plants and plant products other than cereals and their products reported in the present book/article[35]: Fruits (apple), Juices, Juices (pear juice), Puree, Vinegar (apple), PAT, literature[35]

incidence: 2/13, conc. range: 5–10 µg/l, ∅ conc.: 7.5 µg/l, sample origin: retail outlets in Cape Town (city), South Africa, sample year: 1996–1998, country: South Africa[37]

incidence: 2/4*, conc. range: 5–45 µg/l, ∅ conc.: 25 µg/l, sample origin: retail outlets in Cape Town (city), South Africa, sample year: 1996–1998, country: South Africa[37], *apple juices, carbonated

incidence: 2/3*, conc. range: 5 µg/l, ∅ conc.: 5 µg/l, sample origin: retail outlets in Cape Town (city), South Africa, sample year: 1996–1998, country: South Africa[37], *apple juices for infants

- Co-contamination: not reported
- Further contamination of plants and plant products other than cereals and their products reported in the present book/article[37]: Apple cider, Juices, Products (apple product), PAT, literature[37]

incidence: 23/29, conc. range: tr (7 sa.), LOQ-25 µg/l (14 sa.), 25–50 µg/l (2 sa., maximum: 38.8 µg/l), sample origin: supermarkets and handicraft producers, Belgium, sample year: 2001, country: Belgium[38]

incidence: 12/14, conc. range: tr (1 sa.), LOQ-25 µg/l (11 sa., maximum: 10.6 µg/l), sample origin: imports from France, Germany, and Switzerland purchased in supermarkets, Belgium, sample year: 2001, country: Belgium[38]

incidence: 18/21*, conc. range: 2.5–24.2 µg/l, ∅ conc.: 6.5 µg/l, sample origin: industrial, Belgium, sample year: 2001, country: Belgium[38], *apple juice, clear

incidence: 8/11*, conc. range: 6.0–19.8 µg/l, ∅ conc.: 8.1 µg/l, sample origin: industrial, Belgium, sample year: 2001, country: Belgium[38], *apple juice, cloudy

incidence: 3/4*, conc. range: 6.2–38.8 µg/l, ∅ conc.: 15.5 µg/l, sample origin: handicraft-made, Belgium, sample year: 2001, country: Belgium[38], *apple juice, clear

incidence: 6/7*, conc. range: 10.0–26.1 µg/l, ∅ conc.: 14.2 µg/l, sample origin: handicraft-made, Belgium, sample year: 2001, country: Belgium[38], *apple juice, cloudy

- Co-contamination: not reported
- Further contamination of plants and plant products other than cereals and their products reported in the present book/article[38]: Apple cider, PAT, literature[38]

incidence: 31/145* **, conc. range: 8.8–2,700.4 µg/l, sample origin: retail grocery stores, Michigan (state), USA, sample year: 2005/2006, country: USA[41], *apple-based beverages, **conventional

incidence: 6/14* **, conc. range: 17.9–264.8 µg/l, sample origin: retail grocery stores, Michigan (state), USA, sample year: 2005/2006, country: USA[41], *apple-based beverages, **organic

- Co-contamination: not reported
- Further contamination of plants and plant products other than cereals and their products reported in the present book/article[41]: Apple cider, PAT, literature[41]

For detailed information see the article.

incidence: 30/71, conc. range: 6.25–10 µg/kg (25 sa.), 10–25 µg/kg (5 sa., maximum: 15.0 µg/kg), ∅ conc.: 8.1 µg/kg, sample origin: hypermarkets and supermarkets in 12 main cities, Catalonia (autonomous community), Spain, sample year: June/July 2008, country: Spain[42]

- Co-contamination: not reported
- Further contamination of plants and plant products other than cereals and their products reported in the present book/article[42]: Baby food, Products (apple product), PAT, literature[42]

incidence: 5/17, conc. range: 1.5–50.9 µg/l, sample origin: stores and supermarkets, Valencia (city), Spain, sample year: unknown, country: Spain[43], *apple juice and apple-containing beverages

- Co-contamination: not reported
- Further contamination of plants and plant products other than cereals and their products reported in the present book/article[43]: Products (apple product), Puree, PAT, literature[43]

incidence: 10/10*, conc. range: 5.7–26.0 µg/l, sample origin: local store in Munich (city) Germany, sample year: 1996/1997, country: Germany[52], *commercial products

incidence: 2/2*, conc. range: 11.4–23.9 µg/l, ∅ conc.: 17.7 µg/l, sample origin: local garden in Munich (city) Germany, sample year: 1996/1997, country: Germany[52], *apple juice, homemade

incidence: 1/1*, conc.: 0.7 µg/l, sample origin: local store in Munich (city) Germany, sample year: 1996/1997, country: Germany[52], *apple-acerola juice, commercial product

- Co-contamination: not reported
- Further contamination of plants and plant products other than cereals and their products reported in the present book/article[52]: Juices (blackcurrant juice), Juices (cherry juice), Juices (grape juice), Juices (orange juice), Pulp (plum pulp), PAT, literature[52]

incidence: 29/29, conc. range: 0.9–11.8 µg/kg, ∅ conc.: 5.50 µg/kg, sample origin: Greek markets, Greece, sample year: October 2004–June 2006, country: Greece[108]

- Co-contamination: not reported
- Further contamination of plants and plant products other than cereals and their products reported in the present book/article[108]: Baby food, Drinks (apple drink), Juices, Juices (apple concentrate), Juices (apricot concentrate), Juices (orange juice), Juices (pineapple concentrate), Juices (rosehip juice), PAT, literature[108]

incidence: 3/51*, conc. range: ≤152.5 µg/l, ∅ conc.: 140 µg/l, sample origin: retail outlets in Jeddah (city), Saudi Arabia, sample year: 2008, country: Saudi Arabia[444], *local and imported apple juices (only local ones contaminated)

- Co-contamination: not reported
- Further contamination of plants and plant products other than cereals and their products reported in the present book/article[444]: Juices (apple juice), OTA, literature[444]

incidence: 27/45, conc. range: 19.1–47.3 µg/l (7 sa.), 60.2–403.3 µg/l (19 sa.), 732.8 µg/l (1 sa.), ∅ conc.: 139.9 µg/l, sample origin: markets in Istanbul (capital), Turkey, sample year: 1998–2000, country: Turkey[447]

- Co-contamination: not reported
- Further contamination of plants and plant products other than cereals and their products reported in the present book/article[447]: not reported

incidence: 2/2, conc. range: 120–160 µg/l, ∅ conc.: 140 µg/l, sample origin: factory, South Africa, sample year: unknown, country: South Africa[448]

- Co-contamination: not reported
- Further contamination of plants and plant products other than cereals and their products reported in the present book/article[448]: not reported

incidence: 1/1, conc.: 12 µg/l, sample origin: supermarket, USA, sample year: unknown, country: USA[450]

- Co-contamination: not reported
- Further contamination of plants and plant products other than cereals and their products reported in the present book/article[450]: not reported

incidence: 20/20, conc. range: 1.1–107.6 µg/l; ∅ conc.: 41.3 µg/l, sample origin: supermarkets in Navarra (chartered community), Spain, sample year: unknown, country: Spain[452]

- Co-contamination: not reported
- Further contamination of plants and plant products other than cereals and their products reported in the present book/article[452]: not reported

incidence: 9/11, conc. range: 10 µg/l (2 sa.), 51–100 µg/l (2 sa.), 101–200 µg/l (2 sa.), 201–300 µg/l (2 sa.), 350 µg/l (1 sa.), sample origin: roadside stands or directly from producers, Wisconsin (state), USA, sample year: 1976, country: USA[454]

incidence: 14/29, conc. range: 10 µg/l (2 sa.), 11–50 µg/l (8 sa.), 51–100 µg/l (2 sa.), 101–200 µg/l (1 sa.), 204 µg/l (1 sa.), sample origin: roadside stands or directly from producers, Wisconsin (state), USA, sample year: 1977, country: USA[454]

- Co-contamination: not reported
- Further contamination of plants and plant products other than cereals and their products reported in the present book/article[454]: not reported

incidence: 8/13*, conc. range: 5–10 µg/l (5 sa.), >10 µg/l (3 sa., maximum: 30 µg/l), sample origin: retail outlets, Norwich, UK, sample year: unknown, country: UK[455], *conventional

incidence: 1/1*, conc.: 56 µg/l, sample origin: retail outlets, Norwich, UK, sample year: unknown, country: UK[455], *organic

incidence: 1/3*, conc.: 5–10 µg/l, sample origin: retail outlets, Norwich, UK, sample year: unknown, country: UK[455], * apple juice, long life

- Co-contamination: not reported
- Further contamination of plants and plant products other than cereals and their products reported in the present book/article[455]: not reported

incidence: 82/100, conc. range: 0.5–10 µg/l (57 sa.), >10–20 µg/l (10 sa.), >20–30 µg/l (5 sa.), >40–50 µg/l (3 sa.), >50–60 µg/l (4 sa.), >60–70 µg/l (1 sa.), >100 µg/l (2 sa., maximum: 170 µg/l), sample origin: local stores in Madrid (capital), Spain, sample year: April–December 1992, country: Spain[456]

- Co-contamination: not reported
- Further contamination of plants and plant products other than cereals and their products reported in the present book/article[456]: not reported

incidence: 1/1*, conc.: ~1,000 µg/l, sample origin: unknown, sample year: unknown, country: Canada[457], *commercial apple juice

- Co-contamination: not reported
- Further contamination of plants and plant products other than cereals and their products reported in the present book/ article[457]: not reported

incidence: 6/10, conc. range: 9.8–107.2 μg/l, ∅ conc.: 36 μg/l, sample origin: local market in Taipei (capital), Taiwan, sample year: unknown, country: Taiwan[458]

- Co-contamination: not reported
- Further contamination of plants and plant products other than cereals and their products reported in the present book/ article[458]: Vinegar, PAT, literature[458]

incidence: 140/140, conc. range: <1–220 μg/l, sample origin: Norwegian and foreign producers, Norway, sample year: unknown, country: Norway[459]

- Co-contamination: not reported
- Further contamination of plants and plant products other than cereals and their products reported in the present book/ article[459]: not reported

incidence: 8/13, conc. range: 44–309 μg/l, sample origin: stores in Washington D.C. (capital) area, USA, sample year: unknown, country: USA[460]

- Co-contamination: not reported
- Further contamination of plants and plant products other than cereals and their products reported in the present book/ article[460]: not reported

incidence: 74/113, conc. range: ≤50 μg/l (33 sa.), >50–≤300 μg/l (33 sa.), >300 μg/l (8 sa., maximum: 629 μg/l), sample origin: Victoria (state), Australia, sample year: 1988/1989, country: Australia[461]

- Co-contamination: not reported
- Further contamination of plants and plant products other than cereals and their products reported in the present book/ article[461]: not reported

incidence: 3/20, conc. range: 106–216 μg/l, ∅ conc.: 152 μg/l, sample origin: manufactures in Christchurch, Nelson (cities), and Hawkes Bay (region), New Zealand, sample year: 1979/1980, country: New Zealand[462]

- Co-contamination: not reported
- Further contamination of plants and plant products other than cereals and their products reported in the present book/ article[462]: not reported

incidence: 857/1,525, conc. range: 1–50 μg/l (665 sa.), 51–100 μg/l (100 sa.), 101–200 μg/l (57 sa.), >200 μg/l (35 sa.), sample origin: unknown, sample year: 1994–2000, country: USA[464]

- Co-contamination: not reported
- Further contamination of plants and plant products other than cereals and their products reported in the present book/ article[464]: not reported

incidence: 15/76, conc. range: 6.4–77.5 μg/l, ∅ conc.: 15.38 μg/l, sample origin: market in Maringá (municipality), Paraná (state), Brazil, sample year: January 1992–April 1993, country: Brazil[465]

- Co-contamination: not reported
- Further contamination of plants and plant products other than cereals and their products reported in the present book/ article[465]: not reported

incidence: 7/17*, conc. range: 10–49 μg/l (6 sa.), 118 μg/l (1 sa.), sample origin: major retail outlets, health food shops, and farm shops, UK, sample year: 1992, country: UK[466], *many apple juices made from imported apples, respectively their concentrates

incidence: 7/28*, conc. range: 10–49 μg/l, sample origin: major retail outlets, health food shops, and farm shops, UK, sample year: 1993, country: UK[466], sa. imported, *many apple juices made from imported apples, respectively their concentrates

incidence: 11/15*, conc. range: 10–49 μg/l (7 sa.), 50–99 μg/l (2 sa.), 150–199 μg/l (1 sa.), 434 μg/l (1 sa.), sample origin: major retail outlets, health food shops, and farm shops, UK, sample year: 1992, country: UK[466], *apple juices made directly from apples

incidence: 10/34*, conc. range: 10–49 μg/l (7 sa.), 50–99 μg/l (3 sa., maximum: 94 μg/l), sample origin: major retail outlets, health food shops, and farm shops, UK, sample year: 1993, country: UK[466], *apple juices made directly from apples

- Co-contamination: not reported
- Further contamination of plants and plant products other than cereals and their products reported in the present book/ article[466]: not reported

incidence: 26/76*, conc. range: 10–24 μg/l (19 sa.), 50–99 μg/l (5 sa.), >100 μg/l (2 sa., maximum: 118 μg/l), sample origin: large chain stores, small producers, and farm gate shops, UK, sample year: 1993, country: UK[467], *cloudy apple juice

incidence: 13/55*, conc. range: 10–24 μg/l (10 sa.), 24–49 μg/l (2 sa.), 50 μg/l (1 sa.), sample origin: large chain stores, small producers, and farm gate shops, UK, sample year: 1993, country: UK[467], *clear apple juice

incidence: 38/113*, conc. range: 10–24 μg/l (29 sa.), 25–49 μg/l (5 sa.), 50–99 μg/l (2 sa.), >100 μg/l (2 sa., maximum: 497 μg/l), sample origin: large chain stores, small producers, and farm gate shops, UK, sample year: 1994, country: UK[467], *apple juice directly produced equivalent to cloudy

incidence: 27/78*, conc. range: 10–24 μg/l (25 sa.), 25–49 μg/l (2 sa.), sample origin: large chain stores, small producers, and farm gate shops, UK, sample year: 1994, country: UK[467], *apple juice produced from concentrate equivalent to clear

- Co-contamination: not reported
- Further contamination of plants and plant products other than cereals and their products reported in the present book/ article[467]: not reported

incidence: 65/185*, conc. range: 10–24 μg/l (39 sa.), 25–49 μg/l (15 sa.), 50–99 μg/l (7 sa.), >100 μg/l (4 sa., maximum: 490 μg/l), sample origin: retail outlets, small producers, and farm shops, UK, sample year: 1995, country: UK[468], *directly-produced apple juice

- Co-contamination: not reported
- Further contamination of plants and plant products other than cereals and their products reported in the present book/ article[468]: not reported

incidence: 62/173*, conc. range: 10–49 μg/kg (58 sa.), >50 μg/kg (4 sa., maximum: 184 μg/kg), sample origin: retail outlets, small producers, and farm shops, Scotland and England, sample year: 1996, country: UK[469], *directly-produced apple juice and one apple concentrate

- Co-contamination: not reported
- Further contamination of plants and plant products other than cereals and their products reported in the present book/ article[469]: not reported

incidence: 38/101*, conc. range: 5–15 μg/l (35 sa.), >15 μg/l (3 sa., maximum: 30 μg/l), sample origin: retail outlets, small producers, and farm shops, UK, sample year: 1998, country: UK[470], *apple juice made from apple concentrate incidence: 110/199*, conc. range: 5–15 µg/l (55 sa.), 15–50 µg/l (50 sa.), 50–100 µg/l (2 sa.), >100 µg/l (3 sa., maximum: 171 µg/l), sample origin: retail outlets, small producers, and farm shops, UK, sample year: 1998, country: UK[470], *directly-produced apple juice

- Co-contamination: not reported
- Further contamination of plants and plant products other than cereals and their products reported in the present book/article[470]: not reported

incidence: 3/140, conc. range: tr (2 sa), 10–25 µg/l (1 sa.), sample origin: domestic sa. from hand crafters, processing associations, and manufacturers purchased in retail stores in Tohoku (district), Japan, sample year: March–October 2003, country: Japan[471]

incidence: 1/6, conc. range: 10–25 µg/l, sample origin: imported sa. from processing associations and manufacturers purchased in retail stores in Tohoku (district), Japan, sample year: March–October 2003, country: Japan[471]

- Co-contamination: not reported
- Further contamination of plants and plant products other than cereals and their products reported in the present book/article[471]: Juices (apple concentrate), PAT, literature[471]

incidence: 12/90*, conc. range: LOD-LOQ (8 sa.), LOQ-25 µg/l (4 sa.), ∅ conc.: 10.2 µg/l, sample origin: supermarkets, bio shops, health food shops, farmers, small shops, and discounters in Flanders, northern part of Belgium, sample year: unknown, country: Belgium[472], *conventional (production method)

incidence: 8/65*, conc. range: LOD-LOQ (3 sa.), 25–50 µg/l (3 sa.), >50–328.7 µg/l (2 sa.), ∅ conc.: 43.1 µg/l, sample origin: supermarkets, bio shops, health food shops, farmers, small shops, and discounters in Flanders, northern part of Belgium, sample year: unknown, country: Belgium[472], *organic (production method)

incidence: 2/22*, conc. range: LOD-LOQ (1 sa.), LOQ-25 µg/l (1 sa.), ∅ conc.: 10.5 µg/l, sample origin: supermarkets, bio shops, health food shops, farmers, small shops, and discounters in Flanders, northern part of Belgium, sample year: unknown, country: Belgium[472], *handcrafted (production method)

incidence: 11/67*, conc. range: LOD-LOQ (5 sa.), LOQ-25 µg/l (5 sa.), 65.7 µg/l (1 sa.), ∅ conc.: 14.1 µg/l, sample origin: supermarkets, bio shops, health food shops, farmers, small shops, and discounters in Flanders, northern part of Belgium, sample year: unknown, country: Belgium[472], *clear apple juice (clarity)

incidence: 11/110*, conc. range: LOD-LOQ (7 sa.), 25–50 µg/l (3 sa.), 122.6 µg/l (1 sa.), ∅ conc.: 30.3 µg/l, sample origin: supermarkets, bio shops, health food shops, farmers, small shops, and discounters in Flanders, northern part of Belgium, sample year: unknown, country: Belgium[472], *cloudy apple juice (clarity)

- Co-contamination: not reported
- Further contamination of plants and plant products other than cereals and their products reported in the present book/article[472]: not reported

For detailed information see the article.

incidence: 4/4*, conc. range: 80–388 µg/l, ∅ conc.: 228 µg/l, sample origin: local grocery store, USA, sample year: unknown, country: USA[473], *100% pure apple juice

- Co-contamination: not reported
- Further contamination of plants and plant products other than cereals and their products reported in the present book/article[473]: not reported

incidence: 15*/76, conc. range: 1.4–45.6 µg/l, ∅ conc.: 8.91 µg/l, sample origin: supermarkets and convenience stores, Japan, sample year: unknown, country: Japan[475], *5 clear, 1 cloudy, 1 blended, and 8 unspecified apple juices contaminated

- Co-contamination: not reported
- Further contamination of plants and plant products other than cereals and their products reported in the present book/article[475]: not reported

incidence: 29/42, conc. range: >15–50 µg/l (15 sa.), >50 µg/l (14 sa., maximum: 285.3 µg/l), sample origin: retail outlets and producers, Iran, sample year: 2001/2002, country: Iran[476]

- Co-contamination: not reported
- Further contamination of plants and plant products other than cereals and their products reported in the present book/article[476]: Juices (apple concentrate), PAT, literature[476]

incidence: 2/2, conc. range: 38–78 µg/l, ∅ conc.: 58 µg/l, sample origin: unknown, sample year: unknown, country: Canada[477]

- Co-contamination: not reported
- Further contamination of plants and plant products other than cereals and their products reported in the present book/article[477]: not reported

incidence: 3/27, conc. range: 3–7 µg/l, sample origin: 22 cities in São Paulo (state), Brazil, sample year: 2004/2005, country: Brazil[478]

- Co-contamination: not reported
- Further contamination of plants and plant products other than cereals and their products reported in the present book/article[478]: Nectars (apple nectar), PAT, literature[478]

incidence: 54/58*, conc. range: 10–50 µg/l (48 sa.), >50 µg/l (6 sa., maximum: 121.8 µg/l), sample origin: supermarkets in Mashhad (city), east of Iran, sample year: January–June 2006, country: Iran[479], *commercial clear apple juice

- Co-contamination: not reported
- Further contamination of plants and plant products other than cereals and their products reported in the present book/article[479]: not reported

incidence: 4/15*, conc. range: 1.35–6.72 µg/l, ∅ conc.: 3.35 µg/l, sample origin: local supermarket, Okayama (prefecture), Japan, sample year: unknown, country: Japan[480], *apple juice, cloudy

incidence: 5/15*, conc. range: 1.46–14.61 µg/l, ∅ conc.: 5.87 µg/l, sample origin: local supermarket, Okayama (prefecture), Japan, sample year: unknown, country: Japan[480], *apple juice, clear

- Co-contamination: not reported
- Further contamination of plants and plant products other than cereals and their products reported in the present book/article[480]: not reported

incidence: 2/12*, conc. range: 9.1–11.2 µg/l, ∅ conc.: 10.2 µg/l, sample origin: Cesena (city), Italy, sample year: 2005, country: South Africa/Italy[481], *included apple juice, nectar, and infant fruit juice (organic and conventional)

incidence: 5/8*, conc. range: 11.0–75.2 µg/l, ∅ conc.: 31.2 µg/l, sample origin: Belville, Elsies River, and Parow (suburbs) in Tygerberg (district), Cape Town (metropolitan area), South Africa, sample year: 2005, country: South Africa/Italy[481], *included apple juice, apple nectar and infant fruit juice (organic and conventional)

incidence: 7/14*, conc. range: 14.7–166 µg/l, ∅ conc.: 83.9 µg/l, sample origin: retail outlets around commuter taxi and train termini in Tygerberg (district), Cape Town (metropolitan area), South Africa, sample year: 2005?, country: South Africa/Italy[481], *included apple juice, apple nectar, and one orange juice (not contaminated)

- Co-contamination: not reported
- Further contamination of plants and plant products other than cereals and their products reported in the present book/article[481]: not reported

For detailed information see the article.

incidence: 3/52*, conc. range: 63.45–88.44 µg/l, ⌀ conc.: 78.36 µg/l, sample origin: supermarkets and local grocery stores, China, sample year: unknown, country: China/Singapore[482], *domestic apple-based drinks

- Co-contamination: not reported
- Further contamination of plants and plant products other than cereals and their products reported in the present book/article[482]: Juices (hawthorn juice), PAT, literature[482]

For detailed information see the article.

incidence: 3/15, conc. range: 5–6 µg/l, ⌀ conc.: 5.33 µg/l, sample origin: domestic sa. from retail outlets, Tunisia, sample year: unknown, country: Tunisia/France[483]

incidence: 1/10, conc.: 6 µg/l, sample origin: imported sa. from retail outlets, Tunisia, sample year: unknown, country: Tunisia/France[483]

- Co-contamination: not reported
- Further contamination of plants and plant products other than cereals and their products reported in the present book/article[483]: Juices, Juices (apple concentrate), PAT, literature[483]

For detailed information see the article.

incidence: 63/95*, conc. range: 0.47–118.70 µg/l, ⌀ conc.: 30.04 µg/l, sample origin: supermarkets in Navarra (chartered community), Spain, sample year: unknown, country: Spain[484], *conventional

incidence: 3/5*, conc. range: 1.69–17.41 µg/l, ⌀ conc.: 11.4 µg/l, sample origin: supermarkets in Navarra (chartered community), Spain, sample year: unknown, country: Spain[484], *organic

- Co-contamination: not reported
- Further contamination of plants and plant products other than cereals and their products reported in the present book/article[484]: not reported

incidence: 12/71*, conc. range: 15–24 µg/l (8 sa.), 24–49 µg/l (4 sa., maximum: 39.9 µg/l), sample origin: supermarkets, local stores, and salesrooms in schools and organizations in the Taipei (capital) area, Taiwan, sample year: unknown, country: Taiwan[485], *pure apple juice

- Co-contamination: not reported
- Further contamination of plants and plant products other than cereals and their products reported in the present book/article[485]: not reported

incidence: 4/32* **, conc. range: LOD-LOQ (2 sa.), LOQ-10 µg/kg (2 sa., maximum: 5.5 µg/kg), sample origin: supermarkets, organic produce retail outlets, and pharmacies in Lisbon (capital), Portugal, sample year: August 2007–March 2009, country: Portugal[486], *apple juice, clear, **included are ten infant drinks

incidence: 24/36*, conc. range: LOD-LOQ (10 sa.), LOQ-10 µg/kg (5 sa.), 10–25 µg/kg (5 sa.), 25–50 µg/kg (4 sa., maximum: 42 µg/kg), sample origin: supermarkets, organic produce retail outlets, and pharmacies in Lisbon (capital), Portugal, sample year: August 2007–March 2009, country: Portugal[486], *apple juice, cloudy

incidence: 26/109* **, conc. range: LOD-LOQ (9 sa.), LOQ-10 µg/kg (8 sa.), 10–25 µg/kg (5 sa.), 25–50 µg/kg (4 sa., maximum: 42 µg/kg), sample origin: supermarkets, organic product retail outlets, and pharmacies in Lisbon (capital), Portugal, sample year: August 2007–March 2009, country: Portugal[486], *conventional, **included are ten infant drinks

incidence: 7/35*, conc. range: LOD-LOQ (6 sa.), LOQ-10 µg/kg (1 sa., maximum: 9.2 µg/kg), sample origin: supermarkets, organic produce retail outlets, and pharmacies in Lisbon (capital), Portugal, sample year: August 2007–March 2009, country: Portugal[486], *organic

- Co-contamination: not reported
- Further contamination of plants and plant products other than cereals and their products reported in the present book/article[486]: Puree, PAT, literature[486]

incidence: 14/20, conc. range: 3.5–122.5 µg/l, ⌀ conc.: 50.6 µg/l, sample origin: supermarkets in Copenhagen (capital) area, Denmark, sample year: spring 2006, country: Spain/Denmark[547]

- Co-contamination: not reported
- Further contamination of plants and plant products other than cereals and their products reported in the present book/article[547]: not reported

incidence: 7/25*, conc. range: LOQ-10 µg/kg (5 sa.), 10–25 µg/kg (2 sa., maximum: 13.2 µg/kg), ⌀ conc.: 6.1 µg/kg, sample origin: Catalonian (autonomous community) supermarkets and bio shops, Spain, sample year: November 2011–May 2012, country: Spain[594], *conventional

incidence: 14/22*, conc. range: LOQ-10 µg/kg (4 sa.), 10–25 µg/kg (9 sa.), 36.5 µg/kg (1 sa.), ⌀ conc.: 13.9 µg/kg, sample origin: Catalonian (autonomous community) supermarkets and bio shops, Spain, sample year: November 2011–May 2012, country: Spain[594], *organic

incidence: 9/26*, conc. range: LOQ-10 µg/kg (2 sa.), 10–25 µg/kg (6 sa.), 36.5 µg/kg (1 sa.), ⌀ conc.: 14.5 µg/kg, sample origin: Catalonian (autonomous community) supermarkets and bio shops, Spain, sample year: November 2011–May 2012, country: Spain[594], *without vitamin C

incidence: 12/21*, conc. range: LOQ-10 µg/kg (7 sa.), 10–25 µg/kg (5 sa., maximum: 22.1 µg/kg),), ⌀ conc.: 9.6 µg/kg, sample origin: Catalonian (autonomous community) supermarkets and bio shops, Spain, sample year: November 2011–May 2012, country: Spain[594], *with vitamin C

incidence: 10/35*, conc. range: LOQ-10 µg/kg (5 sa.), 10–25 µg/kg (4 sa.), 36.5 µg/kg (1 sa.), ⌀ conc.: 11.7 µg/kg, sample origin: Catalonian (autonomous community) supermarkets and bio shops, Spain, sample year: November 2011–May 2012, country: Spain[594], *apple juice, clear

incidence: 11/12*, conc. range: LOQ-10 µg/kg (4 sa.), 10–25 µg/kg (7 sa., maximum: 22.1 µg/kg),), ⌀ conc.: 11.7 µg/kg, sample origin: Catalonian (autonomous community) supermarkets and bio shops, Spain, sample year: November 2011–May 2012, country: Spain[594], *apple juice, cloudy

- Co-contamination: not reported
- Further contamination of plants and plant products other than cereals and their products reported in the present book/article[594]: Puree, PAT, literature[594]

For detailed information see the article.

incidence: 9/14*, conc. range: 1.7–4.5 µg/l, sample origin: producers and local markets, Italy, sample year: unknown, country: Italy[638], *conventional

incidence: 3/4*, conc. range: 1.2–1.8 µg/l, sample origin: producers and local markets, Italy, sample year: unknown, country: Italy[638], *integrated

incidence: 3/8*, conc. range: 1.0–22 µg/l, sample origin: producers and local markets, Italy, sample year: unknown, country: Italy[638], *organic

- Co-contamination: not reported
- Further contamination of plants and plant products other than cereals and their products reported in the present book/ article[638]: not reported

incidence: 32/49, conc. range: LOD-<10 µg/l (14 sa.), ≤904.4* µg/l (18 sa.), sample origin: handicraft apple presses in Walloon region, south of Belgium, sample year: October/November 2006, country: Belgium[686], *sa. of tree-picked apples after 7 days of storage in a bin

- Co-contamination: not reported
- Further contamination of plants and plant products other than cereals and their products reported in the present book/ article[686]: not reported

incidence: 31/35, conc. range: 6–30 µg/l, sample origin: European and Italian sa. purchased in retail outlets, Italy, sample year: 2008, country: Italy[688]

- Co-contamination: not reported
- Further contamination of plants and plant products other than cereals and their products reported in the present book/ article[688]: Juices, Juices (pear juice), PAT, literature[688]

For detailed information see the article.

incidence: 3/24, conc. range: 2.8–8.9 µg/l, Ø conc.: 6.8 µg/l, sample origin: wholesale stores in Busan, Daegu, Daejeon, Gwangju, Jinju (cities), and Seoul (capital), South Korea, sample year: November 2007, country: South Korea[695]

- Co-contamination: not reported
- Further contamination of plants and plant products other than cereals and their products reported in the present book/ article[695]: Juices (grape juice), Juices (orange juice), PAT, literature[695]

incidence: 1/13, conc.: 26.9 µg/l*, sample origin: purchased in supermarkets imported from Dubai, Indonesia, Japan, and Malaysia*, sample year: November 2012–March 2013, country: Malaysia[773]

- Co-contamination: not reported
- Further contamination of plants and plant products other than cereals and their products reported in the present book/ article[773]: Juices (lychee juice), Juices (pineapple concentrate), PAT, literature[773]

incidence: 2/28, conc. range: 2.5–6.0 µg/kg, Ø conc.: 4.25 µg/kg, sample origin: hypermarkets and supermarkets in Valencia (city), Spain, sample year: 2008, country: Spain[786]

- Co-contamination: not reported
- Further contamination of plants and plant products other than cereals and their products reported in the present book/ article[786]: Juices (apple concentrate), Juices (peach concentrate), Juices (pear concentrate), PAT, literature[786]

incidence: 14/20*, conc. range: 3.18–27.09 µg/l, Ø conc.: 11.5 µg/l, sample origin: supermarkets in Navarra (autonomous community and province), Spain, sample year: 2008, country: Spain[799], *apple juice for infants older than 4 months

- Co-contamination: not reported
- Further contamination of plants and plant products other than cereals and their products reported in the present book/ article[799]: not reported

incidence: 41/50*, conc. range: 0.7–5 µg/l (16 sa.), 5–10 µg/l (7 sa.), 10–50 µg/l (15 sa.), >50 µg/l (3 sa., maximum: 101.9 µg/l), sample origin: local markets, Romania, sample year: unknown, country: Romania[810], *apple juice, clear and cloudy

- Co-contamination: not reported
- Further contamination of plants and plant products other than cereals and their products reported in the present book/ article[810]: not reported

For detailed information see the article.

incidence: 1/30, conc.: 10 µg/l, sample origin: retail markets in Belo Horizonte (city), Minas Gerais (state), Brazil, sample year: March–September 1999, country: Brazil[819]

- Co-contamination: not reported
- Further contamination of plants and plant products other than cereals and their products reported in the present book/ article[819]: not reported

incidence: 12/36, conc. range: 5–15 µg/kg (4 sa.), 30.1–50 µg/kg (4 sa.), ≥50 µg/kg (4 sa., maximum: 190.7 µg/kg), Ø conc.: 52.8 µg/ kg, sample origin: retail outlets and producers of Azerbaijan, Chaharmahal va Bakhtiari, Esfahan, Gillan, Khuzestan, and Tehran (provinces), Iran, sample year: April–November 2013, country: Iran[827]

- Co-contamination: not reported
- Further contamination of plants and plant products other than cereals and their products reported in the present book/ article[827]: Juices (apple concentrate), Juices (grape juice), Juices (peach concentrate), Juices (pear juice), Juices (pomegranate juice), PAT, literature[827]

incidence: 3/15*, conc. range: 100–500 µg/l (2 sa.), 1,839 µg/l (1 sa.), Ø conc.: 845 µg/l, sample origin: local retail markets and juice joints in Lucknow (city), India, sample year: unknown, country: India[838], *branded apple juice

incidence: 7/25*, conc. range: <30 µg/l (1 sa.), 30–50 µg/l (1 sa.), 100–500 µg/l (5 sa., maximum: 325 µg/l), Ø conc.: 191 µg/l, sample origin: local retail markets and juice joints in Lucknow (city), India, sample year: unknown, country: India[838], *apple juice from local vendors

- Co-contamination: not reported
- Further contamination of plants and plant products other than cereals and their products reported in the present book/ article[838]: Juices, PAT, literature[838]

incidence: 10/30*, conc. range: ≤1,650 µg/l, sample origin: retail outlets in Free State, Limpopo, and Western Cape (provinces), South Africa, sample year: latter half of 2006, country: South Africa/Chile[842], *apple juices and apple blends

- Co-contamination: not reported
- Further contamination of plants and plant products other than cereals and their products reported in the present book/ article[842]: not reported

incidence: 2/25, conc. range: 34.05–36.81 µg/kg, Ø conc.: 35.43 µg/kg, sample origin: supermarkets in Beijing (capital), China, sample year: unknown, country: China/Italy[887]

- Co-contamination: not reported
- Further contamination of plants and plant products other than cereals and their products reported in the present book/article[887]: Juices (hawthorn juice), PAT, literature[887]

incidence: 27/42, conc. range: 4–122.36 µg/l, ∅ conc.: 45.71 µg/l, sample origin: local supermarkets, Tunisia, sample year: 2013, country: Tunisia[888]

- Co-contamination: not reported
- Further contamination of plants and plant products other than cereals and their products reported in the present book/article[888]: Jam (apple jam), Jam (pear jam), Juices, Juices (pear juice), Products (fruit product), PAT, literature[888]

incidence: 12/15, conc. range: ≤90.3 µg/kg, sample origin: supermarkets and stores in Changchun (city), China, sample year: 2009, country: China[890]

- Co-contamination: not reported
- Further contamination of plants and plant products other than cereals and their products reported in the present book/article[890]: Baby food, Juices, Juices (apple concentrate), PAT, literature[890]

incidence: 11/30, conc. range: ≤167 µg/l sample origin: major supermarkets and stores, Tunisia, sample year: 2011?, country: Tunisia[895]

- Co-contamination: not reported
- Further contamination of plants and plant products other than cereals and their products reported in the present book/article[895]: Baby food, Juices, PAT, literature[895]

incidence: 7/12* conc. range: 1–5 µg/kg (5 sa.), 5–10 µg/kg (2 sa., maximum: 8.3 µg/kg), ∅ conc.: 4.1 µg/kg, sample origin: supermarkets and retail shops in Novi Sad (city), Vojvodina (province), Serbia, sample year: 2013–2015, country: Serbia/Portugal[904], *apple-based juice for children

incidence: 17/34*, conc. range: 1–15 µg/kg (15 sa.), 10–25 µg/kg (1 sa.), 26.4 µg/kg (1 sa.), ∅ conc.: 5.9 µg/kg, sample origin: supermarkets and retail shops in Novi Sad (city), Vojvodina (province), Serbia, sample year: 2013–2015, country: Serbia/Portugal[904], *apple-based juice (small package juices with straw)

- Co-contamination: not reported
- Further contamination of plants and plant products other than cereals and their products reported in the present book/article[904]: Juices, Puree, PAT, literature[904]

For detailed information see the article.

incidence: 2/2* conc. range: 8.13–13.98 µg/kg, ∅ conc.: 11.6 µg/kg, sample origin: supermarkets and organic shops, Portugal, sample year: unknown, country: Portugal[905], *conventional

incidence: 5/7*, conc. range: <LOQ-45.77 µg/kg, sample origin: supermarkets and organic shops, Portugal, sample year: unknown, country: Portugal[905], *organic

- Co-contamination: not reported
- Further contamination of plants and plant products other than cereals and their products reported in the present book/article[905]: Jellies, Juices (tomato juice), Puree, PAT, literature[905]

For detailed information see the article.

incidence: 16/16*, conc. range: 14.3–46.7 µg/kg, ∅ conc.: 23.1 µg/kg, sample origin: juice producer industry, Rio Grande do Sul (state), Brazil, sample year: 2006/2007, country: Brazil[909], *many damaged apples

- Co-contamination: not reported
- Further contamination of plants and plant products other than cereals and their products reported in the present book/article[909]: Fruits (apple), PAT, literature[909]

incidence: 5/6, conc. range: 3.8–28.4 µg/l, ∅ conc.: 16.7 µg/l, sample origin: Czech retail market, Czech Republic, sample year: unknown, country: Czech Republic[911]

- Co-contamination: not reported
- Further contamination of plants and plant products other than cereals and their products reported in the present book/article[911]: Apple cider, Baby food, Fruit salad, Fruits (apple), Fruits (pear), Juices (pear juice), Pulp (apple pulp), PAT, literature[911]

see also **Juices (apple concentrate)**, and **Products (fruit product)**

Juices (apricot concentrate) may contain the following mycotoxins:

Aspergillus and *Penicillium* Toxins

Patulin

incidence: 2/2, conc. range: 12.2–15.3 µg/kg, ∅ conc.: 13.70 µg/kg, sample origin: imports from China and Greece, sample year: October 2004–June 2006, country: Greece[108]

- Co-contamination: not reported
- Further contamination of plants and plant products other than cereals and their products reported in the present book/article[108]: Baby food, Drinks (apple drink), Juices, Juices (apple concentrate), Juices (apple juice), Juices (orange juice), Juices (pineapple concentrate), Juices (rosehip juice), PAT, literature[108]

Juices (apricot juice) may contain the following mycotoxins:

Alternaria Toxins

Tenuazonic Acid

incidence: 2/2, conc. range: 25–27 µg/kg, ∅ conc.: 26 µg/kg, sample origin: unknown, sample year: unknown, country: Germany[166]

- Co-contamination: not reported
- Further contamination of plants and plant products other than cereals and their products reported in the present book/article[166]: Juices, Juices (apple juice), Juices (banana juice), Juices (blackcurrant juice), Juices (cherry juice), Juices (grape juice), Juices (orange juice), Juices (pear juice), Juices (pomegranate juice), Juices (strawberry juice), Spices, Spices (coriander), Spices (cumin), Spices (curcuma), Spices (curry), Spices (ginger), Spices (paprika), Spices (pepper), TA, literature[166]

Aspergillus and *Penicillium* Toxins

Patulin

incidence: 7/27, conc. range: <10 µg/kg (2 sa.), >10 µg/kg (5 sa., maximum: 32.4 µg/kg), sample origin: supermarkets and organic food shops, Northwestern Italy, sample year: January–September 2007, country: Italy[847]

- Co-contamination: not reported
- Further contamination of plants and plant products other than cereals and their products reported in the present book/article[847]: Juices, Juices (peach juice), Juices (pear juice), PAT, literature[847]

For detailed information see the article.

see also **Products (fruit product)**

Juices (banana juice) may contain the following mycotoxins:

Alternaria Toxins

Tenuazonic Acid

incidence: 2/3, conc. range: 0.5–0.7 μg/kg, ∅ conc.: 0.6 μg/kg, sample origin: unknown, sample year: unknown, country: Germany[166]

- Co-contamination: not reported
- Further contamination of plants and plant products other than cereals and their products reported in the present book/article[166]: Juices, Juices (apple juice), Juices (apricot juice), Juices (blackcurrant juice), Juices (cherry juice), Juices (grape juice), Juices (orange juice), Juices (pear juice), Juices (pomegranate juice), Juices (strawberry juice), Spices, Spices (coriander), Spices (cumin), Spices (curcuma), Spices (curry), Spices (ginger), Spices (paprika), Spices (pepper), TA, literature[166]

Juices (blackcurrant juice) may contain the following mycotoxins:

Alternaria Toxins

Tenuazonic Acid

incidence: 5/5, conc. range: 0.5–2.3 μg/kg, sample origin: unknown, sample year: unknown, country: Germany[166]

- Co-contamination: not reported
- Further contamination of plants and plant products other than cereals and their products reported in the present book/article[166]: Juices, Juices (apple juice), Juices (apricot juice), Juices (banana juice), Juices (cherry juice), Juices (grape juice), Juices (orange juice), Juices (pear juice), Juices (pomegranate juice), Juices (strawberry juice), Spices, Spices (coriander), Spices (cumin), Spices (curcuma), Spices (curry), Spices (ginger), Spices (paprika), Spices (pepper), TA, literature[166]

Aspergillus and *Penicillium* Toxins

Citrinin

incidence: 1/2, conc.: 0.20 μg/l, sample origin: local retail stores, Germany, sample year: unknown, country: Germany[711]

- Co-contamination: not reported
- Further contamination of plants and plant products other than cereals and their products reported in the present book/article[711]: Juices (apple juice), Juices (cherry juice), Juices (tomato juice), CIT, literature[711]

Ochratoxin A

incidence: 3/19, conc. range: ≤0.06 μg/l, sample origin: different origins, Germany, sample year: unknown, country: Germany[110]

- Co-contamination: not reported
- Further contamination of plants and plant products other than cereals and their products reported in the present book/article[110]: Juices (carrot juice), Juices (grape juice), Juices (tomato juice), Sauces (pepper- and barbecue sauce), Sauces (soy sauce), Spices (mustard), Tomato ketchup, Vinegar (apple), Vinegar (balsamico), Vinegar (wine), Wine (red), Wine (rosé), Wine (white), OTA, literature[110]

Patulin

incidence: 1/1, conc.: 0.1 μg/l, sample origin: local store in Munich (city) Germany, sample year: 1996/1997, country: Germany[52], *commercial product

- Co-contamination: not reported
- Further contamination of plants and plant products other than cereals and their products reported in the present book/article[52]: Juices (apple juice), Juices (cherry juice), Juices (grape juice), Juices (orange juice), Pulp (plum pulp), PAT, literature[52]

Juices (carrot juice) may contain the following mycotoxins:

Aspergillus and *Penicillium* Toxins

Ochratoxin A

incidence: 1/18, conc.: 0.01 μg/l, sample origin: different origins, Germany, sample year: unknown, country: Germany[110]

- Co-contamination: not reported
- Further contamination of plants and plant products other than cereals and their products reported in the present book/article[110]: Juices (blackcurrant juice), Juices (grape juice), Juices (tomato juice), Sauces (pepper- and barbecue sauce), Sauces (soy sauce), Spices (mustard), Tomato ketchup, Vinegar (apple), Vinegar (balsamico), Vinegar (wine), Wine (red), Wine (rosé), Wine (white), OTA, literature[110]

Juices (cherry juice) may contain the following mycotoxins:

Alternaria Toxins

Tenuazonic Acid

incidence: 6/6, conc. range: 0.8–54 μg/kg, sample origin: unknown, sample year: unknown, country: Germany[166]

- Co-contamination: not reported
- Further contamination of plants and plant products other than cereals and their products reported in the present book/article[166]: Juices, Juices (apple juice), Juices (apricot juice), Juices (banana juice), Juices (blackcurrant juice), Juices (grape juice), Juices (orange juice), Juices (pear juice), Juices (pomegranate juice), Juices (strawberry juice), Spices, Spices (coriander), Spices (cumin), Spices (curcuma), Spices (curry), Spices (ginger), Spices (paprika), Spices (pepper), TA, literature[166]

Aspergillus and *Penicillium* Toxins

Citrinin

incidence: 1/5, conc.: 0.10 μg/l, sample origin: local retail stores, Germany, sample year: unknown, country: Germany[711]

- Co-contamination: not reported
- Further contamination of plants and plant products other than cereals and their products reported in the present book/article[711]: Juices (apple juice), Juices (blackcurrant juice), Juices (tomato juice), CIT, literature[711]

Patulin

incidence: 1/1*, conc.: 0.2 μg/l, sample origin: local store in Munich (city) Germany, sample year: 1996/1997, country: Germany[52], *sour cherry juice, commercial product

- Co-contamination: not reported

- Further contamination of plants and plant products other than cereals and their products reported in the present book/article[52]: Juices (apple juice), Juices (blackcurrant juice), Juices (grape juice), Juices (orange juice), Pulp (plum pulp), PAT, literature[52]

Juices (citrus juice) may contain the following mycotoxins:

Alternaria Toxins

Alternariol Methyl Ether

incidence: 4/36, conc. range: 0.11–0.20 µg/kg, ∅ conc.: 0.14 µg/kg, sample origin: local supermarkets and agricultural trade markets in Beijing (capital), China, sample year: unknown, country: China[883]

- Co-contamination: For detailed information see the article.
- Further contamination of plants and plant products other than cereals and their products reported in the present book/article[883]: Juices (citrus juice), TA, literature[883]; Juices (tomato juice), AME and TA, literature[883]; Tomato ketchup, AME, AOH and TA, literature[883]; see also Juices (tomato juice) and Tomato ketchup, chapter: Further Mycotoxins and Microbial Metabolites

Tenuazonic Acid

incidence: 9/36, conc. range: 1.21–4.3 µg/kg, ∅ conc.: 1.80 µg/kg, sample origin: local supermarkets and agricultural trade markets in Beijing (capital), China, sample year: unknown, country: China[883]

- Co-contamination: For detailed information see the article.
- Further contamination of plants and plant products other than cereals and their products reported in the present book/article[883]: Juices (citrus juice), AME, literature[883]; Juices (tomato juice), AME and TA, literature[883]; Tomato ketchup, AME, AOH and TA, literature[883]; see also Juices (tomato juice) and Tomato ketchup, chapter: Further Mycotoxins and Microbial Metabolites

Juices (cranberry juice) may contain the following mycotoxins:

Alternaria Toxins

Alternariol

incidence: 1/5, conc.: 0.04 µg/l, sample origin: USA and Canada, sample year: unknown, country: Canada[492]

- Co-contamination: 1 sa. co-contaminated with AME and AOH
- Further contamination of plants and plant products other than cereals and their products reported in the present book/article[492]: Juices (cranberry juice), AME, literature[492]; Juices (grape juice), Wine (red), Wine (white), AME and AOH, literature[492]

Alternariol Methyl Ether

incidence: 1/5, conc.: 0.003 µg/l, sample origin: USA and Canada, sample year: unknown, country: Canada[492]

- Co-contamination: 1 sa. co-contaminated with AME and AOH
- Further contamination of plants and plant products other than cereals and their products reported in the present book/article[492]: Juices (cranberry juice), AOH, literature[492]; Juices (grape juice), Wine (red), Wine (white), AME and AOH, literature[492]

Juices (grape juice) may contain the following mycotoxins:

Alternaria Toxins

Alternariol

incidence: 5/5, conc. range: 0.10–1.05 µg/kg, sample origin: commercially available, sample year: unknown, country: Germany[160]

- Co-contamination: 2 sa. co-contaminated with AME and AOH; 3 sa. contaminated solely with AOH
- Further contamination of plants and plant products other than cereals and their products reported in the present book/article[160]: Beverages, Juices (apple juice), Juices (orange juice), Juices (tomato juice), Juices (vegetable juice), Wine (red), Wine (white), Wine, miscellaneous, AME and AOH, literature[160]; Juices (grape juice), AME, literature[160]

incidence: 1/1, conc.: 1.6 µg/l, sample origin: unknown, sample year: unknown, country: Canada[440]

- Co-contamination: 1 sa. co-contaminated with AME and AOH
- Further contamination of plants and plant products other than cereals and their products reported in the present book/article[440]: Juices (apple juice), Nectars (cranberry nectar), Nectars (prune nectar), AME and AOH, literature[440]; Juices (grape juice), AME, literature[440]; Juices (raspberry juice), Wine (red), AOH, literature[440]

incidence: 3/10*, conc. range: 0.03–0.46 µg/l, ∅ conc.: 0.23 µg/l, sample origin: USA and Canada, sample year: unknown, country: Canada[492], *red grape juice

- Co-contamination: 3 sa. co-contaminated with AME and AOH
- Further contamination of plants and plant products other than cereals and their products reported in the present book/article[492]: Juices (cranberry juice), Wine (red), Wine (white), AME and AOH, literature[492]; Juices (grape juice), AME, literature[492]

Alternariol Methyl Ether

incidence: 2/5, conc. range: pr., sample origin: commercially available, sample year: unknown, country: Germany[160]

- Co-contamination: 2 sa. co-contaminated with AME and AOH
- Further contamination of plants and plant products other than cereals and their products reported in the present book/article[160]: Beverages, Juices (apple juice), Juices (orange juice), Juices (tomato juice), Juices (vegetable juice), Wine (red), Wine (white), Wine, miscellaneous, AME and AOH, literature[160]; Juices (grape juice), AOH, literature[160]

incidence: 1/1, conc.: 0.23 µg/l, sample origin: unknown, sample year: unknown, country: Canada[440]

- Co-contamination: 1 sa. co-contaminated with AME and AOH
- Further contamination of plants and plant products other than cereals and their products reported in the present book/article[440]: Juices (apple juice), Nectars (cranberry nectar), Nectars (prune nectar), AME and AOH, literature[440]; Juices (grape juice), Juices (raspberry juice), Wine (red), AOH, literature[440]

incidence: 5/10*, conc. range: 0.01–39.5 µg/l, ∅ conc.: 7.9 µg/l, sample origin: USA and Canada, sample year: unknown, country: Canada[492], *red grape juice

- Co-contamination: 3 sa. co-contaminated with AME and AOH; 2 sa. contaminated solely with AME
- Further contamination of plants and plant products other than cereals and their products reported in the present book/article[492]: Juices (cranberry juice), Wine (red), Wine

(white), AME and AOH, literature[492]; Juices (grape juice), AOH, literature[492]

Tenuazonic Acid

incidence: 4/4, conc. range: 1.7–7 µg/kg, sample origin: unknown, sample year: unknown, country: Germany[166]

- Co-contamination: not reported
- Further contamination of plants and plant products other than cereals and their products reported in the present book/article[166]: Juices, Juices (apple juice), Juices (apricot juice), Juices (banana juice), Juices (blackcurrant juice), Juices (cherry juice), Juices (orange juice), Juices (pear juice), Juices (pomegranate juice), Juices (strawberry juice), Spices, Spices (coriander), Spices (cumin), Spices (curcuma), Spices (curry), Spices (ginger), Spices (paprika), Spices (pepper), TA, literature[166]

incidence: 11/14, conc. range: <LOQ-19.4 µg/kg, sample origin: local supermarkets, Belgium, sample year: February 2013–February 2015, country: Belgium/Austria/Germany[514]

- Co-contamination: For detailed information see the article.
- Further contamination of plants and plant products other than cereals and their products reported in the present book/article[514]: Juices (apple juice), TA, literature[514]; Juices (tomato juice), Sauces (tomato sauce), Tomato concentrate, ALT, AME, AOH, and TA, literature[514]; see also Juices (tomato juice), Sauces (tomato sauce), Tomato concentrate, chapter: Further Mycotoxins and Microbial Metabolites

For detailed information see the article.

Aspergillus **and** ***Penicillium*** **Toxins**

Ochratoxin A

incidence: 21/27*, conc. range: ≤1.3 µg/l, sample origin: different origins, Germany, sample year: unknown, country: Germany[110], *grape juice, white

incidence: 56/64*, conc. range: ≤5.3 µg/l, sample origin: different origins, Germany, sample year: unknown, country: Germany[110], *grape juice, red

- Co-contamination: not reported
- Further contamination of plants and plant products other than cereals and their products reported in the present book/article[110]: Juices (blackcurrant juice), Juices (carrot juice), Juices (tomato juice), Sauces (pepper- and barbecue sauce), Sauces (soy sauce), Spices (mustard), Tomato ketchup, Vinegar (apple), Vinegar (balsamico), Vinegar (wine), Wine (red), Wine (rosé), Wine (white), OTA, literature[110]

incidence: 19/20*, conc. range: 0.01–0.19 µg/kg (9 sa.), 0.41–1.0 µg/kg (7 sa.), 1.1–2 µg/kg (2 sa.), 2.05 µg/kg (1 sa.), sample origin: 13 regions in the UK, sample year: May 1998, country: UK[263], *red and white grape juice

- Co-contamination: not reported
- Further contamination of plants and plant products other than cereals and their products reported in the present book/article[263]: Chocolate, Currants, Figs, Powder (cocoa powder), Raisins, Sultanas, Wine (red), OTA, literature[263]

incidence: 34/48, conc. range: 0.0212–0.100 µg/l, ∅ conc.: 0.038 µg/l, sample origin: markets, Rio de Janeiro (state), Brazil, sample year: unknown, country: Brazil/Argentina[377]

- Co-contamination: not reported
- Further contamination of plants and plant products other than cereals and their products reported in the present book/article[377]: Pulp (grape pulp), Wine (red), Wine (rosé), Wine (white), OTA, literature[377]

incidence: 7/7*, conc. range: 0.0016–0.0647 µg/l, ∅ conc.: 0.0258 µg/l, sample origin: retail shops in Warsaw (capital), Poland, sample year: unknown, country: Poland[494], *white grape juice and white grape as well as fruit grape drinks

- Co-contamination: not reported
- Further contamination of plants and plant products other than cereals and their products reported in the present book/article[494]: Wine (red), OTA, literature[494]

incidence: 4/25*, conc. range: LOD-0.050 µg/l (3 sa.), 0.071 µg/l (1 sa.), sample origin: USA and Canada, sample year: 1999–2002, country: Canada[495], *white grape juice

incidence: 5/45*, conc. range: LOD-0.050 µg/l (2 sa.), >0.050 µg/l (3 sa., maximum: 0.104 µg/l), sample origin: USA and Canada, sample year: 1999–2002, country: Canada[495], *red grape juice

- Co-contamination: not reported
- Further contamination of plants and plant products other than cereals and their products reported in the present book/article[495]: Wine (red), Wine (white), OTA, literature[495]

incidence: 7?/11*, conc. range: ≤0.311 µg/l, sample origin: partly imported from France and Italy purchased in retail outlets in Berne (capital), Switzerland, sample year: 1994/1995, country: Switzerland[496], *red and white grape juice

- Co-contamination: not reported
- Further contamination of plants and plant products other than cereals and their products reported in the present book/article[496]: Wine (red), Wine (rosé), Wine (white), Wine, miscellaneous, OTA, literature[496]

incidence: 27/37*, conc. range: ≤6.72 µg/l, ∅ conc.: 0.82 µg/l, sample origin: Germany?, sample year: 1999/2000, country: Germany[497], *35 out of them were red juices

- Co-contamination: not reported
- Further contamination of plants and plant products other than cereals and their products reported in the present book/article[497]: Wine (red), OTA, literature[497]

incidence: 2/10, conc. range: 0.90–1.90 µg/kg, ∅ conc.: 1.40 µg/kg, sample origin: different supermarkets and traditional bazaars in Istanbul (capital), Turkey, sample year: 2008/2009, country: Turkey[665]

- Co contamination: not reported
- Further contamination of plants and plant products other than cereals and their products reported in the present book/article[665]: Currants, Pekmez, Sultanas, OTA, literature[665]

incidence: 3/10, conc. range: 0.26–0.54 µg/l, sample origin: local supermarkets in Beijing (capital), China, sample year: unknown, country: China[690]

- Co-contamination: not reported
- Further contamination of plants and plant products other than cereals and their products reported in the present book/article[690]: Wine (red), OTA, literature[690]

incidence: 75/90, conc. range: LOD/LOQ-0.9 µg/kg (50 sa.), 1.0–2.9 µg/kg (20 sa.), 3.0–4.9 µg/kg (4 sa.), 5.261 µg/kg (1 sa.), sample origin: Germany, sample year: 1995–1998, country: EU[698]

incidence: 27/28, conc. range: LOD/LOQ-≤0.56 µg/kg (27 sa.), sample origin: Norway, sample year: 1999, country: EU[698]

incidence: 8/8, conc. range: LOD/LOQ-≤0.176 µg/kg (8 sa.), sample origin: Spain, sample year: 1997, country: EU[698]

- Co-contamination: not reported
- Further contamination of plants and plant products other than cereals and their products reported in the present book/article[698]: Baby food, Beans (soybean), Chocolate, Cocoa, Coffee, Coffee beans, Currants, Fruits, dried, Herbs, Jam, Juices, Legumes, Nuts, Oil seeds, Oils (olive oil), Peas, Peas (chickpea), Products (plant product), Raisins, Spices, Spices (cayenne pepper, chilli), Spices (nutmeg), Spices (paprika), Spices (pepper), Sultanas, Tea, Vinegar, Wine, Wine (red), Wine (rosé), Wine (white), Wine, miscellaneous, OTA, literature[698]

Patulin

incidence: 2/2*, conc. range: 4.9–5.2 µg/l, ∅ conc.: 5.1 µg/l, sample origin: local store in Munich (city) Germany, sample year: 1996/1997, country: Germany[52], *commercial products
- Co-contamination: not reported
- Further contamination of plants and plant products other than cereals and their products reported in the present book/article[52]: Juices (apple juice), Juices (blackcurrant juice), Juices (cherry juice), Juices (orange juice), Pulp (plum pulp), PAT, literature[52]

incidence: 4/24, conc. range: 5.2–14.5 µg/l, ∅ conc.: 11.3 µg/l, sample origin: wholesale stores in Busan, Daegu, Daejeon, Gwangju, Jinju (cities), and Seoul (capital), South Korea, sample year: November 2007, country: South Korea[695]
- Co-contamination: not reported
- Further contamination of plants and plant products other than cereals and their products reported in the present book/article[695]: Juices (apple juice), Juices (orange juice), PAT, literature[695]

incidence: 3/10*, conc. range: 5–15 µg/kg (2 sa.), 16.6 µg/kg (1 sa.), ∅ conc.: 12.3 µg/kg, sample origin: retail outlets and producers of Azerbaijan, Chaharmahal va Bakhtiari, Esfahan, Gillan, Khuzestan, and Tehran (provinces), Iran, sample year: April–November 2013, country: Iran[827], *white grape juice
- Co-contamination: not reported
- Further contamination of plants and plant products other than cereals and their products reported in the present book/article[827]: Juices (apple concentrate), Juices (apple juice), Juices (peach concentrate), Juices (pear juice), Juices (pomegranate juice), PAT, literature[827]

see also **Products (fruit product)**

Juices (grapefruit juice) may contain the following mycotoxins:

Alternaria Toxins

Alternariol

incidence: 1/2, conc.: <LOQ, sample origin: supermarkets, grocery stores, wineries, and small stores in Entre Rios Province, Argentina, sample year: 2011, country: Argentina[498]
- Co-contamination: not reported
- Further contamination of plants and plant products other than cereals and their products reported in the present book/article[498]: Juices (orange juice), Wine (rosé), AOH, literature[498]; Wine (red), Wine (white), AME and AOH, literature[498]

Juices (guava juice) may contain the following mycotoxins:

Aspergillus Toxins

Aflatoxin B_1

incidence: 2/5, conc. range: 12 µg/l, sample origin: shops and markets, Assiut (governorate), Egypt, sample year: unknown, country: Egypt[443]
- Co-contamination: not reported
- Further contamination of plants and plant products other than cereals and their products reported in the present book/article[443]: Beverages, AFB_1 and AFG_1, literature[443]

Juices (hawthorn juice) may contain the following mycotoxins:

Aspergillus and ***Penicillium*** Toxins

Patulin

incidence: 6/43*, conc. range: 19.8–206.88 µg/l, ∅ conc.: 116.34 µg/l, sample origin: supermarkets and local grocery stores, China, sample year: unknown, country: China/Singapore[482], *domestic hawthorn-based drinks
- Co-contamination: not reported
- Further contamination of plants and plant products other than cereals and their products reported in the present book/article[482]: Juices (apple juice), PAT, literature[482]

For detailed information see the article.

incidence: 1/13, conc.: 12.26 µg/kg, sample origin: supermarkets in Beijing (capital), China, sample year: unknown, country: China/Italy[887]
- Co-contamination: not reported
- Further contamination of plants and plant products other than cereals and their products reported in the present book/article[887]: Juices (apple juice), PAT, literature[887]

Juices (lychee juice) may contain the following mycotoxins:

Aspergillus and ***Penicillium*** Toxins

Patulin

incidence: 1/6, conc.: 13.1 µg/l, sample origin: Malaysia, sample year: November 2012–March 2013, country: Malaysia[773]
- Co-contamination: not reported
- Further contamination of plants and plant products other than cereals and their products reported in the present book/article[773]: Juices (apple juice), Juices (pineapple concentrate), PAT, literature[773]

Juices (orange juice) may contain the following mycotoxins:

Alternaria Toxins

Alternariol

incidence: 2/2, conc. range: 0.16–0.24 µg/kg, ∅ conc.: 0.20 µg/kg, sample origin: commercially available, sample year: unknown, country: Germany[160]
- Co-contamination: 2 sa. co-contaminated with AME and AOH
- Further contamination of plants and plant products other than cereals and their products reported in the present book/article[160]: Beverages, Juices (apple juice), Juices (grape juice), Juices (tomato juice), Juices (vegetable juice), Wine (red), Wine (white), Wine, miscellaneous, AME and AOH, literature[160]; Juices (orange juice), AME, literature[160]

incidence: 2/31, conc.: <LOQ, sample origin: supermarkets, grocery stores, wineries, and small stores in Entre Rios and Mendoza (provinces), Argentina, sample year: 2011, country: Argentina[498]
- Co-contamination: not reported

- Further contamination of plants and plant products other than cereals and their products reported in the present book/article[498]: Juices (grapefruit juice), Wine (rosé), AOH, literature[498]; Wine (red), Wine (white), AME and AOH, literature[498]

Alternariol Methyl Ether

incidence: 2/2, conc. range: 0.18–0.27 µg/kg, ∅ conc.: 0.23 µg/kg, sample origin: commercially available, sample year: unknown, country: Germany[160]

- Co-contamination: 2 sa. co-contaminated with AME and AOH
- Further contamination of plants and plant products other than cereals and their products reported in the present book/article[160]: Beverages, Juices (apple juice), Juices (grape juice), Juices (tomato juice), Juices (vegetable juice), Wine (red), Wine (white), Wine, miscellaneous, AME and AOH, literature[160]; Juices (orange juice), AOH, literature[160]

Tenuazonic Acid

incidence: 6/6, conc. range: 1.0–1.9 µg/kg, sample origin: unknown, sample year: unknown, country: Germany[166]

- Co-contamination: not reported
- Further contamination of plants and plant products other than cereals and their products reported in the present book/article[166]: Juices, Juices (apple juice), Juices (apricot juice), Juices (banana juice), Juices (blackcurrant juice), Juices (cherry juice), Juices (grape juice), Juices (pear juice), Juices (pomegranate juice), Juices (strawberry juice), Spices, Spices (coriander), Spices (cumin), Spices (curcuma), Spices (curry), Spices (ginger), Spices (paprika), Spices (pepper), TA, literature[166]

Aspergillus **and** ***Penicillium*** **Toxins**

Patulin

incidence: 1/1, conc.: 0.1 µg/l, sample origin: local store in Munich (city) Germany, sample year: 1996/1997, country: Germany[52], *commercial product

- Co-contamination: not reported
- Further contamination of plants and plant products other than cereals and their products reported in the present book/article[52]: Juices (apple juice), Juices (blackcurrant juice), Juices (cherry juice), Juices (grape juice), Pulp (plum pulp), PAT, literature[52]

incidence: 3/3, conc. range: 3.1–10.8 µg/kg, ∅ conc.: 6.80 µg/kg, sample origin: Greek market, Greece, sample year: October 2004–June 2006, country: Greece[108]

- Co-contamination: not reported
- Further contamination of plants and plant products other than cereals and their products reported in the present book/article[108]: Baby food, Drinks (apple drink), Juices, Juices (apple concentrate), Juices (apple juice), Juices (apricot concentrate), Juices (pineapple concentrate), Juices (rosehip juice), PAT, literature[108]

incidence: 2/24, conc. range: 9.9–30.9 µg/l, ∅ conc.: 20.4 µg/l, sample origin: wholesale stores in Busan, Daegu, Daejeon, Gwangju, Jinju (cities), and Seoul (capital), South Korea, sample year: November 2007, country: South Korea[695]

- Co-contamination: not reported
- Further contamination of plants and plant products other than cereals and their products reported in the present book/article[695]: Juices (apple juice), Juices (grape juice), PAT, literature[695]

Juices (peach concentrate) may contain the following mycotoxins:

Aspergillus **and** ***Penicillium*** **Toxins**

Patulin

incidence: 4/15*, conc. range: LOD-10 µg/kg (1 sa.), 10–50 µg/kg (3 sa., maximum: 21.3 µg/kg), ∅ conc.: 15.7 µg/kg, sample origin: local producers of fruit concentrates in Valencia? (city), Spain, sample year: 2008, country: Spain[786], *non-reconstituted peach concentrate

- Co-contamination: not reported
- Further contamination of plants and plant products other than cereals and their products reported in the present book/article[786]: Juices (apple juice), Juices (apple concentrate), Juices (pear concentrate), PAT, literature[786]

incidence: 3/15, conc. range: 5–15 µg/kg (1 sa.), 30.1–50 µg/kg (2 sa., maximum: 35.2 µg/kg), ∅ conc.: 26.7 µg/kg, sample origin: retail outlets and producers of Azerbaijan, Chaharmahal va Bakhtiari, Esfahan, Gillan, Khuzestan, and Tehran (provinces), Iran, sample year: April–November 2013, country: Iran[827]

- Co-contamination: not reported
- Further contamination of plants and plant products other than cereals and their products reported in the present book/article[827]: Juices (apple concentrate), Juices (apple juice), Juices (grape juice), Juices (pear juice), Juices (pomegranate juice), PAT, literature[827]

Juices (peach juice) may contain the following mycotoxins:

Aspergillus **and** ***Penicillium*** **Toxins**

Patulin

incidence: 2/30, conc. range: <10 µg/kg (2 sa., maximum: 5.0 µg/kg), sample origin: supermarkets and organic food shops, Northwestern Italy, sample year: January–September 2007, country: Italy[847]

- Co-contamination: not reported
- Further contamination of plants and plant products other than cereals and their products reported in the present book/article[847]: Juices, Juices (apricot juice), Juices (pear juice), PAT, literature[847]

For detailed information see the article.

see also **Products (fruit product)**

Juices (pear concentrate) may contain the following mycotoxins:

Aspergillus **and** ***Penicillium*** **Toxins**

Patulin

incidence: 5/10*, conc. range: >50 µg/kg (5 sa., maximum: 126.9 µg/kg), ∅ conc.: 107 µg/kg, sample origin: local producers of fruit concentrates in Valencia? (city), Spain, sample year: 2008, country: Spain[786], *non-reconstituted pear concentrate

- Co-contamination: not reported
- Further contamination of plants and plant products other than cereals and their products reported in the present book/article[786]: Juices (apple juice), Juices (apple concentrate), Juices (peach concentrate), PAT, literature[786]

Juices (pear juice) may contain the following mycotoxins:

Alternaria **Toxins**

Tenuazonic Acid

incidence: 3/3, conc. range: 0.5–23 µg/kg, sample origin: unknown, sample year: unknown, country: Germany[166]

- Co-contamination: not reported
- Further contamination of plants and plant products other than cereals and their products reported in the present book/article[166]: Juices, Juices (apple juice), Juices (apricot juice), Juices (banana juice), Juices (blackcurrant juice), Juices (cherry juice), Juices (grape juice), Juices (orange juice), Juices (pomegranate juice), Juices (strawberry juice), Spices, Spices (coriander), Spices (cumin), Spices (curcuma), Spices (curry), Spices (ginger), Spices (paprika), Spices (pepper), TA, literature[166]

Aspergillus **and** ***Penicillium*** **Toxins**

Patulin

incidence: 1/7*, conc.: 1.1 µg/kg, sample origin: markets, supermarkets, hypermarkets or organic food shops, Italy, sample year: November 2003–February 2004, country: Italy[35], *conventional

incidence: 4/8*, conc. range: >LOQ<10 µg/kg (2 sa.), >10<25 µg/kg (1 sa.), 61.0 µg/kg (1 sa.), sample origin: markets, supermarkets, hypermarkets or organic food shops, Italy, sample year: November 2003–February 2004, country: Italy[35], *organic

- Co-contamination: not reported
- Further contamination of plants and plant products other than cereals and their products reported in the present book/article[35]: Fruits (apple), Juices, Juices (apple juice), Puree, Vinegar (apple), PAT, literature[35]

incidence: 35/35, conc. range: 5–92 µg/l, ∅ conc.: 42.62 µg/kg, sample origin: European and Italian sa. purchased in retail outlets, Italy, sample year: 2008, country: Italy[688]

- Co-contamination: not reported
- Further contamination of plants and plant products other than cereals and their products reported in the present book/article[688]: Juices, Juices (apple juice), PAT, literature[688]

For detailed information see the article.

incidence: 2/15, conc. range: 5–15 µg/kg (1 sa.), 31.3 µg/kg (1 sa.), ∅ conc.: 22.9 µg/kg, sample origin: retail outlets and producers of Azerbaijan, Chaharmahal va Bakhtiari, Esfahan, Gillan, Khuzestan, and Tehran (provinces), Iran, sample year: April–November 2013, country: Iran[827]

- Co-contamination: not reported
- Further contamination of plants and plant products other than cereals and their products reported in the present book/article[827]: Juices (apple concentrate), Juices (apple juice), Juices (grape juice), Juices (peach concentrate), Juices (pomegranate juice), PAT, literature[827]

incidence: 25/39, conc. range: <10 µg/kg (17 sa.), >10 µg/kg (8 sa., maximum: 33.4 µg/kg), sample origin: supermarkets and organic food shops, north-western Italy, sample year: January–September 2007, country: Italy[847]

- Co-contamination: not reported
- Further contamination of plants and plant products other than cereals and their products reported in the present book/article[847]: Juices, Juices (apricot juice), Juices (peach juice), PAT, literature[847]

For detailed information see the article.

incidence: 20/42, conc. range: 5–231 µg/l, ∅ conc.: 62.5 µg/l, sample origin: local supermarkets, Tunisia, sample year: 2013, country: Tunisia[888]

- Co-contamination: not reported
- Further contamination of plants and plant products other than cereals and their products reported in the present book/article[888]: Jam (apple jam), Jam (pear jam), Juices, Juices (apple juice), Products (fruit product), PAT, literature[888]

incidence: 2/3, conc. range: 11.7–38.9 µg/l, ∅ conc.: 25.3 µg/l, sample origin: Czech retail market, Czech Republic, sample year: unknown, country: Czech Republic[911]

- Co-contamination: not reported
- Further contamination of plants and plant products other than cereals and their products reported in the present book/article[911]: Apple cider, Baby food, Fruit salad, Fruits (apple), Fruits (pear), Juices (apple juice), Pulp (apple pulp), PAT, literature[911]

see also **Products (fruit product)**

Juices (pineapple concentrate) may contain the following mycotoxins:

Aspergillus **and** ***Penicillium*** **Toxins**

Patulin

incidence: 1/1*, conc.: 7.7 µg/kg, sample origin: company producing juices, Greece, sample year: October 2004–June 2006, country: Greece[108], *concentrated fruit juice

- Co-contamination: not reported
- Further contamination of plants and plant products other than cereals and their products reported in the present book/article[108]: Baby food, Drinks (apple drink), Juices, Juices (apple concentrate), Juices (apple juice), Juices (apricot concentrate), Juices (orange juice), Juices (rosehip juice), PAT, literature[108]

incidence: 1/6, conc.: 33.1 µg/l*, sample origin: purchased in supermarkets imported from the Philippines* and Malaysia, sample year: November 2012–March 2013, country: Malaysia[773]

- Co-contamination: not reported
- Further contamination of plants and plant products other than cereals and their products reported in the present book/article[773]: Juices (apple juice), Juices (lychee juice), PAT, literature[773]

Juices (pineapple juice) may contain the following mycotoxins:

Fusarium **Toxins**

Fumonisin/s

Fumonisin B_1

incidence: 5/5, conc. range: 8,180–23,730 µg/l, ∅ conc.: 18,000 µg/l, sample origin: commercially available from Costa Rica and Ecuador, sample year: unknown, country: Poland[848]

- Co-contamination: not reported
- Further contamination of plants and plant products other than cereals and their products reported in the present book/article[848]: Juices (pineapple juice), FB_2, FB_3, and MON, literature[848]

Fumonisin B_2

incidence: 5/5, conc. range: 740–1,460 µg/l, ∅ conc.: 1,060 µg/l, sample origin: commercially available from Costa Rica and Ecuador, sample year: unknown, country: Poland[848]

- Co-contamination: not reported
- Further contamination of plants and plant products other than cereals and their products reported in the present book/article[848]: Juices (pineapple juice), FB_1, FB_3, and MON, literature[848]

Fumonisin B_3

incidence: 5/5, conc. range: 170–230 µg/l, ∅ conc.: 200 µg/l, sample origin: commercially available from Costa Rica and Ecuador, sample year: unknown, country: Poland[848]

- Co-contamination: not reported
- Further contamination of plants and plant products other than cereals and their products reported in the present book/article[848]: Juices (pineapple juice), FB_1, FB_2, and MON, literature[848]

Other ***Fusarium*** Toxins

Moniliformin

incidence: 5/5, conc. range: 10–200 µg/l, ∅ conc.: 64 µg/l, sample origin: commercially available from Costa Rica and Ecuador, sample year: unknown, country: Poland[848]

- Co-contamination: not reported
- Further contamination of plants and plant products other than cereals and their products reported in the present book/article[848]: Juices (pineapple juice), FB_1, FB_2, and FB_3, literature[848]

Juices (pomegranate juice) may contain the following mycotoxins:

Alternaria Toxins

Tenuazonic Acid

incidence: 2/2, conc. range: 8–13 µg/kg, ∅ conc.: 10.5 µg/kg, sample origin: unknown, sample year: unknown, country: Germany[166]

- Co-contamination: not reported
- Further contamination of plants and plant products other than cereals and their products reported in the present book/article[166]: Juices, Juices (apple juice), Juices (apricot juice), Juices (banana juice), Juices (blackcurrant juice), Juices (cherry juice), Juices (grape juice), Juices (orange juice), Juices (pear juice), Juices (strawberry juice), Spices, Spices (coriander), Spices (cumin), Spices (curcuma), Spices (curry), Spices (ginger), Spices (paprika), Spices (pepper), TA, literature[166]

Aspergillus and ***Penicillium*** Toxins

Patulin

incidence: 1/12, conc.: 8.2 µg/kg, sample origin: retail outlets and producers of Azerbaijan, Chaharmahal va Bakhtiari, Esfahan, Gillan, Khuzestan, and Tehran (provinces), Iran, sample year: April–November 2013, country: Iran[827]

- Co-contamination: not reported
- Further contamination of plants and plant products other than cereals and their products reported in the present book/article[827]: Juices (apple concentrate), Juices (apple juice), Juices (grape juice), Juices (peach concentrate), Juices (pear juice), PAT, literature[827]

Juices (raspberry juice) may contain the following mycotoxins:

Alternaria Toxins

Alternariol

incidence: 1/1, conc.: 0.84 µg/l, sample origin: unknown, sample year: unknown, country: Canada[440]

- Co-contamination: not reported
- Further contamination of plants and plant products other than cereals and their products reported in the present book/article[440]: Juices (apple juice), Juices (grape juice), Nectars (cranberry nectar), Nectars (prune nectar), AME and AOH, literature[440]; Wine (red), AOH, literature[440]

Juices (rosehip juice) may contain the following mycotoxins:

Aspergillus and ***Penicillium*** Toxins

Patulin

incidence: 2/2, conc. range: 3.5–3.6 µg/kg, ∅ conc.: 3.55 µg/kg, sample origin: Greek markets, Greece, sample year: October 2004–June 2006, country: Greece[108]

- Co-contamination: not reported
- Further contamination of plants and plant products other than cereals and their products reported in the present book/article[108]: Baby food, Drinks (apple drink), Juices, Juices (apple concentrate), Juices (apple juice), Juices (apricot concentrate), Juices (orange juice), Juices (pineapple concentrate), PAT, literature[108]

Juices (strawberry juice) may contain the following mycotoxins:

Alternaria Toxins

Tenuazonic Acid

incidence: 2/2, conc. range: 5–6 µg/kg, ∅ conc.: 5.5 µg/kg, sample origin: unknown, sample year: unknown, country: Germany[166]

- Co-contamination: not reported
- Further contamination of plants and plant products other than cereals and their products reported in the present book/article[166]: Juices, Juices (apple juice), Juices (apricot juice), Juices (banana juice), Juices (blackcurrant juice), Juices (cherry juice), Juices (grape juice), Juices (orange juice), Juices (pear juice), Juices (pomegranate juice), Spices, Spices (coriander), Spices (cumin), Spices (curcuma), Spices (curry), Spices (ginger), Spices (paprika), Spices (pepper), TA, literature[166]

Juices (sugarcane juice) may contain the following mycotoxins:

Aspergillus Toxins

Aflatoxin B_1

incidence: 23/40, conc. range: ≤2.10 µg/kg, ∅ conc.: 0.72 µg/kg, sample origin: juice shops in Assiut (city), Assiut Governorate, Egypt, sample year: January 2016, country: Egypt/Austria[300]

- Co-contamination: For detailed information see the article.
- Further contamination of plants and plant products other than cereals and their products reported in the present book/article[300]: Juices (sugarcane juice), 3-NPA, AFG_1, FA, KA, and PA, literature[300]; Sugarcane, 3-NPA, AFB_1, AFG_1, AME, KA, and PA, literature[300]; see also Juices (sugarcane juice), Sugarcane, chapter: Further Mycotoxins and Microbial Metabolites

Aflatoxin G_1

incidence: 7/40, conc. range: ≤1.34 µg/kg, ∅ conc.: 0.30 µg/kg, sample origin: juice shops in Assiut (city), Assiut Governorate, Egypt, sample year: January 2016, country: Egypt/Austria[300]

- Co-contamination: For detailed information see the article.
- Further contamination of plants and plant products other than cereals and their products reported in the present book/article[300]: Juices (sugarcane juice), 3-NPA, AFB_1, FA, KA, and PA, literature[300]; Sugarcane, 3-NPA, AFB_1, AFG_1, AME, KA, and PA,

literature[300]; see also Juices (sugarcane juice), Sugarcane, chapter: Further Mycotoxins and Microbial Metabolites

Aflatoxin/s

incidence: 10/25, conc. range: 1.0–9.5 µg/kg*, ∅ conc.: 4.1 µg/kg*, sample origin: local industries at the nearby regions of Mandya and Mysore (districts), Karnataka (state), India, sample year: June/July 2012, country: India[453], *AF

incidence: 8/36, conc. range: 0.5–6.5 µg/kg*, sample origin: local vendors at markets in Mysore (district), Karnataka (state), India, sample year: June/July 2012, country: India[453], *AF

incidence: 4/21, conc. range: 0.5–5.5 µg/kg*, sample origin: local vendors at markets in Mandya (district), Karnataka (state), India, sample year: June/July 2012, country: India[453], *AF

- Co-contamination: not reported
- Further contamination of plants and plant products other than cereals and their products reported in the present book/article[453]: Jaggery, AF/S, literature[453]

Aspergillus **and** ***Penicillium*** **Toxins**

3-Nitropropionic Acid

incidence: 12/40, conc. range: ≤13.6 µg/kg, ∅ conc.: 3.58 µg/kg, sample origin: juice shops in Assiut (city), Assiut Governorate, Egypt, sample year: January 2016, country: Egypt/Austria[300]

- Co-contamination: For detailed information see the article.
- Further contamination of plants and plant products other than cereals and their products reported in the present book/article[300]: Juices (sugarcane juice), AFB_1, AFG_1, FA, KA, and PA, literature[300]; Sugarcane, 3-NPA, AFB_1, AFG_1, AME, KA, and PA, literature[300]; see also Juices (sugarcane juice), Sugarcane, chapter: Further Mycotoxins and Microbial Metabolites

Kojic Acid

incidence: 1/40, conc.: 208 µg/kg, sample origin: juice shops in Assiut (city), Assiut Governorate, Egypt, sample year: January 2016, country: Egypt/Austria[300]

- Co-contamination: For detailed information see the article.
- Further contamination of plants and plant products other than cereals and their products reported in the present book/article[300]: Juices (sugarcane juice), 3-NPA, AFB_1, AFG_1, FA, and PA, literature[300]; Sugarcane, 3-NPA, AFB_1, AFG_1, AME, KA, and PA, literature[300]; see also Juices (sugarcane juice), Sugarcane, chapter: Further Mycotoxins and Microbial Metabolites

Penicillic Acid

incidence: 16/40, conc. range: ≤212 µg/kg, ∅ conc.: 45.7 µg/kg, sample origin: juice shops in Assiut (city), Assiut Governorate, Egypt, sample year: January 2016, country: Egypt/Austria[300]

- Co-contamination: For detailed information see the article.
- Further contamination of plants and plant products other than cereals and their products reported in the present book/article[300]: Juices (sugarcane juice), 3-NPA, AFB_1, AFG_1, FA, and KA, literature[300]; Sugarcane, 3-NPA, AFB_1, AFG_1, AME, KA, and PA, literature[300]; see also Juices (sugarcane juice), Sugarcane, chapter: Further Mycotoxins and Microbial Metabolites

Fusarium **Toxins**

Fusaric Acid

incidence: 12/40, conc. range: ≤2,214 µg/kg, ∅ conc.: 258 µg/kg, sample origin: juice shops in Assiut (city), Assiut Governorate, Egypt, sample year: January 2016, country: Egypt/Austria[300]

- Co-contamination: For detailed information see the article.
- Further contamination of plants and plant products other than cereals and their products reported in the present book/article[300]: Juices (sugarcane juice), 3-NPA, AFB_1, AFG_1, KA, and PA, literature[300]; Sugarcane, 3-NPA, AFB_1, AFG_1, AME, KA, and PA, literature[300]; see also Juices (sugarcane juice), Sugarcane, chapter: Further Mycotoxins and Microbial Metabolites

Juices (tomato juice) may contain the following mycotoxins:

Alternaria **Toxins**

Altenuene

incidence: 14/28, conc. range: <LOQ-6.1 µg/kg, sample origin: local supermarkets, Belgium, sample year: February 2013–February 2015, country: Belgium/Austria/Germany[514]

- Co-contamination: For detailed information see the article.
- Further contamination of plants and plant products other than cereals and their products reported in the present book/article[514]: Juices (apple juice), Juices (grape juice), TA, literature[514]; Juices (tomato juice), AME, AOH, and TA, literature[514]; Sauces (tomato sauce), Tomato concentrate, ALT, AME, AOH, and TA, literature[514]; see also Juices (tomato juice), Sauces (tomato sauce), and Tomato concentrate, chapter: Further Mycotoxins and Microbial Metabolites

For detailed information see the article.

Alternariol

incidence: 1/1, conc.: 5.4 µg/kg, sample origin: unknown, sample year: unknown, country: Germany[46]

- Co-contamination: 1 sa. co-contaminated with AME and AOH
- Further contamination of plants and plant products other than cereals and their products reported in the present book/article[46]: Juices (tomato juice), Juices (vegetable juice), AME, literature[46]; Paste (tomato paste), Puree, Tomato ketchup, Wine (red), Wine (white), AME and AOH, literature[46]

incidence: 2/2, conc. range: 0.52–1.99 µg/kg, ∅ conc.: 1.26 µg/kg, sample origin: commercially available, sample year: unknown, country: Germany[160]

- Co-contamination: 2 sa. co-contaminated with AME and AOH
- Further contamination of plants and plant products other than cereals and their products reported in the present book/article[160]: Beverages, Juices (apple juice), Juices (grape juice), Juices (orange juice), Juices (vegetable juice), Wine (red), Wine (white), Wine, miscellaneous, AME and AOH, literature[160]; Juices (tomato juice), AME, literature[160]

incidence: 9/16, conc. range: 1.1–3.1 µg/l, ∅ conc.: 1.9 µg/l, sample origin: retail shops, Hesse (federal state), Germany, sample year: 2009/2010, country: Germany[439]

- Co-contamination: not reported
- Further contamination of plants and plant products other than cereals and their products reported in the present book/article[439]: Juices (apple juice), Paste (tomato paste), Sauces (apple sauce), Tomato ketchup, Wine (white), AOH, literature[439]

incidence: 20/28, conc. range: <LOQ-27.0 µg/kg, sample origin: local supermarkets, Belgium, sample year: February 2013–February 2015, country: Belgium/Austria/Germany[514]

- Co-contamination: For detailed information see the article.
- Further contamination of plants and plant products other than cereals and their products reported in the present book/article[514]: Juices (apple juice), Juices (grape juice), TA, litera-

ture[514]; Juices (tomato juice), ALT, AME, and TA, literature[514]; Sauces (tomato sauce), Tomato concentrate, ALT, AME, AOH, and TA, literature[514]; see also Juices (tomato juice), Sauces (tomato sauce), Tomato concentrate, chapter: Further Mycotoxins and Microbial Metabolites

For detailed information see the article.

incidence: 4/14, conc. range: ≤11 µg/kg, sample origin: retail stores, Netherlands, sample year: autumn 2014, country: Netherlands/Germany[906]

- Co-contamination: not reported
- Further contamination of plants and plant products other than cereals and their products reported in the present book/article[906]: Figs, Fruits (tomato), AOH and TA, literature[906]; Juices (tomato juice), TA, literature[906]; Oils (sunflower oil), AME, literature[906]; Paste (tomato paste), Sauces (tomato sauce), Seeds (sunflower), Tomato ketchup, AME, AOH, and TA, literature[906]

Alternariol Methyl Ether

incidence: 1/1, conc.: 0.9 µg/kg, sample origin: unknown, sample year: unknown, country Germany[46]

- Co-contamination: 1 sa. co-contaminated with AME and AOH
- Further contamination of plants and plant products other than cereals and their products reported in the present book/article[46]: Juices (tomato juice), AOH, literature[46]; Juices (vegetable juice), AME, literature[46]; Paste (tomato paste), Puree, Tomato ketchup, Wine (red), Wine (white), AME and AOH, literature[46]

incidence: 2/2, conc. range: 0.23–0.38 µg/kg, Ø conc.: 0.31 µg/kg, sample origin: commercially available, sample year: unknown, country: Germany[160]

- Co-contamination: 2 sa. co-contaminated with AME and AOH
- Further contamination of plants and plant products other than cereals and their products reported in the present book/article[160]: Beverages, Juices (apple juice), Juices (grape juice), Juices (orange juice), Juices (vegetable juice), Wine (red), Wine (white), Wine, miscellaneous, AME and AOH, literature[160]; Juices (tomato juice), AOH, literature[160]

incidence: 15/28, conc. range: <LOQ-3.3 µg/kg, sample origin: local supermarkets, Belgium, sample year: February 2013–February 2015, country: Belgium/Austria/Germany[514]

- Co-contamination: For detailed information see the article.
- Further contamination of plants and plant products other than cereals and their products reported in the present book/article[514]: Juices (apple juice), Juices (grape juice), TA, literature[514]; Juices (tomato juice), ALT, AOH, and TA, literature[514]; Sauces (tomato sauce), Tomato concentrate, ALT, AME, AOH, and TA, literature[514]; see also Juices (tomato juice), Sauces (tomato sauce), Tomato concentrate, chapter: Further Mycotoxins and Microbial Metabolites

For detailed information see the article.

incidence: 7/9, conc. range: 0.20–5.8 µg/kg, Ø conc.: 1.42 µg/kg, sample origin: local supermarkets and agricultural trade markets in Beijing (capital), China, sample year: unknown, country: China[883]

- Co-contamination: For detailed information see the article.
- Further contamination of plants and plant products other than cereals and their products reported in the present book/article[883]: Juices (citrus juice), AME and TA, literature[883]; Juices (tomato juice), TA, literature[883]; Tomato ketchup, AME, AOH, and TA, literature[883]; see also Juices (tomato juice), Tomato ketchup, chapter: Further Mycotoxins and Microbial Metabolites

For detailed information see the article.

Tenuazonic Acid

incidence: 3/15, conc. range: 61–227 µg/l, Ø conc.: 131 µg/l, sample origin: retail shops in the area of Giessen (town), Germany, sample year: unknown, country: Germany/Indonesia[442]

- Co-contamination: not reported
- Further contamination of plants and plant products other than cereals and their products reported in the present book/article[442]: Juices (apple juice), Tomato ketchup, TA, literature[442]

incidence: 28/28, conc. range: 3.7–333.1 µg/kg, sample origin: local supermarkets, Belgium, sample year: February 2013–February 2015, country: Belgium/Austria/Germany[514]

- Co-contamination: For detailed information see the article.
- Further contamination of plants and plant products other than cereals and their products reported in the present book/article[514]: Juices (apple juice), Juices (grape juice), TA, literature[514]; Juices (tomato juice), ALT, AOH, and AME, literature[514]; Sauces (tomato sauce), Tomato concentrate, ALT, AME, AOH, and TA, literature[514]; see also Juices (tomato juice), Sauces (tomato sauce), Tomato concentrate, chapter: Further Mycotoxins and Microbial Metabolites

For detailed information see the article.

incidence: 9/9, conc. range: 7.4–278 µg/kg, Ø conc.: 87 µg/kg, sample origin: local supermarkets and agricultural trade markets in Beijing (capital), China, sample year: unknown, country: China[883]

- Co-contamination: For detailed information see the article.
- Further contamination of plants and plant products other than cereals and their products reported in the present book/article[883]: Juices (citrus juice), AME and TA, literature[883]; Juices (tomato juice), AME, literature[883]; Tomato ketchup, AME, AOH, and TA, literature[883]; see also Juices (tomato juice), Tomato ketchup, chapter: Further Mycotoxins and Microbial Metabolites

For detailed information see the article.

incidence: 7/14, conc. range: ≤220 µg/kg, sample origin: retail stores, Netherlands, sample year: autumn 2014, country: Netherlands/Germany[906]

- Co-contamination: not reported
- Further contamination of plants and plant products other than cereals and their products reported in the present book/article[906]: Figs, Fruits (tomato), AOH and TA, literature[906]; Juices (tomato juice), AOH, literature[906]; Oils (sunflower oil), AME, literature[906]; Paste (tomato paste), Sauces (tomato sauce), Seeds (sunflower), Tomato ketchup, AME, AOH, and TA, literature[906]

Aspergillus and ***Penicillium*** Toxins

Citrinin

incidence: 1/11, conc.: 0.12 µg/l, sample origin: local retail stores, Germany, sample year: unknown, country: Germany[711]

- Co-contamination: not reported
- Further contamination of plants and plant products other than cereals and their products reported in the present book/article[711]: Juices (apple juice), Juices (blackcurrant juice), Juices (cherry juice), CIT, literature[711]

Ochratoxin A

incidence: 3/30, conc. range: ≤0.032 µg/l, sample origin: different origins, Germany, sample year: unknown, country: Germany[110]

- Co-contamination: not reported
- Further contamination of plants and plant products other than cereals and their products reported in the present book/article[110]: Juices (blackcurrant juice), Juices (carrot juice), Juices (grape juice), Sauces (pepper- and barbecue sauce), Sauces (soy sauce), Spices (mustard), Tomato ketchup, Vinegar (apple), Vinegar (balsamico), Vinegar (wine), Wine (red), Wine (rosé), Wine (white), OTA, literature[110]

Patulin

incidence: 3/4*, conc. range: <LOQ-47.72 µg/kg, sample origin: supermarkets and organic shops, Portugal, sample year: unknown, country: Portugal[905], *conventional

- Co-contamination: not reported
- Further contamination of plants and plant products other than cereals and their products reported in the present book/article[905]: Jellies, Juices (apple juice), Puree, PAT, literature[905]

For detailed information see the article.

Juices (tropical-fruit juice) see Products (fruit product)

Juices (vegetable juice) may contain the following mycotoxins:

Alternaria Toxins

Alternariol

incidence: 1/1*, conc.: 7.82 µg/kg, sample origin: commercially available, sample year: unknown, country: Germany[160], *consisted of tomatoes, carrots, celery, beetroot, and red pepper

- Co-contamination: 1 sa. co-contaminated with AME and AOH
- Further contamination of plants and plant products other than cereals and their products reported in the present book/article[160]: Beverages, Juices (apple juice), Juices (grape juice), Juices (orange juice), Juices (tomato juice), Wine (red), Wine (white), Wine, miscellaneous, AME and AOH, literature[160]; Juices (vegetable juice), AME, literature[160]

Alternariol Methyl Ether

incidence: 1/2, conc.: 0.1 µg/kg, sample origin: unknown, sample year: unknown, country: Germany[46]

- Co-contamination: not reported
- Further contamination of plants and plant products other than cereals and their products reported in the present book/article[46]: Juices (tomato juice), Paste (tomato paste), Puree, Tomato ketchup, Wine (red), Wine (white), AME and AOH, literature[46]

incidence: 1/1*, conc.: 0.79 µg/kg, sample origin: commercially available, sample year: unknown, country: Germany[160], *consisted of tomatoes, carrots, celery, beetroot, and red pepper

- Co-contamination: 1 sa. co-contaminated with AME and AOH
- Further contamination of plants and plant products other than cereals and their products reported in the present book/article[160]: Beverages, Juices (apple juice), Juices (grape juice), Juices (orange juice), Juices (tomato juice), Wine (red), Wine (white), Wine, miscellaneous, AME and AOH, literature[160]; Juices (vegetable juice), AOH, literature[160]

Joven see Wine (red)

"Kabinett" see Wine

Karahi masala see Spices (karahi masala)

Kernels (apricot) may contain the following mycotoxins:

Aspergillus Toxins

Aflatoxin B_1

incidence: 1/8, conc.: 61.05 µg/kg*, sample origin: imported from Algeria and Turkey to Italy, sample year: January 2008–December 2009, country: Italy[320], *contaminated sa. from Algeria

- Co-contamination: 1 sa. co-contaminated with AFB_1 and AFB_2
- Further contamination of plants and plant products other than cereals and their products reported in the present book/article[320]: Coffee beans, Vine fruits, dried, OTA, literature[320]; Kernels (apricot), AFB_2 and AFG_1, literature[320]; Nuts (hazelnut), Paste (hazelnut paste), AFB_1, AFB_2, AFG_1, and AFG_2, literature[320]; Nuts (peanut), Nuts (pistachio nut), AFB_1 and AFB_2, literature[320]

Aflatoxin B_2

incidence: 3/8, conc. range: 0.17–9.64 µg/kg*, Ø conc.: 3.38 µg/kg*, sample origin: imported from Algeria and Turkey to Italy, sample year: January 2008–December 2009, country: Italy[320], *contaminated sa. from Algeria (1) and Turkey (2)

- Co-contamination: 1 sa. co-contaminated with AFB_1 and AFB_2; 2 sa. co-contaminated with AFB_2 and AFG_1
- Further contamination of plants and plant products other than cereals and their products reported in the present book/article[320]: Coffee beans, Vine fruits, dried, OTA, literature[320]; Kernels (apricot), AFB_1 and AFG_1, literature[320]; Nuts (hazelnut), Paste (hazelnut paste), AFB_1, AFB_2, AFG_1, and AFG_2, literature[320]; Nuts (peanut), Nuts (pistachio nut), AFB_1 and AFB_2, literature[320]

Aflatoxin G_1

incidence: 2/8, conc. range: 0.22–0.32 µg/kg*, Ø conc.: 0.27 µg/kg*, sample origin: imported from Algeria and Turkey to Italy, sample year: January 2008–December 2009, country: Italy[320], *contaminated sa. from Turkey

- Co-contamination: 2 sa. co-contaminated with AFB_2 and AFG_1
- Further contamination of plants and plant products other than cereals and their products reported in the present book/article[320]: Coffee beans, Vine fruits, dried, OTA, literature[320]; Kernels (apricot), AFB_1 and AFB_2, literature[320]; Nuts (hazelnut), Paste (hazelnut paste), AFB_1, AFB_2, AFG_1, and AFG_2, literature[320]; Nuts (peanut), Nuts (pistachio nut), AFB_1 and AFB_2, literature[320]

Aflatoxin/s

incidence: 4/15, conc. range: 0.7–5.6 µg/kg*, Ø conc.: 2.65 µg/kg*, sample origin: retail shops and local markets in NWFP and northern areas, Pakistan, sample year: January–November 2009, country: Pakistan[17], *AFB_1, AFB_2, AFG_1, and AFG_2

- Co-contamination: not reported
- Further contamination of plants and plant products other than cereals and their products reported in the present book/article[17]: Dates, Figs, Fruits, dried, Nuts (almond), Nuts (peanut), Nuts (pine nut), Nuts (pistachio nut), Nuts (walnut), Raisins, AF/S, literature[17]

Keskul may contain the following mycotoxins:

***Aspergillus* Toxins**

Aflatoxin/s

incidence: 2/5, conc. range: ≤0.774 µg/kg*, sample origin: groceries and supermarkets, Turkey, sample year: 2003, country: Turkey[119], *AFS total

- Co-contamination: not reported
- Further contamination of plants and plant products other than cereals and their products reported in the present book/article[119]: Asure, Beverages, Cocoa, Cream, Ice cream, Jello, Muhallebi, Pudding, Salep, Sauces (chocolate sauce), Soups, AF/S, literature[119]

Keskul is an almond-based milk pudding. Ingredients were sugar, potato starch, almond flavor, vanilla, and ß-carotene.

Ketchup see **Tomato ketchup**

Kokonte see **Cassava**

Korma masala see **Spices (korma masala)**

Kubeba may contain the following mycotoxins:

***Aspergillus* Toxins**

Aflatoxin B_1

incidence: 1/25, conc.: 150 µg/kg, sample origin: local companies in Cairo (capital), Egypt, sample year: unknown, country: Egypt[104]

- Co-contamination: 1 sa. co-contaminated with AFB_1 and AFB_2
- Further contamination of plants and plant products other than cereals and their products reported in the present book/article[104]: Kubeba, AFB_2, literature[104]; Spices (coriander), Spices (curcuma), AFB_1 and AFG_1, literature[104]; Spices (pepper), AFB_1, literature[104]

Aflatoxin B_2

incidence: 1/25, conc.: 25 µg/kg, sample origin: local companies in Cairo (capital), Egypt, sample year: unknown, country: Egypt[104]

- Co-contamination: 1 sa. co-contaminated with AFB_1 and AFB_2
- Further contamination of plants and plant products other than cereals and their products reported in the present book/article[104]: Kubeba, Spices (pepper), AFB_1, literature[104]; Spices (coriander), Spices (curcuma), AFB_1 and AFG_1, literature[104]

Kubeba is a mixture of black pepper, coriander, garlic, minced meat, rice, and various vegetables used in Egyptian households.

Kucha see **Sauces (leaf sauce)**

Kuru-kuru may contain the following mycotoxins:

***Aspergillus* Toxins**

Aflatoxin/s

Aflatoxin B_1

incidence: 6/6, conc. range: 46–65 µg/kg, ⌀ conc.: 52 µg/kg, sample origin: Bamenda (region, western highlands), Cameroon, sample year: July–September 2011, country: Cameroon/Austria/South Africa[659]

- Co-contamination: For detailed information see the article.
- Further contamination of plants and plant products other than cereals and their products reported in the present book/article[659]: Dagwa, AFB_1, AFB_2, AFG_1, AFG_2, AME, BEA, CIT, DON, ENA, ENA_1, ENB, ENB_1, FA, FB_1, FB_2, FB_3, KA, MA, NIV, OTA, OTB, STG, VIOM, and ZEA, literature[659]; Flour (soybean flour), AFB_1, AFG_1, BEA, DON, ENB, ENB_1, FA, FB_1, FB_2, FB_3, FUS-X, KA, NIV, and ZEA, literature[659]; Kuru-kuru, AFB_2, AFG_1, AFG_2, BEA, ENA, ENA_1, ENB, ENB_1, FA, FB_1, FB_3, KA, MA, OTA, OTB, ROQ-C, STG, VIOM, and ZEA, literature[659]; Paste (peanut paste), AFB_1, AFB_2, AFG_1, AFG_2, AME, BEA, ENA, ENA_1, ENB, ENB_1, FA, FB_1, FB_2, FB_3, KA, MA, OTA, OTB, STG, VIOM, and ZEA, literature[659]; Soups, AFB_1, AFB_2, AFG_1, AFG_2, AME, AOH, BEA, DON, ENA, ENA_1, ENB, ENB_1, FB_1, FB_2, FB_3, KA, OTA, OTB, and STG, literature[659]; see also Dagwa, Flour (soybean flour), Kuru-kuru, Paste (peanut paste), Soups, chapter: Further Mycotoxins and Microbial Metabolites

Aflatoxin B_2

incidence: 6/6, conc. range: 5–9 µg/kg, ⌀ conc.: 7 µg/kg, sample origin: Bamenda (region, western highlands), Cameroon, sample year: July–September 2011, country: Cameroon/Austria/South Africa[659]

- Co-contamination: For detailed information see the article.
- Further contamination of plants and plant products other than cereals and their products reported in the present book/article[659]: Dagwa, AFB_1, AFB_2, AFG_1, AFG_2, AME, BEA, CIT, DON, ENA, ENA_1, ENB, ENB_1, FA, FB_1, FB_2, FB_3, KA, MA, NIV, OTA, OTB, STG, VIOM, and ZEA, literature[659]; Flour (soybean flour), AFB_1, AFG_1, BEA, DON, ENB, ENB_1, FA, FB_1, FB_2, FB_3, FUS-X, KA, NIV, and ZEA, literature[659]; Kuru-kuru, AFB_1, AFG_1, AFG_2, BEA, ENA, ENA_1, ENB, ENB_1, FA, FB_1, FB_3, KA, MA, OTA, OTB, ROQ-C, STG, VIOM, and ZEA, literature[659]; Paste (peanut paste), AFB_1, AFB_2, AFG_1, AFG_2, AME, BEA, ENA, ENA_1, ENB, ENB_1, FA, FB_1, FB_2, FB_3, KA, MA, OTA, OTB, STG, VIOM, and ZEA, literature[659]; Soups, AFB_1, AFB_2, AFG_1, AFG_2, AME, AOH, BEA, DON, ENA, ENA_1, ENB, ENB_1, FB_1, FB_2, FB_3, KA, OTA, OTB, and STG, literature[659]; see also Dagwa, Flour (soybean flour), Kuru-kuru, Paste (peanut paste), Soups, chapter: Further Mycotoxins and Microbial Metabolites

Aflatoxin G_1

incidence: 6/6, conc. range: 8–10 µg/kg, ⌀ conc.: 9 µg/kg, sample origin: Bamenda (region, western highlands), Cameroon, sample year: July–September 2011, country: Cameroon/Austria/South Africa[659]

- Co-contamination: For detailed information see the article.
- Further contamination of plants and plant products other than cereals and their products reported in the present book/article[659]: Dagwa, AFB_1, AFB_2, AFG_1, AFG_2, AME, BEA, CIT, DON, ENA, ENA_1, ENB, ENB_1, FA, FB_1, FB_2, FB_3, KA, MA, NIV, OTA, OTB, STG, VIOM, and ZEA, literature[659]; Flour (soybean flour), AFB_1, AFG_1, BEA, DON, ENB, ENB_1, FA, FB_1, FB_2, FB_3, FUS-X, KA, NIV, and ZEA, literature[659]; Kuru-kuru, AFB_1, AFB_2, AFG_2, BEA, ENA, ENA_1, ENB, ENB_1, FA, FB_1, FB_3, KA, MA, OTA, OTB, ROQ-C, STG, VIOM, and ZEA, literature[659]; Paste (peanut paste), AFB_1, AFB_2, AFG_1, AFG_2, AME, BEA, ENA, ENA_1, ENB, ENB_1, FA, FB_1, FB_2, FB_3, KA, MA, OTA, OTB, STG, VIOM, and

ZEA, literature[659]; Soups, AFB_1, AFB_2, AFG_1, AFG_2, AME, AOH, BEA, DON, ENA, ENA_1, ENB, ENB_1, FB_1, FB_2, FB_3, KA, OTA, OTB, and STG, literature[659]; see also Dagwa, Flour (soybean flour), Kuru-kuru, Paste (peanut paste), Soups, chapter: Further Mycotoxins and Microbial Metabolites

Aflatoxin G_2

incidence: 6/6, conc. range: 5–9 μg/kg, ∅ conc.: 7 μg/kg, sample origin: Bamenda (region, western highlands), Cameroon, sample year: July–September 2011, country: Cameroon/Austria/South Africa[659]

- Co-contamination: For detailed information see the article.
- Further contamination of plants and plant products other than cereals and their products reported in the present book/article[659]: Dagwa, AFB_1, AFB_2, AFG_1, AFG_2, AME, BEA, CIT, DON, ENA, ENA_1, ENB, ENB_1, FA, FB_1, FB_2, FB_3, KA, MA, NIV, OTA, OTB, STG, VIOM, and ZEA, literature[659]; Flour (soybean flour), AFB_1, AFG_1, BEA, DON, ENB, ENB_1, FA, FB_1, FB_2, FB_3, FUS-X, KA, NIV, and ZEA, literature[659]; Kuru-kuru, AFB_1, AFB_2, AFG_1, BEA, ENA, ENA_1, ENB, ENB_1, FA, FB_1, FB_3, KA, MA, OTA, OTB, ROQ-C, STG, VIOM, and ZEA, literature[659]; Paste (peanut paste), AFB_1, AFB_2, AFG_1, AFG_2, AME, BEA, ENA, ENA_1, ENB, ENB_1, FA, FB_1, FB_2, FB_3, KA, MA, OTA, OTB, STG, VIOM, and ZEA, literature[659]; Soups, AFB_1, AFB_2, AFG_1, AFG_2, AME, AOH, BEA, DON, ENA, ENA_1, ENB, ENB_1, FB_1, FB_2, FB_3, KA, OTA, OTB, and STG, literature[659]; see also Dagwa, Flour (soybean flour), Kuru-kuru, Paste (peanut paste), Soups, chapter: Further Mycotoxins and Microbial Metabolites

Other ***Aspergillus*** Toxins

Sterigmatocystin

incidence: 3/6, conc. range: <LOQ-1 μg/kg, ∅ conc.: 1 μg/kg, sample origin: Bamenda (region, western highlands), Cameroon, sample year: July–September 2011, country: Cameroon/Austria/South Africa[659]

- Co-contamination: For detailed information see the article.
- Further contamination of plants and plant products other than cereals and their products reported in the present book/article[659]: Dagwa, AFB_1, AFB_2, AFG_1, AFG_2, AME, BEA, CIT, DON, ENA, ENA_1, ENB, ENB_1, FA, FB_1, FB_2, FB_3, KA, MA, NIV, OTA, OTB, STG, VIOM, and ZEA, literature[659]; Flour (soybean flour), AFB_1, AFG_1, BEA, DON, ENB, ENB_1, FA, FB_1, FB_2, FB_3, FUS-X, KA, NIV, and ZEA, literature[659]; Kuru-kuru, AFB_1, AFB_2, AFG_1, AFG_2, BEA, ENA, ENA_1, ENB, ENB_1, FA, FB_1, FB_3, KA, MA, OTA, OTB, ROQ-C, VIOM, and ZEA, literature[659]; Paste (peanut paste), AFB_1, AFB_2, AFG_1, AFG_2, AME, BEA, ENA, ENA_1, ENB, ENB_1, FA, FB_1, FB_2, FB_3, KA, MA, OTA, OTB, STG, VIOM, and ZEA, literature[659]; Soups, AFB_1, AFB_2, AFG_1, AFG_2, AME, AOH, BEA, DON, ENA, ENA_1, ENB, ENB_1, FB_1, FB_2, FB_3, KA, OTA, OTB, and STG, literature[659]; see also Dagwa, Flour (soybean flour), Kuru-kuru, Paste (peanut paste), Soups, chapter: Further Mycotoxins and Microbial Metabolites

Aspergillus and ***Penicillium*** Toxins

Kojic Acid

incidence: 6/6, conc. range: 1,241–2,043 μg/kg, ∅ conc.: 1,619 μg/kg, sample origin: Bamenda (region, western highlands), Cameroon, sample year: July–September 2011, country: Cameroon/Austria/South Africa[659]

- Co-contamination: For detailed information see the article.
- Further contamination of plants and plant products other than cereals and their products reported in the present book/article[659]: Dagwa, AFB_1, AFB_2, AFG_1, AFG_2, AME, BEA, CIT, DON, ENA, ENA_1, ENB, ENB_1, FA, FB_1, FB_2, FB_3, KA, MA, NIV, OTA, OTB, STG, VIOM, and ZEA, literature[659]; Flour (soybean flour), AFB_1, AFG_1, BEA, DON, ENB, ENB_1, FA, FB_1, FB_2, FB_3, FUS-X, KA, NIV, and ZEA, literature[659]; Kuru-kuru, AFB_1, AFB_2, AFG_1, AFG_2, BEA, ENA, ENA_1, ENB, ENB_1, FA, FB_1, FB_3, MA, OTA, OTB, ROQ-C, STG, VIOM, and ZEA, literature[659]; Paste (peanut paste), AFB_1, AFB_2, AFG_1, AFG_2, AME, BEA, ENA, ENA_1, ENB, ENB_1, FA, FB_1, FB_2, FB_3, KA, MA, OTA, OTB, STG, VIOM, and ZEA, literature[659]; Soups, AFB_1, AFB_2, AFG_1, AFG_2, AME, AOH, BEA, DON, ENA, ENA_1, ENB, ENB_1, FB_1, FB_2, FB_3, KA, OTA, OTB, and STG, literature[659]; see also Dagwa, Flour (soybean flour), Kuru-kuru, Paste (peanut paste), Soups, chapter: Further Mycotoxins and Microbial Metabolites

Ochratoxin A

incidence: 6/6, conc. range: 1.6–2.6 μg/kg, ∅ conc.: 2.2 μg/kg, sample origin: Bamenda (region, western highlands), Cameroon, sample year: July–September 2011, country: Cameroon/Austria/South Africa[659]

- Co-contamination: For detailed information see the article.
- Further contamination of plants and plant products other than cereals and their products reported in the present book/article[659]: Dagwa, AFB_1, AFB_2, AFG_1, AFG_2, AME, BEA, CIT, DON, ENA, ENA_1, ENB, ENB_1, FA, FB_1, FB_2, FB_3, KA, MA, NIV, OTA, OTB, STG, VIOM, and ZEA, literature[659]; Flour (soybean flour), AFB_1, AFG_1, BEA, DON, ENB, ENB_1, FA, FB_1, FB_2, FB_3, FUS-X, KA, NIV, and ZEA, literature[659]; Kuru-kuru, AFB_1, AFB_2, AFG_1, AFG_2, BEA, ENA, ENA_1, ENB, ENB_1, FA, FB_1, FB_3, KA, MA, OTB, ROQ-C, STG, VIOM, and ZEA, literature[659]; Paste (peanut paste), AFB_1, AFB_2, AFG_1, AFG_2, AME, BEA, ENA, ENA_1, ENB, ENB_1, FA, FB_1, FB_2, FB_3, KA, MA, OTA, OTB, STG, VIOM, and ZEA, literature[659]; Soups, AFB_1, AFB_2, AFG_1, AFG_2, AME, AOH, BEA, DON, ENA, ENA_1, ENB, ENB_1, FB_1, FB_2, FB_3, KA, OTA, OTB, and STG, literature[659]; see also Dagwa, Flour (soybean flour), Kuru-kuru, Paste (peanut paste), Soups, chapter: Further Mycotoxins and Microbial Metabolites

Ochratoxin B

incidence: 5/6, conc. range: <LOQ-3 μg/kg, ∅ conc.: 2 μg/kg, sample origin: Bamenda (region, western highlands), Cameroon, sample year: July–September 2011, country: Cameroon/Austria/South Africa[659]

- Co-contamination: For detailed information see the article.
- Further contamination of plants and plant products other than cereals and their products reported in the present book/article[659]: Dagwa, AFB_1, AFB_2, AFG_1, AFG_2, AME, BEA, CIT, DON, ENA, ENA_1, ENB, ENB_1, FA, FB_1, FB_2, FB_3, KA, MA, NIV, OTA, OTB, STG, VIOM, and ZEA, literature[659]; Flour (soybean flour), AFB_1, AFG_1, BEA, DON, ENB, ENB_1, FA, FB_1, FB_2, FB_3, FUS-X, KA, NIV, and ZEA, literature[659]; Kuru-kuru, AFB_1, AFB_2, AFG_1, AFG_2, BEA, ENA, ENA_1, ENB, ENB_1, FA, FB_1, FB_3, KA, MA, OTA, ROQ-C, STG, VIOM, and ZEA, literature[659]; Paste (peanut paste), AFB_1, AFB_2, AFG_1, AFG_2, AME, BEA, ENA, ENA_1, ENB, ENB_1, FA, FB_1, FB_2, FB_3, KA, MA, OTA, OTB, STG, VIOM, and ZEA, literature[659]; Soups, AFB_1, AFB_2, AFG_1, AFG_2, AME, AOH, BEA, DON, ENA, ENA_1, ENB, ENB_1, FB_1, FB_2, FB_3, KA, OTA,

OTB, and STG, literature[659]; see also Dagwa, Flour (soybean flour), Kuru-kuru, Paste (peanut paste), Soups, chapter: Further Mycotoxins and Microbial Metabolites

Viomellein

incidence: 1/6, conc.: <LOQ, sample origin: Bamenda (region, western highlands), Cameroon, sample year: July–September 2011, country: Cameroon/Austria/South Africa[659]

- Co-contamination: For detailed information see the article.
- Further contamination of plants and plant products other than cereals and their products reported in the present book/article[659]: Dagwa, AFB_1, AFB_2, AFG_1, AFG_2, AME, BEA, CIT, DON, ENA, ENA_1, ENB, ENB_1, FA, FB_1, FB_2, FB_3, KA, MA, NIV, OTA, OTB, STG, VIOM, and ZEA, literature[659]; Flour (soybean flour), AFB_1, AFG_1, BEA, DON, ENB, ENB_1, FA, FB_1, FB_2, FB_3, FUS-X, KA, NIV, and ZEA, literature[659]; Kuru-kuru, AFB_1, AFB_2, AFG_1, AFG_2, BEA, ENA, ENA_1, ENB, ENB_1, FA, FB_1, FB_3, KA, MA, OTA, OTB, ROQ-C, STG, and ZEA, literature[659]; Paste (peanut paste), AFB_1, AFB_2, AFG_1, AFG_2, AME, BEA, ENA, ENA_1, ENB, ENB_1, FA, FB_1, FB_2, FB_3, KA, MA, OTA, OTB, STG, VIOM, and ZEA, literature[659]; Soups, AFB_1, AFB_2, AFG_1, AFG_2, AME, AOH, BEA, DON, ENA, ENA_1, ENB, ENB_1, FB_1, FB_2, FB_3, KA, OTA, OTB, and STG, literature[659]; see also Dagwa, Flour (soybean flour), Kuru-kuru, Paste (peanut paste), Soups, chapter: Further Mycotoxins and Microbial Metabolites

Fusarium Toxins

Depsipeptides

Beauvericin

incidence: 6/6, conc. range: 0.6–0.9 µg/kg, ∅ conc.: 0.8 µg/kg, sample origin: Bamenda (region, western highlands), Cameroon, sample year: July–September 2011, country: Cameroon/Austria/South Africa[659]

- Co-contamination: For detailed information see the article.
- Further contamination of plants and plant products other than cereals and their products reported in the present book/article[659]: Dagwa, AFB_1, AFB_2, AFG_1, AFG_2, AME, BEA, CIT, DON, ENA, ENA_1, ENB, ENB_1, FA, FB_1, FB_2, FB_3, KA, MA, NIV, OTA, OTB, STG, VIOM, and ZEA, literature[659]; Flour (soybean flour), AFB_1, AFG_1, BEA, DON, ENB, ENB_1, FA, FB_1, FB_2, FB_3, FUS-X, KA, NIV, and ZEA, literature[659]; Kuru-kuru, AFB_1, AFB_2, AFG_1, AFG_2, ENA, ENA_1, ENB, ENB_1, FA, FB_1, FB_3, KA, MA, OTA, OTB, ROQ-C, STG, VIOM, and ZEA, literature[659]; Paste (peanut paste), AFB_1, AFB_2, AFG_1, AFG_2, AME, BEA, ENA, ENA_1, ENB, ENB_1, FA, FB_1, FB_2, FB_3, KA, MA, OTA, OTB, STG, VIOM, and ZEA, literature[659]; Soups, AFB_1, AFB_2, AFG_1, AFG_2, AME, AOH, BEA, DON, ENA, ENA_1, ENB, ENB_1, FB_1, FB_2, FB_3, KA, OTA, OTB, and STG, literature[659]; see also Dagwa, Flour (soybean flour), Kuru-kuru, Paste (peanut paste), Soups, chapter: Further Mycotoxins and Microbial Metabolites

Enniatin A

incidence: 3/6, conc. range: <LOQ, sample origin: Bamenda (region, western highlands), Cameroon, sample year: July–September 2011, country: Cameroon/Austria/South Africa[659]

- Co-contamination: For detailed information see the article.
- Further contamination of plants and plant products other than cereals and their products reported in the present book/article[659]: Dagwa, AFB_1, AFB_2, AFG_1, AFG_2, AME, BEA, CIT, DON, ENA, ENA_1, ENB, ENB_1, FA, FB_1, FB_2, FB_3, KA, MA, NIV, OTA, OTB, STG, VIOM, and ZEA, literature[659]; Flour (soybean flour), AFB_1, AFG_1, BEA, DON, ENB, ENB_1, FA, FB_1, FB_2, FB_3, FUS-X, KA, NIV, and ZEA, literature[659]; Kuru-kuru, AFB_1, AFB_2, AFG_1, AFG_2, BEA, ENA_1, ENB, ENB_1, FA, FB_1, FB_3, KA, MA, OTA, OTB, ROQ-C, STG, VIOM, and ZEA, literature[659]; Paste (peanut paste), AFB_1, AFB_2, AFG_1, AFG_2, AME, BEA, ENA, ENA_1, ENB, ENB_1, FA, FB_1, FB_2, FB_3, KA, MA, OTA, OTB, STG, VIOM, and ZEA, literature[659]; Soups, AFB_1, AFB_2, AFG_1, AFG_2, AME, AOH, BEA, DON, ENA, ENA_1, ENB, ENB_1, FB_1, FB_2, FB_3, KA, OTA, OTB, and STG, literature[659]; see also Dagwa, Flour (soybean flour), Kuru-kuru, Paste (peanut paste), Soups, chapter: Further Mycotoxins and Microbial Metabolites

Enniatin A_1

incidence: 2/6, conc. range: <LOQ, sample origin: Bamenda (region, western highlands), Cameroon, sample year: July–September 2011, country: Cameroon/Austria/South Africa[659]

- Co-contamination: For detailed information see the article.
- Further contamination of plants and plant products other than cereals and their products reported in the present book/article[659]: Dagwa, AFB_1, AFB_2, AFG_1, AFG_2, AME, BEA, CIT, DON, ENA, ENA_1, ENB, ENB_1, FA, FB_1, FB_2, FB_3, KA, MA, NIV, OTA, OTB, STG, VIOM, and ZEA, literature[659]; Flour (soybean flour), AFB_1, AFG_1, BEA, DON, ENB, ENB_1, FA, FB_1, FB_2, FB_3, FUS-X, KA, NIV, and ZEA, literature[659]; Kuru-kuru, AFB_1, AFB_2, AFG_1, AFG_2, BEA, ENA, ENB, ENB_1, FA, FB_1, FB_3, KA, MA, OTA, OTB, ROQ-C, STG, VIOM, and ZEA, literature[659]; Paste (peanut paste), AFB_1, AFB_2, AFG_1, AFG_2, AME, BEA, ENA, ENA_1, ENB, ENB_1, FA, FB_1, FB_2, FB_3, KA, MA, OTA, OTB, STG, VIOM, and ZEA, literature[659]; Soups, AFB_1, AFB_2, AFG_1, AFG_2, AME, AOH, BEA, DON, ENA, ENA_1, ENB, ENB_1, FB_1, FB_2, FB_3, KA, OTA, OTB, and STG, literature[659]; see also Dagwa, Flour (soybean flour), Kuru-kuru, Paste (peanut paste), Soups, chapter: Further Mycotoxins and Microbial Metabolites

Enniatin B

incidence: 6/6, conc. range: 0.02–0.03 µg/kg, ∅ conc.: 0.02 µg/kg, sample origin: Bamenda (region, western highlands), Cameroon, sample year: July–September 2011, country: Cameroon/Austria/South Africa[659]

- Co-contamination: For detailed information see the article.
- Further contamination of plants and plant products other than cereals and their products reported in the present book/article[659]: Dagwa, AFB_1, AFB_2, AFG_1, AFG_2, AME, BEA, CIT, DON, ENA, ENA_1, ENB, ENB_1, FA, FB_1, FB_2, FB_3, KA, MA, NIV, OTA, OTB, STG, VIOM, and ZEA, literature[659]; Flour (soybean flour), AFB_1, AFG_1, BEA, DON, ENB, ENB_1, FA, FB_1, FB_2, FB_3, FUS-X, KA, NIV, and ZEA, literature[659]; Kuru-kuru, AFB_1, AFB_2, AFG_1, AFG_2, BEA, ENA, ENA_1, ENB_1, FA, FB_1, FB_3, KA, MA, OTA, OTB, ROQ-C, STG, VIOM, and ZEA, literature[659]; Paste (peanut paste), AFB_1, AFB_2, AFG_1, AFG_2, AME, BEA, ENA, ENA_1, ENB, ENB_1, FA, FB_1, FB_2, FB_3, KA, MA, OTA, OTB, STG, VIOM, and ZEA, literature[659]; Soups, AFB_1, AFB_2, AFG_1, AFG_2, AME, AOH, BEA, DON, ENA, ENA_1, ENB, ENB_1, FB_1, FB_2, FB_3, KA, OTA, OTB, and STG, literature[659]; see also Dagwa, Flour (soybean flour), Kuru-kuru, Paste (peanut paste), Soups, chapter: Further Mycotoxins and Microbial Metabolites

Enniatin B_1

incidence: 6/6, conc. range: 0.2–0.4 µg/kg, ∅ conc.: 0.3 µg/kg, sample origin: Bamenda (region, western highlands), Cameroon, sample year: July–September 2011, country: Cameroon/Austria/South Africa[659]

- Co-contamination: For detailed information see the article.
- Further contamination of plants and plant products other than cereals and their products reported in the present book/article[659]: Dagwa, AFB_1, AFB_2, AFG_1, AFG_2, AME, BEA, CIT, DON, ENA, ENA_1, ENB, ENB_1, FA, FB_1, FB_2, FB_3, KA, MA, NIV, OTA, OTB, STG, VIOM, and ZEA, literature[659]; Flour (soybean flour), AFB_1, AFG_1, BEA, DON, ENB, ENB_1, FA, FB_1, FB_2, FB_3, FUS-X, KA, NIV, and ZEA, literature[659]; Kuru-kuru, AFB_1, AFB_2, AFG_1, AFG_2, BEA, ENA, ENA_1, ENB, FA, FB_1, FB_3, KA, MA, OTA, OTB, ROQ-C, STG, VIOM, and ZEA, literature[659]; Paste (peanut paste), AFB_1, AFB_2, AFG_1, AFG_2, AME, BEA, ENA, ENA_1, ENB, ENB_1, FA, FB_1, FB_2, FB_3, KA, MA, OTA, OTB, STG, VIOM, and ZEA, literature[659]; Soups, AFB_1, AFB_2, AFG_1, AFG_2, AME, AOH, BEA, DON, ENA, ENA_1, ENB, ENB_1, FB_1, FB_2, FB_3, KA, OTA, OTB, and STG, literature[659]; see also Dagwa, Flour (soybean flour), Kuru-kuru, Paste (peanut paste), Soups, chapter: Further Mycotoxins and Microbial Metabolites

Fumonisin/s

Fumonisin B_1

incidence: 6/6, conc. range: 2–4.3 µg/kg, ∅ conc.: 3.2 µg/kg, sample origin: Bamenda (region, western highlands), Cameroon, sample year: July–September 2011, country: Cameroon/Austria/South Africa[659]

- Co-contamination: For detailed information see the article.
- Further contamination of plants and plant products other than cereals and their products reported in the present book/article[659]: Dagwa, AFB_1, AFB_2, AFG_1, AFG_2, AME, BEA, CIT, DON, ENA, ENA_1, ENB, ENB_1, FA, FB_1, FB_2, FB_3, KA, MA, NIV, OTA, OTB, STG, VIOM, and ZEA, literature[659]; Flour (soybean flour), AFB_1, AFG_1, BEA, DON, ENB, ENB_1, FA, FB_1, FB_2, FB_3, FUS-X, KA, NIV, and ZEA, literature[659]; Kuru-kuru, AFB_1, AFB_2, AFG_1, AFG_2, BEA, ENA, ENA_1, ENB, ENB_1, FA, FB_3, KA, MA, OTA, OTB, ROQ-C, STG, VIOM, and ZEA, literature[659]; Paste (peanut paste), AFB_1, AFB_2, AFG_1, AFG_2, AME, BEA, ENA, ENA_1, ENB, ENB_1, FA, FB_1, FB_2, FB_3, KA, MA, OTA, OTB, STG, VIOM, and ZEA, literature[659]; Soups, AFB_1, AFB_2, AFG_1, AFG_2, AME, AOH, BEA, DON, ENA, ENA_1, ENB, ENB_1, FB_1, FB_2, FB_3, KA, OTA, OTB, and STG, literature[659]; see also Dagwa, Flour (soybean flour), Kuru-kuru, Paste (peanut paste), Soups, chapter: Further Mycotoxins and Microbial Metabolites

Fumonisin B_3

incidence: 2/6, conc. range: 3.4–4.2 µg/kg, ∅ conc.: 3.8 µg/kg, sample origin: Bamenda (region, western highlands), Cameroon, sample year: July–September 2011, country: Cameroon/Austria/South Africa[659]

- Co-contamination: For detailed information see the article.
- Further contamination of plants and plant products other than cereals and their products reported in the present book/article[659]: Dagwa, AFB_1, AFB_2, AFG_1, AFG_2, AME, BEA, CIT, DON, ENA, ENA_1, ENB, ENB_1, FA, FB_1, FB_2, FB_3, KA, MA, NIV, OTA, OTB, STG, VIOM, and ZEA, literature[659]; Flour (soybean flour), AFB_1, AFG_1, BEA, DON, ENB, ENB_1, FA, FB_1, FB_2, FB_3, FUS-X, KA, NIV, and ZEA, literature[659]; Kuru-kuru, AFB_1, AFB_2, AFG_1, AFG_2, BEA, ENA, ENA_1, ENB, ENB_1, FA, FB_1, KA, MA, OTA, OTB, ROQ-C, STG, VIOM, and ZEA, literature[659]; Paste (peanut paste), AFB_1, AFB_2, AFG_1, AFG_2, AME, BEA, ENA, ENA_1, ENB, ENB_1, FA, FB_1, FB_2, FB_3, KA, MA, OTA, OTB, STG, VIOM, and ZEA, literature[659]; Soups, AFB_1, AFB_2, AFG_1, AFG_2, AME, AOH, BEA, DON, ENA, ENA_1, ENB, ENB_1, FB_1, FB_2, FB_3, KA, OTA, OTB, and STG, literature[659]; see also Dagwa, Flour (soybean flour), Kuru-kuru, Paste (peanut paste), Soups, chapter: Further Mycotoxins and Microbial Metabolites

Other ***Fusarium*** Toxins

Fusaric Acid

incidence: 5/6, conc. range: <LOQ-7.1 µg/kg, ∅ conc.: 5.9 µg/kg, sample origin: Bamenda (region, western highlands), Cameroon, sample year: July–September 2011, country: Cameroon/Austria/South Africa[659]

- Co-contamination: For detailed information see the article.
- Further contamination of plants and plant products other than cereals and their products reported in the present book/article[659]: Dagwa, AFB_1, AFB_2, AFG_1, AFG_2, AME, BEA, CIT, DON, ENA, ENA_1, ENB, ENB_1, FA, FB_1, FB_2, FB_3, KA, MA, NIV, OTA, OTB, STG, VIOM, and ZEA, literature[659]; Flour (soybean flour), AFB_1, AFG_1, BEA, DON, ENB, ENB_1, FA, FB_1, FB_2, FB_3, FUS-X, KA, NIV, and ZEA, literature[659]; Kuru-kuru, AFB_1, AFB_2, AFG_1, AFG_2, BEA, ENA, ENA_1, ENB, ENB_1, FB_1, FB_3, KA, MA, OTA, OTB, ROQ-C, STG, VIOM, and ZEA, literature[659]; Paste (peanut paste), AFB_1, AFB_2, AFG_1, AFG_2, AME, BEA, ENA, ENA_1, ENB, ENB_1, FA, FB_1, FB_2, FB_3, KA, MA, OTA, OTB, STG, VIOM, and ZEA, literature[659]; Soups, AFB_1, AFB_2, AFG_1, AFG_2, AME, AOH, BEA, DON, ENA, ENA_1, ENB, ENB_1, FB_1, FB_2, FB_3, KA, OTA, OTB, and STG, literature[659]; see also Dagwa, Flour (soybean flour), Kuru-kuru, Paste (peanut paste), Soups, chapter: Further Mycotoxins and Microbial Metabolites

Zearalenone

incidence: 1/6, conc.: <LOQ, sample origin: Bamenda (region, western highlands), Cameroon, sample year: July–September 2011, country: Cameroon/Austria/South Africa[659]

- Co-contamination: For detailed information see the article.
- Further contamination of plants and plant products other than cereals and their products reported in the present book/article[659]: Dagwa, AFB_1, AFB_2, AFG_1, AFG_2, AME, BEA, CIT, DON, ENA, ENA_1, ENB, ENB_1, FA, FB_1, FB_2, FB_3, KA, MA, NIV, OTA, OTB, STG, VIOM, and ZEA, literature[659]; Flour (soybean flour), AFB_1, AFG_1, BEA, DON, ENB, ENB_1, FA, FB_1, FB_2, FB_3, FUS-X, KA, NIV, and ZEA, literature[659]; Kuru-kuru, AFB_1, AFB_2, AFG_1, AFG_2, BEA, ENA, ENA_1, ENB, ENB_1, FA, FB_1, FB_3, KA, MA, OTA, OTB, ROQ-C, STG, and VIOM, literature[659]; Paste (peanut paste), AFB_1, AFB_2, AFG_1, AFG_2, AME, BEA, ENA, ENA_1, ENB, ENB_1, FA, FB_1, FB_2, FB_3, KA, MA, OTA, OTB, STG, VIOM, and ZEA, literature[659]; Soups, AFB_1, AFB_2, AFG_1, AFG_2, AME, AOH, BEA, DON, ENA, ENA_1, ENB, ENB_1, FB_1, FB_2, FB_3, KA, OTA, OTB, and STG, literature[659]; see also Dagwa, Flour (soybean flour), Kuru-kuru, Paste (peanut paste), Soups, chapter: Further Mycotoxins and Microbial Metabolites

Penicillium Toxins

Mycophenolic Acid

incidence: 3/6, conc. range: <LOQ-1 µg/kg, ∅ conc.: 1 µg/kg, sample origin: Bamenda (region, western highlands), Cameroon, sample year: July–September 2011, country: Cameroon/Austria/South Africa[659]

- Co-contamination: For detailed information see the article.

- Further contamination of plants and plant products other than cereals and their products reported in the present book/article[659]: Dagwa, AFB_1, AFB_2, AFG_1, AFG_2, AME, BEA, CIT, DON, ENA, ENA_1, ENB, ENB_1, FA, FB_1, FB_2, FB_3, KA, MA, NIV, OTA, OTB, STG, VIOM, and ZEA, literature[659]; Flour (soybean flour), AFB_1, AFG_1, BEA, DON, ENB, ENB_1, FA, FB_1, FB_2, FB_3, FUS-X, KA, NIV, and ZEA, literature[659]; Kuru-kuru, AFB_1, AFB_2, AFG_1, AFG_2, BEA, ENA, ENA_1, ENB, ENB_1, FA, FB_1, FB_3, KA, OTA, OTB, ROQ-C, STG, VIOM, and ZEA, literature[659]; Paste (peanut paste), AFB_1, AFB_2, AFG_1, AFG_2, AME, BEA, ENA, ENA_1, ENB, ENB_1, FA, FB_1, FB_2, FB_3, KA, MA, OTA, OTB, STG, VIOM, and ZEA, literature[659]; Soups, AFB_1, AFB_2, AFG_1, AFG_2, AME, AOH, BEA, DON, ENA, ENA_1, ENB, ENB_1, FB_1, FB_2, FB_3, KA, OTA, OTB, and STG, literature[659]; see also Dagwa, Flour (soybean flour), Kuru-kuru, Paste (peanut paste), Soups, chapter: Further Mycotoxins and Microbial Metabolites

Roquefortine C

incidence: 6/6, conc. range: 0.9–6 µg/kg, ∅ conc.: 3 µg/kg, sample origin: Bamenda (region, western highlands), Cameroon, sample year: July–September 2011, country: Cameroon/Austria/South Africa[659]

- Co-contamination: For detailed information see the article.
- Further contamination of plants and plant products other than cereals and their products reported in the present book/article[659]: Dagwa, AFB_1, AFB_2, AFG_1, AFG_2, AME, BEA, CIT, DON, ENA, ENA_1, ENB, ENB_1, FA, FB_1, FB_2, FB_3, KA, MA, NIV, OTA, OTB, STG, VIOM, and ZEA, literature[659]; Flour (soybean flour), AFB_1, AFG_1, BEA, DON, ENB, ENB_1, FA, FB_1, FB_2, FB_3, FUS-X, KA, NIV, and ZEA, literature[659]; Kuru-kuru, AFB_1, AFB_2, AFG_1, AFG_2, BEA, ENA, ENA_1, ENB, ENB_1, FA, FB_1, FB_3, KA, MA, OTA, OTB, STG, VIOM, and ZEA, literature[659]; Paste (peanut paste), AFB_1, AFB_2, AFG_1, AFG_2, AME, BEA, ENA, ENA_1, ENB, ENB_1, FA, FB_1, FB_2, FB_3, KA, MA, OTA, OTB, STG, VIOM, and ZEA, literature[659]; Soups, AFB_1, AFB_2, AFG_1, AFG_2, AME, AOH, BEA, DON, ENA, ENA_1, ENB, ENB_1, FB_1, FB_2, FB_3, KA, OTA, OTB, and STG, literature[659]; see also Dagwa, Flour (soybean flour), Kuru-kuru, Paste (peanut paste), Soups, chapter: Further Mycotoxins and Microbial Metabolites

Kuru-kuru is a snack consisting of a dry fried paste made from peanuts which is solid.

Legume products see Products (legume product)

Legumes may contain the following mycotoxins:

Aspergillus Toxins

Aflatoxin B_1

incidence: 6/108*, conc. range: ≤3 µg/kg, sample origin: imported to USSR, sample year: 1985–1987, country: USSR[267], *kidney beans, peas, and soybeans

- Co-contamination: not reported
- Further contamination of plants and plant products other than cereals and their products reported in the present book/article[267]: Cocoa beans, Coffee beans, Nuts, Nuts (peanut), AFB_1, literature[267]

incidence: 2/30, conc. range: 1.0–3.6 µg/kg, ∅ conc.: 2.3 µg/kg, sample origin: supermarkets, retail stores, and stock centres, Colombia, sample year: unknown, country: Colombia[499]

- Co-contamination: not reported
- Further contamination of plants and plant products other than cereals and their products reported in the present book/article[499]: not reported

incidence: 5/15*, conc. range: 1–2,672 µg/kg, ∅ conc.: 891 µg/kg, sample origin: Sonoran Desert in Arizona (state), USA, sample year: unknown, country: USA[500], *mesquite, from ground

- Co-contamination: not reported
- Further contamination of plants and plant products other than cereals and their products reported in the present book/article[500]: not reported

For detailed information see the article.

Aspergillus and *Penicillium* Toxins

Ochratoxin A

incidence: 2/64, conc. range: 13.7–15.4 µg/kg, ∅ conc.: 14.6 µg/kg, sample origin: retail outlets across the UK, sample year: August 1996, country: UK[143]

- Co-contamination: not reported
- Further contamination of plants and plant products other than cereals and their products reported in the present book/article[143]: Beans, Powder (cocoa powder), Wine (red), OTA, literature[143]

incidence: 1/103*, conc.: 0.840 µg/kg, sample origin: retail stores in Detmold, Jena, Karlsruhe, Kiel, Kulmbach, and Trier (cities), as well as Berlin (capital), Germany, sample year: 1996–1998, country: Germany[597], *beans, lentils, and peas

- Co-contamination: not reported
- Further contamination of plants and plant products other than cereals and their products reported in the present book/article[597]: Beans (soybean), OTA, literature[597]

incidence: 3/24*, conc. range: ≤0.08 µg/kg, sample origin: supermarkets and fresh provision shops, Hong Kong, China, sample year: August/September 2005, country: China[697], *legumes, pulses and their products

- Co-contamination: not reported
- Further contamination of plants and plant products other than cereals and their products reported in the present book/article[697]: Chocolate, Coffee, Fruits, dried, Juices, Spices, OTA, literature[697]

incidence: 1/104*, conc.: 0.86 µg/kg (1 sa.), sample origin: Germany, sample year: 1995–1998, country: EU[698], *pulses without soya

- Co-contamination: not reported
- Further contamination of plants and plant products other than cereals and their products reported in the present book/article[698]: Baby food, Beans (soybean), Chocolate, Cocoa, Coffee, Coffee beans, Currants, Fruits, dried, Herbs, Jam, Juices, Juices (grape juice), Nuts, Oil seeds, Oils (olive oil), Peas, Peas (chickpea), Products (plant product), Raisins, Spices, Spices (cayenne pepper, chilli), Spices (nutmeg), Spices (paprika), Spices (pepper), Sultanas, Tea, Vinegar, Wine, Wine (red), Wine (rosé), Wine (white), Wine, miscellaneous, OTA, literature[698]

Lentils may contain the following mycotoxins:

Aspergillus Toxins

Aflatoxin B_1

incidence: 1/4, conc.: 72.1 µg/kg, sample origin: different storage facilities, Egypt, sample year: November 1978–April 1979, country: Egypt[4], *sa. moldy with mechanical or insect damage

- Co-contamination: not reported
- Further contamination of plants and plant products other than cereals and their products reported in the present book/article[4]: Beans (broad bean), AFB_1 and AFB_2, literature[4]; Nuts (almond), AFB_1, literature[4]; Nuts (peanut), AFB_1, AFB_2, AFG_1, and AFG_2, literature[4]

incidence: 1/6*, conc.: 3.1 µg/kg**, sample origin: Alexandria, Luxor (cities), and Cairo (capital), Egypt, sample year: unknown, country: Egypt/USA[127], *3 sa. freshly produced and 3 sa. stored for 12 months, **stored sa.

- Co-contamination: not reported
- Further contamination of plants and plant products other than cereals and their products reported in the present book/article[127]: Beans (broad bean), Nuts (peanut), AFB_1 and AFB_2, literature[127]; Spices (fenugreek), AFB_1, AFB_2, and AFG_1, literature[127]

incidence: 13/23, conc. range: 21–100 µg/kg (9 sa.), 101–500 µg/kg (4 sa., maximum: 268 µg/kg), sample origin: villages of Khagaria and Saharsa (districts) of North Bihar (state), India, sample year: August–October 1987, country: India[128]

- Co-contamination: not reported
- Further contamination of plants and plant products other than cereals and their products reported in the present book/article[128]: Beans (cowpea), Beans (mung bean), Field mustard, Seeds (linseed), Peas (chickpea), AFB_1, literature[128]

incidence: 1/4*, conc.: 1.8 µg/kg**, sample origin: retail outlets in Lattakia (city), Syria, sample year: January 1989, country: Italy[347], *lentils, intact and ground, **ground one contaminated

- Co-contamination: not reported
- Further contamination of plants and plant products other than cereals and their products reported in the present book/article[347]: Figs, AFB_1, literature[347]; Nuts (peanut), AFB_1 and AFB_2, literature[347]

Aflatoxin/s

incidence: 1/13, conc.: 14.3 µg/kg*, sample origin: local markets in Khartoum (capital), Sudan, sample year: 1993–1996, country: Sudan[501], *AFB_1, AFB_2, AFG_1, and AFG_2

- Co-contamination: not reported
- Further contamination of plants and plant products other than cereals and their products reported in the present book/article[501]: not reported

Aspergillus and ***Penicillium*** Toxins

Ochratoxin A

incidence: 1/13, conc.: 0.11 µg/kg, sample origin: markets across Lebanon, sample year: unknown, country: Lebanon[502]

- Co-contamination: not reported
- Further contamination of plants and plant products other than cereals and their products reported in the present book/article[502]: not reported

Fusarium Toxins

Type A Trichothecenes

T-2 Toxin

incidence: 1/3*, conc.: 460 µg/kg**, sample origin: market in Istanbul (capital), Turkey, sample year: unknown, country: Turkey[150], *lentils, green and red**

- Co-contamination: not reported
- Further contamination of plants and plant products other than cereals and their products reported in the present book/article[150]: Beans, Peas (chickpea), T-2, literature[150]

Licorice may contain the following mycotoxins:

Alternaria Toxins

Alternariol

incidence: 14/31*, conc. range: LOD-520.60 µg/kg, sample origin: traditional manufactures in Shanghai (city, municipality), China, sample year: unknown, country: China[917], *laminated form of pharmaceutical grade

- Co-contamination: 3 sa. co-contaminated with AOH, FB_1, and ZEA; 3 sa. co-contaminated with AOH and FB_1; 1 sa. co-contaminated with AOH, CIT, FB_1, and OTA; 1 sa. co-contaminated with AOH, DON, FB_1, and ZEA; 1 sa. co-contaminated with AOH, CIT, and FB_1; 5 sa. contaminated solely with AOH
- Further contamination of plants and plant products other than cereals and their products reported in the present book/article[917]: Licorice, CIT, DON, FB_1, OTA, and ZEA, literature[917]

Aspergillus Toxins

Aflatoxin B_1

incidence: 5/28*, conc. range: ≤2.4 µg/kg, Ø conc.: 1.57 µg/kg, sample origin: retail outlets in northern Italy, sample year: 2006/2007, country: Italy[503], *dried licorice extract

- Co-contamination: 5 sa. co-contaminated with AFB_1 and OTA
- Further contamination of plants and plant products other than cereals and their products reported in the present book/article[503]: Licorice, OTA, literature[503]

incidence: 8/54*, conc. range: ≤7.7 µg/kg, Ø conc.: 2.06 µg/kg, sample origin: retail outlets in northern Italy, sample year: 2006/2007, country: Italy[503], *licorice confectionery

- Co-contamination: not reported
- Further contamination of plants and plant products other than cereals and their products reported in the present book/article[503]: Licorice, OTA, literature[503]

incidence: 2/12*, conc. range: 12.56–26.11 µg/kg, Ø conc.: 19.3 µg/kg, sample origin: commercial markets and producing areas, China, sample year: unknown, country: China[872], *licorice roots

- Co-contamination: 1 sa. co-contaminated with AFB_1 and OTA; 1 sa. contaminated solely with AFB_1
- Further contamination of plants and plant products other than cereals and their products reported in the present book/article[872]: Licorice, OTA, literature[872]; Medicinal plants, AFB_1, literature[872]

Aspergillus and ***Penicillium*** Toxins

Citrinin

incidence: 2/31*, conc. range: 6.75–20.44 µg/kg, Ø conc.: 13.6 µg/kg, sample origin: traditional manufactures in Shanghai (city, municipality), China, sample year: unknown, country: China[917], *laminated form of pharmaceutical grade

- Co-contamination: 1 sa. co-contaminated with AOH, CIT, FB_1, and OTA; 1 sa. co-contaminated with AOH, CIT, and FB_1
- Further contamination of plants and plant products other than cereals and their products reported in the present book/article[917]: Licorice, AOH, DON, FB_1, OTA, and ZEA, literature[917]

Ochratoxin A

incidence: 1/2*, conc.: 385 µg/kg, sample origin: CAOBISCO, EU, sample year: unknown, country: Belgium/Russia[192], *licorice powder

- Co-contamination: not reported
- Further contamination of plants and plant products other than cereals and their products reported in the present book/ article[192]: Spices (cayenne pepper, chilli), OTA, literature[192]

incidence: 28/28*, conc. range: : ≤990.1 µg/kg, ∅ conc.: 89.6 µg/kg, sample origin: retail outlets in northern Italy, sample year: 2006/2007, country: Italy[503], *dried licorice extract

- Co-contamination: 5 sa. co-contaminated with AFB_1 and OTA; 23 sa. contaminated solely with OTA
- Further contamination of plants and plant products other than cereals and their products reported in the present book/ article[503]: Licorice, AFB_1, literature[503]

incidence: 33/54*, conc. range: ≤8.3 µg/kg, ∅ conc.: 1.53 µg/kg, sample origin: retail outlets in northern Italy, sample year: 2006/2007, country: Italy[503], *licorice confectionery

- Co-contamination: not reported
- Further contamination of plants and plant products other than cereals and their products reported in the present book/ article[503]: Licorice, AFB_1, literature[503]

incidence: 18/19*, conc. range: 0.4–3.0 µg/kg, ∅ conc.: 1.3 µg/kg, sample origin: drugstores, grocery shops, health food shops, and retail shops in Karlsruhe (city), Baden-Württemberg (state), Germany, sample year: unknown, country: Germany[504], *licorice sweet

incidence: 9/19*, conc. range: 0.3–216.5 µg/kg, ∅ conc.: 29.8 µg/kg, sample origin: drugstores, grocery shops, health food shops, herb shop, and pharmacies in Karlsruhe (city), Baden-Württemberg (state), Germany, sample year: unknown, country: Germany[504], *licorice root

- Co-contamination: not reported
- Further contamination of plants and plant products other than cereals and their products reported in the present book/ article[504]: not reported

incidence: 15/15*, conc. range: 1.4-252.8 µg/kg, ∅ conc.: 63.6 µg/kg, sample origin: herbal stores in Zaragoza (city), Spain, sample year: unknown, country: Spain[505], *dry licorice root (for the preparation of tea)

incidence: 8/8*, conc. range: 3.3–14.7 µg/kg, ∅ conc.: 9.2 µg/kg, sample origin: local growers, Zaragoza (city), Spain, sample year: unknown, country: Spain[505], *fresh licorice root (for sucking or chewing)

incidence: 4/4*, conc. range: 0.5–8.2 µg/kg, ∅ conc.: 3.8 µg/kg, sample origin: local supermarkets in Zaragoza (city), Spain, sample year: unknown, country: Spain[505], *licorice confectionery

incidence: 2/2*, conc. range: 14.6–17.3 µg/kg, ∅ conc.: 16.0 µg/kg, sample origin: herbal stores in Zaragoza (city), Spain, sample year: unknown, country: Spain[505], *liquid licorice extract (as a remedy medicine)

incidence: 1/1*, conc.: 39.5 µg/kg, sample origin: local industry in Zaragoza? (city), Spain, sample year: unknown, country: Spain[505], *solid licorice block (products for domestic, EU, and foreign countries)

- Co-contamination: not reported
- Further contamination of plants and plant products other than cereals and their products reported in the present book/ article[505]: not reported

incidence: 12/16*, conc. range: 0.70–12.0 µg/kg, sample origin: drugstores, supermarkets, health food shops, retail shops in several Spanish cities, Spain, sample year: 2007/2008, country: Spain[506], * licorice confectionery (sweets, pastilles; candy type: hard)

incidence: 11/28*, conc. range: 0.50–1.8 µg/kg, sample origin: partly imported from Italy, Germany, and the UK purchased in drugstores, supermarkets, health food shops, retail shops in several Spanish cities, Spain, sample year: 2007/2008, country: Spain[506], * licorice confectionery (chewy ropes or tubes, gummies; candy type: soft)

- Co-contamination: not reported
- Further contamination of plants and plant products other than cereals and their products reported in the present book/ article[506]: not reported

incidence: 1/1*, conc.: 8.8 µg/kg, sample origin: Beiza (region), Fars (province), Iran, sample year: unknown, country: Iran[764], *licorice root

- Co-contamination: not reported
- Further contamination of plants and plant products other than cereals and their products reported in the present book/ article[764]: not reported

incidence: 12/12*, conc. range: 3.80–36.70 µg/kg, ∅ conc.: 15.80 µg/kg, sample origin: 12 regions, Czech Republic, sample year: 2011–2013, country: Czech Republic[811], *licorice root, dried

- Co-contamination: not reported
- Further contamination of plants and plant products other than cereals and their products reported in the present book/ article[811]: Beans, Chocolate, Coffee, Confectionery, Figs, Herbs and Spices, Juices, Powder (cocoa powder), Products (ginger product), Raisins, Spices (cayenne pepper, chilli), Spices (caraway), Spices (coriander), Spices (ginger), Spices (mustard), Spices (nutmeg), Spices (paprika), Spices (pepper), Tea, Wine (red), Wine (white), OTA, literature[811]

incidence: 1/12*, conc.: 18.73 µg/kg, sample origin: commercial markets and producing areas, China, sample year: unknown, country: China[872], *licorice roots

- Co-contamination: 1 sa. co-contaminated with AFB_1 and OTA
- Further contamination of plants and plant products other than cereals and their products reported in the present book/ article[872]: Licorice, AFB_1, literature[872]; Medicinal plants, AFB_1, literature[872]

incidence: 2/31*, conc. range: LOD-3.93 µg/kg, sample origin: traditional manufactures in Shanghai (city, municipality), China, sample year: unknown, country: China[917], *laminated form of pharmaceutical grade

- Co-contamination: 1 sa. co-contaminated with AOH, CIT, FB_1, and OTA; 1 sa. co-contaminated with FB_1 and OTA
- Further contamination of plants and plant products other than cereals and their products reported in the present book/ article[917]: Licorice, AOH, CIT, DON, FB_1, and ZEA, literature[917]

Fusarium Toxins

Fumonisin/s

Fumonisin B_1
incidence: 12/31*, conc. range: LOD-39.34 µg/kg, sample origin: traditional manufactures in Shanghai (city, municipality), China, sample year: unknown, country: China[917], *laminated form of pharmaceutical grade

- Co-contamination: 3 sa. co-contaminated with AOH, FB_1, and ZEA; 3 sa. co-contaminated with AOH and FB_1; 1 sa. co-contaminated with AOH, CIT, FB_1, and OTA; 1 sa. co-contaminated with AOH, DON, FB_1, and ZEA; 1 sa. co-contaminated with AOH, CIT, and FB_1; 1 sa. co-contaminated with FB_1 and OTA ; 2 sa. contaminated solely with FB_1
- Further contamination of plants and plant products other than cereals and their products reported in the present book/article[917]: Licorice, AOH, CIT, DON, OTA, and ZEA, literature[917]

Type B Trichothecenes

Deoxynivalenol
incidence: 1/31*, conc.: 11.08 µg/kg, sample origin: traditional manufactures in Shanghai (city, municipality), China, sample year: unknown, country: China[917], *laminated form of pharmaceutical grade

- Co-contamination: 1 sa. co-contaminated with AOH, DON, FB_1, and ZEA
- Further contamination of plants and plant products other than cereals and their products reported in the present book/article[917]: Licorice, AOH, CIT, FB_1, OTA, and ZEA, literature[917]

Other ***Fusarium*** Toxins

Zearalenone
incidence: 4/31*, conc. range: 3.37–8.75 µg/kg, ∅ conc.: 5.39 µg/kg, sample origin: traditional manufactures in Shanghai (city, municipality), China, sample year: unknown, country: China[917], *laminated form of pharmaceutical grade

- Co-contamination: 3 sa. co-contaminated with AOH, FB_1, and ZEA; 1 sa. co-contaminated with AOH, DON, FB_1, and ZEA
- Further contamination of plants and plant products other than cereals and their products reported in the present book/article[917]: Licorice, AOH, CIT, DON, FB_1, and OTA, literature[917]

Linseeds see **Seeds (linseed)**

Lupines see **Seeds (lupine)**

Macabeo see **Must**

Mahua seeds may contain the following mycotoxins:

Aspergillus Toxins

Aflatoxin B_1
incidence: 12/15, conc. range: 86.43–322.45 µg/kg, ∅ conc.: 208.26 µg/kg, sample origin: Lucknow (city), India, sample year: July/August 2005, country: India[507]

- Co-contamination: not reported
- Further contamination of plants and plant products other than cereals and their products reported in the present book/article[507]: Mahua seeds, AF/S, literature[507]

Aflatoxin/s
incidence: 12/15, conc. range: 115.35–395.54 µg/kg*, ∅ conc.: 257.40 µg/kg*, sample origin: Lucknow (city), India, sample year: July/August 2005, country: India[507], *AFS total

- Co-contamination: not reported
- Further contamination of plants and plant products other than cereals and their products reported in the present book/article[507]: Mahua seeds, AFB_1, literature[507]

Malaga see **Wine, miscellaneous**

Mandarins see **Fruits (mandarin)**

Manzanilla see **Wine (white)**

Marmalade (apple marmalade) may contain the following mycotoxins:

Aspergillus and ***Penicillium*** Toxins

Patulin
incidence: 6/26, conc. range: 17–39 µg/kg, ∅ conc.: 27.0 µg/kg, sample origin: different Argentinean towns, Argentina, sample year: unknown, country: Argentina[50]

- Co-contamination: not reported
- Further contamination of plants and plant products other than cereals and their products reported in the present book/article[50]: Marmalade (pear marmalade), Puree, PAT, literature[50]

Marmalade (pear marmalade) may contain the following mycotoxins:

Aspergillus and ***Penicillium*** Toxins

Patulin
incidence: 1/6, conc.: 25 µg/kg, sample origin: different Argentinean towns, Argentina, sample year: unknown, country: Argentina[50]

- Co-contamination: not reported
- Further contamination of plants and plant products other than cereals and their products reported in the present book/article[50]: Marmalade (apple marmalade), Puree, PAT, literature[50]

Marsala see **Wine, miscellaneous**

Masala see **Spices (masala)**

Margarine may contain the following mycotoxins:

Fusarium Toxins

Fumonisin/s

Fumonisin B_1
incidence: 2/25*, conc. range: ≤109 µg/kg, sample origin: supermarkets, Spain, sample year: 2011/2012, country: Spain[723], *corn oil margarine

- Co-contamination: not reported
- Further contamination of plants and plant products other than cereals and their products reported in the present book/article[723]: Margarine, ZEA, literature[723]; Oils (maize oil), DON, FB_1, and ZEA, literature[723]

Other ***Fusarium*** Toxins

Zearalenone
incidence: 6/25*, conc. range: ≤86 μg/kg, sample origin: supermarkets, Spain, sample year: 2011/2012, country: Spain[723], *corn oil margarine
- Co-contamination: not reported
- Further contamination of plants and plant products other than cereals and their products reported in the present book/article[723]: Margarine, FB_1, literature[723]; Oils (maize oil), DON, FB_1, and ZEA, literature[723]

Marzipan (almond paste) may contain the following mycotoxins:

Aspergillus Toxins

Aflatoxin B_1
incidence: 1/168, conc.: 39 μg/kg, sample origin: imported to Finland?, sample year: 1974, country: Finland[2]
- Co-contamination: 1 sa. co-contaminated with AFB_1 and AFB_2
- Further contamination of plants and plant products other than cereals and their products reported in the present book/article[2]: Marzipan (almond paste), AFB_2, literature[2]; Nuts, AFB_1, AFB_2, and AFG_1, literature[2]; Nuts (almond), Nuts (hazelnut), AFB_1 and AFB_2, literature[2]; Nuts (peanut), AFB_1, AFB_2, AFG_1, and AFG_2, literature[2]

Aflatoxin B_2
incidence: 1/168, conc.: 7 μg/kg, sample origin: imported to Finland?, sample year: 1974, country: Finland[2]
- Co-contamination: 1 sa. co-contaminated with AFB_1 and AFB_2
- Further contamination of plants and plant products other than cereals and their products reported in the present book/article[2]: Marzipan (almond paste) AFB_1, literature[2]; Nuts, AFB_1, AFB_2, and AFG_1, literature[2]; Nuts (almond), Nuts (hazelnut), AFB_1 and AFB_2, literature[2]; Nuts (peanut), AFB_1, AFB_2, AFG_1, and AFG_2, literature[2]

Meal (copra meal) may contain the following mycotoxins:

Aspergillus Toxins

Aflatoxin B_1
incidence: 20/26, conc. range: 20–500 μg/kg, sample origin: local markets, oil mills, and houses, Andhra Pradesh (state), India, sample year: 1982, country: India[236]
- Co-contamination: not reported
- Further contamination of plants and plant products other than cereals and their products reported in the present book/article[236]: Candies (coconut candy), Copra, AFB_1, AFB_2, AFG_1, and AFG_2, literature[236]; Meal (copra meal), AFB_2, AFG_1, and AFG_2, literature[236]; Oils (coconut oil), AFB_1, AFB_2, and AFG_1, literature[236]

Aflatoxin B_2
incidence: 16/26, conc. range: 20–40 μg/kg, sample origin: local markets, oil mills, and houses, Andhra Pradesh (state), India, sample year: 1982, country: India[236]
- Co-contamination: not reported
- Further contamination of plants and plant products other than cereals and their products reported in the present book/article[236]: Candies (coconut candy), Copra, AFB_1, AFB_2, AFG_1, and AFG_2, literature[236]; Meal (copra meal), AFB_1, AFG_1, and AFG_2, literature[236]; Oils (coconut oil), AFB_1, AFB_2, and AFG_1, literature[236]

Aflatoxin G_1
incidence: 12/26, conc. range: 40–100 μg/kg, sample origin: local markets, oil mills, and houses, Andhra Pradesh (state), India, sample year: 1982, country: India[236]
- Co-contamination: not reported
- Further contamination of plants and plant products other than cereals and their products reported in the present book/article[236]: Candies (coconut candy), Copra, AFB_1, AFB_2, AFG_1, and AFG_2, literature[236]; Meal (copra meal), AFB_1, AFB_2, and AFG_2, literature[236]; Oils (coconut oil), AFB_1, AFB_2, and AFG_1, literature[236]

Aflatoxin G_2
incidence: 10/26, conc. range: 40–100 μg/kg, sample origin: local markets, oil mills, and houses, Andhra Pradesh (state), India, sample year: 1982, country: India[236]
- Co-contamination: not reported
- Further contamination of plants and plant products other than cereals and their products reported in the present book/article[236]: Candies (coconut candy), Copra, AFB_1, AFB_2, AFG_1, and AFG_2, literature[236]; Meal (copra meal), Oils (coconut oil), AFB_1, AFB_2, and AFG_1, literature[236]

Aflatoxin/s
incidence: 1/3, conc.: 2 μg/kg*, sample origin: Fiji, sample year: unknown, country: Fiji/Zambia[321], *AF

incidence: 1/2, conc.: 37 μg/kg*, sample origin: Tonga, sample year: unknown, country: Fiji/Zambia[321], *AF
- Co-contamination: not reported
- Further contamination of plants and plant products other than cereals and their products reported in the present book/article[321]: Copra, Nuts (peanut), AF/S, literature[321]

Meal (cycad meal) may contain the following mycotoxins:

Aspergillus Toxins

Aflatoxin B_1
incidence: 2/4, conc. range: 1.1–2.2 μg/kg, Ø conc.: 1.65 μg/kg, sample origin: unknown, sample year: unknown, country: USA[525]
- Co-contamination: 1 sa. co-contaminated with AFB_1 and AFG; 1 sa. contaminated solely AFB_1
- Further contamination of plants and plant products other than cereals and their products reported in the present book/article[525]: Meal (cycad meal), AF/S, literature[525]; Meal (peanut meal), AFB_1, literature[525]

Aflatoxin/s
incidence: 1/4, conc.: 5–8 μg/kg*, sample origin: unknown, sample year: unknown, country: USA[525], *AFG
- Co-contamination: 1 sa. co-contaminated with AFB_1 and AFG
- Further contamination of plants and plant products other than cereals and their products reported in the present book/article[525]: Meal (cycad meal), Meal (peanut meal), AFB_1, literature[525]

Meal (egusi meal) may contain the following mycotoxins:

Aspergillus Toxins

Aflatoxin/s
incidence: 1/1, conc.: 186 μg/kg*, sample origin: markets in Benin City, Nigeria, sample year: unknown, country: Nigeria[346], *AFB

- Co-contamination: not reported
- Further contamination of plants and plant products other than cereals and their products reported in the present book/article[346]: Emu aran, Ogbono, Ogili-ugba, Ogoro, AF/S, literature[346]

Egusi meal is the preparation of fat- and protein-rich seeds from gourd, melon, and/or squash.

Meal (peanut meal) may contain the following mycotoxins:

Aspergillus Toxins

Aflatoxin B_1

incidence: 3/4*, conc. range: 125–6,250 µg/kg, Ø conc.: 2,217 µg/kg, sample origin: unknown, sample year: unknown, country: USA[525], *ncac

- Co-contamination: not reported
- Further contamination of plants and plant products other than cereals and their products reported in the present book/article[525]: Meal (cycad meal), AFB_1 and AF/S, literature[525]

incidence: 1/1, conc.: 35.7 µg/kg, sample origin: unknown, sample year: unknown, country: USA[526]

incidence: 1/1*, conc.: 269 µg/kg, sample origin: unknown, sample year: unknown, country: USA[526], *de-oiled peanut meal

- Co-contamination: not reported
- Further contamination of plants and plant products other than cereals and their products reported in the present book/article[526]: not reported

Meal (pistachio meal) may contain the following mycotoxins:

Aspergillus Toxins

Aflatoxin B_1

incidence: 1/1, conc.: 61.0 µg/kg, sample origin: unknown, sample year: unknown, country: Denmark[217]

- Co-contamination: 1 sa. co-contaminated with AFB_1 and AFB_2
- Further contamination of plants and plant products other than cereals and their products reported in the present book/article[217]: Butter (peanut butter), Figs, Spices (cayenne pepper, chilli), AFB_1, AFB_2, AFG_1, and AFG_2, literature[217]; Meal (pistachio meal), AFB_2, literature[217]

Aflatoxin B_2

incidence: 1/1, conc.: 5.6 µg/kg, sample origin: unknown, sample year: unknown, country: Denmark[217]

- Co-contamination: 1 sa. co-contaminated with AFB_1 and AFB_2
- Further contamination of plants and plant products other than cereals and their products reported in the present book/article[217]: Butter (peanut butter), Figs, Spices (cayenne pepper, chilli), AFB_1, AFB_2, AFG_1, and AFG_2, literature[217]; Meal (pistachio meal), AFB_1, literature[217]

Meal (soybean meal) may contain the following mycotoxins:

Fusarium Toxins

Type A Trichothecenes

Diacetoxyscirpenol

incidence: 6/17* **, conc. range: <5–130 µg/kg, sample origin: northern Illinois, Iowa, and Michigan (states), USA, sample year: September/October 1986, country: USA[527], *ncac, **discounted or refused by grain merchants

- Co-contamination: 1 sa. co-contaminated with DAS, DON, T-2TET, ZEA, and ZEAOL; 2 sa. co-contaminated with DAS, DON, T-2TET, and ZEA; 3 sa. co-contaminated with DAS, DON, and T-2TET
- Further contamination of plants and plant products other than cereals and their products reported in the present book/article[527]: Meal (soybean meal), DON, T-2TET, and ZEA; Oils (soybean oil), DON, T-2TET, and ZEA, literature[527]; see also Meal (soybean meal), chapter: Further Mycotoxins and Microbial Metabolites

T-2 Tetraol

incidence: 15/17* **, conc. range: <5–1,420 µg/kg***, sample origin: northern Illinois, Iowa, and Michigan (states), USA, sample year: September/October 1986, country: USA[527], *ncac, **discounted or refused by grain merchants, ***T-2 equivalents = T-2TET after base hydrolysis (primarily HT-2)

- Co-contamination: 1 sa. co-contaminated with DAS, DON, T-2TET, ZEA, and ZEAOL; 2 sa. co-contaminated with DAS, DON, T-2TET, and ZEA; 1 sa. co-contaminated with DON,T-2TET, ZEA, and ZEAOL; 3 sa. co-contaminated with DAS, DON, and T-2TET; 8 sa. co-contaminated with DON and T-2TET
- Further contamination of plants and plant products other than cereals and their products reported in the present book/article[527]: Meal (soybean meal), DAS, DON, and ZEA; Oils (soybean oil), DON, T-2TET, and ZEA, literature[527]; see also Meal (soybean meal), chapter: Further Mycotoxins and Microbial Metabolites

Type B Trichothecenes

Deoxynivalenol

incidence: 17/17* **, conc. range: <5–600 µg/kg, sample origin: northern Illinois, Michigan, and Iowa (states), USA, sample year: September/October 1986, country: USA[527], *ncac, **discounted or refused by grain merchants

- Co-contamination: 1 sa. co-contaminated with DAS, DON, T-2TET, ZEA, and ZEAOL; 2 sa. co-contaminated with DAS, DON, T-2TET, and ZEA; 1 sa. co-contaminated with DON, T-2TET, ZEA, and ZEAOL; 3 sa. co-contaminated with DAS, DON, and T-2TET; 1 sa. co-contaminated with DON, T-2TET, and ZEA; 8 sa. co-contaminated with DON and T-2TET; 1 sa. contaminated solely with DON
- Further contamination of plants and plant products other than cereals and their products reported in the present book/article[527]: Meal (soybean meal), DAS, T-2TET, and ZEA; Oils (soybean oil), DON, T-2TET, and ZEA, literature[527]; see also Meal (soybean meal), chapter: Further Mycotoxins and Microbial Metabolites

Other *Fusarium* Toxins

Zearalenone

incidence: 5/17* **, conc. range: 180–760 µg/kg, Ø conc.: 380 µg/kg, sample origin: northern Illinois, Michigan, and Iowa (states), USA, sample year: September/October 1986, country: USA[527], *ncac, **discounted or refused by grain merchants

- Co-contamination: 1 sa. co-contaminated with DAS, DON, T-2TET, ZEA, and ZEAOL; 2 sa. co-contaminated with DAS, DON, T-2TET, and ZEA; 1 sa. co-contaminated with DON, T-2TET, ZEA, and ZEAOL; 1 sa. co-contaminated with DON, T-2TET, and ZEA
- Further contamination of plants and plant products other than cereals and their products reported in the present book/article[527]: Meal (soybean meal), DAS, DON, and T-2TET; Oils (soybean oil), DON, T-2TET, and ZEA, literature[527]; see also Meal (soybean meal), chapter: Further Mycotoxins and Microbial Metabolites

Medicinal herbs see Medicinal plants

Medicinal plants may contain the following mycotoxins:

Aspergillus Toxins

Aflatoxin B_1

incidence: 3/22*, conc. range: 1.58–27.80 µg/kg, Ø conc.: 11.2 µg/kg, sample origin: partly purchased in Beijing (capital), China, sample year: unknown, country: China[59], *traditional Chinese medicines

- Co-contamination: 1 sa. co-contaminated with AFB_1, AFB_2, and AFG_1; 1 sa. co-contaminated with AFB_1 and AFB_2; 1 sa. contaminated solely with AFB_1
- Further contamination of plants and plant products other than cereals and their products reported in the present book/article[59]: Medicinal plants, AFB_2 and AFG_1, literature[59]

For detailed information see the article.

incidence: 1/5*, conc.: 500 µg/kg, sample origin: dried sa. from Sri Lanka, sample year: unknown, country: USA[60], **Aerva lanata* (Linn.)

- Co-contamination: not reported
- Further contamination of plants and plant products other than cereals and their products reported in the present book/article[60]: not reported

incidence: 5/9*, conc. range: 0.03–0.45 µg/kg, Ø conc.: 0.188 µg/kg, sample origin: Penang (state), Malaysia, sample year: 2003, country: Japan/Malaysia[61], *commercial traditional herbal medicines with 3–15 ingredients

- Co-contamination: 3 sa. co-contaminated with AFB_1 and AFB_2; 2 sa. co-contaminated with AFB_1, AFB_2, and AFG_1
- Further contamination of plants and plant products other than cereals and their products reported in the present book/article[61]: Medicinal plants, AFB_2, AFG_1, and AFG_2, literature[61]

incidence: 11/14*, conc. range: 0.02–1.00 µg/kg, Ø conc.: 0.289 µg/kg, sample origin: sa. imported from Indonesia, sample year: 2003, country: Japan/Malaysia[61], *commercial traditional herbal medicines with 3–9 ingredients

- Co-contamination: 1 sa. co-contaminated with AFB_1, AFB_2, AFG_1, and AFG_2; 4 sa. co-contaminated with AFB_1, AFB_2, and AFG_1; 4 sa. co-contaminated with AFB_1 and AFB_2; 2 sa. contaminated solely with AFB_1
- Further contamination of plants and plant products other than cereals and their products reported in the present book/article[61]: Medicinal plants, AFB_2, AFG_1, and AFG_2, literature[61]

For detailed information see the article.

incidence: 14/15*, conc. range: 90–1,200 µg/kg, Ø conc.: 604 µg/kg, sample origin: storage centers of private concerns in southern Bihar (state), India, sample year: unknown, country: India[62], *herbal drugs

- Co-contamination: not reported
- Further contamination of plants and plant products other than cereals and their products reported in the present book/article[62]: not reported

For detailed information see the article.

incidence: 4/10*, Ø conc.: 170 µg/kg, sample origin: store-houses (private concerns) in Bihar (state), India, sample year: unknown, country: India[63], **Glycyrrhiza glabra* Linn.

incidence: 4/10*, Ø conc.: 290 µg/kg, sample origin: store-houses (private concerns) in Bihar (state), India sample year: unknown, country: India[63], **Holarrhena antidysenterica* Wall

incidence: 4/10*, Ø conc.: 510 µg/kg, sample origin: store-houses (private concerns) in Bihar (state), India sample year: unknown, country: India[63], **Ichnocarpus frutescens* (Linn.)

incidence: 8/10*, Ø conc.: 670 µg/kg, sample origin: store-houses (private concerns) in Bihar (state), India sample year: unknown, country: India[63], **Strychnos nux-vomica* Linn.

incidence: 6/10*, Ø conc.: 420 µg/kg, sample origin: store-houses (private concerns) in Bihar (state), India sample year: unknown, country: India[63], **Tribulus terrestris* Linn.

- Co-contamination: not reported
- Further contamination of plants and plant products other than cereals and their products reported in the present book/article[63]: Medicinal plants, CIT, OTA, and ZEA, literature[63]

For detailed information see the article.

incidence: 79/175*, conc. range: 80–910 µg/kg, sample origin: drug centers of Bhagalpur (city) and store houses in southern Bihar (state), India, sample year: January–December 1990, country: India[65], *crude and finished herbal drugs

- Co-contamination: not reported
- Further contamination of plants and plant products other than cereals and their products reported in the present book/article[65]: Medicinal plants, CIT, OTA, and ZEA, literature[65]

For detailed information see the article.

incidence: 4/13*, conc. range: 1.6–7.2 µg/kg, Ø conc.: 3.43 µg/kg, sample origin: enterprises in Zhejiang (province), China, sample year: unknown, country: China[66], *rhizomes and roots

- Co-contamination: 1 sa. co-contaminated with AFB_1, AFB_2, AFG_1, and AFM_1; 1 sa. co-contaminated with AFB_1, AFG_2, and AFM_1; 1 sa. co-contaminated with AFB_1 and AFG_1; 1 sa. co-contaminated with AFB_1 and AFG_2
- Further contamination of plants and plant products other than cereals and their products reported in the present book/article[66]: Medicinal plants, AFB_2, AFG_1, AFG_2, and AFM_1, literature[66]

incidence: 1/4*, conc.: 1.3 µg/kg, sample origin: enterprises in Zhejiang (province), China, sample year: unknown, country: China[66], *flowers

- Co-contamination: 1 sa. co-contaminated with AFB_1 and AFG_1
- Further contamination of plants and plant products other than cereals and their products reported in the present book/article[66]: Medicinal plants, AFB_2, AFG_1, AFG_2, and AFM_1, literature[66]

incidence: 5/8*, conc. range: 0.7–1.6 µg/kg, Ø conc.: 1.0 µg/kg, sample origin: enterprises in Zhejiang (province), China, sample year: unknown, country: China[66], *seeds

- Co-contamination: 1 sa. co-contaminated with AFB_1, AFG_1, and AFG_2; 1 sa. co-contaminated with AFB_1 and AFB_2; 1 sa. co-contaminated with AFB_1 and AFG_2; 2 sa. contaminated solely with AFB_1
- Further contamination of plants and plant products other than cereals and their products reported in the present book/article[66]: Medicinal plants, AFB_2, AFG_1, AFG_2, and AFM_1, literature[66]

incidence: 1/5*, conc.: 2.4 µg/kg, sample origin: enterprises in Zhejiang (province), China, sample year: unknown, country: China[66], *grasses and leaves

- Co-contamination: 1 sa. co-contaminated with AFB_1, AFB_2, and AFG_2
- Further contamination of plants and plant products other than cereals and their products reported in the present book/article[66]: Medicinal plants, AFB_2, AFG_1, AFG_2, and AFM_1, literature[66]

For detailed information see the article.

incidence: 5/28*, conc. range: 1.1–11.3 µg/kg, Ø conc.: 4.06 µg/kg, sample origin: retail suppliers and community hospitals in Khon Kaen (province), north-eastern Thailand, sample year: unknown, country: Thailand/Austria[67], *tablet or capsule (dosage form)

- Co-contamination: 1 sa. co-contaminated with AFB_1, AFB_2, and AFG_2; 1 sa. co-contaminated with AFB_1, AFB_2, and AFG_1; 1 sa. co-contaminated with AFB_1 and AFG_1; 1 sa. co-contaminated with AFB_1 and AFB_2; 1 sa. contaminated solely with AFB_1
- Further contamination of plants and plant products other than cereals and their products reported in the present book/article[67]: Medicinal plants, AFB_2, AFG_1, and AFG_2, literature[67]

For detailed information see the article.

incidence: 7/23*, conc. range: 1.2–9.8 µg/kg, Ø conc.: 3.06 µg/kg, sample origin: unknown, sample year: unknown, country: China[68], *rhizomes and roots

- Co-contamination: 1 sa. co-contaminated with AFB_1, AFB_2, AFM_1, CHAE, CIT, CPA, GT, MA, OTB, STG, and VER; 1 sa. co-contaminated with AFB_1, AFB_2, AFG_1, AFG_2, AFM_1, CHAE, MA, and OTA; 1 sa. co-contaminated with AFB_1, CPA, GT, MA, OTB, STG, T-2, and ZEA; 1 sa. co-contaminated with AFB_1, AFB_2, AFG_2, CHAE, NIV, OTB, and T-2; 1 sa. co-contaminated with AFB_1, CHAE, GT, MA, OTA, OTB, and T-2; 1 sa. co-contaminated with AFB_1, AFB_2, AFG_1, AFM_1, CPA, and NEO; 1 sa. co-contaminated with AFB_1, AFG_1, and AFM_1
- Further contamination of plants and plant products other than cereals and their products reported in the present book/article[68]: Medicinal plants, AFB_2, AFG_1, AFG_2, AFM_1, CIT, CPA, DAS, DON, FUS-X, GT, HT-2, MA, MON, NEO, NIV, OTA, OTB, PAT, STG, T-2, and ZEA, literature[68]; see also Medicinal plants, chapter: Further Mycotoxins and Microbial Metabolites

incidence: 7/15*, conc. range: 0.5–2.3 µg/kg, Ø conc.: 1.46 µg/kg, sample origin: unknown, sample year: unknown, country: China[68], *seeds

- Co-contamination: 1 sa. co-contaminated with AFB_1, AFB_2, AFG_1, AFG_2, AFM_1, CHAE, CPA, GT, HT-2, MA, PEA, STG, VER, and ZEA; 1 sa. co-contaminated with AFB_1, AFB_2, CPA, DAS, GT, MA, MON, NIV, STG, and VER; 1 sa. co-contaminated with AFB_1, AFG_1, AFG_2, DAS, MA, PEA, and STG; 1 sa. co-contaminated with AFB_1, AFB_2, CHAE, CPA, DPDON, and MA; 1 sa. co-contaminated with AFB_1, AFB_2, DAS, GT, STG, and ZEA; 1 sa. co-contaminated with AFB_1, CHAE, STG, ZEA, α-ZEL, and ZEAA; 1 sa. co-contaminated with AFB_1, AFB_2, and MA
- Further contamination of plants and plant products other than cereals and their products reported in the present book/article[68]: Medicinal plants, AFB_2, AFG_1, AFG_2, AFM_1, CIT, CPA, DAS, DON, FUS-X, GT, HT-2, MA, MON, NEO, NIV, OTA, OTB, PAT, STG, T-2, and ZEA, literature[68]; see also Medicinal plants, chapter: Further Mycotoxins and Microbial Metabolites

incidence: 3/11*, conc. range: 0.2–2.4 µg/kg, Ø conc.: 1.17 µg/kg, sample origin: unknown, sample year: unknown, country: China[68], *grasses and leaves

- Co-contamination: 1 sa. co-contaminated with AFB_1, AFG_1, GT, MA, OTB, PEA, and STG; 1 sa. co-contaminated with AFB_1, DAS, OTA, and OTB; 1 sa. co-contaminated with AFB_1, NIV, OTA, and PEA
- Further contamination of plants and plant products other than cereals and their products reported in the present book/article[68]: Medicinal plants, AFB_2, AFG_1, AFG_2, AFM_1, CIT, CPA, DAS, DON, FUS-X, GT, HT-2, MA, MON, NEO, NIV, OTA, OTB, PAT, STG, T-2, and ZEA, literature[68]; see also Medicinal plants, chapter: Further Mycotoxins and Microbial Metabolites

incidence: 1/11*, conc.: 0.3 µg/kg, sample origin: unknown, sample year: unknown, country: China[68], *flowers

- Co-contamination: 1 sa. co-contaminated with AFB_1, AFG_2, CHAE, CPA, DAS, DON3G, PEA, and STG
- Further contamination of plants and plant products other than cereals and their products reported in the present book/article[68]: Medicinal plants, AFB_2, AFG_1, AFG_2, AFM_1, CIT, CPA, DAS, DON, FUS-X, GT, HT-2, MA, MON, NEO, NIV, OTA, OTB, PAT, STG, T-2, and ZEA, literature[68]; see also Medicinal plants, chapter: Further Mycotoxins and Microbial Metabolites

For detailed information see the article.

incidence: 15/58, conc. range: 3.62–73.27 µg/kg, Ø conc.: 16.7 µg/kg, sample origin: commercial herbal medicine markets in Daegu, Daejeon, Gwangju, Jinju, Pusan (cities), and Seoul (capital), South Korea, sample year: unknown, country: Korea[69]

- Co-contamination: 1 sa. co-contaminated with AFB_1, AFB_2, AFG_1, and AFG_2; 1 sa. co-contaminated with AFB_1, AFB_2, and AFG_1; 8 sa. co-contaminated with AFB_1 and AFB_2; 1 sa. co-contaminated with AFB_1 and AFG_1; 4 sa. contaminated solely with AFB_1
- Further contamination of plants and plant products other than cereals and their products reported in the present book/article[69]: Medicinal plants, AFB_2, AFG_1, and AFG_2, literature[69]

For detailed information see the article.

incidence: 21/75*, conc. range: 0.06–239.62 µg/kg, Ø conc.: 22.3 µg/kg, sample origin: Chinese and international suppliers in different provinces, China, sample year: unknown, country: China[70], *seeds and fruits

- Co-contamination: 4 sa. co-contaminated with AFB_1, AFB_2, AFG_1, and AFG_2; 5 sa. co-contaminated with AFB_1, AFB_2, and AFG_1; 10 sa. co-contaminated with AFB_1 and AFB_2; 1 sa. co-contaminated with AFB_1 and AFG_1; 1 sa. contaminated solely with AFB_1
- Further contamination of plants and plant products other than cereals and their products reported in the present book/article[70]: Medicinal plants, AFB_2, AFG_1, and AFG_2, literature[70]

incidence: 6/46*, conc. range: 0.47–68.40 μg/kg, Ø conc.: 12.0 μg/kg, sample origin: Chinese and international suppliers different provinces, China, sample year: unknown, country: China[70], *rhizomes and roots

- Co-contamination: 1 sa. co-contaminated with AFB_1, AFB_2, and AFG_1; 5 sa. co-contaminated with AFB_1 and AFB_2
- Further contamination of plants and plant products other than cereals and their products reported in the present book/article[70]: Medicinal plants, AFB_2, AFG_1, and AFG_2, literature[70]

For detailed information see the article.

incidence: 9/31*, conc. range: 24–105 μg/kg, Ø conc.: 49 μg/kg, sample origin: El Atarin, Dokki Square, and Giza Market (districts) in Cairo (capital), Egypt, sample year: unknown, country: USA/Egypt[344], *herbs and medicinal plants, e.g., helba, kerfa, and karkadie

- Co-contamination: not reported
- Further contamination of plants and plant products other than cereals and their products reported in the present book/article[344]: Food, Spices, Vegetables, dried, AFB_1, literature[344]

For detailed information see the article.

incidence: 10/11*, conc. range: 0.26–22.89 μg/kg, Ø conc.: 4.93 μg/kg, sample origin: Dalian Medical University in Dalian (city), China, sample year: unknown, country: China[474], **Fructus Bruceae*

- Co-contamination: 1 sa. co-contaminated with AFB_1, AFB_2, and AFG_2; 3 sa. co-contaminated with AFB_1 and AFB_2; 1 sa. co-contaminated with AFB_1, AFB_2, AFG_1, and AFG_2; 3 sa. co-contaminated with AFB_1, AFB_2, and AFG_1; 2 sa. contaminated solely with AFB_1
- Further contamination of plants and plant products other than cereals and their products reported in the present book/article[474]: Medicinal plants, AFB_2, AFG_1, and AFG_2, literature[474]

incidence: 6/30, conc. range: 2.27–7.2 μg/kg, Ø conc.: 4.11 μg/kg, sample origin: local markets of different areas in Khyber Pakhtunkhwa (province), Pakistan, sample year: 2011/2012, country: Pakistan[661]

- Co-contamination: 1 sa. co-contaminated with AFB_1, AFG_1, AFG_2, and OTA; 2 sa. co-contaminated with AFB_1 and AFG_1; 2 sa. co-contaminated with AFB_1 and OTA; 1 sa. contaminated solely with AFB_1
- Further contamination of plants and plant products other than cereals and their products reported in the present book/article[661]: Medicinal plants, AFB_2, AFG_1, AFG_2, and OTA, literature[661]

For detailed information see the article.

incidence: 1/12, conc.: 52.0 μg/kg*, sample origin: Anguo (city) herbal market, Hebei (province), China, sample year: unknown, country: China[694], *plactycladi seeds

- Co-contamination: not reported
- Further contamination of plants and plant products other than cereals and their products reported in the present book/article[694]: Medicinal plants, OTA, literature[694]

For detailed information see the article.

incidence: 5/6*, conc. range: 470–1,270 μg/kg**, Ø conc.: 672 μg/kg**, sample origin: drug-stores, India, sample year: 1987/1988, country: India[696], *constituents of drugs, **conc. in plant parts used

incidence: 3/3*, conc. range: 250–1,500 μg/kg, Ø conc.: 800 μg/kg, sample origin: drug-stores, India, sample year: 1987/1988, country: India[696], *powdered herbal drugs

- Co-contamination: not reported
- Further contamination of plants and plant products other than cereals and their products reported in the present book/article[696]: not reported

For detailed information see the article.

incidence: 23?/50*, conc. range: 280–2,330 μg/kg, sample origin: private storehouses and shopkeepers in Santhal Pargana and Chhotanagpur (divisions), Bihar (state), India, sample year: unknown, country: India[772], *included bark, fruit, root, and stem

- Co-contamination: not reported
- Further contamination of plants and plant products other than cereals and their products reported in the present book/article[772]: Medicinal plants, AFB_2 and AFG_1, literature[772]

For detailed information see the article.

incidence: 1/8*, conc.: 1.8 μg/kg, sample origin: supermarkets and traditional Chinese medicine stores in different geographical regions, Taiwan, sample year: unknown, country: Taiwan[776], **Zizyphi fructus*

- Co-contamination: 1 sa. co-contaminated with AFB_1, AFB_2, AFG_1, and AFG_2
- Further contamination of plants and plant products other than cereals and their products reported in the present book/article[776]: Medicinal plants, AFB_2, AFG_1, and AFG_2, literature[776]

incidence: 2/20*, conc. range: 2.07–2.92 μg/kg**, Ø conc.: 2.50 μg/kg, sample origin: markets and stores in Anhui, Gansu, Guangxi, Hebei, and Sichuan (provinces), China, sample year: unknown, country: China[779], **Angelica sinensis*, **only the 2 moldy sa. were contaminated

- Co-contamination: 2 sa. co-contaminated with AFB_1 and AFG_1
- Further contamination of plants and plant products other than cereals and their products reported in the present book/article[779]: Medicinal plants, AFG_1, literature[779]

incidence: 19/20*, conc. range: 0.02–32.0 μg/kg, Ø conc.: 4.38 μg/kg, sample origin: markets and stores in Anhui, Fujian, Hunan, and Zhejiang (provinces), and Beijing (capital), China, sample year: unknown, country: China[780], *lotus seeds (*Nelumbo nucifera* Gaertn.)

- Co-contamination: 1 sa. co-contaminated with AFB_1, AFB_2, AFG_1, and AFG_2; 8 sa. co-contaminated with AFB_1, AFB_2, and AFG_1; 1 sa. co-contaminated with AFB_1, AFG_1, and AFG_2; 1 sa. co-contaminated with AFB_1, AFB_2, and AFG_2; 3 sa. co-contaminated with AFB_1 and AFB_2; 2 sa. co-contaminated with AFB_1 and AFG_1; 3 sa. contaminated solely with AFB_1
- Further contamination of plants and plant products other than cereals and their products reported in the present book/article[780]: Medicinal plants, AFB_2, AFG_1, and AFG_2, literature[780]

incidence: 1/9*, conc.: 10.06 μg/kg, sample origin: commercial markets and producing areas, China, sample year: unknown, country: China[872], *fritillary bulbs

- Co-contamination: not reported
- Further contamination of plants and plant products other than cereals and their products reported in the present book/article[872]: Licorice, AFB_1 and OTA, literature[872]

incidence: 1/17*, conc.: 0.751 μg/kg, sample origin: different provinces, China, sample year: unknown, country: China[873], **Puerariae lobatae radix*

- Co-contamination: not reported

- Further contamination of plants and plant products other than cereals and their products reported in the present book/article[873]: Medicinal plants, T-2, literature[873]

incidence: 6/6*, conc. range: 0.128–1.37 μg/kg, ∅ conc.: 0.61 μg/kg, sample origin: storehouses or markets, China, sample year: unknown, country: China[874], *moldy or heavily moldy sa. of *Glycyrrhiza uralensis*

- Co-contamination: 2 sa. co-contaminated with AFB_1, AFB_2, AFG_1, and OTA; 2 sa. co-contaminated with AFB_1 and OTA; 1 sa. co-contaminated with AFB_1, AFB_2, AFG_1, AFG_2, and OTA; 1 sa. co-contaminated with AFB_1, AFG_1, and OTA
- Further contamination of plants and plant products other than cereals and their products reported in the present book/article[874]: Medicinal plants, AFB_2, AFG_1, AFG_2, and OTA, literature[874]

incidence: 8/9*, conc. range: 0.027–0.267 μg/kg, ∅ conc.: 0.10 μg/kg, sample origin: storehouses or markets, China, sample year: unknown, country: China[874], *non-moldy *Glycyrrhiza uralensis* sa.

- Co-contamination: 4 sa. co-contaminated with AFB_1 and OTA; 2 sa. co-contaminated with AFB_1, AFB_2, and OTA; 1 sa. co-contaminated with AFB_1, AFB_2, AFG_1, and OTA; 1 sa. contaminated solely with AFB_1
- Further contamination of plants and plant products other than cereals and their products reported in the present book/article[874]: Medicinal plants, AFB_2, AFG_1, AFG_2, and OTA, literature[874]

incidence: 4/96*, conc. range: 0.6345–4.2755 μg/kg, ∅ conc.: 2.12 μg/kg, sample origin: Tong Ren Tang crude drugstore, Beijing (capital), China, sample year: unknown, country: China[880], *natural drugs

- Co-contamination: For detailed information see the article.
- Further contamination of plants and plant products other than cereals and their products reported in the present book/article[880]: Medicinal plants, AFB_2, AFG_1, and AFG_2, literature[880]

For detailed information see the article.

incidence: 13/244, conc. range: ≤1,268.6 μg/kg*, sample origin: collected from TCM pharmacies and hospitals in Guangzhou (city), China, sample year: unknown, country: China[886], *AFB_1-concentration of other TCMs significantly lower

- Co-contamination: not reported
- Further contamination of plants and plant products other than cereals and their products reported in the present book/article[886]: Medicinal plants, AFB_2, OTA, and STG, literature[886]

For detailed information see the article.

Aflatoxin B_2

incidence: 3/22*, conc. range: 1.32–5.16 μg/kg, ∅ conc.: 3.62 μg/kg, sample year: unknown, country: China[59], *traditional Chinese medicines

- Co-contamination: 1 sa. co-contaminated with AFB_1, AFB_2, and AFG_1; 1 sa. co-contaminated with AFB_1 and AFB_2; 1 sa. contaminated solely with AFB_2
- Further contamination of plants and plant products other than cereals and their products reported in the present book/article[59]: Medicinal plants, AFB_1 and AFG_1, literature[59]

For detailed information see the article.

incidence: 5/9*, conc. range: 0.01–0.03 μg/kg, ∅ conc.: 0.018 μg/kg, sample origin: Penang (state), Malaysia, sample year: 2003, country: Japan/Malaysia[61], *commercial traditional herbal medicines with 3–15 ingredients

- Co-contamination: 3 sa. co-contaminated with AFB_1 and AFB_2; 2 sa. co-contaminated with AFB_1, AFB_2, and AFG_1
- Further contamination of plants and plant products other than cereals and their products reported in the present book/article[61]: Medicinal plants, AFB_1, AFG_1, and AFG_2, literature[61]

incidence: 11/14*, conc. range: 0.01–0.40 μg/kg, ∅ conc.: 0.096 μg/kg, sample origin: sa. imported from Indonesia, sample year: 2003, country: Japan/Malaysia[61], *commercial traditional herbal medicines with 3–9 ingredients

- Co-contamination: 1 sa. co-contaminated with AFB_1, AFB_2, AFG_1, and AFG_2; 4 sa. co-contaminated with AFB_1, AFB_2, and AFG_1; 4 sa. co-contaminated with AFB_1 and AFB_2; 2 sa. contaminated solely with AFB_2
- Further contamination of plants and plant products other than cereals and their products reported in the present book/article[61]: Medicinal plants, AFB_1, AFG_1, and AFG_2, literature[61]

For detailed information see the article.

incidence: 5/13*, conc. range: 0.8–7.1 μg/kg, ∅ conc.: 3.36 μg/kg, sample origin: enterprises in Zhejiang (province), China, sample year: unknown, country: China[66], *rhizomes and roots

- Co-contamination: 1 sa. co-contaminated with AFB_1, AFB_2, AFG_1, and AFM_1; 1 sa. co-contaminated with AFB_2, AFG_1, and AFG_2; 2 sa. co-contaminated with AFB_2 and AFG_1; 1 sa. co-contaminated with AFB_2 and AFG_2
- Further contamination of plants and plant products other than cereals and their products reported in the present book/article[66]: Medicinal plants, AFB_1, AFG_1, AFG_2, and AFM_1, literature[66]

incidence: 0/4*, conc.: no contamination, sample origin: enterprises in Zhejiang (province), China, sample year: unknown, country: China[66], *flowers

- Co-contamination: not reported
- Further contamination of plants and plant products other than cereals and their products reported in the present book/article[66]: Medicinal plants, AFB_1, AFG_1, AFG_2, and AFM_1, literature[66]

incidence: 2/8*, conc. range: 0.8–0.9 μg/kg, ∅ conc.: 0.85 μg/kg, enterprises in sample origin: Zhejiang (province), China, sample year: unknown, country: China[66], *seeds

- Co-contamination: 1 sa. co-contaminated with AFB_1 and AFB_2; 1 sa. contaminated solely with AFB_2
- Further contamination of plants and plant products other than cereals and their products reported in the present book/article[66]: Medicinal plants, AFB_1, AFG_1, AFG_2, and AFM_1, literature[66]

incidence: 1/5*, conc.: 1.8 μg/kg, sample origin: enterprises in Zhejiang (province), China, sample year: unknown, country: China[66], *grasses and leaves

- Co-contamination: 1 sa. co-contaminated with AFB_1, AFB_2, and AFG_2
- Further contamination of plants and plant products other than cereals and their products reported in the present book/article[66]: Medicinal plants, AFB_1, AFG_1, AFG_2, and AFM_1, literature[66]

For detailed information see the article.

incidence: 3/28*, conc. range: 0.5–1.0 μg/kg, ∅ conc.: 0.7 μg/kg, sample origin: retail suppliers and community hospitals in Khon

Kaen (province), north-eastern Thailand, sample year: unknown, country: Thailand/Austria[67], *tablet or capsule (dosage form)

- Co-contamination: 1 sa. co-contaminated with AFB_1, AFB_2, and AFG_2; 1 sa. co-contaminated with AFB_1, AFB_2, and AFG_1; 1 sa. co-contaminated with AFB_1 and AFB_2
- Further contamination of plants and plant products other than cereals and their products reported in the present book/article[67]: Medicinal plants, AFB_1, AFG_1, and AFG_2, literature[67]

For detailed information see the article.

incidence: 5/23*, conc. range: 0.2–7.1 µg/kg, Ø conc.: 4.18 µg/kg, sample origin: unknown, sample year: unknown, country: China[68], *rhizomes and roots

- Co-contamination: 1 sa. co-contaminated with AFB_1, AFB_2, AFM_1, CHAE, CIT, CPA, GT, MA, OTB, STG, and VER; 1 sa. co-contaminated with AFB_1, AFB_2, AFG_1, AFG_2, AFM_1, CHAE, MA, and OTA; 1 sa. co-contaminated with AFB_1, AFB_2, AFG_2, CHAE, NIV, OTB, and T-2; 1 sa. co-contaminated with AFB_2, AFG_1, AFG_2, MON, MA, NEO, and STG; 1 sa. co-contaminated with AFB_1, AFB_2, AFG_1, AFM_1, CPA, and NEO
- Further contamination of plants and plant products other than cereals and their products reported in the present book/article[68]: Medicinal plants, AFB_1, AFG_1, AFG_2, AFM_1, CIT, CPA, DAS, DON, FUS-X, GT, HT-2, MA, MON, NEO, NIV, OTA, OTB, PAT, STG, T-2, and ZEA, literature[68]; see also Medicinal plants, chapter: Further Mycotoxins and Microbial Metabolites

incidence: 6/15*, conc. range: 0.5–2.3 µg/kg, Ø conc.: 1.15 µg/kg, sample origin: unknown, sample year: unknown, country: China[68], *seeds

- Co-contamination: 1 sa. co-contaminated with AFB_1, AFB_2, AFG_1, AFG_2, AFM_1, CHAE, CPA, GT, HT-2, MA, PEA, STG, VER, and ZEA; 1 sa. co-contaminated with AFB_1, AFB_2, CPA, DAS, GT, MA, MON, NIV, STG, and VER; 1 sa. co-contaminated with AFB_1, AFB_2, CHAE, CPA, DPDON, and MA; 1 sa. co-contaminated with AFB_1, AFB_2, DAS, GT, STG, and ZEA; 1 sa. co-contaminated with AFB_2, DON3G, MA, STG, and VER; 1 sa. co-contaminated with AFB_1, AFB_2, and MA
- Further contamination of plants and plant products other than cereals and their products reported in the present book/article[68]: Medicinal plants, AFB_1, AFG_1, AFG_2, AFM_1, CIT, CPA, DAS, DON, FUS-X, GT, HT-2, MA, MON, NEO, NIV, OTA, OTB, PAT, STG, T-2, and ZEA, literature[68]; see also Medicinal plants, chapter: Further Mycotoxins and Microbial Metabolites

incidence: 1/11*, conc.: 0.2 µg/kg, sample origin: unknown, sample year: unknown, country: China[68], *grasses and leaves

- Co-contamination: 1 sa. co-contaminated with AFB_2, CPA, MA, OTA, PAT, and STG
- Further contamination of plants and plant products other than cereals and their products reported in the present book/article[68]: Medicinal plants, AFB_1, AFG_1, AFG_2, AFM_1, CIT, CPA, DAS, DON, FUS-X, GT, HT-2, MA, MON, NEO, NIV, OTA, OTB, PAT, STG, T-2, and ZEA, literature[68]; see also Medicinal plants, chapter: Further Mycotoxins and Microbial Metabolites

incidence: 0/11*, conc.: no contamination, sample origin: unknown, sample year: unknown, country: China[68], *flowers

- Co-contamination: not reported
- Further contamination of plants and plant products other than cereals and their products reported in the present book/article[68]: Medicinal plants, AFB_1, AFG_1, AFG_2, AFM_1, CIT, CPA, DAS, DON, FUS-X, GT, HT-2, MA, MON, NEO, NIV, OTA, OTB, PAT, STG, T-2, and ZEA, literature[68]; see also Medicinal plants, chapter: Further Mycotoxins and Microbial Metabolites

For detailed information see the article.

incidence: 12/58, conc. range: 5.48–54.98 µg/kg, Ø conc.: 12.0 µg/kg, sample origin: commercial herbal medicine markets in Daegu, Daejeon, Gwangju, Jinju, Pusan (cities), and Seoul (capital), South Korea, sample year: unknown, country: Korea[69]

- Co-contamination: 1 sa. co-contaminated with AFB_1, AFB_2, AFG_1, and AFG_2; 1 sa. co-contaminated with AFB_1, AFB_2, and AFG_1; 8 sa. co-contaminated with AFB_1 and AFB_2; 2 sa. contaminated solely with AFB_2
- Further contamination of plants and plant products other than cereals and their products reported in the present book/article[69]: Medicinal plants, AFB_1, AFG_1, and AFG_2, literature[69]

For detailed information see the article.

incidence: 19/75*, conc. range: 0.05–13.50 µg/kg, Ø conc.: 2.93 µg/kg, sample origin: Chinese and international suppliers different provinces, China, sample year: unknown, country: China[70], *seeds and fruits

- Co-contamination: 4 sa. co-contaminated with AFB_1, AFB_2, AFG_1, and AFG_2; 5 sa. co-contaminated with AFB_1, AFB_2, and AFG_1; 10 sa. co-contaminated with AFB_1 and AFB_2
- Further contamination of plants and plant products other than cereals and their products reported in the present book/article[70]: Medicinal plants, AFB_1, AFG_1, and AFG_2, literature[70]

incidence: 6/46*, conc. range: 0.05–1.71 µg/kg, Ø conc.: 0.59 µg/kg, sample origin: Chinese and international suppliers different provinces, China, sample year: unknown, country: China[70], *rhizomes and roots

- Co-contamination: 1 sa. co-contaminated with AFB_1, AFB_2, and AFG_1; 5 sa. co-contaminated with AFB_1 and AFB_2
- Further contamination of plants and plant products other than cereals and their products reported in the present book/article[70]: Medicinal plants, AFB_1, AFG_1, and AFG_2, literature[70]

For detailed information see the article.

incidence: 8/11*, conc. range: 0.05–4.54 µg/kg, Ø conc.: 0.84 µg/kg, sample origin: Dalian Medical University in Dalian (city), China, sample year: unknown, country: China[474], **Fructus Bruceae*

- Co-contamination: 1 sa. co-contaminated with AFB_1, AFB_2, and AFG_2; 3 sa. co-contaminated with AFB_1 and AFB_2; 1 sa. co-contaminated with AFB_1, AFB_2, AFG_1, and AFG_2; 3 sa. co-contaminated with AFB_1, AFB_2, and AFG_1
- Further contamination of plants and plant products other than cereals and their products reported in the present book/article[474]: Medicinal plants, AFB_1, AFG_1, and AFG_2, literature[474]

incidence: 1/30, conc.: 5.3 µg/kg, sample origin: local markets of different areas in Khyber Pakhtunkhwa (province), Pakistan, sample year: 2011/2012, country: Pakistan[661]

- Co-contamination: not reported
- Further contamination of plants and plant products other than cereals and their products reported in the present book/article[661]: Medicinal plants, AFB_1, AFG_1, AFG_2, and OTA, literature[661]

For detailed information see the article.

incidence: 16?/32*, conc. range: 130–1,340 µg/kg, sample origin: private storehouses and shopkeepers in Santhal Pargana and

Chhotanagpur (divisions), Bihar (state), India, sample year: unknown, country: India[772], *included fruit and root

- Co-contamination: not reported
- Further contamination of plants and plant products other than cereals and their products reported in the present book/article[772]: Medicinal plants, AFB_1 and AFG_1, literature[772]

For detailed information see the article.

incidence: 1/8*, conc.: 1.2 µg/kg, sample origin: supermarkets and traditional Chinese medicine stores in different geographical regions, Taiwan, sample year: unknown, country: Taiwan[776], **Zizyphi fructus*

- Co-contamination: 1 sa. co-contaminated with AFB_1, AFB_2, AFG_1, and AFG_2
- Further contamination of plants and plant products other than cereals and their products reported in the present book/article[776]: Medicinal plants, AFB_1, AFG_1, and AFG_2, literature[776]

incidence: 1/24*, conc.: 2.67 µg/kg, sample origin: commercially available, China, sample year: unknown, country: China[778], *12 medicinal and 12 edible sa. of *Areca catechu* L.

- Co-contamination: not reported
- Further contamination of plants and plant products other than cereals and their products reported in the present book/article[778]: Medicinal plants, ZEA, literature[778]

incidence: 13/20*, conc. range: 0.03–16.4 µg/kg, Ø conc.: 2.17 µg/kg, sample origin: markets and stores in Anhui, Fujian, Hunan, and Zhejiang (provinces), and Beijing (capital), China, sample year: unknown, country: China[780], *lotus seeds (*Nelumbo nucifera* Gaertn.)

- 1 sa. co-contaminated with AFB_1, AFB_2, AFG_1, and AFG_2; 8 sa. co-contaminated with AFB_1, AFB_2, and AFG_1; 1 sa. co-contaminated with AFB_1, AFG_1, and AFG_2; 3 sa. co-contaminated with AFB_1 and AFB_2
- Further contamination of plants and plant products other than cereals and their products reported in the present book/article[780]: Medicinal plants, AFB_1, AFG_1, and AFG_2, literature[780]

incidence: 3/6*, conc. range: 0.057–0.299 µg/kg, Ø conc.: 0.18 µg/kg, sample origin: storehouses or markets, China, sample year: unknown, country: China[874], *moldy or heavily moldy sa. of *Glycyrrhiza uralensis*

- Co-contamination: 2 sa. co-contaminated with AFB_1, AFB_2, AFG_1, and OTA; 1 sa. co-contaminated with AFB_1, AFB_2, AFG_1, AFG_2, and OTA
- Further contamination of plants and plant products other than cereals and their products reported in the present book/article[874]: Medicinal plants, AFB_1, AFG_1, AFG_2, and OTA, literature[874]

incidence: 3/9*, conc. range: 0.032–0.040 µg/kg, Ø conc.: 0.036 µg/kg, sample origin: storehouses or markets, China, sample year: unknown, country: China[874], *non-moldy *Glycyrrhiza uralensis* sa.

- Co-contamination: 2 sa. co-contaminated with AFB_1, AFB_2, and OTA; 1 sa. co-contaminated with AFB_1, AFB_2, AFG_1, and OTA
- Further contamination of plants and plant products other than cereals and their products reported in the present book/article[874]: Medicinal plants, AFB_1, AFG_1, AFG_2, and OTA, literature[874]

incidence: 4/96*, conc. range: 0.1094–1.0462 µg/kg, Ø conc.: 0.51 µg/kg, sample origin: Tong Ren Tang crude drugstore, Beijing (capital), China, sample year: unknown, country: China[880], *natural drugs

- Co-contamination: For detailed information see the article.
- Further contamination of plants and plant products other than cereals and their products reported in the present book/article[880]: Medicinal plants, AFB_1, AFG_1, and AFG_2, literature[880]

For detailed information see the article.

incidence: 5/244, conc. range: 0.3–17.7 µg/kg, Ø conc.: 4.14 µg/kg, sample origin: collected from TCM pharmacies and hospitals in Guangzhou (city), China, sample year: unknown, country: China[886]

- Co-contamination: not reported
- Further contamination of plants and plant products other than cereals and their products reported in the present book/article[886]: Medicinal plants, AFB_1, OTA, and STG, literature[886]

For detailed information see the article.

Aflatoxin G_1

incidence: 1/22*, conc.: 1.31 µg/kg, sample year: unknown, country: China[59], *traditional Chinese medicines

- Co-contamination: 1 sa. co-contaminated with AFB_1, AFB_2, and AFG_1
- Further contamination of plants and plant products other than cereals and their products reported in the present book/article[59]: Medicinal plants, AFB_1 and AFB_2, literature[59]

For detailed information see the article.

incidence: 2/9*, conc. range: 0.02–0.22 µg/kg, Ø conc.: 0.12 µg/kg, sample origin: Penang (state), Malaysia, sample year: 2003, country: Japan/Malaysia[61], *commercial traditional herbal medicines with 3–15 ingredients

- Co-contamination: 2 sa. co-contaminated with AFB_1, AFB_2, and AFG_1
- Further contamination of plants and plant products other than cereals and their products reported in the present book/article[61]: Medicinal plants, AFB_1, AFB_2, and AFG_2, literature[61]

incidence: 5/14*, conc. range: 0.02–0.17 µg/kg, Ø conc.: 0.09 µg/kg, sample origin: sa. imported from Indonesia, sample year: 2003, country: Japan/Malaysia[61], *commercial traditional herbal medicines with 3–9 ingredients

- Co-contamination: 1 sa. co-contaminated with AFB_1, AFB_2, AFG_1, and AFG_2; 4 sa. co-contaminated with AFB_1, AFB_2, and AFG_1
- Further contamination of plants and plant products other than cereals and their products reported in the present book/article[61]: Medicinal plants, AFB_1, AFB_2, and AFG_2, literature[61]

For detailed information see the article.

incidence: 5/13*, conc. range: 0.6–2.5 µg/kg, Ø conc.: 1.3 µg/kg, sample origin: enterprises in Zhejiang (province), China, sample year: unknown, country: China[66], *rhizomes and roots

- Co-contamination: 1 sa. co-contaminated with AFB_1, AFB_2, AFG_1, and AFM_1; 1 sa. co-contaminated with AFB_2, AFG_1, and AFG_2; 1 sa. co-contaminated with AFB_1 and AFG_1; 2 sa. co-contaminated with AFB_2 and AFG_1
- Further contamination of plants and plant products other than cereals and their products reported in the present book/article[66]: Medicinal plants, AFB_1, AFB_2, AFG_2, and AFM_1, literature[66]

incidence: 1/4*, conc.: 0.7 µg/kg, sample origin: enterprises in Zhejiang (province), China, sample year: unknown, country: China[66], *flowers

- Co-contamination: 1 sa. co-contamination with AFB_1 and AFG_1

- Further contamination of plants and plant products other than cereals and their products reported in the present book/ article[66]: Medicinal plants, AFB_1, AFB_2, AFG_2, and AFM_1, literature[66]

incidence: 1/8*, conc.: 0.8 µg/kg, sample origin: enterprises in Zhejiang (province), China, sample year: unknown, country: China[66], *seeds

- Co-contamination: 1 sa. co-contaminated with AFB_1, AFG_1, and AFG_2
- Further contamination of plants and plant products other than cereals and their products reported in the present book/ article[66]: Medicinal plants, AFB_1, AFB_2, AFG_2, and AFM_1, literature[66]

incidence: 0/5*, conc.: no contamination, sample origin: enterprises in Zhejiang (province), China, sample year: unknown, country: China[66], *grasses and leaves

- Co-contamination: not reported
- Further contamination of plants and plant products other than cereals and their products reported in the present book/ article[66]: Medicinal plants, AFB_1, AFB_2, AFG_2, and AFM_1, literature[66]

For detailed information see the article.

incidence: 2/28*, conc. range: 0.6–2.0 µg/kg, Ø conc.: 1.3 µg/kg, sample origin: retail suppliers and community hospitals in Khon Kaen (province), north-eastern Thailand, sample year: unknown, country: Thailand/Austria[67], *tablet or capsule (dosage form)

- Co-contamination: 1 sa. co-contaminated with AFB_1, AFB_2, and AFG_1; 1 sa. co-contaminated with AFB_1 and AFG_1
- Further contamination of plants and plant products other than cereals and their products reported in the present book/ article[67]: Medicinal plants, AFB_1, AFB_2, and AFG_2, literature[67]

For detailed information see the article.

incidence: 4/23*, conc. range: 0.6–2.5 µg/kg, Ø conc.: 1.65 µg/kg, sample origin: unknown, sample year: unknown, country: China[68], *rhizomes and roots

- Co-contamination: 1 sa. co-contaminated with AFB_1, AFB_2, AFG_1, AFG_2, AFM_1, CHAE, MA, and OTA; 1 sa. co-contaminated with AFB_2, AFG_1, AFG_2, MON, MA, NEO, and STG; 1 sa. co-contaminated with AFB_1, AFB_2, AFG_1, AFM_1, CPA, and NEO; 1 sa. co-contaminated with AFB_1, AFG_1, and AFM_1
- Further contamination of plants and plant products other than cereals and their products reported in the present book/ article[68]: Medicinal plants, AFB_1, AFB_2, AFG_2, AFM_1, CIT, CPA, DAS, DON, FUS-X, GT, HT-2, MA, MON, NEO, NIV, OTA, OTB, PAT, STG, T-2, and ZEA, literature[68]; see also Medicinal plants, chapter: Further Mycotoxins and Microbial Metabolites

incidence: 3/15*, conc. range: 0.2–0.8 µg/kg, Ø conc.: 0.43 µg/kg, sample origin: unknown, sample year: unknown, country: China[68], *seeds

- Co-contamination: 1 sa. co-contaminated with AFB_1, AFB_2, AFG_1, AFG_2, AFM_1, CHAE, CPA, GT, HT-2, MA, PEA, STG, VER, and ZEA; 1 sa. co-contaminated with AFB_1, AFG_1, AFG_2, DAS, MA, PEA, and STG; 1 sa. co-contaminated with AFG_1, CHAE, MA, OTA, and OTB
- Further contamination of plants and plant products other than cereals and their products reported in the present book/ article[68]: Medicinal plants, AFB_1, AFB_2, AFG_2, AFM_1, CIT, CPA, DAS, DON, FUS-X, GT, HT-2, MA, MON, NEO, NIV, OTA, OTB, PAT, STG, T-2, and ZEA, literature[68]; see also Medicinal plants, chapter: Further Mycotoxins and Microbial Metabolites

incidence: 1/11*, conc.: 0.3 µg/kg, sample origin: unknown, sample year: unknown, country: China[68], *grasses and leaves

- Co-contamination: 1 sa. co-contaminated with AFB_1, AFG_1, GT, MA, OTB, PEA, and STG
- Further contamination of plants and plant products other than cereals and their products reported in the present book/ article[68]: Medicinal plants, AFB_1, AFB_2, AFG_2, AFM_1, CIT, CPA, DAS, DON, FUS-X, GT, HT-2, MA, MON, NEO, NIV, OTA, OTB, PAT, STG, T-2, and ZEA, literature[68]; see also Medicinal plants, chapter: Further Mycotoxins and Microbial Metabolites

incidence: 0/11*, conc.: no contamination, sample origin: unknown, sample year: unknown, country: China[68], *flowers

- Co-contamination: not reported
- Further contamination of plants and plant products other than cereals and their products reported in the present book/ article[68]: Medicinal plants, AFB_1, AFB_2, AFG_2, AFM_1, CIT, CPA, DAS, DON, FUS-X, GT, HT-2, MA, MON, NEO, NIV, OTA, OTB, PAT, STG, T-2, and ZEA, literature[68]; see also Medicinal plants, chapter: Further Mycotoxins and Microbial Metabolites

For detailed information see the article.

incidence: 3/58, conc. range: 4.86–10.27 µg/kg, Ø conc.: 6.68 µg/kg, sample origin: commercial herbal medicine markets in Daegu, Daejeon, Gwangju, Jinju, Pusan (cities), and Seoul (capital), South Korea, sample year: unknown, country: Korea[69]

- Co-contamination: 1 sa. co-contaminated with AFB_1, AFB_2, AFG_1, and AFG_2; 1 sa. co-contaminated with AFB_1, AFB_2, and AFG_1; 1 sa. co-contaminated with AFB_1 and AFG_1
- Further contamination of plants and plant products other than cereals and their products reported in the present book/ article[69]: Medicinal plants, AFB_1, AFB_2, and AFG_2, literature[69]

For detailed information see the article.

incidence: 10/75*, conc. range: 0.08–34.21 µg/kg, Ø conc.: 6.4 µg/kg, sample origin: Chinese and international suppliers different provinces, China, sample year: unknown, country: China[70], *seeds and fruits

- Co-contamination: 4 sa. co-contaminated with AFB_1, AFB_2, AFG_1, and AFG_2; 5 sa. co-contaminated with AFB_1, AFB_2, and AFG_1; 1 sa. co-contaminated with AFB_1 and AFG_1
- Further contamination of plants and plant products other than cereals and their products reported in the present book/ article[70]: Medicinal plants, AFB_1, AFB_2, and AFG_2, literature[70]

incidence: 1/46*, conc.: 0.95 µg/kg, sample origin: Chinese and international suppliers different provinces, China, sample year: unknown, country: China[70], *rhizomes and roots

- Co-contamination: 1 sa. co-contaminated with AFB_1, AFB_2, and AFG_1
- Further contamination of plants and plant products other than cereals and their products reported in the present book/ article[70]: Medicinal plants, AFB_1, AFB_2, and AFG_2, literature[70]

For detailed information see the article.

incidence: 4/11*, conc. range: 0.20–5.84 µg/kg, Ø conc.: 1.64 µg/kg, sample origin: Dalian Medical University in Dalian (city), China, sample year: unknown, country: China[474], **Fructus Bruceae*

- Co-contamination: 1 sa. co-contaminated with AFB_1, AFB_2, AFG_1, and AFG_2; 3 sa. co-contaminated with AFB_1, AFB_2, and AFG_1

- Further contamination of plants and plant products other than cereals and their products reported in the present book/article[474]: Medicinal plants, AFB_1, AFB_2, and AFG_2, literature[474]

incidence: 4/30, conc. range: 2.61–23.36 µg/kg, Ø conc.: 10.4 µg/kg, sample origin: local markets of different areas in Khyber Pakhtunkhwa (province), Pakistan, sample year: 2011/2012, country: Pakistan[661]

- Co-contamination: 1 sa. co-contaminated with AFB_1, AFG_1, AFG_2, and OTA; 2 sa. co-contaminated with AFB_1 and AFG_1; 1 sa. co-contaminated with AFG_1 and OTA
- Further contamination of plants and plant products other than cereals and their products reported in the present book/article[661]: Medicinal plants, AFB_1, AFB_2, AFG_2, and OTA, literature[661]

For detailed information see the article.

incidence: 6?/13*, conc. range: 230–670 µg/kg, sample origin: private storehouses and shopkeepers in Santhal Pargana and Chhotanagpur (divisions), Bihar (state), India, sample year: unknown, country: India[772], *fruit

- Co-contamination: not reported
- Further contamination of plants and plant products other than cereals and their products reported in the present book/article[772]: Medicinal plants, AFB_1 and AFB_2, literature[772]

For detailed information see the article.

incidence: 1/8*, conc.: 1.0 µg/kg, sample origin: supermarkets and traditional Chinese medicine stores in different geographical regions, Taiwan, sample year: unknown, country: Taiwan[776], **Zizyphi fructus*

- Co-contamination: 1 sa. co-contaminated with AFB_1, AFB_2, AFG_1, and AFG_2
- Further contamination of plants and plant products other than cereals and their products reported in the present book/article[776]: Medicinal plants, AFB_1, AFB_2, and AFG_2, literature[776]

incidence: 2/20*, conc. range: 1.53–2.84 µg/kg**, Ø conc.: 2.19 µg/kg, sample origin: markets and stores in Anhui, Gansu, Guangxi, Hebei, and Sichuan (provinces), China, sample year: unknown, country: China[779], **Angelica sinensis*, **only the 2 moldy sa. were contaminated.

- Co-contamination: 2 sa. co-contaminated with AFB_1 and AFG_1
- Further contamination of plants and plant products other than cereals and their products reported in the present book/article[779]: Medicinal plants, AFB_1, literature[779]

incidence: 12/20*, conc. range: 0.01–0.85 µg/kg, Ø conc.: 0.083 µg/kg, sample origin: markets and stores in Anhui, Fujian, Hunan, and Zhejiang (provinces), and Beijing (capital), China, sample year: unknown, country: China[780], *lotus seeds (*Nelumbo nucifera* Gaertn.)

- 1 sa. co-contaminated with AFB_1, AFB_2, AFG_1, and AFG_2; 8 sa. co-contaminated with AFB_1, AFB_2, and AFG_1; 1 sa. co-contaminated with AFB_1, AFG_1, and AFG_2; 2 sa. co-contaminated with AFB_1 and AFG_1
- Further contamination of plants and plant products other than cereals and their products reported in the present book/article[780]: Medicinal plants, AFB_1, AFB_2, and AFG_2, literature[780]

incidence: 4/6*, conc. range: 0.027–0.785 µg/kg, Ø conc.: 0.29 µg/kg, sample origin: storehouses or markets, China, sample year: unknown, country: China[874], *moldy or heavily moldy sa. of *Glycyrrhiza uralensis*

- Co-contamination: 2 sa. co-contaminated with AFB_1, AFB_2, AFG_1, and OTA; 1 sa. co-contaminated with AFB_1, AFB_2, AFG_1, AFG_2, and OTA; 1 sa. co-contaminated with AFB_1, AFG_1, and OTA
- Further contamination of plants and plant products other than cereals and their products reported in the present book/article[874]: Medicinal plants, AFB_1, AFB_2, AFG_2, and OTA, literature[874]

incidence: 1/9*, conc.: 0.030 µg/kg, sample origin: storehouses or markets, China, sample year: unknown, country: China[874], *non-moldy *Glycyrrhiza uralensis* sa.

- Co-contamination: 1 sa. co-contaminated with AFB_1, AFB_2, AFG_1, and OTA
- Further contamination of plants and plant products other than cereals and their products reported in the present book/article[874]: Medicinal plants, AFB_1, AFB_2, AFG_2, and OTA, literature[874]

incidence: 1/96*, conc.: 1.1536 µg/kg, sample origin: Tong Ren Tang crude drugstore, Beijing (capital), China, sample year: unknown, country: China[880], *natural drugs

- Co-contamination: For detailed information see the article.
- Further contamination of plants and plant products other than cereals and their products reported in the present book/article[880]: Medicinal plants, AFB_1, AFB_2, and AFG_2, literature[880]

For detailed information see the article.

Aflatoxin G_2

incidence: 0/9*, conc.: no contamination, sample origin: Penang (state), Malaysia, sample year: 2003, country: Japan/Malaysia[61], *commercial traditional herbal medicines with 3–15 ingredients

- Co-contamination: not reported
- Further contamination of plants and plant products other than cereals and their products reported in the present book/article[61]: Medicinal plants, AFB_1, AFB_2, and AFG_1, literature[61]

incidence: 1/14*, conc.: 0.03 µg/kg, sample origin: sa. imported from Indonesia, sample year: 2003, country: Japan/Malaysia[61], *commercial traditional herbal medicines with 3–9 ingredients

- Co-contamination: 1 sa. co-contaminated with AFB_1, AFB_2, AFG_1, and AFG_2
- Further contamination of plants and plant products other than cereals and their products reported in the present book/article[61]: Medicinal plants, AFB_1, AFB_2, and AFG_1, literature[61]

For detailed information see the article.

incidence: 4/13*, conc. range: 1.5–4.8 µg/kg, Ø conc.: 2.75 µg/kg, sample origin: enterprises in Zhejiang (province), China, sample year: unknown, country: China[66], *rhizomes and roots

- Co-contamination: 1 sa. co-contaminated with AFB_2, AFG_1, and AFG_2; 1 sa. co-contaminated with AFB_1, AFG_2, AFM_1; 1 sa. co-contaminated with AFB_1 and AFG_2; 1 sa. co-contaminated with AFB_2 and AFG_2
- Further contamination of plants and plant products other than cereals and their products reported in the present book/article[66]: Medicinal plants, AFB_1, AFB_2, AFG_1, and AFM_1, literature[66]

incidence: 0/4*, conc.: no contamination, sample origin: enterprises in Zhejiang (province), China, sample year: unknown, country: China[66], *flowers

- Co-contamination: not reported
- Further contamination of plants and plant products other than cereals and their products reported in the present book/article[66]: Medicinal plants, AFB_1, AFB_2, AFG_1, and AFM_1, literature[66]

incidence: 2/8*, conc. range: 0.9–1.5 μg/kg, Ø conc.: 1.2 μg/kg, sample origin: enterprises in Zhejiang (province), China, sample year: unknown, country: China[66], *seeds

- Co-contamination: 1 sa. co-contaminated with AFB_1, AFG_1, and AFG_2; 1 sa. co-contaminated with AFB_1 and AFG_2
- Further contamination of plants and plant products other than cereals and their products reported in the present book/article[66]: Medicinal plants, AFB_1, AFB_2, AFG_1, and AFM_1, literature[66]

incidence: 1/5*, conc.: 1.6 μg/kg, sample origin: enterprises in Zhejiang (province), China, sample year: unknown, country: China[66], *grasses and leaves

- Co-contamination: 1 sa. co-contaminated with AFB_1, AFB_2, and AFG_2
- Further contamination of plants and plant products other than cereals and their products reported in the present book/article[66]: Medicinal plants, AFB_1, AFB_2, AFG_1, and AFM_1, literature[66]

For detailed information see the article.

incidence: 1/28*, conc.: 0.4 μg/kg, sample origin: retail suppliers and community hospitals in Khon Kaen (province), north-eastern Thailand, sample year: unknown, country: Thailand/Austria[67], *tablet or capsule (dosage form)

- Co-contamination: 1 sa. co-contaminated with AFB_1, AFB_2, and AFG_2
- Further contamination of plants and plant products other than cereals and their products reported in the present book/article[67]: Medicinal plants, AFB_1, AFB_2, and AFG_1, literature[67]

For detailed information see the article.

incidence: 4/23*, conc. range: 0.4–4.8 μg/kg, Ø conc.: 1.48 μg/kg, sample origin: unknown, sample year: unknown, country: China[68], *rhizomes and roots

- Co-contamination: 1 sa. co-contaminated with AFB_1, AFB_2, AFG_1, AFG_2, AFM_1, CHAE, MA, and OTA; 1 sa. co-contaminated with AFB_1, AFB_2, AFG_2, CHAE, NIV, OTB, and T-2; 1 sa. co-contaminated with AFB_2, AFG_1, AFG_2, MON, MA, NEO, and STG; 1 sa. co-contaminated with AFG_2, DPDON, NEO, NIV, OTA, and OTB
- Further contamination of plants and plant products other than cereals and their products reported in the present book/article[68]: Medicinal plants, AFB_1, AFB_2, AFG_1, AFM_1, CIT, CPA, DAS, DON, FUS-X, GT, HT-2, MA, MON, NEO, NIV, OTA, OTB, PAT, STG, T-2, and ZEA, literature[68]; see also Medicinal plants, chapter: Further Mycotoxins and Microbial Metabolites

incidence: 2/15*, conc. range: 0.5–1.5 μg/kg, Ø conc.: 1.00 μg/kg, sample origin: unknown, sample year: unknown, country: China[68], *seeds

- Co-contamination: 1 sa. co-contaminated with AFB_1, AFB_2, AFG_1, AFG_2, AFM_1, CHAE, CPA, GT, HT-2, MA, PEA, STG, VER, and ZEA; 1 sa. co-contaminated with AFB_1, AFG_1, AFG_2, DAS, MA, PEA, and STG
- Further contamination of plants and plant products other than cereals and their products reported in the present book/article[68]: Medicinal plants, AFB_1, AFB_2, AFG_1, AFM_1, CIT, CPA, DAS, DON, FUS-X, GT, HT-2, MA, MON, NEO, NIV, OTA, OTB, PAT, STG, T-2, and ZEA, literature[68]; see also Medicinal plants, chapter: Further Mycotoxins and Microbial Metabolites

incidence: 1/11*, conc.: 0.2 μg/kg, sample origin: unknown, sample year: unknown, country: China[68], *grasses and leaves

- Co-contamination: 1 sa. co-contaminated with AFB_1, AFG_2, CHAE, CPA, DAS, DON3G, PEA, and STG
- Further contamination of plants and plant products other than cereals and their products reported in the present book/article[68]: Medicinal plants, AFB_1, AFB_2, AFG_1, AFM_1, CIT, CPA, DAS, DON, FUS-X, GT, HT-2, MA, MON, NEO, NIV, OTA, OTB, PAT, STG, T-2, and ZEA, literature[68]; see also Medicinal plants, chapter: Further Mycotoxins and Microbial Metabolites

incidence: 0/11*, conc.: no contamination, sample origin: unknown, sample year: unknown, country: China[68], *flowers

- Co-contamination: not reported
- Further contamination of plants and plant products other than cereals and their products reported in the present book/article[68]: Medicinal plants, AFB_1, AFB_2, AFG_1, AFM_1, CIT, CPA, DAS, DON, FUS-X, GT, HT-2, MA, MON, NEO, NIV, OTA, OTB, PAT, STG, T-2, and ZEA, literature[68]; see also Medicinal plants, chapter: Further Mycotoxins and Microbial Metabolites

For detailed information see the article.

incidence: 1/58, conc.: 1.94 μg/kg, sample origin: commercial herbal medicine markets in Daegu, Daejeon, Gwangju, Jinju, Pusan (cities), and Seoul (capital), South Korea, sample year: unknown, country: Korea[69]

- Co-contamination: 1 sa. co-contaminated with AFB_1, AFB_2, AFG_1, and AFG_2
- Further contamination of plants and plant products other than cereals and their products reported in the present book/article[69]: Medicinal plants, AFB_1, AFB_2, and AFG_1, literature[69]

For detailed information see the article.

incidence: 4/75*, conc. range: 0.09–3.50 μg/kg, Ø conc.: 1.09 μg/kg, sample origin: Chinese and international suppliers different provinces, China, sample year: unknown, country: China[70], *seeds and fruits

- Co-contamination: 4 sa. co-contaminated with AFB_1, AFB_2, AFG_1, and AFG_2
- Further contamination of plants and plant products other than cereals and their products reported in the present book/article[70]: Medicinal plants, AFB_1, AFB_2, and AFG_1, literature[70]

incidence: 0/46*, conc.: no contamination, sample origin: Chinese and international suppliers different provinces, China, sample year: unknown, country: China[70], *rhizomes and roots

- Co-contamination: not reported
- Further contamination of plants and plant products other than cereals and their products reported in the present book/article[70]: Medicinal plants, AFB_1, AFB_2, and AFG_1, literature[70]

For detailed information see the article.

incidence: 2/11*, conc. range: 0.09–0.18 μg/kg, Ø conc.: 0.14 μg/kg, sample origin: Dalian Medical University in Dalian (city), China, sample year: unknown, country: China[474], **Fructus Bruceae*

- Co-contamination: 1 sa. co-contaminated with AFB_1, AFB_2, and AFG_2; 1 sa. co-contaminated with AFB_1, AFB_2, AFG_1, and AFG_2
- Further contamination of plants and plant products other than cereals and their products reported in the present book/article[474]: Medicinal plants, AFB_1, AFB_2, and AFG_1, literature[474]

incidence: 2/30, conc. range: 2.5–9.34 μg/kg, ∅ conc.: 5.92 μg/kg, sample origin: local markets of different areas in Khyber Pakhtunkhwa (province), Pakistan, sample year: 2011/2012, country: Pakistan[661]

- Co-contamination: 1 sa. co-contaminated with AFB_1, AFG_1, AFG_2, and OTA; 1 sa. contaminated solely with AFG_2
- Further contamination of plants and plant products other than cereals and their products reported in the present book/article[661]: Medicinal plants, AFB_1, AFB_2, AFG_1, and OTA, literature[661]

For detailed information see the article.

incidence: 1/8*, conc.: 1.4 μg/kg, sample origin: supermarkets and traditional Chinese medicine stores in different geographical regions, Taiwan, sample year: unknown, country: Taiwan[776], **Zizyphi fructus*

- Co-contamination: 1 sa. co-contaminated with AFB_1, AFB_2, AFG_1, and AFG_2
- Further contamination of plants and plant products other than cereals and their products reported in the present book/article[776]: Medicinal plants, AFB_1, AFB_2, and AFG_1, literature[776]

incidence: 3/20*, conc. range: 0.02–0.06 μg/kg, ∅ conc.: 0.04 μg/kg, sample origin: markets and stores in Anhui, Fujian, Hunan, and Zhejiang (provinces), and Beijing (capital), China, sample year: unknown, country: China[780], *lotus seeds (*Nelumbo nucifera* Gaertn.)

- 1 sa. co-contaminated with AFB_1, AFB_2, AFG_1, and AFG_2; 1 sa. co-contaminated with AFB_1, AFG_1, and AFG_2; 1 sa. co-contaminated with AFB_1, AFB_2, and AFG_2
- Further contamination of plants and plant products other than cereals and their products reported in the present book/article[780]: Medicinal plants, AFB_1, AFB_2, and AFG_1, literature[780]

incidence: 1/6*, conc.: 0.107 μg/kg, sample origin: storehouses or markets, China, sample year: unknown, country: China[874], *moldy or heavily moldy sa. of *Glycyrrhiza uralensis*

- Co-contamination: 1 sa. co-contaminated with AFB_1, AFB_2, AFG_1, AFG_2, and OTA
- Further contamination of plants and plant products other than cereals and their products reported in the present book/article[874]: Medicinal plants, AFB_1, AFB_2, AFG_1, and OTA, literature[874]

incidence: 0/9*, conc.: no contamination, sample origin: storehouses or markets, China, sample year: unknown, country: China[874], *non-moldy *Glycyrrhiza uralensis* sa.

- Co-contamination: not reported
- Further contamination of plants and plant products other than cereals and their products reported in the present book/article[874]: Medicinal plants, AFB_1, AFB_2, AFG_1, and OTA, literature[874]

incidence: 1/96*, conc.: 1.1536 μg/kg, sample origin: Tong Ren Tang crude drugstore, Beijing (capital), China, sample year: unknown, country: China[880], *natural drugs

- Co-contamination: For detailed information see the article.
- Further contamination of plants and plant products other than cereals and their products reported in the present book/article[880]: Medicinal plants, AFB_1, AFB_2, and AFG_1, literature[880]

For detailed information see the article.

Aflatoxin M_1

incidence: 2/13*, conc. range: 0.6–0.7 μg/kg, ∅ conc.: 0.65 μg/kg, sample origin: enterprises in Zhejiang (province), China, sample year: unknown, country: China[66], *rhizomes and roots

- Co-contamination: 1 sa. co-contaminated with AFB_1, AFB_2, AFG_1, and AFM_1; 1 sa. co-contaminated with AFB_1, AFG_2, and AFM_1
- Further contamination of plants and plant products other than cereals and their products reported in the present book/article[66]: Medicinal plants, AFB_1, AFB_2, AFG_1, and AFG_2, literature[66]

incidence: 0/4*, conc.: no contamination, sample origin: enterprises in Zhejiang (province), China, sample year: unknown, country: China[66], *flowers

- Co-contamination: not reported
- Further contamination of plants and plant products other than cereals and their products reported in the present book/article[66]: Medicinal plants, AFB_1, AFB_2, AFG_1, and AFG_2, literature[66]

incidence: 0/8*, conc.: no contamination, sample origin: enterprises in Zhejiang (province), China, sample year: unknown, country: China[66], *seeds

- Co-contamination: not reported
- Further contamination of plants and plant products other than cereals and their products reported in the present book/article[66]: Medicinal plants, AFB_1, AFB_2, AFG_1, and AFG_2, literature[66]

incidence: 0/5*, conc.: no contamination, sample origin: enterprises in Zhejiang (province), China, sample year: unknown, country: China[66], *grasses and leaves

- Co-contamination: not reported
- Further contamination of plants and plant products other than cereals and their products reported in the present book/article[66]: Medicinal plants, AFB_1, AFB_2, AFG_1, and AFG_2, literature[66]

For detailed information see the article.

incidence: 5/23*, conc. range: 0.2–1.2 μg/kg, ∅ conc.: 0.68 μg/kg, sample origin: unknown, sample year: unknown, country: China[68], *rhizomes and roots

- Co-contamination: 1 sa. co-contaminated with AFB_1, AFB_2, AFM_1, CHAE, CIT, CPA, GT, MA, OTB, STG, and VER; 1 sa. co-contaminated with AFB_1, AFB_2, AFG_1, AFG_2, AFM_1, CHAE, and OTA; 1 sa. co-contaminated with AFB_1, AFB_2, AFG_1, AFM_1, CPA, and NEO; 1 sa. co-contaminated with AFM_1, CIT, GT, MA, NEO, and STG; 1 sa. co-contaminated with AFB_1, AFG_1, and AFM_1
- Further contamination of plants and plant products other than cereals and their products reported in the present book/article[68]: Medicinal plants, AFB_1, AFB_2, AFG_1, AFG_2, CIT, CPA, DAS, DON, FUS-X, GT, HT-2, MA, MON, NEO, NIV, OTA, OTB, PAT, STG, T-2, and ZEA, literature[68]; see also Medicinal plants, chapter: Further Mycotoxins and Microbial Metabolites

incidence: 1/15*, conc.: 0.4 μg/kg, sample origin: unknown, sample year: unknown, country: China[68], *seeds

- Co-contamination: 1 sa. co-contaminated with AFB_1, AFB_2, AFG_1, AFG_2, AFM_1, CHAE, CPA, GT, HT-2, MA, PEA, STG, VER, and ZEA

- Further contamination of plants and plant products other than cereals and their products reported in the present book/ article[68]: Medicinal plants, AFB_1, AFB_2, AFG_1, AFG_2, CIT, CPA, DAS, DON, FUS-X, GT, HT-2, MA, MON, NEO, NIV, OTA, OTB, PAT, STG, T-2, and ZEA, literature[68]; see also Medicinal plants, chapter: Further Mycotoxins and Microbial Metabolites

incidence: 0/11*, conc.: no contamination, sample origin: unknown, sample year: unknown, country: China[68], *grasses and leaves

- Co-contamination: not reported
- Further contamination of plants and plant products other than cereals and their products reported in the present book/ article[68]: Medicinal plants, AFB_1, AFB_2, AFG_1, AFG_2, CIT, CPA, DAS, DON, FUS-X, GT, HT-2, MA, MON, NEO, NIV, OTA, OTB, PAT, STG, T-2, and ZEA, literature[68]; see also Medicinal plants, chapter: Further Mycotoxins and Microbial Metabolites

incidence: 0/11*, conc.: no contamination, sample origin: unknown, sample year: unknown, country: China[68], *flowers

- Co-contamination: not reported
- Further contamination of plants and plant products other than cereals and their products reported in the present book/ article[68]: Medicinal plants, AFB_1, AFB_2, AFG_1, AFG_2, CIT, CPA, DAS, DON, FUS-X, GT, HT-2, MA, MON, NEO, NIV, OTA, OTB, PAT, STG, T-2, and ZEA, literature[68]; see also Medicinal plants, chapter: Further Mycotoxins and Microbial Metabolites

For detailed information see the article.

Aflatoxin/s

incidence: 2/11*, conc. range: 15.1–15.2 µg/kg**, ∅ conc.: 15.15 µg/kg**, sample origin: Maryland and West Virginia (states) farms, USA, sample year: 2002/2003, country: USA[71], *wild-simulated ginseng roots, **AFB_1, AFB_2, AFG_1, and AFG_2

incidence: 1/1*, conc.: 16 µg/kg**, sample origin: grocery store, USA, sample year: unknown, country: USA[71], *moldy ginseng root, **AFB_1, AFB_2, AFG_1, and AFG_2

- Co-contamination: For detailed information see the article.
- Further contamination of plants and plant products other than cereals and their products reported in the present book/ article[71]: not reported

incidence: 81/84*, conc. range: 1.4–10 µg/kg** (11 sa.), 10.1–20 µg/kg** (16 sa.), 20.1–50 µg/kg** (26 sa.), 50.1–90 µg/kg** (13 sa.), 90.1–100 µg/kg** (4 sa.), ≤853.44 µg/kg** *** (11 sa.), sample origin: herbal stores and a herbal distributor company, Lleida (province), Spain, sample year: unknown, country: Spain[72], *medicinal and aromatic herbs, **AFS total, ***red tea leaves

- Co-contamination: not reported
- Further contamination of plants and plant products other than cereals and their products reported in the present book/ article[72]: Medicinal plants, CIT, DON, FB/S, OTA, T-2, and ZEA, literature[72]

For detailed information see the article.

incidence: 64/83*, conc. range: 4–10 µg/kg** (28 sa.), >10 µg/kg** (36 sa.), sample origin: Tamilnadu (state), India, sample year: unknown, country: Germany[806], *Indian *Cassia senna* L., **AFB_1, AFB_2, AFG_1, and AFG_2

incidence: 34/74*, conc. range: 4–10 µg/kg** (15 sa.), >10 µg/kg** (19 sa.), sample origin: Gujarat (state), India, sample year: unknown, country: Germany[806], *Indian *Cassia senna* L., **AFB_1, AFB_2, AFG_1, and AFG_2

- Co-contamination: not reported
- Further contamination of plants and plant products other than cereals and their products reported in the present book/ article[806]: not reported

For detailed information see the article.

Other *Aspergillus* Toxins

Gliotoxin

incidence: 7/23*, conc. range: 0.3–6.7 µg/kg, ∅ conc.: 3.27 µg/kg, sample origin: unknown, sample year: unknown, country: China[68], *rhizomes and roots

- Co-contamination: 1 sa. co-contaminated with AFB_1, AFB_2, AFM_1, CHAE, CIT, CPA, GT, MA, OTB, STG, and VER; 1 sa. co-contaminated with AFB_1, CPA, GT, MA, OTB, STG, T-2, and ZEA; 1 sa. co-contaminated with AFB_1, CHAE, GT, MA, OTA, OTB, and T-2; 1 sa. co-contaminated with CPA, GT, MA, STG, and VER; 1 sa. co-contaminated with AFM_1, CIT, GT, MA, NEO, and STG; 1 sa. co-contaminated with CHAE, CPA, and GT; 1 sa. co-contaminated with GT, MA, and OTA
- Further contamination of plants and plant products other than cereals and their products reported in the present book/article[68]: Medicinal plants, AFB_1, AFB_2, AFG_1, AFG_2, AFM_1, CIT, CPA, DAS, DON, FUS-X, HT-2, MA, MON, NEO, NIV, OTA, OTB, PAT, STG, T-2, and ZEA, literature[68]; see also Medicinal plants, chapter: Further Mycotoxins and Microbial Metabolites

incidence: 3/15*, conc. range: 0.5–3.2 µg/kg, ∅ conc.: 1.93 µg/kg, sample origin: unknown, sample year: unknown, country: China[68], *seeds

- Co-contamination: 1 sa. co-contaminated with AFB_1, AFB_2, AFG_1, AFG_2, AFM_1, CHAE, CPA, GT, HT-2, MA, PEA, STG, VER, and ZEA; 1 sa. co-contaminated with AFB_1, AFB_2, CPA, DAS, GT, MA, MON, NIV, STG, and VER; 1 sa. co-contaminated with AFB_1, AFB_2, DAS, GT, STG, and ZEA
- Further contamination of plants and plant products other than cereals and their products reported in the present book/ article[68]: Medicinal plants, AFB_1, AFB_2, AFG_1, AFG_2, AFM_1, CIT, CPA, DAS, DON, FUS-X, HT-2, MA, MON, NEO, NIV, OTA, OTB, PAT, STG, T-2, and ZEA, literature[68]; see also Medicinal plants, chapter: Further Mycotoxins and Microbial Metabolites

incidence: 1/11*, conc.: 1.8 µg/kg, sample origin: unknown, sample year: unknown, country: China[68], *grasses and leaves

- Co-contamination: 1 sa. co-contaminated with AFB_1, AFG_1, GT, MA, OTB, PEA, and STG
- Further contamination of plants and plant products other than cereals and their products reported in the present book/ article[68]: Medicinal plants, AFB_1, AFB_2, AFG_1, AFG_2, AFM_1, CIT, CPA, DAS, DON, FUS-X, HT-2, MA, MON, NEO, NIV, OTA, OTB, PAT, STG, T-2, and ZEA, literature[68]; see also Medicinal plants, chapter: Further Mycotoxins and Microbial Metabolites

incidence: 2/11*, conc. range: 2.1–2.5 µg/kg, ∅ conc.: 2.3 µg/kg, sample origin: unknown, sample year: unknown, country: China[68], *flowers

- Co-contamination: 1 sa. co-contaminated with GT, MON, and T-2; 1 sa. contaminated solely with GT
- Further contamination of plants and plant products other than cereals and their products reported in the present book/ article[68]: Medicinal plants, AFB_1, AFB_2, AFG_1, AFG_2, AFM_1, CIT, CPA, DAS, DON, FUS-X, HT-2, MA, MON, NEO, NIV, OTA, OTB,

PAT, STG, T-2, and ZEA, literature[68]; see also Medicinal plants, chapter: Further Mycotoxins and Microbial Metabolites

For detailed information see the article.

Sterigmatocystin

incidence: 10/23*, conc. range: 0.3–10.8 μg/kg, Ø conc.: 2.77 μg/kg, sample origin: unknown, sample year: unknown, country: China[68], *rhizomes and roots

- Co-contamination: 1 sa. co-contaminated with AFB_1, AFB_2, AFM_1, CHAE, CIT, CPA, GT, MA, OTB, STG, and VER; 1 sa. co-contaminated with AFB_1, CPA, GT, MA, OTB, STG, T-2, and ZEA; 1 sa. co-contaminated with AFB_2, AFG_1, AFG_2, MON, MA, NEO, and STG; 1 sa. co-contaminated with AFM_1, CIT, GT, MA, NEO, and STG; 1 sa. co-contaminated with CHAE, MA, and STG; 1 sa. co-contaminated with CIT, DON3G, MON, PAT, and STG; 1 sa. co-contaminated with CIT, DAS, STG, and VER; 1 sa. co-contaminated with CPA, GT, MA, STG, and VER; 1 sa. co-contaminated with CHAE, STG, and ZEA; 1 sa. co-contaminated with CHAE and STG
- Further contamination of plants and plant products other than cereals and their products reported in the present book/article[68]: Medicinal plants, AFB_1, AFB_2, AFG_1, AFG_2, AFM_1, CIT, CPA, DAS, DON, FUS-X, GT, HT-2, MON, NEO, NIV, OTA, OTB, PAT, T-2, and ZEA, literature[68]; see also Medicinal plants, chapter: Further Mycotoxins and Microbial Metabolites

incidence: 8/15*, conc. range: 0.2–7.6 μg/kg, Ø conc.: 2.09 μg/kg, sample origin: unknown, sample year: unknown, country: China[68], *seeds

- Co-contamination: 1 sa. co-contaminated with AFB_1, AFB_2, AFG_1, AFG_2, AFM_1, CHAE, CPA, GT, HT-2, MA, PEA, STG, VER, and ZEA; 1 sa. co-contaminated with AFB_1, AFB_2, CPA, DAS, GT, MA, MON, NIV, STG, and VER; 1 sa. co-contaminated with AFB_1, AFG_1, AFG_2, DAS, MA, PEA, and STG; 1 sa. co-contaminated with FUS-X, MA, NEO, OTA, OTB, PAT, and STG; 1 sa. co-contaminated with AFB_1, AFB_2, DAS, GT, STG, and ZEA; 1 sa. co-contaminated with AFB_1, CHAE, STG, ZEA, α-ZEL, and ZEAA; 1 sa. co-contaminated with AFB_2, DON3G, MA, STG, and VER; 1 sa. co-contaminated with CPA and STG
- Further contamination of plants and plant products other than cereals and their products reported in the present book/article[68]: Medicinal plants, AFB_1, AFB_2, AFG_1, AFG_2, AFM_1, CIT, CPA, DAS, DON, FUS-X, GT, HT-2, MON, NEO, NIV, OTA, OTB, PAT, T-2, and ZEA, literature[68]; see also Medicinal plants, chapter: Further Mycotoxins and Microbial Metabolites

incidence: 6/11*, conc. range: 0.8–9.6 μg/kg, Ø conc.: 4.37 μg/kg, sample origin: unknown, sample year: unknown, country: China[68], *grasses and leaves

- Co-contamination: 1 sa. co-contaminated with AFB_1, AFG_1, GT, MA, OTB, PEA, and STG; 1 sa. co-contaminated with AFB_2, CPA, MA, OTA, PAT, and STG; 1 sa. co-contaminated with CPA, MA, MON, and STG; 1 sa. co-contaminated with DON3G, MA, STG, and ZEA; 1 sa. co-contaminated with CPA, NIV, and STG; 1 sa. co-contaminated with MA, PEA, and STG
- Further contamination of plants and plant products other than cereals and their products reported in the present book/article[68]: Medicinal plants, AFB_1, AFB_2, AFG_1, AFG_2, AFM_1, CIT, CPA, DAS, DON, FUS-X, GT, HT-2, MON, NEO, NIV, OTA, OTB, PAT, T-2, and ZEA, literature[68]; see also Medicinal plants, chapter: Further Mycotoxins and Microbial Metabolites

incidence: 2/11*, conc. range: 0.2–0.6 μg/kg, Ø conc.: 0.4 μg/kg, sample origin: unknown, sample year: unknown, country: China[68], *flowers

- Co-contamination: 1 sa. co-contaminated with AFB_1, AFG_2, CHAE, CPA, DAS, DON3G, PEA, and STG; 1 sa. co-contaminated with MON and STG
- Further contamination of plants and plant products other than cereals and their products reported in the present book/article[68]: Medicinal plants, AFB_1, AFB_2, AFG_1, AFG_2, AFM_1, CIT, CPA, DAS, DON, FUS-X, GT, HT-2, MON, NEO, NIV, OTA, OTB, PAT, T-2, and ZEA, literature[68]; see also Medicinal plants, chapter: Further Mycotoxins and Microbial Metabolites

For detailed information see the article.

incidence: 64/244, conc. range: ≤47.0 μg/kg, sample origin: collected from TCM pharmacies and hospitals in Guangzhou (city), China, sample year: unknown, country: China[886]

- Co-contamination: not reported
- Further contamination of plants and plant products other than cereals and their products reported in the present book/article[886]: Medicinal plants, AFB_1, AFB_2, and OTA, literature[886]

For detailed information see the article.

Aspergillus and *Penicillium* Toxins

Citrinin

incidence: 0/10*, conc.: no contamination, sample origin: store-houses (private concerns) in Bihar (state), India, sample year: unknown, country: India[63], **Glycyrrhiza glabra* Linn.

incidence: 2/10*, Ø conc.: 80 μg/kg, sample origin: store-houses (private concerns) in Bihar (state), India, sample year: unknown, country: India[63], **Holarrhena antidysenterica* Wall

incidence: 0/10*, conc.: no contamination, sample origin: store-houses (private concerns) in Bihar (state), India, sample year: unknown, country: India[63], **Ichnocarpus frutescens* (Linn.)

incidence: 0/10*, conc.: no contamination, sample origin: store-houses (private concerns) in Bihar (state), India, sample year: unknown, country: India[63], **Strychnos nux-vomica* Linn.

incidence: 0/10*, conc.: no contamination, sample origin: store-houses (private concerns) in Bihar (state), India, sample year: unknown, country: India[63], **Tribulus terrestris* Linn.

- Co-contamination: not reported
- Further contamination of plants and plant products other than cereals and their products reported in the present book/article[63]: Medicinal plants, AFB_1, OTA, and ZEA, literature[63]

For detailed information see the article.

incidence: 14/175*, conc. range: 20–160 μg/kg, sample origin: drug centers of Bhagalpur (city) and store houses in southern Bihar (state), India, sample year: January–December 1990, country: India[65], *crude and finished herbal drugs

- Co-contamination: not reported
- Further contamination of plants and plant products other than cereals and their products reported in the present book/article[65]: Medicinal plants, AFB_1, OTA, and ZEA, literature[65]

For detailed information see the article.

incidence: 4/23*, conc. range: 0.3–5.1 μg/kg, Ø conc.: 1.75 μg/kg, sample origin: unknown, sample year: unknown, country: China[68], *rhizomes and roots

- Co-contamination: 1 sa. co-contaminated with AFB_1, AFB_2, AFM_1, CHAE, CIT, CPA, GT, MA, OTB, STG, and VER; 1 sa.

co-contaminated with AFM_1, CIT, GT, MA, NEO, and STG; 1 sa. co-contaminated with CIT, DON3G, MON, PAT, and STG; 1 sa. co-contaminated with CIT, DAS, STG, and VER

- Further contamination of plants and plant products other than cereals and their products reported in the present book/article[68]: Medicinal plants, AFB_1, AFB_2, AFG_1, AFG_2, AFM_1, CPA, DAS, DON, FUS-X, GT, HT-2, MA, MON, NEO, NIV, OTA, OTB, PAT, STG, T-2, and ZEA, literature[68]; see also Medicinal plants, chapter: Further Mycotoxins and Microbial Metabolites

incidence: 0/15*, conc.: no contamination, sample origin: unknown, sample year: unknown, country: China[68], *seeds

- Co-contamination: not reported
- Further contamination of plants and plant products other than cereals and their products reported in the present book/article[68]: Medicinal plants, AFB_1, AFB_2, AFG_1, AFG_2, AFM_1, CPA, DAS, DON, FUS-X, GT, HT-2, MA, MON, NEO, NIV, OTA, OTB, PAT, STG, T-2, and ZEA, literature[68]; see also Medicinal plants, chapter: Further Mycotoxins and Microbial Metabolites

incidence: 1/11*, conc.: 8.5 µg/kg, sample origin: unknown, sample year: unknown, country: China[68], *grasses and leaves

- Co-contamination: 1 sa. co-contaminated with CIT, HT-2, MA, OTA, and OTB
- Further contamination of plants and plant products other than cereals and their products reported in the present book/article[68]: Medicinal plants, AFB_1, AFB_2, AFG_1, AFG_2, AFM_1, CPA, DAS, DON, FUS-X, GT, HT-2, MA, MON, NEO, NIV, OTA, OTB, PAT, STG, T-2, and ZEA, literature[68]; see also Medicinal plants, chapter: Further Mycotoxins and Microbial Metabolites

incidence: 0/11*, conc.: no contamination, sample origin: unknown, sample year: unknown, country: China[68], *flowers

- Co-contamination: not reported
- Further contamination of plants and plant products other than cereals and their products reported in the present book/article[68]: Medicinal plants, AFB_1, AFB_2, AFG_1, AFG_2, AFM_1, CPA, DAS, DON, FUS-X, GT, HT-2, MA, MON, NEO, NIV, OTA, OTB, PAT, STG, T-2, and ZEA, literature[68]; see also Medicinal plants, chapter: Further Mycotoxins and Microbial Metabolites

For detailed information see the article.

incidence: 51/84*, conc. range: 17–30 µg/kg (15 sa.), 31–50 µg/kg (16 sa.), 51–100 µg/kg (12 sa.), 110–200 µg/kg (4 sa.), 250–400 µg/kg (3 sa.), >400 µg/kg (1 sa.), sample origin: herbal stores and a herbal distributor company, Lleida (province), Spain, sample year: unknown, country: Spain[72], *medicinal and aromatic herbs

- Co-contamination: not reported
- Further contamination of plants and plant products other than cereals and their products reported in the present book/article[72]: Medicinal plants, AF/S, DON, FB/S, OTA, T-2, and ZEA, literature[72]

For detailed information see the article.

Cyclopiazonic Acid

incidence: 6/23*, conc. range: 0.5–36.1 µg/kg, ∅ conc.: 11.45 µg/kg, sample origin: unknown, sample year: unknown, country: China[68], *rhizomes and roots

- Co-contamination: 1 sa. co-contaminated with AFB_1, AFB_2, AFM_1, CHAE, CIT, CPA, GT, MA, OTB, STG, and VER; 1 sa. co-contaminated with AFB_1, CPA, GT, MA, OTB, STG, T-2, and ZEA; 1 sa. co-contaminated with AFB_1, AFB_2, AFG_1, AFM_1, CPA, and NEO; 1 sa. co-contaminated with CPA, GT, MA, STG, and VER; 1 sa. co-contaminated with CHAE, CPA, DON3G, and PEA; 1 sa. co-contaminated with CHAE, CPA, and GT
- Further contamination of plants and plant products other than cereals and their products reported in the present book/article[68]: Medicinal plants, AFB_1, AFB_2, AFG_1, AFG_2, AFM_1, CIT, DAS, DON, FUS-X, GT, HT-2, MA, MON, NEO, NIV, OTA, OTB, PAT, STG, T-2, and ZEA, literature[68]; see also Medicinal plants, chapter: Further Mycotoxins and Microbial Metabolites

incidence: 4/15*, conc. range: 2.4–6.8 µg/kg, ∅ conc.: 4.15 µg/kg, sample origin: unknown, sample year: unknown, country: China[68], *seeds

- Co-contamination: 1 sa. co-contaminated with AFB_1, AFB_2, AFG_1, AFG_2, AFM_1, CHAE, CPA, GT, HT-2, MA, PEA, STG, VER, and ZEA; 1 sa. co-contaminated with AFB_1, AFB_2, CPA, DAS, GT, MA, MON, NIV, STG, and VER; 1 sa. co-contaminated with AFB_1, AFB_2, CHAE, CPA, DPDON, and MA; 1 sa. co-contaminated with CPA and STG
- Further contamination of plants and plant products other than cereals and their products reported in the present book/article[68]: Medicinal plants, AFB_1, AFB_2, AFG_1, AFG_2, AFM_1, CIT, DAS, DON, FUS-X, GT, HT-2, MA, MON, NEO, NIV, OTA, OTB, PAT, STG, T-2, and ZEA, literature[68]; see also Medicinal plants, chapter: Further Mycotoxins and Microbial Metabolites

incidence: 3/11*, conc. range: 0.3–9.8 µg/kg, ∅ conc.: 3.5 µg/kg, sample origin: unknown, sample year: unknown, country: China[68], *grasses and leaves

- Co-contamination: 1 sa. co-contaminated with AFB_2, CPA, MA, OTA, PAT, and STG; 1 sa. co-contaminated with CPA, MA, MON, and STG; 1 sa. co-contaminated with CPA, NIV, and STG
- Further contamination of plants and plant products other than cereals and their products reported in the present book/article[68]: Medicinal plants, AFB_1, AFB_2, AFG_1, AFG_2, AFM_1, CIT, DAS, DON, FUS-X, GT, HT-2, MA, MON, NEO, NIV, OTA, OTB, PAT, STG, T-2, and ZEA, literature[68]; see also Medicinal plants, chapter: Further Mycotoxins and Microbial Metabolites

incidence: 2/11*, conc. range: 2.1–2.8 µg/kg, ∅ conc.: 2.45 µg/kg, sample origin: unknown, sample year: unknown, country: China[68], *flowers

- Co-contamination: 1 sa. co-contaminated with AFB_1, AFG_2, CHAE, CPA, DAS, DON3G, PEA, and STG; 1 sa. co-contaminated with CPA and PEA
- Further contamination of plants and plant products other than cereals and their products reported in the present book/article[68]: Medicinal plants, AFB_1, AFB_2, AFG_1, AFG_2, AFM_1, CIT, DAS, DON, FUS-X, GT, HT-2, MA, MON, NEO, NIV, OTA, OTB, PAT, STG, T-2, and ZEA, literature[68]; see also Medicinal plants, chapter: Further Mycotoxins and Microbial Metabolites

For detailed information see the article.

Ochratoxin A

incidence: 2/10*, ∅ conc.: 50 µg/kg, sample origin: store-houses (private concerns) in Bihar (state), India, sample year: unknown, country: India[63], **Glycyrrhiza glabra* Linn.

incidence: 0/10*, conc.: no contamination, sample origin: store-houses (private concerns) in Bihar (state), India, sample year: unknown, country: India[63], **Holarrhena antidysenterica* Wall incidence: 0/10*, conc.: no contamination, sample origin: storehouses (private concerns) in Bihar (state), India, sample year: unknown, country: India[63], **Ichnocarpus frutescens* (Linn.)

incidence: 3/10*, ∅ conc.: 100 µg/kg, sample origin: storehouses (private concerns) in Bihar (state), India, sample year: unknown, country: India[63], **Strychnos nux-vomica* Linn.

incidence: 0/10*, conc.: no contamination, sample origin: storehouses (private concerns) in Bihar (state), India, sample year: unknown, country: India[63], **Tribulus terrestris* Linn.
- Co-contamination: not reported
- Further contamination of plants and plant products other than cereals and their products reported in the present book/article[63]: Medicinal plants, AFB_1, CIT, and ZEA, literature[63]

For detailed information see the article.

incidence: 12/175*, conc. range: 10–130 µg/kg, sample origin: drug centers of Bhagalpur (city) and store houses in southern Bihar (state), India, sample year: January–December 1990, country: India[65], *crude and finished herbal drugs
- Co-contamination: not reported
- Further contamination of plants and plant products other than cereals and their products reported in the present book/article[65]: Medicinal plants, AFB_1, CIT, and ZEA, literature[65]

For detailed information see the article.

incidence: 4/23*, conc. range: 0.2–0.7 µg/kg, ∅ conc.: 0.38 µg/kg, sample origin: unknown, sample year: unknown, country: China[68], *rhizomes and roots
- Co-contamination: 1 sa. co-contaminated with AFB_1, AFB_2, AFG_1, AFG_2, AFM_1, CHAE, MA, and OTA; 1 sa. co-contaminated with AFB_1, CHAE, GT, MA, OTA, OTB, and T-2; 1 sa. co-contaminated with AFG_2, DPDON, NEO, NIV, OTA, and OTB; 1 sa. co-contaminated with GT, MA, and OTA
- Further contamination of plants and plant products other than cereals and their products reported in the present book/article[68]: Medicinal plants, AFB_1, AFB_2, AFG_1, AFG_2, AFM_1, CIT, CPA, DAS, DON, FUS-X, GT, HT-2, MON, NEO, NIV, OTB, PAT, STG, T-2, and ZEA, literature[68]; see also Medicinal plants, chapter: Further Mycotoxins and Microbial Metabolites

incidence: 4/15*, conc. range: 0.2–1.8 µg/kg, ∅ conc.: 0.65 µg/kg, sample origin: unknown, sample year: unknown, country: China[68], *seeds
- Co-contamination: 1 sa. co-contaminated with FUS-X, MA, NEO, OTA, OTB, PAT, and STG; 1 sa. co-contaminated with AFG_1, CHAE, MA, OTA, and OTB; 1 sa. co-contaminated with OTA, ZEA, α-ZEL, ß-ZEL, and ZEAA; 1 sa. co-contaminated with MA and OTA
- Further contamination of plants and plant products other than cereals and their products reported in the present book/article[68]: Medicinal plants, AFB_1, AFB_2, AFG_1, AFG_2, AFM_1, CIT, CPA, DAS, DON, FUS-X, GT, HT-2, MON, NEO, NIV, OTB, PAT, STG, T-2, and ZEA, literature[68]; see also Medicinal plants, chapter: Further Mycotoxins and Microbial Metabolites

incidence: 4/11*, conc. range: 0.2–0.3 µg/kg, ∅ conc.: 0.25 µg/kg, sample origin: unknown, sample year: unknown, country: China[68], *grasses and leaves
- Co-contamination: 1 sa. co-contaminated with AFB_2, CPA, MA, OTA, PAT, and STG; 1 sa. co-contaminated with CIT, HT-2, MA, OTA, and OTB; 1 sa. co-contaminated with AFB_1, DAS, OTA, and OTB; 1 sa. co-contaminated with AFB_1, NIV, OTA, and PEA
- Further contamination of plants and plant products other than cereals and their products reported in the present book/article[68]: Medicinal plants, AFB_1, AFB_2, AFG_1, AFG_2, AFM_1, CIT, CPA, DAS, DON, FUS-X, GT, HT-2, MON, NEO, NIV, OTB, PAT, STG, T-2, and ZEA, literature[68]; see also Medicinal plants, chapter: Further Mycotoxins and Microbial Metabolites

incidence: 1/11*, conc.: 1.1 µg/kg, sample origin: unknown, sample year: unknown, country: China[68], *flowers
- Co-contamination: 1 sa. co-contaminated with HT-2, OTA, and OTB
- Further contamination of plants and plant products other than cereals and their products reported in the present book/article[68]: Medicinal plants, AFB_1, AFB_2, AFG_1, AFG_2, AFM_1, CIT, CPA, DAS, DON, FUS-X, GT, HT-2, MON, NEO, NIV, OTB, PAT, STG, T-2, and ZEA, literature[68]; see also Medicinal plants, chapter: Further Mycotoxins and Microbial Metabolites

For detailed information see the article.

incidence: 53/84*, conc. range: 0.025–2.9 µg/kg (35 sa.), 3–6 µg/kg (12 sa.), 7–9 µg/kg (1 sa.), 10–13 µg/kg (2 sa.), ≤17.3 µg/kg** (3 sa.), sample origin: herbal stores and a herbal distributor company, Lleida (province), Spain, sample year: unknown, country: Spain[72], *medicinal and aromatic herbs, **sage leaves
- Co-contamination: not reported
- Further contamination of plants and plant products other than cereals and their products reported in the present book/article[72]: Medicinal plants, AF/S, CIT, DON, FB/S, T-2, and ZEA, literature[72]

For detailed information see the article.

incidence: 25*/57, conc. range: 1.2–158.7 µg/kg, ∅ conc.: 23.4 µg/kg, sample origin: storehouses or markets in Henan, Jiangxi, and Zhejiang (provinces) as well as Ningxia and Xinjiang (autonomous region) and Beijing (capital), China, sample year: unknown, country: China/Italy[73], *in part moldy
- Co-contamination: not reported
- Further contamination of plants and plant products other than cereals and their products reported in the present book/article[73]: not reported

For detailed information see the article.

incidence: 1/7*, conc.: tr**, sample origin: imported from a tea manufacturing company, Croatia, sample year: unknown, country: Croatia[74], herbal tea and medicinal plants, ***Tillia grandifolia* L.
- Co-contamination: not reported
- Further contamination of plants and plant products other than cereals and their products reported in the present book/article[74]: not reported

For detailed information see the article.

incidence: 2/28*, conc. range: 0.2–0.4 µg/l, ∅ conc.: 0.3 µg/l, sample origin: unknown, sample year: unknown, country: China[75], *rhizomes and roots
- Co-contamination: not reported
- Further contamination of plants and plant products other than cereals and their products reported in the present book/article[75]: Medicinal Plants, OTB, literature[75]

incidence: 1/10*, conc.: 0.3 µg/l, sample origin: unknown, sample year: unknown, country: China[75], *grasses and leaves
- Co-contamination: 1 sa. co-contaminated with OTA and OTB
- Further contamination of plants and plant products other than cereals and their products reported in the present book/article[75]: Medicinal Plants, OTB, literature[75]

incidence: 1/9*, conc.: 1.5 µg/l, sample origin: unknown, sample year: unknown, country: China[75], *seeds

- Co-contamination: 1 sa. co-contaminated with OTA and OTB
- Further contamination of plants and plant products other than cereals and their products reported in the present book/article[75]: Medicinal Plants, OTB, literature[75]

For detailed information see the article.

incidence: 55/129, conc. range: 0.07–2.34* µg/kg, sample origin: storehouses and shopkeepers in Chhotanagpur, Santhal Pargana, and Tirhut (divisions), Bihar (state), India, sample year: unknown, country: India[76], **Holarrhena antidysenterica* Wall (bark)

- Co-contamination: not reported
- Further contamination of plants and plant products other than cereals and their products reported in the present book/article[76]: not reported

For detailed information see the article.

incidence: 1/6*, conc.: 2.5 µg/kg, sample origin: drugstore, Belgium, sample year: unknown, country: Belgium[77], *3 notified and 3 non-notified sa. of maca-plants

- Co-contamination: not reported
- Further contamination of plants and plant products other than cereals and their products reported in the present book/article[77]: Soy isoflavones, FB_1 and OTA, literature[77]; Spices (garlic), FB_1, FB_2, FB_3, and OTA, literature[77]; Vegetables (black radish), FB_1 and FB_2, literature[77]

incidence: 8/30, conc. range: 9.85–125 µg/kg, Ø conc.: 38.8 µg/kg, sample origin: different areas in Khyber Pakhtunkhwa (province), Pakistan, sample year: 2011/2012, country: Pakistan[661]

- Co-contamination: 1 sa. co-contaminated with AFB_1, AFG_1, AFG_2, and OTA; 2 sa. co-contaminated with AFB_1 and OTA; 1 sa. co-contaminated with AFG_1 and OTA; 4 sa. contaminated solely with OTA
- Further contamination of plants and plant products other than cereals and their products reported in the present book/article[661]: Medicinal plants, AFB_1, AFB_2, AFG_1, and AFG_2, literature[661]

For detailed information see the article.

incidence: 1/12, conc.: 92.3 µg/kg*, sample origin: Anguo (city) herbal market, Hebei (province), China, sample year: unknown, country: China[694], *tangerine seeds

- Co-contamination: not reported
- Further contamination of plants and plant products other than cereals and their products reported in the present book/article[694]: Medicinal plants, AFB_1, literature[694]

For detailed information see the article.

incidence: 6/6*, conc. range: 0.049–94.7 µg/kg, Ø conc.: 20.5 µg/kg, sample origin: storehouses or markets, China, sample year: unknown, country: China[874], *moldy or heavily moldy sa. of *Glycyrrhiza uralensis*

- Co-contamination: 2 sa. co-contaminated with AFB_1, AFB_2, AFG_1, and OTA; 2 sa. co-contaminated with AFB_1 and OTA; 1 sa. co-contaminated with AFB_1, AFB_2, AFG_1, AFG_2, and OTA; 1 sa. co-contaminated with AFB_1, AFG_1, and OTA
- Further contamination of plants and plant products other than cereals and their products reported in the present book/article[874]: Medicinal plants, AFB_1, AFB_2, AFG_1, and AFG_2, literature[874]

incidence: 7/9*, conc. range: 0.010–0.106 µg/kg, Ø conc.: 0.038 µg/kg, sample origin: storehouses or markets, China, sample year: unknown, country: China[874], *non-moldy *Glycyrrhiza uralensis* sa.

- Co-contamination: 4 sa. co-contaminated with AFB_1 and OTA; 2 sa. co-contaminated with AFB_1, AFB_2, and OTA; 1 sa. co-contaminated with AFB_1, AFB_2, AFG_1, and OTA
- Further contamination of plants and plant products other than cereals and their products reported in the present book/article[874]: Medicinal plants, AFB_1, AFB_2, AFG_1, and AFG_2, literature[874]

incidence: 13/244, conc. range: 0.5–38.4 µg/kg, Ø conc.: 9.8 µg/kg, sample origin: collected from TCM pharmacies and hospitals in Guangzhou (city), China, sample year: unknown, country: China[886]

- Co-contamination: not reported
- Further contamination of plants and plant products other than cereals and their products reported in the present book/article[886]: Medicinal plants, AFB_1, AFB_2, and STG, literature[886]

For detailed information see the article.

Ochratoxin B

incidence: 5/23*, conc. range: 0.2–1.2 µg/kg, Ø conc.: 0.58 µg/kg, sample origin: unknown, sample year: unknown, country: China[68], *rhizomes and roots

- Co-contamination: 1 sa. co-contaminated with AFB_1, AFB_2, AFM_1, CHAE, CIT, CPA, GT, MA, OTB, STG, and VER; 1 sa. co-contaminated with AFB_1, CPA, GT, MA, OTB, STG, T-2, and ZEA; 1 sa. co-contaminated with AFB_1, AFB_2, AFG_2, CHAE, NIV, OTB, and T-2; 1 sa. co-contaminated with AFB_1, CHAE, GT, MA, OTA, OTB, and T-2; 1 sa. co-contaminated with AFG_2, DPDON, NEO, NIV, OTA, and OTB
- Further contamination of plants and plant products other than cereals and their products reported in the present book/article[68]: Medicinal plants, AFB_1, AFB_2, AFG_1, AFG_2, AFM_1, CIT, CPA, DAS, DON, FUS-X, GT, HT-2, MON, NEO, NIV, OTA, PAT, STG, T-2, and ZEA, literature[68]; see also Medicinal plants, chapter: Further Mycotoxins and Microbial Metabolites

incidence: 2/15*, conc. range: 1.0–1.8 µg/kg, Ø conc.: 1.4 µg/kg, sample origin: unknown, sample year: unknown, country: China[68], *seeds

- Co-contamination: 1 sa. co-contaminated with FUS-X, MA, NEO, OTA, OTB, PAT, and STG; 1 sa. co-contaminated with AFG_1, CHAE, MA, OTA, and OTB
- Further contamination of plants and plant products other than cereals and their products reported in the present book/article[68]: Medicinal plants, AFB_1, AFB_2, AFG_1, AFG_2, AFM_1, CIT, CPA, DAS, DON, FUS-X, GT, HT-2, MON, NEO, NIV, OTA, PAT, STG, T-2, and ZEA, literature[68]; see also Medicinal plants, chapter: Further Mycotoxins and Microbial Metabolites

incidence: 3/11*, conc. range: 0.2–1.5 µg/kg, Ø conc.: 0.7 µg/kg, sample origin: unknown, sample year: unknown, country: China[68], *grasses and leaves

- Co-contamination: 1 sa. co-contaminated with AFB_1, AFG_1, GT, MA, OTB, PEA, and STG; 1 sa. co-contaminated with CIT, HT-2, MA, OTA, and OTB; 1 sa. co-contaminated with AFB_1, DAS, OTA, and OTB

- Further contamination of plants and plant products other than cereals and their products reported in the present book/article[68]: Medicinal plants, AFB_1, AFB_2, AFG_1, AFG_2, AFM_1, CIT, CPA, DAS, DON, FUS-X, GT, HT-2, MON, NEO, NIV, OTA, PAT, STG, T-2, and ZEA, literature[68]; see also Medicinal plants, chapter: Further Mycotoxins and Microbial Metabolites

incidence: 1/11*, conc.: 0.4 µg/kg, sample origin: unknown, sample year: unknown, country: China[68], *flowers
- Co-contamination: 1 sa. co-contaminated with OTA, OTB, and HT-2
- Further contamination of plants and plant products other than cereals and their products reported in the present book/article[68]: Medicinal plants, AFB_1, AFB_2, AFG_1, AFG_2, AFM_1, CIT, CPA, DAS, DON, FUS-X, GT, HT-2, MON, NEO, NIV, OTA, PAT, STG, T-2, and ZEA, literature[68]; see also Medicinal plants, chapter: Further Mycotoxins and Microbial Metabolites

For detailed information see the article.

incidence: 0/28*, conc. no contamination, sample origin: unknown, sample year: unknown, country: China[75], *rhizomes and roots
- Co-contamination: not reported
- Further contamination of plants and plant products other than cereals and their products reported in the present book/article[75]: Medicinal Plants, OTA, literature[75]

incidence: 1/10*, conc.: 0.4 µg/l, sample origin: unknown, sample year: unknown, country: China[75], *grasses and leaves
- Co-contamination: 1 sa. co-contaminated with OTA and OTB
- Further contamination of plants and plant products other than cereals and their products reported in the present book/article[75]: Medicinal Plants, OTA, literature[75]

incidence: 1/9*, conc.: 0.8 µg/l, sample origin: unknown, sample year: unknown, country: China[75], *seeds
- Co-contamination: 1 sa. co-contaminated with OTA and OTB
- Further contamination of plants and plant products other than cereals and their products reported in the present book/article[75]: Medicinal Plants, OTA, literature[75]

For detailed information see the article.

Patulin

incidence: 1/23*, conc.: 8.9 µg/kg, sample origin: unknown, sample year: unknown, country: China[68], *rhizomes and roots
- Co-contamination: 1 sa. co-contaminated with CIT, DON3G, MON, PAT, and STG
- Further contamination of plants and plant products other than cereals and their products reported in the present book/article[68]: Medicinal plants, AFB_1, AFB_2, AFG_1, AFG_2, AFM_1, CIT, CPA, DAS, DON, FUS-X, GT, HT-2, MON, NEO, NIV, OTA, OTB, STG, T-2, and ZEA, literature[68]; see also Medicinal plants, chapter: Further Mycotoxins and Microbial Metabolites

incidence: 1/15*, conc.: 7.1 µg/kg, sample origin: unknown, sample year: unknown, country: China[68], *seeds
- Co-contamination: 1 sa. co-contaminated with FUS-X, MA, NEO, OTA, OTB, PAT, and STG
- Further contamination of plants and plant products other than cereals and their products reported in the present book/article[68]: Medicinal plants, AFB_1, AFB_2, AFG_1, AFG_2, AFM_1, CIT, CPA, DAS, DON, FUS-X, GT, HT-2, MON, NEO, NIV, OTA, OTB, STG, T-2, and ZEA, literature[68]; see also Medicinal plants, chapter: Further Mycotoxins and Microbial Metabolites

incidence: 1/11*, conc.: 3.2 µg/kg, sample origin: unknown, sample year: unknown, country: China[68], *grasses and leaves
- Co-contamination: 1 sa. co-contaminated with AFB_2, CPA, MA, OTA, PAT, and STG
- Further contamination of plants and plant products other than cereals and their products reported in the present book/article[68]: Medicinal plants, AFB_1, AFB_2, AFG_1, AFG_2, AFM_1, CIT, CPA, DAS, DON, FUS-X, GT, HT-2, MON, NEO, NIV, OTA, OTB, STG, T-2, and ZEA, literature[68]; see also Medicinal plants, chapter: Further Mycotoxins and Microbial Metabolites

incidence: 0/11*, conc.: no contamination, sample origin: unknown, sample year: unknown, country: China[68], *flowers
- Co-contamination: not reported
- Further contamination of plants and plant products other than cereals and their products reported in the present book/article[68]: Medicinal plants, AFB_1, AFB_2, AFG_1, AFG_2, AFM_1, CIT, CPA, DAS, DON, FUS-X, GT, HT-2, MON, NEO, NIV, OTA, OTB, STG, T-2, and ZEA, literature[68]; see also Medicinal plants, chapter: Further Mycotoxins and Microbial Metabolites

For detailed information see the article.

Fusarium Toxins

Depsipeptides

Beauvericin

incidence: 15/60, conc. range: >LOD-124.8 µg/kg, Ø conc.: 33.0 µg/kg, sample origin: traditional Chinese medicine stores in Zhejiang (province), China, sample year: unknown, country: Germany[746]
- Co-contamination: 15 sa. co-contaminated with BEA, ENA, ENA_1, ENB, and ENB_1
- Further contamination of plants and plant products other than cereals and their products reported in the present book/article[746]: Medicinal plants, ENA, ENA_1, ENB, and ENB_1, literature[746]; Spices (ginger), BEA, ENA, ENA_1, ENB, and ENB_1, literature[746]

For detailed information see the article.

Enniatin A

incidence: 15/60, conc. range: >LOD-354.6 µg/kg, Ø conc.: 28.9 µg/kg, sample origin: traditional Chinese medicine stores in Zhejiang (province), China, sample year: unknown, country: Germany[746]
- Co-contamination: 15 sa. co-contaminated with BEA, ENA, ENA_1, ENB, and ENB_1
- Further contamination of plants and plant products other than cereals and their products reported in the present book/article[746]: Medicinal plants, BEA, ENA_1, ENB, and ENB_1, literature[746]; Spices (ginger), BEA, ENA, ENA_1, ENB, and ENB_1, literature[746]

For detailed information see the article.

Enniatin A_1

incidence: 15/60, conc. range: >LOD-252.5 µg/kg, Ø conc.: 28.4 µg/kg, sample origin: traditional Chinese medicine stores in Zhejiang (province), China, sample year: unknown, country: Germany[746]
- Co-contamination: 15 sa. co-contaminated with BEA, ENA, ENA_1, ENB, and ENB_1
- Further contamination of plants and plant products other than cereals and their products reported in the present book/article[746]: Medicinal plants, BEA, ENA, ENB, and ENB_1, literature[746]; Spices (ginger), BEA, ENA, ENA_1, ENB, and ENB_1, literature[746]

For detailed information see the article.

Enniatin B

incidence: 15/60, conc. range: >LOD-290.5 µg/kg, ∅ conc.: 32.0 µg/kg, sample origin: traditional Chinese medicine stores in Zhejiang (province), China, sample year: unknown, country: Germany[746]

- Co-contamination: 15 sa. co-contaminated with BEA, ENA, ENA_1, ENB, and ENB_1
- Further contamination of plants and plant products other than cereals and their products reported in the present book/article[746]: Medicinal plants, BEA, ENA, ENA_1, and ENB_1, literature[746]; Spices (ginger), BEA, ENA, ENA_1, ENB, and ENB_1, literature[746]

For detailed information see the article.

Enniatin B_1

incidence: 15/60, conc. range: >LOD-40.2 µg/kg, ∅ conc.: 3.9 µg/kg, sample origin: traditional Chinese medicine stores in Zhejiang (province), China, sample year: unknown, country: Germany[746]

- Co-contamination: 15 sa. co-contaminated with BEA, ENA, ENA_1, ENB, and ENB_1
- Further contamination of plants and plant products other than cereals and their products reported in the present book/article[746]: Medicinal plants, BEA, ENA, ENA_1, and ENB, literature[746]; Spices (ginger), BEA, ENA, ENA_1, ENB, and ENB_1, literature[746]

For detailed information see the article.

Fumonisin/s

Fumonisin B_1

incidence: 12/18*, conc. range: 350–700 µg/kg, sample origin: supermarkets in Lisbon (capital), Portugal, sample year: unknown, country: Portugal[78], *leaves of orange trees

incidence: 12/18*, conc. range: 20–200 µg/kg, sample origin: supermarkets in Lisbon (capital), Portugal, sample year: unknown, country: Portugal[78], *leaves or flowers of linden trees

incidence: 9/15*, conc. range: 50–150 µg/kg, sample origin: supermarkets in Lisbon (capital), Portugal, sample year: unknown, country: Portugal[78], *corn silk

incidence: 8/18*, conc. range: 20–70 µg/kg, sample origin: supermarkets in Lisbon (capital), Portugal, sample year: unknown, country: Portugal[78], *chamomile

- Co-contamination: not reported
- Further contamination of plants and plant products other than cereals and their products reported in the present book/article[78]: Tea, FB_1, literature[78]

incidence: 13/16, conc. range: 14–139 µg/kg, ∅ conc.: 55.5 µg/kg, sample origin: herbal (muthi) markets in Cape Town (city) and Marabastad (business area) in Pretoria (city), South Africa, sample year: 2006, country: South Africa[79]

- Co-contamination: not reported
- Further contamination of plants and plant products other than cereals and their products reported in the present book/article[79]: not reported

For detailed information see the article.

incidence: 6/61*, conc. range: ≤1.487 µg/kg, sample origin: supermarkets and bazaars in Istanbul (capital), Turkey, sample year: unknown, country: Turkey[80]

- Co-contamination: 1 sa. co-contaminated with FB_1 and FB_2; 5 sa. contaminated solely with FB_1
- Further contamination of plants and plant products other than cereals and their products reported in the present book/article[80]: Medicinal plants, FB_2, literature[80]

For detailed information see the article.

incidence: 4/30*, conc. range: 8.6–1,553 µg/kg, ∅ conc.: 410 µg/kg, sample origin: different locations, South Africa, sample year: November 2002, country: South Africa[81], *medicinal wild plants

- Co-contamination: not reported
- Further contamination of plants and plant products other than cereals and their products reported in the present book/article[81]: Dietary plants, FB_1, literature[81]

For detailed information see the article.

incidence: 9/10*, conc. range: 0.58–48.90 µg/kg, ∅ conc.: 14.4 µg/kg, sample origin: regular enterprises, China, sample year: unknown, country: China[739], *seeds

- Co-contamination: 4 sa. co-contaminated with FB_1, FB_2, and FB_3; 4 sa. co-contaminated with FB_1 and FB_2; 1 sa. contaminated solely with FB_1
- Further contamination of plants and plant products other than cereals and their products reported in the present book/article[739]: Medicinal plants, FB_2 and FB_3, literature[739]

incidence: 4/12*, conc. range: 0.69–9.80 µg/kg, ∅ conc.: 3.56 µg/kg, sample origin: regular enterprises, China, sample year: unknown, country: China[739], *rhizomes and roots

- Co-contamination: 1 sa. co-contaminated with FB_1 and FB_2; 3 sa. contaminated solely with FB_1
- Further contamination of plants and plant products other than cereals and their products reported in the present book/article[739]: Medicinal plants, FB_2 and FB_3, literature[739]

incidence: 2/7*, conc. range: 2.50–3.80 µg/kg, ∅ conc.: 3.15 µg/kg, sample origin: regular enterprises, China, sample year: unknown, country: China[739], *grasses and leaves

- Co-contamination: 2 sa. co-contaminated with FB_1, FB_2, and FB_3
- Further contamination of plants and plant products other than cereals and their products reported in the present book/article[739]: Medicinal plants, FB_2 and FB_3, literature[739]

incidence: 2/6*, conc. range: 1.25–2.59 µg/kg, ∅ conc.: 1.92 µg/kg, sample origin: regular enterprises, China, sample year: unknown, country: China[739], *flowers

- Co-contamination: 2 sa. co-contaminated with FB_1 and FB_2
- Further contamination of plants and plant products other than cereals and their products reported in the present book/article[739]: Medicinal plants, FB_2 and FB_3, literature[739]

For detailed information see the article.

Fumonisin B_2

incidence: 1/61*, conc.: pr, sample origin: supermarkets and bazaars in Istanbul (capital), Turkey, sample year: unknown, country: Turkey[80]

- Co-contamination: 1 sa. co-contaminated with FB_1 and FB_2
- Further contamination of plants and plant products other than cereals and their products reported in the present book/article[80]: Medicinal plants, FB_1, literature[80]

For detailed information see the article.

incidence: 8/10*, conc. range: 0.68–45.10 µg/kg, ∅ conc.: 10.4 µg/kg, sample origin: regular enterprises, China, sample year: unknown, country: China[739], *seeds

- Co-contamination: 4 sa. co-contaminated with FB_1, FB_2, and FB_3; 4 sa. co-contaminated with FB_1 and FB_2
- Further contamination of plants and plant products other than cereals and their products reported in the present book/article[739]: Medicinal plants, FB_1 and FB_3, literature[739]

incidence: 1/12*, conc.: 0.82 µg/kg, sample origin: regular enterprises, China, sample year: unknown, country: China[739], *rhizomes and roots

- Co-contamination: 1 sa. co-contaminated with FB_1 and FB_2
- Further contamination of plants and plant products other than cereals and their products reported in the present book/article[739]: Medicinal plants, FB_1 and FB_3, literature[739]

incidence: 3/7*, conc. range: 0.61–1.25 µg/kg, ∅ conc.: 0.88 µg/kg, sample origin: regular enterprises, China, sample year: unknown, country: China[739], *grasses and leaves

- Co-contamination: 2 sa. co-contaminated with FB_1, FB_2, and FB_3; 1 sa. contaminated solely with FB_1
- Further contamination of plants and plant products other than cereals and their products reported in the present book/article[739]: Medicinal plants, FB_1 and FB_3, literature[739]

incidence: 2/6*, conc. range: 0.65–0.72 µg/kg, ∅ conc.: 0.69 µg/kg, sample origin: regular enterprises, China, sample year: unknown, country: China[739], *flowers

- Co-contamination: 2 sa. co-contaminated with FB_1 and FB_2
- Further contamination of plants and plant products other than cereals and their products reported in the present book/article[739]: Medicinal plants, FB_1 and FB_3, literature[739]

For detailed information see the article.

Fumonisin B_3

incidence: 4/10*, conc. range: 0.55–3.75 µg/kg, ∅ conc.: 1.87 µg/kg, sample origin: regular enterprises, China, sample year: unknown, country: China[739], *seeds

- Co-contamination: 4 sa. co-contaminated with FB_1, FB_2, and FB_3
- Further contamination of plants and plant products other than cereals and their products reported in the present book/article[739]: Medicinal plants, FB_1 and FB_2, literature[739]

incidence: 0/12*, conc.: no contamination, sample origin: regular enterprises, China, sample year: unknown, country: China[739], *rhizomes and roots

- Co-contamination: not reported
- Further contamination of plants and plant products other than cereals and their products reported in the present book/article[739]: Medicinal plants, FB_1 and FB_2, literature[739]

incidence: 2/7*, conc. range: 0.87–1.08 µg/kg, ∅ conc.: 0.98 µg/kg, sample origin: regular enterprises, China, sample year: unknown, country: China[739], *grasses and leaves

- Co-contamination: 2 sa. co-contaminated with FB_1, FB_2, and FB_3
- Further contamination of plants and plant products other than cereals and their products reported in the present book/article[739]: Medicinal plants, FB_1 and FB_2, literature[739]

incidence: 0/6*, conc.: no contamination, sample origin: regular enterprises, China, sample year: unknown, country: China[739], *flowers

- Co-contamination: not reported
- Further contamination of plants and plant products other than cereals and their products reported in the present book/article[739]: Medicinal plants, FB_1 and FB_2, literature[739]

For detailed information see the article.

Fumonisin/s

incidence: 11/84*, conc. range: 83–100 µg/kg (3 sa.), 101–200 µg/kg (6 sa.), 201–300 µg/kg (1 sa.), 647 µg/kg** (1 sa.), sample origin: herbal stores and a herbal distributor company, Lleida (province), Spain, sample year: unknown, country: Spain[72], *medicinal and aromatic herbs, **licorice root

- Co-contamination: not reported
- Further contamination of plants and plant products other than cereals and their products reported in the present book/article[72]: Medicinal plants, AF/S, CIT, DON, OTA, T-2, and ZEA, literature[72]

For detailed information see the article.

incidence: 7/16*, conc. range: 124.4–4,503.4 µg/kg**, sample origin: different markets, China, sample year: 2009/2010, country: China[82], *moldy medicinal herbs, **FB_1 and FB_2

incidence: 6/80*, conc. range: 49.0–474.3 µg/kg**, sample origin: different markets, China, sample year: 2009/2010, country: China[82], *non-moldy medicinal herbs, **FB_1 and FB_2

- Co-contamination: not reported
- Further contamination of plants and plant products other than cereals and their products reported in the present book/article[82]: Herbs, FB_1 and FB_2, literature[82]

For detailed information see the article.

Type A Trichothecenes

Diacetoxyscirpenol

incidence: 1/23*, conc.: 0.3 µg/kg, sample origin: unknown, sample year: unknown, country: China[68], *rhizomes and roots

- Co-contamination: 1 sa. co-contaminated with CIT, DAS, STG, and VER
- Further contamination of plants and plant products other than cereals and their products reported in the present book/article[68]: Medicinal plants, AFB_1, AFB_2, AFG_1, AFG_2, AFM_1, CIT, CPA, DON, FUS-X, GT, HT-2, MA, MON, NEO, NIV, OTA, OTB, PAT, STG, T-2, and ZEA, literature[68]; see also Medicinal plants, chapter: Further Mycotoxins and Microbial Metabolites

incidence: 3/15*, conc. range: 0.2–0.3 µg/kg, ∅ conc.: 0.27 µg/kg, sample origin: unknown, sample year: unknown, country: China[68], *seeds

- Co-contamination: 1 sa. co-contaminated with AFB_1, AFB_2, CPA, DAS, GT, MA, MON, NIV, STG, and VER; 1 sa. co-contaminated with AFB_1, AFG_1, AFG_2, DAS, MA, PEA, and STG; 1 sa. co-contaminated with AFB_1, AFB_2, DAS, GT, STG, and ZEA
- Further contamination of plants and plant products other than cereals and their products reported in the present book/article[68]: Medicinal plants, AFB_1, AFB_2, AFG_1, AFG_2, AFM_1, CIT, CPA, DON, FUS-X, GT, HT-2, MA, MON, NEO, NIV, OTA, OTB, PAT, STG, T-2, and ZEA, literature[68]; see also Medicinal plants, chapter: Further Mycotoxins and Microbial Metabolites

incidence: 1/11*, conc.: 1.9 µg/kg, sample origin: unknown, sample year: unknown, country: China[68], *grasses and leaves

- Co-contamination: 1 sa. co-contaminated with AFB_1, DAS, OTA, and OTB
- Further contamination of plants and plant products other than cereals and their products reported in the present book/article[68]: Medicinal plants, AFB_1, AFB_2, AFG_1, AFG_2, AFM_1, CIT, CPA, DON, FUS-X, GT, HT-2, MA, MON, NEO, NIV, OTA, OTB, PAT, STG, T-2, and ZEA, literature[68]; see also Medicinal plants, chapter: Further Mycotoxins and Microbial Metabolites

incidence: 1/11*, conc.: 0.2 μg/kg, sample origin: unknown, sample year: unknown, country: China[68], *flowers

- Co-contamination: 1 sa. co-contaminated with AFB_1, AFG_2, CHAE, CPA, DAS, DON3G, PEA, and STG
- Further contamination of plants and plant products other than cereals and their products reported in the present book/article[68]: Medicinal plants, AFB_1, AFB_2, AFG_1, AFG_2, AFM_1, CIT, CPA, DON, FUS-X, GT, HT-2, MA, MON, NEO, NIV, OTA, OTB, PAT, STG, T-2, and ZEA, literature[68]; see also Medicinal plants, chapter: Further Mycotoxins and Microbial Metabolites

For detailed information see the article.

HT-2 Toxin

incidence: 0/23*, conc.: no contamination, sample origin: unknown, sample year: unknown, country: China[68], *rhizomes and roots

- Co-contamination: not reported
- Further contamination of plants and plant products other than cereals and their products reported in the present book/article[68]: Medicinal plants, AFB_1, AFB_2, AFG_1, AFG_2, AFM_1, CIT, CPA, DAS, DON, FUS-X, GT, MON, NEO, NIV, OTA, OTB, PAT, STG, T-2, and ZEA, literature[68]; see also Medicinal plants, chapter: Further Mycotoxins and Microbial Metabolites

incidence: 1/15*, conc.: 2.9 μg/kg, sample origin: unknown, sample year: unknown, country: China[68], *seeds

- Co-contamination: 1 sa. co-contaminated with AFB_1, AFB_2, AFG_1, AFG_2, AFM_1, CHAE, CPA, GT, HT-2, MA, PEA, STG, VER, and ZEA
- Further contamination of plants and plant products other than cereals and their products reported in the present book/article[68]: Medicinal plants, AFB_1, AFB_2, AFG_1, AFG_2, AFM_1, CIT, CPA, DAS, DON, FUS-X, GT, MON, NEO, NIV, OTA, OTB, PAT, STG, T-2, and ZEA, literature[68]; see also Medicinal plants, chapter: Further Mycotoxins and Microbial Metabolites

incidence: 1/11*, conc.: 2.3 μg/kg, sample origin: unknown, sample year: unknown, country: China[68], *grasses and leaves

- Co-contamination: 1 sa. co-contaminated with CIT, HT-2, MA, OTA, and OTB
- Further contamination of plants and plant products other than cereals and their products reported in the present book/article[68]: Medicinal plants, AFB_1, AFB_2, AFG_1, AFG_2, AFM_1, CIT, CPA, DAS, DON, FUS-X, GT, MON, NEO, NIV, OTA, OTB, PAT, STG, T-2, and ZEA, literature[68]; see also Medicinal plants, chapter: Further Mycotoxins and Microbial Metabolites

incidence: 1/11*, conc.: 0.2 μg/kg, sample origin: unknown, sample year: unknown, country: China[68], *flowers

- Co-contamination: 1 sa. co-contaminated with HT-2, OTA, and OTB
- Further contamination of plants and plant products other than cereals and their products reported in the present book/article[68]: Medicinal plants, AFB_1, AFB_2, AFG_1, AFG_2, AFM_1, CIT, CPA, DAS, DON, FUS-X, GT, MON, NEO, NIV, OTA, OTB, PAT, STG, T-2, and ZEA, literature[68]; see also Medicinal plants, chapter: Further Mycotoxins and Microbial Metabolites

For detailed information see the article.

Neosolaniol

incidence: 5/23*, conc. range: 0.2–0.8 μg/kg, Ø conc.: 0.36 μg/kg, sample origin: unknown, sample year: unknown, country: China[68], *rhizomes and roots

- Co-contamination: 1 sa. co-contaminated with AFB_2, AFG_1, AFG_2, MON, MA, NEO, and STG; 1 sa. co-contaminated with AFB_1, AFB_2, AFG_1, AFM_1, CPA, and NEO; 1 sa. co-contaminated with AFG_2, DPDON, NEO, NIV, OTA, and OTB; 1 sa. co-contaminated with AFM_1, CIT, GT, MA, NEO, and STG; 1 sa. co-contaminated with MA, MON, NEO, and T-2
- Further contamination of plants and plant products other than cereals and their products reported in the present book/article[68]: Medicinal plants, AFB_1, AFB_2, AFG_1, AFG_2, AFM_1, CIT, CPA, DAS, DON, FUS-X, GT, HT-2, MON, NIV, OTA, OTB, PAT, STG, T-2, and ZEA, literature[68]; see also Medicinal plants, chapter: Further Mycotoxins and Microbial Metabolites

incidence: 1/15*, conc.: 0.2 μg/kg, sample origin: unknown, sample year: unknown, country: China[68], *seeds

- Co-contamination: 1 sa. co-contaminated with FUS-X, MA, NEO, OTA, OTB, and PAT
- Further contamination of plants and plant products other than cereals and their products reported in the present book/article[68]: Medicinal plants, AFB_1, AFB_2, AFG_1, AFG_2, AFM_1, CIT, CPA, DAS, DON, FUS-X, GT, HT-2, MON, NIV, OTA, OTB, PAT, STG, T-2, and ZEA, literature[68]; see also Medicinal plants, chapter: Further Mycotoxins and Microbial Metabolites

incidence: 0/11*, conc.: no contamination, sample origin: unknown, sample year: unknown, country: China[68], *grasses and leaves

- Co-contamination: not reported
- Further contamination of plants and plant products other than cereals and their products reported in the present book/article[68]: Medicinal plants, AFB_1, AFB_2, AFG_1, AFG_2, AFM_1, CIT, CPA, DAS, DON, FUS-X, GT, HT-2, MON, NIV, OTA, OTB, PAT, STG, T-2, and ZEA, literature[68]; see also Medicinal plants, chapter: Further Mycotoxins and Microbial Metabolites

incidence: 0/11*, conc.: no contamination, sample origin: unknown, sample year: unknown, country: China[68], *flowers

- Co-contamination: not reported
- Further contamination of plants and plant products other than cereals and their products reported in the present book/article[68]: Medicinal plants, AFB_1, AFB_2, AFG_1, AFG_2, AFM_1, CIT, CPA, DAS, DON, FUS-X, GT, HT-2, MON, NIV, OTA, OTB, PAT, STG, T-2, and ZEA, literature[68]; see also Medicinal plants, chapter: Further Mycotoxins and Microbial Metabolites

For detailed information see the article.

T-2 Toxin

incidence: 4/23*, conc. range: 0.2–0.8 μg/kg, Ø conc.: 0.43 μg/kg, sample origin: unknown, sample year: unknown, country: China[68], *rhizomes and roots

- Co-contamination: 1 sa. co-contaminated with AFB_1, CPA, GT, MA, OTB, STG, T-2, and ZEA; 1 sa. co-contaminated with AFB_1, AFB_2, AFG_2, CHAE, NIV, OTB, and T-2; 1 sa. co-contaminated with AFB_1, CHAE, GT, MA, OTA, OTB, and T-2; 1 sa. co-contaminated with MA, MON, NEO, and T-2
- Further contamination of plants and plant products other than cereals and their products reported in the present book/article[68]: Medicinal plants, AFB_1, AFB_2, AFG_1, AFG_2, AFM_1, CIT, CPA, DAS, DON, FUS-X, GT, HT-2, MON, NEO, NIV, OTA, OTB, PAT, STG, and ZEA, literature[68]; see also Medicinal plants, chapter: Further Mycotoxins and Microbial Metabolites

incidence: 0/15*, conc.: no contamination, sample origin: unknown, sample year: unknown, country: China[68], *seeds

- Co-contamination: not reported
- Further contamination of plants and plant products other than cereals and their products reported in the present book/article[68]: Medicinal plants, AFB_1, AFB_2, AFG_1, AFG_2, AFM_1, CIT, CPA, DAS, DON, FUS-X, GT, HT-2, MON, NEO, NIV, OTA, OTB, PAT, STG, and ZEA, literature[68]; see also Medicinal plants, chapter: Further Mycotoxins and Microbial Metabolites

incidence: 0/11*, conc.: no contamination, sample origin: unknown, sample year: unknown, country: China[68], *grasses and leaves

- Co-contamination: not reported
- Further contamination of plants and plant products other than cereals and their products reported in the present book/article[68]: Medicinal plants, AFB_1, AFB_2, AFG_1, AFG_2, AFM_1, CIT, CPA, DAS, DON, FUS-X, GT, HT-2, MON, NEO, NIV, OTA, OTB, PAT, STG, and ZEA, literature[68]; see also Medicinal plants, chapter: Further Mycotoxins and Microbial Metabolites

incidence: 1/11*, conc.: 0.2 μg/kg, sample origin: unknown, sample year: unknown, country: China[68], *flowers

- Co-contamination: 1 sa. co-contaminated with GT, MON, and T-2
- Further contamination of plants and plant products other than cereals and their products reported in the present book/article[68]: Medicinal plants, AFB_1, AFB_2, AFG_1, AFG_2, AFM_1, CIT, CPA, DAS, DON, FUS-X, GT, HT-2, MON, NEO, NIV, OTA, OTB, PAT, STG, and ZEA, literature[68]; see also Medicinal plants, chapter: Further Mycotoxins and Microbial Metabolites

For detailed information see the article.

incidence: 83/84*, conc. range: 0.3–0.9 μg/kg (8 sa.), 1–9 μg/kg (36 sa.), 10–29 μg/kg (25 sa.), 30–49 μg/kg (11 sa.), 50–99 μg/kg (2 sa.), 256.9 μg/kg** (1 sa.), sample origin: herbal stores and a herbal distributor company, Lleida (province), Spain, sample year: unknown, country: Spain[72], *medicinal and aromatic herbs, **ribgrass leaves

- Co-contamination: not reported
- Further contamination of plants and plant products other than cereals and their products reported in the present book/article[72]: Medicinal plants, AF/S, CIT, DON, FB/S, OTA, and ZEA, literature[72]

For detailed information see the article.

incidence: 1/138, conc.: 64.0 μg/kg*, sample origin: markets, China, sample year: unknown, country: China[854], **Fructus hordei germinatus*

- Co-contamination: not reported
- Further contamination of plants and plant products other than cereals and their products reported in the present book/article[854]: not reported

For detailed information see the article.

incidence: 2/17*, conc. range: 0.853–1.10 μg/kg, ∅ conc.: 0.98 μg/kg, sample origin: different provinces, China, sample year: unknown, country: China[873], **Puerariae lobatae radix*

- Co-contamination: not reported
- Further contamination of plants and plant products other than cereals and their products reported in the present book/article[873]: Medicinal plants, AFB_1, literature[873]

Type B Trichothecenes

Deoxynivalenol

incidence: 0/23*, conc.: no contamination, sample origin: unknown, sample year: unknown, country: China[68], *rhizomes and roots

- Co-contamination: not reported
- Further contamination of plants and plant products other than cereals and their products reported in the present book/article[68]: Medicinal plants, AFB_1, AFB_2, AFG_1, AFG_2, AFM_1, CIT, CPA, DAS, FUS-X, GT, HT-2, MA, MON, NEO, NIV, OTA, OTB, PAT, STG, T-2, and ZEA, literature[68]; see also Medicinal plants, chapter: Further Mycotoxins and Microbial Metabolites

incidence: 1/15*, conc.: 25.4 μg/kg, sample origin: unknown, sample year: unknown, country: China[68], *seeds

- Co-contamination: 1 sa. co-contaminated with DON, DPDON, and NIV
- Further contamination of plants and plant products other than cereals and their products reported in the present book/article[68]: Medicinal plants, AFB_1, AFB_2, AFG_1, AFG_2, AFM_1, CIT, CPA, DAS, FUS-X, GT, HT-2, MA, MON, NEO, NIV, OTA, OTB, PAT, STG, T-2, and ZEA, literature[68]; see also Medicinal plants, chapter: Further Mycotoxins and Microbial Metabolites

incidence: 0/11*, conc.: no contamination, sample origin: unknown, sample year: unknown, country: China[68], *grasses and leaves

- Co-contamination: not reported
- Further contamination of plants and plant products other than cereals and their products reported in the present book/article[68]: Medicinal plants, AFB_1, AFB_2, AFG_1, AFG_2, AFM_1, CIT, CPA, DAS, FUS-X, GT, HT-2, MA, MON, NEO, NIV, OTA, OTB, PAT, STG, T-2, and ZEA, literature[68]; see also Medicinal plants, chapter: Further Mycotoxins and Microbial Metabolites

incidence: 0/11*, conc.: no contamination, sample origin: unknown, sample year: unknown, country: China[68], *flowers

- Co-contamination: not reported
- Further contamination of plants and plant products other than cereals and their products reported in the present book/article[68]: Medicinal plants, AFB_1, AFB_2, AFG_1, AFG_2, AFM_1, CIT, CPA, DAS, FUS-X, GT, HT-2, MA, MON, NEO, NIV, OTA, OTB, PAT, STG, T-2, and ZEA, literature[68]; see also Medicinal plants, chapter: Further Mycotoxins and Microbial Metabolites

For detailed information see the article.

incidence: 52/84*, conc. range: 15–39 μg/kg (8 sa.), 40–100 μg/kg (15 sa.), 101–199 μg/kg (13 sa.), 200–300 μg/kg (9 sa.), 301–500 μg/kg (6 sa.), 5,240 μg/kg** (1 sa.), sample origin: herbal stores and a herbal distributor company, Lleida (province), Spain, sample year: unknown, country: Spain[72], *medicinal and aromatic herbs, **bearberry leaves

- Co-contamination: not reported
- Further contamination of plants and plant products other than cereals and their products reported in the present book/article[72]: Medicinal plants, AF/S, CIT, FB/S, OTA, T-2, and ZEA, literature[72]

For detailed information see the article.

incidence: 1/54*, conc.: 4.4 µg/kg, sample origin: regulated enterprises, China, sample year: unknown, country: China[737], *rhizomes and roots

- Co-contamination: not reported
- Further contamination of plants and plant products other than cereals and their products reported in the present book/ article[737]: Medicinal plants, NIV, literature[737]

incidence: 1/20*, conc.: 26.6 µg/kg, sample origin: regulated enterprises, China, sample year: unknown, country: China[737], *seeds

- Co-contamination: 1 sa. co-contaminated with DON and NIV
- Further contamination of plants and plant products other than cereals and their products reported in the present book/ article[737]: Medicinal plants, NIV, literature[737]

incidence: 2/6*, conc. range: 3.5–5.7 µg/kg, ∅ conc.: 4.6 µg/kg, sample origin: regulated enterprises, China, sample year: unknown, country: China[737], *flowers

- Co-contamination: 1 sa. co-contaminated with DON and NIV; 1 sa. contaminated solely with DON
- Further contamination of plants and plant products other than cereals and their products reported in the present book/ article[737]: Medicinal plants, NIV, literature[737]

incidence: 0/8*, conc.: no contamination, sample origin: regulated enterprises, China, sample year: unknown, country: China[737], *grasses and leaves

- Co-contamination: not reported
- Further contamination of plants and plant products other than cereals and their products reported in the present book/ article[737]: Medicinal plants, NIV, literature[737]

For detailed information see the article.

incidence: 3/58*, conc. range: 17.2–50.5 µg/kg, ∅ conc.: 36.9 µg/kg, sample origin: market in China, sample year: unknown, country: China[894], *medicinal herbs and related products

- Co-contamination: not reported
- Further contamination of plants and plant products other than cereals and their products reported in the present book/ article[894]: not reported

For detailed information see the article.

Fusarenon-X (4-Acetylnivalenol)

incidence: 0/23*, conc.: no contamination, sample origin: unknown, sample year: unknown, country: China[68], *rhizomes and roots

- Co-contamination: not reported
- Further contamination of plants and plant products other than cereals and their products reported in the present book/ article[68]: Medicinal plants, AFB_1, AFB_2, AFG_1, AFG_2, AFM_1, CIT, CPA, DAS, DON, GT, HT-2, MA, MON, NEO, NIV, OTA, OTB, PAT, STG, T-2, and ZEA, literature[68]; see also Medicinal plants, chapter: Further Mycotoxins and Microbial Metabolites

incidence: 1/15*, conc.: 2.0 µg/kg, sample origin: unknown, sample year: unknown, country: China[68], *seeds

- Co-contamination: 1 sa. co-contaminated with FUS-X, MA, NEO, OTA, OTB, PAT, and STG
- Further contamination of plants and plant products other than cereals and their products reported in the present book/ article[68]: Medicinal plants, AFB_1, AFB_2, AFG_1, AFG_2, AFM_1, CIT, CPA, DAS, DON, GT, HT-2, MA, MON, NEO, NIV, OTA, OTB, PAT, STG, T-2, and ZEA, literature[68]; see also Medicinal plants, chapter: Further Mycotoxins and Microbial Metabolites

incidence: 0/11*, conc.: no contamination, sample origin: unknown, sample year: unknown, country: China[68], *grasses and leaves

- Co-contamination: not reported
- Further contamination of plants and plant products other than cereals and their products reported in the present book/ article[68]: Medicinal plants, AFB_1, AFB_2, AFG_1, AFG_2, AFM_1, CIT, CPA, DAS, DON, GT, HT-2, MA, MON, NEO, NIV, OTA, OTB, PAT, STG, T-2, and ZEA, literature[68]; see also Medicinal plants, chapter: Further Mycotoxins and Microbial Metabolites

incidence: 0/11*, conc.: no contamination, sample origin: unknown, sample year: unknown, country: China[68], *flowers

- Co-contamination: not reported
- Further contamination of plants and plant products other than cereals and their products reported in the present book/ article[68]: Medicinal plants, AFB_1, AFB_2, AFG_1, AFG_2, AFM_1, CIT, CPA, DAS, DON, GT, HT-2, MA, MON, NEO, NIV, OTA, OTB, PAT, STG, T-2, and ZEA, literature[68]; see also Medicinal plants, chapter: Further Mycotoxins and Microbial Metabolites

For detailed information see the article.

Nivalenol

incidence: 2/23*, conc. range: 2.9–4.3 µg/kg, ∅ conc.: 3.6 µg/kg, sample origin: unknown, sample year: unknown, country: China[68], *rhizomes and roots

- Co-contamination: 1 sa. co-contaminated with AFB_1, AFB_2, AFG_2, CHAE, NIV, OTB, and T-2; 1 sa. co-contaminated with AFG_2, DPDON, NEO, NIV, OTA, and OTB
- Further contamination of plants and plant products other than cereals and their products reported in the present book/ article[68]: Medicinal plants, AFB_1, AFB_2, AFG_1, AFG_2, AFM_1, CIT, CPA, DAS, DON, FUS-X, GT, HT-2, MON, NEO, OTA, OTB, PAT, STG, T-2, and ZEA, literature[68]; see also Medicinal plants, chapter: Further Mycotoxins and Microbial Metabolites

incidence: 2/15*, conc. range: 3.7–7.9 µg/kg, ∅ conc.: 5.8 µg/kg, sample origin: unknown, sample year: unknown, country: China[68], *seeds

- Co-contamination: 1 sa. co-contaminated with AFB_1, AFB_2, CPA, DAS, GT, MA, MON, NIV, STG, and VER; 1 sa. co-contaminated with DON, DPDON, and NIV
- Further contamination of plants and plant products other than cereals and their products reported in the present book/article[68]: Medicinal plants, AFB_1, AFB_2, AFG_1, AFG_2, AFM_1, CIT, CPA, DAS, DON, FUS-X, GT, HT-2, MON, NEO, OTA, OTB, PAT, STG, T-2, and ZEA, literature[68]; see also Medicinal plants, chapter: Further Mycotoxins and Microbial Metabolites

incidence: 2/11*, conc. range: 2.1–5.8 µg/kg, ∅ conc.: 3.95 µg/kg, sample origin: unknown, sample year: unknown, country: China[68], *grasses and leaves

- Co-contamination: 1 sa. co-contaminated with AFB_1, NIV, OTA, and PEA; 1 sa. co-contaminated with CPA, NIV, and STG
- Further contamination of plants and plant products other than cereals and their products reported in the present book/ article[68]: Medicinal plants, AFB_1, AFB_2, AFG_1, AFG_2, AFM_1, CIT, CPA, DAS, DON, FUS-X, GT, HT-2, MON, NEO, OTA, OTB, PAT, STG, T-2, and ZEA, literature[68]; see also Medicinal plants, chapter: Further Mycotoxins and Microbial Metabolites

incidence: 0/11*, conc.: no contamination, sample origin: unknown, sample year: unknown, country: China[68], *flowers

- Co-contamination: not reported
- Further contamination of plants and plant products other than cereals and their products reported in the present book/article[68]: Medicinal plants, AFB_1, AFB_2, AFG_1, AFG_2, AFM_1, CIT, CPA, DAS, DON, FUS-X, GT, HT-2, MON, NEO, OTA, OTB, PAT, STG, T-2, and ZEA, literature[68]; see also Medicinal plants, chapter: Further Mycotoxins and Microbial Metabolites

For detailed information see the article.

incidence: 2/54*, conc. range: 4.3–4.8 µg/kg, ∅ conc.: 4.55 µg/kg, sample origin: regulated enterprises, China, sample year: unknown, country: China[737], *rhizomes and roots

- Co-contamination: not reported
- Further contamination of plants and plant products other than cereals and their products reported in the present book/article[737]: Medicinal plants, DON, literature[737]

incidence: 2/20*, conc. range: 2.9–3.7 µg/kg, ∅ conc.: 3.3 µg/kg, sample origin: regulated enterprises, China, sample year: unknown, country: China[737], *seeds

- Co-contamination: 1 sa. co-contaminated with DON and NIV; 1 sa. contaminated solely with NIV
- Further contamination of plants and plant products other than cereals and their products reported in the present book/article[737]: Medicinal plants, DON, literature[737]

incidence: 2/6*, conc. range: 3.5–6.6 µg/kg, ∅ conc.: 5.05 µg/kg, sample origin: regulated enterprises, China, sample year: unknown, country: China[737], *flowers

- Co-contamination: 1 sa. co-contaminated with DON and NIV; 1 sa. contaminated solely with NIV
- Further contamination of plants and plant products other than cereals and their products reported in the present book/article[737]: Medicinal plants, DON, literature[737]

incidence: 2/8*, conc. range: 5.9–8.8 µg/kg, ∅ conc.: 7.35 µg/kg, sample origin: regulated enterprises, China, sample year: unknown, country: China[737], *grasses and leaves

- Co-contamination: not reported
- Further contamination of plants and plant products other than cereals and their products reported in the present book/article[737]: Medicinal plants, DON, literature[737]

For detailed information see the article.

Other ***Fusarium*** Toxins

Moniliformin

incidence: 3/23*, conc. range: 2.2–3.8 µg/kg, ∅ conc.: 3.87 µg/kg, sample origin: unknown, sample year: unknown, country: China[68], *rhizomes and roots

- Co-contamination: 1 sa. co-contaminated with AFB_2, AFG_1, AFG_2, MON, MA, NEO, and STG; 1 sa. co-contaminated with CIT, DON3G, MON, PAT, and STG; 1 sa. co-contaminated with MA, MON, NEO, and T-2
- Further contamination of plants and plant products other than cereals and their products reported in the present book/article[68]: Medicinal plants, AFB_1, AFB_2, AFG_1, AFG_2, AFM_1, CIT, CPA, DAS, DON, FUS-X, GT, HT-2, MA, NEO, NIV, OTA, OTB, PAT, STG, T-2, and ZEA, literature[68]; see also Medicinal plants, chapter: Further Mycotoxins and Microbial Metabolites

incidence: 1/15*, conc.: 2.1 µg/kg, sample origin: unknown, sample year: unknown, country: China[68], *seeds

- Co-contamination: 1 sa. co-contaminated with AFB_1, AFB_2, CPA, DAS, GT, MA, MON, NIV, STG, and VER
- Further contamination of plants and plant products other than cereals and their products reported in the present book/article[68]: Medicinal plants, AFB_1, AFB_2, AFG_1, AFG_2, AFM_1, CIT, CPA, DAS, DON, FUS-X, GT, HT-2, MA, NEO, NIV, OTA, OTB, PAT, STG, T-2, and ZEA, literature[68]; see also Medicinal plants, chapter: Further Mycotoxins and Microbial Metabolites

incidence: 1/11*, conc.: 2.7 µg/kg, sample origin: unknown, sample year: unknown, country: China[68], *grasses and leaves

- Co-contamination: 1 sa. co-contaminated with CPA, MA, MON, and STG
- Further contamination of plants and plant products other than cereals and their products reported in the present book/article[68]: Medicinal plants, AFB_1, AFB_2, AFG_1, AFG_2, AFM_1, CIT, CPA, DAS, DON, FUS-X, GT, HT-2, MA, NEO, NIV, OTA, OTB, PAT, STG, T-2, and ZEA, literature[68]; see also Medicinal plants, chapter: Further Mycotoxins and Microbial Metabolites

incidence: 2/11*, conc. range: 2.3–2.9 µg/kg, ∅ conc.: 2.6 µg/kg, sample origin: unknown, sample year: unknown, country: China[68], *flowers

- Co-contamination: 1 sa. co-contaminated with GT, MON, and T-2; 1 sa. co-contaminated with MON and STG
- Further contamination of plants and plant products other than cereals and their products reported in the present book/article[68]: Medicinal plants, AFB_1, AFB_2, AFG_1, AFG_2, AFM_1, CIT, CPA, DAS, DON, FUS-X, GT, HT-2, MA, NEO, NIV, OTA, OTB, PAT, STG, T-2, and ZEA, literature[68]; see also Medicinal plants, chapter: Further Mycotoxins and Microbial Metabolites

For detailed information see the article.

Zearalenone

incidence: 0/10*, conc.: no contamination, sample origin: store-houses (private concerns) in Bihar (state), India, sample year: unknown, country: India[63], **Glycyrrhiza glabra* Linn.

incidence: 0/10*, conc.: no contamination, sample origin: store-houses (private concerns) in Bihar (state), India, sample year: unknown, country: India[63], **Holarrhena antidysenterica* Wall

incidence: 0/10*, conc.: no contamination, sample origin: store-houses (private concerns) in Bihar (state), India, sample year: unknown, country: India[63], **Ichnocarpus frutescens* (Linn.)

incidence: 2/10*, ∅ conc.: 80 µg/kg, sample origin: store-houses (private concerns) in Bihar (state), India sample year: unknown, country: India[63], **Strychnos nux-vomica* Linn.

incidence: 2/10*, ∅ conc.: 20 µg/kg, sample origin: store-houses (private concerns) in Bihar (state), India, sample year: unknown, country: India[63], **Tribulus terrestris* Linn.

- Co-contamination: not reported
- Further contamination of plants and plant products other than cereals and their products reported in the present book/article[63]: Medicinal plants, AFB_1, CIT, and OTA, literature[63]

For detailed information see the article.

incidence: 5/175*, conc. range: 20–110 µg/kg, sample origin: drug centers of Bhagalpur (city) and store houses in southern Bihar (state), India, sample year: January–December 1990, country: India[65], *crude and finished herbal drugs

- Co-contamination: not reported
- Further contamination of plants and plant products other than cereals and their products reported in the present book/ article[65]: Medicinal plants, AFB_1, CIT, and OTA, literature[65]

For detailed information see the article.

incidence: 2/23*, conc. range: 2.1 µg/kg, ∅ conc.: 2.1 µg/kg, sample origin: unknown, sample year: unknown, country: China[68], *rhizomes and roots

- Co-contamination: 1 sa. co-contaminated with AFB_1, CPA, GT, MA, OTB, STG, T-2, and ZEA; 1 sa. co-contaminated with CHAE, STG, and ZEA
- Further contamination of plants and plant products other than cereals and their products reported in the present book/ article[68]: Medicinal plants, AFB_1, AFB_2, AFG_1, AFG_2, AFM_1, CIT, CPA, DAS, DON, FUS-X, GT, HT-2, MON, NEO, NIV, OTA, OTB, PAT, STG, and T-2, literature[68]; see also Medicinal plants, chapter: Further Mycotoxins and Microbial Metabolites

incidence: 4/15*, conc. range: 2.1–10.3 µg/kg, ∅ conc.: 4.88 µg/kg, sample origin: unknown, sample year: unknown, country: China[68], *seeds

- Co-contamination: 1 sa. co-contaminated with AFB_1, AFB_2, AFG_1, AFG_2, AFM_1, CHAE, CPA, GT, HT-2, MA, PEA, STG, VER, and ZEA; 1 sa. co-contaminated with AFB_1, AFB_2, DAS, GT, STG, and ZEA; 1 sa. co-contaminated with AFB_1, CHAE, STG, ZEA, α-ZEL, and ZEAA; 1 sa. co-contaminated with OTA, ZEA, α-ZEL, ß-ZEL, and ZEAA
- Further contamination of plants and plant products other than cereals and their products reported in the present book/ article[68]: Medicinal plants, AFB_1, AFB_2, AFG_1, AFG_2, AFM_1, CIT, CPA, DAS, DON, FUS-X, GT, HT-2, MON, NEO, NIV, OTA, OTB, PAT, STG, and T-2, literature[68]; see also Medicinal plants, chapter: Further Mycotoxins and Microbial Metabolites

incidence: 1/11*, conc.: 2.3 µg/kg, sample origin: unknown, sample year: unknown, country: China[68], *grasses and leaves

- Co-contamination: 1 sa. co-contaminated with DON3G, MA, STG, and ZEA
- Further contamination of plants and plant products other than cereals and their products reported in the present book/article[68]: Medicinal plants, AFB_1, AFB_2, AFG_1, AFG_2, AFM_1, CIT, CPA, DAS, DON, FUS-X, GT, HT-2, MON, NEO, NIV, OTA, OTB, PAT, STG, and T-2, literature[68]; see also Medicinal plants, chapter: Further Mycotoxins and Microbial Metabolites

incidence: 0/11*, conc.: no contamination, sample origin: unknown, sample year: unknown, country: China[68], *flowers

- Co-contamination: not reported
- Further contamination of plants and plant products other than cereals and their products reported in the present book/ article[68]: Medicinal plants, AFB_1, AFB_2, AFG_1, AFG_2, AFM_1, CIT, CPA, DAS, DON, FUS-X, GT, HT-2, MON, NEO, NIV, OTA, OTB, PAT, STG, and T-2, literature[68]; see also Medicinal plants, chapter: Further Mycotoxins and Microbial Metabolites

For detailed information see the article.

incidence: 82/84*, conc. range: 0.3–5 µg/kg (40 sa.), 5.1–10 µg/kg (22 sa.), 10.1–15 µg/kg (12 sa.), 15.1–20 µg/kg (2 sa.), ≤44.1 µg/kg** (6 sa.), sample origin: herbal stores and a herbal distributor company, Lleida (province), Spain, sample year: unknown, country: Spain[72], *medicinal and aromatic herbs, **frangula bark

- Co-contamination: not reported
- Further contamination of plants and plant products other than cereals and their products reported in the present book/ article[72]: Medicinal plants, AF/S, CIT, DON, FB/S, OTA, and T-2, literature[72]

For detailed information see the article.

incidence: 2/10*, conc. range: 1.1–2.1 µg/kg, ∅ conc.: 1.6 µg/kg, sample origin: regulated enterprises, China, sample year: unknown, country: China[740], *rhizomes and roots

- Co-contamination: not reported
- Further contamination of plants and plant products other than cereals and their products reported in the present book/ article[740]: see Medicinal plants, chapter: Further Mycotoxins and Microbial Metabolites

incidence: 4/7*, conc. range: 1.7–10.3 µg/kg, ∅ conc.: 4.83 µg/kg, sample origin: regulated enterprises, China, sample year: unknown, country: China[740], *seeds

- Co-contamination: 1 sa. co-contaminated with ZEA, ZEAA, α-ZEL, and ß-ZEL; 1 sa. co-contaminated with ZEA, ZEAA, and α-ZEL; 2 sa. contaminated solely with ZEA
- Further contamination of plants and plant products other than cereals and their products reported in the present book/ article[740]: see Medicinal plants, chapter: Further Mycotoxins and Microbial Metabolites

For detailed information see the article.

incidence: 1/24*, conc.: 7.09 µg/kg, sample origin: commercially available, China, sample year: unknown, country: China[778], *12 medicinal and 12 edible sa. of *Areca catechu* L.

- Co-contamination: not reported
- Further contamination of plants and plant products other than cereals and their products reported in the present book/ article[778]: Medicinal plants, AFB_2, literature[778]

incidence: 9*/7**, conc. range: 18.7–211.4 µg/kg, ∅ conc.: 92.1 µg/kg, sample origin: Zhangshu (city) in Jiangxi (province), Zhenjian (city) in Jiangsu (province), Zunyi (city in Guizhou (province), and Beijing (capital), China, sample year: unknown, country: China/Italy[898], *coix seeds, **medicinal herbs and related products

- Co-contamination: no reported
- Further contamination of plants and plant products other than cereals and their products reported in the present book/ article[898]: no reported

For detailed information see the article.

Penicillium Toxins

Mycophenolic Acid

incidence: 10/23*, conc. range: 0.2–22.7 µg/kg, ∅ conc.: 5.05 µg/kg, sample origin: unknown, sample year: unknown, country: China[68], *rhizomes and roots

- Co-contamination: 1 sa. co-contaminated with AFB_1, AFB_2, AFM_1, CHAE, CIT, CPA, GT, MA, OTB, STG, and VER; 1 sa. co-contaminated with AFB_1, AFB_2, AFG_1, AFG_2, AFM_1, CHAE, MA, and OTA; 1 sa. co-contaminated with AFB_1, CPA, GT, MA, OTB, STG, T-2, and ZEA; 1 sa. co-contaminated with AFB_1, CHAE, GT, MA, OTA, OTB, and T-2; 1 sa. co-contaminated with AFB_2, AFG_1, AFG_2, MON, MA, NEO, and STG; 1 sa. co-contaminated with AFM_1, CIT, GT, MA, NEO, and STG; 1 sa. co-contaminated with CPA, GT, MA, STG, and VER; 1 sa. co-contaminated with CHAE, MA, and STG; 1 sa. co-contaminated with MA, MON, NEO, and T-2; 1 sa. co-contaminated with GT, MA, and OTA

- Further contamination of plants and plant products other than cereals and their products reported in the present book/article[68]: Medicinal plants, AFB_1, AFB_2, AFG_1, AFG_2, AFM_1, CIT, CPA, DAS, DON, FUS-X, GT, HT-2, MON, NEO, NIV, OTA, OTB, PAT, STG, T-2, and ZEA, literature[68]; see also Medicinal plants, chapter: Further Mycotoxins and Microbial Metabolites

incidence: 9/15*, conc. range: 0.2–10.8 µg/kg, ∅ conc.: 3.6 µg/kg, sample origin: unknown, sample year: unknown, country: China[68], *seeds

- Co-contamination: 1 sa. co-contaminated with AFB_1, AFB_2, AFG_1, AFG_2, AFM_1, CHAE, CPA, GT, HT-2, MA, PEA, STG, VER, and ZEA; 1 sa. co-contaminated with AFB_1, AFB_2, CPA, DAS, GT, MA, MON, NIV, STG, and VER; 1 sa. co-contaminated with AFB_1, AFG_1, AFG_2, DAS, MA, PEA, and STG; 1 sa. co-contaminated with FUS-X, MA, NEO, OTA, OTB, PAT, and STG; 1 sa. co-contaminated with AFB_1, AFB_2, CHAE, CPA, DPDON, and MA; 1 sa. co-contaminated with AFB_2, DON3G, MA, STG, and VER; 1 sa. co-contaminated with AFG_1, CHAE, MA, OTA, and OTB; 1 sa. co-contaminated with AFB_1, AFB_2, and MA; 1 sa. co-contaminated with MA and OTA
- Further contamination of plants and plant products other than cereals and their products reported in the present book/article[68]: Medicinal plants, AFB_1, AFB_2, AFG_1, AFG_2, AFM_1, CIT, CPA, DAS, DON, FUS-X, GT, HT-2, MON, NEO, NIV, OTA, OTB, PAT, STG, T-2, and ZEA, literature[68]; see also Medicinal plants, chapter: Further Mycotoxins and Microbial Metabolites

incidence: 6/11*, conc. range: 0.8–8.1 µg/kg, ∅ conc.: 3.35 µg/kg, sample origin: unknown, sample year: unknown, country: China[68], *grasses and leaves

- Co-contamination: 1 sa. co-contaminated with AFB_1, AFG_1, GT, MA, OTB, PEA, and STG; 1 sa. co-contaminated with AFB_2, CPA, MA, OTA, PAT, and STG; 1 sa. co-contaminated with CIT, HT-2, MA, OTA, and OTB; 1 sa. co-contaminated with CPA, MA, MON, and STG; 1 sa. co-contaminated with DON3G, MA, STG, and ZEA; 1 sa. co-contaminated with MA, PEA, and STG
- Further contamination of plants and plant products other than cereals and their products reported in the present book/article[68]: Medicinal plants, AFB_1, AFB_2, AFG_1, AFG_2, AFM_1, CIT, CPA, DAS, DON, FUS-X, GT, HT-2, MON, NEO, NIV, OTA, OTB, PAT, STG, T-2, and ZEA, literature[68]; see also Medicinal plants, chapter: Further Mycotoxins and Microbial Metabolites

incidence: 0/11*, conc.: no contamination, sample origin: unknown, sample year: unknown, country: China[68], *flowers

- Co-contamination: not reported
- Further contamination of plants and plant products other than cereals and their products reported in the present book/article[68]: Medicinal plants, AFB_1, AFB_2, AFG_1, AFG_2, AFM_1, CIT, CPA, DAS, DON, FUS-X, GT, HT-2, MON, NEO, NIV, OTA, OTB, PAT, STG, T-2, and ZEA, literature[68]; see also Medicinal plants, chapter: Further Mycotoxins and Microbial Metabolites

For detailed information see the article.
see also **Food** and **Fruits (hawthorn)**

Medicinal seeds see **Medicinal plants**

Meju may contain the following mycotoxins:

Aspergillus Toxins

Aflatoxin B_1

incidence: 26/60, conc. range: 2.1–23.5 µg/kg, ∅ conc.: 6.9 µg/kg, sample origin: large supermarkets and small neighbourhood grocery shops in Seoul (capital), Korea, sample year: March 1995–April 1996, country: Korea[528]

- Co-contamination: not reported
- Further contamination of plants and plant products other than cereals and their products reported in the present book/article[528]: not reported

Meju is a crushed fermented soybean cake.

Melon seed products see **Products (melon seed product)**

Melon seeds see **Seeds (melon)**

Melons see **Fruits (melon)**

Merlot see **Wine (red)**

Milk thistle may contain the following mycotoxins:

Fusarium Toxins

Type A Trichothecenes

HT-2 Toxin

incidence: 2*/7**, conc. range: 826.9–943.7 µg/kg, ∅ conc.: 885 µg/kg, sample origin: local markets from Granada (city/province), Spain, sample year: 2013, country: Spain[675], *purchased from a street vendor, **commercial *Silybum marianum*

- Co-contamination: 1 sa. co-contaminated with HT-2, T2, and ZEA; 1 sa. co-contaminated with HT-2 and T-2
- Further contamination of plants and plant products other than cereals and their products reported in the present book/article[675]: Milk thistle, T-2 and ZEA, literature[675]

T-2 Toxin

incidence: 2*/7**, conc. range: 363.0–453.9 µg/kg, ∅ conc.: 409 µg/kg, sample origin: local markets from Granada (city/province), Spain, sample year: 2013, country: Spain[675], *purchased from a street vendor, **commercial *Silybum marianum*

- Co-contamination: 1 sa. co-contaminated with HT-2, T2, and ZEA; 1 sa. co-contaminated with HT-2 and T-2
- Further contamination of plants and plant products other than cereals and their products reported in the present book/article[675]: Milk thistle, HT-2 and ZEA, literature[675]

Other ***Fusarium*** Toxins

Zearalenone

incidence: 1*/7**, conc.: <LOQ, sample origin: local markets from Granada (city/province), Spain, sample year: 2013, country: Spain[675], *purchased from a street vendor, **commercial *Silybum marianum*

- Co-contamination: 1 sa. co-contaminated with HT-2, T2, and ZEA
- Further contamination of plants and plant products other than cereals and their products reported in the present book/article[675]: Milk thistle, HT-2 and T-2, literature[675]

Mistelle see **Wine, miscellaneous**

Moscatel see **Must** and **Wine, miscellaneous**

Muhallebi may contain the following mycotoxins:

Aspergillus **Toxins**

Aflatoxin/s

incidence: 3/4, conc. range: 0.122–1.622 µg/kg*, Ø conc.: 0.628 µg/kg*, sample origin: groceries and supermarkets, Turkey, sample year: 2003, country: Turkey[119], *AFS total

- Co-contamination: not reported
- Further contamination of plants and plant products other than cereals and their products reported in the present book/article[119]: Asure, Beverages, Cocoa, Cream, Ice cream, Jello, Keskul, Pudding, Salep, Sauces (chocolate sauce), and Soups, AF/S, literature[119]

Muhallebi is a creamy pudding. Ingredients were sugar, potato starch, wheat flour, rice flour, natural identical mastic flavor, and vanilla.

Mulberries see **Fruits, dried**

Muscatel see **Wine, miscellaneous**

Must may contain the following mycotoxins:

Aspergillus **Toxins**

Aflatoxin B_1

incidence: 19/47*, conc. range: 0.01–0.08 µg/l (17 sa.), 0.17 and 0.46 µg/l (2 sa.), sample origin: Bekaa valley, Lebanon, sample year: July-October 2005, country: Lebanon/France[533], *handmade must

- Co-contamination: not reported
- Further contamination of plants and plant products other than cereals and their products reported in the present book/article[533]: not reported

Aspergillus **and** ***Penicillium*** **Toxins**

Ochratoxin A

incidence: 20?/22*, conc. range: 0.10–0.50 µg/l (1 sa.), 0.50–1.00 µg/l (4 sa.), 1.00–2.00 µg/l (1 sa.), >2.00–27.29 µg/l (12 sa.), sample origin: unknown, sample year: unknown, country: Spain[158], *fortified must/early stopped fermentation

- Co-contamination: not reported
- Further contamination of plants and plant products other than cereals and their products reported in the present book/article[158]: Berries (grape), Wine, miscellaneous, OTA, literature[158]

incidence: 2/20*, conc. range: 0.06–0.18 µg/l, Ø conc.: 0.13 µg/l, sample origin: supermarkets and small shops, Spain, sample year: 2002, country: Spain[508], *10 white (2 sa. contaminated) and 10 red must sa. (not contaminated)

- Co-contamination: not reported
- Further contamination of plants and plant products other than cereals and their products reported in the present book/article[508]: Wine (red), Wine (white), Wine, miscellaneous, OTA, literature[508]

incidence: 2/8*, conc. range: 0.84–2 µg/l**, Ø conc.: 1.42 µg/l**, sample origin: Andalusia (autonomous community), Spain, sample year: probably 2005, country: Spain[520], *included Matilde, Moscatel, Palomino**, and Sauvignon grape must

- Co-contamination: not reported
- Further contamination of plants and plant products other than cereals and their products reported in the present book/article[520]: Wine, miscellaneous, OTA, literature[520]

incidence: 6/40*, conc. range: 0.071–0.813 µg/l, Ø conc.: 0.255 µg/l, sample origin: important wine-producing regions, Spain, sample year: May–September 2001, country: Spain[534], *white (Chardonnay and Macabeo) and red handmade must (Cabernet Sauvignon, Tempranillo, and Garnacha)

- Co-contamination: not reported
- Further contamination of plants and plant products other than cereals and their products reported in the present book/article[534]: not reported

incidence: 6/6*, conc. range: 0.019–0.050 µg/l, Ø conc.: 0.0315 µg/l, sample origin: coastal Croatia, sample year: 2007, country: Croatia[535], *red must (Crljenak, Dobrinčić, Glavinuša, Ninčuša, Plavac Mali, and Zinfandel)

- Co-contamination: not reported
- Further contamination of plants and plant products other than cereals and their products reported in the present book/article[535]: Wine, OTA, literature[535]

incidence: 27/47*, conc. range: 0.01–0.043 µg/l (15 sa.), >0.05–0.075 µg/l (5 sa.), >0.075–0.221 µg/l (7 sa.), sample origin: vineyards in the Bekaa valley, Lebanon, sample year: July–October 2004, country: Lebanon/France[536], *handmade must

- Co-contamination: not reported
- Further contamination of plants and plant products other than cereals and their products reported in the present book/article[536]: Wine (red), OTA, literature[536]

incidence: 0/4* **, conc.: no contamination, sample origin: vineyard in Cap Bon (peninsula), northeastern Tunisia, sample year: 2004, country: Tunisia/Spain[538], *grape variety: Italia, **sa. taken at early veraison

incidence: 0/4* **, conc.: no contamination, sample origin: vineyard in Cap Bon (peninsula), northeastern Tunisia, sample year: 2004, country: Tunisia/Spain[538], *grape variety: Carignan, **sa. taken at early veraison

incidence: 1/4* **, conc.: 0.59 µg/l, sample origin: vineyard in Cap Bon (peninsula), northeastern Tunisia, sample year: 2004, country: Tunisia/Spain[538], *grape variety: Italia, **sa. taken 2 weeks after veraison

incidence: 2/4* **, conc. range: 0.69–0.94 µg/l, Ø conc.: 0.82 µg/l, sample origin: vineyard in Cap Bon (peninsula), northeastern Tunisia, sample year: 2004, country: Tunisia/Spain[538], *grape variety: Carignan, **sa. taken 2 weeks after veraison

incidence: 2/4* **, conc. range: 0.69–0.73 µg/l, Ø conc.: 0.71 µg/l, sample origin: vineyard in Cap Bon (peninsula), northeastern Tunisia, sample year: 2004, country: Tunisia/Spain[538], *grape variety: Italia, **sa. taken at maturity

incidence: 4/4* **, conc. range: 0.59–2.57 µg/l, Ø conc.: 1.27 µg/l, sample origin: vineyard in Cap Bon (peninsula), northeastern Tunisia, sample year: 2004, country: Tunisia/Spain[538], *grape variety: Carignan, **sa. taken at maturity

- Co-contamination: not reported
- Further contamination of plants and plant products other than cereals and their products reported in the present book/article[538]: not reported

incidence: 11/11, conc. range: 0.003–0.250 µg/l, Ø conc.: 0.043 µg/l, sample origin: Cosenza (province), Calabria (region), southern Italy, sample year: 2007, country: Italy[539] incidence: 8/11*, conc. range: 0.003–0.130 µg/l, ∅ conc.: 0.0345 µg/l, sample origin: Cosenza (province), Calabria (region), southern Italy, sample year: 2007, country: Italy[539], *must after malolactic fermentation

- Co-contamination: not reported
- Further contamination of plants and plant products other than cereals and their products reported in the present book/article[539]: Wine (red), OTA, literature[539]

incidence: 8/11*, conc. range: <0.010–0.461 µg/l, sample origin: Aude (department), southern France, sample year: unknown, country: France[541], *must of red wine

- Co-contamination: not reported
- Further contamination of plants and plant products other than cereals and their products reported in the present book/article[541]: not reported

incidence: 7/8*, conc. range: 0.5–2.5 µg/l, ∅ conc.: 1.03 µg/l, sample origin: wineries, northern Italy, sample year: unknown, country: Italy[674], *fermentable grape must

incidence: 8/8*, conc. range: 0.2–1.3 µg/l, ∅ conc.: 0.71 µg/l, sample origin: wineries, northern Italy, sample year: unknown, country: Italy[674], *mute grape must

- Co-contamination: not reported
- Further contamination of plants and plant products other than cereals and their products reported in the present book/article[674]: Wine (red), Wine (white), OTA, literature[674]

incidence: 3/30*, conc. range: ≤0.1–0.16 µg/l, ∅ conc.: 0.12 µg/l, sample origin: Mendoza (province), Argentina, sample year: harvest season 2008/2009, country: Argentina[816], *red must

incidence: 0/2*, conc.: no contamination, sample origin: Chubut (province), Argentina, sample year: harvest season 2008/2009, country: Argentina[816], *red must

- Co-contamination: not reported
- Further contamination of plants and plant products other than cereals and their products reported in the present book/article[816]: Vine fruits, dried, Wine (red), OTA, literature[816]

Fusarium Toxins

Fumonisin B_2

incidence: 2/12*, conc. range: 10–400 µg/l, ∅ conc.: 205 µg/l, sample origin: Apulian wine cellars, Apulia (region), Italy, sample year: 2007, country: Italy[543], *must of red wine for production of "Primitivo"

- Co-contamination: not reported
- Further contamination of plants and plant products other than cereals and their products reported in the present book/article[543]: not reported

Mustard see Field mustard and Spices (mustard)

Nectars (apple nectar) may contain the following mycotoxins:

Aspergillus* and *Penicillium Toxins

Patulin

incidence: 1/63, conc.: 3–7 µg/l, sample origin: 22 cities in São Paulo (state), Brazil, sample year: 2004/2005, country: Brazil[478]

- Co-contamination: not reported
- Further contamination of plants and plant products other than cereals and their products reported in the present book/article[478]: Juices (apple juice), PAT, literature[478]

Nectars (cranberry nectar) may contain the following mycotoxins:

Alternaria Toxins

Alternariol

incidence: 1/1, conc.: 5.6 µg/l, sample origin: unknown, sample year: unknown, country: Canada[440]

- Co-contamination: 1 sa. co-contaminated with AME and AOH
- Further contamination of plants and plant products other than cereals and their products reported in the present book/article[440]: Juices (apple juice), Juices (grape juice), and Nectars (prune nectar), AME and AOH, literature[440]; Juices (raspberry juice) and Wine (red), AOH, literature[440]; Nectars (cranberry nectar), AME, literature[440]

Alternariol Methyl Ether

incidence: 1/1, conc.: 0.7 µg/l, sample origin: unknown, sample year: unknown, country: Canada[440]

- Co-contamination: 1 sa. co-contaminated with AME and AOH
- Further contamination of plants and plant products other than cereals and their products reported in the present book/article[440]: Juices (apple juice), Juices (grape juice), and Nectars (prune nectar), AME and AOH, literature[440]; Juices (raspberry juice), Nectars (cranberry nectar), and Wine (red), AOH, literature[440]

Nectars (prune nectar) may contain the following mycotoxins:

Alternaria Toxins

Alternariol

incidence: 1/1, conc.: 5.5 µg/l, sample origin: unknown, sample year: unknown, country: Canada[440]

- Co-contamination: 1 sa. co-contaminated with AME and AOH
- Further contamination of plants and plant products other than cereals and their products reported in the present book/article[440]: Juices (apple juice), Juices (grape juice), and Nectars (cranberry nectar), AME and AOH, literature[440]; Nectars (prune nectar), AME, literature[440]; Juices (raspberry juice) and Wine (red), AOH, literature[440]

Alternariol Methyl Ether

incidence: 1/1, conc.: 1.4 µg/l, sample origin: unknown, sample year: unknown, country: Canada[440]

- Co-contamination: 1 sa. co-contaminated with AME and AOH
- Further contamination of plants and plant products other than cereals and their products reported in the present book/article[440]: Juices (apple juice), Juices (grape juice), and Nectars (cranberry nectar), AME and AOH, literature[440]; Juices (raspberry juice), Nectars (prune nectar), and Wine (red), AOH, literature[440]

Nihari masala see Spices (nihari masala)

Nut bar see Bars

Nut cereals may contain the following mycotoxins:

Aspergillus Toxins

Aflatoxin/s

incidence: 1/1, conc.: 2.29 µg/kg*, sample origin: mainly imported from local retail outlets in Lodz (city), Poland, sample year: unknown, country: Poland[18], *AFS total

- Co-contamination: not reported

- Further contamination of plants and plant products other than cereals and their products reported in the present book/article[18]: Nuts, Nuts (almond), Nuts (cashew nut), Nuts (hazelnut), Nuts (peanut), Nuts (pistachio nut), Nuts (walnut), and Seeds (sunflower), AF/S, literature[18]

Nut crème may contain the following mycotoxins:

Aspergillus Toxins

Aflatoxin B_1

incidence: 38/40, conc. range: <1 µg/kg (20 sa.) and 1–5 µg/kg (18 sa.), sample origin: markets in Ankara (capital), Turkey, sample year: September 2002–September 2003, country: Turkey[274]

- Co-contamination: not reported
- Further contamination of plants and plant products other than cereals and their products reported in the present book/article[274]: Nut crème, AF/S, literature[274]; Nuts (hazelnut), AFB_1 and AF/S, literature[274]

Aflatoxin/s

incidence: 39/40, conc. range: <1 µg/kg (1 sa.), 1–5 µg/kg (13 sa.), 5–10 µg/kg (24 sa.), and >10 µg/kg (1 sa.), sample origin: markets in Ankara (capital), Turkey, sample year: September 2002–September 2003, country: Turkey[274]

- Co-contamination: not reported
- Further contamination of plants and plant products other than cereals and their products reported in the present book/article[274]: Nut crème, AFB_1, literature[274]; Nuts (hazelnut), AFB_1 and AF/S, literature[274]

Aspergillus and ***Penicillium*** Toxins

Ochratoxin A

incidence: 31/33*, conc. range: ≤0.270 µg/kg, sample origin: regions around Detmold, Giessen, Hamburg, Jena, Karlsruhe, Kiel, Kulmbach, and Trier (cities), Germany, sample year: unknown, country: Germany[262], *nut nougat crèmes

- Co-contamination: not reported
- Further contamination of plants and plant products other than cereals and their products reported in the present book/article[262]: Bars, Chocolate, Drinks (chocolate drink), Figs, Fruit, dried, Nuts, Nuts (hazelnut), Nuts (peanut), Powder (cocoa powder), Pudding, Raisins, Seeds (linseed), Seeds (sesame), and Seeds (sunflower), OTA, literature[262]

Nut products see Products (nut product)

Nutmeg see Spices (nutmeg)

Nuts and seeds see Food

Nuts may contain the following mycotoxins:

Aspergillus Toxins

Aflatoxin B_1

incidence: 2/10*, conc. range: 10–93 µg/kg, ∅ conc.: 51.5 µg/kg, sample origin: imported to Finland, sample year: 1974, country: Finland[2], *nuts, mixed

- Co-contamination: 1 sa. co-contaminated with AFB_1 and AFB_2; 1 sa. co-contaminated with AFB_1 and AFG_1
- Further contamination of plants and plant products other than cereals and their products reported in the present book/article[2]: Marzipan (almond paste), Nuts (almond), and Nuts (hazelnut), AFB_1 and AFB_2, literature[2]; Nuts, AFB_2, and AFG_1, literature[2]; Nuts (peanut), AFB_1, AFB_2, AFG_1, and AFG_2, literature[2]

incidence: 1/1*, conc.: 6.68 µg/kg, sample origin: grocery markets in Busan, Daegu, Daejeon, Gangneung, and Gwangju (cities) as well as Seoul (capital), Korea, sample year: May 2004–June 2005, country: Korea[10], *nuts, assorted

- Co-contamination: not reported
- Further contamination of plants and plant products other than cereals and their products reported in the present book/article[10]: Butter (peanut butter), Nuts (almond), Nuts (peanut), Nuts, (pistachio nut), Paste (soybean paste), Sauces (soy sauce), Spices (cayenne pepper, chilli), Spices (curry), and Spices (pepper), AFB_1, literature[10]

incidence: 15/76, conc. range: 0.02–7.8* µg/kg, sample origin: Warsaw (capital) market or from food producers, Poland, sample year: 2004, country: Poland[116], *peanuts for direct consumption

incidence: 15/68, conc. range: 0.04–0.78 µg/kg, sample origin: Warsaw (capital) market or from food producers, Poland, sample year: 2005, country: Poland[116]

- Co-contamination: not reported
- Further contamination of plants and plant products other than cereals and their products reported in the present book/article[116]: Coffee and Raisins, OTA, literature[116]; Food, AFB_1 and AFG_1, literature[116]; Nuts, AFB_2, AFG_1, and AFG_2, literature[116]; Spices, AFB_1, AFB_2, AFG_1, and AFG_2, literature[116]

incidence: 2/7*, conc. range: 0.66–1.09 µg/kg, ∅ conc.: 0.87 µg/kg, sample origin: supermarkets in Penang (state), Malaysia, sample year: unknown, country: Malaysia[182], *almond, cashew, hazelnut, and pistachio nuts

- Co-contamination: not reported
- Further contamination of plants and plant products other than cereals and their products reported in the present book/article[182]: Nuts (peanut), Seeds (sesame), Seeds (sunflower), Spices (cayenne pepper, chilli), Spices (cumin), and Spices (pepper), AFB_1, literature[182]

incidence: 1/12*, conc.: 6.68 µg/kg, sample origin: supermarkets, local stores, and markets in Busan, Daejeon, Gwangju (cities), and Seoul (capital), Korea, sample year: June/July 2004, country: Korea[220], *nuts, assorted (seasoned)

- Co-contamination: not reported
- Further contamination of plants and plant products other than cereals and their products reported in the present book/article[220]: Butter (peanut butter), AFB_1 and AFB_2, literature[220]; Nuts, AFB_2, literature[220]; Nuts (peanut), AFB_1, AFB_2, AFG_1, and AFG_2, literature[220]; Nuts (pistachio nut), AFB_1, literature[220]

incidence: 8/167*, conc. range: ≤2.3 µg/kg, sample origin: imported to USSR, sample year: 1985–1987, country: USSR[267], *almonds and cashews

- Co-contamination: not reported
- Further contamination of plants and plant products other than cereals and their products reported in the present book/article[267]: Coffee beans, Cocoa beans, Legumes, and Nuts (peanut), AFB_1, literature[267]

incidence: 12/33, conc. range: 0.3–9.0 μg/kg, ∅ conc.: 1.6 μg/kg, sample origin: available in New Zealand, sample year: 2010, country: New Zealand[277]

- Co-contamination: not reported
- Further contamination of plants and plant products other than cereals and their products reported in the present book/article[277]: Bars, AFB_1, AF/S, DON, NIV, and OTA, literature[277]; Butter (peanut butter), Confectionery, Food, Fruits, dried, Nuts (peanut), Nuts (pistachio nut), and Sauces (peanut sauce), AFB_1 and AF/S, literature[277]; Coffee, Dates, Spices (nutmeg), Spices (pepper), and Vine fruits, dried, OTA, literature[277]; Figs, Spices (cayenne pepper, chilli), Spices (curry), Spices (ginger), and Spices (paprika), AFB_1, AF/S, and OTA, literature[277]; Nuts, AF/S, literature[277]; Nuts (Brazil nut, paranut), AFB_1, literature[277]

incidence: 1/8*, conc.: 0.48 μg/kg, sample origin: local supermarket in Okayama (prefecture), Japan, sample year: unknown, country: Japan[339], *included almonds (raw), cashew nuts (raw and roasted), peanuts**, pecan nuts, pistachio nuts, macadamia nuts (raw), and walnuts (raw), **contaminated

- Co-contamination: 1 sa. co-contaminated with AFB_1, AFG_1, and AFG_2
- Further contamination of plants and plant products other than cereals and their products reported in the present book/article[339]: Coconuts, Fruits, dried, Seeds (sesame), AFB_1 and AFG_2, literature[339]; Figs, AFB_2 and AFG_2, literature[339]; Flour (soybean flour), AFB_1, literature[339]; Nuts, AFB_2, AFG_1, and AFG_2, literature[339]; Raisins, AFG_2, literature[339]; Spices, AFB_1, AFB_2, AFG_1, and AFG_2, literature[339]

For detailed information see the article.

incidence: 1/1*, conc.: 0.29 μg/kg, sample origin: markets and supermarkets in Valencia (city), Spain, sample year: January–December 2002, country: Spain[435], *nut cocktail

- Co-contamination: 1 sa. co-contaminated with AFB_1 and AFG_1
- Further contamination of plants and plant products other than cereals and their products reported in the present book/article[435]: Nuts and Pinhol, AFG_1, literature[435]; Nuts (hazelnut), AFB_1 and AFG_1, literature[435]

Aflatoxin B_2

incidence: 1/10*, conc.: 29 μg/kg, sample origin: imported to Finland, sample year: 1974, country: Finland[2], *nuts, mixed

- Co-contamination: 1 sa. co-contaminated with AFB_1 and AFB_2
- Further contamination of plants and plant products other than cereals and their products reported in the present book/article[2]: Marzipan (almond paste), Nuts (almond), and Nuts (hazelnut), AFB_1 and AFB_2, literature[2]; Nuts, AFB_1, and AFG_1, literature[2]; Nuts (peanut), AFB_1, AFB_2, AFG_1, and AFG_2, literature[2]

incidence: 9/76*, conc. range: 0.02–2.04 μg/kg, sample origin: Warsaw (capital) market or from food producers, Poland, sample year: 2004, country: Poland[116], *peanuts for direct consumption

incidence: 18/68, conc. range: 0.02–0.2 μg/kg, sample origin: Warsaw (capital) market or from food producers, Poland, sample year: 2005, country: Poland[116]

- Co-contamination: not reported
- Further contamination of plants and plant products other than cereals and their products reported in the present book/article[116]: Coffee, Raisins, OTA, literature[116]; Food, AFB_1 and AFG_1, literature[116]; Nuts, AFB_1, AFG_1, and AFG_2, literature[116]; Spices, AFB_1, AFB_2, AFG_1, and AFG_2, literature[116]

incidence: 1/16*, conc.: tr, sample origin: stores in Oslo (capital) area, Norway, sample year: autumn 1973, country: Norway[201], *nuts, mixed

- Co-contamination: 1 sa. co-contaminated with AFB_2, AFG_1, and AFG_2
- Further contamination of plants and plant products other than cereals and their products reported in the present book/article[201]: Cocoa, Nuts (Brazil nut, paranut), Nuts (cashew nut), Nuts (peanut), and Nuts (walnut), AFB_1, AFB_2, AFG_1, and AFG_2, literature[201]; Nuts, AFG_1 and AFG_2, literature[201]

incidence: 1/12*, conc.: 1.21 μg/kg, sample origin: supermarkets, local stores, and markets in Busan, Daejeon, Gwangju (cities), and Seoul (capital), Korea, sample year: June/July 2004, country: Korea[220], *nuts, assorted (seasoned)

- Co-contamination: not reported
- Further contamination of plants and plant products other than cereals and their products reported in the present book/article[220]: Butter (peanut butter), AFB_1 and AFB_2, literature[220]; Nuts and Nuts (pistachio nut), AFB_1, literature[220]; Nuts (peanut), AFB_1, AFB_2, AFG_1, and AFG_2, literature[220]

incidence: 1/8*, conc.: 0.11 μg/kg, sample origin: local supermarket in Okayama (prefecture), Japan, sample year: unknown, country: Japan[339], *included almonds (raw)**, cashew nuts (raw and roasted), peanuts, pecan nuts, pistachio nuts, macadamia nuts (raw), and walnuts (raw), **contaminated

- Co-contamination: 1 sa. co-contaminated with AFB_2 and AFG_2
- Further contamination of plants and plant products other than cereals and their products reported in the present book/article[339]: Coconuts, Fruits, dried, Seeds (sesame), AFB_1 and AFG_2, literature[339]; Figs, AFB_2 and AFG_2, literature[339]; Flour (soybean flour), AFB_1, literature[339]; Nuts, AFB_1, AFG_1, and AFG_2, literature[339]; Raisins, AFG_2, literature[339]; Spices, AFB_1, AFB_2, AFG_1, and AFG_2, literature[339]

For detailed information see the article.

Aflatoxin G_1

incidence: 1/10*, conc.: 5 μg/kg, sample origin: imported to Finland, sample year: 1974, country: Finland[2], *nuts, mixed

- Co-contamination: 1 sa. co-contaminated with AFB_1 and AFG_1
- Further contamination of plants and plant products other than cereals and their products reported in the present book/article[2]: Marzipan (almond paste), Nuts, Nuts (almond), and Nuts (hazelnut), AFB_1 and AFB_2, literature[2]; Nuts (peanut), AFB_1, AFB_2, AFG_1, and AFG_2, literature[2]

incidence: 12/76*, conc. range: 0.02–0.29 μg/kg, sample origin: Warsaw (capital) market or from food producers, Poland, sample year: 2004, country: Poland[116], *peanuts for direct consumption

incidence: 25/68, conc. range: 0.03–1.16 μg/kg, sample origin: Warsaw (capital) market or from food producers, Poland, sample year: 2005, country: Poland[116]

- Co-contamination: not reported
- Further contamination of plants and plant products other than cereals and their products reported in the present book/article[116]: Coffee, Raisins, OTA, literature[116]; Food, AFB_1 and AFG_1, literature[116]; Nuts, AFB_1, AFB_2, and AFG_2, literature[116]; Spices, AFB_1, AFB_2, AFG_1, and AFG_2, literature[116]

incidence: 1/16*, conc.: tr, sample origin: stores in Oslo (capital) area, Norway, sample year: autumn 1973, country: Norway[201], *nuts, mixed

- Co-contamination: 1 sa. co-contaminated with AFB_2, AFG_1, and AFG_2
- Further contamination of plants and plant products other than cereals and their products reported in the present book/article[201]: Cocoa, Nuts (Brazil nut, paranut), Nuts (cashew nut), Nuts (peanut), and Nuts (walnut), AFB_1, AFB_2, AFG_1, and AFG_2, literature[201]; Nuts, AFB_2 and AFG_2, literature[201]

incidence: 1/8*, conc.: 0.84 µg/kg, sample origin: local supermarket in Okayama (prefecture), Japan, sample year: unknown, country: Japan[339], *included almonds (raw), cashew nuts (raw and roasted), peanuts**, pecan nuts, pistachio nuts, macadamia nuts (raw), and walnuts (raw), **contaminated

- Co-contamination: 1 sa. co-contaminated with AFB_1, AFG_1, and AFG_2
- Further contamination of plants and plant products other than cereals and their products reported in the present book/article[339]: Coconuts, Fruits, dried, Seeds (sesame), AFB_1 and AFG_2, literature[339]; Figs, AFB_2 and AFG_2, literature[339]; Flour (soybean flour), AFB_1, literature[339]; Nuts, AFB_1, AFB_2, and AFG_2, literature[339]; Raisins, AFG_2, literature[339]; Spices, AFB_1, AFB_2, AFG_1, and AFG_2, literature[339]

For detailed information see the article.

incidence: 1/1*, conc.: 0.47 µg/kg, sample origin: markets and supermarkets in Valencia (city), Spain, sample year: January–December 2002, country: Spain[435], *nut cocktail

- Co-contamination: 1 sa. co-contaminated with AFB_1 and AFG_1
- Further contamination of plants and plant products other than cereals and their products reported in the present book/article[435]: Nuts, AFB_1, literature[435]; Nuts (hazelnut), AFB_1 and AFG_1, literature[435]; Pinhol, AFG_1, literature[435]

Aflatoxin G_2

incidence: 4/76*, conc. range: 0.02–0.22 µg/kg, sample origin: Warsaw (capital) market or from food producers, Poland, sample year: 2004, country: Poland[116], *peanuts for direct consumption

incidence: 18/68, conc. range: 0.02–0.71 µg/kg, sample origin: Warsaw (capital) market or from food producers, Poland, sample year: 2005, country: Poland[116]

- Co-contamination: not reported
- Further contamination of plants and plant products other than cereals and their products reported in the present book/article[116]: Coffee, Raisins, OTA, literature[116]; Food, AFB_1 and AFG_1, literature[116]; Nuts, AFB_1, AFB_2, and AFG_1, literature[116]; Spices, AFB_1, AFB_2, AFG_1, and AFG_2, literature[116]

incidence: 1/16*, conc.: tr, sample origin: stores in Oslo (capital) area, Norway, sample year: autumn 1973, country: Norway[201], *nuts, mixed

- Co-contamination: 1 sa. co-contaminated with AFB_2, AFG_1, and AFG_2
- Further contamination of plants and plant products other than cereals and their products reported in the present book/article[201]: Cocoa, Nuts (Brazil nut, paranut), Nuts (cashew nut), Nuts (peanut), and Nuts (walnut), AFB_1, AFB_2, AFG_1, and AFG_2, literature[201]; Nuts, AFB_2 and AFG_1, literature[201]

incidence: 2/8*, conc. range: 0.34–1.12 µg/kg, ∅ conc.: 0.73 µg/kg, sample origin: local supermarket in Okayama (prefecture), Japan, sample year: unknown, country: Japan[339], *included almonds (raw)**, cashew nuts (raw and roasted), peanuts**, pecan nuts, pistachio nuts, macadamia nuts (raw), and walnuts (raw), **contaminated

- Co-contamination: 1 sa. co-contaminated with AFB_1, AFG_1, and AFG_2; 1 sa. co-contaminated with AFB_2 and AFG_2
- Further contamination of plants and plant products other than cereals and their products reported in the present book/article[339]: Coconuts, Fruits, dried, Seeds (sesame), AFB_1 and AFG_2, literature[339]; Figs, AFB_2 and AFG_2, literature[339]; Flour (soybean flour), AFB_1, literature[339]; Nuts, AFB_1, AFB_2, and AFG_1, literature[339]; Raisins, AFG_2, literature[339]; Spices, AFB_1, AFB_2, AFG_1, and AFG_2, literature[339]

For detailed information see the article.

Aflatoxin/s

incidence: 1/5*, conc.: 3.91 µg/kg**, sample origin: mainly imported from local retail outlets in Lodz (city), Poland, sample year: unknown, country: Poland[18], *nuts, "chio-felix", **AFS total

- Co-contamination: not reported
- Further contamination of plants and plant products other than cereals and their products reported in the present book/article[18]: Nut cereals, Nuts (almond), Nuts (cashew nut), Nuts (hazelnut), Nuts (peanut), Nuts (pistachio nut), Nuts (walnut), Seeds (sunflower), AF/S, literature[18]

incidence: 1/3*, conc.: 7 µg/kg**, sample origin: imported to the USA, sample year: 1986, country: USA[19], *nuts, mixed, **AFB_1, AFB_2, AFG_1, and AFG_2

- Co-contamination: not reported
- Further contamination of plants and plant products other than cereals and their products reported in the present book/article[19]: Butter (peanut butter), Candies (peanut candy), Candies (pistachio candy), Nuts (almond), Nuts, (Brazil nut, paranut), Nuts (peanut), Nuts (pecan nut), Nuts (pistachio nut), Nuts (walnut), Oils (peanut oil), Paste (peanut paste), Paste (pipian paste), Products (peanut product), Seeds (melon), Spices (cayenne pepper, chilli), Spices (ginger), Spices (nutmeg), AF/S, literature[19]

incidence: 1/82*, conc.: 352.0 µg/kg**, sample origin: local retail stores in the region of Brasilia (capital), Brazil, sample year: 2002–2011, country: Brazil[208], *included almonds, cashew nuts, hazelnuts, nuts, and pistachio nuts (all shelled), **AFB_1, AFB_2, AFG_1, and AFG_2

- Co-contamination: 1 sa. co-contaminated with AFB_1 and AFB_2; no further information available
- Further contamination of plants and plant products other than cereals and their products reported in the present book/article[208]: Nuts (Brazil nut, paranut), Nuts (peanut), Products (peanut product), AF/S, literature[208]

incidence: 3/29*, conc. range: 6–10 µg/kg** (2 sa.) and >50 µg/kg** (1 sa.), sample origin: commercial and imported food from retail outlets and points of entries, Malaysia, sample year: 1995–1999, country: Malaysia[233], *different kinds of nuts, **AF not specified

- Co-contamination: not reported
- Further contamination of plants and plant products other than cereals and their products reported in the present book/article[233]: Butter (peanut butter), Candies (peanut candy), Nuts (peanut), Products (peanut products), Spices (cayenne pepper, chilli), AF/S, literature[233]

incidence: 12/33, conc. range: 0.3–9.7 μg/kg*, ∅ conc.: 1.7 μg/kg*, sample origin: available in New Zealand, sample year: 2010, country: New Zealand[277], *AFS total

- Co-contamination: not reported
- Further contamination of plants and plant products other than cereals and their products reported in the present book/article[277]: Bars, AFB_1, AF/S, DON, NIV, and OTA, literature[277]; Butter (peanut butter), Confectionery, Food, Fruits, dried, Nuts (peanut), Nuts (pistachio nut), Sauces (peanut sauce), AFB_1 and AF/S, literature[277]; Coffee, Dates, Spices (nutmeg), Spices (pepper), Vine fruits, dried, OTA, literature[277]; Figs, Spices (cayenne pepper, chilli), Spices (curry), Spices (ginger), and Spices (paprika), AFB_1, AF/S, and OTA, literature[277]; Nuts, Nuts (Brazil nut, paranut), AFB_1, literature[277]

incidence: 35/35*, conc. range: 2 μg/kg** (23 sa.), 2–5 μg/kg** (6 sa.), 6–15 μg/kg** (1 sa.), 16–50 μg/kg** (1 sa.), 51–200 μg/kg (3 sa.), and 276 μg/kg (1 sa.), sample origin: retail outlets in New South Wales (state), Australia, sample year: 1988/1989, country: Australia[420], *included nuts and nut products, **AFB_1, AFB_2, AFG_1, and AFG_2

- Co-contamination: not reported
- Further contamination of plants and plant products other than cereals and their products reported in the present book/article[420]: Nuts (peanut), AF/S, literature[420]

incidence: 1/5*, conc.: 6 μg/kg**, sample origin: most sa. from Iran, Syria, Turkey, and the USA collected by inspectors at overland and seaport custom offices, Qatar, sample year: July–December 1997, country: Qatar[424], *nuts, mixed, **AFB_1, AFB_2, AFG_1, and AFG_2

- Co-contamination: not reported
- Further contamination of plants and plant products other than cereals and their products reported in the present book/article[424]: Nuts (peanut), Nuts (pistachio nut), AF/S, literature[424]

incidence: 2/29, conc. range: 20.3–23.3 μg/kg, ∅ conc.: 21.8 μg/kg, sample origin: five governates, Kuwait, sample year: unknown, country: Kuwait[704], *AF

- Co-contamination: not reported
- Further contamination of plants and plant products other than cereals and their products reported in the present book/article[704]: Coffee, OTA, literature[704]; Spices, AF/S, literature[704]

incidence: 70/264*, conc. range: 1.0–110 μg/kg**, sample origin: outlets across seven municipalities of Mecca (city), Saudi Arabia, sample year: January/February 2008, country: Saudi Arabia[716], *included almonds, cashew nuts, hazelnuts peanuts, and walnuts, **AFB_1, AFB_2, AFG_1, and AFG_2

- Co-contamination: not reported
- Further contamination of plants and plant products other than cereals and their products reported in the present book/article[716]: not reported

For detailed information see the article.

Aspergillus and ***Penicillium*** Toxins

Ochratoxin A

incidence: 1/1, conc.: 0.02 μg/kg, sample origin: supermarkets and fast-food chains in Quebec (city), Canada, sample year: September/October 2008, country: Canada[90]

incidence: 1/1, conc.: 0.03 μg/kg, sample origin: supermarkets and fast-food chains in Calgary (city), Canada, sample year: September/October 2009, country: Canada[90]

- Co-contamination: not reported
- Further contamination of plants and plant products other than cereals and their products reported in the present book/article[90]: Baby food, Beans, Beverages, Butter (peanut butter), Candies, Chocolate, Coffee, Extracts, Herbs, Ice cream, Pies, Pizza, Products, Pudding, Raisins, Seeds, Wine, OTA, literature[90]

incidence: 26/125*, conc. range: ≤0.270 μg/kg, sample origin: regions around Detmold, Giessen, Hamburg, Jena, Karlsruhe, Kiel, Kulmbach, and Trier (cities), Germany, sample year: unknown, country: Germany[262], *different kinds of nuts

- Co-contamination: not reported
- Further contamination of plants and plant products other than cereals and their products reported in the present book/article[262]: Bars, Chocolate, Drinks (chocolate drink), Figs, Fruits, dried, Nut crème, Nuts (hazelnut), Nuts (peanut), Powder (cocoa powder), Pudding, Raisins, Seeds (linseed), Seeds (sesame), Seeds (sunflower), OTA, literature[262]

incidence: 7/15, conc. range: 2.75–7.42 μg/kg, ∅ conc.: 5.64 μg/kg, sample origin: supermarkets and small shops, Jordan, sample year: September 2008–March 2009, country: Jordan[315]

- Co-contamination: not reported
- Further contamination of plants and plant products other than cereals and their products reported in the present book/article[315]: Coffee beans, Seeds (sunflower), OTA, literature[315]

incidence: 33/142*, conc. range: LOD/LOQ-≤0.27 μg/kg (33 sa.), sample origin: Germany, sample year: 1995–1998, country: EU[698], *tree nuts

- Co-contamination: not reported
- Further contamination of plants and plant products other than cereals and their products reported in the present book/article[698]: Baby food, Beans (soybean), Chocolate, Cocoa, Coffee, Coffee beans, Currants, Fruits, dried, Herbs, Jam, Juices, Juices (grape juice), Legumes, Oil seeds, Oils (olive oil), Peas, Peas (chickpea), Products (plant product), Raisins, Spices, Spices (cayenne pepper, chilli), Spices (nutmeg), Spices (paprika), Spices (pepper), Sultanas, Tea, Vinegar, Wine, Wine (red), Wine (rosé), Wine (white), Wine, (miscellaneous), OTA, literature[698]

Nuts (almond) may contain the following mycotoxins:

Alternaria Toxins

Alternariol

incidence: 1/8, conc.: 1.5 μg/kg, sample origin: KSU, Turkey and stores in Tulln (city) and Vienna (capital), Austria, sample year: unknown, country: Austria[868]

- Co-contamination: For detailed information see the article.
- Further contamination of plants and plant products other than cereals and their products reported in the present book/article[868]: Nuts (almond), AME, BEA, CPA, ENB, ENB_1, and ENB_3, literature[868]; Nuts (hazelnut), 3-NPA, AFB_1, AFB_2, AFG_1, AFG_2, AME, AOH, ATX-I, BEA, ENA, ENA_1, ENB, ENB_2, ENB_3, ENB_4, HT-2, KA, MA, MON, OTA, OTB, STG, T-2, and ZEA, literature[868]; Nuts (peanut), 3-NPA, AFB_1, AFB_2, AME, AOH, ATX-I, BEA, CPA, ENA, ENA_1, ENB, ENB_4, KA, MA, OTA, OTB, STG, and ZEA, literature[868]; Nuts (pistachio nut), BEA, MA, and OTA, literature[868]; see also Nuts (almond), Nuts (hazelnut), Nuts

(peanut), Nuts (pistachio nut), chapter: Further Mycotoxins and Microbial Metabolites
For detailed information see the article.

Alternariol Methyl Ether
incidence: 3/8, conc. range: ≤0.84 μg/kg, ∅ conc.: 0.56 μg/kg, sample origin: KSU, Turkey and stores in Tulln (city) and Vienna (capital), Austria, sample year: unknown, country: Austria[868]

- Co-contamination: For detailed information see the article.
- Further contamination of plants and plant products other than cereals and their products reported in the present book/article[868]: Nuts (almond), AOH, BEA, CPA, ENB, ENB_1, and ENB_3, literature[868]; Nuts (hazelnut), 3-NPA, AFB_1, AFB_2, AFG_1, AFG_2, AME, AOH, ATX-I, BEA, ENA, ENA_1, ENB, ENB_2, ENB_3, ENB_4, HT-2, KA, MA, MON, OTA, OTB, STG, T-2, and ZEA, literature[868]; Nuts (peanut), 3-NPA, AFB_1, AFB_2, AME, AOH, ATX-I, BEA, CPA, ENA, ENA_1, ENB, ENB_4, KA, MA, OTA, OTB, STG, and ZEA, literature[868]; Nuts (pistachio nut), BEA, MA, and OTA, literature[868]; see also Nuts (almond), Nuts (hazelnut), Nuts (peanut), Nuts (pistachio nut), chapter: Further Mycotoxins and Microbial Metabolites

For detailed information see the article.

Aspergillus Toxins

Aflatoxin B_1
incidence: 1/110, conc.: 93 μg/kg, sample origin: imported to Finland, sample year: 1974, country: Finland[2]

- Co-contamination: 1 sa. co-contaminated with AFB_1 and AFB_2
- Further contamination of plants and plant products other than cereals and their products reported in the present book/article[2]: Marzipan (almond paste), Nuts (hazelnut), AFB_1 and AFB_2, literature[2]; Nuts, AFB_1, AFB_2, and AFG_1, literature[2]; Nuts (almond), AFB_2, literature[2]; Nuts (peanut), AFB_1, AFB_2, AFG_1, and AFG_2, literature[2]

incidence: 1/6*, conc.: 67 μg/kg, sample origin: imported to Finland, sample year: 1974, country: Finland[2], *almonds, bitter

- Co-contamination: 1 sa. co-contaminated with AFB_1 and AFB_2
- Further contamination of plants and plant products other than cereals and their products reported in the present book/article[2]: Marzipan (almond paste), Nuts (hazelnut), AFB_1 and AFB_2, literature[2]; Nuts, AFB_1, AFB_2, and AFG_1, literature[2]; Nuts (almond), AFB_2, literature[2]; Nuts (peanut), AFB_1, AFB_2, AFG_1, and AFG_2, literature[2]

incidence: 1/184*, conc.: tr, sample origin: imported to Finland, sample year: 1974, country: Finland[2], *almonds, sliced and crushed

- Co-contamination: not reported
- Further contamination of plants and plant products other than cereals and their products reported in the present book/article[2]: Marzipan (almond paste), Nuts (hazelnut), AFB_1 and AFB_2, literature[2]; Nuts, AFB_1, AFB_2, and AFG_1, literature[2]; Nuts (almond), AFB_2, literature[2]; Nuts (peanut), AFB_1, AFB_2, AFG_1, and AFG_2, literature[2]

incidence: 15/19*, conc. range: 0.5–5 μg/kg (14 sa.) and 6 μg/kg (1 sa.), sample origin: retail outlets in 20 different regions, England, sample year: April 1984, country: UK[3], *almonds, ground

- Co-contamination: not reported
- Further contamination of plants and plant products other than cereals and their products reported in the present book/article[3]: Butter (peanut butter), Nuts (Brazil nut, paranut), Nuts (hazelnut), Nuts (peanut), AFB_1 and AF/S, literature[3]; Nuts (almond), AF/S, literature[3]

incidence: 1/3*, conc.: 5,566.6 μg/kg, sample origin: different storage facilities, Egypt, sample year: November 1978–April 1979, country: Egypt[4], *sa. moldy with mechanical or insect damage

- Co-contamination: not reported
- Further contamination of plants and plant products other than cereals and their products reported in the present book/article[4]: Beans (broad bean), AFB_1 and AFB_2, literature[4]; Lentils, AFB_1, literature[4]; Nuts (peanut), AFB_1, AFB_2, AFG_1, and AFG_2, literature[4]

incidence: 1*/34, conc.: 95 μg/kg, sample origin: Spanish retail market, Spain, sample year: 1986/1987, country: Spain[6], *bulk, unpacked material

- Co-contamination: 1 sa. co-contaminated with AFB_1 and AFB_2
- Further contamination of plants and plant products other than cereals and their products reported in the present book/article[6]: Nuts (almond), AFB_2, literature[6]; Nuts (peanut), AFB_1, literature[6]

incidence: 2/6*, conc. range: 25–71 μg/kg, ∅ conc.: 48 μg/kg, sample origin: major local retail outlets at Nainital (district), India, sample year: 1983–1985, country: India[7], *cuddapah almond

incidence: 3/8, conc. range: 18–194 μg/kg, sample origin: major local retail outlets at Nainital (district), India, sample year: 1983–1985, country: India[7]

- Co-contamination: Often more than one mycotoxin was present in a sa.
- Further contamination of plants and plant products other than cereals and their products reported in the present book/article[7]: Coconuts, AFB_1, AFB_2, AFG_1, AFG_2, and CIT, literature[7]; Dates, AFB_1, AFB_2, AFG_1, and OTA, literature[7]; Grogannuts, AFB_1, AFB_2, AFG_1, AFG_2, and CIT, literature[7]; Nuts (almond), AFB_2, AFG_1, RBT, STG, and ZEA, literature[7]; Nuts (cashew nut), AFB_1, AFB_2, AFG_1, AFG_2, CIT, and OTA, literature[7]; Nuts (peanut), Raisins, AFB_1, AFB_2, AFG_1, AFG_2, and OTA, literature[7]; Nuts (pistachio nut), AFB_1, AFB_2, AFG_1, AFG_2, and STG, literature[7]; Nuts (walnut), AFB_1, AFB_2, AFG_1, AFG_2, and RBT, literature[7]

incidence: 5/25, conc. range: 5.8–140.0 μg/kg, ∅ conc.: 37.5 μg/kg, sample origin: different localities, Pakistan, sample year: unknown, country: Pakistan[8]

- Co-contamination: 2 sa. co-contaminated with AFB_1 and AFB_2; 3 sa. contaminated solely with AFB_1
- Further contamination of plants and plant products other than cereals and their products reported in the present book/article[8]: Nuts (almond), AFB_2, literature[8]

incidence: 1/14, conc.: 0.39 μg/kg, sample origin: unknown, sample year: unknown, country: Korea[9]

- Co-contamination: 1 sa. co-contaminated with AFB_1 and AFB_2
- Further contamination of plants and plant products other than cereals and their products reported in the present book/article[9]: Butter (peanut butter), AFB_1, AFB_2, AFG_1, and AFG_2, literature[9]; Nuts (almond), AFB_2, literature[9]; Nuts (peanut), AFB_1 and AFB_2, literature[9]; Products (nut product) and Tea, AFB_1, literature[9]

incidence: 3/5, conc. range: 0.08–0.55 μg/kg, ∅ conc.: 0.23 μg/kg, sample origin: grocery markets in Busan, Daegu, Daejeon, Gangneung, and Gwangju (cities) as well as Seoul (capital), Korea, sample year: May 2004–June 2005, country: Korea[10]

- Co-contamination: not reported
- Further contamination of plants and plant products other than cereals and their products reported in the present book/

article[10]: Butter (peanut butter), Nuts, Nuts (peanut), Nuts (pistachio nut), Paste (soybean paste), Sauces (soy sauce), Spices (cayenne pepper, chilli), Spices (curry), Spices (pepper), AFB_1, literature[10]

incidence: 1/1*, conc.: 34 µg/kg, sample origin: unknown, sample year: unknown, country: USA[11], *almonds, ground
- Co-contamination: 1 sa. co-contaminated with AFB_1 and AFB_2
- Further contamination of plants and plant products other than cereals and their products reported in the present book/article[11]: Nuts (almond), AFB_2, literature[11]; Nuts (peanut), AFB_1, AFB_2, AFG_1, and AFG_2, literature[11]

incidence: 1*/21**, conc.: 4.8 µg/kg, sample origin: Alfândega, Faro, and Moncorvo (municipalities), Portugal, sample year: 2007/2008, country: Portugal[12], *storage sa., **field and storage sa.
- Co-contamination: 1 sa. co-contaminated with AFB_1 and AFB_2
- Further contamination of plants and plant products other than cereals and their products reported in the present book/article[12]: Nuts (almond), AFB_2, literature[12]

incidence: 11/28*, conc. range: 0.005–2 µg/kg (11 sa., maximum: 0.8 µg/kg), sample origin: ethnic stores, health-food stores, and local stores as well as supermarkets in the four largest cities, Sweden, sample year: autumn 1996–summer 1998, country: Sweden[13], *almonds and almond/nut paste
- Co-contamination: not reported
- Further contamination of plants and plant products other than cereals and their products reported in the present book/article[13]: Beans (brown bean), Peas and Peas (chickpea), OTA, literature[13]; Juices (apple juice), PAT, literature[13]; Nuts (Brazil nut, paranut), Nuts (pistachio nut), Oil seeds, Products (peanut product), AFB_1 and AF/S, literature[13]

incidence: 3/13, conc. range: 1–13 µg/kg, ∅ conc.: 7.4 µg/kg, sample origin: retailers, markets, and dried fruit retail shops in Erzurum (city), Turkey, sample year: March 2001–March 2002, country. Turkey[14]
- Co-contamination: not reported
- Further contamination of plants and plant products other than cereals and their products reported in the present book/article[14]: Nuts (hazelnut), Nuts (peanut), Nuts (walnut), Peas (chickpea), AFB_1, literature[14]

incidence: 19/23*, conc. range: 39–4,000 µg/kg, sample origin: unknown, sample year: unknown, country: Germany[239], *moldy almonds, sweet
- Co-contamination: not reported
- Further contamination of plants and plant products other than cereals and their products reported in the present book/article[239]: Fruits (apple), Fruits (cherry), Fruits (lemon), Fruits (orange), Fruits (peach), Fruits (tomato), Nuts (cashew nut), Nuts (hazelnut), Nuts (walnut), Seeds (sesame), Spice (nutmeg), AFB_1, literature[239]; Nuts (peanut), AFB_1, AFB_2, AFG_1, and AFG_2, literature[239]

incidence: 43/63*, conc. range: ≤45.56 µg/kg, sample origin: retail outlets, cafeterias, and hotels from major districts of Punjab (state), Pakistan, sample year: unknown, country: Pakistan[750], *ingredients for halva
- Co-contamination: not reported
- Further contamination of plants and plant products other than cereals and their products reported in the present book/article[750]: Halva, AFB_1 and AFM_1, literature[750]; Nuts (pistachio nut), Raisins, Spices (cardamom), AFB_1, literature[750]

incidence: 10/74*, conc. range: 0.5–42 µg/kg, ∅ conc.: 7.32 µg/kg, sample origin: California (state), USA, sample year: 1972, country: USA[839], *almonds, unsorted in-shell
- Co-contamination: 6 sa. co-contaminated with AFB_1 and AFB_2; 1 sa. co-contaminated with AFB_1, AFG_1, and AFG_2; 3 sa. contaminated solely with AFB_1
- Further contamination of plants and plant products other than cereals and their products reported in the present book/article[839]: Nuts (almond), AFB_2, AFG_1, AFG_2, and AF/S, literature[839]

Aflatoxin B_2

incidence: 1/110, conc.: 14 µg/kg, sample origin: imported to Finland, sample year: 1974, country: Finland[2]
- Co-contamination: 1 sa. co-contaminated with AFB_1 and AFB_2
- Further contamination of plants and plant products other than cereals and their products reported in the present book/article[2]: Marzipan (almond paste) and Nuts (hazelnut), AFB_1 and AFB_2, literature[2]; Nuts, AFB_1, AFB_2, and AFG_1, literature[2]; Nuts (almond), AFB_1, literature[2]; Nuts (peanut), AFB_1, AFB_2, AFG_1, and AFG_2, literature[2]

incidence: 1/6*, conc.: 5 µg/kg, sample origin: imported to Finland, sample year: 1974, country: Finland[2], *almonds, bitter
- Co-contamination: 1 sa. co-contaminated with AFB_1 and AFB_2
- Further contamination of plants and plant products other than cereals and their products reported in the present book/article[2]: Marzipan (almond paste) and Nuts (hazelnut), AFB_1 and AFB_2, literature[2]; Nuts, AFB_1, AFB_2, and AFG_1, literature[2]; Nuts (almond), AFB_1, literature[2]; Nuts (peanut), AFB_1, AFB_2, AFG_1, and AFG_2, literature[2]

incidence: 0/184*, conc.: no contamination, sample origin: imported to Finland, sample year: 1974, country: Finland[2], *almonds, sliced and crushed
- Co-contamination: not reported
- Further contamination of plants and plant products other than cereals and their products reported in the present book/article[2]: Marzipan (almond paste) and Nuts (hazelnut), AFB_1 and AFB_2, literature[2]; Nuts, AFB_1, AFB_2, and AFG_1, literature[2]; Nuts (almond), AFB_1, literature[2]; Nuts (peanut), AFB_1, AFB_2, AFG_1, and AFG_2, literature[2]

incidence: 1*/34, conc.: 15 µg/kg, sample origin: Spanish retail market, Spain, sample year: 1986/1987, country: Spain[6], *bulk, unpacked material
- Co-contamination: 1 sa. co-contaminated with AFB_1 and AFB_2
- Further contamination of plants and plant products other than cereals and their products reported in the present book/article[6]: Nuts (almond), Nuts (peanut), AFB_1, literature[6]

incidence: 1/6*, conc.: 17 µg/kg, sample origin: major local retail outlets at Nainital (district), India, sample year: 1983–1985, country: India[7], *cuddapah almond

incidence: 2/8, conc. range: 20–125 µg/kg, ∅ conc.: 72.5 µg/kg, sample origin: major local retail outlets at Nainital (district), India, sample year: 1983–1985, country: India[7]
- Co-contamination: Often more than one mycotoxin was present in a sa.
- Further contamination of plants and plant products other than cereals and their products reported in the present book/article[7]: Coconuts, AFB_1, AFB_2, AFG_1, AFG_2, and CIT, literature[7]; Dates, AFB_1, AFB_2, AFG_1, and OTA, literature[7]; Grogannuts, AFB_1, AFB_2, AFG_1, AFG_2, and CIT, literature[7]; Nuts (almond), AFB_1, AFG_1, RBT, STG, and ZEA, literature[7]; Nuts (cashew nut), AFB_1, AFB_2, AFG_1, AFG_2, CIT, and OTA, literature[7]; Nuts (peanut), Raisins,

AFB_1, AFB_2, AFG_1, AFG_2, and OTA, literature[7]; Nuts (pistachio nut), AFB_1, AFB_2, AFG_1, AFG_2, and STG, literature[7]; Nuts (walnut), AFB_1, AFB_2, AFG_1, AFG_2, and RBT, literature[7]

incidence: 2/25, conc. range: 7.0–140.0 µg/kg, ∅ conc.: 73.5 µg/kg, sample origin: different localities, Pakistan, sample year: unknown, country: Pakistan[8]

- Co-contamination: 2 sa. co-contaminated with AFB_1 and AFB_2
- Further contamination of plants and plant products other than cereals and their products reported in the present book/article[8]: Nuts (almond), AFB_1, literature[8]

incidence: 1/14, conc.: 0. 06 µg/kg, sample origin: unknown, sample year: unknown, country: Korea[9]

- Co-contamination: 1 sa. co-contaminated with AFB_1 and AFB_2
- Further contamination of plants and plant products other than cereals and their products reported in the present book/article[9]: Butter (peanut butter), AFB_1, AFB_2, AFG_1, and AFG_2, literature[9]; Nuts (almond), AFB_1, literature[9]; Nuts (peanut), AFB_1 and AFB_2, literature[9]; Products (nut product), Tea, AFB_1, literature[9]

incidence: 1/1*, conc.: 5 µg/kg, sample origin: unknown, sample year: unknown, country: USA[11], *almonds, ground

- Co-contamination: 1 sa. co-contaminated with AFB_1 and AFB_2
- Further contamination of plants and plant products other than cereals and their products reported in the present book/article[11]: Nuts (almond), AFB_1, literature[11]; Nuts (peanut), AFB_1, AFB_2, AFG_1, and AFG_2, literature[11]

incidence: 1*/21**, conc.: 0.17 µg/kg, sample origin: Alfândega, Faro, and Moncorvo (municipalities), Portugal, sample year: 2007/2008, country: Portugal[12], *storage sa., **field and storage sa.

- Co-contamination: 1 sa. co-contaminated with AFB_1 and AFB_2
- Further contamination of plants and plant products other than cereals and their products reported in the present book/article[12]: Nuts (almond), AFB_1, literature[12]

incidence: 6/74*, conc. range: ≤3.8 µg/kg, sample origin: California (state), USA, sample year: 1972, country: USA[839], *almonds, unsorted in-shell

- Co-contamination: 6 sa. co-contaminated with AFB_1 and AFB_2
- Further contamination of plants and plant products other than cereals and their products reported in the present book/article[839]: Nuts (almond), AFB_1, AFG_1, AFG_2, and AF/S, literature[839]

incidence: 2/2*, conc. range: <LOQ, sample origin: supermarkets in Novi Sad (city), Vojvodina (province), Serbia, sample year: February 2013, country: Serbia/Czech Republic[844], *commercial almonds, crude

- Co-contamination: not reported
- Further contamination of plants and plant products other than cereals and their products reported in the present book/article[844]: Nuts (walnut), ZEA, literature[844]

Aflatoxin G_1

incidence: 1/6*, conc.: 25 µg/kg, sample origin: major local retail outlets at Nainital (district), India, sample year: 1983–1985, country: India[7], *cuddapah almond

incidence: 2/8, conc. range: 19–30 µg/kg, ∅ conc.: 24.5 µg/kg, sample origin: major local retail outlets at Nainital (district), India, sample year: 1983–1985, country: India[7]

- Co-contamination: Often more than one mycotoxin was present in a sa.
- Further contamination of plants and plant products other than cereals and their products reported in the present book/article[7]: Coconuts, AFB_1, AFB_2, AFG_1, AFG_2, and CIT, literature[7]; Dates, AFB_1, AFB_2, AFG_1, and OTA, literature[7]; Grogannuts, AFB_1, AFB_2, AFG_1, AFG_2, and CIT, literature[7]; Nuts (almond), AFB_1, AFB_2, RBT, STG, and ZEA, literature[7]; Nuts (cashew nut), AFB_1, AFB_2, AFG_1, AFG_2, CIT, and OTA, literature[7]; Nuts (peanut), Raisins, AFB_1, AFB_2, AFG_1, AFG_2, and OTA, literature[7]; Nuts (pistachio nut), AFB_1, AFB_2, AFG_1, AFG_2, and STG, literature[7]; Nuts (walnut), AFB_1, AFB_2, AFG_1, AFG_2, and RBT, literature[7]

incidence: 1/74*, conc.: 62.5 µg/kg, sample origin: California (state), USA, sample year: 1972, country: USA[839], *almonds, unsorted in-shell

- Co-contamination: 1 sa. co-contaminated with AFB_1, AFG_1, and AFG_2
- Further contamination of plants and plant products other than cereals and their products reported in the present book/article[839]: Nuts (almond), AFB_1, AFB_2, AFG_2, and AF/S, literature[839]

Aflatoxin G_2

incidence: 1/74*, conc.: 2.5 µg/kg, sample origin: California (state), USA, sample year: 1972, country: USA[839], *unsorted in-shell almonds

- Co-contamination: 1 sa. co-contaminated with AFB_1, AFG_1, and AFG_2
- Further contamination of plants and plant products other than cereals and their products reported in the present book/article[839]: Nuts (almond), AFB_1, AFB_2, AFG_1, and AF/S, literature[839]

Aflatoxin/s

incidence: 15/19*, conc. range: 0.5–5 µg/kg** (13 sa.) and 6–10 µg/kg** (2 sa., maximum: 10 µg/kg), sample origin: retail outlets in 20 different regions, England, sample year: April 1984, country: UK[3], *almonds, ground, **AFS total

- Co-contamination: not reported
- Further contamination of plants and plant products other than cereals and their products reported in the present book/article[3]: Butter (peanut butter), Nuts (Brazil nut, paranut), Nuts (hazelnut), Nuts (peanut), AFB_1 and AF/S, literature[3]; Nuts (almond), AFB_1, literature[3]

incidence: 2/78, conc. range: 5.0–9.9 µg/kg* (1 sa.) and >25 µg/kg* (1 sa.), sample origin: brokers, distributors, growers, processors, and retail trade, Canada, sample year: 1970–1975, country: Canada[15], *AFB_1, AFB_2, AFG_1, and AFG_2

- Co-contamination: not reported
- Further contamination of plants and plant products other than cereals and their products reported in the present book/article[15]: Butter (peanut butter), Cocoa beans, Nuts (cashew nut), Nuts (peanut), Nuts (pecan nut), Nuts (pistachio nut), Nuts (walnut), Seeds (sunflower), AF/S, literature[15]

incidence: 1/4, conc.: 0.8 µg/kg*, sample origin: specialist food shops, UK, sample year: unknown, country: UK[16], *AFB_1, AFB_2, AFG_1, and AFG_2

- Co-contamination: not reported

- Further contamination of plants and plant products other than cereals and their products reported in the present book/article[16]: Food, AF/S, DON, FUS-X, NIV, OTA, and ZEA, literature[16]; Oils, Spices (fennel), AF/S and ZEA, literature[16]; Oils (sesame oil), OTA, literature[16]; Paste (curry paste), AF/S, FB/S, NIV, OTA, and ZEA, literature[16]; Pickles, AF/S, FB/S, OTA, and ZEA, literature[16]; Sauces (chilli sauce), NIV, OTA, and ZEA, literature[16]; Spices, AF/S, FB/S, OTA, and ZEA, literature[16]; Spices (cayenne pepper, chilli), AF/S, DAS, DON, HT-2, OTA, and ZEA, literature[16]; Spices (coriander), AF/S, DON, OTA, and ZEA, literature[16]; Spices (curry), AF/S, DAS, FB/S, FUS-X, HT-2, NEO, NIV, OTA, T-2, and ZEA, literature[16]; Spices (garlic), DON, FUS-X, and NIV, literature[16]; Spices (ginger), AF/S, DON, NEO, NIV, OTA, and T-2, literature[16]; Spices (tandoori), AF/S, FB/S, NIV, OTA, and T-2, literature[16]

incidence: 3/10*, conc. range: 1.2–3.4 µg/kg**, ∅ conc.: 2.13 µg/kg**, sample origin: retail shops and local markets in NWFP and northern areas, Pakistan, sample year: January–November 2009, country: Pakistan[17], *almonds without shell, **AFB_1, AFB_2, AFG_1, and AFG_2

- Co-contamination: not reported
- Further contamination of plants and plant products other than cereals and their products reported in the present book/article[17]: Dates, Figs, Fruits, dried, Kernels (apricot), Nuts (peanut), Nuts (pine nut), Nuts (pistachio nut), Nuts (walnut), and Raisins, AF/S, literature[17]

incidence: 1/2, conc.: 2.23 µg/kg*, sample origin: mainly imported from local retail outlets in Lodz (city), Poland, sample year: unknown, country: Poland[18], *AFS total

incidence: 1/1*, conc.: 0.88 µg/kg**, sample origin: mainly imported from local retail outlets in Lodz (city), Poland, sample year: unknown, country: Poland[18], *Californian almonds, **AFS total

- Co-contamination: not reported
- Further contamination of plants and plant products other than cereals and their products reported in the present book/article[18]: Nut cereals, Nuts, Nuts (cashew nut), Nuts (hazelnut), Nuts (peanut), Nuts (pistachio nut), Nuts (walnut), Seeds (sunflower), AF/S, literature[18]

incidence: 1/26, conc.: 6 µg/kg**, sample origin: FDA districts, USA, sample year: 1986, country: USA[19], *AFB_1, AFB_2, AFG_1, and AFG_2

incidence: 1/5, conc.: 10 µg/kg*, sample origin: imported to the USA, sample year: 1986, country: USA[19], *AFB_1, AFB_2, AFG_1, and AFG_2

- Co-contamination: not reported
- Further contamination of plants and plant products other than cereals and their products reported in the present book/article[19]: Butter (peanut butter), Candies (peanut candy), Candies (pistachio candy), Nuts, Nuts (Brazil nut, paranut), Nuts (peanut), Nuts (pecan nut), Nuts (pistachio nut), Nuts (walnut), Oils (peanut oil), Paste (peanut paste), Paste (pipian paste), Products (peanut product), Seeds (melon), Spices (cayenne pepper, chilli), Spices (ginger), Spices (nutmeg), AF/S, literature[19]

incidence: 10/74*, conc. range: ≤107 µg/kg**, ∅ conc.: 14.8 µg/kg**, sample origin: California (state), USA, sample year: 1972, country: USA[839], *unsorted in-shell almonds from growers, **AFB_1, AFB_2, AFG_1, and AFG_2

incidence: 2/16*, conc. range: 10–47 µg/kg**, ∅ conc.: 28.5 µg/kg**, sample origin: California (state), USA, sample year: 1972, country: USA[839], *unsorted almonds, **AFB_1, AFB_2, AFG_1, and AFG_2

incidence: 4/12*, conc. range: 9–272 µg/kg**, ∅ conc.: 132.8 µg/kg**, sample origin: California (state), USA, sample year: 1972, country: USA[839], *ESM-rejected almonds, **AFB_1, AFB_2, AFG_1, and AFG_2

incidence: 1/15*, conc.: 50 µg/kg**, sample origin: California (state), USA, sample year: 1972, country: USA[839], *ESM-rejected almonds, **AFB_1, AFB_2, AFG_1, and AFG_2

incidence: 10/16*, conc. range: 1–217 µg/kg**, ∅ conc.: 109 µg/kg**, sample origin: California (state), USA, sample year: 1972, country: USA[839], *hand-rejected almonds, **AFB_1, AFB_2, AFG_1, and AFG_2

incidence: 0/26*, conc.: no contamination, sample origin: California (state), USA, sample year: 1972, country: USA[839], *hand-rejected almonds (finished product), **AFB_1, AFB_2, AFG_1, and AFG_2

incidence: 1/25*, conc.: 4 µg/kg**, sample origin: California (state), USA, sample year: 1972, country: USA[839], *sliced nutmeats, natural and blanched, **AFB_1, AFB_2, AFG_1, and AFG_2

incidence: 13/27*, conc. range: ≤119 µg/kg**, sample origin: California (state), USA, sample year: 1972, country: USA[839], *diced nutmeats, natural and blanched, **AFB_1, AFB_2, AFG_1, and AFG_2

- Co-contamination: not reported
- Further contamination of plants and plant products other than cereals and their products reported in the present book/article[839]: Nuts (almond), AFB_1, AFB_2, AFG_1, and AFG_2, literature[839]

Other *Aspergillus* Toxins

Sterigmatocystin

incidence: 0/6*, conc.: no contamination, sample origin: major local retail outlets at Nainital (district), India, sample year: 1983–1985, country: India[7], *cuddapah almond

incidence: 2/8, conc. range: 96–130 µg/kg, ∅ conc.: 113 µg/kg, sample origin: major local retail outlets at Nainital (district), India, sample year: 1983–1985, country: India[7]

- Co-contamination: Often more than one mycotoxin was present in a sa.
- Further contamination of plants and plant products other than cereals and their products reported in the present book/article[7]: Coconuts, AFB_1, AFB_2, AFG_1, AFG_2, and CIT, literature[7]; Dates, AFB_1, AFB_2, AFG_1, and OTA, literature[7]; Grogannuts, AFB_1, AFB_2, AFG_1, AFG_2, and CIT, literature[7]; Nuts (almond), AFB_1, AFB_2, AFG_1, RBT, and ZEA, literature[7]; Nuts (cashew nut), AFB_1, AFB_2, AFG_1, AFG_2, CIT, and OTA, literature[7]; Nuts (peanut), Raisins, AFB_1, AFB_2, AFG_1, AFG_2, and OTA, literature[7]; Nuts (pistachio nut), AFB_1, AFB_2, AFG_1, AFG_2, and STG, literature[7]; Nuts (walnut), AFB_1, AFB_2, AFG_1, AFG_2, and RBT, literature[7]

Aspergillus and *Penicillium* Toxins

Cyclopiazonic Acid

incidence: 3/8, conc. range: ≤130 µg/kg, ∅ conc.: 64 µg/kg, sample origin: KSU, Turkey, and stores in Tulln (city) and Vienna (capital), Austria, sample year: unknown, country: Austria[868]

- Co-contamination: For detailed information see the article.
- Further contamination of plants and plant products other than cereals and their products reported in the present book/article[868]: Nuts (almond), AME, AOH, BEA, ENB, ENB_1, and

ENB_3, literature[868]; Nuts (hazelnut), 3-NPA, AFB_1, AFB_2, AFG_1, AFG_2, AME, AOH, ATX-I, BEA, ENA, ENA_1, ENB, ENB_2, ENB_3, ENB_4, HT-2, KA, MA, MON, OTA, OTB, STG, T-2, and ZEA, literature[868]; Nuts (peanut), 3-NPA, AFB_1, AFB_2, AME, AOH, ATX-I, BEA, CPA, ENA, ENA_1, ENB, ENB_4, KA, MA, OTA, OTB, STG, and ZEA, literature[868]; Nuts (pistachio nut), BEA, MA, and OTA, literature[868]; see also Nuts (almond), Nuts (hazelnut), Nuts (peanut), Nuts (pistachio nut), chapter: Further Mycotoxins and Microbial Metabolites

For detailed information see the article.

Ochratoxin A

incidence: 12/44, conc. range: 11–266 µg/kg, Ø conc.: 61 µg/kg, sample origin: retail markets and supermarkets in 24 regions, Tunisia, sample year: unknown, country: Tunisia[20]

- Co-contamination: not reported
- Further contamination of plants and plant products other than cereals and their products reported in the present book/article[20]: Nuts (peanut), Nuts (pistachio nut), Spices (caraway), Spices (coriander), Spices (curcuma), Spices (paprika), Spices (pepper), OTA, literature[20]

incidence: 1/1, conc.: 3.75 µg/kg, sample origin: local supermarket, Japan, sample year: unknown, country: Japan[21]

- Co-contamination: 1 sa. co-contaminated with OTA and OTB
- Further contamination of plants and plant products other than cereals and their products reported in the present book/article[21]: Nuts (almond), OTB, literature[21]; Nuts (cashew nut), Nuts (peanut), OTA and OTB, literature[21]

Ochratoxin B

incidence: 1/1, conc.: 8.83 µg/kg, sample origin: local supermarket, Japan, sample year: unknown, country: Japan[21]

- Co-contamination: 1 sa. co-contaminated with OTA and OTB
- Further contamination of plants and plant products other than cereals and their products reported in the present book/article[21]: Nuts (almond), OTA, literature[21]; Nuts (cashew nut), Nuts (peanut), OTA and OTB, literature[21]

Fusarium Toxins

Depsipeptides

Beauvericin

incidence: 8/8, conc. range: <LOQ, Ø conc.: 0.10 µg/kg, sample origin: KSU, Turkey and stores in Tulln (city) and Vienna (capital), Austria, sample year: unknown, country: Austria[868]

- Co-contamination: For detailed information see the article.
- Further contamination of plants and plant products other than cereals and their products reported in the present book/article[868]: Nuts (almond), AME, AOH, CPA, ENB, ENB_1, and ENB_3, literature[868]; Nuts (hazelnut), 3-NPA, AFB_1, AFB_2, AFG_1, AFG_2, AME, AOH, ATX-I, BEA, ENA, ENA_1, ENB, ENB_2, ENB_3, ENB_4, HT-2, KA, MA, MON, OTA, OTB, STG, T-2, and ZEA, literature[868]; Nuts (peanut), 3-NPA, AFB_1, AFB_2, AME, AOH, ATX-I, BEA, CPA, ENA, ENA_1, ENB, ENB_4, KA, MA, OTA, OTB, STG, and ZEA, literature[868]; Nuts (pistachio nut), BEA, MA, and OTA, literature[868]; see also Nuts (almond), Nuts (hazelnut), Nuts (peanut), Nuts (pistachio nut), chapter: Further Mycotoxins and Microbial Metabolites

For detailed information see the article.

Enniatin B

incidence: 8/8, conc. range: LOQ, Ø conc.: 0.05 µg/kg, sample origin: KSU, Turkey and stores in Tulln (city) and Vienna (capital), Austria, sample year: unknown, country: Austria[868]

- Co-contamination: For detailed information see the article.
- Further contamination of plants and plant products other than cereals and their products reported in the present book/article[868]: Nuts (almond), AME, AOH, BEA, CPA, ENB_1, and ENB_3, literature[868]; Nuts (hazelnut), 3-NPA, AFB_1, AFB_2, AFG_1, AFG_2, AME, AOH, ATX-I, BEA, ENA, ENA_1, ENB, ENB_2, ENB_3, ENB_4, HT-2, KA, MA, MON, OTA, OTB, STG, T-2, and ZEA, literature[868]; Nuts (peanut), 3-NPA, AFB_1, AFB_2, AME, AOH, ATX-I, BEA, CPA, ENA, ENA_1, ENB, ENB_4, KA, MA, OTA, OTB, STG, and ZEA, literature[868]; Nuts (pistachio nut), BEA, MA, and OTA, literature[868]; see also Nuts (almond), Nuts (hazelnut), Nuts (peanut), Nuts (pistachio nut), chapter: Further Mycotoxins and Microbial Metabolites

For detailed information see the article.

Enniatin B_1

incidence: 6/8, conc. range: ≤0.21 µg/kg, Ø conc.: 0.12 µg/kg, sample origin: KSU, Turkey and stores in Tulln (city) and Vienna (capital), Austria, sample year: unknown, country: Austria[868]

- Co-contamination: For detailed information see the article.
- Further contamination of plants and plant products other than cereals and their products reported in the present book/article[868]: Nuts (almond), AME, AOH, BEA, CPA, ENB, and ENB_3, literature[868]; Nuts (hazelnut), 3-NPA, AFB_1, AFB_2, AFG_1, AFG_2, AME, AOH, ATX-I, BEA, ENA, ENA_1, ENB, ENB_2, ENB_3, ENB_4, HT-2, KA, MA, MON, OTA, OTB, STG, T-2, and ZEA, literature[868]; Nuts (peanut), 3-NPA, AFB_1, AFB_2, AME, AOH, ATX-I, BEA, CPA, ENA, ENA_1, ENB, ENB_4, KA, MA, OTA, OTB, STG, and ZEA, literature[868]; Nuts (pistachio nut), BEA, MA, and OTA, literature[868]; see also Nuts (almond), Nuts (hazelnut), Nuts (peanut), Nuts (pistachio nut), chapter: Further Mycotoxins and Microbial Metabolites

For detailed information see the article.

Enniatin B_3

incidence: 1/8, conc.: 0.05 µg/kg, sample origin: KSU, Turkey and stores in Tulln (city) and Vienna (capital), Austria, sample year: unknown, country: Austria[868]

- Co-contamination: For detailed information see the article.
- Further contamination of plants and plant products other than cereals and their products reported in the present book/article[868]: Nuts (almond), AME, AOH, BEA, CPA, ENB, and ENB_1, literature[868]; Nuts (hazelnut), 3-NPA, AFB_1, AFB_2, AFG_1, AFG_2, AME, AOH, ATX-I, BEA, ENA, ENA_1, ENB, ENB_2, ENB_3, ENB_4, HT-2, KA, MA, MON, OTA, OTB, STG, T-2, and ZEA, literature[868]; Nuts (peanut), 3-NPA, AFB_1, AFB_2, AME, AOH, ATX-I, BEA, CPA, ENA, ENA_1, ENB, ENB_4, KA, MA, OTA, OTB, STG, and ZEA, literature[868]; Nuts (pistachio nut), BEA, MA, and OTA, literature[868]; see also Nuts (almond), Nuts (hazelnut), Nuts (peanut), Nuts (pistachio nut), chapter: Further Mycotoxins and Microbial Metabolites

For detailed information see the article.

Other ***Fusarium*** Toxins

Zearalenone

incidence: 1/6*, conc.: pr, sample origin: major local retail outlets at Nainital (district), India, sample year: 1983–1985, country: India[7], *cuddapah almond

incidence: 1/8, conc.: pr, sample origin: major local retail outlets at Nainital (district), India, sample year: 1983–1985, country: India[7]

- Co-contamination: Often more than one mycotoxin was present in a sa.
- Further contamination of plants and plant products other than cereals and their products reported in the present book/article[7]: Coconuts, AFB_1, AFB_2, AFG_1, AFG_2, and CIT, literature[7]; Dates, AFB_1, AFB_2, AFG_1, and OTA, literature[7]; Grogannuts, AFB_1, AFB_2, AFG_1, AFG_2, and CIT, literature[7]; Nuts (almond), AFB_1, AFB_2, AFG_1, RBT, and STG, literature[7]; Nuts (cashew nut), AFB_1, AFB_2, AFG_1, AFG_2, CIT, and OTA, literature[7]; Nuts (peanut), Raisins, AFB_1, AFB_2, AFG_1, AFG_2, and OTA, literature[7]; Nuts (pistachio nut), AFB_1, AFB_2, AFG_1, AFG_2, and STG, literature[7]; Nuts (walnut), AFB_1, AFB_2, AFG_1, AFG_2, and RBT, literature[7]

Penicillium Toxins

Rubratoxin

incidence: 1/6*, conc.: 310 µg/kg, sample origin: major local retail outlets at Nainital (district), India, sample year: 1983–1985, country: India[7], *cuddapah almond

incidence: 0/8, conc.: no contamination, sample origin: major local retail outlets at Nainital (district), India, sample year: 1983–1985, country: India[7]

- Co-contamination: Often more than one mycotoxin was present in a sa.
- Further contamination of plants and plant products other than cereals and their products reported in the present book/article[7]: Coconuts, AFB_1, AFB_2, AFG_1, AFG_2, and CIT, literature[7]; Dates, AFB_1, AFB_2, AFG_1, and OTA, literature[7]; Grogannuts, AFB_1, AFB_2, AFG_1, AFG_2, and CIT, literature[7]; Nuts (almond), AFB_1, AFB_2, AFG_1, STG, and ZEA, literature[7]; Nuts (cashew nut), AFB_1, AFB_2, AFG_1, AFG_2, CIT, and OTA, literature[7]; Nuts (peanut), Raisins, AFB_1, AFB_2, AFG_1, AFG_2, and OTA, literature[7]; Nuts (pistachio nut), AFB_1, AFB_2, AFG_1, AFG_2, and STG, literature[7]; Nuts (walnut), AFB_1, AFB_2, AFG_1, AFG_2, and RBT, literature[7]

Nuts (Brazil nut, paranut) may contain the following mycotoxins:

Aspergillus Toxins

Aflatoxin B_1

incidence: 5/23*, conc. range: 0.5–5 µg/kg (4 sa.) and 33 µg/kg (1 sa.), sample origin: retail outlets in 20 different regions, England, sample year: April 1984, country: UK[3], *Brazil nuts, in-shell

- Co-contamination: not reported
- Further contamination of plants and plant products other than cereals and their products reported in the present book/article[3]: Butter (peanut butter), Nuts (almond), Nuts (hazelnut), Nuts (peanut), AFB_1 and AF/S, literature[3]; Nuts (Brazil nut, paranut), AF/S, literature[3]

incidence: 7/17, conc. range: 0.005–2 µg/kg (3 sa.) and >2 µg/kg (4 sa., maximum: 1,400 µg/kg), sample origin: ethnic stores, health food stores, and local stores as well as supermarkets in the four largest cities, Sweden, sample year: autumn 1996–summer 1998, country: Sweden[13]

- Co-contamination: not reported
- Further contamination of plants and plant products other than cereals and their products reported in the present book/article[13]: Beans (brown bean), Peas, and Peas (chickpea), OTA, literature[13]; Juices (apple juice), PAT, literature[13]; Nuts (almond), AFB_1, literature[13]; Nuts (Brazil nut, paranut), AF/S, literature[13]; Nuts (pistachio nut), Oil seeds, Products (peanut product), AFB_1 and AF/S, literature[13]

incidence: 1/4, conc.: 10.2 µg/kg, sample origin: commercially available market basket sa. of foods and foodstuffs in Tokyo (capital), Japan, sample year: 1986–1990, country: Japan[99]

- Co-contamination: 1 sa. co-contaminated with AFB_1, AFB_2, AFG_1, and AFG_2
- Further contamination of plants and plant products other than cereals and their products reported in the present book/article[99]: Beans, Seeds (sesame), and Sugar, AFB_1 and AFB_2, literature[99]; Food, Jam (bean jam), and Paste (peanut paste), AFB_1, literature[99]; Nuts (Brazil nut, paranut), AFB_2, AFG_1, and AFG_2, literature[99]; Nuts (peanut), Nuts (pistachio nut), and Spices (nutmeg), AFB_1, AFB_2, AFG_1, and AFG_2, literature[99]; Spices, Spices (cayenne pepper, chilli), Spices (pepper), AFB_1, AFB_2, and AFG_1, literature[99]; Spices (paprika), AFB_1 and AFG_1, literature[99]

incidence: 10?/51, conc. range: 8.3–20 µg/kg, Ø conc.: 14.1 µg/kg, sample origin: sites of primary storage or processing (domestic products) and sites of importation, Cyprus, sample year: 1992–1996, country: Cyprus[200]

- Co-contamination: not reported
- Further contamination of plants and plant products other than cereals and their products reported in the present book/article[200]: Butter (peanut butter), Figs, and Nuts (peanut), AFB_1, AFB_2, AFG_1, and AFG_2, literature[200]; Nuts (Brazil nut, paranut), AFB_2 and AFG_1, literature[200]; Nuts (chestnut), Products (sesame product), and Spices, AFB_1, literature[200]; Nuts (pistachio nut), AFB_1 and AFB_2, literature[200]; Nuts (walnut), AFB_1, AFB_2, and AFG_1, literature[200]; Seeds (sesame), AFB_1, AFG_1, and AFG_2, literature[200]

incidence: 16/27, conc. range: 3–4,200 µg/kg, Ø conc.: 457 µg/kg, sample origin: stores in Oslo (capital) area, Norway, sample year: autumn 1973, country: Norway[201]

- Co-contamination: 16 sa. co-contaminated with AFB_1, AFB_2, AFG_1, and AFG_2
- Further contamination of plants and plant products other than cereals and their products reported in the present book/article[201]: Cocoa, Nuts (cashew nut), Nuts (peanut), and Nuts (walnut), AFB_1, AFB_2, AFG_1, and AFG_2, literature[201]; Nuts, Nuts (Brazil nut, paranut), AFB_2, AFG_1, and AFG_2, literature[201]

incidence: 1/2*, conc.: 26 µg/kg, sample origin: food stores and supermarkets in Campinas (city), São Paulo (state), Brazil, sample year: 1991/1995, country: Brazil[202], *Brazil nuts, ground

- Co-contamination: not reported
- Further contamination of plants and plant products other than cereals and their products reported in the present book/article[202]: Nuts (cashew nut), AFB_1 and AFG_1, literature[202]

incidence: 4/12, conc. range: LOQ-2 µg/kg (4 sa., maximum: 0.8 µg/kg), Ø conc.: 0.5 µg/kg, sample origin: retail outlets, UK, sample year: September 2001, country: UK[203]

- Co-contamination: 1 sa. co-contaminated with AFB_1 and AFB_2; 1 sa. co-contaminated with AFB_1 and AFG_1; 2 sa. contaminated solely with AFB_1
- Further contamination of plants and plant products other than cereals and their products reported in the present book/article[203]: Butter (nut butter), AFB_1, AFB_2, AFG_1, AFG_2, and OTA, literature[203]; Butter (peanut butter), AFB_1, AFB_2, AFG_1, AFG_2, and OTA, literature[203]; Dates, AFB_1 and OTA, literature[203]; Figs, AFB_1, AFB_2, AFG_2, and OTA, literature[203]; Fruits, dried, Nuts (pistachio nut), AFB_1, AFB_2, and OTA, literature[203]; Nuts (Brazil nut, paranut), AFB_2 and AFG_1, literature[203]; Nuts (peanut), AFB_1 and AFB_2, literature[203]

incidence: 4/50, conc. range: 11.9–1,058.0 µg/kg, sample origin: cooperatives in Acre (state), northern region in Brazil, sample year: 2009, country: Brazil[204]

incidence: 9/50, conc. range: 6.8–36.6 µg/kg, sample origin: cooperatives in Amazonas (state), northern region in Brazil, sample year: 2009, country: Brazil[204]

incidence: 5/50, conc. range: 2.5–38.7 µg/kg, sample origin: cooperatives in Amapá (state), northern region in Brazil, sample year: 2009, country: Brazil[204]

incidence: 4/50, conc. range: 1.7–6.9 µg/kg, sample origin: cooperatives in Pará (state), northern region in Brazil, sample year: 2009, country: Brazil[204]

- Co-contamination: not reported
- Further contamination of plants and plant products other than cereals and their products reported in the present book/article[204]: Nuts (Brazil nut, paranut), AFB_2, AFG_1, AFG_2, and CPA, literature[204]

incidence: 1/10, conc.: 5.8 µg/kg, sample origin: available in New Zealand, sample year: 2010, country: New Zealand[277]

- Co-contamination: not reported
- Further contamination of plants and plant products other than cereals and their products reported in the present book/article[277]: Bars, AFB_1, AF/S, DON, NIV, and OTA, literature[277]; Butter (peanut butter), Confectionery, Food, Fruits, dried, Nuts, Nuts (peanut), Nuts (pistachio nut), and Sauces (peanut sauce), AFB_1 and AF/S, literature[277]; Coffee, Dates, Spices (pepper), and Vine fruits, dried, OTA, literature[277]; Figs, Spices (cayenne pepper, chilli), Spices (curry), Spices (ginger), Spices (paprika), AFB_1, AF/S, and OTA, literature[277]

incidence: 1/10, conc.: 5.8 µg/kg, sample origin: New Zealand, sample year: unknown, country: New Zealand[701]

- Co-contamination: not reported
- Further contamination of plants and plant products other than cereals and their products reported in the present book/article[701]: Butter (peanut butter), Nuts (peanut), Nuts (pistachio nut), Products (nut product), Products (peanut product), AF/S, literature[701]

Aflatoxin B_2

incidence: 1/4, conc.: 0.8 µg/kg, sample origin: commercially available market basket sa. of foods and foodstuffs in Tokyo (capital), Japan, sample year: 1986–1990, country: Japan[99]

- Co-contamination: 1 sa. co-contaminated with AFB_1, AFB_2, AFG_1, and AFG_2
- Further contamination of plants and plant products other than cereals and their products reported in the present book/article[99]: Beans, Seeds (sesame), and Sugar, AFB_1 and AFB_2, literature[99]; Food, Jam (bean jam), and Paste (peanut paste), AFB_1, literature[99]; Nuts (Brazil nut, paranut), AFB_1, AFG_1, and AFG_2, literature[99]; Nuts (peanut), Nuts (pistachio nut), Spices (nutmeg), AFB_1, AFB_2, AFG_1, and AFG_2, literature[99]; Spices, Spices (cayenne pepper, chilli), Spices (pepper), AFB_1, AFB_2, and AFG_1, literature[99]; Spices (paprika), AFB_1 and AFG_1, literature[99]

incidence: 10?/51, conc. range: 1.1 µg/kg, ∅ conc.: 1.1 µg/kg, sample origin: sites of primary storage or processing (domestic products) and sites of importation, Cyprus, sample year: 1992–1996, country: Cyprus[200]

- Co-contamination: not reported
- Further contamination of plants and plant products other than cereals and their products reported in the present book/article[200]: Butter (peanut butter), Figs, and Nuts (peanut), AFB_1, AFB_2, AFG_1, and AFG_2, literature[200]; Nuts (Brazil nut, paranut), AFB_1 and AFG_1, literature[200]; Nuts (chestnut), Products (sesame product), and Spices, AFB_1, literature[200]; Nuts (pistachio nut), AFB_1 and AFB_2, literature[200]; Nuts (walnut), AFB_1, AFB_2, and AFG_1, literature[200]; Seeds (sesame), AFB_1, AFG_1, and AFG_2, literature[200]

incidence: 16/27, conc. range: tr-1,600 µg/kg, sample origin: stores in Oslo (capital) area, Norway, sample year: autumn 1973, country: Norway[201]

- Co-contamination: 16 sa. co-contaminated with AFB_1, AFB_2, AFG_1, and AFG_2
- Further contamination of plants and plant products other than cereals and their products reported in the present book/article[201]: Cocoa, Nuts (cashew nut), Nuts (peanut), Nuts (walnut), AFB_1, AFB_2, AFG_1, and AFG_2, literature[201]; Nuts, AFB_2, AFG_1, and AFG_2, literature[201]; Nuts (Brazil nut, paranut), AFB_1, AFG_1, and AFG_2, literature[201]

incidence: 1/12, conc.: 0.3 µg/kg, sample origin: retail outlets, UK, sample year: September 2001, country: UK[203]

- Co-contamination: 1 sa. co-contaminated with AFB_1 and AFB_2
- Further contamination of plants and plant products other than cereals and their products reported in the present book/article[203]: Butter (nut butter), AFB_1, AFB_2, AFG_1, AFG_2, and OTA, literature[203]; Butter (peanut butter), AFB_1, AFB_2, AFG_1, AFG_2, and OTA, literature[203]; Dates, AFB_1 and OTA, literature[203]; Figs, AFB_1, AFB_2, AFG_2, and OTA, literature[203]; Fruits, dried, Nuts (pistachio nut), AFB_1, AFB_2, and OTA, literature[203]; Nuts (Brazil nut, paranut), AFB_1 and AFG_1, literature[203]; Nuts (peanut), AFB_1 and AFB_2, literature[203]

incidence: 3/50, conc. range: 13.7–130.4 µg/kg, sample origin: cooperatives in Acre (state), northern region in Brazil, sample year: 2009, country: Brazil[204]

incidence: 2/50, conc. range: 1.1–1.6 µg/kg, ∅ conc.: 1.35 µg/kg, sample origin: cooperatives in Amazonas (state), northern region in Brazil, sample year: 2009, country: Brazil[204]

incidence: 2/50, conc. range: 0.5–7.7 µg/kg, sample origin: cooperatives in Amapá (state), northern region in Brazil, sample year: 2009, country: Brazil[204]

incidence: 0/50, conc.: no contamination, sample origin: cooperatives in Pará (state), northern region in Brazil, sample year: 2009, country: Brazil[204]

- Co-contamination: not reported
- Further contamination of plants and plant products other than cereals and their products reported in the present book/article[204]: Nuts (Brazil nut, paranut), AFB_1, AFG_1, AFG_2, and CPA, literature[204]

Aflatoxin G_1

incidence: 1/4, conc.: 3.2 µg/kg, sample origin: commercially available market basket sa. of foods and foodstuffs in Tokyo (capital), Japan, sample year: 1986–1990, country: Japan[99]

- Co-contamination: 1 sa. co-contaminated with AFB_1, AFB_2, AFG_1, and AFG_2
- Further contamination of plants and plant products other than cereals and their products reported in the present book/article[99]: Beans, Seeds (sesame), and Sugar, AFB_1 and AFB_2, literature[99]; Food, Jam (bean jam), and Paste (peanut paste), AFB_1, literature[99]; Nuts (Brazil nut, paranut), AFB_1, AFB_2, and AFG_2, literature[99]; Nuts (peanut), Nuts (pistachio nut), Spices (nutmeg), AFB_1, AFB_2, AFG_1, and AFG_2, literature[99]; Spices, Spices (cayenne pepper, chilli), Spices (pepper), AFB_1, AFB_2, and AFG_1, literature[99]; Spices (paprika), AFB_1 and AFG_1, literature[99]

incidence: 10?/51, conc. range: 2.3–9.4 µg/kg, ∅ conc.: 5.8 µg/kg, sample origin: sites of primary storage or processing (domestic products) and sites of importation, Cyprus, sample year: 1992–1996, country: Cyprus[200]

- Co-contamination: not reported
- Further contamination of plants and plant products other than cereals and their products reported in the present book/article[200]: Butter (peanut butter), Figs, Nuts (peanut), AFB_1, AFB_2, AFG_1, and AFG_2, literature[200]; Nuts (Brazil nut, paranut), AFB_1 and AFB_2, literature[200]; Nuts (chestnut), Products (sesame product), and Spices, AFB_1, literature[200]; Nuts (pistachio nut), AFB_1 and AFB_2, literature[200]; Nuts (walnut), AFB_1, AFB_2, and AFG_1, literature[200]; Seeds (sesame), AFB_1, AFG_1, and AFG_2, literature[200]

incidence: 16/27, conc. range: 2–3,250 µg/kg, ∅ conc.: 478 µg/kg, sample origin: stores in Oslo (capital) area, Norway, sample year: autumn 1973, country: Norway[201]

- Co-contamination: 16 sa. co-contaminated with AFB_1, AFB_2, AFG_1, and AFG_2
- Further contamination of plants and plant products other than cereals and their products reported in the present book/article[201]: Cocoa, Nuts (cashew nut), Nuts (peanut), Nuts (walnut), AFB_1, AFB_2, AFG_1, and AFG_2, literature[201]; Nuts, AFB_2, AFG_1, and AFG_2, literature[201]; Nuts (Brazil nut, paranut), AFB_1, AFB_2, and AFG_2, literature[201]

incidence: 1/12, conc.: 0.3 µg/kg, sample origin: retail outlets, UK, sample year: September 2001, country: UK[203]

- Co-contamination: 1 sa. co-contaminated with AFB_1 and AFG_1
- Further contamination of plants and plant products other than cereals and their products reported in the present book/article[203]: Butter (nut butter), AFB_1, AFB_2, AFG_1, AFG_2, and OTA, literature[203]; Butter (peanut butter), AFB_1, AFB_2, AFG_1, AFG_2, and OTA, literature[203]; Dates, AFB_1 and OTA, literature[203]; Figs, AFB_1, AFB_2, AFG_2, and OTA, literature[203]; Fruits, dried, Nuts (pistachio nut), AFB_1, AFB_2, and OTA, literature[203]; Nuts (Brazil nut, paranut), Nuts (peanut), AFB_1 and AFB_2, literature[203]

incidence: 3/50, conc. range: 17.9–243.0 µg/kg;, sample origin: cooperatives in Acre (state), northern region in Brazil, sample year: 2009, country: Brazil[204]

incidence: 7/50, conc. range: 2.7–12.6 µg/kg, sample origin: cooperatives in Amazonas (state), northern region in Brazil, sample year: 2009, country: Brazil[204]

incidence: 1/50, conc.: 2.9 µg/kg, sample origin: cooperatives in Amapá (state), northern region in Brazil, sample year: 2009, country: Brazil[204]

incidence: 0/50, conc.: no contamination, sample origin: cooperatives in Pará (state), northern region in Brazil, sample year: 2009, country: Brazil[204]

- Co-contamination: not reported
- Further contamination of plants and plant products other than cereals and their products reported in the present book/article[204]: Nuts (Brazil nut, paranut), AFB_1, AFB_2, AFG_2, and CPA, literature[204]

Aflatoxin G_2

incidence: 1/4, conc.: 0.3 µg/kg, sample origin: commercially available market basket sa. of foods and foodstuffs in Tokyo (capital), Japan, sample year: 1986–1990, country: Japan[99]

- Co-contamination: 1 sa. co-contaminated with AFB_1, AFB_2, AFG_1, and AFG_2
- Further contamination of plants and plant products other than cereals and their products reported in the present book/article[99]: Beans, Seeds (sesame), Sugar, AFB_1 and AFB_2, literature[99]; Food, Jam (bean jam), Paste (peanut paste), AFB_1, literature[99]; Nuts (Brazil nut, paranut), Spices, Spices (cayenne pepper, chilli), Spices (pepper), AFB_1, AFB_2, and AFG_1, literature[99]; Nuts (peanut), Nuts (pistachio nut), Spices (nutmeg), AFB_1, AFB_2, AFG_1, and AFG_2, literature[99]; Spices (paprika), AFB_1 and AFG_1, literature[99]

incidence: 16/27, conc. range: tr-600 µg/kg, sample origin: stores in Oslo (capital) area, Norway, sample year: autumn 1973, country: Norway[201]

- Co-contamination: 16 sa. co-contaminated with AFB_1, AFB_2, AFG_1, and AFG_2
- Further contamination of plants and plant products other than cereals and their products reported in the present book/article[201]: Cocoa, Nuts (cashew nut), Nuts (peanut), and Nuts (walnut), AFB_1, AFB_2, AFG_1, and AFG_2, literature[201]; Nuts, AFB_2, AFG_1, and AFG_2, literature[201]; Nuts (Brazil nut, paranut), AFB_1, AFB_2, and AFG_1, literature[201]

incidence: 2/50, conc. range: 14.7–223.6 µg/kg, ∅ conc.: 119 µg/kg, sample origin: cooperatives in Acre (state), northern region in Brazil, sample year: 2009, country: Brazil[204]

incidence: 1/50, conc.: 1.8 µg/kg, sample origin: cooperatives in Amazonas (state), northern region in Brazil, sample year: 2009, country: Brazil[204]

incidence: 1/50, conc.: 0.8 µg/kg, sample origin: cooperatives in Amapá (state), northern region in Brazil, sample year: 2009, country: Brazil[204]

incidence: 0/50, conc.: no contamination, sample origin: cooperatives in Pará (state), northern region in Brazil, sample year: 2009, country: Brazil[204]

- Co-contamination: not reported
- Further contamination of plants and plant products other than cereals and their products reported in the present book/article[204]: Nuts (Brazil nut, paranut), AFB_1, AFB_2, AFG_1, and CPA, literature[204]

Aflatoxin/s

incidence: 5/23*, conc. range: 0.5–5 µg/kg** (3 sa.), 6–10 µg/kg** (1 sa.), and 60 µg/kg** (1 sa.), sample origin: retail outlets in 20 different regions, England, sample year: April 1984, country: UK[3], *Brazil nuts, in-shell, **AFS total

- Co-contamination: not reported
- Further contamination of plants and plant products other than cereals and their products reported in the present book/article[3]: Butter (peanut butter), Nuts (almond), Nuts (hazelnut), Nuts (peanut), AFB_1 and AFS, literature[3]; Nuts (Brazil nut, paranut), AFB_1, literature[3]

incidence: 7/17, conc. range: 0.01–4 µg/kg* (3 sa.), >4 µg/kg* (4 sa., maximum: 2,500 µg/kg*), sample origin: ethnic stores, health food stores, and local stores as well as supermarkets in the four largest cities, Sweden, sample year: autumn 1996–summer 1998, country: Sweden[13], *AFB_1, AFB_2, AFG_1, and AFG_2

- Co-contamination: not reported
- Further contamination of plants and plant products other than cereals and their products reported in the present book/article[13]: Beans (brown bean), Peas, Peas (chickpea), OTA, literature[13]; Juices (apple juice), PAT, literature[13]; Nuts (almond) and Nuts (Brazil nut, paranut), AFB_1, literature[13]; Nuts (pistachio nut), Oil seeds, Products (peanut product), AFB_1 and AF/S, literature[13]

incidence: 6/12, conc. range: ≤42 µg/kg*, sample origin: imported to the USA, sample year: 1986, country: USA[19], *AFB_1, AFB_2, AFG_1, and AFG_2

- Co-contamination: not reported
- Further contamination of plants and plant products other than cereals and their products reported in the present book/article[19]: Butter (peanut butter), Candies (peanut candy), Candies (pistachio candy), Nuts, Nuts (almond), Nuts (peanut), Nuts (pecan nut), Nuts (pistachio nut), Nuts (walnut), Oils (peanut oil), Paste (peanut paste), Paste (pipian paste), Products (peanut product), Seeds (melon), Spices (cayenne pepper, chilli), Spices (ginger), Spices (nutmeg), AF/S, literature[19]

incidence: 31/69*, conc. range: 6–10 µg/kg** (1 sa.), 11–20 µg/kg** (3 sa.), 21–50 µg/kg** (1 sa.), 51–100 µg/kg** (3 sa.), 101–1,000 µg/kg** (7 sa.), 1,001–10,000 µg/kg** (11 sa.), and >10,000 µg/kg** (5 sa.), sample origin: imported to Sweden, sample year: 1973, country: Sweden[205], *edible, possibly edible, and inedible Brazil nuts, **AFB_1, AFB_2, AFG_1, and AFG_2

- Co-contamination: not reported
- Further contamination of plants and plant products other than cereals and their products reported in the present book/article[205]: Nuts (hazelnut), AF/S, literature[205]

incidence: 3/3, conc. range: 10.1–20 µg/kg* (2 sa.) and 150 µg/kg* (1 sa.), sample origin: port sa. from Felixstowe (town), Manchester (city), and King's Lynn (market town) Port Health Authorities, UK, sample year: unknown, country: UK[206], *AFB_1, AFB_2, AFG_1, and AFG_2

- Co-contamination: not reported
- Further contamination of plants and plant products other than cereals and their products reported in the present book/article[206]: Butter (peanut butter), Figs, Nuts (peanut), Nuts (pistachio nut), Paste (fig paste), AF/S, literature[206]

incidence: 4/60*, conc. range: 10–68 µg/kg**, ⌀ conc.: 46 µg/kg**, sample origin: Brazil nut factory in Manaus (city), Amazonas (state), Brazil, sample year: 2003–2006, country: Brazil[207], *factory stage of the sa.: receiving, **AFB_1, AFB_2, AFG_1, and AFG_2

incidence: 0/30*, conc.: no contamination, sample origin: Brazil nut factory in Manaus (city), Amazonas (state), Brazil, sample year: 2003–2006, country: Brazil[207], *factory stage of the sa.: after drying

incidence: 5/30*, conc. range: 8–686 µg/kg**, ⌀ conc.: 271.7 µg/kg**, sample origin: local markets in Manaus (city), Amazonas (state), Brazil, sample year: 2003–2006, country: Brazil[207], *retail sa., **AFB_1, AFB_2, AFG_1, and AFG_2

- Co-contamination: not reported
- Further contamination of plants and plant products other than cereals and their products reported in the present book/article[207]: not reported

incidence: 4/67*, conc. range: 16.5–1,972 µg/kg**, ⌀ conc.: 553.6 µg/kg**, sample origin: local retail stores in the region of Brasilia (capital), Brazil, sample year: 2002–2011, country: Brazil[208], *Brazil nuts, shelled, **AFB_1, AFB_2, AFG_1, and AFG_2

- Co-contamination: not reported
- Further contamination of plants and plant products other than cereals and their products reported in the present book/article[208]: Nuts, Nuts (peanut), Products (peanut product), AF/S, literature[208]

incidence: 55/80*, conc. range: 1.2–11.5 µg/kg**, sample origin: eastern (oriental) and western (occidental) regions of the Brazilian Amazonas basin, Brazil, sample year: 2006, country: Brazil[209], *Brazil nuts, in-shell, raw, and medium size, **AFB_1, AFB_2, AFG_1, and AFG_2

- Co-contamination: not reported
- Further contamination of plants and plant products other than cereals and their products reported in the present book/article[209]: not reported

incidence: 11/21*, conc. range: ≤0.98 µg/kg**, sample origin: supermarkets in São Paulo (city), São Paulo (state), Brazil, sample year: unknown, country: Brazil[747], *ready-to-eat Brazil nuts, shelled and in-shell, **AFB_1, AFB_2, AFG_1, and AFG_2

incidence: 20/51*, conc. range: ≤7.46 µg/kg**, sample origin: street markets in Manaus (city), Amazonas (state), Brazil, sample year: unknown, country: Brazil[747], *ready-to-eat Brazil nuts, shelled and in-shell, **AFB_1, AFB_2, AFG_1, and AFG_2

incidence: 11/23*, conc. range: ≤140.00 µg/kg**, sample origin: street markets in Belém (city), Pará (state), Brazil, sample year: unknown, country: Brazil[747], *ready-to-eat Brazil nuts, shelled and in-shell, **AFB_1, AFB_2, AFG_1, and AFG_2

- Co-contamination: not reported
- Further contamination of plants and plant products other than cereals and their products reported in the present book/article[747]: not reported

For detailed information see the article.

incidence: 7/40*, conc. range: 1.2–11.5 μg/kg**, sample origin: CIEX factory in Manaus (city), Amazonas (state), Brazil, sample year: unknown, country: Brazil[877], *dry in-shell Brazil nuts for export, **AFB_1, AFB_2, AFG_1, and AFG_2

- Co-contamination: not reported
- Further contamination of plants and plant products other than cereals and their products reported in the present book/article[877]: not reported

Aspergillus and ***Penicillium*** Toxins

Cyclopiazonic Acid

incidence: 3/50, conc. range: 124.7–151.3 μg/kg, sample origin: cooperatives in Acre (state), northern region in Brazil, sample year: 2009, country: Brazil[204]

incidence: 1/50, conc.: 132.8 μg/kg, sample origin: cooperatives in Amazonas (state), northern region in Brazil, sample year: 2009, country: Brazil[204]

incidence: 2/50, conc. range: 143.3–161.2 μg/kg, Ø conc.: 152 μg/kg, sample origin: cooperatives in Amapá (state), northern region in Brazil, sample year: 2009, country: Brazil[204]

incidence: 5/50, conc. range: 98.65–123.31 μg/kg, sample origin: cooperatives in Pará (state), northern region in Brazil, sample year: 2009, country: Brazil[204]

- Co-contamination: not reported
- Further contamination of plants and plant products other than cereals and their products reported in the present book/article[204]: Nuts (Brazil nut, paranut), AFB_1, AFB_2, AFG_1, and AFG_2, literature[204]

Nuts (cashew nut) may contain the following mycotoxins:

Aspergillus Toxins

Aflatoxin B_1

incidence: 3/7, conc. range: 20–190 μg/kg, sample origin: major local retail outlets at Nainital (district), India, sample year: 1983–1985, country: India[7]

- Co-contamination: Often more than one mycotoxin was present in a sa.
- Further contamination of plants and plant products other than cereals and their products reported in the present book/article[7]: Coconuts, AFB_1, AFB_2, AFG_1, AFG_2, and CIT, literature[7]; Dates, AFB_1, AFB_2, AFG_1, and OTA, literature[7]; Grogannuts, AFB_1, AFB_2, AFG_1, AFG_2, and CIT, literature[7]; Nuts (almond), AFB_1, AFB_2, AFG_1, RBT, STG, and ZEA, literature[7]; Nuts (cashew nut), AFB_2, AFG_1, AFG_2, CIT, and OTA, literature[7]; Nuts (peanut), Raisins, AFB_1, AFB_2, AFG_1, AFG_2, and OTA, literature[7]; Nuts (pistachio nut), AFB_1, AFB_2, AFG_1, AFG_2, and STG, literature[7]; Nuts (walnut), AFB_1, AFB_2, AFG_1, AFG_2, and RBT, literature[7]

incidence: 1/3, conc.: tr, sample origin: stores in Oslo (capital) area, Norway, sample year: autumn 1973, country: Norway[201]

- Co-contamination: 1 sa. co-contaminated with AFB_1, AFB_2, AFG_1, and AFG_2
- Further contamination of plants and plant products other than cereals and their products reported in the present book/article[201]: Cocoa, Nuts (Brazil nut, paranut), Nuts (peanut), Nuts (walnut), AFB_1, AFB_2, AFG_1, and AFG_2, literature[201]; Nuts, and Nuts (cashew nut), AFB_2, AFG_1, and AFG_2, literature[201]

incidence: 1/8*, conc.: 10 μg/kg, sample origin: food stores and supermarkets in Campinas (city), São Paulo (state), Brazil, sample year: 1991/1995, country: Brazil[202], *cashew nuts, ground

- Co-contamination: 1 sa. co-contaminated with AFB_1 and AFG_1
- Further contamination of plants and plant products other than cereals and their products reported in the present book/article[202]: Nuts (Brazil nut, paranut nut), AFB_1, literature[202]; Nuts (cashew nut), AFG_1, literature[202]

incidence: 1/6*, conc.: 830 μg/kg, sample origin: unknown, sample year: unknown, country: Germany[239], *cashew nuts, moldy

- Co-contamination: not reported
- Further contamination of plants and plant products other than cereals and their products reported in the present book/article[239]: Fruits (apple), Fruits (cherry), Fruits (lemon), Fruits (orange), Fruits (peach), Fruits (tomato), Nuts (almond), Nuts (hazelnut), Nuts (walnut), Seeds (sesame), Spice (nutmeg), AFB_1, literature[239]; Nuts (peanut), AFB_1, AFB_2, AFG_1, and AFG_2, literature[239]

Aflatoxin B_2

incidence: 3/7, conc. range: 15–161 μg/kg, sample origin: major local retail outlets at Nainital (district), India, sample year: 1983–1985, country: India[7]

- Co-contamination: Often more than one mycotoxin was present in a sa.
- Further contamination of plants and plant products other than cereals and their products reported in the present book/article[7]: Coconuts, AFB_1, AFB_2, AFG_1, AFG_2, and CIT, literature[7]; Dates, AFB_1, AFB_2, AFG_1, and OTA, literature[7]; Grogannuts, AFB_1, AFB_2, AFG_1, AFG_2, and CIT, literature[7]; Nuts (almond), AFB_1, AFB_2, AFG_1, RBT, STG, and ZEA, literature[7]; Nuts (cashew nut), AFB_1, AFG_1, AFG_2, CIT, and OTA, literature[7]; Nuts (peanut), Raisins, AFB_1, AFB_2, AFG_1, AFG_2, and OTA, literature[7]; Nuts (pistachio nut), AFB_1, AFB_2, AFG_1, AFG_2, and STG, literature[7]; Nuts (walnut), AFB_1, AFB_2, AFG_1, AFG_2, and RBT, literature[7]

incidence: 1/3, conc.: tr, sample origin: stores in Oslo (capital) area, Norway, sample year: autumn 1973, country: Norway[201]

- Co-contamination: 1 sa. co-contaminated with AFB_1, AFB_2, AFG_1, and AFG_2
- Further contamination of plants and plant products other than cereals and their products reported in the present book/article[201]: Cocoa, Nuts (Brazil nut, paranut), Nuts (peanut), Nuts (walnut), AFB_1, AFB_2, AFG_1, and AFG_2, literature[201]; Nuts, AFB_2, AFG_1, and AFG_2, literature[201]; Nuts (cashew nut), AFB_1, AFG_1, and AFG_2, literature[201]

Aflatoxin G_1

incidence: 2/7, conc. range: 18–72 μg/kg, Ø conc.: 45 μg/kg, sample origin: major local retail outlets at Nainital (district), India, sample year: 1983–1985, country: India[7]

- Co-contamination: Often more than one mycotoxin was present in a sa.
- Further contamination of plants and plant products other than cereals and their products reported in the present book/article[7]: Coconuts, AFB_1, AFB_2, AFG_1, AFG_2, and CIT, literature[7]; Dates, AFB_1, AFB_2, AFG_1, and OTA, literature[7]; Grogannuts, AFB_1, AFB_2, AFG_1, AFG_2, and CIT, literature[7]; Nuts (almond), AFB_1, AFB_2, AFG_1, RBT, STG, and ZEA, literature[7]; Nuts (cashew nut), AFB_1, AFB_2, AFG_2, CIT, and OTA, literature[7]; Nuts (peanut), Raisins, AFB_1, AFB_2, AFG_1, AFG_2, and OTA, literature[7]; Nuts (pistachio nut), AFB_1, AFB_2, AFG_1, AFG_2, and STG, literature[7]; Nuts (walnut), AFB_1, AFB_2, AFG_1, AFG_2, and RBT, literature[7]

incidence: 1/3, conc.: tr, sample origin: stores in Oslo (capital) area, Norway, sample year: autumn 1973, country: Norway[201]

- Co-contamination: 1 sa. co-contaminated with AFB_1, AFB_2, AFG_1, and AFG_2
- Further contamination of plants and plant products other than cereals and their products reported in the present book/article[201]: Cocoa, Nuts (Brazil nut, paranut), Nuts (peanut), Nuts (walnut), AFB_1, AFB_2, AFG_1, and AFG_2, literature[201]; Nuts, AFB_2, AFG_1, and AFG_2, literature[201]; Nuts (cashew nut), AFB_1, AFB_2, and AFG_2, literature[201]

incidence: 1/8*, conc.: 15 µg/kg, sample origin: food stores and supermarkets in Campinas (city), São Paulo (state), Brazil, sample year: 1991/1995, country: Brazil[202], *Cashew nuts, ground

- Co-contamination: 1 sa. co-contaminated with AFB_1 and AFG_1
- Further contamination of plants and plant products other than cereals and their products reported in the present book/article[202]: Nuts (Brazil nut, paranut nut), Nuts (cashew nut), AFB_1, literature[202]

Aflatoxin G_2

incidence: 1/7, conc.: 20 µg/kg, sample origin: major local retail outlets at Nainital (district), India, sample year: 1983–1985, country: India[7]

- Co-contamination: Often more than one mycotoxin was present in a sa.
- Further contamination of plants and plant products other than cereals and their products reported in the present book/article[7]: Coconuts, AFB_1, AFB_2, AFG_1, AFG_2, and CIT, literature[7]; Dates, AFB_1, AFB_2, AFG_1, and OTA, literature[7]; Grogannuts, AFB_1, AFB_2, AFG_1, AFG_2, and CIT, literature[7]; Nuts (almond), AFB_1, AFB_2, AFG_1, RBT, STG, and ZEA, literature[7]; Nuts (cashew nut), AFB_1, AFB_2, AFG_1, CIT, and OTA, literature[7]; Nuts (peanut), Raisins, AFB_1, AFB_2, AFG_1, AFG_2, and OTA, literature[7]; Nuts (pistachio nut), AFB_1, AFB_2, AFG_1, AFG_2, and STG, literature[7]; Nuts (walnut), AFB_1, AFB_2, AFG_1, AFG_2, and RBT, literature[7]

incidence: 1/3, conc.: tr, sample origin: stores in Oslo (capital) area, Norway, sample year: autumn 1973, country: Norway[201]

- Co-contamination: 1 sa. co-contaminated with AFB_1, AFB_2, AFG_1, and AFG_2
- Further contamination of plants and plant products other than cereals and their products reported in the present book/article[201]: Cocoa, Nuts (Brazil nut, paranut), Nuts (peanut), and Nuts (walnut), AFB_1, AFB_2, AFG_1, and AFG_2, literature[201]; Nuts, AFB_2, AFG_1, and AFG_2, literature[201]; Nuts (cashew nut), AFB_1, AFB_2, and AFG_1, literature[201]

Aflatoxin/s

incidence: 6/20*, Ø conc.: 1.6 µg/kg**, sample origin: greater Manila (capital) area, Philippines, sample year: unknown, country: Philippines[1], *cashew and cashew products, **AFS of pos sa.?

- Co-contamination: not reported
- Further contamination of plants and plant products other than cereals and their products reported in the present book/article[1]: Beans, Butter (peanut butter), Cassava, Cocoa, Coffee, Drinks (alcoholic drink), Nuts (peanut), Products (coconut product), Rootcrops, Spices (cayenne pepper, chilli), Sweet potatoes, Yams, AF/S, literature[1]

incidence: 23/120, conc. range: 5.0–9.9 µg/kg* (10 sa.), 10.0–14.9 µg/kg* (3 sa.), 15.0–19.9 µg/kg* (4 sa.), 20.0–24.9 µg/kg* (2 sa.), and >25 µg/kg* (4 sa.), sample origin: brokers, distributors, growers, processors, and retail trade, Canada, sample year: 1970–1975, country: Canada[15], *AFB_1, AFB_2, AFG_1, and AFG_2

- Co-contamination: not reported
- Further contamination of plants and plant products other than cereals and their products reported in the present book/article[15]: Butter (peanut butter), Cocoa beans, Nuts (almond), Nuts (peanut), Nuts (pecan nut), Nuts (pistachio nut), Nuts (walnut), Seeds (sunflower), AF/S, literature[15]

incidence: 1/3, conc.: 2.95 µg/kg*, sample origin: mainly imported from local retail outlets in Lodz (city), Poland, sample year: unknown, country: Poland[18], *AFS total

- Co-contamination: not reported
- Further contamination of plants and plant products other than cereals and their products reported in the present book/article[18]: Nut cereals, Nuts, Nuts (almond), Nuts (hazelnut), Nuts (peanut), Nuts (pistachio nut), Nuts (walnut), Seeds (sunflower), AF/S, literature[18]

incidence: 24/70, conc. range: 2–4 µg/kg* (22 sa.), 8.7 µg/kg* (1 sa.), and 31.5 µg/kg* (1 sa.), sample origin: Ceara (state), Brazil, sample year: 2010–2012, country: Brazil[572], *AFB_1, AFB_2, AFG_1, and AFG_2

- Co-contamination: not reported
- Further contamination of plants and plant products other than cereals and their products reported in the present book/article[572]: not reported

incidence: 4/20, conc. range: 11.4–23.1 µg/kg, sample origin: imported from India, sample year: before 1995?, country: Korea[643]

- Co-contamination: not reported
- Further contamination of plants and plant products other than cereals and their products reported in the present book/article[643]: not reported

Aspergillus and ***Penicillium*** Toxins

Citrinin

incidence: 1/7, conc.: 40 µg/kg, sample origin: major local retail outlets at Nainital (district), India, sample year: 1983–1985, country: India[7]

- Co-contamination: Often more than one mycotoxin was present in a sa.
- Further contamination of plants and plant products other than cereals and their products reported in the present book/article[7]: Coconuts, AFB_1, AFB_2, AFG_1, AFG_2, and CIT, literature[7]; Dates, AFB_1, AFB_2, AFG_1, and OTA, literature[7]; Grogannuts, AFB_1, AFB_2, AFG_1, AFG_2, and CIT, literature[7]; Nuts (almond), AFB_1, AFB_2, AFG_1, RBT, STG, and ZEA, literature[7]; Nuts (cashew nut), AFB_1, AFB_2, AFG_1, AFG_2, and OTA, literature[7]; Nuts (peanut), Raisins, AFB_1, AFB_2, AFG_1, AFG_2, and OTA, literature[7]; Nuts (pistachio nut), AFB_1, AFB_2, AFG_1, AFG_2, and STG, literature[7]; Nuts (walnut), AFB_1, AFB_2, AFG_1, AFG_2, and RBT, literature[7]

Ochratoxin A

incidence: 2/7, conc. range: pr, sample origin: major local retail outlets at Nainital (district), India, sample year: 1983–1985, country: India[7]

- Co-contamination: Often more than one mycotoxin was present in a sa.
- Further contamination of plants and plant products other than cereals and their products reported in the present book/article[7]: Coconuts, AFB_1, AFB_2, AFG_1, AFG_2, and CIT, literature[7]; Dates, AFB_1, AFB_2, AFG_1, and OTA, literature[7]; Grogannuts, AFB_1, AFB_2, AFG_1, AFG_2, and CIT, literature[7]; Nuts (almond), AFB_1, AFB_2, AFG_1, RBT, STG, and ZEA, literature[7]; Nuts (cashew nut), AFB_1, AFB_2, AFG_1, AFG_2, and CIT, literature[7]; Nuts (peanut), Raisins, AFB_1, AFB_2, AFG_1, AFG_2, and OTA, literature[7]; Nuts (pistachio nut), AFB_1, AFB_2, AFG_1, AFG_2, and STG, literature[7]; Nuts (walnut), AFB_1, AFB_2, AFG_1, AFG_2, and RBT, literature[7]

incidence: 1/1*, conc.: 0.65 µg/kg, sample origin: local supermarket, Japan, sample year: unknown, country: Japan[21], *cashew nuts, raw

- Co-contamination: 1 sa. co-contaminated with OTA and OTB
- Further contamination of plants and plant products other than cereals and their products reported in the present book/article[21]: Nuts (almond), Nuts (peanut), OTA and OTB, literature[21]; Nuts (cashew nut), OTB, literature[21]

incidence: 0/1*, conc.: no contamination, sample origin: local supermarket, Japan, sample year: unknown, country: Japan[21], *cashew nuts, roasted

- Co-contamination: not reported
- Further contamination of plants and plant products other than cereals and their products reported in the present book/article[21]: Nuts (almond), Nuts (peanut), OTA and OTB, literature[21]; Nuts (cashew nut), OTB, literature[21]

Ochratoxin B

incidence: 1/1*, conc.: 7.82 µg/kg, sample origin: local supermarket, Japan, sample year: unknown, country: Japan[21], *cashew nuts, raw

- Co-contamination: 1 sa. co-contaminated with OTA and OTB
- Further contamination of plants and plant products other than cereals and their products reported in the present book/article[21]: Nuts (almond), Nuts (peanut), OTA and OTB, literature[21]; Nuts (cashew nut), OTA, literature[21]

incidence: 1/1*, conc.: 0.91 µg/kg, sample origin: local supermarket, Japan, sample year: unknown, country: Japan[21], *cashew nuts, roasted

- Co-contamination: not reported
- Further contamination of plants and plant products other than cereals and their products reported in the present book/article[21]: Nuts (almond), Nuts (peanut), OTA and OTB, literature[21]; Nuts (cashew nut), OTA, literature[21]

Nuts (chestnut) may contain the following mycotoxins:

Alternaria Toxins

Alternariol

incidence: 1/1*, conc.: 12 µg/kg, sample origin: private households of staff members, Austria, sample year: unknown, country: Austria[850], *moldy

- Co-contamination: 1 sa. co-contaminated with AME, AOH, AVE-Y, CHAE, EMOD, GRIS, SUL, and TEN
- Further contamination of plants and plant products other than cereals and their products reported in the present book/article[850]: Fruits (apple), AME and AOH, literature[850]; Fruits (lemon), Puree, MA, literature[850]; Jam, AOH and MA, literature[850]; Nuts (chestnut), AOH, literature[850]; Spices (garlic), BEA, FB_1, FB_2, and FB_3, literature[850]; Wine (red), AME, AOH, and MA, literature[850]; see also Fruits (lemon), Jam, Nuts (chestnut), Wine (red), chapter: Further Mycotoxins and Microbial Metabolites

Alternariol Methyl Ether

incidence: 1/1*, conc.: 2.0 µg/kg, sample origin: private households of staff members, Austria, sample year: unknown, country: Austria[850], *moldy

- Co-contamination: 1 sa. co-contaminated with AME, AOH, AVE-Y, CHAE, EMOD, GRIS, SUL, and TEN
- Further contamination of plants and plant products other than cereals and their products reported in the present book/article[850]: Fruits (apple), AME and AOH, literature[850]; Fruits (lemon), Puree, MA, literature[850]; Jam, AOH and MA, literature[850]; Nuts (chestnut), AME, literature[850]; Spices (garlic), BEA, FB_1, FB_2, and FB_3, literature[850]; Wine (red), AME, AOH, and MA, literature[850]; see also Fruits (lemon), Jam, Nuts (chestnut), Wine (red), chapter: Further Mycotoxins and Microbial Metabolites

Aspergillus Toxins

Aflatoxin B_1

incidence: 10/118, conc. range: <0.4 µg/kg, sample origin: sites of primary storage or processing (domestic products) and sites of importation, Cyprus, sample year: 1992–1996, country: Cyprus[200]

- Co-contamination: not reported
- Further contamination of plants and plant products other than cereals and their products reported in the present book/article[200]: Butter (peanut butter), Figs, Nuts (peanut), AFB_1, AFB_2, AFG_1, and AFG_2, literature[200]; Nuts (Brazil nut, paranut), Nuts (walnut), AFB_1, AFB_2, and AFG_1, literature[200]; Nuts (pistachio nut), AFB_1 and AFB_2, literature[200]; Products (sesame product), Spices, AFB_1, literature[200]; Seeds (sesame), AFB_1, AFG_1, and AFG_2, literature[200]

incidence: 5/32*, conc. range: ≤1.70 µg/kg, sample origin: retail outlets in Campania, Emilia-Romagna, Piedmont, and Tuscany (regions), northern Italy, sample year: January 2012–June 2013, country: Italy[687], *chestnuts, fresh

incidence: 10/25*, conc. range: ≤2.18 µg/kg, sample origin: retail outlets in Campania, Emilia-Romagna, Piedmont, and Tuscany (regions), northern Italy, sample year: January 2012–June 2013, country: Italy[687], *chestnuts, dried

- Co-contamination: For detailed information see the article.
- Further contamination of plants and plant products other than cereals and their products reported in the present book/article[687]: Flour (chestnut flour), AFB_1, AFB_2, AFG_1, AFG_2, CIT, MA, OTA, and ROQ-C, literature[687]; Nuts (chestnut), AFB_2, AFG_1, CIT, MA, OTA, and ROQ-C, literature[687]

incidence: 3/14*, conc. range: ≤5.25 µg/kg, sample origin: retail outlets principally in Campania, Emilia-Romagna, Piedmont, and Tuscany (regions), northern Italy, sample year: January–June 2011, country: Italy[813], *chestnuts, dried

- Co-contamination: For detailed information see the article.
- Further contamination of plants and plant products other than cereals and their products reported in the present book/article[813]: Flour (chestnut flour), AFB_1, AFB_2, AFG_1, AFG_2, and OTA, literature[813]; Nuts (chestnut), AFB_2, AFG_1, AFG_2, and OTA, literature[813]

Aflatoxin B_2

incidence: 1/32*, conc.: 0.10 µg/kg, sample origin: retail outlets in Campania, Emilia-Romagna, Piedmont, and Tuscany (regions), northern Italy, sample year: January 2012–June 2013, country: Italy[687], *chestnuts, fresh

incidence: 2/25*, conc. range: ≤0.27 µg/kg, sample origin: retail outlets in Campania, Emilia-Romagna, Piedmont, and Tuscany (regions), northern Italy, sample year: January 2012–June 2013, country: Italy[687], *chestnuts, dried

- Co-contamination: For detailed information see the article.
- Further contamination of plants and plant products other than cereals and their products reported in the present book/ article[687]: Flour (chestnut flour), AFB_1, AFB_2, AFG_1, AFG_2, CIT, MA, OTA, and ROQ-C, literature[687]; Nuts (chestnut), AFB_1, AFG_1, CIT, MA, OTA, and ROQ-C, literature[687]

incidence: 1/14*, conc.: 0.29 µg/kg, sample origin: retail outlets principally in Campania, Emilia-Romagna, Piedmont, and Tuscany (regions), northern Italy, sample year: January–June 2011, country: Italy[813], *chestnuts, dried

- Co-contamination: For detailed information see the article.
- Further contamination of plants and plant products other than cereals and their products reported in the present book/ article[813]: Flour (chestnut flour), AFB_1, AFB_2, AFG_1, AFG_2, and OTA, literature[813]; Nuts (chestnut), AFB_1, AFG_1, AFG_2, and OTA, literature[813]

Aflatoxin G_1

incidence: 0/32*, conc.: no contamination, sample origin: retail outlets in Campania, Emilia-Romagna, Piedmont, and Tuscany (regions), northern Italy, sample year: January 2012–June 2013, country: Italy[687], *chestnuts, fresh

incidence: 1/25*, conc.: 0.29 µg/kg, sample origin: retail outlets in Campania, Emilia-Romagna, Piedmont, and Tuscany (regions), northern Italy, sample year: January 2012–June 2013, country: Italy[687], *chestnuts, dried

- Co-contamination: For detailed information see the article.
- Further contamination of plants and plant products other than cereals and their products reported in the present book/ article[687]: Flour (chestnut flour), AFB_1, AFB_2, AFG_1, AFG_2, CIT, MA, OTA, and ROQ-C, literature[687]; Nuts (chestnut), AFB_1, AFB_2, CIT, MA, OTA, and ROQ-C, literature[687]

incidence: 2/14*, conc. range: ≤2.27 µg/kg, sample origin: retail outlets principally in Campania, Emilia-Romagna, Piedmont, and Tuscany (regions), northern Italy, sample year: January–June 2011, country: Italy[813], *chestnuts, dried

- Co-contamination: For detailed information see the article.
- Further contamination of plants and plant products other than cereals and their products reported in the present book/ article[813]: Flour (chestnut flour), AFB_1, AFB_2, AFG_1, AFG_2, and OTA, literature[813]; Nuts (chestnut), AFB_1, AFB_2, AFG_2, and OTA, literature[813]

Aflatoxin G_2

incidence: 1/14*, conc.: 0.28 µg/kg, sample origin: retail outlets principally in Campania, Emilia-Romagna, Piedmont, and Tuscany (regions), northern Italy, sample year: January–June 2011, country: Italy[813], *chestnuts, dried

- Co-contamination: For detailed information see the article.
- Further contamination of plants and plant products other than cereals and their products reported in the present book/ article[813]: Flour (chestnut flour), AFB_1, AFB_2, AFG_1, AFG_2, and OTA, literature[813]; Nuts (chestnut), AFB_1, AFB_2, AFG_1, and OTA, literature[813]

Aflatoxin/s

incidence: 3/5, conc. range: 20–60 µg/kg*, sample origin: markets in Ar´Ar (city), Saudi Arabia, sample year: June 1992, country: Saudi Arabia[658], *AFB_1 and AFG_1

- Co-contamination: not reported
- Further contamination of plants and plant products other than cereals and their products reported in the present book/ article[658]: not reported

***Aspergillus* and *Penicillium* Toxins**

Citrinin

incidence: 0/32*, conc.: no contamination, sample origin: retail outlets in Campania, Emilia-Romagna, Piedmont, and Tuscany (regions), northern Italy, sample year: January 2012–June 2013, country: Italy[687], *chestnuts, fresh

incidence: 1/25*, conc.: 0.97 µg/kg, sample origin: retail outlets in Campania, Emilia-Romagna, Piedmont, and Tuscany (regions), northern Italy, sample year: January 2012–June 2013, country: Italy[687], *chestnuts, dried

- Co-contamination: For detailed information see the article.
- Further contamination of plants and plant products other than cereals and their products reported in the present book/ article[687]: Flour (chestnut flour), AFB_1, AFB_2, AFG_1, AFG_2, CIT, MA, OTA, and ROQ-C, literature[687]; Nuts (chestnut), AFB_1, AFB_2, AFG_1, MA, OTA, and ROQ-C, literature[687]

Ochratoxin A

incidence: 5/32*, conc. range: ≤6.44 µg/kg, sample origin: retail outlets in Campania, Emilia-Romagna, Piedmont, and Tuscany (regions), northern Italy, sample year: January 2012–June 2013, country: Italy[687], *chestnuts, fresh

incidence: 0/25*, conc.: no contamination, sample origin: retail outlets in Campania, Emilia-Romagna, Piedmont, and Tuscany (regions), northern Italy, sample year: January 2012–June 2013, country: Italy[687], *chestnuts, dried

- Co-contamination: For detailed information see the article.
- Further contamination of plants and plant products other than cereals and their products reported in the present book/ article[687]: Flour (chestnut flour), AFB_1, AFB_2, AFG_1, AFG_2, CIT, MA, OTA, and ROQ-C, literature[687]; Nuts (chestnut), AFB_1, AFB_2, AFG_1, CIT, MA, and ROQ-C, literature[687]

incidence: 14/14*, conc. range: 0.47–65.84 µg/kg, Ø conc.: 13.63 µg/kg, sample origin: retail outlets principally in Campania, Emilia-Romagna, Piedmont, and Tuscany (regions), northern Italy, sample year: January–June 2011, country: Italy[813], *chestnuts, dried

- Co-contamination: For detailed information see the article.
- Further contamination of plants and plant products other than cereals and their products reported in the present book/ article[813]: Flour (chestnut flour), AFB_1, AFB_2, AFG_1, AFG_2, and OTA, literature[813]; Nuts (chestnut), AFB_1, AFB_2, AFG_1, and AFG_2, literature[813]

Penicillium Toxins

Mycophenolic Acid

incidence: 4/32*, conc. range: ≤543 µg/kg, sample origin: retail outlets in Campania, Emilia-Romagna, Piedmont, and Tuscany (regions), northern Italy, sample year: January 2012–June 2013, country: Italy[687], *chestnuts, fresh

incidence: 9/25*, conc. range: ≤321 µg/kg, sample origin: retail outlets in Campania, Emilia-Romagna, Piedmont, and Tuscany (regions), northern Italy, sample year: January 2012–June 2013, country: Italy[687], *chestnuts, dried

- Co-contamination: For detailed information see the article.
- Further contamination of plants and plant products other than cereals and their products reported in the present book/article[687]: Flour (chestnut flour), AFB_1, AFB_2, AFG_1, AFG_2, CIT, MA, OTA, and ROQ-C, literature[687]; Nuts (chestnut), AFB_1, AFB_2, AFG_1, CIT, OTA, and ROQ-C, literature[687]

Roquefortine C

incidence: 7/32*, conc. range: ≤57.0 µg/kg, sample origin: retail outlets in Campania, Emilia-Romagna, Piedmont, and Tuscany (regions), northern Italy, sample year: January 2012–June 2013, country: Italy[687], *chestnuts, fresh

incidence: 7/25*, conc. range: ≤12.8 µg/kg, sample origin: retail outlets in Campania, Emilia-Romagna, Piedmont, and Tuscany (regions), northern Italy, sample year: January 2012–June 2013, country: Italy[687], *chestnuts, dried

- Co-contamination: For detailed information see the article.
- Further contamination of plants and plant products other than cereals and their products reported in the present book/article[687]: Flour (chestnut flour), AFB_1, AFB_2, AFG_1, AFG_2, CIT, MA, OTA, and ROQ-C, literature[687]; Nuts (chestnut), AFB_1, AFB_2, AFG_1, CIT, MA, and OTA, literature[687]

Nuts (hazelnut) may contain the following mycotoxins:

Alternaria Toxins

Alternariol

incidence: 20/22, conc. range: ≤650 µg/kg, ∅ conc.: 78 µg/kg, sample origin: markets, orchards, and warehouses in the Black Sea Region, Turkey, sample year: 2007, country: Austria[868]

- Co-contamination: For detailed information see the article.
- Further contamination of plants and plant products other than cereals and their products reported in the present book/article[868]: Nuts (almond), AME, AOH, BEA, CPA, ENB, ENB_1, and ENB_3, literature[868]; Nuts (hazelnut), 3-NPA, AFB_1, AFB_2, AFG_1, AFG_2, AME, ATX-I, BEA, ENA, ENA_1, ENB, ENB_2, ENB_3, ENB_4, HT-2, KA, MA, MON, OTA, OTB, STG, T-2, and ZEA, literature[868]; Nuts (peanut), 3-NPA, AFB_1, AFB_2, AME, AOH, ATX-I, BEA, CPA, ENA, ENA_1, ENB, ENB_4, KA, MA, OTA, OTB, STG, and ZEA, literature[868]; Nuts (pistachio nut), BEA, MA, and OTA, literature[868]; see also Nuts (almond), Nuts (hazelnut), Nuts (peanut), Nuts (pistachio nut), chapter: Further Mycotoxins and Microbial Metabolites

For detailed information see the article.

Alternariol Methyl Ether

incidence: 19/22, conc. range: ≤220 µg/kg, ∅ conc.: 59 µg/kg, sample origin: markets, orchards, and warehouses in the Black Sea Region, Turkey, sample year: 2007, country: Austria[868]

- Co-contamination: For detailed information see the article.
- Further contamination of plants and plant products other than cereals and their products reported in the present book/article[868]: Nuts (almond), AME, AOH, BEA, CPA, ENB, ENB_1, and ENB_3, literature[868]; Nuts (hazelnut), 3-NPA, AFB_1, AFB_2, AFG_1, AFG_2, AOH, ATX-I, BEA, ENA, ENA_1, ENB, ENB_2, ENB_3, ENB_4, HT-2, KA, MA, MON, OTA, OTB, STG, T-2, and ZEA, literature[868]; Nuts (peanut), 3-NPA, AFB_1, AFB_2, AME, AOH, ATX-I, BEA, CPA, ENA, ENA_1, ENB, ENB_4, KA, MA, OTA, OTB, STG, and ZEA, literature[868]; Nuts (pistachio nut), BEA, MA, and OTA, literature[868]; see also Nuts (almond), Nuts (hazelnut), Nuts (peanut), Nuts (pistachio nut), chapter: Further Mycotoxins and Microbial Metabolites

For detailed information see the article.

Altertoxin I

incidence: 2/22, conc. range: <LOQ, ∅ conc.: 7.0 µg/kg, sample origin: markets, orchards, and warehouses in the Black Sea Region, Turkey, sample year: 2007, country: Austria[868]

- Co-contamination: For detailed information see the article.
- Further contamination of plants and plant products other than cereals and their products reported in the present book/article[868]: Nuts (almond), AME, AOH, BEA, CPA, ENB, ENB_1, and ENB_3, literature[868]; Nuts (hazelnut), 3-NPA, AFB_1, AFB_2, AFG_1, AFG_2, AME, AOH, BEA, ENA, ENA_1, ENB, ENB_2, ENB_3, ENB_4, HT-2, KA, MA, MON, OTA, OTB, STG, T-2, and ZEA, literature[868]; Nuts (peanut), 3-NPA, AFB_1, AFB_2, AME, AOH, ATX-I, BEA, CPA, ENA, ENA_1, ENB, ENB_4, KA, MA, OTA, OTB, STG, and ZEA, literature[868]; Nuts (pistachio nut), BEA, MA, and OTA, literature[868]; see also Nuts (almond), Nuts (hazelnut), Nuts (peanut), Nuts (pistachio nut), chapter: Further Mycotoxins and Microbial Metabolites

For detailed information see the article.

Aspergillus Toxins

Aflatoxin B_1

incidence: 1/199, conc.: 325 µg/kg, sample origin: imported to Finland, sample year: 1974, country: Finland[2]

- Co-contamination: 1 sa. co-contaminated with AFB_1 and AFB_2
- Further contamination of plants and plant products other than cereals and their products reported in the present book/article[2]: Marzipan (almond paste), Nuts (almond), AFB_1 and AFB_2, literature[2]; Nuts, AFB_1, AFB_2, and AFG_1, literature[2]; Nuts (hazelnut), AFB_2, literature[2]; Nuts (peanut), AFB_1, AFB_2, AFG_1, and AFG_2, literature[2]

incidence: 1/3*, conc.: 0.6 µg/kg, sample origin: retail outlets in 20 different regions, England, sample year: April 1984, country: UK[3], *hazelnuts, shelled

incidence: 2/18*, conc. range: ≤0.7 µg/kg, sample origin: retail outlets in 20 different regions, England, sample year: April 1984, country. UK[3], *hazelnuts, in-shell

- Co-contamination: not reported
- Further contamination of plants and plant products other than cereals and their products reported in the present book/article[3]: Butter (peanut butter), Nuts (almond), Nuts (Brazil nut, paranut), Nuts (peanut), AFB_1 and AF/S, literature[3]; Nuts (hazelnut), AF/S, literature[3]

incidence: 9/28, conc. range: 1–113 µg/kg, ∅ conc.: 34.4 µg/kg, sample origin: retailers, markets, and dried fruit retail shops in

Erzurum (city), Turkey, sample year: March 2001–March 2002, country. Turkey[14]

- Co-contamination: not reported
- Further contamination of plants and plant products other than cereals and their products reported in the present book/article[14]: Nuts (almond), Nuts (peanut), Nuts (walnut), Peas (chickpea), AFB_1, literature[14]

incidence: 1/1*, conc.: 0.53 μg/kg, sample origin: local stores in Munich (city), Germany, sample year: unknown, country: Germany[181], *hazelnuts, grounded

- Co-contamination: 1 sa. co-contaminated with AFB_1 and AFG_1
- Further contamination of plants and plant products other than cereals and their products reported in the present book/article[181]: Butter (peanut butter), Spices (nutmeg), AFB_1, AFB_2, and AFG_1, literature[181]; Nuts (hazelnut), AFG_1, literature[181]; Nuts (pistachio nut), Spices (pepper), AFB_1, literature[181]; Spices (paprika), AFB_1 and AFG_1, literature[181]

incidence: 18/29*, conc. range: 5–50,000 μg/kg, sample origin: unknown, sample year: unknown, country: Germany[239], *hazelnuts, moldy

- Co-contamination: not reported
- Further contamination of plants and plant products other than cereals and their products reported in the present book/article[239]: Fruits (apple), Fruits (cherry), Fruits (lemon), Fruits (orange), Fruits (peach), Fruits (tomato), Nuts (almond), Nuts (cashew nut), Nuts (walnut), Seeds (sesame), Spice (nutmeg), AFB_1, literature[239]; Nuts (peanut), AFB_1, AFB_2, AFG_1, and AFG_2, literature[239]

incidence: 43/51*, conc. range: <1 μg/kg (41 sa.), 1–5 μg/kg (1 sa.), and 5–10 μg/kg (1 sa.), sample origin: markets in Ankara (capital), Turkey, sample year: September 2002–September 2003, country: Turkey[274], *hazelnuts, dehulled

- Co-contamination: not reported
- Further contamination of plants and plant products other than cereals and their products reported in the present book/article[274]: Nut crème, AFB_1 and AF/S, literature[274]; Nuts (hazelnut), AF/S, literature[274]

incidence: 1/32*, conc.: 0.20 μg/kg**, sample origin: imported from Switzerland and Turkey to Italy, sample year: January 2008–December 2009, country: Italy[320], *hazelnuts without shell, **contaminated sa. from Turkey

- Co-contamination: 1 sa. co-contaminated with AFB_1 and AFG_1
- Further contamination of plants and plant products other than cereals and their products reported in the present book/article[320]: Coffee beans, Vine fruits, dried, OTA, literature[320]; Kernels (apricot), AFB_1, AFB_2, and AFG_1, literature[320]; Nuts (hazelnut), AFB_2 and AFG_1, literature[320]; Nuts (peanut) and Nuts (pistachio nut), AFB_1 and AFB_2, literature[320]; Paste (hazelnut paste), AFB_1, AFB_2, AFG_1, and AFG_2, literature[320]

incidence: 1/9*, conc.: 1.16 μg/kg, sample origin: imported from Turkey to Italy, sample year: January 2008–December 2009, country: Italy[320], *hazelnuts, roasted

- Co-contamination: 1 sa. co-contaminated with AFB_1, AFB_2, AFG_1, and AFG_2
- Further contamination of plants and plant products other than cereals and their products reported in the present book/article[320]: Coffee beans, Vine fruits, dried, OTA, literature[320]; Kernels (apricot), AFB_1, AFB_2, and AFG_1, literature[320]; Nuts (hazelnut), AFB_2, AFG_1, and AFG_2, literature[320]; Nuts (peanut) and Nuts (pistachio nut), AFB_1 and AFB_2, literature[320]; Paste (hazelnut paste), AFB_1, AFB_2, AFG_1, and AFG_2, literature[320]

incidence: 4/72, conc. range: 1.17–1.80 μg/kg, sample origin: public markets/bazaars, supermarkets, retail stores, and wholesale stores in Ankara, Antalya, Isparta, Burdur, and Izmir (provinces), Turkey, sample year: unknown, country: Turkey[409]

- Co-contamination: not reported
- Further contamination of plants and plant products other than cereals and their products reported in the present book/article[409]: Nuts (hazelnut), AFG_1 and AFG_2, literature[409]; Nuts (peanut), AFB_1, AFB_2, AFG_1, and AFG_2, literature[409]; Nuts (pistachio nut), AFB_1 and AFB_2, literature[409]

incidence: 1/5, conc.: 0.42 μg/kg, sample origin: (sa. partly imported) from markets and supermarkets in Valencia (city), Spain, sample year: January–December 2002, country: Spain[435]

- Co-contamination: 1 sa. co-contaminated with AFB_1 and AFG_1
- Further contamination of plants and plant products other than cereals and their products reported in the present book/article[435]: Nuts, AFB_1 and AFG_1, literature[435]; Nuts (hazelnut) and Pinhol, AFG_1, literature[435]

incidence: 3188/3188*, conc. range: 0.02–69.14 μg/kg, sample origin: export companies in Trabzon (city), Turkey, sample year: September 2006–July 2009, country: Turkey[682], *included raw and processed hazelnuts

- Co-contamination: not reported
- Further contamination of plants and plant products other than cereals and their products reported in the present book/article[682]: Nuts (hazelnut), AF/S, literature[682]

For detailed information see the article.

incidence: 24/42, conc. range: ≤4.99 μg/kg, ∅ conc.: 1.37 μg/kg, sample origin: retail stores in Lower Saxony (federal state), Germany, sample year: 2009, country: Germany[829]

- Co-contamination: not reported
- Further contamination of plants and plant products other than cereals and their products reported in the present book/article[829]: Nuts (hazelnut), AF/S, literature[829]; Nuts (pistachio nut), AFB_1, literature[829]; Oils (maize oil), ZEA, literature[829]; Spices (cinnamon), OTA, literature[829]; Spices (curry), Spices (nutmeg), AFB_1, AF/S, and OTA, literature[829]; Spices (ginger), AFB_1 and AF/S, literature[829]

incidence: 10/22, conc. range: ≤15 μg/kg, ∅ conc.: 7.4 μg/kg, sample origin: markets, orchards, and warehouses in the Black Sea Region, Turkey, sample year: 2007, country: Austria[868]

- Co-contamination: For detailed information see the article.
- Further contamination of plants and plant products other than cereals and their products reported in the present book/article[868]: Nuts (almond), AME, AOH, BEA, CPA, ENB, ENB_1, and ENB_3, literature[868]; Nuts (hazelnut), 3-NPA, AFB_2, AFG_1, AFG_2, AME, AOH, ATX-I, BEA, ENA, ENA_1, ENB, ENB_2, ENB_3, ENB_4, HT-2, KA, MA, MON, OTA, OTB, STG, T-2, and ZEA, literature[868]; Nuts (peanut), 3-NPA, AFB_1, AFB_2, AME, AOH, ATX-I, BEA, CPA, ENA, ENA_1, ENB, ENB_4, KA, MA, OTA, OTB, STG, and ZEA, literature[868]; Nuts (pistachio nut), BEA, MA, and OTA, literature[868]; see also Nuts (almond), Nuts (hazelnut), Nuts (peanut), Nuts (pistachio nut), chapter: Further Mycotoxins and Microbial Metabolites

For detailed information see the article.

Aflatoxin B_2

incidence: 1/199, conc.: 29 µg/kg, sample origin: imported to Finland, sample year: 1974, country: Finland[2]

- Co-contamination: 1 sa. co-contaminated with AFB_1 and AFB_2
- Further contamination of plants and plant products other than cereals and their products reported in the present book/article[2]: Marzipan (almond paste), Nuts (almond), AFB_1 and AFB_2, literature[2]; Nuts, AFB_1, AFB_2, and AFG_1, literature[2]; Nuts (hazelnut), AFB_1, literature[2]; Nuts (peanut), AFB_1, AFB_2, AFG_1, and AFG_2, literature[2]

incidence: 0/32*, conc.: no contamination, sample origin: imported from Switzerland and Turkey to Italy, sample year: January 2008–December 2009, country: Italy[320], *hazelnuts without shell

- Co-contamination: not reported
- Further contamination of plants and plant products other than cereals and their products reported in the present book/article[320]: Coffee beans, Vine fruits, dried, OTA, literature[320]; Kernels (apricot), AFB_1, AFB_2, and AFG_1, literature[320]; Nuts (hazelnut), AFB_1 and AFG_1, literature[320]; Nuts (peanut) and Nuts (pistachio nut), AFB_1 and AFB_2, literature[320]; Paste (hazelnut paste), AFB_1, AFB_2, AFG_1, and AFG_2, literature[320]

incidence: 1/9*, conc.: 0.16 µg/kg, sample origin: imported from Turkey to Italy, sample year: January 2008–December 2009, country: Italy[320], *hazelnuts, roasted

- Co-contamination: 1 sa. co-contaminated with AFB_1, AFB_2, AFG_1, and AFG_2
- Further contamination of plants and plant products other than cereals and their products reported in the present book/article[320]: Coffee beans, Vine fruits, dried, OTA, literature[320]; Kernels (apricot), AFB_1, AFB_2, and AFG_1, literature[320]; Nuts (hazelnut), AFB_1, AFG_1, and AFG_2, literature[320]; Nuts (peanut), Nuts (pistachio nut), AFB_1 and AFB_2, literature[320]; Paste (hazelnut paste), AFB_1, AFB_2, AFG_1, and AFG_2, literature[320]

incidence: 4/22, conc. range: <LOQ, ∅ conc.: 5.5 µg/kg, sample origin: markets, orchards, and warehouses in the Black Sea Region, Turkey, sample year: 2007, country: Austria[868]

- Co-contamination: For detailed information see the article.
- Further contamination of plants and plant products other than cereals and their products reported in the present book/article[868]: Nuts (almond), AME, AOH, BEA, CPA, ENB, ENB_1, and ENB_3, literature[868]; Nuts (hazelnut), 3-NPA, AFB_1, AFG_1, AFG_2, AME, AOH, ATX-I, BEA, ENA, ENA_1, ENB, ENB_2, ENB_3, ENB_4, HT-2, KA, MA, MON, OTA, OTB, STG, T-2, and ZEA, literature[868]; Nuts (peanut), 3-NPA, AFB_1, AFB_2, AME, AOH, ATX-I, BEA, CPA, ENA, ENA_1, ENB, ENB_4, KA, MA, OTA, OTB, STG, and ZEA, literature[868]; Nuts (pistachio nut), BEA, MA, and OTA, literature[868]; see also Nuts (almond), Nuts (hazelnut), Nuts (peanut), Nuts (pistachio nut), chapter: Further Mycotoxins and Microbial Metabolites

For detailed information see the article.

Aflatoxin G_1

incidence: 1/1*, conc.: 1.79 µg/kg, sample origin: local stores in Munich (city), Germany, sample year: unknown, country: Germany[181], *hazelnuts, grounded

- Co-contamination: 1 sa. co-contaminated with AFB_1 and AFG_1
- Further contamination of plants and plant products other than cereals and their products reported in the present book/article[181]: Butter (peanut butter), Spices (nutmeg), AFB_1, AFB_2, and AFG_1, literature[181]; Nuts (hazelnut), Nuts (pistachio nut), Spices (pepper), AFB_1, literature[181]; Spices (paprika), AFB_1 and AFG_1, literature[181]

incidence: 1/32*, conc.: 0.31 µg/kg**, sample origin: imported from Switzerland and Turkey to Italy, sample year: January 2008–December 2009, country: Italy[320], *hazelnuts without shell, **contaminated sa. from Turkey

- Co-contamination: 1 sa. co-contaminated with AFB_1 and AFG_1
- Further contamination of plants and plant products other than cereals and their products reported in the present book/article[320]: Coffee beans, Vine fruits, dried, OTA, literature[320]; Kernels (apricot), AFB_1, AFB_2, and AFG_1, literature[320]; Nuts (hazelnut), AFB_1, literature[320]; Nuts (peanut), Nuts (pistachio nut), AFB_1 and AFB_2, literature[320]; Paste (hazelnut paste), AFB_1, AFB_2, AFG_1, and AFG_2, literature[320]

incidence: 1/9*, conc.: 1.82 µg/kg, sample origin: imported from Turkey to Italy, sample year: January 2008–December 2009, country: Italy[320], *hazelnuts, roasted

- Co-contamination: 1 sa. co-contaminated with AFB_1, AFB_2, AFG_1, and AFG_2
- Further contamination of plants and plant products other than cereals and their products reported in the present book/article[320]: Coffee beans, Vine fruits, dried, OTA, literature[320]; Kernels (apricot), AFB_1, AFB_2, and AFG_1, literature[320]; Nuts (hazelnut), AFB_1, AFB_2, and AFG_2, literature[320]; Nuts (peanut), Nuts (pistachio nut), AFB_1 and AFB_2, literature[320]; Paste (hazelnut paste), AFB_1, AFB_2, AFG_1, and AFG_2, literature[320]

incidence: 6/72, conc. range: 0.6–4.75 µg/kg, sample origin: public markets/bazaars, supermarkets, retail stores, and wholesale stores in Ankara, Antalya, Isparta, Burdur, and Izmir (provinces), Turkey, sample year: unknown, country: Turkey[409]

- Co-contamination: not reported
- Further contamination of plants and plant products other than cereals and their products reported in the present book/article[409]: Nuts (hazelnut), AFB_1 and AFG_2, literature[409]; Nuts (peanut), AFB_1, AFB_2, AFG_1, and AFG_2, literature[409]; Nuts (pistachio nut), AFB_1 and AFB_2, literature[409]

incidence: 1/5, conc.: 0.52 µg/kg, sample origin: (sa. partly imported) from markets and supermarkets in Valencia (city), Spain, sample year: January–December 2002, country: Spain[435]

- Co-contamination: 1 sa. co-contaminated with AFB_1 and AFG_1
- Further contamination of plants and plant products other than cereals and their products reported in the present book/article[435]: Nuts, AFB_1 and AFG_1, literature[435]; Nuts (hazelnut), AFB_1, literature[435]; Pinhol, AFG_1, literature[435]

incidence: 12/22, conc. range: ≤28 µg/kg, ∅ conc.: 16 µg/kg, sample origin: markets, orchards, and warehouses in the Black Sea Region, Turkey, sample year: 2007, country: Austria[868]

- Co-contamination: For detailed information see the article.
- Further contamination of plants and plant products other than cereals and their products reported in the present book/article[868]: Nuts (almond), AME, AOH, BEA, CPA, ENB, ENB_1, and ENB_3, literature[868]; Nuts (hazelnut), 3-NPA, AFB_1, AFB_2, AFG_2, AME, AOH, ATX-I, BEA, ENA, ENA_1, ENB, ENB_2, ENB_3, ENB_4, HT-2, KA, MA, MON, OTA, OTB, STG, T-2, and ZEA, literature[868]; Nuts (peanut), 3-NPA, AFB_1, AFB_2, AME, AOH, ATX-

I, BEA, CPA, ENA, ENA_1, ENB, ENB_4, KA, MA, OTA, OTB, STG, and ZEA, literature[868]; Nuts (pistachio nut), BEA, MA, and OTA, literature[868]; see also Nuts (almond), Nuts (hazelnut), Nuts (peanut), Nuts (pistachio nut), chapter: Further Mycotoxins and Microbial Metabolites

For detailed information see the article.

Aflatoxin G_2

incidence: 0/32*, conc.: no contamination, sample origin: imported from Switzerland and Turkey to Italy, sample year: January 2008–December 2009, country: Italy[320], *hazelnuts without shell

- Co-contamination: not reported
- Further contamination of plants and plant products other than cereals and their products reported in the present book/article[320]: Coffee beans, Vine fruits, dried, OTA, literature[320]; Kernels (apricot), AFB_1, AFB_2, and AFG_1, literature[320]; Nuts (hazelnut), AFB_1 and AFG_1, literature[320]; Nuts (peanut) and Nuts (pistachio nut), AFB_1 and AFB_2, literature[320]; Paste (hazelnut paste), AFB_1, AFB_2, AFG_1, and AFG_2, literature[320]

incidence: 1/9*, conc.: 0.31 µg/kg, sample origin: imported from Turkey to Italy, sample year: January 2008–December 2009, country: Italy[320], *hazelnuts, roasted

- Co-contamination: 1 sa. co-contaminated with AFB_1, AFB_2, AFG_1, and AFG_2
- Further contamination of plants and plant products other than cereals and their products reported in the present book/article[320]: Coffee beans, Vine fruits, dried, OTA, literature[320]; Kernels (apricot) and Nuts (hazelnut), AFB_1, AFB_2, and AFG_1, literature[320]; Nuts (peanut), Nuts (pistachio nut), AFB_1 and AFB_2, literature[320]; Paste (hazelnut paste), AFB_1, AFB_2, AFG_1, and AFG_2, literature[320]

incidence: 2/72, conc. range: 0.91 µg/kg, sample origin: public markets/bazaars, supermarkets, retail stores, and wholesale stores in Ankara, Antalya, Isparta, Burdur, and Izmir (provinces), Turkey, sample year: unknown, country: Turkey[409]

- Co-contamination: not reported
- Further contamination of plants and plant products other than cereals and their products reported in the present book/article[409]: Nuts (hazelnut), AFB_1 and AFG_1, literature[409]; Nuts (peanut), AFB_1, AFB_2, AFG_1, and AFG_2, literature[409]; Nuts (pistachio nut), AFB_1 and AFB_2, literature[409]

incidence: 8/22, conc. range: <LOQ, Ø conc.: 5.5 µg/kg, sample origin: markets, orchards, and warehouses in the Black Sea Region, Turkey, sample year: 2007, country: Austria[868]

- Co-contamination: For detailed information see the article.
- Further contamination of plants and plant products other than cereals and their products reported in the present book/article[868]: Nuts (almond), AME, AOH, BEA, CPA, ENB, ENB_1, and ENB_3, literature[868]; Nuts (hazelnut), 3-NPA, AFB_1, AFB_2, AFG_1, AME, AOH, ATX-I, BEA, ENA, ENA_1, ENB, ENB_2, ENB_3, ENB_4, HT-2, KA, MA, MON, OTA, OTB, STG, T-2, and ZEA, literature[868]; Nuts (peanut), 3-NPA, AFB_1, AFB_2, AME, AOH, ATX-I, BEA, CPA, ENA, ENA_1, ENB, ENB_4, KA, MA, OTA, OTB, STG, and ZEA, literature[868]; Nuts (pistachio nut), BEA, MA, and OTA, literature[868]; see also Nuts (almond), Nuts (hazelnut), Nuts (peanut), Nuts (pistachio nut), chapter: Further Mycotoxins and Microbial Metabolites

For detailed information see the article.

Aflatoxin/s

incidence: 1/3*, conc.: 0.6 µg/kg**, sample origin: retail outlets in 20 different regions, England, sample year: April 1984, country: UK[3], *hazelnuts, shelled, **AFS total

incidence: 2/18*, conc. range: ≤0.7 µg/kg**, sample origin: retail outlets in 20 different regions, England, sample year: April 1984, country. UK[3], *hazelnuts, in-shell, **AFS total

- Co-contamination: not reported
- Further contamination of plants and plant products other than cereals and their products reported in the present book/article[3]: Butter (peanut butter), Nuts (almond), Nuts (Brazil nut, paranut), Nuts (peanut), AFB_1 and AF/S, literature[3]; Nuts (hazelnut), AFB_1, literature[3]

incidence: 1/1, conc.: 3.25 µg/kg*, sample origin: mainly imported from local retail outlets in Lodz (city), Poland, sample year: unknown, country: Poland[18], *AFS total

- Co-contamination: not reported
- Further contamination of plants and plant products other than cereals and their products reported in the present book/article[18]: Nut cereals, Nuts, Nuts (almond), Nuts (cashew nut), Nuts (peanut), Nuts (pistachio nut), Nuts (walnut), Seeds (sunflower), AF/S, literature[18]

incidence: 3/35*, conc. range: 6–10 µg/kg**, sample origin: imported to Sweden, sample year: 1973, country: Sweden[205], *edible, possibly edible, and inedible hazelnuts, **AFB_1, AFB_2, AFG_1, and AFG_2

- Co-contamination: not reported
- Further contamination of plants and plant products other than cereals and their products reported in the present book/article[205]: Nuts (Brazil nut, paranut), AF/S, literature[205]

incidence: 47/51*, conc. range: <1 µg/kg (26 sa.), 1–5 µg/kg (17 sa.), 5–10 µg/kg (3 sa.), and >10 µg/kg (1 sa.), sample origin: markets in Ankara (capital), Turkey, sample year: September 2002–September 2003, country: Turkey[274], *hazelnuts, dehulled

- Co-contamination: not reported
- Further contamination of plants and plant products other than cereals and their products reported in the present book/article[274]: Nut crème, AFB_1 and AF/S, literature[274]; Nuts (hazelnut), AFB_1, literature[274]

incidence: 2/80*, conc. range: 5.46–6.55 µg/kg**, Ø conc.: 6.0 µg/kg, sample origin: exporting companies in Aydin (province), Turkey, sample year: January–August 2007, country: Turkey[360], *destined for export, **AFB_1, AFB_2, AFG_1, and AFG_2

- Co-contamination: 1 sa. co-contaminated with AFB_1, AFB_2, and AFG_1; 1 sa. co-contaminated with AFB_1 and AFG_1
- Further contamination of plants and plant products other than cereals and their products reported in the present book/article[360]: Figs, Nuts (peanut), Nuts (pistachio nut), Spices (paprika), AF/S, literature[360]

incidence: 18/20, conc. range: 25–175 µg/kg*, sample origin: markets in Cairo, El-Minya, Assiut, and Sohag (governorates), Egypt, sample year: March 1991, country: Egypt[436], *AFB_1, AFB_2, AFG_1, and AFG_2

- Co-contamination: not reported
- Further contamination of plants and plant products other than cereals and their products reported in the present book/article[436]: Nuts (walnut), AF/S and ZEA, literature[436]

incidence: 3188/3188*, conc. range: 0.07–78.98 μg/kg, sample origin: export companies in Trabzon (city), Turkey, sample year: September 2006–July 2009, country: Turkey[682], *included raw and processed hazelnuts

- Co-contamination: not reported
- Further contamination of plants and plant products other than cereals and their products reported in the present book/article[682]: Nuts (hazelnut), AFB_1, literature[682]

For detailed information see the article.

incidence: 3/8*, conc. range: <LOQ** (2 sa.) and 1.16 μg/kg** (1 sa.), sample origin: Italian, Turkish, and other countries sa. purchased in northwestern Italian stores, Italy, sample year: unknown, country: Italy[821], *hazelnuts, unshelled, **AFB_1, AFB_2, AFG_1, and AFG_2

incidence: 10/32*, conc. range: <LOQ** (4 sa.), LOQ-1 μg/kg** (5 sa.), and 1.22 μg/kg** (1 sa.), ∅ conc.: 0.63 μg/kg**, sample origin: Italian, Turkish, and other countries sa. purchased in northwestern Italian stores, Italy, sample year: unknown, country: Italy[821], *hazelnuts, raw and dehulled, **AFB_1, AFB_2, AFG_1, and AFG_2

incidence: 17/42*, conc. range: <LOQ** (10 sa.) and LOQ-1 μg/kg** (7 sa., maximum: 0.81 μg/kg**), ∅ conc.: 0.62 μg/kg**, sample origin: Italian, Turkish, and other countries sa. purchased in northwestern Italian stores, Italy, sample year: unknown, country: Italy[821], *hazelnuts, roasted and dehulled, **AFB_1, AFB_2, AFG_1, and AFG_2

incidence: 5/11*, conc. range: <LOQ** (3 sa.) and LOQ-1 μg/kg** (2 sa., maximum: 0.45 μg/kg**), ∅ conc.: 0.53 μg/kg**?, sample origin: Italian, Turkish, and other countries sa. purchased in northwestern Italian stores, Italy, sample year: unknown, country: Italy[821], *hazelnuts, grain, and flour, **AFB_1, AFB_2, AFG_1, and AFG_2

First analysis: 93 sa. in total were investigated.

incidence: 28/78*, conc. range: <LOQ** (17 sa.), LOQ-1 μg/kg** (10 sa.), and 1.22 μg/kg** (1 sa.), ∅ conc.: 0.63 μg/kg**, sample origin: Italian sa. purchased in northwestern Italian stores, Italy, sample year: unknown, country: Italy[821], *included unshelled, raw, and roasted dehulled hazelnuts as well as grain and flour, **AFB_1, AFB_2, AFG_1, and AFG_2

incidence: 6/9*, conc. range: <LOQ** (2 sa.), and LOQ-1 μg/kg** (4 sa., maximum: 0.56 μg/kg**), ∅ conc.: 0.57? μg/kg**, sample origin: Turkish sa. purchased in northwestern Italian stores, Italy, sample year: unknown, country: Italy[821], *included unshelled, roasted dehulled hazelnuts as well as grain and flour, **AFB_1, AFB_2, AFG_1, and AFG_2

incidence: 1/6*, conc. range: 1.16 μg/kg**, sample origin: sa. from other countries purchased in northwestern Italian stores, Italy, sample year: unknown, country: Italy[821], *included unshelled and roasted dehulled hazelnuts, **AFB_1, AFB_2, AFG_1, and AFG_2

Second analysis: 93 sa. in total were investigated.

incidence: 33/89*, conc. range: <LOQ** (17 sa.), LOQ-1 μg/kg** (14 sa.), and 1.22 μg/kg** (1 sa.), ∅ conc.: 0.64 μg/kg**, sample origin: Italian, Turkish, and other countries sa. purchased in northwestern Italian stores, Italy, sample year: unknown, country: Italy[821], *included unshelled, raw, and roasted dehulled hazelnuts as well as grain and flour (conventional production), **AFB_1, AFB_2, AFG_1, and AFG_2

incidence: 2/4*, conc. range: <LOQ** (2 sa.), sample origin: Italian, Turkish, and other countries sa. purchased in northwestern Italian stores, Italy, sample year: unknown, country: Italy[821], *included unshelled, raw, and roasted dehulled hazelnuts (organic production), **AFB_1, AFB_2, AFG_1, and AFG_2

Third analysis: 93 sa. in total were investigated.

- Co-contamination: not reported
- Further contamination of plants and plant products other than cereals and their products reported in the present book/article[821]: not reported

For detailed information see the article.

incidence: 24/42, conc. range: ≤23.16 μg/kg*, ∅ conc.: 4.11 μg/kg*, sample origin: retail stores in Lower Saxony (federal state), Germany, sample year: 2009, country: Germany[829], *AFB_1, AFB_2, AFG_1, and AFG_2

- Co-contamination: not reported
- Further contamination of plants and plant products other than cereals and their products reported in the present book/article[829]: Nuts (hazelnut) and Nuts (pistachio nut), AFB_1, literature[829]; Oils (maize oil), ZEA, literature[829]; Spices (cinnamon), OTA, literature[829]; Spices (curry), Spices (nutmeg), AFB_1, AF/S, and OTA, literature[829]; Spices (ginger), AFB_1 and AF/S, literature[829]

Other *Aspergillus* Toxins

Sterigmatocystin

incidence: 21/22, conc. range: ≤5.5 μg/kg, ∅ conc.: 2.3 μg/kg, sample origin: markets, orchards, and warehouses in the Black Sea Region, Turkey, sample year: 2007, country: Austria[868]

- Co-contamination: For detailed information see the article.
- Further contamination of plants and plant products other than cereals and their products reported in the present book/article[868]: Nuts (almond), AME, AOH, BEA, CPA, ENB, ENB_1, and ENB_3, literature[868]; Nuts (hazelnut), 3-NPA, AFB_1, AFB_2, AFG_1, AFG_2, AME, AOH, ATX-I, BEA, ENA, ENA_1, ENB_1, ENB_2, ENB_3, ENB_4, HT-2, KA, MA, MON, OTA, OTB, T-2, and ZEA, literature[868]; Nuts (peanut), 3-NPA, AFB_1, AFB_2, AME, AOH, ATX-I, BEA, CPA, ENA, ENA_1, ENB, ENB_4, KA, MA, OTA, OTB, STG, and ZEA, literature[868]; Nuts (pistachio nut), BEA, MA, and OTA, literature[868]; see also Nuts (almond), Nuts (hazelnut), Nuts (peanut), Nuts (pistachio nut), chapter: Further Mycotoxins and Microbial Metabolites

For detailed information see the article.

Aspergillus and *Penicillium* Toxins

3-Nitropropionic Acid

incidence: 22/22, conc. range: ≤980 μg/kg, ∅ conc.: 440 μg/kg, sample origin: markets, orchards, and warehouses in the Black Sea Region, Turkey, sample year: 2007, country: Austria[868]

- Co-contamination: For detailed information see the article.
- Further contamination of plants and plant products other than cereals and their products reported in the present book/article[868]: Nuts (almond), AME, AOH, BEA, CPA, ENB, ENB_1, and ENB_3, literature[868]; Nuts (hazelnut), AFB_1, AFB_2, AFG_1, AFG_2, AME, AOH, ATX-I, BEA, ENA, ENA_1, ENB_1, ENB_2, ENB_3, ENB_4, HT-2, KA, MA, MON, OTA, OTB, STG, T-2, and ZEA, literature[868]; Nuts (peanut), 3-NPA, AFB_1, AFB_2, AME, AOH, ATX-I, BEA, CPA, ENA, ENA_1, ENB, ENB_4,

KA, MA, OTA, OTB, STG, and ZEA, literature[868]; Nuts (pistachio nut), BEA, MA, and OTA, literature[868]; see also Nuts (almond), Nuts (hazelnut), Nuts (peanut), Nuts (pistachio nut), chapter: Further Mycotoxins and Microbial Metabolites

For detailed information see the article.

Kojic Acid

incidence: 10/22, conc. range: ≤1,400 µg/kg, ∅ conc.: 1,100 µg/kg, sample origin: markets, orchards, and warehouses in the Black Sea Region, Turkey, sample year: 2007, country: Austria[868]

- Co-contamination: For detailed information see the article.
- Further contamination of plants and plant products other than cereals and their products reported in the present book/article[868]: Nuts (almond), AME, AOH, BEA, CPA, ENB, ENB_1, and ENB_3, literature[868]; Nuts (hazelnut), 3-NPA, AFB_1, AFB_2, AFG_1, AFG_2, AME, AOH, ATX-I, BEA, ENA, ENA_1, ENB_1, ENB_2, ENB_3, ENB_4, HT-2, MA, MON, OTA, OTB, STG, T-2, and ZEA, literature[868]; Nuts (peanut), 3-NPA, AFB_1, AFB_2, AME, AOH, ATX-I, BEA, CPA, ENA, ENA_1, ENB, ENB_4, KA, MA, OTA, OTB, STG, and ZEA, literature[868]; Nuts (pistachio nut), BEA, MA, and OTA, literature[868]; see also Nuts (almond), Nuts (hazelnut), Nuts (peanut), Nuts (pistachio nut), chapter: Further Mycotoxins and Microbial Metabolites

For detailed information see the article.

Ochratoxin A

incidence: 19/32, conc. range: ≤0.080 µg/kg, sample origin: regions around Detmold, Giessen, Hamburg, Jena, Karlsruhe, Kiel, Kulmbach, Trier (cities), Germany, sample year: unknown, country: Germany[262]

- Co-contamination: not reported
- Further contamination of plants and plant products other than cereals and their products reported in the present book/article[262]: Bars, Chocolate, Drinks (chocolate drink), Figs, Fruits, dried, Nut crème, Nuts, Nuts (peanut), Powder (cocoa powder), Pudding, Raisins, Seeds (linseed), Seeds (sesame), Seeds (sunflower), OTA, literature[262]

incidence: 1/22, conc.: 220 µg/kg, sample origin: markets, orchards, and warehouses in the Black Sea Region, Turkey, sample year: 2007, country: Austria[868]

- Co-contamination: For detailed information see the article.
- Further contamination of plants and plant products other than cereals and their products reported in the present book/article[868]: Nuts (almond), AME, AOH, BEA, CPA, ENB, ENB_1, and ENB_3, literature[868]; Nuts (hazelnut), 3-NPA, AFB_1, AFB_2, AFG_1, AFG_2, AME, AOH, ATX-I, BEA, ENA, ENA_1, ENB_1, ENB_2, ENB_3, ENB_4, HT-2, KA, MA, MON, OTB, STG, T-2, and ZEA, literature[868]; Nuts (peanut), 3-NPA, AFB_1, AFB_2, AME, AOH, ATX-I, BEA, CPA, ENA, ENA_1, ENB, ENB_4, KA, MA, OTA, OTB, STG, and ZEA, literature[868]; Nuts (pistachio nut), BEA, MA, and OTA, literature[868]; see also Nuts (almond), Nuts (hazelnut), Nuts (peanut), Nuts (pistachio nut), chapter: Further Mycotoxins and Microbial Metabolites

For detailed information see the article.

Ochratoxin B

incidence: 1/22, conc.: 6.9 µg/kg, sample origin: markets, orchards, and warehouses in the Black Sea Region, Turkey, sample year: 2007, country: Austria[868]

- Co-contamination: For detailed information see the article.
- Further contamination of plants and plant products other than cereals and their products reported in the present book/article[868]: Nuts (almond), AME, AOH, BEA, CPA, ENB, ENB_1, and ENB_3, literature[868]; Nuts (hazelnut), 3-NPA, AFB_1, AFB_2, AFG_1, AFG_2, AME, AOH, ATX-I, BEA, ENA, ENA_1, ENB_1, ENB_2, ENB_3, ENB_4, HT-2, KA, MA, MON, OTB, STG, T-2, and ZEA, literature[868]; Nuts (peanut), 3-NPA, AFB_1, AFB_2, AME, AOH, ATX-I, BEA, CPA, ENA, ENA_1, ENB, ENB_4, KA, MA, OTA, OTB, STG, and ZEA, literature[868]; Nuts (pistachio nut), BEA, MA, and OTA, literature[868]; see also Nuts (almond), Nuts (hazelnut), Nuts (peanut), Nuts (pistachio nut), chapter: Further Mycotoxins and Microbial Metabolites

For detailed information see the article.

***Fusarium* Toxins**

Depsipeptides

Beauvericin

incidence: 18/22, conc. range: ≤31 µg/kg, ∅ conc.: 2.4 µg/kg, sample origin: markets, orchards, and warehouses in the Black Sea Region, Turkey, sample year: 2007, country: Austria[868]

- Co-contamination: For detailed information see the article.
- Further contamination of plants and plant products other than cereals and their products reported in the present book/article[868]: Nuts (almond), AME, AOH, BEA, CPA, ENB, ENB_1, and ENB_3, literature[868]; Nuts (hazelnut), 3-NPA, AFB_1, AFB_2, AFG_1, AFG_2, AME, AOH, ATX-I, ENA, ENA_1, ENB, ENB_2, ENB_3, ENB_4, HT-2, KA, MA, MON, OTA, OTB, STG, T-2, and ZEA, literature[868]; Nuts (peanut), 3-NPA, AFB_1, AFB_2, AME, AOH, ATX-I, BEA, CPA, ENA, ENA_1, ENB, ENB_4, KA, MA, OTA, OTB, STG, and ZEA, literature[868]; Nuts (pistachio nut), BEA, MA, and OTA, literature[868]; see also Nuts (almond), Nuts (hazelnut), Nuts (peanut), Nuts (pistachio nut), chapter: Further Mycotoxins and Microbial Metabolites

For detailed information see the article.

Enniatin A

incidence: 8/22, conc. range: ≤170 µg/kg, ∅ conc.: 28 µg/kg, sample origin: markets, orchards, and warehouses in the Black Sea Region, Turkey, sample year: 2007, country: Austria[868]

- Co-contamination: For detailed information see the article.
- Further contamination of plants and plant products other than cereals and their products reported in the present book/article[868]: Nuts (almond), AME, AOH, BEA, CPA, ENB, ENB_1, and ENB_3, literature[868]; Nuts (hazelnut), 3-NPA, AFB_1, AFB_2, AFG_1, AFG_2, AME, AOH, ATX-I, BEA, ENA_1, ENB, ENB_2, ENB_3, ENB_4, HT-2, KA, MA, MON, OTA, OTB, STG, T-2, and ZEA, literature[868]; Nuts (peanut), 3-NPA, AFB_1, AFB_2, AME, AOH, ATX-I, BEA, CPA, ENA, ENA_1, ENB, ENB_4, KA, MA, OTA, OTB, STG, and ZEA, literature[868]; Nuts (pistachio nut), BEA, MA, and OTA, literature[868]; see also Nuts (almond), Nuts (hazelnut), Nuts (peanut), Nuts (pistachio nut), chapter: Further Mycotoxins and Microbial Metabolites

For detailed information see the article.

Enniatin A_1

incidence: 9/22, conc. range: ≤1,100 µg/kg, ∅ conc.: 140 µg/kg, sample origin: markets, orchards, and warehouses in the Black Sea Region, Turkey, sample year: 2007, country: Austria[868]

- Co-contamination: For detailed information see the article.
- Further contamination of plants and plant products other than cereals and their products reported in the present book/article[868]: Nuts (almond), AME, AOH, BEA, CPA, ENB, ENB_1, and ENB_3, literature[868]; Nuts (hazelnut), 3-NPA, AFB_1, AFB_2, AFG_1, AFG_2, AME, AOH, ATX-I, BEA, ENA, ENB, ENB_2, ENB_3, ENB_4, HT-2, KA, MA, MON, OTA, OTB, STG, T-2, and ZEA, literature[868]; Nuts (peanut), 3-NPA, AFB_1, AFB_2, AME, AOH, ATX-I, BEA, CPA, ENA, ENA_1, ENB, ENB_4, KA, MA, OTA, OTB, STG, and ZEA, literature[868]; Nuts (pistachio nut), BEA, MA, and OTA, literature[868]; see also Nuts (almond), Nuts (hazelnut), Nuts (peanut), Nuts (pistachio nut), chapter: Further Mycotoxins and Microbial Metabolites

For detailed information see the article.

Enniatin B

incidence: 17/22, conc. range: ≤540 µg/kg, ∅ conc.: 37 µg/kg, sample origin: markets, orchards, and warehouses in the Black Sea Region, Turkey, sample year: 2007, country: Austria[868]

- Co-contamination: For detailed information see the article.
- Further contamination of plants and plant products other than cereals and their products reported in the present book/article[868]: Nuts (almond), AME, AOH, BEA, CPA, ENB, ENB_1, and ENB_3, literature[868]; Nuts (hazelnut), 3-NPA, AFB_1, AFB_2, AFG_1, AFG_2, AME, AOH, ATX-I, BEA, ENA, ENA_1, ENB_2, ENB_3, ENB_4, HT-2, KA, MA, MON, OTA, OTB, STG, T-2, and ZEA, literature[868]; Nuts (peanut), 3-NPA, AFB_1, AFB_2, AME, AOH, ATX-I, BEA, CPA, ENA, ENA_1, ENB, ENB_4, KA, MA, OTA, OTB, STG, and ZEA, literature[868]; Nuts (pistachio nut), BEA, MA, and OTA, literature[868]; see also Nuts (almond), Nuts (hazelnut), Nuts (peanut), Nuts (pistachio nut), chapter: Further Mycotoxins and Microbial Metabolites

For detailed information see the article.

Enniatin B_2

incidence: 5/22, conc. range: ≤8.9 µg/kg, ∅ conc.: 3.0 µg/kg, sample origin: markets, orchards, and warehouses in the Black Sea Region, Turkey, sample year: 2007, country: Austria[868]

- Co-contamination: For detailed information see the article.
- Further contamination of plants and plant products other than cereals and their products reported in the present book/article[868]: Nuts (almond), AME, AOH, BEA, CPA, ENB, ENB_1, and ENB_3, literature[868]; Nuts (hazelnut), 3-NPA, AFB_1, AFB_2, AFG_1, AFG_2, AME, AOH, ATX-I, BEA, ENA, ENA_1, ENB, ENB_3, ENB_4, HT-2, KA, MA, MON, OTA, OTB, STG, T-2, and ZEA, literature[868]; Nuts (peanut), 3-NPA, AFB_1, AFB_2, AME, AOH, ATX-I, BEA, CPA, ENA, ENA_1, ENB, ENB_4, KA, MA, OTA, OTB, STG, and ZEA, literature[868]; Nuts (pistachio nut), BEA, MA, and OTA, literature[868]; see also Nuts (almond), Nuts (hazelnut), Nuts (peanut), Nuts (pistachio nut), chapter: Further Mycotoxins and Microbial Metabolites

For detailed information see the article.

Enniatin B_3

incidence: 5/22, conc. range: ≤0.11 µg/kg, ∅ conc.: 0.06 µg/kg, sample origin: markets, orchards, and warehouses in the Black Sea Region, Turkey, sample year: 2007, country: Austria[868]

- Co-contamination: For detailed information see the article.
- Further contamination of plants and plant products other than cereals and their products reported in the present book/article[868]: Nuts (almond), AME, AOH, BEA, CPA, ENB, ENB_1, and ENB_3, literature[868]; Nuts (hazelnut), 3-NPA, AFB_1, AFB_2, AFG_1, AFG_2, AME, AOH, ATX-I, BEA, ENA, ENA_1, ENB_1, ENB_2, ENB_4, HT-2, KA, MA, MON, OTA, OTB, STG, T-2, and ZEA, literature[868]; Nuts (peanut), 3-NPA, AFB_1, AFB_2, AME, AOH, ATX-I, BEA, CPA, ENA, ENA_1, ENB, ENB_4, KA, MA, OTA, OTB, STG, and ZEA, literature[868]; Nuts (pistachio nut), BEA, MA, and OTA, literature[868]; see also Nuts (almond), Nuts (hazelnut), Nuts (peanut), Nuts (pistachio nut), chapter: Further Mycotoxins and Microbial Metabolites

For detailed information see the article.

Enniatin B_4

incidence: 10/22, conc. range: ≤190 µg/kg, ∅ conc.: 22 µg/kg, sample origin: markets, orchards, and warehouses in the Black Sea Region, Turkey, sample year: 2007, country: Austria[868]

- Co-contamination: For detailed information see the article.
- Further contamination of plants and plant products other than cereals and their products reported in the present book/article[868]: Nuts (almond), AME, AOH, BEA, CPA, ENB, ENB_1, and ENB_3, literature[868]; Nuts (hazelnut), 3-NPA, AFB_1, AFB_2, AFG_1, AFG_2, AME, AOH, ATX-I, BEA, ENA, ENA_1, ENB_1, ENB_2, ENB_3, HT-2, KA, MA, MON, OTA, OTB, STG, T-2, and ZEA, literature[868]; Nuts (peanut), 3-NPA, AFB_1, AFB_2, AME, AOH, ATX-I, BEA, CPA, ENA, ENA_1, ENB, ENB_4, KA, MA, OTA, OTB, STG, and ZEA, literature[868]; Nuts (pistachio nut), BEA, MA, and OTA, literature[868]; see also Nuts (almond), Nuts (hazelnut), Nuts (peanut), Nuts (pistachio nut), chapter: Further Mycotoxins and Microbial Metabolites

For detailed information see the article.

Type A Trichothecenes

HT-2 Toxin

incidence: 1/5, conc.: 5 µg/kg, sample origin: food stores, Germany, sample year: 2000/2001, country: Germany[151]

- Co-contamination: not reported
- Further contamination of plants and plant products other than cereals and their products reported in the present book/article[151]: Beans and Spices (cayenne pepper, chilli), ZEA, literature[151]; Nuts (hazelnut), T-2 and ZEA, literature[151]; Products (potato product), DAS, HT-2, MAS, SCIRP, and ZEA, literature[151]; Seeds (pumpkin), HT-2 and ZEA, literature[151]; Seeds (sunflower), HT-2, T-2, and ZEA, literature[151]

incidence: 15/22, conc. range: ≤130 µg/kg, ∅ conc.: 39 µg/kg, sample origin: markets, orchards, and warehouses in the Black Sea Region, Turkey, sample year: 2007, country: Austria[868]

- Co-contamination: For detailed information see the article.
- Further contamination of plants and plant products other than cereals and their products reported in the present book/article[868]: Nuts (almond), AME, AOH, BEA, CPA, ENB, ENB_1, and ENB_3, literature[868]; Nuts (hazelnut), 3-NPA, AFB_1, AFB_2, AFG_1, AFG_2, AME, AOH, ATX-I, BEA, ENA, ENA_1, ENB_1, ENB_2, ENB_3, ENB_4, KA, MA, MON, OTA, OTB, STG, T-2, and ZEA, literature[868]; Nuts (peanut), 3-NPA, AFB_1, AFB_2, AME, AOH, ATX-I, BEA, CPA, ENA, ENA_1, ENB, ENB_4, KA, MA, OTA, OTB, STG, and ZEA, literature[868]; Nuts (pistachio nut), BEA, MA, and OTA, literature[868]; see also Nuts (almond), Nuts (hazelnut), Nuts (peanut), Nuts (pistachio nut), chapter: Further Mycotoxins and Microbial Metabolites

For detailed information see the article.

T-2 Toxin

incidence: 1/5, conc.: 6 µg/kg, sample origin: food stores, Germany, sample year: 2000/2001, country: Germany[151]

- Co-contamination: not reported
- Further contamination of plants and plant products other than cereals and their products reported in the present book/article[151]: Beans and Spices (cayenne pepper, chilli), ZEA, literature[151]; Products (potato product), DAS, HT-2, MAS, SCIRP, and ZEA, literature[151]; Seeds (pumpkin), Nuts (hazelnut), HT-2 and ZEA, literature[151]; Seeds (sunflower), HT-2, T-2, and ZEA, literature[151]

incidence: 15/22, conc. range: ≤40 µg/kg, ∅ conc.: 32 µg/kg, sample origin: markets, orchards, and warehouses in the Black Sea Region, Turkey, sample year: 2007, country: Austria[868]

- Co-contamination: For detailed information see the article.
- Further contamination of plants and plant products other than cereals and their products reported in the present book/article[868]: Nuts (almond), AME, AOH, BEA, CPA, ENB, ENB_1, and ENB_3, literature[868]; Nuts (hazelnut), 3-NPA, AFB_1, AFB_2, AFG_1, AFG_2, AME, AOH, ATX-I, BEA, ENA, ENA_1, ENB_1, ENB_2, ENB_3, ENB_4, HT-2, KA, MA, MON, OTA, OTB, STG, and ZEA, literature[868]; Nuts (peanut), 3-NPA, AFB_1, AFB_2, AME, AOH, ATX-I, BEA, CPA, ENA, ENA_1, ENB, ENB_4, KA, MA, OTA, OTB, STG, and ZEA, literature[868]; Nuts (pistachio nut), BEA, MA, and OTA, literature[868]; see also Nuts (almond), Nuts (hazelnut), Nuts (peanut), Nuts (pistachio nut), chapter: Further Mycotoxins and Microbial Metabolites

For detailed information see the article.

Other ***Fusarium*** Toxins

Moniliformin

incidence: 7/22, conc. range: ≤9.2 µg/kg, ∅ conc.: 5.8 µg/kg, sample origin: markets, orchards, and warehouses in the Black Sea Region, Turkey, sample year: 2007, country: Austria[868]

- Co-contamination: For detailed information see the article.
- Further contamination of plants and plant products other than cereals and their products reported in the present book/article[868]: Nuts (almond), AME, AOH, BEA, CPA, ENB, ENB_1, and ENB_3, literature[868]; Nuts (hazelnut), 3-NPA, AFB_1, AFB_2, AFG_1, AFG_2, AME, AOH, ATX-I, BEA, ENA, ENA_1, ENB_1, ENB_2, ENB_3, ENB_4, HT-2, KA, MA, OTA, OTB, STG, T-2, and ZEA, literature[868]; Nuts (peanut), 3-NPA, AFB_1, AFB_2, AME, AOH, ATX-I, BEA, CPA, ENA, ENA_1, ENB, ENB_4, KA, MA, OTA, OTB, STG, and ZEA, literature[868]; Nuts (pistachio nut), BEA, MA, and OTA, literature[868]; see also Nuts (almond), Nuts (hazelnut), Nuts (peanut), Nuts (pistachio nut), chapter: Further Mycotoxins and Microbial Metabolites

For detailed information see the article.

Zearalenone

incidence: 1/5, conc.: 6 µg/kg, sample origin: food stores, Germany, sample year: 2000/2001, country: Germany[151]

- Co-contamination: not reported
- Further contamination of plants and plant products other than cereals and their products reported in the present book/article[151]: Beans and Spices (cayenne pepper, chilli), ZEA, literature[151]; Nuts (hazelnut), HT-2 and T-2, literature[151]; Products (potato product), DAS, HT-2, MAS, SCIRP, and ZEA, literature[151]; Seeds (pumpkin), HT-2 and ZEA, literature[151]; Seeds (sunflower), HT-2, T-2, and ZEA, literature[151]

incidence: 4/22, conc. range: ≤21 µg/kg, ∅ conc.: 7.6 µg/kg, sample origin: markets, orchards, and warehouses in the Black Sea Region, Turkey, sample year: 2007, country: Austria[868]

- Co-contamination: For detailed information see the article.
- Further contamination of plants and plant products other than cereals and their products reported in the present book/article[868]: Nuts (almond), AME, AOH, BEA, CPA, ENB, ENB_1, and ENB_3, literature[868]; Nuts (hazelnut), 3-NPA, AFB_1, AFB_2, AFG_1, AFG_2, AME, AOH, ATX-I, BEA, ENA, ENA_1, ENB_1, ENB_2, ENB_3, ENB_4, HT-2, KA, MA, MON, OTA, OTB, STG, and T-2, literature[868]; Nuts (peanut), 3-NPA, AFB_1, AFB_2, AME, AOH, ATX-I, BEA, CPA, ENA, ENA_1, ENB, ENB_4, KA, MA, OTA, OTB, STG, and ZEA, literature[868]; Nuts (pistachio nut), BEA, MA, and OTA, literature[868]; see also Nuts (almond), Nuts (hazelnut), Nuts (peanut), Nuts (pistachio nut), chapter: Further Mycotoxins and Microbial Metabolites

For detailed information see the article.

Penicillium Toxins

Mycophenolic Acid

incidence: 18/22, conc. range: ≤6,100 µg/kg, ∅ conc.: 700 µg/kg, sample origin: markets, orchards, and warehouses in the Black Sea Region, Turkey, sample year: 2007, country: Austria[868]

- Co-contamination: For detailed information see the article.
- Further contamination of plants and plant products other than cereals and their products reported in the present book/article[868]: Nuts (almond), AME, AOH, BEA, CPA, ENB, ENB_1, and ENB_3, literature[868]; Nuts (hazelnut), 3-NPA, AFB_1, AFB_2, AFG_1, AFG_2, AME, AOH, ATX-I, BEA, ENA, ENA_1, ENB_1, ENB_2, ENB_3, ENB_4, HT-2, KA, MON, OTA, OTB, STG, T-2, and ZEA, literature[868]; Nuts (peanut), 3-NPA, AFB_1, AFB_2, AME, AOH, ATX-I, BEA, CPA, ENA, ENA_1, ENB, ENB_4, KA, MA, OTA, OTB, STG, and ZEA, literature[868]; Nuts (pistachio nut), BEA, MA, and OTA, literature[868]; see also Nuts (almond), Nuts (hazelnut), Nuts (peanut), Nuts (pistachio nut), chapter: Further Mycotoxins and Microbial Metabolites

For detailed information see the article.

Nuts (peanut) may contain the following mycotoxins:

Alternaria Toxins

Alternariol

incidence: 28/84, ∅ conc.: ≤1 µg/kg, sample origin: four agroecological zones, Nigeria, sample year: May–July 2015, country: Nigeria/Austria[148]

- Co-contamination: For detailed information see the article.
- Further contamination of plants and plant products other than cereals and their products reported in the present book/article[148]: Nuts (peanut), AFB_1, AFB_2, AFG_1, AFG_2, AFM_1, AME, BEA, CPA, MON, and NIV, literature[148]; see also Nuts (peanut), chapter: Further Mycotoxins and Microbial Metabolites

For detailed information see the article.

incidence: 3/15, conc.: <LOQ, ∅ conc.: 4.4 µg/kg, sample origin: Osmaniye (province), Turkey, sample year: 2007, country: Austria[868]

- Co-contamination: For detailed information see the article.
- Further contamination of plants and plant products other than cereals and their products reported in the present book/article[868]: Nuts (almond), AME, AOH, BEA, CPA, ENB, ENB_1, and ENB_3, literature[868]; Nuts (hazelnut), 3-NPA, AFB_1, AFB_2,

AFG_1, AFG_2, AME, AOH, ATX-I, BEA, ENA, ENA_1, ENB_1, ENB_2, ENB_3, ENB_4, HT-2, KA, MA, MON, OTA, OTB, STG, T-2, and ZEA, literature[868]; Nuts (peanut), 3-NPA, AFB_1, AFB_2, AME, ATX-I, BEA, CPA, ENA, ENA_1, ENB, ENB_4, KA, MA, OTA, OTB, STG, and ZEA, literature[868]; Nuts (pistachio nut), BEA, MA, and OTA, literature[868]; see also Nuts (almond), Nuts (hazelnut), Nuts (peanut), Nuts (pistachio nut), chapter: Further Mycotoxins and Microbial Metabolites

For detailed information see the article.

Alternariol Methyl Ether

incidence: 40/84, ∅ conc.: ≤1 μg/kg, sample origin: four agroecological zones, Nigeria, sample year: May–July 2015, country: Nigeria/Austria[148]

- Co-contamination: For detailed information see the article.
- Further contamination of plants and plant products other than cereals and their products reported in the present book/article[148]: Nuts (peanut), AFB_1, AFB_2, AFG_1, AFG_2, AFM_1, AOH, BEA, CPA, MON, and NIV, literature[148]; see also Nuts (peanut), chapter: Further Mycotoxins and Microbial Metabolites

For detailed information see the article.

incidence: 3/15, conc.: <LOQ, ∅ conc.: 1.4 μg/kg, sample origin: Osmaniye (province), Turkey, sample year: 2007, country: Austria[868]

- Co-contamination: For detailed information see the article.
- Further contamination of plants and plant products other than cereals and their products reported in the present book/article[868]: Nuts (almond), AME, AOH, BEA, CPA, ENB, ENB_1, and ENB_3, literature[868]; Nuts (hazelnut), 3-NPA, AFB_1, AFB_2, AFG_1, AFG_2, AME, AOH, ATX-I, BEA, ENA, ENA_1, ENB_1, ENB_2, ENB_3, ENB_4, HT-2, KA, MA, MON, OTA, OTB, STG, T-2, and ZEA, literature[868]; Nuts (peanut), 3-NPA, AFB_1, AFB_2, AOH, ATX-I, BEA, CPA, ENA, ENA_1, ENB, ENB_4, KA, MA, OTA, OTB, STG, and ZEA, literature[868]; Nuts (pistachio nut), BEA, MA, and OTA, literature[868]; see also Nuts (almond), Nuts (hazelnut), Nuts (peanut), Nuts (pistachio nut), chapter: Further Mycotoxins and Microbial Metabolites

For detailed information see the article.

Altertoxin I

incidence: ?/15, conc.: <LOQ, sample origin: Osmaniye (province), Turkey, sample year: 2007, country: Austria[868]

- Co-contamination: For detailed information see the article.
- Further contamination of plants and plant products other than cereals and their products reported in the present book/article[868]: Nuts (almond), AME, AOH, BEA, CPA, ENB, ENB_1, and ENB_3, literature[868]; Nuts (hazelnut), 3-NPA, AFB_1, AFB_2, AFG_1, AFG_2, AME, AOH, ATX-I, BEA, ENA, ENA_1, ENB_1, ENB_2, ENB_3, ENB_4, HT-2, KA, MA, MON, OTA, OTB, STG, T-2, and ZEA, literature[868]; Nuts (peanut), 3-NPA, AFB_1, AFB_2, AME, AOH, BEA, CPA, ENA, ENA_1, ENB, ENB_4, KA, MA, OTA, OTB, STG, and ZEA, literature[868]; Nuts (pistachio nut), BEA, MA, and OTA, literature[868]; see also Nuts (almond), Nuts (hazelnut), Nuts (peanut), Nuts (pistachio nut), chapter: Further Mycotoxins and Microbial Metabolites

For detailed information see the article.

Aspergillus Toxins

Aflatoxin B_1

incidence: 9/104, conc. range: tr-954 μg/kg, sample origin: imported to Finland, sample year: 1975/1976, country: Finland[2]

- Co-contamination: 1 sa. co-contaminated with AFB_1, AFB_2, AFG_1, and AFG_2; 1 sa. co-contaminated with AFB_1, AFB_2, and AFG_1; 5 sa. co-contaminated with AFB_1 and AFB_2; 2 sa. contaminated solely with AFB_1
- Further contamination of plants and plant products other than cereals and their products reported in the present book/article[2]: Marzipan (almond paste), Nuts (almond), and Nuts (hazelnut), AFB_1 and AFB_2, literature[2]; Nuts, AFB_1, AFB_2, and AFG_1, literature[2]; Nuts (peanut), AFB_2, AFG_1, and AFG_2, literature[2]

incidence: 8/121*, conc. range: 3–716 μg/kg, ∅ conc.: 160 μg/kg, sample origin: imported to Finland, sample year: 1974–1976, country: Finland[2], *peanuts, roasted

- Co-contamination: 2 sa. co-contaminated with AFB_1, AFB_2, and AFG_1; 6 sa. co-contaminated with AFB_1 and AFB_2
- Further contamination of plants and plant products other than cereals and their products reported in the present book/article[2]: Marzipan (almond paste), Nuts (almond), and Nuts (hazelnut), AFB_1 and AFB_2, literature[2]; Nuts, AFB_1, AFB_2, and AFG_1, literature[2]; Nuts (peanut), AFB_2, AFG_1, and AFG_2, literature[2]

incidence: 20/68*, conc. range: tr-716 μg/kg, sample origin: imported to Finland, sample year: 1974–1976, country: Finland[2], *peanuts, sliced and crushed

- Co-contamination: 4 sa. co-contaminated with AFB_1, AFB_2, AFG_1, and AFG_2; 13 sa. co-contaminated with AFB_1 and AFB_2; 2 sa. co-contaminated with AFB_1 and AFG_1; 1 sa. contaminated solely with AFB_1
- Further contamination of plants and plant products other than cereals and their products reported in the present book/article[2]: Marzipan (almond paste), Nuts (almond), and Nuts (hazelnut), AFB_1 and AFB_2, literature[2]; Nuts, AFB_1, AFB_2, and AFG_1, literature[2]; Nuts (peanut), AFB_2, AFG_1, and AFG_2, literature[2]

incidence: 3/8*, conc. range: 0.5–5 μg/kg (2 sa.) and 88 μg/kg (1 sa.), sample origin: retail outlets in 20 different regions, England, sample year: April 1984, country: UK[3], *peanuts, raw shelled

incidence: 12/24*, conc. range: 0.5–5 μg/kg (9 sa.), 11–30 μg/kg (2 sa.), and 2,520 μg/kg (1 sa), sample origin: retail outlets in 20 different regions, England, sample year: April 1984, country: UK[3], *peanuts, raw in-shell

incidence: 3/17*, conc. range: 0.5–5 μg/kg (2 sa.) and 6 μg/kg (1 sa.), sample origin: retail outlets in 20 different regions, England, sample year: April 1984, country: UK[3], *peanuts, roasted

incidence: 3/14*, conc. range: 0.5 to ≤5 μg/kg, sample origin: retail outlets in 20 different regions, England, sample year: April 1984, country: UK[3], *peanuts, dry roasted

incidence: 5/13*, conc. range: 0.5–5 μg/kg (4 sa.) and 9 μg/kg (1 sa.), sample origin: retail outlets in 20 different regions, England, sample year: April 1984, country: UK[3], *peanuts, roasted in-shell

incidence: 3/17*, conc. range: 0.5 to ≤3 μg/kg, sample origin: retail outlets in 20 different regions, England, sample year: April 1984, country: UK[3], *peanuts, chocolate-coated

incidence: 8/19*, conc. range: 0.5–5 μg/kg (5 sa.), 6–10 μg/kg (1 sa.), 11–30 μg/kg (1 sa.), and 142 μg/kg (1 sa.), sample origin: retail outlets in 20 different regions, England, sample year: April 1984, country: UK[3], *peanut brittle

- Co-contamination: not reported

- Further contamination of plants and plant products other than cereals and their products reported in the present book/article[3]: Butter (peanut butter), Nuts (almond), Nuts (Brazil nut, paranut), and Nuts (hazelnut), AFB_1 and AF/S, literature[3]; Nuts (peanut), AF/S, literature[3]

incidence: 5/45*, conc. range: 90.9–5,871.4 µg/kg, ∅ conc.: 1,877.5 µg/kg, sample origin: different storage facilities, Egypt, sample year: November 1978–April 1979, country: Egypt[4], *sa. moldy with mechanical or insect damage

- Co-contamination: 2 sa. co-contaminated with AFB_1, AFB_2, AFG_1, and AFG_2; 2 sa. co-contaminated with AFB_1, AFB_2, and AFG_1; 1 sa. co-contaminated with AFB_1 and AFG_1
- Further contamination of plants and plant products other than cereals and their products reported in the present book/article[4]: Beans (broad bean), AFB_1 and AFB_2, literature[4]; Lentils and Nuts (almond), AFB_1, literature[4]; Nuts (peanut), AFB_2, AFG_1, and AFG_2, literature[4]

incidence: 1*/38, conc.: <10 µg/kg, sample origin: Spanish retail market, Spain, sample year: 1986/1987, country: Spain[6], *bulk, unpacked material

- Co-contamination: not reported
- Further contamination of plants and plant products other than cereals and their products reported in the present book/article[6]: Nuts (almond), AFB_1 and AFB_2, literature[6]

incidence: 3/6, conc. range: 20–200 µg/kg, sample origin: major local retail outlets at Nainital (district), India, sample year: 1983–1985, country: India[7]

- Co-contamination: Often more than one mycotoxin was present in a sa.
- Further contamination of plants and plant products other than cereals and their products reported in the present book/article[7]: Coconuts, AFB_1, AFB_2, AFG_1, AFG_2, and CIT, literature[7]; Dates, AFB_1, AFB_2, AFG_1, and OTA, literature[7]; Grogannuts, AFB_1, AFB_2, AFG_1, AFG_2, and CIT, literature[7]; Nuts (almond), AFB_1, AFB_2, AFG_1, RBT, STG, and ZEA, literature[7]; Nuts (cashew nut), AFB_1, AFB_2, AFG_1, AFG_2, CIT, and OTA, literature[7]; Nuts (peanut), AFB_2, AFG_1, AFG_2, and OTA, literature[7]; Nuts (pistachio nut), AFB_1, AFB_2, AFG_1, AFG_2, and STG, literature[7]; Nuts (walnut), AFB_1, AFB_2, AFG_1, AFG_2, and RBT, literature[7]; Raisins, AFB_1, AFB_2, AFG_1, AFG_2, and OTA, literature[7]

incidence: 6?/20, conc. range: 0.04–0.31 µg/kg, sample origin: unknown, sample year: unknown, country: Korea[9]

- Co-contamination: not reported
- Further contamination of plants and plant products other than cereals and their products reported in the present book/article[9]: Butter (peanut butter), AFB_1, AFB_2, AFG_1, and AFG_2, literature[9]; Nuts (almond), AFB_1 and AFB_2, literature[9]; Nuts (peanut), AFB_2, literature[9]; Products (nut product), Tea, AFB_1, literature[9]

incidence: 8/15, conc. range: 0.11–18.04 µg/kg, ∅ conc.: 4.07 µg/kg, sample origin: grocery markets in Busan, Daegu, Daejeon, Gangneung, and Gwangju (cities) as well as Seoul (capital), Korea, sample year: May 2004–June 2005, country: Korea[10]

- Co-contamination: not reported
- Further contamination of plants and plant products other than cereals and their products reported in the present book/article[10]: Butter (peanut butter), Nuts, Nuts (almond), Nuts (pistachio nut), Paste (soybean paste), Sauces (soy sauce), Spices (cayenne pepper, chilli), Spices (curry), Spices (pepper), AFB_1, literature[10]

incidence: 3/3*, conc. range: 33–69 µg/kg, ∅ conc.: 45.7 µg/kg, sample origin: unknown, sample year: unknown, country: USA[11], *peanuts, salted-in-shell

- Co-contamination: 1 sa. co-contaminated with AFB_1, AFB_2, AFG_1, and AFG_2; 2 sa. co-contaminated with AFB_1 and AFB_2
- Further contamination of plants and plant products other than cereals and their products reported in the present book/article[11]: Nuts (almond), AFB_1 and AFB_2, literature[11]; Nuts (peanut), AFB_2, AFG_1, and AFG_2, literature[11]

incidence: 1/1*, conc.: 95 µg/kg, sample origin: unknown, sample year: unknown, country: USA[11], *peanuts with cereal coating

- Co-contamination: 1 sa. co-contaminated with AFB_1 and AFB_2
- Further contamination of plants and plant products other than cereals and their products reported in the present book/article[11]: Nuts (almond), AFB_1 and AFB_2, literature[11]; Nuts (peanut), AFB_2, AFG_1, and AFG_2, literature[11]

incidence: 7/18, conc. range: 8–94 µg/kg, ∅ conc.: 43.0 µg/kg, sample origin: retailers, markets, and dried fruit retail shops in Erzurum (city), Turkey, sample year: March 2001–March 2002, country: Turkey[14]

- Co-contamination: not reported
- Further contamination of plants and plant products other than cereals and their products reported in the present book/article[14]: Nuts (almond), Nuts (hazelnut), Nuts (walnut), Peas (chickpea), AFB_1, literature[14]

incidence: 39?/40, conc. range: 1.3–1,600 µg/kg, sample origin: different district hygiene institutes, GDR, sample year: unknown, country: GDR[29]

- Co-contamination: not reported
- Further contamination of plants and plant products other than cereals and their products reported in the present book/article[29]: Butter (peanut butter), Products (peanut product), and Seeds (sunflower), AFB_1, AFB_2, AFG_1, and AFG_2, literature[29]; Coffee, OTA, literature[29]; Fruits, AFB_1, AFG_1, and AFG_2, literature[29]; Fruits (apple), Fruits (peach), and Juices (apple juice), PAT, literature[29]; Nuts (peanut), AFB_2, AFG_1, and AFG_2, literature[29]

incidence: 32*/200, conc. range: 54–1,806 µg/kg, sample origin: 16 districts of eastern Nepal, sample year: 1995–2003, country: Nepal[58], *>30 µg/kg

- Co-contamination: not reported
- Further contamination of plants and plant products other than cereals and their products reported in the present book/article[58]: Betel nuts, AF/S, literature[58]; Butter (peanut butter), AFB_1, literature[58]

incidence: 11?/149, conc. range: 0.4–21.7 µg/kg, ∅ conc.: 4.1 µg/kg, sample origin: commercially available market basket sa. of foods and foodstuffs in Tokyo (capital), Japan, sample year: 1986–1990, country: Japan[99]

incidence: 2/7*, conc. range: 0.7–13.9 µg/kg, ∅ conc.: 7.3 µg/kg, sample origin: commercially available market basket sa. of foods and foodstuffs in Tokyo (capital), Japan, sample year: 1986–1990, country: Japan[99], *peanuts, crushed

incidence: 4/9*, conc. range: 0.5–21.7 µg/kg, sample origin: commercially available market basket sa. of foods and foodstuffs in Tokyo (capital), Japan, sample year: 1986–1990, country: Japan[99], *peanuts, powdered

- Co-contamination: not reported

- Further contamination of plants and plant products other than cereals and their products reported in the present book/article[99]: Beans, Seeds (sesame), and Sugar, AFB_1 and AFB_2, literature[99]; Food, Jam (bean jam), and Paste (peanut paste), AFB_1, literature[99]; Nuts (Brazil nut, paranut), Nuts (pistachio nut), and Spices (nutmeg), AFB_1, AFB_2, AFG_1, and AFG_2, literature[99]; Nuts (peanut), AFB_2, AFG_1, and AFG_2, literature[99]; Spices, Spices (cayenne pepper, chilli), and Spices (pepper), AFB_1, AFB_2, and AFG_1, literature[99]; Spices (paprika), AFB_1 and AFG_1, literature[99]

incidence: 1/9*, conc.: 2.2 μg/kg, sample origin: supermarkets and/or traditional markets in Yogyakarta (city), Indonesia, sample year: 1998/1999, country: Indonesia/Austria/Canada[106], *peanuts, unshelled and roasted

incidence: 5/11*, conc. range: 2.0–4.0 μg/kg (3 sa.) and >4.0 to ≤50.0 μg/kg (2 sa.), sample origin: supermarkets and/or traditional markets in Yogyakarta (city), Indonesia, sample year: 1998/1999, country: Indonesia/Austria/Canada[106], *peanuts, roasted

incidence: 14/20*, conc. range: 2.0–4.0 μg/kg (4 sa.) and >4.0 to ≤61.7 μg/kg (10 sa.), sample origin: supermarkets and/or traditional markets in Yogyakarta (city), Indonesia, sample year: 1998/1999, country: Indonesia/Austria/Canada[106], *peanuts, flour coated (kacang atom)

incidence: 11/13*, conc. range: 2.0–4.0 μg/kg (6 sa.) and >4.0 to ≤7.0 μg/kg (5 sa.), sample origin: supermarkets and/or traditional markets in Yogyakarta (city), Indonesia, sample year: 1998/1999, country: Indonesia/Austria/Canada[106], *peanuts, egg coated (kacang telur)

- Co-contamination: not reported
- Further contamination of plants and plant products other than cereals and their products reported in the present book/article[106]: Butter (peanut butter), Cake (peanut), Food, Products, and Sauces (peanut-chilli sauce), AFB_1, literature[106]

incidence: 2/9, conc. range: 5.6–15.5 μg/kg, ∅ conc.: 10.55 μg/kg, sample origin: local markets, cereal traders, food/feed processing enterprises, grain mills, and small-scale farmers in villages and cities, Burkina Faso, sample year: April 2010, country: Austria/Nigeria[117]

incidence: 3/23, conc. range: 3.4–123 μg/kg, sample origin: local markets, cereal traders, food/feed processing enterprises, grain mills, and small-scale farmers in the local capital and rural villages in Nampula (province), northern Mozambique, sample year: May 2010, country: Austria/Nigeria[117]

- Co-contamination: For detailed information see the article.
- Further contamination of plants and plant products other than cereals and their products reported in the present book/article[117]: Food, 3-NPA, AFB_1, AFB_2, AFG_1, AFG_2, AFM_1, AME, AOH, ATX-I, BEA, CIT, CPA, DON, ENA, ENA_1, ENB, ENB_1, ENB_2, FB_1, FB_2, FB_3, MON, NIV, OTA, STG, and ZEA, literature[117]; Nuts (peanut), 3-NPA, AFB_2, AFG_1, BEA, CPA, ENB_1, and STG, literature[117]; see also Food, chapter: Further Mycotoxins and Microbial Metabolites

incidence: 208/515, conc. range: 1–10 μg/kg (28 sa.), >10–20 μg/kg (85 sa.), >20–50 μg/kg (10 sa.), >50–100 μg/kg (6 sa.), >100–500 μg/kg (23 sa.), >500–2,000 μg/kg (31 sa.), and >2,000 μg/kg (25 sa.), sample origin: unknown, sample year: 1990–1996, country: Cuba[126]

- Co-contamination: not reported
- Further contamination of plants and plant products other than cereals and their products reported in the present book/article[126]: Beans, Beans (soybean), Cocoa beans, Peas, AFB_1, literature[126]

incidence: 2/6*, conc. range: 3.6–5.4 μg/kg**, ∅ conc.: 4.5 μg/kg**, sample origin: Alexandria, Luxor (cities), and Cairo (capital), Egypt, sample year: unknown, country: Egypt/USA[127], *3 sa. freshly produced and 3 sa. stored for 12 months, **stored sa.

- Co-contamination: 2 sa. co-contaminated with AFB_1 and AFB_2
- Further contamination of plants and plant products other than cereals and their products reported in the present book/article[127]: Beans (broad bean), AFB_1 and AFB_2, literature[127]; Lentils, AFB_1, literature[127]; Nuts (peanut), AFB_2, literature[127]; Spices (fenugreek), AFB_1, AFB_2, and AFG_1, literature[127]

incidence: 43/62, conc. range: 0.5–820.0 μg/kg, ∅ conc.: 39.6 μg/kg, sample origin: small markets in several urban Shanghai (city) areas, China, sample year: April 1988–April 1989, country: China/USA[132]

- Co-contamination: not reported
- Further contamination of plants and plant products other than cereals and their products reported in the present book/article[132]: Beans (kidney bean), Butter (peanut butter), Curd (bean curd), Oils (rapeseed oil), Oils (soybean oil), Paste (bean paste), Sauces (soy sauce), Tofu, Wine (miscellaneous), AFB_1, literature[132]

incidence: 13/30*, conc. range: 40–780 μg/kg, ∅ conc.: 180.8 μg/kg, sample origin: markets around Bangkok (capital) area, Thailand, sample year: July 1983–April 1985, country: Thailand[133], *food and sweets derived from peanut

- Co-contamination: 6 sa. co-contaminated with AFB_1 and AFG_1; 7 sa. contaminated solely with AFB_1
- Further contamination of plants and plant products other than cereals and their products reported in the present book/article[133]: Beans and Nuts (peanut), AFG_1, literature[133]; Beans (soybean), Curd (bean curd), Sauces (soy sauce), AFB_1 and AFG_1, literature[133]

incidence: 25/84, conc. range: 0.9–710 μg/kg, ∅ conc.: 117.8 μg/kg, sample origin: four agroecological zones, Nigeria, sample year: May–July 2015, country: Nigeria/Austria[148]

- Co-contamination: For detailed information see the article.
- Further contamination of plants and plant products other than cereals and their products reported in the present book/article[148]: Nuts (peanut), AFB_2, AFG_1, AFG_2, AFM_1, AME, AOH, BEA, CPA, MON, NIV, and OTA, literature[148]; see also Nuts (peanut), chapter: Further Mycotoxins and Microbial Metabolites

For detailed information see the article.

incidence: 1/26*, conc.: 0.1 μg/kg, sample origin: retail stores in Osaka (prefecture), Japan, sample year: 1988–1992, country: Japan[173], *peanuts, shelled and roasted

- Co-contamination: not reported
- Further contamination of plants and plant products other than cereals and their products reported in the present book/article[173]: Butter (peanut butter) and Spices (nutmeg), AFB_1, AFB_2, and AFG_1, literature[173]; Spices (cayenne pepper, chilli), and Spices (pepper), AFB_1, literature[173]; literature[173]

incidence: 11/13, conc. range: 1.47–15.33 μg/kg, ∅ conc.: 4.25 μg/kg, sample origin: supermarkets in Penang (state), Malaysia, sample year: unknown, country: Malaysia[182]

- Co-contamination: not reported
- Further contamination of plants and plant products other than cereals and their products reported in the present book/article[182]: Nuts, Seeds (sesame), Seeds (sunflower), Spices (cayenne pepper, chilli), Spices (cumin), Spices (pepper), AFB_1, literature[182]

incidence: 179?/1,860, conc. range: <0.4–700 µg/kg, Ø conc.: 25.6 µg/kg, sample origin: sites of primary storage or processing (domestic products) and sites of importation, Cyprus, sample year: 1992–1996, country: Cyprus[200]

- Co-contamination: not reported
- Further contamination of plants and plant products other than cereals and their products reported in the present book/article[200]: Butter (peanut butter) and Figs, AFB_1, AFB_2, AFG_1, and AFG_2, literature[200]; Nuts (Brazil nut, paranut) and Nuts (walnut), AFB_1, AFB_2, and AFG_1, literature[200]; Nuts (chestnut), Products (sesame product), and Spices, AFB_1, literature[200]; Nuts (peanut), AFB_2, AFG_1, and AFG_2, literature[200]; Nuts (pistachio nut), AFB_1 and AFB_2, literature[200]; Seeds (sesame), AFB_1, AFG_1, and AFG_2, literature[200]

incidence: 6/34*, conc. range: tr-34 µg/kg, sample origin: stores in Oslo (capital) area, Norway, sample year: autumn 1973, country: Norway[201], *peanuts bought from stores

- Co-contamination: 6 sa. co-contaminated with AFB_1, AFB_2, AFG_1, and AFG_2
- Further contamination of plants and plant products other than cereals and their products reported in the present book/article[201]: Cocoa, Nuts (Brazil nut, paranut), Nuts (cashew nut), and Nuts (walnut), AFB_1, AFB_2, AFG_1, and AFG_2, literature[201]; Nuts, Nuts (peanut), AFB_2, AFG_1, and AFG_2, literature[201]

incidence: 1/3*, conc.: 400 µg/kg, sample origin: stores in Oslo (capital) area, Norway, sample year: autumn 1973, country: Norway[201], *peanuts sorted, pooled sa.

- Co-contamination: 1 sa. co-contaminated with AFB_1, AFB_2, AFG_1, and AFG_2
- Further contamination of plants and plant products other than cereals and their products reported in the present book/article[201]: Cocoa, Nuts (Brazil nut, paranut), Nuts (cashew nut), and Nuts (walnut), AFB_1, AFB_2, AFG_1, and AFG_2, literature[201]; Nuts, Nuts (peanut), AFB_2, AFG_1, and AFG_2, literature[201]

incidence: 8/19, conc. range: LOQ-2 µg/kg (5 sa.), >2–5 µg/kg (1 sa.), and >5 µg/kg (2 sa., maximum: 60.9 µg/kg), sample origin: retail outlets, UK, sample year: September 2001, country: UK[203]

- Co-contamination: 5 sa. co-contaminated with AFB_1 and AFB_2; 3 sa. contaminated solely with AFB_1
- Further contamination of plants and plant products other than cereals and their products reported in the present book/article[203]: Butter (nut butter), AFB_1, AFB_2, AFG_1, AFG_2, and OTA, literature[203]; Butter (peanut butter), AFB_1, AFB_2, AFG_1, AFG_2, and OTA, literature[203]; Dates, AFB_1 and OTA, literature[203]; Figs, AFB_1, AFB_2, AFG_2, and OTA, literature[203]; Fruits, dried, Nuts (pistachio nut), AFB_1, AFB_2, and OTA, literature[203]; Nuts (Brazil nut, paranut), AFB_1, AFB_2, and AFG_1, literature[203]; Nuts (peanut), AFB_2, literature[203]

incidence: 9/9*, conc. range: 74.03–82.12 µg/kg, Ø conc.: 78.21 µg/kg, sample origin: markets in the Niger Delta region, Nigeria, sample year: November 2004–March 2005, country: Nigeria[210]

- Co-contamination: not reported
- Further contamination of plants and plant products other than cereals and their products reported in the present book/article[210]: Breadfruit, AFB_1, literature[210]

incidence: 61/267, conc. range: 5–15,743 µg/kg, Ø conc.: 898 µg/kg, sample origin: markets in 16 areas (cities and counties), Taiwan, sample year: 1977/1978, country: Taiwan[214]

incidence: 5/33*, conc. range: 20–330 µg/kg, Ø conc.: 114 µg/kg, sample origin: markets in 16 areas (cities and counties), Taiwan, sample year: 1977/1978, country: Taiwan[214], *peanuts, fried

incidence: 3/19*, conc. range: 20–2,784 µg/kg, Ø conc.: 1,075 µg/kg, sample origin: markets in 16 areas (cities and counties), Taiwan, sample year: 1977/1978, country: Taiwan[214], *peanuts, roasted

- Co-contamination: not reported
- Further contamination of plants and plant products other than cereals and their products reported in the present book/article[214]: Butter (peanut butter), Cake (peanut), and Oils (peanut oil), AFB_1, AFB_2, AFG_1, and AFG_2, literature[214]; Nuts (peanut), AFB_2, AFG_1, and AFG_2, literature[214]

incidence: 3/8, conc. range: 40–430 µg/kg, Ø conc.: 263 µg/kg, sample origin: peanut storehouses of oil mills, Taiwan, sample year: unknown, country: Taiwan[216]

- Co-contamination: not reported
- Further contamination of plants and plant products other than cereals and their products reported in the present book/article[216]: Butter (peanut butter), Oils (peanut oil), AFB_1, literature[216]

incidence: 15?/29, conc. range: 0.8–16.0 µg/kg, sample origin: storage depots and retail outlets, Botswana, sample year: January 1996–December 1997, country: Botswana[218]

- Co-contamination: not reported
- Further contamination of plants and plant products other than cereals and their products reported in the present book/article[218]: Butter (peanut butter), AFB_1, AFB_2, AFG_1, and AFG_2, literature[218]; Nuts (peanut), AFB_2, AFG_1, and AFG_2, literature[218]

incidence: 1/4 (analyzed sa.)*, conc.: 0.20 µg/kg, sample origin: supermarkets, local stores, and markets in Busan, Daejeon, Gwangju (cities), and Seoul (capital), Korea, sample year: June/July 2004, country: Korea[220], *peanuts, raw

incidence: 4/8 (analyzed sa.)*, conc. range: 1.85–18.04 µg/kg, sample origin: supermarkets, local stores, and markets in Busan, Daejeon, Gwangju (cities), and Seoul (capital), Korea, sample year: June/July 2004, country: Korea[220], *peanuts, roasted

- Co-contamination: not reported
- Further contamination of plants and plant products other than cereals and their products reported in the present book/article[220]: Butter (peanut butter), Nuts, AFB_1 and AFB_2, literature[220]; Nuts (peanut), AFB_2, AFG_1, and AFG_2, literature[220]; Nuts (pistachio nut), AFB_1, literature[220]

For detailed information see the article.

incidence: 1/1, conc.: 59.666 µg/kg, sample origin: market in Khartoum (capital), Sudan, sample year: February 1981–1984?, country: UK/Sudan[223]

- Co-contamination: 1 sa. co-contaminated with AFB_1, AFB_2, and AFG_2
- Further contamination of plants and plant products other than cereals and their products reported in the present book/article[223]: Butter (peanut butter), AFB_1, AFB_2, and AFG_1, literature[223]; Nuts (peanut), AFB_2, AFG_1, and AFG_2, literature[223]; Ochra, AFG_2, literature[223]; Peas (chickpea), AFB_1, literature[223]

incidence: 1/1*, conc.: 3.517 μg/kg, sample origin: market in Khartoum (capital), Sudan, sample year: February 1981–1984?, country: UK/Sudan[223], *peanuts, limed

- Co-contamination: 1 sa. co-contaminated with AFB_1, AFG_1 and AFG_2
- Further contamination of plants and plant products other than cereals and their products reported in the present book/article[223]: Butter (peanut butter), AFB_1, AFB_2, and AFG_1, literature[223]; Nuts (peanut), AFB_2, AFG_1, and AFG_2, literature[223]; Okra, AFG_2, literature[223]; Peas (chickpea), AFB_1, literature[223]

incidence: 6?/14*, conc. range: 15.5–618 μg/kg, sample origin: supermarkets and village stores in five districts of Penang (state) including Penang Island and Wellesley (province), Malaysia, sample year: unknown, country: Malaysia[225], * raw peanuts, shelled

- Co-contamination: not reported
- Further contamination of plants and plant products other than cereals and their products reported in the present book/article[225]: Butter (peanut butter), Confectionery, Nuts (walnut), and Products (nut product), AFB_1 and AFB_2, literature[225]; Cake (peanut), AFB_1, AFB_2, and AFG_2, literature[225]; Nuts (peanut), AFB_2 and AFG_2, literature[225]

incidence: 2/10*, conc. range: 25.3–126 μg/kg, ∅ conc.: 75.7 μg/kg, sample origin: supermarkets and village stores in five districts of Penang (state) including Penang Island and Wellesley (province), Malaysia, sample year: unknown, country: Malaysia[225], *roasted peanuts, in-shell

- Co-contamination: 1 sa. co-contaminated with AFB_1, AFB_2, and AFG_2;1 sa. co-contaminated with AFB_1 and AFB_2
- Further contamination of plants and plant products other than cereals and their products reported in the present book/article[225]: Butter (peanut butter), Confectionery, Nuts (walnut), and Products (nut product), AFB_1 and AFB_2, literature[225]; Cake (peanut), AFB_1, AFB_2, and AFG_2, literature[225]; Nuts (peanut), AFB_2 and AFG_2, literature[225]

incidence: 3?/20*, conc. range: 32.9–37.2 μg/kg, sample origin: supermarkets and village stores in five districts of Penang (state) including Penang Island and Wellesley (province), Malaysia, sample year: unknown, country: Malaysia[225], *roasted peanut, shelled

- Co-contamination: not reported
- Further contamination of plants and plant products other than cereals and their products reported in the present book/article[225]: Butter (peanut butter), Confectionery, Nuts (walnut), Products (nut product), AFB_1 and AFB_2, literature[225]; Cake (peanut), AFB_1, AFB_2, and AFG_2, literature[225]; Nuts (peanut), AFB_2 and AFG_2, literature[225]

incidence: 3?/8*, conc. range: 18.4–56.7 μg/kg, sample origin: supermarkets and village stores in five districts of Penang (state) including Penang Island and Wellesley (province), Malaysia, sample year: unknown, country: Malaysia[225], *pound peanuts

- Co-contamination: not reported
- Further contamination of plants and plant products other than cereals and their products reported in the present book/article[225]: Butter (peanut butter), Confectionery, Nuts (walnut), Products (nut product), AFB_1 and AFB_2, literature[225]; Cake (peanut), AFB_1, AFB_2, and AFG_2, literature[225]; Nuts (peanut), AFB_2 and AFG_2, literature[225]

incidence: 1/2*, conc.: 76.8 μg/kg, sample origin: supermarkets and village stores in five districts of Penang (state) including Penang Island and Wellesley (province), Malaysia, sample year: unknown, country: Malaysia[225], *peanut slices

- Co-contamination: 1 sa. co-contaminated with AFB_1 and AFB_2
- Further contamination of plants and plant products other than cereals and their products reported in the present book/article[225]: Butter (peanut butter), Confectionery, Nuts (walnut), Products (nut product), AFB_1 and AFB_2, literature[225]; Cake (peanut), AFB_1, AFB_2, and AFG_2, literature[225]; Nuts (peanut), AFB_2 and AFG_2, literature[225]

incidence: 1/12*, conc.: 162 μg/kg, sample origin: traditional markets and supermarkets, Indonesia, sample year: 2001/2002, country: Austria/Indonesia/Thailand[226], *peanuts, roasted

incidence: 6/16*, conc. range: 4–357 μg/kg, sample origin: traditional markets and supermarkets, Indonesia, sample year: 2001/2002, country: Austria/Indonesia/Thailand[226], *peanuts, flour/egg coated

incidence: 1/11*, conc.: 13 μg/kg, sample origin: traditional markets and supermarkets, Indonesia, sample year: 2001/2002, country: Austria/Indonesia/Thailand[226], *peanuts, fried

- Co-contamination: not reported
- Further contamination of plants and plant products other than cereals and their products reported in the present book/article[226]: Butter (peanut butter), Cake (peanut), Sauces (peanut-chilli sauce), Snacks, AFB_1 and AF/S, literature[226]; Nuts (peanut), AF/S, literature[226]

incidence: 14?/35*, conc. range: ≤6.4 μg/kg, sample origin: local stores in Hangzhou, Ningbo, and Quzhou (cities), Zhejiang (province), China, sample year: unknown, country: China[227], *peanuts, fresh

incidence: 5/5*, conc. range: 0.2–1,112.8 μg/kg, ∅ conc.: 245.1 μg/kg, sample origin: local stores in Hangzhou, Ningbo, and Quzhou (cities), Zhejiang (province), China, sample year: unknown, country: China[227], *peanuts, musty

- Co-contamination: not reported
- Further contamination of plants and plant products other than cereals and their products reported in the present book/article[227]: Butter (peanut butter), AFB_1, AFB_2, AFG_1, AFG_2, AFM_1, and AFM_2, literature[227]; Nuts (peanut), AFB_2, AFG_1, AFG_2, AFM_1, and AFM_2, literature[227]

incidence: 23/48*, conc. range: 0.55–55.10 μg/kg, ∅ conc.: 6.02 μg/kg, sample origin: Araras, Leme, Pirassununga, and Porto Ferreira (cities), northeast of São Paulo (state), Brazil, sample year: June 2006–May 2007, country: Brazil[237], *peanuts, unprocessed

incidence: 15/48*, conc. range: 0.67–3.09 μg/kg, ∅ conc.: 0.61 μg/kg, sample origin: Araras, Leme, Pirassununga, and Porto Ferreira (cities), northeast of São Paulo (state), Brazil, sample year: June 2006–May 2007, country: Brazil[237], *peanuts, roasted and salty

- Co-contamination: not reported
- Further contamination of plants and plant products other than cereals and their products reported in the present book/article[237]: Candies (peanut candy), AFB_1, AFB_2, AFG_1, and AFG_2, literature[237]; Nuts (peanut), AFB_2, AFG_1, and AFG_2, literature[237]

incidence: 46/62*, conc. range: 20–28,000 μg/kg, sample origin: unknown, sample year: unknown, country: Germany[239], *peanuts, moldy incidence: 10/543, conc. range: 5–20 µg/kg, sample origin: South Africa, sample year: 1970, country: Germany[239]

incidence: 1/258, conc.: 20 µg/kg, sample origin: South Africa, sample year: 1971, country: Germany[239]

incidence: : 2/54, conc. range: 10–125 µg/kg, ∅ conc.: 67.5 µg/kg, sample origin: USA, sample year: 1970, country: Germany[239]

incidence: : 3/35, conc. range: 5–625 µg/kg, ∅ conc.: 213 µg/kg, sample origin: USA, sample year: 1971, country: Germany[239]

incidence: 1/19, conc.: 40 µg/kg, sample origin: Sudan, sample year: 1970, country: Germany[239]

incidence: 2/12, conc. range: 5–20 µg/kg, ∅ conc.: 12.5 µg/kg, sample origin: Sudan, sample year: 1971, country: Germany[239]

- Co-contamination: not reported
- Further contamination of plants and plant products other than cereals and their products reported in the present book/article[239]: Fruits (apple), Fruits (cherry), Fruits (lemmon), Fruits (orange), Fruits (peach), Fruits (tomato), Nuts (almond), Nuts (cashew nut), Nuts (hazelnut), Nuts (walnut), Seeds (sesame), Spice (nutmeg), AFB_1, literature[239]; Nuts (peanut), AFB_2, AFG_1, and AFG_2, literature[239]

incidence: 1/1, conc.: 625 µg/kg, sample origin: Angola, sample year: 1970, country: Germany[239]

- Co-contamination: 1 sa. co-contaminated with AFB_1, AFB_2, AFG_1, and AFG_2
- Further contamination of plants and plant products other than cereals and their products reported in the present book/article[239]: Fruits (apple), Fruits (cherry), Fruits (lemon), Fruits (orange), Fruits (peach), Fruits (tomato), Nuts (almond), Nuts (cashew nut), Nuts (hazelnut), Nuts (walnut), Seeds (sesame), Spice (nutmeg), AFB_1, literature[239]; Nuts (peanut), AFB_2, AFG_1, and AFG_2, literature[239]

incidence: 788/1,559, conc. range: ≤3,650 µg/kg, sample origin: imported to USSR, sample year: 1985–1987, country: USSR[267]

- Co-contamination: not reported
- Further contamination of plants and plant products other than cereals and their products reported in the present book/article[267]: Cocoa beans, Coffee beans, Legumes, Nuts, AFB_1, literature[267]

incidence: 22/26*, conc. range: 1.5–745.7 µg/kg, ∅ conc.: 104.1 µg/kg, sample origin: five markets in Lagos, Ogun, and Oyo (states), Southwest Nigeria, sample year: unknown, country: Nigeria[276], *peanuts, raw

incidence: 21/22*, conc. range: 1.3–59.1 µg/kg, ∅ conc.: 14.1 µg/kg, sample origin: five markets in Lagos, Ogun, and Oyo (states), Southwest Nigeria, sample year: unknown, country: Nigeria[276], *peanuts, roasted

incidence: 25/30*, conc. range: 1.5–745.7 µg/kg, ∅ conc.: 93.3 µg/kg, sample origin: five markets in Lagos, Ogun, and Oyo (states), Southwest Nigeria, sample year: unknown, country: Nigeria[276], *peanuts, coated

incidence: 18/18*, conc. range: 1.3–59.1 µg/kg, ∅ conc.: 14.2 µg/kg, sample origin: five markets in Lagos, Ogun, and Oyo (states), Southwest Nigeria, sample year: unknown, country: Nigeria[276], *peanuts, de-coated

- Co-contamination: not reported
- Further contamination of plants and plant products other than cereals and their products reported in the present book/article[276]: Nuts (peanut), AF/S, literature[276]

For detailed information see the article.

incidence: 10/50, conc. range: 0.1–22.7 µg/kg, ∅ conc.: 5.2 µg/kg, sample origin: available in New Zealand, sample year: 2010, country: New Zealand[277]

- Co-contamination: not reported
- Further contamination of plants and plant products other than cereals and their products reported in the present book/article[277]: Bars, AFB_1, AF/S, DON, NIV, and OTA, literature[277]; Butter (peanut butter), Confectionery, Food, Fruits, dried, Nuts, Nuts (pistachio nut), Sauces (peanut sauce), AFB_1 and AF/S, literature[277]; Coffee, Dates, Spices (nutmeg), Spices (pepper), and Vine fruits, dried, OTA, literature[277]; Figs, Spices (cayenne pepper, chilli), Spices (curry), Spices (ginger), Spices (paprika), AFB_1, AF/S, and OTA, literature[277]; Nuts (peanut), AF/S, literature[277]; Nuts (Brazil nut, paranut), AFB_1, literature[277]

incidence: 1/67*, conc.: 42.06 µg/kg**, sample origin: imported from Argentina, Brazil, Chile, China, Egypt, India, Israel, Turkey, and the USA to Italy, sample year: January 2008–December 2009, country: Italy[320], *peanuts with shell, **contaminated sa. from Egypt

- Co-contamination: 1 sa. co-contaminated with AFB_1 and AFB_2
- Further contamination of plants and plant products other than cereals and their products reported in the present book/article[320]: Coffee beans, Vine fruits, dried, OTA, literature[320]; Kernels (apricot), AFB_1, AFB_2, and AFG_1, literature[320]; Nuts (hazelnut), Paste (hazelnut paste), AFB_1, AFB_2, AFG_1, and AFG_2, literature[320]; Nuts (peanut), AFB_2, literature[320]; Nuts (pistachio nut), AFB_1 and AFB_2, literature[320]

incidence: 0/2*, conc.: no contamination, sample origin: retail outlets in Lattakia (city), Syria, sample year: January 1989, country: Italy[347], *peanuts, in- shell

incidence: 1/2*, conc.: 2.7 µg/kg, sample origin: retail outlets in Lattakia (city), Syria, sample year: January 1989, country: Italy[347], *peanuts, raw and shelled

incidence: 2/3*, conc. range: 0.4–1.9 µg/kg, ∅ conc.: 1.2 µg/kg, sample origin: retail outlets in Lattakia (city), Syria, sample year: January 1989, country: Italy[347], *peanuts, roasted and shelled

- Co-contamination: not reported
- Further contamination of plants and plant products other than cereals and their products reported in the present book/article[347]: Figs, Lentils, AFB_1, literature[347]; Nuts (peanut), AFB_2, literature[347]

incidence: 1/20, conc.: 0.17 µg/kg, sample origin: retail shops and local markets in Rabat-Salé (region), Morocco, sample year: January–October 2006, country: Spain/Morocco[352]

- Co-contamination: not reported
- Further contamination of plants and plant products other than cereals and their products reported in the present book/article[352]: Figs, Nuts (pistachio nut), Nuts (walnut), Raisins, AFB_1 and AF/S, literature[352]; Nuts (peanut), AF/S, literature[352]

incidence: 26/86*, conc. range: 10–2,000 µg/kg, ∅ conc.: 420 µg/kg, sample origin: supermarkets, retail shops, fair, and beach, Recife (city), Pernambuco (state), Brazil, sample year: May–October 1993, country: Brazil[378], *peanuts, raw and roasted

- Co-contamination: 4 sa. co-contaminated with AFB_1, AFB_2, AFG_1, and AFG_2; 16 sa. co-contaminated with AFB_1 and AFB_2; 2 sa. co-contaminated with AFB_1, AFG_1, and AFG_2; 4 sa. contaminated solely with AFB_1
- Further contamination of plants and plant products other than cereals and their products reported in the present book/article[378]: Nuts (peanut), AFB_2, AFG_1, and AFG_2, literature[378]

incidence: 926/2,062, conc. range: >5–833 µg/kg, sample origin: different states in India, sample year: unknown, country: India[379]
- Co-contamination: not reported
- Further contamination of plants and plant products other than cereals and their products reported in the present book/article[379]: not reported

For detailed information see the article.

incidence: 4?/65, conc. range: 6–46 µg/kg, sample origin: storage houses and dealers, Tunisia, sample year: unknown, country: Tunisia/USA[380]
- Co-contamination: not reported
- Further contamination of plants and plant products other than cereals and their products reported in the present book/article[380]: Nuts (peanut), AFG_1, literature[380]; Nuts (pistachio nut), AFB_1, literature[380]; Peas, AFB_1 and AFG_1, literature[380]

incidence: 36/64, conc. range: 6–11 µg/kg (1 sa.), 11–25 µg/kg (6 sa.), 26–50 µg/kg (5 sa.), 51–100 µg/kg (9 sa.), 101–200 µg/kg (11 sa.), 201–300 µg/kg (1 sa.), and 301–400 µg/kg (3 sa.), sample origin: imported from Brazil and the UK, sample year: 1982–1984, country: UK[381]

incidence: 3/40, conc. range: 6–11 µg/kg (1 sa.), 11–25 µg/kg (1 sa.), and 201–300 µg/kg (1 sa.), sample origin: imported from China and the UK, sample year: 1982–1984, country: UK[381]

incidence: 1/1, conc. range: 26–50 µg/kg (1 sa.), sample origin: imported from Gambia and the UK, sample year: 1982–1984, country: UK[381]

incidence: 3/9, conc. range: 11–25 µg/kg (1 sa.), 26–50 µg/kg (1 sa.), and 51–100 µg/kg (1 sa.), sample origin: imported from India and the UK, sample year: 1982–1984, country: UK[381]

incidence: 6/9, conc. range: 11–25 µg/kg (1 sa.), 51–100 µg/kg (1 sa.), and >400 µg/kg (1 sa.), sample origin: imported from Senegal and the UK, sample year: 1982–1984, country: UK[381]

incidence: 2/4, conc. range: 201–300 µg/kg (1 sa.) and 301–400 µg/kg (1 sa.), sample origin: imported from Sudan and the UK, sample year: 1982–1984, country: UK[381]
- Co-contamination: not reported
- Further contamination of plants and plant products other than cereals and their products reported in the present book/article[381]: Nuts (walnut), AFB_1, literature[381]

incidence: 3/40, conc. range: 98–1,056 µg/kg, ∅ conc.: 696 µg/kg, sample origin: different localities in Alexandria (province), Egypt, sample year: unknown, country: Egypt[382]
- Co-contamination: not reported
- Further contamination of plants and plant products other than cereals and their products reported in the present book/article[382]: not reported

incidence: 37/190*, conc. range: 6.5 to <26.0 µg/kg, sample year: unknown, country: Egypt[383], *peanuts, in-shell

incidence: 76/155*, conc. range: 6.5 to <52.0 µg/kg, sample year: unknown, country: Egypt[383], *peanuts, shelled

incidence: 2/58*, conc. range: 52 to <104.0 µg/kg, sample year: unknown, country: Egypt[383], *peanuts, roasted
- Co-contamination: not reported
- Further contamination of plants and plant products other than cereals and their products reported in the present book/article[383]: Nuts (peanut), AFB_2, AFG_1, and AFG_2, literature[383]

incidence: 2/2*, conc. range: 46–498 µg/kg, ∅ conc.: 272 µg/kg, sample origin: unknown, sample year: unknown, country: UK[384], *ncac
- Co-contamination: 2 sa. co-contaminated with AFB_1, AFB_2, AFG_1, and AFG_2
- Further contamination of plants and plant products other than cereals and their products reported in the present book/article[384]: Beans (soybean), AFB_1, AFB_2, AFG_1, and AFG_2, literature[384]; Nuts (peanut), AFB_2, AFG_1, and AFG_2, literature[384]

incidence: 1*/65**, conc.: 120 µg/kg, sample origin: *retail self-service shop, Valencia? (province), Spain, sample year: unknown, country: Spain[385], **peanuts, in-shell
- Co-contamination: 1 sa. co-contaminated with AFB_1 and AFB_2
- Further contamination of plants and plant products other than cereals and their products reported in the present book/article[385]: Nuts (peanut), AFB_2, literature[385]

incidence: 10/20*, conc. range: 1–50 µg/kg (3 sa.), 51–150 µg/kg (3 sa.), 151–300 µg/kg (2 sa.), and 301–700 µg/kg (2 sa.), sample origin: adjoining villages and markets of Calcutta, Dasna, Gwalior, Hapur, Muzaffarnagar, Patna, and Udaipur (cities), India, sample year: 1970–1980, country: India[386], *peanuts, raw

incidence: 5/15*, conc. range: 1–50 µg/kg (3 sa.) and 51–150 µg/kg (2 sa., maximum: 85 µg/kg), sample origin: adjoining villages and markets of Calcutta, Dasna, Gwalior, Hapur, Muzaffarnagar, Patna, and Udaipur (cities), India, sample year: 1970–1980, country: India[386], *peanuts, roasted
- Co-contamination: not reported
- Further contamination of plants and plant products other than cereals and their products reported in the present book/article[386]: not reported

incidence: 1/1*, conc.: 62.2 µg/kg, sample origin: WHO, USA, sample year: unknown, country: USA[387], *ncac
- Co-contamination: not reported
- Further contamination of plants and plant products other than cereals and their products reported in the present book/article[387]: not reported

incidence: 32/72*, conc. range: 1–679 µg/kg, ∅ conc.: 93.7 µg/kg, sample origin: supermarkets and groceries in Maringá (city), Paraná (state), Brazil, sample year: March 1993–February 1994, country: Brazil[388], *peanuts, crude
- Co-contamination: 19 sa. co-contaminated with AFB_1, AFB_2, AFG_1, and AFG_2; 1 sa. co-contaminated with AFB_1, AFB_2, and AFG_1; 10 sa. co-contaminated with AFB_1 and AFB_2; 2 sa. contaminated solely with AFB_1
- Further contamination of plants and plant products other than cereals and their products reported in the present book/article[388]: Nuts (peanut), AFB_2, AFG_1, and AFG_2, literature[388]

incidence: 9/20, conc. range: <4 µg/kg (1 sa.), 4–20 µg/kg (3 sa.), 20–100 µg/kg (2 sa.), and ≤723 µg/kg (3 sa.), sample origin: Babile (district), Ethiopia, sample year: 2013/2014, country: Ethiopia/USA[389]

incidence: 6/20, conc. range: 7.1 µg/kg (1 sa.) and ≤2,526 µg/kg (5 sa.), sample origin: Babile (district), Ethiopia, sample year: 2014/2015, country: Ethiopia/USA[389] incidence: 6/20, conc. range: 4–20 µg/kg (3 sa.) and ≤113 µg/kg (3 sa.), sample origin: Darolabu (district), Ethiopia, sample year: 2013/2014, country: Ethiopia/USA[389]

incidence: 12/20, conc. range: <4 µg/kg (2 sa.), 4–20 µg/kg (2 sa.), 20–100 µg/kg (2 sa.), and ≤1,358 µg/kg (6 sa.), sample origin: Darolabu (district), Ethiopia, sample year: 2014/2015, country: Ethiopia/USA[389]

incidence: 0/20, conc.: no contamination, sample origin: Fedis (district), Ethiopia, sample year: 2013/2014, country: Ethiopia/USA[389]

incidence: 9/20, conc. range: <4 µg/kg (1 sa.), 4–20 µg/kg (2 sa.), 20–100 µg/kg (4 sa.), and ≤2,368 µg/kg (2 sa.), sample origin: Fedis (district), Ethiopia, sample year: 2014/2015, country: Ethiopia/USA[389]

incidence: 0/20, conc.: no contamination, sample origin: Gursum (district), Ethiopia, sample year: 2013/2014, country: Ethiopia/USA[389]

incidence: 5/20, conc. range: 4–20 µg/kg (3 sa.) and ≤682 µg/kg (2 sa.), sample origin: Gursum (district), Ethiopia, sample year: 2014/2015, country: Ethiopia/USA[389]

- Co-contamination: not reported
- Further contamination of plants and plant products other than cereals and their products reported in the present book/article[389]: Cake (peanut), AFB_1 and AFB_2, literature[389]; Nuts (peanut), AFB_2, AFG_1, and AFG_2, literature[389]

incidence: 78/120, conc. range: 1.3–223 µg/kg, Ø conc.: 27 µg/kg, sample origin: retail outlets, Botswana, sample year: July 2001–January 2002, country: Botswana[390]

- Co-contamination: not reported
- Further contamination of plants and plant products other than cereals and their products reported in the present book/article[390]: Nuts (peanut), AFB_2, AFG_1, AFG_2, and CPA, literature[390]

incidence: 3/11*, conc. range: 0.13–0.25 µg/kg, Ø conc.: 0.18 µg/kg, sample origin: Brazil, China, the USA, and unknown, sample year: unknown, country: Spain[391], *included peeled salted, raw, raw in-shell, roasted and salted in-shell, roasted in-shell, and salted peanuts

- Co-contamination: 2 sa. co-contaminated with AFB_1 and AFG_1; 1 sa. contaminated solely with AFB_1
- Further contamination of plants and plant products other than cereals and their products reported in the present book/article[391]: Nuts (peanut), AFG_1, literature[391]

incidence: 10/10, conc. range: 1.5–10 µg/kg, Ø conc.: 4.8 µg/kg, sample origin: markets in Adjamé, Abobo, and Treichville (communes) of Abidjan (city), Ivory Coast, sample year: November 2002, country: France/Ivory Coast[394]

- Co-contamination: 3 sa. co-contaminated with AFB_1, FB_1, OTA, and ZEA; 4 sa. co-contaminated with AFB_1, FB_1, and ZEA; 3 sa. co-contaminated with AFB_1, OTA, and ZEA
- Further contamination of plants and plant products other than cereals and their products reported in the present book/article[394]: Nuts (peanut), FB_1, OTA, and ZEA, literature[394]

incidence: 66/503, conc. range: <100 µg/kg (9 sa.), 100–1,000 µg/kg (37 sa.), and >1,000 µg/kg (20 sa.), sample origin: Córdoba (province), Argentina, sample year: 1973/1974, country: Argentina[395]

- Co-contamination: not reported
- Further contamination of plants and plant products other than cereals and their products reported in the present book/article[395]: not reported

incidence: 50/50*, conc. range: 2–17,203 µg/kg, Ø conc.: 1,377 µg/kg, sample origin: Georgia (state), USA, sample year: 1990, country: USA[396], *peanuts for oil production or feed

- Co-contamination: 14 sa. co-contaminated with AFB_1, AFB_2, AFG_1, AFG_2, and CPA; 2 sa. co-contaminated with AFB_1, AFB_2, AFG_1, and AFG_2; 12 sa. co-contaminated with AFB_1, AFB_2, AFG_1, and CPA; 1 sa. co-contaminated with AFB_1, AFB_2, and AFG_1; 19 sa. co-contaminated with AFB_1, AFB_2, and CPA; 2 sa. co-contaminated with AFB_1 and AFB_2
- Further contamination of plants and plant products other than cereals and their products reported in the present book/article[396]: Nuts (peanut), AFB_2, AFG_1, AFG_2, and CPA, literature[396]

incidence: 19/28, conc. range: 4 to <20 µg/kg (6 sa.), ≥20 µg/kg (13 sa., maximum: 576 µg/kg), sample origin: local markets in the northern part (17 provinces), Thailand, sample year: June 1994–July 1995, country: Japan/Thailand[399]

- Co-contamination: not reported
- Further contamination of plants and plant products other than cereals and their products reported in the present book/article[399]: not reported

incidence: 18/31, conc. range: 300–11,300 µg/kg, Ø conc.: 5,017 µg/kg, sample origin: markets and shops, Uganda, sample year: unknown, country: Uganda/UK[400]

- Co-contamination: not reported
- Further contamination of plants and plant products other than cereals and their products reported in the present book/article[400]: not reported

incidence: 7/10*, conc. range: >LOD-23.59 µg/kg**, sample origin: local supermarkets in Rio de Janeiro (city), Rio de Janeiro (state), Brazil, sample year: unknown, country: Brazil[401], *8 roasted and 2 raw peanut sa., **only roasted peanut sa. contaminated

- Co-contamination: 1 sa. co contaminated with AFB_1, AFB_2, and AFG_1; 1 sa. co-contaminated with AFB_1, AFB_2, AFM_1, and AFM_2; 1 sa. co-contaminated with AFB_1, AFB_2, and AFM_1; 1 sa. co-contaminated with AFB_1, AFB_2, AFG_1, and AFM_1; 1 sa. co-contaminated with AFB_1 and AFB_2; 2 sa. contaminated solely with AFB_1
- Further contamination of plants and plant products other than cereals and their products reported in the present book/article[401]: Nuts (peanut), AFB_2, AFG_1, AFM_1, and AFM_2, literature[401]

incidence: 1/1, conc.: 1,700 µg/kg, sample origin: Nepal, sample year: unknown, country: Japan/Nepal[403]

- Co-contamination: 1 sa. co-contaminated with AFB_1 and AFG_1 as well as other aflatoxins
- Further contamination of plants and plant products other than cereals and their products reported in the present book/article[403]: Beans (soybean), AFB_1 and AFG_1, literature[403]; Nuts (peanut), AFG_1, literature[403]

incidence: 3/24*, conc. range: 2–270 µg/kg, sample origin: villages, India, sample year: unknown, country: UK/France[404], *peanuts and peanut meals

- Co-contamination: not reported

- Further contamination of plants and plant products other than cereals and their products reported in the present book/article[404]: Oils, AFB_1, literature[404]; Oils (peanut oil), AFB_1 and AFB_2, literature[404]

incidence: 9/28*, conc. range: 54–110 µg/kg, sample origin: distributor level in Sitapur (district), India, sample year: November 1995–April 1996, country: India[405], *peanut kernels, unroasted

incidence: 8/24*, conc. range: 56–116 µg/kg, sample origin: distributor level in Shahjahanpur (district), India, sample year: November 1995–April 1996, country: India[405], *peanut kernels, unroasted

incidence: 15/26*, conc. range: 220–442 µg/kg, sample origin: retailor level in Sitapur (district), India, sample year: November 1995–April 1996, country: India[405], *peanut kernels, unroasted

incidence: 14/24*, conc. range: 228–488 µg/kg, sample origin: retailor level in Shahjahanpur (district), India, sample year: November 1995–April 1996, country: India[405], *peanut kernels, unroasted

incidence: 4/25*, conc. range: 24–68 µg/kg, sample origin: distributor level in Sitapur (district), India, sample year: November 1995–April 1996, country: India[405], *peanut kernels, roasted

incidence: 4/24*, conc. range: 27–76 µg/kg, sample origin: distributor level in Shahjahanpur (district), India, sample year: November 1995–April 1996, country: India[405], *peanut kernels, roasted

incidence: 9/32*, conc. range: 108–172 µg/kg, sample origin: retailor level in Sitapur (district), India, sample year: November 1995–April 1996, country: India[405], *peanut kernels, roasted

incidence: 8/28*, conc. range: 110–186 µg/kg, sample origin: retailor level in Shahjahanpur (district), India, sample year: November 1995–April 1996, country: India[405], *peanut kernels, roasted

- Co-contamination: not reported
- Further contamination of plants and plant products other than cereals and their products reported in the present book/article[405]: not reported

incidence: 10/10, conc. range: 0.1–0.5 µg/kg, sample origin: individual households in Huaian area (high-risk area for oesophageal cancer), China, sample year: unknown, country: China[406]

incidence: 7/7, conc. range: 0.1–0.7 µg/kg, sample origin: individual households in Fusui area (high-risk area for HCC and oesophageal cancer), China, sample year: unknown, country: China[406]

- Co-contamination: not reported
- Further contamination of plants and plant products other than cereals and their products reported in the present book/article[406]: Oils, AFB_1, literature[406]

incidence: 20/60*, ∅ conc.: 7.0–116 µg/kg, sample origin: Tupã (region), São Paulo (state), Brazil, sample year: June 2005–May 2006, country: Brazil[408], *evaluated monthly for a storage period of 12 months

- Co-contamination: not reported
- Further contamination of plants and plant products other than cereals and their products reported in the present book/article[408]: Nuts (peanut), AFB_2, literature[408]

For detailed information see the article.

incidence: 12/73, conc. range: 0.22–33.4 µg/kg, sample origin: public markets/bazaars, supermarkets, retail stores, and wholesale stores in Ankara, Antalya, Isparta, Burdur, and Izmir (provinces), Turkey, sample year: unknown, country: Turkey[409]

- Co-contamination: not reported
- Further contamination of plants and plant products other than cereals and their products reported in the present book/article[409]: Nuts (hazelnut), AFB_1, AFG_1, and AFG_2, literature[409]; Nuts (peanut), AFB_2, AFG_1, and AFG_2, literature[409]; Nuts (pistachio nut), AFB_1 and AFB_2, literature[409]

incidence: 259/1040, conc. range: LOD-≤2 µg/kg (220 sa.), 3–5 µg/kg (6 sa.), 6–20 µg/kg (20 sa.), 21–30 µg/kg (3 sa.), and ≥31 µg/kg (10 sa., maximum: 720 µg/kg), sample origin: 12 provinces in eastern China, sample year: 2009/2010, country: China[410]

- Co-contamination: not reported
- Further contamination of plants and plant products other than cereals and their products reported in the present book/article[410]: not reported

For detailed information see the article.

incidence: 9/65, conc. range: 0.15–22.39 µg/kg, sample origin: local markets in Chongqing, Fujian, Guangdong, Guangxi, Hubei, Jiangsu, Shanghai, and Zhejiang (regions), China, sample year: 2003, country: China[411]

- Co-contamination: not reported
- Further contamination of plants and plant products other than cereals and their products reported in the present book/article[411]: Nuts (peanut), AFB_2, AFG_1, and AFG_2, literature[411]; Nuts (pine nut), AFB_1 and AFB_2, literature[411]; Nuts (walnut), AFB_1, AFB_2, AFG_1, and AFG_2, literature[411]

incidence: 40/48*, conc. range: ≤1,197 µg/kg, sample origin: Fushui (county, sa. collected from high incidence area of liver cancer), China, sample year: unknown, country: China[417]

incidence: 24/29*, conc. range: ≤562 µg/kg, sample origin: Fushui (county, sa. collected from low incidence area of liver cancer), China, sample year: unknown, country: China[417]

- Co-contamination: not reported
- Further contamination of plants and plant products other than cereals and their products reported in the present book/article[417]: Oils (peanut oil), AFB_1, literature[417]

incidence: 13/408, conc. range: 0.15–116.64 µg/kg, sample origin: Liaoning (province with temperate monsoon climate), China, sample year: 2010–2013, country: China[441]

incidence: 226/1190, conc. range: 0.06–483.00 µg/kg, sample origin: Henan (province with transitional monsoon climate), China, sample year: 2010–2013, country: China[441]

incidence: 71/455, conc. range: 0.24–1,390 µg/kg, sample origin: Sichuan (province with subtropical temperate monsoon climate), China, sample year: 2010–2013, country: China[441]

incidence: 51/441, conc. range: 0.22–341.41 µg/kg, sample origin: Guangdong (province with subtropical-tropical monsoon climate), China, sample year: 2010–2013, country: China[441]

- Co-contamination: For detailed information see the article.
- Further contamination of plants and plant products other than cereals and their products reported in the present book/article[441]: Nuts (peanuts), AFB_2, AFG_1, and AFG_2, literature[441]

incidence: 63/84, conc. range: ≤92.07 µg/kg, sample origin: supermarkets in Selangor (state), Malaysia, sample year: April–August 2008, country: Malaysia/Indonesia[676]

- Co-contamination: not reported

- Further contamination of plants and plant products other than cereals and their products reported in the present book/article[676]: Nuts (peanut), AFB_2 and AFG_1, literature[676]

For detailed information see the article.

incidence: 29/151, conc. range: 0.16–49.9 µg/kg, ∅ conc.: 4.99 µg/kg, sample origin: markets, supermarkets, groceries, and retail shops in Adana and Osmaniye (provinces), Turkey, sample year: January 2010–December 2012, country: Turkey[741]

- Co-contamination: not reported
- Further contamination of plants and plant products other than cereals and their products reported in the present book/article[741]: Nuts (peanut), AFB_2 and AFG_1, literature[741]; Nuts (pistachio nut), AFB_1, AFB_2, AFG_1, and AFG_2, literature[741]

incidence: 5/58, conc. range: 24.0–87.5 µg/kg, ∅ conc.: 45.3 µg/kg, sample origin: supermarkets and rural fairs in Ijuí, Porto Alegre, Santa Cruz do Sul, Santa Maria, and São Leopoldo (cities), Rio Grande do Sul (state), Brazil, sample year: May 2009–August 2010, country: Brazil[744]

- Co-contamination: not reported
- Further contamination of plants and plant products other than cereals and their products reported in the present book/article[744]: Products (peanut product), AFB_1, literature[744]

incidence: 20/20*, conc. range: 2.19–543.93 µg/kg, ∅ conc.: 97.4 µg/kg, sample origin: street hawkers, local markets, and retail shops in urban areas of Kinshasa (capital), Congo, sample year: December 2010–January 2011, country: South Africa[758], *peanuts, raw

- Co-contamination: 14 sa. co-contaminated with AFB_1, AFB_2, AFG_1, and AFG_2; 1 sa. co-contaminated with AFB_1, AFB_2, and AFG_1; 3 sa. co-contaminated with AFB_1 and AFB_2; 1 sa. co-contaminated with AFB_1 and AFG_2; 1 sa. contaminated solely with AFB_1
- Further contamination of plants and plant products other than cereals and their products reported in the present book/article[758]: Nuts (peanut), AFB_2, AFG_1, and AFG_2, literature[758]

incidence: 18/20*, conc. range: 2.10–35.39 µg/kg, ∅ conc.: 7.67 µg/kg, sample origin: street hawkers, local markets, and retail shops in inner city of Pretoria (capital), South Africa, sample year: December 2010–January 2011, country: South Africa[758], *peanuts, raw

- Co-contamination: 6 sa. co-contaminated with AFB_1, AFB_2, AFG_1, and AFG_2; 2 sa. co-contaminated with AFB_1, AFB_2, and AFG_1; 1 sa. co-contaminated with AFB_1, AFB_2, and AFG_2; 5 sa. co-contaminated with AFB_1 and AFB_2; 4 sa. contaminated solely with AFB_1
- Further contamination of plants and plant products other than cereals and their products reported in the present book/article[758]: Nuts (peanut), AFB_2, AFG_1, and AFG_2, literature[758]

incidence: 27/30, conc. range: 12–937 µg/kg, sample origin: street hawkers, local markets, and retail shops in rural areas of Kinshasa (capital), Congo, sample year: October–December 2009 (rainy season), country: Congo/South Africa[759]

incidence: 16/30, conc. range: 15–390 µg/kg, sample origin: street hawkers, local markets, and retail shops in rural areas of Kinshasa (capital), Congo, sample year: May–July 2009 (dry season), country: Congo/South Africa[759]

- Co-contamination: not reported
- Further contamination of plants and plant products other than cereals and their products reported in the present book/article[759]: not reported

incidence: 2/9, conc. range: 3.16–6.00 µg/kg, ∅ conc.: 4.58 µg/kg, sample origin: supermarkets and night markets in Penang (state), Malaysia, sample year: unknown, country: Malaysia[766]

- Co-contamination: not reported
- Further contamination of plants and plant products other than cereals and their products reported in the present book/article[766]: Nuts (peanut), AFG_1 and AFG_2, literature[766]; Spices (cayenne pepper, chilli), AFB_1, AFB_2, AFG_1, and AFG_2, literature[766]

incidence: 7/8*, conc. range: 22–2,222 µg/kg, ∅ conc.: 682 µg/kg, sample origin: imported to Finland, sample year: unknown, country: Denmark/Finland[769], *peanuts for various purposes

- Co-contamination: 2 sa. co-contaminated with AFB_1, AFB_2, AFG_1, and AFG_2; 1 sa. co-contaminated with AFB_1, AFB_2, and AFG_1; 1 sa. co-contaminated with AFB_1 and AFG_1; 3 sa. contaminated solely with AFB_1
- Further contamination of plants and plant products other than cereals and their products reported in the present book/article[769]: Nuts (peanut), AFB_2, AFG_1, and AFG_2, literature[769]

incidence: 4/4*, conc. range: 3.76–34.8 µg/kg, ∅ conc.: 21.3 µg/kg, sample origin: unknown, sample year: unknown, country: China[775], *ncac

- Co-contamination: 4 sa. co-contaminated with AFB_1, OTA, and ZEA
- Further contamination of plants and plant products other than cereals and their products reported in the present book/article[775]: Nuts (peanut), OTA and ZEA, literature[775]

incidence: 5/10*, conc. range: LOD-20 µg/kg (5 sa., maximum: 19.2 µg/kg), sample origin: retail outlets in Paraná (state), Brazil, sample year: April 2006–May 2007, country: Brazil[784], *peanuts, salted

incidence: 24/45*, conc. range: LOD-20 µg/kg (17 sa.) and >20 µg/kg (7 sa., maximum: 94.6 µg/kg), sample origin: retail outlets in Paraná (state), Brazil, sample year: April 2006–May 2007, country: Brazil[784], *peanuts, sweet

incidence: 9/18*, conc. range: LOD-20 µg/kg (9 sa., maximum: 14.5 µg/kg), sample origin: retail outlets in Paraná (state), Brazil, sample year: April 2006–May 2007, country: Brazil[784], *peanut brittle

- Co-contamination: For detailed information see the article.
- Further contamination of plants and plant products other than cereals and their products reported in the present book/article[784]: Candies (peanut candy), AFB_1, AFB_2, AFG_1, and AFG_2, literatrure[784]; Nuts (peanut), AFB_2, AFG_1, and AFG_2, literature[784]

incidence: 2/9*, conc. range: 0.5 µg/kg, ∅ conc.: 0.5 µg/kg, sample origin: market in Lilongwe (city), Malawi, sample year: December 2012, country: Malawi[790], *peanuts, un-skinned roasted

incidence: 11/15*, conc. range: 0.1–12.3 µg/kg, sample origin: market in Lilongwe (city), Malawi, sample year: December 2012, country: Malawi[790], *peanuts, de-skinned roasted

- Co-contamination: not reported
- Further contamination of plants and plant products other than cereals and their products reported in the present book/article[790]: Butter (peanut butter), Food, AFB_1, AFB_2, AFG_1, and AFG_2, literature[790]; Nuts (peanut), AFB_2, AFG_1, and AFG_2, literature[790]

incidence: 261*/797**, conc. range: >4 µg/kg (195 sa.) and >20 µg/kg (66 sa., maximum: 3,240 µg/kg), sample origin: farmers, traders, supermarkets, processors, and middleman (bought crops from farmers), 11 districts in Malawi, sample year: 2008/2009 harvest, country: Malawi/Kenya/Uganda/Niger/India[796], *>4 µg/kg, **peanuts, fresh and peanut butter?

- Co-contamination: not reported
- Further contamination of plants and plant products other than cereals and their products reported in the present book/article[796]: not reported

incidence: 3/18, conc. range 6.3–528 µg/kg, sample origin: retail shops and vendors at informal markets in Bulawayo metropolitan area, Zimbabwe, sample year: October/November 2011, country: Botswana/South Africa[797]

- Co-contamination: 1 sa. contaminated with AFB_1, AFB_2, and AFG_1, no further information available
- Further contamination of plants and plant products other than cereals and their products reported in the present book/article[797]: Butter (peanut butter), AFB_1, AFB_2, AFG_1, and AFG_2, literature[797]; Nuts (peanut), AF/S, literature[797]

incidence: 26/90, conc. range: 6–125 µg/kg, sample origin: humid forest with monomodal rainfall, humid forest with bimodal rainfall, and western highland, Cameroon, sample year: December 2010–January 2011 (dry season), country: Belgium/Benin[807]

- Co-contamination: For detailed information see the article.
- Further contamination of cereals and cereal products reported in the present book/article[807]: Nuts (peanut), OTA, literature[807]; Products (cassava product), AFB_1 and PA, literature[807]

For detailed information see the article.

incidence: 9/25, conc. range: 0.04–4.74 µg/kg, ∅ conc.: 0.85 µg/kg, sample origin: retail fresh markets or modern department stores in the central region of Thailand, sample year: unknown, country: Thailand/Austria/Germany[857]

- Co-contamination: not reported
- Further contamination of plants and plant products other than cereals and their products reported in the present book/article[857]: Nuts (peanut), AFB_2 and AFG_1, literature[857]; Spices (cayenne pepper, chilli), AFB_1, AFB_2, and AFG_1, literature[857]

incidence: 8/15, conc. range: ≤230 µg/kg, ∅ conc.: 68 µg/kg, sample origin: Osmaniye (province), Turkey, sample year: 2007, country: Austria[868]

- Co-contamination: For detailed information see the article.
- Further contamination of plants and plant products other than cereals and their products reported in the present book/article[868]: Nuts (almond), AME, AOH, BEA, CPA, ENB, ENB_1, and ENB_3, literature[868]; Nuts (hazelnut), 3-NPA, AFB_1, AFB_2, AFG_1, AFG_2, AME, AOH, ATX-I, BEA, ENA, ENA_1, ENB_1, ENB_2, ENB_3, ENB_4, HT-2, KA, MA, MON, OTA, OTB, STG, T-2, and ZEA, literature[868]; Nuts (peanut), 3-NPA, AFB_2, AME, AOH, ATX-I, BEA, CPA, ENA, ENA_1, ENB, ENB_4, KA, MA, OTA, OTB, STG, and ZEA, literature[868]; Nuts (pistachio nut), BEA, MA, and OTA, literature[868]; see also Nuts (almond), Nuts (hazelnut), Nuts (peanut), Nuts (pistachio nut), chapter: Further Mycotoxins and Microbial Metabolites

For detailed information see the article.

incidence: 14/18, conc. range: ≥0.03 to <2 µg/kg (10 sa.), ≥2 to <5 µg/kg (2 sa.), and ≥5 µg/kg (2 sa., maximum: 9.95 µg/kg), ∅ conc.: 1.97 µg/kg, sample origin: sales outlets in nine geographic zones in Tehran (province), Iran, sample year: June 2005, country: Iran[879]

- Co-contamination: not reported
- Further contamination of plants and plant products other than cereals and their products reported in the present book/article[879]: not reported

incidence: 12/30, conc. range: ≤20 µg/kg (11 sa.) and 25.81 µg/kg (1 sa.), sample origin: local markets in Wuhan (city), China, sample year: unknown, country: China[881]

- Co-contamination: not reported
- Further contamination of plants and plant products other than cereals and their products reported in the present book/article[881]: Nuts (peanut), AFB_2, AFG_1, AFG_2, OTA, T-2, and ZEA, literature[881]

Aflatoxin B_2

incidence: 7/104, conc. range: tr-568 µg/kg, sample origin: imported to Finland, sample year: 1975/1976, country: Finland[2]

- Co-contamination: 1 sa. co-contaminated with AFB_1, AFB_2, AFG_1, and AFG_2; 1 sa. co-contaminated with AFB_1, AFB_2, and AFG_1; 5 sa. co-contaminated with AFB_1 and AFB_2
- Further contamination of plants and plant products other than cereals and their products reported in the present book/article[2]: Marzipan (almond paste), Nuts (almond), Nuts (hazelnut), AFB_1 and AFB_2, literature[2]; Nuts, AFB_1, AFB_2, and AFG_1, literature[2]; Nuts (peanut), AFB_1, AFG_1, and AFG_2, literature[2]

incidence: 8/121*, conc. range: 1–89 µg/kg, ∅ conc.: 22 µg/kg, sample origin: imported to Finland, sample year: 1974–1976, country: Finland[2], *peanuts, roasted

- Co-contamination: 2 sa. co-contaminated with AFB_1, AFB_2, and AFG_1; 6 sa. co-contamination with AFB_1 and AFB_2
- Further contamination of plants and plant products other than cereals and their products reported in the present book/article[2]: Marzipan (almond paste), Nuts (almond), and Nuts (hazelnut), AFB_1 and AFB_2, literature[2]; Nuts, AFB_1, AFB_2, and AFG_1, literature[2]; Nuts (peanut), AFB_1, AFG_1, and AFG_2, literature[2]

incidence: 17/68*, conc. range: tr-76 µg/kg, sample origin: imported to Finland, sample year: 1974–1976, country: Finland[2], *peanuts, sliced and crushed

- Co-contamination: 4 sa. co-contaminated with AFB_1, AFB_2, AFG_1, and AFG_2; 13 sa. co-contaminated with AFB_1 and AFB_2
- Further contamination of plants and plant products other than cereals and their products reported in the present book/article[2]: Marzipan (almond paste), Nuts (almond), and Nuts (hazelnut), AFB_1 and AFB_2, literature[2]; Nuts, AFB_1, AFB_2, and AFG_1, literature[2]; Nuts (peanut), AFB_1, AFG_1, and AFG_2, literature[2]

incidence: 4/45*, conc. range: 375–5,567 µg/kg, ∅ conc.: 2,283.5 µg/kg, sample origin: different storage facilities, Egypt, sample year: November 1978–April 1979, country: Egypt[4], *sa. moldy with mechanical or insect damage

- Co-contamination: 2 sa. co-contaminated with AFB_1, AFB_2, AFG_1, and AFG_2; 2 sa. co-contaminated with AFB_1, AFB_2, and AFG_1
- Further contamination of plants and plant products other than cereals and their products reported in the present book/article[4]: Beans (broad bean), AFB_1 and AFB_2, literature[4]; Lentils and Nuts (almond), AFB_1, literature[4]; Nuts (peanut), AFB_1, AFG_1, and AFG_2, literature[4]

incidence: 2/6, conc. range: 10–135 µg/kg, ∅ conc.: 72.5 µg/kg, sample origin: major local retail outlets at Nainital (district), India, sample year: 1983–1985, country: India[7]

- Co-contamination: Often more than one mycotoxin was present in a sa.
- Further contamination of plants and plant products other than cereals and their products reported in the present book/article[7]: Coconuts, AFB_1, AFB_2, AFG_1, AFG_2, and CIT, literature[7]; Dates, AFB_1, AFB_2, AFG_1, and OTA, literature[7]; Grogannuts, AFB_1, AFB_2, AFG_1, AFG_2, and CIT, literature[7]; Nuts (almond), AFB_1, AFB_2, AFG_1, RBT, STG, and ZEA, literature[7]; Nuts (cashew nut), AFB_1, AFB_2, AFG_1, AFG_2, CIT, and OTA, literature[7]; Nuts (peanut), AFB_1, AFG_1, AFG_2, and OTA, literature[7]; Nuts (pistachio nut), AFB_1, AFB_2, AFG_1, AFG_2, and STG, literature[7]; Nuts (walnut), AFB_1, AFB_2, AFG_1, AFG_2, and RBT, literature[7]; Raisins, AFB_1, AFB_2, AFG_1, AFG_2, and OTA, literature[7]

incidence: 6?/20, conc. range: 0.10–0.11 µg/kg, sample origin: unknown, sample year: unknown, country: Korea[9]

- Co-contamination: not reported
- Further contamination of plants and plant products other than cereals and their products reported in the present book/article[9]: Butter (peanut butter), AFB_1, AFB_2, AFG_1, and AFG_2, literature[9]; Nuts (almond), AFB_1 and AFB_2, literature[9]; Nuts (peanut), AFB_1, literature[9]; Products (nut product), Tea, AFB_1, literature[9]

incidence: 3/3*, conc. range: 4–14 µg/kg, ∅ conc.: 10.3 µg/kg, sample origin: unknown, sample year: unknown, country: USA[11], *peanuts, salted in-shell

- Co-contamination: 1 sa. co-contaminated with AFB_1, AFB_2, AFG_1, and AFG_2; 2 sa. co-contaminated with AFB_1 and AFB_2
- Further contamination of plants and plant products other than cereals and their products reported in the present book/article[11]: Nuts (almond), AFB_1 and AFB_2, literature[11]; Nuts (peanut), AFB_1, AFG_1, and AFG_2, literature[11]

incidence: 1/1*, conc.: 20 µg/kg, sample origin: unknown, sample year: unknown, country: USA[11], *peanuts with cereal coating

- Co-contamination: 1 sa. co-contaminated with AFB_1 and AFB_2
- Further contamination of plants and plant products other than cereals and their products reported in the present book/article[11]: Nuts (almond), AFB_1 and AFB_2, literature[11]; Nuts (peanut), AFB_1, AFG_1, and AFG_2, literature[11]

incidence: 39?/40, conc. range: 1.5–744 µg/kg, sample origin: different district hygiene institutes, GDR, sample year: unknown, country: GDR[29]

- Co-contamination: not reported
- Further contamination of plants and plant products other than cereals and their products reported in the present book/article[29]: Butter (peanut butter), Products (peanut product), and Seeds (sunflower), AFB_1, AFB_2, AFG_1, and AFG_2, literature[29]; Coffee, OTA, literature[29]; Fruits, AFB_1, AFG_1, and AFG_2, literature[29]; Fruits (apple), Fruits (peach), Juices (apple juice), PAT, literature[29]; Nuts (peanut), AFB_1, AFG_1, and AFG_2, literature[29]

incidence: 11?/149, conc. range: 0.1–5.3 µg/kg, ∅ conc.: 0.9 µg/kg, sample origin: commercially available market basket sa. of foods and foodstuffs in Tokyo (capital), Japan, sample year: 1986–1990, country: Japan[99]

- Co-contamination: not reported
- Further contamination of plants and plant products other than cereals and their products reported in the present book/article[99]: Beans, Seeds (sesame), Sugar, AFB_1 and AFB_2, literature[99]; Food, Jam (bean jam), Paste (peanut paste), AFB_1, literature[99]; Nuts (Brazil nut, paranut), Nuts (pistachio nut), and Spices (nutmeg), AFB_1, AFB_2, AFG_1, and AFG_2, literature[99]; Nuts (peanut), AFB_1, AFG_1, and AFG_2, literature[99]; Spices, Spices (cayenne pepper, chilli), Spices (pepper), AFB_1, AFB_2, and AFG_1, literature[99]; Spices (paprika), AFB_1 and AFG_1, literature[99]

incidence: 0/9, conc.: no contamination, sample origin: local markets, cereal traders, food/feed processing enterprises, grain mills, and small-scale farmers in villages and cities, Burkina Faso, sample year: April 2010, country: Austria/Nigeria[117]

incidence: 1/23, conc.: 19.5 µg/kg, sample origin: local markets, cereal traders, food/feed processing enterprises, grain mills, and small-scale farmers in the local capital and rural villages in Nampula (province), northern Mozambique, sample year: May 2010, country: Austria/Nigeria[117]

- Co-contamination: For detailed information see the article.
- Further contamination of plants and plant products other than cereals and their products reported in the present book/article[117]: Food, 3-NPA, AFB_1, AFB_2, AFG_1, AFG_2, AFM_1, AME, AOH, ATX-I, BEA, CIT, CPA, DON, ENA, ENA_1, ENB, ENB_1, ENB_2, FB_1, FB_2, FB_3, MON, NIV, OTA, STG, and ZEA, literature[117]; Nuts (peanut), 3-NPA, AFB_1, AFG_1, BEA, CPA, ENB_1, and STG, literature[117]; see also Food, chapter: Further Mycotoxins and Microbial Metabolites

incidence: 2/6*, conc. range: 1.8–2.6 µg/kg**, ∅ conc.: 2.2 µg/kg**, sample origin: Alexandria, Luxor (cities), and Cairo (capital), Egypt, sample year: unknown, country: Egypt/USA[127], *3 sa. freshly produced and 3 sa. stored for 12 months, **stored sa.

- Co-contamination: 2 sa. co-contaminated with AFB_1 and AFB_2
- Further contamination of plants and plant products other than cereals and their products reported in the present book/article[127]: Beans (broad bean), AFB_1 and AFB_2, literature[127]; Lentils, and Nuts (peanut), AFB_1, literature[127]; Spices (fenugreek), AFB_1, AFB_2, and AFG_1, literature[127]

incidence: 15/84, conc. range: 0.4–129 µg/kg, ∅ conc.: 30.8 µg/kg, sample origin: four agroecological zones, Nigeria, sample year: May–July 2015, country: Nigeria/Austria[148]

- Co-contamination: For detailed information see the article.
- Further contamination of plants and plant products other than cereals and their products reported in the present book/article[148]: Nuts (peanut), AFB_1, AFG_1, AFG_2, AFM_1, AME, AOH, BEA, CPA, MON, NIV, and OTA, literature[148]; see also Nuts (peanut), chapter: Further Mycotoxins and Microbial Metabolites

For detailed information see the article.

incidence: 179?/1,860, conc. range: <0.3 to 12.5 µg/kg, ∅ conc.: 1.5 µg/kg, sample origin: sites of primary storage or processing (domestic products) and sites of importation, Cyprus, sample year: 1992–1996, country: Cyprus[200]

- Co-contamination: not reported
- Further contamination of plants and plant products other than cereals and their products reported in the present book/article[200]: Butter (peanut butter) and Figs, AFB_1, AFB_2, AFG_1, and AFG_2, literature[200]; Nuts (Brazil nut, paranut) and Nuts (walnut), AFB_1, AFB_2, and AFG_1, literature[200]; Nuts (chestnut), Products (sesame product), Spices, AFB_1, literature[200]; Nuts (peanut) Seeds (sesame), AFB_1, AFG_1, and AFG_2, literature[200]; Nuts (pistachio nut), AFB_1 and AFB_2, literature[200]

incidence: 6/39, conc. range: tr-4 µg/kg, sample origin: stores in Oslo (capital) area, Norway, sample year: autumn 1973, country: Norway[201]

- Co-contamination: 6 sa. co-contaminated with AFB_1, AFB_2, AFG_1, and AFG_2
- Further contamination of plants and plant products other than cereals and their products reported in the present book/article[201]: Cocoa, Nuts (Brazil nut, paranut), Nuts (cashew nut), Nuts (walnut), AFB_1, AFB_2, AFG_1, and AFG_2, literature[201]; Nuts, AFB_2, AFG_1, and AFG_2, literature[201]; Nuts (peanut), AFB_1, AFG_1, and AFG_2, literature[201]

incidence: 1/1*, conc.: 50 µg/kg, sample origin: stores in Oslo (capital) area, Norway, sample year: autumn 1973, country: Norway[201], *peanuts, sorted

- Co-contamination: 1 sa. co-contaminated with AFB_1, AFB_2, AFG_1, and AFG_2
- Further contamination of plants and plant products other than cereals and their products reported in the present book/article[201]: Cocoa, Nuts (Brazil nut, paranut), Nuts (cashew nut), Nuts (walnut), AFB_1, AFB_2, AFG_1, and AFG_2, literature[201]; Nuts, AFB_2, AFG_1, and AFG_2, literature[201]; Nuts (peanut), AFB_1, AFG_1, and AFG_2, literature[201]

incidence: 5/19, conc. range: LOQ-2 µg/kg (3 sa.) and >5 µg/kg (2 sa., maximum: 10 µg/kg), ∅ conc.: 4.2 µg/kg, sample origin: retail outlets, UK, sample year: September 2001, country: UK[203]

- Co-contamination: 5 sa. co-contaminated with AFB_1 and AFB_2
- Further contamination of plants and plant products other than cereals and their products reported in the present book/article[203]: Butter (nut butter), AFB_1, AFB_2, AFG_1, AFG_2, and OTA, literature[203]; Butter (peanut butter), AFB_1, AFB_2, AFG_1, AFG_2, and OTA, literature[203]; Dates, AFB_1 and OTA, literature[203]; Figs, AFB_1, AFB_2, AFG_2, and OTA, literature[203]; Fruits, dried, Nuts (pistachio nut), AFB_1, AFB_2, and OTA, literature[203]; Nuts (Brazil nut, paranut), AFB_1, AFB_2, and AFG_1, literature[203]; Nuts (peanut), AFB_1, literature[203]

incidence: 45/267, conc. range: 5–1,760 µg/kg, ∅ conc.: 116 µg/kg, sample origin: markets in 16 areas (cities and counties), Taiwan, sample year: 1977/1978, country: Taiwan[214]

incidence: 5/33*, conc. range: 6–56 µg/kg, ∅ conc.: 22 µg/kg, sample origin: markets in 16 areas (cities and counties), Taiwan, sample year: 1977/1978, country: Taiwan[214], *peanuts, fried

incidence: 2/19*, conc. range: 66–711 µg/kg, ∅ conc.: 389 µg/kg, sample origin: markets in 16 areas (cities and counties), Taiwan, sample year: 1977/1978, country: Taiwan[214], *peanuts, roasted

- Co-contamination: not reported
- Further contamination of plants and plant products other than cereals and their products reported in the present book/article[214]: Butter (peanut butter), Cake (peanut), Oils (peanut oil), AFB_1, AFB_2, AFG_1, and AFG_2, literature[214]; Nuts (peanut), AFB_1, AFG_1, and AFG_2, literature[214]

incidence: 15?/29, conc. range: 1.6–16.0 µg/kg, sample origin: storage depots and retail outlets, Botswana, sample year: January 1996–December 1997, country: Botswana[218]

- Co-contamination: not reported
- Further contamination of plants and plant products other than cereals and their products reported in the present book/article[218]: Butter (peanut butter), AFB_1, AFB_2, AFG_1, and AFG_2, literature[218]; Nuts (peanut), AFB_1, AFG_1, and AFG_2, literature[218]

incidence: 0/4 (analyzed sa.)*, conc.: no contamination, sample origin: supermarkets, local stores, and markets in Busan, Daejeon, Gwangju (cities), and Seoul (capital), Korea, sample year: June/July 2004, country: Korea[220], *peanuts, raw

incidence: ?/8 (analyzed sa.)*, conc. range: ≤~2.63 µg/kg, sample origin: supermarkets, local stores, and markets in Busan, Daejeon, Gwangju (cities), and Seoul (capital), Korea, sample year: June/July 2004, country: Korea[220], *peanuts, roasted

- Co-contamination: not reported
- Further contamination of plants and plant products other than cereals and their products reported in the present book/article[220]: Butter (peanut butter) and Nuts, AFB_1 and AFB_2, literature[220]; Nuts (peanut), AFB_1, AFG_1, and AFG_2, literature[220]; Nuts (pistachio nut), AFB_1, literature[220]

For detailed information see the article.

incidence: 1/1, conc.: 0.370 µg/kg, sample origin: market in Khartoum (capital), Sudan, sample year: February 1981–1984?, country: UK/Sudan[223]

- Co-contamination: 1 sa. co-contaminated with AFB_1, AFB_2, and AFG_2
- Further contamination of plants and plant products other than cereals and their products reported in the present book/article[223]: Butter (peanut butter), AFB_1, AFB_2, and AFG_1, literature[223]; Nuts (peanut), AFB_1, AFG_1, and AFG_2, literature[223]; Ochra, AFG_2, literature[223]; Peas (chickpea), AFB_1, literature[223]

incidence: 0/1*, conc.: no contamination, sample origin: market in Khartoum (capital), Sudan, sample year: February 1981–1984?, country: UK/Sudan[223], *peanuts, limed

- Co-contamination: not reported
- Further contamination of plants and plant products other than cereals and their products reported in the present book/article[223]: Butter (peanut butter), AFB_1, AFB_2, and AFG_1, literature[223]; Nuts (peanut), AFB_1, AFG_1, and AFG_2, literature[223]; Okra, AFG_2, literature[223]; Peas (chickpea), AFB_1, literature[223]

incidence: 6?/14*, conc. range: 2.31–92.8 µg/kg, sample origin: supermarkets and village stores in five districts of Penang (state) including Penang Island and Wellesley (province), Malaysia, sample year: unknown, country: Malaysia[225], *raw peanuts, shelled

- Co-contamination: not reported
- Further contamination of plants and plant products other than cereals and their products reported in the present book/article[225]: Butter (peanut butter), Confectionery, Nuts (walnut), Products (nut product), AFB_1 and AFB_2, literature[225]; Cake (peanut), AFB_1, AFB_2, and AFG_2, literature[225]; Nuts (peanut), AFB_1 and AFG_2, literature[225]

incidence: 2/10*, conc. range: 4.46–33.0 µg/kg, ∅ conc.: 17.2 µg/kg, sample origin: supermarkets and village stores in five districts of Penang (state) including Penang Island and Wellesley (province), Malaysia, sample year: unknown, country: Malaysia[225], *roasted peanuts, in-shell

- Co-contamination: 1 sa. co-contaminated with AFB_1, AFB_2 and AFG_2;1 sa. co-contaminated with AFB_1 and AFB_2
- Further contamination of plants and plant products other than cereals and their products reported in the present book/article[225]: Butter (peanut butter), Confectionery, Nuts (walnut), Products (nut product), AFB_1 and AFB_2, literature[225]; Cake (peanut), AFB_1, AFB_2, and AFG_2, literature[225]; Nuts (peanut), AFB_1 and AFG_2, literature[225]

incidence: 3?/20*, conc. range: 7.20–8.77 µg/kg, sample origin: supermarkets and village stores in five districts of Penang (state) including Penang Island and Wellesley (province), Malaysia, sample year: unknown, country: Malaysia[225], *roasted peanuts, shelled

- Co-contamination: not reported
- Further contamination of plants and plant products other than cereals and their products reported in the present book/article[225]: Butter (peanut butter), Confectionery, Nuts (walnut), Products (nut product), AFB$_1$ and AFB$_2$, literature[225]; Cake (peanut), AFB$_1$, AFB$_2$, and AFG$_2$, literature[225]; Nuts (peanut), AFB$_1$ and AFG$_2$, literature[225]

incidence: 3?/8*, conc. range: 3.63–12.9 µg/kg, sample origin: supermarkets and village stores in five districts of Penang (state) including Penang Island and Wellesley (province), Malaysia, sample year: unknown, country: Malaysia[225], *pound peanuts

- Co-contamination: not reported
- Further contamination of plants and plant products other than cereals and their products reported in the present book/article[225]: Butter (peanut butter), Confectionery, Nuts (walnut), Products (nut product), AFB$_1$ and AFB$_2$, literature[225]; Cake (peanut), AFB$_1$, AFB$_2$, and AFG$_2$, literature[225]; Nuts (peanut), AFB$_1$ and AFG$_2$, literature[225]

incidence: 1/2*, conc.: 18.8 µg/kg, sample origin: supermarkets and village stores in five districts of Penang (state) including Penang Island and Wellesley (province), Malaysia, sample year: unknown, country: Malaysia[225], * peanut slices

- Co-contamination: 1 sa. co-contaminated with AFB$_1$ and AFB$_2$
- Further contamination of plants and plant products other than cereals and their products reported in the present book/article[225]: Butter (peanut butter), Confectionery, Nuts (walnut), Products (nut product), AFB$_1$ and AFB$_2$, literature[225]; Cake (peanut), AFB$_1$, AFB$_2$, and AFG$_2$, literature[225]; Nuts (peanut), AFB$_1$ and AFG$_2$, literature[225]

incidence: 14?/35*, conc. range: ≤3.1 µg/kg, sample origin: local stores in Hangzhou, Ningbo, and Quzhou (cities), Zhejiang (province), China, sample year: unknown, country: China[227], *peanuts, fresh

incidence: 5/5*, conc. range: ≤301.2 µg/kg, sample origin: local stores in Hangzhou, Ningbo, and Quzhou (cities), Zhejiang (province), China, sample year: unknown, country: China[227], *peanuts, musty

- Co-contamination: not reported
- Further contamination of plants and plant products other than cereals and their products reported in the present book/article[227]: Butter (peanut butter), AFB$_1$, AFB$_2$, AFG$_1$, AFG$_2$, AFM$_1$, and AFM$_2$, literature[227]; Nuts (peanut), AFB$_1$, AFG$_1$, AFG$_2$, AFM$_1$, and AFM$_2$, literature[227]

incidence: 20/48*, conc. range: 1.60–4.34 µg/kg, ∅ conc.: 0.74 µg/kg, sample origin: Araras, Leme, Pirassununga, and Porto Ferreira (cities), northeast of São Paulo (state), Brazil, sample year: June 2006–May 2007, country: Brazil[237], *peanuts, unprocessed

incidence: 10/48*, conc. range: <LOQ-0.58 µg/kg, ∅ conc.: <LOQ, sample origin: Araras, Leme, Pirassununga, and Porto Ferreira (cities), northeast of São Paulo (state), Brazil, sample year: June 2006–May 2007, country: Brazil[237], *peanuts, roasted and salty

- Co-contamination: not reported
- Further contamination of plants and plant products other than cereals and their products reported in the present book/article[237]: Candies (peanut candy), AFB$_1$, AFB$_2$, AFG$_1$, and AFG$_2$, literature[237]; Nuts (peanut), AFB$_1$, AFG$_1$, and AFG$_2$, literature[237]

incidence: 1/1, conc.: 180 µg/kg, sample origin: Angola, sample year: 1970, country: Germany[239]

- Co-contamination: 1 sa. co-contaminated with AFB$_1$, AFB$_2$, AFG$_1$, and AFG$_2$
- Further contamination of plants and plant products other than cereals and their products reported in the present book/article[239]: Fruits (apple), Fruits (cherry), Fruits (lemmon), Fruits (orange), Fruits (peach), Fruits (tomato), Nuts (almond), Nuts (cashew nut), Nuts (hazelnut), Nuts (walnut), Seeds (sesame), Spice (nutmeg), AFB$_1$, literature[239]; Nuts (peanut), AFB$_1$, AFG$_1$, and AFG$_2$, literature[239]

incidence: 1/67*, conc.: 3.30 µg/kg**, sample origin: imported from Argentina, Brazil, Chile, China, Egypt, India, Israel, Turkey, and the USA to Italy, sample year: January 2008–December 2009, country: Italy[320], *peanuts with shell, **contaminated sa. from Egypt

- Co-contamination: 1 sa. co-contaminated with AFB$_1$ and AFB$_2$
- Further contamination of plants and plant products other than cereals and their products reported in the present book/article[320]: Coffee beans, Vine fruits, dried, OTA, literature[320]; Kernels (apricot), AFB$_1$, AFB$_2$, and AFG$_1$, literature[320]; Nuts Nuts (hazelnut), Paste (hazelnut paste), AFB$_1$, AFB$_2$, AFG$_1$, and AFG$_2$, literature[320]; Nuts (peanut), AFB$_1$, literature[320]; Nuts (pistachio nut), AFB$_1$ and AFB$_2$, literature[320]

incidence: 0/2*, conc.: no contamination, sample origin: retail outlets in Lattakia (city), Syria, sample year: January 1989, country: Italy[347], *peanuts, in-shell

incidence: 1/2*, conc.: no contamination, sample origin: retail outlets in Lattakia (city), Syria, sample year: January 1989, country: Italy[347], *peanuts, raw and shelled

incidence: 2/3*, conc. range: 0.3–0.6 µg/kg, ∅ conc.: 0.45 µg/kg, sample origin: retail outlets in Lattakia (city), Syria, sample year: January 1989, country: Italy[347], *peanuts, roasted and shelled

- Co-contamination: not reported
- Further contamination of plants and plant products other than cereals and their products reported in the present book/article[347]: Figs, Lentils, AFB$_1$, literature[347]; Nuts (peanut), AFB$_1$, literature[347]

incidence: 20/86*, conc. range: 10–400 µg/kg, ∅ conc.: 134 µg/kg, sample origin: supermarkets, retail shops, fair, and beach, Recife (city), Pernambuco (state), Brazil, sample year: May–October 1993, country: Brazil[378], *peanuts, raw and roasted

- Co-contamination: 4 sa. co-contaminated with AFB$_1$, AFB$_2$, AFG$_1$, and AFG$_2$; 16 sa. co-contaminated with AFB$_1$ and AFB$_2$
- Further contamination of plants and plant products other than cereals and their products reported in the present book/article[378]: Nuts (peanut), AFB$_1$, AFG$_1$, and AFG$_2$, literature[378]

incidence: 37/190*, conc. range: ≤6.5 µg/kg, sample year: unknown, country: Egypt[383], *peanuts, in-shell

incidence: 76/155*, conc. range: ≤6.5 µg/kg, sample year: unknown, country: Egypt[383], *peanuts, shelled

incidence: 0/58*, conc.: no contamination, sample year: unknown, country: Egypt[383], *peanuts, roasted

- Co-contamination: not reported
- Further contamination of plants and plant products other than cereals and their products reported in the present book/article[383]: Nuts (peanut), AFB_1, AFG_1, and AFG_2, literature[383]

incidence: 2/2*, conc. range: 11–79 µg/kg, ∅ conc.: 45 µg/kg, sample origin: unknown, sample year: unknown, country: UK[384], *ncac

- Co-contamination: 2 sa. co-contaminated with AFB_1, AFB_2, AFG_1, and AFG_2
- Further contamination of plants and plant products other than cereals and their products reported in the present book/article[384]: Beans (soybean), AFB_1, AFB_2, AFG_1, and AFG_2, literature[384]; Nuts (peanut), AFB_1, AFG_1, and AFG_2, literature[384]

incidence: 1*/65**, conc.: 22 µg/kg, sample origin: *retail self-service shop, Valencia? (province), Spain, sample year: unknown, country: Spain[385], **peanuts, in-shell

- Co-contamination: 1 sa. co-contaminated with AFB_1 and AFB_2
- Further contamination of plants and plant products other than cereals and their products reported in the present book/article[385]: Nuts (peanut), AFB_1, literature[385]

incidence: 30/72*, conc. range: 1–192 µg/kg, ∅ conc.: 51 µg/kg, sample origin: supermarkets and groceries in Maringá (city), Paraná (state), Brazil, sample year: March 1993–February 1994, country: Brazil[388], *peanuts, crude

- Co-contamination: 19 sa. co-contaminated with AFB_1, AFB_2, AFG_1, and AFG_2; 1 sa. co-contaminated with AFB_1, AFB_2, and AFG_1; 10 sa. co-contaminated with AFB_1 and AFB_2
- Further contamination of plants and plant products other than cereals and their products reported in the present book/article[388]: Nuts (peanut), AFB_1, AFG_1, and AFG_2, literature[388]

incidence: 10/20, conc. range: <4 µg/kg (7 sa.), 4–20 µg/kg (1 sa.), and ≤63.5 µg/kg (2 sa.), sample origin: Babile (district), Ethiopia, sample year: 2013/2014, country: Ethiopia/USA[389]

incidence: 6/20, conc. range: 0.3 µg/kg (1 sa.), 4–20 µg/kg (3 sa.), 20–100 µg/kg (1 sa.), and 113 µg/kg (1 sa.), sample origin: Babile (district), Ethiopia, sample year: 2014/2015, country: Ethiopia/USA[389]

incidence: 5/20, conc. range: <4 µg/kg (4 sa.) and 18.8 µg/kg (1 sa.), sample origin: Darolabu (district), Ethiopia, sample year: 2013/2014, country: Ethiopia/USA[389]

incidence: 12/20, conc. range: <4 µg/kg (5 sa.), 4–20 µg/kg (4 sa.), 20–100 µg/kg (2 sa.), and 195 µg/kg (1 sa.), sample origin: Darolabu (district), Ethiopia, sample year: 2014/2015, country: Ethiopia/USA[389]

incidence: 0/20, conc.: no contamination, sample origin: Fedis (district), Ethiopia, sample year: 2013/2014, country: Ethiopia/USA[389]

incidence: 9/20, conc. range: <4 µg/kg (4 sa.), 4–20 µg/kg (2 sa.), 20–100 µg/kg (1 sa.), and ≤237 µg/kg (2 sa.), sample origin: Fedis (district), Ethiopia, sample year: 2014/2015, country: Ethiopia/USA[389]

incidence: 0/20, conc.: no contamination, sample origin: Gursum (district), Ethiopia, sample year: 2013/2014, country: Ethiopia/USA[389]

incidence: 5/20, conc. range: <4 µg/kg (3 sa.) and ≤35.5 µg/kg (2 sa.), sample origin: Gursum (district), Ethiopia, sample year: 2014/2015, country: Ethiopia/USA[389]

- Co-contamination: not reported
- Further contamination of plants and plant products other than cereals and their products reported in the present book/article[389]: Cake (peanut), AFB_1 and AFB_2, literature[389]; Nuts (peanut), AFB_1, AFG_1, and AFG_2, literature[389]

incidence: 48/120, conc. range: 0.5–203 µg/kg, ∅ conc.: 49 µg/kg, sample origin: retail outlets, Botswana, sample year: July 2001–January 2002, country: Botswana[390]

- Co-contamination: not reported
- Further contamination of plants and plant products other than cereals and their products reported in the present book/article[390]: Nuts (peanut), AFB_1, AFG_1, AFG_2, and CPA, literature[390]

incidence: 50/50*, conc. range: 1–2,245 µg/kg, ∅ conc.: 148 µg/kg, sample origin: Georgia (state), USA, sample year: 1990, country: USA[396], *peanuts for oil production or feed

- Co-contamination: 14 sa. co-contaminated with AFB_1, AFB_2, AFG_1, AFG_2, and CPA; 2 sa. co-contaminated with AFB_1, AFB_2, AFG_1, and AFG_2; 12 sa. co-contaminated with AFB_1, AFB_2, AFG_1, and CPA; 1 sa. co-contaminated with AFB_1, AFB_2, and AFG_1; 19 sa. co-contaminated with AFB_1, AFB_2, and CPA; 2 sa. co-contaminated with AFB_1 and AFB_2
- Further contamination of plants and plant products other than cereals and their products reported in the present book/article[396]: Nuts (peanut), AFB_1, AFG_1, AFG_2, and CPA, literature[396]

incidence: 5/10*, conc. range: >LOD-4.08 µg/kg**, sample origin: local supermarkets in Rio de Janeiro (city), Rio de Janeiro (state), Brazil, sample year: unknown, country: Brazil[401], *8 roasted and 2 raw peanut sa., **only roasted peanut sa. contaminated

- Co-contamination: 1 sa. co-contaminated with AFB_1, AFB_2, and AFG_1; 1 sa. co-contaminated with AFB_1, AFB_2, AFM_1, and AFM_2; 1 sa. co-contaminated with AFB_1, AFB_2, and AFM_1; 1 sa. co-contaminated with AFB_1, AFB_2, AFG_1, and AFM_1; 1 sa. co-contaminated with AFB_1 and AFB_2
- Further contamination of plants and plant products other than cereals and their products reported in the present book/article[401]: Nuts (peanut), AFB_1, AFG_1, AFM_1, and AFM_2, literature[401]

incidence: 17/60*, ∅ conc.: 3.3–45.5 µg/kg, sample origin: Tupã (region), São Paulo (state), Brazil, sample year: June 2005–May 2006, country: Brazil[408], *evaluated monthly for a storage period of 12 months

- Co-contamination: not reported
- Further contamination of plants and plant products other than cereals and their products reported in the present book/article[408]: Nuts (peanut), AFB_1, literature[408]

For detailed information see the article.

incidence: 8/73, conc. range: 0.12–28.4 µg/kg, sample origin: public markets/bazaars, supermarkets, retail stores, and wholesale stores in Ankara, Antalya, Isparta, Burdur, and Izmir (provinces), Turkey, sample year: unknown, country: Turkey[409]

- Co-contamination: not reported
- Further contamination of plants and plant products other than cereals and their products reported in the present book/article[409]: Nuts (hazelnut), AFB_1, AFG_1, and AFG_2, literature[409]; Nuts (peanut), AFB_1, AFG_1, and AFG_2, literature[409]; Nuts (pistachio nut), AFB_1 and AFB_2, literature[409]

incidence: 5/65, conc. range: 0.03–6.00 μg/kg, sample origin: local markets in Chongqing, Fujian, Guangdong, Guangxi, Hubei, Jiangsu, Shanghai, and Zhejiang (regions), China, sample year: 2003, country: China[411]

- Co-contamination: not reported
- Further contamination of plants and plant products other than cereals and their products reported in the present book/article[411]: Nuts (peanut), AFB_1, AFG_1, and AFG_2, literature[411]; Nuts (pine nut), AFB_1 and AFB_2, literature[411]; Nuts (walnut), AFB_1, AFB_2, AFG_1, and AFG_2, literature[411]

incidence: 15/408, conc. range: 0.05–27.36 μg/kg, sample origin: Liaoning (province with temperate monsoon climate), China, sample year: 2010–2013, country: China[441]

incidence: 139/1190, conc. range: 0.01–61.50 μg/kg, sample origin: Henan (province with transitional monsoon climate), China, sample year: 2010–2013, country: China[441]

incidence: 60/455, conc. range: 0.08–183.00 μg/kg, sample origin: Sichuan (province with subtropical temperate monsoon climate), China, sample year: 2010–2013, country: China[441]

incidence: 52/441, conc. range: 0.05–30.38 μg/kg, sample origin: Guangdong (province with subtropical-tropical monsoon climate), China, sample year: 2010–2013, country: China[441]

- Co-contamination: For detailed information see the article.
- Further contamination of plants and plant products other than cereals and their products reported in the present book/article[441]: Nuts (peanuts), AFB_1, AFG_1, and AFG_2, literature[441]

incidence: 57/84, conc. range: ≤11.16 μg/kg, sample origin: supermarkets in Selangor (state), Malaysia, sample year: April–August 2008, country: Malaysia/Indonesia[676]

- Co-contamination: not reported
- Further contamination of plants and plant products other than cereals and their products reported in the present book/article[676]: Nuts (peanut), AFB_1 and AFG_1, literature[676]

For detailed information see the article.

incidence: 21/151, conc. range: 0.11–11.0 μg/kg, Ø conc.: 1.43 μg/kg, sample origin: markets, supermarkets, groceries, and retail shops in Adana and Osmaniye (provinces), Turkey, sample year: January 2010–December 2012, country: Turkey[741]

- Co-contamination: not reported
- Further contamination of plants and plant products other than cereals and their products reported in the present book/article[741]: Nuts (peanut), AFB_1 and AFG_1, literature[741]; Nuts (pistachio nut), AFB_1, AFB_2, AFG_1, and AFG_2, literature[741]

incidence: 18/20*, conc. range: 2.01–210.92 μg/kg, Ø conc.: 36.2 μg/kg, sample origin: street hawkers, local markets, and retail shops in urban areas of Kinshasa (capital), Congo, sample year: December 2010–January 2011, country: South Africa[758], *peanuts, raw

- Co-contamination: 14 sa. co-contaminated with AFB_1, AFB_2, AFG_1, and AFG_2; 1 sa. co-contaminated with AFB_1, AFB_2, and AFG_1; 3 sa. co-contaminated with AFB_1 and AFB_2
- Further contamination of plants and plant products other than cereals and their products reported in the present book/article[758]: Nuts (peanut), AFB_1, AFG_1, and AFG_2, literature[758]

incidence: 14/20*, conc. range: 2.00–16.3 μg/kg, Ø conc.: 4.43 μg/kg, sample origin: street hawkers, local markets, and retail shops in inner-city of Pretoria (capital), South Africa, sample year: December 2010–January 2011, country: South Africa[758], *peanuts, raw

- Co-contamination: 6 sa. co-contaminated with AFB_1, AFB_2, AFG_1, and AFG_2; 2 sa. co-contaminated with AFB_1, AFB_2, and AFG_1; 1 sa. co-contaminated with AFB_1, AFB_2, and AFG_2; 5 sa. co-contaminated with AFB_1 and AFB_2
- Further contamination of plants and plant products other than cereals and their products reported in the present book/article[758]: Nuts (peanut), AFB_1, AFG_1, and AFG_2, literature[758]

incidence: 3/8*, conc. range: 167–1,111 μg/kg, Ø conc.: 482 μg/kg, sample origin: imported to Finland, sample year: unknown, country: Denmark/Finland[769], *peanuts for various purposes

- Co-contamination: 2 sa. co-contaminated with AFB_1, AFB_2, AFG_1, and AFG_2; 1 sa. co-contaminated with AFB_1, AFB_2, and AFG_1
- Further contamination of plants and plant products other than cereals and their products reported in the present book/article[769]: Nuts (peanut), AFB_1, AFG_1, and AFG_2, literature[769]

incidence: 3/10*, conc. range: LOD-20 μg/kg (3 sa., maximum: 2.7 μg/kg), sample origin: retail outlets in Paraná (state), Brazil, sample year: April 2006–May 2007, country: Brazil[784], *peanuts, salted

incidence: 24/45*, conc. range: LOD-20 μg/kg (24 sa., maximum: 12.7 μg/kg), sample origin: retail outlets in Paraná (state), Brazil, sample year: April 2006–May 2007, country: Brazil[784], *peanuts, sweet

incidence: 7/18*, conc. range: LOD-20 μg/kg (7 sa., maximum: 2.3 μg/kg), sample origin: retail outlets in Paraná (state), Brazil, sample year: April 2006–May 2007, country: Brazil[784], *peanut brittle

- Co-contamination: For detailed information see the article.
- Further contamination of plants and plant products other than cereals and their products reported in the present book/article[784]: Candies (peanut candy), AFB_1, AFB_2, AFG_1, and AFG_2, literature[784]; Nuts (peanut), AFB_1, AFG_1, and AFG_2, literature[784]

incidence: 2/9*, conc. range: 0.2 μg/kg, Ø conc.: 0.2 μg/kg, sample origin: market in Lilongwe (city), Malawi, sample year: December 2012, country: Malawi[790], *peanuts, un-skinned roasted

incidence: 10/15*, conc. range: 0.2–1.8 μg/kg, sample origin: market in Lilongwe (city), Malawi, sample year: December 2012, country: Malawi[790], *peanuts, de-skinned roasted

- Co-contamination: not reported
- Further contamination of plants and plant products other than cereals and their products reported in the present book/article[790]: Butter (peanut butter) and Food, AFB_1, AFB_2, AFG_1, and AFG_2, literature[790]; Nuts (peanut), AFB_1, AFG_1, and AFG_2, literature[790]

incidence: 6/25, conc. range: 0.01–2.65 μg/kg, Ø conc.: 0.79 μg/kg, sample origin: retail fresh markets or modern department stores in the central region of Thailand, sample year: unknown, country: Thailand/Austria/Germany[857]

- Co-contamination: not reported
- Further contamination of plants and plant products other than cereals and their products reported in the present book/article[857]: Nuts (peanut), AFB_1 and AFG_1, literature[857]; Spices (cayenne pepper, chilli), AFB_1, AFB_2, and AFG_1, literature[857]

incidence: 8/15, conc. range: ≤22 μg/kg, Ø conc.: 11 μg/kg, sample origin: Osmaniye (province), Turkey, sample year: 2007, country: Austria[868]

- Co-contamination: For detailed information see the article.
- Further contamination of plants and plant products other than cereals and their products reported in the present book/

article[868]: Nuts (almond), AME, AOH, BEA, CPA, ENB, ENB_1, and ENB_3, literature[868]; Nuts (hazelnut), 3-NPA, AFB_1, AFB_2, AFG_1, AFG_2, AME, AOH, ATX-I, BEA, ENA, ENA_1, ENB_1, ENB_2, ENB_3, ENB_4, HT-2, KA, MA, MON, OTA, OTB, STG, T-2, and ZEA, literature[868]; Nuts (peanut), 3-NPA, AFB_1, AME, AOH, ATX-I, BEA, CPA, ENA, ENA_1, ENB, ENB_4, KA, MA, OTA, OTB, STG, and ZEA, literature[868]; Nuts (pistachio nut), BEA, MA, and OTA, literature[868]; see also Nuts (almond), Nuts (hazelnut), Nuts (peanut), Nuts (pistachio nut), chapter: Further Mycotoxins and Microbial Metabolites

For detailed information see the article.

incidence: 12/30, conc. range: ≤20 µg/kg (12 sa., maximum: 17.75 µg/kg), sample origin: local markets in Wuhan (city), China, sample year: unknown, country: China[881]

- Co-contamination: not reported
- Further contamination of plants and plant products other than cereals and their products reported in the present book/article[881]: Nuts (peanut), AFB_1, AFG_1, AFG_2, OTA, T-2, and ZEA, literature[881]

Aflatoxin G_1

incidence: 2/104, conc. range: 3–136 µg/kg, ∅ conc.: 69.5 µg/kg, sample origin: imported to Finland, sample year: 1975/1976, country: Finland[2]

- Co-contamination: 1 sa. co-contaminated with AFB_1, AFB_2, AFG_1, and AFG_2; 1 sa. co-contaminated with AFB_1, AFB_2, and AFG_1
- Further contamination of plants and plant products other than cereals and their products reported in the present book/article[2]: Marzipan (almond paste), Nuts (almond), and Nuts (hazelnut), AFB_1 and AFB_2, literature[2]; Nuts, AFB_1, AFB_2, and AFG_1, literature[2]; Nuts (peanut), AFB_1, AFB_2, and AFG_2, literature[2]

incidence: 2/121*, conc. range: 12–20 µg/kg, ∅ conc.: 16 µg/kg, sample origin: imported to Finland, sample year: 1974–1976, country: Finland[2], *peanuts, roasted

- Co-contamination: 2 sa. co-contaminated with AFB_1, AFB_2, and AFG_1
- Further contamination of plants and plant products other than cereals and their products reported in the present book/article[2]: Marzipan (almond paste), Nuts (almond), and Nuts (hazelnut), AFB_1 and AFB_2, literature[2]; Nuts, AFB_1, AFB_2, and AFG_1, literature[2]; Nuts (peanut), AFB_1, AFB_2, and AFG_2, literature[2]

incidence: 6/68*, conc. range: tr-91 µg/kg, sample origin: imported to Finland, sample year: 1974–1976, country: Finland[2], *peanuts, sliced and crushed

- Co-contamination: 4 sa. co-contaminated with AFB_1, AFB_2, AFG_1, and AFG_2; 2 sa. co-contaminated with AFB_1 and AFG_1
- Further contamination of plants and plant products other than cereals and their products reported in the present book/article[2]: Marzipan (almond paste), Nuts (almond), and Nuts (hazelnut), AFB_1 and AFB_2, literature[2]; Nuts, AFB_1, AFB_2, and AFG_1, literature[2]; Nuts (peanut), AFB_1, AFB_2, and AFG_2, literature[2]

incidence: 5/45*, conc. range: 150–10,458 µg/kg, ∅ conc.: 3,561.4 µg/kg, sample origin: different storage facilities, Egypt, sample year: November 1978–April 1979, country: Egypt[4], *sa. moldy with mechanical or insect damage

- Co-contamination: 2 sa. co-contaminated with AFB_1, AFB_2, AFG_1, and AFG_2; 2 sa. co-contaminated with AFB_1, AFB_2, and AFG_1; 1 sa. co-contaminated with AFB_1 and AFG_1
- Further contamination of plants and plant products other than cereals and their products reported in the present book/article[4]: Beans (broad bean), AFB_1 and AFB_2, literature[4]; Lentils and Nuts (almond), AFB_1, literature[4]; Nuts (peanut), AFB_1, AFB_2, and AFG_2, literature[4]

incidence: 1/6, conc.: 96 µg/kg, sample origin: major local retail outlets at Nainital (district), India, sample year: 1983–1985, country: India[7]

- Co-contamination: Often more than one mycotoxin was present in a sa.
- Further contamination of plants and plant products other than cereals and their products reported in the present book/article[7]: Coconuts, AFB_1, AFB_2, AFG_1, AFG_2, and CIT, literature[7]; Dates, AFB_1, AFB_2, AFG_1, and OTA, literature[7]; Grogannuts, AFB_1, AFB_2, AFG_1, AFG_2, and CIT, literature[7]; Nuts (almond), AFB_1, AFB_2, AFG_1, RBT, STG, and ZEA, literature[7]; Nuts (cashew nut), AFB_1, AFB_2, AFG_1, AFG_2, CIT, and OTA, literature[7]; Nuts (peanut), AFB_1, AFB_2, AFG_2, and OTA, literature[7]; Nuts (pistachio nut), AFB_1, AFB_2, AFG_1, AFG_2, and STG, literature[7]; Nuts (walnut), AFB_1, AFB_2, AFG_1, AFG_2, and RBT, literature[7]; Raisins, AFB_1, AFB_2, AFG_1, AFG_2, and OTA, literature[7]

incidence: 1/3*, conc.: 7 µg/kg, sample origin: unknown, sample year: unknown, country: USA[11], *peanuts, salted-in-shell

- Co-contamination: 1 sa. co-contaminated with AFB_1, AFB_2, AFG_1, and AFG_2
- Further contamination of plants and plant products other than cereals and their products reported in the present book/article[11]: Nuts (almond), AFB_1 and AFB_2, literature[11]; Nuts (peanut), AFB_1, AFB_2, and AFG_2, literature[11]

incidence: 0/1*, conc.: no contamination, sample origin: unknown, sample year: unknown, country: USA[11], *peanuts with cereal coating

- Co-contamination: not reported
- Further contamination of plants and plant products other than cereals and their products reported in the present book/article[11]: Nuts (almond), AFB_1 and AFB_2, literature[11]; Nuts (peanut), AFB_1, AFB_2, and AFG_2, literature[11]

incidence: 39?/40, conc. range: 1.0–1,540 µg/kg, sample origin: different district hygiene institutes, GDR, sample year: unknown, country: GDR[29]

- Co-contamination: not reported
- Further contamination of plants and plant products other than cereals and their products reported in the present book/article[29]: Butter (peanut butter), Products (peanut product), and Seeds (sunflower), AFB_1, AFB_2, AFG_1, and AFG_2, literature[29]; Coffee, OTA, literature[29]; Fruits, AFB_1, AFG_1, and AFG_2, literature[29]; Fruits (apple), Fruits (peach), and Juices (apple juice), PAT, literature[29]; Nuts (peanut), AFB_1, AFB_2, and AFG_2, literature[29]

incidence: 11?/149, conc. range: ≤22.1 µg/kg, ∅ conc.: 2.8 µg/kg, sample origin: commercially available market basket sa. of foods and foodstuffs in Tokyo (capital), Japan, sample year: 1986–1990, country: Japan[99]

- Co-contamination: not reported
- Further contamination of plants and plant products other than cereals and their products reported in the present book/arti-

cle[99]: Beans, Seeds (sesame), Sugar, AFB_1 and AFB_2, literature[99]; Food, Jam (bean jam), Paste (peanut paste), AFB_1, literature[99]; Nuts (Brazil nut, paranut), Nuts (pistachio nut), and Spices (nutmeg), AFB_1, AFB_2, AFG_1, and AFG_2, literature[99]; Nuts (peanut), AFB_1, AFB_2, and AFG_2, literature[99]; Spices, Spices (cayenne pepper, chilli), Spices (pepper), AFB_1, AFB_2, and AFG_1, literature[99]; Spices (paprika), AFB_1 and AFG_1, literature[99]

incidence: 0/9, conc.: no contamination, sample origin: local markets, cereal traders, food/feed processing enterprises, grain mills, and small-scale farmers in villages and cities, Burkina Faso, sample year: April 2010, country: Austria/Nigeria[117]

incidence: 1/23, conc.: 30.3 µg/kg, sample origin: local markets, cereal traders, food/feed processing enterprises, grain mills, and small-scale farmers in the local capital and rural villages in Nampula (province), northern Mozambique, sample year: May 2010, country: Austria/Nigeria[117]

- Co-contamination: For detailed information see the article.
- Further contamination of plants and plant products other than cereals and their products reported in the present book/article[117]: Food, 3-NPA, AFB_1, AFB_2, AFG_1, AFG_2, AFM_1, AME, AOH, ATX-I, BEA, CIT, CPA, DON, ENA, ENA_1, ENB, ENB_1, ENB_2, FB_1, FB_2, FB_3, MON, NIV, OTA, STG, and ZEA, literature[117]; Nuts (peanut), 3-NPA, AFB_1, AFB_2, BEA, CPA, ENB_1, and STG, literature[117]; see also Food, chapter: Further Mycotoxins and Microbial Metabolites

incidence: 6/30*, conc. range: 130–160 µg/kg, ∅ conc.: 140 µg/kg, sample origin: markets around Bangkok (capital) area, Thailand, sample year: July 1983–April 1985, country: Thailand[133], *food and sweets derived from peanut

- Co-contamination: 6 sa. co-contaminated with AFB_1 and AFG_1
- Further contamination of plants and plant products other than cereals and their products reported in the present book/article[133]: Beans, AFG_1, literature[133]; Beans (soybean), Curd (bean curd), Sauces (soy sauce), AFB_1 and AFG_1, literature[133]; Nuts (peanut), AFB_1, literature[133]

incidence: 19/84, conc. range: 0.4–1,202 µg/kg, ∅ conc.: 174.3 µg/kg, sample origin: four agroecological zones, Nigeria, sample year: May–July 2015, country: Nigeria/Austria[148]

- Co-contamination: For detailed information see the article.
- Further contamination of plants and plant products other than cereals and their products reported in the present book/article[148]: Nuts (peanut), AFB_1, AFB_2, AFG_2, AFM_1, AME, AOH, BEA, CPA, MON, NIV, and OTA, literature[148]; see also Nuts (peanut), chapter: Further Mycotoxins and Microbial Metabolites

For detailed information see the article.

incidence: 179?/1,860, conc. range: <0.4–72.2 µg/kg, ∅ conc.: 8.9 µg/kg, sample origin: sites of primary storage or processing (domestic products) and sites of importation, Cyprus, sample year: 1992–1996, country: Cyprus[200]

- Co-contamination: not reported
- Further contamination of plants and plant products other than cereals and their products reported in the present book/article[200]: Butter (peanut butter) and Figs, AFB_1, AFB_2, AFG_1, and AFG_2, literature[200]; Nuts (Brazil nut, paranut), Nuts (walnut), AFB_1, AFB_2, and AFG_1, literature[200]; Nuts (chestnut), Products (sesame product), Spices, AFB_1, literature[200]; Nuts (peanut), AFB_1, AFB_2, and AFG_2, literature[200]; Nuts (pistachio nut), AFB_1 and AFB_2, literature[200]; Seeds (sesame), AFB_1, AFG_1, and AFG_2, literature[200]

incidence: 6/34*, conc. range: tr-24 µg/kg, sample origin: stores in Oslo (capital) area, Norway, sample year: autumn 1973, country: Norway[201], *peanuts bought from stores

- Co-contamination: 6 sa. co-contaminated with AFB_1, AFB_2, AFG_1, and AFG_2
- Further contamination of plants and plant products other than cereals and their products reported in the present book/article[201]: Cocoa, Nuts (Brazil nut, paranut), Nuts (cashew nut), and Nuts (walnut), AFB_1, AFB_2, AFG_1, and AFG_2, literature[201]; Nuts, AFB_2, AFG_1, and AFG_2, literature[201]; Nuts (peanut), AFB_1, AFB_2, and AFG_2, literature[201]

incidence: 1/3*, conc.: 350 µg/kg, sample origin: stores in Oslo (capital) area, Norway, sample year: autumn 1973, country: Norway[201], *peanuts sorted, pooled sa.

- Co-contamination: 1 sa. co-contaminated with AFB_1, AFB_2, AFG_1, and AFG_2
- Further contamination of plants and plant products other than cereals and their products reported in the present book/article[201]: Cocoa, Nuts (Brazil nut, paranut), Nuts (cashew nut), and Nuts (walnut), AFB_1, AFB_2, AFG_1, and AFG_2, literature[201]; Nuts, AFB_2, AFG_1, and AFG_2, literature[201]; Nuts (peanut), AFB_1, AFB_2, and AFG_2, literature[201]

incidence: 5/267, conc. range: 12–72 µg/kg, ∅ conc.: 31 µg/kg, sample origin: markets in 16 areas (cities and counties), Taiwan, sample year: 1977/1978, country: Taiwan[214]

incidence: 3/33*, conc. range: 14–1,053 µg/kg, ∅ conc.: 371 µg/kg, sample origin: markets in 16 areas (cities and counties), Taiwan, sample year: 1977/1978, country: Taiwan[214], *peanuts, fried

incidence: 1/19*, conc.: 23 µg/kg, sample origin: markets in 16 areas (cities and counties), Taiwan, sample year: 1977/1978, country: Taiwan[214], *peanuts, roasted

- Co-contamination: not reported
- Further contamination of plants and plant products other than cereals and their products reported in the present book/article[214]: Butter (peanut butter), Cake (peanut), and Oils (peanut oil), AFB_1, AFB_2, AFG_1, and AFG_2, literature[214]; Nuts (peanut), AFB_1, AFB_2, and AFG_2, literature[214]

incidence: 15?/29, conc. range: 1.6–8.0 µg/kg, sample origin: storage depots and retail outlets, Botswana, sample year: January 1996–December 1997, country: Botswana[218]

- Co-contamination: not reported
- Further contamination of plants and plant products other than cereals and their products reported in the present book/article[218]: Butter (peanut butter), AFB_1, AFB_2, AFG_1, and AFG_2, literature[218]; Nuts (peanut), AFB_1, AFB_2, and AFG_2, literature[218]

incidence: 0/4 (analyzed sa.)*, conc.: no contamination, sample origin: supermarkets, local stores, and markets in Busan, Daejeon, Gwangju (cities), and Seoul (capital), Korea, sample year: June/July 2004, country: Korea[220], *peanuts, raw

incidence: ?/8 (analyzed sa.)*, conc. range: ≤~4.07 µg/kg, sample origin: supermarkets, local stores, and markets in Busan, Daejeon, Gwangju (cities), and Seoul (capital), Korea, sample year: June/July 2004, country: Korea[220], *peanuts, roasted

- Co-contamination: not reported
- Further contamination of plants and plant products other than cereals and their products reported in the present book/article[220]: Butter (peanut butter) and Nuts, AFB_1 and AFB_2, literature[220]; Nuts (peanut), AFB_1, AFB_2, and AFG_2, literature[220]; Nuts (pistachio nut), AFB_1, literature[220]

For detailed information see the article.

incidence: 0/1, conc.: no contamination, sample origin: market in Khartoum (capital), Sudan, sample year: February 1981–1984?, country: UK/Sudan[223]

- Co-contamination: not reported
- Further contamination of plants and plant products other than cereals and their products reported in the present book/article[223]: Butter (peanut butter), AFB_1, AFB_2, and AFG_1, literature[223]; Nuts (peanut), AFB_1, AFB_2, and AFG_2, literature[223]; Okra, AFG_2, literature[223]; Peas (chickpea), AFB_1, literature[223]

incidence: 1/1*, conc.: 2.816 μg/kg, sample origin: market in Khartoum (capital), Sudan, sample year: February 1981–1984?, country: UK/Sudan[223], *peanuts, limed

- Co-contamination: 1 sa. co-contaminated with AFB_1, AFG_1 and AFG_2
- Further contamination of plants and plant products other than cereals and their products reported in the present book/article[223]: Butter (peanut butter), AFB_1, AFB_2, and AFG_1, literature[223]; Nuts (peanut), AFB_1, AFB_2, and AFG_2, literature[223]; Okra, AFG_2, literature[223]; Peas (chickpea), AFB_1, literature[223]

incidence: 14?/35*, conc. range: ≤1.3 μg/kg, sample origin: local stores in Hangzhou, Ningbo, and Quzhou (cities), Zhejiang (province), China, sample year: unknown, country: China[227], *peanuts, fresh

incidence: 5/5*, conc. range: ≤0.15 μg/kg, sample origin: local stores in Hangzhou, Ningbo, and Quzhou (cities), Zhejiang (province), China, sample year: unknown, country: China[227], *peanuts, musty

- Co-contamination: not reported
- Further contamination of plants and plant products other than cereals and their products reported in the present book/article[227]: Butter (peanut butter), AFB_1, AFB_2, AFG_1, AFG_2, AFM_1, and AFM_2, literature[227]; Nuts (peanut), AFB_1, AFB_2, AFG_2, AFM_1, and AFM_2, literature[227]

incidence: 22/48*, conc. range: 1.18–36.88 μg/kg, ∅ conc.: 4.39 μg/kg, sample origin: Araras, Leme, Pirassununga, and Porto Ferreira (cities), northeast of São Paulo (state), Brazil, sample year: June 2006–May 2007, country: Brazil[237], *peanuts, unprocessed

incidence: 15/48*, conc. range: 0.58–2.93 μg/kg, ∅ conc.: 0.55 μg/kg, sample origin: Araras, Leme, Pirassununga, and Porto Ferreira (cities), northeast of São Paulo (state), Brazil, sample year: June 2006–May 2007, country: Brazil[237], *peanuts, roasted and salty

- Co-contamination: not reported
- Further contamination of plants and plant products other than cereals and their products reported in the present book/article[237]: Candies (peanut candy), AFB_1, AFB_2, AFG_1, and AFG_2, literature[237]; Nuts (peanut), AFB_1, AFB_2, and AFG_2, literature[237]

incidence: 1/1, conc.: 315 μg/kg, sample origin: Angola, sample year: 1970, country: Germany[239]

- Co-contamination: 1 sa. co-contaminated with AFB_1, AFB_2, AFG_1, and AFG_2
- Further contamination of plants and plant products other than cereals and their products reported in the present book/article[239]: Fruits (apple), Fruits (cherry), Fruits (lemmon), Fruits (orange), Fruits (peach), Fruits (tomato), Nuts (almond), Nuts (cashew nut), Nuts (hazelnut), Nuts (walnut), Seeds (sesame), Spice (nutmeg), AFB_1, literature[239]; Nuts (peanut), AFB_1, AFB_2, and AFG_2, literature[239]

incidence: 6/86*, conc. range: 20–800 μg/kg, ∅ conc.: 207 μg/kg, sample origin: supermarkets, retail shops, fair, and beach, Recife (city), Pernambuco (state), Brazil, sample year: May–October 1993, country: Brazil[378], *peanuts, raw and roasted

- Co-contamination: 4 sa. co-contaminated with AFB_1, AFB_2, AFG_1, and AFG_2; 2 sa. co-contaminated with AFB_1, AFG_1, and AFG_2
- Further contamination of plants and plant products other than cereals and their products reported in the present book/article[378]: Nuts (peanut), AFB_1, AFB_2, and AFG_2, literature[378]

incidence: 1/65, conc.: 0.38 μg/kg, sample origin: storage houses and dealers, Tunisia, sample year: unknown, country: Tunisia/USA[380]

- Co-contamination: not reported
- Further contamination of plants and plant products other than cereals and their products reported in the present book/article[380]: Nuts (peanut), Nuts (pistachio nut), AFB_1, literature[380]; Peas, AFB_1 and AFG_1, literature[380]

incidence: 37/190*, conc. range: 6.5 to <13 μg/kg, sample year: unknown, country: Egypt[383], *peanuts, in-shell

incidence: 76/155*, conc. range: 6.5 to <26 μg/kg, sample year: unknown, country: Egypt[383], *peanuts, shelled

incidence: 0/58*, conc.: no contamination, sample year: unknown, country: Egypt[383], *peanuts, roasted

- Co-contamination: not reported
- Further contamination of plants and plant products other than cereals and their products reported in the present book/article[383]: Nuts (peanut), AFB_1, AFB_2, and AFG_2, literature[383]

incidence: 2/2*, conc. range: 99–113 μg/kg, ∅ conc.: 106 μg/kg, sample origin: unknown, sample year: unknown, country: UK[384], *ncac

- Co-contamination: 2 sa. co-contaminated with AFB_1, AFB_2, AFG_1, and AFG_2
- Further contamination of plants and plant products other than cereals and their products reported in the present book/article[384]: Beans (soybean), AFB_1, AFB_2, AFG_1, and AFG_2, literature[384]; Nuts (peanut), AFB_1, AFB_2, and AFG_2, literature[384]

incidence: 23/72*, conc. range: 1–680 μg/kg, ∅ conc.: 110 μg/kg, sample origin: supermarkets and groceries in Maringá (city), Paraná (state), Brazil, sample year: March 1993–February 1994, country: Brazil[388], *peanuts, crude

- Co-contamination: 19 sa. co-contaminated with AFB_1, AFB_2, AFG_1, and AFG_2; 1 sa. co-contaminated with AFB_1, AFB_2, and AFG_1; 3 sa. co-contaminated with AFG_1 and AFG_2
- Further contamination of plants and plant products other than cereals and their products reported in the present book/article[388]: Nuts (peanut), AFB_1, AFB_2, and AFG_2, literature[388]

incidence: 0/20, conc.: no contamination, sample origin: Babile (district), Ethiopia, sample year: 2013/2014, country: Ethiopia/USA[389]

incidence: 0/20, conc.: no contamination, sample origin: Babile (district), Ethiopia, sample year: 2014/2015, country: Ethiopia/USA[389]

incidence: 4/20, conc. range: 5.99–20 μg/kg (3 sa.) and 61.9 μg/kg (1 sa.);, sample origin: Darolabu (district), Ethiopia, sample year: 2013/2014, country: Ethiopia/USA[389]

incidence: 4/20, conc. range: 199–736 μg/kg (4 sa.), sample origin: Darolabu (district), Ethiopia, sample year: 2014/2015, country: Ethiopia/USA[389] incidence: 0/20, conc.: no contamination, sample origin: Fedis (district), Ethiopia, sample year: 2013/2014, country: Ethiopia/USA[389]

incidence: 2/20, conc. range: 238–477 μg/kg (2 sa.), ∅ conc.: 358 μg/kg, sample origin: Fedis (district), Ethiopia, sample year: 2014/2015, country: Ethiopia/USA[389]

incidence: 0/20, conc.: no contamination, sample origin: Gursum (district), Ethiopia, sample year: 2013/2014, country: Ethiopia/USA[389]

incidence: 1/20, conc. range: 187.1 μg/kg (1 sa.), sample origin: Gursum (district), Ethiopia, sample year: 2014/2015, country: Ethiopia/USA[389]

- Co-contamination: not reported
- Further contamination of plants and plant products other than cereals and their products reported in the present book/article[389]: Cake (peanut), AFB_1 and AFB_2, literature[389]; Nuts (peanut), AFB_1, AFB_2, and AFG_2, literature[389]

incidence: 51/120, conc. range: 0.6–259 μg/kg, ∅ conc.: 91 μg/kg, sample origin: retail outlets, Botswana, sample year: July 2001–January 2002, country: Botswana[390]

- Co-contamination: not reported
- Further contamination of plants and plant products other than cereals and their products reported in the present book/article[390]: Nuts (peanut), AFB_1, AFB_2, AFG_2, and CPA, literature[390]

incidence: 3/11*, conc. range: 0.61–1.68 μg/kg, ∅ conc.: 0.97 μg/kg, sample origin: Brazil, China, the USA, and unknown, sample year: unknown, country: Spain[391], *included peeled salted, raw, raw in-shell, roasted and salted in-shell, roasted in-shell, and salted peanuts

- Co-contamination: 2 sa. co-contaminated with AFB_1 and AFG_1; 1 sa. contaminated solely with AFG_1
- Further contamination of plants and plant products other than cereals and their products reported in the present book/article[391]: Nuts (peanut), AFB_1, literature[391]

incidence: 29/50*, conc. range: 1–2,331 μg/kg, ∅ conc.: 240 μg/kg, sample origin: Georgia (state), USA, sample year: 1990, country: USA[396], *peanuts for oil production or feed

- Co-contamination: 14 sa. co-contaminated with AFB_1, AFB_2, AFG_1, AFG_2, and CPA; 2 sa. co-contaminated with AFB_1, AFB_2, AFG_1, and AFG_2; 12 sa. co-contaminated with AFB_1, AFB_2, AFG_1, and CPA; 1 sa. co-contaminated with AFB_1, AFB_2, and AFG_1
- Further contamination of plants and plant products other than cereals and their products reported in the present book/article[396]: Nuts (peanut), AFB_1, AFB_2, AFG_2, and CPA, literature[396]

incidence: 2/10*, conc. range: >LOD-1.03 μg/kg**, sample origin: local supermarkets in Rio de Janeiro (city), Rio de Janeiro (state), Brazil, sample year: unknown, country: Brazil[401], *8 roasted and 2 raw peanut sa., **only roasted peanut sa. contaminated

- Co-contamination: 1 sa. co-contaminated with AFB_1, AFB_2, and AFG_1; 1 sa. co-contaminated with AFB_1, AFB_2, AFG_1, and AFM_1
- Further contamination of plants and plant products other than cereals and their products reported in the present book/article[401]: Nuts (peanut), AFB_1, AFB_2, AFM_1, and AFM_2, literature[401]

incidence: 1/1, conc.: 2,100 μg/kg, sample origin: Nepal, sample year: unknown, country: Japan/Nepal[403]

- Co-contamination: 1 sa. co-contaminated with AFB_1 and AFG_1 as well as other aflatoxins
- Further contamination of plants and plant products other than cereals and their products reported in the present book/article[403]: Beans (soybean), AFB_1 and AFG_1, literature[403]; Nuts (peanut), AFB_1, literature[403]

incidence: 2/73, conc. range: 20.2 μg/kg, sample origin: public markets/bazaars, supermarkets, retail stores, and wholesale stores in Ankara, Antalya, Isparta, Burdur, and Izmir (provinces), Turkey, sample year: unknown, country: Turkey[409]

- Co-contamination: not reported
- Further contamination of plants and plant products other than cereals and their products reported in the present book/article[409]: Nuts (hazelnut), AFB_1, AFG_1, and AFG_2, literature[409]; Nuts (peanut), AFB_1, AFB_2, and AFG_2, literature[409]; Nuts (pistachio nut), AFB_1 and AFB_2, literature[409]

incidence: 4/65, conc. range: 0.42–11.73 μg/kg, sample origin: local markets in Chongqing, Fujian, Guangdong, Guangxi, Hubei, Jiangsu, Shanghai, and Zhejiang (regions), China, sample year: 2003, country: China[411]

- Co-contamination: not reported
- Further contamination of plants and plant products other than cereals and their products reported in the present book/article[411]: Nuts (peanut), AFB_1, AFB_2, and AFG_2, literature[411]; Nuts (pine nut), AFB_1 and AFB_2, literature[411]; Nuts (walnut), AFB_1, AFB_2, AFG_1, and AFG_2, literature[411]

incidence: 1/408, conc.: 3.61 μg/kg, sample origin: Liaoning (province with temperate monsoon climate), China, sample year: 2010–2013, country: China[441]

incidence: 14/1190, conc. range: 0.33–460.00 μg/kg, sample origin: Henan (province with transitional monsoon climate), China, sample year: 2010–2013, country: China[441]

incidence: 1/455, conc.: 33.40 μg/kg, sample origin: Sichuan (province with subtropical temperate monsoon climate), China, sample year: 2010–2013, country: China[441]

incidence: 4/441, conc. range: 0.50–11.50 μg/kg, sample origin: Guangdong (province with subtropical-tropical monsoon climate), China, sample year: 2010–2013, country: China[441]

- Co-contamination: For detailed information see the article.
- Further contamination of plants and plant products other than cereals and their products reported in the present book/article[441]: Nuts (peanuts), AFB_1, AFB_2, and AFG_2, literature[441]

incidence: 9/84, conc. range: ≤4.36 μg/kg, sample origin: supermarkets in Selangor (state), Malaysia, sample year: April–August 2008, country: Malaysia/Indonesia[676]

- Co-contamination: not reported
- Further contamination of plants and plant products other than cereals and their products reported in the present book/article[676]: Nuts (peanut), AFB_1 and AFB_2, literature[676]

For detailed information see the article.

incidence: 1/151, conc.: 0.59 μg/kg, sample origin: markets, supermarkets, groceries, and retail shops in Adana and Osmaniye (provinces), Turkey, sample year: January 2010–December 2012, country: Turkey[741]

- Co-contamination: not reported
- Further contamination of plants and plant products other than cereals and their products reported in the present book/article[741]: Nuts (peanut), AFB_1 and AFB_2, literature[741]; Nuts (pistachio nut), AFB_1, AFB_2, AFG_1, and AFG_2, literature[741]

incidence: 15/20*, conc. range: 2.17–310.29 μg/kg, ∅ conc.: 48.5 μg/kg, sample origin: street hawkers, local markets, and retail shops in urban areas of Kinshasa (capital), Congo, sample year: December 2010–January 2011, country: South Africa[758], *peanuts, raw

- Co-contamination: 14 sa. co-contaminated with AFB_1, AFB_2, AFG_1, and AFG_2; 1 sa. co-contaminated with AFB_1, AFB_2, and AFG_1
- Further contamination of plants and plant products other than cereals and their products reported in the present book/article[758]: Nuts (peanut), AFB_1, AFB_2, and AFG_2, literature[758]

incidence: 8/20*, conc. range: 2.05–9.77 μg/kg, ∅ conc.: 4.74 μg/kg, sample origin: street hawkers, local markets, and retail shops in inner-city of Pretoria (capital), South Africa, sample year: December 2010–January 2011, country: South Africa[758], *peanuts, raw

- Co-contamination: 6 sa. co-contaminated with AFB_1, AFB_2, AFG_1, and AFG_2; 2 sa. co-contaminated with AFB_1, AFB_2, and AFG_1
- Further contamination of plants and plant products other than cereals and their products reported in the present book/article[758]: Nuts (peanut), AFB_1, AFB_2, and AFG_2, literature[758]

incidence: 2/9, conc. range: 1.05–5.20 μg/kg, ∅ conc.: 3.13 μg/kg, sample origin: supermarkets and night markets in Penang (state), Malaysia, sample year: unknown, country: Malaysia[766]

- Co-contamination: 2 sa. co-contaminated with AFG_1 and AFG_2
- Further contamination of plants and plant products other than cereals and their products reported in the present book/article[766]: Nuts (peanut), AFB_1 and AFG_2, literature[766]; Spices (cayenne pepper, chilli), AFB_1, AFB_2, AFG_1, and AFG_2, literature[766]

incidence: 4/8*, conc. range: 333–556 μg/kg, ∅ conc.: 500 μg/kg, sample origin: imported to Finland, sample year: unknown, country: Denmark/Finland[769], *peanuts for various purposes

- Co-contamination: 2 sa. co-contaminated with AFB_1, AFB_2, AFG_1, and AFG_2; 1 sa. co-contaminated with AFB_1, AFB_2, and AFG_1; 1 sa. co-contaminated with AFB_1 and AFG_1
- Further contamination of plants and plant products other than cereals and their products reported in the present book/article[769]: Nuts (peanut), AFB_1, AFB_2, and AFG_2, literature[769]

incidence: 1/10*, conc.: 1.7 μg/kg, sample origin: retail outlets in Paraná (state), Brazil, sample year: April 2006–May 2007, country: Brazil[784], *peanuts, salted

incidence: 14/45*, conc. range: LOD-20 μg/kg (14 sa., maximum: 10.7 μg/kg), sample origin: retail outlets in Paraná (state), Brazil, sample year: April 2006–May 2007, country: Brazil[784], *peanuts, sweet

incidence: 3/18*, conc. range: LOD-20 μg/kg (3 sa., maximum: 1.1 μg/kg), sample origin: retail outlets in Paraná (state), Brazil, sample year: April 2006–May 2007, country: Brazil[784], *peanut brittle

- Co-contamination: For detailed information see the article.
- Further contamination of plants and plant products other than cereals and their products reported in the present book/article[784]: Candies (peanut candy), AFB_1, AFB_2, AFG_1, and AFG_2, literature[784]; Nuts (peanut), AFB_1, AFB_2, and AFG_2, literature[784]

incidence: 5/9*, conc. range: 0.5–1.6 μg/kg, sample origin: market in Lilongwe (city), Malawi, sample year: December 2012, country: Malawi[790], *peanuts, un-skinned roasted

incidence: 11/15*, conc. range: 0.5–25.1 μg/kg, sample origin: market in Lilongwe (city), Malawi, sample year: December 2012, country: Malawi[790], *peanuts, de-skinned roasted

- Co-contamination: not reported
- Further contamination of plants and plant products other than cereals and their products reported in the present book/article[790]: Butter (peanut butter), Food, AFB_1, AFB_2, AFG_1, and AFG_2, literature[790]; Nuts (peanut), AFB_1, AFB_2, and AFG_2, literature[790]

incidence: 2/25, conc. range: 0.02–0.49 μg/kg, ∅ conc.: 0.26 μg/kg, sample origin: retail fresh markets or modern department stores in the central region of Thailand, sample year: unknown, country: Thailand/Austria/Germany[857]

- Co-contamination: not reported
- Further contamination of plants and plant products other than cereals and their products reported in the present book/article[857]: Nuts (peanut), AFB_1 and AFB_2, literature[857]; Spices (cayenne pepper, chilli), AFB_1, AFB_2, and AFG_1, literature[857]

incidence: 12/30, conc. range: ≤20 μg/kg (12 sa., maximum: 6.22 μg/kg), sample origin: local markets in Wuhan (city), China, sample year: unknown, country: China[881]

- Co-contamination: not reported
- Further contamination of plants and plant products other than cereals and their products reported in the present book/article[881]: Nuts (peanut), AFB_1, AFB_2, AFG_2, OTA, T-2, and ZEA, literature[881]

Aflatoxin G_2

incidence: 1/104, conc.: 34 μg/kg, sample origin: imported to Finland, sample year: 1975/1976, country: Finland[2]

- Co-contamination: 1 sa. co-contaminated with AFB_1, AFB_2, AFG_1, and AFG_2
- Further contamination of plants and plant products other than cereals and their products reported in the present book/article[2]: Marzipan (almond paste), Nuts (almond), Nuts (hazelnut), AFB_1 and AFB_2, literature[2]; Nuts, Nuts (peanut), AFB_1, AFB_2, and AFG_1, literature[2]

incidence: 0/121*, conc.: no contamination, sample origin: imported to Finland, sample year: 1974–1976, country: Finland[2], *peanuts, roasted

- Co-contamination: not reported
- Further contamination of plants and plant products other than cereals and their products reported in the present book/article[2]: Marzipan (almond paste), Nuts (almond), Nuts (hazelnut), AFB_1 and AFB_2, literature[2]; Nuts, Nuts (peanut), AFB_1, AFB_2, and AFG_1, literature[2]

incidence: 4/68*, conc. range: tr-14 μg/kg, sample origin: imported to Finland, sample year: 1974–1976, country: Finland[2], *peanuts, sliced and crushed

- Co-contamination: 4 sa. co-contaminated with AFB_1, AFB_2, AFG_1, and AFG_2
- Further contamination of plants and plant products other than cereals and their products reported in the present book/article[2]: Marzipan (almond paste), Nuts (almond), Nuts (hazelnut), AFB_1 and AFB_2, literature[2]; Nuts, Nuts (peanut), AFB_1, AFB_2, and AFG_1, literature[2]

incidence: 2/45*, conc. range: 148–363 μg/kg, ∅ conc.: 256 μg/kg, sample origin: different storage facilities, Egypt, sample year:

November 1978–April 1979, country: Egypt[4], *sa. moldy with mechanical or insect damage

- Co-contamination: 2 sa. co-contaminated with AFB_1, AFB_2, AFG_1, and AFG_2
- Further contamination of plants and plant products other than cereals and their products reported in the present book/article[4]: Beans (broad bean), AFB_1 and AFB_2, literature[4]; Lentils and Nuts (almond), AFB_1, literature[4]; Nuts (peanut), AFB_1, AFB_2, and AFG_1, literature[4]

incidence: 1/6, conc.: 10 µg/kg, sample origin: major local retail outlets at Nainital (district), India, sample year: 1983–1985, country: India[7]

- Co-contamination: Often more than one mycotoxin was present in a sa.
- Further contamination of plants and plant products other than cereals and their products reported in the present book/article[7]: Coconuts, AFB_1, AFB_2, AFG_1, AFG_2, and CIT, literature[7]; Dates, AFB_1, AFB_2, AFG_1, and OTA, literature[7]; Grogannuts, AFB_1, AFB_2, AFG_1, AFG_2, and CIT, literature[7]; Nuts (almond), AFB_1, AFB_2, AFG_1, RBT, STG, and ZEA, literature[7]; Nuts (cashew nut), AFB_1, AFB_2, AFG_1, AFG_2, CIT, and OTA, literature[7]; Nuts (peanut), AFB_1, AFB_2, AFG_1, and OTA, literature[7]; Nuts (pistachio nut), AFB_1, AFB_2, AFG_1, AFG_2, and STG, literature[7]; Nuts (walnut), AFB_1, AFB_2, AFG_1, AFG_2, and RBT, literature[7]; Raisins, AFB_1, AFB_2, AFG_1, AFG_2, and OTA, literature[7]

incidence: 1/3*, conc.: 3 µg/kg, sample origin: unknown, sample year: unknown, country: USA[11], *peanuts, salted in-shell

- Co-contamination: 1 sa. co-contaminated with AFB_1, AFB_2, AFG_1, and AFG_2
- Further contamination of plants and plant products other than cereals and their products reported in the present book/article[11]: Nuts (almond), AFB_1 and AFB_2, literature[11]; Nuts (peanut), AFB_1, AFB_2, and AFG_1, literature[11]

incidence: 0/1*, conc.: no contamination, sample origin: unknown, sample year: unknown, country: USA[11], *peanuts with cereal coating

- Co-contamination: not reported
- Further contamination of plants and plant products other than cereals and their products reported in the present book/article[11]: Nuts (almond), AFB_1 and AFB_2, literature[11]; Nuts (peanut), AFB_1, AFB_2, and AFG_1, literature[11]

incidence: 39?/40, conc. range: 1.0–548 µg/kg, sample origin: different district hygiene institutes, GDR, sample year: unknown, country: GDR[29]

- Co-contamination: not reported
- Further contamination of plants and plant products other than cereals and their products reported in the present book/article[29]: Butter (peanut butter), Products (peanut product), Seeds (sunflower), AFB_1, AFB_2, AFG_1, and AFG_2, literature[29]; Coffee, OTA, literature[29]; Fruits, AFB_1, AFG_1, and AFG_2, literature[29]; Fruits (apple), Fruits (peach), Juices (apple juice), PAT, literature[29]; Nuts (peanut), AFB_1, AFB_2, and AFG_2, literature[29]

incidence: 11?/149, conc. range: ≤6.7 µg/kg, Ø conc.: 0.8 µg/kg, sample origin: commercially available market basket sa. of foods and foodstuffs in Tokyo (capital), Japan, sample year: 1986–1990, country: Japan[99]

- Co-contamination: not reported
- Further contamination of plants and plant products other than cereals and their products reported in the present book/article[99]: Beans, Seeds (sesame), Sugar, AFB_1 and AFB_2, literature[99]; Food, Jam (bean jam), Paste (peanut paste), AFB_1, literature[99]; Nuts (Brazil nut, paranut), Nuts (pistachio nut), Spices (nutmeg), AFB_1, AFB_2, AFG_1, and AFG_2, literature[99]; Nuts (peanut), Spices, Spices (cayenne pepper, chilli), Spices (pepper), AFB_1, AFB_2, and AFG_1, literature[99]; Spices (paprika), AFB_1 and AFG_1, literature[99]

incidence: 6/84, conc. range: 18.3–123 µg/kg, Ø conc.: 68.3 µg/kg, sample origin: four agroecological zones, Nigeria, sample year: May–July 2015, country: Nigeria/Austria[148]

- Co-contamination: For detailed information see the article.
- Further contamination of plants and plant products other than cereals and their products reported in the present book/article[148]: Nuts (peanut), AFB_1, AFB_2, AFG_1, AFM_1, AME, AOH, BEA, CPA, MON, NIV, and OTA, literature[148]; see also Nuts (peanut), chapter: Further Mycotoxins and Microbial Metabolites

For detailed information see the article.

incidence: 179?/1,860, conc. range: <0.3–3 µg/kg, Ø conc.: 0.6 µg/kg, sample origin: sites of primary storage or processing (domestic products) and sites of importation, Cyprus, sample year: 1992–1996, country: Cyprus[200]

- Co-contamination: not reported
- Further contamination of plants and plant products other than cereals and their products reported in the present book/article[200]: Butter (peanut butter) and Figs, AFB_1, AFB_2, AFG_1, and AFG_2, literature[200]; Nuts (Brazil nut, paranut), Nuts (peanut), and Nuts (walnut), AFB_1, AFB_2, and AFG_1, literature[200]; Nuts (chestnut), Products (sesame product), and Spices, AFB_1, literature[200]; Nuts (pistachio nut), AFB_1 and AFB_2, literature[200]; Seeds (sesame), AFB_1, AFG_1, and AFG_2, literature[200]

incidence: 6/34*, conc. range: tr-2 µg/kg, sample origin: stores in Oslo (capital) area, Norway, sample year: autumn 1973, country: Norway[201], *peanuts bought from stores

- Co-contamination: 6 sa. co-contaminated with AFB_1, AFB_2, AFG_1, and AFG_2
- Further contamination of plants and plant products other than cereals and their products reported in the present book/article[201]: Cocoa, Nuts (Brazil nut, paranut), Nuts (cashew nut), and Nuts (walnut), AFB_1, AFB_2, AFG_1, and AFG_2, literature[201]; Nuts, AFB_2, AFG_1, and AFG_2, literature[201]; Nuts (peanut), AFB_1, AFB_2, and AFG_1, literature[201]

incidence: 1/1*, conc.: 30 µg/kg, sample origin: stores in Oslo (capital) area, Norway, sample year: autumn 1973, country: Norway[201], *peanuts sorted, pooled sa.

- Co-contamination: 1 sa. co-contaminated with AFB_1, AFB_2, AFG_1, and AFG_2
- Further contamination of plants and plant products other than cereals and their products reported in the present book/article[201]: Cocoa, Nuts (Brazil nut, paranut), Nuts (cashew nut), and Nuts (walnut), AFB_1, AFB_2, AFG_1, and AFG_2, literature[201]; Nuts, AFB_2, AFG_1, and AFG_2, literature[201]; Nuts (peanut), AFB_1, AFB_2, and AFG_1, literature[201]

incidence: 3/267, conc. range: 5–80 µg/kg, Ø conc.: 34 µg/kg, sample origin: markets in 16 areas (cities and counties), Taiwan, sample year: 1977/1978, country: Taiwan[214] incidence: 1/33*, conc.: 15 μg/kg, sample origin: markets in 16 areas (cities and counties), Taiwan, sample year: 1977/1978, country: Taiwan[214], *peanuts, fried

incidence: 0/19*, conc.: no contamination, sample origin: markets in 16 areas (cities and counties), Taiwan, sample year: 1977/1978, country: Taiwan[214], *peanuts, roasted

- Co-contamination: not reported
- Further contamination of plants and plant products other than cereals and their products reported in the present book/article[214]: Butter (peanut butter), Cake (peanut), and Oils (peanut oil), AFB_1, AFB_2, AFG_1, and AFG_2, literature[214]; Nuts (peanut), AFB_1, AFB_2, and AFG_1, literature[214]

incidence: 15?/29, conc. range: 1.6–16.0 μg/kg, sample origin: storage depots and retail outlets, Botswana, sample year: January 1996–December 1997, country: Botswana[218]

- Co-contamination: not reported
- Further contamination of plants and plant products other than cereals and their products reported in the present book/article[218]: Butter (peanut butter), AFB_1, AFB_2, AFG_1, and AFG_2, literature[218]; Nuts (peanut), AFB_1, AFB_2, and AFG_1, literature[218]

incidence: 0/4 (analyzed sa.)*, conc.: no contamination, sample origin: supermarkets, local stores, and markets in Busan, Daejeon, Gwangju (cities), and Seoul (capital), Korea, sample year: June/July 2004, country: Korea[220], *peanuts, raw

incidence: ?/8 (analyzed sa.)*, conc. range: ≤~3.50 μg/kg, sample origin: supermarkets, local stores, and markets in Busan, Daejeon, Gwangju (cities), and Seoul (capital), Korea, sample year: June/July 2004, country: Korea[220], *peanuts, roasted

- Co-contamination: not reported
- Further contamination of plants and plant products other than cereals and their products reported in the present book/article[220]: Butter (peanut butter) and Nuts, AFB_1 and AFB_2, literature[220]; Nuts (peanut), AFB_1, AFB_2, and AFG_1, literature[220]; Nuts (pistachio nut), AFB_1, literature[220]

For detailed information see the article.

incidence: 1/1, conc.: 0.023 μg/kg, sample origin: market in Khartoum (capital), Sudan, sample year: February 1981–1984?, country: UK/Sudan[223]

- Co-contamination: 1 sa. co-contaminated with AFB_1, AFB_2, and AFG_2
- Further contamination of plants and plant products other than cereals and their products reported in the present book/article[223]: Butter (peanut butter) and Nuts (peanut), AFB_1, AFB_2, and AFG_1, literature[223]; Okra, AFG_2, literature[223]; Peas (chickpea), AFB_1, literature[223]

incidence: 1/1*, conc.: 0.006 μg/kg, sample origin: market in Khartoum (capital), Sudan, sample year: February 1981–1984?, country: UK/Sudan[223], *peanuts, limed

- Co-contamination: 1 sa. co-contaminated with AFB_1, AFG_1, and AFG_2
- Further contamination of plants and plant products other than cereals and their products reported in the present book/article[223]: Butter (peanut butter), Nuts (peanut), AFB_1, AFB_2, and AFG_1, literature[223]; Ochra, AFG_2, literature[223]; Peas (chickpea), AFB_1, literature[223]

incidence: 0/14*, conc.: no contamination, sample origin: supermarkets and village stores in five districts of Penang (state) including Penang Island and Wellesley (province), Malaysia, sample year: unknown, country: Malaysia[225], *peanut, raw and shelled

- Co-contamination: not reported
- Further contamination of plants and plant products other than cereals and their products reported in the present book/article[225]: Butter (peanut butter), Confectionery, Nuts (peanut), Nuts (walnut), Products (nut product), AFB_1 and AFB_2, literature[225]; Cake (peanut), AFB_1, AFB_2, and AFG_2, literature[225]

incidence: 1/10*, conc.: 20.5 μg/kg, sample origin: supermarkets and village stores in five districts of Penang (state) including Penang Island and Wellesley (province), Malaysia, sample year: unknown, country: Malaysia[225], *roasted peanuts, in-shell

- Co-contamination: 1 sa. co-contaminated with AFB_1, AFB_2, and AFG_2
- Further contamination of plants and plant products other than cereals and their products reported in the present book/article[225]: Butter (peanut butter), Confectionery, Nuts (peanut), Nuts (walnut), Products (nut product), AFB_1 and AFB_2, literature[225]; Cake (peanut), AFB_1, AFB_2, and AFG_2, literature[225]

incidence: 0/20*, conc.: no contamination, sample origin: supermarkets and village stores in five districts of Penang (state) including Penang Island and Wellesley (province), Malaysia, sample year: unknown, country: Malaysia[225], *roasted peanuts, shelled

- Co-contamination: not reported
- Further contamination of plants and plant products other than cereals and their products reported in the present book/article[225]: Butter (peanut butter), Confectionery, Nuts (peanut), Nuts (walnut), Products (nut product), AFB_1 and AFB_2, literature[225]; Cake (peanut), AFB_1, AFB_2, and AFG_2, literature[225]

incidence: 0/8*, conc.: no contamination, sample origin: supermarkets and village stores in five districts of Penang (state) including Penang Island and Wellesley (province), Malaysia, sample year: unknown, country: Malaysia[225], *pound peanut

- Co-contamination: not reported
- Further contamination of plants and plant products other than cereals and their products reported in the present book/article[225]: Butter (peanut butter), Confectionery, Nuts (peanut), Nuts (walnut), Products (nut product), AFB_1 and AFB_2, literature[225]; Cake (peanut), AFB_1, AFB_2, and AFG_2, literature[225]

incidence: 0/2*, conc.: no contamination, sample origin: supermarkets and village stores in five districts of Penang (state) including Penang Island and Wellesley (province), Malaysia, sample year: unknown, country: Malaysia[225], * peanut slices

- Co-contamination: not reported
- Further contamination of plants and plant products other than cereals and their products reported in the present book/article[225]: Butter (peanut butter), Confectionery, Nuts (peanut), Nuts (walnut), Products (nut product), AFB_1 and AFB_2, literature[225]; Cake (peanut), AFB_1, AFB_2, and AFG_2, literature[225]

incidence: 14?/35*, conc. range: ≤4.1 μg/kg, sample origin: local stores in Hangzhou, Ningbo, and Quzhou (cities), Zhejiang (province), China, sample year: unknown, country: China[227], *peanuts, fresh

incidence: 5/5*, conc. range: ≤1.3 μg/kg, sample origin: local stores in Hangzhou, Ningbo, and Quzhou (cities), Zhejiang (province), China, sample year: unknown, country: China[227], *peanuts, musty

- Co-contamination: not reported
- Further contamination of plants and plant products other than cereals and their products reported in the present book/article[227]: Butter (peanut butter), AFB_1, AFB_2, AFG_1, AFG_2, AFM_1, and AFM_2, literature[227]; Nuts (peanut), AFB_1, AFB_2, AFG_1, AFM_1, and AFM_2, literature[227]

incidence: 18/48*, conc. range: 1.33–16.87 μg/kg, ∅ conc.: 1.73 μg/kg, sample origin: Araras, Leme, Pirassununga, and Porto Ferreira (cities), northeast of São Paulo (state), Brazil, sample year: June 2006–May 2007, country: Brazil[237], *peanuts, unprocessed

incidence: 12/48*, conc. range: <LOQ-1.82 μg/kg, ∅ conc.: <LOQ, sample origin: Araras, Leme, Pirassununga, and Porto Ferreira (cities), northeast of São Paulo (state), Brazil, sample year: June 2006–May 2007, country: Brazil[237], *peanuts, roasted and salty

- Co-contamination: not reported
- Further contamination of plants and plant products other than cereals and their products reported in the present book/article[237]: Candies (peanut candy), AFB_1, AFB_2, AFG_1, and AFG_2, literature[237]; Nuts (peanut), AFB_1, AFB_2, and AFG_1, literature[237]

incidence: 1/1, conc.: 40 μg/kg, sample origin: Angola, sample year: 1970, country: Germany[239]

- Co-contamination: 1 sa. co-contaminated with AFB_1, AFB_2, AFG_1, and AFG_2
- Further contamination of plants and plant products other than cereals and their products reported in the present book/article[239]: Fruits (apple), Fruits (cherry), Fruits (lemmon), Fruits (orange), Fruits (peach), Fruits (tomato), Nuts (almond), Nuts (cashew nut), Nuts (hazelnut), Nuts (walnut), Seeds (sesame), Spice (nutmeg), AFB_1, literature[239]; Nuts (peanut), AFB_1, AFB_2, and AFG_1, literature[239]

incidence: 6/86*, conc. range: 20–400 μg/kg, ∅ conc.: 95 μg/kg, sample origin: supermarkets, retail shops, fair, and beach, Recife (city), Pernambuco (state), Brazil, sample year: May–October 1993, country: Brazil[378], *peanuts, raw and roasted

- Co-contamination: 4 sa. co-contaminated with AFB_1, AFB_2, AFG_1, and AFG_2; 2 sa. co-contaminated with AFB_1, AFG_1, and AFG_2
- Further contamination of plants and plant products other than cereals and their products reported in the present book/article[378]: Nuts (peanut), AFB_1, AFB_2, and AFG_1, literature[378]

incidence: 37/190*, conc. range: ≤6.5 μg/kg, sample year: unknown, country: Egypt[383], *peanuts, in-shell

incidence: 76/155*, conc. range: ≤6.5 μg/kg, sample year: unknown, country: Egypt[383], *peanuts, shelled

incidence: 0/58*, conc.: no contamination, sample year: unknown, country: Egypt[383], *peanuts, roasted

- Co-contamination: not reported
- Further contamination of plants and plant products other than cereals and their products reported in the present book/article[383]: Nuts (peanut), AFB_1, AFB_2, and AFG_1, literature[383]

incidence: 2/2*, conc. range: 23–31 μg/kg, ∅ conc.: 27 μg/kg, sample origin: unknown, sample year: unknown, country: UK[384], *ncac

- Co-contamination: 2 sa. co-contaminated with AFB_1, AFB_2, AFG_1, and AFG_2
- Further contamination of plants and plant products other than cereals and their products reported in the present book/article[384]: Beans (soybean), AFB_1, AFB_2, AFG_1, and AFG_2, literature[384]; Nuts (peanut), AFB_1, AFB_2, and AFG_1, literature[384]

incidence: 22/72*, conc. range: 1–320 μg/kg, ∅ conc.: 51 μg/kg, sample origin: supermarkets and groceries in Maringá (city), Paraná (state), Brazil, sample year: March 1993–February 1994, country: Brazil[388], *peanuts, crude

- Co-contamination: 19 sa. co-contaminated with AFB_1, AFB_2, AFG_1, and AFG_2; 3 sa. co-contaminated with AFG_1 and AFG_2
- Further contamination of plants and plant products other than cereals and their products reported in the present book/article[388]: Nuts (peanut), AFB_1, AFB_2, and AFG_1, literature[388]

incidence: 0/20, conc.: no contamination, sample origin: Babile (district), Ethiopia, sample year: 2013/2014, country: Ethiopia/USA[389]

incidence: 0/20, conc.: no contamination, sample origin: Babile (district), Ethiopia, sample year: 2014/2015, country: Ethiopia/USA[389]

incidence: 5/20, conc. range: <4 μg/kg (4 sa.) and 8.2 μg/kg (1 sa.), sample origin: Darolabu (district), Ethiopia, sample year: 2013/2014, country: Ethiopia/USA[389]

incidence: 4/20, conc. range: 15.7–20 μg/kg (2 sa.) and ≤171 μg/kg (2 sa.), sample origin: Darolabu (district), Ethiopia, sample year: 2014/2015, country: Ethiopia/USA[389]

incidence: 0/20, conc.: no contamination, sample origin: Fedis (district), Ethiopia, sample year: 2013/2014, country: Ethiopia/USA[389]

incidence: 2/20, conc. range: 53.0–96.9 μg/kg (2 sa.), ∅ conc.: 75 μg/kg, sample origin: Fedis (district), Ethiopia, sample year: 2014/2015, country: Ethiopia/USA[389]

incidence: 0/20, conc.: no contamination, sample origin: Gursum (district), Ethiopia, sample year: 2013/2014, country: Ethiopia/USA[389]

incidence: 1/20, conc. range: 14.9 μg/kg (1 sa.), sample origin: Gursum (district), Ethiopia, sample year: 2014/2015, country: Ethiopia/USA[389]

- Co-contamination: not reported
- Further contamination of plants and plant products other than cereals and their products reported in the present book/article[389]: Cake (peanut), AFB_1 and AFB_2, literature[389]; Nuts (peanut), AFB_1, AFB_2, and AFG_1, literature[389]

incidence: 45/120, conc. range: 1–164 μg/kg, ∅ conc.: 45 μg/kg, sample origin: retail outlets, Botswana, sample year: July 2001–January 2002, country: Botswana[390]

- Co-contamination: not reported
- Further contamination of plants and plant products other than cereals and their products reported in the present book/article[390]: Nuts (peanut), AFB_1, AFB_2, AFG_1, and CPA, literature[390]

incidence: 16/50*, conc. range: 1–407 μg/kg, ∅ conc.: 63.9 μg/kg, sample origin: Georgia (state), USA, sample year: 1990, country: USA[396], *peanuts for oil production or feed

- Co-contamination: 14 sa. co-contaminated with AFB_1, AFB_2, AFG_1, AFG_2, and CPA; 2 sa. co-contaminated with AFB_1, AFB_2, AFG_1, and AFG_2
- Further contamination of plants and plant products other than cereals and their products reported in the present book/article[396]: Nuts (peanut), AFB_1, AFB_2, AFG_1, and CPA, literature[396]

incidence: 2/73, conc. range: 16.3 μg/kg, sample origin: public markets/bazaars, supermarkets, retail stores, and wholesale stores in Ankara, Antalya, Isparta, Burdur, and Izmir (provinces), Turkey, sample year: unknown, country: Turkey[409]

- Co-contamination: not reported
- Further contamination of plants and plant products other than cereals and their products reported in the present book/article[409]: Nuts (hazelnut), AFB_1, AFG_1, and AFG_2, literature[409]; Nuts (peanut), AFB_1, AFB_2, and AFG_1, literature[409]; Nuts (pistachio nut), AFB_1 and AFB_2, literature[409]

incidence: 3/65, conc. range: 0.12–2.36 µg/kg, sample origin: local markets in Chongqing, Fujian, Guangdong, Guangxi, Hubei, Jiangsu, Shanghai, and Zhejiang (regions), China, sample year: 2003, country: China[411]

- Co-contamination: not reported
- Further contamination of plants and plant products other than cereals and their products reported in the present book/article[411]: Nuts (peanut), AFB_1, AFB_2, and AFG_1, literature[411]; Nuts (pine nut), AFB_1 and AFB_2, literature[411]; Nuts (walnut), AFB_1, AFB_2, AFG_1, and AFG_2, literature[411]

incidence: 3/408, conc. range: 0.27–1.15 µg/kg, sample origin: Liaoning (province with temperate monsoon climate), China, sample year: 2010–2013, country: China[441]

incidence: 48/1190, conc. range: 0.05–104.00 µg/kg, sample origin: Henan (province with transitional monsoon climate), China, sample year: 2010–2013, country: China[441]

incidence: 24/455, conc. range: 06–29.50 µg/kg, sample origin: Sichuan (province with subtropical temperate monsoon climate), China, sample year: 2010–2013, country: China[441]

incidence: 14/441, conc. range: 0.21–5.74 µg/kg, sample origin: Guangdong (province with subtropical-tropical monsoon climate), China, sample year: 2010–2013, country: China[441]

- Co-contamination: For detailed information see the article.
- Further contamination of plants and plant products other than cereals and their products reported in the present book/article[441]: Nuts (peanuts), AFB_1, AFB_2, and AFG_1, literature[441]

incidence: 15/20*, conc. range: 2.38–201.93 µg/kg, ∅ conc.: 53.2 µg/kg, sample origin: street hawkers, local markets, and retail shops in urban areas of Kinshasa (capital), Congo, sample year: December 2010–January 2011, country: South Africa[758], *peanuts, raw

- Co-contamination: 14 sa. co-contaminated with AFB_1, AFB_2, AFG_1, and AFG_2; 1 sa. co-contaminated with AFB_1 and AFG_2
- Further contamination of plants and plant products other than cereals and their products reported in the present book/article[758]: Nuts (peanut), AFB_1, AFB_2, and AFG_1, literature[758]

incidence: 7/20*, conc. range: 2.00–12.31 µg/kg, ∅ conc.: 4.84 µg/kg, sample origin: street hawkers, local markets, and retail shops in inner-city of Pretoria (capital), South Africa, sample year: December 2010–January 2011, country: South Africa[758], *peanuts, raw

- Co-contamination: 6 sa. co-contaminated with AFB_1, AFB_2, AFG_1, and AFG_2; 1 sa. co-contaminated with AFB_1, AFB_2, and AFG_2
- Further contamination of plants and plant products other than cereals and their products reported in the present book/article[758]: Nuts (peanut), AFB_1, AFB_2, and AFG_1, literature[758]

incidence: 2/9, conc. range: 1.10–1.16 µg/kg, ∅ conc.: 1.13 µg/kg, sample origin: supermarkets and night markets in Penang (state), Malaysia, sample year: unknown, country: Malaysia[766]

- Co-contamination: 2 sa. co-contaminated with AFG_1 and AFG_2
- Further contamination of plants and plant products other than cereals and their products reported in the present book/article[766]: Nuts (peanut), AFB_1 and AFG_2, literature[766]; Spices (cayenne pepper, chilli), AFB_1, AFB_2, AFG_1, and AFG_2, literature[766]

incidence: 2/8*, conc. range: 167 µg/kg, ∅ conc.: 167 µg/kg, sample origin: imported to Finland, sample year: unknown, country: Denmark/Finland[769], *peanuts for various purposes

- Co-contamination: 2 sa. co-contaminated with AFB_1, AFB_2, AFG_1, and AFG_2
- Further contamination of plants and plant products other than cereals and their products reported in the present book/article[769]: Nuts (peanut), AFB_1, AFB_2, and AFG_1, literature[769]

incidence: 1/10*, conc.: 1.5 µg/kg, sample origin: retail outlets in Paraná (state), Brazil, sample year: April 2006–May 2007, country: Brazil[784], *peanuts, salted

incidence: 10/45*, conc. range: LOD-20 µg/kg (10 sa., maximum: 2.7 µg/kg), sample origin: retail outlets in Paraná (state), Brazil, sample year: April 2006–May 2007, country: Brazil[784], *peanuts, sweet

incidence: 1/18*, conc.: 1.4 µg/kg, sample origin: retail outlets in Paraná (state), Brazil, sample year: April 2006–May 2007, country: Brazil[784], *peanut brittle

- Co-contamination: For detailed information see the article.
- Further contamination of plants and plant products other than cereals and their products reported in the present book/article[784]: Candies (peanut candy), AFB_1, AFB_2, AFG_1, and AFG_2, literatrure[784]; Nuts (peanut), AFB_1, AFB_2, and AFG_1, literature[784]

incidence: 1/9*, conc.: 0.2 µg/kg, sample origin: market in Lilongwe (city), Malawi, sample year: December 2012, country: Malawi[790], *peanuts, un-skinned roasted

incidence: 9/15*, conc. range: 0.4–2.5 µg/kg, sample origin: market in Lilongwe (city), Malawi, sample year: December 2012, country: Malawi[790], *peanuts, de-skinned roasted

- Co-contamination: not reported
- Further contamination of plants and plant products other than cereals and their products reported in the present book/article[790]: Butter (peanut butter) and Food, AFB_1, AFB_2, AFG_1, and AFG_2, literature[790]; Nuts (peanut), AFB_1, AFB_2, and AFG_1, literature[790]

incidence: 9/30, conc. range: ≤20 µg/kg (9 sa., maximum: 5.54 µg/kg), sample origin: local markets in Wuhan (city), China, sample year: unknown, country: China[881]

- Co-contamination: not reported
- Further contamination of plants and plant products other than cereals and their products reported in the present book/article[881]: Nuts (peanut), AFB_1, AFB_2, AFG_1, OTA, T-2, and ZEA, literature[881]

Aflatoxin M_1

incidence: 10/84, conc. range: 1.5–70.1 µg/kg, ∅ conc.: 19.9 µg/kg, sample origin: four agroecological zones, Nigeria, sample year: May–July 2015, country: Nigeria/Austria[148]

- Co-contamination: For detailed information see the article.
- Further contamination of plants and plant products other than cereals and their products reported in the present book/article[148]: Nuts (peanut), AFB_1, AFB_2, AFG_1, AFG_2, AME, AOH, BEA, CPA, MON, NIV, and OTA, literature[148]; see also Nuts (peanut), chapter: Further Mycotoxins and Microbial Metabolites

For detailed information see the article.

incidence: 0/35*, conc.: no contamination, sample origin: local stores in Hangzhou, Ningbo, and Quzhou (cities), Zhejiang (province), China, sample year: unknown, country: China[227], *peanuts, fresh

incidence: 5/5*, conc. range: ≤64.7 µg/kg, sample origin: local stores in Hangzhou, Ningbo, and Quzhou (cities), Zhejiang (province), China, sample year: unknown, country: China[227], *peanuts, musty

- Co-contamination: not reported
- Further contamination of plants and plant products other than cereals and their products reported in the present book/article[227]: Butter (peanut butter), AFB_1, AFB_2, AFG_1, AFG_2, AFM_1, and AFM_2, literature[227]; Nuts (peanut), AFB_1, AFB_2, AFG_1, AFG_2, and AFM_2, literature[227]

incidence: 3/10*, conc. range: >LOD-0.55 μg/kg**, sample origin: local supermarkets in Rio de Janeiro (city), Rio de Janeiro (state), Brazil, sample year: unknown, country: Brazil[401], *8 roasted and 2 raw peanut sa., **only roasted peanut sa. contaminated

- Co-contamination: 1 sa. co-contaminated with AFB_1, AFB_2, AFM_1, and AFM_2; 1 sa. co-contaminated with AFB_1, AFB_2, and AFM_1; 1 sa. co-contaminated with AFB_1, AFB_2, AFG_1, and AFM_1
- Further contamination of plants and plant products other than cereals and their products reported in the present book/article[401]: Nuts (peanut), AFB_1, AFB_2, AFG_1, and AFM_2, literature[401]

Aflatoxin M_2

incidence: 0/35*, conc.: no contamination, sample origin: local stores in Hangzhou, Ningbo, and Quzhou (cities), Zhejiang (province), China, sample year: unknown, country: China[227], *peanuts, fresh

incidence: 5/5*, conc. range: ≤3.6 μg/kg, sample origin: local stores in Hangzhou, Ningbo, and Quzhou (cities), Zhejiang (province), China, sample year: unknown, country: China[227], *peanuts, musty

- Co-contamination: not reported
- Further contamination of plants and plant products other than cereals and their products reported in the present book/article[227]: Butter (peanut butter), AFB_1, AFB_2, AFG_1, AFG_2, AFM_1, and AFM_2, literature[227]; Nuts (peanut), AFB_1, AFB_2, AFG_1, AFG_2, and AFM_1, literature[227]

incidence: 1/10*, conc.: 0.98 μg/kg**, sample origin: local supermarkets in Rio de Janeiro (city), Rio de Janeiro (state), Brazil, sample year: unknown, country: Brazil[401], *8 roasted and 2 raw peanut sa., **only a roasted peanut sa. contaminated

- Co-contamination: 1 sa. co-contaminated with AFB_1, AFB_2, AFM_1, and AFM_2
- Further contamination of plants and plant products other than cereals and their products reported in the present book/article[401]: Nuts (peanut), AFB_1, AFB_2, AFG_1, and AFM_1, literature[401]

Aflatoxin/s

incidence: 441/630*, Ø conc.: 49.1 μg/kg**, sample origin: greater Manila (capital) area, Philippines, sample year: unknown, country: Philippines[1], *peanut and peanut products except peanut butter, **AF/S of pos sa.?

- Co-contamination: not reported
- Further contamination of plants and plant products other than cereals and their products reported in the present book/article[1]: Beans, Butter (peanut butter), Cassava, Cocoa, Coffee, Drinks (alcoholic drink), Nuts (cashew nut), Products (coconut product), Rootcrops, Spices (cayenne pepper, chilli), Sweet potatoes, Yams, AF/S, literature[1]

incidence: 3/8*, conc. range: 0.5–5 μg/kg** (2 sa.) and 182 μg/kg** (1 sa.), sample origin: retail outlets in 20 different regions, England, sample year: April 1984, country: UK[3], *peanuts, raw shelled, **AFS total

incidence: 13/24*, conc. range: 0.5–5 μg/kg** (8 sa.), 6–10 μg/kg** (1 sa.), 11–30 μg/kg** (1 sa.), 31–100 μg/kg** (2 sa.), and 4,920 μg/kg** (1 sa.), sample origin: retail outlets in 20 different regions, England, sample year: April 1984, country: UK[3], *peanuts, raw in-shell, **AFS total

incidence: 4/17*, conc. range: 0.5–5 μg/kg** (3 sa.) and 7 μg/kg** (1 sa.), sample origin: retail outlets in 20 different regions, England, sample year: April 1984, country: UK[3], *peanuts, roasted, **AFS total

incidence: 3/14*, conc. range: 0.5 to ≤5 μg/kg**, sample origin: retail outlets in 20 different regions, England, sample year: 1984, country: UK[3], *peanuts, dry roasted, **AFS total

incidence: 7/13*, conc. range: 0.5–5 μg/kg** (3 sa.), 6–10 μg/kg** (2 sa.), and 11–30 μg/kg** (2 sa, maximum: 28 μg/kg), sample origin: retail outlets in 20 different regions, England, sample year: April 1984, country: UK[3], *peanuts, roasted in-shell, **AFS total

incidence: 3/17*, conc. range: 0.5 to ≤5 μg/kg**, sample origin: retail outlets in 20 different regions, England, sample year: April 1984, country: UK[3], *peanuts, chocolate-coated, **AFS total

incidence: 8/19*, conc. range: 6–10 μg/kg** (6 sa.), 31–100 μg/kg (1 sa.), and 190 μg/kg (1 sa.), sample origin: retail outlets in 20 different regions, England, sample year: April 1984, country: UK[3], *peanut brittle, **AFS total

- Co-contamination: not reported
- Further contamination of plants and plant products other than cereals and their products reported in the present book/article[3]: Butter (peanut butter), Nuts (almond), Nuts (Brazil nut, paranut), Nuts (hazelnut), AFB_1 and AF/S, literature[3]; Nuts (peanut), AFB_1, literature[3]

incidence: 284/1,679*, conc. range: 5.0–9.9 μg/kg** (62 sa.), 10.0–14.9 μg/kg** (52 sa.), 15.0–19.9 μg/kg** (37 sa.), 20.0–24.9 μg/kg** (35 sa.), and >25.0 μg/kg** (98 sa.), sample origin: brokers, distributors, growers, processors, and retail trade, Canada, sample year: 1970–1975, country: Canada[15], *peanuts, raw, **AFB_1, AFB_2, AFG_1, and AFG_2

incidence: 7/157*, conc. range: 5.0–9.9 μg/kg** (3 sa.), 10.0–14.9 μg/kg** (2 sa.), and >25 μg/kg** (2 sa.), sample origin: brokers, distributors, growers, processors, and retail trade, Canada, sample year: 1970–1975, country: Canada[15], *peanuts, processed, **AFB_1, AFB_2, AFG_1, and AFG_2

- Co-contamination: not reported
- Further contamination of plants and plant products other than cereals and their products reported in the present book/article[15]: Butter (peanut butter), Cocoa beans, Nuts (almond), Nuts (cashew nut), Nuts (pecan nut), Nuts (pistachio nut), Nuts (walnut), Seeds (sunflower), AF/S, literature[15]

incidence: 4/10*, conc. range: 1.5–14.5 μg/kg**, Ø conc.: 5.10 μg/kg**, sample origin: retail shops and local markets in NWFP and northern areas, Pakistan, sample year: January–November 2009, country: Pakistan[17], *peanuts with shell, **AFB_1, AFB_2, AFG_1, and AFG_2

incidence: 5/10*, conc. range: 0.7–12.8 μg/kg**, Ø conc.: 5.20 μg/kg**, sample origin: retail shops and local markets in NWFP and northern areas, Pakistan, sample year: January–November 2009, country: Pakistan[17], *peanuts without shell, **AFB_1, AFB_2, AFG_1, and AFG_2

- Co-contamination: not reported
- Further contamination of plants and plant products other than cereals and their products reported in the present book/article[17]: Dates, Figs, Fruits, dried, Kernels (apricot), Nuts (almond), Nuts (pine nut), Nuts (pistachio nut), Nuts (walnut), Raisins, AF/S, literature[17]

incidence: 1/2, conc.: 1.70 μg/kg*, sample origin: mainly imported from local retail outlets in Lodz (city), Poland, sample year: unknown, country: Poland[18], *AFS total

incidence: 1/6*, conc.: 3.59 μg/kg**, sample origin: mainly imported from local retail outlets in Lodz (city), Poland, sample year: unknown, country: Poland[18], *peanuts, salted, **AFS total

- Co-contamination: not reported
- Further contamination of plants and plant products other than cereals and their products reported in the present book/article[18]: Nut cereals, Nuts, Nuts (almond), Nuts (cashew nut), Nuts (hazelnut), Nuts (pistachio nut), Nuts (walnut), Seeds (sunflower), AF/S, literature[18]

incidence: 6/55*, conc. range: ≤20 μg/kg** (4 sa.) and >20 μg/kg** (2 sa., maximum: 329 μg/kg**,), sample origin: FDA districts, USA, sample year: 1986, country: USA[19], *peanuts, shelled and roasted, **AFB_1, AFB_2, AFG_1, and AFG_2

incidence: 1/4*, conc.: 273 μg/kg**, sample origin: imported to the USA, sample year: 1986, country: USA[19], *peanuts, raw and in-shell**AFB_1, AFB_2, AFG_1, and AFG_2

incidence: 1/1*, conc.: 4 μg/kg**, sample origin: imported to the USA, sample year: 1986, country: USA[19], *peanuts, shelled and roasted, **AFB_1, AFB_2, AFG_1, and AFG_2

incidence: 1/1*, conc.: 302 μg/kg**, sample origin: imported to the USA, sample year: 1986, country: USA[19], *peanut mix, *AFB_1, AFB_2, AFG_1, and AFG_2

- Co-contamination: not reported
- Further contamination of plants and plant products other than cereals and their products reported in the present book/article[19]: Butter (peanut butter), Candies (peanut candy), Candies (pistachio candy), Nuts, Nuts (almond), Nuts (Brazil nut, paranut), Nuts (pecan nut), Nuts (pistachio nut), Nuts (walnut), Oils (peanut oil), Paste (peanut paste), Paste (pipian paste), Products (peanut product), Seeds (melon), Spices (cayenne pepper, chilli), Spices (ginger), Spices (nutmeg), AF/S, literature[19]

incidence: 8/72, conc. range: ≤7.7 μg/kg*, Ø conc.: 2.7 μg/kg*, sample origin: supermarkets and hypermarkets in 12 main cities of Catalonia (autonomous community), Spain, sample year: 2008/2009, country: Spain[93], *AFB_1, AFB_2, AFG_1, and AFG_2

- Co-contamination: not reported
- Further contamination of plants and plant products other than cereals and their products reported in the present book/article[93]: Figs, Nuts (pistachio nut), Spices (cayenne pepper, chilli), AF/S, literature[93]

incidence: 41/47, conc. range: 1–5 μg/kg* (16 sa.), 6–15 μg/kg* (11 sa.), 16–25 μg/kg* (1 sa.), 26–50 μg/kg* (4 sa.), 51–100 μg/kg* (3 sa.), and >100 μg/kg* (6 sa.), sample origin: Keneba (village), Gambia, sample year: October 1988, country: France/England/Germany/USA/Gambia[118], *AF

- Co-contamination: not reported
- Further contamination of plants and plant products other than cereals and their products reported in the present book/article[118]: Food, Sauces (flour sauce), Sauces (leaf sauce), AF/S, literature[118]

incidence: 5/71*, conc.: >30 to ≤100 μg/kg**, sample origin: markets in or near Manila (capital), Philippines, sample year: 1967–1969, country: USA[134], *whole peanuts, **AF

- Co-contamination: not reported
- Further contamination of plants and plant products other than cereals and their products reported in the present book/article[134]: Beans, Butter (peanut butter), Products (coconut product), Products (peanut product), Tubers, AF/S, literature[134]

incidence: 35/170, Ø conc.: 179.8 μg/kg*, sample origin: Highveld, Middleveld, Lowveld, and Lumbombo Region, Swaziland, sample year: July 1982–June 1983, country: South Africa/France/Kenya/Netherlands[135], *AFS total

- Co-contamination: not reported
- Further contamination of plants and plant products other than cereals and their products reported in the present book/article[135]: Beans, Beans (cowpea), Beans (soybean), AF/S, literature[135]

incidence: 27/152, conc. range: 1–100 μg/kg* (11 sa.), 100–1,000 μg/kg* (8 sa.), and >1,000 μg/kg* (8 sa.), sample origin: natives homes and markets, Uganda, sample year: September 1966–June 1967, country: Uganda/USA[137], *AFB_1, AFB_2, AFG_1, and AFG_2

- Co-contamination: 24 sa. contained AFB_1, 16 sa. contained AFB_2, 17 sa. contained AFG_1, 7 sa. contained AFG_2
- Further contamination of plants and plant products other than cereals and their products reported in the present book/article[137]: Beans, Cassava, Peas, AF/S, literature[137]

incidence: 1/10* **, conc.: 72 μg/kg***, sample origin: storage facilities, silos sacks, small shops, and fields in zones with different climatic conditions, Guatemala, sample year: 1975/1976–?, country: Guatemala[138], *ncac, **sa. collected at the end of rainy season, ***AFB_1, AFB_2, and AFG_1

- Co-contamination: not reported
- Further contamination of plants and plant products other than cereals and their products reported in the present book/article[138]: Beans (black beans), Coffee beans, AF/S, literature[138]

incidence: 106/216, conc. range: ≤12,256 μg/kg*, Ø conc.: 1,530 μg/kg*, sample origin: market retailers, street vendors, wholesalers, and distributors, Thailand, sample year: September 1967–July 1969, country: Thailand[139], *AFB_1, AFB_2, AFG_1, and AFG_2

- Co-contamination: not reported
- Further contamination of plants and plant products other than cereals and their products reported in the present book/article[139]: Beans, Beans (mung bean), Food, Seeds (sesame), Spices (cayenne pepper, chilli), Spices (garlic/onion), Starch (cassava starch), Vegetables, AF/S, literature[139]

For detailed information see the article.

incidence: 31/31, conc. range: >6.8–20 μg/kg* (3 sa.), –50 μg/kg* (7 sa.), –100 μg/kg* (2 sa.), –500 μg/kg* (10 sa.), –2,000 μg/kg* (6 sa.), –4,000 μg/kg* (1 sa.), and –6,000 μg/kg* (2 sa.), sample origin: different Ecuadorian regions, Ecuador, sample year: April 1992–October 1994, country: Switzerland[140], *AFB_1, AFB_2, AFG_1, and AFG_2

- Co-contamination: For detailed information see the article.
- Further contamination of plants and plant products other than cereals and their products reported in the present book/article[140]: Beans, AF/S, literature[140]

incidence: 3/11, conc. range: 1.0–3.9 μg/kg* (1 sa) and >50 μg/kg* (2 sa., maximum: 61 μg/kg), sample origin: port sa. from Felixstowe (town), Manchester (city), and King's Lynn (market town) Port Health Authorities, UK, sample year: unknown, country: UK[206], *AFB_1, AFB_2, AFG_1, and AFG_2

incidence: 2/36, conc. range: ≤105 μg/kg*, sample origin: retail sa. from Norwich area and health food stores, market stalls, and supermarkets across the UK, sample year: unknown, country: UK[206], *AFB_1, AFB_2, AFG_1, and AFG_2

- Co-contamination: not reported
- Further contamination of plants and plant products other than cereals and their products reported in the present book/article[206]: Butter (peanut butter), Figs, Nuts (Brazil nut, paranut), Nuts (pistachio nut), Paste (fig paste), AF/S, literature[206]

incidence: 29/359*, conc. range: 7.9–1,496 μg/kg**, Ø conc.: 129.0 μg/kg**, sample origin: local retail stores in the region of Brasilia (capital), Brazil, sample year: 2002–2011, country: Brazil[208], *peanuts, toasted, salted and shelled, **AFB_1, AFB_2, AFG_1, and AFG_2

- Co-contamination: not reported
- Further contamination of plants and plant products other than cereals and their products reported in the present book/article[208]: Nuts, Nuts (Brazil nut, paranut), Products (peanut product), AF/S, literature[208]

incidence: 1/12*, conc.: 204 μg/kg**, sample origin: traditional markets and supermarkets, Indonesia, sample year: 2001/2002, country: Austria/Indonesia/Thailand[226], *peanuts, roasted, **AFB_1, AFB_2, AFG_1, and AFG_2

incidence: 6/16*, conc. range: 5–870 μg/kg**, sample origin: traditional markets and supermarkets, Indonesia, sample year: 2001/2002, country: Austria/Indonesia/Thailand[226], *peanuts, flour/egg coated, **AFB_1, AFB_2, AFG_1, and AFG_2

incidence: 1/11*, conc.: 27 μg/kg**, sample origin: traditional markets and supermarkets, Indonesia, sample year: 2001/2002, country: Austria/Indonesia/Thailand[226], *peanuts, fried, **AFB_1, AFB_2, AFG_1, and AFG_2

- Co-contamination: not reported
- Further contamination of plants and plant products other than cereals and their products reported in the present book/article[226]: Butter (peanut butter), Cake (peanut), Sauces (peanut-chilli sauce), Snacks, AFB_1 and AF/S, literature[226]; Nuts (peanut), AFB_1, literature[226]

incidence: 2/4, conc. range: 0.17–2.13 μg/kg*, Ø conc.: 1.15 μg/kg*, sample origin: sales outlets, Qatar, sample year: October 2002, country: Qatar[229], *AFB_1, AFB_2, AFG_1, and AFG_2

- Co-contamination: not reported
- Further contamination of plants and plant products other than cereals and their products reported in the present book/article[229]: Butter (peanut butter), Figs, and Nuts (pistachio nut), AF/S, literature[229]; Coffee, Raisins, OTA, literature[229]; Spices and Spices (cayenne pepper, chilli), AF/S and OTA, literature[229]

incidence: 8/400, conc. range: 3–8 μg/kg*, Ø conc.: 5.7 μg/kg*, sample origin: local market and processing centers, central region of Sudan, sample year: unknown, country: Sudan/Saudi Arabia[230], *AFB_1, AFB_2, AFG_1, and AFG_2

incidence: 44/400*, conc. range: 4–12 μg/kg**, Ø conc.: 8.5 μg/kg**, sample origin: local market and processing centers, central region of Sudan, sample year: unknown, country: Sudan/Saudi Arabia[230], *peanuts, roasted, **AFB_1, AFB_2, AFG_1, and AFG_2

- Co-contamination: not reported
- Further contamination of plants and plant products other than cereals and their products reported in the present book/article[230]: Butter (peanut butter), AF/S, literature[230]

incidence: 28/86*, conc. range: 0.1–5 μg/kg** (5 sa.), 6–10 μg/kg** (3 sa.), 11–20 μg/kg** (5 sa.), 21–35 μg/kg** (7 sa.), 36–50 μg/kg** (1 sa.), and >50 μg/kg** (7 sa.);, sample origin: commercial and imported food from retail outlets and points of entries, Malaysia, sample year: 1995–1999, country: Malaysia[233], *peanuts, raw, **AF not specified

- Co-contamination: not reported
- Further contamination of plants and plant products other than cereals and their products reported in the present book/article[233]: Butter (peanut butter), Candies (peanut candy), Nuts, Products (peanut product), Spices (cayenne pepper, chilli), AF/S, literature[233]

incidence: 22/26*, conc. range: 1.5–1298.0 μg/kg, Ø conc.: 197.9 μg/kg, sample origin: five markets in Lagos, Ogun, and Oyo (states), Southwest Nigeria, sample year: unknown, country: Nigeria[276], *peanuts, raw

incidence: 21/22*, conc. range: 1.3–134.0 μg/kg, Ø conc.: 23.9 μg/kg, sample origin: five markets in Lagos, Ogun, and Oyo (states), Southwest Nigeria, sample year: unknown, country: Nigeria[276], *peanuts, roasted

incidence: 25/30*, conc. range: 1.5–1298.0 μg/kg, Ø conc.: 176.8 μg/kg, sample origin: five markets in Lagos, Ogun, and Oyo (states), Southwest Nigeria, sample year: unknown, country: Nigeria[276], *peanuts, coated

incidence: 18/18*, conc. range: 1.3–134.0 μg/kg, Ø conc.: 24.2 μg/kg, sample origin: five markets in Lagos, Ogun, and Oyo (states), Southwest Nigeria, sample year: unknown, country: Nigeria[276], *peanuts, de-coated

- Co-contamination: not reported
- Further contamination of plants and plant products other than cereals and their products reported in the present book/article[276]: Nuts (peanut), AFB_1, literature[276]

For detailed information see the article.

incidence: 10/50, conc. range: 0.1–26.6 μg/kg*, Ø conc.: 6.5 μg/kg*, sample origin: available in New Zealand, sample year: 2010, country: New Zealand[277], *AFS total

- Co-contamination: not reported
- Further contamination of plants and plant products other than cereals and their products reported in the present book/article[277]: Bars, AFB_1, AF/S, DON, NIV, and OTA, literature[277]; Butter (peanut butter), Confectionery, Food, Fruits, dried, Nuts, Nuts (pistachio nut), Sauces (peanut sauce), AFB_1 and AF/S, literature[277]; Coffee, Dates, Spices (nutmeg), Spices (pepper), Vine fruits, dried, OTA, literature[277]; Figs, Spices (cayenne pepper, chilli), Spices (curry), Spices (ginger), and Spices (paprika), AFB_1, AF/S, and OTA, literature[277]; Nuts (Brazil nut, paranut), Nuts (peanut), AFB_1, literature[277]

incidence: 11/22, conc. range: 1–18 μg/kg*, sample origin: farmers in Viti Levu and Vanua Levu, main islands of Fiji, sample year: unknown, country: Fiji/Zambia[321], *AF

incidence: 1/11, conc.: 1 μg/kg*, sample origin: Tonga, sample year: unknown, country: Fiji/Zambia[321], *AF

- Co-contamination: not reported
- Further contamination of plants and plant products other than cereals and their products reported in the present book/article[321]: Copra, Meal (copra meal), AF/S, literature[321]

incidence: 1/20, conc.: 0.30 μg/kg*, sample origin: retail shops and local markets in Rabat-Salé (region), Morocco, sample year:

January–October 2006, country: Spain/Morocco[352], *AFB_1, AFB_2, AFG_1, and AFG_2

- Co-contamination: not reported
- Further contamination of plants and plant products other than cereals and their products reported in the present book/article[352]: Figs, Nuts (pistachio nut), Nuts (walnut), and Raisins, AFB_1 and AF/S, literature[352]; Nuts (peanut), AFB_1, literature[352]

incidence: 5/10*, conc. range: 0.75–26.36 μg/kg**, sample origin: exporting companies in Aydin (province), Turkey, sample year: January–August 2007, country: Turkey[360], *destined for export, **AFB_1, AFB_2, AFG_1, and AFG_2

- Co-contamination: 2 sa. co-contaminated with AFB_1 and AFB_2; 3 sa. contaminated solely with AFB_1
- Further contamination of plants and plant products other than cereals and their products reported in the present book/article[360]: Figs, Nuts (hazelnut), Nuts (pistachio nut), Spices (paprika), AF/S, literature[360]

incidence: 3/21*, conc. range: 20–100 μg/kg** (2 sa.) and 787 μg/kg** (1 sa.), sample origin: open-air markets in Port-au-Prince (capital), Haiti, sample year: July 2012, country: USA[376], *ncac, **AFB_1, AFB_2, AFG_1, and AFG_2

- Co-contamination: not reported
- Further contamination of plants and plant products other than cereals and their products reported in the present book/article[376]: Butter (peanut butter), AF/S, literature[376]

incidence: 52/7898*, conc. range: ≥15 μg/kg**, sample origin: China, sample year: 2002–2006, country: Japan[407], *Virginia-type variety, **AFB_1, AFB_2, AFG_1, and AFG_2

incidence: 5/2723*, conc. range: ≥15 μg/kg**, sample origin: China, sample year: 2002–2006, country: Japan[407], *Spanish and runner-type variety, **AFB_1, AFB_2, AFG_1, and AFG_2

incidence: 9/1584, conc. range: ≥15 μg/kg**, sample origin: South Africa, sample year: 2002–2006, country: Japan[407], *AFB_1, AFB_2, AFG_1, and AFG_2

incidence: 17/1005, conc. range: ≥15 μg/kg**, sample origin: USA, sample year: 2002–2006, country: Japan[407], *AFB_1, AFB_2, AFG_1, and AFG_2

incidence: 8/202, conc. range: ≥15 μg/kg**, sample origin: Paraguay, sample year: 2002–2006, country: Japan[407], *AFB_1, AFB_2, AFG_1, and AFG_2

- Co-contamination: not reported
- Further contamination of plants and plant products other than cereals and their products reported in the present book/article[407]: not reported

For detailed information see the article.

incidence: 1/22*, conc.: >1,000 μg/kg** (1 sa.), sample origin: northern Gezira (state), Sudan, sample year: November 1969–March 1970, country: Sudan[412], *irrigated sa., **AF

incidence: 19/20*, conc.: >1,000 μg/kg** (1 sa.), sample origin: central Gezira (state), Sudan, sample year: November 1969–March 1970, country: Sudan[412], *irrigated sa., **AF

incidence: 1/24*, conc.: 50–250 μg/kg** (1 sa.), sample origin: southern Gezira (state), Sudan, sample year: November 1969–March 1970, country: Sudan[412], *irrigated sa., **AF

incidence: 6/44*, conc. range: <50 μg/kg* (5 sa.), and >1,000 μg/kg** (1 sa.), sample origin: Managil, Sudan, sample year: November 1969–March 1970, country: Sudan[412], *irrigated sa., **AF

incidence: 14/15, conc. range: <50 μg/kg* (11 sa.), 50–250 μg/kg* (1 sa.), and 250–1,000 μg/kg* (2 sa.), sample origin: Kosti (city), Sudan, sample year: November 1969–March 1970, country: Sudan[412], *AF

incidence: 48/48, conc. range: <50 μg/kg* (13 sa.), 50–250 μg/kg* (10 sa.), 250–1,000 μg/kg* (14 sa.), and >1,000 μg/kg* (11 sa.), sample origin: El Obeid (state), Sudan, sample year: November 1969–March 1970, country: Sudan[412], *AF

- Co-contamination: not reported
- Further contamination of plants and plant products other than cereals and their products reported in the present book/article[412]: not reported

incidence: 47/150*, conc. range: 6–10 μg/kg** (22 sa.), 14–20 μg/kg** (12 sa.), 25–30 μg/kg** (5 sa.), 40–50 μg/kg** (3 sa.), and 60–125 μg/kg** (5 sa.), sample origin: unknown, sample year: 1967, country: Israel[413], *ncac (stored peanut kernels), **AF

- Co-contamination: not reported
- Further contamination of plants and plant products other than cereals and their products reported in the present book/article[413]: not reported

incidence: 46/84* ** ***, ∅ conc.: 230 μg/kg****, sample origin: farmers' stock peanuts marketed in Georgia (state), USA, sample year: fall 1966, country: USA[416], *ncac, **sa. with visible *Aspergillus flavus* growth (0–2.5 % damage), ***Virginia, Spanish, and runner-type varieties, ****AF

incidence: 23/1,704* ** ***, ∅ conc.: 41 μg/kg****, sample origin: farmers' stock peanuts marketed in Georgia (state), USA, sample year: fall 1966, country: USA[416], *ncac, **sa. with no visible *Aspergillus flavus* growth (0–2.5 % damage), ***Virginia, Spanish, and runner-type varieties, ****AF

incidence: 10/14* ** ***, ∅ conc.: 1,439 μg/kg****, sample origin: farmers' stock peanuts marketed in Georgia (state), USA, sample year: fall 1966, country: USA[416], *ncac, **sa. with visible *Aspergillus flavus* growth (2.5–5.5 % damage), ***Virginia, Spanish, and runner-type varieties, ****AF

incidence: 20/242* ** ***, ∅ conc.: 35 μg/kg****, sample origin: farmers' stock peanuts marketed in Georgia (state), USA, sample year: fall 1966, country: USA[416], *ncac, **sa. with no visible *Aspergillus flavus* growth (2.5–5.5 % damage), ***Virginia, Spanish, and runner-type varieties, ****AF

incidence: 5/6* ** ***, ∅ conc.: 2,951 μg/kg****, sample origin: farmers' stock peanuts marketed in Georgia (state), USA, sample year: fall 1966, country: USA[416], *ncac, **sa. with visible *Aspergillus flavus* growth (> 5.5 % damage), ***Virginia, Spanish, and runner-type varieties, ****AF

incidence: 4/57* ** ***, ∅ conc.: 70 μg/kg****, sample origin: farmers' stock peanuts marketed in Georgia (state), USA, sample year: fall 1966, country: USA[416], *ncac, **sa. with no visible *Aspergillus flavus* growth (> 5.5 % damage), ***Virginia, Spanish, and runner-type varieties, ****AF

- Co-contamination: not reported
- Further contamination of plants and plant products other than cereals and their products reported in the present book/article[416]: not reported

incidence: 88/678, conc. range: ≤4 to <20 μg/kg* (35 sa.) and ≥20 μg/kg* (53 sa., maximum: 7,525.0 μg/kg*), sample origin: households in Busia and Homa Bay (districts), western Kenya, sample year: August 2006, country: South Africa/Kenya/USA[418], *AF

- Co-contamination: not reported
- Further contamination of plants and plant products other than cereals and their products reported in the present book/article[418]: not reported

For detailed information see the article.

incidence: 142/321*, conc. range: 5–30 μg/kg** (26 sa.) and >30 μg/kg** (116 sa., maximum: 2,440 μg/kg**), Ø conc.: 305 μg/kg**, sample origin: markets and milling factories in São Paulo State, Brazil, sample year: January–December 1994, country: Brazil[419], *peanuts and peanut products, **AFB_1 and AFG_1

- Co-contamination: not reported
- Further contamination of plants and plant products other than cereals and their products reported in the present book/article[419]: not reported

incidence: 73/73*, conc. range: 2 μg/kg** (52 sa.), 2–5 μg/kg** (12 sa.), 6–15 μg/kg** (8 sa.), and 61 μg/kg** (1 sa.), sample origin: retail outlets in New South Wales (state), Australia, sample year: 1988/1989, country: Australia[420], *included peanuts and peanut butter, **AFB_1, AFB_2, AFG_1, and AFG_2

- Co-contamination: not reported
- Further contamination of plants and plant products other than cereals and their products reported in the present book/article[420]: Nuts, AF/S, literature[420]

incidence: 11/35, conc. range: >20 to ≤936 μg/kg*, Ø conc.: 440 μg/kg*, sample origin: supermarkets in the region of Campinas (municipality), São Paulo (state), Brazil, sample year: 1995/1996, country: Brazil[421], *AFB_1, AFB_2, AFG_1, and AFG_2

- Co-contamination: not reported
- Further contamination of plants and plant products other than cereals and their products reported in the present book/article[421]: Products (peanut product), AF/S, literature[421]

incidence: 3/13* **, conc. range: 0.1–20 μg/kg*** (3 sa., maximum: 16.3 μg/kg***), sample origin: sa. imported and collected by food inspectors at overland, seaport, and airport health offices, Qatar, sample year: 1998, country: Qatar[423], *ncac, **peanut and peanut products, ***AFB_1, AFB_2, AFG_1, and AFG_2

incidence: 1/8* **, conc.: 4.6 μg/kg***, sample origin: sa. imported and collected by food inspectors at overland, seaport, and airport health offices, Qatar, sample year: 1999, country: Qatar[423], *ncac, **peanut and peanut products, ***AFB_1, AFB_2, AFG_1, and AFG_2

incidence: 5/11* **, conc. range: 0.1–20 μg/kg*** (4 sa.) and 22.7 μg/kg*** (1 sa.), sample origin: sa. imported and collected by food inspectors at overland, seaport, and airport health offices, Qatar, sample year: 2000, country: Qatar[423], *ncac, **peanut and peanut products, ***AFB_1, AFB_2, AFG_1, and AFG_2

- Co-contamination: not reported
- Further contamination of plants and plant products other than cereals and their products reported in the present book/article[423]: Nuts (pistachio nut), AF/S, literature[423]

incidence: 1/3*, conc.: 2.8 μg/kg**, sample origin: most sa. from Iran, Syria, Turkey, and the USA collected by inspectors at overland and seaport custom offices, Qatar, sample year: July–December 1997, country: Qatar[424], *peanuts with shell, **AFB_1, AFB_2, AFG_1, and AFG_2

- Co-contamination: not reported
- Further contamination of plants and plant products other than cereals and their products reported in the present book/article[424]: Nuts, Nuts (pistachio nut), AF/S, literature[424]

incidence: 3/3*, conc. range: 1.7–5.7 μg/kg**, Ø conc.: 3.6 μg/kg**, sample origin: Khartoum, Khartoum North, and Omdurman (towns of the capital Khartoum), Sudan, sample year: unknown, country: Sudan[425], *peanuts, gray roasted, **AFB_1, AFB_2, AFG_1, and AFG_2

incidence: 3/3*, conc. range: 4.3–14.3 μg/kg**, Ø conc.: 9.4 μg/kg**, sample origin: Khartoum, Khartoum North, and Omdurman (towns of the capital Khartoum), Sudan, sample year: unknown, country: Sudan[425], *peanuts, red roasted, **AFB_1, AFB_2, AFG_1, and AFG_2

- Co-contamination: not reported
- Further contamination of plants and plant products other than cereals and their products reported in the present book/article[425]: Paste (peanut paste), AF/S, literature[425]

incidence: 26/27* **, conc. range: 2–5,091 μg/kg***, sample origin: southeastern of the USA, sample year: 1980, country: USA[426], *ncac, **loose-shell kernel fractions, ***AFS

incidence: 20/21* **, conc. range: 3–550 μg/kg***, sample origin: southeastern of the USA, sample year: 1980, country: USA[426], *ncac, **sound mature kernel fractions, ***AFS

- Co-contamination: not reported
- Further contamination of plants and plant products other than cereals and their products reported in the present book/article[426]: Nuts (peanut), CPA, literature[426]

incidence: 109/160*, conc. range: tr-5,850 μg/kg, sample origin: three fields around Tirupati (city), Andhra Pradesh (state), India, sample year: November 1972, country: India[427], *ncac

- Co-contamination: 16 sa. co-contaminated with AFS and CIT; 93 sa. contaminated solely with AFS
- Further contamination of plants and plant products other than cereals and their products reported in the present book/article[427]: Nuts (peanut), CIT, literature[427]

incidence: 20/257, conc. range: 0.3–107.1 μg/kg*, sample origin: traditional markets, supermarkets, grocery stores, wholesale stores, and convenience stores in 25 counties, Taiwan, sample year: 1997–2011 (with exceptions), country: Taiwan[540], *AFB_1, AFB_2, AFG_1, and AFG_2

- Co-contamination: For detailed information see the article.
- Further contamination of plants and plant products other than cereals and their products reported in the present book/article[540]: Butter (peanut butter), Candies (peanut candy), Flour (peanut flour), AF/S, literature[540]

For detailed information see the article.

incidence: 13/22*, conc. range: ≤59.8 μg/kg**, sample origin: markets, superstores, malls, and venders of major districts of Punjab (province), Pakistan, sample year: unknown, country: Pakistan[573], *raw peanuts in-shell, **AFB_1, AFB_2, AFG_1, and AFG_2

incidence: 16/29*, conc. range: ≤82.1 μg/kg**, sample origin: markets, superstores, malls, and venders of major districts of Punjab (province), Pakistan, sample year: unknown, country: Pakistan[573], *raw peanuts without shells, **AFB_1, AFB_2, AFG_1, and AFG_2

incidence: 19/31*, conc. range: ≤71.3 μg/kg**, sample origin: markets, superstores, malls, and venders of major districts of Punjab (province), Pakistan, sample year: unknown, country: Pakistan[573], *roasted peanuts in-shells, **AFB_1, AFB_2, AFG_1, and AFG_2

incidence: 13/19*, conc. range: ≤119.7 μg/kg**, sample origin: markets, superstores, malls, and venders of major districts of

Punjab (province), Pakistan, sample year: unknown, country: Pakistan[573], *roasted peanuts without shells, **AFB_1, AFB_2, AFG_1, and AFG_2
- Co-contamination: not reported
- Further contamination of plants and plant products other than cereals and their products reported in the present book/article[573]: Butter (peanut butter), Cookies, Products (peanut product), AF/S, literature[573]

incidence: 34/39*, conc. range: 15–11,900 µg/kg**, sample origin: farmer's stores and local markets in Babile (district), east Ethiopia, sample year: unknown, country: Ethiopia/Norway[693], *stored and market peanuts, **AFS total

incidence: 35/40*, conc. range: 15–4,940 µg/kg**, sample origin: farmer's stores and local markets in Darolabu (district), east Ethiopia, sample year: unknown, country: Ethiopia/Norway[693], *stored and market peanuts, **AFS total

incidence: 24/41*, conc. range: 15–10,100 µg/kg**, sample origin: farmer´s stores and local markets in Gursum (district), east Ethiopia, sample year: unknown, country: Ethiopia/Norway[693], *stored and market peanuts, **AFS total
- Co-contamination: not reported
- Further contamination of plants and plant products other than cereals and their products reported in the present book/article[693]: not reported

For detailed information see the article.

incidence: 73/949, conc. range: 1–645 µg/kg*, sample origin: imported to New Zealand, sample year: 1996–2004, country: New Zealand[700], *AFB_1, AFB_2, AFG_1, and AFG_2
- Co-contamination: not reported
- Further contamination of plants and plant products other than cereals and their products reported in the present book/article[700]: Butter (peanut butter), Nuts (pistachio nut), AF/S, literature[700]

incidence: 10/50, conc. range: ≤26.6 µg/kg*, sample origin: New Zealand, sample year: 2010, country: New Zealand[701], *AFB_1, AFB_2, AFG_1, and AFG_2
- Co-contamination: not reported
- Further contamination of plants and plant products other than cereals and their products reported in the present book/article[701]: Butter (peanut butter), Nuts (pistachio nut), Products (nut product), and Products (peanut product), AF/S, literature[701]; Nuts (Brazil nut, paranut), AFB_1, literature[701]

incidence: 64/69*, conc. range: ≤501.0 µg/kg**, sample origin: informal local markets across Malawi, sample year: July/August 2012, country: Belgium/Malawi[722], *intended for consumption within the country, **AFB_1, AFB_2, AFG_1, and AFG_2

incidence: 16/27*, conc. range: ≤9.3 µg/kg**, sample origin: aggregated sa. destined for export market, Malawi, sample year: July 2012–January 2013, country: Belgium/Malawi[722], *destined for export market, **AFB_1, AFB_2, AFG_1, and AFG_2
- Co-contamination: not reported
- Further contamination of plants and plant products other than cereals and their products reported in the present book/article[722]: Butter (peanut butter), AF/S, literature[722]

incidence: 3/18, conc. range: 6.6–622.1 µg/kg, sample origin: retail shops and vendors at informal markets in Bulawayo metropolitan area, Zimbabwe, sample year: October/November 2011, country: Botswana/South Africa[797]
- Co-contamination: 1 sa. contaminated with AFB_1, AFB_2, and AFG_1, no further information available
- Further contamination of plants and plant products other than cereals and their products reported in the present book/article[797]: Butter (peanut butter), AFB_1, AFB_2, AFG_1, and AFG_2, literature[797]; Nuts (peanut), AFB_1, literature[797]

incidence: 2/3, conc. range: 0.1–10.6 µg/kg*, Ø conc.: 5.4 µg/kg*, sample origin: purchases and donations in Bamenda (city), Cameroon, sample year: December 2004–April 2005, country: South Africa/Bulgaria[804], *AFB_1, AFB_2, AFG_1, and AFG_2

incidence: 5/5, conc. range: 3.0–10.9 µg/kg*, Ø conc.: 6.8 µg/kg*, sample origin: purchases and donations in Douala (city), Cameroon, sample year: December 2004–April 2005, country: South Africa/Bulgaria[804], *AFB_1, AFB_2, AFG_1, and AFG_2

incidence: 2/4, conc. range: 0.2–10.7 µg/kg*, Ø conc.: 5.5 µg/kg*, sample origin: purchases and donations in Kumbo (city), Cameroon, sample year: December 2004–April 2005, country: South Africa/Bulgaria[804], *AFB_1, AFB_2, AFG_1, and AFG_2

incidence: 3/4, conc. range: 0.1–12.8 µg/kg*, Ø conc.: 8.1 µg/kg*, sample origin: purchases and donations in Yaounde (capital), Cameroon, sample year: 2005, country: South Africa/Bulgaria[804], *AFB_1, AFB_2, AFG_1, and AFG_2
- Co-contamination: For detailed information see the article.
- Further contamination of plants and plant products other than cereals and their products reported in the present book/article[804]: Beans, AF/S, DON, FB_1, and ZEA, literature[804]; Beans (soybean), AF/S, DON, and FB_1, literature[804]; Food, AF/S, DON and ZEA, literature[804]; Nuts (peanut), DON, FB_1, OTA, and ZEA, literature[804]

Other ***Aspergillus*** Toxins

Sterigmatocystin

incidence: 0/26, conc.: no contamination, sample origin: local markets, cereal traders, food/feed processing enterprises, grain mills, and small-scale farmers in villages and cities, Burkina Faso, sample year: April 2010, country: Austria/Nigeria[117]

incidence: 1/23, conc.: 9.7 µg/kg, sample origin: local markets, cereal traders, food/feed processing enterprises, grain mills, and small-scale farmers in the local capital and rural villages in Nampula (province), northern Mozambique, sample year: May 2010, country: Austria/Nigeria[117]
- Co-contamination: For detailed information see the article.
- Further contamination of plants and plant products other than cereals and their products reported in the present book/article[117]: Food, 3-NPA, AFB_1, AFB_2, AFG_1, AFG_2, AFM_1, AME, AOH, ATX-I, BEA, CIT, CPA, DON, ENA, ENA_1, ENB, ENB_1, ENB_2, FB_1, FB_2, FB_3, MON, NIV, OTA, STG, and ZEA, literature[117]; Nuts (peanut), 3-NPA, AFB_1, AFB_2, AFG_1, BEA, CPA, and ENB_1, literature[117]; see also Food, chapter: Further Mycotoxins and Microbial Metabolites

incidence: 5/15, conc. range: <LOQ, Ø conc.: 1.2 µg/kg, sample origin: Osmaniye (province), Turkey, sample year: 2007, country: Austria[868]
- Co-contamination: For detailed information see the article.
- Further contamination of plants and plant products other than cereals and their products reported in the present book/article[868]: Nuts (almond), AME, AOH, BEA, CPA, ENB, ENB_1, and ENB_3, literature[868]; Nuts (hazelnut), 3-NPA, AFB_1, AFB_2,

AFG_1, AFG_2, AME, AOH, ATX-I, BEA, ENA, ENA_1, ENB_1, ENB_2, ENB_3, ENB_4, HT-2, KA, MA, MON, OTA, OTB, STG, T-2, and ZEA, literature[868]; Nuts (peanut), 3-NPA, AFB_1, AFB_2, AME, AOH, BEA, CPA, ENA, ENA_1, ENB, ENB_4, KA, MA, OTA, OTB, and ZEA, literature[868]; Nuts (pistachio nut), BEA, MA, and OTA, literature[868]; see also Nuts (almond), Nuts (hazelnut), Nuts (peanut), Nuts (pistachio nut), chapter: Further Mycotoxins and Microbial Metabolites

For detailed information see the article.

Aspergillus and ***Penicillium*** Toxins

3-Nitropropionic Acid

incidence: 0/9, conc.: no contamination, sample origin: local markets, cereal traders, food/feed processing enterprises, grain mills, and small-scale farmers in villages and cities, Burkina Faso, sample year: April 2010, country: Austria/Nigeria[117]

incidence: 2/23, conc. range: 223–1,349 µg/kg, ∅ conc.: 786 µg/kg, sample origin: local markets, cereal traders, food/feed processing enterprises, grain mills, and small-scale farmers in the local capital and rural villages in Nampula (province), northern Mozambique, sample year: May 2010, country: Austria/Nigeria[117]

- Co-contamination: For detailed information see the article.
- Further contamination of plants and plant products other than cereals and their products reported in the present book/article[117]: Food, 3-NPA, AFB_1, AFB_2, AFG_1, AFG_2, AFM_1, AME, AOH, ATX-I, BEA, CIT, CPA, DON, ENA, ENA_1, ENB, ENB_1, ENB_2, FB_1, FB_2, FB_3, MON, NIV, OTA, STG, and ZEA, literature[117]; Nuts (peanut), AFB_1, AFB_2, AFG_1, BEA, CPA, ENB_1, and STG, literature[117]; see also Food, chapter: Further Mycotoxins and Microbial Metabolites

incidence: 7/15, conc. range: ≤350 µg/kg, ∅ conc.: 82 µg/kg, sample origin: Osmaniye (province), Turkey, sample year: 2007, country: Austria[868]

- Co-contamination: For detailed information see the article.
- Further contamination of plants and plant products other than cereals and their products reported in the present book/article[868]: Nuts (almond), AME, AOH, BEA, CPA, ENB, ENB_1, and ENB_3, literature[868]; Nuts (hazelnut), 3-NPA, AFB_1, AFB_2, AFG_1, AFG_2, AME, AOH, ATX-I, BEA, ENA, ENA_1, ENB, ENB_1, ENB_2, ENB_3, ENB_4, HT-2, KA, MA, MON, OTA, OTB, STG, T-2, and ZEA, literature[868]; Nuts (peanut), AFB_1, AFB_2, AME, AOH, ATX-I, BEA, CPA, ENA, ENA_1, ENB, ENB_4, KA, MA, OTA, OTB, STG, and ZEA, literature[868]; Nuts (pistachio nut), BEA, MA, and OTA, literature[868]; see also Nuts (almond), Nuts (hazelnut), Nuts (peanut), Nuts (pistachio nut), chapter: Further Mycotoxins and Microbial Metabolites

For detailed information see the article.

Citrinin

incidence: 16/160*, conc. range: tr-1,200 µg/kg, sample origin: three fields around Tirupati (city), Andhra Pradesh (state), India, sample year: November 1972, country: India[427], *ncac

- Co-contamination: 16 sa. co-contaminated with AFS and CIT
- Further contamination of plants and plant products other than cereals and their products reported in the present book/article[427]: Nuts (peanut), AF/S, literature[427]

Cyclopiazonic Acid

incidence: 0/9, conc.: no contamination, sample origin: local markets, cereal traders, food/feed processing enterprises, grain mills, and small-scale farmers in villages and cities, Burkina Faso, sample year: April 2010, country: Austria/Nigeria[117]

incidence: 1/23, conc.: 763 µg/kg, sample origin: local markets, cereal traders, food/feed processing enterprises, grain mills, and small-scale farmers in the local capital and rural villages in Nampula (province), northern Mozambique, sample year: May 2010, country: Austria/Nigeria[117]

- Co-contamination: For detailed information see the article.
- Further contamination of plants and plant products other than cereals and their products reported in the present book/article[117]: Food, 3-NPA, AFB_1, AFB_2, AFG_1, AFG_2, AFM_1, AME, AOH, ATX-I, BEA, CIT, CPA, DON, ENA, ENA_1, ENB, ENB_1, ENB_2, FB_1, FB_2, FB_3, MON, NIV, OTA, STG, and ZEA, literature[117]; Nuts (peanut), 3-NPA, AFB_1, AFB_2, AFG_1, BEA, ENB_1, and STG, literature[117]; see also Food, chapter: Further Mycotoxins and Microbial Metabolites

incidence: 5/84, conc. range: 9.6–114 µg/kg, ∅ conc.: 65.8 µg/kg, sample origin: four agroecological zones, Nigeria, sample year: May–July 2015, country: Nigeria/Austria[148]

- Co-contamination: For detailed information see the article.
- Further contamination of plants and plant products other than cereals and their products reported in the present book/article[148]: Nuts (peanut), AFB_1, AFB_2, AFG_1, AFG_2, AFM_1, AME, AOH, BEA, MON, NIV, and OTA, literature[148]; see also Nuts (peanut), chapter: Further Mycotoxins and Microbial Metabolites

For detailed information see the article.

incidence: 18/83, conc. range: 1–10 µg/kg, ∅ conc.: 2 µg/kg, sample origin: retail outlets, Botswana, sample year: July 2001–January 2002, country: Botswana[390]

- Co-contamination: not reported
- Further contamination of plants and plant products other than cereals and their products reported in the present book/article[390]: Nuts (peanut), AFB_1, AFB_2, AFG_1, and AFG_2, literature[390]

incidence: 45/50*, conc. range: tr-2,926 µg/kg, sample origin: Georgia (state), USA, sample year: 1990, country: USA[396], *peanuts for oil production or feed

- Co-contamination: 14 sa. co-contaminated with AFB_1, AFB_2, AFG_1, AFG_2, and CPA; 12 sa. co-contaminated with AFB_1, AFB_2, AFG_1, and CPA; 19 sa. co-contaminated with AFB_1, AFB_2, and CPA
- Further contamination of plants and plant products other than cereals and their products reported in the present book/article[396]: Nuts (peanut), AFB_1, AFB_2, AFG_1, and AFG_2, literature[396]

incidence: 21/27* **, conc. range: 32–6,525 µg/kg, sample origin: southeastern of the USA, sample year: 1980, country: USA[426], *ncac, **loose-shell kernel fractions

incidence: 4/21* **, conc. range: 32–130 µg/kg, sample origin: southeastern of the USA, sample year: 1980, country: USA[426], *ncac, **sound mature kernel fractions

- Co-contamination: not reported
- Further contamination of plants and plant products other than cereals and their products reported in the present book/article[426]: Nuts (peanut), AF/S, literature[426]

incidence: 4/15, conc. range: ≤310 µg/kg, ∅ conc.: 140 µg/kg, sample origin: Osmaniye (province), Turkey, sample year: 2007, country: Austria[868]

- Co-contamination: For detailed information see the article.
- Further contamination of plants and plant products other than cereals and their products reported in the present book/article[868]: Nuts (almond), AME, AOH, BEA, CPA, ENB, ENB_1, and ENB_3, literature[868]; Nuts (hazelnut), 3-NPA, AFB_1, AFB_2, AFG_1, AFG_2, AME, AOH, ATX-I, BEA, ENA, ENA_1, ENB_1, ENB_2, ENB_3, ENB_4, HT-2, KA, MA, MON, OTA, OTB, STG, T-2, and ZEA, literature[868]; Nuts (peanut), 3-NPA, AFB_1, AFB_2, AME, AOH, ATX-I, BEA, ENA, ENA_1, ENB, ENB_4, KA, MA, OTA, OTB, STG, and ZEA, literature[868]; Nuts (pistachio nut), BEA, MA, and OTA, literature[868]; see also Nuts (almond), Nuts (hazelnut), Nuts (peanut), Nuts (pistachio nut), chapter: Further Mycotoxins and Microbial Metabolites

For detailed information see the article.

Kojic Acid

incidence: 6/15, conc. range: ≤40,000 µg/kg, ∅ conc.: 8,900 µg/kg, sample origin: Osmaniye (province), Turkey, sample year: 2007, country: Austria[868]

- Co-contamination: For detailed information see the article.
- Further contamination of plants and plant products other than cereals and their products reported in the present book/article[868]: Nuts (almond), AME, AOH, BEA, CPA, ENB, ENB_1, and ENB_3, literature[868]; Nuts (hazelnut), 3-NPA, AFB_1, AFB_2, AFG_1, AFG_2, AME, AOH, ATX-I, BEA, ENA, ENA_1, ENB_1, ENB_2, ENB_3, ENB_4, HT-2, KA, MA, MON, OTA, OTB, STG, T-2, and ZEA, literature[868]; Nuts (peanut), 3-NPA, AFB_1, AFB_2, AME, AOH, ATX-I, BEA, CPA, ENA, ENA_1, ENB, ENB_4, MA, OTA, OTB, STG, and ZEA, literature[868]; Nuts (pistachio nut), BEA, MA, and OTA, literature[868]; see also Nuts (almond), Nuts (hazelnut), Nuts (peanut), Nuts (pistachio nut), chapter: Further Mycotoxins and Microbial Metabolites

For detailed information see the article.

Ochratoxin A

incidence: 2/6, conc. range: pr, sample origin: major local retail outlets at Nainital (district), India, sample year: 1983–1985, country: India[7]

- Co-contamination: Often more than one mycotoxin was present in a sa.
- Further contamination of plants and plant products other than cereals and their products reported in the present book/article[7]: Coconuts, AFB_1, AFB_2, AFG_1, AFG_2, and CIT, literature[7]; Dates, AFB_1, AFB_2, AFG_1, and OTA, literature[7]; Grogannuts, AFB_1, AFB_2, AFG_1, AFG_2, and CIT, literature[7]; Nuts (almond), AFB_1, AFB_2, AFG_1, RBT, STG, and ZEA, literature[7]; Nuts (cashew nut), AFB_1, AFB_2, AFG_1, AFG_2, CIT, and OTA, literature[7]; Nuts (peanut), AFB_1, AFB_2, AFG_1, and AFG_2, literature[7]; Nuts (pistachio nut), AFB_1, AFB_2, AFG_1, AFG_2, and STG, literature[7]; Nuts (walnut), AFB_1, AFB_2, AFG_1, AFG_2, and RBT, literature[7]; Raisins, AFB_1, AFB_2, AFG_1, AFG_2, and OTA, literature[7]

incidence: 10/42, conc. range: 15–160 µg/kg, ∅ conc.: 60 µg/kg, sample origin: retail markets and supermarkets in 24 regions, Tunisia, sample year: unknown, country: Tunisia[20]

- Co-contamination: not reported
- Further contamination of plants and plant products other than cereals and their products reported in the present book/article[20]: Nuts (almond), Nuts (pistachio nut), Spices (caraway), Spices (coriander), Spices (curcuma), Spices (paprika), and Spices (pepper), OTA, literature[20]

incidence: 1/1, conc.: 7.65 µg/kg, sample origin: local supermarket, Japan, sample year: unknown, country: Japan[21]

- Co-contamination: 1 sa. co-contaminated with OTA and OTB
- Further contamination of plants and plant products other than cereals and their products reported in the present book/article[21]: Nuts (almond), Nuts (cashew nut), OTA and OTB, literature[21]; Nuts (peanut), OTB, literature[21]

incidence: 2/84, conc. range: 1–3 µg/kg, ∅ conc.: 2 µg/kg, sample origin: four agroecological zones, Nigeria, sample year: May–July 2015, country: Nigeria/Austria[148]

- Co-contamination: For detailed information see the article.
- Further contamination of plants and plant products other than cereals and their products reported in the present book/article[148]: Nuts (peanut), AFB_1, AFB_2, AFG_1, AFG_2, AFM_1, AME, AOH, BEA, CPA, MON, and NIV, literature[148]; see also Nuts (peanut), chapter: Further Mycotoxins and Microbial Metabolites

For detailed information see the article.

incidence: 3/31, conc. range: ≤0.080 µg/kg, sample origin: regions around Detmold, Giessen, Hamburg, Jena, Karlsruhe, Kiel, Kulmbach, and Trier (cities), Germany, sample year: unknown, country: Germany[262]

- Co-contamination: not reported
- Further contamination of plants and plant products other than cereals and their products reported in the present book/article[262]: Bars, Chocolate, Drinks (chocolate drink), Figs, Fruits, dried, Nut crème, Nuts, Nuts (hazelnut), Powder (cocoa powder), Pudding, Raisins, Seeds (linseed), Seeds (sesame), Seeds (sunflower), OTA, literature[262]

incidence: 5/20, conc. range: 0.10–2.36 µg/kg, ∅ conc.: 0.68 µg/kg, sample origin: retail shops in Salé (city) and Rabat (capital), Morocco, sample year: July/August 2005, country: Morocco/Spain[364]

- Co-contamination: not reported
- Further contamination of plants and plant products other than cereals and their products reported in the present book/article[364]: Figs, Nuts (walnut), Raisins, OTA, literature[364]

incidence: 6/10, conc. range: 0.20–0.64 µg/kg, ∅ conc.: 0.32 µg/kg, sample origin: markets in Adjamé, Abobo, and Treichville (communes) of Abidjan, Ivory Coast, sample year: November 2002, country: France/Ivory Coast[394]

- Co-contamination: 3 sa. co-contaminated with AFB_1, FB_1, OTA, and ZEA; 3 sa. co-contaminated with AFB_1, OTA, and ZEA
- Further contamination of plants and plant products other than cereals and their products reported in the present book/article[394]: Nuts (peanut), AFB_1, FB_1, and ZEA, literature[394]

incidence: 20/40*, conc. range: 5.5–46.7 µg/kg, ∅ conc.: 27.7 µg/kg, sample origin: storage plant in the south of Córdoba (province), Spain, sample year: May–July 2004, country: Argentina[428], *sa. collected after 1 month of storage

incidence: 20/30*, conc. range: 2.0–11.0 µg/kg, ∅ conc.: 6.63 µg/kg, sample origin: storage plant in the south of Córdoba (province), Spain, sample year: May–July 2004, country: Argentina[428], *sa. collected after 2 months of storage

incidence: 20/30*, conc. range: 4.5–32.0 µg/kg, ∅ conc.: 12.85 µg/kg, sample origin: storage plant in the south of Córdoba (province), Spain, sample year: May–July 2004, country: Argentina[428], *sa. collected after 3 months of storage

- Co-contamination: not reported

- Further contamination of plants and plant products other than cereals and their products reported in the present book/article[428]: not reported

incidence: 6/10, conc. range: 0.6–64 µg/kg, sample origin: markets in Adjamé, Abobo, and Treichville (communes) in Abidjan (city), Ivory Coast, sample year: 2002, country: France/Ivory Coast[429]
- Co-contamination: not reported
- Further contamination of plants and plant products other than cereals and their products reported in the present book/article[429]: not reported

incidence: 2/18, conc. range: 50–2,000 µg/kg, ∅ conc.: 1,025 µg/kg, sample origin: grocery shops in Mysore (city), India, sample year: unknown, country: India[430]
- Co-contamination: not reported
- Further contamination of plants and plant products other than cereals and their products reported in the present book/article[430]: not reported

incidence: 15*/47**, conc. range: ≤170 µg/kg, sample origin: silo of a storage plant in the south of Córdoba (province), Argentina, sample year: unknown, country: Argentina[431], *14 sa. with ≤2.5 µg/kg, **stored "Runner" variety peanut sa.
- Co-contamination: not reported
- Further contamination of plants and plant products other than cereals and their products reported in the present book/article[431]: not reported

incidence: 30/72, conc. range: 0.084–0.774 µg/kg, ∅ conc.: 0.214 µg/kg, sample origin: hypermarkets and supermarkets of 12 cities, Catalonia (autonomous community), Spain, sample year: June–November 2008, country: Spain[432]
- Co-contamination: not reported
- Further contamination of plants and plant products other than cereals and their products reported in the present book/article[432]: Nuts (pistachio nut) OTA, literature[432]

incidence: 4/4*, conc. range: 3.5–21.3 µg/kg, ∅ conc.: 11.2 µg/kg, sample origin: unknown, sample year: unknown, country: China[775], *ncac
- Co-contamination: 4 sa. co-contaminated with AFB_1, OTA, and ZEA
- Further contamination of plants and plant products other than cereals and their products reported in the present book/article[775]: Nuts (peanut), AFB_1 and ZEA, literature[775]

incidence: 0/3, conc.: no contamination, sample origin: purchases and donations in Bamenda (city), Cameroon, sample year: December 2004–April 2005, country: South Africa/Bulgaria[804]

incidence: 1/5, conc.: 2.1 µg/kg, sample origin: purchases and donations in Douala (city), Cameroon, sample year: December 2004–April 2005, country: South Africa/Bulgaria[804]

incidence: 0/4, conc.: no contamination, sample origin: purchases and donations in Kumbo (city), Cameroon, sample year: December 2004–April 2005, country: South Africa/Bulgaria[804]

incidence: 0/4, conc.: no contamination, sample origin: purchases and donations in Yaounde (capital), Cameroon, sample year: 2005, country: South Africa/Bulgaria[804]
- Co-contamination: For detailed information see the article.
- Further contamination of plants and plant products other than cereals and their products reported in the present book/article[804]: Beans, AF/S, DON, FB_1, and ZEA, literature[804]; Beans (soybean), AF/S, DON, and FB_1, literature[804]; Food, AF/S, DON and ZEA, literature[804]; Nuts (peanut), AF/S, DON, FB_1, and ZEA, literature[804]

incidence: 12/90, conc. range: 0.3–12 µg/kg, sample origin: humid forest with monomodal rainfall, humid forest with bimodal rainfall, and western highland, Cameroon, sample year: July/August 2009 (rainy season) and December 2010–January 2011 (dry season), country: Belgium/Benin[807]
- Co-contamination: For detailed information see the article.
- Further contamination of cereals and cereal products reported in the present book/article[807]: Nuts (peanut), AFB_1, literature[807]; Products (cassava product), AFB_1 and PA, literature[807]

For detailed information see the article.

incidence: 6/7, conc. range: ≤5 µg/kg, ∅ conc.: 3.7 µg/kg, sample origin: local markets in Niamey (capital), Niger, sample year: 2011, country: Morocco[858]
- Co-contamination: not reported
- Further contamination of cereals and cereal products reported in the present book/article[858]: not reported

incidence: 6/15, conc. range: ≤260 µg/kg, ∅ conc.: 67 µg/kg, sample origin: Osmaniye (province), Turkey, sample year: 2007, country: Austria[868]
- Co-contamination: For detailed information see the article.
- Further contamination of plants and plant products other than cereals and their products reported in the present book/article[868]: Nuts (almond), AME, AOH, BEA, CPA, ENB, ENB_1, and ENB_3, literature[868]; Nuts (hazelnut), 3-NPA, AFB_1, AFB_2, AFG_1, AFG_2, AME, AOH, ATX-I, BEA, ENA, ENA_1, ENB_1, ENB_2, ENB_3, ENB_4, HT-2, KA, MA, MON, OTA, OTB, STG, T-2, and ZEA, literature[868]; Nuts (peanut), 3-NPA, AFB_1, AFB_2, AME, AOH, ATX-I, BEA, CPA, ENA, ENA_1, ENB, ENB_4, KA, MA, OTB, STG, and ZEA, literature[868]; Nuts (pistachio nut), BEA, MA, and OTA, literature[868]; see also Nuts (almond), Nuts (hazelnut), Nuts (peanut), Nuts (pistachio nut), chapter: Further Mycotoxins and Microbial Metabolites

For detailed information see the article.

incidence: 9/30, conc. range: ≤20 µg/kg (9 sa., maximum: 8.34 µg/kg), sample origin: local markets in Wuhan (city), China, sample year: unknown, country: China[881]
- Co-contamination: not reported
- Further contamination of plants and plant products other than cereals and their products reported in the present book/article[881]: Nuts (peanut), AFB_1, AFB_2, AFG_1, AFG_2, T-2, and ZEA, literature[881]

Ochratoxin B

incidence: 1/1, conc.: 2.56 µg/kg, sample origin: local supermarket, Japan, sample year: unknown, country: Japan[21]
- Co-contamination: 1 sa. co-contaminated with OTA and OTB
- Further contamination of plants and plant products other than cereals and their products reported in the present book/article[21]: Nuts (almond), Nuts (cashew nut), OTA and OTB, literature[21]; Nuts (peanut), OTA, literature[21]

incidence: 5/15, conc. range: ≤82 µg/kg, ∅ conc.: 31 µg/kg, sample origin: Osmaniye (province), Turkey, sample year: 2007, country: Austria[868]
- Co-contamination: For detailed information see the article.
- Further contamination of plants and plant products other than cereals and their products reported in the present book/article[868]: Nuts (almond), AME, AOH, BEA, CPA, ENB, ENB_1,

and ENB_3, literature[868]; Nuts (hazelnut), 3-NPA, AFB_1, AFB_2, AFG_1, AFG_2, AME, AOH, ATX-I, BEA, ENA, ENA_1, ENB_1, ENB_2, ENB_3, ENB_4, HT-2, KA, MA, MON, OTA, OTB, STG, T-2, and ZEA, literature[868]; Nuts (peanut), 3-NPA, AFB_1, AFB_2, AME, AOH, ATX-I, BEA, CPA, ENA, ENA_1, ENB, ENB_4, KA, MA, OTA, STG, and ZEA, literature[868]; Nuts (pistachio nut), BEA, MA, and OTA, literature[868]; see also Nuts (almond), Nuts (hazelnut), Nuts (peanut), Nuts (pistachio nut), chapter: Further Mycotoxins and Microbial Metabolites

For detailed information see the article.

Fusarium Toxins

Depsipeptides

Beauvericin

incidence: 1/26, conc.: 0.1 µg/kg, sample origin: local markets, cereal traders, food/feed processing enterprises, grain mills, and small-scale farmers in villages and cities, Burkina Faso, sample year: April 2010, country: Austria/Nigeria[117]

incidence: 16/23, conc. range: 0.1–24.0 µg/kg, sample origin: local markets, cereal traders, food/feed processing enterprises, grain mills, and small-scale farmers in the local capital and rural villages in Nampula (province), northern Mozambique, sample year: May 2010, country: Austria/Nigeria[117]

- Co-contamination: For detailed information see the article.
- Further contamination of plants and plant products other than cereals and their products reported in the present book/article[117]: Food, 3-NPA, AFB_1, AFB_2, AFG_1, AFG_2, AFM_1, AME, AOH, ATX-I, BEA, CIT, CPA, DON, ENA, ENA_1, ENB, ENB_1, ENB_2, FB_1, FB_2, FB_3, MON, NIV, OTA, STG, and ZEA, literature[117]; Nuts (peanut), 3-NPA, AFB_1, AFB_2, AFG_1, CPA, ENB_1, and STG, literature[117]; see also Food, chapter: Further Mycotoxins and Microbial Metabolites

incidence: 29/84, conc. range: 0.1–6.8 µg/kg, Ø conc.: 1.3 µg/kg, sample origin: four agroecological zones, Nigeria, sample year: May–July 2015, country: Nigeria/Austria[148]

- Co-contamination: For detailed information see the article.
- Further contamination of plants and plant products other than cereals and their products reported in the present book/article[148]: Nuts (peanut), AFB_1, AFB_2, AFG_1, AFG_2, AFM_1, AME, AOH; CPA, MON, NIV, and OTA, literature[148]; see also Nuts (peanut), chapter: Further Mycotoxins and Microbial Metabolites

For detailed information see the article.

incidence: 15/15, conc. range: ≤12 µg/kg, Ø conc.: 1.6 µg/kg, sample origin: Osmaniye (province), Turkey, sample year: 2007, country: Austria[868]

- Co-contamination: For detailed information see the article.
- Further contamination of plants and plant products other than cereals and their products reported in the present book/article[868]: Nuts (almond), AME, AOH, BEA, CPA, ENB, ENB_1, and ENB_3, literature[868]; Nuts (hazelnut), 3-NPA, AFB_1, AFB_2, AFG_1, AFG_2, AME, AOH, ATX-I, BEA, ENA, ENA_1, ENB_1, ENB_2, ENB_3, ENB_4, HT-2, KA, MA, MON, OTA, OTB, STG, T-2, and ZEA, literature[868]; Nuts (peanut), 3-NPA, AFB_1, AFB_2, AME, AOH, ATX-I, CPA, ENA, ENA_1, ENB, ENB_4, KA, MA, OTA, OTB, STG, and ZEA, literature[868]; Nuts (pistachio nut), BEA, MA, and OTA, literature[868]; see also Nuts (almond), Nuts (hazelnut), Nuts (peanut), Nuts (pistachio nut), chapter: Further Mycotoxins and Microbial Metabolites

For detailed information see the article.

Enniatin A

incidence: 3/15, conc. range: <LOQ, Ø conc.: 1.4 µg/kg, sample origin: Osmaniye (province), Turkey, sample year: 2007, country: Austria[868]

- Co-contamination: For detailed information see the article.
- Further contamination of plants and plant products other than cereals and their products reported in the present book/article[868]: Nuts (almond), AME, AOH, BEA, CPA, ENB, ENB_1, and ENB_3, literature[868]; Nuts (hazelnut), 3-NPA, AFB_1, AFB_2, AFG_1, AFG_2, AME, AOH, ATX-I, BEA, ENA, ENA_1, ENB_1, ENB_2, ENB_3, ENB_4, HT-2, KA, MA, MON, OTA, OTB, STG, T-2, and ZEA, literature[868]; Nuts (peanut), 3-NPA, AFB_1, AFB_2, AME, AOH, ATX-I, BEA, CPA, ENA_1, ENB, ENB_4, KA, MA, OTA, OTB, STG, and ZEA, literature[868]; Nuts (pistachio nut), BEA, MA, and OTA, literature[868]; see also Nuts (almond), Nuts (hazelnut), Nuts (peanut), Nuts (pistachio nut), chapter: Further Mycotoxins and Microbial Metabolites

For detailed information see the article.

Enniatin A_1

incidence: 5/15, conc. range: <LOQ, Ø conc.: 0.70 µg/kg, sample origin: Osmaniye (province), Turkey, sample year: 2007, country: Austria[868]

- Co-contamination: For detailed information see the article.
- Further contamination of plants and plant products other than cereals and their products reported in the present book/article[868]: Nuts (almond), AME, AOH, BEA, CPA, ENB, ENB_1, and ENB_3, literature[868]; Nuts (hazelnut), 3-NPA, AFB_1, AFB_2, AFG_1, AFG_2, AME, AOH, ATX-I, BEA, ENA, ENA_1, ENB_1, ENB_2, ENB_3, ENB_4, HT-2, KA, MA, MON, OTA, OTB, STG, T-2, and ZEA, literature[868]; Nuts (peanut), 3-NPA, AFB_1, AFB_2, AME, AOH, ATX-I, BEA, CPA, ENA, ENB, ENB_4, KA, MA, OTA, OTB, STG, and ZEA, literature[868]; Nuts (pistachio nut), BEA, MA, and OTA, literature[868]; see also Nuts (almond), Nuts (hazelnut), Nuts (peanut), Nuts (pistachio nut), chapter: Further Mycotoxins and Microbial Metabolites

For detailed information see the article.

Enniatin B

incidence: 8/15, conc. range: ≤0.45 µg/kg, Ø conc.: 0.23 µg/kg, sample origin: Osmaniye (province), Turkey, sample year: 2007, country: Austria[868]

- Co-contamination: For detailed information see the article.
- Further contamination of plants and plant products other than cereals and their products reported in the present book/article[868]: Nuts (almond), AME, AOH, BEA, CPA, ENB, ENB_1, and ENB_3, literature[868]; Nuts (hazelnut), 3-NPA, AFB_1, AFB_2, AFG_1, AFG_2, AME, AOH, ATX-I, BEA, ENA, ENA_1, ENB_1, ENB_2, ENB_3, ENB_4, HT-2, KA, MA, MON, OTA, OTB, STG, T-2, and ZEA, literature[868]; Nuts (peanut), 3-NPA, AFB_1, AFB_2, AME, AOH, ATX-I, BEA, CPA, ENA, ENA_1, ENB_4, KA, MA, OTA, OTB, STG, and ZEA, literature[868]; Nuts (pistachio nut), BEA, MA, and OTA, literature[868]; see also Nuts (almond), Nuts (hazelnut), Nuts (peanut), Nuts (pistachio nut), chapter: Further Mycotoxins and Microbial Metabolites

For detailed information see the article.

Enniatin B_1

incidence: 0/9, conc.: no contamination, sample origin: local markets, cereal traders, food/feed processing enterprises, grain mills, and small-scale farmers in villages and cities, Burkina Faso, sample year: April 2010, country: Austria/Nigeria[117] incidence: 1/23, conc.: 0.3 µg/kg, sample origin: local markets, cereal traders, food/feed processing enterprises, grain mills, and small-scale farmers in the local capital and rural villages in Nampula (province), northern Mozambique, sample year: May 2010, country: Austria/Nigeria[117]

- Co-contamination: For detailed information see the article.
- Further contamination of plants and plant products other than cereals and their products reported in the present book/article[117]: Food, 3-NPA, AFB_1, AFB_2, AFG_1, AFG_2, AFM_1, AME, AOH, ATX-I, BEA, CIT, CPA, DON, ENA, ENA_1, ENB, ENB_1, ENB_2, FB_1, FB_2, FB_3, MON, NIV, OTA, STG, and ZEA, literature[117]; Nuts (peanut), 3-NPA, AFB_1, AFB_2, AFG_1, BEA, CPA, and STG, literature[117]; see also Food, chapter: Further Mycotoxins and Microbial Metabolites

Enniatin B_4

incidence: 5/15, conc. range: <LOQ, ∅ conc.: 0.1 µg/kg, sample origin: Osmaniye (province), Turkey, sample year: 2007, country: Austria[868]

- Co-contamination: For detailed information see the article.
- Further contamination of plants and plant products other than cereals and their products reported in the present book/article[868]: Nuts (almond), AME, AOH, BEA, CPA, ENB, ENB_1, and ENB_3, literature[868]; Nuts (hazelnut), 3-NPA, AFB_1, AFB_2, AFG_1, AFG_2, AME, AOH, ATX-I, BEA, ENA, ENA_1, ENB, ENB_1, ENB_2, ENB_3, ENB_4, HT-2, KA, MA, MON, OTA, OTB, STG, T-2, and ZEA, literature[868]; Nuts (peanut), 3-NPA, AFB_1, AFB_2, AME, AOH, BEA, CPA, ENA, ENA_1, ENB, KA, MA, OTA, OTB, STG, and ZEA, literature[868]; Nuts (pistachio nut), BEA, MA, and OTA, literature[868]; see also Nuts (almond), Nuts (hazelnut), Nuts (peanut), Nuts (pistachio nut), chapter: Further Mycotoxins and Microbial Metabolites

For detailed information see the article.

Fumonisin/s

Fumonisin B_1

incidence: 7/10, conc. range: <300–600 µg/kg, sample origin: markets in Adjamé, Abobo, and Treichville (communes) of Abidjan, Ivory Coast, sample year: November 2002, country: France/Ivory Coast[394]

- Co-contamination: 3 sa. co-contaminated with AFB_1, FB_1, OTA, and ZEA; 4 sa. co-contaminated with AFB_1, FB_1, and ZEA
- Further contamination of plants and plant products other than cereals and their products reported in the present book/article[394]: Nuts (peanut), AFB_1, OTA, and ZEA, literature[394]

incidence: 1/5, conc.: 1,000 µg/kg, sample origin: throughout Zimbabwe, sample year: April 1995–March 1997, country: Zimbabwe/Belgium[433]

- Co-contamination: not reported
- Further contamination of plants and plant products other than cereals and their products reported in the present book/article[433]: not reported

incidence: 0/3, conc.: no contamination, sample origin: purchases and donations in Bamenda (city), Cameroon, sample year: December 2004–April 2005, country: South Africa/Bulgaria[804]

incidence: 1/5, conc.: 29 µg/kg, sample origin: purchases and donations in Douala (city), Cameroon, sample year: December 2004–April 2005, country: South Africa/Bulgaria[804]

incidence: 1/4, conc.: 25 µg/kg, sample origin: purchases and donations in Kumbo (city), Cameroon, sample year: December 2004–April 2005, country: South Africa/Bulgaria[804]

incidence: 1/4, conc.: 1,498 µg/kg, sample origin: purchases and donations in Yaounde (capital), Cameroon, sample year: 2005, country: South Africa/Bulgaria[804]

- Co-contamination: For detailed information see the article.
- Further contamination of plants and plant products other than cereals and their products reported in the present book/article[804]: Beans, AF/S, DON, FB_1, and ZEA, literature[804]; Beans (soybean), AF/S, DON, and FB_1, literature[804]; Food, AF/S, DON and ZEA, literature[804]; Nuts (peanut), AF/S, DON, OTA, and ZEA, literature[804]

Type A Trichothecenes

Diacetoxyscirpenol

incidence: 2/25*, conc. range: 410–2,030 µg/kg, ∅ conc.: 1,220 µg/kg, sample origin: wholesale market in Mysore (city), India, sample year: February–May 1985, country: India[434], *ncac

- Co-contamination: not reported
- Further contamination of plants and plant products other than cereals and their products reported in the present book/article[434]: Nuts (peanut), T-2, literature[434]

T-2 Toxin

incidence: 3/31*, conc. range: 170–38,890 µg/kg, ∅ conc.: 25,270 µg/kg, sample origin: wholesale market in Mysore (city), India, sample year: October 1984–January 1985, country: India[434], *ncac

incidence: 3/25*, conc. range: 630–1,460 µg/kg, ∅ conc.: 1,043.3 µg/kg, sample origin: wholesale market in Mysore (city), India, sample year: February–May 1985, country: India[434], *ncac

- Co-contamination: not reported
- Further contamination of plants and plant products other than cereals and their products reported in the present book/article[434]: Nuts (peanut), DAS, literature[434]

incidence: 7/30, conc. range: ≤20 µg/kg (7 sa., maximum: 7.82 µg/kg), sample origin: local markets in Wuhan (city), China, sample year: unknown, country: China[881]

- Co-contamination: not reported
- Further contamination of plants and plant products other than cereals and their products reported in the present book/article[881]: Nuts (peanut), AFB_1, AFB_2, AFG_1, AFG_2, OTA, and T-2, literature[881]

Type B Trichothecenes

Deoxynivalenol

incidence: 1/3, conc.: 270 µg/kg, sample origin: purchases and donations in Bamenda (city), Cameroon, sample year: December 2004–April 2005, country: South Africa/Bulgaria[804]

incidence: 5/5, conc. range: 17–138 µg/kg, ∅ conc.: 54 µg/kg, sample origin: purchases and donations in Douala (city), Cameroon, sample year: December 2004–April 2005, country: South Africa/Bulgaria[804]

incidence: 3/4, conc. range: 21–167 µg/kg, ∅ conc.: 59 µg/kg, sample origin: purchases and donations in Kumbo (city), Cameroon, sample year: December 2004–April 2005, country: South Africa/Bulgaria[804] incidence: 3/4, conc. range: 47–217 µg/kg, ∅ conc.: 108 µg/kg, sample origin: purchases and donations in Yaounde (capital), Cameroon, sample year: 2005, country: South Africa/Bulgaria[804]

- Co-contamination: For detailed information see the article.
- Further contamination of plants and plant products other than cereals and their products reported in the present book/ article[804]: Beans, AF/S, DON, FB_1, and ZEA, literature[804]; Beans (soybean), AF/S, DON, and FB_1, literature[804]; Food, AF/S, DON and ZEA, literature[804]; Nuts (peanut), AF/S, FB_1, OTA, and ZEA, literature[804]

Nivalenol

incidence: 3/84, conc. range: 4.5–13.4 µg/kg, ∅ conc.: 8.4 µg/kg, sample origin: four agroecological zones, Nigeria, sample year: May–July 2015, country: Nigeria/Austria[148]

- Co-contamination: For detailed information see the article.
- Further contamination of plants and plant products other than cereals and their products reported in the present book/ article[148]: Nuts (peanut), AFB_1, AFB_2, AFG_1, AFG_2, AFM_1, AME, AOH, BEA, CPA, MON, and OTA, literature[148]; see also Nuts (peanut), chapter: Further Mycotoxins and Microbial Metabolites

For detailed information see the article.

Other ***Fusarium*** Toxins

Moniliformin

incidence: 2/84, conc. range: 7.9–8.8 µg/kg, ∅ conc.: 8.4 µg/kg, sample origin: four agroecological zones, Nigeria, sample year: May–July 2015, country: Nigeria/Austria[148]

- Co-contamination: For detailed information see the article.
- Further contamination of plants and plant products other than cereals and their products reported in the present book/ article[148]: Nuts (peanut), AFB_1, AFB_2, AFG_1, AFG_2, AFM_1, AME, AOH, BEA, CPA, NIV, and OTA, literature[148]; see also Nuts (peanut), chapter: Further Mycotoxins and Microbial Metabolites

For detailed information see the article.

Zearalenone

incidence: 10/10, conc. range: 50–200 µg/kg, ∅ conc.: 95 µg/kg, sample origin: markets in Adjamé, Abobo, and Treichville (communes) of Abidjan, Ivory Coast, sample year: November 2002, country: France/ Ivory Coast[394]

- Co-contamination: 3 sa. co-contaminated with AFB_1, FB_1, OTA, and ZEA; 4 sa. co-contaminated with AFB_1, FB_1, and ZEA; 3 sa. co-contaminated with AFB_1, OTA, and ZEA
- Further contamination of plants and plant products other than cereals and their products reported in the present book/ article[394]: Nuts (peanut), AFB_1, FB_1, and OTA, literature[394]

incidence: 4/4*, conc. range: 28.6–71.7 µg/kg, ∅ conc.: 47.8 µg/kg, sample origin: unknown, sample year: unknown, country: China[775], *ncac

- Co-contamination: 4 sa. co-contaminated with AFB_1, OTA, and ZEA
- Further contamination of plants and plant products other than cereals and their products reported in the present book/ article[775]: Nuts (peanut), AFB_1 and OTA, literature[775]

incidence: 1/3, conc.: 51 µg/kg, sample origin: purchases and donations in Bamenda (city), Cameroon, sample year: December 2004–April 2005, country: South Africa/Bulgaria[804]

incidence: 4/5, conc. range: 40–186 µg/kg, ∅ conc.: 95 µg/kg, sample origin: purchases and donations in Douala (city), Cameroon, sample year: December 2004–April 2005, country: South Africa/Bulgaria[804]

incidence: 3/4, conc. range: 37–160 µg/kg, ∅ conc.: 87 µg/kg, sample origin: purchases and donations in Kumbo (city), Cameroon, sample year: December 2004–April 2005, country: South Africa/Bulgaria[804]

incidence: 2/4, conc. range: 31–59 µg/kg, ∅ conc.: 45 µg/kg, sample origin: purchases and donations in Yaounde (capital), Cameroon, sample year: 2005, country: South Africa/Bulgaria[804]

- Co-contamination: For detailed information see the article.
- Further contamination of plants and plant products other than cereals and their products reported in the present book/ article[804]: Beans, AF/S, DON, FB_1, and ZEA, literature[804]; Beans (soybean), AF/S, DON, and FB_1, literature[804]; Food, AF/S, DON and ZEA, literature[804]; Nuts (peanut), AF/S, DON, FB_1, and OTA, literature[804]

incidence: 2/15, conc. range: <LOQ, ∅ conc.: 2.9 µg/kg, sample origin: Osmaniye (province), Turkey, sample year: 2007, country: Austria[868]

- Co-contamination: For detailed information see the article.
- Further contamination of plants and plant products other than cereals and their products reported in the present book/ article[868]: Nuts (almond), AME, AOH, BEA, CPA, ENB, ENB_1, and ENB_3, literature[868]; Nuts (hazelnut), 3-NPA, AFB_1, AFB_2, AFG_1, AFG_2, AME, AOH, ATX-I, BEA, ENA, ENA_1, ENB_1, ENB_2, ENB_3, ENB_4, HT-2, KA, MA, MON, OTA, OTB, STG, T-2, and ZEA, literature[868]; Nuts (peanut), 3-NPA, AFB_1, AFB_2, AME, AOH, ATX-I, BEA, CPA, ENA, ENA_1, ENB, ENB_4, KA, MA, OTA, OTB, and STG, literature[868]; Nuts (pistachio nut), BEA, MA, and OTA, literature[868]; see also Nuts (almond), Nuts (hazelnut), Nuts (peanut), Nuts (pistachio nut), chapter: Further Mycotoxins and Microbial Metabolites

For detailed information see the article.

incidence: 8/30, conc. range: ≤20 µg/kg (8 sa., maximum: 12.33 µg/kg), sample origin: local markets in Wuhan (city), China, sample year: unknown, country: China[881]

- Co-contamination: not reported
- Further contamination of plants and plant products other than cereals and their products reported in the present book/ article[881]: Nuts (peanut), AFB_1, AFB_2, AFG_1, AFG_2, OTA, and T-2, literature[881]

see also **Food** and **Soups**

Penicillium Toxins

Mycophenolic Acid

incidence: 6/15, conc. range: ≤60 µg/kg, ∅ conc.: 21 µg/kg, sample origin: Osmaniye (province), Turkey, sample year: 2007, country: Austria[868]

- Co-contamination: For detailed information see the article.
- Further contamination of plants and plant products other than cereals and their products reported in the present book/

article[868]: Nuts (almond), AME, AOH, BEA, CPA, ENB, ENB_1, and ENB_3, literature[868]; Nuts (hazelnut), 3-NPA, AFB_1, AFB_2, AFG_1, AFG_2, AME, AOH, ATX-I, BEA, ENA, ENA_1, ENB_1, ENB_2, ENB_3, ENB_4, HT-2, KA, MA, MON, OTA, OTB, STG, T-2, and ZEA, literature[868]; Nuts (peanut), 3-NPA, AFB_1, AFB_2, AME, AOH, ATX-I, BEA, CPA, ENA, ENA_1, ENB, ENB_4, KA, OTA, OTB, STG, and ZEA, literature[868]; Nuts (pistachio nut), BEA, MA, and OTA, literature[868]; see also Nuts (almond), Nuts (hazelnut), Nuts (peanut), Nuts (pistachio nut), chapter: Further Mycotoxins and Microbial Metabolites

For detailed information see the article.

Nuts (pecan nut) may contain the following mycotoxins:

Aspergillus Toxins

Aflatoxin B_1

incidence: 3*/48, conc. range: tr-25 µg/kg, sample origin: Alabama, Georgia, and Louisiana (states), USA, sample year: 1972, country: USA[550], *pecans, blow-outs and sound

- Co-contamination: 1 sa. co-contaminated with AFB_1 and AFG_1; 2 sa. contaminated solely with AFB_1
- Further contamination of plants and plant products other than cereals and their products reported in the present book/article[550]: Nuts (pecan nut), AFG_1, literature[550]

Aflatoxin G_1

incidence: 3*/48, conc. range: tr, sample origin: Alabama, Georgia, and Louisiana (states), USA, sample year: 1972, country: USA[550], *pecans, blow-outs and sound

- Co-contamination: 1 sa. co-contaminated with AFB_1 and AFG_1; 2 sa. contaminated solely with AFG_1
- Further contamination of plants and plant products other than cereals and their products reported in the present book/article[550]: Nuts (pecan nut), AFB_1, literature[550]

Aflatoxin/s

incidence: 1/55, conc. range: 5.0–9.9 µg/kg* (1 sa.), sample origin: brokers, distributors, growers, processors, and retail trade, Canada, sample year: 1970–1975, country: Canada[15], *AFB_1, AFB_2, AFG_1, and AFG_2

- Co-contamination: not reported
- Further contamination of plants and plant products other than cereals and their products reported in the present book/article[15]: Butter (peanut butter), Cocoa beans, Nuts (almond), Nuts (cashew nut), Nuts (peanut), Nuts (pistachio nut), Nuts (walnut), Seeds (sunflower), AF/S, literature[15]

incidence: 3/17, conc. range: ≤334 µg/kg*, sample origin: imported to the USA, sample year: 1986, country: USA[19], *AFB_1, AFB_2, AFG_1, and AFG_2

- Co-contamination: not reported
- Further contamination of plants and plant products other than cereals and their products reported in the present book/article[19]: Butter (peanut butter), Candies (peanut candy), Candies (pistachio candy), Nuts, Nuts (almond), Nuts (Brazil nut, paranut), Nuts (peanut), Nuts (pistachio nut), Nuts (walnut), Oils (peanut oil), Paste (peanut paste), Paste (pipian paste), Products (peanut product), Seeds (melon), Spices (cayenne pepper, chilli), Spices (ginger), Spices (nutmeg), AF/S, literature[19]

incidence: 1*/229, conc.: 40 µg/kg**, sample origin: orchards and shelling plants in Alabama, Florida, and Georgia (states), southeastern USA, sample year: October–December 1974 to January 1975, country: USA[552], *from southern Georgia, weevil-damaged pecan nut sa. collected in January, **AFS

- Co-contamination: not reported
- Further contamination of plants and plant products other than cereals and their products reported in the present book/article[552]: not reported

incidence: 12/185, conc. range: ≤1,900 µg/kg*, ∅ conc.: 156 µg/kg*, sample origin: cracking plants, roasters and repackers, and wholesalers across the USA, sample year: 1969–1971, country: USA[553], *AFS total

incidence: 1/55, conc.: 355 µg/kg*, sample origin: cracking plants, roasters and repackers, and wholesalers across the USA, sample year: 1971/1972, country: USA[553], *AFS total

incidence: 13/95, conc. range: ≤172 µg/kg*, ∅ conc.: 42 µg/kg*, sample origin: cracking plants, roasters and repackers, and wholesalers across the USA, sample year: 1972/1973, country: USA[553], *AFS total

incidence: 11/125, conc. range: ≤125 µg/kg*, ∅ conc.: 34 µg/kg*, sample origin: cracking plants, roasters and repackers, and wholesalers across the USA, sample year: 1973/1974, country: USA[553], *AFS total

incidence: 2/101, conc. range: 10–38 µg/kg*, ∅ conc.: 24 µg/kg*, sample origin: cracking plants, roasters and repackers, and wholesalers across the USA, sample year: 1974/1975, country: USA[553], *AFS total

incidence: 0/26, conc.: no contamination*, sample origin: cracking plants, roasters and repackers, and wholesalers across the USA, sample year: 1975/1976, country: USA[553], *AFS total

incidence: 1/128, conc.: 5 µg/kg*, sample origin: cracking plants, roasters and repackers, and wholesalers across the USA, sample year: 1976/1977, country: USA[553], *AFS total

incidence: 6/214, conc. range: ≤126 µg/kg*, ∅ conc.: 27 µg/kg*, sample origin: cracking plants, roasters and repackers, and wholesalers across the USA, sample year: 1977–1979, country: USA[553], *AFS total

incidence: 3/132, conc. range: ≤9 µg/kg*, ∅ conc.: 8 µg/kg*, sample origin: cracking plants, roasters and repackers, and wholesalers across the USA, sample year: 1979/1980, country: USA[553], *AFS total

- Co-contamination: not reported
- Further contamination of plants and plant products other than cereals and their products reported in the present book/article[553]: not reported

Other ***Aspergillus*** Toxins

Sterigmatocystin

incidence: 1/40*, conc.: ~20,000 µg/kg**, sample origin: probably South Georgia (state) to North Florida (state) area, USA, sample year: November 1975, country: USA[554], *40 halves of in-shell pecans, **in a single half kernel

- Co-contamination: not reported
- Further contamination of plants and plant products other than cereals and their products reported in the present book/article[554]: not reported

Nuts (pine nut) may contain the following mycotoxins:

Aspergillus Toxins

Aflatoxin B_1

incidence: 2/12, conc. range: 0.19–0.23 µg/kg, Ø conc.: 0.21 µg/kg, sample origin: local markets in Chongqing, Fujian, Guangdong, Guangxi, Hubei, Jiangsu, Shanghai, and Zhejiang (regions), China, sample year: 2003, country: China[411]

- Co-contamination: 1 sa. co-contaminated with AFB_1 and AFB_2; 1 sa. contaminated solely with AFB_1
- Further contamination of plants and plant products other than cereals and their products reported in the present book/article[411]: Nuts (peanut) and Nuts (walnut), AFB_1, AFB_2, AFG_1, and AFG_2, literature[411]; Nuts (pine nut), AFB_2, literature[411]

incidence: 26/50*, conc. range: 25–2,080 µg/kg, sample origin: mainly northern and midwestern areas, Tunisia, sample year: unknown, country: Tunisia/USA[555], *Aleppo pine nuts

- Co-contamination: not reported
- Further contamination of plants and plant products other than cereals and their products reported in the present book/article[555]: Nuts (pine nut), AFG_1 and AF/S, literature[555]

Aflatoxin B_2

incidence: 1/12, conc.: 0.02 µg/kg, sample origin: local markets in Chongqing, Fujian, Guangdong, Guangxi, Hubei, Jiangsu, Shanghai, and Zhejiang (regions), China, sample year: 2003, country: China[411]

- Co-contamination: 1 sa. co-contaminated with AFB_1 and AFB_2
- Further contamination of plants and plant products other than cereals and their products reported in the present book/article[411]: Nuts (peanut), Nuts (walnut), AFB_1, AFB_2, AFG_1, and AFG_2, literature[411]; Nuts (pine nut), AFB_1, literature[411]

Aflatoxin G_1

incidence: 26/50*, conc. range: 56–4,570 µg/kg, sample origin: mainly northern and midwestern areas, Tunisia, sample year: unknown, country: Tunisia/USA[555], *Aleppo pine nuts

- Co-contamination: not reported
- Further contamination of plants and plant products other than cereals and their products reported in the present book/article[555]: Nuts (pine nut), AFB_1 and AFS, literature[555]

Aflatoxin/s

incidence: 2/10*, conc. range: 2.6–3.9 µg/kg**, Ø conc.: 3.25 µg/kg**, sample origin: retail shops and local markets in NWFP and northern areas, Pakistan, sample year: January–November 2009, country: Pakistan[17], *pine nuts with shell, **AFB_1, AFB_2, AFG_1, and AFG_2

- Co-contamination: not reported
- Further contamination of plants and plant products other than cereals and their products reported in the present book/article[17]: Dates, Figs, Fruits, dried, Kernels (apricot), Nuts (almond), Nuts (peanut), Nuts (pistachio nut), Nuts (walnut), Raisins, AF/S, literature[17]

incidence: 26/50*, conc. range: 95–7,550 µg/kg**, sample origin: mainly northern and midwestern areas, Tunisia, sample year: unknown, country: Tunisia/USA[555], *Aleppo pine nuts, **AFB_1, AFB_2, AFG_1, and AFG_2

- Co-contamination: not reported
- Further contamination of plants and plant products other than cereals and their products reported in the present book/article[555]: Nuts (pine nut), AFB_1 and AFG_1, literature[555]

Nuts (pistachio nut) may contain the following mycotoxins:

Aspergillus Toxins

Aflatoxin B_1

incidence: 2/6, conc. range: 15–94 µg/kg, Ø conc.: 54.5 µg/kg, sample origin: major local retail outlets at Nainital (district), India, sample year: 1983–1985, country: India[7]

- Co-contamination: Often more than one mycotoxin was present in a sa.
- Further contamination of plants and plant products other than cereals and their products reported in the present book/article[7]: Coconuts, AFB_1, AFB_2, AFG_1, AFG_2, and CIT, literature[7]; Dates, AFB_1, AFB_2, AFG_1, and OTA, literature[7]; Grogannuts, AFB_1, AFB_2, AFG_1, AFG_2, and CIT, literature[7]; Nuts (almond), AFB_1, AFB_2, AFG_1, RBT, STG, and ZEA, literature[7]; Nuts (cashew nut), AFB_1, AFB_2, AFG_1, AFG_2, CIT, and OTA, literature[7]; Nuts (peanut), Raisins, AFB_1, AFB_2, AFG_1, AFG_2, and OTA, literature[7]; Nuts (pistachio nut), AFB_2, AFG_1, AFG_2, and STG, literature[7]; Nuts (walnut), AFB_1, AFB_2, AFG_1, AFG_2, and RBT, literature[7]

incidence: 3/6, conc. range: 3.36–38.66 µg/kg, Ø conc.: 16.22 µg/kg, sample origin: grocery markets in Busan, Daegu, Daejeon, Gangneung, and Gwangju (cities) as well as Seoul (capital), Korea, sample year: May 2004–June 2005, country: Korea[10]

- Co-contamination: not reported
- Further contamination of plants and plant products other than cereals and their products reported in the present book/article[10]: Butter (peanut butter), Nuts, Nuts (almond), Nuts (peanut), Paste (soybean paste), Sauces (soy sauce), Spices (cayenne pepper, chilli), Spices (curry), Spices (pepper), AFB_1, literature[10]

incidence: 7/21, conc. range: 0.005–2 µg/kg (5 sa.) and >2 µg/kg (2 sa., maximum: 1,900 µg/kg), sample origin: ethnic stores, health food stores, and local stores as well as supermarkets in the four largest cities, Sweden, sample year: autumn 1996–summer 1998, country: Sweden[13]

- Co-contamination: not reported
- Further contamination of plants and plant products other than cereals and their products reported in the present book/article[13]: Beans (brown bean), Peas, and Peas (chickpea), OTA, literature[13]; Juices (apple juice), PAT, literature[13]; Nuts (almond), AFB_1, literature[13]; Nuts (Brazil nut, paranut), Oil seeds, Products (peanut product), AFB_1 and AF/S, literature[13]; Nuts (pistachio nut), AF/S, literature[13]

incidence: 5?/165, conc. range: 11.5–1,382 µg/kg, Ø conc.: 323 µg/kg, sample origin: commercially available market basket sa. of foods and foodstuffs in Tokyo (capital), Japan, sample year: 1986–1990, country: Japan[99]

- Co-contamination: not reported
- Further contamination of plants and plant products other than cereals and their products reported in the present book/article[99]: Beans, Seeds (sesame), Sugar, AFB_1 and AFB_2, literature[99]; Food, Jam (bean jam), Paste (peanut paste), AFB_1, literature[99]; Nuts (Brazil nut, paranut), Nuts (peanut), Spices (nutmeg), AFB_1, AFB_2, AFG_1, and AFG_2, literature[99]; Nuts (pistachio nut), AFB_2, AFG_1, and AFG_2, literature[99]; Spices, Spices (cayenne pepper, chilli), Spices (pepper), AFB_1, AFB_2, and AFG_1, literature[99]; Spices (paprika), AFB_1 and AFG_1, literature[99]

incidence: 6/54, conc. range: 7.9–1,830 µg/kg, ∅ conc.: 702 µg/kg, sample origin: commercial sources in Tokyo (capital), Japan, sample year: June–September 1981, country: Japan[129]

- Co-contamination: 5 sa. co-contaminated with AFB_1, AFB_2, AFM_1, and AFL-I; 1 sa. co-contaminated with AFB_1 and AFB_2
- Further contamination of plants and plant products other than cereals and their products reported in the present book/article[129]: Beans, Beans (butter beans), Beans (kidney bean), and Beans (saltani saltapya bean), AFB_1 and AFB_2, literature[129]; Nuts (pistachio nut), AFB_2 and AFM_1, literature[129]; see also Nuts (pistachio nut), chapter: Further Mycotoxins and Microbial Metabolites

incidence: 1/1, conc.: 4.6 µg/kg, sample origin: imported sa. collected by the Mycotoxin Research Association, Japan, sample year: 1985, country: Japan[171]

- Co-contamination: not reported
- Further contamination of plants and plant products other than cereals and their products reported in the present book/article[171]: Cocoa beans, Spices (cayenne pepper, chilli), Spices (pepper), AFB_1, literature[171]

incidence: 1/1*, conc.: 0.36 µg/kg, sample origin: local stores in Munich (city), Germany, sample year: unknown, country: Germany[181], *green pistachios

- Co-contamination: not reported
- Further contamination of plants and plant products other than cereals and their products reported in the present book/article[181]: Butter (peanut butter) and Spices (nutmeg), AFB_1, AFB_2, and AFG_1, literature[181]; Nuts (hazelnut), Spices (paprika), AFB_1 and AFG_1, literature[181]; Spices (pepper), AFB_1, literature[181]

incidence: 53?/856, conc. range: 1.4–206 µg/kg, ∅ conc.: 54.9 µg/kg, sample origin: sites of primary storage or processing (domestic products) and sites of importation, Cyprus, sample year: 1992–1996, country: Cyprus[200]

- Co-contamination: not reported
- Further contamination of plants and plant products other than cereals and their products reported in the present book/article[200]: Butter (peanut butter), Figs, Nuts (peanut), AFB_1, AFB_2, AFG_1, and AFG_2, literature[200]; Nuts (Brazil nut, paranut) and Nuts (walnut), AFB_1, AFB_2, and AFG_1, literature[200]; Nuts (chestnut), Products (sesame product), Spices, AFB_1, literature[200]; Nuts (pistachio nut), AFB_2, literature[200]; Seeds (sesame), AFB_1, AFG_1, and AFG_2, literature[200]

incidence: 9/23, conc. range: LOQ-2 µg/kg (5 sa.), 2–5 µg/kg (2 sa.), and >5 µg/kg (2 sa., maximum: 41.3 µg/kg), ∅ conc.: 9.04 µg/kg, sample origin: retail outlets, UK, sample year: September 2001, country: UK[203]

- Co-contamination: 1 sa. co-contaminated with AFB_1, AFB_2, and OTA; 3 sa. co-contaminated with AFB_1 and AFB_2; 5 sa. contaminated solely with AFB_1
- Further contamination of plants and plant products other than cereals and their products reported in the present book/article[203]: Butter (nut butter), AFB_1, AFB_2, AFG_1, AFG_2, and OTA, literature[203]; Butter (peanut butter), AFB_1, AFB_2, AFG_1, AFG_2, and OTA, literature[203]; Dates, AFB_1 and OTA, literature[203]; Figs, AFB_1, AFB_2, AFG_2, and OTA, literature[203]; Fruits, dried, AFB_1, AFB_2, and OTA, literature[203]; Nuts (Brazil nut, paranut), AFB_1, AFB_2, and AFG_1, literature[203]; Nuts (peanut), AFB_1 and AFB_2, literature[203]; Nuts (pistachio nut), AFB_2 and OTA, literature[203]

incidence: 3/29, conc. range: 8.6–93.3 µg/kg, ∅ conc.: 39.5 µg/kg, sample origin: retail outlets, UK, sample year: September 2001, country: UK[203]

- Co-contamination: 3 sa. co-contaminated with AFB_1 and AFB_2
- Further contamination of plants and plant products other than cereals and their products reported in the present book/article[203]: Butter (nut butter), AFB_1, AFB_2, AFG_1, AFG_2, and OTA, literature[203]; Butter (peanut butter), AFB_1, AFB_2, AFG_1, AFG_2, and OTA, literature[203]; Dates, AFB_1 and OTA, literature[203]; Figs, AFB_1, AFB_2, AFG_2, and OTA, literature[203]; Fruits, dried, AFB_1, AFB_2, and OTA, literature[203]; Nuts (Brazil nut, paranut), AFB_1, AFB_2, and AFG_1, literature[203]; Nuts (peanut), AFB_1 and AFB_2, literature[203]; Nuts (pistachio nut), AFB_2 and OTA, literature[203]

incidence: 1/8, conc.: 3.36 µg/kg, sample origin: supermarkets, local stores, and markets in Busan, Daejeon, Gwangju (cities), and Seoul (capital), Korea, sample year: June/July 2004, country: Korea[220]

- Co-contamination: not reported
- Further contamination of plants and plant products other than cereals and their products reported in the present book/article[220]: Butter (peanut butter) and Nuts, AFB_1 and AFB_2, literature[220]; Nuts (peanut), AFB_1, AFB_2, AFG_1, and AFG_2, literature[220]

incidence: 1/20, conc.: 0.6 µg/kg, sample origin: available in New Zealand, sample year: 2010, country: New Zealand[277]

- Co-contamination: not reported
- Further contamination of plants and plant products other than cereals and their products reported in the present book/article[277]: Bars, AFB_1, AF/S, DON, NIV, and OTA, literature[277]; Butter (peanut butter), Confectionery, Food, Fruits, dried, Nuts, Nuts (peanut), Sauces (peanut sauce), AFB_1 and AF/S, literature[277]; Coffee, Dates, Spices (nutmeg), Spices (pepper), Vine fruits, dried, OTA, literature[277]; Figs, Spices (cayenne pepper, chilli), Spices (curry), Spices (ginger), and Spices (paprika), AFB_1, AF/S, and OTA, literature[277]; Nuts (pistachio nut), AF/S, literature[277]; Nuts (Brazil nut, paranut), AFB_1, literature[277]

incidence: 1/1*, conc.: 1.29 µg/kg, sample origin: imported from Turkey to Italy, sample year: January 2008–December 2009, country: Italy[320], * pistachios without shell

- Co-contamination: 1 sa. co-contaminated with AFB_1 and AFB_2
- Further contamination of plants and plant products other than cereals and their products reported in the present book/article[320]: Coffee beans, Vine fruits, dried, OTA, literature[320]; Kernels (apricot), AFB_1, AFB_2, and AFG_1, literature[320]; Nuts (hazelnut) and Paste (hazelnut paste), AFB_1, AFB_2, AFG_1, and AFG_2, literature[320]; Nuts (peanut), AFB_1 and AFB_2, literature[320]; Nuts (pistachio nut), AFB_2, literature[320]

incidence: 9/20, conc. range: ≤1,430 µg/kg, ∅ conc.: 158 µg/kg, sample origin: retail shops and local markets in Rabat-Salé (region), Morocco, sample year: January–October 2006, country: Spain/Morocco[352]

- Co-contamination: not reported
- Further contamination of plants and plant products other than cereals and their products reported in the present book/article[352]: Figs, Nuts (peanut), Nuts (walnut), Raisins, AFB_1 and AF/S, literature[352]; Nuts (pistachio nut), AF/S, literature[352]

incidence: 1/19, conc.: 22 µg/kg, sample origin: storage houses and dealers, Tunisia, sample year: unknown, country: Tunisia/USA[380]

- Co-contamination: not reported

- Further contamination of plants and plant products other than cereals and their products reported in the present book/article[380]: Nuts (peanut), Peas, AFB_1 and AFG_1, literature[380]

incidence: 8/72, conc. range: 0.48–36.81 μg/kg, sample origin: public markets/bazaars, supermarkets, retail stores, and wholesale stores in Ankara, Antalya, Isparta, Burdur, and Izmir (provinces), Turkey, sample year: unknown, country: Turkey[409]

- Co-contamination: not reported
- Further contamination of plants and plant products other than cereals and their products reported in the present book/article[409]: Nuts (hazelnut), AFB_1, AFG_1, and AFG_2, literature[409]; Nuts (peanut), AFB_1, AFB_2, AFG_1, and AFG_2, literature[409]; Nuts (pistachio nut), AFB_2, literature[409]

incidence: 17/29, conc. range: 2.3–165 μg/kg, ⌀ conc.: 56.4 μg/kg, sample origin: unknown, sample year: unknown, country: Netherlands[556]

- Co-contamination: not reported
- Further contamination of plants and plant products other than cereals and their products reported in the present book/article[556]: not reported

incidence: 3/3*, conc. range: 13.2**–227.2 μg/kg, ⌀ conc.: 115 μg/kg, sample origin: mostly from China or India purchased in retail outlets in Glasgow (city) and surrounding area, UK, sample year: unknown, country: UK[557], *1 roasted** sa. and 2 raw sa. of pistachio nuts

- Co-contamination: 3 sa. co-contaminated with AFB_1 and AFB_2
- Further contamination of plants and plant products other than cereals and their products reported in the present book/article[557]: Nuts (pistachio nut), AFB_2, literature[557]

incidence: 21/40, conc. range: <LOQ-122.4 μg/kg, sample origin: local markets, northern Tunisia, sample year: 2005/2006, country: Tunisia[558]

- Co-contamination: not reported
- Further contamination of plants and plant products other than cereals and their products reported in the present book/article[558]: Nuts (pistachio nut), AFB_2, AFG_1, and AFG_2, literature[558]

incidence: 3/16*, conc. range: 0.12–0.18 μg/kg**, ⌀ conc.: 0.146 μg/kg**, sample origin: imported from Iran, Turkey, the USA, and Spain purchased in commercial outlets of Aragon (autonomous community), northeastern Spain, sample year: January–April 2007, country: Spain[559], *pre-packed pistachio nuts, **from Iran

- Co-contamination: 1 sa. co-contaminated with AFB_1 and AFB_2; 2 sa. contaminated solely with AFB_1
- Further contamination of plants and plant products other than cereals and their products reported in the present book/article[559]: Nuts (pistachio nut), AFB_2, literature[559]

incidence: 8/16*, conc. range: 0.13–0.29 μg/kg**, ⌀ conc.: 0.185 μg/kg**, sample origin: imported from Iran, Turkey, the USA, and Spain purchased in commercial outlets of Aragon (autonomous community), northeastern Spain, sample year: January–April 2007, country: Spain[559], *bulk pistachio nuts, **from Iran

- Co-contamination: 2 sa. co-contaminated with AFB_1 and AFB_2; 6 sa. contaminated solely with AFB_1
- Further contamination of plants and plant products other than cereals and their products reported in the present book/article[559]: Nuts (pistachio nut), AFB_2, literature[559]

incidence: 12/24, conc. range: 0.6–98.5 μg/kg, ⌀ conc.: 27.5 μg/kg, sample origin: community of Madrid (capital), Spain, sample year: December 2003, country: Spain[560]

- Co-contamination: 7 sa. co-contaminated with AFB_1 and AFB_2; 5 sa. contaminated solely with AFB_1
- Further contamination of plants and plant products other than cereals and their products reported in the present book/article[560]: Nuts (pistachio nut), AFB_2, literature[560]

incidence: 5/5, conc. range: 44–213 μg/kg, ⌀ conc.: 145 μg/kg, sample origin: unknown, sample year: unknown, country: Iran[669]

- Co-contamination: 5 sa. co-contaminated with AFB_1 and AFB_2
- Further contamination of plants and plant products other than cereals and their products reported in the present book/article[669]: Nuts (pistachio nut), AFB_2, literature[669]

incidence: 906/3,282, conc. range: LOD-5 μg/kg (678 sa.), 5–8 μg/kg (53 sa.), 8–10 μg/kg (33 sa.), 10–20 μg/kg (58 sa.), 20–50 μg/kg (59 sa.), ≤319.53 μg/kg (25 sa.), sample origin: Rafsanjan (county) and Kerman (province), Iran, sample year: January–December 2009, country: Iran[713]

incidence: 615/2,895, conc. range: LOD-5 μg/kg (418 sa.), 5–8 μg/kg (46 sa.), 8–10 μg/kg (14 sa.), 10–20 μg/kg (50 sa.), 20–50 μg/kg (50 sa.), ≤246.06 μg/kg (37 sa.), sample origin: Rafsanjan (county) and Kerman (province), Iran, sample year: January–December 2010, country: Iran[713]

incidence: 402/2,026, conc. range: LOD-5 μg/kg (270 sa.), 5–8 μg/kg (29 sa.), 8–10 μg/kg (7 sa.), 10–20 μg/kg (41 sa.), 20–50 μg/kg (37 sa.), ≤369.12 μg/kg (18 sa.), sample origin: Rafsanjan (county) and Kerman (province), Iran, sample year: January–December 2011, country: Iran[713]

- Co-contamination: not reported
- Further contamination of plants and plant products other than cereals and their products reported in the present book/article[713]: Nuts (pistachio nut), AFB_2, AFG_1, and AFG_2, literature[713]

incidence: 5/50, conc. range: 2.21–1,037.31 μg/kg, ⌀ conc.: 214 μg/kg, sample origin: imported from California (USA), Iran, and unknown origin purchased in hypermarkets and supermarkets in Catalonia (autonomous community), Spain, sample year: June–December 2008, country: Spain[728]

- Co-contamination: 2 sa. co-contaminated with AFB_1, AFB_2, AFG_1, and AFG_2; 3 sa. co-contaminated with AFB_1 and AFB_2; no further information available
- Further contamination of plants and plant products other than cereals and their products reported in the present book/article[728]: Nuts (pistachio nut), AFB_2, AFG_1, AFG_2, and OTA, literature[728]

incidence: 22/151, conc. range: 0.26–368 μg/kg, ⌀ conc.: 31.2 μg/kg, sample origin: markets, supermarkets, groceries, and retail shops in Adana and Osmaniye (provinces), Turkey, sample year: January 2010–December 2012, country: Turkey[741]

- Co-contamination: not reported
- Further contamination of plants and plant products other than cereals and their products reported in the present book/article[741]: Nuts (peanut), AFB_1, AFB_2, and AFG_1, literature[741]; Nuts (pistachio nut), AFB_2, AFG_1, and AFG_2, literature[741]

incidence: 45/71*, conc. range: ≤19.56 μg/kg, sample origin: retail outlets, cafeterias, and hotels from major districts of Punjab (state), Pakistan, sample year: unknown, country: Pakistan[750], *ingredients for halva

- Co-contamination: not reported
- Further contamination of plants and plant products other than cereals and their products reported in the present book/article[750]: Halva, AFB_1 and AFM_1, literature[750]; Nuts (almond), Raisins, Spices (cardamom), AFB_1, literature[750]

incidence: 35/112*, conc. range: 0.06–5 μg/kg (20 sa.), 5.01–10 μg/kg (3 sa.), 10.01–20 μg/kg (5 sa.), 20.01–50 μg/kg (4 sa.), and >50 μg/kg (3 sa.), sample origin: local factories producing Gaz sweets in Esfahan (city), Iran, sample year: 2007–2012, country: Iran[753], *ingredients for Gaz sweets

- Co-contamination: not reported
- Further contamination of plants and plant products other than cereals and their products reported in the present book/article[753]: not reported

incidence: 2/32, conc. range: ≤0.37 μg/kg, ∅ conc.: 0.3 μg/kg, sample origin: retail stores in Lower Saxony (federal state), Germany, sample year: 2009, country: Germany[829]

- Co-contamination: not reported
- Further contamination of plants and plant products other than cereals and their products reported in the present book/article[829]: Nuts (hazelnut), AFB_1 and AF/S, literature[829]; Oils (maize oil), ZEA, literature[829]; Spices (cinnamon), OTA, literature[829]; Spices (curry), Spices (nutmeg), AFB_1, AF/S, and OTA, literature[829]; Spices (ginger), AFB_1 and AF/S, literature[829]

Aflatoxin B_2

incidence: 1/6, conc.: 75 μg/kg, sample origin: major local retail outlets at Nainital (district), India, sample year: 1983–1985, country: India[7]

- Co-contamination: Often more than one mycotoxin was present in a sa.
- Further contamination of plants and plant products other than cereals and their products reported in the present book/article[7]: Coconuts, AFB_1, AFB_2, AFG_1, AFG_2, and CIT, literature[7]; Dates, AFB_1, AFB_2, AFG_1, and OTA, literature[7]; Grogannuts, AFB_1, AFB_2, AFG_1, AFG_2, and CIT, literature[7]; Nuts (almond), AFB_1, AFB_2, AFG_1, RBT, STG, and ZEA, literature[7]; Nuts (cashew nut), AFB_1, AFB_2, AFG_1, AFG_2, CIT, and OTA, literature[7]; Nuts (peanut), Raisins, AFB_1, AFB_2, AFG_1, AFG_2, and OTA, literature[7]; Nuts (pistachio nut), AFB_1, AFG_1, AFG_2, and STG, literature[7]; Nuts (walnut), AFB_1, AFB_2, AFG_1, AFG_2, and RBT, literature[7]

incidence: 5?/165, conc. range: 0.5–260 μg/kg, ∅ conc.: 58.0 μg/kg, sample origin: commercially available market basket sa. of foods and foodstuffs in Tokyo (capital), Japan, sample year: 1986–1990, country: Japan[99]

- Co-contamination: not reported
- Further contamination of plants and plant products other than cereals and their products reported in the present book/article[99]: Beans, Seeds (sesame), Sugar, AFB_1 and AFB_2, literature[99]; Food, Jam (bean jam), Paste (peanut paste), AFB_1, literature[99]; Nuts (Brazil nut, paranut), Nuts (peanut), and Spices (nutmeg), AFB_1, AFB_2, AFG_1, and AFG_2, literature[99]; Nuts (pistachio nut), AFB_1, AFG_1, and AFG_2, literature[99]; Spices, Spices (cayenne pepper, chilli), Spices (pepper), AFB_1, AFB_2, and AFG_1, literature[99]; Spices (paprika), AFB_1 and AFG_1, literature[99]

incidence: 6/54, conc. range: 1.5–235 μg/kg, ∅ conc.: 104 μg/kg, sample origin: commercial sources in Tokyo (capital), Japan, sample year: June–September 1981, country: Japan[129]

- Co-contamination: 5 sa. co-contaminated with AFB_1, AFB_2, AFM_1, and AFL-I; 1 sa. co-contaminated with AFB_1 and AFB_2
- Further contamination of plants and plant products other than cereals and their products reported in the present book/article[129]: Beans, Beans (butter beans), Beans (kidney bean), and Beans (saltani saltapya bean), AFB_1 and AFB_2, literature[129]; Nuts (pistachio nut), AFB_1 and AFM_1, literature[129]; see also Nuts (pistachio nut), chapter: Further Mycotoxins and Microbial Metabolites

incidence: 53?/856, conc. range: <0.3–2.3 μg/kg, ∅ conc.: 1.2 μg/kg, sample origin: sites of primary storage or processing (domestic products) and sites of importation, Cyprus, sample year: 1992–1996, country: Cyprus[200]

- Co-contamination: not reported
- Further contamination of plants and plant products other than cereals and their products reported in the present book/article[200]: Butter (peanut butter), Figs, Nuts (peanut), AFB_1, AFB_2, AFG_1, and AFG_2, literature[200]; Nuts (Brazil nut, paranut), Nuts (walnut), AFB_1, AFB_2, and AFG_1, literature[200]; Nuts (chestnut), Nuts (pistachio nut), Products (sesame product), Spices, AFB_1, literature[200]; Seeds (sesame), AFB_1, AFG_1, and AFG_2, literature[200]

incidence: 4/23, conc. range: LOQ-2 μg/kg (2 sa.), 2–5 μg/kg (1 sa.), and 6.2 μg/kg (1 sa.), ∅ conc.: 2.48 μg/kg, sample origin: retail outlets, UK, sample year: September 2001, country: UK[203]

- Co-contamination: 1 sa. co-contaminated with AFB1, AFB2, and OTA; 3 sa. co-contaminated with AFB_1 and AFB_2
- Further contamination of plants and plant products other than cereals and their products reported in the present book/article[203]: Butter (nut butter), AFB_1, AFB_2, AFG_1, AFG_2, and OTA, literature[203]; Butter (peanut butter), AFB_1, AFB_2, AFG_1, AFG_2, and OTA, literature[203]; Dates, AFB_1 and OTA, literature[203]; Figs, AFB_1, AFB_2, AFG_2, and OTA, literature[203]; Fruits, dried, AFB_1, AFB_2, and OTA, literature[203]; Nuts (Brazil nut, paranut), AFB_1, AFB_2, and AFG_1, literature[203]; Nuts (peanut), AFB_1 and AFB_2, literature[203]; Nuts (pistachio nut), AFB_1 and OTA, literature[203]

incidence: 3/29, conc. range: LOQ-2 μg/kg (2 sa.) and 13.6 μg/kg (1 sa.), ∅ conc.: 5.27 μg/kg, sample origin: retail outlets, UK, sample year: September 2001, country: UK[203]

- Co-contamination: 3 sa. co-contaminated with AFB_1 and AFB_2
- Further contamination of plants and plant products other than cereals and their products reported in the present book/article[203]: Butter (nut butter), AFB_1, AFB_2, AFG_1, AFG_2, and OTA, literature[203]; Butter (peanut butter), AFB_1, AFB_2, AFG_1, AFG_2, and OTA, literature[203]; Dates, AFB_1 and OTA, literature[203]; Figs, AFB_1, AFB_2, AFG_2, and OTA, literature[203]; Fruits, dried, AFB_1, AFB_2, and OTA, literature[203]; Nuts (Brazil nut, paranut), AFB_1, AFB_2, and AFG_1, literature[203]; Nuts (peanut), AFB_1 and AFB_2, literature[203]; Nuts (pistachio nut), AFB_1 and OTA, literature[203]

incidence: 1/1*, conc.: 0.19 μg/kg, sample origin: imported from Turkey to Italy, sample year: January 2008–December 2009, country: Italy[320], *pistachios without shell

- Co-contamination: 1 sa. co-contaminated with AFB_1 and AFB_2
- Further contamination of plants and plant products other than cereals and their products reported in the present book/article[320]: Coffee beans, Vine fruits, dried, OTA, literature[320]; Kernels (apricot), AFB_1, AFB_2, and AFG_1, literature[320]; Nuts (hazelnut), Paste (hazelnut paste), AFB_1, AFB_2, AFG_1, and AFG_2, literature[320]; Nuts (peanut), AFB_1 and AFB_2, literature[320]; Nuts (pistachio nut), AFB_1, literature[320]

incidence: 4/72, conc. range: 0.076–2.53 μg/kg, sample origin: public markets/bazaars, supermarkets, retail stores, and wholesale stores in Ankara, Antalya, Isparta, Burdur, and Izmir (provinces), Turkey, sample year: unknown, country: Turkey[409]

- Co-contamination: not reported
- Further contamination of plants and plant products other than cereals and their products reported in the present book/ article[409]: Nuts (hazelnut), AFB_1, AFG_1, and AFG_2, literature[409]; Nuts (peanut), AFB_1, AFB_2, AFG_1, and AFG_2, literature[409]; Nuts (pistachio nut), AFB_1, literature[409]

incidence: 3/3*, conc. range: 1.7**–31.7 µg/kg, ∅ conc.: 14.5 µg/kg, sample origin: probably from China or India purchased in retail outlets in Glasgow (city) and surrounding area, UK, sample year: unknown, country: UK[557], *1 roasted** sa. and 2 raw sa. of pistachio nuts

- Co-contamination: 3 sa. co-contaminated with AFB_1 and AFB_2
- Further contamination of plants and plant products other than cereals and their products reported in the present book/ article[557]: Nuts (pistachio nut), AFB_1, literature[557]

incidence: 11/40, conc. range: 0.24 to <LOQ µg/kg, sample origin: local markets, northern Tunisia, sample year: 2005/2006, country: Tunisia[558]

- Co-contamination: not reported
- Further contamination of plants and plant products other than cereals and their products reported in the present book/ article[558]: Nuts (pistachio nut), AFB_1, AFG_1, and AFG_2, literature[558]

incidence: 1/16*, conc.: 0.12 µg/kg**, sample origin: imported from Iran, Turkey, the USA, and Spain purchased in commercial outlets of Aragon (autonomous community), northeastern Spain, sample year: January–April 2007, country: Spain[559], *prepacked pistachio nuts, **from Iran

- Co-contamination: 1 sa. co-contaminated with AFB_1 and AFB_2
- Further contamination of plants and plant products other than cereals and their products reported in the present book/ article[559]: Nuts (pistachio nut), AFB_1, literature[559]

incidence: 2/16*, conc. range: 0.12–0.14 µg/kg**, ∅ conc.: 0.13 µg/kg**, sample origin: imported from Iran, Turkey, the USA, and Spain purchased in commercial outlets of Aragon (autonomous community), northeastern Spain, sample year: January–April 2007, country: Spain[559], *bulk pistachio nuts, **from Iran

- Co-contamination: 2 sa. co-contaminated with AFB_1 and AFB_2
- Further contamination of plants and plant products other than cereals and their products reported in the present book/ article[559]: Nuts (pistachio nut), AFB_1, literature[559]

incidence: 7/24, conc. range: 1.3–19.0 µg/kg, ∅ conc.: 6.87 µg/kg, sample origin: community of Madrid (capital), Spain, sample year: December 2003, country: Spain[560]

- Co-contamination: 7 sa. co-contaminated with AFB_1 and AFB_2
- Further contamination of plants and plant products other than cereals and their products reported in the present book/ article[560]: Nuts (pistachio nut), AFB_1, literature[560]

incidence: 5/5, conc. range: 5.6–21.9 µg/kg, ∅ conc.: 13.1 µg/kg, sample origin: unknown, sample year: unknown, country: Iran[669]

- Co-contamination: 5 sa. co-contaminated with AFB_1 and AFB_2
- Further contamination of plants and plant products other than cereals and their products reported in the present book/ article[560]: Nuts (pistachio nut), AFB_1, literature[669]

incidence: 437/3,282, conc. range: LOD-5 µg/kg (407 sa.), 5–8 µg/kg (18 sa.), 8–10 µg/kg (3 sa.), 10–20 µg/kg (6 sa.), 20–50 µg/kg (2 sa.), 56.51 µg/kg (1 sa.), sample origin: Rafsanjan (county) and Kerman (province), Iran, sample year: 2009, country: Iran[713]

incidence: 313/2,895, conc. range: LOD-5 µg/kg (280 sa.), 5–8 µg/kg (15 sa.), 8–10 µg/kg (4 sa.), 10–20 µg/kg (9 sa.), ≤26.69 µg/ kg (5 sa.), sample origin: Rafsanjan (county) and Kerman (province), Iran, sample year: 2010, country: Iran[713]

incidence: 156/2,026, conc. range: LOD-5 µg/kg (146 sa.), 5–8 µg/kg (6 sa.), 8–10 µg/kg (3 sa.), 8.4 µg/kg (1 sa.), sample origin: Rafsanjan (county) and Kerman (province), Iran, sample year: 2011, country: Iran[713]

- Co-contamination: not reported
- Further contamination of plants and plant products other than cereals and their products reported in the present book/ article[713]: Nuts (pistachio nut), AFB_1, AFG_1, and AFG_2, literature[713]

incidence: 5/50, conc. range: 0.39–97.22 µg/kg, ∅ conc.: 20.3 µg/kg, sample origin: imported from California (state) USA, Iran, and unknown origin purchased in hypermarkets and supermarkets in Catalonia (autonomous community) Spain, sample year: June–December 2008, country: Spain[728]

- Co-contamination: 2 sa. co-contaminated with AFB_1, AFB_2, AFG_1, and AFG_2; 3 sa. co-contaminated with AFB_1 and AFB_2; no further information available
- Further contamination of plants and plant products other than cereals and their products reported in the present book/ article[728]: Nuts (pistachio nut), AFB_1, AFG_1, AFG_2, and OTA, literature[728]

incidence: 17/151, conc. range: 0.11–16.7 µg/kg, ∅ conc.: 2.77 µg/kg, sample origin: markets, supermarkets, groceries, and retail shops in Adana and Osmaniye (provinces), Turkey, sample year: January 2010–December 2012, country: Turkey[741]

- Co-contamination: not reported
- Further contamination of plants and plant products other than cereals and their products reported in the present book/ article[741]: Nuts (peanut), AFB_1, AFB_2, and AFG_1, literature[741]; Nuts (pistachio nut), AFB_1, AFG_1, and AFG_2, literature[741]

Aflatoxin G_1

incidence: 1/6, conc.: 12 µg/kg, sample origin: major local retail outlets at Nainital (district), India, sample year: 1983–1985, country: India[7]

- Co-contamination: Often more than one mycotoxin was present in a sa.
- Further contamination of plants and plant products other than cereals and their products reported in the present book/article[7]: Coconuts, AFB_1, AFB_2, AFG_1, AFG_2, and CIT, literature[7]; Dates, AFB_1, AFB_2, AFG_1, and OTA, literature[7]; Grogannuts, AFB_1, AFB_2, AFG_1, AFG_2, and CIT, literature[7]; Nuts (almond), AFB_1, AFB_2, AFG_1, RBT, STG, and ZEA, literature[7]; Nuts (cashew nut), AFB_1, AFB_2, AFG_1, AFG_2, CIT, and OTA, literature[7]; Nuts (peanut), Raisins, AFB_1, AFB_2, AFG_1, AFG_2, and OTA, literature[7]; Nuts (pistachio nut), AFB_1, AFB_2, AFG_2, and STG, literature[7]; Nuts (walnut), AFB_1, AFB_2, AFG_1, AFG_2, and RBT, literature[7]

incidence: 5?/165, conc. range: ≤306 µg/kg, ∅ conc.: 61.3 µg/kg, sample origin: commercially available market basket sa. of foods and foodstuffs in Tokyo (capital), Japan, sample year: 1986–1990, country: Japan[99]

- Co-contamination: not reported
- Further contamination of plants and plant products other than cereals and their products reported in the present book/article[99]: Beans, Seeds (sesame), Sugar, AFB_1 and AFB_2,

literature[99]; Food, Jam (bean jam), Paste (peanut paste), AFB_1, literature[99]; Nuts (Brazil nut, paranut), Nuts (peanut), Spices (nutmeg), AFB_1, AFB_2, AFG_1, and AFG_2, literature[99]; Nuts (pistachio nut), AFB_1, AFB_2, and AFG_2, literature[99]; Spices, Spices (cayenne pepper, chilli), Spices (pepper), AFB_1, AFB_2, and AFG_1, literature[99]; Spices (paprika), AFB_1 and AFG_1, literature[99]

incidence: 3/40, conc. range: <LOQ, sample origin: local markets, northern Tunisia, sample year: 2005/2006, country: Tunisia[558]

- Co-contamination: not reported
- Further contamination of plants and plant products other than cereals and their products reported in the present book/article[558]: Nuts (pistachio nut), AFB_1, AFB_2, and AFG_2, literature[558]

incidence: 13/3,282, conc. range: LOD-5 μg/kg (10 sa.), 8–10 μg/kg (1 sa.), 10–20 μg/kg (1 sa.), 94.85 μg/kg (1 sa.), sample origin: Rafsanjan (county) and Kerman (province), Iran, sample year: 2009, country: Iran[713]

incidence: 15/2,895, conc. range: LOD-5 μg/kg (11 sa.), 5–8 μg/kg (1 sa.), 20–50 μg/kg (2 sa.), 60.1 μg/kg (1 sa.), sample origin: Rafsanjan (county) and Kerman (province), Iran, sample year: 2010, country: Iran[713]

incidence: 9/2,026, conc. range: LOD-5 μg/kg (5 sa.), 5–8 μg/kg (1 sa.), 10–20 μg/kg (2 sa.), 76.56 μg/kg (1 sa.), sample origin: Rafsanjan (county) and Kerman (province), Iran, sample year: 2011, country: Iran[713]

- Co-contamination: not reported
- Further contamination of plants and plant products other than cereals and their products reported in the present book/article[713]: Nuts (pistachio nut), AFB_1, AFB_2, and AFG_2, literature[713]

incidence: 2/50, conc. range: 0.70–0.71 μg/kg, Ø conc.: 0.705 μg/kg, sample origin: imported from California (state) USA, Iran, and unknown origin purchased in hypermarkets and supermarkets in Catalonia (autonomous community), Spain, sample year: June-December 2008, country: Spain[728]

- Co-contamination: 2 sa. co-contaminated with AFB_1, AFB_2, AFG_1, and AFG_2; no further information available
- Further contamination of plants and plant products other than cereals and their products reported in the present book/article[728]: Nuts (pistachio nut), AFB_1, AFB_2, AFG_2, and OTA, literature[728]

incidence: 3/151, conc. range: 0.79–8.72 μg/kg, Ø conc.: 3.49 μg/kg, sample origin: markets, supermarkets, groceries, and retail shops in Adana and Osmaniye (provinces), Turkey, sample year: January 2010–December 2012, country: Turkey[741]

- Co-contamination: not reported
- Further contamination of plants and plant products other than cereals and their products reported in the present book/article[741]: Nuts (peanut), AFB_1, AFB_2, and AFG_1, literature[741]; Nuts (pistachio nut), AFB_1, AFB_2, and AFG_2, literature[741]

Aflatoxin G_2

incidence: 1/6, conc.: 8 μg/kg, sample origin: major local retail outlets at Nainital (district), India, sample year: 1983–1985, country: India[7]

- Co-contamination: Often more than one mycotoxin was present in a sa.
- Further contamination of plants and plant products other than cereals and their products reported in the present book/article[7]: Coconuts, AFB_1, AFB_2, AFG_1, AFG_2, and CIT, literature[7]; Dates, AFB_1, AFB_2, AFG_1, and OTA, literature[7]; Grogannuts, AFB_1, AFB_2, AFG_1, AFG_2, and CIT, literature[7]; Nuts (almond), AFB_1, AFB_2, AFG_1, RBT, STG, and ZEA, literature[7]; Nuts (cashew nut), AFB_1, AFB_2, AFG_1, AFG_2, CIT, and OTA, literature[7]; Nuts (peanut), Raisins, AFB_1, AFB_2, AFG_1, AFG_2, and OTA, literature[7]; Nuts (pistachio nut), AFB_1, AFB_2, AFG_1, and STG, literature[7]; Nuts (walnut), AFB_1, AFB_2, AFG_1, AFG_2, and RBT, literature[7]

incidence: 5?/165, conc. range: ≤48.3 μg/kg, Ø conc.: 9.7 μg/kg, sample origin: commercially available market basket sa. of foods and foodstuffs in Tokyo (capital), Japan, sample year: 1986–1990, country: Japan[99]

- Co-contamination: not reported
- Further contamination of plants and plant products other than cereals and their products reported in the present book/article[99]: Beans, Seeds (sesame), Sugar, AFB_1 and AFB_2, literature[99]; Food, Jam (bean jam), Paste (peanut paste), AFB_1, literature[99]; Nuts (Brazil nut, paranut), Nuts (peanut), Spices (nutmeg), AFB_1, AFB_2, AFG_1, and AFG_2, literature[99]; Nuts (pistachio nut), Spices, Spices (cayenne pepper, chilli), Spices (pepper), AFB_1, AFB_2, and AFG_1, literature[99]; Spices (paprika), AFB_1 and AFG_1, literature[99]

incidence: 3/40, conc. range: <LOQ, sample origin: local markets, northern Tunisia, sample year: 2005/2006, country: Tunisia[558]

- Co-contamination: not reported
- Further contamination of plants and plant products other than cereals and their products reported in the present book/article[558]: Nuts (pistachio nut), AFB_1, AFB_2, and AFG_1, literature[558]

incidence: 6/3,282, conc. range: LOD-5 μg/kg, (5 sa.), 5.22 μg/kg (1 sa.), sample origin: Rafsanjan (county) and Kerman (province), Iran, sample year: 2009, country: Iran[713]

incidence: 7/2,895, conc. range: LOD to ≤2.66 μg/kg (7 sa.), sample origin: Rafsanjan (county) and Kerman (province), Iran, sample year: 2010, country: Iran[713]

incidence: 7/2,026, conc. range: LOD-≤3.5 μg/kg (7 sa.), sample origin: Rafsanjan (county) and Kerman (province), Iran, sample year: 2011, country: Iran[713]

- Co-contamination: not reported
- Further contamination of plants and plant products other than cereals and their products reported in the present book/article[713]: Nuts (pistachio nut), AFB_1, AFB_2, and AFG_1, literature[713]

incidence: 2/50, conc. range: 0.10–0.23 μg/kg, Ø conc.: 0.165 μg/kg, sample origin: imported from California (USA), Iran, and unknown origin purchased in hypermarkets and supermarkets in Catalonia (autonomous community), Spain, sample year: June–December 2008, country: Spain[728]

- Co-contamination: 2 sa. co-contaminated with AFB_1, AFB_2, AFG_1, and AFG_2; no further information available
- Further contamination of plants and plant products other than cereals and their products reported in the present book/article[728]: Nuts (pistachio nut), AFB_1, AFB_2, AFG_1, and OTA, literature[728]

incidence: 1/151, conc.: 0.42 μg/kg, sample origin: markets, supermarkets, groceries, and retail shops in Adana and Osmaniye (provinces), Turkey, sample year: January 2010–December 2012, country: Turkey[741]

- Co-contamination: not reported

- Further contamination of plants and plant products other than cereals and their products reported in the present book/article[741]: Nuts (peanut), Nuts (pistachio nut), AFB_1, AFB_2, and AFG_1, literature[741]

Aflatoxin M_1

incidence: 5/54, conc. range: 0.9–51.8 µg/kg, ∅ conc.: 21.7 µg/kg, sample origin: commercial sources in Tokyo (capital), Japan, sample year: June–September 1981, country: Japan[129]

- Co-contamination: 5 sa. co-contaminated with AFB_1, AFB_2, AFM_1, and AFL-I
- Further contamination of plants and plant products other than cereals and their products reported in the present book/article[129]: Beans, Beans (butter beans), Beans (kidney bean), Beans (saltani saltapya bean), AFB_1 and AFB_2, literature[129]; Nuts (pistachio nut), AFB_1 and AFB_2, literature[129]; see also Nuts (pistachio nut), chapter: Further Mycotoxins and Microbial Metabolites

Aflatoxin/s

incidence: 7/21, conc. range: 0.01–4 µg/kg* (5 sa.) and >4 µg/kg* (2 sa., maximum: 2,200 µg/kg*), sample origin: ethnic stores, health-food stores, and local stores as well as supermarkets in the four largest cities, Sweden, sample year: autumn 1996–summer 1998, country: Sweden[13], *AFB_1, AFB_2, AFG_1, and AFG_2

- Co-contamination: not reported
- Further contamination of plants and plant products other than cereals and their products reported in the present book/article[13]: Beans (brown bean), Peas, and Peas (chickpea), OTA, literature[13]; Juices (apple juice), PAT, literature[13]; Nuts (almond), Nuts (pistachio nut), AFB_1, literature[13]; Nuts (Brazil nut, paranut), Oil seeds, Products (peanut product), AFB_1 and AF/S, literature[13]

incidence: 17/175, conc. range: 5.0–9.9 µg/kg* (4 sa.), 10.0–14.9 µg/kg* (3 sa.), 15.0–19.9 µg/kg* (1 sa.), 20.0–24.9 µg/kg* (2 sa.), and >25 µg/kg* (7 sa.), sample origin: brokers, distributors, growers, processors, and retail trade, Canada, sample year: 1970–1975, country: Canada[15], *AFB_1, AFB_2, AFG_1, and AFG_2

- Co-contamination: not reported
- Further contamination of plants and plant products other than cereals and their products reported in the present book/article[15]: Butter (peanut butter), Cocoa beans, Nuts (almond), Nuts (cashew nut), Nuts (peanut), Nuts (pecan nut), Nuts (walnut), Seeds (sunflower), AF/S, literature[15]

incidence: 2/10*, conc. range: 1.2–3.0 µg/kg**, ∅ conc.: 2.10 µg/kg**, sample origin: retail shops and local markets in NWFP and northern areas, Pakistan, sample year: January–November 2009, country: Pakistan[17], *pistachios with shell, **AFB_1, AFB_2, AFG_1, and AFG_2

incidence: 5/10*, conc. range: 2.0–14.0 µg/kg**, ∅ conc.: 6.34 µg/kg**, sample origin: retail shops and local markets in NWFP and northern areas, Pakistan, sample year: January–November 2009, country: Pakistan[17], *pistachios without shell, **AFB_1, AFB_2, AFG_1, and AFG_2

- Co-contamination: not reported
- Further contamination of plants and plant products other than cereals and their products reported in the present book/article[17]: Dates, Figs, Fruits, dried, Kernels (apricot), Nuts (almond), Nuts (peanut), Nuts (pine nut), Nuts (walnut), Raisins, AF/S, literature[17]

incidence: 1/4, conc.: 3.87 µg/kg*, sample origin: mainly imported from local retail outlets in Lodz (city), Poland, sample year: unknown, country: Poland[18], *AFS total

- Co-contamination: not reported
- Further contamination of plants and plant products other than cereals and their products reported in the present book/article[18]: Nut cereals, Nuts, Nuts (almond), Nuts (cashew nut), Nuts (hazelnut), Nuts (peanut), Nuts (walnut), Seeds (sunflower), AF/S, literature[18]

incidence: 7/22, conc. range: ≤ 20 µg/kg (4 sa.) and >20 µg/kg* (3 sa., maximum: 252 µg/kg*,), sample origin: FDA districts, USA, sample year: 1986, country: USA[19], *AFB_1, AFB_2, AFG_1, and AFG_2

incidence: 10/21*, conc. range: ≤133 µg/kg**, sample origin: FDA districts, USA, sample year: 1986, country: USA[19], *pistachio nuts, shelled, **AFB_1, AFB_2, AFG_1, and AFG_2

- Co-contamination: not reported
- Further contamination of plants and plant products other than cereals and their products reported in the present book/article[19]: Butter (peanut butter), Candies (peanut candy), Candies (pistachio candy), Nuts, Nuts (almond), Nuts (Brazil nut, paranut), Nuts (peanut), Nuts (pecan nut), Nuts (walnut), Oils (peanut oil), Paste (peanut paste), Paste (pipian paste), Products (peanut product), Seeds (melon), Spices (cayenne pepper, chilli), Spices (ginger), Spices (nutmeg), AF/S, literature[19]

incidence: 14/70, conc. range: ≤108.3 µg/kg*, ∅ conc.: 8.9 µg/kg*, sample origin: supermarkets and hypermarkets in 12 main cities of Catalonia (autonomous community), Spain, sample year: 2008/2009, country: Spain[93], *AFB_1, AFB_2, AFG_1, and AFG_2

- Co-contamination: not reported
- Further contamination of plants and plant products other than cereals and their products reported in the present book/article[93]: Figs, Nuts (peanut), Spices (cayenne pepper, chilli), AF/S, literature[93]

incidence: 8/24, conc. range: 1.0–3.9 µg/kg* (1 sa), 4.0–10.0 µg/kg* (1 sa.), 10.1–20.0 µg/kg* (1 sa.), 20.1–50.0 µg/kg* (2 sa.), and > 50 µg/kg* (3 sa., maximum: 175 µg/kg*), sample origin: port sa. from Felixstowe (town), Manchester (city), and Kings Lynn (market town) Port Health Authorities, UK, sample year: unknown, country: UK[206], *AFB_1, AFB_2, AFG_1, and AFG_2

- Co-contamination: not reported
- Further contamination of plants and plant products other than cereals and their products reported in the present book/article[206]: Butter (peanut butter), Figs, Nuts (Brazil nut, paranut), Nuts (peanut), Paste (fig paste), AF/S, literature[206]

incidence: 3/6, conc. range: 0.23–81.64 µg/kg*, sample origin: sales outlets, Qatar, sample year: October 2002, country: Qatar[229], *AFB_1, AFB_2, AFG_1, and AFG_2

- Co-contamination: not reported
- Further contamination of plants and plant products other than cereals and their products reported in the present book/article[229]: Butter (peanut butter), Figs, and Nuts (peanut), AF/S, literature[229]; Coffee and Raisins, OTA, literature[229]; Spices, Spices (cayenne pepper, chilli), AF/S and OTA, literature[229]

incidence: 1/20, conc.: 0.7 µg/kg*, sample origin: available in New Zealand, sample year: 2010, country: New Zealand[277], *AFS total

- Co-contamination: not reported
- Further contamination of plants and plant products other than cereals and their products reported in the present book/article[277]:

Bars, AFB_1, AF/S, DON, NIV, and OTA, literature[277]; Butter (peanut butter), Confectionery, Food, Fruits, dried, Nuts, Nuts (peanut), Sauces (peanut sauce), AFB_1 and AF/S, literature[277]; Coffee, Dates, Spices (nutmeg), Spices (pepper), Vine fruits, dried, OTA, literature[277]; Figs, Spices (cayenne pepper, chilli), Spices (curry), Spices (ginger), and Spices (paprika), AFB_1, AF/S, and OTA, literature[277]; Nuts (Brazil nut, paranut), Nuts (pistachio nut), AFB_1, literature[277]

incidence: 9/20, conc. range: ≤1,450 µg/kg*, ∅ conc.: 163 µg/kg*, sample origin: retail shops and local markets in Rabat-Salé (region), Morocco, sample year: January–October 2006, country: Spain/Morocco[352], *AFB_1, AFB_2, AFG_1, and AFG_2

- Co-contamination: not reported
- Further contamination of plants and plant products other than cereals and their products reported in the present book/article[352]: Figs, Nuts (peanut), Nuts (walnut), Raisins, AFB_1 and AF/S, literature[352]; Nuts (pistachio nut), AFB_1, literature[352]

incidence: 16/28*, conc. range: 2.31–63.11 µg/kg**, sample origin: exporting companies in Aydin (province), Turkey, sample year: January–August 2007, country: Turkey[360], *destined for export, **AFB_1, AFB_2, AFG_1, and AFG_2

- Co-contamination: 3 sa. co-contaminated with AFB_1, AFB_2, AFG_1, and AFG_2; 1 sa. co-contaminated with AFB_1, AFB_2, and AFG_1; 10 sa. co-contaminated with AFB_1 and AFB_2; 1 sa. co-contaminated with AFB_1 and AFG_1; 1 sa. contaminated solely with AFB_1
- Further contamination of plants and plant products other than cereals and their products reported in the present book/article[360]: Figs, Nuts (hazelnut), Nuts (peanut), Spices (paprika), AF/S, literature[360]

incidence: 16/28* **, conc. range: 0.1–20 µg/kg*** (13 sa.) and >20 µg/kg*** (3 sa., maximum: 55.3 µg/kg***), sample origin: sa. imported and collected by food inspectors at overland, seaport, and airport health offices, Qatar, sample year: 1999, country: Qatar[423], *ncac, **pistachio nut with shell, ***AFB_1, AFB_2, AFG_1, and AFG_2

incidence: 10/18* **, conc. range: 0.1–20 µg/kg*** (7 sa.) and >20 µg/kg*** (3 sa., maximum: 50.4 µg/kg***), sample origin: sa. imported and collected by food inspectors at overland, seaport, and airport health offices, Qatar, sample year: 2000, country: Qatar[423], *ncac, **pistachio nut with shell, ***AFB_1, AFB_2, AFG_1, and AFG_2

incidence: 10/13* **, conc. range: 0.1–20 µg/kg*** (7 sa.) and >20 µg/kg*** (3 sa., maximum: 90.6 µg/kg***), sample origin: sa. imported and collected by food inspectors at overland, seaport, and airport health offices, Qatar, sample year: 1999, country: Qatar[423], *ncac, **pistachio nut without shell, ***AFB_1, AFB_2, AFG_1, and AFG_2

incidence: 5/9* **, conc. range: 0.1–20 µg/kg*** (2 sa.) and >20 µg/kg*** (3 sa., maximum: 117 µg/kg***), sample origin: sa. imported and collected by food inspectors at overland, seaport, and airport health offices, Qatar, sample year: 2000, country: Qatar[423], *ncac, **pistachio nut without shell, ***AFB_1, AFB_2, AFG_1, and AFG_2

- Co-contamination: not reported
- Further contamination of plants and plant products other than cereals and their products reported in the present book/article[423]: Nuts (peanut), AF/S, literature[423]

incidence: 5/23*, conc. range: 0.53–11.9 µg/kg** (3 sa.) and 20.8–23 µg/kg** (2 sa.), sample origin: most sa. from Iran, Syria, Turkey, and the USA collected by inspectors at overland and seaport custom offices, Qatar, sample year: July–December 1997, country: Qatar[424], *pistachio nuts with shell, **AFB_1, AFB_2, AFG_1, and AFG_2

incidence: 12/23*, conc. range: 7.3–18.6 µg/kg** (6 sa.) and 39–289 µg/kg** (6 sa.), sample origin: most sa. from Iran, Syria, Turkey, and the USA collected by inspectors at overland and seaport custom offices, Qatar, sample year: July–December 1997, country: Qatar[424], *pistachio nuts without shell, **AFB_1, AFB_2, AFG_1, and AFG_2

incidence: 8/24*, conc. range: 1.2–18.1 µg/kg** (4 sa.) and 22.3–75 µg/kg** (4 sa.), sample origin: most sa. from Iran, Syria, Turkey, and the USA collected by inspectors at overland and seaport custom offices, Qatar, sample year: 1998, country: Qatar[424], *pistachio nuts with shell, **AFB_1, AFB_2, AFG_1, and AFG_2

incidence: 40/77*, conc. range: 8.3–19.1 µg/kg** (16 sa.) and 30–275 µg/kg** (24 sa.), sample origin: most sa. from Iran, Syria, Turkey, and the USA collected by inspectors at overland and seaport custom offices, Qatar, sample year: 1998, country: Qatar[424], *pistachio nuts without shell, **AFB_1, AFB_2, AFG_1, and AFG_2

- Co-contamination: not reported
- Further contamination of plants and plant products other than cereals and their products reported in the present book/article[424]: Nuts, Nuts (peanut), AF/S, literature[424]

incidence: 46/46, conc. range: 6–97 µg/kg*, ∅ conc.: 24 µg/kg*, sample origin: imported from Iran collected at New York (city) port, USA, sample year: unknown, country: USA[561], *AF

- Co-contamination: not reported
- Further contamination of plants and plant products other than cereals and their products reported in the present book/article[561]: not reported

incidence: 2/31, conc. range: 0.4–0.7 µg/kg, ∅ conc.: 0.55 µg/kg, sample origin: retail outlets in Constantine and Jijel (provinces) as well as Algiers (capital), north Algeria, sample year: 2008, country: Spain[562], *AFS

- Co-contamination: 1 sa. co-contaminated with AFS and OTA; 1 sa. contaminated solely with AFS
- Further contamination of plants and plant products other than cereals and their products reported in the present book/article[562]: Nuts (pistachio nut), OTA, literature[562]

incidence: 11/121, conc. range: 1–305 µg/kg*, sample origin: imported to New Zealand, sample year: 1997–1999 and 2001–2004, country: New Zealand[700], *AFB_1, AFB_2, AFG_1, and AFG_2

- Co-contamination: not reported
- Further contamination of plants and plant products other than cereals and their products reported in the present book/article[700]: Butter (peanut butter), Nuts (peanut), AF/S, literature[700]

incidence: 1/20, conc.: 0.7 µg/kg*, sample origin: New Zealand, sample year: 2010, country: New Zealand[701], *AFB_1 and AFB_2

- Co-contamination: not reported
- Further contamination of plants and plant products other than cereals and their products reported in the present book/article[701]: Butter (peanut butter), Nuts (peanut), Products (nut product), Products (peanut product), AF/S, literature[701]; Nuts (Brazil nut, paranut), AFB_1, literature[701]

Other *Aspergillus* Toxins

Sterigmatocystin

incidence: 2/6, conc. range: 70–100 µg/kg, Ø conc.: 85 µg/kg, sample origin: major local retail outlets at Nainital (district), India, sample year: 1983–1985, country: India[7]

- Co-contamination: Often more than one mycotoxin was present in a sa.
- Further contamination of plants and plant products other than cereals and their products reported in the present book/article[7]: Coconuts, AFB_1, AFB_2, AFG_1, AFG_2, and CIT, literature[7]; Dates, AFB_1, AFB_2, AFG_1, and OTA, literature[7]; Grogannuts, AFB_1, AFB_2, AFG_1, AFG_2, and CIT, literature[7]; Nuts (almond), AFB_1, AFB_2, AFG_1, RBT, STG, and ZEA, literature[7]; Nuts (cashew nut), AFB_1, AFB_2, AFG_1, AFG_2, CIT, and OTA, literature[7]; Nuts (peanut), Raisins, AFB_1, AFB_2, AFG_1, AFG_2, and OTA, literature[7]; Nuts (pistachio nut), AFB_1, AFB_2, AFG_1, and AFG_2, literature[7]; Nuts (walnut), AFB_1, AFB_2, AFG_1, AFG_2, and RBT, literature[7]

Aspergillus and *Penicillium* Toxins

Ochratoxin A

incidence: 6/24, conc. range: 11–203 µg/kg, Ø conc.: 89 µg/kg, sample origin: retail markets and supermarkets in 24 regions, Tunisia, sample year: unknown, country: Tunisia[20]

- Co-contamination: not reported
- Further contamination of plants and plant products other than cereals and their products reported in the present book/article[20]: Nuts (almond), Nuts (peanut), Spices (caraway), Spices (coriander), Spices (curcuma), Spices (paprika), and Spices (pepper), OTA, literature[20]

incidence: 2/23, conc. range: 0.9–46.5 µg/kg, Ø conc.: 23.7 µg/kg, sample origin: retail outlets, UK, sample year: September 2001, country: UK[203]

- Co-contamination: 1 sa. co-contaminated with AFB1, AFB2, and OTA; 1 sa. contaminated solely with OTA
- Further contamination of plants and plant products other than cereals and their products reported in the present book/article[203]: Butter (nut butter), AFB_1, AFB_2, AFG_1, AFG_2, and OTA, literature[203]; Butter (peanut butter), AFB_1, AFB_2, AFG_1, AFG_2, and OTA, literature[203]; Dates, AFB_1 and OTA, literature[203]; Figs, AFB_1, AFB_2, AFG_2, and OTA, literature[203]; Fruits, dried, AFB_1, AFB_2, and OTA, literature[203]; Nuts (Brazil nut, paranut), AFB_1, AFB_2, and AFG_1, literature[203]; Nuts (peanut) and Nuts (pistachio nut), AFB_1 and AFB_2, literature[203]

incidence: 0/29, conc.: no contamination, sample origin: retail outlets, UK, sample year: September 2001, country: UK[203]

- Co-contamination: not reported
- Further contamination of plants and plant products other than cereals and their products reported in the present book/article[203]: Butter (nut butter), AFB_1, AFB_2, AFG_1, AFG_2, and OTA, literature[203]; Butter (peanut butter), AFB_1, AFB_2, AFG_1, AFG_2, and OTA, literature[203]; Dates, AFB_1 and OTA, literature[203]; Figs, AFB_1, AFB_2, AFG_2, and OTA, literature[203]; Fruits, dried, AFB_1, AFB_2, and OTA, literature[203]; Nuts (Brazil nut, paranut), AFB_1, AFB_2, and AFG_1, literature[203]; Nuts (peanut), Nuts (pistachio nut), AFB_1 and AFB_2, literature[203]

incidence: 2/70, conc. range: 0.134–0.321 µg/kg, Ø conc.: 0.228 µg/kg, sample origin: hypermarkets and supermarkets of 12 cities, Catalonia (autonomous community), Spain, sample year: June–November 2008, country: Spain[432]

- Co-contamination: not reported
- Further contamination of plants and plant products other than cereals and their products reported in the present book/article[432]: Nuts (peanut) OTA, literature[432]

incidence: 1/31, conc.: 170 µg/kg, sample origin: retail outlets in Constantine and Jijel (provinces) as well as Algiers (capital), north Algeria, sample year: 2008, country: Spain[562], *AFS

- Co-contamination: 1 sa. co-contaminated with AFS and OTA
- Further contamination of plants and plant products other than cereals and their products reported in the present book/article[562]: Nuts (pistachio nut), AF/S, literature[562]

incidence: 4/59*, conc. range: 0.74–890 µg/kg, Ø conc.: 224 µg/kg, sample origin: supermarkets and stores, USA, sample year: 2012–2014, country: USA[599], *conventional

incidence: 0/26*, conc.: no contamination, sample origin: supermarkets and stores, USA, sample year: 2012–2014, country: USA[599], *organic

- Co-contamination: not reported
- Further contamination of plants and plant products other than cereals and their products reported in the present book/article[599]: Dates, Figs, Raisins, OTA, literature[599]

incidence: 1/50, conc.: 0.67 µg/kg, sample origin: imported from California (USA), Iran, and unknown origin purchased in hypermarkets and supermarkets in Catalonia (autonomous community), Spain, sample year: June–December 2008, country: Spain[728]

- Co-contamination: not reported
- Further contamination of plants and plant products other than cereals and their products reported in the present book/article[728]: Nuts (pistachio nut), AFB_1, AFB_2, AFG_1, and AFG_2, literature[728]

incidence: 1/8, conc.: 9 µg/kg, sample origin: KSU, Turkey, and stores in Tulln (city) and Vienna (capital), Austria, sample year: unknown, country: Austria[868]

- Co-contamination: For detailed information see the article.
- Further contamination of plants and plant products other than cereals and their products reported in the present book/article[868]: Nuts (almond), AME, AOH, BEA, CPA, ENB, ENB_1, and ENB_3, literature[868]; Nuts (hazelnut), 3-NPA, AFB_1, AFB_2, AFG_1, AFG_2, AME, AOH, ATX-I, BEA, ENA, ENA_1, ENB, ENB_1, ENB_2, ENB_3, ENB_4, HT-2, KA, MA, MON, OTA, OTB, STG, T-2, and ZEA, literature[868]; Nuts (peanut), 3-NPA, AFB_1, AFB_2, AME, AOH, ATX-I, BEA, CPA, ENA, ENA_1, ENB, ENB_4, KA, MA, OTA, OTB, STG, and ZEA, literature[868]; Nuts (pistachio nut), BEA and MA, literature[868]; see also Nuts (almond), Nuts (hazelnut), Nuts (peanut), Nuts (pistachio nut), chapter: Further Mycotoxins and Microbial Metabolites

For detailed information see the article.

Fusarium Toxins

Depsipeptides

Beauvericin

incidence: 1/8, conc.: 0.30 µg/kg, sample origin: KSU, Turkey and stores in Tulln (city) and Vienna (capital), Austria, sample year: unknown, country: Austria[868]

- Co-contamination: For detailed information see the article.
- Further contamination of plants and plant products other than cereals and their products reported in the present book/

article[868]: Nuts (almond), AME, AOH, BEA, CPA, ENB, ENB_1, and ENB_3, literature[868]; Nuts (hazelnut), 3-NPA, AFB_1, AFB_2, AFG_1, AFG_2, AME, AOH, ATX-I, BEA, ENA, ENA_1, ENB_1, ENB_2, ENB_3, ENB_4, HT-2, KA, MA, MON, OTA, OTB, STG, T-2, and ZEA, literature[868]; Nuts (peanut), 3-NPA, AFB_1, AFB_2, AME, AOH, ATX-I, BEA, CPA, ENA, ENA_1, ENB, ENB_4, KA, MA, OTA, OTB, STG, and ZEA, literature[868]; Nuts (pistachio nut), MA and OTA, literature[868]; see also Nuts (almond), Nuts (hazelnut), Nuts (peanut), Nuts (pistachio nut), chapter: Further Mycotoxins and Microbial Metabolites

For detailed information see the article.

Penicillium Toxins

Mycophenolic Acid

incidence: 1/8, conc.: 13 µg/kg, sample origin: KSU, Turkey and stores in Tulln (city) and Vienna (capital), Austria, sample year: unknown, country: Austria[868]

- Co-contamination: For detailed information see the article.
- Further contamination of plants and plant products other than cereals and their products reported in the present book/article[868]: Nuts (almond), AME, AOH, BEA, CPA, ENB, ENB_1, and ENB_3, literature[868]; Nuts (hazelnut), 3-NPA, AFB_1, AFB_2, AFG_1, AFG_2, AME, AOH, ATX-I, BEA, ENA, ENA_1, ENB_1, ENB_2, ENB_3, ENB_4, HT-2, KA, MA, MON, OTA, OTB, STG, T-2, and ZEA, literature[868]; Nuts (peanut), 3-NPA, AFB_1, AFB_2, AME, AOH, ATX-I, BEA, CPA, ENA, ENA_1, ENB, ENB_4, KA, MA, OTA, OTB, STG, and ZEA, literature[868]; Nuts (pistachio nut), BEA and OTA, literature[868]; see also Nuts (almond), Nuts (hazelnut), Nuts (peanut), Nuts (pistachio nut), chapter: Further Mycotoxins and Microbial Metabolites

For detailed information see the article.

Nuts (tigernut) may contain the following mycotoxins:

Aspergillus Toxins

Aflatoxin B_1

incidence: 3/37, conc. range: 0.7–4.5 µg/kg, Ø conc.: 2.4 µg/kg, sample origin: supermarkets, street vendors, and candy shops, Valencia (city), Spain, sample year: unknown, country: Spain[162]

- Co-contamination: 1 sa. co-contaminated with AFB_1, AFB_2, AFG_1, and AFG_2; 2 sa. co-contaminated with AFB_1, AFB_2, and AFG_2
- Further contamination of plants and plant products other than cereals and their products reported in the present book/article[162]: Beverages, AFB_1, AFB_2, and AFG_2, literature[162]; Nuts (tigernut), AFB_2, AFG_1, and AFG_2, literature[162]

incidence: 14/48, conc. range: 1.2–2 µg/kg, sample origin: supermarkets, street vendors, and little candy shops, Valencia (city), Spain, sample year: March 2009–March 2010, country: Spain[163]

- Co-contamination: For detailed information see the article.
- Further contamination of plants and plant products other than cereals and their products reported in the present book/article[163]: Beverages, AFB_1, AFB_2, AFG_2, and OTA, literature[163]; Nuts (tigernut), AFB_2, AFG_2, and OTA, literature[163]

For detailed information see the article.

incidence: 27/77, conc. range: 10–120 µg/kg, sample origin: markets in different states of Nigeria, sample year: July–September 1993, country: Nigeria[564]

- Co-contamination: not reported
- Further contamination of plants and plant products other than cereals and their products reported in the present book/article[564]: not reported

For detailed information see the article.

incidence: 0/35*, conc.: no contamination, sample origin: local markets, supermarkets, and cooperatives in the Valencian community, Spain, sample year: 2010/2011, country: Spain[834], *Valencia guaranteed origin

incidence: 14/48*, conc. range: 1.2–1.9 µg/kg, sample origin: local markets, supermarkets, and cooperatives in the Valencian community, Spain, sample year: 2010/2011, country: Spain[834], *unknown origin

- Co-contamination: For detailed information see the article.
- Further contamination of plants and plant products other than cereals and their products reported in the present book/article[834]: Nuts (tigernut), AFB_2, AFG_1, AFG_2, BEA, DON, and OTA, literature[834]

Aflatoxin B_2

incidence: 3/37, conc. range: 2.2–3.5 µg/kg, Ø conc.: 3 µg/kg, sample origin: supermarkets, street vendors, and candy shops, Valencia (city), Spain, sample year: unknown, country: Spain[162]

- Co-contamination: 1 sa. co-contaminated with AFB_1, AFB_2, AFG_1, and AFG_2; 2 sa. co-contaminated with AFB_1, AFB_2, and AFG_2
- Further contamination of plants and plant products other than cereals and their products reported in the present book/article[162]: Beverages, AFB_1, AFB_2, and AFG_2, literature[162]; Nuts (tigernut), AFB_1, AFG_1, and AFG_2, literature[162]

incidence: 3/48, conc. range: 1.5–1.8 µg/kg, sample origin: supermarkets, street vendors, and little candy shops, Valencia (city), Spain, sample year: March 2009–March 2010, country: Spain[163]

- Co-contamination: For detailed information see the article.
- Further contamination of plants and plant products other than cereals and their products reported in the present book/article[163]: Beverages, AFB_1, AFB_2, AFG_2, and OTA, literature[163]; Nuts (tigernut), AFB_1, AFG_2, and OTA, literature[163]

For detailed information see the article.

incidence: 0/35*, conc.: no contamination, sample origin: local markets, supermarkets, and cooperatives in the Valencian community, Spain, sample year: 2010/2011, country: Spain[834], *Valencia guaranteed origin

incidence: 2/48*, conc. range: 1.4–1.8 µg/kg, Ø conc.: 1.6 µg/kg, sample origin: local markets, supermarkets, and cooperatives in the Valencian community, Spain, sample year: 2010/2011, country: Spain[834], *unknown origin

- Co-contamination: For detailed information see the article.
- Further contamination of plants and plant products other than cereals and their products reported in the present book/article[834]: Nuts (tigernut), AFB_1, AFG_1, AFG_2, BEA, DON, and OTA, literature[834]

Aflatoxin G_1

incidence: 1/37, conc.: 3.8 µg/kg, sample origin: supermarkets, street vendors, and candy shops, Valencia (city), Spain, sample year: unknown, country: Spain[162]

- Co-contamination: 1 sa. co-contaminated with AFB_1, AFB_2, AFG_1, and AFG_2

- Further contamination of plants and plant products other than cereals and their products reported in the present book/article[162]: Beverages, AFB_1, AFB_2, and AFG_2, literature[162]; Nuts (tigernut), AFB_1, AFB_2, and AFG_2, literature[162]

incidence: 0/35*, conc.: no contamination, sample origin: local markets, supermarkets, and cooperatives in the Valencian community, Spain, sample year: 2010/2011, country: Spain[834], *Valencia guaranteed origin

incidence: 2/48*, conc. range: 1.3–1.5 µg/kg, ∅ conc.: 1.4 µg/kg, sample origin: local markets, supermarkets, and cooperatives in the Valencian community, Spain, sample year: 2010/2011, country: Spain[834], *unknown origin

- Co-contamination: For detailed information see the article.
- Further contamination of plants and plant products other than cereals and their products reported in the present book/article[834]: Nuts (tigernut), AFB_1, AFB_2, AFG_2, BEA, DON, and OTA, literature[834]

Aflatoxin G_2

incidence: 3/37, conc. range: 1.6–2.7 µg/kg, ∅ conc.: 2.1 µg/kg, sample origin: supermarkets, street vendors, and candy shops, Valencia, sample year: unknown, country: Spain[162]

- Co-contamination: 1 sa. co-contaminated with AFB_1, AFB_2, AFG_1, and AFG_2; 2 sa. co-contaminated with AFB_1, AFB_2, and AFG_2
- Further contamination of plants and plant products other than cereals and their products reported in the present book/article[162]: Beverages, AFB_1, AFB_2, and AFG_2, literature[162]; Nuts (tigernut), AFB_1, AFB_2, and AFG_1, literature[162]

incidence: 8/48, conc. range: 1.3–3.8 µg/kg, sample origin: supermarkets, street vendors, and little candy shops, Valencia (city), Spain, sample year: March 2009–March 2010, country: Spain[163]

- Co-contamination: For detailed information see the article.
- Further contamination of plants and plant products other than cereals and their products reported in the present book/article[163]: Beverages, AFB_1, AFB_2, AFG_2, and OTA, literature[163]; Nuts (tigernut), AFB_1, AFB_2, and OTA, literature[163]

For detailed information see the article.

incidence: 1/35*, conc.: 1.3 µg/kg, sample origin: local markets, supermarkets, and cooperatives in the Valencian community, Spain, sample year: 2010/2011, country: Spain[834], *Valencia guaranteed origin

incidence: 9/48*, conc. range: 1.8–3.1 µg/kg, sample origin: local markets, supermarkets, and cooperatives in the Valencian community, Spain, sample year: 2010/2011, country: Spain[834], *unknown origin

- Co-contamination: For detailed information see the article.
- Further contamination of plants and plant products other than cereals and their products reported in the present book/article[834]: Nuts (tigernut), AFB_1, AFB_2, AFG_1, BEA, DON, and OTA, literature[834]

Aspergillus and ***Penicillium*** Toxins

Ochratoxin A

incidence: 6/48, conc. range: 3.5–11 µg/kg, sample origin: supermarkets, street vendors, and little candy shops, Valencia (city), Spain, sample year: March 2009–March 2010, country: Spain[163]

- Co-contamination: For detailed information see the article.
- Further contamination of plants and plant products other than cereals and their products reported in the present book/article[163]: Beverages, AFB_1, AFB_2, AFG_2, and OTA, literature[163]; Nuts (tigernut), AFB_1, AFB_2, and AFG_2, literature[163]

For detailed information see the article.

incidence: 1/35*, conc.: 1.6 µg/kg, sample origin: local markets, supermarkets, and cooperatives in the Valencian community, Spain, sample year: 2010/2011, country: Spain[834], *Valencia guaranteed origin

incidence: 7/48*, conc. range: 1.2–37 µg/kg, sample origin: local markets, supermarkets, and cooperatives in the Valencian community, Spain, sample year: 2010/2011, country: Spain[834], *unknown origin

- Co-contamination: For detailed information see the article.
- Further contamination of plants and plant products other than cereals and their products reported in the present book/article[834]: Nuts (tigernut), AFB_1, AFB_2, AFG_1, AFG_2, BEA, and DON, literature[834]

Fusarium Toxins

Depsipeptides

Beauvericin

incidence: 5/47, conc. range: 51,600–228,500 µg/kg, sample origin: supermarkets in the Valencian community, Spain, sample year: unknown, country: Spain[565]

- Co-contamination: 5 sa. co-contaminated with BEA and ENB_1; no further information available
- Further contamination of plants and plant products other than cereals and their products reported in the present book/article[565]: Nuts (tigernut), ENA, ENA_1, ENB, and ENB_1, literature[565]

incidence: 3/35*, conc. range: 5–125 µg/kg, sample origin: local markets, supermarkets, and cooperatives in the Valencian community, Spain, sample year: 2010/2011, country: Spain[834], *Valencia guaranteed origin

incidence: 8/48*, conc. range: 2.5–161 µg/kg, sample origin: local markets, supermarkets, and cooperatives in the Valencian community, Spain, sample year: 2010/2011, country: Spain[834], *unknown origin

- Co-contamination: For detailed information see the article.
- Further contamination of plants and plant products other than cereals and their products reported in the present book/article[834]: Nuts (tigernut), AFB_1, AFB_2, AFG_1, AFG_2, DON, and OTA, literature[834]

Enniatin A

incidence: 1/47, conc.: 676,500 µg/kg, sample origin: supermarkets in the Valencian community, Spain, sample year: unknown, country: Spain[565]

- Co-contamination: not reported
- Further contamination of plants and plant products other than cereals and their products reported in the present book/article[565]: Nuts (tigernut), BEA, ENA_1, ENB, and ENB_1, literature[565]

Enniatin A_1

incidence: 8/47, conc. range: 32,200–4,440,00 µg/kg, sample origin: supermarkets in the Valencian community, Spain, sample year: unknown, country: Spain[565]

- Co-contamination: not reported
- Further contamination of plants and plant products other than cereals and their products reported in the present book/article[565]: Nuts (tigernut), BEA, ENA, ENB, and ENB_1, literature[565]

Enniatin B

incidence: 1/47, conc.: 44,800 µg/kg, sample origin: supermarkets in the Valencian community, Spain, sample year: unknown, country: Spain[565]

- Co-contamination: not reported
- Further contamination of plants and plant products other than cereals and their products reported in the present book/article[565]: Nuts (tigernut), BEA, ENA, ENA_1, and ENB_1, literature[565]

Enniatin B_1

incidence: 5/47, conc. range: 21,600–346,000 µg/kg, sample origin: supermarkets in the Valencian community, Spain, sample year: unknown, country: Spain[565]

- Co-contamination: not reported
- Further contamination of plants and plant products other than cereals and their products reported in the present book/article[565]: Nuts (tigernut), BEA, ENA, ENA_1, and ENB, literature[565]

Type B Trichothecenes

Deoxynivalenol

incidence: 2/35*, conc. range: 33–43 µg/kg, ∅ conc.: 38 µg/kg, sample origin: local markets, supermarkets, and cooperatives in the Valencian community, Spain, sample year: 2010/2011, country: Spain[834], *Valencia guaranteed origin

incidence: 3/48*, conc. range: 34–69 µg/kg, sample origin: local markets, supermarkets, and cooperatives in the Valencian community, Spain, sample year: 2010/2011, country: Spain[834], *unknown origin

- Co-contamination: For detailed information see the article.
- Further contamination of plants and plant products other than cereals and their products reported in the present book/article[834]: Nuts (tigernut), AFB_1, AFB_2, AFG_1, AFG_2, BEA, and OTA, literature[834]

Nuts (walnut) may contain the following mycotoxins:

Aspergillus **Toxins**

Aflatoxin B_1

incidence: 3/9, conc. range: 15–110 µg/kg, sample origin: major local retail outlets at Nainital (district), India, sample year: 1983–1985, country: India[7]

- Co-contamination: Often more than one mycotoxin was present in a sa.
- Further contamination of plants and plant products other than cereals and their products reported in the present book/article[7]: Coconuts, AFB_1, AFB_2, AFG_1, AFG_2, and CIT, literature[7]; Dates, AFB_1, AFB_2, AFG_1, and OTA, literature[7]; Grogannuts, AFB_1, AFB_2, AFG_1, AFG_2, and CIT, literature[7]; Nuts (almond), AFB_1, AFB_2, AFG_1, RBT, STG, and ZEA, literature[7]; Nuts (cashew nut), AFB_1, AFB_2, AFG_1, AFG_2, CIT, and OTA, literature[7]; Nuts (peanut), Raisins, AFB_1, AFB_2, AFG_1, AFG_2, and OTA, literature[7]; Nuts (pistachio nut), AFB_1, AFB_2, AFG_1, AFG_2, and STG, literature[7]; Nuts (walnut), AFB_2, AFG_1, AFG_2, and RBT, literature[7]

incidence: 6/24, conc. range: 3–28 µg/kg, ∅ conc.: 22.1 µg/kg, sample origin: retailers, markets, and dried fruit retail shops in Erzurum (city), Turkey, sample year: March 2001–March 2002, country. Turkey[14]

- Co-contamination: not reported
- Further contamination of plants and plant products other than cereals and their products reported in the present book/article[14]: Nuts (almond), Nuts (hazelnut), Nuts (peanut), and Peas (chickpea), AFB_1, literature[14]

incidence: 6?/560, conc. range: <0.4–0.2? µg/kg, ∅ conc.: 0.1 µg/kg?, sample origin: sites of primary storage or processing (domestic products) and sites of importation, Cyprus, sample year: 1992–1996, country: Cyprus[200]

- Co-contamination: not reported
- Further contamination of plants and plant products other than cereals and their products reported in the present book/article[200]: Butter (peanut butter), Figs, Nuts (peanut), AFB_1, AFB_2, AFG_1, and AFG_2, literature[200]; Nuts (Brazil nut, paranut), AFB_1, AFB_2, and AFG_1, literature[200]; Nuts (chestnut), Products (sesame product), Spices, AFB_1, literature[200]; Nuts (pistachio nut), AFB_1 and AFB_2, literature[200]; Nuts (walnut), AFB_2 and AFG_1, literature[200]; Seeds (sesame), AFB_1, AFG_1, and AFG_2, literature[200]

incidence: 1/14, conc.: 8 µg/kg, sample origin: stores in Oslo (capital) area, Norway, sample year: autumn 1973, country: Norway[201]

- Co-contamination: 1 sa. co-contaminated with AFB_1, AFB_2, AFG_1, and AFG_2
- Further contamination of plants and plant products other than cereals and their products reported in the present book/article[201]: Cocoa, Nuts (Brazil nut, paranut), Nuts (cashew nut), and Nuts (peanut), AFB_1, AFB_2, AFG_1, and AFG_2, literature[201]; Nuts, Nuts (walnut), AFB_2, AFG_1, and AFG_2, literature[201]

incidence: 1/3, conc.: 13.5 µg/kg, sample origin: supermarkets and village stores in five districts of Penang (state) including Penang Island and Wellesley (province), Malaysia, sample year: unknown, country: Malaysia[225]

- Co-contamination: 1 sa. co-contaminated with AFB_1 and AFB_2
- Further contamination of plants and plant products other than cereals and their products reported in the present book/article[225]: Butter (peanut butter), Confectionery, Products (nut product), AFB_1 and AFB_2, literature[225]; Cake (peanut), Nuts (peanut), AFB_1, AFB_2, and AFG_2, literature[225]; Nuts (walnut), AFB_2, literature[225]

incidence: 3/12*, conc. range: 5–35,500000 µg/kg?, sample origin: unknown, sample year: unknown, country: Germany[239], *walnuts, moldy

- Co-contamination: not reported
- Further contamination of plants and plant products other than cereals and their products reported in the present book/article[239]: Fruits (apple), Fruits (cherry), Fruits (lemon), Fruits (orange), Fruits (peach), Fruits (tomato), Nuts (almond), Nuts (cashew nut), Nuts (hazelnut), Seeds (sesame), Spice (nutmeg), AFB_1, literature[239]; Nuts (peanut), AFB_1, AFB_2, AFG_1, and AFG_2, literature[239]

incidence: 6/20, conc. range: ≤2,500 µg/kg, ∅ conc.: 360 µg/kg, sample origin: retail shops and local markets in Rabat-Salé (region), Morocco, sample year: January–October 2006, country: Spain/Morocco[352]

- Co-contamination: not reported
- Further contamination of plants and plant products other than cereals and their products reported in the present book/article[352]: Figs, Nuts (peanut), Nuts (pistachio nut), Raisins, AFB_1 and AF/S, literature[352]; Nuts (walnut), AF/S, literature[352]

incidence: 2/12, conc. range: 5 µg/kg, sample origin: imported from Sudan and the UK, sample year: 1982–1984, country: UK[381]

- Co-contamination: not reported
- Further contamination of plants and plant products other than cereals and their products reported in the present book/article[381]: Nuts (peanut), AFB_1, literature[381]

incidence: 21/48, conc. range: 0.14–0.32 µg/kg, sample origin: local markets in Chongqing, Fujian, Guangdong, Guangxi, Hubei, Jiangsu, Shanghai, and Zhejiang (regions), China, sample year: 2003, country: China[411]

- Co-contamination: not reported
- Further contamination of plants and plant products other than cereals and their products reported in the present book/article[411]: Nuts (peanut), AFB_1, AFB_2, AFG_1, and AFG_2, literature[411]; Nuts (pine nut), AFB_1 and AFB_2, literature[411]; Nuts (walnut), AFB_2, AFG_1, and AFG_2, literature[411]

incidence: 7/40*, conc. range: tr-1.3 µg/kg, sample origin: growers and processors in California (state), USA, sample year: September–November 1972 to February–May 1973, country: USA[566], *in-shell walnuts

- Co-contamination: 6 sa. co-contaminated with AFB_1, AFB_2, AFG_1, and AFG_2; 1 sa. co-contaminated with AFB_1 and AFB_2
- Further contamination of plants and plant products other than cereals and their products reported in the present book/article[566]: Nuts (walnut), AFB_2, AFG_1, and AFG_2, literature[566]

incidence: 3/35*, conc. range: 0.12–0.20 µg/kg, ∅ conc.: 0.16 µg/kg, sample origin: retail shops, local markets, and bazaar in different provinces, Turkey, sample year: October 2012–April 2013, country: Turkey[757], *in-shell walnuts

incidence: 5/50*, conc. range: 0.09–12.2 µg/kg, ∅ conc.: 2.65 µg/kg, sample origin: retail shops, local markets, and bazaar in different provinces, Turkey, sample year: October 2012–April 2013, country: Turkey[757], *shelled walnuts

- Co-contamination: not reported
- Further contamination of plants and plant products other than cereals and their products reported in the present book/article[757]: Nuts (walnut), AFB_2, AFG_1, and AFG_2, literature[757]

incidence: 26/35, conc. range: ≤91.90 µg/kg, sample origin: supermarkets in Tehran (capital), Iran, sample year: October/November 2010, country: Iran[802]

- Co-contamination: not reported
- Further contamination of plants and plant products other than cereals and their products reported in the present book/article[802]: Nuts (walnut), AFB_2, AFG_1, and AFG_2, literature[802]

Aflatoxin B_2

incidence: 3/9, conc. range: 12–85 µg/kg, sample origin: major local retail outlets at Nainital (district), India, sample year: 1983–1985, country: India[7]

- Co-contamination: Often more than one mycotoxin was present in a sa.
- Further contamination of plants and plant products other than cereals and their products reported in the present book/article[7]: Coconuts, AFB_1, AFB_2, AFG_1, AFG_2, and CIT, literature[7]; Dates, AFB_1, AFB_2, AFG_1, and OTA, literature[7]; Grogannuts, AFB_1, AFB_2, AFG_1, AFG_2, and CIT, literature[7]; Nuts (almond), AFB_1, AFB_2, AFG_1, RBT, STG, and ZEA, literature[7]; Nuts (cashew nut), AFB_1, AFB_2, AFG_1, AFG_2, CIT, and OTA, literature[7]; Nuts (peanut), Raisins, AFB_1, AFB_2, AFG_1, AFG_2, and OTA, literature[7]; Nuts (pistachio nut), AFB_1, AFB_2, AFG_1, AFG_2, and STG, literature[7]; Nuts (walnut), AFB_1, AFG_1, AFG_2, and RBT, literature[7]

incidence: 6?/560, conc. range: <0.3 µg/kg, sample origin: sites of primary storage or processing (domestic products) and sites of importation, Cyprus, sample year: 1992–1996, country: Cyprus[200]

- Co-contamination: not reported
- Further contamination of plants and plant products other than cereals and their products reported in the present book/article[200]: Butter (peanut butter), Figs, Nuts (peanut), AFB_1, AFB_2, AFG_1, and AFG_2, literature[200]; Nuts (Brazil nut, paranut), AFB_1, AFB_2, and AFG_1, literature[200]; Nuts (chestnut), Products (sesame product), Spices, AFB_1, literature[200]; Nuts (pistachio nut), AFB_1 and AFB_2, literature[200]; Nuts (walnut), AFB_1 and AFG_1, literature[200]; Seeds (sesame), AFB_1, AFG_1, and AFG_2, literature[200]

incidence: 1/14, conc.: tr, sample origin: stores in Oslo (capital) area, Norway, sample year: autumn 1973, country: Norway[201]

- Co-contamination: 1 sa. co-contaminated with AFB_1, AFB_2, AFG_1, and AFG_2
- Further contamination of plants and plant products other than cereals and their products reported in the present book/article[201]: Cocoa, Nuts (Brazil nut, paranut), Nuts (cashew nut), Nuts (peanut), AFB_1, AFB_2, AFG_1, and AFG_2, literature[201]; Nuts, AFB_2, AFG_1, and AFG_2, literature[201]; Nuts (walnut), AFB_1, AFG_1, and AFG_2, literature[201]

incidence: 1/3, conc.: 3.63 µg/kg, sample origin: supermarkets and village stores in five districts of Penang (state) including Penang Island and Wellesley (province), Malaysia, sample year: unknown, country: Malaysia[225]

- Co-contamination: 1 sa. co-contaminated with AFB_1 and AFB_2
- Further contamination of plants and plant products other than cereals and their products reported in the present book/article[225]: Butter (peanut butter), Confectionery, Products (nut product), AFB_1 and AFB_2, literature[225]; Cake (peanut), Nuts (peanut), AFB_1, AFB_2, and AFG_2, literature[225]; Nuts (walnut), AFB_1, literature[225]

incidence: 12/48, conc. range: 0.02–0.70 µg/kg, sample origin: local markets in Chongqing, Fujian, Guangdong, Guangxi, Hubei, Jiangsu, Shanghai, and Zhejiang (regions), China, sample year: 2003, country: China[411]

- Co-contamination: not reported
- Further contamination of plants and plant products other than cereals and their products reported in the present book/article[411]: Nuts (peanut), AFB_1, AFB_2, AFG_1, and AFG_2, literature[411]; Nuts (pine nut), AFB_1 and AFB_2, literature[411]; Nuts (walnut), AFB_1, AFG_1, and AFG_2, literature[411]

incidence: 14/40*, conc. range: <LOQ-10 µg/kg, sample origin: growers and processors in California (state), USA, sample year: September–November 1972 to February–May 1973, country: USA[566], *in-shell walnuts

- Co-contamination: 6 sa. co-contaminated with AFB_1, AFB_2, AFG_1, and AFG_2; 1 sa. co-contaminated with AFB_1 and AFB_2; 7 sa. contaminated solely with AFB_2

- Further contamination of plants and plant products other than cereals and their products reported in the present book/article[566]: Nuts (walnut), AFB_1, AFG_1, and AFG_2, literature[566]

incidence: 0/35*, conc.: no contamination, sample origin: retail shops, local markets, and bazaar in different provinces, Turkey, sample year: October 2012–April 2013, country: Turkey[757], *in-shell walnuts

incidence: 2/50*, conc. range: 0.05–0.61 µg/kg, Ø conc.: 0.33 µg/kg, sample origin: retail shops, local markets, and bazaar in different provinces, Turkey, sample year: October 2012–April 2013, country: Turkey[757], *shelled walnuts

- Co-contamination: not reported
- Further contamination of plants and plant products other than cereals and their products reported in the present book/article[757]: Nuts (walnut), AFB_1, AFG_1, and AFG_2, literature[757]

incidence: 22/35, conc. range: ≤11.05 µg/kg, sample origin: supermarkets in Tehran (capital), Iran, sample year: October/November 2010, country: Iran[802]

- Co-contamination: not reported
- Further contamination of plants and plant products other than cereals and their products reported in the present book/article[802]: Nuts (walnut), AFB_1, AFG_1, and AFG_2, literature[802]

Aflatoxin G_1

incidence: 2/9, conc. range: 10–75 µg/kg, Ø conc.: 42.5 µg/kg, sample origin: major local retail outlets at Nainital (district), India, sample year: 1983–1985, country: India[7]

- Co-contamination: Often more than one mycotoxin was present in a sa.
- Further contamination of plants and plant products other than cereals and their products reported in the present book/article[7]: Coconuts, AFB_1, AFB_2, AFG_1, AFG_2, and CIT, literature[7]; Dates, AFB_1, AFB_2, AFG_1, and OTA, literature[7]; Grogannuts, AFB_1, AFB_2, AFG_1, AFG_2, and CIT, literature[7]; Nuts (almond), AFB_1, AFB_2, AFG_1, RBT, STG, and ZEA, literature[7]; Nuts (cashew nut), AFB_1, AFB_2, AFG_1, AFG_2, CIT, and OTA, literature[7]; Nuts (peanut), Raisins, AFB_1, AFB_2, AFG_1, AFG_2, and OTA, literature[7]; Nuts (pistachio nut), AFB_1, AFB_2, AFG_1, AFG_2, and STG, literature[7]; Nuts (walnut), AFB_1, AFB_2, AFG_2, and RBT, literature[7]

incidence: 6?/560, conc. range: <0.4 µg/kg, sample origin: sites of primary storage or processing (domestic products) and sites of importation, Cyprus, sample year: 1992–1996, country: Cyprus[200]

- Co-contamination: not reported
- Further contamination of plants and plant products other than cereals and their products reported in the present book/article[200]: Butter (peanut butter), Figs, Nuts (peanut), AFB_1, AFB_2, AFG_1, and AFG_2, literature[200]; Nuts (Brazil nut, paranut), AFB_1, AFB_2, and AFG_1, literature[200]; Nuts (chestnut), Products (sesame product), Spices, AFB_1, literature[200]; Nuts (pistachio nut), Nuts (walnut), AFB_1 and AFB_2, literature[200]; Seeds (sesame), AFB_1, AFG_1, and AFG_2, literature[200]

incidence: 1/14, conc.: 4 µg/kg, sample origin: stores in Oslo (capital) area, Norway, sample year: autumn 1973, country: Norway[201]

- Co-contamination: 1 sa. co-contaminated with AFB_1, AFB_2, AFG_1, and AFG_2
- Further contamination of plants and plant products other than cereals and their products reported in the present book/article[201]: Cocoa, Nuts (Brazil nut, paranut), Nuts (cashew nut), and Nuts (peanut), AFB_1, AFB_2, AFG_1, and AFG_2, literature[201]; Nuts, AFB_2, AFG_1, and AFG_2, literature[201]; Nuts (walnut), AFB_1, AFB_2, and AFG_2, literature[201]

incidence: 8/48, conc. range: 0.36–0.83 µg/kg, sample origin: local markets in Chongqing, Fujian, Guangdong, Guangxi, Hubei, Jiangsu, Shanghai, and Zhejiang (regions), China, sample year: 2003, country: China[411]

- Co-contamination: not reported
- Further contamination of plants and plant products other than cereals and their products reported in the present book/article[411]: Nuts (peanut), AFB_1, AFB_2, AFG_1, and AFG_2, literature[411]; Nuts (pine nut), AFB_1 and AFB_2, literature[411]; Nuts (walnut), AFB_1, AFB_2, and AFG_2, literature[411]

incidence: 6/40*, conc. range: tr, sample origin: growers and processors in California (state), USA, sample year: September–November 1972 to February–May 1973, country: USA[566], *in-shell walnuts

- Co-contamination: 6 sa. co-contaminated with AFB_1, AFB_2, AFG_1, and AFG_2
- Further contamination of plants and plant products other than cereals and their products reported in the present book/article[566]: Nuts (walnut), AFB_1, AFB_2, and AFG_2, literature[566]

incidence: 2/35*, conc. range: 0.09–0.10 µg/kg, Ø conc.: 0.09 µg/kg, sample origin: retail shops, local markets, and bazaar in different provinces, Turkey, sample year: October 2012–April 2013, country: Turkey[757], *in-shell walnuts

incidence: 3/50*, conc. range: 0.10–2.52 µg/kg, Ø conc.: 0.92 µg/kg, sample origin: retail shops, local markets, and bazaar in different provinces, Turkey, sample year: October 2012–April 2013, country: Turkey[757], *shelled walnuts

- Co-contamination: not reported
- Further contamination of plants and plant products other than cereals and their products reported in the present book/article[757]: Nuts (walnut), AFB_1, AFB_2, and AFG_2, literature[757]

incidence: 4/35, conc. range: ≤6.41 µg/kg, sample origin: supermarkets in Tehran (capital), Iran, sample year: October/November 2010, country: Iran[802]

- Co-contamination: not reported
- Further contamination of plants and plant products other than cereals and their products reported in the present book/article[802]: Nuts (walnut), AFB_1, AFB_2, and AFG_2, literature[802]

Aflatoxin G_2

incidence: 1/9, conc.: 65 µg/kg, sample origin: major local retail outlets at Nainital (district), India, sample year: 1983–1985, country: India[7]

- Co-contamination: Often more than one mycotoxin was present in a sa.
- Further contamination of plants and plant products other than cereals and their products reported in the present book/article[7]: Coconuts, AFB_1, AFB_2, AFG_1, AFG_2, and CIT, literature[7]; Dates, AFB_1, AFB_2, AFG_1, and OTA, literature[7]; Grogannuts, AFB_1, AFB_2, AFG_1, AFG_2, and CIT, literature[7]; Nuts (almond), AFB_1, AFB_2, AFG_1, RBT, STG, and ZEA, literature[7]; Nuts (cashew nut), AFB_1, AFB_2, AFG_1, AFG_2, CIT, and OTA, literature[7]; Nuts (peanut), Raisins, AFB_1, AFB_2, AFG_1, AFG_2, and OTA, literature[7]; Nuts (pistachio nut), AFB_1, AFB_2,

AFG_1, AFG_2, and STG, literature[7]; Nuts (walnut), AFB_1, AFB_2, AFG_1, and RBT, literature[7]

incidence: 1/14, conc.: tr, sample origin: stores in Oslo (capital) area, Norway, sample year: autumn 1973, country: Norway[201]

- Co-contamination: 1 sa. co-contaminated with AFB_1, AFB_2, AFG_1, and AFG_2
- Further contamination of plants and plant products other than cereals and their products reported in the present book/article[201]: Cocoa, Nuts (Brazil nut, paranut), Nuts (cashew nut), Nuts (peanut), AFB_1, AFB_2, AFG_1, and AFG_2, literature[201]; Nuts, AFB_2, AFG_1, and AFG_2, literature[201]; Nuts (walnut), AFB_1, AFB_2, and AFG_1, literature[201]

incidence: 10/48, conc. range: 0.07–0.12 µg/kg, sample origin: local markets in Chongqing, Fujian, Guangdong, Guangxi, Hubei, Jiangsu, Shanghai, and Zhejiang (regions), China, sample year: 2003, country: China[411]

- Co-contamination: not reported
- Further contamination of plants and plant products other than cereals and their products reported in the present book/article[411]: Nuts (peanut), AFB_1, AFB_2, AFG_1, and AFG_2, literature[411]; Nuts (pine nut), AFB_1 and AFB_2, literature[411]; Nuts (walnut), AFB_1, AFB_2, and AFG_1, literature[411]

incidence: 6/40*, conc. range: tr, sample origin: growers and processors in California (state), USA, sample year: September–November 1972 to February–May 1973, country: USA[566], *in-shell walnuts

- Co-contamination: 6 sa. co-contaminated with AFB_1, AFB_2, AFG_1, and AFG_2
- Further contamination of plants and plant products other than cereals and their products reported in the present book/article[566]: Nuts (walnut), AFB_1, AFB_2, and AFG_1, literature[566]

incidence: 0/35*, conc.: no contamination, sample origin: retail shops, local markets, and bazaar in different provinces, Turkey, sample year: October 2012–April 2013, country: Turkey[757], *in-shell walnuts

incidence: 1/50*, conc.: 0.07 µg/kg, sample origin: retail shops, local markets, and bazaar in different provinces, Turkey, sample year: October 2012–April 2013, country: Turkey[757], *shelled walnuts

- Co-contamination: not reported
- Further contamination of plants and plant products other than cereals and their products reported in the present book/article[757]: Nuts (walnut), AFB_1, AFB_2, and AFG_1, literature[757]

incidence: 3/35, conc. range: ≤3.45 µg/kg, sample origin: supermarkets in Tehran (capital), Iran, sample year: October/November 2010, country: Iran[802]

- Co-contamination: not reported
- Further contamination of plants and plant products other than cereals and their products reported in the present book/article[802]: Nuts (walnut), AFB_1, AFB_2, and AFG_1, literature[802]

Aflatoxin/s

incidence: 10/156, conc. range: 5.0–9.9 µg/kg* (7 sa.), 10.0–14.9 µg/kg* (1 sa.), 20.0–24.9 µg/kg* (1 sa.), and >25 µg/kg* (1 sa.), sample origin: brokers, distributors, growers, processors, and retail trade, Canada, sample year: 1970–1975, country: Canada[15], *AFB_1, AFB_2, AFG_1, and AFG_2

- Co-contamination: not reported
- Further contamination of plants and plant products other than cereals and their products reported in the present book/article[15]: Butter (peanut butter), Cocoa beans, Nuts (almond), Nuts (cashew nut), Nuts (peanut), Nuts (pecan nut), Nuts (pistachio nut), Seeds (sunflower), AF/S, literature[15]

incidence: 4/10*, conc. range: 1.5–13.5 µg/kg**, ∅ conc.: 6.45 µg/kg**, sample origin: retail shops and local markets in NWFP and northern areas, Pakistan, sample year: January–November 2009, country: Pakistan[17], *walnuts with shell, **AFB_1, AFB_2, AFG_1, and AFG_2

incidence: 7/10*, conc. range: 0.8–10.8 µg/kg**, ∅ conc.: 3.43 µg/kg**, sample origin: retail shops and local markets in NWFP and northern areas, Pakistan, sample year: January–November 2009, country: Pakistan[17], *walnuts without shell, **AFB_1, AFB_2, AFG_1, and AFG_2

- Co-contamination: not reported
- Further contamination of plants and plant products other than cereals and their products reported in the present book/article[17]: Dates, Figs, Fruits, dried, Kernels (apricot), Nuts (almond), Nuts (peanut), Nuts (pine nut), Nuts (pistachio nut), Raisins, AF/S, literature[17]

incidence: 1/2, conc.: 5.34 µg/kg*, sample origin: mainly imported from local retail outlets in Lodz (city), Poland, sample year: unknown, country: Poland[18], *AFS total

- Co-contamination: not reported
- Further contamination of plants and plant products other than cereals and their products reported in the present book/article[18]: Nut cereals, Nuts, Nuts (almond), Nuts (cashew nut), Nuts (hazelnut), Nuts (peanut), Nuts (pistachio nut), Seeds (sunflower), AF/S, literature[18]

incidence: 2/27, conc. range: >20 µg/kg* (2 sa., maximum: 41 µg/kg*), sample origin: FDA districts, USA, sample year: 1986, country: USA[19], *AFB_1, AFB_2, AFG_1, and AFG_2

incidence: 2/4, conc. range: ≤8 µg/kg*, sample origin: imported to the USA, sample year: 1986, country: USA[19], *AFB_1, AFB_2, AFG_1, and AFG_2

- Co-contamination: not reported
- Further contamination of plants and plant products other than cereals and their products reported in the present book/article[19]: Butter (peanut butter), Candies (peanut candy), Candies (pistachio candy), Melon seeds, Nuts, Nuts (almond), Nuts (Brazil nut, paranut), Nuts (peanut), Nuts (pecan nut), Nuts (pistachio nut), Oils (peanut oil), Pastes (peanut paste), Pastes (pipian paste), Products (peanut product), Spices (cayenne pepper, chilli), Spices (nutmeg), AF/S, literature[19]

incidence: 6/20, conc. range: ≤4,320 µg/kg*, ∅ conc.: 730 µg/kg*, sample origin: retail shops and local markets in Rabat-Salé (region), Morocco, sample year: January–October 2006, country: Spain/Morocco[352], *AFB_1, AFB_2, AFG_1, and AFG_2

- Co-contamination: not reported
- Further contamination of plants and plant products other than cereals and their products reported in the present book/article[352]: Figs, Nuts (peanut), Nuts (pistachio nut), Raisins, AFB_1 and AF/S, literature[352]; Nuts (walnut), AFB_1, literature[352]

incidence: 15/20, conc. range: 15–25 µg/kg*, sample origin: markets in Cairo, El-Minya, Assiut, and Sohag (governorates), Egypt, sample year: March 1991, country: Egypt[436], *AFB_1, AFB_2, AFG_1, and AFG_2

- Co-contamination: not reported

- Further contamination of plants and plant products other than cereals and their products reported in the present book/article[436]: Nuts (hazelnut), AF/S, literature[436]; Nuts (walnut), ZEA, literature[436]

Aspergillus **and** ***Penicillium*** **Toxins**

Ochratoxin A

incidence: 7/20, conc. range: 0.04–0.23 μg/kg, ⌀ conc.: 0.11 μg/kg, sample origin: retail shops in Salé (city) and Rabat (capital), Morocco, sample year: July/August 2005, country: Morocco/Spain[364]

- Co-contamination: not reported
- Further contamination of plants and plant products other than cereals and their products reported in the present book/article[364]: Figs, Nuts (peanut), Raisins, OTA, literature[364]

Fusarium **Toxins**

Zearalenone

incidence: 1/20, conc.: 125 μg/kg, sample origin: markets in Cairo, El-Minya, Assiut, and Sohag (governorates), Egypt, sample year: March 1991, country: Egypt[436]

- Co-contamination: not reported
- Further contamination of plants and plant products other than cereals and their products reported in the present book/article[436]: Nuts (hazelnut), Nuts (walnut), AF/S, literature[436]

incidence: 2/8*, conc. range: 1.20–3.48 μg/kg, ⌀ conc.: 2.34 μg/kg, sample origin: private resources in Novi Sad (city), Vojvodina (province), Serbia, sample year: February 2013, country: Serbia/Czech Republic[844], *domestic walnuts, stored without shell

- Co-contamination: not reported
- Further contamination of plants and plant products other than cereals and their products reported in the present book/article[844]: Nuts (almond), AFB_2, literature[844]

Penicillium **Toxins**

Rubratoxin

incidence: 1/9, conc.: 210 μg/kg, sample origin: major local retail outlets at Nainital (district), India, sample year: 1983–1985, country: India[7]

- Co-contamination: Often more than one mycotoxin was present in a sample.
- Further contamination of plants and plant products other than cereals and their products reported in the present book/article[7]: Coconuts, AFB_1, AFB_2, AFG_1, AFG_2, and CIT, literature[7]; Dates, AFB_1, AFB_2, AFG_1, and OTA, literature[7]; Grogannuts, AFB_1, AFB_2, AFG_1, AFG_2, and CIT, literature[7]; Nuts (almond), AFB_1, AFB_2, AFG_1, RBT, STG, and ZEA, literature[7]; Nuts (cashew nut), AFB_1, AFB_2, AFG_1, AFG_2, CIT, and OTA, literature[7]; Nuts (peanut), Raisins, AFB_1, AFB_2, AFG_1, AFG_2, and OTA, literature[7]; Nuts (pistachio nut), AFB_1, AFB_2, AFG_1, AFG_2, and STG, literature[7]; Nuts (walnut), AFB_1, AFB_2, AFG_1, and AFG_2, literature[7]

Ogbono may contain the following mycotoxins:

Aspergillus **Toxins**

Aflatoxin/s

incidence: 1/1, conc.: 168 μg/kg*, sample origin: markets in Benin City, Nigeria, sample year: unknown, country: Nigeria[346], *AFB

- Co-contamination: not reported
- Further contamination of plants and plant products other than cereals and their products reported in the present book/article[346]: Emu aran, Meal (egusi meal), Ogili-ugba, Ogoro, AF/S, literature[346]

Ground ogbono seeds served as a thickener giving the prepared soup a black coloration.

see also **Seeds (bush mango)**

Ogiri-ugba may contain the following mycotoxins:

Aspergillus **Toxins**

Aflatoxin/s

incidence: 1/1, conc.: 211 μg/kg*, sample origin: markets in Benin City, Nigeria, sample year: unknown, country: Nigeria[346], *AFB

- Co-contamination: not reported
- Further contamination of plants and plant products other than cereals and their products reported in the present book/article[346]: Emu aran, Meal (egusi meal), Ogbono, Ogoro, AF/S, literature[346]

Ogiri-ugba is a traditional Nigerian foodstuff made from fermented castor and oil bean.

Ogiri may contain the following mycotoxins:

Alternaria **Toxins**

Alternariol Methyl Ether

incidence: 6/31*, conc. range: 64–153 μg/kg, ⌀ conc.: 115 μg/kg, sample origin: fermented food sellers, Southwest Nigeria, sample year: February 2015–July 2016, country: South Africa/Nigeria/Belgium/Kenya[330], *fermented melon

- Co-contamination: For detailed information see the article.
- Further contamination of plants and plant products other than cereals and their products reported in the present book/article[330]: Iru, AFB_1, AFB_2, AFG_1, AFG_2, AME, DON, 3-AcDON, FB_1, FB_2, FB_3, FUS-X, HT-2, OTA, ROQ-C, T-2, and ZEA, literature[330]; Ogiri, AFB_1, AFB_2, AFG_2, DAS, DON, HT-2, NEO, OTA, STG, and ZEA, literature[330]; Ugba, AFB_1, AFB_2, AME, DON, ENB, HT-2, OTA, STG, T-2, and ZEA, literature[330]

Aspergillus **Toxins**

Aflatoxin B_1

incidence: 15/31*, conc. range: 3–4 μg/kg, ⌀ conc.: <LOQ?, sample origin: fermented food sellers, Southwest Nigeria, sample year: February 2015–July 2016, country: South Africa/Nigeria/Belgium/Kenya[330], *fermented melon

- Co-contamination: For detailed information see the article.
- Further contamination of plants and plant products other than cereals and their products reported in the present book/article[330]: Iru, AFB_1, AFB_2, AFG_1, AFG_2, AME, DON, 3-AcDON, FB_1, FB_2, FB_3, FUS-X, HT-2, OTA, ROQ-C, T-2, and ZEA, literature[330]; Ogiri, AFB_2, AFG_2, AME, DAS, DON, HT-2, NEO, OTA, STG, and ZEA, literature[330]; Ugba, AFB_1, AFB_2, AME, DON, ENB, HT-2, OTA, STG, T-2, and ZEA, literature[330]

incidence: 1/9, conc.: 6.99 μg/kg, sample origin: markets in Lagos, Ogun, and Oyo (states), Southwest Nigeria, sample year: March 2016, country: Nigeria/South Africa/Austria/Ireland[361]

- Co-contamination: not reported
- Further contamination of plants and plant products other than cereals and their products reported in the present book/

article[361]: Beans (locust bean), 3-NPA. BEA, and CIT, literature[361]; Iru, CIT, OTA, and STG, literature[361]; Ogiri, 3-NPA, BEA, CIT, and STG, literature[361]; Seeds (melon), 3-NPA, AFB_1, AFB_2, BEA, and STG, literature[361]

Aflatoxin B_2

incidence: 5/31*, conc. range: <LOQ, ∅ conc.: <LOQ, sample origin: fermented food sellers, Southwest Nigeria, sample year: February 2015–July 2016, country: South Africa/Nigeria/Belgium/Kenya[330], *fermented melon

- Co-contamination: For detailed information see the article.
- Further contamination of plants and plant products other than cereals and their products reported in the present book/article[330]: Iru, AFB_1, AFB_2, AFG_1, AFG_2, AME, DON, 3-AcDON, FB_1, FB_2, FB_3, FUS-X, HT-2, OTA, ROQ-C, T-2, and ZEA, literature[330]; Ogiri, AFB_1, AFG_2, AME, DAS, DON, HT-2, NEO, OTA, STG, and ZEA, literature[330]; Ugba, AFB_1, AFB_2, AME, DON, ENB, HT-2, OTA, STG, T-2, and ZEA, literature[330]

Aflatoxin G_2

incidence: 2/31*, conc. range: <LOQ, ∅ conc.: <LOQ, sample origin: fermented food sellers, Southwest Nigeria, sample year: February 2015–July 2016, country: South Africa/Nigeria/Belgium/Kenya[330], *fermented melon

- Co-contamination: For detailed information see the article.
- Further contamination of plants and plant products other than cereals and their products reported in the present book/article[330]: Iru, AFB_1, AFB_2, AFG_1, AFG_2, AME, DON, 3-AcDON, FB_1, FB_2, FB_3, FUS-X, HT-2, OTA, ROQ-C, T-2, and ZEA, literature[330]; Ogiri, AFB_1, AFB_2, AME, DAS, DON, HT-2, NEO, OTA, STG, and ZEA, literature[330]; Ugba, AFB_1, AFB_2, AME, DON, ENB, HT-2, OTA, STG, T-2, and ZEA, literature[330]

Other ***Aspergillus*** Toxins

Sterigmatocystin

incidence: 6/31*, conc. range: <LOQ, ∅ conc.: <LOQ, sample origin: fermented food sellers, Southwest Nigeria, sample year: February 2015–July 2016, country: South Africa/Nigeria/Belgium/Kenya[330], *fermented melon

- Co-contamination: For detailed information see the article.
- Further contamination of plants and plant products other than cereals and their products reported in the present book/article[330]: Iru, AFB_1, AFB_2, AFG_1, AFG_2, AME, DON, 3-AcDON, FB_1, FB_2, FB_3, FUS-X, HT-2, OTA, ROQ-C, T-2, and ZEA, literature[330]; Ogiri, AFB_1, AFB_2, AFG_2, AME, DAS, DON, HT-2, NEO, OTA, and ZEA, literature[330]; Ugba, AFB_1, AFB_2, AME, DON, ENB, HT-2, OTA, STG, T-2, and ZEA, literature[330]

incidence: 5/9, conc. range: 1.5–8.2 μg/kg, ∅ conc.: 3.67 μg/kg, sample origin: markets in Lagos, Ogun, and Oyo (states), Southwest Nigeria, sample year: March 2016, country: Nigeria/South Africa/Austria/Ireland[361]

- Co-contamination: not reported
- Further contamination of plants and plant products other than cereals and their products reported in the present book/article[361]: Beans (locust bean), 3-NPA. BEA, and CIT, literature[361]; Iru, CIT, OTA, and STG, literature[361]; Ogiri, 3-NPA, AFB_1, BEA, and CIT, literature[361]; Seeds (melon), 3-NPA, AFB_1, AFB_2, BEA, and STG, literature[361]

Aspergillus and ***Penicillium*** Toxins

3-Nitropropionic Acid

incidence: 3/9, conc. range: 266.6–2,761.1 μg/kg, ∅ conc.: 1,431.4 μg/kg, sample origin: markets in Lagos, Ogun, and Oyo (states), Southwest Nigeria, sample year: March 2016, country: Nigeria/South Africa/Austria/Ireland[361]

- Co-contamination: not reported
- Further contamination of plants and plant products other than cereals and their products reported in the present book/article[361]: Beans (locust bean), 3-NPA. BEA, and CIT, literature[361]; Iru, CIT, OTA, and STG, literature[361]; Ogiri, AFB_1, BEA, CIT, and STG, literature[361]; Seeds (melon), 3-NPA, AFB_1, AFB_2, BEA, and STG, literature[361]

Citrinin

incidence: 1/9, conc.: 4.49 μg/kg, sample origin: markets in Lagos, Ogun, and Oyo (states), southwest Nigeria, sample year: March 2016, country: Nigeria/South Africa/Austria/Ireland[361]

- Co-contamination: not reported
- Further contamination of plants and plant products other than cereals and their products reported in the present book/article[361]: Beans (locust bean), 3-NPA. BEA, and CIT, literature[361]; Iru, CIT, OTA, and STG, literature[361]; Ogiri, 3-NPA, AFB_1, BEA, and STG, literature[361]; Seeds (melon), 3-NPA, AFB_1, AFB_2, BEA, and STG, literature[361]

Ochratoxin A

incidence: 6/31*, conc. range: ≤27 μg/kg, ∅ conc.: <LOQ?, sample origin: fermented food sellers, Southwest Nigeria, sample year: February 2015–July 2016, country: South Africa/Nigeria/Belgium/Kenya[330], *fermented melon

- Co-contamination: For detailed information see the article.
- Further contamination of plants and plant products other than cereals and their products reported in the present book/article[330]: Iru, AFB_1, AFB_2, AFG_1, AFG_2, AME, DON, 3-AcDON, FB_1, FB_2, FB_3, FUS-X, HT-2, OTA, ROQ-C, T-2, and ZEA, literature[330]; Ogiri, AFB_1, AFB_2, AFG_2, AME, DAS, DON, HT-2, NEO, STG, and ZEA, literature[330]; Ugba, AFB_1, AFB_2, AME, DON, ENB, HT-2, OTA, STG, T-2, and ZEA, literature[330]

Fusarium Toxins

Depsipeptides

Beauvericin

incidence: 2/9, conc. range: 0.24–0.67 μg/kg, ∅ conc.: 0.46 μg/kg, sample origin: markets in Lagos, Ogun, and Oyo (states), Southwest Nigeria, sample year: March 2016, country: Nigeria/South Africa/Austria/Ireland[361]

- Co-contamination: not reported
- Further contamination of plants and plant products other than cereals and their products reported in the present book/article[361]: Beans (locust bean), 3-NPA. BEA, and CIT, literature[361]; Iru, CIT, OTA, and STG, literature[361]; Ogiri, 3-NPA, AFB_1, CIT, and STG, literature[361]; Seeds (melon), 3-NPA, AFB_1, AFB_2, BEA, and STG, literature[361]

Type A Trichothecenes

Diacetoxyscirpenol

incidence: 1/31*, conc.: 1 µg/kg, sample origin: fermented food sellers, Southwest Nigeria, sample year: February 2015–July 2016, country: South Africa/Nigeria/Belgium/Kenya[330], *fermented melon

- Co-contamination: For detailed information see the article.
- Further contamination of plants and plant products other than cereals and their products reported in the present book/article[330]: Iru, AFB_1, AFB_2, AFG_1, AFG_2, AME, DON, 3-AcDON, FB_1, FB_2, FB_3, FUS-X, HT-2, OTA, ROQ-C, T-2, and ZEA, literature[330]; Ogiri, AFB_1, AFB_2, AFG_2, AME, DON, HT-2, NEO, OTA, STG, and ZEA, literature[330]; Ugba, AFB_1, AFB_2, AME, DON, ENB, HT-2, OTA, STG, T-2, and ZEA, literature[330]

HT-2 Toxin

incidence: 4/31*, conc. range: 18–35 µg/kg, ∅ conc.: 27 µg/kg, sample origin: fermented food sellers, Southwest Nigeria, sample year: February 2015–July 2016, country: South Africa/Nigeria/Belgium/Kenya[330], *fermented melon

- Co-contamination: For detailed information see the article.
- Further contamination of plants and plant products other than cereals and their products reported in the present book/article[330]: Iru, AFB_1, AFB_2, AFG_1, AFG_2, AME, DON, 3-AcDON, FB_1, FB_2, FB_3, FUS-X, HT-2, OTA, ROQ-C, T-2, and ZEA, literature[330]; Ogiri, AFB_1, AFB_2, AFG_2, AME, DAS, DON, NEO, OTA, STG, and ZEA, literature[330]; Ugba, AFB_1, AFB_2, AME, DON, ENB, HT-2, OTA, STG, T-2, and ZEA, literature[330]

Neosolaniol

incidence: 1/31*, conc.: <LOQ, sample origin: fermented food sellers, Southwest Nigeria, sample year: February 2015–July 2016, country: South Africa/Nigeria/Belgium/Kenya[330], *fermented melon

- Co-contamination: For detailed information see the article.
- Further contamination of plants and plant products other than cereals and their products reported in the present book/article[330]: Iru, AFB_1, AFB_2, AFG_1, AFG_2, AME, DON, 3-AcDON, FB_1, FB_2, FB_3, FUS-X, HT-2, OTA, ROQ-C, T-2, and ZEA, literature[330]; Ogiri, AFB_1, AFB_2, AFG_2, AME, DAS, DON, HT-2, OTA, STG, and ZEA, literature[330]; Ugba, AFB_1, AFB_2, AME, DON, ENB, HT-2, OTA, STG, T-2, and ZEA, literature[330]

Type B Trichothecenes

Deoxynivalenol

incidence: 3/31*, conc. range: ≤54 µg/kg, ∅ conc.: 31 µg/kg, sample origin: fermented food sellers, Southwest Nigeria, sample year: February 2015–July 2016, country: South Africa/Nigeria/Belgium/Kenya[330], *fermented melon

- Co-contamination: For detailed information see the article.
- Further contamination of plants and plant products other than cereals and their products reported in the present book/article[330]: Iru, AFB_1, AFB_2, AFG_1, AFG_2, AME, DON, 3-AcDON, FB_1, FB_2, FB_3, FUS-X, HT-2, OTA, ROQ-C, T-2, and ZEA, literature[330]; Ogiri, AFB_1, AFB_2, AFG_2, AME, DAS, HT-2, NEO, OTA, STG, and ZEA, literature[330]; Ugba, AFB_1, AFB_2, AME, DON, ENB, HT-2, OTA, STG, T-2, and ZEA, literature[330]

Other *Fusarium* Toxins

Zearalenone

incidence: 8/31*, conc. range: 21–45 µg/kg, ∅ conc.: 33 µg/kg, sample origin: fermented food sellers, Southwest Nigeria, sample year: February 2015–July 2016, country: South Africa/Nigeria/Belgium/Kenya[330], *fermented melon

- Co-contamination: For detailed information see the article.
- Further contamination of plants and plant products other than cereals and their products reported in the present book/article[330]: Iru, AFB_1, AFB_2, AFG_1, AFG_2, AME, DON, 3-AcDON, FB_1, FB_2, FB_3, FUS-X, HT-2, OTA, ROQ-C, T-2, and ZEA, literature[330]; Ogiri, AFB_1, AFB_2, AFG_2, AME, DAS, DON, HT-2, NEO, OTA, and STG, literature[330]; Ugba, AFB_1, AFB_2, AME, DON, ENB, HT-2, OTA, STG, T-2, and ZEA, literature[330]

Ogoro may contain the following mycotoxins:

Aspergillus Toxins

Aflatoxin/s

incidence: 2/2, conc. range: 116*–118** µg/kg***, sample origin: Ugbowo in Benin City, Nigeria, sample year: unknown, country: Nigeria[346], *fresh and **stale ogoro, ***AFB

- Co-contamination: not reported
- Further contamination of plants and plant products other than cereals and their products reported in the present book/article[346]: Emu aran, Meal (egusi meal), Ogbono, Ogili-ugba, AF/S, literature[346]

Ogoro is the sap (fermentable) from the stalk of the male inflorescence or the immature shoot of the oil palm.

Oil seeds may contain the following mycotoxins:

Aspergillus Toxins

Aflatoxin B_1

incidence: 2/21*, conc. range: 0.005–2 µg/kg (2 sa., maximum: 1.0 µg/kg), sample origin: ethnic stores, health food stores, and local stores as well as supermarkets in the four largest cities, Sweden, sample year: autumn 1996–summer 1998, country: Sweden[13], *included pumpkin seeds, poppy seeds, and sunflower seeds

- Co-contamination: not reported
- Further contamination of plants and plant products other than cereals and their products reported in the present book/article[13]: Beans (brown bean), Peas, Peas (chickpea), OTA, literature[13]; Juices (apple juice), PAT, literature[13]; Nuts (almond), AFB_1 literature[13]; Nuts (Brazil nut, paranut), Nuts (pistachio nut), Products (peanut product), AFB_1 and AF/S, literature[13]; Oil seeds, AF/S, literature[13]

incidence: 1/20*, conc.: 0.25 µg/kg, sample origin: Catalonia (autonomous community), Spain, sample year: unknown, country: Spain/Italy[583], *oilseed rape for food and feed

- Co-contamination: not reported
- Further contamination of plants and plant products other than cereals and their products reported in the present book/article[583]: not reported

Aflatoxin/s

incidence: 2/21*, conc. range: 0.005–2 µg/kg** (2 sa., maximum: 1.2 µg/kg**), sample origin: ethnic stores, health food stores, and local stores as well as supermarkets in the four largest cities, Sweden, sample year: autumn 1996–summer 1998, country: Sweden[13], *included pumpkin seeds, poppy seeds, and sunflower seeds, **AFB_1, AFB_2, AFG_1, and AFG_2

- Co-contamination: not reported
- Further contamination of plants and plant products other than cereals and their products reported in the present book/article[13]: Beans (brown bean), Peas, Peas (chickpea), OTA, literature[13]; Juices (apple juice), PAT, literature[13]; Nuts (almond), Oil seeds, AFB_1, literature[13]; Nuts (Brazil nut, paranut), Nuts (pistachio nut), Products (peanut product), AFB_1 and AF/S, literature[13]

incidence: 9/80*, conc. range: 2–20 µg/kg** (7 sa.) and >20 µg/kg** (2 sa.), sample origin: imported and domestic sa. from Uruguay, sample year: 1993–1995, country: Uruguay[269], *ncac, **AFB_1, AFB_2, AFG_1, and AFG_2

- Co-contamination: not reported
- Further contamination of plants and plant products other than cereals and their products reported in the present book/article[269]: Beans (soybean), Oil seeds, ZEA, literature[269]; Cocoa beans, AF/S, literature[269]; Fruits, dried, Vegetables, dried, AF/S and ZEA, literature[269]

Aspergillus **and** ***Penicillium*** **Toxins**

Ochratoxin A

incidence: 29/144, conc. range: LOD/LOQ-0.9 µg/kg (27 sa.), ≤1.79 µg/kg (2 sa.), sample origin: Germany, sample year: 1995–1998, country: EU[698]

- Co-contamination: not reported
- Further contamination of plants and plant products other than cereals and their products reported in the present book/article[698]: Baby food, Beans (soybean), Chocolate, Cocoa, Coffee, Coffee beans, Currants, Fruits, dried, Herbs, Jam, Juices, Juices (grape juice), Legumes, Nuts, Oils (olive oil), Peas, Peas (chickpea), Products (plant product), Raisins, Spices, Spices (cayenne pepper, chilli), Spices (nutmeg), Spices (paprika), Spices (pepper), Sultanas, Tea, Vinegar, Wine, Wine (red), Wine (rosé), Wine (white), Wine, miscellaneous, OTA, literature[698]

Fusarium **Toxins**

Zearalenone

incidence: 5/64*, conc. range: 100–200 µg/kg (2 sa.), >200 µg/kg (3 sa.), sample origin: imported and domestic sa. from Uruguay, sample year: 1993–1995, country: Uruguay[269], *ncac

- Co-contamination: not reported
- Further contamination of plants and plant products other than cereals and their products reported in the present book/article[269]: Beans (soybean), ZEA, literature[269]; Cocoa beans, Oil seeds, AF/S, literature[269]; Fruits, dried, Vegetables, dried, AF/S and ZEA, literature[269]

see also **Kernels (pumpkin)** and **Seeds (sunflower)**

Oils may contain the following mycotoxins:

Aspergillus **Toxins**

Aflatoxin B_1

incidence: 1*/16, conc.: 3 µg/kg, sample origin: villages, India, sample year: unknown, country: UK/France[404], *coconut oil

- Co-contamination: not reported
- Further contamination of plants and plant products other than cereals and their products reported in the present book/article[404]: Nuts (peanut), AFB_1, literature[404]; Oils (peanut oil), AFB_1 and AFB_2, literature[404]

incidence: 12/12*, conc. range: 1.3–22.9 µg/kg, sample origin: individual households in Huantai area (high-risk area for HCC), China, sample year: unknown, country: China[406], *plant oils

incidence: 10/10*, conc. range: 0.2–9.1 µg/kg, sample origin: individual households in Huaian area (high-risk area for oesophageal cancer), China, sample year: unknown, country: China[406], *plant oils

incidence: 17/17*, conc. range: 0.5–114.4 µg/kg, sample origin: individual households in Fusui area (high-risk area for HCC and oesophageal cancer), China, sample year: unknown, country: China[406], *plant oils

- Co-contamination: not reported
- Further contamination of plants and plant products other than cereals and their products reported in the present book/article[406]: Nuts (peanut), AFB_1, literature[406]

incidence: 8*/56, conc. range: 0.2–0.8 µg/kg, sample year: unknown, country: Sudan[569], *7 unrefined sesame oils and 1 unrefined groundnut oil contaminated

- Co-contamination: not reported
- Further contamination of plants and plant products other than cereals and their products reported in the present book/article[569]: not reported

For detailed information see the article.

incidence: 8/15*, conc. range: 1.0–30.3 µg/kg, sample origin: regional factories, Khartoum (state), Sudan, sample year: 2009, country: Sudan[580], *factory A: oils produced from mixed seeds

incidence: 3/12*, conc. range: 0.4–17.6 µg/kg, sample origin: regional factories, Khartoum (state), Sudan, sample year: 2009, country: Sudan[580], *factory B: oils produced from mixed seeds

- Co-contamination: not reported
- Further contamination of plants and plant products other than cereals and their products reported in the present book/article[580]: Oils, AFB_2, AFG_1, and AFG_2, literature[580]; Oils (peanut oil), Oils (sesame oil), Oils (sunflower oil), AFB_1, AFB_2, AFG_1, and AFG_2, literature[580]

incidence: 4/9, conc. range: 1–3.8 µg/kg, ∅ conc.: 2.1 µg/kg, sample origin: supermarkets and wholesale markets of agro-products in Yangtze Delta region, China, sample year: April 2010, country: China[748]

- Co-contamination: 2 sa. co-contaminated with AFB_1, AF/S, and ZEA; 2 sa. co-contaminated with AFB_1 and AF/S
- Further contamination of plants and plant products other than cereals and their products reported in the present book/article[748]: Beans (soybean), AFB_1, AF/S, and ZEA; Oils, AF/S and ZEA, literature[748]

incidence: 6/31*, conc. range: 0.14–0.24 µg/kg, sample origin: local markets in Baoding, Shijiazhuang, and Tangshan (areas), Hebei (province), China, sample year: July/August 2011, country: China[878], *oil, blended

- Co-contamination: not reported
- Further contamination of plants and plant products other than cereals and their products reported in the present book/article[878]: Oils (peanut oil), AFB_1, AFB_2 and AFG_1, literature[878]

Aflatoxin B_2

incidence: 7/15*, conc. range: 0.02–0.8 µg/kg, sample origin: regional factories, Khartoum (state), Sudan, sample year: 2009, country: Sudan[580], *factory A: oils produced from mixed seeds

incidence: 6/12*, conc. range: 0.02–2.3 µg/kg, sample origin: regional factories, Khartoum (state), Sudan, sample year: 2009, country: Sudan[580], *factory B: oils produced from mixed seeds

- Co-contamination: not reported
- Further contamination of plants and plant products other than cereals and their products reported in the present book/article[580]: Oils, AFB_1, AFG_1, and AFG_2, literature[580]; Oils (peanut oil), Oils (sesame oil), Oils (sunflower oil), AFB_1, AFB_2, AFG_1, and AFG_2, literature[580]

Aflatoxin G_1

incidence: 12/15*, conc. range: 0.1–15.1 µg/kg, sample origin: regional factories, Khartoum (state), Sudan, sample year: 2009, country: Sudan[580], *factory A: oils produced from mixed seeds

incidence: 11/12*, conc. range: 0.2–11.8 µg/kg, sample origin: regional factories, Khartoum (state), Sudan, sample year: 2009, country: Sudan[580], *factory B: oils produced from mixed seeds

- Co-contamination: not reported
- Further contamination of plants and plant products other than cereals and their products reported in the present book/article[580]: Oils, AFB_1, AFB_2, and AFG_2, literature[580]; Oils (peanut oil), Oils (sesame oil), Oils (sunflower oil), AFB_1, AFB_2, AFG_1, and AFG_2, literature[580]

Aflatoxin G_2

incidence: 13/15*, conc. range: 0.6–21.7 µg/kg, sample origin: regional factories, Khartoum (state), Sudan, sample year: 2009, country: Sudan[580], *factory A: oils produced from mixed seeds

incidence: 12/12*, conc. range: 0.3–13.2 µg/kg, ∅ conc.: 3.2 µg/kg, sample origin: regional factories, Khartoum (state), Sudan, sample year: 2009, country: Sudan[580], *factory B: oils produced from mixed seeds

- Co-contamination: not reported
- Further contamination of plants and plant products other than cereals and their products reported in the present book/article[580]: Oils, AFB_1, AFB_2, and AFG_1, literature[580]; Oils (peanut oil), Oils (sesame oil), Oils (sunflower oil), AFB_1, AFB_2, AFG_1, and AFG_2, literature[580]

Aflatoxin/s

incidence: 1/4*, conc.: 0.1 µg/kg**, sample origin: specialist food shops, UK, sample year: unknown, country: UK[16], *almond and chilli oils, **AFB_1, AFB_2, AFG_1, and AFG_2

- Co-contamination: not reported
- Further contamination of plants and plant products other than cereals and their products reported in the present book/article[16]: Food, AF/S, DON, FUS-X, NIV, OTA, and ZEA, literature[16]; Nuts (almond), AF/S, literature[16]; Oils, ZEA, literature[16]; Oils (sesame oil), OTA, literature[16]; Paste (curry paste), AF/S, FB/S, NIV, OTA, and ZEA, literature[16]; Pickles, AF/S, FB/S, OTA, and ZEA, literature[16]; Sauces (chilli sauce), NIV, OTA, and ZEA, literature[16]; Spices, AF/S, FB/S, OTA, and ZEA, literature[16]; Spices (cayenne pepper, chilli), AF/S, DAS, DON, HT-2, OTA, and ZEA, literature[16]; Spices (coriander), AF/S, DON, OTA, and ZEA, literature[16]; Spices (curry), AF/S, DAS, FB/S, FUS-X, HT-2, NEO, NIV, OTA, T-2, and ZEA, literature[16]; Spices (fennel), AF/S and ZEA, literature[16]; Spices (garlic), DON, FUS-X, and NIV, literature[16]; Spices (ginger), AF/S, DON, NEO, NIV, OTA, and T-2, literature[16]; Spices (tandoori), AF/S, FB/S, NIV, OTA, and T-2, literature[16]

incidence: 4/9, conc. range: 1.1–6 µg/kg, ∅ conc.: 2.6 µg/kg, sample origin: supermarkets and wholesale markets of agro-products in Yangtze Delta region, China, sample year: April 2010, country: China[748]

- Co-contamination: 2 sa. co-contaminated with AFB_1, AFS, and ZEA; 2 sa. co-contaminated with AFB_1 and AFS
- Further contamination of plants and plant products other than cereals and their products reported in the present book/article[748]: Beans (soybean), AFB_1, AF/S, and ZEA; Oils, AFB_1 and ZEA, literature[748]

Fusarium Toxins

Type A Trichothecenes

Diacetoxyscirpenol

incidence: 1/61*, conc.: 12 µg/kg, sample origin: food stores, Germany, sample year: 2004/2005, country: Germany[840], *refined oil

incidence: 0/49*, conc.: no contamination, sample origin: food stores, Germany, sample year: 2004/2005, country: Germany[840], *non-refined oil

incidence: 1/77*, conc.: 12 µg/kg, sample origin: food stores, Germany, sample year: 2004/2005, country: Germany[840], *conventional

incidence: 0/33*, conc.: no contamination, sample origin: food stores, Germany, sample year: 2004/2005, country: Germany[840], *organic

- Co-contamination: For detailed information see the article.
- Further contamination of plants and plant products other than cereals and their products reported in the present book/article[840]: Oils, DON, 15-AcDON, HT-2, T-2, and ZEA, literature[840]; Oils (maize oil), DAS, DON, 15-AcDON, HT-2, T-2, and ZEA, literature[840]; Oils (soybean oil), 15-AcDON and ZEA, literature[840]; Oils (sunflower oil), HT-2 and T-2, literature[840], see also Oils, Oils (soybean oil), chapter: Further Mycotoxins and Microbial Metabolites

HT-2 Toxin

incidence: 1/61*, conc.: 2 µg/kg, sample origin: food stores, Germany, sample year: 2004/2005, country: Germany[840], *refined oil

incidence: 5/49*, conc. range: 2–6 µg/kg, ∅ conc.: 4 µg/kg, sample origin: food stores, Germany, sample year: 2004/2005, country: Germany[840], *non-refined oil

incidence: 3/77*, conc. range: 2–5 µg/kg, ∅ conc.: 3 µg/kg, sample origin: food stores, Germany, sample year: 2004/2005, country: Germany[840], *conventional

incidence: 3/33*, conc. range: 2–6 µg/kg, ∅ conc.: 4 µg/kg, sample origin: food stores, Germany, sample year: 2004/2005, country: Germany[840], *organic

- Co-contamination: For detailed information see the article.
- Further contamination of plants and plant products other than cereals and their products reported in the present book/article[840]: Oils, DAS, DON, 15-AcDON, T-2, and ZEA, literature[840]; Oils (maize oil), DAS, DON, 15-AcDON, HT-2, T-2, and ZEA, literature[840]; Oils (soybean oil), 15-AcDON and ZEA, literature[840]; Oils (sunflower oil), HT-2 and T-2, literature[840], see also Oils, Oils (soybean oil), chapter: Further Mycotoxins and Microbial Metabolites

T-2 Toxin

incidence: 7/61*, conc. range: 2–11 µg/kg, ∅ conc.: 5 µg/kg, sample origin: food stores, Germany, sample year: 2004/2005, country: Germany[840], *refined oil

incidence: 7/49*, conc. range: 2–13 µg/kg, ∅ conc.: 6 µg/kg, sample origin: food stores, Germany, sample year: 2004/2005, country: Germany[840], *non-refined oil incidence: 11/77*, conc. range: 2–11 µg/kg, ∅ conc.: 5 µg/kg, sample origin: food stores, Germany, sample year: 2004/2005, country: Germany[840], *conventional

incidence: 3/33*, conc. range: 6–13 µg/kg, ∅ conc.: 9 µg/kg, sample origin: food stores, Germany, sample year: 2004/2005, country: Germany[840], *organic

- Co-contamination: For detailed information see the article.
- Further contamination of plants and plant products other than cereals and their products reported in the present book/article[840]: Oils, DAS, DON, 15-AcDON, HT-2, and ZEA, literature[840]; Oils (maize oil), DAS, DON, 15-AcDON, HT-2, T-2, and ZEA, literature[840]; Oils (soybean oil), 15-AcDON and ZEA, literature[840]; Oils (sunflower oil), HT-2 and T-2, literature[840], see also Oils, Oils (soybean oil), chapter: Further Mycotoxins and Microbial Metabolites

Type B Trichothecenes

Deoxynivalenol

incidence: 0/61*, conc.: no contamination, sample origin: food stores, Germany, sample year: 2004/2005, country: Germany[840], *refined oil

incidence: 3/49*, conc. range: 14–110 µg/kg, ∅ conc.: 48 µg/kg, sample origin: food stores, Germany, sample year: 2004/2005, country: Germany[840], *non-refined oil

incidence: 1/77*, conc.: 21 µg/kg, sample origin: food stores, Germany, sample year: 2004/2005, country: Germany[840], *conventional

incidence: 2/33*, conc. range: 14–110 µg/kg, ∅ conc.: 62 µg/kg, sample origin: food stores, Germany, sample year: 2004/2005, country: Germany[840], *organic

- Co-contamination: For detailed information see the article.
- Further contamination of plants and plant products other than cereals and their products reported in the present book/article[840]: Oils, DAS, 15-AcDON, HT-2, T-2, and ZEA, literature[840]; Oils (maize oil), DAS, DON, 15-AcDON, HT-2, T-2, and ZEA, literature[840]; Oils (soybean oil), 15-AcDON and ZEA, literature[840]; Oils (sunflower oil), HT-2 and T-2, literature[840], see also Oils, Oils (soybean oil), chapter: Further Mycotoxins and Microbial Metabolites"

15-Acetyldeoxynivalenol

incidence: 1/61*, conc.: 3 µg/kg, sample origin: food stores, Germany, sample year: 2004/2005, country: Germany[840], *refined oil

incidence: 8/49*, conc. range: 3–116 µg/kg, ∅ conc.: 34 µg/kg, sample origin: food stores, Germany, sample year: 2004/2005, country: Germany[840], *non-refined oil

incidence: 4/77*, conc. range: 3–32 µg/kg, ∅ conc.: 15 µg/kg, sample origin: food stores, Germany, sample year: 2004/2005, country: Germany[840], *conventional

incidence: 5/33*, conc. range: 3–116 µg/kg, ∅ conc.: 43 µg/kg, sample origin: food stores, Germany, sample year: 2004/2005, country: Germany[840], *organic

- Co-contamination: For detailed information see the article.
- Further contamination of plants and plant products other than cereals and their products reported in the present book/article[840]: Oils, DAS, DON, HT-2, T-2, and ZEA, literature[840]; Oils (maize oil), DAS, DON, 15-AcDON, HT-2, T-2, and ZEA, literature[840]; Oils (soybean oil), 15-AcDON and ZEA, literature[840]; Oils (sunflower oil), HT-2 and T-2, literature[840], see also Oils, Oils (soybean oil), chapter: Further Mycotoxins and Microbial Metabolites

Other ***Fusarium*** Toxins

Zearalenone

incidence: 1/4*, conc.: 5.4 µg/kg, sample origin: specialist food shops, UK, sample year: unknown, country: UK[16], *almond and chilli oils

- Co-contamination: not reported
- Further contamination of plants and plant products other than cereals and their products reported in the present book/article[16]: Food, AF/S, DON, FUS-X, NIV, OTA, and ZEA, literature[16]; Nuts (almond), Oils, AF/S, literature[16]; Oils (sesame oil), OTA, literature[16]; Paste (curry paste), AF/S, FB/S, NIV, OTA, and ZEA, literature[16]; Pickles, AF/S, FB/S, OTA, and ZEA, literature[16]; Sauces (chilli sauce), NIV, OTA, and ZEA, literature[16]; Spices, AF/S, FB/S, OTA, and ZEA, literature[16]; Spices (cayenne pepper, chilli), AF/S, DAS, DON, HT-2, OTA, and ZEA, literature[16]; Spices (coriander), AF/S, DON, OTA, and ZEA, literature[16]; Spices (curry), AF/S, DAS, FB/S, FUS-X, HT-2, NEO, NIV, OTA, T-2, and ZEA, literature[16]; Spices (fennel), AF/S and ZEA, literature[16]; Spices (garlic), DON, FUS-X, and NIV, literature[16]; Spices (ginger), AF/S, DON, NEO, NIV, OTA, and T-2, literature[16]; Spices (tandoori), AF/S, FB/S, NIV, OTA, and T-2, literature[16]

incidence: 6/9*, conc. range: 0.21–3.25 µg/kg, ∅ conc.: 0.92 µg/kg, sample origin: market, Korea, sample year: unknown, country: Korea[651], *included maize, soybean, and rice oil (3 sa. in each case)

- Co-contamination: not reported
- Further contamination of plants and plant products other than cereals and their products reported in the present book/article[651]: not reported

incidence: 4/9, conc. range: 12–46 µg/kg, ∅ conc.: 25.3 µg/kg, sample origin: supermarkets and wholesale markets of agro-products in Yangtze Delta region, China, sample year: April 2010, country: China[748]

- Co-contamination: 2 sa. co-contaminated with AFB_1, AF/S, and ZEA; 2 sa. contaminated solely with ZEA
- Further contamination of plants and plant products other than cereals and their products reported in the present book/article[748]: Beans (soybean), AFB_1, AF/S, and ZEA; Oils, AFB_1 and AF/S, literature[748]

incidence: 7/61*, conc. range: 5–173 µg/kg, ∅ conc.: 61 µg/kg, sample origin: food stores, Germany, sample year: 2004/2005, country: Germany[840], *refined oil

incidence: 5/49*, conc. range: 22–1730 µg/kg, ∅ conc.: 839 µg/kg, sample origin: food stores, Germany, sample year: 2004/2005, country: Germany[840], *non-refined oil

incidence: 7/77*, conc. range: 5–173 µg/kg, ∅ conc.: 61 µg/kg, sample origin: food stores, Germany, sample year: 2004/2005, country: Germany[840], *conventional

incidence: 5/33*, conc. range: 22–1,730 µg/kg, ∅ conc.: 839 µg/kg, sample origin: food stores, Germany, sample year: 2004/2005, country: Germany[840], *organic

- Co-contamination: For detailed information see the article.
- Further contamination of plants and plant products other than cereals and their products reported in the present book/article[840]: Oils, DAS, DON, 15-AcDON, HT-2, and T-2, literature[840]; Oils (maize oil), DAS, DON, 15-AcDON, HT-2, T-2,

and ZEA, literature[840]; Oils (soybean oil), 15-AcDON and ZEA, literature[840]; Oils (sunflower oil), HT-2 and T-2, literature[840], see also Oils, Oils (soybean oil), chapter: Further Mycotoxins and Microbial Metabolites

incidence: 4*/44, conc. range: 58–135 µg/kg, ∅ conc.: 94 µg/kg, sample origin: local supermarkets as well as online shops, Germany, sample year: unknown, country: Germany[843], *maize oils

- Co-contamination: not reported
- Further contamination of plants and plant products other than cereals and their products reported in the present book/article[843]: not reported

Oils (coconut oil) may contain the following mycotoxins:

Aspergillus Toxins

Aflatoxin B$_1$

incidence: 10/10, conc. range: 8.33–33.33 µg/kg, sample origin: local markets, oil mills, and houses, Andhra Pradesh (state), India, sample year: 1982, country: India[236]

- Co-contamination: not reported
- Further contamination of plants and plant products other than cereals and their products reported in the present book/article[236]: Candies (coconut candy), Copra, Meal (copra meal), AFB$_1$, AFB$_2$, AFG$_1$, and AFG$_2$, literature[236]; Oils (coconut oil), AFB$_2$ and AFG$_1$, literature[236]

Aflatoxin B$_2$

incidence: 2/10, conc. range: 16.66 µg/kg, sample origin: local markets, oil mills, and houses, Andhra Pradesh (state), India, sample year: 1982, country: India[236]

- Co-contamination: not reported
- Further contamination of plants and plant products other than cereals and their products reported in the present book/article[236]: Candies (coconut candy), Copra, Meal (copra meal), AFB$_1$, AFB$_2$, AFG$_1$, and AFG$_2$, literature[236]; Oils (coconut oil), AFB$_1$ and AFG$_1$, literature[236]

Aflatoxin G$_1$

incidence: 2/10, conc. range: 8.3 µg/kg, sample origin: local markets, oil mills, and houses, Andhra Pradesh (state), India, sample year: 1982, country: India[236]

- Co-contamination: not reported
- Further contamination of plants and plant products other than cereals and their products reported in the present book/article[236]: Candies (coconut candy), Copra, Meal (copra meal), AFB$_1$, AFB$_2$, AFG$_1$, and AFG$_2$, literature[236]; Oils (coconut oil), AFB$_1$ and AFB$_2$, literature[236]

Oils (maize oil) may contain the following mycotoxins:

Aspergillus Toxins

Aflatoxin B$_1$

incidence: 4/7, conc. range: 2.6–3.4 µg/kg, sample origin: local markets, China, sample year: unknown, country: China[853]

- Co-contamination: not reported
- Further contamination of plants and plant products other than cereals and their products reported in the present book/article[853]: Beans, OTA, literature[853]; Oils (maize oil), AFB$_2$ and ZEA, literature[853]; Oils (peanut oil), AFB$_1$ and AFB$_2$, literature[853]

Aflatoxin B$_2$

incidence: 4/7, conc. range: 0.40–13 µg/kg, sample origin: local markets, China, sample year: unknown, country: China[853]

- Co-contamination: not reported
- Further contamination of plants and plant products other than cereals and their products reported in the present book/article[853]: Beans, OTA, literature[853]; Oils (maize oil), AFB$_1$ and ZEA, literature[853]; Oils (peanut oil), AFB$_1$ and AFB$_2$, literature[853]

Fusarium Toxins

Fumonisin/s

Fumonisin B$_1$

incidence: 1/25, conc.: 77 µg/kg, sample origin: supermarkets, Spain, sample year: 2011/2012, country: Spain[723]

- Co-contamination: not reported
- Further contamination of plants and plant products other than cereals and their products reported in the present book/article[723]: Margarine, FB$_1$ and ZEA, literature[723]; Oils (maize oil), DON and ZEA, literature[723]

Type A Trichothecenes

Diacetoxyscirpenol

incidence: 1/17*, conc.: 12 µg/kg, sample origin: food stores, Germany, sample year: 2004/2005, country: Germany[840], *maize germ oil

- Co-contamination: For detailed information see the article.
- Further contamination of plants and plant products other than cereals and their products reported in the present book/article[840]: Oils, DAS, DON, 15-AcDON, HT-2, T-2, and ZEA, literature[840]; Oils (maize oil), DON, 15-AcDON, HT-2, T-2, and ZEA, literature[840]; Oils (soybean oil), 15-AcDON and ZEA, literature[840]; Oils (sunflower oil), HT-2 and T-2, literature[840], see also Oils, Oils (soybean oil), chapter: Further Mycotoxins and Microbial Metabolites

HT-2 Toxin

incidence: 5/17*, conc. range: 2–6 µg/kg, ∅ conc.: 4 µg/kg, sample origin: food stores, Germany, sample year: 2004/2005, country: Germany[840], *maize germ oil

- Co-contamination: For detailed information see the article.
- Further contamination of plants and plant products other than cereals and their products reported in the present book/article[840]: Oils, DAS, DON, 15-AcDON, HT-2, T-2, and ZEA, literature[840]; Oils (maize oil), DAS, DON, 15-AcDON, T-2, and ZEA, literature[840]; Oils (soybean oil), 15-AcDON and ZEA, literature[840]; Oils (sunflower oil), HT-2 and T-2, literature[840], see also Oils, Oils (soybean oil), chapter: Further Mycotoxins and Microbial Metabolites

T-2 Toxin

incidence: 11/17*, conc. range: 2–13 µg/kg, ∅ conc.: 6 µg/kg, sample origin: food stores, Germany, sample year: 2004/2005, country: Germany[840], *maize germ oil

- Co-contamination: For detailed information see the article.
- Further contamination of plants and plant products other than cereals and their products reported in the present book/article[840]: Oils, DAS, DON, 15-AcDON, HT-2, T-2, and ZEA, literature[840]; Oils (maize oil), DAS, DON, 15-AcDON, HT-2, and ZEA, literature[840]; Oils (soybean oil), 15-AcDON and ZEA, literature[840]; Oils (sunflower oil), HT-2 and T-2, literature[840], see

also Oils, Oils (soybean oil), chapter: Further Mycotoxins and Microbial Metabolites

Type B Trichothecenes

Deoxynivalenol

incidence: 2/25, conc. range: ≤216 µg/kg, sample origin: supermarkets, Spain, sample year: 2011/2012, country: Spain[723]

- Co-contamination: not reported
- Further contamination of plants and plant products other than cereals and their products reported in the present book/article[723]: Margarine, Oils (maize oil), FB_1 and ZEA, literature[723]

incidence: 3/17*, conc. range: 14–110 µg/kg, ∅ conc.: 48 µg/kg, sample origin: food stores, Germany, sample year: 2004/2005, country: Germany[840], *maize germ oil

- Co-contamination: For detailed information see the article.
- Further contamination of plants and plant products other than cereals and their products reported in the present book/article[840]: Oils, DAS, DON, 15-AcDON, HT-2, T-2, and ZEA, literature[840]; Oils (maize oil), DAS, 15-AcDON, HT-2, T-2, and ZEA, literature[840]; Oils (soybean oil), 15-AcDON and ZEA, literature[840]; Oils (sunflower oil), HT-2 and T-2, literature[840], see also Oils, Oils (soybean oil), chapter: Further Mycotoxins and Microbial Metabolites

15-Acetyldeoxynivalenol

incidence: 7/17*, conc. range: 3–116 µg/kg, ∅ conc.: 38 µg/kg, sample origin: food stores, Germany, sample year: 2004/2005, country: Germany[840], *maize germ oil

- Co-contamination: For detailed information see the article.
- Further contamination of plants and plant products other than cereals and their products reported in the present book/article[840]: Oils, DAS, DON, 15-AcDON, HT-2, T-2, and ZEA, literature[840]; Oils (maize oil), DAS, DON, HT-2, T-2, and ZEA, literature[840]; Oils (soybean oil), 15-AcDON and ZEA, literature[840]; Oils (sunflower oil), HT-2 and T-2, literature[840], see also Oils, Oils (soybean oil), chapter: Further Mycotoxins and Microbial Metabolites

Other ***Fusarium*** Toxins

Zearalenone

incidence: 8/25, conc. range: ≤67 µg/kg, sample origin: supermarkets, Spain, sample year: 2011/2012, country: Spain[723]

- Co-contamination: not reported
- Further contamination of plants and plant products other than cereals and their products reported in the present book/article[723]: Margarine, FB_1 and ZEA, literature[723]; Oils (maize oil), DON and FB_1, literature[723]

incidence: 1/1, conc.: 111 µg/kg, sample origin: local markets, sample year: unknown, country: China/USA[754]

- Co-contamination: not reported
- Further contamination of plants and plant products other than cereals and their products reported in the present book/article[754]: not reported

incidence: 38/38, conc. range: ≤921 µg/kg, ∅ conc.: 169.7 µg/kg, sample origin: unknown, sample year: unknown, country: Germany[761]

- Co-contamination: not reported
- Further contamination of plants and plant products other than cereals and their products reported in the present book/article[761]: not reported

Other kinds of oils were also contaminated with ZEA but data only in German language.

incidence: 9/16, conc. range: 5.2–184.6 µg/kg, ∅ conc.: 65.1 µg/kg, sample origin: local markets, China, sample year: unknown, country: China[824]

- Co-contamination: not reported
- Further contamination of plants and plant products other than cereals and their products reported in the present book/article[824]: Oils (rapeseed oil), ZEA, literature[824]

incidence: 20/20*, conc. range: ≤97.7 µg/kg, ∅ conc.: 63.9 µg/kg, sample origin: retail stores in Lower Saxony (federal state), Germany, sample year: 2009, country: Germany[829], *maize germ oil

- Co-contamination: not reported
- Further contamination of plants and plant products other than cereals and their products reported in the present book/article[829]: Nuts (hazelnut), AFB_1 and AF/S, literature[829]; Nuts (pistachio nut), AFB_1, literature[829]; Spices (cinnamon), OTA, literature[829]; Spices (curry), Spices (nutmeg), AFB_1, AF/S, and OTA, literature[829]; Spices (ginger), AFB_1 and AF/S, literature[829]

incidence: 9/17*, conc. range: 9–1,730 µg/kg, ∅ conc.: 505 µg/kg, sample origin: food stores, Germany, sample year: 2004/2005, country: Germany[840], *maize germ oil

- Co-contamination: For detailed information see the article.
- Further contamination of plants and plant products other than cereals and their products reported in the present book/article[840]: Oils, DAS, DON, 15-AcDON, HT-2, T-2, and ZEA, literature[840]; Oils (maize oil), DAS, DON, 15-AcDON, HT-2, and T-2, literature[840]; Oils (soybean oil), 15-AcDON and ZEA, literature[840]; Oils (sunflower oil), HT-2 and T-2, literature[840], see also Oils, Oils (soybean oil), chapter: Further Mycotoxins and Microbial Metabolites

incidence: 1/7, conc.: <LOQ, sample origin: local markets, China, sample year: unknown, country: China[853]

- Co-contamination: not reported
- Further contamination of plants and plant products other than cereals and their products reported in the present book/article[853]: Beans, OTA, literature[853]; Oils (maize oil), AFB_1 and AFB_2, literature[853]; Oils (peanut oil), AFB_1 and AFB_2, literature[853]

Oils (mustard oil) may contain the following mycotoxins:

Aspergillus Toxins

Aflatoxin B_1

incidence: 33/100, conc. range: 55–87 µg/kg, sample origin: different zones of Bihar (state), India, sample year: March 1984–February 1985, country: India[544]

- Co-contamination: not reported
- Further contamination of plants and plant products other than cereals and their products reported in the present book/article[544]: Spices (mustard), AFB_1, AFB_2, AFG_1, and AFG_2, literature[544]

Oils (olive oil) may contain the following mycotoxins:

Aspergillus **Toxins**

Aflatoxin B_1

incidence: 36/50, conc. range: 0.0028–0.0047 μg/kg (11 sa.), 0.0052–0.0094 μg/k (20 sa.), 0.0137–0.0157 μg/k (4 sa.), 0.0463 μg/k (1 sa.), sample origin: Greek oil company, Greece, sample year: 1995–1998, country: Greece[570]

- Co-contamination: not reported
- Further contamination of plants and plant products other than cereals and their products reported in the present book/article[570]: not reported

incidence: 14/16*, conc. range: 1–7.5 μg/l, ∅ conc.: 3.6 μg/l, sample origin: Greece, sample year: unknown, country: France[571], *included commercial and crude farm oils

- Co-contamination: 5 sa. co-contaminated with AFB_1, AFB_2, AFG_1, and AFG_2; 5 sa. co-contaminated with AFB_1 and AFB_2; 4 sa. contaminated solely with AFB_1
- Further contamination of plants and plant products other than cereals and their products reported in the present book/article[571]: Oils (olive oil), AFB_2, AFG_1, and AFG_2, literature[571]

incidence: 3/30*, conc. range: 0.5–2.4 μg/kg, ∅ conc.: 1.3 μg/kg, sample origin: olive-press plants and supermarkets in Morocco and Campania, Puglia, and Sicily areas, southern Italy, sample year: unknown, country: Italy[575], *virgin olive oil

- Co-contamination: 3 sa. co-contaminated with AFB_1 and OTA
- Further contamination of plants and plant products other than cereals and their products reported in the present book/article[575]: Oils (olive oil), OTA, literature[575]

incidence: 12/50*, conc. range: tr-0.06** μg/kg, sample origin: producers and from Athens (capital) market, Greece, sample year: 1998–2001, country: Greece[576], *virgin olive oil, **from Crete (island)

- Co-contamination: not reported
- Further contamination of plants and plant products other than cereals and their products reported in the present book/article[576]: Oils (olive oil), OTA, literature[576]

For detailed information see the article.

incidence: 20/20*, conc.: no contamination, sample origin: Institute for Experimental Olive Cultivation, sa. originating from Calabria (region), southern Italy, sample year: unknown, country: Italy[692], *extra virgin, virgin, and virgin lampante olive oil sa.

incidence: 3/15, conc. range: <LOQ*, sample origin: retail markets, Italy, sample year: unknown, country: Italy[692], *all oil sa. of virgin typology

- Co-contamination: not reported
- Further contamination of plants and plant products other than cereals and their products reported in the present book/article[692]: not reported

Aflatoxin B_2

incidence: 10/16*, conc. range: 1–5.5 μg/l, ∅ conc.: 1.8 μg/l, sample origin: Greece, sample year: unknown, country: France[571], *included commercial and crude farm oils

- Co-contamination: 5 sa. co-contaminated with AFB_1, AFB_2, AFG_1, and AFG_2; 5 sa. co-contaminated with AFB_1 and AFB_2
- Further contamination of plants and plant products other than cereals and their products reported in the present book/article[571]: Oils (olive oil), AFB_1, AFG_1, and AFG_2, literature[571]

Aflatoxin G_1

incidence: 5/16*, conc. range: 1–2.5 μg/l, ∅ conc.: 1.6 μg/l, sample origin: Greece, sample year: unknown, country: France[571], *included commercial and crude farm oils

- Co-contamination: 5 sa. co-contaminated with AFB_1, AFB_2, AFG_1, and AFG_2
- Further contamination of plants and plant products other than cereals and their products reported in the present book/article[571]: Oils (olive oil), AFB_1, AFB_2, and AFG_2, literature[571]

Aflatoxin G_2

incidence: 5/16*, conc. range: 1–5 μg/l, ∅ conc.: 2.2 μg/l, sample origin: Greece, sample year: unknown, country: France[571], *included commercial and crude farm oils

- Co-contamination: 5 sa. co-contaminated with AFB_1, AFB_2, AFG_1, and AFG_2
- Further contamination of plants and plant products other than cereals and their products reported in the present book/article[571]: Oils (olive oil), AFB_1, AFB_2, and AFG_1, literature[571]

Aspergillus **and** ***Penicillium*** **Toxins**

Ochratoxin A

incidence: 24/30*, conc. range: 0.1–17.0 μg/kg, ∅ conc.: 2.09 μg/kg, sample origin: olive-press plants and supermarkets in Morocco and Campania, Puglia, and Sicily areas, southern Italy, sample year: unknown, country: Italy[575], *virgin olive oil

- Co-contamination: 3 sa. co-contaminated with AFB_1 and OTA; 21 sa. contaminated solely with OTA
- Further contamination of plants and plant products other than cereals and their products reported in the present book/article[575]: Oils (olive oil) (AFB_1), literature[575]

incidence: 44/50*, conc. range: LOD-0.100 μg/kg (9 sa.), 0.100–0.200 μg/kg (15 sa.), 0.200–0.500 μg/kg (10 sa.), >0.500 μg/kg (10 sa., maximum: 1.03 μg/kg**), sample origin: producers and from Athens (capital) market, Greece, sample year: 2001, country: Greece[576], *virgin olive oil, **from Thessaly (region)

- Co-contamination: not reported
- Further contamination of plants and plant products other than cereals and their products reported in the present book/article[576]: Oils (olive oil), AFB_1, literature[576]

For detailed information see the article.

incidence: 1/12, conc.: 0.6 μg/kg (1 sa.), sample origin: Italy, sample year: 1999, country: EU[698]

- Co-contamination: not reported
- Further contamination of plants and plant products other than cereals and their products reported in the present book/article[698]: Baby food, Beans (soybean), Chocolate, Cocoa, Coffee, Coffee beans, Fruits, dried, Herbs, Jam, Juices, Juices (grape juice), Legumes, Nuts, Oil seeds, Peas, Peas (chickpea), Products (plant product), Raisins, Spices, Spices (cayenne pepper, chilli), Spices (nutmeg), Spices (paprika), Spices (pepper), Sultanas, Tea, Vinegar, Wine, Wine (red), Wine (rosé), Wine (white), Wine, miscellaneous, OTA, literature[698]

Oils (peanut oil) may contain the following mycotoxins:

Aspergillus Toxins

Aflatoxin B$_1$

incidence: 4?/6, conc. range: 0.7 µg/l, sample origin: unknown, sample year: unknown, country: Japan[97]

- Co-contamination: not reported
- Further contamination of plants and plant products other than cereals and their products reported in the present book/article[97]: Beans, Jam (bean jam), Spices, AFB$_1$ and AFB$_2$, literature[97]; Butter (peanut butter), Spices (pepper), AFB$_1$, AFB$_2$, and AFG$_1$, literature[97]; Oils (peanut oil), AFB$_2$ and AFG$_1$, literature[97]; Spices (nutmeg), AFB$_1$, AFB$_2$, AFG$_1$, and AFG$_2$, literature[97]

incidence: 3/19, conc. range: 6–25 µg/kg, ∅ conc.: 16 µg/kg, sample origin: markets in 16 areas (cities and counties), Taiwan, sample year: 1977/1978, country: Taiwan[214]

- Co-contamination: not reported
- Further contamination of plants and plant products other than cereals and their products reported in the present book/article[214]: Butter (peanut butter), Cake (peanut), Nuts (peanut), AFB$_1$, AFB$_2$, AFG$_1$, and AFG$_2$, literature[214]; Oils (peanut oil), AFB$_2$, AFG$_1$, and AFG$_2$, literature[214]

incidence: 8/17, conc. range: 10–70 µg/g, ∅ conc.: 31.3 µg/kg, sample origin: oil mills, Taiwan, sample year: unknown, country: Taiwan[216]

- Co-contamination: not reported
- Further contamination of plants and plant products other than cereals and their products reported in the present book/article[216]: Butter (peanut butter), Nuts (peanut), AFB$_1$, literature[216]

incidence: 15/69*, conc. range: 3–175 µg/kg, sample origin: villages, India, sample year: unknown, country: UK/France[404], *included unbranded, almost certainly unrefined, and possibly refined sa.

- Co-contamination: 7 sa. co-contaminated with AFB$_1$ and AFB$_2$, 8 sa. contaminated solely with AFB$_1$
- Further contamination of plants and plant products other than cereals and their products reported in the present book/article[404]: Nuts (peanut), Oils, AFB$_1$, literature[404]; Oils (peanut oil), AFB$_2$, literature[404]

incidence: 48/50, conc. range: ≤1,009 µg/kg, sample origin: Fushui (county, sa. collected from high-incidence area of liver cancer), China, sample year: unknown, country: China[417]

incidence: 51/51*, conc. range: ≤102 µg/kg, sample origin: Fushui (county, sa. collected from low-incidence area of liver cancer), China, sample year: unknown, country: China[417]

- Co-contamination: not reported
- Further contamination of plants and plant products other than cereals and their products reported in the present book/article[417]: Nuts (peanut), AFB$_1$, literature[417]

incidence: 5/16*, conc. range: 8–16 µg/l, sample origin: Malaysia, sample year: unknown, country: Malaysia[577], *9 refined and 7 unrefined oil sa. (5 of the 7 unrefined contaminated)

- Co-contamination: 5 sa. co-contaminated with AFB$_1$ and AFG$_1$
- Further contamination of plants and plant products other than cereals and their products reported in the present book/article[577]: Oils (peanut oil), AFG$_1$, literature[577]

incidence: 20/30, conc. range: 0.1–52.5 µg/l, ∅ conc.: 7.8 µg/l, sample origin: Zhuqing (village) in Fusui (county), Guangxi (autonomous region), China, sample year: unknown, country: USA/China[578]

- Co-contamination: not reported
- Further contamination of plants and plant products other than cereals and their products reported in the present book/article[578]: not reported

incidence: 20/30*, conc. range: 4.43–2,660 µg/l, sample origin: market of Hapur (city), Uttar Pradesh (state), India, sample year: unknown, country: India[579], *crude peanut oil

- Co-contamination: not reported
- Further contamination of plants and plant products other than cereals and their products reported in the present book/article[579]: not reported

incidence: 11/21, conc. range: 2.9–36.1 µg/kg, sample origin: households, wringers (small-scale oil factories), supermarkets, and grocery shops, Khartoum (state), Sudan, sample year: 2009, country: Sudan[580]

- Co-contamination: not reported
- Further contamination of plants and plant products other than cereals and their products reported in the present book/article[580]: Oils, Oils (sesame oil), Oils (sunflower oil), AFB$_1$, AFB$_2$, AFG$_1$, and AFG$_2$, literature[580]; Oils (peanut oil), AFB$_2$, AFG$_1$, and AFG$_2$, literature[580]

incidence: 104/186, conc. range: 0.5–2.0 µg/kg (3 sa.), 2.1–4.0 µg/kg (16 sa.), 4.1–6.0 µg/kg (6 sa.), 6.1–8.0 µg/kg (8 sa.), 8.1–10 µg/kg (7 sa.), 10.1–15 µg/kg (8 sa.), 15.1–25 µg/kg (16 sa.), 25.1–35 µg/kg (11 sa.), 35.1–70.0 µg/kg (29 sa.), sample origin: Gezira, Khartoum, and Kordofan (states), Sudan, sample year: November 2012, country: Sudan[670]

- Co-contamination: not reported
- Further contamination of plants and plant products other than cereals and their products reported in the present book/article[670]: Oils (sunflower oil), AFB$_1$, literature[670]

incidence: 5/8, conc. range: 0.52–0.72 µg/kg, ∅ conc.: 0.62 µg/kg, sample origin: Japan?, sample year: 1984/1985, country: Japan[760]

- Co-contamination: 3 sa. co-contaminated with AFB$_1$, AFB$_2$, AFG$_1$, and AFG$_2$; 2 sa. co-contaminated with AFB$_1$ and AFB$_2$
- Further contamination of plants and plant products other than cereals and their products reported in the present book/article[760]: Oils (peanut oil), AFB$_2$, AFG$_1$, and AFG$_2$, literature[760]

incidence: 2/4, conc. range: 8.6–8.9 µg/kg, ∅ conc.: 8.75 µg/kg, sample origin: local markets, China, sample year: unknown, country: China[853]

- Co-contamination: not reported
- Further contamination of plants and plant products other than cereals and their products reported in the present book/article[853]: Beans, OTA, literature[853]; Oils (maize oil), AFB$_1$, AFB$_2$, and ZEA, literature[853]; Oils (peanut oil), AFB$_2$, literature[853]

incidence: 15/31, conc. range: 0.15–2.72 µg/kg, sample origin: local markets in Baoding, Shijiazhuang, and Tangshan (areas), Hebei (province), China, sample year: July/August 2011, country: China[878]

- Co-contamination: not reported
- Further contamination of plants and plant products other than cereals and their products reported in the present book/

article[878]: Oils, AFB_1, literature[878]; Oils (peanut oil), AFB_2 and AFG_1, literature[878]

Aflatoxin B₂

incidence: 4?/6, conc. range: 0.1 µg/l, sample origin: unknown, sample year: unknown, country: Japan[97]

- Co-contamination: not reported
- Further contamination of plants and plant products other than cereals and their products reported in the present book/article[97]: Beans, Jams (bean jam), Spices, AFB_1 and AFB_2, literature[97]; Butter (peanut butter), Spices (pepper), AFB_1, AFB_2, and AFG_1, literature[97]; Oils (peanut oil), AFB_1 and AFG_1, literature[97]; Spices (nutmeg), AFB_1, AFB_2, AFG_1, and AFG_2, literature[97]

incidence: 2/19, conc. range: 6 µg/kg, ∅ conc.: 6 µg/kg, sample origin: markets in 16 areas (cities and counties), Taiwan, sample year: 1977/1978, country: Taiwan[214]

- Co-contamination: not reported
- Further contamination of plants and plant products other than cereals and their products reported in the present book/article[214]: Butter (peanut butter), Cake (peanut), Nuts (peanut), AFB_1, AFB_2, AFG_1, and AFG_2, literature[214]; Oils (peanut oil), AFB_1, AFG_1, and AFG_2, literature[214]

incidence: 7/69*, conc. range: nc, sample origin: villages, India, sample year: unknown, country: UK/France[404], *included unbranded, almost certainly unrefined, and possibly refined sa.

- Co-contamination: 7 sa. co-contaminated with AFB_1 and AFB_2
- Further contamination of plants and plant products other than cereals and their products reported in the present book/article[404]: Nuts (peanut), Oils, Oils (peanut oil), AFB_1, literature[404]

incidence: 14/21, conc. range: 0.02–3.6 µg/kg, sample origin: households, wringers (small-scale oil factories), supermarkets, and grocery shops, Khartoum (state), Sudan, sample year: 2009, country: Sudan[580]

- Co-contamination: not reported
- Further contamination of plants and plant products other than cereals and their products reported in the present book/article[580]: Oils, Oils (sesame oil), Oils (sunflower oil), AFB_1, AFB_2, AFG_1, and AFG_2, literature[580]; Oils (peanut oil), AFB_1, AFG_1, and AFG_2, literature[580]

incidence: 5/8, conc. range: 0.09–0.22 µg/kg, ∅ conc.: 0.12 µg/kg, sample origin: Japan?, sample year: 1984/1985, country: Japan[760]

- Co-contamination: 3 sa. co-contaminated with AFB_1, AFB_2, AFG_1, and AFG_2; 2 sa. co-contaminated with AFB_1 and AFB_2
- Further contamination of plants and plant products other than cereals and their products reported in the present book/article[760]: Oils (peanut oil), AFB_1, AFG_1, and AFG_2, literature[760]

incidence: 2/4, conc. range: 1.4–1.7 µg/kg, ∅ conc.: 1.55 µg/kg, sample origin: local markets, China, sample year: unknown, country: China[853]

- Co-contamination: not reported
- Further contamination of plants and plant products other than cereals and their products reported in the present book/article[853]: Beans, OTA, literature[853]; Oils (maize oil), AFB_1, AFB_2, and ZEA, literature[853]; Oils (peanut oil), AFB_1, literature[853]

incidence: 6/31, conc. range: 0.15–0.36 µg/kg, sample origin: local markets in Baoding, Shijiazhuang, and Tangshan (areas), Hebei (province), China, sample year: July/August 2011, country: China[878]

- Co-contamination: not reported
- Further contamination of plants and plant products other than cereals and their products reported in the present book/article[878]: Oils, AFB_1, literature[878]; Oils (peanut oil), AFB_1 and AFG_1, literature[878]

Aflatoxin G₁

incidence: 4?/6, conc. range: 0.1 µg/l, sample origin: unknown, sample year: unknown, country: Japan[97]

- Co-contamination: not reported
- Further contamination of plants and plant products other than cereals and their products reported in the present book/article[97]: Beans, Jams (bean jam), Oils (peanut oil), Spices, AFB_1 and AFB_2, literature[97]; Butter (peanut butter), Spices (pepper), AFB_1, AFB_2, and AFG_1, literature[97]; Spices (nutmeg), AFB_1, AFB_2, AFG_1, and AFG_2, literature[97]

incidence: 2/19, conc. range: 11–19 µg/kg, ∅ conc.: 15 µg/kg, sample origin: markets in 16 areas (cities and counties), Taiwan, sample year: 1977/1978, country: Taiwan[214]

- Co-contamination: not reported
- Further contamination of plants and plant products other than cereals and their products reported in the present book/article[214]: Butter (peanut butter), Cake (peanut), Nuts (peanut), AFB_1, AFB_2, AFG_1, and AFG_2, literature[214]; Oils (peanut oil), AFB_1, AFB_2, and AFG_2, literature[214]

incidence: 5/16*, conc. range: ~8–16 µg/l, sample origin: Malaysia, sample year: unknown, country: Malaysia[577], *9 refined and 7 unrefined oils (5 of the 7 unrefined contaminated)

- Co-contamination: 5 sa. co-contaminated with AFB_1 and AFG_1
- Further contamination of plants and plant products other than cereals and their products reported in the present book/article[577]: Oils (peanut oil), AFB_1, literature[577]

incidence: 20/21, conc. range: 3.1–64.1 µg/kg, sample origin: households, wringers (small-scale oil factories), supermarkets, and grocery shops, Khartoum (state), Sudan, sample year: 2009, country: Sudan[580]

- Co-contamination: not reported
- Further contamination of plants and plant products other than cereals and their products reported in the present book/article[580]: Oils, Oils (sesame oil), Oils (sunflower oil), AFB_1, AFB_2, AFG_1, and AFG_2, literature[580]; Oils (peanut oil), AFB_1, AFB_2, and AFG_2, literature[580]

incidence: 3/8, conc. range: 0.07–0.08 µg/kg, ∅ conc.: 0.08 µg/kg, sample origin: Japan?, sample year: 1984/1985, country: Japan[760]

- Co-contamination: 3 sa. co-contaminated with AFB_1, AFB_2, AFG_1, and AFG_2
- Further contamination of plants and plant products other than cereals and their products reported in the present book/article[760]: Oils (peanut oil), AFB_1, AFB_2, and AFG_2, literature[760]

incidence: 3/31, conc. range: 0.01–0.02 µg/kg, sample origin: local markets in Baoding, Shijiazhuang, and Tangshan (areas), Hebei (province), China, sample year: July/August 2011, country: China[878]

- Co-contamination: not reported
- Further contamination of plants and plant products other than cereals and their products reported in the present book/article[878]: Oils, AFB_1, literature[878]; Oils (peanut oil), AFB_1 and AFB_2, literature[878]

Aflatoxin G₂

incidence: 1/19, conc.: 11 µg/kg, sample origin: markets in 16 areas (cities and counties), Taiwan, sample year: 1977/1978, country: Taiwan[214]

- Co-contamination: not reported
- Further contamination of plants and plant products other than cereals and their products reported in the present book/article[214]: Butter (peanut butter), Cake (peanut), Nuts (peanut), AFB_1, AFB_2, AFG_1, and AFG_2, literature[214]; Oils (peanut oil), AFB_1, AFB_2, and AFG_1, literature[214]

incidence: 19/21, conc. range: 0.5–60.7 µg/kg, sample origin: households, wringers (small-scale oil factories), supermarkets, and grocery shops, Khartoum (state), Sudan, sample year: 2009, country: Sudan[580]

- Co-contamination: not reported
- Further contamination of plants and plant products other than cereals and their products reported in the present book/article[580]: Oils, Oils (sesame oil), Oils (sunflower oil), AFB_1, AFB_2, AFG_1, and AFG_2, literature[580]; Oils (peanut oil), AFB_1, AFB_2, and AFG_1, literature[580]

incidence: 3/8, conc. range: 0.01 µg/kg, ∅ conc.: 0.01 µg/kg, sample origin: Japan?, sample year: 1984/1985, country: Japan[760]

- Co-contamination: 3 sa. co-contaminated with AFB_1, AFB_2, AFG_1, and AFG_2
- Further contamination of plants and plant products other than cereals and their products reported in the present book/article[760]: Oils (peanut oil), AFB_1, AFB_2, and AFG_1, literature[760]

Aflatoxin/s

incidence: 5/8*, conc. range: ≤20 µg/kg (1 sa.), >20 µg/kg** (4 sa., maximum: 310 µg/kg**), sample origin: FDA districts, USA, sample year: 1986, country: USA[19], *peanut oil, crude, **AFB_1, AFB_2, AFG_1, and AFG_2

- Co-contamination: not reported
- Further contamination of plants and plant products other than cereals and their products reported in the present book/article[19]: Butter (peanut butter), Candies (peanut candy), Candies (pistachio candy), Nuts, Nuts (almond), Nuts (Brazil nut, paranut), Nuts (peanut), Nuts (pecan nut), Nuts (pistachio nut), Nuts (walnut), Paste (peanut paste), Paste (pipian paste), Products (peanut product), Seeds (melon), Spices (cayenne pepper, chilli), Spices (ginger), Spices (nutmeg), AF/S, literature[19]

incidence: 43/275*, conc. range: <100 µg/kg** (36 sa.), >100 µg/kg** (7 sa., maximum: 670 µg/kg**), sample origin: Rayalaseema (region, 670 mm rainfall), India, sample year: October 1974–December 1975, country: India[582], *crude peanut oil, **AF

incidence: 196/420*, conc. range: <100 µg/kg** (95 sa.), >100 µg/kg** (101 sa., maximum: 670 µg/kg**), sample origin: Telangana (region, 894 mm rainfall), India, sample year: October 1974–December 1975, country: India[582], *crude peanut oil, **AF

incidence: 304/514*, conc. range: <100 µg/kg** (101 sa.), >100 µg/kg** (203 sa., maximum: 5000 µg/kg**), sample origin: Coastal region, (995 mm rainfall), India, sample year: October 1974–December 1975, country: India[582], *crude peanut oil, **AF

- Co-contamination: not reported
- Further contamination of plants and plant products other than cereals and their products reported in the present book/article[582]: not reported

For detailed information see the article.

Oils (rapeseed oil) may contain the following mycotoxins:

Aspergillus Toxins

Aflatoxin B_1

incidence: 16/70, conc. range: 0.3–3.75 µg/kg, ∅ conc.: 1.53 µg/kg, sample origin: small markets in several urban Shanghai (city) areas, China, sample year: April 1988–April 1989, country: China/USA[132]

- Co-contamination: not reported
- Further contamination of plants and plant products other than cereals and their products reported in the present book/article[132]: Beans (kidney bean), Butter (peanut butter), Curd (bean curd), Nuts (peanut), Oils (soybean oil), Paste (bean paste), Sauces (soy sauce), Tofu, Wine, miscellaneous, AFB_1, literature[132]

Fusarium Toxins

Zearalenone

incidence: 1/6, conc.: 40.7 µg/kg, sample origin: local markets, China, sample year: unknown, country: China[824]

- Co-contamination: not reported
- Further contamination of plants and plant products other than cereals and their products reported in the present book/article[824]: Oils (maize oil), ZEA, literature[824]

Oils (sesame oil) may contain the following mycotoxins:

Aspergillus Toxins

Aflatoxin B_1

incidence: 13/14, conc. range: 1.9–109.2 µg/kg, sample origin: households, wringers (small-scale oil factories), supermarkets, and grocery shops, Khartoum (state), Sudan, sample year: 2009, country: Sudan[580]

- Co-contamination: not reported
- Further contamination of plants and plant products other than cereals and their products reported in the present book/article[580]: Oils, Oils (peanut oil), Oils (sunflower oil), AFB_1, AFB_2, AFG_1, and AFG_2, literature[580]; Oils (sesame oil), AFB_2, AFG_1, and AFG_2, literature[580]

incidence: 70/104, conc. range: 0.5–2.0 µg/kg (15 sa.), 2.1–4.0 µg/kg (36 sa.), 4.1–6.0 µg/kg (11 sa.), 6.1–8.0 µg/kg (4 sa.), 8.1–10.0 µg/kg (4 sa., maximum: 9.8 µg/kg), sample origin: Al-Gadarif, Blue Nile, Khartoum, Kordofan, and Sennar (states), Sudan, sample year: November 2008–January 2010, country: Sudan[902]

- Co-contamination: not reported
- Further contamination of plants and plant products other than cereals and their products reported in the present book/article[902]: Oils (sesame oil), AFB_2, literature[902]

For detailed information see the article.

Aflatoxin B_2

incidence: 10/14, conc. range: 0.2–0.5 µg/kg, sample origin: households, wringers (small-scale oil factories), supermarkets, and grocery shops, Khartoum (state), Sudan, sample year: 2009, country: Sudan[580]

- Co-contamination: not reported
- Further contamination of plants and plant products other than cereals and their products reported in the present book/

article[580]: Oils, Oils (peanut oil), Oils (sunflower oil), AFB_1, AFB_2, AFG_1, and AFG_2, literature[580]; Oils (sesame oil), AFB_1, AFG_1, and AFG_2, literature[580]

incidence: 8/104, conc. range: 0.5–1.3 µg/kg, sample origin: Al-Gadarif, Blue Nile, Khartoum, Kordofan, and Sennar (states), Sudan, sample year: November 2008–January 2010, country: Sudan[902]

- Co-contamination: not reported
- Further contamination of plants and plant products other than cereals and their products reported in the present book/article[902]: Oils (sesame oil), AFB_1, literature[902]

For detailed information see the article.

Aflatoxin G_1

incidence: 13/14, conc. range: 10.6–100.3 µg/kg, sample origin: households, wringers (small-scale oil factories), supermarkets, and grocery shops, Khartoum (state), Sudan, sample year: 2009, country: Sudan[580]

- Co-contamination: not reported
- Further contamination of plants and plant products other than cereals and their products reported in the present book/article[580]: Oils, Oils (peanut oil), Oils (sunflower oil), AFB_1, AFB_2, AFG_1, and AFG_2, literature[580]; Oils (sesame oil), AFB_1, AFB_2, and AFG_2, literature[580]

Aflatoxin G_2

incidence: 14/14, conc. range: 1.8–265.2 µg/kg, ∅ conc.: 102.7 µg/kg, sample origin: households, wringers (small-scale oil factories), supermarkets, and grocery shops, Khartoum (state), Sudan, sample year: 2009, country: Sudan[580]

- Co-contamination: not reported
- Further contamination of plants and plant products other than cereals and their products reported in the present book/article[580]: Oils, Oils (peanut oil), Oils (sunflower oil), AFB_1, AFB_2, AFG_1, and AFG_2, literature[580]; Oils (sesame oil), AFB_1, AFB_2, and AFG_1, literature[580]

Aspergillus and *Penicillium* Toxins

Ochratoxin A

incidence: 1/3, conc.: 0.4 µg/kg, sample origin: specialist food shops, UK, sample year: unknown, country: UK[16]

- Co-contamination: not reported
- Further contamination of plants and plant products other than cereals and their products reported in the present book/article[16]: Food, AF/S, DON, FUS-X, NIV, OTA, and ZEA, literature[16]; Nuts (almond), AF/S, literature[16]; Oils, Spices (fennel), AF/S and ZEA, literature[16]; Paste (curry paste), AF/S, FB/S, NIV, OTA, and ZEA, literature[16]; Pickles, AF/S, FB/S, OTA, and ZEA, literature[16]; Sauces (chilli sauce), NIV, OTA, and ZEA, literature[16]; Spices, AF/S, FB/S, OTA, and ZEA, literature[16]; Spices (cayenne pepper, chilli), AF/S, DAS, DON, HT-2, OTA, and ZEA, literature[16]; Spices (coriander), AF/S, DON, OTA, and ZEA, literature[16]; Spices (curry), AF/S, DAS, FB/S, FUS-X, HT-2, NEO, NIV, OTA, T-2, and ZEA, literature[16]; Spices (garlic), DON, FUS-X, and NIV, literature[16]; Spices (ginger), AF/S, DON, NEO, NIV, OTA, and T-2, literature[16]; Spices (tandoori), AF/S, FB/S, NIV, OTA, and T-2, literature[16]

Oils (soybean oil) may contain the following mycotoxins:

Aspergillus Toxins

Aflatoxin B_1

incidence: 22/73, conc. range: 0.25–4.62 µg/kg, ∅ conc.: 1.10 µg/kg, sample origin: small markets in several urban Shanghai (city) areas, China, sample year: April 1988–April 1989, country: China/USA[132]

- Co-contamination: not reported
- Further contamination of plants and plant products other than cereals and their products reported in the present book/article[132]: Beans (kidney bean), Butter (peanut butter), Curd (bean curd), Nuts (peanut), Oils (rapeseed oil), Paste (bean paste), Sauces (soy sauce), Tofu, Wine, miscellaneous, AFB_1, literature[132]

Fusarium Toxins

Type A Trichothecenes

T-2 Tetraol

incidence: 3/3*, conc. range: 90–1,050 µg/kg**, ∅ conc.: 647 µg/kg**, sample origin: northern Illinois, Michigan, and Iowa (states), USA, sample year: September/October 1986, country: USA[527], *discounted or refused by grain merchants, **T-2 equivalents = T-2TET after base hydrolysis (primarily HT-2)

- Co-contamination: 1 sa. co-contaminated with DON, T-2TET, and ZEA; 1 sa. co-contaminated with T-2TET sand ZEA; 1 sa. contaminated solely with T-2TET
- Further contamination of plants and plant products other than cereals and their products reported in the present book/article[527]: Meal (soybean meal), DAS, DON, T-2TET, and ZEA; Oils (soybean oil), DON and ZEA, literature[527]; see also Meal (soybean meal), chapter: Further Mycotoxins and Microbial Metabolites

Type B Trichothecenes

Deoxynivalenol

incidence: 1/3*, conc.: 30 µg/kg, sample origin: northern Illinois, Michigan, and Iowa (states), USA, sample year: September/October 1986, country: USA[527], *discounted or refused by grain merchants

- Co-contamination: 1 sa. co-contaminated with DON, T-2TET, and ZEA
- Further contamination of plants and plant products other than cereals and their products reported in the present book/article[527]: Meal (soybean meal), DAS, DON, T-2TET, and ZEA; Oils (soybean oil), T-2TET and ZEA, literature[527]; see also Meal (soybean meal), chapter: Further Mycotoxins and Microbial Metabolites

15-Acetyldeoxynivalenol

incidence: 2/14, conc. range: 3–4 µg/kg, ∅ conc.: 4 µg/kg, sample origin: food stores, Germany, sample year: 2004/2005, country: Germany[840]

- Co-contamination: For detailed information see the article.
- Further contamination of plants and plant products other than cereals and their products reported in the present book/article[840]: Oils, DAS, DON, 15-AcDON, HT-2, T-2, and ZEA, litera-

ture[840]; Oils (maize oil), DAS, DON, 15-AcDON, HT-2, T-2, and ZEA, literature[840]; Oils (soybean oil), ZEA, literature[840]; Oils (sunflower oil), HT-2 and T-2, literature[840], see also Oils, Oils (soybean oil), chapter: Further Mycotoxins and Microbial Metabolites

Other ***Fusarium*** Toxins

Zearalenone

incidence: 2/3*, conc. range: 270–590 µg/kg, ∅ conc.: 430 µg/kg, sample origin: northern Illinois, Michigan, and Iowa (states), USA, sample year: September/October 1986, country: USA[527], *discounted or refused by grain merchants

- Co-contamination: 1 sa. co-contaminated with DON, T-2TET, and ZEA; 1 sa. co-contaminated with T-2TET sand ZEA
- Further contamination of plants and plant products other than cereals and their products reported in the present book/article[527]: Meal (soybean meal), DAS, DON, T-2TET, and ZEA; Oils (soybean oil), DON and T-2TET, literature[527]; see also Meal (soybean meal), chapter: Further Mycotoxins and Microbial Metabolites

incidence: 3/14, conc. range: 5–46 µg/kg, ∅ conc.: 24 µg/kg, sample origin: food stores, Germany, sample year: 2004/2005, country: Germany[840]

- Co-contamination: For detailed information see the article.
- Further contamination of plants and plant products other than cereals and their products reported in the present book/article[840]: Oils, DAS, DON, 15-AcDON, HT-2, T-2, and ZEA, literature[840]; Oils (maize oil), DAS, DON, 15-AcDON, HT-2, T-2, and ZEA, literature[840]; Oils (soybean oil), 15-AcDON, literature[840]; Oils (sunflower oil), HT-2 and T-2, literature[840], see also Oils, Oils (soybean oil), chapter: Further Mycotoxins and Microbial Metabolites

Oils (sunflower oil) may contain the following mycotoxins:

Alternaria Toxins

Alternariol Methyl Ether

incidence: 1/11, conc.: 17 µg/kg, sample origin: retail stores, Netherlands, sample year: autumn 2014, country: Netherlands/Germany[906]

- Co-contamination: not reported
- Further contamination of plants and plant products other than cereals and their products reported in the present book/article[906]: Figs, Fruits (tomato), Juices (tomato juice), AOH and TA, literature[906]; Paste (tomato paste), Sauces (tomato sauce), Seeds (sunflower), Tomato ketchup, AME, AOH, and TA, literature[906]

Aspergillus Toxins

Aflatoxin B_1

incidence: 11/19, conc. range: 5.4–71.1 µg/kg, sample origin: households, wringers (small-scale oil factories), supermarkets, and grocery shops, Khartoum (state), Sudan, sample year: 2009, country: Sudan[580]

- Co-contamination: not reported
- Further contamination of plants and plant products other than cereals and their products reported in the present book/article[580]: Oils, Oils (peanut oil), Oils (sesame oil), AFB_1, AFB_2, AFG_1, and AFG_2, literature[580]; Oils (sunflower oil), AFB_2, AFG_1, and AFG_2, literature[580]

incidence: 8/55, conc. range: <10 µg/kg (5 sa.), 15–35 µg/kg (3 sa.), sample origin: Gadarif, Gezira, and Khartoum (states), Sudan, sample year: November 2012, country: Sudan[670]

- Co-contamination: not reported
- Further contamination of plants and plant products other than cereals and their products reported in the present book/article[670]: Oils (peanut oil), AFB_1, literature[670]

Aflatoxin B_2

incidence: 9/19, conc. range: 0.02–1.5 µg/kg, sample origin: households, wringers (small-scale oil factories), supermarkets, and grocery shops, Khartoum (state), Sudan, sample year: 2009, country: Sudan[580]

- Co-contamination: not reported
- Further contamination of plants and plant products other than cereals and their products reported in the present book/article[580]: Oils, Oils (peanut oil), Oils (sesame oil), AFB_1, AFB_2, AFG_1, and AFG_2, literature[580]; Oils (sunflower oil), AFB_1, AFG_1, and AFG_2, literature[580]

Aflatoxin G_1

incidence: 18/19, conc. range: 0.2–173.4 µg/kg, sample origin: households, wringers (small-scale oil factories), supermarkets, and grocery shops, Khartoum (state), Sudan, sample year: 2009, country: Sudan[580]

- Co-contamination: not reported
- Further contamination of plants and plant products other than cereals and their products reported in the present book/article[580]: Oils, Oils (peanut oil), Oils (sesame oil), AFB_1, AFB_2, AFG_1, and AFG_2, literature[580]; Oils (sunflower oil), AFB_1, AFB_2, and AFG_2, literature[580]

Aflatoxin G_2

incidence: 19/19, conc. range: 0.4–66.8 µg/kg, ∅ conc.: 14.8 µg/kg, sample origin: households, wringers (small-scale oil factories), supermarkets, and grocery shops, Khartoum (state), Sudan, sample year: 2009, country: Sudan[580]

- Co-contamination: not reported
- Further contamination of plants and plant products other than cereals and their products reported in the present book/article[580]: Oils, Oils (peanut oil), Oils (sesame oil), AFB_1, AFB_2, AFG_1, and AFG_2, literature[580]; Oils (sunflower oil), AFB_1, AFB_2, and AFG_1, literature[580]

Fusarium Toxins

Type A Trichothecenes

HT-2 Toxin

incidence: 1/16, conc.: 2 µg/kg, sample origin: food stores, Germany, sample year: 2004/2005, country: Germany[840]

- Co-contamination: For detailed information see the article.
- Further contamination of plants and plant products other than cereals and their products reported in the present book/article[840]: Oils, DAS, DON, 15-AcDON, HT-2, T-2, and ZEA, literature[840]; Oils (maize oil), DAS, DON, 15-AcDON, HT-2, T-2, and ZEA, literature[840]; Oils (soybean oil), 15-AcDON and ZEA, literature[840]; Oils (sunflower oil), T-2, literature[840], see also Oils, Oils (soybean oil), chapter: Further Mycotoxins and Microbial Metabolites

T-2 Toxin

incidence: 3/16, conc. range: 2 µg/kg, ∅ conc.: 2 µg/kg, sample origin: food stores, Germany, sample year: 2004/2005, country: Germany[840]

- Co-contamination: For detailed information see the article.
- Further contamination of plants and plant products other than cereals and their products reported in the present book/article[840]: Oils, DAS, DON, 15-AcDON, HT-2, T-2, and ZEA, literature[840]; Oils (maize oil), DAS, DON, 15-AcDON, HT-2, T-2, and ZEA, literature[840]; Oils (soybean oil), 15-AcDON and ZEA, literature[840]; Oils (sunflower oil), HT-2, literature[840], see also Oils, Oils (soybean oil), chapter: Further Mycotoxins and Microbial Metabolites

Oils (vegetable oil) may contain the following mycotoxins:

Alternaria Toxins

Alternariol

incidence: 9/19, conc. range: >LOD (8 sa.), 6.0 µg/kg (1 sa.), sample origin: retail stores in the area of Arnsberg and Münster (cities), Germany, sample year: unknown, country: Germany[743]

- Co-contamination: not reported
- Further contamination of plants and plant products other than cereals and their products reported in the present book/article[743]: Juices, ALT, AME, AOH, and TA, literature[743]; Oils (vegetable oil), AME and TA, literature[743]; Products (tomato product), AME, AOH, and TA, literature[743]; Seeds (sunflower), ALT, AME, AOH, ATX-I, and TA, literature[743]; see also Juices, Oils (vegetable oil), Seeds (sunflower), chapter: Further Mycotoxins and Microbial Metabolites

Alternariol Methyl Ether

incidence: 16/19, conc. range: >LOD (12 sa.), 2.8–14 µg/kg (4 sa.), sample origin: retail stores in the area of Arnsberg and Münster (cities), Germany, sample year: unknown, country: Germany[743]

- Co-contamination: not reported
- Further contamination of plants and plant products other than cereals and their products reported in the present book/article[743]: Juices, ALT, AME, AOH, and TA, literature[743]; Oils (vegetable oil), AOH and TA, literature[743]; Products (tomato product), AME, AOH, and TA, literature[743]; Seeds (sunflower), ALT, AME, AOH, ATX-I, and TA, literature[743]; see also Juices, Oils (vegetable oil), Seeds (sunflower), chapter: Further Mycotoxins and Microbial Metabolites

Tenuazonic Acid

incidence: 4/19, conc. range: >LOD (3 sa.), 15 µg/kg (1 sa.), sample origin: retail stores in the area of Arnsberg and Münster (cities), Germany, sample year: unknown, country: Germany[743]

- Co-contamination: not reported
- Further contamination of plants and plant products other than cereals and their products reported in the present book/article[743]: Juices, ALT, AME, AOH, and TA, literature[743]; Oils (vegetable oil), AME and AOH, literature[743]; Products (tomato product), AME, AOH, and TA, literature[743]; Seeds (sunflower), ALT, AME, AOH, ATX-I, and TA, literature[743]; see also Juices, Oils (vegetable oil), Seeds (sunflower), chapter: Further Mycotoxins and Microbial Metabolites

see also **Butter (peanut butter)**

Okra may contain the following mycotoxins:

Aspergillus Toxins

Aflatoxin G_2

incidence: 1/1*, conc.: 12.675 µg/kg, sample origin: market in Khartoum (capital), Sudan, sample year: February 1981–1984?, country: UK/Sudan[223], *okra dried

- Co-contamination: not reported
- Further contamination of plants and plant products other than cereals and their products reported in the present book/article[223]: Butter (peanut butter), AFB_1, AFB_2, and AFG_1, literature[223]; Nuts (peanut), AFB_1, AFB_2, AFG_1, and AFG_2, literature[223]; Peas (chickpea), AFB_1, literature[223]

Olives may contain the following mycotoxins:

Alternaria Toxins

Altenuene

incidence: 1*/4, conc.: 1,400 µg/kg, sample origin: several localities in Apulia and Basilicata (regions), Italy, sample year: autumn, country: Italy[164], *visibly affected by *Alternaria* rot

- Co-contamination: 1 sa. co-contaminated with ALT, AME, AOH, and TA
- Further contamination of plants and plant products other than cereals and their products reported in the present book/article[164]: Fruits (mandarin), Fruits (tomato), Spices (pepper), AME, AOH, and TA, literature[164]; Fruits (melon), AME and TA, literature[164]; Olives, AME, AOH, and TA, literature[164]; Seeds (sunflower), AME and AOH, literature[164]

Alternariol

incidence: 4*/4, conc. range: 109–2,320 µg/kg, ∅ conc.: 1,120 µg/kg, sample origin: several localities in Apulia and Basilicata (regions), Italy, sample year: autumn, country: Italy[164], *visibly affected by *Alternaria* rot

- Co-contamination: 1 sa. co-contaminated with ALT, AME, AOH, and TA; 1 sa. co-contaminated with AME, AOH, and TA; 2 sa. co-contaminated with AME and AOH
- Further contamination of plants and plant products other than cereals and their products reported in the present book/article[164]: Fruits (mandarin), Fruits (tomato), Spices (pepper), AME, AOH, and TA, literature[164]; Fruits (melon), AME and TA, literature[164]; Olives, ALT, AME, and TA, literature[164]; Seeds (sunflower), AME and AOH, literature[164]

Alternariol Methyl Ether

incidence: 4*/4, conc. range: 30–2,870 µg/kg, ∅ conc.: 818 µg/kg, sample origin: several localities in Apulia and Basilicata (regions), Italy, sample year: autumn, country: Italy[164], *visibly affected by *Alternaria* rot

- Co-contamination: 1 sa. co-contaminated with ALT, AME, AOH, and TA; 1 sa. co-contaminated with AME, AOH, and TA; 2 sa. co-contaminated with AME and AOH
- Further contamination of plants and plant products other than cereals and their products reported in the present book/article[164]: Fruits (mandarin), Fruits (tomato), Spices (pepper), AME, AOH, and TA, literature[164]; Fruits (melon), AME and TA, literature[164]; Olives, ALT, AOH, and TA, literature[164]; Seeds (sunflower), AME and AOH, literature[164]

Tenuazonic Acid

incidence: 2*/4, conc. range: 109–262 µg/kg, ∅ conc. 187 µg/kg, sample origin: several localities in Apulia and Basilicata (regions), Italy, sample year: autumn, country: Italy[164], *visibly affected by *Alternaria* rot

- Co-contamination: 1 sa. co-contaminated with ALT, AME, AOH, and TA; 1 sa. co-contaminated with AME, AOH, and TA
- Further contamination of plants and plant products other than cereals and their products reported in the present book/ article[164]: Fruits (mandarin), Fruits (tomato), Spices (pepper), AME, AOH, and TA, literature[164]; Fruits (melon), AME and TA, literature[164]; Olives, ALT, AME, and AOH, literature[164]; Seeds (sunflower), AME and AOH, literature[164]

incidence: 1/10, conc.: 5.3 µg/kg, sample origin: Dutch survey program (2013) and retail shops across The Netherlands, sample year: 2013?, country: Netherlands[568]

- Co-contamination: not reported
- Further contamination of plants and plant products other than cereals and their products reported in the present book/ article[568]: Figs, Seeds (sunflower), TA, literature[568]; Fruits (apple), AOH, literature[568]; Sauces (tomato sauce), AME, AOH, and TA, literature[568]; Wine (red), AOH and TA, literature[568]; see also Seeds (sunflower), chapter: Further Mycotoxins and Microbial Metabolites

Aspergillus Toxins

Aflatoxin B_1

incidence: 12/103*, conc. range: 5–37 µg/kg, sample origin: Moroccan cities, Morocco, sample year: unknown, country: Morocco[584], *"Greek-style" black olives

- Co-contamination: not reported
- Further contamination of plants and plant products other than cereals and their products reported in the present book/ article[584]: Olives, OTA, literature[584]

incidence: 3/3*, conc. range: 0.6–5 µg/kg, ∅ conc. 2.43 µg/kg, sample origin: retailer, Morocco, sample year: unknown, country: Morocco/France[585], *"Greek-style" black olives

- Co-contamination: 2 sa. co-contaminated with AFB_1, CIT, and OTA; 1 sa. co-contaminated with AFB_1 and OTA
- Further contamination of plants and plant products other than cereals and their products reported in the present book/ article[585]: Olives, CIT and OTA, literature[585]

incidence: 1/1*, conc.: 2.4 µg/kg, sample origin: supermarket, Morocco, sample year: unknown, country: Morocco/France[585], *"Greek-style" black olives

- Co-contamination: 1 sa. co-contaminated with AFB_1, CIT, and OTA
- Further contamination of plants and plant products other than cereals and their products reported in the present book/ article[585]: Olives, CIT and OTA, literature[585]

Aspergillus and ***Penicillium*** Toxins

Citrinin

incidence: 3/6*, conc. range: <LOQ-0.45 µg/kg, sample origin: retailer, Morocco, sample year: unknown, country: Morocco/ France[585], *"Greek-style" black olives

- Co-contamination: 2 sa. co-contaminated with AFB_1, CIT, and OTA; 1 sa. co-contaminated with CIT and OTA
- Further contamination of plants and plant products other than cereals and their products reported in the present book/ article[585]: Olives, AFB_1 and OTA, literature[585]

incidence: 2/4*, conc. range: <LOQ-0.52 µg/kg, sample origin: supermarket, Morocco, sample year: unknown, country: Morocco/ France[585], *"Greek-style" black olives

- Co-contamination: 1 sa. co-contaminated with AFB_1, CIT, and OTA; 1 sa. co-contaminated with CIT and OTA
- Further contamination of plants and plant products other than cereals and their products reported in the present book/ article[585]: Olives, AFB_1 and OTA, literature[585]

Ochratoxin A

incidence: 9/25*, conc. range: 0.62–4.8 µg/kg, ∅ conc.: 1.44 µg/kg, sample origin: popular markets in Rabat (capital), Morocco, sample year: unknown, country: Morocco/France[341], *black olives

- Co-contamination: not reported
- Further contamination of plants and plant products other than cereals and their products reported in the present book/ article[341]: Fruits, dried, Raisins, OTA, literature[341]

incidence: 4/7, conc. range: 0.3–46,830 µg/kg, sample origin: Tunisia, sample year: unknown, country: Tunisia/France[345]

- Co-contamination: not reported
- Further contamination of plants and plant products other than cereals and their products reported in the present book/ article[345]: Bsissa, Vegetables, dried, OTA, literature[345]

incidence: 5/103*, conc. range: 40–80 µg/kg, sample origin: Moroccan cities, Morocco, sample year: unknown, country: Morocco[584], *"Greek-style" black olives

- Co-contamination: not reported
- Further contamination of plants and plant products other than cereals and their products reported in the present book/ article[584]: Olives, AFB_1, literature[584]

incidence: 6/6*, conc. range: <LOQ-1.02 µg/kg, sample origin: retailer, Morocco, sample year: unknown, country: Morocco/ France[585], *"Greek-style" black olives

- Co-contamination: 2 sa. co-contaminated with AFB_1, CIT, and OTA; 1 sa. co-contaminated with AFB_1 and OTA; 1 sa. co-contaminated with CIT and OTA; 2 sa. contaminated solely with OTA
- Further contamination of plants and plant products other than cereals and their products reported in the present book/ article[585]: Olives, AFB_1 and CIT, literature[585]

incidence: 4/4*, conc. range: <LOQ-0.68 µg/kg, sample origin: supermarket, Morocco, sample year: unknown, country: Morocco/ France[585], *"Greek-style" black olives

- Co-contamination: 1 sa. co-contaminated with AFB_1, CIT, and OTA; 1 sa. co-contaminated with CIT and OTA; 2 sa. contaminated solely with OTA
- Further contamination of plants and plant products other than cereals and their products reported in the present book/ article[585]: Olives, AFB_1 and CIT, literature[585]

Oloroso see **Wine (white)**

Palomino see **Must**

Paranuts see **Nuts (Brazil nut, paranut)**

Paste see Dagwa and Kuru-kuru

Paste (bean paste) may contain the following mycotoxins:

Aspergillus Toxins

Aflatoxin B$_1$
incidence: 5/10*, conc. range: 0.2–2.8 µg/kg, ∅ conc.: 1.26 µg/kg, sample origin: small markets in several urban Shanghai (city) areas, China, sample year: April 1988–April 1989, country: China/USA[132], *bean paste, fermented

- Co-contamination: not reported
- Further contamination of plants and plant products other than cereals and their products reported in the present book/article[132]: Beans (kidney bean), Butter (peanut butter), Curd (bean curd), Nuts (peanut), Oils (rapeseed oil), Oils (soybean oil), Sauces (soy sauce), Tofu, Wine, miscellaneous, AFB$_1$, literature[132]

Paste (curry paste) may contain the following mycotoxins:

Aspergillus Toxins

Aflatoxin/s
incidence: 1/4, conc.: 1.2 µg/kg*, sample origin: specialist food shops, UK, sample year: unknown, country: UK[16], *AFB$_1$, AFB$_2$, AFG$_1$, and AFG$_2$

- Co-contamination: not reported
- Further contamination of plants and plant products other than cereals and their products reported in the present book/article[16]: Food, AF/S, DON, FUS-X, NIV, OTA, and ZEA, literature[16]; Nuts (almond), AF/S, literature[16]; Oils, Spices (fennel), AF/S and ZEA, literature[16]; Oils (sesame oil), OTA, literature[16]; Paste (curry paste), FB/S, NIV, OTA, and ZEA, literature[16]; Pickles, AF/S, FB/S, OTA, and ZEA, literature[16]; Sauces (chilli sauce), NIV, OTA, and ZEA, literature[16]; Spices, AF/S, FB/S, OTA, and ZEA, literature[16]; Spices (cayenne pepper, chilli), AF/S, DAS, DON, HT-2, OTA, and ZEA, literature[16]; Spices (coriander), AF/S, DON, OTA, and ZEA, literature[16]; Spices (curry), AF/S, DAS, FB/S, FUS X, HT 2, NEO, NIV, OTA, T-2, and ZEA, literature[16]; Spices (garlic), DON, FUS-X, and NIV, literature[16]; Spices (ginger), AF/S, DON, NEO, NIV, OTA, and T-2, literature[16]; Spices (tandoori), AF/S, FB/S, NIV, OTA, and T-2, literature[16]

Aspergillus and ***Penicillium*** Toxins

Ochratoxin A
incidence: 4?/4, conc. range: 0.6–15.5 µg/kg, sample origin: specialist food shops, UK, sample year: unknown, country: UK[16]

- Co-contamination: not reported
- Further contamination of plants and plant products other than cereals and their products reported in the present book/article[16]: Food, AF/S, DON, FUS-X, NIV, OTA, and ZEA, literature[16]; Nuts (almond), AF/S, literature[16]; Oils and Spices (fennel), AF/S and ZEA, literature[16]; Oils (sesame oil), OTA, literature[16]; Paste (curry paste), AF/S, FB/S, NIV, and ZEA, literature[16]; Pickles, AF/S, FB/S, OTA, and ZEA, literature[16]; Sauces (chilli sauce), NIV, OTA, and ZEA, literature[16]; Spices, AF/S, FB/S, OTA, and ZEA, literature[16]; Spices (cayenne pepper, chilli), AF/S, DAS, DON, HT-2, OTA, and ZEA, literature[16]; Spices (coriander), AF/S, DON, OTA, and ZEA, literature[16]; Spices (curry), AF/S, DAS, FB/S, FUS-X, HT-2, NEO, NIV, OTA, T-2, and ZEA, literature[16]; Spices (garlic), DON, FUS-X, and NIV, literature[16]; Spices (ginger), AF/S, DON, NEO, NIV, OTA, and T-2, literature[16]; Spices (tandoori), AF/S, FB/S, NIV, OTA, and T-2, literature[16]

Fusarium Toxins

Fumonisin/s
incidence: 1/4, conc.: 56 µg/kg*, sample origin: specialist food shops, UK, sample year: unknown, country: UK[16], *FB$_1$ and FB$_2$

- Co-contamination: not reported
- Further contamination of plants and plant products other than cereals and their products reported in the present book/article[16]: Food, AF/S, DON, FUS-X, NIV, OTA, and ZEA, literature[16]; Nuts (almond), AF/S, literature[16]; Oils and Spices (fennel), AF/S and ZEA, literature[16]; Oils (sesame oil), OTA, literature[16]; Paste (curry paste), AF/S, NIV, OTA, and ZEA, literature[16]; Pickles, AF/S, FB/S, OTA, and ZEA, literature[16]; Sauces (chilli sauce), NIV, OTA, and ZEA, literature[16]; Spices, AF/S, FB/S, OTA, and ZEA, literature[16]; Spices (cayenne pepper, chilli), AF/S, DAS, DON, HT-2, OTA, and ZEA, literature[16]; Spices (coriander), AF/S, DON, OTA, and ZEA, literature[16]; Spices (curry), AF/S, DAS, FB/S, FUS-X, HT-2, NEO, NIV, OTA, T-2, and ZEA, literature[16]; Spices (garlic), DON, FUS-X, and NIV, literature[16]; Spices (ginger), AF/S, DON, NEO, NIV, OTA, and T-2, literature[16]; Spices (tandoori), AF/S, FB/S, NIV, OTA, and T-2, literature[16]

Type B Trichothecenes

Nivalenol
incidence: 4?/4, conc. range: 5–16 µg/kg, sample origin: specialist food shops, UK, sample year: unknown, country: UK[16]

- Co-contamination: not reported
- Further contamination of plants and plant products other than cereals and their products reported in the present book/article[16]: Food, AF/S, DON, FUS-X, NIV, OTA, and ZEA, literature[16]; Nuts (almond), AF/S, literature[16]; Oils and Spices (fennel), AF/S and ZEA, literature[16]; Oils (sesame oil), OTA, literature[16]; Paste (curry paste), AF/S, FB/S, OTA, and ZEA, literature[16]; Pickles, AF/S, FB/S, OTA, and ZEA, literature[16]; Sauces (chilli sauce), NIV, OTA, and ZEA, literature[16]; Spices, AF/S, FB/S, OTA, and ZEA, literature[16]; Spices (cayenne pepper, chilli), AF/S, DAS, DON, HT 2, OTA, and ZEA, literature[16]; Spices (coriander), AF/S, DON, OTA, and ZEA, literature[16]; Spices (curry), AF/S, DAS, FB/S, FUS-X, HT-2, NEO, NIV, OTA, T-2, and ZEA, literature[16]; Spices (garlic), DON, FUS-X, and NIV, literature[16]; Spices (ginger), AF/S, DON, NEO, NIV, OTA, and T-2, literature[16]; Spices (tandoori), AF/S, FB/S, NIV, OTA, and T-2, literature[16]

Other ***Fusarium*** Toxins

Zearalenone
incidence: 4?/4, conc. range: 3.1–4.2 µg/kg, sample origin: specialist food shops, UK, sample year: unknown, country: UK[16]

- Co-contamination: not reported
- Further contamination of plants and plant products other than cereals and their products reported in the present book/article[16]: Food, AF/S, DON, FUS-X, NIV, OTA, and ZEA, literature[16]; Nuts (almond), AF/S, literature[16]; Oils and Spices (fennel), AF/S and ZEA, literature[16]; Oils (sesame oil) (OTA), literature[16]; Paste (curry paste), AF/S, FB/S, NIV, and OTA, literature[16]; Pickles, AF/S, FB/S, OTA, and ZEA, literature[16]; Sauces (chilli sauce), NIV, OTA, and ZEA, literature[16]; Spices, AF/S, FB/S, OTA, and ZEA, literature[16]; Spices (cayenne pepper, chilli), AF/S, DAS, DON, HT-2, OTA, and ZEA, literature[16]; Spices (coriander), AF/S, DON, OTA, and ZEA, literature[16]; Spices (curry), AF/S, DAS, FB/S, FUS-X, HT-2, NEO, NIV, OTA,

T-2, and ZEA, literature[16]; Spices (garlic), DON, FUS-X, and NIV, literature[16]; Spices (ginger), AF/S, DON, NEO, NIV, OTA, and T-2, literature[16]; Spices (tandoori), AF/S, FB/S, NIV, OTA, and T-2, literature[16]

Paste (fig paste) may contain the following mycotoxins:

Aspergillus **Toxins**

Aflatoxin B_1

incidence: 3?/5, conc. range: 0.5–2.8 µg/kg, sample origin: Turkey, sample year: 1988, country: Turkey[351]

- Co-contamination: not reported
- Further contamination of plants and plant products other than cereals and their products reported in the present book/article[351]: Figs, AFB_1, AFB_2, AFG_1, AFG_2, and OTA, literature[351]; Paste (fig paste), AFB_2, AFG_1, AFG_2, and OTA, literature[351]

Aflatoxin B_2

incidence: 3?/5, conc. range: 0.5–1.8 µg/kg, sample origin: Turkey, sample year: 1988, country: Turkey[351]

- Co-contamination: not reported
- Further contamination of plants and plant products other than cereals and their products reported in the present book/article[351]: Figs, AFB_1, AFB_2, AFG_1, AFG_2, and OTA, literature[351]; Paste (fig paste), AFB_1, AFG_1, AFG_2, and OTA, literature[351]

Aflatoxin G_1

incidence: 3?/5, conc. range: 0.8–3.2 µg/kg, sample origin: Turkey, sample year: 1988, country: Turkey[351]

- Co-contamination: not reported
- Further contamination of plants and plant products other than cereals and their products reported in the present book/article[351]: Figs, AFB_1, AFB_2, AFG_1, AFG_2, and OTA, literature[351]; Paste (fig paste), AFB_1, AFB_2, AFG_2, and OTA, literature[351]

Aflatoxin G_2

incidence: 3?/5, conc. range: 0.5–1.2 µg/kg, sample origin: Turkey, sample year: 1988, country: Turkey[351]

- Co-contamination: not reported
- Further contamination of plants and plant products other than cereals and their products reported in the present book/article[351]: Figs, AFB_1, AFB_2, AFG_1, AFG_2, and OTA, literature[351]; Paste (fig paste), AFB_1, AFB_2, AFG_1, and OTA, literature[351]

Aflatoxin/s

incidence: 9/10, conc. range: 1.0–3.9 µg/kg* (3 sa.), 4.0–10.0 µg/kg* (5 sa.), and 76 µg/kg* (1 sa.), sample origin: port sa. from Felixstowe (town), Manchester (city), and Kings Lynn (market town) Port Health Authorities, UK, sample year: unknown, country: UK[206], *AFB_1, AFB_2, AFG_1, and AFG_2

- Co-contamination: not reported
- Further contamination of plants and plant products other than cereals and their products reported in the present book/article[206]: Butter (peanut butter), Figs, Nuts (Brazil nut, paranut), Nuts (peanut), and Nuts (pistachio nut), AF/S, literature[206]

incidence: 13/20*, conc. range: 1.0–10.0 µg/kg** (6 sa.), >10.0 µg/kg** (7 sa., maximum: 165 µg/kg**), sample origin: importers and retailers, UK, sample year: November/December 1988, country: UK[359], *Turkish fig paste, **AFS total

incidence: 35/42*, conc. range: 1.0–10.0 µg/kg** (26 sa.), >10.0 µg/kg** (9 sa., maximum: 35 µg/kg**), sample origin: containers at port of entry, UK, sample year: April–December 1989, country: UK[359], *Turkish fig paste, **AFS total

incidence: 57/70*, conc. range: 1.0–10.0 µg/kg** (54 sa.), >10.0 µg/kg** (3 sa., maximum: 40 µg/kg**), sample origin: containers at port of entry, UK, sample year: January–July 1990, country: UK[359], *Turkish fig paste, **AFS total

- Co-contamination: not reported
- Further contamination of plants and plant products other than cereals and their products reported in the present book/article[359]: Figs, AF/S, literature[359]

Aspergillus **and** ***Penicillium*** **Toxins**

Ochratoxin A

incidence: 1/5, conc. range: 5.2 µg/kg, sample origin: Turkey, sample year: 1988, country: Turkey[351]

- Co-contamination: not reported
- Further contamination of plants and plant products other than cereals and their products reported in the present book/article[351]: Figs, AFB_1, AFB_2, AFG_1, AFG_2, and OTA, literature[351]; Paste (fig paste), AFB_1, AFB_2, AFG_1, and AFG_2, literature[351]

Paste (hazelnut paste) may contain the following mycotoxins:

Aspergillus **Toxins**

Aflatoxin B_1

incidence: 5/5, conc. range: 0.26–3.61 µg/kg, ∅ conc.: 1.09 µg/kg, sample origin: imported from Turkey to Italy, sample year: January 2008–December 2009, country: Italy[320]

- Co-contamination: 5 sa. co-contaminated with AFB_1, AFB_2, AFG_1, and AFG_2
- Further contamination of plants and plant products other than cereals and their products reported in the present book/article[320]: Coffee beans, Vine fruits, dried, OTA, literature[320]; Kernels (apricot), AFB_1, AFB_2, and AFG_1, literature[320]; Nuts (hazelnut), AFB_1, AFB_2, AFG_1, and AFG_2, literature[320]; Nuts (peanut), Nuts (pistachio nut), AFB_1 and AFB_2, literature[320]; Paste (hazelnut paste), AFB_2, AFG_1, and AFG_2, literature[320]

Aflatoxin B_2

incidence: 5/5, conc. range: <LOQ-0.55 µg/kg, sample origin: imported from Turkey to Italy, sample year: January 2008–December 2009, country: Italy[320]

- Co-contamination: 5 sa. co-contaminated with AFB_1, AFB_2, AFG_1, and AFG_2
- Further contamination of plants and plant products other than cereals and their products reported in the present book/article[320]: Coffee beans, Vine fruits, dried, OTA, literature[320]; Kernels (apricot), AFB_1, AFB_2, and AFG_1, literature[320]; Nuts (hazelnut), AFB_1, AFB_2, AFG_1, and AFG_2, literature[320]; Nuts (peanut), Nuts (pistachio nut), AFB_1 and AFB_2, literature[320]; Paste (hazelnut paste), AFB_1, AFG_1, and AFG_2, literature[320]

Aflatoxin G_1

incidence: 5/5, conc. range: 0.53–1.84 µg/kg, ∅ conc.: 0.92 µg/kg, sample origin: imported from Turkey to Italy, sample year: January 2008–December 2009, country: Italy[320]

- Co-contamination: 5 sa. co-contaminated with AFB_1, AFB_2, AFG_1, and AFG_2

- Further contamination of plants and plant products other than cereals and their products reported in the present book/article[320]: Coffee beans, Vine fruits, dried, OTA, literature[320]; Kernels (apricot), AFB_1, AFB_2, and AFG_1, literature[320]; Nuts (hazelnut), AFB_1, AFB_2, AFG_1, and AFG_2, literature[320]; Nuts (peanut), Nuts (pistachio nut), AFB_1 and AFB_2, literature[320]; Paste (hazelnut paste), AFB_1, AFB_2, and AFG_2, literature[320]

Aflatoxin G_2
incidence: 5/5, conc. range: <LOQ-0.30 µg/kg, sample origin: imported from Turkey to Italy, sample year: January 2008–December 2009, country: Italy[320]

- Co-contamination: 5 sa. co-contaminated with AFB_1, AFB_2, AFG_1, and AFG_2
- Further contamination of plants and plant products other than cereals and their products reported in the present book/article[320]: Coffee beans, Vine fruits, dried, OTA, literature[320]; Kernels (apricot), Paste (hazelnut paste), AFB_1, AFB_2, and AFG_1, literature[320]; Nuts (hazelnut), AFB_1, AFB_2, AFG_1, and AFG_2, literature[320]; Nuts (peanut), Nuts (pistachio nut), AFB_1 and AFB_2, literature[320]

Paste (peanut paste) may contain the following mycotoxins:

Alternaria Toxins

Alternariol Methyl Ether
incidence: 2/17, conc. range: 0.04–0.1 µg/kg, ∅ conc.: 0.07 µg/kg, sample origin: Yaounde (region, southern tropical forest), Cameroon, sample year: July–September 2011, country: Cameroon/Austria/South Africa[659]

incidence: 5/18, conc. range: 0.004–0.6 µg/kg, ∅ conc.: 0.2 µg/kg, sample origin: Bamenda (region, western highlands), Cameroon, sample year: July–September 2011, country: Cameroon/Austria/South Africa[659]

- Co-contamination: For detailed information see the article.
- Further contamination of plants and plant products other than cereals and their products reported in the present book/article[659]: Dagwa, AFB_1, AFB_2, AFG_1, AFG_2, AME, BEA, CIT, DON, ENA, ENA_1, ENB, ENB_1, FA, FB_1, FB_2, FB_3, KA, MA, NIV, OTA, OTB, STG, VIOM, and ZEA, literature[659]; Flour (soybean flour), AFB_1, AFG_1, BEA, DON, ENB, ENB_1, FA, FB_1, FB_2, FB_3, FUS-X, KA, NIV, and ZEA, literature[659]; Kuru-kuru, AFB_1, AFB_2, AFG_1, AFG_2, BEA, ENA, ENA_1, ENB, ENB_1, FA, FB_1, FB_3, KA, MA, OTA, OTB, ROQ-C, STG, VIOM, and ZEA, literature[659]; Paste (peanut paste), AFB_1, AFB_2, AFG_1, AFG_2, BEA, ENA, ENA_1, ENB, ENB_1, FA, FB_1, FB_2, FB_3, KA, MA, OTA, OTB, STG, VIOM, and ZEA, literature[659]; Soups, AFB_1, AFB_2, AFG_1, AFG_2, AME, AOH, BEA, DON, ENA, ENA_1, ENB, ENB_1, FB_1, FB_2, FB_3, KA, OTA, OTB, and STG, literature[659]; see also Dagwa, Flour (soybean flour), Kuru-kuru, Paste (peanut paste), Soups, chapter: Further Mycotoxins and Microbial Metabolites

Aspergillus Toxins

Aflatoxin B_1
incidence: 5/8, conc. range: 0.4–2.7 µg/kg, sample origin: commercially available market basket sa. of foods and foodstuffs in Tokyo (capital), Japan, sample year: 1986–1990, country: Japan[99]

- Co-contamination: not reported
- Further contamination of plants and plant products other than cereals and their products reported in the present book/article[99]: Beans, Seeds (sesame), Sugar, AFB_1 and AFB_2, literature[99]; Food, Jam (bean jam), AFB_1, literature[99]; Nuts (Brazil nut, paranut), Nuts (peanut), Nuts (pistachio nut), and Spices (nutmeg), AFB_1, AFB_2, AFG_1, and AFG_2, literature[99]; Spices, Spices (cayenne pepper, chilli), Spices (pepper), AFB_1, AFB_2, and AFG_1, literature[99]; Spices (paprika), AFB_1 and AFG_1, literature[99]

incidence: 16/17, conc. range: <LOQ-210 µg/kg, ∅ conc.: 35 µg/kg, sample origin: Yaounde (region, southern tropical forest), Cameroon, sample year: July–September 2011, country: Cameroon/Austria/South Africa[659]

incidence: 18/18, conc. range: <LOQ-198 µg/kg, ∅ conc.: 56 µg/kg, sample origin: Bamenda (region, western highlands), Cameroon, sample year: July–September 2011, country: Cameroon/Austria/South Africa[659]

- Co-contamination: For detailed information see the article.
- Further contamination of plants and plant products other than cereals and their products reported in the present book/article[659]: Dagwa, AFB_1, AFB_2, AFG_1, AFG_2, AME, BEA, CIT, DON, ENA, ENA_1, ENB, ENB_1, FA, FB_1, FB_2, FB_3, KA, MA, NIV, OTA, OTB, STG, VIOM, and ZEA, literature[659]; Flour (soybean flour), AFB_1, AFG_1, BEA, DON, ENB, ENB_1, FA, FB_1, FB_2, FB_3, FUS-X, KA, NIV, and ZEA, literature[659]; Kuru-kuru, AFB_1, AFB_2, AFG_1, AFG_2, BEA, ENA, ENA_1, ENB, ENB_1, FA, FB_1, FB_3, KA, MA, OTA, OTB, ROQ-C, STG, VIOM, and ZEA, literature[659]; Paste (peanut paste), AFB_2, AFG_1, AFG_2, AME, BEA, ENA, ENA_1, ENB, ENB_1, FA, FB_1, FB_2, FB_3, KA, MA, OTA, OTB, STG, VIOM, and ZEA, literature[659]; Soups, AFB_1, AFB_2, AFG_1, AFG_2, AME, AOH, BEA, DON, ENA, ENA_1, ENB, ENB_1, FB_1, FB_2, FB_3, KA, OTA, OTB, and STG, literature[659]; see also Dagwa, Flour (soybean flour), Kuru-kuru, Paste (peanut paste), Soups, chapter: Further Mycotoxins and Microbial Metabolites

Aflatoxin B_2
incidence: 13/17, conc. range: <LOQ-37 µg/kg, ∅ conc.: 19 µg/kg, sample origin: Yaounde (region, southern tropical forest), Cameroon, sample year: July–September 2011, country: Cameroon/Austria/South Africa[659]

incidence: 15/18, conc. range: <LOQ-25 µg/kg, ∅ conc.: 14 µg/kg, sample origin: Bamenda (region, western highlands), Cameroon, sample year: July–September 2011, country: Cameroon/Austria/South Africa[659]

- Co-contamination: For detailed information see the article.
- Further contamination of plants and plant products other than cereals and their products reported in the present book/article[659]: Dagwa, AFB_1, AFB_2, AFG_1, AFG_2, AME, BEA, CIT, DON, ENA, ENA_1, ENB, ENB_1, FA, FB_1, FB_2, FB_3, KA, MA, NIV, OTA, OTB, STG, VIOM, and ZEA, literature[659]; Flour (soybean flour), AFB_1, AFG_1, BEA, DON, ENB, ENB_1, FA, FB_1, FB_2, FB_3, FUS-X, KA, NIV, and ZEA, literature[659]; Kuru-kuru, AFB_1, AFB_2, AFG_1, AFG_2, BEA, ENA, ENA_1, ENB, ENB_1, FA, FB_1, FB_3, KA, MA, OTA, OTB, ROQ-C, STG, VIOM, and ZEA, literature[659]; Paste (peanut paste), AFB_1, AFG_1, AFG_2, AME, BEA, ENA, ENA_1, ENB, ENB_1, FA, FB_1, FB_2, FB_3, KA, MA, OTA, OTB, STG, VIOM, and ZEA, literature[659]; Soups, AFB_1, AFB_2, AFG_1, AFG_2, AME, AOH, BEA, DON, ENA, ENA_1, ENB, ENB_1, FB_1, FB_2, FB_3, KA, OTA, OTB, and STG, literature[659]; see also Dagwa, Flour (soybean flour), Kuru-kuru, Paste (peanut

paste), Soups, chapter: Further Mycotoxins and Microbial Metabolites

Aflatoxin G_1

incidence: 9/17, conc. range: <LOQ-10 µg/kg, ∅ conc.: 6 µg/kg, sample origin: Yaounde (region, southern tropical forest), Cameroon, sample year: July–September 2011, country: Cameroon/Austria/South Africa[659]

incidence: 14/18, conc. range: <LOQ-48 µg/kg, ∅ conc.: 22 µg/kg, sample origin: Bamenda (region, western highlands), Cameroon, sample year: July–September 2011, country: Cameroon/Austria/South Africa[659]

- Co-contamination: For detailed information see the article.
- Further contamination of plants and plant products other than cereals and their products reported in the present book/article[659]: Dagwa, AFB_1, AFB_2, AFG_1, AFG_2, AME, BEA, CIT, DON, ENA, ENA_1, ENB, ENB_1, FA, FB_1, FB_2, FB_3, KA, MA, NIV, OTA, OTB, STG, VIOM, and ZEA, literature[659]; Flour (soybean flour), AFB_1, AFG_1, BEA, DON, ENB, ENB_1, FA, FB_1, FB_2, FB_3, FUS-X, KA, NIV, and ZEA, literature[659]; Kuru-kuru, AFB_1, AFB_2, AFG_1, AFG_2, BEA, ENA, ENA_1, ENB, ENB_1, FA, FB_1, FB_3, KA, MA, OTA, OTB, ROQ-C, STG, VIOM, and ZEA, literature[659]; Paste (peanut paste), AFB_1, AFB_2, AFG_2, AME, BEA, ENA, ENA_1, ENB, ENB_1, FA, FB_1, FB_2, FB_3, KA, MA, OTA, OTB, STG, VIOM, and ZEA, literature[659]; Soups, AFB_1, AFB_2, AFG_1, AFG_2, AME, AOH, BEA, DON, ENA, ENA_1, ENB, ENB_1, FB_1, FB_2, FB_3, KA, OTA, OTB, and STG, literature[659]; see also Dagwa, Flour (soybean flour), Kuru-kuru, Paste (peanut paste), Soups, chapter: Further Mycotoxins and Microbial Metabolites

Aflatoxin G_2

incidence: 9/17, conc. range: <LOQ-7 µg/kg, ∅ conc.: 6 µg/kg, sample origin: Yaounde (region, southern tropical forest), Cameroon, sample year: July–September 2011, country: Cameroon/Austria/South Africa[659]

incidence: 12/18, conc. range: <LOQ-15 µg/kg, ∅ conc.: 10 µg/kg, sample origin: Bamenda (region, western highlands), Cameroon, sample year: July–September 2011, country: Cameroon/Austria/South Africa[659]

- Co-contamination: For detailed information see the article.
- Further contamination of plants and plant products other than cereals and their products reported in the present book/article[659]: Dagwa, AFB_1, AFB_2, AFG_1, AFG_2, AME, BEA, CIT, DON, ENA, ENA_1, ENB, ENB_1, FA, FB_1, FB_2, FB_3, KA, MA, NIV, OTA, OTB, STG, VIOM, and ZEA, literature[659]; Flour (soybean flour), AFB_1, AFG_1, BEA, DON, ENB, ENB_1, FA, FB_1, FB_2, FB_3, FUS-X, KA, NIV, and ZEA, literature[659]; Kuru-kuru, AFB_1, AFB_2, AFG_1, AFG_2, BEA, ENA, ENA_1, ENB, ENB_1, FA, FB_1, FB_3, KA, MA, OTA, OTB, ROQ-C, STG, VIOM, and ZEA, literature[659]; Paste (peanut paste), AFB_1, AFB_2, AFG_1, AME, BEA, ENA, ENA_1, ENB, ENB_1, FA, FB_1, FB_2, FB_3, KA, MA, OTA, OTB, STG, VIOM, and ZEA, literature[659]; Soups, AFB_1, AFB_2, AFG_1, AFG_2, AME, AOH, BEA, DON, ENA, ENA_1, ENB, ENB_1, FB_1, FB_2, FB_3, KA, OTA, OTB, and STG, literature[659]; see also Dagwa, Flour (soybean flour), Kuru-kuru, Paste (peanut paste), Soups, chapter: Further Mycotoxins and Microbial Metabolites

Aflatoxin/s

incidence: 3/4, conc. range: ≤11 µg/kg*, sample origin: imported to the USA, sample year: 1986, country: USA[19], *AFB_1, AFB_2, AFG_1, and AFG_2

- Co-contamination: not reported
- Further contamination of plants and plant products other than cereals and their products reported in the present book/article[19]: Butter (peanut butter), Candies (peanut candy), Candies (pistachio candy), Nuts, Nuts (almond), Nuts (Brazil nut, paranut), Nuts (peanut), Nuts (pecan nut), Nuts (pistachio nut), Nuts (walnut), Oils (peanut oil), Paste (pipian paste), Products (peanut product), Seeds (melon), Spices (cayenne pepper, chilli), Spices (ginger), Spices (nutmeg), AF/S, literature[19]

incidence: 2/2, conc. range: 10.80–29.30 µg/kg*, ∅ conc.: 20.05 µg/kg*, sample origin: unknown, sample year: unknown, country: USA[231], *AFB_1, AFB_2, AFG_1, and AFG_2

- Co-contamination: not reported
- Further contamination of plants and plant products other than cereals and their products reported in the present book/article[231]: Butter (peanut butter), AF/S, literature[231]

incidence: 3/3, conc. range: 1.5–4.7 µg/kg*, ∅ conc.: 3.0 µg/kg*, sample origin: Khartoum, Khartoum North, and Umdorman (towns of the capital Khartoum), Sudan, sample year: unknown, country: Sudan[425], *AFB_1, AFB_2, AFG_1, and AFG_2

- Co-contamination: not reported
- Further contamination of plants and plant products other than cereals and their products reported in the present book/article[425]: Nuts (peanut), AF/S, literature[425]

Other *Aspergillus* Toxins

Sterigmatocystin

incidence: 8/17, conc. range: <LOQ-9.1 µg/kg, ∅ conc.: 6.7 µg/kg, sample origin: Yaounde (region, southern tropical forest), Cameroon, sample year: July–September 2011, country: Cameroon/Austria/South Africa[659]

incidence: 10/18, conc. range: <LOQ-5 µg/kg, ∅ conc.: 3.7 µg/kg, sample origin: Bamenda (region, western highlands), Cameroon, sample year: July–September 2011, country: Cameroon/Austria/South Africa[659]

- Co-contamination: For detailed information see the article.
- Further contamination of plants and plant products other than cereals and their products reported in the present book/article[659]: Dagwa, AFB_1, AFB_2, AFG_1, AFG_2, AME, BEA, CIT, DON, ENA, ENA_1, ENB, ENB_1, FA, FB_1, FB_2, FB_3, KA, MA, NIV, OTA, OTB, STG, VIOM, and ZEA, literature[659]; Flour (soybean flour), AFB_1, AFG_1, BEA, DON, ENB, ENB_1, FA, FB_1, FB_2, FB_3, FUS-X, KA, NIV, and ZEA, literature[659]; Kuru-kuru, AFB_1, AFB_2, AFG_1, AFG_2, BEA, ENA, ENA_1, ENB, ENB_1, FA, FB_1, FB_3, KA, MA, OTA, OTB, ROQ-C, STG, VIOM, and ZEA, literature[659]; Paste (peanut paste), AFB_1, AFB_2, AFG_1, AFG_2, AME, BEA, ENA, ENA_1, ENB, ENB_1, FA, FB_1, FB_2, FB_3, KA, MA, OTA, OTB, VIOM, and ZEA, literature[659]; Soups, AFB_1, AFB_2, AFG_1, AFG_2, AME, AOH, BEA, DON, ENA, ENA_1, ENB, ENB_1, FB_1, FB_2, FB_3, KA, OTA, OTB, and STG, literature[659]; see also Dagwa, Flour (soybean flour), Kuru-kuru, Paste (peanut paste), Soups, chapter: Further Mycotoxins and Microbial Metabolites

Aspergillus and *Penicillium* Toxins

Kojic Acid

incidence: 17/17, conc. range: 35–859 µg/kg, ∅ conc.: 267 µg/kg, sample origin: Yaounde (region, southern tropical forest), Cameroon, sample year: July–September 2011, country: Cameroon/Austria/South Africa[659]

incidence: 18/18, conc. range: 147–1,359 µg/kg, ∅ conc.: 490 µg/kg, sample origin: Bamenda (region, western highlands), Cameroon, sample year: July–September 2011, country: Cameroon/Austria/South Africa[659]

- Co-contamination: For detailed information see the article.
- Further contamination of plants and plant products other than cereals and their products reported in the present book/article[659]: Dagwa, AFB_1, AFB_2, AFG_1, AFG_2, AME, BEA, CIT, DON, ENA, ENA_1, ENB, ENB_1, FA, FB_1, FB_2, FB_3, KA, MA, NIV, OTA, OTB, STG, VIOM, and ZEA, literature[659]; Flour (soybean flour), AFB_1, AFG_1, BEA, DON, ENB, ENB_1, FA, FB_1, FB_2, FB_3, FUS X, KA, NIV, and ZEA, literature[659]; Kuru-kuru, AFB_1, AFB_2, AFG_1, AFG_2, BEA, ENA, ENA_1, ENB, ENB_1, FA, FB_1, FB_3, KA, MA, OTA, OTB, ROQ-C, STG, VIOM, and ZEA, literature[659]; Paste (peanut paste), AFB_1, AFB_2, AFG_1, AFG_2, AME, BEA, ENA, ENA_1, ENB, ENB_1, FA, FB_1, FB_2, FB_3, MA, OTA, OTB, STG, VIOM, and ZEA, literature[659]; Soups, AFB_1, AFB_2, AFG_1, AFG_2, AME, AOH, BEA, DON, ENA, ENA_1, ENB, ENB_1, FB_1, FB_2, FB_3, KA, OTA, OTB, and STG, literature[659]; see also Dagwa, Flour (soybean flour), Kuru-kuru, Paste (peanut paste), Soups, chapter: Further Mycotoxins and Microbial Metabolites

Ochratoxin A

incidence: 2/17, conc. range: <LOQ-0.5 µg/kg, ∅ conc.: 0.5 µg/kg, sample origin: Yaounde (region, southern tropical forest), Cameroon, sample year: July–September 2011, country: Cameroon/Austria/South Africa[659]

incidence: 7/18, conc. range: <LOQ-23 µg/kg, ∅ conc.: 8 µg/kg, sample origin: Bamenda (region, western highlands), Cameroon, sample year: July–September 2011, country: Cameroon/Austria/South Africa[659]

- Co-contamination: For detailed information see the article.
- Further contamination of plants and plant products other than cereals and their products reported in the present book/article[659]: Dagwa, AFB_1, AFB_2, AFG_1, AFG_2, AME, BEA, CIT, DON, ENA, ENA_1, ENB, ENB_1, FA, FB_1, FB_2, FB_3, KA, MA, NIV, OTA, OTB, STG, VIOM, and ZEA, literature[659]; Flour (soybean flour), AFB_1, AFG_1, BEA, DON, ENB, ENB_1, FA, FB_1, FB_2, FB_3, FUS-X, KA, NIV, and ZEA, literature[659]; Kuru-kuru, AFB_1, AFB_2, AFG_1, AFG_2, BEA, ENA, ENA_1, ENB, ENB_1, FA, FB_1, FB_3, KA, MA, OTA, OTB, ROQ-C, STG, VIOM, and ZEA, literature[659]; Paste (peanut paste), AFB_1, AFB_2, AFG_1, AFG_2, AME, BEA, ENA, ENA_1, ENB, ENB_1, FA, FB_1, FB_2, FB_3, KA, MA, OTB, STG, VIOM, and ZEA, literature[659]; Soups, AFB_1, AFB_2, AFG_1, AFG_2, AME, AOH, BEA, DON, ENA, ENA_1, ENB, ENB_1, FB_1, FB_2, FB_3, KA, OTA, OTB, and STG, literature[659]; see also Dagwa, Flour (soybean flour), Kuru-kuru, Paste (peanut paste), Soups, chapter: Further Mycotoxins and Microbial Metabolites

Ochratoxin B

incidence: 2/17, conc. range: <LOQ, sample origin: Yaounde (region, southern tropical forest), Cameroon, sample year: July–September 2011, country: Cameroon/Austria/South Africa[659]

incidence: 9/18, conc. range: <LOQ-16 µg/kg, ∅ conc.: 5 µg/kg, sample origin: Bamenda (region, western highlands), Cameroon, sample year: July–September 2011, country: Cameroon/Austria/South Africa[659]

- Co-contamination: For detailed information see the article.
- Further contamination of plants and plant products other than cereals and their products reported in the present book/article[659]: Dagwa, AFB_1, AFB_2, AFG_1, AFG_2, AME, BEA, CIT, DON, ENA, ENA_1, ENB, ENB_1, FA, FB_1, FB_2, FB_3, KA, MA, NIV, OTA, OTB, STG, VIOM, and ZEA, literature[659]; Flour (soybean flour), AFB_1, AFG_1, BEA, DON, ENB, ENB_1, FA, FB_1, FB_2, FB_3, FUS-X, KA, NIV, and ZEA, literature[659]; Kuru-kuru, AFB_1, AFB_2, AFG_1, AFG_2, BEA, ENA, ENA_1, ENB, ENB_1, FA, FB_1, FB_3, KA, MA, OTA, OTB, ROQ-C, STG, VIOM, and ZEA, literature[659]; Paste (peanut paste), AFB_1, AFB_2, AFG_1, AFG_2, AME, BEA, ENA, ENA_1, ENB, ENB_1, FA, FB_1, FB_2, FB_3, KA, MA, OTA, STG, VIOM, and ZEA, literature[659]; Soups, AFB_1, AFB_2, AFG_1, AFG_2, AME, AOH, BEA, DON, ENA, ENA_1, ENB, ENB_1, FB_1, FB_2, FB_3, KA, OTA, OTB, and STG, literature[659]; see also Dagwa, Flour (soybean flour), Kuru-kuru, Paste (peanut paste), Soups, chapter: Further Mycotoxins and Microbial Metabolites

Viomellein

incidence: 2/17, conc. range: <LOQ, sample origin: Yaounde (region, southern tropical forest), Cameroon, sample year: July–September 2011, country: Cameroon/Austria/South Africa[659]

incidence: 7/18, conc. range: <LOQ, sample origin: Bamenda (region, western highlands), Cameroon, sample year: July–September 2011, country: Cameroon/Austria/South Africa[659]

- Co-contamination: For detailed information see the article.
- Further contamination of plants and plant products other than cereals and their products reported in the present book/article[659]: Dagwa, AFB_1, AFB_2, AFG_1, AFG_2, AME, BEA, CIT, DON, ENA, ENA_1, ENB, ENB_1, FA, FB_1, FB_2, FB_3, KA, MA, NIV, OTA, OTB, STG, VIOM, and ZEA, literature[659]; Flour (soybean flour), AFB_1, AFG_1, BEA, DON, ENB, ENB_1, FA, FB_1, FB_2, FB_3, FUS-X, KA, NIV, and ZEA, literature[659]; Kuru-kuru, AFB_1, AFB_2, AFG_1, AFG_2, BEA, ENA, ENA_1, ENB, ENB_1, FA, FB_1, FB_3, KA, MA, OTA, OTB, ROQ-C, STG, VIOM, and ZEA, literature[659]; Paste (peanut paste), AFB_1, AFB_2, AFG_1, AFG_2, AME, BEA, ENA, ENA_1, ENB, ENB_1, FA, FB_1, FB_2, FB_3, KA, MA, OTA, OTB, STG, and ZEA, literature[659]; Soups, AFB_1, AFB_2, AFG_1, AFG_2, AME, AOH, BEA, DON, ENA, ENA_1, ENB, ENB_1, FB_1, FB_2, FB_3, KA, OTA, OTB, and STG, literature[659]; see also Dagwa, Flour (soybean flour), Kuru-kuru, Paste (peanut paste), Soups, chapter: Further Mycotoxins and Microbial Metabolites

Fusarium Toxins

Depsipeptides

Beauvericin

incidence: 17/17, conc. range: 0.2–2.9 µg/kg, ∅ conc.: 1 µg/kg, sample origin: Yaounde (region, southern tropical forest), Cameroon, sample year: July–September 2011, country: Cameroon/Austria/South Africa[659]

incidence: 18/18, conc. range: 0.5–12.4 µg/kg, ∅ conc.: 6.2 µg/kg, sample origin: Bamenda (region, western highlands), Cameroon, sample year: July–September 2011, country: Cameroon/Austria/South Africa[659]

- Co-contamination: For detailed information see the article.
- Further contamination of plants and plant products other than cereals and their products reported in the present book/article[659]: Dagwa, AFB_1, AFB_2, AFG_1, AFG_2, AME, BEA, CIT, DON, ENA, ENA_1, ENB, ENB_1, FA, FB_1, FB_2, FB_3, KA, MA, NIV, OTA, OTB, STG, VIOM, and ZEA, literature[659]; Flour (soybean flour), AFB_1, AFG_1, BEA, DON, ENB, ENB_1, FA, FB_1, FB_2, FB_3, FUS-X, KA, NIV, and ZEA, literature[659]; Kuru-kuru, AFB_1, AFB_2, AFG_1, AFG_2, BEA, ENA, ENA_1, ENB, ENB_1, FA, FB_1, FB_3, KA, MA, OTA, OTB, ROQ-C, STG, VIOM, and ZEA, literature[659]; Paste (peanut paste), AFB_1, AFB_2, AFG_1, AFG_2, AME, ENA, ENA_1, ENB, ENB_1, FA, FB_1, FB_2, FB_3, KA, MA, OTA, OTB, STG, VIOM, and ZEA, literature[659]; Soups, AFB_1, AFB_2, AFG_1, AFG_2, AME, AOH, BEA, DON, ENA, ENA_1, ENB, ENB_1, FB_1, FB_2, FB_3, KA, OTA, OTB, and STG, literature[659]; see also Dagwa, Flour (soybean flour), Kuru-kuru, Paste (peanut paste), Soups, chapter: Further Mycotoxins and Microbial Metabolites

Enniatin A

incidence: 7/17, conc. range: <LOQ-0.02 µg/kg, Ø conc.: 0.02 µg/kg, sample origin: Yaounde (region, southern tropical forest), Cameroon, sample year: July–September 2011, country: Cameroon/Austria/South Africa[659]

incidence: 13/18, conc. range: <LOQ-0.1 µg/kg, Ø conc.: 0.04 µg/kg, sample origin: Bamenda (region, western highlands), Cameroon, sample year: July–September 2011, country: Cameroon/Austria/South Africa[659]

- Co-contamination: For detailed information see the article.
- Further contamination of plants and plant products other than cereals and their products reported in the present book/article[659]: Dagwa, AFB_1, AFB_2, AFG_1, AFG_2, AME, BEA, CIT, DON, ENA, ENA_1, ENB, ENB_1, FA, FB_1, FB_2, FB_3, KA, MA, NIV, OTA, OTB, STG, VIOM, and ZEA, literature[659]; Flour (soybean flour), AFB_1, AFG_1, BEA, DON, ENB, ENB_1, FA, FB_1, FB_2, FB_3, FUS-X, KA, NIV, and ZEA, literature[659]; Kuru-kuru, AFB_1, AFB_2, AFG_1, AFG_2, BEA, ENA, ENA_1, ENB, ENB_1, FA, FB_1, FB_3, KA, MA, OTA, OTB, ROQ-C, STG, VIOM, and ZEA, literature[659]; Paste (peanut paste), AFB_1, AFB_2, AFG_1, AFG_2, AME, BEA, ENA_1, ENB, ENB_1, FA, FB_1, FB_2, FB_3, KA, MA, OTA, OTB, STG, VIOM, and ZEA, literature[659]; Soups, AFB_1, AFB_2, AFG_1, AFG_2, AME, AOH, BEA, DON, ENA, ENA_1, ENB, ENB_1, FB_1, FB_2, FB_3, KA, OTA, OTB, and STG, literature[659]; see also Dagwa, Flour (soybean flour), Kuru-kuru, Paste (peanut paste), Soups, chapter: Further Mycotoxins and Microbial Metabolites

Enniatin A_1

incidence: 4/17, conc. range: <LOQ, sample origin: Yaounde (region, southern tropical forest), Cameroon, sample year: July–September 2011, country: Cameroon/Austria/South Africa[659]

incidence: 6/18, conc. range: <LOQ-5.8 µg/kg, Ø conc.: 2 µg/kg, sample origin: Bamenda (region, western highlands), Cameroon, sample year: July–September 2011, country: Cameroon/Austria/South Africa[659]

- Co-contamination: For detailed information see the article.
- Further contamination of plants and plant products other than cereals and their products reported in the present book/article[659]: Dagwa, AFB_1, AFB_2, AFG_1, AFG_2, AME, BEA, CIT, DON, ENA, ENA_1, ENB, ENB_1, FA, FB_1, FB_2, FB_3, KA, MA, NIV, OTA, OTB, STG, VIOM, and ZEA, literature[659]; Flour (soybean flour), AFB_1, AFG_1, BEA, DON, ENB, ENB_1, FA, FB_1, FB_2, FB_3, FUS-X, KA, NIV, and ZEA, literature[659]; Kuru-kuru, AFB_1, AFB_2, AFG_1, AFG_2, BEA, ENA, ENA_1, ENB, ENB_1, FA, FB_1, FB_3, KA, MA, OTA, OTB, ROQ-C, STG, VIOM, and ZEA, literature[659]; Paste (peanut paste), AFB_1, AFB_2, AFG_1, AFG_2, AME, BEA, ENA, ENB, ENB_1, FA, FB_1, FB_2, FB_3, KA, MA, OTA, OTB, STG, VIOM, and ZEA, literature[659]; Soups, AFB_1, AFB_2, AFG_1, AFG_2, AME, AOH, BEA, DON, ENA, ENA_1, ENB, ENB_1, FB_1, FB_2, FB_3, KA, OTA, OTB, and STG, literature[659]; see also Dagwa, Flour (soybean flour), Kuru-kuru, Paste (peanut paste), Soups, chapter: Further Mycotoxins and Microbial Metabolites

Enniatin B

incidence: 15/17, conc. range: <LOQ-0.1 µg/kg, Ø conc.: 0.02 µg/kg, sample origin: Yaounde (region, southern tropical forest), Cameroon, sample year: July–September 2011, country: Cameroon/Austria/South Africa[659]

incidence: 17/18, conc. range: 0.001–0.6 µg/kg, Ø conc.: 0.2 µg/kg, sample origin: Bamenda (region, western highlands), Cameroon, sample year: July–September 2011, country: Cameroon/Austria/South Africa[659]

- Co-contamination: For detailed information see the article.
- Further contamination of plants and plant products other than cereals and their products reported in the present book/article[659]: Dagwa, AFB_1, AFB_2, AFG_1, AFG_2, AME, BEA, CIT, DON, ENA, ENA_1, ENB, ENB_1, FA, FB_1, FB_2, FB_3, KA, MA, NIV, OTA, OTB, STG, VIOM, and ZEA, literature[659]; Flour (soybean flour), AFB_1, AFG_1, BEA, DON, ENB, ENB_1, FA, FB_1, FB_2, FB_3, FUS-X, KA, NIV, and ZEA, literature[659]; Kuru-kuru, AFB_1, AFB_2, AFG_1, AFG_2, BEA, ENA, ENA_1, ENB, ENB_1, FA, FB_1, FB_3, KA, MA, OTA, OTB, ROQ-C, STG, VIOM, and ZEA, literature[659]; Paste (peanut paste), AFB_1, AFB_2, AFG_1, AFG_2, AME, BEA, ENA, ENA_1, ENB_1, FA, FB_1, FB_2, FB_3, KA, MA, OTA, OTB, STG, VIOM, and ZEA, literature[659]; Soups, AFB_1, AFB_2, AFG_1, AFG_2, AME, AOH, BEA, DON, ENA, ENA_1, ENB, ENB_1, FB_1, FB_2, FB_3, KA, OTA, OTB, and STG, literature[659]; see also Dagwa, Flour (soybean flour), Kuru-kuru, Paste (peanut paste), Soups, chapter: Further Mycotoxins and Microbial Metabolites

Enniatin B_1

incidence: 14/17, conc. range: 0.03–0.6 µg/kg, Ø conc.: 0.2 µg/kg, sample origin: Yaounde (region, southern tropical forest), Cameroon, sample year: July–September 2011, country: Cameroon/Austria/South Africa[659]

incidence: 18/18, conc. range: 0.02–5 µg/kg, Ø conc.: 2.3 µg/kg, sample origin: Bamenda (region, western highlands), Cameroon, sample year: July–September 2011, country: Cameroon/Austria/South Africa[659]

- Co-contamination: For detailed information see the article.
- Further contamination of plants and plant products other than cereals and their products reported in the present book/article[659]: Dagwa, AFB_1, AFB_2, AFG_1, AFG_2, AME, BEA, CIT, DON, ENA, ENA_1, ENB, ENB_1, FA, FB_1, FB_2, FB_3, KA, MA, NIV, OTA, OTB, STG, VIOM, and ZEA, literature[659]; Flour (soybean flour), AFB_1, AFG_1, BEA, DON, ENB, ENB_1, FA, FB_1, FB_2, FB_3, FUS-X, KA, NIV, and ZEA, literature[659]; Kuru-kuru, AFB_1, AFB_2, AFG_1, AFG_2, BEA, ENA, ENA_1, ENB, ENB_1, FA, FB_1, FB_3, KA, MA, OTA, OTB, ROQ-C, STG, VIOM, and ZEA, literature[659]; Paste (peanut paste), AFB_1, AFB_2, AFG_1, AFG_2, AME, BEA, ENA, ENA_1, ENB, FA, FB_1, FB_2, FB_3, KA, MA, OTA, OTB, STG, VIOM, and ZEA, literature[659]; Soups, AFB_1, AFB_2, AFG_1, AFG_2, AME, AOH,

BEA, DON, ENA, ENA_1, ENB, ENB_1, FB_1, FB_2, FB_3, KA, OTA, OTB, and STG, literature[659]; see also Dagwa, Flour (soybean flour), Kuru-kuru, Paste (peanut paste), Soups, chapter: Further Mycotoxins and Microbial Metabolites

Fumonisin/s

Fumonisin B_1

incidence: 10/17, conc. range: 0.4–11 µg/kg, Ø conc.: 4.8 µg/kg, sample origin: Yaounde (region, southern tropical forest), Cameroon, sample year: July–September 2011, country: Cameroon/Austria/South Africa[659]

incidence: 8/18, conc. range: 1.1–9.6 µg/kg, Ø conc.: 6.3 µg/kg, sample origin: Bamenda (region, western highlands), Cameroon, sample year: July–September 2011, country: Cameroon/Austria/South Africa[659]

- Co-contamination: For detailed information see the article.
- Further contamination of plants and plant products other than cereals and their products reported in the present book/article[659]: Dagwa, AFB_1, AFB_2, AFG_1, AFG_2, AME, BEA, CIT, DON, ENA, ENA_1, ENB, ENB_1, FA, FB_1, FB_2, FB_3, KA, MA, NIV, OTA, OTB, STG, VIOM, and ZEA, literature[659]; Flour (soybean flour), AFB_1, AFG_1, BEA, DON, ENB, ENB_1, FA, FB_1, FB_2, FB_3, FUS-X, KA, NIV, and ZEA, literature[659]; Kuru-kuru, AFB_1, AFB_2, AFG_1, AFG_2, BEA, ENA, ENA_1, ENB, ENB_1, FA, FB_1, FB_3, KA, MA, OTA, OTB, ROQ-C, STG, VIOM, and ZEA, literature[659]; Paste (peanut paste), AFB_1, AFB_2, AFG_1, AFG_2, AME, BEA, ENA, ENA_1, ENB, ENB_1, FA, FB_2, FB_3, KA, MA, OTA, OTB, STG, VIOM, and ZEA, literature[659]; Soups, AFB_1, AFB_2, AFG_1, AFG_2, AME, AOH, BEA, DON, ENA, ENA_1, ENB, ENB_1, FB_1, FB_2, FB_3, KA, OTA, OTB, and STG, literature[659]; see also Dagwa, Flour (soybean flour), Kuru-kuru, Paste (peanut paste), Soups, chapter: Further Mycotoxins and Microbial Metabolites

Fumonisin B_2

incidence: 8/17, conc. range: 0.4–3.5 µg/kg, Ø conc.: 1.8 µg/kg, sample origin: Yaounde (region, southern tropical forest), Cameroon, sample year: July–September 2011, country: Cameroon/Austria/South Africa[659]

incidence: 4/18, conc. range: 1.3–4.2 µg/kg, Ø conc.: 2.5 µg/kg, sample origin: Bamenda (region, western highlands), Cameroon, sample year: July–September 2011, country: Cameroon/Austria/South Africa[659]

- Co-contamination: For detailed information see the article.
- Further contamination of plants and plant products other than cereals and their products reported in the present book/article[659]: Dagwa, AFB_1, AFB_2, AFG_1, AFG_2, AME, BEA, CIT, DON, ENA, ENA_1, ENB, ENB_1, FA, FB_1, FB_2, FB_3, KA, MA, NIV, OTA, OTB, STG, VIOM, and ZEA, literature[659]; Flour (soybean flour), AFB_1, AFG_1, BEA, DON, ENB, ENB_1, FA, FB_1, FB_2, FB_3, FUS-X, KA, NIV, and ZEA, literature[659]; Kuru-kuru, AFB_1, AFB_2, AFG_1, AFG_2, BEA, ENA, ENA_1, ENB, ENB_1, FA, FB_1, FB_3, KA, MA, OTA, OTB, ROQ-C, STG, VIOM, and ZEA, literature[659]; Paste (peanut paste), AFB_1, AFB_2, AFG_1, AFG_2, AME, BEA, ENA, ENA_1, ENB, ENB_1, FA, FB_1, FB_3, KA, MA, OTA, OTB, STG, VIOM, and ZEA, literature[659]; Soups, AFB_1, AFB_2, AFG_1, AFG_2, AME, AOH, BEA, DON, ENA, ENA_1, ENB, ENB_1, FB_1, FB_2, FB_3, KA, OTA, OTB, and STG, literature[659]; see also Dagwa, Flour (soybean flour), Kuru-kuru, Paste (peanut paste), Soups, chapter: Further Mycotoxins and Microbial Metabolites

Fumonisin B_3

incidence: 4/17, conc. range: <LOQ-4.2 µg/kg, Ø conc.: 3 µg/kg, sample origin: Yaounde (region, southern tropical forest), Cameroon, sample year: July–September 2011, country: Cameroon/Austria/South Africa[659]

incidence: 11/18, conc. range: <LOQ-4.6 µg/kg, Ø conc.: 3.3 µg/kg, sample origin: Bamenda (region, western highlands), Cameroon, sample year: July–September 2011, country: Cameroon/Austria/South Africa[659]

- Co-contamination: For detailed information see the article.
- Further contamination of plants and plant products other than cereals and their products reported in the present book/article[659]: Dagwa, AFB_1, AFB_2, AFG_1, AFG_2, AME, BEA, CIT, DON, ENA, ENA_1, ENB, ENB_1, FA, FB_1, FB_2, FB_3, KA, MA, NIV, OTA, OTB, STG, VIOM, and ZEA, literature[659]; Flour (soybean flour), AFB_1, AFG_1, BEA, DON, ENB, ENB_1, FA, FB_1, FB_2, FB_3, FUS-X, KA, NIV, and ZEA, literature[659]; Kuru-kuru, AFB_1, AFB_2, AFG_1, AFG_2, BEA, ENA, ENA_1, ENB, ENB_1, FA, FB_1, FB_3, KA, MA, OTA, OTB, ROQ-C, STG, VIOM, and ZEA, literature[659]; Paste (peanut paste), AFB_1, AFB_2, AFG_1, AFG_2, AME, BEA, ENA, ENA_1, ENB, ENB_1, FA, FB_1, FB_2, KA, MA, OTA, OTB, STG, VIOM, and ZEA, literature[659]; Soups, AFB_1, AFB_2, AFG_1, AFG_2, AME, AOH, BEA, DON, ENA, ENA_1, ENB, ENB_1, FB_1, FB_2, FB_3, KA, OTA, OTB, and STG, literature[659]; see also Dagwa, Flour (soybean flour), Kuru-kuru, Paste (peanut paste), Soups, chapter: Further Mycotoxins and Microbial Metabolites

Other *Fusarium* Toxins

Fusaric Acid

incidence: 13/17, conc. range: <LOQ-19 µg/kg, Ø conc.: 8.3 µg/kg, sample origin: Yaounde (region, southern tropical forest), Cameroon, sample year: July–September 2011, country: Cameroon/Austria/South Africa[659]

incidence: 15/18, conc. range: <LOQ-189 µg/kg, Ø conc.: 42 µg/kg, sample origin: Bamenda (region, western highlands), Cameroon, sample year: July–September 2011, country: Cameroon/Austria/South Africa[659]

- Co-contamination: For detailed information see the article.
- Further contamination of plants and plant products other than cereals and their products reported in the present book/article[659]: Dagwa, AFB_1, AFB_2, AFG_1, AFG_2, AME, BEA, CIT, DON, ENA, ENA_1, ENB, ENB_1, FA, FB_1, FB_2, FB_3, KA, MA, NIV, OTA, OTB, STG, VIOM, and ZEA, literature[659]; Flour (soybean flour), AFB_1, AFG_1, BEA, DON, ENB, ENB_1, FA, FB_1, FB_2, FB_3, FUS-X, KA, NIV, and ZEA, literature[659]; Kuru-kuru, AFB_1, AFB_2, AFG_1, AFG_2, BEA, ENA, ENA_1, ENB, ENB_1, FA, FB_1, FB_3, KA, MA, OTA, OTB, ROQ-C, STG, VIOM, and ZEA, literature[659]; Paste (peanut paste), AFB_1, AFB_2, AFG_1, AFG_2, AME, BEA, ENA, ENA_1, ENB, ENB_1, FB_1, FB_2, FB_3, KA, MA, OTA, OTB, STG, VIOM, and ZEA, literature[659]; Soups, AFB_1, AFB_2, AFG_1, AFG_2, AME, AOH, BEA, DON, ENA, ENA_1, ENB, ENB_1, FB_1, FB_2, FB_3, KA, OTA, OTB, and STG, literature[659]; see also Dagwa, Flour (soybean flour), Kuru-kuru, Paste (peanut paste), Soups, chapter: Further Mycotoxins and Microbial Metabolites

Zearalenone

incidence: 3/17, conc. range: <LOQ-45 µg/kg, Ø conc.: 23 µg/kg, sample origin: Yaounde (region, southern tropical forest), Cameroon, sample year: July–September 2011, country: Cameroon/Austria/South Africa[659]

incidence: 12/18, conc. range: 0.3–2 µg/kg, ∅ conc.: 0.8 µg/kg, sample origin: Bamenda (region, western highlands), Cameroon, sample year: July–September 2011, country: Cameroon/Austria/South Africa[659]

- Co-contamination: For detailed information see the article.
- Further contamination of plants and plant products other than cereals and their products reported in the present book/article[659]: Dagwa, AFB_1, AFB_2, AFG_1, AFG_2, AME, BEA, CIT, DON, ENA, ENA_1, ENB, ENB_1, FA, FB_1, FB_2, FB_3, KA, MA, NIV, OTA, OTB, STG, VIOM, and ZEA, literature[659]; Flour (soybean flour), AFB_1, AFG_1, BEA, DON, ENB, ENB_1, FA, FB_1, FB_2, FB_3, FUS-X, KA, NIV, and ZEA, literature[659]; Kuru-kuru, AFB_1, AFB_2, AFG_1, AFG_2, BEA, ENA, ENA_1, ENB, ENB_1, FA, FB_1, FB_3, KA, MA, OTA, OTB, ROQ-C, STG, VIOM, and ZEA, literature[659]; Paste (peanut paste), AFB_1, AFB_2, AFG_1, AFG_2, AME, BEA, ENA, ENA_1, ENB, ENB_1, FA, FB_1, FB_2, FB_3, KA, MA, OTA, OTB, STG, and VIOM, literature[659]; Soups, AFB_1, AFB_2, AFG_1, AFG_2, AME, AOH, BEA, DON, ENA, ENA_1, ENB, ENB_1, FB_1, FB_2, FB_3, KA, OTA, OTB, and STG, literature[659]; see also Dagwa, Flour (soybean flour), Kuru-kuru, Paste (peanut paste), Soups, chapter: Further Mycotoxins and Microbial Metabolites

Penicillium Toxins

Mycophenolic Acid

incidence: 15/17, conc. range: <LOQ-2 µg/kg, ∅ conc.: 1 µg/kg, sample origin: Yaounde (region, southern tropical forest), Cameroon, sample year: July–September 2011, country: Cameroon/Austria/South Africa[659]

incidence: 8/18, conc. range: <LOQ-1.3 µg/kg, ∅ conc.: 0.8 µg/kg, sample origin: Bamenda (region, western highlands), Cameroon, sample year: July–September 2011, country: Cameroon/Austria/South Africa[659]

- Co-contamination: For detailed information see the article.
- Further contamination of plants and plant products other than cereals and their products reported in the present book/article[659]: Dagwa, AFB_1, AFB_2, AFG_1, AFG_2, AME, BEA, CIT, DON, ENA, ENA_1, ENB, ENB_1, FA, FB_1, FB_2, FB_3, KA, MA, NIV, OTA, OTB, STG, VIOM, and ZEA, literature[659]; Flour (soybean flour), AFB_1, AFG_1, BEA, DON, ENB, ENB_1, FA, FB_1, FB_2, FB_3, FUS-X, KA, NIV, and ZEA, literature[659]; Kuru-kuru, AFB_1, AFB_2, AFG_1, AFG_2, BEA, ENA, ENA_1, ENB, ENB_1, FA, FB_1, FB_3, KA, MA, OTA, OTB, ROQ-C, STG, VIOM, and ZEA, literature[659]; Paste (peanut paste), AFB_1, AFB_2, AFG_1, AFG_2, AME, BEA, ENA, ENA_1, ENB, ENB_1, FA, FB_1, FB_2, FB_3, KA, OTA, OTB, STG, VIOM, and ZEA, literature[659]; Soups, AFB_1, AFB_2, AFG_1, AFG_2, AME, AOH, BEA, DON, ENA, ENA_1, ENB, ENB_1, FB_1, FB_2, FB_3, KA, OTA, OTB, and STG, literature[659]; see also Dagwa, Flour (soybean flour), Kuru-kuru, Paste (peanut paste), Soups, chapter: Further Mycotoxins and Microbial Metabolites

see also **Kuru-kuru**

Paste (pipian paste) may contain the following mycotoxins:

Aspergillus Toxins

Aflatoxin/s

incidence: 3/3, conc. range: ≤78 µg/kg*, sample origin: imported to the USA, sample year: 1986, country: USA[19], *AFB_1, AFB_2, AFG_1, and AFG_2

- Co-contamination: not reported
- Further contamination of plants and plant products other than cereals and their products reported in the present book/article[19]: Butter (peanut butter), Candies (peanut candy), Candies (pistachio candy), Nuts, Nuts (almond), Nuts (Brazil nut, paranut), Nuts (peanut), Nuts (pecan nut), Nuts (pistachio nut), Nuts (walnut), Oils (peanut oil), Paste (peanut paste), Products (peanut product), Seeds (melon), Spices (cayenne pepper, chilli), Spices (ginger), and Spices (nutmeg), AF/S, literature[19]

Paste (red pepper paste) may contain the following mycotoxins:

Aspergillus Toxins

Aflatoxin/s

incidence: 2/15*, conc. range: 0.21–0.55 µg/kg**, ∅ conc.: 0.38 µg/kg**, sample origin: markets in ten cities, Korea, sample year: May–September 2006, country: Korea[188], *red pepper paste (gochujang), **AFB_1, AFB_2, AFG_1, and AFG_2

- Co-contamination: not reported
- Further contamination of plants and plant products other than cereals and their products reported in the present book/article[188]: Spices (cayenne pepper, chilli), Spices (curry), Spices (ginger), AF/S, literature[188]

Paste (sesame paste) may contain the following mycotoxins:

Aspergillus Toxins

Aflatoxin B_1

incidence: 37/100*, conc. range: 0.39–20.45 µg/kg, ∅ conc.: 4.31 µg/kg, sample origin: retail markets in Changchun, Chengdu, Shanghai, Shijiazhuang, Zhengzhou (cities), and Beijing (capital), China, sample year: 2007, country: China[224], *see below

- Co-contamination: 36 sa. co-contaminated with AFB_1 and AFB_2; at least 2 sa. co-contaminated with AFB_1 and AFG_1
- Further contamination of plants and plant products other than cereals and their products reported in the present book/article[224]: Butter (peanut butter), AFB_1, AFB_2, AFG_1, and AFG_2, literature[224]; Paste (sesame paste), AFB_2, AFG_1, and AFG_2, literature[224]

incidence: 8/26*, conc. range: 0.9–1.78 µg/kg, ∅ conc.: 1.29 µg/kg, sample origin: restaurants (in bulk) in Changchun, Chengdu, Shanghai, Shijiazhuang, Zhengzhou (cities), and Beijing (capital), China, sample year: 2007, country: China[224], *freshly prepared

incidence: 7/15*, conc. range: 0.86–14.56 µg/kg, ∅ conc.: 5.32 µg/kg, sample origin: free markets (in bulk) in Changchun, Chengdu, Shanghai, Shijiazhuang, Zhengzhou (cities), and Beijing (capital), China, sample year: 2007, country: China[224], *freshly prepared

incidence: 22/59*, conc. range: 0.39–20.45 µg/kg, ∅ conc.: 5.73 µg/kg, sample origin: supermarkets (bottling) in Changchun, Chengdu, Shanghai, Shijiazhuang, Zhengzhou (cities), and Beijing (capital), China, sample year: 2007, country: China[224], *industrially manufactured and packed in glass bottles with shelf life between 12 and 24 months

- Co-contamination: not reported
- Further contamination of plants and plant products other than cereals and their products reported in the present book/article[224]: Butter (peanut butter), AFB_1, AFB_2, AFG_1, and AFG_2, literature[224]; Paste (sesame paste), AFB_2, AFG_1, and AFG_2, literature[224]

Aflatoxin B_2

incidence: 36/100, conc. range: 0.02–4.92 µg/kg, ∅ conc.: 0.65 µg/kg, sample origin: retail markets in Changchun, Chengdu, Shanghai, Shijiazhuang, Zhengzhou (cities), and Beijing (capital), China, sample year: 2007, country: China[224]

- Co-contamination: 36 sa. co-contaminated with AFB_1 and AFB_2; no further information available
- Further contamination of plants and plant products other than cereals and their products reported in the present book/article[224]: Butter (peanut butter), AFB_1, AFB_2, AFG_1, and AFG_2, literature[224]; Paste (sesame paste), AFB_1, AFG_1, and AFG_2, literature[224]

Aflatoxin G_1

incidence: 29/100, conc. range: 0.05–26.28 µg/kg, ∅ conc.: 2.57 µg/kg, sample origin: retail markets in Changchun, Chengdu, Shanghai, Shijiazhuang, Zhengzhou (cities), and Beijing (capital), China, sample year: 2007, country: China[224]

- Co-contamination: at least 2 sa. co-contamination with AFB_1 and AFG_1; no further information available
- Further contamination of plants and plant products other than cereals and their products reported in the present book/article[224]: Butter (peanut butter), AFB_1, AFB_2, AFG_1, and AFG_2, literature[224]; Paste (sesame paste), AFB_1, AFB_2, and AFG_2, literature[224]

Aflatoxin G_2

incidence: 22/100, conc. range: ≤5.75 µg/kg, sample origin: retail markets in Changchun, Chengdu, Shanghai, Shijiazhuang, Zhengzhou (cities), and Beijing (capital), China, sample year: 2007, country: China[224]

- Co-contamination: not reported
- Further contamination of plants and plant products other than cereals and their products reported in the present book/article[224]: Butter (peanut butter), AFB_1, AFB_2, AFG_1, and AFG_2, literature[224]; Paste (sesame paste), AFB_1, AFB_2, and AFG_1, literature[224]

Paste (soybean paste) may contain the following mycotoxins:

Aspergillus Toxins

Aflatoxin B_1

incidence: 2/13, conc. range: 0.05–0.17 µg/kg, ∅ conc.: 0.11 µg/kg, sample origin: sample origin: grocery markets in Busan, Daegu, Daejeon, Gangneung, and Gwangju (cities) as well as Seoul (capital), Korea, sample year: May 2004–June 2005, country: Korea[10]

- Co-contamination: not reported
- Further contamination of plants and plant products other than cereals and their products reported in the present book/article[10]: Butter (peanut butter), Nuts, Nuts (almond), Nuts (peanut), Nuts (pistachio nut), Sauces (soy sauce), Spices (cayenne pepper, chilli), Spices (curry), Spices (pepper), AFB_1, literature[10]

incidence: 1/80*, conc.: pr., sample origin: markets in Taipei (capital) vicinity, Taiwan, sample year: March–October 1975, country: Taiwan[590], *fermented soybean paste

- Co-contamination: not reported
- Further contamination of plants and plant products other than cereals and their products reported in the present book/article[590]: Paste (soybean paste), AFB_2, AFG_1, and AFG_2, literature[590]; Sauces (soy sauce) AFG_2, literature[590]

Aflatoxin B_2

incidence: 4/80*, conc. range: pr., sample origin: markets in Taipei (capital) vicinity, Taiwan, sample year: March–October 1975, country: Taiwan[590], *fermented soybean paste

- Co-contamination: not reported
- Further contamination of plants and plant products other than cereals and their products reported in the present book/article[590]: Paste (soybean paste), AFB_1, AFG_1, and AFG_2, literature[590]; Sauces (soy sauce) AFG_2, literature[590]

Aflatoxin G_1

incidence: 1/80*, conc.: pr., sample origin: markets in Taipei (capital) vicinity, Taiwan, sample year: March–October 1975, country: Taiwan[590], *fermented soybean paste

- Co-contamination: not reported
- Further contamination of plants and plant products other than cereals and their products reported in the present book/article[590]: Paste (soybean paste), AFB_1, AFB_2, and AFG_2, literature[590]; Sauces (soy sauce) AFG_2, literature[590]

Aflatoxin G_2

incidence: 8/80*, conc. range: pr., sample origin: markets in Taipei (capital) vicinity, Taiwan, sample year: March–October 1975, country: Taiwan[590], *fermented soybean paste

- Co-contamination: not reported
- Further contamination of plants and plant products other than cereals and their products reported in the present book/article[590]: Paste (soybean paste), AFB_1, AFB_2, and AFG_1, literature[590]; Sauces (soy sauce), AFG_2, literature[590]

Aspergillus and *Penicillium* Toxins

Ochratoxin A

incidence: 5/7*, conc. range: 0.97–20.78 µg/kg, ∅ conc.: 11.1 µg/kg, sample origin: homemade, Korea, sample year: unknown, country: Korea[662], *homemade soybean paste (doenjang)

incidence: 5/5*, conc. range: ≤0.90 µg/kg, sample origin: companies, Korea, sample year: unknown, country: Korea[662], *manufactured soybean paste (doenjang)

- Co-contamination: not reported
- Further contamination of plants and plant products other than cereals and their products reported in the present book/article[662]: not reported

Paste (tomato paste) may contain the following mycotoxins:

Alternaria Toxins

Alternariol

incidence: 1/1, conc.: 25 µg/kg, sample origin: unknown, sample year: unknown, country: Germany[46]

- Co-contamination: 1 sa. co-contaminated with AME and AOH
- Further contamination of plants and plant products other than cereals and their products reported in the present book/article[46]: Juices (tomato juice), Puree, Tomato ketchup, Wine (red), Wine (white), AME and AOH, literature[16]; Juices (vegetable juice), Paste (tomato paste), AME, literature[46]

incidence: 10/10, conc. range: 3.1–13 µg/kg, ∅ conc.: 6.6 µg/kg, sample origin: retail shops, Hesse (federal state), Germany, sample year: 2009/2010, country: Germany[439]

- Co-contamination: not reported

- Further contamination of plants and plant products other than cereals and their products reported in the present book/article[439]: Juices (apple juice), Juice (tomato juice), Sauces (apple sauce), Tomato ketchup, Wine (white), AOH, literature[439]

incidence: 3/5, conc. range: ≤26 µg/kg, sample origin: retail stores, the Netherlands, sample year: autumn 2014, country: the Netherlands/Germany[906]

- Co-contamination: not reported
- Further contamination of plants and plant products other than cereals and their products reported in the present book/article[906]: Figs, Fruits (tomato), and Juices (tomato juice), AOH and TA, literature[906]; Oils (sunflower oil), AME, literature[906]; Paste (tomato paste), AME and TA, literature[906]; Sauces (tomato sauce), Seeds (sunflower), Tomato ketchup, AME, AOH, and TA, literature[906]

Alternariol Methyl Ether

incidence: 1/1, conc.: 5.3 µg/kg, sample origin: unknown, sample year: unknown, country: Germany[46]

- Co-contamination: 1 sa. co-contaminated with AME and AOH
- Further contamination of plants and plant products other than cereals and their products reported in the present book/article[46]: Juices (tomato juice), Puree, Tomato ketchup, Wine (red), Wine (white), AME and AOH, literature[46]; Juices (vegetable juice), AME, literature[46]; Paste (tomato paste), AOH, literature[46]

incidence: 1/5, conc.: 5.6 µg/kg, sample origin: retail stores, the Netherlands, sample year: autumn 2014, country: the Netherlands/Germany[906]

- Co-contamination: not reported
- Further contamination of plants and plant products other than cereals and their products reported in the present book/article[906]: Figs, Fruits (tomato), and Juices (tomato juice), AOH and TA, literature[906]; Oils (sunflower oil), AME, literature[906]; Paste (tomato paste), AOH and TA, literature[906]; Sauces (tomato sauce), Seeds (sunflower), Tomato ketchup, AME, AOH, and TA, literature[906]

Tenuazonic Acid

incidence: 4/5, conc. range: ≤344 µg/kg, sample origin: retail stores, Netherlands, sample year: autumn 2014, country: the Netherlands/Germany[906]

- Co-contamination: not reported
- Further contamination of plants and plant products other than cereals and their products reported in the present book/article[906]: Figs, Fruits (tomato), and Juices (tomato juice), AOH and TA, literature[906]; Oils (sunflower oil), AME, literature[906]; Paste (tomato paste), AME and AOH, literature[906]; Sauces (tomato sauce), Seeds (sunflower), Tomato ketchup, AME, AOH, and TA, literature[906]

Peach concentrate
see **Juices (peach concentrate)**

Peanut brittle
see **Nuts (peanut)**

Peanut butter
see **Butter (peanut butter)**

Peanut candies
see **Candies (peanut candy)**

Peanut pastes
see **Paste (peanut paste)**

Peanut products
see **Products (peanut product)**

Peanut sauce
see **Sauces (peanut sauce)**

Peanuts
see **Nuts (peanut)**

Pear concentrate
see **Juices (pear concentrate)**

Pear marmalade
see **Marmalade (pear marmalade)**

Peas
may contain the following mycotoxins:

Aspergillus Toxins

Aflatoxin B_1

incidence: 1/32, conc.: 1–10 µg/kg, sample origin: unknown, sample year: 1990–1996, country: Cuba[126]

- Co-contamination: not reported
- Further contamination of plants and plant products other than cereals and their products reported in the present book/article[126]: Beans, Beans (soybean), Cocoa beans, Nuts (peanut), AFB_1, literature[126]

incidence: 1/35, conc.: 25 µg/kg, sample origin: storage houses and dealers, Tunisia, sample year: unknown, country: Tunisia/USA[380]

- Co-contamination: 1 sa. co-contaminated with AFB_1 and AFG_1
- Further contamination of plants and plant products other than cereals and their products reported in the present book/article[380]: Nuts (peanut), AFB_1 and AFG_1, literature[380]; Nuts (pistachio nut), AFB_1, literature[380]; Peas, AFG_1, literature[380]

Aflatoxin G_1

incidence: 1/35, conc.: 42 µg/kg, sample origin: storage houses and dealers, Tunisia, sample year: unknown, country: Tunisia/USA[380]

- Co-contamination: 1 sa. co-contaminated with AFB_1 and AFG_1
- Further contamination of plants and plant products other than cereals and their products reported in the present book/article[380]: Nuts (peanut), AFB_1 and AFG_1, literature[380]; Nuts (pistachio nut), Peas, AFB_1, literature[380]

Aflatoxin/s

incidence: 3/19, conc. range: 1–100 µg/kg* (3 sa.), sample origin: natives homes and markets, Uganda, sample year: September 1966–June 1967, country: Uganda/USA[137], *AFB_1, AFB_2, AFG_1, and AFG_2

- Co-contamination: 2 sa. contained AFB_1; 1 sa. contained AFB_2; 1 sa. contained AFG_1
- Further contamination of plants and plant products other than cereals and their products reported in the present book/article[137]: Beans, Cassava, Nuts (peanut), AF/S, literature[137]

incidence: 3/13*, conc. range: 18–30 µg/kg**, ⌀ conc.: 24 µg/kg**, sample origin: markets in Khartoum (capital), Sudan, sample year: 1993–1996, country: Sudan/Oman[591], *ncac, **AFB_1, AFB_2, AFG_1, and AFG_2

- Co-contamination: 3 sa. co-contaminated with AFB_1, AFB_2, AFG_1 and AFG_2
- Further contamination of plants and plant products other than cereals and their products reported in the present book/article[591]: not reported

incidence: 4/15*, conc. range: 1.5–9.5 µg/kg**, ⌀ conc.: 4.1 µg/kg**, sample origin: retail shops and local markets, Pakistan, sample year: January–November 2009, country: Pakistan[782], *split pea, **AFB_1, AFB_2, AFG_1, and AFG_2

- Co-contamination: not reported
- Further contamination of plants and plant products other than cereals and their products reported in the present book/article[782]: Beans (cowpea), Beans (kidney bean), Beans (soybean), Peas (chickpea), AF/S, literature[782]

Aspergillus and *Penicillium* Toxins

Ochratoxin A

incidence: 1/20*, conc. range: 0.1–5 μg/kg, sample origin: ethnic stores, health food stores, and local stores as well as supermarkets in the four largest cities, Sweden, sample year: autumn 1996–summer 1998, country: Sweden[13], *yellow peas

- Co-contamination: not reported
- Further contamination of plants and plant products other than cereals and their products reported in the present book/article[13]: Beans (brown bean), Peas (chickpea), OTA, literature[13]; Juices (apple juice), PAT, literature[13]; Nuts (almond), AFB_1, literature[13]; Nuts (Brazil nut, paranut), Nuts (pistachio nut), Oil seeds, Products (peanut product), AFB_1 and AF/S, literature[13]

incidence: 2/27*, conc. range: 10 μg/kg, ∅ conc.: 10 μg/kg, sample origin: retail outlets, Sweden, sample year: 1976, country: Sweden[144], *yellow peas

- Co-contamination: not reported
- Further contamination of plants and plant products other than cereals and their products reported in the present book/article[144]: Beans (brown bean), OTA, literature[144]; Peas, PA and ZEA, literature[144]

incidence: 1/20*, conc.: 0.78 μg/kg (1 sa.), sample origin: Sweden, sample year: 1997/1998, country: EU[698], *peas, yellow

incidence: 2/20*, conc. range: LOD/LOQ-0.9 μg/kg (1 sa.), 1.9 μg/kg (1 sa.), sample origin: Sweden, sample year: 1997/1998, country: EU[698], *peas, brown

- Co-contamination: not reported
- Further contamination of plants and plant products other than cereals and their products reported in the present book/article[698]: Baby food, Beans (soybean), Chocolate, Cocoa, Coffee, Coffee beans, Currants, Fruits, dried, Herbs, Jam, Juices, Juices (grape juice), Legumes, Nuts, Oil seeds, Oils (olive oil), Peas (chickpea), Products (plant product), Raisins, Spices, Spices (cayenne pepper, chilli), Spices (nutmeg), Spices (paprika), Spices (pepper), Sultanas, Tea, Vinegar, Wine, Wine (red), Wine (rosé), Wine (white), Wine, miscellaneous, OTA, literature[698]

Penicillic Acid

incidence: 1/18*, conc.: 24 μg/kg, sample origin: retail outlets, Sweden, sample year: 1976, country: Sweden[144], *green peas

- Co-contamination: not reported
- Further contamination of plants and plant products other than cereals and their products reported in the present book/article[144]: Beans (brown bean), OTA, literature[144]; Peas, OTA and ZEA, literature[144]

Fusarium Toxins

Zearalenone

incidence: 1/27*, conc.: 2 μg/kg, sample origin: retail outlets, Sweden, sample year: 1976, country: Sweden[144], *yellow peas

- Co-contamination: not reported
- Further contamination of plants and plant products other than cereals and their products reported in the present book/article[144]: Beans (brown bean), OTA, literature[144]; Peas, OTA and PA, literature[144]

Peas (chickpea) may contain the following mycotoxins:

Aspergillus Toxins

Aflatoxin B_1

incidence: 1/11*, conc.: 1.7 μg/kg, sample origin: retailers, markets, and dried fruit retail shops in Erzurum (city), Turkey, sample year: March 2001–March 2002, country. Turkey[14], *chickpeas, roasted (Leblebi)

- Co-contamination: not reported
- Further contamination of plants and plant products other than cereals and their products reported in the present book/article[14]: Nuts (almond), Nuts (hazelnut), Nuts (peanut), Nuts (walnut), AFB_1, literature[14]

incidence: 10/28*, conc. range: <20 μg/kg (2 sa.), 21–100 μg/kg (8 sa., maximum: 74 μg/kg), sample origin: villages of Khagaria and Saharsa (districts) of North Bihar (state), India, sample year: August–October 1987, country: India[128], *gram

- Co-contamination: not reported
- Further contamination of plants and plant products other than cereals and their products reported in the present book/article[128]: Beans (cowpea), Beans (mung bean), Field mustard, Lentils, Seeds (linseed), AFB_1, literature[128]

incidence: 1/1, conc.: 0.876 μg/kg, sample origin: market in Khartoum (capital), Sudan, sample year: February 1981–1984?, country: UK/Sudan[223]

- Co-contamination: not reported
- Further contamination of plants and plant products other than cereals and their products reported in the present book/article[223]: Butter (peanut butter), AFB_1, AFB_2, and AFG_1, literature[223]; Nuts (peanut), AFB_1, AFB_2, AFG_1, and AFG_2, literature[223]; Okra, AFG_2, literature[223]

Aflatoxin/s

incidence: 1/10, conc.: 2.5 μg/kg*, sample origin: retail shops and local markets, Pakistan, sample year: January–November 2009, country: Pakistan[782], *AFB_1, AFB_2, AFG_1, and AFG_2

- Co-contamination: not reported
- Further contamination of plants and plant products other than cereals and their products reported in the present book/article[782]: Beans (cowpea), Beans (kidney bean), Beans (soybean), Peas, AF/S, literature[782]

Aspergillus and *Penicillium* Toxins

Ochratoxin A

incidence: 1/10, conc. range: 0.1–5 μg/kg, sample origin: ethnic stores, health-food stores, and local stores as well as supermarkets in the four largest cities, Sweden, sample year: autumn 1996–summer 1998, country: Sweden[13]

- Co-contamination: not reported
- Further contamination of plants and plant products other than cereals and their products reported in the present book/article[13]: Beans (brown bean) and Peas, OTA, literature[13]; Juices (apple juice), PAT, literature[13]; Nuts (almond), AFB_1, literature[13]; Nuts (Brazil nut, paranut), Nuts (pistachio nut), Oil seeds, Products (peanut product), AFB_1 and AF/S, literature[13]

incidence: 1/7, conc.: 0.86 μg/kg (1 sa.), sample origin: Germany, sample year: 1995/1998, country: EU[698]

incidence: 1/10, conc.: 1.2 μg/kg (1 sa.), sample origin: Sweden, sample year: 1997/1998, country: EU[698]

- Co-contamination: not reported
- Further contamination of plants and plant products other than cereals and their products reported in the present book/article[698]: Baby food, Beans (soybean), Chocolate, Cocoa, Coffee, Coffee beans, Currants, Fruits, dried, Herbs, Jam, Juices, Juices (grape juice), Legumes, Nuts, Oil seeds, Oils (olive oil), Peas, Products (plant product), Raisins, Spices, Spices (cayenne pepper, chilli), Spices (nutmeg), Spices (paprika), Spices (pepper), Sultanas, Tea, Vinegar, Wine, Wine (red), Wine (rosé), Wine (white), Wine, miscellaneous, OTA, literature[698]

Fusarium Toxins

Type A Trichothecenes

T-2 Toxin

incidence: 2/5, conc. range: 660–700 μg/kg, ∅ conc.: 680 μg/kg, sample origin: market and bazaar in Istanbul (capital), Turkey, sample year: unknown, country: Turkey[150]

- Co-contamination: not reported
- Further contamination of plants and plant products other than cereals and their products reported in the present book/article[150]: Beans, Lentils,T-2, literature[150]

Pedro Ximenez

see **Wine, miscellaneous**

Pekmez

may contain the following mycotoxins:

Aspergillus and ***Penicillium*** Toxins

Ochratoxin A

incidence: 24/25, conc. range: 0.44–5.32 μg/kg, ∅ conc.: 2.04 μg/kg, sample origin: different supermarkets and traditional bazaars in Istanbul (capital), Turkey, sample year: 2008/2009, country: Turkey[665]

- Co-contamination: not reported
- Further contamination of plants and plant products other than cereals and their products reported in the present book/article[665]: Currants, Juices (grape juice), Sultanas, OTA, literature[665]

incidence: 29/55*, conc. range: 2–5 μg/l (15 sa.), 5–10 μg/l (10 sa.), and 10–20 μg/l (4 sa., maximum: 15 μg/l), sample origin: different provinces, Turkey, sample year: unknown, country: Turkey[862]; *home-made grape pekmez

incidence: 5/20*, conc. range: 2–5 μg/l (3 sa.), 5–10 μg/l (1 sa.), and 12 μg/l (1 sa.), sample origin: different provinces, Turkey, sample year: unknown, country: Turkey[862]; *commercial grape pekmez

incidence: 3/7*, conc. range: 10–20 μg/l (2 sa.) and 31.2 μg/l (1 sa.), sample origin: different provinces, Turkey, sample year: unknown, country: Turkey[862]; *organic grape pekmez

- Co-contamination: not reported
- Further contamination of plants and plant products other than cereals and their products reported in the present book/article[862]: not reported

Pekmez is a Turkish syrup made from grapes or other fruits.

Penedés

see **Wine (red)** and **Wine (white)**

Pepper

see **Spices (pepper)**

Phaseolus vulgaris

see **Beans**

Pickles

may contain the following mycotoxins:

Aspergillus Toxins

Aflatoxin/s

incidence: 4?/4*, conc. range: 1.0–58.5 μg/kg**, sample origin: specialist food shops, UK, sample year: unknown, country: UK[16], *chilli pickle, **AFB_1, AFB_2, AFG_1, and AFG_2

incidence: 4?/4*, conc. range: 0.2–0.6 μg/kg**, sample origin: specialist food shops, UK, sample year: unknown, country: UK[16], *garlic pickle, **AFB_1, AFB_2, AFG_1, and AFG_2

- Co-contamination: not reported
- Further contamination of plants and plant products other than cereals and their products reported in the present book/article[16]: Food, AF/S, DON, FUS-X, NIV, OTA, and ZEA, literature[16]; Nuts (almond), AF/S, literature[16]; Oils and Spices (fennel), AF/S and ZEA, literature[16]; Oils (sesame oil), OTA, literature[16]; Paste (curry paste), AF/S, FB/S, NIV, OTA, and ZEA, literature[16]; Pickles, FB/S, OTA, and ZEA, literature[16]; Sauces (chilli sauce), NIV, OTA, and ZEA, literature[16]; Spices, AF/S, FB/S, OTA, and ZEA, literature[16]; Spices (cayenne pepper, chilli), AF/S, DAS, DON, HT-2, OTA, and ZEA, literature[16]; Spices (coriander), AF/S, DON, OTA, and ZEA, literature[16]; Spices (curry), AF/S, DAS, FB/S, FUS-X, HT-2, NEO, NIV, OTA, T-2, and ZEA, literature[16]; Spices (garlic), DON, FUS-X, and NIV, literature[16]; Spices (ginger), AF/S, DON, NEO, NIV, OTA, and T-2, literature[16]; Spices (tandoori), AF/S, FB/S, NIV, OTA, and T-2, literature[16]

Aspergillus and ***Penicillium*** Toxins

Ochratoxin A

incidence: 4?/4*, conc. range: 0.5–1.2 μg/kg, sample origin: specialist food shops, UK, sample year: unknown, country: UK[16], *chilli pickle

incidence: 4?/4*, conc. range: 0.9–2.5 μg/kg, sample origin: specialist food shops, UK, sample year: unknown, country: UK[16], *garlic pickle

- Co-contamination: not reported
- Further contamination of plants and plant products other than cereals and their products reported in the present book/article[16]: Food, AF/S, DON, FUS-X, NIV, OTA, and ZEA, literature[16]; Nuts (almond), AF/S, literature[16]; Oils and Spices (fennel), AF/S and ZEA, literature[16]; Oils (sesame oil), OTA, literature[16]; Paste (curry paste), AF/S, FB/S, NIV, OTA, and ZEA, literature[16]; Pickles, AF/S, FB/S, and ZEA, literature[16]; Sauces (chilli sauce), NIV, OTA, and ZEA, literature[16]; Spices, AF/S, FB/S, OTA, and ZEA, literature[16]; Spices (cayenne pepper, chilli), AF/S, DAS, DON, HT-2, OTA, and ZEA, literature[16]; Spices (coriander), AF/S, DON, OTA, and ZEA, literature[16]; Spices (curry), AF/S, DAS, FB/S, FUS-X, HT-2, NEO, NIV, OTA, T-2, and ZEA, literature[16]; Spices (garlic), DON, FUS-X, and NIV, literature[16]; Spices (ginger), AF/S, DON, NEO, NIV, OTA,

and T-2, literature[16]; Spices (tandoori), AF/S, FB/S, NIV, OTA, and T-2, literature[16]

Fusarium Toxins

Fumonisin/s

incidence: 1/4*, conc.: 121 µg/kg**, sample origin: specialist food shops, UK, sample year: unknown, country: UK[16], *chilli pickle, **FB_1 and FB_2

incidence: 0/4*, conc.: no contamination**, sample origin: specialist food shops, UK, sample year: unknown, country: UK[16], *garlic pickle, **FB_1 and FB_2

- Co-contamination: not reported
- Further contamination of plants and plant products other than cereals and their products reported in the present book/article[16]: Food, AF/S, DON, FUS-X, NIV, OTA, and ZEA, literature[16]; Nuts (almond), AF/S, literature[16]; Oils and Spices (fennel), AF/S and ZEA, literature[16]; Oils (sesame oil), OTA, literature[16]; Paste (curry paste), AF/S, FB/S, NIV, OTA, and ZEA, literature[16]; Pickles, AF/S, OTA, and ZEA, literature[16]; Sauces (chilli sauce), NIV, OTA, and ZEA, literature[16]; Spices, AF/S, FB/S, OTA, and ZEA, literature[16]; Spices (cayenne pepper, chilli), AF/S, DAS, DON, HT-2, OTA, and ZEA, literature[16]; Spices (coriander), AF/S, DON, OTA, and ZEA, literature[16]; Spices (curry), AF/S, DAS, FB/S, FUS-X, HT-2, NEO, NIV, OTA, T-2, and ZEA, literature[16]; Spices (garlic), DON, FUS-X, and NIV, literature[16]; Spices (ginger), AF/S, DON, NEO, NIV, OTA, and T-2, literature[16]; Spices (tandoori), AF/S, FB/S, NIV, OTA, and T-2, literature[16]

Other ***Fusarium*** Toxins

Zearalenone

incidence: 0/4*, conc.: no contamination, sample origin: specialist food shops, UK, sample year: unknown, country: UK[16], *chilli pickle

incidence: 1/4*, conc.: 3.8 µg/kg, sample origin: specialist food shops, UK, sample year: unknown, country: UK[16], *garlic pickle

- Co-contamination: not reported
- Further contamination of plants and plant products other than cereals and their products reported in the present book/article[16]: Food, AF/S, DON, FUS-X, NIV, OTA, and ZEA, literature[16]; Nuts (almond), AF/S, literature[16]; Oils and Spices (fennel), AF/S and ZEA, literature[16]; Oils (sesame oil), OTA, literature[16]; Paste (curry paste), AF/S, FB/S, NIV, OTA, and ZEA, literature[16]; Pickles, AF/S, FB/S, and OTA, literature[16]; Sauces (chilli sauce), NIV, OTA, and ZEA, literature[16]; Spices, AF/S, FB/S, OTA, and ZEA, literature[16]; Spices (cayenne pepper, chilli), AF/S, DAS, DON, HT-2, OTA, and ZEA, literature[16]; Spices (coriander), AF/S, DON, OTA, and ZEA, literature[16]; Spices (curry), AF/S, DAS, FB/S, FUS-X, HT-2, NEO, NIV, OTA, T-2, and ZEA, literature[16]; Spices (garlic), DON, FUS-X, and NIV, literature[16]; Spices (ginger), AF/S, DON, NEO, NIV, OTA, and T-2, literature[16]; Spices (tandoori), AF/S, FB/S, NIV, OTA, and T-2, literature[16]

Pies may contain the following mycotoxins:

Aspergillus and ***Penicillium*** Toxins

Ochratoxin A

incidence: 2/2*, conc. range: 0.11 µg/kg, ∅ conc.: 0.11 µg/kg, sample origin: supermarkets and fast-food chains in Quebec (city), Canada, sample year: September/October 2008, country: Canada[90], *pies, apple, and others

incidence: 2/2*, conc. range: 0.09–0.13 µg/kg, ∅ conc.: 0.11 µg/kg, sample origin: supermarkets and fast-food chains in Calgary (city), Canada, sample year: September/October 2009, country: Canada[90], *pies, apple and others

- Co-contamination: not reported
- Further contamination of plants and plant products other than cereals and their products reported in the present book/article[90]: Baby food, Beans, Beverages, Butter (peanut butter), Candies, Chocolate, Coffee, Extracts, Herbs, Ice cream, Nuts, Pizza, Products, Pudding, Raisins, Seeds, Wine, OTA, literature[90]

Pili-pili see Spices (cayenne pepper, chilli)

Pinhol may contain the following mycotoxins:

Aspergillus Toxins

Aflatoxin G_1

incidence: 1/2*, conc.: 0.50 µg/kg, sample origin: China and Spain from markets and supermarkets in Valencia (city), Spain, sample year: January–December 2002, country: Spain[435], *nut cocktail

- Co-contamination: not reported
- Further contamination of plants and plant products other than cereals and their products reported in the present book/article[435]: Nuts, Nuts (hazelnut), AFB_1 and AFG_1, literature[435]

Pinotage see Wine (red)

Pipian pastes see Paste (pipian paste)

Pistachio candies see Candies (pistachio candy)

Pistachio nuts see Nuts (pistachio nut)

Pizza may contain the following mycotoxins:

Aspergillus and ***Penicillium*** Toxins

Ochratoxin A

incidence: 1/1, conc.: 0.21 µg/kg, sample origin: supermarkets and fast-food chains in Quebec (city), Canada, sample year: September/October 2008, country: Canada[90]

incidence: 1/1, conc.: 0.22 µg/kg, sample origin: supermarkets and fast-food chains in Calgary (city), Canada, sample year: September/October 2009, country: Canada[90]

- Co-contamination: not reported
- Further contamination of plants and plant products other than cereals and their products reported in the present book/article[90]: Baby food, Beans, Beverages, Butter (peanut butter), Candies, Chocolate, Coffee, Extracts, Herbs, Ice cream, Nuts, Pies, Products, Pudding, Raisins, Seeds, Wine, OTA, literature[90]

Plant food supplements may contain the following mycotoxins:

Aspergillus and ***Penicillium*** Toxins

Ochratoxin A

incidence: 18/24*, conc. range: <LOQ-20.23 µg/kg, sample origin: internet or retail stores in Bari (province), Italy, sample year: unknown, country: Italy[845], *derived from *Vitis vinifera* administered as ampoule, capsule, comfit, powder, solution, syrup, or tablet

- Co-contamination: not reported
- Further contamination of plants and plant products other than cereals and their products reported in the present book/article[845]: Food colorant, Grape pomace and wine lees, OTA, literature[845]

For detailed information, see the article.

Plant products see **Products (plant product)**

Plavac Mali see **Must**

Plums see **Fruits, dried**

Port wine see **Wine, miscellaneous**

Potato products see **Products (potato product)**

Potatoes may contain the following mycotoxins:

Aspergillus and ***Penicillium*** Toxins

Ochratoxin A

incidence: 1/11*, conc.: 4.26 µg/kg, sample origin: households and markets, Germany, sample year: unknown, country: Germany[26], *damaged or moldy

- Co-contamination: not reported
- Further contamination of plants and plant products other than cereals and their products reported in the present book/article[26]: Fruits (apple), Fruits (cherry), Fruits (peach), Fruits (tomato), Strawberries, OTA, literature[26]

Powder (cocoa powder) may contain the following mycotoxins:

Aspergillus Toxins

Aflatoxin B_1

incidence: 5?/12, conc. range: 0.2–0.6 µg/kg, sample origin: unknown, sample year: unknown, country: Japan[112]

- Co-contamination: 3 sa. co-contamination with AFB_1 and OTA; 2 sa. co-contamination with AFB_1, OTA, and OTB; no further information available
- Further contamination of plants and plant products other than cereals and their products reported in the present book/article[112]: Cassis, Juices, Raisins, Wine, OTA, literature[112]; Cocoa, Coffee, Food, Vinegar, OTA and OTB, literature[112]; Powder (cocoa powder), AFB_2, AFG_1, OTA and OTB, literature[112]

incidence: 12/16*, conc. range: ≤1.04 µg/kg, sample origin: different stages of manufacture at processing plants, Brazil, sample year: unknown, country: Brazil[211], *natural cocoa powder

incidence: 18/28*, conc. range: ≤0.92 µg/kg, sample origin: different stages of manufacture at processing plants, Brazil, sample year: unknown, country: Brazil[211], *alkalized cocoa powder

- Co-contamination: not reported
- Further contamination of plants and plant products other than cereals and their products reported in the present book/article[211]: Butter (cocoa butter), AFB_1 and AFB_2, literature[211]; Powder (cocoa powder), AFB_2 and AFG_1, literature[211]; Products (cocoa product), AFB_1, AFB_2, AFG_1, and AFG_2, literature[211]

incidence: 20*/21**, conc. range: ≤0.84 µg/kg, sample origin: Canada, sample year: unknown, country: Canada[864], *n>LOQ, **cocoa, alkalized

- Co-contamination: For detailed information see the article.
- Further contamination of plants and plant products other than cereals and their products reported in the present book/article[864]: Butter (cocoa butter), OTA, literature[864]; Chocolate, Cocoa beans, Products (cocoa product), AFB_1, AFB_2, AFG_1, AFG_2, and OTA, literature[864]; Powder (cocoa powder), AFB_2, AFG_1, and OTA, literature[864]

Aflatoxin B_2

incidence: 1/12, conc.: 0.1 µg/kg, sample origin: unknown, sample year: unknown, country: Japan[112]

- Co-contamination: not reported
- Further contamination of plants and plant products other than cereals and their products reported in the present book/article[112]: Cassis, Juices, Raisins, and Wine, OTA, literature[112]; Cocoa, Coffee, Food, Vinegar, OTA and OTB, literature[112]; Powder (cocoa powder), AFB_1, AFG_1, OTA, and OTB, literature[112]

incidence: 9/16*, conc. range: ≤0.51 µg/kg, sample origin: different stages of manufacture at processing plants, Brazil, sample year: unknown, country: Brazil[211], *natural cocoa powder

incidence: 14/28*, conc. range: ≤0.43 µg/kg, sample origin: different stages of manufacture at processing plants, Brazil, sample year: unknown, country: Brazil[211], *alkalized cocoa powder

- Co-contamination: not reported
- Further contamination of plants and plant products other than cereals and their products reported in the present book/article[211]: Butter (cocoa butter), AFB_1 and AFB_2, literature[211]; Powder (cocoa powder), AFB_1 and AFG_1, literature[211]; Products (cocoa product), AFB_1, AFB_2, AFG_1, and AFG_2, literature[211]

incidence: 15*/21**, conc. range: ≤0.13 µg/kg, sample origin: Canada, sample year: unknown, country: Canada[864], *n>LOQ, **cocoa, alkalized

- Co-contamination: For detailed information see the article.
- Further contamination of plants and plant products other than cereals and their products reported in the present book/article[864]: Butter (cocoa butter), OTA, literature[864]; Chocolate, Cocoa beans, Products (cocoa product), AFB_1, AFB_2, AFG_1, AFG_2, and OTA, literature[864]; Powder (cocoa powder), AFB_1, AFG_1, and OTA, literature[864]

Aflatoxin G_1

incidence: 1/12, conc.: 0.2 µg/kg, sample origin: unknown, sample year: unknown, country: Japan[112]

- Co-contamination: not reported
- Further contamination of plants and plant products other than cereals and their products reported in the present book/article[112]: Cassis, Juices, Raisins, Wine, OTA, literature[112]; Cocoa, Coffee, Food, Vinegar, OTA and OTB, literature[112]; Powder (cocoa powder), AFB_1, AFB_2, OTA, and OTB, literature[112]

incidence: 4/16*, conc. range: ≤0.91 µg/kg, sample origin: different stages of manufacture at processing plants, Brazil, sample year: unknown, country: Brazil[211], *natural cocoa powder incidence: 0/28*, conc.: no contamination, sample origin: different stages of manufacture at processing plants, Brazil, sample year: unknown, country: Brazil[211], *alkalized cocoa powder

- Co-contamination: not reported
- Further contamination of plants and plant products other than cereals and their products reported in the present book/article[211]: Butter (cocoa butter), Powder (cocoa powder), AFB_1 and AFB_2, literature[211]; Products (cocoa product), AFB_1, AFB_2, AFG_1, and AFG_2, literature[211]

incidence: 1*/21**, conc.: 0.24 µg/kg, sample origin: Canada, sample year: unknown, country: Canada[864], *n>LOQ, **cocoa, alkalized

- Co-contamination: For detailed information see the article.
- Further contamination of plants and plant products other than cereals and their products reported in the present book/article[864]: Butter (cocoa butter), OTA, literature[864]; Chocolate, Cocoa beans, Products (cocoa product), AFB_1, AFB_2, AFG_1, AFG_2, and OTA, literature[864]; Powder (cocoa powder), AFB_1, AFB_2, and OTA, literature[864]

Aspergillus and ***Penicillium*** Toxins

Ochratoxin A

incidence: 5?/12, conc. range: 0.25–0.67 µg/kg, sample origin: unknown, sample year: unknown, country: Japan[112]

- Co-contamination: 3 sa. co-contamination with AFB_1 and OTA, 2 sa. co-contamination with AFB_1, OTA, and OTB; no further information available
- Further contamination of plants and plant products other than cereals and their products reported in the present book/article[112]: Cassis, Juices, Raisins, Wine, OTA, literature[112]; Cocoa, Coffee, Food, Vinegar, OTA and OTB, literature[112]; Powder (cocoa powder), AFB_1, AFB_2, AFG_1, and OTB, literature[112]

incidence: 19/20, conc. range: 0.3–1.1 µg/kg, ∅ conc.: 0.71 µg/kg, sample origin: retail outlets across the UK, sample year: August 1996, country: UK[143]

- Co-contamination: not reported
- Further contamination of plants and plant products other than cereals and their products reported in the present book/article[143]: Beans, Legumes, Wine (red), OTA, literature[143]

incidence: 16/16*, conc. range: 0.05–5.13 µg/kg, ∅ conc.: 1.42 µg/kg, sample origin: processing plants in the region of Ilhéus (city) in Bahia (state) and São Paulo (city), São Paulo (state), Brazil, sample year: unknown, country: Brazil[212], *natural cocoa powder

incidence: 28/28*, conc. range: 0.14–3.59 µg/kg, ∅ conc.: 0.90 µg/kg, sample origin: processing plants in the region of Ilhéus (city) in Bahia (state) and São Paulo (city), São Paulo (state), Brazil, sample year: unknown, country: Brazil[212], *alkalized cocoa powder

- Co-contamination: not reported
- Further contamination of plants and plant products other than cereals and their products reported in the present book/article[212]: Butter (cocoa butter), Products (cocoa product), OTA, literature[212]

incidence: 29/31, conc. range: 0.1–4.4 µg/kg, ∅ conc.: 2.41 µg/kg, sample origin: Cameroon, Guinea, Ivory Coast, and Nigeria, sample year: unknown, country: Spain[261]

- Co-contamination: not reported
- Further contamination of plants and plant products other than cereals and their products reported in the present book/article[261]: Chocolate, Cocoa beans, Products (cocoa product), OTA, literature[261]

incidence: 40/40, conc. range: 0.090–1.800 µg/kg, sample origin: regions around Detmold, Giessen, Hamburg, Jena, Karlsruhe, Kiel, Kulmbach, and Trier (cities), Germany, sample year: unknown, country: Germany[262]

- Co-contamination: not reported
- Further contamination of plants and plant products other than cereals and their products reported in the present book/article[262]: Bars, Chocolate, Drinks (chocolate drink), Figs, Fruits, dried, Nut crème, Nuts, Nuts (hazelnut), Nuts (peanut), Pudding, Raisins, Seeds (linseed), Seeds (sesame), Seeds (sunflower), OTA, literature[262]

incidence: 20/20, conc. range: 1.3–2.1 µg/kg, sample origin: 13 regions in the UK, sample year: May 1998, country: UK[263]

- Co-contamination: not reported
- Further contamination of plants and plant products other than cereals and their products reported in the present book/article[263]: Chocolate, Currants, Figs, Juices (grape juice), Raisins, Sultanas, Wine (red), OTA, literature[263]

incidence: 25/26, conc. range: 0.053–0.932 µg/kg, sample origin: different countries, sample year: 2001–2003, country: Spain[264]

- Co-contamination: not reported
- Further contamination of plants and plant products other than cereals and their products reported in the present book/article[264]: Chocolate, OTA, literature[264]

incidence: 16/16*, conc. range: 0.57–7.8 µg/kg, ∅ conc.: 2.0 µg/kg, sample origin: sa. available in Canada, sample year: unknown, country: Canada[265], *cocoa powder, alkalized

incidence: 16/16*, conc. range: 0.25–2.6 µg/kg, ∅ conc.: 0.89 µg/kg, sample origin: sa. available in Canada, sample year: unknown, country: Canada[265], *cocoa powder, natural

- Co-contamination: not reported
- Further contamination of plants and plant products other than cereals and their products reported in the present book/article[265]: Chocolate, OTA, literature[265]

incidence: 40/40, conc. range: 0.18–1.82 µg/kg, ∅ conc.: 0.55 µg/kg, sample origin: discount stores in south, center, and north Italy, sample year: unknown, country: Italy[266]

- Co-contamination: not reported
- Further contamination of plants and plant products other than cereals and their products reported in the present book/article[266]: Chocolate, OTA, literature[266]

incidence: 8/12*, conc. range: 0.23–0.77 µg/kg, ∅ conc.: 0.43 µg/kg, sample origin: supermarkets, local stores, and organic product shops in Naples (city) area, Italy, sample year: unknown, country: Italy[593], *conventional

incidence: 1/6*, conc. range: 0.22 µg/kg, sample origin: supermarkets, local stores, and organic product shops in Naples (city) area, Italy, sample year: unknown, country: Italy[593], *organic

- Co-contamination: not reported
- Further contamination of plants and plant products other than cereals and their products reported in the present book/article[593]: not reported

incidence: 6/12, conc. range: ≤4.10 µg/kg, sample origin: 12 regions, Czech Republic, sample year: 2011–2013, country: Czech Republic[811] incidence: 1/12*, conc.: 0.45 μg/kg, sample origin: 12 regions, Czech Republic, sample year: 2011–2013, country: Czech Republic[811], *cocoa beverage-preparation, powder

- Co-contamination: not reported
- Further contamination of plants and plant products other than cereals and their products reported in the present book/article[811]: Beans, Chocolate, Coffee, Confectionery, Figs, Herbs and Spices, Juices, Licorice, Products (ginger product), Raisins, Spices (caraway), Spices (cayenne pepper, chilli), Spices (coriander), Spices (ginger), Spices (mustard), Spices (nutmeg), Spices (paprika), Spices (pepper), Tea, Wine (red), Wine (white), OTA, literature[811]

incidence: 20*/21**, conc. range: ≤1.88 μg/kg, sample origin: Canada, sample year: unknown, country: Canada[864], *n>LOQ, **cocoa, alkalized

- Co-contamination: For detailed information see the article.
- Further contamination of plants and plant products other than cereals and their products reported in the present book/article[864]: Butter (cocoa butter), OTA, literature[864]; Chocolate, Cocoa beans, Products (cocoa product), AFB_1, AFB_2, AFG_1, AFG_2, and OTA, literature[864]; Powder (cocoa powder), AFB_1, AFB_2, and AFG_1, literature[864]

incidence: 24/36, conc. range: ≤10.55 μg/kg, sample origin: Kumba? (city), Cameroun, sample year: 2005–2007, country: Cameroun/France[907]

- Co-contamination: not reported
- Further contamination of plants and plant products other than cereals and their products reported in the present book/article[907]: Butter (cocoa butter), Chocolate, OTA, literature[907]

Ochratoxin B

incidence: 2?/12, conc. range: 0.10–0.17 μg/kg, ∅ conc.: 0.135 μg/kg?, sample origin: unknown, sample year: unknown, country: Japan[112]

- Co-contamination: 2 sa. co-contamination with AFB_1, OTA, and OTB; no further information available
- Further contamination of plants and plant products other than cereals and their products reported in the present book/article[112]: Cassis, Juices, Raisins, Wine, OTA, literature[112]; Cocoa, Coffee, Food, Vinegar, OTA and OTB, literature[112]; Powder (cocoa powder), AFB_1, AFB_2, AFG_1, and OTA, literature[112]

Primitivo see Must

Products may contain the following mycotoxins:

Aspergillus Toxins

Aflatoxin B1

incidence: 2/4*, conc. range: 2.0–4.0 μg/kg (1 sa.), 6.0 μg/kg (1 sa.), sample origin: supermarkets and/or traditional markets in Yogyakarta (city), Indonesia, sample year: 1998/1999, country: Indonesia/Austria/Canada[106], *fermented products included oncom, tempe, and tape

incidence: 4/4*, conc. range: >4 μg/kg (4 sa., maximum: 26.0 μg/kg), sample origin: supermarkets and/or traditional markets in Yogyakarta (city), Indonesia, sample year: 1998/1999, country: Indonesia/Austria/Canada[106], *processed products included peanut (raw, without shell), red sorghum, green pea flour, and traditional herbal medicine (jamu)

- Co-contamination: not reported
- Further contamination of plants and plant products other than cereals and their products reported in the present book/article[106]: Butter (peanut butter), Cake (peanut), Food, Nuts (peanut), Sauces (peanut-chilli sauce), AFB_1, literature[106]

Aspergillus and *Penicillium* Toxins

Ochratoxin A

incidence: 1/2*, conc.: 0.06 μg/kg**, sample origin: supermarkets and fast-food chains in Quebec (city), Canada, sample year: September/October 2008, country: Canada[90], *dinners, cereal plus vegetable plus meat and * **entrees, frozen

incidence: 2/2*, conc. range: 0.01–0.05 μg/kg**, sample origin: supermarkets and fast-food chains in Calgary (city), Canada, sample year: September/October 2009, country: Canada[90], *dinners, cereal plus vegetable plus meat and * **entrees, frozen

- Co-contamination: not reported
- Further contamination of plants and plant products other than cereals and their products reported in the present book/article[90]: Baby food, Beans, Beverages, Butter (peanut butter), Candies, Chocolate, Coffee, Extracts, Herbs, Ice cream, Nuts, Pies, Pizza, Pudding, Raisins, Seeds, Wine, OTA, literature[90]

Products (almond product) may contain the following mycotoxins:

Aspergillus Toxins

Aflatoxin B1

incidence: 1/9, conc.: 0.3 μg/kg*, sample origin: local markets and online stores, USA, sample year: August 2011–January 2012, country: USA/Taiwan/Canada[777], *almond flour

- Co-contamination: not reported
- Further contamination of plants and plant products other than cereals and their products reported in the present book/article[777]: Products (peanut product), AFB_1 and BEA, literature[777]; Products (pistachio product), AFB_1, AFB_2, AFG_1, BEA, NEO, OTA, and OTB, literature[777]

Products (apple flavor product) may contain the following mycotoxins:

Aspergillus and *Penicillium* Toxins

Patulin

incidence: 3/14, conc. range: 6–1,770 μg/kg, ∅ conc.: 607 μg/kg, sample origin: imported sa. from Finnish customs authorities, Finland, sample year: March–December 1976, country: Finland[33]

- Co-contamination: not reported
- Further contamination of plants and plant products other than cereals and their products reported in the present book/article[33]: Fruits (apple), Jam (apple jam), Juices (apple concentrate), Juices (apple juice), PAT, literature[33]

Products (apple product) may contain the following mycotoxins:

Aspergillus and *Penicillium* Toxins

Patulin

incidence: 18/70*, conc. range: 5–50 μg/kg (18 sa., maximum: 32 μg/kg), sample origin: retail outlets and directly from producers in Sydney (city) and NSW (state), Australia, sample year: August

1989–May 1990, country: Australia[31], *included apples, pears, and mixed fruit sauces, puree and jellies, diced apples, and apple pulps

- Co-contamination: not reported
- Further contamination of plants and plant products other than cereals and their products reported in the present book/article[31]: Juices (apple concentrate), Juices (apple juice), PAT, literature[31]

incidence: 2/6*, conc. range: 10 µg/kg, ∅ conc.: 10 µg/kg, sample origin: retail outlets, Cape Town (city), South Africa, sample year: 1996–1998, country: South Africa[37], *whole apple fruit products

- Co-contamination: not reported
- Further contamination of plants and plant products other than cereals and their products reported in the present book/article[37]: Apple cider, Juices, Juices (apple juice), PAT, literature[37]

incidence: 4/77*, conc. range: 6.25–10 µg/kg (1 sa.), 10–25 µg/kg (3 sa., maximum 17.6 µg/kg), ∅ conc.: 13.5 µg/kg, sample origin: hypermarkets and supermarkets in 12 main cities, Catalonia (autonomous community), Spain, sample year: June/July 2008, country: Spain[42], *solid apple-based products

- Co-contamination: not reported
- Further contamination of plants and plant products other than cereals and their products reported in the present book/article[42]: Baby food, Juices (apple juice), PAT, literature[42]

incidence: 3/6*, conc. range: 4.2–15.2 µg/kg, sample origin: stores and supermarkets, Valencia (city), Spain, sample year: unknown, country: Spain[43], *apple-containing dairy products

- Co-contamination: not reported
- Further contamination of plants and plant products other than cereals and their products reported in the present book/article[43]: Juices (apple juice), Puree, PAT, literature[43]

incidence: 5/20*, conc. range: 5.8–56.4 µg/kg, ∅ conc.: 24.8 µg/kg, sample origin: supermarket, local stores, herbalist, and salesrooms, Naples (city) area, Italy, sample year: unknown, country: Italy[44], *conventional

incidence: 6/17*, conc. range: 1.4–74.2 µg/kg, ∅ conc.: 28.3 µg/kg, sample origin: supermarket, local stores, herbalist, and salesrooms, Naples (city) area, Italy, sample year: unknown, country: Italy[44], *organic

incidence: 0/3*, conc.: no contamination, sample origin: supermarket, local stores, herbalist, and salesrooms, Naples (city) area, Italy, sample year: unknown, country: Italy[44], *integrated

- Co-contamination: not reported
- Further contamination of plants and plant products other than cereals and their products reported in the present book/article[44]: not reported

incidence: 99/105, conc. range: <10 µg/kg (~98 sa.), 42 µg/kg (1 sa.), sample origin: Berlin (capital) market, Germany, sample year: July 1980–February 1981, country: Germany[733]

- Co-contamination: not reported
- Further contamination of plants and plant products other than cereals and their products reported in the present book/article[733]: not reported

For detailed information see the article.

Products (cassava product) may contain the following mycotoxins:

Alternaria **Toxins**

Alternariol Methyl Ether

incidence: ?*/373, conc. range: 0.02–1.49 µg/kg, sample origin: processors and vendors in humid forest, derived savannah, and Southern Guinea Savannah zones, Nigeria, sample year: rainy season, country: Nigeria/Austria[794], *high-quality cassava and fufu flour, lafun, fine yellow, fine white, yellow kpo-kpo, and white kpo-kpo gari sa. contaminated

- Co-contamination: For detailed information see the article.
- Further contamination of plants and plant products other than cereals and their products reported in the present book/article[794]: Products (cassava product), AFB_1, AFG_1, FB_1, FB_2, FB_3, KA, and ZEA, literature[794]; see also Products (cassava product), chapter: Further Mycotoxins and Microbial Metabolites

For detailed information see the article.

Aspergillus **Toxins**

Aflatoxin B_1

incidence: 3*/373, ∅ conc.: 1.16 µg/kg, sample origin: processors and vendors in humid forest, derived savannah, and Southern Guinea Savannah zones, Nigeria, sample year: rainy season, country: Nigeria/Austria[794], *fufu flour (3/36) contaminated

- Co-contamination: For detailed information see the article.
- Further contamination of plants and plant products other than cereals and their products reported in the present book/article[794]: Products (cassava product), AFG_1, AME, FB_1, FB_2, FB_3, KA, and ZEA, literature[794]; see also Products (cassava product), chapter: Further Mycotoxins and Microbial Metabolites

For detailed information see the article.

incidence: 42/165*, conc. range: 6–194 µg/kg, sample origin: humid forest with monomodal rainfall, humid forest with bimodal rainfall, and western highland, Cameroon, sample year: July/August 2009 (rainy season) and December 2010–January 2011 (dry season), country: Belgium/Benin[807], *cassava chips

- Co-contamination: For detailed information see the article.
- Further contamination of cereals and cereal products reported in the present book/article[807]: Nuts (peanut), AFB_1 and OTA, literature[807]; Products (cassava product), PA, literature[807]

For detailed information see the article.

Aflatoxin G_1

incidence: 1*/373, conc.: 2.94 µg/kg, sample origin: processors and vendors in humid forest, derived savannah, and Southern Guinea Savannah zones, Nigeria, sample year: rainy season, country: Nigeria/Austria[794], *high-quality cassava flour (1/29) contaminated

- Co-contamination: For detailed information see the article.
- Further contamination of plants and plant products other than cereals and their products reported in the present book/article[794]: Products (cassava product), AFB_1, AME, FB_1, FB_2, FB_3, KA, and ZEA, literature[794]; see also Products (cassava product), chapter: Further Mycotoxins and Microbial Metabolites

For detailed information see the article.

Aspergillus **and** ***Penicillium*** **Toxins**

Kojic Acid

incidence: ?*/373, conc. range: 8.35–632.68 µg/kg, sample origin: processors and vendors in humid forest, derived savannah, and Southern Guinea Savannah zones, Nigeria, sample year: rainy season, country: Nigeria/Austria[794], *all kinds of sa. contaminated

- Co-contamination: For detailed information see the article.
- Further contamination of plants and plant products other than cereals and their products reported in the present book/article[794]: Products (cassava product), AFB_1, AFG_1, AME, FB_1, FB_2, FB_3, and ZEA, literature[794]; see also Products (cassava product), chapter: Further Mycotoxins and Microbial Metabolites

For detailed information see the article.

Penicillic Acid

incidence: 10/165*, conc. range: 25–184 µg/kg, sample origin: humid forest with monomodal rainfall, humid forest with bimodal rainfall, and western highland, Cameroon, sample year: July/August 2009 (rainy season) and December 2010–January 2011 (dry season), country: Belgium/Benin[807], *cassava flakes

- Co-contamination: For detailed information see the article.
- Further contamination of cereals and cereal products reported in the present book/article[807]: Nuts (peanut), AFB_1 and OTA, literature[807]; Products (cassava product), AFB_1, literature[807]

For detailed information see the article.

Fusarium **Toxins**

Fumonisin/s

Fumonisin B_1

incidence: 2*/373, conc. range: 88.33–102.71 µg/kg, ∅ conc.: 95.5 µg/kg, sample origin: processors and vendors in humid forest, derived savannah, and Southern Guinea Savannah zones, Nigeria, sample year: rainy season, country: Nigeria/Austria[794], *lafun (1/30) and fufu flour (1/36) contaminated

- Co-contamination: For detailed information see the article.
- Further contamination of plants and plant products other than cereals and their products reported in the present book/article[794]: Products (cassava product), AFB_1, AFG_1, AME, FB_2, FB_3, KA, and ZEA, literature[794]; see also Products (cassava product), chapter: Further Mycotoxins and Microbial Metabolites

For detailed information see the article.

Fumonisin B_2

incidence: 5*/373, conc. range: 10.70–218.12 µg/kg, sample origin: processors and vendors in humid forest, derived savannah, and Southern Guinea Savannah zones, Nigeria, sample year: rainy season, country: Nigeria/Austria[794], *lafun (2/30), fufu flour (2/36), and fine white gari (1/113) contaminated

- Co-contamination: For detailed information see the article.
- Further contamination of plants and plant products other than cereals and their products reported in the present book/article[794]: Products (cassava product), AFB_1, AFG_1, AME, FB_1, FB_3, KA, and ZEA, literature[794]; see also Products (cassava product), chapter: Further Mycotoxins and Microbial Metabolites

For detailed information see the article.

Fumonisin B_3

incidence: 1*/373, conc.: 14.49 µg/kg, sample origin: processors and vendors in humid forest, derived savannah, and Southern Guinea Savannah zones, Nigeria, sample year: rainy season, country: Nigeria/Austria[794], *fufu flour (1/36) contaminated

- Co-contamination: For detailed information see the article.
- Further contamination of plants and plant products other than cereals and their products reported in the present book/article[794]: Products (cassava product), AFB_1, AFG_1, AME, FB_1, FB_2, KA, and ZEA, literature[794]; see also Products (cassava product), chapter: Further Mycotoxins and Microbial Metabolites

For detailed information see the article.

Other ***Fusarium*** **Toxins**

Zearalenone

incidence: 16*/373, conc. range: 0.92–90.40 µg/kg, sample origin: processors and vendors in humid forest, derived savannah, and Southern Guinea Savannah zones, Nigeria, sample year: rainy season, country: Nigeria/Austria[794], *high-quality cassava (2/29) and fufu flour (2/36), lafun (6/30), fine white (2/113), fine yellow (1/50), and white kpo-kpo gari (3/52) contaminated

- Co-contamination: For detailed information see the article.
- Further contamination of plants and plant products other than cereals and their products reported in the present book/article[794]: Products (cassava product), AFB_1, AFG_1, AME, FB_1, FB_2, FB_3, and KA, literature[794]; see also Products (cassava product), chapter: Further Mycotoxins and Microbial Metabolites

For detailed information see the article.

Products (cocoa product) may contain the following mycotoxins:

Aspergillus **Toxins**

Aflatoxin B_1

incidence: 10/37*, conc. range: >0.2–2.0 µg/kg (9 sa.), >2.0–10.0 µg/kg (1 sa.), sample origin: different exporting companies, Turkey, sample year: 2007, country: Turkey[115], *commercial Turkish food

incidence: 5/20*, conc. range: >0.2–2.0 µg/kg (5 sa.), sample origin: different exporting companies, Turkey, sample year: 2008, country: Turkey[115], *commercial Turkish food

incidence: 2/9*, conc. range: >0.2–2.0 µg/kg (2 sa.), sample origin: different exporting companies, Turkey, sample year: 2009, country: Turkey[115], *commercial Turkish food

- Co-contamination: not reported
- Further contamination of plants and plant products other than cereals and their products reported in the present book/article[115]: Food, Spices, AFB_1, literature[115]

incidence: 10/25*, conc. range: ≤6.14 µg/kg, sample origin: different stages of manufacture at processing plants, Brazil, sample year: unknown, country: Brazil[211], *cocoa mass

incidence: 13/29*, conc. range: ≤11.21 µg/kg, sample origin: different stages of manufacture at processing plants, Brazil, sample year: unknown, country: Brazil[211], *cocoa nibs

- Co-contamination: not reported
- Further contamination of plants and plant products other than cereals and their products reported in the present book/article[211]: Butter (cocoa butter), AFB_1 and AFB_2, literature[211]; Powder (cocoa powder), AFB_1, AFB_2, and AFG_1, literature[211]; Products (cocoa product), AFB_2, AFG_1, and AFG_2, literature[211]

incidence: 4*/5**, conc.: ≤0.67 µg/kg, sample origin: Canada, sample year: unknown, country: Canada[864], *n>LOQ, **cocoa liquor

- Co-contamination: For detailed information see the article.

- Further contamination of plants and plant products other than cereals and their products reported in the present book/article[864]: Butter (cocoa butter), OTA, literature[864]; Chocolate, Cocoa beans, AFB_1, AFB_2, AFG_1, AFG_2, and OTA, literature[864]; Powder (cocoa powder), AFB_1, AFB_2, AFG_1, and OTA, literature[864]; Products (cocoa product), AFB_2, AFG_1, AFG_2, and OTA, literature[864]

Aflatoxin B_2

incidence: 7/25, conc. range: ≤0.35 µg/kg, sample origin: different stages of manufacture at processing plants, Brazil, sample year: unknown, country: Brazil[211], *cocoa mass

incidence: 9/29, conc. range: ≤1.66 µg/kg, sample origin: different stages of manufacture at processing plants, Brazil, sample year: unknown, country: Brazil[211], *cocoa nibs

- Co-contamination: not reported
- Further contamination of plants and plant products other than cereals and their products reported in the present book/article[211]: Butter (cocoa butter), AFB_1 and AFB_2, literature[211]; Powder (cocoa powder), AFB_1, AFB_2, and AFG_1, literature[211]; Products (cocoa product), AFB_1, AFG_1, and AFG_2, literature[211]

incidence: 3*/5**, conc.: ≤0.09 µg/kg, sample origin: Canada, sample year: unknown, country: Canada[864], *n>LOQ, **cocoa liquor

- Co-contamination: For detailed information see the article.
- Further contamination of plants and plant products other than cereals and their products reported in the present book/article[864]: Butter (cocoa butter), OTA, literature[864]; Chocolate, Cocoa beans, AFB_1, AFB_2, AFG_1, AFG_2, and OTA, literature[864]; Powder (cocoa powder), AFB_1, AFB_2, AFG_1, and OTA, literature[864]; Products (cocoa product), AFB_1, AFG_1, AFG_2, and OTA, literature[864]

Aflatoxin G_1

incidence: 4/25, conc. range: ≤2.58 µg/kg, sample origin: different stages of manufacture at processing plants, Brazil, sample year: unknown, country: Brazil[211], *cocoa mass

incidence: 6/29, conc. range: ≤2.48 µg/kg, sample origin: different stages of manufacture at processing plants, Brazil, sample year: unknown, country: Brazil[211], *cocoa nibs

- Co-contamination: not reported
- Further contamination of plants and plant products other than cereals and their products reported in the present book/article[211]: Butter (cocoa butter), AFB_1 and AFB_2, literature[211]; Powder (cocoa powder), AFB_1, AFB_2, and AFG_1, literature[211]; Products (cocoa product), AFB_1, AFB_2, and AFG_2, literature[211]

incidence: 1*/5**, conc.: 0.13 µg/kg, sample origin: Canada, sample year: unknown, country: Canada[864], *n>LOQ, **cocoa liquor

- Co-contamination: For detailed information see the article.
- Further contamination of plants and plant products other than cereals and their products reported in the present book/article[864]: Butter (cocoa butter), OTA, literature[864]; Chocolate, Cocoa beans, AFB_1, AFB_2, AFG_1, AFG_2, and OTA, literature[864]; Powder (cocoa powder), AFB_1, AFB_2, AFG_1, and OTA, literature[864]; Products (cocoa product), AFB_1, AFB_2, AFG_2, and OTA, literature[864]

Aflatoxin G_2

incidence: 1/25, conc.: 0.73 µg/kg, sample origin: different stages of manufacture at processing plants, Brazil, sample year: unknown, country: Brazil[211], *cocoa mass

incidence: 2/29, conc. range: ≤0.68 µg/kg, sample origin: different stages of manufacture at processing plants, Brazil, sample year: unknown, country: Brazil[211], *cocoa nibs

- Co-contamination: not reported
- Further contamination of plants and plant products other than cereals and their products reported in the present book/article[211]: Butter (cocoa butter), AFB_1 and AFB_2, literature[211]; Powder (cocoa powder), Products (cocoa product), AFB_1, AFB_2, and AFG_1, literature[211]

incidence: 1*/5**, conc.: ≤0.04 µg/kg, sample origin: Canada, sample year: unknown, country: Canada[864], *n>LOQ, **cocoa liquor

- Co-contamination: For detailed information see the article.
- Further contamination of plants and plant products other than cereals and their products reported in the present book/article[864]: Butter (cocoa butter), OTA, literature[864]; Chocolate and Cocoa beans, AFB_1, AFB_2, AFG_1, AFG_2, and OTA, literature[864]; Powder (cocoa powder), Products (cocoa product), AFB_1, AFB_2, AFG_1, and OTA, literature[864]

Aspergillus and ***Penicillium*** Toxins

Ochratoxin A

incidence: 25/25*, conc. range: 0.03–1.09 µg/kg, ∅ conc.: 0.34 µg/kg, sample origin: processing plants in the region of Ilhéus (city) in Bahia (state) and São Paulo (city), São Paulo (state), Brazil, sample year: unknown, country: Brazil[212], *cocoa mass

incidence: 24/29*, conc. range: ≤0.38 µg/kg, sample origin: processing plants in the region of Ilhéus (city) in Bahia (state) and São Paulo (city), São Paulo (state), Brazil, sample year: unknown, country: Brazil[212], *cocoa nibs

- Co-contamination: not reported
- Further contamination of plants and plant products other than cereals and their products reported in the present book/article[212]: Butter (cocoa butter), Powder (cocoa powder), OTA, literature[212]

incidence: 4/8*, conc. range: 0.1–3.5 µg/kg, ∅ conc.: 1.07 µg/kg, sample origin: Cameroon and Ivory Coast, sample year: unknown, country: Spain[261], *cocoa mass

- Co-contamination: not reported
- Further contamination of plants and plant products other than cereals and their products reported in the present book/article[261]: Chocolate, Cocoa beans, Powder (cocoa powder), OTA, literature[261]

incidence: 12/22*, conc. range: 0.043–2.02 µg/kg, ∅ conc.: 0.34 µg/kg, sample origin: cocoa factory, Brazil, sample year: 2003/2004, country: Spain[271], *cocoa nibs

- Co-contamination: not reported
- Further contamination of plants and plant products other than cereals and their products reported in the present book/article[271]: not reported

incidence: 5*/5**, conc.: ≤0.56 µg/kg, sample origin: Canada, sample year: unknown, country: Canada[864], *n>LOQ, **cocoa liquor

- Co-contamination: For detailed information see the article.
- Further contamination of plants and plant products other than cereals and their products reported in the present book/article[864]: Butter (cocoa butter), OTA, literature[864]; Chocolate and Cocoa beans, AFB_1, AFB_2, AFG_1, AFG_2, and OTA, literature[864]; Powder (cocoa powder), AFB_1, AFB_2, AFG_1, and OTA, literature[864]; Products (cocoa product), AFB_1, AFB_2, AFG_1, and AFG_2, literature[864]

see also **Cocoa**

Products (coconut product) may contain the following mycotoxins:

Aspergillus Toxins

Aflatoxin/s

incidence: 57/229, Ø conc.: 2.8 μg/kg*, sample origin: greater Manila (capital) area, Philippines, sample year: unknown, country: the Philippines[1], *AFS of pos sa.?

- Co-contamination: not reported
- Further contamination of plant and plant products other than cereals and their products reported in the present book/article[1]: Beans, Butter (peanut butter), Cassava, Cocoa, Coffee, Drinks (alcoholic drink), Nuts (cashew nut), Nuts (peanut), Rootcrops, Spices (cayenne pepper, chilli), Sweet potatoes, Yams, AF/S, literature[1]

incidence: ?/7, conc. range: ≤26 μg/kg*, sample origin: markets in or near Manila (capital), Philippines, sample year: 1967–1969, country: USA[134], *AF

- Co-contamination: not reported
- Further contamination of plants and plant products other than cereals and their products reported in the present book/article[134]: Beans, Butter (peanut butter), Nuts (peanut), Products (peanut product), Tubers, AF/S, literature[134]

Aspergillus and ***Penicillium*** Toxins

Citrinin

incidence: 4/384*, conc. range: 30–60 μg/kg, Ø conc.: 36.7 μg/kg, sample origin: during early rainy season and late summer from market places, households, oil mills, and cattle-sheds in Eluru, Hyderabad, Kakinada, and Nizamabad (cities), India, sample year: 1982/1983, country: India[278], *dry copra and copra meal

- Co-contamination: not reported
- Further contamination of plants and plant products other than cereals and their products reported in the present book/article[278]: Products (coconut product), OTA, literature[278]

Ochratoxin A

incidence: 1/384*, conc.: 50 μg/kg, sample origin: during early rainy season and late summer from market places, households, oil mills, and cattle-sheds in Eluru, Hyderabad, Kakinada, and Nizamabad (cities), India, sample year: 1982/1983, country: India[278], *dry copra and copra meal

- Co-contamination: not reported
- Further contamination of plants and plant products other than cereals and their products reported in the present book/article[278]: Products (coconut product), CIT, literature[278]

Products (fruit product) may contain the following mycotoxins:

Aspergillus and ***Penicillium*** Toxins

Patulin

incidence: 7/35*, conc. range: 2–76.75 μg/l, Ø conc.: 32.26 μg/l, sample origin: local supermarkets, Tunisia, sample year: 2013, country: Tunisia[888], *compote

- Co-contamination: not reported
- Further contamination of plants and plant products other than cereals and their products reported in the present book/article[888]: Jam (apple jam), Jam (pear jam), Juices, Juices (apple juice), Juices (pear juice), PAT, literature[888]

Products (ginger product) may contain the following mycotoxins:

Aspergillus and ***Penicillium*** Toxins

Ochratoxin A

incidence: 7/12*, conc. range: ≤1.44 μg/kg, sample origin: 12 regions, Czech Republic, sample year: 2011–2013, country: Czech Republic[811], *gingerbread

- Co-contamination: not reported
- Further contamination of plants and plant products other than cereals and their products reported in the present book/article[811]: Beans, Chocolate, Coffee, Confectionery, Figs, Herbs and Spices, Juices, Licorice, Powder (cocoa powder), Raisins, Spices (caraway), Spices (cayenne pepper, chilli), Spices (coriander), Spices (ginger), Spices (mustard), Spices (nutmeg), Spices (paprika), Spices (pepper), Tea, Wine (red), Wine (white), OTA, literature[811]

Products (legume product) may contain the following mycotoxins:

Aspergillus Toxins

Aflatoxin B_1

incidence: 9/12*, conc. range: 90–510 μg/kg, sample origin: market in Agra? (city), India, sample year: unknown, country: India[375], *mangori (black gram)

incidence: 4/15*, conc. range: 170–420 μg/kg, sample origin: market in Agra? (city), India, sample year: unknown, country: India[375], *papad (black gram)

incidence: 5/10*, conc. range: 50–380 μg/kg, sample origin: market in Agra? (city), India, sample year: unknown, country: India[375], *mirchoni (black gram)

incidence: 10/15*, conc. range: 170–1,600 μg/kg, sample origin: market in Agra? (city), India, sample year: unknown, country: India[375], *mangori (green gram)

incidence: 9/14*, conc. range: 114–1,700 μg/kg, sample origin: market in Agra? (city), India, sample year: unknown, country: India[375], *papad (green gram)

- Co-contamination: not reported
- Further contamination of plants and plant products other than cereals and their products reported in the present book/article[375]: Products (legume product), OTA, PAT, STG, and ZEA, literature[375]

Other ***Aspergillus*** Toxins

Sterigmatocystin

incidence: 4/12*, conc. range: 70–310 μg/kg, sample origin: market in Agra? (city), India, sample year: unknown, country: India[375], *mangori (black gram)

incidence: 6/15*, conc. range: 150–820 μg/kg, sample origin: market in Agra? (city), India, sample year: unknown, country: India[375], *papad (black gram)

incidence: 0/10*, conc.: no contamination, sample origin: market in Agra? (city), India, sample year: unknown, country: India[375], *mirchoni (black gram)

incidence: 0/15*, conc.: no contamination, sample origin: market in Agra? (city), India, sample year: unknown, country: India[375], *mangori (green gram) incidence: 0/14*, conc.: no contamination, sample origin: market in Agra? (city), India, sample year: unknown, country: India[375], *papad (green gram)

- Co-contamination: not reported
- Further contamination of plants and plant products other than cereals and their products reported in the present book/article[375]: Products (legume product), AFB_1, OTA, PAT, and ZEA, literature[375]

Aspergillus **and** ***Penicillium*** **Toxins**

Ochratoxin A

incidence: 5/12*, conc. range: 160–1,280 µg/kg, sample origin: market in Agra? (city), India, sample year: unknown, country: India[375], *mangori (black gram)

incidence: 6/15*, conc. range: 90–1,120 µg/kg, sample origin: market in Agra? (city), India, sample year: unknown, country: India[375], *papad (black gram)

incidence: 2/10*, conc. range: 120–580 µg/kg, ∅ conc.: 350 µg/kg, sample origin: market in Agra? (city), India, sample year: unknown, country: India[375], *mirchoni (black gram)

incidence: 7/15*, conc. range: 210–1,360 µg/kg, sample origin: market in Agra? (city), India, sample year: unknown, country: India[375], *mangori (green gram)

incidence: 4/14*, conc. range: 250–1,870 µg/kg, sample origin: market in Agra? (city), India, sample year: unknown, country: India[375], *papad (green gram)

- Co-contamination: not reported
- Further contamination of plants and plant products other than cereals and their products reported in the present book/article[375]: Products (legume product), AFB_1, PAT, STG, and ZEA, literature[375]

Patulin

incidence: 3/12*, conc. range: 130–690 µg/kg, sample origin: market in Agra? (city), India, sample year: unknown, country: India[375], *mangori (black gram)

incidence: 4/15*, conc. range: 220–800 µg/kg, sample origin: market in Agra? (city), India, sample year: unknown, country: India[375], *papad (black gram)

incidence: 0/10*, conc.: no contamination, sample origin: market in Agra? (city), India, sample year: unknown, country: India[375], *mirchoni (black gram)

incidence: 0/15*, conc.: no contamination, sample origin: market in Agra? (city), India, sample year: unknown, country: India[375], *mangori (green gram)

incidence: 0/14*, conc.: no contamination, sample origin: market in Agra? (city), India, sample year: unknown, country: India[375], *papad (green gram)

- Co-contamination: not reported
- Further contamination of plants and plant products other than cereals and their products reported in the present book/article[375]: Products (legume product), AFB_1, OTA, STG, and ZEA, literature[375]

Fusarium **Toxins**

Zearalenone

incidence: 2/12*, conc. range: 240–1,500 µg/kg, ∅ conc.: 870 µg/kg, sample origin: market in Agra? (city), India, sample year: unknown, country: India[375], *mangori (black gram)

incidence: 0/15*, conc.: no contamination, sample origin: market in Agra? (city), India, sample year: unknown, country: India[375], *papad (black gram)

incidence: 0/10*, conc.: no contamination, sample origin: market in Agra? (city), India, sample year: unknown, country: India[375], *mirchoni (black gram)

incidence: 0/15*, conc.: no contamination, sample origin: market in Agra? (city), India, sample year: unknown, country: India[375], *mangori (green gram)

incidence: 0/14*, conc.: no contamination, sample origin: market in Agra? (city), India, sample year: unknown, country: India[375], *papad (green gram)

- Co-contamination: not reported
- Further contamination of plants and plant products other than cereals and their products reported in the present book/article[375]: Products (legume product), AFB_1, OTA, PAT, and STG, literature[375]

Products (melon seed product) may contain the following mycotoxins:

Aspergillus **Toxins**

Aflatoxin B_1

incidence: 10/40*, conc. range: 2–13 µg/kg, sample origin: street hawkers and local markets, southwestern Nigeria, sample year: 2005, country: Nigeria[596], *"ogiri" (fermented "egusi" seed condiment)

incidence: 9/36*, conc. range: 4.1–14.3 µg/kg, sample origin: street hawkers and local markets, southwestern Nigeria, sample year: 2006, country: Nigeria[596], *"ogiri" (fermented "egusi" seed condiment)

incidence: 13/40*, conc. range: 3–15 µg/kg, sample origin: street hawkers and local markets, southwestern Nigeria, sample year: 2005, country: Nigeria[596], *"robo" (melon ball snacks)

incidence: 14/45*, conc. range: 2.5–11.7 µg/kg, sample origin: street hawkers and local markets, southwestern Nigeria, sample year: 2006, country: Nigeria[596], *"robo" (melon ball snacks)

incidence: 7/40*, conc. range: 3–10 µg/kg, sample origin: households and food restaurants within 20 km radius to Olabisi Onabanjo University, Nigeria, sample year: 2005, country: Nigeria[596], *soup from "egusi" melon seed

incidence: 7/32*, conc. range: 4.3–10.7 µg/kg, sample origin: households and food restaurants within 20 km radius to Olabisi Onabanjo University, Nigeria, sample year: 2006, country: Nigeria[596], *soup from "egusi" melon seed

- Co-contamination: not reported
- Further contamination of plants and plant products other than cereals and their products reported in the present book/article[596]: not reported

Products (nut product) may contain the following mycotoxins:

Aspergillus **Toxins**

Aflatoxin B_1

incidence: 1/13, conc.: 0.07 µg/kg, sample origin: unknown, sample year: unknown, country: Korea[9]

- Co-contamination: not reported
- Further contamination of plants and plant products other than cereals and their products reported in the present

book/article[9]: Butter (peanut butter), AFB_1, AFB_2, AFG_1, and AFG_2, literature[9]; Nuts (almond) and Nuts (peanut), AFB_1 and AFB_2, literature[9]; Tea, AFB_1, literature[9]

incidence: 4?/20*, conc. range: 97.8–453 µg/kg, sample origin: supermarkets and village stores in five districts of Penang (state) including Penang Island and Wellesley (province), Malaysia, sample year: unknown, country: Malaysia[225], *nut products, coated

- Co-contamination: not reported
- Further contamination of plants and plant products other than cereals and their products reported in the present book/article[225]: Butter (peanut butter), Confectionery, Nuts (walnut), AFB_1 and AFB_2, literature[225]; Cake (peanut), Nuts (peanut), AFB_1, AFB_2, and AFG_2, literature[225]; Products (nut product), AFB_2, literature[225]

Aflatoxin B$_2$

incidence: 4?/20*, conc. range: 15.3–61.7 µg/kg, sample origin: supermarkets and village stores in five districts of Penang (state) including Penang Island and Wellesley (province), Malaysia, sample year: unknown, country: Malaysia[225], *nut products, coated

- Co-contamination: not reported
- Further contamination of plants and plant products other than cereals and their products reported in the present book/article[225]: Butter (peanut butter), Confectionery, Nuts (walnut), AFB_1 and AFB_2, literature[225]; Cake (peanut), Nuts (peanut), AFB_1, AFB_2, and AFG_2, literature[225]; Products (nut product), AFB_1, literature[225]

Aflatoxin/s

incidence: 16/59*, conc. range: ≤9.7 µg/kg**, sample origin: New Zealand, sample year: unknown, country: New Zealand[701], *mixed nut products, **AFB_1, AFB_2, AFG_1, and AFG_2

- Co-contamination: not reported
- Further contamination of plants and plant products other than cereals and their products reported in the present book/article[701]: Butter (peanut butter), Nuts (peanut), Nuts (pistachio nut), Products (peanut product), AF/S, literature[701]; Nuts (Brazil nut, paranut), AFB_1, literature[701]

see also **Nuts**

Products (peanut product) may contain the following mycotoxins:

Aspergillus Toxins

Aflatoxin B$_1$

incidence: 22/74, conc. range: 0.005–2 µg/kg (18 sa.), >2 µg/kg (4 sa., maximum: 15 µg/kg), sample origin: ethnic stores, health food stores, and local stores as well as supermarkets in the four largest cities, Sweden, sample year: autumn 1996–summer 1998, country: Sweden[13]

- Co-contamination: not reported
- Further contamination of plants and plant products other than cereals and their products reported in the present book/article[13]: Beans (brown bean), Peas, and Peas (chickpea), OTA, literature[13]; Juices (apple juice), PAT, literature[13]; Nuts (almond), AFB_1, literature[13]; Nuts (Brazil nut, paranut), Nuts (pistachio nut), and Oil seeds, AFB_1 and AF/S, literature[13]; Products (peanut product), AF/S, literature[13]

incidence: 5/11*, conc. range: 0.3–0.6 µg/kg, Ø conc.: 0.42 µg/kg, sample origin: local markets and online stores, USA, sample year: August 2011–January 2012, country: USA/Taiwan[28], *included peanut butter, peanut flour, and raw peanut

- Co-contamination: 2 sa. co-contaminated with AFB_1 and BEA; 3 sa. contaminated solely with AFB_1
- Further contamination of plants and plant products other than cereals and their products reported in the present book/article[28]: Products (peanut product), BEA, literature[28]; Products (pistachio product), AFB_1, AFB_2, AFG_1, BEA, NEO, OTA, and OTB, literature[28]

incidence: 13/20, conc. range: 15–138 µg/kg, Ø conc.: 64.9 µg/kg, sample origin: different district hygiene institutes, GDR, sample year: unknown, country: GDR[29]

- Co-contamination: not reported
- Further contamination of plants and plant products other than cereals and their products reported in the present book/article[29]: Butter (peanut butter), Nuts (peanut), Seeds (sunflower), AFB_1, AFB_2, AFG_1, and AFG_2, literature[29]; Coffee, OTA, literature[29]; Fruits, AFB_1, AFG_1, and AFG_2, literature[29]; Fruits (apple), Fruits (peach), Juices (apple juice), PAT, literature[29]; Products (peanut product), AFB_2, AFG_1, and AFG_2, literature[29]

incidence: 11/17, conc. range: 3–144 µg/kg, Ø conc.: 39.1 µg/kg, sample origin: marketed food from Malaysia, sample year: September 1996 and February 1997, country: Japan/Malaysia[567]

- Co-contamination: 10 sa. co-contaminated with AFB_1 and AFB_2; 1 sa. contaminated solely with AFB_1
- Further contamination of plants and plant products other than cereals and their products reported in the present book/article[567]: Products (peanut product), AFB_2, AFG_1, and AFG_2, literature[567]

incidence: 3/5, conc. range: 1–244 µg/kg, Ø conc.: 101 µg/kg, sample origin: marketed food from the Philippines, sample year: December 1996, country: Japan/Malaysia[567]

- Co-contamination: 1 sa. co-contaminated with AFB_1, AFB_2, AFG_1, and AFG_2; 1 sa. co-contaminated with AFB_1, AFB_2, and AFG_1; 1 sa. contaminated solely with AFB_1
- Further contamination of plants and plant products other than cereals and their products reported in the present book/article[567]: Products (peanut product), AFB_2, AFG_1, and AFG_2, literature[567]

incidence: 9/43*, conc. range: 22.0–84.6 µg/kg, Ø conc.: 49.8 µg/kg, sample origin: supermarkets and rural fairs in Ijuí, Porto Alegre, Santa Cruz do Sul, Santa Maria, and São Leopoldo (cities), Rio Grande do Sul (State), Brazil, sample year: May 2009–August 2010, country: Brazil[744]

- Co-contamination: not reported
- Further contamination of plants and plant products other than cereals and their products reported in the present book/article[744]: Nuts (peanut), AFB_1, literature[744]

incidence: 5/11*, conc. range: 0.3–0.6 µg/kg, sample origin: local markets and online stores, USA, sample year: August 2011–January 2012, country: USA/Taiwan/Canada[777], *included peanut butter, peanut flour, and raw peanut

- Co-contamination: 2 sa. co-contaminated with AFB_1 and BEA; 3 sa. contaminated solely with AFB_1

- Further contamination of plants and plant products other than cereals and their products reported in the present book/article[777]: Products (almond product), AFB_1, literature[777]; Products (peanut product), BEA, literature[777]; Products (pistachio product), AFB_1, AFB_2, AFG_1, BEA, NEO, OTA, and OTB, literature[777]

Aflatoxin B$_2$

incidence: 2/20, conc. range: 3–24 µg/kg, ∅ conc.: 13.5 µg/kg, sample origin: different district hygiene institutes, GDR, sample year: unknown, country: GDR[29]

- Co-contamination: not reported
- Further contamination of plants and plant products other than cereals and their products reported in the present book/article[29]: Butter (peanut butter), Nuts (peanut), Seeds (sunflower), AFB_1, AFB_2, AFG_1, and AFG_2, literature[29]; Coffee, OTA, literature[29]; Fruits, AFB_1, AFG_1, and AFG_2, literature[29]; Fruits (apple), Fruits (peach), Juices (apple juice), PAT, literature[29]; Products (peanut product), AFB_1, AFG_1, and AFG_2, literature[29]

incidence: 10/17, conc. range: 3–36 µg/kg, ∅ conc.: 12.2 µg/kg, sample origin: marketed food from Malaysia, sample year: September 1996 and February 1997, country: Japan/Malaysia[567]

- Co-contamination: 10 sa. co-contaminated with AFB_1 and AFB_2
- Further contamination of plants and plant products other than cereals and their products reported in the present book/article[567]: Products (peanut product), AFB_1, AFG_1, and AFG_2, literature[567]

incidence: 2/5, conc. range: 17–125 µg/kg, ∅ conc.: 71 µg/kg, sample origin: marketed food from the Philippines, sample year: December 1996, country: Japan/Malaysia[567]

- Co-contamination: 1 sa. co-contaminated with AFB_1, AFB_2, AFG_1, and AFG_2; 1 sa. co-contaminated with AFB_1, AFB_2, and AFG_1
- Further contamination of plants and plant products other than cereals and their products reported in the present book/article[567]: Products (peanut product), AFB_1, AFG_1, and AFG_2, literature[567]

Aflatoxin G$_1$

incidence: 8/20, conc. range: 9–44 µg/kg, ∅ conc.: 28 µg/kg, sample origin: different district hygiene institutes, GDR, sample year: unknown, country: GDR[29]

- Co-contamination: not reported
- Further contamination of plants and plant products other than cereals and their products reported in the present book/article[29]: Butter (peanut butter), Nuts (peanut), Seeds (sunflower), AFB_1, AFB_2, AFG_1, and AFG_2, literature[29]; Coffee, OTA, literature[29]; Fruits, AFB_1, AFG_1, and AFG_2, literature[29]; Fruits (apple), Fruits (peach), Juices (apple juice), PAT, literature[29]; Products (peanut product), AFB_1, AFB_2, and AFG_2, literature[29]

incidence: 0/17, conc.: no contamination, sample origin: marketed food from Malaysia, sample year: September 1996 and February 1997, country: Japan/Malaysia[567]

- Co-contamination: not reported
- Further contamination of plants and plant products other than cereals and their products reported in the present book/article[567]: Products (peanut product), AFB_1, AFB_2, and AFG_2, literature[567]

incidence: 2/5, conc. range: 6–68 µg/kg, ∅ conc.: 37 µg/kg, sample origin: marketed food from the Philippines, sample year: December 1996, country: Japan/Malaysia[567]

- Co-contamination: 1 sa. co-contaminated with AFB_1, AFB_2, AFG_1, and AFG_2; 1 sa. co-contaminated with AFB_1, AFB_2, and AFG_1
- Further contamination of plants and plant products other than cereals and their products reported in the present book/article[567]: Products (peanut product), AFB_1, AFB_2, and AFG_2, literature[567]

Aflatoxin G$_2$

incidence: 2/20, conc. range: 4–18 µg/kg, ∅ conc.: 11 µg/kg, sample origin: different district hygiene institutes, GDR, sample year: unknown, country: GDR[29]

- Co-contamination: not reported
- Further contamination of plants and plant products other than cereals and their products reported in the present book/article[29]: Butter (peanut butter), Nuts (peanut), Sunflower seeds, AFB_1, AFB_2, AFG_1, and AFG_2, literature[29]; Coffee, OTA, literature[29]; Fruits, AFB_1, AFG_1, and AFG_2, literature[29]; Fruits (apple), Fruits (peach), Juices (apple juice), PAT, literature[29]; Products (peanut product), AFB_1, AFB_2, and AFG_1, literature[29]

incidence: 0/17, conc.: no contamination, sample origin: marketed food from Malaysia, sample year: September 1996 and February 1997, country: Japan/Malaysia[567]

- Co-contamination: not reported
- Further contamination of plants and plant products other than cereals and their products reported in the present book/article[567]: Products (peanut product), AFB_1, AFB_2, and AFG_1, literature[567]

incidence: 1/5, conc.: 33 µg/kg, sample origin: marketed food from the Philippines, sample year: December 1996, country: Japan/Malaysia[567]

- Co-contamination: 1 sa. co-contaminated with AFB_1, AFB_2, AFG_1, and AFG_2
- Further contamination of plants and plant products other than cereals and their products reported in the present book/article[567]: Products (peanut product), AFB_1, AFB_2, and AFG_1, literature[567]

Aflatoxin/s

incidence: 22/74, conc. range: 0.01–4 µg/kg* (19 sa.), >4 µg/kg* (3 sa., maximum: 19 µg/kg*), sample origin: ethnic, health-food, and local stores as well as supermarkets in the four largest cities, Sweden, sample year: autumn 1996–summer 1998, country: Sweden[13], *AFB_1, AFB_2, AFG_1, and AFG_2

- Co-contamination: not reported
- Further contamination of plants and plant products other than cereals and their products reported in the present book/article[13]: Beans (brown bean), Peas, Peas (chickpea), OTA, literature[13]; Juices (apple juice), PAT, literature[13]; Nuts (almond), Products (peanut product), AFB_1, literature[13]; Nuts (Brazil nut, paranut), Nuts (pistachio nut), Oil seeds, AFB_1 and AF/S, literature[13]

incidence: 1/6, conc.: 2 µg/kg*, sample origin: FDA districts, USA, sample year: 1986, country: USA[19], *AFB_1, AFB_2, AFG_1, and AFG_2

- Co-contamination: not reported
- Further contamination of plants and plant products other than cereals and their products reported in the present book/

article[19]: Butter (peanut butter), Candies (peanut candy), Candies (pistachio candy), Nuts, Nuts (almond), Nuts (Brazil nut, paranut), Nuts (peanut), Nuts (pecan nut), Nuts (pistachio nut), Nuts (walnut), Oils (peanut oil), Paste (peanut paste), Paste (pipian paste), Seeds (melon), Spices (cayenne pepper, chilli), Spices (ginger), Spices (nutmeg), AF/S, literature[19]

incidence: 11/32, conc. range: >30–220 μg/kg*, sample origin: markets in or near Manila (capital), Philippines, sample year: 1967–1969, country: USA[134], *AF

- Co-contamination: not reported
- Further contamination of plants and plant products other than cereals and their products reported in the present book/article[134]: Beans, Butter (peanut butter), Nuts (peanut), Products (coconut product), Tubers, AF/S, literature[134]

incidence: 8/295*, conc. range: 15.0–340.0 μg/kg**, ∅ conc.: 101.5 μg/kg**, sample origin: local retail stores in the region of Brasilia (capital), Brazil, sample year: 2002–2011, country: Brazil[208], *peanut butter and peanut candies, **AFB_1, AFB_2, AFG_1, and AFG_2

- Co-contamination: not reported
- Further contamination of plants and plant products other than cereals and their products reported in the present book/article[208]: Nuts, Nuts (Brazil nut, paranut), Nuts (peanut), AF/S, literature[208]

incidence: 49/100, conc. range: 0.1–5 μg/k* (14 sa.), 6–10 μg/kg* (10 sa.), 11–20 μg/kg* (3 sa.), 21–35 μg/kg* (17 sa.), and >50 μg/kg* (5 sa.), sample origin: commercial and imported food from retail outlets and points of entries, Malaysia, sample year: 1995–1999, country: Malaysia[233], *AF not specified

- Co-contamination: not reported
- Further contamination of plants and plant products other than cereals and their products reported in the present book/article[233]: Butter (peanut butter), Candies (peanut candy), Nuts, Nuts (peanut), Spices (cayenne pepper, chilli), AF/S, literature[233]

incidence: 30/45, conc. range: >20 to ≤1,789 μg/kg*, ∅ conc.: 384 μg/kg*, sample origin: supermarkets in the region of Campinas (municipality), São Paulo (state), Brazil, sample year: 1995/1996, country: Brazil[421], *AFB_1, AFB_2, AFG_1, and AFG_2

- Co-contamination: not reported
- Further contamination of plants and plant products other than cereals and their products reported in the present book/article[421]: Nuts (peanut), AF/S, literature[421]

incidence: 8/41*, conc. range: ≤11.4 μg/kg*, sample origin: markets, superstores, malls, and venders of major districts of Punjab (province), Pakistan, sample year: unknown, country: Pakistan[573], *packed peanut nimko, **AFB_1, AFB_2, AFG_1, and AFG_2

- Co-contamination: not reported
- Further contamination of plants and plant products other than cereals and their products reported in the present book/article[573]: Butter (peanut butter), Cookies, Nuts (peanut), AF/S, literature[573]

incidence: 14/71, conc. range: ≤10.5 μg/kg*, sample origin: New Zealand, sample year: unknown, country: New Zealand[701], *AFB_1, AFB_2, AFG_1, and AFG_2

- Co-contamination: not reported
- Further contamination of plants and plant products other than cereals and their products reported in the present book/article[701]: Butter (peanut butter), Nuts (peanut), Nuts (pistachio nut), Products (nut product), AF/S, literature[701]; Nuts (Brazil nut, paranut), AFB_1, literature[701]

Fusarium Toxins

Depsipeptides

Beauvericin

incidence: 2/11*, conc. range: 0.9–5.0 μg/kg, ∅ conc.: 2.95 μg/kg, sample origin: local markets and online stores, USA, sample year: August 2011–January 2012, country: USA/Taiwan[28], *included peanut butter, peanut flour, and raw peanut

- Co-contamination: 2 sa. co-contaminated with AFB_1 and BEA
- Further contamination of plants and plant products other than cereals and their products reported in the present book/article[28]: Products (peanut product), AFB_1, literature[28]; Products (pistachio product), AFB_1, AFB_2, AFG_1, BEA, NEO, OTA, and OTB, literature[28]

incidence: 2/11*, conc. range: 0.8–4.5 μg/kg, ∅ conc.: 2.65 μg/kg, sample origin: local markets and online stores, USA, sample year: August 2011–January 2012, country: USA/Taiwan/Canada[777], *included peanut butter, peanut flour, and raw peanut

- Co-contamination: 2 sa. co-contaminated with AFB_1 and BEA
- Further contamination of plants and plant products other than cereals and their products reported in the present book/article[777]: Products (almond product), Products (peanut product), AFB_1, literature[777]; Products (pistachio product), AFB_1, AFB_2, AFG_1, BEA, NEO, OTA, and OTB, literature[777]

Products (pistachio product) may contain the following mycotoxins:

Aspergillus Toxins

Aflatoxin B_1

incidence: 2/10*, conc. range: 0.7–1.4 μg/kg, ∅ conc.: 1.05 μg/kg, sample origin: local markets and online stores, USA, sample year: August 2011–January 2012, country: USA/Taiwan[28], *included pistachio butter, pistachio flour, and roasted pistachio

- Co-contamination: 1 sa. co-contaminated with AFB_1, AFB_2, AFG_1, OTA, and OTB; no further information available
- Further contamination of plants and plant products other than cereals and their products reported in the present book/article[28]: Products (peanut product), AFB_1 and BEA literature[28]; Products (pistachio product), AFB_2, AFG_1, BEA, NEO, OTA, and OTB, literature[28]

incidence: 2/10*, conc. range: 0.5–1.2 μg/kg, ∅ conc.: 0.85 μg/kg, sample origin: local markets and online stores, USA, sample year: August 2011–January 2012, country: USA/Taiwan/Canada[777], *included pistachio butter, pistachio flour, and roasted pistachio

- Co-contamination: 1 sa. co-contaminated with AFB_1, AFB_2, AFG_1, OTA, and OTB; 1 sa. contaminated solely with AFB_1
- Further contamination of plants and plant products other than cereals and their products reported in the present book/article[777]: Products (almond product), AFB_1, literature[777]; Products (peanut product), AFB_1 and BEA, literature[777]; Products (pistachio product), AFB_2, AFG_1, BEA, NEO, OTA, and OTB, literature[777]

Aflatoxin B_2
incidence: 1/10*, conc.: 0.6 µg/kg, sample origin: local markets and online stores, USA, sample year: August 2011–January 2012, country: USA/Taiwan[28], *included pistachio butter, pistachio flour, and roasted pistachio
- Co-contamination: 1 sa. co-contaminated with AFB_1, AFB_2, AFG_1, OTA, and OTB; no further information available
- Further contamination of plants and plant products other than cereals and their products reported in the present book/article[28]: Products (peanut product), AFB_1 and BEA literature[28]; Products (pistachio product), AFB_1, AFG_1, BEA, NEO, OTA, and OTB, literature[28]

incidence: 1/10*, conc.: 0.9 µg/kg, sample origin: local markets and online stores, USA, sample year: August 2011–January 2012, country: USA/Taiwan/Canada[777], *included pistachio butter, pistachio flour, and roasted pistachio
- Co-contamination: 1 sa. co-contaminated with AFB_1, AFB_2, AFG_1, OTA, and OTB
- Further contamination of plants and plant products other than cereals and their products reported in the present book/article[777]: Products (almond product), AFB_1, literature[777]; Products (peanut product), AFB_1 and BEA, literature[777]; Products (pistachio product), AFB_1, AFG_1, BEA, NEO, OTA, and OTB, literature[777]

Aflatoxin G_1
incidence: 1/10*, conc.: 0.4 µg/kg, sample origin: local markets and online stores, USA, sample year: August 2011–January 2012, country: USA/Taiwan[28], *included pistachio butter, pistachio flour, and roasted pistachio
- Co-contamination: 1 sa. co-contaminated with AFB_1, AFB_2, AFG_1, OTA, and OTB; no further information available
- Further contamination of plants and plant products other than cereals and their products reported in the present book/article[28]: Products (peanut product), AFB_1 and BEA literature[28]; Products (pistachio product), AFB_1, AFB_2, BEA, NEO, OTA, and OTB, literature[28]

incidence: 1/10*, conc.: 0.5 µg/kg, sample origin: local markets and online stores, USA, sample year: August 2011–January 2012, country: USA/Taiwan/Canada[777], *included pistachio butter, pistachio flour, and roasted pistachio
- Co-contamination: 1 sa. co-contaminated with AFB_1, AFB_2, AFG_1, OTA, and OTB
- Further contamination of plants and plant products other than cereals and their products reported in the present book/article[777]: Products (almond product), AFB_1, literature[777]; Products (peanut product), AFB_1 and BEA, literature[777]; Products (pistachio product), AFB_1, AFB_2, BEA, NEO, OTA, and OTB, literature[777]

Aspergillus **and** ***Penicillium*** **Toxins**

Ochratoxin A
incidence: 3/10*, conc. range: 1.1–7.1 µg/kg, sample origin: local markets and online stores, USA, sample year: August 2011–January 2012, country: USA/Taiwan[28], *included pistachio butter, pistachio flour, and roasted pistachio
- Co-contamination: 1 sa. co-contaminated with AFB_1, AFB_2, AFG_1, OTA, and OTB; no further information available
- Further contamination of plants and plant products other than cereals and their products reported in the present book/article[28]: Products (peanut product), AFB_1 and BEA literature[28]; Products (pistachio product), AFB_1, AFB_2, AFG_1, BEA, NEO, and OTB, literature[28]

incidence: 3/10*, conc. range: 1.0–6.6 µg/kg, sample origin: local markets and online stores, USA, sample year: August 2011–January 2012, country: USA/Taiwan/Canada[777], *included pistachio butter, pistachio flour, and roasted pistachio
- Co-contamination: 1 sa. co-contaminated with AFB_1, AFB_2, AFG_1, OTA, and OTB
- Further contamination of plants and plant products other than cereals and their products reported in the present book/article[777]: Products (almond product), AFB_1, literature[777]; Products (peanut product), AFB_1 and BEA, literature[777]; Products (pistachio product), AFB_1, AFB_2, AFG_1, BEA, NEO, and OTB, literature[777]

Ochratoxin B
incidence: 2/10*, conc. range: 0.6–2.8 µg/kg, Ø conc.: 1.7 µg/kg, sample origin: local markets and online stores, USA, sample year: August 2011–January 2012, country: USA/Taiwan[28], *included pistachio butter, pistachio flour, and roasted pistachio
- Co-contamination: 1 sa. co-contaminated with AFB_1, AFB_2, AFG_1, OTA, and OTB; no further information available
- Further contamination of plants and plant products other than cereals and their products reported in the present book/article[28]: Products (peanut product), AFB_1 and BEA literature[28]; Products (pistachio product), AFB_1, AFB_2, AFG_1, BEA, NEO, and OTA, literature[28]

incidence: 2/10*, conc. range: 0.5–2.3 µg/kg, Ø conc.: 1.4 µg/kg, sample origin: local markets and online stores, USA, sample year: August 2011–January 2012, country: USA/Taiwan/Canada[777], *included pistachio butter, pistachio flour, and roasted pistachio
- Co-contamination: 1 sa. co-contaminated with AFB_1, AFB_2, AFG_1, OTA, and OTB
- Further contamination of plants and plant products other than cereals and their products reported in the present book/article[777]: Products (almond product), AFB_1, literature[777]; Products (peanut product), AFB_1 and BEA, literature[777]; Products (pistachio product), AFB_1, AFB_2, AFG_1, BEA, NEO, and OTA, literature[777]

Fusarium Toxins

Depsipeptides

Beauvericin
incidence: 1/10*, conc.: 1.4 µg/kg, sample origin: local markets and online stores, USA, sample year: August 2011–January 2012, country: USA/Taiwan[28], *included pistachio butter, pistachio flour, and roasted pistachio
- Co-contamination: not reported
- Further contamination of plants and plant products other than cereals and their products reported in the present book/article[28]: Products (peanut product), AFB_1 and BEA literature[28]; Products (pistachio product), AFB_1, AFB_2, AFG_1, NEO, OTA, and OTB, literature[28]

incidence: 1/10*, conc.: 1.9 µg/kg, sample origin: local markets and online stores, USA, sample year: August 2011–January 2012, country: USA/Taiwan/Canada[777], *included pistachio butter, pistachio flour, and roasted pistachio

- Co-contamination: 1 sa. co-contaminated with AFB_1, AFB_2, AFG_1, OTA, and OTB
- Further contamination of plants and plant products other than cereals and their products reported in the present book/article[777]: Products (almond product), AFB_1, literature[777]; Products (peanut product), AFB_1 and BEA, literature[777]; Products (pistachio product), AFB_1, AFB_2, AFG_1, NEO, OTA, and OTB, literature[777]

Type A Trichothecenes

Neosolaniol

incidence: 1/10*, conc.: 58 µg/kg, sample origin: local markets and online stores, USA, sample year: August 2011–January 2012, country: USA/Taiwan[28], *included pistachio butter, pistachio flour, and roasted pistachio

- Co-contamination: not reported
- Further contamination of plants and plant products other than cereals and their products reported in the present book/article[28]: Products (peanut product), AFB_1 and BEA literature[28]; Products (pistachio product), AFB_1, AFB_2, AFG_1, BEA, OTA, and OTB, literature[28]

incidence: 1/10*, conc.: 46 µg/kg, sample origin: local markets and online stores, USA, sample year: August 2011–January 2012, country: USA/Taiwan/Canada[777], *included pistachio butter, pistachio flour, and roasted pistachio

- Co-contamination: not reported
- Further contamination of plants and plant products other than cereals and their products reported in the present book/article[777]: Products (almond product), AFB_1, literature[777]; Products (peanut product), AFB_1 and BEA, literature[777]; Products (pistachio product), AFB_1, AFB_2, AFG_1, BEA, OTA, and OTB, literature[777]

Products (plant product) may contain the following mycotoxins:

Aspergillus and ***Penicillium*** Toxins

Ochratoxin A

incidence: 2/9, conc. range: LOD/LOQ-0.9 µg/kg (1 sa.), 1.0 µg/kg (1 sa.), sample origin: Finland, sample year: 1998, country: EU[698]

incidence: 4/15*, conc. range: 1.0–2.9 µg/kg (2 sa.), ≤7.0 µg/kg (2 sa.), sample origin: Finland, sample year: 1998, country: EU[698], *miscellaneous commodities of plant origin

incidence: 0/3*, conc.: no contamination, sample origin: Finland, sample year: 1999, country: EU[698], *miscellaneous commodities of plant origin

- Co-contamination: not reported
- Further contamination of plants and plant products other than cereals and their products reported in the present book/article[698]: Baby food, Beans (soybean), Chocolate, Cocoa, Coffee, Coffee beans, Currants, Fruits, dried, Herbs, Jam, Juices, Juices (grape juice), Legumes, Nuts, Oil seeds, Oils (olive oil), Peas, Peas (chickpea), Raisins, Spices, Spices (cayenne pepper, chilli), Spices (nutmeg), Spices (paprika), Spices (pepper), Sultanas, Tea, Vinegar, Wine, Wine (red), Wine (rosé), Wine (white), Wine, miscellaneous, OTA, literature[698]

Products (potato product) may contain the following mycotoxins:

Fusarium Toxins

Type A Trichothecenes

Diacetoxyscirpenol

incidence: 1/21, conc.: 21 µg/kg, sample origin: food stores, Germany, sample year: 2000/2001, country: Germany[151]

- Co-contamination: not reported
- Further contamination of plants and plant products other than cereals and their products reported in the present book/article[151]: Beans, Spices (cayenne pepper, chilli), ZEA, literature[151]; Nuts (hazelnut), Seeds (sunflower), HT-2, T-2, and ZEA, literature[151]; Products (potato product), HT-2, MAS, SCIRP, and ZEA, literature[151]; Seeds (pumpkin), HT-2 and ZEA, literature[151]

HT-2 Toxin

incidence: 1/21, conc.: 5 µg/kg, sample origin: food stores, Germany, sample year: 2000/2001, country: Germany[151]

- Co-contamination: not reported
- Further contamination of plants and plant products other than cereals and their products reported in the present book/article[151]: Beans, Spices (cayenne pepper, chilli), ZEA, literature[151]; Nuts (hazelnut), Seeds (sunflower), HT-2, T-2, and ZEA, literature[151]; Products (potato product), DAS, MAS, SCIRP, and ZEA, literature[151]; Seeds (pumpkin), HT-2 and ZEA, literature[151]

Monoacetoxyscirpenol

incidence: 3/21, conc.: 5–26 µg/kg, sample origin: food stores, Germany, sample year: 2000/2001, country: Germany[151]

- Co-contamination: not reported
- Further contamination of plants and plant products other than cereals and their products reported in the present book/article[151]: Beans, Spices (cayenne pepper, chilli), ZEA, literature[151]; Nuts (hazelnut), Seeds (sunflower), HT-2, T-2, and ZEA, literature[151]; Products (potato product), DAS, HT-2, SCIRP, and ZEA, literature[151]; Seeds (pumpkin), HT-2 and ZEA, literature[151]

Scirpentriol

incidence: 2/21, conc. range: 23–35 µg/kg, Ø conc.: 29 µg/kg, sample origin: food stores, Germany, sample year: 2000/2001, country: Germany[151]

- Co-contamination: not reported
- Further contamination of plants and plant products other than cereals and their products reported in the present book/article[151]: Beans, Spices (cayenne pepper, chilli), ZEA, literature[151]; Nuts (hazelnut), Seeds (sunflower), HT-2, T-2, and ZEA, literature[151]; Products (potato product), DAS, HT-2, MAS, and ZEA, literature[151]; Seeds (pumpkin), HT-2 and ZEA, literature[151]

Other ***Fusarium*** Toxins

Zearalenone

incidence: 1/21, conc.: 2 µg/kg, sample origin: food stores, Germany, sample year: 2000/2001, country: Germany[151]

- Co-contamination: not reported

- Further contamination of plants and plant products other than cereals and their products reported in the present book/article[151]: Beans, Spices (cayenne pepper, chilli), ZEA, literature[151]; Nuts (hazelnut), Seeds (sunflower), HT-2, T-2, and ZEA, literature[151]; Products (potato product), DAS, HT-2, MAS, and SCIRP, literature[151]; Seeds (pumpkin), HT-2 and ZEA, literature[151]

Products (sesame product) may contain the following mycotoxins:

Aspergillus Toxins

Aflatoxin B$_1$

incidence: 5/130, conc. range: 1 µg/kg, sample origin: sites of primary storage or processing (domestic products) and sites of importation, Cyprus, sample year: 1992–1996, country: Cyprus[200]

- Co-contamination: not reported
- Further contamination of plants and plant products other than cereals and their products reported in the present book/article[200]: Butter (peanut butter), Figs, Nuts (peanut), AFB$_1$, AFB$_2$, AFG$_1$, and AFG$_2$, literature[200]; Nuts (Brazil nut, paranut) and Nuts (walnut), AFB$_1$, AFB$_2$, and AFG$_1$, literature[200]; Nuts (chestnut), Spices, AFB$_1$, literature[200]; Nuts (pistachio nut), AFB$_1$ and AFB$_2$, literature[200]; Seeds (sesame), AFB$_1$, AFG$_1$, and AFG$_2$, literature[200]

see also **Seeds (sesame)**

Products (soybean product) may contain the following mycotoxins:

Fusarium Toxins

Fumonisin/s

incidence: 3/35, conc. range: ≤21 µg/kg, sample origin: German market, Germany, sample year: 1999–2001, country: Germany[510]

- Co-contamination: not reported
- Further contamination of plants and plant products other than cereals and their products reported in the present book/article[510]: Sauces, Vegetables (asparagus), FB/S, literature[510]

Products (tobacco product) may contain the following mycotoxins:

Aspergillus Toxins

Aflatoxin B$_1$

incidence: 2/5*, conc. range: 0.016–0.021 µg/kg (dry mass), ∅ conc.: 0.019 µg/kg (dry mass), sample origin: Washington D.C. area, USA, sample year: unknown, country: USA[725], *chew tobacco

incidence: 6/6*, conc. range: 0.013–0.271 µg/kg (dry mass), ∅ conc.: 0.091 µg/kg (dry mass), sample origin: Washington D.C. area, USA, sample year: unknown, country: USA[725], *dry snuff tobacco

- Co-contamination: not reported
- Further contamination of plants and plant products other than cereals and their products reported in the present book/article[725]: not reported

Products (tomato product) may contain the following mycotoxins:

Alternaria Toxins

Altenuene

incidence: 8/10, conc. range: 3.8–4.82 µg/kg, sample origin: Italian supermarkets, Italy, sample year: unknown, country: Italy[822]

- Co-contamination: not reported
- Further contamination of plants and plant products other than cereals and their products reported in the present book/article[822]: Juices (apple juice), ALT and TA, literature[822]; Products (tomato product), AOH, literature[822]; see also Products (tomato product), chapter: Further Mycotoxins and Microbial Metabolites

Alternariol

incidence: 24/34, conc. range: >LOD (14 sa.), 6.1–25 µg/kg (10 sa.), sample origin: retail stores in the area of Arnsberg and Münster (cities), Germany, sample year: unknown, country: Germany[743]

- Co-contamination: not reported
- Further contamination of plants and plant products other than cereals and their products reported in the present book/article[743]: Juices, ALT, AME, AOH, and TA, literature[743]; Oils (vegetable oil), AME, AOH, and TA, literature[743]; Products (tomato product), AME and TA, literature[743]; Seeds (sunflower), ALT, AME, AOH, ATX-I, and TA, literature[743]; see also Juices, Oils (vegetable oil), Seeds (sunflower), chapter: Further Mycotoxins and Microbial Metabolites

incidence: 5/10, conc. range: 4.0–6.8 µg/kg, sample origin: Italian supermarkets, Italy, sample year: unknown, country: Italy[822]

- Co-contamination: not reported
- Further contamination of plants and plant products other than cereals and their products reported in the present book/article[822]: Juices (apple juice), ALT and TA, literature[822]; Products (tomato product), ALT, literature[822]; see also Products (tomato product), chapter: Further Mycotoxins and Microbial Metabolites

Alternariol Methyl Ether

incidence: 27/34, conc. range: >LOD (19 sa.), 1.2–7.4 µg/kg (8 sa.), sample origin: retail stores in the area of Arnsberg and Münster (cities), Germany, sample year: unknown, country: Germany[743]

- Co-contamination: not reported
- Further contamination of plants and plant products other than cereals and their products reported in the present book/article[743]: Juices, ALT, AME, AOH, and TA, literature[743]; Oils (vegetable oil), AME, AOH, and TA, literature[743]; Products (tomato product), AOH and TA, literature[743]; Seeds (sunflower), ALT, AME, AOH, ATX-I, and TA, literature[743]; see also Juices, Oils (vegetable oil), Seeds (sunflower), chapter: Further Mycotoxins and Microbial Metabolites

Tenuazonic Acid

incidence: 31/34, conc. range: >LOD (16 sa.), 52–460 µg/kg (15 sa.), sample origin: retail stores in the area of Arnsberg and Münster (cities), Germany, sample year: unknown, country: Germany[743]

- Co-contamination: not reported

- Further contamination of plants and plant products other than cereals and their products reported in the present book/article[743]: Juices, ALT, AME, AOH, and TA, literature[743]; Oils (vegetable oil), AME, AOH, and TA, literature[743]; Products (tomato product), AME and AOH, literature[743]; Seeds (sunflower), ALT, AME, AOH, ATX-I, and TA, literature[743]; see also Juices, Oils (vegetable oil), Seeds (sunflower), chapter: Further Mycotoxins and Microbial Metabolites

incidence: 26/26*, conc. range: 7–2,330 μg/kg, ∅ conc.: 187 μg/kg, sample origin: German market, Germany, sample year: unknown, country: Germany[781], *included different tomato products
- Co-contamination: not reported
- Further contamination of plants and plant products other than cereals and their products reported in the present book/article[781]: Sauces (pepper sauce), TA, literature[781]

For detailed information see the article.

Prunes see **Fruits (dried)**

Pudding may contain the following mycotoxins:

Aspergillus Toxins

Aflatoxin/s
incidence: 6/8, conc. range: ≤3.585 μg/kg*, sample origin: groceries and supermarkets, Turkey, sample year: 2003, country: Turkey[119], *AFS total
- Co-contamination: not reported
- Further contamination of plants and plant products other than cereals and their products reported in the present book/article[119]: Asure, Beverages, Cocoa, Cream, Ice cream, Jello, Keskul, Muhallebi, Salep, Sauces (chocolate sauce), Soups, AF/S, literature[119]

Pudding ingredients were sugar, cornstarch, instant cocoa, thickener agent, chocolate flavor, and vanilla.

Aspergillus and ***Penicillium*** Toxins

Ochratoxin A
incidence: 1/1, conc.: 0.03 μg/kg, sample origin: supermarkets and fast-food chains in Quebec (city), Canada, sample year: September/October 2008, country: Canada[90]

incidence: 1/1, conc.: 0.02 μg/kg, sample origin: supermarkets and fast-food chains in Calgary (city), Canada, sample year: September/October 2009, country: Canada[90]
- Co-contamination: not reported
- Further contamination of plants and plant products other than cereals and their products reported in the present book/article[90]: Baby food, Beans, Beverages, Butter (peanut butter), Candies, Chocolate, Coffee, Extracts, Herbs, Ice cream, Nuts, Pies, Pizza, Products, Raisins, Seeds, Wine, OTA, literature[90]

incidence: 27/32*, conc. range: ≤0.080 μg/kg, sample origin: regions around Detmold, Giessen, Hamburg, Jena, Karlsruhe, Kiel, Kulmbach, and Trier (cities), Germany, sample year: unknown, country: Germany[262], *choco crème

incidence: 7/32*, conc. range: ≤0.090 μg/kg, sample origin: regions around Detmold, Giessen, Hamburg, Jena, Karlsruhe, Kiel, Kulmbach, and Trier (cities), Germany, sample year: unknown, country: Germany[262], *different kinds of puddings and crèmes
- Co-contamination: not reported
- Further contamination of plants and plant products other than cereals and their products reported in the present book/article[262]: Bars, Chocolate, Drinks (chocolate drink), Figs, Fruits, dried, Nut crème, Nuts, Nuts (hazelnut), Nuts (peanut), Powder (cocoa powder), Raisins, Seeds (linseed), Seeds (sesame), Seeds (sunflower), OTA, literature[262]

Pulp (apple pulp) may contain the following mycotoxins:

Aspergillus and ***Penicillium*** Toxins

Patulin
incidence: 25/32, conc. range: 1.3–102.1 μg/kg, ∅ conc.: 24.1 μg/kg, sample origin: Czech baby food producer, Czech Republic, sample year: unknown, country: Czech Republic[911]
- Co-contamination: not reported
- Further contamination of plants and plant products other than cereals and their products reported in the present book/article[911]: Apple cider, Baby food, Fruit salad, Fruits (apple), Fruits (pear), Juices (apple juice), Juices (pear juice), PAT, literature[911]

see also **Fruits (apple)**

Pulp (grape pulp) may contain the following mycotoxins:

Aspergillus and ***Penicillium*** Toxins

Ochratoxin A
incidence: 14/16*, conc. range: 0.0212–0.0354 μg/l, ∅ conc.: 0.0283 μg/l, sample origin: markets, Rio de Janeiro (state), Brazil, sample year: unknown, country: Brazil/Argentina[377], *frozen grape pulp
- Co-contamination: not reported
- Further contamination of plants and plant products other than cereals and their products reported in the present book/article[377]: Juices (grape juice), Wine (red), Wine (rosé), Wine (white), OTA, literature[377]

Pulp (plum pulp) may contain the following mycotoxins:

Aspergillus and ***Penicillium*** Toxins

Patulin
incidence: 1/1, conc.: 0.8 μg/l, sample origin: local store in Munich (city) Germany, sample year: 1996/1997, country: Germany[52], *commercial product
- Co-contamination: not reported
- Further contamination of plants and plant products other than cereals and their products reported in the present book/article[52]: Juices (apple juice), Juices (blackcurrant juice), Juices (cherry juice), Juices (grape juice), Juices (orange juice), PAT, literature[52]

Pulp (tomato pulp) may contain the following mycotoxins:

Alternaria Toxins

Tenuazonic Acid
incidence: 7/22, conc. range: 39–111 μg/kg, ∅ conc.: 73.6 μg/kg, sample origin: local markets and supermarkets in Campinas (city), São Paulo (state), Brazil, sample year: July 1994–December 1995, country: Brazil[48]
- Co-contamination: 1 sa. co-contaminated with CA and TA; 6 sa. contaminated solely with TA
- Further contamination of plants and plant products other than cereals and their products reported in the present book/article[48]: Pulp (tomato pulp), CPA, literature[48]; Puree, CPA and TA, literature[48]

Aspergillus **and** ***Penicillium*** **Toxins**

Cyclopiazonic Acid

incidence: 6/22, conc. range: 64–178 µg/kg, ∅ conc.: 109 µg/kg, sample origin: local markets and supermarkets in Campinas (city), São Paulo (state), Brazil, sample year: July 1994–December 1995, country: Brazil[48]

- Co-contamination: 1 sa. co-contaminated with CA and TA; 6 sa. contaminated solely with CPA
- Further contamination of plants and plant products other than cereals and their products reported in the present book/article[48]: Pulp (tomato pulp), TA, literature[48]; Puree, CPA and TA, literature[48]

Pulses see Legumes

Pumpkin kernels see Seeds (pumpkin)

Puree may contain the following mycotoxins:

Alternaria **Toxins**

Alternariol

incidence: 5/80*, conc. range: 187–8,756 µg/kg, ∅ conc.: 2,219 µg/kg, sample origin: local markets and supermarkets Buenos Aires (city), Argentina, sample year: March/April 2006, country: Argentina[45], *tomato puree

- Co-contamination: 2 sa. co-contaminated with AOH and TA; 2 sa. co-contaminated with AOH and AME; 1 sa. contaminated solely with AOH
- Further contamination of plants and plant products other than cereals and their products reported in the present book/article[45]: Puree, AME and TA, literature[45]

incidence: 2/3*, conc. range: 3.8–5.9 µg/kg, ∅ conc.: 4.85 µg/kg, sample origin: unknown, sample year: unknown, country: Germany[46], *tomato puree

- Co-contamination: 2 sa. co-contaminated with AME and AOH
- Further contamination of plants and plant products other than cereals and their products reported in the present book/article[46]: Juices (tomato juice), Paste (tomato paste), Tomato ketchup, Wine (red), Wine (white), AME and AOH, literature[46]; Juices (vegetable juice), Puree, AME, literature[46]

incidence: 11/17*, conc. range: 4–33 µg/kg, sample origin: Swiss market, Switzerland, sample year: autumn and winter 2010, country: Switzerland[47], *tomato puree and concentrates

- Co-contamination: For detailed information see the article.
- Further contamination of plants and plant products other than cereals and their products reported in the present book/article[47]: Fruits (tomato), Sauces (tomato sauce), Tomato ketchup, AME, AOH, and TA, literature[47]; Fruits, dried, ALT, AME, AOH, and TA, literature[47]; Puree, AME and TA, literature[47]; see also Fruits (tomato), Fruits, dried, Puree, chapter: Further Mycotoxins and Microbial Metabolites

Alternariol Methyl Ether

incidence: 21/80*, conc. range: 84–1,734 µg/kg, ∅ conc.: 593 µg/kg, sample origin: local markets and supermarkets Buenos Aires (city), Argentina, sample year: March/April 2006, country: Argentina[45], *tomato puree

- Co-contamination: 6 sa. co-contamination with AME and TA; 2 sa. co-contamination with AME and AOH; 13 sa. contamination solely with AME
- Further contamination of plants and plant products other than cereals and their products reported in the present book/article[45]: Puree, AOH and TA, literature[45]

incidence: 2/3*, conc. range: 0.2–0.5 µg µg/kg, ∅ conc.: 0.35 µg/kg, sample origin: unknown, sample year: unknown, country: Germany[46], *tomato puree

- Co-contamination: 2 sa. co-contaminated with AME and AOH
- Further contamination of plants and plant products other than cereals and their products reported in the present book/article[46]: Juices (tomato juice), Paste (tomato paste), Tomato ketchup, Wine (red), Wine (white), AME and AOH, literature[46]; Juices (vegetable juice), AME, literature[46]; Puree, AOH, literature[46]

incidence: 12/17*, conc. range: 1–9 µg/kg, sample origin: Swiss market, Switzerland, sample year: autumn and winter 2010, country: Switzerland[47], *tomato puree and concentrates

- Co-contamination: For detailed information see the article.
- Further contamination of plants and plant products other than cereals and their products reported in the present book/article[47]: Fruits (tomato), Sauces (tomato sauce), Tomato ketchup, AME, AOH, and TA, literature[47]; Fruits, dried, ALT, AME, AOH, and TA, literature[47]; Puree, AOH and TA, literature[47]; see also Fruits (tomato), Fruits, dried, Puree, chapter: Further Mycotoxins and Microbial Metabolites

Tenuazonic Acid

incidence: 23/80*, conc. range: 39–4,021 µg/kg, ∅ conc.: 775 µg/kg, sample origin: local markets and supermarkets Buenos Aires (city), Argentina, sample year: March/April 2006, country: Argentina[45], *tomato puree

- Co-contamination: 6 sa. co-contamination with AME and TA; 2 sa. co-contamination with AOH and TA; 15 sa. contamination solely with TA
- Further contamination of plants and plant products other than cereals and their products reported in the present book/article[45]: Puree, AME and AOH, literature[45]

incidence: 17/17*, conc. range: 2–790 µg/kg, ∅ conc.: 165 µg/kg, sample origin: Swiss market, Switzerland, sample year: autumn and winter 2010, country: Switzerland[47], *tomato puree and concentrates

- Co-contamination: For detailed information see the article.
- Further contamination of plants and plant products other than cereals and their products reported in the present book/article[47]: Fruits (tomato), Sauces (tomato sauce), Tomato ketchup, AME, AOH, and TA, literature[47]; Fruits, dried, ALT, AME, AOH, and TA, literature[47]; Puree, AME and AOH, literature[47]; see also Fruits (tomato), Fruits, dried, Puree, chapter: Further Mycotoxins and Microbial Metabolites

incidence: 4/22*, conc. range: 29–76 µg/kg, ∅ conc.: 53 µg/kg, sample origin: local markets and supermarkets in Campinas (city), São Paulo (state), Brazil, sample year: July 1994–December 1995, country: Brazil[48], *tomato puree

- Co-contamination: 2 sa. co-contaminated with CA and TA; 2 sa. contaminated solely with TA
- Further contamination of plants and plant products other than cereals and their products reported in the present book/article[48]: Pulp (tomato pulp), CPA and TA, literature[48]; Puree, CPA, literature[48]

incidence: 12/12*, conc. range: 1.0–78 µg/kg, ∅ conc.: 14.3 µg/kg, sample origin: retail stores, Germany, sample year: unknown, country: Germany[581], *for composition of each single puree see the article

- Co-contamination: not reported
- Further contamination of plants and plant products other than cereals and their products reported in the present book/ article[581]: Tea, TA, literature[581]

Aspergillus and ***Penicillium*** Toxins

Cyclopiazonic Acid

incidence: 2/22*, conc. range: 36–117 µg/kg, ∅ conc.: 76.5 µg/kg, sample origin: local markets and supermarkets in Campinas (city), São Paulo (state), Brazil, sample year: July 1994–December 1995, country: Brazil[48], *tomato puree

- Co-contamination: 2 sa. co-contaminated with CA and TA
- Further contamination of plants and plant products other than cereals and their products reported in the present book/ article[48]: Pulp (tomato pulp), CPA and TA, literature[48]; Puree, TA, literature[48]

Ochratoxin A

incidence: 5/5*, conc. range: 2.3–13.9 µg/kg, ∅ conc.: 9.1 µg/kg, sample origin: grapes collected in a field in Apulia (region), southern Italy, sample year: unknown, country: Italy[49], *red grape berry puree

- Co-contamination: not reported
- Further contamination of plants and plant products other than cereals and their products reported in the present book/ article[49]: Sultanas, OTA, literature[49]

Patulin

incidence: 1/15* **, conc.: 0.7 µg/kg, sample origin: markets, supermarkets, hypermarkets or organic food shops, Italy, sample year: November 2003–February 2004, country: Italy[35], *fruit puree, **conventional

incidence: 9/25* **, conc. range: >LOQ<10 µg/kg (8 sa.), 13.0 µg/kg (1 sa.), sample origin: markets, supermarkets, hypermarkets or organic food shops, Italy, sample year: November 2003–February 2004, country: Italy[35], *fruit puree, **organic

- Co-contamination: not reported
- Further contamination of plants and plant products other than cereals and their products reported in the present book/ article[35]: Fruits (apple), Juices (apple juice), Juices (pear juice), Vinegar (apple), PAT, literature[35]

Conventional and organic purees included 23 baby food samples (homogenized fruits, 11 conventional and 12 organic sa.).

incidence: 6/18*, conc. range: 7.7–28.4 µg/kg, sample origin: stores and supermarkets, Valencia (city), Spain, sample year: unknown, country: Spain[43], *apple compote and puree

- Co-contamination: not reported
- Further contamination of plants and plant products other than cereals and their products reported in the present book/ article[43]: Juices (apple juice), Products (apple product), PAT, literature[43]

incidence: 4/8*, conc. range: 22–221 µg/kg, ∅ conc.: 123.0 µg/kg, sample origin: different Argentinean towns, Argentina, sample year: unknown, country: Argentina[50], *apple puree

- Co-contamination: not reported
- Further contamination of plants and plant products other than cereals and their products reported in the present book/ article[50]: Marmalade (apple marmalade), Marmalade (pear marmalade), PAT, literature[50]

incidence: 5/76*, conc. range: LOD-LOQ (3 sa.), LOQ-10 µg/kg (2 sa., maximum: 5.7 µg/kg), sample origin: supermarkets, organic produce retail outlets, and pharmacies in Lisbon (capital), Portugal, sample year: August 2007–March 2009, country: Portugal[486], *apple purees for infants and young children, homogenized

- Co-contamination: not reported
- Further contamination of plants and plant products other than cereals and their products reported in the present book/ article[486]: Juices (apple juice), PAT, literature[486]

incidence: 0/24* **, conc.: no contamination, sample origin: Catalonian (autonomous community) supermarkets and bio shops, Spain, sample year: November 2011–May 2012, country: Spain[594], *apple puree included baby foods, **conventional

incidence: 6/22* **, conc. range: LOQ-10 µg/kg (2 sa.), 10–25 µg/kg (2 sa.), 25–50 µg/kg (1 sa.), and 50.2 µg/kg (1 sa.), ∅ conc.: 22.5 µg/kg, sample origin: Catalonian (autonomous community) supermarkets and bio shops, Spain, sample year: November 2011–May 2012, country: Spain[594], *apple puree included baby foods, **organic

incidence: 5/17* **, conc. range: LOQ-10 µg/kg (1 sa.), 10–25 µg/kg (2 sa.), 25–50 µg/kg (1 sa.), and 50.2 µg/kg (1 sa.), ∅ conc.: 25.3 µg/kg, sample origin: Catalonian (autonomous community) supermarkets and bio-shops, Spain, sample year: November 2011–May 2012, country: Spain[594], *apple puree included baby foods, **without vitamin C

incidence: 1/29* **, conc.: 8.3 µg/kg (1 sa.), sample origin: Catalonian (autonomous community) supermarkets and bio-shops, Spain, sample year: November 2011–May 2012, country: Spain[594], *apple puree included baby foods, **with vitamin C

- Co-contamination: not reported
- Further contamination of plants and plant products other than cereals and their products reported in the present book/ article[594]: Juices (apple juice), PAT, literature[594]

For detailed information see the article.

incidence: 1/16* conc.: 2.9 µg/kg, sample origin: supermarkets and retail shops in Novi Sad (city), Vojvodina (province), Serbia, sample year: 2013–2015, country: Serbia/Portugal[904], *apple puree

incidence: 10/50*, conc. range: 1–5 µg/kg (8 sa.), 5–10 µg/kg (2 sa., maximum: 7.7 µg/kg), ∅ conc.: 3.5 µg/kg, sample origin: supermarkets and retail shops in Novi Sad (city), Vojvodina (province), Serbia, sample year: 2013–2015, country: Serbia/ Portugal[904], *multifruit puree

- Co-contamination: not reported
- Further contamination of plants and plant products other than cereals and their products reported in the present book/ article[904]: Juices, Juices (apple juice), PAT, literature[904]

For detailed information see the article.

incidence: 6/18* conc. range: 7.59–41.10 µg/kg, ∅ conc.: 16.2 µg/kg, sample origin: supermarkets and organic shops, Portugal, sample year: unknown, country: Portugal[905], *conventional tomato puree

incidence: 1/2*, conc.: 13.15 µg/kg, sample origin: supermarkets and organic shops, Portugal, sample year: unknown, country: Portugal[905], *organic tomato puree

- Co-contamination: not reported
- Further contamination of plants and plant products other than cereals and their products reported in the present book/ article[905]: Jellies and Juices (apple juice), Juices (tomato juice), PAT, literature[905]

For detailed information see the article.

Penicillium Toxins

Mycophenolic Acid

incidence: 1/1*, conc.: 31,000 µg/kg, sample origin: private households of staff members, Austria, sample year: unknown, country: Austria[850], *moldy apple puree

- Co-contamination: not reported
- Further contamination of plants and plant products other than cereals and their products reported in the present book/article[850]: Fruits (apple), Nuts (chestnut), AME and AOH, literature[850]; Fruits (lemon), MA, literature[850]; Jam, AOH and MA, literature[850]; Spices (garlic), BEA, FB_1, FB_2, and FB_3, literature[850]; Wine (red), AME, AOH, and MA, literature[850]; see also Fruits (lemon), Jam, Nuts (chestnut), Wine (red), chapter: Further Mycotoxins and Microbial Metabolites

see also **Fruits (apple)**

"QbA" see **Wine**

Quinoa see **Flour**

Raisins may contained the following mycotoxins:

Alternaria Toxins

Alternariol

incidence: 3/57, conc. range: 3.5–15.6 µg/kg, ∅ conc.: 8.9 µg/kg, sample origin: principally Xinjiang (province), purchased in supermarkets and local markets in Beijing (capital), China, sample year: 2016/2017, country: China[815]

- Co-contamination: For detailed information see the article.
- Further contamination of plants and plant products other than cereals and their products reported in the present book/article[815]: Dates, MA, PA, and TA, literature[815]; Fruits, dried, AME, AOH, MA, PAT, and TA, literature[815]; Raisins, AME, MA, OTA, OTB, and TA, literature[815]; see also Dates, Fruits, dried, chapter: Further Mycotoxins and Microbial Metabolites

Alternariol Methyl Ether

incidence: 11/57, conc. range: 0.3–13.5 µg/kg, ∅ conc.: 3.1 µg/kg, sample origin: principally Xinjiang (province), purchased in supermarkets and local markets in Beijing (capital), China, sample year: 2016/2017, country: China[815]

- Co-contamination: For detailed information see the article.
- Further contamination of plants and plant products other than cereals and their products reported in the present book/article[815]: Dates, MA, PA, and TA, literature[815]; Fruits, dried, AME, AOH, MA, PAT, and TA, literature[815]; Raisins, AOH, MA, OTA, OTB, and TA, literature[815]; see also Dates, Fruits, dried, chapter: Further Mycotoxins and Microbial Metabolites

Tenuazonic Acid

incidence: 20/57, conc. range: 6.9–594.4 µg/kg, ∅ conc.: 104.8 µg/kg, sample origin: principally Xinjiang (province), purchased in supermarkets and local markets in Beijing (capital), China, sample year: 2016/2017, country: China[815]

- Co-contamination: For detailed information see the article.
- Further contamination of plants and plant products other than cereals and their products reported in the present book/article[815]: Dates, MA, PA, and TA, literature[815]; Fruits, dried, AME, AOH, MA, PAT, and TA, literature[815]; Raisins, AME, AOH, MA, OTA, and OTB, literature[815]; see also Dates, Fruits, dried, chapter: Further Mycotoxins and Microbial Metabolites

Aspergillus Toxins

Aflatoxin B_1

incidence: 3/8, conc. range: 15–180 µg/kg, sample origin: major local retail outlets at Nainital (district), India, sample year: 1983–1985, country: India[7]

- Co-contamination: Often more than one mycotoxin was present in a sa.
- Further contamination of plants and plant products other than cereals and their products reported in the present book/article[7]: Coconuts, AFB_1, AFB_2, AFG_1, AFG_2, and CIT, literature[7]; Dates, AFB_1, AFB_2, AFG_1, and OTA, literature[7]; Grogannuts, AFB_1, AFB_2, AFG_1, AFG_2, and CIT, literature[7]; Nuts (almond), AFB_1, AFB_2, AFG_1, RBT, STG, and ZEA, literature[7]; Nuts (cashew nut), AFB_1, AFB_2, AFG_1, AFG_2, CIT, and OTA, literature[7]; Nuts (peanut), AFB_1, AFB_2, AFG_1, AFG_2, and OTA, literature[7]; Nuts (pistachio nut), AFB_1, AFB_2, AFG_1, AFG_2, and STG, literature[7]; Nuts (walnut), AFB_1, AFB_2, AFG_1, AFG_2, and RBT, literature[7]; Raisins, AFB_2, AFG_1, AFG_2, and OTA, literature[7]

incidence: 4/20, conc. range: ≤13.9 µg/kg, ∅ conc.: 10.7 µg/kg, sample origin: retail shops and local markets in Rabat-Salé (region), Morocco, sample year: January–October 2006, country: Spain/Morocco[352]

- Co-contamination: not reported
- Further contamination of plants and plant products other than cereals and their products reported in the present book/article[352]: Figs, Nuts (peanut), Nuts (pistachio nut), Nuts (walnut), AFB_1 and AF/S, literature[352]; Raisins, AF/S, literature[352]

incidence: 2/5*, conc. range: 1.68–2.35 µg/kg, ∅ conc.: 2.02 µg/kg, sample origin: Chile, Iran, and Turkey, sa. purchased in local shops in Poland, sample year: unknown, country: Poland[664], *organic sa. from Turkey without any mycotoxins

- Co-contamination: 2 sa. co-contaminated with AFB_1 and OTA
- Further contamination of plants and plant products other than cereals and their products reported in the present book/article[664]: Fruits (goji), Raisins, OTA, literature[664]

incidence: 1/22, conc.: 0.64 µg/kg, sample origin: supermarkets, shops, and market stalls in Mashhad (city), Khorasan (province), Iran, sample year: 2009–2011, country: Iran[726]

- Co-contamination: not reported
- Further contamination of plants and plant products other than cereals and their products reported in the present book/article[726]: Currants, AFB_1, AFB_2, AFG_1, and OTA, literature[726]; Raisins, OTA, literature[726]

incidence: 19/46*, conc. range: ≤9.67 µg/kg, sample origin: retail outlets, cafeterias, and hotels from major districts of Punjab (state), Pakistan, sample year: unknown, country: Pakistan[750], *ingredients for halva

- Co-contamination: not reported
- Further contamination of plants and plant products other than cereals and their products reported in the present book/article[750]: Halva, AFB_1 and AFM_1, literature[750]; Nuts (almond), Nuts (pistachio nut), Spices (cardamom), AFB_1, literature[750]

incidence: 1/6, conc.: 0.6 µg/kg, sample origin: Corinth (city), Peloponnese (peninsula), Greece, sample year: 2012, country: Greece[767]

- Co-contamination: 1 sa. (Corinth) co-contaminated with AFB_1 and OTA
- Further contamination of plants and plant products other than cereals and their products reported in the present book/article[767]: Currants, Sultanas, AFB_1 and OTA, literature[767]; Raisins, OTA, literature[767]

Aflotoxin B_2

incidence: 2/8, conc. range: 16–150 µg/kg, Ø conc.: 83 µg/kg, sample origin: major local retail outlets at Nainital (district), India, sample year: 1983–1985, country: India[7]

- Co-contamination: Often more than one mycotoxin was present in a sa.
- Further contamination of plants and plant products other than cereals and their products reported in the present book/article[7]: Coconuts, AFB_1, AFB_2, AFG_1, AFG_2, and CIT, literature[7]; Dates, AFB_1, AFB_2, AFG_1, and OTA, literature[7]; Grogannuts, AFB_1, AFB_2, AFG_1, AFG_2, and CIT, literature[7]; Nuts (almond), AFB_1, AFB_2, AFG_1, RBT, STG, and ZEA, literature[7]; Nuts (cashew nut) (AFB_1, AFB_2, AFG_1, AFG_2, CIT, and OTA, literature[7]; Nuts (peanut), AFB_1, AFB_2, AFG_1, AFG_2, and OTA, literature[7]; Nuts (pistachio nut), AFB_1, AFB_2, AFG_1, AFG_2, and STG, literature[7]; Nuts (walnut), AFB_1, AFB_2, AFG_1, AFG_2, and RBT, literature[7]; Raisins, AFB_1, AFG_1, AFG_2, and OTA, literature[7]

Aflotoxin G_1

incidence: 2/8, conc. range: 15–80 µg/kg, Ø conc.: 47.5 µg/kg, sample origin: major local retail outlets at Nainital (district), India, sample year: 1983–1985, country: India[7]

- Co-contamination: Often more than one mycotoxin was present in a sa.
- Further contamination of plants and plant products other than cereals and their products reported in the present book/article[7]: Coconuts, AFB_1, AFB_2, AFG_1, AFG_2, and CIT, literature[7]; Dates, AFB_1, AFB_2, AFG_1, and OTA, literature[7]; Grogannuts, AFB_1, AFB_2, AFG_1, AFG_2, and CIT, literature[7]; Nuts (almond), AFB_1, AFB_2, AFG_1, RBT, STG, and ZEA, literature[7]; Nuts (cashew nut), AFB_1, AFB_2, AFG_1, AFG_2, CIT, and OTA, literature[7]; Nuts (peanut), AFB_1, AFB_2, AFG_1, AFG_2, and OTA, literature[7]; Nuts (pistachio nut), AFB_1, AFB_2, AFG_1, AFG_2, and STG, literature[7]; Nuts (walnut), AFB_1, AFB_2, AFG_1, AFG_2, and RBT, literature[7]; Raisins, AFB_1, AFB_2, AFG_2, and OTA, literature[7]

Aflotoxin G_2

incidence: 1/8, conc.: 15 µg/kg, sample origin: major local retail outlets at Nainital (district), India, sample year: 1983–1985, country: India[7]

- Co-contamination: Often more than one mycotoxin was present in a sa.
- Further contamination of plants and plant products other than cereals and their products reported in the present book/article[7]: Coconuts, AFB_1, AFB_2, AFG_1, AFG_2, and CIT, literature[7]; Dates, AFB_1, AFB_2, AFG_1, and OTA, literature[7]; Grogannuts, AFB_1, AFB_2, AFG_1, AFG_2, and CIT, literature[7]; Nuts (almond), AFB_1, AFB_2, AFG_1, RBT, STG, and ZEA, literature[7]; Nuts (cashew nut), AFB_1, AFB_2, AFG_1, AFG_2, CIT, and OTA, literature[7]; Nuts (peanut), AFB_1, AFB_2, AFG_1, AFG_2, and OTA, literature[7]; Nuts (pistachio nut), AFB_1, AFB_2, AFG_1, AFG_2, and STG, literature[7]; Nuts (walnut), AFB_1, AFB_2, AFG_1, AFG_2, and RBT, literature[7]; Raisins, AFB_1, AFB_2, AFG_1, and OTA, literature[7]

incidence: 1/1, conc.: 0.19 µg/kg, sample origin: local supermarket in Okayama (prefecture), Japan, sample year: unknown, country: Japan[339]

- Co-contamination: not reported
- Further contamination of plants and plant products other than cereals and their products reported in the present book/article[339]: Coconuts, Fruits, dried, Seeds (sesame), AFB_1 and AFG_2, literature[339]; Figs, AFB_2 and AFG_2, literature[339]; Flour (soybean flour), AFB_1, literature[339]; Nuts and Spices, AFB_1, AFB_2, AFG_1, and AFG_2, literature[339]

Aflatoxin/s

incidence: 2/10, conc. range: 2.6–7.5 µg/kg*, Ø conc.: 5.05 µg/kg*, sample origin: retail shops and local markets in NWFP and northern areas, Pakistan, sample year: January–November 2009, country: Pakistan[17], *AFB_1, AFB_2, AFG_1, and AFG_2

- Co-contamination: not reported
- Further contamination of plants and plant products other than cereals and their products reported in the present book/article[17]: Dates, Figs, Fruits, dried, Kernels (apricot), Nuts (almond), Nuts (peanut), Nuts (pine nut), Nuts (pistachio nut), Nuts (walnut), AF/S, literature[17]

incidence: 4/20, conc. range: ≤13.9 µg/kg*, Ø conc.: 10.7 µg/kg*, sample origin: retail shops and local markets in Rabat-Salé (region), Morocco, sample year: January–October 2006, country: Spain/Morocco[352], *AFB_1, AFB_2, AFG_1, and AFG_2

- Co-contamination: not reported
- Further contamination of plants and plant products other than cereals and their products reported in the present book/article[352]: Figs, Nuts (peanut), Nuts (pistachio nut), Nuts (walnut), AFB_1 and AF/S, literature[352]; Raisins, AFB_1, literature[352]

incidence: 3/53, conc. range: ≤1 µg/kg* (2 sa.) and 1.64 µg/kg* (1 sa.), sample origin: vendors, Pakistan, sample year: 2012, country: Pakistan/Ireland[680], *AFB_1, AFB_2, AFG_1, and AFG_2

incidence: 2/51, conc. range: ≤1 µg/kg* (1 sa.) and 1.54 µg/kg* (1 sa.), Ø conc.: 0.81 µg/kg*, sample origin: vendors, Pakistan, sample year: 2013, country: Pakistan/Ireland[680], *AFB_1, AFB_2, AFG_1, and AFG_2

incidence: 3/66, conc. range: ≤1 µg/kg* (1 sa.) and 1–4 µg/kg* (2 sa., maximum 2.23 µg/kg*), sample origin: vendors, Pakistan, sample year: 2014, country: Pakistan/Ireland[680], *AFB_1, AFB_2, AFG_1, and AFG_2

- Co-contamination: not reported
- Further contamination of plants and plant products other than cereals and their products reported in the present book/article[680]: Raisins, OTA, literature[680]

incidence: 3/16, conc. range: 0.85–2.14 µg/kg*, sample origin: Tehran (capital), Iran, sample year: unknown, country: Iran[828], *total AFS

- Co-contamination: not reported
- Further contamination of plants and plant products other than cereals and their products reported in the present book/article[828]: Fruits, dried, AF/S, literature[828]

For detailed information see the article.

Aspergillus and ***Penicillium*** Toxins

Citrinin

incidence: 1/1, conc.: 0.5 µg/kg, sample origin: unknown, sample year: unknown, country: Germany[673]

- Co-contamination: 1 sa. co-contaminated with CIT and OTA
- Further contamination of plants and plant products other than cereals and their products reported in the present book/article[673]: Cocoa, AF/S and OTA, literature[673]; Raisins, OTA, literature[673]; Sultanas, CIT and OTA, literature[673]

Ochratoxin A

incidence: 1/8, conc.: pr, sample origin: major local retail outlets at Nainital (district), India, sample year: 1983–1985, country: India[7]

- Co-contamination: Often more than one mycotoxin was present in a sa.
- Further contamination of plants and plant products other than cereals and their products reported in the present book/article[7]: Coconuts, AFB_1, AFB_2, AFG_1, AFG_2, and CIT, literature[7]; Dates, AFB_1, AFB_2, AFG_1, and OTA, literature[7]; Grogannuts, AFB_1, AFB_2, AFG_1, AFG_2, and CIT, literature[7]; Nuts (almond), AFB_1, AFB_2, AFG_1, RBT, STG, and ZEA, literature[7]; Nuts (cashew nut), AFB_1, AFB_2, AFG_1, AFG_2, CIT, and OTA, literature[7]; Nuts (peanut), AFB_1, AFB_2, AFG_1, AFG_2, and OTA, literature[7]; Nuts (pistachio nut), AFB_1, AFB_2, AFG_1, AFG_2, and STG, literature[7]; Nuts (walnut), AFB_1, AFB_2, AFG_1, AFG_2, and RBT, literature[7]; Raisins, AFB_1, AFB_2, AFG_1, and AFG_2, literature[7]

incidence: 31/52, conc. range: ≤12.5 µg/kg, ∅ conc.: 0.93 µg/kg, sample origin: local supermarkets and small retail shops in seven major cities, Japan, sample year: summer 2004–winter 2007, country: Japan[85]

- Co-contamination: not reported
- Further contamination of plants and plant products other than cereals and their products reported in the present book/article[85]: Beans (soybean), Vegetables (asparagus), FB_1 and FB_2, literature[85]; Beverages, Chocolate, Cocoa, Coffee, Coffee beans, Spices (coriander), Wine, OTA, literature[85]

incidence: 1/1, conc.: 2.3 µg/kg, sample origin: supermarkets and fast-food chains in Quebec (city), Canada, sample year: September/October 2008, country: Canada[90]

incidence: 1/1, conc.: 0.17 µg/kg, sample origin: supermarkets and fast-food chains in Calgary (city), Canada, sample year: September/October 2009, country: Canada[90]

- Co-contamination: not reported
- Further contamination of plants and plant products other than cereals and their products reported in the present book/article[90]: Baby food, Beans, Beverages, Butter (peanut butter), Candies, Chocolate, Coffee, Extracts, Herbs, Ice cream, Nuts, Pies, Pizza, Products, Pudding, Seeds, Wine, OTA, literature[90]

incidence: 6/13, conc. range: 0.14–0.70 µg/kg, sample origin: unknown, sample year: unknown, country: Japan[112]

- Co-contamination: not reported
- Further contamination of plants and plant products other than cereals and their products reported in the present book/article[112]: Cassis, Juices, Wine, OTA, literature[112]; Cocoa, Coffee, Food, and Vinegar, OTA and OTB, literature[112]; Powder (cocoa powder), AFB_1, AFB_2, AFG_1, OTA, and OTB, literature[112]

incidence: 4/4, conc. range: 0.02–5.7 µg/kg, ∅ conc.: 1.56 µg/kg, sample origin: Warsaw (capital) market or from food producers, Poland, sample year: 2004, country: Poland[116]

incidence: 1/6*, conc. range: 0.02–0.08 µg/kg, sample origin: Warsaw (capital) market or from food producers, Poland, sample year: 2005, country: Poland[116], *raisins and fruit, dried

- Co-contamination: not reported
- Further contamination of plants and plant products other than cereals and their products reported in the present book/article[116]: Coffee, OTA, literature[116]; Food, AFB_1 and AFG_1, literature[116]; Nuts, Spices, AFB_1, AFB_2, AFG_1, and AFG_2, literature[116]

incidence: 10*/11, conc. range: 0.02–12.5 µg/kg, ∅ conc.: 1.73 µg/kg, sample origin: local supermarkets and small retail shops across Japan, sample year: summer 2004–winter 2005, country: Japan[146], *imported products

- Co-contamination: not reported
- Further contamination of plants and plant products other than cereals and their products reported in the present book/article[146]: Beans (green bean), Coffee, Wine, OTA, literature[146]; Butter (peanut butter), AFB_1, AFB_2, AFG_1, and AFG_2, literature[146]; Chocolate, AFB_1 and OTA, literature[146]

incidence: 1/1*, conc.: 4.2 µg/kg**, sample origin: stores, Taiwan, sample year: unknown, country: Taiwan/Russia[147], *black raisins, **CL-ELISA

- Co-contamination: not reported
- Further contamination of plants and plant products other than cereals and their products reported in the present book/article[147]: Beans (kidney beans), Berries (grape), Coffee, OTA, literature[147]

incidence: 2/7*, conc. range: 0.93–1.20 µg/kg, ∅ conc.: 1.07 µg/kg, sample origin: sales outles, Qatar, sample year: October 2002, country: Qatar[229], *grape raisins, dried

- Co-contamination: not reported
- Further contamination of plants and plant products other than cereals and their products reported in the present book/article[229]: Butter (peanut butter), Figs, Nuts (peanut), Nuts (pistachio nut), AF/S, literature[229]; Coffee, OTA, literature[229]; Spices, Spices (cayenne pepper, chilli), AF/S and OTA, literature[229]

incidence: 101/106*, conc. range: ≤21.40 µg/kg, sample origin: regions around Detmold, Giessen, Hamburg, Jena, Karlsruhe, Kiel, Kulmbach, Trier (cities), Germany, sample year: unknown, country: Germany[262], *raisins and currants

- Co-contamination: not reported
- Further contamination of plants and plant products other than cereals and their products reported in the present book/article[262]: Bars, Chocolate, Drinks (chocolate drink), Figs, Fruits, dried, Nut crème, Nuts, Nuts (hazelnut), Nuts (peanut), Powder (cocoa powder), Pudding, Seeds (linseed), Seeds (sesame), Seeds (sunflower), OTA, literature[262]

incidence: 98/101, conc. range: 0.1 µg/kg (5 sa.), 0.2–4.0 µg/kg (73 sa.), 4.1–10 µg/kg (13 sa.), 10.1–20 µg/kg (6 sa.), and 29.8 µg/kg (1 sa.), sample origin: 13 regions in UK, sample year: October/November 1998, country: UK[263]

- Co-contamination: not reported
- Further contamination of plants and plant products other than cereals and their products reported in the present book/article[263]: Chocolate, Currants, Figs, Juices (grape juice), Powder (cocoa powder), Sultanas, Wine (red), OTA, literature[263]

incidence: 18/20, conc. range: 0.125–6.2* µg/kg, ∅ conc.: 1.48 µg/kg, sample origin: imported from Greece, Iran, and Turkey purchased in Hungarian outlets, Hungary, sample year: unknown, country: Hungary[273], *Iranian sa.

- Co-contamination: not reported
- Further contamination of plants and plant products other than cereals and their products reported in the present book/article[273]: not reported

incidence: 4/20, conc. range: 4.6–9.9 µg/kg, ∅ conc.: 6.95 µg/kg, sample origin: imported from Australia, Chile, Greece, and the USA and purchased in unknown shops in Norwich and supermarkets, small shops, etc. across UK, sample year: unknown, country: UK[328]

- Co-contamination: not reported
- Further contamination of plants and plant products other than cereals and their products reported in the present book/ article[328]: Currants and Sultanas, OTA, literature[328]

incidence: 67/85, conc. range: ≤26.6 µg/kg, ∅ conc.: 2.29 µg/kg, sample origin: different countries and unknown collected in unknown retail locations across Canada, sample year: 1998–2000, country: Canada[329]

- Co-contamination: not reported
- Further contamination of plants and plant products other than cereals and their products reported in the present book/ article[329]: Currants, Sultanas, OTA, literature[329]

For detailed information see the article.

incidence: 10/11, conc. range: 0.02–0.25 µg/kg, ∅ conc.: 0.069 µg/kg, sample origin: popular markets in Rabat (capital), Morocco, sample year: unknown, country: Morocco/France[341]

- Co-contamination: not reported
- Further contamination of plants and plant products other than cereals and their products reported in the present book/ article[341]: Fruits, dried, Olives, OTA, literature[341]

incidence: 6/20*, conc. range: 0.05–4.95 µg/kg, ∅ conc.: 0.96 µg/kg, sample origin: retail shops in Salé (city) and Rabat (capital), Morocco, sample year: July/August 2005, country: Morocco/ Spain[364], *raisins, dried

- Co-contamination: not reported
- Further contamination of plants and plant products other than cereals and their products reported in the present book/ article[364]: Figs, Nuts (peanut), Nuts (walnut), OTA, literature[364]

incidence: 96/118*, conc. range: >0.1 µg/kg (91 sa.) and >10 µg/kg (5 sa., maximum 34.6 µg/kg), sample origin: local stores, Sweden, sample year: winter seasons 1999/2000 and 2001/2002, country: Sweden[598], *raisins and currants

- Co-contamination: not reported
- Further contamination of plants and plant products other than cereals and their products reported in the present book/ article[598]: not reported

For detailed information see the article.

incidence: 30/56*, conc. range: 0.34–15.34 µg/kg, ∅ conc.: 2.43 µg/kg, sample origin: supermarkets and stores, USA, sample year: 2012–2014, country: USA[599], *conventional

incidence: 18/53*, conc. range: 0.28–8.21 µg/kg, ∅ conc.: 2.09 µg/kg, sample origin: supermarkets and stores, USA, sample year: 2012–2014, country: USA[599], *organic

- Co-contamination: not reported
- Further contamination of plants and plant products other than cereals and their products reported in the present book/ article[599]: Dates, Figs, Nuts (pistachio nut), OTA, literature[599]

incidence: 2/5*, conc. range: 0.44–1.27 µg/kg, ∅ conc.: 0.86 µg/kg, sample origin: Chile, Iran, and Turkey, sa. purchased in local shops in Poland, sample year: unknown, country: Poland[664], *organic sa. from Turkey without any mycotoxins

- Co-contamination: 2 sa. co-contaminated with AFB_1 and OTA
- Further contamination of plants and plant products other than cereals and their products reported in the present book/ article[664]: Fruits (goji), OTA, literature[664]; Raisins, AFB_1, literature[664]

incidence: 1/1, conc.: 1.7 µg/kg, sample origin: unknown, sample year: unknown, country: Germany[673]

- Co-contamination: 1 sa. co-contaminated with CIT and OTA
- Further contamination of plants and plant products other than cereals and their products reported in the present book/ article[673]: Cocoa, AF/S and OTA, literature[673]; Raisins, CIT, literature[673]; Sultanas, CIT and OTA, literature[673]

incidence: 38/53, conc. range: ≤1 µg/kg (9 sa.), 1–10 µg/kg (26 sa.), and ≥10 µg/kg (3 sa., maximum 12.75 µg/kg), sample origin: vendors, Pakistan, sample year: 2012, country: Pakistan/Ireland[680]

incidence: 38/51, conc. range: ≤1 µg/kg (12 sa.), 1–10 µg/kg (24 sa.), and ≥10 µg/kg (2 sa., maximum 11.16 µg/kg), sample origin: vendors, Pakistan, sample year: 2013, country: Pakistan/Ireland[680]

incidence: 46/66, conc. range: ≤1 µg/kg (15 sa.), 1–10 µg/kg (28 sa.), and ≥10 µg/kg (3 sa., maximum 10.88 µg/kg), sample origin: vendors, Pakistan, sample year: 2014, country: Pakistan/Ireland[680]

- Co-contamination: not reported
- Further contamination of plants and plant products other than cereals and their products reported in the present book/ article[680]: Raisins, AF/S, literature[680]

incidence: 2/2, conc. range: LOD/LOQ-0.9 µg/kg (1 sa.) and 5.0 µg/kg (1 sa.), sample origin: Finland, sample year: 1998, country: EU[698]

incidence: 20/29, conc. range: LOD/LOQ-0.9 µg/kg (9 sa.), 1.0–2.9 µg/kg (9 sa.), 3.0–4.9 µg/kg (1 sa.), and 7.0 µg/kg (1 sa.), sample origin: Finland, sample year: 1999, country: EU[698]

incidence: 6/13, conc. range: LOD/LOQ-0.9 µg/kg (3 sa.), 1.0–2.9 µg/kg (2 sa.), and 4.3 µg/kg (1 sa.), sample origin: France, sample year: 1998/1999, country: EU[698]

- Co-contamination: not reported
- Further contamination of plants and plant products other than cereals and their products reported in the present book/ article[698]: Baby food, Beans (soybean), Chocolate, Coffee, Coffee beans, Cocoa, Currants, Herbs, Jam, Juices, Juices (grape juice), Fruits, dried, Legumes, Nuts, Oil seeds, Oils (olive oil), Peas, Peas (chickpea), Products (plant product), Spices, Spices (cayenne pepper, chilli), Spices (nutmeg), Spices (paprika), Spices (pepper), Sultanas, Tea, Vinegar, Wine, Wine (red), Wine (rosé), Wine (white), Wine, miscellaneous, OTA, literature[698]

incidence: 3/22, conc.: 0.8–4.9 µg/kg, ∅ conc.: 2.21 µg/kg, sample origin: supermarkets, shops, and market stalls in Mashhad (city), Khorasan (province), Iran, sample year: 2009–2011, country: Iran[726]

- Co-contamination: not reported
- Further contamination of plants and plant products other than cereals and their products reported in the present book/ article[726]: Currants, AFB_1, AFB_2, AFG_1, and OTA, literature[726]; Raisins, AFB_1, literature[726]

incidence: 6/10, conc. range: 0.16–5 µg/kg (6 sa., maximum: 4.10 µg/kg), sample origin: dried grape factories in Hamadan (province), Iran, sample year: October 2012–March 2013, country: Iran[742]

- Co-contamination: not reported

- Further contamination of plants and plant products other than cereals and their products reported in the present book/article[742]: Currants, Sultanas, OTA literature[742]

incidence: 4/4, conc. range: <0.8–10 µg/kg (3 sa.) and 103.3 µg/kg (1 sa.), sample origin: Crete (island, administrative unit), Greece, sample year: 2012, country: Greece[767]

incidence: 6/6, conc. range: <0.8–10 µg/kg (3 sa.) and 10–100 µg/kg (3 sa.), sample origin: Aigio and Corinth (town and city), Peloponnese (peninsula), Greece, sample year: 2012, country: Greece[767]

- Co-contamination: 1 sa. (Corinth) co-contaminated with AFB_1 and OTA; no further information available
- Further contamination of plants and plant products other than cereals and their products reported in the present book/article[767]: Currants, Sultanas, AFB_1 and OTA, literature[767]; Raisins, AFB_1, literature[767]

incidence: 2/10, conc. range: 0.75–6.48 µg/kg, Ø conc.: 3.62 µg/kg, sample origin: dried fruit store, Iran, sample year: unknown, country: Iran[783]

- Co-contamination: not reported
- Further contamination of plants and plant products other than cereals and their products reported in the present book/article[783]: not reported

incidence: 5/12, conc. range: ≤2.17 µg/kg, sample origin: 12 regions, Czech Republic, sample year: 2011–2013, country: Czech Republic[811]

- Co-contamination: not reported
- Further contamination of plants and plant products other than cereals and their products reported in the present book/article[811]: Beans, Chocolate, Coffee, Confectionery, Figs, Herbs and Spices, Juices, Licorice, Powder (cocoa powder), Products (ginger product), Spices (caraway), Spices (cayenne pepper, chilli), Spices (coriander), Spices (ginger), Spices (mustard), Spices (nutmeg), Spices (paprika), Spices (pepper), Tea, Wine (red), Wine (white), OTA, literature[811]

incidence: 4/4, conc. range: 1.3–31.2 µg/kg, Ø conc.: 9.6 µg/kg, sample origin: Achaia (Peloponnese region), Greece, sample year: September 2011, country: Italy/Greece[812]

incidence: 4/4, conc. range: 1.1–98.2 µg/kg, Ø conc.: 35.9 µg/kg, sample origin: Ilia (Peloponnese region), Greece, sample year: September 2011, country: Italy/Greece[812]

incidence: 2/6, conc. range: ≤1.2 µg/kg, sample origin: Corinthia (Peloponnese region), Greece, sample year: September 2011, country: Italy/Greece[812]

incidence: 13/18, conc. range: ≤22.3 µg/kg, sample origin: Messinia (Peloponnese region), Greece, sample year: September 2011, country: Italy/Greece[812]

incidence: 7/9, conc. range: ≤66.7 µg/kg, sample origin: Zante Island, Greece, sample year: September 2011, country: Italy/Greece[812]

- Co-contamination: not reported
- Further contamination of plants and plant products other than cereals and their products reported in the present book/article[812]: Raisins, FB_2, literature[812]

For detailed information see the article.

incidence: 11/57, conc. range: 0.2–8.8 µg/kg, Ø conc.: 1.9 µg/kg, sample origin: principally Xinjiang (province), purchased in supermarkets and local markets in Beijing (capital), China, sample year: 2016/2017, country: China[815]

- Co-contamination: For detailed information see the article.
- Further contamination of plants and plant products other than cereals and their products reported in the present book/article[815]: Dates, MA, PA, and TA, literature[815]; Fruits, dried, AME, AOH, MA, PAT, and TA, literature[815]; Raisins, AME, AOH, MA, OTB, and TA, literature[815]; see also Dates, Fruits, dried, chapter: Further Mycotoxins and Microbial Metabolites

see also **Currants, Fruits, dried,** and **Sultanas**

Ochratoxin B

incidence: 2/57, conc. range: 0.1–0.3 µg/kg, Ø conc.: 0.2 µg/kg, sample origin: principally Xinjiang (province), purchased in supermarkets and local markets in Beijing (capital), China, sample year: 2016/2017, country: China[815]

- Co-contamination: For detailed information see the article.
- Further contamination of plants and plant products other than cereals and their products reported in the present book/article[815]: Dates, MA, PA, and TA, literature[815]; Fruits, dried, AME, AOH, MA, PAT, and TA, literature[815]; Raisins, AME, AOH, MA, OTA, and TA, literature[815]; see also Dates, Fruits, dried, chapter: Further Mycotoxins and Microbial Metabolites

Fusarium Toxins

Fumonisin/s

Fumonisin B_1

incidence: 7/7, conc. range: 243.42–17,321.51 µg/kg, Ø conc.: 2924 µg/kg, sample origin: imported from China, India, Iran, South Africa, Turkey, USA, and unknown origin, sample year: unknown, country: Hungary[600]

- Co-contamination: 1 sa. co-contaminated with FB_1, iso-FB_1, FB_2, iso-$FB_{2,3}$, FB_3, 3-epi-FB_3, FB_4, 3-epi-FB_4, FB_5, and iso-FB_5, 1 sa. co-contaminated with FB_1, FB_2, FB_3, 3-epi-FB_3, FB_4, and 3-epi-FB_4; 3 sa. co-contaminated with FB_1, FB_2, FB_3, FB_4, and 3-epi-FB_4; 1 sa. co-contaminated with FB_1, FB_2, FB_3, and FB_4; 1 sa. co-contaminated with FB_1, FB_2, and FB_3
- Further contamination of plants and plant products other than cereals and their products reported in the present book/article[600]: Raisins, FB_2 and FB_3, literature[600]; see also Raisins, chapter: Further Mycotoxins and Microbial Metabolites

Fumonisin B_2

incidence: 7/7, conc. range: 157.14–11,515.78 µg/kg, Ø conc.: 2279 µg/kg, sample origin: imported from China, India, Iran, South Africa, Turkey, the USA, and unknown origin, sample year: unknown, country: Hungary[600]

- Co-contamination: 1 sa. co-contaminated with FB_1, iso-FB_1, FB_2, iso-$FB_{2,3}$, FB_3, 3-epi-FB_3, FB_4, 3-epi-FB_4, FB_5, and iso-FB_5; 1 sa. co-contaminated with FB_1, FB_2, FB_3, 3-epi-FB_3, FB_4, and 3-epi-FB_4; 3 sa. co-contaminated with FB_1, FB_2, FB_3, FB_4, and 3-epi-FB_4; 1 sa. co-contaminated with FB_1, FB_2, FB_3, and FB_4; 1 sa. co-contaminated with FB_1, FB_2, and FB_3
- Further contamination of plants and plant products other than cereals and their products reported in the present book/article[600]: Raisins, FB_1 and FB_3, literature[600]; see also Raisins, chapter: Further Mycotoxins and Microbial Metabolites

incidence: 10/21, conc. range: <LOQ-7.5 µg/kg, sample origin: imported from Chile, China, Greece, South Africa, Turkey, and the USA sample year: unknown, country: Denmark[601]

- Co-contamination: 9 sa. co-contaminated with FB_2 and FB_4 and 1 sa. contaminated solely with FB_2
- Further contamination of plants and plant products other than cereals and their products reported in the present book/article[601]: see also Raisins, chapter: Further Mycotoxins and Microbial Metabolites

For detailed information see the article.

incidence: 0/4, conc.: no contamination, sample origin: Achaia (Peloponnese region), Greece, sample year: September 2011, country: Italy/Greece[812]

incidence: 0/4, conc.: no contamination, sample origin: Ilia (Peloponnese region), Greece, sample year: September 2011, country: Italy/Greece[812]

incidence: 2/6, conc. range: ≤9.9 µg/kg, sample origin: Corinthia (Peloponnese region), Greece, sample year: September 2011, country: Italy/Greece[812]

incidence: 7/18, conc. range: ≤17.7 µg/kg, sample origin: Messinia (Peloponnese region), Greece, sample year: September 2011, country: Italy/Greece[812]

incidence: 3/9, conc. range: ≤25.5 µg/kg, sample origin: Zante Island, Greece, sample year: September 2011, country: Italy/Greece[812]

- Co-contamination: not reported
- Further contamination of plants and plant products other than cereals and their products reported in the present book/article[812]: Raisins, OTA, literature[812]

For detailed information see the article.

Fumonisin B_3

incidence: 7/7, conc. range: 50.27–3,716.03 µg/kg, Ø conc.: 684 µg/kg, sample origin: imported from China, India, Iran, South Africa, Turkey, USA, and unknown origin, sample year: unknown, country: Hungary[600]

- Co-contamination: 1 sa. co-contaminated with FB_1, iso-FB_1, FB_2, iso-$FB_{2,3}$, FB_3, 3-epi-FB_3, FB_4, 3-epi-FB_4, FB_5, and iso-FB_5; 1 sa. co-contaminated with FB_1, FB_2, FB_3, 3-epi-FB_3, FB_4, and 3-epi-FB_4; 3 sa. co-contaminated with FB_1, FB_2, FB_3, FB_4, and 3-epi-FB_4; 1 sa. co-contaminated with FB_1, FB_2, FB_3, and FB_4; 1 sa. co-contaminated with FB_1, FB_2, and FB_3
- Further contamination of plants and plant products other than cereals and their products reported in the present book/article[600]: Raisins, FB_1 and FB_2, literature[600]; see also Raisins, chapter: Further Mycotoxins and Microbial Metabolites

Penicillium Toxins

Mycophenolic Acid

incidence: 27/57, conc. range: 0.3–2647.3 µg/kg, Ø conc.: 130.7 µg/kg, sample origin: principally Xinjiang (province), purchased in supermarkets and local markets in Beijing (capital), China, sample year: 2016/2017, country: China[815]

- Co-contamination: For detailed information see the article.
- Further contamination of plants and plant products other than cereals and their products reported in the present book/article[815]: Dates, MA, PA, and TA, literature[815]; Fruits, dried, AME, AOH, MA, PAT, and TA, literature[815]; Raisins, AME, AOH, OTA, OTB, and TA, literature[815]; see also Dates, Fruits, dried, chapter: Further Mycotoxins and Microbial Metabolites

Rapeseeds see Oil seeds

Red kidney beans see Beans (kidney bean)

Red pepper see Spices (cayenne pepper, chilli)

Reserva see Wine (red)

Retsina see Wine, miscellaneous

Rhine Riesling see Wine (white)

Rice wine see Wine, miscellaneous

Riesling see Wine (white)

Rioja see Wine (red) and Wine (white)

Roero Arneis see Wine

Rootcrops may contained the following mycotoxins:

Aspergillus Toxins

Aflotoxin/s

incidence: 43/129*, Ø conc.: 3.8 µg/kg**, sample origin: greater Manila (capital) area, Philippines, sample year: unknown, country: Philippines[1], *rootcrops and products, **AF of pos sa.?

- Co-contamination: not reported
- Further contamination of plants and plant products other than cereals and their products reported in the present book/article[1]: Beans, Butter (peanut butter), Cassava, Cocoa, Coffee, Drinks (alcoholic drink), Nuts (cashew nut), Nuts (peanut), Products (coconut product), Spices (cayenne pepper, chilli), Sweet potatoes, Yams, AF/S, literature[1]

Roero Arneis see Wine

Safflower seeds see Seeds (safflower)

Sago see Starch (cassava starch)

Salep may contain the following mycotoxins:

Aspergillus Toxins

Aflatoxin/s

incidence: 6/7, conc. range: ≤1.562 µg/kg*, sample origin: groceries and supermarkets, Turkey, sample year: 2003, country: Turkey[119], *AFS total

- Co-contamination: not reported
- Further contamination of plants and plant products other than cereals and their products reported in the present book/article[119]: Asure, Beverages, Cocoa, Cream, Ice cream, Jello, Keskul, Muhallebi, Pudding, Sauces (chocolate sauce), Soups, AF/S, literature[119]

Salep is a flour made from tubers. Ingredients were sugar, thickener agent, salep flavor, and cinnamon.

Sauces may contain the following mycotoxins:

Fusarium Toxins

Fumonisin/s

incidence: 8/53*, conc. range: ≤17 µg/kg, sample origin: German market, Germany, sample year: 1999–2001, country: Germany[510], *seasoning sauces and pastes

- Co-contamination: not reported
- Further contamination of plants and plant products other than cereals and their products reported in the present book/article[510]: Products (soybean product), Vegetables (asparagus), FB/S, literature[510]

Sauces (apple sauce) may contain the following mycotoxins:

Alternaria Toxins

Alternariol

incidence: 6/10, conc. range: 1.2–2.0 µg/l, ∅ conc.: 1.5 µg/l, sample origin: retail shops, Hesse (federal state), Germany, sample year: 2009/2010, country: Germany[439]

- Co-contamination: not reported
- Further contamination of plants and plant products other than cereals and their products reported in the present book/article[439]: Juices (apple juice), Juice (tomato juice), Paste (tomato paste), Tomato ketchup, Wine (white), AOH, literature[439]

Sauces (chilli sauce) may contain the following mycotoxins:

Aspergillus Toxins

Aflatoxin/s

incidence: 10/26, conc. range: ≤54.5 µg/kg*, sample origin: retail markets of Faisalabad, Jhang, Lahore, and Toba Tek Singh (districts) in Punjab (state), Pakistan, sample year: unknown, country: Pakistan[751], *AFB_1, AFB_2, AFG_1, and AFG_2

incidence: 7/24, conc. range: ≤33.9 µg/kg*, sample origin: food restaurants in Faisalabad, Jhang, Lahore, and Toba Tek Singh (districts) in Punjab (state), Pakistan, sample year: unknown, country: Pakistan[751], *AFB_1, AFB_2, AFG_1, and AFG_2

- Co-contamination: not reported
- Further contamination of plants and plant products other than cereals and their products reported in the present book/article[751]: Sauces (chilli sauce), OTA, literature[751]; Spices (cayenne pepper, chilli), AF/S and OTA, literature[751]

Aspergillus and ***Penicillium*** Toxins

Ochratoxin A

incidence: 1/4, conc.: 3.3 µg/kg, sample origin: specialist food shops, UK, sample year: unknown, country: UK[16]

- Co-contamination: not reported
- Further contamination of plants and plant products other than cereals and their products reported in the present book/article[16]: Food, AF/S, DON, FUS-X, NIV, OTA, and ZEA, literature[16]; Nuts (almond), AF/S, literature[16]; Oils, Spices (fennel), AF/S and ZEA, literature[16]; Oils (sesame oil), OTA, literature[16]; Paste (curry paste), AF/S, FB/S, NIV, OTA, and ZEA, literature[16]; Pickles, AF/S, FB/S, OTA, and ZEA, literature[16]; Sauces (chilli sauce), NIV and ZEA, literature[16]; Spices, AF/S, FB/S, OTA, and ZEA, literature[16]; Spices (cayenne pepper, chilli), AF/S, DAS, DON, HT-2, OTA, and ZEA, literature[16]; Spices (coriander), AF/S, DON, OTA, and ZEA, literature[16]; Spices (curry), AF/S, DAS, FB/S, FUS-X, HT-2, NEO, NIV, OTA, T-2, and ZEA, literature[16]; Spices (garlic), DON, FUS-X, and NIV, literature[16]; Spices (ginger), AF/S, DON, NEO, NIV, OTA, and T-2, literature[16]; Spices (tandoori), AF/S, FB/S, NIV, OTA, and T-2, literature[16]

incidence: 5/26, conc. range: ≤24.1 µg/kg*, sample origin: retail markets of Faisalabad, Jhang, Lahore, and Toba Tek Singh (districts) in Punjab (state), Pakistan, sample year: unknown, country: Pakistan[751], *AFB_1, AFB_2, AFG_1, and AFG_2

incidence: 4/24, conc. range: ≤20.9 µg/kg*, sample origin: food restaurants in Faisalabad, Jhang, Lahore, and Toba Tek Singh (districts) in Punjab (state), Pakistan, sample year: unknown, country: Pakistan[751], *AFB_1, AFB_2, AFG_1, and AFG_2

- Co-contamination: not reported
- Further contamination of plants and plant products other than cereals and their products reported in the present book/article[751]: Sauces (chilli sauce), AF/S, literature[751]; Spices (cayenne pepper, chilli), AF/S and OTA, literature[751]

Fusarium Toxins

Type B Trichothecenes

Nivalenol

incidence: 1/4, conc.: 15 µg/kg, sample origin: specialist food shops, UK, sample year: unknown, country: UK[16]

- Co-contamination: not reported
- Further contamination of plants and plant products other than cereals and their products reported in the present book/article[16]: Food, AF/S, DON, FUS-X, NIV, OTA, and ZEA, literature[16]; Nuts (almond), AF/S, literature[16]; Oils, Spices (fennel), AF/S and ZEA, literature[16]; Oils (sesame oil), OTA, literature[16]; Paste (curry paste), AF/S, FB/S, NIV, OTA, and ZEA, literature[16]; Pickles, AF/S, FB/S, OTA, and ZEA, literature[16]; Sauces (chilli sauce), OTA and ZEA, literature[16]; Spices, AF/S, FB/S, OTA, and ZEA, literature[16]; Spices (cayenne pepper, chilli), AF/S, DAS, DON, HT-2, OTA, and ZEA, literature[16]; Spices (coriander), AF/S, DON, OTA, and ZEA, literature[16]; Spices (curry), AF/S, DAS, FB/S, FUS-X, HT-2, NEO, NIV, OTA, T-2, and ZEA, literature[16]; Spices (garlic), DON, FUS-X, and NIV, literature[16]; Spices (ginger), AF/S, DON, NEO, NIV, OTA, and T-2, literature[16]; Spices (tandoori), AF/S, FB/S, NIV, OTA, and T-2, literature[16]

Other ***Fusarium*** Toxins

Zearalenone

incidence: 1/4, conc.: 7.1 µg/kg, sample origin: specialist food shops, UK, sample year: unknown, country: UK[16]

- Co-contamination: not reported
- Further contamination of plants and plant products other than cereals and their products reported in the present book/article[16]: Food, AF/S, DON, FUS-X, NIV, OTA, and ZEA, literature[16]; Nuts (almond), AF/S, literature[16]; Oils, Spices (fennel), AF/S and ZEA, literature[16]; Oils (sesame oil), OTA, literature[16]; Paste (curry paste), AF/S, FB/S, NIV, OTA, and ZEA, literature[16]; Pickles, AF/S, FB/S, OTA, and ZEA, literature[16]; Sauces (chilli sauce), NIV and OTA, literature[16]; Spices, AF/S, FB/S, OTA, and ZEA, literature[16]; Spices (cayenne pepper, chilli), AF/S, DAS, DON, HT-2, OTA, and ZEA, literature[16]; Spices (coriander), AF/S, DON, OTA, and ZEA, literature[16]; Spices (curry), AF/S, DAS, FB/S, FUS-X, HT-2, NEO, NIV, OTA, T-2, and ZEA, literature[16]; Spices (garlic), DON, FUS-X, and NIV, literature[16]; Spices (ginger), AF/S, DON, NEO, NIV, OTA, and T-2, literature[16]; Spices (tandoori), AF/S, FB/S, NIV, OTA, and T-2, literature[16]

Sauces (chocolate sauce) may contain the following mycotoxins:

Aspergillus **Toxins**

Aflatoxin/s

incidence: 3/3, conc. range: 0.575–8.672 µg/kg*, Ø conc.: 4.131 µg/kg, sample origin: groceries and supermarkets, Turkey, sample year: 2003, country: Turkey[119], *AFS total

- Co-contamination: not reported
- Further contamination of plants and plant products other than cereals and their products reported in the present book/article[119]: Asure, Beverages, Cocoa, Cream, Ice cream, Jello, Keskul, Muhallebi, Pudding, Salep, Soups, AF/S, literature[119]

Chocolate sauce ingredients were sugar, cocoa, and milk powder.

Sauces (flour sauce) may contain the following mycotoxins:

Aspergillus **Toxins**

Aflatoxin/s

incidence: 13/22*, conc. range: 1–5 µg/kg** (1 sa.), 6–15 µg/kg** (8 sa.), 16–25 µg/kg** (3 sa.), >100 µg/kg** (1 sa.), sample origin: Keneba (village), Gambia, sample year: October 1988, country: France/England/Germany/USA/Gambia[118], *bukolo, **AF

- Co-contamination: not reported
- Further contamination of plants and plant products other than cereals and their products reported in the present book/article[118]: Food, Nuts (peanut), Sauces (leaf sauce), AF/S, literature[118]

Sauces (fruit sauce) see **Fruits (apple)**

Sauces (leaf-sauce) may contain the following mycotoxins:

Aspergillus **Toxins**

Aflatoxin/s

incidence: 3/3, conc. range: 21–34 µg/kg*, Ø conc.: 27 µg/kg, sample origin: Keneba (village), Gambia, sample year: September (end of rainy season) 1989 or 1992, country: UK/France/USA[101], *AFB_1, AFB_2, AFG_1, and AFG_2

- Co-contamination: not reported
- Further contamination of plants and plant products other than cereals and their products reported in the present book/article[101]: Sauces (peanut sauce), AF/S, literature[101]

incidence: 56/87*, conc. range: 1–5 µg/kg** (3 sa.), 6–15 µg/kg** (34 sa.), 16–25 µg/kg** (9 sa.), 26–50 µg/kg** (5 sa.), 51–100 µg/kg** (4 sa.), >100 µg/kg** (1 sa.), sample origin: Keneba (village), Gambia, sample year: October 1988, country: France/England/Germany/USA/Gambia[118], *kucha, **AF

- Co-contamination: not reported
- Further contamination of plants and plant products other than cereals and their products reported in the present book/article[118]: Food, Nuts (peanut), Sauces (flour sauce), AF/S, literature[118]

Sauces (peanut sauce) may contain the following mycotoxins:

Aspergillus **Toxins**

Aflatoxin B_1

incidence: 8/24, conc. range: 0.2–8.3 µg/kg, Ø conc.: 2.0 µg/kg, sample origin: available in New Zealand, sample year: 20010, country: New Zealand[277]

- Co-contamination: not reported
- Further contamination of plants and plant products other than cereals and their products reported in the present book/article[277]: Bars, AFB_1, AF/S, DON, NIV, and OTA, literature[277]; Butter (peanut butter), Confectionery, Food, Fruits, dried, Nuts, Nuts (peanut), Nuts (pistachio nut), AFB_1 and AF/S, literature[277]; Coffee, Dates, Spices (nutmeg), Spices (pepper), Vine fruits, dried, OTA, literature[277]; Figs, Spices (cayenne pepper, chilli), Spices (curry), Spices (ginger), Spices (paprika), AFB_1, AF/S, and OTA, literature[277]; Nuts (Brazil nut, paranut), AFB_1, literature[277]; Sauces (peanut sauce), AF/S, literature[277]

Aflatoxin/s

incidence: 18/20, conc. range: 19–943 µg/kg*, sample origin: Keneba (village), Gambia, sample year: September (end of rainy season) 1989 or 1992, country: UK/France/USA[101], *AFB_1, AFB_2, AFG_1, and AFG_2

- Co-contamination: not reported
- Further contamination of plants and plant products other than cereals and their products reported in the present book/article[101]: Sauces (leaf sauce), AF/S, literature[101]

incidence: 8/24, conc. range: 0.2–10.5 µg/kg*, Ø conc.: 2.7 µg/kg*, sample origin: available in New Zealand, sample year: 2010, country: New Zealand[277], *AFS total

- Co-contamination: not reported
- Further contamination of plants and plant products other than cereals and their products reported in the present book/article[277]: Bars, AFB_1, AF/S, DON, NIV, and OTA, literature[277]; Butter (peanut butter), Confectionery, Food, Fruits, dried, Nuts, Nuts (peanut), Nuts (pistachio nut), AFB_1 and AF/S, literature[277]; Coffee, Dates, Spices (nutmeg), Spices (pepper), Vine fruits, dried, OTA, literature[277]; Figs, Spices (cayenne pepper, chilli), Spices (curry), Spices (ginger), Spices (paprika), AFB_1, AF/S, and OTA, literature[277]; Nuts (Brazil nut, paranut), Sauces (peanut sauce), AFB_1, literature[277]

Sauces (peanut-chilli sauce) may contain the following mycotoxins:

Aspergillus **Toxins**

Aflatoxin B_1

incidence: 16/16, conc. range: 2.0–4.0 µg/kg (4 sa.), >4.0–≤230.0 µg/kg (12 sa.), sample origin: supermarkets and/or traditional markets in Yogyakarta (city), Indonesia, sample year: 1998/1999, country: Indonesia/Austria/Canada[106], *sambal pecel/sambal sate

- Co-contamination: not reported
- Further contamination of plants and plant products other than cereals and their products reported in the present book/article[106]: Butter (peanut butter), Cake (peanut), Food, Nuts (peanut), Products, AFB_1, literature[106]

incidence: 7/12, conc. range: 8–207 µg/kg, sample origin: traditional markets and supermarkets, Indonesia, sample year: 2001/2002, country: Austria/Indonesia/Thailand[226]

- Co-contamination: not reported
- Further contamination of plants and plant products other than cereals and their products reported in the present book/article[226]: Butter (peanut butter), Cake (peanut), Nuts (peanut), Snacks, AFB_1 and AF/S, literature[226]; Sauces (peanut chilli sauce), AF/S, literature[226]

Aflatoxin/s
incidence: 9/12, conc. range: 7–613 μg/kg, sample origin: traditional markets and supermarkets, Indonesia, sample year: 2001/2002, country: Austria/Indonesia/Thailand[226]
- Co-contamination: not reported
- Further contamination of plants and plant products other than cereals and their products reported in the present book/article[226]: Butter (peanut butter), Cake (peanut), Nuts (peanut), Snacks, AFB_1 and AF/S, literature[226]; Sauces (peanut chilli sauce), AFB_1, literature[226]

Sauces (pepper sauce) may contain the following mycotoxins:

Alternaria Toxins

Tenuazonic Acid
incidence: 4/4*, conc. range: 3–712 μg/kg, ∅ conc.: 218 μg/kg, sample origin: German market, Germany, sample year: unknown, country: Germany[781], *included normal, Ajvar and Calabrian pepper sauce
- Co-contamination: not reported
- Further contamination of plants and plant products other than cereals and their products reported in the present book/article[781]: Products (tomato product), TA, literature[781]

Sauces (pepper and barbecue sauce) may contain the following mycotoxins:

Aspergillus and *Penicillium* Toxins

Ochratoxin A
incidence: 7/50*, conc. range: ≤0.72 μg/l sample origin: different origins, Germany, sample year: unknown, country: Germany[110]
- Co-contamination: not reported
- Further contamination of plants and plant products other than cereals and their products reported in the present book/article[110]: Juices (blackcurrant juice), Juices (carrot juice), Juices (grape juice), Juices (tomato juice), Sauces (soy sauce), Spices (mustard), Tomato ketchup, Vinegar (apple), Vinegar (balsamico), Vinegar (wine), Wine (red), Wine (rosé), Wine (white), OTA, literature[110]

Sauces (soy sauce) may contain the following mycotoxins:

Aspergillus Toxins

Aflatoxin B_1
incidence: 1/1, conc.: 1.81 μg/kg, sample origin: sample origin: grocery markets in Busan, Daegu, Daejeon, Gangneung, and Gwangju (cities) as well as Seoul (capital), Korea, sample year: May 2004–June 2005, country: Korea[10]
- Co-contamination: not reported
- Further contamination of plants and plant products other than cereals and their products reported in the present book/article[10]: Butter (peanut butter), Nuts, Nuts (almond), Nuts (peanut), Nuts (pistachio nut), Paste (soybean paste), Spices (cayenne pepper, chilli), Spices (curry), Spices (pepper), AFB_1, literature[10]

incidence: 32/56, conc. range: 0.4–188 μg/kg, ∅ conc.: 16.6 μg/kg, sample origin: small markets in several urban Shanghai (city) areas, China, sample year: April 1988–April 1989, country: China/USA[132]
- Co-contamination: not reported
- Further contamination of plants and plant products other than cereals and their products reported in the present book/article[132]: Beans (kidney bean), Butter (peanut butter), Curd (bean curd), Nuts (peanut), Oils (rapeseed oil), Oils (soybean oil), Paste (bean paste), Tofu, Wine, miscellaneous, AFB_1, literature[132]

incidence: 2/10, conc. range: 20–170 μg/kg, ∅ conc.: 95 μg/kg, sample origin: markets around Bangkok (capital) area, Thailand, sample year: July 1983–April 1985, country: Thailand[133]
- Co-contamination: 1 sa. co-contaminated with AFB_1 and AFG_1; 1 sa. contaminated solely with AFB_1
- Further contamination of plants and plant products other than cereals and their products reported in the present book/article[133]: Beans, Sauces (soy sauce), AFG_1, literature[133]; Beans (soybean), Curd (bean curd), Nuts (peanut), AFB_1 and AFG_1, literature[133]

incidence: 34/53, conc. range: 0.31–12.5 μg/kg*, sample origin: unknown, sample year: unknown, country: China[402] (measured by TRFIA, only 3 sa. exceed 5.0 μg AFB_1/kg
- Co-contamination: not reported
- Further contamination of plants and plant products other than cereals and their products reported in the present book/article[402]: not reported

Aflatoxin G_1
incidence: 3/10, conc. range: 20–70 μg/kg, ∅ conc.: 36.7 μg/kg, sample origin: markets around Bangkok (capital) area, Thailand, sample year: July 1983–April 1985, country: Thailand[133]
- Co-contamination: 1 sa. co-contaminated with AFB_1 and AFG_1; 2 sa. contaminated solely with AFG_1
- Further contamination of plants and plant products other than cereals and their products reported in the present book/article[133]: Beans, AFG_1, literature[133]; Beans (soybean), Curd (bean curd), Nuts (peanut), AFB_1 and AFG_1, literature[133]; Sauces (soy sauce), AFB_1, literature[133]

Aflatoxin G_2
incidence: 1/149, conc.: pr, sample origin: factories, Taiwan, sample year: April–December 1974, country: Taiwan[590]
- Co-contamination: not reported
- Further contamination of plants and plant products other than cereals and their products reported in the present book/article[590]: Paste (soybean paste), AFB_1, AFB_2, AFG_1, and AFG_2, literature[590]

Aspergillus and *Penicillium* Toxins

Ochratoxin A
incidence: 2/15*, conc. range: ≤0.25 μg/l sample origin: different origins, Germany, sample year: unknown, country: Germany[110], *soja- and herbal sauce
- Co-contamination: not reported
- Further contamination of plants and plant products other than cereals and their products reported in the present book/article[110]: Juices (blackcurrant juice), Juices (carrot juice), Juices (grape juice), Juices (tomato juice), Sauces (pepper and barbecue sauce), Spices (mustard), Tomato ketchup, Vinegar (apple), Vinegar (balsamico), Vinegar (wine), Wine (red), Wine (rosé), Wine (white), OTA, literature[110]

incidence: 5/5, ∅ conc.: 6.82 μg/kg, sample origin: Japan, sample year: 1996, country: Japan[592]

- Co-contamination: not reported
- Further contamination of plants and plant products other than cereals and their products reported in the present book/article[592]: not reported

Fusarium Toxins

Type B Trichothecenes

Deoxynivalenol

incidence: 34/35, conc. range: 4.5–1,245.6 µg/kg, ∅ conc.: 142 µg/kg, sample origin: local markets in Guangzhou (city), Guangdong (province), China, sample year: unknown, country: China[882]

incidence: 5/5, conc. range: 30.5–238.3 µg/kg, ∅ conc.: 114 µg/kg, sample origin: imports from Japan purchased on local markets in Guangzhou (city), Guangdong (province), China, sample year: unknown, country: China[882]
- Co-contamination: not reported
- Further contamination of plants and plant products other than cereals and their products reported in the present book/article[882]: not reported

For detailed information see the article.

Sauces (tomato sauce) may contain the following mycotoxins:

Alternaria Toxins

Altenuene

incidence: 9/28, conc. range: <LOQ-12.1 µg/kg, sample origin: local supermarkets, Belgium, sample year: February 2013–February 2015, country: Belgium/Austria/Germany[514]
- Co-contamination: For detailed information see the article.
- Further contamination of plants and plant products other than cereals and their products reported in the present book/article[514]: Juices (apple juice), Juices (grape juice), TA, literature[514]; Juices (tomato juice), Tomato concentrate, ALT, AME, AOH, and TA, literature[514]; Sauces (tomato sauce), AME, AOH, and TA, literature[514]; see also Juices (tomato juice), Sauces (tomato sauce), and Tomato concentrate, chapter: Further Mycotoxins and Microbial Metabolites

For detailed information see the article.

Alternariol

incidence: 8/24*, conc. range: 4–10 µg/kg, sample origin: Swiss market, Switzerland, sample year: autumn and winter 2010, country: Switzerland[47], *tomato sauces and soup
- Co-contamination: For detailed information see the article.
- Further contamination of plants and plant products other than cereals and their products reported in the present book/article[47]: Fruits (tomato), Puree, Tomato ketchup, AME, AOH, and TA, literature[47]; Fruits, dried, ALT, AME, AOH, and TA, literature[47]; Sauces (tomato sauce), AME and TA, literature[47]; see also Fruits (tomato), Fruits, dried, Puree, chapter: Further Mycotoxins and Microbial Metabolites

incidence: 24/28, conc. range: <LOQ-41.6 µg/kg, sample origin: local supermarkets, Belgium, sample year: February 2013–February 2015, country: Belgium/Austria/Germany[514]
- Co-contamination: For detailed information see the article.
- Further contamination of plants and plant products other than cereals and their products reported in the present book/article[514]: Juices (apple juice), Juices (grape juice), TA, literature[514]; Juices (tomato juice), Tomato concentrate, ALT, AME, AOH, and TA, literature[514]; Sauces (tomato sauce), ALT, AME, and TA, literature[514]; see also Juices (tomato juice), Sauces (tomato sauce), Tomato concentrate, chapter: Further Mycotoxins and Microbial Metabolites

For detailed information see the article.

incidence: 4/8, conc. range: ≤25 µg/kg, ∅ conc.: 16 µg/kg, sample origin: Dutch survey program (2013) and retail shops across The Netherlands, sample year: 2013?, country: Netherlands[568]
- Co-contamination: 4 sa. co-contaminated with AME, AOH, and TA.
- Further contamination of plants and plant products other than cereals and their products reported in the present book/article[568]: Figs, Olives, Seeds (sunflower), TA, literature[568]; Fruits (apple), AOH, literature[568]; Sauces (tomato sauce), AME and TA, literature[568]; Wine (red), AOH and TA, literature[568]; see also Seeds (sunflower), chapter: Further Mycotoxins and Microbial Metabolites

incidence: 3/9, conc. range: ≤16 µg/kg, sample origin: retail stores, Netherlands, sample year: autumn 2014, country: Netherlands/Germany[906]
- Co-contamination: not reported
- Further contamination of plants and plant products other than cereals and their products reported in the present book/article[906]: Figs, Fruits (tomato), Juices (tomato juice), AOH and TA, literature[906]; Oils (sunflower oil), AME, literature[906]; Paste (tomato paste), Seeds (sunflower), Tomato ketchup, AME, AOH, and TA, literature[906]; Sauces (tomato sauce), AME and TA, literature[906]

Alternariol Methyl Ether

incidence: 7/24*, conc. range: 1–4 µg/kg, sample origin: Swiss market, Switzerland, sample year: autumn and winter 2010, country: Switzerland[47], *tomato sauces and soup
- Co-contamination: For detailed information see the article.
- Further contamination of plants and plant products other than cereals and their products reported in the present book/article[47]: Fruits (tomato), Puree, Tomato ketchup, AME, AOH, and TA, literature[47]; Fruits, dried, ALT, AME, AOH, and TA, literature[47]; Sauces (tomato sauce), AOH and TA, literature[47]; see also Fruits (tomato), Fruits, dried, Puree, chapter: Further Mycotoxins and Microbial Metabolites

incidence: 22/28, conc. range: <LOQ-3.8 µg/kg, sample origin: local supermarkets, Belgium, sample year: February 2013–February 2015, country: Belgium/Austria/Germany[514]
- Co-contamination: For detailed information see the article.
- Further contamination of plants and plant products other than cereals and their products reported in the present book/article[514]: Juices (apple juice), Juices (grape juice), TA, literature[514]; Juices (tomato juice), Tomato concentrate, ALT, AME, AOH, and TA, literature[514]; Sauces (tomato sauce), ALT, AOH, and TA, literature[514]; see also Juices (tomato juice), Sauces (tomato sauce), Tomato concentrate, chapter: Further Mycotoxins and Microbial Metabolites

For detailed information see the article.

incidence: 4/8, conc. range: ≤7.8 µg/kg, ∅ conc.: 3.8 µg/kg, sample origin: Dutch survey program (2013) and retail shops across The Netherlands, sample year: 2013?, country: Netherlands[568]
- Co-contamination: 4 sa. co-contaminated with AME, AOH, and TA.

- Further contamination of plants and plant products other than cereals and their products reported in the present book/article[568]: Figs, Olives, Seeds (sunflower), TA, literature[568]; Fruits (apple), AOH, literature[568]; Sauces (tomato sauce), Wine (red), AOH and TA, literature[568]; see also Seeds (sunflower), chapter: Further Mycotoxins and Microbial Metabolites

incidence: 2/9, conc. range: ≤1.6 μg/kg, sample origin: retail stores, Netherlands, sample year: autumn 2014, country: Netherlands/Germany[906]

- Co-contamination: not reported
- Further contamination of plants and plant products other than cereals and their products reported in the present book/article[906]: Figs, Fruits (tomato), Juices (tomato juice), Sauces (tomato sauce), AOH and TA, literature[906]; Oils (sunflower oil), AME, literature[906]; Paste (tomato paste), Seeds (sunflower), Tomato ketchup, AME, AOH, and TA, literature[906]

Tenuazonic Acid

incidence: 24/24*, conc. range: 4–144 μg/kg, ∅ conc.: 40 μg/kg, sample origin: Swiss market, Switzerland, sample year: autumn and winter 2010, country: Switzerland[47], *tomato sauces and soup

- Co-contamination: For detailed information see the article.
- Further contamination of plants and plant products other than cereals and their products reported in the present book/article[47]: Fruits (tomato), Puree, Tomato ketchup, AME, AOH, and TA, literature[47]; Fruits, dried, ALT, AME, AOH, and TA, literature[47]; Sauces (tomato sauce), AME and AOH, literature[47]; see also Fruits (tomato), Fruits, dried, Puree, chapter: Further Mycotoxins and Microbial Metabolites

incidence: 28/28, conc. range: 7.7–330.6 μg/kg, sample origin: local supermarkets, Belgium, sample year: February 2013–February 2015, country: Belgium/Austria/Germany[514]

- Co-contamination: For detailed information see the article.
- Further contamination of plants and plant products other than cereals and their products reported in the present book/article[514]: Juices (apple juice), Juices (grape juice), TA, literature[514]; Juices (tomato juice), Tomato concentrate, ALT, AME, AOH, and TA, literature[514]; Sauces (tomato sauce), ALT, AME, and AOH, literature[514]; see also Juices (tomato juice), Sauces (tomato sauce), Tomato concentrate, chapter: Further Mycotoxins and Microbial Metabolites

For detailed information see the article.

incidence: 8/8, conc. range: 66–462 μg/kg, ∅ conc.: 202 μg/kg, sample origin: Dutch survey program (2013) and retail shops across The Netherlands, sample year: 2013?, country: Netherlands[568]

- Co-contamination: 4 sa. co-contaminated with AME, AOH, and TA; 4 sa. contaminated solely with TA
- Further contamination of plants and plant products other than cereals and their products reported in the present book/article[568]: Figs, Olives, Seeds (sunflower), TA, literature[568]; Fruits (apple), AOH, literature[568]; Sauces (tomato sauce), AME and AOH, literature[568]; Wine (red), AOH and TA, literature[568]; see also Seeds (sunflower), chapter: Further Mycotoxins and Microbial Metabolites

incidence: 7/9, conc. range: ≤111 μg/kg, sample origin: retail stores, Netherlands, sample year: autumn 2014, country: Netherlands/Germany[906]

- Co-contamination: not reported
- Further contamination of plants and plant products other than cereals and their products reported in the present book/article[906]: Figs, Fruits (tomato), Juices (tomato juice), AOH and TA, literature[906]; Oils (sunflower oil), AME, literature[906]; Paste (tomato paste), Seeds (sunflower), Tomato ketchup, AME, AOH, and TA, literature[906]; Sauces (tomato sauce), AME and AOH, literature[906]

Sauvignon see Must

Sauvignon Blanc see Wine (white)

Seasonings see Spices

Seeds may contain the following mycotoxins:

Aspergillus and *Penicillium* Toxins

Ochratoxin A

incidence: 1/1*, conc.: 0.07 μg/kg, sample origin: supermarkets and fast-food chains in Quebec (city), Canada, sample year: September/October 2008, country: Canada[90], *seeds, shelled

incidence: 1/1*, conc.: 0.35 μg/kg, sample origin: supermarkets and fast-food chains in Calgary (city), Canada, sample year: September/October 2009, country: Canada[90], *seeds, shelled

- Co-contamination: not reported
- Further contamination of plants and plant products other than cereals and their products reported in the present book/article[90]: Baby food, Beans, Beverages, Butter (peanut butter), Candies, Chocolate, Coffee, Extracts, Herbs, Ice cream, Nuts, Pies, Pizza, Products, Pudding, Raisins, Wine, OTA, literature[90]

Seeds (bush mango) may contain the following mycotoxins:

Alternaria Toxins

Tenuazonic Acid

incidence: 26/40, conc. range: ≤80.0 μg/kg, ∅ conc.: 22.3 μg/kg, sample origin: local markets in Lagos (state), Nigeria, sample year: September/October 2014, country: Nigeria/Austria/Ireland[603]

- Co-contamination: For detailed information see the article.
- Further contamination of plants and plant products other than cereals and their products reported in the present book/article[603]: Seeds (bush mango), 3-NPA, AFB_1, AFB_2, AFG_1, AFM_1, BEA, CPA, HT-2, KA, MA, MON, NIV, OTA, OTB, STG, and T-2, literature[603]; Seeds (melon), 3-NPA, AFB_1, AFB_2, AFG_1, AFG_2, AFM_1, CPA, KA, OTA, and STG, literature[603]; see also Seeds (melon), Seeds (bush mango), chapter: Further Mycotoxins and Microbial Metabolites

For detailed information see the article.

Aspergillus Toxins

Aflatoxin B_1

incidence: 23/40, conc. range: ≤713.2 μg/kg, ∅ conc.: 68.1 μg/kg, sample origin: local markets in Lagos (state), Nigeria, sample year: September/October 2014, country: Nigeria/Austria/Ireland[603]

- Co-contamination: For detailed information see the article.
- Further contamination of plants and plant products other than cereals and their products reported in the present book/article[603]: Seeds (bush mango), 3-NPA, AFB_2, AFG_1, AFM_1, BEA, CPA, HT-2, KA, MA, MON, NIV, OTA, OTB, STG, T-2, and TA, literature[603]; Seeds (melon), 3-NPA, AFB_1, AFB_2, AFG_1, AFG_2, AFM_1, CPA, KA, OTA, and STG, literature[603]; see also Seeds (melon), Seeds (bush mango), chapter: Further Mycotoxins and Microbial Metabolites

For detailed information see the article.

incidence: 1/1, conc.: 11.71 μg/kg, sample origin: shops in Watt Market, Calabar (city), Nigeria, sample year: January 2013, country: Nigeria[639]

- Co-contamination: 1 sa. co-contaminated with AFB_1, AFB_2, AFG_1, and AFG_2
- Further contamination of plants and plant products other than cereals and their products reported in the present book/article[639]: Seeds (bush mango), AFB_2, AFG_1, and AFG_2, literature[639]; Seeds (melon), Spices (cayenne pepper, chilli), AFB_1, AFB_2, AFG_1, and AFG_2, literature[639]

Aflatoxin B_2

incidence: 10/40, conc. range: ≤105.7 μg/kg, ∅ conc.: 22.8 μg/kg, sample origin: local markets in Lagos (state), Nigeria, sample year: September/October 2014, country: Nigeria/Austria/Ireland[603]

- Co-contamination: For detailed information see the article.
- Further contamination of plants and plant products other than cereals and their products reported in the present book/article[603]: Seeds (bush mango), 3-NPA, AFB_1, AFG_1, AFM_1, BEA, CPA, HT-2, KA, MA, MON, NIV, OTA, OTB, STG, T-2, and TA, literature[603]; Seeds (melon), 3-NPA, AFB_1, AFB_2, AFG_1, AFG_2, AFM_1, CPA, KA, OTA, and STG, literature[603]; see also Seeds (melon), Seeds (bush mango), chapter: Further Mycotoxins and Microbial Metabolites

For detailed information see the article.

incidence: 1/1, conc.: 2.43 μg/kg, sample origin: shops in Watt Market, Calabar (city), Nigeria, sample year: January 2013, country: Nigeria[639]

- Co-contamination: 1 sa. co-contaminated with AFB_1, AFB_2, AFG_1, and AFG_2
- Further contamination of plants and plant products other than cereals and their products reported in the present book/article[639]: Seeds (bush mango), AFB_1, AFG_1, and AFG_2, literature[639]; Seeds (melon), Spices (cayenne pepper, chilli), AFB_1, AFB_2, AFG_1, and AFG_2, literature[639]

Aflatoxin G_1

incidence: 19/40, conc. range: ≤6.8 μg/kg, ∅ conc.: 3.0 μg/kg, sample origin: local markets in Lagos (state), Nigeria, sample year: September/October 2014, country: Nigeria/Austria/Ireland[603]

- Co-contamination: For detailed information see the article.
- Further contamination of plants and plant products other than cereals and their products reported in the present book/article[603]: Seeds (bush mango), 3-NPA, AFB_1, AFB_2, AFM_1, BEA, CPA, HT-2, KA, MA, MON, NIV, OTA, OTB, STG, T-2, and TA, literature[603]; Seeds (melon), 3-NPA, AFB_1, AFB_2, AFG_1, AFG_2, AFM_1, CPA, KA, OTA, and STG, literature[603]; see also Seeds (melon), Seeds (bush mango), chapter: Further Mycotoxins and Microbial Metabolites

For detailed information see the article.

incidence: 1/1, conc.: 1.85 μg/kg, sample origin: shops in Watt Market, Calabar (city), Nigeria, sample year: January 2013, country: Nigeria[639]

- Co-contamination: 1 sa. co-contaminated with AFB_1, AFB_2, AFG_1, and AFG_2
- Further contamination of plants and plant products other than cereals and their products reported in the present book/article[639]: Seeds (bush mango), AFB_1, AFB_2, and AFG_2, literature[639]; Seeds (melon), Spices (cayenne pepper, chilli), AFB_1, AFB_2, AFG_1, and AFG_2, literature[639]

Aflatoxin G_2

incidence: 1/1, conc.: 0.14 μg/kg, sample origin: shops in Watt Market, Calabar (city), Nigeria, sample year: January 2013, country: Nigeria[639]

- Co-contamination: 1 sa. co-contaminated with AFB_1, AFB_2, AFG_1, and AFG_2
- Further contamination of plants and plant products other than cereals and their products reported in the present book/article[639]: Seeds (bush mango), AFB_1, AFB_2, and AFG_1, literature[639]; Seeds (melon), Spices (cayenne pepper, chilli), AFB_1, AFB_2, AFG_1, and AFG_2, literature[639]

Aflatoxin M_1

incidence: 6/40, conc. range: ≤25.5 μg/kg, ∅ conc.: 8.2 μg/kg, sample origin: local markets in Lagos (state), Nigeria, sample year: September/October 2014, country: Nigeria/Austria/Ireland[603]

- Co-contamination: For detailed information see the article.
- Further contamination of plants and plant products other than cereals and their products reported in the present book/article[603]: Seeds (bush mango), 3-NPA, AFB_1, AFB_2, AFG_1, BEA, CPA, HT-2, KA, MA, MON, NIV, OTA, OTB, STG, T-2, and TA, literature[603]; Seeds (melon), 3-NPA, AFB_1, AFB_2, AFG_1, AFG_2, AFM_1, CPA, KA, OTA, and STG, literature[603]; see also Seeds (melon), Seeds (bush mango), chapter: Further Mycotoxins and Microbial Metabolites

For detailed information see the article.

Other *Aspergillus* Toxins

Sterigmatocystin

incidence: 16/40, conc. range: ≤8.2 μg/kg, ∅ conc.: 1.4 μg/kg, sample origin: local markets in Lagos (state), Nigeria, sample year: September/October 2014, country: Nigeria/Austria/Ireland[603]

- Co-contamination: For detailed information see the article.
- Further contamination of plants and plant products other than cereals and their products reported in the present book/article[603]: Seeds (bush mango), 3-NPA, AFB_1, AFB_2, AFG_1, AFM_1, BEA, CPA, HT-2, KA, MA, MON, NIV, OTA, OTB, T-2, and TA, literature[603]; Seeds (melon), 3-NPA, AFB_1, AFB_2, AFG_1, AFG_2, AFM_1, CPA, KA, OTA, and STG, literature[603]; see also Seeds (melon), Seeds (bush mango), chapter: Further Mycotoxins and Microbial Metabolites

For detailed information see the article.

Aspergillus and *Penicillium* Toxins

3-Nitropropionic Acid

incidence: 29/40, conc. range: ≤117.2 μg/kg, ∅ conc.: 20.9 μg/kg, sample origin: local markets in Lagos (state), Nigeria, sample year: September/October 2014, country: Nigeria/Austria/Ireland[603]

- Co-contamination: For detailed information see the article.
- Further contamination of plants and plant products other than cereals and their products reported in the present book/article[603]: Seeds (bush mango), AFB_1, AFB_2, AFG_1, AFM_1, BEA, CPA, HT-2, KA, MA, MON, NIV, OTA, OTB, STG, T-2, and TA, literature[603]; Seeds (melon), 3-NPA, AFB_1, AFB_2, AFG_1, AFG_2, AFM_1, CPA, KA, OTA, and STG, literature[603]; see also Seeds (melon), Seeds (bush mango), chapter: Further Mycotoxins and Microbial Metabolites

For detailed information see the article.

Cyclopiazonic Acid

incidence: 22/40, conc. range: ≤533.0 μg/kg, ∅ conc.: 59.0 μg/kg, sample origin: local markets in Lagos (state), Nigeria, sample year: September/October 2014, country: Nigeria/Austria/Ireland[603]

- Co-contamination: For detailed information see the article.
- Further contamination of plants and plant products other than cereals and their products reported in the present book/article[603]: Seeds (bush mango), 3-NPA, AFB_1, AFB_2, AFG_1, AFM_1, BEA, HT-2, KA, MA, MON, NIV, OTA, OTB, STG, T-2, and TA, literature[603]; Seeds (melon), 3-NPA, AFB_1, AFB_2, AFG_1, AFG_2, AFM_1, CPA, KA, OTA, and STG, literature[603]; see also Seeds (melon), Seeds (bush mango), chapter: Further Mycotoxins and Microbial Metabolites

For detailed information see the article.

Kojic Acid

incidence: 40/40, conc. range: ≤488,800.0 μg/kg, ∅ conc.: 46,160.9 μg/kg, sample origin: local markets in Lagos (state), Nigeria, sample year: September/October 2014, country: Nigeria/Austria/Ireland[603]

- Co-contamination: For detailed information see the article.
- Further contamination of plants and plant products other than cereals and their products reported in the present book/article[603]: Seeds (bush mango), 3-NPA, AFB_1, AFB_2, AFG_1, AFM_1, BEA, CPA, HT-2, MA, MON, NIV, OTA, OTB, STG, T-2, and TA, literature[603]; Seeds (melon), 3-NPA, AFB_1, AFB_2, AFG_1, AFG_2, AFM_1, CPA, KA, OTA, and STG, literature[603]; see also Seeds (melon), Seeds (bush mango), chapter: Further Mycotoxins and Microbial Metabolites

For detailed information see the article.

Ochratoxin A

incidence: 28/40, conc. range: ≤831.4 μg/kg, ∅ conc.: 43.4 μg/kg, sample origin: local markets in Lagos (state), Nigeria, sample year: September/October 2014, country: Nigeria/Austria/Ireland[603]

- Co-contamination: For detailed information see the article.
- Further contamination of plants and plant products other than cereals and their products reported in the present book/article[603]: Seeds (bush mango), 3-NPA, AFB_1, AFB_2, AFG_1, AFM_1, BEA, CPA, HT-2, KA, MA, MON, NIV, OTB, STG, T-2, and TA, literature[603]; Seeds (melon), 3-NPA, AFB_1, AFB_2, AFG_1, AFG_2, AFM_1, CPA, KA, OTA, and STG, literature[603]; see also Seeds (melon), Seeds (bush mango), chapter: Further Mycotoxins and Microbial Metabolites

For detailed information see the article.

Ochratoxin B

incidence: 7/40, conc. range: ≤129.7 μg/kg, ∅ conc.: 23.5 μg/kg, sample origin: local markets in Lagos (state), Nigeria, sample year: September/October 2014, country: Nigeria/Austria/Ireland[603]

- Co-contamination: For detailed information see the article.
- Further contamination of plants and plant products other than cereals and their products reported in the present book/article[603]: Seeds (bush mango), 3-NPA, AFB_1, AFB_2, AFG_1, AFM_1, BEA, CPA, HT-2, KA, MA, MON, NIV, OTA, STG, T-2, and TA, literature[603]; Seeds (melon), 3-NPA, AFB_1, AFB_2, AFG_1, AFG_2, AFM_1, CPA, KA, OTA, and STG, literature[603]; see also Seeds (melon), Seeds (bush mango), chapter: Further Mycotoxins and Microbial Metabolites

For detailed information see the article.

Fusarium Toxins

Depsipeptides

Beauvericin

incidence: 8/40, conc. range: ≤7.1 μg/kg, ∅ conc.: 3.7 μg/kg, sample origin: local markets in Lagos (state), Nigeria, sample year: September/October 2014, country: Nigeria/Austria/Ireland[603]

- Co-contamination: For detailed information see the article.
- Further contamination of plants and plant products other than cereals and their products reported in the present book/article[603]: Seeds (bush mango), 3-NPA, AFB_1, AFB_2, AFG_1, AFM_1, CPA, HT-2, KA, MA, MON, NIV, OTA, OTB, STG, T-2, and TA, literature[603]; Seeds (melon), 3-NPA, AFB_1, AFB_2, AFG_1, AFG_2, AFM_1, CPA, KA, OTA, and STG, literature[603]; see also Seeds (melon), Seeds (bush mango), chapter: Further Mycotoxins and Microbial Metabolites

For detailed information see the article.

Type A Trichothecenes

HT-2 Toxin

incidence: 7/40, conc. range: ≤104.4 μg/kg, ∅ conc.: 43.4 μg/kg, sample origin: local markets in Lagos (state), Nigeria, sample year: September/October 2014, country: Nigeria/Austria/Ireland[603]

- Co-contamination: For detailed information see the article.
- Further contamination of plants and plant products other than cereals and their products reported in the present book/article[603]: Seeds (bush mango), 3-NPA, AFB_1, AFB_2, AFG_1, AFM_1, BEA, CPA, KA, MA, MON, NIV, OTA, OTB, STG, T-2, and TA, literature[603]; Seeds (melon), 3-NPA, AFB_1, AFB_2, AFG_1, AFG_2, AFM_1, CPA, KA, OTA, and STG, literature[603]; see also Seeds (melon), Seeds (bush mango), chapter: Further Mycotoxins and Microbial Metabolites

For detailed information see the article.

T-2 Toxin

incidence: 18/40, conc. range: ≤112.6 μg/kg, ∅ conc.: 22.8 μg/kg, sample origin: local markets in Lagos (state), Nigeria, sample year: September/October 2014, country: Nigeria/Austria/Ireland[603]

- Co-contamination: For detailed information see the article.
- Further contamination of plants and plant products other than cereals and their products reported in the present book/article[603]: Seeds (bush mango), 3-NPA, AFB_1, AFB_2, AFG_1, AFM_1, BEA, CPA, HT-2, KA, MA, MON, NIV, OTA, OTB, STG, and TA, literature[603]; Seeds (melon), 3-NPA, AFB_1, AFB_2, AFG_1, AFG_2, AFM_1, CPA, KA, OTA, and STG, literature[603]; see also Seeds (melon), Seeds (bush mango), chapter: Further Mycotoxins and Microbial Metabolites

For detailed information see the article.

Type B Trichothecenes

Nivalenol

incidence: 5/40, conc. range: ≤3.8 μg/kg, ∅ conc.: 3.8 μg/kg, sample origin: local markets in Lagos (state), Nigeria, sample year: September/October 2014, country: Nigeria/Austria/Ireland[603]

- Co-contamination: For detailed information see the article.
- Further contamination of plants and plant products other than cereals and their products reported in the present book/article[603]: Seeds (bush mango), 3-NPA, AFB_1, AFB_2, AFG_1, AFM_1, BEA, CPA, HT-2, KA, MA, MON, OTA, OTB, STG, T-2, and TA,

literature[603]; Seeds (melon), 3-NPA, AFB_1, AFB_2, AFG_1, AFG_2, AFM_1, CPA, KA, OTA, and STG, literature[603]; see also Seeds (melon), Seeds (bush mango), chapter: Further Mycotoxins and Microbial Metabolites
For detailed information see the article.

Other ***Fusarium*** Toxins

Moniliformin

incidence: 9/40, conc. range: ≤602.2 µg/kg, Ø conc.: 73.6 µg/kg, sample origin: local markets in Lagos (state), Nigeria, sample year: September/October 2014, country: Nigeria/Austria/Ireland[603]

- Co-contamination: For detailed information see the article.
- Further contamination of plants and plant products other than cereals and their products reported in the present book/article[603]: Seeds (bush mango), 3-NPA, AFB_1, AFB_2, AFG_1, AFM_1, BEA, CPA, HT-2, MA, KA, NIV, OTA, OTB, STG, T-2, and TA, literature[603]; Seeds (melon), 3-NPA, AFB_1, AFB_2, AFG_1, AFG_2, AFM_1, CPA, KA, OTA, and STG, literature[603]; see also Seeds (melon), Seeds (bush mango), chapter: Further Mycotoxins and Microbial Metabolites

For detailed information see the article.

Penicillium Toxins

Mycophenolic Acid

incidence: 22/40, conc. range: ≤291.6 µg/kg, Ø conc.: 27.6 µg/kg, sample origin: local markets in Lagos (state), Nigeria, sample year: September/October 2014, country: Nigeria/Austria/Ireland[603]

- Co-contamination: For detailed information see the article.
- Further contamination of plants and plant products other than cereals and their products reported in the present book/article[603]: Seeds (bush mango), 3-NPA, AFB_1, AFB_2, AFG_1, AFM_1, BEA, CPA, HT-2, KA, MON, NIV, OTA, OTB, STG, T-2, and TA, literature[603]; Seeds (melon), 3-NPA, AFB_1, AFB_2, AFG_1, AFG_2, AFM_1, CPA, KA, OTA, and STG, literature[603]; see also Seeds (melon), Seeds (bush mango), chapter: Further Mycotoxins and Microbial Metabolites

For detailed information see the article.
see also **Ogbono**

Seeds (linseed) may contain the following mycotoxins:

Aspergillus Toxins

Aflatoxin B_1

incidence: 15/48, conc. range: <20 µg/kg (2 sa.), 21–100 µg/kg (7 sa.), 101–500 µg/kg (3 sa.), 501–1,000 µg/kg (3 sa., maximum: 875 µg/kg), sample origin: villages of Khagaria and Saharsa (districts) of North Bihar (state), India, sample year: August-October 1987, country: India[128]

- Co-contamination: not reported
- Further contamination of plants and plant products other than cereals and their products reported in the present book/article[128]: Beans (cowpea), Beans (mung bean), Field mustard, Lentils, Peas (chickpea), AFB_1, literature[128]

incidence: 13/198, conc. range: 18–460 µg/kg, sample origin: 16 districts of Bihar (state), India, sample year: February/March 1987, country: India[463]

- Co-contamination: 4 sa. co-contaminated with AFB_1 and AFB_2, 9 sa. contaminated solely with AFB_1
- Further contamination of plants and plant products other than cereals and their products reported in the present book/article[463]: Seeds (linseed), AFB_2, literature[463]

Aflatoxin B_2

incidence: 4/198, conc. range: pr, sample origin: 16 districts of Bihar (state), India, sample year: February/March 1987, country: India[463]

- Co-contamination: 4 sa. co-contaminated with AFB_1 and AFB_2
- Further contamination of plants and plant products other than cereals and their products reported in the present book/article[463]: Seeds (linseed), AFB_1, literature[463]

Aflatoxin/s

incidence: 46/105, conc. range: 120–810 µg/kg, sample origin: different areas of Bihar (state), India, sample year: March 1984–February 1985, country: India[524]

- Co-contamination: For detailed information see the article.
- Further contamination of plants and plant products other than cereals and their products reported in the present book/article[524]: For detailed information see the article.

Aspergillus and ***Penicillium*** Toxins

Ochratoxin A

incidence: 12/24, conc. range: ≤1.79 µg/kg, sample origin: regions around Detmold, Giessen, Hamburg, Jena, Karlsruhe, Kiel, Kulmbach, and Trier (cities), Germany, sample year: unknown, country: Germany[262]

- Co-contamination: not reported
- Further contamination of plants and plant products other than cereals and their products reported in the present book/article[262]: Bars, Chocolate, Drinks (chocolate drink), Figs, Fruits, dried, Nut crème, Nuts, Nuts (hazelnut), Nuts (peanut), Powder (cocoa powder), Pudding, Raisins, Seeds (sesame), Seeds (sunflower), OTA, literature[262]

Seeds (lupine) may contain the following mycotoxins:

Aspergillus Toxins

Aflatoxin/s

incidence: 2/13*, conc. range: 4.5–6.5 µg/kg**, sample origin: local markets in Karthoum (capital), Sudan, sample year: 1993–1996, country: Yemen/Oman[715], *for food and feed, **AFB_1 and AFB_2

- Co-contamination: not reported
- Further contamination of plants and plant products other than cereals and their products reported in the present book/article[715]: not reported

Seeds (melon) may contain the following mycotoxins:

Aspergillus Toxins

Aflatoxin B_1

incidence: 6/9, conc. range: 1.12–22.4 µg/kg, Ø conc.: 5.6 µg/kg, sample origin: markets in Lagos, Ogun, and Oyo (states), southwest Nigeria, sample year: March 2016, country: Nigeria/South Africa/Austria/Ireland[361]

- Co-contamination: not reported
- Further contamination of plants and plant products other than cereals and their products reported in the present book/article[361]: Beans (locust bean), 3-NPA. BEA, and CIT, literature[361]; Iru, CIT, OTA, and STG, literature[361]; Ogiri, 3-NPA, AFB_1, BEA, CIT, and STG, literature[361]; Seeds (melon), 3-NPA, AFB_2, BEA, and STG, literature[361]

incidence: 9/32, conc. range: 5 to <10 µg/kg (3 sa.), 10 to <15 µg/kg (4 sa.), >20 µg/kg (2 sa., maximum: 36 µg/kg), sample origin: open markets in villages in Ogun (state in the rain forest), Nigeria, sample year: March–June 2001, country: Nigeria[531]

incidence: 8/28, conc. range: 5 to <10 µg/kg (2 sa.), 10 to <15 µg/kg (3 sa.), 15–20 µg/kg (3 sa., maximum: 18 µg/kg), sample origin: open markets in towns in Ogun (state in the rain forest), Nigeria, sample year: March–June 2001, country: Nigeria[531]

incidence: 10/26, conc. range: 5 to <10 µg/kg (6 sa.), 10 to <15 µg/kg (2 sa.), >20 µg/kg (2 sa., maximum: 50 µg/kg), sample origin: open markets in villages in Oyo (state in the rain forest), Nigeria, sample year: March–June 2001, country: Nigeria[531]

incidence: 7/29, conc. range: 10 to <15 µg/kg (4 sa.), 15–20 µg/kg (2 sa.), 23 µg/kg (1 sa.), sample origin: open markets in towns in Oyo (state in the rain forest), Nigeria, sample year: March–June 2001, country: Nigeria[531]

incidence: 9/22, conc. range: 5 to <10 µg/kg (4 sa.), 10 to <15 µg/kg (2 sa.), 15–20 µg/kg (2 sa.), 26 µg/kg (1 sa.), sample origin: open markets in villages in Osun (state in the rain forest), Nigeria, sample year: March–June 2001, country: Nigeria[531]

incidence: 14/24, conc. range: 5 to <10 µg/kg (6 sa.), 10 to <15 µg/kg (4 sa.), 15–20 µg/kg (3 sa.), 26 µg/kg (1 sa.), sample origin: open markets in towns in Osun (state in the rain forest), Nigeria, sample year: March–June 2001, country: Nigeria[531]

incidence: 9/35, conc. range: 5 to <10 µg/kg (4 sa.), 10 to <15 µg/kg (2 sa.), 15–20 µg/kg (3 sa., maximum: 16 µg/kg), sample origin: open markets in villages in Kaduna (state in the savanna), Nigeria, sample year: March–June 2001, country: Nigeria[531]

incidence: 6/26, conc. range: 5 to <10 µg/kg (2 sa.), 10 to <15 µg/kg (3 sa.), 15 µg/kg (1 sa.), sample origin: open markets in towns in Kaduna (state in the savanna), Nigeria, sample year: March–June 2001, country: Nigeria[531]

incidence: 11/28, conc. range: 5 to <10 µg/kg (5 sa.), 10 to <15 µg/kg (3 sa.), 15–20 µg/kg (2 sa.), 25 µg/kg (1 sa.), sample origin: open markets in villages in Niger (state in the savanna), Nigeria, sample year: March–June 2001, country: Nigeria[531]

incidence: 5/25, conc. range: 5 to <10 µg/kg (1 sa.), 10 to <15 µg/kg (2 sa.), 15–20 µg/kg (2 sa., maximum: 17 µg/kg), sample origin: open markets in towns in Niger (state in the savanna), Nigeria, sample year: March–June 2001, country: Nigeria[531]

incidence: 8/23, conc. range: 5 to <10 µg/kg (3 sa.), 10 to <15 µg/kg (1 sa.), 15–20 µg/kg (2 sa.), >20 µg/kg (2 sa., maximum: 30 µg/kg), sample origin: open markets in villages in Bauchi (state in the savanna), Nigeria, sample year: March–June 2001, country: Nigeria[531]

incidence: 6/19, conc. range: 5 to <10 µg/kg (2 sa.), 10 to <15 µg/kg (2 sa.), 15–20 µg/kg (1 sa.), 28 µg/kg (1 sa.), sample origin: open markets in towns in Bauchi (state in the savanna), Nigeria, sample year: March–June 2001, country: Nigeria[531]

- Co-contamination: not reported
- Further contamination of plants and plant products other than cereals and their products reported in the present book/article[531]: not reported

incidence: 24/75, conc. range: 2.3–47.7 µg/kg, ∅ conc.: 15.3 µg/kg, sample origin: farmers' stores in humid forest, Nigeria, sample year: July–December 2000, country: Nigeria[532]

incidence: 13/62, conc. range: 2.3–25.1 µg/kg, ∅ conc.: 10.5 µg/kg, sample origin: farmers' stores in the Northern Guinea savanna, Nigeria, sample year: July–December 2000, country: Nigeria[532]

- Co-contamination: not reported
- Further contamination of plants and plant products other than cereals and their products reported in the present book/article[532]: not reported

For detailed information see the article.

incidence: 13/16, conc. range: ≤127.5 µg/kg, ∅ conc.: 37.5 µg/kg, sample origin: local markets in Lagos (state), Nigeria, sample year: September/October 2014, country: Nigeria/Austria/Ireland[603]

- Co-contamination: For detailed information see the article.
- Further contamination of plants and plant products other than cereals and their products reported in the present book/article[603]: Seeds (bush mango), 3-NPA, AFB_1, AFB_2, AFG_1, AFM_1, BEA, CPA, HT-2, KA, MA, MON, NIV, OTA, OTB, STG, T-2, and TA, literature[603]; Seeds (melon), 3-NPA, AFB_2, AFG_1, AFG_2, AFM_1, CPA, KA, OTA, and STG, literature[603]; see also Seeds (melon), Seeds (bush mango), chapter: Further Mycotoxins and Microbial Metabolites

For detailed information see the article.

incidence: 1/1, conc.: 0.82 µg/kg, sample origin: shops in Watt Market, Calabar (city), Nigeria, sample year: January 2013, country: Nigeria[639]

- Co-contamination: 1 sa. co-contaminated with AFB_1, AFB_2, AFG_1, and AFG_2
- Further contamination of plants and plant products other than cereals and their products reported in the present book/article[639]: Seeds (bush mango), Spices (cayenne pepper, chilli), AFB_1, AFB_2, AFG_1, and AFG_2, literature[639]; Seeds (melon), AFB_2, AFG_1, and AFG_2, literature[639]

incidence: 64/182*, conc. range: ≤55 µg/kg**, sample origin: Ogun, Osun, and Oyo (states) in the humid forest, Nigeria, sample year: unknown, country: Nigeria[668], *storage time for "egusi" melon seeds in different facilities 0–10 months; **after 6–8-month storage time

incidence: 49/179*, conc. range: ≤45 µg/kg**, sample origin: Bauchi, Kaduna, and Niger (states) in the savanna, Nigeria, sample year: unknown, country: Nigeria[668], *storage time for "egusi" melon seeds in different facilities 0–12 months, **after 10–12-month storage time

- Co-contamination: not reported
- Further contamination of plants and plant products other than cereals and their products reported in the present book/article[668]: not reported

For detailed information see the article.

incidence: 58/65, conc. range: 0.4–2 µg/kg (38 sa.), 2–5 µg/kg (8 sa.), 5–15 µg/kg (5 sa.), 15–50 µg/kg (5 sa.), 50–200 µg/kg (2 sa.), sample origin: Khorasan (province), Iran, sample year: unknown, country: Iran[727]

- Co-contamination: not reported
- Further contamination of plants and plant products other than cereals and their products reported in the present book/article[727]: Seeds (melon), AFB_2, AFG_1, and AFG_2, literature[727]

incidence: 22/22, conc. range: 0.1–2 µg/kg (7 sa.), 2.1–10 µg/kg (7 sa.), >10 µg/kg (8 sa., maximum: 66.5 µg/kg), ∅ conc.: 9.7 µg/kg, sample origin: retailers and households, Ireland and UK, sample year: September 2015, country: Ireland/Nigeria/UK/Italy[846]

- Co-contamination: For detailed information see the article.
- Further contamination of plants and plant products other than cereals and their products reported in the present book/article[846]: Seeds (melon), AFB_2, AFG_1, AFG_2, CIT, and OTA, literature[846]

For detailed information see the article.

Aflatoxin B_2

incidence: 2/9, conc. range: 0.75–5.30 µg/kg, ∅ conc.: 3 µg/kg, sample origin: markets in Lagos, Ogun, and Oyo (states), southwest Nigeria, sample year: March 2016, country: Nigeria/South Africa/Austria/Ireland[361]

- Co-contamination: not reported
- Further contamination of plants and plant products other than cereals and their products reported in the present book/article[361]: Beans (locust bean), 3-NPA. BEA, and CIT, literature[361]; Iru, CIT, OTA, and STG, literature[361]; Ogiri, 3-NPA, AFB_1, BEA, CIT, and STG, literature[361]; Seeds (melon), 3-NPA, AFB_1, BEA, and STG, literature[361]

incidence: 13/16, conc. range: ≤14.8 µg/kg, ∅ conc.: 4.7 µg/kg, sample origin: local markets in Lagos (state), Nigeria, sample year: September/October 2014, country: Nigeria/Austria/Ireland[603]

- Co-contamination: For detailed information see the article.
- Further contamination of plants and plant products other than cereals and their products reported in the present book/article[603]: Seeds (bush mango), 3-NPA, AFB_1, AFB_2, AFG_1, AFM_1, BEA, CPA, HT-2, KA, MA, MON, NIV, OTA, OTB, STG, T-2, and TA, literature[603]; Seeds (melon), 3-NPA, AFB_1, AFG_1, AFG_2, AFM_1, CPA, KA, OTA, and STG, literature[603]; see also Seeds (melon), Seeds (bush mango), chapter: Further Mycotoxins and Microbial Metabolites

For detailed information see the article.

incidence: 1/1, conc.: 0.05 µg/kg, sample origin: shops in Watt Market, Calabar (city), Nigeria, sample year: January 2013, country: Nigeria[639]

- Co-contamination: 1 sa. co-contaminated with AFB_1, AFB_2, AFG_1, and AFG_2
- Further contamination of plants and plant products other than cereals and their products reported in the present book/article[639]: Seeds (bush mango), Spices (cayenne pepper, chilli), AFB_1, AFB_2, AFG_1, and AFG_2, literature[639]; Seeds (melon), AFB_1, AFG_1, and AFG_2, literature[639]

incidence: 14/65, conc. range: 0.4–2 (7 sa.) µg/kg, 2–5 (5 sa.) µg/kg, 5–15 (2 sa.) µg/kg, sample origin: Khorasan (province), Iran, sample year: unknown, country: Iran[727]

- Co-contamination: not reported
- Further contamination of plants and plant products other than cereals and their products reported in the present book/article[727]: Seeds (melon), AFB_1, AFG_1, and AFG_2, literature[727]

incidence: 22/22, conc. range: 0.1–2 µg/kg (17 sa.), 2.1–10 µg/kg (4 sa.), 13.9 µg/kg (1 sa.), ∅ conc.: 1.5 µg/kg, sample origin: retailers and households, Ireland and UK, sample year: September 2015, country: Ireland/Nigeria/UK/Italy[846]

- Co-contamination: For detailed information see the article.
- Further contamination of plants and plant products other than cereals and their products reported in the present book/article[846]: Seeds (melon), AFB_1, AFG_1, AFG_2, CIT, and OTA, literature[846]

For detailed information see the article.

Aflatoxin G_1

incidence: 7/16, conc. range: ≤12.7 µg/kg, ∅ conc.: 9.9 µg/kg, sample origin: local markets in Lagos (state), Nigeria, sample year: September/October 2014, country: Nigeria/Austria/Ireland[603]

- Co-contamination: For detailed information see the article.
- Further contamination of plants and plant products other than cereals and their products reported in the present book/article[603]: Seeds (bush mango), 3-NPA, AFB_1, AFB_2, AFG_1, AFM_1, BEA, CPA, HT-2, KA, MA, MON, NIV, OTA, OTB, STG, T-2, and TA, literature[603]; Seeds (melon), 3-NPA, AFB_1, AFB_2, AFG_2, AFM_1, CPA, KA, OTA, and STG, literature[603]; see also Seeds (melon), Seeds (bush mango), chapter: Further Mycotoxins and Microbial Metabolites

For detailed information see the article.

incidence: 1/1, conc.: 0.39 µg/kg, sample origin: shops in Watt Market, Calabar (city), Nigeria, sample year: January 2013, country: Nigeria[639]

- Co-contamination: 1 sa. co-contaminated with AFB_1, AFB_2, AFG_1, and AFG_2
- Further contamination of plants and plant products other than cereals and their products reported in the present book/article[639]: Seeds (bush mango), Spices (cayenne pepper, chilli), AFB_1, AFB_2, AFG_1, and AFG_2, literature[639]; Seeds (melon), AFB_1, AFB_2, and AFG_2, literature[639]

incidence: 6/65, conc. range: 0.4–2 (4 sa.) µg/kg, 2–5 (2 sa.) µg/kg, sample origin: Khorasan (province), Iran, sample year: unknown, country: Iran[727]

- Co-contamination: not reported
- Further contamination of plants and plant products other than cereals and their products reported in the present book/article[727]: Seeds (melon), AFB_1, AFB_2, and AFG_2, literature[727]

incidence: 11/22, conc. range: 0.1–2 µg/kg (9 sa.), 2.1–10 µg/kg (2 sa.), ∅ conc.: 1.3 µg/kg, sample origin: retailers and households, Ireland and UK, sample year: September 2015, country: Ireland/Nigeria/UK/Italy[846]

- Co-contamination: For detailed information see the article.
- Further contamination of plants and plant products other than cereals and their products reported in the present book/article[846]: Seeds (melon), AFB_1, AFB_2, AFG_2, CIT, and OTA, literature[846]

For detailed information see the article.

Aflatoxin G_2

incidence: 3/16, conc. range: ≤6.7 µg/kg, ∅ conc.: 4.7 µg/kg, sample origin: local markets in Lagos (state), Nigeria, sample year: September/October 2014, country: Nigeria/Austria/Ireland[603]

- Co-contamination: For detailed information see the article.
- Further contamination of plants and plant products other than cereals and their products reported in the present book/article[603]: Seeds (bush mango), 3-NPA, AFB_1, AFB_2, AFG_1, AFM_1, BEA, CPA, HT-2, KA, MA, MON, NIV, OTA, OTB, STG, T-2, and TA, literature[603]; Seeds (melon), 3-NPA, AFB_1, AFB_2, AFG_1, AFM_1, CPA, KA, OTA, and STG, literature[603]; see also Seeds (melon), Seeds (bush mango), chapter: Further Mycotoxins and Microbial Metabolites

For detailed information see the article.

incidence: 1/1, conc.: 0.09 µg/kg, sample origin: shops in Watt Market, Calabar (city), Nigeria, sample year: January 2013, country: Nigeria[639]

- Co-contamination: 1 sa. co-contaminated with AFB_1, AFB_2, AFG_1, and AFG_2
- Further contamination of plants and plant products other than cereals and their products reported in the present book/article[639]: Seeds (bush mango), Spices (cayenne pepper, chilli), AFB_1, AFB_2, AFG_1, and AFG_2, literature[639]; Seeds (melon), AFB_1, AFB_2, and AFG_1, literature[639]

incidence: 1/65, conc.: 0.4–2 µg/kg, sample origin: Khorasan (province), Iran, sample year: unknown, country: Iran[727]

- Co-contamination: not reported
- Further contamination of plants and plant products other than cereals and their products reported in the present book/ article[727]: Seeds (melon), AFB_1, AFG_1, and AFG_1, literature[727]

incidence: 6/22, conc. range: 0.1–2 µg/kg (6 sa.), ∅ conc.: 0.2 µg/kg, sample origin: retailers and households, Ireland and UK, sample year: September 2015, country: Ireland/Nigeria/UK/Italy[846]

- Co-contamination: For detailed information see the article.
- Further contamination of plants and plant products other than cereals and their products reported in the present book/ article[846]: Seeds (melon), AFB_1, AFB_2, AFG_1, CIT, and OTA, literature[846]

For detailed information see the article.

Aflatoxin M_1

incidence: 4/16, conc. range: ≤2.0 µg/kg, ∅ conc.: 1.3 µg/kg, sample origin: local markets in Lagos (state), Nigeria, sample year: September/October 2014, country: Nigeria/Austria/Ireland[603]

- Co contamination: For detailed information see the article.
- Further contamination of plants and plant products other than cereals and their products reported in the present book/ article[603]: Seeds (bush mango), 3-NPA, AFB_1, AFB_2, AFG_1, AFM_1, BEA, CPA, HT-2, KA, MA, MON, NIV, OTA, OTB, STG, T-2, and TA, literature[603]; Seeds (melon), 3-NPA, AFB_1, AFB_2, AFG_1, AFG_2, CPA, KA, OTA, and STG, literature[603]; see also Seeds (melon), Seeds (bush mango), chapter: Further Mycotoxins and Microbial Metabolites

For detailed information see the article.

Aflatoxin/s

incidence: 2/4, conc. range: ≤29 µg/kg*, sample origin: imported to USA, sample year: 1986, country: USA[19], *AFB_1, AFB_2, AFG_1, and AFG_2

- Co-contamination: not reported
- Further contamination of plants and plant products other than cereals and their products reported in the present book/ article[19]: Butter (peanut butter), Candies (peanut candy), Candies (pistachio candy), Nuts, Nuts (almond), Nuts (Brazil nut, paranut), Nuts (peanut), Nuts (pecan nut), Nuts (pistachio nut), Nuts (walnut), Oils (peanut oil), Paste (peanut paste), Paste (pipian paste), Products (peanut product), Spices (cayenne pepper, chilli), Spices (ginger), Spices (nutmeg), AF/S, literature[19]

Other *Aspergillus* Toxins

Sterigmatocystin

incidence: 4/9, conc. range: 0.64–12.8 µg/kg, ∅ conc.: 3.98 µg/kg, sample origin: markets in Lagos, Ogun, and Oyo (states), Southwest Nigeria, sample year: March 2016, country: Nigeria/ South Africa/Austria/Ireland[361]

- Co-contamination: not reported
- Further contamination of plants and plant products other than cereals and their products reported in the present book/ article[361]: Beans (locust bean), 3-NPA. BEA, and CIT, literature[361]; Iru, CIT, OTA, and STG, literature[361]; Ogiri, 3-NPA, AFB_1, BEA, CIT, and STG, literature[361]; Seeds (melon), 3-NPA, AFB_1, AFB_2, and BEA, literature[361]

incidence: 13/16, conc. range: ≤2.8 µg/kg, ∅ conc.: 1.4 µg/kg, sample origin: local markets in Lagos (state), Nigeria, sample year: September/October 2014, country: Nigeria/Austria/Ireland[603]

- Co-contamination: For detailed information see the article.
- Further contamination of plants and plant products other than cereals and their products reported in the present book/ article[603]: Seeds (bush mango), 3-NPA, AFB_1, AFB_2, AFG_1, AFM_1, BEA, CPA, HT-2, KA, MA, MON, NIV, OTA, OTB, STG, T-2, and TA, literature[603]; Seeds (melon), 3-NPA, AFB_1, AFB_2, AFG_1, AFG_2, AFM_1, CPA, KA, and OTA, literature[603]; see also Seeds (melon), Seeds (bush mango), chapter: Further Mycotoxins and Microbial Metabolites

For detailed information see the article.

Aspergillus and *Penicillium* Toxins

3-Nitropropionic Acid

incidence: 2/9, conc. range: 177.9–1,472.4 µg/kg, ∅ conc.: 825.2 µg/kg, sample origin: markets in Lagos, Ogun, and Oyo (states), Southwest Nigeria, sample year: March 2016, country: Nigeria/ South Africa/Austria/Ireland[361]

- Co-contamination: not reported
- Further contamination of plants and plant products other than cereals and their products reported in the present book/ article[361]: Beans (locust bean), 3-NPA. BEA, and CIT, literature[361]; Iru, CIT, OTA, and STG, literature[361]; Ogiri, 3-NPA, AFB_1, BEA, CIT, and STG, literature[361]; Seeds (melon), AFB_1, AFB_2, BEA, and STG, literature[361]

incidence: 7/16, conc. range: ≤40.2 µg/kg, ∅ conc.: 14.2 µg/kg, sample origin: local markets in Lagos (state), Nigeria, sample year: September/October 2014, country: Nigeria/Austria/Ireland[603]

- Co-contamination: For detailed information see the article.
- Further contamination of plants and plant products other than cereals and their products reported in the present book/ article[603]: Seeds (bush mango), 3-NPA, AFB_1, AFB_2, AFG_1, AFM_1, BEA, CPA, HT-2, KA, MA, MON, NIV, OTA, OTB, STG, T-2, and TA, literature[603]; Seeds (melon), AFB_1, AFB_2, AFG_1, AFG_2, AFM_1, CPA, KA, OTA, and STG, literature[603]; see also Seeds (melon), Seeds (bush mango), chapter: Further Mycotoxins and Microbial Metabolites

For detailed information see the article.

Citrinin

incidence: 4/22, conc. range: 0.2–2 µg/kg (1 sa.), 2.1–10 µg/kg (3 sa., maximum: 3.6 µg/kg), ∅ conc.: 3.0 µg/kg, sample origin: retailers and households, Ireland and UK, sample year: September 2015, country: Ireland/Nigeria/UK/Italy[846]

- Co-contamination: For detailed information see the article.
- Further contamination of plants and plant products other than cereals and their products reported in the present book/ article[846]: Seeds (melon), AFB_1, AFB_2, AFG_1, AFG_2, and OTA, literature[846]

For detailed information see the article.

Cyclopiazonic Acid

incidence: 3/16, conc. range: ≤36.4 µg/kg, ∅ conc.: 26.9 µg/kg, sample origin: local markets in Lagos (state), Nigeria, sample year: September/October 2014, country: Nigeria/Austria/Ireland[603]

- Co-contamination: For detailed information see the article.
- Further contamination of plants and plant products other than cereals and their products reported in the present book/ article[603]: Seeds (bush mango), 3-NPA, AFB_1, AFB_2, AFG_1, AFM_1, BEA, CPA, HT-2, KA, MA, MON, NIV, OTA, OTB, STG, T-2, and TA, literature[603]; Seeds (melon), 3-NPA, AFB_1, AFB_2, AFG_1, AFG_2,

AFM$_1$, KA, OTA, and STG, literature[603]; see also Seeds (melon), Seeds (bush mango), chapter: Further Mycotoxins and Microbial Metabolites

For detailed information see the article.

Kojic Acid

incidence: 16/16, conc. range: ≤3,942.0 µg/kg, ∅ conc.: 1,288.5 µg/kg, sample origin: local markets in Lagos (state), Nigeria, sample year: September/October 2014, country: Nigeria/Austria/Ireland[603]

- Co-contamination: For detailed information see the article.
- Further contamination of plants and plant products other than cereals and their products reported in the present book/article[603]: Seeds (bush mango), 3-NPA, AFB$_1$, AFB$_2$, AFG$_1$, AFM$_1$, BEA, CPA, HT-2, KA, MA, MON, NIV, OTA, OTB, STG, T-2, and TA, literature[603]; Seeds (melon), 3-NPA, AFB$_1$, AFB$_2$, AFG$_1$, AFG$_2$, AFM$_1$, CPA, OTA, and STG, literature[603]; see also Seeds (melon), Seeds (bush mango), chapter: Further Mycotoxins and Microbial Metabolites

For detailed information see the article.

Ochratoxin A

incidence: 1/16, conc.: 0.6 µg/kg, sample origin: local markets in Lagos (state), Nigeria, sample year: September/October 2014, country: Nigeria/Austria/Ireland[603]

- Co-contamination: For detailed information see the article.
- Further contamination of plants and plant products other than cereals and their products reported in the present book/article[603]: Seeds (bush mango), 3-NPA, AFB$_1$, AFB$_2$, AFG$_1$, AFM$_1$, BEA, CPA, HT-2, KA, MA, MON, NIV, OTA, OTB, STG, T-2, and TA, literature[603]; Seeds (melon), 3-NPA, AFB$_1$, AFB$_2$, AFG$_1$, AFG$_2$, AFM$_1$, CPA, KA, and STG, literature[603]; see also Seeds (melon), Seeds (bush mango), chapter: Further Mycotoxins and Microbial Metabolites

For detailed information see the article.

incidence: 3/22, conc. range: 0.2–2 µg/kg (3 sa., maximum: 0.95 µg/kg), ∅ conc.: 0.6 µg/kg, sample origin: retailers and households, Ireland and UK, sample year: September 2015, country: Ireland/Nigeria/UK/Italy[846]

- Co-contamination: For detailed information see the article.
- Further contamination of plants and plant products other than cereals and their products reported in the present book/article[846]: Seeds (melon), AFB$_1$, AFB$_2$, AFG$_1$, AFG$_2$, and CIT, literature[846]

For detailed information see the article.

Fusarium Toxins

Depsipeptides

Beauvericin

incidence: 7/9, conc. range: 0.19–2.1 µg/kg, ∅ conc.: 1.09 µg/kg, sample origin: markets in Lagos, Ogun, and Oyo (states), Southwest Nigeria, sample year: March 2016, country: Nigeria/South Africa/Austria/Ireland[361]

- Co-contamination: not reported
- Further contamination of plants and plant products other than cereals and their products reported in the present book/article[361]: Beans (locust bean), 3-NPA. BEA, and CIT, literature[361]; Iru, CIT, OTA, and STG, literature[361]; Ogiri, 3-NPA, AFB$_1$, BEA, CIT, and STG, literature[361]; Seeds (melon), 3-NPA, AFB$_1$, AFB$_2$, and STG, literature[361]

Seeds (pumpkin) may contain the following mycotoxins:

Fusarium Toxins

Type A Trichothecenes

HT-2 Toxin

incidence: 1/5, conc.: 5 µg/kg, sample origin: food stores, Germany, sample year: 2000/2001, country: Germany[151]

- Co-contamination: not reported
- Further contamination of plants and plant products other than cereals and their products reported in the present book/article[151]: Beans, Seeds (pumpkin), Spices (cayenne pepper, chilli), ZEA, literature[151]; Nuts (hazelnut), Seeds (sunflower), HT-2, T-2, and ZEA, literature[151]; Products (potato product), DAS, HT-2, MAS, SCIRP, and ZEA, literature[151]

Other *Fusarium* Toxins

Zearalenone

incidence: 1/5, conc.: 4 µg/kg, sample origin: food stores, Germany, sample year: 2000/2001, country: Germany[151]

- Co-contamination: not reported
- Further contamination of plants and plant products other than cereals and their products reported in the present book/article[151]: Beans, Spices (cayenne pepper, chilli), ZEA, literature[151]; Nuts (hazelnut), Seeds (sunflower), HT-2, T-2, and ZEA, literature[151]; Products (potato product), DAS, HT-2, MAS, SCIRP, and ZEA, literature[151]; Seeds (pumpkin), HT-2, literature[151]

see also **Oil seeds**

Seeds (sesame) may contain the following mycotoxins:

Alternaria Toxins

Alternariol

incidence: 15/17, conc. range: 0.3–14 µg/kg, ∅ conc.: 3 µg/kg, sample origin: B'Ladi, Bokkos, Mekera, and Vwang (districts), Plateau State, Nigeria, sample year: April/May 2011, country: Nigeria/Austria[629]

- Co-contamination: For detailed information see the article.
- Further contamination of plants and plant products other than cereals and their products reported in the present book/article[629]: Seeds (sesame), 3-NPA, AME, BEA, CIT, DON, FA, KA, STG, TA, and ZEA, literature[629]; see also Seeds (sesame), chapter: Further Mycotoxins and Microbial Metabolites

Alternariol Methyl Ether

incidence: 16/17, conc. range: 0.2–4 µg/kg, ∅ conc.: 2 µg/kg, sample origin: B'Ladi, Bokkos, Mekera, and Vwang (districts), Plateau State, Nigeria, sample year: April/May 2011, country: Nigeria/Austria[629]

- Co-contamination: For detailed information see the article.
- Further contamination of plants and plant products other than cereals and their products reported in the present book/article[629]: Seeds (sesame), 3-NPA, AOH, BEA, CIT, DON, FA, KA, STG, TA, and ZEA, literature[629]; see also Seeds (sesame), chapter: Further Mycotoxins and Microbial Metabolites

Tenuazonic Acid

incidence: 16/17, conc. range: 2–40 µg/kg, ∅ conc.: 16 µg/kg, sample origin: B'Ladi, Bokkos, Mekera, and Vwang (districts), Plateau State, Nigeria, sample year: April/May 2011, country: Nigeria/Austria[629]

- Co-contamination: For detailed information see the article.
- Further contamination of plants and plant products other than cereals and their products reported in the present book/article[629]: Seeds (sesame), 3-NPA, AME, AOH, BEA, CIT, DON, FA, KA, STG, and ZEA, literature[629]; see also Seeds (sesame), chapter: Further Mycotoxins and Microbial Metabolites

Aspergillus Toxins

Aflatoxin B_1

incidence: 2/19, conc. range: 0.6–2.4 µg/kg, ∅ conc.: 1.5 µg/kg, sample origin: commercially available market basket sa. of foods and foodstuffs in Tokyo (capital), Japan, sample year: 1986–1990, country: Japan[99]

- Co contamination: not reported
- Further contamination of plants and plant products other than cereals and their products reported in the present book/article[99]: Beans, Sugar, AFB_1 and AFB_2, literature[99]; Food, Jam (bean jam), Paste (peanut paste), AFB_1, literature[99]; Nuts (Brazil nut, paranut), Nuts (peanut), Nuts (pistachio nut), Spices (nutmeg), AFB_1, AFB_2, AFG_1, and AFG_2, literature[99]; Seeds (sesame), AFB_2, literature[99]; Spices, Spices (cayenne pepper, chilli), Spices (pepper), AFB_1, AFB_2, and AFG_1, literature[99]; Spices (paprika), AFB_1 and AFG_1, literature[99]

incidence: 1/6*, conc.: 5 µg/kg, sample origin: silos, homes, and small shops in different villages on the southern coast, Guatemala, sample year: 1976, country: Guatemala[136], *sa. stored for 6 months during dry season

- Co-contamination: not reported
- Further contamination of plants and plant products other than cereals and their products reported in the present book/article[136]: Beans (black bean), AF/S, literature[136]; Coffee, AFB_1, literature[136]

incidence: 7/8, conc. range: 0.54–1.82 µg/kg, ∅ conc.: 0.90 µg/kg, sample origin: supermarkets in Penang (state), Malaysia, sample year: unknown, country: Malaysia[182]

- Co-contamination: not reported
- Further contamination of plants and plant products other than cereals and their products reported in the present book/article[182]: Nuts, Nuts (peanut), Seeds (sunflower), Spices (cayenne pepper, chilli), Spices (cumin), Spices (pepper), AFB_1, literature[182]

incidence: 4?/211, conc. range: 2 µg/kg, sample origin: sites of primary storage or processing (domestic products) and sites of importation, Cyprus, sample year: 1992–1996, country: Cyprus[200]

- Co-contamination: not reported
- Further contamination of plants and plant products other than cereals and their products reported in the present book/article[200]: Butter (peanut butter), Figs, Nuts (peanut), AFB_1, AFB_2, AFG_1, and AFG_2, literature[200]; Nuts (Brazil nut, paranut), Nuts (walnut), AFB_1, AFB_2, and AFG_1, literature[200]; Nuts (chestnut), Products (sesame product), Spices, AFB_1, literature[200]; Nuts (pistachio nut), AFB_1 and AFB_2, literature[200]; Seeds (sesame), AFG_1 and AFG_2, literature[200]

incidence: 4/19*, conc. range: 4–10 µg/kg, sample origin: unknown, sample year: unknown, country: Germany[239], *sesame seeds, moldy

- Co-contamination: not reported
- Further contamination of plants and plant products other than cereals and their products reported in the present book/article[239]: Fruits (apple), Fruits (cherry), Fruits (lemon), Fruits (orange), Fruits (peach), Fruits (tomato), Nuts (almond), Nuts (cashew nut), Nuts (hazelnut), Nuts (walnut), Spice (nutmeg), AFB_1, literature[239]; Nuts (peanut), AFB_1, AFB_2, AFG_1, and AFG_2, literature[239]

incidence: 1/1*, conc.: 0.54 µg/kg, sample origin: local supermarket in Okayama (prefecture), Japan, sample year: unknown, country: Japan[339], *white sesame

- Co-contamination: 1 sa. co-contaminated with AFB_1 and AFG_2
- Further contamination of plants and plant products other than cereals and their products reported in the present book/article[339]: Coconuts, Fruits, dried, AFB_1 and AFG_2, literature[339]; Figs, AFB_2 and AFG_2, literature[339]; Flour (soybean flour), AFB_1, literature[339]; Nuts, Spices, AFB_1, AFB_2, AFG_1, and AFG_2, literature[339]; Seeds (sesame), Raisins, AFG_2, literature[339]

incidence: 23/30*, conc. range: 0.1–14.5 µg/kg, sample origin: mostly imported from other countries and purchased on the Athens (capital) market, Greece, sample year: autumn 2014, country: Greece[397], *sesame seeds and sesame products

- Co-contamination: not reported
- Further contamination of plants and plant products other than cereals and their products reported in the present book/article[397]: not reported

incidence: 33?/182*, conc. range: 0.1–0.5 µg/kg (31 sa.), 0.5–1 µg/kg (16 sa.), 1–2 µg/kg (7 sa.), 2–6 µg/kg (9 sa., maximum: 5.54 µg/kg), sample origin: Khorasan (province), Iran, sample year: September 2009–August 2010, country: Iran[677], *sesame seeds, unprocessed

- Co-contamination: not reported
- Further contamination of plants and plant products other than cereals and their products reported in the present book/article[677]: Seeds (sesame), AFB_2, AFG_1, and AFG_2, literature[677]

incidence: 91*/269, conc. range: 1–2 µg/kg (58 sa.), 2–3 µg/kg (17 sa.), 3–4 µg/kg (7 sa.), 4–5 µg/kg (4 sa.), >5 µg/kg (5 sa., maximum: 40.57 µg/kg), sample origin: imported from Afghanistan to Khorasan (province), Iran, sample year: April 2008–January 2009, country: Iran[745], *only sa. contaminated >1 µg/kg are considered

- Co-contamination: not reported
- Further contamination of plants and plant products other than cereals and their products reported in the present book/article[745]: Seeds (sesame), AFB_2 and AFG_1, literature[745]

Aflatoxin B_2

incidence: 2/19, conc. range: 0.2–0.5 µg/kg, ∅ conc.: 0.4 µg/kg, sample origin: commercially available market basket sa. of foods and foodstuffs in Tokyo (capital), Japan, sample year: 1986–1990, country: Japan[99]

- Co-contamination: not reported
- Further contamination of plants and plant products other than cereals and their products reported in the present book/article[99]: Beans, Sugar, AFB_1 and AFB_2, literature[99]; Food, Jam (bean jam), Paste (peanut paste), AFB_1, literature[99]; Nuts (Brazil nut, paranut), Nuts (peanut), Nuts (pistachio nut), Spices (nutmeg), AFB_1, AFB_2, AFG_1, and AFG_2, literature[99]; Seeds (sesame), AFB_1, literature[99]; Spices, Spices (cayenne pepper, chilli), Spices (pepper), AFB_1, AFB_2, and AFG_1, literature[99]; Spices (paprika), AFB_1 and AFG_1, literature[99]

incidence: 8?/182*, conc. range: 0.1–0.5 µg/kg (13 sa., maximum: 0.42 µg/kg), sample origin: Khorasan (province), Iran, sample year: September 2009–August 2010, country: Iran[677], *sesame seeds, unprocessed

- Co-contamination: not reported

- Further contamination of plants and plant products other than cereals and their products reported in the present book/article[677]: Seeds (sesame), AFB_1, AFG_1, and AFG_2, literature[677]

incidence: 4*/269, conc. range: 1–2 µg/kg (1 sa.), 2–3 µg/kg (1 sa.), >5 µg/kg (2 sa., maximum: 5.71 µg/kg), sample origin: imported from Afghanistan to Khorasan (province), Iran, sample year: April 2008–January 2009, country: Iran[745], *only sa. contaminated >1 µg/kg are considered

- Co-contamination: not reported
- Further contamination of plants and plant products other than cereals and their products reported in the present book/article[745]: Seeds (sesame), AFB_1 and AFG_1, literature[745]

Aflatoxin G_1

incidence: 4?/211, conc. range: <0.4 µg/kg, sample origin: sites of primary storage or processing (domestic products) and sites of importation, Cyprus, sample year: 1992–1996, country: Cyprus[200]

- Co-contamination: not reported
- Further contamination of plants and plant products other than cereals and their products reported in the present book/article[200]: Butter (peanut butter), Figs, Nuts (peanut), AFB_1, AFB_2, AFG_1, and AFG_2, literature[200]; Nuts (Brazil nut, paranut), Nuts (walnut), AFB_1, AFB_2, and AFG_1, literature[200]; Nuts (chestnut), Products (sesame product), Spices, AFB_1, literature[200]; Nuts (pistachio nut), AFB_1 and AFB_2, literature[200]; Seeds (sesame), AFB_1 and AFG_2, literature[200]

incidence: 20/20, conc. range: 0.06–2.04 µg/kg, ∅ conc.: 0.75 µg/kg, sample origin: markets in Ankara (capital), Turkey, sample year: unknown, country: Turkey[221]

- Co-contamination: not reported
- Further contamination of plants and plant products other than cereals and their products reported in the present book/article[221]: Butter (peanut butter), AFB_1, AFB_2, and AFG_1, literature[221]

incidence: 1/182*, conc.: 1–2 µg/kg, sample origin: Khorasan (province), Iran, sample year: September 2009–August 2010, country: Iran[677], *sesame seeds, unprocessed

- Co-contamination: not reported
- Further contamination of plants and plant products other than cereals and their products reported in the present book/article[677]: Seeds (sesame), AFB_1, AFB_2, and AFG_2, literature[677]

incidence: 5*/269, conc. range: 1–2 µg/kg (3 sa.), 3–4 µg/kg (2 sa., maximum: 3.86 µg/kg), sample origin: imported from Afghanistan to Khorasan (province), Iran, sample year: April 2008–January 2009, country: Iran[745], *only sa. contaminated >1 µg/kg are considered

- Co-contamination: not reported
- Further contamination of plants and plant products other than cereals and their products reported in the present book/article[745]: Seeds (sesame), AFB_1 and AFB_2, literature[745]

Aflatoxin G_2

incidence: 4?/211, conc. range: <0.3 µg/kg, sample origin: sites of primary storage or processing (domestic products) and sites of importation, Cyprus, sample year: 1992–1996, country: Cyprus[200]

- Co-contamination: not reported
- Further contamination of plants and plant products other than cereals and their products reported in the present book/article[200]: Butter (peanut butter), Figs, Nuts (peanut), AFB_1, AFB_2, AFG_1, and AFG_2, literature[200]; Nuts (Brazil nut, paranut), Nuts (walnut), AFB_1, AFB_2, and AFG_1, literature[200]; Nuts (chestnut), Products (sesame product), Spices, AFB_1, literature[200]; Nuts (pistachio nut), AFB_1 and AFB_2, literature[200]; Seeds (sesame), AFB_1 and AFG_1, literature[200]

incidence: 1/1*, conc.: 7.05 µg/kg, sample origin: local supermarket in Okayama (prefecture), Japan, sample year: unknown, country: Japan[339], *white sesame

- Co-contamination: 1 sa. co-contaminated with AFB_1 and AFG_2
- Further contamination of plants and plant products other than cereals and their products reported in the present book/article[339]: Coconuts, Fruits, dried, AFB_1 and AFG_2, literature[339]; Figs, AFB_2 and AFG_2, literature[339]; Flour (soybean flour), Seeds (sesame), AFB_1, literature[339]; Nuts, Spices, AFB_1, AFB_2, AFG_1, and AFG_2, literature[339]; Raisins, AFG_2, literature[339]

incidence: 1/182*, conc.: 0.1–0.5 µg/kg, sample origin: Khorasan (province), Iran, sample year: September 2009–August 2010, country: Iran[677], *sesame seeds, unprocessed

- Co-contamination: not reported
- Further contamination of plants and plant products other than cereals and their products reported in the present book/article[677]: Seeds (sesame), AFB_1, AFB_2, and AFG_1, literature[677]

Aflatoxin/s

incidence: 2/75, conc. range: <10 µg/kg*, ∅ conc.: <1 µg/kg*, sample origin: market retailers, street vendors, wholesalers, and distributors, Thailand, sample year: September 1967–July 1969, country: Thailand[139], *AFB_1, AFB_2, AFG_1, and AFG_2

- Co-contamination: not reported
- Further contamination of plants and plant products other than cereals and their products reported in the present book/article[139]: Beans, Beans (mung bean), Food, Nuts (peanut), Spices (cayenne pepper, chilli), Spices (garlic/onion), Starch (cassava starch), Vegetables, AF/S, literature[139]

For detailed information see the article.

incidence: 4/10, conc. range: 0.10–1.01 µg/kg*, ∅ conc.: 0.46 µg/kg*, sample origin: "Casamance" region in southern Guinea, Senegal, sample year: unknown, country: Senegal/Nigeria/USA[710], *B-AF/S

incidence: 9/15, conc. range: 0.10–1.23 µg/kg*, ∅ conc.: 0.33 µg/kg*, sample origin: "Bassin arachidier" and "Senegal Oriental" regions, Senegal, sample year: unknown, country: Senegal/Nigeria/USA[710], *B-AF/S

- Co-contamination: not reported
- Further contamination of plants and plant products other than cereals and their products reported in the present book/article[710]: not reported

Other ***Aspergillus*** Toxins

Sterigmatocystin

incidence: 1/17, conc.: 25 µg/kg, sample origin: B'Ladi, Bokkos, Mekera, and Vwang (districts), Plateau State, Nigeria, sample year: April/May 2011, country: Nigeria/Austria[629]

- Co-contamination: For detailed information see the article.
- Further contamination of plants and plant products other than cereals and their products reported in the present book/article[629]: Seeds (sesame), 3-NPA, AME, AOH, BEA, CIT, DON, FA, KA, TA, and ZEA, literature[629]; see also Seeds (sesame), chapter: Further Mycotoxins and Microbial Metabolites

Aspergillus and ***Penicillium*** Toxins

3-Nitropropionic Acid

incidence: 3/17, conc. range: 3–95 µg/kg, Ø conc.: 35 µg/kg, sample origin: B'Ladi, Bokkos, Mekera, and Vwang (districts), Plateau State, Nigeria, sample year: April/May 2011, country: Nigeria/Austria[629]

- Co-contamination: For detailed information see the article.
- Further contamination of plants and plant products other than cereals and their products reported in the present book/ article[629]: Seeds (sesame), AME, AOH, BEA, CIT, DON, FA, KA, STG, TA, and ZEA, literature[629]; see also Seeds (sesame), chapter: Further Mycotoxins and Microbial Metabolites"

Citrinin

incidence: 3/17, conc. range: 13–17 µg/kg; Ø conc.: 15 µg/kg, sample origin: B'Ladi, Bokkos, Mekera, and Vwang (districts), Plateau State, Nigeria, sample year: April/May 2011, country: Nigeria/Austria[629]

- Co-contamination: For detailed information see the article.
- Further contamination of plants and plant products other than cereals and their products reported in the present book/ article[629]: Seeds (sesame), 3-NPA, AME, AOH, BEA, DON, FA, KA, STG, TA, and ZEA, literature[629]; see also Seeds (sesame), chapter: Further Mycotoxins and Microbial Metabolites

Kojic Acid

incidence: 9/17, conc. range: 50–64,600 µg/kg, Ø conc.: 7,449 µg/kg, sample origin: B'Ladi, Bokkos, Mekera, and Vwang (districts), Plateau State, Nigeria, sample year: April/May 2011, country: Nigeria/Austria[629]

- Co-contamination: For detailed information see the article.
- Further contamination of plants and plant products other than cereals and their products reported in the present book/ article[629]: Seeds (sesame), 3-NPA, AME, AOH, BEA, CIT, DON, FA, STG, TA, and ZEA, literature[629]; see also Seeds (sesame), chapter: Further Mycotoxins and Microbial Metabolites

Ochratoxin A

incidence: 9/24, conc. range: ≤0.860 µg/kg, sample origin: regions around Detmold, Giessen, Hamburg, Jena, Karlsruhe, Kiel, Kulmbach, and Trier (cities), Germany, sample year: unknown, country: Germany[262]

- Co-contamination: not reported
- Further contamination of plants and plant products other than cereals and their products reported in the present book/ article[262]: Bars, Chocolate, Drinks (chocolate drink), Figs, Fruits, dried, Nut crème, Nuts, Nuts (hazelnut), Nuts (peanut), Powder (cocoa powder), Pudding, Raisins, Seeds (linseed), Seeds (sunflower), OTA, literature[262]

incidence: 19/19, conc. range: 1.90–15.66 µg/kg, Ø conc.: 8.14 µg/kg, sample origin: markets in Kaffin Koro, Minna, and Pasiko (towns), Niger State, Nigeria, sample year: 2010, country: Nigeria[785]

- Co-contamination: not reported
- Further contamination of plants and plant products other than cereals and their products reported in the present book/ article[785]: Flakes (cassava), OTA, literature[785]

Fusarium Toxins

Depsipeptides

Beauvericin

incidence: 16/17, conc. range: 3–82 µg/kg, Ø conc.: 25 µg/kg, sample origin: B'Ladi, Bokkos, Mekera, and Vwang (districts), Plateau State, Nigeria, sample year: April/May 2011, country: Nigeria/Austria[629]

- Co-contamination: For detailed information see the article.
- Further contamination of plants and plant products other than cereals and their products reported in the present book/ article[629]: Seeds (sesame), 3-NPA, AME, AOH, CIT, DON, FA, KA, STG, TA, and ZEA, literature[629]; see also Seeds (sesame), chapter: Further Mycotoxins and Microbial Metabolites

Type A Trichothecenes

T-2 Toxin

incidence: 1/30, conc.: 0.1 µg/kg, sample origin: local supermarkets and small retail shops, Japan, sample year: summer 2010–winter 2012, country: Japan[892]

- Co-contamination: not reported
- Further contamination of plants and plant products other than cereals and their products reported in the present book/ article[892]: Beans (mung bean), DON, HT-2, T-2, and ZEA, literature[892]; Beans (soybean), HT-2 and T-2, literature[892]; Seeds (sesame), ZEA, literature[892]

Type B Trichothecenes

Deoxynivalenol

incidence: 15/17, conc. range: 8–76 µg/kg, Ø conc.: 28 µg/kg, sample origin: B'Ladi, Bokkos, Mekera, and Vwang (districts), Plateau State, Nigeria, sample year: April/May 2011, country: Nigeria/Austria[629]

- Co-contamination: For detailed information see the article.
- Further contamination of plants and plant products other than cereals and their products reported in the present book/ article[629]: Seeds (sesame), 3-NPA, AME, AOH, BEA, CIT, FA, KA, STG, TA, and ZEA, literature[629]; see also Seeds (sesame), chapter: Further Mycotoxins and Microbial Metabolites

Other ***Fusarium*** Toxins

Fusaric Acid

incidence: 1/17, conc.: 9 µg/kg, sample origin: B'Ladi, Bokkos, Mekera, and Vwang (districts), Plateau State, Nigeria, sample year: April/May 2011, country: Nigeria/Austria[629]

- Co-contamination: For detailed information see the article.
- Further contamination of plants and plant products other than cereals and their products reported in the present book/ article[629]: Seeds (sesame), 3-NPA, AME, AOH, BEA, CIT, DON, KA, STG, TA, and ZEA, literature[629]; see also Seeds (sesame), chapter: Further Mycotoxins and Microbial Metabolites

Zearalenone

incidence: 15/17, conc.: range: 0.7–38 µg/kg, Ø conc.: 7 µg/kg, sample origin: B'Ladi, Bokkos, Mekera, and Vwang (districts), Plateau State, Nigeria, sample year: April/May 2011, country: Nigeria/Austria[629]

- Co-contamination: For detailed information see the article.
- Further contamination of plants and plant products other than cereals and their products reported in the present book/ article[629]: Seeds (sesame), 3-NPA, AME, AOH, BEA, CIT, DON, FA, KA, STG, and TA, literature[629]; see also Seeds (sesame), chapter: Further Mycotoxins and Microbial Metabolites

incidence: 2/7*, Ø conc.: 10 µg/kg, sample origin: Yemen, sample year: 1984, country: Japan[630], *ncac

- Co-contamination: not reported
- Further contamination of plants and plant products other than cereals and their products reported in the present book/ article[630]: not reported

incidence: 20/30, conc. range: ≤21.3 µg/kg, ∅ conc.: 4.5 µg/kg, sample origin: local supermarkets and small retail shops, Japan, sample year: summer 2010–winter 2012, country: Japan[892]
- Co-contamination: not reported
- Further contamination of plants and plant products other than cereals and their products reported in the present book/article[892]: Beans (mung bean), DON, HT-2, T-2, and ZEA, literature[892]; Beans (soybean), HT-2 and T-2, literature[892]; Seeds (sesame), T-2, literature[892]

Seeds (sunflower) may contain the following mycotoxins:

Alternaria Toxins

Altenuene

incidence: 1/11, conc.: >LOD, sample origin: retail stores in the area of Arnsberg and Münster (cities), Germany, sample year: unknown, country: Germany[743]
- Co-contamination: not reported
- Further contamination of plants and plant products other than cereals and their products reported in the present book/article[743]: Juices, ALT, AME, AOH, and TA, literature[743]; Oils (vegetable oil), Products (tomato product), AME, AOH, and TA, literature[743]; Seeds (sunflower), AME, AOH, ATX-I, and TA, literature[743]; see also Juices, Oils (vegetable oil), Seeds (sunflower), chapter: Further Mycotoxins and Microbial Metabolites

Alternariol

incidence: 2*/2, conc. range: 357–1,840 µg/kg, ∅ conc.: 1,090 µg/kg, sample origin: several localities in Apulia and Basilicata (regions), Italy, sample year: late summer, country: Italy[164], *visibly affected by *Alternaria* rot
- Co-contamination: 1 sa. co-contaminated with AME and AOH; 1 sa. contaminated solely with AOH
- Further contamination of plants and plant products other than cereals and their products reported in the present book/article[164]: Fruits (mandarin), Fruits (tomato), Spices (pepper), AME, AOH, and TA, literature[164]; Fruits (melon), AME and TA, literature[164]; Olives, ALT, AME, and TA, literature[164]; Seeds (sunflower), AME, literature[164]

incidence: 9/79*, conc. range: 4.6–246 µg/kg, ∅ conc.: 62.4 µg/kg, sample origin: farmer's fields and cultivar trials, South Africa, sample year: 2016, country: Germany/Brazil/South Africa[414], *sunflower seeds, unshelled

incidence: 4/80*, conc. range: 7.7–119 µg/kg, ∅ conc.: 63.2 µg/kg, sample origin: farmer's fields and cultivar trials, South Africa, sample year: 2016, country: Germany/Brazil/South Africa[414], *sunflower seeds, shelled
- Co-contamination: For detailed information see the article.
- Further contamination of plants and plant products other than cereals and their products reported in the present book/article[414]: Seeds (sunflower), AME and TA, literature[414]; see also Seeds (sunflower), chapter: Further Mycotoxins and Microbial Metabolites

incidence: 128/150*, conc. range: 50–676 µg/kg, ∅ conc.: 189 µg/kg, sample origin: Buenos Aires, Cordoba, La Pampa, San Luis, and Santa Fe (provinces), Argentina, sample year: 1991/1992, country: Argentina[648], *ncac
- Co-contamination: 46 sa. co-contaminated with AME, AOH, and TA; 44 sa. co-contaminated solely with AOH and TA; 18 sa. co-contaminated solely with AME and AOH; 20 sa. contaminated solely with AOH
- Further contamination of plants and plant products other than cereals and their products reported in the present book/article[648]: Seeds (sunflower), AME and TA, literature[648]

incidence: 20/20*, conc. range: 250–980 µg/kg**, ∅ conc.: 661 µg/kg**, sample origin: silo in Cordoba (province), Argentina, sample year: 1994, country: Argentina[649], *ncac, **first sampling
- Co-contamination: 16 sa. co-contaminated with AOH and TA; 4 sa. contaminated solely with AOH
- Further contamination of plants and plant products other than cereals and their products reported in the present book/article[649]: Seeds (sunflower), AME and TA, literature[649]

incidence: 17/20*, conc. range: 60–800 µg/kg**, ∅ conc.: 360 µg/kg**, sample origin: silo in Cordoba (province), Argentina, sample year: 1994, country: Argentina[649], *ncac, **second sampling
- Co-contamination: 1 sa. co-contaminated with AME, AOH and TA; 12 sa. co-contaminated with AOH and TA; 4 sa. contaminated solely with AOH
- Further contamination of plants and plant products other than cereals and their products reported in the present book/article[649]: Seeds (sunflower), AME and TA, literature[649]

incidence: 4/20*, conc. range: 800–1,600 µg/kg**, ∅ conc.: 1,070 µg/kg**, sample origin: silo in Cordoba (province), Argentina, sample year: 1994, country: Argentina[649], *ncac, **third sampling
- Co-contamination: 1 sa. co-contaminated with AME, AOH and TA; 3 sa. co-contaminated with AOH and TA
- Further contamination of plants and plant products other than cereals and their products reported in the present book/article[649]: Seeds (sunflower), AME and TA, literature[649]

incidence: 38/50, conc. range: 35–792 µg/kg, ∅ conc.: 166 µg/kg, sample origin: oil seed factory, Argentina, sample year: unknown, country: Argentina[650]
- Co-contamination: 31 sa. co-contaminated with AME and AOH; 7 sa. contaminated solely with AOH
- Further contamination of plants and plant products other than cereals and their products reported in the present book/article[650]: Seeds (sunflower), AME, literature[650]

incidence: 6/11, conc. range: >LOD (3 sa.), 16–39 µg/kg (3 sa.), sample origin: retail stores in the area of Arnsberg and Münster (cities), Germany, sample year: unknown, country: Germany[743]
- Co-contamination: not reported
- Further contamination of plants and plant products other than cereals and their products reported in the present book/article[743]: Juices, ALT, AME, AOH, and TA, literature[743]; Oils (vegetable oil), Products (tomato product), AME, AOH, and TA, literature[743]; Seeds (sunflower), ALT, AME, ATX-I, and TA, literature[743]; see also Juices, Oils (vegetable oil), Seeds (sunflower), chapter: Further Mycotoxins and Microbial Metabolites

incidence: 1/10, conc.: 36 µg/kg, sample origin: retail stores, Netherlands, sample year: autumn 2014, country: Netherlands/Germany[906]
- Co-contamination: 1 sa. co-contaminated with AME and AOH; no further information available
- Further contamination of plants and plant products other than cereals and their products reported in the present book/article[906]: Figs, Fruits (tomato), Juices (tomato juice), AOH and TA, literature[906]; Oils (sunflower oil), AME, literature[906]; Paste (tomato paste), Sauces (tomato sauce), Tomato ketchup, AME, AOH, and TA, literature[906]; Seeds (sunflower), AME and TA, literature[906]

Alternariol Methyl Ether

incidence: 1*/2, conc.: 129 µg/kg, sample origin: several localities in Apulia and Basilicata (regions), Italy, sample year: late summer, country: Italy[164], *visibly affected by *Alternaria* rot

- Co-contamination: 1 sa. co-contaminated with AME and AOH
- Further contamination of plants and plant products other than cereals and their products reported in the present book/article[164]: Fruits (mandarin), Fruits (tomato), Spices (pepper), AME, AOH, and TA, literature[164]; Fruits (melon), AME and TA, literature[164]; Olives, ALT, AME, and TA, literature[164]; Seeds (sunflower), AOH, literature[164]

incidence: 13/79*, conc. range: 0.9–197 µg/kg, ∅ conc.: 24.1 µg/kg, sample origin: farmer's fields and cultivar trials, South Africa, sample year: 2016, country: Germany/Brazil/South Africa[414], *sunflower seeds, unshelled

incidence: 5/80*, conc. range: 0.7–27.8 µg/kg, ∅ conc.: 8.1 µg/kg, sample origin: farmer's fields and cultivar trials, South Africa, sample year: 2016, country: Germany/Brazil/South Africa[414], *sunflower seeds, shelled

- Co-contamination: For detailed information see the article.
- Further contamination of plants and plant products other than cereals and their products reported in the present book/article[414]: Seeds (sunflower), AOH and TA, literature[414]; see also Seeds (sunflower), chapter: Further Mycotoxins and Microbial Metabolites

incidence: 68/150*, conc. range: 30–836 µg/kg, ∅ conc.: 202 µg/kg, sample origin: Buenos Aires, Cordoba, La Pampa, San Luis, and Santa Fe (provinces), Argentina, sample year: 1991/1992, country: Argentina[648], *ncac

- Co-contamination: 46 sa. co-contaminated with AME, AOH, and TA; 18 sa. co-contaminated solely with AME and AOH; 4 sa. co-contaminated solely with AME and TA; 2 sa. contaminated solely with AME
- Further contamination of plants and plant products other than cereals and their products reported in the present book/article[648]: Seeds (sunflower), AOH and TA, literature[648]

incidence: 0/20*, conc.: no contamination**, sample origin: silo in Cordoba (province), Argentina, sample year: 1994, country: Argentina[649], *ncac, **first sampling

- Co-contamination: not reported
- Further contamination of plants and plant products other than cereals and their products reported in the present book/article[649]: Seeds (sunflower), AOH and TA, literature[649]

incidence: 1/20*, conc.: 600 µg/kg**, sample origin: silo in Cordoba (province), Argentina, sample year: 1994, country: Argentina[649], *ncac, **second sampling

- Co-contamination: 1 sa. co-contaminated with AME, AOH and TA
- Further contamination of plants and plant products other than cereals and their products reported in the present book/article[649]: Seeds (sunflower), AOH and TA, literature[649]

incidence: 1/20*, conc.: 800 µg/kg**, sample origin: silo in Cordoba (province), Argentina, sample year: 1994, country: Argentina[649], *ncac, **third sampling

- Co-contamination: 1 sa. co-contaminated with AME, AOH and TA
- Further contamination of plants and plant products other than cereals and their products reported in the present book/article[649]: Seeds (sunflower), AOH and TA, literature[649]

incidence: 31/50, conc. range: 90–630 µg/kg, ∅ conc.: 114 µg/kg, sample origin: oil seed factory, Argentina, sample year: unknown, country: Argentina[650]

- Co-contamination: 31 sa. co-contaminated with AME and AOH
- Further contamination of plants and plant products other than cereals and their products reported in the present book/article[650]: Seeds (sunflower), AOH, literature[650]

incidence: 7?/11, conc. range: >LOD (2 sa.), 0.64–21 µg/kg (4 sa.), sample origin: retail stores in the area of Arnsberg and Münster (cities), Germany, sample year: unknown, country: Germany[743]

- Co-contamination: not reported
- Further contamination of plants and plant products other than cereals and their products reported in the present book/article[743]: Juices, ALT, AME, AOH, and TA, literature[743]; Oils (vegetable oil), Products (tomato product), AME, AOH, and TA, literature[743]; Seeds (sunflower), ALT, AME, ATX-I, and TA, literature[743]; see also Juices, Oils (vegetable oil), Seeds (sunflower), chapter: Further Mycotoxins and Microbial Metabolites

incidence: 1/10, conc.: 1.7 µg/kg, sample origin: retail stores, Netherlands, sample year: autumn 2014, country: Netherlands/Germany[906]

- Co-contamination: 1 sa. co-contaminated with AME and AOH; no further information available
- Further contamination of plants and plant products other than cereals and their products reported in the present book/article[906]: Figs, Fruits (tomato), Juices (tomato juice), AOH and TA, literature[906]; Oils (sunflower oil), AME, literature[906]; Paste (tomato paste), Sauces (tomato sauce), Tomato ketchup, AME, AOH, and TA, literature[906]; Seeds (sunflower), AOH and TA, literature[906]

Altertoxin I

incidence: 1/11, conc.: >LOD, sample origin: retail stores in the area of Arnsberg and Münster (cities), Germany, sample year: unknown, country: Germany[743]

- Co-contamination: not reported
- Further contamination of plants and plant products other than cereals and their products reported in the present book/article[743]: Juices, ALT, AME, AOH, and TA, literature[743]; Oils (vegetable oil), Products (tomato product), AME, AOH, and TA, literature[743]; Seeds (sunflower), ALT, AME, OH, and TA, literature[743]; see also Juices, Oils (vegetable oil), Seeds (sunflower), chapter: Further Mycotoxins and Microbial Metabolites

Tenuazonic Acid

incidence: 53/79*, conc. range: 82.7–5,160 µg/kg, ∅ conc.: 510 µg/kg, sample origin: farmer's fields and cultivar trials, South Africa, sample year: 2016, country: Germany/Brazil/South Africa[414], *sunflower seeds, unshelled

incidence: 28/80*, conc. range: 65.7–6,260 µg/kg, ∅ conc.: 7521 µg/kg, sample origin: farmer's fields and cultivar trials, South Africa, sample year: 2016, country: Germany/Brazil/South Africa[414], *sunflower seeds, shelled

- Co-contamination: For detailed information see the article.
- Further contamination of plants and plant products other than cereals and their products reported in the present book/article[414]: Seeds (sunflower), AME and AOH, literature[414]; see also Seeds (sunflower), chapter: Further Mycotoxins and Microbial Metabolites

incidence: 5/5, conc. range: 85–449 µg/kg, ∅ conc.: 223 µg/kg, sample origin: Dutch survey program (2013) and retail shops across The Netherlands, sample year: 2013?, country: Netherlands[568]

- Co-contamination: 1 sa. co-contaminated with TA and TEN; 4 sa. contaminated solely with TA
- Further contamination of plants and plant products other than cereals and their products reported in the present book/article[568]: Figs, Olives, TA, literature[568]; Fruits (apple), AOH, literature[568]; Sauces (tomato sauce), AME, AOH, and TA, literature[568]; Wine (red), AOH and TA, literature[568]; see also Seeds (sunflower), chapter: Further Mycotoxins and Microbial Metabolites

incidence: 98/150*, conc. range: 2,500–15,796 µg/kg, Ø conc.: 6,459 µg/kg, sample origin: Buenos Aires, Cordoba, La Pampa, San Luis, and Santa Fe (provinces), Argentina, sample year: 1991/1992, country: Argentina[648], *ncac

- Co-contamination: 46 sa. co-contaminated with AME, AOH, and TA; 44 sa. co-contaminated solely with AOH and TA; 4 sa. co-contaminated solely with AME and TA; 4 sa. contaminated solely with TA
- Further contamination of plants and plant products other than cereals and their products reported in the present book/article[648]: Seeds (sunflower), AME and AOH, literature[648]

incidence: 16/20*, conc. range: 3,900–31,600 µg/kg**, Ø conc.: 16,000 µg/kg**, sample origin: silo in Cordoba (province), Argentina, sample year: 1994, country: Argentina[649], *ncac, **first sampling

- Co-contamination: 16 sa. co-contaminated with AOH and TA
- Further contamination of plants and plant products other than cereals and their products reported in the present book/article[649]: Seeds (sunflower), AME and AOH, literature[649]

incidence: 13/20*, conc. range: 3,600–15,800 µg/kg**, Ø conc.: 8,000 µg/kg**, sample origin: silo in Cordoba (province), Argentina, sample year: 1994, country: Argentina[649], *ncac, **second sampling

- Co-contamination: 1 sa. co-contaminated with AME, AOH and TA; 12 sa. co-contaminated with AOH and TA
- Further contamination of plants and plant products other than cereals and their products reported in the present book/article[649]: Seeds (sunflower), AME and AOH, literature[649]

incidence: 9/20*, conc. range: 3,120–6,240 µg/kg**, Ø conc.: 4,500 µg/kg**, sample origin: silo in Cordoba (province), Argentina, sample year: 1994, country: Argentina[649], *ncac, **third sampling

- Co-contamination: 1 sa. co-contaminated with AME, AOH and TA; 3 sa. co-contaminated with AOH and TA; 5 sa. contaminated solely with TA
- Further contamination of plants and plant products other than cereals and their products reported in the present book/article[649]: Seeds (sunflower), AME and AOH, literature[649]

incidence: 11/11, conc. range: >LOD (9 sa.), 350–490 µg/kg (2 sa.), Ø conc.: 420 µg/kg, sample origin: retail stores in the area of Arnsberg and Münster (cities), Germany, sample year: unknown, country: Germany[743]

- Co-contamination: not reported
- Further contamination of plants and plant products other than cereals and their products reported in the present book/article[743]: Juices, ALT, AME, AOH, and TA, literature[743]; Oils (vegetable oil), Products (tomato product), AME, AOH, and TA, literature[743]; Seeds (sunflower), ALT, AME, AOH, and ATX-I, literature[743]; see also Juices, Oils (vegetable oil), Seeds (sunflower), chapter: Further Mycotoxins and Microbial Metabolites

incidence: 8/10, conc. range: ≤1,350 µg/kg, sample origin: retail stores, Netherlands, sample year: autumn 2014, country: Netherlands/Germany[906]

- Co-contamination: not reported
- Further contamination of plants and plant products other than cereals,and their products reported in the present book/article[906]: Figs, Fruits (tomato), Juices (tomato juice), AOH and TA, literature[906]; Oils (sunflower oil), AME, literature[906]; Paste (tomato paste), Sauces (tomato sauce), Tomato ketchup, AME, AOH, and TA, literature[906]; Seeds (sunflower), AME and AOH, literature[906]

Aspergillus Toxins

Aflatoxin B_1

incidence: 1/4*, conc.: 10.5 µg/kg, sample origin: local marketing institutions, GDR, sample year: unknown, country: GDR[29], *moldy

- Co-contamination: not reported
- Further contamination of plants and plant products other than cereals and their products reported in the present book/article[29]: Butter (peanut butter), Nuts (peanut), Products (peanut product), AFB_1, AFB_2, AFG_1, and AFG_2, literature[29]; Coffee, OTA, literature[29]; Fruits, AFB_1, AFG_1, and AFG_2, literature[29]; Fruits (apple), Fruits (peach), Juices (apple juice), PAT, literature[29]; Seeds (sunflower), AFB_2, AFG_1, and AFG_2, literature[29]

incidence: 6/7, conc. range: 1.14–5.33 µg/kg, Ø conc.: 2.55 µg/kg, sample origin: supermarkets in Penang (state), Malaysia, sample year: unknown, country: Malaysia[182]

- Co-contamination: not reported
- Further contamination of plants and plant products other than cereals and their products reported in the present book/article[182]: Nuts, Nuts (peanut), Seeds (sesame), Spices (cayenne pepper, chilli), Spices (cumin), Spices (pepper), AFB_1, literature[182]

Aflatoxin B_2

incidence: 1/4*, conc.: 0.5 µg/kg, sample origin: local marketing institutions, GDR, sample year: unknown, country: GDR[29], *moldy

- Co-contamination: not reported
- Further contamination of plants and plant products other than cereals and their products reported in the present book/article[29]: Butter (peanut butter), Nuts (peanut), Products (peanut product), AFB_1, AFB_2, AFG_1, and AFG_2, literature[29]; Coffee, OTA, literature[29]; Fruits, AFB_1, AFG_1, and AFG_2, literature[29]; Fruits (apple), Fruits (peach), Juices (apple juice), PAT, literature[29]; Seeds (sunflower), AFB_1, AFG_1, and AFG_2, literature[29]

Aflatoxin G_1

incidence: 1/4*, conc.: 0.4 µg/kg, sample origin: local marketing institutions, GDR, sample year: unknown, country: GDR[29], *moldy

- Co-contamination: not reported
- Further contamination of plants and plant products other than cereals and their products reported in the present book/article[29]: Butter (peanut butter, Nuts (peanut), Products (peanut product), AFB_1, AFB_2, AFG_1, and AFG_2, literature[29]; Coffee, OTA, literature[29]; Fruits, AFB_1, AFG_1, and AFG_2, literature[29]; Fruits (apple), Fruits (peach), Juices (apple juice), PAT, literature[29]; Seeds (sunflower), AFB_1, AFB_2, and AFG_2, literature[29]

Aflatoxin G_2

incidence: 1/4*, conc.: 0.03 µg/kg, sample origin: local marketing institutions, GDR, sample year: unknown, country: GDR[29], *moldy

- Co-contamination: not reported

- Further contamination of plants and plant products other than cereals and their products reported in the present book/article[29]: Butter (peanut butter), Nuts (peanut), Products (peanut product), AFB_1, AFB_2, AFG_1, and AFG_2, literature[29]; Coffee, OTA, literature[29]; Fruits, AFB_1, AFG_1, and AFG_2, literature[29]; Fruits (apple), Fruits (peach), Juices (apple juice), PAT, literature[29]; Seeds (sunflower), AFB_1, AFB_2, and AFG_1, literature[29]

Aflatoxin/s

incidence: 7/136, conc. range: 5.0–9.9 µg/kg* (5 sa.), 10.0–14.9 µg/kg* (1 sa.), 15.0–19.9 µg/kg* (1 sa.), sample origin: brokers, distributors, growers, processors, and retail trade, Canada, sample year: 1970–1975, country: Canada[15], *AFB_1, AFB_2, AFG_1, and AFG_2

- Co-contamination: not reported
- Further contamination of plants and plant products other than cereals and their products reported in the present book/article[15]: Butter (peanut butter), Cocoa beans, Nuts (almond), Nuts (cashew nut), Nuts (peanut), Nuts (pecan nut), Nuts (pistachio nut), Nuts (walnut), AF/S, literature[15]

incidence: 1/1, conc.: 1.92 µg/kg*, sample origin: mainly imported from local retail outlets in Lodz (city), Poland, sample year: unknown, country: Poland[18], *AFS total

- Co-contamination: not reported
- Further contamination of plants and plant products other than cereals and their products reported in the present book/article[18]: Nut cereals, Nuts, Nuts (almond), Nuts (cashew nut), Nuts (hazelnut), Nuts (peanut), Nuts (pistachio nut), Nuts (walnut), AF/S, literature[18]

Aspergillus and ***Penicillium*** Toxins

Ochratoxin A

incidence: 10/34, conc. range: ≤0.100 µg/kg, sample origin: regions around Detmold, Giessen, Hamburg, Jena, Karlsruhe, Kiel, Kulmbach, and Trier (cities), Germany, sample year: unknown, country: Germany[262]

- Co-contamination: not reported
- Further contamination of plants and plant products other than cereals and their products reported in the present book/article[262]: Bars, Chocolate, Drinks (chocolate drink), Figs, Fruits, dried, Nut crème, Nuts, Nuts (hazelnut), Nuts (peanut), Powder (cocoa powder), Pudding, Raisins, Seeds (linseed), Seeds (sesame), OTA, literature[262]

incidence: 1/3, conc.: 4.34 µg/kg, sample origin: supermarkets and small shops, Jordan, sample year: September 2008–March 2009, country: Jordan[315]

- Co-contamination: not reported
- Further contamination of plants and plant products other than cereals and their products reported in the present book/article[315]: Coffee beans, Nuts, OTA, literature[315]

Fusarium Toxins

Type A Trichothecenes

HT-2 Toxin

incidence: 2/5, conc. range: 5 µg/kg, sample origin: food stores, Germany, sample year: 2000/2001, country: Germany[151]

- Co-contamination: not reported
- Further contamination of plants and plant products other than cereals and their products reported in the present book/article[151]: Beans, Spices (cayenne pepper, chilli), ZEA, literature[151]; Nuts (hazelnut), HT-2, T-2, and ZEA, literature[151]; Products (potato product), DAS, HT-2, MAS, SCIRP, and ZEA, literature[151]; Seeds (pumpkin), HT-2 and ZEA, literature[151]; Seeds (sunflower), T-2 and ZEA, literature[151]

T-2 Toxin

incidence: 2/5, conc. range: 20–21 µg/kg, ∅ conc.: 20.5 µg/kg, sample origin: food stores, Germany, sample year: 2000/2001, country: Germany[151]

- Co-contamination: not reported
- Further contamination of plants and plant products other than cereals and their products reported in the present book/article[151]: Beans, Spices (cayenne pepper, chilli), ZEA, literature[151]; Nuts (hazelnut), HT-2, T-2, and ZEA, literature[151]; Products (potato product), DAS, HT-2, MAS, SCIRP, and ZEA, literature[151]; Seeds (pumpkin), Seeds (sunflower), HT-2 and ZEA, literature[151]

Type B Trichothecenes

Deoxynivalenol

incidence: 4/9* **, conc. range: 40–304 µg/kg, ∅ conc.: 155 µg/kg, sample origin: different locations, Serbia, sample year: 2004, country: Serbia[634], *ncac, **sunflower and sunflower meal

incidence: 5/10* **, conc. range: 114–788 µg/kg, ∅ conc.: 447 µg/kg, sample origin: different locations, Serbia, sample year: 2005, country: Serbia[634], *ncac, **sunflower and sunflower meal

- Co-contamination: not reported
- Further contamination of plants and plant products other than cereals and their products reported in the present book/article[633]: Beans (soybean), DON, literature[634]

incidence: 1/8*, conc.: 100–500 µg/kg (1 sa.), sample origin: home-grown as well as imported and purchased in retail outlets and from Portuguese cereal traders, Portugal, sample year: unknown, country: Portugal[787], *for food and feed

- Co-contamination: For detailed information see the article.
- Further contamination of plants and plant products other than cereals and their products reported in the present book/article[787]: Beans (soybean), DON and ZEA, literature[787]; Seeds (sunflower), ZEA, literature[787]

Other ***Fusarium*** Toxins

Zearalenone

incidence: 2/5, conc. range: 2–4 µg/kg, ∅ conc.: 3 µg/kg, sample origin: food stores, Germany, sample year: 2000/2001, country: Germany[151]

- Co-contamination: not reported
- Further contamination of plants and plant products other than cereals and their products reported in the present book/article[151]: Beans, Spices (cayenne pepper, chilli), ZEA, literature[151]; Products (potato product), DAS, HT-2, MAS, SCIRP, and ZEA, literature[151]; Nuts (hazelnut), HT-2, T-2, and ZEA, literature[151]; Seeds (pumpkin), HT-2 and ZEA, literature[151]; Seeds (sunflower), HT-2 and T-2, literature[151]

incidence: 1/8*, conc.: 5–100 µg/kg (1 sa.), sample origin: home-grown as well as imported and purchased in retail outlets and from Portuguese cereal traders, Portugal, sample year: unknown, country: Portugal[787], *for food and feed

- Co-contamination: For detailed information see the article.
- Further contamination of plants and plant products other than cereals and their products reported in the present book/ article[787]: Beans (soybean), DON and ZEA, literature[787]; Seeds (sunflower), DON, literature[787]

see also **Oil seeds**

Sesame oils see **Oils (sesame oil)**

Sesame products see **Products (sesame product)**

Sesame seeds see **Seeds (sesame)**

Shea nuts may contain the following mycotoxins:

Aspergillus **Toxins**

Aflatoxin B_1

incidence: 2/21*, conc. range: <5 µg/kg (1 sa.**), 15 µg/kg (1 sa.**), sample origin: Ghana, Ivory Coast, Mali, and Upper Volta, sample year: May–September 1981, country: England[563], *providing cocoa-butter substitutes, **moldy shea nut sa.

- Co-contamination: 1 sa. co-contaminated with AFB_1 and AFB_2; 1 sa. contaminated solely with AFB_1
- Further contamination of plants and plant products other than cereals and their products reported in the present book/ article[563]: Shea nuts, AFB_2, literature[563]

Aflatoxin B_2

incidence: 1/21*, conc.: pr**, sample origin: Ghana, Ivory Coast, Mali, and Upper Volta, sample year: May–September 1981, country: England[563], *providing cocoa-butter substitutes, **moldy shea nut sa.

- Co-contamination: 1 sa. co-contaminated with AFB_1 and AFB_2
- Further contamination of plants and plant products other than cereals and their products reported in the present book/ article[563]: Shea nuts, AFB_1, literature[563]

Sherry see **Wine, miscellaneous**

Shiitake may contain the following mycotoxins:

Fusarium **Toxins**

Fumonisin/s

Fumonisin B_1

incidence: 6/30*, conc. range: 8.2–78.7 µg/kg, Ø conc.: 35.1 µg/kg, sample origin: supermarkets in Shanghai (city), China, sample year: unknown, country: China[736]; **Lentinula edodes*

- Co-contamination: 3 sa. co-contaminated with FB_1 and MA; 3 sa. contaminated solely with FB_1
- Further contamination of plants and plant products other than cereals and their products reported in the present book/ article[736]: Shiitake, DON, FB_2, and MA, literature[736]

Fumonisin B_2

incidence: 5/30*, conc. range: 8.2–71.5 µg/kg, Ø conc.: 31.4 µg/kg, sample origin: supermarkets in Shanghai (city), China, sample year: unknown, country: China[736]; **Lentinula edodes*

- Co-contamination: 1 sa. co-contaminated with FB_1 and FB_2; 1 sa. co-contaminated with FB_2 and MA; 3 sa. contaminated solely with FB_1
- Further contamination of plants and plant products other than cereals and their products reported in the present book/ article[736]: Shiitake, DON, FB_1, and MA, literature[736]

Type B Trichothecenes

Deoxynivalenol

incidence: 5/30*, conc. range: 6.3–33.2 µg/kg, Ø conc.: 12.6 µg/kg, sample origin: supermarkets in Shanghai (city), China, sample year: unknown, country: China[736]; **Lentinula edodes*

- Co-contamination: 1 sa. co-contaminated with DON and MA; 4 sa. contaminated solely with DON
- Further contamination of plants and plant products other than cereals and their products reported in the present book/ article[736]: Shiitake, FB_1, FB_2, and MA, literature[736]

Penicillium **Toxins**

Mycophenolic Acid

incidence: 12/30*, conc. range: 3.3–28,819.5 µg/kg, Ø conc.: 3,378 µg/kg, sample origin: supermarkets in Shanghai (city), China, sample year: unknown, country: China[736]; **Lentinula edodes*

- Co-contamination: 1 sa. co-contaminated with DON and MA; 3 sa. co-contaminated with FB_1 and MA; 1 sa. co-contaminated with FB_2 and MA; 7 sa. contaminated solely with MA
- Further contamination of plants and plant products other than cereals and their products reported in the present book/ article[736]: Shiitake, DON, FB_1, and FB_2, literature[736]

Shiraz see **Wine (red)**

Shiro may contain the following mycotoxins:

Aspergillus **Toxins**

Aflatoxin B_1

incidence: 5/60, conc. range: 100–500 µg/kg, Ø conc.: 271 µg/kg, sample origin: government-owned stores, retail shops, and open markets in Addis Ababa (capital), Ethiopia, sample year: unknown, country: Ethiopia[168]

- Co-contamination: not reported
- Further contamination of plants and plant products other than cereals and their products reported in the present book/ article[168]: Spices (cayenne pepper, chilli), AFB_1, literature[168]

For detailed information see the article.

Shiro is a homogenous stew. It consisted primarily of powdered chickpeas or broad bean meal and is often prepared with spices.

Singkamas see **Tubers**

Snack bar see **Bars**

Snacks may contain the following mycotoxins:

Alternaria **Toxins**

Alternariol Methyl Ether

incidence: 1/10*, conc.: 47 µg/kg, sample origin: street traders in Lagos (city), Nigeria, sample year: unknown, country: Nigeria/ Austria[763], *snacks, groundnut-based incidence: 0/2*, conc.: no contamination, sample origin: street traders in Lagos (city), Nigeria, sample year: unknown, country: Nigeria/Austria[763], *snacks, groundnut-/maize-based

- Co-contamination: For detailed information see the article.
- Further contamination of plants and plant products other than cereals and their products reported in the present book/ article[763]: Snacks, 3-NPA, AFB_1, AFB_2, AFM_1, BEA, FB_1, FB_2, FB_3, KA, and MON, literature[763]; see also Snacks, chapter: Further Mycotoxins and Microbial Metabolites

Aspergillus Toxins

Aflatoxin B_1

incidence: 5/11*, conc. range: 8–75 µg/kg, sample origin: traditional markets and supermarkets, Indonesia, sample year: 2001/2002, country: Austria/Indonesia/Thailand[226], *snacks, traditional (fried floured peanut, peanuts with brown sugar, and sweet-grounded peanuts)

- Co-contamination: not reported
- Further contamination of plants and plant products other than cereals and their products reported in the present book/ article[226]: Butter (peanut butter), Cake (peanut), Nuts (peanut), Sauces (peanut chilli sauce), AFB_1 and AF/S, literature[226]; Snacks, AF/S, literature[226]

incidence: 68/106*, conc. range: 5–165 µg/kg, Ø conc.: 25.5 µg/kg, sample origin: street hawkers, markets, and retail shops, Southwest Nigeria, sample year: unknown, country: Nigeria[398], *groundnut snacks, dry-roasted

- Co-contamination: not reported
- Further contamination of plants and plant products other than cereals and their products reported in the present book/ article[398]: Snacks, AFB_2, AFG_1, and AFG_2, literature[398]

incidence: 3/10*, conc. range: 52–362 µg/kg, sample origin: street traders in Lagos (city), Nigeria, sample year: unknown, country: Nigeria/Austria[763], *snacks, groundnut-based

incidence: 2/2*, conc. range: 9.3–12 µg/kg, Ø conc.: 10.7 µg/kg, sample origin: street traders in Lagos (city), Nigeria, sample year: unknown, country: Nigeria/Austria[763], *snacks, groundnut-/ maize-based

- Co-contamination: For detailed information see the article.
- Further contamination of plants and plant products other than cereals and their products reported in the present book/ article[763]: Snacks, 3-NPA, AFB_2, AFM_1, AME, BEA, FB_1, FB_2, FB_3, KA, and MON, literature[763]; see also Snacks, chapter: Further Mycotoxins and Microbial Metabolites

incidence: 0/1*, conc.: no contamination, sample origin: street vendors in Lagos (metropolis), Nigeria, sample year: unknown, country: Spain/Nigeria[833], *snacks, cassava-based

incidence: 1/3*, conc.: 23 µg/kg, sample origin: street vendors in Lagos (metropolis), Nigeria, sample year: unknown, country: Spain/Nigeria[833], *snacks, coconut-based (baked coconut and coconut candy)

incidence: 2/8*, conc. range: ≤19 µg/kg, sample origin: street vendors in Lagos (metropolis), Nigeria, sample year: unknown, country: Spain/Nigeria[833], *snacks, groundnut-based

- Co-contamination: not reported
- Further contamination of plants and plant products other than cereals and their products reported in the present book/ article[833]: Snacks, AFB_2, literature[833]; see also Snacks, chapter: Further Mycotoxins and Microbial Metabolites

Aflatoxin B_2

incidence: 28/106*, conc. range: 6–26 µg/kg, Ø conc.: 10.7 µg/kg, sample origin: street hawkers, markets, and retail shops, Southwest Nigeria, sample year: unknown, country: Nigeria[398], *groundnut snacks, dry-roasted

- Co-contamination: not reported
- Further contamination of plants and plant products other than cereals and their products reported in the present book/ article[398]: Snacks, AFB_1, AFG_1, and AFG_2, literature[398]

incidence: 2/10*, conc. range: 559–679 µg/kg, Ø conc.: 619 µg/kg, sample origin: street traders in Lagos (city), Nigeria, sample year: unknown, country: Nigeria/Austria[763], *snacks, groundnut-based

incidence: 2/2*, conc. range: 0.8–1.7 µg/kg, Ø conc.: 1.3 µg/kg, sample origin: street traders in Lagos (city), Nigeria, sample year: unknown, country: Nigeria/Austria[763], *snacks, groundnut-/maize-based

- Co-contamination: For detailed information see the article.
- Further contamination of plants and plant products other than cereals and their products reported in the present book/ article[763]: Snacks, 3-NPA, AFB_1, AFM_1, AME, BEA, FB_1, FB_2, FB_3, KA, and MON, literature[763]; see also Snacks, chapter: Further Mycotoxins and Microbial Metabolites

incidence: 0/1*, conc.: no contamination, sample origin: street vendors in Lagos (metropolis), Nigeria, sample year: unknown, country: Spain/Nigeria[833], *snacks, cassava-based

incidence: 1/3*, conc.: 3 µg/kg, sample origin: street vendors in Lagos (metropolis), Nigeria, sample year: unknown, country: Spain/Nigeria[833], *snacks, coconut-based (baked coconut and coconut candy)

incidence: 1/8*, conc.: 2 µg/kg, sample origin: street vendors in Lagos (metropolis), Nigeria, sample year: unknown, country: Spain/Nigeria[833], *groundnut-based snacks

- Co-contamination: not reported
- Further contamination of plants and plant products other than cereals and their products reported in the present book/ article[833]: Snacks, AFB_1, literature[833]; see also Snacks, chapter: Further Mycotoxins and Microbial Metabolites

Aflatoxin G_1

incidence: 12/106*, conc. range: 5–20 µg/kg, Ø conc.: 7.2 µg/kg, sample origin: street hawkers, markets, and retail shops, Southwest Nigeria, sample year: unknown, country: Nigeria[398], *groundnut snacks, dry-roasted

- Co-contamination: not reported
- Further contamination of plants and plant products other than cereals and their products reported in the present book/ article[398]: Snacks, AFB_1, AFB_2, and AFG_2, literature[398]

Aflatoxin G_2

incidence: 3/106*, conc. range: 7–10 µg/kg, Ø conc.: 8.0 µg/kg, sample origin: street hawkers, markets, and retail shops, Southwest Nigeria, sample year: unknown, country: Nigeria[398], *groundnut snacks, dry-roasted

- Co-contamination: not reported
- Further contamination of plants and plant products other than cereals and their products reported in the present book/ article[398]: Snacks, AFB_1, AFB_2, and AFG_1, literature[398]

Aflatoxin M_1

incidence: 2/10*, conc. range: 256–410 µg/kg, ∅ conc.: 333 µg/kg, sample origin: street traders in Lagos (city), Nigeria, sample year: unknown, country: Nigeria/Austria[763], *snacks, groundnut-based

incidence: 0/2*, conc.: no contamination, sample origin: street traders in Lagos (city), Nigeria, sample year: unknown, country: Nigeria/Austria[763], *snacks, groundnut-/maize-based

- Co-contamination: For detailed information see the article.
- Further contamination of plants and plant products other than cereals and their products reported in the present book/article[763]: Snacks, 3-NPA, AFB_1, AFB_2, AME, BEA, FB_1, FB_2, FB_3, KA, and MON, literature[763]; see also Snacks, chapter: Further Mycotoxins and Microbial Metabolites

Aflatoxin/s

incidence: 5/11*, conc. range: 7–112 µg/kg, sample origin: traditional markets and supermarkets, Indonesia, sample year: 2001/2002, country: Austria/Indonesia/Thailand[226], *snacks, traditional (fried floured peanut, peanuts with brown sugar, and sweet-grounded peanuts)

- Co-contamination: not reported
- Further contamination of plants and plant products other than cereals and their products reported in the present book/article[226]: Butter (peanut butter), Cake (peanut), Nuts (peanut), Sauces (peanut chilli sauce), AFB_1 and AF/S, literature[226]; Snacks, AFB_1, literature[226]

incidence: 4/40*, conc. range: 10–40 µg/kg**, sample origin: markets in Idadan (city), Southwest Nigeria, sample year: unknown, country: Nigeria[529], *"robo" = fried melon balls, **AFB_1, AFB_2, and AFG_1

- Co-contamination: 1 sa. co-contaminated with AFB_1 and AFB_2; 2 sa. co-contaminated with AFB_1 and AFG_1; 1 sa. co-contaminated with AFB_1, AFB_2, and AFG_1
- Further contamination of plants and plant products other than cereals and their products reported in the present book/article[529]: not reported

incidence: 9/12*, conc. range: ≤43.8 µg/kg**, sample origin: markets and street corners in Lagos metropolis, Nigeria, sample year: unknown, country: Nigeria[721], *snacks, groundnut-based (groundnut cake, peanut candy, roasted and boiled groundnut), **AFB_1, AFB_2, and AFG_1

incidence: 1/5*, conc.: 6.0 µg/kg**, sample origin: markets and street corners in Lagos metropolis, Nigeria, sample year: unknown, country: Nigeria[721], *snacks, nut-based (tiger nut, gurundi, coconut chips, and walnut), **AFB_1, AFB_2, and AFG_1

- Co-contamination: not reported
- Further contamination of plants and plant products other than cereals and their products reported in the present book/article[721]: not reported

Aspergillus and *Penicillium* Toxins

3-Nitropropionic Acid

incidence: 3/10*, conc. range: 104–150 µg/kg, sample origin: street traders in Lagos (city), Nigeria, sample year: unknown, country: Nigeria/Austria[763], *snacks, groundnut-based

incidence: 2/2*, conc. range: 2.9–4.8 µg/kg, ∅ conc.: 3.85 µg/kg, sample origin: street traders in Lagos (city), Nigeria, sample year: unknown, country: Nigeria/Austria[763], *snacks, groundnut-/maize-based

- Co-contamination: For detailed information see the article.
- Further contamination of plants and plant products other than cereals and their products reported in the present book/article[763]: Snacks, AFB_1, AFB_2, AFM_1, AME, BEA, FB_1, FB_2, FB_3, KA, and MON, literature[763]; see also Snacks, chapter: Further Mycotoxins and Microbial Metabolites

Kojic Acid

incidence: 6/10*, conc. range: 23–485 µg/kg, sample origin: street traders in Lagos (city), Nigeria, sample year: unknown, country: Nigeria/Austria[763], *snacks, groundnut-based

incidence: 2/2*, conc. range: 194.4–218 µg/kg, ∅ conc.: 206 µg/kg, sample origin: street traders in Lagos (city), Nigeria, sample year: unknown, country: Nigeria/Austria[763], *snacks, groundnut-/maize-based

- Co-contamination: For detailed information see the article.
- Further contamination of plants and plant products other than cereals and their products reported in the present book/article[763]: Snacks, 3-NPA, AFB_1, AFB_2, AFM_1, AME, BEA, FB_1, FB_2, FB_3, and MON, literature[763]; see also Snacks, chapter: Further Mycotoxins and Microbial Metabolites

Fusarium Toxins

Depsipeptides

Beauvericin

incidence: 6/10*, conc. range: 2–84 µg/kg, sample origin: street traders in Lagos (city), Nigeria, sample year: unknown, country: Nigeria/Austria[763], *snacks, groundnut-based

incidence: 2/2*, conc. range: 1.8–1.9 µg/kg, ∅ conc.: 1.85 µg/kg, sample origin: street traders in Lagos (city), Nigeria, sample year: unknown, country: Nigeria/Austria[763], *snacks, groundnut-/maize-based

- Co-contamination: For detailed information see the article.
- Further contamination of plants and plant products other than cereals and their products reported in the present book/article[763]: Snacks, 3-NPA, AFB_1, AFB_2, AFM_1, AME, FB_1, FB_2, FB_3, KA, and MON, literature[763]; see also Snacks, chapter: Further Mycotoxins and Microbial Metabolites

Fumonisin/s

Fumonisin B_1

incidence: 0/10*, conc.: no contamination, sample origin: street traders in Lagos (city), Nigeria, sample year: unknown, country: Nigeria/Austria[763], *snacks, groundnut-based

incidence: 2/2*, conc. range: 124–130 µg/kg, ∅ conc.: 127 µg/kg, sample origin: street traders in Lagos (city), Nigeria, sample year: unknown, country: Nigeria/Austria[763], *snacks, groundnut-/maize-based

- Co-contamination: For detailed information see the article.
- Further contamination of plants and plant products other than cereals and their products reported in the present book/article[763]: Snacks, 3-NPA, AFB_1, AFB_2, AFM_1, AME, BEA, FB_2, FB_3, KA, and MON, literature[763]; see also Snacks, chapter: Further Mycotoxins and Microbial Metabolites

Fumonisin B_2

incidence: 0/10*, conc.: no contamination, sample origin: street traders in Lagos (city), Nigeria, sample year: unknown, country:

Nigeria/Austria[763], *snacks, groundnut-based

incidence: 2/2*, conc. range: 35–40 μg/kg, Ø conc.: 37.5 μg/kg, sample origin: street traders in Lagos (city), Nigeria, sample year: unknown, country: Nigeria/Austria[763], *snacks, groundnut-/maize-based

- Co-contamination: For detailed information see the article.
- Further contamination of plants and plant products other than cereals and their products reported in the present book/article[763]: Snacks, 3-NPA, AFB_1, AFB_2, AFM_1, AME, BEA, FB_1, FB_3, KA, and MON, literature[763]; see also Snacks, chapter: Further Mycotoxins and Microbial Metabolites

Fumonisin B_3
incidence: 0/10*, conc.: no contamination, sample origin: street traders in Lagos (city), Nigeria, sample year: unknown, country: Nigeria/Austria[763], *snacks, groundnut-based

incidence: 2/2*, conc. range: 12.5–14.4 μg/kg, Ø conc.: 13.5 μg/kg, sample origin: street traders in Lagos (city), Nigeria, sample year: unknown, country: Nigeria/Austria[763], *snacks, groundnut-/maize-based

- Co-contamination: For detailed information see the article.
- Further contamination of plants and plant products other than cereals and their products reported in the present book/article[763]: Snacks, 3-NPA, AFB_1, AFB_2, AFM_1, AME, BEA, FB_1, FB_2, KA, and MON, literature[763]; see also Snacks, chapter: Further Mycotoxins and Microbial Metabolites

Other ***Fusarium*** Toxins

Moniliformin
incidence: 0/10*, conc.: no contamination, sample origin: street traders in Lagos (city), Nigeria, sample year: unknown, country: Nigeria/Austria[763], *snacks, groundnut-based

incidence: 2/2*, conc. range: 68–72 μg/kg, Ø conc.: 70 μg/kg, sample origin: street-traders in Lagos (city), Nigeria, sample year: unknown, country: Nigeria/Austria[763], *snacks, groundnut-/maize-based

- Co-contamination: For detailed information see the article.
- Further contamination of plants and plant products other than cereals and their products reported in the present book/article[763]: Snacks, 3-NPA, AFB_1, AFB_2, AFM_1, AME, BEA, FB_1, FB_2, FB_3, and KA, literature[763]; see also Snacks, chapter: Further Mycotoxins and Microbial Metabolites

Soft drink see Juices (apple juice)

Soups may contain the following mycotoxins:

Alternaria Toxins

Alternariol
incidence: 4/7*, conc. range: 0.04–3 μg/kg, Ø conc.: 0.8 μg/kg, sample origin: Yaounde (region, southern tropical forest), Cameroon, sample year: July–September 2011, country: Cameroon/Austria/South Africa[659], *peanut soup

incidence: 3/8*, conc. range: 0.1–2 μg/kg, Ø conc.: 0.7 μg/kg, sample origin: Bamenda (region, western highlands), Cameroon, sample year: July–September 2011, country: Cameroon/Austria/South Africa[659], *peanut soup

- Co-contamination: For detailed information see the article.
- Further contamination of plants and plant products other than cereals and their products reported in the present book/article[659]: Dagwa, AFB_1, AFB_2, AFG_1, AFG_2, AME, BEA, CIT, DON, ENA, ENA_1, ENB, ENB_1, FA, FB_1, FB_2, FB_3, KA, MA, NIV, OTA, OTB, STG, VIOM, and ZEA, literature[659]; Flour (soybean flour), AFB_1, AFG_1, BEA, DON, ENB, ENB_1, FA, FB_1, FB_2, FB_3, FUS-X, KA, NIV, and ZEA, literature[659]; Kuru-kuru, AFB_1, AFB_2, AFG_1, AFG_2, BEA, ENA, ENA_1, ENB, ENB_1, FA, FB_1, FB_3, KA, MA, OTA, OTB, ROQ-C, STG, VIOM, and ZEA, literature[659]; Paste (peanut paste), AFB_1, AFB_2, AFG_1, AFG_2, AME, BEA, ENA, ENA_1, ENB, ENB_1, FA, FB_1, FB_2, FB_3, KA, MA, OTA, OTB, STG, VIOM, and ZEA, literature[659]; Soups, AFB_1, AFB_2, AFG_1, AFG_2, AME, BEA, DON, ENA, ENA_1, ENB, ENB_1, FB_1, FB_2, FB_3, KA, OTA, OTB, and STG, literature[659]; see also Dagwa, Flour (soybean flour), Kuru-kuru, Paste (peanut paste), Soups, chapter: Further Mycotoxins and Microbial Metabolites

Alternariol Methyl Ether
incidence: 6/7*, conc. range: 0.2–23.6 μg/kg, Ø conc.: 6.4 μg/kg, sample origin: Yaounde (region, southern tropical forest), Cameroon, sample year: July–September 2011, country: Cameroon/Austria/South Africa[659], *peanut soup

incidence: 7/8*, conc. range: 0.2–1.8 μg/kg, Ø conc.: 0.9 μg/kg, sample origin: Bamenda (region, western highlands), Cameroon, sample year: July–September 2011, country: Cameroon/Austria/South Africa[659], *peanut soup

- Co-contamination: For detailed information see the article.
- Further contamination of plants and plant products other than cereals and their products reported in the present book/article[659]: Dagwa, AFB_1, AFB_2, AFG_1, AFG_2, AME, BEA, CIT, DON, ENA, ENA_1, ENB, ENB_1, FA, FB_1, FB_2, FB_3, KA, MA, NIV, OTA, OTB, STG, VIOM, and ZEA, literature[659]; Flour (soybean flour), AFB_1, AFG_1, BEA, DON, ENB, ENB_1, FA, FB_1, FB_2, FB_3, FUS-X, KA, NIV, and ZEA, literature[659]; Kuru-kuru, AFB_1, AFB_2, AFG_1, AFG_2, BEA, ENA, ENA_1, ENB, ENB_1, FA, FB_1, FB_3, KA, MA, OTA, OTB, ROQ-C, STG, VIOM, and ZEA, literature[659]; Paste (peanut paste), AFB_1, AFB_2, AFG_1, AFG_2, AME, BEA, ENA, ENA_1, ENB, ENB_1, FA, FB_1, FB_2, FB_3, KA, MA, OTA, OTB, STG, VIOM, and ZEA, literature[659]; Soups, AFB_1, AFB_2, AFG_1, AFG_2, AOH, BEA, DON, ENA, ENA_1, ENB, ENB_1, FB_1, FB_2, FB_3, KA, OTA, OTB, and STG, literature[659]; see also Dagwa, Flour (soybean flour), Kuru-kuru, Paste (peanut paste), Soups, chapter: Further Mycotoxins and Microbial Metabolites

Aspergillus Toxins

Aflatoxin B_1
incidence: 5/7*, conc. range: <LOQ-37 μg/kg, Ø conc.: 16 μg/kg, sample origin: Yaounde (region, southern tropical forest), Cameroon, sample year: July–September 2011, country: Cameroon/Austria/South Africa[659], *peanut soup

incidence: 6/8*, conc. range: 1–36 μg/kg, Ø conc.: 14 μg/kg, sample origin: Bamenda (region, western highlands), Cameroon, sample year: July–September 2011, country: Cameroon/Austria/South Africa[659], *peanut soup

- Co-contamination: For detailed information see the article.
- Further contamination of plants and plant products other than cereals and their products reported in the present book/article[659]: Dagwa, AFB_1, AFB_2, AFG_1, AFG_2, AME, BEA, CIT, DON, ENA, ENA_1, ENB, ENB_1, FA, FB_1, FB_2, FB_3, KA, MA, NIV,

OTA, OTB, STG, VIOM, and ZEA, literature[659]; Flour (soybean flour), AFB_1, AFG_1, BEA, DON, ENB, ENB_1, FA, FB_1, FB_2, FB_3, FUS-X, KA, NIV, and ZEA, literature[659]; Kuru-kuru, AFB_1, AFB_2, AFG_1, AFG_2, BEA, ENA, ENA_1, ENB, ENB_1, FA, FB_1, FB_3, KA, MA, OTA, OTB, ROQ-C, STG, VIOM, and ZEA, literature[659]; Paste (peanut paste), AFB_1, AFB_2, AFG_1, AFG_2, AME, BEA, ENA, ENA_1, ENB, ENB_1, FA, FB_1, FB_2, FB_3, KA, MA, OTA, OTB, STG, VIOM, and ZEA, literature[659]; Soups, AFB_2, AFG_1, AFG_2, AME, AOH, BEA, DON, ENA, ENA_1, ENB, ENB_1, FB_1, FB_2, FB_3, KA, OTA, OTB, and STG, literature[659]; see also Dagwa, Flour (soybean flour), Kuru-kuru, Paste (peanut paste), Soups, chapter: Further Mycotoxins and Microbial Metabolites

Aflatoxin B_2

incidence: 4/7*, conc. range: <LOQ-8 µg/kg, ∅ conc.: 3 µg/kg, sample origin: Yaounde (region, southern tropical forest), Cameroon, sample year: July–September 2011, country: Cameroon/Austria/South Africa[659], *peanut soup

incidence: 5/8*, conc. range: 0.4–7 µg/kg, ∅ conc.: 3 µg/kg, sample origin: Bamenda (region, western highlands), Cameroon, sample year: July/September 2011, country: Cameroon/Austria/South Africa[659], *peanut soup

- Co-contamination: For detailed information see the article.
- Further contamination of plants and plant products other than cereals and their products reported in the present book/article[659]: Dagwa, AFB_1, AFB_2, AFG_1, AFG_2, AME, BEA, CIT, DON, ENA, ENA_1, ENB, ENB_1, FA, FB_1, FB_2, FB_3, KA, MA, NIV, OTA, OTB, STG, VIOM, and ZEA, literature[659]; Flour (soybean flour), AFB_1, AFG_1, BEA, DON, ENB, ENB_1, FA, FB_1, FB_2, FB_3, FUS-X, KA, NIV, and ZEA, literature[659]; Kuru-kuru, AFB_1, AFB_2, AFG_1, AFG_2, BEA, ENA, ENA_1, ENB, ENB_1, FA, FB_1, FB_3, KA, MA, OTA, OTB, ROQ-C, STG, VIOM, and ZEA, literature[659]; Paste (peanut paste), AFB_1, AFB_2, AFG_1, AFG_2, AME, BEA, ENA, ENA_1, ENB, ENB_1, FA, FB_1, FB_2, FB_3, KA, MA, OTA, OTB, STG, VIOM, and ZEA, literature[659]; Soups, AFB_1, AFG_1, AFG_2, AME, AOH, BEA, DON, ENA, ENA_1, ENB, ENB_1, FB_1, FB_2, FB_3, KA, OTA, OTB, and STG, literature[659]; see also Dagwa, Flour (soybean flour), Kuru-kuru, Paste (peanut paste), Soups, chapter: Further Mycotoxins and Microbial Metabolites

Aflatoxin G_1

incidence: 3/7*, conc. range: 0.6–44 µg/kg, ∅ conc.: 19 µg/kg, sample origin: Yaounde (region, southern tropical forest), Cameroon, sample year: July/September 2011, country: Cameroon/Austria/South Africa[659], *peanut soup

incidence: 3/8*, conc. range: 2.4–6 µg/kg, ∅ conc.: 5 µg/kg, sample origin: Bamenda (region, western highlands), Cameroon, sample year: July–September 2011, country: Cameroon/Austria/South Africa[659], *peanut soup

- Co-contamination: For detailed information see the article.
- Further contamination of plants and plant products other than cereals and their products reported in the present book/article[659]: Dagwa, AFB_1, AFB_2, AFG_1, AFG_2, AME, BEA, CIT, DON, ENA, ENA_1, ENB, ENB_1, FA, FB_1, FB_2, FB_3, KA, MA, NIV, OTA, OTB, STG, VIOM, and ZEA, literature[659]; Flour (soybean flour), AFB_1, AFG_1, BEA, DON, ENB, ENB_1, FA, FB_1, FB_2, FB_3, FUS-X, KA, NIV, and ZEA, literature[659]; Kuru-kuru, AFB_1, AFB_2, AFG_1, AFG_2, BEA, ENA, ENA_1, ENB, ENB_1, FA, FB_1, FB_3, KA, MA, OTA, OTB, ROQ-C, STG, VIOM, and ZEA, literature[659]; Paste (peanut paste), AFB_1, AFB_2, AFG_1, AFG_2, AME, BEA, ENA, ENA_1, ENB, ENB_1, FA, FB_1, FB_2, FB_3, KA, MA, OTA, OTB, STG, VIOM, and ZEA, literature[659]; Soups, AFB_1, AFB_2, AFG_2, AME, AOH, BEA, DON, ENA, ENA_1, ENB, ENB_1, FB_1, FB_2, FB_3, KA, OTA, OTB, and STG, literature[659]; see also Dagwa, Flour (soybean flour), Kuru-kuru, Paste (peanut paste), Soups, chapter: Further Mycotoxins and Microbial Metabolites

Aflatoxin G_2

incidence: 3/7*, conc. range: 10–36 µg/kg, ∅ conc.: 23 µg/kg, sample origin: Yaounde (region, southern tropical forest), Cameroon, sample year: July–September 2011, country: Cameroon/Austria/South Africa[659], *peanut soup

incidence: 7/8*, conc. range: 10.8–22.9 µg/kg, ∅ conc.: 17.6 µg/kg, sample origin: Bamenda (region, western highlands), Cameroon, sample year: July/September 2011, country: Cameroon/Austria/South Africa[659], *peanut soup

- Co-contamination: For detailed information see the article.
- Further contamination of plants and plant products other than cereals and their products reported in the present book/article[659]: Dagwa, AFB_1, AFB_2, AFG_1, AFG_2, AME, BEA, CIT, DON, ENA, ENA_1, ENB, ENB_1, FA, FB_1, FB_2, FB_3, KA, MA, NIV, OTA, OTB, STG, VIOM, and ZEA, literature[659]; Flour (soybean flour), AFB_1, AFG_1, BEA, DON, ENB, ENB_1, FA, FB_1, FB_2, FB_3, FUS-X, KA, NIV, and ZEA, literature[659]; Kuru-kuru, AFB_1, AFB_2, AFG_1, AFG_2, BEA, ENA, ENA_1, ENB, ENB_1, FA, FB_1, FB_3, KA, MA, OTA, OTB, ROQ-C, STG, VIOM, and ZEA, literature[659]; Paste (peanut paste), AFB_1, AFB_2, AFG_1, AFG_2, AME, BEA, ENA, ENA_1, ENB, ENB_1, FA, FB_1, FB_2, FB_3, KA, MA, OTA, OTB, STG, VIOM, and ZEA, literature[659]; Soups, AFB_1, AFB_2, AFG_1, AME, AOH, BEA, DON, ENA, ENA_1, ENB, ENB_1, FB_1, FB_2, FB_3, KA, OTA, OTB, and STG, literature[659]; see also Dagwa, Flour (soybean flour), Kuru-kuru, Paste (peanut paste), Soups, chapter: Further Mycotoxins and Microbial Metabolites

Aflatoxin/s

incidence: 9/10*, conc. range: ≤3.073 µg/kg**, sample origin: groceries and supermarkets, Turkey, sample year: 2003, country: Turkey[119], *instant soup, **AFS total

- Co-contamination: not reported
- Further contamination of plants and plant products other than cereals and their products reported in the present book/article[119]: Asure, Beverages, Cocoa, Cream, Ice cream, Jello, Keskul, Muhallebi, Pudding, Salep, Sauces (chocolate sauce), AF/S, literature[119]

In this case instant soup ingredients were lentil flour, corn starch, salt, wheat flour, vegetable oil, various spices, flavor enhancer monosodium glutamate, and acidity regulator citric acid.

Other ***Aspergillus*** Toxins

Sterigmatocystin

incidence: 2/7*, conc. range: <LOQ-0.5 µg/kg, ∅ conc.: 0.5 µg/kg, sample origin: Yaounde (region, southern tropical forest), Cameroon, sample year: July–September 2011, country: Cameroon/Austria/South Africa[659], *peanut soup

incidence: 3/8*, conc. range: <LOQ-0.6 µg/kg, ∅ conc.: 0.6 µg/kg, sample origin: Bamenda (region, western highlands), Cameroon, sample year: July–September 2011, country: Cameroon/Austria/South Africa[659], *peanut soup

- Co-contamination: For detailed information see the article.

- Further contamination of plants and plant products other than cereals and their products reported in the present book/article[659]: Dagwa, AFB_1, AFB_2, AFG_1, AFG_2, AME, BEA, CIT, DON, ENA, ENA_1, ENB, ENB_1, FA, FB_1, FB_2, FB_3, KA, MA, NIV, OTA, OTB, STG, VIOM, and ZEA, literature[659]; Flour (soybean flour), AFB_1, AFG_1, BEA, DON, ENB, ENB_1, FA, FB_1, FB_2, FB_3, FUS-X, KA, NIV, and ZEA, literature[659]; Kuru-kuru, AFB_1, AFB_2, AFG_1, AFG_2, BEA, ENA, ENA_1, ENB, ENB_1, FA, FB_1, FB_3, KA, MA, OTA, OTB, ROQ-C, STG, VIOM, and ZEA, literature[659]; Paste (peanut paste), AFB_1, AFB_2, AFG_1, AFG_2, AME, BEA, ENA, ENA_1, ENB, ENB_1, FA, FB_1, FB_2, FB_3, KA, MA, OTA, OTB, STG, VIOM, and ZEA, literature[659]; Soups, AFB_1, AFB_2, AFG_1, AFG_2, AME, AOH, DON, BEA, ENA, ENA_1, ENB, ENB_1, FB_1, FB_2, FB_3, KA, OTA, and OTB, literature[659]; see also Dagwa, Flour (soybean flour), Kuru-kuru, Paste (peanut paste), Soups, chapter: Further Mycotoxins and Microbial Metabolites

Aspergillus and ***Penicillium*** Toxins

Kojic Acid

incidence: 7/7*, conc. range: 34–329 µg/kg, Ø conc.: 130 µg/kg, sample origin: Yaounde (region, southern tropical forest), Cameroon, sample year: July–September 2011, country: Cameroon/Austria/South Africa[659], *peanut soup

incidence: 8/8*, conc. range: 72–509 µg/kg, Ø conc.: 268 µg/kg, sample origin: Bamenda (region, western highlands), Cameroon, sample year: July/September 2011, country: Cameroon/Austria/South Africa[659], *peanut soup

- Co-contamination: For detailed information see the article.
- Further contamination of plants and plant products other than cereals and their products reported in the present book/article[659]: Dagwa, AFB_1, AFB_2, AFG_1, AFG_2, AME, BEA, CIT, DON, ENA, ENA_1, ENB, ENB_1, FA, FB_1, FB_2, FB_3, KA, MA, NIV, OTA, OTB, STG, VIOM, and ZEA, literature[659]; Flour (soybean flour), AFB_1, AFG_1, BEA, DON, ENB, ENB_1, FA, FB_1, FB_2, FB_3, FUS-X, KA, NIV, and ZEA, literature[659]; Kuru-kuru, AFB_1, AFB_2, AFG_1, AFG_2, BEA, ENA, ENA_1, ENB, ENB_1, FA, FB_1, FB_3, KA, MA, OTA, OTB, ROQ-C, STG, VIOM, and ZEA, literature[659]; Paste (peanut paste), AFB_1, AFB_2, AFG_1, AFG_2, AME, BEA, ENA, ENA_1, ENB, ENB_1, FA, FB_1, FB_2, FB_3, KA, MA, OTA, OTB, STG, VIOM, and ZEA, literature[659]; Soups, AFB_1, AFB_2, AFG_1, AFG_2, AME, AOH, DON, BEA, ENA, ENA_1, ENB, ENB_1, FB_1, FB_2, FB_3, OTA, OTB, and STG, literature[659]; see also Dagwa, Flour (soybean flour), Kuru-kuru, Paste (peanut paste), Soups, chapter: Further Mycotoxins and Microbial Metabolites

Ochratoxin A

incidence: 1/7*, conc.: <LOQ, sample origin: Yaounde (region, southern tropical forest), Cameroon, sample year: July–September 2011, country: Cameroon/Austria/South Africa[659], *peanut soup

incidence: 5/8*, conc. range: <LOQ-2.7 µg/kg, Ø conc.: 1.7 µg/kg, sample origin: Bamenda (region, western highlands), Cameroon, sample year: July/September 2011, country: Cameroon/Austria/South Africa[659], *peanut soup

- Co-contamination: For detailed information see the article.
- Further contamination of plants and plant products other than cereals and their products reported in the present book/article[659]: Dagwa, AFB_1, AFB_2, AFG_1, AFG_2, AME, BEA, CIT, DON, ENA, ENA_1, ENB, ENB_1, FA, FB_1, FB_2, FB_3, KA, MA, NIV, OTA, OTB, STG, VIOM, and ZEA, literature[659]; Flour (soybean flour), AFB_1, AFG_1, BEA, DON, ENB, ENB_1, FA, FB_1, FB_2, FB_3, FUS-X, KA, NIV, and ZEA, literature[659]; Kuru-kuru, AFB_1, AFB_2, AFG_1, AFG_2, BEA, ENA, ENA_1, ENB, ENB_1, FA, FB_1, FB_3, KA, MA, OTA, OTB, ROQ-C, STG, VIOM, and ZEA, literature[659]; Paste (peanut paste), AFB_1, AFB_2, AFG_1, AFG_2, AME, BEA, ENA, ENA_1, ENB, ENB_1, FA, FB_1, FB_2, FB_3, KA, MA, OTA, OTB, STG, VIOM, and ZEA, literature[659]; Soups, AFB_1, AFB_2, AFG_1, AFG_2, AME, AOH, DON, BEA, ENA, ENA_1, ENB, ENB_1, FB_1, FB_2, FB_3, KA, OTB, and STG, literature[659]; see also Dagwa, Flour (soybean flour), Kuru-kuru, Paste (peanut paste), Soups, chapter: Further Mycotoxins and Microbial Metabolites

Ochratoxin B

incidence: 2/7*, conc. range: <LOQ, sample origin: Yaounde (region, southern tropical forest), Cameroon, sample year: July–September 2011, country: Cameroon/Austria/South Africa[659], *peanut soup

incidence: 4/8*, conc. range: <LOQ-1 µg/kg, Ø conc.: 1 µg/kg, sample origin: Bamenda (region, western highlands), Cameroon, sample year: July/September 2011, country: Cameroon/Austria/South Africa[659], *peanut soup

- Co-contamination: For detailed information see the article.
- Further contamination of plants and plant products other than cereals and their products reported in the present book/article[659]: Dagwa, AFB_1, AFB_2, AFG_1, AFG_2, AME, BEA, CIT, DON, ENA, ENA_1, ENB, ENB_1, FA, FB_1, FB_2, FB_3, KA, MA, NIV, OTA, OTB, STG, VIOM, and ZEA, literature[659]; Flour (soybean flour), AFB_1, AFG_1, BEA, DON, ENB, ENB_1, FA, FB_1, FB_2, FB_3, FUS-X, KA, NIV, and ZEA, literature[659]; Kuru-kuru, AFB_1, AFB_2, AFG_1, AFG_2, BEA, ENA, ENA_1, ENB, ENB_1, FA, FB_1, FB_3, KA, MA, OTA, OTB, ROQ-C, STG, VIOM, and ZEA, literature[659]; Paste (peanut paste), AFB_1, AFB_2, AFG_1, AFG_2, AME, BEA, ENA, ENA_1, ENB, ENB_1, FA, FB_1, FB_2, FB_3, KA, MA, OTA, OTB, STG, VIOM, and ZEA, literature[659]; Soups, AFB_1, AFB_2, AFG_1, AFG_2, AME, AOH, DON, BEA, ENA, ENA_1, ENB, ENB_1, FB_1, FB_2, FB_3, KA, OTA, and STG, literature[659]; see also Dagwa, Flour (soybean flour), Kuru-kuru, Paste (peanut paste), Soups, chapter: Further Mycotoxins and Microbial Metabolites

Fusarium Toxins

Depsipeptides

Beauvericin

incidence: 7/7*, conc. range: 0.04–0.7 µg/kg, Ø conc.: 0.4 µg/kg, sample origin: Yaounde (region, southern tropical forest), Cameroon, sample year: July/September 2011, country: Cameroon/Austria/South Africa[659], *peanut soup

incidence: 8/8*, conc. range: 0.1–1.2 µg/kg, Ø conc.: 0.6 µg/kg, sample origin: Bamenda (region, western highlands), Cameroon, sample year: July/September 2011, country: Cameroon/Austria/South Africa[659], *peanut soup

- Co-contamination: For detailed information see the article.
- Further contamination of plants and plant products other than cereals and their products reported in the present book/article[659]: Dagwa, AFB_1, AFB_2, AFG_1, AFG_2, AME, BEA, CIT, DON, ENA, ENA_1, ENB, ENB_1, FA, FB_1, FB_2, FB_3, KA, MA, NIV, OTA, OTB, STG, VIOM, and ZEA, literature[659]; Flour (soybean flour), AFB_1, AFG_1, BEA, DON, ENB, ENB_1, FA, FB_1, FB_2, FB_3, FUS-X, KA, NIV, and ZEA, literature[659]; Kuru-kuru, AFB_1, AFB_2, AFG_1, AFG_2, BEA, ENA, ENA_1, ENB, ENB_1, FA, FB_1, FB_3, KA, MA, OTA, OTB, ROQ-C, STG, VIOM, and ZEA, literature[659]; Paste (peanut paste), AFB_1, AFB_2, AFG_1, AFG_2, AME, BEA, ENA, ENA_1,

ENB, ENB_1, FA, FB_1, FB_2, FB_3, KA, MA, OTA, OTB, STG, VIOM, and ZEA, literature[659]; Soups, AFB_1, AFB_2, AFG_1, AFG_2, AME, AOH, DON, ENA, ENA_1, ENB, ENB_1, FB_1, FB_2, FB_3, KA, OTA, OTB, and STG, literature[659]; see also Dagwa, Flour (soybean flour), Kuru-kuru, Paste (peanut paste), Soups, chapter: Further Mycotoxins and Microbial Metabolites

Enniatin A

incidence: 6/7*, conc. range: 0.02–0.03 µg/kg, Ø conc.: 0.02 µg/kg, sample origin: Yaounde (region, southern tropical forest), Cameroon, sample year: July–September 2011, country: Cameroon/Austria/South Africa[659], *peanut soup

incidence: 7/8*, conc. range: <LOQ-0.1 µg/kg, Ø conc.: 0.04 µg/kg, sample origin: Bamenda (region, western highlands), Cameroon, sample year: July–September 2011, country: Cameroon/Austria/South Africa[659], *peanut soup

- Co-contamination: For detailed information see the article.
- Further contamination of plants and plant products other than cereals and their products reported in the present book/article[659]: Dagwa, AFB_1, AFB_2, AFG_1, AFG_2, AME, BEA, CIT, DON, ENA, ENA_1, ENB, ENB_1, FA, FB_1, FB_2, FB_3, KA, MA, NIV, OTA, OTB, STG, VIOM, and ZEA, literature[659]; Flour (soybean flour), AFB_1, AFG_1, BEA, DON, ENB, ENB_1, FA, FB_1, FB_2, FB_3, FUS-X, KA, NIV, and ZEA, literature[659]; Kuru-kuru, AFB_1, AFB_2, AFG_1, AFG_2, BEA, ENA, ENA_1, ENB, ENB_1, FA, FB_1, FB_3, KA, MA, OTA, OTB, ROQ-C, STG, VIOM, and ZEA, literature[659]; Paste (peanut paste), AFB_1, AFB_2, AFG_1, AFG_2, AME, BEA, ENA, ENA_1, ENB, ENB_1, FA, FB_1, FB_2, FB_3, KA, MA, OTA, OTB, STG, VIOM, and ZEA, literature[659]; Soups, AFB_1, AFB_2, AFG_1, AFG_2, AME, AOH, DON, BEA, ENA_1, ENB, ENB_1, FB_1, FB_2, FB_3, KA, OTA, OTB, and STG, literature[659]; see also Dagwa, Flour (soybean flour), Kuru-kuru, Paste (peanut paste), Soups, chapter: Further Mycotoxins and Microbial Metabolites

Enniatin A_1

incidence: 7/7*, conc. range: 0.02–0.2 µg/kg, Ø conc.: 0.1 µg/kg, sample origin: Yaounde (region, southern tropical forest), Cameroon, sample year: July–September 2011, country: Cameroon/Austria/South Africa[659], *peanut soup

incidence: 8/8*, conc. range: <LOQ-0.1 µg/kg, Ø conc.: 0.1 µg/kg, sample origin: Bamenda (region, western highlands), Cameroon, sample year: July/September 2011, country: Cameroon/Austria/South Africa[659], *peanut soup

- Co-contamination: For detailed information see the article.
- Further contamination of plants and plant products other than cereals and their products reported in the present book/article[659]: Dagwa, AFB_1, AFB_2, AFG_1, AFG_2, AME, BEA, CIT, DON, ENA, ENA_1, ENB, ENB_1, FA, FB_1, FB_2, FB_3, KA, MA, NIV, OTA, OTB, STG, VIOM, and ZEA, literature[659]; Flour (soybean flour), AFB_1, AFG_1, BEA, DON, ENB, ENB_1, FA, FB_1, FB_2, FB_3, FUS-X, KA, NIV, and ZEA, literature[659]; Kuru-kuru, AFB_1, AFB_2, AFG_1, AFG_2, BEA, ENA, ENA_1, ENB, ENB_1, FA, FB_1, FB_3, KA, MA, OTA, OTB, ROQ-C, STG, VIOM, and ZEA, literature[659]; Paste (peanut paste), AFB_1, AFB_2, AFG_1, AFG_2, AME, BEA, ENA, ENA_1, ENB, ENB_1, FA, FB_1, FB_2, FB_3, KA, MA, OTA, OTB, STG, VIOM, and ZEA, literature[659]; Soups, AFB_1, AFB_2, AFG_1, AFG_2, AME, AOH, DON, BEA, ENA, ENB, ENB_1, FB_1, FB_2, FB_3, KA, OTA, OTB, and STG, literature[659]; see also Dagwa, Flour (soybean flour), Kuru-kuru, Paste (peanut paste), Soups, chapter: Further Mycotoxins and Microbial Metabolites

Enniatin B

incidence: 7/7*, conc. range: <LOQ-0.2 µg/kg, Ø conc.: 0.1 µg/kg, sample origin: Yaounde (region, southern tropical forest), Cameroon, sample year: July/September 2011, country: Cameroon/Austria/South Africa[659], *peanut soup

incidence: 8/8*, conc. range: 0.01–0.1 µg/kg, Ø conc.: 0.04 µg/kg, sample origin: Bamenda (region, western highlands), Cameroon, sample year: July/September 2011, country: Cameroon/Austria/South Africa[659], *peanut soup

- Co-contamination: For detailed information see the article.
- Further contamination of plants and plant products other than cereals and their products reported in the present book/article[659]: Dagwa, AFB_1, AFB_2, AFG_1, AFG_2, AME, BEA, CIT, DON, ENA, ENA_1, ENB, ENB_1, FA, FB_1, FB_2, FB_3, KA, MA, NIV, OTA, OTB, STG, VIOM, and ZEA, literature[659]; Flour (soybean flour), AFB_1, AFG_1, BEA, DON, ENB, ENB_1, FA, FB_1, FB_2, FB_3, FUS-X, KA, NIV, and ZEA, literature[659]; Kuru-kuru, AFB_1, AFB_2, AFG_1, AFG_2, BEA, ENA, ENA_1, ENB, ENB_1, FA, FB_1, FB_3, KA, MA, OTA, OTB, ROQ-C, STG, VIOM, and ZEA, literature[659]; Paste (peanut paste), AFB_1, AFB_2, AFG_1, AFG_2, AME, BEA, ENA, ENA_1, ENB, ENB_1, FA, FB_1, FB_2, FB_3, KA, MA, OTA, OTB, STG, VIOM, and ZEA, literature[659]; Soups, AFB_1, AFB_2, AFG_1, AFG_2, AME, AOH, DON, BEA, ENA, ENA_1, ENB_1, FB_1, FB_2, FB_3, KA, OTA, OTB, and STG, literature[659]; see also Dagwa, Flour (soybean flour), Kuru-kuru, Paste (peanut paste), Soups, chapter: Further Mycotoxins and Microbial Metabolites

Enniatin B_1

incidence: 6/7*, conc. range: 0.03–0.3 µg/kg, Ø conc.: 0.2 µg/kg, sample origin: Yaounde (region, southern tropical forest), Cameroon, sample year: July/September 2011, country: Cameroon/Austria/South Africa[659], *peanut soup

incidence: 8/8*, conc. range: 0.01–0.3 µg/kg, Ø conc.: 0.1 µg/kg, sample origin: Bamenda (region, western highlands), Cameroon, sample year: July/September 2011, country: Cameroon/Austria/South Africa[659], *peanut soup

- Co-contamination: For detailed information see the article.
- Further contamination of plants and plant products other than cereals and their products reported in the present book/article[659]: Dagwa, AFB_1, AFB_2, AFG_1, AFG_2, AME, BEA, CIT, DON, ENA, ENA_1, ENB, ENB_1, FA, FB_1, FB_2, FB_3, KA, MA, NIV, OTA, OTB, STG, VIOM, and ZEA, literature[659]; Flour (soybean flour), AFB_1, AFG_1, BEA, DON, ENB, ENB_1, FA, FB_1, FB_2, FB_3, FUS-X, KA, NIV, and ZEA, literature[659]; Kuru-kuru, AFB_1, AFB_2, AFG_1, AFG_2, BEA, ENA, ENA_1, ENB, ENB_1, FA, FB_1, FB_3, KA, MA, OTA, OTB, ROQ-C, STG, VIOM, and ZEA, literature[659]; Paste (peanut paste), AFB_1, AFB_2, AFG_1, AFG_2, AME, BEA, ENA, ENA_1, ENB, ENB_1, FA, FB_1, FB_2, FB_3, KA, MA, OTA, OTB, STG, VIOM, and ZEA, literature[659]; Soups, AFB_1, AFB_2, AFG_1, AFG_2, AME, AOH, DON, BEA, ENA, ENA_1, ENB, FB_1, FB_2, FB_3, KA, OTA, OTB, and STG, literature[659]; see also Dagwa, Flour (soybean flour), Kuru-kuru, Paste (peanut paste), Soups, chapter: Further Mycotoxins and Microbial Metabolites

Fumonisin/s

Fumonisin B_1

incidence: 5/7*, conc. range: 0.6–8 µg/kg, Ø conc.: 2.6 µg/kg, sample origin: Yaounde (region, southern tropical forest), Cameroon, sample year: July/September 2011, country: Cameroon/Austria/South Africa[659], *peanut soup incidence: 6/8*, conc. range: 0.9–17 µg/kg, ∅ conc.: 6 µg/kg, sample origin: Bamenda (region, western highlands), Cameroon, sample year: July/September 2011, country: Cameroon/Austria/South Africa[659], *peanut soup

- Co-contamination: For detailed information see the article.
- Further contamination of plants and plant products other than cereals and their products reported in the present book/article[659]: Dagwa, AFB_1, AFB_2, AFG_1, AFG_2, AME, BEA, CIT, DON, ENA, ENA_1, ENB, ENB_1, FA, FB_1, FB_2, FB_3, KA, MA, NIV, OTA, OTB, STG, VIOM, and ZEA, literature[659]; Flour (soybean flour), AFB_1, AFG_1, BEA, DON, ENB, ENB_1, FA, FB_1, FB_2, FB_3, FUS-X, KA, NIV, and ZEA, literature[659]; Kuru-kuru, AFB_1, AFB_2, AFG_1, AFG_2, BEA, ENA, ENA_1, ENB, ENB_1, FA, FB_1, FB_3, KA, MA, OTA, OTB, ROQ-C, STG, VIOM, and ZEA, literature[659]; Paste (peanut paste), AFB_1, AFB_2, AFG_1, AFG_2, AME, BEA, ENA, ENA_1, ENB, ENB_1, FA, FB_1, FB_2, FB_3, KA, MA, OTA, OTB, STG, VIOM, and ZEA, literature[659]; Soups, AFB_1, AFB_2, AFG_1, AFG_2, AME, AOH, DON, BEA, ENA, ENA_1, ENB, ENB_1, FB_2, FB_3, KA, OTA, OTB, and STG, literature[659]; see also Dagwa, Flour (soybean flour), Kuru-kuru, Paste (peanut paste), Soups, chapter: Further Mycotoxins and Microbial Metabolites

Fumonisin B_2

incidence: 1/7*, conc.: 5.8 µg/kg, sample origin: Yaounde (region, southern tropical forest), Cameroon, sample year: July–September 2011, country: Cameroon/Austria/South Africa[659], *peanut soup

incidence: 4/8*, conc. range: <LOQ-2.4 µg/kg, ∅ conc.: 2.1 µg/kg, sample origin: Bamenda (region, western highlands), Cameroon, sample year: July/September 2011, country: Cameroon/Austria/South Africa[659], *peanut soup

- Co-contamination: For detailed information see the article.
- Further contamination of plants and plant products other than cereals and their products reported in the present book/article[659]: Dagwa, AFB_1, AFB_2, AFG_1, AFG_2, AME, BEA, CIT, DON, ENA, ENA_1, ENB, ENB_1, FA, FB_1, FB_2, FB_3, KA, MA, NIV, OTA, OTB, STG, VIOM, and ZEA, literature[659]; Flour (soybean flour), AFB_1, AFG_1, BEA, DON, ENB, ENB_1, FA, FB_1, FB_2, FB_3, FUS-X, KA, NIV, and ZEA, literature[659]; Kuru-kuru, AFB_1, AFB_2, AFG_1, AFG_2, BEA, ENA, ENA_1, ENB, ENB_1, FA, FB_1, FB_3, KA, MA, OTA, OTB, ROQ-C, STG, VIOM, and ZEA, literature[659]; Paste (peanut paste), AFB_1, AFB_2, AFG_1, AFG_2, AME, BEA, ENA, ENA_1, ENB, ENB_1, FA, FB_1, FB_2, FB_3, KA, MA, OTA, OTB, STG, VIOM, and ZEA, literature[659]; Soups, AFB_1, AFB_2, AFG_1, AFG_2, AME, AOH, DON, BEA, ENA, ENA_1, ENB, ENB_1, FB_1, FB_3, KA, OTA, OTB, and STG, literature[659]; see also Dagwa, Flour (soybean flour), Kuru-kuru, Paste (peanut paste), Soups, chapter: Further Mycotoxins and Microbial Metabolites

Fumonisin B_3

incidence: 1/7*, conc.: 1.9 µg/kg, sample origin: Yaounde (region, southern tropical forest), Cameroon, sample year: July–September 2011, country: Cameroon/Austria/South Africa[659], *peanut soup

incidence: 0/8*, conc.: no contamination, sample origin: Bamenda (region, western highlands), Cameroon, sample year: July–September 2011, country: Cameroon/Austria/South Africa[659], *peanut soup

- Co-contamination: For detailed information see the article.
- Further contamination of plants and plant products other than cereals and their products reported in the present book/article[659]: Dagwa, AFB_1, AFB_2, AFG_1, AFG_2, AME, BEA, CIT, DON, ENA, ENA_1, ENB, ENB_1, FA, FB_1, FB_2, FB_3, KA, MA, NIV, OTA, OTB, STG, VIOM, and ZEA, literature[659]; Flour (soybean flour), AFB_1, AFG_1, BEA, DON, ENB, ENB_1, FA, FB_1, FB_2, FB_3, FUS-X, KA, NIV, and ZEA, literature[659]; Kuru-kuru, AFB_1, AFB_2, AFG_1, AFG_2, BEA, ENA, ENA_1, ENB, ENB_1, FA, FB_1, FB_3, KA, MA, OTA, OTB, ROQ-C, STG, VIOM, and ZEA, literature[659]; Paste (peanut paste), AFB_1, AFB_2, AFG_1, AFG_2, AME, BEA, ENA, ENA_1, ENB, ENB_1, FA, FB_1, FB_2, FB_3, KA, MA, OTA, OTB, STG, VIOM, and ZEA, literature[659]; Soups, AFB_1, AFB_2, AFG_1, AFG_2, AME, AOH, DON, BEA, ENA, ENA_1, ENB, ENB_1, FB_1, FB_2, KA, OTA, OTB, and STG, literature[659]; see also Dagwa, Flour (soybean flour), Kuru-kuru, Paste (peanut paste), Soups, chapter: Further Mycotoxins and Microbial Metabolites

Type B Trichothecenes

Deoxynivalenol

incidence: 1/7*, conc.: 1.5 µg/kg, sample origin: Yaounde (region, southern tropical forest), Cameroon, sample year: July–September 2011, country: Cameroon/Austria/South Africa[659], *peanut soup

incidence: 3/8*, conc. range: 1–1.8 µg/kg, ∅ conc.: 1.3 µg/kg, sample origin: Bamenda (region, western highlands), Cameroon, sample year: July/September 2011, country: Cameroon/Austria/South Africa[659], *peanut soup

- Co-contamination: For detailed information see the article.
- Further contamination of plants and plant products other than cereals and their products reported in the present book/article[659]: Dagwa, AFB_1, AFB_2, AFG_1, AFG_2, AME, BEA, CIT, DON, ENA, ENA_1, ENB, ENB_1, FA, FB_1, FB_2, FB_3, KA, MA, NIV, OTA, OTB, STG, VIOM, and ZEA, literature[659]; Flour (soybean flour), AFB_1, AFG_1, BEA, DON, ENB, ENB_1, FA, FB_1, FB_2, FB_3, FUS-X, KA, NIV, and ZEA, literature[659]; Kuru-kuru, AFB_1, AFB_2, AFG_1, AFG_2, BEA, ENA, ENA_1, ENB, ENB_1, FA, FB_1, FB_3, KA, MA, OTA, OTB, ROQ-C, STG, VIOM, and ZEA, literature[659]; Paste (peanut paste), AFB_1, AFB_2, AFG_1, AFG_2, AME, BEA, ENA, ENA_1, ENB, ENB_1, FA, FB_1, FB_2, FB_3, KA, MA, OTA, OTB, STG, VIOM, and ZEA, literature[659]; Soups, AFB_1, AFB_2, AFG_1, AFG_2, AME, AOH, BEA, ENA, ENA_1, ENB, ENB_1, FB_1, FB_2, FB_3, KA, OTA, OTB, and STG, literature[659]; see also Dagwa, Flour (soybean flour), Kuru-kuru, Paste (peanut paste), Soups, chapter: Further Mycotoxins and Microbial Metabolites

see also **Sauces (tomato sauce)**

Soy isoflavones may contain the following mycotoxins:

Aspergillus and ***Penicillium*** Toxins

Ochratoxin A

incidence: 1/12*, conc.: 1 µg/kg, sample origin: drugstore and the internet, Belgium, sample year: unknown, country: Belgium[77], *5 notified and 7 non-notified sa.

- Co-contamination: not reported
- Further contamination of plants and plant products other than cereals and their products reported in the present book/article[77]: Medicinal plants, OTA, literature[77]; Soy isoflavones, FB_1, literature[77]; Spices (garlic), FB_1, FB_2, FB_3, and OTA, literature[77]; Vegetables (black radish), FB_1 and FB_2, literature[77]

***Fusarium* Toxins**

Fumonisin/s

Fumonisin B_1

incidence: 1/12*, conc.: 4 µg/kg, sample origin: drugstore and the internet, Belgium, sample year: unknown, country: Belgium[77], *5 notified and 7 non-notified sa.

- Co-contamination: not reported
- Further contamination of plants and plant products other than cereals and their products reported in the present book/article[77]: Medicinal plants, Soy isoflavones, OTA, literature[77]; Spices (garlic), FB_1, FB_2, FB_3, and OTA, literature[77]; Vegetables (black radish), FB_1 and FB_2, literature[77]

Soy sauce see **Sauces (soy sauce)**

Soybean paste see **Paste (soybean paste)**

Soybean products see **Products (soybean product)**

Soybeans see **Beans (soybean)**

Sparkling wine see **Wine, miscellaneous**

"Spätlese" see **Wine** and **Wine (white)**

Spices may contain the following mycotoxins:

***Alternaria* Toxins**

Tenuazonic Acid

incidence: 5/7*, conc. range: 60–420 µg/kg, sample origin: unknown, sample year: unknown, country: Germany[166], *included allspice, caraway, cardamom, marjoram, and nutmeg (contaminated); not contaminated: cinnamon and clove

- Co-contamination: not reported
- Further contamination of plants and plant products other than cereals and their products reported in the present book/article[166]: Juices, Juices (apple juice), Juices (apricot juice), Juices (banana juice), Juices (blackcurrant juice), Juices (cherry juice), Juices (grape juice), Juices (orange juice), Juices (pear juice), Juices (pomegranate juice), Juices (strawberry juice), Spices (coriander), Spices (cumin), Spices (curcuma), Spices (curry), Spices (ginger), Spices (paprika), Spices (pepper), TA, literature[166]

***Aspergillus* Toxins**

Aflatoxin B_1

incidence: 5?/37*, conc. range: 0.2–0.8 µg/kg, sample origin: unknown, sample year: unknown, country: Japan[97], *spices, mixed

- Co-contamination: not reported
- Further contamination of plants and plant products other than cereals and their products reported in the present book/article[97]: Beans, Jam (bean jam), AFB_1 and AFB_2, literature[97]; Butter (peanut butter), Oils (peanut oil), Spices (pepper), AFB_1, AFB_2, and AFG_1, literature[97]; Spices, AFB_2, literature[97]; Spices (nutmeg), AFB_1, AFB_2, AFG_1, and AFG_2, literature[97]

incidence: 16?/78*, conc. range: ≤1.9 µg/kg, ∅ conc.: 0.8 µg/kg, sample origin: commercially available market basket sa. of foods and foodstuffs in Tokyo (capital), Japan, sample year: 1986–1990, country: Japan[99], *spices, mixed

- Co-contamination: not reported
- Further contamination of plants and plant products other than cereals and their products reported in the present book/article[99]: Beans, Seeds (sesame), Sugar, AFB_1 and AFB_2, literature[99]; Food, Jam (bean jam), Paste (peanut paste), AFB_1, literature[99]; Nuts (Brazil nut, paranut), Nuts (peanut), Nuts (pistachio nut), Spices (nutmeg), AFB_1, AFB_2, AFG_1, and AFG_2, literature[99]; Spices, AFB_2 and AFG_1, literature[99]; Spices (cayenne pepper, chilli), Spices (pepper), AFB_1, AFB_2, and AFG_1, literature[99]; Spices (paprika), AFB_1 and AFG_1, literature[99]

incidence: 58?/136*, conc. range: >0.2–2.0 µg/kg (34 sa.), >2.0–10.0 µg/kg (22 sa.), sample origin: different exporting companies, Turkey, sample year: 2007, country: Turkey[115], *commercial Turkish food

incidence: 99/248*, conc. range: >0.2–2.0 µg/kg (70 sa.), >2.0–10.0 µg/kg (26 sa.), >10.0–20.0 µg/kg (2 sa.), >20.0–50.0 µg/kg (1 sa.), sample origin: different exporting companies, Turkey, sample year: 2008, country: Turkey[115], *commercial Turkish food

incidence: 35/57*, conc. range: >0.2–2.0 µg/kg (30 sa.), >2.0–10.0 µg/kg (4 sa.), >10.0–20.0 µg/kg (1 sa.), sample origin: different exporting companies, Turkey, sample year: 2009, country: Turkey[115], *commercial Turkish food

- Co-contamination: not reported
- Further contamination of plants and plant products other than cereals and their products reported in the present book/article[115]: Food, Products (cocoa products), AFB_1, literature[115]

incidence: 3/4*, conc. range: 0.04–0.41 µg/kg, sample origin: Warsaw (capital) market or from food producers, Poland, sample year: 2004, country: Poland[116], *culinary mixes

incidence: 12/15*, conc. range: 0.05–6.35 µg/kg, sample origin: Warsaw (capital) market or from food producers, Poland, sample year: 2005, country: Poland[116], *spices (mixes)

- Co-contamination: not reported
- Further contamination of plants and plant products other than cereals and their products reported in the present book/article[116]: Coffee, Raisins, OTA, literature[116]; Food, AFB_1 and AFG_1, literature[116]; Nuts, AFB_1, AFB_2, AFG_1, and AFG_2, literature[116]; Spices, AFB_2, AFG_1, and AFG_2, literature[116]

incidence: 2/5*, conc. range: 0.16–0.91 µg/kg, sample origin: food shops in Debrecen (town), Hungary, sample year: October 2004, country: Hungary[170], *spice mixture

- Co-contamination: For detailed information see the article.
- Further contamination of plants and plant products other than cereals and their products reported in the present book/article[170]: Spices, AFB_2, literature[170]; Spices (cayenne pepper, chilli), AFB_1, AFB_2, and OTA, literature[170]; Spices (pepper), AFB_1, literature[170]

incidence: 1/6*, conc.: <0.4 µg/kg, sample origin: sites of primary storage or processing (domestic products) and sites of importation, Cyprus, sample year: 1992–1996, country: Cyprus[200], *pepper, turmeric and others

- Co-contamination: not reported
- Further contamination of plants and plant products other than cereals and their products reported in the present book/article[200]: Butter (peanut butter), Figs, Nuts (peanut), AFB_1, AFB_2, AFG_1, and AFG_2, literature[200]; Nuts (Brazil nut, paranut), Nuts (walnut), AFB_1, AFB_2, and AFG_1, literature[200]; Nuts (chestnut), Products (sesame product), AFB_1, literature[200]; Nuts

(pistachio nut), AFB_1 and AFB_2, literature[200]; Seeds (sesame), AFB_1, AFG_1, and AFG_2, literature[200]

incidence: 9/13*, conc. range: 0.15–14.9 µg/kg, sample origin: local markets, Tunisia, sample year: 2005/2006, country: Tunisia[338], *included black pepper, cumin, and red pepper

- Co-contamination: 2 sa. co-contaminated with AFB_1 and AFG_2; no further information available
- Further contamination of plants and plant products other than cereals and their products reported in the present book/article[338]: Fruits, dried, AFB_1, AFB_2, and AFG_1, literature[338]; Spices, AFB_2, AFG_1, and AFG_2, literature[338]

incidence: 2/6*, conc. range: 0.79–0.95 µg/kg, Ø conc.: 0.87 µg/kg, sample origin: local supermarket in Okayama (prefecture), Japan, sample year: unknown, country: Japan[339], *included cardamom**, clove**, cumin, red pepper, turmeric (curcuma), and white pepper, **contaminated

- Co-contamination: 1 sa. co-contaminated with AFB_1, AFB_2, and AFG_1; 1 sa. contaminated solely with AFB_1
- Further contamination of plants and plant products other than cereals and their products reported in the present book/article[339]: Coconuts, Fruits, dried, Seeds (sesame), AFB_1 and AFG_2, literature[339]; Figs, AFB_2 and AFG_2, literature[339]; Flour (soybean flour), AFB_1, literature[339]; Nuts, AFB_1, AFB_2, AFG_1, and AFG_2, literature[339]; Raisins, AFG_2, literature[339]; Spices, AFB_2, AFG_1, and AFG_2, literature[339]

For detailed information see the article.

incidence: 4/10*, conc. range: 10–46 µg/kg, Ø conc.: 25 µg/kg, sample origin: El Atarin, Dokki Square, and Giza Market (districts) in Cairo (capital), Egypt, sample year: unknown, country: USA/Egypt[344], *included cinnamon = 10 and 46 µg AFB_1/kg, coriander = no contamination, black pepper = 33 µg AFB_1/kg, and red pepper = 10 µg AFB_1/kg,

- Co-contamination: not reported
- Further contamination of plants and plant products other than cereals and their products reported in the present book/article[344]: Food, Medicinal plants, Vegetables, dried, AFB_1, literature[344]

incidence: 8/10*, conc. range: 0.05–3.32 µg/kg, Ø conc.: 0.85 µg/kg, sample origin: Penang (state, island), Malaysia, sample year: unknown (first study), country: Malaysia[530], *included mixed spices for "biryani," "dalca," "kerutuk," "kurma," "soto," "rendang," seafood, and vegetables

- Co-contamination: 4 sa. co-contaminated with AFB_1 and OTA; 2 sa. co-contaminated with AFB_1, AFB_2, and OTA; 1 sa. co-contaminated with AFB_1, AFB_2, AFG_1, AFG_2, and OTA; 1 sa. co-contaminated with AFB_1 and AFB_2
- Further contamination of plants and plant products other than cereals and their products reported in the present book/article[530]: Spices, AFB_2, AFG_1, AFG_2, and OTA, literature[530]; Spices (cayenne pepper, chilli), Spices (coriander), Spices (cumin), Spices (curry), Spices (fennel), Spices (masala), AFB_1, AFB_2, AFG_1, AFG_2, and OTA, literature[530]; Spices (curcuma), AFB_1, AFB_2, AFG_1, and OTA, literature[530]; Spices (pepper), AFB_1 and AFB_2, literature[530]; Spices (poppy seed), AFB_1, literature[530]

incidence: 6/6*, conc. range: 2.44–11.03 µg/kg, Ø conc.: 6.68 µg/kg, sample origin: Penang (state, island), Malaysia, sample year: unknown (second study), country: Malaysia[530], *included mixed spices for "biryani," "kurma," "masala," "soto," "rendang," and barbeque

- Co-contamination: 3 sa. co-contaminated with AFB_1, AFB_2, and AFG_1; 2 sa. co-contaminated with AFB_1 and AFB_2; 1 sa. co-contaminated with AFB_1, AFB_2, AFG_1, and AFG_2
- Further contamination of plants and plant products other than cereals and their products reported in the present book/article[530]: Spices, AFB_2, AFG_1, AFG_2, and OTA, literature[530]; Spices (cayenne pepper, chilli), Spices (coriander), Spices (cumin), Spices (curry), Spices (fennel), Spices (masala), AFB_1, AFB_2, AFG_1, AFG_2, and OTA, literature[530]; Spices (curcuma), AFB_1, AFB_2, AFG_1, and OTA, literature[530]; Spices (pepper), AFB_1 and AFB_2, literature[530]; Spices (poppy seed), AFB_1, literature[530]

incidence: 1*/50, conc.: 5.1 µg/kg*, sample origin: Netherlands, sample year: unknown, country: Netherlands[546], *bay leaf

- Co-contamination: not reported
- Further contamination of plants and plant products other than cereals and their products reported in the present book/article[546]: Spices (nutmeg), AFB_1 and AF/S, literature[546]

incidence: 1/10*, conc.: <LOQ, sample origin: Chinese markets in Beijing (capital), China, sample year: unknown, country: China/Italy[768], **Pelargonium graveolens* L.

- Co-contamination: not reported
- Further contamination of plants and plant products other than cereals and their products reported in the present book/article[768]: Food, AFB_1, AFB_2, and OTA, literature[768]; Spices (fennel), Spices (star anise), AFG_2, literature[768]

incidence: 11/130, conc. range: <LOQ (8 sa.), LOQ-5.0 µg/kg (3 sa., maximum: 1.95 µg/kg), Ø conc.: 0.31 µg/kg, sample origin: imports from Europe, India, Northern Africa, South America, Sub-Saharan area, and USA purchased in stores in northwestern Italy, sample year: September 2011–February 2012, country: Italy[884]

- Co-contamination: 6 sa. co-contaminated with AFS and OTA
- Further contamination of plants and plant products other than cereals and their products reported in the present book/article[884]: Spices, AFB_2, AFG_1, AFG_2, and OTA, literature[884]

For detailed information see the article.

Aflatoxin B_2

incidence: 1/37*, conc.: 0.2 µg/kg, sample origin: unknown, sample year: unknown, country: Japan[97], *spices, mixed

- Co-contamination: not reported
- Further contamination of plants and plant products other than cereals and their products reported in the present book/article[97]: Beans, Jam (bean jam), AFB_1 and AFB_2, literature[97]; Butter (peanut butter), Oils (peanut oil), Spices (pepper), AFB_1, AFB_2, and AFG_1, literature[97]; Spices, AFB_1, literature[97]; Spices (nutmeg), AFB_1, AFB_2, AFG_1, and AFG_2, literature[97]

incidence: 16?/78*, conc. range: ≤0.4 µg/kg, Ø conc.: <0.1 µg/kg, sample origin: commercially available market basket sa. of foods and foodstuffs in Tokyo (capital), Japan, sample year: 1986–1990, country: Japan[99], *spices, mixed

- Co-contamination: not reported
- Further contamination of plants and plant products other than cereals and their products reported in the present book/article[99]: Beans, Seeds (sesame), Sugar, AFB_1 and AFB_2, literature[99]; Food, Jam (bean jam), Paste (peanut paste), AFB_1, literature[99]; Nuts (Brazil nut, paranut), Nuts (peanut), Nuts (pistachio nut), Spices (nutmeg), AFB_1, AFB_2, AFG_1, and AFG_2, literature[99]; Spices, Spices (paprika), AFB_1 and AFG_1, literature[99]; Spices (cayenne pepper, chilli), Spices (pepper), AFB_1, AFB_2, and AFG_1, literature[99]

incidence: 3/4*, conc. range: 0.02–0.08 μg/kg, sample origin: Warsaw (capital) market or from food producers, Poland, sample year: 2004, country: Poland[116], *culinary mixes

incidence: 4/15*, conc. range: 0.06–0.32 μg/kg, sample origin: Warsaw (capital) market or from food producers, Poland, sample year: 2005, country: Poland[116], *spices (mixes)

- Co-contamination: not reported
- Further contamination of plants and plant products other than cereals and their products reported in the present book/article[116]: Coffee, Raisins, OTA, literature[116]; Food, AFB_1 and AFG_1, literature[116]; Nuts, AFB_1, AFB_2, AFG_1, and AFG_2, literature[116]; Spices, AFB_1, AFG_1, and AFG_2, literature[116]

incidence: 1/5*, conc.: 0.25 μg/kg, sample origin: food shops in Debrecen (town), Hungary, sample year: October 2004, country: Hungary[170], *spice mixture

- Co-contamination: For detailed information see the article.
- Further contamination of plants and plant products other than cereals and their products reported in the present book/article[170]: Spices, Spices (pepper), AFB_1, literature[170]; Spices (cayenne pepper, chilli), AFB_1, AFB_2, and OTA, literature[170]

incidence: 7/13*, conc. range: 0.05–1.2 μg/kg, sample origin: local markets, Tunisia, sample year: 2005/2006, country: Tunisia[338], *included black pepper, cumin, and red pepper

- Co-contamination: 7 sa. co-contaminated with AFB_1 and AFB_2; no further information available
- Further contamination of plants and plant products other than cereals and their products reported in the present book/article[338]: Fruits, dried, AFB_1, AFB_2, and AFG_1, literature[338]; Spices, AFB_1, AFG_1, and AFG_2, literature[338]

incidence: 2/6*, conc. range: 0.41–0.70 μg/kg, ∅ conc.: 0.56 μg/kg, sample origin: local supermarket in Okayama (prefecture), Japan, sample year: unknown, country: Japan[339], *included cardamom**, clove, cumin, red pepper**, turmeric (curcuma), and white pepper, **contaminated

- Co-contamination: 1 sa. co-contaminated with AFB_1, AFB_2, and AFG_1; 1 sa. co-contaminated with AFB_2, AFG_1, and AFG_2
- Further contamination of plants and plant products other than cereals and their products reported in the present book/article[339]: Coconuts, Fruits, dried, Seeds (sesame), AFB_1 and AFG_2, literature[339]; Figs, AFB_2 and AFG_2, literature[339]; Flour (soybean flour), AFB_1, literature[339]; Nuts, AFB_1, AFB_2, AFG_1, and AFG_2, literature[339]; Raisins, AFG_2, literature[339]; Spices, AFB_1, AFG_1, and AFG_2, literature[339]

For detailed information see the article.

incidence: 4/10*, conc. range: 0.09–0.67 μg/kg, ∅ conc.: 0.27 μg/kg, sample origin: Penang (state, island), Malaysia, sample year: unknown (first study), country: Malaysia[530], *included mixed spices for "biryani," "dalca," "kerutuk," "kurma," "soto," "rendang," seafood, and vegetables

- Co-contamination: 2 sa. co-contaminated with AFB_1, AFB_2, and OTA; 1 sa. co-contaminated with AFB_1, AFB_2, AFG_1, AFG_2, and OTA; 1 sa. co-contaminated with AFB_1 and AFB_2
- Further contamination of plants and plant products other than cereals and their products reported in the present book/article[530]: Spices, AFB_1, AFG_1, AFG_2, and OTA, literature[530]; Spices (cayenne pepper, chilli), Spices (coriander), Spices (cumin), Spices (curry), Spices (fennel), Spices (masala), AFB_1, AFB_2, AFG_1, AFG_2, and OTA, literature[530]; Spices (curcuma), AFB_1, AFB_2, AFG_1, and OTA, literature[530]; Spices (pepper), AFB_1 and AFB_2, literature[530]; Spices (poppy seed), AFB_1, literature[530]

incidence: 6/6*, conc. range: 0.41–1.54 μg/kg, ∅ conc.: 1.03 μg/kg, sample origin: Penang (state, island), Malaysia, sample year: unknown (second study), country: Malaysia[530], *included mixed spices for "biryani," "kurma," "masala," "soto," "rendang," and barbeque

- Co-contamination: 3 sa. co-contaminated with AFB_1, AFB_2, and AFG_1; 2 sa. co-contaminated with AFB_1 and AFB_2; 1 sa. co-contaminated with AFB_1, AFB_2, AFG_1, and AFG_2
- Further contamination of plants and plant products other than cereals and their products reported in the present book/article[530]: Spices, AFB_1, AFG_1, AFG_2, and OTA, literature[530]; Spices (cayenne pepper, chilli), Spices (coriander), Spices (cumin), Spices (curry), Spices (fennel), Spices (masala), AFB_1, AFB_2, AFG_1, AFG_2, and OTA, literature[530]; Spices (curcuma), AFB_1, AFB_2, AFG_1, and OTA, literature[530]; Spices (pepper), AFB_1 and AFB_2, literature[530]; Spices (poppy seed), AFB_1, literature[530]

incidence: 6/130, conc. range: <LOQ (4 sa.), LOQ-5.0 μg/kg (2 sa.), ∅ conc.: 1.07 μg/kg, sample origin: imports from Europe, India, Northern Africa, South America, Sub-Saharan area, and USA purchased in stores in north-western Italy, sample year: September 2011–February 2012, country: Italy[884]

- Co-contamination: 6 sa. co-contaminated with AFS and OTA
- Further contamination of plants and plant products other than cereals and their products reported in the present book/article[884]: Spices, AFB_1, AFG_1, AFG_2, and OTA, literature[884]

For detailed information see the article.

Aflatoxin G_1

incidence: 16?/78*, conc. range: ≤0.5 μg/kg, ∅ conc.: <0.1 μg/kg, sample origin: commercially available market basket sa. of foods and foodstuffs in Tokyo (capital), Japan, sample year: 1986–1990, country: Japan[99], *spices, mixed

- Co-contamination: not reported
- Further contamination of plants and plant products other than cereals and their products reported in the present book/article[99]: Beans, Seeds (sesame), Spices, Sugar, AFB_1 and AFB_2, literature[99]; Food, Jam (bean jam), Paste (peanut paste), AFB_1, literature[99]; Nuts (Brazil nut, paranut), Nuts (peanut), Nuts (pistachio nut), Spices (nutmeg), AFB_1, AFB_2, AFG_1, and AFG_2, literature[99]; Spices (paprika), AFB_1 and AFG_1, literature[99]; Spices (cayenne pepper, chilli), Spices (pepper), AFB_1, AFB_2, and AFG_1, literature[99]

incidence: 1/4*, conc.: 0.02 μg/kg, sample origin: Warsaw (capital) market or from food producers, Poland, sample year: 2004, country: Poland[116], *culinary mixes

incidence: 8/15*, conc. range: 0.04–1.12 μg/kg, sample origin: Warsaw (capital) market or from food producers, Poland, sample year: 2005, country: Poland[116], *spices (mixes)

- Co-contamination: not reported
- Further contamination of plants and plant products other than cereals and their products reported in the present book/article[116]: Coffee, Raisins, OTA, literature[116]; Food, AFB_1 and AFG_1, literature[116]; Nuts, AFB_1, AFB_2, AFG_1, and AFG_2, literature[116]; Spices, AFB_1, AFB_2, and AFG_2, literature[116]

incidence: 6/13*, conc. range: 0.07–2.0 μg/kg, sample origin: local markets, Tunisia, sample year: 2005/2006, country: Tunisia[338], *included black pepper, cumin, and red pepper

- Co-contamination: 6 sa. co-contaminated with AFB_1 and AFG_1; no further information available
- Further contamination of plants and plant products other than cereals and their products reported in the present book/article[338]: Fruits, dried, AFB_1, AFB_2, and AFG_1, literature[338]; Spices, AFB_1, AFB_2, and AFG_2, literature[338]

incidence: 4/6*, conc. range: 0.69–7.09 μg/kg, Ø conc.: 2.64 μg/kg, sample origin: local supermarket in Okayama (prefecture), Japan, sample year: unknown, country: Japan[339], *included cardamom**, clove, cumin**, red pepper**, turmeric (curcuma)**, and white pepper, **contaminated

- Co-contamination: 1 sa. co-contaminated with AFB_1, AFB_2, and AFG_1; 1 sa. co-contaminated with AFB_2, AFG_1, and AFG_2; 1 sa. co-contaminated with AFG_1 and AFG_2; 1 sa. contaminated solely with AFG_1
- Further contamination of plants and plant products other than cereals and their products reported in the present book/article[339]: Coconuts, Fruits, dried, Seeds (sesame), AFB_1 and AFG_2, literature[339]; Figs, AFB_2 and AFG_2, literature[339]; Flour (soybean flour), AFB_1, literature[339]; Nuts, AFB_1, AFB_2, AFG_1, and AFG_2, literature[339]; Raisins, AFG_2, literature[339]; Spices, AFB_1, AFB_2, and AFG_2, literature[339]

For detailed information see the article.

incidence: 1/10*, conc.: 0.17 μg/kg, sample origin: Penang (state, island), Malaysia, sample year: unknown (first study), country: Malaysia[530], *included mixed spices for "biryani," "dalca," "kerutuk," "kurma," "soto," "rendang," seafood, and vegetables

- Co-contamination: 1 sa. co-contaminated with AFB_1, AFB_2, AFG_1, AFG_2, and OTA
- Further contamination of plants and plant products other than cereals and their products reported in the present book/article[530]: Spices, AFB_1, AFB_2, AFG_2, and OTA, literature[530]; Spices (cayenne pepper, chilli), Spices (coriander), Spices (cumin), Spices (curry), Spices (fennel), Spices (masala), AFB_1, AFB_2, AFG_1, AFG_2, and OTA, literature[530]; Spices (curcuma), AFB_1, AFB_2, AFG_1, and OTA, literature[530]; Spices (pepper), AFB_1 and AFB_2, literature[530]; Spices (poppy seed), AFB_1, literature[530]

incidence: 4/6*, conc. range: 0.75–2.74 μg/kg, Ø conc.: 1.63 μg/kg, sample origin: Penang (state, island), Malaysia, sample year: unknown (second study), country: Malaysia[530], *included mixed spices for "biryani," "kurma," "masala," "soto," "rendang," and barbeque

- Co-contamination: 3 sa. co-contaminated with AFB_1, AFB_2, and AFG_1; 1 sa. co-contaminated with AFB_1, AFB_2, AFG_1, and AFG_2
- Further contamination of plants and plant products other than cereals and their products reported in the present book/article[530]: Spices, AFB_1, AFB_2, AFG_2, and OTA, literature[530]; Spices (cayenne pepper, chilli), Spices (coriander), Spices (cumin), Spices (curry), Spices (fennel), Spices (masala), AFB_1, AFB_2, AFG_1, AFG_2, and OTA, literature[530]; Spices (curcuma), AFB_1, AFB_2, AFG_1, and OTA, literature[530]; Spices (pepper), AFB_1 and AFB_2, literature[530]; Spices (poppy seed), AFB_1, literature[530]

incidence: 3/130, conc. range: <LOQ (3 sa.), Ø conc.: 0.42 μg/kg, sample origin: imports from Europe, India, Northern Africa, South America, Sub-Saharan area, and USA purchased in stores in north-western Italy, sample year: September 2011–February 2012, country: Italy[884]

- Co-contamination: 6 sa. co-contaminated with AFS and OTA
- Further contamination of plants and plant products other than cereals and their products reported in the present book/article[884]: Spices, AFB_1, AFB_2, AFG_2, and OTA, literature[884]

For detailed information see the article.

Aflatoxin G_2

incidence: 0/4*, conc.: no contamination, sample origin: Warsaw (capital) market or from food producers, Poland, sample year: 2004, country: Poland[116], *culinary mixes

incidence: 1/15*, conc.: 0.04 μg/kg, sample origin: Warsaw (capital) market or from food producers, Poland, sample year: 2005, country: Poland[116], *spices (mixes)

- Co-contamination: not reported
- Further contamination of plants and plant products other than cereals and their products reported in the present book/article[116]: Coffee, Raisins, OTA, literature[116]; Food, AFB_1 and AFG_1, literature[116]; Nuts, AFB_1, AFB_2, AFG_1, and AFG_2, literature[116]; Spices, AFB_1, AFB_2, and AFG_1, literature[116]

incidence: 2/13*, conc. range: 0.07–0.09 μg/kg, Ø conc.: 0.08 μg/kg, sample origin: local markets, Tunisia, sample year: 2005/2006, country: Tunisia[338], *included black pepper, cumin, and red pepper

- Co-contamination: 2 sa. co-contaminated with AFB_1 and AFG_2; no further information available
- Further contamination of plants and plant products other than cereals and their products reported in the present book/article[338]: Fruits, dried, Spices, AFB_1, AFB_2, and AFG_1, literature[338]

incidence: 3/6*, conc. range: 3.93–6.94 μg/kg, Ø conc.: 5.56 μg/kg, sample origin: local supermarket in Okayama (prefecture), Japan, sample year: unknown, country: Japan[339], *included cardamom, clove, cumin, red pepper**, turmeric (curcuma)**, and white pepper**, **contaminated

- Co-contamination: 1 sa. co-contaminated with AFB_2, AFG_1, and AFG_2, 1 sa. co-contaminated with AFG_1 and AFG_2; 1 sa. contaminated solely with AFG_2
- Further contamination of plants and plant products other than cereals and their products reported in the present book/article[339]: Coconuts, Fruits, dried, Seeds (sesame), AFB_1 and AFG_2, literature[339]; Figs, AFB_2 and AFG_2, literature[339]; Flour (soybean flour), AFB_1, literature[339]; Nuts, AFB_1, AFB_2, AFG_1, and AFG_2, literature[339]; Raisins, AFG_2, literature[339]; Spices, AFB_1, AFB_2, and AFG_1, literature[339]

For detailed information see the article.

incidence: 1/10*, conc.: 0.01 μg/kg, sample origin: Penang (state, island), Malaysia, sample year: unknown (first study), country: Malaysia[530], *included mixed spices for "biryani," "dalca," "kerutuk," "kurma," "soto," "rendang," seafood, and vegetables

- Co-contamination: 1 sa. co-contaminated with AFB_1, AFB_2, AFG_1, AFG_2, and OTA
- Further contamination of plants and plant products other than cereals and their products reported in the present book/article[530]: Spices, AFB_1, AFB_2, AFG_1, and OTA, literature[530]; Spices (cayenne pepper, chilli), Spices (coriander), Spices (cumin), Spices (curry), Spices (fennel), Spices (masala), AFB_1, AFB_2, AFG_1, AFG_2, and OTA, literature[530]; Spices (curcuma), AFB_1, AFB_2, AFG_1, and OTA, literature[530]; Spices (pepper), AFB_1 and AFB_2, literature[530]; Spices (poppy seed), AFB_1, literature[530]

incidence: 1/6*, conc.: 0.22 μg/kg, sample origin: Penang (state, island), Malaysia, sample year: unknown (second study), country:

Malaysia[530], *included mixed spices for "biryani," "kurma," "masala," "soto," "rendang," and barbeque

- Co-contamination: 1 sa. co-contaminated with AFB_1, AFB_2, AFG_1, and AFG_2
- Further contamination of plants and plant products other than cereals and their products reported in the present book/article[530]: Spices, AFB_1, AFB_2, AFG_1, and OTA, literature[530]; Spices (cayenne pepper, chilli), Spices (coriander), Spices (cumin), Spices (curry), Spices (fennel), Spices (masala), AFB_1, AFB_2, AFG_1, AFG_2, and OTA, literature[530]; Spices (curcuma), AFB_1, AFB_2, AFG_1, and OTA, literature[530]; Spices (pepper), AFB_1 and AFB_2, literature[530]; Spices (poppy seed), AFB_1, literature[530]

incidence: 7/130, conc. range: <LOQ (3 sa.), LOQ-5.0 μg/kg (4 sa.), ∅ conc.: 2.26 μg/kg, sample origin: imports from Europe, India, Northern Africa, South America, Sub-Saharan area, and USA purchased in stores in north-western Italy, sample year: September 2011–February 2012, country: Italy[884]

- Co-contamination: 6 sa. co-contaminated with AFS and OTA
- Further contamination of plants and plant products other than cereals and their products reported in the present book/article[884]: Spices, AFB_1, AFB_2, AFG_1, and OTA, literature[884]

For detailed information see the article.

Aflatoxin/s

incidence: 1/4*, conc.: 0.9 μg/kg**, sample origin: specialist food shops, UK, sample year: unknown, country: UK[16], *five spice powder, **AFB_1, AFB_2, AFG_1, and AFG_2

- Co-contamination: not reported
- Further contamination of plants and plant products other than cereals and their products reported in the present book/article[16]: Food, AF/S, DON, FUS-X, NIV, OTA, and ZEA, literature[16]; Nuts (almond), AF/S, literature[16]; Oils, Spices (fennel), AF/S and ZEA, literature[16]; Oils (sesame oil), OTA, literature[16]; Paste (curry paste), AF/S, FB/S, NIV, OTA, and ZEA, literature[16]; Pickles, AF/S, FB/S, OTA, and ZEA, literature[16]; Sauces (chilli sauce), NIV, OTA, and ZEA, literature[16]; Spices, FB/S, OTA, and ZEA, literature[16]; Spices (cayenne pepper, chilli), AF/S, DAS, DON, HT-2, OTA, and ZEA, literature[16]; Spices (coriander), AF/S, DON, OTA, and ZEA, literature[16]; Spices (curry), AF/S, DAS, FB/S, FUS-X, HT-2, NEO, NIV, OTA, T-2, and ZEA, literature[16]; Spices (garlic), DON, FUS-X, and NIV, literature[16]; Spices (ginger), AF/S, DON, NEO, NIV, OTA, and T-2, literature[16]; Spices (tandoori), AF/S, FB/S, NIV, OTA, and T-2, literature[16]

incidence: 5/6*, conc. range: 0.16–5.12 μg/kg**, sample origin: sales outlets, Qatar, sample year: October 2002, country: Qatar[229], *spice powder, mixed, **AFB_1, AFB_2, AFG_1, and AFG_2

- Co-contamination: not reported
- Further contamination of plants and plant products other than cereals and their products reported in the present book/article[229]: Butter (peanut butter), Figs, Nuts (peanut), Nuts (pistachio nut), AF/S, literature[229]; Coffee, Raisins, Spices, OTA, literature[229]; Spices (cayenne pepper, chilli), AF/S and OTA, literature[229]

incidence: 5/41*, conc. range: 1–1.9 μg/kg** (1 sa.), 2–3.9 μg/kg** (3 sa.), 4–10 μg/kg** (1 sa.), sample origin: supermarkets, local and ethnic shops in Norwich (city) area, UK, sample year: unknown, country: UK[250], *herbs and spices, **AFB_1, AFB_2, AFG_1, and AFG_2, (garlic powder: 3.3 μg AFS**/kg; ginger: 1.3, 2.0, 8.4 μg AFS**/ kg; fenugreek: 2.5 μg AFS**/ kg)

- Co-contamination: not reported
- Further contamination of plants and plant products other than cereals and their products reported in the present book/article[250]: Spices (cayenne pepper, chilli), Spices (curry), AF/S, literature[250]

incidence: 5/12, conc. range: 8.1–84.09 μg/kg* **, ∅ conc.: 39.0 μg/kg* **, sample origin: local markets, Qatar, sample year: unknown, country: Qatar/Italy[445], *AFB_1, AFB_2, AFG_1, and AFG_2, **garam masala: 8.1 μg AFS*/kg; turmeric/curcuma: 13.21 μg AFS*/kg; tandoori: 18.76 μg AFS*/kg; chilli powder: 71.01 μg AFS*/kg; black pepper: 84.09 μg AFS*/kg

- Co-contamination: not reported
- Further contamination of plants and plant products other than cereals and their products reported in the present book/article[445]: not reported

For detailed information see the article.

incidence: 43/56, conc. range: 0.48–94.96 μg/kg, sample origin: 5 governorates, Kuwait, sample year: unknown, country: Kuwait[704], *AF

- Co-contamination: not reported
- Further contamination of plants and plant products other than cereals and their products reported in the present book/article[704]: Coffee, OTA, literature[704]; Nuts, AF/S, literature[704]

Aspergillus and ***Penicillium*** Toxins

Ochratoxin A

incidence: 1/4*, conc.: 2.6 μg/kg, sample origin: specialist food shops, UK, sample year: unknown, country: UK[16], *five spice powder

- Co-contamination: not reported
- Further contamination of plants and plant products other than cereals and their products reported in the present book/article[16]: Food, AF/S, DON, FUS-X, NIV, OTA, and ZEA, literature[16]; Nuts (almond), AF/S, literature[16]; Oils, Spices (fennel), AF/S and ZEA, literature[16]; Oils (sesame oil), OTA, literature[16]; Paste (curry paste), AF/S, FB/S, NIV, OTA, and ZEA, literature[16]; Pickles, AF/S, FB/S, OTA, and ZEA, literature[16]; Sauces (chilli sauce), NIV, OTA, and ZEA, literature[16]; Spices, AF/S, FB/S, and ZEA, literature[16]; Spices (cayenne pepper, chilli), AF/S, DAS, DON, HT-2, OTA, and ZEA, literature[16]; Spices (coriander), AF/S, DON, OTA, and ZEA, literature[16]; Spices (curry), AF/S, DAS, FB/S, FUS-X, HT-2, NEO, NIV, OTA, T-2, and ZEA, literature[16]; Spices (garlic), DON, FUS-X, and NIV, literature[16]; Spices (ginger), AF/S, DON, NEO, NIV, OTA, and T-2, literature[16]; Spices (tandoori), AF/S, FB/S, NIV, OTA, and T-2, literature[16]

incidence: 1/6*, conc.: 0.86 μg/kg, sample origin: sales outlets, Qatar, sample year: October 2002, country: Qatar[229], *spice powder, mixed

- Co-contamination: not reported
- Further contamination of plants and plant products other than cereals and their products reported in the present book/article[229]: Butter (peanut butter), Figs, Nuts (peanut), Nuts (pistachio nut), Spices, AF/S, literature[229]; Coffee, Raisins, OTA, literature[229]; Spices (cayenne pepper, chilli), AF/S and OTA, literature[229]

incidence: 4/13*, conc. range: 0.75–5.5 μg/kg, sample origin: local markets and traditional family reserves, Tunisia, sample year: 2005/2006, country: Tunisia[340], *included cumin, black pepper, and red pepper

- Co-contamination: not reported

- Further contamination of plants and plant products other than cereals and their products reported in the present book/article[340]: Fruits, dried, OTA, literature[340]

incidence: 9/10*, conc. range: 0.30–3.65 µg/kg, Ø conc.: 1.36 µg/kg, sample origin: Penang (state, island), Malaysia, sample year: unknown (first study), country: Malaysia[530], *included mixed spices for "biryani," "dalca," "kerutuk," "kurma," "soto," "rendang," seafood, and vegetables

- Co-contamination: 4 sa. co-contaminated with AFB_1 and OTA; 2 sa. co-contaminated with AFB_1, AFB_2, and OTA; 1 sa. co-contaminated with AFB_1, AFB_2, AFG_1, AFG_2, and OTA; 2 sa. contaminated solely with OTA
- Further contamination of plants and plant products other than cereals and their products reported in the present book/article[530]: Spices, AFB_1, AFB_2, AFG_1, and AFG_2, literature[530]; Spices (cayenne pepper, chilli), Spices (coriander), Spices (cumin), Spices (curry), Spices (fennel), Spices (masala), AFB_1, AFB_2, AFG_1, AFG_2, and OTA, literature[530]; Spices (curcuma), AFB_1, AFB_2, AFG_1, and OTA, literature[530]; Spices (pepper), AFB_1 and AFB_2, literature[530]; Spices (poppy seed), AFB_1, literature[530]

incidence: 11/22*, conc. range: ≤0.82 µg/kg, sample origin: supermarkets and fresh provision shops, Hong Kong, China, sample year: August/September 2005, country: China[697]

incidence: 5/30*, conc. range: ≤0.99 µg/kg, sample origin: supermarkets and fresh provision shops, Hong Kong, China, sample year: August/September 2005, country: China[697], *seasonings

- Co-contamination: not reported
- Further contamination of plants and plant products other than cereals and their products reported in the present book/article[697]: Chocolate, Coffee, Fruits, dried, Juices, Legumes, OTA, literature[697]

incidence: 5/5, conc. range: LOD/LOQ-0.9 µg/kg (1 sa.), 1.0–2.9 µg/kg (2 sa.), ≤23.8 µg/kg (2 sa.), sample origin: Italy, sample year: 1999, country: EU[698]

- Co-contamination: not reported
- Further contamination of plants and plant products other than cereals and their products reported in the present book/article[698]: Baby food, Beans (soybean), Chocolate, Cocoa, Coffee, Coffee beans, Currants, Fruits, dried, Herbs, Jam, Juices, Juices (grape juice), Legumes, Nuts, Oil seeds, Oils (olive oil), Peas, Peas (chick-pea), Products (plant product), Raisins, Spices (cayenne pepper, chilli), Spices (nutmeg), Spices (paprika), Spices (pepper), Sultanas, Tea, Vinegar, Wine, Wine (red), Wine (rosé), Wine (white), Wine, miscellaneous, OTA, literature[698]

incidence: 143/681*, conc. range: ≤41.83 µg/kg**, sample origin: German spice company, Germany, sample year: unknown, country: Germany[712], *more than 50 different spice commodities, **chilli

- Co-contamination: not reported
- Further contamination of plants and plant products other than cereals and their products reported in the present book/article[712]: Spices, OTB, literature[712]

incidence: 31/130, conc. range: LOQ-5.0 µg/kg (15 sa.), 5.0–10.0 µg/kg (11 sa.), 10.0–15.0 µg/kg (2 sa.), >15.0 µg/kg (3 sa., maximum: 19.06 µg/kg), Ø conc.: 6.18 µg/kg, sample origin: imports from Europe, India, Northern Africa, South America, Sub-Saharan area, and USA purchased in stores in north-western Italy, sample year: September 2011–February 2012, country: Italy[884]

- Co-contamination: 6 sa. co-contaminated with AFS and OTA
- Further contamination of plants and plant products other than cereals and their products reported in the present book/article[884]: Spices, AFB_1, AFB_2, AFG_1, and AFG_2, literature[884]

For detailed information see the article.

Ochratoxin B

incidence: 68/681*, conc. range: ≤15.07 µg/kg**, sample origin: German spice company, Germany, sample year: unknown, country: Germany[712], *more than 50 different spice commodities, **vanilla extract

- Co-contamination: not reported
- Further contamination of plants and plant products other than cereals and their products reported in the present book/article[712]: Spices, OTA, literature[712]

Fusarium Toxins

Fumonisin/s

Fumonisin B_1

incidence: 2/13, conc. range: 70–130 µg/kg, Ø conc.: 100.0 µg/kg, sample origin: local markets and traditional home reserves, Tunisia, sample year: 2005/2006, country: Tunisia[642]

- Co-contamination: not reported
- Further contamination of plants and plant products other than cereals and their products reported in the present book/article[642]: not reported

Fumonisin/s

incidence: 4?/4*, conc. range: 13–17 µg/kg**, sample origin: specialist food shops, UK, sample year: unknown, country: UK[16], *five spice powder, **FB_1 and FB_2

- Co-contamination: not reported
- Further contamination of plants and plant products other than cereals and their products reported in the present book/article[16]: Food, AF/S, DON, FUS-X, NIV, OTA, and ZEA, literature[16]; Nuts (almond), AF/S, literature[16]; Oils, Spices (fennel), AF/S and ZEA, literature[16]; Oils (sesame oil), OTA, literature[16]; Paste (curry paste), AF/S, FB/S, NIV, OTA, and ZEA, literature[16]; Pickles, AF/S, FB/S, OTA, and ZEA, literature[16]; Sauces (chilli sauce), NIV, OTA, and ZEA, literature[16]; Spices, AF/S, OTA, and ZEA, literature[16]; Spices (cayenne pepper, chilli), AF/S, DAS, DON, HT-2, OTA, and ZEA, literature[16]; Spices (coriander), AF/S, DON, OTA, and ZEA, literature[16]; Spices (curry), AF/S, DAS, FB/S, FUS-X, HT-2, NEO, NIV, OTA, T-2, and ZEA, literature[16]; Spices (garlic), DON, FUS-X, and NIV, literature[16]; Spices (ginger), AF/S, DON, NEO, NIV, OTA, and T-2, literature[16]; Spices (tandoori), AF/S, FB/S, NIV, OTA, and T-2, literature[16]

Other *Fusarium* Toxins

Zearalenone

incidence: 4?/4*, conc. range: 3.2–5.2 µg/kg, sample origin: specialist food shops, UK, sample year: unknown, country: UK[16], *five spice powder

- Co-contamination: not reported
- Further contamination of plants and plant products other than cereals and their products reported in the present book/article[16]: Food, AF/S, DON, FUS-X, NIV, OTA, and ZEA, literature[16]; Nuts (almond), AF/S, literature[16]; Oils, Spices (fennel), AF/S and ZEA, literature[16]; Oils (sesame oil), OTA, literature[16];

Paste (curry paste), AF/S, FB/S, NIV, OTA, and ZEA, literature[16]; Pickles, AF/S, FB/S, OTA, and ZEA, literature[16]; Sauces chilli sauce), NIV, OTA, and ZEA, literature[16]; Spices, AF/S, FB/S, and OTA, literature[16]; Spices (cayenne pepper, chilli), AF/S, DAS, DON, HT-2, OTA, and ZEA, literature[16]; Spices (coriander), AF/S, DON, OTA, and ZEA, literature[16]; Spices (curry), AF/S, DAS, FB/S, FUS-X, HT-2, NEO, NIV, OTA, T-2, and ZEA, literature[16]; Spices (garlic), DON, FUS-X, and NIV, literature[16]; Spices (ginger), AF/S, DON, NEO, NIV, OTA, and T-2, literature[16]; Spices (tandoori), AF/S, FB/S, NIV, OTA, and T-2, literature[16]

see also **Herbs**

Spices (ammi) may contain the following mycotoxins:

Aspergillus Toxins

Aflatoxin B_1

incidence: 1/7, conc.: 60 µg/kg, sample origin: major retail outlets in Nainital (district), India, sample year: January 1984–November 1986, country: India[22]

- Co-contamination: 1 sa. co-contaminated with AFB_1, AFB_2, and AFG_1
- Further contamination of plants and plant products other than cereals and their products reported in the present book/article[22]: Spices (ammi), AFB_2 and AFG_1, literature[22]; Spices (cardamom), AFB_1, AFB_2, AFG_1, AFG_2, CIT, and OTA, literature[22]; Spices (cassia), AFB_1, AFB_2, AFG_1, and AFG_2, literature[22]; Spices (cayenne pepper, chilli), AFB_1, AFB_2, AFG_1, AFG_2, and ZEA, literature[22]; Spices (coriander), AFB_1, AFB_2, AFG_1, AFG_2, CIT, OTA, and ZEA, literature[22]; Spices (cumin), AFB_1, AFB_2, AFG_1, AFG_2, and CIT, literature[22]; Spices (curcuma), AFB_1, AFB_2, AFG_1, AFG_2, CIT, and RBT, literature[22]; Spices (fennel), AFB_1, AFB_2, AFG_1, AFG_2, CIT, OTA, and STG, literature[22]; Spices (garlic), AFB_1, AFB_2, and AFG_1, literature[22]; Spices (pepper), AFB_1, AFB_2, AFG_1, AFG_2, CIT, STG, and ZEA, literature[22]

Aflatoxin B_2

incidence: 1/7, conc.: 34 µg/kg, sample origin: major retail outlets in Nainital (district), India, sample year: January 1984–November 1986, country: India[22]

- Co-contamination: 1 sa. co-contaminated with AFB_1, AFB_2, and AFG_1
- Further contamination of plants and plant products other than cereals and their products reported in the present book/article[22]: Spices (ammi), AFB_1 and AFG_1, literature[22]; Spices (cardamom), AFB_1, AFB_2, AFG_1, AFG_2, CIT, and OTA, literature[22]; Spices (cassia), AFB_1, AFB_2, AFG_1, and AFG_2, literature[22]; Spices (cayenne pepper, chilli), AFB_1, AFB_2, AFG_1, AFG_2, and ZEA, literature[22]; Spices (coriander), AFB_1, AFB_2, AFG_1, AFG_2, CIT, OTA, and ZEA, literature[22]; Spices (cumin), AFB_1, AFB_2, AFG_1, AFG_2, and CIT, literature[22]; Spices (curcuma), AFB_1, AFB_2, AFG_1, AFG_2, CIT, and RBT, literature[22]; Spices (fennel), AFB_1, AFB_2, AFG_1, AFG_2, CIT, OTA, and STG, literature[22]; Spices (garlic), AFB_1, AFB_2, and AFG_1, literature[22]; Spices (pepper), AFB_1, AFB_2, AFG_1, AFG_2, CIT, STG, and ZEA, literature[22]

Aflatoxin G_1

incidence: 1/7, conc.: 32 µg/kg, sample origin: major retail outlets in Nainital (district), India, sample year: January 1984–November 1986, country: India[22]

- Co-contamination: 1 sa. co-contaminated with AFB_1, AFB_2, and AFG_1
- Further contamination of plants and plant products other than cereals and their products reported in the present book/article[22]: Spices (ammi), AFB_1 and AFB_2, literature[22]; Spices (cardamom), AFB_1, AFB_2, AFG_1, AFG_2, CIT, and OTA, literature[22]; Spices (cassia), AFB_1, AFB_2, AFG_1, and AFG_2, literature[22]; Spices (cayenne pepper, chilli), AFB_1, AFB_2, AFG_1, AFG_2, and ZEA, literature[22]; Spices (coriander), AFB_1, AFB_2, AFG_1, AFG_2, CIT, OTA, and ZEA, literature[22]; Spices (cumin), AFB_1, AFB_2, AFG_1, AFG_2, and CIT, literature[22]; Spices (curcuma), AFB_1, AFB_2, AFG_1, AFG_2, CIT, and RBT, literature[22]; Spices (fennel), AFB_1, AFB_2, AFG_1, AFG_2, CIT, OTA, and STG, literature[22]; Spices (garlic), AFB_1, AFB_2, and AFG_1, literature[22]; Spices (pepper), AFB_1, AFB_2, AFG_1, AFG_2, CIT, STG, and ZEA, literature[22]

Spices (anise) may contain the following mycotoxins:

Aspergillus Toxins

Aflatoxin B_1

incidence: 4/6*, conc. range: 4.9–8.4 µg/kg, Ø conc.: 7.1 µg/kg, sample origin: organic markets and organic shops, Turkey, sample year: May 2010–May 2011, country: Turkey[859], *organic anise sa.

- Co-contamination: not reported
- Further contamination of plants and plant products other than cereals and their products reported in the present book/article[859]: Herbs (camomile), Herbs (linden), Herbs (rosehip), Herbs (sage), Spices (basil), Spices (cayenne pepper, chilli), Spices (cinnamon), Spices (coriander), Spices (cumin), Spices (fennel), Spices (ginger), Spices (laurel leave), Spices (pepper), Spices (peppermint), Spices (poppy seed), Spices (rosemary), Spices (sumac), AFB_1, literature[859]

Aflatoxin/S

incidence: 3/5, conc. range: 30–35 µg/kg*, sample origin: Assiut (governorate), Egypt, sample year: unknown, country: Egypt[641], *AFB_1 and AFB_2

- Co-contamination: not reported
- Further contamination of plants and plant products other than cereals and their products reported in the present book/article[641]: Spices (black cumin), AF/S and CIT, literature[641]; Spices (caraway), Spices (marjoram), AF/S and STG, literature[641]; Spices (cayenne pepper, chilli), Spices (cumin), STG, literature[641]; Spices (coriander), Spices (fennel), Spices (pepper), Spices (peppermint), AF/S, literature[641]

Aspergillus and *Penicillium* Toxins

Ochratoxin

incidence: 6/80*, conc. range: 0.41–0.98 µg/kg, Ø conc.: 0.73 µg/kg, sample origin: supermarkets, shops, and market stalls in eight cities, China, sample year: April–July 2009, country: China[885], *aniseed

- Co-contamination: not reported
- Further contamination of plants and plant products other than cereals and their products reported in the present book/article[885]: Spices (cayenne pepper, chilli), Spices (cinnamon), Spices (cumin), Spices (curry), Spices (prickly ash), OTA, literature[885]

Spices (basil) may contain the following mycotoxins:

Aspergillus Toxins

Aflatoxin B_1

incidence: 6/6*, conc. range: 0.8–18.1 µg/kg, ∅ conc.: 7.5 µg/kg, sample origin: organic markets and organic shops, Turkey, sample year: May 2010–May 2011, country: Turkey[859], *organic basil sa.

- Co-contamination: not reported
- Further contamination of plants and plant products other than cereals and their products reported in the present book/article[859]: Herbs (camomile), Herbs (linden), Herbs (rosehip), Herbs (sage), Spices (anise), Spices (cayenne pepper, chilli), Spices (cinnamon), Spices (coriander), Spices (cumin), Spices (fennel), Spices (ginger), Spices (laurel leaves), Spices (pepper), Spices (peppermint), Spices (poppy seed), Spices (rosemary), Spices (sumac), AFB_1, literature[859]

Spices (black cumin) may contain the following mycotoxins:

Aspergillus Toxins

Aflatoxin/s

incidence: 2/5, conc. range: 25–31 µg/kg*, ∅ conc.: 28 µg/kg*, sample origin: Assiut (governorate), Egypt, sample year: unknown, country: Egypt[641], *AFB_1 and AFG_1

- Co-contamination: not reported
- Further contamination of plants and plant products other than cereals and their products reported in the present book/article[641]: Spices (anise), Spices (coriander), Spices (fennel), Spices (pepper), Spices (peppermint), AF/S, literature[641]; Spices (black cumin), CIT, literature[641]; Spices (caraway), Spices (marjoram), AF/S and STG, literature[641]; Spices (cayenne pepper, chilli), Spices (cumin), STG, literature[641]

Aspergillus and *Penicillium* Toxins

Citrinin

incidence: 2/5, conc. range: 8–12 µg/kg, ∅ conc.: 10 µg/kg, sample origin: Assiut (governorate), Egypt, sample year: unknown, country: Egypt[641]

- Co-contamination: not reported
- Further contamination of plants and plant products other than cereals and their products reported in the present book/article[22]: Spices (anise), Spices (black cumin), Spices (coriander), Spices (fennel), Spices (pepper), Spices (peppermint), AF/S, literature[641]; Spices (caraway), Spices (marjoram), AF/S and STG, literature[641]; Spices (cayenne pepper, chilli), Spices (cumin), STG, literature[641]

Spices (caraway) may contain the following mycotoxins:

Aspergillus Toxins

Aflatoxin/s

incidence: 3/5, conc. range: 12–14 µg/kg*, sample origin: Assiut (governorate), Egypt, sample year: unknown, country: Egypt[641], *AFB_1 and AFB_2

- Co-contamination: not reported
- Further contamination of plants and plant products other than cereals and their products reported in the present book/article[641]: Spices (anise), Spices (coriander), Spices (fennel), Spices (pepper), AFS, literature[641]; Spices (black cumin), AF/S and CIT, literature[641]; Spices (caraway), Spices (cayenne pepper, chilli), Spices (cumin), STG, literature[641]; Spices (marjoram), AF/S and STG, literature[641]

Other *Aspergillus* Toxins

Sterigmatocystin

incidence: 3/5, conc. range: 14–18 µg/kg, sample origin: Assiut (governorate), Egypt, sample year: unknown, country: Egypt[641]

- Co-contamination: not reported
- Further contamination of plants and plant products other than cereals and their products reported in the present book/article[641]: Spices (anise), Spices (caraway), Spices (coriander), Spices (fennel), Spices (pepper), AF/S, literature[641]; Spices (black cumin), AF/S and CIT, literature[641]; Spices (cayenne pepper, chilli), Spices (cumin), STG, literature[641]; Spices (marjoram), AF/S and STG, literature[641]

Aspergillus and *Penicillium* Toxins

Ochratoxin A

incidence: 13/24, conc. range: 28–510 µg/kg, ∅ conc.: 244 µg/kg, sample origin: retail markets and supermarkets in 24 regions, Tunisia, sample year: unknown, country: Tunisia[20]

- Co-contamination: not reported
- Further contamination of plants and plant products other than cereals and their products reported in the present book/article[20]: Nuts (almond), Nuts (peanut), Nuts (pistachio nut), Spices (coriander), Spices (curcuma), Spices (paprika), Spices (pepper), OTA, literature[20]

incidence: 2/12, conc. range: ≤0.71 µg/kg, sample origin: 12 regions, Czech Republic, sample year: 2011–2013, country: Czech Republic[811]

- Co-contamination: not reported
- Further contamination of plants and plant products other than cereals and their products reported in the present book/article[811]: Beans, Chocolate, Coffee, Confectionery, Figs, Herbs and Spices, Juices, Licorice, Powder (cocoa powder), Products (ginger product), Raisins, Spices (cayenne pepper, chilli), Spices (coriander), Spices (ginger), Spices (mustard), Spices (nutmeg), Spices (paprika), Spices (pepper), Tea, Wine (red), Wine (white), OTA, literature[811]

Spices (cardamom) may contain the following mycotoxins:

Aspergillus Toxins

Aflatoxin B_1

incidence: 1/6, conc.: 20 µg/kg, sample origin: major retail outlets in Nainital (district), India, sample year: February 1984–October 1986, country: India[22]

- Co-contamination: 1 sa. co-contaminated with AFB_1, AFB_2, and AFG_1
- Further contamination of plants and plant products other than cereals and their products reported in the present book/article[22]: Spices (ammi), AFB_1, AFB_2, and AFG_1, literature[22]; Spices (cardamom), AFB_2, AFG_1, AFG_2, CIT, and OTA, literature[22]; Spices (cassia), AFB_1, AFB_2, AFG_1, and AFG_2, literature[22]; Spices (cayenne pepper, chilli), AFB_1, AFB_2, AFG_1, AFG_2, and ZEA, literature[22]; Spices (coriander), AFB_1, AFB_2, AFG_1, AFG_2, CIT, OTA, and ZEA, literature[22]; Spices (cumin), AFB_1, AFB_2, AFG_1, AFG_2, and CIT, literature[22]; Spices (curcuma), AFB_1, AFB_2, AFG_1, AFG_2, CIT, and RBT, literature[22]; Spices (fennel), AFB_1, AFB_2, AFG_1, AFG_2, CIT, OTA, and STG, literature[22]; Spices (garlic), AFB_1, AFB_2, and AFG_1, literature[22]; Spices (pepper), AFB_1, AFB_2, AFG_1, AFG_2, CIT, STG, and ZEA, literature[22]

incidence: 3?/6*, conc. range: 18–129 µg/kg, sample origin: major retail outlets in Nainital (district), India, sample year: May 1984–December 1986, country: India[22], *greater cardamom

- Co-contamination: not reported
- Further contamination of plants and plant products other than cereals and their products reported in the present book/article[22]: Spices (ammi), AFB_1, AFB_2, and AFG_1, literature[22]; Spices (cardamom), AFB_2, AFG_1, AFG_2, CIT, and OTA, literature[22]; Spices (cassia), AFB_1, AFB_2, AFG_1, and AFG_2, literature[22]; Spices (cayenne pepper, chilli), AFB_1, AFB_2, AFG_1, AFG_2, and ZEA, literature[22]; Spices (coriander), AFB_1, AFB_2, AFG_1, AFG_2, CIT, OTA, and ZEA, literature[22]; Spices (cumin), AFB_1, AFB_2, AFG_1, AFG_2, and CIT, literature[22]; Spices (curcuma), AFB_1, AFB_2, AFG_1, AFG_2, CIT, and RBT, literature[22]; Spices (fennel), AFB_1, AFB_2, AFG_1, AFG_2, CIT, OTA, and STG, literature[22]; Spices (garlic), AFB_1, AFB_2, and AFG_1, literature[22]; Spices (pepper), AFB_1, AFB_2, AFG_1, AFG_2, CIT, STG, and ZEA, literature[22]

incidence: 21/34*, conc. range: ≤7.34 µg/kg, sample origin: retail outlets, cafeterias, and hotels from major districts of Punjab (state), Pakistan, sample year: unknown, country: Pakistan[750], *ingredients for halva

- Co-contamination: not reported
- Further contamination of plants and plant products other than cereals and their products reported in the present book/article[750]: Halva, AFB_1 and AFM_1, literature[750]; Nuts (almond), Nuts (pistachio nut), Raisins, AFB_1, literature[750]

Aflatoxin B_2

incidence: 1/6, conc.: 15 µg/kg, sample origin: major retail outlets in Nainital (district), India, sample year: February 1984–October 1986, country: India[22]

- Co-contamination: 1 sa. co-contaminated with AFB_1, AFB_2, and AFG_1
- Further contamination of plants and plant products other than cereals and their products reported in the present book/article[22]: Spices (ammi), AFB_1, AFB_2, and AFG_1, literature[22]; Spices (cardamom), AFB_1, AFG_1, AFG_2, CIT, and OTA, literature[22]; Spices (cassia), AFB_1, AFB_2, AFG_1, and AFG_2, literature[22]; Spices (cayenne pepper, chilli), AFB_1, AFB_2, AFG_1, AFG_2, and ZEA, literature[22]; Spices (coriander), AFB_1, AFB_2, AFG_1, AFG_2, CIT, OTA, and ZEA, literature[22]; Spices (cumin), AFB_1, AFB_2, AFG_1, AFG_2, and CIT, literature[22]; Spices (curcuma), AFB_1, AFB_2, AFG_1, AFG_2, CIT, and RBT, literature[22]; Spices (fennel), AFB_1, AFB_2, AFG_1, AFG_2, CIT, OTA, and STG, literature[22]; Spices (garlic), AFB_1, AFB_2, and AFG_1, literature[22]; Spices (pepper), AFB_1, AFB_2, AFG_1, AFG_2, CIT, STG, and ZEA, literature[22]

incidence: 3?/6*, conc. range: 14–108 µg/kg, sample origin: major retail outlets in Nainital (district), India, sample year: May 1984–December 1986, country: India[22], *greater cardamom

- Co-contamination: not reported
- Further contamination of plants and plant products other than cereals and their products reported in the present book/article[22]: Spices (ammi), AFB_1, AFB_2, and AFG_1, literature[22]; Spices (cardamom), AFB_1, AFG_1, AFG_2, CIT, and OTA, literature[22]; Spices (cassia), AFB_1, AFB_2, AFG_1, and AFG_2, literature[22]; Spices (cayenne pepper, chilli), AFB_1, AFB_2, AFG_1, AFG_2, and ZEA, literature[22]; Spices (coriander), AFB_1, AFB_2, AFG_1, AFG_2, CIT, OTA, and ZEA, literature[22]; Spices (cumin), AFB_1, AFB_2, AFG_1, AFG_2, and CIT, literature[22]; Spices (curcuma), AFB_1, AFB_2, AFG_1, AFG_2, CIT, and RBT, literature[22]; Spices (fennel), AFB_1, AFB_2, AFG_1, AFG_2, CIT, OTA, and STG, literature[22]; Spices (garlic), AFB_1, AFB_2, and AFG_1, literature[22]; Spices (pepper), AFB_1, AFB_2, AFG_1, AFG_2, CIT, STG, and ZEA, literature[22]

Aflatoxin G_1

incidence: 1/6, conc.: 12 µg/kg, sample origin: major retail outlets in Nainital (district), India, sample year: February 1984–October 1986, country: India[22]

- Co-contamination: 1 sa. co-contaminated with AFB_1, AFB_2, and AFG_1
- Further contamination of plants and plant products other than cereals and their products reported in the present book/article[22]: Spices (ammi), AFB_1, AFB_2, and AFG_1, literature[22]; Spices (cardamom), AFB_1, AFB_2, AFG_2, CIT, and OTA, literature[22]; Spices (cassia), AFB_1, AFB_2, AFG_1, and AFG_2, literature[22]; Spices (cayenne pepper, chilli), AFB_1, AFB_2, AFG_1, AFG_2, and ZEA, literature[22]; Spices (coriander), AFB_1, AFB_2, AFG_1, AFG_2, CIT, OTA, and ZEA, literature[22]; Spices (cumin), AFB_1, AFB_2, AFG_1, AFG_2, and CIT, literature[22]; Spices (curcuma), AFB_1, AFB_2, AFG_1, AFG_2, CIT, and RBT, literature[22]; Spices (fennel), AFB_1, AFB_2, AFG_1, AFG_2, CIT, OTA, and STG, literature[22]; Spices (garlic), AFB_1, AFB_2, and AFG_1, literature[22]; Spices (pepper), AFB_1, AFB_2, AFG_1, AFG_2, CIT, STG, and ZEA, literature[22]

incidence: 3?/6*, conc. range: 17–78 µg/kg, sample origin: major retail outlets in Nainital (district), India, sample year: May 1984–December 1986, country: India[22], *greater cardamom

- Co-contamination: not reported
- Further contamination of plants and plant products other than cereals and their products reported in the present book/article[22]: Spices (ammi), AFB_1, AFB_2, and AFG_1, literature[22]; Spices (cardamom), AFB_1, AFB_2, AFG_2, CIT, and OTA, literature[22]; Spices (cassia), AFB_1, AFB_2, AFG_1, and AFG_2, literature[22]; Spices (cayenne pepper, chilli), AFB_1, AFB_2, AFG_1, AFG_2, and ZEA, literature[22]; Spices (coriander), AFB_1, AFB_2, AFG_1, AFG_2, CIT, OTA, and ZEA, literature[22]; Spices (cumin), AFB_1, AFB_2, AFG_1, AFG_2, and CIT, literature[22]; Spices (curcuma), AFB_1, AFB_2, AFG_1, AFG_2, CIT, and RBT, literature[22]; Spices (fennel), AFB_1, AFB_2, AFG_1, AFG_2, CIT, OTA, and STG, literature[22]; Spices (garlic), AFB_1, AFB_2, and AFG_1, literature[22]; Spices (pepper), AFB_1, AFB_2, AFG_1, AFG_2, CIT, STG, and ZEA, literature[22]

Aflatoxin G_2

incidence: 0/6, conc.: no contamination, sample origin: major retail outlets in Nainital (district), India, sample year: February 1984–October 1986, country: India[22]

- Co-contamination: not reported
- Further contamination of plants and plant products other than cereals and their products reported in the present book/article[22]: Spices (ammi), AFB_1, AFB_2, and AFG_1, literature[22]; Spices (cardamom), AFB_1, AFB_2, AFG_1, CIT, and OTA, literature[22]; Spices (cassia), AFB_1, AFB_2, AFG_1, and AFG_2, literature[22]; Spices (cayenne pepper, chilli), AFB_1, AFB_2, AFG_1, AFG_2, and ZEA, literature[22]; Spices (coriander), AFB_1, AFB_2, AFG_1, AFG_2, CIT, OTA, and ZEA, literature[22]; Spices (cumin), AFB_1, AFB_2, AFG_1, AFG_2, and CIT, literature[22]; Spices (curcuma), AFB_1, AFB_2, AFG_1, AFG_2, CIT, and RBT, literature[22]; Spices (fennel), AFB_1, AFB_2, AFG_1, AFG_2, CIT, OTA, and STG, literature[22]; Spices (garlic), AFB_1, AFB_2, and AFG_1, literature[22]; Spices (pepper), AFB_1, AFB_2, AFG_1, AFG_2, CIT, STG, and ZEA, literature[22]

incidence: 3?/6*, conc. range: 5–76 µg/kg, sample origin: major retail outlets in Nainital (district), India, sample year: May 1984–December 1986, country: India[22], *greater cardamom

- Co-contamination: not reported
- Further contamination of plants and plant products other than cereals and their products reported in the present book/article[22]: Spices (ammi), AFB_1, AFB_2, and AFG_1, literature[22]; Spices (cardamom), AFB_1, AFB_2, AFG_1, CIT, and OTA, literature[22]; Spices (cassia), AFB_1, AFB_2, AFG_1, and AFG_2, literature[22]; Spices (cayenne pepper, chilli), AFB_1, AFB_2, AFG_1, AFG_2, and ZEA, literature[22]; Spices (coriander), AFB_1, AFB_2, AFG_1, AFG_2, CIT, OTA, and ZEA, literature[22]; Spices (cumin), AFB_1, AFB_2, AFG_1, AFG_2, and CIT, literature[22]; Spices (curcuma), AFB_1, AFB_2, AFG_1, AFG_2, CIT, and RBT, literature[22]; Spices (fennel), AFB_1, AFB_2, AFG_1, AFG_2, CIT, OTA, and STG, literature[22]; Spices (garlic), AFB_1, AFB_2, and AFG_1, literature[22]; Spices (pepper), AFB_1, AFB_2, AFG_1, AFG_2, CIT, STG, and ZEA, literature[22]

Aspergillus **and** ***Penicillium*** **Toxins**

Citrinin

incidence: 1/6, conc.: 25 µg/kg, sample origin: major retail outlets in Nainital (district), India, sample year: February 1984–October 1986, country: India[22]

- Co-contamination: not reported
- Further contamination of plants and plant products other than cereals and their products reported in the present book/article[22]: Spices (ammi), AFB_1, AFB_2, and AFG_1, literature[22]; Spices (cardamom), AFB_1, AFB_2, AFG_1, AFG_2, and OTA, literature[22]; Spices (cassia), AFB_1, AFB_2, AFG_1, and AFG_2, literature[22]; Spices (cayenne pepper, chilli), AFB_1, AFB_2, AFG_1, AFG_2, and ZEA, literature[22]; Spices (coriander), AFB_1, AFB_2, AFG_1, AFG_2, CIT, OTA, and ZEA, literature[22]; Spices (cumin), AFB_1, AFB_2, AFG_1, AFG_2, and CIT, literature[22]; Spices (curcuma), AFB_1, AFB_2, AFG_1, AFG_2, CIT, and RBT, literature[22]; Spices (fennel), AFB_1, AFB_2, AFG_1, AFG_2, CIT, OTA, and STG, literature[22]; Spices (garlic), AFB_1, AFB_2, and AFG_1, literature[22]; Spices (pepper), AFB_1, AFB_2, AFG_1, AFG_2, CIT, STG, and ZEA, literature[22]

incidence: 0/6*, conc.: no contamination, sample origin: major retail outlets in Nainital (district), India, sample year: May 1984–December 1986, country: India[22], *greater cardamom

- Co-contamination: not reported
- Further contamination of plants and plant products other than cereals and their products reported in the present book/article[22]: Spices (ammi), AFB_1, AFB_2, and AFG_1, literature[22]; Spices (cardamom), AFB_1, AFB_2, AFG_1, AFG_2, and OTA, literature[22]; Spices (cassia), AFB_1, AFB_2, AFG_1, and AFG_2, literature[22]; Spices (cayenne pepper, chilli), AFB_1, AFB_2, AFG_1, AFG_2, and ZEA, literature[22]; Spices (coriander), AFB_1, AFB_2, AFG_1, AFG_2, CIT, OTA, and ZEA, literature[22]; Spices (cumin), AFB_1, AFB_2, AFG_1, AFG_2, and CIT, literature[22]; Spices (curcuma), AFB_1, AFB_2, AFG_1, AFG_2, CIT, and RBT, literature[22]; Spices (fennel), AFB_1, AFB_2, AFG_1, AFG_2, CIT, OTA, and STG, literature[22]; Spices (garlic), AFB_1, AFB_2, and AFG_1, literature[22]; Spices (pepper), AFB_1, AFB_2, AFG_1, AFG_2, CIT, STG, and ZEA, literature[22]

Ochratoxin A

incidence: 0/6, conc.: no contamination, sample origin: major retail outlets in Nainital (district), India, sample year: February 1984–October 1986, country: India[22]

- Co-contamination: not reported
- Further contamination of plants and plant products other than cereals and their products reported in the present book/article[22]: Spices (ammi), AFB_1, AFB_2, and AFG_1, literature[22]; Spices (cardamom), AFB_1, AFB_2, AFG_1, AFG_2, and CIT, literature[22]; Spices (cassia), AFB_1, AFB_2, AFG_1, and AFG_2, literature[22]; Spices (cayenne pepper, chilli), AFB_1, AFB_2, AFG_1, AFG_2, and ZEA, literature[22]; Spices (coriander), AFB_1, AFB_2, AFG_1, AFG_2, CIT, OTA, and ZEA, literature[22]; Spices (cumin), AFB_1, AFB_2, AFG_1, AFG_2, and CIT, literature[22]; Spices (curcuma), AFB_1, AFB_2, AFG_1, AFG_2, CIT, and RBT, literature[22]; Spices (fennel), AFB_1, AFB_2, AFG_1, AFG_2, CIT, OTA, and STG, literature[22]; Spices (garlic), AFB_1, AFB_2, and AFG_1, literature[22]; Spices (pepper), AFB_1, AFB_2, AFG_1, AFG_2, CIT, STG, and ZEA, literature[22]

incidence: 2/6*, conc. range: pr, sample origin: major retail outlets in Nainital (district), India, sample year: May 1984–December 1986, country: India[22], *greater cardamom

- Co-contamination: not reported
- Further contamination of plants and plant products other than cereals and their products reported in the present book/article[22]: Spices (ammi), AFB_1, AFB_2, and AFG_1, literature[22]; Spices (cardamom), AFB_1, AFB_2, AFG_1, AFG_2, and CIT, literature[22]; Spices (cassia), AFB_1, AFB_2, AFG_1, and AFG_2, literature[22]; Spices (cayenne pepper, chilli), AFB_1, AFB_2, AFG_1, AFG_2, and ZEA, literature[22]; Spices (coriander), AFB_1, AFB_2, AFG_1, AFG_2, CIT, OTA, and ZEA, literature[22]; Spices (cumin), AFB_1, AFB_2, AFG_1, AFG_2, and CIT, literature[22]; Spices (curcuma), AFB_1, AFB_2, AFG_1, AFG_2, CIT, and RBT, literature[22]; Spices (fennel), AFB_1, AFB_2, AFG_1, AFG_2, CIT, OTA, and STG, literature[22]; Spices (garlic), AFB_1, AFB_2, and AFG_1, literature[22]; Spices (pepper), AFB_1, AFB_2, AFG_1, AFG_2, CIT, STG, and ZEA, literature[22]

Spices (cassia) may contain the following mycotoxins:

Aspergillus **Toxins**

Aflatoxin B_1

incidence: 1/6*, conc.: 13 µg/kg, sample origin: major retail outlets in Nainital (district), India, sample year: February 1984–November 1986, country: India[22], *Indian cassia

- Co-contamination: 1 sa. co-contaminated with AFB_1, AFB_2, AFG_1, and AFG_2
- Further contamination of plants and plant products other than cereals and their products reported in the present book/article[22]: Spices (ammi), AFB_1, AFB_2, and AFG_1, literature[22]; Spices (cardamom), AFB_1, AFB_2, AFG_1, AFG_2, CIT, and OTA, literature[22]; Spices (cassia), AFB_2, AFG_1, and AFG_2, literature[22]; Spices (cayenne pepper, chilli), AFB_1, AFB_2, AFG_1, AFG_2, and ZEA, literature[22]; Spices (coriander), AFB_1, AFB_2, AFG_1, AFG_2, CIT, OTA, and ZEA, literature[22]; Spices (cumin), AFB_1, AFB_2, AFG_1, AFG_2, and CIT, literature[22]; Spices (curcuma), AFB_1, AFB_2, AFG_1, AFG_2, CIT, and RBT, literature[22]; Spices (fennel), AFB_1, AFB_2, AFG_1, AFG_2, CIT, OTA, and STG, literature[22]; Spices (garlic), AFB_1, AFB_2, and AFG_1, literature[22]; Spices (pepper), AFB_1, AFB_2, AFG_1, AFG_2, CIT, STG, and ZEA, literature[22]

Aflatoxin B_2

incidence: 1/6*, conc.: 11 µg/kg, sample origin: major retail outlets in Nainital (district), India, sample year: February 1984–November 1986, country: India[22], *Indian cassia

- Co-contamination: 1 sa. co-contaminated with AFB_1, AFB_2, AFG_1, and AFG_2
- Further contamination of plants and plant products other than cereals and their products reported in the present book/article[22]: Spices (ammi), AFB_1, AFB_2, and AFG_1, literature[22];

Spices (cardamom), AFB_1, AFB_2, AFG_1, AFG_2, CIT, and OTA, literature[22]; Spices (cassia), AFB_1, AFG_1, and AFG_2, literature[22]; Spices (cayenne pepper, chilli), AFB_1, AFB_2, AFG_1, AFG_2, and ZEA, literature[22]; Spices (coriander), AFB_1, AFB_2, AFG_1, AFG_2, CIT, OTA, and ZEA, literature[22]; Spices (cumin), AFB_1, AFB_2, AFG_1, AFG_2, and CIT, literature[22]; Spices (curcuma), AFB_1, AFB_2, AFG_1, AFG_2, CIT, and RBT, literature[22]; Spices (fennel), AFB_1, AFB_2, AFG_1, AFG_2, CIT, OTA, and STG, literature[22]; Spices (garlic), AFB_1, AFB_2, and AFG_1, literature[22]; Spices (pepper), AFB_1, AFB_2, AFG_1, AFG_2, CIT, STG, and ZEA, literature[22]

Aflatoxin G_1

incidence: 1/6*, conc.: 8 µg/kg, sample origin: major retail outlets in Nainital (district), India, sample year: February 1984–November 1986, country: India[22], *Indian cassia

- Co-contamination: 1 sa. co-contaminated with AFB_1, AFB_2, AFG_1, and AFG_2
- Further contamination of plants and plant products other than cereals and their products reported in the present book/article[22]: Spices (ammi), AFB_1, AFB_2, and AFG_1, literature[22]; Spices (cardamom), AFB_1, AFB_2, AFG_1, AFG_2, CIT, and OTA, literature[22]; Spices (cassia), AFB_1, AFB_2, and AFG_2, literature[22]; Spices (cayenne pepper, chilli), AFB_1, AFB_2, AFG_1, AFG_2, and ZEA, literature[22]; Spices (coriander), AFB_1, AFB_2, AFG_1, AFG_2, CIT, OTA, and ZEA, literature[22]; Spices (cumin), AFB_1, AFB_2, AFG_1, AFG_2, and CIT, literature[22]; Spices (curcuma), AFB_1, AFB_2, AFG_1, AFG_2, CIT, and RBT, literature[22]; Spices (fennel), AFB_1, AFB_2, AFG_1, AFG_2, CIT, OTA, and STG, literature[22]; Spices (garlic), AFB_1, AFB_2, and AFG_1, literature[22]; Spices (pepper), AFB_1, AFB_2, AFG_1, AFG_2, CIT, STG, and ZEA, literature[22]

Aflatoxin G_2

incidence: 1/6*, conc.: 4 µg/kg, sample origin: major retail outlets in Nainital (district), India, sample year: February 1984–November 1986, country: India[22], *Indian cassia

- Co-contamination: 1 sa. co-contaminated with AFB_1, AFB_2, AFG_1, and AFG_2
- Further contamination of plants and plant products other than cereals and their products reported in the present book/article[22]: Spices (ammi), AFB_1, AFB_2, and AFG_1, literature[22]; Spices (cardamom), AFB_1, AFB_2, AFG_1, AFG_2, CIT, and OTA, literature[22]; Spices (cassia), AFB_1, AFB_2, and AFG_1, literature[22]; Spices (cayenne pepper, chilli), AFB_1, AFB_2, AFG_1, AFG_2, and ZEA, literature[22]; Spices (coriander), AFB_1, AFB_2, AFG_1, AFG_2, CIT, OTA, and ZEA, literature[22]; Spices (cumin), AFB_1, AFB_2, AFG_1, AFG_2, and CIT, literature[22]; Spices (curcuma), AFB_1, AFB_2, AFG_1, AFG_2, CIT, and RBT, literature[22]; Spices (fennel), AFB_1, AFB_2, AFG_1, AFG_2, CIT, OTA, and STG, literature[22]; Spices (garlic), AFB_1, AFB_2, and AFG_1, literature[22]; Spices (pepper), AFB_1, AFB_2, AFG_1, AFG_2, CIT, STG, and ZEA, literature[22]

Spices (cayenne pepper, chilli) may contain the following mycotoxins:

Alternaria Toxins

Alternariol Methyl Ether

incidence: 0/18*, conc.: no contamination, sample origin: open markets, Sri Lanka, sample year: 2012/2013, country: Belgium[889], *chilli pods, whole

incidence: 0/26*, conc.: no contamination, sample origin: open markets, Sri Lanka, sample year: 2012/2013, country: Belgium[889], *chilli flakes

incidence: 0/42*, conc.: no contamination, sample origin: open markets, Sri Lanka, sample year: 2012/2013, country: Belgium[889], *chilli powder

incidence: 2/35*, conc. range: 69.72–222 µg/kg, Ø conc.: 145.9 µg/kg, sample origin: exotic shops and supermarkets in Antwerp and Ghent (cities), Belgium, sample year: 2012/2013, country: Belgium[889], *included whole chilli pods, chilli powders, and chilli flakes

- Co-contamination: For detailed information see the article.
- Further contamination of plants and plant products other than cereals and their products reported in the present book/article[889]: Spices (cayenne pepper, chilli), AFB_1, AFB_2, AFG_1, AFG_2, CIT, FB_2, OTA, and STG, literature[889]

For detailed information see the article.

Aspergillus Toxins

Aflatoxin B_1

incidence: 4/4*, conc. range: 0.08–0.63 µg/kg, Ø conc.: 0.40 µg/kg, sample origin: grocery markets in Busan, Daegu, Daejeon, Gangneung, and Gwangju (cities) as well as Seoul (capital), Korea, sample year: May 2004–June 2005, country: Korea[10], *hot pepper powder

- Co-contamination: not reported
- Further contamination of plants and plant products other than cereals and their products reported in the present book/article[10]: Butter (peanut butter), Nuts, Nuts (almond), Nuts (peanut), Nuts (pistachio nut), Paste (soybean paste), Sauces (soy sauce), Spices (curry), Spices (pepper), AFB_1, literature[10]

incidence: 4?/9*, conc. range: 15–146 µg/kg, sample origin: major retail outlets in Nainital (district), India, sample year: February 1984–December 1986, country: India[22], *chillies

- Co-contamination: not reported
- Further contamination of plants and plant products other than cereals and their products reported in the present book/article[22]: Spices (ammi), AFB_1, AFB_2, and AFG_1, literature[22]; Spices (cardamom), AFB_1, AFB_2, AFG_1, AFG_2, CIT, and OTA, literature[22]; Spices (cassia), AFB_1, AFB_2, AFG_1, and AFG_2, literature[22]; Spices (cayenne pepper, chilli), AFB_2, AFG_1, AFG_2, and ZEA, literature[22]; Spices (coriander), AFB_1, AFB_2, AFG_1, AFG_2, CIT, OTA, and ZEA, literature[22]; Spices (cumin), AFB_1, AFB_2, AFG_1, AFG_2, and CIT, literature[22]; Spices (curcuma), AFB_1, AFB_2, AFG_1, AFG_2, CIT, and RBT, literature[22]; Spices (fennel), AFB_1, AFB_2, AFG_1, AFG_2, CIT, OTA, and STG, literature[22]; Spices (garlic), AFB_1, AFB_2, and AFG_1, literature[22]; Spices (pepper), AFB_1, AFB_2, AFG_1, AFG_2, CIT, STG, and ZEA, literature[22]

incidence: 2/10*, conc. range: 2.6–9.1 µg/kg, Ø conc.: 5.9 µg/kg, sample origin: commercially available market basket sa. of foods and foodstuffs in Tokyo (capital), Japan, sample year: 1986–1990, country: Japan[99], *red pepper

- Co-contamination: not reported
- Further contamination of plants and plant products other than cereals and their products reported in the present book/article[99]: Beans, Seeds (sesame), Sugar, AFB_1 and AFB_2, literature[99]; Food, Jam (bean jam), Paste (peanut paste), AFB_1, literature[99]; Nuts (Brazil nut, paranut), Nuts (peanut), Nuts (pistachio nut), Spices (nutmeg), AFB_1, AFB_2, AFG_1, and AFG_2, literature[99]; Spices, Spices (pepper), AFB_1, AFB_2, and AFG_1, literature[99]; Spices (paprika), AFB_1 and AFG_1, literature[99]; Spices (cayenne pepper, chilli), AFB_2 and AFG_1, literature[99]

incidence: 8/15*, conc. range: 0.2–32.9 µg/kg, Ø conc.: 8.1 µg/kg, sample origin: unknown, sample year: unknown, country: USA[167], *red pepper

- Co-contamination: 1 sa. co-contaminated with AFB_1, AFB_2, AFG_1, and AFG_2; 3 sa. co-contaminated with AFB_1, AFB_2, and AFG_1; 1 sa. co-contaminated with AFB_1, AFB_2, and AFG_2; 3 sa. co-contaminated with AFB_1 and AFB_2
- Further contamination of plants and plant products other than cereals and their products reported in the present book/article[167]: Spices (cayenne pepper, chilli), AFB_2, AFG_1, and AFG_2, literature[167]; Spices (ginger), AFB_1, AFB_2, AFG_1, and AFG_2, literature[167]; Spices (pepper), AFB_1, literature[167]

incidence: 8/60*, conc. range: 250–525 µg/kg, Ø conc.: 426 µg/kg, sample origin: government-owned stores, retail shops, and open markets in Addis Ababa (capital), Ethiopia, sample year: unknown, country: Ethiopia[168], *red pepper, grounded

- Co-contamination: not reported
- Further contamination of plants and plant products other than cereals and their products reported in the present book/article[168]: Shiro, AFB_1, literature[168]

incidence: 5/5*, conc. range: 2.3–31.8 µg/kg, Ø conc.: 8.87 µg/kg, sample origin: supermarkets and ethnic shops in Lisbon (capital) area, Portugal, sample year: unknown, country: Portugal[169], *cayenne pepper, packaged

incidence: 3/8*, conc. range: 1.5–2.2 µg/kg, Ø conc.: 1.97 µg/kg, sample origin: supermarkets and ethnic shops in Lisbon (capital) area, Portugal, sample year: unknown, country: Portugal[169], *chilli, packaged

- Co-contamination: not reported
- Further contamination of plants and plant products other than cereals and their products reported in the present book/article[169]: Spices (cumin), Spices (curry), Spices (nutmeg), Spices (paprika), Spices (pepper), Spices (saffron), AFB_1, literature[169]

incidence: 10/22*, conc. range: 0.14–15.7 µg/kg, sample origin: food shops in Hajdú-Bihar (county), Hungary, sample year: October 2004, country: Hungary[170], *red pepper, grounded

incidence: 7/16*, conc. range: 0.24–9.4 µg/kg, sample origin: shops in/around Kaposvár (town), Somogy (county), Hungary, sample year: October 2004, country: Hungary[170], *red pepper, grounded and originally packed

incidence: 1/13*, conc.: 0.28 µg/kg, sample origin: food shops in Hajdú-Bihar (county), Hungary, sample year: November 2004, country: Hungary[170], *red pepper, grounded

incidence: 0/19*, conc.: no contamination, sample origin: 10 Hungarian counties, Hungary, sample year: October/November 2004, country: Hungary[170], *red pepper, home-made

incidence: 2/5*, conc. range: 0.75–8.1 µg/kg, Ø conc.: 4.43 µg/kg, sample origin: food shops in Debrecen (town), Hungary, sample year: October 2004, country: Hungary[170], *chilli

- Co-contamination: For detailed information see the article.
- Further contamination of plants and plant products other than cereals and their products reported in the present book/article[170]: Spices, AFB_1 and AFB_2, literature[170]; Spices (cayenne pepper, chilli), AFB_2 and OTA, literature[170], Spices (pepper), AFB_1, literature[170]

incidence: 1/1*, conc.: 2.0 µg/kg, sample origin: imported sa. collected by the Mycotoxin Research Association, Japan, sample year: 1985, country: Japan[171], *red pepper

- Co-contamination: not reported
- Further contamination of plants and plant products other than cereals and their products reported in the present book/article[171]: Cocoa beans, Nuts (pistachio nut), Spices (pepper), AFB_1, literature[171]

incidence: 3/6*, conc. range: 2.5–40 µg/kg**, Ø conc.: 20.8 µg/kg**, sample origin: markets (marked sa. of unknown origin), Turkey, sample year: unknown, country: Turkey[172], *red pepper, **TLC

- Co-contamination: not reported
- Further contamination of plants and plant products other than cereals and their products reported in the present book/article[172]: Spices (cayenne pepper, chilli), AFB_2, AFG_1 and AFG_2, literature[172]

incidence: 10/15*, conc. range: 1.7–100 µg/kg**, Ø conc.: 36.3 µg/kg**, sample origin: spice-sellers (unmarked sa. from unknown origin, India, and Turkey), Turkey, sample year: unknown, country: Turkey[172], *red pepper, **TLC and HPLC

- Co-contamination: 3 sa. co-contaminated with AFB_1, AFB_2, and AFG_1; 2 sa. co-contaminated with AFB_1 and AFB_2; 5 sa. contaminated solely with AFB_1
- Further contamination of plants and plant products other than cereals and their products reported in the present book/article[172]: Spices (cayenne pepper, chilli), AFB_2, AFG_1 and AFG_2, literature[172]

incidence: 7/11*, conc. range: 2.4–60 µg/kg**, Ø conc.: 21.7 µg/kg**, sample origin: street bazaars (unmarked sa. from unknown origin and Turkey), Turkey, sample year: unknown, country: Turkey[172], *red pepper, **TLC and HPLC

- Co-contamination: 2 sa. co-contaminated with AFB_1, AFB_2, and AFG_1; 1 sa. co-contaminated with AFB_1, AFB_2, and AFG_2; 4 sa. contaminated solely with AFB_1
- Further contamination of plants and plant products other than cereals and their products reported in the present book/article[172]: Spices (cayenne pepper, chilli), AFB_2, AFG_1 and AFG_2, literature[172]

For detailed information see the article.

incidence: 1/2*, conc.: 0.8 µg/kg, sample origin: retail stores in Osaka (prefecture), Japan, sample year: 1988–1992, country: Japan[173], *red pepper

- Co-contamination: not reported
- Further contamination of plants and plant products other than cereals and their products reported in the present book/article[173]: Butter (peanut butter), Spices (nutmeg), AFB_1, AFB_2, and AFG_1, literature[173]; Nuts (peanut), Spices (pepper), AFB_1, literature[173]

incidence: 14/14*, conc. range: ≤5.40 µg/kg, Ø conc.: 2.88 µg/kg, sample origin: popular markets in Salé (city) and Rabat (capital), Morocco, sample year: unknown, country: Morocco[174], *red paprika

- Co-contamination: not reported
- Further contamination of plants and plant products other than cereals and their products reported in the present book/article[174]: Spice (cayenne pepper, chilli), AF/S, literature[174]; Spices (cumin), Spices (ginger), Spices (pepper), AFB_1 and AF/S, literature[174]

incidence: 17/26*, conc. range: 0.6–56 µg/kg, Ø conc.: 17.5 µg/kg, sample origin: various cities, Turkey, sample year: unknown, country: Turkey[175], *red pepper

- Co-contamination: not reported

- Further contamination of plants and plant products other than cereals and their products reported in the present book/ article[175]: Spices (cayenne pepper, chilli), AFB_2, literature[175]

incidence: 5/11*, conc. range: 0.57–26.90 µg/kg, Ø conc.: 7.16 µg/kg, sample origin: public markets, supermarkets, shops, and bonded warehouse in Emilia Romagna (administrative region), Italy, sample year: 2000–2005, country: Italy[176], *hot pepper, whole/smashed

- Co-contamination: 1 sa. co-contaminated with AFB_1, AFB_2, AFG_1, and AFG_2; 1 sa. co-contaminated with AFB_1 and AFB_2; 3 sa. contaminated solely with AFB_1
- Further contamination of plants and plant products other than cereals and their products reported in the present book/ article[176]: Spices (cayenne pepper, chilli), AFB_2, AFG_1, and AFG_2, literature[176]; Spices (cinnamon), AFB_1, literature[176]; Spices (nutmeg), AFB_1 and AFB_2, literature[176]

incidence: 72/75*, conc. range: 0.025–5 µg/kg (61 sa.), >5 µg/kg (11 sa., maximum: 24.7 µg/kg), Ø conc.: 1.9 µg/kg, sample origin: bazaars and herbal shops in Sanliurfa (town), Turkey, sample year: April, country: Turkey[177], *pepper, deep-red ground

- Co-contamination: not reported
- Further contamination of plants and plant products other than cereals and their products reported in the present book/ article[177]: not reported

incidence: 15/30*, conc. range: 0.2–32.3 µg/kg, Ø conc.: 13.4 µg/kg, sample origin: bazaars and markets in Istanbul (capital), Turkey, sample year: unknown, country: Turkey[178], *red-scaled pepper

incidence: 7/30*, conc. range: 0.2–10.7 µg/kg, Ø conc.: 4.7 µg/kg, sample origin: bazaars and markets in Istanbul (capital), Turkey, sample year: unknown, country: Turkey[178], *red pepper

- Co-contamination: not reported
- Further contamination of plants and plant products other than cereals and their products reported in the present book/ article[178]: Spices (cayenne pepper, chilli), AF/S, literature[178]; Spices (pepper), AFB_1 and AF/S, literature[178]

incidence: 68/100*, conc. range: 0.025–5 µg/kg (50 sa), >5 µg/kg (18 sa., maximum: 40.9 µg/kg), sample origin: retail sales points in Istanbul (capital), Turkey, sample year: April/May 2004, country: Turkey[179], *red pepper, powdered

- Co-contamination: not reported
- Further contamination of plants and plant products other than cereals and their products reported in the present book/ article[179]: not reported

incidence: 5/5*, conc. range: 2.3–70.4 µg/kg, sample origin: bazaars, Turkey, sample year: unknown, country: Turkey[180], *chilli powder

incidence: 5/5*, conc. range: 1.6–80.4 µg/kg, sample origin: herbal shops, Turkey, sample year: unknown, country: Turkey[180], *chilli powder

incidence: 5/5*, conc. range: 2.1–4.2 µg/kg, sample origin: supermarkets, Turkey, sample year: unknown, country: Turkey[180], *chilli powder

- Co-contamination: not reported
- Further contamination of plants and plant products other than cereals and their products reported in the present book/ article[180]: Spices (cayenne pepper, chilli), AF/S, literature[180]; Spices (paprika), Spices (pepper), AFB_1 and AF/S, literature[180]

incidence: 8/8, conc. range: 0.58–3.5 µg/kg, Ø conc.: 2.62 µg/kg, sample origin: supermarkets in Penang (state), Malaysia, sample year: unknown, country: Malaysia[182]

- Co-contamination: not reported
- Further contamination of plants and plant products other than cereals and their products reported in the present book/ article[182]: Nuts, Nuts (peanut), Seeds (sesame), Seeds (sunflower), Spices (cumin), Spices (pepper), AFB_1, literature[182]

incidence: 19/24*, conc. range: 0.13–11.45 µg/kg, Ø conc.: 2.61 µg/kg, sample origin: markets, herbal shops, and bazaars in Corum and Ankara (cities), Turkey, sample year: November 2010–August 2011, country: Turkey[183], *red chilli flakes

- Co-contamination: 3 sa. co-contaminated with AFB_1, AFB_2, AFG_1, and OTA; 11 sa. co-contaminated with AFB_1, AFB_2, and OTA; 1 sa. co-contaminated with AFB_1 and OTA; 4 sa. contaminated solely with AFB_1
- Further contamination of plants and plant products other than cereals and their products reported in the present book/ article[183]: Spices (cayenne pepper, chilli), AFB_2, AFG_1, and OTA, literature[183]; Spices (cumin), Spices (pepper), AFB_1, AFB_2, and OTA, literature[183]

incidence: 14/22*, conc. range: 0.20–35.77 µg/kg, Ø conc.: 5.10 µg/kg, sample origin: markets, herbal shops, and bazaars in Corum and Ankara (cities), Turkey, sample year: November 2010–August 2011, country: Turkey[183], *red chilli powder

- Co-contamination: 3 sa. co-contaminated with AFB_1, AFB_2, AFG_1, and OTA; 5 sa. co-contaminated with AFB_1, AFB_2, and OTA; 1 sa. co-contaminated with AFB_1 and OTA; 5 sa. contaminated solely with AFB_1
- Further contamination of plants and plant products other than cereals and their products reported in the present book/ article[183]: Spices (cayenne pepper, chilli), AFB_2, AFG_1, and OTA, literature[183]; Spices (cumin), Spices (pepper), AFB_1, AFB_2, and OTA, literature[183]

incidence: 2/2*, conc. range: 2.5–2.7 µg/kg, Ø conc.: 2.6 µg/kg, sample origin: unknown, sample year: unknown, country: Denmark[217], *chilli

incidence: 2/2*, conc. range: 2.1–5.0 µg/kg, Ø conc.: 3.6 µg/kg, sample origin: unknown, sample year: unknown, country: Denmark[217], *cayenne pepper

- Co-contamination: 2 sa. co-contaminated with AFB_1, AFB_2, AFG_1, and AFG_2
- Further contamination of plants and plant products other than cereals and their products reported in the present book/ article[217]: Butter (peanut butter), Figs, AFB_1, AFB_2, AFG_1, and AFG_2, literature[217]; Spices (cayenne pepper, chilli), AFB_2, AFG_1, and AFG_2, literature[217]; Meal (pistachio meal), AFB_1 and AFB_2, literature[217]

incidence: 21/42*, conc. range: <10 µg/kg (16 sa.), 11–30 µg/kg (3 sa.), 31–50 µg/kg (1 sa.), 51100 µg/kg (1 sa.), sample origin: regulated market yards and cold storage facilities, Andhra Pradesh (state), India, sample year: 1998/1999, country: India[241], *chilli grade 1

incidence: 25/38*, conc. range: <10 µg/kg (10 sa.), 11–30 µg/kg (6 sa.), 31–50 µg/kg (3 sa.), 51–100 µg/kg (4 sa.), >100 µg/kg (2 sa.), sample origin: regulated market yards and cold storage facilities, Andhra Pradesh (state), India, sample year: 1998/1999, country: India[241], *chilli grade 2 incidence: 41/44*, conc. range: <10 μg/kg (21 sa.), 11–30 μg/kg (4 sa.), 31–50 μg/kg (3 sa.), 51–100 μg/kg (2 sa.), >100 μg/kg (11 sa., maximum: 969 μg/kg), sample origin: regulated market yards and cold storage facilities, Andhra Pradesh (state), India, sample year: 1998/1999, country: India[241], *chilli grade 3

incidence: 3/12*, conc. range: <10 μg/kg (2 sa.), 31–50 μg/kg (1 sa.), sample origin: regulated market yards and cold storage facilities, Andhra Pradesh (state), India, sample year: 1998/1999, country: India[241], *chillies from cold store

incidence: 17/43*, conc. range: <10 μg/kg (12 sa.), 11–30 μg/kg (1 sa.), 51–100 μg/kg (3 sa.), >100 μg/kg (1 sa.), sample origin: supermarkets in Hyderabad (city), India, sample year: 1998/1999, country: India[241], *chilli powder

- Co-contamination: not reported
- Further contamination of plants and plant products other than cereals and their products reported in the present book/article[241]: not reported

incidence: 10/33*, conc. range: tr-8 μg/kg, sample origin: India, sample year: unknown, country: Canada[242], *cayenne pepper

- Co-contamination: 5 sa. co-contamination with AFB_1 and AFB_2; 5 sa. contaminated solely with AFB_1
- Further contamination of plants and plant products other than cereals and their products reported in the present book/article[242]: Spices (cayenne pepper, chilli), AFB_2 and AFG_1, literature[242]

incidence: 6/6*, conc. range: tr-6 μg/kg, sample origin: India, sample year: unknown, country: Canada[242], *Indian chilli powder

- Co-contamination: 1 sa. co-contaminated with AFB_1, AFB_2, and AFG_1; 1 sa. co-contamination with AFB_1 and AFB_2; 2 sa. co-contaminated with AFB_1 and AFG_1; 2 sa. contaminated solely with AFB_1
- Further contamination of plants and plant products other than cereals and their products reported in the present book/article[242]: Spices (cayenne pepper, chilli), AFB_2 and AFG_1, literature[242]

incidence: 13/24*, conc. range: 9.5–211.3 μg/kg, ∅ conc.: 62.3 μg/kg, sample origin: retail and wholesale, Nordwürttemberg (administrative region), Germany, sample year: unknown, country: Germany[243], *cayenne pepper powder and ground chilli

- Co-contamination: 11 sa. co-contaminated with AFB_1, AFB_2, and AFG_1, 2 sa. co-contaminated with AFB_1 and AFB_2
- Further contamination of plants and plant products other than cereals and their products reported in the present book/article[243]: Spices (cayenne pepper, chilli), AFB_2 and AFG_1, literature[243]; Spices (paprika), AFB_1 and AFB_2, literature[243]

incidence: 11/22*, conc. range: ≤5 μg/kg (7 sa.), >5 μg/kg (4 sa., maximum: 24 μg/kg), sample origin: unknown, sample year: unknown, country: Germany/Thailand[244], *chilli and cayenne pepper

- Co-contamination: not reported
- Further contamination of plants and plant products other than cereals and their products reported in the present book/article[244]: Spices (coriander), Spices (nutmeg), AFB_1, literature[244]

incidence: 29/31*, conc. range: 0.6–13.9 μg/kg, sample origin: warehouses, packing establishments, supermarkets, and smaller shops, UK, sample year: unknown, country: UK[245], *chilli

- Co-contamination: 1 sa. co-contaminated with AFB_1, AFB_2, AFG_1, AFG_2, and OTA; 1 sa. co-contaminated with AFB_1, AFB_2, AFG_1, and OTA; 8 sa. co-contaminated with AFB_1, AFB_2 and OTA; 1 sa. co-contaminated with AFB_1, AFG_1 and OTA; 1 sa. co-contaminated with AFB_1, AFG_2 and OTA; 17 sa. co-contaminated with AFB_1 and OTA
- Further contamination of plants and plant products other than cereals and their products reported in the present book/article[245]: Spices (cayenne pepper, chilli), AFB_2, AFG_1, AFG_2, and OTA, literature[245]; Spices (paprika), AFB_1, AFB_2, AFG_1, AFG_2, and OTA, literature[245]

incidence: 4/4*, conc. range: 0.2–6.8 μg/kg, ∅ conc.: 2.26 μg/kg, sample origin: warehouses, packing establishments, supermarkets, and smaller shops, UK, sample year: unknown, country: UK[245], *cayenne pepper

- Co-contamination: 1 sa. co-contaminated with AFB_1, AFB_2, and OTA; 1 sa. co-contaminated with AFB_1, AFG_1 and OTA; 2 sa. co-contaminated with AFB_1 and OTA
- Further contamination of plants and plant products other than cereals and their products reported in the present book/article[245]: Spices (cayenne pepper, chilli), AFB_2, AFG_1, AFG_2, and OTA, literature[245]; Spices (paprika), AFB_1, AFB_2, AFG_1, AFG_2, and OTA, literature[245]

incidence: 18/29*, conc. range: 1–12.4 μg/kg, sample origin: different sources, Pakistan, sample year: unknown, country: Pakistan[246], *red chilli, whole

incidence: 17/23*, conc. range: 1–13.9 μg/kg, sample origin: different sources, Pakistan, sample year: unknown, country: Pakistan[246], *red chilli, crushed

incidence: 81/124*, conc. range: 1–79.9 μg/kg, sample origin: different sources, Pakistan, sample year: unknown, country: Pakistan[246], *red chilli powder

- Co-contamination: not reported
- Further contamination of plants and plant products other than cereals and their products reported in the present book/article[246]: not reported

incidence: 16/22*, conc. range: ≤96.3 μg/kg, sample origin: markets, herbal shops, and chilli-growing areas, Faisalabad (city), Punjab (state), Pakistan, sample year: June–December 2008, country: Pakistan/Portugal/USA[247], *chilli, whole

incidence: 19/22*, conc. range: ≤89.6 μg/kg, sample origin: markets, herbal shops, and chilli-growing areas, Faisalabad (city), Punjab (state), Pakistan, sample year: June December 2008, country: Pakistan/Portugal/USA[247], *chilli, grounded

- Co-contamination: not reported
- Further contamination of plants and plant products other than cereals and their products reported in the present book/article[247]: not reported

incidence: 2/16, conc. range: 2.3–9.0 μg/kg, ∅ conc.: 5.65 μg/kg, sample origin: retail stores in and around Calcutta (city), India, sample year: unknown, country: India[248]

- Co-contamination: not reported
- Further contamination of plants and plant products other than cereals and their products reported in the present book/article[248]: not reported

incidence: 52/80*, conc. range: <1.0 μg/kg (9 sa.), 1–5 μg/kg (34 sa.), 5–10 μg/kg (3 sa.), >10 μg/kg (6 sa., maximum: 56.61 μg/kg), sample origin: local markets and open markets in Kuala Lumpur (capital), Putrajaya (city), and Selangor (state) areas, Malaysia, sample year: May/June 2009, country: Malaysia[249], *chilli, dried

- Co-contamination: not reported
- Further contamination of plants and plant products other than cereals and their products reported in the present book/article[249]: Spice (cayenne pepper, chilli), AFB_2, AFG_1, AFG_2, and OTA, literature[249]

incidence: 4/5*, conc. range: 2.9–4.8 µg/kg, Ø conc.: 3.7 µg/kg, sample origin: available in New Zealand, sample year: 2009, country: New Zealand[277], *cayenne pepper

incidence: 4/5*, conc. range: 3.5–7.7 µg/kg, Ø conc.: 5.1 µg/kg, sample origin: available in New Zealand, sample year: 2009, country: New Zealand[277];*chilli powder

- Co-contamination: not reported
- Further contamination of plants and plant products other than cereals and their products reported in the present book/article[277]: Bars, AFB_1, AF/S, DON, NIV, and OTA, literature[277]; Butter (peanut butter), Confectionery, Food, Fruits, dried, Nuts, Nuts (peanut), Nuts (pistachio nut), Sauces (peanut sauce), AFB_1 and AF/S, literature[277]; Coffee, Dates, Spices (nutmeg), Spices (pepper), Vine fruits, dried, OTA, literature[277]; Figs, Spices (curry), Spices (ginger), Spices (paprika), AFB_1, AF/S, and OTA, literature[277]; Nuts (Brazil nut, paranut), AFB_1, literature[277]; Spices (cayenne pepper, chilli), AF/S and OTA, literature[277]

incidence: 1/1*, conc.: 10.7 µg/kg, sample origin: imported from Asian countries purchased from major wholesale/retail store of ethnic foods, West of Scotland, sample year: unknown, country: UK[314], *chilli, dried and crushed

- Co-contamination: 1 sa. co-contaminated with AFB_1 and AFB_2
- Further contamination of plants and plant products other than cereals and their products reported in the present book/article[314]: Spice (cayenne pepper, chilli), AFB_2, literature[314]

incidence: 10/10, conc. range: 15.5–57.5 µg/kg, sample origin: retail sales points in different regions, Iran, sample year: March–May 2015, country: Iran[491]

- Co-contamination: not reported
- Further contamination of plants and plant products other than cereals and their products reported in the present book/article[491]: Herbs (camomile), Herbs (sage), AFB_1, literature[491]; Spices (curcuma), Spices (ginger), Spices (pepper), Spices (thyme), AFB_1, literature[491]

incidence: 1/2*, conc.: 3.18 µg/kg, sample origin: Penang (state, island), Malaysia, sample year: unknown (first study), country: Malaysia[530], *chilli, dried or coarse dried

- Co-contamination: 1 sa. co-contaminated with AFB_1, AFB_2, AFG_1, AFG_2, and OTA
- Further contamination of plants and plant products other than cereals and their products reported in the present book/article[530]: Spices, Spices (coriander), Spices (cumin), Spices (curry), Spices (fennel), Spices (masala), AFB_1, AFB_2, AFG_1, AFG_2, and OTA, literature[530]; Spices (cayenne pepper, chilli), AFB_2, AFG_1, AFG_2, and OTA, literature[530]; Spices (curcuma), AFB_1, AFB_2, AFG_1, and OTA, literature[530]; Spices (pepper), AFB_1 and AFB_2, literature[530]; Spices (poppy seed), AFB_1, literature[530]

incidence: 3/3*, conc. range: 6.44–28.43 µg/kg, Ø conc.: 16.2 µg/kg, sample origin: Penang (state, island), Malaysia, sample year: unknown (second study), country: Malaysia[530], *chilli, dried

- Co-contamination: 3 sa. co-contaminated with AFB_1 and AFB_2
- Further contamination of plants and plant products other than cereals and their products reported in the present book/article[530]: Spices, Spices (coriander), Spices (cumin), Spices (curry), Spices (fennel), Spices (masala), AFB_1, AFB_2, AFG_1, AFG_2, and OTA, literature[530]; Spices (cayenne pepper, chilli), AFB_2, AFG_1, AFG_2, and OTA, literature[530]; Spices (curcuma), AFB_1, AFB_2, AFG_1, and OTA, literature[530]; Spices (pepper), AFB_1 and AFB_2, literature[530]; Spices (poppy seed), AFB_1, literature[530]

incidence: 38/42*, conc. range: 0.34–77.13 µg/kg, sample origin: apiaries, supermarkets, open bazaars, and retail shops in Sanliurfa (city, province) region, Turkey, sample year: 2013, country: Turkey[595], *isot red pepper

- Co-contamination: not reported
- Further contamination of plants and plant products other than cereals and their products reported in the present book/article[595]: Spices (cayenne pepper, chilli), AFB_2, AFG_1, and AFG_2, literature[595]

incidence: 4/10*, conc. range: 0.126–1.424 µg/kg, Ø conc.: 0.527 µg/kg, sample origin: local fresh market in Sanliurfa (city), Turkey, sample year: 2013, country: Turkey[595], *isot red pepper dried on concrete surface

- Co-contamination: 2 sa. co-contaminated with AFB_1 and AFB_2; 1 sa. co-contaminated with AFB_1, AFB_2, AFG_1, and AFG_2; 1 sa. co-contaminated with AFB_1 and AFG_1
- Further contamination of plants and plant products other than cereals and their products reported in the present book/article[595]: Spices (cayenne pepper, chilli), AFB_2, AFG_1, and AFG_2, literature[595]

incidence: 10/10*, conc. range: 1.926–9.822 µg/kg, Ø conc.: 4.208 µg/kg, sample origin: local fresh market in Sanliurfa (city), Turkey, sample year: 2013, country: Turkey[595], *isot red pepper dried on soil surface

- Co-contamination: 6 sa. co-contaminated with AFB_1 and AFB_2; 2 sa. co-contaminated with AFB_1, AFB_2, AFG_1, and AFG_2; 2 sa. co-contaminated with AFB_1, AFB_2, and AFG_2
- Further contamination of plants and plant products other than cereals and their products reported in the present book/article[595]: Spices (cayenne pepper, chilli), AFB_2, AFG_1, and AFG_2, literature[595]

incidence: 1/1, conc.: 1.21 µg/kg, sample origin: shops in Watt Market, Calabar (city), Nigeria, sample year: January 2013, country: Nigeria[639]

- Co-contamination: 1 sa. co-contaminated with AFB_1, AFB_2, AFG_1, and AFG_2
- Further contamination of plants and plant products other than cereals and their products reported in the present book/article[639]: Seeds (bush mango), Seeds (melon), AFB_1, AFB_2, AFG_1, and AFG_2, literature[639]; Spices (cayenne pepper, chilli), AFB_2, AFG_1, and AFG_2, literature[639]

incidence: 1/3*, conc.: 11.3 µg/kg, sample origin: supermarkets in La Paz (capital), Bolivia, sample year: unknown, country: Japan/Peru/Bolivia[678], *crushed or powdered red chilli pepper

incidence: 2/4*, conc. range: 0.7–0.9 µg/kg, Ø conc.: 0.8 µg/kg, sample origin: supermarkets in Lima (capital), Peru, sample year: unknown, country: Japan/Peru/Bolivia[678], *red chilli pepper, crushed or powdered incidence: 1/2*, conc.: 2.3 μg/kg, sample origin: supermarkets, China, sample year: unknown, country: Japan/Peru/Bolivia[678], *red chilli pepper, crushed or powdered

incidence: 0/1*, conc.: no contamination, sample origin: supermarket, Japan, sample year: unknown, country: Japan/Peru/Bolivia[678], *red chilli pepper, crushed or powdered

- Co-contamination: not reported
- Further contamination of plants and plant products other than cereals and their products reported in the present book/article[678]: not reported

incidence: 150/182*, conc. range: LOQ-5 μg/kg (100 sa.), 5–10 μg/kg (15 sa.), 10–50 μg/kg (25 sa.), >50 μg/kg (10 sa., maximum: 165 μg/kg), Ø conc.: 8.89 μg/kg, sample origin: markets, supermarkets, groceries, and herbal shops in Adana and Osmaniye (provinces), Turkey, sample year: January 2011–December 2012, country: Turkey[734], *chilli

- Co-contamination: 84 sa. co-contaminated with AFB_1 and AFB_2; no further information available
- Further contamination of plants and plant products other than cereals and their products reported in the present book/article[734]: Spices (cayenne pepper, chilli), AFB_2 and AFG_1, literature[734]

incidence: 3/13, conc. range: 0.9–1.6 μg/kg, Ø conc.: 1.3 μg/kg, sample origin: central markets in Cusco and Trujillo (cities) and Lima (capital), Peru, sample year: unknown, country: Japan[749]

- Co-contamination: 3 sa. co-contaminated with AFB_1 and OTA
- Further contamination of plants and plant products other than cereals and their products reported in the present book/article[749]: Spices (cayenne pepper, chilli), OTA, literature[749]

incidence: 2/10, conc. range: 10.8–33.2 μg/kg, Ø conc.: 22 μg/kg, sample origin: supermarkets and night markets in Penang (state), Malaysia, sample year: unknown, country: Malaysia[766]

- Co-contamination: 1 sa. co-contaminated with AFB_1 and AFB_2; 1 sa. co-contaminated with AFB_1, AFG_1, and AFG_2
- Further contamination of plants and plant products other than cereals and their products reported in the present book/article[766]: Nuts (peanut), AFB_1 AFG_1, and AFG_2, literature[766]; Spices (cayenne pepper, chilli), AFB_2, AFG_1, and AFG_2, literature[766]

incidence: 6/6*, conc. range: 0.03355–0.04149 μg/kg, Ø conc.: 0.03886 μg/kg, sample origin: popular markets in Hyderabad (city), Andhra Pradesh (province), India, sample year: June 2012, country: Iran[800], *chilli, whole

- Co-contamination: not reported
- Further contamination of plants and plant products other than cereals and their products reported in the present book/article[800]: Spices (pepper), AFB_1, literature[800]

incidence: 6/6*, conc. range: 0.25552–0.62681 μg/kg, Ø conc.: 0.41427 μg/kg, sample origin: popular herbal shops in Qaemshahr (city), Mazandaran (province), Iran, sample year: June 2012, country: Iran[801], *chilli powder

incidence: 6/6*, conc. range: 0.06463–0.24594 μg/kg, Ø conc.: 0.15311 μg/kg, sample origin: popular markets in Hyderabad (city), Andhra Pradesh (province), India, sample year: June 2012, country: Iran[801], *chilli powder

- Co-contamination: not reported
- Further contamination of plants and plant products other than cereals and their products reported in the present book/article[801]: Spices (pepper), AFB_1, literature[801]

incidence: 29/30*, conc. range: 0.51–46.28 μg/kg, Ø conc.: 13.03 μg/kg, sample origin: retail fresh markets or modern department stores in the central region of Thailand, sample year: unknown, country: Thailand/Austria/Germany[857], *chilli powder

incidence: 11/30*, conc. range: 0.1–53.62 μg/kg, Ø conc.: 6.09 μg/kg, sample origin: retail fresh markets or modern department stores in the central region of Thailand, sample year: unknown, country: Thailand/Austria/Germany[857], *chilli pods

- Co-contamination: not reported
- Further contamination of plants and plant products other than cereals and their products reported in the present book/article[857]: Nuts (peanut), AFB_1, AFB_2, and AFG_1, literature[857]; Spices (cayenne pepper, chilli), AFB_2 and AFG_1, literature[857]

incidence: 6/8*, conc. range: 23.4–46.6 μg/kg, Ø conc.: 41.5 μg/kg, sample origin: organic markets and organic shops, Turkey, sample year: May 2010–May 2011, country: Turkey[859], *organic red pepper sa.

incidence: 4/7*, conc. range: 3.5–30.3 μg/kg, Ø conc.: 23.4 μg/kg, sample origin: organic markets and organic shops, Turkey, sample year: May 2010–May 2011, country: Turkey[859], *organic red pepper flake sa.

- Co-contamination: not reported
- Further contamination of plants and plant products other than cereals and their products reported in the present book/article[859]: Herbs (camomile), Herbs (linden), Herbs (rosehip), Herbs (sage), Spices (anise), Spices (basil), Spices (cinnamon), Spices (coriander), Spices (cumin), Spices (fennel), Spices (ginger), Spices (laurel leave), Spices (pepper), Spices (peppermint), Spices (poppy seed), Spices (rosemary), Spices (sumac), AFB_1, literature[859]

incidence: 65/65*, conc. range: 15.75–500 μg/kg, Ø conc.: 182 μg/kg, sample origin: market, India, sample year: unknown, country: India[863], *red chilli powder

- Co-contamination: not reported
- Further contamination of plants and plant products other than cereals and their products reported in the present book/article[863]: not reported

incidence: 7/18*, conc. range: <LOQ-687 μg/kg, Ø conc.: 133.1 μg/kg, sample origin: open markets, Sri Lanka, sample year: 2012/2013, country: Belgium[889], *chilli pods, whole

incidence: 24/26*, conc. range: <LOQ-91 μg/kg, Ø conc.: 20.5 μg/kg, sample origin: open markets, Sri Lanka, sample year: 2012/2013, country: Belgium[889], *chilli flakes

incidence: 35/42*, conc. range: <LOQ-31.2 μg/kg, Ø conc.: 10.4 μg/kg, sample origin: open markets, Sri Lanka, sample year: 2012/2013, country: Belgium[889], *chilli powder

incidence: 11/35*, conc. range: <LOQ-11.9 μg/kg, Ø conc.: 8.07 μg/kg, sample origin: exotic shops and supermarkets in Antwerp and Ghent (cities), Belgium, sample year: 2012/2013, country: Belgium[889], *included whole chilli pods, chilli powders, and chilli flakes

- Co-contamination: For detailed information see the article.
- Further contamination of plants and plant products other than cereals and their products reported in the present book/article[889]: Spices (cayenne pepper, chilli), AFB_2, AFG_1, AFG_2, AME, CIT, FB_2, OTA, and STG, literature[889]

For detailed information see the article.

incidence: 9/10*, conc. range: <LOQ (3 sa.), 5.1–35 µg/kg (6 sa.), sample origin: markets, Sri Lanka, sample year: unknown, country: Belgium[891], *red chilli

- Co-contamination: not reported
- Further contamination of plants and plant products other than cereals and their products reported in the present book/article[891]: Spices (cayenne pepper, chilli), AFB_2, FB_2, OTA, and STG, literature[891]; Spices (pepper), AFB_1, AFB_2, AFG_2, CIT, FB_1, OTA, and STG, literature[891]

incidence: 4/56*, conc. range: <LOQ-0.89 µg/kg, ∅ conc.: 0.64 µg/kg, sample origin: imported from China and Vietnam as well as Korea and purchased in local markets, Korea, sample year: unknown, country: Korea[901], *red pepper powder

- Co-contamination: not reported
- Further contamination of plants and plant products other than cereals and their products reported in the present book/article[901]: Spices (cayenne pepper, chilli), OTA, literature[901]

Aflatoxin B_2

incidence: 4?/9*, conc. range: 11–88 µg/kg, sample origin: major retail outlets in Nainital (district), India, sample year: February 1984–December 1986, country: India[22], *chillies

- Co-contamination: not reported
- Further contamination of plants and plant products other than cereals and their products reported in the present book/article[22]: Spices (ammi), AFB_1, AFB_2, and AFG_1, literature[22]; Spices (cardamom), AFB_1, AFB_2, AFG_1, AFG_2, CIT, and OTA, literature[22]; Spices (cassia), AFB_1, AFB_2, AFG_1, and AFG_2, literature[22]; Spices (cayenne pepper, chilli), AFB_1, AFG_1, AFG_2, and ZEA, literature[22]; Spices (coriander), AFB_1, AFB_2, AFG_1, AFG_2, CIT, OTA, and ZEA, literature[22]; Spices (cumin), AFB_1, AFB_2, AFG_1, AFG_2, and CIT, literature[22]; Spices (curcuma), AFB_1, AFB_2, AFG_1, AFG_2, CIT, and RBT, literature[22]; Spices (fennel), AFB_1, AFB_2, AFG_1, AFG_2, CIT, OTA, and STG, literature[22]; Spices (garlic), AFB_1, AFB_2, and AFG_1, literature[22]; Spices (pepper), AFB_1, AFB_2, AFG_1, AFG_2, CIT, STG, and ZEA, literature[22]

incidence: 2/10*, conc. range: ≤0.4 µg/kg, ∅ conc.: 0.2 µg/kg, sample origin: commercially available market basket sa. of foods and foodstuffs in Tokyo (capital), Japan, sample year: 1986–1990, country: Japan[99], *red pepper

- Co-contamination: not reported
- Further contamination of plants and plant products other than cereals and their products reported in the present book/article[99]: Beans, Seeds (sesame), Sugar, AFB_1 and AFB_2, literature[99]; Food, Jam (bean jam), Paste (peanut paste), AFB_1, literature[99]; Nuts (Brazil nut, paranut), Nuts (peanut), Nuts (pistachio nut), Spices (nutmeg), AFB_1, AFB_2, AFG_1, and AFG_2, literature[99]; Spices, Spices (pepper), AFB_1, AFB_2, and AFG_1, literature[99]; Spices (paprika), AFB_1 and AFG_1, literature[99]; Spices (cayenne pepper, chilli), AFB_1 and AFG_1, literature[99]

incidence: 8/15*, conc. range: 0.1–1.5 µg/ kg, ∅ conc.: 0.38 µg/kg, sample origin: unknown, sample year: unknown, country: USA[167], *red pepper

- Co-contamination: 1 sa. co-contaminated with AFB_1, AFB_2, AFG_1, and AFG_2; 3 sa. co-contaminated with AFB_1, AFB_2, and AFG_1; 1 sa. co-contaminated with AFB_1, AFB_2, and AFG_2; 3 sa. co-contaminated with AFB_1 and AFB_2
- Further contamination of plants and plant products other than cereals and their products reported in the present book/article[167]: Spices (cayenne pepper, chilli), AFB_1, AFG_1, and AFG_2, literature[167]; Spices (ginger), AFB_1, AFB_2, AFG_1, and AFG_2, literature[167]; Spices (pepper), AFB_1, literature[167]

incidence: 7/22*, conc. range: 0.22–1.25 µg/kg, sample origin: food shops in Hajdú-Bihar (county), Hungary, sample year: October 2004, country: Hungary[170], *red pepper, grounded

incidence: 3/16*, conc. range: 0.52–0.73 µg/kg, sample origin: shops in/around Kaposvár (town), Somogy (county), Hungary, sample year: October 2004, country: Hungary[170], *red pepper, grounded and originally packed

incidence: 0/13*, conc.: no contamination, sample origin: food shops in Hajdú-Bihar (county), Hungary, sample year: November 2004, country: Hungary[170], *red pepper, grounded

incidence: 0/19*, conc.: no contamination, sample origin: 10 Hungarian counties, Hungary, sample year: October/November 2004, country: Hungary[170], *red pepper, home-made

incidence: 1/5*, conc.: 0.71 µg/kg, sample origin: food shops in Debrecen (town), Hungary, sample year: October 2004, country: Hungary[170], *chilli

- Co-contamination: For detailed information see the article.
- Further contamination of plants and plant products other than cereals and their products reported in the present book/article[170]: Spices, AFB_1 and AFB_2, literature[170]; Spices (cayenne pepper, chilli), AFB_1 and OTA, literature[170]; Spices (pepper), AFB_1, literature[170]

incidence: 0/6*, conc.: no contamination, sample origin: markets (marked sa. of unknown origin), Turkey, sample year: unknown, country: Turkey[172], *red pepper

- Co-contamination: not reported
- Further contamination of plants and plant products other than cereals and their products reported in the present book/article[172]: Spices (cayenne pepper, chilli), AFB_1, AFG_1 and AFG_2, literature[172]

incidence: 5/15*, conc. range: 3.2–40 µg/kg**, ∅ conc.: 12.4 µg/kg**, sample origin: spice sellers (unmarked sa. from unknown origin, India, and Turkey), Turkey, sample year: unknown, country: Turkey[172], *red pepper, **TLC and HPLC

- Co-contamination: 3 sa. co-contaminated with AFB_1, AFB_2, and AFG_1; 2 sa. co-contaminated with AFB_1 and AFB_2
- Further contamination of plants and plant products other than cereals and their products reported in the present book/article[172]: Spices (cayenne pepper, chilli), AFB_1, AFG_1 and AFG_2, literature[172]

incidence: 3/11*, conc. range: 3.4–7.8 µg/kg**, ∅ conc.: 5.06 µg/kg**, sample origin: street bazaars (unmarked sa. from unknown origin and Turkey), Turkey, sample year: unknown, country: Turkey[172], *red pepper, **TLC and HPLC

- Co-contamination: 2 sa. co-contaminated with AFB_1, AFB_2, and AFG_1; 1 sa. co-contaminated with AFB_1, AFB_2, and AFG_2
- Further contamination of plants and plant products other than cereals and their products reported in the present book/article[172]: Spices (cayenne pepper, chilli), AFB_1, AFG_1 and AFG_2, literature[172]

For detailed information see the article.

incidence: 5/26*, conc. range: 0.4–2.3 μg/kg, ∅ conc.: 1.46 μg/kg, sample origin: various cities, Turkey, sample year: unknown, country: Turkey[175], *red pepper

- Co-contamination: not reported
- Further contamination of plants and plant products other than cereals and their products reported in the present book/article[175]: Spices (cayenne pepper, chilli), AFB_1, literature[175]

incidence: 2/11*, conc. range: 0.81–1.20 μg/kg, ∅ conc.: 1.01 μg/kg, sample origin: public markets, supermarkets, shops, and bonded warehouse in Emilia Romagna (administrative region), Italy, sample year: 2000–2005, country: Italy[176], *hot pepper, whole/smashed

- Co contamination: 1 sa. co-contaminated with AFB_1, AFB_2, AFG_1, and AFG_2; 1 sa. co-contaminated with AFB_1 and AFB_2
- Further contamination of plants and plant products other than cereals and their products reported in the present book/article[176]: Spices (cayenne pepper, chilli), AFB_1, AFG_1, and AFG_2, literature[176]; Spices (cinnamon), AFB_1, literature[176]; Spices (nutmeg), AFB_1 and AFB_2, literature[176]

incidence: 17/24*, conc. range: 0.04–1.28 μg/kg, ∅ conc.: 0.24 μg/kg, sample origin: markets, herbal shops, and bazaars in Corum and Ankara (cities), Turkey, sample year: November 2010–August 2011, country: Turkey[183], *red chilli flakes

- Co-contamination: 3 sa. co-contaminated with AFB_1, AFB_2, AFG_1, and OTA; 11 sa. co-contaminated with AFB_1, AFB_2, and OTA; 3 sa. contaminated solely with AFB_2
- Further contamination of plants and plant products other than cereals and their products reported in the present book/article[183]: Spices (cayenne pepper, chilli), AFB_1, AFG_1, and OTA, literature[183]; Spices (cumin), Spices (pepper), AFB_1, AFB_2, and OTA, literature[183]

incidence: 11/22*, conc. range: 0.05–1.07 μg/kg, ∅ conc.: 0.32 μg/kg, sample origin: markets, herbal shops, and bazaars in Corum and Ankara (cities), Turkey, sample year: November 2010–August 2011, country: Turkey[183], *red chilli powder

- Co-contamination: 3 sa. co-contaminated with AFB_1, AFB_2, AFG_1, and OTA; 5 sa. co-contaminated with AFB_1, AFB_2, and OTA; 3 sa. contaminated solely with AFB_2
- Further contamination of plants and plant products other than cereals and their products reported in the present book/article[183]: Spices (cayenne pepper, chilli), AFB_1, AFG_1, and OTA, literature[183]; Spices (cumin), Spices (pepper), AFB_1, AFB_2, and OTA, literature[183]

incidence: 2/2*, conc. range: 1.7–2.2 μg/kg, ∅ conc.: 2 μg/kg, sample origin: unknown, sample year: unknown, country: Denmark[217], *chilli

incidence: 2/2*, conc. range: 2.1–3.9 μg/kg, ∅ conc.: 3 μg/kg, sample origin: unknown, sample year: unknown, country: Denmark[217], *cayenne pepper

- Co-contamination: 2 sa. co-contaminated with AFB_1, AFB_2, AFG_1, and AFG_2
- Further contamination of plants and plant products other than cereals and their products reported in the present book/article[217]: Butter (peanut butter), Figs, AFB_1, AFB_2, AFG_1, and AFG_2, literature[217]; Spices (cayenne pepper, chilli), AFB_1, AFG_1, and AFG_2, literature[217]; Meal (pistachio meal), AFB_1 and AFB_2, literature[217]

incidence: 5/33*, conc. range: tr, sample origin: India, sample year: unknown, country: Canada[242], *cayenne pepper

- Co-contamination: 5 sa. co-contamination with AFB_1 and AFB_2
- Further contamination of plants and plant products other than cereals and their products reported in the present book/article[242]: Spices (cayenne pepper, chilli), AFB_1 and AFG_1, literature[242]

incidence: 2/6*, conc. range: tr, sample origin: India, sample year: unknown, country: Canada[242], *Indian chilli powder

- Co-contamination: 1 sa. co-contaminated with AFB_1, AFB_2, and AFG_1; 1 sa. co-contamination with AFB_1 and AFB_2
- Further contamination of plants and plant products other than cereals and their products reported in the present book/article[242]: Spices (cayenne pepper, chilli), AFB_1 and AFG_1, literature[242]

incidence: 13/24*, conc. range: 0.3–7.1 μg/kg, ∅ conc.: 1.8 μg/kg, sample origin: retail and wholesale, Nordwürttemberg (administrative region), Germany, sample year: unknown, country: Germany[243], *cayenne pepper powder and ground chilli

- Co-contamination: 11 sa. co-contaminated with AFB_1, AFB_2, and AFG_1; 2 sa. co-contaminated with AFB_1 and AFB_2
- Further contamination of plants and plant products other than cereals and their products reported in the present book/article[243]: Spices (cayenne pepper, chilli), AFB_1 and AFG_1, literature[243]; Spices (paprika), AFB_1 and AFB_2, literature[243]

incidence: 10/31*, conc. range: 0.2–0.7 μg/kg, sample origin: warehouses, packing establishments, supermarkets, and smaller shops, UK, sample year: unknown, country: UK[245], *chilli

- Co-contamination: 1 sa. co-contaminated with AFB_1, AFB_2, AFG_1, AFG_2, and OTA; 1 sa. co-contaminated with AFB_1, AFB_2, AFG_1, and OTA; 8 sa. co-contaminated with AFB_1, AFB_2, and OTA
- Further contamination of plants and plant products other than cereals and their products reported in the present book/article[245]: Spices (cayenne pepper, chilli), AFB_1, AFG_1, AFG_2, and OTA, literature[245]; Spices (paprika), AFB_1, AFB_2, AFG_1, AFG_2, and OTA, literature[245]

incidence: 1/4*, conc.: 0.2 μg/kg, sample origin: warehouses, packing establishments, supermarkets, and smaller shops, UK, sample year: unknown, country: UK[245], *cayenne pepper

- Co-contamination: 1 sa. co-contaminated with AFB_1, AFB_2, and OTA
- Further contamination of plants and plant products other than cereals and their products reported in the present book/article[245]: Spices (cayenne pepper, chilli), AFB_1, AFG_1, AFG_2, and OTA, literature[245]; Spices (paprika), AFB_1, AFB_2, AFG_1, AFG_2, and OTA, literature[245]

incidence: 14/80*, conc. range: <1.0 μg/kg (8 sa.), 1–5 μg/kg (3 sa.), 5–10 μg/kg (2 sa.), 11.45 μg/kg (1 sa.), sample origin: local markets and open markets in Kuala Lumpur (capital), Putrajaya (city), and Selangor (state) areas, Malaysia, sample year: May/June 2009, country: Malaysia[249], *chilli, dried

- Co-contamination: not reported
- Further contamination of plants and plant products other than cereals and their products reported in the present book/article[249]: Spice (cayenne pepper, chilli), AFB_1, AFG_1, AFG_2, and OTA, literature[249]

incidence: 1/1*, conc.: 0.7 µg/kg, sample origin: imported from Asian countries purchased from major wholesale/retail store of ethnic foods, West of Scotland, sample year: unknown, country: UK[314], *chilli, dried and crushed

- Co-contamination: 1 sa. co-contaminated with AFB_1 and AFB_2
- Further contamination of plants and plant products other than cereals and their products reported in the present book/article[314]: Spice (cayenne pepper, chilli), AFB_1, literature[314]

incidence: 1/2*, conc.: 0.46 µg/kg, sample origin: Penang (state, island), Malaysia, sample year: unknown (first study), country: Malaysia[530], *chilli, dried or coarse dried

- Co-contamination: 1 sa. co-contaminated with AFB_1, AFB_2, AFG_1, AFG_2, and OTA
- Further contamination of plants and plant products other than cereals and their products reported in the present book/article[530]: Spices, Spices (coriander), Spices (cumin), Spices (curry), Spices (fennel), Spices (masala), AFB_1, AFB_2, AFG_1, AFG_2, and OTA, literature[530]; Spices (cayenne pepper, chilli), AFB_1, AFG_1, AFG_2, and OTA, literature[530]; Spices (curcuma), AFB_1, AFB_2, AFG_1, and OTA, literature[530]; Spices (pepper), AFB_1 and AFB_2, literature[530]; Spices (poppy seed), AFB_1, literature[530]

incidence: 3/3*, conc. range: 0.63–2.74 µg/kg, Ø conc.: 1.36 µg/kg, sample origin: Penang (state, island), Malaysia, sample year: unknown (second study), country: Malaysia[530], *chilli, dried

- Co-contamination: 3 sa. co-contaminated with AFB_1 and AFB_2
- Further contamination of plants and plant products other than cereals and their products reported in the present book/article[530]: Spices, Spices (coriander), Spices (cumin), Spices (curry), Spices (fennel), Spices (masala), AFB_1, AFB_2, AFG_1, AFG_2, and OTA, literature[530]; Spices (cayenne pepper, chilli), AFB_1, AFG_1, AFG_2, and OTA, literature[530]; Spices (curcuma), AFB_1, AFB_2, AFG_1, and OTA, literature[530]; Spices (pepper), AFB_1 and AFB_2, literature[530]; Spices (poppy seed), AFB_1, literature[530]

incidence: 36/42*, conc. range: 0.031–7.72 µg/kg, sample origin: apiaries, supermarkets, open bazaars, and retail shops in Sanliurfa (city, province) region, Turkey, sample year: 2013, country: Turkey[595], *isot red pepper

- Co-contamination: not reported
- Further contamination of plants and plant products other than cereals and their products reported in the present book/article[595]: Spices (cayenne pepper, chilli), AFB_1, AFG_1, and AFG_2, literature[595]

incidence: 7/10*, conc. range: 0.125–1.705 µg/kg, Ø conc.: 0.357 µg/kg, sample origin: local fresh market in Sanliurfa (city), Turkey, sample year: 2013, country: Turkey[595], *isot red pepper dried on concrete surface

- Co-contamination: 3 sa. co-contaminated with AFB_2 and AFG_2; 2 sa. co-contaminated with AFB_1 and AFB_2; 1 sa. co-contaminated with AFB_1, AFB_2, AFG_1, and AFG_2; 1 sa. contaminated solely with AFB_2
- Further contamination of plants and plant products other than cereals and their products reported in the present book/article[595]: Spices (cayenne pepper, chilli), AFB_1, AFG_1, and AFG_2, literature[595]

incidence: 10/10*, conc. range: 0.128–0.478 µg/kg, Ø conc.: 0.278 µg/kg, sample origin: local fresh market in Sanliurfa (city), Turkey, sample year: 2013, country: Turkey[595], *isot red pepper dried on soil surface

- Co-contamination: 6 sa. co-contaminated with AFB_1 and AFB_2; 2 sa. co-contaminated with AFB_1, AFB_2, AFG_1, and AFG_2; 2 sa. co-contaminated with AFB_1, AFB_2, and AFG_2
- Further contamination of plants and plant products other than cereals and their products reported in the present book/article[595]: Spices (cayenne pepper, chilli), AFB_1, AFG_1, and AFG_2, literature[595]

incidence: 1/1, conc.: 0.34 µg/kg, sample origin: shops in Watt Market, Calabar (city), Nigeria, sample year: January 2013, country: Nigeria[639]

- Co-contamination: 1 sa. co-contaminated with AFB_1, AFB_2, AFG_1, and AFG_2
- Further contamination of plants and plant products other than cereals and their products reported in the present book/article[639]: Seeds (bush mango), Seeds (melon), AFB_1, AFB_2, AFG_1, and AFG_2, literature[639]; Spices (cayenne pepper, chilli), AFB_1, AFG_1, and AFG_2, literature[639]

incidence: 84/182*, conc. range: LOQ-5 µg/kg (82 sa.), 5–10 µg/kg (1 sa.), 11.3 µg/kg (1 sa.), Ø conc.: 0.88 µg/kg, sample origin: markets, supermarkets, groceries, and herbal shops in Adana and Osmaniye (provinces), Turkey, sample year: January 2011–December 2012, country: Turkey[734], *chilli

- Co-contamination: 84 sa. co-contaminated with AFB_1 and AFB_2; no further information available
- Further contamination of plants and plant products other than cereals and their products reported in the present book/article[734]: Spices (cayenne pepper, chilli), AFB_1 and AFG_1, literature[734]

incidence: 1/10, conc.: 5.0 µg/kg, sample origin: supermarkets and night markets in Penang (state), Malaysia, sample year: unknown, country: Malaysia[766]

- Co-contamination: 1 sa. co-contaminated with AFB_1 and AFB_2
- Further contamination of plants and plant products other than cereals and their products reported in the present book/article[766]: Nuts (peanut), Spices (cayenne pepper, chilli), AFB_1, AFG_1, and AFG_2, literature[766]

incidence: 26/30*, conc. range: 0.09–6.12 µg/kg, Ø conc.: 1.27 µg/kg, sample origin: retail fresh markets or modern department stores in the central region of Thailand, sample year: unknown, country: Thailand/Austria/Germany[857], *chilli powder

incidence: 4/30*, conc. range: 0.12–4.67 µg/kg, Ø conc.: 1.44 µg/kg, sample origin: retail fresh markets or modern department stores in the central region of Thailand, sample year: unknown, country: Thailand/Austria/Germany[857], *chilli pods

- Co-contamination: not reported
- Further contamination of plants and plant products other than cereals and their products reported in the present book/article[857]: Nuts (peanut), AFB_1, AFB_2, and AFG_1, literature[857]; Spices (cayenne pepper, chilli), AFB_1 and AFG_1, literature[857]

incidence: 1/18*, conc.: 31.5 µg/kg, sample origin: open markets, Sri Lanka, sample year: 2012/2013, country: Belgium[889], *chilli pods, whole

incidence: 6/26*, conc. range: <LOQ-6.1 µg/kg, Ø conc.: 6.1 µg/kg, sample origin: open markets, Sri Lanka, sample year: 2012/2013, country: Belgium[889], *chilli flakes

incidence: 2/42*, conc. range: <LOQ, Ø conc.: LOQ, sample origin: open markets, Sri Lanka, sample year: 2012/2013, country: Belgium[889], *chilli powder

incidence: 0/35*, conc.: no contamination, sample origin: exotic shops and supermarkets in Antwerp and Ghent (cities), Belgium, sample year: 2012/2013, country: Belgium[889], *included whole chilli pods, chilli powders, and chilli flakes

- Co-contamination: For detailed information see the article.
- Further contamination of plants and plant products other than cereals and their products reported in the present book/article[889]: Spices (cayenne pepper, chilli), AFB_1, AFG_1, AFG_2, AME, CIT, FB_2, OTA, and STG, literature[889]

For detailed information see the article.

incidence: 2/10*, conc. range: <LOQ, sample origin: markets, Sri Lanka, sample year: unknown, country: Belgium[891], *red chilli

- Co-contamination: not reported
- Further contamination of plants and plant products other than cereals and their products reported in the present book/article[891]: Spices (cayenne pepper, chilli), AFB_1, FB_2, OTA, and STG, literature[891]; Spices (pepper), AFB_1, AFB_2, AFG_2, CIT, FB_1, OTA, and STG, literature[891]

Aflatoxin G_1

incidence: 4?/9*, conc. range: 8–58 µg/kg, sample origin: major retail outlets in Nainital (district), India, sample year: February 1984–December 1986, country: India[22], *chillies

- Co-contamination: not reported
- Further contamination of plants and plant products other than cereals and their products reported in the present book/article[22]: Spices (ammi), AFB_1, AFB_2, and AFG_1, literature[22]; Spices (cardamom), AFB_1, AFB_2, AFG_1, AFG_2, CIT, and OTA, literature[22]; Spices (cassia), AFB_1, AFB_2, AFG_1, and AFG_2, literature[22]; Spices (cayenne pepper, chilli), AFB_1, AFB_2, AFG_2, and ZEA, literature[22]; Spices (coriander), AFB_1, AFB_2, AFG_1, AFG_2, CIT, OTA, and ZEA, literature[22]; Spices (cumin), AFB_1, AFB_2, AFG_1, AFG_2, and CIT, literature[22]; Spices (curcuma), AFB_1, AFB_2, AFG_1, AFG_2, CIT, and RBT, literature[22]; Spices (fennel), AFB_1, AFB_2, AFG_1, AFG_2, CIT, OTA, and STG, literature[22]; Spices (garlic), AFB_1, AFB_2, and AFG_1, literature[22]; Spices (pepper), AFB_1, AFB_2, AFG_1, AFG_2, CIT, STG, and ZEA, literature[22]

incidence: 2/10*, conc. range: ≤1.9 µg/kg, Ø conc.: 1.0 µg/kg, sample origin: commercially available market basket sa. of foods and foodstuffs in Tokyo (capital), Japan, sample year: 1986–1990, country: Japan[99], *red pepper

- Co-contamination: not reported
- Further contamination of plants and plant products other than cereals and their products reported in the present book/article[99]: Beans, Seeds (sesame), Spices (cayenne pepper, chilli), Sugar, AFB_1 and AFB_2, literature[99]; Food, Jam (bean jam), Paste (peanut paste), AFB_1, literature[99]; Nuts (Brazil nut, paranut), Nuts (peanut), Nuts (pistachio nut), Spices (nutmeg), AFB_1, AFB_2, AFG_1, and AFG_2, literature[99]; Spices, Spices (pepper), AFB_1, AFB_2, and AFG_1, literature[99]; Spices (paprika), AFB_1 and AFG_1, literature[99]

incidence: 4/15*, conc. range: 0.7–20.4 µg/kg, Ø conc.: 9.1 µg/kg, sample origin: unknown, sample year: unknown, country: USA[167], *red pepper

- Co-contamination: 1 sa. co-contaminated with AFB_1, AFB_2, AFG_1, and AFG_2; 3 sa. co-contaminated with AFB_1, AFB_2, and AFG_1
- Further contamination of plants and plant products other than cereals and their products reported in the present book/article[167]: Spices (cayenne pepper, chilli), AFB_1, AFB_2, and AFG_2, literature[167]; Spices (ginger), AFB_1, AFB_2, AFG_1, and AFG_2, literature[167]; Spices (pepper), AFB_1, literature[167]

incidence: 0/6*, conc.: no contamination, sample origin: markets (marked sa. of unknown origin), Turkey, sample year: unknown, country: Turkey[172], *red pepper

- Co-contamination: not reported
- Further contamination of plants and plant products other than cereals and their products reported in the present book/article[172]: Spices (cayenne pepper, chilli), AFB_1, AFB_2, and AFG_2, literature[172]

incidence: 3/15*, conc. range: 4–20 µg/kg**, Ø conc.: 11.3 µg/kg**, sample origin: spice-sellers (unmarked sa. from unknown origin, India, and Turkey), Turkey, sample year: unknown, country: Turkey[172], *red pepper, **TLC and HPLC

- Co-contamination: 3 sa. co-contaminated with AFB_1, AFB_2, and AFG_1
- Further contamination of plants and plant products other than cereals and their products reported in the present book/article[172]: Spices (cayenne pepper, chilli), AFB_1, AFB_2, and AFG_2, literature[172]

incidence: 2/11*, conc. range: 0.9–4.1 µg/kg**, Ø conc.: 2.5 µg/kg**, sample origin: street bazaars (unmarked sa. from unknown origin and Turkey), Turkey, sample year: unknown, country: Turkey[172], *red pepper, **TLC and HPLC

- Co-contamination: 2 sa. co-contaminated with AFB_1, AFB_2, and AFG_1
- Further contamination of plants and plant products other than cereals and their products reported in the present book/article[172]: Spices (cayenne pepper, chilli), AFB_1, AFB_2, and AFG_2, literature[172]

For detailed information see the article.

incidence: 1/11*, conc.: 1.40 µg/kg, sample origin: public markets, supermarkets, shops, and bonded warehouse in Emilia Romagna (administrative region), Italy, sample year: 2000–2005, country: Italy[176], *hot pepper, whole/smashed

- Co-contamination: 1 sa. co-contaminated with AFB_1, AFB_2, AFG_1, and AFG_2
- Further contamination of plants and plant products other than cereals and their products reported in the present book/article[176]: Spices (cayenne pepper, chilli), AFB_1, AFB_2, and AFG_2, literature[176]; Spices (cinnamon), AFB_1, literature[176]; Spices (nutmeg), AFB_1 and AFB_2, literature[176]

incidence: 4/24*, conc. range: 0.24–2.05 µg/kg, Ø conc.: 1.27 µg/kg, sample origin: markets, herbal shops, and bazaars in Corum and Ankara (cities), Turkey, sample year: November 2010–August 2011, country: Turkey[183], *red chilli flakes

incidence: 4/22*, conc. range: 0.25–1.05 µg/kg, Ø conc.: 0.68 µg/kg, sample origin: markets, herbal shops, and bazaars in Corum and Ankara (cities), Turkey, sample year: November 2010–August 2011, country: Turkey[183], *red chilli powder

- Co-contamination: 3 sa. co-contaminated with AFB_1, AFB_2, AFG_1, and OTA; 1 sa. contaminated solely with AFG_1
- Further contamination of plants and plant products other than cereals and their products reported in the present book/article[183]: Spices (cayenne pepper, chilli), Spices (cumin), Spices (pepper), AFB_1, AFB_2, and OTA, literature[183]

incidence: 2/2*, conc. range: 0.6–1.2 µg/kg, Ø conc.: 0.9 µg/kg, sample origin: unknown, sample year: unknown, country: Denmark[217], *chilli incidence: 2/2*, conc. range: 1.0–2.4 μg/kg, ∅ conc.: 1.7 μg/kg, sample origin: unknown, sample year: unknown, country: Denmark[217], *cayenne pepper

- Co-contamination: 2 sa. co-contaminated with AFB_1, AFB_2, AFG_1, and AFG_2
- Further contamination of plants and plant products other than cereals and their products reported in the present book/article[217]: Butter (peanut butter), Figs, AFB_1, AFB_2, AFG_1, and AFG_2, literature[217]; Spices (cayenne pepper, chilli), AFB_1, AFB_2, and AFG_2, literature[217]; Meal (pistachio meal), AFB_1 and AFB_2, literature[217]

incidence: 0/33*, conc.: no contamination, sample origin: India, sample year: unknown, country: Canada[242], *cayenne pepper

- Co-contamination: not reported
- Further contamination of plants and plant products other than cereals and their products reported in the present book/article[242]: Spices (cayenne pepper, chilli), AFB_1 and AFB_2, literature[242]

incidence: 3/6*, conc. range: tr-3 μg/kg, sample origin: India, sample year: unknown, country: Canada[242], *Indian chilli powder

- Co-contamination: 1 sa. co-contaminated with AFB_1, AFB_2, and AFG_1; 2 sa. co-contaminated with AFB_1 and AFG_1
- Further contamination of plants and plant products other than cereals and their products reported in the present book/article[242]: Spices (cayenne pepper, chilli), AFB_1 and AFB_2, literature[242]

incidence: 11/24*, conc. range: 0.2–18.3 μg/kg, ∅ conc.: 5.9 μg/kg, sample origin: retail and wholesale, Nordwürttemberg (administrative region), Germany, sample year: unknown, country: Germany[243], *cayenne pepper powder and ground chilli

- Co-contamination: 11 sa. co-contaminated with AFB_1, AFB_2, and AFG_1
- Further contamination of plants and plant products other than cereals and their products reported in the present book/article[243]: Spices (cayenne pepper, chilli), Spices (paprika), AFB_1 and AFB_2, literature[243]

incidence: 3/31*, conc. range: 0.2–1.1 μg/kg, sample origin: warehouses, packing establishments, supermarkets, and smaller shops, UK, sample year: unknown, country: UK[245], *chilli

- Co-contamination: 1 sa. co-contaminated with AFB_1, AFB_2, AFG_1, AFG_2, and OTA; 1 sa. co-contaminated with AFB_1, AFB_2, AFG_1, and OTA; 1 sa. co-contaminated with AFB_1, AFG_1 and OTA
- Further contamination of plants and plant products other than cereals and their products reported in the present book/article[245]: Spices (cayenne pepper, chilli), AFB_1, AFB_2, AFG_2, and OTA, literature[245]; Spices (paprika), AFB_1, AFB_2, AFG_1, AFG_2, and OTA, literature[245]

incidence: 1/4*, conc.: 0.8 μg/kg, sample origin: warehouses, packing establishments, supermarkets, and smaller shops, UK, sample year: unknown, country: UK[245], *cayenne pepper

- Co-contamination: 1 sa. co-contaminated with AFB_1, AFG_1 and OTA
- Further contamination of plants and plant products other than cereals and their products reported in the present book/article[245]: Spices (cayenne pepper, chilli), AFB_1, AFB_2, AFG_2, and OTA, literature[245]; Spices (paprika), AFB_1, AFB_2, AFG_1, AFG_2, and OTA, literature[245]

incidence: 21/80*, conc. range: <1.0 μg/kg (14 sa.), 1–5 μg/kg (3 sa.), 5–10 μg/kg (2 sa.), >10 μg/kg (2 sa., maximum: 17.65 μg/kg), sample origin: local markets and open markets in Kuala Lumpur (capital), Putrajaya (city), and Selangor (state) areas, Malaysia, sample year: May/June 2009, country: Malaysia[249], *chilli, dried

- Co-contamination: not reported
- Further contamination of plants and plant products other than cereals and their products reported in the present book/article[249]: Spice (cayenne pepper, chilli), AFB_1, AFB_2, AFG_2, and OTA, literature[249]

incidence: 1/2*, conc.: 0.26 μg/kg, sample origin: Penang (state, island), Malaysia, sample year: unknown (first study), country: Malaysia[530], *chilli, dried or coarse dried

- Co-contamination: 1 sa. co-contaminated with AFB_1, AFB_2, AFG_1, AFG_2, and OTA
- Further contamination of plants and plant products other than cereals and their products reported in the present book/article[530]: Spices, Spices (coriander), Spices (cumin), Spices (curry), Spices (fennel), Spices (masala), AFB_1, AFB_2, AFG_1, AFG_2, and OTA, literature[530]; Spices (cayenne pepper, chilli), AFB_1, AFB_2, AFG_2, and OTA, literature[530]; Spices (curcuma), AFB_1, AFB_2, AFG_1, and OTA, literature[530]; Spices (pepper), AFB_1 and AFB_2, literature[530]; Spices (poppy seed), AFB_1, literature[530]

incidence: 0/3*, conc.: no contamination, sample origin: Penang (state, island), Malaysia, sample year: unknown (second study), country: Malaysia[530], *chilli, dried

- Co-contamination: not reported
- Further contamination of plants and plant products other than cereals and their products reported in the present book/article[530]: Spices, Spices (coriander), Spices (cumin), Spices (curry), Spices (fennel), Spices (masala), AFB_1, AFB_2, AFG_1, AFG_2, and OTA, literature[530]; Spices (cayenne pepper, chilli), AFB_1, AFB_2, AFG_2, and OTA, literature[530]; Spices (curcuma), AFB_1, AFB_2, AFG_1, and OTA, literature[530]; Spices (pepper), AFB_1 and AFB_2, literature[530]; Spices (poppy seed), AFB_1, literature[530]

incidence: 21/42*, conc. range: 0.01–4.423 μg/kg, sample origin: apiaries, supermarkets, open bazaars, and retail shops in Sanliurfa (city, province) region, Turkey, sample year: 2013, country: Turkey[595], *isot red pepper

- Co-contamination: not reported
- Further contamination of plants and plant products other than cereals and their products reported in the present book/article[595]: Spices (cayenne pepper, chilli), AFB_1, AFB_2, and AFG_2, literature[595]

incidence: 2/10*, conc. range: 0.123–0.134 μg/kg, ∅ conc.: 0.129 μg/kg, sample origin: local fresh market in Sanliurfa (city), Turkey, sample year: 2013, country: Turkey[595], *isot red pepper dried on concrete surface

- Co-contamination: 1 sa. co-contaminated with AFB_1, AFB_2, AFG_1, and AFG_2; 1 sa. co-contaminated with AFB_1 and AFG_1
- Further contamination of plants and plant products other than cereals and their products reported in the present book/article[595]: Spices (cayenne pepper, chilli), AFB_1, AFB_2, and AFG_2, literature[595]

incidence: 2/10*, conc. range: 0.148–0.160 μg/kg, ∅ conc.: 0.154 μg/kg, sample origin: local fresh market in Sanliurfa (city), Turkey, sample year: 2013, country: Turkey[595], *isot red pepper dried on soil surface

- Co-contamination: 2 sa. co-contaminated with AFB_1, AFB_2, AFG_1, and AFG_2
- Further contamination of plants and plant products other than cereals and their products reported in the present book/article[595]: Spices (cayenne pepper, chilli), AFB_1, AFB_2, and AFG_2, literature[595]

incidence: 1/1, conc.: 3.73 µg/kg, sample origin: shops in Watt Market, Calabar (city), Nigeria, sample year: January 2013, country: Nigeria[639]

- Co-contamination: 1 sa. co-contaminated with AFB_1, AFB_2, AFG_1, and AFG_2
- Further contamination of plants and plant products other than cereals and their products reported in the present book/article[639]: Seeds (bush mango), Seeds (melon), AFB_1, AFB_2, AFG_1, and AFG_2, literature[639]; Spices (cayenne pepper, chilli), AFB_1, AFB_2, and AFG_2, literature[639]

incidence: 32/182*, conc. range: LOQ-5 µg/kg (32 sa., maximum: 3.88 µg/kg), ∅ conc.: 0.77 µg/kg, sample origin: markets, supermarkets, groceries, and herbal shops in Adana and Osmaniye (provinces), Turkey, sample year: January 2011–December 2012, country: Turkey[734], *chilli

- Co-contamination: 84 sa. co-contaminated with AFB_1 and AFB_2; no further information available
- Further contamination of plants and plant products other than cereals and their products reported in the present book/article[734]: Spices (cayenne pepper, chilli), AFB_1 and AFB_2, literature[734]

incidence: 2/10, conc. range: 5.80–14.0 µg/kg, ∅ conc.: 9.9 µg/kg, sample origin: supermarkets and night markets in Penang (state), Malaysia, sample year: unknown, country: Malaysia[766]

- Co-contamination: 1 sa. co-contaminated with AFB_1, AFG_1, and AFG_2; 1 sa. co-contaminated with AFG_1 and AFG_2
- Further contamination of plants and plant products other than cereals and their products reported in the present book/article[766]: Nuts (peanut), AFB_1 AFG_1, and AFG_2, literature[766]; Spices (cayenne pepper, chilli), AFB_1, AFB_2, and AFG_2, literature[766]

incidence: 2/30*, conc. range: 0.07–0.1 µg/kg, ∅ conc.: 0.08 µg/kg, sample origin: retail fresh markets or modern department stores in the central region of Thailand, sample year: unknown, country: Thailand/Austria/Germany[857], *chilli powder

incidence: 0/30*, conc.: no contamination, sample origin: retail fresh markets or modern department stores in the central region of Thailand, sample year: unknown, country: Thailand/Austria/Germany[857], *chilli pods

- Co-contamination: not reported
- Further contamination of plants and plant products other than cereals and their products reported in the present book/article[857]: Nuts (peanut), AFB_1, AFB_2, and AFG_1, literature[857]; Spices (cayenne pepper, chilli), AFB_1 and AFB_2, literature[857]

incidence: 0/18*, conc.: no contamination, sample origin: open markets, Sri Lanka, sample year: 2012/2013, country: Belgium[889], *chilli pods, whole

incidence: 1/26*, conc.: <LOQ, sample origin: open markets, Sri Lanka, sample year: 2012/2013, country: Belgium[889], *chilli flakes

incidence: 0/42*, conc.: no contamination, sample origin: open markets, Sri Lanka, sample year: 2012/2013, country: Belgium[889], *chilli powder

incidence: 0/35*, conc.: no contamination, sample origin: exotic shops and supermarkets in Antwerp and Ghent (cities), Belgium, sample year: 2012/2013, country: Belgium[889], *included whole chilli pods, chilli powders, and chilli flakes

- Co-contamination: For detailed information see the article.
- Further contamination of plants and plant products other than cereals and their products reported in the present book/article[889]: Spices (cayenne pepper, chilli), AFB_1, AFB_2, AFG_2, AME, CIT, FB_2, OTA, and STG, literature[889]

For detailed information see the article.

Aflatoxin G_2

incidence: 4?/9*, conc. range: 6–40 µg/kg, sample origin: major retail outlets in Nainital (district), India, sample year: February 1984–December 1986, country: India[22], *chillies

- Co-contamination: not reported
- Further contamination of plants and plant products other than cereals and their products reported in the present book/article[22]: Spices (ammi), AFB_1, AFB_2, and AFG_1, literature[22]; Spices (cardamom), AFB_1, AFB_2, AFG_1, AFG_2, CIT, and OTA, literature[22]; Spices (cassia), AFB_1, AFB_2, AFG_1, and AFG_2, literature[22]; Spices (cayenne pepper, chilli), AFB_1, AFB_2, AFG_1, and ZEA, literature[22]; Spices (coriander), AFB_1, AFB_2, AFG_1, AFG_2, CIT, OTA, and ZEA, literature[22]; Spices (cumin), AFB_1, AFB_2, AFG_1, AFG_2, and CIT, literature[22]; Spices (curcuma), AFB_1, AFB_2, AFG_1, AFG_2, CIT, and RBT, literature[22]; Spices (fennel), AFB_1, AFB_2, AFG_1, AFG_2, CIT, OTA, and STG, literature[22]; Spices (garlic), AFB_1, AFB_2, and AFG_1, literature[22]; Spices (pepper), AFB_1, AFB_2, AFG_1, AFG_2, CIT, STG, and ZEA, literature[22]

incidence: 2/15*, conc. range: 0.5–1.1 µg/kg, ∅ conc.: 0.8 µg/kg, sample origin: unknown, sample year: unknown, country: USA[167], *red pepper

- Co-contamination: 1 sa. co-contaminated with AFB_1, AFB_2, AFG_1, and AFG_2; 1 sa. co-contaminated with AFB_1, AFB_2, and AFG_2
- Further contamination of plants and plant products other than cereals and their products reported in the present book/article[167]: Spices (cayenne pepper, chilli), AFB_1, AFB_2, and AFG_2, literature[167]; Spices (ginger), AFB_1, AFB_2, AFG_1, and AFG_2, literature[167]; Spices (pepper), AFB_1, literature[167]

incidence: 0/6*, conc.: no contamination, sample origin: markets (marked sa. of unknown origin), Turkey, sample year: unknown, country: Turkey[172], *red pepper

- Co-contamination: not reported
- Further contamination of plants and plant products other than cereals and their products reported in the present book/article[172]: Spices (cayenne pepper, chilli), AFB_1, AFB_2, and AFG_1, literature[172]

incidence: 0/15*, conc.: no contamination, sample origin: spice-sellers (unmarked sa. from unknown origin, India, and Turkey), Turkey, sample year: unknown, country: Turkey[172], *red pepper

- Co-contamination: not reported
- Further contamination of plants and plant products other than cereals and their products reported in the present book/article[172]: Spices (cayenne pepper, chilli), AFB_1, AFB_2, and AFG_1, literature[172]

incidence: 1/11*, conc.: 2 µg/kg**, sample origin: street bazaars (unmarked sa. from unknown origin and Turkey), Turkey, sample year: unknown, country: Turkey[172], *red pepper, **TLC and HPLC

- Co-contamination: 1 sa. co-contaminated with AFB_1, AFB_2, and AFG_1
- Further contamination of plants and plant products other than cereals and their products reported in the present book/article[172]: Spices (cayenne pepper, chilli), AFB_1, AFB_2, and AFG_1, literature[172]

For detailed information see the article.

incidence: 1/11*, conc.: 1.20 µg/kg, sample origin: public markets, supermarkets, shops, and bonded warehouse in Emilia Romagna (administrative region), Italy, sample year: 2000–2005, country: Italy[176], *hot pepper, whole/smashed

- Co-contamination: 1 sa. co-contaminated with AFB_1, AFB_2, AFG_1, and AFG_2

- Further contamination of plants and plant products other than cereals and their products reported in the present book/ article[176]: Spices (cayenne pepper, chilli), AFB_1, AFB_2, and AFG_1, literature[176]; Spices (cinnamon), AFB_1, literature[176]; Spices (nutmeg), AFB_1 and AFB_2, literature[176]

incidence: 2/2*, conc. range: 0.7–1.0 µg/kg, ∅ conc.: 0.85 µg/kg, sample origin: unknown, sample year: unknown, country: Denmark[217], *chilli

incidence: 2/2*, conc. range: 0.6–1.5 µg/kg, ∅ conc.: 1.1 µg/kg, sample origin: unknown, sample year: unknown, country: Denmark[217], *cayenne pepper

- Co-contamination: 2 sa. co-contaminated with AFB_1, AFB_2, AFG_1, and AFG_2
- Further contamination of plants and plant products other than cereals and their products reported in the present book/ article[217]: Butter (peanut butter), Figs, AFB_1, AFB_2, AFG_1, and AFG_2, literature[217]; Spices (cayenne pepper, chilli), AFB_1, AFB_2, and AFG_1, literature[217]; Meal (pistachio meal), AFB_1 and AFB_2, literature[217]

incidence: 2/31*, conc. range: 0.3–0.4 µg/kg, ∅ conc.: 0.35 µg/kg, sample origin: warehouses, packing establishments, supermarkets, and smaller shops, UK, sample year: unknown, country: UK[245], *chilli

- Co-contamination: 1 sa. co-contaminated with AFB_1, AFB_2, AFG_1, AFG_2, and OTA; 1 sa. co-contaminated with AFB_1, AFG_2 and OTA
- Further contamination of plants and plant products other than cereals and their products reported in the present book/ article[245]: Spices (cayenne pepper, chilli), AFB_1, AFB_2, AFG_1, and OTA, literature[245]; Spices (paprika), AFB_1, AFB_2, AFG_1, AFG_2, and OTA, literature[245]

incidence: 0/4*, conc.: no contamination, sample origin: warehouses, packing establishments, supermarkets, and smaller shops, UK, sample year: unknown, country: UK[245], *cayenne pepper

- Co-contamination: not reported
- Further contamination of plants and plant products other than cereals and their products reported in the present book/ article[245]: Spices (cayenne pepper, chilli), AFB_1, AFB_2, AFG_1, and OTA, literature[245]; Spices (paprika), AFB_1, AFB_2, AFG_1, AFG_2, and OTA, literature[245]

incidence: 3/80*, conc. range: <1.0 µg/kg (1 sa.), 1–5 µg/kg (2 sa., maximum: 3.2 µg/kg), sample origin: local markets and open markets in Kuala Lumpur (capital), Putrajaya (city), and Selangor (state) areas, Malaysia, sample year: May/June 2009, country: Malaysia[249], *chilli, dried

- Co-contamination: not reported
- Further contamination of plants and plant products other than cereals and their products reported in the present book/ article[249]: Spice (cayenne pepper, chilli), AFB_1, AFB_2, AFG_1, and OTA, literature[249]

incidence: 1/2*, conc.: 0.04 µg/kg, sample origin: Penang (state, island), Malaysia, sample year: unknown (first study), country: Malaysia[530], *chilli, dried or coarse dried

- Co-contamination: 1 sa. co-contaminated with AFB_1, AFB_2, AFG_1, AFG_2, and OTA
- Further contamination of plants and plant products other than cereals and their products reported in the present book/ article[530]: Spices, Spices (coriander), Spices (cumin), Spices (curry), Spices (fennel), Spices (masala), AFB_1, AFB_2, AFG_1, AFG_2, and OTA, literature[530]; Spices (cayenne pepper, chilli), AFB_1, AFB_2, AFG_1, and OTA, literature[530]; Spices (curcuma), AFB_1, AFB_2, AFG_1, and OTA, literature[530]; Spices (pepper), AFB_1 and AFB_2, literature[530]; Spices (poppy seed), AFB_1, literature[530]

incidence: 0/3*, conc.: no contamination, sample origin: Penang (state, island), Malaysia, sample year: unknown (second study), country: Malaysia[530], *chilli, dried

- Co-contamination: not reported
- Further contamination of plants and plant products other than cereals and their products reported in the present book/ article[530]: Spices, Spices (coriander), Spices (cumin), Spices (curry), Spices (fennel), Spices (masala), AFB_1, AFB_2, AFG_1, AFG_2, and OTA, literature[530]; Spices (cayenne pepper, chilli), AFB_1, AFB_2, AFG_1, and OTA, literature[530]; Spices (curcuma), AFB_1, AFB_2, AFG_1, and OTA, literature[530]; Spices (pepper), AFB_1 and AFB_2, literature[530]; Spices (poppy seed), AFB_1, literature[530]

incidence: 7/42*, conc. range: 0.001–1.542 µg/kg, sample origin: apiaries, supermarkets, open bazaars, and retail shops in Sanliurfa (city, province) region, Turkey, sample year: 2013, country: Turkey[595], *isot red pepper

- Co-contamination: not reported
- Further contamination of plants and plant products other than cereals and their products reported in the present book/ article[595]: Spices (cayenne pepper, chilli), AFB_1, AFB_2, and AFG_1, literature[595]

incidence: 5/10*, conc. range: 0.125–0.229 µg/kg, ∅ conc.: 0.180 µg/kg, sample origin: local fresh market in Sanliurfa (city), Turkey, sample year: 2013, country: Turkey[595], *isot red pepper dried on concrete surface

- Co-contamination: 3 sa. co-contaminated with AFB_2 and AFG_2; 1 sa. co-contaminated with AFB_1, AFB_2, AFG_1, and AFG_2; 1 sa. contaminated solely with AFG_2
- Further contamination of plants and plant products other than cereals and their products reported in the present book/ article[595]: Spices (cayenne pepper, chilli), AFB_1, AFB_2, and AFG_1, literature[595]

incidence: 4/10*, conc. range: 0.127–0.451 µg/kg, ∅ conc.: 0.220 µg/kg, sample origin: local fresh market in Sanliurfa (city), Turkey, sample year: 2013, country: Turkey[595], *isot red pepper dried on soil surface

- Co-contamination: 2 sa. co-contaminated with AFB_1, AFB_2, AFG_1, and AFG_2; 2 sa. co-contaminated with AFB_1, AFB_2, and AFG_2
- Further contamination of plants and plant products other than cereals and their products reported in the present book/ article[595]: Spices (cayenne pepper, chilli), AFB_1, AFB_2, and AFG_1, literature[595]

incidence: 1/1, conc.: 0.20 µg/kg, sample origin: shops in Watt Market, Calabar (city), Nigeria, sample year: January 2013, country: Nigeria[639]

- Co-contamination: 1 sa. co-contaminated with AFB_1, AFB_2, AFG_1, and AFG_2
- Further contamination of plants and plant products other than cereals and their products reported in the present book/ article[639]: Seeds (bush mango), Seeds (melon), AFB_1, AFB_2, AFG_1, and AFG_2, literature[639]; Spices (cayenne pepper, chilli), AFB_1, AFB_2, and AFG_1, literature[639]

incidence: 8/10, conc. range: 5.20–8.82 μg/kg, ∅ conc.: 6.7 μg/kg, sample origin: supermarkets and night markets in Penang (state), Malaysia, sample year: unknown, country: Malaysia[766]

- Co-contamination: 1 sa. co-contaminated with AFB_1, AFG_1, and AFG_2; 1 sa. co-contaminated with AFG_1 and AFG_2; 6 sa. contaminated solely with AFG_2
- Further contamination of plants and plant products other than cereals and their products reported in the present book/article[766]: Nuts (peanut), AFB_1 AFG_1, and AFG_2, literature[766]; Spices (cayenne pepper, chilli), AFB_1, AFB_2, and AFG_1, literature[766]

incidence: 0/18*, conc.: no contamination, sample origin: open markets, Sri Lanka, sample year: 2012/2013, country: Belgium[889], *chilli pods, whole

incidence: 0/26*, conc.: no contamination, sample origin: open markets, Sri Lanka, sample year: 2012/2013, country: Belgium[889], *chilli flakes

incidence: 1/42*, conc.: <LOQ, sample origin: open markets, Sri Lanka, sample year: 2012/2013, country: Belgium[889], *chilli powder

incidence: 0/35*, conc.: no contamination, sample origin: exotic shops and supermarkets in Antwerp and Ghent (cities), Belgium, sample year: 2012/2013, country: Belgium[889], *included whole chilli pods, chilli powders, and chilli flakes

- Co-contamination: For detailed information see the article.
- Further contamination of plants and plant products other than cereals and their products reported in the present book/article[889]: Spices (cayenne pepper, chilli), AFB_1, AFB_2, AFG_1, AME, CIT, FB_2, OTA, and STG, literature[889]

For detailed information see the article.

Aflatoxin/s

incidence: 14/23*, ∅ conc.: 11.6 μg/kg**, sample origin: greater Manila (capital) area, Philippines, sample year: unknown, country: Philippines[1], *pili and pili products, **AFS of pos sa.?

- Co-contamination: not reported
- Further contamination of plants and plant products other than cereals and their products reported in the present book/article[1]: Beans, Butter (peanut butter), Cassava, Cocoa, Coffee, Drinks (alcoholic drink), Nuts (cashew nut), Nuts (peanut), Products (coconut product), Rootcrops, Sweet potatoes, Yams, AF, literature[1]

incidence: 4?/4*, conc. range: 1.1–5.4 μg/kg**, sample origin: specialist food shops, UK, sample year: unknown, country: UK[16], *chilli powder, **AFB_1, AFB_2, AFG_1, and AFG_2

- Co-contamination: not reported
- Further contamination of plants and plant products other than cereals and their products reported in the present book/article[16]: Food, AF/S, DON, FUS-X, NIV, OTA, and ZEA, literature[16]; Nuts (almond), AF/S, literature[16]; Oils, Spices (fennel), AF/S and ZEA, literature[16]; Oils (sesame oil), OTA, literature[16]; Paste (curry paste), AF/S, FB/S, NIV, OTA, and ZEA, literature[16]; Pickles, AF/S, FB/S, OTA, and ZEA, literature[16]; Sauces (chilli sauce), NIV, OTA, and ZEA, literature[16]; Spices, AF/S, FB/S, OTA, and ZEA, literature[16]; Spices (cayenne pepper, chilli), DAS, DON, HT-2, OTA, and ZEA, literature[16]; Spices (coriander), AF/S, DON, OTA, and ZEA, literature[16]; Spices (curry), AF/S, DAS, FB/S, FUS-X, HT-2, NEO, NIV, OTA, T-2, and ZEA, literature[16]; Spices (garlic), DON, FUS-X, and NIV, literature[16]; Spices (ginger), AF/S, DON, NEO, NIV, OTA, and T-2, literature[16]; Spices (tandoori), AF/S, FB/S, NIV, OTA, and T-2, literature[16]

incidence: 9/12*, conc. range: ≤30 μg/kg**, sample origin: imported to USA, sample year: 1986, country: USA[19], *capsicum, **AFB_1, AFB_2, AFG_1, and AFG_2

- Co-contamination: not reported
- Further contamination of plants and plant products other than cereals and their products reported in the present book/article[19]: Butter (peanut butter), Candies (peanut candy), Candies (pistachio candy), Nuts, Nuts (almond), Nuts (Brazil nut, paranut), Nuts (peanut), Nuts (pecan nut), Nuts (pistachio nut), Nuts (walnut), Oils (peanut oil), Paste (peanut paste), Paste (pipian paste), Products (peanut product), Seeds (melon), Spices (ginger), Spices (nutmeg), AF/S, literature[19]

incidence: 41/72*, conc. range: ≤8.8 μg/kg**, ∅ conc.: 2.2 μg/kg**, sample origin: supermarkets and hypermarkets in 12 main cities of Catalonia (autonomous community), Spain, sample year: 2008/2009, country: Spain[93], *red pepper, **AFB_1, AFB_2, AFG_1, and AFG_2

- Co-contamination: not reported
- Further contamination of plants and plant products other than cereals and their products reported in the present book/article[93]: Figs, Nuts (peanut), Nuts (pistachio nut), AF/S, literature[93]

incidence: 12/106, conc. range: ≤966 μg/kg*, ∅ conc.: 125 μg/kg*, sample origin: market retailers, street vendors, wholesalers, and distributors, Thailand, sample year: September 1967–July 1969, country: Thailand[139], *AFB_1, AFB_2, AFG_1, and AFG_2

- Co-contamination: not reported
- Further contamination of plants and plant products other than cereals and their products reported in the present book/article[139]: Beans, Beans (mung bean), Food, Nuts (peanut), Seeds (sesame), Spices (garlic/onion), Starch (cassava starch), Vegetables, AF/S, literature[139]

For detailed information see the article.

incidence: 14/14*, conc. range: ≤9.68 μg/kg**, ∅ conc.: 5.23 μg/kg**, sample origin: popular markets in Salé (city) and Rabat (capital), Morocco, sample year: unknown, country: Morocco[174], *red paprika, **AFS total

- Co-contamination: not reported
- Further contamination of plants and plant products other than cereals and their products reported in the present book/article[174]: Spice (cayenne pepper, chilli), AFB_1, literature[174]; Spices (cumin), Spices (ginger), Spices (pepper), AFB_1 and AF/S, literature[174]

incidence: 17/30*, conc. range: 0.3–44.7 μg/kg**, ∅ conc.: 16.8 μg/kg**, sample origin: bazaars and markets in Istanbul (capital), Turkey, sample year: unknown, country: Turkey[178], *red-scaled pepper, **AFB_1, AFB_2, AFG_1, and AFG_2

incidence: 10/30*, conc. range: 0.3–14.2 μg/kg**, ∅ conc.: 5.6 μg/kg**, sample origin: bazaars and markets in Istanbul (capital), Turkey, sample year: unknown, country: Turkey[178], *red pepper, **AFB_1, AFB_2, AFG_1, and AFG_2

- Co-contamination: not reported
- Further contamination of plants and plant products other than cereals and their products reported in the present book/article[178]: Spices (cayenne pepper, chilli), AFB_1, literature[178]; Spices (pepper), AFB_1 and AF/S, literature[178]

incidence: 5/5*, conc. range: 2.7–76.8 μg/kg, sample origin: bazaars, Turkey, sample year: unknown, country: Turkey[180], *chilli powder incidence: 5/5*, conc. range: 1.8–85.9 μg/kg, sample origin: herbal shops, Turkey, sample year: unknown, country: Turkey[180], *chilli powder

incidence: 5/5*, conc. range: 2.1–4.9 μg/kg, sample origin: supermarkets, Turkey, sample year: unknown, country: Turkey[180], *chilli powder

- Co-contamination: not reported
- Further contamination of plants and plant products other than cereals and their products reported in the present book/article[180]: Spices (cayenne pepper, chilli), AFB_1, literature[180]; Spices (paprika), Spices (pepper), AFB_1 and AF/S, literature[180]

incidence: 8/44*, conc. range: 1.1–97.5 μg/kg**, ∅ conc.: 20.5 μg/kg**, sample origin: retailers and spice shops, Sanliurfa (province), Turkey, sample year: unknown, country: Turkey[186], *red-scaled pepper, **AFB and AFG

incidence: 3/26*, conc. range: 1.8–16.4 μg/kg**, ∅ conc.: 7.3 μg/kg**, sample origin: retailers and spice shops, Sanliurfa (province), Turkey, sample year: unknown, country: Turkey[186], *red pepper powder, **AFB and AFG

- Co-contamination: not reported
- Further contamination of plants and plant products other than cereals and their products reported in the present book/article[186]: Spices (pepper), AF/S, literature[186]

incidence: 10/30*, conc. range: 0.35–27.50 μg/kg**, sample origin: market stalls, shops, and supermarkets, Ireland, sample year: unknown, country: Ireland[187], *chilli powder, **AFB_1, AFB_2, AFG_1, and AFG_2

incidence: 2/8*, conc. range: 0.72–18.50 μg/kg**, ∅ conc.: 9.61 μg/kg**, sample origin: market stalls, shops, and supermarkets, Ireland, sample year: unknown, country: Ireland[187], *cayenne pepper, **AFB_1, AFB_2, AFG_1, and AFG_2

- Co-contamination: not reported
- Further contamination of plants and plant products other than cereals and their products reported in the present book/article[187]: Spices (coriander), Spices (curcuma), Spices (curry), Spices (paprika), Spices (pepper), AF/S, literature[187]

incidence: 7/41*, conc. range: 0.08–4.66 μg/kg**, sample origin: markets in ten cities, Korea, sample year: May–September 2006, country: Korea[188], *red pepper flour, **AFB_1, AFB_2, AFG_1, and AFG_2

- Co-contamination: not reported
- Further contamination of plants and plant products other than cereals and their products reported in the present book/article[188]: Paste (red pepper paste), Spices (curry), Spices (ginger), AF/S, literature[188]

incidence: 34/34*, conc. range: 3.55–9.55 μg/kg**, ∅ conc.: 6.36 μg/kg**, sample origin: bazaars and markets in Burdur, Isparta, and Antalya city centers, Turkey, sample year: unknown, country: Turkey[190], *red pepper, **AFS total

incidence: ?/38*, conc. range: ≤9.46 μg/kg**, sample origin: bazaars and markets in Burdur, Isparta, and Antalya city centers, Turkey, sample year: unknown, country: Turkey[190], *red pepper, flaked, **AFS total

- Co-contamination: not reported
- Further contamination of plants and plant products other than cereals and their products reported in the present book/article[190]: Spices (cumin), Spices (pepper), AF/S, literature[190]

incidence: 4/6*, conc. range: 5.60–69.28 μg/kg**, sample origin: sales outlets, Qatar, sample year: October 2002, country: Qatar[229], *chilli powder, **AFB_1, AFB_2, AFG_1, and AFG_2

- Co-contamination: not reported
- Further contamination of plants and plant products other than cereals and their products reported in the present book/article[229]: Butter (peanut butter), Figs, Nuts (peanut), Nuts (pistachio nut), AF/S, literature[229]; Coffee, Raisins, Spices (cayenne pepper, chilli), OTA, literature[229]; Spices, AF/S and OTA, literature[229]

incidence: 5/16*, conc. range: 0.1–5 μg/kg** (3 sa.), 6–10 μg/kg** (1 sa.), >50 μg/kg** (1 sa.), sample origin: commercial and imported food from retail outlets and points of entries, Malaysia, sample year: 1995–1999, country: Malaysia[233], *chilli powder etc., **AF not specified

- Co-contamination: not reported
- Further contamination of plants and plant products other than cereals and their products reported in the present book/article[233]: Butter (peanut butter), Candies (peanut candy), Nuts, Nuts (peanut), Products (peanut products), AF/S, literature[233]

incidence: 18/50*, conc. range: 1.0–1.9 μg/kg** (2 sa.), 2.0–3.9 μg/kg** (5 sa.), 4.0–10 μg/kg** (2 sa.), 10–50 μg/kg** (9 sa., maximum: 14.8 μg/kg** ***), sample origin: supermarkets, local and ethnic shops in Norwich (city) area, UK, sample year: unknown, country: UK[250], *included cayenne pepper, chilli, and paprika, **AFB_1, AFB_2, AFG_1, and AFG_2, ***cayenne pepper sa.

incidence: 9/14*, conc. range: 1–1.9 μg/kg** (2 sa.), 2–3.9 μg/kg** (3 sa.), 4–10 μg/kg** (1 sa.), 10–50 μg/kg** (2 sa.), >50 μg/kg** (1 sa.), sample origin: Port Health Authorities at Felixstowe (town), UK, sample year: unknown, country: UK[250], *included whole chilli and chilli powder, **AFB_1, AFB_2, AFG_1, and AFG_2

- Co-contamination: not reported
- Further contamination of plants and plant products other than cereals and their products reported in the present book/article[250]: Spices, Spices (curry), AF/S, literature[250]

incidence: 9/9*, conc. range: 6.8–96.2 μg/kg**, ∅ conc.: 47.3 μg/kg**, sample origin: directly from markets in Pakistan as sold by vendors, sample year: unknown, country: Portugal[251], *chilli, ground, **AFS total

incidence: 4/4*, conc. range: 0.1–6.6 μg/kg**, ∅ conc.: 2.15 μg/kg**, sample origin: directly from markets in Pakistan as sold by vendors, sample year: unknown, country: Portugal[251], *pod chilli, **AFS total

- Co-contamination: not reported
- Further contamination of plants and plant products other than cereals and their products reported in the present book/article[251]: not reported

incidence: 14/35*, conc. range: ≤2.49 μg/kg**, sample origin: retail shops, Catalonia (autonomous community), Spain, sample year: June–August 2008, country: Spain[252], *chilli, **AFB_1, AFB_2, AFG_1, and AFG_2

- Co-contamination: For detailed information see the article.
- Further contamination of plants and plant products other than cereals and their products reported in the present book/article[252]: Spices (cayenne pepper, chilli), OTA and ZEA, literature[252]; Spices (paprika), AF/S, OTA, and ZEA, literature[252]

incidence: 11/11*, conc. range: 1.9–65.7 μg/kg**, sample origin: distributor companies, Spain, sample year: 2008–2010, country: Spain[253], *chilli, **AFB_1, AFB_2, AFG_1, and AFG_2

- Co-contamination: 9 sa. co-contaminated with AFS, DON, OTA, T-2, and ZEA; 2 sa. co-contaminated with AFS, OTA, and ZEA
- Further contamination of plants and plant products other than cereals and their products reported in the present book/article[253]: Spices (cayenne pepper, chilli), DON, OTA, T-2, and ZEA, literature[253]; Spices (paprika), AF/S, DON, OTA, and ZEA, literature[253]

incidence: 29/42* **, conc. range: ≤95.9 μg/kg***, sample origin: markets, herbal shops, and chilli-growing areas, Faisalabad (city), Punjab (state), Pakistan, sample year: 2009, country: Pakistan/Portugal/USA[254], *whole and ground chilli, **collected in summer season, ***AFB_1, AFB_2, AFG_1, and AFG_2

incidence: 32/45* **, conc. range: ≤74.6 μg/kg***, sample origin: markets, herbal shops, and chilli growing areas, Faisalabad (city), Punjab (state), Pakistan, sample year: 2009/2010, country: Pakistan/Portugal/USA[254], *whole and ground chilli, **collected in winter season, ***AFB_1, AFB_2, AFG_1, and AFG_2

- Co-contamination: not reported
- Further contamination of plants and plant products other than cereals and their products reported in the present book/article[254]: not reported

incidence: 4/5*, conc. range: 2.9–5.0 μg/kg**, ∅ conc.: 3.8 μg/kg**, sample origin: available in New Zealand, sample year: 2009, country: New Zealand[277], *cayenne pepper, **AFS total

incidence: 4/5*, conc. range: 3.5–8.5 μg/kg**, ∅ conc.: 5.5 μg/kg**, sample origin: available in New Zealand, sample year: 2009, country: New Zealand[277];*chilli powder, **AFS total

- Co-contamination: not reported
- Further contamination of plants and plant products other than cereals and their products reported in the present book/article[277]: Bars, AFB_1, AF/S, DON, NIV, and OTA, literature[277]; Butter (peanut butter), Confectionery, Food, Fruits, dried, Nuts, Nuts (peanut), Nuts (pistachio nut), Sauces (peanut sauce), AFB_1 and AF/S, literature[277]; Coffee, Dates, Spices (nutmeg), Spices (pepper), Vine fruits, dried, OTA, literature[277]; Figs, Spices (curry), Spices (ginger), Spices (paprika), AFB_1, AF/S, and OTA, literature[277]; Nuts (Brazil nut, paranut), AFB_1, literature[277]; Spices (cayenne pepper, chilli), AFB_1 and OTA, literature[277]

incidence: 4/5*, conc. range: 3.5–8.5 μg/kg**, sample origin: New Zealand, sample year: 2009, country: New Zealand[702], *chilli powder, **AFB_1, AFB_2, AFG_1, and AFG_2

incidence: 4/5*, conc. range: 2.9–5.0 μg/kg**, sample origin: New Zealand, sample year: 2009, country: New Zealand[702], *cayenne pepper, **AFB_1 and AFB_2

- Co-contamination: not reported
- Further contamination of plants and plant products other than cereals and their products reported in the present book/article[702]: Figs, Spices (curry), Spices (ginger), Spices (paprika), AF/S and OTA, literature[702]; Fruits, dried, AF/S, literature[702]; Spices (cayenne pepper, chilli), Spices (pepper), Vine fruits, dried, OTA, literature[702]

incidence: 17/29*, conc. range: ≤94.1 μg/kg**, sample origin: retail markets of Faisalabad, Jhang, Lahore, and Toba Tek Singh (districts) in Punjab (state), Pakistan, sample year: unknown, country: Pakistan[751], *chilli, crushed, **AFB_1, AFB_2, AFG_1, and AFG_2

incidence: 15/28*, conc. range: ≤99.3 μg/kg**, sample origin: food restaurants in Faisalabad, Jhang, Lahore, and Toba Tek Singh (districts) in Punjab (state), Pakistan, sample year: unknown, country: Pakistan[751], *chilli, crushed, **AFB_1, AFB_2, AFG_1, and AFG_2

incidence: 19/34*, conc. range: ≤104.4 μg/kg**, sample origin: retail markets of Faisalabad, Jhang, Lahore, and Toba Tek Singh (districts) in Punjab (state), Pakistan, sample year: unknown, country: Pakistan[751], *chilli powder, **AFB_1, AFB_2, AFG_1, and AFG_2

incidence: 17/29*, conc. range: ≤92.9 μg/kg**, sample origin: food restaurants in Faisalabad, Jhang, Lahore, and Toba Tek Singh (districts) in Punjab (state), Pakistan, sample year: unknown, country: Pakistan[751], *chilli powder, **AFB_1, AFB_2, AFG_1, and AFG_2

- Co-contamination: not reported
- Further contamination of plants and plant products other than cereals and their products reported in the present book/article[751]: Sauces (chilli sauce), AF/S and OTA, literature[751]; Spices (cayenne pepper, chilli), OTA, literature[751]

incidence: 223/226*, conc. range: 1–10 μg/kg** (138 sa.), 10–20 μg/kg** (63 sa.), >20 μg/kg** (22 sa., maximum: 91.4 μg/kg**), sample origin: warehouses, packing establishments, supermarkets, and smaller shops across Pakistan, sample year: July 2006–June 2011, country: Pakistan[765], *red chilli, whole, **AFB_1, AFB_2, AFG_1, and AFG_2

incidence: 64/69*, conc. range: 1–10 μg/kg** (13 sa.), 10–20 μg/kg** (19 sa.), >20 μg/kg** (32 sa., maximum: 77.9 μg/kg**), sample origin: warehouses, packing establishments, supermarkets, and smaller shops across Pakistan, sample year: July 2006–June 2011, country: Pakistan[765], *red chilli, powder, **AFB_1, AFB_2, AFG_1, and AFG_2

incidence: 34/36*, conc. range: 1–10 μg/kg** (5 sa.), 10–20 μg/kg** (10 sa.), >20 μg/kg** (19 sa., maximum: 93.7 μg/kg**), sample origin: warehouses, packing establishments, supermarkets, and smaller shops across Pakistan, sample year: July 2006–June 2011, country: Pakistan[765], *red chilli, crushed, **AFB_1, AFB_2, AFG_1, and AFG_2

- Co-contamination: not reported
- Further contamination of plants and plant products other than cereals and their products reported in the present book/article[765]: not reported

For detailed information see the article.

incidence: 3/13*, conc. range: 0.04–0.21 μg/kg**, sample origin: (domestic) the internet shopping mall, supermarket, and wholesale market, South Korea, sample year: unknown, country: Korea[809], *packed, dried, and ground red pepper sa., **AFB_1, AFB_2, AFG_1, and AFG_2

incidence: 0/45*, conc.: no contamination, sample origin: (domestic) the internet shopping mall, supermarket, and wholesale market, South Korea, sample year: unknown, country: Korea[809], *non-packed, dried, and ground red pepper sa.

incidence: 2/4*, conc. range: 0.05–0.11 μg/kg**, sample origin: imported from China purchased in the internet shopping mall, supermarket, and wholesale market, South Korea, sample year: unknown, country: Korea[809], *packed, dried, and ground red pepper sa., **AFB_1, AFB_2, AFG_1, and AFG_2

incidence: 0/6*, conc.: no contamination, sample origin: imported from China purchased in the internet shopping mall, supermarket, and wholesale market, South Korea, sample year: unknown, country: Korea[809], *non-packed, dried, and ground red pepper sa.

incidence: 0/1*, conc.: no contamination, sample origin: imported from Vietnam purchased in the internet shopping mall, supermarket, and wholesale market, South Korea, sample year: unknown, country: Korea[809], *packed, dried, and ground red pepper sa.

incidence: 1/4*, conc.: 0.90 µg/kg**, sample origin: imported from Vietnam purchased in the internet shopping mall, supermarket, and wholesale market, South Korea, sample year: unknown, country: Korea[809], *non-packed, dried, and ground red pepper sa., **AFB_1, AFB_2, AFG_1, and AFG_2

incidence: 5/5*, conc. range: 4.22–38.03 µg/kg**, Ø conc.: 18.98 µg/kg**, sample origin: imported from India purchased in the internet shopping mall, supermarket, and wholesale market, South Korea, sample year: unknown, country: Korea[809], *packed, dried, and ground red pepper sa., **AFB_1, AFB_2, AFG_1, and AFG_2

incidence: 0/0*, conc.: no sa. available, sample origin: imported from India purchased in the internet shopping mall, supermarket, and wholesale market, South Korea, sample year: unknown, country: Korea[809], *non-packed, dried, and ground red pepper sa.

- Co-contamination: not reported
- Further contamination of plants and plant products other than cereals and their products reported in the present book/article[809]: Spices (cayenne pepper, chilli), OTA, literature[809]

incidence: 116/116*, conc. range: 0.15–70.3 µg/kg**, Ø conc.: 9.48 µg/kg**, sample origin: markets of Karachi (city), Pakistan, sample year: 2007–2009, country: Pakistan[836], *red chilli sa., **AF

- Co-contamination: not reported
- Further contamination of plants and plant products other than cereals and their products reported in the present book/article[836]: not reported

incidence: 10/10* **, conc. range: 0.23–1.46 µg/kg***, Ø conc.: 0.90 µg/kg***, sample origin: local supermarket, sample year: unknown, country: China/Germany[855], *red paprika sa., **storage time: 0 days, ***AFS

incidence: 10/10* **, conc. range: 7.93–9.32 µg/kg***, Ø conc.: 8.50 µg/kg***, sample origin: local supermarket, sample year: unknown, country: China/Germany[855], *red paprika sa., **storage time: 90 days, ***AFS

- Co-contamination: not reported
- Further contamination of plants and plant products other than cereals and their products reported in the present book/article[855]: not reported

For detailed information see the article.

Other ***Aspergillus*** Toxins

Sterigmatocystin

incidence: 3/5*, conc. range: 10–23 µg/kg, sample origin: Assiut (governorate), Egypt, sample year: unknown, country: Egypt[641], *red pepper

- Co-contamination: not reported
- Further contamination of plants and plant products other than cereals and their products reported in the present book/article[641]: Spices (anise), Spices (coriander), Spices (fennel), Spices (pepper), Spices (peppermint), AF/S, literature[641]; Spices (black cumin), AF/S and CIT, literature[641]; Spices (caraway), Spices (marjoram), AF/S and STG, literature[641]; Spices (cumin), STG, literature[641]

incidence: 5/18*, conc. range: <LOQ-31.8 µg/kg, Ø conc.: 34.3 µg/kg?, sample origin: open markets, Sri Lanka, sample year: 2012/2013, country: Belgium[889], *chilli pods, whole

incidence: 10/26*, conc. range: <LOQ, Ø conc.: <LOQ, sample origin: open markets, Sri Lanka, sample year: 2012/2013, country: Belgium[889], *chilli flakes

incidence: 18/42*, conc. range: <LOQ-13.6 µg/kg, Ø conc.: 13.0 µg/kg, sample origin: open markets, Sri Lanka, sample year: 2012/2013, country: Belgium[889], *chilli powder

incidence: 10/35*, conc. range: <LOQ, sample origin: exotic shops and supermarkets in Antwerp and Ghent (cities), Belgium, sample year: 2012/2013, country: Belgium[889], *included whole chilli pods, chilli powders, and chilli flakes

- Co-contamination: For detailed information see the article.
- Further contamination of plants and plant products other than cereals and their products reported in the present book/article[889]: Spices (cayenne pepper, chilli), AFB_1, AFB_2, AFG_1, AFG_2, AME, CIT, FB_2, and OTA, literature[889]

For detailed information see the article.

incidence: 4/10*, conc. range: <LOQ, sample origin: markets, Sri Lanka, sample year: unknown, country: Belgium[891], *red chilli

- Co-contamination: not reported
- Further contamination of plants and plant products other than cereals and their products reported in the present book/article[891]: Spices (cayenne pepper, chilli), AFB_1, AFB_2, FB_2, and OTA, literature[891]; Spices (pepper), AFB_1, AFB_2, AFG_2, CIT, FB_1, OTA, and STG, literature[891]

Aspergillus and ***Penicillium*** Toxins

Citrinin

incidence: 4/18*, conc. range: <LOQ-2,107 µg/kg, Ø conc.: 1,363 µg/kg, sample origin: open markets, Sri Lanka, sample year: 2012/2013, country: Belgium[889], *chilli pods, whole

incidence: 1/26*, conc.: <LOQ, sample origin: open markets, Sri Lanka, sample year: 2012/2013, country: Belgium[889], *chilli flakes

incidence: 2/42*, conc. range: <LOQ, Ø conc.: LOQ, sample origin: open markets, Sri Lanka, sample year: 2012/2013, country: Belgium[889], *chilli powder

incidence: 2/35*, conc. range: <LOQ, Ø conc.: LOQ, sample origin: exotic shops and supermarkets in Antwerp and Ghent (cities), Belgium, sample year: 2012/2013, country: Belgium[889], *included whole chilli pods, chilli powders, and chilli flakes

- Co-contamination: For detailed information see the article.
- Further contamination of plants and plant products other than cereals and their products reported in the present book/article[889]: Spices (cayenne pepper, chilli), AFB_1, AFB_2, AFG_1, AFG_2, AME, FB_2, OTA, and STG, literature[889]

For detailed information see the article.

Ochratoxin A

incidence: 4?/4*, conc. range: 1.6–50.4 µg/kg, sample origin: specialist food shops, UK, sample year: unknown, country: UK[16], *chilli powder

- Co-contamination: not reported
- Further contamination of plants and plant products other than cereals and their products reported in the present book/article[16]: Food, AF/S, DON, FUS-X, NIV, OTA, and ZEA, literature[16]; Nuts (almond), AF/S, literature[16]; Oils, Spices (fennel), AF/S and ZEA, literature[16]; Oils (sesame oil), OTA, literature[16]; Paste (curry paste), AF/S, FB/S, NIV, OTA, and ZEA, literature[16]; Pickles, AF/S, FB/S, OTA, and ZEA, literature[16]; Sauces (chilli sauce), NIV, OTA, and ZEA, literature[16]; Spices, AF/S, FB/S, OTA, and ZEA, literature[16]; Spices (cayenne pepper, chilli), AF/S, DAS, DON, HT-2, and ZEA, literature[16]; Spices (coriander), AF/S, DON, OTA, and ZEA, literature[16]; Spices (curry), AF/S, DAS, FB/S, FUS-X, HT-2, NEO, NIV, OTA, T-2, and ZEA, literature[16]; Spices (garlic), DON, FUS-X, and NIV, literature[16]; Spices (ginger), AF/S, DON, NEO, NIV, OTA, and T-2, literature[16]; Spices (tandoori), AF/S, FB/S, NIV, OTA, and T-2, literature[16]

incidence: 14/22*, conc. range: 0.4–26.8 µg/kg, sample origin: food shops in Hajdú-Bihar (county), Hungary, sample year: October 2004, country: Hungary[170], *red pepper, grounded

incidence: 9/16*, conc. range: 0.4–66.2 µg/kg, sample origin: shops in/around Kaposvár (town), Somogy (county), Hungary, sample year: October 2004, country: Hungary[170], *red pepper, grounded and originally packed

incidence: 3/13*, conc.: 2.5 µg/kg, sample origin: food shops in Hajdú-Bihar (county), Hungary, sample year: November 2004, country: Hungary[170], *red pepper, grounded

incidence: 6/19*, conc. range: 0.4–3.5 µg/kg, sample origin: 10 Hungarian counties, Hungary, sample year: October/November 2004, country: Hungary[170], *red pepper, home-made

incidence: 1/5*, conc.: 2.1 µg/kg, sample origin: food shops in Debrecen (town), Hungary, sample year: October 2004, country: Hungary[170], *chilli

- Co-contamination: For detailed information see the article.
- Further contamination of plants and plant products other than cereals and their products reported in the present book/article[170]: Spices, Spices (cayenne pepper, chilli), AFB_1 and AFB_2, literature[170]; Spices (pepper), AFB_1, literature[170]

incidence: 18/24*, conc. range: 0.46–53.04 µg/kg, ∅ conc.: 16.44 µg/kg, sample origin: markets, herbal shops, and bazaars in Corum and Ankara (cities), Turkey, sample year: November 2010–August 2011, country: Turkey[183], *red chilli flakes

- Co-contamination: 3 sa. co-contaminated with AFB_1, AFB_2, AFG_1, and OTA; 11 sa. co-contaminated with AFB_1, AFB_2, and OTA; 1 sa. co-contaminated with AFB_1 and OTA; 3 sa. contaminated solely with OTA
- Further contamination of plants and plant products other than cereals and their products reported in the present book/article[183]: Spices (cayenne pepper, chilli), AFB_1, AFB_2, and AFG_1, literature[183]; Spices (cumin), Spices (pepper), AFB_1, AFB_2, and OTA, literature[183]

incidence: 12/22*, conc. range: 0.78–98.2 µg/g, ∅ conc.: 24.65 µg/kg, sample origin: markets, herbal shops, and bazaars in Corum and Ankara (cities), Turkey, sample year: November 2010–August 2011, country: Turkey[183], *red chilli powder

- Co-contamination: 3 sa. co-contaminated with AFB_1, AFB_2, AFG_1, and OTA; 5 sa. co-contaminated with AFB_1, AFB_2, and OTA; 1 sa. co-contaminated with AFB_1 and OTA; 3 sa. contaminated solely with OTA
- Further contamination of plants and plant products other than cereals and their products reported in the present book/article[183]: Spices (cayenne pepper, chilli), AFB_1, AFB_2, and AFG_1, literature[183]; Spices (cumin), Spices (pepper), AFB_1, AFB_2, and OTA, literature[183]

incidence: 1/2*, conc.: 12 µg/kg, sample origin: retail shops, Belgium, Netherlands, and Russia, sample year: unknown, country: Belgium/Russia[192], *pili-pili

incidence: 2/2*, conc. range: 20–30 µg/kg, ∅ conc.: 25 µg/kg, sample origin: retail shops, Belgium, Netherlands, and Russia, sample year: unknown, country: Belgium/Russia[192], *red pepper

incidence: 1/3*, conc.: 31 µg/kg, sample origin: retail shops, Belgium, Netherlands, and Russia, sample year: unknown, country: Belgium/Russia[192], *chilli

incidence: 1/2*, conc.: 12 µg/kg, sample origin: retail shops, Belgium, Netherlands, and Russia, sample year: unknown, country: Belgium/Russia[192], *cayenne

- Co-contamination: not reported
- Further contamination of plants and plant products other than cereals and their products reported in the present book/article[192]: Licorice, OTA, literature[192]

incidence: 2/3*, conc. range: 1.4–13 µg/kg, ∅ conc.: 7.2 µg/kg, sample origin: retail shops in Belgium, Netherlands, and Russia, sample year: autumn–winter 2005/2006, country: Belgium/Russia/Hungary[193], *cayenne

incidence: 2/4*, conc. range: 1.6–22 µg/kg, ∅ conc.: 11.8 µg/kg, sample origin: retail shops in Belgium, Netherlands, and Russia, sample year: autumn–winter 2005/2006, country: Belgium/Russia/Hungary[193], *included chilli and chilli pods

incidence: 2/2*, conc. range: 18–30 µg/kg, ∅ conc.: 24 µg/kg, sample origin: retail shops in Belgium, Netherlands, and Russia, sample year: autumn–winter 2005/2006, country: Belgium/Russia/Hungary[193], *red pepper

incidence: 1/3*, conc.: 19 µg/kg, sample origin: retail shops in Belgium, Netherlands, and Russia, sample year: autumn–winter 2005/2006, country: Belgium/Russia/Hungary[193], *included pili-pili and pili-pili pods

- Co-contamination: not reported
- Further contamination of plants and plant products other than cereals and their products reported in the present book/article[193]: Spices (ginger), Spices (nutmeg), Spices (paprika), Spices (pepper), OTA, literature[193]

incidence: 1/1*, conc.: 160 µg/kg (1 sa.), sample origin: Austria, sample year: 1994, country: EU[194], *cayenne pepper

incidence: 13/18*, conc. range: LOD/LOQ-4.9 µg/kg (4 sa.), 5.0–9.9 µg/kg (3 sa.), 10.0–24.9 µg/kg (5 sa.), 38 µg/kg (1 sa.), sample origin: Austria, sample year: 1994, country: EU[194], *red pepper

- Co contamination: not reported
- Further contamination of plants and plant products other than cereals and their products reported in the present book/article[194]: Coffee, Coffee beans, Spices (curry), Spices (marjoram), Spices (pepper), OTA, literature[194]

incidence: 2/6*, conc. range: 2.34–4.91 µg/kg, ∅ conc.: 3.63 µg/kg, sample origin: sales outlets, Qatar, sample year: October 2002, country: Qatar[229], *chilli powder

- Co-contamination: not reported

- Further contamination of plants and plant products other than cereals and their products reported in the present book/article[229]: Butter (peanut butter), Figs, Nuts (peanut), Nuts (pistachio nut), Spices (cayenne pepper, chilli), AF/S, literature[229]; Coffee, Raisins, OTA, literature[229]; Spices, AF/S and OTA, literature[229]

incidence: 31/31*, conc. range: 0.2–152.2 µg/kg, sample origin: warehouses, packing establishments, supermarkets, and smaller shops, UK, sample year: unknown, country: UK[245], *chilli

- Co-contamination: 1 sa. co-contaminated with AFB_1, AFB_2, AFG_1, AFG_2, and OTA; 1 sa. co-contaminated with AFB_1, AFB_2, AFG_1, and OTA; 8 sa. co-contaminated with AFB_1, AFB_2 and OTA; 1 sa. co-contaminated with AFB_1, AFG_1 and OTA; 1 sa. co-contaminated with AFB_1, AFG_2 and OTA; 17 sa. co-contaminated with AFB_1 and OTA; 2 sa. contaminated solely with OTA
- Further contamination of plants and plant products other than cereals and their products reported in the present book/article[245]: Spices (cayenne pepper, chilli), AFB_1, AFB_2, AFG_1, and AFG_2, literature[245]; Spices (paprika), AFB_1, AFB_2, AFG_1, AFG_2, and OTA, literature[245]

incidence: 4/4*, conc. range: 2.4–16.5 µg/kg, ∅ conc.: 6.98 µg/kg, sample origin: warehouses, packing establishments, supermarkets, and smaller shops, UK, sample year: unknown, country: UK[245], *cayenne pepper

- Co-contamination: 1 sa. co-contaminated with AFB_1, AFB_2, and OTA; 1 sa. co-contaminated with AFB_1, AFG_1 and OTA; 2 sa. co-contaminated with AFB_1 and OTA
- Further contamination of plants and plant products other than cereals and their products reported in the present book/article[245]: Spices (cayenne pepper, chilli), AFB_1, AFB_2, AFG_1, and AFG_2, literature[245]; Spices (paprika), AFB_1, AFB_2, AFG_1, AFG_2, and OTA, literature[245]

incidence: 65/80*, conc. range: <1.0 µg/kg (18 sa.), 1–5 µg/kg (26 sa.), 5–10 µg/kg (9 sa.), >10 µg/kg (12 sa., maximum: 101.42 µg/kg), sample origin: local markets and open markets in Kuala Lumpur (capital), Putrajaya (city), and Selangor (state) areas, Malaysia, sample year: May/June 2009, country: Malaysia[249], *chilli, dried

- Co-contamination: not reported
- Further contamination of plants and plant products other than cereals and their products reported in the present book/article[249]: Spice (cayenne pepper, chilli), AFB_1, AFB_2, AFG_1, and AFG_2, literature[249]

incidence: 35/35*, conc. range: 0.62–44.6 µg/kg**, sample origin: retail shops, Catalonia (autonomous community), Spain, sample year: June–August 2008, country: Spain[252], *chilli, **AFB_1, AFB_2, AFG_1, and AFG_2

- Co-contamination: For detailed information see the article.
- Further contamination of plants and plant products other than cereals and their products reported in the present book/article[252]: Spices (cayenne pepper, chilli), AF/S and ZEA, literature[252]; Spices (paprika), AF/S, OTA, and ZEA, literature[252]

incidence: 11/11*, conc. range: 4.3–19.1 µg/kg, sample origin: distributor companies, Spain, sample year: 2008–2010, country: Spain[253], *chilli

- Co-contamination: 9 sa. co-contaminated with AFS, DON, OTA, T-2, and ZEA; 2 sa. co-contaminated with AFS, OTA, and ZEA
- Further contamination of plants and plant products other than cereals and their products reported in the present book/article[253]: Spices (cayenne pepper, chilli), AF/S, DON, T-2, and ZEA, literature[253]; Spices (paprika), AF/S, DON, OTA, and ZEA, literature[253]

incidence: 2/32*, conc. range: 47 µg/kg (1 sa.), 93 µg/kg (1 sa.), sample origin: markets and cold storage units, Guntur and Khammam (districts), Andhra Pradesh (state), India, sample year: unknown, country: UK/India[255], *dried chilli pods (grade 1)

incidence: 2/14*, conc. range: 10–29 µg/kg (1 sa.), 50–100 µg/kg (1 sa.), sample origin: markets and cold storage units, Guntur and Khammam (districts), Andhra Pradesh (state), India, sample year: unknown, country: UK/India[255], *dried chilli pods (grade 2)

incidence: 8/23*, conc. range: 10–29 µg/kg (3 sa.), 30–49 µg/kg (5 sa.), sample origin: markets and cold storage units, Guntur and Khammam (districts), Andhra Pradesh (state), India, sample year: unknown, country: UK/India[255], *dried chilli pods (grade 3)

incidence: 14/21*, conc. range: 10–29 µg/kg (8 sa.), 30–49 µg/kg (4 sa.), 50–100 µg/kg (1 sa.), 120 µg/kg (1 sa.), sample origin: markets and cold storage units, Guntur and Khammam (districts), Andhra Pradesh (state), India, sample year: unknown, country: UK/India[255], * chilli powder from supermarkets

- Co-contamination: not reported
- Further contamination of plants and plant products other than cereals and their products reported in the present book/article[255]: not reported

incidence: 2/2*, conc. range: 0.83–2.74 µg/kg, ∅ conc.: 1.79 µg/kg, sample origin: available in New Zealand, sample year: 2007, country: New Zealand[277], *cayenne pepper

incidence: 5/5*, conc. range: 1.8–7.3 µg/kg, sample origin: available in New Zealand, sample year: 2008/2009, country: New Zealand[277], *cayenne pepper

incidence: 4/4*, conc. range: 0.23–39.9 µg/kg, sample origin: available in New Zealand, sample year: 2007, country: New Zealand[277];*chilli powder

incidence: 4/5*, conc. range: 1.3–6.3 µg/kg, sample origin: available in New Zealand, sample year: 2008/2009, country: New Zealand[277];*chilli powder

- Co-contamination: not reported
- Further contamination of plants and plant products other than cereals and their products reported in the present book/article[277]: Bars, AFB_1, AF/S, DON, NIV, and OTA, literature[277]; Butter (peanut butter), Confectionery, Food, Fruits, dried, Nuts, Nuts (peanut), Nuts (pistachio nut), Sauces (peanut sauce), Spices (cayenne pepper, chilli), AFB_1 and AF/S, literature[277]; Coffee, Dates, Spices (nutmeg), Spices (pepper), Vine fruits, dried, OTA, literature[277]; Figs, Spices (curry), Spices (ginger), Spices (paprika), AFB_1, AF/S, and OTA, literature[277]; Nuts (Brazil nut, paranut), AFB_1, literature[277]

incidence: 1/2*, conc.: 0.62 µg/kg, sample origin: Penang (state, island), Malaysia, sample year: unknown (first study), country: Malaysia[530], *chilli, dried or coarse dried

- Co-contamination: 1 sa. co-contaminated with AFB_1, AFB_2, AFG_1, AFG_2, and OTA
- Further contamination of plants and plant products other than cereals and their products reported in the present book/article[530]: Spices, Spices (coriander), Spices (cumin), Spices (curry), Spices (fennel), Spices (masala), AFB_1, AFB_2, AFG_1, AFG_2, and OTA, literature[530]; Spices (cayenne pepper, chilli), AFB_1, AFB_2, AFG_1, and AFG_2, literature[530]; Spices (curcuma), AFB_1, AFB_2, AFG_1, and OTA, literature[530]; Spices (pepper), AFB_1 and AFB_2, literature[530]; Spices (poppy seed), AFB_1, literature[530]

incidence: 0/3*, conc.: no contamination, sample origin: Penang (state, island), Malaysia, sample year: unknown (second study), country: Malaysia[530], *chilli, dried

- Co-contamination: not reported
- Further contamination of plants and plant products other than cereals and their products reported in the present book/article[530]: Spices, Spices (coriander), Spices (cumin), Spices (curry), Spices (fennel), Spices (masala), AFB_1, AFB_2, AFG_1, AFG_2, and OTA, literature[530]; Spices (cayenne pepper, chilli), AFB_1, AFB_2, AFG_1, and AFG_2, literature[530]; Spices (curcuma), AFB_1, AFB_2, AFG_1, and OTA, literature[530]; Spices (pepper), AFB_1 and AFB_2, literature[530]; Spices (poppy seed), AFB_1, literature[530]

incidence: 0/41* **, conc.: no contamination, sample origin: supermarkets, online shopping mall, small stakeholder mills, HACCP-implemented factories, and an important company, Korea, sample year: August 2008, country: Korea[588], *red paprika/red peppers, **pre-harvest stages

incidence: 11/68* **, conc. range: ≤34.96 μg/kg, sample origin: supermarkets, online shopping mall, small stakeholder mills, HACCP-implemented factories, and an important company, Korea, sample year: August 2008, country: Korea[588], *red paprika/red peppers, **post-harvest-and processing stages

incidence: 31/83* **, conc. range: ≤33.59 μg/kg, sample origin: supermarkets, online shopping mall, small stakeholder mills, HACCP-implemented factories, and an important company, Korea, sample year: August 2008, country: Korea[588], *red paprika/red peppers, **distribution stages

- Co-contamination: not reported
- Further contamination of plants and plant products other than cereals and their products reported in the present book/article[588]: not reported

incidence: 6/8*, conc. range: LOD/LOQ-0.9 μg/kg (1 sa.), 1.0–2.9 μg/kg (1 sa.), ≤14.5 μg/kg (4 sa.), sample origin: Netherlands, sample year: 1996, country: EU[698], *pepper powder

- Co-contamination: not reported
- Further contamination of plants and plant products other than cereals and their products reported in the present book/article[698]: Baby food, Beans (soybean), Chocolate, Cocoa, Coffee, Coffee beans, Currants, Fruits, dried, Herbs, Jam, Juices, Juices (grape juice), Legumes, Nuts, Oil seeds, Oils (olive oil), Peas, Peas (chickpea), Products (plant product), Raisins, Spices, Spices (nutmeg), Spices (paprika), Spices (pepper), Sultanas, Tea, Vinegar, Wine, Wine (red), Wine (rosé), Wine (white), Wine, miscellaneous, OTA, literature[698]

incidence: 4/5*, conc. range: 1.3–6.3 μg/kg, sample origin: New Zealand, sample year: 2009, country: New Zealand[702], *chilli powder

incidence: 5/5*, conc. range: 1.8–7.3 μg/kg, sample origin: New Zealand, sample year: 2009, country: New Zealand[702], *cayenne pepper

- Co-contamination: not reported
- Further contamination of plants and plant products other than cereals and their products reported in the present book/article[702]: Figs, Spices (curry), Spices (ginger), Spices (paprika), AF/S and OTA, literature[702]; Fruits, dried, Spices (cayenne pepper, chilli), AF/S, literature[702]; Spices (pepper), Vine fruits, dried, OTA, literature[702]

incidence: 5/13, conc. range: 0.6–17.1 μg/kg, Ø conc.: 18.9 μg/kg, sample origin: central markets in Cusco and Trujillo (cities), and Lima (capital), Peru, sample year: unknown, country: Japan[749]

- Co-contamination: 3 sa. co-contaminated with AFB_1 and OTA; 2 sa. contaminated solely with OTA
- Further contamination of plants and plant products other than cereals and their products reported in the present book/article[749]: Spices (cayenne pepper, chilli), AFB_1, literature[749]

incidence: 3/4, conc. range: 163.4–1,059.1 μg/kg, Ø conc.: 474 μg/kg, sample origin: central markets in Santiago de Chile (capital), Chile, sample year: unknown, country: Japan[749]

incidence: 4/4, conc. range: 0.8–628.3 μg/kg, Ø conc.: 207 μg/kg, sample origin: central markets in La Paz (capital), Bolivia, sample year: unknown, country: Japan[749]

- Co-contamination: not reported
- Further contamination of plants and plant products other than cereals and their products reported in the present book/article[749]: Spices (cayenne pepper, chilli), AFB_1, literature[749]

incidence: 11/29*, conc. range: ≤54.3 μg/kg, sample origin: retail markets of Faisalabad, Jhang, Lahore, and Toba Tek Singh (districts) in Punjab (state), Pakistan, sample year: unknown, country: Pakistan[751], *chilli, crushed

incidence: 14/28*, conc. range: ≤48.7 μg/kg, sample origin: food restaurants in Faisalabad, Jhang, Lahore, and Toba Tek Singh (districts) in Punjab (state), Pakistan, sample year: unknown, country: Pakistan[751], *chilli, crushed

incidence: 13/34, conc. range: ≤64.5 μg/kg, sample origin: retail markets of Faisalabad, Jhang, Lahore, and Toba Tek Singh (districts) in Punjab (state), Pakistan, sample year: unknown, country: Pakistan[751]

incidence: 12/29, conc. range: ≤58.1 μg/kg, sample origin: food restaurants in Faisalabad, Jhang, Lahore, and Toba Tek Singh (districts) in Punjab (state), Pakistan, sample year: unknown, country: Pakistan[751]

- Co-contamination: not reported
- Further contamination of plants and plant products other than cereals and their products reported in the present book/article[751]: Sauces (chilli sauce), AF/S and OTA, literature[751]; Spices (cayenne pepper, chilli), AF/S, literature[751]

incidence: 5/13*, conc. range: 0.64–3.85 μg/kg, sample origin: (domestic) the internet shopping mall, supermarket, and wholesale market, South Korea, sample year: unknown, country: Korea[809], *red pepper, packed, dried, and ground

incidence: 0/45*, conc.: no contamination, sample origin: (domestic) the internet shopping mall, supermarket, and wholesale market, South Korea, sample year: unknown, country: Korea[809], *red pepper, non-packed, dried, and ground

incidence: 1/4*, conc.: 0.23 μg/kg, sample origin: imported from China purchased in the internet shopping mall, supermarket, and wholesale market, South Korea, sample year: unknown, country: Korea[809], *red pepper, packed, dried, and ground

incidence: 1/6*, conc.: 0.15 μg/kg, sample origin: imported from China purchased in the internet shopping mall, supermarket, and wholesale market, South Korea, sample year: unknown, country: Korea[809], *red pepper, non-packed, dried, and ground

incidence: 0/1*, conc.: no contamination, sample origin: imported from Vietnam purchased in the internet shopping mall, supermarket, and wholesale market, South Korea, sample year: unknown, country: Korea[809], *red pepper, packed, dried, and ground incidence: 1/4*, conc.: 0. 20 µg/kg, sample origin: imported from Vietnam purchased in the internet shopping mall, supermarket, and wholesale market, South Korea, sample year: unknown, country: Korea[809], *red pepper, non-packed, dried, and ground

incidence: 5/5*, conc. range: 3.75–56.30 µg/kg, ∅ conc.: 17.31 µg/kg, sample origin: imported from India purchased in the internet shopping mall, supermarket, and wholesale market, South Korea, sample year: unknown, country: Korea[809], *red pepper, packed, dried, and ground

incidence: 0/0*, conc.: no sa. available, sample origin: imported from India purchased in the internet shopping mall, supermarket, and wholesale market, South Korea, sample year: unknown, country: Korea[809], *red pepper, non-packed, dried, and ground

- Co-contamination: not reported
- Further contamination of plants and plant products other than cereals and their products reported in the present book/article[809]: Spices (cayenne pepper, chilli), AF/S, literature[809]

incidence: 11/12*, conc. range: ≤32.7 µg/kg, sample origin: 12 regions, Czech Republic, sample year: 2011–2013, country: Czech Republic[811], *chilli pepper, dried

- Co-contamination: not reported
- Further contamination of plants and plant products other than cereals and their products reported in the present book/article[811]: Beans, Chocolate, Coffee, Confectionery, Figs, Herbs and Spices, Juices, Licorice, Powder (cocoa powder), Products (ginger product), Raisins, Spices (caraway), Spices (coriander), Spices (ginger), Spices (mustard), Spices (nutmeg), Spices (paprika), Spices (pepper), Tea, Wine (red), Wine (white), OTA, literature[811]

incidence: 12/30*, conc. range: LOD-2.0 µg/kg (4 sa.), 2.0–5.0 µg/kg (5 sa.), 5.0–15.0 µg/kg (2 sa.), >15.0 µg/kg (1 sa.), ∅ conc.: 8.45 µg/kg, sample origin: marketed in Italy, sample year: unknown, country: Italy[820], *chilli

- Co-contamination: not reported
- Further contamination of plants and plant products other than cereals and their products reported in the present book/article[820]: Coffee, Wine (red), OTA, literature[820]

incidence: 44/44*, conc. range: 1–5 µg/kg (31 sa.), 5–15 µg/kg (9 sa.), 15–30 µg/kg (2 sa.), >30 µg/kg (2 sa., maximum: 31.7 µg/kg), ∅ conc.: 5.3 µg/kg, sample origin: supermarkets, free markets, traditional bazaars, and traditional family reserves in Istanbul (capital), Turkey, sample year: 2012/2013, country: Turkey[861], *red pepper flakes, unpacked

incidence: 27/31*, conc. range: 0.05–1 µg/kg, sample origin: supermarkets, free markets, traditional bazaars, and traditional family reserves in Istanbul (capital), Turkey, sample year: 2012/2013, country: Turkey[861], *red pepper flakes, packed

- Co-contamination: not reported
- Further contamination of plants and plant products other than cereals and their products reported in the present book/article[861]: not reported

incidence: 16/80*, conc. range: 0.41–30.73 µg/kg, ∅ conc.: 6.77 µg/kg, sample origin: supermarkets, shops, and market stalls in eight cities, China, sample year: April–July 2009, country: China[885], *chilli

- Co-contamination: not reported
- Further contamination of plants and plant products other than cereals and their products reported in the present book/article[885]: Spices (anise), Spices (cinnamon), Spices (cumin), Spices (curry), Spices (prickly ash), OTA, literature[885]

incidence: 2/18*, conc. range: <LOQ-5.3 µg/kg, sample origin: open markets, Sri Lanka, sample year: 2012/2013, country: Belgium[889], *chilli pods, whole

incidence: 13/26*, conc. range: <LOQ-15 µg/kg, ∅ conc.: 12.6 µg/kg, sample origin: open markets, Sri Lanka, sample year: 2012/2013, country: Belgium[889], *chilli flakes

incidence: 20/42*, conc. range: <LOQ-282 µg/kg, ∅ conc.: 37.8 µg/kg, sample origin: open markets, Sri Lanka, sample year: 2012/2013, country: Belgium[889], *chilli powder

incidence: 16/35*, conc. range: <LOQ-17.9 µg/kg, ∅ conc.: 9.93 µg/kg, sample origin: exotic shops and supermarkets in Antwerp and Ghent (cities), Belgium, sample year: 2012/2013, country: Belgium[889], *included whole chilli pods, chilli powders, and chilli flakes

- Co-contamination: For detailed information see the article.
- Further contamination of plants and plant products other than cereals and their products reported in the present book/article[889]: Spices (cayenne pepper, chilli), AFB_1, AFB_2, AFG_1, AFG_2, AME, CIT, FB_2, and STG, literature[889]

For detailed information see the article.

incidence: 6/10*, conc. range: <LOQ (2 sa.), 7.0–27 µg/kg (4 sa.), sample origin: markets, Sri Lanka, sample year: unknown, country: Belgium[891], *red chilli

- Co-contamination: not reported
- Further contamination of plants and plant products other than cereals and their products reported in the present book/article[891]: Spices (cayenne pepper, chilli), AFB_1, AFB_2, FB_2, and STG, literature[891]; Spices (pepper), AFB_1, AFB_2, AFG_2, CIT, FB_1, OTA, and STG, literature[891]

incidence: 6/56*, conc. range: <LOQ-4.51 µg/kg, ∅ conc.: 2.38 µg/kg, sample origin: imported from China and Vietnam as well as Korea and purchased in local markets, Korea, sample year: unknown, country: Korea[901], *red pepper powder

- Co-contamination: not reported
- Further contamination of plants and plant products other than cereals and their products reported in the present book/article[901]: Spices (cayenne pepper, chilli), AFB_1, literature[901]

Fusarium Toxins

Fumonisin/s

Fumonisin B_1

incidence: 7/14*, conc. range: 20.2–140 µg/kg, ∅ conc.: 99.9 µg/kg, sample origin: FUCHS Gewürze GmbH, Latvia, sample year: 2014, country: Latvia[679], *hot paprika, ground

- Co-contamination: not reported
- Further contamination of plants and plant products other than cereals and their products reported in the present book/article[679]: Spices (nutmeg), FB_1 and OTA, literature[679]; Spices (oregano), STG, literature[679]; Spices (paprika), FB_1, OTA, and STG, literature[679]; Spices (thyme), DON, STG, and ZEA, literature[679]

Fumonisin B_2

incidence: 1/18*, conc.: 87.1 µg/kg, sample origin: open markets, Sri Lanka, sample year: 2012/2013, country: Belgium[889], *chilli pods, whole

incidence: 9/26*, conc. range: <LOQ, ∅ conc.: <LOQ, sample origin: open markets, Sri Lanka, sample year: 2012/2013, country: Belgium[889], *chilli flakes

incidence: 3/42*, conc. range: <LOQ, ∅ conc.: LOQ, sample origin: open markets, Sri Lanka, sample year: 2012/2013, country: Belgium[889], *chilli powder

incidence: 4/35*, conc. range: <LOQ-53.8 µg/kg, sample origin: exotic shops and supermarkets in Antwerp and Ghent (cities), Belgium, sample year: 2012/2013, country: Belgium[889], *included whole chilli pods, chilli powders, and chilli flakes

- Co-contamination: For detailed information see the article.
- Further contamination of plants and plant products other than cereals and their products reported in the present book/article[889]: Spices (cayenne pepper, chilli), AFB_1, AFB_2, AFG_1, AFG_2, AME, CIT, OTA, and STG, literature[889]

For detailed information see the article.

incidence: 5/10*, conc. range: <LOQ, sample origin: markets, Sri Lanka, sample year: unknown, country: Belgium[891], *red chilli

- Co-contamination: not reported
- Further contamination of plants and plant products other than cereals and their products reported in the present book/article[891]: Spices (cayenne pepper, chilli), AFB_1, AFB_2, FB_2, and STG, literature[891]; Spices (pepper), AFB_1, AFB_2, AFG_2, CIT, FB_1, OTA, and STG, literature[891]

Type A Trichothecenes

Diacetoxyscirpenol

incidence: 4?/4*, conc. range: 47–81 µg/kg, sample origin: specialist food shops, UK, sample year: unknown, country: UK[16], *chilli powder

- Co-contamination: not reported
- Further contamination of plants and plant products other than cereals and their products reported in the present book/article[16]: Food, AF/S, DON, FUS-X, NIV, OTA, and ZEA, literature[16]; Nuts (almond), AF/S, literature[16]; Oils, Spices (fennel), AF/S and ZEA, literature[16]; Oils (sesame oil), OTA, literature[16]; Paste (curry paste), AF/S, FB/S, NIV, OTA, and ZEA, literature[16]; Pickles, AF/S, FB/S, OTA, and ZEA, literature[16]; Sauces (chilli sauce), NIV, OTA, and ZEA, literature[16]; Spices, AF/S, FB/S, OTA, and ZEA, literature[16]; Spices (cayenne pepper, chilli), AF/S, DON, HT-2, OTA, and ZEA, literature[16]; Spices (coriander), AF/S, DON, OTA, and ZEA, literature[16]; Spices (curry), AF/S, DAS, FB/S, FUS-X, HT-2, NEO, NIV, OTA, T-2, and ZEA, literature[16]; Spices (garlic), DON, FUS-X, and NIV, literature[16]; Spices (ginger), AF/S, DON, NEO, NIV, OTA, and T-2, literature[16]; Spices (tandoori), AF/S, FB/S, NIV, OTA, and T-2, literature[16]

HT-2 Toxin

incidence: 1/4*, conc.: 24 µg/kg, sample origin: specialist food shops, UK, sample year: unknown, country: UK[16], *chilli powder

- Co-contamination: not reported
- Further contamination of plants and plant products other than cereals and their products reported in the present book/article[16]: Food, AF/S, DON, FUS-X, NIV, OTA, and ZEA, literature[16]; Nuts (almond), AF/S, literature[16]; Oils, Spices (fennel), AF/S and ZEA, literature[16]; Oils (sesame oil), OTA, literature[16]; Paste (curry paste), AF/S, FB/S, NIV, OTA, and ZEA, literature[16]; Pickles, AF/S, FB/S, OTA, and ZEA, literature[16]; Sauces (chilli sauce), NIV, OTA, and ZEA, literature[16]; Spices, AF/S, FB/S, OTA, and ZEA, literature[16]; Spices (cayenne pepper, chilli), AF/S, DAS, DON, OTA, and ZEA, literature[16]; Spices (coriander), AF/S, DON, OTA, and ZEA, literature[16]; Spices (curry), AF/S, DAS, FB/S, FUS-X, HT-2, NEO, NIV, OTA, T-2, and ZEA, literature[16]; Spices (garlic), DON, FUS-X, and NIV, literature[16]; Spices (ginger), AF/S, DON, NEO, NIV, OTA, and T-2, literature[16]; Spices (tandoori), AF/S, FB/S, NIV, OTA, and T-2, literature[16]

T-2 Toxin

incidence: 2/11*, conc. range: ≤0.2 µg/kg, sample origin: distributor companies, Spain, sample year: 2008–2010, country: Spain[253], *chilli

- Co-contamination: 2 sa. co-contaminated with AFS, DON, OTA, T-2, and ZEA
- Further contamination of plants and plant products other than cereals and their products reported in the present book/article[253]: Spices (cayenne pepper, chilli), Spices (paprika), AF/S, DON, OTA, and ZEA, literature[253]

Type B Trichothecenes

Deoxynivalenol

incidence: 1/4*, conc.: 8 µg/kg, sample origin: specialist food shops, UK, sample year: unknown, country: UK[16], *chilli powder

- Co-contamination: not reported
- Further contamination of plants and plant products other than cereals and their products reported in the present book/article[16]: Food, AF/S, DON, FUS-X, NIV, OTA, and ZEA, literature[16]; Nuts (almond), AF/S, literature[16]; Oils, Spices (fennel), AF/S and ZEA, literature[16]; Oils (sesame oil), OTA, literature[16]; Paste (curry paste), AF/S, FB/S, NIV, OTA, and ZEA, literature[16]; Pickles, AF/S, FB/S, OTA, and ZEA, literature[16]; Sauces (chilli sauce), NIV, OTA, and ZEA, literature[16]; Spices, AF/S, FB/S, OTA, and ZEA, literature[16]; Spices (cayenne pepper, chilli), AF/S, DAS, HT-2, OTA, and ZEA, literature[16]; Spices (coriander), AF/S, DON, OTA, and ZEA, literature[16]; Spices (curry), AF/S, DAS, FB/S, FUS-X, HT-2, NEO, NIV, OTA, T-2, and ZEA, literature[16]; Spices (garlic), DON, FUS-X, and NIV, literature[16]; Spices (ginger), AF/S, DON, NEO, NIV, OTA, and T-2, literature[16]; Spices (tandoori), AF/S, FB/S, NIV, OTA, and T-2, literature[16]

incidence: 2/11*, conc. range: ≤0.3 µg/kg, sample origin: distributor companies, Spain, sample year: 2008–2010, country: Spain[253], *chilli

- Co-contamination: 2 sa. co-contaminated with AFS, DON, OTA, T-2, and ZEA
- Further contamination of plants and plant products other than cereals and their products reported in the present book/article[253]: Spices (cayenne pepper, chilli), AF/S, OTA, T-2, and ZEA, literature[253]; Spices (paprika), AF/S, DON, OTA, and ZEA, literature[253]

Other ***Fusarium*** Toxins

Zearalenone

incidence: 4?/4*, conc. range: 1.5–15.4 µg/kg, sample origin: specialist food shops, UK, sample year: unknown, country: UK[16], *chilli powder

- Co-contamination: not reported
- Further contamination of plants and plant products other than cereals and their products reported in the present book/article[16]: Food, AF/S, DON, FUS-X, NIV, OTA, and ZEA, literature[16]; Nuts (almond), AF/S, literature[16]; Oils, Spices (fennel), AF/S and ZEA, literature[16]; Oils (sesame oil), OTA, literature[16]; Paste (curry paste), AF/S, FB/S, NIV, OTA, and ZEA, literature[16];

Pickles, AF/S, FB/S, OTA, and ZEA, literature[16]; Sauces (chilli sauce), NIV, OTA, and ZEA, literature[16]; Spices, AF/S, FB/S, OTA, and ZEA, literature[16]; Spices (cayenne pepper, chilli), AF/S, DAS, DON, HT-2, and OTA, literature[16]; Spices (coriander), AF/S, DON, OTA, and ZEA, literature[16]; Spices (curry), AF/S, DAS, FB/S, FUS-X, HT-2, NEO, NIV, OTA, T-2, and ZEA, literature[16]; Spices (garlic), DON, FUS-X, and NIV, literature[16]; Spices (ginger), AF/S, DON, NEO, NIV, OTA, and T-2, literature[16]; Spices (tandoori), AF/S, FB/S, NIV, OTA, and T-2, literature[16]

incidence: 2/9, conc. range: pr, sample origin: major retail outlets in Nainital (district), India, sample year: February 1984–December 1986, country: India[22], *chillies

- Co-contamination: not reported
- Further contamination of plants and plant products other than cereals and their products reported in the present book/article[22]: Spices (ammi), AFB_1, AFB_2, and AFG_1, literature[22]; Spices (cardamom), AFB_1, AFB_2, AFG_1, AFG_2, CIT, and OTA, literature[22]; Spices (cassia), AFB_1, AFB_2, AFG_1, and AFG_2, literature[22]; Spices (cayenne pepper, chilli), AFB_1, AFB_2, AFG_1, and AFG_2, literature[22]; Spices (coriander), AFB_1, AFB_2, AFG_1, AFG_2, CIT, OTA, and ZEA, literature[22]; Spices (cumin), AFB_1, AFB_2, AFG_1, AFG_2, and CIT, literature[22]; Spices (curcuma), AFB_1, AFB_2, AFG_1, AFG_2, CIT, and RBT, literature[22]; Spices (fennel), AFB_1, AFB_2, AFG_1, AFG_2, CIT, OTA, and STG, literature[22]; Spices (garlic), AFB_1, AFB_2, and AFG_1, literature[22]; Spices (pepper), AFB_1, AFB_2, AFG_1, AFG_2, CIT, STG, and ZEA, literature[22]

incidence: 5/10*, conc. range: 2–17 μg/kg, sample origin: food stores, Germany, sample year: 2000/2001, country: Germany[151], *red pepper

- Co-contamination: not reported
- Further contamination of plants and plant products other than cereals and their products reported in the present book/article[151]: Beans, ZEA, literature[151]; Nuts (hazelnut), Seeds (sunflower), HT-2, T-2, and ZEA, literature[151]; Products (potato product), DAS, HT-2, MAS, SCIRP, and ZEA, literature[151]; Seeds (pumpkin), HT-2 and ZEA, literature[151]

incidence: 15/35*, conc. range: ≤129 μg/kg**, sample origin: retail shops, Catalonia (autonomous community), Spain, sample year: June–August 2008, country: Spain[252], *chilli, **AFB_1, AFB_2, AFG_1, and AFG_2

- Co-contamination: For detailed information see the article.
- Further contamination of plants and plant products other than cereals and their products reported in the present book/article[252]: Spices (cayenne pepper, chilli), AF/S and OTA, literature[252]; Spices (paprika), AF/S, OTA, and ZEA, literature[252]

incidence: 4/11*, conc. range: ≤114.3 μg/kg, sample origin: distributor companies, Spain, sample year: 2008–2010, country: Spain[253], *chilli

- Co-contamination: 2 sa. co-contaminated with AFS, DON, OTA, T-2, and ZEA; 2 sa. co-contaminated with AFS, OTA, and ZEA
- Further contamination of plants and plant products other than cereals and their products reported in the present book/article[253]: Spices (cayenne pepper, chilli), AF/S, DON, OTA, and T-2, literature[253]; Spices (paprika), AF/S, DON, OTA, and ZEA, literature[253]

see also **Spices, Spices (paprika), Spices (pepper)**

Spices (celery) may contain the following mycotoxins:

Aspergillus Toxins

Aflatoxin G_1

incidence: 1/9*, conc.: 3.7 μg/kg, sample origin: imported from India, Indonesia, and Malaysia, USA, sample year: unknown, country: USA[185], *celery seeds

- Co-contamination: not reported
- Further contamination of plants and plant products other than cereals and their products reported in the present book/article[185]: Spices (nutmeg), AFB_1 and AFB_2, literature[185]; Spices (pepper), AFG_1 and AFG_2, literature[185]

Spices (cinnamon) may contain the following mycotoxins:

Aspergillus Toxins

Aflatoxin B_1

incidence: 1/1, conc.: 0.98 μg/kg, sample origin: public markets, supermarkets, shops, and bonded warehouse in Emilia Romagna (administrative region), Italy, sample year: 2000–2005, country: Italy[176]

- Co-contamination: not reported
- Further contamination of plants and plant products other than cereals and their products reported in the present book/article[176]: Spices (cayenne pepper, chilli), AFB_1, AFB_2, AFG_1, and AFG_2, literature[176]; Spices (nutmeg), AFB_1 and AFB_2, literature[176]

incidence: 5/5*, conc. range: 49.4–53 μg/kg, ∅ conc.: 51.6 μg/kg, sample origin: organic markets and organic shops, Turkey, sample year: May 2010–May 2011, country: Turkey[859], *organic cinnamon sa.

- Co-contamination: not reported
- Further contamination of plants and plant products other than cereals and their products reported in the present book/article[859]: Herbs (camomile), Herbs (linden), Herbs (rosehip), Herbs (sage), Spices (anise), Spices (basil), Spices (cayenne pepper, chilli), Spices (coriander), Spices (cumin), Spices (fennel), Spices (ginger), Spices (laurel leaves), Spices (pepper), Spices (peppermint), Spices (poppy seed), Spices (rosemary), Spices (sumac), AFB_1, literature[859]

Aspergillus and ***Penicillium*** Toxins

Ochratoxin A

incidence: 7/13, conc. range: ≤2.21 μg/kg, ∅ conc.: 0.60 μg/kg, sample origin: retail stores in Lower Saxony (federal state), Germany, sample year: 2009, country: Germany[829]

- Co-contamination: not reported
- Further contamination of plants and plant products other than cereals and their products reported in the present book/article[829]: Nuts (hazelnut), AFB_1 and AF/S, literature[829]; Nuts (pistachio nut), AFB_1, literature[829]; Oils (maize oil), ZEA, literature[829]; Spices (curry), Spices (nutmeg), AFB_1, AF/S, and OTA, literature[829]; Spices (ginger), AFB_1 and AF/S, literature[829]

incidence: 4/80, conc. range: 0.55–2.02 μg/kg, ∅ conc.: 1.10 μg/kg, sample origin: supermarkets, shops, and market stalls in eight cities, China, sample year: April–July 2009, country: China[885]

- Co-contamination: not reported
- Further contamination of plants and plant products other than cereals and their products reported in the present book/article[885]: Spices (anise), Spices (cayenne pepper, chilli), Spices (cumin), Spices (curry), Spices (prickly ash), OTA, literature[885]

Spices (coriander) may contain the following mycotoxins:

Alternaria Toxins

Tenuazonic Acid

incidence: 2/2, conc. range: 1,500–1,700 μg/kg, ⌀ conc.: 1,600 μg/kg, sample origin: unknown, sample year: unknown, country: Germany[166]

- Co-contamination: not reported
- Further contamination of plants and plant products other than cereals and their products reported in the present book/article[166]: Juices, Juices (apple juice), Juices (apricot juice), Juices (banana juice), Juices (blackcurrant juice), Juices (cherry juice), Juices (grape juice), Juices (orange juice), Juices (pear juice), Juices (pomegranate juice), Juices (strawberry juice), Spices, Spices (cumin), Spices (curcuma), Spices (curry), Spices (ginger), Spices (paprika), Spices (pepper), TA, literature[166]

Aspergillus Toxins

Aflatoxin B_1

incidence: 6?/9, conc. range: 25–230 μg/kg, sample origin: major retail outlets in Nainital (district), India, sample year: January 1984–December 1986, country: India[22]

- Co-contamination: not reported
- Further contamination of plants and plant products other than cereals and their products reported in the present book/article[22]: Spices (ammi), AFB_1, AFB_2, and AFG_1, literature[22]; Spices (cardamom), AFB_1, AFB_2, AFG_1, AFG_2, CIT, and OTA, literature[22]; Spices (cassia), AFB_1, AFB_2, AFG_1, and AFG_2, literature[22]; Spices (cayenne pepper, chilli), AFB_1, AFB_2, AFG_1, AFG_2, and ZEA, literature[22]; Spices (coriander), AFB_2, AFG_1, AFG_2, CIT, OTA, and ZEA, literature[22]; Spices (cumin), AFB_1, AFB_2, AFG_1, AFG_2, and CIT, literature[22]; Spices (curcuma), AFB_1, AFB_2, AFG_1, AFG_2, CIT, and RBT, literature[22]; Spices (fennel), AFB_1, AFB_2, AFG_1, AFG_2, CIT, OTA, and STG, literature[22]; Spices (garlic), AFB_1, AFB_2, and AFG_1, literature[22]; Spices (pepper), AFB_1, AFB_2, AFG_1, AFG_2, CIT, STG, and ZEA, literature[22]

incidence: 2?/15, ⌀ conc.: 8 μg/kg, sample origin: local companies in Cairo (capital), Egypt, sample year: unknown, country: Egypt[104]

- Co-contamination: not reported
- Further contamination of plants and plant products other than cereals and their products reported in the present book/article[104]: Kubeba, AFB_1 and AFB_2, literature[104]; Spices (coriander), AFG_1, literature[104]; Spices (pepper), AFB_1, literature[104]; Spices (curcuma), AFB_1 and AFG_1, literature[104]

incidence: 2/12, conc. range: ≤5 μg/kg (1 sa.), 5.2 μg/kg (1 sa.), sample origin: unknown, sample year: unknown, country: Germany/Thailand[244]

- Co-contamination: not reported
- Further contamination of plants and plant products other than cereals and their products reported in the present book/article[244]: Spices (cayenne pepper, chilli), Spices (nutmeg), AFB_1, literature[244]

incidence: 3/10*, conc. range: 19–37 μg/kg, ⌀ conc.: 25.7 μg/kg, sample origin: unknown, sample year: unknown, country: India[322], *not stored

- Co-contamination: 1 sa. co-contaminated with AFB_1, AFB_2, and AFG_1; 2 sa. co-contaminated with AFB_1 and AFG_1
- Further contamination of plants and plant products other than cereals and their products reported in the present book/article[322]: Spices (coriander), AFB_2 and AFG_1, literature[322]

For detailed information see the article.

incidence: 1/9, conc.: 45.5 μg/kg*, sample origin: processors located in six Canadian cities, Canada, sample year: unknown, country: Canada[323], *approximate conc.

- Co-contamination: 1 sa. co-contaminated with AFB_1 and AFB_2
- Further contamination of plants and plant products other than cereals and their products reported in the present book/article[323]: Spices (coriander), AFB_2, literature[323]; Spices (curcuma), Spices (nutmeg), AFB_1 and AFB_2, literature[323]; Spices (ginger), AFB_1, AFB_2, AFG_1, and AFG_2, literature[323]

incidence: 1/1*, conc.: 7.68 μg/kg, sample origin: Penang (state, island), Malaysia, sample year: unknown (first study), country: Malaysia[530], *"ketumbar"

- Co-contamination: 1 sa. co-contaminated with AFB_1, AFB_2, AFG_1, AFG_2, and OTA
- Further contamination of plants and plant products other than cereals and their products reported in the present book/article[530]: Spices, Spices (cayenne pepper, chilli), Spices (cumin), Spices (curry), Spices (fennel), Spices (masala), AFB_1, AFB_2, AFG_1, AFG_2, and OTA, literature[530]; Spices (coriander), AFB_2, AFG_1, AFG_2, and OTA, literature[530]; Spices (curcuma), AFB_1, AFB_2, AFG_1, and OTA, literature[530]; Spices (pepper), AFB_1 and AFB_2, literature[530]; Spices (poppy seed), AFB_1, literature[530]

incidence: 1/1, conc.: 10.97 μg/kg, sample origin: Penang (state, island), Malaysia, sample year: unknown (second study), country: Malaysia[530]

- Co-contamination: 1 sa. co-contaminated with AFB_1, AFB_2, and AFG_1
- Further contamination of plants and plant products other than cereals and their products reported in the present book/article[530]: Spices, Spices (cayenne pepper, chilli), Spices (cumin), Spices (curry), Spices (fennel), Spices (masala), AFB_1, AFB_2, AFG_1, AFG_2, and OTA, literature[530]; Spices (coriander), AFB_2, AFG_1, AFG_2, and OTA, literature[530]; Spices (curcuma), AFB_1, AFB_2, AFG_1, and OTA, literature[530]; Spices (pepper), AFB_1 and AFB_2, literature[530]; Spices (poppy seed), AFB_1, literature[530]

incidence: 1/2*, conc.: 15.6 μg/kg, sample origin: organic markets and organic shops, Turkey, sample year: May 2010–May 2011, country: Turkey[859], *organic coriander sa.

- Co-contamination: not reported
- Further contamination of plants and plant products other than cereals and their products reported in the present book/article[859]: Herbs (camomile), Herbs (linden), Herbs (rosehip), Herbs (sage), Spices (anise), Spices (basil), Spices (cayenne pepper, chilli), Spices (cinnamon), Spices (cumin), Spices (fennel), Spices (ginger), Spices (laurel leave), Spices (pepper), Spices (peppermint), Spices (poppy seed), Spices (rosemary), Spices (sumac), AFB_1, literature[859]

Aflatoxin B_2

incidence: 6?/9, conc. range: 20–72 μg/kg, sample origin: major retail outlets in Nainital (district), India, sample year: January 1984–December 1986, country: India[22]

- Co-contamination: not reported

- Further contamination of plants and plant products other than cereals and their products reported in the present book/article[22]: Spices (ammi), AFB_1, AFB_2, and AFG_1, literature[22]; Spices (cardamom), AFB_1, AFB_2, AFG_1, AFG_2, CIT, and OTA, literature[22]; Spices (cassia), AFB_1, AFB_2, AFG_1, and AFG_2, literature[22]; Spices (cayenne pepper, chilli), AFB_1, AFB_2, AFG_1, AFG_2, and ZEA, literature[22]; Spices (coriander), AFB_1, AFG_1, AFG_2, CIT, OTA, and ZEA, literature[22]; Spices (cumin), AFB_1, AFB_2, AFG_1, AFG_2, and CIT, literature[22]; Spices (curcuma), AFB_1, AFB_2, AFG_1, AFG_2, CIT, and RBT, literature[22]; Spices (fennel), AFB_1, AFB_2, AFG_1, AFG_2, CIT, OTA, and STG, literature[22]; Spices (garlic), AFB_1, AFB_2, and AFG_1, literature[22]; Spices (pepper), AFB_1, AFB_2, AFG_1, AFG_2, CIT, STG, and ZEA, literature[22]

incidence: 1/10*, conc.: 5 µg/kg, sample origin: unknown, sample year: unknown, country: India[322], *not stored

- Co-contamination: 1 sa. co-contaminated with AFB_1, AFB_2, and AFG_1
- Further contamination of plants and plant products other than cereals and their products reported in the present book/article[322]: Spices (coriander), AFB_1 and AFG_1, literature[322]

For detailed information see the article.

incidence: 1/9, conc.: 16 µg/kg*, sample origin: processors located in six Canadian cities, Canada, sample year: unknown, country: Canada[323], *approximate conc.

- Co-contamination: 1 sa. co-contaminated with AFB_1 and AFB_2
- Further contamination of plants and plant products other than cereals and their products reported in the present book/article[323]: Spices (coriander), AFB_1, literature[323]; Spices (curcuma), Spices (nutmeg), AFB_1 and AFB_2, literature[323]; Spices (ginger), AFB_1, AFB_2, AFG_1, and AFG_2, literature[323]

incidence: 1/1*, conc.: 1.16 µg/kg, sample origin: Penang (state, island), Malaysia, sample year: unknown (first study), country: Malaysia[530], *"ketumbar"

- Co-contamination: 1 sa. co-contaminated with AFB_1, AFB_2, AFG_1, AFG_2, and OTA
- Further contamination of plants and plant products other than cereals and their products reported in the present book/article[530]: Spices, Spices (cayenne pepper, chilli), Spices (cumin), Spices (curry), Spices (fennel), Spices (masala), AFB_1, AFB_2, AFG_1, AFG_2, and OTA, literature[530]; Spices (coriander), AFB_1, AFG_1, AFG_2, and OTA, literature[530]; Spices (curcuma), AFB_1, AFB_2, AFG_1, and OTA, literature[530]; Spices (pepper), AFB_1 and AFB_2, literature[530]; Spices (poppy seed), AFB_1, literature[530]

incidence: 1/1, conc.: 1.57 µg/kg, sample origin: Penang (state, island), Malaysia, sample year: unknown (second study), country: Malaysia[530]

- Co-contamination: 1 sa. co-contaminated with AFB_1, AFB_2, and AFG_1
- Further contamination of plants and plant products other than cereals and their products reported in the present book/article[530]: Spices, Spices (cayenne pepper, chilli), Spices (cumin), Spices (curry), Spices (fennel), Spices (masala), AFB_1, AFB_2, AFG_1, AFG_2, and OTA, literature[530]; Spices (coriander), AFB_1, AFG_1, AFG_2, and OTA, literature[530]; Spices (curcuma), AFB_1, AFB_2, AFG_1, and OTA, literature[530]; Spices (pepper), AFB_1 and AFB_2, literature[530]; Spices (poppy seed), AFB_1, literature[530]

Aflatoxin G_1

incidence: 6?/9, conc. range: 13–40 µg/kg, sample origin: major retail outlets in Nainital (district), India, sample year: January 1984–December 1986, country: India[22]

- Co-contamination: not reported
- Further contamination of plants and plant products other than cereals and their products reported in the present book/article[22]: Spices (ammi), AFB_1, AFB_2, and AFG_1, literature[22]; Spices (cardamom), AFB_1, AFB_2, AFG_1, AFG_2, CIT, and OTA, literature[22]; Spices (cassia), AFB_1, AFB_2, AFG_1, and AFG_2, literature[22]; Spices (cayenne pepper, chilli), AFB_1, AFB_2, AFG_1, AFG_2, and ZEA, literature[22]; Spices (coriander), AFB_1, AFB_2, AFG_2, CIT, OTA, and ZEA, literature[22]; Spices (cumin), AFB_1, AFB_2, AFG_1, AFG_2, and CIT, literature[22]; Spices (curcuma), AFB_1, AFB_2, AFG_1, AFG_2, CIT, and RBT, literature[22]; Spices (fennel), AFB_1, AFB_2, AFG_1, AFG_2, CIT, OTA, and STG, literature[22]; Spices (garlic), AFB_1, AFB_2, and AFG_1, literature[22]; Spices (pepper), AFB_1, AFB_2, AFG_1, AFG_2, CIT, STG, and ZEA, literature[22]

incidence: 2?/15, ∅ conc.: 2 µg/kg, sample origin: local companies in Cairo (capital), Egypt, sample year: unknown, country: Egypt[104]

- Co-contamination: not reported
- Further contamination of plants and plant products other than cereals and their products reported in the present book/article[104]: Kubeba, AFB_1 and AFB_2, literature[104]; Spices (coriander), Spices (pepper), AFB_1, literature[104]; Spices (curcuma), AFB_1 and AFG_1, literature[104]

incidence: 3/10*, conc. range: 3–4 µg/kg, ∅ conc.: 3.7 µg/kg, sample origin: unknown, sample year: unknown, country: India[322], *not stored

- Co-contamination: 1 sa. co-contaminated with AFB_1, AFB_2, and AFG_1; 2 sa. co-contaminated with AFB_1 and AFG_1
- Further contamination of plants and plant products other than cereals and their products reported in the present book/article[322]: Spices (coriander), AFB_1 and AFB_2, literature[322]

For detailed information see the article.

incidence: 1/1*, conc.: 0.43 µg/kg, sample origin: Penang (state, island), Malaysia, sample year: unknown (first study), country: Malaysia[530], *"ketumbar"

- Co-contamination: 1 sa. co-contaminated with AFB_1, AFB_2, AFG_1, AFG_2, and OTA
- Further contamination of plants and plant products other than cereals and their products reported in the present book/article[530]: Spices, Spices (cayenne pepper, chilli), Spices (cumin), Spices (curry), Spices (fennel), Spices (masala), AFB_1, AFB_2, AFG_1, AFG_2, and OTA, literature[530]; Spices (coriander), AFB_1, AFB_2, AFG_2, and OTA, literature[530]; Spices (curcuma), AFB_1, AFB_2, AFG_1, and OTA, literature[530]; Spices (pepper), AFB_1 and AFB_2, literature[530]; Spices (poppy seed), AFB_1, literature[530]

incidence: 1/1, conc.: 0.84 µg/kg, sample origin: Penang (state, island), Malaysia, sample year: unknown (second study), country: Malaysia[530]

- Co-contamination: 1 sa. co-contaminated with AFB_1, AFB_2, and AFG_1
- Further contamination of plants and plant products other than cereals and their products reported in the present book/article[530]: Spices, Spices (cayenne pepper, chilli), Spices (cumin), Spices (curry), Spices (fennel), Spices (masala), AFB_1, AFB_2, AFG_1, AFG_2, and OTA, literature[530]; Spices (coriander), AFB_1,

AFB_2, AFG_2, and OTA, literature[530]; Spices (curcuma), AFB_1, AFB_2, AFG_1, and OTA, literature[530]; Spices (pepper), AFB_1 and AFB_2, literature[530]; Spices (poppy seed), AFB_1, literature[530]

Aflatoxin G_2

incidence: 6?/9, conc. range: 14–35 µg/kg, sample origin: major retail outlets in Nainital (district), India, sample year: January 1984–December 1986, country: India[22]

- Co-contamination: not reported
- Further contamination of plants and plant products other than cereals and their products reported in the present book/article[22]: Spices (ammi), AFB_1, AFB_2, and AFG_1, literature[22]; Spices (cardamom), AFB_1, AFB_2, AFG_1, AFG_2, CIT, and OTA, literature[22]; Spices (cassia), AFB_1, AFB_2, AFG_1, and AFG_2, literature[22]; Spices (cayenne pepper, chilli), AFB_1, AFB_2, AFG_1, AFG_2, and ZEA, literature[22]; Spices (coriander), AFB_1, AFB_2, AFG_1, CIT, OTA, and ZEA, literature[22]; Spices (cumin), AFB_1, AFB_2, AFG_1, AFG_2, and CIT, literature[22]; Spices (curcuma), AFB_1, AFB_2, AFG_1, AFG_2, CIT, and RBT, literature[22]; Spices (fennel), AFB_1, AFB_2, AFG_1, AFG_2, CIT, OTA, and STG, literature[22]; Spices (garlic), AFB_1, AFB_2, and AFG_1, literature[22]; Spices (pepper), AFB_1, AFB_2, AFG_1, AFG_2, CIT, STG, and ZEA, literature[22]

incidence: 1/1*, conc.: 0.07 µg/kg, sample origin: Penang (state, island), Malaysia, sample year: unknown (first study), country: Malaysia[530], *"ketumbar"

- Co-contamination: 1 sa. co-contaminated with AFB_1, AFB_2, AFG_1, AFG_2, and OTA
- Further contamination of plants and plant products other than cereals and their products reported in the present book/article[530]: Spices, Spices (cayenne pepper, chilli), Spices (cumin), Spices (curry), Spices (fennel), Spices (masala), AFB_1, AFB_2, AFG_1, AFG_2, and OTA, literature[530]; Spices (coriander), AFB_1, AFB_2, AFG_1, and OTA, literature[530]; Spices (curcuma), AFB_1, AFB_2, AFG_1, and OTA, literature[530]; Spices (pepper), AFB_1 and AFB_2, literature[530]; Spices (poppy seed), AFB_1, literature[530]

incidence: 0/1, conc.: no contamination, sample origin: Penang (state, island), Malaysia, sample year: unknown (second study), country: Malaysia[530]

- Co-contamination: not reported
- Further contamination of plants and plant products other than cereals and their products reported in the present book/article[530]: Spices, Spices (cayenne pepper, chilli), Spices (cumin), Spices (curry), Spices (fennel), Spices (masala), AFB_1, AFB_2, AFG_1, AFG_2, and OTA, literature[530]; Spices (coriander), AFB_1, AFB_2, AFG_1, and OTA, literature[530]; Spices (curcuma), AFB_1, AFB_2, AFG_1, and OTA, literature[530]; Spices (pepper), AFB_1 and AFB_2, literature[530]; Spices (poppy seed), AFB_1, literature[530]

Aflatoxin/s

incidence: 1/3, conc.: 0.7 µg/kg*, sample origin: specialist food shops, UK, sample year: unknown, country: UK[16], *AFB_1, AFB_2, AFG_1, and AFG_2

- Co-contamination: not reported
- Further contamination of plants and plant products other than cereals and their products reported in the present book/article[16]: Food, AF/S, DON, FUS-X, NIV, OTA, and ZEA, literature[16]; Nuts (almond), AF/S, literature[16]; Oils, Spices (fennel), AF/S and ZEA, literature[16]; Oils (sesame oil), OTA, literature[16]; Paste (curry paste), AF/S, FB/S, NIV, OTA, and ZEA, literature[16]; Pickles, AF/S, FB/S, OTA, and ZEA, literature[16]; Sauces (chilli sauce), NIV, OTA, and ZEA, literature[16]; Spices, AF/S, FB/S, OTA, and ZEA, literature[16]; Spices (cayenne pepper, chilli), AF/S, DAS, DON, HT-2, OTA, and ZEA, literature[16]; Spices (coriander), DON, OTA, and ZEA, literature[16]; Spices (curry), AF/S, DAS, FB/S, FUS-X, HT-2, NEO, NIV, OTA, T-2, and ZEA, literature[16]; Spices (garlic), DON, FUS-X, and NIV, literature[16]; Spices (ginger), AF/S, DON, NEO, NIV, OTA, and T-2, literature[16]; Spices (tandoori), AF/S, FB/S, NIV, OTA, and T-2, literature[16]

incidence: 1/9, conc. range: 2.90*–12.80** µg/kg, sample origin: market stalls, shops, and supermarkets, Ireland, sample year: unknown, country: Ireland[187], *AFB_2, **AFB_1

- Co-contamination: not reported
- Further contamination of plants and plant products other than cereals and their products reported in the present book/article[187]: Spices (cayenne pepper, chilli), Spices (curcuma), Spices (curry), Spices (paprika), Spices (pepper), AF/S, literature[187]

incidence: 1/5, conc.: 8 µg/kg*, sample origin: Assiut (governorate), Egypt, sample year: unknown, country: Egypt[641], *AFB_1 and AFG_1

- Co-contamination: not reported
- Further contamination of plants and plant products other than cereals and their products reported in the present book/article[641]: Spices (anise), Spices (fennel), Spices (pepper), AF/S, literature[641]; Spices (black cumin), AF/S and CIT, literature[641]; Spices (caraway), Spices (marjoram), AF/S and STG, literature[641]; Spices (cayenne pepper, chilli), Spices (cumin), STG, literature[641]

Aspergillus and *Penicillium* Toxins

Citrinin

incidence: 1/9, conc.: 34 µg/kg, sample origin: major retail outlets in Nainital (district), India, sample year: January 1984–December 1986, country: India[22]

- Co-contamination: not reported
- Further contamination of plants and plant products other than cereals and their products reported in the present book/article[22]: Spices (ammi), AFB_1, AFB_2, and AFG_1, literature[22]; Spices (cardamom), AFB_1, AFB_2, AFG_1, AFG_2, CIT, and OTA, literature[22]; Spices (cassia), AFB_1, AFB_2, AFG_1, and AFG_2, literature[22]; Spices (cayenne pepper, chilli), AFB_1, AFB_2, AFG_1, AFG_2, and ZEA, literature[22]; Spices (coriander), AFB_1, AFB_2, AFG_1, AFG_2, OTA, and ZEA, literature[22]; Spices (cumin), AFB_1, AFB_2, AFG_1, AFG_2, and CIT, literature[22]; Spices (curcuma), AFB_1, AFB_2, AFG_1, AFG_2, CIT, and RBT, literature[22]; Spices (fennel), AFB_1, AFB_2, AFG_1, AFG_2, CIT, OTA, and STG, literature[22]; Spices (garlic), AFB_1, AFB_2, and AFG_1, literature[22]; Spices (pepper), AFB_1, AFB_2, AFG_1, AFG_2, CIT, STG, and ZEA, literature[22]

Ochratoxin A

incidence: 1/3, conc.: 4.0 µg/kg, sample origin: specialist food shops, UK, sample year: unknown, country: UK[16]

- Co-contamination: not reported
- Further contamination of plants and plant products other than cereals and their products reported in the present book/article[16]: Food, AF/S, DON, FUS-X, NIV, OTA, and ZEA, literature[16]; Nuts (almond), AF/S, literature[16]; Oils, Spices (fennel), AF/S and ZEA, literature[16]; Oils (sesame oil), OTA, literature[16]; Paste (curry paste), AF/S, FB/S, NIV, OTA, and ZEA, literature[16];

Pickles, AF/S, FB/S, OTA, and ZEA, literature[16]; Sauces (chilli sauce), NIV, OTA, and ZEA, literature[16]; Spices, AF/S, FB/S, OTA, and ZEA, literature[16]; Spices (cayenne pepper, chilli), AF/S, DAS, DON, HT-2, OTA, and ZEA, literature[16]; Spices (coriander), AF/S, DON, and ZEA, literature[16]; Spices (curry), AF/S, DAS, FB/S, FUS-X, HT-2, NEO, NIV, OTA, T-2, and ZEA, literature[16]; Spices (garlic), DON, FUS-X, and NIV, literature[16]; Spices (ginger), AF/S, DON, NEO, NIV, OTA, and T-2, literature[16]; Spices (tandoori), AF/S, FB/S, NIV, OTA, and T-2, literature[16]

incidence: 10/20, conc. range: 74–576 µg/kg, ∅ conc.: 206 µg/kg, sample origin: retail markets and supermarkets in 24 regions, Tunisia, sample year: unknown, country: Tunisia[20]
- Co-contamination: not reported
- Further contamination of plants and plant products other than cereals and their products reported in the present book/article[20]: Nuts (almond), Nuts (peanut), Nuts (pistachio nut), Spices (caraway), Spices (curcuma), Spices (paprika), Spices (pepper), OTA, literature[20]

incidence: 1/9, conc.: pr, sample origin: major retail outlets in Nainital (district), India, sample year: January 1984–December 1986, country: India[22]
- Co-contamination: not reported
- Further contamination of plants and plant products other than cereals and their products reported in the present book/article[22]: Spices (ammi), AFB_1, AFB_2, and AFG_1, literature[22]; Spices (cardamom), AFB_1, AFB_2, AFG_1, AFG_2, CIT, and OTA, literature[22]; Spices (cassia), AFB_1, AFB_2, AFG_1, and AFG_2, literature[22]; Spices (cayenne pepper, chilli), AFB_1, AFB_2, AFG_1, AFG_2, and ZEA, literature[22]; Spices (coriander), AFB_1, AFB_2, AFG_1, AFG_2, CIT, and ZEA, literature[22]; Spices (cumin), AFB_1, AFB_2, AFG_1, AFG_2, and CIT, literature[22]; Spices (curcuma), AFB_1, AFB_2, AFG_1, AFG_2, CIT, and RBT, literature[22]; Spices (fennel), AFB_1, AFB_2, AFG_1, AFG_2, CIT, OTA, and STG, literature[22]; Spices (garlic), AFB_1, AFB_2, and AFG_1, literature[22]; Spices (pepper), AFB_1, AFB_2, AFG_1, AFG_2, CIT, STG, and ZEA, literature[22]

incidence: 1/5, conc.: 0.93 µg/kg, sample origin: local supermarkets and small retail shops in seven major cities, Japan, sample year: winter 2007, country: Japan[85]
- Co-contamination: not reported
- Further contamination of plants and plant products other than cereals and their products reported in the present book/article[85]: Beans (soybean), Vegetables (asparagus), FB_1 and FB_2, literature[85]; Beverages, Chocolate, Cocoa, Coffee, Coffee beans, Raisins, Wine, OTA, literature[85]

incidence: 20/50*, conc. range: 10–29 µg/kg (16 sa.), 30–49 µg/kg (3 sa.), 50–100 µg/kg (1 sa.), sample origin: retail shops, India, sample year: unknown, country: India/UK/Belgium[191], *coriander seeds
- Co-contamination: not reported
- Further contamination of plants and plant products other than cereals and their products reported in the present book/article[191]: Spices (curcuma), Spices (ginger), Spices (pepper), OTA, literature[191]

incidence: 1/1*, conc.: 0.91 µg/kg, sample origin: Penang (state, island), Malaysia, sample year: unknown (first study), country: Malaysia[530], *"ketumbar"
- Co-contamination: 1 sa. co-contaminated with AFB_1, AFB_2, AFG_1, AFG_2, and OTA
- Further contamination of plants and plant products other than cereals and their products reported in the present book/article[530]: Spices, Spices (cayenne pepper, chilli), Spices (cumin), Spices (curry), Spices (fennel), Spices (masala), AFB_1, AFB_2, AFG_1, AFG_2, and OTA, literature[530]; Spices (coriander), AFB_1, AFB_2, AFG_1, and AFG_2, literature[530]; Spices (curcuma), AFB_1, AFB_2, AFG_1, and OTA, literature[530]; Spices (pepper), AFB_1 and AFB_2, literature[530]; Spices (poppy seed), AFB_1, literature[530]

incidence: 4/12*, conc. range: ≤1.96 µg/kg, sample origin: 12 regions, Czech Republic, sample year: 2011–2013, country: Czech Republic[811], *coriander seeds
- Co-contamination: not reported
- Further contamination of plants and plant products other than cereals and their products reported in the present book/article[811]: Beans, Chocolate, Coffee, Confectionery, Figs, Herbs and Spices, Juices, Licorice, Powder (cocoa powder), Products (ginger product), Raisins, Spices (caraway), Spices (cayenne pepper, chilli), Spices (ginger), Spices (mustard), Spices (nutmeg), Spices (paprika), Spices (pepper), Tea, Wine (red), Wine (white), OTA, literature[811]

Fusarium Toxins

Type B Trichothecenes

Deoxynivalenol

incidence: 1/3, conc.: 21 µg/kg, sample origin: specialist food shops, UK, sample year: unknown, country: UK[16]
- Co-contamination: not reported
- Further contamination of plants and plant products other than cereals and their products reported in the present book/article[16]: Food, AF/S, DON, FUS-X, NIV, OTA, and ZEA, literature[16]; Nuts (almond), AF/S, literature[16]; Oils, Spices (fennel), AF/S and ZEA, literature[16]; Oils (sesame oil), OTA, literature[16]; Paste (curry paste), AF/S, FB/S, NIV, OTA, and ZEA, literature[16]; Pickles, AF/S, FB/S, OTA, and ZEA, literature[16]; Sauces (chilli sauce), NIV, OTA, and ZEA, literature[16]; Spices, AF/S, FB/S, OTA, and ZEA, literature[16]; Spices (cayenne pepper, chilli), AF/S, DAS, DON, HT-2, OTA, and ZEA, literature[16]; Spices (coriander), AF/S, OTA, and ZEA, literature[16]; Spices (curry), AF/S, DAS, FB/S, FUS-X, HT-2, NEO, NIV, OTA, T-2, and ZEA, literature[16]; Spices (garlic), DON, FUS-X, and NIV, literature[16]; Spices (ginger), AF/S, DON, NEO, NIV, OTA, and T-2, literature[16]; Spices (tandoori), AF/S, FB/S, NIV, OTA, and T-2, literature[16]

Other ***Fusarium*** Toxins

Zearalenone

incidence: 3?/3, conc. range: 3.6–6.7 µg/kg, sample origin: specialist food shops, UK, sample year: unknown, country: UK[16]
- Co-contamination: not reported
- Further contamination of plants and plant products other than cereals and their products reported in the present book/article[16]: Food, AF/S, DON, FUS-X, NIV, OTA, and ZEA, literature[16]; Nuts (almond), AF/S, literature[16]; Oils, Spices (fennel), AF/S and ZEA, literature[16]; Oils (sesame oil), OTA, literature[16]; Paste (curry paste), AF/S, FB/S, NIV, OTA, and ZEA, literature[16]; Pickles, AF/S, FB/S, OTA, and ZEA, literature[16]; Sauces (chilli

sauce), NIV, OTA, and ZEA, literature[16]; Spices, AF/S, FB/S, OTA, and ZEA, literature[16]; Spices (cayenne pepper, chilli), AF/S, DAS, DON, HT-2, OTA, and ZEA, literature[16]; Spices (coriander), AF/S, DON, and OTA, literature[16]; Spices (curry), AF/S, DAS, FB/S, FUS-X, HT-2, NEO, NIV, OTA, T-2, and ZEA, literature[16]; Spices (garlic), DON, FUS-X, and NIV, literature[16]; Spices (ginger), AF/S, DON, NEO, NIV, OTA, and T-2, literature[16]; Spices (tandoori), AF/S, FB/S, NIV, OTA, and T-2, literature[16]

incidence: 2/9, conc. range: pr, sample origin: major retail outlets in Nainital (district), India, sample year: January 1984–December 1986, country: India[22]

- Co-contamination: not reported
- Further contamination of plants and plant products other than cereals and their products reported in the present book/article[22]: Spices (ammi), AFB_1, AFB_2, and AFG_1, literature[22]; Spices (cardamom), AFB_1, AFB_2, AFG_1, AFG_2, CIT, and OTA, literature[22]; Spices (cassia), AFB_1, AFB_2, AFG_1, and AFG_2, literature[22]; Spices (cayenne pepper, chilli), AFB_1, AFB_2, AFG_1, AFG_2, and ZEA, literature[22], Spices (coriander), AFB_1, AFB_2, AFG_1, AFG_2, CIT, and OTA, literature[22]; Spices (cumin), AFB_1, AFB_2, AFG_1, AFG_2, and CIT, literature[22]; Spices (curcuma), AFB_1, AFB_2, AFG_1, AFG_2, CIT, and RBT, literature[22]; Spices (fennel), AFB_1, AFB_2, AFG_1, AFG_2, CIT, OTA, and STG, literature[22]; Spices (garlic), AFB_1, AFB_2, and AFG_1, literature[22]; Spices (pepper), AFB_1, AFB_2, AFG_1, AFG_2, CIT, STG, and ZEA, literature[22]

Spices (cumin) may contain the following mycotoxins:

Alternaria Toxins

Tenuazonic Acid

incidence: 2/2, conc. range: 530–1,350 µg/kg, ∅ conc.: 940 µg/kg, sample origin: unknown, sample year: unknown, country: Germany[166]

- Co-contamination: not reported
- Further contamination of plants and plant products other than cereals and their products reported in the present book/article[166]: Juices, Juices (apple juice), Juices (apricot juice), Juices (banana juice), Juices (blackcurrant juice), Juices (cherry juice), Juices (grape juice), Juices (orange juice), Juices (pear juice), Juices (pomegranate juice), Juices (strawberry juice), Spices, Spices (coriander), Spices (curcuma), Spices (curry), Spices (ginger), Spices (paprika), Spices (pepper), TA, literature[166]

Aspergillus Toxins

Aflatoxin B_1

incidence: 2/8, conc. range: 24–104 µg/kg, ∅ conc.: 64 µg/kg, sample origin: major retail outlets in Nainital (district), India, sample year: July 1984–December 1986, country: India[22]

- Co contamination: not reported
- Further contamination of plants and plant products other than cereals and their products reported in the present book/article[22]: Spices (ammi), AFB_1, AFB_2, and AFG_1, literature[22]; Spices (cardamom), AFB_1, AFB_2, AFG_1, AFG_2, CIT, and OTA, literature[22]; Spices (cassia), AFB_1, AFB_2, AFG_1, and AFG_2, literature[22]; Spices (cayenne pepper, chilli), AFB_1, AFB_2, AFG_1, AFG_2, and ZEA, literature[22]; Spices (coriander), AFB_1, AFB_2, AFG_1, AFG_2, CIT, OTA, and ZEA, literature[22]; Spices (cumin), AFB_2, AFG_1, AFG_2, and CIT, literature[22]; Spices (curcuma), AFB_1, AFB_2, AFG_1, AFG_2, CIT, and RBT, literature[22]; Spices (fennel), AFB_1, AFB_2, AFG_1, AFG_2, CIT, OTA, and STG, literature[22]; Spices (garlic), AFB_1, AFB_2, and AFG_1, literature[22]; Spices (pepper), AFB_1, AFB_2, AFG_1, AFG_2, CIT, STG, and ZEA, literature[22]

incidence: 3/7*, conc. range: 1.25–2.3 µg/kg, ∅ conc.: 1.85 µg/kg, sample origin: supermarkets and ethnic shops in Lisbon (capital) area, Portugal, sample year: unknown, country: Portugal[169], *cumin, packaged

- Co-contamination: not reported
- Further contamination of plants and plant products other than cereals and their products reported in the present book/article[169]: Spices (cayenne pepper, chilli), Spices (curry), Spices (nutmeg), Spices (paprika), Spices (pepper), Spices (saffron), AFB_1, literature[169]

incidence: 8/14, conc. range: ≤0.08 µg/kg, ∅ conc.: 0.03 µg/kg, sample origin: popular markets in Salé (city) and Rabat (capital), Morocco, sample year: unknown, country: Morocco[174]

- Co-contamination: not reported
- Further contamination of plants and plant products other than cereals and their products reported in the present book/article[174]: Spice (cayenne pepper, chilli), Spices (ginger), Spices (pepper), AFB_1 and AF/S, literature[174]; Spices (cumin), AF/S, literature[174]

incidence: 2/3*, conc. range: 1.89–4.64 µg/kg, ∅ conc.: 3.26 µg/kg, sample origin: supermarkets in Penang (state), Malaysia, sample year: unknown, country: Malaysia[182], *cumin powder

- Co-contamination: not reported
- Further contamination of plants and plant products other than cereals and their products reported in the present book/article[182]: Nuts, Nuts (peanut), Seeds (sesame), Seeds (sunflower), Spices (cayenne pepper, chilli), Spices (pepper), AFB_1, literature[182]

incidence: 4/19, conc. range: 0.32–0.88 µg/kg, ∅ conc.: 0.58 µg/kg, sample origin: markets, herbal shops, and bazaars in Corum and Ankara (cities), Turkey, sample year: November 2010–August 2011, country: Turkey[183]

- Co-contamination: not reported
- Further contamination of plants and plant products other than cereals and their products reported in the present book/article[183]: Spices (cayenne pepper, chilli), AFB_1, AFB_2, AFG_1, and OTA, literature[183]; Spices (cumin), AFB_2 and OTA, literature[183]; Spices (pepper), AFB_1, AFB_2, and OTA, literature[183]

incidence: 1/2*, conc.: 2.45 µg/kg, sample origin: Penang (state, island), Malaysia, sample year: unknown (first study), country: Malaysia[530], *"jintan putih"

- Co-contamination: 1 sa. co-contaminated with AFB_1, AFB_2, AFG_1, AFG_2, and OTA
- Further contamination of plants and plant products other than cereals and their products reported in the present book/article[530]: Spices, Spices (cayenne pepper, chilli), Spices (coriander), Spices (curry), Spices (fennel), Spices (masala), AFB_1, AFB_2, AFG_1, AFG_2, and OTA, literature[530]; Spices (cumin), AFB_2, AFG_1, AFG_2, and OTA, literature[530]; Spices (curcuma), AFB_1, AFB_2, AFG_1, and OTA, literature[530]; Spices (pepper), AFB_1 and AFB_2, literature[530]; Spices (poppy seed), AFB_1, literature[530]

incidence: 0/1, conc.: no contamination, sample origin: Penang (state, island), Malaysia, sample year: unknown (second study), country: Malaysia[530]

- Co-contamination: not reported
- Further contamination of plants and plant products other than cereals and their products reported in the present book/article[530]: Spices, Spices (cayenne pepper, chilli), Spices (coriander), Spices (curry), Spices (fennel), Spices (masala), AFB_1, AFB_2, AFG_1, AFG_2, and OTA, literature[530]; Spices (cumin), AFB_2, AFG_1, AFG_2, and OTA, literature[530]; Spices (curcuma), AFB_1, AFB_2, AFG_1, and OTA, literature[530]; Spices (pepper), AFB_1 and AFB_2, literature[530]; Spices (poppy seed), AFB_1, literature[530]

incidence: 5/8*, conc. range: 0.5–26.3 µg/kg, Ø conc.: 10.7 µg/kg, sample origin: organic markets and organic shops, Turkey, sample year: May 2010–May 2011, country: Turkey[859], *organic cumin sa.

- Co-contamination: not reported
- Further contamination of plants and plant products other than cereals and their products reported in the present book/article[859]: Herbs (camomile), Herbs (linden), Herbs (rosehip), Herbs (sage), Spices (anise), Spices (basil), Spices (cayenne pepper, chilli), Spices (cinnamon), Spices (coriander), Spices (fennel), Spices (ginger), Spices (laurel leave), Spices (pepper), Spices (peppermint), Spices (poppy seed), Spices (rosemary), Spices (sumac), AFB_1, literature[859]

Aflatoxin B_2

incidence: 2/8, conc. range: 12–78 µg/kg, Ø conc.: 45 µg/kg, sample origin: major retail outlets in Nainital (district), India, sample year: July 1984–December 1986, country: India[22]

- Co-contamination: not reported
- Further contamination of plants and plant products other than cereals and their products reported in the present book/article[22]: Spices (ammi), AFB_1, AFB_2, and AFG_1, literature[22]; Spices (cardamom), AFB_1, AFB_2, AFG_1, AFG_2, CIT, and OTA, literature[22]; Spices (cassia), AFB_1, AFB_2, AFG_1, and AFG_2, literature[22]; Spices (cayenne pepper, chilli), AFB_1, AFB_2, AFG_1, AFG_2, and ZEA, literature[22]; Spices (coriander), AFB_1, AFB_2, AFG_1, AFG_2, CIT, OTA, and ZEA, literature[22]; Spices (cumin), AFB_1, AFG_1, AFG_2, and CIT, literature[22]; Spices (curcuma), AFB_1, AFB_2, AFG_1, AFG_2, CIT, and RBT, literature[22]; Spices (fennel), AFB_1, AFB_2, AFG_1, AFG_2, CIT, OTA, and STG, literature[22]; Spices (garlic), AFB_1, AFB_2, and AFG_1, literature[22]; Spices (pepper), AFB_1, AFB_2, AFG_1, AFG_2, CIT, STG, and ZEA, literature[22]

incidence: 2/19, conc. range: 0.07–0.10 µg/kg, Ø conc.: 0.08 µg/kg, sample origin: markets, herbal shops, and bazaars in Corum and Ankara (cities), Turkey, sample year: November 2010–August 2011, country: Turkey[183]

- Co-contamination: not reported
- Further contamination of plants and plant products other than cereals and their products reported in the present book/article[183]: Spices (cayenne pepper, chilli), AFB_1, AFB_2, AFG_1, and OTA, literature[183]; Spices (cumin), AFB_1 and OTA, literature[183]; Spices (pepper), AFB_1, AFB_2, and OTA, literature[183]

incidence: 1/2*, conc.: 0.35 µg/kg, sample origin: Penang (state, island), Malaysia, sample year: unknown (first study), country: Malaysia[530], *"jintan putih"

- Co-contamination: 1 sa. co-contaminated with AFB_1, AFB_2, AFG_1, AFG_2, and OTA
- Further contamination of plants and plant products other than cereals and their products reported in the present book/article[530]: Spices, Spices (cayenne pepper, chilli), Spices (coriander), Spices (curry), Spices (fennel), Spices (masala), AFB_1, AFB_2, AFG_1, AFG_2, and OTA, literature[530]; Spices (cumin), AFB_1, AFG_1, AFG_2, and OTA, literature[530]; Spices (curcuma), AFB_1, AFB_2, AFG_1, and OTA, literature[530]; Spices (pepper), AFB_1 and AFB_2, literature[530]; Spices (poppy seed), AFB_1, literature[530]

incidence: 0/1, conc.: no contamination, sample origin: Penang (state, island), Malaysia, sample year: unknown (second study), country: Malaysia[530]

- Co-contamination: not reported
- Further contamination of plants and plant products other than cereals and their products reported in the present book/article[530]: Spices, Spices (cayenne pepper, chilli), Spices (coriander), Spices (curry), Spices (fennel), Spices (masala), AFB_1, AFB_2, AFG_1, AFG_2, and OTA, literature[530]; Spices (cumin), AFB_1, AFG_1, AFG_2, and OTA, literature[530]; Spices (curcuma), AFB_1, AFB_2, AFG_1, and OTA, literature[530]; Spices (pepper), AFB_1 and AFB_2, literature[530]; Spices (poppy seed), AFB_1, literature[530]

Aflatoxin G_1

incidence: 2/8, conc. range: 8–45 µg/kg, Ø conc.: 26.5 µg/kg, sample origin: major retail outlets in Nainital (district), India, sample year: July 1984–December 1986, country: India[22]

- Co-contamination: not reported
- Further contamination of plants and plant products other than cereals and their products reported in the present book/article[22]: Spices (ammi), AFB_1, AFB_2, and AFG_1, literature[22]; Spices (cardamom), AFB_1, AFB_2, AFG_1, AFG_2, CIT, and OTA, literature[22]; Spices (cassia), AFB_1, AFB_2, AFG_1, and AFG_2, literature[22]; Spices (cayenne pepper, chilli), AFB_1, AFB_2, AFG_1, AFG_2, and ZEA, literature[22]; Spices (coriander), AFB_1, AFB_2, AFG_1, AFG_2, CIT, OTA, and ZEA, literature[22]; Spices (cumin), AFB_1, AFB_2, AFG_2, and CIT, literature[22]; Spices (curcuma), AFB_1, AFB_2, AFG_1, AFG_2, CIT, and RBT, literature[22]; Spices (fennel), AFB_1, AFB_2, AFG_1, AFG_2, CIT, OTA, and STG, literature[22]; Spices (garlic), AFB_1, AFB_2, and AFG_1, literature[22]; Spices (pepper), AFB_1, AFB_2, AFG_1, AFG_2, CIT, STG, and ZEA, literature[22]

incidence: 1/2*, conc.: 0.35 µg/kg, sample origin: Penang (state, island), Malaysia, sample year: unknown (first study), country: Malaysia[530], *"jintan putih"

- Co-contamination: 1 sa. co-contaminated with AFB_1, AFB_2, AFG_1, AFG_2, and OTA
- Further contamination of plants and plant products other than cereals and their products reported in the present book/article[530]: Spices, Spices (cayenne pepper, chilli), Spices (coriander), Spices (curry), Spices (fennel), Spices (masala), AFB_1, AFB_2, AFG_1, AFG_2, and OTA, literature[530]; Spices (cumin), AFB_1, AFB_2, AFG_2, and OTA, literature[530]; Spices (curcuma), AFB_1, AFB_2, AFG_1, and OTA, literature[530]; Spices (pepper), AFB_1 and AFB_2, literature[530]; Spices (poppy seed), AFB_1, literature[530]

incidence: 1/1, conc.: 0.91 µg/kg, sample origin: Penang (state, island), Malaysia, sample year: unknown (second study), country: Malaysia[530]

- Co-contamination: not reported
- Further contamination of plants and plant products other than cereals and their products reported in the present book/article[530]: Spices, Spices (cayenne pepper, chilli), Spices (coriander), Spices (curry), Spices (fennel), Spices (masala), AFB_1,

AFB_2, AFG_1, AFG_2, and OTA, literature[530]; Spices (cumin), AFB_1, AFB_2, AFG_2, and OTA, literature[530]; Spices (curcuma), AFB_1, AFB_2, AFG_1, and OTA, literature[530]; Spices (pepper), AFB_1 and AFB_2, literature[530]; Spices (poppy seed), AFB_1, literature[530]

Aflatoxin G_2

incidence: 1/8, conc.: 30 µg/kg, sample origin: major retail outlets in Nainital (district), India, sample year: July 1984–December 1986, country: India[22]

- Co-contamination: not reported
- Further contamination of plants and plant products other than cereals and their products reported in the present book/article[22]: Spices (ammi), AFB_1, AFB_2, and AFG_1, literature[22]; Spices (cardamom), AFB_1, AFB_2, AFG_1, AFG_2, CIT, and OTA, literature[22]; Spices (cassia), AFB_1, AFB_2, AFG_1, and AFG_2, literature[22]; Spices (cayenne pepper, chilli), AFB_1, AFB_2, AFG_1, AFG_2, and ZEA, literature[22]; Spices (coriander), AFB_1, AFB_2, AFG_1, AFG_2, CIT, OTA, and ZEA, literature[22]; Spices (cumin), AFB_1, AFB_2, AFG_1, and CIT, literature[22]; Spices (curcuma), AFB_1, AFB_2, AFG_1, AFG_2, CIT, and RBT, literature[22]; Spices (fennel), AFB_1, AFB_2, AFG_1, AFG_2, CIT, OTA, and STG, literature[22]; Spices (garlic), AFB_1, AFB_2, and AFG_1, literature[22]; Spices (pepper), AFB_1, AFB_2, AFG_1, AFG_2, CIT, STG, and ZEA, literature[22]

incidence: 1/2*, conc.: 0.05 µg/kg, sample origin: Penang (state, island), Malaysia, sample year: unknown (first study), country: Malaysia[530], *"jintan putih"

- Co-contamination: 1 sa. co-contaminated with AFB_1, AFB_2, AFG_1, AFG_2, and OTA
- Further contamination of plants and plant products other than cereals and their products reported in the present book/article[530]: Spices, Spices (cayenne pepper, chilli), Spices (coriander), Spices (curry), Spices (fennel), Spices (masala), AFB_1, AFB_2, AFG_1, AFG_2, and OTA, literature[530]; Spices (cumin), AFB_1, AFB_2, AFG_1, and OTA, literature[530]; Spices (curcuma), AFB_1, AFB_2, AFG_1, and OTA, literature[530]; Spices (pepper), AFB_1 and AFB_2, literature[530]; Spices (poppy seed), AFB_1, literature[530]

incidence: 0/1, conc.: no contamination, sample origin: Penang (state, island), Malaysia, sample year: unknown (second study), country: Malaysia[530]

- Co-contamination: not reported
- Further contamination of plants and plant products other than cereals and their products reported in the present book/article[530]: Spices, Spices (cayenne pepper, chilli), Spices (coriander), Spices (curry), Spices (fennel), Spices (masala), AFB_1, AFB_2, AFG_1, AFG_2, and OTA, literature[530]; Spices (cumin), AFB_1, AFB_2, AFG_1, and OTA, literature[530]; Spices (curcuma), AFB_1, AFB_2, AFG_1, and OTA, literature[530]; Spices (pepper), AFB_1 and AFB_2, literature[530]; Spices (poppy seed), AFB_1, literature[530]

Aflatoxin/S

incidence: 8/14, conc. range: ≤0.18 µg/kg*, ⌀ conc.: 0.05 µg/kg*, sample origin: popular markets in Salé (city) and Rabat (capital), Morocco, sample year: unknown, country: Morocco[174], *AFS total

- Co-contamination: not reported
- Further contamination of plants and plant products other than cereals and their products reported in the present book/article[174]: Spice (cayenne pepper, chilli), Spices (ginger), Spices (pepper), AFB_1 and AF/S, literature[174]; Spices (cumin), AFB_1, literature[174]

incidence: 28/28, conc. range: 4.55–8.57 µg/kg*, ⌀ conc.: 7.03 µg/kg*, sample year: unknown, country: Turkey[190], *AFS total

- Co-contamination: not reported
- Further contamination of plants and plant products other than cereals and their products reported in the present book/article[190]: Spices (cayenne pepper, chilli), Spices (pepper), AF/S, literature[190]

Other *Aspergillus* Toxins

Sterigmatocystin

incidence: 3/5, conc. range: 11 µg/kg, sample origin: Assiut (governorate), Egypt, sample year: unknown, country: Egypt[641]

- Co-contamination: not reported
- Further contamination of plants and plant products other than cereals and their products reported in the present book/article[641]: Spices (anise), Spices (coriander), Spices (fennel), Spices (pepper), AF/S, literature[641]; Spices (black cumin), AF/S and CIT, literature[641]; Spices (caraway), Spices (marjoram), AF/S and STG, literature[641]; Spices (cayenne pepper, chilli), STG, literature[641]

Aspergillus and *Penicillium* Toxins

Citrinin

incidence: 1/8, conc.: 22 µg/kg, sample origin: major retail outlets in Nainital (district), India, sample year: July 1984–December 1986, country: India[22]

- Co-contamination: not reported
- Further contamination of plants and plant products other than cereals and their products reported in the present book/article[22]: Spices (ammi), AFB_1, AFB_2, and AFG_1, literature[22]; Spices (cardamom), AFB_1, AFB_2, AFG_1, AFG_2, CIT, and OTA, literature[22]; Spices (cassia), AFB_1, AFB_2, AFG_1, and AFG_2, literature[22]; Spices (cayenne pepper, chilli), AFB_1, AFB_2, AFG_1, AFG_2, and ZEA, literature[22]; Spices (coriander), AFB_1, AFB_2, AFG_1, AFG_2, CIT, OTA, and ZEA, literature[22]; Spices (cumin), AFB_1, AFB_2, AFG_1, and AFG_2, literature[22]; Spices (curcuma), AFB_1, AFB_2, AFG_1, AFG_2, CIT, and RBT, literature[22]; Spices (fennel), AFB_1, AFB_2, AFG_1, AFG_2, CIT, OTA, and STG, literature[22]; Spices (garlic), AFB_1, AFB_2, and AFG_1, literature[22]; Spices (pepper), AFB_1, AFB_2, AFG_1, AFG_2, CIT, STG, and ZEA, literature[22]

Ochratoxin A

incidence: 1/19, conc.: 0.63 µg/kg, sample origin: markets, herbal shops, and bazaars in Corum and Ankara (cities), Turkey, sample year: November 2010–August 2011, country: Turkey[183]

- Co-contamination: not reported
- Further contamination of plants and plant products other than cereals and their products reported in the present book/article[183]: Spices (cayenne pepper, chilli), AFB_1, AFB_2, AFG_1, and OTA, literature[183]; Spices (cumin), AFB_1 and AFB_2, literature[183]; Spices (pepper), AFB_1, AFB_2, and OTA, literature[183]

incidence: 1/2*, conc.: 20.40 µg/kg, sample origin: Penang (state, island), Malaysia, sample year: unknown (first study), country: Malaysia[530], *"jintan putih"

- Co-contamination: 1 sa. co-contaminated with AFB_1, AFB_2, AFG_1, AFG_2, and OTA
- Further contamination of plants and plant products other than cereals and their products reported in the present book/article[530]: Spices, Spices (cayenne pepper, chilli), Spices (coriander), Spices (curry), Spices (fennel), Spices (masala), AFB_1, AFB_2, AFG_1, AFG_2, and OTA, literature[530]; Spices (cumin), AFB_1, AFB_2, AFG_1, and AFG_2, literature[530]; Spices (curcuma), AFB_1, AFB_2, AFG_1, and OTA, literature[530]; Spices (pepper), AFB_1 and AFB_2, literature[530]; Spices (poppy seed), AFB_1, literature[530]

incidence: 7/29, conc. range: 0.31–3.25 µg/kg, ∅ conc.: 1.46 µg/kg, sample origin: supermarkets, shops, and market stalls in eight cities, China, sample year: April–July 2009, country: China[885]

- Co-contamination: not reported
- Further contamination of plants and plant products other than cereals and their products reported in the present book/article[885]: Spices (anise), Spices (cayenne pepper, chilli), Spices (cinnamon), Spices (curry), Spices (prickly ash), OTA, literature[885]

see also **Spices**

Spices (curcuma) may contain the following mycotoxins:

Alternaria Toxins

Tenuazonic Acid

incidence: 2/2, conc. range: 60–100 µg/kg, ∅ conc.: 80 µg/kg, sample origin: unknown, sample year: unknown, country: Germany[166]

- Co-contamination: not reported
- Further contamination of plants and plant products other than cereals and their products reported in the present book/article[166]: Juices, Juices (apple juice), Juices (apricot juice), Juices (banana juice), Juices (blackcurrant juice), Juices (cherry juice), Juices (grape juice), Juices (orange juice), Juices (pear juice), Juices (pomegranate juice), Juices (strawberry juice), Spices, Spices (coriander), Spices (cumin), Spices (curry), Spices (ginger), Spices (paprika), Spices (pepper), TA, literature[166]

Aspergillus Toxins

Aflatoxin B_1

incidence: 5?/9, conc. range: 21–165 µg/kg, sample origin: major retail outlets in Nainital (district), India, sample year: March 1984–December 1986, country: India[22]

- Co-contamination: not reported
- Further contamination of plants and plant products other than cereals and their products reported in the present book/article[22]: Spices (ammi), AFB_1, AFB_2, and AFG_1, literature[22]; Spices (cardamom), AFB_1, AFB_2, AFG_1, AFG_2, CIT, and OTA, literature[22]; Spices (cassia), AFB_1, AFB_2, AFG_1, and AFG_2, literature[22]; Spices (cayenne pepper, chilli), AFB_1, AFB_2, AFG_1, AFG_2, and ZEA, literature[22]; Spices (coriander), AFB_1, AFB_2, AFG_1, AFG_2, CIT, OTA, and ZEA, literature[22]; Spices (cumin), AFB_1, AFB_2, AFG_1, AFG_2, and CIT, literature[22]; Spices (curcuma), AFB_2, AFG_1, AFG_2, CIT, and RBT, literature[22]; Spices (fennel), AFB_1, AFB_2, AFG_1, AFG_2, CIT, OTA, and STG, literature[22]; Spices (garlic), AFB_1, AFB_2, and AFG_1, literature[22]; Spices (pepper), AFB_1, AFB_2, AFG_1, AFG_2, CIT, STG, and ZEA, literature[22]

incidence: 2?/15, ∅ conc.: 12 µg/kg, sample origin: local companies in Cairo (capital), Egypt, sample year: unknown, country: Egypt[104]

- Co-contamination: not reported
- Further contamination of plants and plant products other than cereals and their products reported in the present book/article[104]: Kubeba, AFB_1 and AFB_2, literature[104]; Spices (coriander), AFB_1 and AFG_1, literature[104]; Spices (pepper), AFB_1, literature[104]; Spices (curcuma), AFG_1, literature[104]

incidence: 6/6, conc. range: tr-3.8 µg/kg*, sample origin: processors located in six Canadian cities, Canada, sample year: unknown, country: Canada[323], *approximate conc.

- Co-contamination: 3 sa. co-contaminated with AFB_1 and AFB_2; 3 sa. contaminated solely with AFB_1
- Further contamination of plants and plant products other than cereals and their products reported in the present book/article[323]: Spices (coriander), Spices (nutmeg), AFB_1 and AFB_2, literature[323]; Spices (curcuma), AFB_2, literature[323]; Spices (ginger), AFB_1, AFB_2, AFG_1, and AFG_2, literature[323]

incidence: 8/10, conc. range: 1.3–4.2 µg/kg, sample origin: retail sales points in different regions, Iran, sample year: March–May 2015, country: Iran[491]

- Co-contamination: not reported
- Further contamination of plants and plant products other than cereals and their products reported in the present book/article[491]: Herbs (camomile), Herbs (sage), AFB_1, literature[491]; Spices (cayenne pepper, chilli), Spices (ginger), Spices (pepper), Spices (thyme), AFB_1, literature[491]

incidence: 2/2*, conc. range: 0.05–0.18 µg/kg, ∅ conc.: 0.12 µg/kg, sample origin: Penang (state, island), Malaysia, sample year: unknown (first study), country: Malaysia[530], *"kunyit"

- Co-contamination: 1 sa. co-contaminated with AFB_1, AFB_2, and OTA; 1 sa. co-contaminated with AFB_1 and OTA
- Further contamination of plants and plant products other than cereals and their products reported in the present book/article[530]: Spices, Spices (cayenne pepper, chilli), Spices (coriander), Spices (cumin), Spices (curry), Spices (fennel), Spices (masala), AFB_1, AFB_2, AFG_1, AFG_2, and OTA, literature[530]; Spices (curcuma), AFB_2, AFG_1, and OTA, literature[530]; Spices (pepper), AFB_1 and AFB_2, literature[530]; Spices (poppy seed), AFB_1, literature[530]

incidence: 2/2, conc. range: 4.79–7.28 µg/kg, ∅ conc.: 6.01 µg/kg, sample origin: Penang (state, island), Malaysia, sample year: unknown (second study), country: Malaysia[530]

- Co-contamination: 1 sa. co-contaminated with AFB_1, AFB_2, and AFG_1; 1 sa. co-contaminated with AFB_1 and AFG_1
- Further contamination of plants and plant products other than cereals and their products reported in the present book/article[530]: Spices, Spices (cayenne pepper, chilli), Spices (coriander), Spices (curry), Spices (fennel), Spices (masala), AFB_1, AFB_2, AFG_1, AFG_2, and OTA, literature[530]; Spices (cumin), AFB_2, AFG_1, AFG_2, and OTA, literature[530]; Spices (curcuma), AFB_2, AFG_1, and OTA, literature[530]; Spices (pepper), AFB_1 and AFB_2, literature[530]; Spices (poppy seed), AFB_1, literature[530]

Aflatoxin B_2

incidence: 5?/9, conc. range: 12–150 µg/kg, sample origin: major retail outlets in Nainital (district), India, sample year: March 1984–December 1986, country: India[22]

- Co-contamination: not reported
- Further contamination of plants and plant products other than cereals and their products reported in the present book/article[22]: Spices (ammi), AFB_1, AFB_2, and AFG_1, literature[22]; Spices (cardamom), AFB_1, AFB_2, AFG_1, AFG_2, CIT, and OTA, literature[22]; Spices (cassia), AFB_1, AFB_2, AFG_1, and AFG_2, literature[22]; Spices (cayenne pepper, chilli), AFB_1, AFB_2, AFG_1, AFG_2, and ZEA, literature[22]; Spices (coriander), AFB_1, AFB_2, AFG_1, AFG_2, CIT, OTA, and ZEA, literature[22]; Spices (cumin), AFB_1, AFB_2, AFG_1, AFG_2, and CIT, literature[22]; Spices (curcuma), AFB_1, AFG_1, AFG_2, CIT, and RBT, literature[22]; Spices (fennel),

AFB_1, AFB_2, AFG_1, AFG_2, CIT, OTA, and STG, literature[22]; Spices (garlic), AFB_1, AFB_2, and AFG_1, literature[22]; Spices (pepper), AFB_1, AFB_2, AFG_1, AFG_2, CIT, STG, and ZEA, literature[22]

incidence: 3/6, conc. range: tr-1.3 µg/kg*, sample origin: processors located in six Canadian cities, Canada, sample year: unknown, country: Canada[323], *approximate conc.

- Co-contamination: 3 sa. co-contaminated with AFB_1 and AFB_2
- Further contamination of plants and plant products other than cereals and their products reported in the present book/article[323]: Spices (coriander), Spices (nutmeg), AFB_1 and AFB_2, literature[323]; Spices (curcuma), AFB_1, literature[323]; Spices (ginger), AFB_1, AFB_2, AFG_1, and AFG_2, literature[323]

incidence: 1/2*, conc.: 0.03 µg/kg, sample origin: Penang (state, island), Malaysia, sample year: unknown (first study), country: Malaysia[530], *"kunyit"

- Co-contamination: 1 sa. co-contaminated with AFB_1, AFB_2, and OTA
- Further contamination of plants and plant products other than cereals and their products reported in the present book/article[530]: Spices, Spices (cayenne pepper, chilli), Spices (coriander), Spices (cumin), Spices (curry), Spices (fennel), Spices (masala), AFB_1, AFB_2, AFG_1, AFG_2, and OTA, literature[530]; Spices (curcuma), AFB_1, AFG_1, and OTA, literature[530]; Spices (pepper), AFB_1 and AFB_2, literature[530]; Spices (poppy seed), AFB_1, literature[530]

incidence: 1/2, conc.: 1.68 µg/kg, sample origin: Penang (state, island), Malaysia, sample year: unknown (second study), country: Malaysia[530]

- Co-contamination: 1 sa. co-contaminated with AFB_1, AFB_2, and AFG_1; 1 sa. co-contaminated with AFB_1 and AFG_1
- Further contamination of plants and plant products other than cereals and their products reported in the present book/article[530]: Spices, Spices (cayenne pepper, chilli), Spices (coriander), Spices (curry), Spices (fennel), Spices (masala), AFB_1, AFB_2, AFG_1, AFG_2, and OTA, literature[530]; Spices (cumin), AFB_2, AFG_1, AFG_2, and OTA, literature[530]; Spices (curcuma), AFB_1, AFG_1, and OTA, literature[530]; Spices (pepper), AFB_1 and AFB_2, literature[530]; Spices (poppy seed), AFB_1, literature[530]

Aflatoxin G_1

incidence: 5?/9, conc. range: 20–125 µg/kg, sample origin: major retail outlets in Nainital (district), India, sample year: March 1984–December 1986, country: India[22]

- Co-contamination: not reported
- Further contamination of plants and plant products other than cereals and their products reported in the present book/article[22]: Spices (ammi), AFB_1, AFB_2, and AFG_1, literature[22]; Spices (cardamom), AFB_1, AFB_2, AFG_1, AFG_2, CIT, and OTA, literature[22]; Spices (cassia), AFB_1, AFB_2, AFG_1, and AFG_2, literature[22]; Spices (cayenne pepper, chilli), AFB_1, AFB_2, AFG_1, AFG_2, and ZEA, literature[22]; Spices (coriander), AFB_1, AFB_2, AFG_1, AFG_2, CIT, OTA, and ZEA, literature[22]; Spices (cumin), AFB_1, AFB_2, AFG_1, AFG_2, and CIT, literature[22]; Spices (curcuma), AFB_1, AFB_2, AFG_2, CIT, and RBT, literature[22]; Spices (fennel), AFB_1, AFB_2, AFG_1, AFG_2, CIT, OTA, and STG, literature[22]; Spices (garlic), AFB_1, AFB_2, and AFG_1, literature[22]; Spices (pepper), AFB_1, AFB_2, AFG_1, AFG_2, CIT, STG, and ZEA, literature[22]

incidence: 2?/15, ∅ conc.: 8 µg/kg, sample origin: local companies in Cairo (capital), Egypt, sample year: unknown, country: Egypt[104]

- Co-contamination: not reported
- Further contamination of plants and plant products other than cereals and their products reported in the present book/article[104]: Kubeba, AFB_1 and AFB_2, literature[104]; Spices (coriander), AFB_1 and AFG_1, literature[104]; Spices (pepper), Spices (curcuma), AFB_1, literature[104]

incidence: 0/2*, conc.: no contamination, sample origin: Penang (state, island), Malaysia, sample year: unknown (first study), country: Malaysia[530], *"kunyit"

- Co-contamination: not reported
- Further contamination of plants and plant products other than cereals and their products reported in the present book/article[530]: Spices, Spices (cayenne pepper, chilli), Spices (coriander), Spices (cumin), Spices (curry), Spices (fennel), Spices (masala), AFB_1, AFB_2, AFG_1, AFG_2, and OTA, literature[530]; Spices (curcuma), AFB_1, AFB_2, and OTA, literature[530]; Spices (pepper), AFB_1 and AFB_2, literature[530]; Spices (poppy seed), AFB_1, literature[530]

incidence: 2/2, conc. range: 0.56–1.26 µg/kg, ∅ conc.: 0.91 µg/kg, sample origin: Penang (state, island), Malaysia, sample year: unknown (second study), country: Malaysia[530]

- Co-contamination: 1 sa. co-contaminated with AFB_1, AFB_2, and AFG_1; 1 sa. co-contaminated with AFB_1 and AFG_1
- Further contamination of plants and plant products other than cereals and their products reported in the present book/article[530]: Spices, Spices (cayenne pepper, chilli), Spices (coriander), Spices (curry), Spices (fennel), Spices (masala), AFB_1, AFB_2, AFG_1, AFG_2, and OTA, literature[530]; Spices (cumin), AFB_2, AFG_1, AFG_2, and OTA, literature[530]; Spices (curcuma), AFB_1, AFB_2, and OTA, literature[530]; Spices (pepper), AFB_1 and AFB_2, literature[530]; Spices (poppy seed), AFB_1, literature[530]

Aflatoxin G_2

incidence: 5?/9, conc. range: 14–125 µg/kg, sample origin: major retail outlets in Nainital (district), India, sample year: March 1984–December 1986, country: India[22]

- Co-contamination: not reported
- Further contamination of plants and plant products other than cereals and their products reported in the present book/article[22]: Spices (ammi), AFB_1, AFB_2, and AFG_1, literature[22]; Spices (cardamom), AFB_1, AFB_2, AFG_1, AFG_2, CIT, and OTA, literature[22]; Spices (cassia), AFB_1, AFB_2, AFG_1, and AFG_2, literature[22]; Spices (cayenne pepper, chilli), AFB_1, AFB_2, AFG_1, AFG_2, and ZEA, literature[22]; Spices (coriander), AFB_1, AFB_2, AFG_1, AFG_2, CIT, OTA, and ZEA, literature[22]; Spices (cumin), AFB_1, AFB_2, AFG_1, AFG_2, and CIT, literature[22]; Spices (curcuma), AFB_1, AFB_2, AFG_1, CIT, and RBT, literature[22]; Spices (fennel), AFB_1, AFB_2, AFG_1, AFG_2, CIT, OTA, and STG, literature[22]; Spices (garlic), AFB_1, AFB_2, and AFG_1, literature[22]; Spices (pepper), AFB_1, AFB_2, AFG_1, AFG_2, CIT, STG, and ZEA, literature[22]

Aflatoxin/s

incidence: 4/10, conc. range: 0.81–16.40 µg/kg*, sample origin: market stalls, shops, and supermarkets, Ireland, sample year: unknown, country: Ireland[187], *AFB_1, AFB_2, AFG_1, and AFG_2

- Co-contamination: not reported
- Further contamination of plants and plant products other than cereals and their products reported in the present book/ article[187]: Spices (cayenne pepper, chilli), Spices (coriander), Spices (curry), Spices (paprika), Spices (pepper), AF/S, literature[187]

Aspergillus **and** ***Penicillium*** **Toxins**

Citrinin

incidence: 2/9, conc. range: 48–52 µg/kg, ∅ conc.: 50 µg/kg, sample origin: major retail outlets in Nainital (district), India, sample year: March 1984–December 1986, country: India[22]

- Co-contamination: not reported
- Further contamination of plants and plant products other than cereals and their products reported in the present book/ article[22]: Spices (ammi), AFB_1, AFB_2, and AFG_1, literature[22]; Spices (cardamom), AFB_1, AFB_2, AFG_1, AFG_2, CIT, and OTA, literature[22]; Spices (cassia), AFB_1, AFB_2, AFG_1, and AFG_2, literature[22]; Spices (cayenne pepper, chilli), AFB_1, AFB_2, AFG_1, AFG_2, and ZEA, literature[22]; Spices (coriander), AFB_1, AFB_2, AFG_1, AFG_2, CIT, OTA, and ZEA, literature[22]; Spices (cumin), AFB_1, AFB_2, AFG_1, AFG_2, and CIT, literature[22]; Spices (curcuma), AFB_1, AFB_2, AFG_1, AFG_2, and RBT, literature[22]; Spices (fennel), AFB_1, AFB_2, AFG_1, AFG_2, CIT, OTA, and STG, literature[22]; Spices (garlic), AFB_1, AFB_2, and AFG_1, literature[22]; Spices (pepper), AFB_1, AFB_2, AFG_1, AFG_2, CIT, STG, and ZEA, literature[22]

Ochratoxin A

incidence: 13/25, conc. range: 39–925 µg/kg, ∅ conc.: 290 µg/kg, sample origin: retail markets and supermarkets in 24 regions, Tunisia, sample year: unknown, country: Tunisia[20]

- Co-contamination: not reported
- Further contamination of plants and plant products other than cereals and their products reported in the present book/ article[20]: Nuts (almond), Nuts (peanut), Nuts (pistachio nut), Spices (caraway), Spices (coriander), Spices (paprika), Spices (pepper), OTA, literature[20]

incidence: 9/25*, conc. range: 10–29 µg/kg (3 sa.), 30–49 µg/kg (2 sa.), 50–100 µg/kg (3 sa.), 110 µg/kg (1 sa.), sample origin: retail shops, India, sample year: unknown, country: India/UK/ Belgium[191], *curcuma powder

- Co-contamination: not reported
- Further contamination of plants and plant products other than cereals and their products reported in the present book/ article[191]: Spices (coriander), Spices (ginger), Spices (pepper), OTA, literature[191]

incidence: 2/2*, conc. range: conc. range: 0.20–3.58 µg/kg, ∅ conc.: 1.89 µg/kg, sample origin: Penang (state, island), Malaysia, sample year: unknown (first study), country: Malaysia[530], *"kunyit"

- Co-contamination: not reported
- Further contamination of plants and plant products other than cereals and their products reported in the present book/ article[530]: Spices, Spices (cayenne pepper, chilli), Spices (coriander), Spices (cumin), Spices (curry), Spices (fennel), Spices (masala), AFB_1, AFB_2, AFG_1, AFG_2, and OTA, literature[530]; Spices (curcuma), AFB_1, AFB_2, and AFG_1, literature[530]; Spices (pepper), AFB_1 and AFB_2, literature[530]; Spices (poppy seed), AFB_1, literature[530]

Penicillium **Toxins**

Rubratoxin

incidence: 1/9, conc.: 375 µg/kg, sample origin: major retail outlets in Nainital (district), India, sample year: March 1984–December 1986, country: India[22]

- Co-contamination: not reported
- Further contamination of plants and plant products other than cereals and their products reported in the present book/ article[22]: Spices (ammi), AFB_1, AFB_2, and AFG_1, literature[22]; Spices (cardamom), AFB_1, AFB_2, AFG_1, AFG_2, CIT, and OTA, literature[22]; Spices (cassia), AFB_1, AFB_2, AFG_1, and AFG_2, literature[22]; Spices (cayenne pepper, chilli), AFB_1, AFB_2, AFG_1, AFG_2, and ZEA, literature[22]; Spices (coriander), AFB_1, AFB_2, AFG_1, AFG_2, CIT, OTA, and ZEA, literature[22]; Spices (cumin), AFB_1, AFB_2, AFG_1, AFG_2, and CIT, literature[22]; Spices (curcuma), AFB_1, AFB_2, AFG_1, AFG_2, and CIT, literature[22]; Spices (fennel), AFB_1, AFB_2, AFG_1, AFG_2, CIT, OTA, and STG, literature[22]; Spices (garlic), AFB_1, AFB_2, and AFG_1, literature[22]; Spices (pepper), AFB_1, AFB_2, AFG_1, AFG_2, CIT, STG, and ZEA, literature[22]

see also **Spices**

Spices (curry) may contain the following mycotoxins:

Alternaria **Toxins**

Tenuazonic Acid

incidence: 5/5, conc. range: 190–2,100 µg/kg, sample origin: unknown, sample year: unknown, country: Germany[166]

- Co-contamination: not reported
- Further contamination of plants and plant products other than cereals and their products reported in the present book/ article[166]: Juices, Juices (apple juice), Juices (apricot juice), Juices (banana juice), Juices (blackcurrant juice), Juices (cherry juice), Juices (grape juice), Juices (orange juice), Juices (pear juice), Juices (pomegranate juice), Juices (strawberry juice), Spices, Spices (coriander), Spices (cumin), Spices (curcuma), Spices (ginger), Spices (paprika), Spices (pepper), TA, literature[166]

Aspergillus **Toxins**

Aflatoxin B_1

incidence: 2/2*, conc. range: 3.64–4.12 µg/kg, ∅ conc.: 3.88 µg/kg, sample origin: grocery markets in Busan, Daegu, Daejeon, Gangneung, and Gwangju (cities) as well as Seoul (capital), Korea, sample year: May 2004–June 2005, country: Korea[10], *curry powder

- Co-contamination: not reported
- Further contamination of plants and plant products other than cereals and their products reported in the present book/ article[10]: Butter (peanut butter), Nuts, Nuts (almond), Nuts (peanut), Nuts (pistachio nut), Paste (soybean paste), Sauces (soy sauce), Spices (cayenne pepper, chilli), Spices (pepper), AFB_1, literature[10]

incidence: 2/5*, conc. range: 1.75–2.5 µg/kg, ∅ conc.: 2.13 µg/kg, sample origin: supermarkets and ethnic shops in Lisbon (capital) area, Portugal, sample year: unknown, country: Portugal[169], *curry powder, packaged

- Co-contamination: not reported
- Further contamination of plants and plant products other than cereals and their products reported in the present book/

article[169]: Spices (cayenne pepper, chilli), Spices (cumin), Spices (nutmeg), Spices (paprika), Spices (pepper), Spices (saffron), AFB_1, literature[169]

incidence: 5/5*, conc. range: 0.2–202 µg/kg, ∅ conc.: 41.4 µg/kg, sample origin: available in New Zealand, sample year: 2009, country: New Zealand[277], *curry powder

- Co-contamination: not reported
- Further contamination of plants and plant products other than cereals and their products reported in the present book/article[277]: Bars, AFB_1, AF/S, DON, NIV, and OTA, literature[277]; Butter (peanut butter), Confectionery, Food, Fruits, dried, Nuts, Nuts (peanut), Nuts (pistachio nut), Sauces (peanut sauce), AFB_1 and AF/S, literature[277]; Coffee, Dates, Spices (nutmeg), Spices (pepper), Vine fruits, dried, OTA, literature[277]; Figs, Spices (cayenne pepper, chilli), Spices (ginger), Spices (paprika), AFB_1, AF/S, and OTA, literature[277]; Nuts (Brazil nut, paranut), AFB_1, literature[277]; Spices (curry), AF/S and OTA, literature[277]

incidence: 5/5*, conc. range: 0.03–2.21 µg/kg, ∅ conc.: 0.64 µg/kg, sample origin: Penang (state, island), Malaysia, sample year: unknown (first study), country: Malaysia[530], *mixed spices for fish curry

- Co-contamination: 2 sa. co-contaminated with AFB_1 and OTA; 1 sa. co-contaminated with AFB_1, AFB_2, AFG_1, AFG_2, and OTA; 1 sa. co-contaminated with AFB_1, AFB_2, AFG_1, and OTA; 1 sa. co-contaminated with AFB_1, AFB_2, and OTA
- Further contamination of plants and plant products other than cereals and their products reported in the present book/article[530]: Spices, Spices (cayenne pepper, chilli), Spices (coriander), Spices (cumin), Spices (fennel), Spices (masala), AFB_1, AFB_2, AFG_1, AFG_2, and OTA, literature[530]; Spices (curcuma), AFB_1, AFB_2, AFG_1, and OTA, literature[530]; Spices (curry), AFB_2, AFG_1, AFG_2, and OTA, literature[530]; Spices (pepper), AFB_1 and AFB_2, literature[530]; Spices (poppy seed), AFB_1, literature[530]

incidence: 3/3*, conc. range: 0.16–2.66 µg/kg, ∅ conc.: 1.14 µg/kg, sample origin: Penang (state, island), Malaysia, sample year: unknown (first study), country: Malaysia[530], *mixed spices for meat curry

- Co-contamination: 2 sa. co-contaminated with AFB_1, AFB_2, and OTA; 1 sa. co-contaminated with AFB_1 and OTA
- Further contamination of plants and plant products other than cereals and their products reported in the present book/article[530]: Spices, Spices (cayenne pepper, chilli), Spices (coriander), Spices (cumin), Spices (fennel), Spices (masala), AFB_1, AFB_2, AFG_1, AFG_2, and OTA, literature[530]; Spices (curcuma), AFB_1, AFB_2, AFG_1, and OTA, literature[530]; Spices (curry), AFB_2, AFG_1, AFG_2, and OTA, literature[530]; Spices (pepper), AFB_1 and AFB_2, literature[530]; Spices (poppy seed), AFB_1, literature[530]

incidence: 1/2*, conc.: 4.38 µg/kg, sample origin: Penang (state, island), Malaysia, sample year: unknown (second study), country: Malaysia[530], *mixed spices for fish curry

- Co-contamination: 1 sa. co-contaminated with AFB_1, AFB_2, AFG_1, and OTA
- Further contamination of plants and plant products other than cereals and their products reported in the present book/article[530]: Spices, Spices (cayenne pepper, chilli), Spices (coriander), Spices (cumin), Spices (fennel), Spices (masala), AFB_1, AFB_2, AFG_1, AFG_2, and OTA, literature[530]; Spices (curcuma), AFB_1, AFB_2, AFG_1, and OTA, literature[530]; Spices (curry), AFB_2, AFG_1, AFG_2, and OTA, literature[530]; Spices (pepper), AFB_1 and AFB_2, literature[530]; Spices (poppy seed), AFB_1, literature[530]

incidence: 2/2*, conc. range: 3.15–3.89 µg/kg, ∅ conc.: 3.52 µg/kg, sample origin: Penang (state, island), Malaysia, sample year: unknown (second study), country: Malaysia[530], *mixed spices for chicken curry

- Co-contamination: 2 sa. co-contaminated with AFB_1 and AFB_2
- Further contamination of plants and plant products other than cereals and their products reported in the present book/article[530]: Spices, Spices (cayenne pepper, chilli), Spices (coriander), Spices (cumin), Spices (fennel), Spices (masala), AFB_1, AFB_2, AFG_1, AFG_2, and OTA, literature[530]; Spices (curcuma), AFB_1, AFB_2, AFG_1, and OTA, literature[530]; Spices (curry), AFB_2, AFG_1, AFG_2, and OTA, literature[530]; Spices (pepper), AFB_1 and AFB_2, literature[530]; Spices (poppy seed), AFB_1, literature[530]

incidence: 0/1*, conc.: no contamination, sample origin: Penang (state, island), Malaysia, sample year: unknown (second study), country: Malaysia[530], *mixed spices for meat curry

- Co-contamination: not reported
- Further contamination of plants and plant products other than cereals and their products reported in the present book/article[530]: Spices, Spices (cayenne pepper, chilli), Spices (coriander), Spices (cumin), Spices (fennel), Spices (masala), AFB_1, AFB_2, AFG_1, AFG_2, and OTA, literature[530]; Spices (curcuma), AFB_1, AFB_2, AFG_1, and OTA, literature[530]; Spices (curry), AFB_2, AFG_1, AFG_2, and OTA, literature[530]; Spices (pepper), AFB_1 and AFB_2, literature[530]; Spices (poppy seed), AFB_1, literature[530]

incidence: 6/6, conc. range: ≤1.03 µg/kg, ∅ conc.: 0.41 µg/kg, sample origin: retail stores in Lower Saxony (federal state), Germany, sample year: 2009, country: Germany[829]

- Co-contamination: not reported
- Further contamination of plants and plant products other than cereals and their products reported in the present book/article[829]: Nuts (hazelnut), AFB_1 and AF/S, literature[829]; Nuts (pistachio nut), AFB_1, literature[829]; Oils (maize oil), ZEA, literature[829]; Spices (cinnamon), OTA, literature[829]; Spices (curry), AF/S and OTA, literature[829]; Spices (ginger), AFB_1 and AF/S, literature[829]; Spices (nutmeg), AFB_1, AF/S, and OTA, literature[829]

Aflatoxin B_2

incidence: 3/5*, conc. range: 0.03–0.28 µg/kg, ∅ conc.: 0.12 µg/kg, sample origin: Penang (state, island), Malaysia, sample year: unknown (first study), country: Malaysia[530], *mixed spices for fish curry

- Co-contamination: 1 sa. co-contaminated with AFB_1, AFB_2, AFG_1, AFG_2, and OTA; 1 sa. co-contaminated with AFB_1, AFB_2, AFG_1, and OTA; 1 sa. co-contaminated with AFB_1, AFB_2, and OTA
- Further contamination of plants and plant products other than cereals and their products reported in the present book/article[530]: Spices, Spices (cayenne pepper, chilli), Spices (coriander), Spices (cumin), Spices (fennel), Spices (masala), AFB_1, AFB_2, AFG_1, AFG_2, and OTA, literature[530]; Spices (curcuma), AFB_1, AFB_2, AFG_1, and OTA, literature[530]; Spices (curry), AFB_1, AFG_1, AFG_2, and OTA, literature[530]; Spices (pepper), AFB_1 and AFB_2, literature[530]; Spices (poppy seed), AFB_1, literature[530]

incidence: 2/3*, conc. range: 0.04–0.38 µg/kg, ∅ conc.: 0.21 µg/kg, sample origin: Penang (state, island), Malaysia, sample year: unknown (first study), country: Malaysia[530], *mixed spices for meat curry

- Co-contamination: 2 sa. co-contaminated with AFB_1, AFB_2, and OTA

- Further contamination of plants and plant products other than cereals and their products reported in the present book/ article[530]: Spices, Spices (cayenne pepper, chilli), Spices (coriander), Spices (cumin), Spices (fennel), Spices (masala), AFB_1, AFB_2, AFG_1, AFG_2, and OTA, literature[530]; Spices (curcuma), AFB_1, AFB_2, AFG_1, and OTA, literature[530]; Spices (curry), AFB_1, AFG_1, AFG_2, and OTA, literature[530]; Spices (pepper), AFB_1 and AFB_2, literature[530]; Spices (poppy seed), AFB_1, literature[530]

incidence: 1/2*, conc.: 0.92 μg/kg, sample origin: Penang (state, island), Malaysia, sample year: unknown (second study), country: Malaysia[530], *mixed spices for fish curry

- Co-contamination: 1 sa. co-contaminated with AFB_1, AFB_2, AFG_1, and OTA
- Further contamination of plants and plant products other than cereals and their products reported in the present book/ article[530]: Spices, Spices (cayenne pepper, chilli), Spices (coriander), Spices (cumin), Spices (fennel), Spices (masala), AFB_1, AFB_2, AFG_1, AFG_2, and OTA, literature[530]; Spices (curcuma), AFB_1, AFB_2, AFG_1, and OTA, literature[530]; Spices (curry), AFB_1, AFG_1, AFG_2, and OTA, literature[530]; Spices (pepper), AFB_1 and AFB_2, literature[530]; Spices (poppy seed), AFB_1, literature[530]

incidence: 2/2*, conc. range: 0.57–0.63 μg/kg, ∅ conc.: 0.60 μg/kg, sample origin: Penang (state, island), Malaysia, sample year: unknown (second study), country: Malaysia[530], *mixed spices for chicken curry

- Co-contamination: 2 sa. co-contaminated with AFB_1 and AFB_2
- Further contamination of plants and plant products other than cereals and their products reported in the present book/ article[530]: Spices, Spices (cayenne pepper, chilli), Spices (coriander), Spices (cumin), Spices (fennel), Spices (masala), AFB_1, AFB_2, AFG_1, AFG_2, and OTA, literature[530]; Spices (curcuma), AFB_1, AFB_2, AFG_1, and OTA, literature[530]; Spices (curry), AFB_1, AFG_1, AFG_2, and OTA, literature[530]; Spices (pepper), AFB_1 and AFB_2, literature[530]; Spices (poppy seed), AFB_1, literature[530]

incidence: 0/1*, conc.: no contamination, sample origin: Penang (state, island), Malaysia, sample year: unknown (second study), country: Malaysia[530], *mixed spices for meat curry

- Co-contamination: not reported
- Further contamination of plants and plant products other than cereals and their products reported in the present book/ article[530]: Spices, Spices (cayenne pepper, chilli), Spices (coriander), Spices (cumin), Spices (fennel), Spices (masala), AFB_1, AFB_2, AFG_1, AFG_2, and OTA, literature[530]; Spices (curcuma), AFB_1, AFB_2, AFG_1, and OTA, literature[530]; Spices (curry), AFB_1, AFG_1, AFG_2, and OTA, literature[530]; Spices (pepper), AFB_1 and AFB_2, literature[530]; Spices (poppy seed), AFB_1, literature[530]

Aflatoxin G_1

incidence: 2/5*, conc. range: 0.01–0.16 μg/kg, ∅ conc.: 0.09 μg/kg, sample origin: Penang (state, island), Malaysia, sample year: unknown (first study), country: Malaysia[530], *mixed spices for fish curry

- Co-contamination: 1 sa. co-contaminated with AFB_1, AFB_2, AFG_1, AFG_2, and OTA; 1 sa. co-contaminated with AFB_1, AFB_2, AFG_1, and OTA
- Further contamination of plants and plant products other than cereals and their products reported in the present book/ article[530]: Spices, Spices (cayenne pepper, chilli), Spices (coriander), Spices (cumin), Spices (fennel), Spices (masala), AFB_1, AFB_2, AFG_1, AFG_2, and OTA, literature[530]; Spices (curcuma), AFB_1, AFB_2, AFG_1, and OTA, literature[530]; Spices (curry), AFB_1, AFB_2, AFG_2, and OTA, literature[530]; Spices (pepper), AFB_1 and AFB_2, literature[530]; Spices (poppy seed), AFB_1, literature[530]

incidence: 0/3*, conc.: no contamination, sample origin: Penang (state, island), Malaysia, sample year: unknown (first study), country: Malaysia[530], *mixed spices for meat curry

- Co-contamination: not reported
- Further contamination of plants and plant products other than cereals and their products reported in the present book/ article[530]: Spices, Spices (cayenne pepper, chilli), Spices (coriander), Spices (cumin), Spices (fennel), Spices (masala), AFB_1, AFB_2, AFG_1, AFG_2, and OTA, literature[530]; Spices (curcuma), AFB_1, AFB_2, AFG_1, and OTA, literature[530]; Spices (curry), AFB_1, AFB_2, AFG_2, and OTA, literature[530]; Spices (pepper), AFB_1 and AFB_2, literature[530]; Spices (poppy seed), AFB_1, literature[530]

incidence: 1/2*, conc.: 0.62 μg/kg, sample origin: Penang (state, island), Malaysia, sample year: unknown (second study), country: Malaysia[530], *mixed spices for fish curry

- Co-contamination: not reported
- Further contamination of plants and plant products other than cereals and their products reported in the present book/ article[530]: Spices, Spices (cayenne pepper, chilli), Spices (coriander), Spices (cumin), Spices (fennel), Spices (masala), AFB_1, AFB_2, AFG_1, AFG_2, and OTA, literature[530]; Spices (curcuma), AFB_1, AFB_2, AFG_1, and OTA, literature[530]; Spices (curry), AFB_1, AFB_2, AFG_2, and OTA, literature[530]; Spices (pepper), AFB_1 and AFB_2, literature[530]; Spices (poppy seed), AFB_1, literature[530]

incidence: 0/2*, conc.: no contamination, sample origin: Penang (state, island), Malaysia, sample year: unknown (second study), country: Malaysia[530], *mixed spices for chicken curry

- Co-contamination: not reported
- Further contamination of plants and plant products other than cereals and their products reported in the present book/ article[530]: Spices, Spices (cayenne pepper, chilli), Spices (coriander), Spices (cumin), Spices (fennel), Spices (masala), AFB_1, AFB_2, AFG_1, AFG_2, and OTA, literature[530]; Spices (curcuma), AFB_1, AFB_2, AFG_1, and OTA, literature[530]; Spices (curry), AFB_1, AFB_2, AFG_2, and OTA, literature[530]; Spices (pepper), AFB_1 and AFB_2, literature[530]; Spices (poppy seed), AFB_1, literature[530]

incidence: 0/1*, conc.: no contamination, sample origin: Penang (state, island), Malaysia, sample year: unknown (second study), country: Malaysia[530], *mixed spices for meat curry

- Co-contamination: not reported
- Further contamination of plants and plant products other than cereals and their products reported in the present book/ article[530]: Spices, Spices (cayenne pepper, chilli), Spices (coriander), Spices (cumin), Spices (fennel), Spices (masala), AFB_1, AFB_2, AFG_1, AFG_2, and OTA, literature[530]; Spices (curcuma), AFB_1, AFB_2, AFG_1, and OTA, literature[530]; Spices (curry), AFB_1, AFB_2, AFG_2, and OTA, literature[530]; Spices (pepper), AFB_1 and AFB_2, literature[530]; Spices (poppy seed), AFB_1, literature[530]

Aflatoxin G_2

incidence: 1/5*, conc.: 0.02 μg/kg, sample origin: Penang (state, island), Malaysia, sample year: unknown (first study), country: Malaysia[530], *mixed spices for fish curry

- Co-contamination: not reported

- Further contamination of plants and plant products other than cereals and their products reported in the present book/article[530]: Spices, Spices (cayenne pepper, chilli), Spices (coriander), Spices (cumin), Spices (fennel), Spices (masala), AFB_1, AFB_2, AFG_1, AFG_2, and OTA, literature[530]; Spices (curcuma), AFB_1, AFB_2, AFG_1, and OTA, literature[530]; Spices (curry), AFB_1, AFB_2, AFG_1, and OTA, literature[530]; Spices (pepper), AFB_1 and AFB_2, literature[530]; Spices (poppy seed), AFB_1, literature[530]

incidence: 0/3*, conc.: no contamination, sample origin: Penang (state, island), Malaysia, sample year: unknown (first study), country: Malaysia[530], *mixed spices for meat curry
- Co-contamination: not reported
- Further contamination of plants and plant products other than cereals and their products reported in the present book/article[530]: Spices, Spices (cayenne pepper, chilli), Spices (coriander), Spices (cumin), Spices (fennel), Spices (masala), AFB_1, AFB_2, AFG_1, AFG_2, and OTA, literature[530]; Spices (curcuma), AFB_1, AFB_2, AFG_1, and OTA, literature[530]; Spices (curry), AFB_1, AFB_2, AFG_1, and OTA, literature[530]; Spices (pepper), AFB_1 and AFB_2, literature[530]; Spices (poppy seed), AFB_1, literature[530]

incidence: 0/2*, conc.: no contamination, sample origin: Penang (state, island), Malaysia, sample year: unknown (second study), country: Malaysia[530], *mixed spices for fish curry
- Co-contamination: not reported
- Further contamination of plants and plant products other than cereals and their products reported in the present book/article[530]: Spices, Spices (cayenne pepper, chilli), Spices (coriander), Spices (cumin), Spices (fennel), Spices (masala), AFB_1, AFB_2, AFG_1, AFG_2, and OTA, literature[530]; Spices (curcuma), AFB_1, AFB_2, AFG_1, and OTA, literature[530]; Spices (curry), AFB_1, AFB_2, AFG_1, and OTA, literature[530]; Spices (pepper), AFB_1 and AFB_2, literature[530]; Spices (poppy seed), AFB_1, literature[530]

incidence: 0/2*, conc.: no contamination, sample origin: Penang (state, island), Malaysia, sample year: unknown (second study), country: Malaysia[530], *mixed spices for chicken curry
- Co-contamination: not reported
- Further contamination of plants and plant products other than cereals and their products reported in the present book/article[530]: Spices, Spices (cayenne pepper, chilli), Spices (coriander), Spices (cumin), Spices (fennel), Spices (masala), AFB_1, AFB_2, AFG_1, AFG_2, and OTA, literature[530]; Spices (curcuma), AFB_1, AFB_2, AFG_1, and OTA, literature[530]; Spices (curry), AFB_1, AFB_2, AFG_1, and OTA, literature[530]; Spices (pepper), AFB_1 and AFB_2, literature[530]; Spices (poppy seed), AFB_1, literature[530]

incidence: 0/1*, conc.: no contamination, sample origin: Penang (state, island), Malaysia, sample year: unknown (second study), country: Malaysia[530], *mixed spices for meat curry
- Co-contamination: not reported
- Further contamination of plants and plant products other than cereals and their products reported in the present book/article[530]: Spices, Spices (cayenne pepper, chilli), Spices (coriander), Spices (cumin), Spices (fennel), Spices (masala), AFB_1, AFB_2, AFG_1, AFG_2, and OTA, literature[530]; Spices (curcuma), AFB_1, AFB_2, AFG_1, and OTA, literature[530]; Spices (curry), AFB_1, AFB_2, AFG_1, and OTA, literature[530]; Spices (pepper), AFB_1 and AFB_2, literature[530]; Spices (poppy seed), AFB_1, literature[530]

Aflatoxin/s

incidence: 3?/3*, conc. range: 0.8–61.2 µg/kg**, sample origin: specialist food shops, UK, sample year: unknown, country: UK[16], *curry powder, "hot," **AFB_1, AFB_2, AFG_1, and AFG_2

incidence: 1/3*, conc.: 0.4 µg/kg**, sample origin: specialist food shops, UK, sample year: unknown, country: UK[16], *curry powder, "mild," **AFB_1, AFB_2, AFG_1, and AFG_2

incidence: 0/4*, conc.: no contamination**, sample origin: specialist food shops, UK, sample year: unknown, country: UK[16], *curry mixes, dry, **AFB_1, AFB_2, AFG_1, and AFG_2
- Co-contamination: not reported
- Further contamination of plants and plant products other than cereals and their products reported in the present book/article[16]: Food, AF/S, DON, FUS-X, NIV, OTA, and ZEA, literature[16]; Nuts (almond), AF/S, literature[16]; Oils, Spices (fennel), AF/S and ZEA, literature[16]; Oils (sesame oil), OTA, literature[16]; Paste (curry paste), AF/S, FB/S, NIV, OTA, and ZEA, literature[16]; Pickles, AF/S, FB/S, OTA, and ZEA, literature[16]; Sauces (chilli sauce), NIV, OTA, and ZEA, literature[16]; Spices, AF/S, FB/S, OTA, and ZEA, literature[16]; Spices (cayenne pepper, chilli), AF/S, DAS, DON, HT-2, OTA, and ZEA, literature[16]; Spices (coriander), AF/S, DON, OTA, and ZEA, literature[16]; Spices (curry), DAS, FB/S, FUS-X, HT-2, NEO, NIV, OTA, T-2, and ZEA, literature[16]; Spices (garlic), DON, FUS-X, and NIV, literature[16]; Spices (ginger), AF/S, DON, NEO, NIV, OTA, and T-2, literature[16]; Spices (tandoori), AF/S, FB/S, NIV, OTA, and T-2, literature[16]

incidence: 3/20*, conc. range: 0.50–9.10 µg/kg**, sample origin: market stalls, shops, and supermarkets, Ireland, sample year: unknown, country: Ireland[187], *curry powder, **AFB_1, AFB_2, AFG_1, and AFG_2
- Co-contamination: not reported
- Further contamination of plants and plant products other than cereals and their products reported in the present book/article[187]: Spices (cayenne pepper, chilli), Spices (coriander), Spices (curcuma), Spices (paprika), Spices (pepper), AF/S, literature[187]

incidence: 2/20, conc. range: 0.13–0.46 µg/kg*, Ø conc.: 0.295 µg/kg*, sample origin: markets in ten cities, Korea, sample year: May–September 2006, country: Korea[188], *AFB_1, AFB_2, AFG_1, and AFG_2
- Co-contamination: not reported
- Further contamination of plants and plant products other than cereals and their products reported in the present book/article[188]: Paste (red pepper paste), Spices (cayenne pepper, chilli), Spices (ginger), AF/S, literature[188]

incidence: 10/29, conc. range: 1–1.9 µg/kg* (6 sa.), 2–3.9 µg/kg* (2 sa.), 4–10 µg/kg* (2 sa., maximum: 5.2 µg/kg*), sample origin: supermarkets, local and ethnic shops in Norwich (city) area, UK, sample year: unknown, country: UK[250], *AFB_1, AFB_2, AFG_1, and AFG_2
- Co-contamination: not reported
- Further contamination of plants and plant products other than cereals and their products reported in the present book/article[250]: Spices, Spices (cayenne pepper, chilli), AF/S, literature[250]

incidence: 5/5*, conc. range: 0.2–225 µg/kg**, Ø conc.: 46.1 µg/kg**, sample origin: available in New Zealand, sample year: 2009, country: New Zealand[277], *curry powder, **AFS total

- Co-contamination: not reported
- Further contamination of plants and plant products other than cereals and their products reported in the present book/article[277]: Bars, AFB_1, AF/S, DON, NIV, and OTA, literature[277]; Butter (peanut butter), Confectionery, Food, Fruits, dried, Nuts, Nuts (peanut), Nuts (pistachio nut), Sauces (peanut sauce), AFB_1 and AF/S, literature[277]; Coffee, Dates, Spices (nutmeg), Spices (pepper), Vine fruits, dried, OTA, literature[277]; Figs, Spices (cayenne pepper, chilli), Spices (ginger), Spices (paprika), AFB_1, AF/S, and OTA, literature[277]; Nuts (Brazil nut, paranut), AFB_1, literature[277]; Spices (curry), AFB_1 and OTA, literature[277]

incidence: 5/5*, conc. range: ≤225 µg/kg**, sample origin: New Zealand, sample year: 2009, country: New Zealand[702], *curry powder, **AFB_1, AFB_2, and AFG_1

- Co-contamination: not reported
- Further contamination of plants and plant products other than cereals and their products reported in the present book/article[702]: Figs, Spices (cayenne pepper, chilli), Spices (ginger), Spices (paprika), AF/S and OTA, literature[702]; Fruits, dried, AF/S, literature[702]; Spices (curry), Spices (pepper), Vine fruits, dried, OTA, literature[702]

incidence: 6/6, conc. range: ≤1.17 µg/kg*, ∅ conc.: 0.45 µg/kg*, sample origin: retail stores in Lower Saxony (federal state), Germany, sample year: 2009, country: Germany[829], *AFB_1, AFB_2, AFG_1, and AFG_2

- Co-contamination: not reported
- Further contamination of plants and plant products other than cereals and their products reported in the present book/article[829]: Nuts (hazelnut), AFB_1 and AF/S, literature[829]; Nuts (pistachio nut), AFB_1, literature[829]; Oils (maize oil), ZEA, literature[829]; Spices (cinnamon), OTA, literature[829]; Spices (curry), AFB_1 and OTA, literature[829]; Spices (ginger), AFB_1 and AF/S, literature[829]; Spices (nutmeg), AFB_1, AF/S, and OTA, literature[829]

Aspergillus and ***Penicillium*** Toxins

Ochratoxin A

incidence: 3?/3*, conc. range: 2.3–21.3 µg/kg, sample origin: specialist food shops, UK, sample year: unknown, country: UK[16], *curry powder, "hot"

incidence: 3?/3*, conc. range: 1.8–9.4 µg/kg, sample origin: specialist food shops, UK, sample year: unknown, country: UK[16], *curry powder, "mild"

incidence: 4?/ 4*, conc. range: 1.2–5.4 µg/kg, sample origin: specialist food shops, UK, sample year: unknown, country: UK[16], *curry mixes, dry

- Co-contamination: not reported
- Further contamination of plants and plant products other than cereals and their products reported in the present book/article[16]: Food, AF/S, DON, FUS-X, NIV, OTA, and ZEA, literature[16]; Nuts (almond), AF/S, literature[16]; Oils, Spices (fennel), AF/S and ZEA, literature[16]; Oils (sesame oil), OTA, literature[16]; Paste (curry paste), AF/S, FB/S, NIV, OTA, and ZEA, literature[16]; Pickles, AF/S, FB/S, OTA, and ZEA, literature[16]; Sauces (chilli sauce), NIV, OTA, and ZEA, literature[16]; Spices, AF/S, FB/S, OTA, and ZEA, literature[16]; Spices (cayenne pepper, chilli), AF/S, DAS, DON, HT-2, OTA, and ZEA, literature[16]; Spices (coriander), AF/S, DON, OTA, and ZEA, literature[16]; Spices (curry), AF/S, DAS, FB/S, FUS-X, HT-2, NEO, NIV, T-2, and ZEA, literature[16]; Spices (garlic), DON, FUS-X, and NIV, literature[16]; Spices (ginger), AF/S, DON, NEO, NIV, OTA, and T-2, literature[16]; Spices (tandoori), AF/S, FB/S, NIV, OTA, and T-2, literature[16]

incidence: 3/3*, conc. range: 5.0–9.9 µg/kg (1 sa.), >25 µg/kg (2 sa., maximum: 33 µg/kg), sample origin: Austria, sample year: 1994, country: EU[194], *curry powder

- Co-contamination: not reported
- Further contamination of plants and plant products other than cereals and their products reported in the present book/article[194]: Coffee, Coffee beans, Spices (cayenne pepper, chilli), Spices (marjoram), Spices (pepper), OTA, literature[194]

incidence: 5/5*, conc. range: 0.2–3.5 µg/kg, sample origin: available in New Zealand, sample year: 2008/2009, country: New Zealand[277], *curry powder

- Co-contamination: not reported
- Further contamination of plants and plant products other than cereals and their products reported in the present book/article[277]: Bars, AFB_1, AF/S, DON, NIV, and OTA, literature[277]; Butter (peanut butter), Confectionery, Food, Fruits, dried, Nuts, Nuts (peanut), Nuts (pistachio nut), Sauces (peanut sauce), Spices (curry), AFB_1 and AF/S, literature[277]; Coffee, Dates, Spices (nutmeg), Spices (pepper), Vine fruits, dried, OTA, literature[277]; Figs, Spices (cayenne pepper, chilli), Spices (ginger), Spices (paprika), AFB_1, AF/S, and OTA, literature[277]; Nuts (Brazil nut, paranut), AFB_1, literature[277]

incidence: 5/5*, conc. range: 0.28–4.53 µg/kg, ∅ conc.: 1.72 µg/kg, sample origin: Penang (state, island), Malaysia, sample year: unknown, country: Malaysia[530], *mixed spices for fish curry

- Co-contamination: not reported
- Further contamination of plants and plant products other than cereals and their products reported in the present book/article[530]: Spices, Spices (cayenne pepper, chilli), Spices (coriander), Spices (cumin), Spices (fennel), Spices (masala), AFB_1, AFB_2, AFG_1, AFG_2, and OTA, literature[530]; Spices (curcuma), AFB_1, AFB_2, AFG_1, and OTA, literature[530]; Spices (curry), AFB_1, AFB_2, AFG_1, and AFG_2, literature[530]; Spices (pepper), AFB_1 and AFB_2, literature[530]; Spices (poppy seed), AFB_1, literature[530]

incidence: 3/3*, conc. range: 0.14–9.59 µg/kg, ∅ conc.: 3.42 µg/kg, sample origin: Penang (state, island), Malaysia, sample year: unknown, country: Malaysia[530], *mixed spices for meat curry

- Co-contamination: not reported
- Further contamination of plants and plant products other than cereals and their products reported in the present book/article[530]: Spices, Spices (cayenne pepper, chilli), Spices (coriander), Spices (cumin), Spices (fennel), Spices (masala), AFB_1, AFB_2, AFG_1, AFG_2, and OTA, literature[530]; Spices (curcuma), AFB_1, AFB_2, AFG_1, and OTA, literature[530]; Spices (curry), AFB_1, AFB_2, AFG_1, and AFG_2, literature[530]; Spices (pepper), AFB_1 and AFB_2, literature[530]; Spices (poppy seed), AFB_1, literature[530]

incidence: 5/5*, conc. range: ≤3.5 µg/kg, sample origin: New Zealand, sample year: 2009, country: New Zealand[702], *curry powder

- Co-contamination: not reported
- Further contamination of plants and plant products other than cereals and their products reported in the present book/article[702]: Figs, Spices (cayenne pepper, chilli), Spices (ginger), Spices (paprika), AF/S and OTA, literature[702]; Fruits, dried,

Spices (curry), AF/S, literature[702]; Spices (pepper), Vine fruits, dried, OTA, literature[702]

incidence: 6/6, conc. range: ≤2.74 μg/kg*, ∅ conc.: 1.44 μg/kg*, sample origin: retail stores in Lower Saxony (federal state), Germany, sample year: 2009, country: Germany[829]

- Co-contamination: not reported
- Further contamination of plants and plant products other than cereals and their products reported in the present book/article[829]: Nuts (hazelnut), AFB_1 and AF/S, literature[829]; Nuts (pistachio nut), AFB_1, literature[829]; Oils (maize oil), ZEA, literature[829]; Spices (cinnamon), OTA, literature[829]; Spices (curry), AFB_1 and AF/S, literature[829]; Spices (ginger), AFB_1 and AF/S, literature[829]; Spices (nutmeg), AFB_1, AF/S, and OTA, literature[829]

incidence: 5/11*, conc. range: 1.04–6.24 μg/kg, ∅ conc.: 2.44 μg/kg, sample origin: supermarkets, shops, and market stalls in eight cities, China, sample year: April–July 2009, country: China[885], *curry powder

- Co-contamination: not reported
- Further contamination of plants and plant products other than cereals and their products reported in the present book/article[885]: Spices (anise), Spices (cayenne pepper, chilli), Spices (cinnamon), Spices (cumin), Spices (prickly ash), OTA, literature[885]

Fusarium Toxins

Fumonisin/s

incidence: 3?/3*, conc. range: 15–16 μg/kg**, sample origin: specialist food shops, UK, sample year: unknown, country: UK[16], *curry powder, "hot," **FB_1 and FB_2

incidence: 1/3*, conc.: 21 μg/kg**, sample origin: specialist food shops, UK, sample year: unknown, country: UK[16], *curry powder, "mild, **FB_1 and FB_2

incidence: 1/4*, conc.: 230 μg/kg**, sample origin: specialist food shops, UK, sample year: unknown, country: UK[16], *curry mixes, dry, **FB_1 and FB_2

- Co-contamination: not reported
- Further contamination of plants and plant products other than cereals and their products reported in the present book/article[16]: Food, AF/S, DON, FUS-X, NIV, OTA, and ZEA, literature[16]; Nuts (almond), AF/S, literature[16]; Oils, Spices (fennel), AF/S and ZEA, literature[16]; Oils (sesame oil), OTA, literature[16]; Paste (curry paste), AF/S, FB/S, NIV, OTA, and ZEA, literature[16]; Pickles, AF/S, FB/S, OTA, and ZEA, literature[16]; Sauces (chilli sauce), NIV, OTA, and ZEA, literature[16]; Spices, AF/S, FB/S, OTA, and ZEA, literature[16]; Spices (cayenne pepper, chilli), AF/S, DAS, DON, HT-2, OTA, and ZEA, literature[16]; Spices (coriander), AF/S, DON, OTA, and ZEA, literature[16]; Spices (curry), AF/S, DAS, FUS-X, HT-2, NEO, NIV, OTA, T-2, and ZEA, literature[16]; Spices (garlic), DON, FUS-X, and NIV, literature[16]; Spices (ginger), AF/S, DON, NEO, NIV, OTA, and T-2, literature[16]; Spices (tandoori), AF/S, FB/S, NIV, OTA, and T-2, literature[16]

Type A Trichothecenes

Diacetoxyscirpenol

incidence: 0/3*, conc.: no contamination, sample origin: specialist food shops, UK, sample year: unknown, country: UK[16], *curry powder, "hot"

incidence: 0/3*, conc.: no contamination, sample origin: specialist food shops, UK, sample year: unknown, country: UK[16], *curry powder, "mild"

incidence: 1/4*, conc.: 25 μg/kg, sample origin: specialist food shops, UK, sample year: unknown, country: UK[16], *curry mixes, dry

- Co-contamination: not reported
- Further contamination of plants and plant products other than cereals and their products reported in the present book/article[16]: Food, AF/S, DON, FUS-X, NIV, OTA, and ZEA, literature[16]; Nuts (almond), AF/S, literature[16]; Oils, Spices (fennel), AF/S and ZEA, literature[16]; Oils (sesame oil), OTA, literature[16]; Paste (curry paste), AF/S, FB/S, NIV, OTA, and ZEA, literature[16]; Pickles, AF/S, FB/S, OTA, and ZEA, literature[16]; Sauces (chilli sauce), NIV, OTA, and ZEA, literature[16]; Spices, AF/S, FB/S, OTA, and ZEA, literature[16]; Spices (cayenne pepper, chilli), AF/S, DAS, DON, HT-2, OTA, and ZEA, literature[16]; Spices (coriander), AF/S, DON, OTA, and ZEA, literature[16]; Spices (curry), AF/S, FB/S, FUS-X, HT-2, NEO, NIV, OTA, T-2, and ZEA, literature[16]; Spices (garlic), DON, FUS-X, and NIV, literature[16]; Spices (ginger), AF/S, DON, NEO, NIV, OTA, and T-2, literature[16]; Spices (tandoori), AF/S, FB/S, NIV, OTA, and T-2, literature[16]

HT-2 Toxin

incidence: 0/3*, conc.: no contamination, sample origin: specialist food shops, UK, sample year: unknown, country: UK[16], *curry powder, "hot"

incidence: 0/3*, conc.: no contamination, sample origin: specialist food shops, UK, sample year: unknown, country: UK[16], *curry powder, "mild"

incidence: 1/4*, conc.: 17 μg/kg, sample origin: specialist food shops, UK, sample year: unknown, country: UK[16], *curry mixes, dry

- Co-contamination: not reported
- Further contamination of plants and plant products other than cereals and their products reported in the present book/article[16]: Food, AF/S, DON, FUS-X, NIV, OTA, and ZEA, literature[16]; Nuts (almond), AF/S, literature[16]; Oils, Spices (fennel), AF/S and ZEA, literature[16]; Oils (sesame oil), OTA, literature[16]; Paste (curry paste), AF/S, FB/S, NIV, OTA, and ZEA, literature[16]; Pickles, AF/S, FB/S, OTA, and ZEA, literature[16]; Sauces (chilli sauce), NIV, OTA, and ZEA, literature[16]; Spices, AF/S, FB/S, OTA, and ZEA, literature[16]; Spices (cayenne pepper, chilli), AF/S, DAS, DON, HT-2, OTA, and ZEA, literature[16]; Spices (coriander), AF/S, DON, OTA, and ZEA, literature[16]; Spices (curry), AF/S, DAS, FB/S, FUS-X, NEO, NIV, OTA, T-2, and ZEA, literature[16]; Spices (garlic), DON, FUS-X, and NIV, literature[16]; Spices (ginger), AF/S, DON, NEO, NIV, OTA, and T-2, literature[16]; Spices (tandoori), AF/S, FB/S, NIV, OTA, and T-2, literature[16]

Neosolaniol

incidence: 0/3*, conc.: no contamination, sample origin: specialist food shops, UK, sample year: unknown, country: UK[16], *curry powder, "hot"

incidence: 0/3*, conc.: no contamination, sample origin: specialist food shops, UK, sample year: unknown, country: UK[16], *curry powder, "mild"

incidence: 1/4*, conc.: 9 μg/kg, sample origin: specialist food shops, UK, sample year: unknown, country: UK[16], *curry mixes, dry

- Co-contamination: not reported

- Further contamination of plants and plant products other than cereals and their products reported in the present book/article[16]: Food, AF/S, DON, FUS-X, NIV, OTA, and ZEA, literature[16]; nuts (almond), AF/S, literature[16]; Oils, Spices (fennel), AF/S and ZEA, literature[16]; Oils (sesame oil), OTA, literature[16]; Paste (curry paste), AF/S, FB/S, NIV, OTA, and ZEA, literature[16]; Pickles, AF/S, FB/S, OTA, and ZEA, literature[16]; Sauces (chilli sauce), NIV, OTA, and ZEA, literature[16]; Spices, AF/S, FB/S, OTA, and ZEA, literature[16]; Spices (cayenne pepper, chilli), AF/S, DAS, DON, HT-2, OTA, and ZEA, literature[16]; Spices (coriander), AF/S, DON, OTA, and ZEA, literature[16]; Spices (curry), AF/S, DAS, FB/S, FUS-X, HT-2, NIV, OTA, T-2, and ZEA, literature[16]; Spices (garlic), DON, FUS-X, and NIV, literature[16]; Spices (ginger), AF/S, DON, NEO, NIV, OTA, and T-2, literature[16]; Spices (tandoori), AF/S, FB/S, NIV, OTA, and T-2, literature[16]

T-2 Toxin

incidence: 0/3*, conc.: no contamination, sample origin: specialist food shops, UK, sample year: unknown, country: UK[16], *curry powder, "hot

incidence: 0/3*, conc.: no contamination, sample origin: specialist food shops, UK, sample year: unknown, country: UK[16], *curry powder, "mild

incidence: 1/4*, conc.: 13 µg/kg, sample origin: specialist food shops, UK, sample year: unknown, country: UK[16], *curry mixes, dry

- Co-contamination: not reported
- Further contamination of plants and plant products other than cereals and their products reported in the present book/article[16]: Food, AF/S, DON, FUS-X, NIV, OTA, and ZEA, literature[16]; Nuts (almond), AF/S, literature[16]; Oils, Spices (fennel), AF/S and ZEA, literature[16]; Oils (sesame oil), OTA, literature[16]; Paste (curry paste), AF/S, FB/S, NIV, OTA, and ZEA, literature[16]; Pickles, AF/S, FB/S, OTA, and ZEA, literature[16]; Sauces (chilli sauce), NIV, OTA, and ZEA, literature[16]; Spices, AF/S, FB/S, OTA, and ZEA, literature[16]; Spices (cayenne pepper, chilli), AF/S, DAS, DON, HT-2, OTA, and ZEA, literature[16]; Spices (coriander), AF/S, DON, OTA, and ZEA, literature[16]; Spices (curry), AF/S, DAS, FB/S, FUS-X, HT-2, NEO, NIV, OTA, and ZEA, literature[16]; Spices (garlic), DON, FUS-X, and NIV, literature[16]; Spices (ginger), AF/S, DON, NEO, NIV, OTA, and T-2, literature[16]; Spices (tandoori), AF/S, FB/S, NIV, OTA, and T-2, literature[16]

Type B Trichothecenes

Fusarenon-X (4-Acetylnivalenol)

incidence: 0/3*, conc.: no contamination, sample origin: specialist food shops, UK, sample year: unknown, country: UK[16], *curry powder, "hot"

incidence: 0/3*, conc.: no contamination, sample origin: specialist food shops, UK, sample year: unknown, country: UK[16], *curry powder, "mild"

incidence: 1/4*, conc.: 7 µg/kg, sample origin: specialist food shops, UK, sample year: unknown, country: UK[16], *curry mixes, dry

- Co-contamination: not reported
- Further contamination of plants and plant products other than cereals and their products reported in the present book/article[16]: Food, AF/S, DON, FUS-X, NIV, OTA, and ZEA, literature[16]; Nuts (almond), AF/S, literature[16]; Oils, Spices (fennel), AF/S and ZEA, literature[16]; Oils (sesame oil), OTA, literature[16]; Paste (curry paste), AF/S, FB/S, NIV, OTA, and ZEA, literature[16]; Pickles, AF/S, FB/S, OTA, and ZEA, literature[16]; Sauces (chilli sauce), NIV, OTA, and ZEA, literature[16]; Spices, AF/S, FB/S, OTA, and ZEA, literature[16]; Spices (cayenne pepper, chilli), AF/S, DAS, DON, HT-2, OTA, and ZEA, literature[16]; Spices (coriander), AF/S, DON, OTA, and ZEA, literature[16]; Spices (curry), AF/S, DAS, FB/S, HT-2, NEO, NIV, OTA, T-2, and ZEA, literature[16]; Spices (garlic), DON, FUS-X, and NIV, literature[16]; Spices (ginger), AF/S, DON, NEO, NIV, OTA, and T-2, literature[16]; Spices (tandoori), AF/S, FB/S, NIV, OTA, and T-2, literature[16]

Nivalenol

incidence: 3?/3*, conc. range: 15–50 µg/kg, sample origin: specialist food shops, UK, sample year: unknown, country: UK[16], *curry powder, "hot"

incidence: 3?/3*, conc. range: 9–67 µg/kg, sample origin: specialist food shops, UK, sample year: unknown, country: UK[16], *curry powder, "mild"

incidence: 1/4*, conc.: 14 µg/kg, sample origin: specialist food shops, UK, sample year: unknown, country: UK[16], *curry mixes, dry

- Co-contamination: not reported
- Further contamination of plants and plant products other than cereals and their products reported in the present book/article[16]: Food, AF/S, DON, FUS-X, NIV, OTA, and ZEA, literature[16]; Nuts (almond), AF/S, literature[16]; Oils, Spices (fennel), AF/S and ZEA, literature[16]; Oils (sesame oil), OTA, literature[16]; Paste (curry paste), AF/S, FB/S, NIV, OTA, and ZEA, literature[16]; Pickles, AF/S, FB/S, OTA, and ZEA, literature[16]; Sauces (chilli sauce), NIV, OTA, and ZEA, literature[16]; Spices, AF/S, FB/S, OTA, and ZEA, literature[16]; Spices (cayenne pepper, chilli), AF/S, DAS, DON, HT-2, OTA, and ZEA, literature[16]; Spices (coriander), AF/S, DON, OTA, and ZEA, literature[16]; Spices (curry), AF/S, DAS, FB/S, FUS-X, HT-2, NEO, OTA, T-2, and ZEA, literature[16]; Spices (garlic), DON, FUS-X, and NIV, literature[16]; Spices (ginger), AF/S, DON, NEO, NIV, OTA, and T-2, literature[16]; Spices (tandoori), AF/S, FB/S, NIV, OTA, and T-2, literature[16]

Other *Fusarium* Toxins

Zearalenone

incidence: 3?/3*, conc. range: 1.2–10.8 µg/kg, sample origin: specialist food shops, UK, sample year: unknown, country: UK[16], *curry powder, "hot"

incidence: 0/3*, conc.: no contamination, sample origin: specialist food shops, UK, sample year: unknown, country: UK[16], *curry powder, "mild"

incidence: 1/4*, conc.: 5.2 µg/kg, sample origin: specialist food shops, UK, sample year: unknown, country: UK[16], *curry mixes, dry

- Co-contamination: not reported
- Further contamination of plants and plant products other than cereals and their products reported in the present book/article[16]: Food, AF/S, DON, FUS-X, NIV, OTA, and ZEA, literature[16]; Nuts (almond), AF/S, literature[16]; Oils, Spices (fennel), AF/S and ZEA, literature[16]; Oils (sesame oil), OTA, literature[16]; Paste (curry paste), AF/S, FB/S, NIV, OTA, and ZEA, literature[16]; Pickles, AF/S, FB/S, OTA, and ZEA, literature[16]; Sauces (chilli sauce), NIV, OTA, and ZEA, literature[16]; Spices, AF/S, FB/S, OTA, and ZEA, literature[16]; Spices (cayenne pepper, chilli), AF/S, DAS, DON, HT-2, OTA, and ZEA, literature[16]; Spices

(coriander), AF/S, DON, OTA, and ZEA, literature[16]; Spices (curry), AF/S, DAS, FB/S, FUS-X, HT-2, NEO, NIV, OTA, and T-2, literature[16]; Spices (garlic), DON, FUS-X, and NIV, literature[16]; Spices (ginger), AF/S, DON, NEO, NIV, OTA, and T-2, literature[16]; Spices (tandoori), AF/S, FB/S, NIV, OTA, and T-2, literature[16]

Spices (fennel) may contain the following mycotoxins:

Aspergillus **Toxins**

Aflatoxin B_1

incidence: 6?/9, conc. range: 30–275 µg/kg, sample origin: major retail outlets in Nainital (district), India, sample year: January 1984–September 1986, country: India[22]

- Co-contamination: not reported
- Further contamination of plants and plant products other than cereals and their products reported in the present book/article[22]: Spices (ammi), AFB_1, AFB_2, and AFG_1, literature[22]; Spices (cardamom), AFB_1, AFB_2, AFG_1, AFG_2, CIT, and OTA, literature[22] Spices (cassia), AFB_1, AFB_2, AFG_1, and AFG_2, literature[22]; Spices (cayenne pepper, chilli), AFB_1, AFB_2, AFG_1, AFG_2, and ZEA, literature[22]; Spices (coriander), AFB_1, AFB_2, AFG_1, AFG_2, CIT, OTA, and ZEA, literature[22]; Spices (cumin), AFB_1, AFB_2, AFG_1, AFG_2, and CIT, literature[22]; Spices (curcuma), AFB_1, AFB_2, AFG_1, AFG_2, CIT, and RBT, literature[22]; Spices (fennel), AFB_2, AFG_1, AFG_2, CIT, OTA, and STG, literature[22]; Spices (garlic), AFB_1, AFB_2, and AFG_1, literature[22]; Spices (pepper), AFB_1, AFB_2, AFG_1, AFG_2, CIT, STG, and ZEA, literature[22]

incidence: 2/2*, conc. range: 0.01–4.37 µg/kg, ∅ conc.: 2.19 µg/kg, sample origin: Penang (state, island), Malaysia, sample year: unknown (first study), country: Malaysia[530], *"jintan manis"

- Co-contamination: 1 sa. co-contaminated with AFB_1, AFB_2, AFG_1, AFG_2, and OTA; 1 sa. contaminated solely with AFB_1
- Further contamination of plants and plant products other than cereals and their products reported in the present book/article[530]: Spices, Spices (cayenne pepper, chilli), Spices (coriander), Spices (cumin), Spices (curry), Spices (masala), AFB_1, AFB_2, AFG_1, AFG_2, and OTA, literature[530]; Spices (curcuma), AFB_1, AFB_2, AFG_1, and OTA, literature[530]; Spices (fennel), AFB_2, AFG_1, AFG_2, and OTA, literature[530]; Spices (pepper), AFB_1 and AFB_2, literature[530]; Spices (poppy seed), AFB_1, literature[530]

incidence: 2/2, conc. range: 0.38–21.74 µg/kg, ∅ conc.: 11.1 µg/kg, sample origin: Penang (state, island), Malaysia, sample year: unknown (second study), country: Malaysia[530]

- Co-contamination: 1 sa. co-contaminated with AFB_1 and AFB_2; 1 sa. co-contaminated with AFB_1 and AFG_1
- Further contamination of plants and plant products other than cereals and their products reported in the present book/article[530]: Spices, Spices (cayenne pepper, chilli), Spices (coriander), Spices (cumin), Spices (curry), Spices (masala), AFB_1, AFB_2, AFG_1, AFG_2, and OTA, literature[530]; Spices (curcuma), AFB_1, AFB_2, AFG_1, and OTA, literature[530]; Spices (fennel), AFB_2, AFG_1, AFG_2, and OTA, literature[530]; Spices (pepper), AFB_1 and AFB_2, literature[530]; Spices (poppy seed), AFB_1, literature[530]

incidence: 6/7*, conc. range: 1.1–11 µg/kg, ∅ conc.: 5.7 µg/kg, sample origin: organic markets and organic shops, Turkey, sample year: May 2010–May 2011, country: Turkey[859], *organic fennel seed sa.

- Co-contamination: not reported
- Further contamination of plants and plant products other than cereals and their products reported in the present book/article[859]: Herbs (camomile), Herbs (linden), Herbs (rosehip), Herbs (sage), Spices (anise), Spices (basil), Spices (cayenne pepper, chilli), Spices (cinnamon), Spices (coriander), Spices (cumin), Spices (ginger), Spices (laurel leaves), Spices (pepper), Spices (peppermint), Spices (poppy seed), Spices (rosemary), Spices (sumac), AFB_1, literature[859]

Aflatoxin B_2

incidence: 6?/9, conc. range: 28–173 µg/kg, sample origin: major retail outlets in Nainital (district), India, sample year: January 1984–September 1986, country: India[22]

- Co-contamination: not reported
- Further contamination of plants and plant products other than cereals and their products reported in the present book/article[22]: Spices (ammi), AFB_1, AFB_2, and AFG_1, literature[22]; Spices (cardamom), AFB_1, AFB_2, AFG_1, AFG_2, CIT, and OTA, literature[22]; Spices (cassia), AFB_1, AFB_2, AFG_1, and AFG_2, literature[22]; Spices (cayenne pepper, chilli), AFB_1, AFB_2, AFG_1, AFG_2, and ZEA, literature[22]; Spices (coriander), AFB_1, AFB_2, AFG_1, AFG_2, CIT, OTA, and ZEA, literature[22]; Spices (cumin), AFB_1, AFB_2, AFG_1, AFG_2, and CIT, literature[22]; Spices (curcuma), AFB_1, AFB_2, AFG_1, AFG_2, CIT, and RBT, literature[22]; Spices (fennel), AFB_1, AFG_1, AFG_2, CIT, OTA, and STG, literature[22]; Spices (garlic), AFB_1, AFB_2, and AFG_1, literature[22]; Spices (pepper), AFB_1, AFB_2, AFG_1, AFG_2, CIT, STG, and ZEA, literature[22]

incidence: 1/2*, conc.: 0.67 µg/kg, sample origin: Penang (state, island), Malaysia, sample year: unknown (first study), country: Malaysia[530], *"jintan manis"

- Co-contamination: 1 sa. co-contaminated with AFB_1, AFB_2, AFG_1, AFG_2, and OTA
- Further contamination of plants and plant products other than cereals and their products reported in the present book/article[530]: Spices, Spices (cayenne pepper, chilli), Spices (coriander), Spices (cumin), Spices (curry), Spices (masala), AFB_1, AFB_2, AFG_1, AFG_2, and OTA, literature[530]; Spices (curcuma), AFB_1, AFB_2, AFG_1, and OTA, literature[530]; Spices (fennel), AFB_1, AFG_1, AFG_2, and OTA, literature[530]; Spices (pepper), AFB_1 and AFB_2, literature[530]; Spices (poppy seed), AFB_1, literature[530]

incidence: 1/2, conc.: 2.31 µg/kg, sample origin: Penang (state, island), Malaysia, sample year: unknown (second study), country: Malaysia[530]

- Co-contamination: 1 sa. co-contaminated with AFB_1 and AFB_2
- Further contamination of plants and plant products other than cereals and their products reported in the present book/article[530]: Spices, Spices (cayenne pepper, chilli), Spices (coriander), Spices (cumin), Spices (curry), Spices (masala), AFB_1, AFB_2, AFG_1, AFG_2, and OTA, literature[530]; Spices (curcuma), AFB_1, AFB_2, AFG_1, and OTA, literature[530]; Spices (fennel), AFB_1, AFG_1, AFG_2, and OTA, literature[530]; Spices (pepper), AFB_1 and AFB_2, literature[530]; Spices (poppy seed), AFB_1, literature[530]

Aflatoxin G_1

incidence: 6?/9, conc. range: 15–76 µg/kg, sample origin: major retail outlets in Nainital (district), India, sample year: January 1984–September 1986, country: India[22]

- Co-contamination: not reported
- Further contamination of plants and plant products other than cereals and their products reported in the present book/

article[22]: Spices (ammi), AFB_1, AFB_2, and AFG_1, literature[22]; Spices (cardamom), AFB_1, AFB_2, AFG_1, AFG_2, CIT, and OTA, literature[22]; Spices (cassia), AFB_1, AFB_2, AFG_1, and AFG_2, literature[22]; Spices (cayenne pepper, chilli), AFB_1, AFB_2, AFG_1, AFG_2, and ZEA, literature[22]; Spices (coriander), AFB_1, AFB_2, AFG_1, AFG_2, CIT, OTA, and ZEA, literature[22]; Spices (cumin), AFB_1, AFB_2, AFG_1, AFG_2, and CIT, literature[22]; Spices (curcuma), AFB_1, AFB_2, AFG_1, AFG_2, CIT, and RBT, literature[22]; Spices (fennel), AFB_1, AFB_2, AFG_2, CIT, OTA, and STG, literature[22]; Spices (garlic), AFB_1, AFB_2, and AFG_1, literature[22]; Spices (pepper), AFB_1, AFB_2, AFG_1, AFG_2, CIT, STG, and ZEA, literature[22]

incidence: 1/2*, conc.: 0.22 µg/kg, sample origin: Penang (state, island), Malaysia, sample year: unknown (first study), country: Malaysia[530], *"jintan manis"

- Co-contamination: 1 sa. co-contaminated with AFB_1, AFB_2, AFG_1, AFG_2, and OTA
- Further contamination of plants and plant products other than cereals and their products reported in the present book/article[530]: Spices, Spices (cayenne pepper, chilli), Spices (coriander), Spices (cumin), Spices (curry), Spices (masala), AFB_1, AFB_2, AFG_1, AFG_2, and OTA, literature[530]; Spices (curcuma), AFB_1, AFB_2, AFG_1, and OTA, literature[530]; Spices (fennel), AFB_1, AFB_2, AFG_2, and OTA, literature[530]; Spices (pepper), AFB_1 and AFB_2, literature[530]; Spices (poppy seed), AFB_1, literature[530]

incidence: 1/2, conc.: 0.85 µg/kg, sample origin: Penang (state, island), Malaysia, sample year: unknown (second study), country: Malaysia[530]

- Co-contamination: 1 sa. co-contaminated with AFB_1 and AFG_1
- Further contamination of plants and plant products other than cereals and their products reported in the present book/article[530]: Spices, Spices (cayenne pepper, chilli), Spices (coriander), Spices (cumin), Spices (curry), Spices (masala), AFB_1, AFB_2, AFG_1, AFG_2, and OTA, literature[530]; Spices (curcuma), AFB_1, AFB_2, AFG_1, and OTA, literature[530]; Spices (fennel), AFB_1, AFB_2, AFG_2, and OTA, literature[530]; Spices (pepper), AFB_1 and AFB_2, literature[530]; Spices (poppy seed), AFB_1, literature[530]

Aflatoxin G_2

incidence: 6?/9, conc. range: 9–69 µg/kg, sample origin: major retail outlets in Nainital (district), India, sample year: January 1984–September 1986, country: India[22]

- Co-contamination: not reported
- Further contamination of plants and plant products other than cereals and their products reported in the present book/article[22]: Spices (ammi), AFB_1, AFB_2, and AFG_1, literature[22]; Spices (cardamom), AFB_1, AFB_2, AFG_1, AFG_2, CIT, and OTA, literature[22]; Spices (cassia), AFB_1, AFB_2, AFG_1, and AFG_2, literature[22]; Spices (cayenne pepper, chilli), AFB_1, AFB_2, AFG_1, AFG_2, and ZEA, literature[22]; Spices (coriander), AFB_1, AFB_2, AFG_1, AFG_2, CIT, OTA, and ZEA, literature[22]; Spices (cumin), AFB_1, AFB_2, AFG_1, AFG_2, and CIT, literature[22]; Spices (curcuma), AFB_1, AFB_2, AFG_1, AFG_2, CIT, and RBT, literature[22]; Spices (fennel), AFB_1, AFB_2, AFG_1, CIT, OTA, and STG, literature[22]; Spices (garlic), AFB_1, AFB_2, and AFG_1, literature[22]; Spices (pepper), AFB_1, AFB_2, AFG_1, AFG_2, CIT, STG, and ZEA, literature[22]

incidence: 1/2*, conc.: 0.03 µg/kg, sample origin: Penang (state, island), Malaysia, sample year: unknown (first study), country: Malaysia[530], *"jintan manis"

- Co-contamination: 1 sa. co-contaminated with AFB_1, AFB_2, AFG_1, AFG_2, and OTA
- Further contamination of plants and plant products other than cereals and their products reported in the present book/article[530]: Spices, Spices (cayenne pepper, chilli), Spices (coriander), Spices (cumin), Spices (curry), Spices (masala), AFB_1, AFB_2, AFG_1, AFG_2, and OTA, literature[530]; Spices (curcuma), AFB_1, AFB_2, AFG_1, and OTA, literature[530]; Spices (fennel), AFB_1, AFB_2, AFG_1, and OTA, literature[530]; Spices (pepper), AFB_1 and AFB_2, literature[530]; Spices (poppy seed), AFB_1, literature[530]

incidence: 0/2, conc.: no contamination, sample origin: Penang (state, island), Malaysia, sample year: unknown (second study), country: Malaysia[530]

- Co-contamination: not reported
- Further contamination of plants and plant products other than cereals and their products reported in the present book/article[530]: Spices, Spices (cayenne pepper, chilli), Spices (coriander), Spices (cumin), Spices (curry), Spices (masala), AFB_1, AFB_2, AFG_1, AFG_2, and OTA, literature[530]; Spices (curcuma), AFB_1, AFB_2, AFG_1, and OTA, literature[530]; Spices (fennel), AFB_1, AFB_2, AFG_1, and OTA, literature[530]; Spices (pepper), AFB_1 and AFB_2, literature[530]; Spices (poppy seed), AFB_1, literature[530]

incidence: 1/2, conc.: <LOQ, sample origin: Chinese markets in Hebei (province), China, sample year: unknown, country: China/Italy[768]

- Co-contamination: not reported
- Further contamination of plants and plant products other than cereals and their products reported in the present book/article[768]: Food, AFB_1, AFB_2, and OTA, literature[768]; Spices, AFB_1, literature[768]; Spices (star anise), AFG_2, literature[768]

Aflatoxin/s

incidence: 1/3, conc.: 1.2 µg/kg*, sample origin: specialist food shops, UK, sample year: unknown, country: UK[16], *AFB_1, AFB_2, AFG_1, and AFG_2

- Co-contamination: not reported
- Further contamination of plants and plant products other than cereals and their products reported in the present book/article[16]: Food, AF/S, DON, FUS-X, NIV, OTA, and ZEA, literature[16]; Nuts (almond), AFS, literature[16]; Oils, AF/S and ZEA, literature[16]; Oils (sesame oil), OTA, literature[16]; Paste (curry paste), AF/S, FB/S, NIV, OTA, and ZEA, literature[16]; Pickles, AF/S, FB/S, OTA, and ZEA, literature[16]; Sauces (chilli sauce), NIV, OTA, and ZEA, literature[16]; Spices, AF/S, FB/S, OTA, and ZEA, literature[16]; Spices (cayenne pepper, chilli), AF/S, DAS, DON, HT-2, OTA, and ZEA, literature[16]; Spices (coriander), AF/S, DON, OTA, and ZEA, literature[16]; Spices (curry), AF/S, DAS, FB/S, FUS-X, HT-2, NEO, NIV, OTA, T-2, and ZEA, literature[16]; Spices (fennel), ZEA, literature[16]; Spices (garlic), DON, FUS-X, and NIV, literature[16]; Spices (ginger), AF/S, DON, NEO, NIV, OTA, and T-2, literature[16]; Spices (tandoori), AF/S, FB/S, NIV, OTA, and T-2, literature[16]

incidence: 2/5, conc. range: 20–25 µg/kg*, Ø conc.: 22.5 µg/kg*, sample origin: Assiut (governorate), Egypt, sample year: unknown, country: Egypt[641], *AFB_1 and AFG_1

- Co-contamination: not reported
- Further contamination of plants and plant products other than cereals and their products reported in the present book/article[22]: Spices (anise), Spices (coriander), Spices (pepper), AF/S, literature[641]; Spices (black cumin), AF/S and CIT, literature[641]; Spices (caraway), Spices (marjoram), AF/S and STG, literature[641]; Spices (cayenne pepper, chilli), Spices (cumin), STG, literature[641]

Other ***Aspergillus*** Toxins

STERIGMATOCYSTIN

incidence: 1/9, conc.: 142 μg/kg, sample origin: major retail outlets in Nainital (district), India, sample year: January 1984–September 1986, country: India[22]

- Co-contamination: not reported
- Further contamination of plants and plant products other than cereals and their products reported in the present book/article[22]: Spices (ammi), AFB_1, AFB_2, and AFG_1, literature[22]; Spices (cardamom), AFB_1, AFB_2, AFG_1, AFG_2, CIT, and OTA, literature[22]; Spices (cassia), AFB_1, AFB_2, AFG_1, and AFG_2, literature[22]; Spices (cayenne pepper, chilli), AFB_1, AFB_2, AFG_1, AFG_2, and ZEA, literature[22]; Spices (coriander), AFB_1, AFB_2, AFG_1, AFG_2, CIT, OTA, and ZEA, literature[22]; Spices (cumin), AFB_1, AFB_2, AFG_1, AFG_2, and CIT, literature[22]; Spices (curcuma), AFB_1, AFB_2, AFG_1, AFG_2, CIT, and RBT, literature[22]; Spices (fennel), AFB_1, AFB_2, AFG_1, AFG_2, CIT, and OTA, literature[22]; Spices (garlic), AFB_1, AFB_2, and AFG_1, literature[22]; Spices (pepper), AFB_1, AFB_2, AFG_1, AFG_2, CIT, STG, and ZEA, literature[22]

Aspergillus and ***Penicillium*** Toxins

CITRININ

incidence: 2/9, conc. range: 28–59 μg/kg, ∅ conc.: 43.5 μg/kg, sample origin: major retail outlets in Nainital (district), India, sample year: January 1984–September 1986, country: India[22]

- Co-contamination: not reported
- Further contamination of plants and plant products other than cereals and their products reported in the present book/article[22]: Spices (ammi), AFB_1, AFB_2, and AFG_1, literature[22]; Spices (cardamom), AFB_1, AFB_2, AFG_1, AFG_2, CIT, and OTA, literature[22]; Spices (cassia), AFB_1, AFB_2, AFG_1, and AFG_2, literature[22]; Spices (cayenne pepper, chilli), AFB_1, AFB_2, AFG_1, AFG_2, and ZEA, literature[22]; Spices (coriander), AFB_1, AFB_2, AFG_1, AFG_2, CIT, OTA, and ZEA, literature[22]; Spices (cumin), AFB_1, AFB_2, AFG_1, AFG_2, and CIT, literature[22]; Spices (curcuma), AFB_1, AFB_2, AFG_1, AFG_2, CIT, and RBT, literature[22]; Spices (fennel), AFB_1, AFB_2, AFG_1, AFG_2, OTA, and STG, literature[22]; Spices (garlic), AFB_1, AFB_2, and AFG_1, literature[22]; Spices (pepper), AFB_1, AFB_2, AFG_1, AFG_2, CIT, STG, and ZEA, literature[22]

OCHRATOXIN A

incidence: 2/9, conc. range: pr, sample origin: major retail outlets in Nainital (district), India, sample year: January 1984–September 1986, country: India[22]

- Co-contamination: not reported
- Further contamination of plants and plant products other than cereals and their products reported in the present book/article[22]: Spices (ammi), AFB_1, AFB_2, and AFG_1, literature[22]; Spices (cardamom), AFB_1, AFB_2, AFG_1, AFG_2, CIT, and OTA, literature[22]; Spices (cassia), AFB_1, AFB_2, AFG_1, and AFG_2, literature[22]; Spices (cayenne pepper, chilli), AFB_1, AFB_2, AFG_1, AFG_2, and ZEA, literature[22]; Spices (coriander), AFB_1, AFB_2, AFG_1, AFG_2, CIT, OTA, and ZEA, literature[22]; Spices (cumin), AFB_1, AFB_2, AFG_1, AFG_2, and CIT, literature[22]; Spices (curcuma), AFB_1, AFB_2, AFG_1, AFG_2, CIT, and RBT, literature[22]; Spices (fennel), AFB_1, AFB_2, AFG_1, AFG_2, CIT, and STG, literature[22]; Spices (garlic), AFB_1, AFB_2, and AFG_1, literature[22]; Spices (pepper), AFB_1, AFB_2, AFG_1, AFG_2, CIT, STG, and ZEA, literature[22]

incidence: 1/2*, conc.: 1.26 μg/kg, sample origin: Penang (state, island), Malaysia, sample year: unknown (first study), country: Malaysia[530], *"jintan manis"

- Co-contamination: 1 sa. co-contaminated with AFB_1, AFB_2, AFG_1, AFG_2, and OTA
- Further contamination of plants and plant products other than cereals and their products reported in the present book/article[530]: Spices, Spices (cayenne pepper, chilli), Spices (coriander), Spices (cumin), Spices (curry), Spices (masala), AFB_1, AFB_2, AFG_1, AFG_2, and OTA, literature[530]; Spices (curcuma), AFB_1, AFB_2, AFG_1, and OTA, literature[530]; Spices (fennel), AFB_1, AFB_2, AFG_1, and AFG_2, literature[530]; Spices (pepper), AFB_1 and AFB_2, literature[530]; Spices (poppy seed), AFB_1, literature[530]

Fusarium Toxins

ZEARALENONE

incidence: 1/3, conc.: 7.0 μg/kg, sample origin: specialist food shops, UK, sample year: unknown, country: UK[16]

- Co-contamination: not reported
- Further contamination of plants and plant products other than cereals and their products reported in the present book/article[16]: Food, AF/S, DON, FUS-X, NIV, OTA, and ZEA, literature[16]; Nuts (almond), AF/S, literature[16]; Oils, AF/S and ZEA, literature[16]; Oils (sesame oil), OTA, literature[16]; Paste (curry paste), AF/S, FB/S, NIV, OTA, and ZEA, literature[16]; Pickles, AF/S, FB/S, OTA, and ZEA, literature[16]; Sauces (chilli sauce), NIV, OTA, and ZEA, literature[16]; Spices, AF/S, FB/S, OTA, and ZEA, literature[16]; Spices (cayenne pepper, chilli), AF/S, DAS, DON, HT-2, OTA, and ZEA, literature[16]; Spices (coriander), AF/S, DON, OTA, and ZEA, literature[16]; Spices (curry), AF/S, DAS, FB/S, FUS-X, HT-2, NEO, NIV, OTA, T-2, and ZEA, literature[16]; Spices (fennel), AF/S, literature[16]; Spices (garlic), DON, FUS-X, and NIV, literature[16]; Spices (ginger), AF/S, DON, NEO, NIV, OTA, and T-2, literature[16]; Spices (tandoori), AF/S, FB/S, NIV, OTA, and T-2, literature[16]

Spices (fenugreek) may contain the following mycotoxins:

Aspergillus Toxins

AFLATOXIN B_1

incidence: 3/6*, conc. range: 2.0–4.3 μg/kg**, ∅ conc.: 4.15 μg/kg, sample origin: Alexandria, Luxor (cities), and Cairo (capital), Egypt, sample year: unknown, country: Egypt/USA[127], *3 sa. freshly produced and 3 sa. stored for 12 months, **1 freshly produced sa. and 2 stored sa.

- Co-contamination: 1 sa. co-contaminated with AFB_1, AFB_2 and AFG_1; 1 sa. co-contaminated with AFB_1 and AFB_2 (2 sa. were the stored sa.)
- Further contamination of plants and plant products other than cereals and their products reported in the present book/article[127]: Beans (broad bean), Nuts (peanut), AFB_1 and AFB_2, literature[127]; Lentils, AFB_1, literature[127]; Spices (fenugreek), AFB_2 and AFG_1, literature[127]

AFLATOXIN B_2

incidence: 2/6*, conc. range: 2.5–3.0 μg/kg**, ∅ conc.: 2.75 μg/kg**, sample origin: Alexandria, Luxor (cities), and Cairo (capital), Egypt, sample year: unknown, country: Egypt/USA[127], *3 sa. freshly produced and 3 sa. stored for 12 months, **stored sa.

- Co-contamination: 1 sa. co-contaminated with AFB_1, AFB_2 and AFG_1; 1 sa. co-contaminated with AFB_1 and AFB_2
- Further contamination of plants and plant products other than cereals and their products reported in the present book/article[127]: Beans (broad bean), Nuts (peanut), AFB_1 and AFB_2,

literature[127]; Lentils, AFB_1, literature[127]; Spices (fenugreek), AFB_1 and AFG_1, literature[127]

Aflatoxin G_1

incidence: 1/6*, conc.: 1.8 µg/kg**, sample origin: Alexandria, Luxor (cities), and Cairo (capital), Egypt, sample year: unknown, country: Egypt/USA[127], *3 sa. freshly produced and 3 sa. stored for 12 months, **stored sa.

- Co-contamination: 1 sa. co-contaminated with AFB_1, AFB_2 and AFG_1
- Further contamination of plants and plant products other than cereals and their products reported in the present book/article[127]: Beans (broad bean), Nuts (peanut), Spices (fenugreek), AFB_1 and AFB_2, literature[127]; Lentils, AFB_1, literature[127]

Spices (garlic) may contain the following mycotoxins:

Aspergillus Toxins

Aflatoxin B_1

incidence: 1/6, conc.: 12 µg/kg, sample origin: major retail outlets in Nainital (district), India, sample year: January 1984–November 1986, country: India[22]

- Co-contamination: 1 sa. co-contaminated with AFB_1, AFB_2, and AFG_1
- Further contamination of plants and plant products other than cereals and their products reported in the present book/article[22]: Spices (ammi), AFB_1, AFB_2, and AFG_1, literature[22]; Spices (cardamom), AFB_1, AFB_2, AFG_1, AFG_2, CIT, and OTA, literature[22]; Spices (cassia), AFB_1, AFB_2, AFG_1, and AFG_2, literature[22]; Spices (cayenne pepper, chilli), AFB_1, AFB_2, AFG_1, and AFG_2, literature[22]; Spices (coriander), AFB_1, AFB_2, AFG_1, AFG_2, CIT, OTA, and ZEA, literature[22]; Spices (cumin), AFB_1, AFB_2, AFG_1, AFG_2, and CIT, literature[22]; Spices (curcuma), AFB_1, AFB_2, AFG_1, AFG_2, CIT, and RBT, literature[22]; Spices (fennel), AFB_1, AFB_2, AFG_1, AFG_2, CIT, OTA, and STG, literature[22]; Spices (garlic), AFB_2 and AFG_1, literature[22]; Spices (pepper), AFB_1, AFB_2, AFG_1, AFG_2, CIT, STG, and ZEA, literature[22]

Aflatoxin B_2

incidence: 1/6, conc.: 15 µg/kg, sample origin: major retail outlets in Nainital (district), India, sample year: January 1984–November 1986, country: India[22]

- Co-contamination: 1 sa. co-contaminated with AFB_1, AFB_2, and AFG_1
- Further contamination of plants and plant products other than cereals and their products reported in the present book/article[22]: Spices (ammi), AFB_1, AFB_2, and AFG_1, literature[22]; Spices (cardamom), AFB_1, AFB_2, AFG_1, AFG_2, CIT, and OTA, literature[22]; Spices (cassia), AFB_1, AFB_2, AFG_1, and AFG_2, literature[22]; Spices (cayenne pepper, chilli), AFB_1, AFB_2, AFG_1, and AFG_2, literature[22]; Spices (coriander), AFB_1, AFB_2, AFG_1, AFG_2, CIT, OTA, and ZEA, literature[22]; Spices (cumin), AFB_1, AFB_2, AFG_1, AFG_2, and CIT, literature[22]; Spices (curcuma), AFB_1, AFB_2, AFG_1, AFG_2, CIT, and RBT, literature[22]; Spices (fennel), AFB_1, AFB_2, AFG_1, AFG_2, CIT, OTA, and STG, literature[22]; Spices (garlic), AFB_1 and AFG_1, literature[22]; Spices (pepper), AFB_1, AFB_2, AFG_1, AFG_2, CIT, STG, and ZEA, literature[22]

Aflatoxin G_1

incidence: 1/6, conc.: 10 µg/kg, sample origin: major retail outlets in Nainital (district), India, sample year: January 1984–November 1986, country: India[22]

- Co-contamination: 1 sa. co-contaminated with AFB_1, AFB_2, and AFG_1
- Further contamination of plants and plant products other than cereals and their products reported in the present book/article[22]: Spices (ammi), AFB_1, AFB_2, and AFG_1, literature[22]; Spices (cardamom), AFB_1, AFB_2, AFG_1, AFG_2, CIT, and OTA, literature[22]; Spices (cassia), AFB_1, AFB_2, AFG_1, and AFG_2, literature[22]; Spices (cayenne pepper, chilli), AFB_1, AFB_2, AFG_1, and AFG_2, literature[22]; Spices (coriander), AFB_1, AFB_2, AFG_1, AFG_2, CIT, OTA, and ZEA, literature[22]; Spices (cumin), AFB_1, AFB_2, AFG_1, AFG_2, and CIT, literature[22]; Spices (curcuma), AFB_1, AFB_2, AFG_1, AFG_2, CIT, and RBT, literature[22]; Spices (fennel), AFB_1, AFB_2, AFG_1, AFG_2, CIT, OTA, and STG, literature[22]; Spices (garlic), AFB_1 and AFB_2, literature[22]; Spices (pepper), AFB_1, AFB_2, AFG_1, AFG_2, CIT, STG, and ZEA, literature[22]

Aspergillus and *Penicillium* Toxins

Ochratoxin A

incidence: 2/12*, conc. range: 1–6 µg/kg, ∅ conc.: 3.5 µg/kg, sample origin: drugstore and specialized shop, Belgium, sample year: unknown, country: Belgium[77], *five notified and seven non-notified sa.

- Co-contamination: 1 sa. co-contaminated with FB_1, FB_2, FB_3, and OTA; 1 sa. contaminated solely with OTA
- Further contamination of plants and plant products other than cereals and their products reported in the present book/article[77]: Medicinal plants, OTA, literature[77]; Soy isoflavones, FB_1 and OTA, literature[77]; Spices (garlic), FB_1, FB_2, and FB_3, literature[77]; Vegetables (black radish), FB_1 and FB_2, literature[77]

Fusarium Toxins

Depsipeptides

Beauvericin

incidence: 1/1*, conc.: 4,900 µg/kg, sample origin: private households of staff members, Austria, sample year: unknown, country: Austria[850], *moldy

- Co-contamination: 1 sa. co-contaminated with BEA, FB_1, FB_2, and FB_3.
- Further contamination of plants and plant products other than cereals and their products reported in the present book/article[850]: Fruits (apple), Nuts (chestnut), AME and AOH, literature[850]; Fruits (lemon), Puree, MA, literature[850]; Jam, AOH and MA, literature[850]; Spices (garlic), FB_1, FB_2, and FB_3, literature[850]; Wine (red), AME, AOH, and MA, literature[850]; see also Fruits (lemon), Jam, Nuts (chestnut), Wine (red), chapter: Further Mycotoxins and Microbial Metabolites

Fumonisin/s

Fumonisin B_1

incidence: 1/12*, conc.: 10 µg/kg, sample origin: drugstore and specialized shop, Belgium, sample year: unknown, country: Belgium[77], *five notified and seven non-notified sa.

- Co-contamination: 1 sa. co-contaminated with FB_1, FB_2, FB_3, and OTA

- Further contamination of plants and plant products other than cereals and their products reported in the present book/article[77]: Medicinal plants, OTA, literature[77]; Soy isoflavones, FB_1 and OTA, literature[77]; Spices (garlic), FB_2, FB_3, and OTA, literature[77]; Vegetables (black radish), FB_1 and FB_2, literature[77]

incidence: 2/3*, conc. range: ~46–440 µg/kg, sample origin: commercially available for industrial use, Netherlands, sample year: unknown, country: Netherlands[184], *garlic powder

- Co-contamination: not reported
- Further contamination of plants and plant products other than cereals and their products reported in the present book/article[184]: Spices (onion), FB_1, literature[184]; Spices (paprika), AFB_1 and OTA, literature[184]; Spices (pepper), AFB_2, literature[184]

incidence: 1/1*, conc.: 6,400 µg/kg, sample origin: private households of staff members, Austria, sample year: unknown, country: Austria[850], *moldy

- Co-contamination: 1 sa. co-contaminated with BEA, FB_1, FB_2, and FB_3
- Further contamination of plants and plant products other than cereals and their products reported in the present book/article[850]: Fruits (apple), Nuts (chestnut), AME and AOH, literature[850]; Fruits (lemon), Puree, MA, literature[850]; Jam, AOH and MA, literature[850]; Spices (garlic), BEA, FB_2, and FB_3, literature[850]; Wine (red), AME, AOH, and MA, literature[850]; see also Fruits (lemon), Jam, Nuts (chestnut), Wine (red), chapter: Further Mycotoxins and Microbial Metabolites

Fumonisin B2

incidence: 1/12*, conc.: 8 µg/kg, sample origin: drugstore and specialized shop, Belgium, sample year: unknown, country: Belgium[77], *five notified and seven non-notified sa.

- Co-contamination: 1 sa. co-contaminated with FB_1, FB_2, FB_3, and OTA
- Further contamination of plants and plant products other than cereals and their products reported in the present book/article[77]: Medicinal plants, OTA, literature[77]; Soy isoflavones, FB_1 and OTA, literature[77]; Spices (garlic), FB_1, FB_3, and OTA, literature[77]; Vegetables (black radish), FB_1 and FB_2, literature[77]

incidence: 1/1*, conc.: 130 µg/kg, sample origin: private households of staff members, Austria, sample year: unknown, country: Austria[850], *moldy

- Co-contamination: 1 sa. co-contaminated with BEA, FB_1, FB_2, and FB_3
- Further contamination of plants and plant products other than cereals and their products reported in the present book/article[850]: Fruits (apple), Nuts (chestnut), AME and AOH, literature[850]; Fruits (lemon), Puree, MA, literature[850]; Jam, AOH and MA, literature[850]; Spices (garlic), BEA, FB_1, and FB_3, literature[850]; Wine (red), AME, AOH, and MA, literature[850]; see also Fruits (lemon), Jam, Nuts (chestnut), Wine (red), chapter: Further Mycotoxins and Microbial Metabolites

Fumonisin B3

incidence: 1/12*, conc.: 3 µg/kg, sample origin: drugstore and specialized shop, Belgium, sample year: unknown, country: Belgium[77], *five notified and seven non-notified sa.

- Co-contamination: 1 sa. co-contaminated with FB_1, FB_2, FB_3, and OTA
- Further contamination of plants and plant products other than cereals and their products reported in the present book/article[77]: Medicinal plants, OTA, literature[77]; Soy isoflavones, FB_1 and OTA, literature[77]; Spices (garlic), FB_1, FB_2, and OTA, literature[77]; Vegetables (black radish), FB_1 and FB_2, literature[77]

incidence: 1/1*, conc.: 660 µg/kg, sample origin: private households of staff members, Austria, sample year: unknown, country: Austria[850], *moldy

- Co-contamination: 1 sa. co-contaminated with BEA, FB_1, FB_2, and FB_3
- Further contamination of plants and plant products other than cereals and their products reported in the present book/article[850]: Fruits (apple), Nuts (chestnut), AME and AOH, literature[850]; Fruits (lemon), Puree, MA, literature[850]; Jam, AOH and MA, literature[850]; Spices (garlic), BEA, FB_1, and FB_2, literature[850]; Wine (red), AME, AOH, and MA, literature[850]; see also Fruits (lemon), Jam, Nuts (chestnut), Wine (red), chapter: Further Mycotoxins and Microbial Metabolites

Type B Trichothecenes

Deoxynivalenol

incidence: 1/4, conc.: 14 µg/kg, sample origin: specialist food shops, UK, sample year: unknown, country: UK[16]

- Co-contamination: not reported
- Further contamination of plants and plant products other than cereals and their products reported in the present book/article[16]: Food, AF/S, DON, FUS-X, NIV, OTA, and ZEA, literature[16]; Nuts (almond), AF/S, literature[16]; Oils, Spices (fennel), AF/S and ZEA, literature[16]; Oils (sesame oil), OTA, literature[16]; Paste (curry paste), AF/S, FB/S, NIV, OTA, and ZEA, literature[16]; Pickles, AF/S, FB/S, OTA, and ZEA, literature[16]; Sauces (chilli sauce), NIV, OTA, and ZEA, literature[16]; Spices, AF/S, FB/S, OTA, and ZEA, literature[16]; Spices (cayenne pepper, chilli), AF/S, DAS, DON, HT-2, OTA, and ZEA, literature[16]; Spices (coriander), AF/S, DON, OTA, and ZEA, literature[16]; Spices (curry), AF/S, DAS, FB/S, FUS-X, HT-2, NEO, NIV, OTA, T-2, and ZEA, literature[16]; Spices (garlic), FUS-X and NIV, literature[16]; Spices (ginger), AF/S, DON, NEO, NIV, OTA, and T-2, literature[16]; Spices (tandoori), AF/S, FB/S, NIV, OTA, and T-2, literature[16]

Fusarenon-X (4-Acetylnivalenol)

incidence: 1/4, conc.: 5 µg/kg, sample origin: specialist food shops, UK, sample year: unknown, country: UK[16]

- Co-contamination: not reported
- Further contamination of plants and plant products other than cereals and their products reported in the present book/article[16]: Food, AF/S, DON, FUS-X, NIV, OTA, and ZEA, literature[16]; Nuts (almond) AF/S, literature[16]; Oils, Spices (fennel), AF/S and ZEA, literature[16]; Oils (sesame oil), OTA, literature[16]; Paste (curry pastes), AF/S, FB/S, NIV, OTA, and ZEA, literature[16]; Pickles, AF/S, FB/S, OTA, and ZEA, literature[16]; Sauces (chilli sauce), NIV, OTA, and ZEA, literature[16]; Spices, AF/S, FB/S, OTA, and ZEA, literature[16]; Spices (cayenne pepper, chilli), AF/S, DAS, DON, HT-2, OTA, and ZEA, literature[16]; Spices (coriander), AF/S, DON, OTA, and ZEA, literature[16]; Spices (curry), AF/S, DAS, FB/S, FUS-X, HT-2, NEO, NIV, OTA, T-2, and ZEA, literature[16]; Spices (garlic), DON and NIV, literature[16]; Spices (ginger), AF/S, DON, NEO, NIV, OTA, and T-2, literature[16]; Spices (tandoori), AF/S, FB/S, NIV, OTA, and T-2, literature[16]

Nivalenol

incidence: 1/4, conc.: 21 µg/kg, sample origin: specialist food shops, UK, sample year: unknown, country: UK[16]

- Co-contamination: not reported
- Further contamination of plants and plant products other than cereals and their products reported in the present book/article[16]: Food, AF/S, DON, FUS-X, NIV, OTA, and ZEA, literature[16]; Nuts (almond), AF/S, literature[16]; Oils, Spices (fennel), AF/S and ZEA, literature[16]; Oils (sesame oil), OTA, literature[16]; Paste (curry paste), AF/S, FB/S, NIV, OTA, and ZEA, literature[16]; Pickles, AF/S, FB/S, OTA, and ZEA, literature[16]; Sauces (chilli sauce), NIV, OTA, and ZEA, literature[16]; Spices, AF/S, FB/S, OTA, and ZEA, literature[16]; Spices (cayenne pepper, chilli), AF/S, DAS, DON, HT-2, OTA, and ZEA, literature[16]; Spices (coriander), AF/S, DON, OTA, and ZEA, literature[16]; Spices (curry), AF/S, DAS, FB/S, FUS-X, HT-2, NEO, NIV, OTA, T-2, and ZEA, literature[16]; Spices (garlic), DON and FUS-X, literature[16]; Spices (ginger), AF/S, DON, NEO, NIV, OTA, and T-2, literature[16]; Spices (tandoori), AF/S, FB/S, NIV, OTA, and T-2, literature[16]

Spices (garlic/onion) may contain the following mycotoxins:

Aspergillus Toxins

Aflatoxin/s

incidence: 2/58, conc. range: ≤60 µg/kg*?, ∅ conc.: 67 µg/kg*?, sample origin: market retailers, street vendors, wholesalers, and distributors, Hong Kong (British colony) and Thailand, sample year: September 1967–July 1969, country: Thailand[139], *AFB_1, AFB_2, AFG_1, and AFG_2

- Co-contamination: not reported
- Further contamination of plants and plant products other than cereals and their products reported in the present book/article[139]: Beans, Beans (mung bean), Food, Nuts (peanut), Seeds (sesame), Spices (cayenne pepper, chilli), Starch (cassava starch), Vegetables, AF/S, literature[139]

For detailed information see the article.

Spices (ginger) may contain the following mycotoxins:

Alternaria Toxins

Tenuazonic Acid

incidence: 2/2, conc. range: 50–70 µg/kg, ∅ conc.: 60 µg/kg, sample origin: unknown, sample year: unknown, country: Germany[166]

- Co-contamination: not reported
- Further contamination of plants and plant products other than cereals and their products reported in the present book/article[166]: Juices, Juices (apple juice), Juices (apricot juice), Juices (banana juice), Juices (blackcurrant juice), Juices (cherry juice), Juices (grape juice), Juices (orange juice), Juices (pear juice), Juices (pomegranate juice), Juices (strawberry juice), Spices, Spices (coriander), Spices (cumin), Spices (curcuma), Spices (curry), Spices (paprika), Spices (pepper), TA, literature[166]

Aspergillus Toxins

Aflatoxin B_1

incidence: 3/5, conc. range: 1.4–6.5 µg/kg, ∅ conc.: 4 µg/kg, sample origin: unknown, sample year: unknown, country: USA[167]

- Co-contamination: 1 sa. co-contaminated with AFB_1, AFB_2, AFG_1, and AFG_2; 1 sa. co-contaminated with AFB_1 and AFB_2; 1 sa. contaminated solely with AFB_1
- Further contamination of plants and plant products other than cereals and their products reported in the present book/article[167]: Spices (cayenne pepper, chilli), AFB_1, AFB_2, AFG_1, and AFG_2, literature[167]; Spices (ginger), AFB_2, AFG_1, and AFG_2, literature[167]; Spices (pepper), AFB_1, literature[167]

incidence: 10/12, conc. range: ≤3.50 µg/kg, ∅ conc.: 0.63 µg/kg, sample origin: popular markets in Salé (city) and Rabat (capital), Morocco, sample year: unknown, country: Morocco[174]

- Co-contamination: not reported
- Further contamination of plants and plant products other than cereals and their products reported in the present book/article[174]: Spice (cayenne pepper, chilli), Spices (cumin), Spices (pepper), AFB_1 and AF/S, literature[174]; Spices (ginger), AF/S, literature[174]

incidence: 4/5*, conc. range: 0.3–2.7 µg/kg, ∅ conc.: 1.4 µg/kg, sample origin: available in New Zealand, sample year: 2009, country: New Zealand[277], *ginger, ground

- Co-contamination: not reported
- Further contamination of plants and plant products other than cereals and their products reported in the present book/article[277]: Bars, AFB_1, AF/S, DON, NIV, and OTA, literature[277]; Butter (peanut butter), Confectionery, Food, Fruits, dried, Nuts, Nuts (peanut), Nuts (pistachio nut), Sauces (peanut sauce), AFB_1 and AF/S, literature[277]; Coffee, Dates, Spices (pepper), Vine fruits, dried, OTA, literature[277]; Figs, Spices (cayenne pepper, chilli), Spices (curry), Spices (paprika), AFB_1, AF/S, and OTA, literature[277]; Nuts (Brazil nut, paranut), AFB_1, literature[277]; Spices (ginger), AF/S and OTA, literature[277]

incidence: 8/15, conc. range: tr-25 µg/kg*, sample origin: processors located in six Canadian cities, Canada, sample year: unknown, country: Canada[323], *approximate conc.

- Co-contamination: 2 sa. co-contaminated with AFB_1, AFB_2, AFG_1, and AFG_2; 1 sa. co-contaminated with AFB_1, AFB_2, and AFG_1; 4 sa. co-contamination with AFB_1 and AFB_2; 1 sa. contamination solely with AFB_1
- Further contamination of plants and plant products other than cereals and their products reported in the present book/article[323]: Spices (coriander), Spices (curcuma), Spices (nutmeg), AFB_1 and AFB_2, literature[323]; Spices (ginger), AFB_2, AFG_1, and AFG_2, literature[323]

incidence: 5/9, conc. range: 1.9–39.8 µg/kg, sample origin: retail sales points in different regions, Iran, sample year: March–May 2015, country: Iran[491]

- Co-contamination: not reported
- Further contamination of plants and plant products other than cereals and their products reported in the present book/article[491]: Herbs (camomile), Herbs (sage), AFB_1, literature[491]; Spices (cayenne pepper, chilli), Spices (curcuma), Spices (pepper), Spices (thyme), AFB_1, literature[491]

incidence: 2/7, conc. range: ≤0.2 µg/kg, ∅ conc.: 0.16 µg/kg, sample origin: retail stores in Lower Saxony (federal state), Germany, sample year: 2009, country: Germany[829]

- Co-contamination: not reported
- Further contamination of plants and plant products other than cereals and their products reported in the present book/article[829]: Nuts (hazelnut), AFB_1 and AF/S, literature[829]; Nuts (pistachio nut), AFB_1, literature[829]; Oils (maize oil), ZEA, literature[829]; Spices (cinnamon), OTA, literature[829]; Spices (curry), Spices (nutmeg), AFB_1, AF/S, and OTA, literature[829]; Spices (ginger), AF/S, literature[829]

incidence: 3/4*, conc. range: 3.8–23.1 µg/kg, ∅ conc.: 16.5 µg/kg, sample origin: organic markets and organic shops, Turkey, sample year: May 2010–May 2011, country: Turkey[859], *organic ginger sa.

- Co-contamination: not reported
- Further contamination of plants and plant products other than cereals and their products reported in the present book/article[859]: Herbs (camomile), Herbs (linden), Herbs (rosehip), Herbs (sage), Spices (anise), Spices (basil), Spices (cayenne pepper, chilli), Spices (cinnamon), Spices (coriander), Spices (cumin), Spices (fennel), Spices (laurel leave), Spices (pepper), Spices (peppermint), Spices (poppy seed), Spices (rosemary), Spices (sumac), AFB_1, literature[859]

incidence: 1/5*, conc.: <LOQ, sample origin: local markets, Beijing (capital), China, sample year: unknown, country: China[875], *ginger powder

- Co-contamination: 1 sa. co-contaminated with AFB_1 and AFG_2
- Further contamination of plants and plant products other than cereals and their products reported in the present book/article[875]: Spices (ginger), AFB_2, AFG_1, AFG_2, and OTA, literature[875]; Tea, AFB_1, AFG_1, AFG_2, and OTA, literature[875]

incidence: 0/10*, conc.: no contamination, sample origin: local markets, Beijing (capital), China, sample year: unknown, country: China[875], *moldy and normal fresh ginger

- Co-contamination: not reported
- Further contamination of plants and plant products other than cereals and their products reported in the present book/article[875]: Spices (ginger), AFB_2, AFG_1, AFG_2, and OTA, literature[875]; Tea, AFB_1, AFG_1, AFG_2, and OTA, literature[875]

Aflatoxin B_2

incidence: 2/5, conc. range: 0.2 µg/kg, ∅ conc.: 0.2 µg/kg, sample origin: unknown, sample year: unknown, country: USA[167]

- Co-contamination: 1 sa. co-contaminated with AFB_1, AFB_2, AFG_1, and AFG_2; 1 sa. co-contaminated with AFB_1 and AFB_2
- Further contamination of plants and plant products other than cereals and their products reported in the present book/article[167]: Spices (cayenne pepper, chilli), AFB_1, AFB_2, AFG_1, and AFG_2, literature[167]; Spices (ginger), AFB_1, AFG_1, and AFG_2, literature[167]; Spices (pepper), AFB_1, literature[167]

incidence: 7/15, conc. range: tr-15 µg/kg*, sample origin: processors located in six Canadian cities, Canada, sample year: unknown, country: Canada[323], *approximate conc.

- Co-contamination: 2 sa. co-contaminated with AFB_1, AFB_2, AFG_1, and AFG_2; 1 sa. co-contaminated with AFB_1, AFB_2, and AFG_1; 4 sa. co-contamination with AFB_1 and AFB_2
- Further contamination of plants and plant products other than cereals and their products reported in the present book/article[323]: Spices (coriander), Spices (curcuma), Spices (nutmeg), AFB_1 and AFB_2, literature[323]; Spices (ginger), AFB_1, AFG_1, and AFG_2, literature[323]

incidence: 0/5*, conc.: no contamination, sample origin: local markets, Beijing (capital), China, sample year: unknown, country: China[875], *ginger powder

- Co-contamination: not reported
- Further contamination of plants and plant products other than cereals and their products reported in the present book/article[875]: Spices (ginger), AFB_1, AFB_2, AFG_2, and OTA, literature[875]; Tea, AFB_1, AFG_1, AFG_2, and OTA, literature[875]

incidence: 2/10*, conc. range: <LOQ**, sample origin: local markets, Beijing (capital), China, sample year: unknown, country: China[875], *moldy and normal fresh ginger, **moldy ginger sa.

- Co-contamination: 1 sa. co-contaminated with AFB_2, AFG_1, and AFG_2; 1 sa. co-contaminated with AFB_2 and AFG_1
- Further contamination of plants and plant products other than cereals and their products reported in the present book/article[875]: Spices (ginger), AFB_1, AFG_1, AFG_2, and OTA, literature[875]; Tea, AFB_1, AFG_1, AFG_2, and OTA, literature[875]

Aflatoxin G_1

incidence: 1/5, conc.: 2.5 µg/kg, sample origin: unknown, sample year: unknown, country: USA[167]

- Co-contamination: 1 sa. co-contaminated with AFB_1, AFB_2, AFG_1, and AFG_2
- Further contamination of plants and plant products other than cereals and their products reported in the present book/article[167]: Spices (cayenne pepper, chilli), AFB_1, AFB_2, AFG_1, and AFG_2, literature[167]; Spices (ginger), AFB_1, AFB_2, and AFG_2, literature[167]; Spices (pepper), AFB_1, literature[167]

incidence: 3/15, conc. range: tr-15 µg/kg*, sample origin: processors located in six Canadian cities, Canada, sample year: unknown, country: Canada[323], *approximate conc.

- Co-contamination: 2 sa. co-contaminated with AFB_1, AFB_2, AFG_1, and AFG_2; 1 sa. co-contaminated with AFB_1, AFB_2, and AFG_1
- Further contamination of plants and plant products other than cereals and their products reported in the present book/article[323]: Spices (coriander), Spices (curcuma), Spices (nutmeg), AFB_1 and AFB_2, literature[323]; Spices (ginger), AFB_1, AFB_2, and AFG_2, literature[323]

incidence: 0/5*, conc.: no contamination, sample origin: local markets, Beijing (capital), China, sample year: unknown, country: China[875], *ginger powder

- Co-contamination: not reported
- Further contamination of plants and plant products other than cereals and their products reported in the present book/article[875]: Spices (ginger), AFB_1, AFB_2, AFG_2, and OTA, literature[875]; Tea, AFB_1, AFG_1, AFG_2, and OTA, literature[875]

incidence: 2/10*, conc. range: 0.70–0.89 µg/kg**, ∅ conc.: 0.80 µg/kg**, sample origin: local markets, Beijing (capital), China, sample year: unknown, country: China[875], *moldy and normal fresh ginger, **moldy ginger sa.

- Co-contamination: 1 sa. co-contaminated with AFB_2, AFG_1, and AFG_2; 1 sa. co-contaminated with AFB_2 and AFG_1
- Further contamination of plants and plant products other than cereals and their products reported in the present book/article[875]: Spices (ginger), AFB_1, AFB_2, AFG_2, and OTA, literature[875]; Tea, AFB_1, AFG_1, AFG_2, and OTA, literature[875]

Aflatoxin G_2

incidence: 1/5, conc.: 0.2 µg/kg, sample origin: unknown, sample year: unknown, country: USA[167]

- Co-contamination: 1 sa. co-contaminated with AFB_1, AFB_2, AFG_1, and AFG_2
- Further contamination of plants and plant products other than cereals and their products reported in the present book/article[167]: Spices (cayenne pepper, chilli), AFB_1, AFB_2, AFG_1, and AFG_2, literature[167]; Spices (ginger), AFB_1, AFB_2, and AFG_1, literature[167]; Spices (pepper), AFB_1, literature[167]

incidence: 2/15, conc. range: tr-4 μg/kg*, sample origin: processors located in six Canadian cities, Canada, sample year: unknown, country: Canada[323], *approximate conc.

- Co-contamination: 2 sa. co-contaminated with AFB_1, AFB_2, AFG_1, and AFG_2
- Further contamination of plants and plant products other than cereals and their products reported in the present book/article[323]: Spices (coriander), Spices (curcuma), Spices (nutmeg), AFB_1 and AFB_2, literature[323]; Spices (ginger), AFB_1, AFB_2, and AFG_1, literature[323]

incidence: 3/5*, conc. range: <LOQ, sample origin: local markets, Beijing (capital), China, sample year: unknown, country: China[875], *ginger powder

- Co-contamination: 1 sa. co-contaminated with AFG_1 and AFG_2; 2 sa. contaminated solely with AFG_2
- Further contamination of plants and plant products other than cereals and their products reported in the present book/article[875]: Spices (ginger), AFB_1, AFB_2, AFG_1, and OTA, literature[875]; Tea, AFB_1, AFG_1, AFG_2, and OTA, literature[875]

incidence: 2/10*, conc. range: 0.25–0.65 μg/kg**, Ø conc.: 0.45 μg/kg**, sample origin: local markets, Beijing (capital), China, sample year: unknown, country: China[875], *moldy and normal fresh ginger, **moldy ginger sa.

- Co-contamination: 1 sa. co-contaminated with AFB_2, AFG_1, and AFG_2; 1 sa. contaminated solely with AFG_2
- Further contamination of plants and plant products other than cereals and their products reported in the present book/article[875]: Spices (ginger), AFB_1, AFB_2, AFG_1, and OTA, literature[875]; Tea, AFB_1, AFG_1, AFG_2, and OTA, literature[875]

Aflatoxin/s

incidence: 4?/4*, conc. range: 4.2–13.5 μg/kg, sample origin: specialist food shops, UK, sample year: unknown, country: UK[16], *AFB_1, AFB_2, AFG_1, and AFG_2

- Co-contamination: not reported
- Further contamination of plants and plant products other than cereals and their products reported in the present book/article[16]: Food, AF/S, DON, FUS-X, NIV, OTA, and ZEA, literature[16]; Nuts (almond), AF/S, literature[16]; Oils, Spices (fennel), AF/S and ZEA, literature[16]; Oils (sesame oil), OTA, literature[16]; Paste (curry paste), AF/S, FB/S, NIV, OTA, and ZEA, literature[16]; Pickles, AF/S, FB/S, OTA, and ZEA, literature[16]; Sauces (chilli sauce), NIV, OTA, and ZEA, literature[16]; Spices, AF/S, FB/S, OTA, and ZEA, literature[16]; Spices (cayenne pepper, chilli), AF/S, DAS, DON, HT-2, OTA, and ZEA, literature[16]; Spices (coriander), AF/S, DON, OTA, and ZEA, literature[16]; Spices (curry), AF/S, DAS, FB/S, FUS-X, HT-2, NEO, NIV, OTA, T-2, and ZEA, literature[16]; Spices (garlic), DON, FUS-X, and NIV, literature[16]; Spices (ginger), DON, NEO, NIV, OTA, and T-2, literature[16]; Spices (tandoori), AF/S, FB/S, NIV, OTA, and T-2, literature[16]

incidence: 1/3, conc.: 2 μg/kg*, sample origin: imported to USA, sample year: 1986, country: USA[19], *AFB_1, AFB_2, AFG_1, and AFG_2

- Co-contamination: not reported
- Further contamination of plants and plant products other than cereals and their products reported in the present book/article[19]: Butter (peanut butter), Candies (peanut candy), Candies (pistachio candy), Nuts, Nuts (almond), Nuts (Brazil nut, paranut), Nuts (peanut), Nuts (pecan nut), Nuts (pistachio nut), Nuts (walnut), Oils (peanut oil), Paste (peanut paste), Paste (pipian paste), Products (peanut product), Seeds (melon), Spices (cayenne pepper, chilli), Spices (nutmeg), AF/S, literature[19]

incidence: 10/12, conc. range: ≤9.10 μg/kg*, Ø conc.: 1.47 μg/kg*, sample origin: popular markets in Salé (city) and Rabat (capital), Morocco, sample year: unknown, country: Morocco[174], *AFS total

- Co-contamination: not reported
- Further contamination of plants and plant products other than cereals and their products reported in the present book/article[174]: Spice (cayenne pepper, chilli), Spices (cumin), Spices (pepper), AFB_1 and AF/S, literature[174]; Spices (ginger), AFB_1, literature[174]

incidence: 1/7*, conc.: 0.18 μg/kg**, sample origin: markets in ten cities, Korea, sample year: May–September 2006, country: Korea[188], *ginger product, **AFB_1, AFB_2, AFG_1, and AFG_2

- Co-contamination: not reported
- Further contamination of plants and plant products other than cereals and their products reported in the present book/article[188]: Paste (red pepper paste), Spices (cayenne pepper, chilli), Spices (curry), AF/S, literature[188]

incidence: 4/5*, conc. range: 0.3–3.6 μg/kg**, Ø conc.: 2.0 μg/kg**, sample origin: available in New Zealand, sample year: 2009, country: New Zealand[277], *ginger, ground, **AFS total

- Co-contamination: not reported
- Further contamination of plants and plant products other than cereals and their products reported in the present book/article[277]: Bars, AFB_1, AF/S, DON, NIV, and OTA, literature[277]; Butter (peanut butter), Confectionery, Food, Fruits, dried, Nuts, Nuts (peanut), Nuts (pistachio nut), Sauces (peanut sauce), AFB_1 and AF/S, literature[277]; Coffee, Dates, Spices (pepper), Vine fruits, dried, OTA, literature[277]; Figs, Spices (cayenne pepper, chilli), Spices (curry), Spices (paprika), AFB_1, AF/S, and OTA, literature[277]; Nuts (Brazil nut, paranut), AFB_1, literature[277]; Spices (ginger), AFB_1 and OTA, literature[277]

incidence: 4/5*, conc. range: 0.3–3.6 μg/kg**, sample origin: New Zealand, sample year: 2009, country: New Zealand[702], *ginger, ground, **AFB_1, AFB_2, AFG_1, and AFG_2

- Co-contamination: not reported
- Further contamination of plants and plant products other than cereals and their products reported in the present book/article[702]: Figs, Spices (cayenne pepper, chilli), Spices (curry), Spices (paprika), AF/S and OTA, literature[702]; Fruits, dried, AF/S, literature[702]; Spices (ginger), Spices (pepper), Vine fruits, dried, OTA, literature[702]

incidence: 2/7, conc. range: ≤0.39 μg/kg*, Ø conc.: 0.35 μg/kg*, sample origin: retail stores in Lower Saxony (federal state), Germany, sample year: 2009, country: Germany[829], *AFB_1, AFB_2, AFG_1, and AFG_2

- Co-contamination: not reported
- Further contamination of plants and plant products other than cereals and their products reported in the present book/article[829]: Nuts (hazelnut), AFB_1 and AF/S, literature[829]; Nuts (pistachio nut), Spices (ginger), AFB_1, literature[829]; Oils (maize oil), ZEA, literature[829]; Spices (cinnamon), OTA, literature[829]; Spices (curry), Spices (nutmeg), AFB_1, AF/S, and OTA, literature[829]

Aspergillus **and** ***Penicillium*** **Toxins**

Ochratoxin A

incidence: 4?/4, conc. range: 2.1–7.5 μg/kg, sample origin: specialist food shops, UK, sample year: unknown, country: UK[16]

- Co-contamination: not reported
- Further contamination of plants and plant products other than cereals and their products reported in the present book/article[16]: Food, AF/S, DON, FUS-X, NIV, OTA, and ZEA, literature[16]; Nuts (almond), AF/S, literature[16]; Oils, Spices (fennel), AF/S and ZEA, literature[16]; Oils (sesame oil), OTA, literature[16]; Paste (curry paste), AF/S, FB/S, NIV, OTA, and ZEA, literature[16]; Pickles, AF/S, FB/S, OTA, and ZEA, literature[16]; Sauces (chilli sauce), NIV, OTA, and ZEA, literature[16]; Spices, AF/S, FB/S, OTA, and ZEA, literature[16]; Spices (cayenne pepper, chilli), AF/S, DAS, DON, HT-2, OTA, and ZEA, literature[16]; Spices (coriander), AF/S, DON, OTA, and ZEA, literature[16]; Spices (curry), AF/S, DAS, FB/S, FUS-X, HT-2, NEO, NIV, OTA, T-2, and ZEA, literature[16]; Spices (garlic), DON, FUS-X, and NIV, literature[16]; Spices (ginger), AF/S, DON, NEO, NIV, and T-2, literature[16]; Spices (tandoori), AF/S, FB/S, NIV, OTA, and T-2, literature[16]

incidence: 2/25*, conc. range: 10–29 μg/kg (1 sa.), 30–49 μg/kg (1 sa.), sample origin: retail shops, India, sample year: unknown, country: India/UK/Belgium[191], *ginger powder

- Co-contamination: not reported
- Further contamination of plants and plant products other than cereals and their products reported in the present book/article[191]: Spices (coriander), Spices (curcuma), Spices (pepper), OTA, literature[191]

incidence: 1/5, conc.: 2.5 μg/kg, sample origin: retail shops in Belgium, Netherlands, and Russia, sample year: autumn-winter 2005/2006, country: Belgium/Russia/Hungary[193]

- Co-contamination: not reported
- Further contamination of plants and plant products other than cereals and their products reported in the present book/article[193]: Spices (cayenne pepper, chilli), Spices (nutmeg), Spices (paprika), Spices (pepper), OTA, literature[193]

incidence: 2/2*, conc. range: 0.6–4.5 μg/kg, ∅ conc.: 2.55 μg/kg, sample origin: available in New Zealand, sample year: 2007, country: New Zealand[277], *ginger, ground

incidence: 4/5*, conc. range: 0.5–2.1 μg/kg, sample origin: available in New Zealand, sample year: 2008/2009, country: New Zealand[277], *ginger, ground

- Co-contamination: not reported
- Further contamination of plants and plant products other than cereals and their products reported in the present book/article[277]: Bars, AFB_1, AF/S, DON, NIV, and OTA, literature[277]; Butter (peanut butter), Confectionery, Food, Fruits, dried, Nuts, Nuts (peanut), Nuts (pistachio nut), Sauces (peanut sauce), Spices (ginger), AFB_1 and AF/S, literature[277]; Coffee, Dates, Spices (pepper), Vine fruits, dried, OTA, literature[277]; Figs, Spices (cayenne pepper, chilli), Spices (curry), Spices (paprika), AFB_1, AF/S, and OTA, literature[277]; Nuts (Brazil nut, paranut), AFB_1, literature[277]

incidence: 4/5*, conc. range: ≤2.1 μg/kg, sample origin: New Zealand, sample year: 2009, country: New Zealand[702], *ginger, ground

- Co-contamination: not reported
- Further contamination of plants and plant products other than cereals and their products reported in the present book/article[702]: Figs, Spices (cayenne pepper, chilli), Spices (curry), Spices (paprika), AF/S and OTA, literature[702]; Fruits, dried, Spices (ginger), AF/S, literature[702]; Spices (pepper), Vine fruits, dried, OTA, literature[702]

incidence: 7/12*, conc. range: ≤12.7 μg/kg, sample origin: 12 regions, Czech Republic, sample year: 2011–2013, country: Czech Republic[811], *ginger roots, dried

- Co-contamination: not reported
- Further contamination of plants and plant products other than cereals and their products reported in the present book/article[811]: Beans, Chocolate, Coffee, Confectionery, Figs, Herbs and Spices, Juices, Licorice, Powder (cocoa powder), Products (ginger product), Raisins, Spices (caraway), Spices (cayenne pepper, chilli), Spices (coriander), Spices (mustard), Spices (nutmeg), Spices (paprika), Spices (pepper), Tea, Wine (red), Wine (white), OTA, literature[811]

incidence: 0/5*, conc.: no contamination, sample origin: local markets, Beijing (capital), China, sample year: unknown, country: China[875], *ginger powder

incidence: 1/10*, conc.: <LOQ**, sample origin: local markets, Beijing (capital), China, sample year: unknown, country: China[875], *moldy and normal fresh ginger, **moldy ginger sa.

- Co-contamination: not reported
- Further contamination of plants and plant products other than cereals and their products reported in the present book/article[875]: Spices (ginger), AFB_1, AFB_2, AFG_1, and AFG_2, literature[875]; Tea, AFB_1, AFG_1, AFG_2, and OTA, literature[875]

Fusarium **Toxins**

Depsipeptides

Beauvericin

incidence: 3/3*, conc. range: 73.7–117.9 μg/kg, ∅ conc.: 97.4 μg/kg, sample origin: Bavarian retail stores, Germany, sample year: unknown, country: Germany[746], *ginger spice, dried

- Co-contamination: 2 sa. co-contaminated with BEA, ENA, ENA_1, ENB, and ENB_1; 1 sa. co-contaminated with BEA and ENB
- Further contamination of plants and plant products other than cereals and their products reported in the present book/article[746]: Medicinal plants, BEA, ENA, ENA_1, ENB, and ENB_1, literature[746]; Spices (ginger), ENA, ENA_1, ENB, and ENB_1, literature[746]

incidence: 2/2*, conc. range: 29.3–59.2 μg/kg, ∅ conc.: 44.3 μg/kg, sample origin: Bavarian retail stores, Germany, sample year: unknown, country: Germany[746], *ginger herb, dried for infusions

- Co-contamination: 1 sa. co-contaminated with BEA, ENB, and ENB_1, 1 sa. contaminated solely with BEA
- Further contamination of plants and plant products other than cereals and their products reported in the present book/article[746]: Medicinal plants, BEA, ENA, ENA_1, ENB, and ENB_1, literature[746]; Spices (ginger), ENA, ENA_1, ENB, and ENB_1, literature[746]

incidence: 0/1*, conc.: no contamination, sample origin: Bavarian retail stores, Germany, sample year: unknown, country: Germany[746], *ginger root, fresh

- Co-contamination: not reported

- Further contamination of plants and plant products other than cereals and their products reported in the present book/article[746]: Medicinal plants, BEA, ENA, ENA_1, ENB, and ENB_1, literature[746]; Spices (ginger), ENA, ENA_1, ENB, and ENB_1, literature[746]

Enniatin A

incidence: 2/3*, conc. range: 22.4–90.2 µg/kg, ∅ conc.: 56.3 µg/kg, sample origin: Bavarian retail stores, Germany, sample year: unknown, country: Germany[746], *ginger spice, dried

- Co-contamination: 2 sa. co-contaminated with BEA, ENA, ENA_1, ENB, and ENB_1
- Further contamination of plants and plant products other than cereals and their products reported in the present book/article[746]: Medicinal plants, BEA, ENA, ENA_1, ENB, and ENB_1, literature[746]; Spices (ginger), BEA, ENA_1, ENB, and ENB_1, literature[746]

incidence: 0/2*, conc.: no contamination, sample origin: Bavarian retail stores, Germany, sample year: unknown, country: Germany[746], *ginger herb, dried for infusions

- Co-contamination: not reported
- Further contamination of plants and plant products other than cereals and their products reported in the present book/article[746]: Medicinal plants, BEA, ENA, ENA_1, ENB, and ENB_1, literature[746]; Spices (ginger), BEA, ENA_1, ENB, and ENB_1, literature[746]

incidence: 0/1*, conc.: no contamination, sample origin: Bavarian retail stores, Germany, sample year: unknown, country: Germany[746], *ginger root, fresh

- Co-contamination: not reported
- Further contamination of plants and plant products other than cereals and their products reported in the present book/article[746]: Medicinal plants, BEA, ENA, ENA_1, ENB, and ENB_1, literature[746]; Spices (ginger), BEA, ENA_1, ENB, and ENB_1, literature[746]

Enniatin A_1

incidence: 2/3*, conc. range: 20.6–65.0 µg/kg, ∅ conc.: 42.8 µg/kg, sample origin: Bavarian retail stores, Germany, sample year: unknown, country: Germany[746], *ginger spice, dried

- Co-contamination: 2 sa. co-contaminated with BEA, ENA, ENA_1, ENB, and ENB_1
- Further contamination of plants and plant products other than cereals and their products reported in the present book/article[746]: Medicinal plants, BEA, ENA, ENA_1, ENB, and ENB_1, literature[746]; Spices (ginger), BEA, ENA, ENB, and ENB_1, literature[746]

incidence: 0/2*, conc.: no contamination, sample origin: Bavarian retail stores, Germany, sample year: unknown, country: Germany[746], *ginger herb, dried for infusions

- Co-contamination: not reported
- Further contamination of plants and plant products other than cereals and their products reported in the present book/article[746]: Medicinal plants, BEA, ENA, ENA_1, ENB, and ENB_1, literature[746]; Spices (ginger), BEA, ENA, ENB, and ENB_1, literature[746]

incidence: 0/1*, conc.: no contamination, sample origin: Bavarian retail stores, Germany, sample year: unknown, country: Germany[746], *ginger root, fresh

- Co-contamination: not reported
- Further contamination of plants and plant products other than cereals and their products reported in the present book/article[746]: Medicinal plants, BEA, ENA, ENA_1, ENB, and ENB_1, literature[746]; Spices (ginger), BEA, ENA, ENB, and ENB_1, literature[746]

Enniatin B

incidence: 3/3*, conc. range: >LOD-6.2 µg/kg, sample origin: Bavarian retail stores, Germany, sample year: unknown, country: Germany[746], *ginger spice, dried

- Co-contamination: 2 sa. co-contaminated with BEA, ENA, ENA_1, ENB, and ENB_1; 1 sa. co-contaminated with BEA and ENB
- Further contamination of plants and plant products other than cereals and their products reported in the present book/article[746]: Medicinal plants, BEA, ENA, ENA_1, ENB, and ENB_1, literature[746]; Spices (ginger), BEA, ENA, ENA_1, and ENB_1, literature[746]

incidence: 1/2*, conc.: 6.0 µg/kg, sample origin: Bavarian retail stores, Germany, sample year: unknown, country: Germany[746], *ginger herb, dried for infusions

- Co-contamination: 1 sa. co-contaminated with BEA, ENB, and ENB_1
- Further contamination of plants and plant products other than cereals and their products reported in the present book/article[746]: Medicinal plants, BEA, ENA, ENA_1, ENB, and ENB_1, literature[746]; Spices (ginger), BEA, ENA, ENA_1, and ENB_1, literature[746]

incidence: 0/1*, conc.: no contamination, sample origin: Bavarian retail stores, Germany, sample year: unknown, country: Germany[746], *ginger root, fresh

- Co-contamination: not reported
- Further contamination of plants and plant products other than cereals and their products reported in the present book/article[746]: Medicinal plants, BEA, ENA, ENA_1, ENB, and ENB_1, literature[746]; Spices (ginger), BEA, ENA, ENA_1, and ENB_1, literature[746]

Enniatin B_1

incidence: 2/3*, conc. range: 11.5–30.9 µg/kg, ∅ conc.: 20.7 µg/kg, sample origin: Bavarian retail stores, Germany, sample year: unknown, country: Germany[746], *ginger spice, dried

- Co-contamination: 2 sa. co-contaminated with BEA, ENA, ENA_1, ENB, and ENB_1
- Further contamination of plants and plant products other than cereals and their products reported in the present book/article[746]: Medicinal plants, BEA, ENA, ENA_1, ENB, and ENB_1, literature[746]; Spices (ginger), BEA, ENA, ENA_1, and ENB, literature[746]

incidence: 1/2*, conc.: 6.4 µg/kg, sample origin: Bavarian retail stores, Germany, sample year: unknown, country: Germany[746], *ginger herb, dried for infusions

- Co-contamination: 1 sa. co-contaminated with BEA, ENB, and ENB_1
- Further contamination of plants and plant products other than cereals and their products reported in the present book/article[746]: Medicinal plants, BEA, ENA, ENA_1, ENB, and ENB_1, literature[746]; Spices (ginger), BEA, ENA, ENA_1, and ENB, literature[746]

incidence: 0/1*, conc.: no contamination, sample origin: Bavarian retail stores, Germany, sample year: unknown, country: Germany[746], *ginger root, fresh

- Co-contamination: not reported
- Further contamination of plants and plant products other than cereals and their products reported in the present book/article[746]: Medicinal plants, BEA, ENA, ENA_1, ENB, and ENB_1, literature[746]; Spices (ginger), BEA, ENA, ENA_1, and ENB, literature[746]

Type A Trichothecenes

Neosolaniol

incidence: 1/4, conc.: 23 µg/kg, sample origin: specialist food shops, UK, sample year: unknown, country: UK[16]

- Co-contamination: not reported
- Further contamination of plants and plant products other than cereals and their products reported in the present book/article[16]: Food, AF/S, DON, FUS-X, NIV, OTA, and ZEA, literature[16]; Nuts (almond), AF/S, literature[16]; Oils, Spices (fennel), AF/S and ZEA, literature[16]; Oils (sesame oil), OTA, literature[16]; Paste (curry paste), AF/S, FB/S, NIV, OTA, and ZEA, literature[16]; Pickles, AF/S, FB/S, OTA, and ZEA, literature[16]; Sauces (chilli sauce), NIV, OTA, and ZEA, literature[16]; Spices, AF/S, FB/S, OTA, and ZEA, literature[16]; Spices (cayenne pepper, chilli), AF/S, DAS, DON, HT-2, OTA, and ZEA, literature[16]; Spices (coriander), AF/S, DON, OTA, and ZEA, literature[16]; Spices (curry), AF/S, DAS, FB/S, FUS-X, HT-2, NEO, NIV, OTA, T-2, and ZEA, literature[16]; Spices (garlic), DON, FUS-X, and NIV, literature[16]; Spices (ginger), AF/S, DON, NIV, OTA, and T-2, literature[16]; Spices (tandoori), AF/S, FB/S, NIV, OTA, and T-2, literature[16]

T-2 Toxin

incidence: 1/4, conc.: 18 µg/kg, sample origin: specialist food shops, UK, sample year: unknown, country: UK[16]

- Co-contamination: not reported
- Further contamination of plants and plant products other than cereals and their products reported in the present book/article[16]: Food, AF/S, DON, FUS-X, NIV, OTA, and ZEA, literature[16]; Nuts (almond), AF/S, literature[16]; Oils, Spices (fennel), AF/S and ZEA, literature[16]; Oils (sesame oil), OTA, literature[16]; Paste (curry paste), AF/S, FB/S, NIV, OTA, and ZEA, literature[16]; Pickles, AF/S, FB/S, OTA, and ZEA, literature[16]; Sauces (chilli sauce), NIV, OTA, and ZEA, literature[16]; Spices, AF/S, FB/S, OTA, and ZEA, literature[16]; Spices (cayenne pepper, chilli), AF/S, DAS, DON, HT-2, OTA, and ZEA, literature[16]; Spices (coriander), AF/S, DON, OTA, and ZEA, literature[16]; Spices (curry), AF/S, DAS, FB/S, FUS-X, HT-2, NEO, NIV, OTA, T-2, and ZEA, literature[16]; Spices (garlic), DON, FUS-X, and NIV, literature[16]; Spices (ginger), AF/S, DON, NEO, NIV, and OTA, literature[16]; Spices (tandoori), AF/S, FB/S, NIV, OTA, and T-2, literature[16]

Type B Trichothecenes

Deoxynivalenol

incidence: 1/4, conc.: 9 µg/kg, sample origin: specialist food shops, UK, sample year: unknown, country: UK[16]

- Co-contamination: not reported
- Further contamination of plants and plant products other than cereals and their products reported in the present book/article[16]: Food, AF/S, DON, FUS-X, NIV, OTA, and ZEA, literature[16]; Nuts (almond), AF/S, literature[16]; Oils, Spices (fennel), AF/S and ZEA, literature[16]; Oils (sesame oil), OTA, literature[16]; Paste (curry paste), AF/S, FB/S, NIV, OTA, and ZEA, literature[16]; Pickles, AF/S, FB/S, OTA, and ZEA, literature[16]; Sauces (chilli sauce), NIV, OTA, and ZEA, literature[16]; Spices, AF/S, FB/S, OTA, and ZEA, literature[16]; Spices (cayenne pepper, chilli), AF/S, DAS, DON, HT-2, OTA, and ZEA, literature[16]; Spices (coriander), AF/S, DON, OTA, and ZEA, literature[16]; Spices (curry), AF/S, DAS, FB/S, FUS-X, HT-2, NEO, NIV, OTA, T-2, and ZEA, literature[16]; Spices (garlic), DON, FUS-X, and NIV, literature[16]; Spices (ginger), AF/S, NEO, NIV, OTA, and T-2, literature[16]; Spices (tandoori), AF/S, FB/S, NIV, OTA, and T-2, literature[16]

Nivalenol

incidence: 1/4, conc.: 34 µg/kg, sample origin: specialist food shops, UK, sample year: unknown, country: UK[16]

- Co-contamination: not reported
- Further contamination of plants and plant products other than cereals and their products reported in the present book/article[16]: Food, AF/S, DON, FUS-X, NIV, OTA, and ZEA, literature[16]; Nuts (almond), AF/S, literature[16]; Oils, Spices (fennel), AF/S and ZEA, literature[16]; Oils (sesame oil), OTA, literature[16]; Paste (curry paste), AF/S, FB/S, NIV, OTA, and ZEA, literature[16]; Pickles, AF/S, FB/S, OTA, and ZEA, literature[16]; Sauces (chilli sauce), NIV, OTA, and ZEA, literature[16]; Spices, AF/S, FB/S, OTA, and ZEA, literature[16]; Spices (cayenne pepper, chilli), AF/S, DAS, DON, HT-2, OTA, and ZEA, literature[16]; Spices (coriander), AF/S, DON, OTA, and ZEA, literature[16]; Spices (curry), AF/S, DAS, FB/S, FUS-X, HT-2, NEO, NIV, OTA, T-2, and ZEA, literature[16]; Spices (garlic), DON, FUS-X, and NIV, literature[16]; Spices (ginger), AF/S, DON, NEO, OTA, and T-2, literature[16]; Spices (tandoori), AF/S, FB/S, NIV, OTA, and T-2, literature[16]

Spices (laurel leave) may contain the following mycotoxins:

Aspergillus Toxins

Aflatoxin B_1

incidence: 2/7*, conc. range: 16.5–20.3 µg/kg, Ø conc.: 18.4 µg/kg, sample origin: organic markets and organic shops, Turkey, sample year: May 2010–May 2011, country: Turkey[859], *organic laurel leaves sa.

- Co-contamination: not reported
- Further contamination of plants and plant products other than cereals and their products reported in the present book/article[859]: Herbs (camomile), Herbs (linden), Herbs (rosehip), Herbs (sage), Spices (anise), Spices (basil), Spices (cayenne pepper, chilli), Spices (cinnamon), Spices (coriander), Spices (cumin), Spices (fennel), Spices (ginger), Spices (pepper), Spices (peppermint), Spices (poppy seed), Spices (rosemary), Spices (sumac), AFB_1, literature[859]

Spice (majoran) may contain the following mycotoxins:

Aspergillus Toxins

Aflatoxin/s

incidence: 1/5, conc.: 9 µg/kg*, sample origin: Assiut (governorate), Egypt, sample year: unknown, country: Egypt[641], *AFB_1 and AFB_2

- Co-contamination: not reported

- Further contamination of plants and plant products other than cereals and their products reported in the present book/article[641]: Spices (anise), Spices (coriander), Spices (fennel), Spices (pepper), Spices (peppermint), AF/S, literature[641]; Spices (black cumin), AF/S and CIT, literature[641]; Spices (caraway), AF/S and STG, literature[641]; Spices (cayenne pepper, chilli), Spices (cumin), Spices (marjoram), STG, literature[641]

Other ***Aspergillus*** Toxins

Sterigmatocystin

incidence: 1/5, conc.: 17 μg/kg, sample origin: Assiut (governorate), Egypt, sample year: unknown, country: Egypt[641]

- Co-contamination: not reported
- Further contamination of plants and plant products other than cereals and their products reported in the present book/article[641]: Spices (anise), Spices (coriander), Spices (fennel), Spices (marjoram), Spices (pepper), Spices (peppermint), AF/S, literature[641]; Spices (black cumin), AF/S and CIT, literature[641]; Spices (caraway), AF/S and STG, literature[641]; Spices (cayenne pepper, chilli), Spices (cumin), STG, literature[641]

Aspergillus and ***Penicillium*** Toxins

Ochratoxin A

incidence: 1/13, conc.: 28 μg/kg (1 sa.), sample origin: Austria, sample year: 1994, country: EU[194]

- Co-contamination: not reported
- Further contamination of plants and plant products other than cereals and their products reported in the present book/article[194]: Coffee, Coffee beans, Spices (cayenne pepper, chilli), Spices (curry), Spices (pepper), OTA, literature[194]

Spices (masala) may contain the following mycotoxins:

Aspergillus Toxins

Aflatoxin B_1

incidence: 4/4, conc. range: 0.46–3.88 μg/kg, ∅ conc.: 1.66 μg/kg, sample origin: Penang (state, island), Malaysia, sample year: unknown, country: Malaysia[530]

- Co-contamination: 1 sa. co-contaminated with AFB_1, AFB_2, AFG_1, AFG_2, and OTA; 3 sa. co-contaminated with AFB_1, AFB_2, and OTA
- Further contamination of plants and plant products other than cereals and their products reported in the present book/article[530]: Spices, Spices (cayenne pepper, chilli), Spices (coriander), Spices (cumin), Spices (curry), Spices (fennel), AFB_1, AFB_2, AFG_1, AFG_2, and OTA, literature[530]; Spices (curcuma), AFB_1, AFB_2, AFG_1, and OTA, literature[530]; Spices (masala), AFB_2, AFG_1, AFG_2, and OTA, literature[530]; Spices (pepper), AFB_1 and AFB_2, literature[530]; Spices (poppy seed), AFB_1, literature[530]

Aflatoxin B_2

incidence: 4/4, conc. range: 0.05–0.34 μg/kg, ∅ conc.: 0.21 μg/kg, sample origin: Penang (state, island), Malaysia, sample year: unknown, country: Malaysia[530]

- Co-contamination: 1 sa. co-contaminated with AFB_1, AFB_2, AFG_1, AFG_2, and OTA; 3 sa. co-contaminated with AFB_1, AFB_2, and OTA
- Further contamination of plants and plant products other than cereals and their products reported in the present book/article[530]: Spices, Spices (cayenne pepper, chilli), Spices (coriander), Spices (cumin), Spices (curry), Spices (fennel), AFB_1, AFB_2, AFG_1, AFG_2, and OTA, literature[530]; Spices (curcuma), AFB_1, AFB_2, AFG_1, and OTA, literature[530]; Spices (masala), AFB_1, AFG_1, AFG_2, and OTA, literature[530]; Spices (pepper), AFB_1 and AFB_2, literature[530]; Spices (poppy seed), AFB_1, literature[530]

Aflatoxin G_1

incidence: 1/4, conc.: 0.18 μg/kg, sample origin: Penang (state, island), Malaysia, sample year: unknown, country: Malaysia[530]

- Co-contamination: 1 sa. co-contaminated with AFB_1, AFB_2, AFG_1, AFG_2, and OTA
- Further contamination of plants and plant products other than cereals and their products reported in the present book/article[530]: Spices, Spices (cayenne pepper, chilli), Spices (coriander), Spices (cumin), Spices (curry), Spices (fennel), AFB_1, AFB_2, AFG_1, AFG_2, and OTA, literature[530]; Spices (curcuma), AFB_1, AFB_2, AFG_1, and OTA, literature[530]; Spices (masala), AFB_1, AFB_2, AFG_2, and OTA, literature[530]; Spices (pepper), AFB_1 and AFB_2, literature[530]; Spices (poppy seed), AFB_1, literature[530]

Aflatoxin G_2

incidence: 1/4, conc.: 0.01 μg/kg, sample origin: Penang (state, island), Malaysia, sample year: unknown, country: Malaysia[530]

- Co-contamination: 1 sa. co-contaminated with AFB_1, AFB_2, AFG_1, AFG_2, and OTA
- Further contamination of plants and plant products other than cereals and their products reported in the present book/article[530]: Spices, Spices (cayenne pepper, chilli), Spices (coriander), Spices (cumin), Spices (curry), Spices (fennel), AFB_1, AFB_2, AFG_1, AFG_2, and OTA, literature[530]; Spices (curcuma), AFB_1, AFB_2, AFG_1, and OTA, literature[530]; Spices (masala), AFB_1, AFB_2, AFG_1, and OTA, literature[530]; Spices (pepper), AFB_1 and AFB_2, literature[530]; Spices (poppy seed), AFB_1, literature[530]

Aflatoxin/s

incidence: 10/15*, conc. range: 1–10 μg/kg** (8 sa.), ≤14.01 μg/kg** (2 sa.), sample origin: retail shops, open bazaars, and local markets in Karachi (city), Pakistan, sample year: January–July 2015, country: Pakistan/Ireland[393], *composite spices (nihari masala), unpacked, **AFB_1, AFB_2, AFG_1, and AFG_2

incidence: 12?/15*, conc. range: 1–10 μg/kg** (9 sa.), ≤16.57 μg/kg** (2 sa.), sample origin: retail shops, open bazaars, and local markets in Karachi (city), Pakistan, sample year: January–July 2015, country: Pakistan/Ireland[393], *composite spices (biryani masala), unpacked, **AFB_1, AFB_2, AFG_1, and AFG_2

incidence: 12/15*, conc. range: ≤7.20 μg/kg**, sample origin: retail shops, open bazaars, and local markets in Karachi (city), Pakistan, sample year: January–July 2015, country: Pakistan/Ireland[393], *composite spices (karhai masala), unpacked, **AFB_1, AFB_2, AFG_1, and AFG_2

incidence: 11/15*, conc. range: 1–10 μg/kg** (8 sa.), 10–20 μg/kg** (2 sa.), 25.74 μg/kg** (1 sa.), sample origin: retail shops, open bazaars, and local markets in Karachi (city), Pakistan, sample year: January–July 2015, country: Pakistan/Ireland[393], *composite spices (korma masala), unpacked, **AFB_1, AFB_2, AFG_1, and AFG_2

incidence: 13?/15, conc. range: 1–10 μg/kg* (9 sa.), 10–20 μg/kg* (1 sa.), 22.10 μg/kg* (1 sa.), sample origin: retail shops, open bazaars, and local markets in Karachi (city), Pakistan, sample year: January–July 2015, country: Pakistan/Ireland[393], *composite spices (tikka masala), unpacked, **AFB_1, AFB_2, AFG_1, and AFG_2

- Co-contamination: not reported
- Further contamination of plants and plant products other than cereals and their products reported in the present book/article[393]: not reported

Aspergillus and ***Penicillium*** Toxins

Ochratoxin A

incidence: 4/4, conc. range: 0.26–0.56 μg/kg, ∅ conc.: 0.38 μg/kg, sample origin: Penang (state, island), Malaysia, sample year: unknown, country: Malaysia[530]

- Co-contamination: 1 sa. co-contaminated with AFB_1, AFB_2, AFG_1, AFG_2, and OTA; 3 sa. co-contaminated with AFB_1, AFB_2, and OTA
- Further contamination of plants and plant products other than cereals and their products reported in the present book/article[530]: Spices, Spices (cayenne pepper, chilli), Spices (coriander), Spices (cumin), Spices (curry), Spices (fennel), AFB_1, AFB_2, AFG_1, AFG_2, and OTA, literature[530]; Spices (curcuma), AFB_1, AFB_2, AFG_1, and OTA, literature[530]; Spices (masala), AFB_1, AFB_2, AFG_1, and AFG_2, literature[530]; Spices (pepper), AFB_1 and AFB_2, literature[530]; Spices (poppy seed), AFB_1, literature[530]

Spices (mustard) may contain the following mycotoxins:

Aspergillus Toxins

Aflatoxin B_1

incidence: 44/100, conc. range: tr-750 μg/kg, sample origin: different zones of Bihar (state), India, sample year: March 1984–February 1985, country: India[544]

- Co-contamination: 9 sa. co-contaminated with AFB_1, AFB_2; 3 sa. co-contaminated with AFB_1, AFB_2, and AFG_1; 1 sa. co-contaminated with AFB_1, AFB_2, AFG_1, and AFG_2; 31 sa. contaminated solely with AFB_1
- Further contamination of plants and plant products other than cereals and their products reported in the present book/article[544]: Oils (mustard oil), AFB_1, literature[544]; Spices (mustard), AFB_2, AFG_1, and AFG_2, literature[544]

incidence: 14/125*, conc. range: 15–640 μg/kg, sample origin: mustard growing areas (16 districts) of Bihar (state), India, sample year: February/March 1987, country: India[545], **Brassica juncea* Coss. (brown-seeded, *Rai*)

- Co-contamination: 9 sa. co-contaminated with AFB_1 and AFB_2; 2 sa. co-contaminated with AFB_1, AFB_2 and AFG_1; 3 sa. contaminated solely with AFB_1
- Further contamination of plants and plant products other than cereals and their products reported in the present book/article[545]: Field mustard, AFB_1 and AFB_2, literature[545]; Spices (mustard), AFB_2 and AFG_1, literature[545]

Aflatoxin B_2

incidence: 13/100, conc. range: pr, sample origin: different zones of Bihar (state), India, sample year: March 1984–February 1985, country: India[544]

- Co-contamination: 9 sa. co-contaminated with AFB_1, AFB_2; 3 sa. co-contaminated with AFB_1, AFB_2, and AFG_1; 1 sa. co-contaminated with AFB_1, AFB_2, AFG_1, and AFG_2
- Further contamination of plants and plant products other than cereals and their products reported in the present book/article[544]: Oils (mustard oil), AFB_1, literature[544]; Spices (mustard), AFB_1, AFG_1, and AFG_2, literature[544]

incidence: 11/125*, conc. range: pr, sample origin: mustard growing areas (16 districts) of Bihar (state), India, sample year: February/March 1987, country: India[545], **Brassica juncea* Coss. (brown-seeded, *Rai*)

- Co-contamination: 9 sa. co-contaminated with AFB_1 and AFB_2; 2 sa. co-contaminated with AFB_1, AFB_2 and AFG_1
- Further contamination of plants and plant products other than cereals and their products reported in the present book/article[545]: Field mustard, AFB_1 and AFB_2, literature[545]; Spices (mustard), AFB_1 and AFG_1, literature[545]

Aflatoxin G_1

incidence: 4/100, conc. range: pr, sample origin: different zones of Bihar (state), India, sample year: March 1984–February 1985, country: India[544]

- Co-contamination: 3 sa. co-contaminated with AFB_1, AFB_2, and AFG_1; 1 sa. co-contaminated with AFB_1, AFB_2, AFG_1, and AFG_2
- Further contamination of plants and plant products other than cereals and their products reported in the present book/article[544]: Oils (mustard oil), AFB_1, literature[544]; Spices (mustard), AFB_1, AFB_2, and AFG_2, literature[544]

incidence: 2/125*, conc. range: pr, sample origin: mustard growing areas (16 districts) of Bihar (state), India, sample year: February/March 1987, country: India[545], **Brassica juncea* Coss. (brown-seeded, *Rai*)

- Co-contamination: 2 sa. co-contaminated with AFB_1, AFB_2 and AFG_1
- Further contamination of plants and plant products other than cereals and their products reported in the present book/article[545]: Field mustard, AFB_1 and AFB_2, literature[545]; Spices (mustard), AFB_1 and AFB_2, literature[545]

Aflatoxin G_2

incidence: 1/100, conc.: pr, sample origin: different zones of Bihar (state), India, sample year: March 1984–February 1985, country: India[544]

- Co-contamination: 1 sa. co-contaminated with AFB_1, AFB_2, AFG_1, and AFG_2
- Further contamination of plants and plant products other than cereals and their products reported in the present book/article[544]: Oils (mustard oil), AFB_1, literature[544]; Spices (mustard), AFB_1, AFB_2, and AFG_1, literature[544]

Aspergillus and ***Penicillium*** Toxins

Ochratoxin A

incidence: 3/4, conc. range: ≤0.34 μg/l sample origin: different origins, Germany, sample year: unknown, country: Germany[110]

- Co-contamination: not reported
- Further contamination of plants and plant products other than cereals and their products reported in the present book/article[110]: Juices (blackcurrant juice), Juices (carrot juice), Juices (grape juice), Juices (tomato juice), Sauces (pepper and barbecue sauce), Sauces (soy sauce), Tomato ketchup, Vinegar (apple), Vinegar (balsamico), Vinegar (wine), Wine (red), Wine (rosé), Wine (white), OTA, literature[110]

incidence: 3/12, conc. range: ≤0.37 μg/kg, sample origin: 12 regions, Czech Republic, sample year: 2011–2013, country: Czech Republic[811]

- Co-contamination: not reported

- Further contamination of plants and plant products other than cereals and their products reported in the present book/article[811]: Beans, Chocolate, Coffee, Confectionery, Figs, Herbs and Spices, Juices, Licorice, Powder (cocoa powder), Products (ginger product), Raisins, Spices (caraway), Spices (cayenne pepper, chilli), Spices (coriander), Spices (ginger), Spices (nutmeg), Spices (paprika), Spices (pepper), Tea, Wine (red), Wine (white), OTA, literature[811]

Spices (nutmeg) may contain the following mycotoxins:

Aspergillus Toxins

Aflatoxin B_1

incidence: 5/5*, ∅ conc.: 23.8 µg/kg, sample origin: shops, supermarkets and wholesalers, France, sample year: 1989/1990, country: France[5], *nutmeg, powdered

- Co-contamination: not reported
- Further contamination of plants and plant products other than cereals and their products reported in the present book/article[5]: Butter (peanut butter), AFB_1, literature[5]

incidence: 25?/56, conc. range: 0.2–60.3 µg/kg, sample origin: unknown, sample year: unknown, country: Japan[97]

- Co-contamination: not reported
- Further contamination of plants and plant products other than cereals and their products reported in the present book/article[97]: Beans, Jam (bean jam), Spices, AFB_1 and AFB_2, literature[97]; Butter (peanut butter), Oils (peanut oil), Spices (pepper), AFB_1, AFB_2, and AFG_1, literature[97]; Spices (nutmeg), AFB_2, AFG_1, and AFG_2, literature[97]

incidence: 84?/108, conc. range: 0.2–13.4 µg/kg, ∅ conc.: 3.6 µg/kg, sample origin: commercially available market basket sa. of foods and foodstuffs in Tokyo (capital), Japan, sample year: 1986–1990, country: Japan[99]

- Co-contamination: not reported
- Further contamination of plants and plant products other than cereals and their products reported in the present book/article[99]: Beans, Seeds (sesame), Sugar, AFB_1 and AFB_2, literature[99]; Food, Jam (bean jam), Paste (peanut paste), AFB_1, literature[99]; Nuts (Brazil nut, paranut), Nuts (peanut), Nuts (pistachio nut), AFB_1, AFB_2, AFG_1, and AFG_2, literature[99]; Spices, Spices (cayenne pepper, chilli), Spices (pepper), AFB_1, AFB_2, and AFG_1, literature[99]; Spices (nutmeg), AFB_2, AFG_1, and AFG_2, literature[99]; Spices (paprika), AFB_1 and AFG_1, literature[99]

incidence: 8/10*, conc. range: 1.25–58.0 µg/kg, ∅ conc.: 20.85 µg/kg, sample origin: supermarkets and ethnic shops in Lisbon (capital) area, Portugal, sample year: unknown, country: Portugal[169], *nutmeg, packaged

- Co-contamination: not reported
- Further contamination of plants and plant products other than cereals and their products reported in the present book/article[169]: Spices (cayenne pepper, chilli), Spices (cumin), Spices (curry), Spices (paprika), Spices (pepper), Spices (saffron), AFB_1, literature[169]

incidence: 2/3, conc. range: 0.4–0.6 µg/kg, ∅ conc.: 0.5 µg/kg, sample origin: retail stores in Osaka (prefecture), Japan, sample year: 1988–1992, country: Japan[173]

- Co-contamination: 1 sa. co-contaminated with AFB_1, AFB_2, and AFG_1; 1 sa. contaminated solely with AFB_1
- Further contamination of plants and plant products other than cereals and their products reported in the present book/article[173]: Butter (peanut butter), AFB_1, AFB_2, and AFG_1, literature[173]; Nuts (peanuts), Spices (cayenne pepper, chilli), Spices (pepper), AFB_1, literature[173]; Spices (nutmeg), AFB_2 and AFG_1, literature[173]

incidence: 1/3, conc.: 2.27 µg/kg, sample origin: public markets, supermarkets, shops, and bonded warehouse in Emilia Romagna (administrative region), Italy, sample year: 2000–2005, country: Italy[176]

- Co-contamination: 1 sa. co-contaminated with AFB_1 and AFB_2
- Further contamination of plants and plant products other than cereals and their products reported in the present book/article[176]: Spices (cayenne pepper, chilli), AFB_1, AFB_2, AFG_1, and AFG_2, literature[176]; Spices (cinnamon), AFB_1, literature[176]; Spices (nutmeg), AFB_2, literature[176]

incidence: 1/1*, conc.: 1.77 µg/kg, sample origin: local stores in Munich (city), Germany, sample year: unknown, country: Germany[181], *nutmeg powder

- Co-contamination: 1 sa. co-contaminated with AFB_1, AFB_2, and AFG_1
- Further contamination of plants and plant products other than cereals and their products reported in the present book/article[181]: Butter (peanut butter), AFB_1, AFB_2, and AFG_1, literature[181]; Nuts (hazelnut), Spices (paprika), AFB_1 and AFG_1, literature[181]; Nuts (pistachio nut), Spices (pepper), AFB_1, literature[181]; Spices (nutmeg), AFB_2 and AFG_1, literature[181]

incidence: 3?/5, conc. range: 2.5–5.5 µg/kg, sample origin: imported from India, Indonesia, and Malaysia, USA, sample year: unknown, country: USA[185]

- Co-contamination: not reported
- Further contamination of plants and plant products other than cereals and their products reported in the present book/article[185]: Spices (celery), AFG_1, literature[185]; Spices (nutmeg), AFB_2, literature[185]; Spices (pepper), AFG_1 and AFG_2, literature[185]

incidence: 3/13*, conc. range: 5–15 µg/kg, sample origin: unknown, sample year: unknown, country: Germany[239], *nutmeg, moldy

- Co-contamination: not reported
- Further contamination of plants and plant products other than cereals and their products reported in the present book/article[239]: Fruits (apple), Fruits (cherry), Fruits (lemon), Fruits (orange), Fruits (peach), Fruits (tomato), Nuts (almond), Nuts (cashew nut), Nuts (hazelnut), Nuts (walnut), Seeds (sesame), AFB_1, literature[239]; Nuts (peanut), AFB_1, AFB_2, AFG_1, and AFG_2, literature[239]

incidence: 11/28, conc. range: ≤5 µg/kg (8 sa.), >5 µg/kg (3 sa., maximum: 7.7 µg/kg), sample origin: unknown, sample year: unknown, country: Germany/Thailand[244]

- Co-contamination: not reported
- Further contamination of plants and plant products other than cereals and their products reported in the present book/article[244]: Spices (cayenne pepper, chilli), Spices (coriander), AFB_1, literature[244]

incidence: 4/13, conc. range: 5–37.5 µg/kg*, sample origin: processors located in six Canadian cities, Canada, sample year: unknown, country: Canada[323], *approximate conc.

- Co-contamination: 3 sa. co-contaminated with AFB_1 and AFB_2; 1 sa. contaminated solely with AFB_1
- Further contamination of plants and plant products other than cereals and their products reported in the present book/article[323]: Spices (coriander), Spices (curcuma), AFB_1 and AFB_2,

literature[323]; Spices (ginger), AFB_1, AFB_2, AFG_1, and AFG_2, literature[323]; Spices (nutmeg), AFB_2, literature[323]

incidence: 30/32*, conc. range: 1.0–23.2 µg/kg, ∅ conc.: 12.5 µg/kg, sample origin: Netherlands, sample year: unknown, country: Netherlands[546], *nutmeg powder

- Co-contamination: not reported
- Further contamination of plants and plant products other than cereals and their products reported in the present book/article[546]: Spices, AFB_1, literature[546]; Spices (nutmeg), AF/S, literature[546]

incidence: 25/25, conc. range: 1.0–378 µg/kg, ∅ conc.: 52.3 µg/kg, sample origin: imported from Indonesia, Japan, sample year: unknown, country: Japan[548]

- Co-contamination: 2 sa. co-contaminated with AFB_1, AFB_2, AFG_1, and AFG_2; 8 sa. co-contaminated with AFB_1, AFB_2, and AFG_1; 10 sa. co-contaminated with AFB_1 and AFB_2; 5 sa. contaminated solely with AFB_1
- Further contamination of plants and plant products other than cereals and their products reported in the present book/article[548]: Spices (nutmeg), AFB_2, AFG_1, and AFG_2, literature[548]

incidence: 2/4, conc. range: 0.73–0.82 µg/kg, ∅ conc.: 0.78 µg/kg, sample origin: sa. from Hainan (province) purchased in markets, herbal shops, and bazaars in Beijing (capital), China, sample year: unknown, country: China[549]

- Co-contamination: 1 sa. co-contaminated with AFB_1 and AFB_2; 1 sa. contaminated solely with AFB_1
- Further contamination of plants and plant products other than cereals and their products reported in the present book/article[549]: Spices (nutmeg), AFB_2, AFG_2, and OTA, literature[549]

incidence: 2/6, conc. range: 1.62–16.31 µg/kg, ∅ conc.: 9.0 µg/kg, sample origin: sa. from Yunnan (province) purchased in markets, herbal shops, and bazaars in Beijing (capital), China, sample year: unknown, country: China[549]

- Co-contamination: 1 sa. co-contaminated with AFB_1, AFB_2, and AFG_2; 1 sa. co-contaminated with AFB_1 and AFB_2
- Further contamination of plants and plant products other than cereals and their products reported in the present book/article[549]: Spices (nutmeg), AFB_2, AFG_2, and OTA, literature[549]

incidence: 0/3, conc.: no contamination, sample origin: sa. from Guangdong (province) purchased in markets, herbal shops, and bazaars in Beijing (capital), China, sample year: unknown, country: China[549]

- Co-contamination: not reported
- Further contamination of plants and plant products other than cereals and their products reported in the present book/article[549]: Spices (nutmeg), AFB_2, AFG_2, and OTA, literature[549]

incidence: 20/20, conc. range: ≤19.69 µg/kg, ∅ conc.: 4.11 µg/kg, sample origin: retail stores in Lower Saxony (federal state), Germany, sample year: 2009, country: Germany[829]

- Co-contamination: not reported
- Further contamination of plants and plant products other than cereals and their products reported in the present book/article[829]: Nuts (hazelnut), AFB_1 and AF/S, literature[829]; Nuts (pistachio nut), AFB_1, literature[829]; Oils (maize oil), ZEA, literature[829]; Spices (cinnamon), OTA, literature[829]; Spices (curry), AFB_1, AF/S, and OTA, literature[829]; Spices (ginger), AFB_1 and AF/S, literature[829]; Spices (nutmeg), AF/S and OTA, literature[829]

Aflatoxin B_2

incidence: 25?/56, conc. range: 0.1–6.5 µg/kg, sample origin: unknown, sample year: unknown, country: Japan[97]

- Co-contamination: not reported
- Further contamination of plants and plant products other than cereals and their products reported in the present book/article[97]: Beans, Jam (bean jam), Spices, AFB_1 and AFB_2, literature[97]; Butter (peanut butter), Oils (peanut oil), Spices (pepper), AFB_1, AFB_2, and AFG_1, literature[97]; Spices (nutmeg), AFB_1, AFG_1, and AFG_2, literature[97]

incidence: 84?/108, conc. range: ≤2.5 µg/kg, ∅ conc.: 0.5 µg/kg, sample origin: commercially available market basket sa. of foods and foodstuffs in Tokyo (capital), Japan, sample year: 1986–1990, country: Japan[99]

- Co-contamination: not reported
- Further contamination of plants and plant products other than cereals and their products reported in the present book/article[99]: Beans, Seeds (sesame), Sugar, AFB_1 and AFB_2, literature[99]; Food, Jam (bean jam), Paste (peanut paste), AFB_1, literature[99]; Nuts (Brazil nut, paranut), Nuts (peanut), Nuts (pistachio nut), AFB_1, AFB_2, AFG_1, and AFG_2, literature[99]; Spices, Spices (cayenne pepper, chilli), Spices (pepper), AFB_1, AFB_2, and AFG_1, literature[99]; Spices (nutmeg), AFB_1, AFG_1, and AFG_2, literature[99]; Spices (paprika), AFB_1 and AFG_1, literature[99]

incidence: 1/3, conc.: 0.2 µg/kg, sample origin: retail stores in Osaka (prefecture), Japan, sample year: 1988–1992, country: Japan[173]

- Co-contamination: 1 sa. co-contaminated with AFB_1, AFB_2, and AFG_1
- Further contamination of plants and plant products other than cereals and their products reported in the present book/article[173]: Butter (peanut butter), AFB_1, AFB_2, and AFG_1, literature[173]; Nuts (peanuts), Spices (cayenne pepper, chilli), Spices (pepper), AFB_1, literature[173]; Spices (nutmeg), AFB_1 and AFG_1, literature[173]

incidence: 1/3, conc.: 0.47 µg/kg, sample origin: public markets, supermarkets, shops, and bonded warehouse in Emilia Romagna (administrative region), Italy, sample year: 2000–2005, country: Italy[176]

- Co-contamination: 1 sa. co-contaminated with AFB_1 and AFB_2
- Further contamination of plants and plant products other than cereals and their products reported in the present book/article[176]: Spices (cayenne pepper, chilli), AFB_1, AFB_2, AFG_1, and AFG_2, literature[176]; Spices (cinnamon), Spices (nutmeg), AFB_1, literature[176]

incidence: 1/1*, conc.: 0.40 µg/kg, sample origin: local stores in Munich (city), Germany, sample year: unknown, country: Germany[181], *nutmeg powder

- Co-contamination: 1 sa. co-contaminated with AFB_1, AFB_2, and AFG_1
- Further contamination of plants and plant products other than cereals and their products reported in the present book/article[181]: Butter (peanut butter), AFB_1, AFB_2, and AFG_1, literature[181]; Nuts (hazelnut), Spices (nutmeg), Spices (paprika), AFB_1 and AFG_1, literature[181]; Nuts (pistachio nut), Spices (pepper), AFB_1, literature[181]

incidence: 3?/5, conc. range: 0.75–1.1 µg/kg, sample origin: imported from India, Indonesia, and Malaysia, USA, sample year: unknown, country: USA[185]

- Co-contamination: not reported

- Further contamination of plants and plant products other than cereals and their products reported in the present book/article[185]: Spices (celery), AFG_1, literature[185]; Spices (nutmeg), AFB_1, literature[185]; Spices (pepper), AFG_1 and AFG_2, literature[185]

incidence: 3/13, conc. range: tr-15 µg/kg*, sample origin: processors located in six Canadian cities, Canada, sample year: unknown, country: Canada[323], *approximate conc.

- Co-contamination: 3 sa. co-contaminated with AFB_1 and AFB_2
- Further contamination of plants and plant products other than cereals and their products reported in the present book/article[323]: Spices (coriander), Spices (curcuma), AFB_1 and AFB_2, literature[323]; Spices (ginger), AFB_1, AFB_2, AFG_1, and AFG_2, literature[323]; Spices (nutmeg), AFB_1, literature[323]

incidence: 20/25, conc. range: 0.2–43 µg/kg, ∅ conc.: 5.4 µg/kg, sample origin: imported from Indonesia, Japan, sample year: unknown, country: Japan[548]

- Co-contamination: 2 sa. co-contaminated with AFB_1, AFB_2, AFG_1, and AFG_2; 8 sa. co-contaminated with AFB_1, AFB_2, and AFG_1; 10 sa. co-contaminated with AFB_1 and AFB_2
- Further contamination of plants and plant products other than cereals and their products reported in the present book/article[548]: Spices (nutmeg), AFB_1, AFG_1, and AFG_2, literature[548]

incidence: 1/4, conc.: 0.17 µg/kg, sample origin: sa. from Hainan (province) purchased in markets, herbal shops, and bazaars in Beijing (capital), China, sample year: unknown, country: China[549]

- Co-contamination: 1 sa. co-contaminated with AFB_1 and AFB_2
- Further contamination of plants and plant products other than cereals and their products reported in the present book/article[549]: Spices (nutmeg), AFB_1, AFG_2, and OTA, literature[549]

incidence: 2/6, conc. range: 0.10–1.55 µg/kg, ∅ conc.: 0.83 µg/kg, sample origin: sa. from Yunnan (province) purchased in markets, herbal shops, and bazaars in Beijing (capital), China, sample year: unknown, country: China[549]

- Co-contamination: 1 sa. co-contaminated with AFB_1, AFB_2, and AFG_2; 1 sa. co-contaminated with AFB_1 and AFB_2
- Further contamination of plants and plant products other than cereals and their products reported in the present book/article[549]: Spices (nutmeg), AFB_1, AFG_2, and OTA, literature[549]

incidence: 0/3, conc.: no contamination, sample origin: sa. from Guangdong (province) purchased in markets, herbal shops, and bazaars in Beijing (capital), China, sample year: unknown, country: China[549]

- Co-contamination: not reported
- Further contamination of plants and plant products other than cereals and their products reported in the present book/article[549]: Spices (nutmeg), AFB_1, AFG_2, and OTA, literature[549]

Aflatoxin G_1

incidence: 25?/56, conc. range: 0.4–15.8 µg/kg, sample origin: unknown, sample year: unknown, country: Japan[97]

- Co-contamination: not reported
- Further contamination of plants and plant products other than cereals and their products reported in the present book/article[97]: Beans, Jam (bean jam), Spices, AFB_1 and AFB_2, literature[97]; Butter (peanut butter), Oils (peanut oil), Spices (pepper), AFB_1, AFB_2, and AFG_1, literature[97]; Spices (nutmeg), AFB_1, AFB_2, and AFG_2, literature[97]

incidence: 84?/108, conc. range: ≤2.8 µg/kg, ∅ conc.: 0.1 µg/kg, sample origin: commercially available market basket sa. of foods and foodstuffs in Tokyo (capital), Japan, sample year: 1986–1990, country: Japan[99]

- Co-contamination: not reported
- Further contamination of plants and plant products other than cereals and their products reported in the present book/article[99]: Beans, Seeds (sesame), Sugar, AFB_1 and AFB_2, literature[99]; Food, Jam (bean jam), Paste (peanut paste), AFB_1, literature[99]; Nuts (Brazil nut, paranut), Nuts (peanut), Nuts (pistachio nut), AFB_1, AFB_2, AFG_1, and AFG_2, literature[99]; Spices, Spices (cayenne pepper, chilli), Spices (pepper), AFB_1, AFB_2, and AFG_1, literature[99]; Spices (nutmeg), AFB_1, AFB_2, and AFG_2, literature[99]; Spices (paprika), AFB_1 and AFG_1, literature[99]

incidence: 1/3, conc.: 0.2 µg/kg, sample origin: retail stores in Osaka (prefecture), Japan, sample year: 1988–1992, country: Japan[173]

- Co-contamination: 1 sa. co-contaminated with AFB_1, AFB_2, and AFG_1
- Further contamination of plants and plant products other than cereals and their products reported in the present book/article[173]: Butter (peanut butter), AFB_1, AFB_2, and AFG_1, literature[173]; Nuts (peanuts), Spices (cayenne pepper, chilli), Spices (pepper), AFB_1, literature[173]; Spices (nutmeg), AFB_1 and AFB_2, literature[173]

incidence: 1/1*, conc.: 0.63 µg/kg, sample origin: local stores in Munich (city), Germany, sample year: unknown, country: Germany[181], *nutmeg powder

- Co-contamination: 1 sa. co-contaminated with AFB_1, AFB_2, and AFG_1
- Further contamination of plants and plant products other than cereals and their products reported in the present book/article[181]: Butter (peanut butter), AFB_1, AFB_2, and AFG_1, literature[181]; Nuts (hazelnut), Spices (paprika), AFB_1 and AFG_1, literature[181]; Nuts (pistachio nut), Spices (pepper), AFB_1, literature[181]; Spices (nutmeg), AFB_1 and AFB_2, literature[181]

incidence: 10/25, conc. range: 0.6–78 µg/kg, ∅ conc.: 14.4 µg/kg, sample origin: imported from Indonesia, Japan, sample year: unknown, country: Japan[548]

- Co-contamination: 2 sa. co-contaminated with AFB_1, AFB_2, AFG_1, and AFG_2; 8 sa. co-contaminated with AFB_1, AFB_2, and AFG_1
- Further contamination of plants and plant products other than cereals and their products reported in the present book/article[548]: Spices (nutmeg), AFB_1, AFB_2, and AFG_2, literature[548]

Aflatoxin G_2

incidence: 1/56, conc.: 0.3 µg/kg, sample origin: unknown, sample year: unknown, country: Japan[97]

- Co-contamination: not reported
- Further contamination of plants and plant products other than cereals and their products reported in the present book/article[97]: Beans, Jam (bean jam), Spices, AFB_1 and AFB_2, literature[97]; Butter (peanut butter), AFB_1, AFB_2, and AFG_1, literature[97]; Oils (peanut oil), Spices (pepper), Spices (nutmeg), AFB_1, AFB_2, and AFG_1, literature[97]

incidence: 84?/108, conc. range: ≤0.4 µg/kg, ∅ conc.: <0.1 µg/kg, sample origin: commercially available market basket sa. of foods and foodstuffs in Tokyo (capital), Japan, sample year: 1986–1990, country: Japan[99]

- Co-contamination: not reported
- Further contamination of plants and plant products other than cereals and their products reported in the present book/article[99]: Beans, Seeds (sesame), Sugar, AFB_1 and AFB_2, literature[99]; Food, Jam (bean jam), Paste (peanut paste), AFB_1, literature[99]; Nuts (Brazil nut, paranut), Nuts (peanut), Nuts (pistachio nut), AFB_1, AFB_2, AFG_1, and AFG_2, literature[99]; Spices, Spices (cayenne pepper, chilli), Spices (nutmeg), Spices (pepper), AFB_1, AFB_2, and AFG_1, literature[99]; Spices (paprika), AFB_1 and AFG_1, literature[99]

incidence: 2/25, conc. range: 0.4–4.4 µg/kg, ∅ conc.: 2.4 µg/kg, sample origin: imported from Indonesia, Japan, sample year: unknown, country: Japan[548]
- Co-contamination: 2 sa. co-contaminated with AFB_1, AFB_2, AFG_1, and AFG_2
- Further contamination of plants and plant products other than cereals and their products reported in the present book/article[548]: Spices (nutmeg), AFB_1, AFB_2, and AFG_1, literature[548]

incidence: 0/4, conc.: no contamination, sample origin: sa. from Hainan (province) purchased in markets, herbal shops, and bazaars in Beijing (capital), China, sample year: unknown, country: China[549]
- Co-contamination: not reported
- Further contamination of plants and plant products other than cereals and their products reported in the present book/article[549]: Spices (nutmeg), AFB_1, AFB_2, and OTA, literature[549]

incidence: 1/6, conc.: 0.29 µg/kg, sample origin: sa. from Yunnan (province) purchased in markets, herbal shops, and bazaars in Beijing (capital), China, sample year: unknown, country: China[549]
- Co-contamination: 1 sa. co-contaminated with AFB_1, AFB_2, and AFG_2
- Further contamination of plants and plant products other than cereals and their products reported in the present book/article[549]: Spices (nutmeg), AFB_1, AFB_2, and OTA, literature[549]

incidence: 0/3, conc.: no contamination, sample origin: sa. from Guangdong (province) purchased in markets, herbal shops, and bazaars in Beijing (capital), China, sample year: unknown, country: China[549]
- Co-contamination: not reported
- Further contamination of plants and plant products other than cereals and their products reported in the present book/article[549]: Spices (nutmeg), AFB_1, AFB_2, and OTA, literature[549]

Aflatoxin/s

incidence: 5/5, conc. range: ≤2 µg/kg*, sample origin: imported to USA, sample year: 1986, country: USA[19], *AFB_1, AFB_2, AFG_1, and AFG_2
- Co-contamination: not reported
- Further contamination of plants and plant products other than cereals and their products reported in the present book/article[19]: Butter (peanut butter), Candies (peanut candy), Candies (pistachio candy), Nuts, Nuts (almond), Nuts (Brazil nut, paranut), Nuts (peanut), Nuts (pecan nut), Nuts (pistachio nut), Nuts (walnut), Oils (peanut oil), Paste (peanut paste), Paste (pipian paste), Products (peanut product), Seeds (melon), Spices (cayenne pepper, chilli), Spices (ginger), AF/S, literature[19]

incidence: 30/32*, conc. range: 2.3–36.5 µg/kg**, ∅ conc.: 15.8 µg/kg**, sample origin: Netherlands, sample year: unknown, country: Netherlands[546], *nutmeg powder, **AFB_1, AFB_2, AFG_1, and AFG_2
- Co-contamination: not reported
- Further contamination of plants and plant products other than cereals and their products reported in the present book/article[546]: Spices, Spices (nutmeg), AFB_1, literature[546]

incidence: 20/20, conc. range: ≤23.74 µg/kg*, ∅ conc.: 5.24 µg/kg*, sample origin: retail stores in Lower Saxony (federal state), Germany, sample year: 2009, country: Germany[829], *AFB_1, AFB_2, AFG_1, and AFG_2
- Co-contamination: not reported
- Further contamination of plants and plant products other than cereals and their products reported in the present book/article[829]: Nuts (hazelnut), AFB_1 and AF/S, literature[829]; Nuts (pistachio nut), AFB_1, literature[829]; Oils (maize oil), ZEA, literature[829]; Spices (cinnamon), OTA, literature[829]; Spices (curry), AFB_1, AF/S, and OTA, literature[829]; Spices (ginger), AFB_1 and AF/S, literature[829]; Spices (nutmeg), AFB_1 and OTA, literature[829]

Aspergillus and ***Penicillium*** Toxins

Ochratoxin A

incidence: 4/8, conc. range: 1.1–2.1 µg/kg, ∅ conc.: 1.5 µg/kg, sample origin: retail shops in Belgium, Netherlands, and Russia, sample year: autumn-winter 2005/2006, country: Belgium/Russia/Hungary[193]
- Co-contamination: not reported
- Further contamination of plants and plant products other than cereals and their products reported in the present book/article[193]: Spices (cayenne pepper, chilli), Spices (ginger), Spices (paprika), Spices (pepper), OTA, literature[193]

incidence: 2/2, conc. range: 4.3–23.5 µg/kg, ∅ conc.: 13.9 µg/kg, sample origin: available in New Zealand, sample year: 2007, country: New Zealand[277]
- Co-contamination: not reported
- Further contamination of plants and plant products other than cereals and their products reported in the present book/article[277]: Bars, AFB_1, AF/S, DON, NIV, and OTA, literature[277]; Butter (peanut butter), Confectionery, Food, Fruits, dried, Nuts, Nuts (peanut), Nuts (pistachio nut), Sauces (peanut sauce), AFB_1 and AF/S, literature[277]; Coffee, Dates, Spices (pepper), Vine fruits, dried, OTA, literature[277]; Figs, Spices (cayenne pepper, chilli), Spices (curry), Spices (ginger), Spices (paprika), AFB_1, AF/S, and OTA, literature[277]; Nuts (Brazil nut, paranut), AFB_1, literature[277]

incidence: 0/4, conc.: no contamination, sample origin: sa. from Hainan (province) purchased in markets, herbal shops, and bazaars in Beijing (capital), China, sample year: unknown, country: China[549]
- Co-contamination: not reported
- Further contamination of plants and plant products other than cereals and their products reported in the present book/article[549]: Spices (nutmeg), AFB_1, AFB_2, and AFG_2, literature[549]

incidence: 1/6, conc.: <LOQ, sample origin: sa. from Yunnan (province) purchased in markets, herbal shops, and bazaars in Beijing (capital), China, sample year: unknown, country: China[549]
- Co-contamination: not reported
- Further contamination of plants and plant products other than cereals and their products reported in the present book/article[549]: Spices (nutmeg), AFB_1, AFB_2, and AFG_2, literature[549]

incidence: 0/3, conc.: no contamination, sample origin: sa. from Guangdong (province) purchased in markets, herbal shops, and bazaars in Beijing (capital), China, sample year: unknown, country: China[549]

- Co-contamination: not reported
- Further contamination of plants and plant products other than cereals and their products reported in the present book/article[549]: Spices (nutmeg), AFB_1, AFB_2, and AFG_2, literature[549]

incidence: 1/50, conc.: 14 µg/kg, sample origin: FUCHS Gewürze GmbH, Latvia, sample year: 2014, country: Latvia[679]

- Co-contamination: not reported
- Further contamination of plants and plant products other than cereals and their products reported in the present book/article[679]: Spices (cayenne pepper, chilli), Spices (nutmeg), FB_1, literature[679]; Spices (oregano), STG, literature[679]; Spices (paprika), FB_1, OTA, and STG, literature[679]; Spices (thyme), DON, STG, and ZEA, literature[679]

incidence: 3/3, conc. range: LOD/LOQ-0.9 µg/kg (1 sa.), ≤8.5 µg/kg (2 sa.), sample origin: Portugal, sample year: 1999, country: EU[698]

- Co-contamination: not reported
- Further contamination of plants and plant products other than cereals and their products reported in the present book/article[698]: Baby food, Beans (soybean), Chocolate, Cocoa, Coffee, Coffee beans, Currants, Fruits, dried, Herbs, Jam, Juices, Juices (grape juice), Legumes, Nuts, Oil seeds, Oils (olive oil), Peas, Peas (chickpea), Products (plant product), Raisins, Spices, Spices (cayenne pepper, chilli), Spices (paprika), Spices (pepper), Sultanas, Tea, Vinegar, Wine, Wine (red), Wine (rosé), Wine (white), Wine, miscellaneous, OTA, literature[698]

incidence: 12/12, conc. range: 0.30–60.70 µg/kg, Ø conc.: 8.70 µg/kg, sample origin: 12 regions, Czech Republic, sample year: 2011–2013, country: Czech Republic[811]

- Co-contamination: not reported
- Further contamination of plants and plant products other than cereals and their products reported in the present book/article[811]: Beans, Chocolate, Coffee, Confectionery, Figs, Herbs and Spices, Juices, Licorice, Powder (cocoa powder), Products (ginger product), Raisins, Spices (cayenne pepper, chilli), Spices (caraway), Spices (coriander), Spices (ginger), Spices (mustard), Spices (paprika), Spices (pepper), Tea, Wine (red), Wine (white), OTA, literature[811]

incidence: 20/20, conc. range: ≤10.3 µg/kg, Ø conc.: 2.56 µg/kg, sample origin: retail stores in Lower Saxony (federal state), Germany, sample year: 2009, country: Germany[829]

- Co-contamination: not reported
- Further contamination of plants and plant products other than cereals and their products reported in the present book/article[829]: Nuts (hazelnut), AFB_1 and AF/S, literature[829]; Nuts (pistachio nut), AFB_1, literature[829]; Oils (maize oil), ZEA, literature[829]; Spices (cinnamon), OTA, literature[829]; Spices (curry), AFB_1, AF/S, and OTA, literature[829]; Spices (ginger), Spices (nutmeg), AFB_1 and AF/S, literature[829]

Fusarium Toxins

Fumonisin/s

Fumonisin B_1

incidence: 1/50, conc.: 25 µg/kg, sample origin: FUCHS Gewürze GmbH, Latvia, sample year: 2014, country: Latvia[679]

- Co-contamination: not reported
- Further contamination of plants and plant products other than cereals and their products reported in the present book/article[679]: Spices (cayenne pepper, chilli), FB_1, literature[679]; Spices (nutmeg), OTA, literature[679]; Spices (oregano), STG, literature[679]; Spices (paprika), FB_1, OTA, and STG, literature[679]; Spices (thyme), DON, STG, and ZEA, literature[679]

Spices (onion) may contain the following mycotoxins:

Fusarium Toxins

Fumonisin/s

Fumonisin B_1

incidence: 3/3*, conc. range: ~5–135 µg/kg, sample origin: commercially available for industrial use, Netherlands, sample year: unknown, country: Netherlands[184], *onion powder

- Co-contamination: not reported
- Further contamination of plants and plant products other than cereals and their products reported in the present book/article[184]: Spices (garlic), FB_1, literature[184]; Spices (paprika), AFB_1 and OTA, literature[184]; Spices (pepper), AFB_2, literature[184]

Spices (oregano) may contain the following mycotoxins:

Aspergillus Toxins

Sterigmatocystin

incidence: 5/50, conc. range: 15–28 µg/kg, sample origin: FUCHS Gewürze GmbH, Latvia, sample year: 2014, country: Latvia[679]

- Co-contamination: not reported
- Further contamination of plants and plant products other than cereals and their products reported in the present book/article[679]: Spices (cayenne pepper, chilli), FB_1, literature[679]; Spices (nutmeg), FB_1 and OTA, literature[679]; Spices (paprika), FB_1, OTA, and STG, literature[679]; Spices (thyme), DON, STG, and ZEA, literature[679]

Spices (paprika) may contain the following mycotoxins:

Alternaria Toxins

Alternariol

incidence: 6/20*, conc. range: <6.6–101 µg/kg, sample origin: grocery in Oostkamp (municipality), Belgium, sample year: June–October 2008, country: Belgium[165], *sweet pepper with moldy or brown spots

- Co-contamination: not reported
- Further contamination of plants and plant products other than cereals and their products reported in the present book/article[165]: Spices (paprika), BEA, literature[165]

Tenuazonic Acid

incidence: 9/9, conc. range: 880–37,300 µg/kg, sample origin: unknown, sample year: unknown, country: Germany[166]

- Co-contamination: not reported
- Further contamination of plants and plant products other than cereals and their products reported in the present book/article[166]: Juices, Juices (apple juice), Juices (apricot juice), Juices (banana juice), Juices (blackcurrant juice), Juices (cherry juice), Juices (grape juice), Juices (orange juice), Juices (pear juice), Juices

(pomegranate juice), Juices (strawberry juice), Spices, Spices (coriander), Spices (cumin), Spices (curcuma), Spices (curry), Spices (ginger), Spices (pepper), TA, literature[166]

Aspergillus Toxins

Aflatoxin B_1

incidence: 5?/12, conc. range: 0.2–1.3 µg/kg, Ø conc.: 0.6 µg/kg, sample origin: commercially available market basket sa. of foods and foodstuffs in Tokyo (capital), Japan, sample year: 1986–1990, country: Japan[99]

- Co-contamination: not reported
- Further contamination of plants and plant products other than cereals and their products reported in the present book/article[99]: Beans, Seeds (sesame), Sugar, AFB_1 and AFB_2, literature[99]; Food, Jam (bean jam), Paste (peanut paste), AFB_1, literature[99]; Nuts (Brazil nut, paranut), Nuts (peanut), Nuts (pistachio nut), Spices (nutmeg), AFB_1, AFB_2, AFG_1, and AFG_2, literature[99]; Spices, Spices (cayenne pepper, chilli), Spices (pepper), AFB_1, AFB_2, and AFG_1, literature[99]; Spices (paprika), AFG_1, literature[99]

incidence: 8/12, conc. range: 1.25–17.75 µg/kg, Ø conc.: 6.98 µg/kg, sample origin: supermarkets and ethnic shops in Lisbon (capital) area, Portugal, sample year: unknown, country: Portugal[169], *paprika, packaged

- Co-contamination: not reported
- Further contamination of plants and plant products other than cereals and their products reported in the present book/article[169]: Spices (cayenne pepper, chilli), Spices (cumin), Spices (curry), Spices (nutmeg), Spices (pepper), Spices (saffron), AFB_1, literature[169]

incidence: 10/10, conc. range: 8.2–116.4 µg/kg, sample origin: bazaars, Turkey, sample year: unknown, country: Turkey[180]

incidence: 10/10, conc. range: 1.7–59.3 µg/kg, sample origin: herbal shops, Turkey, sample year: unknown, country: Turkey[180]

incidence: 7/10, conc. range: 0.5–1.8 µg/kg, sample origin: supermarkets, Turkey, sample year: unknown, country: Turkey[180]

- Co-contamination: not reported
- Further contamination of plants and plant products other than cereals and their products reported in the present book/article[180]: Spices (cayenne pepper, chilli), Spices (pepper), AFB_1 and AF/S, literature[180]; Spices (paprika), AF/S, literature[180]

incidence: 1/1*, conc.: 1.76 µg/kg, sample origin: local stores in Munich (city), Germany, sample year: unknown, country: Germany[181], *paprika powder

- Co-contamination: 1 sa. co-contaminated with AFB_1 and AFG_1
- Further contamination of plants and plant products other than cereals and their products reported in the present book/article[181]: Butter (peanut butter), Spices (nutmeg), AFB_1, AFB_2, and AFG_1, literature[181]; Nuts (hazelnut), AFB_1 and AFG_1, literature[181]; Nuts (pistachio nut), Spices (pepper), AFB_1, literature[181]; Spices (paprika), AFG_1, literature[181]

incidence: 1/5*, conc.: 0.6 µg/kg, sample origin: commercially available for industrial use, Netherlands, sample year: unknown, country: Netherlands[184], *paprika powder

- Co-contamination: not reported
- Further contamination of plants and plant products other than cereals and their products reported in the present book/article[184]: Spices (garlic), Spices (onion), FB_1, literature[184]; Spices (paprika), OTA, literature[184]; Spices (pepper), AFB_2, literature[184]

incidence: 7/15*, conc. range: 2.5–14.5 µg/kg, Ø conc.: 7.6 µg/kg, sample origin: retail and wholesale, Nordwürttemberg (administrative region), Germany, sample year: unknown, country: Germany[243], *different kinds of paprika powder

- Co-contamination: 7 sa. co-contaminated with AFB_1 and AFB_2
- Further contamination of plants and plant products other than cereals and their products reported in the present book/article[243]: Spices (cayenne pepper, chilli), AFB_1, AFB_2, and AFG_1, literature[243]; Spices (paprika), AFB_2, literature[243]

incidence: 25/25, conc. range: 0.2–3.4 µg/kg, sample origin: warehouses, packing establishments, supermarkets, and smaller shops, UK, sample year: unknown, country: UK[245]

- Co-contamination: 1 sa. co-contaminated with AFB_1, AFB_2, AFG_1, AFG_2, and OTA; 2 sa. co-contaminated with AFB_1, AFB_2, and AFG_1 and OTA; 9 sa. co-contaminated with AFB_1, AFG_1, AFG_2, and OTA; 11 sa. co-contaminated with AFB_1, AFG_1 and OTA; 1 sa. co-contaminated with AFB_1, AFG_2 and OTA; 1 sa. co-contaminated with AFB_1 and OTA
- Further contamination of plants and plant products other than cereals and their products reported in the present book/article[245]: Spices (cayenne pepper, chilli), AFB_1, AFB_2, AFG_1, AFG_2, and OTA, literature[245]; Spices (paprika), AFB_2, AFG_1, AFG_2, and OTA, literature[245]

incidence: 5/5, conc. range: 0.2–3.5 µg/kg, Ø conc.: 1.5 µg/kg, sample origin: available in New Zealand, sample year: 2009, country: New Zealand[277]

- Co-contamination: not reported
- Further contamination of plants and plant products other than cereals and their products reported in the present book/article[277]: Bars, AFB_1, AF/S, DON, NIV, and OTA, literature[277]; Butter (peanut butter), Confectionery, Food, Fruits, dried, Nuts, Nuts (peanut), Nuts (pistachio nut), Sauces (peanut sauce), AFB_1 and AF/S, literature[277]; Coffee, Dates, Spices (pepper), Vine fruits, dried, OTA, literature[277]; Figs, Spices (cayenne pepper, chilli), Spices (curry), Spices (ginger), AFB_1, AF/S, and OTA, literature[277]; Nuts (Brazil nut, paranut), AFB_1, literature[277]; Spices (paprika), AF/S and OTA, literature[277]

incidence: 38/50*, conc. range: ≤0.91 µg/kg, sample origin: local markets in Izmir (province), Turkey, sample year: February/March 2015, country: Turkey[449], *bell pepper, dried green

- Co-contamination: not reported
- Further contamination of plants and plant products other than cereals and their products reported in the present book/article[449]: Eggplant, AFB_1, AFB_2, AFG_1, AFG_2, and OTA, literature[449]; Spices (paprika), AFB_2, AFG_1, AFG_2, and OTA, literature[449]

incidence: 19/21*, conc. range: 0.7–3.8 µg/kg, Ø conc.: 1.1 µg/kg, sample origin: local retail outlets, Spain, sample year: 2004–2007, country: Spain/Nicaragua[586], *red paprika powder

- Co-contamination: 1 sa. co-contaminated with AFB_1, AFB_2, AFG_1, AFG_2, and OTA; 1 sa. co-contaminated with AFB_1, AFB_2, AFG_1, and OTA; 2 sa. co-contaminated with AFB_1, AFB_2, AFG_2, and OTA; 1 sa. co-contaminated with AFB_1, AFB_2, and AFG_2; 2 sa. co-contaminated with AFB_1, AFB_2, and OTA; 2 sa. co-contaminated with AFB_1, AFG_2, and OTA; 1 sa. co-contaminated with AFB_1 and AFG_2; 5 sa. co-contaminated with AFB_1 and OTA; 4 sa. contaminated solely with AFB_1
- Further contamination of plants and plant products other than cereals and their products reported in the present book/article[586]: Spices (paprika), AFB_2, AFG_1, AFG_2, and OTA, literature[586]

incidence: 58/70, conc. range: LOD-5.0 µg/kg (45 sa.), 5.0–10.0 µg/kg (13 sa., maximum: 7.3 µg/kg), ∅ conc.: 3.4 µg/kg, sample origin: local supermarkets and free markets, São Paulo (city), Brazil, sample year: January–April 2006, country: Brazil[587]

- Co-contamination: not reported
- Further contamination of plants and plant products other than cereals and their products reported in the present book/article[587]: Spices (paprika), AFB_2, AFG_1, and OTA, literature[587]

Aflatoxin B_2

incidence: 7/15*, conc. range: 0.1–1.7 µg/kg, ∅ conc.: 0.81 µg/kg, sample origin: retail and wholesale, Nordwürttemberg (administrative region), Germany, sample year: unknown, country: Germany[243], *different kinds of paprika powder

- Co-contamination: 7 sa. co-contaminated with AFB_1 and AFB_2
- Further contamination of plants and plant products other than cereals and their products reported in the present book/article[243]: Spices (cayenne pepper, chilli), AFB_1, AFB_2, and AFG_1, literature[243]; Spices (paprika), AFB_1, literature[243]

incidence: 3/25, conc. range: 0.2–0.3 µg/kg, ∅ conc.: 0.23 µg/kg, sample origin: warehouses, packing establishments, supermarkets, and smaller shops, UK, sample year: unknown, country: UK[245]

- Co-contamination: 1 sa. co-contaminated with AFB_1, AFB_2, AFG_1, AFG_2, and OTA; 2 sa. co-contaminated with AFB_1, AFB_2, and AFG_1 and OTA
- Further contamination of plants and plant products other than cereals and their products reported in the present book/article[245]: Spices (cayenne pepper, chilli), AFB_1, AFB_2, AFG_1, AFG_2, and OTA, literature[245]; Spices (paprika), AFB_1, AFG_1, AFG_2, and OTA, literature[245]

incidence: 50/50*, conc. range: 0.13–0.66 µg/kg, ∅ conc.: 0.26 µg/kg, sample origin: local markets in Izmir (province), Turkey, sample year: February/March 2015, country: Turkey[449], *bell pepper, dried green

- Co-contamination: not reported
- Further contamination of plants and plant products other than cereals and their products reported in the present book/article[449]: Eggplant, AFB_1, AFB_2, AFG_1, AFG_2, and OTA, literature[449]; Spices (paprika), AFB_1, AFG_1, AFG_2, and OTA, literature[449]

incidence: 7/21*, conc. range: 0.6–0.7 µg/kg, ∅ conc.: 0.6 µg/kg, sample origin: local retail outlets, Spain, sample year: 2004–2007, country: Spain/Nicaragua[586], *red paprika powder

- Co-contamination: 1 sa. co-contaminated with AFB_1, AFB_2, AFG_1, AFG_2, and OTA; 1 sa. co-contaminated with AFB_1, AFB_2, AFG_1, and OTA; 2 sa. co-contaminated with AFB_1, AFB_2, AFG_2, and OTA; 1 sa. co-contaminated with AFB_1, AFB_2, and AFG_2; 2 sa. co-contaminated with AFB_1, AFB_2, and OTA
- Further contamination of plants and plant products other than cereals and their products reported in the present book/article[586]: Spices (paprika), AFB_1, AFG_1, AFG_2, and OTA, literature[586]

incidence: 9/70, conc. range: LOD-5.0 µg/kg (9 sa., maximum: 0.5 µg/kg), ∅ conc.: 1.0 µg/kg, sample origin: local supermarkets and free markets, São Paulo (city), Brazil, sample year: January–April 2006, country: Brazil[587]

- Co-contamination: not reported
- Further contamination of plants and plant products other than cereals and their products reported in the present book/article[587]: Spices (paprika), AFB_1, AFG_1, and OTA, literature[587]

Aflatoxin G_1

incidence: 5?/12, conc. range: ≤0.2 µg/kg, ∅ conc.: <0.1 µg/kg, sample origin: commercially available market basket sa. of foods and foodstuffs in Tokyo (capital), Japan, sample year: 1986–1990, country: Japan[99]

- Co-contamination: not reported
- Further contamination of plants and plant products other than cereals and their products reported in the present book/article[99]: Beans, Seeds (sesame), Sugar, AFB_1 and AFB_2, literature[99]; Food, Jam (bean jam), Paste (peanut paste), Spices (paprika), AFB_1, literature[99]; Nuts (Brazil nut, paranut), Nuts (peanut), Nuts (pistachio nut), Spices (nutmeg), AFB_1, AFB_2, AFG_1, and AFG_2, literature[99]; Spices, Spices (cayenne pepper, chilli), Spices (pepper), AFB_1, AFB_2, and AFG_1, literature[99]

incidence: 1/1*, conc.: 1.56 µg/kg, sample origin: local stores in Munich (city), Germany, sample year: unknown, country: Germany[181], *paprika powder

- Co-contamination: 1 sa. co-contaminated with AFB_1 and AFG_1
- Further contamination of plants and plant products other than cereals and their products reported in the present book/article[181]: Butter (peanut butter), Spices (nutmeg), AFB_1, AFB_2, and AFG_1, literature[181]; Nuts (hazelnut), AFB_1 and AFG_1, literature[181]; Nuts (pistachio nut), Spices (paprika), Spices (pepper), AFB_1, literature[181]

incidence: 23/25, conc. range: 0.2–2.2 µg/kg, sample origin: warehouses, packing establishments, supermarkets, and smaller shops, UK, sample year: unknown, country: UK[245]

- Co-contamination: 1 sa. co-contaminated with AFB_1, AFB_2, AFG_1, AFG_2, and OTA; 2 sa. co-contaminated with AFB_1, AFB_2, and AFG_1 and OTA; 9 sa. co-contaminated with AFB_1, AFG_1, AFG_2, and OTA; 11 sa. co-contaminated with AFB_1, AFG_1 and OTA
- Further contamination of plants and plant products other than cereals and their products reported in the present book/article[245]: Spices (cayenne pepper, chilli), AFB_1, AFB_2, AFG_1, AFG_2, and OTA, literature[245]; Spices (paprika), AFB_1, AFB_2, AFG_2, and OTA, literature[245]

incidence: 50/50*, conc. range: 0.32–1.38 µg/kg, ∅ conc.: 0.61 µg/kg, sample origin: local markets in Izmir (province), Turkey, sample year: February/March 2015, country: Turkey[449], *bell pepper, dried green

- Co-contamination: not reported
- Further contamination of plants and plant products other than cereals and their products reported in the present book/article[449]: Eggplant, AFB_1, AFB_2, AFG_1, AFG_2, and OTA, literature[449]; Spices (paprika), AFB_1, AFB_2, AFG_2, and OTA, literature[449]

incidence: 2/21*, conc. range: 0.9–1.1 µg/kg, ∅ conc.: 1.0 µg/kg, sample origin: local retail outlets, Spain, sample year: 2004–2007, country: Spain/Nicaragua[586], *red paprika powder

- Co-contamination: 1 sa. co-contaminated with AFB_1, AFB_2, AFG_1, AFG_2, and OTA; 1 sa. co-contaminated with AFB_1, AFB_2, AFG_1, and OTA
- Further contamination of plants and plant products other than cereals and their products reported in the present book/article[586]: Spices (paprika), AFB_1, AFB_2, AFG_2, and OTA, literature[586]

incidence: 15/70, conc. range: LOD-5.0 µg/kg (15 sa., maximum: 1.6 µg/kg), ∅ conc.: 0.4 µg/kg, sample origin: local supermarkets and free markets, São Paulo (city), Brazil, sample year: January–April 2006, country: Brazil[587]

- Co-contamination: not reported
- Further contamination of plants and plant products other than cereals and their products reported in the present book/article[587]: Spices (paprika), AFB_1, AFB_2, and OTA, literature[587]

Aflatoxin G_2

incidence: 11/25, conc. range: 0.2–0.7 μg/kg, sample origin: warehouses, packing establishments, supermarkets, and smaller shops, UK, sample year: unknown, country: UK[245]

- Co-contamination: 1 sa. co-contaminated with AFB_1, AFB_2, AFG_1, AFG_2, and OTA; 9 sa. co-contaminated with AFB_1, AFG_1, AFG_2, and OTA; 1 sa. co-contaminated with AFB_1, AFG_2 and OTA
- Further contamination of plants and plant products other than cereals and their products reported in the present book/article[245]: Spices (cayenne pepper, chilli), AFB_1, AFB_2, AFG_1, AFG_2, and OTA, literature[245]; Spices (paprika), AFB_1, AFB_2, AFG_1, and OTA, literature[245]

incidence: 50/50*, conc. range: 0.11–0.22 μg/kg, ∅ conc.: 0.19 μg/kg, sample origin: local markets in Izmir (province), Turkey, sample year: February/March 2015, country: Turkey[449], *bell pepper, dried green

- Co-contamination: not reported
- Further contamination of plants and plant products other than cereals and their products reported in the present book/article[449]: Eggplant, AFB_1, AFB_2, AFG_1, AFG_2, and OTA, literature[449]; Spices (paprika), AFB_1, AFB_2, AFG_1, and OTA, literature[449]

incidence: 8/21*, conc. range: 0.5–0.8 μg/kg, ∅ conc.: 0.6 μg/kg, sample origin: local retail outlets, Spain, sample year: 2004–2007, country: Spain/Nicaragua[586], *red paprika powder

- Co-contamination: 1 sa. co-contaminated with AFB_1, AFB_2, AFG_1, AFG_2, and OTA; 2 sa. co-contaminated with AFB_1, AFB_2, AFG_2, and OTA; 1 sa. co-contaminated with AFB_1, AFB_2, and AFG_2; 2 sa. co-contaminated with AFB_1, AFG_2, and OTA; 1 sa. co-contaminated with AFB_1 and AFG_2; 1 sa. co-contaminated with AFG_2 and OTA
- Further contamination of plants and plant products other than cereals and their products reported in the present book/article[586]: Spices (paprika), AFB_1, AFB_2, AFG_1, and OTA, literature[586]

Aflatoxin/s

incidence: 10/10, conc. range: 9.7–124.6 μg/kg*, sample origin: bazaars, Turkey, sample year: unknown, country: Turkey[180], *AFB_1, AFB_2, AFG_1, and AFG_2

incidence: 10/10, conc. range: 1.8–62.4 μg/kg*, sample origin: herbal shops, Turkey, sample year: unknown, country: Turkey[180], *AFB_1, AFB_2, AFG_1, and AFG_2

incidence: 7/10, conc. range: 0.5–1.8 μg/kg*, sample origin: supermarkets, Turkey, sample year: unknown, country: Turkey[180], *AFB_1, AFB_2, AFG_1, and AFG_2

- Co-contamination: not reported
- Further contamination of plants and plant products other than cereals and their products reported in the present book/article[180]: Spices (cayenne pepper, chilli), Spices (pepper), AFB_1 and AF/S, literature[180]; Spices (paprika), AFB_1, literature[180]

incidence: 2/10, conc. range: 0.40–6.40 μg/kg*, ∅ conc.: 3.40 μg/kg*, sample origin: market stalls, shops, and supermarkets, Ireland, sample year: unknown, country: Ireland[187], *AFB_1, AFB_2, AFG_1, and AFG_2

- Co-contamination: not reported
- Further contamination of plants and plant products other than cereals and their products reported in the present book/article[187]: Spices (cayenne pepper, chilli), Spices (coriander), Spices (curcuma), Spices (curry), Spices (pepper), AF/S, literature[187]

incidence: 38/64, conc. range: ≤7.25 μg/kg*, sample origin: retail shops, Catalonia (autonomous community), Spain, sample year: June–August 2008, country: Spain[252], *AFB_1, AFB_2, AFG_1, and AFG_2

- Co-contamination: For detailed information see the article.
- Further contamination of plants and plant products other than cereals and their products reported in the present book/article[252]: Spices (cayenne pepper, chilli), AF/S, OTA, and ZEA, literature[252]; Spices (paprika), OTA and ZEA, literature[252]

incidence: 17/17, conc. range: 1.8–50.4 μg/kg*, sample origin: distributor companies, Spain, sample year: 2008–2010, country: Spain[253], *AFB_1, AFB_2, AFG_1, and AFG_2

- Co-contamination: 15 sa. co-contaminated with AFS, DON, OTA, and ZEA; 2 sa. co-contaminated with AFS, OTA, and ZEA
- Further contamination of plants and plant products other than cereals and their products reported in the present book/article[253]: Spices (cayenne pepper, chilli), AF/S, DON, OTA, T-2, and ZEA, literature[253]; Spices (paprika), DON, OTA, and, ZEA, literature[253]

incidence: 4/4*, conc. range: 22.3–83.7 μg/kg**, sample origin: distributor companies, Spain, sample year: 2008–2010, country: Spain[253], *paprika, smoked, **AFB_1, AFB_2, AFG_1, and AFG_2

- Co-contamination: 4 sa. co-contaminated with AFS and OTA
- Further contamination of plants and plant products other than cereals and their products reported in the present book/article[253]: Spices (cayenne pepper, chilli), AF/S, DON, OTA, T-2, and ZEA, literature[253]; Spices (paprika), DON, OTA, and, ZEA, literature[253]

incidence: 5/5, conc. range: 0.2–3.5 μg/kg*, ∅ conc.: 1.7 μg/kg*, sample origin: available in New Zealand, sample year: 2009, country: New Zealand[277], *AFS total

- Co-contamination: not reported
- Further contamination of plants and plant products other than cereals and their products reported in the present book/article[277]: Bars, AFB_1, AF/S, DON, NIV, and OTA, literature[277]; Butter (peanut butter), Confectionery, Food, Fruits, dried, Nuts, Nuts (peanut), Nuts (pistachio nut), Sauces (peanut sauce), AFB_1 and AF/S, literature[277]; Coffee, Dates, Spices (pepper), Vine fruits, dried, OTA, literature[277]; Figs, Spices (cayenne pepper, chilli), Spices (curry), Spices (ginger), AFB_1, AF/S, and OTA, literature[277]; Nuts (Brazil nut, paranut), AFB_1, literature[277]; Spices (paprika), AFB_1 and OTA, literature[277]

incidence: 19/23*, conc. range: 0.38–14.71 μg/kg**, sample origin: exporting companies in Aydin (province), Turkey, sample year: January–August 2007, country: Turkey[360], *destined for export, **AFB_1, AFB_2, AFG_1, and AFG_2

- Co-contamination: 1 sa. co-contaminated with AFB_1, AFB_2, and AFG_1; 2 sa. co-contaminated with AFB_1 and AFB_2; 16 sa. contaminated solely with AFB_1

- Further contamination of plants and plant products other than cereals and their products reported in the present book/article[360]: Figs, Nuts (hazelnut), Nuts (peanut), Nuts (pistachio nut), AF/S, literature[360]

incidence: 5/5, conc. range: ≤3.5 µg/kg*, sample origin: New Zealand, sample year: 2009, country: New Zealand[702], *AFB_1 and AFG_1

- Co-contamination: not reported
- Further contamination of plants and plant products other than cereals and their products reported in the present book/article[702]: Figs, Spices (cayenne pepper, chilli), Spices (curry), Spices (ginger), AF/S and OTA, literature[702]; Fruits, dried, AF/S, literature[702]; Spices (paprika), Spices (pepper), Vine fruits, dried, OTA, literature[702]

Other ***Aspergillus*** Toxins

Sterigmatocystin

incidence: 1/36*, conc.: 3.5 µg/kg, sample origin: FUCHS Gewürze GmbH, Latvia, sample year: 2014, country: Latvia[679], *sweet paprika, ground

- Co-contamination: not reported
- Further contamination of plants and plant products other than cereals and their products reported in the present book/article[679]: Spices (cayenne pepper, chilli), FB_1, literature[679]; Spices (nutmeg), Spices (paprika), FB_1 and OTA, literature[679]; Spices (oregano), STG, literature[679]; Spices (thyme), DON, STG, and ZEA, literature[679]

Aspergillus and ***Penicillium*** Toxins

Ochratoxin A

incidence: 16/23*, conc. range: 31–528 µg/kg, ∅ conc.: 203 µg/kg, sample origin: retail markets and supermarkets in 24 regions, Tunisia, sample year: unknown, country: Tunisia[20], *red paprika

- Co-contamination: not reported
- Further contamination of plants and plant products other than cereals and their products reported in the present book/article[20]: Nuts (almond), Nuts (peanut), Nuts (pistachio nut), Spices (caraway), Spices (coriander), Spices (curcuma), Spices (pepper), OTA, literature[20]

incidence: 3/3*, conc. range: 2.1–8.0 µg/kg, ∅ conc.: 4.5 µg/kg, sample origin: commercially available for industrial use, Netherlands, sample year: unknown, country: Netherlands[184], *paprika powder

- Co-contamination: not reported
- Further contamination of plants and plant products other than cereals and their products reported in the present book/article[184]: Spices (garlic), Spices (onion), FB_1, literature[184]; Spices (paprika), AFB_1, literature[184]; Spices (pepper), AFB_2, literature[184]

incidence: 2/5, conc. range: 2.6–35 µg/kg, ∅ conc.: 18.8 µg/kg, sample origin: retail shops in Belgium, Netherlands, and Russia, sample year: autumn-winter 2005/2006, country: Belgium/Russia/Hungary[193]

- Co-contamination: not reported
- Further contamination of plants and plant products other than cereals and their products reported in the present book/article[193]: Spices (cayenne pepper, chilli), Spices (ginger), Spices (nutmeg), Spices (pepper), OTA, literature[193]

incidence: 25/25, conc. range: 0.3–47.7 µg/kg, sample origin: warehouses, packing establishments, supermarkets, and smaller shops, UK, sample year: unknown, country: UK[245]

- Co-contamination: 1 sa. co-contaminated with AFB_1, AFB_2, AFG_1, AFG_2, and OTA; 2 sa. co-contaminated with AFB_1, AFB_2, and AFG_1 and OTA; 9 sa. co-contaminated with AFB_1, AFG_1, AFG_2, and OTA; 11 sa. co-contaminated with AFB_1, AFG_1 and OTA; 1 sa. co-contaminated with AFB_1, AFG_2 and OTA; 1 sa. co-contaminated with AFB_1 and OTA
- Further contamination of plants and plant products other than cereals and their products reported in the present book/article[245]: Spices (cayenne pepper, chilli), AFB_1, AFB_2, AFG_1, AFG_2, and OTA, literature[245]; Spices (paprika), AFB_1, AFB_2, AFG_1, and AFG_2, literature[245]

incidence: 63/64, conc. range: ≤281 µg/kg*, sample origin: retail shops, Catalonia (autonomous community), Spain, sample year: June–August 2008, country: Spain[252], *AFB_1, AFB_2, AFG_1, and AFG_2

- Co-contamination: For detailed information see the article.
- Further contamination of plants and plant products other than cereals and their products reported in the present book/article[252]: Spices (cayenne pepper, chilli), AF/S, OTA, and ZEA, literature[252]; Spices (paprika), AF/S and ZEA, literature[252]

incidence: 17/17, conc. range: 4.3–118.7 µg/kg, sample origin: distributor companies, Spain, sample year: 2008–2010, country: Spain[253]

- Co-contamination: 15 sa. co-contaminated with AFS, DON, OTA, and ZEA; 2 sa. co-contaminated with AFS, OTA, and ZEA
- Further contamination of plants and plant products other than cereals and their products reported in the present book/article[253]: Spices (cayenne pepper, chilli), AF/S, DON, OTA, T-2, and ZEA, literature[253]; Spices (paprika), AF/S, DON, and ZEA, literature[253]

incidence: 4/4*, conc. range: 9.0–474.7 µg/kg, sample origin: distributor companies, Spain, sample year: 2008–2010, country: Spain[253], *paprika, smoked

- Co-contamination: 4 sa. co-contaminated with AFS and OTA
- Further contamination of plants and plant products other than cereals and their products reported in the present book/article[253]: Spices (cayenne pepper, chilli), AF/S, DON, OTA, T-2, and ZEA, literature[253]; Spices (paprika), AF/S, DON, and ZEA, literature[253]

incidence: 4/4, conc. range: 13.3–50.6 µg/kg, sample origin: available in New Zealand, sample year: 2007, country: New Zealand[277]

incidence: 5/5, conc. range: 15–103 µg/kg, sample origin: available in New Zealand, sample year: 2008/2009, country: New Zealand[277]

- Co-contamination: not reported
- Further contamination of plants and plant products other than cereals and their products reported in the present book/article[277]: Bars, AFB_1, AF/S, DON, NIV, and OTA, literature[277]; Butter (peanut butter), Confectionery, Food, Fruits, dried, Nuts, Nuts (peanut), Nuts (pistachio nut), Sauces (peanut sauce), Spices (paprika), AFB_1 and AF/S, literature[277]; Coffee, Dates, Spices (pepper), Vine fruits, dried, OTA, literature[277]; Figs, Spices (cayenne pepper, chilli), Spices (curry), Spices (ginger), AFB_1, AF/S, and OTA, literature[277]; Nuts (Brazil nut, paranut), AFB_1, literature[277]

incidence: 50/50*, conc. range: 15.38–24.70 µg/kg, ∅ conc.: 20.53 µg/kg, sample origin: local markets in Izmir (province), Turkey, sample year: February/March 2015, country: Turkey[449], *bell pepper, dried green

- Co-contamination: not reported

- Further contamination of plants and plant products other than cereals and their products reported in the present book/article[449]: Eggplant, AFB_1, AFB_2, AFG_1, AFG_2, and OTA, literature[449]; Spices (paprika), AFB_1, AFB_2, AFG_1, and AFG_2, literature[449]

incidence: 14/21*, conc. range: 0.7–73.8 µg/kg, ∅ conc.: 11.9 µg/kg, sample origin: local retail outlets, Spain, sample year: 2004–2007, country: Spain/Nicaragua[586], *red paprika powder

- Co-contamination: 1 sa. co-contaminated with AFB_1, AFB_2, AFG_1, AFG_2, and OTA; 1 sa. co-contaminated with AFB_1, AFB_2, AFG_1, and OTA; 2 sa. co-contaminated with AFB_1, AFB_2, AFG_2, and OTA; 2 sa. co-contaminated with AFB_1, AFB_2, and OTA; 2 sa. co-contaminated with AFB_1, AFG_2, and OTA; 5 sa. co-contaminated with AFB_1 and OTA; 1 sa. co-contaminated with AFG_2 and OTA
- Further contamination of plants and plant products other than cereals and their products reported in the present book/article[586]: Spices (paprika), AFB_1, AFB_2, AFG_1, and AFG_2, literature[586]

incidence: 60/70, conc. range: LOD-5.0 µg/kg (31 sa), 5.0–10.0 µg/kg (20 sa.), ≥10.0 µg/kg (9 sa., maximum: 97.3 µg/kg), ∅ conc.: 7.0 µg/kg, sample origin: local supermarkets and free markets, São Paulo (city), Brazil, sample year: January–April 2006, country: Brazil[587]

- Co-contamination: not reported
- Further contamination of plants and plant products other than cereals and their products reported in the present book/article[587]: Spices (paprika), AFB_1, AFB_2, and AFG_1, literature[587]

incidence: 2/36*, conc. range: 3.0–12 µg/kg, ∅ conc.: 7.5 µg/kg, sample origin: FUCHS Gewürze GmbH, Latvia, sample year: 2010–2013, country: Latvia[679], *sweet paprika, ground

- Co-contamination: not reported
- Further contamination of plants and plant products other than cereals and their products reported in the present book/article[679]: Spices (cayenne pepper, chilli), FB_1, literature[679]; Spices (nutmeg), FB_1 and OTA, literature[679]; Spices (oregano), STG, literature[679]; Spices (paprika), FB_1 and STG, literature[679]; Spices (thyme), DON, STG, and ZEA, literature[679]

incidence: 3/6*, conc. range: 1.0–2.9 µg/kg (2 sa.), 4.3 µg/kg (1 sa.), sample origin: Portugal, sample year: 1999, country: EU[698], *sweet pepper

incidence: 9/12*, conc. range: LOD/LOQ 0.9 µg/kg (1 sa.), 1.0–2.9 µg/kg (3 sa.), 3.0–4.9 µg/kg (1 sa.), 9.8 µg/kg (1 sa.), sample origin: Netherlands, sample year: 1996–1998, country: EU[698], *paprika powder

- Co-contamination: not reported
- Further contamination of plants and plant products other than cereals and their products reported in the present book/article[698]: Baby food, Beans (soybean), Chocolate, Cocoa, Coffee, Coffee beans, Currants, Fruits, dried, Herbs, Jam, Juices, Juices (grape juice), Legumes, Nuts, Oil seeds, Oils (olive oil), Peas, Peas (chickpea), Products (plant product), Raisins, Spices, Spices (cayenne pepper, chilli), Spices (nutmeg), Spices (pepper), Sultanas, Tea, Vinegar, Wine, Wine (red), Wine (rosé), Wine (white), Wine, miscellaneous, OTA, literature[698]

incidence: 5/5, conc. range: 15–103 µg/kg, sample origin: New Zealand, sample year: 2009, country: New Zealand[702]

- Co-contamination: not reported
- Further contamination of plants and plant products other than cereals and their products reported in the present book/article[702]: Figs, Spices (cayenne pepper, chilli), Spices (curry), Spices (ginger), AF/S and OTA, literature[702]; Fruits, dried, Spices (paprika), AF/S, literature[702]; Spices (pepper), Vine fruits, dried, OTA, literature[702]

incidence: 12/12*, conc. range: 5.50–91.80 µg/kg, ∅ conc.: 19.00 µg/kg, sample origin: 12 regions, Czech Republic, sample year: 2011–2013, country: Czech Republic[811], *paprika powder, fiery

incidence: 12/12*, conc. range: 1.10–38.40 µg/kg, ∅ conc.: 16.00 µg/kg, sample origin: 12 regions, Czech Republic, sample year: 2011–2013, country: Czech Republic[811], *paprika powder, sweet

- Co-contamination: not reported
- Further contamination of plants and plant products other than cereals and their products reported in the present book/article[811]: Beans, Chocolate, Coffee, Confectionery, Figs, Herbs and Spices, Juices, Licorice, Powder (cocoa powder), Products (ginger product), Raisins, Spices (cayenne pepper, chilli), Spices (caraway), Spices (coriander), Spices (ginger), Spices (mustard), Spices (nutmeg), Spices (pepper), Tea, Wine (red), Wine (white), OTA, literature[811]

Fusarium Toxins

Depsipeptides

Beauvericin

incidence: 1/20*, conc.: 124 µg/kg, sample origin: grocery in Oostkamp (municipality), Belgium, sample year: June–October 2008, country: Belgium[165], *sweet pepper with moldy or brown spots

- Co-contamination: not reported
- Further contamination of plants and plant products other than cereals and their products reported in the present book/article[165]: Spices (paprika), AOH, literature[165]

Fumonisin/s

Fumonisin B_1

incidence: 4/36*, conc. range: 9–40 µg/kg, ∅ conc.: 23.3 µg/kg, sample origin: FUCHS Gewürze GmbH, Latvia, sample year: 2014, country: Latvia[679], *sweet paprika, ground

- Co-contamination: not reported
- Further contamination of plants and plant products other than cereals and their products reported in the present book/article[679]: Spices (cayenne pepper, chilli), FB_1, literature[679]; Spices (nutmeg), FB_1 and OTA, literature[679]; Spices (oregano), STG, literature[679]; Spices (paprika), OTA and STG, literature[679]; Spices (thyme), DON, STG, and ZEA, literature[679]

Type A Trichothecenes

HT-2 Toxin

incidence: 3/32*, conc. range: ≤199 µg/kg, ∅ conc.: 82 µg/kg, sample orogin: different suppliers, Spain, sample year: 2008–2010, country: Spain[589], *paprika and chilli pepper

- Co-contamination: For detailed information see the article.
- Further contamination of plants and plant products other than cereals and their products reported in the present book/article[589]: Spices (paprika), DON and T-2, literature[589]

T-2 Toxin

incidence: 2/32*, conc. range: 9–11 µg/kg, ∅ conc.: 10 µg/kg, sample orogin: different suppliers, Spain, sample year: 2008–2010, country: Spain[589], *paprika and chilli pepper

- Co-contamination: For detailed information see the article.
- Further contamination of plants and plant products other than cereals and their products reported in the present book/article[589]: Spices (paprika), DON and HT-2, literature[589]

Type B Trichothecenes

Deoxynivalenol

incidence: 1/17, conc.: 0.1 µg/kg, sample origin: distributor companies, Spain, sample year: 2008–2010, country: Spain[253]

- Co-contamination: 1 sa. co-contaminated with AFS, DON, OTA, and ZEA
- Further contamination of plants and plant products other than cereals and their products reported in the present book/article[253]: Spices (cayenne pepper, chilli), AF/S, DON, OTA, T-2, and ZEA, literature[253]; Spices (paprika), AF/S, OTA, and ZEA, literature[253]

incidence: 0/4*, conc.: no contamination, sample origin: distributor companies, Spain, sample year: 2008–2010, country: Spain[253], *paprika, smoked

- Co-contamination: not reported
- Further contamination of plants and plant products other than cereals and their products reported in the present book/article[253]: Spices (cayenne pepper, chilli), AF/S, DON, OTA, T-2, and ZEA, literature[253]; Spices (paprika), AF/S, OTA, and ZEA, literature[253]

incidence: 4/32*, conc. range: ≤269 µg/kg, Ø conc.: 125 µg/kg, sample orogin: different suppliers, Spain, sample year: 2008–2010, country: Spain[589], *paprika and chilli pepper

- Co-contamination: For detailed information see the article.
- Further contamination of plants and plant products other than cereals and their products reported in the present book/article[589]: Spices (paprika), HT-2 and T-2, literature[589].

Other ***Fusarium*** Toxins

Zearalenone

incidence: 25/64, conc. range: ≤131 µg/kg*, sample origin: retail shops, Catalonia (autonomous community), Spain, sample year: June–August 2008, country: Spain[252], *AFB_1, AFB_2, AFG_1, and AFG_2

- Co-contamination: For detailed information see the article.
- Further contamination of plants and plant products other than cereals and their products reported in the present book/article[252]: Spices (cayenne pepper, chilli), AF/S, OTA, and ZEA, literature[252]; Spices (paprika), AF/S and OTA, literature[252]

incidence: 3/17, conc. range: ≤78.2 µg/kg, sample origin: distributor companies, Spain, sample year: 2008–2010, country: Spain[253]

- Co-contamination: 1 sa. co-contaminated with AFS, DON, OTA, and ZEA; 2 sa. co-contaminated with AFS, OTA, and ZEA
- Further contamination of plants and plant products other than cereals and their products reported in the present book/article[253]: Spices (cayenne pepper, chilli), AF/S, DON, OTA, T-2, and ZEA, literature[253]; Spices (paprika), AF/S, DON, and, OTA, literature[253]

incidence: 0/4*, conc.: no contamination, sample origin: distributor companies, Spain, sample year: 2008–2010, country: Spain[253], *paprika, smoked

- Co-contamination: not reported
- Further contamination of plants and plant products other than cereals and their products reported in the present book/article[253]: Spices (cayenne pepper, chilli), AF/S, DON, OTA, T-2, and ZEA, literature[253]; Spices (paprika), AF/S, DON, and, OTA, literature[253]

Spices (pepper) may contain the following mycotoxins:

Alternaria Toxins

Alternariol

incidence: 1*/1, conc.: 640 µg/kg, sample origin: several localities in Apulia and Basilicata (regions), Italy, sample year: late summer, country: Italy[164], *visibly affected by *Alternaria* rot

- Co-contamination: 1 sa. co-contaminated with AME, AOH, and TA
- Further contamination of plants and plant products other than cereals and their products reported in the present book/article[164]: Fruits (mandarin), Fruits (tomato), AME, AOH, and TA, literature[164]; Fruits (melon), AME and TA, literature[164]; Olives, ALT, AME, AOH, and TA, literature[164]; Seeds (sunflower), AME and AOH, literature[164]; Spices (pepper), AME and TA, literature[164]

Alternariol Methyl Ether

incidence: 1*/1, conc.: 49 µg/kg, sample origin: several localities in Apulia and Basilicata (regions), Italy, sample year: late summer, country: Italy[164], *visibly affected by *Alternaria* rot

- Co-contamination: 1 sa. co-contaminated with AME, AOH, and TA
- Further contamination of plants and plant products other than cereals and their products reported in the present book/article[164]: Fruits (mandarin), Fruits (tomato), AME, AOH, and TA, literature[164]; Fruits (melon), AME and TA, literature[164]; Olives, ALT, AME, AOH, and TA, literature[164]; Seeds (sunflower), AME and AOH, literature[164]; Spices (pepper), AOH and TA, literature[164]

Tenuazonic Acid

incidence: 1*/1, conc.: 54 µg/kg, sample origin: several localities in Apulia and Basilicata (regions), Italy, sample year: late summer, country: Italy[164], *visibly affected by *Alternaria* rot

- Co-contamination: 1 sa. co-contaminated with AME, AOH, and TA
- Further contamination of plants and plant products other than cereals and their products reported in the present book/article[164]: Fruits (mandarin), Fruits (tomato), AME, AOH, and TA, literature[164]; Fruits (melon), AME and TA, literature[164]; Olives, ALT, AME, AOH, and TA, literature[164]; Seeds (sunflower), AME and AOH, literature[164]; Spices (pepper), AME and AOH, literature[164]

incidence: 6/6, conc. range: 60–1,000 µg/kg, sample origin: unknown, sample year: unknown, country: Germany[166]

- Co-contamination: not reported
- Further contamination of plants and plant products other than cereals and their products reported in the present book/article[166]: Juices, Juices (apple juice), Juices (apricot juice), Juices (banana juice), Juices (blackcurrant juice), Juices (cherry juice), Juices (grape juice), Juices (orange juice), Juices (pear juice), Juices (pomegranate juice), Juices (strawberry juice), Spices, Spices (coriander), Spices (cumin), Spices (curcuma), Spices (curry), Spices (ginger), Spices (paprika), TA, literature[166]

Aspergillus Toxins

Aflatoxin B_1

incidence: 1/1*, conc.: 0.08 µg/kg, sample origin: sample origin: grocery markets in Busan, Daegu, Daejeon, Gangneung, and Gwangju (cities) as well as Seoul (capital), Korea, sample year: May 2004–June 2005, country: Korea[10], *pepper powder

- Co-contamination: not reported
- Further contamination of plants and plant products other than cereals and their products reported in the present book/article[10]: Butter (peanut butter), Nuts, Nuts (almond), Nuts (peanut), Nuts (pistachio nut), Paste (soybean paste), Sauces (soy sauce), Spices (cayenne pepper, chilli), Spices (curry), AFB_1, literature[10]

incidence: 5?/8*, conc. range: 17–190 µg/kg, sample origin: major retail outlets in Nainital (district), India, sample year: April 1984–September 1986, country: India[22], *black pepper

- Co-contamination: not reported
- Further contamination of plants and plant products other than cereals and their products reported in the present book/article[22]: Spices (ammi), AFB_1, AFB_2, and AFG_1, literature[22]; Spices (cardamom), AFB_1, AFB_2, AFG_1, AFG_2, CIT, and OTA, literature[22]; Spices (cassia), AFB_1, AFB_2, AFG_1, and AFG_2, literature[22]; Spices (cayenne pepper, chilli), AFB_1, AFB_2, AFG_1, AFG_2, and ZEA, literature[22]; Spices (coriander), AFB_1, AFB_2, AFG_1, AFG_2, CIT, OTA, and ZEA, literature[22]; Spices (cumin), AFB_1, AFB_2, AFG_1, AFG_2, and CIT, literature[22]; Spices (curcuma), AFB_1, AFB_2, AFG_1, AFG_2, CIT, and RBT, literature[22]; Spices (fennel), AFB_1, AFB_2, AFG_1, AFG_2, CIT, OTA, and STG, literature[22]; Spices (garlic), AFB_1, AFB_2, and AFG_1, literature[22]; Spices (pepper), AFB_2, AFG_1, AFG_2, CIT, STG, and ZEA, literature[22]

incidence: 7?/24*, conc. range: 0.6–2.3 µg/kg, sample origin: unknown, sample year: unknown, country: Japan[97], *white pepper

- Co-contamination: not reported
- Further contamination of plants and plant products other than cereals and their products reported in the present book/article[97]: Beans, Jam (bean jam), Spices, AFB_1 and AFB_2, literature[97]; Butter (peanut butter), Oils (peanut oil), AFB_1, AFB_2, and AFG_1, literature[97]; Spices (pepper), AFB_2 and AFG_1, literature[97]; Spices (nutmeg), AFB_1, AFB_2, AFG_1, and AFG_2, literature[97]

incidence: 11?/84*, conc. range: 0.1–1.8 µg/kg, ∅ conc.: 0.8 µg/kg, sample origin: commercially available market basket sa. of foods and foodstuffs in Tokyo (capital), Japan, sample year: 1986–1990, country: Japan[99], *white pepper

- Co-contamination: not reported
- Further contamination of plants and plant products other than cereals and their products reported in the present book/article[99]: Beans, Seeds (sesame), Sugar, AFB_1 and AFB_2, literature[99]; Food, Jam (bean jam), Paste (peanut paste), AFB_1, literature[99]; Nuts (Brazil nut, paranut), Nuts (peanut), Nuts (pistachio nut), Spices (nutmeg), AFB_1, AFB_2, AFG_1, and AFG_2, literature[99]; Spices, Spices (cayenne pepper, chilli), AFB_1, AFB_2, and AFG_1, literature[99]; Spices (paprika), AFB_1 and AFG_1, literature[99]; Spices (pepper), AFB_2 and AFG_1, literature[99]

incidence: 4/15*, ∅ conc.: 35 µg/kg, sample origin: local companies in Cairo (capital), Egypt, sample year: unknown, country: Egypt[104], *black pepper

incidence: 4/15*, ∅ conc.: 22 µg/kg, sample origin: local companies in Cairo (capital), Egypt, sample year: unknown, country: Egypt[104], *white pepper

- Co-contamination: not reported
- Further contamination of plants and plant products other than cereals and their products reported in the present book/article[104]: Kubeba, AFB_1 and AFB_2, literature[104]; Spices (coriander), Spices (curcuma), AFB_1 and AFG_1, literature[104]

incidence: 1/7*, conc.: 0.3 µg/kg, sample origin: unknown, sample year: unknown, country: USA[167], *white pepper

- Co-contamination: not reported
- Further contamination of plants and plant products other than cereals and their products reported in the present book/article[167]: Spices (cayenne pepper, chilli), Spices (ginger), AFB_1, AFB_2, AFG_1, and AFG_2, literature[167]

incidence: 3/7*, conc. range: 1.25–5.0 µg/kg, ∅ conc.: 3.67 µg/kg, sample origin: supermarkets and ethnic shops in Lisbon (capital) area, Portugal, sample year: unknown, country: Portugal[169], *white pepper, packaged

- Co-contamination: not reported
- Further contamination of plants and plant products other than cereals and their products reported in the present book/article[169]: Spices (cayenne pepper, chilli), Spices (cumin), Spices (curry), Spices (nutmeg), Spices (paprika), Spices (saffron), AFB_1, literature[169]

incidence: 1/6*, conc.: 0.46 µg/kg, sample origin: food shops in Debrecen (town), Hungary, sample year: October 2004, country: Hungary[170], *black pepper

- Co-contamination: For detailed information see the article.
- Further contamination of plants and plant products other than cereals and their products reported in the present book/article[170]: Spices, AFB_1 and AFB_2, literature[170]; Spices (cayenne pepper, chilli), AFB_1, AFB_2, and OTA, literature[170]

incidence: 1/1*, conc.: 10.8 µg/kg, sample origin: imported sa. collected by the Mycotoxin Research Association, Japan, sample year: 1985, country: Japan[171], *black pepper

- Co-contamination: not reported
- Further contamination of plants and plant products other than cereals and their products reported in the present book/article[171]: Cocoa beans, Nuts (pistachio nut), Spices (cayenne pepper, chilli), AFB_1, literature[171]

incidence: 1/13*, conc.: 0.6 µg/kg, sample origin: retail stores in Osaka (prefecture), Japan, sample year: 1988–1992, country: Japan[173], *white pepper

- Co-contamination: not reported
- Further contamination of plants and plant products other than cereals and their products reported in the present book/article[173]: Butter (peanut butter), Spices (nutmeg), AFB_1, AFB_2, and AFG_1, literature[173]; Nuts (peanuts), Spices (cayenne pepper, chilli), AFB_1, literature[173]

incidence: 7/15*, conc. range: ≤0.30 µg/kg, ∅ conc.: 0.09 µg/kg, sample origin: popular markets in Salé (city) and Rabat (capital), Morocco, sample year: unknown, country: Morocco[174], *black pepper

- Co-contamination: not reported
- Further contamination of plants and plant products other than cereals and their products reported in the present book/article[174]: Spice (cayenne pepper, chilli), Spices (cumin), Spices (ginger), AFB_1 and AF/S, literature[174]; Spices (pepper), AF/S, literature[174]

incidence: 2/24*, conc. range: 8.6–9.5 µg/kg**, ∅ conc.: 9.05 µg/kg**, sample origin: bazaars and markets in Istanbul (capital),

Turkey, sample year: unknown, country: Turkey[178], *black pepper, **AFB_1, AFB_2, AFG_1, and AFG_2

- Co-contamination: not reported
- Further contamination of plants and plant products other than cereals and their products reported in the present book/article[178]: Spices (cayenne pepper, chilli), AFB_1 and AF/S, literature[178]; Spices (pepper), AF/S, literature[178]

incidence: 2/5*, conc. range: 0.9–1.2 μg/kg, Ø conc.: 1.05 μg/kg, sample origin: bazaars, Turkey, sample year: unknown, country: Turkey[180], *black pepper, grounded

incidence: 1/5*, conc.: 0.8 μg/kg, sample origin: herbal shops, Turkey, sample year: unknown, country: Turkey[180], *black pepper, grounded

incidence: 1/5*, conc.: 0.3 μg/kg, sample origin: supermarkets, Turkey, sample year: unknown, country: Turkey[180], *black pepper, grounded

- Co-contamination: not reported
- Further contamination of plants and plant products other than cereals and their products reported in the present book/article[180]: Spices (cayenne pepper, chilli), Spices (paprika), AFB_1 and AF/S, literature[180]; Spices (pepper), AF/S, literature[180]

incidence: 1/1*, conc.: 1.00 μg/kg, sample origin: local stores in Munich (city), Germany, sample year: unknown, country: Germany[181], *black pepper powder

- Co-contamination: not reported
- Further contamination of plants and plant products other than cereals and their products reported in the present book/article[181]: Butter (peanut butter), Spices (nutmeg), AFB_1, AFB_2, and AFG_1, literature[181]; Nuts (hazelnut), AFB_1 and AFG_1, literature[181]; Nuts (pistachio nut), Spices (paprika), AFB_1, literature[181]

incidence: 4/4*, conc. range: 0.65–2.1 μg/kg, Ø conc.: 1.20 μg/kg, sample origin: supermarkets in Penang (state), Malaysia, sample year: unknown, country: Malaysia[182], *black pepper

- Co-contamination: not reported
- Further contamination of plants and plant products other than cereals and their products reported in the present book/article[182]: Nuts, Nuts (peanut), Seeds (sesame), Seeds (sunflower), Spices (cayenne pepper, chilli), Spices (cumin), AFB_1, literature[182]

incidence: 7/23*, conc. range: 0.13–0.42 μg/kg, Ø conc.: 0.20 μg/kg, sample origin: markets, herbal shops, and bazaars in Corum and Ankara (cities), Turkey, sample year: November 2010–August 2011, country: Turkey[183], *black pepper powder

- Co-contamination: 1 sa. co-contaminated with AFB_1 and OTA; no further information available
- Further contamination of plants and plant products other than cereals and their products reported in the present book/article[183]: Spices (cayenne pepper, chilli), AFB_1, AFB_2, AFG_1, and OTA, literature[183]; Spices (cumin), AFB_1, AFB_2, and OTA, literature[183]; Spices (pepper), AFB_2 and OTA, literature[183]

incidence: 5/7, conc. range: 0.7–8.4 μg/kg, sample origin: retail sales points in different regions, Iran, sample year: March–May 2015, country: Iran[491]

- Co-contamination: not reported
- Further contamination of plants and plant products other than cereals and their products reported in the present book/article[491]: Herbs (camomile), Herbs (sage), AFB_1, literature[491]; Spices (cayenne pepper, chilli), Spices (curcuma), Spices (ginger), Spices (thyme), AFB_1, literature[491]

incidence: 1/2*, conc.: 0.25 μg/kg, sample origin: Penang (state, island), Malaysia, sample year: unknown (first study), country: Malaysia[530], *black ("lada hitam") and white pepper ("lada putih")

- Co-contamination: 1 sa. co-contaminated with AFB_1 and AFB_2
- Further contamination of plants and plant products other than cereals and their products reported in the present book/article[530]: Spices, Spices (cayenne pepper, chilli), Spices (coriander), Spices (cumin), Spices (curry), Spices (fennel), Spices (masala), AFB_1, AFB_2, AFG_1, AFG_2, and OTA, literature[530]; Spices (curcuma), AFB_1, AFB_2, AFG_1, and OTA, literature[530]; Spices (pepper), AFB_2, literature[530]; Spices (poppy seed), AFB_1, literature[530]

incidence: 2/2*, conc. range: 0.32–0.39 μg/kg, Ø conc.: 0.35 μg/kg, sample origin: Penang (state, island), Malaysia, sample year: unknown (second study), country: Malaysia[530], *white pepper

- Co-contamination: 1 sa. co-contaminated with AFB_1 and AFB_2; 1 sa. contaminated solely with AFB_1
- Further contamination of plants and plant products other than cereals and their products reported in the present book/article[530]: Spices, Spices (cayenne pepper, chilli), Spices (coriander), Spices (cumin), Spices (curry), Spices (fennel), Spices (masala), AFB_1, AFB_2, AFG_1, AFG_2, and OTA, literature[530]; Spices (curcuma), AFB_1, AFB_2, AFG_1, and OTA, literature[530]; Spices (pepper), AFB_2, literature[530]; Spices (poppy seed), AFB_1, literature[530]

incidence: 7/7*, conc. range: 1–5 μg/kg (1 sa.), 5–10 μg/kg (4 sa.), 10–15 μg/kg (2 sa.), sample origin: different shops, France, sample year: unknown, country: France[706], *black pepper

incidence: 4/4*, conc. range: 1–5 μg/kg (1 sa.), 5–10 μg/kg (1 sa.), 15–20 μg/kg (2 sa.), sample origin: different shops, France, sample year: unknown, country: France[706], *white pepper

incidence: 0/1*, conc.: no contamination, sample origin: different shops, France, sample year: unknown, country: France[706], *green pepper

- Co-contamination: not reported
- Further contamination of plants and plant products other than cereals and their products reported in the present book/article[706]: not reported

incidence: 6/6*, conc. range: 0.04835–0.17468 μg/kg, Ø conc.: 0.12370 μg/kg, sample origin: popular markets in Hyderabad (city), Andhra Pradesh (province), India, sample year: June 2012, country: Iran[800], *black pepper powder

incidence: 6/6*, conc. range: 0.03115–0.04412 μg/kg, Ø conc.: 0.03652 μg/kg, sample origin: popular markets in Hyderabad (city), Andhra Pradesh (province), India, sample year: June 2012, country: Iran[800], *black pepper, whole

- Co-contamination: not reported
- Further contamination of plants and plant products other than cereals and their products reported in the present book/article[800]: Spices (cayenne pepper, chilli), AFB_1, literature[800]

incidence: 6/6*, conc. range: 0.10075–0.53584 μg/kg, Ø conc.: 0.24334 μg/kg, sample origin: popular herbal shops in Qaemshahr (city), Mazandaran (province), Iran, sample year: June 2012, country: Iran[801], *black pepper powder

incidence: 6/6*, conc. range: 0.06316–0.26679 μg/kg, Ø conc.: 0.18757 μg/kg, sample origin: popular herbal shops in Qaemshahr (city), Mazandaran (province), Iran, sample year: June 2012, country: Iran[801], *black pepper, whole

- Co-contamination: not reported

- Further contamination of plants and plant products other than cereals and their products reported in the present book/article[801]: Spices (cayenne pepper, chilli), AFB_1, literature[801]

incidence: 4/6*, conc. range: 24.6–30.0 µg/kg, ∅ conc.: 27.6 µg/kg, sample origin: organic markets and organic shops, Turkey, sample year: May 2010–May 2011, country: Turkey[859], *organic black pepper sa.

- Co-contamination: not reported
- Further contamination of plants and plant products other than cereals and their products reported in the present book/article[859]: Herbs (camomile), Herbs (linden), Herbs (rosehip), Herbs (sage), Spices (anise), Spices (basil), Spices (cayenne pepper, chilli), Spices (cinnamon), Spices (coriander), Spices (cumin), Spices (fennel), Spices (ginger), Spices (laurel leave), Spices (peppermint), Spices (poppy seed), Spices (rosemary), Spices (sumac), AFB_1, literature[859]

incidence: 1/10*, conc.: 11 µg/kg, sample origin: markets, Sri Lanka, sample year: unknown, country: Belgium[891], *black pepper

incidence: 1/10*, conc.: 4.9 µg/kg, sample origin: markets, Sri Lanka, sample year: unknown, country: Belgium[891], *white pepper

- Co-contamination: not reported
- Further contamination of plants and plant products other than cereals and their products reported in the present book/article[891]: Spices (cayenne pepper, chilli), AFB_1, AFB_2, FB_2, OTA, and STG, literature[891]; Spices (pepper), AFB_2, AFG_2, CIT, FB_1, OTA, and STG, literature[891]

Aflatoxin B_2

incidence: 5?/8*, conc. range: 12–150 µg/kg, sample origin: major retail outlets in Nainital (district), India, sample year: April 1984–September 1986, country: India[22], *black pepper

- Co-contamination: not reported
- Further contamination of plants and plant products other than cereals and their products reported in the present book/article[22]: Spices (ammi), AFB_1, AFB_2, and AFG_1, literature[22]; Spices (cardamom), AFB_1, AFB_2, AFG_1, AFG_2, CIT, and OTA, literature[22]; Spices (cassia), AFB_1, AFB_2, AFG_1, and AFG_2, literature[22]; Spices (cayenne pepper, chilli), AFB_1, AFB_2, AFG_1, AFG_2, and ZEA, literature[22]; Spices (coriander), AFB_1, AFB_2, AFG_1, AFG_2, CIT, OTA, and ZEA, literature[22]; Spices (cumin), AFB_1, AFB_2, AFG_1, AFG_2, and CIT, literature[22]; Spices (curcuma), AFB_1, AFB_2, AFG_1, AFG_2, CIT, and RBT, literature[22]; Spices (fennel), AFB_1, AFB_2, AFG_1, AFG_2, CIT, OTA, and STG, literature[22]; Spices (garlic), AFB_1, AFB_2, and AFG_1, literature[22]; Spices (pepper), AFB_1, AFG_1, AFG_2, CIT, STG, and ZEA, literature[22]

incidence: 7?/24*, conc. range: 0.1–0.2 µg/kg, sample origin: unknown, sample year: unknown, country: Japan[97], *white pepper

- Co-contamination: not reported
- Further contamination of plants and plant products other than cereals and their products reported in the present book/article[97]: Beans, Jam (bean jam), Spices, AFB_1 and AFB_2, literature[97]; Butter (peanut butter), Oils (peanut oil), AFB_1, AFB_2, and AFG_1, literature[97]; Spices (pepper), AFB_1 and AFG_1, literature[97]; Spices (nutmeg), AFB_1, AFB_2, AFG_1, and AFG_2, literature[97]

incidence: 11?/84*, conc. range: 0.3 µg/kg, ∅ conc.: <0.1 µg/kg, sample origin: commercially available market basket sa. of foods and foodstuffs in Tokyo (capital), Japan, sample year: 1986–1990, country: Japan[99], *white pepper

- Co-contamination: not reported
- Further contamination of plants and plant products other than cereals and their products reported in the present book/article[99]: Beans, Seeds (sesame), Sugar, AFB_1 and AFB_2, literature[99]; Food, Jam (bean jam), Paste (peanut paste), AFB_1, literature[99]; Nuts (Brazil nut, paranut), Nuts (peanut), Nuts (pistachio nut), Spices (nutmeg), AFB_1, AFB_2, AFG_1, and AFG_2, literature[99]; Spices, Spices (cayenne pepper, chilli), AFB_1, AFB_2, and AFG_1, literature[99]; Spices (paprika), AFB_1 and AFG_1, literature[99]; Spices (pepper), AFB_1 and AFG_1, literature[99]

incidence: 6/23*, conc. range: 0.04–0.05 µg/kg, ∅ conc.: 0.05 µg/kg, sample origin: markets, herbal shops, and bazaars in Corum and Ankara (cities), Turkey, sample year: November 2010–August 2011, country: Turkey[183], *black pepper powder

- Co-contamination: not reported
- Further contamination of plants and plant products other than cereals and their products reported in the present book/article[183]: Spices (cayenne pepper, chilli), AFB_1, AFB_2, AFG_1, and OTA, literature[183]; Spices (cumin), AFB_1, AFB_2, and OTA, literature[183]; Spices (pepper), AFB_1 and OTA, literature[183]

incidence: 1/5, conc.: 0.6 µg/kg, sample origin: commercially available for industrial use, Netherlands, sample year: unknown, country: Netherlands[184]

- Co-contamination: not reported
- Further contamination of plants and plant products other than cereals and their products reported in the present book/article[184]: Spices (garlic), Spices (onion), FB_1, literature[184]; Spices (paprika), AFB_1 and OTA, literature[184]

incidence: 1/2*, conc.: 0.03 µg/kg, sample origin: Penang (state, island), Malaysia, sample year: unknown (first study), country: Malaysia[530], *black ("lada hitam") and white pepper ("lada putih")

- Co-contamination: 1 sa. co-contaminated with AFB_1 and AFB_2
- Further contamination of plants and plant products other than cereals and their products reported in the present book/article[530]: Spices, Spices (cayenne pepper, chilli), Spices (coriander), Spices (cumin), Spices (curry), Spices (fennel), Spices (masala), AFB_1, AFB_2, AFG_1, AFG_2, and OTA, literature[530]; Spices (curcuma), AFB_1, AFB_2, AFG_1, and OTA, literature[530]; Spices (pepper), Spices (poppy seed), AFB_1, literature[530]

incidence: 1/2*, conc.: 0.11 µg/kg, sample origin: Penang (state, island), Malaysia, sample year: unknown (second study), country: Malaysia[530], *white pepper

- Co-contamination: 1 sa. co-contaminated with AFB_1 and AFB_2
- Further contamination of plants and plant products other than cereals and their products reported in the present book/article[530]: Spices, Spices (cayenne pepper, chilli), Spices (coriander), Spices (cumin), Spices (curry), Spices (fennel), Spices (masala), AFB_1, AFB_2, AFG_1, AFG_2, and OTA, literature[530]; Spices (curcuma), AFB_1, AFB_2, AFG_1, and OTA, literature[530]; Spices (pepper), Spices (poppy seed), AFB_1, literature[530]

incidence: 0/10*, conc.: no contamination, sample origin: markets, Sri Lanka, sample year: unknown, country: Belgium[891], *black pepper

incidence: 1/10*, conc.: <LOQ, sample origin: markets, Sri Lanka, sample year: unknown, country: Belgium[891], *white pepper

- Co-contamination: not reported
- Further contamination of plants and plant products other than cereals and their products reported in the present book/article[891]: Spices (cayenne pepper, chilli), AFB_1, AFB_2, FB_2, OTA, and STG, literature[891]; Spices (pepper), AFB_1, AFG_2, CIT, FB_1, OTA, and STG, literature[891]

Aflatoxin G_1

incidence: 5?/8*, conc. range: 15–75 µg/kg, sample origin: major retail outlets in Nainital (district), India, sample year: April 1984–September 1986, country: India[22], *black pepper

- Co-contamination: not reported
- Further contamination of plants and plant products other than cereals and their products reported in the present book/article[22]: Spices (ammi), AFB_1, AFB_2, and AFG_1, literature[22]; Spices (cardamom), AFB_1, AFB_2, AFG_1, AFG_2, CIT, and OTA, literature[22]; Spices (cassia), AFB_1, AFB_2, AFG_1, and AFG_2, literature[22]; Spices (cayenne pepper, chilli), AFB_1, AFB_2, AFG_1, AFG_2, and ZEA, literature[22]; Spices (coriander), AFB_1, AFB_2, AFG_1, AFG_2, CIT, OTA, and ZEA, literature[22]; Spices (cumin), AFB_1, AFB_2, AFG_1, AFG_2, and CIT, literature[22]; Spices (curcuma), AFB_1, AFB_2, AFG_1, AFG_2, CIT, and RBT, literature[22]; Spices (fennel), AFB_1, AFB_2, AFG_1, AFG_2, CIT, OTA, and STG, literature[22]; Spices (garlic), AFB_1, AFB_2, and AFG_1, literature[22]; Spices (pepper), AFB_1, AFB_2, AFG_2, CIT, STG, and ZEA, literature[22]

incidence: 7?/24*, conc. range: 0.2–1.4 µg/kg, sample origin: unknown, sample year: unknown, country: Japan[97], *white pepper

- Co-contamination: not reported
- Further contamination of plants and plant products other than cereals and their products reported in the present book/article[97]: Beans, Jam (bean jam), Spices, Spices (pepper), AFB_1 and AFB_2, literature[97]; Butter (peanut butter), Oils (peanut oil), AFB_1, AFB_2, and AFG_1, literature[97]; Spices (nutmeg), AFB_1, AFB_2, AFG_1, and AFG_2, literature[97]

incidence: 11?/84*, conc. range: 0.9 µg/kg, ∅ conc.: 0.2 µg/kg, sample origin: commercially available market basket sa. of foods and foodstuffs in Tokyo (capital), Japan, sample year: 1986–1990, country: Japan[99], *white pepper

- Co-contamination: not reported
- Further contamination of plants and plant products other than cereals and their products reported in the present book/article[99]: Beans, Seeds (sesame), Spices (pepper), Sugar, AFB_1 and AFB_2, literature[99]; Food, Jam (bean jam), Paste (peanut paste), AFB_1, literature[99]; Nuts (Brazil nut, paranut), Nuts (peanut), Nuts (pistachio nut), Spices (nutmeg), AFB_1, AFB_2, AFG_1, and AFG_2, literature[99]; Spices, Spices (cayenne pepper, chilli), AFB_1, AFB_2, and AFG_1, literature[99]; Spices (paprika), AFB_1 and AFG_1, literature[99]

incidence: 3?/7*, conc. range: 1.8–3.7 µg/kg, sample origin: imported from India, Indonesia, and Malaysia, USA, sample year: unknown, country: USA[185], *black pepper

- Co-contamination: not reported
- Further contamination of plants and plant products other than cereals and their products reported in the present book/article[185]: Spices (celery), AFG_1, literature[185]; Spices (nutmeg), AFB_1 and AFB_2, literature[185]; Spices (pepper), AFG_2, literature[185]

Aflatoxin G_2

incidence: 5?/8*, conc. range: 12–76 µg/kg, sample origin: major retail outlets in Nainital (district), India, sample year: April 1984–September 1986, country: India[22], *black pepper

- Co-contamination: not reported
- Further contamination of plants and plant products other than cereals and their products reported in the present book/article[22]: Spices (ammi), AFB_1, AFB_2, and AFG_1, literature[22]; Spices (cardamom), AFB_1, AFB_2, AFG_1, AFG_2, CIT, and OTA, literature[22]; Spices (cassia), AFB_1, AFB_2, AFG_1, and AFG_2, literature[22]; Spices (cayenne pepper, chilli), AFB_1, AFB_2, AFG_1, AFG_2, and ZEA, literature[22]; Spices (coriander), AFB_1, AFB_2, AFG_1, AFG_2, CIT, OTA, and ZEA, literature[22]; Spices (cumin), AFB_1, AFB_2, AFG_1, AFG_2, and CIT, literature[22]; Spices (curcuma), AFB_1, AFB_2, AFG_1, AFG_2, CIT, and RBT, literature[22]; Spices (fennel), AFB_1, AFB_2, AFG_1, AFG_2, CIT, OTA, and STG, literature[22]; Spices (garlic), AFB_1, AFB_2, and AFG_1, literature[22]; Spices (pepper), AFB_1, AFB_2, AFG_1, CIT, STG, and ZEA, literature[22]

incidence: 1/7*, conc.: 1.1 µg/kg, sample origin: imported from India, Indonesia, and Malaysia, USA, sample year: unknown, country: USA[185], *black pepper

- Co-contamination: not reported
- Further contamination of plants and plant products other than cereals and their products reported in the present book/article[185]: Spices (celery), AFG_1, literature[185]; Spices (nutmeg), AFB_1 and AFB_2, literature[185]; Spices (pepper), AFG_1, literature[185]

incidence: 1/10*, conc.: 5.7 µg/kg, sample origin: markets, Sri Lanka, sample year: unknown, country: Belgium[891], *black pepper

incidence: 1/10*, conc.: 2.6 µg/kg, sample origin: markets, Sri Lanka, sample year: unknown, country: Belgium[891], *white pepper

- Co-contamination: not reported
- Further contamination of plants and plant products other than cereals and their products reported in the present book/article[891]: Spices (cayenne pepper, chilli), AFB_1, AFB_2, FB_2, OTA, and STG, literature[891]; Spices (pepper), AFB_1, AFB_2, CIT, FB_1, OTA, and STG, literature[891]

Aflatoxin/s

incidence: 14/15*, conc. range: ≤0.55 µg/kg**, ∅ conc.: 0.21 µg/kg**, sample origin: popular markets in Salé (city) and Rabat (capital), Morocco, sample year: unknown, country: Morocco[174], *black pepper, **AFS total

- Co-contamination: not reported
- Further contamination of plants and plant products other than cereals and their products reported in the present book/article[174]: Spice (cayenne pepper, chilli), Spices (cumin), Spices (ginger), AFB_1 and AF/S, literature[174]; Spices (pepper), AFB_1, literature[174]

incidence: 8/24*, conc. range: 0.1–15.3 µg/kg**, ∅ conc.: 3.6 µg/kg**, sample origin: bazaars and markets in Istanbul (capital), Turkey, sample year: unknown, country: Turkey[178], *black pepper, **AFB_1, AFB_2, AFG_1, and AFG_2

- Co-contamination: not reported
- Further contamination of plants and plant products other than cereals and their products reported in the present book/article[178]: Spices (cayenne pepper, chilli), AFB_1 and AF/S, literature[178]; Spices (pepper), AFB_1, literature[178]

incidence: 2/5*, conc. range: 1.2–2.3 µg/kg**, ∅ conc.: 1.75 µg/kg**, sample origin: bazaars, Turkey, sample year: unknown, country: Turkey[180], *black pepper, ground, **AFB_1, AFB_2, AFG_1, and AFG_2

incidence: 1/5*, conc.: 1.1 µg/kg, sample origin: herbal shops, Turkey, sample year: unknown, country: Turkey[180], *black pepper, ground, **AFB_1, AFB_2, AFG_1, and AFG_2

incidence: 1/5*, conc.: 0.3 µg/kg, sample origin: supermarkets, Turkey, sample year: unknown, country: Turkey[180], *black pepper, ground, **AFB_1, AFB_2, AFG_1, and AFG_2

- Co-contamination: not reported

- Further contamination of plants and plant products other than cereals and their products reported in the present book/article[180]: Spices (cayenne pepper, chilli), Spices (paprika), AFB_1 and AF/S, literature[180]; Spices (pepper), AFB_1, literature[180]

incidence: 1/20*, conc.: 13.8 µg/kg**, sample origin: retailers and spice shops, Sanliurfa (province), Turkey, sample year: unknown, country: Turkey[186], *isot pepper, **AFB and AFG

- Co-contamination: not reported
- Further contamination of plants and plant products other than cereals and their products reported in the present book/article[186]: Spices (cayenne pepper, chilli), AF/S, literature[186]

incidence: 4/30, conc. range: 0.42–3.24 µg/kg*, sample origin: market stalls, shops, and supermarkets, Ireland, sample year: unknown, country: Ireland[107], *AFB_1, AFB_2, AFG_1, and AFG_2

- Co-contamination: not reported
- Further contamination of plants and plant products other than cereals and their products reported in the present book/article[187]: Spices (cayenne pepper, chilli), Spices (coriander), Spices (curcuma), Spices (curry), Spices (paprika), AF/S, literature[187]

incidence: 10/24*, conc. range: 0.2–4.5 µg/kg, sample origin: imported from Australia and Singapore as well as from supermarkets in Kuala Lumpur, Putrajaya, and Selangor areas, Malaysia, sample year: unknown, country: Malaysia[189], *white pepper seed

incidence: 15/33*, conc. range: 0.1–4.6 µg/kg, sample origin: imported from Australia and Singapore as well as from supermarkets in Kuala Lumpur, Putrajaya, and Selangor areas, Malaysia, sample year: unknown, country: Malaysia[189], *white pepper powder

incidence: 18/30*, conc. range: 0.1–4.8 µg/kg, sample origin: imported from Australia and Singapore as well as from supermarkets in Kuala Lumpur, Putrajaya, and Selangor areas, Malaysia, sample year: unknown, country: Malaysia[189], *black pepper seed

incidence: 27/39*, conc. range: 0.7–4.9 µg/kg, sample origin: imported from Australia and Singapore as well as from supermarkets in Kuala Lumpur, Putrajaya, and Selangor areas, Malaysia, sample year: unknown, country: Malaysia[189], *black pepper powder

- Co-contamination: not reported
- Further contamination of plants and plant products other than cereals and their products reported in the present book/article[189]: not reported

For detailed information see the article.

incidence: 34/34*, conc. range: 0.67–6.15 µg/kg**, ∅ conc.: 3.44 µg/kg**, sample year: unknown, country: Turkey[190], *black pepper, **AFS total

- Co-contamination: not reported
- Further contamination of plants and plant products other than cereals and their products reported in the present book/article[190]: Spices (cayenne pepper, chilli), Spices (cumin), AF/S, literature[190]

incidence: 3/5*, conc. range: 28–35 µg/kg**, sample origin: Assiut (governorate), Egypt, sample year: unknown, country: Egypt[641], *black pepper, **AFB_1 and AFG_1

- Co-contamination: not reported
- Further contamination of plants and plant products other than cereals and their products reported in the present book/article[641]: Spices (anise), Spices (coriander), Spices (fennel), Spices (peppermint), AF/S, literature[641]; Spices (black cumin), AF/S and CIT, literature[641]; Spices (caraway), Spices (marjoran), AF/S and STG, literature[641]; Spices (cayenne pepper, chilli), Spices (cumin), STG, literature[641]

Other ***Aspergillus*** Toxins

Sterigmatocystin

incidence: 2/8*, conc. range: 105–125 µg/kg, ∅ conc.: 115 µg/kg, sample origin: major retail outlets in Nainital (district), India, sample year: April 1984–September 1986, country: India[22], *black pepper

- Co-contamination: not reported
- Further contamination of plants and plant products other than cereals and their products reported in the present book/article[22]: Spices (ammi), AFB_1, AFB_2, and AFG_1, literature[22]; Spices (cardamom), AFB_1, AFB_2, AFG_1, AFG_2, CIT, and OTA, literature[22]; Spices (cassia), AFB_1, AFB_2, AFG_1, and AFG_2, literature[22]; Spices (cayenne pepper, chilli), AFB_1, AFB_2, AFG_1, AFG_2, and ZEA, literature[22]; Spices (coriander), AFB_1, AFB_2, AFG_1, AFG_2, CIT, OTA, and ZEA, literature[22]; Spices (cumin), AFB_1, AFB_2, AFG_1, AFG_2, and CIT, literature[22]; Spices (curcuma), AFB_1, AFB_2, AFG_1, AFG_2, CIT, and RBT, literature[22]; Spices (fennel), AFB_1, AFB_2, AFG_1, AFG_2, CIT, OTA, and STG, literature[22]; Spices (garlic), AFB_1, AFB_2, and AFG_1, literature[22]; Spices (pepper), AFB_1, AFB_2, AFG_1, AFG_2, CIT, and ZEA, literature[22]

incidence: 5/10*, conc. range: <LOQ, sample origin: markets, Sri Lanka, sample year: unknown, country: Belgium[891], *black pepper

incidence: 8/10*, conc. range: <LOQ (4 sa.), 15–36 µg/kg (4 sa.), sample origin: markets, Sri Lanka, sample year: unknown, country: Belgium[891], *white pepper

- Co-contamination: not reported
- Further contamination of plants and plant products other than cereals and their products reported in the present book/article[891]: Spices (cayenne pepper, chilli), AFB_1, AFB_2, FB_2, OTA, and STG, literature[891]; Spices (pepper), AFB_1, AFB_2, AFG_2, CIT, FB_1, and OTA, literature[891]

Aspergillus and ***Penicillium*** Toxins

Citrinin

incidence: 1/8*, conc.: 50 µg/kg, sample origin: major retail outlets in Nainital (district), India, sample year: April 1984–September 1986, country: India[22], *black pepper

- Co-contamination: not reported
- Further contamination of plants and plant products other than cereals and their products reported in the present book/article[22]: Spices (ammi), AFB_1, AFB_2, and AFG_1, literature[22]; Spices (cardamom), AFB_1, AFB_2, AFG_1, AFG_2, CIT, and OTA, literature[22]; Spices (cassia), AFB_1, AFB_2, AFG_1, and AFG_2, literature[22]; Spices (cayenne pepper, chilli), AFB_1, AFB_2, AFG_1, AFG_2, and ZEA, literature[22]; Spices (coriander), AFB_1, AFB_2, AFG_1, AFG_2, CIT, OTA, and ZEA, literature[22]; Spices (cumin), AFB_1, AFB_2, AFG_1, AFG_2, and CIT, literature[22]; Spices (curcuma), AFB_1, AFB_2, AFG_1, AFG_2, CIT, and RBT, literature[22]; Spices (fennel), AFB_1,

AFB_2, AFG_1, AFG_2, CIT, OTA, and STG, literature[22]; Spices (garlic), AFB_1, AFB_2, and AFG_1, literature[22]; Spices (pepper), AFB_1, AFB_2, AFG_1, AFG_2, STG, and ZEA, literature[22]

incidence: 1/10*, conc.: <LOQ, sample origin: markets, Sri Lanka, sample year: unknown, country: Belgium[891], *black pepper

incidence: 0/10*, conc.: no contamination, sample origin: markets, Sri Lanka, sample year: unknown, country: Belgium[891], *white pepper

- Co-contamination: not reported
- Further contamination of plants and plant products other than cereals and their products reported in the present book/article[891]: Spices (cayenne pepper, chilli), AFB_1, AFB_2, FB_2, OTA, and STG, literature[891]; Spices (pepper), AFB_1, AFB_2, AFG_2, FB_1, OTA, and STG, literature[891]

Ochratoxin A

incidence: 13/20*, conc. range: 26–643 µg/kg, ∅ conc.: 274 µg/kg, sample origin: retail markets and supermarkets in 24 regions, Tunisia, sample year: unknown, country: Tunisia[20], *black pepper

- Co-contamination: not reported
- Further contamination of plants and plant products other than cereals and their products reported in the present book/article[20]: Nuts (almond), Nuts (peanut), Nuts (pistachio nut), Spices (caraway), Spices (coriander), Spices (curcuma), Spices (paprika), OTA, literature[20]

incidence: 4/23*, conc. range: 0.87–3.48 µg/kg, ∅ conc.: 1.82 µg/kg, sample origin: markets, herbal shops, and bazaars in Corum and Ankara (cities), Turkey, sample year: November 2010–August 2011, country: Turkey[183], *black pepper powder

- Co-contamination: 1 sa. co-contaminated with AFB_1 and OTA; no further information available
- Further contamination of plants and plant products other than cereals and their products reported in the present book/article[183]: Spices (cayenne pepper, chilli), AFB_1, AFB_2, AFG_1, and OTA, literature[183]; Spices (cumin), AFB_1, AFB_2, and OTA, literature[183]; Spices (pepper), AFB_1 and AFB_2, literature[183]

incidence: 14/26*, conc. range: 10–29 µg/kg (8 sa.), 30–49 µg/kg (5 sa.), 50–100 µg/kg (1 sa.), sample origin: retail shops, India, sample year: unknown, country: India/UK/Belgium[191], *black pepper seeds

- Co-contamination: not reported
- Further contamination of plants and plant products other than cereals and their products reported in the present book/article[191]: Spices (coriander), Spices (curcuma), Spices (ginger), OTA, literature[191]

incidence: 1/6*, conc.: 0.6 µg/kg, sample origin: retail shops in Belgium, Netherlands, and Russia, sample year: autumn-winter 2005/2006, country: Belgium/Russia/Hungary[193], *included black pepper and black pepper seeds

incidence: 0/7*, conc.: no contamination, sample origin: retail shops in Belgium, Netherlands, and Russia, sample year: autumn-winter 2005/2006, country: Belgium/Russia/Hungary[193], *included white pepper and white pepper seeds

- Co-contamination: not reported
- Further contamination of plants and plant products other than cereals and their products reported in the present book/article[193]: Spices (cayenne pepper, chilli), Spices (ginger), Spices (nutmeg), Spices (paprika), OTA, literature[193]

incidence: 11/11, conc. range: LOD/LOQ-4.9 µg/kg (6 sa.), 5.0–9.9 µg/kg (5 sa., maximum: 8 µg/kg), sample origin: Austria, sample year: 1994, country: EU[194]

- Co-contamination: not reported
- Further contamination of plants and plant products other than cereals and their products reported in the present book/article[194]: Coffee, Coffee beans, Spices (cayenne pepper, chilli), Spices (curry), Spices (marjoram), OTA, literature[194]

incidence: 22/60, conc. range: 0.15–2.20 µg/kg, ∅ conc.: 0.72 µg/kg, sample origin: supermarkets and open markets in Kuala Lumpur (capital), Putrajaya (planned city), Selangor (state) areas, Malaysia, sample year: October–December 2008, country: Malaysia[195], *included black and white pepper seeds as well as pepper powder

incidence: 3/12, conc. range: 0.20–0.45 µg/kg, ∅ conc.: 0.32 µg/kg, sample origin: imported from Australia, Malaysia, sample year: October–December 2008, country: Malaysia[195], *included black and white pepper seeds as well as pepper powder

incidence: 5/12, conc. range: 0.20–1.76 µg/kg, ∅ conc.: 1.19 µg/kg, sample origin: imported from Singapore, Malaysia, sample year: October–December 2008, country: Malaysia[195], *included black and white pepper seeds as well as pepper powder

incidence: 27/36, conc. range: 0.5–13.58 µg/kg, ∅ conc.: 5.21 µg/kg, sample origin: market, Malaysia, sample year: October–December 2008, country: Malaysia[195], *included black and white pepper seeds as well as pepper powder

- Co-contamination: not reported
- Further contamination of plants and plant products other than cereals and their products reported in the present book/article[195]: not reported

incidence: 4/5*, conc. range: 0.7–8.3 µg/kg, sample origin: available in New Zealand, sample year: 2008/2009, country: New Zealand[277], *black and white pepper

- Co-contamination: not reported
- Further contamination of plants and plant products other than cereals and their products reported in the present book/article[277]: Bars, AFB_1, AF/S, DON, NIV, and OTA, literature[277]; Butter (peanut butter), Confectionery, Food, Fruits, dried, Nuts, Nuts (peanut), Nuts (pistachio nut), Sauces (peanut sauce), AFB_1 and AF/S, literature[277]; Coffee, Dates, Spices (nutmeg), Vine fruits, dried, OTA, literature[277]; Figs, Spices (cayenne pepper, chilli), Spices (curry), Spices (ginger), Spices (paprika), AFB_1, AF/S, and OTA, literature[277]; Nuts (Brazil nut, paranut), AFB_1, literature[277]

incidence: 1/6, conc.: 0.80 µg/kg (1 sa.), sample origin: Netherlands, sample year: 1996/1997, country: EU[698]

- Co-contamination: not reported
- Further contamination of plants and plant products other than cereals and their products reported in the present book/article[698]: Baby food, Beans (soybean), Chocolate, Cocoa, Coffee, Coffee beans, Currants, Fruits, dried, Herbs, Jam, Juices, Juices (grape juice), Legumes, Nuts, Oil seeds, Oils (olive oil), Peas, Peas (chickpea), Products (plant product), Raisins, Spices, Spices (cayenne pepper, chilli), Spices (nutmeg), Spices

(paprika), Sultanas, Tea, Vinegar, Wine, Wine (red), Wine (rosé), Wine (white), Wine, miscellaneous, OTA, literature[698]

incidence: 4/5*, conc. range: ≤8.3 μg/kg, sample origin: New Zealand, sample year: 2009, country: New Zealand[702], *black and white pepper

- Co-contamination: not reported
- Further contamination of plants and plant products other than cereals and their products reported in the present book/article[702]: Figs, Spices (cayenne pepper, chilli), Spices (curry), Spices (ginger), Spices (paprika), AF/S and OTA, literature[702]; Fruits, dried, AF/S, literature[702]; Vine fruits, dried, OTA, literature[702]

incidence: 11/12*, conc. range: ≤2.82 μg/kg, sample origin: 12 regions, Czech Republic, sample year: 2011–2013, country: Czech Republic[811], *black pepper

- Co-contamination: not reported
- Further contamination of plants and plant products other than cereals and their products reported in the present book/article[811]: Beans, Chocolate, Coffee, Confectionery, Figs, Herbs and Spices, Juices, Licorice, Powder (cocoa powder), Products (ginger product), Raisins, Spices (cayenne pepper, chilli), Spices (caraway), Spices (coriander), Spices (ginger), Spices (mustard), Spices (nutmeg), Spices (paprika), Tea, Wine (red), Wine (white), OTA, literature[811]

incidence: 4/10*, conc. range: <LOQ (3 sa.), 51 μg/kg (1 sa.), sample origin: markets, Sri Lanka, sample year: unknown, country: Belgium[891], *black pepper

incidence: 0/10*, conc.: no contamination, sample origin: markets, Sri Lanka, sample year: unknown, country: Belgium[891], *white pepper

- Co-contamination: not reported
- Further contamination of plants and plant products other than cereals and their products reported in the present book/article[891]: Spices (cayenne pepper, chilli), AFB_1, AFB_2, FB_2, OTA, and STG, literature[891]; Spices (pepper), AFB_1, AFB_2, AFG_2, CIT, FB_1, and STG, literature[891]

Fusarium Toxins

Fumonisin/s

Fumonisin B_1

incidence: 1/10*, conc.: 134.5 μg/kg, sample origin: markets, Sri Lanka, sample year: unknown, country: Belgium[891], *black pepper

incidence: 0/10*, conc.: no contamination, sample origin: markets, Sri Lanka, sample year: unknown, country: Belgium[891], *white pepper

- Co-contamination: not reported
- Further contamination of plants and plant products other than cereals and their products reported in the present book/article[891]: Spices (cayenne pepper, chilli), AFB_1, AFB_2, FB_2, OTA, and STG, literature[891]; Spices (pepper), AFB_1, AFB_2, AFG_2, CIT, OTA, and STG, literature[891]

Other ***Fusarium*** Toxins

Zearalenone

incidence: 2/8*, conc. range: pr, sample origin: major retail outlets in Nainital (district), India, sample year: April 1984–September 1986, country: India[22], *black pepper

- Co-contamination: not reported
- Further contamination of plants and plant products other than cereals and their products reported in the present book/article[22]: Spices (ammi), AFB_1, AFB_2, and AFG_1, literature[22]; Spices (cardamom), AFB_1, AFB_2, AFG_1, AFG_2, CIT, and OTA, literature[22]; Spices (cassia), AFB_1, AFB_2, AFG_1, and AFG_2, literature[22]; Spices (cayenne pepper, chilli), AFB_1, AFB_2, AFG_1, AFG_2, and ZEA, literature[22]; Spices (coriander), AFB_1, AFB_2, AFG_1, AFG_2, CIT, OTA, and ZEA, literature[22]; Spices (cumin), AFB_1, AFB_2, AFG_1, AFG_2, and CIT, literature[22]; Spices (curcuma), AFB_1, AFB_2, AFG_1, AFG_2, CIT, and RBT, literature[22]; Spices (fennel), AFB_1, AFB_2, AFG_1, AFG_2, CIT, OTA, and STG, literature[22]; Spices (garlic), AFB_1, AFB_2, and AFG_1, literature[22]; Spices (pepper), AFB_1, AFB_2, AFG_1, AFG_2, CIT, and STG, literature[22]

see also **Spices, Spices (cayenne pepper, chilli)**

Spices (peppermint) may contain the following mycotoxins:

Aspergillus Toxins

Aflatoxin B_1

incidence: 4/5*, conc. range: 4.2–26.7 μg/kg, Ø conc.: 14.7 μg/kg, sample origin: organic markets and organic shops, Turkey, sample year: May 2010–May 2011, country: Turkey[859], *organic peppermint sa.

- Co-contamination: not reported
- Further contamination of plants and plant products other than cereals and their products reported in the present book/article[859]: Herbs (camomile), Herbs (linden), Herbs (rosehip), Herbs (sage), Spices (anise), Spices (basil), Spices (cayenne pepper, chilli), Spices (cinnamon), Spices (coriander), Spices (cumin), Spices (fennel), Spices (ginger), Spices (laurel leave), Spices (pepper), Spices (poppy seed), Spices (rosemary), Spices (sumac), AFB_1, literature[859]

Aflatoxin/s

incidence: 1/5, conc.: 15 μg/kg, sample origin: Assiut (governorate), Egypt, sample year: unknown, country: Egypt[641], *AFB_1 and AFB_2

- Co-contamination: not reported
- Further contamination of plants and plant products other than cereals and their products reported in the present book/article[641]: Spices (anise), Spices (coriander), Spices (fennel), Spices (pepper), AF/S, literature[641]; Spices (black cumin), AF/S and CIT, literature[641]; Spices (caraway), Spices (marjoram), AF/S and STG, literature[641]; Spices (cayenne pepper, chilli), Spices (cumin), STG, literature[641]

Spices (pili-pili) see Spices (cayenne pepper, chilli)

Spices (poppy seed) may contain the following mycotoxins:

Aspergillus Toxins

Aflatoxin B_1

incidence: 1/1*, conc.: 0.05 μg/kg, sample origin: Penang (state, island), Malaysia, sample year: unknown (first study), country: Malaysia[530], *"kas-kas"

- Co-contamination: not reported
- Further contamination of plants and plant products other than cereals and their products reported in the present book/article[530]: Spices, Spices (cayenne pepper, chilli), Spices (coriander), Spices (cumin), Spices (curry), Spices (fennel), Spices

(masala), AFB_1, AFB_2, AFG_1, AFG_2, and OTA, literature[530]; Spices (curcuma), AFB_1, AFB_2, AFG_1, and OTA, literature[530]; Spices (pepper), AFB_1 and AFB_2, literature[530]

incidence: 3/7*, conc. range: 0.98–3.2 µg/kg, ∅ conc.: 2.4 µg/kg, sample origin: organic markets and organic shops, Turkey, sample year: May 2010–May 2011, country: Turkey[859], *organic poppy seed sa.

- Co-contamination: not reported
- Further contamination of plants and plant products other than cereals and their products reported in the present book/article[859]: Herbs (camomile), Herbs (linden), Herbs (rosehip), Herbs (sage), Spices (anise), Spices (basil), Spices (cayenne pepper, chilli), Spices (cinnamon), Spices (coriander), Spices (cumin), Spices (fennel), Spices (ginger), Spices (laurel leave), Spices (pepper), Spices (peppermint), Spices (rosemary), Spices (sumac), AFB_1, literature[859]

Spices (prickly ash) may contain the following mycotoxins:

Aspergillus and *Penicillium* Toxins

Ochratoxin A

incidence: 8/80, conc. range: 0.57–9.82 µg/kg, ∅ conc.: 3.17 µg/kg, sample origin: supermarkets, shops, and market stalls in eight cities, China, sample year: April–July 2009, country: China[885]

- Co-contamination: not reported
- Further contamination of plants and plant products other than cereals and their products reported in the present book/article[885]: Spices (anise), Spices (cayenne pepper, chilli), Spices (cinnamon), Spices (cumin), Spices (curry), OTA, literature[885]

Spices (rosemary) may contain the following mycotoxins:

Aspergillus Toxins

Aflatoxin B_1

incidence: 3/6*, conc. range: 3.3–10 µg/kg, ∅ conc.: 6.7 µg/kg, sample origin: organic markets and organic shops, Turkey, sample year: May 2010–May 2011, country: Turkey[859], *organic rosemary sa.

- Co-contamination: not reported
- Further contamination of plants and plant products other than cereals and their products reported in the present book/article[859]: Herbs (camomile), Herbs (linden), Herbs (rosehip), Herbs (sage), Spices (anise), Spices (basil), Spices (cayenne pepper, chilli), Spices (cinnamon), Spices (coriander), Spices (cumin), Spices (fennel), Spices (ginger), Spices (laurel leaves), Spices (pepper), Spices (peppermint), Spices (poppy seed), Spices (sumac), AFB_1, literature[859]

Spices (saffron) may contain the following mycotoxins:

Aspergillus Toxins

Aflatoxin B_1

incidence: 2/5*, conc. range: 2.0–2.75 µg/kg, ∅ conc.: 2.38 µg/kg, sample origin: supermarkets and ethnic shops in Lisbon (capital) area, Portugal, sample year: unknown, country: Portugal[169], *saffron, packaged

- Co-contamination: not reported
- Further contamination of plants and plant products other than cereals and their products reported in the present book/article[169]: Spices (cayenne pepper, chilli), Spices (cumin), Spices (curry), Spices (nutmeg), Spices (paprika), Spices (pepper), AFB_1, literature[169]

Spices (star anise) may contain the following mycotoxins:

Aspergillus Toxins

Aflatoxin G_2

incidence: 1/3, conc.: <LOQ, sample origin: Chinese markets in Beijing (capital) city and Fujian (province), China, sample year: unknown, country: China/Italy[768]

- Co-contamination: not reported
- Further contamination of plants and plant products other than cereals and their products reported in the present book/article[768]: Food, AFB_1, AFB_2, and OTA, literature[768]; Spices, AFB_1, literature[768]; Spices (fennel), AFG_2, literature[768]

Spices (sumac) may contain the following mycotoxins:

Aspergillus Toxins

Aflatoxin B_1

incidence: 8/10*, conc. range: 51.2–52.5 µg/kg, ∅ conc.: 45.8 µg/kg?, sample origin: organic markets and organic shops, Turkey, sample year: May 2010–May 2011, country: Turkey[859], *organic sumac

- Co-contamination: not reported
- Further contamination of plants and plant products other than cereals and their products reported in the present book/article[859]: Herbs (camomile), Herbs (linden), Herbs (rosehip), Herbs (sage), Spices (anise), Spices (basil), Spices (cayenne pepper, chilli), Spices (cinnamon), Spices (coriander), Spices (cumin), Spices (fennel), Spices (ginger), Spices (laurel leaves), Spices (pepper), Spices (peppermint), Spices (poppy seed), Spices (rosemary), AFB_1, literature[859]

Spices (tandoori) may contain the following mycotoxins:

Aspergillus Toxins

Aflatoxin/s

incidence: 3?/3, conc. range: 1.9–6.8 µg/kg*, sample origin: specialist food shops, UK, sample year: unknown, country: UK[16], *AFB_1, AFB_2, AFG_1, and AFG_2

- Co-contamination: not reported
- Further contamination of plants and plant products other than cereals and their products reported in the present book/article[16]: Food, AF/S, DON, FUS-X, NIV, OTA, and ZEA, literature[16]; Nuts (almond), AF/S, literature[16]; Oils, Spices (fennel), AF/S and ZEA, literature[16]; Oils (sesame oil), OTA, literature[16]; Paste (curry paste), AF/S, FB/S, NIV, OTA, and ZEA, literature[16]; Pickles, AF/S, FB/S, OTA, and ZEA, literature[16]; Sauces (chilli sauce), NIV, OTA, and ZEA, literature[16]; Spices, AF/S, FB/S, OTA, and ZEA, literature[16]; Spices (cayenne pepper, chilli), AF/S, DAS, DON, HT-2, OTA, and ZEA, literature[16]; Spices (coriander), AF/S, DON, OTA, and ZEA, literature[16]; Spices (curry), AF/S, DAS, FB/S, FUS-X, HT-2, NEO, NIV, OTA, T-2, and ZEA, literature[16]; Spices (garlic), DON, FUS-X, and NIV, literature[16]; Spices (ginger), AF/S, DON, NEO, NIV, OTA, and T-2, literature[16]; Spices (tandoori), FB/S, NIV, OTA, and T-2, literature[16]

Aspergillus and *Penicillium* Toxins

Ochratoxin A

incidence: 3?/3, conc. range: 2.2–23.9 µg/kg, sample origin: specialist food shops, UK, sample year: unknown, country: UK[16]

- Co-contamination: not reported

- Further contamination of plants and plant products other than cereals and their products reported in the present book/article[16]: Food, AF/S, DON, FUS-X, NIV, OTA, and ZEA, literature[16]; Nuts (almond), AF/S, literature[16]; Oils, Spices (fennel), AF/S and ZEA, literature[16]; Oils (sesame oil), OTA, literature[16]; Paste (curry paste), AF/S, FB/S, NIV, OTA, and ZEA, literature[16]; Pickles, AF/S, FB/S, OTA, and ZEA, literature[16]; Sauces (chilli sauce), NIV, OTA, and ZEA, literature[16]; Spices, AF/S, FB/S, OTA, and ZEA, literature[16]; Spices (cayenne pepper, chilli), AF/S, DAS, DON, HT-2, OTA, and ZEA, literature[16]; Spices (coriander), AF/S, DON, OTA, and ZEA, literature[16]; Spices (curry), AF/S, DAS, FB/S, FUS-X, HT-2, NEO, NIV, OTA, T-2, and ZEA, literature[16]; Spices (garlic), DON, FUS-X, and NIV, literature[16]; Spices (ginger), AF/S, DON, NEO, NIV, OTA, and T-2, literature[16]; Spices (tandoori), AF/S, FB/S, NIV, and T-2, literature[16]

Fusarium Toxins

Fumonisin/s

incidence: 1/3, conc.: 19 µg/kg*, sample origin: specialist food shops, UK, sample year: unknown, country: UK[16], *FB_1 and FB_2

- Co-contamination: not reported
- Further contamination of plants and plant products other than cereals and their products reported in the present book/article[16]: Food, AF/S, DON, FUS-X, NIV, OTA, and ZEA, literature[16]; Nuts (almond), AF/S, literature[16]; Oils, Spices (fennel), AF/S and ZEA, literature[16]; Oils (sesame oil), OTA, literature[16]; Paste (curry paste), AF/S, FB/S, NIV, OTA, and ZEA, literature[16]; Pickles, AF/S, FB/S, OTA, and ZEA, literature[16]; Sauces (chilli sauce), NIV, OTA, and ZEA, literature[16]; Spices, AF/S, FB/S, OTA, and ZEA, literature[16]; Spices (cayenne pepper, chilli), AF/S, DAS, DON, HT-2, OTA, and ZEA, literature[16]; Spices (coriander), AF/S, DON, OTA, and ZEA, literature[16]; Spices (curry), AF/S, DAS, FB/S, FUS-X, HT-2, NEO, NIV, OTA, T-2, and ZEA, literature[16]; Spices (garlic), DON, FUS-X, and NIV, literature[16]; Spices (ginger), AF/S, DON, NEO, NIV, OTA, and T-2, literature[16]; Spices (tandoori), AF/S, NIV, OTA, and T-2, literature[16]

Type A Trichothecenes

T-2 Toxin

incidence: 3?/3, conc. range: 23–281 µg/kg, sample origin: specialist food shops, UK, sample year: unknown, country: UK[16]

- Co-contamination: not reported
- Further contamination of plants and plant products other than cereals and their products reported in the present book/article[16]: Food, AF/S, DON, FUS-X, NIV, OTA, and ZEA, literature[16]; Nuts (almond), AF/S, literature[16]; Oils, Spices (fennel), AF/S and ZEA, literature[16]; Oils (sesame oil), OTA, literature[16]; Paste (curry paste), AF/S, FB/S, NIV, OTA, and ZEA, literature[16]; Pickles, AF/S, FB/S, OTA, and ZEA, literature[16]; Sauces (chilli sauce), NIV, OTA, and ZEA, literature[16]; Spices, AF/S, FB/S, OTA, and ZEA, literature[16]; Spices (cayenne pepper, chilli), AF/S, DAS, DON, HT-2, OTA, and ZEA, literature[16]; Spices (coriander), AF/S, DON, OTA, and ZEA, literature[16]; Spices (curry), AF/S, DAS, FB/S, FUS-X, HT-2, NEO, NIV, OTA, T-2, and ZEA, literature[16]; Spices (garlic), DON, FUS-X, and NIV, literature[16]; Spices (ginger), AF/S, DON, NEO, NIV, OTA, and T-2, literature[16]; Spices (tandoori), AF/S, FB/S, NIV, and OTA, literature[16]

Type B Trichothecenes

Nivalenol

incidence: 3?/3, conc. range: 60–126 µg/kg, sample origin: specialist food shops, UK, sample year: unknown, country: UK[16]

- Co-contamination: not reported
- Further contamination of plants and plant products other than cereals and their products reported in the present book/article[16]: Food, AF/S, DON, FUS-X, NIV, OTA, and ZEA, literature[16]; Nuts (almond), AF/S, literature[16]; Oils, Spices (fennel), AF/S and ZEA, literature[16]; Oils (sesame oil), OTA, literature[16]; Paste (curry paste), AF/S, FB/S, NIV, OTA, and ZEA, literature[16]; Pickles, AF/S, FB/S, OTA, and ZEA, literature[16]; Sauces (chilli sauce), NIV, OTA, and ZEA, literature[16]; Spices, AF/S, FB/S, OTA, and ZEA, literature[16]; Spices (cayenne pepper, chilli), AF/S, DAS, DON, HT-2, OTA, and ZEA, literature[16]; Spices (coriander), AF/S, DON, OTA, and ZEA, literature[16]; Spices (curry), AF/S, DAS, FB/S, FUS-X, HT-2, NEO, NIV, OTA, T-2, and ZEA, literature[16]; Spices (garlic), DON, FUS-X, and NIV, literature[16]; Spices (ginger), AF/S, DON, NEO, NIV, OTA, and T-2, literature[16]; Spices (tandoori), AF/S, FB/S, OTA, and T-2, literature[16]

see also **Spices**

Spices (thyme) may contain the following mycotoxins:

Aspergillus Toxins

Aflatoxin B_1

incidence: 1/2, conc.: 2.2 µg/kg, sample origin: retail sales points in different regions, Iran, sample year: March–May 2015, country: Iran[491]

- Co-contamination: not reported
- Further contamination of plants and plant products other than cereals and their products reported in the present book/article[491]: Herbs (camomile), Herbs (sage), AFB_1, literature[491]; Spices (cayenne pepper, chilli), Spices (curcuma), Spices (ginger), Spices (pepper), AFB_1, literature[491]

Other *Aspergillus* Toxins

Sterigmatocystin

incidence: 2/50, conc. range: 10–14 µg/kg, Ø conc.: 12 µg/kg, sample origin: FUCHS Gewürze GmbH, Latvia, sample year: 2014, country: Latvia[679]

- Co-contamination: not reported
- Further contamination of plants and plant products other than cereals and their products reported in the present book/article[679]: Spices (cayenne pepper, chilli), FB_1, literature[679]; Spices (nutmeg), FB_1 and OTA, literature[679]; Spices (oregano), STG, literature[679]; Spices (paprika), FB_1, OTA, and STG, literature[679]; Spices (thyme), DON and ZEA, literature[679]

Fusarium Toxins

Type B Trichothecenes

Deoxynivalenol

incidence: 7/50, conc. range: 10–142 µg/kg, sample origin: FUCHS Gewürze GmbH, Latvia, sample year: 2014, country: Latvia[679]

- Co-contamination: not reported
- Further contamination of plants and plant products other than cereals and their products reported in the present book/

article[679]: Spices (cayenne pepper, chilli), FB_1, literature[679]; Spices (nutmeg), FB_1 and OTA, literature[679]; Spices (oregano), STG, literature[679]; Spices (paprika), FB_1, OTA, and STG, literature[679]; Spices (thyme), STG and ZEA, literature[679]

Other ***Fusarium*** Toxins

Zearalenone

incidence: 9/50, conc. range: 40–209 µg/kg, sample origin: FUCHS Gewürze GmbH, Latvia, sample year: 2014, country: Latvia[679]

- Co-contamination: not reported
- Further contamination of plants and plant products other than cereals and their products reported in the present book/article[679]: Spices (cayenne pepper, chilli), FB_1, literature[679]; Spices (nutmeg), FB_1 and OTA, literature[679]; Spices (oregano), STG, literature[679]; Spices (paprika), FB_1, OTA, and STG, literature[679]; Spices (thyme), DON and STG, literature[679]

Spices (turmeric) see Spices (curcuma)

Starch (cassava starch) may contain the following mycotoxins:

Aspergillus Toxins

Aflatoxin/s

incidence: 2/65*, conc. range: ≤294 µg/kg**, ∅ conc.: 150 µg/kg**, sample origin: market retailers, street vendors, wholesalers, and distributors, Thailand, sample year: September 1967–July 1969, country: Thailand[139], *sago, **AFB_1, AFB_2, AFG_1, and AFG_2

- Co-contamination: not reported
- Further contamination of plants and plant products other than cereals and their products reported in the present book/article[139]: Beans, Beans (mung bean), Food, Nuts (peanut), Seeds (sesame), Spices (cayenne pepper, chilli), Spices (garlic/onion), Vegetables, AF/S, literature[139]

For detailed information see the article.

Strawberries may contain the following mycotoxins:

Alternaria Toxins

Alternariol

incidence: 7/21*, conc. range: 13–724 µg/kg, sample origin: local markets in Valencia (city), Spain, sample year: unknown, country: Spain[755], *stored for <7 days at 6 ± 2 °C

incidence: 8/13*, conc. range: 26–752 µg/kg, sample origin: local markets in Valencia (city), Spain, sample year: unknown, country: Spain[755], *stored for <5 days at 22 ± 2 °C

- Co-contamination: not reported
- Further contamination of plants and plant products other than cereals and their products reported in the present book/article[755]: Strawberries, AME, literature[755]

For detailed information see the article.

Alternariol Methyl Ether

incidence: 11/21*, conc. range: 10–26 µg/kg, sample origin: local markets in Valencia (city), Spain, sample year: unknown, country: Spain[755], *stored for <7 days at 6 ± 2 °C

incidence: 4/13*, conc. range: 11–137 µg/kg, sample origin: local markets in Valencia (city), Spain, sample year: unknown, country: Spain[755], *stored for <5 days at 22 ± 2 °C

- Co-contamination: not reported
- Further contamination of plants and plant products other than cereals and their products reported in the present book/article[755]: Strawberries, AOH, literature[755]

For detailed information see the article.

Aspergillus and ***Penicillium*** Toxins

Ochratoxin A

incidence: 4/10*, conc. range: 0.22–1.44 µg/kg, sample origin: households and markets, Germany, sample year: unknown, country: Germany[26], *damaged or moldy

- Co-contamination: not reported
- Further contamination of plants and plant products other than cereals and their products reported in the present book/article[26]: Fruits (apple), Fruits (cherry), Fruits (peach), Fruits (tomato), Potatoes, OTA, literature[26]

Sugar may contain the following mycotoxins:

Aspergillus Toxins

Aflatoxin B_1

incidence: 3?/9, conc. range: 1.0–1.5 µg/kg, ∅ conc.: 1.2 µg/kg, sample origin: commercially available market basket sa. of foods and foodstuffs in Tokyo (capital), Japan, sample year: 1986–1990, country: Japan[99]

- Co-contamination: For detailed information see the article.
- Further contamination of plants and plant products other than cereals and their products reported in the present book/article[99]: Beans, Seeds (sesame), AFB_1 and AFB_2, literature[99]; Food, Jam (bean jam), Paste (peanut paste), AFB_1, literature[99]; Nuts (Brazil nut, paranut), Nuts (peanut), Nuts (pistachio nut), Spices (nutmeg), AFB_1, AFB_2, AFG_1, and AFG_2, literature[99]; Spices, Spices (cayenne pepper, chilli), Spices (pepper), AFB_1, AFB_2, and AFG_1, literature[99]; Spices (paprika), AFB_1 and AFG_1, literature[99]; Sugar, AFB_2, literature[99]

Aflatoxin B_2

incidence: 3?/9, conc. range: 0.1–0.2 µg/kg, ∅ conc.: 0.1 µg/kg, sample origin: commercially available market basket sa. of foods and foodstuffs in Tokyo (capital), Japan, sample year: 1986–1990, country: Japan[99]

- Co-contamination: For detailed information see the article.
- Further contamination of plants and plant products other than cereals and their products reported in the present book/article[99]: Beans, Seeds (sesame), AFB_1 and AFB_2, literature[99]; Food, Jam (bean jam), Paste (peanut paste), AFB_1, literature[99]; Nuts (Brazil nut, paranut), Nuts (peanut), Nuts (pistachio nut), Spices (nutmeg), AFB_1, AFB_2, AFG_1, and AFG_2, literature[99]; Spices, Spices (cayenne pepper, chilli), Spices (pepper), AFB_1, AFB_2, and AFG_1, literature[99]; Spices (paprika), AFB_1 and AFG_1, literature[99]; Sugar, AFB_1, literature[99]

Sugar beets may contain the following mycotoxins:

Fusarium Toxins

Zearalenone

incidence: 6/25*, conc. range: 12–391 µg/kg, ∅ conc.: 154 µg/kg, sample origin: northwestern Minnesota (state), USA, sample year: September 1988, country: USA[689], *moldy sugar beet roots (*Fusarium* spp.)

incidence: 31/75*, conc. range: 13–4,650 µg/kg, Ø conc.: 281 µg/kg, sample origin: outdoor stockpiles of a sugar-processing company in northwestern Minnesota (state), USA, sample year: 1987/1988, country: USA[689], *sugar beet fibers (high incidence of *Fusarium equiseti* on sugar beets before harvest and on stored sugar beets)

- Co-contamination: For detailed information see the article.
- Further contamination of plants and plant products other than cereals and their products reported in the present book/article[689]: not reported; see also Sugar beets, chapter: Further Mycotoxins and Microbial Metabolites

Sugarcane may contain the following mycotoxins:

Alternaria Toxins

Alternariol Methyl Ether

incidence: 9/21*, conc. range: ≤1.26 µg/kg, Ø conc.: 0.45 µg/kg, sample origin: local vendors in Assiut (city), Assiut (governorate), Egypt, sample year: January 2016, country: Egypt/Austria[300], *sugarcane grass

- Co-contamination: For detailed information see the article.
- Further contamination of plants and plant products other than cereals and their products reported in the present book/article[300]: Juices (sugarcane juice), 3-NPA, AFB_1, AFG_1, FA, KA, and PA, literature[300]; Sugarcane, 3-NPA, AFB_1, AFG_1, KA, and PA, literature[300]; see also Juices (sugarcane juice), Sugarcane, chapter: Further Mycotoxins and Microbial Metabolites

Aspergillus Toxins

Aflatoxin B_1

incidence: 10/21*, conc. range: ≤30.6 µg/kg, Ø conc.: 13.6 µg/kg, sample origin: local vendors in Assiut (city), Assiut (governorate), Egypt, sample year: January 2016, country: Egypt/Austria[300], *sugarcane grass

- Co-contamination: For detailed information see the article.
- Further contamination of plants and plant products other than cereals and their products reported in the present book/article[300]: Juices (sugarcane juice), 3-NPA, AFB_1, AFG_1, FA, KA, and PA, literature[300]; Sugarcane, 3-NPA, AFG_1, AME, KA, and PA, literature[300]; see also Juices (sugarcane juice), Sugarcane, chapter: Further Mycotoxins and Microbial Metabolites

Aflatoxin G_1

incidence: 2/21*, conc. range: ≤7.76 µg/kg, Ø conc.: 5.10 µg/kg, sample origin: local vendors in Assiut (city), Assiut (governorate), Egypt, sample year: January 2016, country: Egypt/Austria[300], *sugarcane grass

- Co-contamination: For detailed information see the article.
- Further contamination of plants and plant products other than cereals and their products reported in the present book/article[300]: Juices (sugarcane juice), 3-NPA, AFB_1, AFG_1, FA, KA, and PA, literature[300]; Sugarcane, 3-NPA, AFB_1, AME, KA, and PA, literature[300]; see also Juices (sugarcane juice), Sugarcane, chapter: Further Mycotoxins and Microbial Metabolites

Aspergillus and ***Penicillium*** Toxins

3-Nitropropionic Acid

incidence: 13/21*, conc. range: ≤193 µg/kg, Ø conc.: 27.5 µg/kg, sample origin: local vendors in Assiut (city), Assiut (governorate), Egypt, sample year: January 2016, country: Egypt/Austria[300], *sugarcane grass

- Co-contamination: For detailed information see the article.
- Further contamination of plants and plant products other than cereals and their products reported in the present book/article[300]: Juices (sugarcane juice), 3-NPA, AFB_1, AFG_1, FA, KA, and PA, literature[300]; Sugarcane, AFB_1, AFG_1, AME, KA, and PA, literature[300]; see also Juices (sugarcane juice), Sugarcane, chapter: Further Mycotoxins and Microbial Metabolites

Kojic Acid

incidence: 16/21*, conc. range: ≤117,815 µg/kg, Ø conc.: 20,176 µg/kg, sample origin: local vendors in Assiut (city), Assiut (governorate), Egypt, sample year: January 2016, country: Egypt/Austria[300], *sugarcane grass

- Co-contamination: For detailed information see the article.
- Further contamination of plants and plant products other than cereals and their products reported in the present book/article[300]: Juices (sugarcane juice), 3-NPA, AFB_1, AFG_1, FA, KA, and PA, literature[300]; Sugarcane, 3-NPA, AFB_1, AFG_1, AME, and PA, literature[300]; see also Juices (sugarcane juice), Sugarcane, chapter: Further Mycotoxins and Microbial Metabolites

Penicillic Acid

incidence: 10/21*, conc. range: ≤2,683 µg/kg, Ø conc.: 333 µg/kg, sample origin: local vendors in Assiut (city), Assiut (governorate), Egypt, sample year: January 2016, country: Egypt/Austria[300], *sugarcane grass

- Co-contamination: For detailed information see the article.
- Further contamination of plants and plant products other than cereals and their products reported in the present book/article[300]: Juices (sugarcane juice), 3-NPA, AFB_1, AFG_1, FA, KA, and PA, literature[300]; Sugarcane, 3-NPA, AFB_1, AFG_1, AME, and KA, literature[300]; see also Juices (sugarcane juice), Sugarcane, chapter: Further Mycotoxins and Microbial Metabolites

Sugarcane juice see Juices (sugarcane juice)

Sultanas may contain the following mycotoxins:

Aspergillus Toxins

Aflatoxin B_1

incidence: 1/1, conc.: 0.05 µg/kg, sample origin: Aigio (town), Peloponnese (peninsula), Greece, sample year: 2012, country: Greece[767]

incidence: 1/1, conc.: 0.05 µg/kg, sample origin: Crete (island, administrative unit), Greece, sample year: 2012, country: Greece[767]

- Co-contamination: 2 sa. co-contaminated with AFB_1 and OTA
- Further contamination of plants and plant products other than cereals and their products reported in the present book/article[767]: Currants, Raisins, AFB_1 and OTA, literature[767]; Sultanas, OTA, literature[767]

Aflatoxin/s

incidence: 0/24*, conc.: no contamination, sample origin: imported from Argentina, Chile, Iran, and Turkey purchased from markets in Campinas (city) and São Paulo (city, state), Brazil, sample year: 2002/2003, country: Brazil[356], *black sultanas

incidence: 3/19*, conc. range: 0.3–2.0 µg/kg**, sample origin: imported from Argentina, Chile, Iran, and Turkey purchased from markets in Campinas (city) and São Paulo (city, state), Brazil, sample year: 2002/2003, country: Brazil[356], *white sultanas, **AFB_1 and AFB_2

- Co-contamination: not reported

- Further contamination of plants and plant products other than cereals and their products reported in the present book/ article[356]: Figs, AF/S, literature[356]

Aspergillus and ***Penicillium*** Toxins

Citrinin

incidence: 1/2, conc.: tr, sample origin: unknown, sample year: unknown, country: Germany[673]

- Co-contamination: 1 sa. co-contaminated with CIT and OTA
- Further contamination of plants and plant products other than cereals and their products reported in the present book/ article[673]: Cocoa, AF/S and OTA, literature[673]; Raisins, CIT and OTA, literature[673]; Sultanas, OTA, literature[673]

Ochratoxin A

incidence: 2/2, conc. range: 1.0–3.7 µg/kg, ∅ conc.: 2.3 µg/kg, sample origin: local market, southern Italy, sample year: unknown, country: Italy[49]

- Co-contamination: not reported
- Further contamination of plants and plant products other than cereals and their products reported in the present book/ article[49]: Puree, OTA, literature[49]

incidence: 28/53, conc. range: 0.51–58.04 µg/kg, sample origin: different export companies, Turkey, sample year: unknown, country: Turkey[55]

- Co-contamination: not reported
- Further contamination of plants and plant products other than cereals and their products reported in the present book/ article[55]: Figs, AFB_1, AF/S, and OTA, literature[55]; Fruits, dried, OTA, literature[55]

incidence: 92/100, conc. range: 0.1 µg/kg (5 sa.), 0.2–4.0 µg/kg (67 sa.), 4.1–10 µg/kg (11 sa.), 10.1–20 µg/kg (7 sa.), 20.1–30 µg/kg (2 sa., maximum: 21.5 µg/kg), sample origin: 13 regions in UK, sample year: October/November 1998, country: UK[263]

- Co-contamination: not reported
- Further contamination of plants and plant products other than cereals and their products reported in the present book/ article[263]: Chocolate, Currants, Figs, Juices (grape juice), Powder (cocoa powder), Raisins, Wine (red), OTA, literature[263]

incidence: 1/3, conc.: 3.6 µg/kg, sample origin: Peloponnese (peninsula), Greece, sample year: 1998, country: Greece[327]

incidence: 0/1, conc.: no contamination, sample origin: Peloponnese (peninsula), Greece, sample year: 1999, country: Greece[327]

incidence: 7/9, conc. range: ≤13.2 µg/kg, sample origin: Crete (island), Greece, sample year: 1998, country: Greece[327]

incidence: 5/6, conc. range: ≤1.8 µg/kg, sample origin: Crete (island), Greece, sample year: 1999, country: Greece[327]

incidence: 3/8, conc. range: ≤10.3 µg/kg, sample origin: Crete (island), Greece, sample year: 2000, country: Greece[327]

- Co-contamination: not reported
- Further contamination of plants and plant products other than cereals and their products reported in the present book/ article[327]: Currants, Wine (red), Wine (rosé), Wine (white), Wine, miscellaneous, OTA, literature[327]

incidence: 8/20, conc. range: 3.7–15.2 µg/kg, ∅ conc.: 8 µg/kg, sample origin: imported from Greece, Turkey, and unknown purchased in shops in Norwich and supermarkets, small shops, etc. across the UK, sample year: unknown, country: UK[328]

- Co-contamination: not reported
- Further contamination of plants and plant products other than cereals and their products reported in the present book/ article[328]: Currants, Raisins, OTA, literature[328]

incidence: 39/66, conc. range: ≤26.0 µg/kg, ∅ conc.: 3.11 µg/kg, sample origin: different countries and unknown collected in retail locations across Canada, sample year: 1998–2000, country: Canada[329]

- Co-contamination: not reported
- Further contamination of plants and plant products other than cereals and their products reported in the present book/ article[329]: Currants, Raisins, OTA, literature[329]

For detailed information see the article.

incidence: 19/24*, conc. range: 0.1–5.0 µg/kg (11 sa.), 5.1–10.0 µg/kg (4 sa.), 10.1–20.0 µg/kg (3 sa.), 33.9 µg/kg (1 sa.), sample origin: different markets in Campinas (city) and São Paulo (city/state), Brazil, sample year: 2002/2003, country: Brazil[331], *black sultanas

incidence: 10/19*, conc. range: 0.1–5.0 µg/kg, sample origin: different markets in Campinas (city) and São Paulo (city/state), Brazil, sample year: 2002/2003, country: Brazil[331], *white sultanas

- Co-contamination: not reported
- Further contamination of plants and plant products other than cereals and their products reported in the present book/ article[331]: Dates, Figs, Fruits, dried, OTA, literature[331]

incidence: 20/62*, conc. range: 0.026–1.0 µg/kg (12 sa.), 1.01–2.0 µg/kg (2 sa.), 2.01–5.0 µg/kg (3 sa.), 5.01–10 µg/kg (2 sa.), 25.5 µg/kg (1 sa.), sample origin: unprocessed sultanas from vineyards in Aegean (region), Turkey, sample year: 1998, country: Turkey[645]

incidence: 16/20*, conc. range: 0.026–1.0 µg/kg (1 sa.), 1.01–2.0 µg/kg (2 sa.), 2.01–5.0 µg/kg (4 sa.), 5.01–10 µg/kg (3 sa.), >10 µg/kg (6 sa., maximum: 33.6 µg/kg), sample origin: unprocessed sultanas from vineyards in Aegean (region), Turkey, sample year: 1999, country: Turkey[645]

incidence: 4/10*, conc. range: 0.026–1.0 µg/kg (1 sa.), 2.01–5.0 µg/kg (1 sa.), 5.01–10 µg/kg (1 sa.), 18.5 µg/kg (1 sa.), sample origin: unprocessed sultanas from vineyards in Aegean (region), Turkey, sample year: 2000, country: Turkey[645]

incidence: 60/65*, conc. range: 0.026–1.0 µg/kg (4 sa.), 1.01–2.0 µg/kg (6 sa.), 2.01–5.0 µg/kg (20 sa.), 5.01–10 µg/kg (18 sa.), >10 µg/kg (12 sa., maximum: 40.4 µg/kg), sample origin: unprocessed sultanas from packing-houses in Aegean (region), Turkey, sample year: 2002, country: Turkey[645]

incidence: 43/64*, conc. range: 0.026–1.0 µg/kg (18 sa.), 1.01–2.0 µg/kg (8 sa.), 2.01–5.0 µg/kg (12 sa.), 5.01–10 µg/kg (3 sa.), >10 µg/kg (2 sa., maximum: 54.0 µg/kg), sample origin: unprocessed sultanas from packing-houses in Aegean (region), Turkey, sample year: 2003, country: Turkey[645]

incidence: 36/39*, conc. range: 0.026–1.0 µg/kg (15 sa.), 1.01–2.0 µg/kg (6 sa.), 2.01–5.0 µg/kg (9 sa.), 5.01–10 µg/kg (2 sa.), >10 µg/kg (4 sa., maximum: 18.43), sample origin: unprocessed sultanas from packing-houses in Aegean (region), Turkey, sample year: 2004, country: Turkey[645]

- Co-contamination: not reported
- Further contamination of plants and plant products other than cereals and their products reported in the present book/article[645]: not reported

incidence: 10/16*, conc. range: 0.30–1.00 μg/kg (1 sa.), 1.01–2.00 μg/kg (6 sa.), 2.01–5.00 μg/kg (1 sa.), 5.01–10.00 μg/kg (2 sa., maximum: 5.90 μg/kg), sample origin: Aegean (region), Turkey, sample year: 1999, country: Turkey[646], *processed sultanas

incidence: 343/418*, conc. range: 0.30–1.00 μg/kg (245 sa.), 1.01–2.00 μg/kg (63 sa.), 2.01–5.00 μg/kg (30 sa.), 5.01–10.00 μg/kg (3 sa.), >10.01 μg/kg (2 sa., maximum: 100.00 μg/kg), sample origin: Aegean (region), Turkey, sample year: 2000, country: Turkey[646], *processed sultanas

incidence: 752/805*, conc. range: 0.30–1.00 μg/kg (336 sa.), 1.01–2.00 μg/kg (228 sa.), 2.01–5.00 μg/kg (158 sa.), 5.01–10.00 μg/kg (22 sa.), >10.01 μg/kg (8 sa., maximum: 47.81 μg/kg), sample origin: Aegean (region), Turkey, sample year: 2001, country: Turkey[646], *processed sultanas

incidence: 414/414*, conc. range: 0.30–1.00 μg/kg (132 sa.), 1.01–2.00 μg/kg (156 sa.), 2.01–5.00 μg/kg (120 sa.), 5.01–10.00 μg/kg (5 sa.), 11.31 μg/kg (1 sa.), sample origin: Aegean (region), Turkey, sample year: 2002, country: Turkey[646], *processed sultanas

incidence: 194/232*, conc. range: 0.30–1.00 μg/kg (160 sa.), 1.01–2.00 μg/kg (23 sa.), 2.01–5.00 μg/kg (11 sa., maximum: 4.10 μg/kg), sample origin: Aegean (region), Turkey, sample year: 2003, country: Turkey[646], *processed sultanas

- Co-contamination: not reported
- Further contamination of plants and plant products other than cereals and their products reported in the present book/article[646]: not reported

incidence: 2/28, conc. range: 0.19–1.07 μg/kg, Ø conc.: 0.63 μg/kg, sample origin: different supermarkets and traditional bazaars in Istanbul (capital), Turkey, sample year: 2008/2009, country: Turkey[665]

- Co-contamination: not reported
- Further contamination of plants and plant products other than cereals and their products reported in the present book/article[665]: Currants, Juices (grape juice), Pekmez, OTA, literature[665]

incidence: 2/2, conc. range: 3.5–3.9 μg/kg, Ø conc.: 3.7 μg/kg, sample origin: unknown, sample year: unknown, country: Germany[673]

- Co-contamination: 1 sa. co-contaminated with CIT and OTA; 1 sa. contaminated solely with OTA
- Further contamination of plants and plant products other than cereals and their products reported in the present book/article[673]: Cocoa, AF/S and OTA, literature[673]; Raisins, CIT and OTA, literature[673]; Sultanas, CIT, literature[673]

incidence: 100/106, conc. range: LOD/LOQ-0.9 μg/kg (69 sa.), 1.0–2.9 μg/kg (19 sa.), 3.0–4.9 μg/kg (4 sa.), 5.0–9.9 μg/kg (6 sa.), ≤21.40 μg/kg (2 sa.), sample origin: Germany, sample year: 1995–1998, country: EU[698]

incidence: 0/1, conc.: no contamination, sample origin: Greece, sample year: 1997, country: EU[698]

incidence: 11/15, conc. range: 1.0–2.9 μg/kg (7 sa.), 3.0–4.9 μg/kg (3 sa.), 16.5 μg/kg (1 sa.), sample origin: Greece, sample year: 1998, country: EU[698]

incidence: 11/19, conc. range: LOD/LOQ-0.9 μg/kg (2 sa.), 1.0–2.9 μg/kg (7 sa.), 3.0–4.9 μg/kg (1 sa.), 13.17 μg/kg (1 sa.), sample origin: Greece, sample year: 1999, country: EU[698]

- Co-contamination: not reported
- Further contamination of plants and plant products other than cereals and their products reported in the present book/article[698]: Baby food, Beans (soybean), Chocolate, Cocoa, Coffee, Coffee beans, Currants, Fruits, dried, Herbs, Jam, Juices, Juices (grape juice), Legumes, Nuts, Oil seeds, Oils (olive oil), Peas, Peas (chickpea), Products (plant product), Raisins, Spices, Spices (cayenne pepper, chilli), Spices (nutmeg), Spices (paprika), Spices (pepper), Tea, Vinegar, Wine, Wine (red), Wine (rosé), Wine (white), Wine, miscellaneous, OTA, literature[698]

incidence: 10/16, conc. range: 0.16–5 μg/kg (8 sa.), 5–10 μg/kg (2 sa., maximum: 8.40 μg/kg), sample origin: dried grape factories in Hamadan (province), Iran, sample year: October 2012–March 2013, country: Iran[742]

- Co-contamination: not reported
- Further contamination of plants and plant products other than cereals and their products reported in the present book/article[742]: Currants, Raisins, OTA literature[742]

incidence: 2/2, conc. range: 10–100 μg/kg (2 sa.), sample origin: Turkish sa. imported to Greece, sample year: 2012, country: Greece[767]

incidence: 1/1, conc.: 117.8 μg/kg, sample origin: Aigio (town), Peloponnese (peninsula), Greece, sample year: 2012, country: Greece[767]

incidence: 1/1, conc.: <0.8–10 μg/kg, sample origin: Crete (island, administrative unit), Greece, sample year: 2012, country: Greece[767]

- Co-contamination: 2 sa. (Aigio and Crete) co-contaminated with AFB_1 and OTA; 2 sa. contaminated solely with OTA
- Further contamination of plants and plant products other than cereals and their products reported in the present book/article[767]: Currants, Raisins, AFB_1 and OTA, literature[767]; Sultanas, AFB_1, literature[767]

incidence: 19/29, conc. range: ≤12.83 μg/kg, sample origin: retail markets, China, sample year: 2012, country: China[897]

- Co-contamination: not reported
- Further contamination of plants and plant products other than cereals and their products reported in the present book/article[897]: Currants, OTA, literature[897]

For detailed information see the article.

see also **Currants, Fruits, dried,** and **Raisins**

Sunflower seeds see **Seeds (sunflower)**

Sweet potatoes may contain the following mycotoxins:

Aspergillus Toxins

Aflatoxin/s

incidence: 31/37*, Ø conc.: 60.6 μg/kg**, sample origin: greater Manila (capital) area, Philippines, sample year: unknown, country: Philippines[1], *sweet potato and sweet potato products, **AFS of pos sa.?

- Co-contamination: not reported
- Further contamination of plants and plant products other than cereals and their products reported in the present book/article[1]: Beans, Butter (peanut butter), Cassava, Cocoa, Coffee, Drinks (alcoholic drink), Nuts (cashew nut), Nuts (peanut), Products (coconut product), Root crops, Spices (cayenne pepper, chilli), Yams, AF/S, literature[1]

see also **Tubers**

Sweets may contain the following mycotoxins:

Aspergillus Toxins

Aflatoxin M_1

incidence: 134/138*, conc. range: 0.01–1.5 µg/kg, Ø conc.: 0.48 µg/kg, sample origin: 36 districts of Punjab (province), Pakistan, sample year: unknown, country: Pakistan/USA[835], *burfi (mithai or a type of sweets with main ingredients condensed milk and sugar)

- Co-contamination: not reported
- Further contamination of plants and plant products other than cereals and their products reported in the present book/article[835]: not reported

see also **Chocolate**

Tahini see **Butter (sesame butter)**

Tandoori see **Spices (tandoori)**

Tea may contain the following mycotoxins:

Alternaria Toxins

Tenuazonic Acid

incidence: 3/3*, conc. range: 1.3–2.4 µg/l, Ø conc.: 1.87 µg/l, sample origin: retail stores, Germany, sample year: unknown, country: Germany[581], *infant fruit tea

incidence: 2/2*, conc. range: 1.0–1.5 µg/l, Ø conc.: 1.25 µg/l, sample origin: retail stores, Germany, sample year: unknown, country: Germany[581], *fruit tea

incidence: 2/2*, conc. range: 0.7–2.7 µg/l, Ø conc.: 1.7 µg/l, sample origin: retail stores, Germany, sample year: unknown, country: Germany[581], *infant herbal tea

incidence: 1/1*, conc.: 3.8 µg/l, sample origin: retail stores, Germany, sample year: unknown, country: Germany[581], *herbal tea

incidence: 7/7*, conc. range: 0.8–17.5 µg/l, Ø conc.: 4.73 µg/l, sample origin: retail stores, Germany, sample year: unknown, country: Germany[581], *infant fennel tea

incidence: 3/3*, conc. range: 6.1–13.7 µg/l, Ø conc.: 10.6 µg/l, sample origin: retail stores, Germany, sample year: unknown, country: Germany[581], *fennel tea

- Co-contamination: not reported
- Further contamination of plants and plant products other than cereals and their products reported in the present book/article[581]: Puree, TA, literature[581]

Aspergillus Toxins

Aflatoxin B_1

incidence: 1/9*, conc.: 1.45 µg/kg, sample origin: unknown, sample year: unknown, country: Korea[9], *adlay tea

- Co-contamination: not reported
- Further contamination of plants and plant products other than cereals and their products reported in the present book/article[9]: Butter (peanut butter), AFB_1, AFB_2, AFG_1, and AFG_2, literature[9]; Nuts (almond), Nuts (peanut), AFB_1 and AFB_2, literature[9]; Products (nut product), AFB_1, literature[9]

incidence: 1/10*, conc.: 5.4 µg/kg, sample origin: supermarkets in Almería (city), Spain, sample year: unknown, country: Spain[788], *green tea

- Co-contamination: not reported
- Further contamination of plants and plant products other than cereals and their products reported in the present book/article[788]: not reported

incidence: 1/37*, conc.: <1 µg/kg**, sample origin: Belo Horizonte (city) in Minas Gerais (state), Brazil, sample year: 2010, country: Brazil[818], *vegetable drugs, dried plant extracts, and industrialized products, **green tea (*Camellia sinensis*)

- Co-contamination: not reported
- Further contamination of plants and plant products other than cereals and their products reported in the present book/article[818]: not reported

For detailed information see the article.

incidence: 2/5*, conc. range: <LOQ, sample origin: local markets, Beijing (capital), China, sample year: unknown, country: China[875], *ginger tea bags

incidence: 0/5*, conc.: no contamination, sample origin: local markets, Beijing (capital), China, sample year: unknown, country: China[875], *ginger black tea bags

- Co-contamination: not reported
- Further contamination of plants and plant products other than cereals and their products reported in the present book/article[875]: Spices (ginger), AFB_1, AFB_2, AFG_1, AFG_2, and OTA, literature[875]; Tea, AFG_1, AFG_2, and OTA, literature[875]

Aflatoxin B_2

incidence: 0/12*, conc.: no contamination, sample origin: supermarkets in Valencia (city), Spain, sample year: unknown, country: Spain[373], *black tea bags

incidence: 0/14*, conc.: no contamination, sample origin: supermarkets in Valencia (city), Spain, sample year: unknown, country: Spain[373], *red tea bags

incidence: 0/10*, conc.: no contamination, sample origin: supermarkets in Valencia (city), Spain, sample year: unknown, country: Spain[373], *green tea bags

incidence: 6/8*, conc. range: 14.4–32.2 µg/l, Ø conc.: 26 µg/l, sample origin: supermarkets in Valencia (city), Spain, sample year: unknown, country: Spain[373], *green mint tea bags

- Co-contamination: not reported
- Further contamination of plants and plant products other than cereals and their products reported in the present book/article[373]: Tea, AFG_1, AFG_2, 15-AcDON, and ENB, literature[373]

Aflatoxin G_1

incidence: 0/12*, conc.: no contamination, sample origin: supermarkets in Valencia (city), Spain, sample year: unknown, country: Spain[373], *black tea bags

incidence: 0/14*, conc.: no contamination, sample origin: supermarkets in Valencia (city), Spain, sample year: unknown, country: Spain[373], *red tea bags

incidence: 0/10*, conc.: no contamination, sample origin: supermarkets in Valencia (city), Spain, sample year: unknown, country: Spain[373], *green tea bags

incidence: 4/8*, conc. range: <LOQ, sample origin: supermarkets in Valencia (city), Spain, sample year: unknown, country: Spain[373], *green mint tea bags

- Co-contamination: not reported
- Further contamination of plants and plant products other than cereals and their products reported in the present book/article[373]: Tea, AFB_2, AFG_2, 15-AcDON, and ENB, literature[373]

incidence: 0/5*, conc.: no contamination, sample origin: local markets, Beijing (capital), China, sample year: unknown, country: China[875], *ginger tea bags

incidence: 2/5*, conc. range: 1.32–2.67 µg/kg, ∅ conc.: 2.0 µg/kg, sample origin: local markets, Beijing (capital), China, sample year: unknown, country: China[875], *ginger black tea bags

- Co-contamination: not reported
- Further contamination of plants and plant products other than cereals and their products reported in the present book/article[875]: Spices (ginger), AFB_1, AFB_2, AFG_1, AFG_2, and OTA, literature[875]; Tea, AFB_1, AFG_2, and OTA, literature[875]

Aflatoxin G_2

incidence: 2/12*, conc. range: <LOQ, sample origin: supermarkets in Valencia (city), Spain, sample year: unknown, country: Spain[373], *black tea bags

incidence: 2/14*, conc. range: <LOQ, sample origin: supermarkets in Valencia (city), Spain, sample year: unknown, country: Spain[373], *red tea bags

incidence: 0/10*, conc.: no contamination, sample origin: supermarkets in Valencia (city), Spain, sample year: unknown, country: Spain[373], *green tea bags

incidence: 4/8*, conc. range: 1.9–2.6 µg/l, ∅ conc.: 2.3 µg/l, sample origin: supermarkets in Valencia (city), Spain, sample year: unknown, country: Spain[373], *green mint tea bags

- Co-contamination: not reported
- Further contamination of plants and plant products other than cereals and their products reported in the present book/article[373]: Tea, AFB_2, AFG_1, 15-AcDON, and ENB, literature[373]

incidence: 1/5*, conc.: <LOQ, sample origin: local markets, Beijing (capital), China, sample year: unknown, country: China[875], *ginger tea bags

incidence: 3/5*, conc. range: <LOQ-3.09 µg/kg, sample origin: local markets, Beijing (capital), China, sample year: unknown, country: China[875], *ginger black tea bags

- Co-contamination: not reported
- Further contamination of plants and plant products other than cereals and their products reported in the present book/article[875]: Spices (ginger), AFB_1, AFB_2, AFG_1, AFG_2, and OTA, literature[875]; Tea, AFB_1, AFG_1, and OTA, literature[875]

Aflatoxin/s

incidence: 4/20, conc. range: 2.8–18.4 µg/kg*, ∅ conc.: 7.25 µg/kg*, sample origin: markets from Aswan, Assiut, El-Minya, Qena, and Sohag, (governorates), Egypt, sample year: unknown, country: Egypt[196], *AFB_1 and AFB_2

- Co-contamination: not reported
- Further contamination of plants and plant products other than cereals and their products reported in the present book/article[196]: Cocoa, AF/S, literature[196]

incidence: 21/21, conc. range: 2.1–14.5 µg/kg*, ∅ conc.: 7.1 µg/kg*, sample origin: domestic sa. purchased in retail stores in Amol (city), north of Iran, sample year: autumn 2010, country: Iran[683], *AFS total

incidence: 21/21, conc. range: 1.5–16.5 µg/kg*, ∅ conc.: 6.11 µg/kg*, sample origin: imported sa. purchased in retail stores in Amol (city), north of Iran, sample year: autumn 2010, country: Iran[683], *AFS total

- Co-contamination: not reported
- Further contamination of plants and plant products other than cereals and their products reported in the present book/article[683]: not reported

incidence: 11/40*, conc. range: ~190 µg/kg**, ∅ conc.: 12.07 µg/kg**, sample origin: Iranian and imported sa. purchased from Tehran (capital) markets, Iran, sample year: July 2010, country: Iran[817], *black tea, **AFB_1, AFB_2, AFG_1, and AFG_2

- Co-contamination: not reported
- Further contamination of plants and plant products other than cereals and their products reported in the present book/article[817]: not reported

Aspergillus and ***Penicillium*** Toxins

Ochratoxin A

incidence: 1/32*, conc.: 1.33 µg/kg, sample origin: different regions, Germany, sample year: unknown, country: Germany[197], *green tea

incidence: 3/34*, conc. range: ≤1.78 µg/kg, ∅ conc.: 1.14 µg/kg, sample origin: different regions, Germany, sample year: unknown, country: Germany[197], *herb tea

incidence: 13/31*, conc. range: ≤10.3 µg/kg, ∅ conc.: 2.78 µg/kg, sample origin: different regions, Germany, sample year: unknown, country: Germany[197], *children´s herb tea

- Co-contamination: not reported
- Further contamination of plants and plant products other than cereals and their products reported in the present book/article[197]: Coffee, OTA, literature[197]

incidence: 8/139, conc. range: LOD/LOQ-0.9 µg/kg (2 sa.), 1.0–2.9 µg/kg (5 sa.), 10.30 µg/kg (1 sa.), sample origin: Germany, sample year: 1995–1998, country: EU[698]

- Co-contamination: not reported
- Further contamination of plants and plant products other than cereals and their products reported in the present book/article[698]: Baby food, Beans (soybean), Chocolate, Cocoa, Coffee, Coffee beans, Currants, Fruits, dried, Herbs, Jam, Juices, Juices (grape juice), Legumes, Nuts, Oil seeds, Oils (olive oil), Peas, Peas (chickpea), Products (plant product), Raisins, Spices, Spices (cayenne pepper, chilli), Spices (nutmeg), Spices (paprika), Spices (pepper), Sultanas, Vinegar, Wine, Wine (red), Wine (rosé), Wine (white), Wine, miscellaneous, OTA, literature[698]

incidence: 2/16*, conc. range: 0.65 µg/kg, ∅ conc.: 0.65 µg/kg, sample origin: Yunnan (province) in south-western China, sa. purchased in tea shops, Austria/Germany, sample year: unknown, country: Austria/Germany[735], *loose Pu-erh tea, conventional

incidence: 1/9*, conc.: 14.8 µg/kg, sample origin: Yunnan (province) in south-western China, sa. purchased in tea shops, Austria/Germany, sample year: unknown, country: Austria/Germany[735], *loose Pu-erh tea, organic

incidence: 1/11*, conc.: 94.7 µg/kg, sample origin: Yunnan (province) in south-western China, sa. purchased in tea shops, Austria/Germany, sample year: unknown, country: Austria/Germany[735], *compressed Pu-erh tea

- Co-contamination: not reported
- Further contamination of plants and plant products other than cereals and their products reported in the present book/article[735]: not reported

incidence: 8/12*, conc. range: ≤8.50 µg/kg, sample origin: 12 regions, Czech Republic, sample year: 2011–2013, country: Czech Republic[811], *green tea leaves and stalks, dried incidence: 4/12*, conc. range: ≤250 μg/kg, sample origin: 12 regions, Czech Republic, sample year: 2011–2013, country: Czech Republic[811], *tea leaves and stalks, fermented

incidence: 4/12*, conc. range: ≤104 μg/kg, sample origin: 12 regions, Czech Republic, sample year: 2011–2013, country: Czech Republic[811], *herbs and vegetables for infusions—fruit tea

incidence: 1/12*, conc.: 1.62 μg/kg, sample origin: 12 regions, Czech Republic, sample year: 2011–2013, country: Czech Republic[811], *herbs and vegetables for infusions—herbal tea

- Co-contamination: not reported
- Further contamination of plants and plant products other than cereals and their products reported in the present book/article[811]: Beans, Chocolate, Coffee, Confectionery, Figs, Herbs and Spices, Juices, Licorice, Powder (cocoa powder), Products (ginger product), Raisins, Spices (caraway), Spices (cayenne pepper, chilli), Spices (coriander), Spices (ginger), Spices (mustard), Spices (nutmeg), Spices (paprika), Spices (pepper), Wine (red), Wine (white), OTA, literature[811]

incidence: 2/5*, conc. range: 1.05–1.19 μg/kg, ∅ conc.: 1.12 μg/kg, sample origin: local markets, Beijing (capital), China, sample year: unknown, country: China[875], *ginger tea bags

incidence: 0/5*, conc.: no contamination, sample origin: local markets, Beijing (capital), China, sample year: unknown, country: China[875], *ginger black tea bags

- Co-contamination: not reported
- Further contamination of plants and plant products other than cereals and their products reported in the present book/article[875]: Spices (ginger), AFB_1, AFB_2, AFG_1, AFG_2, and OTA, literature[875]; Tea, AFB_1, AFG_1, and AFG_2, literature[875]

***Fusarium* Toxins**

Depsipeptides

Enniatin B

incidence: 0/12*, conc.: no contamination, sample origin: supermarkets in Valencia (city), Spain, sample year: unknown, country: Spain[373], *black tea bags

incidence: 0/14*, conc.: no contamination, sample origin: supermarkets in Valencia (city), Spain, sample year: unknown, country: Spain[373], *red tea bags

incidence: 2/10*, conc. range: <LOQ, sample origin: supermarkets in Valencia (city), Spain, sample year: unknown, country: Spain[373], *green tea bags

incidence: 0/8*, conc.: no contamination, sample origin: supermarkets in Valencia (city), Spain, sample year: unknown, country: Spain[373], *green mint tea bags

- Co-contamination: not reported
- Further contamination of plants and plant products other than cereals and their products reported in the present book/article[373]: Tea, AFB_2, AFG_1, AFG_2, and 15-AcDON, literature[373]

Fumonisin/s

Fumonisin B_1

incidence: 16/18*, conc. range: 80–280 μg/kg, sample origin: supermarkets in Lisbon (capital), Portugal, sample year: unknown, country: Portugal[78], *black tea

- Co-contamination: not reported
- Further contamination of plants and plant products other than cereals and their products reported in the present book/article[78]: Medicinal plants, FB_1, literature[78]

incidence: 1/91*, conc.: 76 μg/kg**, sample origin: supermarkets, pharmacies, and private collections in China and Belgium, sample year: 2008–2010, country: Belgium/China[198], *tea raw material, **Ceylon mélange (no mycotoxins were detected in the derived drinkable product)

- Co-contamination: not reported
- Further contamination of plants and plant products other than cereals and their products reported in the present book/article[198]: not reported

Type B Trichothecenes

15-Acetyldeoxynivalenol

incidence: 0/12*, conc.: no contamination, sample origin: supermarkets in Valencia (city), Spain, sample year: unknown, country: Spain[373], *black tea bags

incidence: 0/14*, conc.: no contamination, sample origin: supermarkets in Valencia (city), Spain, sample year: unknown, country: Spain[373], *red tea bags

incidence: 0/10*, conc.: no contamination, sample origin: supermarkets in Valencia (city), Spain, sample year: unknown, country: Spain[373], *green tea bags

incidence: 2/8*, conc. range: 60.5–61 μg/l, ∅ conc.: 60.7 μg/l, sample origin: supermarkets in Valencia (city), Spain, sample year: unknown, country: Spain[373], *green mint tea bags

- Co-contamination: not reported
- Further contamination of plants and plant products other than cereals and their products reported in the present book/article[373]: Tea, AFB_2, AFG_1, AFG_2, and ENB, literature[373]

see also **Coffee** and **Medicinal plants**

Tempranillo see **Must**

Tigernuts see **Nuts (tigernut)**

Tikka masala see **Spices (tikka masala)**

Tobacco may contain the following mycotoxins:

***Aspergillus* Toxins**

Aflatoxin B_1

incidence: 1/40*, conc.: 15.5 μg/kg**, sample origin: Assiut and Sohag (governorates), Egypt, sample year: unknown, country: Egypt[654], *Cleopatra and Rothman cigarettes, **Cleopatra cigarette

- Co-contamination: not reported
- Further contamination of plants and plant products other than cereals and their products reported in the present book/article[654]: Tobacco, AFB_2, T-2, and ZEA, literature[654]

Aflatoxin B_2

incidence: 1/40*, conc.: 20.7 μg/kg**, sample origin: Assiut and Sohag (governorates), Egypt, sample year: unknown, country: Egypt[654], *Cleopatra and Rothman cigarettes, **Cleopatra cigarette

- Co-contamination: not reported
- Further contamination of plants and plant products other than cereals and their products reported in the present book/article[654]: Tobacco, AFB_1, T-2, and ZEA, literature[654]

Aspergillus and *Penicillium* Toxins

Penicillic Acid

incidence: 2/2*, conc. range: ~110–230 μg/kg, sample origin: commercial tobacco company, USA, sample year: unknown, country: USA[103], *commercial moldy tobacco

- Co-contamination: not reported
- Further contamination of plants and plant products other than cereals and their products reported in the present book/article[103]: not reported

Fusarium Toxins

Type A Trichothecenes

T-2 Toxin

incidence: 1/40*, conc.: 2.8 μg/kg**, sample origin: Assiut and Sohag (governorates), Egypt, sample year: unknown, country: Egypt[654], *Cleopatra and Rothman cigarettes, **Rothman cigarette

- Co-contamination: not reported
- Further contamination of plants and plant products other than cereals and their products reported in the present book/article[654]: Tobacco, AFB_1, AFB_2, and ZEA, literature[654]

Other *Fusarium* Toxins

Zearalenone

incidence: 1/40*, conc.: 5.5 μg/kg**, sample origin: Assiut and Sohag (governorates), Egypt, sample year: unknown, country: Egypt[654], *Cleopatra and Rothman cigarettes, **Cleopatra cigarette

- Co-contamination: not reported
- Further contamination of plants and plant products other than cereals and their products reported in the present book/article[654]: Tobacco, AFB_1, AFB_2, and T-2, literature[654]

Tobacco products see Products (tobacco product)

Toffees may contain the following mycotoxins:

Aspergillus Toxins

Aflatoxin B_1

incidence: 19/67*, conc. range: 2–400 μg/kg, sample origin: Mysore (city) market, India, sample year: October 1991–January 1992, country: India[199], *groundnut toffee (chikki), sweet snack

- Co-contamination: not reported
- Further contamination of plants and plant products other than cereals and their products reported in the present book/article[199]: Bondakadledkai, Congressbele, AFB_1 and AFB_2, literature[199]; Toffees, AFB_2, literature[199]

Aflatoxin B_2

incidence: incidence: 3/67*, conc. range: 3–120 μg/kg, sample origin: Mysore (city) market, India, sample year: October 1991–January 1992, country: India[199], *groundnut toffee (chikki), sweet snack

- Co-contamination: not reported
- Further contamination of plants and plant products other than cereals and their products reported in the present book/article[199]: Bondakadledkai, Congressbele, AFB_1 and AFB_2, literature[199]; Toffees, AFB_1, literature[199]

Tofu may contain the following mycotoxins:

Aspergillus Toxins

Aflatoxin B_1

incidence: 2/24*, conc. range: 1.7–6.3 μg/kg, ∅ conc.: 4.03 μg/kg, sample origin: small markets in several urban Shanghai (city) areas, China, sample year: April 1988–April 1989, country: China/USA[132], *tofu, partially dried

- Co-contamination: not reported
- Further contamination of plants and plant products other than cereals and their products reported in the present book/article[132]: Beans (kidney bean), Butter (peanut butter), Curd (bean curd), Nuts (peanut), Oils (rapeseed oil), Oils (soybean oil), Paste (bean paste), Sauces (soy sauce), Wine, miscellaneous, AFB_1, literature[132]

Tomato concentrate may contain the following mycotoxins:

Alternaria Toxins

Altenuene

incidence: 15/27, conc. range: 18.7–62.0 μg/kg, sample origin: local supermarkets, Belgium, sample year: February 2013–February 2015, country: Belgium/Austria/Germany[514]

- Co-contamination: For detailed information see the article.
- Further contamination of plants and plant products other than cereals and their products reported in the present book/article[514]: Juices (apple juice), Juices (grape juice), TA, literature[514]; Juices (tomato juice), Sauces (tomato sauce), ALT, AME, AOH, and TA, literature[514]; Tomato concentrate, AME, AOH, and TA, literature[514]; see also Juices (tomato juice), Sauces (tomato sauce), Tomato concentrate, chapter: Further Mycotoxins and Microbial Metabolites

For detailed information see the article.

Alternariol

incidence: 23/27, conc. range: <LOQ-31.0 μg/kg, sample origin: local supermarkets, Belgium, sample year: February 2013–February 2015, country: Belgium/Austria/Germany[514]

- Co-contamination: For detailed information see the article.
- Further contamination of plants and plant products other than cereals and their products reported in the present book/article[514]: Juices (apple juice), Juices (grape juice), TA, literature[514]; Juices (tomato juice), Sauces (tomato sauce), ALT, AME, AOH, and TA, literature[514]; Tomato concentrate, ALT, AME, and TA, literature[514]; see also Juices (tomato juice), Sauces (tomato sauce), Tomato concentrate, chapter: Further Mycotoxins and Microbial Metabolites

For detailed information see the article.

Alternariol Methyl Ether

incidence: 18/27, conc. range: <LOQ-6.10 μg/kg, sample origin: local supermarkets, Belgium, sample year: February 2013–February 2015, country: Belgium/Austria/Germany[514]

- Co-contamination: For detailed information see the article.
- Further contamination of plants and plant products other than cereals and their products reported in the present book/article[514]: Juices (apple juice), Juices (grape juice), TA, literature[514]; Juices (tomato juice), Sauces (tomato sauce), ALT, AME,

AOH, and TA, literature[514]; Tomato concentrate, ALT, AOH, and TA, literature[514]; see also Juices (tomato juice), Sauces (tomato sauce), Tomato concentrate, chapter: Further Mycotoxins and Microbial Metabolites

For detailed information see the article.

Tenuazonic Acid

incidence: 27/27, conc. range: <LOQ-174.3 µg/kg, sample origin: local supermarkets, Belgium, sample year: February 2013–February 2015, country: Belgium/Austria/Germany[514]

- Co-contamination: For detailed information see the article.
- Further contamination of plants and plant products other than cereals and their products reported in the present book/article[514]: Juices (apple juice), Juices (grape juice), TA, literature[514]; Juices (tomato juice), Sauces (tomato sauce), ALT, AME, AOH, and TA, literature[514]; Tomato concentrate, ALT, AME, and AOH, literature[514]; see also Juices (tomato juice), Sauces (tomato sauce), Tomato concentrate, chapter: Further Mycotoxins and Microbial Metabolites

For detailed information see the article.

see also **Puree**

Tomato juice see Juices (tomato juice)

Tomato ketchup may contain the following mycotoxins:

Alternaria Toxins

Alternariol

incidence: 1/1, conc.: 2.6 µg/kg, sample origin: unknown, sample year: unknown, country: Germany[46]

- Co-contamination: 1 sa. co-contaminated with AME and AOH
- Further contamination of plants and plant products other than cereals and their products reported in the present book/article[46]: Juices (tomato juice), Paste (tomato paste), Puree, Wine (red), Wine (white), AME and AOH, literature[46]; Juices (vegetable juice), Tomato ketchup, AME, literature[46]

incidence: 3/19, conc. range: 4–5 µg/kg, sample origin: Swiss market, Switzerland, sample year: autumn and winter 2010, country: Switzerland[47]

- Co-contamination: For detailed information see the article.
- Further contamination of plants and plant products other than cereals and their products reported in the present book/article[47]: Fruits (tomato), Puree, Sauces (tomato sauce), AME, AOH, and TA, literature[47]; Fruits, dried, ALT, AME, AOH, and TA, literature[47]; Tomato ketchup, AME and TA, literature[47]; see also Fruits (tomato), Fruits, dried, Puree, chapter: Further Mycotoxins and Microbial Metabolites

incidence: 18/18, conc. range: 1.0–5.0 µg/kg, ∅ conc.: 2.5 µg/kg, sample origin: retail shops, Hesse (federal state), Germany, sample year: 2009/2010, country: Germany[439]

- Co-contamination: not reported
- Further contamination of plants and plant products other than cereals and their products reported in the present book/article[439]: Juices (apple juice), Juice (tomato juice), Paste (tomato paste), Sauces (apple sauce), Wine (white), AOH, literature[439]

incidence: 14/31, conc. range: 2.5–300 µg/kg, ∅ conc.: 80 µg/kg, sample origin: local supermarkets and agricultural trade markets in Beijing (capital), China, sample year: unknown, country: China[883]

- Co-contamination: For detailed information see the article.
- Further contamination of plants and plant products other than cereals and their products reported in the present book/article[883]: Juices (citrus juice), Juices (tomato juice), Tomato ketchup, AME and TA, literature[883]; see also Juices (tomato juice), Tomato ketchup, chapter: Further Mycotoxins and Microbial Metabolites

For detailed information see the article.

incidence: 3/5, conc. range: ≤18 µg/kg, sample origin: retail stores, Netherlands, sample year: autumn 2014, country: Netherlands/Germany[906]

- Co-contamination: not reported
- Further contamination of plants and plant products other than cereals and their products reported in the present book/article[906]: Figs, Fruits (tomato), Juices (tomato juice), AOH and TA, literature[906]; Oils (sunflower oil), AME, literature[906]; Paste (tomato paste), Sauces (tomato sauce), Seeds (sunflower), AME, AOH, and TA, literature[906]; Tomato ketchup, AME and TA, literature[906]

Alternariol Methyl Ether

incidence: 1/1, conc.: 0.4 µg/kg, sample origin: unknown, sample year: unknown, country: Germany[46]

- Co-contamination: 1 sa. co-contaminated with AME and AOH
- Further contamination of plants and plant products other than cereals and their products reported in the present book/article[46]: Juices (tomato juice), Paste (tomato paste), Puree, Wine (red), Wine (white), AME and AOH, literature[46]; Juices (vegetable juice), AME, literature[46]; Tomato ketchup, AOH, literature[46]

incidence: 3/19, conc. range: 1 µg/kg, sample origin: Swiss market, Switzerland, sample year: autumn and winter 2010, country: Switzerland[47]

- Co-contamination: For detailed information see the article.
- Further contamination of plants and plant products other than cereals and their products reported in the present book/article[47]: Fruits (tomato), Puree, Sauces (tomato sauce), AME, AOH, and TA, literature[47]; Fruits, dried, ALT, AME, AOH, and TA, literature[47]; Tomato ketchup, AOH and TA, literature[47]; see also Fruits (tomato), Fruits, dried, Puree, chapter: Further Mycotoxins and Microbial Metabolites.

incidence: 28/31, conc. range: 0.32–38 µg/kg, ∅ conc.: 8.0 µg/kg, sample origin: local supermarkets and agricultural trade markets in Beijing (capital), China, sample year: unknown, country: China[883]

- Co-contamination: For detailed information see the article.
- Further contamination of plants and plant products other than cereals and their products reported in the present book/article[883]: Juices (citrus juice), Juices (tomato juice), AME and TA, literature[883]; Tomato ketchup, AOH and TA, literature[883]; see also Juices (tomato juice), Tomato ketchup, chapter: Further Mycotoxins and Microbial Metabolites

For detailed information see the article.

incidence: 1/5, conc.: 5.6 µg/kg, sample origin: retail stores, Netherlands, sample year: autumn 2014, country: Netherlands/Germany[906]

- Co-contamination: not reported
- Further contamination of plants and plant products other than cereals and their products reported in the present book/article[906]: Figs, Fruits (tomato), Juices (tomato juice), AOH and TA, literature[906]; Oils (sunflower oil), AME, literature[906]; Paste (tomato paste), Sauces (tomato sauce), Seeds (sunflower), AME, AOH, and TA, literature[906]; Tomato ketchup, AOH and TA, literature[906]

Tenuazonic Acid

incidence: 19/19, conc. range: 3–141 µg/kg, Ø conc.: 37 µg/kg, sample origin: Swiss market, Switzerland, sample year: autumn and winter 2010, country: Switzerland[47]

- Co-contamination: For detailed information see the article.
- Further contamination of plants and plant products other than cereals and their products reported in the present book/article[47]: Fruits (tomato), Puree, Sauces (tomato sauce), AME, AOH, and TA, literature[47]; Fruits, dried, ALT, AME, AOH, and TA, literature[47]; Tomato ketchup, AME and AOH, literature[47]; see also Fruits (tomato), Fruits, dried, Puree, chapter: Further Mycotoxins and Microbial Metabolites.

incidence: 2/18, conc. range: 55–67 µg/kg, Ø conc.: 61 µg/kg, sample origin: retail shops in the area of Giessen (town), Germany, sample year: unknown, country: Germany/Indonesia[442]

- Co-contamination: not reported
- Further contamination of plants and plant products other than cereals and their products reported in the present book/article[442]: Juices (apple juice), Juices (tomato juice), TA, literature[442]

incidence: 31/31, conc. range: 10.2–1,787 µg/kg, Ø conc.: 338 µg/kg, sample origin: local supermarkets and agricultural trade markets in Beijing (capital), China, sample year: unknown, country: China[883]

- Co-contamination: For detailed information see the article.
- Further contamination of plants and plant products other than cereals and their products reported in the present book/article[883]: Juices (citrus juice), Juices (tomato juice), AME and TA, literature[883]; Tomato ketchup, AME and AOH, literature[883]; see also Juices (tomato juice), Tomato ketchup, chapter: Further Mycotoxins and Microbial Metabolites

For detailed information see the article.

incidence: 2/5, conc. range: ≤138 µg/kg, sample origin: retail stores, Netherlands, sample year: autumn 2014, country: Netherlands/Germany[906]

- Co-contamination: not reported
- Further contamination of plants and plant products other than cereals and their products reported in the present book/article[906]: Figs, Fruits (tomato), Juices (tomato juice), AOH and TA, literature[906]; Oils (sunflower oil), AME, literature[906]; Paste (tomato paste), Sauces (tomato sauce), Seeds (sunflower), AME, AOH, and TA, literature[906]; Tomato ketchup, AME and AOH, literature[906]

Aspergillus **and** ***Penicillium*** **Toxins**

Ochratoxin A

incidence: 18/57, conc. range: ≤3.8 µg/l sample origin: different origins, Germany, sample year: unknown, country: Germany[110]

- Co-contamination: not reported
- Further contamination of plants and plant products other than cereals and their products reported in the present book/article[110]: Juices (blackcurrant juice), Juices (carrot juice), Juices (grape juice), Juices (tomato juice), Sauces (pepper- and barbecue sauce), Sauces (soy sauce), Spices (mustard), Vinegar (apple), Vinegar (balsamico), Vinegar (wine), Wine (red), Wine (rosé), Wine (white), OTA, literature[110]

Tomato puree see Puree

Tomato sauce see Sauces (tomato sauce)

Tomatoes see Fruits (tomato) and Fruits, dried

Traditional Chinese medicine see Medicinal plants

"Trockenbeerenauslese" see Wine and Wine (white)

Tubers may contain the following mycotoxins:

Aspergillus **Toxins**

Aflatoxin/s

incidence: 6/59*, conc. range: >30 to ≤440 µg/kg**, sample origin: markets in or near Manila (capital), Philippines, sample year: 1967–1969, country: USA[134], *included ubi, gabi, tugi, singkamas, and sweet potatoes, **AF

- Co-contamination: not reported
- Further contamination of plants and plant products other than cereals and their products reported in the present book/article[134]: Beans, Butter (peanut butter), Nuts (peanut), Products (coconut product), Products (peanut products), AF/S, literature[134]

Tugi see Tubers

Turmeric see Spices (curcuma)

Ubi see Tubers

Ugba may contain the following mycotoxins:

Alternaria **Toxins**

Alternariol Methyl Ether

incidence: 4/30*, conc. range: 25–193 µg/kg, Ø conc.: 109 µg/kg, sample origin: fermented food sellers, Southwest Nigeria, sample year: February 2015–July 2016, country: South Africa/Nigeria/Belgium/Kenya[330], *fermented African oil bean

- Co-contamination: For detailed information see the article.
- Further contamination of plants and plant products other than cereals and their products reported in the present book/article[330]: Iru, AFB_1, AFG_1, AFG_1, AFG_2, AME, DON, 3-AcDON, FB_1, FB_2, FB_3, FUS-X, HT-2, OTA, ROQ-C, T-2, and ZEA, literature[330]; Ogiri, AFB_1, AFB_2, AFG_2, AME, DAS, DON, HT-2, NEO, OTA, STG, and ZEA, literature[330]; Ugba, AFB_1, AFB_2, DON, ENB, HT-2, OTA, STG, T-2, and ZEA, literature[330]

Aspergillus **Toxins**

Aflatoxin B_1

incidence: 6/30*, conc. range: 3–36 µg/kg, Ø conc.: 20 µg/kg, sample origin: fermented food sellers, Southwest Nigeria, sample year: February 2015–July 2016, country: South Africa/Nigeria/Belgium/Kenya[330], *fermented African oil bean

- Co-contamination: For detailed information see the article.
- Further contamination of plants and plant products other than cereals and their products reported in the present book/article[330]: Iru, AFB_1, AFG_1, AFG_1, AFG_2, AME, DON, 3-AcDON, FB_1, FB_2, FB_3, FUS-X, HT-2, OTA, ROQ-C, T-2, and ZEA, literature[330]; Ogiri, AFB_1, AFB_2, AFG_2, AME, DAS, DON, HT-2, NEO, OTA, STG, and ZEA, literature[330]; Ugba, AFB_2, AME, DON, ENB, HT-2, OTA, STG, T-2, and ZEA, literature[330]

Aflatoxin B_2

incidence: 1/30*, conc.: <LOQ, sample origin: fermented food sellers, Southwest Nigeria, sample year: February 2015–July 2016, country: South Africa/Nigeria/Belgium/Kenya[330], *fermented African oil bean

- Co-contamination: For detailed information see the article.
- Further contamination of plants and plant products other than cereals and their products reported in the present book/article[330]: Iru, AFB_1, AFG_1, AFG_1, AFG_2, AME, DON, 3-AcDON, FB_1, FB_2, FB_3, FUS-X, HT-2, OTA, ROQ-C, T-2, and ZEA, literature[330]; Ogiri, AFB_1, AFB_2, AFG_2, AME, DAS, DON, HT-2, NEO, OTA, STG, and ZEA, literature[330]; Ugba, AFB_1, AME, DON, ENB, HT-2, OTA, STG, T-2, and ZEA, literature[330]

Other *Aspergillus* Toxins

Sterigmatocystin

incidence: 2/30*, conc. range: 22–27 µg/kg, Ø conc.: 25 µg/kg, sample origin: fermented food sellers, Southwest Nigeria, sample year: February 2015–July 2016, country: South Africa/Nigeria/Belgium/Kenya[330], *fermented African oil bean

- Co-contamination: For detailed information see the article.
- Further contamination of plants and plant products other than cereals and their products reported in the present book/article[330]: Iru, AFB_1, AFG_1, AFG_1, AFG_2, AME, DON, 3-AcDON, FB_1, FB_2, FB_3, FUS-X, HT-2, OTA, ROQ-C, T-2, and ZEA, literature[330]; Ogiri, AFB_1, AFB_2, AFG_2, AME, DAS, DON, HT-2, NEO, OTA, STG, and ZEA, literature[330]; Ugba, AFB_1, AFB_2, AME, DON, ENB, HT-2, OTA, T-2, and ZEA, literature[330]

Aspergillus and *Penicillium* Toxins

Ochratoxin A

incidence: 1/30*, conc.: 9 µg/kg, sample origin: fermented food sellers, Southwest Nigeria, sample year: February 2015–July 2016, country: South Africa/Nigeria/Belgium/Kenya[330], *fermented African oil bean

- Co-contamination: For detailed information see the article.
- Further contamination of plants and plant products other than cereals and their products reported in the present book/article[330]: Iru, AFB_1, AFG_1, AFG_1, AFG_2, AME, DON, 3-AcDON, FB_1, FB_2, FB_3, FUS-X, HT-2, OTA, ROQ-C, T-2, and ZEA, literature[330]; Ogiri, AFB_1, AFB_2, AFG_2, AME, DAS, DON, HT-2, NEO, OTA, STG, and ZEA, literature[330]; Ugba, AFB_1, AFB_2, AME, DON, ENB, HT-2, STG, T-2, and ZEA, literature[330]

Fusarium Toxins

Depsipeptides

Enniatin B

incidence: 4/30*, conc. range: <LOQ, sample origin: fermented food sellers, Southwest Nigeria, sample year: February 2015–July 2016, country: South Africa/Nigeria/Belgium/Kenya[330], *fermented African oil bean

- Co-contamination: For detailed information see the article.
- Further contamination of plants and plant products other than cereals and their products reported in the present book/article[330]: Iru, AFB_1, AFG_1, AFG_1, AFG_2, AME, DON, 3-AcDON, FB_1, FB_2, FB_3, FUS-X, HT-2, OTA, ROQ-C, T-2, and ZEA, literature[330]; Ogiri, AFB_1, AFB_2, AFG_2, AME, DAS, DON, HT-2, NEO, OTA, STG, and ZEA, literature[330]; Ugba, AFB_1, AFB_2, AME, DON, HT-2, OTA, STG, T-2, and ZEA, literature[330]

Type A Trichothecenes

HT-2 Toxin

incidence: 1/30*, conc.: 17 µg/kg, sample origin: fermented food sellers, Southwest Nigeria, sample year: February 2015–July 2016, country: South Africa/Nigeria/Belgium/Kenya[330], *fermented African oil bean

- Co-contamination: For detailed information see the article.
- Further contamination of plants and plant products other than cereals and their products reported in the present book/article[330]: Iru, AFB_1, AFG_1, AFG_1, AFG_2, AME, DON, 3-AcDON, FB_1, FB_2, FB_3, FUS-X, HT-2, OTA, ROQ-C, T-2, and ZEA, literature[330]; Ogiri, AFB_1, AFB_2, AFG_2, AME, DAS, DON, HT-2, NEO, OTA, STG, and ZEA, literature[330]; Ugba, AFB_1, AFB_2, AME, DON, ENB, OTA, STG, T-2, and ZEA, literature[330]

T-2 Toxin

incidence: 1/30*, conc.: 15 µg/kg, sample origin: fermented food sellers, Southwest Nigeria, sample year: February 2015–July 2016, country: South Africa/Nigeria/Belgium/Kenya[330], *fermented African oil bean

- Co-contamination: For detailed information see the article.
- Further contamination of plants and plant products other than cereals and their products reported in the present book/article[330]: Iru, AFB_1, AFG_1, AFG_1, AFG_2, AME, DON, 3-AcDON, FB_1, FB_2, FB_3, FUS-X, HT-2, OTA, ROQ-C, T-2, and ZEA, literature[330]; Ogiri, AFB_1, AFB_2, AFG_2, AME, DAS, DON, HT-2, NEO, OTA, STG, and ZEA, literature[330]; Ugba, AFB_1, AFB_2, AME, DON, ENB, HT-2, OTA, STG, and ZEA, literature[330]

Type B Trichothecenes

Deoxynivalenol

incidence: 4/30*, conc. range: 36–38 µg/kg, Ø conc.: 37 µg/kg, sample origin: fermented food sellers, Southwest Nigeria, sample year: February 2015–July 2016, country: South Africa/Nigeria/Belgium/Kenya[330], *fermented African oil bean

- Co-contamination: For detailed information see the article.
- Further contamination of plants and plant products other than cereals and their products reported in the present book/article[330]: Iru, AFB_1, AFG_1, AFG_1, AFG_2, AME, DON, 3-AcDON, FB_1, FB_2, FB_3, FUS-X, HT-2, OTA, ROQ-C, T-2, and ZEA, literature[330]; Ogiri, AFB_1, AFB_2, AFG_2, AME, DAS, DON, HT-2, NEO, OTA, STG, and ZEA, literature[330]; Ugba, AFB_1, AFB_2, AME, ENB, HT-2, OTA, STG, T-2, and ZEA, literature[330]

Other *Fusarium* Toxins

Zearalenone

incidence: 4/30*, conc. range: 39–117 µg/kg, Ø conc.: 72 µg/kg, sample origin: fermented food sellers, Southwest Nigeria, sample year: February 2015–July 2016, country: South Africa/Nigeria/Belgium/Kenya[330], *fermented African oil bean

- Co-contamination: For detailed information see the article.
- Further contamination of plants and plant products other than cereals and their products reported in the present book/article[330]: Iru, AFB_1, AFG_1, AFG_1, AFG_2, AME, DON, 3-AcDON, FB_1, FB_2, FB_3, FUS-X, HT-2, OTA, ROQ-C, T-2, and ZEA, literature[330]; Ogiri, AFB_1, AFB_2, AFG_2, AME, DAS, DON, HT-2, NEO, OTA, STG, and ZEA, literature[330]; Ugba, AFB_1, AFB_2, AME, DON, ENB, HT-2, OTA, STG, and T-2, literature[330]

Utiel-Requena see **Wine (red)** and **Wine (white)**

Vanilla extract see Extracts

Vegetables may contain the following mycotoxins:

Aspergillus **Toxins**

Aflatoxin/s

incidence: 1/140*, conc.: 46 μg/kg**, sample origin: market retailers, street vendors, wholesalers, and distributors, Thailand, sample year: September 1967–July 1969, country: Thailand[139], *fresh vegetables, **AFB_1, AFB_2, AFG_1, and AFG_2

- Co-contamination: not reported
- Further contamination of plants and plant products other than cereals and their products reported in the present book/article[139]: Beans, Beans (mung bean), Nuts (peanut), Food, Seeds (sesame), Spices (cayenne pepper, chilli), Spices (garlic/onion), Starch (cassava starch), AF/S, literature[139]

For detailed information see the article.

Vegetables (asparagus) may contain the following mycotoxins:

Fusarium **Toxins**

Depsipeptides

Beauvericin

incidence: 1/5*, conc: 24.77 μg/kg, sample origin: wasteland in Poznań (city), central Poland, sample year: May 2013, country: Poland[849], *spears of fresh asparagus plants

- Co-contamination: 1 sa. co-contaminated with BEA and FB_1
- Further contamination of plants and plant products other than cereals and their products reported in the present book/article[849]: Vegetables (asparagus), ENA_1, ENB, FB_1, FB_2, FB_3, and MON, literature[849]

incidence: 1/3*, conc: 103.64 μg/kg, sample origin: wasteland in Poznań (city), central Poland, sample year: May 2013, country: Poland[849], *top of fresh asparagus plants

- Co-contamination: 1 sa. co-contaminated with BEA, FB_1, FB_2, and MON
- Further contamination of plants and plant products other than cereals and their products reported in the present book/article[849]: Vegetables (asparagus), ENA_1, ENB, FB_1, FB_2, FB_3, and MON, literature[849]

incidence: 2/3*, conc. range: 42.54–47.57 μg/kg, Ø conc.: 45.1 μg/kg, sample origin: wasteland in Poznań (city), central Poland, sample year: May 2013, country: Poland[849], *base of fresh asparagus plants

- Co-contamination: 1 sa. co-contaminated with BEA, ENB, and FB_1; 1 sa. co-contaminated with BEA and FB_1
- Further contamination of plants and plant products other than cereals and their products reported in the present book/article[849]: Vegetables (asparagus), ENA_1, ENB, FB_1, FB_2, FB_3, and MON, literature[849]

For detailed information see the article.

Enniatin A_1

incidence: 0/5*, conc: no contamination, sample origin: wasteland in Poznań (city), central Poland, sample year: May 2013, country: Poland[849], *spears of fresh asparagus plants

- Co-contamination: not reported
- Further contamination of plants and plant products other than cereals and their products reported in the present book/article[849]: Vegetables (asparagus), BEA, ENB, FB_1, FB_2, FB_3, and MON, literature[849]

incidence: 1/3*, conc.: 43.49 μg/kg, sample origin: wasteland in Poznań (city), central Poland, sample year: May 2013, country: Poland[849], *top of fresh asparagus plants

- Co-contamination: 1 sa. co-contaminated with ENA_1, FB_1, and FB_2
- Further contamination of plants and plant products other than cereals and their products reported in the present book/article[849]: Vegetables (asparagus), BEA, ENB, FB_1, FB_2, FB_3, and MON, literature[849]

incidence: 0/3*, conc: no contamination, sample origin: wasteland in Poznań (city), central Poland, sample year: May 2013, country: Poland[849], *base of fresh asparagus plants

- Co-contamination: not reported
- Further contamination of plants and plant products other than cereals and their products reported in the present book/article[849]: Vegetables (asparagus), BEA, ENB, FB_1, FB_2, FB_3, and MON, literature[849]

For detailed information see the article.

Enniatin B

incidence: 0/5*, conc: no contamination, sample origin: wasteland in Poznań (city), central Poland, sample year: May 2013, country: Poland[849], *spears of fresh asparagus plants

- Co-contamination: not reported
- Further contamination of plants and plant products other than cereals and their products reported in the present book/article[849]: Vegetables (asparagus), BEA, ENA_1, FB_1, FB_2, FB_3, and MON, literature[849]

incidence: 0/3*, conc.: no contamination, sample origin: wasteland in Poznań (city), central Poland, sample year: May 2013, country: Poland[849], *top of fresh asparagus plants

- Co-contamination: not reported
- Further contamination of plants and plant products other than cereals and their products reported in the present book/article[849]: Vegetables (asparagus), BEA, ENA_1, FB_1, FB_2, FB_3, and MON, literature[849]

incidence: 2/3*, conc. range: 20.16–40.33 μg/kg, Ø conc.: 30.3 μg/kg, sample origin: wasteland in Poznań (city), central Poland, sample year: May 2013, country: Poland[849], *base of fresh asparagus plants

- Co-contamination: 1 sa. co-contaminated with BEA, ENB, and FB_1; 1 sa. co-contaminated with ENB and FB_1
- Further contamination of plants and plant products other than cereals and their products reported in the present book/article[849]: Vegetables (asparagus), BEA, ENA_1, FB_1, FB_2, FB_3, and MON, literature[849]

For detailed information see the article.

Fumonisin/s

Fumonisin B_1

incidence: 2/25*, conc. range: 460 μg/kg** and 7,400 μg/kg***, sample origin: asparagus producing area in Potenza (city, commune), southern Italy, sample year: unknown, country: Italy[83], *affected by crown rot, **stem, ***crown

- Co-contamination: 2 sa. co-contaminated with FB_1 and FB_2

- Further contamination of plants and plant products other than cereals and their products reported in the present book/article[83]: Vegetables (asparagus), FB_2, literature[83]

incidence: 9*/20?, conc. range: 0.5–5.6 µg/kg, ∅ conc.: 2.8 µg/kg, sample origin: area of Poznan (city), central Poland, sample year: 2002, country: Poland/Germany[84], *base of asparagus spears

- Co-contamination: 6 sa. co-contaminated with FB_1 and MON; 3 sa. contaminated solely with FB_1
- Further contamination of plants and plant products other than cereals and their products reported in the present book/article[84]: Vegetables (asparagus), MON, literature[84]

incidence: 4*/20?, conc. range: tr-2.1 µg/kg, sample origin: area of Poznan (city), central Poland, sample year: 2002, country: Poland/Germany[84], *center of asparagus spears

- Co-contamination: 4 sa. co-contaminated with FB_1 and MON
- Further contamination of plants and plant products other than cereals and their products reported in the present book/article[84]: Vegetables (asparagus), MON, literature[84]

incidence: 3*/20?, conc. range: 0.5–0.9 µg/kg, ∅ conc.: 0.73 µg/kg, sample origin: area of Poznan (city), central Poland, sample year: 2002, country: Poland/Germany[84], *top of asparagus spears

- Co-contamination: 3 sa. co-contaminated with FB_1 and MON
- Further contamination of plants and plant products other than cereals and their products reported in the present book/article[84]: Vegetables (asparagus), MON, literature[84]

incidence: 13*/20?, conc. range: 0.6–3.8 µg/kg, ∅ conc.: 1.57 µg/kg, sample origin: area of Poznan (city), central Poland, sample year: 2003, country: Poland/Germany[84], *base of asparagus spears

- Co-contamination: 8 sa. co-contaminated with FB_1 and MON; 5 sa. contaminated solely with FB_1
- Further contamination of plants and plant products other than cereals and their products reported in the present book/article[84]: Vegetables (asparagus), MON, literature[84]

incidence: 15*/20?, conc. range: tr-4.9 µg/kg, sample origin: area of Poznan (city), central Poland, sample year: 2003, country: Poland/Germany[84], *top of asparagus spears

- Co-contamination: 12 sa. co-contaminated with FB_1 and MON; 3 sa. contaminated solely with FB_1
- Further contamination of plants and plant products other than cereals and their products reported in the present book/article[84]: Vegetables (asparagus), MON, literature[84]

For detailed information see the article.

incidence: 2/20, conc. range: 2.6–2.8 µg/kg, ∅ conc.: 2.7 µg/kg, sample origin: local supermarkets and small retail shops in 7 major cities, Japan, sample year: winter 2007, country: Japan[85]

- Co-contamination: not reported
- Further contamination of plants and plant products other than cereals and their products reported in the present book/article[85]: beans (soybean) (FB_1 and FB_2), literature[85]; Beverages, Chocolate, Cocoa, Coffee, Coffee beans, Raisins, Spices (coriander), and Wine (OTA), literature[85]; Vegetables (asparagus), FB_2, literature[85]

incidence: 9/10*, conc. range: 36.4–366.5 µg/kg**, ∅ conc.: 178 µg/kg**, sample origin: fields of an asparagus-producing area in Rhineland-Palatinate (state), western Germany, sample year: July 2000, country: Germany[86], *asparagus spears damaged by rot and growth depression, **data of subsamples A based on dry weight

incidence: 9/10*, conc. range: 43.2–4,513.7 µg/kg**, ∅ conc.: 1,223 µg/kg**, sample origin: fields of an asparagus-producing area in Rhineland-Palatinate (state), western Germany, sample year: 2000, country: Germany[86], *asparagus spears damaged by rot and growth depression, **data of subsamples B based on dry weight

- Co-contamination: not reported
- Further contamination of plants and plant products other than cereals and their products reported in the present book/article[86]: not reported

Besides FB_1, FB_2, and FB_3 were found in 6 sa. but no declaration of concentration.

incidence: 24/30*, conc. range: 24–670 µg/kg**, ∅ conc.: 149 µg/kg**, sample origin: fields in CaizhuangWeiji, Kangzhuang, and Shangwacun (areas), Shandong (province), China, sample year: June 2004, country: China[87], *healthy asparagus spears, **based on dry weight

- Co-contamination: 24 sa. co-contaminated with FB_1 and FB_2
- Further contamination of plants and plant products other than cereals and their products reported in the present book/article[87]: Vegetables (asparagus), FB_2, literature[87]

incidence: 5/5*, conc. range: 2.25–19.52 µg/kg, ∅ conc.: 7.44 µg/kg, sample origin: wasteland in Poznań (city), central Poland, sample year: May 2013, country: Poland[849], *spears of fresh asparagus plants

- Co-contamination: 3 sa. co-contaminated with FB_1, FB_2, and FB_3; 1 sa. co-contaminated with FB_1 and FB_2; 1 sa. co-contaminated with BEA and FB_1
- Further contamination of plants and plant products other than cereals and their products reported in the present book/article[849]: Vegetables (asparagus), BEA ENA_1, ENB,, FB_2, FB_3, and MON, literature[849]

incidence: 3/3*, conc. range: 2.83–9.33 µg/kg, ∅ conc.: 6.22 µg/kg, sample origin: wasteland in Poznań (city), central Poland, sample year: May 2013, country: Poland[849], *top of fresh asparagus plants

- Co-contamination: 1 sa. co-contaminated with BEA, FB_1, FB_2, and MON; 1 sa. co-contaminated with ENA_1, FB_1, and FB_2; 1 sa. co-contaminated with FB_1 and MON
- Further contamination of plants and plant products other than cereals and their products reported in the present book/article[849]: Vegetables (asparagus), BEA, ENA_1, ENB, FB_2, FB_3, and MON, literature[849]

incidence: 3/3*, conc. range: 4.02–5.38 µg/kg, ∅ conc.: 4.87 µg/kg, sample origin: wasteland in Poznań (city), central Poland, sample year: May 2013, country: Poland[849], *base of fresh asparagus plants

- Co-contamination: 1 sa. co-contaminated with BEA, ENB, and FB_1; 1 sa. co-contaminated with BEA and FB_1; 1 sa. co-contaminated with ENB and FB_1
- Further contamination of plants and plant products other than cereals and their products reported in the present book/article[849]: Vegetables (asparagus), BEA, ENA_1, ENB, FB_2, FB_3, and MON, literature[849]

For detailed information see the article.

Fumonisin B_2

incidence: 2/25*, conc. range: 60 µg/kg** and 830 µg/kg***, sample origin: asparagus producing area in Potenza (city, commune), southern Italy, sample year: unknown, country: Italy[83], *affected by crown rot, **stem, ***crown

- Co-contamination: 2 sa. co-contaminated with FB_1 and FB_2
- Further contamination of plants and plant products other than cereals and their products reported in the present book/article[83]: Vegetables (asparagus), FB_1, literature[83]

incidence: 2/20, conc. range: 2.5 µg/kg, ∅ conc.: 2.5 µg/kg, sample origin: local supermarkets and small retail shops in 7 major cities, Japan, sample year: winter 2007, country: Japan[85]

- Co-contamination: not reported
- Further contamination of plants and plant products other than cereals and their products reported in the present book/article[85]: Beans (soybean), FB_1 and FB_2, literature[85]; Beverages, Chocolate, Cocoa, Coffee, Coffee beans, Raisins, Spices (coriander), Wine, OTA, literature[85]; Vegetables (asparagus), FB_1, literature[85]

incidence: 24/30*, conc. range: 17–138 µg/kg**, ∅ conc.: 43.9 µg/kg**, sample origin: fields in CaizhuangWeiji, Kangzhuang, and Shangwacun (areas), Shandong (province), China, sample year: June 2004, country: China[87], *healthy asparagus spears, **based on dry weight

- Co-contamination: 24 sa. co-contaminated with FB_1 and FB_2
- Further contamination of plants and plant products other than cereals and their products reported in the present book/article[87]: Vegetables (asparagus), FB_1, literature[87]

incidence: 4/5*, conc. range: 0.14–1.08 µg/kg, ∅ conc.: 0.64 µg/kg, sample origin: wasteland in Poznań (city), central Poland, sample year: May 2013, country: Poland[849], *spears of fresh asparagus plants

- Co-contamination: 3 sa. co-contaminated with FB_1, FB_2, and FB_3; 1 sa. co-contaminated with FB_1 and FB_2
- Further contamination of plants and plant products other than cereals and their products reported in the present book/article[849]: Vegetables (asparagus), BEA, ENA_1, ENB, FB_1, FB_3, and MON, literature[849]

incidence: 2/3*, conc. range: 0.38–1.05 µg/kg, ∅ conc.: 0.72 µg/kg, sample origin: wasteland in Poznań (city), central Poland, sample year: May 2013, country: Poland[849], *top of fresh asparagus plants

- Co-contamination: 1 sa. co-contaminated with ENA_1, FB_1, and FB_2; 1 sa. co-contaminated with FB_1 and FB_2
- Further contamination of plants and plant products other than cereals and their products reported in the present book/article[849]: Vegetables (asparagus), BEA, ENA_1, ENB, FB_1, FB_3, and MON, literature[849]

incidence: 0/3*, conc.: no contamination, sample origin: wasteland in Poznań (city), central Poland, sample year: May 2013, country: Poland[849], *base of fresh asparagus plants

- Co-contamination: not reported
- Further contamination of plants and plant products other than cereals and their products reported in the present book/article[849]: Vegetables (asparagus), BEA, ENA_1, ENB, FB_1, FB_3, and MON, literature[849]

For detailed information see the article.

Fumonisin B3

incidence: 3/5*, conc. range: 0.06–0.76 µg/kg, ∅ conc.: 0.34 µg/kg, sample origin: wasteland in Poznań (city), central Poland, sample year: May 2013, country: Poland[849], *spears of fresh asparagus plants

- Co-contamination: 3 sa. co-contaminated with FB_1, FB_2, and FB_3
- Further contamination of plants and plant products other than cereals and their products reported in the present book/article[849]: Vegetables (asparagus), BEA, ENA_1, ENB, FB_1, FB_2, and MON, literature[849]

incidence: 0/3*, conc.: no contamination, sample origin: wasteland in Poznań (city), central Poland, sample year: May 2013, country: Poland[849], *top of fresh asparagus plants

- Co-contamination: not reported
- Further contamination of plants and plant products other than cereals and their products reported in the present book/article[849]: Vegetables (asparagus), BEA, ENA_1, ENB, FB_1, FB_2, and MON, literature[849]

incidence: 0/3*, conc.: no contamination, sample origin: wasteland in Poznań (city), central Poland, sample year: May 2013, country: Poland[849], *base of fresh asparagus plants

- Co-contamination: not reported
- Further contamination of plants and plant products other than cereals and their products reported in the present book/article[849]: Vegetables (asparagus), BEA, ENA_1, ENB, FB_1, FB_2, and MON, literature[849]

For detailed information see the article.

Fumonisin/s

incidence: 6/51, conc. range: ≤49 µg/kg, sample origin: German market, Germany, sample year: 1999–2001, country: Germany[510]

- Co-contamination: not reported
- Further contamination of plants and plant products other than cereals and their products reported in the present book/article[510]: Products (soybean product), Sauces, FB/S, literature[510]

Other ***Fusarium*** Toxins

Moniliformin

incidence: 12*/20?, conc. range: tr-1,350 µg/kg, sample origin: area of Poznan (city), central Poland, sample year: 2002, country: Poland/Germany[84], *base of asparagus spears

- Co-contamination: 5 sa. co-contaminated with FB_1 and MON; 7 sa. contaminated solely with MON
- Further contamination of plants and plant products other than cereals and their products reported in the present book/article[84]: Vegetables (asparagus), FB_1, literature[84]

incidence: 8*/19, conc. range: 70–290 µg/kg, ∅ conc.: 139 µg/kg, sample origin: area of Poznan (city), central Poland, sample year: 2003, country: Poland/Germany[84], *base of asparagus spears

- Co-contamination: 8 sa. co-contaminated with FB_1 and MON
- Further contamination of plants and plant products other than cereals and their products reported in the present book/article[84]: Vegetables (asparagus), FB_1, literature[84]

incidence: 13*/19, conc. range: tr-360 µg/kg, sample origin: area of Poznan (city), central Poland, sample year: 2003, country: Poland/Germany[84], *top of asparagus spears

- Co-contamination: 12 sa. co-contaminated with FB_1 and MON; 1 sa. contaminated solely with MON)
- Further contamination of plants and plant products other than cereals and their products reported in the present book/article[84]: Vegetables (asparagus), FB_1, literature[84]

For detailed information see the article.

incidence: 0/5*, conc: no contamination, sample origin: wasteland in Poznań (city), central Poland, sample year: May 2013, country: Poland[849], *spears of fresh asparagus plants

- Co-contamination: not reported
- Further contamination of plants and plant products other than cereals and their products reported in the present book/article[849]: Vegetables (asparagus), BEA, ENA_1, ENB, FB_1, FB_2, and FB_3, literature[849]

incidence: 2/3*, conc. range: 10.54–15.96 μg/kg, Ø conc.: 13.25 μg/kg, sample origin: wasteland in Poznań (city), central Poland, sample year: May 2013, country: Poland[849], *top of fresh asparagus plants

- Co-contamination: 1 sa. co-contaminated with BEA, FB_1, FB_2, and MON; 1 sa. co-contaminated with FB_1 and MON
- Further contamination of plants and plant products other than cereals and their products reported in the present book/article[849]: Vegetables (asparagus), BEA, ENA_1, ENB, FB_1, FB_2, and FB_3, literature[849]

incidence: 0/3*, conc: no contamination, sample origin: wasteland in Poznań (city), central Poland, sample year: May 2013, country: Poland[849], *base of fresh asparagus plants

- Co-contamination: not reported
- Further contamination of plants and plant products other than cereals and their products reported in the present book/article[849]: Vegetables (asparagus), BEA, ENA_1, ENB, FB_1, FB_2, and FB_3, literature[849]

For detailed information see the article.

Vegetables (black radish) may contain the following mycotoxins:

Fusarium Toxins

Fumonisin/s

Fumonisin B_1

incidence: 1/6*, conc.: 4 μg/kg, sample origin: internet, Belgium, sample year: unknown, country: Belgium[77], *2 notified and 4 non-notified sa.

- Co-contamination: 1 sa. co-contaminated with FB_1 and FB_2,
- Further contamination of plants and plant products other than cereals and their products reported in the present book/article[77]: Medicinal plants, OTA, literature[77]; Soy isoflavones, FB_1 and OTA, literature[77]; Spices (garlic), FB_1, FB_2, FB_3, and OTA, literature[77]; Vegetables (black radish), FB_2, literature[77]

Fumonisin B_2

incidence: 1/6*, conc.: 2 μg/kg, sample origin: internet, Belgium, sample year: unknown, country: Belgium[77], *2 notified and 4 non-notified sa.

- Co-contamination: 1 sa. co-contaminated with FB_1 and FB_2
- Further contamination of plants and plant products other than cereals and their products reported in the present book/article[77]: Medicinal plants, OTA, literature[77]; Soy isoflavones, FB_1 and OTA, literature[77]; Spices (garlic), FB_1, FB_2, FB_3, and OTA, literature[77]; Vegetables (black radish), FB_1, literature[77]

Vegetables, dried may contain the following mycotoxins:

Aspergillus Toxins

Aflatoxin B_1

incidence: 3/12*, conc. range: 7–28 μg/kg, Ø conc.: 20 μg/kg, sample origin: El Atarin, Dokki Square, and Giza Market (districts) in Cairo (capital), Egypt, sample year: unknown, country: USA/Egypt[344], *included beans, freek**, and molokhia**, **contaminated

- Co-contamination: not reported
- Further contamination of plants and plant products other than cereals and their products reported in the present book/article[344]: Food, Medicinal plants, Spices, AFB_1, literature[344]

Aflatoxin/s

incidence: 3/100, conc. range: 2–20 μg/kg* (2 sa.), >20 μg/kg* (1 sa.), sample origin: imported and domestic sa. from Uruguay, sample year: 1993–1995, country: Uruguay[269], *AFB_1, AFB_2, AFG_1, and AFG_2

- Co-contamination: not reported
- Further contamination of plants and plant products other than cereals and their products reported in the present book/article[269]: Beans (soybean), Vegetables, dried, ZEA, literature[269]; Cocoa beans, AF/S, literature[269]; Fruits, dried, Oil seeds, AF/S and ZEA, literature[269]

Aspergillus and *Penicillium* Toxins

Ochratoxin A

incidence: 6/7*, conc. range: 245–7,444 μg/kg, sample origin: Tunisia, sample year: unknown, country: Tunisia/France[345], *chickpea, bean, and lentil?

- Co-contamination: not reported
- Further contamination of plants and plant products other than cereals and their products reported in the present book/article[345]: Bsissa, Olives, OTA, literature[345]

Fusarium Toxins

Zearalenone

incidence: 1/99, conc. range: >200 μg/kg, sample origin: imported and domestic sa. from Uruguay, sample year: 1993–1995, country: Uruguay[269], *vegetables (dried)

- Co-contamination: not reported
- Further contamination of plants and plant products other than cereals and their products reported in the present book/article[269]: Beans (soybean), ZEA, literature[269]; Cocoa beans, Vegetables, dried, AF/S, literature[269]; Fruits, dried, Oil seeds, AF/S and ZEA, literature[269]

Vermouth see Wine, miscellaneous

Vine fruits, dried may contain the following mycotoxins:

Aspergillus Toxins

Aflatoxin B_1

incidence: 10/71, conc. range: <LOQ-17.5 μg/kg*, Ø conc.: 16.5 μg/kg*, sample origin: local and of Algeria, Chile, Iran, Israel, and Turkey purchased from markets in Valencia (city), Spain, and Tunis (capital), Tunisia, sample year: 2012/2013, country: Spain[720], *purchased in Tunisia

- Co-contamination: For detailed information see the article.
- Further contamination of plants and plant products other than cereals and their products reported in the present book/article[720]: Dates, AFB_2, AFG_1, AFG_2, DAS, ENA_1, ENB, ENB_1, and OTA, literature[720]; Figs, AFG_1, BEA, DAS, ENA_1, ENB, ENB_1, HT-2, and OTA, literature[720]; Fruits, dried, AFG_2, DAS, ENA_1, ENB, ENB_1, HT-2, and OTA, literature[720]; Vine fruits, dried, AFG_2, ENA, ENA_1, ENB, ENB_1, HT-2, and OTA, literature[720]

For detailed information see the article.

Aflatoxin G_2

incidence: 6/71, conc. range: <LOQ-3.1 µg/kg*, Ø conc.: 2.35 µg/kg*, sample origin: local and of Algeria, Chile, Iran, Israel, and Turkey purchased from markets in Valencia (city), Spain, and Tunis (capital), Tunisia, sample year: 2012/2013, country: Spain[720], *purchased in Spain and Tunisia

- Co-contamination: For detailed information see the article.
- Further contamination of plants and plant products other than cereals and their products reported in the present book/article[720]: Dates, AFB_2, AFG_1, AFG_2, DAS, ENA_1, ENB, ENB_1, and OTA, literature[720]; Figs, AFG_1, BEA, DAS, ENA_1, ENB, ENB_1, HT-2, and OTA, literature[720]; Fruits, dried, AFG_2, DAS, ENA_1, ENB, ENB_1, HT-2, and OTA, literature[720]; Vine fruits, dried, AFB_1, ENA, ENA_1, ENB, ENB_1, HT-2, and OTA, literature[720]

For detailed information see the article.

Aspergillus and *Penicillium* Toxins

Ochratoxin A

incidence: 2/3*, conc. range: 0.28–0.74 µg/kg, sample origin: available in New Zealand, sample year: 2007, country: New Zealand[277], *raisins, sultanas, and currants

incidence: 5/10*, conc. range: 0.2–0.6 µg/kg, sample origin: available in New Zealand, sample year: 2008/2009, country: New Zealand[277];*raisins, sultanas, and currants

- Co-contamination: not reported
- Further contamination of plants and plant products other than cereals and their products reported in the present book/article[277]: Bars, AFB_1, AF/S, DON, NIV, and OTA, literature[277]; Butter (peanut butter), Confectionery, Food, Fruits, dried, Nuts, Nuts (peanut), Nuts (pistachio nut), Sauces (peanut sauce), AFB_1 and AF/S, literature[277]; Coffee, Dates, Spices (nutmeg), Spices (pepper), OTA, literature[277]; Figs, Spices (cayenne pepper, chilli), Spices (curry), Spices (ginger), Spices (paprika), AFB_1, AF/S, and OTA, literature[277]; Nuts (Brazil nut, paranut), AFB_1, literature[277]

incidence: 6*/27, conc. range: <LOQ-9.87 µg/kg*, sample origin: imported from Chile and Turkey to Italy, sample year: January 2008–December 2009, country: Italy[320], *contaminated sa. from Chile (1) and Turkey (5)

- Co-contamination: not reported
- Further contamination of plants and plant products other than cereals and their products reported in the present book/article[320]: Coffee beans, OTA, literature[320]; Kernels (apricot), AFB_1, AFB_2, and AFG_1, literature[320]; Nuts (hazelnut), Paste (hazelnut paste), AFB_1, AFB_2, AFG_1, and AFG_2, literature[320]; Nuts (peanut), Nuts (pistachio nut), AFB_1 and AFB_2, literature[320]

incidence: 21/31*, conc. range: 1.5–14 µg/kg, Ø conc.: 2.29 µg/kg, sample origin: Mendoza and San Juan (provinces), Argentina, sample year: unknown, country: Argentina/Brazil[647], *dried black vine fruits without seeds

incidence: 16/19*, conc. range: 1.4–7.5 µg/kg, Ø conc.: 3.23 µg/kg, sample origin: Mendoza and San Juan (provinces), Argentina, sample year: unknown, country: Argentina/Brazil[647], *dried white vine fruits without seeds

- Co-contamination: not reported
- Further contamination of plants and plant products other than cereals and their products reported in the present book/article[647]: not reported

incidence: 5/10*, conc. range: 0.2–0.6 µg/kg, sample origin: New Zealand, sample year: 2009, country: New Zealand[702], *raisins, sultanas, and currants

- Co-contamination: not reported
- Further contamination of plants and plant products other than cereals and their products reported in the present book/article[702]: Figs, Spices (cayenne pepper, chilli), Spices (curry), Spices (ginger), Spices (paprika), AF/S and OTA, literature[702]; Spices (pepper), OTA, literature[702]; Fruits, dried, AF/S, literature[702]

incidence: 18/71, conc. range: <LOQ-5.5 µg/kg*, Ø conc.: 3.03 µg/kg*, sample origin: local and of Algeria, Chile, Iran, Israel, and Turkey purchased from markets in Valencia (city), Spain, and Tunis (capital), Tunisia, sample year: 2012/2013, country: Spain[720], *purchased in Spain and Tunisia

- Co-contamination: For detailed information see the article.
- Further contamination of plants and plant products other than cereals and their products reported in the present book/article[720]: Dates, AFB_2, AFG_1, AFG_2, DAS, ENA_1, ENB, ENB_1, and OTA, literature[720]; Figs, AFG_1, BEA, DAS, ENA_1, ENB, ENB_1, HT-2, and OTA, literature[720]; Fruits, dried, AFG_2, DAS, ENA_1, ENB, ENB_1, HT-2, and OTA, literature[720]; Vine fruits, dried, AFB_1, AFG_2, ENA, ENA_1, ENB, ENB_1, and HT-2, literature[720]

For detailed information see the article.

incidence: 10/48*, conc. range: 0.26–20.28 µg/kg, Ø conc.: 3.75 µg/kg, sample origin: San Juan Province, Argentina, sample year: unknown, country: Argentina[816], *black dried vine fruits of muscatel variety without seed grapes and respected export quality criteria

- Co-contamination: not reported
- Further contamination of plants and plant products other than cereals and their products reported in the present book/article[816]: Must, Wine (red), OTA, literature[816]

Fusarium Toxins

Depsipeptides

Enniatin A

incidence: 6/71, conc. range: 19.4–60.8 µg/kg*, Ø conc.: 27.2 µg/kg*, sample origin: local and of Algeria, Chile, Iran, Israel, and Turkey purchased from markets in Valencia (city), Spain, and Tunis (capital), Tunisia, sample year: 2012/2013, country: Spain[720], *purchased in Tunisia

- Co-contamination: For detailed information see the article.
- Further contamination of plants and plant products other than cereals and their products reported in the present book/article[720]: Dates, AFB_2, AFG_1, AFG_2, DAS, ENA_1, ENB, ENB_1, and OTA, literature[720]; Figs, AFG_1, BEA, DAS, ENA_1, ENB, ENB_1, HT-2, and OTA, literature[720]; Fruits, dried, AFG_2, DAS, ENA_1, ENB, ENB_1, HT-2, and OTA, literature[720]; Vine fruits, dried, AFB_1, AFG_2, ENA_1, ENB, ENB_1, HT-2, and OTA, literature[720]

For detailed information see the article.

Enniatin A_1

incidence: 34/71, conc. range: <LOQ-155 µg/kg*, Ø conc.: 12.2 µg/kg*, sample origin: local and of Algeria, Chile, Iran, Israel, and Turkey purchased from markets in Valencia (city), Spain, and Tunis (capital), Tunisia, sample year: 2012/2013, country: Spain[720], *purchased in Spain and Tunisia

- Co-contamination: For detailed information see the article.

- Further contamination of plants and plant products other than cereals and their products reported in the present book/article[720]: Dates, AFB_2, AFG_1, AFG_2, DAS, ENA_1, ENB, ENB_1, and OTA, literature[720]; Figs, AFG_1, BEA, DAS, ENA_1, ENB, ENB_1, HT-2, and OTA, literature[720]; Fruits, dried, AFG_2, DAS, ENA_1, ENB, ENB_1, HT-2, and OTA, literature[720]; Vine fruits, dried, AFB_1, AFG_2, ENA, ENB, ENB_1, HT-2, and OTA, literature[720]

For detailed information see the article.

Enniatin B

incidence: 42/71, conc. range: <LOQ-4,251 µg/kg*, ∅ conc.: 273 µg/kg*, sample origin: local and of Algeria, Chile, Iran, Israel, and Turkey purchased from markets in Valencia (city), Spain, and Tunis (capital), Tunisia, sample year: 2012/2013, country: Spain[720], *purchased in Spain and Tunisia

- Co-contamination: For detailed information see the article.
- Further contamination of plants and plant products other than cereals and their products reported in the present book/article[720]: Dates, AFB_2, AFG_1, AFG_2, DAS, ENA_1, ENB, ENB_1, and OTA, literature[720]; Figs, AFG_1, BEA, DAS, ENA_1, ENB, ENB_1, HT-2, and OTA, literature[720]; Fruits, dried, AFG_2, DAS, ENA_1, ENB, ENB_1, HT-2, and OTA, literature[720]; Vine fruits, dried, AFB_1, AFG_2, ENA, ENA_1, ENB_1, HT-2, and OTA, literature[720]

For detailed information see the article.

Enniatin B_1

incidence: 21/71, conc. range: <LOQ-454 µg/kg*, ∅ conc.: 53.9 µg/kg*, sample origin: local and of Algeria, Chile, Iran, Israel, and Turkey purchased from markets in Valencia (city), Spain, and Tunis (capital), Tunisia, sample year: 2012/2013, country: Spain[720], *purchased in Spain and Tunisia

- Co-contamination: For detailed information see the article.
- Further contamination of plants and plant products other than cereals and their products reported in the present book/article[720]: Dates, AFB_2, AFG_1, AFG_2, DAS, ENA_1, ENB, ENB_1, and OTA, literature[720]; Figs, AFG_1, BEA, DAS, ENA_1, ENB, ENB_1, HT-2, and OTA, literature[720]; Fruits, dried, AFG_2, DAS, ENA_1, ENB, ENB_1, HT-2, and OTA, literature[720]; Vine fruits, dried, AFB_1, AFG_2, ENA, ENA_1, ENB, HT-2, and OTA, literature[720]

For detailed information see the article.

Type A Trichothecenes

HT-2 Toxin

incidence: 5/71, conc. range: <LOQ-8.5 µg/kg*, ∅ conc.: 6.45 µg/kg*, sample origin: local and of Algeria, Chile, Iran, Israel, and Turkey purchased from markets in Valencia (city), Spain, and Tunis (capital), Tunisia, sample year: 2012/2013, country: Spain[720], *purchased in Spain and Tunisia

- Co-contamination: For detailed information see the article.
- Further contamination of plants and plant products other than cereals and their products reported in the present book/article[720]: Dates, AFB_2, AFG_1, AFG_2, DAS, ENA_1, ENB, ENB_1, and OTA, literature[720]; Figs, AFG_1, BEA, DAS, ENA_1, ENB, ENB_1, HT-2, and OTA, literature[720]; Fruits, dried, AFG_2, DAS, ENA_1, ENB, ENB_1, HT-2, and OTA, literature[720]; Vine fruits, dried, AFB_1, AFG_2, ENA, ENA_1, ENB, ENB_1, and OTA, literature[720]

For detailed information see the article.

see also **Currants, Fruits, dried, Raisins** and **Sultanas**

Vinegar may contain the following mycotoxins:

Aspergillus and ***Penicillium*** Toxins

Ochratoxin A

incidence: 1/6, conc.: 0.80 µg/kg, sample origin: unknown, sample year: unknown, country: Japan[112]

- Co-contamination: not reported
- Further contamination of plants and plant products other than cereals and their products reported in the present book/article[112]: Cassis, Juices, Raisins, Wine, OTA, literature[112]; Cocoa, Coffee, Food, OTA and OTB, literature[112]; Powder (cocoa powder), AFB_1, AFB_2, AFG_1, OTA, and OTB, literature[112]; Vinegar, OTB, literature[112]

incidence: 44/87, conc. range: LOD/LOQ-0.9 µg/kg (33 sa.), 1.0–2.9 µg/kg (8 sa.), ≤4.35 µg/kg (3 sa.), sample origin: Germany, sample year: 1995–1998, country: EU[698]

- Co-contamination: not reported
- Further contamination of plants and plant products other than cereals and their products reported in the present book/article[698]: Baby food, Beans (soybean), Chocolate, Cocoa, Coffee, Coffee beans, Currants, Fruits, dried, Herbs, Jam, Juices, Juices (grape juice), Legumes, Nuts, Oil seeds, Oils (olive oil), Peas, Peas (chickpea), Products (plant product), Raisins, Spices, Spices (cayenne pepper, chilli), Spices (nutmeg), Spices (paprika), Spices (pepper), Sultanas, Tea, Wine, Wine (red), Wine (rosé), Wine (white), Wine, miscellaneous, OTA, literature[698]

Ochratoxin B

incidence: 1/6, conc.: 0.29 µg/kg, sample origin: unknown, sample year: unknown, country: Japan[112]

- Co-contamination: not reported
- Further contamination of plants and plant products other than cereals and their products reported in the present book/article[112]: Cassis, Juices, Raisins, Vinegar, Wine, OTA, literature[112]; Cocoa, Coffee, Food, OTA and OTB, literature[112]; Powder (cocoa powder), AFB_1, AFB_2, AFG_1, OTA, and OTB, literature[112]

Patulin

incidence: 1/2, conc.: 14.2 µg/l, sample origin: local market in Taipei (capital), Taiwan, sample year: unknown, country: Taiwan[458]

- Co-contamination: not reported
- Further contamination of plants and plant products other than cereals and their products reported in the present book/article[458]: Juices (apple juice), PAT, literature[458]

Vinegar (apple) may contain the following mycotoxins:

Aspergillus and ***Penicillium*** Toxins

Ochratoxin A

incidence: 1/18*, conc.: <0.01 µg/l, sample origin: different origins, Germany, sample year: unknown, country: Germany[110], *apple and fruit vinegar

- Co-contamination: not reported
- Further contamination of plants and plant products other than cereals and their products reported in the present book/article[110]: Juices (blackcurrant juice), Juices (carrot juice), Juices (grape juice), Juices (tomato juice), Sauces (pepper- and barbecue sauce), Sauces (soy sauce), Spices (mustard), Tomato ketchup, Vinegar (balsamico), Vinegar (wine), Wine (red), Wine (rosé), Wine (white), OTA, literature[110]

Patulin
incidence: 0/2* **, conc.: no contamination, sample origin: markets, supermarkets, hypermarkets, or organic food shops, Italy, sample year: November 2003–February 2004, country: Italy[35], *conventional, **apple vinegar

incidence: 2/6* **, conc. range: >LOQ<10 μg/kg (2 sa., maximum: 4.2 μg/kg), sample origin: markets, supermarkets, hypermarkets, or organic food shops, Italy, sample year: November 2003–February 2004, country: Italy[35], *organic, **apple vinegar
- Co-contamination: not reported
- Further contamination of plants and plant products other than cereals and their products reported in the present book/article[35]: Fruits (apple), Juices (apple juice), Juices (pear juice), Puree, PAT, literature[35]

Vinegar (balsamico) may contain the following mycotoxins:

Aspergillus **and *Penicillium*** Toxins

Ochratoxin A
incidence: 21/29, conc. range: ≤4.35 μg/l, sample origin: different origins, Germany, sample year: unknown, country: Germany[110]
- Co-contamination: not reported
- Further contamination of plants and plant products other than cereals and their products reported in the present book/article[110]: Juices (blackcurrant juice), Juices (carrot juice), Juices (grape juice), Juices (tomato juice), Sauces (pepper- and barbecue sauce), Sauces (soy sauce), Spices (mustard), Tomato ketchup, Vinegar (apple), Vinegar (wine), Wine (red), Wine (rosé), Wine (white), OTA, literature[110]

incidence: 15/15, conc. range: 0.008–0.046 μg/l (12 sa.), 0.102–0.252 μg/l (3 sa.*), sample origin: imported from southern European countries and Morocco bought in local supermarkets and small shops, France, sample year: summer 1999, country: Greece/France[111], *balsamico vinegar sa.
- Co-contamination: not reported
- Further contamination of plants and plant products other than cereals and their products reported in the present book/article[111]: Wine (red), OTA, literature[111]

Vinegar (wine) may contain the following mycotoxins:

Aspergillus **and *Penicillium*** Toxins

Ochratoxin A
incidence: 19/38, conc. range: ≤1.9 μg/l, sample origin: different origins, Germany, sample year: unknown, country: Germany[110]
- Co-contamination: not reported
- Further contamination of plants and plant products other than cereals and their products reported in the present book/article[110]: Juices (blackcurrant juice), Juices (carrot juice), Juices (grape juice), Juices (tomato juice), Sauces (pepper- and barbecue sauce), Sauces (soy sauce), Spices (mustard), Tomato ketchup, Vinegar (apple), Vinegar (balsamico), Wine (red), Wine (rosé), Wine (white), OTA, literature[110]

Vinho Verde see Wine, miscellaneous

"Vollreiflese" see Wine (red)

Walnuts see Nuts (walnut)

Weaning food see Baby food

"Weisser Riesling" see Wine (white)

Wine may contain the following mycotoxins:

Aspergillus Toxins

Aflatoxin B_1
incidence: 5/22*, conc. range: 0.4–1 μg/l**, ∅ conc.: 0.78 μg/l, sample origin: different countries, sample year: unknown, country: Finland[332], *different kinds of wine, **contaminated sa. from Algeria, Greece, Portugal, Spain, and Yugoslavia (only >LOQ concentrations mentioned)
- Co-contamination: 5 sa. co-contaminated with AFB_1, AFB_2, AFG_1, and AFG_2
- Further contamination of plants and plant products other than cereals and their products reported in the present book/article[332]: Wine, AFB_2, AFG_1, and AFG_2, literature[332]

For detailed information see the article.

Aflatoxin B_2
incidence: 5/22*, conc. range: 0.1–0.3 μg/l**, ∅ conc.: 0.2 μg/l, sample origin: different countries, sample year: unknown, country: Finland[332], *different kinds of wine, **contaminated sa. from Algeria, Greece, Portugal, Spain, and Yugoslavia (only >LOQ concentrations mentioned)
- Co-contamination: 5 sa. co-contaminated with AFB_1, AFB_2, AFG_1, and AFG_2
- Further contamination of plants and plant products other than cereals and their products reported in the present book/article[332]: Wine, AFB_1, AFG_1, and AFG_2, literature[332]

For detailed information see the article.

Aflatoxin G_1
incidence: 5/22*, conc. range: 0.3–1 μg/l**, ∅ conc.: 0.66 μg/l, sample origin: different countries, sample year: unknown, country: Finland[332], *different kinds of wine, **contaminated sa. from Algeria, Greece, Portugal, Spain, and Yugoslavia (only >LOQ concentrations mentioned)
- Co-contamination: 5 sa. co-contaminated with AFB_1, AFB_2, AFG_1, and AFG_2
- Further contamination of plants and plant products other than cereals and their products reported in the present book/article[332]: Wine, AFB_1, AFB_2, and AFG_2, literature[332]

For detailed information see the article.

Aflatoxin G_2
incidence: 5/22*, conc. range: 0.1–0.3 μg/l**, ∅ conc.: 0. 18 μg/l, sample origin: different countries, sample year: unknown, country: Finland[332], *different kinds of wine, **contaminated sa. from Algeria, Greece, Portugal, Spain, and Yugoslavia (only >LOQ concentrations mentioned)
- Co-contamination: 5 sa. co-contaminated with AFB_1, AFB_2, AFG_1, and AFG_2,
- Further contamination of plants and plant products other than cereals and their products reported in the present book/article[332]: Wine, AFB_1, AFB_2, and AFG_1, literature[332]

For detailed information see the article.

Aspergillus **and *Penicillium*** Toxins

Ochratoxin A
incidence: 15/83, conc. range: ≤1.29 μg/kg, ∅ conc.: 0.29 μg/kg, sample origin: local supermarkets and small retail shops in 7 major cities, Japan, sample year: summer 2004–winter 2007, country: Japan[85]

- Co-contamination: not reported
- Further contamination of plants and plant products other than cereals and their products reported in the present book/article[85]: Beans (soybean), Vegetables (asparagus), FB_1 and FB_2, literature[85]; Beverages, Chocolate, Cocoa, Coffee, Coffee beans, Spices (coriander), Raisins, OTA, literature[85]

incidence: 1/1, conc.: 0.02 μg/l, sample origin: supermarkets and fast-food chains in Quebec (city), Canada, sample year: September/October 2008, country: Canada[90]

incidence: 0/1, conc.: no contamination, sample origin: supermarkets and fast-food chains in Calgary (city), Canada, sample year: September/October 2009, country: Canada[90]

- Co-contamination: not reported
- Further contamination of plants and plant products other than cereals and their products reported in the present book/article[90]: Baby food, Beans, Beverages, Butter (peanut butter), Candies, Chocolate, Coffee, Extracts, Herbs, Ice cream, Nuts, Pies, Pizza, Products, Pudding, Raisins, Seeds, OTA, literature[90]

incidence: 1/14, conc.: 0.15 μg/kg, sample origin: unknown, sample year: unknown, country: Japan[112]

- Co-contamination: not reported
- Further contamination of plants and plant products other than cereals and their products reported in the present book/article[112]: Cassis, Juices, Raisins, OTA, literature[112]; Cocoa, Coffee, Food, Vinegar, OTA and OTB, literature[112]; Powder (cocoa powder), AFB_1, AFB_2, AFG_1, OTA, and OTB, literature[112]

incidence: 8*/10, conc. range: 0.02–0.72 μg/kg, ∅ conc.: 0.26 μg/kg, sample origin: local supermarkets and small retail shops across Japan, sample year: summer 2004–winter 2005, country: Japan[146], *domestic and imported products

- Co-contamination: not reported
- Further contamination of plants and plant products other than cereals and their products reported in the present book/article[146]: Beans (green bean), Coffee, Raisins, OTA, literature[146]; Butter (peanut butter), AFB_1, AFB_2, AFG_1, and AFG_2, literature[146]; Chocolate, AFB_1 and OTA, literature[146]

incidence: 6/18*, conc. range: ≤0.11 μg/l, sample origin: unknown, sample year: unknown, country: Switzerland[333], *white, red (mainly contaminated), and rosé wines

- Co-contamination: not reported
- Further contamination of plants and plant products other than cereals and their products reported in the present book/article[333]: Drinks, OTA, literature[333]

incidence: 1/117*, conc.: 0.02 μg/l, sample origin: Austrian wines, Austria, sample year: unknown, country: Austria[334], *red and white wine and "Prädikatswein"

- Co-contamination: not reported
- Further contamination of plants and plant products other than cereals and their products reported in the present book/article[334]: not reported

incidence: 77/107*, conc. range: ≤1.10 μg/l, sample origin: supermarkets and wine shops in Piedmont (region), Italy, sample year: 2003/2004 (production year), country: Italy[336], *red and white wine (included Barbera, Cortese, Dolcetto, Erbaluce, and Roero Arneis)

- Co-contamination: not reported
- Further contamination of plants and plant products other than cereals and their products reported in the present book/article[336]: not reported

For detailed information see the article.

incidence: 14/18*, conc. range: LOQ-0.1 μg/l (7 sa.), 0.1–0.5 μg/l (7 sa., maximum: 0.420 μg/l), ∅ conc.: 0.088 μg/l, sample origin: wineries in Piedmont (region), Italy, sample year: 2000, country: Italy[337], *red and white wine

incidence: 17/33*, conc. range: LOQ-0.1 μg/l (8 sa.), 0.1–0.5 μg/l (8 sa.), 0.870 μg/l (1 sa.), ∅ conc.: 0.071 μg/l, sample origin: wineries in Piedmont (region), Italy, sample year: 2001, country: Italy[337], *red and white wine

incidence: 28/46*, conc. range: LOQ-0.1 μg/l (13 sa.), 0.1–0.5 μg/l (13 sa.), 0.5–2 μg/l (2 sa., maximum: 0.810 μg/l), ∅ conc.: 0.089 μg/l, sample origin: wineries in Piedmont (region), Italy, sample year: 2002, country: Italy[337], *red and white wine

incidence: 122/162*, conc. range: LOQ-0.1 μg/l (87 sa.), 0.1–0.5 μg/l (33 sa.), 0.5–2 μg/l (2 sa., maximum: 1.390 μg/l), ∅ conc.: 0.069 μg/l, sample origin: wineries in Piedmont (region), Italy, sample year: 2003, country: Italy[337], *red and white wine

incidence: 104/151*, conc. range: LOQ-0.1 μg/l (63 sa.), 0.1–0.5 μg/l (36 sa.), 0.5–2 μg/l (5 sa., maximum: 0.736 μg/l), ∅ conc.: 0.081 μg/l, sample origin: wineries in Piedmont (region), Italy, sample year: 2004, country: Italy[337], *red and white wine

incidence: 166/305*, conc. range: LOQ-0.1 μg/l (103 sa.), 0.1–0.5 μg/l (57 sa.), 0.5–2 μg/l (5 sa.), 2.630 μg/l (1 sa.), ∅ conc.: 0.068 μg/l, sample origin: wineries in Piedmont (region), Italy, sample origin: wineries in Piedmont (region), Italy, sample year: 2005, country: Italy[337], *red and white wine

incidence: 298/396*, conc. range: LOQ-0.1 μg/l (197 sa.), 0.1–0.5 μg/l (88 sa.), 0.5–2 μg/l (12 sa.), 2.010 μg/l (1 sa.), ∅ conc.: 0.096 μg/l, sample origin: wineries in Piedmont (region), Italy, sample year: 2006, country: Italy[337], *red and white wine

incidence: 71/95*, conc. range: LOQ-0.1 μg/l (68 sa.), 0.1–0.5 μg/l (1 sa.), 0.5–2 μg/l (2 sa., maximum: 1.360 μg/l), ∅ conc.: 0.048 μg/l, sample origin: wineries in Piedmont (region), Italy, sample year: 2007, country: Italy[337], *red and white wine

- Co-contamination: not reported
- Further contamination of plants and plant products other than cereals and their products reported in the present book/article[337]: Wine (red), Wine (white), OTA, literature[337]

For detailed information see the article.

incidence: 8/10*, conc. range: 0.005–0.021 μg/l, ∅ conc.: 0.0148 μg/l, sample origin: coastal Croatia, sample year: 2007, country: Croatia[535], *red wine, partly bottled

- Co-contamination: not reported
- Further contamination of plants and plant products other than cereals and their products reported in the present book/article[535]: Must, OTA, literature[535]

incidence: 83/117*, conc. range: ≤2.00 μg/l, sample origin: different winemaking regions of Greece, sample year: 1999–2006, country: Greece/the UK[621], *conventional

incidence: 21/33*, conc. range: ≤1.48 μg/l, sample origin: different winemaking regions of Greece, sample year: 1999–2006, country: Greece/the UK[621], *organic

- Co-contamination: not reported
- Further contamination of plants and plant products other than cereals and their products reported in the present book/article[621]: Wine (red), Wine (rosé), Wine (white), OTA, literature[621]

For detailed information see the article.

incidence: 1/9*, conc.: 1.1 μg/l**, sample origin: home-made wine and from local supermarkets in Almeria (city), Spain, sample year: unknown, country: Spain[627], *included red, rosé, and white wine, **rosé wine

- Co-contamination: not reported
- Further contamination of plants and plant products other than cereals and their products reported in the present book/article[627]: not reported

incidence: 14/34*, conc. range: LOD/LOQ-≤0.36 µg/kg (14 sa.), sample origin: France, sample year: 1998/1999, country: EU[698], *domestic wine (all types)

incidence: 50/70*, conc. range: LOD/LOQ-0.9 µg/kg (47 sa.), ≤1.64 µg/kg (3 sa.), sample origin: France, sample year: 1998/1999, country: EU[698], *imported wine (all types)

- Co-contamination: not reported
- Further contamination of plants and plant products other than cereals and their products reported in the present book/article[698]: Baby food, Beans (soybean), Chocolate, Cocoa, Coffee, Coffee beans, Currants, Fruits, dried, Herbs, Jam, Juices, Juices (grape juice), Legumes, Nuts, Oil seeds, Oils (olive oil), Peas, Peas (chickpea), Products (plant product), Raisins, Spices, Spices (cayenne pepper, chilli), Spices (nutmeg), Spices (paprika), Spices (pepper), Sultanas, Tea, Vinegar, Wine (red), Wine (rosé), Wine (white), Wine, miscellaneous, OTA, literature[698]

incidence: 7/30*, conc. range: 0.092–0.446 µg/l, ∅ conc.: 0.235 µg/l, sample origin: Romanian viticultural regions, Romania, sample year: 2009/2010, country: Romania[732], *included red, white, dry and sweet wine

- Co-contamination: not reported
- Further contamination of plants and plant products other than cereals and their products reported in the present book/article[732]: not reported

For detailed information see the article.

incidence: 25/80, conc. range: 0.021–0.030 µg/l (14 sa.), 0.031–0.040 µg/l (8 sa.), 0.041–0.050 µg/l (3 sa.), sample origin: Argentinian, Brazilian, and Chilean sa. purchased in supermarkets, Rio de Janeiro (state), Brazil, sample year: 2002–2008, country: Brazil[770]

- Co-contamination: not reported
- Further contamination of plants and plant products other than cereals and their products reported in the present book/article[770]: not reported

incidence: 34/1188*, conc. range: LOD-LOQ (13 sa.), ≤0.35 µg/l (21 sa.), sample origin: minor stores, wine shops, wineries, and supermarkets, Chile, sample year: 2007–2009, country: Chile[870],*red and white wine

- Co-contamination: not reported
- Further contamination of plants and plant products other than cereals and their products reported in the present book/article[870]: not reported

For detailed information see the article.

Patulin

incidence: 5/7*, conc. range: 4.6–31.4 µg/l, sample origin: Mosel-Saar-Ruwer (region), Germany, sample year: 2006 vintage, country: Germany[916], *"QbA"

incidence: 3/6*, conc. range: 7.2–48.5 µg/l, sample origin: Mosel-Saar-Ruwer (region), Germany, sample year: 2006 vintage, country: Germany[916], *"Kabinett"

incidence: 6/19*, conc. range: 4.6–31.8 µg/l, sample origin: Mosel-Saar-Ruwer (region), Germany, sample year: 2006 vintage, country: Germany[916], * "Spätlese"

incidence: 14/28*, conc. range: 3.5–80 µg/l, sample origin: Mosel-Saar-Ruwer (region), Germany, sample year: 2006 vintage, country: Germany[916], * "Auslese"

incidence: 18/27*, conc. range: 5.7–20.3 µg/l, sample origin: Mosel-Saar-Ruwer (region), Germany, sample year: 2006 vintage, country: Germany[916], *"Beerenauslese"

incidence: 4/5*, conc. range: 17.8–35.3 µg/l, sample origin: Mosel-Saar-Ruwer (region), Germany, sample year: 2006 vintage, country: Germany[916], *"Trockenbeerenauslese"

incidence: 2/4*, conc. range: 5.3–12.1 µg/l, sample origin: Mosel-Saar-Ruwer (region), Germany, sample year: 2006 vintage, country: Germany[916],*unknown quality level

- Co-contamination: not reported
- Further contamination of plants and plant products other than cereals and their products reported in the present book/article[916]: not reported

For detailed information see the article.

Wine (red) may contain the following mycotoxins:

Alternaria Toxins

Alternariol

incidence: 1/1*, conc.: 4.9 µg/kg, sample origin: unknown, sample year: unknown, country: Germany[46], * "Vollreiflese"

- Co-contamination: 1 sa. co-contaminated with AOH and AME
- Further contamination of plants and plant products other than cereals and their products reported in the present book/article[46]: Juices (tomato juice), Paste (tomato paste), Puree, Tomato ketchup, Wine (white), AME and AOH, literature[46]; Juices (vegetable juice), Wine (red), AME, literature[46]

incidence: 5/5, conc. range: 0.36–7.50 µg/kg, sample origin: commercially available, sample year: unknown, country: Germany[160]

- Co-contamination: 5 sa. co-contaminated with AME and AOH
- Further contamination of plants and plant products other than cereals and their products reported in the present book/article[160]: Beverages, Juices (apple juice), Juices (grape juice), Juices (orange juice), Juices (tomato juice), Juices (vegetable juice), Wine (white), Wine, miscellaneous, AME and AOH, literature[160]; Wine (red), AME, literature[160]

incidence: 1/1, conc.: 1.9 µg/l, sample origin: unknown, sample year: unknown, country: Canada[440]

- Co-contamination: not reported
- Further contamination of plants and plant products other than cereals and their products reported in the present book/article[440]: Juices (apple juice), Juices (grape juice), Nectars (cranberry nectar), Nectars (prune nectar), AME and AOH, literature[440], Juices (raspberry juice), AOH, literature[440]

incidence: 13/17, conc. range: 0.03–5.02 µg/l, ∅ conc.: 1.3 µg/l, sample origin: domestic, sample year: unknown, country: Canada[492]

- Co-contamination: 13 sa. co-contaminated with AME and AOH
- Further contamination of plants and plant products other than cereals and their products reported in the present book/article[492]: Juices (cranberry juice), Juices (grape juice), Wine (white), AME and AOH, literature[492]; Wine (red), AME, literature[492]

incidence: 7/7, conc. range: 0.27–19.4 µg/l, ∅ conc.: 4.7 µg/l, sample origin: imported from Argentina, Chile, Italy, and the USA, sample year: unknown, country: Canada[492]
- Co-contamination: 6 sa. co-contaminated with AME and AOH; 1 sa. contaminated solely with AOH
- Further contamination of plants and plant products other than cereals and their products reported in the present book/article[492]: Juices (cranberry juice), Juices (grape juice), Wine (white), AME and AOH, literature[492]; Wine (red), AME, literature[492]

incidence: 6/56, conc. range: <LOQ-13 µg/l, sample origin: supermarkets, grocery stores, wineries, and small stores in Entre Rios, Mendoza, and San Juan (provinces), Argentina, sample year: 2011, country: Argentina[498]
- Co-contamination: not reported
- Further contamination of plants and plant products other than cereals and their products reported in the present book/article[498]: Juices (grapefruit juice), Juices (orange juice), Wine (rosé), AOH, literature[498]; Wine (red), AME, literature[498]; Wine (white), AME and AOH, literature[498]

incidence: 1*/5, conc.: 11 µg/kg, sample origin: Dutch survey program (2013) and retail shops across the Netherlands, sample year: 2013?, country: Netherlands[568], *bottled in 2011
- Co-contamination: not reported
- Further contamination of plants and plant products other than cereals and their products reported in the present book/article[568]: Figs, Olives, Seeds (sunflower), Wine (red), TA, literature[568]; Fruits (apple), AOH, literature[568]; Sauces (tomato sauce), AME, AOH, and TA, literature[568]; see also Seeds (sunflower), chapter: Further Mycotoxins and Microbial Metabolites

incidence: 3/62*, conc. range: 6.4–12 µg/kg, sample origin: imported from South Africa purchased in supermarkets, wine stores, and importers, Netherlands, sample year: 2009, country: Brazil/Netherlands[814], *Cabernet Sauvignon
- Co-contamination: not reported
- Further contamination of plants and plant products other than cereals and their products reported in the present book/article[814]: Wine (red), CPA, MA, OTA, and ZEA, literature[814]

incidence: 1/1*, conc.: 34 µg/kg, sample origin: private households of staff members, Austria, sample year: unknown, country: Austria[850], *moldy
- Co-contamination: 1 sa. co-contaminated with AOH, AME, MA, and TEN.
- Further contamination of plants and plant products other than cereals and their products reported in the present book/article[850]: Fruits (apple), Nuts (chestnut), AME and AOH, literature[850]; Fruits (lemon), Puree, MA, literature[850]; Jam, AOH and MA, literature[850]; Spices (garlic), BEA, FB_1, FB_2, and FB_3, literature[850]; Wine (red), AME and MA, literature[850]; see also Fruits (lemon), Jam, Nuts (chestnut), Wine (red), chapter: Further Mycotoxins and Microbial Metabolites

Alternariol Methyl Ether

incidence: 1/1*, conc.: 0.3 µg/kg, sample origin: unknown, sample year: unknown, country: Germany[46], * "Vollreiflese"
- Co-contamination: 1 sa. co-contaminated with AOH and AME
- Further contamination of plants and plant products other than cereals and their products reported in the present book/article[46]: Juices (tomato juice), Paste (tomato paste), Puree, Tomato ketchup, Wine (white), AME and AOH, literature[46]; Juices (vegetable juice), AME, literature[46]; Wine (red), AOH, literature[46]

incidence: 5/5, conc. range: pr.-0.15 µg/kg, sample origin: commercially available, sample year: unknown, country: Germany[160]
- Co-contamination: 5 sa. co-contaminated with AME and AOH
- Further contamination of plants and plant products other than cereals and their products reported in the present book/article[160]: Beverages, Juices (apple juice), Juices (grape juice), Juices (orange juice), Juices (tomato juice), Juices (vegetable juice), Wine (white), Wine, miscellaneous, AME and AOH, literature[160]; Wine (red), AOH, literature[160]

incidence: 13/17, conc. range: 0.01–0.23 µg/l, ∅ conc.: 0.119 µg/l, sample origin: domestic, sample year: unknown, country: Canada[492]
- Co-contamination: 13 sa. co-contaminated with AME and AOH
- Further contamination of plants and plant products other than cereals and their products reported in the present book/article[492]: Juices (cranberry juice), Juices (grape juice), Wine (white), AME and AOH, literature[492]; Wine (red), AOH, literature[492]

incidence: 6/7, conc. range: 0.01–0.19 µg/l, ∅ conc.: 0.067 µg/l, sample origin: imported from Argentina, Chile, Italy, and the USA, sample year: unknown, country: Canada[492]
- Co-contamination: 6 sa. co-contaminated with AME and AOH
- Further contamination of plants and plant products other than cereals and their products reported in the present book/article[492]: Juices (cranberry juice), Juices (grape juice), Wine (white), AME and AOH, literature[492]; Wine (red), AOH, literature[492]

incidence: 1/56, conc.: <LOQ, sample origin: supermarkets, grocery stores, wineries, and small stores in Entre Rios, Mendoza, and San Juan (provinces), Argentina, sample year: 2011, country: Argentina[498]
- Co-contamination: not reported
- Further contamination of plants and plant products other than cereals and their products reported in the present book/article[498]: Juices (grapefruit juice), Juices (orange juice), Wine (red), Wine (rosé), AOH, literature[498]; Wine (white), AME and AOH, literature[498]

incidence: 1/1*, conc.: 0.5 µg/kg, sample origin: private households of staff members, Austria, sample year: unknown, country: Austria[850], *moldy
- Co-contamination: 1 sa. co-contaminated with AOH, AME, MA, and TEN
- Further contamination of plants and plant products other than cereals and their products reported in the present book/article[850]: Fruits (apple), Nuts (chestnut), AME and AOH, literature[850]; Fruits (lemon), Puree, MA, literature[850]; Jam, Wine (red), AOH and MA, literature[850]; Spices (garlic), BEA, FB_1, FB_2, and FB_3, literature[850]; see also Fruits (lemon), Jam, Nuts (chestnut), Wine (red), chapter: Further Mycotoxins and Microbial Metabolites

Tenuazonic Acid

incidence: 3/5, conc. range: ≤46 µg/kg, ∅ conc.: 25 µg/kg, sample origin: Dutch survey program (2013) and retail shops across the Netherlands, sample year: 2013?, country: Netherlands[568]

- Co-contamination: not reported
- Further contamination of plants and plant products other than cereals and their products reported in the present book/article[568]: Figs, Olives, Seeds (sunflower), TA, literature[568]; Fruits (apple), Wine (red), AOH, literature[568]; Sauces (tomato sauce), AME, AOH, and TA, literature[568]; see also Seeds (sunflower), chapter: Further Mycotoxins and Microbial Metabolites

Aspergillus Toxins

Aflatoxin B_1

incidence: 1/57*, conc.: 0.15 µg/l, sample origin: local retails, Sicily (island, autonomous region), Italy, sample year: 2009–2013 vintages, country: Italy[708], *red wine, bottled

- Co-contamination: 1 sa. co-contaminated with AFB_1, AFG_1, and OTA
- Further contamination of plants and plant products other than cereals and their products reported in the present book/article[708]: Wine (red), AFB_2, AFG_1, AFG_2, OTA, and OTB, literature[708]

Aflatoxin B_2

incidence: 21/24, conc. range: 1.25–25.73 µg/l, Ø conc.: 7.89 µg/l, sample origin: local markets, Spain, sample year: unknown, country: Spain[602]

- Co-contamination: not reported
- Further contamination of plants and plant products other than cereals and their products reported in the present book/article[602]: not reported

incidence: 4/57*, conc. range: 0.02–0.08 µg/l, Ø conc.: 0.04 µg/l, sample origin: local retails, Sicily (island, autonomous region), Italy, sample year: 2009–2013 vintages, country: Italy[708], *red wine, bottled

- Co-contamination: 2 sa. co-contaminated with AFB_2, OTA, and OTB; 1 sa. co-contaminated with AFB_2, AFG_1, OTA, and OTB; 1 sa. co-contaminated with AFB_2 and OTA
- Further contamination of plants and plant products other than cereals and their products reported in the present book/article[708]: Wine (red), AFB_1, AFG_1, AFG_2, OTA, and OTB, literature[708]

Aflatoxin G_1

incidence: 21/57*, conc. range: 0.02–0.13 µg/l, Ø conc.: 0.05 µg/l, sample origin: local retails, Sicily (island, autonomous region), Italy, sample year: 2009–2013 vintages, country: Italy[708], *red wine, bottled

- Co-contamination: 4 sa. co-contaminated with AFG_1 and OTA; 3 sa. co-contaminated with AFG_1 and AFG_2; 2 sa. co-contaminated with AFG_1, AFG_2, OTA, and OTB; 2 sa. co-contaminated with AFG_1, OTA, and OTB; 1 sa. co-contaminated with AFB_1, AFG_1, and OTA; 1 sa. co-contaminated with AFB_2, AFG_1, OTA, and OTB; 8 sa. contaminated solely with AFG_1
- Further contamination of plants and plant products other than cereals and their products reported in the present book/article[708]: Wine (red), AFB_1, AFB_2, AFG_2, OTA, and OTB, literature[708]

Aflatoxin G_2

incidence: 8/57*, conc. range: 0.04–0.06 µg/l, Ø conc.: 0.05 µg/l, sample origin: local retails, Sicily (island, autonomous region), Italy, sample year: 2009–2013 vintages, country: Italy[708], *red wine, bottled

- Co-contamination: 3 sa. co-contaminated with AFG_1 and AFG_2; 3 sa. co-contaminated with AFG_2, OTA, and OTB; 2 sa. co-contaminated with AFG_1, AFG_2, OTA, and OTB
- Further contamination of plants and plant products other than cereals and their products reported in the present book/article[708]: Wine (red), AFB_1, AFB_2, AFG_1, OTA, and OTB, literature[708]

Aspergillus and *Penicillium* Toxins

Cyclopiazonic Acid

incidence: 8/68*, conc. range: 1.2–2.7 µg/kg, sample origin: imported from South Africa purchased in supermarkets, wine stores, and importers, the Netherlands, sample year: 2008, country: Brazil/the Netherlands[814], *Cabernet Sauvignon

- Co-contamination: not reported
- Further contamination of plants and plant products other than cereals and their products reported in the present book/article[814]: Wine (red), AOH, MA, OTA, and ZEA, literature[814]

Ochratoxin A

incidence: 79/172, conc. range: ≤7.0 µg/l, sample origin: importers, producers, stores, wine and sparkling wine cellars, Germany, sample year: unknown, country: Germany[110]

- Co-contamination: not reported
- Further contamination of plants and plant products other than cereals and their products reported in the present book/article[110]: Juices (blackcurrant juice), Juices (carrot juice), Juices (grape juice), Juices (tomato juice), Sauces (pepper- and barbecue sauce), Sauces (soy sauce), Spices (mustard), Tomato ketchup, Vinegar (apple), Vinegar (balsamico), Vinegar (wine), Wine (rosé), Wine (white), OTA, literature[110]

incidence: 31/31, conc. range: 0.009–0.100 µg/l (22 sa.*), 0.551–0.892 µg/l (3 sa.**), 1.54 µg/l (1 sa.***), 1.59–3.40 µg/l (5 sa.****), sample origin: imported from southern European countries and Morocco bought in local supermarkets and small shops, France, sample year: summer 1999, country: Greece/France[111], *sa. from Mediterranean sea countries, **sa. from Italy and Morocco, ***sa. from Communauté Europeène, ****sa. from Greece and France

- Co-contamination: not reported
- Further contamination of plants and plant products other than cereals and their products reported in the present book/article[111]: Vinegar (balsamico), OTA, literature[111]

incidence: 4/10, conc. range: 0.7–1.1 µg/kg, Ø conc.: 0.95 µg/kg, sample origin: retail outlets across the UK, sample year: August 1996, country: the UK[143]

- Co-contamination: not reported
- Further contamination of plants and plant products other than cereals and their products reported in the present book/article[143]: Beans, Legumes, Powder (cocoa powder), OTA, literature[143]

incidence: 28/50, conc. range: 0.01–0.19 µg/kg (22 sa.), 0.2–0.4 µg/kg (4 sa.), 0.41–1.0 µg/kg (2 sa., maximum: 0.8 µg/kg), sample origin: 13 regions in the UK, sample year: May 1998, country: the UK[263]

- Co-contamination: not reported
- Further contamination of plants and plant products other than cereals and their products reported in the present book/article[263]: Chocolate, Currants, Figs, Juices (grape juice), Powder (cocoa powder), Raisins, Sultanas, OTA, literature[263]

incidence: 1/2, conc.: 0.1–0.3 µg µg/l (1 sa.), sample origin: domestic, purchased from supermarkets in 8 counties, Taiwan, sample year: spring and fall 2003, country: Taiwan[284]

incidence: 4/8, conc. range: 0.1–0.3 µg/l (2 sa.), 0.3–0.5 µg/l (2 sa.), sample origin: imported, purchased from supermarkets in 8 counties, Taiwan, sample year: spring and fall 2003, country: Taiwan[284]

- Co-contamination: not reported
- Further contamination of plants and plant products other than cereals and their products reported in the present book/article[284]: Coffee, OTA, literature[284]

incidence: 3/10, conc. range: 0.02–0.16 µg/l, ∅ conc.: 0.08 µg/l, sample origin: supermarkets and small shops in the northeast of China, sample year: January–June 2010, country: China[302]

- Co-contamination: not reported
- Further contamination of plants and plant products other than cereals and their products reported in the present book/article[302]: Berries (grape), Coffee, OTA, literature[302]

incidence: 71/104*, conc. range: ≤2.69 µg/l, sample origin: manufacturers' stock and Greek retail market, Greece, sample year: 1995–1999, country: Greece[327], *red wine, dry

- Co-contamination: not reported
- Further contamination of plants and plant products other than cereals and their products reported in the present book/article[327]: Currants, Sultanas, Wine (rosé), Wine (white), Wine, miscellaneous, OTA, literature[327]

For detailed information see the article.

incidence: 9/35*, conc. range: tr (8 sa.), 1.23 µg/l (1 sa.), sample origin: homemade wines and from retail shops, Portugal, sample year: unknown, country: Portugal[335]

- Co-contamination: not reported
- Further contamination of plants and plant products other than cereals and their products reported in the present book/article[335]: Wine (white), OTA, literature[335]

incidence: 695/1002, conc. range: LOQ-0.1 µg/l (449 sa.), 0.1–0.5 µg/l (218 sa.), 0.5–2 µg/l (26 sa.), >2 µg/l (2 sa., maximum: 2.630 µg/l), ∅ conc.: 0.121 µg/l, sample origin: wineries in Piedmont (region), Italy, sample year: 2004–2008, country: Italy[337]

- Co-contamination: not reported
- Further contamination of plants and plant products other than cereals and their products reported in the present book/article[337]: Wine, Wine (white), OTA, literature[337]

incidence: 3/10*, conc. range: 0.0283–0.0424 µg/l, ∅ conc.: 0.03536 µg/l, sample origin: from Brazil at markets in Rio de Janeiro (state), Brazil, sample year: unknown, country: Brazil/Argentina[377], *table wines

incidence: 2/7*, conc. range: 0.0283–0.0424 µg/l, ∅ conc.: 0.03535 µg/l, sample origin: imported from Argentina at markets in Rio de Janeiro (state), Brazil, sample year: unknown, country: Brazil/Argentina[377], *table wines

incidence: 2/5*, conc. range: 0.0283–0.0707 µg/l, ∅ conc.: 0.0495 µg/l, sample origin: imported from Chile at markets in Rio de Janeiro (state), Brazil, sample year: unknown, country: Brazil/Argentina[377], *table wines

incidence: 1/4*, conc.: 0.0424 µg/l, sample origin: imported from Spain at markets in Rio de Janeiro (state), Brazil, sample year: unknown, country: Brazil/Argentina[377], *table wines

incidence: 2/5*, conc. range: 0.0283 µg/l, ∅ conc.: 0.0283 µg/l, sample origin: imported from France at markets in Rio de Janeiro (state), Brazil, sample year: unknown, country: Brazil/Argentina[377], *table wines

incidence: 3/5*, conc. range: 0.0283–0.0424 µg/l, ∅ conc.: 0.033 µg/l, sample origin: imported from Italy at markets in Rio de Janeiro (state), Brazil, sample year: unknown, country: Brazil/Argentina[377], *table wines

incidence: 3/7*, conc. range: 0.0283–0.0567 µg/l, ∅ conc.: 0.0401 µg/l, sample origin: imported from Portugal at markets in Rio de Janeiro (state), Brazil, sample year: unknown, country: Brazil/Argentina[377], *table wines

- Co-contamination: not reported
- Further contamination of plants and plant products other than cereals and their products reported in the present book/article[377]: Juices (grape juice), Pulp (grape pulp), Wine (rosé), Wine (white), OTA, literature[377]

incidence: 20/20, conc. range: 0.04–3.24 µg/l, ∅ conc.: 0.912 µg/l, sample origin: supermarkets in Rabat (capital), Morocco, sample year: unknown, country: Morocco/France[493]

- Co-contamination: not reported
- Further contamination of plants and plant products other than cereals and their products reported in the present book/article[493]: Juices, Wine (rosé), Wine (white), OTA, literature[493]

incidence: 49/53, conc. range: 0.0022–6.71 µg/l, ∅ conc.: 0.039 µg/l, sample origin: imported from retail shops in Warsaw (capital), Poland, sample year: unknown, country: Poland[494]

- Co-contamination: not reported
- Further contamination of plants and plant products other than cereals and their products reported in the present book/article[494]: Juices (grape juice), OTA, literature[494]

For detailed information see the article.

incidence: 5/36, conc. range: LOD-0.049 µg/l (3 sa.), 0.050–0.500 µg/l (2 sa., maximum: 0.393 µg/l), sample origin: Canada, sample year: 1999–2002, country: Canada[495]

incidence: 26/48, conc. range: LOD-0.049 µg/l (6 sa.), 0.050–0.500 µg/l (15 sa.), >0.500 µg/l (5 sa., maximum: 2.320 µg/l), sample origin: imported, Canada, sample year: 1999–2002, country: Canada[495]

- Co-contamination: not reported
- Further contamination of plants and plant products other than cereals and their products reported in the present book/article[495]: Juices (grape juice), Wine (white), OTA, literature[495]

incidence: 60/77?, conc. range: ≤0.388 µg/l, sample origin: partly imported from retail outlets in Berne (capital), Switzerland, sample year: 1994/1995, country: Switzerland[496]

- Co-contamination: not reported
- Further contamination of plants and plant products other than cereals and their products reported in the present book/article[496]: Juices (grape juice), Wine (rosé), Wine (white), Wine, miscellaneous, OTA, literature[496]

incidence: 21/21, conc. range: ≤1.30 µg/l*, ∅ conc.: 0.39 µg/l, sample origin: imported from Bulgaria (1 sa.), Italy (9 sa.), and Tunisia* (11 sa.), Germany, sample year: 1999/2000, country: Germany[497]

- Co-contamination: not reported
- Further contamination of plants and plant products other than cereals and their products reported in the present book/article[497]: Juices (grape juice), OTA, literature[497]

incidence: 24/130*, conc. range: 0.06–4.24 µg/l, Ø conc.: 0.25 µg/l, sample origin: supermarkets and small shops, Spain, sample year: 2002, country: Spain[508], *included Costers del Segre, Penedés, Rioja, Utiel-Requena, and ordinary wine

- Co-contamination: not reported
- Further contamination of plants and plant products other than cereals and their products reported in the present book/article[508]: Must, Wine (white), Wine, miscellaneous, OTA, literature[508]

incidence: 9/14*, conc. range: 0.02–0.5 µg/l (7 sa.), >2.0 µg/l (2 sa., maximum: 2.51 µg/l), sample origin: viticultural regions, Greece, sample year: 1995–1997, 1999, country: Greece/France[509], *red wine, dry

incidence: 2/3*, conc. range: 0.02–0.5 µg/l (2 sa.), sample origin: viticultural regions, Greece, sample year: 1995–1997, 1999, country: Greece/France[509], *red wine, sweet

- Co-contamination: not reported
- Further contamination of plants and plant products other than cereals and their products reported in the present book/article[509]: Wine (white), OTA, literature[509]

For detailed information see the article.

incidence: 84/91, conc. range: ≤0.603 µg/l, sample origin: partly from different European countries and Spain, sample year: 1997, country: Spain[511]

- Co-contamination: not reported
- Further contamination of plants and plant products other than cereals and their products reported in the present book/article[511]: Wine (rosé), Wine (white), Wine, miscellaneous, OTA, literature[511]

incidence: 1/14*, conc.: 0.021 µg/l, sample origin: wine-producing regions of South Africa, sample year: unknown, country: South Africa[512], *Cabernet Sauvignon

incidence: 1/5*, conc.: 0.031 µg/l, sample origin: wine-producing regions of South Africa, sample year: unknown, country: South Africa[512], *Merlot

incidence: 4/7*, conc. range: ≤0.100 µg/l, sample origin: wine-producing regions of South Africa, sample year: unknown, country: South Africa[512], *Shiraz

incidence: 2/11*, conc. range: ≤0.016 µg/l, sample origin: wine-producing regions of South Africa, sample year: unknown, country: South Africa[512], *Pinotage

incidence: 1/3*, conc.: 0.030 µg/l, sample origin: wine-producing regions of South Africa, sample year: unknown, country: South Africa[512], *Blend (red)

- Co-contamination: not reported
- Further contamination of plants and plant products other than cereals and their products reported in the present book/article[512]: Wine (white), Wine, miscellaneous, OTA, literature[512]

incidence: 15/23, conc. range: ≤0.079 µg/l, sample origin: north western Italy, sample year: 1998/1999, country: Italy[513]

incidence: 17/19, conc. range: ≤0.227 µg/l, sample origin: north eastern Italy, sample year: 1998/1999, country: Italy[513]

incidence: 26/30, conc. range: ≤1.450 µg/l, sample origin: central Italy, sample year: 1998/1999, country: Italy[513]

incidence: 24/24, conc. range: ≤3.177 µg/l, sample origin: southern Italy, sample year: 1998/1999, country: Italy[513]

- Co-contamination: not reported
- Further contamination of plants and plant products other than cereals and their products reported in the present book/article[513]: Wine miscellaneous, OTA, literature[513]

incidence: 535/773, conc. range: ≤7.50 µg/l, sample origin: food stores in Rome (capital), Italy, sample year: 2002–2004, country: Italy[516]

- Co-contamination: not reported
- Further contamination of plants and plant products other than cereals and their products reported in the present book/article[516]: Wine (rosé), Wine (white), Wine, miscellaneous, OTA, literature[516]

For detailed information see the article.

incidence: 2/18, conc. range: LOD-LOQ (1 sa.), 0.06 µg/l (1 sa.), sample origin: supermarkets in Alicante (city) and Utiel Requena (wine production area), Valencia (community), Spain, sample year: July 2003 (production year: 1999), country: Spain[517]

incidence: 6/19, conc. range: LOD-LOQ (2 sa.), 0.20–0.40 µg/l (4 sa.), sample origin: supermarkets in Alicante (city) and Utiel Requena (wine production area), Valencia (community), Spain, sample year: July 2003 (production year: 2000), country: Spain[517]

incidence: 13/24, conc. range: LOD-LOQ (6 sa.), 0.10–0.53 µg/l (7 sa.), sample origin: supermarkets in Alicante (city), Valencia (city), and Utiel Requena (wine production area), Valencia (community), Spain, sample year: July 2003 (production year: 2001), country: Spain[517]

- Co-contamination: not reported
- Further contamination of plants and plant products other than cereals and their products reported in the present book/article[517]: Wine (rosé), Wine (white), Wine, miscellaneous, OTA, literature[517]

incidence: 1/5, conc. range: 0.06 µg/l, sample origin: produced in Apulia (region) purchased in markets and local retail stores, Italy, sample year: 2003 (production year: 1999), country: Italy/Spain[518]

incidence: 11/11, conc. range: LOD-0.10 µg/l (4 sa.), >0.10–0.50 µg/l (6 sa.), 2.37 µg/l (1 sa.), sample origin: produced in Apulia (region) purchased in markets and local retail stores, Italy, sample year: 2003 (production year: 2000), country: Italy/Spain[518]

incidence: 18/18, conc. range: LOD-0.10 µg/l (3 sa.), >0.10–0.50 µg/l (9 sa.), >0.50–1 µg/l (4 sa.), >0.1–2 µg/l (2 sa., maximum: 1.62 µg/l), sample origin: produced in Apulia (region) purchased in markets and local retail stores, Italy, sample year: 2003 (production year: 2001), country: Italy/Spain[518]

incidence: 3/10, conc. range: >0.10–0.50 µg/l (1 sa.), >0.50–1 µg/l (1 sa.), 1.00 µg/l (1 sa.), sample origin: produced in Apulia (region) purchased in markets and local retail stores, Italy, sample year: 2003 (production year: 2002), country: Italy/Spain[518]

incidence: 49/59, conc. range: LOD-0.10 µg/l (22 sa.), >0.10–0.50 µg/l (23 sa.), >0.50–1 µg/l (3 sa.), 1.39 µg/l (1 sa.), sample origin: produced in Sicily (autonomous region, island) purchased in markets and local retail stores, Italy, sample year: 2003 (production year: 2000), country: Italy/Spain[518]

incidence: 42/46, conc. range: LOD-0.10 µg/l (18 sa.), >0.10–0.50 µg/l (22 sa.), >0.50–1 µg/l (1 sa.), 1.10 µg/l (1 sa.), sample origin: produced in Sicily (autonomous region, island) purchased in markets and local retail stores, Italy, sample year: 2003 (production year: 2001), country: Italy/Spain[518] incidence: 10/10, conc. range: LOD-0.10 μg/l (6 sa.), >0.10-0.50 μg/l (4 sa., maximum: 0.41 μg/l), sample origin: produced in Sicily (autonomous region, island) purchased in markets and local retail stores, Italy, sample year: 2003 (production year: 2002), country: Italy/Spain[518]

- Co-contamination: not reported
- Further contamination of plants and plant products other than cereals and their products reported in the present book/article[518]: Wine (rosé), Wine (white), Wine miscellaneous, OTA, literature[518]

incidence: 26/27*, conc. range: ≤7.63 μg/l, ∅ conc.: 1.269 μg/l, sample origin: bottled wine from local retails, Italy, sample year: unknown, country: Italy[522], *commercial red wine

incidence: 11/11*, conc. range: 0.46–4.72 μg/l, ∅ conc.: 1.185 μg/l, sample origin: home-made wine from small farmers, Italy, sample year: unknown, country: Italy[522], *home-made red wine

- Co-contamination: not reported
- Further contamination of plants and plant products other than cereals and their products reported in the present book/article[522]: Wine (rosé), Wine (white), Wine, miscellaneous, OTA, literature[522]

incidence: 18/54, conc. range: <0.01–≤0.03 μg/l (10 sa.), 0.03≤0.20 μg/l (7 sa.), 1.68 μg/l (1 sa.), sample origin: different states of the USA, sample year: 1995–2001, country: the USA[523]

- Co-contamination: not reported
- Further contamination of plants and plant products other than cereals and their products reported in the present book/article[523]: Wine (white), Wine, miscellaneous, OTA, literature[523]

incidence: 42/70*, conc. range: 0.01–0.048 μg/l (33 sa.), >0.05-0.1 μg/l (7 sa.), 0.1–0.126 μg/l (2 sa.), sample origin: Bekaa valley (sa. purchased from supermarkets), Lebanon, sample year: July-October 2004, country: Lebanon/France[536], *red wine, finished

- Co-contamination: not reported
- Further contamination of plants and plant products other than cereals and their products reported in the present book/article[536]: Must, OTA, literature[536]

incidence: 10/11*, conc. range: 0.002–0.100 μg/l, ∅ conc.: 0.0282 μg/l, sample origin: Cosenza (province), Calabria (region), southern Italy, sample year: 2007, country: Italy[539], *commercial red wine

- Co-contamination: not reported
- Further contamination of plants and plant products other than cereals and their products reported in the present book/article[539]: Must, OTA, literature[539]

incidence: 2/2, conc. range: 1.50–3.78 μg/l, ∅ conc.: 2.64 μg/l, sample origin: Experimental Winemaking Station, University of Bologna (city), Italy, sample year: unknown, country: Italy[604]

- Co-contamination: not reported
- Further contamination of plants and plant products other than cereals and their products reported in the present book/article[604]: not reported

incidence: 5/31, conc. range: ≤0.030 μg/l, ∅ conc.: 0.027 μg/l, sample origin: unknown, sample year: unknown, country: Japan[605]

- Co-contamination: not reported
- Further contamination of plants and plant products other than cereals and their products reported in the present book/article[605]: Wine (white), OTA, literature[605]

incidence: 33/33*, conc. range: 0.03–0.533 μg/l, ∅ conc.: 0.117 μg/l, sample origin: wine-producing areas, Hungary, sample year: vintages 2000–2002, country: Hungary[606], *sa. of red bottled quality wine and directly from wineries

- Co-contamination: not reported
- Further contamination of plants and plant products other than cereals and their products reported in the present book/article[606]: Wine (white), OTA, literature[606]

incidence: 11/14, conc. range: 0.056–0.316 μg/l, ∅ conc.: 0.160 μg/l, sample origin: EVENA, Spain, sample year: 1997, country: Spain[607]

incidence: 2/14, conc. range: 0.074–0.193 μg/l, ∅ conc.: 0.133 μg/l, sample origin: EVENA, Spain, sample year: 1998, country: Spain[607]

- Co-contamination: not reported
- Further contamination of plants and plant products other than cereals and their products reported in the present book/article[607]: Wine (white), OTA, literature[607]

incidence: 4/4*, conc. range: 0.18–0.38 μg/l, ∅ conc.: 0.29 μg/l, sample origin: local retail stores in Cape Town (city), South Africa, sample year: 2000/2001, country: South Africa/Italy[608], *red wine, bottled

incidence: 5/5*, conc. range: 0.07–0.39 μg/l, ∅ conc.: 0.21 μg/l, sample origin: local retail stores in Cape Town (city), South Africa, sample year: 2000/2001, country: South Africa/Italy[608], *red wine, boxed

incidence: 5/5, conc. range: 0.23–0.91 μg/l, ∅ conc.: 0.58 μg/l, sample origin: local retail stores in Cesena (city), Italy, sample year: 2000/2001, country: South Africa/Italy[608]

- Co-contamination: not reported
- Further contamination of plants and plant products other than cereals and their products reported in the present book/article[608]: Wine (white), OTA, literature[608]

incidence: 96/580*, conc. range: 0.051–0.100 μg/l (94 sa.), >0.100-<0.200 μg/l (2 sa.), sample origin: from 21 countries, sample year: unknown, country: Canada[609], *red wine, bottled

- Co-contamination: not reported
- Further contamination of plants and plant products other than cereals and their products reported in the present book/article[609]: Wine (white), OTA, literature[609]

For detailed information see the article.

incidence: 33/267*, conc. range: <0.1 μg/l (19 sa.), 0.1–<0.2 μg/l (12 sa.), ≥0.2 μg/l (2 sa., maximum: 0.62 μg/l), sample origin: wine producers and retail outlets in Sydney (city), New South Wales (state), Australia, sample year: 1984–1986, 1997–2000, country: Australia[610], *sa. in bottles included sparkling wine

incidence: 0/45*, conc.: no contamination, sample origin: wine producers and retail outlets in Sydney (city), New South Wales (state), Australia, sample year: 1984–1986, 1997–2000, country: Australia[610], *sa. in tanks

incidence: 16/32*, conc. range: <0.1 μg/l (10 sa.), 0.1–<0.2 μg/l (4 sa.), ≥0.2 μg/l (2 sa., maximum: 0.215 μg/l), sample origin: wine producers and retail outlets in Sydney (city), New South Wales (state), Australia, sample year: 1984–1986, 1997–2000, country: Australia[610], *sa. in casks

- Co-contamination: not reported
- Further contamination of plants and plant products other than cereals and their products reported in the present book/article[610]: Wine (white), OTA, literature[610]

For detailed information see the article.

incidence: 6/6, conc. range: 0.06–1.06 μg/l, ∅ conc.: 0.24 μg/l, sample origin: wine producers in Central Anatolia (region), Turkey, sample year: 1999–2001, country: Turkey[611]

incidence: 11/11, conc. range: 0.04–1.32 μg/l, ∅ conc.: 0.52 μg/l, sample origin: wine producers in East Anatolia (region), Turkey, sample year: 1999–2001, country: Turkey[611]

incidence: 6/6, conc. range: 0.05–1.92 μg/l, ∅ conc.: 0.87 μg/l, sample origin: wine producers in Aegean (region), Turkey, sample year: 2000/2001, country: Turkey[611]

incidence: 12/12, conc. range: 0.04–1.05 μg/l, ∅ conc.: 0.24 μg/l, sample origin: wine producers in Thrace (region), Turkey, sample year: 1999–2001, country: Turkey[611]

- Co-contamination: not reported
- Further contamination of plants and plant products other than cereals and their products reported in the present book/article[611]: Wine (rosé), Wine (white), OTA, literature[611]

For detailed information see the article.

incidence: 16/29, conc. range: 0.31–2.92 μg/l, sample origin: retail shops and home-made wine in Apulia (region), Italy, sample year: unknown, country: Italy[612]

- Co-contamination: not reported
- Further contamination of plants and plant products other than cereals and their products reported in the present book/article[612]: Wine (rosé), Wine (white), OTA, literature[612]

incidence: 31/38*, conc. range: 0.4–1.44 μg/l, ∅ conc.: 0.29 μg/l, sample origin: local stores, Italy, sample year: April–June 2004, country: Italy[613], *red wine, bottled

incidence: 5/5*, conc. range: 0.17–0.87 μg/l, ∅ conc.: 0.33 μg/l, sample origin: local stores, Italy, sample year: April–June 2004, country: Italy[613], *red wine, boxed (glass or Tetrapak packaging)

- Co-contamination: not reported
- Further contamination of plants and plant products other than cereals and their products reported in the present book/article[613]: Wine (rosé), Wine (white), OTA, literature[613]

incidence: 13/16, conc. range: tr-1.530 μg/l, sample origin: local stores in Naples and Salerno (cities), Italy, sample year: 2009, country: Italy[614]

- Co-contamination: not reported
- Further contamination of plants and plant products other than cereals and their products reported in the present book/article[614]: Wine (white), OTA, literature[614]

incidence: 7/7*, conc. range: 0.012–0.047 μg/l, ∅ conc.: 0.022 μg/l, sample origin: local store, Croatia, sample year: 2002/2003, country: Croatia[616], *red wine, bottled

- Co-contamination: not reported
- Further contamination of plants and plant products other than cereals and their products reported in the present book/article[616]: Wine (white), OTA, literature[616]

For detailed information see the article.

incidence: 5/11*, conc. range: 0.12–0.30 μg/l**, ∅ conc.: 0.21 μg/l**, sample origin: local supermarket in Vienna (capital), Austria, sample year: 2000–2002 (production year), country: Austria[617], *wine, bottled, **contaminated sa. from Italy (4) and Croatia (1)

- Co-contamination: not reported
- Further contamination of plants and plant products other than cereals and their products reported in the present book/article[617]: not reported

For detailed information see the article.

incidence: 15/15, conc. range: ≤0.26 μg/l, ∅ conc.: 0.029 μg/l, sample origin: domestic wines from local supermarket and retail stores, Japan, sample year: summer 2006, country: Japan[618]

incidence: 20/23, conc. range: ≤0.82 μg/l, sample origin: imported wines from local supermarket and retail stores, Japan, sample year: summer 2006, country: Japan[618]

- Co-contamination: not reported
- Further contamination of plants and plant products other than cereals and their products reported in the present book/article[618]: Wine (white), OTA, literature[618]

incidence: 3/7*, conc. range: 1.8**–4.4 μg/l, ∅ conc.: 2.93 μg/l, sample origin: retail markets in Kuban (region), Russia, sample year: unknown, country: Russia/Belgium[619], *thereof 2 sa. home-made red wine 1 of them contaminated**

- Co-contamination: not reported
- Further contamination of plants and plant products other than cereals and their products reported in the present book/article[619]: not reported

incidence: 9/29, conc. range: 0.10–1.33 μg/l, sample origin: domestic wines purchased in supermarkets and retail stores in São Paulo (city), Brazil, sample year: summer 2005, country: Brazil[620]

incidence: 1/6, conc.: <0.03 μg/l, sample origin: imported wines from Argentina purchased in supermarkets and retail stores in São Paulo (city), Brazil, sample year: summer 2005, country: Brazil[620]

incidence: 0/7, conc.: no contamination, sample origin: imported wines from Chile purchased in supermarkets and retail stores in São Paulo (city), Brazil, sample year: summer 2005, country: Brazil[620]

incidence: 3/5, conc. range: 0.20–0.29 μg/l, sample origin: imported wines from France purchased in supermarkets and retail stores in São Paulo (city), Brazil, sample year: summer 2005, country: Brazil[620]

incidence: 5/5, conc. range: 0.03–0.32 μg/l, sample origin: imported wines from Italy purchased in supermarkets and retail stores in São Paulo (city), Brazil, sample year: summer 2005, country: Brazil[620]

incidence: 5/5, conc. range: 0.03–0.25 μg/l, sample origin: imported wines from Portugal purchased in supermarkets and retail stores in São Paulo (city), Brazil, sample year: summer 2005, country: Brazil[620]

incidence: 1/1, conc.: <0.03 μg/l, sample origin: imported wines from South Africa purchased in supermarkets and retail stores in São Paulo (city), Brazil, sample year: summer 2005, country: Brazil[620]

incidence: 2/2, conc. range: 0.07–0.12 μg/l, sample origin: imported wines from Spain purchased in supermarkets and retail stores in São Paulo (city), Brazil, sample year: summer 2005, country: Brazil[620]

incidence: 1/3, conc.: <0.03 μg/l, sample origin: imported wines from Uruguay purchased in supermarkets and retail stores in São Paulo (city), Brazil, sample year: summer 2005, country: Brazil[620]

- Co-contamination: not reported
- Further contamination of plants and plant products other than cereals and their products reported in the present book/article[620]: not reported

incidence: 45/64*, conc. range: ≤1.31 μg/l, sample origin: different winemaking regions of Greece, sample year: 1999–2006, country: Greece/the UK[621], *red dry wine

incidence: 13/14*, conc. range: ≤2.00 μg/l, sample origin: different winemaking regions of Greece, sample year: 1999–2006, country: Greece/the UK[621], *red dessert wine

- Co-contamination: not reported

- Further contamination of plants and plant products other than cereals and their products reported in the present book/ article[621]: Wine, Wine (rosé), Wine (white), OTA, literature[621]

For detailed information see the article.

incidence: 4/14, conc. range: ≤0.20 µg/l, sample origin: domestic and imported wines purchased in local supermarkets, Japan, sample year: 2010, country: Japan[622]

- Co-contamination: 1 sa. co-contaminated with FB_1, FB_2, FB_3, and OTA: 2 sa. co-contaminated with FB_1 and OTA; 1 sa. contaminated solely with OTA
- Further contamination of plants and plant products other than cereals and their products reported in the present book/ article[622]: Wine red, FB_1, FB_2, and FB_3, literature[622]; Wine (white), FB_1 and OTA, literature[622]

incidence: 18/63, conc. range: LOD-0.2 µg/l (6 sa.), 0.2–0.5 µg/l (9 sa.), >0.5 µg/l (3 sa., maximum: 1.18 µg/l), ∅ conc.: 0.32 µg/l, sample origin: supermarkets, China, sample year: 2008/2009, country: China[623]

- Co-contamination: not reported
- Further contamination of plants and plant products other than cereals and their products reported in the present book/ article[623]: Wine (white), OTA, literature[623]

incidence: 12/16, conc. range: 0.03–0.62 µg/l, sample origin: tropical winemaking region, Brazil, sample year: 2009, country: Brazil[624]

- Co-contamination: not reported
- Further contamination of plants and plant products other than cereals and their products reported in the present book/ article[624]: Wine (white), OTA, literature[624]

incidence: 88/112, conc. range: ≤4.93 µg/l, sample origin: local retailers in Lazio and Apulia (regions), Italy, sample year: 1997–2002 (production year), country: Italy[626]

- Co-contamination: not reported
- Further contamination of plants and plant products other than cereals and their products reported in the present book/ article[626]: not reported

incidence: 9/9*, conc. range: 0.03–4.90 µg/l, ∅ conc.: 2.34 µg/l, sample origin: imported from France, Spain, the USA, and Italy and purchased in local stores, Italy, sample year: unknown, country: Italy[644], *red wine, bottled

- Co-contamination: not reported
- Further contamination of plants and plant products other than cereals and their products reported in the present book/ article[644]: Wine (white), OTA, literature[644]

incidence: 20/20, conc. range: 0.39–7.96 µg/l, ∅ conc.: 2.85 µg/l, sample origin: mostly produced in different regions in Anatolia (region), Turkey, sample year: 1987–2005, country: Turkey/Egypt[671]

- Co-contamination: not reported
- Further contamination of plants and plant products other than cereals and their products reported in the present book/ article[671]: Wine (white), OTA, literature[671]

incidence: 6/27, conc. range: <LOQ-0.26 µg/l, sample origin: supermarkets and specialized stores, Chile, sample year: unknown, country: Chile[672]

incidence: 5/16, conc. range: 0.15–0.25 µg/l, ∅ conc.: 0.20 µg/l, sample origin: imported from Australia, France, Germany, Italy, and Spain and purchased in supermarkets and specialized stores, Chile, sample year: unknown, country: Chile[672]

- Co-contamination: not reported
- Further contamination of plants and plant products other than cereals and their products reported in the present book/ article[672]: Wine (white), OTA, literature[672]

incidence: 16/22, conc. range: 0.2–1.9 µg/l, ∅ conc.: 0.56 µg/l, sample origin: wineries, northern Italy, sample year: unknown, country: Italy[674]

- Co-contamination: not reported
- Further contamination of plants and plant products other than cereals and their products reported in the present book/ article[674]: Must, Wine (white), OTA, literature[674]

incidence: 19/25*, conc. range: ~0.011 µg/l (5 sa.), 0.011–0.033 µg/l (11 sa.), ≤0.463 µg/l (3 sa.), sample origin: retail stores and wholesale trade, Slovakia, sample year: 2005, country: Slovakia[684], *red wine, bottled

- Co-contamination: not reported
- Further contamination of plants and plant products other than cereals and their products reported in the present book/ article[684]: Wine (white), OTA, literature[684]

For detailed information see the article.

incidence: 3/7, conc. range: 1.8–3.4 µg/l, ∅ conc.: 2.6 µg/l, sample origin: retail markets in Belgium and Russia, sample year: unknown, country: Russia/Spain/Belgium[685]

- Co-contamination: not reported
- Further contamination of plants and plant products other than cereals and their products reported in the present book/ article[685]: not reported

incidence: 1/10, conc.: 0.63 µg/l, sample origin: local supermarkets in Beijing (capital), China, sample year: unknown, country: China[690]

- Co-contamination: not reported
- Further contamination of plants and plant products other than cereals and their products reported in the present book/ article[690]: Juices (grape juice), OTA, literature[690]

incidence: 37/54, conc. range: LOD/LOQ-0.9 µg/kg (35 sa.), ≤1.9 µg/kg (2 sa.), sample origin: Finland, sample year: 1998, country: EU[698]

incidence: 72/112?, conc. range: LOD/LOQ-0.9 µg/kg (69 sa.), ≤1.29 µg/kg (3 sa.), sample origin: Finland, sample year: 1999, country: EU[698]

incidence: 79/172, conc. range: LOD/LOQ-0.9 µg/kg (69 sa.), 1.0–2.9 µg/kg (8 sa.), 3.0–4.9 µg/kg (1 sa.), 7.0 µg/kg (1 sa.), sample origin: Germany, sample year: 1995–1998, country: EU[698]

incidence: 21/38, conc. range: LOD/LOQ-0.9 µg/kg (20 sa.), 2.61 µg/kg (1 sa.), sample origin: Greece, sample year: 1999, country: EU[698]

incidence: 159/184, conc. range: LOD/LOQ-0.9 µg/kg (89 sa.), 1.0–2.9 µg/kg (42 sa.), 3.0–4.9 µg/kg (9 sa.), 5.0–9.9 µg/kg (12 sa.), ≤15.6 µg/kg (7 sa.), sample origin: Italy, sample year: 1992–1999, country: EU[698]

incidence: 51/60, conc. range: LOD/LOQ-0.9 µg/kg (43 sa.), ≤1.5 µg/kg (8 sa.), sample origin: Italy, sample year: 1999, country: EU[698]

incidence: 66/72, conc. range: LOD/LOQ-≤0.603 µg/kg (66 sa.), sample origin: Spain, sample year: 1997, country: EU[698]

incidence: 29/84, conc. range: LOD/LOQ-0.9 µg/kg (28 sa.), 1.20 µg/kg (1 sa.), sample origin: the Netherlands, sample year: 1999, country: EU[698] incidence: 32/66, conc. range: LOD/LOQ-0.9 µg/kg (23 sa.), 1.0–2.9 µg/kg (8 sa.), 3.10 µg/kg (1 sa.), sample origin: the Netherlands, sample year: 1999, country: EU[698]

- Co-contamination: not reported
- Further contamination of plants and plant products other than cereals and their products reported in the present book/article[698]: Baby food, Beans (soybean), Chocolate, Cocoa, Coffee, Coffee beans, Currants, Fruits, dried, Herbs, Jam, Juices, Juices (grape juice), Legumes, Nuts, Oil seeds, Oils (olive oil), Peas, Peas (chickpea), Products (plant product), Raisins, Spices, Spices (cayenne pepper, chilli), Spices (nutmeg), Spices (paprika), Spices (pepper), Sultanas, Tea, Vinegar, Wine, Wine (rosé), Wine (white), Wine, miscellaneous, OTA, literature[698]

incidence: 72/131*, conc. range: tr (33 sa.), >0.5 µg/l (22 sa.), >1.00–6.00 µg/l (17 sa.), sample origin: France, Germany, and Italy, sample year: 1999–2007, country: Italy/Germany[699], *organic

- Co-contamination: not reported
- Further contamination of plants and plant products other than cereals and their products reported in the present book/article[699]: Wine (rosé), Wine (white), OTA, literature[699]

incidence: 31/57*, conc. range: 0.01–0.73 µg/l, ∅ conc.: 0.14 µg/l, sample origin: local retails, Sicily (island, autonomous region), Italy, sample year: 2009–2013 vintages, country: Italy[708], *red wine, bottled

- Co-contamination: 10 sa. co-contaminated with OTA and OTB; 4 sa. co-contaminated with AFG_1 and OTA; 3 sa. co-contaminated with AFG_2, OTA, and OTB; 2 sa. co-contaminated with AFB_2, OTA, and OTB; 2 sa. co-contaminated with AFG_1, AFG_2, OTA, and OTB; 2 sa. co-contaminated with AFG_1, OTA, and OTB; 1 sa. co-contaminated with AFB_1, AFG_1, and OTA; 1 sa. co-contaminated with AFB_2, AFG_1, OTA, and OTB; 1 sa. co-contaminated with AFB_2 and OTA; 5 sa. contaminated solely with OTA
- Further contamination of plants and plant products other than cereals and their products reported in the present book/article[708]: Wine (red), AFB_1, AFB_2, AFG_1, AFG_2, and OTB, literature[708]

incidence: 5/5*, conc. range: 0.05–1.28 µg/l, ∅ conc.: 0.38 µg/l, sample origin: Italian market, Italy, sample year: unknown, country: Italy[791], *red table wine

incidence: 4/4*, conc. range: 0.09–0.80 µg/l, ∅ conc.: 0.45 µg/l, sample origin: Italian market, Italy, sample year: unknown, country: Italy[791], *Apulian Denominazione di origine controllata (DOC) wines

incidence: 6/6*, conc. range: 0.06–0.26 µg/l, ∅ conc.: 0.14 µg/l, sample origin: local farm in Salento (region in southern Apulia), Italy, sample year: unknown, country: Italy[791], *organic

- Co-contamination: not reported
- Further contamination of plants and plant products other than cereals and their products reported in the present book/article[791]: not reported

incidence: 2/13, conc. range: 0.0012–0.0712 µg/l, ∅ conc.: 0.0362 µg/l, sample origin: retail stores, specialized shops delivered directly by a producer in Moravian subregions, historical country, Czech Republic, sample year: 2006–2008 purchased in 2009, country: Czech Republic[792]

incidence: 6/10, conc. range: 0.0046–0.1273 µg/l, ∅ conc.: 0.0611 µg/l, sample origin: imported from Chile*, Italy, France, Greece*, and Spain*, Czech Republic, sample year: 2000, 2005–2007 purchased 2009, country: Czech Republic[792], *contaminated sa.

- Co-contamination: not reported
- Further contamination of plants and plant products other than cereals and their products reported in the present book/article[792]: Wine (rosé), Wine (white), OTA, literature[792]

incidence: 3/12, conc. range: ≤0.695 µg/kg, sample origin: 12 regions, Czech Republic, sample year: 2011–2013, country: Czech Republic[811]

- Co-contamination: not reported
- Further contamination of plants and plant products other than cereals and their products reported in the present book/article[811]: Beans, Chocolate, Coffee, Confectionery, Figs, Herbs and Spices, Juices, Licorice, Powder (cocoa powder), Products (ginger product), Raisins, Spices (caraway), Spices (cayenne pepper, chilli), Spices (coriander), Spices (ginger), Spices (mustard), Spices (nutmeg), Spices (paprika), Spices (pepper), Tea, Wine (white), OTA, literature[811]

incidence: 6/68*, conc. range: 0.2–0.6 µg/kg, sample origin: imported from South Africa purchased in supermarkets, wine stores, and importers, Netherlands, sample year: 2008, country: Brazil/Netherlands[814], *Cabernet Sauvignon

- Co-contamination: not reported
- Further contamination of plants and plant products other than cereals and their products reported in the present book/article[814]: Wine (red), AOH, CPA, MA, and ZEA, literature[814]

incidence: 4/47, conc. range: 0.02–4.82 µg/l, sample origin: 7 provinces, Argentina, sample year: harvest season vintage 2008, country: Argentina[816]

- Co-contamination: not reported
- Further contamination of plants and plant products other than cereals and their products reported in the present book/article[816]: Must, Vine fruits, dried, OTA, literature[816]

For detailed information see the article.

incidence: 2/30, conc. range: 2.0–5.0 µg/kg (2 sa.), ∅ conc.: 2.34 µg/kg, sample origin: southern, central, and northern Italy, sample year: unknown, country: Italy[820]

- Co-contamination: not reported
- Further contamination of plants and plant products other than cereals and their products reported in the present book/article[820]: Coffee, Spices (cayenne pepper, chilli), OTA, literature[820]

incidence: 14/16*, conc. range: 0.010–0.237 µg/l, ∅ conc.: 0.051 µg/l, sample origin: retail market, sa. sold in the UK, sample year: 2009, country: Spain[825], *DOCa Rioja exportation wine market

incidence: 7/8*, conc. range: 0.060–0.138 µg/l, ∅ conc.: 0.085 µg/l, sample origin: retail market, sa. sold in Germany, sample year: 2009, country: Spain[825], *DOCa Rioja exportation wine market

incidence: 3/3*, conc. range: 0.030–0.353 µg/l, ∅ conc.: 0.149 µg/l, sample origin: retail market, sa. sold in Switzerland, sample year: 2009, country: Spain[825], *DOCa Rioja exportation wine market

incidence: 2/3*, conc. range: 0.063–0.309 µg/l, ∅ conc.: 0.125 µg/l, sample origin: retail market, sa. sold in the USA, sample year: 2009, country: Spain[825], *DOCa Rioja exportation wine market

- Co-contamination: not reported

- Further contamination of plants and plant products other than cereals and their products reported in the present book/ article[825]: not reported

For detailed information see the article.

incidence: 17/32*, conc. range: 0.007–0.179 µg/l, ∅ conc.: 0.037 µg/l, sample origin: semiarid and dry area of Rioja Alavesa, Spain, sample year: produced 1995–2008, country: Spain[826], * "Joven" red wine

incidence: 26/41*, conc. range: 0.004–0.169 µg/l, ∅ conc.: 0.040 µg/l, sample origin: semiarid and dry area of Rioja Alavesa, Spain, sample year: produced 1995–2008, country: Spain[826], * "Crianza" red wine

incidence: 10/18*, conc. range: 0.004–0.060 µg/l, ∅ conc.: 0.022 µg/l, sample origin: semiarid and dry area of Rioja Alavesa, Spain, sample year: produced 1995–2008, country: Spain[826], * "Reserva" red wine

incidence: 1/3*, conc.: 0.048 µg/l, sample origin: semiarid and dry area of Rioja Alavesa, Spain, sample year: produced 1995–2008, country: Spain[826], * "Gran reserva" red wine

- Co-contamination: not reported
- Further contamination of plants and plant products other than cereals and their products reported in the present book/ article[826]: Wine (rosé), Wine (white), OTA, literature[826]

For detailed information see the article.

incidence: 35/35*, conc. range: 0.00049–0.0949 µg/l, ∅ conc.: 0.0127 µg/l, sample origin: supermarkets in Navarra (chartered community, province), Spain, sample year: vintages 2004, 2006/2007, country: Spain[830], *conventional

- Co-contamination: 35 sa. co-contaminated with OTA and OTB; no further information available
- Further contamination of plants and plant products other than cereals and their products reported in the present book/ article[830]: Wine (red), OTB, literature[830]; see also Wine (red), chapter: Further Mycotoxins and Microbial Metabolites

incidence: 16/16*, conc. range: 0.00060–0.142 µg/l, ∅ conc.: 0.0229 µg/l, sample origin: wineries in Navarra (chartered community, province), Spain, sample year: vintages 2007/2008, country: Spain[830], *organic

- Co-contamination: 16 sa. co-contaminated with OTA, OTB, and MeOTB; no further information available.
- Further contamination of plants and plant products other than cereals and their products reported in the present book/ article[830]: Wine (red), OTB, literature[830]; see also Wine (red), chapter: Further Mycotoxins and Microbial Metabolites

incidence: 20/20*, conc. range: 0.0005–0.0448 µg/l, ∅ conc.: 0.0099 µg/l, sample origin: wineries in Navarra (chartered community, province), Spain, sample year: vintages 2006/2007, country: Spain[831], *red wine, bottled

- Co-contamination: 8 sa. co-contaminated with OTA, OTB, and OTC; 6 sa. co-contaminated with OTA, OTB, OTC, and MeOTA; 4 sa. co-contaminated with OTA, OTB, and MeOTA; 2 sa. co-contaminated with OTA and OTB
- Further contamination of plants and plant products other than cereals and their products reported in the present book/ article[831]: Wine (red), OTB, literature[831]; see also Wine (red), chapter: Further Mycotoxins and Microbial Metabolites

incidence: 95/96, conc. range: 0.00020–0.455 µg/l, ∅ conc.: 0.0562 µg/l, sample origin: Mediterranean countries, sa. purchased on-line as well as in supermarkets in Greece and Spain, sample year: vintages 1998–2009, country: Spain[832]

- Co-contamination: For detailed information see the article.
- Further contamination of plants and plant products other than cereals and their products reported in the present book/ article[832]: Wine (red), OTB, literature[832]; see also Wine (red), chapter: Further Mycotoxins and Microbial Metabolites

For detailed information see the article.

incidence: 16/21*, conc. range: LOD-0.50 µg/l (15 sa.), 0.71 µg/l (1 sa.), sample origin: Thessaly (region) as well as western and central Greece, sample year produced 2007–2009, country: Greece[837], *red wine, dry

incidence: 1/1*, conc.: 0.1 µg/l, sample origin: Thessaly (region) as well as western and central Greece, sample year: produced 2007–2009, country: Greece[837], *red wine, sweet

- Co-contamination: not reported
- Further contamination of plants and plant products other than cereals and their products reported in the present book/ article[837]: Wine (rosé), Wine (white), OTA, literature[837]

For detailed information see the article.

incidence: 14/18, conc. range: 0.09–0.94 µg/l, ∅ conc.: 0.40 µg/l, sample origin: supermarkets, Tunisia, sample year: produced 2004–2006, country: Tunisia[852]

- Co-contamination: not reported
- Further contamination of plants and plant products other than cereals and their products reported in the present book/ article[852]: Wine (rosé), Wine (white), OTA, literature[852]

For detailed information see the article.

incidence: 30/33, conc. range: LOD-0.1 µg/l (20 sa.), 0.1–0.5 µg/l (8 sa.), >0.5 µg/l (2 sa., maximum: 0.815 µg/l), ∅ conc.: 0.134 µg/l, sample origin: Thrace (region), Turkey, sample year: 2005/2006, country: Turkey[867]

incidence: 8/9, conc. range: LOD-0.1 µg/l (6 sa.), 0.1–0.5 µg/l (2 sa., maximum: 0.252 µg/l), ∅ conc.: 0.088 µg/l, sample origin: Aegean (region), Turkey, sample year: 2005/2006, country: Turkey[867]

incidence: 1/3, conc.: 0.023 µg/l, sample origin: central Anatolia (region), Turkey, sample year: 2005/2006, country: Turkey[867]

incidence: 5/6, conc. range: LOD-0.1 µg/l (5 sa., maximum: 0.026 µg/l), ∅ conc.: 0.026 µg/l, sample origin: east Anatolia (region), Turkey, sample year: 2005/2006, country: Turkey[867]

- Co-contamination: not reported
- Further contamination of plants and plant products other than cereals and their products reported in the present book/ article[867]: Wine (rosé), Wine (white), OTA, literature[867]

incidence: 44/77*, conc. range: <0.1 µg/l (3 sa.), 0.1–0.5 µg/l (23 sa.), 0.5–1.0 µg/l (10 sa.), 1.0–1.5 µg/l (2 sa.), 1.5–2.0 µg/l (3 sa.), >2.0 µg/l (3 sa., maximum: 5.65 µg/l), ∅ conc.: 0.80 µg/l, sample origin: four major wine-producing regions, China, sample year: 1999–2009, country: China[899], *red wine, dry

- Co-contamination: not reported
- Further contamination of plants and plant products other than cereals and their products reported in the present book/ article[899]: Wine (rosé), Wine (white), Wine, miscellaneous, OTA, literature[899]

For detailed information see the article.

incidence: 1/34, conc.: 4.5 µg/l, sample origin: retail stores, produced in Rio Grande do Sul (state), Brazil, sample year: unknown, country: Brazil[908]

- Co-contamination: not reported
- Further contamination of plants and plant products other than cereals and their products reported in the present book/article[908]: not reported

incidence: 16/17, conc. range: 0.165–0.886 µg/l, Ø conc.: 0.36 µg/l, sample origin: different Italian regions, Italy, sample year: 2005–2012, country: Italy[915]

- Co-contamination: not reported
- Further contamination of plants and plant products other than cereals and their products reported in the present book/article[915]: not reported

Ochratoxin B

incidence: 21/57*, conc. range: 0.02–0.66 µg/l, Ø conc.: 0.16 µg/l, sample origin: Sicily (island, autonomous region), Italy, sample year: 2009–2013 vintages, country: Italy[708], *red wine, bottled

- Co-contamination: 10 sa. co-contaminated with OTA and OTB; 2 sa. co-contaminated with AFG_1, AFG_2, OTA, and OTB; 2 sa. co-contaminated with AFG_1, OTA, and OTB; 3 sa. co-contaminated with AFG_2, OTA, and OTB; 2 sa. co-contaminated with AFB_2, OTA, and OTB; 1 sa. co-contaminated with AFB_2, AFG_1, OTA, and OTB; 1 sa. contaminated solely with OTB
- Further contamination of plants and plant products other than cereals and their products reported in the present book/article[708]: Wine (red), AFB_1, AFB_2, AFG_1, AFG_2, and OTA, literature[708]

incidence: 35/35*, conc. range: 0.00321–0.0700 µg/l, Ø conc.: 0.0110 µg/l, sample origin: supermarkets in Navarra (chartered community, province), Spain, sample year: vintages 2004, 2006/2007, country: Spain[830], *conventional

- Co-contamination: 35 sa. co-contaminated with OTA and OTB; no further information available
- Further contamination of plants and plant products other than cereals and their products reported in the present book/article[830]: Wine (red), OTA, literature[830]; see also Wine (red), chapter: Further Mycotoxins and Microbial Metabolites

incidence: 16/16*, conc. range: 0.00308–0.0443 µg/l, Ø conc.: 0.0114 µg/l, sample origin: wineries in Navarra (chartered community, province), Spain, sample year: vintages 2007/2008, country: Spain[830], *organic

- Co-contamination: 16 sa. co-contaminated with OTA, OTB, and MeOTB; no further information available.
- Further contamination of plants and plant products other than cereals and their products reported in the present book/article[830]: Wine (red), OTA, literature[830]; see also Wine (red), chapter: Further Mycotoxins and Microbial Metabolites

incidence: 20/20*, conc. range: 0.0033–0.0228 µg/l, Ø conc.: 0.0096 µg/l, sample origin: wineries in Navarra (chartered community, province), Spain, sample year: vintages 2006/2007, country: Spain[831], *red wine, bottled

- Co-contamination: 8 sa. co-contaminated with OTA, OTB, and OTC; 6 sa. co-contaminated with OTA, OTB, OTC, and MeOTA; 4 sa. co-contaminated with OTA, OTB, and MeOTA; 2 sa. co-contaminated with OTA and OTB.
- Further contamination of plants and plant products other than cereals and their products reported in the present book/article[831]: Wine (red), OTA, literature[831]; see also Wine (red), chapter: Further Mycotoxins and Microbial Metabolites

incidence: 96/96, conc. range: 0.00205–0.119 µg/l, Ø conc.: 0.0236 µg/l, sample origin: Mediterranean countries, sa. purchased on-line as well as in supermarkets in Greece and Spain, sample year: vintages 1998–2009, country: Spain[832]

- Co-contamination: For detailed information see the article.
- Further contamination of plants and plant products other than cereals and their products reported in the present book/article[832]: Wine (red), OTA, literature[832]; see also Wine (red), chapter: Further Mycotoxins and Microbial Metabolites

For detailed information see the article.

Fusarium Toxins

Fumonisin/s

Fumonisin B_1

incidence: 4/14, conc. range: <1.0 µg/l, sample origin: domestic and imported wines purchased in local supermarkets, Japan, sample year: 2010, country: Japan[622]

- Co-contamination: 1 sa. co-contaminated with FB_1, FB_2, FB_3, and OTA; 2 sa. co-contaminated with FB_1 and OTA; 1 sa. contaminated solely with FB_1
- Further contamination of plants and plant products other than cereals and their products reported in the present book/article[622]: Wine (red), FB_2, FB_3, and OTA, literature[622]; Wine (white), FB_1 and OTA, literature[622]

Fumonisin B_2

incidence: 9/45, conc. range: 0.5–2.4 µg/l, Ø conc.: 0.93 µg/l, sample origin: sa. originated from different wine-producing regions and purchased in a market in Apulia (region), Italy, sample year: 2004–2008, country: Italy[542]

- Co-contamination: not reported
- Further contamination of plants and plant products other than cereals and their products reported in the present book/article[542]: not reported

incidence: 2/14, conc. range: <1.0 µg/l, sample origin: domestic and imported wines purchased in local supermarkets, Japan, sample year: 2010, country: Japan[622]

- Co-contamination: 1 sa. co-contaminated with FB_1, FB_2, FB_3, and OTA; 1 sa. contaminated solely with FB_2
- Further contamination of plants and plant products other than cereals and their products reported in the present book/article[622]: Wine (red), FB_1, FB_3, and OTA, literature[622]; Wine (white), FB_1 and OTA, literature[622]

incidence: 16/56, conc. range: 1.0–25 µg/l, Ø conc.: 5.69 µg/l, sample origin: different countries, sample year: 1991, 1994, 1996, 1998–2008, country: Denmark[625]

- Co-contamination: not reported
- Further contamination of plants and plant products other than cereals and their products reported in the present book/article[625]: Wine (white), Wine, miscellaneous, FB_2, literature[625]

Fumonisin B_3

incidence: 1/14, conc.: <1.0 µg/l, sample origin: domestic and imported wines purchased in local supermarkets, Japan, sample year: 2010, country: Japan[622]

- Co-contamination: 1 sa. co-contaminated with FB_1, FB_2, FB_3, and OTA
- Further contamination of plants and plant products other than cereals and their products reported in the present book/article[622]: Wine (red), FB_1, FB_2, and OTA, literature[622]; Wine (white), FB_1 and OTA, literature[622]

Type A Trichothecenes

T-2 Toxin

incidence: 6/72, conc. range: 0.11–0.43 μg/l, sample origin: imported from Australia, Europe, and South America as well as domestic, the USA, sample year: 2003–2011, country: the USA[666]

- Co-contamination: not reported
- Further contamination of plants and plant products other than cereals and their products reported in the present book/article[666]: Wine (white), T-2, literature[666]

For detailed information see the article.

Other ***Fusarium*** Toxins

Zearalenone

incidence: 2/68*, conc. range: 3.1–3.5 μg/kg, ∅ conc.: 3.3 μg/kg, sample origin: imported from South Africa and purchased in supermarkets, wine stores, and importers, Netherlands, sample year: 2008, country: Brazil/Netherlands[814], *Cabernet Sauvignon

- Co-contamination: not reported
- Further contamination of plants and plant products other than cereals and their products reported in the present book/article[814]: Wine (red), AOH, CPA, MA, and OTA, literature[814]

Penicillium Toxins

Mycophenolic Acid

incidence: 40/44, conc. range: ≤130 μg/kg (most of the sa. 3–20 μg/kg), sample origin: e.g. Greece and Spain, sample year: unknown, country: Germany[422]

- Co-contamination: not reported
- Further contamination of plants and plant products other than cereals and their products reported in the present book/article[422]: not reported

incidence: 1/68*, conc.: 6.8 μg/kg, sample origin: imported from South Africa and purchased in supermarkets, wine stores, and importers, Netherlands, sample year: 2008, country: Brazil/Netherlands[814], *Cabernet Sauvignon

incidence: 7/62*, conc. range: 6.2–35 μg/kg, sample origin: imported from South Africa and purchased in supermarkets, wine stores, and importers, Netherlands, sample year: 2009, country: Brazil/Netherlands[814], *Cabernet Sauvignon

incidence: 1/50*, conc.: 54 μg/kg, sample origin: imported from South Africa and purchased in supermarkets, wine stores, and importers, Netherlands, sample year: 2010, country: Brazil/Netherlands[814], *Cabernet Sauvignon

incidence: 3/60*, conc. range: 11–14 μg/kg, sample origin: imported from South Africa and purchased in supermarkets, wine stores, and importers, Netherlands, sample year: 2011, country: Brazil/Netherlands[814], *Cabernet Sauvignon

- Co-contamination: not reported
- Further contamination of plants and plant products other than cereals and their products reported in the present book/article[814]: Wine (red), AOH, CPA, OTA, and ZEA, literature[814]

incidence: 1/1*, conc.: 38 μg/kg, sample origin: private households of staff members, Austria, sample year: unknown, country: Austria[850], *moldy

- Co-contamination: 1 sa. co-contaminated with AOH, AME, MA, and TEN
- Further contamination of plants and plant products other than cereals and their products reported in the present book/article[850]: Fruits (apple), Nuts (chestnut), Wine (red), AME and AOH, literature[850]; Fruits (lemon), Puree, MA, literature[850]; Jam, AOH and MA, literature[850]; Spices (garlic), BEA, FB_1, FB_2, and FB_3, literature[850]; see also Fruits (lemon), Jam, Nuts (chestnut), Wine (red), chapter: Further Mycotoxins and Microbial Metabolites

Wine (rosé) may contain the following mycotoxins:

Alternaria Toxins

Alternariol

incidence: 1/4, conc.: <LOQ, sample origin: supermarkets, grocery stores, wineries, and small stores in Entre Rios, Mendoza, and San Juan (provinces), Argentina, sample year: 2011, country: Argentina[498]

- Co-contamination: not reported
- Further contamination of plants and plant products other than cereals and their products reported in the present book/article[498]: Juices (grapefruit juice), Juices (orange juice), AOH, literature[498]; Wine (red), Wine (white), AME and AOH, literature[498]

Aspergillus and ***Penicillium*** Toxins

Ochratoxin A

incidence: 18/51, conc. range: ≤2.4 μg/l, sample origin: importers, producers, stores, wine and sparkling wine cellars, Germany, sample year: unknown, country: Germany[110]

- Co-contamination: not reported
- Further contamination of plants and plant products other than cereals and their products reported in the present book/article[110]: Juices (blackcurrant juice), Juices (carrot juice), Juices (grape juice), Juices (tomato juice), Sauces (pepper- and barbecue sauce), Sauces (soy sauce), Spices (mustard), Tomato ketchup, Vinegar (apple), Vinegar (balsamico), Vinegar (wine), Wine (red), Wine (white), OTA, literature[110]

incidence: 13/20*, conc. range: ≤1.16 μg/l, sample origin: manufacturers´ stock and Greek retail market, Greece, sample year: 1995–1999, country: Greece[327], *rosé wine, dry

- Co-contamination: not reported
- Further contamination of plants and plant products other than cereals and their products reported in the present book/article[327]: Currants, Sultanas, Wine (red), Wine (white), Wine, miscellaneous, OTA, literature[327]

For detailed information see the article.

incidence: 1/5*, conc.: 0.0354 μg/l, sample origin: from Brazil at markets in Rio de Janeiro (state), Brazil, sample year: unknown, country: Brazil/Argentina[377], *table wines

- Co-contamination: not reported
- Further contamination of plants and plant products other than cereals and their products reported in the present book/article[377]: Juices (grape juice), Pulp (grape pulp), Wine (red), Wine (white), OTA, literature[377]

incidence: 3/3, conc. range: 0.04–0.54 μg/l, ∅ conc.: 0.223 μg/l, sample origin: supermarkets in Rabat (capital), Morocco, sample year: unknown, country: Morocco/France[493]

- Co-contamination: not reported
- Further contamination of plants and plant products other than cereals and their products reported in the present book/article[493]: Juices, Wine (red), Wine (white), OTA, literature[493]

incidence: 12/13?, conc. range: ≤0.123 µg/l, sample origin: imported from retail outlets in Berne (capital), Switzerland, sample year: 1994/1995, country: Switzerland[496]
- Co-contamination: not reported
- Further contamination of plants and plant products other than cereals and their products reported in the present book/article[496]: Juices (grape juice), Wine (red), Wine (white), Wine, miscellaneous, OTA, literature[496]

incidence: 29/32, conc. range: ≤0.161 µg/l, sample origin: partly from different European countries and Spain, sample year: 1997, country: Spain[511]
- Co-contamination: not reported
- Further contamination of plants and plant products other than cereals and their products reported in the present book/article[511]: Wine (red), Wine (white), Wine, miscellaneous, OTA, literature[511]

incidence: 69/75, conc. range: ≤4.07 µg/l, sample origin: food stores in Rome (capital), Italy, sample year: 2002–2004, country: Italy[516]
- Co-contamination: not reported
- Further contamination of plants and plant products other than cereals and their products reported in the present book/article[516]: Wine (red), Wine (white), Wine, miscellaneous, OTA, literature[516]

For detailed information see the article.

incidence: 5/11, conc. range: LOD-LOQ (2 sa.), 0.25–0.40 µg/l (3 sa.), sample origin: supermarkets in Utiel Requena (wine production area), Valencia (community), Spain, sample year: July 2003 (production year: 2000), country: Spain[517]

incidence: 7/10, conc. range: LOD-LOQ (2 sa.), 0.11–0.46 µg/l (5 sa.), sample origin: supermarkets in Utiel Requena (wine production area), Valencia (community), Spain, sample year: July 2003 (production year: 2001), country: Spain[517]
- Co-contamination: not reported
- Further contamination of plants and plant products other than cereals and their products reported in the present book/article[517]: Wine (red), Wine (white), Wine, miscellaneous, OTA, literature[517]

incidence: 1/5, conc.: 0.28 µg/l, sample origin: produced in Apulia (region) purchased in markets and local retail stores, Italy, sample year: 2003 (production year: 2000), country: Italy/Spain[518]

incidence: 4/4, conc. range: >0.10–0.50 µg/l (2 sa.), >0.50–1 µg/l (1 sa.), 1.04 µg/l (1 sa.), sample origin: produced in Apulia (region) purchased in markets and local retail stores, Italy, sample year: 2003 (production year: 2002), country: Italy/Spain[518]
- Co-contamination: not reported
- Further contamination of plants and plant products other than cereals and their products reported in the present book/article[518]: Wine (red), Wine (white), Wine miscellaneous, OTA, literature[518]

incidence: 5/6*, conc. range: ≤1.15 µg/l, ∅ conc.: 0.804 µg/l, sample origin: bottled wine from local retails, Italy, sample year: unknown, country: Italy[522], *commercial red wine

incidence: 2/2*, conc. range: 0.41–0.64 µg/l, ∅ conc.: 0.525 µg/l, sample origin: home-made wine from small farmers, Italy, sample year: unknown, country: Italy[522], *homemade red wine
- Co-contamination: not reported
- Further contamination of plants and plant products other than cereals and their products reported in the present book/article[522]: Wine (red), Wine (white), Wine, miscellaneous, OTA, literature[522]

incidence: 4/4, conc. range: 0.03–2.23 µg/l, ∅ conc.: 0.86 µg/l, sample origin: wine producers in Aegean (region), Turkey, sample year: 2000/2001, country: Turkey[611]
- Co-contamination: not reported
- Further contamination of plants and plant products other than cereals and their products reported in the present book/article[611]: Wine (red), Wine (white), OTA, literature[611]

For detailed information see the article.

incidence: 2/4, conc. range: 0.28–0.40 µg/l, sample origin: retail shops and homemade wine in Apulia (region), Italy, sample year: unknown, country: Italy[612]
- Co-contamination: not reported
- Further contamination of plants and plant products other than cereals and their products reported in the present book/article[612]: Wine (red), Wine (white), OTA, literature[612]

incidence: 5/7*, conc. range: 0.10–0.82 µg/l, ∅ conc.: 0.32 µg/l, sample origin: local stores, Italy, sample year: April–June 2004, country: Italy[613], *rosé wine, bottled
- Co-contamination: not reported
- Further contamination of plants and plant products other than cereals and their products reported in the present book/article[613]: Wine (red), Wine (white), OTA, literature[613]

incidence: 6/10*, conc. range: ≤0.38 µg/l, sample origin: different winemaking regions of Greece, sample year: 1999–2006, country: Greece/the UK[621], *rosé dry wine
- Co-contamination: not reported
- Further contamination of plants and plant products other than cereals and their products reported in the present book/article[621]: Wine, Wine (red), Wine (white), OTA, literature[621]

incidence: 18/51, conc. range: LOD/LOQ-0.9 µg/kg (16 sa.), ≤2.38 µg/kg (2 sa.), sample origin: Germany, sample year: 1995–1998, country: EU[698]

incidence: 3/5, conc. range: LOD/LOQ-≤0.13 µg/kg (3 sa.), sample origin: Greece, sample year: 1999, country: EU[698]

incidence: 2/4, conc. range: LOD/LOQ-≤0.283 µg/kg (2 sa.), sample origin: Italy, sample year: 1999, country: EU[698]

incidence: 24/26, conc. range: LOD/LOQ-≤0.155 µg/kg (24 sa.), sample origin: Spain, sample year: 1997, country: EU[698]
- Co-contamination: not reported
- Further contamination of plants and plant products other than cereals and their products reported in the present book/article[698]: Baby food, Beans (soybean), Chocolate, Cocoa, Coffee, Coffee beans, Currants, Fruits, dried, Herbs, Jam, Juices, Juices (grape juice), Legumes, Nuts, Oil seeds, Oils (olive oil), Peas, Peas (chickpea), Products (plant product), Raisins, Spices, Spices (cayenne pepper, chilli), Spices (nutmeg), Spices (paprika), Spices (pepper), Sultanas, Tea, Vinegar, Wine, Wine (red), Wine (white), Wine, miscellaneous, OTA, literature[698]

incidence: 1/10*, conc.: 2.36 µg/l, sample origin: France, Germany, and Italy, sample year: 1999–2007, country: Italy/Germany[699], *organic
- Co-contamination: not reported
- Further contamination of plants and plant products other than cereals and their products reported in the present book/article[699]: Wine (red), Wine (white), OTA, literature[699]

incidence: 2/3, conc. range: 0.002–0.0034 µg/l, ∅ conc.: 0.0027 µg/l, sample origin: imported from France* and Portugal*, Czech Republic, sample year: 2006/2007 purchased in 2009, country: Czech Republic[792], *contaminated sa.

- Co-contamination: not reported
- Further contamination of plants and plant products other than cereals and their products reported in the present book/article[792]: Wine (red), Wine (white), OTA, literature[792]

incidence: 1/3*, conc.: 0.007 µg/l, sample origin: semiarid and dry area of Rioja Alavesa, Spain, sample year: produced 1995–2008, country: Spain[826]

- Co-contamination: not reported
- Further contamination of plants and plant products other than cereals and their products reported in the present book/article[826]: Wine (red), Wine (white), OTA, literature[826]

For detailed information see the article.

incidence: 8/8*, conc. range: LOD-0.50 µg/l (2 sa.), 0.50–1.00 µg/l (1 sa.), >2.00 µg/l (4 sa., maximum: 2.52 µg/l), sample origin: Thessaly (region) as well as western and central Greece, sample year: produced 2007–2009, country: Greece[837], *róse wine, dry

- Co-contamination: not reported
- Further contamination of plants and plant products other than cereals and their products reported in the present book/article[837]: Wine (red), Wine (white), OTA, literature[837]

For detailed information see the article.

incidence: 8/9, conc. range: 0.09–0.64 µg/l, Ø conc.: 0.22 µg/l, sample origin: Tunisia, sample year: produced 2004–2006, country: Tunisia[852]

- Co-contamination: not reported
- Further contamination of plants and plant products other than cereals and their products reported in the present book/article[852]: Wine (red), Wine (white), OTA, literature[852]

For detailed information see the article.

incidence: 2/2, conc. range: LOD-0.1 µg/l (1 sa.), 0.161 µg/l (1 sa.), Ø conc.: 0.103 µg/l, sample origin: Thrace (region), Turkey, sample year: 2005/2006, country: Turkey[867]

incidence: 5/5, conc. range: LOD-0.1 µg/l (5 sa., maximum: 0.086 µg/l), Ø conc.: 0.041 µg/l, sample origin: Aegean (region), Turkey, sample year: 2005/2006, country: Turkey[867]

incidence: 1/1, conc.: 0.020 µg/l, sample origin: central Anatolia (region), Turkey, sample year: 2005/2006, country: Turkey[867]

incidence: 1/2, conc.: 0.035 µg/l, sample origin: east Anatolia (region), Turkey, sample year: 2005/2006, country: Turkey[867]

- Co-contamination: not reported
- Further contamination of plants and plant products other than cereals and their products reported in the present book/article[867]: Wine (red), Wine (white), OTA, literature[867]

incidence: 1/4, conc.: 0.22 µg/l, sample origin: 4 major wine-producing regions, China, sample year: 1999–2009, country: China[899]

- Co-contamination: not reported
- Further contamination of plants and plant products other than cereals and their products reported in the present book/article[899]: Wine (red), Wine (white), Wine, miscellaneous, OTA, literature[899]

For detailed information see the article.

Wine (white) may contain the following mycotoxins:

Alternaria Toxins

Alternariol

incidence: 3/5*, conc. range: 1.2–4.9 µg/kg, Ø conc.: 2.73 µg/kg, sample origin: unknown, sample year: unknown, country: Germany[46], *included "Spätlese," "Beerenauslese," and "Trockenbeerenauslese"

- Co-contamination: 2 sa. co-contaminated with AME and AOH; 1 sa. contaminated solely with AOH
- Further contamination of plants and plant products other than cereals and their products reported in the present book/article[46]: Juices (tomato juice), Paste (tomato paste), Puree, Tomato ketchup, Wine (red), AME and AOH, literature[46]; Juices (vegetable juice), Wine (white), AME, literature[46]

For detailed information see the article.

incidence: 6/6, conc. range: 0.10–7.59* µg/kg, sample origin: commercially available, sample year: unknown, country: Germany[160], *variety Riesling

- Co-contamination: 1 sa. co-contaminated with AME and AOH; 5 sa. contaminated solely with AOH
- Further contamination of plants and plant products other than cereals and their products reported in the present book/article[160]: Beverages, Juices (apple juice), Juices (grape juice), Juices (orange juice), Juices (tomato juice), Juices (vegetable juice), Wine (red), Wine, miscellaneous, AME and AOH, literature[160]; Wine (white), AME, literature[160]

incidence: 2/18, conc. range: 1.2–2.0 µg/l, Ø conc.: 1.6 µg/l, sample origin: retail shops, Hesse (federal state), Germany, sample year: 2009/2010, country: Germany[439]

- Co-contamination: not reported
- Further contamination of plants and plant products other than cereals and their products reported in the present book/article[439]: Juices (apple juice), Juice (tomato juice), Paste (tomato paste), Sauces (apple sauce), Tomato ketchup, AOH, literature[439]

incidence: 1/19, conc.: 1.48 µg/l, sample origin: domestic, sample year: unknown, country: Canada[492]

- Co-contamination: 1 sa. co-contaminated with AME and AOH
- Further contamination of plants and plant products other than cereals and their products reported in the present book/article[492]: Juices (cranberry juice), Juices (grape juice), Wine (red), AME and AOH, literature[492]; Wine (white), AME, literature[492]

incidence: 1/4, conc.: 0.67 µg/l, sample origin: imported from Argentina, Italy, and the USA, sample year: unknown, country: Canada[492]

- Co-contamination: not reported
- Further contamination of plants and plant products other than cereals and their products reported in the present book/article[492]: Juices (cranberry juice), Juices (grape juice), Wine (red), AME and AOH, literature[492]; Wine (white), AME, literature[492]

incidence: 4/53, conc. range: <LOQ-18 µg/l, sample origin: supermarkets, grocery stores, wineries, and small stores in Entre Rios, Mendoza, and San Juan (provinces), Argentina, sample year: 2011, country: Argentina[498]

- Co-contamination: not reported
- Further contamination of plants and plant products other than cereals and their products reported in the present book/article[498]: Juices (grapefruit juice), Juices (orange juice), Wine (rosé), AOH, literature[498]; Wine (red), AME and AOH, literature[498]; Wine (white), AME, literature[498]

Alternariol Methyl Ether

incidence: 3/5*, conc. range: 0.1–0.2 µg/kg, Ø conc.: 0.17 µg/kg, sample origin: unknown, sample year: unknown, country: Germany[46], *included the varieties "Spätlese," "Beerenauslese," and "Trockenbeerenauslese"

- Co-contamination: 2 sa. co-contaminated with AME and AOH; 1 sa. contaminated solely with AME
- Further contamination of plants and plant products other than cereals and their products reported in the present book/article[46]: Juices (tomato juice), Paste (tomato paste), Puree, Tomato ketchup, Wine (red), AME and AOH, literature[46]; Juices (vegetable juice), AME, literature[46]; Wine (white), AOH, literature[46]

For detailed information see the article.

incidence: 1/6, conc.: pr., sample origin: commercially available, sample year: unknown, country: Germany[160]

- Co-contamination: 1 sa. co-contaminated with AME and AOH
- Further contamination of plants and plant products other than cereals and their products reported in the present book/article[160]: Beverages, Juices (apple juice), Juices (grape juice), Juices (orange juice), Juices (tomato juice), Juices (vegetable juice), Wine (red), Wine, miscellaneous, AME and AOH, literature[160]; Wine (white), AOH, literature[160]

incidence: 2/19, conc. range: 0.02–0.059 μg/l, ∅ conc.: 0.0395 μg/l, sample origin: domestic, sample year: unknown, country: Canada[492]

- Co-contamination: 1 sa. co-contaminated with AME and AOH; 1 sa. contaminated solely with AME
- Further contamination of plants and plant products other than cereals and their products reported in the present book/article[492]: Juices (cranberry juice), Juices (grape juice), Wine (red), AME and AOH, literature[492]; Wine (white), AOH, literature[492]

incidence: 0/4, conc.: no contamination, sample origin: imported from Argentina, Italy, and the USA, Canada, sample year: unknown, country: Canada[492]

- Co-contamination: not reported
- Further contamination of plants and plant products other than cereals and their products reported in the present book/article[492]: Juices (cranberry juice), Juices (grape juice), Wine (red), AME and AOH, literature[492]; Wine (white), AOH, literature[492]

incidence: 3/53, conc. range: <LOQ-225 μg/l, sample origin: supermarkets, grocery stores, wineries, and small stores in Entre Rios, Mendoza, and San Juan (provinces), Argentina, sample year: 2011, country: Argentina[498]

- Co-contamination: not reported
- Further contamination of plants and plant products other than cereals and their products reported in the present book/article[498]: Juices (grapefruit juice), Juices (orange juice), Wine (rosé), Wine (white), AOH, literature[498]; Wine (red), AME and AOH, literature[498]

Aspergillus and ***Penicillium*** Toxins

Ochratoxin A

incidence: 14/58, conc. range: ≤1.4 μg/l, sample origin: importers, producers, stores, wine and sparkling wine cellars, Germany, sample year: unknown, country: Germany[110]

- Co-contamination: not reported
- Further contamination of plants and plant products other than cereals and their products reported in the present book/article[110]: Juices (blackcurrant juice), Juices (carrot juice), Juices (grape juice), Juices (tomato juice), Sauces (pepper- and barbecue sauce), Sauces (soy sauce), Spices (mustard), Tomato ketchup, Vinegar (apple), Vinegar (balsamico), Vinegar (wine), Wine (red), Wine (rosé), OTA, literature[110]

incidence: 63/118*, conc. range: ≤1.72 μg/l, sample origin: manufacturers' stock and Greek retail market, Greece, sample year: 1995–1999, country: Greece[327], *white wine, dry

- Co-contamination: not reported
- Further contamination of plants and plant products other than cereals and their products reported in the present book/article[327]: Currants, Sultanas, Wine (red), Wine (rosé), Wine, miscellaneous, OTA, literature[327]

For detailed information see the article.

incidence: 3/25*, conc. range: tr (2 sa.), 2.4 μg/l (1 sa.), sample origin: homemade wines and from retail shops, Portugal, sample year: unknown, country: Portugal[335]

- Co-contamination: not reported
- Further contamination of plants and plant products other than cereals and their products reported in the present book/article[335]: Wine (red), OTA, literature[335]

incidence: 125/204, conc. range: LOQ-0.1 μg/l (97 sa.), 0.1–0.5 μg/l (25 sa.), 0.5–2 μg/l (3 sa., maximum: 1.360 μg/l), ∅ conc.: 0.086 μg/l, sample origin: wineries in Piedmont (region), Italy, sample year: 2004–2008, country: Italy[337]

- Co-contamination: not reported
- Further contamination of plants and plant products other than cereals and their products reported in the present book/article[337]: Wine, Wine (red), OTA, literature[337]

incidence: 2/10*, conc. range: 0.0282 μg/l, ∅ conc.: 0.0283 μg/l?, sample origin: from Brazil at markets in Rio de Janeiro (state), Brazil, sample year: unknown, country: Brazil/Argentina[377], *table wines

incidence: 0/5*, conc.: no contamination, sample origin: imported from Argentina at markets in Rio de Janeiro (state), Brazil, sample year: unknown, country: Brazil/Argentina[377], *table wines

incidence: 0/4*, conc.: no contamination, sample origin: imported from Spain at markets in Rio de Janeiro (state), Brazil, sample year: unknown, country: Brazil/Argentina[377], *table wines

incidence: 1/5*, conc.: 0.0213 μg/l, sample origin: imported from France at markets in Rio de Janeiro (state), Brazil, sample year: unknown, country: Brazil/Argentina[377], *table wines

incidence: 1/4*, conc.: 0.0212, sample origin: imported from Italy at markets in Rio de Janeiro (state), Brazil, sample year: unknown, country: Brazil/Argentina[377], *table wines

incidence: 1/5*, conc.: 0.0283, sample origin: imported from Portugal at markets in Rio de Janeiro (state), Brazil, sample year: unknown, country: Brazil/Argentina[377], *table wines

- Co-contamination: not reported
- Further contamination of plants and plant products other than cereals and their products reported in the present book/article[377]: Juices (grape juice), Pulp (grape pulp), Wine (red), Wine (rosé), OTA, literature[377]

incidence: 7/7, conc. range: 0.028–0.18 μg/l, ∅ conc.: 0.073 μg/l, sample origin: supermarkets in Rabat (capital), Morocco, sample year: unknown, country: Morocco/France[493]

- Co-contamination: not reported
- Further contamination of plants and plant products other than cereals and their products reported in the present book/article[493]: Juices, Wine (red), Wine (rosé), OTA, literature[493]

incidence: 10/43, conc. range: LOD-0.049 μg/l (8 sa.), 0.050–0.500 μg/l (2 sa., maximum: 0.156 μg/l), sample origin: Canada, sample year: 1999–2002, country: Canada[495] incidence: 22/53, conc. range: LOD-0.049 µg/l (10 sa.), 0.050–0.500 µg/l (10 sa.), >0.500 µg/l (2 sa., maximum: 3.720 µg/l), sample origin: imported, Canada, sample year: 1999–2002, country: Canada[495]

- Co-contamination: not reported
- Further contamination of plants and plant products other than cereals and their products reported in the present book/article[495]: Juices (grape juice), Wine (red), OTA, literature[495]

incidence: 6/22?, conc. range: ≤0.178 µg/l, sample origin: partly imported purchased in retail outlets in Berne (capital), Switzerland, sample year: 1994/1995, country: Switzerland[496]

- Co-contamination: not reported
- Further contamination of plants and plant products other than cereals and their products reported in the present book/article[496]: Juices (grape juice), Wine (red), Wine (rosé), Wine, miscellaneous, OTA, literature[496]

incidence: 4/50*, conc. range: 0.11–1.13 µg/l, ∅ conc.: 0.37 µg/l, sample origin: supermarkets and small shops, Spain, sample year: 2002, country: Spain[508], *included Costers del Segre, Penedés, Rioja, Utiel Requena, and ordinary wine

- Co-contamination: not reported
- Further contamination of plants and plant products other than cereals and their products reported in the present book/article[508]: Must, Wine (red), Wine, miscellaneous, OTA, literature[508]

incidence: 7/13*, conc. range: 0.02–0.5 µg/l (5 sa.), 0.5–1.0 µg/l (2 sa., maximum: 0.87 µg/l), sample origin: viticultural regions, Greece, sample year: 1995–1997, 1999, country: Greece/France[509], *white wine, dry

incidence: 4/4*, conc. range: 0.02–0.5 µg/l (2 sa.), 0.5–1.0 µg/l (1 sa.), 3.20 µg/l (1 sa.), sample origin: viticultural regions, Greece, sample year: 1995–1997, 1999, country: Greece/France[509], *white wine, sweet

- Co-contamination: not reported
- Further contamination of plants and plant products other than cereals and their products reported in the present book/article[509]: Wine (red), OTA, literature[509]

For detailed information see the article.

incidence: 45/69, conc. range: ≤0.267 µg/l, sample origin: partly from different European countries, California, and Spain, sample year: 1997, country: Spain[511]

- Co-contamination: not reported
- Further contamination of plants and plant products other than cereals and their products reported in the present book/article[511]: Wine (red), Wine (rosé), Wine, miscellaneous, OTA, literature[511]

incidence: 4/10*, conc. range: ≤0.126 µg/l, sample origin: wine-producing regions of South Africa, sample year: unknown, country: South Africa[512], *Chardonnay

incidence: 3/6*, conc. range: ≤0.057 µg/l, sample origin: wine-producing regions of South Africa, sample year: unknown, country: South Africa[512], *Chenin Blanc

incidence: 2/6*, conc. range: ≤0.110 µg/l, sample origin: wine-producing regions of South Africa, sample year: unknown, country: South Africa[512], *Sauvignon Blanc

incidence: 3/3*, conc. range: ≤0.026 µg/l, sample origin: wine-producing regions of South Africa, sample year: unknown, country: South Africa[512], *Gewürztraminer

incidence: 1/1*, conc.: 0.017 µg/l, sample origin: wine-producing regions of South Africa, sample year: unknown, country: South Africa[512], *Rhine Riesling

incidence: 1/1*, conc.: 0.027 µg/l, sample origin: wine-producing regions of South Africa, sample year: unknown, country: South Africa[512], *"Weisser Riesling"

- Co-contamination: not reported
- Further contamination of plants and plant products other than cereals and their products reported in the present book/article[512]: Wine (red), Wine, miscellaneous, OTA, literature[512]

incidence: 128/290, conc. range: ≤1.95 µg/l, sample origin: food stores in Rome (capital), Italy, sample year: 2002–2004, country: Italy[516]

- Co-contamination: not reported
- Further contamination of plants and plant products other than cereals and their products reported in the present book/article[516]: Wine (red), Wine (rosé), Wine, miscellaneous, OTA, literature[516]

For detailed information see the article.

incidence: 0/11, conc.: no contamination, sample origin: supermarkets in Valencia (city), Valencia (community), Spain, sample year: July 2003 (production year: 2000), country: Spain[517]

incidence: 1/10, conc.: 0.09 µg/l, sample origin: supermarkets in Valencia (city), Valencia (community), Spain, sample year: July 2003 (production year: 2001), country: Spain[517]

incidence: 3/3, conc. range: 0.32–0.76 µg/l (5 sa.), sample origin: food stores in Valencia (city), Valencia (community), Spain, sample year: July 2003 (production year: 2003), country: Spain[517]

- Co-contamination: not reported
- Further contamination of plants and plant products other than cereals and their products reported in the present book/article[517]: Wine (red), Wine (rosé), Wine, miscellaneous, OTA, literature[517]

incidence: 0/5, conc.: no contamination, sample origin: produced in Apulia (region) purchased in markets and local retail stores, Italy, sample year: 2003 (production year: 2000), country: Italy/Spain[518]

incidence: 0/6, conc.: no contamination, sample origin: produced in Apulia (region) purchased in markets and local retail stores, Italy, sample year: 2003 (production year: 2001), country: Italy/Spain[518]

incidence: 4/10, conc. range: >LOD-0.10 µg/l (3 sa.), 0.21 µg/l (1 sa.), sample origin: produced in Apulia (region) purchased in markets and local retail stores, Italy, sample year: 2003 (production year: 2002), country: Italy/Spain[518]

- Co-contamination: not reported
- Further contamination of plants and plant products other than cereals and their products reported in the present book/article[518]: Wine (red), Wine (rosé), Wine miscellaneous, OTA, literature[518]

incidence: 2/7*, conc. range: ≤0.06 µg/l, ∅ conc.: 0.045 µg/l, sample origin: bottled wine from local retails, Italy, sample year: unknown, country: Italy[522], *commercial white wine

incidence: 2/2*, conc. range: 0.10–0.97 µg/l, ∅ conc.: 0.535 µg/l, sample origin: home-made wine from small farmers, Italy, sample year: unknown, country: Italy[522], *homemade white wine

- Co-contamination: not reported

- Further contamination of plants and plant products other than cereals and their products reported in the present book/article[522]: Wine (red), Wine (rosé), Wine, miscellaneous, OTA, literature[522]

incidence: 7/27, conc. range: <0.01–≤0.03 µg/l (3 sa.), 0.03 ≤0.20 µg/l (4 sa., maximum: 0.08 µg/l), sample origin: different states of the USA, sample year: 1995–2001, country: the USA[523]

- Co-contamination: not reported
- Further contamination of plants and plant products other than cereals and their products reported in the present book/article[523]: Wine (red), Wine, miscellaneous, OTA, literature[523]

incidence: 5/28, conc. range: ≤0.022 µg/l, ∅ conc.: 0.020 µg/l, sample origin: unknown, sample year: unknown, country: Japan[605]

- Co-contamination: not reported
- Further contamination of plants and plant products other than cereals and their products reported in the present book/article[605]: Wine (red), OTA, literature[605]

incidence: 31/32*, conc. range: ≤0.156 µg/l, ∅ conc.: 0.097 µg/l, sample origin: wine-producing areas, Hungary, sample year: vintages 2000–2002, country: Hungary[606], *sa. of white bottled quality wine and directly from wineries

- Co-contamination: not reported
- Further contamination of plants and plant products other than cereals and their products reported in the present book/article[606]: Wine (red), OTA, literature[606]

incidence: 6/6, conc. range: 0.154–0.208 µg/l, ∅ conc.: 0.185 µg/l, sample origin: EVENA, Spain, sample year: 1997, country: Spain[607]

incidence: 1/6, conc.: 0.192 µg/l, sample origin: EVENA, Spain, sample year: 1998, country: Spain[607]

- Co-contamination: not reported
- Further contamination of plants and plant products other than cereals and their products reported in the present book/article[607]: Wine (red), OTA, literature[607]

incidence: 6/6*, conc. range: 0.05–0.30 µg/l, ∅ conc.: 0.19 µg/l, sample origin: local retail stores in Cape Town (city), South Africa, sample year: 2000/2001, country: South Africa/Italy[608], *white wine, bottled

incidence: 2/2*, conc. range: 0.04–0.33 µg/l, ∅ conc.: 0.19 µg/l, sample origin: local retail stores in Cape Town (city), South Africa, sample year: 2000/2001, country: South Africa/Italy[608], *white wine, boxed

incidence: 7/7*, conc. range: 0.06–0.18 µg/l, ∅ conc.: 0.12 µg/l, sample origin: local retail stores in Cape Town (city), South Africa, sample year: 2000/2001, country: South Africa/Italy[608], *white wine, late harvest

incidence: 3/3, conc. range: 0.01–0.08 µg/l, ∅ conc.: 0.05 µg/l, sample origin: local retail stores in Cesena (city), Italy, sample year: 2000/2001, country: South Africa/Italy[608]

- Co-contamination: not reported
- Further contamination of plants and plant products other than cereals and their products reported in the present book/article[608]: Wine (red), OTA, literature[608]

incidence: 14/362*, conc. range: 0.051–0.100 µg/l, sample origin: from 21 countries, sample year: unknown, country: Canada[609], *white wine, bottled

- Co-contamination: not reported
- Further contamination of plants and plant products other than cereals and their products reported in the present book/article[609]: Wine (white), OTA, literature[609]

For detailed information see the article.

incidence: 33/219*, conc. range: <0.1 µg/l (15 sa.), 0.1–<0.2 µg/l (11 sa.), ≥0.2 µg/l (7 sa., maximum: 0.26 µg/l), sample origin: wine producers and retail outlets in Sydney (city), New South Wales (state), Australia, sample year: 1984–1986, 1997–2000, country: Australia[610], *sa. in bottles included sparkling, dessert, and fortified wine, and Chardonnay juices

incidence: 2/7*, conc. range: 0.05 µg/l, sample origin: wine producers and retail outlets in Sydney (city), New South Wales (state), Australia, sample year: 1984–1986, 1997–2000, country: Australia[610], *sa. in tanks

incidence: 8/31*, conc. range: <0.1 µg/l (4 sa.), 0.1–<0.2 µg/l (2 sa.), ≥0.2 µg/l (2 sa., maximum: 0.50 µg/l), sample origin: wine producers and retail outlets in Sydney (city), New South Wales (state), Australia, sample year: 1984–1986, 1997–2000, country: Australia[610], *sa. in casks

- Co-contamination: not reported
- Further contamination of plants and plant products other than cereals and their products reported in the present book/article[610]: Wine (red), OTA, literature[610]

For detailed information see the article.

incidence: 4/4, conc. range: 0.02–0.11 µg/l, ∅ conc.: 0.048 µg/l, sample origin: wine producers in Central Anatolia (region), Turkey, sample year: 2000, country: Turkey[611]

incidence: 4/4, conc. range: 0.09–0.34 µg/l, ∅ conc.: 0.21 µg/l, sample origin: wine producers in Aegean (region), Turkey, sample year: 2001, country: Turkey[611]

- Co-contamination: not reported
- Further contamination of plants and plant products other than cereals and their products reported in the present book/article[611]: Wine (red), Wine (rosé), OTA, literature[611]

For detailed information see the article.

incidence: 4/10, conc. range: 0.60–1.95 µg/l, sample origin: retail shops and home-made wine in Apulia (region), Italy, sample year: unknown, country: Italy[612]

- Co-contamination: not reported
- Further contamination of plants and plant products other than cereals and their products reported in the present book/article[612]: Wine (red), Wine (rosé), OTA, literature[612]

incidence: 8/13*, conc. range: 0.03–0.42 µg/l, ∅ conc.: 0.16 µg/l, sample origin: local stores, Italy, sample year: April–June 2004, country: Italy[613], *white wine, bottled

incidence: 2/3*, conc. range: 0.11–0.17 µg/l, ∅ conc.: 0.14 µg/l, sample origin: local stores, Italy, sample year: April–June 2004, country: Italy[613], *white wine, boxed (glass or Tetrapak packaging)

- Co-contamination: not reported
- Further contamination of plants and plant products other than cereals and their products reported in the present book/article[613]: Wine (red), Wine (rosé), OTA, literature[613]

incidence: 11/13, conc. range: tr-0.263 µg/l, sample origin: local stores in Naples and Salerno (cities), Italy, sample year: 2009, country: Italy[614]

- Co-contamination: not reported

- Further contamination of plants and plant products other than cereals and their products reported in the present book/article[614]: Wine (red), OTA, literature[614]

incidence: 4/7*, conc. range: 0.015–0.022 µg/l, sample origin: local store, Croatia, sample year: 2002/2003, country: Croatia[616], *white wine, bottled
- Co-contamination: not reported
- Further contamination of plants and plant products other than cereals and their products reported in the present book/article[616]: Wine (red), OTA, literature[616]

For detailed information see the article.

incidence: 12/14, conc. range: ≤0.010 µg/l, sample origin: domestic wines from local supermarket and retail stores, Japan, sample year: summer 2006, country: Japan[618]

incidence: 16/22, conc. range: ≤0.51 µg/l, sample origin: imported wines from local supermarket and retail stores, Japan, sample year: summer 2006, country: Japan[618]
- Co-contamination: not reported
- Further contamination of plants and plant products other than cereals and their products reported in the present book/article[618]: Wine (red), OTA, literature[618]

incidence: 31/49*, conc. range: ≤0.51 µg/l, sample origin: different winemaking regions of Greece, sample year: 1999–2006, country: Greece/the UK[621], *white wine, dry

incidence: 9/13*, conc. range: ≤1.16 µg/l, sample origin: different winemaking regions of Greece, sample year: 1999–2006, country: Greece/the UK[621], *white dessert wine
- Co-contamination: not reported
- Further contamination of plants and plant products other than cereals and their products reported in the present book/article[621]: Wine, Wine (red), Wine (rosé), OTA, literature[621]

For detailed information see the article.

incidence: 1/13, conc.: 0.42 µg/l, sample origin: domestic and imported wines purchased in local supermarkets, Japan, sample year: 2010, country: Japan[622]
- Co-contamination: not reported
- Further contamination of plants and plant products other than cereals and their products reported in the present book/article[622]: Wine (red), FB_1, FB_2, FB_3, and OTA, literature[622]; Wine (white), FB_1, literature[622]

incidence: 4/42, conc. range: LOD-0.2 µg/l (3 sa.), 0.53 µg/l (1 sa.), ∅ conc.: 0.23 µg/l, sample origin: supermarkets, China, sample year: 2008/2009, country: China[623]
- Co-contamination: not reported
- Further contamination of plants and plant products other than cereals and their products reported in the present book/article[623]: Wine (red), OTA, literature[623]

incidence: 1/7, conc.: 0.03 µg/l, sample origin: tropical winemaking region, Brazil, sample year: 2009, country: Brazil[624]
- Co-contamination: not reported
- Further contamination of plants and plant products other than cereals and their products reported in the present book/article[624]: Wine (red), OTA, literature[624]

incidence: 2/2*, conc. range: 0.10 µg/l, ∅ conc.: 0.10 µg/l, sample origin: imported from France, Spain, the USA, and Italy purchased in local stores, Italy, sample year: unknown, country: Italy[644], *white wine, bottled
- Co-contamination: not reported
- Further contamination of plants and plant products other than cereals and their products reported in the present book/article[644]: Wine (red), OTA, literature[644]

incidence: 5/5, conc. range: 0.25–1.80 µg/l, ∅ conc.: 0.87 µg/l, sample origin: mostly produced in different regions in Anatolia (region), Turkey, sample year: 2000–2004, country: Turkey/Egypt[671]
- Co-contamination: not reported
- Further contamination of plants and plant products other than cereals and their products reported in the present book/article[671]: Wine (red), OTA, literature[671]

incidence: 2/8, conc. range: <LOQ, sample origin: supermarkets and specialized stores, Chile, sample year: unknown, country: Chile[672]

incidence: 0/2, conc.: no contamination, sample origin: imported from Australia, France, Germany, Italy, and Spain and purchased in supermarkets and specialized stores, Chile, sample year: unknown, country: Chile[672]
- Co-contamination: not reported
- Further contamination of plants and plant products other than cereals and their products reported in the present book/article[672]: Wine (red), OTA, literature[672]

incidence: 8/9, conc. range: 0.2–1.6 µg/l, ∅ conc.: 0. 60 µg/l, sample origin: wineries, northern Italy, sample year: unknown, country: Italy[674]
- Co-contamination: not reported
- Further contamination of plants and plant products other than cereals and their products reported in the present book/article[674]: Must, Wine (red), OTA, literature[674]

incidence: 9/41, conc. range: ~0.011 µg/l (5 sa.), 0.011–0.033 µg/l (4 sa.), sample origin: retail stores and wholesale trade, Slovakia, sample year: 2005, country: Slovakia[684]
- Co-contamination: not reported
- Further contamination of plants and plant products other than cereals and their products reported in the present book/article[684]: Wine (red), OTA, literature[684]

For detailed information see the article.

incidence: 7/10, conc. range: LOD/LOQ-≤0.39 µg/kg (7 sa.), sample origin: Finland, sample year: 1999, country: EU[698]

incidence: 12/56, conc. range: LOD/LOQ-0.9 µg/kg (10 sa.), ≤1.36 µg/kg (2 sa.), sample origin: Germany, sample year: 1995–1998, country: EU[698]

incidence: 23/45, conc. range: LOD/LOQ-0.9 µg/kg (22 sa.), 1.17 µg/kg (1 sa.), sample origin: Greece, sample year: 1999, country: EU[698]

incidence: 7/20, conc. range: LOD/LOQ-0.9 µg/kg (5 sa.), 1.0–2.9 µg/kg (1 sa.), 8.864 µg/kg (1 sa.), sample origin: Italy, sample year: 1999, country: EU[698]

incidence: 36/44, conc. range: LOD/LOQ-≤0.267 µg/kg (36 sa.), sample origin: Spain, sample year: 1999, country: EU[698]

incidence: 2/16, conc. range: LOD/LOQ-0.9 µg/kg (1 sa.), 2.10 µg/kg (1 sa.), sample origin: the Netherlands, sample year: 1999, country: EU[698]

incidence: 0/4, conc.: no contamination, sample origin: the Netherlands, sample year: 1999, country: EU[698]
- Co-contamination: not reported
- Further contamination of plants and plant products other than cereals and their products reported in the present book/

article[698]: Baby food, Beans (soybean), Chocolate, Cocoa, Coffee, Coffee beans, Currants, Fruits, dried, Herbs, Jam, Juices, Juices (grape juice), Legumes, Nuts, Oil seeds, Oils (olive oil), Peas, Peas (chickpea), Products (plant product), Raisins, Spices, Spices (cayenne pepper, chilli), Spices (nutmeg), Spices (paprika), Spices (pepper), Sultanas, Tea, Vinegar, Wine, Wine (red), Wine (rosé), Wine, miscellaneous, OTA, literature[698]

incidence: 4/58*, conc. range: <1.00 µg/l (2 sa.), <2.00 µg/l (2 sa.), sample origin: France, Germany, and Italy, sample year: 1999–2007, country: Italy/Germany[699], *organic

- Co-contamination: not reported
- Further contamination of plants and plant products other than cereals and their products reported in the present book/article[699]: Wine (red), Wine (rosé), OTA, literature[699]

incidence: 2/33, conc. range: 0.0035–0.0173 µg/l, ∅ conc.: 0.00827 µg/l, sample origin: retail stores, specialized shops and delivered by a producer in Moravian subregions, historical country, Czech Republic, sample year: 2003–2009 purchased 2009, country: Czech Republic[792]

incidence: 8/13, conc. range: 0.0016–0.227 µg/l, ∅ conc.:. 0.00432 µg/l, sample origin: imported from Australia (south)*, France*, Greece*, Hungary, Italy*, Portugal*, and Spain*, Czech Republic, sample year: 2000, 2003–2008 purchased 2009, country: Czech Republic[792], *contaminated sa.

- Co-contamination: not reported
- Further contamination of plants and plant products other than cereals and their products reported in the present book/article[792]: Wine (red), Wine (rosé), OTA, literature[792]

incidence: 7/10*, conc. range: ≤0.086 µg/l, sample origin: supermarkets and specialized shops, Spain, sample year: different years, country: Spain[798], *Manzanilla (white grapes used mainly Palomino and Pedro Ximenez)

incidence: 5/10*, conc. range: ≤0.110 µg/l, sample origin: supermarkets and specialized shops, Spain, sample year: different years, country: Spain[798], *Fino (white grapes used mainly Palomino and Pedro Ximenez)

incidence: 10/10*, conc. range: 0.042–0.413 µg/l, ∅ conc.: 0.144 µg/l, sample origin: supermarkets and specialized shops, Spain, sample year: different years, country: Spain[798], *Amontillado (white grapes used mainly Palomino and Pedro Ximenez)

incidence: 10/10*, conc. range: 0.117–0.642 µg/l, ∅ conc.: 0.319 µg/l, sample origin: supermarkets and specialized shops, Spain, sample year: different years, country: Spain[798], *Oloroso (white grapes used mainly Palomino and Pedro Ximenez)

- Co-contamination: not reported
- Further contamination of plants and plant products other than cereals and their products reported in the present book/article[798]: not reported

For detailed information see the article.

incidence: 5/12, conc. range: ≤0.036 µg/kg, sample origin: 12 regions, Czech Republic, sample year: 2011–2013, country: Czech Republic[811]

- Co-contamination: not reported
- Further contamination of plants and plant products other than cereals and their products reported in the present book/article[811]: Beans, Chocolate, Coffee, Confectionery, Figs, Herbs and Spices, Juices, Licorice, Powder (cocoa powder), Products (ginger product), Raisins, Spices (caraway), Spices (cayenne pepper, chilli), Spices (coriander), Spices (ginger), Spices (mustard), Spices (nutmeg), Spices (paprika), Spices (pepper), Tea, Wine (red), OTA, literature[811]

incidence: 2/3, conc. range: 0.013–0.015 µg/l, ∅ conc.: 0.014 µg/l, sample origin: semiarid and dry area of Rioja Alavesa, Spain, sample year: produced 1995–2008, country: Spain[826]

- Co-contamination: not reported
- Further contamination of plants and plant products other than cereals and their products reported in the present book/article[826]: Wine (rosé), Wine (red), OTA, literature[826]

For detailed information see the article.

incidence: 23/26*, conc. range: LOD-0.50 µg/l (21 sa.), 0.50–1.00 µg/l (2 sa., maximum: 0.56 µg/l), sample origin: Thessaly (region) as well as western and central Greece, sample year: produced 2007–2009, country: Greece[837], *white wine, dry

incidence: 4/4*, conc. range: LOD-0.50 µg/l (3 sa.), 2.30 µg/l (1 sa.), sample origin: Thessaly (region) as well as western and central Greece, sample year: produced 2007–2009, country: Greece[837], *white wine, sweet

- Co-contamination: not reported
- Further contamination of plants and plant products other than cereals and their products reported in the present book/article[837]: Wine (rosé), Wine (red), OTA, literature[837]

For detailed information see the article.

incidence: 6/7, conc. range: 0.11–1.5 µg/l, ∅ conc.: 0.59 µg/l, sample origin: Tunisia, sample year: produced 2004–2006, country: Tunisia[852]

- Co-contamination: not reported
- Further contamination of plants and plant products other than cereals and their products reported in the present book/article[852]: Wine (red), Wine (rosé), OTA, literature[852]

For detailed information see the article.

incidence: 9/9, conc. range: LOD-0.1 µg/l (2 sa.), 0.1–0.5 µg/l (6 sa.), 0.618 µg/l (1 sa.), ∅ conc.: 0.249 µg/l, sample origin: Thrace (region), Turkey, sample year: 2005/2006, country: Turkey[867]

incidence: 14/14, conc. range: LOD-0.1 µg/l (11 sa.), 0.1–0.5 µg/l (3 sa., maximum: 0.192 µg/l), ∅ conc.: 0.051 µg/l, sample origin: Aegean (region), Turkey, sample year: 2005/2006, country: Turkey[867]

incidence: 6/11, conc. range: LOD-0.1 µg/l (6 sa., maximum: 0.072 µg/l), ∅ conc.: 0.029 µg/l, sample origin: central Anatolia (region), Turkey, sample year: 2005/2006, country: Turkey[867]

- Co-contamination: not reported
- Further contamination of plants and plant products other than cereals and their products reported in the present book/article[867]: Wine (red), Wine (rosé), OTA, literature[867]

incidence: 1/34*, conc.: 0.07 µg/l, sample origin: 4 major wine-producing regions, China, sample year: 1999–2009, country: China[899], *white wine, dry

- Co-contamination: not reported
- Further contamination of plants and plant products other than cereals and their products reported in the present book/article[899]: Wine (red), Wine (rosé), Wine, miscellaneous, OTA, literature[899]

For detailed information see the article.

Fusarium Toxins

Fumonisin/s

Fumonisin B_1

incidence: 2/13, conc. range: <1.0 µg/l, sample origin: domestic and imported wines purchased in local supermarkets, Japan, sample year: 2010, country: Japan[622]

- Co-contamination: not reported
- Further contamination of plants and plant products other than cereals and their products reported in the present book/article[622]: Wine (red), FB_1, FB_2, FB_3, and OTA, literature[622]; Wine (white), OTA, literature[622]

Fumonisin B_2

incidence: 1/13, conc.: 1.0 µg/l, sample origin: different countries, sample year: 2002/2003, 2008, country: Denmark[625]

- Co-contamination: not reported
- Further contamination of plants and plant products other than cereals and their products reported in the present book/article[625]: Wine (red), Wine, miscellaneous, FB_2, literature[625]

Type A Trichothecenes

T-2 Toxin

incidence: 16/71, conc. range: 0.23–0.76 µg/l, sample origin: imported from Australia, Europe, and South America as well as domestic, the USA, sample year: 2008–2010, country: USA[666]

- Co-contamination: not reported
- Further contamination of plants and plant products other than cereals and their products reported in the present book/article[666]: Wine (red), OTA, literature[666]

For detailed information see the article.

Wine, miscellaneous may contain the following mycotoxins:

Alternaria Toxins

Alternariol

incidence: 2/2*, conc. range: 2.04–2.70 µg/kg, ∅ conc.: 2.37 µg/kg, sample origin: commercially available, sample year: unknown, country: Germany[160], *mulled wine

- Co-contamination: 1 sa. co-contaminated with AME and AOH; 1 sa. contaminated solely with AOH
- Further contamination of plants and plant products other than cereals and their products reported in the present book/article[160]: Beverages, Juices (apple juice), Juices (grape juice), Juices (orange juice), Juices (tomato juice), Juices (vegetable juice), Wine (red), Wine (white), AME and AOH, literature[160]; Wine, miscellaneous, AME, literature[160]

incidence: 2/2*, conc. range: tr, sample origin: local market, China, sample year: unknown, country: China[724], *sorghum wine

incidence: 1/1*, conc.: 0.04 µg/l, sample origin: local market, China, sample year: unknown, country: China[724], *rice wine

incidence: 1/1*, conc.: 0.06 µg/l, sample origin: local market, China, sample year: unknown, country: China[724], *pine nuts wine

incidence: 2/2*, conc. range: 0.04–0.13 µg/l, ∅ conc.: 0.085 µg/l, sample origin: local market, China, sample year: unknown, country: China[724], *buckwheat wine

incidence: 2/2*, conc. range: tr, sample origin: local market, China, sample year: unknown, country: China[724], *barley wine

incidence: 1/1*, conc.: 0.05 µg/l, sample origin: local market, China, sample year: unknown, country: China[724], *medicinal wine

incidence: 1/1*, conc.: 0.04 µg/l, sample origin: local market, China, sample year: unknown, country: China[724], *walnut wine

incidence: 1/1*, conc.: 0.70 µg/l, sample origin: local market, China, sample year: unknown, country: China[724], *millet wine

- Co-contamination: 10 sa. co-contaminated with AME and AOH; 1 sa. co-contaminated with AME, AOH, and TA (walnut wine)
- Further contamination of plants and plant products other than cereals and their products reported in the present book/article[724]: Juices (apple juice), AME, AOH, and TA, literature[724]; Wine, miscellaneous, AME and TA, literature[724]

Alternariol Methyl Ether

incidence: 1/2*, conc.: pr., sample origin: commercially available, sample year: unknown, country: Germany[160], *mulled wine

- Co-contamination: 1 sa. contaminated solely with AOH
- Further contamination of plants and plant products other than cereals and their products reported in the present book/article[160]: Beverages, Juices (apple juice), Juices (grape juice), Juices (orange juice), Juices (tomato juice), Juices (vegetable juice), Wine (red), Wine (white), AME and AOH, literature[160]; Wine, miscellaneous, AOH, literature[160]

incidence: 2/2*, conc. range: 0.03–0.06 µg/l, ∅ conc.: 0.045 µg/l, sample origin: local market, China, sample year: unknown, country: China[724], *sorghum wine

incidence: 1/1*, conc.: 0.05 µg/l, sample origin: local market, China, sample year: unknown, country: China[724], *rice wine

incidence: 1/1*, conc.: 0.03 µg/l, sample origin: local market, China, sample year: unknown, country: China[724], *pine nuts wine

incidence: 2/2*, conc. range: tr, sample origin: local market, China, sample year: unknown, country: China[724], *buckwheat wine

incidence: 2/2*, conc. range: tr, sample origin: local market, China, sample year: unknown, country: China[724], *barley wine

incidence: 1/1*, conc.: 0.08 µg/l, sample origin: local market, China, sample year: unknown, country: China[724], *medicinal wine

incidence: 1/1*, conc.: 0.03 µg/l, sample origin: local market, China, sample year: unknown, country: China[724], *walnut wine

incidence: 1/1*, conc.: tr, sample origin: local market, China, sample year: unknown, country: China[724], *millet wine

- Co-contamination: 10 sa. co-contaminated with AME and AOH; 1 sa. co-contaminated with AME, AOH, and TA (walnut wine)
- Further contamination of plants and plant products other than cereals and their products reported in the present book/article[724]: Juices (apple juice), AME, AOH, and TA, literature[724]; Wine, miscellaneous, AME and AOH, literature[724]

Tenuazonic Acid

incidence: 1/1*, conc.: 0.31 µg/l, sample origin: local market, China, sample year: unknown, country: China[724], *walnut wine

- Co-contamination: 1 sa. co-contaminated with AME, AOH, and TA
- Further contamination of plants and plant products other than cereals and their products reported in the present book/article[724]: Juices (apple juice), AME, AOH, and TA, literature[724]; Wine, miscellaneous, AME and AOH, literature[724]

Aspergillus Toxins

Aflatoxin B_1

incidence: 2/56*, conc. range: 5–10 µg/kg, Ø conc.: 7.50 µg/kg, sample origin: small markets in several urban Shanghai (city) areas, China, sample year: April 1988–April 1989, country: China/the USA[132], *yellow rice wine

- Co-contamination: not reported
- Further contamination of plants and plant products other than cereals and their products reported in the present book/article[132]: Beans (kidney bean), Butter (peanut butter), Curd (bean curd), Nuts (peanut), Oils (rapeseed oil), Oils (soybean oil), Paste (bean paste), Sauces (soy sauce), Tofu, AFB_1, literature[132]

Aspergillus and ***Penicillium*** Toxins

Ochratoxin A

incidence: 1/2*, conc.: 0.13 µg/l, sample origin: Spanish and Portuguese markets as well as Italian and Spanish distributors, Spain, sample year: unknown, country: Spain[158], *sparkling wine

incidence: 13/25*, conc. range: 0.10–0.50 µg/l (8 sa.), 0.50 to ≤0.96 µg/l (5 sa.), sample origin: Spanish and Portuguese markets as well as Italian and Spanish distributors, Spain, sample year: unknown, country: Spain[158], *fortified wines long aging in wooden casks

incidence: 1/12*, conc.: 0.08 µg/l, sample origin: Spanish and Portuguese markets as well as Italian and Spanish distributors, Spain, sample year: unknown, country: Spain[158], *fortified wines, second fermentation with Flor yeast, long aging in wooden casks

- Co-contamination: not reported
- Further contamination of plants and plant products other than cereals and their products reported in the present book/article[158]: Berries (grape), Must, OTA, literature[158]

incidence: 15/18*, conc. range: ≤2.82 µg/l, sample origin: manufacturers´ stock and Greek retail market, Greece, sample year: 1995–1999, country: Greece[327], *dessert wine

incidence: 6/8*, conc. range: ≤1.75 µg/l, sample origin: manufacturers´ stock and Greek retail market, Greece, sample year: 1995–1999, country: Greece[327], *retsina

- Co-contamination: not reported
- Further contamination of plants and plant products other than cereals and their products reported in the present book/article[327]: Currants, Sultanas, Wine (red), Wine (rosé), Wine (white), OTA, literature[327]

For detailed information see the article.

incidence: ?/6*, conc. range: ≤0.017 µg/l, sample origin: imported from Portugal purchased in retail outlets in Berne (capital), Switzerland, sample year: 1994/1995, country: Switzerland[496], *Port wine

incidence: 2/2*, conc. range: 0.029–0.054 µg/l, Ø conc.: 0.042 µg/l, sample origin: imported from Spain purchased in retail outlets in Berne (capital), Switzerland, sample year: 1994/1995, country: Switzerland[496], *Sherry

incidence: 2/2*, conc. range: 0.044–0.337 µg/l, Ø conc.: 0.191 µg/l, sample origin: imported from Italy purchased in retail outlets in Berne (capital), Switzerland, sample year: 1994/1995, country: Switzerland[496], *Marsala

incidence: 3/3*, conc. range: 0.049–0.451 µg/l, Ø conc.: 0.290 µg/l, sample origin: imported from Italy purchased in retail outlets in Berne (capital), Switzerland, sample year: 1994/1995, country: Switzerland[496], *Malaga

incidence: 0/2*, conc.: no contamination, sample origin: imported from Italy purchased in retail outlets in Berne (capital), Switzerland, sample year: 1994/1995, country: Switzerland[496], *Vermouth

- Co-contamination: not reported
- Further contamination of plants and plant products other than cereals and their products reported in the present book/article[496]: Juices (grape juice), Wine (red), Wine (rosé), Wine (white), OTA, literature[496]

incidence: 4/10*, conc. range: 0.14–0.71 µg/l, Ø conc.: 0.44 µg/l, sample origin: supermarkets and small shops, Spain, sample year: 2002, country: Spain[508], *sparkling wine

incidence: 9/20*, conc. range: 0.09–15.25 µg/l, Ø conc.: 4.47 µg/l, sample origin: supermarkets and small shops, Spain, sample year: 2002, country: Spain[508], *included Malaga, Mistelle, Muscatel, Sherry, and Vermouth

- Co-contamination: not reported
- Further contamination of plants and plant products other than cereals and their products reported in the present book/article[508]: Must, Wine (red), Wine (white), OTA, literature[508]

incidence: 35/47*, conc. range: ≤0.254 µg/l, sample origin: partly from different European countries and Spain, sample year: 1997, country: Spain[511], *Sherry type wine (aperitif)

incidence: 15/16*, conc. range: ≤2.540 µg/l, sample origin: Italy and Spain, sample year: 1997, country: Spain[511], *dessert wine

incidence: 10/12*, conc. range: ≤0.037 µg/l, sample origin: Spain, sample year: 1997, country: Spain[511], *sparkling wine

- Co-contamination: not reported
- Further contamination of plants and plant products other than cereals and their products reported in the present book/article[511]: Wine (red), Wine (rosé), Wine (white), OTA, literature[511]

incidence: 3/3*, conc. range: ≤2.672 µg/l, sample origin: wine-producing regions of South Africa, sample year: unknown, country: South Africa[512], *noble late harvest wine

incidence: 4/7*, conc. range: ≤0.168 µg/l, sample origin: wine-producing regions of South Africa, sample year: unknown, country: South Africa[512], *fortified wine

- Co-contamination: not reported
- Further contamination of plants and plant products other than cereals and their products reported in the present book/article[512]: Wine (red), Wine (white), OTA, literature[512]

incidence: 0/2*, conc.: no contamination, sample origin: northern Italy, sample year: 1998/1999, country: Italy[513], *special dessert wine

incidence: 5/8*, conc. range: ≤0.041 µg/l, sample origin: central Italy, sample year: 1998/1999, country: Italy[513], *special dessert wine

incidence: 4/5*, conc. range: ≤3.856 µg/l, sample origin: south Italy, sample year: 1998/1999, country: Italy[513], *special dessert wine

- Co-contamination: not reported
- Further contamination of plants and plant products other than cereals and their products reported in the present book/article[513]: Wine (red), OTA, literature[513]

incidence: 3/64*, conc. range: 0.05–0.08 μg/l**, ∅ conc.: 0.063 μg/l**, sample origin: northern region of Portugal, sample year: unknown, country: Portugal[515], *included standard Port wine (19 sa.), outstanding Port wine (12 sa.), adulteration of white Port wine (3 sa.**), and Vinho Verde (30 sa.)

- Co-contamination: not reported
- Further contamination of plants and plant products other than cereals and their products reported in the present book/article[515]: not reported

incidence: 18/28*, conc. range: ≤1.90 μg/l, sample origin: food stores in Rome (capital), Italy, sample year: 2002–2004, country: Italy[516], *dessert wine

- Co-contamination: not reported
- Further contamination of plants and plant products other than cereals and their products reported in the present book/article[516]: Wine (red), Wine (rosé), Wine (white), OTA, literature[516]

For detailed information see the article.

incidence: 8/13*, conc. range: LOD-LOQ (2 sa.), 0.10–0.40 μg/l (6 sa.), sample origin: supermarkets in Alicante and Valencia (cities), Valencia (community), Spain, sample year: July 2003 (production year: 2001), country: Spain[517], *dessert wines included Fondillón (2 sa. contaminated) and Moscatel (3 sa. contaminated)

- Co-contamination: not reported
- Further contamination of plants and plant products other than cereals and their products reported in the present book/article[517]: Wine (red), Wine (rosé), Wine (white), OTA, literature[517]

incidence: 7/9*, conc. range: LOD-0.10 μg/l (1 sa.), >0.10–0.50 μg/l (4 sa.), >0.50–1 μg/l (2 sa., maximum: 0.82 μg/l), sample origin: produced in Sicily (autonomous region, island) purchased in markets and local retail stores, Italy, sample year: 2003 (production year: 2000), country: Italy/Spain[518], *dessert wine

incidence: 5/10*, conc. range: LOD-0.10 μg/l (1 sa.), >0.10–0.50 μg/l (1 sa.), >0.50–1 μg/l (1 sa.), >1–2 μg/l (2 sa., maximum: 1.64 μg/l), sample origin: produced in Sicily (autonomous region, island) purchased in markets and local retail stores, Italy, sample year: 2003 (production year: 2001), country: Italy/Spain[518], *dessert wine

- Co-contamination: not reported
- Further contamination of plants and plant products other than cereals and their products reported in the present book/article[518]: Wine (red), Wine (rosé), Wine (white), OTA, literature[518]

incidence: 186/188*, conc. range: LOD-<LOQ (17 sa.), LOQ-0.500 μg/l (102 sa.), >0.500–2.000 μg/l (49 sa.), >2.000 μg/l (18 sa., maximum: 4.630 μg/l), sample origin: Spain[1], sample year: 2002–2005, country: Spain[519], *sweet wine

incidence: 47/49*, conc. range: LOD-<LOQ (12 sa.), LOQ-0.500 μg/l (29 sa.), >0.500–2.000 μg/l (6 sa., maximum, 1.218 μg/l), sample origin: France[519], sample year: 2002–2005, country: Spain[519], *sweet wine

incidence: 2/3*, conc. range: LOD-<LOQ, sample origin: Germany, sample year: 2002–2005, country: Spain[519], *sweet wine

incidence: 3/5*, conc. range: LOD-<LOQ (2 sa.), 0.029 μg/l (1 sa.), sample origin: Italy, sample year: 2002–2005, country: Spain[519], *sweet wine

incidence: 1/1*, conc.: 0.026 μg/l, sample origin: Japan, sample year: 2002–2005, country: Spain[519], *sweet wine

incidence: 9/9*, conc. range: LOD-<LOQ (7 sa.), LOQ-0.500 μg/l (2 sa., maximum: 0.090 μg/l), sample origin: Austria, sample year: 2002–2005, country: Spain[519], *sweet wine

incidence: 1/1*, conc.: <0.020 μg/l, sample origin: Canada, sample year: 2002–2005, country: Spain[519], *sweet wine

incidence: 9/9*, conc. range: LOD-<LOQ (3 sa.), LOQ-0.500 μg/l (6 sa., maximum: 0.070 μg/l), sample origin: Chile, sample year: 2002–2005, country: Spain[519], *sweet wine

incidence: 2/2*, conc. range: 0.020–0.047 μg/l, sample origin: Slovenia, sample year: 2002–2005, country: Spain[519], *sweet wine

incidence: 6/6*, conc. range: LOQ-0.500 μg/l (4 sa.), > 0.500–2.000 μg/l (1 sa.), 2.947 μg/l (1 sa.), sample origin: Greece, sample year: 2002–2005, country: Spain[519], *sweet wine

incidence: 2/2*, conc. range: LOD-<LOQ (1 sa.), 0.030 μg/l (1 sa.), sample origin: Hungary, sample year: 2002–2005, country: Spain[519], *sweet wine

incidence: 8/9*, conc. range: LOD-<LOQ (1 sa.), LOQ-0.500 μg/l (7 sa., maximum: 0.139 μg/l), sample origin: Portugal, sample year: 2002–2005, country: Spain[519], *sweet wine

incidence: 1/1*, conc.: 0.029 μg/l, sample origin: New Zealand, sample year: 2002–2005, country: Spain[519], *sweet wine

incidence: 1/1*, conc.: <0.020 μg/l, sample origin: South Africa, sample year: 2002–2005, country: Spain[519], *sweet wine

incidence: 2/2*, conc. range: <0.020 μg/l, sample origin: Switzerland, sample year: 2002–2005, country: Spain[519], *sweet wine

incidence: 1/1*, conc.: 0.090 μg/l, sample origin: the USA, sample year: 2002–2005, country: Spain[519], *sweet wine

- Co-contamination: not reported
- Further contamination of plants and plant products other than cereals and their products reported in the present book/article[519]: not reported

[1]For detailed information see the article.

incidence: 6/8*, conc. range: 0.56-3.42 μg/l, ∅ conc.: 2.42 μg/l, sample origin: Andalusia (autonomous community), Spain, sample year: probably 2005, country: Spain[520], *Moscatel

incidence: 8/8*, conc. range: 0.39-7.30 μg/l, ∅ conc.: 2.86 μg/l, sample origin: Spain, sample year: probably 2005, country: Spain[520], *Pedro Ximenez

- Co-contamination: not reported
- Further contamination of plants and plant products other than cereals and their products reported in the present book/article[520]: Must, OTA, literature[520]

incidence: 69/340*, conc. range: ≤2.1 μg/l (only 2 >0.5 μg/l), sample origin: Portuguese Demarcated Regions, Portugal, sample year: unknown, country: Portugal[521], *included Port wine, Vinho Verde, and wines from other regions of Portugal

- Co-contamination: not reported
- Further contamination of plants and plant products other than cereals and their products reported in the present book/article[521]: not reported

incidence: 1/1*, conc.: 0.29 μg/l, sample origin: bottled and homemade wine from local retails and farmers, Italy, sample year: unknown, country: Italy[522], *Marsala

- Co-contamination: not reported
- Further contamination of plants and plant products other than cereals and their products reported in the present book/article[522]: Wine (red), Wine (rosé), Wine (white), OTA, literature[522]

incidence: 1/2*, conc.: 0.014 µg/l, sample origin: different states of the USA, sample year: 1995–2001, country: the USA[523], *fruit/berry blended wines

- Co-contamination: not reported
- Further contamination of plants and plant products other than cereals and their products reported in the present book/article[523]: Wine (red), Wine (white), OTA, literature[523]

incidence: 0/2*, conc.: no contamination, sample origin: Finland, sample year: 1999, country: EU[698], *red fruit wine

incidence: 1/1*, conc.: LOD/LOQ-0.9 µg/kg (1 sa.), sample origin: Finland, sample year: 1999, country: EU[698], *white fruit wine

incidence: 1/1*, conc.: 1.04 µg/kg (1 sa.), sample origin: Germany, sample year: 1995–1998, country: EU[698], *sweet wine

incidence: 6/7*, conc. range: LOD/LOQ-0.9 µg/kg (4 sa.), ≤1.68 µg/kg (2 sa.), sample origin: Greece, sample year: 1999, country: EU[698], *sweet wine

incidence: 9/15*, conc. range: LOD/LOQ-0.9 µg/kg (5 sa.), 1.0–2.9 µg/kg (2 sa.), ≤3.856 µg/kg (2 sa.), sample origin: Italy, sample year: 1988–1997, country: EU[698], *sweet wine

incidence: 15/16*, conc. range: LOD/LOQ-0.9 µg/kg (7 sa.), ≤2.54 µg/kg (8 sa.), sample origin: Spain, sample year: 1997, country: EU[698], *sweet wine

incidence: 23/27*, conc. range: LOD/LOQ ≤0.254 µg/kg (23 sa.), sample origin: Spain, sample year: 1997, country: EU[698], *aperitif wine

incidence: 10/12*, conc. range: LOD/LOQ-≤0.037 µg/kg (10 sa.), sample origin: Spain, sample year: 1997, country: EU[698], *sparkling wine

- Co-contamination: not reported
- Further contamination of plants and plant products other than cereals and their products reported in the present book/article[698]: Baby food, Beans (soybean), Chocolate, Cocoa, Coffee, Coffee beans, Currants, Fruits, dried, Herbs, Jam, Juices, Juices (grape juice), Legumes, Nuts, Oil seeds, Oils (olive oil), Peas, Peas (chickpea), Products (plant product), Raisins, Spices, Spices (cayenne pepper, chilli), Spices (nutmeg), Spices (paprika), Spices (pepper), Sultanas, Tea, Vinegar, Wine, Wine (red), Wine (rosé), Wine (white), OTA, literature[698]

incidence: 16/36*, conc. range: <LOQ (15 sa.), 0.24 µg/l (1 sa.), sample origin: Ningxiahong Wolfberry Industrial Group Co., Ltd. purchased from retail markets in Beijing (capital), China, sample year: 2005–2010, country: China[771], *wolfberry fruit wine

- Co-contamination: not reported
- Further contamination of plants and plant products other than cereals and their products reported in the present book/article[771]: not reported

incidence: 1/4*, conc.: 0.10 µg/l (1 sa.), sample origin: 4 major wine-producing regions, China, sample year: 1999–2009, country: China[899], *ice wine

- Co-contamination: not reported
- Further contamination of plants and plant products other than cereals and their products reported in the present book/article[899]: Wine (red), Wine (rosé), Wine (white), OTA, literature[899]

For detailed information see the article.

incidence: 1/10*, conc.: 0.107 µg/l, sample origin: local stores and the supermarket in Guangzhou (city), Guangdong (province), China, sample year: unknown, country: China[912], *rice wine with 28%–45% alcohol content

- Co-contamination: not reported
- Further contamination of plants and plant products other than cereals and their products reported in the present book/article[912]: not reported

Fusarium Toxins

Fumonisin B_1

incidence: 1/15*, conc.: 17.3 µg/l, sample origin: markets in Beijing (capital) markets, China, sample year: unknown, country: China[628], *Chinese rice wine

- Co-contamination: not reported
- Further contamination of plants and plant products other than cereals and their products reported in the present book/article[628]: not reported

Fumonisin B_2

incidence: 1/6*, conc.: 2.8 µg/l, sample origin: different countries, sample year: 1998, 2002, 2007, country: Denmark[625], *Port wine

- Co-contamination: not reported
- Further contamination of plants and plant products other than cereals and their products reported in the present book/article[625]: Wine (red), Wine (white), FB_2, literature[625]

Yams may contain the following mycotoxins:

Aspergillus Toxins

Aflatoxin B_1

incidence: 0/18*, conc.: no contamination, sample origin: markets and pharmacies in different provinces, China, sample year: unknown, country: China[774], *non-moldy sa.

- Co-contamination: not reported
- Further contamination of plants and plant products other than cereals and their products reported in the present book/article[774]: Yams, FB_1, literature[774]

incidence: 3/4*, conc. range: 1.305–9.688 µg/kg, Ø conc.: 6.44 µg/kg, sample origin: markets and pharmacies in different provinces, China, sample year: unknown, country: China[774], *moldy sa.

- Co-contamination: 1 sa. co-contaminated with AFB_1, FB_1, and FB_2; 1 sa. co-contaminated with AFB_1 and FB_1; 1 sa. contaminated solely with AFB_1
- Further contamination of plants and plant products other than cereals and their products reported in the present book/article[774]: Yams, FB_1 and FB_2, literature[774]

Aflatoxin/s

incidence: 22/57*, Ø conc.: 88.8 µg/kg**, sample origin: greater Manila (capital) area, Philippines, sample year: unknown, country: Philippines[1], *yam and yam products, **AFS of pos sa.?

- Co-contamination: not reported
- Further contamination of plants and plant products other than cereals and their products reported in the present book/article[1]: Beans, Butter (peanut butter), Cassava, Cocoa, Coffee, Drinks (alcoholic drink), Nuts (cashew nut), Nuts (peanut), Products (coconut product), Rootcrops, Spices (cayenne pepper, chilli), Sweet potatoes, AF/S, literature[1]

Fusarium Toxins

Fumonisin/s

Fumonisin B_1

incidence: 1/18*, conc.: 0.878 µg/kg, sample origin: markets and pharmacies in different provinces, China, sample year: unknown, country: China[774], *non-moldy sa.

- Co-contamination: not reported
- Further contamination of plants and plant products other than cereals and their products reported in the present book/article[774]: not reported

incidence: 3/4*, conc. range: <LOQ-3.727 µg/kg, sample origin: markets and pharmacies in different provinces, China, sample year: unknown, country: China[774], *moldy sa.

- Co-contamination: 1 sa. co-contaminated with AFB_1, FB_1, and FB_2; 1 sa. co-contaminated with AFB_1 and FB_1; 1 sa. contaminated solely with FB_1
- Further contamination of plants and plant products other than cereals and their products reported in the present book/article[774]: Yams, AFB_1 and FB_2, literature[774]

Fumonisin B_2

incidence: 0/18*, conc.: no contamination, sample origin: markets and pharmacies in different provinces, China, sample year: unknown, countr y: China[774], *non-moldy sa.

- Co-contamination: not reported
- Further contamination of plants and plant products other than cereals and their products reported in the present book/article[774]: Yams, FB_1, literature[774]

incidence: 1/4*, conc.: <LOQ, sample origin: markets and pharmacies in different provinces, China, sample year: unknown, country: China[774], *moldy sa.

- Co-contamination: 1 sa. co-contaminated with AFB_1, FB_1, and FB_2
- Further contamination of plants and plant products other than cereals and their products reported in the present book/article[774]: Yams, AFB_1 and FB_1, literature[774]

Zinfandel see **Must**

Further Mycotoxins and Microbial Metabolites

Cake (peanut) (literature[738]) contained further mycotoxins/microbial metabolites:
Averufin, averufanin, cyclosporin A, emodin, equisetin, nidurufin, norsolorinic acid, O-methylsterigmatocystin, versicolorin A, and versicolorin C

Cassava (literature[823]) contained further mycotoxins/microbial metabolites:
Agistatin D, agroclavine, alamethicin, andrastin A, andrastin B, andrastin C, andrastin D, apicidin Ia, aspergillimide, asperloxine A, aspinolide B, aspterric acid, asterric acid, aurasperone B, aurasperone C, aurasperone G, aurofusarin, avenacein Y, averantin, averufanin, averufin, bikaverin, brevianamide F, butenolide, chanoclavine, chlamydosporol, chloramphenicol, chrysophanol, CJ 21058, clonostachydiol, curvularin, cycloaspeptide A, cycloechinullin, cyclosporin A, cyclosporin C, cyclosporin D, cyclosporin H, cytochalasin B, cytochalasin D, cytochalasin H, cytochalasin J, dechlorogriseofulvin, dihydrogriseofulvin, dinactin, emodin, equisetin, festuclavine, fumigaclavine A, fumonisin B_4, fusarinolic acid, geldanamycin, griseofulvin, hydrolysed fumonisin B_1, linamarin, lotaustralin, macrosporin, macrosporin A, malformin A2, malformin C, marcfortine A, marcfortine B, marcfortine C, methylsterigmatocystin, methylsulochrin, monactin, monocerin, nidurufin, nigericin, nonactin, norsolorinic acid, ophiobolin A, paspalin, paspalitrem A, paxillin, physicon, pseurotin A, pyranonigrin, radicicol, rubrofusarin, rugulosine, secalonic acid D, semi-vioxanth, semi-xanthomegnin, siccanol, skyrin, stachybotrylactam, sulochrin, tentoxin, terrein, trichodermin, tryptophol, valinomycin, versicolorin A, versicolorin C, zearalenone-4 sulfate, α-zearalenol, and ß-zearalenol

Coffee beans (literature[003]) contained further mycotoxins/microbial metabolites:
Fumonisin B_4 and fumonisin B_6

Dagwa (literature[659]) contained further mycotoxins/microbial metabolites:
15-Hydroxyculmorin, averantin, averufin, brevianamide F, butenolide, chlamydosporol, culmorin, cytochalasin J, cytochalasin H, deoxynivalenol-3-glucoside, equisetin, emodin, equisetin derivative (CJ21058), griseofulvin, macrosporin, marcfortine A, O-methylsterigmatocystin, monocerin, nigericin, physicon, radicicol, skyrin, tryptophol, versicolorin A, and versicolorin C

Dates (literature[815]) contained further mycotoxins/microbial metabolites:
Tentoxin

Dietary supplements (literature[286]) contained further mycotoxins/microbial metabolites:
Tentoxin

Flour (soybean flour) (literature[659]) contained further mycotoxins/microbial metabolites:
Aurofusarin, averantin, averufin, brevianamide F, butenolide, chlamydosporol, culmorin, cytochalasin H, deoxynivalenol-3-glucoside, emodin, equisetin, griseofulvin, macrosporin, O-methylsterigmatocystin, physicon, and tryptophol

Food (literature[117]) contained further mycotoxins/microbial metabolites:
Agroclavine, bacitracin, deoxynivalenol-3-glucoside, and elymoclavine

Food (literature[121]) contained further mycotoxins/microbial metabolites:
α-Zearalenol and ß-zearalenol

Fruits (apple) (literature[36]) contained further mycotoxins/microbial metabolites:
Acuminatopyrone, aurofusarin, and chlamydosporols

Fruits (lemon) (literature[850]) contained further mycotoxins/microbial metabolites:
Roridin A

Fruits (tomato) (literature[47]) contained further mycotoxins/microbial metabolites:
Tentoxin

Fruits, dried (literature[47]) contained further mycotoxins/microbial metabolites:
Tentoxin

Fruits, dried (literature[815]) contained further mycotoxins/microbial metabolites:
Tentoxin

Jam (literature[850]) contained further mycotoxins/microbial metabolites:
Chaetoglobosin A and emodin

Juices (literature[743]) contained further mycotoxins/microbial metabolites:
Tentoxin

Juices (sugarcane juice) (literature[300]) contained further mycotoxins/microbial metabolites:
Agroclavine, ascochlorin, asperglaucide, asperphenamate, aspinolid B, averfurin, berkedrimane B, chlorocitreorosein, citreorosein, emodin, fusapyron, gibberellic acid, griseofulvin, illicicolin B, integracin A, integracin B, iso rhodoptilometrin, monocerin, nidurufin, norlichexanthone, oxaline, quinolactacin A, skyrin, tryptophol, versicolorin A, versicolorin C, and xanthotoxin

M. Weidenbörner, *Mycotoxins in Plants and Plant Products*, https://doi.org/10.1007/978-3-319-92850-0_2

Juices (tomato juice) (literature[514]) contained further mycotoxins/microbial metabolites:
Alternariol-3-sulfate, alternariol monomethyl ether-3-glucoside, and tentoxin

Juices (tomato juice) (literature[883]) contained further mycotoxins/microbial metabolites:
Tentoxin

Kuru-kuru (literature[659]) contained further mycotoxins/microbial metabolites:
15-Hydroxyculmorin, averantin, averufin, brevianamide F, chlamydosporol, culmorin, cytochalasin H, cytochalasin J, emodin, equisetin, griseofulvin, marcfortine A, O-methylsterigmatocystin, monocerin, physicon, skyrin, tryptophol, versicolorin A, and versicolorin C

Meal (soybean meal) (literature[527]) contained further mycotoxins/microbial metabolites:
Zearalenol

Medicinal plants (literature[68]) contained further mycotoxins/microbial metabolites:
Chaetoglobosin A, deepoxydeoxynivalenol, deoxynivalenol-3-glucoside, penitrem A, verruculogen, zearalanone, α-zearalenol, and ß-zearalenol

Medicinal plants (literature[740]) contained further mycotoxins/microbial metabolites:
Zearalanone, α-zearalenol, and ß-zearalenol

Nuts (almond) (literature[868]) contained further mycotoxins/microbial metabolites:
Emodin, equisetin, macrosporin, secalonic acid D, tentoxin, and viridicatin

Nuts (chestnut) (literature[850]) contained further mycotoxins/microbial metabolites:
Avenacein Y, chaetoglobosin A, emodin, griseofulvin, sulochrin, and tentoxin

Nuts (hazelnut) (literature[868]) contained further mycotoxins/microbial metabolites:
Alamethicin F30, apicidin, curvularin, emodin, equisetin, macrosporin, 3-*O*-methylsterigmatocystin, pestalotin, physicon, tentoxin, viridicatin, and zearalenone-14-sulphate

Nuts (peanut) (literature[148]) contained further mycotoxins/microbial metabolites:
Andrastatin A, asperphenamate, curvularin, cyclo (L-Pro-L-Tyr), cyclo (L-Pro-L-Val), epiequisetin, equisetin, fonsecin, neoechinulin A, nigragillin, and further 36 microbial metabolites

Nuts (peanut) (literature[868]) contained further mycotoxins/microbial metabolites:
Curvularin, emodin, equisetin, macrosporin, 3-*O*-methylsterigmatocystin, pestalotin, tentoxin, viridicatin, and zearalenone-14-sulphate

Nuts (pistachio nut) (literature[129]) contained further mycotoxins/microbial metabolites:
Aflatoxicol I

Nuts (pistachio nut) (literature[868]) contained further mycotoxins/microbial metabolites:
Macrosporin and viridicatin

Oils (literature[840]) contained further mycotoxins/microbial metabolites:
α-Zearalenol

Oils (soybean oil) (literature[840]) contained further mycotoxins/microbial metabolites:
α-Zearalenol

Oils (vegetable oil) (literature[743]) contained further mycotoxins/microbial metabolites:
Tentoxin

Paste (peanut paste) (literature[659]) contained further mycotoxins/microbial metabolites:
15-Hydroxyculmorin, aurofusarin, averantin, averufin, brevianamide F, chlamydosporol, CJ21058, culmorin, curvularin, cytochalasin C, cytochalasin J, emodin, equisetin, griseofulvin, marcfortine A, macrosporin, O-methylsterigmatocystin, monocerin, physicon, skyrin, tryptophol, versicolorin A, and versicolorin C

Products (cassava product) (literature[794]) contained further mycotoxins/microbial metabolites:
Asperglaucide, asperphenamate, brevianamide F, cyclo (L-Pro-L-Val), cyclo (L-Pro-L-Tyr), emodin, fellutanine A, N-benzol-phenylalanine, rugulusovin, and tryptophol

Products (tomato product) (literature[743]) contained further mycotoxins/microbial metabolites:
Tentoxin

Products (tomato product) (literature[822]) contained further mycotoxins/microbial metabolites:
Tentoxin

Puree (literature[47]) contained further mycotoxins/microbial metabolites:
Tentoxin

Raisins (literature[600]) contained further mycotoxins/microbial metabolites:
Iso-fumonisin B_1, iso-fumonisin $B_{2,3}$, 3-epi-fumonisin B_3, fumonisin B_4, 3-epi-fumonisin B_4, fumonisin B_5, and iso-fumonisin B_5

Raisins (literature[601]) contained further mycotoxins/microbial metabolites:
Fumonisin B_4

Sauces (tomato sauce) (literature[514]) contained further mycotoxins/microbial metabolites:
Alternariol-3-sulfate, alternariol monomethyl ether-3-sulfate, and tentoxin

Seeds (bush mango) (literature[603]) contained further mycotoxins/microbial metabolites:
Alamethicin, andrastatin A, ascochlorin, asperfuran, asperglaucide, aspinolid B, averantin, averufin, bikaverin,

brevianamide F, butenolide, chlorocitreorosein, citreorosein, curvularin, cyclo (L-Pro-L-Tyr), cytochalasin D, demethylsulochrin, dihydrocitrinone, dihydrotrichotetronine, emodin, epiequisetin, equisetin, integracin A, integracin B, malformin A2, malformin C, neoechinulin A, nidurufin, *O*-methylsterigmatocystin, pestalotin, physicon, puromycin, questiomycin A, quinolactacin A, radicicol, rugulosin, rugulusovin, skyrin, trichodimerol, tryptophol, versicolorin A, and versicolorin C

Seeds (melon) (literature[603]) contained further mycotoxins/microbial metabolites:
Ascochlorin, asperfuran, asperglaucide, averantin, averufin, bikaverin, butenolide, citreorosein, cyclo (L-Pro-L-Tyr), emodin, epiequisetin, equisetin, malformin A2, malformin C, neoechinulin A, nidurufin, *O*-methylsterigmatocystin, physicon, questiomycin A, quinolactacin A, rugulusovin, tryptophol, versicolorin A, and versicolorin C

Seeds (sesame) (literature[629]) contained further mycotoxins/microbial metabolites:
Asperlacton, aurofusarin, avenacein, averufin, brefeldin A, brevianamide F, chloramphenicol, curvularin, dechlorogriseofulvin, emodin, equisetin, griseofulvin, macrosporin A, monocerin, nonactin, pestalotin, physicon, versicolorin C, and zearalenone-sulfate

Seeds (sunflower) (literature[414]) contained further mycotoxins/microbial metabolites:
Altenuisol, tentoxin, and valine-tenuazonic acid

Seeds (sunflower) (literature[568]) contained further mycotoxins/microbial metabolites:
Iso-altenuene and tentoxin

Seeds (sunflower) (literature[743]) contained further mycotoxins/microbial metabolites:
Tentoxin

Snacks (literature[763]) contained further mycotoxins/microbial metabolites:
Averufin, brevianamide F, emodin, equisetin, macrosporin A, monocerin, O-methylsterigmatocystin, nidurufin, norsolorinic acid, pestalotin, tryptophol, versicolorin A, and versicolorin C

Snacks (literature[833]) contained further mycotoxins/microbial metabolites:
α-Zearalenol

Soups (literature[659]) contained further mycotoxins/microbial metabolites:
Andrastin A, averantin, averufanin, averufin, brevianamide F, chlamydosporol, CJ21058, chrysophanol, curvularin, emodin, equisetin, fumonisin B_6, macrosporin, O-methylsterigmatocystin, monocerin, nonactin, norsolorinic acid, physicon, skyrin, tentoxin, tryptophol, versicolorin A, versicolorin C, zearalenone-sulfate, α-zearalenol, and ß-zearalenol

Sugar beets (literature[689]) contained further mycotoxins/microbial metabolites:
cis-Zearalenone, *trans*-α-zearalenol, and *trans*-ß-zearalenol

Sugarcane (literature[300]) contained further mycotoxins/microbial metabolites:
Agroclavine, ascochlorin, asperglaucide, asperphenamate, averufin, berkedrimane B, brevianamide F, citreorosein, cyclo (L-Pro-L-Tyr), cyclo (L-Pro-L-Val), cytochalasin D, emodin, illicicolin B, illicicolin E, iso-rhodoptilometrin, macrosporin, *N*-benzoyl-phenylalanine, norlichexanthone, oxaline, physicon, quinolactacin A, skyrin, and tryptophol

Tomato concentrate (literature[514]) contained further mycotoxins/microbial metabolites:
Alternariol-3-sulfate, alternariol monomethyl ether-3-sulfate, and tentoxin

Tomato ketchup (literature[883]) contained further mycotoxins/microbial metabolites:
Tentoxin

Wine (red) (literature[830]) contained further mycotoxins/microbial metabolites:
Ethylochratoxin B, methylochratoxin A, methylochratoxin B, and ochratoxin C

Wine (red) (literature[831]) contained further mycotoxins/microbial metabolites:
Methylochratoxin A and ochratoxin C

Wine (red) (literature[832]) contained further mycotoxins/microbial metabolites:
Ethylochratoxin B, methylochratoxin A, methylochratoxin B, and ochratoxin C

Wine (red) (literature[850]) contained further mycotoxins/microbial metabolites:
Tentoxin

For incidence and concentration as well as possible co-contamination of these compounds see the corresponding article.

Tables

Mycotoxins and Their Plant and Plant Product Spectrum (Except Cereals and Cereal Products)

Table 1. ***Alternaria*** **toxins in plants and plant products**

Altenuene	
	Fruits (apple); fruits (tomato); fruits, dried; juices; juices (apple juice); juices (tomato juice); olives; products (tomato product); sauces (tomato sauce); seeds (sunflower); tomato concentrate
Alternariol	
	Beans (soybean); beverages; cassava; dietary supplements; figs; food; fruits (apple); fruits (mandarin); fruits (tangerine); fruits (tomato); fruits, dried; jam; juices; juices (apple concentrate); juices (apple juice); juices (cranberry juice); juices (grape juice); juices (grapefruit juice); juices (orange juice); juices (raspberry juice); juices (tomato juice); juices (vegetable juice); licorice; nectars (cranberry nectar); nectars (prune nectar); nuts (almond); nuts (chestnut); nuts (hazelnut); nuts (peanut); oils (vegetable oil); olives; paste (tomato paste); products (tomato product); puree; raisins; sauces (apple sauce); sauces (tomato sauce); seeds (sesame); seeds (sunflower); soups; spices (paprika); spices (pepper); strawberries; tomato concentrate; tomato ketchup; wine (red); wine (rosé); wine (white); wine, miscellaneous
Alternariol methyl ether	
	Beans (soybean); beverages; cassava; dagwa; dietary supplements; food; fruits (apple); fruits (mandarin); fruits (melon); fruits (tangerine); fruits (tomato); fruits, dried; iru; juices; juices (apple concentrate); juices (apple juice); juices (citrus juice); juices (cranberry juice); juices (grape juice); juices (orange juice); juices (tomato juice); juices (vegetable juice); nectars (cranberry nectar); nectars (prune nectar); nuts (almond); nuts (chestnut); nuts (hazelnut); nuts (peanut); ogiri; oils (sunflower oil); oils (vegetable oil); olives; paste (peanut paste); paste (tomato paste); products (cassava product); products (tomato product); puree; raisins; sauces (tomato sauce); seeds (sesame); seeds (sunflower); snacks; soups; spices (cayenne pepper, chilli); spices (pepper); strawberries; sugarcane; tomato concentrate; tomato ketchup; ugba; wine (red); wine (white); wine, miscellaneous
Altertoxin I	
	Cassava; food; fruits (apple); nuts (hazelnut); nuts (peanut); seeds (sunflower)
Tenuazonic acid	
	Cassava; dates; dietary supplements; figs; fruits (apple); fruits (mandarin); fruits (melon); fruits (tomato); fruits, dried; juices; juices (apple juice); juices (apricot juice); juices (banana juice); juices (blackcurrant juice); juices (cherry juice); juices (citrus juice); juices (grape juice); juices (orange juice); juices (pear juice); juices (pomegranate juice); juices (strawberry juice); juices (tomato juice); oils (vegetable oil); olives; paste (tomato paste); products (tomato product); pulp (tomato pulp); puree; raisins; sauces (pepper sauce); sauces (tomato sauce); seeds (bush mango); seeds (sesame); seeds (sunflower); spices; spices (coriander); spices (cumin); spices (curcuma); spices (curry); spices (ginger); spices (paprika); spices (pepper); tea; tomato concentrate; tomato ketchup; wine (red); wine, miscellaneous

M. Weidenbörner, *Mycotoxins in Plants and Plant Products*, https://doi.org/10.1007/978-3-319-92850-0_3

Table 2. ***Aspergillus*** **toxins in plants and plant products**

Aflatoxin B_1

Baby food; bars; beans; beans (broad bean); beans (butter bean); beans (cowpea); beans (kidney bean); beans (mung bean); beans (pinto bean); beans (saltani saltapya bean); beans (soybean); betel nuts; beverages; biscuits; bondakadledkai; breadfruit; breakfast cereals; butter (cocoa butter); butter (nut butter); butter (peanut butter); cake (peanut); candies (coconut candy); candies (peanut candy); cassava; chips (cassava chips); chips (yam chips); chocolate; cocoa; cocoa beans; coconuts; coffee; coffee beans; confectionery; congressbele; copra; curd (bean curd); currants; dagwa; dates; drinking water; drinks (tiger nut drink); eggplant; field mustard; figs; flour; flour (cassava flour); flour (chestnut flour); flour (gram flour); flour (lentil flour); flour (soybean flour); food; fruits; fruits (apple); fruits (cherry); fruits (lemon); fruits (mango); fruits (orange); fruits (peach); fruits (quince); fruits (tomato); fruits, dried; fruits and vegetables; ginkgo; grogannuts; gums, food-grade; halva; herbs (camomile); herbs (linden); herbs (rosehip); herbs (sage); infant food; iru; jam (bean jam); jam (fig jam); juices (guava juice); juices (sugarcane juice); kernels (apricot); kubeba; kuru-kuru; legumes; lentils; licorice; mahua seeds; marzipan (almond paste); meal (copra meal); meal (cycad meal); meal (peanut meal); meal (pistachio meal); medicinal plants; meju; must; nut crème; nuts; nuts (almond); nuts (Brazil nut, paranut); nuts (cashew nut); nuts (chestnut); nuts (hazelnut); nuts (peanut); nuts (pecan nut); nuts (pine nut); nuts (pistachio nut); nuts (tigernut); nuts (walnut); ogiri; oil seeds; oils; oils (coconut oil); oils (maize oil); oils (mustard oil); oils (olive oil); oils (peanut oil); oils (rapeseed oil); oils (sesame oil); oils (soybean oil); oils (sunflower oil); olives; paste (bean paste); paste (fig paste); paste (hazelnut paste); paste (peanut paste); paste (sesame paste); paste (soybean paste); peas; peas (chickpea); powder (cocoa powder); products; products (almond product); products (cassava product); products (cocoa product); products (legume product); products (melon seed product); products (nut product); products (peanut product); products (pistachio product); products (sesame product); products (tobacco product); raisins; sauces (peanut sauce); sauces (peanut-chilli sauce); sauces (soy sauce); seeds (bush mango); seeds (linseed); seeds (melon); seeds (sesame); seeds (sunflower); shea-nuts; shiro; snacks; soups; spices; spices (ammi); spices (anise); spices (basil); spices (cardamom); spices (cassia); spices (cayenne pepper, chilli); spices (cinnamon); spices (coriander); spices (cumin); spices (curcuma); spices (curry); spices (fennel); spices (fenugreek); spices (garlic); spices (ginger); spices (laurel leaves); spices (masala); spices (mustard); spices (nutmeg); spices (paprika); spices (pepper); spices (peppermint); spices (poppy seed); spices (rosemary); spices (saffron); spices (sumac); spices (thyme); sugar; sugarcane; sultanas; tea; tobacco; toffees; tofu; ugba; vegetables, dried; vine fruits, dried; wine; wine (red); wine, miscellaneous; yams

Aflatoxin B_2

Baby food; beans; beans (broad bean); beans (butter bean); beans (kidney bean); beans (mung bean); beans (saltani saltapya bean); beans (soybean); betel nuts; beverages; bondakadledkai; breakfast cereals; butter (cocoa butter); butter (nut butter); butter (peanut butter); cake (peanut); candies (coconut candy); candies (peanut candy); cassava; chips (yam chips); chocolate; cocoa; cocoa beans; coconuts; coffee; coffee beans; confectionery; congressbele; copra; currants; dagwa; dates; drinking water; eggplant; field mustard; figs; flour (cassava flour); flour (chestnut flour); flour (gram flour); flour (lentil flour); flour (potato flour); flour (soybean flour); food; fruits (lemon); fruits (mango); fruits, dried; ginkgo; grogannuts; infant food; iru; jam (bean jam); jam (fig jam); kernels (apricot); kubeba; kuru-kuru; marzipan (almond paste); meal (copra meal); meal (pistachio meal); medicinal plants; nuts; nuts (almond); nuts (Brazil nut, paranut); nuts (cashew nut); nuts (chestnut); nuts (hazelnut); nuts (peanut); nuts (pine nut); nuts (pistachio nut); nuts (tigernut); nuts (walnut); ogiri; oils; oils (coconut oil); oils (maize oil); oils (olive oil); oils (peanut oil); oils (sesame oil); oils (sunflower oil); paste (fig paste); paste (hazelnut paste); paste (peanut paste); paste (sesame paste); paste (soybean paste); powder (cocoa powder); products (cocoa product); products (nut product); products (peanut product); products (pistachio product); raisins; seeds (bush mango); seeds (linseed); seeds (melon); seeds (sesame); seeds (sunflower); shea-nuts; snacks; soups; spices; spices (ammi); spices (cardamom); spices (cassia); spices (cayenne pepper, chilli); spices (coriander); spices (cumin); spices (curcuma); spices (curry); spices (fennel); spices (fenugreek); spices (garlic); spices (ginger); spices (masala); spices (mustard); spices (nutmeg); spices (paprika); spices (pepper); sugar; tea; tobacco; toffees; ugba; wine; wine (red)

Aflatoxin G_1

Baby food; beans; beans (kidney bean); beans (pinto bean); beans (soybean); betel nuts; beverages; breakfast cereals; butter (nut butter); butter (peanut butter); cake (peanut); candies (coconut candy); candies (peanut candy); cassava; chips (yam chips); chocolate; cocoa; cocoa beans; coconuts; coffee; coffee beans; copra; curd (bean curd); currants; dagwa; dates; drinking water; eggplant; figs; flour (chestnut flour); flour (lentil flour); flour (potato flour); flour (soybean flour); flour (vetch flour); food; fruits; fruits (lemon); fruits (mango); fruits, dried; grogannuts; infant food; iru; juices (sugarcane juice); kernels (apricot); kuru-kuru; meal (copra meal); medicinal plants; nuts; nuts (almond); nuts (Brazil nut, paranut); nuts (cashew nut); nuts (chestnut); nuts (hazelnut); nuts (peanut); nuts (pecan nut); nuts (pine nut); nuts (pistachio nut); nuts (tigernut); nuts (walnut); oils; oils (coconut oil); oils (olive oil); oils (peanut oil); oils (sesame oil); oils (sunflower oil); paste (fig paste); paste (hazelnut paste); paste (peanut paste); paste (sesame paste); paste (soybean paste); peas; pinhol; powder (cocoa powder); products (cassava product); products (cocoa product); products (peanut product); products (pistachio product); raisins; sauces (soy sauce); seeds (bush mango); seeds (melon); seeds (sesame); seeds (sunflower); snacks; soups; spices; spices (ammi); spices (cardamom); spices (cassia); spices (cayenne pepper, chilli); spices (celery); spices (coriander); spices (cumin); spices (curcuma); spices (curry); spices (fennel); spices (fenugreek); spices (garlic); spices (ginger); spices (masala); spices (mustard); spices (nutmeg); spices (paprika); spices (pepper); sugarcane; tea; wine; wine (red)

Aflatoxin G_2

Baby food; beans (kidney bean); beans (pinto bean); beans (soybean); betel nuts; beverages; butter (nut butter); butter (peanut butter); cake (peanut); candies (coconut candy); candies (peanut candy); cassava; chips (yam chips); chocolate; cocoa; cocoa beans; coconuts; coffee; coffee beans; copra; dagwa; dates; eggplant; figs; flour (chestnut flour); flour (lentil flour); food; fruits; fruits (lemon); fruits (mango); fruits, dried; grogannuts; infant food; iru; kuru-kuru; meal (copra meal); medicinal plants; nuts; nuts (almond); nuts (Brazil nut, paranut); nuts (cashew nut); nuts (chestnut); nuts (hazelnut); nuts (peanut); nuts (pistachio nut); nuts (tigernut); nuts (walnut); ochra; ogiri; oils; oils (olive oil); oils (peanut oil); oils (sesame oil); oils (sunflower oil); paste (fig paste); paste (hazelnut paste); paste (peanut paste); paste (sesame paste); paste (soybean paste); products (cocoa product); products (peanut product); raisins; sauces (soy sauce); seeds (bush mango); seeds (melon); seeds (sesame); seeds (sunflower); snacks; soups; spices; spices (cardamom); spices (cassia); spices (cayenne pepper, chilli); spices (coriander); spices (cumin); spices (curcuma); spices (curry); spices (fennel); spices (ginger); spices (masala); spices (mustard); spices (nutmeg); spices (paprika); spices (pepper); spices (star anise); tea; vine fruits, dried; wine; wine (red)

(continued)

Table 2. (continued)

Aflatoxin M_1
Baby food; breakfast cereals; butter (peanut butter); cake (peanut); cassava; food; halva; medicinal plants; nuts (peanut); nuts (pistachio nut); seeds (bush mango); seeds (melon); snacks; sweets

Aflatoxin M_2
Butter (peanut butter); nuts (peanut)

Aflatoxin/s
Asure; baby food; bars; beans; beans (black bean); beans (cowpea); beans (kidney bean); beans (mung bean); beans (soybean); bee pollen; betel nuts; beverages; butter (peanut butter); butter (sesame butter); cake (date); cake (peanut); candies (peanut candy); candies (pistachio candy); cassava; chips (cassava chips); chips (yam chips); cocoa; cocoa beans; coffee; coffee beans; confectionery; cookies; copra; cream; dates; drinks (alcoholic drink); emu aran; figs; flour (peanut flour); food; fruits, dried; halva; honey; ice cream; infant food; jaggery; jello; juices (sugarcane juice); kernels (apricot); keskul; lentils; mahua seeds; meal (copra meal); meal (cycad meal); meal (egusi meal); medicinal plants; muhallebi; nut cereals; nut crème; nuts; nuts (almond); nuts (Brazil nut, paranut); nuts (cashew nut); nuts (chestnut); nuts (hazelnut); nuts (peanut); nuts (pecan nut); nuts (pine nut); nuts (pistachio nut); nuts (walnut); ogbono; ogili-ugba; ogoro; oil seeds; oils; oils (peanut oil); paste (curry paste); paste (fig paste); paste (peanut paste); paste (pipian paste); paste (red pepper paste); peas; peas (chickpea); pickles; products (coconut product); products (nut product); products (peanut product); pudding; raisins; rootcrops; salep; sauces (chilli sauce); sauces (chocolate sauce); sauces (flour sauce); sauces (leaf-sauce); sauces (peanut sauce); sauces (peanut-chilli sauce); seeds (linseed); seeds (lupine); seeds (melon); seeds (sesame); seeds (sunflower); snacks; soups; spices; spices (anise); spices (black cumin); spices (caraway); spices (cayenne pepper, chilli); spices (coriander); spices (cumin); spices (curcuma); spices (curry); spices (fennel); spices (garlic/onion); spices (ginger); spices (majoran); spices (masala); spices (nutmeg); spices (paprika); spices (pepper); spices (peppermint); spices (tandoori); starch (cassava starch); sultanas; sweet potatoes; tea; tubers; vegetables; vegetables, dried; yams

Other *Aspergillus* Toxins

Gliotoxin
Medicinal plants

Sterigmatocystin
Cake (peanut); cassava; coffee; coffee beans; dagwa; dietary supplements; food; iru; kuru-kuru; medicinal plants; nuts (almond); nuts (hazelnut); nuts (peanut); nuts (pecan nut); nuts (pistachio nut); ogiri; paste (peanut paste); products (legume product); seeds (bush mango); seeds (melon); seeds (sesame); soups; spices (caraway); spices (cayenne pepper, chilli); spices (cumin); spices (fennel); spices (majoran); spices (oregano); spices (paprika); spices (pepper); spices (thyme); ugba

Table 3. ***Aspergillus*** **and** ***Penicillium*** **toxins in plants and plant products**

3-Nitropropionic acid
Beans (locust bean); cake (peanut); cassava; food; juices (sugarcane juice); nuts (hazelnut); nuts (peanut); ogiri; seeds (bush mango); seeds (melon); seeds (sesame); snacks; sugarcane

Citreoviridin
Cassava

Citrinin
Beans (locust bean); breakfast cereals; cassava; coconuts; dagwa; flour (chestnut flour); food; food colorant; fruits (apple); grogannuts; iru; juices (apple juice); juices (blackcurrant juice); juices (cherry juice); juices (tomato juice); licorice; medicinal plants; nuts (cashew nut); nuts (chestnut); nuts (peanut); ogiri; olives; products (coconut product); raisins; seeds (melon); seeds (sesame); spices (black cumin); spices (cardamom); spices (cayenne pepper, chilli); spices (coriander); spices (cumin); spices (curcuma); spices (fennel); spices (pepper); sultanas

Cyclopiazonic acid
Cassava; figs; food; medicinal plants; nuts (almond); nuts (Brazil nut, paranut); nuts (peanut); pulp (tomato pulp); puree; seeds (bush mango); seeds (melon); wine (red)

Kojic acid
Cake (peanut); cassava; dagwa; figs; flour (soybean flour); juices (sugarcane juice); kuru-kuru; nuts (hazelnut); nuts (peanut); paste (peanut paste); products (cassava product); seeds (bush mango); seeds (melon); seeds (sesame); snacks; soups; sugarcane

(continued)

Table 3. (continued)

Ochratoxin A

Baby food; bars; beans; beans (brown bean); beans (green bean); beans (kidney bean); beans (pinto bean); beans (soybean); bee pollen; berries (grape); beverages; biscuits; breakfast cereals; bsissa; butter (cocoa butter); butter (nut butter); butter (peanut butter); cake (peanut); candies; cassava; cassis; chocolate; cocoa; cocoa beans; coconuts; coffee; coffee beans; confectionery; currants; dagwa; dates; dietary supplements; drinking water; drinks; drinks (chocolate drink); drinks (coffee drink); eggplant; extracts; figs; flakes (cassava); flour; flour (cassava flour); flour (chestnut flour); flour (lentil flour); flour (potato flour); flour (soybean flour); flour (vetch flour); food; food colorant; fruits (apple); fruits (cherry); fruits (goji); fruits (peach); fruits (quince); fruits (tomato); fruits, dried; grape extract; grape pomace and wine lees; herbs; herbs and spices; ice cream; infant food; iru; jam; jam (fig jam); juices; juices (apple juice); juices (blackcurrant juice); juices (carrot juice); juices (grape juice); juices (tomato juice); kuru-kuru; legumes; lentils; licorice; medicinal plants; must; nut crème; nuts; nuts (almond); nuts (cashew nut); nuts (chestnut); nuts (hazelnut); nuts (peanut); nuts (pistachio nut); nuts (tigernut); nuts (walnut); ogiri; oil seeds; oils (olive oil); oils (sesame oil); olives; paste (curry paste); paste (fig paste); paste (peanut paste); paste (soybean paste); peas; peas (chickpea); pekmez; pickles; pies; pizza; plant food supplements; potatoes; powder (cocoa powder); products; products (cocoa product); products (coconut product); products (ginger product); products (legume product); products (pistachio product); products (plant product); pudding; pulp (grape pulp); puree; raisins; sauces (chilli sauce); sauces (pepper- and barbecue sauce); sauces (soy sauce); seeds; seeds (bush mango); seeds (linseed); seeds (melon); seeds (sesame); seeds (sunflower); soups; soy isoflavones; spices; spices (anise); spices (caraway); spices (cardamom); spices (cayenne pepper, chilli); spices (cinnamon); spices (coriander); spices (cumin); spices (curcuma); spices (curry); spices (fennel); spices (garlic); spices (ginger); spices (majoran); spices (masala); spices (mustard); spices (nutmeg); spices (paprika); spices (pepper); spices (prickly ash); spices (tandoori); strawberries; sultanas; tea; tomato ketchup; ugba; vegetables, dried; vine fruits, dried; vinegar; vinegar (apple); vinegar (balsamico); vinegar (wine); wine; wine (red); wine (rosé); wine (white); wine, miscellaneous

Ochratoxin B

Cassava; cocoa; dagwa; dietary supplements; food; kuru-kuru; medicinal plants; nuts (almond); nuts (cashew nut); nuts (hazelnut); nuts (peanut); powder (cocoa powder); products (pistachio product); raisins; seeds (bush mango); soups; spices; vinegar; wine (red)

Patulin

Apple cider; apple leather; apple rings; baby food; berries (blueberry); berries (lingonberry); beverages; cassava; coffee; dietary supplements; drinks (apple drink); food; fruit salad; fruits (apple); fruits (hawthorn); fruits (peach); fruits (pear); fruits (plum); fruits (quince); fruits, dried; jellies; jam (apple jam); jam (pear jam); jam (quince jam); juices; juices (apple concentrate); juices (apple juice); juices (apricot concentrate); juices (apricot juice); juices (blackcurrant juice); juices (cherry juice); juices (grape juice); juices (hawthorn juice); juices (lychee juice); juices (orange juice); juices (peach concentrate); juices (peach juice); juices (pear concentrate); juices (peach juice); juices (pineapple concentrate); juices (pomegranate juice); juices (rosehip juice); juices (tomato juice); marmalade (apple marmalade); marmalade (pear marmalade); medicinal plants; nectars (apple nectar); paste (peanut paste); products (apple flavor product); products (apple product); products (fruit product); products (legume product); pulp (apple pulp); pulp (plum pulp); puree; vinegar; vinegar (apple); wine

Penicillic Acid

Beans; cassava; dates; dietary supplements; juices (sugarcane juice); peas; products (cassava product); sugarcane; tobacco

Viomellein

Dagwa; kuru-kuru; paste (peanut paste)

Table 4. ***Fusarium*** **toxins in plants and plant products**

Depsipeptides

Beauvericin

Beans (locust bean); cake (peanut); cassava; coffee; dagwa; dietary supplements; figs; flour (soybean flour); food; kuru-kuru; medicinal plants; nuts (almond); nuts (hazelnut); nuts (peanut); nuts (pistachio nut); nuts (tigernut); ogiri; paste (peanut paste); products (peanut product); products (pistachio product); seeds (bush mango); seeds (melon); seeds (sesame); snacks; soups; spices (garlic); spices (ginger); spices (paprika); vegetables (asparagus)

Enniatin A

Cassava; coffee; dagwa; dietary supplements; food; kuru-kuru; medicinal plants; nuts (hazelnut); nuts (peanut); nuts (tigernut); paste (peanut paste); soups; spices (ginger); vine fruits, dried

Enniatin A_1

Cassava; coffee; dagwa; dates; dietary supplements; figs; food; fruits, dried; kuru-kuru; medicinal plants; nuts (hazelnut); nuts (peanut); nuts (tigernut); paste (peanut paste); soups; spices (ginger); vegetables (asparagus); vine fruits, dried

Enniatin B

Cassava; coffee; dagwa; dates; dietary supplements; figs; flour (soybean flour); food; fruits, dried; kuru-kuru; medicinal plants; nuts (almond); nuts (hazelnut); nuts (peanut); nuts (tigernut); paste (peanut paste); soups; spices (ginger); tea; ugba; vegetables (asparagus); vine fruits, dried

Enniatin B_1

Cassava; coffee; dagwa; dates; dietary supplements; figs; flour (soybean flour); food; fruits, dried; kuru-kuru; medicinal plants; nuts (almond); nuts (peanut); nuts (tigernut); paste (peanut paste); soups; spices (ginger); vine fruits, dried

(continued)

Table 4. (continued)

Enniatin B_2
Cassava; food; nuts (hazelnut)

Enniatin B_3
Cassava; nuts (almond); nuts (hazelnut)

Enniatin B_4
Nuts (hazelnut); nuts (peanut)

Enniatin/s
Fruits (apple)

Fumonisin/s

Fumonisin B_1
Beans; beans (cowpea); beans (soybean); breakfast cereals; cassava; coffee; dagwa; dietary plants; dietary supplements; figs; flour (cassava flour); flour (soybean flour); food; herbs; iru; juices (pineapple juice); kuru-kuru; licorice; margarine; medicinal plants; nuts (peanut); oils (maize oil); paste (peanut paste); products (cassava product); raisins; shiitake; snacks; soups; soy isoflavones; spices; spices (cayenne pepper, chilli); spices (garlic); spices (nutmeg); spices (onion); spices (paprika); spices (pepper); tea; vegetables (asparagus); vegetables (black radish); wine (red); wine (white); wine, miscellaneous; yams

Fumonisin B_2
Beans (soybean); breakfast cereals; cassava; coffee; coffee beans; dagwa; dietary supplements; figs; flour (soybean flour); food; herbs; iru; juices (pineapple juice); medicinal plants; must; paste (peanut paste); products (cassava product); raisins; shiitake; snacks; soups; spices (cayenne pepper, chilli); spices (garlic); vegetables (asparagus); vegetables (black radish); wine (red); wine (white); wine, miscellaneous; yams

Fumonisin B_3
Breakfast cereals; cassava; dagwa; dietary supplements; flour (soybean flour); food; iru; juices (pineapple juice); kuru-kuru; medicinal plants; paste (peanut paste); products (cassava product); raisins; snacks; soups; spices (garlic); vegetables (asparagus)

Fumonisin/s
Bee pollen; infant food; medicinal plants; paste (curry paste); pickles; products (soybean product); sauces; spices; spices (curry); spices (tandoori); vegetables (asparagus); wine (red)

Type A trichothecenes

Diacetoxyscirpenol
Cassava; coffee; dates; dietary supplements; figs; flour (cassava flour); food; fruits, dried; meal (soybean meal); medicinal plants; nuts (peanut); ogiri; oils; oils (maize oil); products (potato product); spices (cayenne pepper, chilli); spices (curry)

HT-2 toxin
Beans (black bean); beans (mung bean); beans (soybean); cassava; coffee; dietary supplements; figs; food; fruits, dried; iru; medicinal plants; milk thistle; nuts (hazelnut); ogiri; oils; oils (maize oil); oils (sunflower oil); products (potato product); seeds (bush mango); seeds (pumpkin); seeds (sunflower); spices (cayenne pepper, chilli); spices (curry); spices (paprika); ugba; vine fruits, dried

Monoacetoxyscirpenol
Cassava; food; products (potato product)

Neosolaniol
Bee pollen; cassava; coffee; dietary supplements; medicinal plants; ogiri; products (pistachio product); spices (curry); spices (ginger)

Scirpentriol
Food; products (potato product)

T-2 toxin
Beans; beans (black bean); beans (mung bean); beans (soybean); bee pollen; biscuits; cassava; coffee; dietary supplements; ginkgo; iru; jam (fig jam); lentils; medicinal plants; milk thistle; nuts (hazelnut); nuts (peanut); oils; oils (maize oil); oils (sunflower oil); peas (chickpea); seeds (bush mango); seeds (sesame); seeds (sunflower); spices (cayenne pepper, chilli); spices (curry); spices (ginger); spices (paprika); spices (tandoori); tobacco; ugba; wine (red); wine (white)

T-2 tetraol
Cassava; food; meal (soybean meal); oils (soybean oil)

Type B trichothecenes

Deoxynivalenol
Bars; beans; beans (mung bean); beans (soybean); bee pollen; breakfast cereals; cassava; coffee; dagwa; dietary supplements; flour (cassava flour); flour (soybean flour); food; infant food; iru; licorice; meal (soybean meal); medicinal plants; nuts (peanut); nuts (tigernut); ogiri; oils; oils (maize oil); oils (soybean oil); sauces (soy sauce); seeds (sesame); seeds (sunflower); shiitake; soups; spices (cayenne pepper, chilli); spices (coriander); spices (garlic); spices (ginger); spices (paprika); spices (thyme); ugba

3-Acetyldeoxynivalenol
Coffee; dietary supplements; food; iru

(continued)

Table 4. (continued)

15-Acetyldeoxynivalenol
Coffee; food; oils; oils (maize oil); oils (soybean oil); tea

Fusarenon-X
Dietary supplements; flour (soybean flour); food; iru; medicinal plants; spices (curry); spices (garlic)

Nivalenol
Bars; beans; beans (soybean); bee pollen; breakfast cereals; cassava; coffee; dagwa; flour (soybean flour); food; medicinal plants; nuts (peanut); paste (curry paste); sauces (chilli sauce); seeds (bush mango); spices (curry); spices (garlic); spices (ginger); spices (tandoori)

Other *Fusarium* Toxins

Fusaric acid
Cassava; dagwa; flour (soybean flour); juices (sugarcane juice); kuru-kuru; paste (peanut paste); seeds (sesame)

Fusaproliferin
Cassava

Moniliformin
Cassava; food; fruits (apple); juices (pineapple juice); medicinal plants; nuts (hazelnut); nuts (peanut); seeds (bush mango); snacks; vegetables (asparagus)

Zearalenone
Beans; beans (mung bean); beans (soybean); bee pollen; beverages; biscuits; breakfast cereals; cassava; dagwa; dietary supplements; flour (cassava flour); flour (soybean flour); food; fruits, dried; food-grade; infant food; iru; kuru-kuru; licorice; margarine; meal (soybean meal); medicinal plants; milk thistle; nuts (almond); nuts (hazelnut); nuts (peanut); nuts (walnut); ogiri; oil seeds; oils; oils (maize oil); oils (rapeseed oil); oils (soybean oil); paste (curry paste); paste (peanut paste); peas; pickles; products (cassava product); products (legume product); products (potato product); sauces (chilli sauce); seeds (pumpkin); seeds (sesame); seeds (sunflower); spices; spices (cayenne pepper, chilli); spices (coriander); spices (curry); spices (fennel); spices (paprika); spices (pepper); spices (thyme); sugar beets; tobacco; ugba; vegetables, dried; wine (red)

Table 5. *Penicillium* toxins in plants and plant products

Mycophenolic acid
Cassava; dagwa; dates; dietary supplements; flour (chestnut flour); fruits (lemon); fruits, dried; jam; kuru-kuru; medicinal plants; nuts (chestnut); nuts (hazelnut); nuts (peanut); nuts (pistachio nut); paste (peanut paste); puree; raisins; seeds (bush mango); shiitake; wine (red)

Roquefortine C
Cassava; flour (chestnut flour); iru; kuru-kuru; nuts (chestnut)

Rubratoxin
Nuts (almond); nuts (walnut); spices (curcuma)

Plants and Plant Products and Their Mycotoxins (Except Cereals and Cereal Products)

Apple cider

Aspergillus and ***Penicillium*** toxins
Patulin

Apple leather

Aspergillus and ***Penicillium*** toxins
Patulin

Apple rings

Aspergillus and ***Penicillium*** toxins
Patulin

Asure

Aspergillus toxins
Aflatoxin/s

Baby food

Aspergillus toxins
Aflatoxin/s: aflatoxin B_1, aflatoxin B_2, aflatoxin G_1, aflatoxin G_2, aflatoxin M_1, aflatoxin/s

Aspergillus and ***Penicillium*** toxins
Ochratoxin A, patulin

Bars

Aspergillus toxins
Aflatoxin/s: aflatoxin B_1, aflatoxin/s

Aspergillus and ***Penicillium*** toxins
Ochratoxin A

Fusarium toxins
Type B trichothecenes: deoxynivalenol, nivalenol

Beans

Aspergillus toxins
Aflatoxin/s: aflatoxin B_1, aflatoxin B_2, aflatoxin G_1, aflatoxin/s

Aspergillus and ***Penicillium*** toxins
Ochratoxin A, penicillic acid

Fusarium toxins
Fumonisin/s: fumonisin B_1
Type A trichothecenes: T-2 toxin
Type B trichothecenes: deoxynivalenol, nivalenol
Other *Fusarium* toxins: zearalenone

Beans (black bean)

Aspergillus toxins
Aflatoxin/s

Fusarium toxins
Type A trichothecenes: HT-2 toxin, T-2 toxin

Beans (broad bean)

Aspergillus toxins
Aflatoxin/s: aflatoxin B_1, aflatoxin B_2

Beans (brown bean)

Aspergillus and ***Penicillium*** toxins
Ochratoxin A

Beans (butter bean)

Aspergillus toxins
Aflatoxin/s: aflatoxin B_1, aflatoxin B_2

Beans (cowpea)

Aspergillus toxins
Aflatoxin/s: aflatoxin B_1, aflatoxin/s

Fusarium toxins
Fumonisin/s: fumonisin B_1

Beans (green bean)

Aspergillus and ***Penicillium*** toxins
Ochratoxin A

Beans (kidney bean)

Aspergillus toxins
Aflatoxin/s: aflatoxin B_1, aflatoxin B_2, aflatoxin G_1, aflatoxin G_2, aflatoxin/s

Aspergillus and ***Penicillium*** toxins
Ochratoxin A

Beans (locust bean)

Aspergillus and ***Penicillium*** toxins
3-Nitropropionic acid, citrinin

Fusarium toxins
Depsipeptides: beauvericin

Beans (mung bean)

Aspergillus toxins
Aflatoxin/s: aflatoxin B_1, aflatoxin B_2, aflatoxin/s

Fusarium toxins
Type A trichothecenes: HT-2 toxin, T-2 toxin
Type B trichothecenes: deoxynivalenol
Other *Fusarium* toxins: zearalenone

Beans (pinto bean)

Aspergillus toxins
Aflatoxin/s: aflatoxin B_1, aflatoxin G_1, aflatoxin G_2

Aspergillus and ***Penicillium*** toxins
Ochratoxin A

M. Weidenbörner, *Mycotoxins in Plants and Plant Products*, https://doi.org/10.1007/978-3-319-92850-0_4

Beans (saltani saltapya bean)

Aspergillus toxins
Aflatoxin/s: aflatoxin B_1, aflatoxin B_2

Beans (soybean)

Alternaria toxins
Alternariol, alternariol methyl ether

Aspergillus toxins
Aflatoxin/s: aflatoxin B_1, aflatoxin B_2, aflatoxin G_1, aflatoxin G_2, aflatoxin/s

Aspergillus and ***Penicillium*** toxins
Ochratoxin A

Fusarium toxins
Fumonisin/s: fumonisin B_1, fumonisin B_2
Type A trichothecenes: HT-2 toxin, T-2 toxin
Type B trichothecenes: deoxynivalenol, nivalenol
Other *Fusarium* toxins: zearalenone

Bee pollen

Aspergillus toxins
Aflatoxin/s

Aspergillus and ***Penicillium*** toxins
Ochratoxin A

Fusarium toxins
Fumonisin/s
Type A trichothecenes: neosolaniol, T-2 toxin
Type B trichothecenes: deoxynivalenol, nivalenol
Other *Fusarium* toxins: zearalenone

Berries (blueberry)

Aspergillus and ***Pernicillium*** toxins
Patulin

Berries (grape)

Aspergillus and ***Pernicillium*** toxins
Ochratoxin A

Berries (lingonberry)

Aspergillus and ***Pernicillium*** toxins
Patulin

Betel nuts

Aspergillus toxins
Aflatoxin/s: aflatoxin B_1, aflatoxin B_2, aflatoxin G_1, aflatoxin G_2, aflatoxin/s

Beverages

Alternaria toxins
Alternariol, alternariol methyl ether

Aspergillus toxins
Aflatoxin/s: aflatoxin B_1, aflatoxin B_2, aflatoxin G_1, aflatoxin G_2, aflatoxin/s

Aspergillus and ***Penicillium*** toxins
Ochratoxin A, patulin

Fusarium toxins: zearalenone

Biscuits

Aspergillus toxins
Aflatoxin/s: aflatoxin B_1

Aspergillus and ***Penicillium*** toxins
Ochratoxin A

Fusarium toxins
Type A trichothecenes: T-2 toxin
Other *Fusarium* toxins: zearalenone

Bondakadledkai

Aspergillus toxins
Aflatoxin/s: aflatoxin B_1, aflatoxin B_2

Breadfruit

Aspergillus toxins
Aflatoxin/s: aflatoxin B_1

Breakfast cereals

Aspergillus toxins
Aflatoxin/s: aflatoxin B_1, aflatoxin B_2, aflatoxin G_1, aflatoxin M_1

Aspergillus and ***Penicillium*** toxins
Citrinin, ochratoxin A

Fusarium toxins
Fumonisin/s: fumonisin B_1, fumonisin B_2, fumonisin B_3
Type B trichothecenes: deoxynivalenol, nivalenol
Other *Fusarium* toxins: zearalenone

Bsissa

Aspergillus and ***Penicillium*** toxins
Ochratoxin A

Butter (cocoa butter)

Aspergillus toxins
Aflatoxin/s: aflatoxin B_1, aflatoxin B_2

Aspergillus and ***Penicillium*** toxins
Ochratoxin A

Butter (nut butter)

Aspergillus toxins
Aflatoxin/s: aflatoxin B_1, aflatoxin B_2, aflatoxin G_1, aflatoxin G_2

Aspergillus and ***Penicillium*** toxins
Ochratoxin A

Butter (peanut butter)

Aspergillus toxins
Aflatoxin/s: aflatoxin B_1, aflatoxin B_2, aflatoxin G_1, aflatoxin G_2, aflatoxin M_1, aflatoxin M_2, aflatoxin/s

Aspergillus and ***Penicillium*** toxins
Ochratoxin A

Butter (sesame butter)

Aspergillus toxins
Aflatoxin/s

Cake (date)

Aspergillus toxins
Aflatoxin/s

Cake (peanut)

Aspergillus toxins
Aflatoxin/s: aflatoxin B_1, aflatoxin B_2, aflatoxin G_1, aflatoxin G_2, aflatoxin M_1, aflatoxin/s
Other *Aspergillus* toxins: sterigmatocystin

Aspergillus and ***Penicillium*** toxins
3-Nitropropionic acid, kojic acid, ochratoxin A

Fusarium toxins
Depsipeptides: beauvericin

Candies

Aspergillus and ***Penicillium*** toxins
Ochratoxin A

Candies (coconut candy)

Aspergillus toxins
Aflatoxin/s: aflatoxin B_1, aflatoxin B_2, aflatoxin G_1, aflatoxin G_2

Candies (peanut candy)

Aspergillus toxins
Aflatoxin/s: aflatoxin B_1, aflatoxin B_2, aflatoxin G_1, aflatoxin G_2, aflatoxin/s

Candies (pistachio candy)

Aspergillus toxins
Aflatoxin/s

Cassava

Alternaria toxins
Alternariol, alternariol methyl ether, altertoxin I, tenuazonic acid

Aspergillus toxins
Aflatoxin/s: aflatoxin B_1, aflatoxin B_2, aflatoxin G_1, aflatoxin G_2, aflatoxin M_1, aflatoxin/s
Other *Aspergillus* toxins: sterigmatocystin

Aspergillus and ***Penicillium*** toxins
3-Nitropropionic acid, citreoviridin, citrinin, cyclopiazonic acid, kojic acid, ochratoxin A, ochratoxin B, patulin, penicillic acid

Fusarium toxins
Depsipeptides: beauvericin, enniatin A, enniatin A_1, enniatin B, enniatin B_1, enniatin B_2, enniatin B_3
Fumonisin/s: fumonisin B_1, fumonisin B_2, fumonisin B_3
Type A trichothecenes: diacetoxyscirpenol, HT-2 toxin, monoacetoxyscirpenol, neosolaniol, T-2 toxin, T-2 tetraol
Type B trichothecenes: deoxynivalenol, nivalenol
Other *Fusarium* toxins: fusaric acid, fusaproliferin, moniliformin, zearalenone

Penicillium toxins
Mycophenolic acid, roquefortine C

Cassis

Aspergillus and ***Penicillium*** toxins
Ochratoxin A

Chips (cassava chips)

Aspergillus toxins
Aflatoxin/s: aflatoxin B_1, aflatoxin/s

Chips (yam chips)

Aspergillus toxins
Aflatoxin/s: aflatoxin B_1, aflatoxin B_2, aflatoxin G_1, aflatoxin G_2, aflatoxin/s

Chocolate

Aspergillus toxins
Aflatoxin/s: aflatoxin B_1, aflatoxin B_2, aflatoxin G_1, aflatoxin G_2

Aspergillus and ***Penicillium*** toxins
Ochratoxin A

Cocoa

Aspergillus toxins
Aflatoxin/s: aflatoxin B_1, aflatoxin B_2, aflatoxin G_1, aflatoxin G_2, aflatoxin/s

Aspergillus and ***Penicillium*** toxins
Ochratoxin A, ochratoxin B

Cocoa beans

Aspergillus toxins
Aflatoxin/s: aflatoxin B_1, aflatoxin B_2, aflatoxin G_1, aflatoxin G_2, aflatoxin/s

Aspergillus and ***Penicillium*** toxins
Ochratoxin A

Coconuts

Aspergillus toxins
Aflatoxin/s: aflatoxin B_1, aflatoxin B_2, aflatoxin G_1, aflatoxin G_2

Aspergillus and ***Penicillium*** toxins
Citrinin, ochratoxin A

Coffee

Aspergillus toxins
Aflatoxin/s: aflatoxin B_1, aflatoxin B_2, aflatoxin G_1, aflatoxin G_2, aflatoxin/s
Other *Aspergillus* toxins: sterigmatocystin

Aspergillus and ***Penicillium*** toxins
Ochratoxin A, ochratoxin B

Fusarium toxins
Depsipeptides: beauvericin, enniatin A, enniatin A_1, enniatin B, enniatin B_1
Fumonisin/s: fumonisin B_1, fumonisin B_2
Type A trichothecenes: diacetoxyscirpenol, HT-2 toxin, neosolaniol, T-2 toxin
Type B trichothecenes: deoxynivalenol, 3-acetyldeoxynivalenol, 15-acetyldeoxynivalenol, nivalenol

Coffee beans

Aspergillus toxins
Aflatoxin/s: aflatoxin B_1, aflatoxin B_2, aflatoxin G_1, aflatoxin G_2, aflatoxin/s
Other *Aspergillus* toxins: sterigmatocystin

Aspergillus and ***Penicillium*** toxins
Ochratoxin A

Fusarium toxins
Fumonisin/s: fumonisin B_1

Confectionery

Aspergillus toxins
Aflatoxin/s: aflatoxin B_1, aflatoxin B_2, aflatoxin/s

Aspergillus and ***Penicillium*** toxins
Ochratoxin A

Congressbele

Aspergillus toxins
Aflatoxin/s: aflatoxin B_1, aflatoxin B_2

Cookies

Aspergillus toxins
Aflatoxin/s

Copra

Aspergillus toxins
Aflatoxin/s: aflatoxin B_1, aflatoxin B_2, aflatoxin G_1, aflatoxin G_2, aflatoxin/s

Cream

Aspergillus toxins
Aflatoxin/s

Curd (bean curd)

Aspergillus toxins
Aflatoxin/s: aflatoxin B_1, aflatoxin G_1

Currants

Aspergillus toxins
Aflatoxin/s: aflatoxin B_1, aflatoxin B_2, aflatoxin G_1

Aspergillus and ***Penicillium*** toxins
Ochratoxin A

Dagwa

Alternaria toxins
Alternariol methyl ether

Aspergillus toxins
Aflatoxin/s: aflatoxin B_1, aflatoxin B_2, aflatoxin G_1, aflatoxin G_2
Other *Aspergillus* toxins: sterigmatocystin

Aspergillus and ***Penicillium*** toxins
Citrinin, kojic acid, ochratoxin A, ochratoxin B, viomellein

Fusarium toxins
Depsipeptides: beauvericin, enniatin A, enniatin A_1, enniatin B, enniatin B_1
Fumonisin/s: fumonisin B_1, fumonisin B_2, fumonisin B_3
Type B trichothecenes: deoxynivalenol, nivalenol
Other *Fusarium* toxins: fusaric acid, zearalenone

Penicillium toxins
Mycophenolic acid

Dates

Alternaria toxins
Tenuazonic acid

Aspergillus toxins
Aflatoxin/s: aflatoxin B_1, aflatoxin B_2, aflatoxin G_1, aflatoxin G_2, aflatoxin/s

Aspergillus and ***Penicillium*** toxins
Ochratoxin A, penicillic acid

Fusarium toxins
Depsipeptides: enniatin A_1, enniatin B, enniatin B_1
Type A trichothecenes: diacetoxyscirpenol

Penicillium toxins
Mycophenolic acid

Dietary plants

Fusarium toxins
Fumonisin/s: fumonisin B_1

Dietary supplements

Alternaria toxins
Alternariol, alternariol methyl ether, tenuazonic acid

Aspergillus toxins
Sterigmatocystin

Aspergillus and ***Penicillium*** toxins
Ochratoxin A, ochratoxin B, patulin, penicillic acid

Fusarium toxins
Depsipeptides: enniatin A, enniatin A_1, enniatin B, enniatin B_1
Fumonisin/s: fumonisin B_1, fumonisin B_2, fumonisin B_3
Type A trichothecenes: diacetoxyscirpenol, HT-2 toxin, neosolaniol, T-2 toxin
Type B trichothecenes: deoxynivalenol, 3-acetyldeoxynivalenol, fusarenon-X (4-acetylnivalenol)
Other *Fusarium* toxins: zearalenone

Penicillium toxins
Mycophenolic acid

Dates

Aspergillus toxins
Aflatoxin/s: aflatoxin B_1, aflatoxin B_2, aflatoxin G_1

Aspergillus and ***Penicillium*** toxins
Ochratoxin A

Drinks

Aspergillus and ***Penicillium*** toxins
Ochratoxin A

Drinks (alcoholic drink)

Aspergillus toxins
Aflatoxin/s

Drinks (apple drink)

Aspergillus and ***Penicillium*** toxins
Patulin

Drinks (chocolate drink)

Aspergillus and ***Penicillium*** toxins
Ochratoxin A

Drinks (coffee drink)

Aspergillus and ***Penicillium*** toxins
Ochratoxin A

Drinks (tiger nut drink)

Aspergillus toxins
Aflatoxin/s: aflatoxin B_1

Eggplant

Aspergillus toxins
Aflatoxin/s: aflatoxin B_1, aflatoxin B_2, aflatoxin G_1, aflatoxin G_2

Aspergillus and ***Penicillium*** toxins
Ochratoxin A

Emu aran

Aspergillus toxins
Aflatoxin/s

Extracts

Aspergillus and ***Penicillium*** toxins
Ochratoxin A

Field mustard

Aspergillus toxins
Aflatoxin/s: aflatoxin B_1, aflatoxin B_2

Figs

Alternaria toxins
Alternariol, tenuazonic acid

Aspergillus toxins
Aflatoxin/s: aflatoxin B_1, aflatoxin B_2, aflatoxin G_1, aflatoxin G_2, aflatoxin/s

Aspergillus and ***Penicillium*** toxins
Cyclopiazonic acid, kojic acid, ochratoxin A

Fusarium toxins
Depsipeptides: beauvericin, enniatin A_1, enniatin B, enniatin B_1
Fumonisin/s: fumonisin B_1, fumonisin B_2
Type A trichothecenes: diacetoxyscirpenol, HT-2 toxin

Flakes (cassava)

Aspergillus and ***Penicillium*** toxins
Ochratoxin A

Flour

Aspergillus toxins
Aflatoxin/s: aflatoxin B_1

Aspergillus and ***Penicillium*** toxins
Ochratoxin A

Flour (cassava flour)

Aspergillus toxins
Aflatoxin/s: aflatoxin B_1, aflatoxin B_2

Aspergillus and ***Penicillium*** toxins
Ochratoxin A

Fusarium toxins
Fumonisin/s: fumonisin B_1
Type A trichothecenes: diacetoxyscirpenol, HT-2 toxin
Type B trichothecenes: deoxynivalenol
Other *Fusarium* toxins: zearalenone

Flour (chestnut flour)

Aspergillus toxins
Aflatoxin/s: aflatoxin B_1, aflatoxin B_2, aflatoxin G_1, aflatoxin G_2

Aspergillus and ***Penicillium*** toxins
Citrinin, ochratoxin A

Penicillium toxins
Mycophenolic acid, roquefortine C

Flour (gram flour)

Aspergillus toxins
Aflatoxin/s: aflatoxin B_1, aflatoxin B_2

Flour (lentil flour)

Aspergillus toxins
Aflatoxin/s: aflatoxin B_1, aflatoxin B_2, aflatoxin G_1, aflatoxin G_2

Aspergillus and ***Penicillium*** toxins
Ochratoxin A

Flour (peanut flour)

Aspergillus toxins
Aflatoxin/s

Flour (lentil flour)

Aspergillus toxins
Aflatoxin/s: aflatoxin B_2, aflatoxin G_1

Aspergillus and ***Penicillium*** toxins
Ochratoxin A

Flour (soybean flour)

Aspergillus toxins
Aflatoxin/s: aflatoxin B_1, aflatoxin B_2, aflatoxin G_1

Aspergillus and ***Penicillium*** toxins
Kojic acid, ochratoxin A

Fusarium toxins
Depsipeptides: beauvericin, enniatin B, enniatin B_1
Fumonisin/s: fumonisin B_1, fumonisin B_2, fumonisin B_3
Type B trichothecenes: deoxynivalenol, fusarenon-X (4-acetylnivalenol), nivalenol
Other *Fusarium* toxins: fusaric acid, zearalenone

Flour (vetch flour)

Aspergillus toxins
Aflatoxin/s: aflatoxin G_1

Aspergillus and ***Penicillium*** toxins
Ochratoxin A

Food

Alternaria toxins
Alternariol, alternariol methyl ether, altertoxin I

Aspergillus toxins
Aflatoxin/s: aflatoxin B_1, aflatoxin B_2, aflatoxin G_1, aflatoxin G_2, aflatoxin M_1, aflatoxin/s
Other *Aspergillus* toxins: sterigmatocystin

Aspergillus and ***Penicillium*** toxins
3-Nitropropionic acid, citrinin, cyclopiazonic acid, ochratoxin A, ochratoxin B, patulin

Fusarium toxins
Depsipeptides: beauvericin, enniatin A, enniatin A_1, enniatin B, enniatin B_1, enniatin B_2
Fumonisin/s: fumonisin B_1, fumonisin B_2, fumonisin B_3
Type A trichothecenes: diacetoxyscirpenol, HT-2 toxin, monoacetoxyscirpenol, scirpentriol, T-2 tetraol
Type B trichothecenes: deoxynivalenol, 15-acetydeoxynivalenol, 3-acetyldeoxynivalenol, fusarenon-X (4-acetylnivalenol), nivalenol
Other *Fusarium* toxins: moniliformin, zearalenone

Food colorant

Aspergillus and ***Penicillium*** toxins
Citrinin, ochratoxin A

Fruit salad

Aspergillus and ***Penicillium*** toxins
Patulin

Fruits

Aspergillus toxins
Aflatoxin/s: aflatoxin B_1, aflatoxin G_1, aflatoxin G_2

Fruits (apple)

Alternaria toxins
Altenuene, alternariol, alternariol methyl ether, altertoxin I, tenuazonic acid

Aspergillus toxins
Aflatoxin/s: aflatoxin B_1

Aspergillus and ***Penicillium*** toxins
Citrinin, ochratoxin A, patulin

Fusarium toxins
Depsipeptides: enniatin/s (A, A_1, B, B_1)
Other *Fusarium* toxins: moniliformin

Fruits (cherry)

Aspergillus toxins
Aflatoxin/s: aflatoxin B_1

Aspergillus and ***Penicillium*** toxins
Ochratoxin A

Fruits (goji)

Aspergillus and ***Penicillium*** toxins
Ochratoxin A

Fruits (hawthorn)

Aspergillus and ***Penicillium*** toxins
Patulin

Fruits (lemon)

Aspergillus toxins
Aflatoxin/s: aflatoxin B_1, aflatoxin B_2, aflatoxin G_1, aflatoxin G_2

Penicillium toxins
Mycophenolic acid

Fruits (mandarin)

Alternaria toxins
Alternariol, alternariol methyl ether, tenuazonic acid

Fruits (mango)

Aspergillus toxins
Aflatoxin/s: aflatoxin B_1, aflatoxin B_2, aflatoxin G_1, aflatoxin G_2

Fruits (melon)

Alternaria toxins
Alternariol methyl ether, tenuazonic acid

Fruits (orange)

Aspergillus toxins
Aflatoxin/s: aflatoxin B_1

Fruits (peach)

Aspergillus toxins
Aflatoxin/s: aflatoxin B_1

Aspergillus and ***Penicillium*** toxins
Ochratoxin A, patulin

Fruits (pear)

Aspergillus and ***Penicillium*** toxins
Patulin

Fruits (plum)

Aspergillus and ***Penicillium*** toxins
Patulin

Fruits (quince)

Aspergillus toxins
Aflatoxin/s: aflatoxin B_1

Aspergillus and ***Penicillium*** toxins
Ochratoxin A, patulin

Fruits (tangerine)

Alternaria toxins
Alternariol, alternariol methyl ether

Fruits (tomato)

Alternaria toxins
Atenuene, alternariol, alternariol methyl ether, tenuazonic acid

Aspergillus toxins
Aflatoxin/s: aflatoxin B_1

Aspergillus and ***Penicillium*** toxins
Ochratoxin A

Fruits and vegetables

Aspergillus toxins
Aflatoxin/s: aflatoxin B_1

Fruits, dried

Alternaria toxins
Atenuene, alternariol, alternariol methyl ether, tenuazonic acid

Aspergillus toxins
Aflatoxin/s: aflatoxin B_1, aflatoxin B_2, aflatoxin G_1, aflatoxin G_2, aflatoxin/s

Aspergillus and ***Penicillium*** toxins
Ochratoxin A, patulin

Fusarium toxins
Depsipeptides: enniatin A_1, enniatin B, enniatin B_1
Type A trichothecenes: diacetoxyscirpenol, HT-2 toxin
Other *Fusarium* toxins: zearalenone

Penicillium toxins
Mycophenolic acid

Ginkgo

Aspergillus toxins
Aflatoxin/s: aflatoxin B_1, aflatoxin B_2

Fusarium toxins
Type A trichothecenes: T-2 toxin

Grape extract

Aspergillus and ***Penicillium*** toxins
Ochratoxin A

Grape pomace and wine lees

Aspergillus and ***Penicillium*** toxins
Ochratoxin A

Grogannuts

Aspergillus toxins
Aflatoxin/s: aflatoxin B_1, aflatoxin B_2, aflatoxin G_1, aflatoxin G_2

Aspergillus and ***Penicillium*** toxins
Citrinin

Gums, food-grade

Aspergillus toxins
Aflatoxin/s: aflatoxin B_1

Fusarium toxins
Zearalenone

Halva

Aspergillus toxins
Aflatoxin/s: aflatoxin B_1, aflatoxin M_1, aflatoxin/s

Herbs

Aspergillus and ***Penicillium*** toxins
Ochratoxin A

Fusarium toxins
Fumonisin/s: fumonisin B_1, fumonisin B_2

Herbs (camomile)

Aspergillus toxins
Aflatoxin/s: aflatoxin B_1

Herbs (linden)

Aspergillus toxins
Aflatoxin/s: aflatoxin B_1

Herbs (roseship)

Aspergillus toxins
Aflatoxin/s: aflatoxin B_1

Herbs (sage)

Aspergillus toxins
Aflatoxin/s: aflatoxin B_1

Herbs and spices

Aspergillus and ***Penicillium*** toxins
Ochratoxin A

Honey

Aspergillus toxins
Aflatoxin/s

Ice cream

Aspergillus toxins
Aflatoxin/s

Aspergillus and ***Penicillium*** toxins
Ochratoxin A

Infant food

Aspergillus toxins
Aflatoxin/s: aflatoxin B_1, aflatoxin B_2, aflatoxin G_1, aflatoxin G_2, aflatoxin/s

Aspergillus and ***Penicillium*** toxins
Ochratoxin A

Fusarium toxins
Fumonisin/s
Type B trichothecenes: deoxynivalenol
Other *Fusarium* toxins: zearalenone

Iru

Alternaria toxins
Alternariol methyl ether

Aspergillus toxins
Aflatoxin/s: aflatoxin B_1, aflatoxin B_2, aflatoxin G_1, aflatoxin G_2
Other *Aspergillus* toxins: sterigmatocystin

Aspergillus and ***Penicillium*** toxins
Citrinin, ochratoxin A

Fusarium toxins
Fumonisin/s: fumonisin B_1, fumonisin B_2, fumonisin B_3
Type A trichothecenes: HT-2 toxin, T-2 toxin
Type B trichothecenes: deoxynivalenol, 3-acetyldeoxynivalenol, fusarenon-X (4-acetylnivalenol)
Other *Fusarium* toxins: zearalenone

Penicillium toxins
Roquefortine C

Jaggery

Aspergillus toxins
Aflatoxin/s

Jam

Alternaria toxins
Alternariol

Aspergillus and ***Penicillium*** toxins
Ochratoxin A

Penicillium toxins
Mycophenolic acid

Jam (apple jam)

Aspergillus and ***Penicillium*** toxins
Patulin

Jam (bean jam)

Aspergillus toxins
Aflatoxin/s: aflatoxin B_1, aflatoxin B_2

Jam (fig jam)

Aspergillus toxins
Aflatoxin/s: aflatoxin B_1, aflatoxin B_2

Aspergillus and ***Penicillium*** toxins
Ochratoxin A

Fusarium toxins
Type A trichothecenes: T-2 toxin

Jam (pear jam)

Aspergillus and ***Penicillium*** toxins
Patulin

Jam (quince jam)

Aspergillus and ***Penicillium*** toxins
Patulin

Jellies

Aspergillus and ***Penicillium*** toxins
Patulin

Jello

Aspergillus toxins
Aflatoxin/s

Juices

Alternaria toxins
Atenuene, alternariol, alternariol methyl ether, tenuazonic acid

Aspergillus and ***Penicillium*** toxins
Ochratoxin A, patulin

Juices (apple concentrate)

Alternaria toxins
Alternariol, alternariol methyl ether

Aspergillus and ***Penicillium*** toxins
Patulin

Juices (apple juice)

Alternaria toxins
Altenuene, alternariol, alternariol methyl ether, tenuazonic acid

Aspergillus and ***Penicillium*** toxins
Citrinin, ochratoxin A, patulin

Juices (apricot concentrate)

Aspergillus and ***Penicillium*** toxins
Patulin

Juices (apricot juice)

Alternaria toxins
Tenuazonic acid

Aspergillus and ***Penicillium*** toxins
Patulin

Juices (banana juice)

Alternaria toxins
Tenuazonic acid

Juices (blackcurrant juice)

Alternaria toxins
Tenuazonic acid

Aspergillus and ***Penicillium*** toxins
Citrinin, ochratoxin A, patulin

Juices (carrot juice)

Aspergillus and ***Penicillium*** toxins
Ochratoxin A

Juices (cherry juice)

Alternaria toxins
Tenuazonic acid

Aspergillus and ***Penicillium*** toxins
Citrinin, patulin

Juices (citrus juice)

Alternaria toxins
Alternariol methyl ether, tenuazonic acid

Juices (cranberry juice)

Alternaria toxins
Alternariol, alternariol methyl ether

Juices (grape juice)

Alternaria toxins
Alternariol, alternariol methyl ether, tenuazonic acid

Aspergillus and ***Penicillium*** toxins
Ochratoxin A, patulin

Juices (grapefruit juice)

Alternaria toxins
Alternariol

Juices (guava juice)

Aspergillus toxins
Aflatoxin/s: aflatoxin B_1

Juices (hawthorn juice)

Aspergillus and ***Penicillium*** toxins
Patulin

Juices (lychee juice)

Aspergillus and ***Penicillium*** toxins
Patulin

Juices (orange juice)

Alternaria toxins
Alternariol, alternariol methyl ether, tenuazonic acid

Aspergillus and ***Penicillium*** toxins
Patulin

Juices (peach concentrate)

Aspergillus and ***Penicillium*** toxins
Patulin

Juices (peach juice)

Aspergillus and ***Penicillium*** toxins
Patulin

Juices (pear concentrate)

Aspergillus and ***Penicillium*** toxins
Patulin

Juices (pear juice)

Alternaria toxins
Tenuazonic acid

Aspergillus and ***Penicillium*** toxins
Patulin

Juices (pineapple concentrate)

Aspergillus and ***Penicillium*** toxins
Patulin

Juices (pineapple juice)

Fusarium toxins
Fumonisin/s: fumonisin B_1, fumonisin B_2, fumonisin B_3
Other *Fusarium* toxins: moniliformin

Juices (pomegranate juice)

Alternaria toxins
Tenuazonic acid

Aspergillus and ***Penicillium*** toxins
Patulin

Juices (rasberry juice)

Alternaria toxins
Alternariol

Juices (rosehip juice)

Aspergillus and ***Penicillium*** toxins
Patulin

Juices (strawberry juice)

Alternaria toxins
Tenuazonic acid

Juices (sugarcane juice)

Aspergillus toxins
Aflatoxin/s: aflatoxin B_1, aflatoxin G_1, aflatoxin/s

Aspergillus and ***Penicillium*** toxins
3-Nitropropionic acid, kojic acid, penicillic acid

Fusarium toxins
Fusaric acid

Juices (tomato juice)

Alternaria toxins
Altenuene, alternariol, alternariol methyl ether, tenuazonic acid

Aspergillus and ***Penicillium*** toxins
Citrinin, ochratoxin A, patulin

Juices (vegetable juice)

Alternaria toxins
Alternariol, alternariol methyl ether

Kernels (apricot)

Aspergillus toxins
Aflatoxin/s: aflatoxin B_1, aflatoxin B_2, aflatoxin G_1, aflatoxin/s

Keskul

Aspergillus toxins
Aflatoxin/s

Kubeba

Aspergillus toxins
Aflatoxin/s: aflatoxin B_1, aflatoxin B_2

Kuru-kuru

Aspergillus toxins
Aflatoxin/s: aflatoxin B_1, aflatoxin B_2, aflatoxin G_1, aflatoxin G_2
Other *Aspergillus* toxins: sterigmatocystin

Aspergillus and ***Penicillium*** toxins
Kojic acid, ochratoxin A, ochratoxin B, viomellein

Fusarium toxins
Depsipeptides: beauvericin, enniatin A, enniatin A_1, enniatin B, enniatin B_1
Fumonisin/s: fumonisin B_1, fumonisin B_3
Other *Fusarium* toxins: fusaric acid, zearalenone

Penicillium toxins
Mycophenolic acid, roquefortine C

Legumes

Aspergillus toxins
Aflatoxin/s: aflatoxin B_1

Aspergillus and ***Penicillium*** toxins
Ochratoxin A

Lentils

Aspergillus toxins
Aflatoxin/s: aflatoxin B_1, aflatoxin/s

Aspergillus and ***Penicillium*** toxins
Ochratoxin A

Fusarium toxins
Type A trichothecenes: T-2 toxin

Licorice

Alternaria toxins
Alternariol

Aspergillus toxins
Aflatoxin/s: aflatoxin B_1

Aspergillus and ***Penicillium*** toxins
Citrinin, ochratoxin A

Fusarium toxins
Fumonisin/s: fumonisin B_1
Type B trichothecenes: deoxynivalenol
Other *Fusarium* toxins: zearalenone

Mahua seeds

Aspergillus toxins
Aflatoxin/s: aflatoxin B_1, aflatoxin/s

Marmalade (apple marmalade)

Aspergillus and ***Penicillium*** toxins
Patulin

Marmalade (pear marmalade)

Aspergillus and ***Penicillium*** toxins
Patulin

Margarine

Fusarium toxins
Fumonisin/s: fumonisin B_1
Other *Fusarium* toxins: zearalenone

Marzipan (almond paste)

Aspergillus toxins
Aflatoxin/s: aflatoxin B_1, aflatoxin B_2

Meal (copra meal)

Aspergillus toxins
Aflatoxin/s: aflatoxin B_1, aflatoxin B_2, aflatoxin G_1, aflatoxin G_2, aflatoxin/s

Meal (cycad meal)

Aspergillus toxins
Aflatoxin/s: aflatoxin B_1, aflatoxin/s

Meal (egusi meal)

Aspergillus toxins
Aflatoxin/s

Meal (peanut meal)

Aspergillus toxins
Aflatoxin/s: aflatoxin B_1

Meal (pistachio meal)

Aspergillus toxins
Aflatoxin/s: aflatoxin B_1, aflatoxin B_2

Meal (soybean meal)

Fusarium toxins
Type A trichothecenes: diacetoxyscirpenol, T-2 tetraol
Type B trichothecenes: deoxynivalenol
Other *Fusarium* toxins: zearalenone

Medicinal plants

Aspergillus toxins
Aflatoxin/s: aflatoxin B_1, aflatoxin B_2, aflatoxin G_1, aflatoxin G_2, aflatoxin M_1, aflatoxin/s
Other *Aspergillus* toxins: gliotoxin, sterigmatocystin

Aspergillus **and** ***Penicillium*** toxins
Citrinin, cyclopiazonic acid, ochratoxin A, ochratoxin B, patulin

Fusarium toxins
Depsipeptides: beauvericin, enniatin A, enniatin A_1, enniatin B, enniatin B_1
Fumonisin/s: fumonisin B_1, fumonisin B_2, fumonisin B_3, fumonisin/s
Type A trichothecenes: diacetoxyscirpenol, HT-2 toxin, neosolaniol, T-2 toxin
Type B trichothecenes: deoxynivalenol, fusarenon-X (4-acetylnivalenol), nivalenol
Other *Fusarium* toxins: moniliformin, zearalenone

Penicillium toxins
Mycophenolic acid

Meju

Aspergillus toxins
Aflatoxin/s: aflatoxin B_1

Milk thistle

Fusarium toxins
Type A trichothecenes: HT-2 toxin, T-2 toxin
Other *Fusarium* toxins: zearalenone

Muhallebi

Aspergillus toxins
Aflatoxin/s

Must

Aspergillus toxins
Aflatoxin/s: aflatoxin B_1

Aspergillus **and** ***Penicillium*** toxins
Ochratoxin A

Fusarium toxins
Fumonisin/s: fumonisin B_2

Nectars (apple nectar)

Aspergillus **and** ***Penicillium*** toxins
Patulin

Nectars (cranberry nectar)

Alternaria toxins
Alternariol, alternariol methyl ether

Nectars (prune nectar)

Alternaria toxins
Alternariol, alternariol methyl ether

Nut cereals

Aspergillus toxins
Aflatoxin/s

Nut crème

Aspergillus toxins
Aflatoxin/s: aflatoxin B_1, aflatoxin/s

Aspergillus **and** ***Penicillium*** toxins
Ochratoxin A

Nuts

Aspergillus toxins
Aflatoxin/s: aflatoxin B_1, aflatoxin B_2, aflatoxin G_1, aflatoxin G_2, aflatoxin/s

Aspergillus **and** ***Penicillium*** toxins
Ochratoxin A

Nuts (almond)

Alternaria toxins
Alternariol, alternariol methyl ether

Aspergillus toxins
Aflatoxin/s: aflatoxin B_1, aflatoxin B_2, aflatoxin G_1, aflatoxin G_2, aflatoxin/s
Other *Aspergillus* toxins: sterigmatocystin

Aspergillus **and** ***Penicillium*** toxins
Cyclopiazonic acid, ochratoxin A, ochratoxin B

Fusarium toxins
Depsipeptides: beauvericin, enniatin B, enniatin B_1, enniatin B_3
Other *Fusarium* toxins: zearalenone

Penicillium toxins
Rubratoxin

Nuts (Brazil nut, paranut)

Aspergillus toxins
Aflatoxin/s: aflatoxin B_1, aflatoxin B_2, aflatoxin G_1, aflatoxin G_2, aflatoxin/s

Aspergillus and ***Penicillium*** toxins
Cyclopiazonic acid

Nuts (cashew nut)

Aspergillus toxins
Aflatoxin/s: aflatoxin B_1, aflatoxin B_2, aflatoxin G_1, aflatoxin G_2, aflatoxin/s

Aspergillus and ***Penicillium*** toxins
Citrinin, ochratoxin A, ochratoxin B

Nuts (chestnut)

Alternaria toxins
Alternariol, alternariol methyl ether

Aspergillus toxins
Aflatoxin/s: aflatoxin B_1, aflatoxin B_2, aflatoxin G_1, aflatoxin G_2, aflatoxin/s

Aspergillus and ***Penicillium*** toxins
Citrinin, ochratoxin A

Penicillium toxins
Mycophenolic acid, roquefortine C

Nuts (hazelnut)

Alternaria toxins
Alternariol, alternariol methyl ether, altertoxin I

Aspergillus toxins
Aflatoxin/s: aflatoxin B_1, aflatoxin B_2, aflatoxin G_1, aflatoxin G_2, aflatoxin/s
Other *Aspergillus* toxins: sterigmatocystin

Aspergillus and ***Penicillium*** toxins
3-Nitropropionic acid, kojic acid, ochratoxin A, ochratoxin B

Fusarium toxins
Depsipeptides: beauvericin, enniatin A, enniatin A_1, enniatin B, enniatin B_2, enniatin B_3, enniatin B_4
Type A trichothecenes: HT-2 toxin, T-2 toxin
Other *Fusarium* toxins: moniliformin, zearalenone

Penicillium toxins
Mycophenolic acid

Nuts (peanut)

Alternaria toxins
Alternariol, alternariol methyl ether, altertoxin I

Aspergillus toxins
Aflatoxin/s: aflatoxin B_1, aflatoxin B_2, aflatoxin G_1, aflatoxin G_2, aflatoxin M_1, aflatoxin M_2, aflatoxin/s
Other *Aspergillus* toxins: sterigmatocystin

Aspergillus and ***Penicillium*** toxins
3-Nitropropionic acid, citrinin, cyclopiazonic acid, kojic acid, ochratoxin A, ochratoxin B

Fusarium toxins
Depsipeptides: beauvericin, enniatin A, enniatin A_1, enniatin B, enniatin B_1, enniatin B_4
Fumonisin/s: fumonisin B_1
Type A trichothecenes: diacetoxyscirpenol, T-2 toxin
Type B trichothecenes: deoxynivalenol, nivalenol
Other *Fusarium* toxins: moniliformin, zearalenone

Penicillium toxins
Mycophenolic acid

Nuts (pecan nut)

Aspergillus toxins
Aflatoxin/s: aflatoxin B_1, aflatoxin G_1, aflatoxin/s
Other *Aspergillus* toxins: sterigmatocystin

Nuts (pine nut)

Aspergillus toxins
Aflatoxin/s: aflatoxin B_1, aflatoxin B_2, aflatoxin G_1, aflatoxin/s

Nuts (pistachio nut)

Aspergillus toxins
Aflatoxin/s: aflatoxin B_1, aflatoxin B_2, aflatoxin G_1, aflatoxin G_2, aflatoxin M_1, aflatoxin/s
Other *Aspergillus* toxins: sterigmatocystin

Aspergillus and ***Penicillium*** toxins
Ochratoxin A

Fusarium toxins
Depsipeptides: beauvericin

Penicillium toxins
Mycophenolic acid

Nuts (tiger nut)

Aspergillus toxins
Aflatoxin/s: aflatoxin B_1, aflatoxin B_2, aflatoxin G_1, aflatoxin G_2

Aspergillus and ***Penicillium*** toxins
Ochratoxin A

Fusarium toxins
Depsipeptides: beauvericin, enniatin A, enniatin A_1, enniatin B, enniatin B_1
Type B trichothecenes: deoxynivalenol

Nuts (walnut)

Aspergillus toxins
Aflatoxin/s: aflatoxin B_1, aflatoxin B_2, aflatoxin G_1, aflatoxin G_2, aflatoxin/s

Aspergillus and ***Penicillium*** toxins
Ochratoxin A

Fusarium toxins
Zearalenone

Penicillium toxins
Rubratoxin

Ogbono

Aspergillus toxins
Aflatoxin/s

Ogili-ugba

Aspergillus toxins
Aflatoxin/s

Ogiri

Alternaria toxins
Alternariol methyl ether

Aspergillus toxins
Aflatoxin/s: aflatoxin B_1, aflatoxin B_2, aflatoxin G_2
Other *Aspergillus* toxins: sterigmatocystin

Aspergillus and ***Penicillium*** toxins
3-Nitropropionic acid, citrinin, ochratoxin A

Fusarium toxins
Depsipeptides: beauvericin
Type A trichothecenes: diacetoxyscirpenol, HT-2 toxin, neosolaniol
Type B trichothecenes: deoxynivalenol
Other *Fusarium* toxins: zearalenone

Ogoro

Aspergillus toxins
Aflatoxin/s

Oil seeds

Aspergillus toxins
Aflatoxin/s: aflatoxin B_1, aflatoxin/s

Aspergillus and ***Penicillium*** toxins
Ochratoxin A

Fusarium toxins
Zearalenone

Oils

Aspergillus toxins
Aflatoxin/s: aflatoxin B_1, aflatoxin B_2, aflatoxin G_1, aflatoxin G_2, aflatoxin/s

Fusarium toxins
Type A trichothecenes: diacetoxyscirpenol, HT-2 toxin, T-2 toxin
Type B trichothecenes: deoxynivalenol, 15-acetyldeoxynivalenol
Other *Fusarium* toxins: zearalenone

Oils (coconut oil)

Aspergillus toxins
Aflatoxin/s: aflatoxin B_1, aflatoxin B_2, aflatoxin G_1

Oils (maize oil)

Aspergillus toxins
Aflatoxin/s: aflatoxin B_1, aflatoxin B_2

Fusarium toxins
Fumonisin/s: fumonisin B_1
Type A trichothecenes: diacetoxyscirpenol, HT-2 toxin, T-2 toxin
Type B trichothecenes: deoxynivalenol, 15-acetyldeoxynivalenol
Other *Fusarium* toxins: zearalenone

Oils (mustard oil)

Aspergillus toxins
Aflatoxin/s: aflatoxin B_1

Oils (olive oil)

Aspergillus toxins
Aflatoxin/s: aflatoxin B_1, aflatoxin B_2, aflatoxin G_1, aflatoxin G_2

Aspergillus and ***Penicillium*** toxins
Ochratoxin A

Oils (peanut oil)

Aspergillus toxins
Aflatoxin/s: aflatoxin B_1, aflatoxin B_2, aflatoxin G_1, aflatoxin G_2, aflatoxin/s

Oils (rapeseed oil)

Aspergillus toxins
Aflatoxin/s: aflatoxin B_1

Fusarium toxins
Zearalenone

Oils (sesame oil)

Aspergillus toxins
Aflatoxin/s: aflatoxin B_1, aflatoxin B_2, aflatoxin G_1, aflatoxin G_2,

Aspergillus and ***Penicillium*** toxins
Ochratoxin A

Oils (soybean oil)

Aspergillus toxins
Aflatoxin/s: aflatoxin B_1

Fusarium toxins
Type A trichothecenes: T-2 tetraol
Type B trichothecenes: deoxynivalenol, 15-acetyldeoxynivalenol
Other *Fusarium* toxins: zearalenone

Oils (sunflower oil)

Alternaria toxins
Alternariol methyl ether

Aspergillus toxins
Aflatoxin/s: aflatoxin B_1, aflatoxin B_2, aflatoxin G_1, aflatoxin G_2

Fusarium toxins
Type A trichothecenes: HT-2 toxin, T-2 toxin

Oils (vegetable oil)

Alternaria toxins
Alternariol, alternariol methyl ether, tenuazonic acid

Okra

Aspergillus toxins
Aflatoxin/s: aflatoxin G_2

Olives

Alternaria toxins
Altenuene, alternariol, alternariol methyl ether, tenuazonic acid

Aspergillus toxins
Aflatoxin/s: aflatoxin B_1

Aspergillus and ***Penicillium*** toxins
Citrinin, ochratoxin A

Paste (bean paste)

Aspergillus toxins
Aflatoxin/s: aflatoxin B_1

Paste (curry paste)

Aspergillus toxins
Aflatoxin/s

Aspergillus and ***Penicillium*** toxins
Ochratoxin A

Fusarium toxins
Fumonisin/s
Type B trichothecenes: nivalenol
Other *Fusarium* toxins: zearalenone

Paste (fig paste)

Aspergillus toxins
Aflatoxin/s: aflatoxin B_1, aflatoxin B_2, aflatoxin G_1, aflatoxin G_2, aflatoxin/s

Aspergillus and ***Penicillium*** toxins
Ochratoxin A

Paste (hazelnut paste)

Aspergillus toxins
Aflatoxin/s: aflatoxin B_1, aflatoxin B_2, aflatoxin G_1, aflatoxin G_2

Paste (peanut paste)

Alternaria toxins
Alternariol methyl ether

Aspergillus toxins
Aflatoxin/s: aflatoxin B_1, aflatoxin B_2, aflatoxin G_1, aflatoxin G_2, aflatoxin/s
Other *Aspergillus* toxins: sterigmatocystin

Aspergillus and ***Penicillium*** toxins
Kojic acid, ochratoxin A, ochratoxin B, viomellein

Fusarium toxins
Depsipeptides: beauvericin, enniatin A, enniatin A_1, enniatin B, enniatin B_1
Fumonisin/s: fumonisin B_1, fumonisin B_2, fumonisin B_3
Other *Fusarium* toxins: fusaric acid, zearalenone

Penicillium toxins
Mycophenolic acid

Paste (pipian paste)

Aspergillus toxins
Aflatoxin/s

Paste (red pepper paste)

Aspergillus toxins
Aflatoxin/s

Paste (sesame paste)

Aspergillus toxins
Aflatoxin/s: aflatoxin B_1, aflatoxin B_2, aflatoxin G_1, aflatoxin G_2

Paste (soybean paste)

Aspergillus toxins
Aflatoxin/s: aflatoxin B_1, aflatoxin B_2, aflatoxin G_1, aflatoxin G_2

Aspergillus and ***Penicillium*** toxins
Ochratoxin A

Paste (peanut paste)

Alternaria toxins
Alternariol, alternariol methyl ether, tenuazonic acid

Peas

Aspergillus toxins
Aflatoxin/s: aflatoxin B_1, aflatoxin G_1, aflatoxin/s

Aspergillus and ***Penicillium*** toxins
Ochratoxin A, penicillic acid

Fusarium toxins
Zearalenone

Peas (chickpea)

Aspergillus toxins
Aflatoxin/s: aflatoxin B_1, aflatoxin/s

Aspergillus and ***Penicillium*** toxins
Ochratoxin A

Fusarium toxins
Type A trichothecenes: T-2 toxin

Pekmez

Aspergillus and ***Penicillium*** toxins
Ochratoxin A

Pickles

Aspergillus toxins
Aflatoxin/s

Aspergillus and ***Penicillium*** toxins
Ochratoxin A

Fusarium toxins
Fumonisin/s
Other *Fusarium* toxins: zearalenone

Pies

Aspergillus and ***Penicillium*** toxins
Ochratoxin A

Pinhol

Aspergillus toxins
Aflatoxin/s: aflatoxin G_1

Pizza

Aspergillus and ***Penicillium*** toxins
Ochratoxin A

Plant food supplements

Aspergillus and ***Penicillium*** toxins
Ochratoxin A

Potatoes

Aspergillus and ***Penicillium*** toxins
Ochratoxin A

Powder (cocoa powder)

Aspergillus toxins
Aflatoxin/s: aflatoxin B_1, aflatoxin B_2, aflatoxin G_1

Aspergillus and ***Penicillium*** toxins
Ochratoxin A, ochratoxin B

Products

Aspergillus toxins
Aflatoxin/s: aflatoxin B_1

Aspergillus and ***Penicillium*** toxins
Ochratoxin A

Products (almond product)

Aspergillus toxins
Aflatoxin/s: aflatoxin B_1

Products (apple flavor product)

Aspergillus and ***Penicillium*** toxins
Patulin

Products (apple product)

Aspergillus and ***Penicillium*** toxins
Patulin

Products (cassava product)

Alternaria toxins
Alternariol methyl ether

Aspergillus toxins
Aflatoxin/s: aflatoxin B_1, aflatoxin G_1

Aspergillus and ***Penicillium*** toxins
Kojic acid, penicillic acid

Fusarium toxins
Fumonisin/s: fumonisin B_1, fumonisin B_2, fumonisin B_3
Other *Fusarium* toxins: zearalenone

Products (cocoa product)

Aspergillus toxins
Aflatoxin/s: aflatoxin B_1, aflatoxin B_2, aflatoxin G_1, aflatoxin G_2

Aspergillus and ***Penicillium*** toxins
Ochratoxin A

Products (coconut product)

Aspergillus toxins
Aflatoxin/s

Aspergillus and ***Penicillium*** toxins
Citrinin, ochratoxin A

Products (fruit product)

Aspergillus and ***Penicillium*** toxins
Patulin

Products (ginger product)

Aspergillus and ***Penicillium*** toxins
Ochratoxin A

Products (legume product)

Aspergillus toxins
Aflatoxin/s: aflatoxin B_1
Other *Aspergillus* toxins: sterigmatocystin

Aspergillus and ***Penicillium*** toxins
Ochratoxin A, patulin

Fusarium toxins
Zearalenone

Products (melon seed product)

Aspergillus toxins
Aflatoxin/s: aflatoxin B_1

Products (nut product)

Aspergillus toxins
Aflatoxin/s: aflatoxin B_1, aflatoxin B_2, aflatoxin/s

Products (peanut product)

Aspergillus toxins
Aflatoxin/s: aflatoxin B_1, aflatoxin B_2, aflatoxin G_1, aflatoxin G_2, aflatoxin/s

Fusarium toxins
Depsipeptides: beauvericin

Products (pistachio product)

Aspergillus toxins
Aflatoxin/s: aflatoxin B_1, aflatoxin B_2, aflatoxin G_1

Aspergillus and ***Penicillium*** toxins
Ochratoxin A, ochratoxin B

Fusarium toxins
Depsipeptides: beauvericin
Type A trichothecenes: neosolaniol

Products (plant product)

Aspergillus and ***Penicillium*** toxins
Ochratoxin A

Products (potato product)

Fusarium toxins
Type A trichothecenes: diacetoxyscirpenol, HT-2 toxin, monoacetoxyscirpenol, scirpentriol
Other *Fusarium* toxins: zearalenone

Products (sesame product)

Aspergillus toxins
Aflatoxin/s: aflatoxin B_1

Products (soybean product)

Fusarium toxins
Fumonisin/s

Products (tobacco product)

Aspergillus toxins
Aflatoxin/s: aflatoxin B_1

Products (tomato product)

Alternaria toxins
Altenuene, alternariol, alternariol methyl ether, tenuazonic acid

Pudding

Aspergillus toxins
Aflatoxin/s

Aspergillus and ***Penicillium*** toxins
Ochratoxin A

Pulp (apple pulp)

Aspergillus and ***Penicillium*** toxins
Patulin

Pulp (grape pulp)

Aspergillus and ***Penicillium*** toxins
Ochratoxin A

Pulp (plum pulp)

Aspergillus and ***Penicillium*** toxins
Patulin

Pulp (tomato pulp)

Alternaria toxins
Tenuazonic acid

Aspergillus and ***Penicillium*** toxins
Cyclopiazonic acid

Puree

Alternaria toxins
Alternariol, alternariol methyl ether, tenuazonic acid

Aspergillus and ***Penicillium*** toxins
Cyclopiazonic acid, ochratoxin A, patulin

Penicillium toxins
Mycophenolic acid

Raisins

Alternaria toxins
Alternariol, alternariol methyl ether, tenuazonic acid

Aspergillus toxins
Aflatoxin/s: aflatoxin B_1, aflatoxin B_2, aflatoxin G_1, aflatoxin G_2, aflatoxin/s

Aspergillus and ***Penicillium*** toxins
Citrinin, ochratoxin A, ochratoxin B

Fusarium toxins
Fumonisin/s: fumonisin B_1, fumonisin B_2, fumonisin B_3

Penicillium toxins
Mycophenolic acid

Rootcrops

Aspergillus toxins
Aflatoxin/s

Salep

Aspergillus toxins
Aflatoxin/s

Sauces

Fusarium toxins
Fumonisin/s

Sauces (apple sauce)

Alternaria toxins
Alternariol

Sauces (chilli sauce)

Aspergillus toxins
Aflatoxin/s

Aspergillus and ***Penicillium*** toxins
Ochratoxin A

Fusarium toxins
Type B trichothecenes: nivalenol
Other *Fusarium* toxins: zearalenone

Sauces (chocolate sauce)

Aspergillus toxins
Aflatoxin/s

Sauces (flour sauce)

Aspergillus toxins
Aflatoxin/s

Sauces (leaf-sauce)

Aspergillus toxins
Aflatoxin/s

Sauces (peanut sauce)

Aspergillus toxins
Aflatoxin/s: aflatoxin B_1, aflatoxin/s

Sauces (peanut-chilli sauce)

Aspergillus toxins
Aflatoxin/s: aflatoxin B_1, aflatoxin/s

Sauces (pepper sauce)

Alternaria toxins
Tenuazonic acid

Sauces (pepper- and barbecue sauce)

Aspergillus and ***Penicillium*** toxins
Ochratoxin A

Sauces (soy sauce)

Aspergillus toxins
Aflatoxin/s: aflatoxin B_1, aflatoxin G_1, aflatoxin G_2

Aspergillus and ***Penicillium*** toxins
Ochratoxin A

Fusarium toxins
Type B trichothecenes: deoxynivalenol

Sauces (tomato sauce)

Alternaria toxins
Altenuene, alternariol, alternariol methyl ether, tenuazonic acid

Seeds

Aspergillus and ***Penicillium*** toxins
Ochratoxin A

Seeds (bush mango)

Alternaria toxins
Tenuazonic acid

Aspergillus toxins
Aflatoxin/s: aflatoxin B_1, aflatoxin B_2, aflatoxin G_1, aflatoxin G_2, aflatoxin M_1
Other *Aspergillus* toxins: sterigmatocystin

Aspergillus and ***Penicillium*** toxins
3-Nitropropionic acid, cyclopiazonic acid, kojic acid, ochratoxin A, ochratoxin B

Fusarium toxins
Depsipeptides: beauvericin
Type A trichothecenes: HT-2 toxin, T-2 toxin
Type B trichothecenes: nivalenol
Other *Fusarium* toxins: moniliformin

Penicillium toxins
Mycophenolic acid

Seeds (linseed)

Aspergillus toxins
Aflatoxin/s: aflatoxin B_1, aflatoxin B_2, aflatoxin/s

Aspergillus and ***Penicillium*** toxins
Ochratoxin A

Seeds (lupine)

Aspergillus toxins
Aflatoxin/s

Seeds (melon)

Aspergillus toxins
Aflatoxin/s: aflatoxin B_1, aflatoxin B_2, aflatoxin G_1, aflatoxin G_2, aflatoxin M_1, aflatoxin/s
Other *Aspergillus* toxins: sterigmatocystin

Aspergillus and ***Penicillium*** toxins
3-Nitropropionic acid, citrinin, cyclopiazonic acid, kojic acid, ochratoxin A

Fusarium toxins
Depsipeptides: beauvericin

Seeds (pumpkin)

Fusarium toxins
Type A trichothecenes: HT-2 toxin
Other *Fusarium* toxins: zearalenone

Seeds (sesame)

Alternaria toxins
Alternariol, alternariol methyl ether, tenuazonic acid

Aspergillus toxins
Aflatoxin/s: aflatoxin B_1, aflatoxin B_2, aflatoxin G_1, aflatoxin G_2, aflatoxin/s
Other *Aspergillus* toxins: sterigmatocystin

Aspergillus and ***Penicillium*** toxins
3-Nitropropionic acid, citrinin, kojic acid, ochratoxin A

Fusarium toxins
Depsipeptides: beauvericin
Type A trichothecenes: T-2 toxin
Type B trichothecenes: deoxynivalenol
Other *Fusarium* toxins: fusaric acid, zearalenone

Seeds (sunflower)

Alternaria toxins
Altenuene, alternariol, alternariol methyl ether, altertoxin I, tenuazonic acid

Aspergillus toxins
Aflatoxin/s: aflatoxin B_1, aflatoxin B_2, aflatoxin G_1, aflatoxin G_2, aflatoxin/s

Aspergillus and ***Penicillium*** toxins
Ochratoxin A

Fusarium toxins
Type A trichothecenes: HT-2 toxin, T-2 toxin
Type B trichothecenes: deoxynivalenol
Other *Fusarium* toxins: zearalenone

Shea-nuts

Aspergillus toxins
Aflatoxin/s: aflatoxin B_1, aflatoxin B_2

Shiitake

Fusarium toxins
Fumonisin/s: fumonisin B_1, fumonisin B_2
Type B trichothecenes: deoxynivalenol

Penicillium toxins
Mycophenolic acid

Shiro

Aspergillus toxins
Aflatoxin/s: aflatoxin B_1

Snacks

Alternaria toxins
Alternariol methyl ether

Aspergillus toxins
Aflatoxin/s: aflatoxin B_1, aflatoxin B_2, aflatoxin G_1, aflatoxin G_2, aflatoxin M_1, aflatoxin/s

Aspergillus and ***Penicillium*** toxins
3-Nitropropionic acid, kojic acid

Fusarium toxins
Depsipeptides: beauvericin
Fumonisin/s: fumonisin B_1, fumonisin B_2, fumonisin B_3
Other *Fusarium* toxins: moniliformin

Soups

Alternaria toxins
Alternariol, alternariol methyl ether

Aspergillus toxins
Aflatoxin/s: aflatoxin B_1, aflatoxin B_2, aflatoxin G_1, aflatoxin G_2, aflatoxin/s
Other *Aspergillus* toxins: sterigmatocystin

Aspergillus and ***Penicillium*** toxins
Kojic acid, ochratoxin A, ochratoxin B

Fusarium toxins
Depsipeptides: beauvericin, enniatin A, enniatin A_1, enniatin B, enniatin B_1
Fumonisin/s: fumonisin B_1, fumonisin B_2, fumonisin B_3
Type B trichothecenes: deoxynivalenol

Soy isoflavones

Aspergillus and ***Penicillium*** toxins
Ochratoxin A

Fusarium toxins
Fumonisin/s: fumonisin B_1

Spices

Alternaria toxins
Tenuazonic acid

Aspergillus toxins
Aflatoxin/s: aflatoxin B_1, aflatoxin B_2, aflatoxin G_1, aflatoxin G_2, aflatoxin/s

Aspergillus and ***Penicillium*** toxins
Ochratoxin A, ochratoxin B

Fusarium toxins
Fumonisin/s: fumonisin B_1, fumonisin/s
Other *Fusarium* toxins: zearalenone

Spices (ammi)

Aspergillus toxins
Aflatoxin/s: aflatoxin B_1, aflatoxin B_2, aflatoxin G_1

Spices (anise)

Aspergillus toxins
Aflatoxin/s: aflatoxin B_1, aflatoxin/s

Aspergillus and ***Penicillium*** toxins
Ochratoxin A

Spices (basil)

Aspergillus toxins
Aflatoxin/s: aflatoxin B_1

Spices (black cumin)

Aspergillus toxins
Aflatoxin/s

Aspergillus and ***Penicillium*** toxins
Citrinin

Spices (caraway)

Aspergillus toxins
Aflatoxin/s
Other *Aspergillus* toxins: sterigmatocystin

Aspergillus and ***Penicillium*** toxins
Ochratoxin A

Spices (cardamom)

Aspergillus toxins
Aflatoxin/s: aflatoxin B_1, aflatoxin B_2, aflatoxin G_1, aflatoxin G_2,

Aspergillus and ***Penicillium*** toxins
Citrinin, ochratoxin A

Spices (cassia)

Aspergillus toxins
Aflatoxin/s: aflatoxin B_1, aflatoxin B_2, aflatoxin G_1, aflatoxin G_2

Fusarium toxins
Fumonisin/s: fumonisin B_1

Spices (cayenne pepper, chilli)

Alternaria toxins
Alternariol methyl ether

Aspergillus toxins
Aflatoxin/s: aflatoxin B_1, aflatoxin B_2, aflatoxin G_1, aflatoxin G_2, aflatoxin/s
Other *Aspergillus* toxins: sterigmatocystin

Aspergillus and ***Penicillium*** toxins
Citrinin, ochratoxin A

Fusarium toxins
Fumonisin/s: fumonisin B_1, fumonisin B_2
Type A trichothecenes: diacetoxyscirpenol, HT-2 toxin, T-2 toxin
Type B trichothecenes: deoxynivalenol
Other *Fusarium* toxins: zearalenone

Spices (celery)

Aspergillus toxins
Aflatoxin/s: aflatoxin G_1

Spices (cinnamon)

Aspergillus toxins
Aflatoxin/s: aflatoxin B_1

Aspergillus and ***Penicillium*** toxins
Ochratoxin A

Spices (coriander)

Alternaria toxins
Tenuazonic acid

Aspergillus toxins
Aflatoxin/s: aflatoxin B_1, aflatoxin B_2, aflatoxin G_1, aflatoxin G_2, aflatoxin/s

Aspergillus and ***Penicillium*** toxins
Citrinin, ochratoxin A

Fusarium toxins
Type B trichothecenes: deoxynivalenol
Other *Fusarium* toxins: zearalenone

Spices (cumin)

Alternaria toxins
Tenuazonic acid

Aspergillus toxins
Aflatoxin/s: aflatoxin B_1, aflatoxin B_2, aflatoxin G_1, aflatoxin G_2, aflatoxin/s
Other *Aspergillus* toxins: sterigmatocystin

Aspergillus and ***Penicillium*** toxins
Citrinin, ochratoxin A

Spices (curcuma)

Alternaria toxins
Tenuazonic acid

Aspergillus toxins
Aflatoxin/s: aflatoxin B_1, aflatoxin B_2, aflatoxin G_1, aflatoxin G_2, aflatoxin/s

Aspergillus and ***Penicillium*** toxins
Citrinin, ochratoxin A

Penicillium toxins
Rubratoxin

Spices (curry)

Alternaria toxins
Tenuazonic acid

Aspergillus toxins
Aflatoxin/s: aflatoxin B_1, aflatoxin B_2, aflatoxin G_1, aflatoxin G_2, aflatoxin/s

Aspergillus and ***Penicillium*** toxins
Ochratoxin A

Fusarium toxins
Fumonisin/s: fumonisin/s
Type A trichothecenes: diacetoxyscirpenol, HT-2 toxin, neosolaniol, T-2 toxin
Type B trichothecenes: fusarenon-X (4-acetylnivalenol), nivalenol,
Other *Fusarium* toxins: zearalenone

Spices (fennel)

Aspergillus toxins
Aflatoxin/s: aflatoxin B_1, aflatoxin B_2, aflatoxin G_1, aflatoxin G_2, aflatoxin/s
Other *Aspergillus* toxins: sterigmatocystin

Aspergillus and ***Penicillium*** toxins
Citrinin, ochratoxin A

Fusarium toxins
Zearalenone

Spices (fenugreek)

Aspergillus toxins
Aflatoxin/s: aflatoxin B_1, aflatoxin B_2, aflatoxin G_1

Spices (garlic)

Aspergillus toxins
Aflatoxin/s: aflatoxin B_1, aflatoxin B_2, aflatoxin G_1

Aspergillus and ***Penicillium*** toxins
Ochratoxin A

Fusarium toxins
Depsipeptides: beauvericin
Fumonisin/s: fumonisin B_1, fumonisin B_2, fumonisin B_3
Type B trichothecenes: deoxynivalenol, fusarenon-X (4-acetylnivalenol), nivalenol

Spices (garlic/onion)

Aspergillus toxins
Aflatoxin/s

Spices (ginger)

Alternaria toxins
Tenuazonic acid

Aspergillus toxins
Aflatoxin/s: aflatoxin B_1, aflatoxin B_2, aflatoxin G_1, aflatoxin G_2, aflatoxin/s

Aspergillus and ***Penicillium*** toxins
Ochratoxin A

Fusarium toxins
Depsipeptides: beauvericin, enniatin A, enniatin A_1, enniatin B, enniatin B_1
Type A trichothecenes: neosolaniol, T-2 toxin
Type B trichothecenes: deoxynivalenol, nivalenol

Spices (laurel leave)

Aspergillus toxins
Aflatoxin/s: aflatoxin B_1

Spices (majoran)

Aspergillus toxins
Aflatoxin/s
Other *Aspergillus* toxins: sterigmatocystin

Aspergillus and ***Penicillium*** toxins
Ochratoxin A

Spices (masala)

Aspergillus toxins
Aflatoxin/s: aflatoxin B_1, aflatoxin B_2, aflatoxin G_1, aflatoxin G_2, aflatoxin/s

Aspergillus and ***Penicillium*** toxins
Ochratoxin A

Spices (mustard)

Aspergillus toxins
Aflatoxin/s: aflatoxin B_1, aflatoxin B_2, aflatoxin G_1, aflatoxin G_2

Aspergillus and ***Penicillium*** toxins
Ochratoxin A

Spices (nutmeg)

Aspergillus toxins
Aflatoxin/s: aflatoxin B_1, aflatoxin B_2, aflatoxin G_1, aflatoxin G_2, aflatoxin/s

Aspergillus and ***Penicillium*** toxins
Ochratoxin A

Fusarium toxins
Fumonisin/s: fumonisin B_1

Spices (onion)

Fusarium toxins
Fumonisin/s: fumonisin B_1

Spices (oregano)

Aspergillus toxins
Sterigmatocystin

Spices (paprika)

Alternaria toxins
Alternariol, tenuazonic acid

Aspergillus toxins
Aflatoxin/s: aflatoxin B_1, aflatoxin B_2, aflatoxin G_1, aflatoxin G_2, aflatoxin/s
Other *Aspergillus* toxins: sterigmatocystin

Aspergillus and ***Penicillium*** toxins
Ochratoxin A

Fusarium toxins
Depsipeptides: beauvericin
Fumonisin/s: fumonisin B_1
Type A trichothecenes: HT-2 toxin, T-2 toxin
Type B trichothecenes: deoxynivalenol
Other *Fusarium* toxins: zearalenone

Spices (pepper)

Alternaria toxins
Alternariol, alternariol methyl ether, tenuazonic acid

Aspergillus toxins
Aflatoxin/s: aflatoxin B_1, aflatoxin B_2, aflatoxin G_1, aflatoxin G_2, aflatoxin/s
Other *Aspergillus* toxins: sterigmatocystin

Aspergillus and ***Penicillium*** toxins
Citrinin, ochratoxin A

Fusarium toxins
Fumonisin/s: fumonisin B_1
Other *Fusarium* toxins: zearalenone

Spices (peppermint)

Aspergillus toxins
Aflatoxin/s: aflatoxin B_1, aflatoxin/s

Spices (poppy seed)

Aspergillus toxins
Aflatoxin/s: aflatoxin B_1

Spices (prickly ash)

Aspergillus and ***Penicillium*** toxins
Ochratoxin A

Spices (rosemary)

Aspergillus toxins
Aflatoxin/s: aflatoxin B_1

Spices (saffron)

Aspergillus toxins
Aflatoxin/s: aflatoxin B_1

Spices (star anise)

Aspergillus toxins
Aflatoxin/s: aflatoxin G_2

Spices (sumac)

Aspergillus toxins
Aflatoxin/s: aflatoxin B_1

Spices (tandoori)

Aspergillus toxins
Aflatoxin/s

Aspergillus and ***Penicillium*** toxins
Ochratoxin A

Fusarium toxins
Fumonisin/s
Type A trichothecenes: T-2 toxin
Type B trichothecenes: nivalenol

Spices (thyme)

Aspergillus toxins
Aflatoxin/s: aflatoxin B_1
Other *Aspergillus* toxins: sterigmatocystin

Fusarium toxins
Type B trichothecenes: deoxynivalenol
Other *Fusarium* toxins: zearalenone

Starch (cassava starch)

Aspergillus toxins
Aflatoxin/s

Strawberries

Alternaria toxins
Alternariol, alternariol methyl ether

Aspergillus and ***Penicillium*** toxins
Ochratoxin A

Sugar

Aspergillus toxins
Aflatoxin/s: aflatoxin B_1, aflatoxin B_2

Sugar beets

Fusarium toxins
Zearalenone

Sugarcane

Alternaria toxins
Alternariol methyl ether

Aspergillus toxins
Aflatoxin/s: aflatoxin B_1, aflatoxin G_1

Aspergillus and ***Penicillium*** toxins
3-Nitropropionic acid, kojic acid, penicillic acid

Sultanas

Aspergillus toxins
Aflatoxin/s: aflatoxin B_1, aflatoxin/s

Aspergillus and ***Penicillium*** toxins
Citrinin, ochratoxin A

Sweet potatoes

Aspergillus toxins
Aflatoxin/s

Sweets

Aspergillus toxins
Aflatoxin/s: aflatoxin M_1

Tea

Alternaria toxins
Tenuazonic acid

Aspergillus toxins
Aflatoxin/s: aflatoxin B_1, aflatoxin B_2, aflatoxin G_1, aflatoxin G_2, aflatoxin/s

Aspergillus and ***Penicillium*** toxins
Ochratoxin A

Fusarium toxins
Depsipeptides: enniatin B
Fumonisin/s: fumonisin B_1
Type B trichothecenes: 15-acetyldeoxynivalenol

Tobacco

Aspergillus toxins
Aflatoxin/s: aflatoxin B_1, aflatoxin B_2

Aspergillus and ***Penicillium*** toxins
Penicillic acid

Fusarium toxins
Type A trichothecenes: T-2 toxin
Other *Fusarium* toxins: zearalenone

Toffees

Aspergillus toxins
Aflatoxin/s: aflatoxin B_1, aflatoxin B_2

Tofu

Aspergillus toxins
Aflatoxin/s: aflatoxin B_1

Tomato concentrate

Alternaria toxins
Altenuene, alternariol, alternariol methyl ether, tenuazonic acid

Tomato ketchup

Alternaria toxins
Alternariol, alternariol methyl ether, tenuazonic acid

Aspergillus and ***Penicillium*** toxins
Ochratoxin A

Tubers

Aspergillus toxins
Aflatoxin/s

Ugba

Alternaria toxins
Alternariol methyl ether

Aspergillus toxins
Aflatoxin/s: aflatoxin B_1, aflatoxin B_2
Other *Aspergillus* toxins: sterigmatocystin

Aspergillus and ***Penicillium*** toxins
Ochratoxin A

Fusarium toxins
Depsipeptides: enniatin B
Type A trichothecenes: HT-2 toxin, T-2 toxin
Type B trichothecenes: deoxynivalenol
Other *Fusarium* toxins: zearalenone

Vegetables

Aspergillus toxins
Aflatoxin/s

Vegetables (asparagus)

Fusarium toxins
Depsipeptides: beauvericin, enniatin A_1, enniatin B
Fumonisin/s: fumonisin B_1, fumonisin B_2, fumonisin B_3, fumonisin/s
Other *Fusarium* toxins: moniliformin

Vegetables (black radish)

Fusarium toxins
Fumonisin/s: fumonisin B_1, fumonisin B_2

Vegetables, dried

Aspergillus toxins
Aflatoxin/s: aflatoxin B_1, aflatoxin/s

Aspergillus and ***Penicillium*** toxins
Ochratoxin A

Fusarium toxins
Zearalenone

Vine fruits, dried

Aspergillus toxins
Aflatoxin/s: aflatoxin B_1, aflatoxin G_1

Aspergillus and ***Penicillium*** toxins
Ochratoxin A

Fusarium toxins
Depsipeptides: enniatin A, enniatin A_1, enniatin B, enniatin B_1
Type A trichothecenes: HT-2 toxin

Vinegar

Aspergillus and ***Penicillium*** toxins
Ochratoxin A, ochratoxin B, patulin

Vinegar (apple)

Aspergillus and ***Penicillium*** toxins
Ochratoxin A, patulin

Vinegar (balsamico)

Aspergillus and ***Penicillium*** toxins
Ochratoxin A

Vinegar (wine)

Aspergillus and ***Penicillium*** toxins
Ochratoxin A

Wine

Aspergillus toxins
Aflatoxin/s: aflatoxin B_1, aflatoxin B_2, aflatoxin G_1 aflatoxin G_2

Aspergillus and ***Penicillium*** toxins
Ochratoxin A, patulin

Wine (red)

Alternaria toxins
Alternariol, alternariol methyl ether, tenuazonic acid

Aspergillus toxins
Aflatoxin/s: aflatoxin B_1, aflatoxin B_2, aflatoxin G_1, aflatoxin G_2

Aspergillus and ***Penicillium*** toxins
Cyclopiazonic acid, ochratoxin A, ochratoxin B

Fusarium toxins
Fumonisin/s: fumonisin B_1, fumonisin B_2, fumonisin B_3
Type A trichothecenes: T-2 toxin
Other *Fusarium* toxins: zearalenone

Penicillium toxins
Mycophenolic acid

Wine (rosé)

Alternaria toxins
Alternariol

Aspergillus and ***Penicillium*** toxins
Ochratoxin A

Wine (white)

Alternaria toxins
Alternariol, alternariol methyl ether

Aspergillus and ***Penicillium*** toxins
Ochratoxin A

Fusarium toxins
Fumonisin/s: fumonisin B_1, fumonisin B_2
Type A trichothecenes: T-2 toxin

Wine, miscellaneous

Alternaria toxins
Alternariol, alternariol methyl ether, tenuazonic acid

Aspergillus toxins
Aflatoxin/s: aflatoxin B_1

Aspergillus and ***Penicillium*** toxins
Ochratoxin A

Fusarium toxins
Fumonisin/s: fumonisin B_1, fumonisin B_2

Yams

Aspergillus toxins
Aflatoxin/s: aflatoxin B_1, aflatoxin/s

Fusarium toxins
Fumonisin/s: fumonisin B_1, fumonisin B_2

Mycotoxin Contamination in Conventional and Organic Farming

Ahn, J., Kim, D., Jang, H.-s., Kim, Y., Shim, W.-B., Chung, D.-H., 2010: Occurrence of ochratoxin A in Korean red paprika and factors to be considered in prevention strategy. Mycotoxin Research 26, 279-286. doi: https://doi.org/10.1007/s12550-010-0067-2 (588).

Alvito, P.C., Sizoo, E.A., Almeida, C.M.M., Van Egmond, H.P., 2010: Occurrence of aflatoxins and ochratoxin A in baby foods in Portugal. Food Analytical Methods 3, 22-30. doi: https://doi.org/10.1007/s12161-008-9064-x (88).

Baert, K., De Meulenaer, B., Kamala, A., Kasase, C., Devlieghere, F., 2006: Occurrence of patulin in organic, conventional, and handcrafted apple juices marketed in Belgium. Journal of Food Protection 69, 1371-1378 (472).

Barreira, M.J., Alvito, P.C., Almeida, C.M.M., 2010: Occurrence of patulin in apple-based-foods in Portugal. Food Chemistry 121, 653-658. doi: https://doi.org/10.1016/j.foodchem.2009.12.085 (486).

Beretta, B., Gaiaschi, A., Galli, C.L., Restani, P., 2000: Patulin in apple-based foods: occurrence and safety evaluation. Food Additives and Contaminants 17, 399-406 (30).

Cunha, S.C., Faria, M.A., Fernandes, J.O., 2009: Determination of patulin in apple and quince products by GC-MS using $^{13}C_{5-7}$ patulin as internal standard. Food Chemistry 115, 352-359. doi: https://doi.org/10.1016/j.foodchem.2008.11.074 (27).

Cunha, S.C., Faria, M.A., Pereira, V.L., Oliveira, T.,M., Lima, A.C., Pinto, E., 2014: Patulin assessment and fungi identification in organic and conventional fruits and derived products. Food Control 44, 185-190. doi: https://doi.org/10.1016/j.foodcont.2014.03.043 (905).

De Magalhães, J.T., Sodré, G.A., Viscogliosi, H., Grenier-Loustalot, M.-F., 2011: Occurrence of ochratoxin A in Brazilian cocoa beans. Food Control 22, 744-748. doi: https://doi.org/10.1016/j.foodcont.2010.11.006 (705).

Food Standards Agency, 2002: Survey of nuts, nut products and dried tree fruits for mycotoxins. Food-Survey-Information-Sheet No. 21/02 (203).

Gilbert, J., Shepherd, M.J., 1985: A survey of aflatoxins in peanut butter, nuts and nut confectionery products by HPLC with fluorescence detection. Food Additives and Contaminants 2, 171-183 (3).

Gopinandhan, T.N., Velmourrougane, K., Panneerselvam, P., Keshamma, E., Raghuramulu, Y., 2007: Occurrence of ochratoxin-A (OT-A) in green and commercial coffee samples. Journal of Food Science and Technology 44, 247-249 (298).

Haas, D., Pfeifer, B., Reiterich, C., Partenheimer, R., Reck, B., Buzina, W., 2013: Identification and quantification of fungi and mycotoxins from Pu-erh tea. International Journal of Food Microbiology 166, 316-322. doi: https://doi.org/10.1016/j.ijfoodmicro.2013.07.024 (735).

Harris, K.L., Bobe, G., Bourquin, L.D., 2009: Patulin surveillance in apple cider and juice marketed in Michigan. Journal of Food Protection 72, 1255-1261 (41).

Hernández-Martínez, R., Navarro-Blasco, I., 2010: Aflatoxin levels and exposure assessment of Spanish infant cereals. Food Additives and Contaminants: Part B 3, 275-288. doi: https://doi.org/10.1080/19393210.2010.531402 (98).

Jeszka-Skowron, M., Zgoła-Grześkowiak, A., Stanisz, E., Waśkiewicz, A., 2017: Potential health benefits and quality of dried fruits: goji fruits, cranberries and raisins. Food Chemistry 221, 228-236. doi: https://doi.org/10.1016/j.foodchem.2016.10.049 (664).

Katerere, D.R., Stockenström, S., Thembo, K.M., Balducci, G., Shephard, G.S., 2007: Investigation of patulin contamination in apple juice sold in retail outlets in Italy and South Africa. Food Additives and Contaminants 24, 630-634. doi: https://doi.org/10.1080/02652030601137668 (481).

Knudsen, P.B., Mogensen, J.M., Larsen, T.O., Nielsen, K.F., 2011: Occurrence of fumonisin B_2 and B_4 in retail raisins. Journal of Agricultural and Food Chemistry 59, 772-776. doi: https://doi.org/10.1021/jf103855x (601).

Labrinea, E.P., Natskoulis, P.I., Spiropoulos, A.E., Magan, N., Tassou, C.C., 2011: A survey of ochratoxin A occurrence in Greek wines. Food Additives and Contaminants: Part B 4, 61-66. doi: https://doi.org/10.1080/19393210.2010.539707 (621).

Larcher, R., Nicolini, G., 2001: Survey of ochratoxin A in musts, concentrated musts and wines produced or marketed in Trentino (Italy). Journal of Commodity Science 40, 69-78 (537).

Malmauret, L., Parent-Massin, D., Hardy, J.-L., Verger, P., 2002: Contaminants in organic and conventional foodstuffs in France. Food Additives and Contaminants 19, 524-532. doi: https://doi.org/10.1080/02652030210123878 (34).

Meletis, K., Meniades-Meimaroglou, S., Markaki, P., 2007. Determination of ochratoxin A in grapes of Greek origin by immunoaffinity and high-performance liquid chromatography. Food Additives and Contaminants 24, 1275-1282. doi: https://doi.org/10.1080/02652030701413852 (159).

Miceli, A., Negro, C., Tommasi, L., De Leo, P., 2003: Polyphenols, resveratrol, antioxidant activity and ochratoxin A contamination in red table wines, controlled denomination of origin (DOC) wines and wines obtained from organic farming. Journal of Wine Research 14, 115-120. doi: https://doi.org/10.1080/09571260410001678030 (791).

Mortimer, D.N., Parker, I., Shepherd, M.J., Gilbert, J., 1985: A limited survey of retail apple and grape juices for the mycotoxin patulin. Food Additives and Contaminants 2, 165-170 (455).

M. Weidenbörner, *Mycotoxins in Plants and Plant Products*, https://doi.org/10.1007/978-3-319-92850-0_5

Murillo-Arbizu, M., Amězqueta, S., González-Peñas, E., López De Cerain, A., 2009: Occurrence of patulin and its dietary intake through apple juice consumption by the Spanish population. Food Chemistry 113, 420-423. doi: https://doi.org/10.1016/j.foodchem.2008.07.054 (484).

Palumbo, J.D., O´Keeffe, T.L., Ho, Y.S., Santillan, C.J., 2015: Occurrence of ochratoxin A contamination and detection of ochratoxigenic *Aspergillus* species in retail samples of dried fruits and nuts. Journal of Food Protection 78, 836-842. doi: https://doi.org/10.4315/0362-028X.JFP-14-471 (599).

Piemontese, L., Solfrizzo, M., Visconti, A., 2005: Occurrence of patulin in conventional and organic fruit products in Italy and subsequent exposure assessment. Food Additives and Contaminants 22, 437-442. doi: https://doi.org/10.1080/02652030500073550 (35).

Piqué, E., Vargas-Murga, L., Gómez-Catalán, J., De Lapuente, J., Llobet, J.M., 2013: Occurrence of patulin in organic and conventional apple-based food marketed in Catalonia and exposure assessment. Food and Chemical Toxicology 60, 190-204. doi: https://doi.org/10.1016/j.fct.2013.07.052 (594).

Prelle, A., Spadaro, D., Garibaldi, A., Gullino, M.L., 2012: Aflatoxin monitoring in Italian hazelnut products by LC-MS. Food Additives and Contaminants: Part B 5, 279-285 doi: https://doi.org/10.1080/19393210.2012.711371 and personal communication (821).

Ritieni, A., 2003: Patulin in Italian commercial apple products. Journal of Agricultural and Food Chemistry 51, 6086-6090. doi: https://doi.org/10.1021/jf034523c (44).

Remiro, R., González-Peñas, E., Lizarraga, E., López de Cerain, A., 2012: Quantification of ochratoxin A and five analogs in Navarra red wines. Food Control 27, 139-145 doi: https://doi.org/10.1016/j.foodcont.2012.03.006 (830).

Rubert, J., Sebastià, N., Soriano, J.M., Soler, C., Mañes, J., 2011: One-year monitoring of aflatoxins and ochratoxin A in tiger-nuts and their beverages. Food Chemistry 127, 822-826. doi: https://doi.org/10.1016/j.foodchem.2011.01.016 (163).

Schollenberger, M., Müller, H.-M., Rüfle, M., Drochner, W., 2008: Natural occurrence of 16 *Fusarium* toxins in edible oil marketed in Germany. Food Control 19, 475-482. doi: https://doi.org/10.1016/j.foodcont.2007.05.012 (840).

Spadaro, D., Ciavorella, A., Frati, S., Garibaldi, A., Gullino, M.L., 2007: Incidence and level of patulin contamination in pure and mixed apple juices marketed in Italy. Food Control 18, 1098-1102. doi: https://doi.org/10.1016/j.foodcont.2006.07.007 (438).

Spadaro, D., Garibaldi, A., Gullino, M.L., 2008: Occurrence of patulin and its dietary intake through pear, peach, and apricot juices in Italy. Food Additives and Contaminants: Part B 1, 134-139. doi: https://doi.org/10.1080/02652030802363790 (847).

Tafuri, A., Ferracane, R., Ritieni, A., 2004: Ochratoxin A in Italian marketed cocoa products. Food Chemistry 88, 487-494. doi: https://doi.org/10.1016/j.foodchem.2004.01.061 (593).

Versari, A., Parpinello, G.P., Mattioli, A.U., 2007: Survey of patulin contamination in Italian apple juices from organic and conventional agriculture. Journal of Food Technology 5, 143-146 (638).

Villa, P., Markaki, P., 2009: Aflatoxin B_1 and ochratoxin A in breakfast cereals from Athens market: occurrence and risk assessment. Food Control 20, 455-461. doi: https://doi.org/10.1016/j.foodcont.2008.07.012 (91).

Wilson, D.M., Nuovo, G.J., 1973: Patulin production in apples decayed by *Penicillium expansum*. Applied Microbiology 26,124-125 (40).

Articles dealing with the mycotoxin contamination of organic products

Comuzzo, P., Rauhut, D., Werner, M., Lagazio, C., Zironi, R., 2013: A survey on wines from organic viticulture from different European countries. Food Control 34, 274-282. doi: https://doi.org/10.1016/j.foodcont.2013.04.039 (699).

Mortimer, D.N., Shepherd, M.J., Gilbert, J., Morgan, M.R.A., 1987: A survey of the occurrence of aflatoxin B_1 in peanut butters by enzyme-linked immunosorbent assay. Food Additives and Contaminants 5, 127-132 (213).

Tosun, H., Arslan, R., 2013: Determination of aflatoxin B_1 levels in organic spices and herbs. Scientific World Journal 2013, article ID 874093. doi: https://doi.org/10.1155/2013/974093 (859).

Tosun, H., Yildiz, H., Obuz, E., Seçkin, A.K., 2014: Ochratoxin A in grape pekmez (grape molasses) consumed in Turkey. Food Additives and Contaminants: Part B 7, 37-39. doi: https://doi.org/10.1080/19393210.2013.833298 (862).

Turcotte, A.-M., Scott, P.M., Tague, B., 2013: Analysis of cocoa products for ochratoxin A and aflatoxins. Mycotoxin Research 29, 193-201. doi: https://doi.org/10.1007/s12550-013-0167-x (864).

Literature

Numerical Bibliography

1. Bulato-Jayme, J., Almero, E.A., Castro, M.C.A., Jardeleza, M.T., Salamat, L.A., 1982: A case-control dietary study of primary liver cancer risk from aflatoxin exposure. International Journal of Epidemiology 11, 112-119.
2. Pensala, O., Niskanen, A., Lahtinen, S., 1977: The occurrence of aflatoxin in nuts and nut products imported to Finland for human consumption during the years 1974-1976. Nordisk Veterinaer Medicin 29, 347-355.
3. Gilbert, J., Shepherd, M.J., 1985: A survey of aflatoxins in peanut butters, nuts and nut confectionery products by HPLC with fluorescence detection. Food Additives and Contaminants 2, 171-183.
4. Qutet, S.M., El-Tabey Shehata, A.M., Mesallam, A.S., 1983: Occurrence of aflatoxins in some Egyptian food crops collected from two coastal regions. Food Chemistry 10, 149-153.
5. Herry, M.-P., Lemetayer, N., 1992: Aflatoxin B_1 contamination in oilseeds, dried fruits and spices. Microbiologie—Aliments—Nutrition 10, 261-266.
6. Jiménez, M., Mateo, R., Querol, A., Huerta, T., Hernández, E., 1991: Mycotoxins and mycotoxigenic moulds in nuts and sunflower seeds for human consumption. Mycopathologia 115, 121-127.
7. Saxena, J., Mehrota, B.S., 1990: The occurrence of mycotoxins in some dry fruits retail marketed in Nainital district of India. Acta Alimentaria 19, 221-224.
8. Bilgrami, Z., Ghaffar, A., 1999: Detection of aflatoxin in almond seeds. Pakistan Journal of Botany 31, 227-230.
9. Jang, M.-R., Lee, C.-H., Cho, S.-H., Park, J.-S., Kwon, E.-Y., Lee, E.-J., Kim, S.-H., Kim, D.-B., 2007: A survey of total aflatoxins in food using high performance liquid chromatography-fluorescence detector (HPLC-FLD) and liquid chromatography tandem mass spectrometry (LC-MS/MS). Korean Journal of Food Science and Technology 39, 488-493.
10. Ok, H.E., Kim, H.J., Shim, W.B., Lee, H., Bae, D.-H., Chung, D.-H., Chun, H.S., 2007: Natural occurrence of aflatoxin B_1 in marketed foods and risk estimates of dietary exposure in Koreans. Journal of Food Protection 70, 2824-2828.
11. Beebe, R.M., 1978: Reverse phase high pressure liquid chromatographic determination of aflatoxins in foods. Journal of the Association of Analytical Chemists 61, 1347-1352.
12. Rodriguez, P., Venâncio, A., Lima, M., 2012: Aflatoxigenic fungi and aflatoxins in Portuguese almonds. Scientific World Journal, doi: https://doi.org/10.1100/2012/471926.
13. Thuvander, A., Möller, T., Enghardt Barbieri, H., Jansson, A., Salomonsson, A.-C., Olsen, M., 2001: Dietary intake of some important mycotoxins by the Swedish population. Food Additives and Contaminants 18, 696-706. doi: https://doi.org/10.1080/02652030110047775.
14. Gürses, M., 2006: Mycoflora and aflatoxin content of hazelnuts, walnuts, peanuts, almonds and roasted chickpeas (Leblebi) sold in Turkey. International Journal of Food Properties 9, 395-399. doi: https://doi.org/10.1080/10942910600596597.
15. Gelda, C.S., Luyt, L.J., 1977: Survey of total aflatoxin content in peanuts, peanut butter, and other foodstuffs. Annales de la Nutrition et de l´Alimentation 31, 477-483.
16. Patel, S., Hazel, C.M., Winterton, A.G.M., Mortby, E., 1996: Survey of ethnic foods for mycotoxins. Food Additives and Contaminants 13, 833-841.
17. Luttfullah, G., Hussain, A., 2011: Studies on contamination level of aflatoxins in some dried fruits and nuts of Pakistan. Food Control 22, 426-429. doi: https://doi.org/10.1016/j.foodcont.2010.09.015.
18. Leszcynska, J., Kucharska, U., Żegota, H., 2000: Aflatoxins in nuts assayed by immunological methods. European Food Research and Technology 210, 213-215.
19. Wood, G.E., 1989: Aflatoxins in domestic and imported foods and feeds. Journal of the Association of Official Analytical Chemists 72, 543-548.
20. Zaied, C., Abid, S., Bouaziz, C., Chouchane, S., Jomaa, M., Bacha, H., 2010: Ochratoxin A levels in spices and dried nuts consumed in Tunisia. Food Additives and Contaminants: Part B 3, 52-57. doi: https://doi.org/10.1080/19440041003587302.
21. Saito, K., Ikeuchi, R., Kataoka, H., 2012: Determination of ochratoxins in nuts and grain samples by in-tube solid-phase microextraction coupled with liquid chromatography-mass spectrometry. Journal of Chromatography A 1220, 1-6. doi: https://doi.org/10.1016/j.chroma.2011.11.008.
22. Saxena, J., Mehrotra, B.S., 1989: Screening of spices commonly marketed in India for natural occurrence of mycotoxins. Journal of Food Composition and Analysis 2, 286-292.

M. Weidenbörner, *Mycotoxins in Plants and Plant Products*, https://doi.org/10.1007/978-3-319-92850-0

23. Stinson, E.E., Osman, S.F., Heisler, E.G., Siciliano, J., Bills, D.D., 1981: Mycotoxin production in whole tomatoes, apples, oranges, and lemons. Journal of Agricultural and Food Chemistry 29, 790-792.
24. Wittkowski, M., Baltes, W., Krönert, W., Weber, R., 1983: Determination of *Alternaria* toxins in fruit and vegetable products. Zeitschrift für Lebensmittel-Untersuchung und -Forschung 177, 447-453.
25. Martins, M.L., Gimeno, A., Martins, H.M., Bernardo, F., 2002: Co-occurrence of patulin and citrinin in Portuguese apples with rotten spots. Food Additives and Contaminants 19, 568-574. doi: https://doi.org/10.1080/02652030210121320.
26. Engelhardt, G., Ruhland, M., Wallnöfer, P.R., 1999: Occurrence of ochratoxin A in moldy vegetables and fruits analysed after removal of rotten tissue parts. Advances in Food Science 21, 88-92.
27. Cunha, S.C., Faria, M.A., Fernandes, J.O., 2009: Determination of patulin in apple and quince products by GC-MS using $^{13}C_{5-7}$ patulin as internal standard. Food Chemistry 115, 352-359. doi: https://doi.org/10.1016/j.foodchem.2008.11.074.
28. Liao, C.-D., Wong, J.W., Zhang, K., Hayward, D.G., Lee, N.S., Trucksess, M.W., 2013: Multi-mycotoxin analysis of finished grain and nut products using high-performance liquid chromatography-triple-quadrupole mass spectrometry. Journal of Agricultural and Food Chemistry 61, 4771-4782. doi: https://doi.org/10.1021/4000677.
29. Fritz, W., Engst, R., 1981: Survey of selected mycotoxins in food. Journal of Environmental Science and Health B16, 193-210.
30. Beretta, B., Gaiaschi, A., Galli, C.L., Restani, P., 2000: Patulin in apple-based foods: occurrence and safety evaluation. Food Additives and Contaminants 17, 399-406.
31. Burda, K., 1992: Incidence of patulin in apple, pear, and mixed fruit products marketed in New South Wales. Journal of Food Protection 55, 796-798.
32. De Sylos, C.M., Rodriguez-Amaya, D.B., 1999: Incidence of patulin in fruits and fruit juices marketed in Campinas, Brazil. Food Additives and Contaminants 16, 71-74.
33. Lindroth, S., Niskanen, A., 1978: Comparison of potential patulin hazard in home-made and commercial apple products. Journal of Food Science 43, 446-448.
34. Malmauret, L., Parent-Massin, D., Hardy, J.-L., Verger, P., 2002: Contaminants in organic and conventional foodstuffs in France. Food Additives and Contaminants 19, 524-532. doi: https://doi.org/10.1080/02652030210123878.
35. Piemontese, L., Solfrizzo, M., Visconti, A., 2005: Occurrence of patulin in conventional and organic fruit products in Italy and subsequent exposure assessment. Food Additives and Contaminants 22, 437-442. doi: https://doi.org/10.1080/02652030500073550.
36. Sørensen, J.L., Phipps, R.K., Nielsen, K.F., Schroers, H.-J., Frank, J., Thrane, U., 2009: Analysis of *Fusarium avenaceum* metabolites produced during wet apple core rot. Journal of Agricultural and Food Chemistry 57, 1632-1639. doi: https://doi.org/10.1021/jf802926u.
37. Leggott, N.L., Shephard, G.S., 2001: Patulin in South African commercial apple products. Food Control 12, 73-76.
38. Tangni, E.K., Theys, R., Mignolet, E., Maudoux, M., Michelet, J.Y., Larondelle, Y., 2003: Patulin in domestic and imported apple-based drinks in Belgium: occurrence and exposure assessment. Food Additives and Contaminants 20, 482-489. doi: https://doi.org/10.1080/026520303 1000093204.
39. Wheeler, J.L., Harrison, M.A., Koehler, P.E., 1987: Presence and stability of patulin in pasteurized apple cider. Journal of Food Science 52, 479-480.
40. Wilson, D.M., Nuovo, G.J., 1973: Patulin production in apples decayed by *Penicillium expansum*. Applied Microbiology 26,124-125.
41. Harris, K.L., Bobe, G., Bourquin, L.D., 2009: Patulin surveillance in apple cider and juice marketed in Michigan. Journal of Food Protection 72, 1255-1261.
42. Cano-Sancho, G., Marin, S., Ramos, A.J., Sanchis, V., 2009: Survey of patulin occurrence in apple juice and apple products in Catalonia, Spain, and an estimate of dietary intake. Food Additives and Contaminants: Part B 2, 59-65. doi: https://doi.org/10.1080/02652030902897747.
43. González-Osnaya, L., Soriano, J.M., Moltó, J.C., Mañes, J., 2007: Exposure to patulin from consumption of apple-based products. Food Additives and Contaminants 24, 1268-1274. doi: https://doi.org/10.1080/02652030701361556.
44. Ritieni, A., 2003: Patulin in Italian commercial apple products. Journal of Agricultural and Food Chemistry 51, 6086-6090. doi: https://doi.org/10.1021/jf034523c.
45. Terminiello, L., Patriarca, A., Pose, G., Pinto, V.F., 2006: Occurrence of alternariol, alternariol monomethyl ether and tenuazonic acid in Argentinean tomato puree. Mycotoxin Research 22, 236-240.
46. Asam, S., Konitzer, K., Rychlik, M., 2011: Precise determination of the *Alternaria* mycotoxins alternariol and alternariol monomethyl ether in cereal, fruit and vegetable products using stable isotope dilution assays. Mycotoxin Research 27, 23-28. doi: https://doi.org/10.1007/s12550-010-0071-6.
47. Noser, J., Schneider, P., Rother, M., Schmutz, H., 2011: Determination of six *Alternaria* toxins with UPLC-MS/MS and their occurrence in tomatoes and tomato products from the Swiss market. Mycotoxin Research 27, 265-271. doi: https://doi.org/10.1007/s12550-011-0103-x.
48. Da Motta, S., Valente Soares, L.M., 2001: Survey of Brazilian tomato products for alternariol, alternariol monomethyl ether, tenuazonic acid and cyclopiazonic acid. Food Additives and Contaminants 18, 630-634. doi: https://doi.org/10.1080/02652030110035093.
49. Solfrizzo, M., Panzarini, G., Visconti, A., 2008: Determination of ochratoxin A in grapes, dried vine fruits, and winery byproducts by high performance liquid chromatography with fluorometric detection (HPLC-FLD) and immunoaffinity cleanup. Journal of Agricultural and Food Chemistry 56, 11081-11086. doi: https://doi.org/10.1021/jf802380d.
50. Funes, G.J., Resnik, S.L., 2009: Determination of patulin in solid and semisolid apple and pear products marketed in Argentina. Food Control 20, 277-280. doi: https://doi.org/10.1016/j.foodcont.2008.05.010.
51. Katerere, D.R., Stockenström, S., Shephard, G.S., 2008: HPLC-DAD method for the determination of patulin in dried apple rings. Food Control 19, 389-392. doi: https://doi.org/10.1016/j.foodcont.2007.04.015.
52. Rychlik, M., Schieberle, P., 1999: Quantification of the mycotoxin patulin by a stable isotope dilution assay. Journal of Agricultural and Food Chemistry 47, 3749-3755. doi: https://doi.org/10.1021/jf990198a.
53. Fakoor Janati, S.S., Beheshti, H.R., Asadi, M., Mihanparast, S., Feizy, J., 2012: Preliminary survey of aflatoxins and ochratoxin A in dried fruits from Iran. Bulletin of Environmental Contamination and Toxicology 88, 391-395. doi: https://doi.org/10.1007/s00128-011-0477-7.
54. Zohri, A.A., Abdel-Gawad, K.M., 1993: Survey of mycoflora and mycotoxins of some dried fruits in Egypt. Journal of Basic Microbiology 33, 279-288.

55. Bircan, C., 2009: Incidence of ochratoxin A in dried fruits and co-occurrence with aflatoxins in dried figs. Food and Chemical Toxicology 47, 1996-2001. doi: https://doi.org/10.1016/j.fct.2009.05.008.
56. Van Der Bijl, P., Stockenström, S., Vismer, H.F., Van Wyk, C.W., 1996: Incidence of fungi and aflatoxins in imported areca nut samples. South African Journal of Science 92, 154-156.
57. Raisuddin, S., Misra, J.K., 1991: Aflatoxin in betel nut and its control by use of food preservatives. Food Additives and Contaminants 8, 707-712.
58. Koirala, P., Kumar, S., Yadav, B.K., Premarajan, K.C., 2005: Occurrence of aflatoxin in some of the food and feed in Nepal. Indian Journal of Medical Sciences 59, 331-336.
59. Yang, M.-H., Chen, J.-M., Zhang, X.-H., 2005: Immunoaffinity column clean-up and liquid chromatography with post-column derivatization for analysis of aflatoxins in traditional Chinese medicine. Chromatographia 62, 499-504. doi: https://doi.org/10.1365/s10337-005-0647-z.
60. Abeywickrama, K., Bean, G.A., 1991: Toxigenic *Aspergillus flavus* and aflatoxins in Sri Lanka medicinal plant material. Mycopathologia 113, 187-190.
61. Ali, N., Hashim, N.H., Saad, B., Safan, K., Nakajima, M., Yoshizawa, T., 2005: Evaluation of a method to determine the natural occurrence of aflatoxins in commercial traditional herbal medicines from Malaysia and Indonesia. Food and Chemical Toxicology 43, 1763-1772. doi: https://doi.org/10.1016/j.fct.2005.05.019.
62. Roy, A.K., Sinha, K.K., Chourasia, H.K., 1988: Aflatoxin contamination of some common drug plants. Applied and Environmental Microbiology 54, 842-843.
63. Roy, A.K., Chourasia, H.K., 1990: Mycoflora, mycotoxin producibility and mycotoxins in traditional herbal drugs from India. Journal of General and Applied Microbiology 36, 295-302.
64. Martinéz-Domínguez, G., Romero-González, R., Frenich, A.G., 2015: Determination of toxic substances, pesticides and mycotoxins, in ginkgo biloba nutraceutical products by liquid chromatography Orbitrap-mass spectrometry. Microchemical Journal 118, 124-130. doi: https://doi.org/10.1016/j.microc.2014.09.002.
65. Chourasia, H.K., 1995: Mycobiota and mycotoxins in herbal drugs of Indian pharmaceutical industries. Mycological Research 99, 697-703.
66. Han, Z., Zheng, Y., Luan, L., Cai, Z., Ren, Y., Wu, Y., 2010: An ultra-performance liquid chromatography-tandem mass spectrometry method for simultaneous determination of aflatoxins B_1, B_2, G_1, G_2, M_1 and M_2 in traditional Chinese medicines. Analytica Chimica Acta 664, 165-171. doi: https://doi.org/10.1016/j.aca.2010.02.009.
67. Tassaneeyakul, W., Razzazi-Fazeli, E., Porasuphatana, S., Bohm, J., 2004: Contamination of aflatoxins in herbal medicinal products in Thailand. Mycopathologia 158, 239-244.
68. Han, Z., Ren, Y., Zhu, J., Cai, Z., Chen, Y., Luan, L., Wu, Y., 2012: Multianalysis of 35 mycotoxins in traditional Chinese medicines by ultra-high-performance liquid chromatography-tandem mass spectrometry coupled with accelerated solvent extraction. Journal of Agricultural and Food Chemistry 60, 8233-8247. doi: https://doi.org/10.1021/jf301928r.
69. Shim, W.-B., Kim, K., Ofori, J.A., Chung, Y.-C., Chung, D.-H., 2012: Occurrence of aflatoxins in herbal medicine distributed in South Korea. Journal of Food Protection 75, 1991-1999. doi: https://doi.org/10.4315/0362-028X.JFP-12-190.
70. Liu, L., Jin, H., Sun, L., Ma, S., Lin, R., 2012: Determination of aflatoxins in medicinal herbs by high-performance liquid chromatography-tandem mass spectrometry. Phytochemical Analysis 23, 469-476. doi: https://doi.org/10.1002/pca.2343.
71. D'Ovidio, K., Trucksess, M., Weaver, C., Horn, E., McIntosh, M., Bean, G., 2006: Aflatoxins in ginseng roots. Food Additives and Contaminants 23, 174-180. doi: https://doi.org/10.1080/02652030500442524.
72. Santos, L., Marín, S., Sanchis, V., Ramos, A.J., 2009: Screening of mycotoxin multicontamination in medicinal and aromatic herbs sampled in Spain. Journal of the Science of the Food and Agriculture 89, 1802-1807. doi: https://doi.org/10.1002/jsfa.3647.
73. Yang, L., Wang, L., Pan, J., Xiang, L., Yang, M., Logrieco, A.F., 2010: Determination of ochratoxin A in traditional Chinese medicinal plants by HPLC-FLD. Food Additives and Contaminants: Part A 27, 989-997. doi: https://doi.org/10.1080/19440041003647122.
74. Halt, M., 1998: Moulds and mycotoxins in herb tea and medicinal plants. European Journal of Epidemiology 14, 269-274.
75. Han, Z., Zheng, Y., Luan, L., Ren, Y., Wu, Y., 2010: Analysis of ochratoxin A and ochratoxin B in traditional Chinese medicines by ultra-high-performance liquid chromatography-tandem mass spectrometry using [$^{13}C_{20}$]-ochratoxin A as an internal standard. Journal of Chromatography A 1217, 4365-4374. doi: https://doi.org/10.1016/j.chroma.2010.04.052.
76. Roy, A.K., Kumar, S., 1993: Occurrence of ochratoxin A in herbal drugs of Indian origin—a report. Mycotoxin Research 9, 94-98.
77. Di Mavungu, J.D., Monbaliu, S., Scippo, M.-L., Maghuin-Rogister, G., Schneider, Y.-J., Larondelle, Y., Callebaut, A., Robbens, J., Van Petheghem, C., De Saeger, S., 2009: LC-MS/MS multi-analyte method for mycotoxin determination in food supplements. Food Additives and Contaminants: Part A 26, 885-895. doi: https://doi.org/10.1080/02652030902774649 (and personal communication).
78. Martins, M.L., Martins, H.M., Bernado, F., 2001: Fumonisins B_1 and B_2 in black tea and medicinal plants. Journal of Food Protection 64, 1268-1270.
79. Katerere, D.R., Stockenström, S., Thembo, K.M., Rheeder, J.P., Shephard, G.S., Vismer, H.F., 2008: A preliminary survey of mycological and fumonisin and aflatoxin contamination of African traditional herbal medicines sold in South Africa. Human & Experimental Toxicology 27, 793-798. doi: https://doi.org/10.1177/0960327108099535.
80. Omurtag, G.Z., Yazicioğlu, D., 2004: Determination of fumonisins B_1 and B_2 in herbal tea and medicinal plants in Turkey by high-performance liquid chromatography. Journal of Food Protection 67, 1782-1786.
81. Sewram, V., Shephard, G.S., Van Der Merwe, L., Jacobs, T.V., 2006: Mycotoxin contamination of dietary and medicinal wild plants in the Eastern Cape Province of South Africa. Journal of Agricultural and Food Chemistry 54, 5688-5693. doi: https://doi.org/10.1021/jf060483b.
82. Kong, W., Xie, T., Li, J., Wei, J., Qiu, F., Qi, A., Zheng, Y., Yang, M., 2012: Analysis of fumonisins B_1 and B_2 in spices and aromatic and medicinal herbs by HPLC-FLD with on-line post-column derivatization and positive confirmation by LC-MS/MS. Analyst 137, 3166-3174. doi: https://doi.org/10.1039/c2an35164a.
83. Logrieco, A., Doko, B., Moretti, A., Frisullo, S., Visconti, A., 1998: Occurrence of fumonisins B_1 and B_2 in *Fusarium proliferatum* infected asparagus plants. Journal of Agricultural and Food Chemistry 46, 5201-5204. doi: https://doi.org/10.1021/jf9804903.

84. Weber, Z., Kostecki, M., Von Bargen, S., Gossmann, M., Waskiewicz, A., Bocianowski, J., Knaflewski, M., Büttner, C., Golinski, P., 2006: *Fusarium* species colonizing spears and forming mycotoxins in field samples of asparagus from Germany and Poland. Journal of Phytopathology 154, 209-216.
85. Aoyama, K., Nakajima, M., Tabata, S., Ishikuro, E., Tanaka, T., Norizuki, H., Itoh, Y., Fujita, K., Kai, S., Tsutsumi, T., Takahashi, M., Tanaka, H., IIzuka, S., Ogiso, M., Maeda, M., Yamaguchi, S., Sugiyama, K.-I., Sugita-Konishi, Y., Kumagai, S., 2010: Four-year surveillance for ochratoxin A and fumonisins in retail foods in Japan. Journal of Food Protection 73, 344-352.
86. Seefelder, W., Gossmann, M., Humpf, H.-U., 2002: Analysis of fumonisin B_1 in *Fusarium proliferatum*-infected asparagus spears and garlic bulbs from Germany by liquid chromatography-electrospray ionization mass spectrometry. Journal of Agricultural and Food Chemistry 50, 2778-2781. doi: https://doi.org/10.1021/jf0115037.
87. Liu, C., Liu, F., Xu, W., Kofoet, A., Humpf, H.-U., Jiang S., 2005: Occurrence of fumonisins B_1 and B_2 in asparagus from Shandong province, P.R. China. Food Additives and Contaminants 22, 673-676. doi: https://doi.org/10.1080/02652030500133016.
88. Alvito, P.C., Sizoo, E.A., Almeida, C.M.M., Van Egmond, H.P., 2010: Occurrence of aflatoxins and ochratoxin A in baby foods in Portugal. Food Analytical Methods 3, 22-30. doi: https://doi.org/10.1007/s12161-008-9064-x.
89. Kabak, B., 2012: Aflatoxin M_1 and ochratoxin A in baby formulae in Turkey: occurrence and safety evaluation. Food Control 26, 182-187. doi: https://doi.org/10.1016/j.foodcont.2012.01.032.
90. Tam, J., Pantazopoulos, P., Scott, P.M., Moisey, J., Dabeka, R.W., Richard, I.D.K., 2011: Application of isotope dilution mass spectrometry: determination of ochratoxin A in the Canadian total diet study. Food Additives and Contaminants: Part A 28, 754-761. doi: https://doi.org/10.1080/19440049.2010.504750.
91. Villa, P., Markaki, P., 2009: Aflatoxin B_1 and ochratoxin A in breakfast cereals from Athens market: occurrence and risk assessment. Food Control 20, 455-461. doi: https://doi.org/10.1016/j.foodcont.2008.07.012.
92. Tam, J., Mankotia, M., Malby, M., Pantazopoulos, P., Neil, R.J., Calway, P., Scott, P.M., 2006: Survey of breakfast and infant cereals for aflatoxins B_1, B_2, G_1 and G_2. Food Additives and Contaminants 23, 693-699. doi: https://doi.org/10.1080/02652030600627230.
93. Cano-Sancho, G., Sanchis, V., Marín, S., Ramos, A.J., 2013: Occurrence and exposure assessment of aflatoxins in Catalonia (Spain). Food and Chemical Toxicology 51, 188-193. doi: https://doi.org/10.1016/j.fct.2012.09.032.
94. Molinié, A., Faucet, V., Castegnaro, M., Pfohl-Leszkowicz, A., 2005: Analysis of some breakfast cereals on the French market for their contents of ochratoxin A, citrinin and fumonisin B_1: development of a method for simultaneous extraction of ochratoxin A and citrinin. Food Chemistry 92, 391-400. doi: https://doi.org/10.1016/j.foodchem.2004.06.035.
95. Zinedine, A., Blesa, J., Mahnine, N., El Abidi, A., Montesano, D., Mañes, J., 2010: Pressurized liquid extraction coupled to liquid chromatography for the analysis of ochratoxin A in breakfast and infants cereals from Morocco. Food Control 21, 132-135. doi: https://doi.org/10.1016/j.foodcont.2009.04.009.
96. Cunha, S.C., Fernandes, J.O., 2010: Development and validation of a method based on a QuEChERS procedure and heart-cutting GC-MS for determination of five mycotoxins in cereal products. Journal of Separation Science 33, 600-609. doi: https://doi.org/10.1002/jssc.200900695.
97. Tabata, S., Kamimura, H., Tamura, Y., Yasuda, K., Ushiyama, H., Hashimoto, H., Nishijima, M., Nishima, T., 1987: Investigation of aflatoxins contamination in foods and foodstuffs. Journal of Food Hygienic Society of Japan 28, 395-401.
98. Hernández-Martínez, R., Navarro-Blasco, I., 2010: Aflatoxin levels and exposure assessment of Spanish infant cereals. Food Additives and Contaminants: Part B 3, 275-288. doi: https://doi.org/10.1080/19393210.2010.531402.
99. Tabata, S., Kamimura, H., Ibe, A., Hashimoto, H., Iida, M., Tamura, Y., Nishima, T., 1993: Aflatoxin contamination in foods and foodstuffs in Tokyo: 1986-1990. Journal of the Association of Official Analytical Chemists International 76, 32-35.
100. Sinha, K.K., Sinha, A.K., 1991: Monitoring and identification of aflatoxins in wheat, gram and maize flours in Bihar state (India). Food Additives and Contaminants 8, 453-457.
101. Hudson, G.J., Wild, C.P., Zarba, A., Groopman, J.D., 1992: Aflatoxins isolated by immunoaffinity chromatography from foods consumed in The Gambia, West Africa. Natural Toxins 1, 100-105.
102. Thorpe, C.W., Johnson, R.L., 1974: Analysis of penicillic acid by gas-liquid chromatography. Journal of the Association of Official Analytical Chemists 57, 861-865.
103. Snow, J.P., Lucas, G.R., Harvan, D., Pero, N.W., Owens, R.G., 1972: Analysis of tobacco and smoke condensate for penicillic acid. Applied Microbiology 24, 34-36.
104. Aziz, N.H., Youssef, Y.A., 1991: Occurrence of aflatoxins and aflatoxin-producing moulds in fresh and processed meat in Egypt. Food Additives and Contaminants 8, 321-33.
105. Oluwafemi, F., Nnanna Ibeh, I., 2011: Microbial contamination of seven major weaning foods in Nigeria. Journal of Health, Population and Nutrition 29, 415-419.
106. Noviandi, C.T., Razzazi, E., Agus, A., Böhm, J., Hulan, H.W., Wedhastri, S., Maryudhani, Y.B., Nuryono, Sardjono, Leibetseder, J., 2001: Natural occurrence of aflatoxin B_1 in some Indonesian food and feed products in Yogyakarta in year 1998-1999. Mycotoxin Research 17A, 174-178.
107. MAFF—UK 1994: Survey of apple products for patulin. Food-Surveillance-Information-Sheet No. 29.
108. Moukas, A., Panagiotopoulou, V., Markaki, P., 2008: Determination of patulin in fruit juices using HPLC-DAD and GC-MSD techniques. Food Chemistry 109, 860-867. doi: https://doi.org/10.1016/j.foodchem.2008.01.015.
109. Bonerba, E., Conte, R., Ceci, E., Tantillo, G., 2010: Assessment of dietary intake of patulin from baby foods. Journal of Food Science 75, T123-T125. doi: https://doi.org/10.1111/j.1750-3841.2010.01743.x.
110. Majerus, P., Bresch, H., Otteneder, H., 2000: Ochratoxin A in wines, fruit juices and seasonings. Archiv für Lebensmittelhygiene 51, 95-97.
111. Markaki, P., Delpont-Binet, C., Grosso, F., Dragacci, S., 2001: Determination of ochratoxin A in red wine and vinegar by immunoaffinity high-pressure liquid chromatography. Journal of Food Protection 64, 533-537.
112. Tabata, S., Iida, K., Kimura, K., Iwasaki, Y., Nakazato, M., Kamata, K., Hirokado, M., 2008: Investigation of ochratoxin A, B and citrinin contamination in various commercial foods. Journal of Food Hygienic Society of Japan 49, 111-115.
113. Lohiya, G., Nichols, L., Hsieh, D., Lohiya, S., Nguyen, H., 1987: Aflatoxin content of foods served to a population with a high incidence of hepatocellular carcinoma. Hepatology 7, 750-752.

114. Van Rensburg, S.J., Cook-Mozaffari, P., Van Schalkwyk, D.J., Van Der Watt, J.J., Vincent, T.J., Purchase, I.F., 1985: Hepatocellular carcinoma and dietary aflatoxin in Mozambique and Transkei. British Journal of Cancer 51, 713-726.
115. Ulca, P., Evcimen, M.K., Senyuva, H.Z., 2010: Surveys of aflatoxin B_1 contamination of retail Turkish foods and of products intended for export between 2007 and 2009. Food Additives and Contaminants: Part B 3, 120-125. doi: https://doi.org/10.1080/19393210.2010.493619.
116. Czerwiecki, L., Wilczyńska, G., Kwiecień, A., 2006: Mycotoxins in several Polish food products in 2004-2005. Mycotoxin Research 22, 159-162.
117. Warth, B., Parich, A., Atehnkeng, J., Bandyopadhyay, R., Schumacher, R., Sulyok, M., Krska, R., 2012: Quantitation of mycotoxins in food and feed from Burkina Faso and Mozambique using a modern LC-MS/MS multitoxin method. Journal of Agricultural and Food Chemistry 60, 9352-9363. doi: https://doi.org/10.1021/jf302003n.
118. Wild, C.P., Hudson, G.J., Sabbioni, G., Chapot, B., Hall, A.J., Wogan, G.N., Whittle, H., Montesano, R., Groopman, J.D., 1992: Dietary intake of aflatoxins and the level of albumin-bound aflatoxin in peripheral blood in The Gambia, West Africa. Cancer, Epidemiology, Biomarkers & Prevention 1, 229-234.
119. Dogan, M., Liman, B.C., Sagdic, O., 2006: Total aflatoxin levels of instant foods in Turkey. Archiv für Lebensmittelhygiene 57, 55-58.
120. Plestina, R., Ceović, S., Gatenbeck, S., Habazin-Novak, V., Hult, K., Hökby, E., Krogh, P., Radić, B., 1990: Human exposure to ochratoxin A in areas of Yugoslavia with endemic nephropathy. Journal of Environmental Pathology, Toxicology, Oncology 10, 145-148.
121. Schollenberger, M., Müller, H.-M., Rüfle, M., Terry-Jara, H., Suchy, S., Plank, S., Drochner, W., 2007: Natural occurrence of *Fusarium* toxins in soy food marketed in Germany. International Journal of Food Microbiology 113, 142-146. doi: https://doi.org/10.1016/j.ijfoodmicro.2006.06.022.
122. Kong, W.-J., Shen, H.-H., Zhang, X.-F., Yang, X.-L., Qiu, F., Ou-yang, Z., Yang, M.-H., 2013: Analysis of zearalenone and α-zearalenol in 100 foods and medicinal plants determined by HPLC-FLD and positive confirmation by LC-MS-MS. Journal of the Science of Food and Agriculture 93, 1584-1590. doi: https://doi.org/10.1002/jsfa.5926.
123. Lombaert, G.A., Pellaers, P., Roscoe, V., Mankotia, M., Neil, R., Scott, P.M., 2003: Mycotoxins in infant cereal foods from the Canadian retail market. Food Additives and Contaminants 20, 494-504. doi: https://doi.org/10.1080/0265203010094645.
124. Dietrich, R., Usleber, E., Märtlbauer, E., Gareis, M., 1999: Nachweis des nephrotoxischen Mykotoxins Citrinin in Lebensmitteln und mit *Monascus* spp. hergestellten Lebensmittelfarbstoffen. Archiv für Lebensmittelhygiene 50, 17-21.
125. Kumari, P., Sinha, K.K., 1987: Aflatoxin elaboration in mungbean seeds. Indian Phytopathology 40, 201-203.
126. Escobar, A., Sánchez Regueiro, O., 2002: Determination of aflatoxin B_1 in food and feedstuffs in Cuba (1990 through 1996) using an immunoenzymatic reagent kit (aflacen). Journal of Food Protection 65, 219-221.
127. Girgis, A.N., El-Sherif, S., Rofael, N., Nesheim, S., 1977: Aflatoxins in Egyptian foodstuffs. Journal of the Association of Official Analytical Chemists 60, 746-747.
128. Sinha, A.K., Ranjan, K.S., 1990: Incidence of aflatoxins in food grains during flood and post flood period at North Bihar. Proceedings of the Indian National Science Academy B56, 295-298.
129. Nishijima, M., 1983: Survey for mycotoxins in commercial foods. In: Kurata, H., Ueno, Y. (eds.), Developments in Food Science 7. Toxigenic Fungi—Their Toxins and Health Hazard. Proceedings of the Mycotoxin Symposia held in the Third International Mycological Congress, Tokyo, August 30-September 3, pp. 172-181.
130. Soares, L.M.V., Rodriguez-Amaya, D.B., 1989: Survey of aflatoxins, ochratoxin A, zearalenone, and sterigmatocystin in some Brazilian foods by using multi-toxin thin layer chromatographic method. Journal of the Association of Official Analytical Chemists 72, 22-26.
131. Fakoor Janati, S.S., Beheshti, H.R., Koshbakht Fahim, N., Feizy, J., 2011: Aflatoxins and ochratoxin A in bean from Iran. Bulletin of Environmental Contamination and Toxicology 87, 194-197. doi: https://doi.org/10.1007/s00128-011-0311-2.
132. Qian, G.-S., Ross, R.K., Yu, M.C., Yuan, J.-M., Gao, Y.-T., Henderson, B.E., Wogan, G.N., Groopman, J.D., 1994: A follow-up study of urinary markers of aflatoxin exposure and liver cancer risk in Shanghai, People's Republic of China. Cancer Epidemiology, Biomarkers & Prevention 3, 3-10.
133. Imwidthaya, S., Anukarahanonta, T., Komolpis, P., 1987: Bacterial, fungal and aflatoxin contamination of cereals and cereal products in Bangkok. Journal of the Medical Association of Thailand 70, 390-396.
134. Campbell, T.C., Salamat, L., 1971: Aflatoxin ingestion and excretion by humans. In: Purchase, I.F.H. (ed.) Symposium on Mycotoxins in Human Health pp. 271-280. London.
135. Peers, F., Bosch, X., Kaldor, J., Linsell, A., Pluijmen, M., 1987: Aflatoxin exposure, hepatitis B virus infection and liver cancer in Swaziland. International Journal of Cancer 39, 545-553.
136. De Campos, M., Santos, C.J., Olszyna-Marzys, A.E., 1980: Aflatoxin contamination in grains from the Pacific coast in Guatemala and the effect of storage upon contamination. Bulletin of Environmental Contamination and Toxicology 24, 789-795.
137. Alpert, M.E., Hutt, M.S.R., Wogan, G.N., Davidson, C.S., 1971: Association between aflatoxin content of food and hepatoma frequency in Uganda. Cancer 28, 253-260.
138. De Campos, M., Olszyna-Marzys, A.E., 1979: Aflatoxin contamination in grains and grain products during dry season in Guatemala. Bulletin of Environmental Contamination and Toxicology 22, 350-356.
139. Shank, R.C., Wogan, G.N., Gibson, J.B., Nondasuta, A., 1972: Dietary aflatoxins and human liver cancer. II. Aflatoxins in marketed foods and foodstuffs of Thailand and Hong Kong. Food Cosmetics and Toxicology 10, 61-69.
140. Mühlemann, M., Lüthy, J., Hübner, P., 1997: Mycotoxin contamination of food in Ecuador. A: aflatoxins. Mitteilungen aus dem Gebiete der Lebensmitteluntersuchung und Hygiene 88, 474-496.
141. Petkova-Bocharova, T., Castegnaro, M., 1985: Ochratoxin A contamination of cereals in an area of high incidence of Balkan endemic nephropathy in Bulgaria. Food Additives and Contaminants 2, 267-270.
142. El-Sayed, A.A.A., 1996: Natural occurrence of ochratoxin A and citrinin in food stuffs in Egypt. Mycotoxin Research 12, 41-44.
143. MAFF—UK 1997: Survey of aflatoxins and ochratoxin A in cereals and retail products. Food-Surveillance-Information-Sheet No. 130.
144. Åkerstrand, A.K., Josefsson, E., 1979: Fungi and mycotoxins in beans and peas. Vår Föda 31, 405-414.
145. Domijan, A.-M., Peraica, M., Žlener, V., Cvetković, B., Jurjević, Ž., Topolovec-Pintarić, S., Ivić, D., 2005: Seed-borne fungi and ochratoxin A contamination of dry beans (*Phaseolus vulgaris* L.) in the Republic of Croatia. Food and Chemical Toxicology 43, 427-432. doi: https://doi.org/10.1016/j.fct.2004.12.002.

146. Kumagai, S., Nakajima, M., Tabata, S., Ishikuro, E., Tanaka, T., Norizuki, H., Itoh, Y., Aoyama, K., Fujita, K., Kai, S., Sato, T., Saito, S., Yoshiike, N., Sugita-Konishi, Y., 2008: Aflatoxin and ochratoxin A contamination of retail foods and intake of these mycotoxins in Japan. Food Additives and Contaminants: Part A 25, 1101-1106. doi: https://doi.org/10.1080/02652030802226187.
147. Yu, F.-Y., Vdovenko, M.M., Wang, J.-J., Sakharov, I.Y., 2011: Comparison of enzyme-linked immunosorbent assays with chemiluminescent and colorimetric detection for the determination of ochratoxin A in food. Journal of Agricultural and Food Chemistry 59, 809-813. doi: https://doi.org/10.1021/jf103261u.
148. Oyedele, O.A., Ezekiel, C.N., Sulyok, M., Adetjuni, M.C., Warth, B., Atanda, O.O., Krska, R., 2017: Mycotoxin risk assessment for consumers of groundnut in domestic markets in Nigeria. International Journal of Food Microbiology 251, 24-32. doi: https://doi.org/10.1016/j.ijfoodmicro.2017.03.020.
149. Pepeljnjak, S., Ueno, Y., Tanaka, T., 1992: Mycological situation and occurrence of fusariotoxins in cereals from endemic nephropathy region in Croatia. Microbiologie - Aliments - Nutrition 10, 191-197.
150. Omurtag, G.Z., Yazicioğlu, D., 2001: Occurrence of T-2 toxin in processed cereals and pulses in Turkey determined by HPLC and TLC. Food Additives and Contaminants 18, 844-849. doi: https://doi.org/10.1080/02652030110046901.
151. Schollenberger, M., Müller, H.-M., Rüfle, M., Suchy, S., Planck, S., Drochner, W., 2005: Survey of *Fusarium* toxins in foodstuffs of plant origin marketed in Germany. International Journal of Food Microbiology 97, 317-326. doi: https://doi.org/10.1016/j.ijfoodmicro.2004.05.001.
152. Ruhland, M., Engelhardt, G., Wallnöfer, P.R., 1998: Production of ochratoxin A on artificially and naturally contaminated jam. Advances in Food Science (CMTL) 20, 13-16.
153. Kačániová, M., Juráček, M., Chlebo, R., Kňazovická, V., Kadasi-Horáková, M., Kunová, S., Lejková, J.,Haščík, P., Mareček, J., Šimko, M., 2011: Mycobiota and mycotoxins in bee pollen collected from different areas of Slovakia. Journal of Environmental Science and Health, Part B 46, 623-629. doi: https://doi.org/10.1080/03601234.2011.589322.
154. Rodríguez-Carrasco, Y., Font, G., Mañes, J., Berrada, H., 2013: Determination of mycotoxins in bee pollen by gas chromatography-tandem mass spectrometry. Journal of Agricultural and Food Chemistry 61, 1999-2005. doi: https://doi.org/10.1021/jf400256f.
155. Andersson, A., Josefsson, E., Nilsson, G., Åkerstrand, K., 1977: Fungi and patulin in fruits and berries. Vår Föda 29, 292-298.
156. Åkerstrand, A.K., Molander, A., Andersson, A., Nilsson, G., 1976: Occurrence of moulds and mycotoxins in frozen blueberries. Vår Föda 28, 197-200.
157. Cozzi, G., Pascale, M., Perrone, G., Visconti, A., Logrieco, A., 2006: Effect of *Lobesia botrana* on black aspergilli rot and ochratoxin A content in grapes. International Journal of Food Microbiology 111, S88-S92. doi: https://doi.org/10.1016/j.ijfoodmicro.2006.03.012.
158. Valero, A., Marín, S., Ramos, A.J., Sanchis, V., 2008: Survey: ochratoxin A in European special wines. Food Chemistry 108, 593-599. doi: https://doi.org/10.1016/j.foodchem.2007.11.040.
159. Meletis, K., Meniades-Meimaroglou, S., Markaki, P., 2007: Determination of ochratoxin A in grapes of Greek origin by immunoaffinity and high-performance liquid chromatography. Food Additives and Contaminants 24, 1275-1282. doi: https://doi.org/10.1080/02652030701413852.
160. Asam, S., Konitzer, K., Schieberle, P., Rychlik, M., 2009: Stable isotope dilution assays of alternariol and alternariol monomethyl ether in beverages. Journal of Agricultural and Food Chemistry 57, 5152-5160. doi: https://doi.org/10.1021/jf900450w.
161. Nikander, P., Seppälä, T., Kilonzo, G.P., Huttunen, P., Saarinen, L., Kilima, E., Pitkänen, T., 1991: Ingredients and contaminants of traditional alcoholic beverages in Tanzania. Transactions of the Royal Society of Tropical Medicine and Hygiene 85, 133-135.
162. Sebastià, N., Soler, C., Soriano, J.M., Mañes, J., 2010: Occurrence of aflatoxins in tigernuts and their beverages commercialized in Spain. Journal of Agricultural and Food Chemistry 58, 2609-2612. doi: https://doi.org/10.1021/jf903818x.
163. Rubert, J., Sebastià, N., Soriano, J.M., Soler, C., Mañes, J., 2011: One-year monitoring of aflatoxins and ochratoxin A in tiger-nuts and their beverages. Food Chemistry 127, 822-826. doi: https://doi.org/10.1016/j.foodchem.2011.01.016.
164. Logrieco, A., Bottalico, A., Visconti, A., Vurro, M., 1988: Natural occurrence of *Alternaria*-mycotoxins in some plant products. Microbologie - Aliments - Nutrition 6, 13-17.
165. Monbaliu, S., Van Poucke, K., Heungens, K., Van Peteghem, C., De Saeger, S., 2010: Production and migration of mycotoxins in sweet pepper analyzed by multimycotoxin LC-MS/MS. Journal of Agricultural and Food Chemistry 58, 10475-10479. doi: https://doi.org/10.1021/jf102722k.
166. Asam, S., Lichtenegger, M., Liu, Y., Rychlik, M., 2012: Content of the *Alternaria* mycotoxin tenuazonic acid in food commodities determined by a staple isotope dilution assay. Mycotoxin Research 28, 9-15. doi: https://doi.org/10.1007/s12550-011-0111-x.
167. Awe, M.J., Schranz, J.L., 1981: High pressure liquid chromatographic determination of aflatoxins in spices. Journal of the Association of Official Analytical Chemists 64, 1377-1382.
168. Fufa, H., Urga, K., 1996: Screening of aflatoxins in Shiro and ground red pepper in Addis Ababa. Ethiopian Medical Journal 34, 243-249.
169. Martins, M.L., Martins, H.M., Bernardo, F., 2001: Aflatoxins in spices marketed in Portugal. Food Additives and Contaminants 18, 315-319. doi: https://doi.org/10.1080/02652030010021558.
170. Fazekas, B., Tar, A., Kovács, M., 2005: Aflatoxin and ochratoxin A content of spices in Hungary. Food Additives and Contaminants 22, 856-863. doi: https://doi.org/10.1080/02652030500198027.
171. Adachi, Y., Hara, M., Kumazawa, N.H., Hirano, K., Ueno, I., Egawa, K., 1991: Detection of aflatoxin B_1 in imported food products into Japan by enzyme-linked immunosorbent assay and high performance liquid chromatography. Journal of Veterinary Medical Science 53, 49-52.
172. Omurtag, G.Z., Atak, G., Yurdun, T., Ersoy, Ö., 1998: Aflatoxin analysis in dried red pepper samples by TLC and HPLC. Acta Pharmaceutica Turcica XXXX, 125-130.
173. Taguchi, S., Fukushima, S., Sumimoto, T., Yoshida, S., Nishimune, T., 1995: Aflatoxins in foods collected in Osaka, Japan, from 1988 to 1992. Journal of the Association of Official Analytical Chemists International 78, 325-327.
174. Zinedine, A., Brera, C., Elakhdari, S., Catano, C., Debegnach, F., Angelini, S., De Santis, B., Faid, M., Benlemlih, M., Minardi, V., Miraglia, M., 2006: Natural occurrence of mycotoxins in cereals and spices commercialized in Morocco. Food Control 17, 868-874. doi: https://doi.org/10.1016/j.foodcont.2005.06.001.
175. Omurtag, G.Z., Atak, G., Keskin., G., Ersoy, Ö., 2002: HPLC assay for aflatoxins in dried red peppers and feedstuffs in Turkey. Acta Pharmaceutica Turcica 44, 11-22.

176. Romagnoli, B., Menna, V., Gruppioni, N., Bergamini, C., 2007: Aflatoxins in spices, aromatic herbs, herb-teas and medicinal plants marketed in Italy. Food Control 18, 697-701. doi: https://doi.org/10.1016/j.foodcont.2006.02.020.
177. Ardic, M., Karakaya, Y., Atasever, M., Durmaz, H., 2008: Determination of aflatoxin B_1 levels in deep-red ground pepper (isot) using immunoaffinity column combined with ELISA. Food and Chemical Toxicology 46, 1596-1599. doi: https://doi.org/10.1016/j.fct.2007.12.025.
178. Colak, H., Bingol, E.B., Hampikyan, H., Nazli, B., 2006: Determination of aflatoxin contamination in red-scaled, red and black pepper by ELISA and HPLC. Journal of Food and Drug Analysis 14, 292-296.
179. Aydin, A., Emin Erkan, M., Başkaya, R., Ciftcioglu, G., 2007: Determination of aflatoxin B_1 levels in powdered red pepper. Food Control 18, 1015-1018. doi: https://doi.org/10.1016/j.foodcont.2006.03.013.
180. Bircan, C., 2005: The determination of aflatoxins in spices by immunoaffinity extraction using HPLC. International Journal of Food Science and Technology 40, 929-934. doi: https://doi.org/10.1111/j.1365-2621.2005.01025x.
181. Cervino, C., Asam, S., Knopp, D., Rychlik, M., Niessner, R., 2008: Use of isotope-labeled aflatoxins for LC-MS/MS stable isotope dilution analysis of foods. Journal of Agricultural and Food Chemistry 56, 1873-1879. doi: https://doi.org/10.1021/jf073231z.
182. Reddy, K.R.N., Farhana, N.I., Salleh, B., 2011: Occurrence of *Aspergillus* spp. and aflatoxin B_1 in Malaysian foods used for human consumption. Journal of Food Science 76, T99-T104. doi: https://doi.org/10.1111/j.1750-3841.2011.02133.x.
183. Ozbey, F., Kabak, B., 2012: Natural co-occurrence of aflatoxins and ochratoxin A in spices. Food Control 28, 354-361. doi: https://doi.org/10.1016/j.foodcont.2012.05.039.
184. Boonzaaijer, G., Van Osenbruggen, W.A., Kleinnijenhuis, A.J., Van Dongen, W.D., 2008: An exploratory investigation of several mycotoxins and their natural occurrence in flavor ingredients and spices, using a multi-mycotoxin LC-MS/MS method. World Mycotoxin Journal 1, 167-174.
185. Suzuki, J.I., Dainius, B., Kilbuck, J.H., 1973: A modified method for aflatoxin determination in spices. Journal of Food Science 38, 949-950.
186. Erdogan, A., 2004: The aflatoxin contamination of some pepper types sold in Turkey. Chemosphere 56, 321-325. doi: https://doi.org/10.1016/j.chemosphere.2004.02.020.
187. O'Riordan, M.J., Wilkinson, M.G., 2008: A survey of the incidence and level of aflatoxin contamination in a range of imported spice preparations on the Irish retail market. Food Chemistry 107, 1429-1435. doi: https://doi.org/10.1016/j.foodchem.2007.09.073.
188. Cho, S.-H., Lee, C.-H., Jang, M.-R., Son, Y.-W., Lee, S.-M., Choi, I.-S., Kim, S.-H., Kim, D.-B., 2008: Aflatoxins contamination in spices and processed spice products commercialized in Korea. Food Chemistry 107, 1283-1288. doi: https://doi.org/10.1016/j.foodchem.2007.08.049.
189. Jalili, M., Jinap, S., Adzahan, N., 2009: Survey of aflatoxins in retail samples of whole and ground black and white peppercorns. Food Additives and Contaminants: Part B 2, 178-182. doi: https://doi.org/10.1080/19440040903384190.
190. Kursun, O., Mutlu, A.G., 2010: Aflatoxin in spices marketed in the West Mediterranean region of Turkey. Journal of Animal and Veterinary Advances 9, 2979-2981.
191. Thirumala-Devi, K., Mayo, M.A., Reddy, G., Emmanuel, K.E., Larondelle, Y., Reddy, D.V.R., 2001: Occurrence of ochratoxin A in black pepper, coriander, ginger and turmeric in India. Food Additives and Contaminants 18, 830-835. doi: https://doi.org/10.1080/02652030110044921.
192. Goryacheva, I.Y., De Saeger, S., Nesterenko, I.S., Eremin, S.A., Van Peteghem, C., 2007: Rapid all-in-one three-step immunoassay for non-instrumental detection of ochratoxin A in high-coloured herbs and spices. Talanta 72, 1230–1234 (and personal communication). doi: https://doi.org/10.1016/j.talanta.2006.12.049.
193. Goryacheva, I.Yu., De Saeger, S., Lobeau, M., Eremin, S.A., Barna-Vetró, I., Van Peteghem, C., 2006: Approach for ochratoxin A fast screening in spices using clean-up tandem immunoassay columns with confirmation by high performance liquid chromatography-tandem mass spectrometry (HPLC-MS/MS). Analytica Chimica Acta 577, 38-45. doi: https://doi.org/10.1016/j.aca.2006.06.033.
194. Commission of the European Communities, 1997. Reports on tasks for scientific cooperation. Report of experts participating in task 3.2.2. assessment of dietary intake of ochratoxin A by the population of EU member states. Report EUR 17523 EN.
195. Jalili, M., Jinap, S., Radu, S., 2010: Natural occurrence of ochratoxin A contamination in commercial black and white pepper products. Mycopathologia 170, 251-258. doi: https://doi.org/10.1007/s11046-010-9320-7.
196. Abdel-Hafez, A., El-Maghraby, O.M.O., 1992: Fungal flora and aflatoxin associated with cocoa, roasted coffee and tea powders in Egypt. Cryptogamie, Mycologie, 13, 31-45.
197. Bresch, H., Urbanek, M., Hell, K., 2000: Ochratoxin A in coffee, tea and beer. Archiv für Lebensmittelhygiene 51, 89-94.
198. Monbaliu, S., Wu, A., Zhang, D., Van Peteghem, C., De Saeger, S., 2010: Multimycotoxin UPLC-MS/MS for tea, herbal infusions and the derived drinkable products. Journal of Agricultural and Food Chemistry 58, 12664-12671. doi: https://doi.org/10.1021/jf1033043.
199. Ratt, E.R., Shantha, T., 1994: Incidence of aflatoxin in groundnut-based snack products. Journal of Food Science and Technology 31, 327-329.
200. Ioannou-Kakouri, E., Aletrari, M., Christou, E., Hadjioannou-Ralli, A., Koliou, A., Akkelidou, D., 1999: Surveillance and control of aflatoxins B_1, B_2, G_1, G_2, and M_1 in foodstuffs in the Republic of Cyprus: 1992-1996. Journal of the Association of Official Analytical Chemists International 82, 883-892.
201. Yndestad, M., Underdal, B., 1975: Aflatoxin in foods on the Norwegian market. Nordisk Veterinaer Medicin 27, 42-48.
202. Furlani, R.P.Z., Valente Soares, L.M., 1996: Survey of aflatoxins, ochratoxin A, zearalenone, and sterigmatocystin in tree nuts commercialized in Campinas, São Paulo. Ciência e Tecnologia de Alimentos. 16, 238-240.
203. Food Standards Agency, 2002: Survey of nuts, nut products and dried tree fruits for mycotoxins. Food-Survey-Information-Sheet No. 21/02.
204. Reis, T.A., Oliveira, T.D., Baquião, A.C., Gonçalves, S.S., Zorzete, P., Corrêa, B., 2012: Mycobiota and mycotoxins in Brazil nut samples from different states of the Brazilian Amazon region. International Journal of Food Microbiology 159, 61-68. doi: https://doi.org/10.1016/j.ijfoodmicro.2012.08.005.
205. Nilsson, G., Åkerstrand, K., Slorach, S., 1974: Aflatoxins in nuts. Var Foda 26, 56-62.
206. MAFF—UK, 1996: Aflatoxin surveillance of retail and imported nuts, nut products, dried figs and fig products. Food-Surveillance-Information-Sheet No. 81.

207. Pacheco, A.M., Lucas, A., Parente, R., Pacheco, N., 2010: Association between aflatoxin and aflatoxigenic fungi in Brazil nut (*Bertholletia excelsa* H.B.K.). Ciência e Tecnologia de Alimentos 30, 330-334.
208. Andrade, P.D., Homem de Mello, M., França, J.A., Caldas, E.D., 2013: Aflatoxins in food products consumed in Brazil: a preliminary dietary risk assessment. Food Additives and Contaminants: Part A 30, 127-136. doi: https://doi.org/10.1080/19440049.2012.720037.
209. Pacheco, A.M., Scussel, V.M., 2007: Selenium and aflatoxin levels in raw Brazil nuts from the Amazon Basin. Journal of Agricultural and Food Chemistry 55, 11087-11092. doi: https://doi.org/10.1021/jf072434k.
210. Odoemelam, S.A., Osu, C.I., 2009: Aflatoxin B_1 contamination of some edible grains marketed in Nigeria. E-Journal of Chemistry 6, 308-314.
211. Copetti, M.V., Iamanaka, B.T., Pereira, J.L., Lemes, D.P., Nakano, F., Taniwaki, M.H., 2012: Determination of aflatoxins in by-products of industrial processing of cocoa beans. Food Additives and Contaminants: Part A 29, 972-978. doi: https://doi.org/10.1080/19440049.2012.6 60657.
212. Copetti, M.V., Iamanaka, B.T., Nester, M.A., Efraim, P., Taniwaki, M.H., 2013: Occurrence of ochratoxin A in cocoa by-products and determination of its reduction during chocolate manufacture. Food Chemistry 136, 100-104. doi: https://doi.org/10.1016/j.foodchem.2012.07.093.
213. Mortimer, D.N., Shepherd, M.J., Gilbert, J., Morgan, M.R.A., 1987: A survey of the occurrence of aflatoxin B_1 in peanut butters by enzyme-linked immunosorbent assay. Food Additives and Contaminants 5, 127-132.
214. Wei, D.-L., Wei, R.-D., 1980: High pressure liquid chromatographic determination of aflatoxins in peanut and peanut products of Taiwan. Proceedings of the National Science Council, Republic of China 4, 152-155.
215. Ram, B.P., Hart, L.P., Cole, R.J., Pestka, J.J., 1986: Application of ELISA to retail survey of aflatoxin B_1 in peanut butter. Journal of Food Protection 49, 792-795.
216. Tung, T.C., Ling, K.H., 1968: Study on aflatoxin of foodstuffs in Taiwan. The Journal of Vitaminology 14, 48-52.
217. Vahl, M., Jørgensen, K., 1998: Determination of aflatoxins in food using LC/MS/MS. Zeitschrift für Lebensmittel-Untersuchung und–Forschung 206, 243-245.
218. Siame, B.A., Mpuchane, S.F., Gashe, B.A., Allotey, J., Teffera, G., 1998: Occurrence of aflatoxins, fumonisin B_1, and zearalenone in foods and feeds in Botswana. Journal of Food Protection 61, 1670-1673.
219. Elshafie, S.Z.B., ElMubarak, A., El-Nagerabi, S.A.F., Elshafie, A.E., 2011: Aflatoxin B_1 contamination of traditionally processed peanuts butter for human consumption in Sudan. Mycopathologia 171, 435-439. doi: https://doi.org/10.1007/s11046-010-9378-2.
220. Chun, H. S., Kim, H. J., Ok, H. E., Hwang, J.-B., Chung, D.-H., 2007: Determination of aflatoxin levels in nuts and their products consumed in South Korea. Food Chemistry 102, 385-391. doi: https://doi.org/10.1016/j.foodchem.2006.05.031.
221. Yentür, G., Er, B., Özkan, M.G., Öktem, A.B., 2006: Determination of aflatoxins in peanut butter and sesame samples using high-performance liquid chromatography method. European Food Research and Technology 224, 167-170. doi: https://doi.org/10.1007/s00217-006-0310-4.
222. Francis, O.J., Jr., Lipinski, L.J., Gaul, J.A., Campbell, A.D., 1982: High pressure liquid chromatographic determination of aflatoxins in peanut butter using a silicia gel-packed flowcell for fluorescence detection. Journal of the Association of Official Analytical Chemists 65, 672-676.
223. Hendrickse, R.G., Coulter, J.B.S., Lamplugh, S.M., MacFarlane, S.B.J., Williams, T.E., Omer, M.I.A., Suliman, G.I., 1982: Aflatoxins and kwashiorkor: a study in Sudanese children. British Medical Journal 285, 843-846.
224. Li, F.-Q., Li, Y.-W., Wang, Y.-R., Luo, X.-Y., 2009: Natural occurrence of aflatoxins in Chinese peanut butter and sesame paste. Journal of Agricultural and Food Chemistry 57, 3519-3524. doi: https://doi.org/10.1021/jf804055n.
225. Leong, Y.-H., Ismail, N., Latif, A.A., Ahmad, R., 2010: Aflatoxin occurrence in nuts and commercial nutty products in Malaysia. Food Control 21, 334-338. doi: https://doi.org/10.1016/j.foodcont.2009.06.002.
226. Razzazi-Fazeli, E., Noviandi, C.T., Porasuphatana, S., Agus, A., Böhm, J., 2004: A survey of aflatoxin B_1 and total aflatoxin contamination in baby food, peanut and corn products sold at retail in Indonesia analysed by ELISA and HPLC. Mycotoxin Research 20, 51-58.
227. Huang, B., Han, Z., Cai, Z., Wu, Y., Ren, Y., 2010: Simultaneous determination of aflatoxin B_1, B_2, G_1, G_2, M_1 and M_2 in peanuts and their derivative products by ultra-high-performance liquid chromatography-tandem mass spectrometry. Analytica Chimica Acta 662, 62-68. doi: https://doi.org/10.1016/j.aca.2010.01.002.
228. Morgan, M.R.A., Kang, A.S., Chan, H.W.-S., 1986: Aflatoxin determination in peanut butter by enzyme-linked immunosorbent assay. Journal of the Science of Food and Agriculture 37, 908-914.
229. Abdulkadar, A.H.W., Al-Ali, A.A., Al-Kildi, A.M., Al-Jedah, J.H., 2004: Mycotoxins in food products available in Qatar. Food Control 15, 543-548. doi: https://doi.org/10.1016/j.foodcont.2003.08.008.
230. Younis, Y.M.H., Malik, K.M., 2003: TLC and HPLC assays of aflatoxin contamination in Sudanese peanuts and peanut products. Kuwait Journal of Science & Engineering 30, 79-93.
231. Nasir, M.S., Jolley, M.E., 2002: Development of a fluorescence polarization assay for the determination of aflatoxins in grains. Journal of Agricultural and Food Chemistry 50, 3116-3121. doi: https://doi.org/10.1021/jf011638c.
232. Gagliardi, S.J., Cheatle, T.F., Mooney, R.L., Llewellyn, G.C., O'Rear, C.E., Llewellyn, G.C., 1991: The occurrence of aflatoxin in peanut butter from 1982 to 1989. Journal of Food Protection 54, 627-631.
233. Ali, N., 2000: Aflatoxins in Malaysian food. Mycotoxins 50, 31-35.
234. Nilüfer, D., Boyacioğlu, D., 2002: Comparative study of three different methods for the determination of aflatoxins in Tahini. Journal of Agricultural and Food Chemistry 50, 3375-3379. doi: https://doi.org/10.1021/jf020005a.
235. Njumbe Ediage, E., Di Mavungu, J.D., Monbaliu, S., Van Peteghem, C., De Saeger, S., 2011: A validated multianalyte LC-MS/MS method for quantification of 25 mycotoxins in cassava flour, peanut cake and maize samples. Journal of Agricultural and Food Chemistry 59, 5173-5180. doi: https://doi.org/10.1021/jf2009364.
236. Krishna Kumari, C., Nusrath, M., Reddy, B.N., 1984: Aflatoxin contamination of coconut products from Andhra Pradesh. Indian Phytopathology 37, 284-287.
237. Oliveira, C.A.F., Gonçalves, N., Rosim, R.E., Fernandez, A.M., 2009: Determination of aflatoxins in peanut products in the northeast region of São Paulo, Brazil. International Journal of Molecular Sciences 10, 174-183. doi: https://doi.org/10.3390/ijms10010174.

238. Wareing, P.W., Westby, A., Gibbs, J.A., Allotey, L.T., Halm, M., 2001: Consumer preferences and fungal and mycotoxin contamination of dried cassava products from Ghana. International Journal of Food Science and Technology 36, 1-10.
239. Hanssen, E., Jung, M., 1973: Control of aflatoxins in the food industry. Pure and Applied Chemistry 35, 239-250
240. Shank, R.C., Bourgeois, C.H., Keschamras, N., Chandavimol, P., 1971: Aflatoxins in autopsy specimens from Thai children with an acute disease of unknown aetiology. Food and Cosmetics Toxicology 9, 501-507
241. Reddy, S.V., Kiran Mayi, D., Uma Reddy, M., Thirumala-Devi, K., Reddy, D.V.R., 2001: Aflatoxins B_1 in different grades of chillies (*Capsicum annum* L.) in India as determined by indirect competitive-ELISA. Food Additives and Contaminants 18, 553-558. doi: https://doi.org/10.1080/02652030010025383.
242. Scott, P.M., Kennedy, B.P.C., 1973: Analysis and survey of ground black, white, and Capsicum peppers for aflatoxins. Journal of the Association of Official Analytical Chemists 56, 1452-1457.
243. El-Dessouki, S., 1992: Aflatoxine in Cayenne-Pfeffer- und Paprika-Pulver. Deutsche Lebensmittel-Rundschau 88, 78.
244. Majerus, P., Woller, R., Leevivat, P., Klintrimas, T., 1985: Spices: Mould attack and content of aflatoxins, ochratoxin A and sterigmatocystine. Fleischwirtschaft 65, 1155-1158.
245. Food Standards Agency, 2005: Survey of spices for aflatoxins and ochratoxin A. Food-Survey-Information-Sheet No. 73/05.
246. Shamsuddin, Z.A., Khan, M.A., Khan, B.A., Ahmad, M.A., Ahmed, A., 1995: Contamination of red chilli with aflatoxin B_1 in Pakistan. Mycotoxin Research 11, 21-24.
247. Iqbal, S.Z., Paterson, R.R.M., Bhatti, I.A., Asi, M.R., Sheikh, M.A., Bhatti, H.N., 2010: Aflatoxin B_1 in chilies from the Punjab region, Pakistan. Mycotoxin Research 26, 205-209. doi: https://doi.org/10.1007/s12550-010-0055-6.
248. Saha, D., Acharya, D., Roy, D., Shrestha, D., Dhar, T.K., 2007: Simultaneous enzyme immunoassay for the screening of aflatoxin B_1 and ochratoxin A in chili samples. Analytica Chimica Acta 584, 343-349. doi: https://doi.org/10.1016/j.aca.2006.11.042.
249. Jalili, M., Jinap, S., 2012: Natural occurrence of aflatoxins and ochratoxin A in commercial dried chili. Food Control 24, 160-164. doi: https://doi.org/10.1016/j.foodcont.2011.09.020.
250. MacDonald, S., Castle, L., 1996: A UK retail survey of aflatoxins in herbs and spices and their fate during cooking. Food Additives and Contaminants 13, 120-128.
251. Paterson, R.R.M., 2007: Aflatoxins contamination in chilli samples from Pakistan. Food Control 18, 817-820. doi: https://doi.org/10.1016/j.foodcont.2006.04.005.
252. Santos, L., Marín, S., Sanchis, V., Ramos, A.J., 2010: Co-occurrence of aflatoxins, ochratoxin A and zearalenone in *Capsicum* powder samples available on the Spanish market. Food Chemistry 122, 826-830. doi: https://doi.org/10.1016/j.foodchem.2010.03.070.
253. Santos, L., Marín, S., Mateo, E.M., Gil-Serna, J., Valle-Algarra, F.M., Patiño, B., Ramos, A.J., 2011: Mycobiota and co-occurrence of mycotoxins in *Capsicum* powder. International Journal of Food Microbiology 151, 270-276. doi: https://doi.org/10.1016/j.ijfoodmicro.2011.09.011.
254. Iqbal, S.Z., Paterson, R.R.M., Bhatti, I.A., Asi, M.R., 2011: Comparing aflatoxin contamination in chilies from Punjab, Pakistan produced in summer and winter. Mycotoxin Research 27, 75-80. doi: https://doi.org/10.1007/s12550-010-0078-z.
255. Thirumala-Devi, K., Mayo, M.A., Reddy, G., Reddy, S.V., Delfosse, P., Reddy, D.V.R., 2000: Production of polyclonal antibodies against ochratoxin A and its detection in chilies by ELISA. Journal of Agricultural and Food Chemistry 48, 5079-5082. doi: https://doi.org/10.1021/jf000049o.
256. Manjula, K., Hell, K., Fandohan, P., Abass, A., Bandyopadhyay, R., 2009: Aflatoxin and fumonisin contamination of cassava products and maize grain from markets in Tanzania and Republic of the Congo. Toxin Reviews 28, 63-69. doi: https://doi.org/10.1080/15569540802462214.
257. Bankole, S.A., Mabekoje, O.O., 2004: Mycoflora and occurrence of aflatoxin B_1 in dried yam chips from markets in Ogun and Oyo States, Nigeria. Mycopathologia 157, 111-115.
258. Bankole, S.A., Adebanjo, A., 2003: Aflatoxin contamination of dried yam chips marketed in Nigeria. Tropical Science 43, 201-203.
259. Mestres, C., Bassa, S., Fagbohoun, E., Nago, M., Hell, K., Vernier, P., Champiat, D., Hounhouigan, J., Cardwell, K.F., 2004: Yam chip food subsector: hazardous practices and presence of aflatoxins in Benin. Journal of Stored Products Research 40, 575-585. doi: https://doi.org/10.1016/j.jspr.2003.11.003.
260. Copetti, M.V., Iamanaka, B.T., Pereira, J.L., Lemes, D.P., Nakano, F., Taniwaki, M.H., 2012: Co-occurrence of ochratoxin A and aflatoxins in chocolate marketed in Brazil. Food Control 26, 36-41. doi: https://doi.org/10.1016/j.foodcont.2011.12.023.
261. Serra Bonvehi, J., 2004: Occurrence of ochratoxin A in cocoa products and chocolate. Journal of Agricultural and Food Chemistry 52, 6347-6352. doi: https://doi.org/10.1021/jf040153w.
262. Engel, G., 2000: Ochratoxin A in sweets, oil seeds and dairy products. Archiv für Lebensmittelhygiene 51, 98-101.
263. MAFF—UK 1999: 1998 Survey of retail products for ochratoxin A. Food-Surveillance-Information-Sheet No. 185.
264. Burdaspal, P.A., Legarda, T.M., 2003: Occurrence of ochratoxin A in samples of different types of chocolate and cacao powder, marketed in Spain and fifteen foreign countries. Alimentaria 40, 143-153.
265. Turcotte, A.-M., Scott, P.M., 2011: Ochratoxin A in cocoa and chocolate sampled in Canada. Food Additives and Contaminants: Part A 28, 762-766. doi: https://doi.org/10.1080/19440049.2010.508796.
266. Brera, C., Debegnach, F., De Santis, B., Iafrate, E., Pannunzi, E., Berdini, C., Prantera, E., Gregori, E., Miraglia, M., 2011: Ochratoxin A in cocoa and chocolate products from the Italian market: occurrence and exposure assessment. Food Control 22, 1663-1667. doi: https://doi.org/10.1016/j.foodcont.2011.03.026.
267. Tutelyan, V.A., Eller, K.I., Sobolev, V.S., 1989: A survey using normal-phase high-performance liquid chromatography of aflatoxins in domestic and imported foods in the USSR. Food Additives and Contaminants 6, 159-165.
268. Copetti, M.V., Iamanaka, B.T., Pereira, J.L., Fungaro, M.H., Taniwaki, M.H., 2011: Aflatoxigenic fungi and aflatoxin in cocoa. International Journal of Food Microbiology 148, 141-144. doi: https://doi.org/10.1016/j.ijfoodmicro.2011.05.020.
269. Piñeiro, M., Dawson, R., Costarrica, M.L., 1996: Monitoring program for mycotoxin contamination in Uruguayan food and feeds. Natural Toxins 4, 242-245.
270. Amézqueta, S., González-Peñas, E., Murillo, M., López De Cerain, A., 2004: Validation of a high-performance liquid chromatography analytical method for ochratoxin A quantification in cocoa beans. Food Additives and Contaminants 21, 1096-1106. doi: https://doi.org/10.1080/02652030400019422.

271. Amézqueta, S., González-Peñas, E., Murillo, M., López De Cerain, A., 2005: Occurrence of ochratoxin A in cocoa beans: effect of shelling. Food Additives and Contaminants 22, 590-596. doi: https://doi.org/10.1080/02652030500130160.
272. Copetti, M.V., Pereira, J.L., Iamanaka, B.T., Pitt, J.I., Taniwaki, M.H., 2010: Ochratoxigenic fungi and ochratoxin A in cocoa during farm processing. International Journal of Food Microbiology 143, 67-70. doi: https://doi.org/10.1016/j.ijfoodmicro.2010.07.031.
273. Varga, J., Kocsubé, S., Koncz, Z., Téren, J., 2006: Mycobiota and ochratoxin A in raisins purchased in Hungary. Acta Alimentaria 35, 289-294. doi: https://doi.org/10.1556/AAlim.35.2006.3.6.
274. Aycicek, H., Aksoy, A., Saygi, S., 2005: Determination of aflatoxin levels in some dairy and food products which consumed in Ankara, Turkey. Food Control 16, 263-266. doi: https://doi.org/10.1016/j.foodcont.2004.03.004.
275. Zohri, A.A., Saber, S.M., 1993: Filamentous fungi and mycotoxin detected in coconut. Zentralblatt für Mikrobiologie 148, 325-332.
276. Afolabi, C.G., Ezekiel, C.N., Kehinde, I.A., Olaolu, A.W., Ogunsanya, O.M., 2015: Contamination of groundnut in south-western Nigeria by aflatoxigenic fungi and aflatoxins in relation to processing. Journal of Phytopathology 163, 279-286. doi: https://doi.org/10.1111/jph.12317.
277. Cressey, P., Pearson, A., 2014: Risk Profile: Mycotoxins in the New Zealand Food Supply. ESR Client Report FW 14005. ESR, Christchurch, New Zealand.
278. Krishna Kumari, C., Nusrath, M., 1987: Natural occurrence of citrinin and ochratoxin A in coconut products. National Academy Science Letters 10, 303-305.
279. Soliman, K.M., 2002: Incidence, level, and behavior of aflatoxins during coffee bean roasting and decaffeination. Journal of Agricultural and Food Chemistry 50, 7477-7481. doi: https://doi.org/10.1021/jf011338v.
280. Purchase, I.F.H., Pretorius, M.E., 1973: Sterigmatocystin in coffee beans. Journal of the Association of Official Analytical Chemists 56, 225-226.
281. Leoni, L.A.B., Valente Soares, L.M., Oliveira, P.L.C., 2000: Ochratoxin A in Brazilian roasted and instant coffee. Food Additives and Contaminants 17, 867-870.
282. Cantàfora, A., Grossi, M., Miraglia, M., Benelli, L., 1983: Determination of ochratoxin A in coffee beans using reversed-phase high performance liquid chromatography. La Rivista della Società Italiana di Scienza dell´ Alimentazione 12, 103-108.
283. Fazekas, B., Tar, A.K., Zomborsky-Kovács, M., 2002: Ochratoxin A contamination of cereal grains and coffee in Hungary in the year 2001. Acta Veterinaria Hungarica 50, 177-188.
284. Lin, L.-C., Chen, P.-C., Fu, Y.-M., Shih, D.Y.-C., 2005: Ochratoxin A contamination in coffees, cereals, red wines and beers in Taiwan. Journal of Food and Drug Analysis 13, 84-92.
285. Jørgensen, K., 1998: Survey of pork, poultry, coffee, beer and pulses for ochratoxin A. Food Additives and Contaminants 15, 550-554.
286. Veprikova, Z., Zachariasova, M., Dzuman, Z., Zachariasova, A., Fenclova, M., Slavikova, P., Vaclavikova, M., Mastovska, K., Hengst, D., Hajslova, J., 2015: Mycotoxins in plant-based dietary supplements: hidden health risk for consumers. Journal of Agricultural and Food Chemistry 63, 6633-6643. doi: https://doi.org/10.1021/acs.jafc.5b02105.
287. Lombaert, G.A., Pellaers, P., Chettiar, M., Lavalee, D., Scott, P.M., Lau, B.P.-Y., 2002: Survey of Canadian retail coffees for ochratoxin A. Food Additives and Contaminants 19, 869-877. doi: https://doi.org/10.1080/02652030210145027.
288. Otteneder, H., Majerus, P., 2001: Ochratoxin A (OTA) in coffee: nation-wide evaluation of data collected by German Food Control 1995-1999. Food Additives and Contaminants 18, 431-435. doi: https://doi.org/10.1080/02652030010021567.
289. Patel, S., Hazel, C.M., Winterton, A.G.M., Gleadle, A.E., 1997: Survey of ochratoxin A in UK retail coffees. Food Additives and Contaminants 14, 217-222.
290. Pittet, A., Tornare, D., Huggett, A., Viani, R., 1996: Liquid chromatographic determination of ochratoxin A in pure and adulterated soluble coffee using an immunoaffinity column cleanup procedure. Journal of Agricultural and Food Chemistry 44, 3564-3569.
291. Van Der Stegen, G., Jörissen, U., Pittet, A., Saccon, M., Steiner, W., Vincenzi, M., Winkler, M., Zapp, J., Schlatter, Chr. 1997: Screening of European coffee final products for occurrence of ochratoxin A (OTA). Food Additives and Contaminants 14, 211-216.
292. Trucksess, M.W., Giler, J., Young, K., White, K.D., Page, S.W., 1999: Determination and survey of ochratoxin A in wheat, barley and coffee—1997. Journal of the Association of Official Analytical Chemists International 82, 85-89.
293. Tsubouchi, H., Terada, H., Yamamoto, K., Hisada, K., Sakabe, Y., 1988: Ochratoxin A found in commercial roast coffee. Journal of Agricultural and Food Chemistry 36, 540-542.
294. De Almeida, A.P., Alaburda, J., Shundo, L., Ruvieri, V., Navas, S.A., Lamardo, L.C.A., Sabino, M., 2007: Ochratoxin A in Brazilian instant coffee. Brazilian Journal of Microbiology 38, 300-303. doi: https://doi.org/10.1590/S1517-83822007000200022.
295. Mounjouenpou, P., Durand, N., Guyot, B., Guiraud, J.-P., 2007: Effect of operating conditions on ochratoxin A extraction from roasted coffee. Food Additives and Contaminants 24, 730-734. doi: https://doi.org/10.1080/02652030701199972.
296. Burdaspal, P.A., Legarda, T.M., 1998: Ochratoxin A in roasted and soluble coffees marketed in Spain. Alimentaria 35, 31-35.
297. Coronel, M.B., Marin, S., Cano, G., Ramos, A.J., Sanchis, V., 2011: Ochratoxin A in Spanish retail ground roasted coffee: occurrence and assessment of the exposure in Catalonia. Food Control 22, 414-419. doi: https://doi.org/10.1016/j.foodcont.2010.09.012.
298. Gopinandhan, T.N., Velmourougane, K., Panneerselvam, P., Keshamma, E., Raghuramulu, Y., 2007: Occurrence of ochratoxin-A (OT-A) in green and commercial coffee samples. Journal of Food Science and Technology 44, 247-249.
299. Tozlovanu, M., Pfohl-Leszkowicz, A., 2010: Ochratoxin A in roasted coffee from French supermarkets and transfer in coffee beverages: comparison of analysis methods. Toxins 2, 1928-1942. doi: https://doi.org/10.3390/toxins2081928.
300. Abdallah, M.F., Krska, R., Sulyok, M., 2016: Mycotoxin contamination in sugarcane grass and juice: first report on detection of multiple mycotoxins and exposure assessment for aflatoxin B_1 and G_1 in humans. Toxins (Basel), 8 (11), 343. doi: https://doi.org/10.3390/toxins8110343.
301. Bhat, R.V., Vasanthi, S., 2005: Natural occurrence of ochratoxins in Indian coffee. Indian Journal of Nutrition and Dietetics 42, 106-113.
302. Wu, J., Tan, Y., Wang, Y., Xu, R., 2012: Occurrence of ochratoxin A in grain and manufactured food products in China detected by HPLC with fluorescence detection and confirmed by LC-ESI-MS/MS. Mycopathologia 173, 199-205. doi: https://doi.org/10.1007/s11046-011-9393-y.
303. Nakajima, M., Tsubouchi, H., Miyabe, M., Ueno, Y., 1997: Survey of aflatoxin B_1 and ochratoxin A in commercial green coffee beans by high-performance liquid chromatography linked with immunoaffinity chromatography. Food and Agricultural Immunology 9, 77-83.

304. Schmidt, H., Bannier, M., Vogel, R.F., Niessen, L., 2004: Detection and quantification of *Aspergillus ochraceus* in green coffee by PCR. Letters in Applied Microbiology 38, 464-469. doi: https://doi.org/10.1111/j.1472-765X.2004.01524.x.
305. De Palo, D., Gabucci, G., Valussi, S., 1979: Study on the possible presence of aflatoxine, sterigmatocystine and ochratoxine in green coffee. VIIIe Colloque Scientifique International sur le Café, Abidjan, 28 nov.-3 déc. 1977. ASIC (Paris) 1979, 539-543.
306. Leoni, L.A.B., Furlani, R.P.Z., Valente Soares, L.M., Oliveira, P.L.C., 2001: Ochratoxin A in Brazilian green coffee. Ciência e Tecnologia de Alimentos 21, 105-107. doi: https://doi.org/10.1590/S0101-20612001000100022.
307. Studer-Rohr, I., Dietrich, D.R., Schlatter, J., Schlatter, C., 1995: The occurrence of ochratoxin A in coffee. Food Chemical and Toxicology 33, 341-355.
308. Sibanda, L., De Saeger, S., Bauters, T.G.M., Nelis, H.J., Van Peteghem, C., 2001: Development of a flow-through enzyme immunoassay and application in screening green coffee samples for ochratoxin A with confirmation by high-performance liquid chromatography. Journal of Food Protection 64, 1597-1602.
309. Pardo, E., Marín, S., Ramos, A.J., Sanchis, V., 2004: Occurrence of ochratoxigenic fungi and ochratoxin A in green coffee from different origins. Food Science and Technology International 10, 45-49. doi: https://doi.org/10.1177/1082013204041509.
310. Levi, C.P., Trenk, H.L., Mohr, H.K., 1974: Study of the occurrence of ochratoxin A in green coffee beans. Journal of the Association of Official Analytical Chemists 57, 866-870.
311. Micco, C., Grossi, M., Miraglia, M., Brera, C., 1989: A study of the contamination by ochratoxin A of green and roasted coffee beans. Food Additives and Contaminants 6, 333-339.
312. Romani, S., Sacchetti, G., Chaves López, C., Pinnavaia, G.G., Dalla Rosa, M., 2000: Screening on the occurrence of ochratoxin A in green coffee beans of different origins and types. Journal of Agricultural and Food Chemistry 48, 3616-3619. doi: https://doi.org/10.1021/jf990783b.
313. MAFF—UK, 1996: Surveillance of ochratoxin A in green (unroasted) coffee beans. Food-Surveillance-Information-Sheet No. 80.
314. Ruadrew, S., Craft, J., Aidoo, K., 2013: Occurrence of toxigenic *Aspergillus* spp. and aflatoxins in selected food commodities of Asian origin sourced in the West of Scotland. Food and Chemical Toxicology 55, 653-658. doi: https://doi.org/10.1016/j.fct.2013.02.001.
315. Salem, N.M., Ahmad, R., 2010: Mycotoxins in food from Jordan: preliminary survey. Food Control 21, 1099-1103. doi: https://doi.org/10.1016/j.foodcont.2010.01.002.
316. Duris, D., Mburu, J.K., Durand, N., Clarke, R., Frank, J.M., Guyot, B., 2010: Ochratoxin A contamination of coffee batches from Kenya in relation to cultivation methods and post-harvest processing treatments. Food Additives and Contaminants: Part A 27, 836-841. doi: https://doi.org/10.1080/19440040903571770.
317. Martins, M.L., Martins, H.M., Gimeno, A., 2003: Incidence of microflora and of ochratoxin A in green coffee beans (*Coffea arabica*). Food Additives and Contaminants 20, 1127-1131. doi: https://doi.org/10.1080/02652030310001620405.
318. Paulino De Moraes, M.H., Luchese, R.H., 2003: Ochratoxin A on green coffee: influence of harvest and drying processing procedures. Journal of Agricultural and Food Chemistry 51, 5824-5828. doi: https://doi.org/10.1021/jf026248k.
319. Gopinandhan, T.N., Kannan, G.S., Panneerselvam, P., Velmourougane, K., Raghuramulu, Y., Jayarama J, 2008: Survey on ochratoxin A in Indian green coffee destined for export. Food Additives and Contaminants: Part B 1, 51-57. doi: https://doi.org/10.1080/19393210802236984.
320. Imperato, R., Campone, L., Piccinelli, A.L., Veneziano, A., Rastrelli, L., 2011: Survey of aflatoxins and ochratoxin A contamination in food products imported in Italy. Food Control 22, 1905-1910. doi: https://doi.org/10.1016/j.foodcont.2011.05.002.
321. Lovelace, C.E.A., Aalbersberg, W.G.L., 1989: Aflatoxin levels in foodstuffs in Fiji and Tonga islands. Plant Foods for Human Nutrition 39, 393-399.
322. Misra, N., 1987: Mycotoxins in spices. III. Investigation on the natural occurrence of aflatoxins in *Coriandrum sativum* L. Journal of Food Science and Technology 24, 324-325.
323. Scott, P.M., Kennedy, B.P.C., 1975: The analysis of spices and herbs for aflatoxins. Canadian Institute of Food Science and Technology Journal 8, 124-125.
324. Houssou, P.A., Ahohuendo, B.C., Fandohan, P., Kpodo, K., Hounhouigan, D.J., Jakobsen, M., 2009: Natural infection of cowpea (*Vigna unguiculata* (L.) Walp.) by toxigenic fungi and mycotoxin contamination in Benin, West Africa. Journal of Stored Products Research 45, 40-44. doi: https://doi.org/10.1016/j.jspr.2008.07.002.
325. Zohri, A.A., Khayria, M., Gawad, A.El., 1992: Studies on mycoflora and mycotoxins of cowpea cultivars. Korean Mycology 20, 252-258.
326. Kritzinger, Q., Aveling, T.A.S., Marasas, W.F.O., Rheeder, J.P., Van Der Westhuizen, L., Shephard, G.S., 2003: Mycoflora and fumonisin mycotoxins associated with cowpea [*Vigna unguiculata* (L.) Walp] seeds. Journal of Agricultural and Food Chemistry 51, 2188-2192. doi: https://doi.org/10.1021/jf026121v.
327. Stefanaki, I., Foufa, E., Tsatsou-Dritsa, A., Dais, P., 2003: Ochratoxin A concentrations in Greek domestic wines and dried vine fruits. Food Additives and Contaminants 20, 74-83. doi: https://doi.org/10.1080/0265203021000031537.
328. MacDonald, S., Wilson, P., Barnes, K., Damant, A., Massey, R., Mortby, E., Shepherd, M.J., 1999: Ochratoxin A in dried vine fruit: method development and survey. Food Additives and Contaminants 16, 253-260.
329. Lombaert, G.A., Pellaers, P., Neumann, G., Kitchen, D., Huzel, V., Trelka, R., Kotello, S., Scott, P.M., 2004: Ochratoxin A in dried vine fruits on the Canadian retail market. Food Additives and Contaminants 21, 578-585. doi: https://doi.org/10.1080/02652030410001687681.
330. Adekoya, I., Njobeh, P., Obadina, A., Chilaka, C., Okoth, S., De Boevre, M., De Saeger, S., 2017: Awareness and prevalence of mycotoxin contamination in selected Nigerian fermented foods. Toxins (Basel), 9(11), 363. doi: https://doi.org/10.3390/toxins9110363.
331. Iamanaka, B.T., Taniwaki, M.H., Menezes, H.C., Vicente, E., Fungaro, M.H.P., 2005: Incidence of toxigenic fungi and ochratoxin A in dried fruits sold in Brazil. Food Additives and Contaminants 22, 1258-1263. doi: https://doi.org/10.1080/02652030500260447
332. Lehtonen, M., 1973: Detection of aflatoxines in wines. Chemie, Mikrobiologie & Technologie der Lebensmittel 2, 161-164
333. Zimmerli, B., Dick, R., 1995: Determination of ochratoxin A at the ppt level in human blood, serum, milk and some foodstuffs by high performance liquid chromatography with enhanced fluorescence detection and immunoaffinity column cleanup: methodology and Swiss data. Journal of Chromatography B 666, 85-99.
334. Eder, R., Paar, E., Edinger, W., Lew, H., 2002: Investigations into the ochratoxin A contents of Austrian wines with special respect to wines of the "Prädikatswein"-category. Mitteilungen Klosterneuburg 52, 125-131.

335. Pena, A., Cerejo, F., Silva, L.J.G., Lino, C.M., 2010: Ochratoxin A survey in Portuguese wine by LC-FD with direct injection. Talanta 82, 1556-1561. doi: https://doi.org/10.1016/j.talanta.2010.07.041.

336. Spadaro, D., Ciavorella, A., Lore, A., Garibaldi, A., Gullino, M.L., 2007: Low levels of ochratoxin A in wines from Piedmont. Communications in Agricultural and Applied Biological Science 72, 327-332.

337. Spadaro, D., Lorè, A., Garibaldi, A., Gullino, M.L., 2010: Occurrence of ochratoxin A before bottling in DOC and DOCG wines produced in Piedmont (northern Italy). Food Control 21, 1294-1297. doi: https://doi.org/10.1016/j.foodcont.2010.02.

338. Ghali, R., Khlifa, K.H., Ghorbel, H., Maaroufi, K., Hedilli, A., 2010: Aflatoxin determination in commonly consumed foods in Tunisia. Journal of the Science of Food and Agriculture 90, 2347-2351. doi: https://doi.org/10.1002/jsfa.4069.

339. Nonaka, Y., Saito, K., Hanioka, N., Narimatsu, S., Kataoka, H., 2009: Determination of aflatoxins in food samples by automated on-line in-tube solid-phase microextraction coupled with liquid chromatography-mass spectrometry. Journal of Chromatography A 1216, 4416-4422. doi: https://doi.org/10.1016/j.chroma.2009.03.035.

340. Ghali, R., Hmaissia-khlifa, K., Ghorbel, H., Maaroufi, K., Hedili, A., 2009: HPLC determination of ochratoxin A in high consumption Tunisian foods. Food Control 20, 716-720. doi: https://doi.org/10.1016/j.foodcont.2008.09.004.

341. Zinedine, A., Betbeder, A.-M., Faid, M., Benlemlih, M., Idrissi, L., Creppy, E.E., 2004: Ochratoxin A, determination in dried fruits and black olives from Morocco. Alimentaria 359, 73-76.

342. Arranz, I., Stroka, J., Neugebauer, M., 2006: Determination of aflatoxin B_1 in tiger nut-based soft drinks. Food Additives and Contaminants 23, 305-308. doi: https://doi.org/10.1080/02652030500415652.

343. Noba, S., Uyama, A., Mochizuki, N., 2009: Determination of ochratoxin A in ready-to-drink coffee by immunoaffinity cleanup and liquid chromatography-tandem mass spectrometry. Journal of Agricultural and Food Chemistry 57, 6036-6040. doi: https://doi.org/10.1021/jf900546p.

344. Selim, M.I., Popendorf, W., Ibrahim, M.S., El Sharkawy, S., El Kashory, E.S., 1996: Aflatoxin B_1 in common Egyptian foods. Journal of the Association of Official Analytical Chemists International 79, 1124-1129.

345. Maaroufi, K., Achour, A., Betbeder, A.M., Hammami, M., Ellouz, F., Creppy, E.E., Bacha, H., 1995: Foodstuffs and human blood contamination by the mycotoxin ochratoxin A: correlation with chronic interstitial nephropathy in Tunisia. Archives of Toxicology 69, 552-558.

346. Alozie, T.C., Rotimi, C.N., Oyibo, B.B., 1980: Production of aflatoxin by *Aspergillus flavus* (UBMI) in some Nigerian indigenous beverages and foodstuffs. Mycopathologia 70, 125-128.

347. Haydar, M., Benelli, L., Brera, C., 1990: Occurrence of aflatoxins in Syrian foods and foodstuffs: a preliminary study. Food Chemistry 37, 261-268.

348. Steiner, E.W., Rieker, R.H., Battaglia, R., 1988: Aflatoxin contamination in dried figs: distribution and association with fluorescence. Journal of Agricultural and Food Chemistry 36, 88-91.

349. Senyuva, H.Z., Gilbert, J., Ozcan, S., Ulken, U., 2005: Survey for co-occurrence of ochratoxin A and aflatoxin B_1 in dried figs in Turkey by using a single laboratory-validated alkaline extraction method for ochratoxin A. Journal of Food Protection 68, 1512-1515 .

350. Steiner, W., Brunschweiler, K., Leimbacher, E., Schneider, R., 1993: Aflatoxin B_1 and G_1, cyclopiazonic acid, kojic acid and ochratoxin A in dried figs showing BGY-fluorescence. Mitteilungen aus dem Gebiete der Lebensmitteluntersuchung und Hygiene 84, 523-536.

351. Özay, G., Alperden, I., 1991: Aflatoxin and ochratoxin—a contamination of dried figs (*Ficus carica* L.) from the 1988 crop. Mycotoxin Research 7, 85-91.

352. Juan, C., Zinedine, A., Moltó, J.C., Idrissi, L., Mañes, J., 2008: Aflatoxin levels in dried fruits and nuts from Rabat-Salé area, Morocco. Food Control 19, 849-853. doi: https://doi.org/10.1016/j.foodcont.2007.08.010.

353. Heperkan, D., Somuncuoglu, S., Karbancioglu-Güler, F., Mecik, N., 2012: Natural contamination of cyclopiazonic acid in dried figs and co-occurrence of aflatoxin. Food Control 23, 82-86. doi: https://doi.org/10.1016/j.foodcont.2011.06.015.

354. Şenyuva, H.Z., Gilbert, J., Samson, R.A., Özcan, S., Öztürkoğlu, Ş., Önal, D., 2008: Occurrence of fungi and their mycotoxins in individual Turkish dried figs. World Mycotoxin Journal 1, 79-86.

355. Senyuva, H.Z., Gilbert, J., Ulken, U., 2007: Aflatoxins in Turkish dried figs intended for export to the European Union. Journal of Food Protection 70, 1029-1032.

356. Iamanaka, B.T., De Menezes, H.C., Vicente, E., Leite, R.S.F., Taniwaki, M.H., 2007: Aflatoxigenic fungi and aflatoxins occurrence in sultanas and dried figs commercialized in Brazil. Food Control 18, 454-457. doi: https://doi.org/10.1016/j.foodcont.2005.12.002.

357. Roy, R., 1990: Aflatoxins in imported figs. Acta Horticulturae 269, 521-525.

358. Bircan, C., Barringer, S.A., Ulken, U., Pehlivan, R., 2008: Increased aflatoxin contamination of dried figs in a drought year. Food Additives and Contaminants 25, 1400-1408. doi: https://doi.org/10.1080/02652030802163414.

359. Sharman, M., Patey, A.L., Bloomfield, D.A., Gilbert, J., 1991: Surveillance and control of aflatoxin contamination of dried figs and fig paste imported into the United Kingdom. Food Additives and Contaminants 8, 299-304.

360. Bircan, C., Barringer, S.A., Ulken, U., Pehlivan, R., 2008: Aflatoxin levels in dried figs, nuts and paprika for export from Turkey. International Journal of Food Science and Technology 43, 1492-1498. doi: https://doi.org/10.1111/j.1365-2621.2008.01726.x.

361. Adedeji, B.S., Ezeokoli, O.T., Ezekiel, C.N., Obadina, A.O., Somorin, Y.M., Sulyok, M., Adeleke, R.A., Warth, B., Nwangburuka, C.C., Omemu, A.M., Oyewole, O.B., Krska, R., 2017: Bacterial species and mycotoxin contamination associated with locust bean, melon and their fermented products in south-western Nigeria. International Journal of Food Microbiology 258, 73-80. doi:https://doi.org/10.1016/j.ijfoodmicro.2017.07.014.

362. Heperkan, D., Karbancioglu Güler, F., Oktay, H.I., 2012: Mycoflora and natural occurrence of aflatoxin, cyclopiazonic acid, fumonisin and ochratoxin in dried figs. Food Additives and Contaminants Part: A 29, 277-286. doi: https://doi.org/10.1080/19440049.2011.597037.

363. Baumann, U., Zimmerli, B., 1988: A simple determination of ochratoxin A in foods. Mitteilungen aus dem Gebiete der Lebensmitteluntersuchung und Hygiene 79, 151-158.

364. Zinedine, A., Soriano, J.M., Juan, C., Mojemmi, B., Moltó, J.C., Bouklouze, A., Cherrah, Y., Idrissi, L., El Aouad, R., Mañes, J., 2007: Incidence of ochratoxin A in rice and dried fruits from Rabat and Salé area, Morocco. Food Additives and Contaminants 24, 285-291. doi: https://doi.org/10.1080/02652030600967230.

365. Karbancioğlu-Güler, F., Heperkan, D., 2008: Natural occurrence of ochratoxin A in dried figs. Analytica Chimica Acta 617, 32-36. doi: https://doi.org/10.1016/j.aca.2008.01.009.

366. Pavón, M.Á., González, I., De La Cruz, S., Martín, R., Lacarra, T.G., 2012: The use of high-performance liquid chromatography to detect ochratoxin A in dried figs from the Spanish market. Journal of the Science of Food and Agriculture 92, 74-77. doi: https://doi.org/10.1002/jsfa.4543.
367. Kosoglu, I., Aksoy, U., Pehlivan, R., 2011: Fumonisin B_1 and B_2 occurrence in dried fig fruits (*Ficus carica* L.) under Meander Valley's climatic conditions and relationship with fruit quality. Food Additives and Contaminants: Part A 28, 1569-1577. doi: https://doi.org/10.1080/19440049.2011.593190.
368. Baydar, T., Engin, A.B., Girgin, G., Aydin, S., Şahin, G., 2005: Aflatoxin and ochratoxin in various types of commonly consumed retail ground samples in Ankara, Turkey. Annals of Agricultural and Environmental Medicine 12, 193-197.
369. Xiang, L., Gao, Y.H., Liu, D.Y., Yang, M.H., 2012: Rapid determination of patulin in medicinal hawthorn fruits by HPLC-MS/MS. World Mycotoxin Journal 5, 31-36. doi: https://doi.org/10.3920/WMJ2011.1336.
370. Subrahmanyam, P., Rao, A.S., 1973: Occurrence of aflatoxins in some pickled foods. Indian Journal of Microbiology 13, 157-159.
371. Sharma, Y.P., Sumbali, G., 1999: Incidence of aflatoxin producing strains and aflatoxin contamination in dry fruit slices of quinces (*Cydonia oblonga* Mill.) from the Indian state of Jammu and Kashmir. Mycopathologia 148, 103-107.
372. Sharma, Y.P., Sumbali, G., 1999: Natural occurrence of ochratoxin A and toxigenic *Aspergillus ochraceus* strains in dry fruit slices of quinces from Jammu and Kashmir. Indian Phytopathology 52, 148-150.
373. Pallarés, N., Font, G., Mañes, J., Ferrer, E., 2017: Multimycotoxin LC-MS/MS analysis in tea beverages after dispersive liquid-liquid microextraction (DLLME). Journal of Agricultural and Food Chemistry 65, 10282-10289. doi: https://doi.org/10.1021/acs.jafc.7b03507.
374. Sinha, K.K., Singh, A., 1982: Aflatoxin producing potentials of *Aspergillus flavus* and natural occurrences of aflatoxins in some fruits and vegetables. National Academy Science Letters 5, 213-214.
375. Rani, N., Singh, S., 1996: Natural occurrence of mycoflora and mycotoxins in some pulse products of human use. Journal of the Indian Botanical Society 75, 283-285.
376. Schwartzbord, J.R., Brown, D.L., 2015: Aflatoxin contamination in Haitian peanut products and maize and the safety of oils processed from contaminated peanuts. Food Control 56, 114-118. doi: https://doi.org/10.1016/j.foodcont.2015.03.014.
377. Rosa, C.A.R., Magnoli, C.E., Fraga, M.E., Dalcero, A.M., Santana, D.M.N., 2004: Occurrence of ochratoxin A in wine and grape juice marketed in Rio de Janeiro, Brazil. Food Additives and Contaminants 21, 358-364. doi: https://doi.org/10.1080/02652030310001639549.
378. Araújo, A.C.P., Telles, D.L., Das Graças Almeida, M., Nascimento, E.S., 1994: Determination of aflatoxins in peanuts marketed in Recife-PE, Brazil. Alimentaria 249, 35-37 .
379. Bhat, R.V., Vasanthi, S., Sashidhar Rao, B., Nageswara Rao, R., Sudershan Rao, V., Nagaraja, K.V., Girija Bai, R., Krishna Prasad, C.A., Vanchinathan, S., Rajat Roy, Subrata Saha, Mukherjee, A., Ghosh, P.K., Toteja, G.S., Saxena, B.N., 1996: Aflatoxin B_1 contamination in groundnut samples collected from different geographical regions of India: a multicentre study. Food Additives and Contaminants 13, 325-331.
380. Boutrif, E., Jemmali, M., Campbell, A.D., Pohland, A.E., 1977: Aflatoxin in Tunisian foods and foodstuffs. Annales de la Nutrition et de l'Alimentation 31, 431-434.
381. Kershaw, S.J., 1985: Aflatoxins in imported edible nuts: some data 1982-1984. Journal of Food Technology 20, 647-649
382. El-Gohary, A.H., 1995: Study on aflatoxins in some foodstuffs with special reference to public health hazard in Egypt. Asian-Australasian Journal of Animal Science 8, 571-575
383. El-Khadem, M., 1990: Aflatoxins in Egyptian peanut. An overview. Journal of Plant Diseases and Protection 97, 233-236.
384. Hetmanski, M.T., Scudamore, K.A., 1989: A simple quantitative HPLC method for determination of aflatoxins in cereals and animal feedstuffs using gel permeation chromatography clean-up. Food Additives and Contaminants 6, 35-48.
385. Sanchis, V., Sala, N., Palomes, A., Santamarina, P., Burdaspal, P.A., 1986: Occurrence of aflatoxin and aflatoxigenic molds in foods and feed in Spain. Journal of Food Protection 49, 445-448, 454.
386. Singh, T., Tyagi, R.P.S., Varma, B.K., 1982: Study of the occurrence of aflatoxin B_1 in foodgrains. Journal of Food Science and Technology 19, 35-37.
387. Chu, F.S., Fan, T.S.L., Zhang, G.-S., Xu, Y.-C., Faust, S., McMahon, P.L., 1987: Improved enzyme-linked immunosorbent assay for aflatoxin B_1 in agricultural commodities. Journal of the Association of Official Analytical Chemists 70, 854-857.
388. Martins-Maciel, E.R., Machinski, M., Jr., Pereira, C.R.S., Takahachi, G., Kemmelmeier, C., Nishiyama, P., 1996: Incidence of aflatoxins and *Aspergillus flavus* in peanuts consumed in Maringá City, Brazil. Arquivos de Biologia e Technologia 39, 807-813.
389. Mohammed, A., Chala, A., Dejene, M., Fininsa, C., Hoisington, D.A., Sobolev, V.S., Arias, R.S., 2016: *Aspergillus* and aflatoxin in roundnut (*Arachis hypogaea* L.) and groundnut cake in eastern Ethiopia. Food Additives and Contaminants: Part B 9, 290-298. doi: https://doi.org/10.1080/19393210.2016.1216468.
390. Mphande, F.A., Siame, B.A., Taylor, J.E., 2004: Fungi, aflatoxins, and cyclopiazonic acid associated with peanut retailing in Botswana. Journal of Food Protection 67, 96-102.
391. Blesa, J., Soriano, J.M., Moltó, J.C., Marín, R., Mañes, J., 2003: Determination of aflatoxins in peanuts by matrix solid-phase dispersion and liquid chromatography. Journal of Chromatography A 1011, 49-54. doi: https://doi.org/10.1016/S0021-9673(03)01102-6.
392. Garcia, L.P., Savi, G.D., Santos, K., Scussel, V.M., 2016: Fumonisins and fungi in dry soybeans (*Glycine max* L.) for human consumption. Food Additives and Contaminants: Part B 9, 79-84. doi: https://doi.org/10.1080/19393210.2015.1135481.
393. Asghar, M.A., Zahir, E., Rantilal, S., Ahmed, A., Iqbal, J., 2016: Aflatoxins in composite spices collected from local markets of Karachi, Pakistan. Food Additives and Contaminants: Part B 9, 113-119. doi: https://doi.org/10.1080/19393210.2016.1150356.
394. Sangare-Tigori, B., Moukha, S., Kouadio, H.J., Betbeder, A.-M., Dano, D.S., Creppy, E.E., 2006: Co-occurrence of aflatoxin B_1, fumonisin B_1, ochratoxin A and zearalenone in cereals and peanuts from Côte d'Ivoire. Food Additives and Contaminants 23, 1000-1007. doi: https://doi.org/10.1080/02652030500415686.
395. Varsavsky, E., Sommer, S.E., 1977: Determination of aflatoxins in peanuts. Annales de la Nutrition et de l'Alimentation 31, 539-544.
396. Urano, T., Trucksess, M.W., Beaver, R.W., Wilson, D.M., Dorner, J.W., Dowell, F.E., 1992. Co-occurrence of cyclopiazonic acid and aflatoxins in corn and peanuts. Journal of the Association of Official Analytical Chemists International 75, 838-841.
397. Kollia, E., Tsourouflis, T., Markaki, P., 2016: Aflatoxin B_1 in sesame seeds and sesame products from the Greek market. Food Additives and Contaminants: Part B 9, 217-222. doi: https://doi.org/10.1080/19393210.2016.1179349.

398. Bankole, S.A., Ogunsanwo, B.M., Eseigbe, D.A., 2005: Aflatoxins in Nigerian dry-roasted groundnuts. Food Chemistry 89, 503-506. doi: https://doi.org/10.1016/j.foodchem.2004.03.004.
399. Lipigorngoson, S., Limtrakul, P., Suttajit, M., Yoshizawa, T., 2003: In-house direct cELISA for determining aflatoxin B_1 in Thai corn and peanuts. Food Additives and Contaminants 20, 838-845. doi: https://doi.org/10.1080/0265203031000156060.
400. Lopez, A., Crawford, M.A., 1967: Aflatoxin content of groundnuts sold for human consumption in Uganda. Lancet 2, 1351-1354.
401. Sartori, A.,V., Swensson de Mattos, J., Pereira Souza, Y., Pereira dos Santos, R., De Moares, M.H.P., Wanderley da Nóbrega, A., 2015: Determination of aflatoxins M_1, M_2, B_1, B_2, G_1 and G_2 in peanut by modified QuEChERS method and ultra-high performance liquid chromatography-tandem mass spectrometry. Vigilância Sanitária em Debate: Sociedade, Ciência & Tecnologia 3, 115-121. doi: https://doi.org/10.3395/2317-269X.00406.
402. Wang, D., Zhang, Z., Peiwu, L., Zhang, Q., Zhang, W., 2016: Time-resolved fluorescent immunochromatography of aflatoxin B_1 in soybean sauce: a rapid and sensitive quantitative analysis. Sensors 16, 1094, doi: https://doi.org/10.3390/s16071094.
403. Okumura, H., Kawamura, O., Kishimoto, S., Hasegawa, A., Shrestha, S.M., Okuda, K., Obata, H., Okuda, H., Haruki, K., Uchida, T., Ogasawara, Y., Ueno, Y., 1993: Aflatoxin M_1 in Nepalese sera, quantified by combination of monoclonal antibody immunoaffinity chromatography and enzyme-linked immunosorbent assay. Carcinogenesis 14, 1233-1235.
404. Peers, F.G., Linsell, C.A., 1975: Aflatoxin contamination and its heat stability in Indian cooking oils. Tropical Science 17, 229-232.
405. Verma, R.K., Agarwal, R.K., 2000: Occurrence of aflatoxin in groundnut and groundnut cake available at distributors and retailers of Uttar Pradesh. Indian Journal of Animal Nutrition 17, 156-159.
406. Sun, G., Wang, S., Hu, X., Su, J., Zhang, Y., Xie, Y., Zhang, H., Tang, L., Wang, J.-S., 2011: Co-contamination of aflatoxin B_1 and fumonisin B_1 in food and human dietary exposure in three areas of China. Food Additives and Contaminants: Part A 28, 461-470. doi: https://doi.org/10.1080/19440049.2010.544678.
407. Okano, K., Tomita, T., Kumeda, Y., Matsumaru, K., Ichinoe, M., 2008: Aflatoxin B and G contamination in imported peanuts and *Aspergillus* section *Flavi* of the causing fungi. Mycotoxins 58, 107-114.
408. Nakai, V.K., De Oliveira Rocha, L., Gonçalez, E., Fonseca, H., Ortega, E.M.M., Corrêa, B., 2008: Distribution of fungi and aflatoxins in a stored peanut variety. Food Chemistry 106, 285-290. doi: https://doi.org/10.1016/j.foodchem.2007.05.087.
409. Basaran, P., Ozcan, M., 2009: Occurrence of aflatoxins in various nuts commercialized in Turkey. Journal of Food Safety 29, 95-105.
410. Ding, X., Li, P., Bai, Y., Zhou, H., 2012: Aflatoxin B_1 in post-harvest peanuts and dietary risk in China. Food Control 23, 143-148. doi:https://doi.org/10.1016/j.foodcont.2011.06.026.
411. Wang, J., Liu, X.-M., 2007: Contamination of aflatoxins in different kinds of foods in China. Biomedical and Environmental Sciences 20, 483-487.
412. Habish, H.A., Abdulla, M.H., Broadbent, J.H., 1972: The incidence of aflatoxin in Sudanese groundnuts. Tropical Science 13, 279-287.
413. Joffe, A.Z., 1970: The presence of aflatoxin in kernels from five years groundnut crops and of *Aspergillus flavus* isolates from kernels and soils. Plant and Soil 33, 91-96.
414. Hickert, S., Hermes, L., Marques, L.M.M., Focke, C., Cramer, B., Lopes, N.P., Flett, B., Humpf, H.-U., 2017: *Alternaria* toxins in South African sunflower seeds: cooperative study. Mycotoxin Research 33, 309-321. doi: https://doi.org/10.1007/s12550-017-0290-1.
415. Rodríguez-Carrasco, Y., Mañes, J., Berrada, H., Juan, C., 2016: Development and validation of a LC-ESI-MS/MS method for the determination of *Alternaria* toxins alternariol, alternariol methyl-ether and tentotoxin in tomato and tomato-based products. Toxins (Basel) 8 (11), 328. doi: https://doi.org/10.3390/toxins8110328.
416. Dickens, J.W., Welty, R.E., 1969: Detecting farmers´ stock peanuts containing aflatoxin by examination for visible growth of *Aspergillus flavus*. Mycopathologia et Mycologia Applicata 37, 65-69.
417. Liu, Z.-h., Tu, W.-s., Li, D.-r., Li, Y.-d., Xie, C.-h., Yang, Y.-z., Qin, B.-b., 1990: A new method for the quantitation of aflatoxin M_1 in urine by high performance liquid chromatography and its application to the etiologic study of hepatoma. Biomedical Chromatography 4, 83-86.
418. Mutegi, C.K., Ngugi, H.K., Hendriks, S.L., Jones, R.B., 2009: Prevalence and factors associated with aflatoxin contamination of peanuts from western Kenya. International Journal of Food Microbiology 130, 27-34. doi: https://doi.org/10.1016/j.ijfoodmicro.2008.12.030.
419. Sabino, M., Inomata, E.I., Lamardo, L.C.A., Milanez, T.V., Navas, S.A., Zorzetto, M.A.P., 1999: A survey of the occurrence of aflatoxins in groundnuts (peanuts) and groundnut products in São Paulo State/Brazil in 1994. Revista do Instituto Adolfo Luz 58, 53-57.
420. Azer, M., Cooper, C., 1991: Determination of aflatoxins in foods using HPLC and a commercial ELISA system. Journal of Food Protection 54, 291-294.
421. Freitas, V.P.S., Brigido, B.M., 1998: Occurrence of aflatoxins B_1, B_2, G_1, and G_2 in peanuts and their products marketed in the region of Campinas, Brazil in 1995 and 1996. Food Additives and Contaminants 15, 807-811.
422. Meyer, K., Müller, T., Dietrich, R., Märtlbauer, E., Bauer, J., 2001: Nachweis und Vorkommen von Mykophenolsäure und Citrinin in Rotwein. Mycotoxin Research, 17 (Suppl. 2) 160-164.
423. Abdulkadar, A.H.W., Al-Ali, A., Al-Jedah, J.H., 2002: Occurrence of aflatoxin in commodities imported into Qatar, 1997-2000. Food Additives and Contaminants 19, 666-670. doi: https://doi.org/10.1080/02652030210121339.
424. Abdulkadar, A.H.W., Al-Ali, A., Al-Jedah, J., 2000: Aflatoxin contamination in edible nuts imported in Qatar. Food Control 1, 157-160.
425. Hag Elamin, N.H., Abdel-Rahim, A.M., Khalid, A.E., 1988: Aflatoxin contamination of groundnuts in Sudan. Mycopathologia 104, 25-31.
426. Landsen, J.A., Davidson, J.I., 1983: Occurrence of cyclopiazonic acid in peanuts. Applied and Environmental Microbiology 45, 766-769.
427. Subrahmanyam, P., Rao, A.S., 1974: Occurrence of aflatoxins and citrinin in groundnut (*Arachis hypogaea* L.) at harvest in relation to pod condition and kernel moisture content. Current Science 43, 707-710.
428. Magnoli, C., Astoreca, A., Ponsone, M.L., Fernández-Juri, M.G., Barberis C., Dalcero, A.M., 2007: Ochratoxin A and *Aspergillus* section *Nigri* in peanut seeds at different months of storage in Córdoba, Argentina. International Journal of Food Microbiology 119, 213-218. doi: https://doi.org/10.1016/j.ijfoodmicro.2007.07.056.
429. Sangare-Tigori, B., Dem, A.A., Kouadio, H.J., Betbeder, A.-M., Dano, D.S., Moukha, S., Creppy, E.E., 2006: Preliminary survey of ochratoxin A in millet, maize, rice and peanuts in Côte d´Ivoire from 1998 to 2002. Human & Experimental Toxicology 25, 211-216. doi: https://doi.org/10.1191/0960327106ht605oa.
430. Rati Rao, E., Basappa, S.C., Sreenivasa Murthy, V., 1979: Studies on the occurrence of ochratoxins in foodgrains. Journal of Food Science and Technology 16, 113-114.

431. Magnoli, C., Astoreca, A., Ponsone, L., Fernández-Juri, M.G., Chiacchiera, S., Dalcero, A., 2006: Ochratoxin A and the occurrence of ochratoxin A-producing black Aspergilli in stored peanut seeds from Córdoba, Argentina. Journal of the Science of Food and Agriculture 86, 2369-2373. doi: https://doi.org/10.1002/jsfa.2625.
432. Coronel, M.B., Marín, S., Cano-Sancho, G., Ramos, A.J., Sanchis, V., 2012: Exposure assessment to ochratoxin A in Catalonia (Spain) based on the consumption of cereals, nuts, coffee, wine, and beer. Food Additives and Contaminants: Part A 29, 979-993. doi: https://doi.org/10.1080/19440049.2012.660708.
433. Gamanya, R., Sibanda, L., 2001: Survey of *Fusarium moniliforme* (*F. verticillioides*) and production of fumonisin B_1 in cereal grains and oilseeds in Zimbabwe. International Journal of Food Microbiology 71, 145-149.
434. Bhavanishankar, T.N., Shantha, T., 1987: Natural occurrence of *Fusarium* toxins in peanut, sorghum and maize from Mysore (India). Journal of the Science of Food and Agriculture 40, 327-332.
435. Blesa, J., Soriano, J.M., Moltó, J.C., Mañes, J., 2004: Limited survey for the presence of aflatoxins in foods from local markets and supermarkets in Valencia, Spain. Food Additives and Contaminants 21, 165-171. doi: https://doi.org/10.1080/0265203031000.
436. Abdel-Hafez, A.I.I., Saber, S.M., 1993: Mycoflora and mycotoxins of hazelnut (*Corylus avellana* L.) and walnut (*Juglans regia* L.) seeds in Egypt. Zentralblatt für Mikrobiologie 148, 137-147.
437. Var, I., Kabak, B., Gök, F., 2007: Survey of aflatoxin B_1 in helva, a traditional Turkish food, by TLC. Food Control 18, 59-62. doi: https://doi.org/10.1016/j.foodcont.2005.08.008.
438. Spadaro, D., Ciavorella, A., Frati, S., Garibaldi, A., Gullino, M.L., 2007: Incidence and level of patulin contamination in pure and mixed apple juices marketed in Italy. Food Control 18, 1098-1102. doi: https://doi.org/10.1016/j.foodcont.2006.07.007.
439. Ackermann, Y., Curtui, V., Dietrich, R., Gross, M., Latif, H., Märtlbauer, E., Usleber, E., 2011: Widespread occurrence of low levels of alternariol in apple and tomato products, as determined by comparative immunochemical assessment using monoclonal and polyclonal antibodies. Journal of Agricultural and Food Chemistry 59, 6360-6368. doi: https://doi.org/10.1021/jf201516f.
440. Lau, B. P.-Y., Scott, P.M., Lewis, D.A., Kanhere, S.R., Cléroux, C., Roscoe, V.A., 2003: Liquid chromatography-mass spectrometry and liquid chromatography-tandem mass spectrometry of the *Alternaria* mycotoxins alternariol and alternariol monomethyl ether in fruit juices and beverages. Journal of Chromatography A 998, 119-131. doi: https://doi.org/10.1016/S0021-9673(03)00606-X.
441. Wu, L.X, Ding, X.X., Li, P.W., Du, X.H., Zhou, H.Y., Bai, Y.Zh., Zhang, L.X., 2016: Aflatoxin contamination of peanuts at harvest in China from 2010 to 2013 and its relationship with climatic conditions. Food Control 60, 117-123. doi: https://doi.org/10.1016/j.foodcont.2015.06.029.
442. Gross, M., Curtui, V., Ackermann, Y., Latif, H., Usleber, E., 2011: Enzyme immunoassay for tenuazonic acid in apple and tomato products. Journal of Agricultural and Food Chemistry 59, 12317-12322. doi: https://doi.org/10.1021/jf203540y.
443. Abdel-Sater, M.A., Zohri, A.A., Ismail, M.A., 2001: Natural contamination of some Egyptian fruit juices and beverages by mycoflora and mycotoxins. Journal of Food Science and Technology 38, 407-411.
444. Al-Hazmi, N.A., 2010: Determination of patulin and ochratoxin A using HPLC in apple juice samples in Saudi Arabia. Saudi Journal of Biological Sciences 17, 353-359. doi: https://doi.org/10.1016/j.sjbs.2010.06.006.
445. Hammami, W., Fiori, S., Al Thani, R., Kali, N.A., Balmas, V., Migheli, Q., Jaoua, S., 2014: Fungal and aflatoxin contamination of marketed spices. Food Control 37, 177-181. doi: https://doi.org/10.1016/j.foodcont.2013.09.027.
446. Gökmen, V., Acar, J., 2000: Long-term survey of patulin in apple juice concentrates produced in Turkey. Food Additives and Contaminants 17, 933-936.
447. Yurdun, T., Omurtag, G.Z., Ersoy, Ö., 2001: Incidence of patulin in apple juices marketed in Turkey. Journal of Food Protection 64, 1851-1853.
448. Leggott, N.L., Shephard, G.S., Stockenström, S., Staal, E., Van Schalkwyk, D.J., 2001: The reduction of patulin in apple juice by three different types of activated carbon. Food Additives and Contaminants 18, 825-829. doi: https://doi.org/10.1080/02652030110044903.
449. Çağındı, Ö., Gürhayta, O.F., 2016: Aflatoxins and ochratoxin A in dried eggplant and green bell pepper. Food Control 70, 216-220. doi: https://doi.org/10.1016/j.foodcont.2016.05.060.
450. Rosen, J.D., Pareles, S.R., 1974: Quantitative analysis of patulin in apple juice. Journal of Agricultural and Food Chemistry 22, 1024-1026.
451. Josefsson, E., Andersson, A., 1976: Analysis of patulin in apple beverages sold in Sweden. Vår Föda 8, 189-196.
452. Murillo, M., Gonzaléz-Peñas, E., Amézqueta, S., 2008: Determination of patulin in commercial apple juice by micellar electrokinetic chromatography. Food and Chemical Toxicology 46, 57-64. doi: https://doi.org/10.1016/j.fct.2007.06.024.
453. Hariprasad, P., Vipin, A.V., Karuna, S., Raksha, R.K., Venkateswaran, G., 2015: Natural aflatoxin uptake by sugarcane (*Saccharum officinaurum* L.) and its persistence in jaggery. Environmental Science and Pollution Research 22, 6246-6253. doi: https://doi.org/10.1007/s11356-014-3851-2.
454. Brackett, R.E., Marth, E.H., 1979: Patulin in apple juice from roadside stands in Wisconsin. Journal of Food Protection 42, 862-863.
455. Mortimer, D.N., Parker, I., Shepherd, M.J., Gilbert, J., 1985: A limited survey of retail apple and grape juices for the mycotoxin patulin. Food Additives and Contaminants 2, 165-170.
456. Prieta, J., Moreno, M.A., Díaz, S., Suárez, G., Domínguez, L., 1994: Survey of patulin in apple juice and children´s apple food by the diphasic dialysis membrane procedure. Journal of Agricultural and Food Chemistry 42, 1701-1703.
457. Scott, P.M., Miles, W.F., Toft, P., Dubé, J.G., 1972: Occurrence of patulin in apple juice. Journal of Agricultural and Food Chemistry 20, 450-451.
458. Sheu, F., Shyu, Y.T., 1999: Analysis of patulin in apple juice by diphasic dialysis extraction with in situ acylation and mass spectrometric determination. Journal of Agricultural and Food Chemistry 47, 2711-2714. doi: https://doi.org/10.1021/jf9812329.
459. Stray, H., 1978: High pressure liquid chromatographic determination of patulin in apple juice. Journal of the Association of Official Analytical Chemists 61, 1359-1362.
460. Ware, G. M., Thorpe, C.W., Pohland, A.E., 1974: Liquid chromatographic method for determination of patulin in apple juice. Journal of the Association of Official Analytical Chemists 57, 1111-1113.
461. Watkins, K.L., Fazekas, G., Palmer, M.V., 1990: Patulin in Australian apple juice. Food Australia 42, 438-439.
462. Wilson, R.D., 1981: Surveying some apple juice for patulin. Food Technology in New Zealand 6, 27-31.

463. Sinha, A.K., Ranjan, K.S., 1989: Preharvest aflatoxin problem in linseed (*Linum usitatissimum*) in Bihar. Journal of the Indian Botanical Society 68, 19-20.
464. Roach, J.A.G., Brause, A.R., Eisele, T.A., Rupp, H.S., 2002: HPLC detection of patulin in apple juice with GC/MS confirmation of patulin identity. Advances in Experimental Medicine and Biology 504, 135-140.
465. Manchisky, M., Jr., Midio, A.F., 1996: Incidencia de patulina en jugo de manzana industrializado. Alimentaria 276, 61-64.
466. MAFF—UK, 1993: Patulin in apple juice: 1993 survey. Food-Surveillance-Information-Sheet No. 16.
467. MAFF—UK, 1994: 1994 Survey of apple juice for patulin. Food-Surveillance-Information-Sheet No. 36.
468. MAFF—UK, 1995: 1995 Survey of apple juice for patulin. Food-Surveillance-Information-Sheet No. 74.
469. MAFF—UK, 1996: 1996 Survey of apple juice for patulin. Food-Surveillance-Information-Sheet No. 97.
470. MAFF—UK, 1999: 1998 survey of apple juice for patulin. Food-Surveillance-Information-Sheet No. 173.
471. Watanabe, M., Shimizu, H., 2005: Detection of patulin in apple juice marketed in the Tohoku district, Japan. Journal of Food Protection 68, 610-612.
472. Baert, K., De Meulenaer, B., Kamala, A., Kasase, C., Devlieghere, F., 2006: Occurrence of patulin in organic, conventional, and handcrafted apple juices marketed in Belgium. Journal of Food Protection 69, 1371-1378.
473. Rupp, H.S., Turnipseed, S.B., 2000: Confirmation of patulin and 5-hydroxymethylfurfural in apple juice by gas chromatography/mass spectrometry. Journal of the Association of Analytical Chemists International 83, 612-620.
474. Cao, J., Zhou, S., Kong, W., Ma, X., Yang, M., Wan, L., Yang, S., 2014: Simultaneous determination of aflatoxins B_1, B_2, G_1, G_2 in *Fructus Bruceae* by high-performance liquid chromatography with online postcolumn photochemical derivatization. Journal of Separation Science 37, 2771-2778. doi: https://doi.org/10.1002/jssc.201400501.
475. Ito, R., Yamazaki, H., Inoue, K., Yoshimura, Y., Kawaguchi, M., Nakazawa, H., 2004: Development of liquid chromatography-electrospray mass spectrometry for the determination of patulin in apple juice: investigation of its contamination levels in Japan. Journal of Agricultural and Food Chemistry 52, 7464-7468. doi: https://doi.org/10.1021/jf049264l.
476. Majid Cheraghali, A., Mohammadi, H.R., Amirahmadi, M., Yazdanpanah, H., Abouhossain, G., Zamanian, F., Khansari, M.G., Afshar, M., 2005: Incidence of patulin contamination in apple juice produced in Iran. Food Control 16, 165-167. doi: https://doi.org/10.1016/j.foodcont.2004.01.006.
477. Tarter, E.J., Scott, P.M., 1991: Determination of patulin by capillary gas chromatography of the heptafluorobutyrate derivative. Journal of Chromatography 538, 441-446.
478. Iha, M.H., Sabino, M., 2008: Incidence of patulin in Brazilian apple-based drinks. Food Control 19, 417-422. doi: https://doi.org/10.1016/j.foodcont.2007.05.001.
479. Karimi, G., Hassanzadeh, M., Yazdanpanah, H., Nazari, F., Iranshahi, M., Nili, A., 2008: Contamination of patulin in clear apple juice in Mashhad, Iran. Journal of Food Safety 28, 413-421.
480. Kataoka, H., Itano, M., Ishizaki, A., Saito, K., 2009: Determination of patulin in fruit juice and dried fruit samples by in-tube solid-phase microextraction coupled with liquid chromatography-mass spectrometry. Journal of Chromatography A 1216, 3746-3750. doi: https://doi.org/10.1016/j.chroma.2009.03.017.
481. Katerere, D.R., Stockenström, S., Thembo, K.M., Balducci, G., Shephard, G.S., 2007: Investigation of patulin contamination in apple juice sold in retail outlets in Italy and South Africa. Food Additives and Contaminants 24, 630-634. doi: https://doi.org/10.1080/02652030601137668.
482. Li, F., Zhao, S., Chin, L., Li, Y., Wu, D., Zhao, X., Han, C., Zhang, H., Ji, R., 2007: Determination of patulin in apple and hawthorn beverages by solid-phase filtration column and liquid chromatography. Journal of the Association of Official Analytical Chemists International 90, 167-172.
483. Mhadhbi, H., Bouzouita, N., Martel, A., Zarrouk, H., 2007: Occurrence of mycotoxin patulin in apple-based products marketed in Tunisia. Journal of Food Protection 70, 2642-2645.
484. Murillo-Arbizu, M., Amĕzqueta, S., González-Peñas, E., López De Cerain, A., 2009: Occurrence of patulin and its dietary intake through apple juice consumption by the Spanish population. Food Chemistry 113, 420-423. doi: https://doi.org/10.1016/j.foodchem.2008.07.054.
485. Lai, C.-L., Fuh, Y.-M., Shih, D.Y.-C., 2000: Detection of mycotoxin patulin in apple juice. Journal of Food and Drug Analysis 8, 85-96.
486. Barreira, M.J., Alvito, P.C., Almeida, C.M.M., 2010: Occurrence of patulin in apple-based-foods in Portugal. Food Chemistry 121, 653-658. doi: https://doi.org/10.1016/j.foodchem.2009.12.085.
487. Delgado, T., Gómez-Cordovés, C., Scott, P.M., 1996: Determination of alternariol and alternariol methyl ether in apple juice using solid-phase extraction and high-performance liquid chromatography. Journal of Chromatography A 731, 109-114.
488. Delgado, T., Gómez-Cordovéz, C., 1998: Natural occurrence of alternariol and alternariol methyl ether in Spanish apple juice concentrates. Journal of Chromatography A 815, 93-97.
489. Welke, J.E., Hoeltz, M., Dottori, H.A., Noll, I.B., 2009: Quantitative analysis of patulin in apple juice by thin-layer chromatography using a charge coupled device detector. Food Additives and Contaminants: Part A 26, 754-758. doi: https://doi.org/10.1080/02652030802662746.
490. Gökmen, V., Acar, J., 1998: Incidence of patulin in apple juice concentrates produced in Turkey. Journal of Chromatography A 815, 99-102.
491. Khazaeli, P., Mehrabani, M., Heidari, M.R., Asadikaram, G., Lari Najafi, M., 2017: Prevalence of aflatoxin contamination in herbs and spices in different regions of Iran. Iranian Journal of Public Health 46, 1540-1545.
492. Scott, P.M., Lawrence, G.A., Lau, B.P.-Y., 2006: Analysis of wines, grape juices and cranberry juices for *Alternaria* toxins. Mycotoxin Research 22, 142-147
493. Filali, A., Ouammi, L., Betbeder, A.M., Baudrimont, I., Soulaymani, R., Benayada, A., Creppy, E.E., 2001: Ochratoxin A in beverages from Morocco: a preliminary survey. Food Additives and Contaminants 18, 565-568. doi: https://doi.org/10.1080/02652030110035138.
494. Czerwiecki, L., Wilczyńska, G., Kwiecień, A., 2005: Ochratoxin A: an improvement clean up and HPLC method used to investigate wine and grape juice on the Polish market. Food Additives and Contaminants 22, 158-162. doi: https://doi.org/10.1080/02652030500038066.
495. Ng, W., Mankotia, M., Pantazopoulos, P., Neil, R.J., Scott, P.M., 2004: Ochratoxin in wine and grape juice sold in Canada. Food Additives and Contaminants 21, 971-981. doi: https://doi.org/10.1080/02652030400000653.
496. Zimmerli, B., Dick, R., 1996: Ochratoxin A in table wine and grape-juice: occurrence and risk assessment. Food Additives and Contaminants 13, 655-668.
497. Woese, K., 2000: Ochratoxin A in grape juice and wine. Mycotoxin Research 16, 132-135.

498. Broggi, L., Reynoso, C., Resnik, S., Martinez, F., Drunday, V., Bernal, Á.R., 2013: Occurrence of alternariol and alternariol monomethyl ether in beverages from the Entre Rios Province market, Argentina. Mycotoxin Research 29, 17-22. doi: https://doi.org/10.1007/s12550-012-0147-6.
499. Diaz, G.J., Perilla, N.S., Rojas, Y., 2001: Occurrence of aflatoxins in selected Colombian foods. Mycotoxin Research 17, 15-20.
500. Boyd, M.L., Cotty, P.J., 2001: *Aspergillus flavus* and aflatoxin contamination of leguminous trees of the Sonoran Desert in Arizona. Phytopathology 91, 913-919.
501. El-Nagerabi, S.A.F., Elshafie, A.E., 2000: Incidence of seed-borne fungi and aflatoxins in Sudanese lentil seeds. Mycopathologia 149, 151-156.
502. Assaf, H., Betbeder, A.-M., Creppy, E.E., Pallardy, M., Azouri, H., 2004: Ochratoxin A levels in human plasma and foods in Lebanon. Human & Experimental Toxicology 23, 495-501. doi: https://doi.org/10.1191/0960327104ht481oa.
503. Pietri, A., Rastelli, S., Bertuzzi, T., 2010: Ochratoxin A and aflatoxins in liquorice products. Toxins (Basel) 2, 758-770. doi: https://doi.org/10.3390/toxins2040758.
504. Bresch, H., Urbanek, M., Nusser, M., 2000: Ochratoxin A in food containing liquorice. Nahrung 44, 276-278.
505. Ariño, A., Herrera, M., Estopañan, G., Juan, T., 2007: High levels of ochratoxin A in licorice and derived products. International Journal of Food Microbiology 114, 366-369. doi: https://doi.org/10.1016/j.ijfoodmicro.2006.08.005.
506. Herrera, M., Herrera, A., Ariño, A., 2009: Estimation of dietary intake of ochratoxin A liquorice confectionery. Food and Chemical Toxicology 47, 2002-2006. doi: https://doi.org/10.1016/j.fct.2009.05.009.
507. Sidhu, O.P, Chandra, H., Behl, H.M., 2009: Occurrence of aflatoxins in mahua (*Madhuca indica* Gmel.) seeds: synergistic effect of plant extracts on inhibition of *Aspergillus flavus* growth and aflatoxin production. Food and Chemical Toxicology 47, 774-777. doi: https://doi.org/10.1016/j.fct.2009.01.001.
508. Bellí, N., Marín, S., Duaigües, A., Ramos, A.J., Sanchis, V., 2004: Ochratoxin A in wines, musts and grape juices from Spain. Journal of the Science of Food and Agriculture 84, 591-594. doi: https://doi.org/10.1002/jsfa.1702.
509. Soufleros, E.H., Tricard, C., Bouloumpasi, E.C., 2003: Occurrence of ochratoxin A in Greek wines. Journal of the Science of Food and Agriculture 83, 173-179. doi: https://doi.org/10.1002/jsfa.1300.
510. Zimmer, I., Usleber, E., Klaffke, H., Weber, R., Majerus, P., Otteneder, H., Gareis, M., Dietrich, R., Märtlbauer, E., 2008: Fumonisin intake of the German consumer. Mycotoxin Research 24, 40-52.
511. Burdaspal, P.A., Legarda, T.M., 1999: Ochratoxin A in wines and grape products originated from Spain and other European countries. Alimentaria (Madrid) 299, 107-113.
512. Stander, M.A., Steyn, P.S., 2002: Survey of ochratoxin A in South African wines. South African Journal of Enology and Viticulture 23, 9-13.
513. Pietri, A., Bertuzzi, T., Pallaroni, L., Piva, G., 2001: Occurrence of ochratoxin A in Italian wines. Food Additives and Contaminants 18, 647-654. doi: https://doi.org/10.1080/02652030110028353.
514. Walravens, J., Mikula, H., Rychlik, M., Asam, S., Devos, T., Njumbe Ediage, E., Diana Di Mavungu, J., Jacxsens, L., Van Landschoot, A., Vanhaecke, L., De Saeger, S., 2016: Validated UPLC-MS/MS methods to quantitate free and conjugated *Alternaria* toxins in commercially available tomato products and fruit and vegetable juices in Belgium. Journal of Agricultural and Food Chemistry 64, 5101-5109. doi: https://doi.org/10.1021/acs.jafc.6b01029.
515. Festas, I., Herbert, P., Santos, L., Cabral, M., Barros, P., Alves, A., 2000: Ochratoxin A in some Portuguese wines: method validation and screening in Port Wine and Vinho Verde. American Journal of Enology and Viticulture 51, 150-154.
516. Brera, C., Debegnach, F., Minardi, V., Prantera, E., Pannunzi, E., Faleo, S., De Santis, B., Miraglia, M., 2008: Ochratoxin A contamination in Italian wine samples and evaluation of the exposure in the Italian population. Journal of Agricultural and Food Chemistry 56, 10611-10618. doi: https://doi.org/10.1021/jf8016282.
517. Blesa, J., Soriano, J.M., Moltó, J.C., Mañes, J., 2004: Concentration of ochratoxin A in wines from supermarkets and stores of Valencian community (Spain). Journal of Chromatography A 1054, 397-401. doi: https://doi.org/10.1016/j.chroma.2004.04.058.
518. Brera, C., Soriano, J.M., Debegnach, F., Miraglia, M., 2005: Exposure assessment to ochratoxin A from the consumption of Italian and Hungarian wines. Microchemical Journal 79, 109-113. doi: https://doi.org/10.1016/j.microc.2004.10.004.
519. Burdaspal, P., Legarda, T., 2007: Occurrence of ochratoxin A in sweet wines in Spain and other countries. Food Additives and Contaminants 24, 976-986. doi: https://doi.org/10.1080/02652030701311155.
520. Hernández, M.J., García-Moreno, M.V., Durán, E., Guillén, D., Barosso, C.G., 2006: Validation of two analytical methods for the determination of ochratoxin A by reversed-phase high-performance liquid chromatography coupled to fluorescence detection in musts and sweet wines from Andalusia. Analytica Chimica Acta 566, 117-121. doi: https://doi.org/10.1016/j.aca.2006.02.002.
521. Ratola, N., Martins, L., Alves, A., 2004: Ochratoxin A in wines-assessing global uncertainty associated with the results. Analytica Chimica Acta 513, 319-324. doi: https://doi.org/10.1016/j.aca.2003.11.063.
522. Visconti, A., Pascale, M., Centonze, G., 1999: Determination of ochratoxin A in wine by means of immunoaffinity column clean-up and high-performance liquid chromatography. Journal of Chromatography A 864, 89-101.
523. Siantar, D.P., Halverson, C.A., Kirmiz, C., Peterson, G.F., Hill, N.R., Dugar, S.M., 2003: Ochratoxin A in wine: survey by antibody-and polymeric-based SPE columns using HPLC/fluorescence detection. American Journal of Enology and Viticulture 54, 170-177.
524. Sahay, S.S., Prasad, T., Sinha, K.K., 1990: Postharvest incidence of aflatoxins in *Linum usitatissimum* seeds. Journal of the Science of Food and Agriculture 53, 169-174.
525. Chen, S.-C., Friedman, L., 1966: Aflatoxin determination in seed meal. Journal of the Association of Official Analytical Chemists 49, 28-33.
526. El-Nakib, O., Pestka, J.J., Chu, F.S., 1981: Determination of aflatoxin B_1 in corn, wheat, and peanut butter by enzyme-linked immunosorbent assay and solid phase radioimmunoassay. Journal of the Association of Official Analytical Chemists 64, 1077-1082.
527. Jacobsen, B.J., Harlin, K.S., Swanson, S.P., Lambert, R.J., Beasley, V.R., Sinclair, J.B., Wei, L.S., 1995: Occurrence of fungi and mycotoxins associated with field mold damaged soybeans in the Midwest. Plant Disease 79, 86-88.
528. Kim, E.-K., Shon, D.-H., Yoo, J.Y., Ryu, D., Lee, C., Kim, Y.-B., 2001: Natural occurrence of aflatoxins in Korean meju. Food Additives and Contaminants 18, 151-156. doi: https://doi.org/10.1080/02652030010006104.
529. Bankole, S.A., 1995: Mycoflora, mycotoxin producing fungi and mycotoxins from melon ball snacks in Nigeria. Chemie, Mikrobiologie & Technologie der Lebensmittel 17, 164-168.

530. Ali, N., Hashim, N.H., Shuib, N.S., 2015: Natural occurrence of aflatoxins and ochratoxin A in processed spices marketed in Malaysia. Food Additives and Contaminants: Part A 32, 518-532. doi: https://doi.org/10.1080/19440049.2015.1011712.
531. Bankole, S.A., Ogunsanwo, B.M., Mabekoje, O.O., 2004: Natural occurrence of moulds and aflatoxin B_1 in melon seeds from markets in Nigeria. Food and Chemical Toxicology 42, 1309-1314. doi: https://doi.org/10.1016/j.fct.2004.03.015.
532. Bankole, S.A., Ogunsanwo, B.M., Osho, A., Adewuyi, G.O., 2006: Fungal contamination and aflatoxin B_1 of "egusi" melon seeds in Nigeria. Food Control 17, 814-818. doi: https://doi.org/10.1016/j.foodcont.2005.05.008.
533. El Khoury, A., Rizk, T., Lteif, R., Azouri, H., Delia, M.-L., Lebrihi, A., 2008: Fungal contamination and aflatoxin B_1 and ochratoxin A in Lebanese wine-grapes and musts. Food and Chemical Toxicology 46, 2244-2250. doi: https://doi.org/10.1016/j.fct.2008.02.026.
534. Bellí, N., Pardo, E., Marín, S., Farré, G., Ramos, A.J., Sanchis, V., 2004: Occurrence of ochratoxin A and toxigenic potential of fungal isolates from Spanish grapes. Journal of the Science of Food and Agriculture 84, 541-546. doi: https://doi.org/10.1002/jsfa.1658.
535. Flajs, D., Domijan, A.-M., Ivić, D., Cvjetković, B., Peraica, M., 2009: ELISA and HPLC analysis of ochratoxin A in red wines of Croatia. Food Control 20, 590-592. doi: https://doi.org/10.1016/j.foodcont.2008.08.021.
536. El Khoury, A., Rizk, T., Lteif, R., Azouri, H., Delia, M.-L., Lebrihi, A., 2006: Occurrence of ochratoxin A- and aflatoxin B_1-producing fungi in Lebanese grapes and ochratoxin A content in musts and finished wines during 2004. Journal of Agricultural and Food Chemistry 54, 8977-8982. doi: https://doi.org/10.1021/jf062085e.
537. Martins, C., Assunção, R., Cunha, S.C., Fernandes, J.O., Jager, A., Petta, T., Oliveira, C.A., Alvito, P., 2018: Assessment of multiple mycotoxins in breakfast cereals available in the Portuguese market. Food Chemistry 239, 132-140. doi: https://doi.org/10.1016/j.foodchem.2017.06.088.
538. Lasram, S., Bellí, N., Chebil, S., Nahla, Z., Ahmed, M., Sanchis, V., Ghorbel, A., 2007: Occurrence of ochratoxigenic fungi and ochratoxin A in grapes from a Tunisian vineyard. International Journal of Food Microbiology 114, 376-379. doi: https://doi.org/10.1016/j.ijfoodmicro.2006.09.027.
539. Loizzo, M.R., Nigro, S., De Luca, D., Menichini, F., 2011: Detection of ochratoxin A and *cis*- and *trans*-resveratrol in red wines and their musts from Calabria (Italy). Food Additives and Contaminants: Part A 28, 1561-1568. doi: https://doi.org/10.1080/19440049.2011.590454.
540. Chen, Y.-C., Liao, C.-D., Lin, H.-Y., Chiueh, L.-C., Shih, D.Y.-C., 2013: Survey of aflatoxin contamination in peanut products in Taiwan from 1997 to 2011. Journal of Food and Drug Analysis 21, 247-252. doi: https://doi.org/10.1016/j.jfda.2013.07.001.
541. Sage, L., Krivobok, S., Delbos, E., Seigle-Murandi, F., Creppy, E.E., 2002: Fungal flora and ochratoxin A production in grapes and musts from France. Journal of Agricultural and Food Chemistry 50, 1306-1311. doi: https://doi.org/10.1021/jf011015z.
542. Logrieco, A., Ferracane, R., Visconti, A., Ritieni, A., 2010: Natural occurrence of fumonisin B_2 in red wine from Italy. Food Additives and Contaminants: Part A 27, 1136-1141. doi: https://doi.org/10.1080/19440041003716547.
543. Logrieco, A., Ferracane, R., Haidukowsky, M., Cozzi, G., Visconti, A., Ritieni, A., 2009: Fumonisin B_2 production by *Aspergillus niger* from grapes and natural occurrence in must. Food Additives and Contaminants: Part A 26, 1495-1500. doi: https://doi.org/10.1080/02652030903148322.
544. Sahay, S.S., Prasad, T., 1990: The occurrence of aflatoxins in mustard and mustard products. Food Additives and Contaminants 7, 509-513.
545. Sinha, A.K., Bilgrami, K.S., Ranjan, K.S., Prasad, T., 1988: Incidence of aflatoxins in mustard crop in Bihar. Indian Phytopathology 41, 434-437.
546. Beljaars, P.R., Schumans, J.C.H.M., Koken, P.J., 1975: Quantitative fluorodensitometric determination and survey of aflatoxins in nutmeg. Journal of the Association of Official Analytical Chemists 58, 263-271.
547. Murillo-Arbizu, M., González-Peñas, E., Hansen, S.H., Amézqueta, S., Østergaard, J., 2008: Development and validation of a microemulsion electrokinetic chromatography method for patulin quantification in commercial apple juice. Food and Chemical Toxicology 46, 2251-2257. doi: https://doi.org/10.1016/j.fct.2008.02.027.
548. Okano, K., Tomita, T., Ohzu, Y., Takai, M., Ose, A., Kotsuka, A., Ikeda, N., Sakata, J., Kumeda, Y., Nakamura, N., Ichinoe, M., 2012: Aflatoxins B and G contamination and aflatoxigenic fungi in nutmeg. Shokuhin Eiseigaku Zasshi 53, 211-216.
549. Kong, W.-J., Liu, S.-Y., Qiu, F., Xiao, X.-H., Yang, M.-H., 2013: Simultaneous multi-mycotoxin determination in nutmeg by ultrasound-assisted solid-liquid extraction and immunoaffinity column clean-up coupled with liquid chromatography and on-line post-column photochemical derivatization-fluorescence detection. Analyst 138, 2729-2739. doi: https://doi.org/10.1039/c3an00059a.
550. Escher, F.E., Koehler, P.E., Ayres, J.C., 1974: A study on aflatoxin and mold contaminations in improved variety pecans. Journal of Food Science 39, 1127-1129.
551. Assunção, R., Martins, C., Dupont, D., Alvito, P., 2016: Patulin and ochratoxin A co-occurrence and their bioaccessibility in processed cereal-based foods: a contribution for Portuguese children risk assessment. Food and Chemical Toxicology 96, 205-214. doi: https://doi.org/10.1016/j.fct.2016.08.004.
552. Wells, J.M., Payne, J.A., 1976: Incidence of aflatoxin contamination on a sampling of southeastern pecans. Proceedings of the Florida State Horticultural Society 89, 256-257.
553. Wells, J.M., Payne, J.A., Stoloff, L., 1983: Weather-related incidence of aflatoxin contamination in late-harvested pecans. Plant Disease 67, 751-753.
554. Schroeder, H.W., Hein, H., Jr., 1977: Natural occurrence of sterigmatocystin in in-shell pecans. Canadian Journal of Microbiology 23, 639-641.
555. Boutrif, E., Jemmali, M., Pohland, A.E., Campbell, A.D., 1977: Aflatoxin in Tunisian Aleppo pine nuts. Journal of the Association of Official Analytical Chemists 60, 747-748.
556. Scholten, J.M., Spanjer, M.C., 1996: Determination of aflatoxin B_1 in pistachio kernels and shells. Journal of the Association of Official Analytical Chemists International 79, 1360-1364.
557. Candlish, A.A.G., Pearson, S.M., Aidoo, K.E., Smith, J.E., Kelly, B., Irvine, H., 2001: A survey of ethnic foods for microbial quality and aflatoxin content. Food Additives and Contaminants 18, 129-136. doi: https://doi.org/10.1080/02652030010021404.
558. Ghali, R., Belouaer, I., Hdiri, S., Ghorbel, H., Maaroufi, K., Hedilli, A., 2009: Simultaneous HPLC determination of aflatoxins B_1, B_2, G_1 and G_2 in Tunisian sorghum and pistachios. Journal of Food Composition and Analysis 22, 751-755. doi: https://doi.org/10.1016/j.jfca.2009.04.009.

559. Ariño, A., Herrera, M., Estopañan, G., Rota, M.C., Carramiñana, J.J., Juan, T., Herrera, A., 2009: Aflatoxins in bulk and pre-packed pistachios sold in Spain and effect of roasting. Food Control 20, 811-814. doi: https://doi.org/10.1016/j.foodcont.2008.10.013.
560. Burdaspal, P.A., Legarda, T.M., Iñigo, S., 1998: A survey of aflatoxins in pistachios in the community of Madrid, Spain. Alimentaria 364, 106-113.
561. Dickens, J.W., Welty, R.E., 1975: Fluorescence in pistachio nuts contaminated with aflatoxin. Journal of the American Oil Chemists´ Society 52, 448-450.
562. Fernane, F., Sanchis, V., Marín, S., Ramos, A.J., 2010: First report on mould and mycotoxin contamination of pistachios sampled in Algeria. Mycopathologia 170, 423-429. doi: https://doi.org/10.1007/s11046-010-9332-3.
563. Kershaw, S.J., 1982: Occurrence of aflatoxins in oilseeds providing cocoa-butter substitutes. Applied and Environmental Microbiology 43, 1210-1212.
564. Bankole, S.A., Eseigbe, D.A., 1996: Occurrence of mycoflora and aflatoxins in marketed tiger nut in Nigeria. Crop Research 11, 219-223.
565. Sebastià, N., Meca, G., Soriano, J.M., Mañes, J., 2012: Presence of *Fusarium* emerging mycotoxins in tiger-nuts commercialized in Spain. Food Control 25, 631-635. doi: https://doi.org/10.1016/j.foodcont.2011.11.040.
566. Fuller, G., Spooncer, W.W., King, A.D. Jr., Schade, J., Mackey, B., 1977: Survey of aflatoxins in California tree nuts. Journal of the American Oil Chemists´ Society 54, 231A-234A.
567. Ali, N., Hashim, N.H., Yoshizawa, T., 1999: Evaluation and application of a simple and rapid method for the analysis of aflatoxins in commercial foods from Malaysia and the Philippines. Food Additives and Contaminants 16, 273-280.
568. López, P., Venema, D., De Rijk, T., De Kok, A., Scholten, J.M., Mol, H.G.J., De Nijs, M., 2016: Occurrence of *Alternaria* toxins in food products in The Netherlands. Food Control, 60, 196-204. doi: https://doi.org/10.1016/j.foodcont.2015.07.032.
569. Idris, Y.M.A., Mariod, A.A., Elnour, I.A., Mohamed, A.A., 2010: Determination of aflatoxin levels in Sudanese edible oils. Food and Chemical Toxicology 48, 2539-2541. doi: https://doi.org/10.1016/j.fct.2010.05.021.
570. Daradimos, E., Marcaki, P., Koupparis, M., 2000: Evaluation and validation of two fluorometric HPLC methods for the determination of aflatoxin B_1 in olive oil. Food Additives and Contaminants 17, 65-73.
571. Toussaint, G., Lafaverges, F., Walker, E.A., 1977: The use of high pressure liquid chromatography for determination of aflatoxins in olive oil. Archives de l´Institut Pasteur de Tunis 3-4, 325-335.
572. Milhome, M.A.L., Lima, C.G., De Lima, L.K., Lima, F.A.F., Sousa, D.O.B., Nascimento, R.F., 2014: Occurrence of aflatoxins in cashew nuts produced in northeastern Brazil. Food Control, 42, 34-37. doi: https://doi.org/10.1016/j.foodcont.2014.01.033.
573. Iqbal, S.Z., Asi, M.R., Zuber, M., Akram, N., Batool, N., 2013: Aflatoxins contamination in peanut and peanut products commercially available in retail markets of Punjab, Pakistan. Food Control 32, 83-86. doi: https://doi.org/10.1016/j.foodcont.2012.11.024.
574. Iqbal, S.Z., Asi, M.R., Jinap, S., 2014: Aflatoxins in dates and dates products. Food Control, 43, 163-166. doi: https://doi.org/10.1016/j.foodcont.2014.03.010.
575. Ferracane, R., Tafuri, A., Logieco, A., Galvano, F., Balzano, D., Ritieni, A., 2007: Simultaneous determination of aflatoxin B_1 and ochratoxin A and their natural occurrence in Mediterranean virgin olive oil. Food Additives and Contaminants 24, 173-180. doi: https://doi.org/10.1080/02652030600986040.
576. Papachristou, A., Markaki, P., 2004: Determination of ochratoxin A in virgin olive oils of Greek origin by immunoaffinity column clean-up and high-performance liquid chromatography. Food Additives and Contaminants 21, 85-92. doi: https://doi.org/10.1080/02652030310001632547.
577. Chong, Y.H., Beng, C.G., 1965: Aflatoxins in unrefined groundnut oil. Medical Journal of Malaya XX, 49-50.
578. Wang, J.-S., Huang, T., Su, J., Liang, F., Wei, Z., Liang, Y., Luo, H., Kuang, S.-Y., Qian, G.-S., Sun, G., He, X., Kensler, T.W., Groopman, J.D., 2001: Hepatocellular carcinoma and aflatoxin exposure in Zhuqing Village, Fusui County, People´s Republic of China. Cancer Epidemiology, Biomarkers & Prevention 10, 143-146.
579. Pal, R., Varma, B.K., Srivastava, D.D., 1979: Aflatoxin in groundnut oil, groundnut cake and hydrogenated oil in Hapur (Uttar Pradesh, India) market. Journal of Food Science and Technology 16, 160-170.
580. Elzupir, A.O., Suliman, M.A., Ibrahim, I.A., Fadul, M.H., Elhussein, A.M., 2010: Aflatoxin levels in vegetable oils in Khartoum State, Sudan. Mycotoxin Research 26, 69-73. doi: https://doi.org/10.1007/s12550-010-0041-z.
581. Asam, S., Rychlik, M., 2013: Potential health hazards due to the occurrence of the mycotoxin tenuazonic acid in infant food. European Food Research and Technology 236, 491-497 doi: https://doi.org/10.1007/s00217-012-1901-x.
582. Giridhar, N., Krishnamurthy, G.V., 1977: Studies on aflatoxin content of groundnut oil in Andhra Pradesh with reference to climatic conditions and seasonal variations. Journal of Food Science and Technology, India 14, 84-85.
583. Viñas, I., Palma, J., Garza, S., Sibilia, A., Sanchis, V., Visconti, A., 1994: Natural occurrence of aflatoxin and *Alternaria* mycotoxins in oilseed rape from Catalonia (Spain): incidence of toxigenic strains. Mycopathologia 128, 175-179.
584. Tantaoui-Elaraki, A., Le Tutour, B., Bouzid, M., Keddani, M.J., 1983: Contamination des olives noires "façon Grèce" par les spores d´*Aspergillus* toxinogènes et leurs toxines. Industries Alimentaires et Agricoles 100, 997-1000.
585. El Adlouni, C., Tozlovanu, M., Naman, F., Faid, M., Pfohl-Leszkowicz, A., 2006: Preliminary data on the presence of mycotoxins (ochratoxin A, citrinin and aflatoxin B_1) in black table olives "Greek style" of Moroccan origin. Molecular Nutrition & Food Research 50, 507-512. doi: https://doi.org/10.1002/mnfr.200600055.
586. Hernández Hierro, J.M., Garcia-Villanova, R.J., Torrero, P.R., Toruño Fonseca, I.M., 2008: Aflatoxins and ochratoxin A in red paprika for retail sale in Spain: occurrence and evaluation of a simultaneous analytical method. Journal of Agricultural and Food Chemistry 56, 751-756. doi: https://doi.org/10.1021/jf073002c.
587. Shundo, L., De Almeida, A.P., Alaburda, J., Lamardo, L.C.A., Navas, S.A., Ruvieri, V., Sabino, M., 2009: Aflatoxins and ochratoxin A in Brazilian paprika. Food Control 20, 1099-1102. doi: https://doi.org/10.1016/j.foodcont.2009.02.008.
588. Ahn, J., Kim, D., Jang, H.-s., Kim, Y., Shim, W.-B., Chung, D.-H., 2010: Occurrence of ochratoxin A in Korean red paprika and factors to be considered in prevention strategy. Mycotoxin Research 26, 279-286. doi: https://doi.org/10.1007/s12550-010-0067-2.
589. Valle-Algarra, F.M., Mateo, E.M., Mateo, R., Gimeno-Adelantado, J.V., Jiménez, M., 2011: Determination of type A and type B trichothecenes in paprika and chili pepper using LC-triple quadrupole-MS and GC-ECD. Talanta 84, 1112-1117. doi: https://doi.org/10.1016/j.talanta.2011.03.017.

590. Tang, S.-Y., Ling, K.-H., 1977: Mycotoxins in soy sauce and fermented soybean paste of Taiwan. Journal of the Formosan Medical Association 76, 482-487.
591. El-Nagerabi, S.A.F., Elshafie, A.E., Abdalla, A.H., 2000: Composition of mycoflora and aflatoxins in pea seeds from the Sudan. Kuwait Journal of Science & Engineering 27, 109-120.
592. Ueno, Y., 1998: Residue and risk of ochratoxin A in human plasma and beverages in Japan. Mycotoxins 47, 25-32.
593. Tafuri, A., Ferracane, R., Ritieni, A., 2004: Ochratoxin A in Italian marketed cocoa products. Food Chemistry 88, 487-494. doi: https://doi.org/10.1016/j.foodchem.2004.01.061.
594. Piqué, E., Vargas-Murga, L., Gómez-Catalán, J., De Lapuente, J., Llobet, J.M., 2013: Occurrence of patulin in organic and conventional apple-based food marketed in Catalonia and exposure assessment. Food and Chemical Toxicology 60, 190-204. doi: https://doi.org/10.1016/j.fct.2013.07.052.
595. Karaaslan, M., Arslanğray, Y., 2015: Aflatoxins B_1, B_2, G_1, and G_2 contamination in ground red peppers commercialized in Sanliurfa, Turkey. Environmental Monitoring and Assessment 187 (4), 184. doi: https://doi.org/10.1007/s10661-015-4402-0.
596. Bankole, S.A., Adenusi, A.A., Lawal, O.S., Adesanya, O.O., 2010: Occurrence of aflatoxin B_1 in food products derivable from "egusi" melon seeds consumed in southwestern Nigeria. Food Control 21, 974-976. doi: https://doi.org/10.1016/j.foodcont.2009.11.014.
597. Wolff, J., 2000: Ochratoxin A in cereals and cereal products. Archiv für Lebensmittelhygiene 51, 85-88.
598. Möller, T.E., Nyberg, M., 2003: Ochratoxin A in raisins and currants: basic extraction procedure used in two small marketing surveys of the occurrence and control of the heterogeneity of the toxins in samples. Food Additives and Contaminants 20, 1072-1076. doi: https://doi.org/10.1080/02652030310001615212.
599. Palumbo, J.D., O´Keeffe, T.L., Ho, Y.S., Santillan, C.J., 2015: Occurrence of ochratoxin A contamination and detection of ochratoxigenic *Aspergillus* species in retail samples of dried fruits and nuts. Journal of Food Protection 78, 836-842. doi: https://doi.org/10.4315/0362-028X.JFP-14-471.
600. Varga, J., Kocsubé, S., Suri, K., Szigeti, Gy., Szekeres, A., Varga, M., Tóth, B., Bartók, T., 2010: Fumonisin contamination and fumonisin producing black Aspergilli in dried vine fruits of different origin. International Journal of Food Microbiology 143, 143-149. doi: https://doi.org/10.1016/j.ijfoodmicro.2010.08.008.
601. Knudsen, P.B., Mogensen, J.M., Larsen, T.O., Nielsen, K.F., 2011: Occurrence of fumonisin B_2 and B_4 in retail raisins. Journal of Agricultural and Food Chemistry 59, 772-776. doi: https://doi.org/10.1021/jf103855x.
602. Pérez-Ortega, P., Gilbert-López, B., García-Reyes, J.F., Ramos-Martos, N., Molina-Díaz, A., 2012: Generic sample treatment method for simultaneous determination of multiclass pesticides and mycotoxins in wines by liquid chromatography-mass spectrometry. Journal of Chromatography A 1249, 32-40. doi: https://doi.org/10.1016/j.chroma.2012.06.020.
603. Ezekiel, C.N., Sulyok, M., Somorin, Y., Odutayo, F.I., Nwabekee, S.U., Balogun, A.T., Krska, R., 2016: Mould and mycotoxin exposure assessment of melon and bush mango seeds, two common soup thickeners consumed in Nigeria. International Journal of Food Microbiology 237, 83-91 doi: https://doi.org/10.1016/j.ijfoodmicro.2016.08.019.
604. Castellari, M., Versari, A., Fabiani, A., Parpinello, G.P., Galassi, S., 2001: Removal of ochratoxin A in red wines by means of adsorption treatments with commercial fining agents. Journal of Agricultural and Food Chemistry 49, 3917-3921. doi: https://doi.org/10.1021/jf010137o.
605. Horii, S., Hashiguchi, T., Igi, Y., Sudou, S., 2010: Investigation of ochratoxin A in Japanese wine by LC/MC/MS. Journal of the American Society for Enology and Viticulture, Japan 21, 3-7.
606. Varga, J., Kiss, R., Mátrai, T., Mátrai, T., Téren, J., 2005: Detection of ochratoxin A in Hungarian wines and beers. Acta Alimentaria 34, 381-392.
607. Lopez De Cerain, A., González-Peñas, E., Jiménez, A.M., Bello, J., 2002: Contribution to the study of ochratoxin A in Spanish wines. Food Additives and Contaminants 19, 1058-1064. doi: https://doi.org/10.1080/02652030210145928.
608. Shephard, G.S., Fabiani, A., Stockenström, S., Mshicileli, N., Sewram, V., 2003: Quantitation of ochratoxin A in South African wines. Journal of Agricultural and Food Chemistry 51, 1102-1106. doi: https://doi.org/10.1021/jf0259866.
609. Soleas, G.J., Yan, J., Goldberg, D.M., 2001: Assay of ochratoxin A in wine and beer by high-pressure liquid chromatography photodiode array and gas chromatography mass selective detection. Journal of Agricultural and Food Chemistry 49, 2733-2740. doi: https://doi.org/10.1021/jf0100651.
610. Hocking, A.D., Varelis, P., Pitt, J.I., Cameron, S.F., Leong, S.-L.L., 2003: Occurrence of ochratoxin A in Australian wine. Australian Journal of Grape and Wine Research 9, 72-78.
611. Anli, E., Çabuk, B., Vural, N., Başpinar, E., 2005: Ochratoxin A in Turkish wines. Journal of Food Biochemistry 29, 611-623.
612. Aresta, A., Vatinno, R., Palmisano, F., Zambonin, C.G., 2006: Determination of ochratoxin A in wine at sub ng/mL levels by solid-phase microextraction coupled to liquid chromatography with fluorescence detection. Journal of Chromatography A 1115, 196-201. doi: https://doi.org/10.1016/j.chroma.2006.02.092.
613. Bacaloni, A., Cavaliere, C., Faberi, A., Pastorini, E., Samperi, R., Laganà, A., 2005: Automated on-line solid-phase extraction-liquid chromatography-electrospray tandem mass spectrometry method for the determination of ochratoxin A in wine and beer. Journal of Agricultural and Food Chemistry 53, 5518-5525. doi: https://doi.org/10.1021/jf050254+.
614. Campone, L., Piccinelli, A.L., Rastrelli, L., 2011: Dispersive liquid-liquid microextraction combined with high-performance liquid chromatography-tandem mass spectrometry for the identification and the accurate quantification by isotope dilution assay of ochratoxin A in wine samples. Analytical and Bioanalytical Chemistry 399, 1279-1286. doi: https://doi.org/10.1007/s00216-010-4347-7.
615. Maciel, L.F., Felício, A.L.S.M., Miranda, L.C.R., Pires, T.C., Bispo, E.D.S., Hirooka, E.Y., 2018: Aflatoxins and ochratoxin A in different cocoa clones (*Theobroma cacao* L.) developed in the southern region of Bahia, Brazil. Food Additives and Contaminants: Part A 35, 134-143. doi: https://doi.org/10.1080/19440049.2017.1397293.
616. Domijan, A.-M., Peraica, M., 2005: Ochratoxin A in wine. Arhir za Higijena Rada I Toksikologija 56, 17-20.
617. Maier, N.M., Buttinger, G., Welhartizki, S., Gavioli, E., Lindner, W., 2004: Molecularly imprinted polymer-assisted sample clean-up of ochratoxin A from red wine: merits and limitations. Journal of Chromatography B 804, 103-111. doi: https://doi.org/10.1016/j.jchromb.2004.01.014.

618. Noba, S., Omote, M., Kitagawa, Y., Mochizuki, N., 2008: Determination of ochratoxin A in wine by immunoaffinity cleanup and liquid chromatography tandem mass spectrometry. Journal of Food Protection 71, 1038-1042.
619. Rusanova, T.Yu., Beloglazova, N.V., Goryacheva, I.Yu., Lobeau, M., Van Peteghem, C., De Saeger, S., 2009: Non-instrumental immunochemical tests for rapid ochratoxin A detection in red wine. Analytica Chimica Acta 653, 97-102. doi: https://doi.org/10.1016/j.aca.2009.08.036.
620. Shundo, L., De Almeida, A.P., Alaburda, J., Ruvieri, V., Navas, S.A., Lamardo, L.C.A., Sabino, M., 2006: Ochratoxin A in wines and grape juices commercialized in the city of São Paulo, Brazil. Brazilian Journal of Microbiology 37, 533-537. doi:https://doi.org/10.1590/S1517-83822006000400024.
621. Labrinea, E.P., Natskoulis, P.I., Spiropoulos, A.E., Magan, N., Tassou, C.C., 2011: A survey of ochratoxin A occurrence in Greek wines. Food Additives and Contaminants: Part B 4, 61-66. doi: https://doi.org/10.1080/19393210.2010.539707.
622. Tamura, M., Takahashi, A., Uyama, A., Mochizuki, N., 2012: A method for multiple mycotoxin analysis in wines by solid phase extraction and multifunctional cartridge purification, and ultra-high-performance liquid chromatography coupled to tandem mass spectrometry. Toxins 4, 476-486. doi: https://doi.org/10.3390/toxins4060476.
623. Wu, J., Tan, Y., Wang, Y., Xu, R., 2011: Occurrence of ochratoxin A in wine and beer samples from China. Food Additives and Contaminants: Part B 4, 52-56. doi: https://doi.org/10.1080/19393210.2010.535909.
624. Terra, M.F., Prado, G., Pereira, G.E., Ematné, H.J., Batista, L.R., 2013: Detection of ochratoxin A in tropical wine and grape juice from Brazil. Journal of the Science of Food and Agriculture 93, 890-894. doi: https://doi.org/10.1002/jsfa.5817.
625. Mogensen, J.M., Larsen, T.O., Nielsen, K.F., 2010: Widespread occurrence of the mycotoxin fumonisin B_2 in wine. Journal of Agricultural and Food Chemistry 58, 4853-4857. doi: https://doi.org/10.1021/jf904520t.
626. Perrone, G., Nicoletti, I., Pascale, M., De Rossi, A., De Girolamo, A., Visconti, A., 2007: Positive correlation between high levels of ochratoxin A and resveratol-related compounds in red wines. Journal of Agricultural and Food Chemistry 55, 6807-6812. doi: https://doi.org/10.1021/jf070993I.
627. Romero-González, R., Garrido Frenich, A., Martínez Vidal, J.L., Aguilera-Luiz, M.M., 2010: Determination of ochratoxin A and T-2 toxin in alcoholic beverages by hollow fiber liquid phase microextraction and ultra high-pressure liquid chromatography coupled to tandem mass spectrometry. Talanta 82, 171-176. doi: https://doi.org/10.1016/j.talanta.2010.04.016.
628. Ma, L., Xu, W., He, X., Huang, K., Wang, Y., Luo, Y., 2013: Determination of fumonisins B_1 and B_2 in Chinese rice wine by HPLC using AQC precolumn derivatisation. Journal of the Science of Food and Agriculture 93, 1128-1133. doi: https://doi.org/10.1002/jsfa.5862.
629. Ezekiel, C.N., Sulyok, M., Warth, B., Krska, R., 2012: Multi-microbial metabolites in fonio millet (acha) and sesame seeds in Plateau State, Nigeria. European Food Research and Technology 235, 285-293. doi: https://doi.org/10.1007/s00217-012-1755-2.
630. Tanaka, T., Hasegawa, A., Yamamoto, S., Lee, U.-S., Sugiura, Y., Ueno, Y., 1988: Worldwide contamination of cereals by the *Fusarium* mycotoxins nivalenol, deoxynivalenol, and zearalenone. 1. Survey of 19 countries. Journal of Agricultural and Food Chemistry 36, 979-983.
631. Oviedo, M.S., Barros, G.G., Chulze, S.N., Ramirez, M.L., 2012: Natural occurrence of alternariol and alternariol monomethyl ether in soya beans. Mycotoxin Research 28, 169-174. doi: https://doi.org/10.1007/s12550-012-0132-0.
632. Misra, R.S., Singh, R.S., Tripathi, R.K., 1980: Aflatoxin contamination in some food commodities in and around Tarai area of Nainital, Uttar Pradesh. Indian Phytopathology 33, 217-220.
633. Yuwai, K.E., Rao, K.S., Singh, K., Tanaka, T., Ueno, Y., 1994: Occurrence of nivalenol, deoxynivalenol, and zearalenone in imported cereals in Papua, New Guinea. Natural Toxins 2, 19-21.
634. Jajić, I., Jurić, V., Abramović, B., 2008: First survey of deoxynivalenol occurrence in crops in Serbia. Food Control 19, 545-500. doi: https://doi.org/10.1016/j.foodcont.2007.05.009.
635. Castellá, G., Bragulat, M.R., Cabañes, F.J., 1999: Surveillance of fumonisins in maize-based feeds and cereals from Spain. Journal of Agriculture and Food Chemistry 47, 4707-4710. doi: https://doi.org/10.1021/jf981236d.
636. Barros, G., García, D., Oviedo, M.S., Ramirez, M.L., Torres, A., Lattanzio, V., Pascale, M., Chulze, S., 2011: Survey of T-2 and HT-2 toxins in soybean and soy meal from Argentina using immunoaffinity clean-up and high performance liquid chromatography. World Mycotoxin Journal 4, 189-197. doi: https://doi.org/10.3920/WMJ2010.1272.
637. Lee, Y.-W., Kim, J.-G., Chung, D.-H., Roh, P.-U., Pestka, J.J., 1991: Natural occurrence of zearalenone in rice and soybean produced in Korea. Mycotoxin Research 7, 69-72.
638. Versari, A., Parpinello, G.P., Mattioli, A.U., 2007: Survey of patulin contamination in Italian apple juices from organic and conventional agriculture. Journal of Food Technology 5, 143-146.
639. Williams, I.O, Ugbaje, S.A., Igile, G.O., Ekpe, O.O., 2015: Occurrence of aflatoxin in some food commodities commonly consumed in Nigeria. Journal of Food Research 4, 81-88 doi: https://doi.org/10.5539/jfr.v4n5p81.
640. Isohata, E., Toyoda, M., Saito, Y., 1986: Studies on chemical analysis of mycotoxin (XV). Analysis of nivalenol and deoxynivalenol in imported cereals, baby foods and processed foods (confectioneries). Bulletin of National Institute of Health Science 104, 113-118.
641. El-Kady, I.A., El-Maraghy, S.S.M., Eman Mostafa, M., 1995: Natural occurrence of mycotoxins in different spices in Egypt. Folia Microbiologica 40, 297-300.
642. Ghali, R., Ghorbel, H., Hedilli, A., 2009: Fumonisin determination in Tunisian foods and feeds. ELISA and HPLC methods comparison. Journal of Agricultural and Food Chemistry 57, 3955-3960. doi: https://doi.org/10.1021/jf803786h.
643. Shon, D.-H., Cho, M.-H., Lee, M.-H., 1997: Aflatoxin residues in agricultural commodities determined by direct ELISA. Journal of Food Hygiene and Safety 12, 281-287.
644. Castellari, M., Fabbri, S., Fabiani, A., Amati, A., Galassi, S., 2000: Comparison of different immunoaffinity clean-up procedures for high-performance liquid chromatographic analysis of ochratoxin A in wines. Journal of Chromatography A 888, 129-136.
645. Meyvaci, K.B., Altindisli, A., Aksoy, U., Eltem, R., Turgut, H., Arasiler, Z., Kartal, N., 2005: Ochratoxin A in sultanas from Turkey I: survey of unprocessed sultanas from vineyards and packing-houses. Food Additives and Contaminants 22, 1138-1143. doi: https://doi.org/10.1080/02652030500199132.
646. Aksoy, U., Eltem, R., Meyvaci, K.B., Altındıslı, A., Karabat, S., 2007: Five-year survey of ochratoxin A in processed sultanas from Turkey. Food Additives and Contaminants 24, 292-296. doi: https://doi.org/10.1080/02652030601039021.
647. Magnoli, C., Astoreca, A., Ponsone, L., Combina, M., Palacio, G., Rosa, C.A.R., Dalcero, A.M., 2004: Survey of mycoflora and ochratoxin A in dried vine fruits from Argentina markets. Letters in Applied Microbiology 39, 326-331. doi:https://doi.org/10.1111/j.1472-765X.2004.01583.x.

648. Chulze, S.N., Torres, A.M., Dalcero, A.M., Etcheverry, M.G., Ramirez, M.L., Farnochi, M.C., 1995: *Alternaria* mycotoxins in sunflower seeds: incidence and distribution of the toxins in oil and meal. Journal of Food Protection 58, 1133-1135.
649. Dalcero, A.M., Combina, M., Etcheverry, M., Varsavsky, E., Rodriguez, M.I., 1997: Evaluation of *Alternaria* and its mycotoxins during ensiling of sunflower seeds. Natural Toxins 5, 20-23.
650. Torres, A., Chulze, S., Varsavasky, E., Rodriguez, M., 1993: *Alternaria* metabolites in sunflower seeds. Mycopathologia 121, 17-20.
651. Eom, T., Cho, H.-D., Kim, J., Park, M., An, J., Kim, M., Kim, S.-H., Han, S.B., 2017: Multiclass mycotoxin analysis in edible oils using a simple solvent extraction method and liquid chromatography with tandem mass spectrometry. Food Additives and Contaminants: Part A 34, 2011-2022. doi: https://doi.org/10.1080/19440049.2017.1363416.
652. Magnani, R.F., De Souza, G.D., Rodriguez-Filho, E., 2007: Analysis of alternariol and alternariol monomethyl ether on flavedo and albedo tissues of tangerines (*Citrus reticulata*) with symptoms of *Alternaria* brown spot. Journal of Agricultural and Food Chemistry 55, 4980-4986. doi: https://doi.org/10.1021/jf0704256.
653. Wongworapat, K., Tu Ho, M.H., Soontornjanagit, M., Kawamura, O., 2016: Occurrence of ochratoxin A and ochratoxin B in commercial coffee in Vietnam and Thailand. JSM Mycotoxins 66, 1-6. doi: https://doi.org/10.2520/myco.66.1.
654. El-Maghraby, O.M.O., Abdel-Sater, M.A., 1993: Mycoflora and natural occurrence of mycotoxins in tobacco from cigarettes in Egypt. Zentralblatt für Mikrobiologie 148, 253-264.
655. Mislivec, P.B., Bruce, V.R., Stack, M.E., Bandler, R., 1987: Molds and tenuazonic acid in fresh tomatoes used for catsup production. Journal of Food Protection 50, 38-41.
656. Mahnine, N., Meca, G., Fernández-Franzón, M., Mañes, J., Zinedine, A., 2012: Occurrence of fumonisins B_1, B_2 and B_3 in breakfast and infant cereals from Morocco. Phytopathologia Mediterranea 51, 193-197.
657. Hashimoto, R., Nakagawa, H., Onji, Y., Asano, K., Yokoyama, K., Takahashi, H., 2015: Mycotoxin contamination of Vietnamese coffee beans caused by *Aspergillus Nigri* and *Circumdati*. JSM Mycotoxins 65, 1-6. doi: https://doi.org/10.2520/myco.65.1.
658. Abdel-Gawad, K.M., Zohri, A.A., 1993: Fungal flora and mycotoxins of six kinds of nut seeds for human consumption in Saudi Arabia. Mycopathologia 124, 55-64.
659. Abia, W.A., Warth, B., Sulyok, M., Krska, R., Tchana, A.N., Njobeh, P.B., Dutton, M.F., Moundipa, P.F., 2013: Determination of multi-mycotoxin occurrence in cereals, nuts and their product in Cameroon by liquid chromatography tandem mass spectrometry (LC-MS/MS). Food Control 31, 438-453. doi: https://doi.org/10.1016/j.foodcont.2012.10.006.
660. Abiala, M.A., Ezekiel, C.N., Chukwura, N.I., Odebode, A.C., 2011: Toxigenic *Aspergillus* section *Flavi* and aflatoxins in dried yam chips in Oyo State, Nigeria. Academia Arena 3, 42-49.
661. Ahmad, B., Ashiq, S., Hussain, A., Bashir, S., Hussain, M., 2014: Evaluation of mycotoxins, mycobiota, and toxigenic fungi in selected medicinal plants of Khyber Pakhtunkhwa, Pakistan. Fungal Biology 118, 776-784. doi: https://doi.org/10.1016/j.funbio.2014.06.002.
662. Ahn, S., Lee, S., Lee, J., Kim, B., 2016: Accurate determination of ochratoxin A in Korean fermented soybean paste by isotope dilution-liquid chromatography tandem mass spectrometry. Food Chemistry 190, 368-373. doi: https://doi.org/10.1016/j.foodchem.2015.05.114.
663. Aidoo, K.E., Mohamed, S.M., Candlish, A.A., Tester, R.F., Elgerbi, A.M., 2011: Occurrence of fungi and mycotoxins in some commercial baby foods in North Africa. Food and Nutrition Sciences 2, 751-758. doi: https://doi.org/10.4236/fns.2011.27103.
664. Jeszka-Skowron, M., Zgoła-Grześkowiak, A., Stanisz, E., Waśkiewicz, A., 2017: Potential health benefits and quality of dried fruits: goji fruits, cranberries and raisins. Food Chemistry 221, 228-236. doi: https://doi.org/10.1016/j.foodchem.2016.10.049.
665. Akdeniz, A.S., Ozden, S., Alpertunga, B., 2013: Ochratoxin A in dried grapes and grape-derived products in Turkey. Food Additives and Contaminants: Part B 6, 265-269. doi: https://doi.org/10.1080/19393210.2013.814719.
666. Al-Taher, F., Banaszewski, K., Jackson, L., Zweigenbaum, J., Ryu, D., Cappozzo, J., 2013: Rapid method for the determination of multiple mycotoxins in wines and beers by LC-MS/MS using a stable isotope dilution assay. Journal of Agricultural and Food Chemistry 61, 2378-2384. doi: https://doi.org/10.1021/jf304729f.
667. Chompurat, S., Sato, K., Mahakarnchanakul, W., Sirison, J., Kladpan, S., 2006: Determination of ochratoxin A in roasted coffee and corn by immunoaffinity column cleanup with high-performance thin layer chromatography. Mycotoxins Suppl 4, P 197-201. doi: https://doi.org/10.2520/myco1975.2006.Suppl4_197.
668. Bankole, S.A., Lawal, O.A., Adebanjo, A., 2004: Storage practices and aflatoxin B_1 contamination of "egusi" melon seeds in Nigeria. Tropical Science 44, 150-153.
669. Yazdanpanah, H., Mohammadi, T., Abouhossain, G., Cheraghali, A.M., 2005: Effect of roasting on degradation of aflatoxins in contaminated pistachio nuts. Food and Chemical Toxicology 43, 1135-1139. doi: https://doi.org/10.1016/j.fct.2005.03.004.
670. Mariod, A.A., Idris, Y.M.A., 2015: Aflatoxin B_1 levels in groundnut and sunflower oils in different Sudanese states. Food Additives and Contaminants: Part B 8, 266-270. doi: https://doi.org/10.1080/19393210.2015.1082511.
671. Altiokka, G., Can, N.O., Atkoşar, Z., Aboul-Enein, H.Y., 2009: Determination of ochratoxin A in Turkish wines. Journal of Food and Drug Analysis 17, 467-473.
672. Alvarado, M., Galarce-Bustos, O., Vega, M., Aranda, M., 2013: Fast and selective determination of ochratoxin A in wines using an optimized and validated liquid chromatographic method. Food Analytical Methods 6, 621-629. doi: https://doi.org/10.1007/s12161-012-9446-y.
673. Meister, U., 2003 : Detection of citrinin in ochratoxin A-containing products by a new HPLC method. Mycotoxin Research 19, 27-30.
674. Anfossi, L., Giovannoli, C., Giraudi, G., Biagioli, F., Passini, C., Baggiani, C., 2012: A lateral flow immunoassay for the rapid detection of ochratoxin A in wine and grape must. Journal of Agricultural and Food Chemistry 60, 11491-11497. doi: https://doi.org/10.1021/j3031666.
675. Arroyo-Manzanares, N., García-Campaña, A.M., Gámiz-Gracia, L., 2013: Multiclass mycotoxin analysis in *Silybum marianum* by ultra high performance liquid chromatography-tandem mass spectrometry using a procedure based on QuEChERS and dispersive liquid-liquid microextraction. Journal of Chromatography A 1282 11-19. doi: https://doi.org/10.1016/j.chroma.2013.01.072.
676. Arzandeh, S., Selamat, J., Lioe, H., 2010: Aflatoxin in raw peanut kernels marketed in Malaysia. Journal of Food and Drug Analysis 18, 44-50.
677. Asadi, M., Beheshti, H.R., Feizy, J., 2011: A survey of aflatoxins in sesame in Iran. Mycotoxin Research 27, 259-263. doi: https://doi.org/10.1007/s12550-011-0102-y.
678. Asai, T., Tsuchiya, Y., Okano, K., Piscoya, A., Nishi, C.Y., Ikoma, T., Oyama, T., Ikegami, K., Yamamoto, M., 2012: Aflatoxin contamination of red chili pepper from Bolivia and Peru, countries with high gallbladder cancer incidence rates. Asian Pacific Journal of Cancer Prevention 13, 5167-5170. doi: https://doi.org/10.7314/APJCP.2012.13.10.5167.

679. Reinholds, I., Pugajeva, I., Bavrins, K., Kuchovska, G., Bartkevics, V., 2017: Mycotoxins, pesticides and toxic metals in commercial spices and herbs. Food Additives and Contaminants: Part B 10, 5-14. doi: https://doi.org/10.1080/19393210.2016.1210244 (and personal communication).
680. Asghar, M.A., Ahmed, A., Iqbal, J., 2016: Aflatoxins and ochratoxin A in export quality raisins collected from different areas of Pakistan. Food Additives and Contaminants: Part B 9, 51-58. doi: https://doi.org/10.1080/19393210.2015.1127293.
681. Asghar, M.A., Iqbal, J., Ahmed, A., Khan, M.A., Shamsuddin, Z.A., 2014: Aflatoxin B_1 in betel nuts (*Areca catechu* L.) imported to Pakistan from different regions of South Asia. Food Additives and Contaminants: Part B 7, 176-181. doi: https://doi.org/10.1080/19393210.2013.869771.
682. Baltaci, C., İlyasoğlu, H., Cavrar, S., 2012: Aflatoxin levels in raw and processed hazelnuts in Turkey. Food Additives and Contaminants: Part B 5, 83-86. doi: https://doi.org/10.1080/19393210.2012.656146.
683. Sani, A.M., Azizi, E.G., Naeejic, Z., 2012: Aflatoxins level in tea samples in Amol (north of Iran). Nutrition & Food Science 42, 422-427. doi: https://doi.org/10.1108/00346651211277672.
684. Belajová, E., Rauová, D., 2007: Determination of ochratoxin A and its occurrence in wines of Slovakian retail. Journal of Food and Nutrition Research 46, 68-74.
685. Beloglazova, N.V., Goryacheva, I.Yu., Rusanova, T.Yu., Yurasov, N.A., Galve, R., Marco, M.-P., De Saeger, S., 2010: Gel-based immunotest for simultaneous detection of 2,4,6-trichorophenol and ochratoxin A in red wine. Analytica Chimica Acta 672, 3-8. doi: https://doi.org/10.1016/j.aca.2010.05.024.
686. Gillard, N., Agneessens, R., Dubois, M.L., Delahaut, Ph., 2009: Quantification of patulin in Belgian handicraft-made apple juices. World Mycotoxin Journal 2, 95-104. doi: https://doi.org/10.3920/WMJ2008.1054.
687. Bertuzzi, T., Rastelli, S., Pietri, A., 2015: *Aspergillus* and *Penicillium* toxins in chestnuts and derived products produced in Italy. Food Control 50, 876-880. doi: https://doi.org/10.1016/j.foodcont.2014.10.047.
688. Bonerba, E., Ceci, E., Conte, R., Tantillo, G., 2010: Survey of the presence of patulin in fruit juices. Food Additives and Contaminants: Part B 3, 114-119. doi: https://doi.org/10.1080/19393210.2010.490882.
689. Bosch, U., Mirocha, C.J., 1992: Toxin production by *Fusarium* species from sugar beets and natural occurrence of zearalenone in beets and beet fibers. Applied and Environmental Microbiology 58, 3233-3239.
690. Cao, J., Kong, W., Zhou, S., Yin, L., Wan, L., Yang, M., 2013: Molecularly imprinted polymer-based solid phase clean-up for analysis of ochratoxin A in beer, red wine, and grape juice. Journal of Separation Science 36, 1291-1297. doi: https://doi.org/10.1002/jssc.201201055.
691. Casal, S., Vieira, T., Cruz, R., Cunha, S.C., 2014: Ochratoxin A in commercial soluble coffee and coffee substitutes. Food Research International 61, 56-60 doi: https://doi.org/10.1016/j.foodres.2014.04.045.
692. Cavaliere, C., Foglia, P., Guarino, C., Nazzari, M., Samperi, R., Laganà, A., 2007: Determination of aflatoxins in olive oil by liquid chromatography-tandem mass spectrometry. Analytica Chimica Acta 596, 141-148. doi: https://doi.org/10.1016/j.aca.2007.05.055.
693. Chala, A., Mohammed, A., Ayalew, A., Skinnes, H., 2013: Natural occurrence of aflatoxins in groundnut (*Arachis hypogaea* L.) from eastern Ethiopia. Food Control 30, 602-605. doi: https://doi.org/10.1016/j.foodcont.2012.08.023.
694. Chen, A.J., Jiao, X., Hu, Y., Lu, X., Gao, W., 2015: Mycobiota and mycotoxins in traditional medicinal seeds from China. Toxins (Basel) 7, 3858-3875. doi: https://doi.org/10.3390/toxins7103858.
695. Cho, M.S., Kim, K., Seo, E., Kassim, N., Mtenga, A.B., Shim, W.-B., Lee, S.-H., Chung, D.-H., 2010: Occurrence of patulin in various fruit juices from South Korea: an exposure assessment. Food Science and Biotechnology 19, 1-5. doi: https://doi.org/10.1007/s10068-010-0001-6.
696. Chourasia, H.K., 1990: Aflatoxin contamination in drug yielding plants. Journal of the Indian Botanical Society 69, 281-283.
697. Chung, S. W.-c., Kwong, K.P., Tang, A.S.P., Yeung, S.T.K., 2009: Ochratoxin A levels in foodstuffs marketed in Hong Kong. Journal of Food Composition and Analysis. 22, 756-761. doi: https://doi.org/10.1016/j.jfca.2009.02.014.
698. Commission of the European Communities, January 2002. Reports on tasks for scientific cooperation. Report of experts participating in task 3.2.7: assessment of dietary intake of ochratoxin A by the population of EU member states.
699. Comuzzo, P., Rauhut, D., Werner, M., Lagazio, C., Zironi, R., 2013: A survey on wines from organic viticulture from different European countries. Food Control 34, 274-282. doi: https://doi.org/10.1016/j.foodcont.2013.04.039 (and personal communication).
700. Cressey, P., Thomson, B., 2006: Risk Profile: Mycotoxins in the New Zealand Food Supply. ESR Client Report FW 0617. ESR, Christchurch, New Zealand.
701. Cressey, P., Jones, S., 2010: Mycotoxin surveillance programme 2009-2010. Aflatoxins in nuts and nut products. ESR Client Report FW 10036. ESR, Christchurch, New Zealand.
702. Cressey, P., Jones, S., 2009: Mycotoxin surveillance programme 2008-09. Aflatoxins and ochratoxin A in dried fruits and spices. ESR Client Report FW 09042. ESR, Christchurch, New Zealand.
703. Cressey, P., Jones, S., 2011: Mycotoxin surveillance programme 2011. Ochratoxin A in cereal products, wine, beer and coffee. ESR Client Report FW 11075. ESR, Christchurch, New Zealand.
704. Dashti, B.H.R., Al-Hamli, S.F., 2010: Investigation of mycotoxins in selected food commodities in Kuwait: a case study. Journal of the American Dietetic Association 110 (Suppl.), A 92.
705. De Magalhães, J.T., Sodré, G.A., Viscogliosi, H., Grenier-Loustalot, M.-F., 2011. Occurrence of ochratoxin A in Brazilian cocoa beans. Food Control 22, 744-748. doi: https://doi.org/10.1016/j.foodcont.2010.11.006.
706. Delcourt, A., Rousset, A., Lemaître, J.P., 1994: Microbial and mycotoxic contamination of peppers and food safety. Bottelino Chimico Farmaceutico 133, 235-238.
707. Demirel, R., Sariozlu, N.Y., 2014: Mycotoxigenic moulds and mycotoxins in flours consumed in Turkey. Journal of the Science of Food and Agriculture 94, 1577-1584. doi: https://doi.org/10.1002/jsfa.6460.
708. Di Stefano, V., Avellone, G., Pitonzo, R., Capocchiano, V.G., Mazza, A., Cicero, N., Dugo, G., 2015: Natural co-occurrence of ochratoxin A, ochratoxin B and aflatoxins in Sicilian red wines. Food Additives and Contaminants. Part A 32, 1343-1351. doi: https://doi.org/10.1080/19440049.2015.1055521.
709. Diaz, G.J., Krska, R., Sulyok, M., 2015: Mycotoxins and cyanogenic glycosides in staple foods of three indigenous people of the Colombian Amazon. Food Additives and Contaminants: Part B 8, 291-297. doi: https://doi.org/10.1080/19393210.2015.1089948.

710. Diedhiou, P.M., Bandyopadhyay, R., Atehnkeng, J., Ojiambo, P.S., 2011: *Aspergillus* colonization and aflatoxin contamination of maize and sesame kernels in two agro-ecological zones in Senegal. Journal of Phytopathology 159, 268-275. doi: https://doi.org/10.1111/j.1439-0434.2010.01761.x (and personal communication).

711. Dietrich, R., Schmid, A., Märtlbauer, E., 2001: Citrinin in fruit juices. Mycotoxin Research 17, 156-159.

712. Scheuer, R., Gareis, M., 2002: Occurrence of ochratoxin A and B in spices. Mycotoxin Research 18, 62-66.

713. Dini, A., Khazaeli, P., Roohbakhsh, A., Madadlou, A., Pourenamdari, M., Setoodeh, L., Askarian, A., Doraki, N., Farrokhi, H., Moradi, H., Khodadadi, E., 2013: Aflatoxin contamination level in Iran´s pistachio nut during years 2009-2011. Food Control 30, 540-544. doi: https://doi.org/10.1016/j.foodcont.2012.08.012.

714. El-Kady, I.A., Youssef, M.S., 1993: Survey of mycoflora and mycotoxins in Egyptian soybean seeds. Journal of Basic Microbiology 33, 371-378.

715. El-Nagerabi, S.A.F., Elshafie, A.E., 2000: Composition of mycoflora and aflatoxins in lupine seeds from the Sudan. Phytopathologia Mediterranea 39, 257-262.

716. El tawila, M.M., Neamatallah, A., Serdar, S.A., 2013: Incidence of aflatoxins in commercial nuts in the holy city of Mekkah. Food Control 29, 121-124. doi: https://doi.org/10.1016/j.foodcont.2012.06.004.

717. Boyacioğlu, D., Gönül, M., 1990: Survey of aflatoxin contamination of dried figs grown in Turkey in 1986. Food Additives and Contaminants 7, 235-237.

718. Elzupir, A.O., Salih, A.O.A., Suliman, S.A., Adam, A.A., Elhussein, A.M., 2011: Aflatoxins in peanut butter in Khartoum State, Sudan. Mycotoxin Research 27, 183-186. doi: https://doi.org/10.1007/s12550-011-0094-7.

719. Essono, G., Ayodele, M., Akoa, A., Foko, J., Filtenborg, O., Olembo, S., 2009: Aflatoxin-producing *Aspergillus* spp. and aflatoxin levels in stored cassava chips as affected by processing practices. Food Control 20, 648-654. doi: https://doi.org/10.1016/j.foodcont.2008.09.018.

720. Azaiez, I., Font, G., Mañes J., Fernández-Franzón, M., 2015: Survey of mycotoxins in dates and dried fruits from Tunisian and Spanish markets. Food Control 51, 340-346. doi: https://doi.org/10.1016/j.foodcont.2014.11.033.

721. Ezekiel, C.N., Kayode, F.O., Fapohunda, S.O., Olorunfemi, M.F., Kponi, B.T., 2012: Aflatoxigenic moulds and aflatoxins in street-vended snacks in Lagos, Nigeria. Internet Journal of Food Safety 14, 83-88.

722. Matumba, L., Van Poucke, C., Monjerezi, M., Njumbe Ediage, E., De Saeger, S., 2015: Concentrating aflatoxins on the domestic market through groundnut export: a focus on Malawian groundnut value and supply chain. Food Control 51, 236-239. doi: https://doi.org/10.1016/j.foodcont.2014.11.035.

723. Escobar, J., Lorán, S., Giménez, I., Ferruz, E., Herrera, M., Herrera, A., Ariño, A., 2013: Occurrence and exposure assessment of *Fusarium* mycotoxins in maize germ, refined corn oil and margarine. Food and Chemical Toxicology 62, 514-520. doi: https://doi.org/10.1016/j.fct.2013.09.020.

724. Fan, C., Cao, X., Liu, M., Wang, W., 2016: Determination of *Alternaria* mycotoxins in wine and juice using ionic liquid modified countercurrent chromatography as a pretreatment method followed by high-performance liquid chromatography. Journal of Chromatography A 1436, 133-140. doi: https://doi.org/10.1016/j.chroma.2016.01.069.

725. Zitomer, N., Rybak, M.E., Li, Z., Walters, M.J., Holman, M.R., 2015: Determination of aflatoxin B_1 in smokeless tobacco products by use of UHPLC-MS/MS. Journal of Agricultural and Food Chemistry 63, 9131-9138. doi: https://doi.org/10.1021/acs.jafc.5b02622.

726. Feizy, J., Beheshti, H.R., Asadi, M., 2012: Ochratoxin A and aflatoxins in dried vine fruits from the Iranian market. Mycotoxin Research 28, 237-242. doi: https://doi.org/10.1007/s12550-012-0145-8.

727. Feizy, J., Beheshti, H.R., Fakoor Janati, S.S., Fahim, N.K., 2011: Survey of aflatoxins in watermelon seeds from Iran using immunoaffinity column cleanup and HPLC with fluorescence detection. Food Additives and Contaminants: Part B 4, 106-109. doi: https://doi.org/10.1080/19393210.2011.561932.

728. Fernane, F., Cano-Sancho, G., Sanchis, V., Marin, S., Ramos, A.J., 2010: Aflatoxins and ochratoxin A in pistachios sampled in Spain: occurrence and presence of mycotoxigenic fungi. Food Additives and Contaminants: Part B 3, 185-192. doi: https://doi.org/10.1080/19440049.2010.497257.

729. Flamini, R., Dalla Vedova, A., De Rosso, M., Panighel, A., 2007: A new sensitive and selective method for analysis of ochratoxin A in grape and wine by direct liquid chromatography/surface-activated chemical ionization tandem mass spectrometry. Rapid Communications in Mass Spectrometry 21, 3737-3742. doi: https://doi.org/10.1002/rcm.3275.

730. Galarce-Bustos, O., Alvarado, M., Vega, M., Aranda, M., 2014: Occurrence of ochratoxin A in roasted and instant coffees in Chilean market. Food Control 46, 102-107. doi: https://doi.org/10.1016/j.foodcont.2014.05.014.

731. García-Moraleja, A., Font, G., Mañes, J., Ferrer, E., 2015: Analysis of mycotoxins in coffee and risk assessment in Spanish adolescents and adults. Food and Chemical Toxicology 86, 225-233 doi: https://doi.org/10.1016/j.fct.2015.10.014.

732. Geana, E.-I., Iordache, A.-M., Ionete, R.E., 2012: Ochratoxin A occurrence in Romanian wines. Revista de Chimie 63, 883-886.

733. Geipel, M., Baltes, W., Krönert, W., Weber, R., 1981: Determination of patulin in apple products by high performance liquid chromatography. Chemie, Mikrobiologie, Technologie der Lebensmittel 7, 93-96.

734. Golge, O., Hepsag, F., Kabak, B., 2013: Incidence and level of aflatoxin contamination in chilli commercialised in Turkey. Food Control 33, 514-520. doi: https://doi.org/10.1016/j.foodcont.2013.03.048.

735. Haas, D., Pfeifer, B., Reiterich, C., Partenheimer, R., Reck, B., Buzina, W., 2013: Identification and quantification of fungi and mycotoxins from Pu-erh tea. International Journal of Food Microbiology 166, 316-322. doi: https://doi.org/10.1016/j.ijfoodmicro.2013.07.024.

736. Han, Z., Feng, Z., Shi, W., Zhao, Z., Wu, Y., Wu, A., 2014: A quick, easy, cheap, effective, rugged, and safe sample pretreatment and liquid chromatography with tandem mass spectrometry method for the simultaneous quantification of 33 mycotoxins in *Lentinula edodes*. Journal of Separation Science 37, 1957-1966. doi: https://doi.org/10.1002/jssc.201400329.

737. Han, Z., Liu, X., Ren, Y., Luan, L., Wu, Y., 2010: A rapid method with ultra-high-performance liquid chromatography-tandem mass spectrometry for simultaneous determination of five type B trichothecenes in traditional Chinese medicines. Journal of Separation Science 33, 1923-1932. doi: https://doi.org/10.1002/jssc.201000094.

738. Ezekiel, C.N., Sulyok, M., Warth, B., Odebode, A.C., Krska, R., 2012: Natural occurrence of mycotoxins in peanut cake from Nigeria. Food Control 27, 338-342. doi: https://doi.org/10.1016/j.foodcont.2012.04.010 (and personal communication).

739. Han, Z., Ren, Y., Liu, X., Luan, L., Wu, Y., 2010: A reliable isotope dilution method for simultaneous determination of fumonisins B_1, B_2 and B_3 in traditional Chinese medicines by ultra-high-performance liquid chromatography-tandem mass spectrometry. Journal of Separation Science 33, 2723-2733. doi: https://doi.org/10.1002/jssc.201000423.
740. Han, Z., Ren, Y., Zhou, H., Luan, L., Cai, Z., Wu, Y., 2011: A rapid method for simultaneous determination of zearalenone, α-zearalenol, ß-zearalenol, zearalanone, α-zearalanol and ß-zearalanol in traditional Chinese medicines by ultra-high-performance liquid chromatography-tandem mass spectrometry. Journal of Chromatography B 879, 411-420. doi: https://doi.org/10.1016/j.jchromb.2010.12.028.
741. Hepsag, F., Golge, O., Kabak, B., 2014: Quantitation of aflatoxins in pistachios and groundnuts using HPLC-FLD method. Food Control 38, 75-81. doi: https://doi.org/10.1016/j.foodcont.2013.10.005.
742. Heshmati, A., Mozaffari Nejad, A.S., 2015: Ochratoxin A in dried grapes in Hamadan province, Iran. Food Additives and Contaminants: Part B 8, 255-259. doi: https://doi.org/10.1080/19393210.2015.1074945.
743. Hickert, S., Bergmann, M., Ersen, S., Cramer, B., Humpf, H.-U., 2016: Survey of *Alternaria* toxin contamination in food from the German market, using a rapid HPLC-MS/MS approach. Mycotoxin Research 32, 7-18. doi: https://doi.org/10.1007/s12550-015-0233-7.
744. Hoeltz, M., Einloft, T.C., Oldoni, V.P., Dottori, H.A., Noll, I.B., 2012: The occurrence of aflatoxin B_1 contamination in peanuts and peanut products marketed in southern Brazil. Brazilian Archives of Biology and Technology 55, 313-317. doi: https://doi.org/10.1590/S1516-89132012000200019.
745. Hosseininia, A.R., Vahabzadeh, M., Rashedinia, M., Riahi-Zanjani, B., Karimi, G., 2014: A survey of aflatoxins in sesame seeds imported into Khorasan Province, Iran. Mycotoxin Research 30, 43-46. doi: https://doi.org/10.1007/s12550-013-0186-7.
746. Hu, L., Rychlik, M., 2014: Occurrence of enniatins and beauvericin in 60 Chinese medicinal herbs. Food Additives and Contaminants: Part A 31, 1240-1245. doi: https://doi.org/10.1080/19440049.2014.913813.
747. Iamanaka, B.T., Nakano, F., Lemes, D.P., Ferranti, L.S., Taniwaki, M.H., 2014: Aflatoxin evaluation in ready-to-eat brazil nuts using reversed-phase liquid chromatography and post-column derivatisation. Food Additives and Contaminants: Part A 31, 917-923. doi: https://doi.org/10.1080/19440049.2014.895857.
748. Li, R., Wang, X., Zhou, T., Yang, D., Wang, Q., Zhou, Y., 2014: Occurrence of four mycotoxins in cereal and oil products in Yangtze Delta region of China and their food safety risks. Food Control 35, 117-122. doi: https://doi.org/10.1016/j.foodcont.2013.06.042 (and personal communication).
749. Ikoma, T., Tsuchiya, Y., Asai, T., Okano, K., Ito, N., Endoh, K., Yamamoto, M., Nakamura, K., 2015: Ochratoxin A contamination of red chili peppers from Chile, Bolivia and Peru, countries with a high incidence of gallbladder cancer. Asian Pacific Journal of Cancer Prevention 16, 5987-5991. doi: https://doi.org/10.7314/APJCP.2015.16.14.5987.
750. Iqbal, S.Z., Asi, M.R., Jinap, S., 2013: Natural occurrence of aflatoxin B_1 and aflatoxin M_1 in "halva" and its ingredients. Food Control 34, 404-407. doi: https://doi.org/10.1016/j.foodcont.2013.04.038.
751. Iqbal, S.Z., Asi, M.R., Zuber, M., Akhtar, J., Saif, M.J., 2013: Natural occurrence of aflatoxins and ochratoxin A in commercial chilli and chilli sauce samples. Food Control 30, 621-625. doi: https://doi.org/10.1016/j.foodcont.2012.09.003.
752. Jager, A.V., Tedesco, M.P., Souto, P.C.M.C., Oliveira, C.A.F., 2013: Assessment of aflatoxin intake in São Paulo, Brazil. Food Control 33, 87-92 doi: https://doi.org/10.1016/j.foodcont.2013.02.016.
753. Jahanmard, E., Azarani, F., Sharifi, M., Esfandiari, Z., 2014: Aflatoxin in pistachio nuts used as ingredients in Gaz sweets produced in Isfahan, Iran. Food Additives and Contaminants: Part B 7, 70-73. doi: https://doi.org/10.1080/19393210.2013.846942.
754. Zhao, Y., Wan, L.-H., Bai, X.-L., Liu, Y.-M., Zhang, F.-P., Liu, Y.-M., Liao, X., 2017: Quantification of mycotoxins in vegetable oil by UPLC-MS/MS after magnetic solid-phase extraction. Food Additives and Contaminants: Part A 34, 1201-1210. doi: https://doi.org/10.1080/19440049.2017.1319074.
755. Juan, C., Oueslati, S., Mañes, J., 2016: Evaluation of *Alternaria* mycotoxins in strawberries: quantification and storage condition. Food Additives and Contaminants: Part A 33, 861-868. doi: https://doi.org/10.1080/19440049.2016.1177375.
756. Kaaya, A.N., Eboku, D., 2010: Mould and aflatoxin contamination of dried cassava chips in eastern Uganda: association with traditional processing and storage practices. Journal of Biological Sciences 10, 718-729.
757. Kabak, B., 2014: Quantitation of aflatoxins in walnut kernels by high-performance liquid chromatography with fluorescence detection. Food Additives and Contaminants: Part B 7, 288-294. doi: https://doi.org/10.1080/19393210.2014.928372.
758. Kamika, I., Mngqawa, P., Rheeder, J.P., Teffo, S.L., Katerere, D.R., 2014: Mycological and aflatoxin contamination of peanuts sold at markets in Kinshasa, Democratic Republic of Congo, and Pretoria, South Africa. Food Additives and Contaminants: Part B 7, 120-126. doi: https://doi.org/10.1080/19393210.2013.858187.
759. Kamika, I., Takoy, L.L., 2011: Natural occurrence of aflatoxin B_1 in peanut collected from Kinshasa, Democratic Republic of Congo. Food Control 22, 1760-1764. doi: https://doi.org/10.1016/j.foodcont.2011.04.010.
760. Kamimura, H., Nishijima, M., Tabata, S., Yasuda, K., Ushiyama, H., Nishima, T., 1986: Survey of mycotoxins contamination in edible oil and fate of mycotoxins during oil-refining processes. Journal of the Food Hygienic Society of Japan 27, 59-63.
761. Kappenstein, O., Klaffke, H.St., Mehlitz, I., Tiebach, R., Weber, R., Lepschy, J., Wittkowski, R., 2005: Determination of zearalenone in edible oils with SEC and LC-ESI-MS/MS. Mycotoxin Research 21, 3-6.
762. Kassim, N., Kim, K., Mtenga, A.B., Song, J.-E., Liu, Q., Shim, W.-B., Chung, D.-H., 2011: A preliminary study of T-2 and HT-2 toxins in cereals sold in traditional market in South Korea. Food Control 22, 1408-1412. doi: https://doi.org/10.1016/j.foodcont.2011.02.023.
763. Kayode, O.F., Sulyok, M., Fapohunda, S.O, Ezekiel, C.N., Krska, R., Oguntona, C.R.B., 2013: Mycotoxins and fungal metabolites in groundnut- and maize-based snacks from Nigeria. Food Additives and Contaminants: Part B 6, 294-300. doi: https://doi.org/10.1080/19393210.2013.823626.
764. Khalesi, M., Sheikh-Zeinoddin, M., Tabrizchi, M., 2011: Determination of ochratoxin A in licorice root using inverse ion mobility spectrometry. Talanta 83, 988-993. doi: https://doi.org/10.1016/j.talanta.2010.11.004.
765. Khan, M.A., Asghar, M.A., Iqbal, J., Ahmed, A., Shamsuddin, Z.A., 2014: Aflatoxins contamination and prevention in red chillies (*Capsicum annuum* L.) in Pakistan. Food Additives and Contaminants: Part B 7, 1-6. doi: https://doi.org/10.1080/19393210.2013.825330
766. Khayoon, W.S., Saad, B., Lee, T.P., Salleh, B., 2012: High performance liquid chromatographic determination of aflatoxins in chilli, peanut and rice using silica based monolithic column. Food Chemistry 133, 489-496. doi: https://doi.org/10.1016/j.foodchem.2012.01.010.

767. Kollia, E., Kanapitsas, A., Markaki, P., 2014: Occurrence of aflatoxin B_1 and ochratoxin A in dried vine fruits from Greek market. Food Additives and Contaminants: Part B 7, 11-16. doi: https://doi.org/10.1080/19393210.2013.825647.
768. Kong, W., Wei, R., Logrieco, A.F., Wei, J., Wen, J., Xiao, X., Yang, M., 2014: Occurrence of toxigenic fungi and determination of mycotoxins by HPLC-FLD in functional foods and spices in China markets. Food Chemistry 146, 320-326. doi: https://doi.org/10.1016/j.foodchem.2013.09.005.
769. Krogh, P., Hald, B., Korpinen, E.-L., 1970: Occurrence of aflatoxin in groundnut- and copra-products imported to Finland. Nordisk Veterinaer Medicin 22, 584-589.
770. Krüger, C.D., Fernandes, A.M., Rosa, C.A.R., 2012: Ochratoxin A in wines from 2002 to 2008 harvest marketed in Rio de Janeiro, Brazil. Food Additives and Contaminants: Part B 5, 204-207. doi: https://doi.org/10.1080/19393210.2012.697485.
771. Kuang, Y., Qiu, F., Kong, W., Yang, M., 2012: Natural occurrence of ochratoxin A in wolfberry fruit wine marketed in China. Food Additives and Contaminants: Part 5, 70-74. doi: https://doi.org/10.1080/19393210.2012.660198.
772. Kumar, S., Roy, A.K., 1993: Occurrence of aflatoxin in some liver curative herbal medicines. Letters in Applied Microbiology 17, 112-114.
773. Lee, T.P., Sakai, R., Manaf, N.A., Rodhi, A.M., Saad, B., 2014: High performance liquid chromatography method for the determination of patulin and 5-hydroxymethylfurfural in fruit juices marketed in Malaysia. Food Control 38, 142-149. doi: https://doi.org/10.1016/j.foodcont.2013.10.018.
774. Li, M., Kong, W., Li, Y., Liu, H., Liu, Q., Dou, X., Ou-yang, Z., Yang, M., 2016: High-throughput determination of multi-mycotoxins in Chinese yam and related products by ultra fast liquid chromatography coupled with tandem mass spectrometry after one-step extraction. Journal of Chromatography B 1022, 118-125. doi: https://doi.org/10.1016/j.jchromb.2016.04.014.
775. Li, X., Li, P., Zhang, Q., Li, R., Zhang, W., Zhang, Z., Ding, X., Tang, X., 2013: Multi-component immunochromatographic assay for simultaneous detection of aflatoxin B_1, ochratoxin A and zearalenone in agro-food. Biosensors and Bioelectronics 49, 426-432. doi: https://doi.org/10.1016/j.bios.2013.05.039.
776. Liau, B.-C., Jong, T.-T., Lee, M.-R., Chang, C.-M.J., 2007: Supercritical fluid extraction and quantification of aflatoxins in Zizyphi Fructus by liquid chromatography/atmospheric pressure chemical ionization tandem mass spectrometry. Rapid Communications in Mass Spectrometry 21, 667-673. doi: https://doi.org/10.1002/rcm.2870.
777. Liao, C.-D., Wong, J.W., Zhang, K., Yang, P., Wittenberg, J.B., Trucksess, M.W., Hayward, D.G., Lee, N.S., Chang, J.S., 2015: Multi-mycotoxin analysis of finished grain and nut products using ultrahigh-performance liquid chromatography and positive electrospray ionization-quadrupole orbital ion trap high-resolution mass spectrometry. Journal of Agricultural and Food Chemistry 63, 8314-8332. doi: https://doi.org/10.1021/jf505049a.
778. Liu, H., Luo, J., Kong, W., Liu, Q., Hu, Y., Yang, M., 2016: UFLC-ESI-MS/MS analysis of multiple mycotoxins in medicinal and edible *Areca catechu*. Chemosphere 150, 176-183. doi: https://doi.org/10.1016/j.chemosphere.2016.02.032.
779. Liu, Q., Kong, W., Guo, W., Yang, M., 2015: Multi-class mycotoxins analysis in *Angelica sinensis* by ultra fast liquid chromatography coupled with tandem mass spectrometry. Journal of Chromatography B 988, 175-181. doi: https://doi.org/10.1016/j.jchromb.2015.02.022.
780. Liu, S., Qiu, F., Kong, W., Wei, J., Xiao, X., Yang, M., 2013: Development and validation of an accurate and rapid LC-ESI-MS/MS method for the simultaneous quantification of aflatoxin B_1, B_2, G_1 and G_2 in lotus seeds. Food Control 29, 156-161. doi: https://doi.org/10.1016/j.foodcont.2012.05.069.
781. Lohrey, L., Marschik, S., Cramer, B., Humpf, H.-U., 2013: Large scale synthesis of isotopically labeled $^{13}C_2$-tenuazonic acid and development of a rapid HPLC-MS/MS method for the analysis of tenuazonic acid in tomato and pepper products. Journal of Agricultural and Food Chemistry 61, 114-120. doi: https://doi.org/10.1021/jf305138k.
782. Lutfullah, G., Hussain, A., 2012: Studies on contamination level of aflatoxins in some cereals and beans of Pakistan. Food Control 23, 32-36. doi: https://doi.org/10.1016/j.foodcont.2011.06.004.
783. Karami-Osboo, R., Miri, R., Javidnia, K., Kobarfard, F., AliAbadi, M.H.S., Maham, M., 2015: A validated dispersive liquid-liquid microextraction method for extraction of ochratoxin A from raisin samples. Journal of Food Science and Technology 52, 2440-2445. doi: https://doi.org/10.1007/s13197-013-1215-4.
784. Magrine, I.O.C., Ferrari, S.S.C., Souza, G.F., Minamihara, L., Kemmelmeier, C., Bando, E., Machinski, M. Jr., 2011: Intake of aflatoxins through the consumption of peanut products in Brazil. Food Additives and Contaminants: Part B 4, 99-105. doi: https://doi.org/10.1080/19393210.2011.561931.
785. Makun, H.A., Adeniran, A.L., Mailafiya, S.C., Ayanda, I.S., Mudashiru, A.T., Ojukwu, U.J., Jagaba, A.S., Usman, Z., Salihu, D.A., 2013: Natural occurrence of ochratoxin A in some marketed Nigerian foods. Food Control 31, 566-571. doi: https://doi.org/10.1016/j.foodcont.2012.09.043.
786. Marín, S., Mateo, E.M., Sanchis, V., Valle-Algarra, F.M., Ramos, A.J., Jiménez, M., 2011: Patulin contamination in fruit derivatives, including baby food, from the Spanish market. Food Chemistry 124, 563-568. doi: https://doi.org/10.1016/j.foodchem.2010.06.072.
787. Marques, M.F., Martins, H.M., Costa, J.M., Bernardo, F., 2008: Co-occurrence of deoxynivalenol and zearalenone in crops marketed in Portugal. Food Additives and Contaminants: Part B 1, 130-133. doi: https://doi.org/10.1080/02652030802253983.
788. Martinéz-Domínguez, G., Romero-González, R., Frenich, A.G., 2016: Multi-class methodology to determine pesticides and mycotoxins in green tea and royal jelly supplements by liquid chromatography coupled to Orbitrap high resolution mass spectrometry. Food Chemistry 197, 907-915. doi: https://doi.org/10.1016/j.foodchem.2015.11.070.
789. Mata, A.T., Ferreira, J.P., Oliveira, B.R., Batoréu, M.C., Barreto Crespo, M.T., Pereira, V.J., Bronze, M.R., 2015: Bottled water: analysis of mycotoxins by LC-MS/MS. Food Chemistry 176, 455-464. doi: https://doi.org/10.1016/j.foodchem.2014.12.088.
790. Matumba, L., Monjerezi, M., Biswick, T., Mwatseteza, J., Makumba, W., Kamangira, D., Mtukuso, A., 2014: A survey of the incidence and level of aflatoxin contamination in a range of locally and imported processed foods on Malawian retail market. Food Control 39, 87-91. doi: https://doi.org/10.1016/j.foodcont.2013.09.068.
791. Miceli, A., Negro, C., Tommasi, L., De Leo, P., 2003: Polyphenols, resveratrol, antioxidant activity and ochratoxin A contamination in red table wines, controlled denomination of origin (DOC) wines and wines obtained from organic farming. Journal of Wine Research 14, 115-120. doi: https://doi.org/10.1080/09571260410001678030.
792. Mikulíková, R., Běláková, S., Benešová, K., Svoboda, Z., 2012: Study of ochratoxin A content in south Moravian and foreign wines by the UPLC method with fluorescence detection. Food Chemistry 133, 55-59 doi: https://doi.org/10.1016/j.foodchem.2011.12.061.

793. Miró-Abella, E., Herrero, P., Canela, N., Arola, L., Borrull, F., Ras, R., Fontanals, N., 2017: Determination of mycotoxins in plant-based beverages using QuEChERS and liquid chromatography-tandem mass spectrometry. Food Chemistry 229, 366-372. doi: https://doi.org/10.1016/j.foodchem.2017.02.078.
794. Abass, A.B., Awoyale, W., Sulyok, M., 2017: Occurrence of regulated mycotoxins and other microbial metabolites in dried cassava products from Nigeria. Toxins (Basel), 9 (7), 207. doi: https://doi.org/10.3390/toxins.9070207.
795. Montaseri, H., Eskandari, M.H., Yeganeh, A.T., Karami, S., Javidnia, K., Dehghanzadeh, G.R., Mesbahi, G.R., Niakousari, M., 2014: Patulin in apple leather in Iran. Food Additives and Contaminants: Part B 7, 106-109. doi: https://doi.org/10.1080/19393210.2013.855825.
796. Monyo, E.S., Njoroge, S.M.C., Coe, R., Osiru, M., Madinda, F., Waliyar, F., Thakur, R.P., Chilunjika, T., Anitha, S., 2012: Occurrence and distribution of aflatoxin contamination in groundnuts (*Arachis hypogaea* L) and population density of *Aflatoxigenic Aspergilli* in Malawi. Crop Protection 42, 149-155 doi: https://doi.org/10.1016/j.cropro.2012.07.004.
797. Mupunga, I., Lebelo, S.L., Mngqawa, P., Rheeder, J.P., Katerere, D.R., 2014: Natural occurrence of aflatoxins in peanuts and peanut butter from Bulawayo, Zimbabwe. Journal of Food Protection 77, 1814-1818. doi: https://doi.org/10.4315/0362-028X.JPF-14-129.
798. Murillo-Arbizu, M.T., Amézqueta, S., González-Peñas, E., López de Cerain, A., 2010: Occurrence of ochratoxin A in southern Spanish generous wines under the denomination of origin "Jerez-Xérès-Sherry and 'Manzanilla' Sanlúcar de Barrameda". Toxins (Basel) 2, 1054-1064. doi: https://doi.org/10.3390/toxins2051054.
799. Murillo-Arbizu, M., González-Peñas, E., Amézqueta, S., 2010: Comparison between capillary electrophoresis and high performance liquid chromatography for the study of the occurrence of patulin in apple juice intended for infants. Food and Chemical Toxicology 48, 2429-2434. doi: https://doi.org/10.1016/j.fct.2010.06.003.
800. Nejad, A.S.M., Bayat, M., Ahmadi, A.A., 2013: Investigation of aflatoxin B_1 in spices marketed in Hyderabad, India by ELISA method. Journal of Pure and Applied Microbiology 7, 3219-3223.
801. Nejad, A.S.M., Ghannad, M.S., Kamkar, A., 2014: Determination of aflatoxin B_1 levels in Iranian and Indian spices by ELISA method. Toxin Reviews 33, 151-154. doi: https://doi.org/10.3109/15569543.2014.942319.
802. Nejad, M.I., Farahani, A.D., 2012: Aflatoxin in raw walnut kernels marketed in Tehran, Iran. Food Additives and Contaminants: Part B 5, 8-10. doi: https://doi.org/10.1080/19393210.2011.637239.
803. Nielsen, K.F., Ngemela, A.F., Jensen, L.B., Soman de Medeiros, L., Rasmussen, P.H., 2015: UHPLC-MS/MS determination of ochratoxin A and fumonisins in coffee using QuEChERS extraction combined with mixed-mode SPE purification. Journal of Agricultural and Food Chemistry 63, 1029-1034. doi: https://doi.org/10.1021/jf504254q.
804. Njobeh, P.B., Dutton, M.F., Koch, S.H., Chuturgoon, A.A., Stoev, S.D., Mosonik, J.S., 2010: Simultaneous occurrence of mycotoxins in human food commodities from Cameroon. Mycotoxin Research 26, 47-57. doi: https://doi.org/10.1007/s12550-009-0039-6.
805. Njumbe Ediage, E., Di Mavungu, J. D., Goryacheva, I.Y., Van Peteghem, C., De Saeger, S., 2012: Multiplex flow-through immunoassay formats for screening of mycotoxins in a variety of food matrices. Analytical and Bioanalytical Chemistry 403, 265-278. doi: https://doi.org/10.1007/s00216-012-5803-3.
806. Müller, P., Basedow, T., 2007: Aflatoxin contamination of pods of Indian *Cassia senna* L. (Caesalpiniaceae) before harvest, during drying and in storage: reasons and possible methods of reduction. Journal of Stored Products Research 43, 323-329. doi: https://doi.org/10.1016/j.jspr.2006.08.005.
807. Njumbe Ediage, E., Hell, K., De Saeger, S., 2014: A comprehensive study to explore differences in mycotoxin patterns from agro-ecological regions through maize, peanut, and cassava products: a case study, Cameroon. Journal of Agricultural and Food Chemistry 62, 4789-4797. doi: https://doi.org/10.1021/jf501710u.
808. Nyikal, J., Misore, A., Nzioka, C., Njuguna, C., Muchiri, E., Njau, J., Maingi, S., Njoroge, J., Mutiso, J., Onteri, J., Langat, A., Kilei, I.K, Nyamongo, J., Ogana, G., Muture, B., Tukei, P., Onyango, C., Ochieng, W., Tetteh, C., Likimani, S., Nguku, P., Galgalo, T., Kibet, S., Manya, A., Dahiye, A., Mwihia, J., Mugoya, I., Onsongo, J., Ngindu, A., DeCock, K.M., Lindblade, K., Slutsker, L., Amornkul, P., Rosen, D., Feiken, D., Thomas, T., Mensah, P., Eseko, N., Nejjar, A., Onsongo, M., Kessel, F., Njapau, H., Park, D.L., Lewis, L., Luber, G., Rogers, H., Backer, L., Rubin, C., Gieseker, K.E., Azziz-Baumgartner, E., Chege, W., Bowen, A., 2004: Outbreak of aflatoxin poisoning - eastern and central provinces, Kenya, January-July 2004. Morbidity and Mortality Weekly Report 53, 790-793.
809. Ok, H.E., Chung, S.H., Lee, N., Chun, H.S., 2015: Simple high-performance liquid chromatography method for the simultaneous analysis of aflatoxins, ochratoxin A, and zearalenone in dried and ground red pepper. Journal of Food Protection 78, 1226-1231. doi: https://doi.org/10.4315/0362-028X.JFP-14-451.
810. Oroian, M., Amariei, S., Gutt, G., 2014: Patulin in apple juices from the Romanian market. Food Additives and Contaminants 7, 147-150. doi: https://doi.org/10.1080/19393210.2013.861518.
811. Ostry, V., Malir, F., Dofkova, M., Skarkova, J., Pfohl-Leszkowicz, A., Ruprich, J., 2015: Ochratoxin A dietary exposure of ten population groups in the Czech Republic: comparison with data over the world. Toxins (Basel) 7, 3608-3635. doi: https://doi.org/10.3390/toxins7093608.
812. Perrone, G., De Girolamo, A., Sarigiannis, Y., Haidukowski, M.E., Visconti, A., 2013: Occurrence of ochratoxin A, fumonisin B_2 and black aspergilli in raisins from Western Greece regions in relation to environmental and geographical factors. Food Additives and Contaminants: Part A 30, 1339-1347. doi: https://doi.org/10.1080/19440049.2013.796594.
813. Pietri, A., Rastelli, S., Mulazzi, A., Bertuzzi, T., 2012: Aflatoxins and ochratoxin A in dried chestnuts and chestnut flour produced in Italy. Food Control 25, 601-606. doi: https://doi.org/10.1016/j.foodcont.2011.11.042.
814. Pizzutti, I.R., De Kok, A., Scholten, J., Righi, L.W., Cardoso, C.D., Rohers, G.N., Da Silva, R.C., 2014: Development, optimization and validation of a multimethod for the determination of 36 mycotoxins in wines by liquid chromatography-tandem mass spectrometry. Talanta 129, 352-363. doi: https://doi.org/10.1016/j.talanta.2014.05.017.
815. Wei, D., Wang, Y., Jiang, D., Feng, X., Li, J., Wang, M., 2017: Survey of *Alternaria* toxins and other mycotoxins in dried fruits in China. Toxins (Basel) 9 (7), 200. doi: https://doi.org/10.3390/toxins9070200.
816. Ponsone, M.L., Chiotta, M.L., Combina, M., Torres, A., Knass, P., Dalcero, A., Chulze, S., 2010: Natural occurrence of ochratoxin A in musts, wines and grape vine fruits from grapes harvested in Argentina. Toxins (Basel) 2(8), 1984-1996. doi: https://doi.org/10.3390/toxins2081984.
817. Pouretedal, Z., Mazaheri, M., 2013: Aflatoxins in black tea in Iran. Food Additives and Contaminants: Part B 6, 127-129. doi: https://doi.org/10.1080/19393210.2013.764551.

818. Prado, G., Altoé, A.F., Gomes, T.C.B., Leal, A.S., Morais, V.A.D., Oliveira, M.S., Ferreira, M.B., Gomes, M.B., Paschoal, F.N., Von S. Souza, R., Silva, D.A., Cruz Madeira, J.E.G., 2012: Occurrence of aflatoxin B_1 in natural products. Brazilian Journal of Microbiology 43, 1428-1435.
819. Prado, G., De Oliveira, M.S., Da Cunha, M.R., De Fátima Gomides, M., Abrantes, F.M., Dos Santos, L.G., Veloso, T., De S. Barroso, R.E., 2000: Occurrence of patulin in apple juice by high-performance liquid chromatography. Revista do Instituto Adolfo Lutz 59, 21-25.
820. Prelle, A., Spadaro, D., Denca, A., Garibaldi, A., Gullino, M.L., 2013: Comparison of clean-up methods for ochratoxin A on wine, beer, roasted coffee and chili commercialized in Italy. Toxins (Basel) 5, 1827-1844. doi: https://doi.org/10.3390/toxins5101827.
821. Prelle, A., Spadaro, D., Garibaldi, A., Gullino, M.L., 2012: Aflatoxin monitoring in Italian hazelnut products by LC-MS. Food Additives and Contaminants: Part B 5, 279-285 doi: https://doi.org/10.1080/19393210.2012.711371 (and personal communication).
822. Prelle, A., Spadaro, D., Garibaldi, A., Gullino, M.L., 2013: A new method for detection of five *Alternaria* toxins in food matrices based on LC-APCI-MS. Food Chemistry 140, 161-167. doi: https://doi.org/10.1016/j.foodchem.2012.12.065.
823. Sulyok, M., Beed, F., Boni, S., Abass, A., Mukunzi, A., Krska, R., 2015: Quantitation of multiple mycotoxins and cyanogenic glucosides in cassava samples from Tanzania and Rwanda by an LC-MS/MS-based multi-toxin method. Food Additives and Contaminants: Part A 32, 488-502. doi: https://doi.org/10.1080/19440049.2014.975752.
824. Qian, M., Zhang, H., Wu, L., Jin, N., Wang, J., Jiang, K., 2015: Simultaneous determination of zearalenone and its derivatives in edible vegetable oil by gel permeation chromatography and gas chromatography-triple quadrupole mass spectrometry. Food Chemistry 166, 23-28. doi: https://doi.org/10.1016/j.foodchem.2014.05.133.
825. Quintela, S., Villarán, M.C., López de Armentia, I., Elejalde, E., 2012: Ochratoxin A in Spanish exportation wine market. Food Control 25, 501-504. doi: https://doi.org/10.1016/j.foodcont.2011.11.023.
826. Quintela, S., Villarán, M.C., López de Armentia, I., Elejalde, E., 2011: Occurrence of ochratoxin A in Rioja Alavesa wines. Food Chemistry 126, 302-305. doi: https://doi.org/10.1016/j.foodchem.2010.09.096.
827. Rahimi, E., Jeiran, M.R., 2015: Patulin and its dietary intake by fruit juice consumption in Iran. Food Additives and Contaminants: Part B 8, 40-43. doi: https://doi.org/10.1080/19393210.2014.967814.
828. Reazai, M., Sayadi, M., Akbarpour, B., Mohammadpourfard, I., Behzadi, A., Teimoory, H., 2014: Safety of dried fruits marketed in Tehran as assessed by aflatoxin contamination. International Journal of Food Nutrition and Safety 5, 24-30.
829. Reinhold, L., Reinhardt, K., 2011: Mycotoxins in foods in Lower Saxony (Germany): results of official control analyses performed in 2009. Mycotoxin Research 27, 137-143. doi: https://doi.org/10.1007/s12550-011-0086-7.
830. Remiro, R., González-Peñas, E., Lizarraga, E., López de Cerain, A., 2012: Quantification of ochratoxin A and five analogs in Navarra red wines. Food Control 27, 139-145. doi: https://doi.org/10.1016/j.foodcont.2012.03.006.
831. Remiro, R., Ibáñez-Vea, M., González-Peñas, E., Lizarraga, E., 2010: Validation of a liquid chromatography method for the simultaneous quantification of ochratoxin A and its analogues in red wines. Journal of Chromatography A 1217, 8249-8256. doi: https://doi.org/10.1016/j.chroma.2010.11.004.
832. Remiro, R., Irigoyen, A., González-Peñas, E., Lizarraga, E., López de Cerain, A., 2013: Levels of ochratoxins in Mediterranean red wines. Food Control 32, 63-68. doi: https://doi.org/10.1016/j.foodcont.2012.11.040.
833. Rubert, J., Fapohunda, S.O., Soler, C., Ezekiel, C.N., Mañes, J., Kayode, F., 2013: A survey of mycotoxins in random street-vended snacks from Lagos, Nigeria, using QuEChERS-HPLC-MS/MS. Food Control 32, 673-677. doi: https://doi.org/10.1016/j.foodcont.2013.01.017.
834. Rubert, J., Soler, C., Mañes, J., 2012: Occurrence of fourteen mycotoxins in tiger-nuts. Food Control 25, 374-379. doi: https://doi.org/10.1016/j.foodcont.2011.11.004.
835. Sadia, A., Jabbar, M.A., Deng, Y., Hussain, E.A., Riffat, S., Naveed, S., Arif, M., 2012: A survey of aflatoxin M_1 in milk and sweets of Punjab, Pakistan. Food Control 26, 235-240. doi: https://doi.org/10.1016/j.foodcont.2012.01.055.
836. Sahar, N., Arif, S., Iqbal, S., Ul Ain Afzal, Q., Aman, S., Ara, J., Ahmed, M., 2015: Moisture content and its impact on aflatoxin levels in ready-to-use red chillies. Food Additives and Contaminants: Part B 8, 67-72. doi: https://doi.org/10.1080/19393210.2014.978395.
837. Sarigiannis, Y., Kapolos, J., Koliadima, A., Tsegenidis, T., Karaiskakis, G., 2014: Ochratoxin A levels in Greek retail wines. Food Control 42, 139-142. doi: https://doi.org/10.1016/j.foodcont.2014.02.003.
838. Saxena, N., Dwivedi, P.D., Ansari, K.M., Das, M., 2008: Patulin in apple juices: incidence and likely intake in an Indian population. Food Additives and Contaminants: Part B 1, 140-146. doi: https://doi.org/10.1080/02652030802378848.
839. Schade, J.E., McGreevy, K., King, A.D.Jr., Mackey, B., Fuller, G., 1975: Incidence of aflatoxin in California almonds. Applied Microbiology 29, 48-53.
840. Schollenberger, M., Müller, H.-M., Rüfle, M., Drochner, W., 2008: Natural occurrence of 16 *Fusarium* toxins in edible oil marketed in Germany. Food Control 19, 475-482. doi: https://doi.org/10.1016/j.foodcont.2007.05.012.
841. Serrat, R., Mendoça, C., Venâncio, A, 2006: Fungi and ochratoxin A detected in healthy grapes for wine production. Letters in Applied Microbiology 42, 42-47.
842. Shepard, G.S., Van Der Westhuizen, L., Katerere, D.R., Herbst, M., Pineiro, M., 2010: Preliminary exposure assessment of deoxynivalenol and patulin in South Africa. Mycotoxin Research 26, 181-185. doi: https://doi.org/10.1007/s12550-010-0052-9.
843. Siegel, D., Andrae, K., Proske, M., Kochan, C., Koch, M., Weber, M., Nehls, I., 2010: Dynamic covalent hydrazine chemistry as a selective extraction and cleanup technique for the quantification of the *Fusarium* mycotoxin zearalenone in edible oils. Journal of Chromatography A 1217, 2206-2215. doi: https://doi.org/10.1016/j.chroma.2010.02.019.
844. Škrbić, B., Živančev, J., Godula, M., 2014: Multimycotoxin analysis of crude extracts of nuts with ultra-high performance liquid chromatography/tandem mass spectrometry. Journal of Food Composition and Analysis 34, 171-177. doi: https://doi.org/10.1016/j.jfca.2014.03.002.
845. Solfrizzo, M., Piemontese, L., Gambacorta, L., Zivoli, R., Longobardi, F., 2015: Food coloring agents and plant food supplements derived from *Vitis vinifera*: a new source of human exposure to ochratoxin A. Journal of Agricultural and Food Chemistry 63, 3609-3614. doi: https://doi.org/10.1021/acs.jafc.5b00326.
846. Somorin, Y., Akinyemi, A., Bertuzzi, T., Pietri, A., 2016: Co-occurrence of aflatoxins, ochratoxin A and citrinin in "egusi" melon (*Colocynthis citrullus* L.) seeds consumed in Ireland and the United Kingdom. Food Additives and Contaminants: Part B 9, 230-235. doi: https://doi.org/10.1080/19393210.2016.1183051.
847. Spadaro, D., Garibaldi, A., Gullino, M.L., 2008: Occurrence of patulin and its dietary intake through pear, peach, and apricot juices in Italy. Food Additives and Contaminants: Part B 1, 134-139. doi: https://doi.org/10.1080/02652030802363790.

848. Stępień, L., Koczyk, G., Waśkiewicz, A., 2013: Diversity of *Fusarium* species and mycotoxins contaminating pineapple. Journal of Applied Genetics 54, 367-380. doi: https://doi.org/10.1007/s13353-013-0146-0.

849. Stępień, L., Waśkiewicz, A., Urbaniak, M., 2016: Wildly growing asparagus (*Asparagus officinalis* L.) hosts pathogenic *Fusarium* species and accumulates their mycotoxins. Microbial Ecology 71: 927-937. doi: https://doi.org/10.1007/s00248-015-0717-1.

850. Sulyok, M., Krska, R., Schuhmacher, R., 2007: A liquid chromatography/tandem mass spectrometric multi-mycotoxin method for the quantification of 87 analytes and its application to semi-quantitative screening of moldy food samples. Analytical and Bioanalytical Chemistry 389, 1505-1523. doi: https://doi.org/10.1007/s00216-007-1542-2.

851. Swaileh, K.M., Abdulkhaliq, A., 2013: Analysis of aflatoxins, caffeine, nicotine and heavy metals in Palestinian multifloral honey from different geographic regions. Journal of the Science of Food and Agriculture 93, 2116-2120. doi: https://doi.org/10.1002/jsfa.6014.

852. Lasram, S., Oueslati, S., Chebil, S., Mliki, A., Ghorbel, A., 2013: Occurrence of ochratoxin A in domestic beers and wines from Tunisia by immunoaffinity clean-up and liquid chromatography. Food Additives and Contaminants: Part B 6, 1-5. doi: https://doi.org/10.1080/19393210.2012.716453.

853. Zhou, J., Xu, J.-J., Huang, B.-F., Cai, Z.-X., Ren, Y.-P., 2017: High-performance liquid chromatographic determination of multi-mycotoxin in cereals and bean foodstuffs using interference-removal solid-phase extraction combined with optimized dispersive liquid-liquid microextraction. Journal of Separation Science 40, 2141-2150. doi: https://doi.org/10.1002/jssc.201601326.

854. Tan, Y., Kuang, Y., Zhao, R., Chen, B., Wu, J., 2011: Determination of T-2 and HT-2 toxins in traditional Chinese medicine marketed in China by LC-ELSD after sample clean-up by two solid-phase extractions. Chromatographia 73, 407-410. doi: https://doi.org/10.1007/s10337-010-1890-5.

855. Tang, D., Zhong, Z., Niessner, R., Knopp, D., 2009: Multifunctional magnetic bead-based electrochemical immunoassay for the detection of aflatoxin B_1 in food. Analyst 134, 1554-1560. doi: https://doi.org/10.1039/b902401h.

856. Taniwaki, M.H., Pitt, J.I., Teixeira, A.A., Iamanaka, B.T., 2003: The source of ochratoxin A in Brazilian coffee and its formation in relation to processing methods. International Journal of Food Microbiology 82, 173-179.

857. Tansakul, N., Limsuwan, S., Böhm, J., Hollmann, M., Razzazi-Fazeli, E., 2013: Aflatoxins in selected Thai commodities. Food Additives and Contaminants: Part B 6, 254-259. doi: https://doi.org/10.1080/19393210.2013.812148.

858. Toffa, D.D., Mahnine, N., Ouaffak, L., El Abidi, A., El Alaoui Faris, F.Z., Zinedine, A., 2013: First survey on the presence of ochratoxin A and fungi in raw cereals and peanut available in the Republic of Niger. Food Control 32, 558-562. doi: https://doi.org/10.1016/j.foodcont.2013.01.028.

859. Tosun, H., Arslan, R., 2013: Determination of aflatoxin B_1 levels in organic spices and herbs. Scientific World Journal 2013, article ID 874093. doi: https://doi.org/10.1155/2013/874093.

860. Levi, C., 1980: Mycotoxins in coffee. Journal of the Association of Official Analytical Chemists 63, 1282-1285.

861. Tosun, A., Ozden, S., 2016: Ochratoxin A in red pepper flakes commercialised in Turkey. Food Additives and Contaminants: Part B 9, 46-50. doi: https://doi.org/10.1080/19393210.2015.1121929.

862. Tosun, H., Yildiz, H., Obuz, E., Seçkin, A.K., 2014: Ochratoxin A in grape pekmez (grape molasses) consumed in Turkey. Food Additives and Contaminants: Part B 7, 37-39. doi: https://doi.org/10.1080/19393210.2013.833298.

863. Tripathi, S., Mishra, H.N., 2009: A rapid FT-NIR method for estimation of aflatoxin B_1 in red chilli powder. Food Control 20, 840-846. doi: https://doi.org/10.1016/j.foodcont.2008.11.003.

864. Turcotte, A.-M., Scott, P.M., Tague, B., 2013: Analysis of cocoa products for ochratoxin A and aflatoxins. Mycotoxin Research 29, 193-201. doi: https://doi.org/10.1007/s12550-013-0167-x.

865. Vaclavik, L., Vaclavikova, M., Begley, T.H., Krynitsky, A.J., Rader, J.L., 2013: Determination of multiple mycotoxins in dietary supplements containing green coffee bean extracts using ultrahigh-performance liquid chromatography-tandem mass spectrometry (UHPLC-MC/MS). Journal of Agricultural and Food Chemistry 61, 4822-4830. doi: https://doi.org/10.1021/jf401139u.

866. Vanesa, D., Ana, P., 2013: Occurrence of ochratoxin A in coffee beans, ground roasted coffee and soluble coffee and method validation. Food Control 30, 675-678. doi: https://doi.org/10.1016/j.foodcont.2012.09.004.

867. Var, I., Kabak, B., 2007: Occurrence of ochratoxin A in Turkish wines. Microchemical Journal 86, 241-247. doi: https://doi.org/10.1016/j.microc.2007.04.002.

868. Varga, E., Glauner, T., Berthiller, F., Krska, R., Schumacher, R., Sulyok, M., 2013: Development and validation of a (semi-)quantitative UHPLC-MS/MS method for the determination of 191 mycotoxins and other fungal metabolites in almonds, hazelnuts, peanuts and pistachios. Analytical and Bioanalytical Chemistry 405, 5087-5104. doi: https://doi.org/10.1007/s00216-013-6831-3.

869. Vecchio, A., Mineo, V., Planeta, D., 2012: Ochratoxin A in instant coffee in Italy. Food Control 28, 220-223. doi: https://doi.org/10.1016/j.foodcont.2012.04.029.

870. Vega, M., Ríos, G., Von Baer, D., Mardones, C., Tessini, C., Herlitz, E., Saelzer, R., Ruiz, M.A., 2012: Ochratoxin A occurrence in wines produced in Chile. Food Control 28, 147-150. doi: https://doi.org/10.1016/j.foodcont.2012.04.032.

871. Vijayasamundeeswari, A., Mohankumar, M., Karthikeyan, M., Vijayanandraj, S., Paranidharan, V., Velazhahan, R., 2009: Prevalence of aflatoxin B_1 contamination in pre- and post-harvest maize kernels, food products, poultry and livestock feeds in Tamil Nadu, India. Journal of Plant Protection Research 49, 221-224. doi: https://doi.org/10.2478/v10045-009-0033-1.

872. Wang, L., Wang, Z., Gao, W., Chen, J., Yang, M., Kuang, Y., Huang, L., Chen, S., 2013: Simultaneous determination of aflatoxin B_1 and ochratoxin A in licorice roots and fritillary bulbs by solid-phase extraction coupled with high-performance liquid chromatography-tandem mass spectrometry. Food Chemistry 138, 1048-1054. doi: https://doi.org/10.1016/j.foodchem.2012.11.066.

873. Wang, S., Cheng, L., Ji, S., Wang, K., 2014: Simultaneous determination of seventeen mycotoxin residues in *Puerariae lobatae radix* by liquid chromatography-tandem mass spectrometry. Journal of Pharmaceutical and Biomedical Analysis 98, 201-209. doi: https://doi.org/10.1016/j.jpba.2014.05.037.

874. Wei, R., Qiu, F., Kong, W., Wei, J., Yang, M., Luo, Z., Qin, J., Ma, X., 2013: Co-occurrence of aflatoxin B_1, B_2, G_1, G_2 and ochrotoxin A in *Glycyrrhiza uralensis* analyzed by HPLC-MS/MS. Food Control 32, 216-221. doi: https://doi.org/10.1016/j.foodcont.2012.11.028.

875. Wen, J., Kong, W., Wang, J., Yang, M., 2013: Simultaneous determination of four aflatoxins and ochratoxin A in ginger and related products by HPLC with fluorescence detection after immunoaffinity column clean-up and postcolumn photochemical derivatization. Journal of Separation Science 36, 3709-3716. doi: https://doi.org/10.1002/jssc.201300885.

876. Wu, J., Zhao, R., Chen, B., Yang, M., 2011: Determination of zearalenone in barley by high-performance liquid chromatography coupled with evaporative light scattering detection and natural occurrence of zearalenone in functional food. Food Chemistry 126, 1508-1511. doi: https://doi.org/10.1016/j.foodchem.2010.11.159.
877. Xavier, J.J.M., Scussel, V.M., 2008: Development of an LC-MS/MS method for the determination of aflatoxins B_1, B_2, G_1, and G_2 in Brazil nut. International Journal of Environmental Analytical Chemistry 88, 425-433. doi: https://doi.org/10.1080/03067310701836816.
878. Yang, L.-x., Liu, Y.-p., Miao, H., Dong, B., Yang, N.-j., Chang, F.-q., Sun, J.-b., 2011: Determination of aflatoxins in edible oil from markets in Hebei Province of China by liquid chromatography-tandem mass spectrometry. Food Additives and Contaminants: Part B 4, 244-247. doi: https://doi.org/10.1080/19393210.2011.632694.
879. Yazdanpanah, H., Zarghi, A., Shafaati, A.R., Foroutan, S.M., Aboul-Fathi, F., Khoddam, A., Nazari, F., Shaki, F., 2013: Analysis of aflatoxin B_1 in Iranian foods using HPLC and a monolithic column and estimation of its dietary intake. Iranian Journal of Pharmaceutical Research 12 (Suppl) 83-89.
880. Zhang, X., Liu, H., Chen, J., 2005: Immunoaffinity column cleanup with liquid chromatography using post-column bromination for aflatoxins in medicinal herbs and plant extracts. Journal of Chromatographic Science 43, 47-51. doi: https://doi.org/10.1093/chromsci/43.1.47.
881. Zhang, Z., Hu, X., Zhang, Q., Li, P., 2016: Determination for multiple mycotoxins in agricultural products using HPLC-MS/MS via a multiple antibody immunoaffinity column. Journal of Chromatography B 1021, 145-152. doi: https://doi.org/10.1016/j.jchromb.2016.02.035.
882. Zhao, H., Wang, Y., Zou, Y., Zhao, M., 2013: Natural occurrence of deoxynivalenol in soy sauces consumed in China. Food Control 29, 71-75. doi: https://doi.org/10.1016/j.foodcont.2012.05.066.
883. Zhao, K., Shao, B., Yang, D., Li, F., 2015: Natural occurrence of four *Alternaria* mycotoxins in tomato- and citrus-based foods in China. Journal of Agricultural and Food Chemistry 63, 343-348. doi: https://doi.org/10.1021/jf5052738.
884. Prelle, A., Spadaro, D., Garibaldi, A., Gullino, M.L., 2014: Co-occurrence of aflatoxins and ochratoxin A in spices commercialized in Italy. Food Control 39, 192-197. doi: https://doi.org/10.1016/j.foodcont.2013.11.013 (and personal communication).
885. Zhao, X., Yuan, Y., Zhang, X., Yue, T., 2014: Identification of ochratoxin A in Chinese spices using HPLC fluorescent detectors with immunoaffinity column cleanup. Food Control 46, 332-337. doi: https://doi.org/10.1016/j.foodcont.2014.05.052.
886. Zheng, R., Xu, H., Wang, W., Zhan, R., Chen, W., 2014: Simultaneous determination of aflatoxin B_1, B_2, G_1, G_2, ochratoxin A, and sterigmatocystin in traditional Chinese medicines by LC-MS-MS. Analytical and Bioanalytical Chemistry 406, 3031-3039. doi: https://doi.org/10.1007/s00216-014-7750-7.
887. Zhou, Y., Kong, W., Li, Y., Logrieco, A.F., Xu, J., Yang, M., 2012: A new solid-phase extraction and HPLC method for determination of patulin in apple products and hawthorn juice in China. Journal of Separation Science 35, 641-649. doi: https://doi.org/10.1002/jssc.201100919.
888. Zouaoui, N., Sbaii, N., Bacha, H., Abid-Essefi, S., 2015: Occurrence of patulin in various fruit juice marketed in Tunisia. Food Control 51, 356-360. doi: https://doi.org/10.1016/j.foodcont.2014.09.048.
889. Yogendrarajah, P., Jacxsens, L., De Saeger, S., De Meulenaer, B., 2014: Co-occurrence of multiple mycotoxins in dry chilli (*Capsicum annum* L.) samples from the markets of Sri Lanka and Belgium. Food Control 46, 26-34. doi: https://doi.org/10.1016/j.foodcont.2014.04.043.
890. Yuan, Y., Zhuang, H., Zhang, T., Liu, J., 2010: Patulin content in apple products marketed in Northeast China. Food Control 21, 1488-1491. doi: https://doi.org/10.1016/j.foodcont.2010.04.019 (and personal communication).
891. Yogendrarajah, P., Van Poucke, C., De Meulenaer, B., De Saeger, S., 2013: Development and validation of a QuEChERS based liquid chromatography tandem mass spectrometry method for the determination of multiple mycotoxins in spices. Journal of Chromatography A 1297, 1-11. doi: https://doi.org/10.1016/j.chroma.2013.04.075.
892. Yoshinari, T., Takeuchi, H., Aoyama, K., Taniguchi, M., Hashiguchi, S., Kai, S., Ogiso, M., Sato, T., Akiyama, Y., Nakajima, M., Tabata, S., Tanaka, T., Ishikuro, E., Sugita-Konishi, Y., 2014: Occurrence of four *Fusarium* mycotoxins, deoxynivalenol, zearalenone, T-2 toxin, and HT-2 toxin, in wheat, barley, and Japanese retail food. Journal of Food Protection 77, 1940-1946 doi: https://doi.org/10.4315/0362-028X.JFP-14-185.
893. Yu, F.-Y., Gribas, A.V., Vdovenko, M.M., Sakharov, I.Y., 2013: Development of ultrasensitive direct chemiluminescent enzyme immunoassay for the determination of aflatoxin B_1 in food products. Talanta 107, 25-29. doi: https://doi.org/10.1016/j.talanta.2012.12.047.
894. Yue, Y.-T., Zhang, X.-F., Pan, J., Ou-Yang, Z., Wu, J., Yang, M.-H., 2010: Determination of deoxynivalenol in medicinal herbs and related products by GC-ECD and confirmation by GC-MS. Chromatographia 71, 533-538. doi: https://doi.org/10.1365/s10337-010-1477-1.
895. Zaied, C., Abid, S., Hlel, W., Bacha, H., 2013: Occurrence of patulin in apple-based-foods largely consumed in Tunisia. Food Control 31, 263-267. doi: https://doi.org/10.1016/j.foodcont.2012.10.005.
896. Zhang, K., Wong, J.W., Jia, Z., Vaclavikova, M., Trucksess, M.W., Begley, T.H., 2014: Screening multimycotoxins in food-grade gums by stable isotope dilution and liquid chromatography/tandem mass spectrometry. Journal of the Association of Official Analytical Chemists International 97, 889-895.
897. Zhang, X., Li, J., Zong, N., Zhou, Z., Ma, L., 2014: Ochratoxin A in dried vine fruits from Chinese markets. Food Additives and Contaminants: Part B 7, 157-161. doi: https://doi.org/10.1080/19393210.2013.867365.
898. Zhang, X., Liu, W., Logrieco, A.F., Yang, M., Ou-yang, Z., Wang, X., Guo, Q., 2011: Determination of zearalenone in traditional Chinese medicinal plants and related products by HPLC-FLD. Food Additives and Contaminants: Part A 28, 885-893. doi: https://doi.org/10.1080/19440049.2011.563429.
899. Zhang, X., Chen, L., Li, J., Zhu, BQ., Ma, L., 2013: Occurrence of ochratoxin A in Chinese wines: influence of local meteorological parameters. European Food Research and Technology 236, 277-283. doi: https://doi.org/10.1007/s00217-012-1886-5.
900. Torović, L., Trajković Pavlović, L., Popović, M., 2017: Ochratoxin A and aflatoxin B_1 in breakfast cereals marketed in Serbia - occurrence and health risk characterisation. Food Additives and Contaminants: Part B 10, 176-184. doi: https://doi.org/10.1080/19393210.2017.1285358.
901. Kim, S., Lee, S., Nam, T.-G., Seo, D., Yoo, M., 2017: Comparison of a newly developed liquid chromatography with tandem mass spectrometry method and enzyme-linked immunosorbent assay for detection of multiple mycotoxins in red pepper powder. Journal of Food Protection 80, 1347-1354. doi: https://doi.org/10.4315/0362-028X.JFP-17-006.
902. Idris, Y.M.A., Hassan, S.A., Mariod, A.A., 2013: Physicochemical characteristics and aflatoxin levels in two types of Sudanese sesame oil. Journal of the American Oil Chemists´ Society 90, 989-998. doi: https://doi.org/10.1007/s11746-013-2241-0.

903. Škrbić, B., Antić, I., Cvejanov, J., 2017: Determination of mycotoxins in biscuits, dried fruits and fruit jams: an assessment of human exposure. Food Additives and Contaminants: Part A, 34, 1012-1025. doi: https://doi.org/10.1080/19440049.2017.1303195 (and personal communication).
904. Torović, L., Dimitrov, N., Assunção, R., Alvito, P., 2017: Risk assessment of patulin intake through apple-based food by infants and preschool children in Serbia. Food Additives and Contaminants: Part A 34, 2023-2032. doi: https://doi.org/10.1080/19440049.2017.1364434.
905. Cunha, S.C., Faria, M.A., Pereira, V.L., Oliveira, T,M., Lima, A.C., Pinto, E., 2014: Patulin assessment and fungi identification in organic and conventional fruits and derived products. Food Control 44, 185-190. doi: https://doi.org/10.1016/j.foodcont.2014.03.043.
906. López, P., Venema, D., Mol, H., Spanjer, M., De Stoppelaar, J., Pfeiffer, E., De Nijs, M., 2016: *Alternaria* toxins and conjugates in selected foods in The Netherlands. Food Control, 60, 153-159. doi: https://doi.org/10.1016/j.foodcont.2016.04.001.
907. Mounjouenpou, P., Mbang, J.A.A.., Guyot, B., Guiraud, J.P., 20127: Traditional procedures of cocoa processing and occurrence of ochratoxin - A in the derived products. Journal of Chemical and Pharmaceutical Research 4, 1332-1339.
908. Welke, J.E., Hoeltz, M., Dottori, H.A., Noll, I.B., 2010: Determination of ochratoxin A in wine by high-performance thin-layer chromatography using charged coupled device. Journal of the Brazilian Chemical Society 21, 441-446.
909. Welke, J.E., Hoeltz, M., Dottori, H.A., Noll, I.B., 2010: Fungi and patulin in apples and the role of processing on patulin levels in juices: a study on naturally contaminated apples. Journal of Food Safety 30, 276-287.
910. Guo, Y., Zhou, Z., Yuan, Y., Yue, T., 2013: Survey of patulin in apple juice concentrates in Shaanxi (China) and its dietary intake. Food Control 34, 570-572. doi: https://doi.org/10.1016/j.foodcont.2013.05.029.
911. Vaclavikova, M., Dzuman, Z., Lacina, O., Fenclova, M., Veprikova, Z., Zachariasova, M., Hajslova, J., 2015: Monitoring survey of patulin in a variety of fruit-based products using a sensitive UHPLC-MS/MS analytical procedure. Food Control 47, 577-584. doi: https://doi.org/10.1016/j.foodcont.2014.07.064.
912. Lai, X., Ruan, C., Liu, R., Liu, C., 2014: Application of ionic liquid-based dispersive liquid-liquid microextraction for the analysis of ochratoxin A in rice wines. Food Chemistry 161, 317-322. doi: https://doi.org/10.1016/j.foodchem.2014.04.033.
913. Yacine Ware, L., Durand, N., Nikiema, P.A., Alter, P., Fontana, A., Montet, D., Barro, N., 2017: Occurrence of mycotoxins in commercial infant formulas locally produced in Ouagadougou (Burkina Faso). Food Control 73, 518-523. doi: https://doi.org/10.1016/j.foodcont.2016.08.047.
914. Lucchetta, G., Bazzo, I., Dal Cortivo, G., Stringher, L., Bellotto, D., Borgo, M., Angelini, E., 2010: Occurrence of black aspergilli and ochratoxin A on grapes in Italy. Toxins (Basel) 2, 840-855. doi: https://doi.org/10.3390/toxins2040840.
915. Giovannoli, C., Passini, C., Di Nardo, F., Anfossi, L., Baggiani, C., 2014: Determination of ochratoxin A in Italian red wines by molecularly imprinted solid phase extraction and HPLC analysis. Journal of Agricultural and Food Chemistry 62, 5220-5225. doi: https://doi.org/10.1021/jf5010995.
916. Majerus, P., Hain, J., Kölb, C., 2008: Patulin in grape must, and new, still fermenting wine (Federweißer). Mycotoxin Research 24, 135-139.
917. Huang, X., Wang, S., Mao, D., Miao, S., Hu, Q., Ji, S., 2017: Optimized QuEChERS method combined with UHPLC-MS/MS for the simultaneous determination of 15 mycotoxins in liquorice. Journal of the Association of Official Analytical Chemists International 101, 1-10. doi: https://doi.org/10.5740/jaoacint.17-0365.

Alphabetical Bibliography

Abass, A.B., Awoyale, W., Sulyok, M., 2017: Occurrence of regulated mycotoxins and other microbial metabolites in dried cassava products from Nigeria. Toxins (Basel), 9 (7), 207. doi: https://doi.org/10.3390/toxins.9070207 (794).

Abdallah, M.F., Krska, R., Sulyok, M., 2016: Mycotoxin contamination in sugarcane grass and juice: first report on detection of multiple mycotoxins and exposure assessment for aflatoxin B_1 and G_1 in humans. Toxins (Basel), 8 (11), 343. doi: https://doi.org/10.3390/toxins8110343 (300).

Abdel-Gawad, K.M., Zohri, A.A., 1993: Fungal flora and mycotoxins of six kinds of nut seeds for human consumption in Saudi Arabia. Mycopathologia 124, 55–64 (658).

Abdel-Hafez, A., El-Maghraby, O.M.O., 1992: Fungal flora and aflatoxin associated with cocoa, roasted coffee and tea powders in Egypt. Cryptogamie, Mycologie, 13, 31–45 (196).

Abdel-Hafez, A.I.I., Saber, S.M., 1993: Mycoflora and mycotoxins of hazelnut (*Corylus avellana* L.) and walnut (*Juglans regia* L.) seeds in Egypt. Zentralblatt für Mikrobiologie 148, 137–147 (436).

Abdel-Sater, M.A., Zohri, A.A., Ismail, M.A., 2001: Natural contamination of some Egyptian fruit juices and beverages by mycoflora and mycotoxins. Journal of Food Science and Technology 38, 407–411 (443).

Abdulkadar, A.H.W., Al-Ali, A., Al-Jedah, J., 2000: Aflatoxin contamination in edible nuts imported in Qatar. Food Control 1, 157–160 (424).

Abdulkadar, A.H.W., Al-Ali, A., Al-Jedah, J.H., 2002: Occurrence of aflatoxin in commodities imported into Qatar, 1997–2000. Food Additives and Contaminants 19, 666-670. doi: https://doi.org/10.1080/02652030210121339 (423).

Abdulkadar, A.H.W., Al-Ali, A.A., Al-Kildi, A.M., Al-Jedah, J.H., 2004: Mycotoxins in food products available in Qatar. Food Control 15, 543-548. doi: https://doi.org/10.1016/j.foodcont.2003.08.008 (229).

Abeywickrama, K., Bean, G.A., 1991: Toxigenic *Aspergillus flavus* and aflatoxins in Sri Lanka medicinal plant material. Mycopathologia 113, 187-190 (60).

Abia, W.A., Warth, B., Sulyok, M., Krska, R., Tchana, A.N., Njobeh, P.B., Dutton, M.F., Moundipa, P.F., 2013: Determination of multi-mycotoxin occurrence in cereals, nuts and their product in Cameroon by liquid chromatography tandem mass spectrometry (LC-MS/MS). Food Control 31, 438-453. doi: https://doi.org/10.1016/j.foodcont.2012.10.006 (659).

Abiala, M.A., Ezekiel, C.N., Chukwura, N.I., Odebode, A.C., 2011: Toxigenic *Aspergillus* section *Flavi* and aflatoxins in dried yam chips in Oyo State, Nigeria. Academia Arena 3, 42-49 (660).

Ackermann, Y., Curtui, V., Dietrich, R., Gross, M., Latif, H., Märtlbauer, E., Usleber, E., 2011: Widespread occurrence of low levels of alternariol in apple and tomato products, as determined by comparative immunochemical assessment using monoclonal and polyclonal antibodies. Journal of Agricultural and Food Chemistry 59, 6360-6368. doi: https://doi.org/10.1021/jf201516f (439).

Adachi, Y., Hara, M., Kumazawa, N. H., Hirano, K., Ueno, I., Egawa, K., 1991: Detection of aflatoxin B_1 in imported food products into Japan by enzyme-linked immunosorbent assay and high performance liquid chromatography. Journal of Veterinary Medical Science 53, 49-52 (171).

Adedeji, B.S., Ezeokoli, O.T., Ezekiel, C.N., Obadina, A.O., Somorin, Y.M., Sulyok, M., Adeleke, R.A., Warth, B., Nwangburuka, C.C., Omemu, A.M., Oyewole, O.B., Krska, R., 2017: Bacterial species and mycotoxin contamination associated with locust bean, melon and their fermented products in south-western Nigeria. International Journal of Food Microbiology 258, 73-80. doi:https://doi.org/10.1016/j.ijfoodmicro.2017.07.014 (361).

Adekoya, I., Njobeh, P., Obadina, A., Chilaka, C., Okoth, S., De Boevre, M., De Saeger, S., 2017: Awareness and prevalence of mycotoxin contamination in selected Nigerian fermented foods. Toxins (Basel), 9(11), 363. doi: https://doi.org/10.3390/toxins9110363 (330).

Afolabi, C.G., Ezekiel, C.N., Kehinde, I.A., Olaolu, A.W., Ogunsanya, O.M., 2015: Contamination of groundnut in south-western Nigeria by aflatoxigenic fungi and aflatoxins in relation to processing. Journal of Phytopathology 163, 279-286. doi: https://doi.org/10.1111/jph.12317 (276).

Ahmad, B., Ashiq, S., Hussain, A., Bashir, S., Hussain, M., 2014: Evaluation of mycotoxins, mycobiota, and toxigenic fungi in selected medicinal plants of Khyber Pakhtunkhwa, Pakistan. Fungal Biology 118, 776-784. doi: https://doi.org/10.1016/j.funbio.2014.06.002 (661).

Ahn, J., Kim, D., Jang, H.-S., Kim, Y., Shim, W.-B., Chung, D.-H., 2010: Occurrence of ochratoxin A in Korean red paprika and factors to be considered in prevention strategy. Mycotoxin Research 26, 279-286. doi: https://doi.org/10.1007/s12550-010-0067-2 (588).

Ahn, S., Lee, S., Lee, J., Kim, B., 2016: Accurate determination of ochratoxin A in Korean fermented soybean paste by isotope dilution-liquid chromatography tandem mass spectrometry. Food Chemistry 190, 368-373. doi: https://doi.org/10.1016/j.foodchem.2015.05.114 (662).

Aidoo, K.E., Mohamed, S.M., Candlish, A.A., Tester, R.F., Elgerbi, A.M., 2011: Occurrence of fungi and mycotoxins in some commercial baby foods in North Africa. Food and Nutrition Sciences 2, 751-758. doi: https://doi.org/10.4236/fns.2011.27103 (663).

Akdeniz, A.S., Ozden, S., Alpertunga, B., 2013: Ochratoxin A in dried grapes and grape-derived products in Turkey. Food Additives and Contaminants: Part B 6, 265-269. doi: https://doi.org/10.1080/19393210.2013.814719 (665).

Åkerstrand, A.K., Josefsson, E., 1979: Fungi and mycotoxins in beans and peas. Vår Föda 31, 405-414 (144).

Åkerstrand, A.K., Molander, A., Andersson, A., Nilsson, G., 1976: Occurrence of moulds and mycotoxins in frozen blueberries. Vår Föda 28, 197-200 (156).

Aksoy, U., Eltem, R., Meyvaci, K.B., Altindisli, A., Karabat, S., 2007: Five-year survey of ochratoxin A in processed sultanas from Turkey. Food Additives and Contaminants 24, 292-296. doi: https://doi.org/10.1080/02652030601039021 (646).

Al-Hazmi, N.A., 2010: Determination of patulin and ochratoxin A using HPLC in apple juice samples in Saudi Arabia. Saudi Journal of Biological Sciences 17, 353-359. doi: https://doi.org/10.1016/j.sjbs.2010.06.006 (444).

Ali, N., 2000: Aflatoxins in Malaysian food. Mycotoxins 50, 31-35 (233).

Ali, N., Hashim, N.H., Saad, B., Safan, K., Nakajima, M., Yoshizawa, T., 2005: Evaluation of a method to determine the natural occurrence of aflatoxins in commercial traditional herbal medicines from Malaysia and Indonesia. Food and Chemical Toxicology 43, 1763-1772. doi: https://doi.org/10.1016/j.fct.2005.05.019 (61).

Ali, N., Hashim, N.H., Shuib, N.S., 2015: Natural occurrence of aflatoxins and ochratoxin A in processed spices marketed in Malaysia. Food Additives and Contaminants: Part A 32, 518-532. doi: https://doi.org/10.1080/19440049.2015.1011712 (530).

Ali, N., Hashim, N.H., Yoshizawa, T., 1999: Evaluation and application of a simple and rapid method for the analysis of aflatoxins in commercial foods from Malaysia and the Philippines. Food Additives and Contaminants 16, 273-280 (567).

Alozie, T.C., Rotimi, C.N., Oyibo, B.B., 1980: Production of aflatoxin by *Aspergillus flavus* (UBMI) in some Nigerian indigenous beverages and foodstuffs. Mycopathologia 70, 125-128 (346).

Alpert, M.E., Hutt, M.S.R., Wogan, G.N., Davidson, C.S., 1971: Association between aflatoxin content of food and hepatoma frequency in Uganda. Cancer 28, 253-260 (137).

Al-Taher, F., Banaszewski, K., Jackson, L., Zweigenbaum, J., Ryu, D., Cappozzo, J., 2013: Rapid method for the determination of multiple mycotoxins in wines and beers by LC-MS/MS using a stable isotope dilution assay. Journal of Agricultural and Food Chemistry 61, 2378-2384. doi: https://doi.org/10.1021/jf304729f (666).

Altiokka, G., Can, N.O., Atkoşar, Z., Aboul-Enein, H.Y., 2009: Determination of ochratoxin A in Turkish wines. Journal of Food and Drug Analysis 17, 467-473 (671).

Alvarado, M., Galarce-Bustos, O., Vega, M., Aranda, M., 2013: Fast and selective determination of ochratoxin A in wines using an optimized and validated liquid chromatographic method. Food Analytical Methods 6, 621-629. doi: https://doi.org/10.1007/s12161-012-9446-y (672).

Alvito, P.C., Sizoo, E.A., Almeida, C.M.M., Van Egmond, H.P., 2010: Occurrence of aflatoxins and ochratoxin A in baby foods in Portugal. Food Analytical Methods 3, 22-30. doi: https://doi.org/10.1007/s12161-008-9064-x (88).

Amézqueta, S., González-Peñas, E., Murillo, M., López De Cerain, A., 2004: Validation of a high-performance liquid chromatography analytical method for ochratoxin A quantification in cocoa beans. Food Additives and Contaminants 21, 1096-1106. doi: https://doi.org/10.1080/02652030400019422 (270).

Amézqueta, S., González-Peñas, E., Murillo, M., López De Cerain, A., 2005: Occurrence of ochratoxin A in cocoa beans: effect of shelling. Food Additives and Contaminants 22, 590-596. doi: https://doi.org/10.1080/02652030500130160 (271).

Andersson, A., Josefsson, E., Nilsson, G., Åkerstrand, K., 1977: Fungi and patulin in fruits and berries. Vår Föda 29, 292-298 (155).

Andrade, P.D., Homem de Mello, M., França, J.A., Caldas, E.D., 2013: Aflatoxins in food products consumed in Brazil: a preliminary dietary risk assessment. Food Additives and Contaminants: Part A 30, 127-136. doi: https://doi.org/10.1080/19440049.2012.720037 (208).

Anfossi, L., Giovannoli, C., Giraudi, G., Biagioli, F., Passini, C., Baggiani, C., 2012: A lateral flow immunoassay for the rapid detection of ochratoxin A in wine and grape must. Journal of Agricultural and Food Chemistry 60, 11491-11497. doi: https://doi.org/10.1021/j3031666 (674).

Anli, E., Çabuk, B., Vural, N., Başpinar, E., 2005: Ochratoxin A in Turkish wines. Journal of Food Biochemistry 29, 611-623 (611).

Aoyama, K., Nakajima, M., Tabata, S., Ishikuro, E., Tanaka, T., Norizuki, H., Itoh, Y., Fujita, K., Kai, S., Tsutsumi, T., Takahashi, M., Tanaka, H., IIzuka, S., Ogiso, M., Maeda, M., Yamaguchi, S., Sugiyama, K.-I., Sugita-Konishi, Y., Kumagai, S., 2010: Four-year surveillance for ochratoxin A and fumonisins in retail foods in Japan. Journal of Food Protection 73, 344-352 (85).

Araújo, A.C.P., Telles, D.L., Das Graças Almeida, M., Nascimento, E.S., 1994: Determination of aflatoxins in peanuts marketed in Recife-PE, Brazil. Alimentaria 249, 35-37 (378).

Ardic, M., Karakaya, Y., Atasever, M., Durmaz, H., 2008: Determination of aflatoxin B_1 levels in deep-red ground pepper (isot) using immunoaffinity column combined with ELISA. Food and Chemical Toxicology 46, 1596-1599. doi: https://doi.org/10.1016/j.fct.2007.12.025 (177).

Aresta, A., Vatinno, R., Palmisano, F., Zambonin, C.G., 2006: Determination of ochratoxin A in wine at sub ng/mL levels by solid-phase microextraction coupled to liquid chromatography with fluorescence detection. Journal of Chromatography A 1115, 196-201. doi: https://doi.org/10.1016/j.chroma.2006.02.092 (612).

Ariño, A., Herrera, M., Estopañan, G., Juan, T., 2007: High levels of ochratoxin A in licorice and derived products. International Journal of Food Microbiology 114, 366-369. doi: https://doi.org/10.1016/j.ijfoodmicro.2006.08.005 (505).

Ariño, A., Herrera, M., Estopañan, G., Rota, M.C., Carramiñana, J.J., Juan, T., Herrera, A., 2009: Aflatoxins in bulk and pre-packed pistachios sold in Spain and effect of roasting. Food Control 20, 811-814. doi: https://doi.org/10.1016/j.foodcont.2008.10.013 (559).

Arranz, I., Stroka, J., Neugebauer, M., 2006: Determination of aflatoxin B_1 in tiger nut-based soft drinks. Food Additives and Contaminants 23, 305-308. doi: https://doi.org/10.1080/02652030500415652 (342).

Arroyo-Manzanares, N., García-Campaña, A.M., Gámiz-Gracia, L., 2013: Multiclass mycotoxin analysis in *Silybum marianum* by ultra high performance liquid chromatography-tandem mass spectrometry using a procedure based on QuEChERS and dispersive liquid-liquid microextraction. Journal of Chromatography A 1282. 11-19. doi: https://doi.org/10.1016/j.chroma.2013.01.072 (675).

Arzandeh, S., Selamat, J., Lioe, H., 2010: Aflatoxin in raw peanut kernels marketed in Malaysia. Journal of Food and Drug Analysis 18, 44-50 (676).

Asadi, M., Beheshti, H.R., Feizy, J., 2011: A survey of aflatoxins in sesame in Iran. Mycotoxin Research 27, 259-263. doi: https://doi.org/10.1007/s12550-011-0102-y (677).

Asai, T., Tsuchiya, Y., Okano, K., Piscoya, A., Nishi, C.Y., Ikoma, T., Oyama, T., Ikegami, K., Yamamoto, M., 2012: Aflatoxin contamination of red chili pepper from Bolivia and Peru, countries with high gallbladder cancer incidence rates. Asian Pacific Journal of Cancer Prevention 13, 5167-5170. doi: https://doi.org/10.7314/APJCP.2012.13.10.5167 (678).

Asam, S., Konitzer, K., Rychlik, M., 2011: Precise determination of the *Alternaria* mycotoxins alternariol and alternariol monomethyl ether in cereal, fruit and vegetable products using stable isotope dilution assays. Mycotoxin Research 27, 23-28. doi: https://doi.org/10.1007/s12550-010-0071-6 (46).

Asam, S., Konitzer, K., Schieberle, P., Rychlik, M., 2009: Stable isotope dilution assays of alternariol and alternariol monomethyl ether in beverages. Journal of Agricultural and Food Chemistry 57, 5152-5160. doi: https://doi.org/10.1021/jf900450w (160).

Asam, S., Lichtenegger, M., Liu, Y., Rychlik, M., 2012: Content of the *Alternaria* mycotoxin tenuazonic acid in food commodities determined by a staple isotope dilution assay. Mycotoxin Research 28, 9-15. doi: https://doi.org/10.1007/s12550-011-0111-y (166).

Asam, S., Rychlik, M., 2013: Potential health hazards due to the occurrence of the mycotoxin tenuazonic acid in infant food. European Food Research and Technology 236, 491-497 doi: https://doi.org/10.1007/s00217-012-1901-x (581).

Asghar, M.A., Ahmed, A., Iqbal, J., 2016a: Aflatoxins and ochratoxin A in export quality raisins collected from different areas of Pakistan. Food Additives and Contaminants: Part B 9, 51-58. doi: https://doi.org/10.1080/19393210.2015.1127293 (680).

Asghar, M.A., Iqbal, J., Ahmed, A., Khan, M.A., Shamsuddin, Z.A., 2014: Aflatoxin B_1 in betel nuts (*Areca catechu* L.) imported to Pakistan from different regions of South Asia. Food Additives and Contaminants: Part B 7, 176-181. doi: https://doi.org/10.1080/19393210.2013.869771 (681).

Asghar, M.A., Zahir, E., Rantilal, S., Ahmed, A., Iqbal, J., 2016b: Aflatoxins in composite spices collected from local markets of Karachi, Pakistan. Food Additives and Contaminants: Part B 9, 113-119. doi: https://doi.org/10.1080/19393210.2016.1150356 (393).

Assaf, H., Betbeder, A.-M., Creppy, E.E., Pallardy, M., Azouri, H., 2004: Ochratoxin A levels in human plasma and foods in Lebanon. Human & Experimental Toxicology 23, 495-501. doi: https://doi.org/10.1191/0960327104ht481oa (502).

Assunção, R., Martins, C., Dupont, D., Alvito, P., 2016: Patulin and ochratoxin A co-occurrence and their bioaccessibility in processed cereal-based foods: a contribution for Portuguese children risk assessment. Food and Chemical Toxicology 96, 205-214. doi: https://doi.org/10.1016/j.fct.2016.08.004 (551).

Awe, M.J., Schranz, J.L., 1981: High pressure liquid chromatographic determination of aflatoxins in spices. Journal of the Association of Official Analytical Chemists 64, 1377-1382 (167).

Aycicek, H., Aksoy, A., Saygi, S., 2005: Determination of aflatoxin levels in some dairy and food products which consumed in Ankara, Turkey. Food Control 16, 263-266. doi: https://doi.org/10.1016/j.foodcont.2004.03.004 (274).

Aydin, A., Emin Erkan, M., Başkaya, R., Ciftcioglu, G., 2007: Determination of aflatoxin B_1 levels in powdered red pepper. Food Control 18, 1015-1018. doi: https://doi.org/10.1016/j.foodcont.2006.03.013 (179).

Azaiez, I., Font, G., Mañes J., Fernández-Franzón, M., 2015: Survey of mycotoxins in dates and dried fruits from Tunisian and Spanish markets. Food Control 51, 340-346. doi: https://doi.org/10.1016/j.foodcont.2014.11.033 (720).

Azer, M., Cooper, C., 1991: Determination of aflatoxins in foods using HPLC and a commercial ELISA system. Journal of Food Protection 54, 291-294 (420).

Aziz, N.H., Youssef, Y.A., 1991: Occurrence of aflatoxins and aflatoxin-producing moulds in fresh and processed meat in Egypt. Food Additives and Contaminants 8, 321-331 (104).

Bacaloni, A., Cavaliere, C., Faberi, A., Pastorini, E., Samperi, R., Laganà, A., 2005: Automated on-line solid-phase extraction-liquid chromatography-electrospray tandem mass spectrometry method for the determination of ochratoxin A in wine and beer. Journal of Agricultural and Food Chemistry 53, 5518-5525. doi: https://doi.org/10.1021/jf050254+ (613).

Baert, K., De Meulenaer, B., Kamala, A., Kasase, C., Devlieghere, F., 2006: Occurrence of patulin in organic, conventional, and handcrafted apple juices marketed in Belgium. Journal of Food Protection 69, 1371-1378 (472).

Baltaci, C., İlyasoğlu, H., Cavrar, S., 2012: Aflatoxin levels in raw and processed hazelnuts in Turkey. Food Additives and Contaminants: Part B 5, 83-86. doi: https://doi.org/10.1080/19393210.2012.656146 (682).

Bankole, S.A., 1995: Mycoflora, mycotoxin producing fungi and mycotoxins from melon ball snacks in Nigeria. Chemie, Mikrobiologie & Technologie der Lebensmittel 17, 164-168 (529).

Bankole, S.A., Adebanjo, A., 2003: Aflatoxin contamination of dried yam chips marketed in Nigeria. Tropical Science 43, 201-203 (258).

Bankole, S.A., Adenusi, A.A., Lawal, O.S., Adesanya, O.O., 2010: Occurrence of aflatoxin B_1 in food products derivable from "egusi" melon seeds consumed in southwestern Nigeria. Food Control 21, 974-976. doi: https://doi.org/10.1016/j.foodcont.2009.11.014 (596).

Bankole, S.A., Eseigbe, D.A., 1996: Occurrence of mycoflora and aflatoxins in marketed tiger nut in Nigeria. Crop Research 11, 219-223 (564).

Bankole, S.A., Lawal, O.A., Adebanjo, A., 2004a: Storage practices and aflatoxin B_1 contamination of "egusi" melon seeds in Nigeria. Tropical Science 44, 150-153 (668).

Bankole, S.A., Mabekoje, O.O., 2004: Mycoflora and occurrence of aflatoxin B_1 in dried yam chips from markets in Ogun and Oyo States, Nigeria. Mycopathologia 157, 111-115 (257).

Bankole, S.A., Ogunsanwo, B.M., Eseigbe, D.A., 2005: Aflatoxins in Nigerian dry-roasted groundnuts. Food Chemistry 89, 503-506. doi: https://doi.org/10.1016/j.foodchem.2004.03.004 (398).

Bankole, S.A., Ogunsanwo, B.M., Mabekoje, O.O., 2004b: Natural occurrence of moulds and aflatoxin B_1 in melon seeds from markets in Nigeria. Food and Chemical Toxicology 42, 1309-1314. doi: https://doi.org/10.1016/j.fct.2004.03.015 (531).

Bankole, S.A., Ogunsanwo, B.M., Osho, A., Adewuyi, G.O., 2006: Fungal contamination and aflatoxin B_1 of "egusi" melon seeds in Nigeria. Food Control 17, 814-818. doi: https://doi.org/10.1016/j.foodcont.2005.05.008 (532).

Barreira, M.J., Alvito, P.C., Almeida, C.M.M., 2010: Occurrence of patulin in apple-based-foods in Portugal. Food Chemistry 121, 653-658. doi: https://doi.org/10.1016/j.foodchem.2009.12.085 (486).

Barros, G., García, D., Oviedo, M.S., Ramirez, M.L., Torres, A., Lattanzio, V., Pascale, M., Chulze, S., 2011: Survey of T-2 and HT-2 toxins in soybean and soy meal from Argentina using immunoaffinity clean-up and high performance liquid chromatography. World Mycotoxin Journal 4, 189-197. doi: https://doi.org/10.3920/WMJ2010.1272 (636).

Basaran, P., Ozcan, M., 2009: Occurrence of aflatoxins in various nuts commercialized in Turkey. Journal of Food Safety 29, 95-105 (409).

Baumann, U., Zimmerli, B., 1988: A simple determination of ochratoxin A in foods. Mitteilungen aus dem Gebiete der Lebensmitteluntersuchung und Hygiene 79, 151-158 (363).

Baydar, T., Engin, A.B., Girgin, G., Aydin, S., Şahin, G., 2005: Aflatoxin and ochratoxin in various types of commonly consumed retail ground samples in Ankara, Turkey. Annals of Agricultural and Environmental Medicine 12, 193-197 (368).

Beebe, R.M., 1978: Reverse phase high pressure liquid chromatographic determination of aflatoxins in foods. Journal of the Association of Official Analytical Chemists 61, 1347-1352 (11).

Belajová, E., Rauová, D., 2007: Determination of ochratoxin A and its occurrence in wines of Slovakian retail. Journal of Food and Nutrition Research 46, 68-74 (684).

Beljaars, P.R., Schumans, J.C.H.M., Koken, P.J., 1975: Quantitative fluorodensitometric determination and survey of aflatoxins in nutmeg. Journal of the Association of Official Analytical Chemists 58, 263-271 (546).

Bellí, N., Marín, S., Duaigües, A., Ramos, A.J., Sanchis, V., 2004a: Ochratoxin A in wines, musts and grape juices from Spain. Journal of the Science of Food and Agriculture 84, 591-594. doi: https://doi.org/10.1002/jsfa.1702 (508).

Bellí, N., Pardo, E., Marín, S., Farré, G., Ramos, A.J., Sanchis, V., 2004b: Occurrence of ochratoxin A and toxigenic potential of fungal isolates from Spanish grapes. Journal of the Science of Food and Agriculture 84, 541-546. doi: https://doi.org/10.1002/jsfa.1658 (534).

Beloglazova, N.V., Goryacheva, I.Yu., Rusanova, T.Yu., Yurasov, N.A., Galve, R., Marco, M.-P., De Saeger, S., 2010: Gel-based immunotest for simultaneous detection of 2,4,6-trichorophenol and ochratoxin A in red wine. Analytica Chimica Acta 672, 3-8. doi: https://doi.org/10.1016/j.aca.2010.05.024 (685).

Beretta, B., Gaiaschi, A., Galli, C.L., Restani, P., 2000: Patulin in apple-based foods: occurrence and safety evaluation. Food Additives and Contaminants 17, 399-406 (30).

Bertuzzi, T., Rastelli, S., Pietri, A., 2015: *Aspergillus* and *Penicillium* toxins in chestnuts and derived products produced in Italy. Food Control 50, 876-880. doi: https://doi.org/10.1016/j.foodcont.2014.10.047 (687).

Bhat, R.V., Vasanthi, S., 2005: Natural occurrence of ochratoxins in Indian coffee. Indian Journal of Nutrition and Dietetics 42, 106-113 (301).

Bhat, R.V., Vasanthi, S., Sashidhar Rao, B., Nageswara Rao, R., Sudershan Rao, V., Nagaraja, K.V., Girija Bai, R., Krishna Prasad, C.A., Vanchinathan, S., Rajat Roy, Subrata Saha, Mukherjee, A., Ghosh, P.K., Toteja, G.S., Saxena, B.N., 1996: Aflatoxin B_1 contamination in groundnut samples collected from different geographical regions of India: a multicentre study. Food Additives and Contaminants 13, 325-331 (379).

Bhavanishankar, T.N., Shantha, T., 1987: Natural occurrence of *Fusarium* toxins in peanut, sorghum and maize from Mysore (India). Journal of the Science of Food and Agriculture 40, 327-332 (434).

Bilgrami, Z., Ghaffar, A., 1999: Detection of aflatoxin in almond seeds. Pakistan Journal of Botany 31, 227-230 (8).

Bircan, C., 2005: The determination of aflatoxins in spices by immunoaffinity extraction using HPLC. International Journal of Food Science and Technology 40, 929-934. doi: https://doi.org/10.1111/j.1365-2621.2005.01025x (180).

Bircan, C., 2009: Incidence of ochratoxin A in dried fruits and co-occurrence with aflatoxins in dried figs. Food and Chemical Toxicology 47, 1996-2001. doi: https://doi.org/10.1016/j.fct.2009.05.008 (55).

Bircan, C., Barringer, S.A., Ulken, U., Pehlivan, R., 2008a: Increased aflatoxin contamination of dried figs in a drought year. Food Additives and Contaminants 25, 1400-1408. doi: https://doi.org/10.1080/02652030802163414 (358).

Bircan, C., Barringer, S.A., Ulken, U., Pehlivan, R., 2008b: Aflatoxin levels in dried figs, nuts and paprika for export from Turkey. International Journal of Food Science and Technology 43, 1492-1498. doi: https://doi.org/10.1111/j.1365-2621.2008.01726.x (360).

Blesa, J., Soriano, J.M., Moltó, J.C., Mañes, J., 2004a: Concentration of ochratoxin A in wines from supermarkets and stores of Valencian community (Spain). Journal of Chromatography A 1054, 397-401. doi: https://doi.org/10.1016/j.chroma.2004.04.058 (517).

Blesa, J., Soriano, J.M., Moltó, J.C., Mañes, J., 2004b: Limited survey for the presence of aflatoxins in foods from local markets and supermarkets in Valencia, Spain. Food Additives and Contaminants 21, 165-171. doi: https://doi.org/10.1080/0265203031000 (435).

Blesa, J., Soriano, J.M., Moltó, J.C., Marín, R., Mañes, J., 2003: Determination of aflatoxins in peanuts by matrix solid-phase dispersion and liquid chromatography. Journal of Chromatography A 1011, 49-54. doi: https://doi.org/10.1016/S0021-9673(03)01102-6 (391).

Bonerba, E., Ceci, E., Conte, R., Tantillo, G., 2010a: Survey of the presence of patulin in fruit juices. Food Additives and Contaminants: Part B 3, 114-119. doi: https://doi.org/10.1080/19393210.2010.490882 (688).

Bonerba, E., Conte, R., Ceci, E., Tantillo, G., 2010b: Assessment of dietary intake of patulin from baby foods. Journal of Food Science 75, T123-T125. doi: https://doi.org/10.1111/j.1750-3841.2010.01743.x (109).

Boonzaaijer, G., Van Osenbruggen, W.A., Kleinnijenhuis, A.J., Van Dongen, W.D., 2008: An exploratory investigation of several mycotoxins and their natural occurrence in flavor ingredients and spices, using a multi-mycotoxin LC-MS/MS method. World Mycotoxin Journal 1, 167-174 (184).

Bosch, U., Mirocha, C.J., 1992: Toxin production by *Fusarium* species from sugar beets and natural occurrence of zearalenone in beets and beet fibers. Applied and Environmental Microbiology 58, 3233-3239 (689).

Boutrif, E., Jemmali, M., Campbell, A.D., Pohland, A.E., 1977a: Aflatoxin in Tunisian foods and foodstuffs. Annales de la Nutrition et de l´Alimentation 31, 431-434 (380).

Boutrif, E., Jemmali, M., Pohland, A.E., Campbell, A.D., 1977b: Aflatoxin in Tunisian Aleppo pine nuts. Journal of the Association of Official Analytical Chemists 60, 747-748 (555).

Boyacioğlu, D., Gönül, M., 1990: Survey of aflatoxin contamination of dried figs grown in Turkey in 1986. Food Additives and Contaminants 7, 235-237 (717).

Boyd, M.L., Cotty, P.J., 2001: *Aspergillus flavus* and aflatoxin contamination of leguminous trees of the Sonoran Desert in Arizona. Phytopathology 91, 913-919 (500).

Brackett, R.E., Marth, E.H., 1979: Patulin in apple juice from roadside stands in Wisconsin. Journal of Food Protection 42, 862-863 (454).

Brera, C., Debegnach, F., De Santis, B., Iafrate, E., Pannunzi, E., Berdini, C., Prantera, E., Gregori, E., Miraglia, M., 2011: Ochratoxin A in cocoa and chocolate products from the Italian market: occurrence and exposure assessment. Food Control 22, 1663-1667. doi: https://doi.org/10.1016/j.foodcont.2011.03.026 (266).

Brera, C., Debegnach, F., Minardi, V., Prantera, E., Pannunzi, E., Faleo, S., De Santis, B., Miraglia, M., 2008: Ochratoxin A contamination in Italian wine samples and evaluation of the exposure in the Italian population. Journal of Agricultural and Food Chemistry 56, 10611-10618. doi: https://doi.org/10.1021/jf8016282 (516).

Brera, C., Soriano, J.M., Debegnach, F., Miraglia, M., 2005: Exposure assessment to ochratoxin A from the consumption of Italian and Hungarian wines. Microchemical Journal 79, 109-113. doi: https://doi.org/10.1016/j.microc.2004.10.004 (518).

Bresch, H., Urbanek, M., Hell, K., 2000a: Ochratoxin A in coffee, tea and beer. Archiv für Lebensmittelhygiene 51, 89-94 (197).

Bresch, H., Urbanek, M., Nusser, M., 2000b: Ochratoxin A in food containing liquorice. Nahrung 44, 276-278 (504).

Broggi, L., Reynoso, C., Resnik, S., Martinez, F., Drunday, V., Bernal, Á.R., 2013: Occurrence of alternariol and alternariol monomethyl ether in beverages from the Entre Rios Province market, Argentina. Mycotoxin Research 29, 17-22. doi: https://doi.org/10.1007/s12550-012-0147-6 (498).

Bulato-Jayme, J., Almero, E.A., Castro, M.C.A., Jardeleza, M.T., Salamat, L.A., 1982: A case-control dietary study of primary liver cancer risk from aflatoxin exposure. International Journal of Epidemiology 11, 112-119 (1).

Burda, K., 1992: Incidence of patulin in apple, pear, and mixed fruit products marketed in New South Wales. Journal of Food Protection 55, 796-798 (31).

Burdaspal, P., Legarda, T., 2007: Occurrence of ochratoxin A in sweet wines in Spain and other countries. Food Additives and Contaminants 24, 976-986. doi: https://doi.org/10.1080/02652030701311155 (519).

Burdaspal, P.A., Legarda, T.M., 1998: Ochratoxin A in roasted and soluble coffees marketed in Spain. Alimentaria 35, 31-35 (296).

Burdaspal, P.A., Legarda, T.M., 1999: Ochratoxin A in wines and grape products originated from Spain and other European countries. Alimentaria (Madrid) 299, 107-113 (511).

Burdaspal, P.A., Legarda, T.M., 2003: Occurrence of ochratoxin A in samples of different types of chocolate and cacao powder, marketed in Spain and fifteen foreign countries. Alimentaria 40, 143-153 (264).

Burdaspal, P.A., Legarda, T.M., Iñigo, S., 1998: A survey of aflatoxins in pistachios in the community of Madrid, Spain. Alimentaria 364, 106-113 (560).

Çağındı, Ö., Gürhayta, O.F., 2016: Aflatoxins and ochratoxin A in dried eggplant and green bell pepper. Food Control 70, 216-220. doi: https://doi.org/10.1016/j.foodcont.2016.05.060 (449).

Campbell, T.C., Salamat, L., 1971: Aflatoxin ingestion and excretion by humans. In: Purchase, I.F.H. (ed.) Symposium on Mycotoxins in Human Health pp. 271-280. London (134).

Campone, L., Piccinelli, A.L., Rastrelli, L., 2011: Dispersive liquid-liquid microextraction combined with high-performance liquid chromatography-tandem mass spectrometry for the identification and the accurate quantification by isotope dilution assay of ochratoxin A in wine samples. Analytical and Bioanalytical Chemistry 399, 1279-1286. doi: https://doi.org/10.1007/s00216-010-4347-7 (614).

Candlish, A.A.G., Pearson, S.M., Aidoo, K.E., Smith, J.E., Kelly, B., Irvine, H., 2001: A survey of ethnic foods for microbial quality and aflatoxin content. Food Additives and Contaminants 18, 129-136. doi: https://doi.org/10.1080/02652030010021404 (557).

Cano-Sancho, G., Marin, S., Ramos, A.J., Sanchis, V., 2009: Survey of patulin occurrence in apple juice and apple products in Catalonia, Spain, and an estimate of dietary intake. Food Additives and Contaminants: Part B 2, 59-65. doi: https://doi.org/10.1080/02652030902897747 (42).

Cano-Sancho, G., Sanchis, V., Marín, S., Ramos, A.J., 2013: Occurrence and exposure assessment of aflatoxins in Catalonia (Spain). Food and Chemical Toxicology 51, 188-193. doi: https://doi.org/10.1016/j.fct.2012.09.032 (93).

Cantàfora, A., Grossi, M., Miraglia, M., Benelli, L., 1983: Determination of ochratoxin A in coffee beans using reversed-phase high performance liquid chromatography. La Rivista della Società Italiana di Scienza dell′ Alimentazione 12, 103-108 (282).

Cao, J., Kong, W., Zhou, S., Yin, L., Wan, L., Yang, M., 2013: Molecularly imprinted polymer-based solid phase clean-up for analysis of ochratoxin A in beer, red wine, and grape juice. Journal of Separation Science 36, 1291-1297. doi: https://doi.org/10.1002/jssc.201201055 (690).

Cao, J., Zhou, S., Kong, W., Ma, X., Yang, M., Wan, L., Yang, S., 2014: Simultaneous determination of aflatoxins B_1, B_2, G_1, G_2 in *Fructus Bruceae* by high-performance liquid chromatography with online postcolumn photochemical derivatization. Journal of Separation Science 37, 2771-2778. doi: https://doi.org/10.1002/jssc.201400501 (474).

Casal, S., Vieira, T., Cruz, R., Cunha, S.C., 2014: Ochratoxin A in commercial soluble coffee and coffee substitutes. Food Research International 61, 56-60 doi: https://doi.org/10.1016/j.foodres.2014.04.045 (691).

Castellá, G., Bragulat, M.R., Cabañes, F.J., 1999: Surveillance of fumonisins in maize-based feeds and cereals from Spain. Journal of Agriculture and Food Chemistry 47, 4707-4710. doi: https://doi.org/10.1021/jf981236d (635).

Castellari, M., Fabri, S., Fabiani, A., Amati, A., Galassi, S., 2000: Comparison of different immunoaffinity clean-up procedures for high-performance liquid chromatographic analysis of ochratoxin A in wines. Journal of Chromatography A 888, 129-136 (644).

Castellari, M., Versari, A., Fabiani, A., Parpinello, G.P., Galassi, S., 2001: Removal of ochratoxin A in red wines by means of adsorption treatments with commercial fining agents. Journal of Agricultural and Food Chemistry 49, 3917-3921. doi: https://doi.org/10.1021/jf010137o (604).

Cavaliere, C., Foglia, P., Guarino, C., Nazzari, M., Samperi, R., Laganà, A., 2007: Determination of aflatoxins in olive oil by liquid chromatography-tandem mass spectrometry. Analytica Chimica Acta 596, 141-148. doi: https://doi.org/10.1016/j.aca.2007.05.055 (692).

Cervino, C., Asam, S., Knopp, D., Rychlik, M., Niessner, R., 2008: Use of isotope-labeled aflatoxins for LC-MS/MS stable isotope dilution analysis of foods. Journal of Agricultural and Food Chemistry 56, 1873-1879. doi: https://doi.org/10.1021/jf073231z (181).

Chala, A., Mohammed, A., Ayalew, A., Skinnes, H., 2013: Natural occurrence of aflatoxins in groundnut (*Arachis hypogaea* L.) from eastern Ethiopia. Food Control 30, 602-605. doi: https://doi.org/10.1016/j.foodcont.2012.08.023 (693).

Chen, A.J., Jiao, X., Hu, Y., Lu, X., Gao, W., 2015: Mycobiota and mycotoxins in traditional medicinal seeds from China. Toxins (Basel) 7, 3858-3875. doi: https://doi.org/10.3390/toxins7103858 (694).

Chen, S.-C., Friedman, L., 1966: Aflatoxin determination in seed meal. Journal of the Association of Official Analytical Chemists 49, 28–33 (525).

Chen, Y.-C., Liao, C.-D., Lin, H.-Y., Chiueh, L.-C., Shih, D.Y.-C., 2013: Survey of aflatoxin contamination in peanut products in Taiwan from 1997 to 2011. Journal of Food and Drug Analysis 21, 247-252. doi: https://doi.org/10.1016/j.jfda.2013.07.001 (540).

Cho, M.S., Kim, K., Seo, E., Kassim, N., Mtenga, A.B., Shim, W.-B., Lee, S.-H., Chung, D.-H., 2010: Occurrence of patulin in various fruit juices from South Korea: an exposure assessment. Food Science and Biotechnology 19, 1-5. doi: https://doi.org/10.1007/s10068-010-0001-6 (695).

Cho, S.-H., Lee, C.-H., Jang, M.-R., Son, Y.-W., Lee, S.-M., Choi, I.-S., Kim, S.-H., Kim, D.-B., 2008: Aflatoxins contamination in spices and processed spice products commercialized in Korea. Food Chemistry 107, 1283-1288. doi: https://doi.org/10.1016/j.foodchem.2007.08.049 (188).

Chompurat, S., Sato, K., Mahakarnchanakul, W., Sirison, J., Kladpan, S., 2006: Determination of ochratoxin A in roasted coffee and corn by immunoaffinity column cleanup with high-performance thin layer chromatography. Mycotoxins Suppl 4, P 197-201. doi: https://doi.org/10.2520/myco1975.2006 (667).

Chong, Y.H., Beng, C.G., 1965: Aflatoxins in unrefined groundnut oil. Medical Journal of Malaya XX, 49-50 (577).

Chourasia, H.K., 1990: Aflatoxin contamination in drug yielding plants. Journal of the Indian Botanical Society 69, 281–283 (696).

Chourasia, H.K., 1995: Mycobiota and mycotoxins in herbal drugs of Indian pharmaceutical industries. Mycological Research 99, 697–703 (65).

Chu, F.S., Fan, T.S.L., Zhang, G.-S., Xu, Y.-C., Faust, S., McMahon, P.L., 1987: Improved enzyme-linked immunosorbent assay for aflatoxin B_1 in agricultural commodities. Journal of the Association of Official Analytical Chemists 70, 854–857 (387).

Chulze, S.N., Torres, A.M., Dalcero, A.M., Etcheverry, M.G., Ramirez, M.L., Farnochi, M.C., 1995: *Alternaria* mycotoxins in sunflower seeds: incidence and distribution of the toxins in oil and meal. Journal of Food Protection 58, 1133–1135 (648).

Chun, H. S., Kim, H. J., Ok, H. E., Hwang, J.-B., Chung, D.-H., 2007: Determination of aflatoxin levels in nuts and their products consumed in South Korea. Food Chemistry 102, 385-391. doi: https://doi.org/10.1016/j.foodchem.2006.05.031 (220).

Chung, S. W.-C., Kwong, K.P., Tang, A.S.P., Yeung, S.T.K., 2009: Ochratoxin A levels in foodstuffs marketed in Hong Kong. Journal of Food Composition and Analysis. 22, 756-761. doi: https://doi.org/10.1016/j.jfca.2009.02.014 (697).

Colak, H., Bingol, E.B., Hampikyan, H., Nazli, B., 2006: Determination of aflatoxin contamination in red-scaled, red and black pepper by ELISA and HPLC. Journal of Food and Drug Analysis 14, 292-296 (178).

Commission of the European Communities, 1997. Reports on tasks for scientific cooperation. Report of experts participating in task 3.2.2: assessment of dietary intake of ochratoxin A by the population of EU member states. Report EUR 17523 EN (194).

Commission of the European Communities, 2002. Reports on tasks for scientific cooperation. Report of experts participating in task 3.2.7: assessment of dietary intake of ochratoxin A by the population of EU member states (698).

Comuzzo, P., Rauhut, D., Werner, M., Lagazio, C., Zironi, R., 2013: A survey on wines from organic viticulture from different European countries. Food Control 34, 274-282. doi: https://doi.org/10.1016/j.foodcont.2013.04.039 (699). (and personal communication).

Copetti, M.V., Iamanaka, B.T., Nester, M.A., Efraim, P., Taniwaki, M.H., 2013: Occurrence of ochratoxin A in cocoa by-products and determination of its reduction during chocolate manufacture. Food Chemistry 136, 100-104. doi: https://doi.org/10.1016/j.foodchem.2012.07.093 (212).

Copetti, M.V., Iamanaka, B.T., Pereira, J.L., Fungaro, M.H., Taniwaki, M.H., 2011: Aflatoxigenic fungi and aflatoxin in cocoa. International Journal of Food Microbiology 148, 141-144. doi: https://doi.org/10.1016/j.ijfoodmicro.2011.05.020 (268).

Copetti, M.V., Iamanaka, B.T., Pereira, J.L., Lemes, D.P., Nakano, F., Taniwaki, M.H., 2012a: Co-occurrence of ochratoxin A and aflatoxins in chocolate marketed in Brazil. Food Control 26, 36-41. doi: https://doi.org/10.1016/j.foodcont.2011.12.023 (260).

Copetti, M.V., Iamanaka, B.T., Pereira, J.L., Lemes, D.P., Nakano, F., Taniwaki, M.H., 2012b: Determination of aflatoxins in by-products of industrial processing of cocoa beans. Food Additives and Contaminants: Part A 29, 972-978. doi: https://doi.org/10.1080/19440049.2012.660657 (211).

Copetti, M.V., Pereira, J.L., Iamanaka, B.T., Pitt, J.I., Taniwaki, M.H., 2010: Ochratoxigenic fungi and ochratoxin A in cocoa during farm processing. International Journal of Food Microbiology 143, 67-70. doi: https://doi.org/10.1016/j.ijfoodmicro.2010.07.031 (272).

Coronel, M.B., Marin, S., Cano, G., Ramos, A.J., Sanchis, V., 2011: Ochratoxin A in Spanish retail ground roasted coffee: occurrence and assessment of the exposure in Catalonia. Food Control 22, 414-419. doi: https://doi.org/10.1016/j.foodcont.2010.09.012 (297).

Coronel, M.B., Marín, S., Cano-Sancho, G., Ramos, A.J., Sanchis, V., 2012: Exposure assessment to ochratoxin A in Catalonia (Spain) based on the consumption of cereals, nuts, coffee, wine, and beer. Food Additives and Contaminants: Part A 29, 979-993. doi: https://doi.org/10.1080/19440049.2012.660708 (432).

Cozzi, G., Pascale, M., Perrone, G., Visconti, A., Logrieco, A., 2006: Effect of *Lobesia botrana* on black aspergilli rot and ochratoxin A content in grapes. International Journal of Food Microbiology 111, S88-S92. doi: https://doi.org/10.1016/j.ijfoodmicro.2006.03.012 (157).

Cressey, P., Jones, S., 2009: Mycotoxin surveillance programme 2008-09. Aflatoxins and ochratoxin A in dried fruits and spices. ESR Client Report FW 09042. ESR, Christchurch, New Zealand (702).

Cressey, P., Jones, S., 2010: Mycotoxin surveillance programme 2009-2010. Aflatoxins in nuts and nut products. ESR Client Report FW 10036. ESR, Christchurch, New Zealand (701).

Cressey, P., Jones, S., 2011: Mycotoxin surveillance programme 2011. Ochratoxin A in cereal products, wine, beer and coffee. ESR Client Report FW 11075. ESR, Christchurch, New Zealand (703).

Cressey, P., Pearson, A., 2014: Risk Profile: Mycotoxins in the New Zealand Food Supply. ESR Client Report FW 14005. ESR, Christchurch, New Zealand (277).

Cressey, P., Thomson, B., 2006: Risk Profile: Mycotoxins in the New Zealand Food Supply. ESR Client Report FW 0617. ESR, Christchurch, New Zealand (700).

Cunha, S.C., Faria, M.A., Fernandes, J.O., 2009: Determination of patulin in apple and quince products by GC-MS using $^{13}C_{5-7}$ patulin as internal standard. Food Chemistry 115, 352-359. doi: https://doi.org/10.1016/j.foodchem.2008.11.074 (27).

Cunha, S.C., Faria, M.A., Pereira, V.L., Oliveira, T.M., Lima, A.C., Pinto, E., 2014: Patulin assessment and fungi identification in organic and conventional fruits and derived products. Food Control 44, 185-190. doi: https://doi.org/10.1016/j.foodcont.2014.03.043 (905).

Cunha, S.C., Fernandes, J.O., 2010: Development and validation of a method based on a QuEChERS procedure and heart-cutting GC-MS for determination of five mycotoxins in cereal products. Journal of Separation Science 33, 600-609. doi: https://doi.org/10.1002/jssc.200900695 (96).

Czerwiecki, L., Wilczyńska, G., Kwiecień, A., 2005: Ochratoxin A: an improvement clean-up and HPLC method used to investigate wine and grape juice on the Polish market. Food Additives and Contaminants 22, 158-162. doi: https://doi.org/10.1080/02652030500038066 (494).

Czerwiecki, L., Wilczyńska, G., Kwiecień, A., 2006: Mycotoxins in several Polish food products in 2004–2005. Mycotoxin Research 22, 159–162 (116).

D'Ovidio, K., Trucksess, M., Weaver, C., Horn, E., McIntosh, M., Bean, G., 2006: Aflatoxins in ginseng roots. Food Additives and Contaminants 23, 174-180. doi: https://doi.org/10.1080/02652030500442524 (71).

Da Motta, S., Valente Soares, L.M., 2001. Survey of Brazilian tomato products for alternariol, alternariol monomethyl ether, tenuazonic acid and cyclopiazonic acid. Food Additives and Contaminants 18, 630-634. doi: https://doi.org/10.1080/02652030110035093 (48).

Dalcero, A.M., Combina, M., Etcheverry, M., Varsavsky, E., Rodriguez, M.I., 1997: Evaluation of *Alternaria* and its mycotoxins during ensiling of sunflower seeds. Natural Toxins 5, 20–23 (649).

Daradimos, E., Marcaki, P., Koupparis, M., 2000: Evaluation and validation of two fluometric HPLC methods for the determination of aflatoxin B_1 in olive oil. Food Additives and Contaminants 17, 65–73 (570).

Dashti, B.H.R., Al-Hamli, S.F., 2010: Investigation of mycotoxins in selected food commodities in Kuwait: a case study. Journal of the American Dietetic Association 110 (Suppl.), A 92 (704).

De Almeida, A.P., Alaburda, J., Shundo, L., Ruvieri, V., Navas, S.A., Lamardo, L.C.A., Sabino, M., 2007: Ochratoxin A in Brazilian instant coffee. Brazilian Journal of Microbiology 38, 300-303. doi: https://doi.org/10.1590/S1517-83822007000200022 (294).

De Campos, M., Olszyna-Marzys, A.E., 1979: Aflatoxin contamination in grains and grain products during dry season in Guatemala. Bulletin of Environmental Contamination and Toxicology 22, 350–356 (138).

De Campos, M., Santos, C.J., Olszyna-Marzys, A.E., 1980: Aflatoxin contamination in grains from the Pacific coast in Guatemala and the effect of storage upon contamination. Bulletin of Environmental Contamination and Toxicology 24, 789–795 (136).

De Magalhães, J.T., Sodré, G.A., Viscogliosi, H., Grenier-Loustalot, M.-F., 2011: Occurrence of ochratoxin A in Brazilian cocoa beans. Food Control 22, 744-748. doi: https://doi.org/10.1016/j.foodcont.2010.11.006 (705).

De Palo, D., Gabucci, G., Valussi, S., 1979: Study on the possible presence of aflatoxine, sterigmatocystine and ochratoxine in green coffee. VIIIe Colloque Scientifique International sur le Café, Abidjan, 28 nov.-3 déc. 1977. ASIC (Paris) 1979, 539–543 (305).

De Sylos, C.M., Rodriguez-Amaya, D.B., 1999: Incidence of patulin in fruits and fruit juices marketed in Campinas, Brazil. Food Additives and Contaminants 16, 71–74 (32).

Delcourt, A., Rousset, A., Lemaître, J.P., 1994: Microbial and mycotoxic contamination of peppers and food safety. Bottelino Chimico Farmaceutico 133, 235–238 (706).

Delgado, T., Gómez-Cordovéz, C., 1998: Natural occurrence of alternariol and alternariol methyl ether in Spanish apple juice concentrates. Journal of Chromatography A 815, 93–97 (488).

Delgado, T., Gómez-Cordovés, C., Scott, P.M., 1996: Determination of alternariol and alternariol methyl ether in apple juice using solid-phase extraction and high-performance liquid chromatography. Journal of Chromatography A 731, 109–114 (487).

Demirel, R., Sariozlu, N.Y., 2014: Mycotoxigenic moulds and mycotoxins in flours consumed in Turkey. Journal of the Science of Food and Agriculture 94, 1577-1584. doi: https://doi.org/10.1002/jsfa.6460 (707).

Di Mavungu, J.D., Monbaliu, S., Scippo, M.-L., Maghuin-Rogister, G., Schneider, Y.-J., Larondelle, Y., Callebaut, A., Robbens, J., Van Petheghem, C., De Saeger, S., 2009: LC-MS/MS multi-analyte method for mycotoxin determination in food supplements. Food Additives and Contaminants: Part A 26, 885–895. doi: https://doi.org/10.1080/02652030902774649 (and personal communication) (77).

Di Stefano, V., Avellone, G., Pitonzo, R., Capocchiano, V.G., Mazza, A., Cicero, N., Dugo, G., 2015: Natural co-occurrence of ochratoxin A, ochratoxin B and aflatoxins in Sicilian red wines. Food Additives and Contaminants: Part A 32, 1343-1351 doi: https://doi.org/10.1080/19440049.2015.1055521 (708).

Diaz, G.J., Krska, R., Sulyok, M., 2015: Mycotoxins and cyanogenic glycosides in staple foods of three indigenous people of the Colombian Amazon. Food Additives and Contaminants: Part B 8, 291-297. doi: https://doi.org/10.1080/19393210.2015.1089948 (709).

Diaz, G.J., Perilla, N.S., Rojas, Y., 2001: Occurrence of aflatoxins in selected Colombian foods. Mycotoxin Research 17, 15–20 (499).

Dickens, J.W., Welty, R.E., 1969: Detecting farmers´ stock peanuts containing aflatoxin by examination for visible growth of *Aspergillus flavus*. Mycopathologia et Mycologia Applicata 37, 65–69 (416).

Dickens, J.W., Welty, R.E., 1975: Fluorescence in pistachio nuts contaminated with aflatoxin. Journal of the American Oil Chemists´ Society 52, 448–450 (561).

Diedhiou, P.M., Bandyopadhyay, R., Atehnkeng, J., Ojiambo, P.S., 2011: *Aspergillus* colonization and aflatoxin contamination of maize and sesame kernels in two agro-ecological zones in Senegal. Journal of Phytopathology 159, 268–275. doi: https://doi.org/10.1111/j.1439-0434.2010.01761.x (710). (and personal communication).

Dietrich, R., Schmid, A., Märtlbauer, E., 2001: Citrinin in fruit juices. Mycotoxin Research 17, 156–159 (711).

Dietrich, R., Usleber, E., Märtlbauer, E., Gareis, M., 1999: Nachweis des nephrotoxischen Mykotoxins Citrinin in Lebensmitteln und mit *Monascus* spp. hergestellten Lebensmittelfarbstoffen. Archiv für Lebensmittelhygiene 50, 17–21 (124).

Ding, X., Li, P., Bai, Y., Zhou, H., 2012: Aflatoxin B_1 in post-harvest peanuts and dietary risk in China. Food Control 23, 143-148. doi: https://doi.org/10.1016/j.foodcont.2011.06.026 (410).

Dini, A., Khazaeli, P., Roohbakhsh, A., Madadlou, A., Pourenamdari, M., Setoodeh, L., Askarian, A., Doraki, N., Farrokhi, H., Moradi, H., Khodadadi, E., 2013: Aflatoxin contamination level in Iran's pistachio nut during years 2009–2011. Food Control 30, 540-544. doi: https://doi.org/10.1016/j.foodcont.2012.08.012 (713).

Dogan, M., Liman, B.C., Sagdic, O., 2006: Total aflatoxin levels of instant foods in Turkey. Archiv für Lebensmittelhygiene 57, 55–58 (119).

Domijan, A.-M., Peraica, M., 2005: Ochratoxin A in wine. Arhir za Higijena Rada I Toksikologija 56, 17–20 (616).

Domijan, A.-M., Peraica, M., Žlener, V., Cvetković, B., Jurjević, Ž., Topolovec-Pintarić, S., Ivić, D., 2005: Seed-borne fungi and ochratoxin A contamination of dry beans (*Phaseolus vulgaris* L.) in the Republic of Croatia. Food and Chemical Toxicology 43, 427-432. doi: https://doi.org/10.1016/j.fct.2004.12.002 (145).

Duris, D., Mburu, J.K., Durand, N., Clarke, R., Frank, J.M., Guyot, B., 2010: Ochratoxin A contamination of coffee batches from Kenya in relation to cultivation methods and post-harvest processing treatments. Food Additives and Contaminants: Part A 27, 836-841. doi: https://doi.org/10.1080/19440040903571770 (316).

Eder, R., Paar, E., Edinger, W., Lew, H., 2002: Investigations into the ochratoxin A contents of Austrian wines with special respect to wines of the "Prädikatswein"-category. Mitteilungen Klosterneuburg 52, 125–131 (334).

El Adlouni, C., Tozlovanu, M., Naman, F., Faid, M., Pfohl-Leszkowicz, A., 2006: Preliminary data on the presence of mycotoxins (ochratoxin A, citrinin and aflatoxin B_1) in black table olives "Greek style" of Moroccan origin. Molecular Nutrition & Food Research 50, 507-512. doi: https://doi.org/10.1002/mnfr.200600055 (585).

El Khoury, A., Rizk, T., Lteif, R., Azouri, H., Delia, M.-L., Lebrihi, A., 2006: Occurrence of ochratoxin A- and aflatoxin B_1-producing fungi in Lebanese grapes and ochratoxin A content in musts and finished wines during 2004. Journal of Agricultural and Food Chemistry 54, 8977-8982. doi: https://doi.org/10.1021/jf062085e (536).

El Khoury , A., Rizk, T., Lteif, R., Azouri, H., Delia, M.-L., Lebrihi, A., 2008: Fungal contamination and aflatoxin B_1 and ochratoxin A in Lebanese wine-grapes and musts. Food and Chemical Toxicology 46, 2244-2250. doi: https://doi.org/10.1016/j.fct.2008.02.026 (533).

El-Dessouki, S., 1992: Aflatoxine in Cayenne-Pfeffer- und Paprika-Pulver. Deutsche Lebensmittel-Rundschau 88, 78 (243).

El-Gohary, A.H., 1995: Study on aflatoxins in some foodstuffs with special reference to public health hazard in Egypt. Asian-Australasian Journal of Animal Science 8, 571–575 (382).

El-Kady, I.A., El-Maraghy, S.S.M., Eman Mostafa, M., 1995: Natural occurrence of mycotoxins in different spices in Egypt. Folia Microbiologica 40, 297–300 (641).

El-Kady, I.A., Youssef, M.S., 1993: Survey of mycoflora and mycotoxins in Egyptian soybean seeds. Journal of Basic Microbiology 33, 371–378 (714).

El-Khadem, M., 1990: Aflatoxins in Egyptian peanut. An overview. Journal of Plant Diseases and Protection 97, 233–236 (383).

El-Maghraby, O.M.O., Abdel-Sater, M.A., 1993: Mycoflora and natural occurrence of mycotoxins in tobacco from cigarettes in Egypt. Zentralblatt für Mikrobiologie 148, 253–264 (654).

El-Nagerabi, S.A.F., Elshafie, A.E., 2000a: Composition of mycoflora and aflatoxins in lupine seeds from the Sudan. Phytopathologia Mediterranea 39, 257–262 (715).

El-Nagerabi, S.A.F., Elshafie, A.E., 2000b: Incidence of seed-borne fungi and aflatoxins in Sudanese lentil seeds. Mycopathologia 149, 151–156 (501).

El-Nagerabi, S.A.F., Elshafie, A.E., Abdalla, A.H., 2000: Composition of mycoflora and aflatoxins in pea seeds from the Sudan. Kuwait Journal of Science & Engineering 27, 109–120 (591).

El-Nakib, O., Pestka, J.J., Chu, F.S., 1981: Determination of aflatoxin B_1 in corn, wheat, and peanut butter by enzyme-linked immunosorbent assay and solid phase radioimmunoassay. Journal of the Association of Official Analytical Chemists 64, 1077–1082 (526).

El-Sayed, A.A.A., 1996: Natural occurrence of ochratoxin A and citrinin in food stuffs in Egypt. Mycotoxin Research 12, 41–44 (142).

Elshafie, S.Z.B., ElMubarak, A., El-Nagerabi, S.A.F., Elshafie, A.E., 2011: Aflatoxin B_1 contamination of traditionally processed peanuts butter for human consumption in Sudan. Mycopathologia 171, 435-439. doi: https://doi.org/10.1007/s11046-010-9378-2 (219).

El tawila, M.M., Neamatallah, A., Serdar, S.A., 2013: Incidence of aflatoxins in commercial nuts in the holy city of Mekkah. Food Control 29, 121-124. doi: https://doi.org/10.1016/j.foodcont.2012.06.004 (716).

Elzupir, A.O., Salih, A.O.A., Suliman, S.A., Adam, A.A., Elhussein, A.M., 2011: Aflatoxins in peanut butter in Khartoum State, Sudan. Mycotoxin Research 27, 183-186. doi: https://doi.org/10.1007/s12550-011-0094-7 (718).

Elzupir, A.O., Suliman, M.A., Ibrahim, I.A., Fadul, M.H., Elhussein, A.M., 2010: Aflatoxin levels in vegetable oils in Khartoum State, Sudan. Mycotoxin Research 26, 69-73. doi: https://doi.org/10.1007/s12550-010-0041-z (580).

Engel, G., 2000: Ochratoxin A in sweets, oil seeds and dairy products. Archiv für Lebensmittelhygiene 51, 98–101 (262).

Engelhardt, G., Ruhland, M., Wallnöfer, P.R., 1999: Occurrence of ochratoxin A in moldy vegetables and fruits analysed after removal of rotten tissue parts. Advances in Food Science 21, 88–92 (26).

Eom, T., Cho, H.-D., Kim, J., Park, M., An, J., Kim, M., Kim, S.-H., Han, S.B., 2017: Multiclass mycotoxin analysis in edible oils using a simple solvent extraction method and liquid chromatography with tandem mass spectrometry. Food Additives and Contaminants: Part A 34, 2011–2022. doi: https://doi.org/10.1080/19440049.2017.1363416 (651).

Erdogan, A., 2004: The aflatoxin contamination of some pepper types sold in Turkey. Chemosphere 56, 321-325. doi: https://doi.org/10.1016/j.chemosphere.2004.02.020 (186).

Escher, F.E., Koehler, P.E., Ayres, J.C., 1974: A study on aflatoxin and mold contaminations in improved variety pecans. Journal of Food Science 39, 1127–1129 (550).

Escobar, J., Lorán, S., Giménez, I., Ferruz, E., Herrera, M., Herrera, A., Ariño, A., 2013: Occurrence and exposure assessment of *Fusarium* mycotoxins in maize germ, refined corn oil and margarine. Food and Chemical Toxicology 62, 514-520. doi: https://doi.org/10.1016/j.fct.2013.09.020 (723).

Escobar, A., Sánchez Regueiro, O., 2002: Determination of aflatoxin B_1 in food and feedstuffs in Cuba (1990 through 1996) using an immunoenzymatic reagent kit (aflacen). Journal of Food Protection 65, 219–221 (126).

Essono, G., Ayodele, M., Akoa, A., Foko, J., Filtenborg, O., Olembo, S., 2009: Aflatoxin-producing *Aspergillus* spp. and aflatoxin levels in stored cassava chips as affected by processing practices. Food Control 20, 648-654. doi: https://doi.org/10.1016/j.foodcont.2008.09.018 (719).

Ezekiel, C.N., Kayode, F.O., Fapohunda, S.O., Olorunfemi, M.F., Kponi, B.T., 2012a: Aflatoxigenic moulds and aflatoxins in street-vended snacks in Lagos, Nigeria. Internet Journal of Food Safety 14, 83–88 (721).

Ezekiel, C.N., Sulyok, M., Somorin, Y., Odutayo, F.I., Nwabekee, S.U., Balogun, A.T., Krska, R., 2016: Mould and mycotoxin exposure assessment of melon and bush mango seeds, two common soup thickeners consumed in Nigeria. International Journal of Food Microbiology 237, 83-91 doi: https://doi.org/10.1016/j.ijfoodmicro.2016.08.019 (603).

Ezekiel, C.N., Sulyok, M., Warth, B., Krska, R., 2012b: Multi-microbial metabolites in fonio millet (acha) and sesame seeds in Plateau State, Nigeria. European Food Research and Technology 235, 285-293. doi: https://doi.org/10.1007/s00217-012-1755-2 (629).

Ezekiel, C.N., Sulyok, M., Warth, B., Odebode, A.C., Krska, R., 2012c: Natural occurrence of mycotoxins in peanut cake from Nigeria. Food Control 27, 338–342. doi: https://doi.org/10.1016/j.foodcont.2012.04.010 (and personal communication) (738).

Fakoor Janati, S.S., Beheshti, H.R., Asadi, M., Mihanparast, S., Feizy, J., 2012: Preliminary survey of aflatoxins and ochratoxin A in dried fruits from Iran. Bulletin of Environmental Contamination and Toxicology 88, 391-395. doi: https://doi.org/10.1007/s00128-011-0477-7 (53).

Fakoor Janati, S.S., Beheshti, H.R., Koshbakht Fahim, N., Feizy, J., 2011: Aflatoxins and ochratoxin A in bean from Iran. Bulletin of Environmental Contamination and Toxicology 87, 194-197. doi: https://doi.org/10.1007/s00128-011-0311-2 (131).

Fan, C., Cao, X., Liu, M., Wang, W., 2016: Determination of *Alternaria* mycotoxins in wine and juice using ionic liquid modified countercurrent chromatography as a pretreatment method followed by high-performance liquid chromatography. Journal of Chromatography A 1436, 133-140. doi: https://doi.org/10.1016/j.chroma.2016.01.069 (724).

Fazekas, B., Tar, A., Kovács, M., 2005: Aflatoxin and ochratoxin A content of spices in Hungary. Food Additives and Contaminants 22, 856-863. doi: https://doi.org/10.1080/02652030500198027 (170).

Fazekas, B., Tar, A.K., Zomborsky-Kovács, M., 2002: Ochratoxin A contamination of cereal grains and coffee in Hungary in the year 2001. Acta Veterinaria Hungarica 50, 177–188 (283).

Feizy, J., Beheshti, H.R., Asadi, M., 2012: Ochratoxin A and aflatoxins in dried vine fruits from the Iranian market. Mycotoxin Research 28, 237-242. doi: https://doi.org/10.1007/s12550-012-0145-8 (726).

Feizy, J., Beheshti, H.R., Fakoor Janati, S.S., Fahim, N.K., 2011: Survey of aflatoxins in watermelon seeds from Iran using immunoaffinity column cleanup and HPLC with fluorescence detection. Food Additives and Contaminants: Part B 4, 106-109. doi: https://doi.org/10.1080/19393210.2011.561932 (727).

Fernane, F., Cano-Sancho, G., Sanchis, V., Marin, S., Ramos, A.J., 2010a: Aflatoxins and ochratoxin A in pistachios sampled in Spain: occurrence and presence of mycotoxigenic fungi. Food Additives and Contaminants: Part B 3, 185-192. doi: https://doi.org/10.1080/19440049.2010.497257 (728).

Fernane, F., Sanchis, V., Marín, S., Ramos, A.J., 2010b: First report on mould and mycotoxin contamination of pistachios sampled in Algeria. Mycopathologia 170, 423-429. doi: https://doi.org/10.1007/s11046-010-9332-3 (562).

Ferracane, R., Tafuri, A., Logieco, A., Galvano, F., Balzano, D., Ritieni, A., 2007: Simultaneous determination of aflatoxin B_1 and ochratoxin A and their natural occurrence in Mediterranean virgin olive oil. Food Additives and Contaminants 24, 173–180. doi: https://doi.org/10.1080/02652030600986040 (575).

Festas, I., Herbert, P., Santos, L., Cabral, M., Barros, P., Alves, A., 2000: Ochratoxin A in some Portuguese wines: method validation and screening in Port Wine and Vinho Verde. American Journal of Enology and Viticulture 51, 150–154 (515).

Filali, A., Ouammi, L., Betbeder, A.M., Baudrimont, I., Soulaymani, R., Benayada, A., Creppy, E.E., 2001: Ochratoxin A in beverages from Morocco: a preliminary survey. Food Additives and Contaminants 18, 565-568. doi: https://doi.org/10.1080/02652030110035138 (493).

Flajs, D., Domijan, A.-M., Ivić, D., Cvjetković, B., Peraica, M., 2009: ELISA and HPLC analysis of ochratoxin A in red wines of Croatia. Food Control 20, 590-592. doi: https://doi.org/10.1016/j.foodcont.2008.08.021 (535).

Flamini, R., Dalla Vedova, A., De Rosso, M., Panighel, A., 2007: A new sensitive and selective method for analysis of ochratoxin A in grape and wine by direct liquid chromatography/surface-activated chemical ionization tandem mass spectrometry. Rapid Communications in Mass Spectrometry 21, 3737-3742. doi: https://doi.org/10.1002/rcm.3275 (729).

Food Standards Agency, 2002: Survey of nuts, nut products and dried tree fruits for mycotoxins. Food-Survey-Information-Sheet No. 21/02 (203).

Food Standards Agency, 2005: Survey of spices for aflatoxins and ochratoxin A. Food-Survey-Information-Sheet No. 73/05 (245).

Francis, O.J., Jr., Lipinski, L.J., Gaul, J.A., Campbell, A.D., 1982: High pressure liquid chromatographic determination of aflatoxins in peanut butter using a silica gel-packed flowcell for fluorescence detection. Journal of the Association of Official Analytical Chemists 65, 672-676 (222).

Freitas, V.P.S., Brigido, B.M., 1998: Occurrence of aflatoxins B_1, B_2, G_1, and G_2 in peanuts and their products marketed in the region of Campinas, Brazil in 1995 and 1996. Food Additives and Contaminants 15, 807–811 (421).

Fritz, W., Engst, R., 1981: Survey of selected mycotoxins in food. Journal of Environmental Science and Health B16, 193–210 (29).

Fufa, H., Urga, K., 1996: Screening of aflatoxins in Shiro and ground red pepper in Addis Ababa. Ethiopian Medical Journal 34, 243–249 (168).

Fuller, G., Spooncer, W.W., King, A.D. Jr., Schade, J., Mackey, B., 1977: Survey of aflatoxins in California tree nuts. Journal of the American Oil Chemists′ Society 54, 231A–234A (566).

Funes, G.J., Resnik, S.L., 2009: Determination of patulin in solid and semisolid apple and pear products marketed in Argentina. Food Control 20, 277-280. doi: https://doi.org/10.1016/j.foodcont.2008.05.010 (50).

Furlani, R.P.Z., Valente Soares, L.M., 1996: Survey of aflatoxins, ochratoxin A, zearalenone, and sterigmatocystin in tree nuts commercialized in Campinas, São Paulo. Ciência e Tecnologia de Alimentos. 16, 238–240 (202).

Gagliardi, S.J., Cheatle, T.F., Mooney, R.L., Llewellyn, G.C., O'Rear, C.E., Llewellyn, G.C., 1991: The occurrence of aflatoxin in peanut butter from 1982 to 1989. Journal of Food Protection 54, 627–631 (232).

Galarce-Bustos, O., Alvarado, M., Vega, M., Aranda, M., 2014: Occurrence of ochratoxin A in roasted and instant coffees in Chilean market. Food Control 46, 102-107. doi: https://doi.org/10.1016/j.foodcont.2014.05.014 (730).

Gamanya, R., Sibanda, L., 2001: Survey of *Fusarium moniliforme* (*F. verticillioides*) and production of fumonisin B_1 in cereal grains and oilseeds in Zimbabwe. International Journal of Food Microbiology 71, 145–149 (433).

Garcia, L.P., Savi, G.D., Santos, K., Scussel, V.M., 2016: Fumonisins and fungi in dry soybeans (*Glycine max* L.) for human consumption. Food Additives and Contaminants: Part B 9, 79-84. doi: https://doi.org/10.1080/19393210.2015.1135484 (392).

García-Moraleja, A., Font, G., Mañes, J., Ferrer, E., 2015: Analysis of mycotoxins in coffee and risk assessment in Spanish adolescents and adults. Food and Chemical Toxicology 86, 225-233 doi: https://doi.org/10.1016/j.fct.2015.10.014 (731).

Geana, E.-I., Iordache, A.-M., Ionete, R.E., 2012: Ochratoxin A occurrence in Romanian wines. Revista de Chimie 63, 883–886 (732).

Geipel, M., Baltes, W., Krönert, W., Weber, R., 1981: Determination of patulin in apple products by high performance liquid chromatography. Chemie, Mikrobiologie, Technologie der Lebensmittel 7, 93–96 (733).

Gelda, C.S., Luyt, L.J., 1977: Survey of total aflatoxin content in peanuts, peanut butter, and other foodstuffs. Annales de la Nutrition et de l′Alimentation 31, 477–483 (15).

Ghali, R., Belouaer, I., Hdiri, S., Ghorbel, H., Maaroufi, K., Hedilli, A., 2009a: Simultaneous HPLC determination of aflatoxins B_1, B_2, G_1 and G_2 in Tunisian sorghum and pistachios. Journal of Food Composition and Analysis 22, 751–755. doi: https://doi.org/10.1016/j.jfca.2009.04.009 (558).

Ghali, R., Ghorbel, H., Hedilli, A., 2009b: Fumonisin determination in Tunisian foods and feeds. ELISA and HPLC methods comparison. Journal of Agricultural and Food Chemistry 57, 3955-3960. doi: https://doi.org/10.1021/jf803786h (642).

Ghali, R., Hmaissia-khlifa, K., Ghorbel, H., Maaroufi, K., Hedili, A., 2009c: HPLC determination of ochratoxin A in high consumption Tunisian foods. Food Control 20, 716-720. doi: https://doi.org/10.1016/j.foodcont.2008.09.004 (340).

Ghali, R., Khlifa, K.H., Ghorbel, H., Maaroufi, K., Hedilli, A., 2010: Aflatoxin determination in commonly consumed foods in Tunisia. Journal of the Science of Food and Agriculture 90, 2347-2351. doi: https://doi.org/10.1002/jsfa.4069 (338).

Gilbert, J., Shepherd, M.J., 1985: A survey of aflatoxins in peanut butters, nuts and nut confectionery products by HPLC with fluorescence detection. Food Additives and Contaminants 2, 171–183 (3).

Gillard, N., Agneessens, R., Dubois, M.L., Delahaut, Ph., 2009: Quantification of patulin in Belgian handicraft-made apple juices. World Mycotoxin Journal 2, 95-104. doi: https://doi.org/10.3920/WMJ2008.1054 (686).

Giovannoli, C., Passini, C., Di Nardo, F., Anfossi, L., Baggiani, C., 2014: Determination of ochratoxin A in Italian red wines by molecularly imprinted solid phase extraction and HPLC analysis. Journal of Agricultural and Food Chemistry 62, 5220-5225. doi: https://doi.org/10.1021/jf5010995 (915).

Girgis, A.N., El-Sherif, S., Rofael, N., Nesheim, S., 1977: Aflatoxins in Egyptian foodstuffs. Journal of the Association of Official Analytical Chemists 60, 746–747 (127).

Giridhar, N., Krishnamurthy, G.V., 1977: Studies on aflatoxin content of groundnut oil in Andhra Pradesh with reference to climatic conditions and seasonal variations. Journal of Food Science and Technology, India 14, 84–85 (582).

Gökmen, V., Acar, J., 1998: Incidence of patulin in apple juice concentrates produced in Turkey. Journal of Chromatography A 815, 99–102 (490).

Gökmen, V., Acar, J., 2000: Long-term survey of patulin in apple juice concentrates produced in Turkey. Food Additives and Contaminants 17, 933–936 (446).

Golge, O., Hepsag, F., Kabak, B., 2013: Incidence and level of aflatoxin contamination in chilli commercialised in Turkey. Food Control 33, 514-520. doi: https://doi.org/10.1016/j.foodcont.2013.03.048 (734).

González-Osnaya, L., Soriano, J.M., Moltó, J.C., Mañes, J., 2007: Exposure to patulin from consumption of apple-based products. Food Additives and Contaminants 24, 1268-1274. doi: https://doi.org/10.1080/02652030701361556 (43).

Gopinandhan, T.N., Kannan, G.s., Panneerselvam, P., Velmourougane, K., Raghuramulu, Y., Jayarama J 2008: Survey on ochratoxin A in Indian green coffee destined for export. Food Additives and Contaminants: Part B 1, 51-57. doi: https://doi.org/10.1080/19393210802236984 (319).

Gopinandhan, T.N., Velmourougane, K., Panneerselvam, P., Keshamma, E., Raghuramulu, Y., 2007: Occurrence of ochratoxin-A (OT-A) in green and commercial coffee samples. Journal of Food Science and Technology 44, 247–249 (298).

Goryacheva, I.Yu., De Saeger, S., Lobeau, M., Eremin, S.A., Barna-Vetró, I., Van Peteghem, C., 2006: Approach for ochratoxin A fast screening in spices using clean-up tandem immunoassay columns with confirmation by high performance liquid chromatography-tandem mass spectrometry (HPLC-MS/MS). Analytica Chimica Acta 577, 38-45. doi: https://doi.org/10.1016/j.aca.2006.06.033 (193).

Goryacheva, I.Y., De Saeger, S., Nesterenko, I.S., Eremin, S.A., Van Peteghem, C., 2007: Rapid all-in-one three-step immunoassay for non-instrumental detection of ochratoxin A in high-coloured herbs and spices. Talanta 72, 1230–1234 (and personal communication). doi: https://doi.org/10.1016/j.talanta.2006.12.049 (192).

Gross, M., Curtui, V., Ackermann, Y., Latif, H., Usleber, E., 2011: Enzyme immunoassay for tenuazonic acid in apple and tomato products. Journal of Agricultural and Food Chemistry 59, 12317-12322. doi: https://doi.org/10.1021/jf203540y (442).

Guo, Y., Zhou, Z., Yuan, Y., Yue, T., 2013: Survey of patulin in apple juice concentrates in Shaanxi (China) and its dietary intake. Food Control 34, 570-572. doi: https://doi.org/10.1016/j.foodcont.2013.05.029 (910).

Gürses, M., 2006: Mycoflora and aflatoxin content of hazelnuts, walnuts, peanuts, almonds and roasted chickpeas (Leblebi) sold in Turkey. International Journal of Food Properties 9, 395-399. doi: https://doi.org/10.1080/10942910600596597 (14).

Haas, D., Pfeifer, B., Reiterich, C., Partenheimer, R., Reck, B., Buzina, W., 2013: Identification and quantification of fungi and mycotoxins from Pu-erh tea. International Journal of Food Microbiology 166, 316-322. doi: https://doi.org/10.1016/j.ijfoodmicro.2013.07.024 (735).

Habish, H.A., Abdulla, M.H., Broadbent, J.H., 1972: The incidence of aflatoxin in Sudanese groundnuts. Tropical Science 13, 279–287 (412).

Hag Elamin, N.H., Abdel-Rahim, A.M., Khalid, A.E., 1988: Aflatoxin contamination of groundnuts in Sudan. Mycopathologia 104, 25–31 (425).

Halt, M., 1998: Moulds and mycotoxins in herb tea and medicinal plants. European Journal of Epidemiology 14, 269–274 (74).

Hammami, W., Fiori, S., Al Thani, R., Kali, N.A., Balmas, V., Migheli, Q., Jaoua, S., 2014: Fungal and aflatoxin contamination of marketed spices. Food Control 37, 177-181. doi: https://doi.org/10.1016/j.foodcont.2013.09.027 (445).

Han, Z., Feng, Z., Shi, W., Zhao, Z., Wu, Y., Wu, A., 2014: A quick, easy, cheap, effective, rugged, and safe sample pretreatment and liquid chromatography with tandem mass spectrometry method for the simultaneous quantification of 33 mycotoxins in *Lentinula edodes*. Journal of Separation Science 37, 1957-1966. doi: https://doi.org/10.1002/jssc.201400329 (736).

Han, Z., Liu, X., Ren, Y., Luan, L., Wu, Y., 2010a: A rapid method with ultra-high-performance liquid chromatography-tandem mass spectrometry for simultaneous determination of five type B trichothecenes in traditional Chinese medicines. Journal of Separation Science 33, 1923-1932. doi: https://doi.org/10.1002/jssc.201000094 (737).

Han, Z., Ren, Y., Liu, X., Luan, L., Wu, Y., 2010b: A reliable isotope dilution method for simultaneous determination of fumonisins B_1, B_2 and B_3 in traditional Chinese medicines by ultra-high-performance liquid chromatography-tandem mass spectrometry. Journal of Separation Science 33, 2723-2733. doi: https://doi.org/10.1002/jssc.201000423 (739).

Han, Z., Ren, Y., Zhou, H., Luan, L., Cai, Z., Wu, Y., 2011: A rapid method for simultaneous determination of zearalenone, α-zearalenol, ß-zearalenol, zearalanone, α-zearalanol and ß-zearalanol in traditional Chinese medicines by ultra-high-performance liquid chromatography-tandem mass spectrometry. Journal of Chromatography B 879, 411-420. doi: https://doi.org/10.1016/j.jchromb.2010.12.028 (740).

Han, Z., Ren, Y., Zhu, J., Cai, Z., Chen, Y., Luan, L., Wu, Y., 2012: Multianalysis of 35 mycotoxins in traditional Chinese medicines by ultra-high-performance liquid chromatography-tandem mass spectrometry coupled with accelerated solvent extraction. Journal of Agricultural and Food Chemistry 60, 8233-8247. doi: https://doi.org/10.1021/jf301928r (68).

Han, Z., Zheng, Y., Luan, L., Cai, Z., Ren, Y., Wu, Y., 2010c: An ultra-performance liquid chromatography-tandem mass spectrometry method for simultaneous determination of aflatoxins B_1, B_2, G_1, G_2, M_1 and M_2 in traditional Chinese medicines. Analytica Chimica Acta 664, 165-171. doi: https://doi.org/10.1016/j.aca.2010.02.009 (66).

Han, Z., Zheng, Y., Luan, L., Ren, Y., Wu, Y., 2010d: Analysis of ochratoxin A and ochratoxin B in traditional Chinese medicines by ultra-high-performance liquid chromatography-tandem mass spectrometry using [$^{13}C_{20}$]-ochratoxin A as an internal standard. Journal of Chromatography A 1217, 4365-4374. doi: https://doi.org/10.1016/j.chroma.2010.04.052 (75).

Hanssen, E., Jung, M., 1973: Control of aflatoxins in the food industry. Pure and Applied Chemistry 35, 239–250 (239).

Hariprasad, P., Vipin, A.V., Karuna, S., Raksha, R.K., Venkateswaran, G., 2015: Natural aflatoxin uptake by sugarcane (*Saccharum officinaurum* L.) and its persistence in jaggery. Environmental Science and Pollution Research 22, 6246-6253. doi: https://doi.org/10.1007/s11356-014-3851-2 (453).

Harris, K.L., Bobe, G., Bourquin, L.D., 2009: Patulin surveillance in apple cider and juice marketed in Michigan. Journal of Food Protection 72, 1255–1261 (41).

Hashimoto, R., Nakagawa, H., Onji, Y., Asano, K., Yokoyama, K., Takahashi, H., 2015: Mycotoxin contamination of Vietnamese coffee beans caused by *Aspergillus Nigri* and *Circumdati*. JSM Mycotoxins 65, 1-6. doi: https://doi.org/10.2520/myco.65.1 (657).

Haydar, M., Benelli, L., Brera, C., 1990: Occurrence of aflatoxins in Syrian foods and foodstuffs: a preliminary study. Food Chemistry 37, 261–268 (347).

Hendrickse, R.G., Coulter, J.B.S., Lamplugh, S.M., MacFarlane, S.B.J., Williams, T.E., Omer, M.I.A., Suliman, G.I., 1982: Aflatoxins and kwashiorkor: a study in Sudanese children. British Medical Journal 285, 843–846 (223).

Heperkan, D., Karbancioglu Güler, F., Oktay, H.I., 2012a: Mycoflora and natural occurrence of aflatoxin, cyclopiazonic acid, fumonisin and ochratoxin in dried figs. Food Additives and Contaminants Part: A 29, 277-286. doi: https://doi.org/10.1080/19440049.2011.597037 (362).

Heperkan, D., Somuncuoglu, S., Karbancioglu-Güler, F., Mecik, N., 2012b: Natural contamination of cyclopiazonic acid in dried figs and co-occurrence of aflatoxin. Food Control 23, 82-86. doi: https://doi.org/10.1016/j.foodcont.2011.06.015 (353).

Hepsag, F., Golge, O., Kabak, B., 2014: Quantitation of aflatoxins in pistachios and groundnuts using HPLC-FLD method. Food Control 38, 75-81. doi: https://doi.org/10.1016/j.foodcont.2013.10.005 (741).

Hernández, M.J., García-Moreno, M.V., Durán, E., Guillén, D., Barosso, C.G., 2006: Validation of two analytical methods for the determination of ochratoxin A by reversed-phase high-performance liquid chromatography coupled to fluorescence detection in musts and sweet wines from Andalusia. Analytica Chimica Acta 566, 117–121. doi: https://doi.org/10.1016/j.aca.2006.02.002 (520).

Hernández Hierro, J.M., Garcia-Villanova, R.J., Torrero, P.R., Toruño Fonseca, I.M., 2008: Aflatoxins and ochratoxin A in red paprika for retail sale in Spain: occurrence and evaluation of a simultaneous analytical method. Journal of Agricultural and Food Chemistry 56, 751-756. doi: https://doi.org/10.1021/jf073002c (586).

Hernández-Martínez, R., Navarro-Blasco, I., 2010: Aflatoxin levels and exposure assessment of Spanish infant cereals. Food Additives and Contaminants: Part B 3, 275-288. doi: https://doi.org/10.1080/19393210.2010.531402 (98).

Herrera, M., Herrera, A., Ariño, A., 2009: Estimation of dietary intake of ochratoxin A liquorice confectionery. Food and Chemical Toxicology 47, 2002-2006. doi: https://doi.org/10.1016/j.fct.2009.05.009 (506).

Herry, M.-P., Lemetayer, N., 1992: Aflatoxin B_1 contamination in oilseeds, dried fruits and spices. Microbiologie – Aliments – Nutrition 10, 261–266 (5).

Heshmati, A., Mozaffari Nejad, A.S., 2015: Ochratoxin A in dried grapes in Hamadan province, Iran. Food Additives and Contaminants: Part B 8, 255-259. doi: https://doi.org/10.1080/19393210.2015.1074945 (742).

Hetmanski, M.T., Scudamore, K.A., 1989: A simple quantitative HPLC method for determination of aflatoxins in cereals and animal feedstuffs using gel permeation chromatography clean-up. Food Additives and Contaminants 6, 35–48 (384).

Hickert, S., Bergmann, M., Ersen, S., Cramer, B., Humpf, H.-U., 2016: Survey of *Alternaria* toxin contamination in food from the German market, using a rapid HPLC-MS/MS approach. Mycotoxin Research 32, 7-18. doi: https://doi.org/10.1007/s12550-015-0233-7 (743).

Hickert, S., Hermes, L., Marques, L.M.M., Focke, C., Cramer, B., Lopes, N.P., Flett, B., Humpf, H.-U., 2017: *Alternaria* toxins in South African sunflower seeds: cooperative study. Mycotoxin Research 33, 309-321. doi: https://doi.org/10.1007/s12550-017-0290-1 (414).

Hocking, A.D., Varelis, P., Pitt, J.I., Cameron, S.F., Leong, S.-L.L., 2003: Occurrence of ochratoxin A in Australian wine. Australian Journal of Grape and Wine Research 9, 72–78 (610).

Hoeltz, M., Einloft, T.C., Oldoni, V.P., Dottori, H.A., Noll, I.B., 2012: The occurrence of aflatoxin B_1 contamination in peanuts and peanut products marketed in southern Brazil. Brazilian Archives of Biology and Technology 55, 313-317. doi: https://doi.org/10.1590/S1516-89132012000200019 (744).

Horii, S., Hashiguchi, T., Igi, Y., Sudou, S., 2010: Investigation of ochratoxin A in Japanese wine by LC/MC/MS. Journal of the American Society for Enology and Viticulture, Japan 21, 3–7 (605).

Hosseininia, A.R., Vahabzadeh, M., Rashedinia, M., Riahi-Zanjani, B., Karimi, G., 2014: A survey of aflatoxins in sesame seeds imported into Khorasan Province, Iran. Mycotoxin Research 30, 43-46. doi: https://doi.org/10.1007/s12550-013-0186-7 (745).

Houssou, P.A., Ahohuendo, B.C., Fandohan, P., Kpodo, K., Hounhouigan, D.J., Jakobsen, M., 2009: Natural infection of cowpea (*Vigna unguiculata* (L.) Walp.) by toxigenic fungi and mycotoxin contamination in Benin, West Africa. Journal of Stored Products Research 45, 40-44. doi: https://doi.org/10.1016/j.jspr.2008.07.002 (324).

Hu, L., Rychlik, M., 2014: Occurrence of enniatins and beauvericin in 60 Chinese medicinal herbs. Food Additives and Contaminants: Part A 31, 1240-1245. doi: https://doi.org/10.1080/19440049.2014.913813 (746).

Huang, B., Han, Z., Cai, Z., Wu, Y., Ren, Y., 2010: Simultaneous determination of aflatoxin B_1, B_2, G_1, G_2, M_1 and M_2 in peanuts and their derivative products by ultra-high-performance liquid chromatography-tandem mass spectrometry. Analytica Chimica Acta 662, 62-68. doi: https://doi.org/10.1016/j.aca.2010.01.002 (227).

Huang, X., Wang, S., Mao, D., Miao, S., Hu, Q., Ji, S., 2017: Optimized QuEChERS method combined with UHPLC-MS/MS for the simultaneous determination of 15 mycotoxins in liquorice. Journal of the Association of Official Analytical Chemists International 101, 1-10. doi: https://doi.org/10.5740/jaoacint.17-0365 (917).

Hudson, G.J., Wild, C.P., Zarba, A., Groopman, J.D., 1992: Aflatoxins isolated by immunoaffinity chromatography from foods consumed in The Gambia, West Africa. Natural Toxins 1, 100–105 (101).

Iamanaka, B.T., De Menezes, H.C., Vicente, E., Leite, R.S.F., Taniwaki, M.H., 2007: Aflatoxigenic fungi and aflatoxins occurrence in sultanas and dried figs commercialized in Brazil. Food Control 18, 454-457. doi: https://doi.org/10.1016/j.foodcont.2005.12.002 (356).

Iamanaka, B.T., Nakano, F., Lemes, D.P., Ferranti, L.S., Taniwaki, M.H., 2014: Aflatoxin evaluation in ready-to-eat Brazil nuts using reversed-phase liquid chromatography and post-column derivatisation. Food Additives and Contaminants: Part A 31, 917-923. doi: https://doi.org/10.1080/19440049.2014.895857 (747).

Iamanaka, B.T., Taniwaki, M.H., Menezes, H.C., Vicente, E., Fungaro, M.H.P., 2005: Incidence of toxigenic fungi and ochratoxin A in dried fruits sold in Brazil. Food Additives and Contaminants 22, 1258-1263. doi: https://doi.org/10.1080/02652030500260447 (331).

Idris, Y.M.A., Hassan, S.A., Mariod, A.A., 2013: Physicochemical characteristics and aflatoxin levels in two types of Sudanese sesame oil. Journal of the American Oil Chemists´ Society 90, 989–998. doi: https://doi.org/10.1007/s11746-013-2241-0 (902).

Idris, Y.M.A., Mariod, A.A., Elnour, I.A., Mohamed, A.A., 2010: Determination of aflatoxin levels in Sudanese edible oils. Food and Chemical Toxicology 48, 2539-2541. doi: https://doi.org/10.1016/j.fct.2010.05.021 (569).

Iha, M.H., Sabino, M., 2008: Incidence of patulin in Brazilian apple-based drinks. Food Control 19, 417-422. doi: https://doi.org/10.1016/j.foodcont.2007.05.001 (478).

Ikoma, T., Tsuchiya, Y., Asai, T., Okano, K., Ito, N., Endoh, K., Yamamoto, M., Nakamura, K., 2015: Ochratoxin A contamination of red chili peppers from Chile, Bolivia and Peru, countries with a high incidence of gallbladder cancer. Asian Pacific Journal of Cancer Prevention 16, 5987-5991. doi: https://doi.org/10.7314/APJCP.2015.16.14.5987 (749).

Imperato, R., Campone, L., Piccinelli, A.L., Veneziano, A., Rastrelli, L., 2011: Survey of aflatoxins and ochratoxin A contamination in food products imported in Italy. Food Control 22, 1905-1910. doi: https://doi.org/10.1016/j.foodcont.2011.05.002 (320).

Imwidthaya, S., Anukarahanonta, T., Komolpis, P., 1987: Bacterial, fungal and aflatoxin contamination of cereals and cereal products in Bangkok. Journal of the Medical Association of Thailand 70, 390–396 (133).

Ioannou-Kakouri, E., Aletrari, M., Christou, E., Hadjioannou-Ralli, A., Koliou, A., Akkelidou, D., 1999: Surveillance and control of aflatoxins B_1, B_2, G_1, G_2, and M_1 in foodstuffs in the Republic of Cyprus: 1992-1996. Journal of the Association of Official Analytical Chemists International 82, 883–892 (200).

Iqbal, S.Z., Asi, M.R., Jinap, S., 2013a: Natural occurrence of aflatoxin B_1 and aflatoxin M_1 in "halva" and its ingredients. Food Control 34, 404-407. doi: https://doi.org/10.1016/j.foodcont.2013.04.038 (750).

Iqbal, S.Z., Asi, M.R., Jinap, S., 2014: Aflatoxins in dates and dates products. Food Control, 43, 163-166. doi: https://doi.org/10.1016/j.foodcont.2014.03.010 (574).

Iqbal, S.Z., Asi, M.R., Zuber, M., Akhtar, J., Saif, M.J., 2013b: Natural occurrence of aflatoxins and ochratoxin A in commercial chilli and chilli sauce samples. Food Control 30, 621-625. doi: https://doi.org/10.1016/j.foodcont.2012.09.003 (751).

Iqbal, S.Z., Asi, M.R., Zuber, M., Akram, N., Batool, N., 2013c: Aflatoxins contamination in peanut and peanut products commercially available in retail markets of Punjab, Pakistan. Food Control 32, 83-86. doi: https://doi.org/10.1016/j.foodcont.2012.11.024 (573).

Iqbal, S.Z., Paterson, R.R.M., Bhatti, I.A., Asi, M.R., 2011: Comparing aflatoxin contamination in chilies from Punjab, Pakistan produced in summer and winter. Mycotoxin Research 27, 75-80. doi: https://doi.org/10.1007/s12550-010-0078-z (254).

Iqbal, S.Z., Paterson, R.R.M., Bhatti, I.A., Asi, M.R., Sheikh, M.A., Bhatti, H.N., 2010: Aflatoxin B_1 in chilies from the Punjab region, Pakistan. Mycotoxin Research 26, 205-209. doi: https://doi.org/10.1007/s12550-010-0055-6 (247).

Isohata, E., Toyoda, M., Saito, Y., 1986: Studies on chemical analysis of mycotoxin (XV). Analysis of nivalenol and deoxynivalenol in imported cereals, baby foods and processed foods (confectioneries). Bulletin of National Institute of Health Science 104, 113–118 (640).

Ito, R., Yamazaki, H., Inoue, K., Yoshimura, Y., Kawaguchi, M., Nakazawa, H., 2004: Development of liquid chromatography-electrospray mass spectrometry for the determination of patulin in apple juice: investigation of its contamination levels in Japan. Journal of Agricultural and Food Chemistry 52, 7464-7468. doi: https://doi.org/10.1021/jf049264l (475).

Jacobsen, B.J., Harlin, K.S., Swanson, S.P., Lambert, R.J., Beasley, V.R., Sinclair, J.B., Wei, L.S., 1995: Occurrence of fungi and mycotoxins associated with field mold damaged soybeans in the Midwest. Plant Disease 79, 86–88 (527).

Jager, A.V., Tedesco, M.P., Souto, P.C.M.C., Oliveira, C.A.F., 2013: Assessment of aflatoxin intake in São Paulo, Brazil. Food Control 33, 87-92 doi: https://doi.org/10.1016/j.foodcont.2013.02.016 (752).

Jahanmard, E., Azarani, F., Sharifi, M., Esfandiari, Z., 2014: Aflatoxin in pistachio nuts used as ingredients in Gaz sweets produced in Isfahan, Iran. Food Additives and Contaminants: Part B 7, 70-73. doi: https://doi.org/10.1080/19393210.2013.846942 (753).

Jajić, I., Jurić, V., Abramović, B., 2008: First survey of deoxynivalenol occurrence in crops in Serbia. Food Control 19, 545-500. doi: https://doi.org/10.1016/j.foodcont.2007.05.009 (634).

Jalili, M., Jinap, S., 2012: Natural occurrence of aflatoxins and ochratoxin A in commercial dried chili. Food Control 24, 160-164. doi: https://doi.org/10.1016/j.foodcont.2011.09.020 (294).

Jalili, M., Jinap, S., Adzahan, N., 2009: Survey of aflatoxins in retail samples of whole and ground black and white peppercorns. Food Additives and Contaminants: Part B 2, 178-182. doi: https://doi.org/10.1080/19440040903384190 (189).

Jalili, M., Jinap, S., Radu, S., 2010: Natural occurrence of ochratoxin A contamination in commercial black and white pepper products. Mycopathologia 170, 251-258. doi: https://doi.org/10.1007/s11046-010-9320-7 (195).

Jang, M.-R., Lee, C.-H., Cho, S.-H., Park, J.-S., Kwon, E.-Y., Lee, E.-J., Kim, S.-H., Kim, D.-B., 2007: A survey of total aflatoxins in food using high performance liquid chromatography-fluorescence detector (HPLC-FLD) and liquid chromatography tandem mass spectrometry (LC-MS/MS). Korean Journal of Food Science and Technology 39, 488–493 (9).

Jeszka-Skowron, M., Zgoła-Grześkowiak, A., Stanisz, E., Waśkiewicz, A., 2017: Potential health benefits and quality of dried fruits: goji fruits, cranberries and raisins. Food Chemistry 221, 228-236. doi: https://doi.org/10.1016/j.foodchem.2016.10.049 (664).

Jiménez, M., Mateo, R., Querol, A., Huerta, T., Hernández, E., 1991: Mycotoxins and mycotoxigenic moulds in nuts and sunflower seeds for human consumption. Mycopathologia 115, 121–127 (6).

Joffe, A.Z., 1970: The presence of aflatoxin in kernels from five years groundnut crops and of *Aspergillus flavus* isolates from kernels and soils. Plant and Soil 33, 91–96 (413).

Jørgensen, K., 1998: Survey of pork, poultry, coffee, beer and pulses for ochratoxin A. Food Additives and Contaminants 15, 550–554 (285).

Josefsson, E., Andersson, A., 1976: Analysis of patulin in apple beverages sold in Sweden. Vår Föda 8, 189–196 (451).

Juan, C., Oueslati, S., Mañes, J., 2016: Evaluation of *Alternaria* mycotoxins in strawberries: quantification and storage condition. Food Additives and Contaminants: Part A 33, 861-868. doi: https://doi.org/10.1080/19440049.2016.1177375 (755).

Juan, C., Zinedine, A., Moltó, J.C., Idrissi, L., Mañes, J., 2008: Aflatoxin levels in dried fruits and nuts from Rabat-Salé area, Morocco. Food Control 19, 849-853. doi: https://doi.org/10.1016/j.foodcont.2007.08.010 (352).

Kaaya, A.N., Eboku, D., 2010: Mould and aflatoxin contamination of dried cassava chips in eastern Uganda: association with traditional processing and storage practices. Journal of Biological Sciences 10, 718–729 (756).

Kabak, B., 2012: Aflatoxin M_1 and ochratoxin A in baby formulae in Turkey: occurrence and safety evaluation. Food Control 26, 182-187. doi: https://doi.org/10.1016/j.foodcont.2012.01.032 (89).

Kabak, B., 2014: Quantitation of aflatoxins in walnut kernels by high-performance liquid chromatography with fluorescence detection. Food Additives and Contaminants. Part B 7, 288-294. doi: https://doi.org/10.1080/19393210.2014.928372 (757).

Kačániová, M., Juráček, M., Chlebo, R., Kňazovická, V., Kadasi-Horáková, M., Kunová, S., Lejková, J.,Haščík, P., Mareček, J., Šimko, M., 2011: Mycobiota and mycotoxins in bee pollen collected from different areas of Slovakia. Journal of Environmental Science and Health, Part B 46, 623-629. doi: https://doi.org/10.1080/03601234.2011.589322 (153).

Kamika, I., Mngqawa, P., Rheeder, J.P., Teffo, S.L., Katerere, D.R., 2014: Mycological and aflatoxin contamination of peanuts sold at markets in Kinshasa, Democratic Republic of Congo, and Pretoria, South Africa. Food Additives and Contaminants: Part B 7, 120-126. doi: https://doi.org/10.1080/19393210.2013.858187 (758).

Kamika, I., Takoy, L.L., 2011: Natural occurrence of aflatoxin B_1 in peanut collected from Kinshasa, Democratic Republic of Congo. Food Control 22, 1760-1764. doi: https://doi.org/10.1016/j.foodcont.2011.04.010 (759).

Kamimura, H., Nishijima, M., Tabata, S., Yasuda, K., Ushiyama, H., Nishima, T., 1986: Survey of mycotoxins contamination in edible oil and fate of mycotoxins during oil refining processes. Journal of the Food Hygienic Society of Japan 27, 59–63 (760).

Kappenstein, O., Klaffke, H.St., Mehlitz, I., Tiebach, R., Weber, R., Lepschy, J., Wittkowski, R., 2005: Determination of zearalenone in edible oils with SEC and LC-ESI-MS/MS. Mycotoxin Research 21, 3–6 (761).

Karaaslan, M., Arslanğray, Y., 2015: Aflatoxins B_1, B_2, G_1, and G_2 contamination in ground red peppers commercialized in Sanliurfa, Turkey. Environmental Monitoring and Assessment 187 (4), 184. doi: https://doi.org/10.1007/s10661-015-4402-0 (595).

Karbancioğlu-Güler, F., Heperkan, D., 2008: Natural occurrence of ochratoxin A in dried figs. Analytica Chimica Acta 617, 32-36. doi: https://doi.org/10.1016/j.aca.2008.01.009 (365).

Karami-Osboo, R., Miri, R., Javidnia, K., Kobarfard, F., AliAbadi, M.H.S., Maham, M., 2015: A validated dispersive liquid-liquid microextraction method for extraction of ochratoxin A from raisin samples. Journal of Food Science and Technology 52, 2440-2445. doi: https://doi.org/10.1007/s13197-013-1215-4 (783).

Karimi, G., Hassanzadeh, M., Yazdanpanah, H., Nazari, F., Iranshahi, M., Nili, A., 2008: Contamination of patulin in clear apple juice in Mashhad, Iran. Journal of Food Safety 28, 413–421 (479).

Kassim, N., Kim, K., Mtenga, A.B., Song, J.-E., Liu, Q., Shim, W.-B., Chung, D.-H., 2011: A preliminary study of T-2 and HT-2 toxins in cereals sold in traditional market in South Korea. Food Control 22, 1408-1412. doi: https://doi.org/10.1016/j.foodcont.2011.02.023 (762).

Kataoka, H., Itano, M., Ishizaki, A., Saito, K., 2009: Determination of patulin in fruit juice and dried fruit samples by in-tube solid-phase microextraction coupled with liquid chromatography-mass spectrometry. Journal of Chromatography A 1216, 3746-3750. doi: https://doi.org/10.1016/j.chroma.2009.03.017 (480).

Katerere, D.R., Stockenström, S., Shephard, G.S., 2008a: HPLC-DAD method for the determination of patulin in dried apple rings. Food Control 19, 389-392. doi: https://doi.org/10.1016/j.foodcont.2007.04.015 (51).

Katerere, D.R., Stockenström, S., Thembo, K.M., Balducci, G., Shephard, G.S., 2007: Investigation of patulin contamination in apple juice sold in retail outlets in Italy and South Africa. Food Additives and Contaminants 24, 630-634. doi: https://doi.org/10.1080/02652030601137668 (481).

Katerere, D.R., Stockenström, S., Thembo, K.M., Rheeder, J.P., Shephard, G.S., Vismer, H.F., 2008b: A preliminary survey of mycological and fumonisin and aflatoxin contamination of African traditional herbal medicines sold in South Africa. Human & Experimental Toxicology 27, 793-798. doi: https://doi.org/10.1177/0960327108099535 (79).

Kayode, O.F., Sulyok, M., Fapohunda, S.O, Ezekiel, C.N., Krska, R., Oguntona, C.R.B., 2013: Mycotoxins and fungal metabolites in groundnut- and maize-based snacks from Nigeria. Food Additives and Contaminants: Part B 6, 294-300. doi: https://doi.org/10.1080/19393210.2013.823626 (763).

Kershaw, S.J., 1982: Occurrence of aflatoxins in oilseeds providing cocoa-butter substitutes. Applied and Environmental Microbiology 43, 1210–1212 (563).

Kershaw, S.J., 1985: Aflatoxins in imported edible nuts: some data 1982-1984. Journal of Food Technology 20, 647–649 (381).

Khalesi, M., Sheikh-Zeinoddin, M., Tabrizchi, M., 2011: Determination of ochratoxin A in licorice root using inverse ion mobility spectrometry. Talanta 83, 988-993. doi: https://doi.org/10.1016/j.talanta.2010.11.004 (764).

Khan, M.A., Asghar, M.A., Iqbal, J., Ahmed, A., Shamsuddin, Z.A., 2014: Aflatoxins contamination and prevention in red chillies (*Capsicum annuum* L.) in Pakistan. Food Additives and Contaminants: Part B 7, 1-6. doi: https://doi.org/10.1080/19393210.2013.825330 (765).

Khayoon, W.S., Saad, B., Lee, T.P., Salleh, B., 2012: High performance liquid chromatographic determination of aflatoxins in chilli, peanut and rice using silica based monolithic column. Food Chemistry 133, 489-496. doi: https://doi.org/10.1016/j.foodchem.2012.01.010 (766).

Khazaeli, P., Mehrabani, M., Heidari, M.R., Asadikaram, G., Lari Najafi, M., 2017: Prevalence of aflatoxin contamination in herbs and spices in different regions of Iran. Iranian Journal of Public Health 46, 1540–1545 (491).

Kim, E.-K., Shon, D.-H., Yoo, J.Y., Ryu, D., Lee, C., Kim, Y.-B., 2001: Natural occurrence of aflatoxins in Korean meju. Food Additives and Contaminants 18, 151-156. doi: https://doi.org/10.1080/02652030010006104 (528).

Kim, S., Lee, S., Nam, T.-G., Seo, D., Yoo, M., 2017: Comparison of a newly developed liquid chromatography with tandem mass spectrometry method and enzyme-linked immunosorbent assay for detection of multiple mycotoxins in red pepper powder. Journal of Food Protection 80, 1347-1354. doi: https://doi.org/10.4315/0362-028X.JFP-17-006 (901).

Knudsen, P.B., Mogensen, J.M., Larsen, T.O., Nielsen, K.F., 2011: Occurrence of fumonisin B_2 and B_4 in retail raisins. Journal of Agricultural and Food Chemistry 59, 772-776. doi: https://doi.org/10.1021/jf103855x (601).

Koirala, P., Kumar, S., Yadav, B.K., Premarajan, K.C., 2005: Occurrence of aflatoxin in some of the food and feed in Nepal. Indian Journal of Medical Sciences 59, 331–336 (58).

Kollia, E., Kanapitsas, A., Markaki, P., 2014: Occurrence of aflatoxin B_1 and ochratoxin A in dried vine fruits from Greek market. Food Additives and Contaminants: Part B 7, 11-16. doi: https://doi.org/10.1080/19393210.2013.825647 (767).

Kollia, E., Tsourouflis, T., Markaki, P., 2016: Aflatoxin B_1 in sesame seeds and sesame products from the Greek market. Food Additives and Contaminants: Part B 9, 217-222. doi: https://doi.org/10.1080/19393210.2016.1179349 (397).

Kong, W., Wei, R., Logrieco, A.F., Wei, J., Wen, J., Xiao, X., Yang, M., 2014: Occurrence of toxigenic fungi and determination of mycotoxins by HPLC-FLD in functional foods and spices in China markets. Food Chemistry 146, 320-326. doi: https://doi.org/10.1016/j.foodchem.2013.09.005 (768).

Kong, W., Xie, T., Li, J., Wei, J., Qiu, F., Qi, A., Zheng, Y., Yang, M., 2012: Analysis of fumonisins B_1 and B_2 in spices and aromatic and medicinal herbs by HPLC-FLD with on-line post-column derivatization and positive confirmation by LC-MS/MS. Analyst 137, 3166-3174. doi: https://doi.org/10.1039/c2an35164a (82).

Kong, W.-J., Liu, S.-Y., Qiu, F., Xiao, X.-H., Yang, M.-H., 2013a: Simultaneous multi-mycotoxin determination in nutmeg by ultrasound-assisted solid-liquid extraction and immunoaffinity column clean-up coupled with liquid chromatography and on-line post-column photochemical derivatization-fluorescence detection. Analyst 138, 2729-2739. doi: https://doi.org/10.1039/c3an00059a (549).

Kong, W.-J., Shen, H.-H., Zhang, X.-F., Yang, X.-L., Qiu, F., Ou-yang, Z., Yang, M.-H., 2013b: Analysis of zearalenone and α-zearalenol in 100 foods and medicinal plants determined by HPLC-FLD and positive confirmation by LC-MS-MS. Journal of the Science of Food and Agriculture 93, 1584-1590. doi: https://doi.org/10.1002/jsfa.5926 (122).

Kosoglu, I., Aksoy, U., Pehlivan, R., 2011: Fumonisin B_1 and B_2 occurrence in dried fig fruits (*Ficus carica* L.) under Meander Valley's climatic conditions and relationship with fruit quality. Food Additives and Contaminants: Part A 28, 1569-1577. doi: https://doi.org/10.1080/19440049.2011.593190 (367).

Krishna Kumari, C., Nusrath, M., 1987: Natural occurrence of citrinin and ochratoxin A in coconut products. National Academy Science Letters 10, 303–305 (278).

Krishna Kumari, C., Nusrath, M., Reddy, B.N., 1984: Aflatoxin contamination of coconut products from Andhra Pradesh. Indian Phytopathology 37, 284–287 (236).

Kritzinger, Q., Aveling, T.A.S., Marasas, W.F.O., Rheeder, J.P., Van Der Westhuizen, L., Shephard, G.S., 2003: Mycoflora and fumonisin mycotoxins associated with cowpea [*Vigna unguiculata* (L.) Walp] seeds. Journal of Agricultural and Food Chemistry 51, 2188-2192. doi: https://doi.org/10.1021/jf026121v (326).

Krogh, P., Hald, B., Korpinen, E.-L., 1970: Occurrence of aflatoxin in groundnut- and copra-products imported to Finland. Nordisk Veterinaer Medicin 22, 584–589 (769).

Krüger, C.D., Fernandes, A.M., Rosa, C.A.R., 2012: Ochratoxin A in wines from 2002 to 2008 harvest marketed in Rio de Janeiro, Brazil. Food Additives and Contaminants: Part 5, 204-207. doi: https://doi.org/10.1080/19393210.2012.697485 (770).

Kuang, Y., Qiu, F., Kong, W., Yang, M., 2012: Natural occurrence of ochratoxin A in wolfberry fruit wine marketed in China. Food Additives and Contaminants: Part 5, 70-74. doi: https://doi.org/10.1080/19393210.2012.660198 (771).

Kumagai, S., Nakajima, M., Tabata, S., Ishikuro, E., Tanaka, T., Norizuki, H., Itoh, Y., Aoyama, K., Fujita, K., Kai, S., Sato, T., Saito, S., Yoshiike, N., Sugita-Konishi, Y., 2008: Aflatoxin and ochratoxin A contamination of retail foods and intake of these mycotoxins in Japan. Food Additives and Contaminants: Part A 25, 1101-1106. doi: https://doi.org/10.1080/02652030802226187 (146).

Kumar, S., Roy, A.K., 1993: Occurrence of aflatoxin in some liver curative herbal medicines. Letters in Applied Microbiology 17, 112–114 (772).

Kumari, P., Sinha, K.K., 1987: Aflatoxin elaboration in mungbean seeds. Indian Phytopathology 40, 201–203 (125).

Kursun, O., Mutlu, A.G., 2010: Aflatoxin in spices marketed in the West Mediterranean region of Turkey. Journal of Animal and Veterinary Advances 9, 2979–2981 (190).

Labrinea, E.P., Natskoulis, P.I., Spiropoulos, A.E., Magan, N., Tassou, C.C., 2011: A survey of ochratoxin A occurrence in Greek wines. Food Additives and Contaminants: Part B 4, 61-66. doi: https://doi.org/10.1080/19393210.2010.539707 (621).

Lai, C.-L., Fuh, Y.-M., Shih, D.Y.-C., 2000: Detection of mycotoxin patulin in apple juice. Journal of Food and Drug Analysis 8, 85–96 (485).

Lai, X., Ruan, C., Liu, R., Liu, C., 2014: Application of ionic liquid-based dispersive liquid-liquid microextraction for the analysis of ochratoxin A in rice wines. Food Chemistry 161, 317-322. doi: https://doi.org/10.1016/j.foodchem.2014.04.033 (912).

Landsen, J.A., Davidson, J.I., 1983: Occurrence of cyclopiazonic acid in peanuts. Applied and Environmental Microbiology 45, 766–769 (426).

Lasram, S., Bellí, N., Chebil, S., Nahla, Z., Ahmed, M., Sanchis, V., Ghorbel, A., 2007: Occurrence of ochratoxigenic fungi and ochratoxin A in grapes from a Tunisian vineyard. International Journal of Food Microbiology 114, 376-379. doi: https://doi.org/10.1016/j.ijfoodmicro.2006.09.027 (538).

Lasram, S., Oueslati, S., Chebil, S., Mliki, A., Ghorbel, A., 2013: Occurrence of ochratoxin A in domestic beers and wines from Tunisia by immunoaffinity clean-up and liquid chromatography. Food Additives and Contaminants: Part B 6, 1–5. doi: https://doi.org/10.1080/19393210.2012.716453 (852).

Lau, B. P.-Y., Scott, P.M., Lewis, D.A., Kanhere, S.R., Cléroux, C., Roscoe, V.A., 2003: Liquid chromatography-mass spectrometry and liquid chromatography-tandem mass spectrometry of the *Alternaria* mycotoxins alternariol and alternariol monomethyl ether in fruit juices and beverages. Journal of Chromatography A 998, 119-131. doi: https://doi.org/10.1016/S0021-9673(03)00606-X (440).

Lee, T.P., Sakai, R., Manaf, N.A., Rodhi, A.M., Saad, B., 2014: High performance liquid chromatography method for the determination of patulin and 5-hydroxymethylfurfural in fruit juices marketed in Malaysia. Food Control 38, 142-149. doi: https://doi.org/10.1016/j.foodcont.2013.10.018 (773).

Lee, Y.-W., Kim, J.-G., Chung, D.-H., Roh, P.-U., Pestka, J.J., 1991: Natural occurrence of zearalenone in rice and soybean produced in Korea. Mycotoxin Research 7, 69–72 (637).

Leggott, N.L., Shephard, G.S., 2001: Patulin in South African commercial apple products. Food Control 12, 73–76 (37).

Leggott, N.L., Shephard, G.S., Stockenström, S., Staal, E., Van Schalkwyk, D.J., 2001: The reduction of patulin in apple juice by three different types of activated carbon. Food Additives and Contaminants 18, 825-829. doi: https://doi.org/10.1080/02652030110044903 (448).

Leong, Y.-H., Ismail, N., Latif, A.A., Ahmad, R., 2010: Aflatoxin occurrence in nuts and commercial nutty products in Malaysia. Food Control 21, 334-338. doi: https://doi.org/10.1016/j.foodcont.2009.06.002 (225).

Leoni, L.A.B., Furlani, R.P.Z., Valente Soares, L.M., Oliveira, P.L.C., 2001: Ochratoxin A in Brazilian green coffee. Ciência e Tecnologia de Alimentos 21, 105-107. doi: https://doi.org/10.1590/S0101-20612001000100022 (306).

Leoni, L.A.B., Valente Soares, L.M., Oliveira, P.L.C., 2000: Ochratoxin A in Brazilian roasted and instant coffee. Food Additives and Contaminants 17, 867–870 (281).

Leszcynska, J., Kucharska, U., Żegota, H., 2000: Aflatoxins in nuts assayed by immunological methods. European Food Research and Technology 210, 213–215 (18).

Lehtonen, M., 1973: Detection of aflatoxines in wines. Chemie, Mikrobiologie & Technologie der Lebensmittel 2, 161–164 (332).

Levi, C., 1980: Mycotoxins in coffee. Journal of the Association of Official Analytical Chemists 63, 1282–1285 (860).

Levi, C.P., Trenk, H.L., Mohr, H.K., 1974: Study of the occurrence of ochratoxin A in green coffee beans. Journal of the Association of Official Analytical Chemists 57, 866–870 (310).

Li, F., Zhao, S., Chin, L., Li, Y., Wu, D., Zhao, X., Han, C., Zhang, H., Ji, R., 2007: Determination of patulin in apple and hawthorn beverages by solid-phase filtration column and liquid chromatography. Journal of the Association of Official Analytical Chemists International 90, 167–172 (482).

Li, F.-Q., Li, Y.-W., Wang, Y.-R., Luo, X.-Y., 2009: Natural occurrence of aflatoxins in Chinese peanut butter and sesame paste. Journal of Agricultural and Food Chemistry 57, 3519-3524. doi: https://doi.org/10.1021/jf804055n (224).

Li, M., Kong, W., Li, Y., Liu, H., Liu, Q., Dou, X., Ou-yang, Z., Yang, M., 2016: High-throughput determination of multi-mycotoxins in Chinese yam and related products by ultra fast liquid chromatography coupled with tandem mass spectrometry after one-step extraction. Journal of Chromatography B 1022, 118-125. doi: https://doi.org/10.1016/j.jchromb.2016.04.014 (774).

Li, R., Wang, X., Zhou, T., Yang, D., Wang, Q., Zhou, Y., 2014: Occurrence of four mycotoxins in cereal and oil products in Yangtze Delta region of China and their food safety risks. Food Control 35, 117–122. doi: https://doi.org/10.1016/j.foodcont.2013.06.042 (and personal communication) (748).

Li, X., Li, P., Zhang, Q., Li, R., Zhang, W., Zhang, Z., Ding, X., Tang, X., 2013: Multi-component immunochromatographic assay for simultaneous detection of aflatoxin B_1, ochratoxin A and zearalenone in agro-food. Biosensors and Bioelectronics 49, 426-432. doi: https://doi.org/10.1016/j.bios.2013.05.039 (775).

Liao, C. D., Wong, J.W., Zhang, K., Hayward, D.G., Lee, N.S., Trucksess, M.W., 2013: Multi-mycotoxin analysis of finished grain and nut products using high-performance liquid chromatography-triple-quadrupole mass spectrometry. Journal of Agricultural and Food Chemistry 61, 4771-4782. doi: https://doi.org/10.1021/4000677 (28).

Liao, C.-D., Wong, J.W., Zhang, K., Yang, P., Wittenberg, J.B., Trucksess, M.W., Hayward, D.G., Lee, N.S., Chang, J.S., 2015: Multi-mycotoxin analysis of finished grain and nut products using ultrahigh performance liquid chromatography and positive electrospray ionization-quadrupole orbital ion trap high-resolution mass spectrometry. Journal of Agricultural and Food Chemistry 63, 8314-8332. doi: https://doi.org/10.1021/jf505049a (777).

Liau, B.-C., Jong, T.-T., Lee, M.-R., Chang, C.-M.J., 2007: Supercritical fluid extraction and quantification of aflatoxins in Zizyphi Fructus by liquid chromatography/atmospheric pressure chemical ionization tandem mass spectrometry. Rapid Communications in Mass Spectrometry 21, 667-673. doi: https://doi.org/10.1002/rcm.2870 (776).

Lin, L.-C., Chen, P.-C., Fu, Y.-M., Shih, D.Y.-C., 2005: Ochratoxin A contamination in coffees, cereals, red wines and beers in Taiwan. Journal of Food and Drug Analysis 13, 84–92 (284).

Lindroth, S., Niskanen, A., 1978: Comparison of potential patulin hazard in home-made and commercial apple products. Journal of Food Science 43, 446–448 (33).

Lipigorngoson, S., Limtrakul, P., Suttajit, M., Yoshizawa, T., 2003: In-house direct cELISA for determining aflatoxin B_1 in Thai corn and peanuts. Food Additives and Contaminants 20, 838-845. doi: https://doi.org/10.1080/0265203031000156060 (399).

Liu, C., Liu, F., Xu, W., Kofoet, A., Humpf, H.-U., Jiang S., 2005: Occurrence of fumonisins B_1 and B_2 in asparagus from Shandong province, P.R. China. Food Additives and Contaminants 22, 673-676. doi: https://doi.org/10.1080/02652030500133016 (87).

Liu, H., Luo, J., Kong, W., Liu, Q., Hu, Y., Yang, M., 2016: UFLC-ESI-MS/MS analysis of multiple mycotoxins in medicinal and edible *Areca catechu*. Chemosphere 150, 176-183. doi: https://doi.org/10.1016/j.chemosphere.2016.02.032 (778).

Liu, L., Jin, H., Sun, L., Ma, S., Lin, R., 2012: Determination of aflatoxins in medicinal herbs by high-performance liquid chromatography-tandem mass spectrometry. Phytochemical Analysis 23, 469-476. doi: https://doi.org/10.1002/pca.2343 (70).

Liu, Q., Kong, W., Guo, W., Yang, M., 2015: Multi-class mycotoxins analysis in *Angelica sinensis* by ultra fast liquid chromatography coupled with tandem mass spectrometry. Journal of Chromatography B 988, 175-181. doi: https://doi.org/10.1016/j.jchromb.2015.02.022 (779).

Liu, S., Qiu, F., Kong, W., Wei, J., Xiao, X., Yang, M., 2013: Development and validation of an accurate and rapid LC-ESI-MS/MS method for the simultaneous quantification of aflatoxin B_1, B_2, G_1 and G_2 in lotus seeds. Food Control 29, 156-161. doi: https://doi.org/10.1016/j.foodcont.2012.05.069 (780).

Liu, Z-H., Tu, W-S., Li, D-R., Li, Y-D., Xie, C-H., Yang, Y-Z., Qin, B-B., 1990: A new method for the quantitation of aflatoxin M_1 in urine by high performance liquid chromatography and its application to the etiologic study of hepatoma. Biomedical Chromatography 4, 83–86 (417).

Logrieco, A., Bottalico, A., Visconti, A., Vurro, M., 1988: Natural occurrence of *Alternaria*-mycotoxins in some plant products. Microbiologie—Aliments—Nutrition 6, 13-17 (164).

Logrieco, A., Doko, B., Moretti, A., Frisullo, S., Visconti, A., 1998: Occurrence of fumonisins B_1 and B_2 in *Fusarium proliferatum* infected asparagus plants. Journal of Agricultural and Food Chemistry 46, 5201-5204. doi: https://doi.org/10.1021/jf9804903 (83).

Logrieco, A., Ferracane, R., Haidukowsky, M., Cozzi, G., Visconti, A., Ritieni, A., 2009: Fumonisin B_2 production by *Aspergillus niger* from grapes and natural occurrence in must. Food Additives and Contaminants: Part A 26, 1495-1500. doi: https://doi.org/10.1080/02652030903148322 (543).

Logrieco, A., Ferracane, R., Visconti, A., Ritieni, A., 2010: Natural occurrence of fumonisin B_2 in red wine from Italy. Food Additives and Contaminants: Part A 27, 1136-1141. doi: https://doi.org/10.1080/19440041003716547 (542).

Lohiya, G., Nichols, L., Hsieh, D., Lohiya, S., Nguyen, H., 1987: Aflatoxin content of foods served to a population with a high incidence of hepatocellular carcinoma. Hepatology 7, 750–752 (113).

Lohrey, L., Marschik, S., Cramer, B., Humpf, H.-U., 2013: Large scale synthesis of isotopically labeled $^{13}C_2$-tenuazonic acid and development of a rapid HPLC-MS/MS method for the analysis of tenuazonic acid in tomato and pepper products. Journal of Agricultural and Food Chemistry 61, 114-120. doi: https://doi.org/10.1021/jf305138k (781).

Loizzo, M.R., Nigro, S., De Luca, D., Menichini, F., 2011: Detection of ochratoxin A and *cis-* and *trans*-resveratrol in red wines and their musts from Calabria (Italy). Food Additives and Contaminants: Part A 28, 1561-1568. doi: https://doi.org/10.1080/19440049.2011.590454 (539).

Lombaert, G.A., Pellaers, P., Chettiar, M., Lavalee, D., Scott, P.M., Lau, B.P.-Y., 2002: Survey of Canadian retail coffees for ochratoxin Food Additives and Contaminants 19, 869–877. doi: https://doi.org/10.1080/02652030210145027 (287).

Lombaert, G.A., Pellaers, P., Neumann, G., Kitchen, D., Huzel, V., Trelka, R., Kotello, S., Scott, P.M., 2004: Ochratoxin A in dried vine fruits on the Canadian retail market. Food Additives and Contaminants 21, 578-585. doi: https://doi.org/10.1080/02652030410001687681 (329).

Lombaert, G.A., Pellaers, P., Roscoe, V., Mankotia, M., Neil, R., Scott, P.M., 2003: Mycotoxins in infant cereal foods from the Canadian retail market. Food Additives and Contaminants 20, 494-504. doi: https://doi.org/10.1080/0265203010094645 (123).

Lopez, A., Crawford, M.A., 1967: Aflatoxin content of groundnuts sold for human consumption in Uganda. Lancet 2, 1351–1354 (400).

López, P., Venema, D., De Rijk, T., De Kok, A., Scholten, J.M., Mol, H.G.J., De Nijs, M., 2016a: Occurrence of *Alternaria* toxins in food products in The Netherlands. Food Control, 60, 196-204. doi: https://doi.org/10.1016/j.foodcont.2015.07.032 (568).

López, P., Venema, D., Mol, H., Spanjer, M., De Stoppelaar, J., Pfeiffer, E., De Nijs, M., 2016b: *Alternaria* toxins and conjugates in selected foods in The Netherlands. Food Control, 60, 153-159. doi: https://doi.org/10.1016/j.foodcont.2016.04.001 (906).

Lopez De Cerain, A., González-Peñas, E., Jiménez, A.M., Bello, J., 2002: Contribution to the study of ochratoxin A in Spanish wines. Food Additives and Contaminants 19, 1058-1064. doi: https://doi.org/10.1080/02652030210145928 (607).

Lovelace, C.E.A., Aalbersberg, W.G.L., 1989: Aflatoxin levels in foodstuffs in Fiji and Tonga islands. Plant Foods for Human Nutrition 39, 393–399 (321).

Lucchetta, G., Bazzo, I., Dal Cortivo, G., Stringher, L., Bellotto, D., Borgo, M., Angelini, E., 2010: Occurrence of black aspergilli and ochratoxin A on grapes in Italy. Toxins (Basel) 2, 840-855. doi: https://doi.org/10.3390/toxins2040840 (914).

Lutfullah, G., Hussain, A., 2012: Studies on contamination level of aflatoxins in some cereals and beans of Pakistan. Food Control 23, 32-36. doi: https://doi.org/10.1016/j.foodcont.2011.06.004 (782).

Luttfullah, G., Hussain, A., 2011: Studies on contamination level of aflatoxins in some dried fruits and nuts of Pakistan. Food Control 22, 426-429. doi: https://doi.org/10.1016/j.foodcont.2010.09.015 (17).

Ma, L., Xu, W., He, X., Huang, K., Wang, Y., Luo, Y., 2013: Determination of fumonisins B_1 and B_2 in Chinese rice wine by HPLC using AQC pre-column derivatisation. Journal of the Science of Food and Agriculture 93, 1128-1133. doi: https://doi.org/10.1002/jsfa.5862 (628).

Maaroufi, K., Achour, A., Betbeder, A.M., Hammami, M., Ellouz, F., Creppy, E.E., Bacha, H., 1995: Foodstuffs and human blood contamination by the mycotoxin ochratoxin A: correlation with chronic interstitial nephropathy in Tunisia. Archives of Toxicology 69, 552–558 (345).

MacDonald, S., Castle, L., 1996: A UK retail survey of aflatoxins in herbs and spices and their fate during cooking. Food Additives and Contaminants 13, 120–128 (250).

MacDonald, S., Wilson, P., Barnes, K., Damant, A., Massey, R., Mortby, E., Shepherd, M.J., 1999: Ochratoxin A in dried vine fruit: method development and survey. Food Additives and Contaminants 16, 253–260 (328).

Maciel, L.F., Felício, A.L.d.S.M., Miranda, L.C.R., Pires, T.C., Bispo, E.d.S., Hirooka, E.Y., 2018: Aflatoxins and ochratoxin A in different cocoa clones (*Theobroma cacao* L.) developed in the southern region of Bahia, Brazil. Food Additives and Contaminants: Part A 35, 134-143. doi: https://doi.org/10.1080/19440049.2017.1397293 (615).

MAFF—UK, 1993: Patulin in apple juice: 1993 survey. Food-Surveillance-Information-Sheet No. 16 (466).

MAFF—UK, 1994: Survey of apple products for patulin. Food-Surveillance-Information-Sheet No. 29 (107).

MAFF—UK, 1994: 1994 Survey of apple juice for patulin. Food-Surveillance-Information-Sheet No. 36 (467).

MAFF—UK, 1995: 1995 Survey of apple juice for patulin. Food-Surveillance-Information-Sheet No. 74 (468).

MAFF—UK, 1996: 1996 Survey of apple juice for patulin. Food-Surveillance-Information-Sheet No. 97 (469).

MAFF—UK, 1996: Surveillance of ochratoxin A in green (unroasted) coffee beans. Food-Surveillance-Information-Sheet No. 80 (313).

MAFF—UK, 1996: Aflatoxin surveillance of retail and imported nuts, nut products, dried figs and fig products. Food-Surveillance-Information-Sheet No. 81 (206).

MAFF—UK, 1997: Survey of aflatoxins and ochratoxin A in cereals and retail products. Food-Surveillance-Information-Sheet No. 130 (143).

MAFF—UK, 1999: 1998 Survey of retail products for ochratoxin A. Food-Surveillance-Information-Sheet No. 185 (263).

MAFF—UK, 1999: 1998 survey of apple juice for patulin. Food-Surveillance-Information-Sheet No. 173 (470).

Magnani, R.F., De Souza, G.D., Rodriguez-Filho, E., 2007: Analysis of alternariol and alternariol monomethyl ether on flavedo and albedo tissues of tangerines (*Citrus reticulata*) with symptoms of *Alternaria* brown spot. Journal of Agricultural and Food Chemistry 55, 4980-4986. doi: https://doi.org/10.1021/jf0704256 (652).

Magnoli, C., Astoreca, A., Ponsone, L., Combina, M., Palacio, G., Rosa, C.A.R., Dalcero, A.M., 2004: Survey of mycoflora and ochratoxin A in dried vine fruits from Argentina markets. Letters in Applied Microbiology 39, 326-331. doi: https://doi.org/10.1111/j.1472-765X.2004.01583.x (647).

Magnoli, C., Astoreca, A., Ponsone, M.L., Fernández-Juri, M.G., Barberis C., Dalcero, A.M., 2007: Ochratoxin A and *Aspergillus* section *Nigri* in peanut seeds at different months of storage in Córdoba, Argentina. International Journal of Food Microbiology 119, 213-218. doi: https://doi.org/10.1016/j.ijfoodmicro.2007.07.056 (428).

Magnoli, C., Astoreca, A., Ponsone, L., Fernández-Juri, M.G., Chiacchiera, S., Dalcero, A., 2006: Ochratoxin A and the occurrence of ochratoxin A-producing black Aspergilli in stored peanut seeds from Córdoba, Argentina. Journal of the Science of Food and Agriculture 86, 2369-2373. doi: https://doi.org/10.1002/jsfa.2625 (431).

Magrine, I.O.C., Ferrari, S.S.C., Souza, G.F., Minamihara, L., Kemmelmeier, C., Bando, E., Machinski, M. Jr., 2011: Intake of aflatoxins through the consumption of peanut products in Brazil. Food Additives and Contaminants: Part B 4, 99-105. doi: https://doi.org/10.1080/19393210.2011.561931 (784).

Mahnine, N., Meca, G., Fernández-Franzón, M., Mañes, J., Zinedine, A., 2012: Occurrence of fumonisins B_1, B_2 and B_3 in breakfast and infant cereals from Morocco. Phytopathologia Mediterranea 51, 193–197 (656).

Maier, N.M., Buttinger, G., Welhartizki, S., Gavioli, E., Lindner, W., 2004: Molecularly imprinted polymer-assisted sample clean-up of ochratoxin A from red wine: merits and limitations. Journal of Chromatography B 804, 103-111. doi: https://doi.org/10.1016/j.jchromb.2004.01.014 (617).

Majerus, P., Bresch, H., Otteneder, H., 2000: Ochratoxin A in wines, fruit juices and seasonings. Archiv für Lebensmittelhygiene 51, 95–97 (110).

Majerus, P., Hain, J., Kölb, C., 2008: Patulin in grape must, and new, still fermenting wine (Federweißer). Mycotoxin Research 24, 135–139 (916).

Majerus, P., Woller, R., Leevivat, P., Klintrimas, T., 1985: Spices: Mould attack and content of aflatoxins, ochratoxin A and sterigmatocystine. Fleischwirtschaft 65, 1155–1158 (244).

Majid Cheraghali, A., Mohammadi, H.R., Amirahmadi, M., Yazdanpanah, H., Abouhossain, G., Zamanian, F., Khansari, M.G., Afshar, M., 2005: Incidence of patulin contamination in apple juice produced in Iran. Food Control 16, 165-167. doi: https://doi.org/10.1016/j.foodcont.2004.01.006 (476).

Makun, H.A., Adeniran, A.L., Mailafiya, S.C., Ayanda, I.S., Mudashiru, A.T., Ojukwu, U.J., Jagaba, A.S., Usman, Z., Salihu, D.A., 2013: Natural occurrence of ochratoxin A in some marketed Nigerian foods. Food Control 31, 566-571. doi: https://doi.org/10.1016/j.foodcont.2012.09.043 (785).

Malmauret, L., Parent Massin, D., Hardy, J. L., Verger, P., 2002: Contaminants in organic and conventional foodstuffs in France. Food Additives and Contaminants 19, 524-532. doi: https://doi.org/10.1080/02652030210123878 (34).

Manchisky, M., Jr., Midio, A.F., 1996: Incidencia de patulina en jugo de manzana industrializado. Alimentaria 276, 61–64 (465).

Manjula, K., Hell, K., Fandohan, P., Abass, A., Bandyopadhyay, R., 2009: Aflatoxin and fumonisin contamination of cassava products and maize grain from markets in Tanzania and Republic of the Congo. Toxin Reviews 28, 63-69. doi: https://doi.org/10.1080/15569540802462214 (256).

Marín, S., Mateo, E.M., Sanchis, V., Valle-Algarra, F.M., Ramos, A.J., Jiménez, M., 2011: Patulin contamination in fruit derivatives, including baby food, from the Spanish market. Food Chemistry 124, 563-568. doi: https://doi.org/10.1016/j.foodchem.2010.06.072 (786).

Mariod, A.A., Idris, Y.M.A., 2015: Aflatoxin B_1 levels in groundnut and sunflower oils in different Sudanese states. Food Additives and Contaminants: Part B 8, 266-270. doi: https://doi.org/10.1080/19393210.2015.1082511 (670).

Markaki, P., Delpont-Binet, C., Grosso, F., Dragacci, S., 2001: Determination of ochratoxin A in red wine and vinegar by immunoaffinity high-pressure liquid chromatography. Journal of Food Protection 64, 533–537 (111).

Marques, M.F., Martins, H.M., Costa, J.M., Bernardo, F., 2008: Co-occurrence of deoxynivalenol and zearalenone in crops marketed in Portugal. Food Additives and Contaminants: Part B 1, 130-133. doi: https://doi.org/10.1080/02652030802253983 (787).

Martinéz-Domínguez, G., Romero-González, R., Frenich, A.G., 2015: Determination of toxic substances, pesticides and mycotoxins, in ginkgo biloba nutraceutical products by liquid chromatography Orbitrap-mass spectrometry. Microchemical Journal 118, 124-130. doi: https://doi.org/10.1016/j.microc.2014.09.002 (64).

Martinéz-Domínguez, G., Romero-González, R., Frenich, A.G., 2016: Multi-class methodology to determine pesticides and mycotoxins in green tea and royal jelly supplements by liquid chromatography coupled to Orbitrap high resolution mass spectrometry. Food Chemistry 197, 907-915. doi: https://doi.org/10.1016/j.foodchem.2015.11.070 (788).

Martins, C., Assunção, R., Cunha, S.C., Fernandes, J.O., Jager, A., Petta, T., Oliveira, C.A., Alvito, P., 2018: Assessment of multiple mycotoxins in breakfast cereals available in the Portuguese market. Food Chemistry 239, 132-140. doi: https://doi.org/10.1016/j.foodchem.2017.06.088 (537).

Martins, M.L., Gimeno, A., Martins, H.M., Bernardo, F., 2002: Co-occurrence of patulin and citrinin in Portuguese apples with rotten spots. Food Additives and Contaminants 19, 568-574. doi: https://doi.org/10.1080/02652030210121320 (25).

Martins, M.L., Martins, H.M., Bernardo, F., 2001a: Aflatoxins in spices marketed in Portugal. Food Additives and Contaminants 18, 315-319. doi: https://doi.org/10.1080/02652030010021558 (169).

Martins, M.L., Martins, H.M., Bernado, F., 2001b: Fumonisins B_1 and B_2 in black tea and medicinal plants. Journal of Food Protection 64, 1268–1270 (78).

Martins, M.L., Martins, H.M., Gimeno, A., 2003: Incidence of microflora and of ochratoxin A in green coffee beans (*Coffea arabica*). Food Additives and Contaminants 20, 1127-1131. doi: https://doi.org/10.1080/02652030310001620405 (317).

Martins-Maciel, E.R., Machinski, M., Jr., Pereira, C.R.S., Takahachi, G., Kemmelmeier, C., Nishiyama, P., 1996: Incidence of aflatoxins and *Aspergillus flavus* in peanuts consumed in Maringá City, Brazil. Arquivos de Biologia e Technologia 39, 807–813 (388).

Mata, A.T., Ferreira, J.P., Oliveira, B.R., Batoréu, M.C., Barreto Crespo, M.T., Pereira, V.J., Bronze, M.R., 2015 : Bottled water: analysis of mycotoxins by LC-MS/MS. Food Chemistry 176, 455–464. doi: https://doi.org/10.1016/j.foodchem.2014.12.088 (789).

Matumba, L., Monjerezi, M., Biswick, T., Mwatseteza, J., Makumba, W., Kamangira, D., Mtukuso, A., 2014: A survey of the incidence and level of aflatoxin contamination in a range of locally and imported processed foods on Malawian retail market. Food Control 39, 87–91. doi : https://doi.org/10.1016/j.foodcont.2013.09.068 (790).

Matumba, L., Van Poucke, C., Monjerezi, M., Njumbe Ediage, E., De Saeger, S., 2015: Concentrating aflatoxins on the domestic market through groundnut export: a focus on Malawian groundnut value and supply chain. Food Control 51, 236-239. doi: https://doi.org/10.1016/j.foodcont.2014.11.035 (722).

Meister, U., 2003 : Detection of citrinin in ochratoxin A-containing products by a new HPLC method. Mycotoxin Research 19, 27–30 (673).

Meletis, K., Meniades-Meimaroglou, S., Markaki, P., 2007: Determination of ochratoxin A in grapes of Greek origin by immunoaffinity and high-performance liquid chromatography. Food Additives and Contaminants 24, 1275-1282. doi: https://doi.org/10.1080/02652030701413852 (159).

Mestres, C., Bassa, S., Fagbohoun, E., Nago, M., Hell, K., Vernier, P., Champiat, D., Hounhouigan, J., Cardwell, K.F., 2004: Yam chip food sub-sector: hazardous practices and presence of aflatoxins in Benin. Journal of Stored Products Research 40, 575-585. doi: https://doi.org/10.1016/j.jspr.2003.11.003 (259).

Meyer, K., Müller, T., Dietrich, R., Märtlbauer, E., Bauer, J., 2001: Nachweis und Vorkommen von Mykophenolsäure und Citrinin in Rotwein. Mycotoxin Research, 17 (Suppl. 2) 160–164 (422).

Meyvaci, K.B., Altindisli, A., Aksoy, U., Eltem, R., Turgut, H., Arasiler, Z., Kartal, N., 2005: Ochratoxin A in sultanas from Turkey I: survey of unprocessed sultanas from vineyards and packing-houses. Food Additives and Contaminants 22, 1138-1143. doi: https://doi.org/10.1080/02652030500199132 (645).

Mhadhbi, H., Bouzouita, N., Martel, A., Zarrouk, H., 2007: Occurrence of mycotoxin patulin in apple-based products marketed in Tunisia. Journal of Food Protection 70, 2642–2645 (483).

Micco, C., Grossi, M., Miraglia, M., Brera, C., 1989: A study of the contamination by ochratoxin A of green and roasted coffee beans. Food Additives and Contaminants 6, 333–339 (311).

Miceli, A., Negro, C., Tommasi, L., De Leo, P., 2003: Polyphenols, resveratrol, antioxidant activity and ochratoxin A contamination in red table wines, controlled denomination of origin (DOC) wines and wines obtained from organic farming. Journal of Wine Research 14, 115-120. doi: https://doi.org/10.1080/0957126041000l678030 (791).

Mikulíková, R., Běláková, S., Benešová, K., Svoboda, Z., 2012: Study of ochratoxin A content in south Moravian and foreign wines by the UPLC method with fluorescence detection. Food Chemistry 133, 55-59 doi: https://doi.org/10.1016/j.foodchem.2011.12.061 (792).

Milhome, M.A.L., Lima, C.G., De Lima, L.K., Lima, F.A.F., Sousa, D.O.B., Nascimento, R.F., 2014: Occurrence of aflatoxins in cashew nuts produced in northeastern Brazil. Food Control, 42, 34-37. doi: https://doi.org/10.1016/j.foodcont.2014.01.033 (572).

Miró-Abella, E., Herrero, P., Canela, N., Arola, L., Borrull, F., Ras, R., Fontanals, N., 2017: Determination of mycotoxins in plant-based beverages using QuEChERS and liquid chromatography-tandem mass spectrometry. Food Chemistry 229, 366-372. doi: https://doi.org/10.1016/j.foodchem.2017.02.078 (793).

Mislivec, P.B., Bruce, V.R., Stack, M.E., Bandler, R., 1987: Molds and tenuazonic acid in fresh tomatoes used for catsup production. Journal of Food Protection 50, 38–41 (655).

Misra, N., 1987: Mycotoxins in spices. III. Investigation on the natural occurrence of aflatoxins in *Coriandrum sativum* L. Journal of Food Science and Technology 24, 324–325 (322).

Misra, R.S., Singh, R.S., Tripathi, R.K., 1980: Aflatoxin contamination in some food commodities in and around Tarai area of Nainital, Uttar Pradesh. Indian Phytopathology 33, 217–220 (632).

Mogensen, J.M., Larsen, T.O., Nielsen, K.F., 2010: Widespread occurrence of the mycotoxin fumonisin B_2 in wine. Journal of Agricultural and Food Chemistry 58, 4853-4857. doi: https://doi.org/10.1021/jf904520t (625).

Mohammed, A., Chala, A., Dejene, M., Fininsa, C., Hoisington, D.A., Sobolev, V.S., Arias, R.S., 2016: *Aspergillus* and aflatoxin in groundnut (*Arachis hypogaea* L.) and groundnut cake in eastern Ethiopia. Food Additives and Contaminants: Part B 9, 290-298. doi: https://doi.org/10.1080/19393210.2016.1216468 (389).

Möller, T.E., Nyberg, M., 2003: Ochratoxin A in raisins and currants: basic extraction procedure used in two small marketing surveys of the occurrence and control of the heterogeneity of the toxins in samples. Food Additives and Contaminants 20, 1072-1076. doi: https://doi.org/10.1080/02652030310001615212 (598).

Molinié, A., Faucet, V., Castegnaro, M., Pfohl-Leszkowicz, A., 2005: Analysis of some breakfast cereals on the French market for their contents of ochratoxin A, citrinin and fumonisin B_1: development of a method for simultaneous extraction of ochratoxin A and citrinin. Food Chemistry 92, 391-400. doi: https://doi.org/10.1016/j.foodchem.2004.06.035 (94).

Monbaliu, S., Van Poucke, K., Heungens, K., Van Peteghem, C., De Saeger, S., 2010a: Production and migration of mycotoxins in sweet pepper analyzed by multimycotoxin LC-MS/MS. Journal of Agricultural and Food Chemistry 58, 10475-10479. doi: https://doi.org/10.1021/jf102722k (165).

Monbaliu, S., Wu, A., Zhang, D., Van Peteghem, C., De Saeger, S., 2010b: Multimycotoxin UPLC-MS/MS for tea, herbal infusions and the derived drinkable products. Journal of Agricultural and Food Chemistry 58, 12664-12671. doi: https://doi.org/10.1021/jf1033043 (198).

Montaseri, H., Eskandari, M.H., Yeganeh, A.T., Karami, S., Javidnia, K., Dehghanzadeh, G.R., Mesbahi, G.R., Niakousari, M., 2014: Patulin in apple leather in Iran. Food Additives and Contaminants: Part B 7, 106-109. doi: https://doi.org/10.1080/19393210.2013.855825 (795).

Monyo, E.S., Njoroge, S.M.C., Coe, R., Osiru, M., Madinda, F., Waliyar, F., Thakur, R.P., Chilunjika, T., Anitha, S., 2012: Occurrence and distribution of aflatoxin contamination in groundnuts (*Arachis hypogaea* L) and population density of *Aflatoxigenic Aspergilli* in Malawi. Crop Protection 42, 149-155 doi: https://doi.org/10.1016/j.cropro.2012.07.004 (796).

Morgan, M.R.A., Kang, A.S., Chan, H.W.-S., 1986: Aflatoxin determination in peanut butter by enzyme-linked immunosorbent assay. Journal of the Science of Food and Agriculture 37, 908–914 (228).

Mortimer, D.N., Parker, I., Shepherd, M.J., Gilbert, J., 1985: A limited survey of retail apple and grape juices for the mycotoxin patulin. Food Additives and Contaminants 2, 165–170 (455).

Mortimer, D.N., Shepherd, M.J., Gilbert, J., Morgan, M.R.A., 1987: A survey of the occurrence of aflatoxin B_1 in peanut butters by enzyme-linked immunosorbent assay. Food Additives and Contaminants 5, 127–132 (213).

Moukas, A., Panagiotopoulou, V., Markaki, P., 2008: Determination of patulin in fruit juices using HPLC-DAD and GC-MSD techniques. Food Chemistry 109, 860-867. doi: https://doi.org/10.1016/j.foodchem.2008.01.015 (108).

Mounjouenpou, P., Durand, N., Guyot, B., Guiraud, J.-P., 2007: Effect of operating conditions on ochratoxin A extraction from roasted coffee. Food Additives and Contaminants 24, 730-734. doi: https://doi.org/10.1080/02652030701199972 (295).

Mounjouenpou, P., Mbang, J.A.A.., Guyot, B., Guiraud, J.P., 2017: Traditional procedures of cocoa processing and occurrence of ochratoxin-A in the derived products. Journal of Chemical and Pharmaceutical Research 4, 1332–1339 (907).

Mphande, F.A., Siame, B.A., Taylor, J.E., 2004: Fungi, aflatoxins, and cyclopiazonic acid associated with peanut retailing in Botswana. Journal of Food Protection 67, 96–102 (390).

Mühlemann, M., Lüthy, J., Hübner, P., 1997: Mycotoxin contamination of food in Ecuador. A: aflatoxins. Mitteilungen aus dem Gebiete der Lebensmitteluntersuchung und Hygiene 88, 474–496 (140).

Müller, P., Basedow, T., 2007: Aflatoxin contamination of pods of Indian *Cassia senna* L. (Caesalpinaceae) before harvest, during drying and in storage: reasons and possible methods of reduction. Journal of Stored Products Research 43, 323-329. doi: https://doi.org/10.1016/j.jspr.2006.08.005 (806).

Mupunga, I., Lebelo, S.L., Mngqawa, P., Rheeder, J.P., Katerere, D.R., 2014: Natural occurrence of aflatoxins in peanuts and peanut butter from Bulawayo, Zimbabwe. Journal of Food Protection 77, 1814-1818. doi: https://doi.org/10.4315/0362-028X.JPF-14-129 (797).

Murillo, M., Gonzaléz-Peñas, E., Amézqueta, S., 2008: Determination of patulin in commercial apple juice by micellar electrokinetic chromatography. Food and Chemical Toxicology 46, 57-64. doi: https://doi.org/10.1016/j.fct.2007.06.024 (452).

Murillo-Arbizu, M., Amĕzqueta, S., González-Peñas, E., López De Cerain, A., 2009: Occurrence of patulin and its dietary intake through apple juice consumption by the Spanish population. Food Chemistry 113, 420-423. doi: https://doi.org/10.1016/j.foodchem.2008.07.054 (484).

Murillo-Arbizu, M.T., Amézqueta, S., González-Peñas, E., López de Cerain, A., 2010a: Occurrence of ochratoxin A in southern Spanish generous wines under the denomination of origin, Jerez-Xérès-Sherry and 'Manzanilla' Sanlúcar de Barrameda. Toxins (Basel) 2, 1054-1064. doi: https://doi.org/10.3390/toxins2051054 (798).

Murillo-Arbizu, M., González-Peñas, E., Amézqueta, S., 2010b: Comparison between capillary electrophoresis and high performance liquid chromatography for the study of the occurrence of patulin in apple juice intended for infants. Food and Chemical Toxicology 48, 2429-2434. doi: https://doi.org/10.1016/j.fct.2010.06.003 (799).

Murillo-Arbizu, M., González-Peñas, E., Hansen, S.H., Amézqueta, S., Østergaard, J., 2008: Development and validation of a microemulsion electrokinetic chromatography method for patulin quantification in commercial apple juice. Food and Chemical Toxicology 46, 2251-2257. doi: https://doi.org/10.1016/j.fct.2008.02.027 (547).

Mutegi, C.K., Ngugi, H.K., Hendriks, S.L., Jones, R.B., 2009: Prevalence and factors associated with aflatoxin contamination of peanuts from western Kenya. International Journal of Food Microbiology 130, 27-34. doi: https://doi.org/10.1016/j.ijfoodmicro.2008.12.030 (418).

Nakai, V.K., De Oliveira Rocha, L., Gonçalez, E., Fonseca, H., Ortega, E.M.M., Corrêa, B., 2008: Distribution of fungi and aflatoxins in a stored peanut variety. Food Chemistry 106, 285-290. doi: https://doi.org/10.1016/j.foodchem.2007.05.087 (408).

Nakajima, M., Tsubouchi, H., Miyabe, M., Ueno, Y., 1997: Survey of aflatoxin B_1 and ochratoxin A in commercial green coffee beans by high-performance liquid chromatography linked with immunoaffinity chromatography. Food and Agricultural Immunology 9, 77–83 (303).

Nasir, M.S., Jolley, M.E., 2002: Development of a fluorescence polarization assay for the determination of aflatoxins in grains. Journal of Agricultural and Food Chemistry 50, 3116-3121. doi: https://doi.org/10.1021/jf011638c (231).

Nejad, A.S.M., Bayat, M., Ahmadi, A.A., 2013: Investigation of aflatoxin B_1 in spices marketed in Hyderabad, India by ELISA method. Journal of Pure and Applied Microbiology 7, 3219–3223 (800).

Nejad, A.S.M., Ghannad, M.S., Kamkar, A., 2014: Determination of aflatoxin B_1 levels in Iranian and Indian spices by ELISA method. Toxin Reviews 33, 151-154. doi: https://doi.org/10.3109/15569543.2014.942319 (801).

Nejad, M.I., Farahani, A.D., 2012: Aflatoxin in raw walnut kernels marketed in Tehran, Iran. Food Additives and Contaminants: Part B 5, 8-10. doi: https://doi.org/10.1080/19393210.2011.637239 (802).

Ng, W., Mankotia, M., Pantazopoulos, P., Neil, R.J., Scott, P.M., 2004: Ochratoxin in wine and grape juice sold in Canada. Food Additives and Contaminants 21, 971-981. doi: https://doi.org/10.1080/02652030400000653 (495).

Nielsen, K.F., Ngemela, A.F., Jensen, L.B., Soman de Medeiros, L., Rasmussen, P.H., 2015: UHPLC-MS/MS determination of ochratoxin A and fumonisins in coffee using QuEChERS extraction combined with mixed-mode SPE purification. Journal of Agricultural and Food Chemistry 63, 1029-1034. doi: https://doi.org/10.1021/jf504254q (803).

Nikander, P., Seppälä, T., Kilonzo, G.P., Huttunen, P., Saarinen, L., Kilima, E., Pitkänen, T., 1991: Ingredients and contaminants of traditional alcoholic beverages in Tanzania. Transactions of the Royal Society of Tropical Medicine and Hygiene 85, 133–135 (161).

Nilsson, G., Åkerstrand, K., Slorach, S., 1974: Aflatoxins in nuts. Vår Föda 26, 56–62 (205).

Nilüfer, D., Boyacioğlu, D., 2002: Comparative study of three different methods for the determination of aflatoxins in Tahini. Journal of Agricultural and Food Chemistry 50, 3375-3379. doi: https://doi.org/10.1021/jf020005a (234).

Nishijima, M., 1983: Survey for mycotoxins in commercial foods. In: Kurata, H., Ueno, Y. (eds.), Developments in Food Science 7. Toxigenic Fungi – Their Toxins and Health Hazard. Proceedings of the Mycotoxin Symposia held in the Third International Mycological Congress, Tokyo, August 30-September 3, pp. 172–181 (129).

Njobeh, P.B., Dutton, M.F., Koch, S.H., Chuturgoon, A.A., Stoev, S.D., Mosonik, J.S., 2010: Simultaneous occurrence of mycotoxins in human food commodities from Cameroon. Mycotoxin Research 26, 47-57. doi: https://doi.org/10.1007/s12550-009-0039-6 (804).

Njumbe Ediage, E., Di Mavungu, J. D., Goryacheva, I.Y., Van Peteghem, C., De Saeger, S., 2012: Multiplex flow-through immunoassay formats for screening of mycotoxins in a variety of food matrices. Analytical and Bioanalytical Chemistry 403, 265-278. doi: https://doi.org/10.1007/s00216-012-5803-3 (805).

Njumbe Ediage, E., Di Mavungu, J.D., Monbaliu, S., Van Peteghem, C., De Saeger, S., 2011: A validated multianalyte LC-MS/MS method for quantification of 25 mycotoxins in cassava flour, peanut cake and maize samples. Journal of Agricultural and Food Chemistry 59, 5173-5180. doi: https://doi.org/10.1021/jf2009364 (235).

Njumbe Ediage, E., Hell, K., De Saeger, S., 2014: A comprehensive study to explore differences in mycotoxin patterns from agro-ecological regions through maize, peanut, and cassava products: a case study, Cameroon. Journal of Agricultural and Food Chemistry 62, 4789-4797. doi: https://doi.org/10.1021/jf501710u (807).

Noba, S., Omote, M., Kitagawa, Y., Mochizuki, N., 2008: Determination of ochratoxin A in wine by immunoaffinity cleanup and liquid chromatography tandem mass spectrometry. Journal of Food Protection 71, 1038–1042 (618).

Noba, S., Uyama, A., Mochizuki, N., 2009: Determination of ochratoxin A in ready-to-drink coffee by immunoaffinity cleanup and liquid chromatography-tandem mass spectrometry. Journal of Agricultural and Food Chemistry 57, 6036-6040. doi: https://doi.org/10.1021/jf900546p (343).

Nonaka, Y., Saito, K., Hanioka, N., Narimatsu, S., Kataoka, H., 2009: Determination of aflatoxins in food samples by automated on-line in-tube solid-phase microextraction coupled with liquid chromatography-mass spectrometry. Journal of Chromatography A 1216, 4416-4422. doi: https://doi.org/10.1016/j.chroma.2009.03.035 (339).

Noser, J., Schneider, P., Rother, M., Schmutz, H., 2011: Determination of six *Alternaria* toxins with UPLC- MS/MS and their occurrence in tomatoes and tomato products from the Swiss market. Mycotoxin Research 27, 265-271. doi: https://doi.org/10.1007/s12550-011-0103-x (47).

Noviandi, C.T., Razzazi, E., Agus, A., Böhm, J., Hulan, H.W., Wedhastri, S., Maryudhani, Y.B., Nuryono, Sardjono, Leibetseder, J., 2001: Natural occurrence of aflatoxin B_1 in some Indonesian food and feed products in Yogyakarta in year 1998–1999. Mycotoxin Research 17A, 174–178 (106).

Nyikal, J., Misore, A., Nzioka, C., Njuguna, C., Muchiri, E., Njau, J., Maingi, S., Njoroge, J., Mutiso, J., Onteri, J., Langat, A., Kilei, I.K, Nyamongo, J., Ogana, G., Muture, B., Tukei, P., Onyango, C., Ochieng, W., Tetteh C., Likimani, S., Nguku, P., Galgalo, T., Kibet, S., Manya, A., Dahiye, A., Mwihia, J., Mugoya, I., Onsongo, J., Ngindu, A., DeCock, K.M., Lindblade, K., Slutsker, L., Amornkul, P., Rosen, D., Feiken, D., Thomas, T., Mensah, P., Eseko, N., Nejjar, A., Onsongo, M., Kessel, F., Njapau, H., Park, D.L., Lewis, L., Luber, G., Rogers, H., Backer, L., Rubin, C., Gieseker, K.E., Azziz-Baumgartner, E., Chege, W., Bowen, A., 2004: Outbreak of aflatoxin poisoning—eastern and central provinces, Kenya, January–July 2004. Morbidity and Mortality Weekly Report 53, 790–793 (808).

O'Riordan, M.J., Wilkinson, M.G., 2008: A survey of the incidence and level of aflatoxin contamination in a range of imported spice preparations on the Irish retail market. Food Chemistry 107, 1429-1435. doi: https://doi.org/10.1016/j.foodchem.2007.09.073 (187).

Odoemelam, S.A., Osu, C.I., 2009: Aflatoxin B_1 contamination of some edible grains marketed in Nigeria. E-Journal of Chemistry 6, 308–314 (210).

Ok, H.E., Chung, S.H., Lee, N., Chun, H.S., 2015: Simple high-performance liquid chromatography method for the simultaneous analysis of aflatoxins, ochratoxin A, and zearalenone in dried and ground red pepper. Journal of Food Protection 78, 1226-1231. doi: https://doi.org/10.4315/0362-028X.JFP-14-451 (809).

Ok, H.E., Kim, H.J., Shim, W.B., Lee, H., Bae, D.-H., Chung, D.-H., Chun, H.S., 2007: Natural occurrence of aflatoxin B_1 in marketed foods and risk estimates of dietary exposure in Koreans. Journal of Food Protection 70, 2824–2828 (10).

Okano, K., Tomita, T., Kumeda, Y., Matsumaru, K., Ichinoe, M., 2008: Aflatoxin B and G contamination in imported peanuts and *Aspergillus* section *Flavi* of the causing fungi. Mycotoxins 58, 107–114 (407).

Okano, K., Tomita, T., Ohzu, Y., Takai, M., Ose, A., Kotsuka, A., Ikeda, N., Sakata, J., Kumeda, Y., Nakamura, N., Ichinoe, M., 2012: Aflatoxins B and G contamination and aflatoxigenic fungi in nutmeg. Shokuhin Eiseigaku Zasshi 53, 211–216 (548).

Okumura, H., Kawamura, O., Kishimoto, S., Hasegawa, A., Shrestha, S.M., Okuda, K., Obata, H., Okuda, H., Haruki, K., Uchida, T., Ogasawara, Y., Ueno, Y., 1993: Aflatoxin M_1 in Nepalese sera, quantified by combination of monoclonal antibody immunoaffinity chromatography and enzyme-linked immunosorbent assay. Carcinogenesis 14, 1233–1235 (403).

Oliveira, C.A.F., Gonçalves, N., Rosim, R.E., Fernandez, A.M., 2009: Determination of aflatoxins in peanut products in the northeast region of São Paulo, Brazil. International Journal of Molecular Sciences 10, 174-183. doi: https://doi.org/10.3390/ijms10010174 (237).

Oluwafemi, F., Nnanna Ibeh, I., 2011: Microbial contamination of seven major weaning foods in Nigeria. Journal of Health, Population and Nutrition 29, 415–419 (105).

Omurtag, G.Z., Atak, G., Keskin, G., Ersoy, Ö., 2002: HPLC assay for aflatoxins in dried red peppers and feedstuffs in Turkey. Acta Pharmaceutica Turcica 44, 11–22 (175).

Omurtag, G.Z., Atak, G., Yurdun, T., Ersoy, Ö., 1998: Aflatoxin analysis in dried red pepper samples by TLC and HPLC. Acta Pharmaceutica Turcica XXXX, 125-130 (172).

Omurtag, G.Z., Yazicioğlu, D., 2001: Occurrence of T-2 toxin in processed cereals and pulses in Turkey determined by HPLC and TLC. Food Additives and Contaminants 18, 844-849. doi: https://doi.org/10.1080/02652030110046901 (150).

Omurtag, G.Z., Yazicioğlu, D., 2004: Determination of fumonisins B_1 and B_2 in herbal tea and medicinal plants in Turkey by high-performance liquid chromatography. Journal of Food Protection 67, 1782–1786 (80).

Oroian, M., Amariei, S., Gutt, G., 2014: Patulin in apple juices from the Romanian market. Food Additives and Contaminants 7, 147–150. doi: https://doi.org/10.1080/19393210.2013.861518 (810).

Ostry, V., Malir, F., Dofkova, M., Skarkova, J., Pfohl-Leszkowicz, A., Ruprich, J., 2015: Ochratoxin A dietary exposure of ten population groups in the Czech Republic: comparison with data over the world. Toxins (Basel) 7, 3608-3635. doi: https://doi.org/10.3390/toxins7093608 (811).

Otteneder, H., Majerus, P., 2001: Ochratoxin A (OTA) in coffee: nation-wide evaluation of data collected by German Food Control 1995–1999. Food Additives and Contaminants 18, 431-435. doi: https://doi.org/10.1080/02652030010021567 (288).

Oviedo, M.S., Barros, G.G., Chulze, S.N., Ramirez, M.L., 2012: Natural occurrence of alternariol and alternariol monomethyl ether in soya beans. Mycotoxin Research 28, 169-174. doi: https://doi.org/10.1007/s12550-012-0132-0 (631).

Oyedele, O.A., Ezekiel, C.N., Sulyok, M., Adetjuni, M.C., Warth, B., Atanda, O.O., Krska, R., 2017: Mycotoxin risk assessment for consumers of groundnut in domestic markets in Nigeria. International Journal of Food Microbiology 251, 24-32. doi: https://doi.org/10.1016/j.ijfoodmicro.2017.03.020 (148).

Özay, G., Alperden, I., 1991: Aflatoxin and ochratoxin—a contamination of dried figs (*Ficus carina* L) from the 1988 crop. Mycotoxin Research 7, 85–91 (351).

Ozbey, F., Kabak, B., 2012: Natural co-occurrence of aflatoxins and ochratoxin A in spices. Food Control 28, 354-361. doi: https://doi.org/10.1016/j.foodcont.2012.05.039 (183).

Pacheco, A.M., Lucas, A., Parente, R., Pacheco, N., 2010: Association between aflatoxin and aflatoxigenic fungi in Brazil nut (*Bertholletia excelsa* H.B.K.). Ciência e Tecnologia de Alimentos 30, 330–334 (207).

Pacheco, A.M., Scussel, V.M., 2007: Selenium and aflatoxin levels in raw Brazil nuts from the Amazon Basin. Journal of Agricultural and Food Chemistry 55, 11087-11092. doi: https://doi.org/10.1021/jf072434k (209).

Pal, R., Varma, B.K., Srivastava, D.D., 1979: Aflatoxin in groundnut oil, groundnut cake and hydrogenated oil in Hapur (Uttar Pradesh, India) market. Journal of Food Science and Technology 16, 169–170 (579).

Pallarés, N., Font, G., Mañes, J., Ferrer, E., 2017: Multimycotoxin LC-MS/MS analysis in tea beverages after dispersive liquid-liquid microextraction (DLLME). Journal of Agricultural and Food Chemistry 65, 10282-10289. doi: https://doi.org/10.1021/acs.jafc.7b03507 (373).

Palumbo, J.D., O'Keeffe, T.L., Ho, Y.S., Santillan, C.J., 2015: Occurrence of ochratoxin A contamination and detection of ochratoxigenic *Aspergillus* species in retail samples of dried fruits and nuts. Journal of Food Protection 78, 836-842. doi: https://doi.org/10.4315/0362-028X.JFP-14-471 (599).

Papachristou, A., Markaki, P., 2004: Determination of ochratoxin A in virgin olive oils of Greek origin by immunoaffinity column clean-up and high-performance liquid chromatography. Food Additives and Contaminants 21, 85-92. doi: https://doi.org/10.1080/02652030310001632547 (576).

Pardo, E., Marín, S., Ramos, A.J., Sanchis, V., 2004: Occurrence of ochratoxigenic fungi and ochratoxin A in green coffee from different origins. Food Science and Technology International 10, 45-49. doi: https://doi.org/10.1177/1082013204041509 (309).

Patel, S., Hazel, C.M., Winterton, A.G.M., Gleadle, A.E., 1997: Survey of ochratoxin A in UK retail coffees. Food Additives and Contaminants 14, 217–222 (289).

Patel, S., Hazel, C.M., Winterton, A.G.M., Mortby, E., 1996: Survey of ethnic foods for mycotoxins. Food Additives and Contaminants 13, 833–841 (16).

Paterson, R.R.M., 2007: Aflatoxins contamination in chilli samples from Pakistan. Food Control 18, 817-820. doi: https://doi.org/10.1016/j.foodcont.2006.04.005 (251).

Pavón, M.Á., González, I., De La Cruz, S., Martín, R., Lacarra, T.G., 2012: The use of high-performance liquid chromatography to detect ochratoxin A in dried figs from the Spanish market. Journal of the Science of Food and Agriculture 92, 74–77. doi: https://doi.org/10.1002/jsfa.4543 (366).

Paulino De Moraes, M.H., Luchese, R.H., 2003: Ochratoxin A on green coffee: influence of harvest and drying processing procedures. Journal of Agricultural and Food Chemistry 51, 5824-5828. doi: https://doi.org/10.1021/jf026248k (318).

Peers, F., Bosch, X., Kaldor, J., Linsell, A., Pluijmen, M., 1987: Aflatoxin exposure, hepatitis B virus infection and liver cancer in Swaziland. International Journal of Cancer 39, 545–553 (135).

Peers, F.G., Linsell, C.A., 1975: Aflatoxin contamination and its heat stability in Indian cooking oils. Tropical Science 17, 229–232 (404).

Pena, A., Cerejo, F., Silva, L.J.G., Lino, C.M., 2010: Ochratoxin A survey in Portuguese wine by LC-FD with direct injection. Talanta 82, 1556-1561. doi: https://doi.org/10.1016/j.talanta.2010.07.041 (335).

Pensala, O., Niskanen, A., Lahtinen, S., 1977: The occurrence of aflatoxin in nuts and nut products imported to Finland for human consumption during the years 1974–1976. Nordisk Veterinaer Medicin 29, 347–355 (2).

Pepeljnjak, S., Ueno, Y., Tanaka, T., 1992: Mycological situation and occurrence of fusariotoxins in cereals from endemic nephropathy region in Croatia. Microbiologie—Aliments—Nutrition 10, 191–197 (149).

Pérez-Ortega, P., Gilbert-López, B., García-Reyes, J.F., Ramos-Martos, N., Molina-Díaz, A., 2012: Generic sample treatment method for simultaneous determination of multiclass pesticides and mycotoxins in wines by liquid chromatography-mass spectrometry. Journal of Chromatography A 1249, 32-40. doi: https://doi.org/10.1016/j.chroma.2012.06.020 (602).

Perrone, G., De Girolamo, A., Sarigiannis, Y., Haidukowski, M.E., Visconti, A., 2013: Occurrence of ochratoxin A, fumonisin B_2 and black aspergilli in raisins from Western Greece regions in relation to environmental and geographical factors. Food Additives and Contaminants: Part A 30, 1339-1347. doi: https://doi.org/10.1080/19440049.2013.796594 (812).

Perrone, G., Nicoletti, I., Pascale, M., De Rossi, A., De Girolamo, A., Visconti, A., 2007: Positive correlation between high levels of ochratoxin A and resveratrol-related compounds in red wines. Journal of Agricultural and Food Chemistry 55, 6807-6812. doi: https://doi.org/10.1021/jf0709931 (626).

Petkova-Bocharova, T., Castegnaro, M., 1985: Ochratoxin A contamination of cereals in an area of high incidence of Balkan endemic nephropathy in Bulgaria. Food Additives and Contaminants 2, 267–270 (141).

Piemontese, L., Solfrizzo, M., Visconti, A., 2005: Occurrence of patulin in conventional and organic fruit products in Italy and subsequent exposure assessment. Food Additives and Contaminants 22, 437-442. doi: https://doi.org/10.1080/02652030500073550 (35).

Pietri, A., Bertuzzi, T., Pallaroni, L., Piva, G., 2001: Occurrence of ochratoxin A in Italian wines. Food Additives and Contaminants 18, 647-654. doi: https://doi.org/10.1080/02652030110028353 (513).

Pietri, A., Rastelli, S., Bertuzzi, T., 2010. Ochratoxin A and aflatoxins in liquorice products. Toxins (Basel) 2, 758-770. doi: https://doi.org/10.3390/toxins2040758 (503).

Pietri, A., Rastelli, S., Mulazzi, A., Bertuzzi, T., 2012: Aflatoxins and ochratoxin A in dried chestnuts and chestnut flour produced in Italy. Food Control 25, 601-606. doi: https://doi.org/10.1016/j.foodcont.2011.11.042 (813).

Piñeiro, M., Dawson, R., Costarrica, M.L., 1996: Monitoring program for mycotoxin contamination in Uruguayan food and feeds. Natural Toxins 4, 242–245 (269).

Piqué, E., Vargas-Murga, L., Gómez-Catalán, J., De Lapuente, J., Llobet, J.M., 2013: Occurrence of patulin in organic and conventional apple-based food marketed in Catalonia and exposure assessment. Food and Chemical Toxicology 60, 190-204. doi: https://doi.org/10.1016/j.fct.2013.07.052 (594).

Pittet, A., Tornare, D., Huggett, A., Viani, R., 1996: Liquid chromatographic determination of ochratoxin A in pure and adulterated soluble coffee using an immunoaffinity column cleanup procedure. Journal of Agricultural and Food Chemistry 44, 3564–3569 (290).

Pizzutti, I.R., De Kok, A., Scholten, J., Righi, L.W., Cardoso, C.D., Rohers, G.N., Da Silva, R.C., 2014: Development, optimization and validation of a multimethod for the determination of 36 mycotoxins in wines by liquid chromatography-tandem mass spectrometry. Talanta 129, 352-363. doi: https://doi.org/10.1016/j.talanta.2014.05.017 (814).

Plestina, R., Ceović, S., Gatenbeck, S., Habazin-Novak, V., Hult, K., Hökby, E., Krogh, P., Radić, B., 1990: Human exposure to ochratoxin A in areas of Yugoslavia with endemic nephropathy. Journal of Environmental Pathology, Toxicology, Oncology 10, 145–148 (120).

Ponsone, M.L., Chiotta, M.L., Combina, M., Torres, A., Knass, P., Dalcero, A., Chulze, S., 2010: Natural occurrence of ochratoxin A in musts, wines and grape vine fruits from grapes harvested in Argentina. Toxins (Basel) 2(8), 1984-1996. doi: https://doi.org/10.3390/toxins2081984 (817).

Pouretedal, Z., Mazaheri, M., 2013: Aflatoxins in black tea in Iran. Food Additives and Contaminants: Part B 6, 127-129. doi: https://doi.org/10.1080/19393210.2013.764551 (817).

Prado, G., Altoé, A.F., Gomes, T.C.B., Leal, A.S., Morais, V.A.D., Oliveira, M.S., Ferreira, M.B., Gomes, M.B., Paschoal, F.N., Von S. Souza, R., Silva, D.A., Cruz Madeira, J.E.G., 2012: Occurrence of aflatoxin B_1 in natural products. Brazilian Journal of Microbiology 43, 1428–1435 (818).

Prado, G., De Oliveira, M.S., Da Cunha, M.R., De Fátima Gomides, M., Abrantes, F.M., Dos Santos, L.G., Veloso, T., De S. Barroseo, R.E., 2000: Occurrence of patulin in apple juice by high-performance liquid chromatography. Revista do Instituto Adolfo Lutz 59, 21–25 (819).

Prelle, A., Spadaro, D., Denca, A., Garibaldi, A., Gullino, M.L., 2013a: Comparison of clean-up methods for ochratoxin A on wine, beer, roasted coffee and chili commercialized in Italy. Toxins (Basel) 5, 1827-1844. doi: https://doi.org/10.3390/toxins5101827 (820).

Prelle, A., Spadaro, D., Garibaldi, A., Gullino, M.L., 2012: Aflatoxin monitoring in Italian hazelnut products by LC-MS. Food Additives and Contaminants: Part B 5, 279–285 doi: 10.1080/19393210.2012.711371 (and personal communication) (821).

Prelle, A., Spadaro, D., Garibaldi, A., Gullino, M.L., 2013b: A new method for detection of five *Alternaria* toxins in food matrices based on LC-APCI-MS. Food Chemistry 140, 161-167. doi: https://doi.org/10.1016/j.foodchem.2012.12.065 (822).

Prelle, A., Spadaro, D., Garibaldi, A., Gullino, M.L., 2014: Co-occurrence of aflatoxins and ochratoxin A in spices commercialized in Italy. Food Control 39, 192–197. doi: https://doi.org/10.1016/j.foodcont.2013.11.013 (and personal communication) (884).

Prieta, J., Moreno, M.A., Díaz, S., Suárez, G., Domínguez, L., 1994: Survey of patulin in apple juice and children's apple food by the diphasic dialysis membrane procedure. Journal of Agricultural and Food Chemistry 42, 1701–1703 (456).

Purchase, I.F.H., Pretorius, M.E., 1973: Sterigmatocystin in coffee beans. Journal of the Association of Official Analytical Chemists 56, 225–226 (280).

Qian, M., Zhang, H., Wu, L., Jin, N., Wang, J., Jiang, K., 2015: Simultaneous determination of zearalenone and its derivatives in edible vegetable oil by gel permeation chromatography and gas chromatography-triple quadrupole mass spectrometry. Food Chemistry 166, 23-28. doi: https://doi.org/10.1016/j.foodchem.2014.05.133 (824).

Qian, G.-S., Ross, R.K., Yu, M.C., Yuan, J.-M., Gao, Y.-T., Henderson, B.E., Wogan, G.N., Groopman, J.D., 1994: A follow-up study of urinary markers of aflatoxin exposure and liver cancer risk in Shanghai, People´s Republic of China. Cancer Epidemiology, Biomarkers & Prevention 3, 3–10 (132).

Quintela, S., Villarán, M.C., López de Armentia, I., Elejalde, E., 2011: Occurrence of ochratoxin A in Rioja Alavesa wines. Food Chemistry 126, 302-305. doi: https://doi.org/10.1016/j.foodchem.2010.09.096 (826).

Quintela, S., Villarán, M.C., López de Armentia, I., Elejalde, E., 2012: Ochratoxin A in Spanish exportation wine market. Food Control 25, 501-504. doi: https://doi.org/10.1016/j.foodcont.2011.11.023 (825).

Qutet, S.M., El-Tabey Shehata, A.M., Mesallam, A.S., 1983: Occurrence of aflatoxins in some Egyptian food crops collected from two coastal regions. Food Chemistry 10, 149–153 (4).

Rahimi, E., Jeiran, M.R., 2015: Patulin and its dietary intake by fruit juice consumption in Iran. Food Additives and Contaminants: Part B 8, 40-43. doi: https://doi.org/10.1080/19393210.2014.967814 (827).

Raisuddin, S., Misra, J.K., 1991: Aflatoxin in betel nut and its control by use of food preservatives. Food Additives and Contaminants 8, 707–712 (57).

Ram, B.P., Hart, L.P., Cole, R.J., Pestka, J.J., 1986: Application of ELISA to retail survey of aflatoxin B_1 in peanut butter. Journal of Food Protection 49, 792–795 (215).

Rani, N., Singh, S., 1996: Natural occurrence of mycoflora and mycotoxins in some pulse products of human use. Journal of the Indian Botanical Society 75, 283–285 (375).

Rati Rao, E., Basappa, S.C., Sreenivasa Murthy, V., 1979: Studies on the occurrence of ochratoxins in foodgrains. Journal of Food Science and Technology 16, 113–114 (430).

Ratola, N., Martins, L., Alves, A., 2004: Ochratoxin A in wines-assessing global uncertainty associated with the results. Analytica Chimica Acta 513, 319-324. doi: https://doi.org/10.1016/j.aca.2003.11.063 (521).

Ratt, E.R., Shantha, T., 1994: Incidence of aflatoxin in groundnut-based snack products. Journal of Food Science and Technology 31, 327–329 (199).

Razzazi-Fazeli, E., Noviandi, C.T., Porasuphatana, S., Agus, A., Böhm, J., 2004: A survey of aflatoxin B_1 and total aflatoxin contamination in baby food, peanut and corn products sold at retail in Indonesia analysed by ELISA and HPLC. Mycotoxin Research 20, 51–58 (226).

Reazai, M., Sayadi, M., Akbarpour, B., Mohammadpourfard, I., Behzadi, A., Teimoory, H., 2014: Safety of dried fruits marketed in Tehran as assessed by aflatoxin contamination. International Journal of Food Nutrition and Safety 5, 24–30 (828).

Reddy, K.R.N., Farhana, N.I., Salleh, B., 2011: Occurrence of *Aspergillus* spp. and aflatoxin B_1 in Malaysian foods used for human consumption. Journal of Food Science 76, T99-T104. doi: https://doi.org/10.1111/j.1750-3841.2011.02133.x (182).

Reddy, S.V., Kiran Mayi, D., Uma Reddy, M., Thirumala-Devi, K., Reddy, D.V.R., 2001: Aflatoxins B_1 in different grades of chillies (*Capsicum annum* L.) in India as determined by indirect competitive-ELISA. Food Additives and Contaminants 18, 553-558. doi: https://doi.org/10.1080/02652030010025383 (241).

Reinhold, L., Reinhardt, K., 2011: Mycotoxins in foods in Lower Saxony (Germany): results of official control analyses performed in 2009. Mycotoxin Research 27, 137-143. doi: https://doi.org/10.1007/s12550-011-0086-7 (829).

Reinholds, I., Pugajeva, I., Bavrins, K., Kuchovska, G., Bartkevics, V., 2017: Mycotoxins, pesticides and toxic metals in commercial spices and herbs. Food Additives and Contaminants: Part B 10, 5–14. doi: https://doi.org/10.1080/19393210.2016.1210244 (and personal communication) (679).

Reis, T.A., Oliveira, T.D., Baquião, A.C., Gonçalves, S.S., Zorzete, P., Corrêa, B., 2012: Mycobiota and mycotoxins in Brazil nut samples from different states of the Brazilian Amazon region. International Journal of Food Microbiology 159, 61-68. doi: https://doi.org/10.1016/j.ijfoodmicro.2012.08.005 (204).

Remiro, R., González-Peñas, E., Lizarraga, E., López de Cerain, A., 2012: Quantification of ochratoxin A and five analogs in Navarra red wines. Food Control 27, 139-145. doi: https://doi.org/10.1016/j.foodcont.2012.03.006 (830).

Remiro, R., Ibáñez-Vea, M., González-Peñas, E., Lizarraga, E., 2010: Validation of a liquid chromatography method for the simultaneous quantification of ochratoxin A and its analogues in red wines. Journal of Chromatography A 1217, 8249-8256. doi: https://doi.org/10.1016/j.chroma.2010.11.004 (831).

Remiro, R., Irigoyen, A., González-Peñas, E., Lizarraga, E., López de Cerain, A., 2013: Levels of ochratoxins in Mediterranean red wines. Food Control 32, 63-68. doi: https://doi.org/10.1016/j.foodcont.2012.11.040 (832).

Ritieni, A., 2003: Patulin in Italian commercial apple products. Journal of Agricultural and Food Chemistry 51, 6086-6090. doi: https://doi.org/10.1021/jf034523c (44).

Roach, J.A.G., Brause, A.R., Eisele, T.A., Rupp, H.S., 2002: HPLC detection of patulin in apple juice with GC/MS confirmation of patulin identity. Advances in Experimental Medicine and Biology 504, 135–140 (464).

Rodriguez, P., Venâncio, A., Lima, M., 2012: Aflatoxigenic fungi and aflatoxins in Portuguese almonds. Scientific World Journal, doi: https://doi.org/10.1100/2012/471926 (12).

Rodríguez-Carrasco, Y., Font, G., Mañes, J., Berrada, H., 2013: Determination of mycotoxins in bee pollen by gas chromatography-tandem mass spectrometry. Journal of Agricultural and Food Chemistry 61, 1999-2005. doi: https://doi.org/10.1021/jf400256f (154).

Rodríguez-Carrasco, Y., Mañes, J., Berrada, H., Juan, C., 2016: Development and validation of a LC-ESI-MS/MS method for the determination of *Alternaria* toxins alternariol, alternariol methyl-ether and tentotoxin in tomato and tomato-based products. Toxins (Basel) 8 (11), 328. doi: https://doi.org/10.3390/toxins8110328 (415).

Romagnoli, B., Menna, V., Gruppioni, N., Bergamini, C., 2007: Aflatoxins in spices, aromatic herbs, herb-teas and medicinal plants marketed in Italy. Food Control 18, 697-701. doi: https://doi.org/10.1016/j.foodcont.2006.02.020 (176).

Romani, S., Sacchetti, G., Chaves López, C., Pinnavaia, G.G., Dalla Rosa, M., 2000: Screening on the occurrence of ochratoxin A in green coffee beans of different origins and types. Journal of Agricultural and Food Chemistry 48, 3616-3619. doi: https://doi.org/10.1021/jf990783b (312).

Romero-González, R., Garrido Frenich, A., Martínez Vidal, J.L., Aguilera-Luiz, M.M., 2010: Determination of ochratoxin A and T-2 toxin in alcoholic beverages by hollow fiber liquid phase microextraction and ultra high-pressure liquid chromatography coupled to tandem mass spectrometry. Talanta 82, 171-176. doi: https://doi.org/10.1016/j.talanta.2010.04.016 (627).

Rosa, C.A.R., Magnoli, C.E., Fraga, M.E., Dalcero, A.M., Santana, D.M.N., 2004: Occurrence of ochratoxin A in wine and grape juice marketed in Rio de Janeiro, Brazil. Food Additives and Contaminants 21, 358-364. doi: https://doi.org/10.1080/02652030310001639549 (377).

Rosen, J.D., Pareles, S.R., 1974: Quantitative analysis of patulin in apple juice. Journal of Agricultural and Food Chemistry 22, 1024–1026 (450).

Roy, A.K., Chourasia, H.K., 1990: Mycoflora, mycotoxin producibility and mycotoxins in traditional herbal drugs from India. Journal of General and Applied Microbiology 36, 295–302 (63).

Roy, A.K., Kumar, S., 1993: Occurrence of ochratoxin A in herbal drugs of Indian origin—a report. Mycotoxin Research 9, 94–98 (76).

Roy, A.K., Sinha, K.K., Chourasia, H.K., 1988: Aflatoxin contamination of some common drug plants. Applied and Environmental Microbiology 54, 842–843 (62).

Roy, R., 1990: Aflatoxins in imported figs. Acta Horticulturae 269, 521–525 (357).

Ruadrew, S., Craft, J., Aidoo, K., 2013: Occurrence of toxigenic *Aspergillus* spp. and aflatoxins in selected food commodities of Asian origin sourced in the West of Scotland. Food and Chemical Toxicology 55, 653-658. doi: https://doi.org/10.1016/j.fct.2013.02.001 (314).

Rubert, J., Fapohunda, S.O., Soler, C., Ezekiel, C.N., Mañes, J., Kayode, F., 2013: A survey of mycotoxins in random street-vended snacks from Lagos, Nigeria, using QuEChERS-HPLC-MS/MS. Food Control 32, 673-677. doi: https://doi.org/10.1016/j.foodcont.2013.01.017 (833).

Rubert, J., Sebastià, N., Soriano, J.M., Soler, C., Mañes, J., 2011: One-year monitoring of aflatoxins and ochratoxin A in tiger-nuts and their beverages. Food Chemistry 127, 822-826. doi: https://doi.org/10.1016/j.foodchem.2011.01.016 (163).

Rubert, J., Soler, C., Mañes, J., 2012: Occurrence of fourteen mycotoxins in tiger-nuts. Food Control 25, 374-379. doi: https://doi.org/10.1016/j.foodcont.2011.11.004 (834).

Ruhland, M., Engelhardt, G., Wallnöfer, P.R., 1998: Production of ochratoxin A on artificially and naturally contaminated jam. Advances in Food Science (CMTL) 20, 13–16 (152).

Rupp, H.S., Turnipseed, S.B., 2000: Confirmation of patulin and 5-hydroxymethylfurfural in apple juice by gas chromatography/mass spectrometry. Journal of the Association of Analytical Chemists International 83, 612–620 (473).

Rusanova, T.Yu., Beloglazova, N.V., Goryacheva, I.Yu., Lobeau, M., Van Peteghem, C., De Saeger, S., 2009: Non-instrumental immunochemical tests for rapid ochratoxin A detection in red wine. Analytica Chimica Acta 653, 97-102. doi: https://doi.org/10.1016/j.aca.2009.08.036 (619).

Rychlik, M., Schieberle, P., 1999: Quantification of the mycotoxin patulin by a stable isotope dilution assay. Journal of Agricultural and Food Chemistry 47, 3749-3755. doi: https://doi.org/10.1021/jf990198a (52).

Sabino, M., Inomata, E.I., Lamardo, L.C.A., Milanez, T.V., Navas, S.A., Zorzetto, M.A.P., 1999: A survey of the occurrence of aflatoxins in groundnuts (peanuts) and groundnut products in São Paulo State/Brazil in 1994. Revista do Instituto Adolfo Luz 58, 53–57 (419).

Sadia, A., Jabbar, M.A., Deng, Y., Hussain, E.A., Riffat, S., Naveed, S., Arif, M., 2012: A survey of aflatoxin M_1 in milk and sweets of Punjab, Pakistan. Food Control 26, 235-240. doi: https://doi.org/10.1016/j.foodcont.2012.01.055 (835).

Sage, L., Krivobok, S., Delbos, E., Seigle-Murandi, F., Creppy, E.E., 2002: Fungal flora and ochratoxin A production in grapes and musts from France. Journal of Agricultural and Food Chemistry 50, 1306-1311. doi: https://doi.org/10.1021/jf011015z (541).

Saha, D., Acharya, D., Roy, D., Shrestha, D., Dhar, T.K., 2007: Simultaneous enzyme immunoassay for the screening of aflatoxin B_1 and ochratoxin A in chili samples. Analytica Chimica Acta 584, 343-349. doi: https://doi.org/10.1016/j.aca.2006.11.042 (248).

Sahar, N., Arif, S., Iqbal, S., Ul Ain Afzal, Q., Aman, S., Ara, J., Ahmed, M., 2015: Moisture content and its impact on aflatoxin levels in ready-to-use red chillies. Food Additives and Contaminants: Part B 8, 67-72. doi: https://doi.org/10.1080/19393210.2014.978395 (836).

Sahay, S.S., Prasad, T., 1990: The occurrence of aflatoxins in mustard and mustard products. Food Additives and Contaminants 7, 509–513 (544).

Sahay, S.S., Prasad, T., Sinha, K.K., 1990: Postharvest incidence of aflatoxins in *Linum usitatissimum* seeds. Journal of the Science of Food and Agriculture 53, 169–174 (524).

Saito, K., Ikeuchi, R., Kataoka, H., 2012: Determination of ochratoxins in nuts and grain samples by in-tube solid-phase microextraction coupled with liquid chromatography-mass spectrometry. Journal of Chromatography A 1220, 1-6. doi: https://doi.org/10.1016/j.chroma.2011.11.008 (21).

Salem, N.M., Ahmad, R., 2010: Mycotoxins in food from Jordan: preliminary survey. Food Control 21, 1099-1103. doi: https://doi.org/10.1016/j.foodcont.2010.01.002 (315).

Sanchis, V., Sala, N., Palomes, A., Santamarina, P., Burdaspal, P.A., 1986: Occurrence of aflatoxin and aflatoxigenic molds in foods and feed in Spain. Journal of Food Protection 49, 445–448, 454 (385).

Sangare-Tigori, B., Dem, A.A., Kouadio, H.J., Betbeder, A.-M., Dano, D.S., Moukha, S., Creppy, E.E., 2006a: Preliminary survey of ochratoxin A in millet, maize, rice and peanuts in Côte d'Ivoire from 1998 to 2002. Human & Experimental Toxicology 25, 211-216. doi: https://doi.org/10.1191/0960327106ht605oa (429).

Sangare-Tigori, B., Moukha, S., Kouadio, H.J., Betbeder, A.-M., Dano, D.S., Creppy, E.E., 2006b: Co-occurrence of aflatoxin B_1, fumonisin B_1, ochratoxin A and zearalenone in cereals and peanuts from Côte d'Ivoire. Food Additives and Contaminants 23, 1000-1007. doi: https://doi.org/10.1080/02652030500415686 (394).

Sani, A.M., Azizi, E.G., Naeejic, Z., 2012: Aflatoxins level in tea samples in Amol (north of Iran). Nutrition & Food Science 42, 422-427. doi: https://doi.org/10.1108/00346651211277672 (683).

Santos, L., Marín, S., Mateo, E.M., Gil-Serna, J., Valle-Algarra, F.M., Patiño, B., Ramos, A.J., 2011: Mycobiota and co-occurrence of mycotoxins in *Capsicum* powder. International Journal of Food Microbiology 151, 270-276. doi: https://doi.org/10.1016/j.ijfoodmicro.2011.09.011 (253).

Santos, L., Marín, S., Sanchis, V., Ramos, A.J., 2009: Screening of mycotoxin multicontamination in medicinal and aromatic herbs sampled in Spain. Journal of the Science of the Food and Agriculture 89, 1802-1807. doi: https://doi.org/10.1002/jsfa.3647 (72).

Santos, L., Marín, S., Sanchis, V., Ramos, A.J., 2010: Co-occurrence of aflatoxins, ochratoxin A and zearalenone in *Capsicum* powder samples available on the Spanish market. Food Chemistry 122, 826-830. doi: https://doi.org/10.1016/j.foodchem.2010.03.070 (252).

Sarigiannis, Y., Kapolos, J., Koliadima, A., Tsegenidis, T., Karaiskakis, G., 2014: Ochratoxin A levels in Greek retail wines. Food Control 42, 139-142. doi: https://doi.org/10.1016/j.foodcont.2014.02.003 (837).

Sartori, A.V., Swensson de Mattos, J., Pereira Souza, Y., Pereira dos Santos, R., De Moares, M.H.P., Wanderley da Nóbrega, A., 2015: Determination of aflatoxins M_1, M_2, B_1, B_2, G_1 and G_2 in peanut by modified QuEChERS method and ultra-high performance liquid chromatography-tandem mass spectrometry. Vigilância Sanitária em Debate: Sociedade, Ciência & Tecnologia 3, 115-121. doi: 10.3395/2317-269X.00406 (401).

Saxena, N., Dwivedi, P.D., Ansari, K.M., Das, M., 2008: Patulin in apple juices: incidence and likely intake in an Indian population. Food Additives and Contaminants: Part B 1, 140-146. doi: https://doi.org/10.1080/02652030802378848 (838).

Saxena, J., Mehrotra, B.S., 1989: Screening of spices commonly marketed in India for natural occurrence of mycotoxins. Journal of Food Composition and Analysis 2, 286–292 (22).

Saxena, J., Mehrota, B.S., 1990: The occurrence of mycotoxins in some dry fruits retail marketed in Nainital district of India. Acta Alimentaria 19, 221–224 (7).

Schade, J.E., McGreevy, K., King, A.D., Jr., Mackey, B., Fuller, G., 1975: Incidence of aflatoxin in California almonds. Applied Microbiology 29, 48–53 (839).

Scheuer, R., Gareis, M., 2002: Occurrence of ochratoxin A and B in spices. Mycotoxin Research 18, 62–66 (712).

Schmidt, H., Bannier, M., Vogel, R.F., Niessen, L., 2004: Detection and quantification of *Aspergillus ochraceus* in green coffee by PCR. Letters in Applied Microbiology 38, 464-469. doi: https://doi.org/10.1111/j.1472-765X.2004.01524.x (304).

Schollenberger, M., Müller, H.-M., Rüfle, M., Drochner, W., 2008: Natural occurrence of 16 *Fusarium* toxins in edible oil marketed in Germany. Food Control 19, 475-482. doi: https://doi.org/10.1016/j.foodcont.2007.05.012 (840).

Schollenberger, M., Müller, H.-M., Rüfle, M., Suchy, S., Planck, S., Drochner, W., 2005: Survey of *Fusarium* toxins in foodstuffs of plant origin marketed in Germany. International Journal of Food Microbiology 97, 317-326. doi: https://doi.org/10.1016/j.ijfoodmicro.2004.05.001 (151).

Schollenberger, M., Müller, H.-M., Rüfle, M., Terry-Jara, H., Suchy, S., Plank, S., Drochner, W., 2007: Natural occurrence of *Fusarium* toxins in soy food marketed in Germany. International Journal of Food Microbiology 113, 142-146. doi: https://doi.org/10.1016/j.ijfoodmicro.2006.06.022 (121).

Scholten, J.M., Spanjer, M.C., 1996: Determination of aflatoxin B_1 in pistachio kernels and shells. Journal of the Association of Official Analytical Chemists International 79, 1360–1364 (556).

Schroeder, H.W., Hein, H., Jr., 1977: Natural occurrence of sterigmatocystin in in-shell pecans. Canadian Journal of Microbiology 23, 639–641 (554).

Schwartzbord, J.R., Brown, D.L., 2015: Aflatoxin contamination in Haitian peanut products and maize and the safety of oils processed from contaminated peanuts. Food Control 56, 114-118. doi: https://doi.org/10.1016/j.foodcont.2015.03.014 (376).

Scott, P.M., Kennedy, B.P.C., 1973: Analysis and survey of ground black, white, and *Capsicum* peppers for aflatoxins. Journal of the Association of Official Analytical Chemists 56, 1452–1457 (242).

Scott, P.M., Kennedy, B.P.C., 1975: The analysis of spices and herbs for aflatoxins. Canadian Institute of Food Science and Technology Journal 8, 124–125 (323).

Scott, P.M., Lawrence, G.A., Lau, B.P.-Y., 2006: Analysis of wines, grape juices and cranberry juices for *Alternaria* toxins. Mycotoxin Research 22, 142–147 (492).

Scott, P.M., Miles, W.F., Toft, P., Dubé, J.G., 1972: Occurrence of patulin in apple juice. Journal of Agricultural and Food Chemistry 20, 450–451 (457).

Sebastià, N., Soler, C., Soriano, J.M., Mañes, J., 2010: Occurrence of aflatoxins in tigernuts and their beverages commercialized in Spain. Journal of Agricultural and Food Chemistry 58, 2609-2612. doi: https://doi.org/10.1021/jf903818x (162).

Sebastià, N., Meca, G., Soriano, J.M., Mañes, J., 2012: Presence of *Fusarium* emerging mycotoxins in tiger-nuts commercialized in Spain. Food Control 25, 631-635. doi: https://doi.org/10.1016/j.foodcont.2011.11.040 (565).

Seefelder, W., Gossmann, M., Humpf, H.-U., 2002: Analysis of fumonisin B_1 in *Fusarium proliferatum*-infected asparagus spears and garlic bulbs from Germany by liquid chromatography-electrospray ionization mass spectrometry. Journal of Agricultural and Food Chemistry 50, 2778-2781. doi: https://doi.org/10.1021/jf0115037 (86).

Selim, M.I., Popendorf, W., Ibrahim, M.S., El Sharkawy, S., El Kashory, E.S., 1996: Aflatoxin B_1 in common Egyptian foods. Journal of the Association of Official Analytical Chemists International 79, 1124–1129 (344).

Senyuva, H.Z., Gilbert, J., Ozcan, S., Ulken, U., 2005: Survey for co-occurrence of ochratoxin A and aflatoxin B_1 in dried figs in Turkey by using a single laboratory-validated alkaline extraction method for ochratoxin A. Journal of Food Protection 68, 1512–1515 (349).

Şenyuva, H.Z., Gilbert, J., Samson, R.A., Özcan, S., Öztürkoğlu, Ş., Önal, D., 2008: Occurrence of fungi and their mycotoxins in individual Turkish dried figs. World Mycotoxin Journal 1, 79–86 (354).

Senyuva, H.Z., Gilbert, J., Ulken, U., 2007: Aflatoxins in Turkish dried figs intended for export to the European Union. Journal of Food Protection 70, 1029–1032 (355).

Serra Bonvehi, J., 2004: Occurrence of ochratoxin A in cocoa products and chocolate. Journal of Agricultural and Food Chemistry 52, 6347-6352. doi: https://doi.org/10.1021/jf040153w (261).

Serrat, R., Mendoça, C., Venâncio, A, 2006: Fungi and ochratoxin A detected in healthy grapes for wine production. Letters in Applied Microbiology 42, 42–47 (841).

Sewram, V., Shephard, G.S., Van Der Merve, L., Jacobs, T.V., 2006: Mycotoxin contamination of dietary and medicinal wild plants in the Eastern Cape Province of South Africa. Journal of Agricultural and Food Chemistry 54, 5688-5693. doi: https://doi.org/10.1021/jf060483b (81).

Shamsuddin, Z.A., Khan, M.A., Khan, B.A., Ahmad, M.A., Ahmed, A., 1995: Contamination of red chilli with aflatoxin B_1 in Pakistan. Mycotoxin Research 11, 21–24 (246).

Shank, R.C., Bourgeois, C.H., Keschamras, N., Chandavimol, P., 1971: Aflatoxins in autopsy specimens from Thai children with an acute disease of unknown aetiology. Food and Cosmetics Toxicology 9, 501–507 (240).

Shank, R.C., Wogan, G.N., Gibson, J.B., Nondasuta, A., 1972: Dietary aflatoxins and human liver cancer. II. Aflatoxins in marketed foods and foodstuffs of Thailand and Hong Kong. Food Cosmetics and Toxicology 10, 61–69 (139).

Sharma, Y.P., Sumbali, G., 1999a: Incidence of aflatoxin producing strains and aflatoxin contamination in dry fruit slices of quinces (*Cydonia oblonga* Mill.) from the Indian state of Jammu and Kashmir. Mycopathologia 148, 103–107 (371).

Sharma, Y.P., Sumbali, G., 1999b: Natural occurrence of ochratoxin A and toxigenic *Aspergillus ochraceus* strains in dry fruit slices of quinces from Jammu and Kashmir. Indian Phytopathology 52, 148–150 (372).

Sharman, M., Patey, A.L., Bloomfield, D.A., Gilbert, J., 1991: Surveillance and control of aflatoxin contamination of dried figs and fig paste imported into the United Kingdom. Food Additives and Contaminants 8, 299–304 (359).

Shephard, G.S., Fabiani, A., Stockenström, S., Mshicileli, N., Sewram, V., 2003: Quantitation of ochratoxin A in South African wines. Journal of Agricultural and Food Chemistry 51, 1102-1106. doi: https://doi.org/10.1021/jf0259866 (608).

Shepard, G.S., Van der Westhuizen, L., Katerere, D.R., Herbst, M., Pineiro, M., 2010: Preliminary exposure assessment of deoxynivalenol and patulin in South Africa. Mycotoxin Research 26, 181-185. doi: https://doi.org/10.1007/s12550-010-0052-9 (842).

Sheu, F., Shyu, Y.T., 1999: Analysis of patulin in apple juice by diphasic dialysis extraction with in situ acylation and mass spectrometric determination. Journal of Agricultural and Food Chemistry 47, 2711-2714. doi: https://doi.org/10.1021/jf9812329 (458).

Shim, W.-B., Kim, K., Ofori, J.A., Chung, Y.-C., Chung, D.-H., 2012: Occurrence of aflatoxins in herbal medicine distributed in South Korea. Journal of Food Protection 75, 1991-1999. doi: https://doi.org/10.4315/0362-028X.JFP-12-190 (69).

Shon, D.-H., Cho, M.-H., Lee, M.-H., 1997: Aflatoxin residues in agricultural commodities determined by direct ELISA. Journal of Food Hygiene and Safety 12, 281–287 (643).

Shundo, L., De Almeida, A.P., Alaburda, J., Lamardo, L.C.A., Navas, S.A., Ruvieri, V., Sabino, M., 2009: Aflatoxins and ochratoxin A in Brazilian paprika. Food Control 20, 1099-1102. doi: https://doi.org/10.1016/j.foodcont.2009.02.008 (587).

Shundo, L., De Almeida, A.P., Alaburda, J., Ruvieri, V., Navas, S.A., Lamardo, L.C.A., Sabino, M., 2006: Ochratoxin A in wines and grape juices commercialized in the city of São Paulo, Brazil. Brazilian Journal of Microbiology 37, 533-537. doi:https://doi.org/10.1590/S1517-83822006000400021 (620).

Siame, B.A., Mpuchane, S.F., Gashe, B.A., Allotey, J., Teffera, G., 1998: Occurrence of aflatoxins, fumonisin B_1, and zearalenone in foods and feeds in Botswana. Journal of Food Protection 61, 1670–1673 (218).

Siantar, D.P., Halverson, C.A., Kirmiz, C., Peterson, G.F., Hill, N.R., Dugar, S.M., 2003: Ochratoxin A in wine: survey by antibody-and polymeric-based SPE columns using HPLC/fluorescence detection. American Journal of Enology and Viticulture 54, 170–177 (523).

Sibanda, L., De Saeger, S., Bauters, T.G.M., Nelis, H.J., Van Peteghem, C., 2001: Development of a flow through enzyme immunoassay and application in screening green coffee samples for ochratoxin A with confirmation by high-performance liquid chromatography. Journal of Food Protection 64, 1597–1602 (308).

Sidhu, O.P., Chandra, H., Behl, H.M., 2009: Occurrence of aflatoxins in mahua (*Madhuca indica* Gmel.) seeds: synergistic effect of plant extracts on inhibition of *Aspergillus flavus* growth and aflatoxin production. Food and Chemical Toxicology 47, 774-777. doi: https://doi.org/10.1016/j.fct.2009.01.001 (507).

Siegel, D., Andrae, K., Proske, M., Kochan, C., Koch, M., Weber, M., Nehls, I., 2010: Dynamic covalent hydrazine chemistry as a selective extraction and cleanup technique for the quantification of the *Fusarium* mycotoxin zearalenone in edible oils. Journal of Chromatography A 1217, 2206-2215. doi: https://doi.org/10.1016/j.chroma.2010.02.019 (843).

Singh, T., Tyagi, R.P.S., Varma, B.K., 1982: Study of the occurrence of aflatoxin B_1 in foodgrains. Journal of Food Science and Technology 19, 35–37 (386).

Sinha, A.K., Bilgrami, K.S., Ranjan, K.S., Prasad, T., 1988: Incidence of aflatoxins in mustard crop in Bihar. Indian Phytopathology 41, 434–437 (545).

Sinha, A.K., Ranjan, K.S., 1989: Preharvest aflatoxin problem in linseed (*Linum usitatissimum*) in Bihar. Journal of the Indian Botanical Society 68, 19–20 (463).

Sinha, A.K., Ranjan, K.S., 1990: Incidence of aflatoxins in food grains during flood and post flood period at North Bihar. Proceedings of the Indian National Science Academy B 56, 295–298 (128).

Sinha, K.K., Singh, A., 1982: Aflatoxin producing potentials of *Aspergillus flavus* and natural occurrences of aflatoxins in some fruits and vegetables. National Academy Science Letters 5, 213–214 (374).

Sinha, K.K., Sinha, A.K., 1991: Monitoring and identification of aflatoxins in wheat, gram and maize flours in Bihar state (India). Food Additives and Contaminants 8, 453–457 (100).

Škrbić, B., Antić, I., Cvejanov, J., 2017: Determination of mycotoxins in biscuits, dried fruits and fruit jams: an assessment of human exposure. Food Additives and Contaminants: Part A, 34, 1012–1025. doi: 10.1080/19440049.2017.1303195 (and personal communication) (903).

Škrbić, B., Živančev, J., Godula, M., 2014: Multimycotoxin analysis of crude extracts of nuts with ultra-high performance liquid chromatography/tandem mass spectrometry. Journal of Food Composition and Analysis 34, 171-177. doi: https://doi.org/10.1016/j.jfca.2014.03.002 (844).

Snow, J.P., Lucas, G.R., Harvan, D., Pero, N.W., Owens, R.G., 1972: Analysis of tobacco and smoke condensate for penicillic acid. Applied Microbiology 24, 34–36 (103).

Soares, L.M.V., Rodriguez-Amaya, D.B., 1989: Survey of aflatoxins, ochratoxin A, zearalenone, and sterigmatocystin in some Brazilian foods by using multi-toxin thin layer chromatographic method. Journal of the Association of Official Analytical Chemists 72, 22–26 (130).

Soleas, G.J., Yan, J., Goldberg, D.M., 2001: Assay of ochratoxin A in wine and beer by high-pressure liquid chromatography photodiode array and gas chromatography mass selective detection. Journal of Agricultural and Food Chemistry 49, 2733-2740. doi: https://doi.org/10.1021/jf0100651 (609).

Solfrizzo, M., Panzarini, G., Visconti, A., 2008: Determination of ochratoxin A in grapes, dried vine fruits, and winery byproducts by high performance liquid chromatography with fluorometric detection (HPLC-FLD) and immunoaffinity cleanup. Journal of Agricultural and Food Chemistry 56, 11081-11086. doi: https://doi.org/10.1021/jf802380d (49).

Solfrizzo, M., Piemontese, L., Gambacorta, L., Zivoli, R., Longobardi, F., 2015: Food coloring agents and plant food supplements derived from *Vitis vinifera*: a new source of human exposure to ochratoxin A. Journal of Agricultural and Food Chemistry 63, 3609-3614. doi: https://doi.org/10.1021/acs.jafc.5b00326 (845).

Soliman, K.M., 2002: Incidence, level, and behavior of aflatoxins during coffee bean roasting and decaffeination. Journal of Agricultural and Food Chemistry 50, 7477-7481. doi: https://doi.org/10.1021/jf011338v (279).

Somorin, Y., Akinyemi, A., Bertuzzi, T., Pietri, A., 2016: Co-occurrence of aflatoxins, ochratoxin A and citrinin in "egusi" melon (*Colocynthis citrullus* L.) seeds consumed in Ireland and the United Kingdom. Food Additives and Contaminants: Part B 9, 230-235. doi: https://doi.org/10.1080/19393210.2016.1183051 (846).

Sørensen, J.L., Phipps, R.K., Nielsen, K.F., Schroers, H.-J., Frank, J., Thrane, U., 2009: Analysis of *Fusarium avenaceum* metabolites produced during wet apple core rot. Journal of Agricultural and Food Chemistry 57, 1632-1639. doi: https://doi.org/10.1021/jf802926u (36).

Soufleros, E.H., Tricard, C., Bouloumpasi, E.C., 2003: Occurrence of ochratoxin A in Greek wines. Journal of the Science of Food and Agriculture 83, 173-179. doi: https://doi.org/10.1002/jsfa.1300 (509).

Spadaro, D., Ciavorella, A., Frati, S., Garibaldi, A., Gullino, M.L., 2007a: Incidence and level of patulin contamination in pure and mixed apple juices marketed in Italy. Food Control 18, 1098-1102. doi: https://doi.org/10.1016/j.foodcont.2006.07.007 (438).

Spadaro, D., Ciavorella, A., Lore, A., Garibaldi, A., Gullino, M.L., 2007b: Low levels of ochratoxin A in wines from Piedmont. Communications in Agricultural and Applied Biological Science 72, 327–332 (336).

Spadaro, D., Garibaldi, A., Gullino, M.L., 2008: Occurrence of patulin and its dietary intake through pear, peach, and apricot juices in Italy. Food Additives and Contaminants: Part B 1, 134-139. doi: https://doi.org/10.1080/02652030802363790 (847).

Spadaro, D., Lorè, A., Garibaldi, A., Gullino, M.L., 2010: Occurrence of ochratoxin A before bottling in DOC and DOCG wines produced in Piedmont (northern Italy). Food Control 21, 1294-1297. doi: https://doi.org/10.1016/j.foodcont.2010.02.017 (337).

Stander, M.A., Steyn, P.S., 2002: Survey of ochratoxin A in South African wines. South African Journal of Enology and Viticulture 23, 9–13 (512).

Stefanaki, I., Foufa, E., Tsatsou-Dritsa, A., Dais, P., 2003: Ochratoxin A concentrations in Greek domestic wines and dried vine fruits. Food Additives and Contaminants 20, 74-83. doi: https://doi.org/10.1080/0265203021000031537 (327).

Steiner, E.W., Rieker, R.H., Battaglia, R., 1988: Aflatoxin contamination in dried figs: distribution and association with fluorescence. Journal of Agricultural and Food Chemistry 36, 88–91 (348).

Steiner, W., Brunschweiler, K., Leimbacher, E., Schneider, R., 1993: Aflatoxin B_1 and G_1, cyclopiazonic acid, kojic acid and ochratoxin A in dried figs showing BGY-fluorescence. Mitteilungen aus dem Gebiete der Lebensmitteluntersuchung und Hygiene 84, 523–536 (350).

Stępień, L., Koczyk, G., Waśkiewicz, A., 2013: Diversity of *Fusarium* species and mycotoxins contaminating pineapple. Journal of Applied Genetics 54, 367-380. doi: https://doi.org/10.1007/s13353-013-0146-0 (848).

Stępień, L., Waśkiewicz, A., Urbaniak, M., 2016: Wildly growing asparagus (*Asparagus officinalis* L.) hosts pathogenic *Fusarium* species and accumulates their mycotoxins. Microbial Ecology 71: 927-937. doi: https://doi.org/10.1007/s00248-015-0717-1 (849).

Stinson, E.E., Osman, S.F., Heisler, E.G., Siciliano, J., Bills, D.D., 1981: Mycotoxin production in whole tomatoes, apples, oranges, and lemons. Journal of Agricultural and Food Chemistry 29, 790–792 (23).

Stray, H., 1978: High pressure liquid chromatographic determination of patulin in apple juice. Journal of the Association of Official Analytical Chemists 61, 1359–1362 (459).

Studer-Rohr, I., Dietrich, D.R., Schlatter, J., Schlatter, C., 1995: The occurrence of ochratoxin A in coffee. Food Chemical and Toxicology 33, 341–355 (307).

Subrahmanyam, P., Rao, A.S., 1973: Occurrence of aflatoxins in some pickled foods. Indian Journal of Microbiology 13, 157–159 (370).

Subrahmanyam, P., Rao, A.S., 1974: Occurrence of aflatoxins and citrinin in groundnut (*Arachis hypogaea* L.) at harvest in relation to pod condition and kernel moisture content. Current Science 43, 707–710 (427).

Sulyok, M., Beed, F., Boni, S., Abass, A., Mukunzi, A., Krska, R., 2015: Quantitation of multiple mycotoxins and cyanogenic glucosides in cassava samples from Tanzania and Rwanda by an LC-MS/MS-based multi-toxin method. Food Additives and Contaminants: Part A 32, 488-502. doi: https://doi.org/10.1080/19440049.2014.975752 (823).

Sulyok, M., Krska, R., Schuhmacher, R., 2007: A liquid chromatography/tandem mass spectrometric multi-mycotoxin method for the quantification of 87 analytes and its application to semi-quantitative screening of moldy food samples. Analytical and Bioanalytical Chemistry 389, 1505-1523. doi: https://doi.org/10.1007/s00216-007-1542-2 (850).

Sun, G., Wang, S., Hu, X., Su, J., Zhang, Y., Xie, Y., Zhang, H., Tang, L., Wang, J.-S., 2011: Co-contamination of aflatoxin B_1 and fumonisin B_1 in food and human dietary exposure in three areas of China. Food Additives and Contaminants: Part A 28, 461-470. doi: https://doi.org/10.1080/19440049.2010.544678 (406).

Suzuki, J.I., Dainius, B., Kilbuck, J.H., 1973: A modified method for aflatoxin determination in spices. Journal of Food Science 38, 949–950 (185).

Swaileh, K.M., Abdulkhaliq, A., 2013: Analysis of aflatoxins, caffeine, nicotine and heavy metals in Palestinian multifloral honey from different geographic regions. Journal of the Science of Food and Agriculture 93, 2116-2120. doi: https://doi.org/10.1002/jsfa.6014 (851).

Tabata, S., Iida, K., Kimura, K., Iwasaki, Y., Nakazato, M., Kamata, K., Hirokado, M., 2008: Investigation of ochratoxin A, B and citrinin contamination in various commercial foods. Journal of Food Hygienic Society of Japan 49, 111–115 (112).

Tabata, S., Kamimura, H., Ibe, A., Hashimoto, H., Iida, M., Tamura, Y., Nishima, T., 1993: Aflatoxin contamination in foods and foodstuffs in Tokyo: 1986–1990. Journal of the Association of Official Analytical Chemists International 76, 32–35 (99).

Tabata, S., Kamimura, H., Tamura, Y., Yasuda, K., Ushiyama, H., Hashimoto, H., Nishijima, M., Nishima, T., 1987: Investigation of aflatoxins contamination in foods and foodstuffs. Journal of Food Hygienic Society of Japan 28, 395–401 (97).

Tafuri, A., Ferracane, R., Ritieni, A., 2004: Ochratoxin A in Italian marketed cocoa products. Food Chemistry 88, 487-494. doi: https://doi.org/10.1016/j.foodchem.2004.01.061 (593).

Taguchi, S., Fukushima, S., Sumimoto, T., Yoshida, S., Nishimune, T., 1995: Aflatoxins in foods collected in Osaka, Japan, from 1988 to 1992. Journal of the Association of Official Analytical Chemists International 78, 325–327 (173).

Tam, J., Mankotia, M., Malby, M., Pantazopoulos, P., Neil, R.J., Calway, P., Scott, P.M., 2006: Survey of breakfast and infant cereals for aflatoxins B_1, B_2, G_1 and G_2. Food Additives and Contaminants 23, 693-699. doi: https://doi.org/10.1080/02652030600627230 (92).

Tam, J., Pantazopoulos, P., Scott, P.M., Moisey, J., Dabeka, R.W., Richard, I.D.K., 2011: Application of isotope dilution mass spectrometry: determination of ochratoxin A in the Canadian total diet study. Food Additives and Contaminants: Part A 28, 754-761. doi: https://doi.org/10.1080/19440049.2010.504750 (90).

Tamura, M., Takahashi, A., Uyama, A., Mochizuki, N., 2012: A method for multiple mycotoxin analysis in wines by solid phase extraction and multifunctional cartridge purification, and ultra-high-performance liquid chromatography coupled to tandem mass spectrometry. Toxins 4, 476-486. doi: https://doi.org/10.3390/toxins4060476 (622).

Tan, Y., Kuang, Y., Zhao, R., Chen, B., Wu, J., 2011: Determination of T-2 and HT-2 toxins in traditional Chinese medicine marketed in China by LC-ELSD after sample clean-up by two solid-phase extractions. Chromatographia 73, 407-410. doi: https://doi.org/10.1007/s10337-010-1890-5 (854).

Tanaka, T., Hasegawa, A., Yamamoto, S., Lee, U.-S., Sugiura, Y., Ueno, Y., 1988: Worldwide contamination of cereals by the *Fusarium* mycotoxins nivalenol, deoxynivalenol, and zearalenone. 1. Survey of 19 countries. Journal of Agricultural and Food Chemistry 36, 979–983 (630).

Tang, D., Zhong, Z., Niessner, R., Knopp, D., 2009: Multifunctional magnetic bead-based electrochemical immunoassay for the detection of aflatoxin B_1 in food. Analyst 134, 1554-1560. doi: https://doi.org/10.1039/b902401h (855).

Tang, S.-Y., Ling, K.-H., 1977: Mycotoxins in soy sauce and fermented soybean paste of Taiwan. Journal of the Formosan Medical Association 76, 482–487 (590).

Tangni, E.K., Theys, R., Mignolet, E., Maudoux, M., Michelet, J.Y., Larondelle, Y., 2003: Patulin in domestic and imported apple-based drinks in Belgium: occurrence and exposure assessment. Food Additives and Contaminants 20, 482-489. doi: https://doi.org/10.1080/0265203031000093204 (38).

Taniwaki, M.H., Pitt, J.I., Teixeira, A.A., Iamanaka, B.T., 2003: The source of ochratoxin A in Brazilian coffee and its formation in relation to processing methods. International Journal of Food Microbiology 82, 173–179 (856).

Tansakul, N., Limsuwan, S., Böhm, J., Hollmann, M., Razzazi-Fazeli, E., 2013: Aflatoxins in selected Thai commodities. Food Additives and Contaminants: Part B 6, 254-259. doi: https://doi.org/10.1080/19393210.2013.812148 (857).

Tantaoui-Elaraki, A., Le Tutour, B., Bouzid, M., Keddani, M.J., 1983: Contamination des olives noires "façon Grèce" par les spores d´*Aspergillus* toxinogènes et leurs toxines. Industries Alimentaires et Agricoles 100, 997–1000 (584).

Tarter, E.J., Scott, P.M., 1991: Determination of patulin by capillary gas chromatography of the heptafluorobutyrate derivative. Journal of Chromatography 538, 441–446 (477).

Tassaneeyakul, W., Razzazi-Fazeli, E., Porasuphatana, S., Bohm, J., 2004: Contamination of aflatoxins in herbal medicinal products in Thailand. Mycopathologia 158, 239–244 (67).

Terminiello, L., Patriarca, A., Pose, G., Pinto, V.F., 2006: Occurrence of alternariol, alternariol monomethyl ether and tenuazonic acid in Argentinean tomato puree. Mycotoxin Research 22, 236–240 (45).

Terra, M.F., Prado, G., Pereira, G.E., Ematné, H.J., Batista, L.R., 2013: Detection of ochratoxin A in tropical wine and grape juice from Brazil. Journal of the Science of Food and Agriculture 93, 890-894. doi: https://doi.org/10.1002/jsfa.5817 (624).

Thirumala-Devi, K., Mayo, M.A., Reddy, G., Emmanuel, K.E., Larondelle, Y., Reddy, D.V.R., 2001: Occurrence of ochratoxin A in black pepper, coriander, ginger and turmeric in India. Food Additives and Contaminants 18, 830 835. doi: https://doi.org/10.1080/02652030110044921 (191).

Thirumala-Devi, K., Mayo, M.A., Reddy, G., Reddy, S.V., Delfosse, P., Reddy, D.V.R., 2000: Production of polyclonal antibodies against ochratoxin A and its detection in chilies by ELISA. Journal of Agricultural and Food Chemistry 48, 5079-5082. doi: https://doi.org/10.1021/jf000049o (255).

Thorpe, C.W., Johnson, R.L., 1974: Analysis of penicillic acid by gas-liquid chromatography. Journal of the Association of Official Analytical Chemists 57, 861–865 (102).

Thuvander, A., Möller, T., Enghardt Barbieri, H., Jansson, A., Salomonsson, A.-C., Olsen, M., 2001: Dietary intake of some important mycotoxins by the Swedish population. Food Additives and Contaminants 18, 696-706. doi: https://doi.org/10.1080/02652030110047775 (13).

Toffa, D.D., Mahnine, N., Ouaffak, L., El Abidi, A., El Alaoui Faris, F.Z., Zinedine, A., 2013: First survey on the presence of ochratoxin A and fungi in raw cereals and peanut available in the Republic of Niger. Food Control 32, 558-562. doi: https://doi.org/10.1016/j.foodcont.2013.01.028 (858).

Torović, L., Dimitrov, N., Assunção, R., Alvito, P., 2017a: Risk assessment of patulin intake through apple-based food by infants and preschool children in Serbia. Food Additives and Contaminants: Part A 34, 2023–2032. doi: https://doi.org/10.1080/19440049.2017.1364434 (904).

Torović, L., Trajković Pavlović, L., Popović, M., 2017b: Ochratoxin A and aflatoxin B_1 in breakfast cereals marketed in Serbia - occurrence and health risk characterisation. Food Additives and Contaminants: Part B 10, 176-184. doi: https://doi.org/10.1080/19393210.2017.1285358 (900).

Torres, A., Chulze, S., Varsavasky, E., Rodriguez, M., 1993: *Alternaria* metabolites in sunflower seeds. Mycopathologia 121, 17–20 (650).

Tosun, H., Arslan, R., 2013: Determination of aflatoxin B_1 levels in organic spices and herbs. Scientific World Journal 2013, article ID 874093. doi: https://doi.org/10.1155/2013/874093 (859).

Tosun, A., Ozden, S., 2016: Ochratoxin A in red pepper flakes commercialised in Turkey. Food Additives and Contaminants: Part B 9, 46-50. doi: https://doi.org/10.1080/19393210.2015.1121929 (861).

Tosun, H., Yildiz, H., Obuz, E., Seçkin, A.K., 2014: Ochratoxin A in grape pekmez (grape molasses) consumed in Turkey. Food Additives and Contaminants: Part B 7, 37-39. doi: https://doi.org/10.1080/19393210.2013.833298 (862).

Toussaint, G., Lafaverges, F., Walker, E.A., 1977: The use of high pressure liquid chromatography for determination of aflatoxins in olive oil. Archives de l´Institut Pasteur de Tunis 3–4, 325–335 (571).

Tozlovanu, M., Pfohl-Leszkowicz, A., 2010: Ochratoxin A in roasted coffee from French supermarkets and transfer in coffee beverages: comparison of analysis methods. Toxins 2, 1928-1942. doi: https://doi.org/10.3390/toxins2081928 (299).

Tripathi, S., Mishra, H.N., 2009: A rapid FT-NIR method for estimation of aflatoxin B_1 in red chilli powder. Food Control 20, 840-846. doi: https://doi.org/10.1016/j.foodcont.2008.11.003 (863).

Trucksess, M.W., Giler, J., Young, K., White, K.D., Page, S.W., 1999: Determination and survey of ochratoxin A in wheat, barley and coffee—1997. Journal of the Association of Official Analytical Chemists International 82, 85–89 (292).

Tsubouchi, H., Terada, H., Yamamoto, K., Hisada, K., Sakabe, Y., 1988: Ochratoxin A found in commercial roast coffee. Journal of Agricultural and Food Chemistry 36, 540–542 (293).

Tung, T.C., Ling, K.H., 1968: Study on aflatoxin of foodstuffs in Taiwan. The Journal of Vitaminology 14, 48–52 (216).

Turcotte, A.-M., Scott, P.M., 2011: Ochratoxin A in cocoa and chocolate sampled in Canada. Food Additives and Contaminants: Part A 28, 762-766. doi: https://doi.org/10.1080/19440049.2010.508796 (265).

Turcotte, A.-M., Scott, P.M., Tague, B., 2013: Analysis of cocoa products for ochratoxin A and aflatoxins. Mycotoxin Research 29, 193-201. doi: https://doi.org/10.1007/s12550-013-0167-x (864).

Tutelyan, V.A., Eller, K.I., Sobolev, V.S., 1989: A survey using normal-phase high-performance liquid chromatography of aflatoxins in domestic and imported foods in the USSR. Food Additives and Contaminants 6, 459–465 (267).

Ueno, Y., 1998: Residue and risk of ochratoxin A in human plasma and beverages in Japan. Mycotoxins 47, 25–32 (592).

Ulca, P., Evcimen, M.K., Senyuva, H.Z., 2010: Surveys of aflatoxin B_1 contamination of retail Turkish foods and of products intended for export between 2007 and 2009. Food Additives and Contaminants: Part B 3, 120-125. doi: https://doi.org/10.1080/19393210.2010.493619 (115).

Urano, T., Trucksess, M.W., Beaver, R.W., Wilson, D.M., Dorner, J.W., Dowell, F.E., 1992: Co-occurrence of cyclopiazonic acid and aflatoxins in corn and peanuts. Journal of the Association of Official Analytical Chemists International 75, 838–841 (396).

Vaclavik, L., Vaclavikova, M., Begley, T.H., Krynitsky, A.J., Rader, J.L., 2013: Determination of multiple mycotoxins in dietary supplements containing green coffee bean extracts using ultrahigh-performance liquid chromatography-tandem mass spectrometry (UHPLC-MC/MS). Journal of Agricultural and Food Chemistry 61, 4822-4830. doi: https://doi.org/10.1021/jf401139u (865).

Vaclavikova, M., Dzuman, Z., Lacina, O., Fenclova, M., Veprikova, Z., Zachariasova, M., Hajslova, J., 2015: Monitoring survey of patulin in a variety of fruit-based products using a sensitive UHPLC-MS/MS analytical procedure. Food Control 47, 577-584. doi: https://doi.org/10.1016/j.foodcont.2014.07.064 (911).

Vahl, M., Jørgensen, K., 1998: Determination of aflatoxins in food using LC/MS/MS. Zeitschrift für Lebensmittel-Untersuchung und – Forschung 206, 243–245 (217).

Valero, A., Marín, S., Ramos, A.J., Sanchis, V., 2008: Survey: ochratoxin A in European special wines. Food Chemistry 108, 593-599. doi: https://doi.org/10.1016/j.foodchem.2007.11.040 (158).

Valle-Algarra, F.M., Mateo, E.M., Mateo, R., Gimeno-Adelantado, J.V., Jiménez, M., 2011: Determination of type A and type B trichothecenes in paprika and chili pepper using LC-triple quadrupole-MS and GC-ECD. Talanta 84, 1112-1117. doi: https://doi.org/10.1016/j.talanta.2011.03.017 (589).

Van Der Bijl, P., Stockenström, S., Vismer, H.F., Van Wyk, C.W., 1996: Incidence of fungi and aflatoxins in imported areca nut samples. South African Journal of Science 92, 154–156 (56).

Van Der Stegen, G., Jörissen, U., Pittet, A., Saccon, M., Steiner, W., Vincenzi, M., Winkler, M., Zapp, J., Schlatter, Chr. 1997: Screening of European coffee final products for occurrence of ochratoxin A (OTA). Food Additives and Contaminants 14, 211–216 (291).

Van Rensburg, S.J., Cook-Mozaffari, P., Van Schalkwyk, D.J., Van Der Watt, J.J., Vincent, T.J., Purchase, I.F., 1985: Hepatocellular carcinoma and dietary aflatoxin in Mozambique and Transkei. British Journal of Cancer 51, 713–726 (114).

Vanesa, D., Ana, P., 2013: Occurrence of ochratoxin A in coffee beans, ground roasted coffee and soluble coffee and method validation. Food Control 30, 675-678. doi: https://doi.org/10.1016/j.foodcont.2012.09.004 (866).

Var, I., Kabak, B., 2007: Occurrence of ochratoxin A in Turkish wines. Microchemical Journal 86, 241-247. doi: https://doi.org/10.1016/j.microc.2007.04.002 (867).

Var, I., Kabak, B., Gök, F., 2007: Survey of aflatoxin B_1 in helva, a traditional Turkish food, by TLC. Food Control 18, 59-62. doi: https://doi.org/10.1016/j.foodcont.2005.08.008 (437).

Varga, E., Glauner, T., Berthiller, F., Krska, R., Schumacher, R., Sulyok, M., 2013: Development and validation of a (semi-)quantitative UHPLC-MS/MS method for the determination of 191 mycotoxins and other fungal metabolites in almonds, hazelnuts, peanuts and pistachios. Analytical and Bioanalytical Chemistry 405, 5087-5104. doi: https://doi.org/10.1007/s00216-013-6831-3 (868).

Varga, J., Kiss, R., Mátrai, T., Mátrai, T., Téren, J., 2005: Detection of ochratoxin A in Hungarian wines and beers. Acta Alimentaria 34, 381–392 (606).

Varga, J., Kocsubé, S., Koncz, Z., Téren, J., 2006: Mycobiota and ochratoxin A in raisins purchased in Hungary. Acta Alimentaria 35, 289-294. doi: https://doi.org/10.1556/AAlim.35.2006.3.6 (273).

Varga, J., Kocsubé, S., Suri, K., Szigeti, Gy., Szekeres, A., Varga, M., Tóth, B., Bartók, T., 2010: Fumonisin contamination and fumonisin producing black Aspergilli in dried vine fruits of different origin. International Journal of Food Microbiology 143, 143-149. doi: https://doi.org/10.1016/j.ijfoodmicro.2010.08.008 (600).

Varsavsky, E., Sommer, S.E., 1977: Determination of aflatoxins in peanuts. Annales de la Nutrition et de l´Alimentation 31, 539–544 (395).

Vecchio, A., Mineo, V., Planeta, D., 2012: Ochratoxin A in instant coffee in Italy. Food Control 28, 220-223. doi: https://doi.org/10.1016/j.foodcont.2012.04.029 (869).

Vega, M., Ríos, G., Von Baer, D., Mardones, C., Tessini, C., Herlitz, E., Saelzer, R., Ruiz, M.A., 2012: Ochratoxin A occurrence in wines produced in Chile. Food Control 28, 147-150. doi: https://doi.org/10.1016/j.foodcont.2012.04.032 (870).

Veprikova, Z., Zachariasova, M., Dzuman, Z., Zachariasova, A., Fenclova, M., Slavikova, P., Vaclavikova, M., Mastovska, K., Hengst, D., Hajslova, J., 2015: Mycotoxins in plant-based dietary supplements: hidden health risk for consumers. Journal of Agricultural and Food Chemistry 63, 6633-6643. doi: https://doi.org/10.1021/acs.jafc.5b02105 (286).

Verma, R.K., Agarwal, R.K., 2000: Occurrence of aflatoxin in groundnut and groundnut cake available at distributors and retailers of Utar Pradesh. Indian Journal of Animal Nutrition 17, 156–159 (405).

Versari, A., Parpinello, G.P., Mattioli, A.U., 2007: Survey of patulin contamination in Italian apple juices from organic and conventional agriculture. Journal of Food Technology 5, 143–146 (638).

Vijayasamundeeswari, A., Mohankumar, M., Karthikeyan, M., Vijayanandraj, S., Paranidharan, V., Velazhahan, R., 2009: Prevalence of aflatoxin B_1 contamination in pre- and post-harvest maize kernels, food products, poultry and livestock feeds in Tamil Nadu, India. Journal of Plant Protection Research 49, 221-224. doi: https://doi.org/10.2478/v10045-009-0033-1 (871).

Villa, P., Markaki, P., 2009: Aflatoxin B_1 and ochratoxin A in breakfast cereals from Athens market: occurrence and risk assessment. Food Control 20, 455-461. doi: https://doi.org/10.1016/j.foodcont.2008.07.012 (91).

Viñas, I., Palma, J., Garza, S., Sibilia, A., Sanchis, V., Visconti, A., 1994: Natural occurrence of aflatoxin and *Alternaria* mycotoxins in oilseed rape from Catalonia (Spain): incidence of toxigenic strains. Mycopathologia 128, 175–179 (583).

Visconti, A., Pascale, M., Centonze, G., 1999: Determination of ochratoxin A in wine by means of immunoaffinity column clean-up and high-performance liquid chromatography. Journal of Chromatography A 864, 89–101 (522).

Walravens, J., Mikula, H., Rychlik, M., Asam, S., Devos, T., Njumbe Ediage, E., Diana Di Mavungu, J., Jacxsens, L., Van Landschoot, A., Vanhaecke, L., De Saeger, S., 2016: Validated UPLC-MS/MS methods to quantitate free and conjugated *Alternaria* toxins in commercially available tomato products and fruit and vegetable juices in Belgium. Journal of Agricultural and Food Chemistry 64, 5101-5109. doi: https://doi.org/10.1021/acs.jafc.6b01029 (514).

Wang, D., Zhang, Z., Peiwu, L., Zhang, Q., Zhang, W., 2016: Time-resolved fluorescent immunochromatography of aflatoxin B_1 in soybean sauce: a rapid and sensitive quantitative analysis. Sensors 16, 1094, doi: https://doi.org/10.3390/s16071094 (402).

Wang, J., Liu, X.-M., 2007: Contamination of aflatoxins in different kinds of foods in China. Biomedical and Environmental Sciences 20, 483–487 (411).

Wang, J.-S., Huang, T., Su, J., Liang, F., Wie, Z., Liang, Y., Luo, H., Kuang, S.-Y., Qian, G.-S., Sun, G., He, X., Kensler, T.W., Groopman, J.D., 2001: Hepatocellular carcinoma and aflatoxin exposure in Zhuqing Village, Fusui County, People's Republic of China. Cancer Epidemiology, Biomarkers & Prevention 10, 143–146 (578).

Wang, L., Wang, Z., Gao, W., Chen, J., Yang, M., Kuang, Y., Huang, L., Chen, S., 2013: Simultaneous determination of aflatoxin B_1 and ochratoxin A in licorice roots and fritillary bulbs by solid-phase extraction coupled with high-performance liquid chromatography-tandem mass spectrometry. Food Chemistry 138, 1048-1054. doi: https://doi.org/10.1016/j.foodchem.2012.11.066 (872).

Wang, S., Cheng, L., Ji, S., Wang, K., 2014: Simultaneous determination of seventeen mycotoxin residues in *Puerariae lobatae radix* by liquid chromatography-tandem mass spectrometry. Journal of Pharmaceutical and Biomedical Analysis 98, 201-209. doi: https://doi.org/10.1016/j.jpba.2014.05.037 (873).

Ware, G. M., Thorpe, C.W., Pohland, A.E., 1974: Liquid chromatographic method for determination of patulin in apple juice. Journal of the Association of Official Analytical Chemists 57, 1111–1113 (460).

Wareing, P.W., Westby, A., Gibbs, J.A., Allotey, L.T., Halm, M., 2001: Consumer preferences and fungal and mycotoxin contamination of dried cassava products from Ghana. International Journal of Food Science and Technology 36, 1–10 (238).

Warth, B., Parich, A., Atehnkeng, J., Bandyopadhyay, R., Schumacher, R., Sulyok, M., Krska, R., 2012: Quantitation of mycotoxins in food and feed from Burkina Faso and Mozambique using a modern LC-MS/MS multitoxin method. Journal of Agricultural and Food Chemistry 60, 9352-9363. doi: https://doi.org/10.1021/jf302003n (117).

Watanabe, M., Shimizu, H., 2005: Detection of patulin in apple juice marketed in the Tohoku district, Japan. Journal of Food Protection 68, 610–612 (471).

Watkins, K.L., Fazekas, G., Palmer, M.V., 1990: Patulin in Australian apple juice. Food Australia 42, 438–439 (461).

Weber, Z., Kostecki, M., Von Bargen, S., Gossmann, M., Waskiewicz, A., Bocianowski, J., Knaflewski, M., Büttner, C., Golinski, P., 2006: *Fusarium* species colonizing spears and forming mycotoxins in field samples of asparagus from Germany and Poland. Journal of Phytopathology 154, 209–216 (84).

Wei, D., Wang, Y., Jiang, D., Feng, X., Li, J., Wang, M., 2017: Survey of *Alternaria* toxins and other mycotoxins in dried fruits in China. Toxins (Basel) 9 (7), 200. doi: https://doi.org/10.3390/toxins9070200 (815).

Wei, D. L., Wei, R. D., 1980: High pressure liquid chromatographic determination of aflatoxins in peanut and peanut products of Taiwan. Proceedings of the National Science Council, Republic of China 4, 152–155 (214).

Wei, R., Qiu, F., Kong, W., Wei, J., Yang, M., Luo, Z., Qin, J., Ma, X., 2013: Co-occurrence of aflatoxin B_1, B_2, G_1, G_2 and ochrotoxin A in *Glycyrrhiza uralensis* analyzed by HPLC-MS/MS. Food Control 32, 216-221. doi: https://doi.org/10.1016/j.foodcont.2012.11.028 (874).

Welke, J.E., Hoeltz, M., Dottori, H.A., Noll, I.B., 2009: Quantitative analysis of patulin in apple juice by thin-layer chromatography using a charge coupled device detector. Food Additives and Contaminants: Part A 26, 754-758. doi: https://doi.org/10.1080/02652030802662746 (489).

Welke, J.E., Hoeltz, M., Dottori, H.A., Noll, I.B., 2010a: Determination of ochratoxin A in wine by high-performance thin-layer chromatography using charged coupled device. Journal of the Brazilian Chemical Society 21, 441-446 (908).

Welke, J.E., Hoeltz, M., Dottori, H.A., Noll, I.B., 2010b: Fungi and patulin in apples and the role of processing on patulin levels in juices: a study on naturally contaminated apples. Journal of Food Safety 30, 276–287 (909).

Wells, J.M., Payne, J.A., 1976: Incidence of aflatoxin contamination on a sampling of southeastern pecans. Proceedings of the Florida State Horticultural Society 89, 256–257 (552).

Wells, J.M., Payne, J.A., Stoloff, L., 1983: Weather-related incidence of aflatoxin contamination in late-harvested pecans. Plant Disease 67, 751–753 (553).

Wen, J., Kong, W., Wang, J., Yang, M., 2013: Simultaneous determination of four aflatoxins and ochratoxin A in ginger and related products by HPLC with fluorescence detection after immunoaffinity column clean-up and postcolumn photochemical derivatization. Journal of Separation Science 36, 3709-3716. doi: https://doi.org/10.1002/jssc.201300885 (875).

Wheeler, J.L., Harrison, M.A., Koehler, P.E., 1987: Presence and stability of patulin in pasteurized apple cider. Journal of Food Science 52, 479–480 (39).

Wild, C.P., Hudson, G.J., Sabbioni, G., Chapot, B., Hall, A.J., Wogan, G.N., Whittle, H., Montesano, R., Groopman, J.D., 1992: Dietary intake of aflatoxins and the level of albumin-bound aflatoxin in peripheral blood in The Gambia, West Africa. Cancer, Epidemiology, Biomarkers & Prevention 1, 229–234 (118).

Williams, I.O, Ugbaje, S.A., Igile, G.O., Ekpe, O.O., 2015: Occurrence of aflatoxin in some food commodities commonly consumed in Nigeria. Journal of Food Research 4, 81-88 doi: https://doi.org/10.5539/jfr.v4n5p81 (639).

Wilson, D.M., Nuovo, G.J., 1973: Patulin production in apples decayed by *Penicillium expansum*. Applied Microbiology 26,124–125 (40).

Wilson, R.D., 1981: Surveying some apple juice for patulin. Food Technology in New Zealand 6, 27–31 (462).

Wittkowski, M., Baltes, W., Krönert, W., Weber, R., 1983: Determination of *Alternaria* toxins in fruit and vegetable products. Zeitschrift für Lebensmittel-Untersuchung und -Forschung 177, 447–453 (24).

Woese, K., 2000: Ochratoxin A in grape juice and wine. Mycotoxin Research 16, 132–135 (497).

Wolff, J., 2000: Ochratoxin A in cereals and cereal products. Archiv für Lebensmittelhygiene 51, 85–88 (597).

Wongworapat, K., Tu Ho, M.H., Soontornjanagit, M., Kawamura, O., 2016: Occurrence of ochratoxin A and ochratoxin B in commercial coffee in Vietnam and Thailand. JSM Mycotoxins 66, 1-6. doi: https://doi.org/10.2520/myco.66.1 (653).

Wood, G.E., 1989: Aflatoxins in domestic and imported foods and feeds. Journal of the Association of Official Analytical Chemists 72, 543–548 (19).

Wu, J., Tan, Y., Wang, Y., Xu, R., 2011a: Occurrence of ochratoxin A in wine and beer samples from China. Food Additives and Contaminants: Part B 4, 52-56. doi: https://doi.org/10.1080/19393210.2010.535909 (623).

Wu, J., Tan, Y., Wang, Y., Xu, R., 2012: Occurrence of ochratoxin A in grain and manufactured food products in China detected by HPLC with fluorescence detection and confirmed by LC-ESI-MS/MS. Mycopathologia 173, 199-205. doi: https://doi.org/10.1007/s11046-011-9393-y (302).

Wu, J., Zhao, R., Chen, B., Yang, M., 2011b: Determination of zearalenone in barley by high-performance liquid chromatography coupled with evaporative light scattering detection and natural occurrence of zearalenone in functional food. Food Chemistry 126, 1508-1511. doi: https://doi.org/10.1016/j.foodchem.2010.11.159 (876).

Wu, L.X., Ding, X.X., Li, P.W., Du, X.H., Zhou, H.Y., Bai, Y.Zh., Zhang, L.X., 2016: Aflatoxin contamination of peanuts at harvest in China from 2010 to 2013 and its relationship with climatic conditions. Food Control 60, 117-123. doi: https://doi.org/10.1016/j.foodcont.2015.06.029 (441).

Xavier, J.J.M., Scussel, V.M., 2008: Development of an LC-MS/MS method for the determination of aflatoxins B_1, B_2, G_1, and G_2 in Brazil nut. International Journal of Environmental Analytical Chemistry 88, 425-433. doi: https://doi.org/10.1080/03067310701836816 (877).

Xiang, L., Gao, Y.H., Liu, D.Y., Yang, M.H., 2012: Rapid determination of patulin in medicinal hawthorn fruits by HPLC-MS/MS. World Mycotoxin Journal 5, 31-36. doi: https://doi.org/10.3920/WMJ2011.1336 (369).

Yacine Ware, L., Durand, N., Nikiema, P.A., Alter, P., Fontana, A., Montet, D., Barro, N., 2017: Occurrence of mycotoxins in commercial infant formulas locally produced in Ouagadougou (Burkina Faso). Food Control 73, 518-523. doi: https://doi.org/10.1016/j.foodcont.2016.08.047 (913).

Yang, L., Wang, L., Pan, J., Xiang, L., Yang, M., Logrieco, A.F., 2010: Determination of ochratoxin A in traditional Chinese medicinal plants by HPLC-FLD. Food Additives and Contaminants: Part A 27, 989-997. doi: https://doi.org/10.1080/19440041003647122 (73).

Yang, L.-X., Liu, Y.-P., Miao, H., Dong, B., Yang, N.-J., Chang, F.-Q., Sun, J.-B., 2011: Determination of aflatoxins in edible oil from markets in Hebei Province of China by liquid chromatography-tandem mass spectrometry. Food Additives and Contaminants: Part B 4, 244-247. doi: https://doi.org/10.1080/19393210.2011.632694 (878).

Yang, M.-H., Chen, J.-M., Zhang, X.-H., 2005: Immunoaffinity column clean-up and liquid chromatography with post-column derivatization for analysis of aflatoxins in traditional Chinese medicine. Chromatographia 62, 499-504. doi: https://doi.org/10.1365/s10337-005-0647-z (59).

Yazdanpanah, H., Mohammadi, T., Abouhossain, G., Cheraghali, A.M., 2005: Effect of roasting on degradation of aflatoxins in contaminated pistachio nuts. Food and Chemical Toxicology 43, 1135-1139. doi: https://doi.org/10.1016/j.fct.2005.03.004 (669).

Yazdanpanah, H., Zarghi, A., Shafaati, A.R., Foroutan, S.M., Aboul-Fathi, F., Khoddam, A., Nazari, F., Shaki, F., 2013: Analysis of aflatoxin B_1 in Iranian foods using HPLC and a monolithic column and estimation of its dietary intake. Iranian Journal of Pharmaceutical Research 12 (Suppl) 83–89 (879).

Yndestad, M., Underdal, B., 1975: Aflatoxin in foods on the Norwegian market. Nordisk Veterinaer Medicin 27, 42–48 (201).

Yentür, G., Er, B., Özkan, M.G., Öktem, A.B., 2006: Determination of aflatoxins in peanut butter and sesame samples using high-performance liquid chromatography method. European Food Research and Technology 224, 167-170. doi: https://doi.org/10.1007/s00217-006-0310-4 (221).

Yogendrarajah, P., Jacxsens, L., De Saeger, S., De Meulenaer, B., 2014: Co-occurrence of multiple mycotoxins in dry chilli (*Capsicum annum* L.) samples from the markets of Sri Lanka and Belgium. Food Control 46, 26-34. doi: https://doi.org/10.1016/j.foodcont.2014.04.043 (889).

Yogendrarajah, P., Van Poucke, C., De Meulenaer, B., De Saeger, S., 2013: Development and validation of a QuEChERS based liquid chromatography tandem mass spectrometry method for the determination of multiple mycotoxins in spices. Journal of Chromatography A 1297, 1-11. doi: https://doi.org/10.1016/j.chroma.2013.04.075 (891).

Yoshinari, T., Takeuchi, H., Aoyama, K., Taniguchi, M., Hashiguchi, S., Kai, S., Ogiso, M., Sato, T., Akiyama, Y., Nakajima, M., Tabata, S., Tanaka, T., Ishikuro, E., Sugita-Konishi, Y., 2014: Occurrence of four *Fusarium* mycotoxins, deoxynivalenol, zearalenone, T-2 toxin, and HT-2 toxin, in wheat, barley, and Japanese retail food. Journal of Food Protection 77, 1940-1946 doi: https://doi.org/10.4315/0362-028X.JFP-14-185 (892).

Younis, Y.M.H., Malik, K.M., 2003: TLC and HPLC assays of aflatoxin contamination in Sudanese peanuts and peanut products. Kuwait Journal of Science & Engineering 30, 79–93 (230).

Yu, F.-Y., Gribas, A.V., Vdovenko, M.M., Sakharov, I.Y., 2013: Development of ultrasensitive direct chemiluminescent enzyme immunoassay for the determination of aflatoxin B_1 in food products. Talanta 107, 25-29. doi: https://doi.org/10.1016/j.talanta.2012.12.047 (893).

Yu, F.-Y., Vdovenko, M.M., Wang, J.-J., Sakharov, I.Y., 2011: Comparison of enzyme-linked immunosorbent assays with chemiluminescent and colorimetric detection for the determination of ochratoxin A in food. Journal of Agricultural and Food Chemistry 59, 809-813. doi: https://doi.org/10.1021/jf103261u (147).

Yuan, Y., Zhuang, H., Zhang, T., Liu, J., 2010: Patulin content in apple products marketed in Northeast China. Food Control 21, 1488–1491. doi: https://doi.org/10.1016/j.foodcont.2010.04.019 (and personal communication) (890).

Yue, Y.-T., Zhang, X.-F., Pan, J., Ou-Yang, Z., Wu, J., Yang, M.-H., 2010: Determination of deoxynivalenol in medicinal herbs and related products by GC-ECD and confirmation by GC-MS. Chromatographia 71, 533-538. doi: https://doi.org/10.1365/s10337-010-1477-1 (894).

Yurdun, T., Omurtag, G.Z., Ersoy, Ö., 2001: Incidence of patulin in apple juices marketed in Turkey. Journal of Food Protection 64, 1851–1853 (447).

Yuwai, K.E., Rao, K.S., Singh, K., Tanaka, T., Ueno, Y., 1994: Occurrence of nivalenol, deoxynivalenol, and zearalenone in imported cereals in Papua, New Guinea. Natural Toxins 2, 19–21 (633).

Zaied, C., Abid, S., Bouaziz, C., Chouchane, S., Jomaa, M., Bacha, H., 2010: Ochratoxin A levels in spices and dried nuts consumed in Tunisia. Food Additives and Contaminants: Part B 3, 52-57. doi: https://doi.org/10.1080/19440041003587302 (20).

Zaied, C., Abid, S., Hlel, W., Bacha, H., 2013: Occurrence of patulin in apple-based-foods largely consumed in Tunisia. Food Control 31, 263-267. doi: https://doi.org/10.1016/j.foodcont.2012.10.005 (895).

Zhang, K., Wong, J.W., Jia, Z., Vaclavikova, M., Trucksess, M.W., Begley, T.H., 2014a: Screening multimycotoxins in food-grade gums by stable isotope dilution and liquid chromatography/tandem mass spectrometry. Journal of the Association of Official Analytical Chemists International 97, 889–895 (896).

Zhang, X., Chen, L., Li, J., Zhu, BQ., Ma, L., 2013: Occurrence of ochratoxin A in Chinese wines: influence of local meteorological parameters. European Food Research and Technology 236, 277-283. doi: https://doi.org/10.1007/s00217-012-1886-5 (899).

Zhang, X., Li, J., Zong, N., Zhou, Z., Ma, L., 2014b: Ochratoxin A in dried vine fruits from Chinese markets. Food Additives and Contaminants: Part B 7, 157-161. doi: https://doi.org/10.1080/19393210.2013.867365 (897).

Zhang, X., Liu, H., Chen, J., 2005: Immunoaffinity column cleanup with liquid chromatography using post-column bromination for aflatoxins in medicinal herbs and plant extracts. Journal of Chromatographic Science 43, 47-51. doi: https://doi.org/10.1093/chromsci/43.1.47 (880).

Zhang, X., Liu, W., Logrieco, A.F., Yang, M., Ou-yang, Z., Wang, X., Guo, Q., 2011: Determination of zearalenone in traditional Chinese medicinal plants and related products by HPLC-FLD. Food Additives and Contaminants: Part A 28, 885-893. doi: https://doi.org/10.1080/19440049.2011.563429 (898).

Zhang, Z., Hu, X., Zhang, Q., Li, P., 2016: Determination for multiple mycotoxins in agricultural products using HPLC-MS/MS via a multiple antibody immunoaffinity column. Journal of Chromatography B 1021, 145-152. doi: https://doi.org/10.1016/j.jchromb.2016.02.035 (881).

Zhao, H., Wang, Y., Zou, Y., Zhao, M., 2013: Natural occurrence of deoxynivalenol in soy sauces consumed in China. Food Control 29, 71-75. doi: https://doi.org/10.1016/j.foodcont.2012.05.066 (882).

Zhao, K., Shao, B., Yang, D., Li, F., 2015: Natural occurrence of four *Alternaria* mycotoxins in tomato- and citrus-based foods in China. Journal of Agricultural and Food Chemistry 63, 343-348. doi: https://doi.org/10.1021/jf5052738 (883).

Zhao, X., Yuan, Y., Zhang, X., Yue, T., 2014: Identification of ochratoxin A in Chinese spices using HPLC fluorescent detectors with immunoaffinity column cleanup. Food Control 46, 332-337. doi: https://doi.org/10.1016/j.foodcont.2014.05.052 (885).

Zhao, Y., Wan, L.-H., Bai, X.-L., Liu, Y.-M., Zhang, F.-P., Liu, Y.-M., Liao, X., 2017: Quantification of mycotoxins in vegetable oil by UPLC-MS/MS after magnetic solid-phase extraction. Food Additives and Contaminants: Part A 34, 1201-1210. doi: https://doi.org/10.1080/19440049.2017.1319074 (754).

Zheng, R., Xu, H., Wang, W., Zhan, R., Chen, W., 2014: Simultaneous determination of aflatoxin B_1, B_2, G_1, G_2, ochratoxin A, and sterigmatocystin in traditional Chinese medicines by LC-MS-MS. Analytical and Bioanalytical Chemistry 406, 3031-3039. doi: https://doi.org/10.1007/s00216-014-7750-7 (886).

Zhou, J., Xu, J.-J., Huang, B.-F., Cai, Z.-X., Ren, Y.-P., 2017: High-performance liquid chromatographic determination of multi-mycotoxin in cereals and bean foodstuffs using interference-removal solid-phase extraction combined with optimized dispersive liquid-liquid microextraction. Journal of Separation Science 40, 2141-2150. doi: https://doi.org/10.1002/jssc.201601326 (853).

Zhou, Y., Kong, W., Li, Y., Logrieco, A.F., Xu, J., Yang, M., 2012: A new solid-phase extraction and HPLC method for determination of patulin in apple products and hawthorn juice in China. Journal of Separation Science 35, 641-649. doi: https://doi.org/10.1002/jssc.201100919 (887).

Zimmer, I., Usleber, E., Klaffke, H., Weber, R., Majerus, P., Otteneder, H., Gareis, M., Dietrich, R., Märtlbauer, E., 2008: Fumonisin intake of the German consumer. Mycotoxin Research 24, 40–52 (510).

Zimmerli, B., Dick, R., 1995: Determination of ochratoxin A at the ppt level in human blood, serum, milk and some foodstuffs by high performance liquid chromatography with enhanced fluorescence detection and immunoaffinity column cleanup: methodology and Swiss data. Journal of Chromatography B 666, 85–99 (333).

Zimmerli, B., Dick, R., 1996: Ochratoxin A in table wine and grape-juice: occurrence and risk assessment. Food Additives and Contaminants 13, 655–668 (496).

Zinedine, A., Betbeder, A.-M., Faid, M., Benlemlih, M., Idrissi, L., Creppy, E.E., 2004: Ochratoxin A, determination in dried fruits and black olives from Morocco. Alimentaria 359, 73–76 (341).

Zinedine, A., Blesa, J., Mahnine, N., El Abidi, A., Montesano, D., Mañes, J., 2010: Pressurized liquid extraction coupled to liquid chromatography for the analysis of ochratoxin A in breakfast and infants cereals from Morocco. Food Control 21, 132-135. doi: https://doi.org/10.1016/j.foodcont.2009.04.009 (95).

Zinedine, A., Brera, C., Elakhdari, S., Catano, C., Debegnach, F., Angelini, S., De Santis, B., Faid, M., Benlemlih, M., Minardi, V., Miraglia, M., 2006: Natural occurrence of mycotoxins in cereals and spices commercialized in Morocco. Food Control 17, 868-874. doi: https://doi.org/10.1016/j.foodcont.2005.06.001 (174).

Zinedine, A., Soriano, J.M., Juan, C., Mojemmi, B., Moltó, J.C., Bouklouze, A., Cherrah, Y., Idrissi, L., El Aouad, R., Mañes, J., 2007: Incidence of ochratoxin A in rice and dried fruits from Rabat and Salé area, Morocco. Food Additives and Contaminants 24, 285-291. doi: https://doi.org/10.1080/02652030600967230 (364).

Zitomer, N., Rybak, M.E., Li, Z., Walters, M.J., Holman, M.R., 2015: Determination of aflatoxin B_1 in smokeless tobacco products by use of UHPLC-MS/MS. Journal of Agricultural and Food Chemistry 63, 9131-9138. doi: https://doi.org/10.1021/acs.jafc.5b02622 (725).

Zohri, A.A., Abdel-Gawad, K.M., 1993: Survey of mycoflora and mycotoxins of some dried fruits in Egypt. Journal of Basic Microbiology 33, 279–288 (54).

Zohri, A.A., Khayria, M., Gawad, A.El., 1992: Studies on mycoflora and mycotoxins of cowpea cultivars. Korean Mycology 20, 252–258 (325).

Zohri, A.A., Saber, S.M., 1993: Filamentous fungi and mycotoxin detected in coconut. Zentralblatt für Mikrobiologie 148, 325–332 (275).

Zouaoui, N., Sbaii, N., Bacha, H., Abid-Essefi, S., 2015: Occurrence of patulin in various fruit juice marketed in Tunisia. Food Control 51, 356-360. doi: https://doi.org/10.1016/j.foodcont.2014.09.048 (888).

The manufacturer's authorised representative in the EU is Springer Nature Customer Service Centre GmbH, Europaplatz 3, 69115 Heidelberg, Germany. If you have any concerns regarding our products, please contact ProductSafety@springernature.com

Printed and bound by CPI Group (UK) Ltd, Croydon, CR0 4YY
01/10/2026
02225888-0001